Dale Carmichael

PLASTICS MATERIALS AND PROCESSES

PLASTICS MATERIALS AND PROCESSES

Seymour S. Schwartz
Plastics Consultant

Senior Staff Engineer (Retired)
Technology Support Div.
Hughes Aircraft Co.

Lecturer-Polymer Science & Technology
Los Angeles Trade Technical College and U.C.L.A.

Sidney H. Goodman
Manager, Materials Products Department
Technology Support Div.
Hughes Aircraft Co.

Senior Lecturer
Department of Chemical Engineering
University of Southern California

VNR VAN NOSTRAND REINHOLD COMPANY
NEW YORK CINCINNATI TORONTO LONDON MELBOURNE

Library of Congress Catalog Card Number: 81-1887
ISBN: 0-442-22777-9

Manufactured in the United States of America

Published by Van Nostrand Reinhold Company Inc.
135 West 50th Street, New York, N.Y. 10020

Van Nostrand Reinhold Limited
1410 Birchmount Road
Scarborough, Ontario MIP 2E7, Canada

Van Nostrand Reinhold Australia Pty. Ltd.
17 Queen Street
Mircham, Victoria 3132, Australia

Van Nostrand Reinhold Company Limited
Molly Millars Lane
Wokingham, Berkshire, England

15 14 13 12 11 10 9 8 7 6 5 4 2 1

Library of Congress Cataloging in Publication Data

Schwartz, Seymour S., 1915–
Plastics materials and processes.

Includes index.
1. Plastics. I. Goodman, Sidney H. II. Title.
TP1120.S32 668.4 81-1887
ISBN 0-442-22777-9 AACR2

To Esther and Carol for their devotion, encouragement, and tolerance; and to Rose, Susan, Stephen, and Leslie who agreed to wait until this task was ended.

Preface

We have often asked ourselves during the course of this book's creation, why bother generating "another book on plastics"? The early motivation came from the difficulties we encountered in prescribing a text for our students. The needs of students in the trade technical level paralleled, yet differed from, those at the university graduate level. Available literature for the former stressed processing, fabrication, and the "mechanics" of plastics. The latter had to depend primarily on polymer science (chemistry and physics) texts and occasionally on encyclopedic reference manuals. The texts were usually theoretically biased and the encyclopedias continuously updated with little chronological continuity of plastics technology from its early days.

In the nonacademic environment as well, we often found that a comprehensive overview of the technology of plastics was not available. Highly specific or theoretical material could be found; for example, entire books devoted to one class of materials or one particular processing method. Alternatively, the so-called survey texts often treated specific subjects in a cursory fashion, because size limitations precluded a truly in-depth presentation of each one of hundreds of topics. Of particular concern was the rather sparse treatment always given plastics materials and properties relative to processing and fabrication.

It was our intent, therefore, to bridge these gaps and provide an in-depth survey of the plastics field. We have tried not to become overly theoretical, preserving this endeavor for those qualified academics. We have strived not to gloss over subjects, even where a paucity of information was available. We have tried to compile and collate data that would be useful to the designer as well as to the plastics engineer. We would like the "expert" to use the book to broaden his scope of knowledge in other plastics areas. We have attempted to include sufficient detailed discussions which provide the background and technological support for the student pursuing, or just beginning his pursuit, into the broad plastics arena.

The one problem area which faces all authors of technical manuscripts is when to stop. Rapid obsolescence of information and/or advancement of the art are facts of life in all high technology. The temptation to prolong final manuscript submission arose every time a new piece of trade literature or a monthly journal crossed our desks. In each, "the hottest thing since sliced bread" was energetically displayed. We considered the deletion of such critical advances to be an abrogation of duty to our readers to provide the very latest developments.

At the same time, we recognized that to cover every facet of the plastics industry would be impossible. The decision to include a particular subject or to delete another ultimately was based on our best judgment of the relevancy of the subject for most readers. It is a certainty that there are areas some will feel have not been sufficiently addressed; still others will feel we have overdone a particular treatment. From those readers we request their indulgence and understanding. Our sincere desire was to present a comprehensive survey for as large an audience as practical.

Although the materials chapters begin with the chemistry of each plastics class, the trade school student or nonchemist should have no trouble in bypassing these portions and going directly to the discussions of properties and applications. The chapters on processes are designed to give the lay reader or beginning student the historical background of the process and the later developments, and at the same time to be of value to the professional by providing a discussion of the capabilities and limitations of each process. It should be noted that the chapters are on processes—not *processing*; that is, there is very little "button pushing" information. However, "trouble shooting" information for many processes is included so the volume can be used as a desk or shop reference for the student, beginning professional, and the experienced operator.

A chapter has been included on safety considerations which we believe is the first attempt at presenting this topic in a relatively brief but specific manner. The chapter on materials and process selection is also a unique feature. In this chapter we have summarized the various topics discussed throughout the book in a chartlike fashion. The intent is to provide a series of concise reference guides that will be of particular value to the designer or engineer who requires selection of the optimum materials and/or process for any given application.

We have taken great pains to avoid using trade names for generic types of plastics in the belief that no one manufacturer should be singled out as the exclusive source of the plastic. At the same time, we recognize that many plastics are commonly recognized by their trade designations and so have included a list of trade names after each discussion of the particular class of plastics, except where five or less exist. In such a case the trade names are listed at the beginning of the specific discussion.

Seymour S. Schwartz
Sidney H. Goodman

Acknowledgments

To acknowledge and thank every individual who, directly or indirectly, has contributed to this venture would require an inordinate amount of time and space. We are deeply indebted to many individuals, companies, and organizations whose forbearance, cooperation, and encouragement have made this task considerably easier.

There are those whose special efforts must be recognized. Frank Evans of Noland Paper Company; Jack Boswell and Dr. George Firth of Applied Plastics Company; Brian Keller, Robert Johnson, Arnold Tuckerman, and many others of the Technology Support Division of Hughes Aircraft Company; Dr. Ronald Salovey and Professor Emeritus Edward Partridge of the University of Southern California; R. M. Milby of Los Angeles Trade-Technical College; The Lady Davis Foundation; and the Department of Chemical Engineering of the Technion, Israel Institute of Technology have all been generous in supplying material, support, and assistance. M. Scott Moulton gave early support and encouragement and germinated the beginnings of this project.

To Ashak M. Rawji, Susan Munger, Alberta Gordon, and Linda Kucan, our editors, for their guidance, prodding, and leadership through the publication maze; and to Lorraine Freedman, who somehow found time between all her studies and school activities to handle the myriad details of a research assistant, conscientiously, carefully, and cheerfully —our special thanks.

To those whom we have inadvertently overlooked—our apologies. We know who you are and will always be grateful for your contributions. To one and all, our sincere appreciation and thanks.

Contents

1
Introduction to Plastics

BRIEF HISTORY OF PLASTICS

As a material of construction, plastics are the "babies" of the world. They are, for the most part, produced through the combination of man-made chemicals and, as such, do not come directly "from nature" like wood, ceramic, or metal. There are some plastics derived from naturally occurring materials that, like wood, ceramic, or metal, man has modified for his use. But they comprise a small percentage of all plastics. If one considers the beginning of plastics as occurring at the onset of the evolution of organic chemistry, then one might argue that their "birth" occurred circa 1820.

Practically, though, there were three events which can better be described as the beginning of plastics. The first occured in 1839 when Goodyear discovered the vulcanization of rubber. This was the first successful attempt at synthetically modifying a "natural" substance into a more practical and useful product.

The second event was Hyatt's synthesization of the first pyroxylin plastic, "Celluloid," in 1868. He discovered a method of not only chemically modifying a naturally occurring substance, cellulose to nitrated cellulose, but was also able to improve this chemical modification, through the addition of camphor, to produce a product with a specifically predetermined set of properties needed to guarantee its successful commercialization. Figures 1-1 to 1-4 show early cellulosic products made from the first "plastic."

The third event was Baekeland's introduction in 1909 of the first truly synthetic plastic; that is, produced entirely from basic chemicals, phenol, and formaldehyde, and combined in a controlled manner to yield a moldable powder. He also developed the procedures and equipment needed to convert this powder into useful end products.

Table 1-1 gives a more detailed history of many of the plastics industry's milestones. The growth of this "baby" during the period of 1909–1967 was quite obviously phenomenal and, in fact, continues to this day.

We will now attempt to survey this "child," to see what he is made of, how he is made, what he can and, importantly, cannot do, and how he has adjusted to life with his older brothers and sisters—metals, ceramics, glasses, and woods.

Fig. 1-1. Billiard ball, circa 1868, the year of Hyatt's invention of Celluloid. (*Courtesy: Celanese Plastics Co.*)

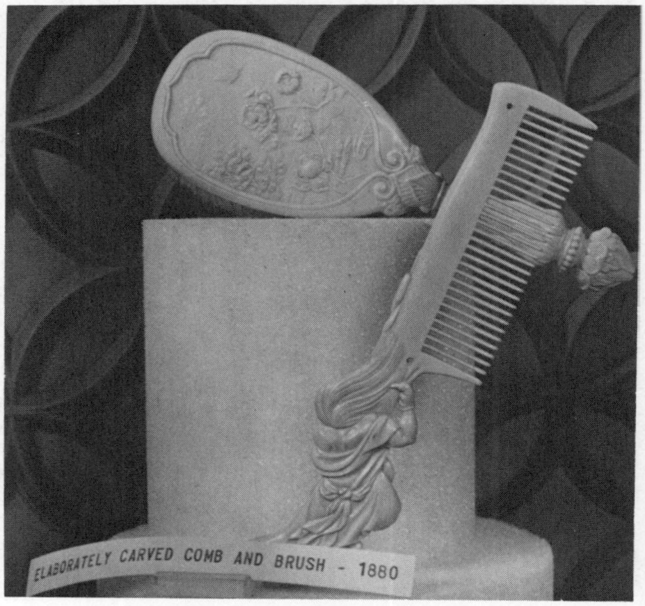

Fig. 1-2. Example of an early plastics application. (*Courtesy: Celanese Plastics Co.*)

ADVANTAGES AND DISADVANTAGES

Plastics must be considered as materials of construction—a raw source of material that can be adapted to the needs

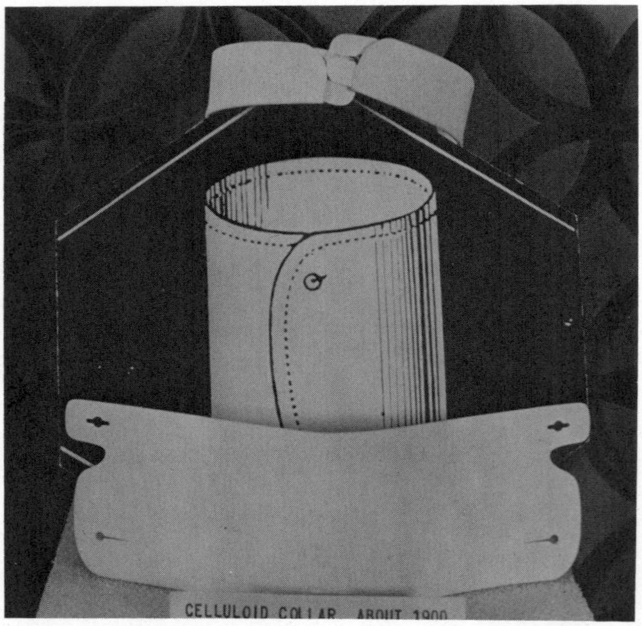

Fig. 1-3. Celluloid collars. (*Courtesy: Celanese Plastics Co.*)

of man, just as iron, birch, copper, glass, and brick are used for the improvement of man's existence. They have significant positive values and properties, but they also have deficiencies, both minor and major, as is the case for any other materials. The successful application of their strengths and an understanding of their weaknesses allow engineers to transform all the older materials into useful end products, and engineers and fabricators apply these same principles to plastics.

Plastics are no longer just useful materials for the fabrication of toys and kitchenware, nor are they alternatives or substitutes. They now represent a very important and highly versatile group of engineering materials. They are used in all fields of man's endeavors, from prefabricated homes to cable insulation, from gears and bearings to truck and car bodies, from boat hulls to printed circuits, from sports equipment to satellites, from paints and coatings to prosthetic devices (see Figures 1-5 and 1-6).

There is a wide variation in properties among the materials which are classified as plastics. This makes it difficult to discuss the advantages and disadvantages when comparing certain plastics to other materials or even with other plastics. We can only generalize, and for every advantage cited, a counter disadvantage can probably be found, and vice versa. However, let us itemize a few of these generalizations for descriptive purposes.

Most plastics can be easily fabricated into items with highly intricate shapes, using special techniques which are suitable for mass production. Many plastics can also be worked using common shop techniques. Plastics are, on the other hand, often difficult to repair if broken. Plastics are

Fig. 1-4. Early plastics applications in automobiles. (*Courtesy: Celanese Plastics Co.*)

Table 1-1. Significant Milestones in Plastics Development: 1839–1975.*

1839	Goodyear discovered vulcanization of rubber.
1868	"Celluloid," first pyroxylin plastic (Hyatt).
1909	Baekeland granted his "Heat and Pressure" patent for phenolic resins.
1921	Rayon introduced commercially.
1922	Staudinger began work on macromolecules.
1923	Polysulfide introduced commercially.
1925	Nitrocellulose lacquers introduced commercially.
1926	Alkyd introduced.
	Aniline-formaldehyde introduced in U.S.
	Eckert and Zeigler marketed plastics injection molding machine.
	L. E. Shaw initiated transfer molding.
1927	Cellulose acetate introduced commercially.
	Polyacrylate introduced commercially.
	Polyvinyl chloride introduced in U.S.
1928	Carothers started his research on polymers and polymerization.
	Urea-formaldehyde introduced commercially.
	Vinyl acetate and vinyl chloride copolymerized.
1930	Semon plasticized polyvinyl chloride.
1931	Carothers described condensation of polymers.
	Carothers discovered neoprene.
	Hyde began research on organo-silicon polymers.
1932	Screw preplasticization in injection molding patented.
1933	Crawford devised commercial synthesis for methyl methacrylate.
	Ellis patented unsaturated polyester resins.
	Ethyl cellulose introduced commercially.
1934	Fawcett and Gibson polymerized polyethylene under high pressure.
	Polyvinyl chloride (PVC) introduced commercially.
1935	Henkel made melamine-formaldehyde resins.
	Plasticized PVC introduced in U.S.
1936	Polystyrene introduced commercially.
1937	Automatic compression molding introduced commercially.
	Imperial Chemical Industries, Ltd. Patent 471,590 on high-pressure polyethylene.
	Polyamide introduced commercially.
	Polyurethanes first produced.
1938	Cellulose acetate butyrate introduced.
	Melamine introduced commercially.
	Nylon (polyamide).
	Plax started commercial blow molding of acetate and styrene.
	Polymerization of caprolactam (Nylon 6).
	Polystyrene.
1939	First patent (in Germany) on epoxy.
	Styrene-based plastics became commercial.
1940	Polyethylene introduced commercially.
1941	Rubber Reserve Co. (U.S. Government) initiated synthetic rubber industry in U.S.
	Urethane–polyester type–introduced in Germany.
	Whinfield and Dickson invented polyethylene glycol terephthalate.
1942	Dow Corning made silicone industrially.
	Polyester introduced commercially.
	Production of polyethylene started.
1943	Castan's patent issued on epoxy.
	Fluorocarbon introduced.
1945	Cellulose propionate introduced.
	Tetrafluoroethylene introduced commercially.
1946	Organosols and plastisols introduced commercially.
	Polyurethane elastomers introduced.
1947	Epoxy introduced commercially.
1949	Keaton brothers developed automatic screw transfer press.
1950	First large-scale production of (poly tetra fluoroethylene (PTFE).
1953	Staudinger received Nobel Prize for his work on macromolecules.
	Ziegler made polyethylene using organo-metallic catalyst.
1954	Cis-polyisoprene (synthetic natural rubber) was synthesized.
	Natta made high molecular weight stereoregular polypropylene.
	Polyurethane introduced in U.S.
	Ziegler produced polyethylene without pressure.
1955	Polypropylene introduced commercially.
1957	Polycarbonates developed commercially.
	Polypropylene introduced in U.S.
	Urethane–polyether type–introduced in U.S.
1959	Chlorinated polyether introduced commercially.
	Polycarbonates introduced commercially.
	Polyformaldehyde (acetal) introduced commercially.
1960	Isotactic polymer developments occur.
1961	Vinylidene fluoride introduced commercially.
1962	Phenoxy introduced commercially.
	Polyallomers introduced commercially.
	Ethylene propylene rubber introduced commercially.
1964	Ethylene vinyl acetate introduced.
	Ionomers introduced commercially.
	Polyimides introduced as a fabricated product.
	Polyphenylene oxide introduced commercially.
1965	Methyl pentene introduced commercially.
	Parylene introduced commercially.
	Polysulfone introduced commercially.
1970	Thermoplastic elastomers.
1972	Injection moldable polyesters.
1974	Aromatic polyamides.

*Extracted from *SPE Journal*, 1967.

not necessarily cheap, as is commonly thought. In fact, some are so expensive they are limited to those products used by the most sophisticated technology, high-cost enterprises such as the space program. Plastics are not typically as strong as metals, and they are more prone to dimensional changes, especially under load. They cannot survive extreme high and low temperatures as well as other materials. However, they are generally less dense and their strength-versus-weight ratios are highly favorable. Newer composites of plastics with such materials as glass or graphite fiber yield strengths comparable to and occasionally greater than metals, however, at a sacrifice in ease of fabrication (see Figure 1-7).

Plastics are good thermal insulators. They show high

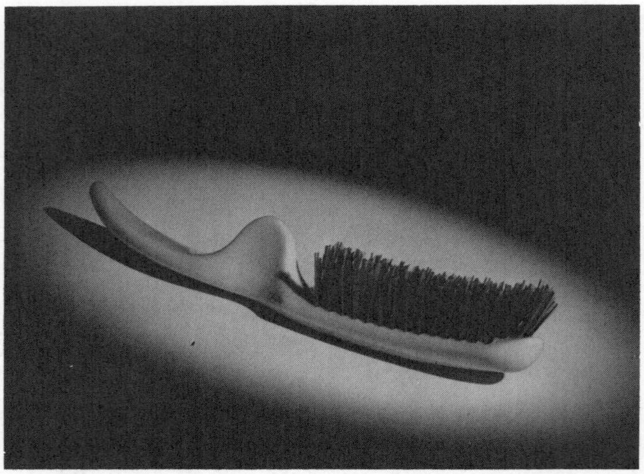

Fig. 1-5. Plastics brush handle of Tenite polypropylene. (*Courtesy: Eastman Chemical Products, Inc.*)

Fig. 1-6. Reinforced-nylon chain saw handle. (*Courtesy: Fiberfil Division, Dart Industries, Inc.*)

mechanical dampening. They are much more resistant to corrosion and chemical attack. They can be produced transparent, translucent, or opaque in a wide range of colors and tints. Many, however, deteriorate when exposed to sunlight, weather, or ultraviolet light. Plastics have exceptional electrical resistance characteristics with correspondingly adequate mechanical properties and ease of fabrication.

These generalizations indicate some of the advantages and disadvantages of plastics. Manv of their end uses depend less on these general considerations than on a specific and important characteristic of an individual plastic. For example, low coefficient of friction, complete chemical inertness, excellent dielectric loss properties, and ultralow temperature resistance may be of such significance and specificity as to warrant the use of only one plastic above all other material choices (see Figure 1-8).

Designing with plastics is not easy. The older materials

Fig. 1-7. Reinforced plastics truck cab. (*Courtesy: Ferro Corporation*)

Fig. 1-8. Plastics drums for chemicals. (*Courtesy: U.S. Industrial Chemicals Co.*)

have generated, design data—in some cases, over centuries, even millenia. Plastics have only a 100-year track record. Even so, what data is available is not always necessarily predictive, since changes in production methods or fabrication techniques can lead to variable product performance.

There have been many unsatisfactory applications of plastics. Some of them because plastics were not the best type of material, but others, more often than not, because the wrong plastic was chosen, the wrong fabricating technique was used, or quality was sacrificied to cost.

DEFINITIONS

Plastics are polymeric materials as are proteins, rubber, and many other naturally occuring materials. All are high molecular weight organic compounds whose structure can be graphically represented by a repeating, small constituent unit, the "mer." Whittington (1968, p. 179) provides a more rigorous definition of plastics as: "Materials that in their finished state contain as an essential ingredient a synthetic polymer of high molecular weight, are flexible or rigid solids but not elastomers in their finished state, and at some stage in their manufacture or processing into finished articles can be shaped by flow or by *in situ* polymerization or curing." He adds, "It has been suggested . . . that the plural form always should be used to distinguish 'plastics' as known in the plastics industry from other plastic materials such as molten glass, modeling wax and clay compounds in the unfired state."

The general classification of polymeric materials includes many rubbers, synthetic fibers, coatings, adhesives, chemical intermediates, as well as those commonly termed plastics. Except for the so-called inorganic polymers and silicones, the repeating units are based on carbon.

For the purposes of this survey, we shall consider only those polymers which are of use as materials of design and construction, are not bioactive (proteins), and may only incidentally have rubberlike characteristics.

REFERENCES AND BIBLIOGRAPHY

Celanese Plastics Magazine. Celanese Plastics Co., March 1965, **23**(1).

Masson, D., ed. *The Story of the Plastics Industry.* 12th Rev. Ed. The Society of the Plastics Industry, Inc., New York, January 1972.

Phillips Petroleum Company. *Glossary of Plastics Terms.* 3rd Ed. Bulletin 2108, Phillips Petroleum Co., Chemical Dept./Plastics Division, Bartlesville, Oklahoma.

Whittington, L. R. *Whittington's Dictionary of Plastics.* Technomic Publishing Co., Inc., Stamford, Connecticut, 1968.

Winding, C. C., and Hiatt, G. D. *Polymeric Materials.* McGraw-Hill Book Co., New York, 1961.

2
Structure and Characteristics of Polymers

TYPES OF POLYMERS

There are a number of ways to typify a carbon-based polymer. Three are based on the arrangement of the atoms in the polymer. Two are based on the nature of the chemical reaction used to generated the polymer.

Straight-Chain or Linear Polymer

If we draw a picture of elemental carbon in its 4-valence state, we see that we can connect one carbon to another in a long continuous chain through two of the available bonding sites.

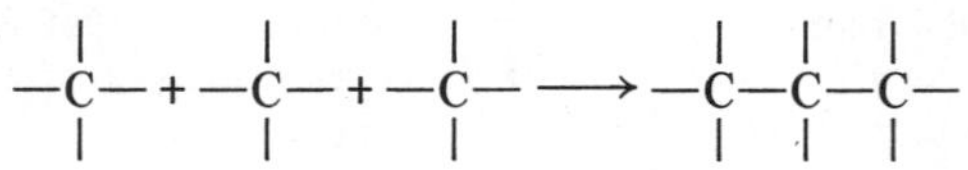

Structurally, the chain would be more apt to look like

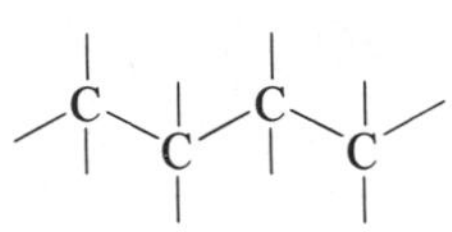

or, for that matter, any of a number of other spatial configurations, especially when we are talking about chains many thousands of units long. In fact, these various configurations will be constantly changing due to molecular motions such as vibration, stretching, or bending. The nearest visual representation of such a condition would be a bead key chain on a vibrating glass plate (see Figure 2-1a, b, and c.) However, since we can hypothetically represent the structure on a chalkboard as a straight length, these types of polymers are referred to as "straight-chain" or "linear" polymers.

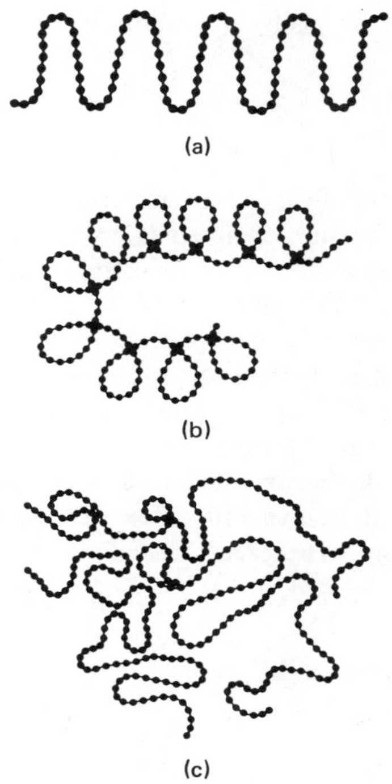

Fig. 2-1. Spatial distributions of key chains on vibrating glass plate.

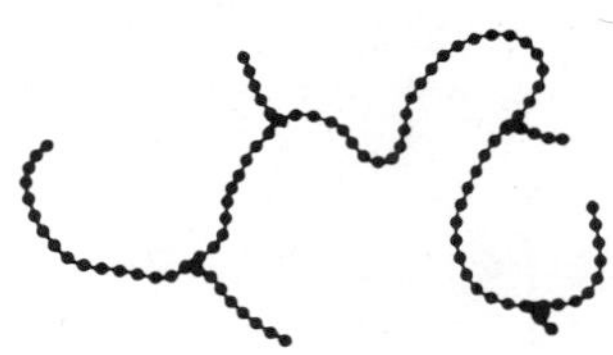

Fig. 2-2. Branched polymer.

Branched Chain Polymer

The remaining two bonds on each carbon atom may or may not have other substituent groups attached. These substituent groups may be from one to a few carbons in length. Such substituted structures are still referred to as linear polymers. These "branches" are generally few in number, occur irregularly along the main chain, and can only be joined to one main chain. The term "branch-chain" is often used to describe these linear polymers. (See Figure 2-2.)

Network Polymer

Some of the remaining bond sites on a linear chain can be interconnected with bond sites on other chains. This may occur as the growth of the chains proceeds from inception, or may be induced via reaction of specific reactive sites associated with the chains. When such interconnection occurs we have formed a "network" polymer. The process of forming these interconnections is called "crosslinking" and, depending on the frequency of occurrence and spacing of the interconnections, we can have a "lightly crosslinked" or "highly crosslinked" polymer (see Figures 2-3 and 2-4.)

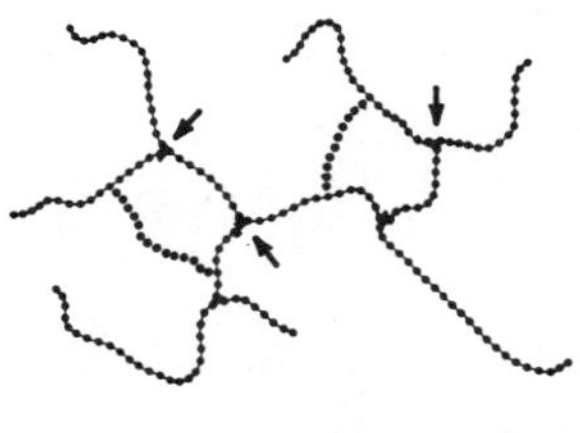

Fig. 2-3. Lightly crosslinked polymer.

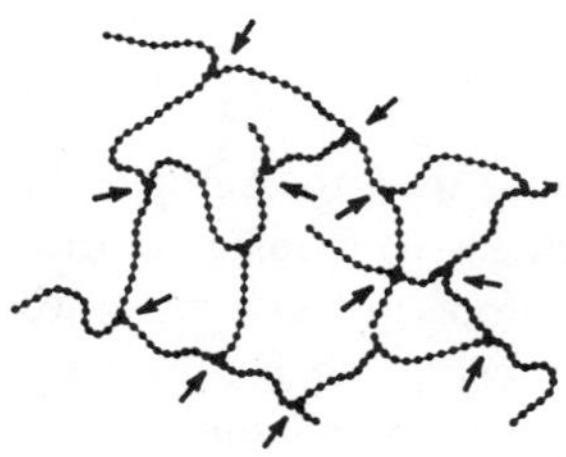

Fig. 2-4. Highly crosslinked polymer.

Addition Polymer

Polymers are also classified by referring to the chemical reaction mechanism used to form the basic structure. A double bond between two carbon atoms can be broken and caused to react with other reactive materials.

$$\begin{array}{c} H\quad H \\ |\quad\ | \\ C=C \\ |\quad\ | \\ H\quad H \end{array} + Cl_2 \longrightarrow Cl-\begin{array}{c} H\quad H \\ |\quad\ | \\ C-C \\ |\quad\ | \\ H\quad H \end{array}-Cl$$

This type of reaction is called an "addition" reaction. When we react such double bond-containing molecules with each other, we can generate consecutive addition reactions until all the molecules have linked together to form a long chain-like molecule (i.e., a polymer) of significantly increased molecular weight.

$$nCH_2{=}CH_2 \longrightarrow -\!\!(CH_2CH_2)_{\overline{n}}$$

Values of n may vary from 10 to 10,000 and more. Such a polymer formed as a result of an addition reaction is called an "addition" polymer.

Where more than one double bond exists in a molecule (e.g., butadiene, $CH_2{=}CH{-}CH{=}CH_2$), we can first generate a linear polymer chain, $-CH_2{-}CH{=}CH{-}CH_2-$, which has as part of its backbone remaining double bonds. These double bonds may then participate in addition reactions with double bonds from other chains and form points of cross-linking, thus forming a network polymer.

$$\begin{array}{l} \qquad\quad\ | \\ -CH_2-CH-CH-CH_2- \\ \qquad\qquad\qquad\ | \\ \qquad -CH_2-CH-CH-CH_2- \\ \qquad\qquad\qquad\quad\ | \end{array}$$

Condensation Polymer

Many organic molecules contain highly reactive functional groups. Reactions between these functional groups can occur and, when such reactions are constructed so as to yield high molecular weight materials, we can generate polymers.

For example, if we react an organic acid with a primary amine, we can generate an amide and water.

$$R-\overset{O}{\overset{\|}{C}}-OH + H_2N-R' \longrightarrow R-\overset{O}{\overset{\|}{C}}-\overset{H}{\overset{|}{N}}-R' + H_2O$$

This type of reaction is called a "condensation" reaction and is characterized by the reaction of the two functional groups to form a product and the elimination of a simple molecule.

The above reaction between a monofunctional acid and amine gives an amide, which contains neither acid nor amine functional groups. If now, a dibasic acid is reacted with a diamine, a series of condensation reactions can be generated. The product is a new dibasic acid, or a new diamine, of increased size.

$$HO\overset{O}{\overset{\|}{C}}R\overset{O}{\overset{\|}{C}}OH + H_2NR\,NH_2 + HO\overset{O}{\overset{\|}{C}}R\overset{O}{\overset{\|}{C}}OH \longrightarrow$$
$$HO\overset{O}{\overset{\|}{C}}R\overset{O}{\overset{\|}{C}}\overset{H}{\overset{|}{N}}R'\overset{H}{\overset{|}{N}}\overset{O}{\overset{\|}{C}}R\overset{O}{\overset{\|}{C}}OH + 2H_2O$$

The reaction sequence can continue until a polymer molecule is created. Such a polymer formed as a result of a condensation reaction is called a "condensation" polymer.

It should not be too difficult to visualize that if we react molecules that have more than two functional groups per molecule, we can generate network polymers via the condensation route.

TERMINOLOGY

Monomers and Polymers

The term *monomer* is used to define the small reactant molecules from which the polymer is made. If we have an addition polymer, then both the monomer and the repeating unit within the polymer have the same composition,

$$nCH_2{=}CH_2 \longrightarrow -\!\!(CH_2CH_2)_{\overline{n}}$$

If the polymer is a condensation polymer made by reaction of a single substance, the repeating unit will have a different, but similar composition to the monomer. If it is a condensation polymer made by the reaction of two substances, both these substances are called monomers and the repeating unit will contain the basic structure of both.

$$n\mathrm{HO\overset{O}{\overset{\|}{C}}R\overset{O}{\overset{\|}{C}}OH} + n\mathrm{H_2NR'NH_2} \longrightarrow -\!\!\left(\mathrm{O\overset{O}{\overset{\|}{C}}R\overset{O}{\overset{\|}{C}}\overset{H}{\overset{|}{N}}R'}\right)_{\!n} + 2n\mathrm{H_2O}$$

Polymers with a single repeating structure are called *homopolymers.*

Copolymers

Polymers with more than one repeating structure are called *copolymers*. These can be made by the addition polymerization of two ethylenic monomers or by the reaction of three different multifunctional monomers.

Example 1:

$$n\mathrm{CH_2{=}CH_2} + m\mathrm{\overset{CH_3}{\overset{|}{C}}H{=}CH_2} \longrightarrow -\!\!\left(\mathrm{CH_2CH_2}\right)_{\!n}\!-\!\!\left(\mathrm{\overset{CH_3}{\overset{|}{C}}H{-}CH_2}\right)_{\!m}$$

Example 2:

$$\mathrm{HO\overset{O}{\overset{\|}{C}}R\overset{O}{\overset{\|}{C}}OH} + \mathrm{H_2NR'NH_2} + \mathrm{HO\overset{O}{\overset{\|}{C}}R''\overset{O}{\overset{\|}{C}}OH} \longrightarrow -\!\!\left(\mathrm{\overset{O}{\overset{\|}{C}}R\overset{O}{\overset{\|}{C}}\overset{H}{\overset{|}{N}}R'}\right)\!- \text{ and } -\!\!\left(\mathrm{\overset{O}{\overset{\|}{C}}R''\overset{O}{\overset{\|}{C}}\overset{H}{\overset{|}{N}}R'}\right)\!-$$

In Example 1 above, let us designate that part of the copolymer $-\!\!\left(\mathrm{CH_2CH_2}\right)_{\!n}$ by the letter **A** and the other part of the copolymer $-\!\!\left(\mathrm{\overset{CH_3}{\overset{|}{C}}H{-}CH_2}\right)_{\!m}$ by the letter **B**. In a given chain, the relationship between **A** and **B** can take a number of forms.

If **A** and **B** are spaced randomly along the chain (i.e., **AABBBBABBAAAA**), we call such a copolymer a *random* copolymer. If, on the other hand, each repeating unit is spaced alternately along the chain (i.e., **ABABABAB**), we call such a copolymer an *alternating* copolymer.

Similarly, if we construct the polymer in the following ways we can describe their construction as:

a. **AAABBBBAAABBBB** — *block* copolymer

b. **AAAAAAA** — *graft* copolymer
    B
    B
    B

c. **AAABBBB** or **AAABBBBCCC** — *terpolymer*
    C
    C
    C

It is possible then, using the above definitions, to construct an "alternating graft copolymer" or a "random block copolymer."

Alloys

The term alloying is used to represent the physical admixture of two polymers. The object is to combine resins with widely differing properties into a homogeneous mass, and at the same time overcome phase-separation problems associated with blending of polymers that differ significantly in molecular weight.

Alloying occurs by establishing sites of compatibility along molecular chains such that a degree of physical cross-linking occurs. This is not a true chemical process, but is achieved by addition of special "wetting" agents or compounding under high shear conditions. Figure 2-5 compares the various combining techniques of polymers.

Tacticity

When we encounter a monomer of the general type $\underset{H}{\underset{|}{\overset{H}{\overset{|}{C}}}}{=}\underset{H}{\underset{|}{\overset{R}{\overset{|}{C}}}}$, we can generate three types of polymer structure. In most cases, polymerization will yield a random arrangement of the R group along the chain. Such a polymer is *atactic*. When, however, certain stereospecific catalysts and reaction conditions are chosen, we can generate a polymer where the R groups are all on the same side of the chain,

$$-C-\overset{R}{\overset{|}{C}}-C-\overset{R}{\overset{|}{C}}-C-\overset{R}{\overset{|}{C}}-C-\overset{R}{\overset{|}{C}}-.$$

Such a polymer is *isotactic*. We can also generate a polymer where the R group regularly alternates on either side of the chain, $-\overset{R}{\overset{|}{C}}-C-\underset{R}{\underset{|}{C}}-C-\overset{R}{\overset{|}{C}}-C-\underset{R}{\underset{|}{C}}-C-$. Such a polymer is *syndiotactic*.

The catalysts used to promote specific tacticity were developed by Karl Ziegler in Germany around 1953. Dr. Giulio Natta in Italy shared a Nobel prize with Ziegler for his pioneering work in stereospecific polymerization (principally of polypropylene) using Ziegler catalysts.

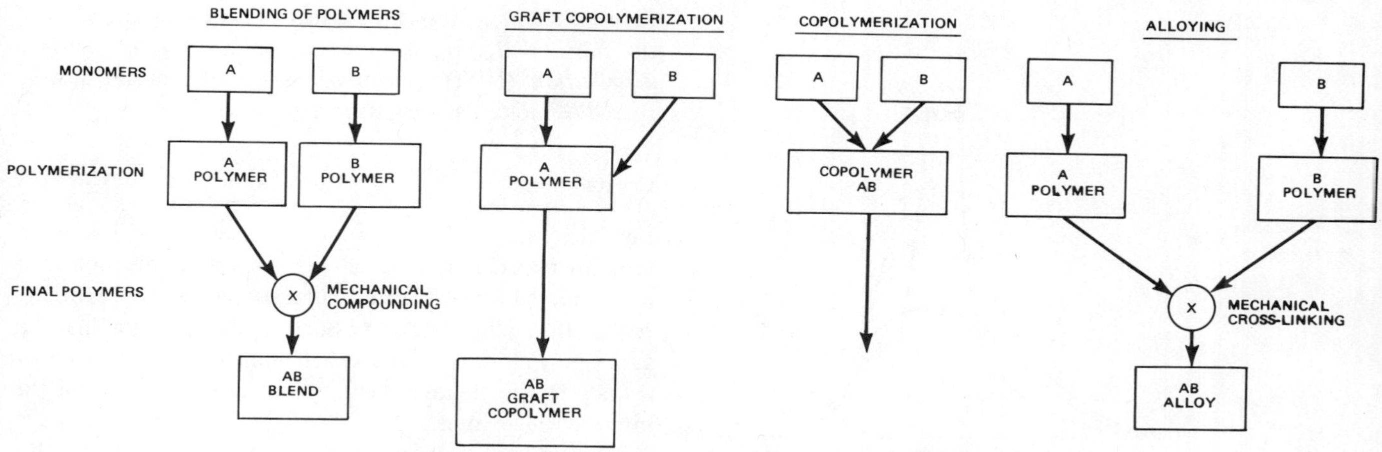

Fig. 2-5. Methods of producing combinations of polymers. (Mack, 1971)

Thermoplastic and Thermoset

From a user's point of view, an important classification of plastics is that of thermoplastic and thermoset. Thermoplastic materials are those materials which undergo no permanent change on heating. They flow and may be molded into a shape which they retain on cooling. The material will flow again when reheated and can be remolded. Common nonplastic examples of this phenomenon are ice and wax. Thermoplastic polymers contain linear molecules that are not crosslinked. These linear molecules can be crosslinked under some circumstances, at which time the material loses its thermoplastic character and becomes thermoset.

Thermoset materials are capable of a high degree of crosslinking. They are generally molded in a partially polymerized state so that they can flow when heated. During the molding process further polymerization occurs, and the plastic becomes highly crosslinked and can no longer flow. Common nonplastic examples are cooking a raw egg or baking bread dough. Thermoset materials in the partially polymerized state are often called *resins*. The conversion to the final state is known as *curing* (in rubber technology, this conversion is known as vulcanization).

To summarize: *a thermoplastic material is a polymeric material that can be formed upon the application of heat and pressure and can be continually reformed with further application of heat and pressure. A thermoset material is a polymeric material which can be formed by the application of heat and pressure, but cannot be reformed upon further application of heat and pressure.*

STRUCTURE OF POLYMERIC MATERIALS

Molecular Weight

For a crosslinked polymer, the term molecular weight has no significance. The material does not exist in the form of discrete molecules but, theoretically at least, as one huge three-dimensional fish net; that is, one massive interconnected molecule of indeterminate size. Linear polymers, on the other hand, are discrete, although tangled, molecules, and their size can be expressed in terms of molecular weight. More precisely, the molecular weight is a function of the degree of polymerization, which is the number of structurally recurring units in the chain, each with its own specific molecular weight.

In a substance which consists of small discrete molecules, the molecular weight of each molecule is the same, because each one has the same composition. One molecule of aniline has a molecular weight of 93 and a formula, $C_6H_5NH_2$. This does not change, regardless of the sample size containing this molecule.

Molecular Weight Distribution (MWD). The generalized formula for a linear polymer is $-\!\!(M)\!\!-_n$ where M is the repeating unit and n is the degree of polymerization. In a given sample of a polymeric material, molecules with varying n values will be present and n can typically range in the thousands. The spread of these n values will depend on the nature of the polymer, the method of polymerization, and the processing history of the polymer. In a sample with an average n value of 5,000, there will be a percentage of molecules with n values of 50 and of 50,000. (See Figure 2-6.) Thus, a molecular weight measurement of a given sample of polymer will only represent an average value of the weights of the individual molecules.

Determination Methods. There are many methods of determining molecular weight of polymers, and different methods give different values. Thus, comparisons must be made only when those methods yielding the same type of average result are used.

Where a wide range of molecular sizes exist, the different methods yield different results because the average molecular weight is actually based on two different methods of

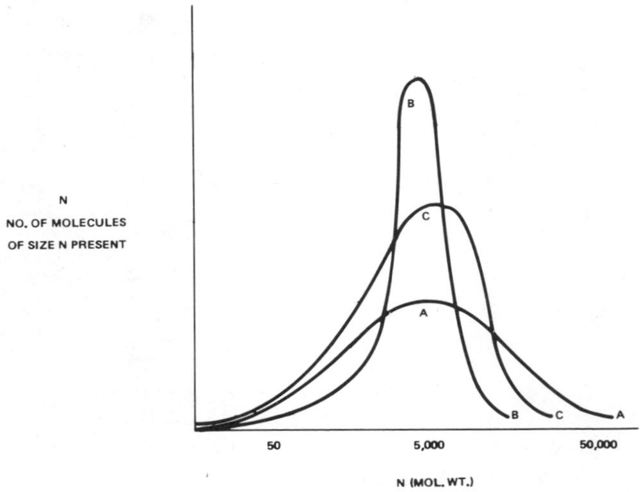

Fig. 2-6. Molecular weight distribution curves for various polymers. Sample A represents a polymer with a typical bell-shaped distribution of molecular sizes with an average value of 5,000. Sample B also has an average value of 5,000; its spectrum of molecular sizes is, however, considerably narrower. Sample C represents a polymer having an average molecular weight of 5,000, but is molecular distribution is skewed toward the large molecular sizes.

averaging. In end-group and osmotic-pressure methods, the results are determined by the number of molecules present. In viscosity, light-scattering and size-exclusion (chromatographic), methods, the values obtained are influenced by the molecular size, shape, and flexibility, as well as by the actual number of molecules.

When the molecular weight of a substance composed of simple molecules of identical size is measured by both of these method types, the resultant molecular weight values are usually the same, since the size and shape factors are reduced to an experimentally evaluated constant. For a widely variant polymer, the molecular weights obtained by the two methods may be significantly different.

Number and Weight Average Molecular Weights. The *number average molecular weight* is defined as

$$\overline{M}_n = \Sigma N_i M_i$$

where N_i is the mole fraction of the fraction that has a molecular weight M_i. This is the molecular weight in the commonly accepted sense.

The *weight average molecular weight* is defined as

$$\overline{M}_w = \Sigma W_i M_i$$

where W_i is the weight fraction of the fraction that has a molecular weight M_i.

For molecules of the same size, the two methods give identical results; but when there is a size distribution, $\overline{M}_w$ is greater than $\overline{M}_n$, and this ratio increases as the distribution broadens. $\overline{M}_n$ imparts greater importance to small molecules, while $\overline{M}_w$ stresses the effect of large molecules. Thus, the ratio of $\overline{M}_w/\overline{M}_n$ is a measure of the *molecular weight distribution* (MWD). High values would indicate a wide spread of molecular weights in a given sample.

Crystallinity

For solid materials consisting of small molecules, intermolecular forces are large enough to restrain the molecules in an ordered array. The molecular arrangement is essentially regular throughout the mass. Such solids are crystalline and, as such, will melt at the exact temperature which corresponds to the thermal energy input needed to disrupt the intermolecular forces.

Polymers, by their nature extremely large and complex molecules, show no sharp melting point. Even though a polymer is a solid because it is hard and rigid, it may in fact be completely amorphous with no regular molecular arrangement. Such a material is often thought of as an ultrahigh-viscosity liquid, whose viscosity decreases with increases in temperature.

Crystallites and Spherulites. Because of their length and complex configuration, polymer molecules can align themselves in an amorphous matrix such that select areas become more ordered than the surrounding environment. Such ordered regions are called *crystallites*. The crystallinity of a polymer refers to the degree that such ordered crystallite regions are present.

A typical crystallite has dimensions less than the length of the extended polymer molecules so that a given molecule can traverse more than one crystallite region, as well as the amorphous regions in between. The transition from crystallite to amorphous area is not sharp and delineated, and semi-ordered transition regions exist as well.

An accepted feature of most polymer crystals is that the covalent axis of the polymer chains is perpendicular to the longer dimensions of the crystals. It is believed that the polymer folds over on itself after about a hundred chain atoms have entered the lattice (see Figure 2-7).

A number of hypotheses have been proposed to explain the configuration of the chains within a crystalline area.

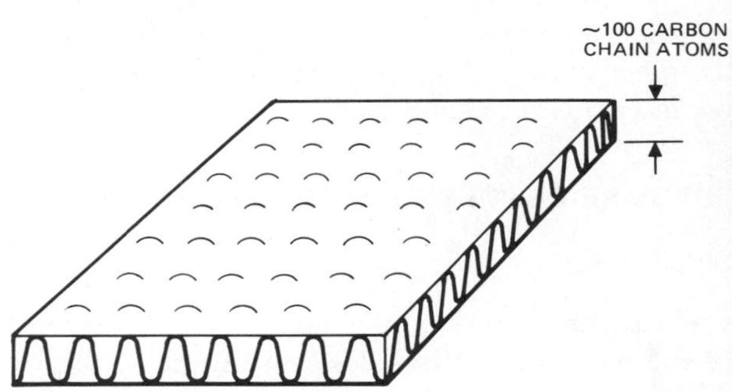

Fig. 2-7. Sketch of the folded-chain description of polymer crystals.

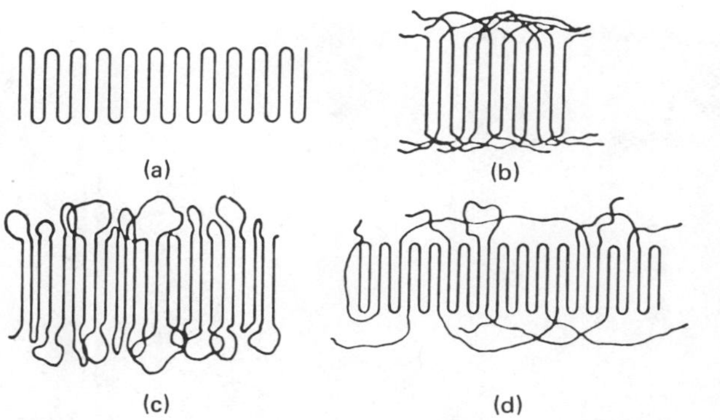

Fig. 2-8. Schematic two-dimensional representations of models of the fold surface in polymer crystals: (a) sharp folds, (b) "switchboard" model, (c) loose loops with adjacent reentry, (d) a combination of several features. (From *Textbook of Polymer Science* by Billmeyer. Used with permission of A. Peterlin, contributor/author.)

Figure 2-8 shows a two-dimensional schematic of four proposed models, the last being currently most favored. Figure 2-9 shows a representation of how the chains align in an orderly fashion to form a single polyethylene crystal. Figure 2-10 shows some photomicrographs of a series of high polymer crystals.

The relative percentages of these crystalline regions vary widely with the nature of the polymer and the history of the sample being examined. There are no known commercial polymers that are 100% crystalline.

Often, a number of crystallites will order themselves further into larger groups. Such groups are called *spherulites*.

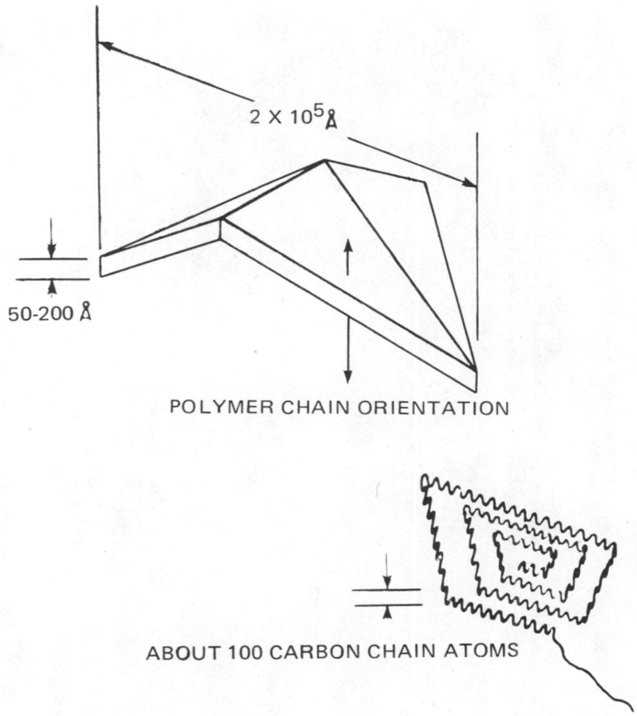

Fig. 2-9. A large single crystal of polyethylene might have the dimensions indicated here. The growth mechanism involves the folding over of the planar zigzag chain on itself at intervals of about every 100 chain atoms. A single crystal may contain many individual molecules. (From *Principles of Polymer Systems* by Rodriguez. Copyright 1970 by McGraw-Hill, Inc. Used with permission of McGraw-Hill Book Co.)

Fig. 2-10. High polymer crystals. (Wunderlich, 1963)

Fig. 2-11. Spherulites are dendritic crystals that have formed in solid polyethylene as it cooled from the molten state. (Wunderlich, 1964)

Spherulites are observed in a polarizing microscope. They appear as almost circular birefringent areas possessing a dark Maltese cross pattern (see Figure 2-8.) The polymer chains are tangentially ordered around the nuclei despite the fact that the spheroidal area consists of a collection of crystallites. Spherulites rarely exceed 100 microns in diameter.

Phase Transitions. There is a marked change in such physical properties as specific volume, heat capacity, and viscosity at the melting point of a low molecular weight crystalline solid. This temperature is usually very clear-cut and sharp. With a polymer, however, these changes occur over a temperature range. Essentially, the melting point (T_m) of a polymer is that temperature at which the crystallite regions lose their order. Since there is a spectrum of crystallite regions which will disperse, the energy required for such dispersal is spread over a wider range.

Glass Transition. The properties of a polymer may change over a temperature range due to another mechanism as well—*glass transition* (T_g). This point manifests itself in the material as a change in properties from those of a hard, brittle, glassy substance to those of a soft, flexible material. If we plot density or specific volume versus temperature (Figure 2-12), a change in slope will occur at T_g. *Below the T_g, the polymer is stiff and dimensionally stable; above it, it becomes more rubberlike.*

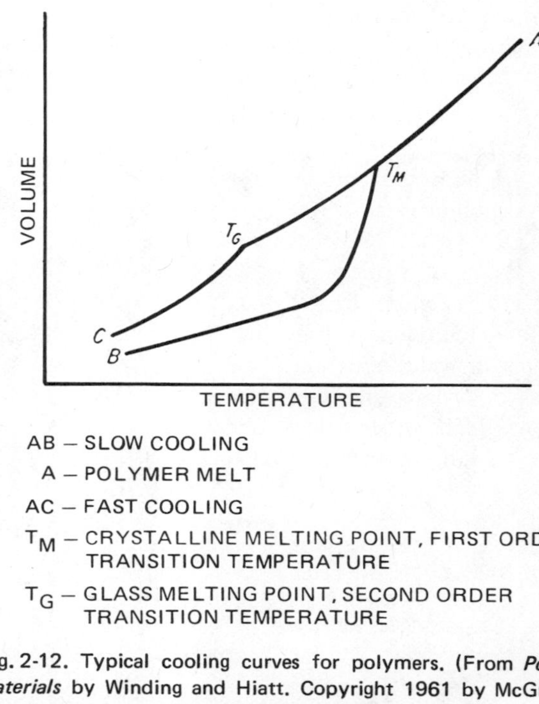

AB – SLOW COOLING
A – POLYMER MELT
AC – FAST COOLING
T_M – CRYSTALLINE MELTING POINT, FIRST ORDER TRANSITION TEMPERATURE
T_G – GLASS MELTING POINT, SECOND ORDER TRANSITION TEMPERATURE

Fig. 2-12. Typical cooling curves for polymers. (From *Polymeric Materials* by Winding and Hiatt. Copyright 1961 by McGraw-Hill, Inc. Used with permission of McGraw-Hill Book Co.)

Glass transition effects are greatest with amorphous polymers. Below the T_g, thermal energy is insufficient to allow chain atoms to do more than vibrate about their equilibrium positions. The polymer chains are "fixed" in position and the polymer is glasslike and brittle. More and more movement in the chains is generated as the temperature passes up, through, and beyond T_g; and the polymer properties reflect this molecular motion by appearing, first, tougher and more flexible and then, finally, more rubbery in nature.

The temperature of these changes relative to room temperature is an effective determinant for the end use of a polymer. If T_m is below room temperature (i.e., its T_g is considerably below room temperature), the polymer will probably show rubbery or elastomeric characteristics under normal use conditions. Conversely, polymers used as rigid molding plastics or tough fibers generally have a T_g at room temperature or above, and a considerably elevated temperature is required to soften these materials.

There are some polymers, primarily highly crystalline ones, which show additional transition points at temperatures below T_g. These transitions are often referred to as α, β, γ transitions. These transitions are attributed to motion effects of short segments of the main polymer chain.

Chain-Straightening and Molecular Slippage. Much of the significance of glass transition relates to the interrelationships between the molecular chains and their segments. Two facets of these associations that are of specific interest as they concern plastics properties (see section on testing and properties) are chain-straightening and molecular slippage. Chain-straightening refers to the elongation of a given molecular chain from its equilibrium position at rest (unstressed) to a new dimension under stress. Molecular slippage refers to the movement of molecules (or molecular segments) past adjacent molecules due to the application of stress. As the polymer matrix is stressed, both phenomena may occur depending on the degree of crystallinity, or cross-linking, if any.

To illustrate these effects, we can imagine grabbing an entangled pile of string and pulling the ends (see Figure 2-13a,b and c). Those strings which are being held at both ends will extend (chain-straighten) until they reach their full length, at which time motion of these strings will stop. Those strings which are being pulled at one end only will slide by both the other strings being pulled in the opposite direction and the strings that are not being gripped at all (i.e., molecular slippage).

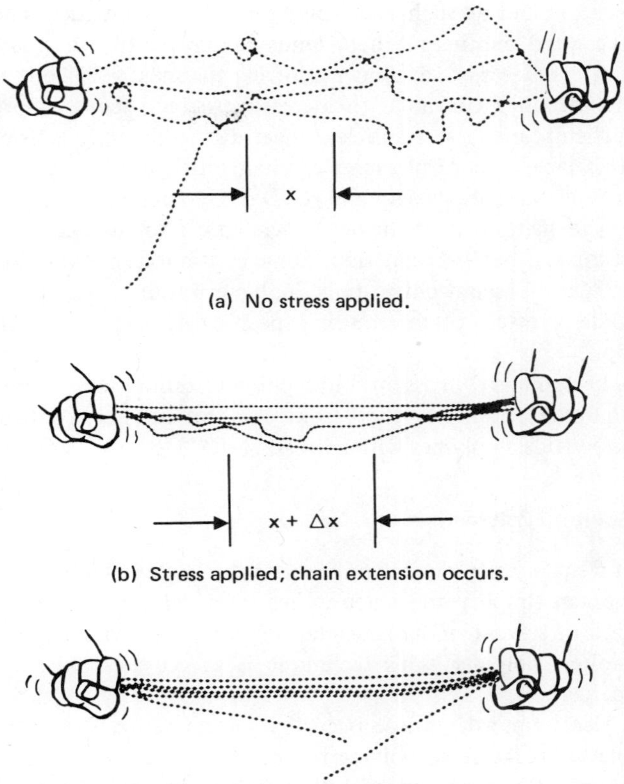

(a) No stress applied.

(b) Stress applied; chain extension occurs.

(c) Continued stress; molecular slippage occurs.

Fig. 2-13. Effect of tensile stress on molecular configuration.

POLYMERIZATION METHODS

There are four basic methods of producing a polymer. Many factors influence the choice of a particular method to produce a polymer. In many instances, the nature of the reaction chemistry dictates the specific method to be used. In other instances, the resultant polymer (low or semiviscous liquid, friable or rigid solid) may limit one's choice.

It is not within the scope of this survey to detail the various methods used for each plastic we are to discuss, except for those occasional situations that are sufficiently unique to warrant more vigorous treatment (e.g., interfacial condensation of nylon). Rather, we shall examine the basic procedures in general and illustrate the key advantages and disadvantages of each, irrespective of any plastic that might be made via these procedures. The interested reader is referred to any basic organic chemistry text for more detailed descriptions.

Bulk Polymerization

From the point of view of equipment, complexity, and economics, the simplest method is *mass or bulk* polymerization. This procedure merely allows the monomer to react at a predetermined reaction temperature, with or without catalysts, to form the polymer.

Theoretically, the monomer can be a gas, liquid, or solid, but, practically, almost all mass polymerizations take place in a liquid phase.

The polymer may be either soluble or insoluble in the monomer. If the former, then the mass viscosity continually increases until the final degree of polymerization is obtained. If the latter, the polymer will precipitate from the remaining unreacted monomer and can be subsequently separated.

A serious drawback to bulk polymerization is control of heat of reaction. Both monomer and polymer are heat insulators and as such have low thermal conductivities. The generated exothermic heat tends to stay within the mass and is not easily withdrawn. Stirring the mass helps, but as the viscosity continues to increase, stirring becomes more difficult and a less efficient heat-dissipating mechanism. This lack of control causes resultant difficulty in the control of molecular weight and MWD of the final polymer.

The method does, however, lend itself for use in small casting or batch production. Some continuous methods are in use. Gaseous-phase bulk polymerization takes place under pressure often requiring specific catalysts for conversion.

In summary, mass or bulk polymerization uses simple equipment, is highly exothermic with difficult heat control, and yields a polymer with a broad MWD.

Solution Polymerization

If we now carry out the polymerization in a suitable solvent, we can simplify the reaction heat removal, since the solution of solvent, monomer, and polymer is less viscous than molten polymer. This technique is called *solution* polymerization.

Heat removal can also occur by solvent reflux. However, the temperature of polymerization is restricted to the boiling point of the solvent, and, in many cases, this can cause the time for reaction completion to increase. Other drawbacks to this procedure include: production of a lower molecular weight polymer, separation and recovery of the solvent may be complex and expensive (especially with recent environmental restrictions), and the solvent may be flammable and/or toxic.

If a solvent can be found in which the monomer is soluble but the polymer insoluble, the resultant polymer precipitation facilitates the separation steps.

In summary, one can control heat more readily in solution polymerization, although higher molecular weight polymers are difficult to produce. A solution of the polymer itself may be marketable, but the purification of solid polymer involves complex and potentially hazardous procedures.

Emulsion Polymerization

If the monomer can be polymerized in a water emulsion, then we can retain the low viscosity needed for good heat control without the hazards associated with the handling of solvents. Such a procedure is called *emulsion* polymerization.

Reaction rates and molecular weights are usually higher with this method than with mass or solution polymerization. The MWD is often quite narrow, water is cheaper and less hazardous than solvent, and recovery steps are not as complex. However, ingredients must be added to aid emulsification (emulsifying and stabilizing agents). This added contamination and the requirement of a drying step for the polymer constitute significant disadvantages of the process.

Suspension Polymerization

Finally, there is *suspension* polymerization in which the monomer and forming polymer globules are maintained in suspension by agitation without the use of an emulsifying agent. The polymer beads are formed by coalescence and their size regulated by suspension stabilizers and the amount and intensity of agitation. The final beads must be screened out of the liquid phase and washed and dried before they can be used, although suspensions can be, and are, marketable. Control of exothermic heat is good, and high molecular weight polymers with relatively narrow MWD's are possible.

REFERENCES AND BIBLIOGRAPHY

Billmeyer, F. W. Jr. *Textbook of Polymer Science*. 2nd Ed. Wiley-Interscience, New York, 1971.

Brydson, J. A. *Plastics Materials*. 3rd Ed. Robert E. Krieger Publishing Co., Inc., Huntington, New York, 1975.

Cowie, J. M. G. *Polymers: Chemistry and Physics of Modern Materials.* Intext Educational Publishing, New York, 1973.

Goodyear Chemical Division. *Molecular Weight Distribution Analysis*. Tech Book Facts R-1 Methods, April 1965.

Mack, W. A. Resins mix it up in new combinations. *Packaging Digest*, November 1971, pp. 40–45.

Mark, H. F. Tailor-making plastics. *International Science & Technology*, March 1964, **27**:72–79.

Moore, W. R. *An Introduction to Polymer Chemistry*. Aldine Publishing Co., Chicago, 1963.

Rodriguez, F. *Principles of Polymer Systems*. TATA McGraw-Hill Publishing Co., Ltd., New Delhi, 1974.

Tsuruta, T., and Inoue, S. Well-ordered polymers. *Science & Technology,* November 1967, pp. 66–76.

Winding, C. C., and Hiatt, G. D. *Polymeric Materials*, McGraw-Hill Book Co., New York, 1961.

Wunderlich, B. The solid state of polyethylene. *Scientific American*, November 1964, **211**(5):80–94.

——. *J. Poly. Sci*., Part C, No. 1, pp. 41–64, 1963.

3
Testing and Properties of Plastics

How hard is steel? How brittle is glass? How flexible is bamboo? Will ceramic resist salt water and retain its insulation characteristics? These are typical of the questions asked by people who are going to use materials of construction. Since we have based our study of plastics on the fact that they too are materials of construction, we must be in a position to answer these questions as they relate to plastics.

The answers come from testing and are not always indisputable. Consequently, it would be well to become familiar with the test methods used in the plastics industry, their significance, and shortcomings.

Table 3-1 describes many of the tests used with plastics materials. (This list by no means covers the complete range of tests used. The Government currently lists over 3,000 tests for plastics alone.) We will discuss some of the more common and generally used ones.

MECHANICAL PROPERTIES

Tensile Properties

Tensile properties are one of the most important single indications of the strength of a material. The force necessary to pull the specimen apart is determined, along with the amount of stretching that occurs before breaking. Figure 3-1 shows a universal testing machine with a typical tensile "dumbbell" test specimen. Figure 3-2 shows typical tensile test specimens. Figure 3-3 shows various kinds of test jaws and fixtures used to run tensile tests on different plastics configurations.

Figure 3-4 shows a typical stress-strain plot of a plastics specimen when stressed in tension. This plot provides five descriptive parameters: (1) *tensile strength*—the maximum stress the material withstands at point of rupture; (2) *yield strength*—the stress at which nonelastic deformation begins; (3) *ultimate elongation*—the total amount of extension the sample undergoes; (4) *modulus of elasticity*—the slope of the initial steep portion of the curve, this represents the stress-strain ratio in the elastic region in which Hooke's law holds; and (5) *toughness*—or more precisely, the energy to break per unit volume, this is the integrated area under the curve and is a rough measure of the plastics' toughness.

Translating the data from such a diagram provides a terminology useful in describing various plastics. A hard plastic is one with a high modulus. A strong plastic would be represented by high tensile yield. A brittle plastic would generate a curve that terminates before reaching the yield point. A tough plastic is one with a high energy to break per unit volume; that is, a large area under the curve.

Examples of stress-strain curves for various types of plastics are shown in Figure 3-5.

Correlation between Tensile Strength and Polymer Structure. Certain generalizations can be made about the relationship between polymer structure and tensile properties, although these generalizations are by no means predictive. Highly crosslinked polymers and thermoplastics below T_g are hard and brittle. They can be strong, since the stress at break is very high even before a yield occurs. A thermoplastic above T_g is softer, since chain-straightening and molecular slippage will allow more extension. Such a material will yield, the yield being determined by the relative time dependencies of the straightening and slipping processes. If the rate of slip is slow compared to straightening, then considerable extension can occur before the nonelastic effect of molecular slip manifests itself. If the rates are equivalent, permanent set will occur at low stresses.

Permanent set is the change in dimension, expressed as a percentage of the original dimension, by which an elastic material fails to return to the original dimension after being stressed for a standard period of time.

If a plastic is crystalline, it will be harder than a completely amorphous plastic. The existence of crystallite regions increases the yield stress and modulus, but molecular slippage can still occur in the surrounding amorphous regions. In a lightly crosslinked polymer, chain-straightening can occur, but molecular slipping cannot. Such a material above its T_g is soft and strong, since large elastic extensions can occur.

Compressive Strength

The stress-strain curves of plastics in compression are the same as in tension, and the terminology is the same. However, the nature of the test piece now becomes significant. Comparisons between values are not always valid because some are taken at fracture (if fracture occurs) where they are precisely defined, but others are arbitrarily chosen at some degree of unacceptable distortion. In many plastics, the sample will not break but merely flatten to a disc shape. (See Figures 3-6 and 3-7.) There are some plastics whose tensile and compressive characteristics differ. An example is polystyrene, which is tough under compressive load but very brittle in tension.

Table 3-1. SPE Guide for Uniform Testing and Reporting of Identification Properties of Plastics Materials: Updated in 1972 and Issued by Engineering Properties and Structure Division.*

Property	ASTM test method[a]	Recommendations[b]	U.S. customary units	SI units	X[c]	s[c]
Apparent density	D1895					
Free flowing	Method A	—	—	g/cu cm		
Non-pouring	Method B	—	—	g/cu cm		
Bulk factor	D1895	—		Dimensionless		
Specific gravity	D792 Method A or B	—		Dimensionless		
Density	D1505	—	—	G/cu cm		
Mold shrinkage	D955	—	in. /in.	mm/mm		
Flow temperature Rossi-Peakes	D569 Procedure A	Used to classify cellulose acetate and cellulose acetate butyrate compounds	°C	°C		
Melt-flow rate, thermoplastics	D1238	State condition employed	—	g/10 min		
Molding index	D731	—	1 bf (at close time), sec	kgf (at close time), s		
Dielectric constant; dissipation factor at 60 Hz, 1 kHz, 1 MHz	D150	⅛-in.-thick specimens		Dimensionless		
Volume resistivity 1 min at 500 v	D257	⅛-in.-thick specimens	—	ohm-cm		
Arc resistance high voltage, low current	D495 Stainless steel electrodes	⅛-in.-thick specimens	sec	s		
Dielectric strength	D149	⅛-in.-thick specimens	v/mil	v/mm		
Short time	Sec. 6.1.1					
Step by step	Sec. 6.1.3					
Tensile strength	D638	⅛-in.-thick type 1 specimens; state testing speed used	psi State whether at yield or break	Pa		
Elongation	D638	⅛-in.-thick type I specimens; state testing speed used	% State whether at yield or break	%		
Modulus of elasticity, tensile	D638	⅛-in.-thick type I specimens; state testing speed used	psi	Pa		
Flexural strength	D790	State procedure employed	psi State whether at yield, at rupture, or 5% strain	Pa		
Tangent modulus of elasticity, flexural	D790	State procedure employed	psi	Pa		
Compressive strength	D695	—	psi State whether at yield, at failure, or at given strain	Pa		
Modulus of elasticity, compressive	D695	—	psi	Pa		
Impact resistance, Izod	D256 Method A	Specimen width, as tested, ⅛ in.	ft-1 bf/in. of notch	joule/m of notch		
Notch sensitivity	D256 Method D	Specimen width, as tested, ⅛ in.	ft-1 bf/in. (notch)-in. (rad)	joule/m (notch)-m (rad)		
Hardness, durometer	D2240	—	Scale reading; state durometer type and time under pressure			
Hardness, Rockwell	D785 Procedure A	—	Scale reading; state scale			
Haze	D1003	⅛-in.-thick specimens; state method (within D1003) used	%	%		
Luminous transmittance	D1003	⅛-in.-thick specimens; state method (within D1003) used	%	%		
Index of refraction	D542	⅛-in.-thick specimens; state method (within D542) used		Dimensionless		
Water absorption	D570	—	%	%		
24 hr immersion	Sec. 6.1					
Long-term immersion	Sec. 6.4					
Brittleness temperature	D746	—	°C at 50% failure point	°C at 50% failure point		
Coefficient of linear thermal expansion	D696	—	$°C^{-1}$	$°C^{-1}$		
Deflection temperature	D648	¼-in.-thick specimens	°C at 264 and 66 psi fibers stress	°C at 18.5 and 4.6 kgf/sq cm stress		

Table 3-1. (*Continued*)

Property	ASTM test method[a]	Recommendations[b]	U.S. customary units	SI units	X[c]	s[c]
Vicat softening point	D1525	—	°C	°C		
Flammability	D635	⅛-in.-thick specimens	in./min (if burning) in., sec (if SE)	cm/min (if burning) cm, s (if SE)		
Oxygen index	D2863	—	%	%		
Deformation under load	D621	State temperature and stress	%	%		
Rigid plastics	Method A					
Non-rigid plastics	Method B					
Dynamic mechanical properties	D2236	Report values over range of temperature of show plot	Dimensionless			
Logarithmic decrement						
Elastic shear modulus			psi	Pa		
Creep	D2990	⅛-in.-thick type II specimens; report stress to produce 1% creep strain in 1000 hr	psi	Pa		
Creep rupture	D2990	⅛-in.-thick type II specimens; report stress to produce rupture in 1000 hr	psi	Pa		

[a] Use the latest revision of any ASTM method referenced.
[b] Applicable material specification should be the primary guide to test specimen dimensions and methods of preparing and conditioning specimens. (See Part 26 of the Annual Book of ASTM Standards.) If specifications are lacking, test method recommendations should be followed. In their absence, specimens should be prepared by compression molding, fully positive. Conditioning in accordince with Procedure A of ASTM Methods D618 is usually appropriate for materials which absorb little moisture. Hygroscopic plastics are routinely tested in the dry or "as molded" state; specimens conditioned to an "equilibrium moisture content" (at 23°C and 50% RH) may exhibit properties more closely related to those which obtain in use.
[c] Characterization of a given grade of material should be based on tests of representative samples from at least five batches. The arithmetic average "X" and the estimated standard deviations "s" for each property of that material would be computed from the average values for all batches evaluated.
Note: The reporting of all significant test parameters is strongly recommended. (See the Report Section of each test method used.) All departures from standard or recommended procedures should be noted clearly on the report or data sheet. The use of International SI units is encouraged.

*Courtesy: Society of Plastics Engineers, Inc.

Fig. 3-1. Typical universal tensile testing machine. (*Courtesy: Instron Corporation*)

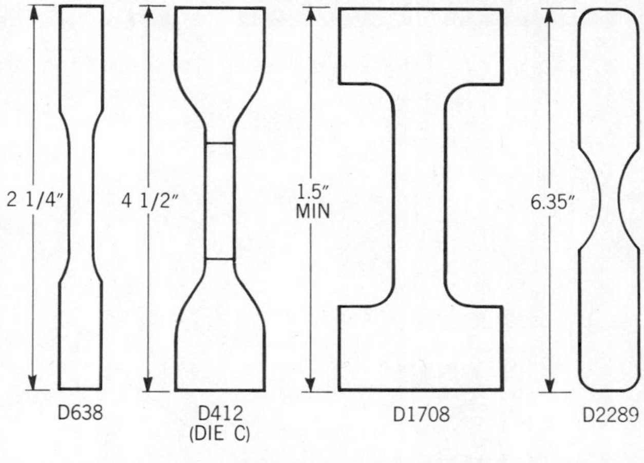

Fig. 3-2. Approximate shapes of flat plastics test specimens for some of the ASTM tensile test methods. (Supnick, 1966)

Flexural Strength

In flexure, a bar is supported at its outer ends and loaded steadily at the center, as shown in Figure 3-8. The distortion is measured as the load is increased. Flexural strength is the maximum stress in the outer fiber of the test specimen at rupture. If a plastic does not break, then the results may be reported at some given unacceptable level of distortion, as with the compressive test.

Fig. 3-3. Clamping jaws for tensile testing of plastics materials. (*Courtesy: Instron Corporation*)

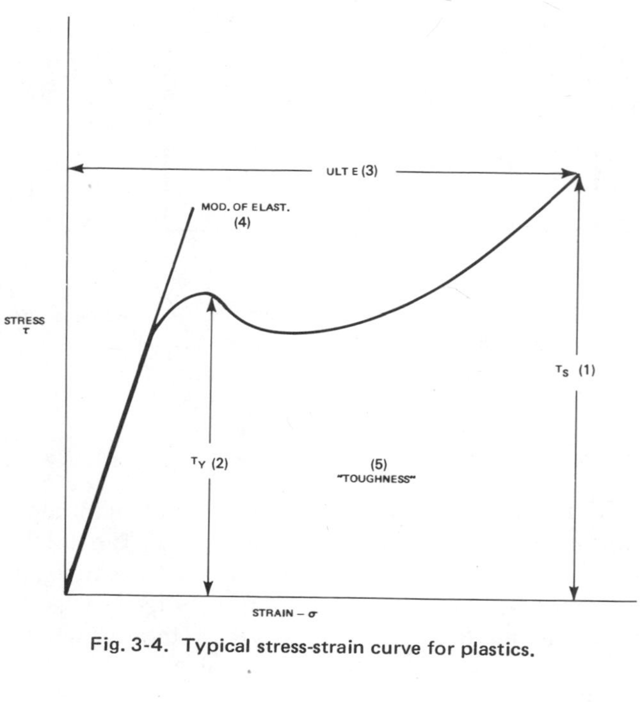

Fig. 3-4. Typical stress-strain curve for plastics.

Impact Strength

The ability of a plastic to withstand an impact load depends on its toughness. As mentioned earlier, toughness can be approximated from the area under the tensile stress-strain curve. The major difference is that in an impact test the rate of applied load is significantly faster than that typically used in a tensile test. Measured impact strength is the measure of the work done in breaking a test sample. Figure 3-9 compares test impact velocities with more common examples.

The most common method used for measuring impact strength is the Izod notched bar test. A notch is machined or molded into a rectangular test piece which is supported vertically at one end (see Figures 3-10 and 3-11).

A free-swinging pendulum impacts the test piece at the notch and breaks it. The energy absorbed is found from the reduction in swing of the pendulum. Other commonly used tests are the Izod unnotched; Charpy (pendulum hits a horizontally supported, unclamped, unnotched rectangular specimen, Figure 3-12); falling ball (ball bearing or metal

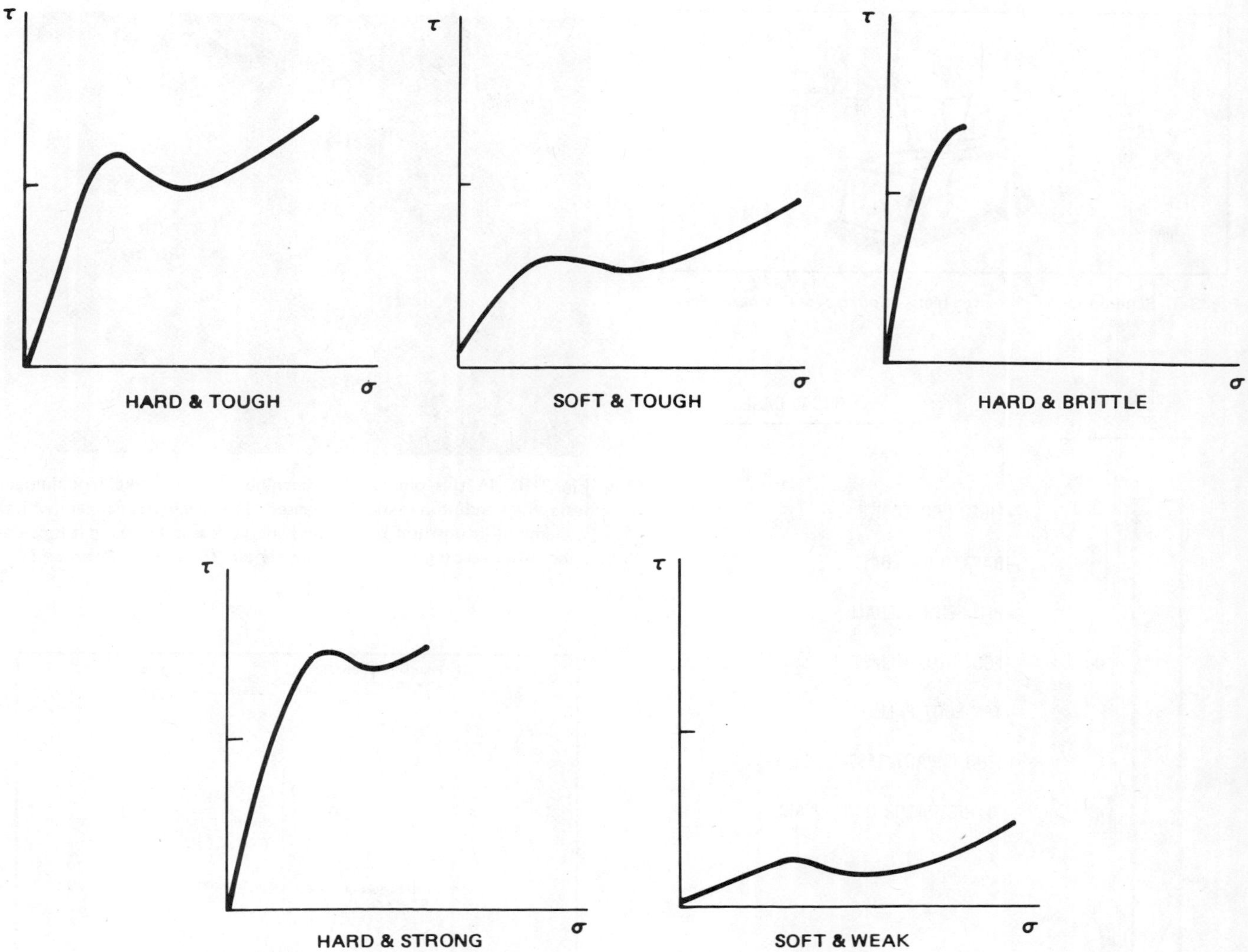

Fig. 3-5. Stress-strain curves of various types of plastics.

Fig. 3-6. Epoxy compression specimen showing brittle fracture. (*Courtesy: Delsen Corporation*)

Fig. 3-7. Fluorocarbon compression specimen showing distortion but no break. (*Courtesy: Delsen Corporation*)

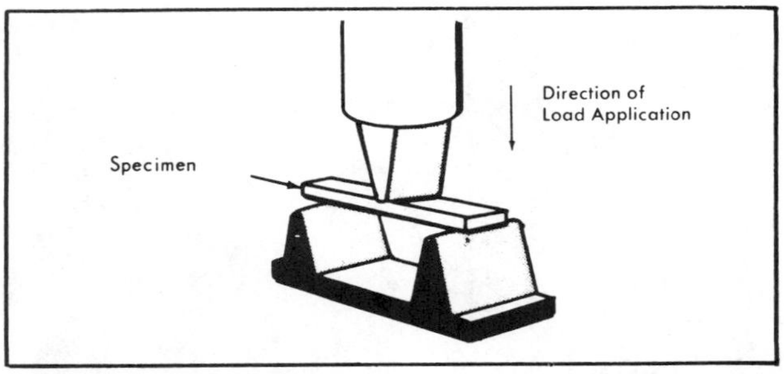

Fig. 3-8. Schematic of flexure test. (*Courtesy: Celanese Plastics Co.*)

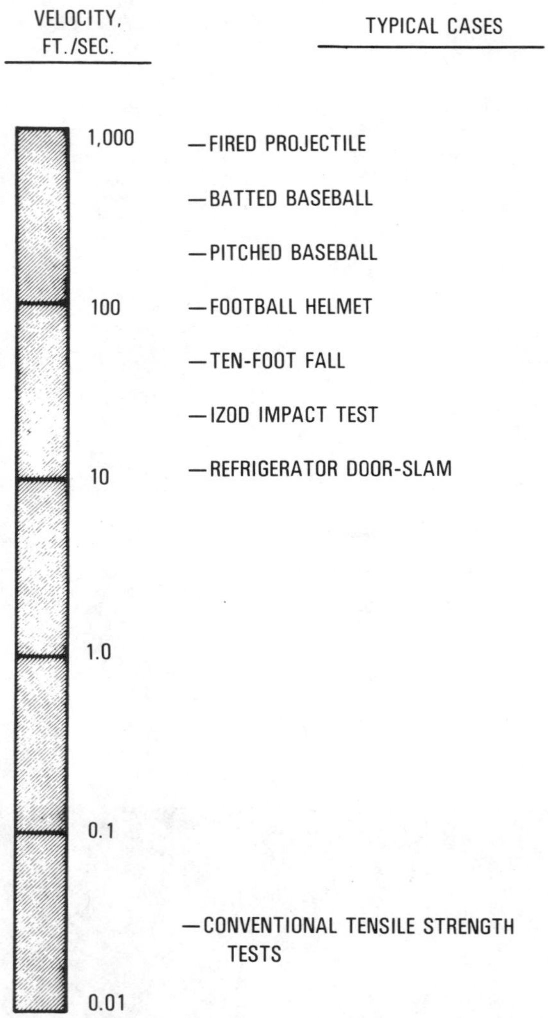

Fig. 3-9. Some typical velocities. (Printed with permission of Dow Chemical Co.)

dart of fixed weight dropped from varying heights onto a disc, Figure 3-13); tensile impact (dogbone specimen rapidly pulled in tension), as well as many variations of these basic methods.

By itself, an impact strength value can have little significance. An extreme example is trying to compare a hard

Fig. 3-10. In this photo of a machine used to make Izod impact tests, the pendulum is shown released. The notched specimen (white) is clamped in position below, and energy loss in breaking it is calculated from reading the dial in the center. (*Courtesy: Celanese Plastics Co.*)

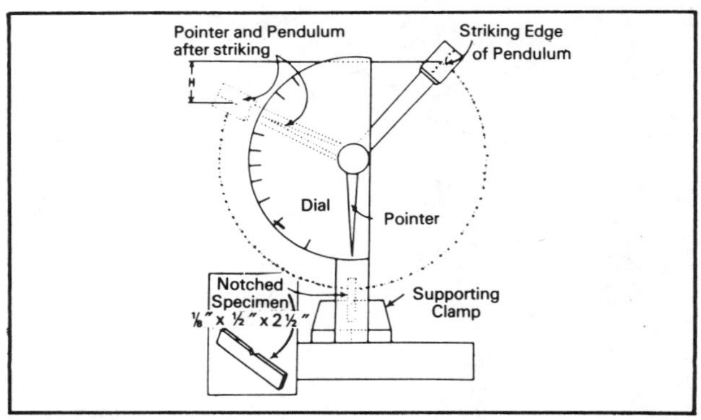

Fig. 3-11. Schematic Izod impact tester. (*Courtesy: Celanese Plastics Co.*)

rubber and extremely tough plastic, both of which stop a pendulum but don't break. However, when an impact strength measurement is used in conjunction with other tests, such as tensile and hardness, a more descriptive picture of the sample's character emerges.

Creep Resistance

When physical properties such as tension and compression are analyzed at various loading rates, the time scale involved is always relatively short (seconds and minutes). These results are not sufficient to predict the effect of such loadings under continuous application over a longer time scale (days and years). The deformations that can, and do, occur over such a time span can be so large that they become a most serious limiting factor to the plastics' use in load-bearing applications. In fact, it is only in recent years that sufficient

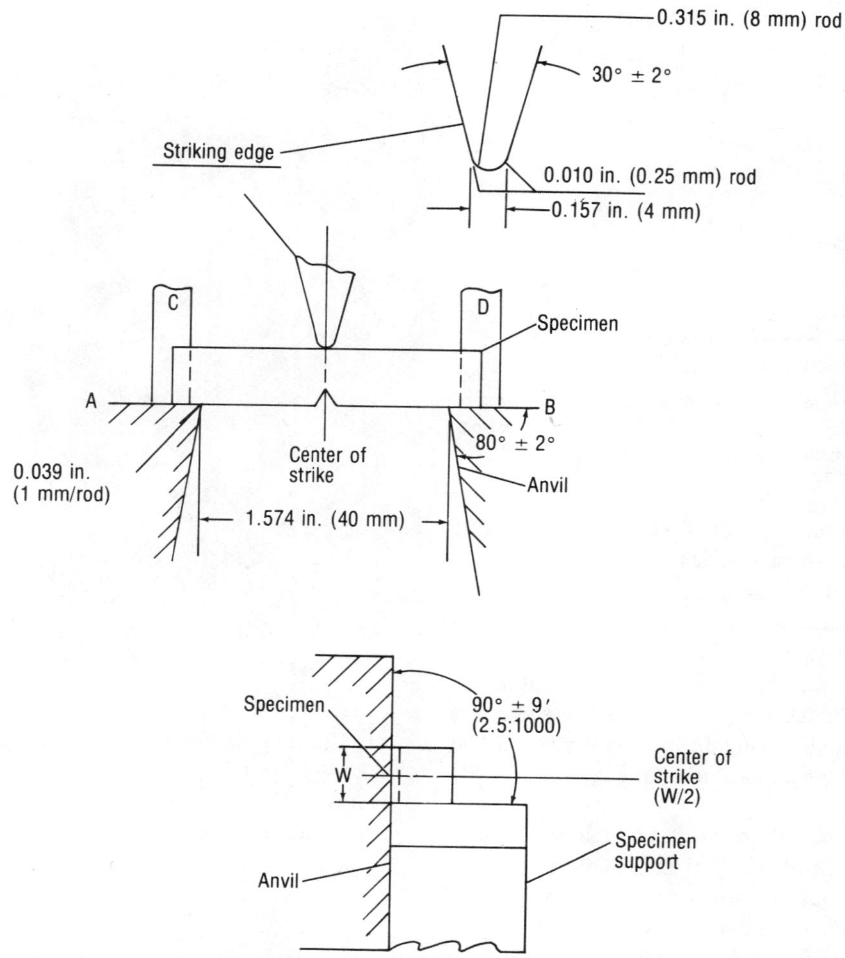

Fig. 3-12. Charpy impact test uses a specimen which is supported as a horizontal simple beam. (Mock, 1977)

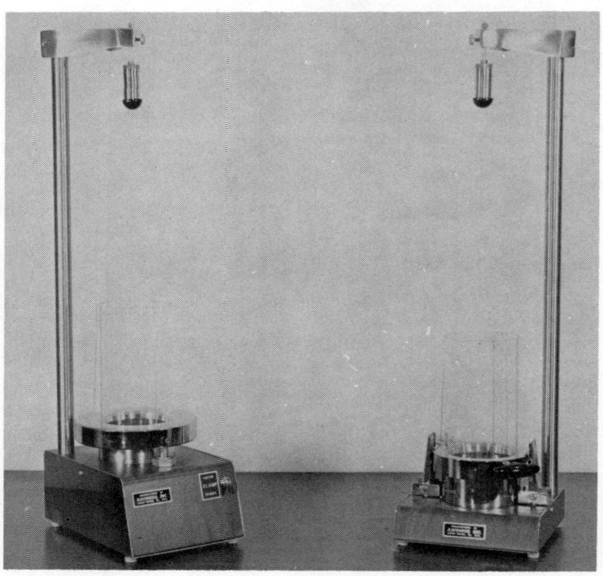

Fig. 3-13. Views of pneumatic and manual dart-drop impact testers. (*Courtesy: Kayeness, Inc., Honeybrook, Pa.*)

data on plastics' creep has been accumulated and published for design purposes.

The standard method of studying this phenomenon is to apply a fixed load to a standard test piece (often, but not necessarily, a tensile bar) and examine the strain as a function of time. The creep characteristics of a plastic do depend on structure; the more the allowable molecular movement, the greater the creep. Thermoplastics creep more than thermosets, amorphous plastics more than crystalline ones, and thermoplastics above their T_g more than those below. Because creep is the result of a viscous movement in the polymer, the effect of a small stress over a long period can be just as severe as a large stress for a short time (see Figure 3-14).

Hardness

Although hardness is a somewhat nebulous term, it has already been defined in terms of the tensile modulus of elasticity. From a more practical side, it can be further

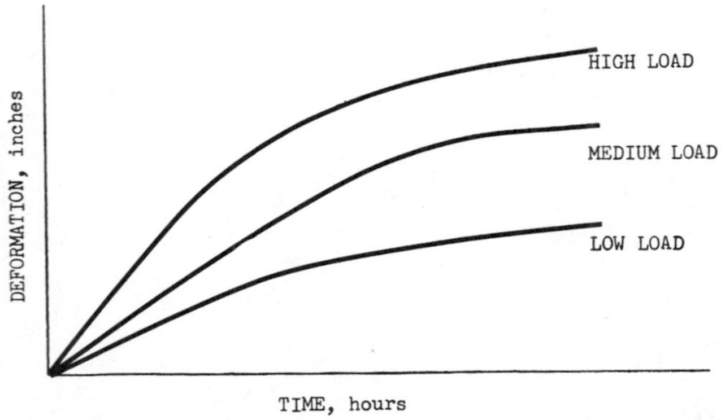

Fig. 3-14. Creep test data—deformation versus time. (Richardson, 1963. Reprinted by permission from *Plastics Design & Processing.* Copyright 1963, Lake Publishing Corporation, Libertyville, Ill.)

characterized by a combination of three other measurable parameters: (1) scratch resistance, (2) abrasion and/or mar resistance, and (3) indentation under load.

To measure scratch hardness, a specimen is moved laterally under a loaded diamond point (see Figure 3-15). The hardness value is expressed as load/width of scratch. In other tests (especially in the paint industry), the surface is scratched with lead pencils having different hardnesses. The hardness is defined as the lead hardness first causing a visible scratch.

Abrasion resistance is usually measured by a loss in weight or change in optical transmission/reflectance after a sample has been exposed to an abrasive surface, usually under load, for a predetermined number of cycles or time period. Figure 3-16 shows a typical abrasion tester for plastics.

Indentation under load is performed similarly to the tests used to measure the hardness of other materials, such as metals or ceramics.

There are at least four popular hardness scales in use. They are Shore A and Shore D, for soft to relatively hard plastics and elastomers, respectively (see Chapter 7, Figures 7-8, 7-9, and 7-10); Barcol, for values around midrange to just over the high range of Shore D; and Rockwell M (Figure 3-17), for very hard plastics. Figure 3-18 describes the test

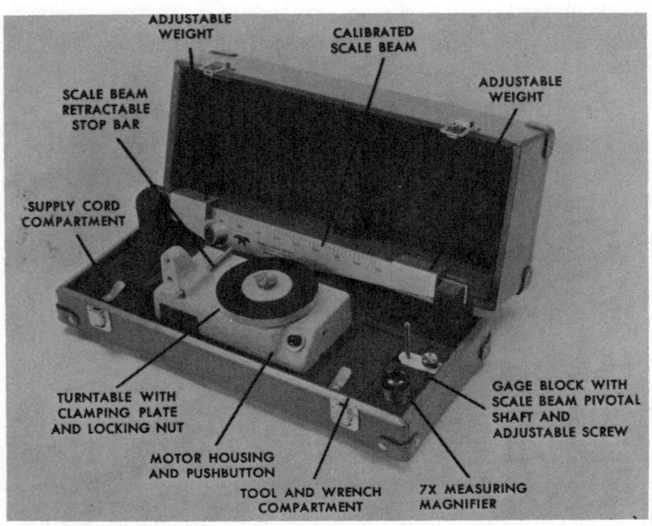

Fig. 3-15. Scratch resistance tester. (*Courtesy: Teledyne Taber*)

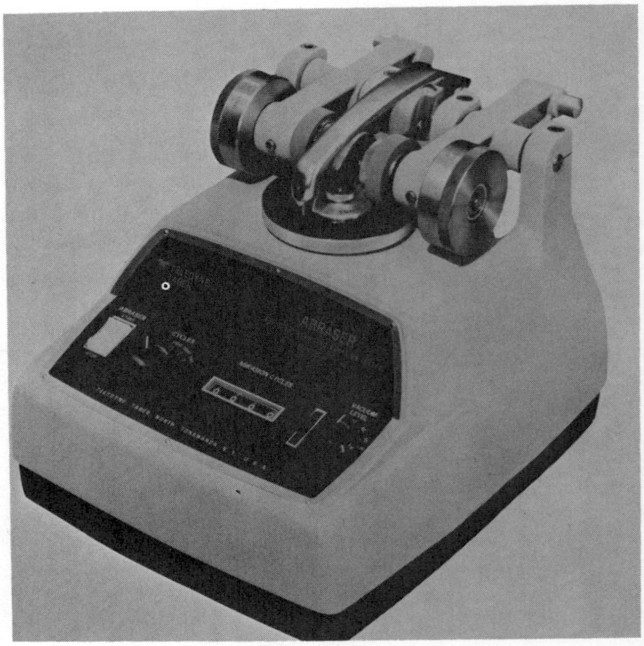

Fig. 3-16. Abrasion tester. (*Courtesy: Teledyne Taber*)

Fig. 3-17. Wilson® Rockwell® digital hardness tester used for testing plastics. (*Courtesy: Acco Industries, Inc., Wilson Instrument Division, Bridgeport, Conn.*)

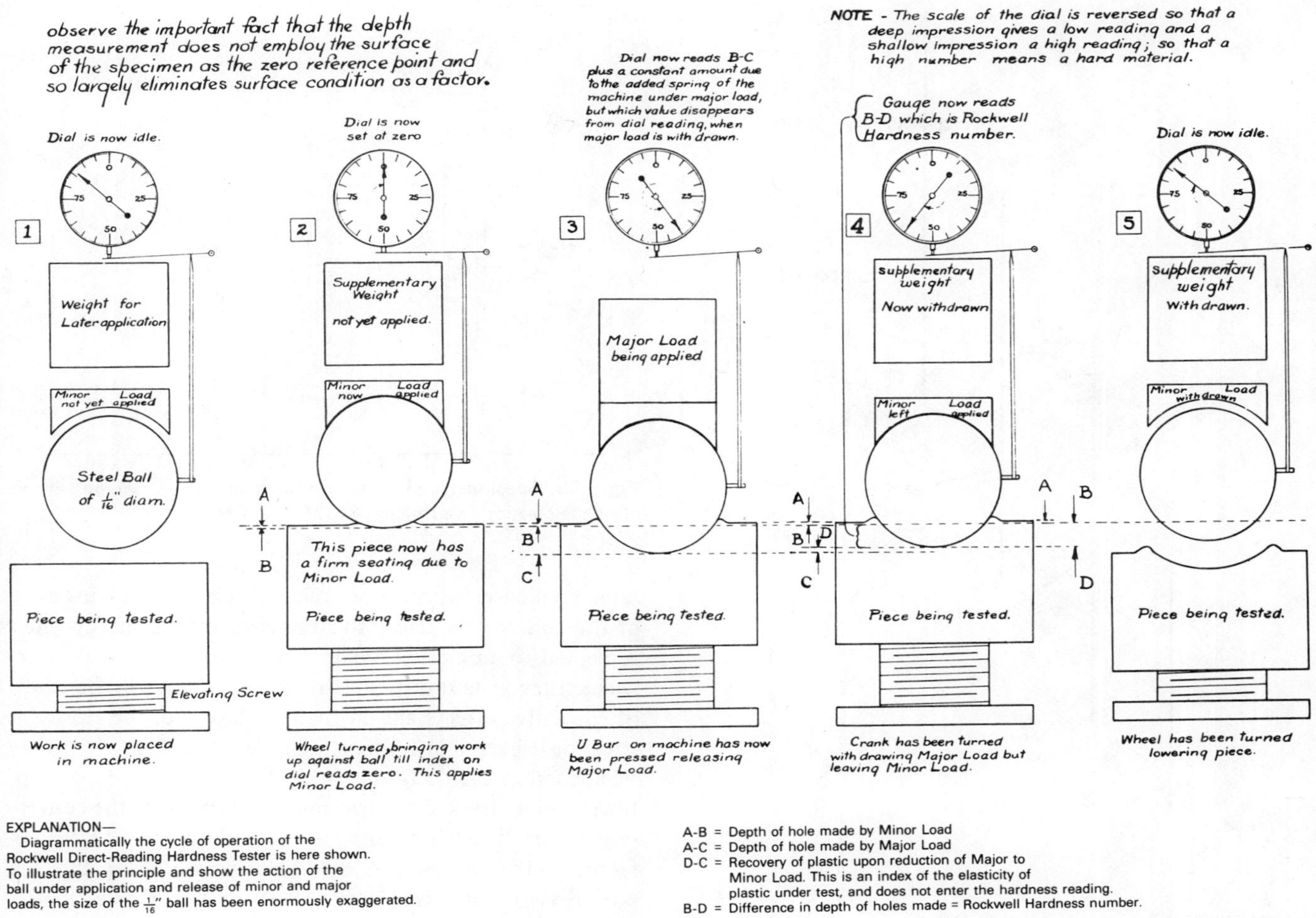

Fig. 3-18. The Rockwell hardness tester accomplishes a test which in principle is represented by this series of sketches. (*Courtesy: Acco Industries, Inc., Wilson Instrument Division, Bridgeport, Conn.*)

procedure for measuring hardness using the Rockwell instrument. Figure 3-19 shows the relative hardness ranges covered by these durometers. It should be noted, however, that there is no correlation whatsoever between these scales.

Stress-Cracking

As with metals, stress-cracking is the apparent premature failure of a plastic under stresses considerably less than those predicted to produce such failure. Figure 3-20 shows the test apparatus used to evaluate stress-crack resistance of plastics. The failure can manifest itself as a series of fine surface cracks (crazing) at one extreme, to catastrophic rupture at the other.

As with metals, two basic conditions are required for a stress-crack related failure: (1) a stress and (2) an external initiating agent. However, for plastics, in addition to the above two prerequisites, the nature and structure of the plastic and its process history can influence the extent of stress-crack resistance.

Also, the external initiating agent is not limited to chemical exposure as with metals, but may also include a thermal or other energy-intensive environment. The range of initiating agents is wide and includes surface-active agents, oils, greases, solvents (particularly ones which cause softening and swelling of the plastics' surface), heat (due to structural changes within the polymer), and oxidation (due to surface degradation).

The classic example of the effects of stress-cracking is the case of the polyethylene milk bottle. It was only after many years that a polyethylene polymer and a process to make the bottles were integrated to the point that the lactic acid in the milk would not cause a premature split in the highly stressed neck area of the bottle.

ELECTRICAL PROPERTIES

Figure 3-21 illustrates the test configurations used for the tests described below.

Dielectric Constant, Dissipation, Loss, and Power Factors

The dielectric constant is the ratio of the capacitance of a condenser with plastic between the plates to that of the

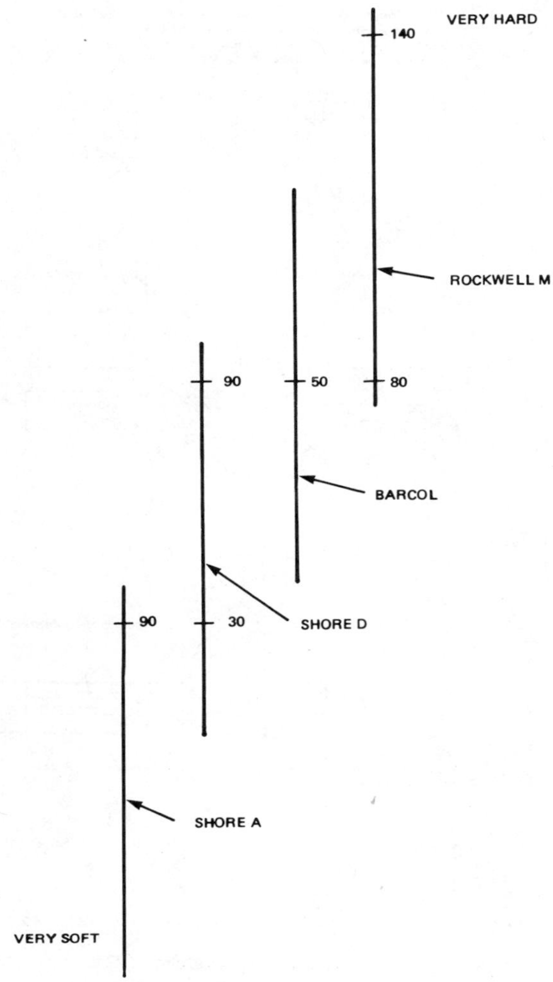

Fig. 3-19. Range of hardnesses common to plastics.

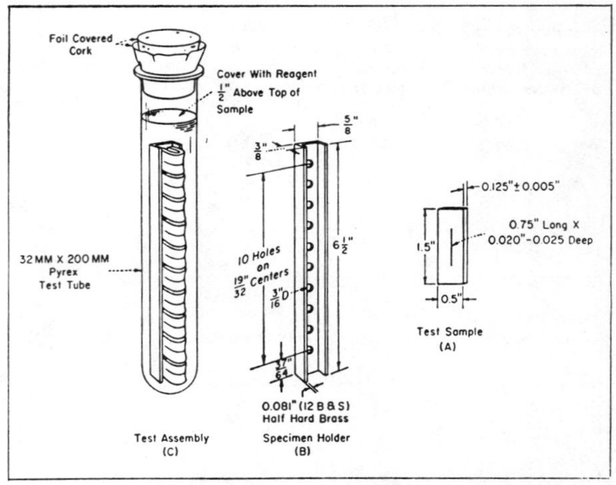

Fig. 3-20. Specimen and apparatus used in stress-cracking test of ethylene plastics. (Kaufmann, 1957)

same condenser with air between the plates. It is a measure of the ability of a plastic to store electrostatic energy when subjected to an electric field; that is , to act as a capacitor. Dielectric constant or permittivity is important in the study of capacitive effects and in the calculation of the electrical stress distribution in composite insulators.

Dissipation factor is the ratio of conductance to susceptibility of a plastic dielectric and is a measure of the conversion of reactive power to real power, showing as heat. Loss factor is the product of the dielectric constant and the power factor and is a measure of total AC losses in the dielectric.

Power factor is the measurement of electrical power loss in a dielectric material. In a perfect condenser, the current

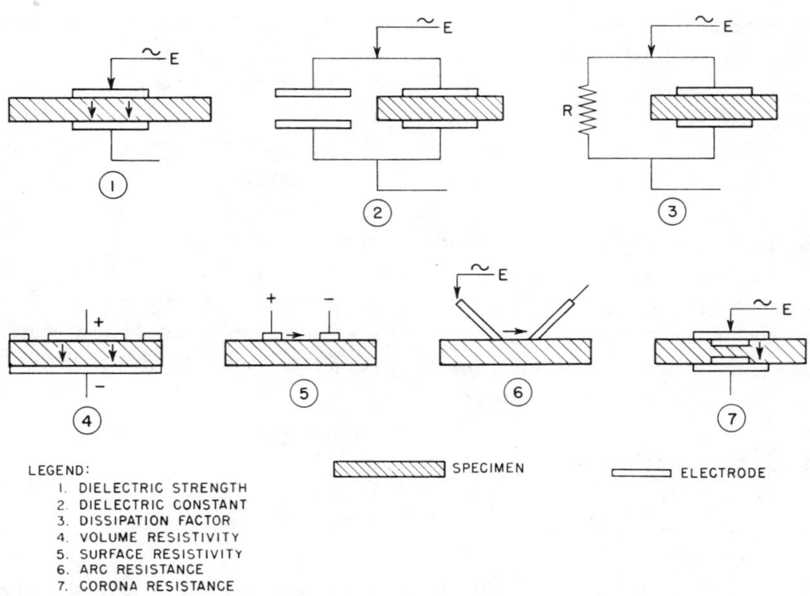

Fig. 3-21. Schematic illustration of electrical properties. (Craig, 1962. Reprinted by permission from *Plastics Design & Processing*. Copyright 1962, Lake Publishing Corporation, Libertyville, Ill.)

leads the voltage by 90°. This angle is proportionately changed with the absorbed current when the dielectric absorbs power to produce heat. The power factor then is the ratio of power dissipated to the product of the sinusoidal voltage and current.

Dielectric Strength

Dielectric strength is defined as the voltage gradient at which a plastic fails as a dielectric due to electrical breakdown. This occurs when the voltage across an insulator reaches such a value that there is a sudden onrush of current.

There are two tests commonly used in practice. In both, the specimen is placed between two high-voltage electrodes (usually immersed in a suitable insulative liquid, air for room temperature studies or silicone oil for high-temperature studies). In the first test, a voltage is applied from zero to breakdown at a uniform and rapid rate. This is known as the *short-time* test and the voltage recorded at breakdown is the short-time dielectric strength. In the second method, the voltage is applied in increasing increments (often 100 or 500 volts) and held at each increment for a specified time. This is known as the *step-by-step* test. In general, higher values of dielectric strength are obtained from the short-time test.

As the thickness of a sample increases, so does the voltage necessary to cause breakdown. However, this relationship is not linear and, more importantly, the voltage *gradient* at breakdown (volts/thickness) *DECREASES* as the *THICKNESS OF THE SAMPLE INCREASES.* Thus, a $\frac{1}{8}''$ thick sample of plastic may have a dielectric strength of 400 v/mil (0.001 inch), but a 3-mil film of the same material may record a value of 1,000 v/mil. This variation of voltage breakdown gradient with thickness accounts for the remarkably good insulating properties of thin films of plastics compared to their performance in bulk form. A more in-depth technical discussion of this phenomenon can be found in Table 3-1, ASTM test method D-149.

Insulation Resistance

The measure of the resistance a plastic provides to current flow when two terminals are placed in contact with it is called the insulation resistance. It is the ratio of the applied d.c. voltage on the electrodes to the total current between them. Insulation resistance is often reported as the ohms resistance generated between the two electrodes, but is in fact a composite of volume and surface resistivity.

Volume resistivity represents the extent to which current can pass through the bulk of the plastic, and is dependent on the nature of the plastic itself. Typical values of resistivity are shown in Figure 3-22.

Surface resistivity is more a function of the nature of the plastics' surface, its cleanliness, purity, and freedom from contamination. Surface resistivity is the ratio of the potential gradient parallel to the current along the surface, to the current per unit width of surface. The most important factor in this measurement is the presence of moisture (a relatively good conductor), which will significantly lower the surface resistance.

Arc Resistance

If an electric arc passes over the surface of a plastic, there is a tendency for tracking to occur. The arc initially passes through the air, but the surface of the plastic decomposes along the line of the arc. The result of the decomposition is a carbon track which can conduct the current. This causes further decomposition and increases the size of the current-

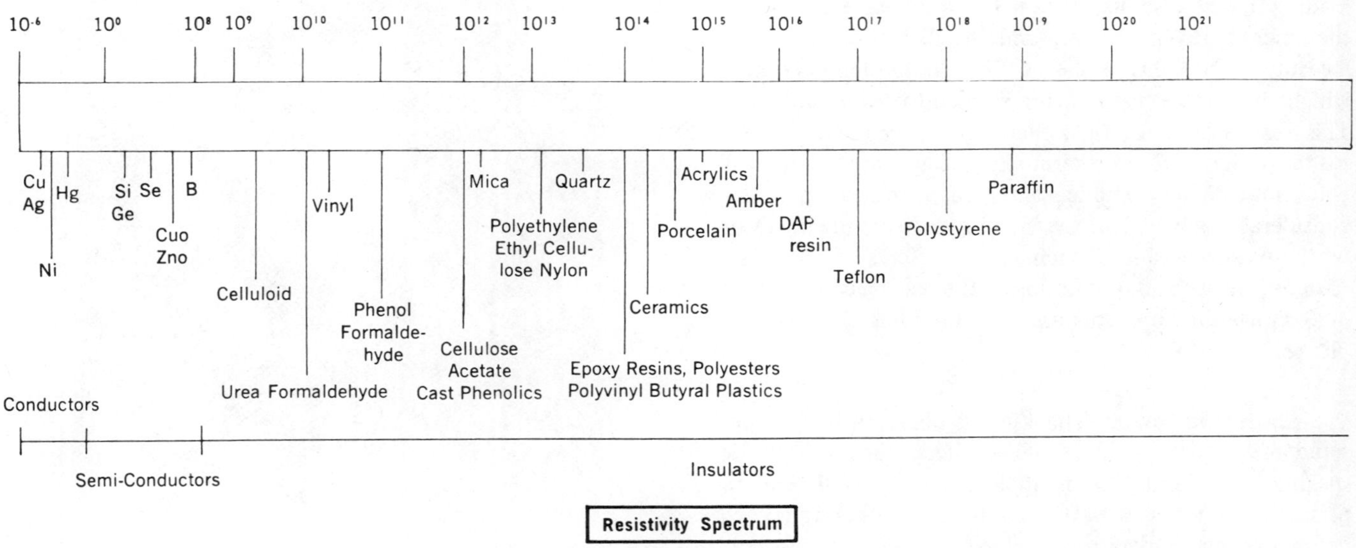

Fig. 3-22. Resistivity spectrum. (*Courtesy: FMC Corporation*)

carrying track until the current passes completely along the surface. The value of arc resistance depends on the ease of decomposition of the plastic and the nature of the decomposition products.

In recent years, concern has been raised about the flammable nature of a plastic that has arc-tracked. A test gaining prominence is one in which an arc is purposely generated (regardless of the level of resistance to arcing the plastic has) and then observations are made as to whether or not the arcing causes combustion and, if so, to what degree the plastic continues to burn.

THERMAL PROPERTIES

Useful Temperature Limits

Compared to metals, ceramics, and glass, plastics are restricted in their tolerance to upper-temperature exposure. On the lower end of the temperature scale, plastics generally perform as well, or better than, most other materials, including metals and ceramics.

A guide to the potential usability of a given plastic at a given temperature can be obtained from studies of its T_m and T_g. The ideal determination of a plastic's ability to withstand a given temperature extreme is to run one or more of the specific tests discussed earlier over the temperature range of interest.

For screening purposes, or where it is impractical to run the other tests, there are two additional methods that have found wide utility as reasonably accurate predictors of a plastic's performance at elevated temperatures. These are the heat-deflection temperature (HDT) and the Vicat softening point (Vicat).

Heat-Deflection Temperature. The HDT is obtained by supporting a sample bar as in the flexural strength test and loading it with a standard load. The ambient temperature is then raised at a fixed rate and the distortion of the bar measured. The HDT is then defined as the temperature at which the sample distorts to a standard level of deflection. This test is used for rigid thermoplastics and thermosets. In fact, the HDT of thermoplastics in the "as-molded" condition is a very practical and realistic measurement. It is considerably lower than the intrinsic softening temperature of the plastic because it includes the effects of processing, such as stresses and orientation in the test specimen. Figure 3-23 shows the apparatus used to determine the HDT of a plastic.

Vicat Softening Point. The Vicat is obtained by placing a standardized flat-ended needle on the surface of a plastics specimen. A fixed load is applied to the needle and the plastic is heated at a uniform rate. The softening point is defined as the temperature at which the penetration reaches a standard value. Vicats are usually higher than HDT's and are run primarily on thermoplastics. Vicat readings are not affected by sample size and molded-in stresses. Figures 3-24 and 3-25 illustrate this test.

Fig. 3-23. Apparatus for determining deflection temperature of plastics under load. (*Courtesy: Tinius Olsen Testing Machine Co., Inc.*)

Melting Point Tests. The melting point is a rough indication of ultimate use temperature and is used to characterize and identify polymers. Melting point tests measure either melting point (in the case of crystalline polymers) or soft-

Fig. 3-24. Close-up of Vicat softening point test frame. (*Courtesy: Tinius Olsen Testing Machine Co., Inc.*)

Fig. 3-25. Test frame being removed after Vicat softening point determination. (*Courtesy: Celanese Plastics Co.*)

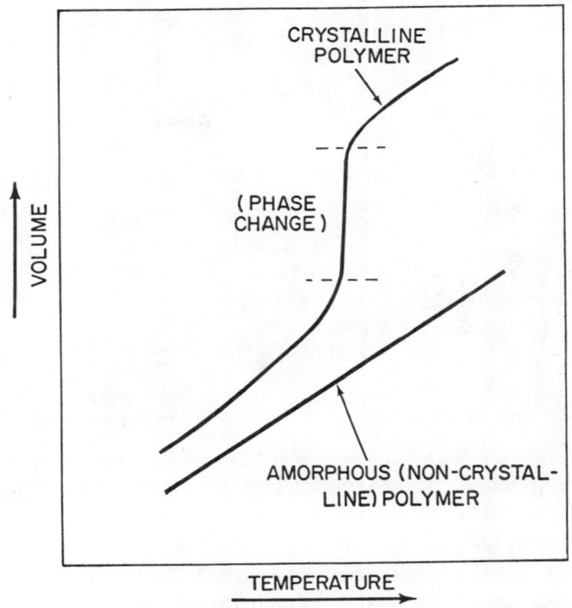

Fig. 3-26. Typical melting curves for crystalline and amorphous polymers. (Black, 1964. Reprinted by permission from *Plastics Design & Processing*. Copyright 1964, Lake Publishing Corporation, Libertyville, Ill.)

ening point (in the case of noncrystalline or amorphous polymers).

The melting point of a highly crystalline thermoplastic, such as a nylon or acetal, occurs over a very narrow temperature range. Less crystalline thermoplastics, such as cellulosics or polystyrene, soften or melt over wider temperature ranges. Thermosets do not melt, but sinter and decompose as the temperature is raised. A graphical representation of the volume of thermoplastics on heating (or cooling) is given in Figure 3-26.

Three tests are commonly used to determine melting point. Each consists essentially of heating a sample and noting the temperature at which it melts. Though they vary in complexity and accuracy, these tests provide a means for observing closely the precise temperature of transition. Equipment is shown in Figure 3-27 for melting point tests.

Other Thermal Tests. *Flow tests*, which measure temperature or rate of flow under heat and pressure, relate material characteristics to processability.

The *torsion pendulum test* (schematically pictured in

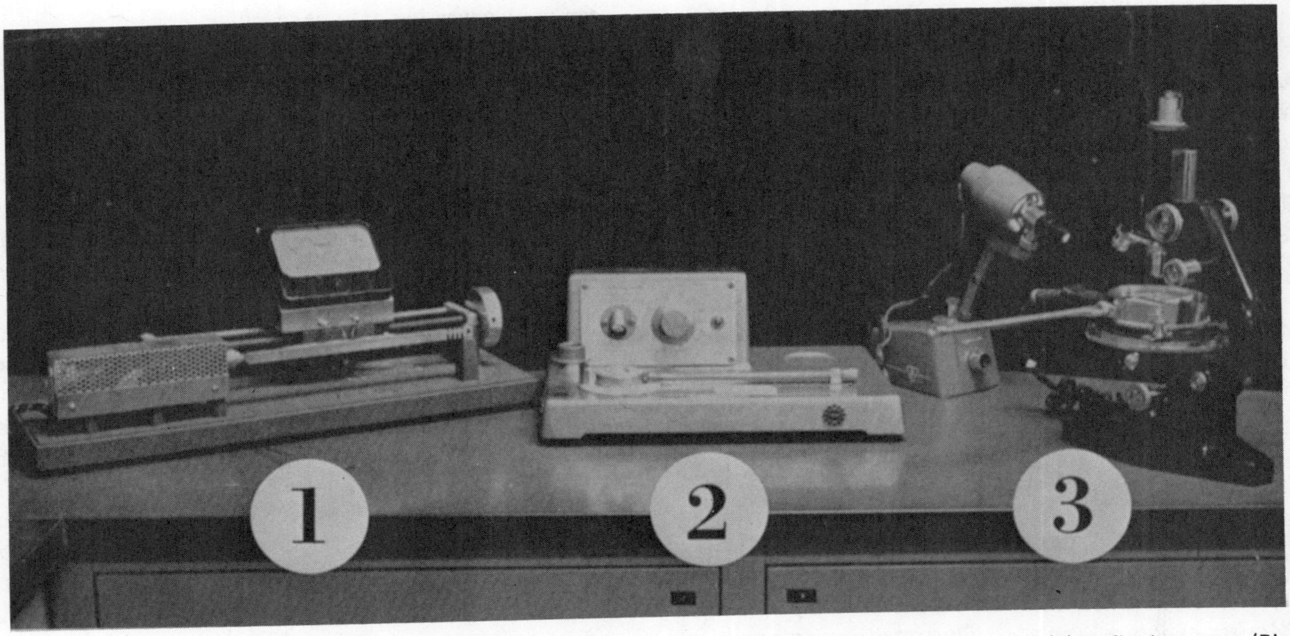

Fig. 3-27. Equipment used for melting point determinations: (1) heated bar, (2) Fisher-Johns apparatus, and (3) Kofler hot stage. (Black, 1964. Reprinted by permission from *Plastics Design & Processing*. Copyright 1964, Lake Publishing Corporation, Libertyville, Ill.)

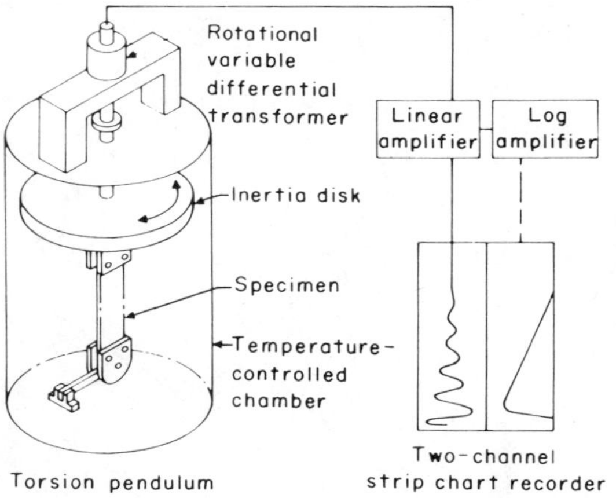

Fig. 3-28. Torsion pendulum spectrograph for measuring shear modulus and log decrement (damping factor) of plastics. (Riddell, 1974)

Figure 3-28) directly measures two dynamic properties of plastics as a function of temperature: shear or torsion modulus, and damping factor. When these two properties are plotted against changing temperature, they reveal the polymer transitions, particularly glass transition, of plastics. The shear modulus also indicates the relative change in rigidity due to transitions.

The mechanical damping of a polymer is related to impact, vibration, and similar dynamic stresses. Although damping has been correlated with fatigue performance, neither it nor torsion modulus can be used to direct design data, because creep effects are not included.

PERMANENCE PROPERTIES

Water Absorption

Although most plastics are insoluble in water, they will absorb it. Depending on the degree of absorption, the properties of a plastic may be affected to a greater or lesser degree. The greatest deleterious effect of water on plastics is in the area of electrical resistance. There are many plastics as well whose susceptibility to water causes significant deterioration in mechanical properties.

If a more accurate assessment of plastics' water resistance is required, mechanical and/or electrical tests may be run and compared before and after water immersion or exposure. There are also a few plastics which are water soluble and their end uses are designed to take advantage of this characteristic.

Resistance to Solvents, Chemicals, and Radiation

Radiation resistance has only recently become of marked interest and data is scarce. Visual observations, as well as standard mechanical tests, are made on specimens before and after exposure. Table 3-2 illustrates some general resistance data.

Resistance of plastics to solvents and chemicals is extremely difficult to generalize. Each class of plastics has its own peculiar resistance properties. The test methods used to evaluate these characteristics are patterned after the water absorption tests, with the specific chemical or solvent replacing water.

Because of the wide variety of potential solvents and chemicals that contact plastics, it is best to specify their effect as each individual class of plastics is discussed.

Weather Resistance

Successful laboratory simulation of the effects of weather on plastics has been elusive. Most investigations have had only limited success and conclude that there is no substitute for natural weathering. Plastics deterioration from weathering is a process of slow oxidation in which the ultraviolet (UV) radiation furnishes the energy, and the atmosphere (including moisture) furnishes the oxygen.

In polymeric materials, the UV energy causes scissioning which exposes the material to oxidation from oxygen in the air and, more rapidly, from dissolved oxygen in rain or dew. The UV energy in sunlight that degrades plastics varies in wavelength distribution and intensity with time of day and year, as well as latitude and elevation of exposure. The presence of water on an outdoor plastic is a transient, erratic, local occurrence. It depends on color and radiation characteristics of the surface, amount of insulation behind the surface, angle of inclination from horizontal, proximity of water sources, wind velocity, and temperature and relative humidity of the ambient air. The photodegradation process of polymers is significantly influenced by temperature. As with many chemical reactions, an increase of 10°C can double weathering-type reactions.

Laboratory weathering equipment must not duplicate weather conditions so closely that the tests become long-term rather than accelerated tests. They also must not be so severe that the degradation is very rapid, but different from natural weathering. Most widely used of the laboratory exposure instruments is the Atlas Weather-Ometer (see Figure 3-29). It consists of an enclosure with a high-intensity light source mounted inside a rotating cylindrical rack on which are mounted the test specimens.

These units also are supplied with water-cooled, xenon-arc lamps. Xenon-arc emission is closer to natural sunlight spectral distribution and intensity. The instrument is equipped for temperature and humidity in addition to the energy source and a programmed spray of water. The effects of SO_2, ozone, or nitrous oxides can also be studied.

Outdoor exposure is the oldest method of determining weather resistance. The standard procedure is to expose samples vertically, facing south. So-called accelerated tests are those that use inclined racks (5-45° from horizontal) facing south, in southern locations where sunlight is strongest and available for more hours per day.

Table 3-2. Radiation Damage to Common Insulating and Packaging Materials.

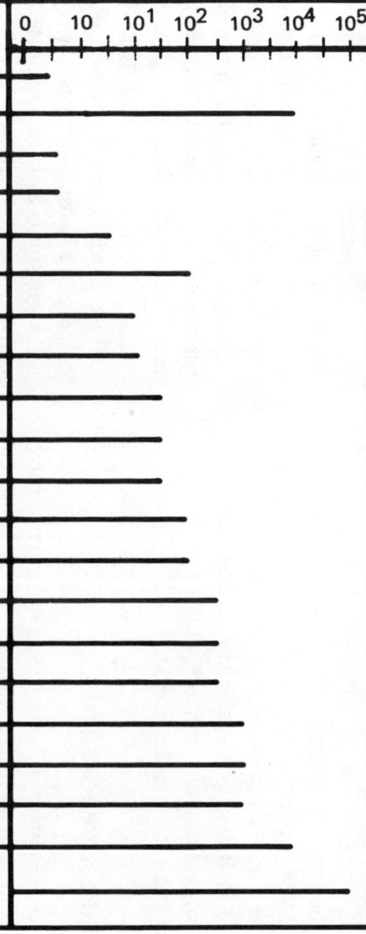

MATERIALS	DOSE IN MEGARADS	COMMENTS
PTFE		LOSES 40% OF ITS TENSILE STRENGTH, CRUMBLES AT HIGHER DOSES
NYLON		LOSES ELONGATION PROPERTIES RAPIDLY, TENSILE STRENGTH REMAINS
METHYLMETHACRYLATE		SERIOUS DETERIORATION
BUTYL & FLUOROCARBON PLASTICS		DETERIORATION BEGINNING
SILICONE RUBBER		LOSES ITS ELASTICITY
POLYESTER FILM		THRESHOLD MECHANICAL DAMAGE: BECOMES WEAK AND BRITTLE
PHENOLIC-UREA		LITTLE EFFECT
CELLULOSE ACETATE		LOSES 25% MECHANICAL PROPERTIES: ELECTRICAL PROPERTIES UNAFFECTED
VINYLS		LOSES UP TO HALF OF ITS TENSILE STRENGTH AND ELONGATION
POLYETHYLENE		LOSES 75% ELONGATION, 15% TENSILE STRENGTH
NATURAL RUBBER		LOSES 25% OF ITS PROPERTIES
MELAMINE (GENERAL PURPOSE)		LOSES 25% OR MORE OF ITS STRENGTH
POLYSTYRENE		RESISTIVITY DECREASES, MECHANICAL PROPERTIES REMAIN
UNFILLED SEMI-FLEXIBLE EPOXY		LITTLE EVIDENCE OF CHANGE
FILLED SEMI-FLEXIBLE EPOXY		PRODUCES STIFFENING AND SOME BLISTERS
RIGID EPOXY (GENERAL PURPOSE)		MAY CRACK, BECOME WEAK AND BRITTLE
SILICONE GLASS LAMINATE		LITTLE OR NO EFFECT
MINERAL FILLED PHENOLIC		LITTLE EFFECT
HEAT RESISTANT, FILLED EPOXY		NO DETERIORATION
MICA-GLASS LAMINATE		VIRTUALLY UNAFFECTED
INORGANICS: MICA, CERAMICS GLASS		RESISTIVITY TEMPORARILY CHANGED, BUT LITTLE OTHER EFFECTS NOTED

Fig. 3-29. Xenon-arc Weather-Ometer, with automatic humidity control and monitors for ozone, sulfur dioxide, and oxides of nitrogen. Access to test chamber is on right side. (*Courtesy: Atlas Electric Devices Co., Chicago, Ill.*)

Maximum acceleration in sunlight is achieved with EMMA (Equatorial Mount with Mirrors for Acceleration) and EMMAQUA (EMMA plus water). These machines automatically follow the sun so that rays approximate a normal exposure to the surface of test specimens. In addition, they are equipped with aluminum mirrors which multiply the intensity of the sun's rays on the plastic.

"SPECIAL-PURPOSE" PROPERTIES

The foregoing tests represent many of the basic descriptive parameters common to nearly all plastics. There are other properties which by themselves may have only limited value for design purposes. Yet, under a given set of circumstances, any one of these properties may distinguish a particular plastic to such an extent that, in conjunction with its more general (and perhaps average or lower) physical characteristics, this plastic becomes a highly desirable (or even the only) material applicable for use. Among these "special-purpose" properties are:

Friction. Many plastics have coefficients of friction around 0.2, compared with 0.1 for graphite and 0.05 for diamond. A coefficient-of-friction tester is used to determine the dynamic and static coefficients of friction. Figure 3-30 shows a typical instrument.

A force gauge measures the force needed to pull the plastic over various surfaces with a fixed weight normal to the direction of pull. This weight is typically 200 grams. Standard tests measure the frictional characteristics of the plastics sample on stainless steel; however, many combinations of plastics, metals, ceramics, and woods are used.

Thermal Conductivity. We have already categorized plastics as thermal insulators and, as such, they have very low $[(2\text{–}10) \times 10^{-4}$ cal-cm/cm^2-°C-sec$]$ thermal conductivity. In many instances, where additional heat transfer is required, proper incorporation of other ingredients, "formulating," can raise these values to a more useful level.

Thermal conductivity measurements are made using a guarded hot plate technique. An example of such apparatus is shown in Figure 3-31. The heat for the apparatus is supplied by a central heater plate surrounded by a guard heater. Specimens are sandwiched between the central heater and cold sinks on either end. The unit is designed to transmit heat normal to the specimen surface.

Specific Heat. Specific heat determinations are made using a drop-type adiabatic calorimeter. This instrument, seen in schematic form in Figure 3-32, consists of an inner cup to hold the plastics sample, a guard with heating and cooling devices to maintain adiabatic conditions, and temperature-sensing devices. Specific heat of plastics remains constant with temperature but will vary if fillers or reinforcements are added.

Fig. 3-30. A coefficient-of-friction tester. (*Courtesy: Kayeness, Inc., Honeybrook, Pa.*)

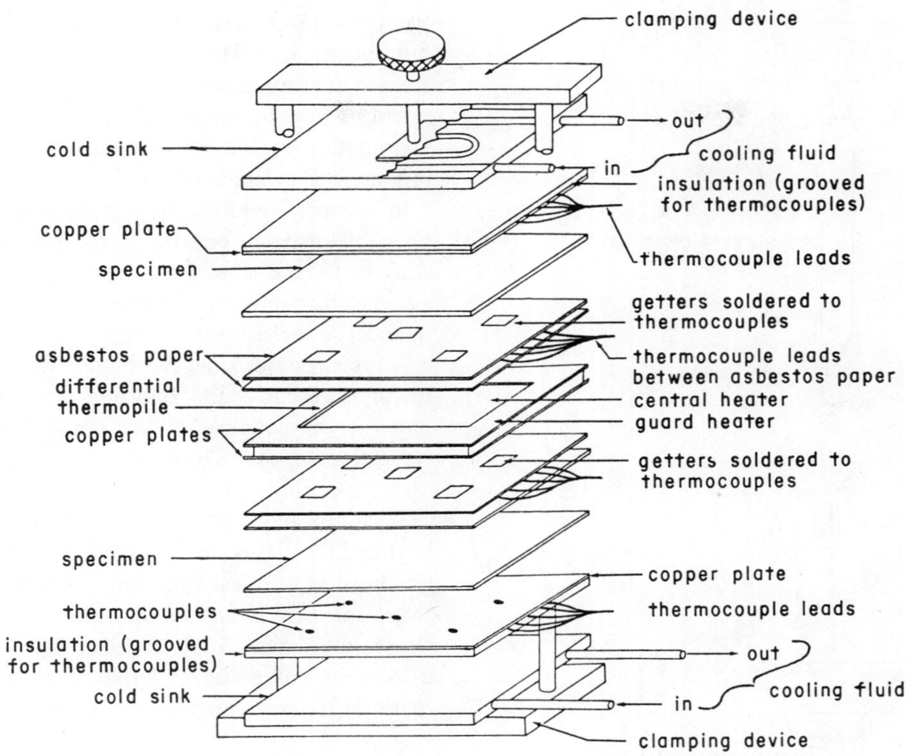

Fig. 3-31. Thermal conductivity apparatus. (Howes and Pears, 1961. Reprinted by permission of *Modern Plastics*, McGraw-Hill, Inc.)

Coefficient of Expansion. The values for plastics are typically ten times higher than for metals. Although such high values can, and do, cause problems of shrinkage and thermally induced stresses and strains, careful formulation often minimizes these effects.

Thermal expansion measurements are based on National Bureau of Standards designs. The sample is fitted within quartz tube dilatometers which are heated and/or cooled. The motion of the specimen with change in temperature is measured by the dilatometer dial gauge (see Figure 3-33).

Thermal Analysis. Many of the thermal properties of plastics can be determined using more sophisticated methods of thermal analysis. These techniques include differential scanning calorimetry (DSC), thermal and differential

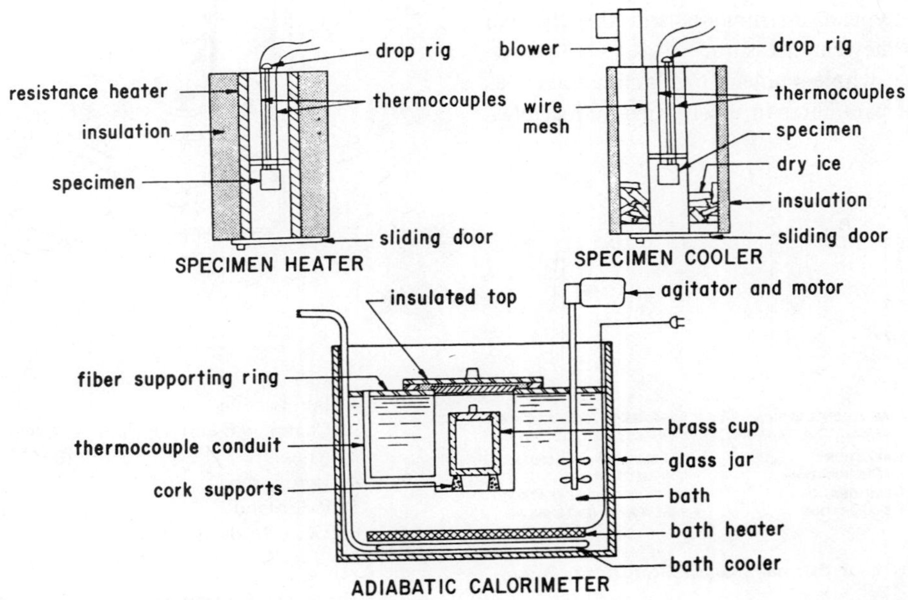

Fig. 3-32. Specific heat apparatus. (Howes and Pears, 1961. Reprinted by permission of *Modern Plastics*, McGraw-Hill, Inc.)

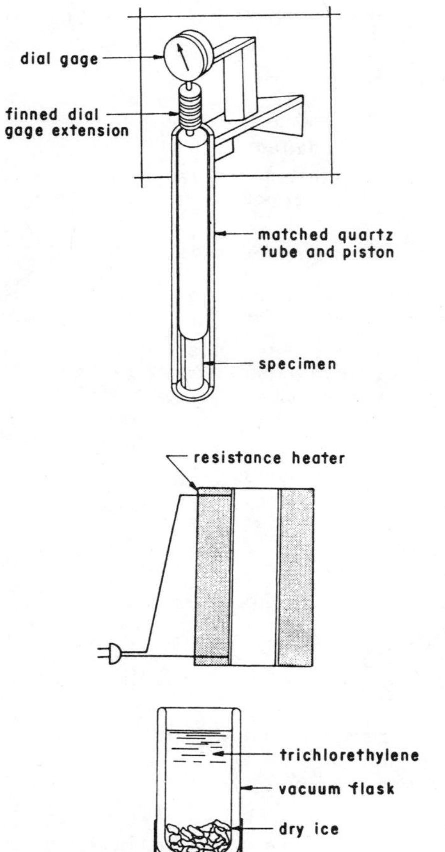

Fig. 3-33. Sketches of thermal expansion apparatus. (Howes and Pears, 1961. Reprinted by permission of *Modern Plastics*, McGraw-Hill, Inc.)

thermal gravimetry (TG and DTG), and thermomechanical (TMA) and differential thermomechanical analyses (DTMA). Figure 3-34 shows typical instruments and the thermal properties of plastics they can measure.

Flammability. The whole subject of flammability has emerged as a critical parameter in recent years. Most plastics burn. (Some are self-extinguishing and a select few do not burn.) When plastics are foamed, the added exposed surface area increases flammability potential. Most work in restricting the burning of plastics has been achieved through formulating methods, although there are instances where inflammability has been "built in to" the polymer structure.

In the past, ignition temperature and burning rate have been of greatest concern. More recently, the emphasis has shifted to the amount and nature of smoke generated, gaseous by-products and their toxicology, and the consequences of incineration during disposal operations.

A system used to assess the relative flammability of polymeric materials is the Limiting Oxygen Index (LOI). This index is a measure of the percentage of oxygen in an oxygen-nitrogen mixture which will just support burning under specific test conditions.

Since the oxygen content of air is about 21%, materials having LOI values greater than 21 (in reality, 23–25) will not burn in air. Test conditions must be duplicated exactly; thus, the LOI is not an absolute measure of a polymer's flame retardance. Table 3-3 lists representative LOI values of various polymeric materials. Figure 3-35 shows a schematic of the oxyen index test apparatus.

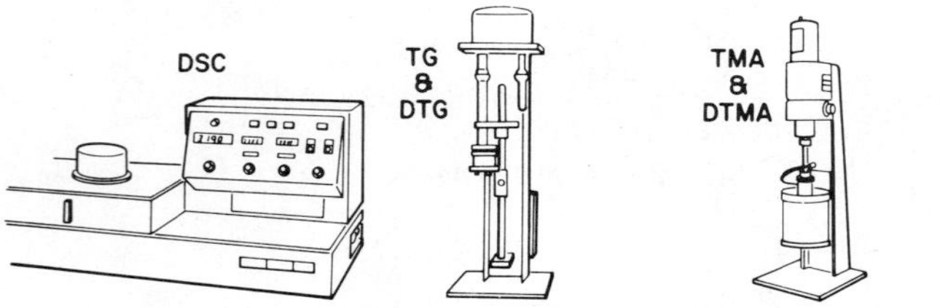

MEASURES	MEASURES WEIGHT CHANGES ATTENDING:	MEASURES DIMENSIONAL CHANGES ATTENDING:
• SPECIFIC HEAT	• REACTION	• THERMAL EXPANSION
• HEATS OF TRANSITION	• DETERIORATION	• CONTRACTION
• ONSET TEMPERATURES	• DEHYDRATION	• SOFTENING (PENETRATION)
• RATES OF REACTION	• EVAPORATION	• STRETCHING (EXTENSION)
• PURITY		

Fig. 3-34. Primary methods of thermal analysis. (*Courtesy: Perkin-Elmer Corporation*)

GLASS COLUMN (MINIMUM DIMENSION 45CM H x 7.5CM I.D.)

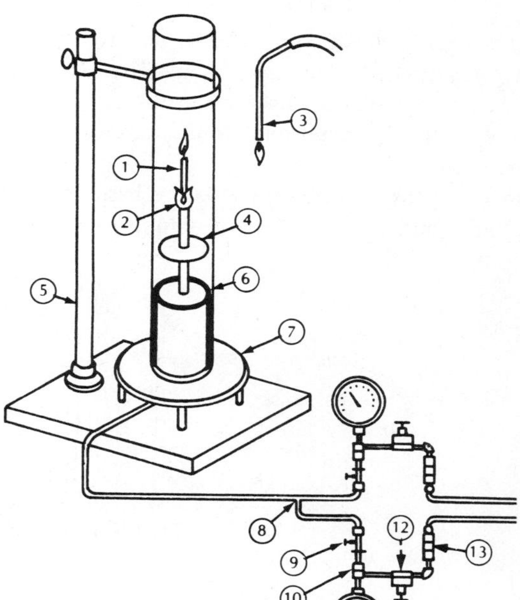

1. Burning Specimen
2. Clamp with Rod Support
3. Igniter
4. Wire Screen
5. Ring Stand
6. Glass Beads in a Bed
7. Brass Base
8. Tee
9. Cutoff Valve
10. Orifice in Holder
11. Pressure Gauge
12. Precision Pressure Regulator
13. Filter

Fig. 3-35. Oxygen index apparatus. (ASTM D2863).

Table 3-3. Limiting Oxygen Index Values of Various Polymeric Materials.*

LOI of Various Plastics

Plastic	LOI, %O_2
Polyacetal	15.0
Poly(methyl methacrylate)	17.3
Polyethylene	17.4
Polypropylene	17.5
Polystyrene	17.8
Poly(4-methylpentene)	18.0
Filter paper (cellulose)	18.2
ABS resin	18.8
Cellulose acetate	19.0
Styrene-acrylonitrile	19.1
Poly(ethylene terephthalate)	20.0
Birch wood	20.5
Poly(vinyl fluoride)	22.6
Chlorinated polyether (Penton)	23.2
Noryl 731	24.3
Nylon 66	24.3
Polycarbonate	24.9
Nylon 6	26.4
Poly(phenylene oxide)	30.0
Polysulfone, P-1700	30.0
Polyimides	31 > 45
Polyimide (Kapton film)	36.5
Polysulfone (PES)	38.0
Poly(phenylene sulfide)	>40.0
Poly(vinyl chloride)	40.3
Poly(vinylidene fluoride)	43.7
Chlorinated PVC	45.0
Polysulfone (Astrel)	>50.0
Poly(vinylidene chloride)	60.0
Polytetrafluoroethylene	95.0

*Vaccari, 1977.

LOI of High Performance Fabrics

Fabric[a]	LOI, %O_2
Untreated cotton	16–17
FR treated cotton	31–32
Natural Nomex	27–28
Dyed Nomex	25–27
Kynol	29–30
Fypro	29–30
Durette	35–38
PBI (polybenzimidazole)	38–43
Nomex HT-4	42–46
PBI-S	42–49

[a]Weight: 4.5–7.0 oz/yd^2.

LOI of Spun Fabrics

Fabric[a]	LOI, %O_2
Acrilan (acrylic)	18.2
Arnel triacetate (cellulose ester)	18.4
Acetate (cellulose ester)	18.6
Polypropylene	18.6
Rayon	19.7
Cotton (greige)	20.1
Nylon	20.1
Polyester	20.6
Wool (dry cleaned)	25.2
Dynel (modacrylic)	26.7
Rhovyl 55 (polyvinyl chloride)	27.1
Nomex N-4272 (aromatic polyamide)	28.2

[a]Weight: 4.8–7.0 oz/yd^2.

The basic mechanism of burning is the same for plastics as for other materials. Plastics burn when they are subjected to sufficient heat to break down the polymer structure into volatile gases. The gases burn and radiate heat from the flame back to the plastic, causing further pyrolysis and evolution of flammable gases. With flame-retardant plastics, the gases either do not ignite, or they ignite and then extinguish if there is no continued external heat input. Flame retardants can be added to polymeric materials. Their mechanisms for rendering the plastic non-burning are shown in Figure 3-36.

Light Transmission and Color. Plastics range from opaque to transparent. Many have greater transmission than glass. In plastics, the extent of transmission and wave scattering is independent of wavelength, and plastics in their purest form have no color of their own. As impurities are introduced (whether from added ingredients, variable molecular weight homologs, or extremely high crystallinity), a plastic becomes more translucent than opaque, and begins to acquire a yellowish tinge. Most plastics transmit at ultraviolet wavelengths, and these same plastics are often affected by UV-caused degradation. This degradation almost always manifests itself in color deterioration, but in many plastics, the deterioration can extend to a significant loss in mechanical and other properties. Figure 3-37 shows an apparatus used for measurement of haze in clear plastics films.

Adhesive Properties. Adhesive properties of polymers are determined by a variety of mechanical tests. The most popular involve bonding standardized aluminum coupons in various configurations and pulling them apart in tension. Obviously, any adherend can be used, and special test techniques exist to accommodate those that are particularly brittle, rubbery, or otherwise nonconforming to the standard aluminum specimen.

The tensile lap-shear test is the most common of all adhesive tests (see Figure 3-38), because it is easily fabricated and simple to test. Many of the lap-shear tests used cannot account for peel forces imposed during stressing. Two types of peel tests are common, the T-peel (Figure 3-39) and the

The mechanisms of flame retardants are generally classified into five categories, or modes, of operation. In most flame-retardant systems, two or more of these modes operate simultaneously, by

1

. . . **altering** the chemistry of the thermal degradation (pyrolysis) process so that more nonvolatile tars and char are formed and smaller amounts of flammable gases are produced.

Phosphorus-based flame retardants generally function in this way.

2

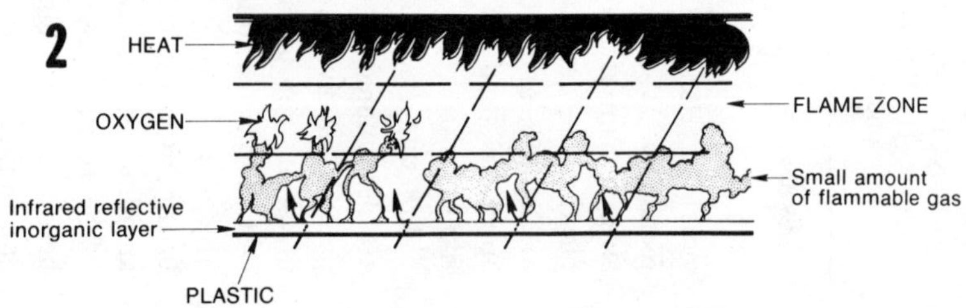

. . . **reducing** the heat flow from the solid plastic to decrease the rate of pyrolysis and the production of flammable gases.

This mode requries the use of inorganic fillers in the plastic, which become exposed as the plastic burns away, leaving a layer of filler as a barrier to heat flow into the polymer. In other systems, a glassy coating, formed at high temperatures from fillers and additives such as phosphates and borates, may act as the barrier. A surface layer of char resulting from Mode 1 may also act as a barrier to heat flow in the polymer.

3

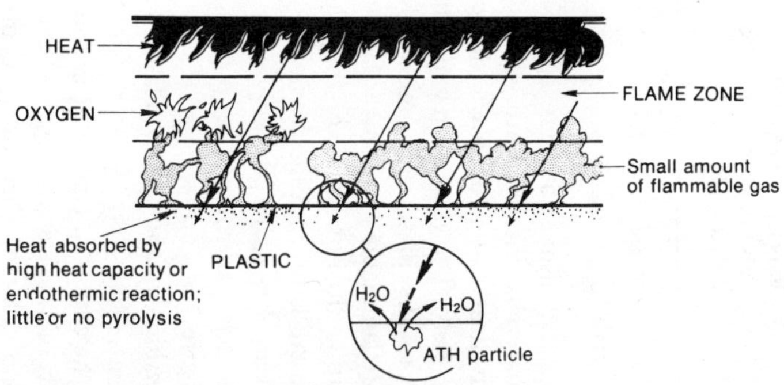

. . . **absorbing** heat through endothermic reactions, to reduce the amount of heat available to degrade the polymer.

Hydrated additives and fillers such as alumina trihydrate (ATH) function in this way. When exposed to high temperatures, the ATH absorbs heat and gives off water. Some hydrated borates react similarly.

4

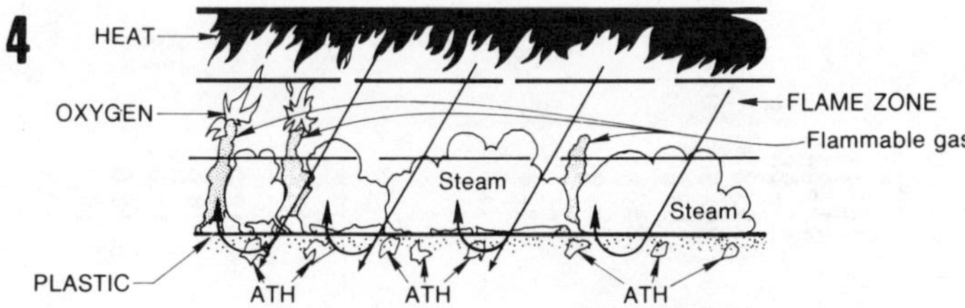

. . . **reducing** the rate of diffusion of pyrolysis gases from the solid to the flame and/or generating nonflammable gases which reduce the concentration of flammable gases.

This mode usually operates in conjunction with Mode 2 since a barrier to heat flow also acts as a barrier to pyrolysis diffusion, or with Mode 3 since water (or other released gas) dilutes the flammable gases.

5

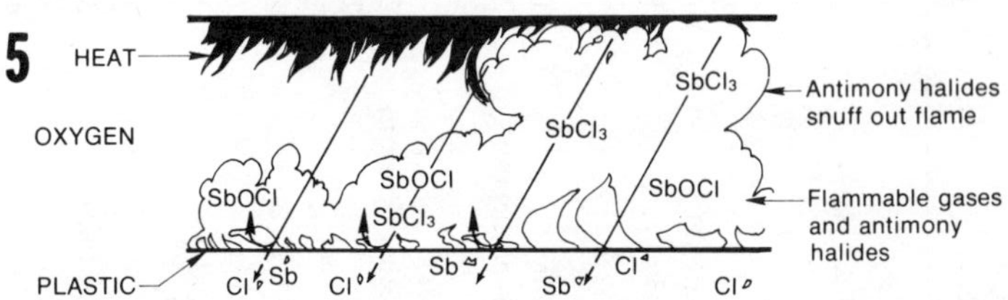

. . . **generating** chemicals that act as flame "snuffers" through a direct effect on the combustion reaction.

The exact mechanisms involved are not fully understood but, because this mode is very efficient, it is the most widely used of the five. Although other elements such as nitrogen and phosphorus may function by this mode, it most commonly utilizes a halogen (chlorine or bromine). In most cases, antimony is also used because it provides a synergistic effect that significantly improves the flame retardancy of the halogen. When the plastic is heated, the antimony and the halogen combine to form a volatile compound that mixes with the flammable gases and chemically inhibits the burning reaction.

Fig. 3-36. How flame retardants do their job. (Lindstrom et al., 1977)

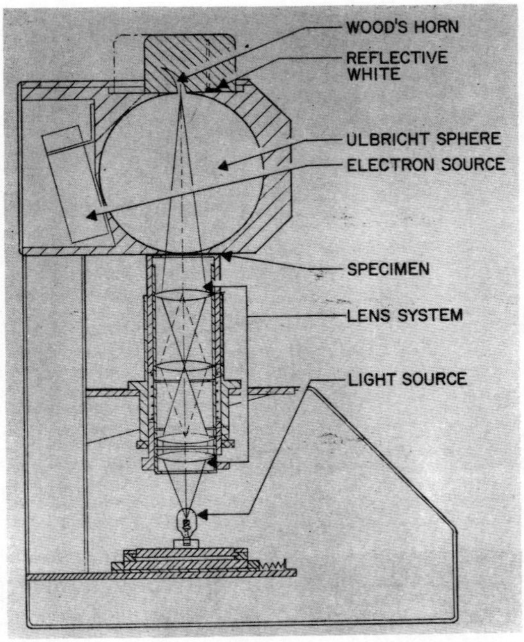

Fig. 3-37. Schematic of haze measurement instrument. (Reprinted by permission from *Plastics Design & Processing*: December, 1968. Copyright 1968, Lake Publishing Corporation, Libertyville, Ill.)

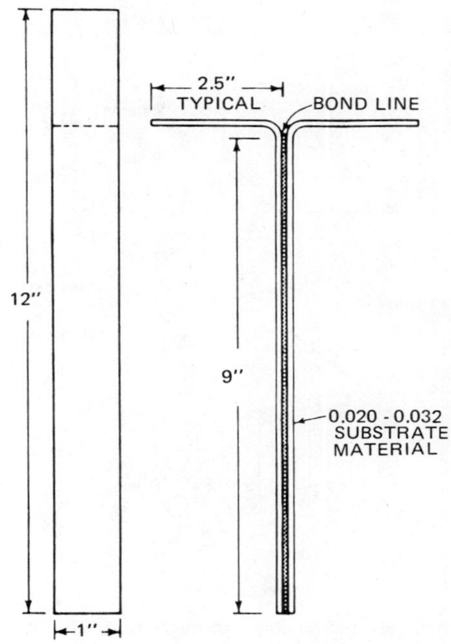

Fig. 3-39. T-peel specimen. (From *Adhesive Bonding Techniques and Applications* by Cagle. Copyright 1968 by McGraw-Hill, Inc. Used with permission of McGraw-Hill Book Co.)

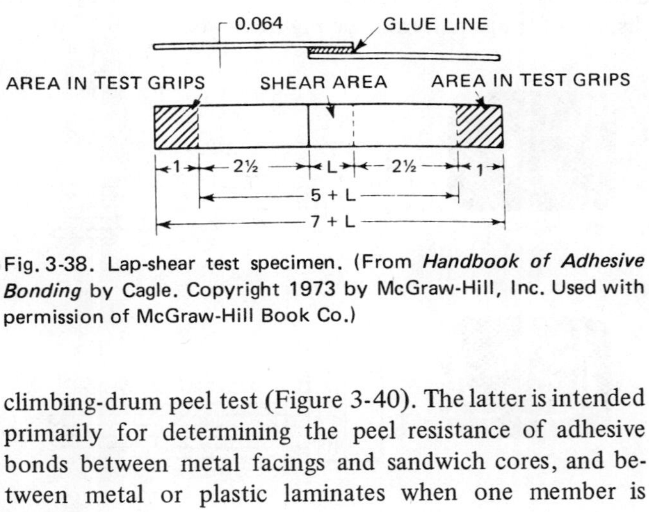

Fig. 3-38. Lap-shear test specimen. (From *Handbook of Adhesive Bonding* by Cagle. Copyright 1973 by McGraw-Hill, Inc. Used with permission of McGraw-Hill Book Co.)

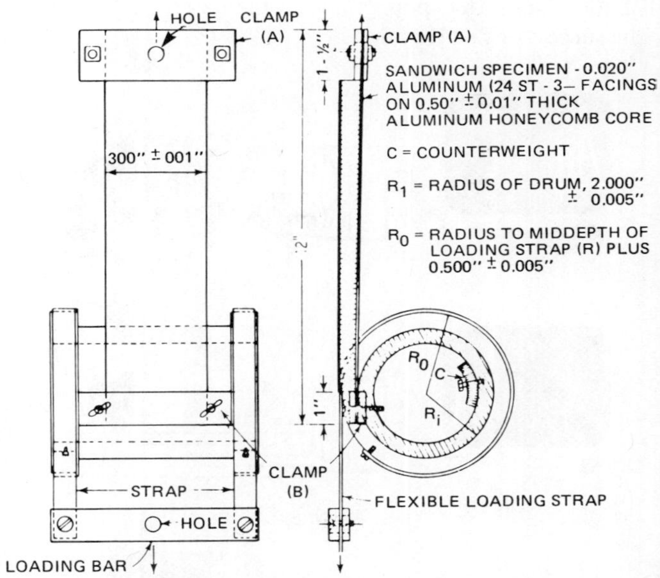

Fig. 3-40. Climbing-drum peel specimen and test fixture. (From *Adhesive Bonding Techniques and Applications* by Cagle. Copyright 1968 by McGraw-Hill, Inc. Used with permission of McGraw-Hill Book Co.)

climbing-drum peel test (Figure 3-40). The latter is intended primarily for determining the peel resistance of adhesive bonds between metal facings and sandwich cores, and between metal or plastic laminates when one member is flexible.

Nondestructive Testing. Plastics are susceptible to flaws which cannot always be seen; thus, premature failure can only be prevented if methods are available that will disclose the presence and extent of such flaws without destroying the plastic part. Three of the more common methods are ultrasonic, gamma-radiometric, and holographic testing.

The *through-transmission gamma radiation gauging technique* (Figure 3-41) is the most direct approach to nondestructively determine local volume fractions of reinforcement, resin, and porosity in composite systems.

Ultrasonic testing is sensitive enough to permit the detection of small defects and penetrates to a degree that makes the examination of thick sections possible. It can pinpoint flaws, distinguish among them, and measure their extent. The basis of ultrasonics in flaw detection (see Figure 3-42) is that any type of wave motion is affected by the interface between areas of dissimilar density. Such interfaces reflect a portion of the energy back to its starting point and reduce

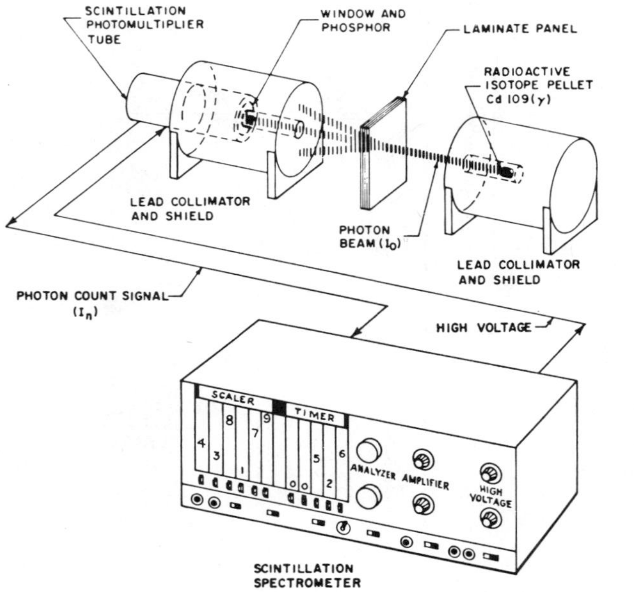

Fig. 3-41. Gamma radiometry through-transmission technique. (Zubrick, 1968)

the level of that which passes through. Proper interpretation of the defect, reflected pulse, or amplitude reduction of the transmitted pulse through the defect, is the key to the success of this technique.

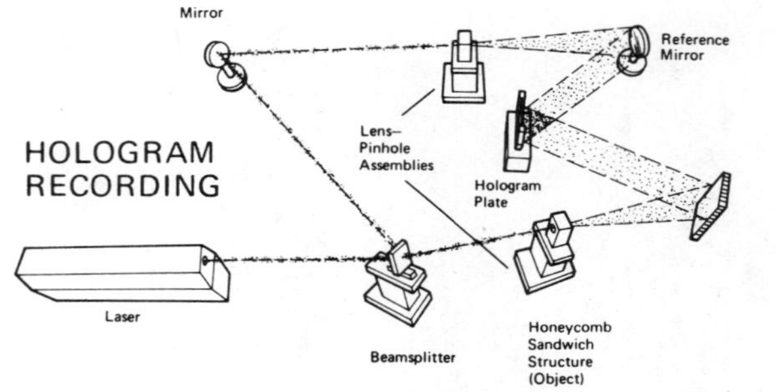

Fig. 3-43. Hologram recording technique. *(Courtesy: G. C. Optronics, Inc.)*

In *holographic analysis*, a beam splitter divides a narrow laser beam into two controllable beams; one goes through the plastics sample (object) and the other is used as a reference (Figure 3-43). Through interferometric recombination, displacements on the level of microinches can be monitored. Low-level stresses can be applied either statically (vacuum, pressure, or heat) or dynamically (variable-frequency ultrasonic transducer).

A hologram is taken of the test object and developed before stressing. If the hologram is then viewed so that the reconstructed virtual image is superimposed on the object, any stress-induced deformation will appear as a dynamic display of fringe patterns (see Figure 3-44).

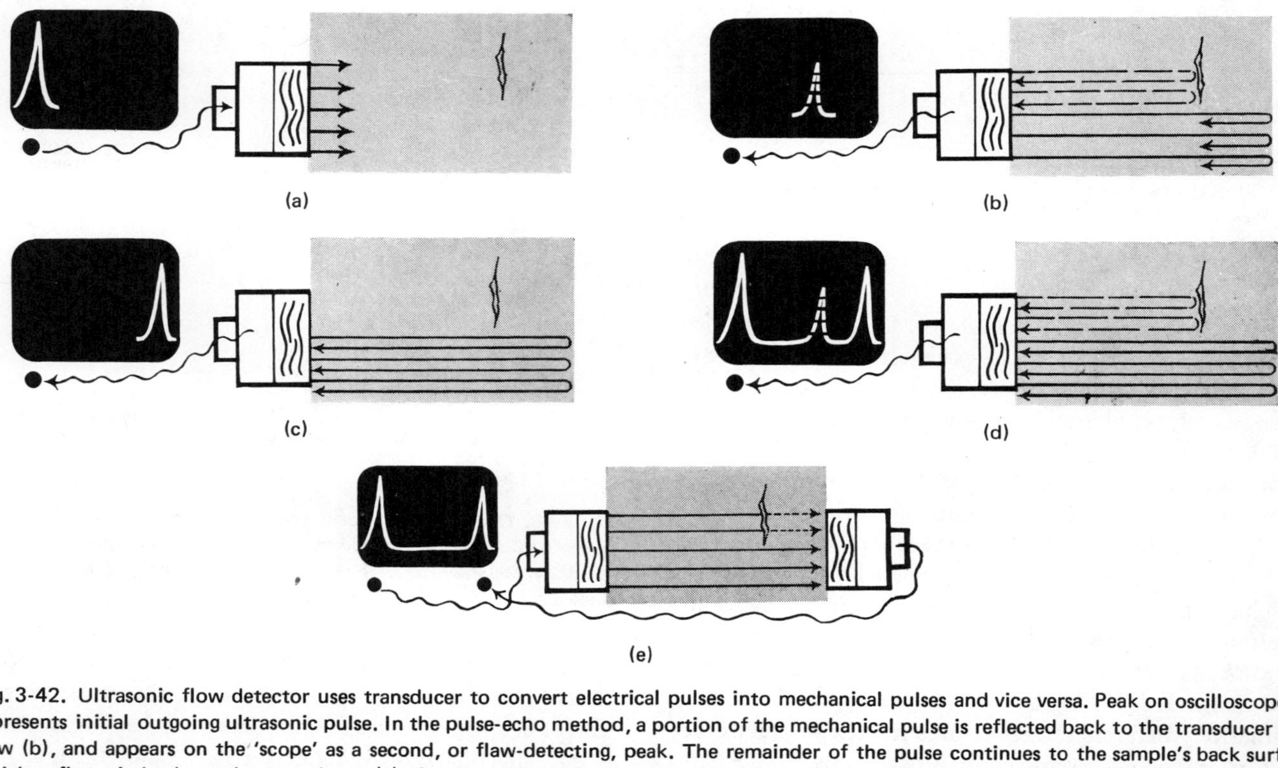

Fig. 3-42. Ultrasonic flow detector uses transducer to convert electrical pulses into mechanical pulses and vice versa. Peak on oscilloscope (a) represents initial outgoing ultrasonic pulse. In the pulse-echo method, a portion of the mechanical pulse is reflected back to the transducer by a flaw (b), and appears on the 'scope' as a second, or flaw-detecting, peak. The remainder of the pulse continues to the sample's back surface, which reflects it back to the transducer (c), forming the third, or back surface, peak. All three peaks are generated in rapid succession but appear simultaneously on the 'scope' (d) because of persistence of vision and lag in response of CRT phosphor. In through-transmission testing (d), a second transducer receives the pulse emitted by the first transducer and generates a second peak on the oscilloscope. The reduced amplitude of the second peak indicates the presence and extent of a defect. (Hand, 1962)

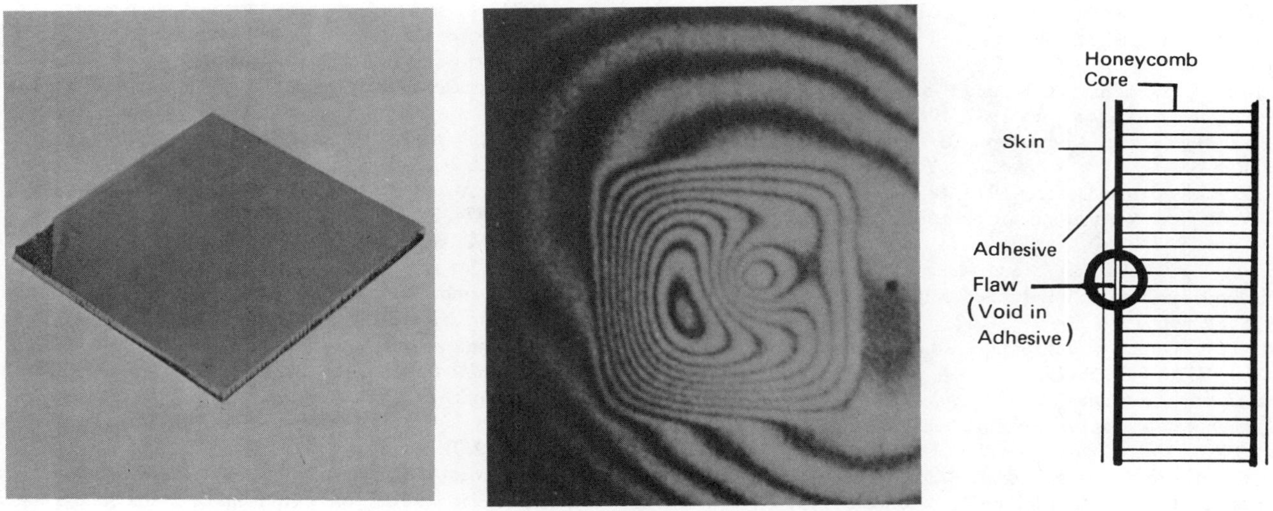

Fig. 3-44. Location, size, and shape of a subsurface defect (adhesive void between skin and core) in an aluminum honeycomb-sandwich panel is clearly revealed in this real-time photograph of a holographic fringe pattern. *(Courtesy: G. C. Optronics, Inc.)*

REFERENCES AND BIBLIOGRAPHY

Abolins, V. Softening point of thermoplastics: deflection temperature vs. Vicat. *Materials Engineering*, April 1975, **81**(4):52–54.

——. Gardner impact versus Izod–which is better for plastics? *Materials Engineering*, November 1973, 78(6):52–54.

ACCO, Wilson Instrument Division. *Rockwell Digital Readout Hardness Testers*. Bulletin 18-11-366, 1978.

——. *Rockwell Hardness Testers*. Bulletin 18-11-341, May 1978.

Black, M. K. Guide to standard tests for plastics: Vicat softening point. *Plastics Design & Processing*, January 1964, **4**(1):20–21.

——. Guide to standard tests for plastics: melting point determinations. *Plastics Design & Processing*, March 1964, **4**(3):22–23.

Borg-Warner Chemicals. *Measurement Reporting and Interpretation of Thermoplastics Properties*. Bulletin P-127.

Bourgault, C. V. Test method characterizes crosslinked, heat-shrinkable polymers. *Plastics Design & Processing*, July 1974, **14**(7):17–19.

Brady, D. G., and Hill, H. W. Jr. How some high-perfomance resins stack up in chemical resistance. *Modern Plastics*, May 1974, **51**(5):60–62.

Cagle, C. V. *Handbook of Adhesive Bonding*. McGraw-Hill Book Co., New York, 1973.

——. *Adhesive Bonding Techniques and Applications*. McGraw-Hill Book Co., New York, 1968.

Caryl, C. R. "Accelerated Outdoor Exposure Testing." Paper presentation, *Regional Technical Conference, Society of Plastics Engineers, Inc*., June 1966.

Cassel, B. "Characterization of Thermosets." Paper presentation, *28th Pittsburgh Conference*, Cleveland, Ohio, March 1977.

Celanese Plastics Co. *Standard Tests on Plastics*. 6th Ed. Bulletin GIC, June 1973.

Chastain, C. E. In designing with plastics, how much should you trust ASTM test data? *Machine Design*, January 23, 1975, **47**(2):107–111.

Craig, C. L. Sr. Electrical properties of plastics–what they mean to the designer. *Plastics Design & Processing*, February 1962, **2**(2):12–16.

Dow Chemical Co. *Impact Resistance and Toughness of Plastic Molding Materials.* Bulletins S-302-19 and 0-306-19, 1973.

——. *Electrical Properties of Plastic Molding Materials*. Bulletin 171-616, 1968.

——. *Permanence Properties of Plastic Molding Materials*. Bulletin 171-639, 1968.

Dreger, D. D. How dependable are accelerated weathering tests for plastics and finishes? *Machine Design*, November 1973, **45**(29):61–67.

FMC Corporation. *Dapon Diallyl Phthalate Resin for Molding Materials*. Bulletin 2500, March 1964.

Gardner Laboratory Inc. *Plastics Testing*. Bulletin B 11, June 1977.

GC Optronics, Inc. *GCO Series 300 Sandwich Structure Analyzers*.

General Electric Co. Foam flammability update. *Foam*, Vol. 2, Issue 2.

Gruber, W. Heat distortion test instrument provides automatic resin quality analysis. *Plastics Design & Processing*, July 1975, **15**(7):16.

Hand, W. Testing reinforced plastics with ultrasonics. *Plastics Technology*, February 1962, **18**(2):35–38.

Houston, A. M. Thermosets and thermoplastics vie on performance, moldability, cost. *Materials Engineering*, March 1976, **83**(3):40–46.

——. Consider variables in choosing plastics for impact resistance. *Materials Engineering*, April 1976, **83**(4):38–40.

Howes, P. T. Jr., and Pears, C. D. Thermal properties of reinforced plastics. *Modern Plastics*, September 1961, **38**(1):140–157, 2–252.

Instron Engineering Corporation. *Microcon I, A Breakthrough in Materials Testing Instrumentation*. Bulletin IC-3-124, 1978.

——. *Servohydraulic Systems for Materials Testing*. Bulletin IC-6-47(A), 1978.

——. *Opening a New Era in Materials Testing*. Bulletins IC-3-65(A) and IC-3-66(A), 1977.

——. *Condensed Catalog*. April 1962.

——. *Jaws and Testing Fixtures*, Catalog 1-5-1.

Johnston, G. C. Izod impact testing. *Durez Molder*, December 1962–March 1963.

Jost, H. D. Quick control tests for strength and rigidity of phenolics. *SPE Journal*, February 1963, **19**(2):208–210.

Kaufman, K. A. Environmental stress cracking of ethylene plastics. *Modern Plastics*, February 1957, **34**(6):146–148.

Kirchner, R. F., and Smith D. J. Melt temperature uniformity and profile control in extrusion coating. *TAPPI, The Journal of the Technical Association of the Pulp and Paper Industry*, July 1974, **57**(7):90–93.

Lamond, K. Right perspective on test data matches materials to application criteria. *Plastics Design & Processing*, March 1976, **16**(3):6–10.

Lindstrom, R. S., Sidman, K. R., and Lanzon, K. S. Flame retardant plastics. *Machine Design*, July 1977, **49**(16):83–87.

Materials Engineering. Comparison of materials. February 1975, **81**(2):18–58.

——. Plastics vs. time. October 1973, **78**(5):29–33.

May, W. P. New test method for plastisol foams. *Plastics Technology*, June 1977, **23**(6).

Miller, D. P., Petrella, R. V., and Menca, A. Smoke and toxic-gas emission from burning unsaturated polyester resins. *Modern Plastics*, September 1976, **53**(9):95–103.

Mock, J. A. Guide to destructive testing. *Materials Engineering*, August 1977, **86**(2):18–29.

Owens-Corning Fiberglas. *Recommended Testing Procedures for Polyester Resins, Control Molding*. Bulletin No. 1-RC-6525, October 1974.

——. *Owens-Corning Press Molding Resins–Recommended Testing Procedures*. Bulletin No. 1-RC-6679, December 1974.

PLASTEC Note 17, *Guide to Test Methods for Plastics and Related Materials*. Plastics Technical Evaluation Center, Picatinny Arsenal, Dover, New Jersey, August 1977.

Plastics Design & Processing. A new method for haze measurement of plastic films. December 1968, pp. 29–30.

Plastics Design Forum. Protecting your product from mother nature. November/December 1977.

Plastics World. Chemical resistance of many common plastics. February 17, 1975, **33**(1):27.

——. Data on glass-reinforced and unmodified thermoplastics. October 21, 1974.

Richardson, P. N. Guide to standard physical tests for plastics–deflection temperature under load. *Plastics Design & Processing*, January 1963, **3**(1):33–35.

——. Guide to standard physical tests for plastics–hardness. *Plastics Design & Processing*, March 1963, **3**(3):30–31.

——. Guide to standard physical tests for plastics–Izod impact strength. *Plastics Design & Processing*, March 1963, **3**(3):30–31.

——. Guide to standard physical tests for plastics–fatigue and creep. *Plastics Design & Processing*, April 1963, **3**(4):27–29.

——. Guide to standard physical tests for plastics–compressive properties and shear strength. *Plastics Design & Processing*, November 1962, **2**(11):21–23.

Riddell, M. N. A guide to better testing of plastics. *Plastics Engineering*, April 1974, **30**(4):71–78.

Robinson, A. K. A statistical evaluation of the portable Barcol Hardness Tester for measuring polyester cure. *Plastics Design & Processing*, November 1962, **2**(11):28–31.

Scavuzzo, J. J., and McNally, C. U. Dimensional stability of thermoplastics. *Plastics World*, March 1961, **19**(3):26–27.

Skiest, I. *Handbook of Adhesives*. 2nd Ed. Van Nostrand Reinhold Co., New York, 1977.

Supnick, R. H. Mechanical testing of plastics. *SPE Journal*, September 1966, **22**(9):21–27.

Taber Instrument Corporation. *Model 139-B Shear Hardness and Diamond Scratch Tester for Plastics and Organic Coatings*. Instruction Bulletin 5601.

Theberge, J. E., Arkles, B., and Cloud, P. How time and heat affect properties of plastics. *Machine Design*, March 20, 1975, **47**(6):79–81.

Tinius Olsen Testing Machine Company. *Testing Machines for Plastics*. Bulletin 71-D, July 1978.

——. *Impact Testing Machines*. Bulletin 99, November 1976.

Vaccari, J. A. NMAB questions fire safety of polymeric materials in aircraft. *Materials Engineering*, June 1977, **85**(6):31–40.

Whittington, L. R. *Whittington's Dictionary of Plastics*. Technomic Publishing Company, Inc., Stamford, Connecticut, 1968.

Winding, C. C., and Hiatt, C. G. *Polymeric Materials*. McGraw-Hill Book Co., New York, 1961.

Zubrick, J. R. Nondestructive testing of glass fiber-RP: key to composition characterization and design properties prediction. *SPE Journal*, September 1968, **24**:56–60.

4
Rheology

Rheology is the study of flow characteristics of fluids. In idealized fluids, Newton postulated that resistance to flow is proportional to the rate of shear. Elastic solids were also described by Hooke as having a stress proportional to strain.

Earlier we stated that plastics can be thought of as extremely high-viscosity fluids (i.e., they are solid). We might then surmise that the application of a strain (shear) to a plastic will produce a corresponding stress (resistance to flow). Such in fact occurs in many plastics processing operations. However, rarely do plastics behave ideally (i.e., obey precisely Newtonian behavior).

Measuring Equipment

In order to visualize the types of rheological flow normally associated with plastics, it is necessary first to examine how rheological properties are obtained. To generate stress-strain data, an instrument must be capable of shearing a fluid at a known rate, and must be capable of producing a value of the stress associated with this shear rate. There are many such instruments, often proprietary in nature, with complex configurations that make it extremely difficult to establish rigorous stress-strain relationships. Figure 4-1 shows one example, a Plasti-Corder. These devices usually provide data in the form of units specific to the individual instrument, and the units are often untransferable to common rheological units such as centipoise.

The extrusion rheometer and the rotational viscometer are two types of instruments that can yield rheological values which generate rate-stress relationships.

The extrusion rheometer (see Figure 4-2) consists of a large cylinder with a cylindrical orifice attached to the bottom. The material in the large cylinder is forced through the orifice by a controlled pressure. The pressure exerted is a measure of the shearing stress, and the amount of extruded material per unit time is a measure of the shear rate.

The rotational viscometer (see Figure 4-3) consists of a small cylinder rotating in a larger cylinder. The material fills the annular space between the two cylinders. It is not necessary for the small cylinder to rotate; the large one may rotate while the small cylinder remains in position. The difference in torque required for the rotation is a measure of the shearing stress, and the speed of rotation is a measure of the rate of shear.

Two types of data now can be generated. We can plot

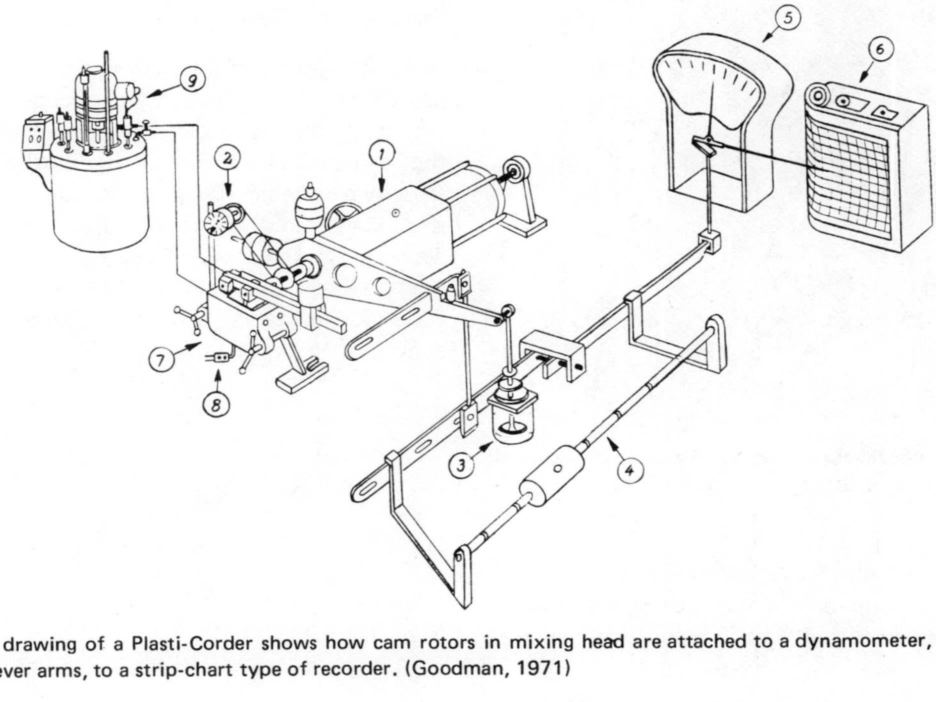

Fig. 4-1. Schematic drawing of a Plasti-Corder shows how cam rotors in mixing head are attached to a dynamometer, which is linked in turn, through a series of lever arms, to a strip-chart type of recorder. (Goodman, 1971)

Fig. 4-2. Melt indexer with optional digit-timer automatic actuating switch, and in-line weight support and lowering device. (*Courtesy: Tinius Olsen Testing Machine Co., Inc.*)

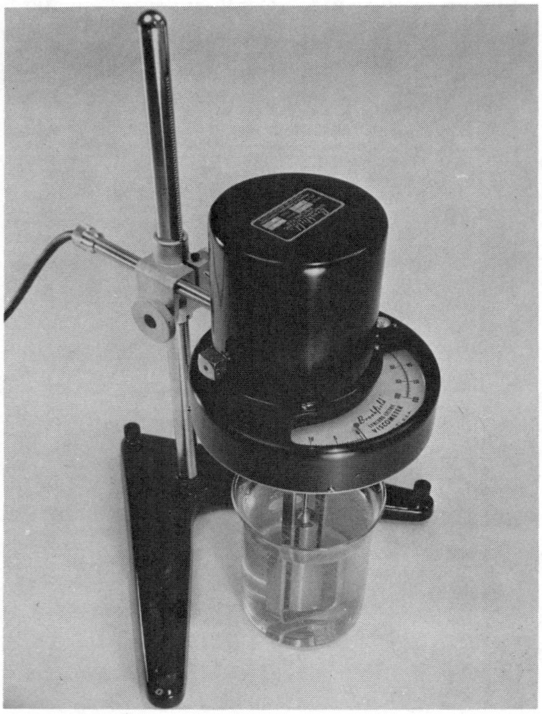

Fig. 4-3. A Brookfield viscometer. (With permission of Brookfield Engineering Laboratories, Inc., Stoughton, Mass.)

extrusion rheometer pressure versus material flow rate, and we can plot viscosity versus rate of shear. Using both these methods lets us examine the five basic rheological parameters common to plastics.

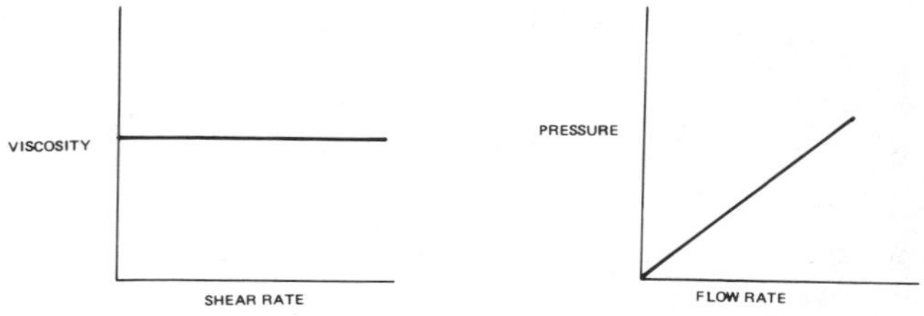

Fig. 4-4. Rheological character of a Newtonian fluid.

TYPES OF FLOW

Newtonian Fluid

A *Newtonian* fluid is one in which a doubling of pressure will double the flow rate (see Figure 4-4). (Increasing shear rate does not change the viscosity.) Water is a good example of such a fluid wherein the stress-strain relationship is proportionately linear.

Bingham Plastics

Grease, on the other hand, acts differently. It does not flow until a certain minimum force has been applied. Once this force is exceeded, flow occurs. This minimum force is referred to as the *yield value* and arises as a result of the strength of the gel bonds in the grease structure. Once the bonds are broken, the grease flows easily.

This type of flow can be represented by the curves in Figure 4-5. Notice that the stress-strain relationship is proportionately linear once flow begins. This type of rheological behavior is called *Bingham plastic* flow, and the type of material which exhibits such flow, a *Bingham plastic*.

Pseudoplastic Flow

A dilute solution of a polymer in a solvent will closely approach Newtonian flow. A highly concentrated solution of the same polymer in the same solvent may form a gel and thus show the characteristics of a Bingham plastic. Between these two extremes, the solution does not have a yield value, since it will flow, yet upon stirring, the viscosity appears to decrease. In a rheometer, the solution extrudes slowly at low pressures, but a doubling of the pressure may cause a fourfold increase in flow rate. Figure 4-6 illustrates this behavior.

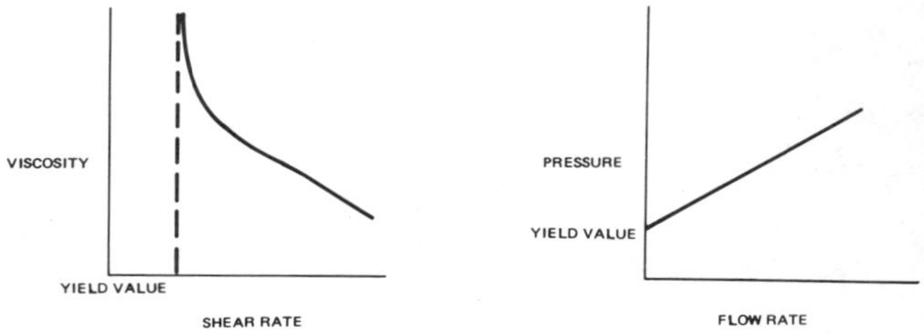

Fig. 4-5. Rheological character of a Bingham plastic.

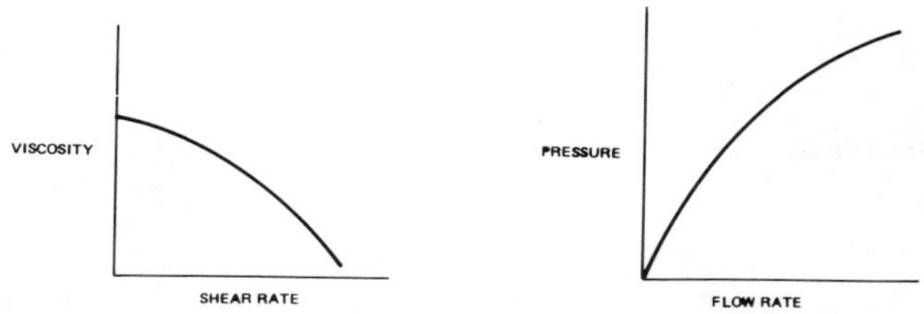

Fig. 4-6. Rheological character of a pseudoplastic fluid.

Occasionally, the term *shear thinning* is used to describe such *pseudoplastic* flow. A polymer that exhibits pseudoplastic flow is called a *pseudoplastic polymer*. Such a polymer approaches Newtonian flow at low shear rates but behaves more like a Bingham plastic at high rates. This behavior is believed to be due to the molecular randomness of the polymer at low flow rates; while at higher flow rates, the molecules tend to orient themselves in the direction of flow and thus offer less resistance.

Dilatancy

The opposite of shear thinning is *shear thickening* or, in rheological terms, *dilatancy*. Solid polymers rarely exhibit dilatant flow, but polymer dispersions or solutions (paint, coatings, adhesives) do occasionally exhibit this phenomenon. In such a case, doubling the pressure may only cause a 50% increase in flow rate. Figure 4-7 illustrates this.

Note that the viscosity/shear-rate curve shows that if greater shear is applied, the viscosity actually increases. This can be helpful for the student who gets caught in quicksand, which is a dilatant fluid. Instead of standing still (applying no shear), the victim is advised to run rapidly with a flat slapping motion of the feet (applying high lateral shear). This will cause the quicksand to thicken sufficiently for support. The theoretical basis for this condition lies in the fact that, while standing still, the sand particles can glide past each other and allow the victim's body to displace the moving sand. At the high shear rate, however, the sand particles interfere with each other's motion, making it necessary for a faster-flowing particle to jump an adjacent particle, thus causing an expansion effect.

As a rule, unfortunately, the victim does not discover he's dealing with a dilatant fluid until his body has begun to submerge, in which case other remedial measures are called for.

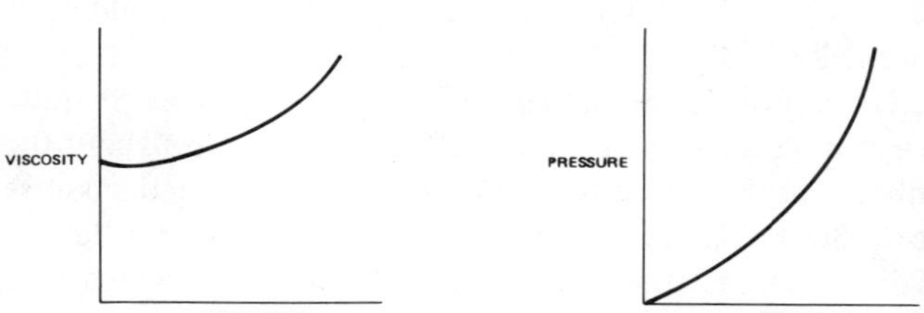

Fig. 4-7. Rheological character of a dilatant fluid.

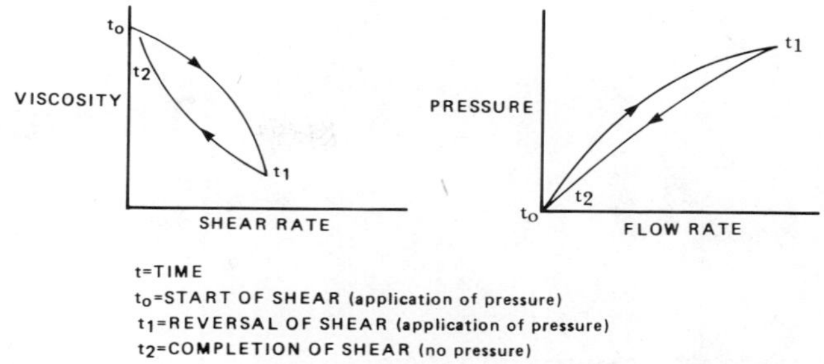

Fig. 4-8. Rheological character of a thixotropic fluid.

Thixotropy

The final rheological characteristic of interest is *thixotropy*. This term is commonly misused for Bingham plastic and pseudoplastic flow. Although thixotropy resembles them, it involves, additionally, a hysteresis, or time delay, effect. Figure 4-8 illustrates this effect.

Practical Applications

Dilatancy is rarely desirable in polymers or their formulations. Pseudoplasticity is useful in many processing methods of solid plastics which involve shear-induced flow (e.g., extrusion). Thixotropy is often required where sag or flow control is needed (e.g., vertical surface adhesives, caulks, dip coatings, non-sag paints).

Rheological principles are also used to determine molecular weights of polymers, to describe and determine flow properties of polymers during processing, to examine viscosity/temperature behavior, and to correlate with time and degree of crosslinking in thermosets.

REFERENCES AND BIBLIOGRAPHY

Goodman, S. H. Measuring elastomer flow behavior. *Rubber World*, May 1971, **164**(2):61–64.

Haake Brabender. *C. W. Brabender Instruments*. Bulletin GP-75, June 1975.

——. *Haake Viskosimeter*. Bulletin N100,9076, October 1974.

——. *35th Anniversary Symposium–A Compendium of Technical Papers*, 1973.

——. *Rheocord Torque Rheometer Viscosity Measuring System.* Catalog No. RC-100.

Jemal, R. Viscosity measurements in adhesive and sealant systems. *Adhesives Age*, May 1974.

NL Industries. *Rheology Handbook, A Practical Guide to Rheological Additives*. PB 113.

——. *Rheological Additives Selection Chart*. PB 123.

Park, R. A. *Dispersion Resin Handbook*. Firestone Plastics Co.

Severs, E. T. *Rheology of Polymers*. Reinhold Publishing Corporation, New York, 1962.

Smoluk, G. R. (ed.) Rheology, study of flow. *SPE Journal*, December 1971, **27**(12):20–30.

Swanson, J. W. Testing of plastic materials and plastic products. *Western Plastics*, Technical Report No. 40, June 1958.

Tinius Olsen Testing Machine Co. *Testing Machines for Plastics*. Bulletin 71-C, June 1976.

5
Thermoplastics

GENERAL CONSIDERATIONS

There are at the present time six main classes of plastics belonging to the broad category of thermoplastics. In each of these classes, there are many subclasses which include over 50 individual polymer types, copolymers, and alloys. A broad spectrum of properties and characteristics is generated by these materials. Some of the included materials perform more as thermosets rather than thermoplastics. Their chemical structure, method of polymerization, primary field of usage, and/or competitive property status makes it more convenient to classify them as thermoplastics. Many of these polymers can even be crosslinked, yet the crosslinked versions represent such a small percentage of volume that the polymers are still classed as thermoplastics.

It is interesting to note the historical sales trends of these materials. After Hyatt, the only "plastics" commercially available were the "Celluloids"–a thermoplastic group. Then Baekeland's discovery of phenolics presaged the ascendency of thermosets to the point that they dominated sales volume well into World War II. The development of new thermoplastics began in the 1930s but didn't achieve gross commercial success until after 1945. Since that time, thermoplastics have outgrown thermosets to the point where three of the main classes of thermoplastics alone (PE/PP, PVC, and PS and alloys) outsell all other plastics combined by a ratio of about 3:1. In 1977, sales of two of these thermoplastic polymers (polyethylene and polypropylene) by themselves totalled 6.25 million pounds, roughly 40% of the total United States sales (16 million pounds) for that year. Table 5-1 shows the projected sales patterns of principal plastics until 1985 as well as the sales for the 1976–1977 period. Table 5-2 shows the relative costs of the principal plastics as of January 1978.

We will begin our consideration of the thermoplastics by examining the so-called general-purpose classes, and then proceed to the more specialized class of "engineering" thermoplastics. As an aid to the use of abbreviations we may be using in the following discussion, an abbreviated list of ISO-approved plastics abbreviations is included in Table 5-3.

A standard practice for the materials chapters of this book will be to list all trade names at the end of each appropriate section. Should, at the time of writing, there be five or less commercial trade names for a given type of plastic, then they will be listed next to the appropriate section heading.

GENERAL-PURPOSE THERMOPLASTICS

Cellulosics

Cellulose is the chief constituent of nearly all forms of plant life and, as such, is one of the most widely distributed and available chemical raw materials. Theoretically, the supply is unlimited in almost all parts of the world. There are two important sources for chemical processing: cotton linters and wood pulp.

Cotton linters are the short fuzz fibers left on the cotton seed after ginning. Cotton is the purest naturally occurring form of cellulose, and the linters are as pure chemically as the staple fiber used by the textile industry. The chemical purification of linters involves a caustic soda treatment, followed by bleaching. Cellulose is obtained from wood by various pulping processes, but the sulfite process, in which the non-cellulosic constituents are dissolved in a cooking liquor of sulfurous acid and a calcium or magnesium base, provides the largest amount. It is necessary to increase the alpha-cellulose content to 90–96% for acetylation grades. Table 5-4 shows the variation in molecular size for the various types of cellulose.

The best description of the cellulose molecule chain is that of a number of beta-glucose anhydride units linked together by primary valence through oxygen bridges at the 1- and 4- positions, each unit containing three free hydroxyl groups in positions 2, 3, and 6. One of these hydroxyl groups is primary and the other two are secondary.

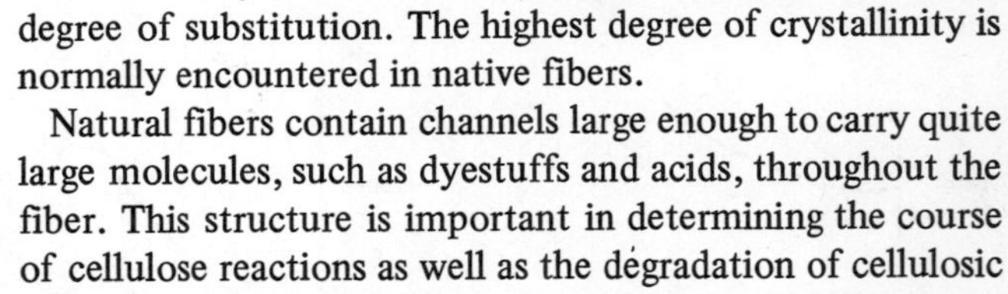

Cellulose and its derivatives may be obtained in states of greater or lesser crystallinity, depending on the treatment to which they have been subjected and the existence and degree of substitution. The highest degree of crystallinity is normally encountered in native fibers.

Natural fibers contain channels large enough to carry quite large molecules, such as dyestuffs and acids, throughout the fiber. This structure is important in determining the course of cellulose reactions as well as the dégradation of cellulosic fabrics. The relatively large channels serve to bring reagents quickly to within a short distance of any given point in the fiber. From this point, the nature of attack is governed by

Table 5-1. Projected Sales Patterns of Principal Plastics until 1985.[a,*]

	1976	1977	1980	1985	Projected Growth Rate, %/yr 1976–1980	1980–1985
Polyethylene						
Low density	5765 (2615)	6350 (2880)	7840 (3556)	10976 (4979)	8.0	7.0
High density	3126 (1418)	3450 (1565)	4412 (2001)	6481 (2940)	9.0	8.0
Polypropylene	2535 (1150)	2700 (1225)	3809 (1728)	5181 (2350)	9.0	8.0
Polyvinyl chloride	4638 (2104)	5078 (2303)	6194 (2810)	8672 (3934)	7.5	7.0
Polystyrene	3144 (1426)	3450 (1565)	4496 (2039)	6294 (2855)	7.5	7.0
ABS, SAN	1098 (498)	1250 (567)	1490 (676)	2139 (970)	8.0	7.5
Other styrenes	740 (336)	777 (352)	899 (408)	1120 (508)	5.0	4.5
Acrylic	488 (221)	525 (238)	659 (299)	946 (429)	7.8	7.5
Cellulosics	154 (70)	160 (73)	177 (80)	211 (96)	3.6	3.6
Epoxy	243 (110)	260 (118)	306 (139)	388 (176)	6.0	5.0
Nylon	247 (112)	282 (128)	348 (158)	522 (237)	9.0	8.5
Phenolic	1339 (607)	1405 (637)	1627 (738)	2077 (942)	5.0	5.0
Polyacetal	88 (40)	100 (45)	126 (57)	189 (86)	9.5	8.5
Polycarbonate	120 (54)	137 (62)	182 (83)	280 (127)	11.0	9.0
Polyesters[c]	947 (430)	1022 (464)	1336 (606)	2004 (909)	9.0	8.5
Thermoplastic polyesters	32 (15)	40 (18)	73 (33)	146 (66)	23.0	15.0
Polyurethane	1623 (736)	1750 (794)	2207 (1001)	3168 (1437)	8.0	7.5
Alkyd, other vinyls	1883 (854)	1977 (892)	2287 (1037)	2919 (1324)	5.0	5.0
Urea, melamine	988 (448)	1060 (481)	1200 (544)	1532 (695)	5.0	5.0
Others[d]	145 (66)	170 (77)	290 (132)	580 (263)	15.0	15.0
Total	20,343 (13,310)	31,943 (14,489)	39,958 (18,125)	55,825 (25,322)	8.0[b]	6.9[b]

[a]Figures in million lb (million kg).
[b]Average.
[c]Including reinforcing.
[d]Fluoroplastic, polybutylene, silicones, thermoplastic urethanes, phenyleneoxide based resins and others.
*Reprinted by permission from *Materials in Design Engineering*, November 1977.

Table 5-2. The Resin-Price Picture as of January 1, 1978.*

Material	Price, ¢/lb.[a]	Material	Price, ¢/lb.[a]
ABS, medium impact	57	Polyethylene, low density	
high impact	61	Liner grade	31½
Acrylic molding powder	56	Clarity film	38½
Cellulosics		Injection molding	31½
Acetate	83	Polypropylene	
Butyrate	85	G-P homopolymer	30
Propionate	85	Medium-impact copolymer	33–35½
Epoxy	76	High-impact copolymer	33–35½
Melamine molding powder	55–59	Polystyrene	
Nylon 6	116	Crystal	28
Nylon 6/6	116	High impact	29
Nylon 11	207	Polyurethane	
Phenolic molding compound	47	Polymeric MDI	58
Polyacetal	94	Polymeric TDI	54
Polycarbonate	109	Flexible slabstock polyol	35
Polyesters		Thermoplastic PUR	135
Unsaturated (thermoset)	36	Polyvinyl chloride	
Thermoplastic (PBT)	98	Homopolymer, dispersion	39
Polyethylene, high density		Homopolymer, suspension	27
G-P blow molding	81¾	Styrene acrylonitrile	40
Injection molding	81¾	Urea molding powder	52–55

[a]Prevailing list price, in largest bulk quantity. General-purpose (G-P) unless otherwise specified.
*Reprinted by permission from *Modern Plastics*, January 1978, McGraw-Hill, Inc.

Table 5-3. Recommended Abbreviations for Some Commonly Used Plastics.

ABS	Acrylonitrile-butadiene-styrene
CA	Cellulose acetate
CAB	Cellulose acetate butyrate
EP	Epoxy
FEP	Fluorinated ethylene propylene
MF	Melamine-formaldehyde
PA	Polyamide (nylon)
PC	Polycarbonate
PE	Polyethylene
PETP	Polyethylene terephthalate
PF	Phenol-formaldehyde
PMMA	Polymethyl methacrylate
POM	Polyoxymethylene (polyformaldehyde; acetals)
PP	Polypropylene
PS	Polystyrene
PTFE	Polytetrafluoroethylene
PUR	Polyurethane
PVAC	Polyvinyl acetate
PVAL	Polyvinyl alcohol
PVC	Polyvinyl chloride
PVDC	Polyvinylidene chloride (saran)
SAN	Styrene-acrylonitrile
UF	Urea-formaldehyde

Table 5-4. Molecular Weights of Cellulose.*

Type of Cellulose	Molecular Weight (weight average)	Number of Glucose Units
Native cellulose	570,000	3,500
Purified cotton linters	150,000–500,000	1,000–3,000
Wood pulps	90,000–150,000	600–1,000

*From *Polymeric Materials* by Winding and Hiatt. Copyright 1961 by McGraw-Hill, Inc. Used with permission of McGraw-Hill Book Co.

the degree of swelling of the fiber; the greater the swelling, the more rapid the attack. Degradation of fibers by dilute aqueous acids, heat, oxygen, and other reagents, producing relatively little swelling, leads primarily to attack on the amorphous, uncrystallized portions of the fiber.

Reactions of Cellulose. There are two general types of cellulose reactions: those involving the hydroxyl groups and those involving the degradation of the cellulose chain. We shall begin our discussion with a consideration of the cellulose-chain degradation reactions, proceed to esterification of the hydroxyl groups, and, finally, consider the etherification of the hydroxyl groups.

Regenerated Cellulose

Viscose Rayon. In the latter part of the 19th century, three processes were competing for the newly created synthetic fiber market: (1) cellulose nitrate (Chardonnet) rayon, (2) the cuprammonium process, and (3) the xanthate process, discovered by Cross and Bevan in 1892. In the beginning, the xanthate (or viscose) process was not very important, but owing to changes in the economics of the processes, most of the regenerated cellulose rayon is now produced by this process.

When cellulose is treated with NaOH solutions, an addition compound is formed that has some of the features of an alcoholate. One of these characteristics is the ability to react with CS_2 to form an ester of dithiocarbonic acid (xanthogenic acid).

$$(C_6H_9O_4)ONa + CS_2 \longrightarrow \begin{array}{l} SNa \\ | \\ C{=}S \\ | \\ O(C_6H_9O_4) \end{array} \qquad (1)$$

This is an idealized representation of the actual reaction, as only one of the anhydroglucose units is shown, and the actual ratio of the $C_6H_{10}O_5$ groups to Na is nearer to 2:1. Also, there are many possible side reactions.

The xanthate ester from (1) dissolves in dilute alkalis to give a so-called viscose solution, which is extruded into a H_2SO_4-Na_2SO_4 bath for coagulation to regenerated cellulose. The regeneration reaction can be represented as follows:

$$\begin{array}{l} SNa \\ | \\ C{=}S \\ | \\ OR \end{array} + H_2SO_4 \longrightarrow \begin{array}{l} SH \\ | \\ C{=}S \\ | \\ OR \end{array} + NaHSO_4 \qquad (2)$$

$$\begin{array}{l} SH \\ | \\ C{=}S \\ | \\ OR \end{array} \longrightarrow HOR + CS_2 \qquad (3)$$

In rayon manufacture, the spinning operation consists of forcing the solution through very fine orifices. In the viscose process, the spinnerettes are usually made of platinum and gold alloys and may contain 3,000–15,000 orifices of 0.002–0.01 inch in diameter.

As the solution is forced through the spinnerette, it contacts a coagulating bath such that the filaments do not stick together, but are hardened and can be removed from the bath and collected. The fibers are then washed and formed into skeins, following which they are desulfurized, bleached, treated with dilute acid, washed and finally dried. Special surface-conditioning agents are also added.

For staple manufacture, the extruded rope, or "tow," of filaments is not twisted, but is washed and cut into 1–8 inch lengths. The product is baled and, when used, it is fluffed and generally blended, combed, and twisted with other noncontinuous fibers such as wool, cotton, or staple

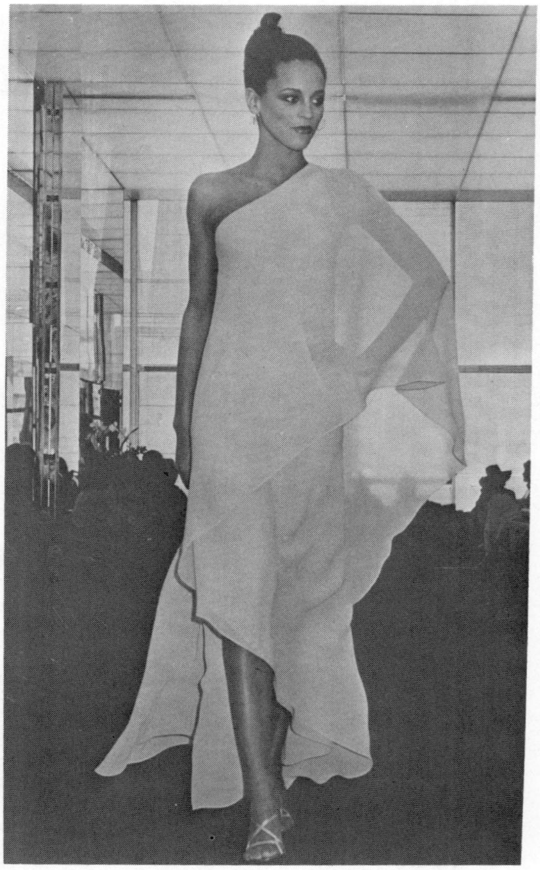

Fig. 5-1. Evening dress of 100% rayon Georgette. (*Courtesy: ITT Rayonier Inc.*)

Fig. 5-2. Dress of cotton/rayon jersey. (*Courtesy: ITT Rayonier Inc.*)

synthetic fibers. (See Figures 5-1 and 5-2.) These blends give attractive dyeing and handling combinations.

A high-tenacity tire cord is made from high-denier rayon. This fiber is obtained by maintaining a high degree of polymerization in the cellulose and by keeping filaments in an incompletely precipitated form so that a high degree of drawing and orientation is obtained. The newer synthetic fibers have almost eliminated rayon as a commercial tire cord today.

Cuprammonium Rayon. This process for the production of regenerated cellulose yields the finest filament of any synthetic fiber. Of all rayon processes it is the most complicated, and the production costs are relatively high. However, it is still manufactured for premium-grade fabrics.

Schweitzer, in 1857, discovered that cellulose can be generated from an ammoniacal solution of copper oxide by dilution with water or weak acids and bases. This solution is filtered and passed through nickel spinnerettes. The fibers are coagulated and stretched, then decopperized, washed, and dried.

Cellophane. A special sheet-forming machine is used to produce cellophane. It consists of a hopper, the bottom of which is a nozzle with a long slit as an orifice from which the viscose solution is extruded under pressure in the form of a sheet. The film is exposed for a short time to air and then is carried into a solution of $(NH_4)_2SO_4$ of carefully controlled concentration and temperature, which coagulates and hardens the deposited film. It then passes through a number of Na_2SO_4, H_2SO_4, and water baths for washing the film.

The regenerated cellulose appears as a clear transparent sheet. It then passes through a bath, wherein it absorbs approximately 7% by weight of glycerol, rendering the sheet flexible. Glucose and polyhydroxy alcohols are also used as plasticizing agents. Most of these plasticizers are chosen because they are hydroscopic and hold moisture in the film. This property is desirable since water is also an effective plasticizing agent. The film is then dried.

Since most food packaging requires a wrapping that maintains the initial moisture content of the food, a sheet that has high moisture impermeability is desirable. The most recent method for obtaining such a low vapor transmission involves coating the cellophane film with a polyvinylidene chloride latex. This coating also reduces gas permeability and affords increased grease and chemical resistance.

Even cellophane is disappearing from the marketplace. Production peaked in 1960 at 440 million pounds. Demand for this film has dropped at a 28% annual rate from 1974–1977, while its price increased by 70% in the same time period.

Vulcanized Fiber. Cellulose sheets which have been gelatinized in aqueous $ZnCl_2$, washed free of chemicals, formed while still moist, and dried into sheets, rods and tubes are used to produce vulcanized fiber.

Vulcanized fiber is one of the strongest, most rigid, and lightweight materials available (about half as heavy as aluminum). It is oilproof but will absorb 20–60% water on long immersion. It has high impact resistance and good machinability. Standard grades are available for punching and forming for dielectric applications. Bone grade has greater hardness and stiffness, and semiflexible grades are made for use as gaskets and packing. An intermediate grade serves for manufacture of items such as suitcases.

Major limitations include lack of flexibility, dimensional instability caused by poor resistance to water absorption, and exceptionally long processing time ($ZnCl_2$ extraction).

Cellulose Esters

Cellulose Nitrate. Bracconot first prepared cellulose nitrate in 1833. The modern method of preparation was introduced in 1845. Hyatt prepared collodion in 1861 and obtained patents on celluloid in 1869.

The esterification between cellulose and HNO_3 can be controlled to yield varying degrees of nitration, depending on the number of hydroxyl groups that have been replaced by nitrate groups. The completely nitrated product is the trinitrate with a nitrogen content of 14.16%. The dinitrate contains 11.3% nitrogen. The various possible products can be classified according to their nitrogen content as follows:

1. Gun cotton: 13.0–13.7% nitrogen, with a degree of polymerization of approximately 1500. This is military gun cotton and is used as a propellant powder.
2. Smokeless powder and dynamite: 12.1–12.8% nitrogen, the degree of polymerization ranges from 800 to 1200. This product is used for rifle and smokeless powders, and gelatin dynamite. It is more soluble than the higher nitrates, but not sufficiently so to use in lacquers.
3. Soluble pyroxylin: 11.5–12.2% nitrogen, degree of polymerization varies from 100 to 500. This material is soluble in all common lacquer solvents and can be produced in almost any viscosity.
4. Plastic grade: 10.5–11.5% nitrogen, degree of polymerization approximately 300–400. This material is used for plastics and dispersed by camphor into a plastic gel.

Although cellulose nitrate did not survive as a textile fiber because it burned so explosively, it became the first synthetic resin. "Celluloid" was registered as a trademark in 1873 and was readily converted into such articles as combs, toothbrush handles, ornaments, collars, and cuffs (see Figures 1-2 and 1-3). To this day, it is made by gelantinizing cellulose nitrate with a solution of camphor in alcohol, mixing in the desired fillers, working it between slightly heated rolls, and forming it with presses (see Figure 5-3).

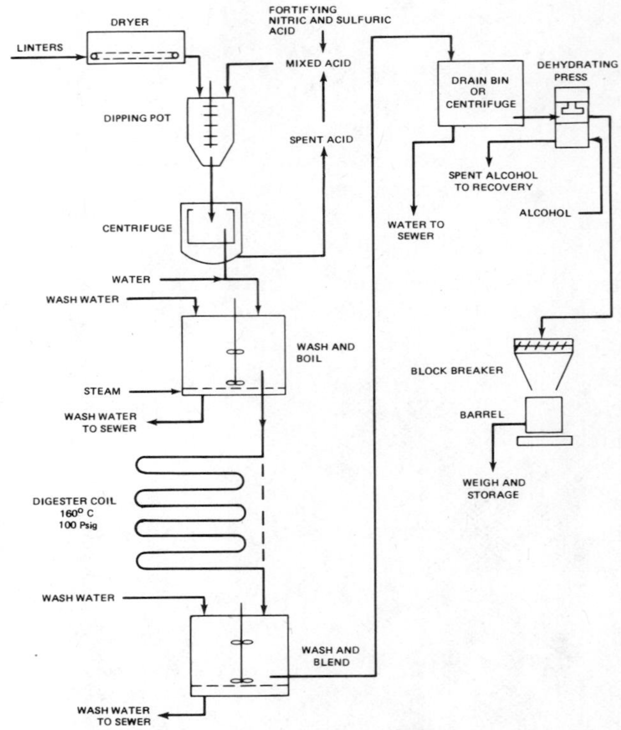

Fig. 5-3. Cellulose nitrate manufacture flow sheet.

Cellulose nitrate is not suited to compression molding because its easy thermoplasticity makes necessary the chilling of the mold before the removal of the molded article. It cannot be used in injection-molding presses because of its sensitivity to high temperature. In addition, the fire hazard caused by the much greater surface area of granules and powder, has almost completely limited its use to the blocking process.

The blocking process consists of filtration of an alcoholic solution of the ester, reduction of the alcohol content through vacuum kneading, and milling to approximately 0.5 inch-thick sheets. The sheets are then pressed under controlled conditions (excess heating often causes explosions) into blocks, cooled, and sliced into sheets from 0.005 to 1.0 inch in thickness. The sheets must then be "aged" from 3 to 56 days to reduce the volatile content to about 2%.

It is possible to extrude alcohol-containing cellulose nitrate under carefully controlled conditions, but such processes are obsolete. Table 5-5 shows typical properties of cellulosic materials.

Cellulose nitrate plastics products are inexpensive, easily worked, readily machined, cemented, swaged, blown, and formed. Practically all colors are available in transparent, translucent, mottled, and opaque effects. The average specific gravity is about 1.41. Cutting, sawing, punching, drilling, drawing, turning, printing, embossing, and polishing may be done with ordinary tools, provided there is a stream of cold water running over the material while it is being worked.

Cellulose nitrate is practically unaffected by hydrocarbons, animal, vegetable and mineral oils, and low concentrations of mineral acids at normal temperatures, except that leaching of plasticizer may occur. It is decomposed by strong alkalis and acids, and dissolved or softened by many alcohols, ketones, and esters. Its resistance to water absorption is fair; however, it is superior to the acetate in this respect.

Cellulose nitrate discolors on exposure to sunlight and embrittles on aging because of loss of plasticizers. This difficulty has been partially overcome through the use of special plasticizers.

Cellulose nitrate in a suitable solvent is used as a household cement and for airplane covering or "dope." If this type of cement contains aluminum powder, it is known as liquid solder. The solvents that are used, as well as the nitrate itself, are highly flammable.

Principal uses of cellulose nitrate are knife handles and table-tennis balls. The latter continue to be made from cellulose nitrate because it has been difficult to match the "bounce" and hand of the ball using newer materials (see Figure 5-4).

Cellulose Acetate. During World War I, cellulose acetate was developed as the result of a search for a less flammable airplane dope. The major method of preparation of cellulose acetate is homogeneous acetylation in which the acetylated cellulose dissolves into solvent as it forms. The cellulose must be first acetylated completely to the triacetate. The triacetate is then hydrolyzed to the resultant desired level of substitution.

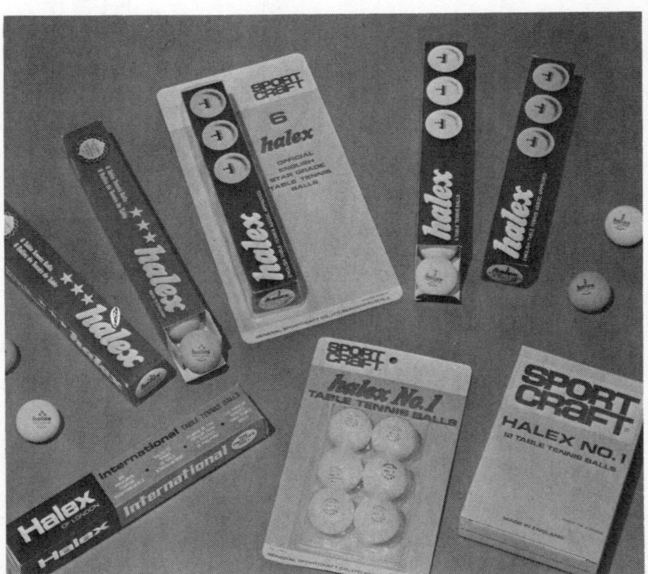

Fig. 5-4. Ping-Pong balls made from cellulose nitrate. *(Courtesy: General Sportcraft Co., Ltd.)*

The cellulose is pretreated, typically with glacial acetic acid. The acetylation is then carried out in a mixture of acetylating agent (typically acetic anhydride), a catalyst (H_2SO_4), and a diluent (acetic acid/methylene chloride). Esterification is complete in 5–6 hours at temperatures below 50°C (122°F), and the product at the end of this stage is called *primary cellulose acetate.*

Following acetylation, the primary acetate must be "ripened." Acetic acid or water is added and hydrolyzation occurs over a 72-hour period. When the required degree of substitution is obtained, the cellulose acetate is precipitated out of a water solution. It is then washed and dried. The product at this stage is called *secondary cellulose acetate.*

Compounding cellulose acetate is achieved through the addition of plasticizers, stabilizers, extenders, lubricants, and color. The prime function of the plasticizers is to provide for easy processing of the polymer as well as to modify such properties as melt viscosity and toughness. Stabilizers are added to improve weather resistance. Compounding is typically done using dry processes such as two-roll mills, internal mixers, and extruders.

Cellulose acetate is the first, truly general-purpose plastic with no one outstanding property (see Table 5-5). It provides toughness and good appearance at a reasonable cost and with low flammability. It is available in translucent, transparent (including water-white) and opaque colors, in mottled and pearled effects. It can be extruded and injection or compression molded.

Cellulose acetates have a high rate of water absorption, poor electrical resistance, and limited resistance to aging, heat, and chemicals. Water resistance is improved by increasing the degree of acetylation. This approach, however, can decrease hardness and impact strength and increase flow temperature.

Table 5-6 and Figure 5-5 show the effect of degree of acetylation on resin properties.

Primary areas of application of cellulose acetate are film and sheeting. Because of its clarity, the film is extensively used in photographic film and packaging of all types. Thin sheet is used in display boxes, and spectacle frames are made from thick sheet. Other major markets include toothbrushes, combs, and lacquers.

Most of the cellulose triacetate produced at the present time is used in the manufacture of textile fiber. It is soluble in many solvents, but this is not too significant a disadvantage since in normal use there are few opportunities for contact with solvents. However, should liquid fingernail polish and polish removers contact the cellulose acetate rayon, it will dissolve it, because these items contain such solvents.

The chief disadvantage of cellulose acetate is its sensitivity to heat. Acetate rayons should not be washed in hot water or subjected to much heat when being pressed. The triace-

Table 5-5. Cellulosic Molding Compounds and Sheet.*

		ASTM test method	Ethyl cellulose molding compounds and sheet	Cellulose acetate		Cellulose propionate molding compound	Cellulose acetate butyrate sheet	Cellulose acetate butyrate, molding	Cellulose nitrate
				Sheet	Molding				
Processing	1. Molding qualities	—	Excellent	Excellent	Excellent	Excellent	Excellent	Excellent	Good
	2. Compression molding temp., °F.	—	250-390	—	260-420	265-400	—	265-390	185-250
	3. Comp. molding pressure, p.s.i.	—	500-5000	—	100-5000	100-5000	—	100-5000	2000-5000
	4. Injection molding temp., °F.	—	350-500	—	335-490	335-515	—	335-480	—
	5. Injection molding pressure, p.s.i.	—	8000-32000	—	8000-32000	8000-32000	—	8000-32000	—
	6. Compression ratio	—	1.8-2.4	—	1.8-2.6	1.8-2.4	—	1.8-2.4	—
	7. Mold (linear) shrinkage, in./in.	—	0.005-0.009	—	0.003-0.010(C) 0.003-0.008(I)	0.003-0.009(C) 0.003-0.006(I)	—	0.003-0.009(C) 0.003-0.006(I)	—
	8. Specific gravity (density)	D792	1.09-1.17	1.28-1.32	1.22-1.34	1.17-1.24	1.15-1.22	1.15-1.22	1.35-1.40
	9. Specific volume, cu. in./lb.	D792	25.5-23.6	21.7-21.0	22.7-20.6	23.4-22.4	24.1-22.7	24.1-22.7	20.5-19.8
	10. Machining qualities	—	Good	Excellent	Excellent	Excellent	Excellent	Excellent	Excellent
Mechanical	11. Tensile strength, p.s.i.	D638	2000-8000	4500-8000	1900-9000	2000-7800	2600-6900	2600-6900	7000-8000
	12. Elongation, %	D638	5.0-40.0	20.0-50.0	6.0-70.0	29.0-100.0	50.0-100.0	40.0-88.0	40.0-45.0
	13. Tensile elastic modulus, 10^5 p.s.i.	D638	1.0-3.0	3.0-6.0	0.65-4.0	0.6-2.15	2.0-2.5	0.5-2.0	1.9-2.2
	14. Compressive strength, p.s.i.	D695	10000-35000	—	2000-36000	2400-22000	—	2100-22000	22000-35000
	15. Flexural yield strength, p.s.i.	D790	4000-12000	6000-10000	2000-16000	2900-11400	4000-9000	1800-9300	9000-11000
	16. Impact strength, ft. lb./in. of notch (½ × ½ in. notched bar, Izod test)	D256	2.0-8.5 0.3-1.7 at −40°F.	—	0.4-5.2	0.5-11.5	—	0.8-6.3	5.0-7.0
	17. Hardness, Rockwell	D785	R50-R115	R85-R120	R34-R125	R10-R122	R30-R115	R31-R116	R95-R115
	18. Flexural elastic modulus, p.s.i. × 10^5	D790	—	—	—	1.5-3.4	—	—	—
	19. Compressive modulus, p.s.i. × 10^5	D695	—	—	—	—	—	—	—
Thermal	20. Thermal conductivity, 10^{-4} cal./sec./sq. cm./1 (°C./cm.)	C177	3.8-7.0	4.0-8.0	4.0-8.0	4.0-8.0	4.0-8.0	4.0-8.0	5.5
	21. Specific heat, cal./°C./gm.	—	0.3-0.75	0.3-0.5	0.3-0.42	0.30-0.40	0.3-0.4	0.3-0.4	0.3-0.4
	22. Thermal expansion, 10^{-5} in./in./°C.	D696	10-20	10-15	8-18	11-17	11-17	11-17	8-12
	23. Resistance to heat, °F. (continuous)	—	115-185	140-220	140-220	155-220	140-220	140-220	ca. 140
	24. Deflection temp., °F. at 264 p.s.i. fiber stress	D648	115-190	—	111-195	111-228	—	113-202	140-160
	at 66 p.s.i. fiber stress		—	—	120-209	147-250	—	130-227	—
Electrical	25. Volume resistivity, ohm-cm. (50% RH and 23° C.)	D257	10^{12}-10^{14}	10^{11}-10^{15}	10^{10}-10^{14}	10^{12}-10^{16}	10^{11}-10^{15}	10^{11}-10^{15}	10-15 × 10^{10}
	26. Dielectric strength, short-time, ⅛-in. thickness, volts/mil	D149	350-500	250-600	250-600	300-450	250-400	250-400	300-600
	27. Dielectric strength, step-by-step, ⅛-in. thickness, volts/mil	D149	300-500	200-520	200-520	230-385	225-320	225-320	250-550
	28. Dielectric constant, 60 cyc.	D150	3.0-4.2	4.0-5.0	3.5-7.5	3.7-4.3	4.0-5.0	3.5-6.4	7.0-7.5
	29. Dielectric constant, 10^3 cyc.	D150	3.0-4.1	4.0-5.0	3.4-7.0	3.6-4.3	4.0-5.0	3.4-6.4	7.0
	30. Dielectric constant, 10^6 cyc.	D150	2.8-3.9	4.0-5.0	3.2-7.0	3.3-3.8	4.0-5.0	3.2-6.2	6.4
	31. Dissipation (power) factor, 60 cyc.	D150	0.005-0.020	0.01-0.02	0.01-0.06	0.01-0.04	0.01-0.02	0.01-0.04	0.09-0.12
	32. Dissipation (power) factor, 10^3 cyc.	D150	0.002-0.020	0.01-0.03	0.01-0.07	0.01-0.04	0.01-0.03	0.01-0.04	0.03
	33. Dissipation (power) factor, 10^6 cyc.	D150	0.010-0.060	0.03-0.04	0.01-0.10	0.01-0.05	0.02-0.04	0.01-0.04	0.06-0.09
	34. Arc resistance, sec.	D495	60-80	180-200	50-310	175-190	—	—	—
Optical	35. Refractive index, n_D	D542	1.47	1.49-1.50	1.46-1.50	1.46-1.49	1.46-1.49	1.46-1.49	1.49-1.51
	36. Clarity	—	Transparent, translucent, opaque					—	—
	37. Transmittance, %	—	—	88 (⅛-in. thick)	—	88 (⅛-in. thick)	88 (⅛-in. thick)	—	—
	38. Haze, %	—	—	<1 (film, thin sheet)	—	<1 (film, thin sheet)	<1 (film, thin sheet)	—	—

Resistance characteristics

39. Water absorp., 24 hr., ⅛-in. thick, %	D570	0.8-1.8	2.0-7.0	1.7-6.5	1.2-2.8	0.9-2.2	0.9-2.2	1.0-2.0
40. Flammability [b] Burning rate, in./min.	D635	—	—	—	1.0-1.3	—	—	—
AEB, in./ATB, sec. [c]		—	—	—	—	—	—	—
41. Effect of sunlight	—	Slight when properly stabilized	Slight	Slight	Slight	Slight	Slight	Discolors and becomes brittle
42. Effect of weak acids	D543	Slight	Slight	Slight	Slight	Slight	Slight	Slight
43. Effect of strong acids	D543	Decomposes	Decomposes	Decomposes	Decomposes	Decomposes	Decomposes	Decomposes
44. Effect of weak alkalies	D543	None	Slight	Slight	Slight	Slight	Slight	Slight
45. Effect of strong alkalies	D543	Slight	Swells	Decomposes	Decomposes	Decomposes	Decomposes	Decomposes
46. Effect of organic solvents	D543	Widely soluble		Soluble in liquid ketones and esters of lower alcohols; softened or dissolved by chlorinated hydrocarbons and aromatic hydrocarbons; little affected by aliphatic hydrocarbons			—	Widely soluble

[a] This numerical flame spread rating is not intended to reflect hazards presented by this or any other material under actual fire conditions.
[b] AEB = Avg. extent of burning in inches; ATB = Avg. time of burning in seconds.
*Reprinted by permission from *Modern Plastics Encyclopedia*, McGraw-Hill, Inc.

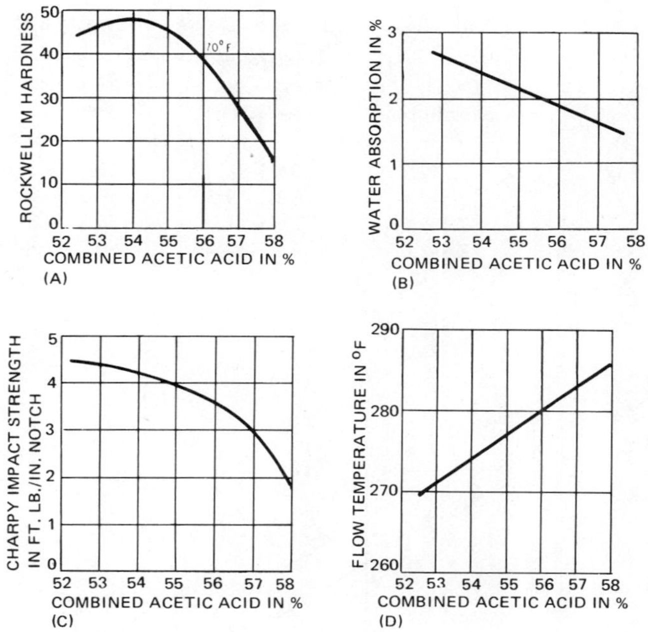

Fig. 5-5. Effect of degree of acetylation on (a) hardness, (b) water absorption, (c) impact strength, and (d) flow temperature (31% plasticizer content). (Brydson, 1975)

tate fiber, however, is still more resistant to elevated temperatures than many other fibers.

An interesting use of continuous-tow acetate is in the manufacture of filter-tip cigarettes. Thousands of separate filaments held parallel are compacted and cut into short plugs which offer low pressure-drop, and yet a great absorbing surface for the collection of tars.

Cellulose Acetate Butyrate. The appearance and uses of cellulose acetate butyrate are similar to those of cellulose acetate. Manufacturing techniques are equivalent to those used for cellulose acetate, using a mixture of butyric and acetic anhydrides. The properties of the resulting product can be varied by altering the relative proportions of the two acids and anhydrides used, as well as the degree of hydrolysis, and the nature and amount of plasticizer used. In general, the mixed ester has a greater solubility in a wider range of solvents and is compatible with a greater number of plasticizers than is the straight acetate, the moisture absorption is lower, and the impact strength is higher. Molded articles made from cellulose acetate butyrate have excellent dimensional stability. (See Table 5-5 for properties of cellulose esters.)

A typical cellulose acetate butyrate used in plastics contains about 13% acetyl and 37% butyryl. A much smaller quantity of plasticizer is required with cellulose-mixed esters than with cellulose acetate. As the butyryl content is raised, less plasticizer is required and compatibility with various plasticizers is greater. The resulting product is thus more stable because of the reduced plasticizer loss. See Figure 5-6 for the effect of butyryl content on properties.

Table 5-6. Influence of Degree of Substitution on the Properties and Uses of Cellulose Acetate.

Degree of Substitution	Acetyl Content (%–$COCH_3$)	Acetic Acid Yield (%)	Solubility	Uses
2.2–2.3	36.5–38.0	52.1–54.3	Soluble in acetone	Injection moulding
2.3–2.4	38.0–39.5	54.3–56.4		Film
2.4–2.6	39.5–41.5	56.4–59.3		Lacquers
2.8–3.0	4.25–44.0	60.7–62.8	Insoluble in acetone, soluble in chloroform	Triacetate film and fibre

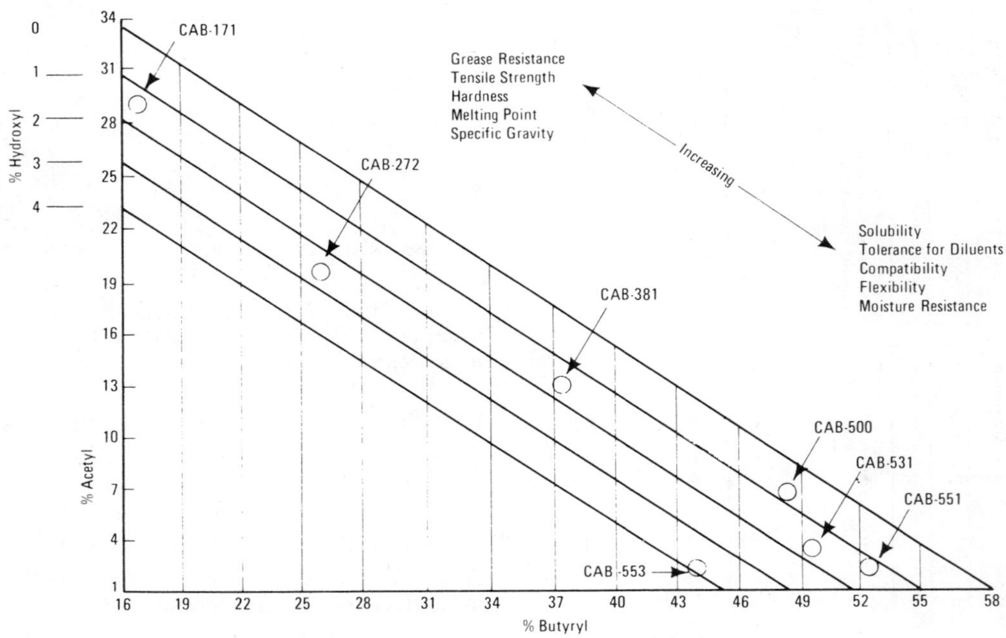

Fig. 5-6. Effect of butyryl content on properties of Eastman cellulose acetate butyrate. (*Courtesy: Eastman Chemical Products, Inc.*)

Cellulose acetate butyrate compounds are available in granular form for injection and compression molding and extrusion, and in sheet form for thermoforming. The molding and thermoforming compositions are suited to applications requiring resistance to humidity and exposure to weather (see Figures 5-7 and 5-8). The butyrates are exceptionally well suited to dry extrusion, which permits production of various shapes in any desired length.

Fig. 5-7. Signal light lenses made from Tenite cellulose acetate butyrate. (*Courtesy: Eastman Chemical Products, Inc.*)

Acetate butyrate compositions are formulated into gel dipping lacquers. Solvents are used to give clear, fluid solutions at around 49°C (120°F) which set to a clear, firm gel on cooling to room temperature. Relatively thick coatings result. The process is especially applicable for coating shaped articles of metal, wood, or cardboard. A low-viscosity type, termed half-second butyrate, is available for compounding into lacquers with excellent weathering properties. Compositions are recommended for cloth, paper, and wood, and melted coatings for adhesives and for special lacquers for plastics.

A higher butyryl-type ester is the base material in strippable protective coatings. Hot-melt compositions which are fluid at 121–177°C (250–350°F) allow dip-coating of tools and gauges for protection during temporary storage. No volatile solvent is used (see Figure 5-9).

Cellulose Propionate. Efforts to produce a cellulosic resin with improved processing characteristics led to the develop-

Fig. 5-8. Large outdoor signs made of Uvex cellulose acetate butyrate sheet. (*Courtesy: Eastman Chemical Products, Inc.*)

ment of cellulose propionate. The same techniques and equipment are used for the manufacture of this ester as are used for cellulose acetate and cellulose acetate butyrate. The successful commercialization of this ester (as well as the mixed acetate-propionate ester) was retarded due to the cost and availability of propionic acid.

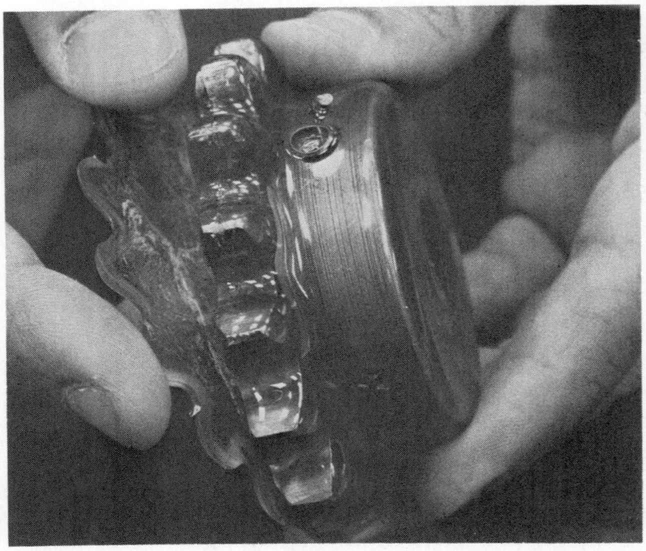

Fig. 5-9. Hot-melt cellulose acetate butyrate coatings are easy to apply and remove. After the part is dipped in a molten bath at about 350°F for less than a minute and allowed to dry, the coating is punctured with a fingernail or wooden stick and peeled off. The coating can then be reused after melting. (*Courtesy: Eastman Chemical Products, Inc.*)

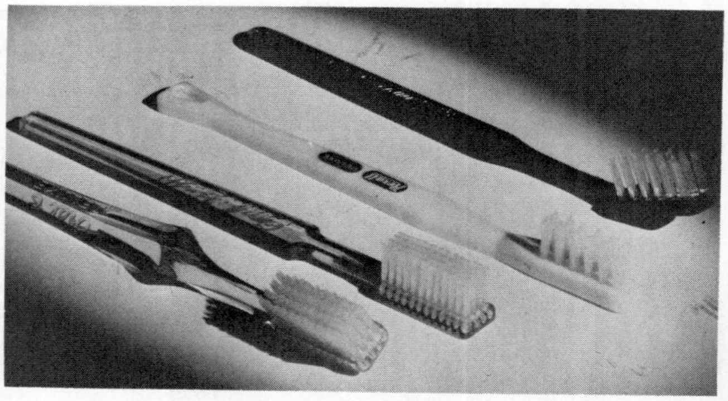

Fig. 5-10. Tenite cellulose propionate toothbrush handles. (*Courtesy: Eastman Chemical Products, Inc.*)

The composition of propionate formulations are also similar to the other acetate esters, except that the level of required plasticizer for equivalent flow properties lies between the acetate and the acetate-butyrate types. (See Table 5-5 for properties of cellulose esters.)

Compared to cellulose acetate, the propionates have approximately 6% lower specific gravity, higher resistance to plasticizer exudation and migration during processing, and, at elevated temperature and humidity conditions, lower moisture absorption and improved dimensional stability.

Typical applications include filter bowls, telephone knobs and dials, safety flasher lenses, film cases, handles (see Figure 5-10), safety goggles, steering wheels, and extruded sheet and thin wall tubing. Both the propionate and butyrate esters are used extensively for photographic film. They provide the excellent wet and dry dimensional stability, clarity, and slow burning characteristics needed for safety film.

Blister packages made from propionate cost the same as packages made from rigid polyvinyl chloride. However, the propionate provides better visibility, product protection, and is more resistant to distortion during storage at elevated temperatures (see Table 5-7). The propionate esters also vacuum-form better and are easier to effect proper trim.

Cellulose Ethers

Manufacture. All commercial ethers are prepared via the Williamson synthesis. First the cellulose is treated with caustic at elevated temperature. The alkali cellulose thus produced is pressed free of excess alkali and dried to a 10–25% moisture content.

The alkali cellulose is then reacted under pressure with the appropriate alkyl chloride at temperatures from 60°C to 140°C (140–284°F) for about 12 hours, to produce ethyl or methyl cellulose. Ethylene oxide is the coreactant used to produce hydroxyethyl cellulose, and the sodium salt of chloracetic acid is used for the production of carboxymethyl cellulose. All the ethers are washed and dried after reaction.

Table 5-7. Properties of Blister Packaging Film Extruded from TENITE Butyrate, Propionate, and Acetate.*

		10-Mil Sheet			
			TENITE Propionate		
Property	ASTM Method	TENITE Butyrate MH Flow	H3 Flow	H2 Flow	TENITE Acetate H3 Flow
Density, g/cm^3	D 1505	1.20	1.20	1.20	1.30
Haze, %	D 1003	0.5	0.5	0.5	0.5
Gloss at 45°, Gardner Rating	D 2457	85	85	85	85
Transparency, %	D 1746	85	85	85	75
Elmendorf tear strength, M.D.	D 1922	200	190	190	230
grams T.D.		200	190	190	250
Dart impact strength, $\frac{1}{2}''$ diameter	D 1709				
Dart at 26 in. at 73°F(23°C), grams	modified	550	480	525	500
Tensile strength					
At yield, psi (MPa) M.D.	D 882	4700(32.4)	4800(33.1)	4500(31.0)	5500(37.9)
T.D.		4500(31.2)	4600(31.7)	4500(31.0)	5200(35.9)
At fracture, psi (MPa) M.D.	D 882	5400(37.2)	5500(37.9)	4900(33.8)	5600(38.6)
T.D.		5200(35.9)	5300(36.5)	5000(34.5)	5300(36.5)
Ultimate elongation, % M.D.	D 882	100	90	90	50
T.D.		100	105	105	50
Tensile modulus of elasticity, M.D.	D 882	200,000(1379)	210,000(1448)	191,000(1317)	225,000(1551)
psi (MPa) T.D.		200,000(1379)	210,000(1448)	187,000(1289)	225,000(1551)
Water vapor transmission rate (grams/24 h)/100 sq in. at 23°C from 100% R.H. to 50% R.H.	E 96-B	4	4	4	5
Gas transmission rate					
$cm^3/m^2 \cdot 24\ h \cdot atm$	D 1434-V				
CO_2		15,000	14,000	–	1,600
O_2		1,500	1,200	–	350
N_2		500	400	–	100

*Courtesy: Eastman Chemical Products, Inc.

Ethyl Cellulose. In the ether ethyl cellulose, the ethyl groups have replaced the hydrogens in the hydroxyl groups of the glucose residues. The standard ether cellulose has a substitution of 2.4–2.5 ethoxyl groups per glucose unit.

Ethyl cellulose plastics containing this proportion of ethoxyl groups (47–48%) provide a wide range of useful properties. They contain a combination of desirable characteristics which include unusual toughness, thermoplasticity, and low flammability. The unplasticized material is flexible at temperatures as low as -40°C (-40°F), and it is fairly extensible. It is inert to alkalis of all strengths and has excellent solubility and compatibility with a wide variety of solvents, resins, waxes, oils, and plasticizers. Its moisture absorption of 2% is somewhat better than cellulose acetate, but slightly poorer than cellulose nitrate. Ethyl cellulose has a specific gravity of 1.14, compared to 1.32 for cellulose acetate and 1.47 for cellulose nitrate. Ethyl cellulose dissolves readily in hot resins, oil, waxes, plasticizers, and in just about all organic liquids except aliphatic hydrocarbons. Formulations can be applied as molten mixtures at elevated temperatures (i.e., hot-melts) and form the basis for some adhesives and coatings. Table 5-8 shows examples of ethyl cellulose hot-melt formulations and their properties.

Ethyl cellulose mechanical properties are adequate over a wide range of temperatures and humidity conditions. Its electrical insulating properties are satisfactory for many purposes, particularly in circumstances where its ability to retain high flexibility at low temperatures is useful. In applications such as gasketing for extreme low-temperature conditions, many resins embrittle. Ethyl cellulose, however, retains the necessary flexibility for dependable use.

One of the most important applications of ethyl cellulose is its use with other resins in lacquers, varnishes, and adhesives to improve toughness and flexibility. It is added to waxes to increase strength and melting point. A wide variety of films and foils are made for general-purpose use. Other uses include molded objects of all sorts, fabric coatings, insulation, and extruded tubing.

Methyl and Hydroxyethyl Cellulose. The typical degree of substitution for methyl cellulose is 1.7–1.9. These materials, having a methoxy range of 27–31%, are water soluble. They are more soluble in cold water than in hot, so that solutions become viscous and finally gel as the temperature is raised. This combination of water solubility and the ability to produce viscous solutions has made methyl cellulose useful as

Table 5-8. Examples of Ethyl Cellulose Hot-Melt Formulations and Their Properties.*

	Formula					
	1	2	3	4	5	6
Ingredients						
Ethocel 10 CPS standard ethyl cellulose (1)	15	10	20	20	10	20
Super Beckacite 2000 resin (2)	15	18	–	–	–	–
Staybelite ester #10 (3)	20	–				
Staybelite resin (3)	–	–			63	
Dow resin 276-V9 (1)	–	–	20	40		
Dioctyl phthalate (6)	5	–				
Paraffin wax (135°F)	25	28	32	14	27	32
Microcrystalline wax (4)				12		
Hydrogenated castor oil (5)	20	44	28	14		28
Santicizer B-16 plasticizer (7)						20
Properties						
Specific gravity	1.02	0.99	0.98	1.01	1.01	1.01
Mar resistance	Good	Fair	Fair	Fair	Good	Fair
Water vapor transmission grams/100 in^2/24 hrs	0.70	1.46	1.34	2.40	0.46	2.34
Color	← NEARLY COLORLESS →					
Odor	← VERY SLIGHT →				← SLIGHT →	← NONE →
Heat sealability	Excellent	Good	← Excellent →			
Low temperature flexibility	Good	Fair	Fair	Good	Poor	Excellent
Suggested uses	Soap Wraps, Shelf Paper, Label Coating	Bread Wrapper Adhesive	Non-Contact Food Packages	Frozen Food Packages	Low Cost Paper Adhesive	Drinking Cup Adhesive

(1) Dow Chemical Co.
(2) Reichhold Chemicals Inc.
(3) Hercules Inc.
(4) Socony Vacuum Oil Co.
(5) Emery Industries Inc.
(6) Ohio Apex Co.
(7) Monsanto Co.
*Data extracted from Dow Chemical Co. Bulletin 170-108. Reprinted with permission of the Dow Chemical Co.

a thickening agent in the textile, food, and adhesives industries. Since it is edible, it can be used as a thickener for mayonnaise, salad dressings, sauces, and other thickened food products. It is also employed as an emulsifying and dispersing agent in the textile, cosmetic, and paint industries, and has been used as a greaseproof paper size and coating. Hydroxyethyl cellulose is used in similar applications for essentially the same reasons.

Carboxymethyl Celluiose. The carboxymethyl cellulose ether has a relatively low degree of substitution. As the degree of substitution increases from 1.0 to 0.5 carboxymethyl group/glucose unit, the solubility of the polymer in water or dilute (6%) alkali increases.

Commercial grades have about 0.7-0.9 degree of substitution and are not pure materials, but contain residual sodium salts. For most uses, these inorganic salts are physiologically inert.

Water solutions are good protective colloids and find application in fabric-sizing systems, resin-emulsion paints, drilling muds, food thickeners, and ice cream and cosmetic texturizers. A main market for this ether is as a soil-suspension additive in detergents to facilitate the carrying away of dirt.

Trade Names

Regenerated Cellulose		
Arbonier	Cellulose for plastic filler	ITT Rayonier, Inc.
Aristocrat	Rayon fiber	Beaunit Corporation
Avicel	Purified cellulose	American Viscose Corporation
Avril	Rayon fiber	FMC Corporation
Avron	Rayon fiber	FMC Corporation
Belastraw	Rayon fiber	Akzona, Inc.
Bemberg	Rayon fiber	Beaunit Corporation
Bemberg Sarophan	Polymer-coated cellophane	Bemberg Folien Gambit

Briglo	Rayon fiber	Akzona, Inc.
Briphane	Cellulose copolymer film	British Cellophane, Ltd.
Bubblefil	Regenerated cellulose	E.I. Du Pont de Nemours & Co., Inc., Polychemicals Division
Cello	Cellulose film	British Cellophane, Ltd.
Cellokote	Wax-coated cellulose film	UCB-Sidac Corporation
Cellolam	Wax-laminated cellulose film	UCB-Sidac Corporation
Cellophan	Cellulose film	Calle & Co., A.G.
Cellsy	Cellulose film	Darcel, Ltd.
Cellufoil	Transparent cellulose	Alcan Canada Foils
Civona	Rayon fiber	Beaunit Corporation
Coloray	Rayon fiber	Courtaulds
Comiso	Rayon fiber	Beaunit Corporation
Cord fil	Random-cut rayon cord	Rayon Processing Co. of Rhode Island
Cordura	Viscose tire cord	E.I. Du Pont de Nemours & Co., Inc.
Cowal	Rayon fiber	Courtaulds
Crinofil	Rayon fiber	Beaunit Corporation
Crystaton	Rayon fiber	Akzona, Inc.
Cupersil	Rayon fiber	Beaunit Corporation
Cupioni	Rayon fiber	Beaunit Corporation
Cupracolor	Rayon fiber	Beaunit Corporation
Delovisca	Rayon fiber	Delaware Rayon
Delray	Rayon fiber	Delaware Rayon
Diamond	Vulcanized fiber	Continental-Diamond Fibre Corporation
Dream Slub	Rayon fiber	Beaunit Corporation
Dulesco	Rayon fiber	FMC Corporation
Dycel	Cellophane lamination	La Cellophane, S.A. Groupe Rhone-Poulenc
Dylok	Rayon fiber	IRC
Dynopas	Vulcanized fiber sheet and rod	Dynamit Nobel, A.G.
Dynos	Vulcanized fiber sheet and rod	Dynamit Nobel, A.G.
Englo	Rayon fiber	Akzona, Inc.
Fairlite	Cellulose derivative	Shoe Form Co., Inc.
Fairylite	Cellulose derivative	Shoe Form Co., Inc.
Featheray	Rayon fiber	Akzona, Inc.
Fibro	Rayon fiber	Courtaulds
Flackona	Rayon fiber	Beaunit Corporation
Floramer	Cellulose for cellulose esters	ITT Rayonier, Inc.
Forbon	Vulcanized fiber	NVF Co.
Fortisan	Saponified acetate–rayon yarn	Celanese Corp. of America
Glitter	Rayon fiber	Beaunit Corporation
Hasta	Wax-coated cellulose film	UCB-Sidac Corporation
Isal	Coated cellulose film	UCB-Sidac Corporation
Jetspun	Rayon fiber	Akzona, Inc.
"K"	Cellophane	E.I. Du Pont de Nemours & Co., Inc.
Kolorbon	Rayon fiber	Akzona, Inc.
Laktroset	Rayon fiber	IRC
Lastex	Rayon fiber	Uniroyal, Inc.
Mateso	Rayon fiber	Beaunit Mills Corporation
Multi-Shato	Rayon fiber	Beaunit Corporation
Narco	Viscose rayon	Beaunit Corporation
Narcon	Rayon fiber	Beaunit Corporation
Narene	Rayon fiber	Beaunit Corporation
Newbray	Rayon fiber	New Bedford Rayon
New-Dull	Rayon fiber	New Bedford Rayon
Newlow	Rayon fiber	New Bedford Rayon
Nojax	Regenerated cellulose tubing	Union Carbide Corporation
Non-Flam	Cellulose derivative	Shoe Form Co., Inc.
Nub-Lit	Rayon fabric	Beaunit Corporation
Nupron	Rayon fiber	IRC
Nymph Wrap	Moisture-proof cellophane	Sylvania, Inc.
Perglo	Rayon fiber	Akzona, Inc.
Placetate	High-purity cellulose	ITT Rayonier, Inc.
Raybrite	Cellulose reinforcement	ITT Rayonier, Inc.
Raymix	Cut rayon fibers	Rayon Processing Co. of Rhode Island; also Microfibres, Inc.
Rayocord (X)	Cellulose for cellulose esters	ITT Rayonier, Inc.

Sibon	Rayon paper	Kahjin Co.
Sicofane	Regenerated cellulose film	Mazzucchelli Celluloide, S.p.A.
Sicophane	Regenerated cellulose film	Mazzucchelli Celluloide, S.p.A.
Sidac	Transparent cellulose film	UCB-Sidac Corporation
Sidaphane	Regenerated cellulose film	UCB-Sidac Corporation
Skybloom	Rayon fiber	Akzona, Inc.
Skyloft	Rayon fiber	Akzona, Inc.
Softglo	Rayon fiber	Akzona, Inc.
Spaulding	Vulcanized fiber	Spaulding Fiber Co.
Strata	Rayon fiber	Beaunit Corporation
Stratella	Rayon fiber	Beaunit Corporation
Suprenka	Rayon fiber	Akzona, Inc.
Survival	Moisture-proof cellophane	American Viscose Corporation
Survival	Moisture-proof cellophane	FMC Corporation
Sylphwrap	Regenerated cellulose wrapping	FMC Corporation
Taylorite	Vulcanized fiber	Synthane-Taylor Corporation
Taylorite	Vulcanized fiber	Taylor Fibre Co.
Tempra	Rayon fiber	Akzona, Inc.
Terek	Cellulose-coated fabrics	Athol Mfg. Co.
Terekan	Cellulose-coated paper	Athol Mfg. Co.
Tinselfil	Rayon fiber	Beaunit Corporation
Topel	Rayon fiber	Courtaulds
Tusson	Rayon fiber	Beaunit Corporation
Tyret	Cellulose monofilament	E.I. Du Pont de Nemours & Co., Inc.
Veri-Dull	Rayon fiber	Beaunit Corporation
Viscacelle	Cellulose film	British Cellophane, Ltd.
Vitrocelle	Cellophane	Papeterics Dalle & Lecomte
Vul-Cot	Vulcanized fiber	NVF Co.
Wilmington Fibre	Vulcanized fiber	Wilmington Fiber Specialty Co.
Zantrel (Polynosic)	High wet-modulus rayon fiber	Akzona, Inc.
Zephyr	Regenerated cellulose film	Union Carbide Corporation
Cellulose Esters		
Accobond	Film resins	American Cyanamid Co.
Acele	Cellulose diacetate	E.I. Du Pont de Nemours & Co., Inc.
Acelon	Acetate sheet and film	M & B Plastics, Ltd.
Acelor	Acetate yarn	Amcel Europe, S.A.
Acetophane	Acetate film	UCB-Sidac Corporation
Acetyloid	Cellulose acetate	Dainippon Ink & Chemicals America, Inc.
Airloft	Acetate nylon filament yarn	Celanese Corporation
Amerith	Cellulose nitrate	Celanese Corp. of America
Ampol	Cellulose acetate	Americal Polymers, Inc.
Arnel	Cellulose triacetate fiber	Celanese Corp. of America
Astracet	Cellulose acetate butyrate sheets	Dynamit Nobel, A.G.
Bexfilm	Cellulose acetate film	Bakelite Xylonite, Ltd.
Bexoid	Cellulose acetate sheet	Bakelite Xylonite, Ltd.
Brasslyfe	Cellulose lacquers	Bee Chemical Co.
Cabulite	Cellulose acetate butyrate sheet	May & Baker, Ltd.
Cal-Stix	Acetate film	Topflight Corporation
Celacloud	Acetate fiber	Celanese Fibers Co., Division of Celanese Corp. of America
Celacol	Water-soluble cellulose esters	British Celanese, Ltd.
Celacrimp	Acetate fiber	Celanese Corp. of America
Celadex	Cellulose propionate	British Celanese, Ltd.
Celafil	Acetate fiber	Celanese Corp. of America
Celaire	Acetate fiber	Celanese Corp. of America
Celairese	Acetate fiber	Celanese Corp. of America
Celaloft	Acetate fiber	Celanese Corp. of America
Celanese	Acetate film, sheeting, molding material	Celanese Plastics Co., Division of Celanese Corp. of America
Celaperm	Acetate fiber	Celanese Fibers Co., Division of Celanese Corp. of America
Celarادem	Acetate fiber	Celanese Corp. of America
Celaspun	Acetate fiber	Celanese Corp. of America
Celastoid	Cellulose acetate sheet	British Celanese, Ltd.
Celastoid Wireweld	Cellulose acetate sheet	British Celanese, Ltd.
Celastraw	Acetate fiber	Celanese Corp. of America

Celatow	Acetate fiber	Celanese Corp. of America
Celatress	Acetate fiber	Celanese Corp. of America
Celaweb	Acetate roving	Celanese Corp. of America
Celawrap	Cellulose acetate film	British Celanese, Ltd.
Celenamel	Cellulose acetate coating	Belder Mfg. Co.
Cellastine	Cellulose acetate sheet, film, rod, and tube	British Celanese, Ltd.
Cellider	Plasticized cellulose acetate	Bayer, A.G.
Cellit	Unplasticized cellulose esters	Bayer, A.G.
Cellon	Plasticized cellulose acetate	Dynamit Nobel, A.G.
Cellonex	Cellulose acetate sheet	Dynamit Nobel, A.G.
Charmour	Acetate sheets	Celanese Corp. of America
Chromspun	Colored cellulose acetate fiber	Eastman Chemical Products, Inc.
Cinelin	Cellulose acetate and nitrate	Cinelin Co.
Cinemoid	Cellulose acetate sheet	British Celanese, Ltd.
Clairfoil	Cellulose acetate film	British Celanese, Ltd.
Clarifoil	Cellulose acetate film	British Celanese, Ltd.
Claritex	Wire-reinforced cellulose acetate sheet	British Celanese, Ltd.
Clearsite	Cellulose acetate and acetate butyrate	Celluplastic Corporation
Cobocell	Cellulose acetate butyrate tubing	Cobon Plastics Corporation
Col-O-Tex	Cellulose nitrate–coated fabric	Columbus Coated Fabrics Corporation
Cover Seal	Acetate sheet for packaging	SLM Plastics, Inc.
Crystal	Acetate fiber	Tennessee Eastman Co.
Cycloset	Acetate fiber	E.I. Du Pont de Nemours & Co., Inc.
Dexel	Cellulose acetate	British Celanese, Ltd.
Dup-Pak	Cellulose acetate butyrate	Fidelity Chemical Products Co.
Dynal	Acetate rods and sheets	Dynamit Nobel, A.G.
Eanplast	Acetate molding powders	East Anglia Plastic, Ltd.
Edge-Glow	Fluorescent acetate	Monsanto Chemical Co.
Embafilm	Cellulose acetate–coated films	M & B Plastics, Ltd.
Embatex	Cellulose acetate–coated fabrics	M & B Plastics, Ltd.
Estran	Acetate filament	Eastman Chemical Products, Inc.
Fabrikoid	Cellulose nitrate–coated material	E.I. Du Pont de Nemours & Co., Inc.
Fedroid	Cellulose nitrate–coated fabric	Federal Industries, Division of Textron, Inc.
Fiberlac	Cellulose nitrate lacquer	Monsanto Chemical Co., Plastics Division
Fibestos	Cellulose acetate	Monsanto Chemical Co., Plastics Division
Forticel	Cellulose propionate sheet, films, and Molding Powders	Plastics Division, Celanese Corp. of America
Fortisan	Saponified cellulose acetate	Celanese Plastics Co.
Genco	Cellulose acetate butyrate	General Plastics Corporation
Glass-O-Net	Scrim-reinforced butyrate	Warp Brothers
Hercocel	Cellulose acetate molding powder	Hercules Powder Co., Inc.
Hercose AP	Cellulose acetate propionate	Hercules Powder Co., Inc.
Herculoid	Cellulose nitrate	Hercules Powder Co., Inc.
Hycoloid	Cellulose nitrate	Hygienic Tube & Container Co.
Iridon	Cellulose acetate sheet and film, acetate butyrate sheet	Commercial Plastics, Ltd.
Joda	Acetate and acetate butyrate sheet, cellulose nitrate	Joseph Davis Plastics Co.
Kleer Cite	Extruded acetate sheet and film	Kleer Kast, Inc.
Kleer Kast	Cast acetate sheet and film	Kleer Kast, Inc.
Kodacel	Cellulose acetate film and sheet	Eastman Chemical Products, Inc.
Kodapak	Cellulose acetate	Eastman Kodak Co.
Kodapak Sheet	Cellulose acetate butyrate	Eastman Kodak Co.
Laticel	Cellulose acetate and acetate butyrate	Lati, S.p.A.
Loftura	Acetate filament	Eastman Chemical Products, Inc.
Lumaline	Metallized cellulose acetate film	La Cellophane, S.A.
Lumarith	Cellulose acetate sheet, film, and molding powders	Celanese Corp. of America
Macoid	Cellulosic dip coatings	Detroit Macoid Corporation
Micro-Brite	Metallized acetate and butyrate sheet	Coating Products
Mirralon	Metallized cellulose acetate film	M & B Plastics, Ltd.

Mystikleer	Mat finish acetate	Borden, Inc.
Nacrolaque	Cellulose acetate sheet	M & B Plastics, Ltd.
Naxoid	Cellulose acetate sheet	Convert, Ets.G.
Nitrocol	Nitrocellulose base pigment dispersion	C.J. Osborn Co.
Nitrol	Cellulose nitrate	Monsanto Chemical Co., Plastics Division
Nitron	Cellulose nitrate	Monsanto Chemical Co., Plastics Division
Nixon C/A	Cellulose acetate sheets, tubing, and rods	Nixon Nitration Works, Inc.
Novoid	Cellulose nitrate sheet	Convert, Ets.G.
Plastacele	Cellulose acetate flake	E.I. Du Pont de Nemours & Co., Inc.
Plastigen	Nitrocellulose lacquer resins	BASF Wyandotte Corporation
Plastocele	Cellulose acetate	E.I. Du Pont de Nemours & Co., Inc.
Poly-Cell	Cellulose casting resins	Poly-Cell Plastics Co.
Proxcote	Cellulose nitrate coatings	Pyroxylin Products Inc.
Prox-Mesh	Wire-reinforced cellulose nitrate	Pyroxylin Products Inc.
Prox-Peel	Strippable coatings	Pyroxylin Products Inc.
Pyralin	Cellulose nitrate rods and sheets	E.I. Du Pont de Nemours & Co., Inc.
Prya-Shell	Cellulose nitrate and acetate	Shoe Form Co.
Quiltacel	Acetate film	Celanese Corp. of America
Rhodex	Cellulose acetate film and extrusion	May & Baker, Ltd.
Rhodialine	Cast cellulose acetate film	May & Baker, Ltd.
Rhodialite	Cellulose acetate molding powders	May & Baker, Ltd.
Rhodiod	Cellulose acetate sheet	M & B Plastics, Ltd.
Rhodoglass	Cellulose acetate sheet	M & B Plastics, Ltd.
Rhodophane	Cast cellulose acetate film	May & Baker, Ltd.
Rocel	Cellulose acetate	British Celanese, Ltd.
Santaloid	Cellulose esters	Monsanto Chemical Co.
Sclas	Cellulose foil laminate	UCB-Sidac Corporation
Scrim-Glass	Scrim-reinforced butyrate	Warp Brothers
Sculptured	Acetate fiber	Chemical Products, Inc.
Setilithe	Cellulose acetate granules	Akzo Plastics, N.V.
Sicalit	Cellulose acetate molding powder	Mazzucchelli Celluloide, S.p.A.
Sicaloid	Cellulose acetate sheets	Mazzucchelli Celluloide, S.p.A.
Sicofoil	Cellulose acetate and triacetate film	Mazzucchelli Celluloide, S.p.A.
Sicograph	Triacetate film	Mazzucchelli Celluloide, S.p.A.
Sicoplast	Extruded acetate	Mazzucchelli Celluloide, S.p.A.
Strux	Cellular cellulose acetate	Aircraft Specialties
Tan-O-Tex	Cellulose nitrate–coated fabric	Columbia Mills, Inc.
Tec	Cellulose acetate	Tennessee Eastman Co.
Telcocell	Cellulose acetate butyrate powders	Telcon Plastics, Ltd.
Tenex	Acetate resin	Teijin, Ltd.
Tenite	Cellulose acetate, acetate butyrate, and acetate propionate	Eastman Chemical Products, Inc.
Textileather	Cellulose nitrate–coated material and leather cloth	General Tire & Rubber Co., Textileather Division
Textiloid	Cellulose nitrate–coated, resin-bonded composition	General Tire & Rubber Co., Textileather Division
Trolit	Cellulose acetate molding	Dynamit Nobel, A.G.
Tubiceta	Cellulose acetate flakes	Akzo Plastics, N.V.
Utex	Cellulose acetate film	Utilex, Ltd.
Uvex	Cellulose acetate butyrate sheet	Eastman Chemical Products, Inc.
Vondafol	Cellulose acetate film	Vondeling Enplatet, N.V.
Vuepak	Cellulose acetate film	Monsanto Co.
Wopalite	Acetate powders	Worbla, Ltd.
Wopaloid	Cellulose nitrate sheet, tubes, and rods	Worbla, Ltd.
Wopalon	Cellulose acetate sheets, rods, and tubes	Worbla, Ltd.
Worbla	Nitrocellulose resins for lacquers, bleached cotton linters	Worbla, Ltd.
Xelox	Cellulose nitrate sheet	Mazzucchelli Celluloide, S.p.A.

Xylonite	Cellulose nitrate	Bakelite Xylonite, Ltd.
Zapon	Cellulose nitrate lacquer	Atlas Powder Co.
Cellulose Ethers		
Cellosize	Hydroxyethyl cellulose	Union Carbide Chemical Co.
Cliderite	Ethyl cellulose potting compound	Bischoff Chemical Co.
Courlose	Water-soluble cellulose ethers	British Celanese, Ltd.
Covinax	Hydroxyethyl cellulose	Franklin Chemical Co.
Cyanocel	Cyanoethylated cellulose	American Cyanamid Co.
Dip-Pak	Ethyl cellulose	Fidelity Chemical Products Corporation
Edisol-M	Methyl cellulose film	Polymer Films, Ltd.
Ethocel	Ethyl cellulose molding compound	Dow Chemical Co.
Ethomulsion	Ethyl cellulose lacquer emulsion	Dow Chemical Co.
Ethulon	Ethyl cellulose films	May & Baker, Ltd.
Ethycol	Ethyl cellulose pigment dispersions	C.J. Osborne Chemicals, Inc.
Hercocel	Ethyl cellulose molding powder	Hercules Powder Co., Inc.
Joda	Ethyl cellulose	Joseph Davis Plastics Co.
Klucel	Hydroxylpropyl cellulose	Hercules, Inc.
Macoid	Cellulosic dip coatings	Detroit Macoid Corporation
Methocel	Methyl cellulose	Dow Chemical Co.
Metolose	Methyl cellulose	Shin-Etsu Chemical Co.
Natrosol	Hydroxyethyl cellulose	Hercules, Inc.
Terucello	Carboxymethyl cellulose	Showa High Polymer Co.
Thermocast	Ethyl cellulose hot-melt compounds	Bischoff Chemical Co.
Thermocote	Ethyl cellulose strippable coating	Bischoff Chemical Co.
Tylose	Methyl cellulose and carboxymethyl cellulose	Hoechst, A.G.

REFERENCES AND BIBLIOGRAPHY

Arnold, L. K. *Introduction to Plastics*. Iowa State University Press, Ames, Iowa, 1968.

Brydson, J. A. *Plastics Materials*. 3rd Ed. Robert E. Krieger Publishing Co., Inc., Huntington, New York, 1975.

Celanese Plastics Co. *Forticel Cellulose Propionate Sheet–Properties and Processing*. Bulletin B1B, June 1972.

——. *Forticel Cellulose Propionate–Properties*. Bulletin A1A, June 1969.

——. *Celanese Cellulose Acetate Film–Properties and Processing*. Bulletin F1A, February 1967.

——. *Celanese Cellulose Acetate–Properties*. Bulletin A1B, September 1967.

——. *Celanese Cellulose Acetate Sheet–Properties and Processing*. Bulletin B1A, April 1963.

Dow Chemical Co. *Ethocel Ethyl Cellulose Resins*. Bulletin 192-662, 1978.

——. *Ethocel for Coatings Applications*. Bulletin 190-028, 1972.

——. *Hotmelts of Ethocel*. Bulletin 170–108, 1968.

DuBois, H. J., and John, F. W. *Plastics*. 5th Ed. Van Nostrand Reinhold Co., New York, 1974.

E.I. Du Pont de Nemours & Co., Inc. *Du Pont Cellophane–Properties and Uses*. Bulletin A-6166.

Eastman Chemical Products, Inc. *Blister Packaging with Tenite Cellulosic Plastics*. Publication No. PC-17A, ECP 3612, March 1977.

——. *Eastman Cellulose Acetate Butyrate for Coatings*. Publication No. E-101B, ECP 2739, April 1977.

——. *Effect of Various Chemicals & Reagents on Tenite Butyrate*. Publication No. PC-3, ECP 1083, September 1969.

——. *Effect of Various Chemicals & Reagents on Tenite Acetate*. Publication No. PC-6, ECP 1226, December 1969.

——. *Properties and Characteristics of Tenite Butyrate*. Materials Bulletin MB-33, ECP 463, May 1968.

——. *Properties and Characteristics of Tenite Acetate*. Materials Bulletin MB-34, ECP 463, July 1968.

——. *Injection Molding Tenite Cellulosics Plastics*. Publication No. PC-2, December 1968.

——. *A Comparison of Tenite Propionate and Rigid Poly(Vinyl Chloride) for Blister Packaging*. Technical Report TR-27, ECP 465, February 1966.

——. *Cellulose Acetate Butyrate for Protective Coatings*. 2nd Ed., Bulletin No. E-101, 1962.

——. *A Guide to the Use of Tenite Butyrate for Colorful Outdoor Signs*. 1960.

——. *Tenite Cellulosic Plastics–Properties and Reference Data*.

Fabian, R. J. Strippable coatings keep products bright, clean, safe. . . . *Materials in Design Engineering*, May 1959, **49**(5):110–115.

Gross, S. (exec. ed.) *Modern Plastics Encyclopedia*. **45**(14A). McGraw-Hill, Inc., New York, 1968–69.

Materials Engineering. One coat butyrate paint protects plastic shutters. October 1976, **84**(4):72.

Modern Plastics. The resin-price picture as of Jan. 1, 1978. January 1978, **55**(1):55.

Morris, W. Cellophane is wrapping it up. *Los Angeles Times*, Times-Mirror Corporation, Los Angeles, April 1978, pp. 11–12.

National Electrical Manufacturers Association. *Electrical Insulation–Terms and Definitions*.

PLASTEC Note 19, *Trade Designations of Plastics and Related Materials*, (revised). Plastics Technical Evaluation Center, Picatinny Arsenal, Dover, New Jersey, May 1978.

Plastics World. Transparent plastics: new & improved. January 1977, **35**(1):24–26.

Reidell, A. W. The nitrocellulose dilemma. *Modern Paint & Coatings*. September 1978, **68**(9):51–52.

Scales, R. E. Cellulosics, from Plastics in Packaging. *Plastics Engineering*. October 1978, **34**(10):31–33.
Simonds, H. R., and Church, J. M. *A Concise Guide to Plastics*. 2nd Ed. Van Nostrand Reinhold Co., New York, 1963.
Swanson, R. S. *Plastics Technology–Basic Materials and Processes*. McKnight & McKnight Publishing Co., Bloomington, Illinois, 1965.
Winding, C. C., and Hiatt, G. D. *Polymeric Materials*. McGraw-Hill Book Co., New York, 1961.

OLEFINICS

There is a large group of polymers formed by polymerizing monomers containing the C=C group. Since ethylene, containing the olefinic double bond $CH_2{=}CH_2$, is the simplest monomer that contains this structure, all other polymers having a substituted ethenic base could be called ethenic polymers.

It is easier to visualize the polymerization chemistry of C=C-containing monomers and polymers if one thinks of them as substituted ethylenes, even though this is not consistent with organic chemistry practices.

If we take a basic C=C structure and generally substitute as follows: $\begin{matrix} w & & x \\ | & & | \\ C & = & C \\ | & & | \\ y & & z \end{matrix}$, we can visualize a given polymeric structure by replacing the w, x, y, and z's with the appropriate substituent atom or molecule. For example, in polyethylene, w, x, y, and z would be H's. Polystyrene's "x" would be a benzene ring, and in monochlorotrifluoroethylene, w would be Cl and x, y, and z would be F's.

It is possible to alter the polymer after its chain formation if the substituent groups lend themselves to further chemical reaction. For example, chlorine may replace hydrogen through chlorination or a free radical formed by displacement of a hydrogen through irradiation.

The substituents on adjacent carbon atoms (w and x, or y and z) do not have to be separate, but may be part of a ring structure which connects to the two carbon atoms.

The entire class of ethenic polymers tends to have a linear structure. Some branching is possible, but highly branched polymers are not usually formed. Crosslinkages between chains may also be formed, but such materials are a small part of the total production. Thus, these polymers are usually thermoplastics. The general linear structure of ethenic polymers also indicates suitability for fiber and film formation. Such applications are common in this class. Since the hydrocarbon monomers used to generate these polymers are olefins, there is a tendency to group the polymers together and call them polyolefins or olefinics. It is more practical, however, to break this large agglomeration of polymers into subclasses based on similarities of use, chemical nature of precursor monomers, and/or properties. Thus, polyethylene, polypropylene, polybutene, and their copolymers are grouped into one subclass; polystyrene, its copolymers and alloys into a second subclass; vinyls and copolymers into a third, and so on. In common trade usage, the first subclass (consisting of polyethylene, polypropylene, etc.) is often referred to as the "olefinics" or "ethenics." The other subclasses are then referred to by a title referencing the base monomer (e.g., vinyls–from vinyl chloride; halocarbons–from haloethylenes, etc.). Even though this usage is rigorously incorrect, we will continue the tradition and further discussions will be based on this classification system.

Polyethylenes

Although the chemical notation for the polymerization of ethylene indicates a simple reaction,

$$nCH_2{=}CH_2 \longrightarrow \text{-}\!\!\left(CH_2CH_2\right)\!\!\text{-}_n,$$

the properties of the polymer vary over a wide range, depending on molecular weight, the method of manufacture, and differences in structure. At one time these variations in structure were thought to be minor, but they are actually of such importance that entire areas of application are influenced.

The production of oils by the polymerization of ethylene was described as early as 1879, and a considerable amount of research on the production of wax-free oils from ethylene was done in the 1920s. The first polyethylene was produced in England in the early 1930s by Imperial Chemical Industries, Ltd., during studies of the effects of very high pressures on chemical reactions. The resulting patent described the polymerization of ethylene at pressures of 15,000-30,000 psi and temperatures of 200-400°C in the presence of very small amounts of O_2 as a catalyst. In spite of some difficulties (e.g., major explosions), commercial production started in England in 1940 and in the United States in 1943. The commercial processes that were developed from this early work are known as "high-pressure" or "conventional" and also "low-density," and the same words are used to describe the polyethylene produced by these processes.

High-Pressure Polyethylenes. The high-pressure process requires ultrahigh-purity ethylene (less than 0.2% CO, acetylene, O_2, and moisture). Oxygen and peroxides act as catalysts. The ethylene is compressed to 20,000 psi and heated to around 204°C (400°F). It is fed to a bank of autoclaves where the stagewise reaction takes place. Because catalyst concentration directly affects the average molecular weight, it is added at carefully controlled rates. The reaction is highly exothermic; thus, extremely good heat transfer is required to prevent runaway reactions.

From the reaction stages, the polymer/monomer mixture is sent to a series of separators, from which the monomer is recycled back to the compressors, and the overall system pressure is reduced to below atmospheric. The polyethylene is extruded into a continuous rod which is chopped to size for blending, compounding, and bagging.

Low-Pressure Polyethylenes. In the early 1950s, considerable interest in low-pressure (under 500 psi) processes was stimulated by the development of the Ziegler catalysts in Germany. The low-pressure methods, pioneered by Phillips and Du Pont in the U.S., are called Ziegler or "linear" processes, in contrast to the high-pressure, or conventional, processes. Low-pressure polymerization produces a polyethylene having remarkably different properties, due to the greater linearity of the polymer and differences in molecular weight distribution.

There are two basic low-pressure processes: the Ziegler (suspension) method and the Phillips (solution) method. In the former process, catalysts made from the reaction of titanium tri- or tetrachloride with aluminum alkyls are used. The monomer is added to a well-stirred dispersion of catalyst in a liquid hydrocarbon. Reaction conditions are less than 100°C and 100 atms. As the polymer forms from the catalyst, it produces a slurry which is sent to a flash drum from which the remaining monomer is removed for recycling. The polymer is freed of solvent and catalyst through a series of filtrations, extractions, and washings, after which it is dried, extruded, chopped, compounded, and bagged.

In the solution process, a solvent such as cyclohexane is added to a reactor. Then monomer and catalyst are added. The reaction conditions range from 130°C to 160°C (266–320°F) and 200 to 500 psi. A typical catalyst is a composition of chromium oxides on an activated silica-alumina base. The reactor is a two-zone continuous unit which has a well agitated reaction zone and a quiescent separation zone for catalyst settling. The polymer solution still has some remaining catalyst; therefore, the solution is passed through a flash drum (removes and recycles monomer), centrifuge (removes catalyst), and cooler/precipitator (removes polymer). The polymer is then dried, extruded, chopped, compounded, and bagged as in the other processes.

Both processes require exceptionally efficient catalyst removal, particularly if the polymer is to be used in electrical applications.

In 1977 a low-pressure gas-phase processs for the manufacture of low-density polyethylene was announced. The process is reported to require half the capital investment, one quarter of the energy, and 10% of the plant space now needed to produce equal amounts of conventional polyethylene by high-pressure processes.

In this new process, gaseous ethylene, comonomer (when required), and catalyst are continuously fed to a fluidized-bed reactor. The reaction pressure is 100–300 psi, which is about one-hundredth the pressure used in high-pressure production. The reaction temperature is controlled at less than 100°C (212°F), which is roughly a third of that used in a high-pressure operation.

Because of the low-pressure reaction zone, energy use is cut by 75%. Less and much thinner piping is required. The first reactor, a 110,000 ton/yr unit, consists of a single 14-foot-diameter tank. The reactor stage required to produce an equivalent amount of polymer in a conventional plant requires one mile of heavy wall pipe. The new process eliminates or sharply reduces complex pumping and recovery systems, refrigerated ethylene streams, and heat- and noise-producing rotary and reciprocating equipment. Simplified instrumentation also makes possible servicing without reactor shutdowns.

The low-density polyethylene produced under the process is polymerized directly as a granular solid, requiring no monomer stripping or pelletizing. The material is purported to heat faster than pellets, thus lowering energy requirements at the user level.

Polyethylene Branching. Because of the intensity of polymerization conditions, the high-pressure process tends to form molecules with side chains. These branches vary from one to five or more carbons in length and may be one to three in number for every 100 carbons in the main chain.

Any changes in the amount of branching in a polymer will affect the tendency to form crystallites and spherulites. In general, for a polymer which tends to crystallize, greater linearity (less branching) will increase the ratio of crystalline to amorphous regions. Figure 5-11 shows a polyethylene

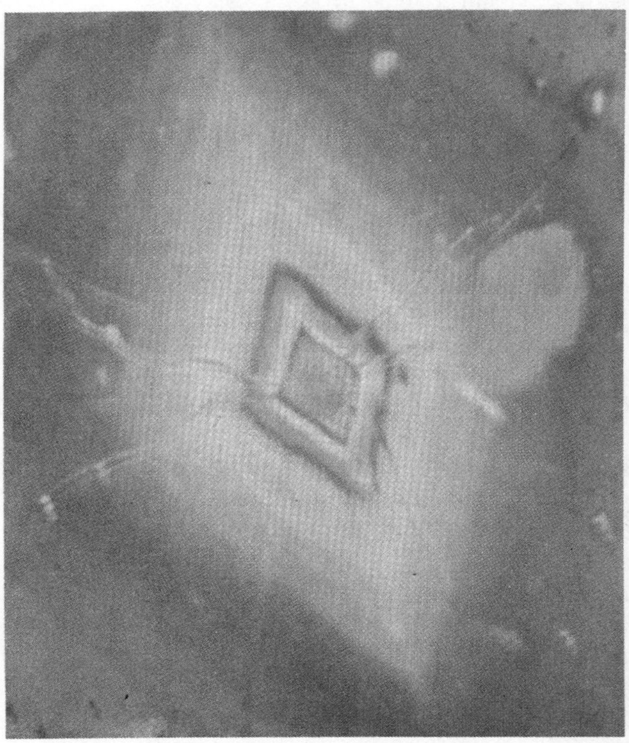

Fig. 5-11. Growth spiral in a crystal of polyethylene is revealed by the interference microscope. Such a spiral occurs when the polyethylene molecules in a dilute solution form a thin plate that, as it grows, regularly turns back on itself like a winding staircase. (Wunderlich, 1964)

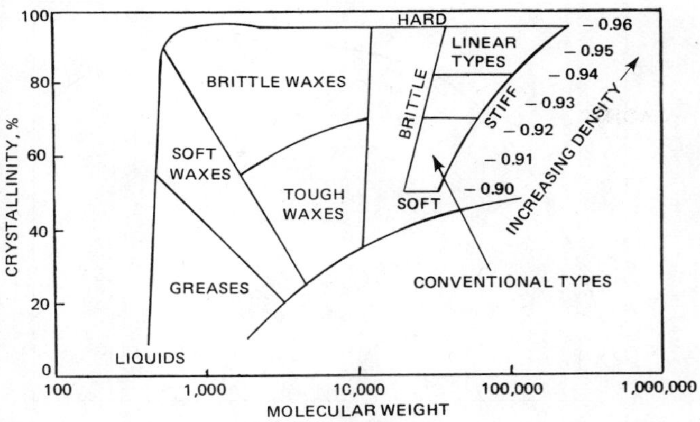

Fig. 5-12. Relationships between crystallinity, molecular weight, and properties of possible polyethylene. (Adapted from Kirk/Othmer, *Encyclopedia of Chemical Technology*, first supplemental volume, copyright 1957. From *Polymeric Materials* by Winding and Hiatt, copyright 1961. Used with permission of Interscience Publishers, Inc. and McGraw-Hill Book Co.)

crystal. The percentage crystallinity will have a significant effect on properties. Greater crystallinity causes higher density, so more linear polymers are known as high-density polyethylene. Figure 5-12 indicates how properties are affected by changes in linearity involved in the different commercial polymerization processes.

Properties. The processes now used to produce polyethylene for plastics by no means cover the entire range of possible products. Oils and waxes may also be made by modifications of these polymerization processes. The molecular weight/crystallinity relationships allow the generation of a wide range of properties in both waxes and plastics.

The most important intrinsic properties of commercial polyethylenes indicating performance in principal applications are: (1) density, (2) melt index, and (3) molecular weight distribution.

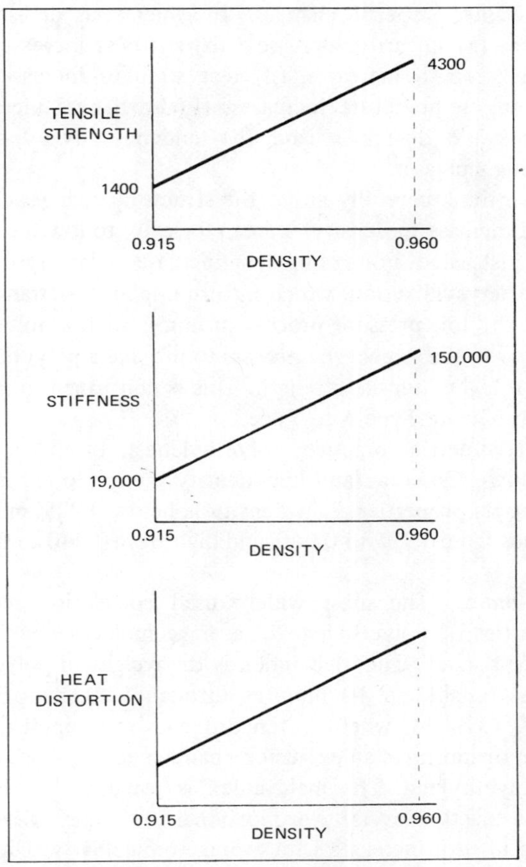

Fig. 5-14. Typical density effects. (*Courtesy: Soltex Polymer Corporation*)

Density. The complete range of densities encountered in polyethylene extends from 0.910 to 0.965 (see Figures 5-13 and 5-14). The tremendous change of properties with density is reflected in the need to express densities to three decimal places. A change of three in the third place produces

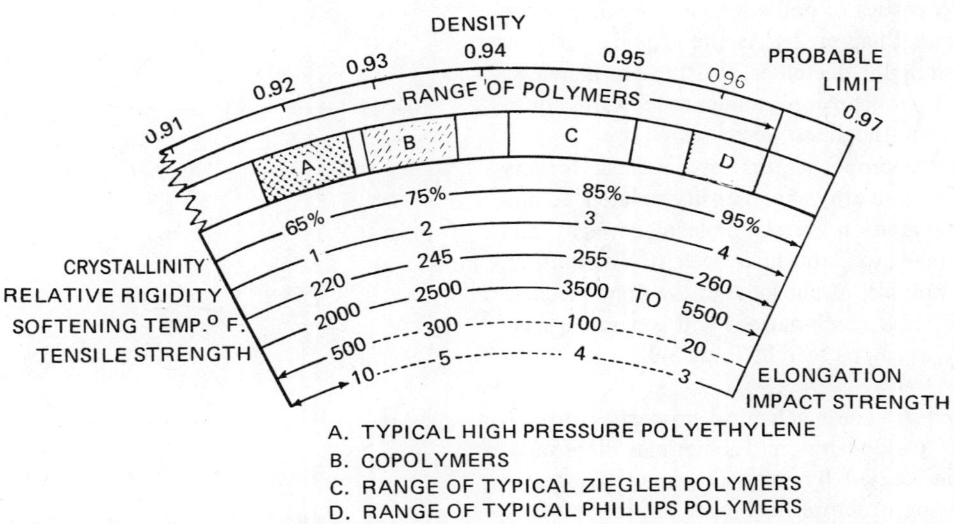

Fig. 5-13. Range of properties obtained in various polyethylene polymerization processes. (From *Polymeric Materials* by Winding and Hiatt. Copyright 1961 by McGraw-Hill, Inc. Used with permission of McGraw-Hill Book Co.)

appreciable property changes. In general, as density increases: (a) linearity increases, (b) stiffness increases, (c) tensile strength increases, (d) tear strength increases, (e) softening temperature increases, (f) brittleness increases, (g) flex life decreases, and (h) tendency toward stress-cracking increases.

If we think carefully about the structural changes caused by an increase in density, we can logically follow the variations just listed. For example, an increase in density implies greater crystallization, which in turn implies less branching. Since the low-pressure process produces such a polyethylene, we would expect the process to provide a polyethylene with a higher tensile strength. This is confirmed in Figure 5-13 (compare Type A to Type D).

In commercial practice, polyethylene is often classified into low-, medium-, and high-density groups to aid in predicting its properties. Low-density is below 0.925, medium extends from 0.925 to 0.940, and high from 0.940 to 0.965.

Melt index. The most widely used correlation relating properties of polyethylene to average molecular weight is the melt index. The melt index is the weight of polyethylene extrudable in 10 minutes through a fixed orifice at 190°C (374°F), when a standard weight is applied to a piston operating in an extrusion chamber containing 3 grams of polyethylene. The melt index is roughly, but by no means exactly, inversely proportional to the melt viscosity, which in turn increases with average molecular weight (see Figure 5-15).

Thus, a high melt index indicates greater fluidity at processing temperatures. Melt index was developed originally to indicate flow properties as a measure of extrudability. In addition, as melt index increases, tensile strength decreases, tear strength decreases, softening temperature decreases, and toughness decreases (see Figure 5-16).

Molecular Weight Distribution. Molecular weight distribution affects the properties of polyethylene as well. (See the general discussion in Chapter 2.) As the $\overline{M}_w/\overline{M}_n$ ratio increases, tensile strength decreases, brittleness increases, softening temperature decreases, tendency toward stress-cracking increases, and toughness decreases.

We now have three properties that are intrinsic for any particular grade of polyethylene: density related to linearity, melt index based on average molecular weight, and $\overline{M}_w/\overline{M}_n$ proportional to molecular weight distribution. Since all of these variables affect some of the same mechanical properties, different combinations will generate a spectrum of properties, as can be seen in Table 5-9.

Mechanical Properties. The mechanical properties of polyethylene vary over a wide range, and sometimes these variations appear to be caused by relatively small changes in physical properties or structure. Extremes in properties are usually encompassed by conventional (low-density) polyethylene at one end of the scale and linear (high-density)

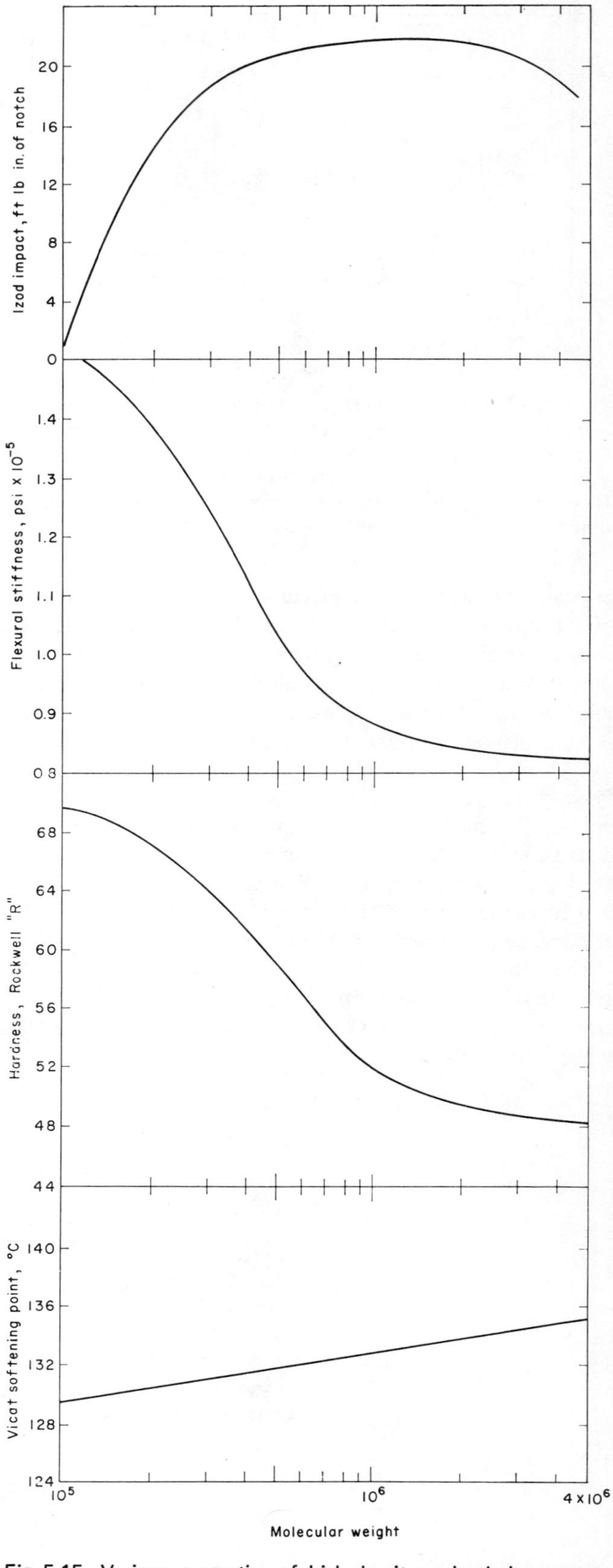

Fig. 5-15. Various properties of high-density polyethylene versus molecular weight. (Margolis, 1971)

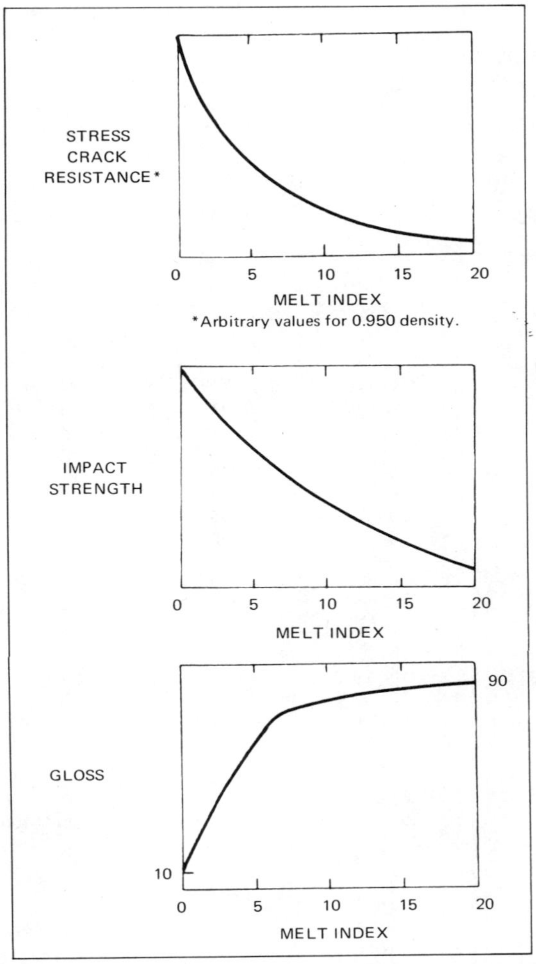

Fig. 5-16. Typical melt index effects. (*Courtesy: Soltex Polymer Corporation*)

at the other. However, polyethylene types with intermediate properties are available. They can be made by variations in both the high- and low-pressure processes. Table 5-10 shows typical polyethylene properties.

Polyethylene deteriorates when exposed to sunlight, but the incorporation of about 3% carbon black greatly reduces attack. Such compounds are considered to have sufficient stability to be used for outdoor electrical insulation. Under certain conditions, O_2 causes some degradation. Some antioxidants have been shown to retard such degradation, although the effectiveness seems to be dependent on the type of polyethylene. High-density polyethylene appears to benefit more from the use of antioxidants.

Polyethylene has excellent resistance to both corrosive chemicals and solvents at room temperature. However, its resistance to solvents changes rapidly with temperature and exposed surface conditions, especially where elevated temperatures may be encountered.

Polyethylene is subject to environmental stress-cracking. Some formulations are much more resistant than others (see Figure 5-17). In addition to variations between polymer types, the resistance of any polyethylene to stress-cracking increases rapidly as the melt index of the material decreases below about 1. Whether or not a stress-cracking hazard exists in any particular situation involving contact of a liquid with polyethylene, therefore, depends upon the degree and type of stress on the polymer, the particular liquid involved, and the particular formulation.

Water and water solutions of inorganic salts normally do not exhibit activity as stress-cracking agents, but water solutions of detergents are very active in this regard. Some of the concentrated mineral and organic acids and concentrated aqueous solutions of the alkali hydroxides

Table 5-9. How Intrinsic Properties Affect Polyethylene Resins.

Physical Properties	If Density Increases	If Melt Index Increases	If MWD Broadens
Melt viscosity	Higher	Lower	-
Vicat softening point	Much higher	Lower	Lower
Surface hardness	Higher	Slightly lower	Lower
Tensile strength			
Yield	Much higher	Slightly lower	Lower
Break	Slightly higher	Lower	Lower
Elongation	Lower	Lower	Higher
Creep resistance	Higher	Slightly lower	Higher
Flexural stiffness	Much higher	Slightly lower	-
Flexibility	Lower	-	-
Toughness	Lower	Lower	Lower
Low temperature brittleness	Lower	Lower	Higher
Stress crack resistance	Lower	Lower	Higher
Optical properties			
Transparency	Higher*	Higher	-
Freedom from haze	Higher*	Higher	-
Gloss	Higher*	Higher	-
Barrier properties			
MVT rate	Lower	-	-
Gas and liquid			
Transmission	Much lower	-	-
Grease proofness	Much higher	Slightly lower	-
Electrical properties	Slightly higher	No effect	-

* Not true in the high-density range.

Table 5-10. Polyethylene

Grade (Melt Index, gm/10 min)	ASTM Method	Polyethylene: Type I (0.910–0.925 gm/cu cm) 1(>25)	2(>10–25)	3(>1.0–10)	4(>0.3–1.0)	5(0.3 max)
1. Processing Suitability[a]		M, RM, PC	EC, M, RM, PC, E	B, M, EC, EW	B, EW, E, M	B, EW, E
Special Grades Available						
2. UV Stabilized		Yes	Yes	Yes	Yes	Yes
3. Heat Stabilized		Yes	Yes	Yes	Yes	Yes
4. FDA Approved		Yes	Yes	Yes	Yes	Yes
Processing Characteristics						
5. Bulk Factor	D1895	1.6–2.4	1.6–2.4	1.6–2.4	1.6–2.4	1.6–2.4
6. Vicat Softening Point, °F	D1525	154–218	178–218	180–210	186–215	195–216
7. Mold Shrinkage, in./in.	D955	0.01–0.05	0.01–0.05	0.015–0.05	0.015–0.05	0.015–0.050
Injection Molding Temperatures, °F	–					
8. Rec. Stock Temp		250–500	280–500	320–550	375–550	–
9. Feed Zone		250–450	280–450	320–500	350–500	–
10. Middle Zone		280–550	310–600	360–600	375–600	–
11. Nozzle		280–500	310–500	360–550	375–550	–
12. Mold		40–150	40–150	50–150	50–150	–
Extrusion Temperatures, °F	–					
13. Feed Zone		–	280–450	250–600	250–350	250–450
14. Transition Zone		–	300–550	250–600	250–400	250–500
15. Metering Zone		–	300–620	300–650	350–550	350–550
General Physical Properties						
16. Specific Gravity	D792	0.910–0.933	0.910–0.933	0.910–0.933	0.910–0.933	0.910–0.933
Water Absorption, %	D570					
17. Method A (24 hr)		<0.01–0.02	<0.01–0.02	<0.01–0.02	<0.01–0.02	<0.01–0.03
18. Method D (Long Term)		0.01–0.15	0.01–0.15	0.01–0.15	0.01	0.01
Mechanical Properties						
19. Tensile Yield Strength, 1000 psi	D638	1–2[e]	1–2[e]	1.1–2[e]	1.4–2.3	1.15–2.3
20. Ultimate Elongation, %	D638	50–600[e]	60–600[e]	200–740[e]	400–675	400–700
21. Flexural Strength at Yield, 1000 psi	D790	–	–	–	–	–
22. Flexural Modulus of Elasticity, 10^5 psi	D790	4.5[e]	6[e]	4.5[e]	–	–
23. Compressive Strength at Yield, 1000 psi	D695	–	–	–	–	–
24. Rockwell Hardness (Method A)	D785	10[e]	15[e]	10[e]	44–48[f]	44–48[f]
25. 1/8 in. specimen		No break	No break	No break	No break	No break
26. 1/4 in. specimen		>16[e]	>16[e]	>16[e]	10–>16	>16
27. Deformation Under Load	D621	–	–	–	–	–
Thermal Properties						
Deflection Temperature, °F[d]	D648					
28. At 66 psi Fiber Stress		100–120	100–125	95–120	100–120	100–120
29. At 264 psi Fiber Stress		90–105	90–105	90–105	90–105	–
30. Coeff of Lin Ther Expansion, in./in./°F	D696	$8–20 \times 10^{-5}$	$8–20 \times 10^{-5}$	$8–20 \times 10^{-5}$	$10–20 \times 10^{-5}$	$10–20 \times 10^{-5}$
31. Flam. ($\frac{1}{8} \times \frac{1}{2} \times 5$ in. spec.), in./min	D635	1–1.04	1–1.04	1–1.04	1–1.04	1–1.04
Electrical Properties						
32. Dielectric Strength, v/mil, short time	D149	500–1000[g]	500–1000[g]	425–1000[g]	425–1000	425–1000
Dielectric Constant	D150					
33. At 60 cps		2.3–2.35[g]	2.3–2.35[g]	2.2–2.4[g]	2.2–2.4	2.2–2.4
34. At 10^6 cps		2.3–2.35[g]	2.3–2.35[g]	2.2–2.4[g]	2.2–2.4	2.2–2.4
Dissipation Factor	D150					
35. At 60 cps		<0.0005[g]	<0.0005[g]	<0.0005[g]	<0.0005	0.0001–<0.0005
36. At 10^6 cps		<0.0005[g]	<0.0005[g]	<0.0005[g]	<0.0005	0.0002–<0.0005
37. Volume Resistivity, ohm-cm	D257	$>10^{15}–>10^{16}$[g]	$>10^{15}–>10^{16}$[g]	$>10^{15}–>10^{16}$[g]	$>10^{15}–>10^{16}$	$>10^{15}–>10^{16}$
38. Arc Resistance (SS Electrode), sec	D495	135–160[g]	135–160[g]	135–160[g]	135–160	135–>200
Cost (Natural)	–					
39. Cents/lb		15–25	15–19.25	15.5–19.25	18–23.5	17–35
40. Cents/cu in.		0.50–0.82	0.50–0.64	0.52–0.64	0.60	0.78–1.15

[a]E = Extrusion; EC = Extrusion coating; EW = Extrusion wire coating; M = Injection molding; B = Blow molding; RM = Rotational molding; PC = Power coating.
[b]USI has grade with rec. stock temp down to 280°F.
[c]UC range broader, see column 1.
[d]Unannealed, except where noted.
[e]Compression tested at 20 in./min.
[f]Shore D hardness.
[g]Compression molded.
*Extracted from "Plastics Technology's Buyer's Guide to Polyolefins and Cellulosics."

Properties.*

Polyethylene: Type II (0.940 gm/cu cm)					Polyethylene: Type III (0.941–0.965 gm/cu cm)			
1(>25)	2(>10–25)	3(>1.0–10)	4(>1.0–10)	5(0.3 max)	1(>25)	2(>10–25)	3(>1.0–10)	4(>0.3–1.0)
M	M, E, EC	M, B, E, EC, EW	B, E, EW	B, E	M	M, EC	RM, M, B, E, EC	M, B, E
No	No	No	Yes	Yes	No	Yes	Yes	Yes
No	No	No	Yes	Yes	No	Yes	Yes	Yes
Yes	Yes	Yes	Yes	Yes	Yes	Yes	Yes	Yes
1.6	1.6–2.4	1.6–2.4	1.7–2.4	1.6	1.6–1.8	1.6–2.2	1.6–2.2	1.6–2.2
100–200	195–235	195–235	195–245	245	260	255–260	215–260	250–265
0.01–0.05	0.015–0.050	0.02–0.05	0.02–0.05	–	0.01–0.05	0.01–0.05	0.015–0.05	0.01–0.05
375–500[b]	340–420[b,c]	375–550[b]	–	–	350–550	300–600	300–600	250–600
325–450	325–420[c]	350–500	–	–	350–400	250–70	250–450	350–530
375–550	350–450	375–600	–	–	400–450	300–550	300–500	400–530
375–500	370–450	375–550	–	–	400–450	300–550	300–600	400–600
60–150	70–150	40–150	–	–	50–120	40–180	40–180	70–150
250–300	250–450	250–600[e]	300–600	300–450	–	350–600	280–450	280–450
–	320–625[e]	325–625[e]	325–625	350–450	–	350–600	350–550	350–400
–	350–620[e]	350–620[e]	350–600	375–450	–	350–600	400–620	375–550
0.926–0.927	0.926–0.940	0.926–0.940	0.930–0.936	0.93–0.94	0.945–0.965	0.945–0.965	0.942–0.965	0.945–0.965
<0.01	<0.01	<0.01–<0.1	<0.01	<0.01	0.01–0.03	0.01–0.03	0.01–0.03	0.01–0.03
<0.01	–	–	–	<0.01	<0.01	<0.01	<0.01	<0.01
1.3–2.2	1.4–3.0	1.4–2.6	1.4–2.5	1.5–2.5	3–5.5[i]	3–5.5[i]	2.3–5.5[i]	3–5.5[i]
50–125	74–500	50–500	100–500	>600	30–1000	30–1000	100–1000	100–1000
–	–	–	–	–	1	1	1	1
0.4	0.36–1.15	0.36–1.15	0.4–0.65	0.5–0.75	0.8–1.6	0.8–2.6	0.5–2.6	0.8–2.6
–	–	–	–	–	3.2	3.2	3.2	3.2
54[f]	48–60[f]	48–60[f]	48–55[f]	48–55[f]	55–70[j]	55–70[j]	55–70[j]	48–70[j]
0.5–No break	0.5–No break	0.5–No break	No break	No break	0.3–0.6	0.5–2.0[m]	0.6–5[m]	1.5–7[m]
No break	No break	No break	No break	No break	1.5	0.6–1.5	0.4–1.5	0.4–1.5
–	–	–	–	–	–	–	–	–
120	120–165	120–165	120–140	120–140	140–180	140–180	140–185	140–185
–	105–120	105–120	–	–	110–125	110–125	110–125	110–125
10×10^{-5}	10×10^{-5}	10×10^{-5}	10×10^{-5}	10×10^{-5}	$6\text{–}7.2 \times 10^{-5}$[k]	$6\text{–}7.2 \times 10^{-5}$[n]	$6\text{–}7.2 \times 10^{-5}$[n]	$4\text{–}7.2 \times 10^{-5}$[n]
<1	<1–<2.5	<1–<2.5	<1	<1	1–1.04	1.0–<2.5	1.0–<2.5	1.0–<2.5
550	450–1000	450–1000	550–1000	550	450–550[i]	450–500[o]	450–550[o]	450–550[o]
2.3–2.35	2.3–2.35	2.3–2.35	2.3–2.35	2.3–2.35	2.30–2.35	2.30–2.35	2.30–2.35	2.30–2.35
2.3–2.35	2.3–2.35	2.3–2.35	2.3–2.35	2.3–2.35	2.30–2.35	2.30–2.35	2.30–2.35	2.30–2.35
0.0004	0.0004–<0.0005	0.004–<0.0005	0.0004–<0.0005	0.0004	<0.0005	<0.0005	<0.0005	<0.0005
0.0003	0.0003	0.0003–<0.0005	0.0003–<0.0005	0.0003	<0.0005	<0.0005	<0.0005	<0.0005
$>10^{16}$	$>10^{16}$	$>10^{15}\text{–}>10^{16}$	$>10^{15}\text{–}>10^{-16}$	$>10^{16}$	$>10^{16}$	$>10^{16}$	$>10^{16}$	$1\text{–}5 \times 10^{17}$
160–240	150–240	150–240	160–240	160–240	120–250	120–250	120–250	100–250
16	16–19.25	16–29.5	19–24	18–22	17–19	16–18	16–28	18–25
–	–	–	–	.74	0.59–0.65	0.62	0.57–0.97	0.60–1.00

[h] Eastman says it make special heat resist grade.
[i] Spec. compression molded; ratio of loading: 2 in./min.
[j] Shore D.
[k] Value cited by Phillips Petroleum, Union Carbide reports range of $11\text{–}13 \times 10^{-15}$ in./in./°C. Shell reports 1.3×10^{-4} in./in./° F.
[l] In./in./°C.
[m] Izod values of 1–3 ft lb/in. of notch given for $\frac{1}{2} \times \frac{1}{2}$ in. spec.
[n] Union Carbide reports $11\text{–}13 \times 10^{-5}$ in./in./°C. Shell reports 1.3×10^{-4} in./in./° F.
[o] Eastman Chemical lists a diel. str. of 1000 v/mil (short time).

Table 5-10. *(Continued)*

Polyethylene: Type III (Continued)				Polyethylene: Special Types			
Grade (Melt Index, gm/10 min)	ASTM Method	5 (0.3 max)	6 (Glass-Filled)	0.2	0.1	8.0	7.0
1. Processing Suitability[a]		M, B, E, EW	M	E, W	E	M	M
Special Grades Available							
2. UV Stabilized		Yes	No	Yes	No	No	No
3. Heat Stabilized		Yes	No	Yes	No	No	No
4. FDA Approved		Yes	No	No	Yes	No	No
Processing Characteristics							
5. Bulk Factor	D1895	1.6–2.2	–	1.6	1.8	1.8	1.8
6. Vicat Softening Point, °F	D1525	230–265	–	255	250	260	250
7. Mold Shrinkage, in./in.	D955	0.01–0.05	0.002–0.003	–	–	0.015–0.030	0.015–0.030
Injection Molding Temperatures, °F	–						
8. Rec. Stock Temp		250–530	480	–	–	400	400
9. Feed Zone		480–520	430	–	–	–	–
10. Middle Zone		510–550	450	–	–	–	–
11. Nozzle		510–550	380	–	–	–	–
12. Mold		70–150	110	–	–	50–100	50–100
Extrusion Temperatures, °F	–						
13. Feed Zone		280–450	–	280–550	350–390	–	–
14. Transition Zone		350–400	–	280–550	350–390	–	–
15. Metering Zone		400–550	–	280–550	350–390	–	–
General Physical Properties							
16. Specific Gravity	D792	0.945–0.965	1.10–1.26	0.960	0.935	1.05	1.03
Water Absorption, %	D750						
17. Method A (24 hr)		0.01–0.03	0.01–0.05	0.1	<0.01	0.01	<0.01
18. Method D (Long Term)		<0.01	0.1–0.2	–	<0.01	–	–
Mechanical Properties							
19. Tensile Yield Strength, 1000 psi	D638	3–5.5[i]	6–11	3.5	2.2	4.6	4.0
20. Ultimate Elongation, %	D638	100–1000	3–3.5	800	150	40	50
21. Flexural Strength at Yield, 1000 psi	D790	1	7–12	3.5–4	3.5–4	3.5–4	3.5–4
22. Flexural Modulus of Elasticity, 10^5 psi	D790	0.8–2.6	3–6	1.0	0.75	1.75	1.2
23. Compressive Strength at Yield, 1000 psi	D695	3.2	4–6	–	–	4.5	3.3
24. Rockwell Hardness (Method A)	D785	45–70[j]	R50–65	68D[m]	60D[m]	R60	R50
25. 1/8 in. specimen		1.5–12[m]	–	6	does not break	0.7	0.8
26. 1/4 in. specimen		0.4–1.5	1.1–4	–	–	–	–
27. Deformation Under Load	D621	–	0.4–1.6	–	–	–	–
Thermal Properties							
Deflection Temperture, °F[d]	D648						
28. At 66 psi Fiber Stress		140–185	260–265	158	–	160	140
29. At 264 psi Fiber Stress		110–125	225–260	–	–	–	–
30. Coeff of Lin Ther Expansion, in./in./°F	D696	6–7.2 × 10^{-5}[n]	1.7–2.3 × 10^{-5}	7 × 10^{-5}	–	3.9 × 10^{-5}	4.4 × 10^{-5}
31. Flam. ($\frac{1}{8}$ × $\frac{1}{2}$ × 5 in. spec.), in./min	D635	1.0–<2.5	–	–	1.0	No burn	No burn
Electrical Properties							
32. Dielectric Strength, v/mil, short time	D149	350–550[o]	–	475	–	500	500
Dielectric Constant	D150						
33. At 60 cps		2.30–2.7	–	2.6	–	–	–
34. At 10^6 cps		2.30–2.6	–	2.6	–	–	–
Dissipation Factor	D150						
35. At 60 cps		<0.0005	–	0.002	–	–	–
36. At 10^6 cps		<0.0005	–	0.002	–	–	–
37. Volume Resistivity, ohm-cm	D257	>10^{16}	2.9 × 10^{16}	6 × 10^{16}	–	–	–
38. Arc Resistance (SS Electrode), sec	D495	60–250	–	60	–	60	60
Cost (Natural)	–						
39. Cents/lb		19–50	62–67	–	–	–	–
40. Cents/cu in.		0.60–1.70	2.4–3.1	–	–	–	–

[a]E = Extrusion; EC = Extrusion coating; EW = Extrusion wire coating; M = Injection molding; B = Blow molding; RM = Rotational molding; PC = Power coating.
[b]USI has grade with rec. stock temp down to 280°F.
[c]UC range broader, see column 1.
[d]Unannealed, except where noted.
[e]Compression tested at 20 in./min.
[f]Shore D hardness.
[g]Compression molded.
[h]Eastman says it makes special heat resist grade.
[i]Spec. compression molded; ratio of loading: 2 in./min.
[j]Shore D.
[k]Value cited by Phillips Petroleum, Union Carbide reports range of 11–13 × 10^{-15} in./in./°C. Shell reports 1.3 × 10^{-4} in./in./°F.
[l]In./in./°C.
[m]Izod values of 1–3 ft lb/in. of notch given for $\frac{1}{2}$ × $\frac{1}{2}$ in. spec.
[n]Union Carbide reports 11–13 × 10^{-5} in./in./°C. Shell reports 1.3 × 10^{-4} in./in./°F.
[o]Eastman Chemical lists a diel. str. of 1000 v/mil (short time).

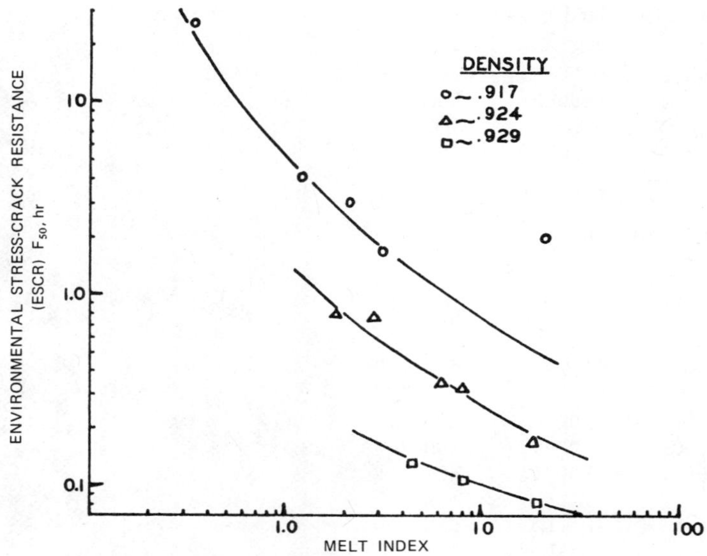

Fig. 5-17. Influence of melt index and density on environmental stress-crack resistance (ESCR) of low-density polyethylene. (Wild et al., 1967)

exhibit stress-cracking activity. Most organic liquids, under the proper conditions, are stress-cracking agents.

High-density polyethylenes seem to be more susceptible to this type of deterioration. Small amounts of rubbery polymers aid greatly in reducing this problem. High-density polyethylene employed by the cable industry often contains about 5% crude butyl rubber to protect against stress-crack failure.

Although the mechanical properties of polyethylene vary widely, as discussed earlier, they are considered average or below average when compared to other plastics materials. The electrical properties of polyethylene, on the other hand, are outstanding. Polyethylene is free from taste, odor, and toxicity and, of course, is very lightweight. It is perhaps the easiest of the thermoplastics to process, being adaptable to all types of high-volume molding and extrusion. It can be molded in a wide range of sizes, thicknesses, and under a broad range of molding conditions.

This combination, along with its excellent chemical resistance and formulatability, as well as its low cost, makes polyethylene production the largest in volume for plastics manufactured and used in the world. Properties of the new low-pressure, low-density polyethylenes are equal to or better than those obtained with conventional high-pressure resins. Pipe-burst strengths have been running 25% higher than the upper limit of conventional polyethylene, modulus values are higher, stress-crack resistance and low-temperature properties are good, and extractables meet FDA requirements.

Applications. The production of films consumes by far the greatest proportion of polyethylene. Slip agents are formulated into the polymer to prevent sticking ("blocking") of the thin sheets together. Film thicknesses vary from under 0.05 mil to 8 mils, and may run as high as 20 mils. (Note: In the trade, any film with a thickness of greater than 10 mils is often considered sheet stock.) Variations in clarity are possible, maximum transparency being desired for display wrapping. Because of the amount of crystallization that occurs, with its resultant opacity, films having a thickness of greater than 10 mils are never transparent.

Flat-die extrusion into water baths or onto chilled rolls produces higher clarity in the film and, coupled with variations in the polymers and copolymers, makes polyethylene competitive with traditional cellophane as well as many newer films. The principal uses for film are food, soft goods (sheets, shirts, etc.) and hardware packaging, garment bags, drum liners, temporary building closures for protection during construction, moisture barriers under concrete slabs and on foundations, soil mulching, greenhouses, automobile seat covers, tarpaulins, and temporary decoration displays. Table 5-11 shows the gas permeability of two types of polyethylene.

Electrical insulation was the first, and to this day, still is

Table 5-11. Gas Permeability (cm^3-mil/hr-in^2).*

Gas	Low Density	High Density
Carbon dioxide	0.820	0.143
Ethane	0.827	0.098
Freon	0.582	0.039
Helium	0.363	0.103
Hydrogen	0.535	0.134
Natural gas	0.231	0.047
Nitrogen	0.074	0.022

*Courtesy: Soltex Polymer Corporation.

Table 5-12. Comparison of Cellular Insulations.*

Property	ASTM	Cellular Type III Polyethylene[a]	Cellular Polypropylene	Solid Low-Density Polyethylene
Density, lb/cu ft	–	40	40	57.4
Tensile strength, psi	D412	2800	3100	2200
Elongation, %	D412	300+	300+	300+
Mandrel bend (2D), F[b]	–	<–94	–40	<–70
Power factor				
At 1 kc	D150	0.0012	0.0023	0.00018
At 50 kc	D150	0.0009	0.0025	0.00020
At 1 mc	D150	0.0006	0.0021	0.00023
Dielectric constant				
At 1 kc	D150	1.5–1.7	1.5–1.7	2.28
At 50 kc	D150	1.5–1.7	1.5–1.7	2.28
At 1 mc	D150	1.5–1.7	1.5–1.7	2.28

[a]DGDA-2580.
[b]Lowest temperature passed.
*Reprinted by permission from *Materials in Design Engineering*, August 1961.

Fig. 5-18. Because of its superior toughness and chemical resistance, polyethylene is widely used in industrial containers. (*Courtesy: Soltex Polymer Corporation*)

a major use of polyethylene. Low power factor and dielectric constant at high frequencies make it particularly valuable for communication wire and TV lead-ins. Table 5-12 compares the electrical properties of a foamed polyethylene wire insulation to unfoamed polyethylene and foamed polypropylene. Because of its excellent electrical resistance and its high chemical inertness, it finds prime use as buried power-cable jacketing.

Housewares such as pails, pans, pitchers, measures, refrigerator and freezer dishes, and closures are common items in homes, farms, and laboratories.

Polyethylene pipe is another major application for this polymer. Drums and drum liners, containers for chemicals (see Figure 5-18), labware, and tubing are other important industrial applications. The much greater stiffness of high-density polyethylene greatly increases its use for bottles (see Figure 5-19), particularly squeeze bottles for cosmetics, household products, and pharmaceuticals.

Polyethylene powder has been added to thermoplastic and thermoset resins to improve various properties of the matrix resin. In vinyls, the polyethylene powder improves flow and impact strength. In polystyrenes, a slight improvement in flow and apparent improvement in processing smoothness is achieved. However, impact strength and clarity decrease.

Despite system incompatibility, polyethylene powder in FEP fluorocarbon resins improves melt flow at higher shear rates. In thermosetting polyesters, the powder is effective for improving the polyester resin-filler distribution and surface smoothness and appearance of moldings.

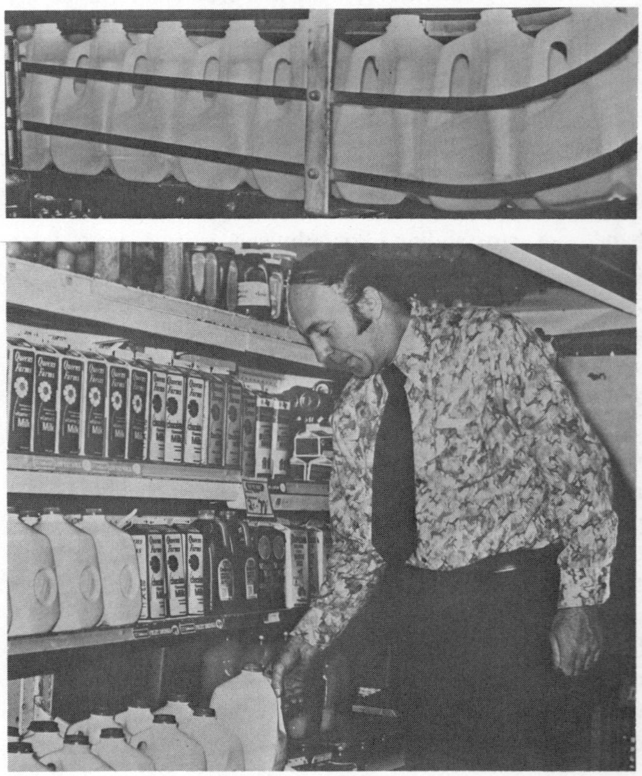

Fig. 5-19. Blow-molded gallon milk containers of low-density polyethylene are produced at a rate of 32 per minute (a). A dairy manager checks the stocking of the day's delivery (b). (*Courtesy: Phillips Chemical Co., a division of Phillips Petroleum Co.*)

Trade Names

Aerazate	Expanded sheet	Bakelite Xylonite, Ltd.
Aeroflex	Extrusions	Anchor Plastics Company
Aerozote	Expanded sheet	Bakelite Xylonite, Ltd.
Agiline	Resin	American Agile Corporation
Alathon	Resin	E.I. Du Pont de Nemours & Co., Inc.
Aldyl	Pipe	E.I. Du Pont de Nemours & Co., Inc.
Alkathene	Resins and compounds	Imperial Chemical Industries, Ltd.
Amcel	Semi-fabric foam	American Excelsior
Amerifilm	Film and sheeting	Ross & Roberts, Inc.
Ameriplast	Film and sheeting	Ross & Roberts, Inc.
Ameripol	Resin	Goodrich-Gulf Chemicals, Inc.
Amoco	Resin	Amoco Chemical Corporation
Anar	Resin	Rohm & Haas, Inc.
Apex	Semi-rigid compound	Apex Tire & Rubber Co.
Arlin	Film	Poly-Version, Inc.
Arman	Film and sheeting	Ross & Roberts, Inc.
Arnar	Film and sheeting	Ross & Roberts, Inc.
Bakelite	Resin	Union Carbide Corporation
Baylon	Low-density resin	Mobay Chemical Co.
Bexcourse	Resin	Bakelite Xylonite, Ltd.
Bexthene	High-density sheet and film	Bakelite Xylonite, Ltd.
Bexyarn	High-density filament	Bakelite Xylonite, Ltd.
Black Poly	Pipe and fittings	Anesile Co.
Blapol	Compounds	Reichhold Chemicals, Inc.
Borinal	Resin	Allied Resinous Products, Inc.
Brea	Resin	Brea Chemicals, Inc.
Caragy	Resin	Shell Chemical Co.
Carflex	Tubing	Carlon Products, Inc.
Carlona	Resins	Shell Chemical Co.
Celene	Low-density resin	Montedison, S.p.A.
Cell-O-Air	Expanded	Superior Cable Corporation
Chasekote P.S.	Film	Chase & Sons, Inc.
Chemplex	High-density film	Chemplex Co.
Chesfoil	Foil laminate	Chester Packaging Products Co.
Cheselene	Film tubing	Chester Packaging Products Co.
Chespak	Glassine laminate	Chester Packaging Products Co.
Cities Service	Resin	Cities Service Corporation
Clearsite	Rods and tubes	Celluplastic, Inc.
Clearstream	Flexible pipe	Yardley Plastics Co.
Clysar	Resin	E.I. Du Pont de Nemours & Co., Inc.
Cobonol	Tubing	Cobon Plastics Corporation
Compact	High-density	Stamicarbon, N.V.
Conolene	Low-density film	Continental Can Co.
Conolex	High-density film	Continental Can Co.
Corrofoam	Foamed coaxial cable	Phelps Dodge Copper Products Corporation
Crinothene	Sheet	Canadian Industries, Ltd.
Crown Poly	Coated paper	Crown Zellerbach Corporation
Crown-Seal	Film	Crown Zellerbach Corporation
Crown-Tango	Tinted low-density film	Crown Zellerbach Corporation
Crystalene	Foam	Crystal-X Corporation
Cyclothane	Tubing	Polykote, Inc.
Dapco	Resin	Davidson Products Co.
Deltaplast	Low-density pipe and fittings	Foster Brothers Plastics, Ltd.
Deltathene	High-density pipe and fittings	Foster Brothers Plastics, Ltd.
Densothene	Film	Metal Box Co.
Diothene	Film	Metal Box Co.
DLP	Resin	Dawbarn Brothers, Ltd.
Dow	Resin	Dow Chemical Co.
DP Strongwall	Pipe	Diversified Parkway Industries, Inc.
Dupont	Resin	E.I. Du Pont de Nemours & Co., Inc.
Dura-Clear	Film	Harwid Co.
Duraflo	Resin	Daymond
Duraphane UVI	Films	British Cellophane, Ltd.
Durethane	Thin wall tubing and sheeting	ARCO Polymers, Inc.

Durethene	Film	ARCO Polymers, Inc.
Dylan	Resins and fiber	ARCO Polymers, Inc.
Dynalon	Tubes	Dynamit Nobel, A.G.
Dynapol	Resin	Dynamit Nobel, A.G.
Eanplast	Molding compounds	East Anglia Plastic, Ltd.
Econoklear	Shrinkable film	Phillips Film Co., Inc.
Eftlon	Closed-cell foam	High Voltage Energy Corporation, Sokisui Chemical Co.
Elrex	Resin	Rexene Polyolefins
Elrexene	Resin	Rexene Polyolefins
El Rey	Low-density resin	Dart Industries, Inc.
Ende	Naturally degradable	Akerlund & Rausing
Enthofil	Reinforced	Fiberfil, Inc.
Enthoglas	Reinforced	Fiberfil, Inc.
Epolene	Low melt for casting and slush molding	Eastman Chemical Products, Inc.
Epotal	Dispersions	BASF, A.G.
Erinold	Molding compounds	Mobil Chemicals, Ltd.
Ethafoam	Expanded	Dow Chemical Co.
Ethofil	Reinforced	Dart Industries, Inc.
Ethoglas	Reinforced	Dart Industries, Inc.
Ethosar	Reinforced	Dart Industries, Inc.
Ethron	Film and resin	Dow Chemical Co.
Ethylon	Filament	Kureha Chemical Industry Co., Ltd.
Ethylux	Extrusion	Westlake Plastics Co.
Evatane	Resins	Imperial Chemical Industries, Ltd.
Evatate	Resins	Sumitomo Chemical Co.
Everex	Tube	Evered & Co., Ltd.
Excelite	Tubing	Thermoplastic Processes, Inc.
Ex-O-Line	Pipe	Carlon Products, Inc.
Extrel	Film	Exxon Chemical Co.
Extvel	Film	Extrudo Film Corporation
Exxon	Resin	Exxon Chemical Co.
E-Z Flow	Film	Extrudo Film Corporation
Fablothene	Film	Commercial Plastics, Ltd.
Fairlite	Film	Chippewa Plastics, Inc.
Fairlon		Chippewa Plastics, Inc.
Fertene	Low-density compounds	Montedison, S.p.A.
Fiberlene	Glass-filled compounds	Service Color Corporation
Firestone Polyethylene	Resin	Firestone Plastics Co.
Foam-cel	Foam insulation	Simplex Wire & Cable Co.
Foamflex	Foam cable	Phelps Dodge Copper Products Corporation
Formid	High-density	Polymon Developments, Ltd.
Fortiflex	Resins	Solex Polymer Co.
Fortilene	High-density	Celanese Canada, Ltd.
Full Flo	Pipe	Carlon Products, Inc.
Futron	Powder	Rubbermaid Industrial Products Co.
Garmil	Film	Canadian Industries, Ltd.
Genco	Resin	General Plastics, Inc.
Gen-Pak	Film	Monsanto Co.
Ger-Pak	Film and sheet	Monsanto Co.
Ger-Tube	Tubing	Monsanto Co.
Gilco	Tubing and sheet	Gilman Brothers Co.
Gil-Fold	Sheet	Gilman Brothers Co.
Glass-Weave	Scrim-reinforced	Warp Brothers
Grex	High-density resins	W.R. Grace & Co.
Guardex	Barrier-coated	Allied Chemical Corporation
Gulf	Resin	Gulf Oil Chemicals Co.
Habirlene	Insulation	Phelps Dodge Copper Products Corporation
Helex	High-density sheet	British Celanese, Ltd.
Hi-Fax	High-density	Hercules, Inc.
Hi-Fax	Resin	Canadian Industries, Inc.
Hilex	High-density sheet	British Celanese, Ltd.
Hi-Mol	High-density	Carlon Products Corporation
Hipack	Film	Showa High-Polymer Co.
Hitalex	Film	Hitachi Chemical Co.

Hi-Ten	High-density pipe and fittings	P.U. Muntz & Barwell, Ltd.
Hi-Zex	High-density	Mitsui Petrochemical Industries, Ltd.
Hostalen	Resin	American Hoechst Corporation
Inter-Fuse	Coating	M.J. Fassler & Co.
Iridon	Resin	Commercial Plastics, Ltd.
Irvethene	Resins and compounds	Great American Plastics Co.
Isoflex	Tubing	Keystone Plastics, Inc.
Ivithene	Film and flexible sheet	3M Co.
Katharon	Film and sheet	Ross & Roberts, Inc.
Kelon	Resin	W.S. Shamban & Co.
Klearcor	Resin	Western Plastics Corporation
Lacqtene	Resin	ATO Chimie of Paris
Lamabond	Reinforced	Lamex, Inc.
Lightlon	Foamed sheet	Sekisui Chemical Co., Ltd.
Lorival	Tube	United Ebonite & Lorival, Ltd.
Lotrene	Low-density	CdF Chimie
Lupolen	Resin	Lupoline Automatic Polishing
Marlex	Resins	Phillips Petroleum Co.
Marleythene	Low-density film, pipe, and fittings	Marley Extrusions, Ltd.
Mayflex	Resin	Maynard Plastics Co.
Measure-Marked	Pipe	Vistron Corporation
Medi-Gard	Antibacterial sheet	Medical Plastics Corp. of America
Metafoam	Medium-density foam	Ludlow Plastics
Metaplene	Medium-density film	Ludlow Plastics
Micropol Powder	Powder	Belgrave Industries, Ltd.
Microthene	Powder	U.S. Industrial Chemicals Co.
Milpac	Film	Canadian Industries, Ltd.
Milrol	Film	Canadian Industries, Ltd.
Miltite	Shrink film	Canadian Industries, Ltd.
Milwrap	Film	Canadian Industries, Ltd.
Mini-Vaps	Expanded	Nalge Co.
Mirason	Low-density	Mitsui Polychemicals
Mobil	Resin	Mobil Oil Co.
Modulene	Resin and molding compounds	Muehlstein & Co.
Moplen-RO	High-density	Montedison, S.p.A.
Moseley	Pipe and fittings	David Moseley & Sons, Ltd.
Muntz	Tubing	Glynwed Plastics, Ltd.
Naltene	Sheet	Convert, Ets.G.
Natene	High-density	Rhone Progil, Direction Commerciale Matieres Plastiques
Nazarene	Resin	H & R Plastics Industries, Inc.
Neopolen	Foam	BASF Wyandotte Corporation
Neo Zex	Medium-density	Mitsui Petrochemical Industries, Ltd.
Nipolon	Low-density	Nippon Polychemicals Co.
Nissan Vinyl	Compounds	Nissan Chemical Co.
Norchem	Low-density	Northern Petrochemical Co.
Novatec	High-density	Mitsubishi Petrochemical Industries, Ltd.
Okolene	Insulated wire	Okonite Co.
Optithene	Extruded film	Gulf Oil Corporation
Orthoplene	Film	Ludlow Corporation
Pactene	Resin	Schmidt Industries, Inc.
Paraplene	High-density film	Ludlow Corporation
Paraweb	Polyester-reinforced	ICI America, Inc.
Paxon	High-density	Allied Chemical Corporation
Pearlon	Film	Visking Corporation
PEP-1	Foam	Soviet Origin
Perfaqueen	Perforated film	Ethyl Corporation
Perflex	Film	Union Carbide Corporation
Perflex S	Shrinkable film	Union Carbide Corporation
Petrothene	Resin	U.S. Industrial Chemicals Co.
Plaskon	Resin	Allied Chemical Corporation
Plaspac	Extruded film	PPC Industries, Inc.
Plastazote	Cellular	Bakelite Xylonite, Ltd.
Plastiglass	Film	Warp Brothers
Plastiglaze	Coating powders	Plastic Coatings, Ltd.

Plastissue	High-density film	C.K. Addison
Plastylene	Low-density	Rhone-Progil Direction Commerciale Matieres Plastiques
Plexar	High-density film	Chemplex Co.
Plicose	Resin	Harte & Co.
Pliothane	Modified	Ametek/Westchester Plastics, Inc.
Polyane	Low-density film	La Cellophane S.A., Rhodia, Inc.
Polychromatic	Film and sheeting	Ross & Roberts, Inc.
Poly-Con	Resin	Norton Co.
Poly-Cover	Resin	Flex-O-Glass, Inc.
Polyethylene Omni		Omni Products Corporation
Polyfilm	Film	Dow Chemical Co.
Polyfilm	Film	Extruders, Inc.
Polyflexon	Film	Plymouth Industrial Products, Inc.
Poly-Flo	Tubing	Imperial Eastman Corporation
Poly-Fresh	Wrap	Crown Zellerbach Corporation
Polyfusion	Pipe	Nalge Co.
Polyigen	Film	BASF Wyandotte Corporation
Polyken	Film	Kendall Co.
Polymist	Powdered	Allied Chemical Corporation
Polymix	Colored compounds	Monsanto Co.
Polymul	Emulsions	Diamond Shamrock Co.
Polyorc	Low-density pipe and fittings	Yorkshire Imperial Metals, Ltd.
Poly Paper	High-density paper	Nalge Co.
Poly-Plant	Film	Texas Plastics Inc.
Polyprene Amalgam	Film, pipe, and fittings	G & M Plastics, Ltd.
Polysol	Suspension dispersions	Showa High Polymer Co.
Poly-T	Film	Tupper Corporation
Poly-The	Low- and medium-density	Gulf Oil Chemicals Co.
Polython	Film, tubing, and sheet	Poly Plastic Products, Inc.
Polytreat	Film	Extruders, Inc.
Portex	Low-density rod and tube	Portland Plastics, Ltd.
Prestolene	Film, tube, and sheet	J.P. Frank Chemical & Plastic Co.
Prohi	High-density	Protective Lining Corporation
Python	Tube, profile, and rod	Plastic Tube & Conduit Co., Ltd.
Ravinal	Low-density	ANIC, S.p.A.
Redithene	Low-density sheet	Rediweld, Ltd.
Reet	Film and sheeting	Ross & Roberts, Inc.
Reevon	Filaments and fabrics	Reeves Brothers, Inc.
Resinite	Resin	Resin Industries, Inc.
Resinol	Resin	Allied Resin Products Co.
Rexlon	Low-density	Nippon Petrochemicals Co.
Reynolon	Resin	Reynolds Metal Co.
Rhino	Reinforced	Cities Service Co.
Riblene	Resins	ANIC, S.p.A.
Ribofil	Fibers	Plasticisers, Ltd.
Rigidene	Resin	Heil Process Equipment Corporation
Rigidex	High-density	BP Plastics, Ltd.
Roll-Tex	Film	Texas Plastics, Inc.
Rotene	High-density compounds	Montedison, S.p.A.
Roto Flame	Resin	Rototron Corporation
Rotothene	Resin	Rototron Corporation
Rumiten	Resin	Rumianca, S.p.A.
San-D-Lite	Sheet	Sandee Mfg. Co.
Sarolene	High-density sheet	Saro Products, Ltd.
Sclair	Resin	E.I. Du Pont de Nemours & Co., Inc.
Sclairfilm	Film	E.I. Du Pont de Nemours & Co., Inc.
Sclairpipe	Pipe	E.I. Du Pont de Nemours & Co., Inc.
Seilon		Seiberling Rubber Co., Plastics Division
Sholex	High-density	Showa Yuka, K.K.
Sicalene	Sheet	Mazzucchelli Celluloide, S.p.A.
Sirtene	Low-density	Societa Italiana Resine
Sisalite	Film	American Sisalkraft Co.
Softon	Foam	Sekisui Chemical Co.
Solidur	Resin	Pennekamp & Huesker, A.G.
Soltex	Resin	Soltex Corporation

Spex-Film	Film	Griffolyn Co.
Stamylon	Low- and high-density	DSM
Sumikathene	Resin	Sumitomo Chemical Co., Ltd.
Suntec	High-density	Asahi Chemical Industry Co.
Superaero-flex	Extrusions	Anchor Plastics Co., Inc.
Super Dylan	High-density	ARCO Polymers, Inc.
Super Modulene	Resin and molding compounds	H. Muehlstein & Co.
Supernaltene	Sheet	Convert, Ets.G.
Super Seam	Foam	Charles Industries
Suprane	High-density film	Rhodia Corporation
Suprathen	Low-density film	Kalle & Co., A.G.
Suprel	Resin	Allied Chemical Corporation
Suprex	Pipe	Vistron Corporation
Surfaseptic	Film and resin	Dow Chemical Co.
Tarpo	Textile-reinforced film	UCB-Sidac Corporation
Telcothene	Powder and sheet	Telcon Plastics, Ltd.
Tenite	Resin	Eastman Chemical Products, Inc.
Tetralene	Resin	Royal Industries
Three Crown	Low-density pipe and fittings	P.H. Muntz & Barwell, Ltd.
Toray	Foam sheet	Toray Industries, Inc.
Toray PEF	Foam	Toray Industries, Inc.
Torolon	Fabrics	Polymer International
Touchstone	Film and sheeting	Ross & Roberts, Inc.
Tower Sof	Film	Tower Products, Inc.
Trocal	Film and sheet	Dynamit Nobel, A.G.
Trofil	Monofilaments	Dynamit Nobel, A.G.
Trojan	Pipe	Vistron Corporation
Trolen	Sheet	Dynamit Nobel, A.G.
Trovilon	Film	Dynamit Nobel, A.G.
Twin-Tex	Pipe	Vistron Corporation
Tygothene	Resin	Norton Co.
Tyvek	Spunbonded high-density sheet	E.I. Du Pont de Nemours & Co., Inc.
Ultra-Ethylux	Extrusions	Westlake Plastics Co.,
Union Carbide	Resin	Union Carbide Corporation
Valsof	Emulsions	United Merchants, Inc.
Verolene	Film and sheeting	J.P. Frank Chemical & Plastic Co.
Vestolen	High-density	Chemische-Werke Huls, A.G.
Vispose	Film	Ethyl Corporation
VisQueen	Low-density film	Imperial Chemical Industries, Ltd.
Vistal	Film	UCB-Sidac Corporation
Vitral	High-density film	Papeteries Dalle & Lecomte
Vitrathene	Sheet, rod, tube	Stanley Smith & Co.
Vivelle	Resin	Imperial Chemicals Industries, Ltd.
Volara	Low-density foam	Voltek, Inc.
Volasta	Medium-density foam	Voltek, Inc.
Volite	High-density foam	Voltek, Inc.
Vyon	Porous high-density	Atlas Mineral & Chemical Division, ESB, Inc.
Wascofilm	Film	Wasco Products, Inc.
Yukalon	Low-density	Mitsubishi Petrochemical Co.
Zee	Wrap	Crown Zellerbach Corporation
Zelite	Resin	Newage Industries, Inc.
Zendel	Film	Union Carbide Corporation

REFERENCES AND BIBLIOGRAPHY

Birks, A. M., and Dowd, L. E. "Polyethylene Film." Paper presentation, *SPE ANTEC*, New Orleans, May 1979.

Bloyer, S. In-line systems cut costs in HDPE bag making. *Modern Plastics*, April, 1978, **55**(4):63–65.

Brydson, J. A. *Plastics Materials*. 3rd Ed. Robert E. Kreiger Publishing Co., Inc., Huntington, New York, 1975.

Celanese Plastics Co. *Fortiflex Polyethylene–Properties*. Bulletin P1A, March 1972.

Chimura, L. J. Polyolefins are definitely in Ford's future. *Plastics Engineering*, August 1978, **34**(8):18–21.

Concha, M. Low-density polyethylene, from Plastics in Packaging. *Plastics Engineering*, October 1978, **34**(10):39–40.

Deanin, R. D., and Kivong, P. C. "Expansion Coefficients of Filled High-Density Polyethylenes." Paper presentation, *SPE ANTEC*, New Orleans, May 1979.

Dubois, J. H., and John, F. W. *Plastics*. 5th Ed. Van Nostrand Reinhold Co., New York, 1974.

E.I. Du Pont de Nemours & Co., Inc. *Alathon Polyethylene Resin–Many-Purpose Plastic*. School Service Bulletin A-70068.

Eastman Chemical Products, Inc. *The Effect of Various Chemicals on Tenite Polyethylene*. Technical Report 17B, ECP 3125, March 1974.

Hardesty, G. H., and Camp, J. R. High-density polyethylene, from Plastics in Packaging. *Plastics Engineering*, October 1968, **34**(10): 37–38.

Hercules Powder Co., Inc. *Hi-Fax Hercules High-Density Polyethylene: What It Is... Where to Use It... How to Use It....* Bulletin No. 500-273A, December 1957.

Houston, A. M. Polyethylenes combine toughness with processing versatility. *Materials Engineering*, May 1976, **83**(5):32–34.

Joyner, R. S., Edmunds, N. J., and Gilbert, P. G. Troubleshooting HDPE processing problems. *Plastics Design & Processing*, April 1979, **19**(4):45–49.

Kaufmann, K. A., and Imig, G. S. Effect of low-M.W. polyethylene waxes on PE injection moldings. *Modern Plastics*, February 1959, **36**(6):137–140.

Longarzo, P. J., and Margolies, A. F. The many faces of polyethylene. *Machine Design*, January 11, 1973, **45**(1):113–117.

Lustiger, A., and Corneliussen, R. D. "Stress Cracking in Polyethylene." Paper presentation, *SPE ANTEC*, New Orleans, May 1979.

MacBride, R. R. Carbide has new LDPE technology—and backs it with vast expansion. *Plastics World*, June 1971, **29**(6).

Margolis, A. F. Effect of molecular weight on properties of HDPE. *SPE Journal*, June 1971, **27**(6):44–48.

Martino, R. Polyethylene, from Materials '79. *Modern Plastics*, January 1979, **56**:46–47.

——. Volume thermoplastics: there will be plenty around, but it will cost more. *Modern Plastics*, May 1979, **56**(5):92–94.

——. That new LDPE technology: what does it mean to users? *Modern Plastics*, February 1978, **55**(2):38–40.

——. Chaos in LDPE pricing. *Modern Plastics*, April 1978, **55**(4):46–47.

Materials Engineering. Rotomolding HDPE is flame retardant. January 1977, **85**(1):19.

——. Modified polyethylene insulates cable better. July 1977, **86**(1):42–43.

Materials in Design Engineering. Tougher cellular polyethylene made from type III resin for wire and cable insulation. August 1961, **54**(2):19.

Modern Plastics. New Polyethylenes for wire and cable. September 1978, **55**(9):98.

——. New use for an old technique makes a pipe-insulation "package." July 1978, **55**(7):30–32.

Phillips Chemical Co. *Marlex 50 Polyethylene*. Bulletin, 1956.

Pelagatti, U., and Baretta, G. Stress cracking of polyethylene. *Modern Plastics*, June 1959, **36**(10):140–146, 212.

Plastics Technology. PT's Buyer's Guide to Polyolefins and Cellulosics, March 1967, **13**(3).

Simonds, H. R., and Church, J. M. *A Concise Guide to Plastics*. 2nd Ed. Van Nostrand Reinhold Co., New York, 1963.

Sneller, J. New high-flow polyolefins make thin-wall molding a better bet. *Modern Plastics*, January 1979, **56**(1):75–77.

Swanson, R. S. *Plastics Technology—Basic Materials and Processes*. McKnight & McKnight Publishing Co., Bloomington, Illinois, 1965.

Tanaka, M., Otsuki, F., Kinugasa, F., Nagasawa, T., and Tsurutani, I. "The rheology of polyethylene in pipe coating." Paper presentation, *SPE ANTEC*, New Orleans, May 1979.

Titus, J. B. *Trade Designations of Plastics and Related Materials* (revised). PLASTEC Note N9C, Plastics Technical Evaluation Center, Picatinny Arsenal, Dover, New Jersey, May 1978.

Turley, R. R. The Future for LDPE. Paper Presentation *SPE ANTEC*, New Orleans, May 1979.

U.S. Industrial Chemicals Co. *Petrothene Polyolefins...A Processing Guide*. Bulletin PL 3 471, 4th Ed., 1971.

——. *Evaluation of Microthene F Powdered Polyethylene in Thermoplastic and Thermoset Resins*. Bulletin PTD 55, May 1968.

——. *Molding Microthene Polyethylene Powder*. Bulletin PTD 36-964, 1st Ed., 1963.

Wild, L., Woldering, J. F., and Guliana, R. T. "Number-Average Molecular Weights of Low Density Polyethylenes: Relationship to Physical Properties." Paper presentation, *SPE ANTEC*, May 1967.

Winding, C. C., and Hiatt, G. D. *Polymeric Materials*. McGraw-Hill Book Co., New York, 1961.

Wunderlich, B. The solid state of polyethylene, *Scientific American*, November 1964, **211**(5):80–94.

Ultrahigh Molecular Weight Polyethylene

Ultrahigh molecular weight (UHMWPE) polyethylene is a low-pressure, highly polymerized polyethylene produced with Ziegler catalysts. The molecular weight of UHMWPE varies between 1 million and 10 million compared to 100,000–500,000 for conventional polyethylene. Figure 5-20 compares the relationship of intrinsic viscosity and molecular weight between UHMWPE and high-density polyethylene. The increased molecular weight causes an improvement in impact strength (particularly at low temperatures), energy absorption under high loading rates (see Figure 5-21), tensile strength at elevated temperatures, and resistance to stress-cracking. Table 5-13 compares the properties of UHMWPE with standard polyethylenes and Table 5-14 compares the properties of three commercial grades of UHMWPE.

Tables 5-15 and 5-16 illustrate the exceptional tensile impact strength and excellent abrasion resistance of UHMWPE. As an example of its low-temperature toughness, 1-inch-

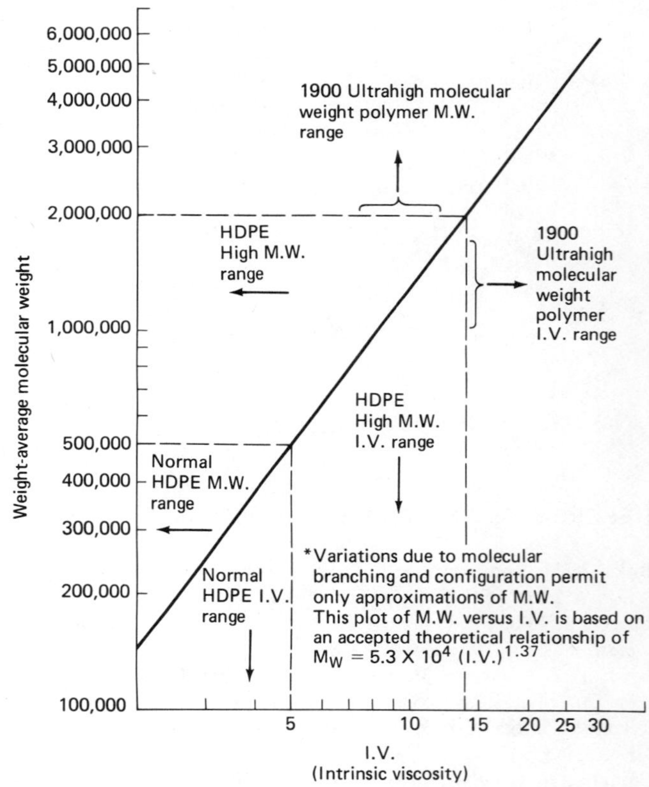

Fig. 5-20. Relationship of intrinsic viscosity and molecular weight for ultrahigh molecular weight polymers. (*Courtesy: Hercules, Inc.*)

Table 5-13. Comparison of Polyethylene Types.*

Properties	Branched Low Density	Linear High Density	UHMW
Molecular weight (weight average)	300,000	126,000	1,500,000
Density, gm/cc	0.92	0.96	0.945
Crystalline melting point, °C (°F)	113 (235.4)	136 (276.8)	135 (275)
Izod impact, notched, ft-lbs/in	16	4	>20[a]
Stiffness modulus, psi	14,000	150,000	91,000
Tensile strength, psi	1,600	4,400	5,400
Heat deformation °C (°F) @66 psi	48.9 (120)	71.1 (160)	68.9 (156)

[a]Specimen deflects, but does not break.
*Courtesy: Specialty Chemicals Division, Allied Chemical Corporation.

Table 5-14. Molecular Weight Determines Property Variations.*

Property	Allied 1220	Hercules 1900	Allied 260–100
Molecular weight	2.8×10^6	$4\text{–}5 \times 10^6$	$8\text{–}10 \times 10^6$
Specific gravity, gm/cc	0.936	0.938	0.930
Tensile strength, 73 F, psi	5600	6400	6000
Yield strength, psi	3100	3500	3000
Ultimate tensile elongation 73 F, %	450	350	200
−100 F, %	190	–	–
−200 F, %	110	–	–
Izod impact (notched), ft-lb/in. notch	>20	no break	>20
Stiffness, cantilever beam, psi	80,000	80,000	80,000
Rockwell hardness, R. scale	50	–	50
Tear strength, gm/Mil	550	–	–
Heat deflection temperature, 264 psi, F	113	117	115
Brittleness temperature, F	<−94	–	–
Coefficient of friction, static	0.11	0.14	–
Abrasion resistance (Armstrong), cm^3	0.355	–	–
Abrasion resistance (Taber), gm/10,000 cycles	0.008	–	–
Dielectric constant 100 Hz	2.34	2.34	2.34
10^6 Hz	2.30	2.30	2.30
Dissipation factor 100 Hz	0.0003	0.0003	0.0003
10^6 Hz	0.0002	0.0002	0.0002
Dielectric strength short time, volts/Mil	710	710	710
stepwise	680	680	680
Volume resistivity, ohm/cm	$>10^{16}$	$>10^{16}$	$>10^{16}$
Burning rate	Very slow	Very slow	Very slow

*Reprinted by permission from *Product Engineering*, January 1973.

Table 5-15. Comparative Impact Strengths of Various Plastics.*

Material	D-1822-61T Type L Specimen Tensile Impact Strength ft-lbs/sq in
Polyacetal	30
Polypropylene (normal impact)	45
High-density polyethylene	60
Polypropylene (high impact)	120
Low-density polyethylene	180
Nylon	270
Polycarbonate	310
UHMW polyethylene	>1,000

*Courtesy: Specialty Chemicals Division, Allied Chemical Corporation.

thick slabs of UHMWPE were subjected to small bore ballistics tests at 0°C (32°F) and -54°C (-65°F). The slabs stopped .45 caliber, .357 magnum, and .44 magnum pellets at a distance of 25 feet. The first two projectiles actually bounced out of the specimen, while the latter pellet remained embedded in the slab.

UHMWPE has very good dielectric properties which remain constant over a wide temperature range. Like all polyethylenes, UHMWPE offers good resistance to corrosive media, except strong oxidizing acids. It shows slight swelling in aromatic and halogenated hydrocarbons, a significant improvement over traditional polyethylenes.

Because the wear resistance and physical properties of UHMWPE are, in many areas, similar to the engineering thermoplastics, its primary uses have been directed at the same markets: gears, bushings, textile machinery compo-

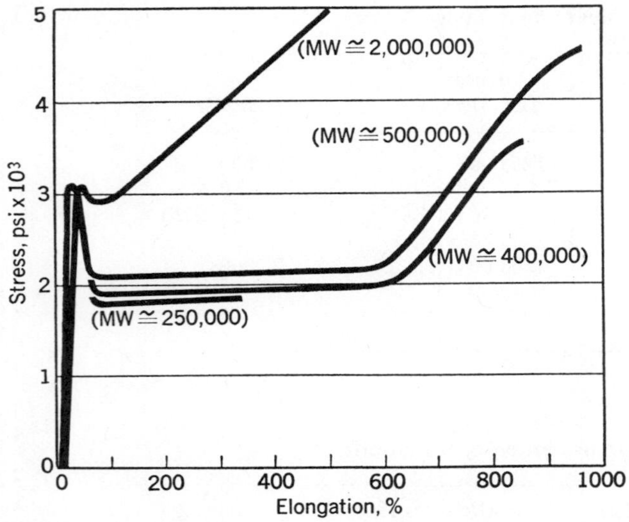

Fig. 5-21. Polyethylene stress-strain curves show why ultrahigh molecular weight type is high-energy absorber. (*Courtesy: Product Engineering*)

Table 5-16. Relative Abrasion Resistance of Various Materials (Carbon Steel Set = 100).*

UHMW Polymer	15	Polypropylene	190
Nylon 6-6	31	Phosphor bronze	190
Polyurethane (D-70)	37	Phenolic laminate L.E.	200
TFE	72	Polysulfone	300
304 Stainless steel	84	Yellow brass	400
High-density polyethylene	86	Low-density polyethylene	530
Polycarbonate	96	Maple	690
Carbon steel	100	Hard neoprene rubber	800
Polyacetal	110	Hickory wood	950

*Courtesy: Hercules, Inc.

nents; linings for chutes and conveyors in the mining industry; guide and deflector rollers, and star wheels for bottling and labelling machines in the food and packaging industries (see Figure 5-22); sporting goods (golf club faces, helmets, etc.); mallet heads, door stops, and vibration damper pads.

Fig. 5-22. Conveyor star wheel made of ultrahigh molecular weight polyethylene; structural component of a bottling plant. (*Courtesy: American Hoechst Corporation*)

REFERENCES AND BIBLIOGRAPHY

Allied Chemical Corporation. *A-C Polyethylene 1220.* Technical Data Bulletin No. C-1A, October 1968.

American Hoechst Corporation. *Product Data Sheet—Hostalen GM 9255F.* February 1978.

——. *Hoechst Plastics—Hostalen GUR.* Bulletin KR112e-7097, September 1977.

——. *Hostalen GUR in Mining, Refining, Materials Handling.* Bulletin PT-76-1, 1974.

——. *New Pathways in Engineering Hostalen GUR (In Industry).* Bulletin.

Fabian, R. J. Engineer's guide to UHMW polyethylene. *Materials Engineering*, September 1971, 74(3):34–39.

Hercules, Inc. *Abrasion Resistant 1900 UHMW Polymer.* Bulletin 500-693, 1972.

——. *Properties, Uses and Fabrication of the Hi-Fax 1900 Series.* Bulletin Hi-Fax No. 30, January 1964.

——. *1900 UHMW Polymer Relationship I.V. to M.W.* Bulletin 51865H.

Trade Names

AC		Allied Chemical Corporation
Aycee	Pipe	Triangle
Ceram P		Solidur Plastics
Driscopipe	Pipe	Phillips Products
Embafort	Plate	M & B Plastics, Ltd.
Gar-dur Plastic		Garland Mfg. Co.
Hi-Fax 1900		Hercules Inc., Polymers Department
Hi-Mo		Ireland Alloys, Ltd.
Hostalen GUR		American Hoechst Corporation
Impax		Impact Plastics, Inc.
Jaytrex		Johnston Industrial Plastics
Perplas		Performance Plastics, Ltd.
Polyceram		Impact Plastics, Inc.
Polypenco 19		Polymer Corporation
Ultra Bluejet	Pipe	Celanese Plastics Co.
Ultrex	Extrusions	Spriratex Co.
Xelon		International Silver Co.

Longarzo, P. J., and Margolies, A. F. The many faces of polyethylene. *Machine Design*, January 11, 1973, **45**(1):113–117.
Lowery, J. H. Properties and applications of ultra high molecular weight linear polyethylene. *Plastics Design & Processing*, April 1963, 3(4):30–35.
Modern Plastics. New job for HDPE: it protects steel. May 1978, **55**(5):32.
——. UHMWPE–a sleeping giant stirs. April 1975, **52**(4):38–40.
——. HMWPE–the tough olefin is starting to move. July 1965, **42**(7):84–88, 150–160.
Plastics Design & Processing. Ultra high molecular weight polyethylene offers ultra high resistance to abrasion. June 1972, **12**(6):22–23.
The Polymer Corporation. *Polypenco 19 Ultra-High Molecular Weight Polyolefin.* Bulletin BR-44S, July 1972.
Product Engineering. New processing twists and materials add to UHMW plastics versatility. January 1973, **44** (1):20–23.
——. Long chain plastic bids for tough industrial jobs. February 13, 1967, **38**(4):81–83.

Crosslinked Polyethylene

Although polyethylene is a highly saturated polymer, it is possible to crosslink it. The first work done on crosslinking polyethylene was done in England in the 1940s. In the 1950s, workers at General Electric succeeded in crosslinking polyethylene with dicumyl peroxide. About the same time the use of gamma radiation and high-speed electrons for crosslinking polymers was described.

The basic process of crosslinking polyethylene is via a free-radical mechanism. Either a peroxide degrades thermally to provide the free radical (chemical crosslinked polyethylene), or the free radical is generated within the polymer chain by the displacement of hydrogen from the chain by high-energy irradiation (irradiated crosslinked polyethylene). The latter is typically carried out in an inert atmosphere to prevent oxidative degradation. Also, the irradiation process is limited to thicknesses compatible with the depth of penetration of the radiation.

Although the crosslinking of polyethylene markedly increases its properties, it does so at the expense of processing ease and cost (see Table 5-17). Flow properties of the crosslinked polymer are restricted (see Figure 5-23), and the polymer becomes insoluble in many of the solvents that would dissolve normal polyethylene. Mechanical properties, particularly high-temperature modulus, increase. The material is stiffer over its entire temperature range and much more fatigue-resistant. Figure 5-24 compares the fatigue resistance of crosslinked polyethylene to that of some other thermoplastics. Crystallinity and stress-crack susceptibility decrease. The electrical properties remain excellent and are only affected by such secondary factors as the nature of the peroxide used, its decomposition products, and/or added fillers and modifiers. Dielectric strength at high temperatures is improved. Figure 5-25 shows the comparative effect of temperature on the tensile strength of polyethylene and crosslinked polyethylene.

Crosslinked polyethylene can be formulated with fillers and modifiers to a greater degree than normal polyethylene.

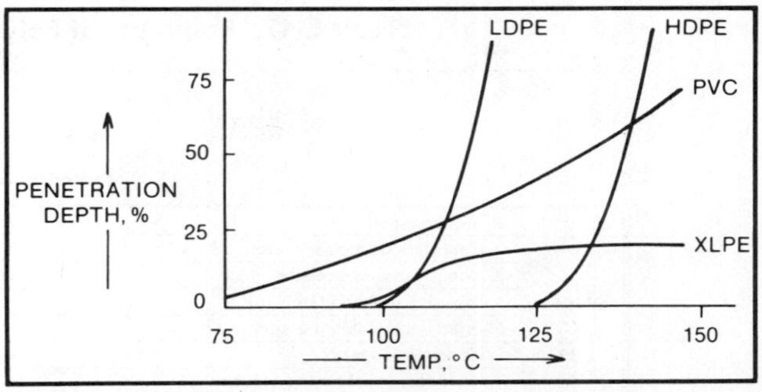

Fig. 5-23. Thermomechanical deformation resistance. (Thomas and Bowrey, 1977)

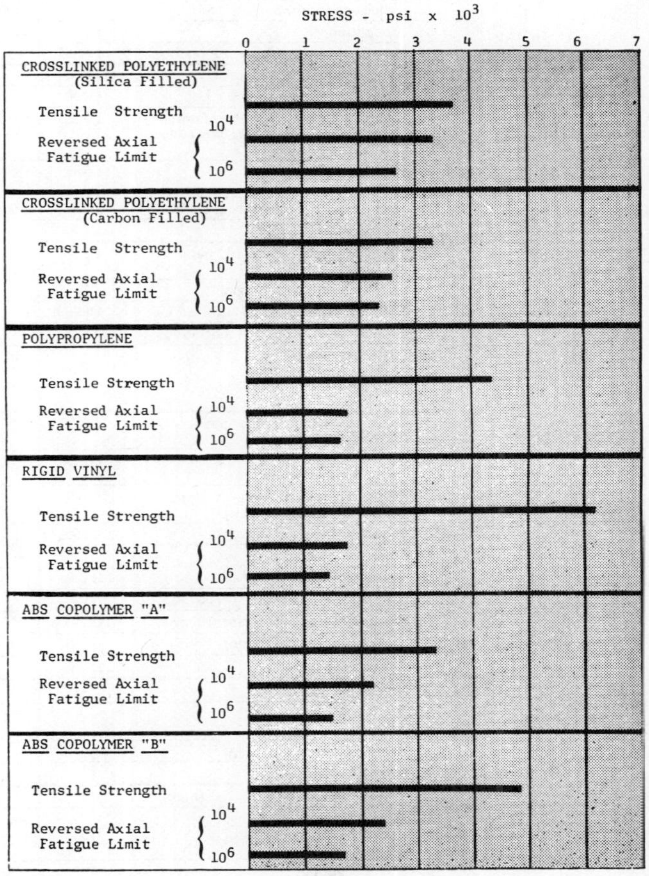

Fig. 5-24. Resistance to fatigue shown by crosslinked polyethylene and some thermoplastics. (Boot and Jordan, 1963. Reprinted by permission from *Plastics Design & Processing*. Copyright 1963, Lake Publishing Corporation, Libertyville, Ill.)

This helps to vary properties and reduce costs. It is produced in three basic forms: molding powder for rotational molding, pellets for injection molding (and in Japan for foam production), and thin film for irradiation.

The powder used for rotational molding has found its largest use in the manufacture of automotive gas tanks, although the material and procedure is adaptable to hollow items ranging from small toy dolls to containers of several hundred gallons.

Table 5-17. Properties of Polyethylene Crosslinkable Compounds.*

		ASTM test method	Polyethylene cross-linkable compounds: Rotational molding grades	Molding grades	Wire and cable grades	Ultra high molecular weight
Processing	1. Molding qualities	—	Good	Excellent	—	Fair
	2. Compression molding temp., °F.	—	390-450	240-450	—	400-500
	3. Comp. molding pressure, p.s.i.	—	500-800	100-800	—	800-1200
	4. Injection molding temp., °F.	—	—	250-300	—	—
	5. Injection molding pressure, p.s.i.	—	—	—	—	—
	6. Compression ratio	—	—	—	—	—
	7. Mold (linear) shrinkage, in./in.	—	0.02-0.05	0.007 0.090	0.020-0.050	—
	8. Specific gravity (density)	D792	0.930-0.939	0.95-1.45	0.91-1.40	0.94
	9. Specific volume, cu. in./lb.	D792	—	—	—	29.4
	10. Machining qualities	—	—	Good to excellent	Good to excellent	Good to excellent
Mechanical	11. Tensile strength, p.s.i.	D638	2500-2700	1600-4600	1500-3100	2500-3500
	12. Elongation, %	D638	450->660	10.0-440.0	180-600	300-500
	13. Tensile elastic modulus, 10^5 p.s.i.	D638	—	0.5-5.0	—	0.20-1.10
	14. Compressive strength, p.s.i.	D695	—	2000-5500	—	—
	15. Flexural yield strength, p.s.i.	D790	—	2000-6500	—	—
	16. Impact strength, ft. lb./in. of notch (½ × ½ in. notched bar, Izod test)	D256	No break	1.0-20.0	—	No break
	17. Hardness, Rockwell	D785	D60-63 (Shore)	55-80 (Shore 1)	33-57 (Shore D)	D60-70 (Shore)
	18. Flexural elastic modulus, p.s.i. × 10^5	D790	1.0	0.7-3.5	—	1.30-1.40
	19. Compressive modulus, p.s.i. × 10^5	D695	—	0.5-1.5	—	—
Thermal	20. Thermal conductivity, 10^{-4} cal./sec./sq. cm./1 (°C./cm.)	C177	—	—	—	—
	21. Specific heat, cal./°C./gm.	—	—	—	—	—
	22. Thermal expansion, 10^{-5} in./in./°C.	D696	—	10.0-35.0	10.0-35.0	7.2
	23. Resistance to heat, °F. (continuous)	—	—	275	275	—
	24. Deflection temp., °F. at 264 p.s.i. fiber stress	D648	140	105-145	100-175	105-120
	at 66 p.s.i. fiber stress	D648	—	130-225	100-175	155-180
Electrical	25. Volume resistivity, ohm-cm. (50% RH and 23° C.)	D257	—	<200->5 × 10^8	—	>10^{16}
	26. Dielectric strength, short-time, ⅛-in. thickness, volts/mil	D149	—	—	230-1420	710
	27. Dielectric strength, step-by-step, ⅛-in. thickness, volts/mil	D149	—	—	—	680
	28. Dielectric constant, 60 cyc.	D150	—	—	2.28-7.60	—
	29. Dielectric constant, 10^3 cyc.	D150	—	—	2.27-7.40	—
	30. Dielectric constant, 10^6 cyc.	D150	—	2.63-3.11	2.27-7.50	2.30
	31. Dissipation (power) factor, 60 cyc.	D150	—	—	0.003-0.044	—
	32. Dissipation (power) factor, 10^3 cyc.	D150	—	—	0.00048-0.04900	—
	33. Dissipation (power) factor, 10^6 cyc.	D150	—	0.001-0.002	0.0004-0.0050	0.0002
	34. Arc resistance, sec.	D495	—	—	—	—
Optical	35. Refractive index, n_D	D542	—	—	—	—
	36. Clarity	—	Transparent to translucent		Transparent to opaque	Transl.-opaque
	37. Transmittance, %	—	—	—	—	—
	38. Haze, %	—	—	—	—	—
Resistance characteristics	39. Water absorp., 24 hr., ⅛-in. thick., %	D570	—	<0.01-0.06	<0.01-0.06	<0.01
	40. Flammability[b] Burning rate, in./min.	D635	—	—	—	—
	AEB, in./ATB, sec.[c]	D635	—	—	—	—
	41. Effect of sunlight	—	Unprotected material crazes rapidly; requires black for complete protection but weather resistant grades available in natural and colors		Same as for polyethylene	—
	42. Effect of weak acids	D543		Very resistant		Very resistant
	43. Effect of strong acids	D543		Attacked slowly by oxidizing acids		Attacked by oxidizing acids
	44. Effect of weak alkalies	D543		Very resistant		Very resistant
	45. Effect of strong alkalies	D543		Very resistant		Very resistant
	46. Effect of organic solvents	D543		Resistant below 80° C.		Resists many solvents below 80° C.

*Reprinted by permission from *Modern Plastics Encyclopedia*, McGraw-Hill, Inc.

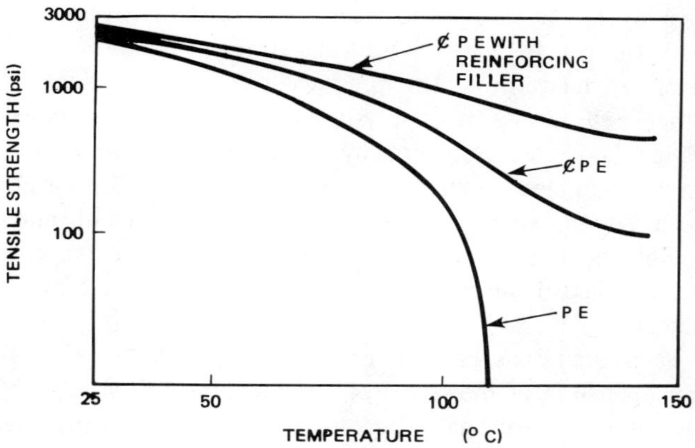

Fig. 5-25. Effect of temperature on strength of crosslinked polyethylene. (Scrima, 1976)

The major market for extruded grades is wire and cable insulation. Its outstanding weatherability and stability at high-exposure temperatures along with tough mechanical properties make the material ideal for this use.

A recent application for crosslinked polyethylene is as a molded union sleeve for concrete conduit. Here, the plastic is molded into a sleeve and simply slipped over the conduit at the connection point. In this case, it is the resistance to stress-cracking, moisture, chemicals, and creep (see Figure 5-26) that specifies the use of crosslinked polyethylene.

A foam is made from crosslinked polyethylene in Japan. Parts are cut from preblown sheet and are blended with polyethylene co vinyl acetate for the soles and heels of leisure and sport shoes.

There are two major markets for irradiated film. The first is as a packaging film for food and small parts. The film has improved strength and reduced permeability. When oriented, the film shrinks to conform to the contours of the product and has crystal clarity. The second market is shrink tubing for electrical connectors. Here, polyethylene is used together with crosslinked polyethylene to form a protective sleeve. The interlayer is polyethylene which, when heated, flows to fill voids and completely encapsulates the wire. The outer layer is crosslinked polyethylene, which shrinks to form a compact protective shell over the connector.

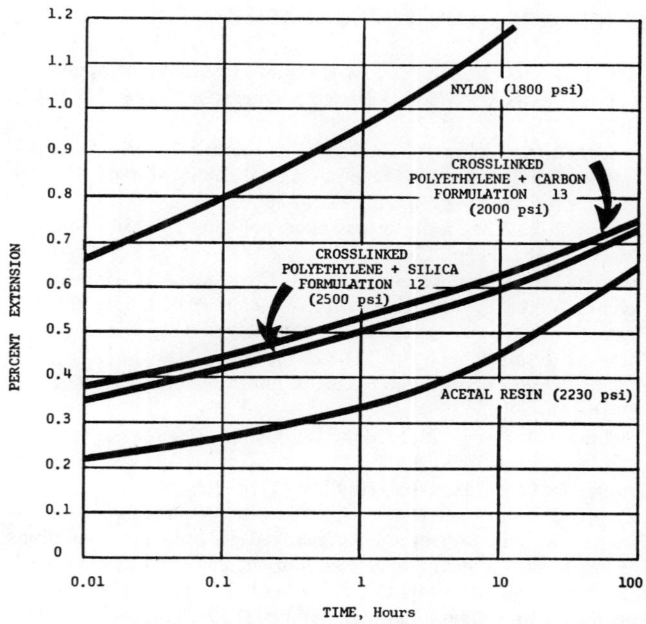

Fig. 5-26. Tensile creep behavior of crosslinked polyethylene compounds is less than that of nylon and about equal to acetal resin. (Reprinted by permission from *Plastics Design & Processing*, September, 1963. Copyright 1963, Lake Publishing Corporation, Libertyville, Ill.)

Trade Names

Alveolit	Foam	Sekisui
Blanex	Compounds	Reichhold Chemicals, Inc.
Cab	Compounds	Cabot Corporation
Enrad	Pre-irradiated	Enflo Corporation
Exar	Insulated wire	Haveg Industries, Inc.
Frelen	Foam	Pellan Corporation
Genkene	Insulation	General Cable Corporation
Hiray	Irradiated film and tapes	Hitachi Chemical Co.
Insulrad	Irradiated sheet and rod	Electronized Chemical Corporation
Irracure		Cooke Division, Reichhold Chemicals, Inc.
Irrathene	Irradiated	General Electric Co.
Kablend	Blended with EPR	Kaiser Aluminum & Chemical Co.
Kathene	Resins	Kaiser Aluminum & Chemical Co.
Marlex	Resins	Phillips Petroleum Co.
Petrothene	Compounds	U.S.I. Industrial Chemicals Co.
Polycure	Compounds	Reichhold Chemicals, Inc.
Rome	Compounds	Alcoa
Softion	Foam sheet	Sekisui Chemical Co.
Thermolene	Film	Continental Can Co.
Trocellen	Foam	Dynamit Nobel, A.G.
Trolen-Dur	Rods	Dynamit Nobel, A.G.
Velvotex	Foam	Leslies, Ltd.
Vulkene	Compounds	General Electric Co.

REFERENCES AND BIBLIOGRAPHY

Boot, R. J., and Jordan, T. J. A study of crosslinked compounded polyolefin systems. *Plastics Design & Processing*, September 1963, 3(9):11–16.

Botham, R. A., Jenkins, G. H., Ball, G. L. III, and Salyer, I. O. Cross-linked HDPE makes the grade in thermal energy storage. *Modern Plastics*, December 1978, **55**(12):54–58.

Carlson, B. C. Chemically cross-linked polyethylene. *SPE Journal*, March 1961, **17**(3):265–270.

Pastorino, R., Lewis, R., and Halle, R. Crosslinking HDPE with cyclic peroxyketals. *Modern Plastics*, September 1978, **55**(9):86–88.

Plastics Technology. Crosslinkable polyolefins: now you can mold them. November 1968, **14**(11):11.

Plastics World. Crosslinking extends limits of PE. August 1966, **24**(8):48–49.

Reichhold Chemicals, Inc., Cooke Div. *Compounds for the Wire and Cable Industry*. Bulletin.

Scrima, D. Crosslinked Polyethylene. Term Report, Spring, 1976, Dept. of Chem. Eng., University of Southern California.

Thomas, B., and Bowrey, M. Crosslinked polyethylene insulations using the sioplas technology. *Wire Journal*, May 1977, **10**(5):88–92.

U.S. Industrial Chemicals Co. *Petrothene XL–Crosslinkable Polyolefin Molding Compounds*. Bulletin PTD 59-1068, 1st Ed., 1968.

——. *Petrothene XL–Crosslinkable Polyethylene Compounds . . . For Molding and Extrusion*. Bulletin PTD 46-366, 1st Ed., 1966.

Vaughan, D. J. "Extrusion of Crosslinkable Polyethylene." Paper presentation, *SPE ANTEC*, New Orleans, May 1979.

Chlorinated Polyethylene

Chlorinated linear polyethylene is produced by a simple chemical substitution of chlorine on the linear polyethylene backbone. The chlorination can occur by directly exposing the polymer to gaseous chlorine either in solution or in finely divided form. The presence of ultraviolet light or other activator is necessary for practical reaction rates. Materials offered commercially are in the 30–45% chlorine range and include amorphous as well as semicrystalline products. The materials are softer, more elastomeric, and more soluble than the unmodified polymer. Harder versions can be obtained by increasing the chlorine content to 50% or higher.

These materials have inherent oxygen and ozone resistance because of the saturated nature of the polymer. They also show exceptionally high tear strength and heat/aging characteristics, and excellent oil and chemical resistance. Table 5-18 shows properties of five different grades of chlorinated polyethylene. To reduce cost, the base polymer can be heavily extended with processing oils and inert fillers. Figure 5-27 shows the effect of fillers on chlorinated polyethylene resins.

The significant advantages of products produced from chlorinated polyethylene are their improved resistance to chemical extraction, plasticizer volatility, and weathering. Products made from chlorinated polyethylene do not fog at higher use temperatures and can be made completely flame retardant. They do, however, exhibit chemical instability similar to polyvinyl chloride. They may be used as primary compounding materials or as blending resins with PVC, high- and low-density polyethylene, and other polymers. They are crosslinkable by irradiation or chemical means.

Table 5-18. Dow CPE Chlorinated Polyethylene Resins: Comparative Properties.*

Property	Test Method	Resin Types				
		CPE 3614A	CPE 4211	CPE 4213	CPE 3623A	CPE 2552
Bulk density lb/ft^3 (g/cm^3)	CPE-B2[a]	26 (.42)	35 (.56)	25 (.56)	26 (.38)	18 (.29)
Melt viscosity, poise/1,000	CPE-D3a[a]	21.5	9.0	18.5	17.5	13
Chlorine content %	CPE-D6a[a]	36	42	42	36	25
Residual crystallinity %	CPE-D-14a[a]	2	2	<2	5	20
Volatiles % max.	CPE-B4[a]	0.2	0.2	0.2	0.2	0.2
Typical screen analysis % retained on 20 mesh	ASTM D1921	5	5	3	5	10
Tensile strength psi (kg/cm^2)	ASTM D412	1200 (84.4)	1800 (126.7)	1800 (126.7)	1900 (133.9)	1500 (105.7)
Ultimate elongation %	ASTM D412	700	500	600	700	800
100% modulus, psi (kg/cm^2)	ASTM D412	150 (19.9)	250 (17.6)	175 (12.3)	200 (14.1)	500 (35.2)
Hardness, Shore A	ASTM D1706	60	70	65	65	90
Specific gravity	ASTM D792	1.16	1.22	1.22	1.16	1.08
Low temperature brittleness, °F (°C)	ASTM D746	−90 (−68)	−35 (−37)	−70 (−57)	−80 (−60)	−76 (−60)

[a] Dow test methods available upon request.
*Reprinted with permission of Dow Chemical Co.

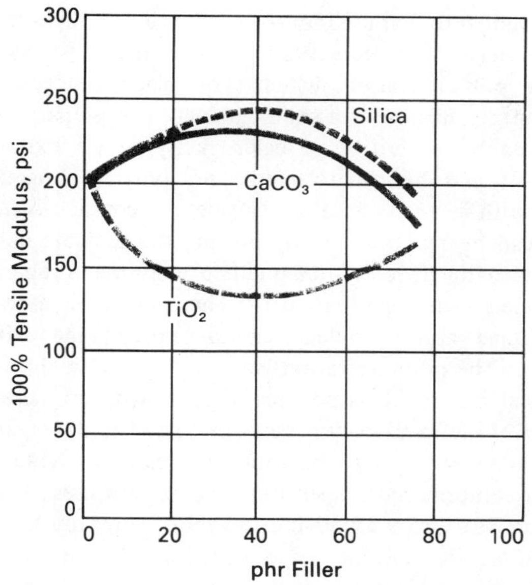

Fig. 5-27. Effect of filler on modulus of Dow CPE resin compound (ATM D412). (Reprinted with permission of The Dow Chemical Co.)

Applications include film and sheeting via standard extrusion, calendering, and molding techniques. Extruded sheet, supported by foams, is being used for automobile dashboards, upholstery, and door liners. Flexible film is used for nursery, agricultural, and weather protection.

Chlorinated polyethylene resins are blended with PVC for extruded pipe and building materials to serve as impact modifiers and processing aids. In vinyl-asbestos tile, their use aids in reducing the amount of asbestos reinforcement and enables higher filler loadings.

Trade Names

Daisolac	Osaka Soda Co.
Hostapren	Hoechst, A. G.
Lutrigen	BASF Wyandotte Corporation
Plaskon CPE	Allied Chemical Corporation
Polyliner	Goodyear Tire & Rubber Co.
Solarflex	Pantasote Co.
Tyrin	Dow Chemical Co.

REFERENCES AND BIBLIOGRAPHY

Brydson, J. A. *Plastic Materials.* 3rd Ed. Robert E. Krieger Publishing Co., Inc., Huntington, New York, 1975.

Dow Chemical Co. *Dow CPE Resins, Versatility–Function–Performance.* Bulletin 305-371, 1976.

——. Product Information Sheets CPE 3614A, CPE 4211, CPE 4213, CPE 3623A, and CPE 2552. Form Nos. 305-591, 305-594, 305-378, 305-593, and 305-590, 1979.

——. *Dow Chlorinated Polyethylene Resins.* Bulletin 301-93, 1971.

——. *Blending or Compounding With Dow CPE Resins.* Bulletin 301-108, 1971.

Modern Plastics. New chlorinated polyethylenes bring new application potential. August 1967, 44(8):87–88, 144.

Winding, C. C., and Hiatt, G. C. *Polymeric Materials.* McGraw-Hill Book Co., New York, 1961.

Polypropylene

Polypropylene is the fastest growing general-purpose thermoplastic in existence today. Copolymer and modified polypropylenes are responsible for the growth of polypropylene as an engineering thermoplastic as well as a general-purpose plastic. Polypropylene is the only high-volume thermoplastic that is processable by all four major fabrication methods: molding, extrusion, film, and fibers. Because of polypropylene's excellent processability and unusually good properties, its market has grown rapidly and steadily since its first commercial production in 1957 (except for a slight slump in 1974 caused by feedstock shortages).

A major reason for polypropylene's success is its low density (0.90 for unmodified polymer), good mechanical properties, and high temperature resistance. This makes it a cheaper material to use than most general-purpose plastics. Many new polypropylenes with improved engineering properties have been developed. The most promising are the glass-reinforced polypropylenes, foamed polypropylenes (including glass-reinforced foams), heat-resistant polypropylenes, and oriented sheeting. Polypropylene has found wide use in the textile industry, where it is successfully competing with rayon and other synthetic fibers as well as the natural fibers in many textile applications. This is particularly true in the carpet industry, where polypropylene is becoming increasingly popular for use in attractive, long-wearing carpets.

Polypropylene was not developed in the isotactic form until Ziegler discovered catalysts that permitted the synthesis of isotactic polymers (see discussion in Chapter 2). Atactic polypropylene is a soft, rubbery material of little commerical value, while the isotactic form (because of its increased crystallinity) is hard and tough, exhibiting properties very similar to those of high-density polyethylene.

The first synthesis of crystalline polypropylene is credited to Natta in the mid-1950s. The discovery of stereospecific catalysts enabled stereoregulated polymerization to occur, and polypropylene benefited from this perhaps more than any other thermoplastic. The addition polymerization of propylene is based on dehydrogenation with these stereospecific catalysts. Because ethylene is symmetrical, propylene is the simplest monomer which can be polymerized into an isotactic chain. The basic reaction is:

$$n\mathrm{CH_2{=}\underset{\displaystyle CH}{\overset{\displaystyle CH_3}{|}}} \longrightarrow \left(\mathrm{CH_2\underset{\displaystyle CH}{\overset{\displaystyle CH_3}{|}}}\right)_n$$

Crystalline titanium trichloride mixed with aluminum trimethyl are the initiators used in this process.

In 1974, Solvay (of Belgium) developed a new form of the metal-alkyl catalyst. The older method had several faults that were difficult and expensive to overcome. The old catalysts could not be used in any crystal form small enough to disperse properly; thus, relatively large amounts of the catalyst would be retained, unreacted, in the polymer. This residue caused degradation of thermal and other physical properties. Solvay was able to increase the catalyst crystal area, which not only solved the previously mentioned problem, but also had another advantage—the increased surface area made the catalyst four to five times more active. Since much less catalyst could be used, the catalyst residue was reduced from the old level of 100–300 ppm to only 15-20 ppm. This eliminated an expensive alcohol washing and purifying step which was used to bring the residual catalyst level down below 100 ppm.

The polypropylene formed by these processes is highly isotactic, as shown in the schematic representation below:

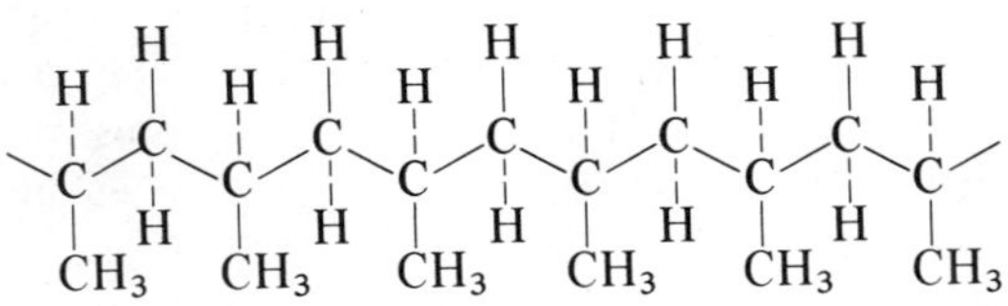

Being isotactic, all the methyl groups in the polymer are aligned, protruding from the chain on the same side. The chain has rotational flexibility and will allow the methyl groups to seek positions as far from each other as possible. The tetrahedral bond angles in the carbon chains allow three positions for any radical group; thus, the final chain structure is actually a spiral or helical shape with a repeating cycle of approximately three monomeric units. This spiral molecule interlocks with other similar chains to form the crystallite structure which gives polypropylene its physical properties. The entwined helix holds the chains closely together, resulting in a polypropylene with better heat and creep resistance than polyethylene.

The low-pressure process used in making polypropylene is very similar to that used for the production of high-density polyethylene. Very pure (95.5%) monomer must be used for the polymerization, which takes place at active locations on the surface of a solid catalyst. The preparation of this catalyst is critical. Ziegler catalysts are extremely dangerous and difficult to handle, not only reacting explosively with water, but also igniting on contact with air. They can be made *in situ* by putting the separate components into the reactor, but one is usually made separately and added as a single ingredient. The catalyst is suspended in heptane and propylene passed through the agitated mixture. The polymerization is rather slow, and allows for easy heat control. It is possible to get nearly complete conversion of olefin to polymer in one pass. The slow reaction rate also allows copolymerization to be easily carried out.

Both solution and suspension methods are used. For the solution process, a liquid hydrocarbon, typically heptane, is used as the solvent. Hydrogen gas may be added to reduce the molecular weight of the product, which improves the flow characteristics of the polymer. Temperature control is very important. Too high a temperature will produce atactic polymer. The solution must be warm enough, however, to keep the polymer in solution. Typical process temperatures for polypropylene range from 70°C to 80°C (158-176°F). The isotactic polymer separates out into a slurry consisting of solvent/polymer/catalyst. By centrifuging with various solvents, and distillation, the polypropylene is purified. All of these processes must be done in an inert atmosphere to prevent oxidation of the polymer. Additives such as stabilizers, colorants, and antioxidants can be added at this stage. The polymer is then dried, extruded, and chopped into pellets.

Polypropylenes, being long-chain aliphatic hydrocarbons, exhibit many of the properties of polyethylene. (Table 5-19 and Figures 5-28 through 5-32 show how polypropylene

Table 5-19. How Molecular Properties Affect Physical Properties of Polypropylenes.*

Physical Properties	As Density (Crystallinity) Increases	As Average Molecular Weight Increases	As Molecular Weight Distribution Is Narrowed
Tensile strength			
Yield	Higher	Lower	–
Break	Lower	Lower	Higher
Elongation	Lower	Lower	–
Flexural stiffness	Higher	Lower	–
Flexibility	Lower	–	–
Creep resistance	Higher	Lower	Higher
Abrasion resistance	Higher	Lower	–
Low-temperature brittleness	Lower	Lower	Higher
Environmental stress cracking res.	Lower	Lower	Higher
Softening temperature	Higher	Lower	Higher
Transmission rate of film	Lower	–	–
Electrical properties	Higher	–	–
Shrinkage	Higher	Lower	–
Warpage	Higher	Lower	Higher

*Hauck, 1966.

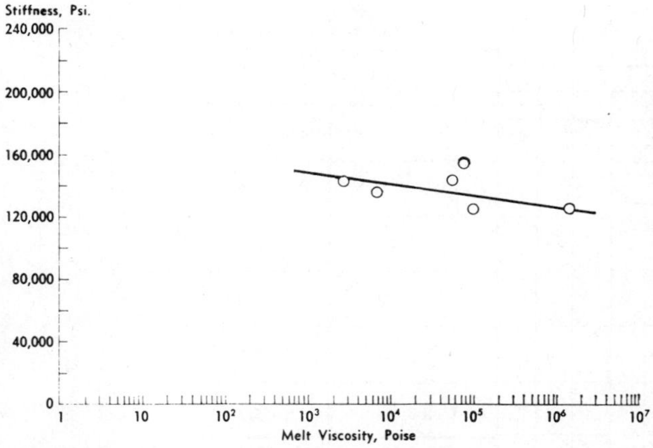

Fig. 5-28. Variation in stiffness with molecular weight. (Shearer, Guillet, and Coover, 1961)

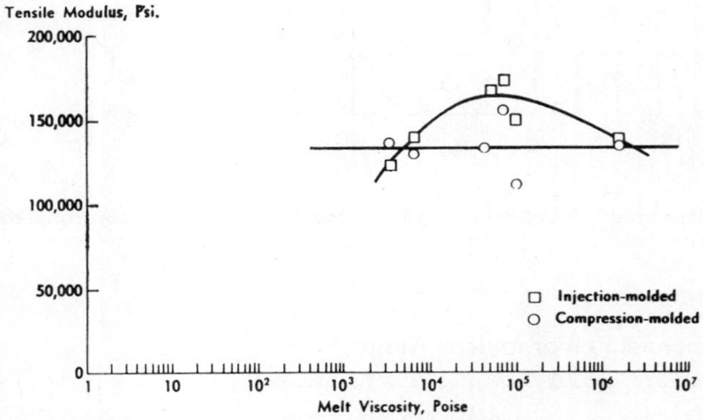

Fig. 5-29. Variation in tensile modulus with molecular weight. (Shearer, Guillet, and Coover, 1961)

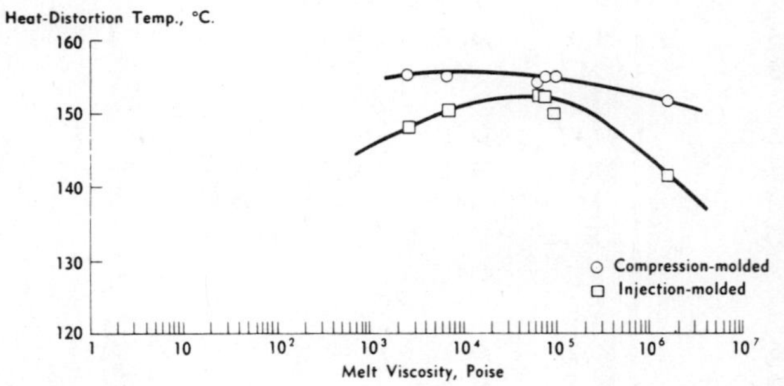

Fig. 5-30. Variation in heat-distortion temperature with molecular weight. (Shearer, Guillet, and Coover, 1961)

properties vary with change in density, molecular weight, and molecular weight distribution.) They have moderate tensile and high impact strengths and are resistant to many inorganic solvents. Polypropylenes have excellent electrical properties, are extremely processable, highly water-resistant, and absorb very little moisture (which means they remain dimensionally stable in humid environments). They are not very good barriers to oxygen and other gases. Polypropylene is the fastest burning of the olefins, but successful flame-retardant treatments are possible.

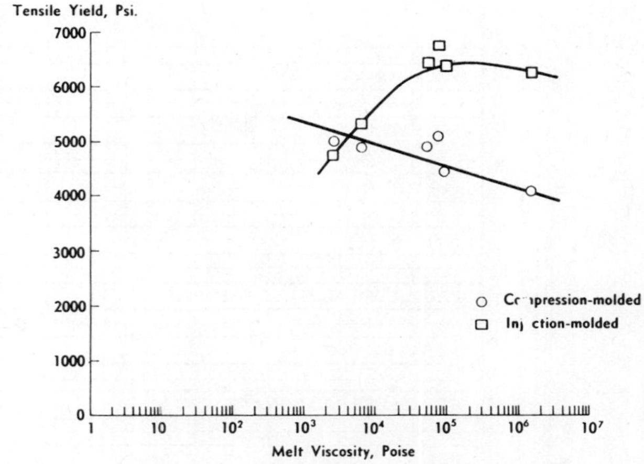

Fig. 5-31. Variation in tensile yield with molecular weight. (Shearer, Guillet, and Coover, 1961)

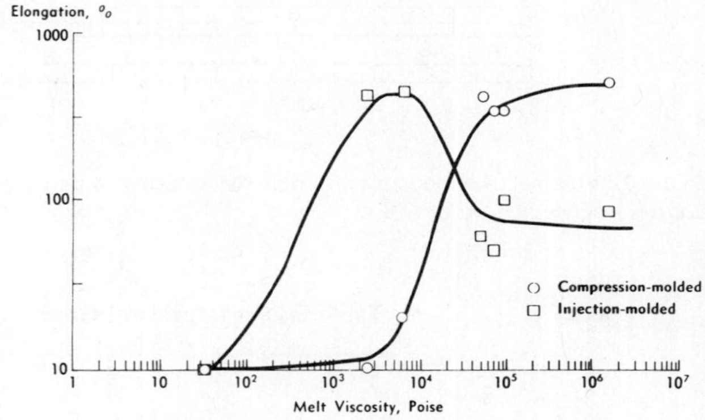

Fig. 5-32. Variation in elongation with molecular weight. (Shearer, Guillet, and Coover, 1961)

Perhaps polypropylene's most outstanding features are its very low density and its rigidity. These qualities make the polymer extremely light-weight, yet the most rigid of any polyolefin (see Figure 5-33). Flexural modulus of unfilled polypropylene ranges from 170,000 to 250,000 psi. Unfilled polypropylene generally is not recommended for use in load-bearing applications, since the plastic creeps rapidly even under moderate stresses. Glass-reinforced versions, however, can be used in high-load applications. Table 5-20 lists the types and grades of commercial polypropylene. Table 5-21 shows typical properties of various polypropylene resin types. Table 5-22 compares polypropylene properties to those of other thermoplastics.

Polypropylene has outstanding resistance to low-frequency flexing. The plastic is often used as a "living hinge" for items like small containers. Initial flexing of oriented polymer further aligns the molecules, allowing very good fatigue resistance. (Figure 5-34 shows the basic polypropylene-hinge design.)

Polypropylene has good unnotched and excellent notched impact strength, but is notch-dependent as is seen in Figures 5-35 through 5-37. Impact strength can be im-

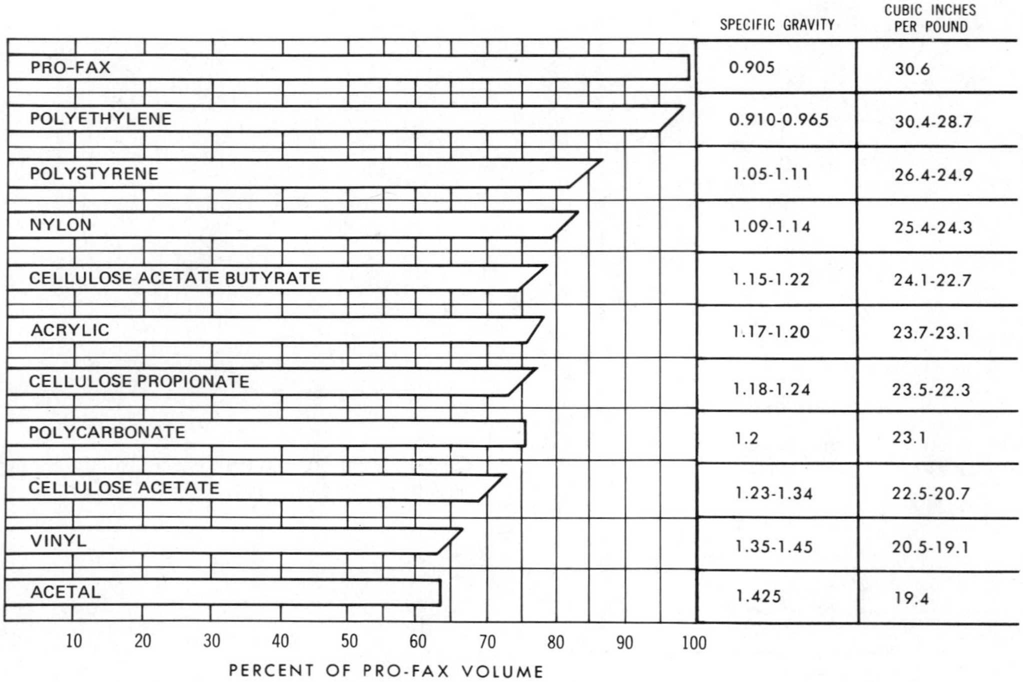

Fig. 5-33. Volume of various plastics materials plotted as percentages of the volume occupied by an equal weight of Pro-fax polypropylene. (*Courtesy: Hercules, Inc.*)

Table 5-20. Types and Grades of Commercial Polypropylene Resins.*

Types and Grades[a]	Specific Gravity	Melt Flow Rates[c] Low	Medium	High
Homopolymers				
General purpose	0.905	M, E	M, E	M, E
Long-term heat aging (dry)	0.905	M	M	M
Long-term heat aging (wet)	0.905	M	–	–
Weather resistant (UV stabilized)	0.905	M, E	M, E	–
FDA grades	0.905	M, E	M, E	M, E
UL grades[b]	0.910	M	M	–
NSF grades	0.905–0.920	E	–	–
Flame-retardant grades	1.05–1.41	–	M	–
Copolymers				
General purpose	0.905	M, E	M, E	–
Medium impact	0.900	M, E	M, E	–
High impact	0.900	M, E	M, E	–
Polyblends				
General purpose	0.905	M, E	M, E	–
Medium impact	0.905	M, E	M, E	–
High impact	0.905	M, E	–	–
Filled and/or reinforced				
Asbestos	1.05–1.25	–	M	–
Glass fibers	1.06–1.25	M	–	–
Talc	0.985–1.032	M	M	–

[a]Basic colors are translucent white, black, blue-gray and opaque gray. [b]Recognized for 203 F service. [c]M = molding resins, which can include injection and blow molding types; E = extrusion resins, which can include film, sheet, profile, and fiber extrusion types.

*Hauck 1966.

Table 5-21. Polypropylene Properties.*

		ASTM test method	Unmodified	Copolymer	Inert filled	Glass reinforced	Impact (rubber modified)
Processing	1. Molding qualities	—	Excellent	Excellent	Excellent	Excellent	Excellent
	2. Compression molding temp., °F.	—	340-450	340-550	340-550	340-550	340-450
	3. Comp. molding pressure, p.s.i.	—	500-1000	500-1000	500-1000	500-1000	500-1000
	4. Injection molding temp., °F.	—	400-550	400-550	375-550	400-550	400-550
	5. Injection molding pressure, p.s.i.	—	10000-20000	10000-20000	10000-20000	10000-20000	15000
	6. Compression ratio	—	2.0-2.4	2.0-2.4	—	—	2.0-2.4
	7. Mold (linear) shrinkage, in./in.	—	0.010-0.025	0.010-0.025	0.005-0.015	0.002-0.008	0.01-0.025
	8. Specific gravity (density)	D792	0.902-0.910	0.890-0.905	1.0-1.3	1.05-1.24	0.890-0.91
	9. Specific volume, cu. in./lb.	D792	30.8-30.4	31.2-30.5	26.0	24.5	30.8-30.5
	10. Machining qualities	—	Good	Good	Fair-good	Fair	Good
Mechanical	11. Tensile strength, p.s.i.	D638	4300-5500	2900-4500	4500-8200	6000-14500	2800-4400
	12. Elongation, %	D638	200.0-700.0	200-700(at break)	3.0-20.0(at break)	2.0-3.6	350.0->500.0
	13. Tensile elastic modulus, 10^5 p.s.i.	D638	1.60-2.25	1.0-1.7	4.0-8.0	4.5-9.0	1.0-1.7
	14. Compressive strength, p.s.i.	D695	5500-8000	<3700-8000	—	5500-7000	4000-6500
	15. Flexural yield strength, p.s.i.	D790	6000-8000	5000-7000	8000-9000	7000-11000	—
	16. Impact strength, ft. lb./in. of notch (½ × ½ in. notched bar, Izod test)	D256	0.5-2.2 at 73°F. (⅛ × ½ in. bars)	1.1-20.0 at 73°F. (⅛ × ½ in. bars)	0.4-3.0 at 73°F. (⅛ × ½ in. bars)	1.0-5.0 at 73° F. (⅛ × ½ in. bars)	1.0-15.0 73° F. (⅛ × ½ in. bars)
	17. Hardness, Rockwell	D785	R80-110	R50-96	R94-100	R110	R50-85
	18. Flexural elastic modulus, p.s.i. × 10^5	D790	1.7-2.5	1.3-2.0	3.0-6.5	3.8-8.5	1.2-1.8
	19. Compressive modulus, p.s.i. × 10^5	D695	1.5-3.0	—	2.4	—	—
Thermal	20. Thermal conductivity, 10^{-4} cal./sec./sq. cm./1 (°C./cm.)	C177	2.8	2.0-4.0	—	—	3.0-4.0
	21. Specific heat, cal./°C./gm.	—	0.46	0.5	—	—	0.5
	22. Thermal expansion, 10^{-5} in./in./°C.	D696	5.8-10.2	8.0-9.5	2.9	2.9-5.2	6.0-8.5
	23. Resistance to heat, °F. (continuous)	—	225-260	190-240	250-280	270-290	200-250
	24. Deflection temp., °F. at 264 p.s.i. fiber stress	D648	125-140	115-140	140-200	230-300	120-135
	at 66 p.s.i. fiber stress	D648	200-250	185-235	210-290	305-310	160-210
Electrical	25. Volume resistivity, ohm-cm. (50% RH and 23° C.)	D257	$>10^{16}$	10^{17}	1.5-2.8 × 10^{15}	—	$>10^{15}$
	26. Dielectric strength, short-time, ⅛-in. thickness, volts/mil	D149	500-660	500-660	450-650	—	500-650
	27. Dielectric strength, step-by-step, ⅛-in. thickness, volts/mil	D149	450-650	450-600	—	—	450-600
	28. Dielectric constant, 60 cyc.	D150	2.2-2.6	2.25-2.30	2.50-2.75	2.37	2.3
	29. Dielectric constant, 10^3 cyc.	D150	2.2-2.6	2.24-2.30	2.50-2.65	2.36	2.3
	30. Dielectric constant, 10^6 cyc.	D150	2.2-2.6	2.24-2.30	2.50-2.60	2.38	2.3
	31. Dissipation (power) factor, 60 cyc.	D150	<0.0005	<0.0001-0.0005	0.0054-0.0070	0.0022	>0.0003
	32. Dissipation (power) factor, 10^3 cyc.	D150	<0.0005-0.0018	<0.0001-0.0008	0.0036-0.0050	0.0017	>0.0003
	33. Dissipation (power) factor, 10^6 cyc.	D150	<0.0005-0.0018	<0.0001-0.0018	0.0021	0.0035	>0.0003
	34. Arc resistance. sec.	D495	136-185	136	121	74	—
Optical	35. Refractive index, n_D	D542	1.49	—	—	—	—
	36. Clarity	—	Transparent, translucent, opaque		Opaque	Opaque	Translucent
	37. Transmittance, %	—	55-90	—	—	—	—
	38. Haze, %	—	1.0-3.5	—	—	—	—
Resistance characteristics	39. Water absorp., 24 hr., ⅛-in. thick., %	D570	<0.01-0.03	<0.01-0.03	0.02-0.10	0.01-0.05	<0.01-0.03
	40. Flammability[a] Burning rate, in./min.	D635	0.75-0.83	—	1.5-2.5	0.75-0.83	—
	AEB, in./ATB, sec.[b]	D635	0.2-0.8/4-16 (FR gr.)	—	—	—	—
	41. Effect of sunlight	—	Unprotected material crazes rapidly; requires black for complete protection, but weather-resistant grades available in natural and colors				
	42. Effect of weak acids	D543	None				
	43. Effect of strong acids	D543	Attacked slowly by oxidizing acids				Attacked slowly by oxidizing acids
	44. Effect of weak alkalies	D543	None	None	None	None	None
	45. Effect of strong alkalies	D543	Very resistant		Very resistant	Resistant	Very resistant
	46. Effect of organic solvents	D543	Resistant below 80° C.				Attacked by hydrocarbons and chlorinated hydrocarbons

[a]This numerical flame spread rating is not intended to reflect hazards presented by this or any other material under actual fire conditions.
[b]AEB = Avg. extent of burning in inches; ATB = Avg. time of burning in seconds.
*Reprinted by permission from *Modern Plastics Encyclopedia*, McGraw-Hill, Inc.

Table 5-22. Polypropylene Possesses a Unique Combination of Properties*

Material	Polypropylene	Acrylics	Cellulosics	Nylon	High-density Polyethylene	Polystyrene	Low-density Polyethylene
Sp. gravity	0.89–0.92	1.18–1.19	1.09–1.22	1.09–1.14	0.93–0.96	1.04–1.65	0.92–0.94
Tensile strength, psi	3,500–5,000	7,000–9,000	2,000–8,000	7,000–10,900	2,800–5,000	5,000–9,000	2,000–2,500
Heat distortion temp., °F (66 psi)	210–230	160–195	110–235	300–360	130–165	160–210	105–180
Notched impact, ft. lb./in. of notch, 77°F	0.5–2	0.4–0.5	0.4–6.0	1–3	0.5–5.5	0.05–0.50	1.5–20
Molding temp., °F	300–600	380–390	350–490	470–720	300–600	325–360	330–600
Molding pressure, psi	8,000–20,000	10,000–20,000	8,000–32,000	10,000–25,000	8,000–15,000	10,000–30,000	8,000–20,000
Cost ¢/lb.	42–55	70	41–72	140–150	38–50	25	35–38

*Dunkel and Westlund, 1960.

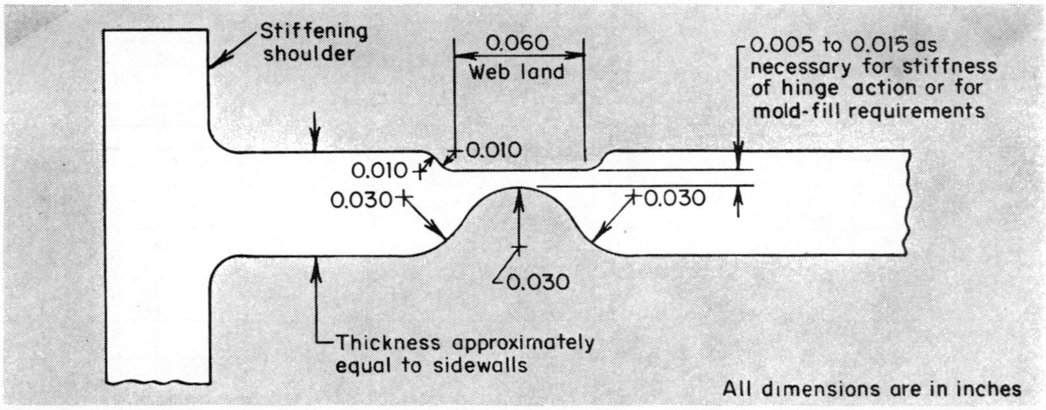

Fig. 5-34. Basic polypropylene-hinge design. When a hinge of this type is used in a boxlike structure, the upper plane of the hinge should be dropped up to 0.005 inch, to ensure a snug fit between lid and box. (Hanna, 1962)

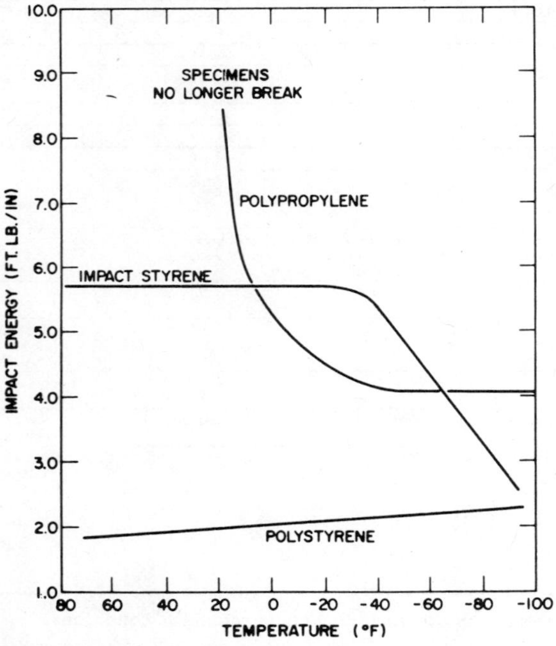

Fig. 5-35. Polypropylene has good unnotched impact resistance. (Dunkel and Westlund, 1960)

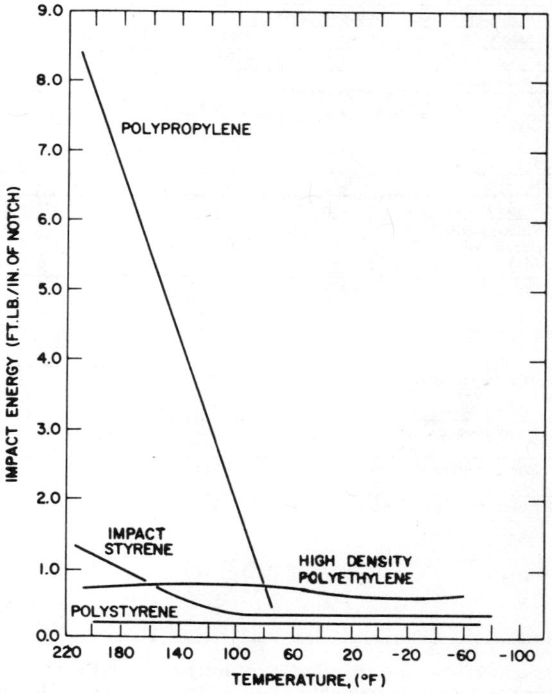

Fig. 5-36. Polypropylene has excellent notched impact resistance at elevated temperatures. (Dunkel and Westlund, 1960)

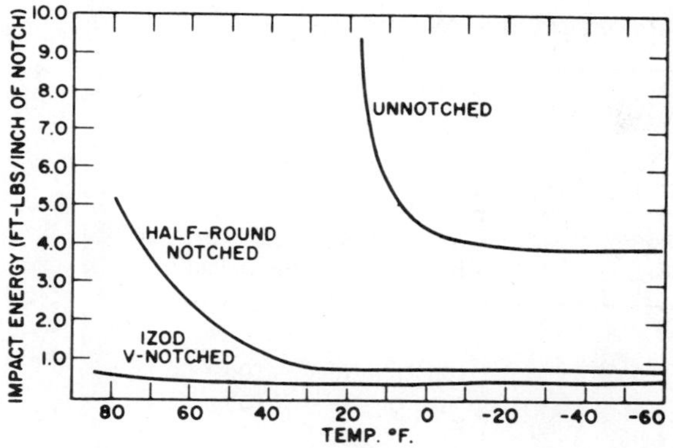

Fig. 5-37. Impact properties of polypropylene are notch-dependent. (Dunkel and Westlund, 1960)

proved by adding glass reinforcement, blending with nitrile rubber, or copolymerizing with ethylene. Impact strength can be raised from 1 ft-lb/in (notched samples) to 20 ft-lb/in, using these modifications. Figure 5-38 compares the effects of blending and copolymerization on impact properties.

Polypropylene's heat resistance is much superior to polyethylene. Thermal stability can be further increased by adding reinforcing fillers. Fire-retardancy can be imparted to polypropylenes through the use of additives such as halogen-containing compounds (like chlorinated paraffins), antimony oxide, or phosphorous.

Being a nonpolar hydrocarbon polymer, polypropylene has excellent electrical properties. Outstanding features are its use as a high-frequency dielectric even under moist, hot conditions. Oil-impregnated polypropylene paper is used as a cable wrap.

Polypropylene's stress-crack resistance is excellent. It is virtually insusceptible to environmental stress-cracking. Reinforcement with glass fiber improves this characteristic even further.

Like other polyolefins, polypropylene has very good resistance to nonoxidizing inorganic chemicals, detergents, low-boiling hydrocarbons, and alcohols at elevated temperatures as well as at room temperature. Polypropylenes are

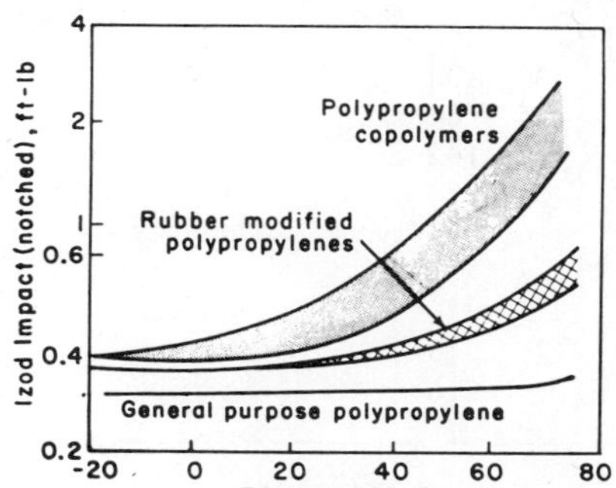

Fig. 5-38. Regardless of test used, rigid polyolefin copolymers have much better impact strength than general-purpose polypropylenes. (Reprinted by permission from *Materials in Design Engineering*, July 1965)

vulnerable to oxidizing agents, such as nitric acid and hydrogen peroxide, as well as some ketones, like methyl ethyl ketone and methyl isobutyl ketone.

Polypropylenes are easily machinable and can be electroplated using special processes. Because of their chemical inertness and waxlike properties, polypropylenes are very difficult to bond. Polyolefin-based adhesives are sometimes useful, but often exotic coating or finishing methods must be used.

Many modifications are being used to tailor-make these polymers. By controlling molecular weight, using additives (stabilizers, reinforcements, etc.), and through copolymerization, it is possible to accentuate and improve many of the properties of polypropylene. Table 5-23 shows the effect of fillers on some properties of polypropylene. Higher molecular weight resins tend to be softer and less stiff, but tougher. Lower molecular weights produce products that are harder and stiffer and tend to be brittle. Molecular weight distribution greatly affects the melt-flow characteristics of polypropylene; hence, processing is reliant upon these factors.

Polypropylene is suitable for nearly any type of plastics processing technique. Injection molding still accounts for the largest percentage of resin use, but both fiber and film production are rapidly gaining ground. Polypropylene is an easy material to extrude and will perform well in any machine designed to extrude polyethylene. As with other polyolefins, a relatively long extruder produces a better-quality melt. Multiple screw arrangements have been used successfully with polypropylene. Because of the heat-sensitive nature of polypropylene, long exposure to the heat of the screw extruder cannot be tolerated.

Because polypropylene does not absorb moisture, drying is not required. The polymer is sensitive to oxygen at elevated temperatures, so in certain processes inert gas is fed into the extruder along with the pellets. The polymer does not show the high degree of specific heat that the polyethylenes do. This means that it can be heated and cooled more rapidly and easily. Polypropylene exhibits a high degree of thixotropic behavior, which causes superior mold-flow characteristics.

Polypropylene has been well adapted to blow molding. Bottles made from polypropylene can be made thinner than polyethylene bottles because of polypropylene's greater strength; thus, even though polypropylenes cost more per pound, the cost per bottle is less.

The largest use of extruded polypropylene is in thin film packaging. The chill-roll film-casting method is generally used. In this process, the polypropylene is extruded from a straight slit die, and dropped onto a highly polished, plated roll, which is cooled by internal fluid circulation. The film cools and solidifies on the roll, is stripped off, trimmed, treated, and rolled. This is more expensive than the blown film process, but good transparent polypropylene film cannot be produced by blown film extrusion.

Table 5-23. Physical Properties Comparison: Talc versus Other Mineral Fillers.*

General-purpose extrusion grade polypropylene (GWE 21 Propathene by ICI)

50% filler levels by weight
All figures based on talc filler = 100

Filler	Tensile Strength	Tensile Modulus	Impact Strength[a] Notched	Impact Strength[a] Unnotched
Silicates				
Talc	100	100	100	100
Coarse kaolin clay	90	86	81	155
Calcined kaolin clay	91	96	75	158
Calcium carbonates				
Ultrafine calcium carbonate	81	89	111	281
Ultrafine calcium carbonate, stearate coated	77	80	176	280
Ground chalk	74	68	97	231
Ground chalk, stearate coated	75	74	108	205
Unfilled	121	47	156	340+

Polypropylene homopolymer (PXC 3906 Propathene by ICI)

50% filler levels by weight
All figures based on talc filler = 100

Filler	Tensile Strength	Tensile Modulus	Impact Strength[a] Notched	Impact Strength[a] Unnotched
Silicates				
Talc	100	100	100	100
Coarse kaolin clay	92	119	68	59
Calcined kaolin clay	105	127	54	70
Calcium carbonates				
Ultrafine calcium carbonate	83	96	47	No break
Ultrafine calcium carbonate, stearate coated	68	76	214	230
Ground chalk	77	90	110	No break
Ground chalk, stearate coated	76	76	142	No break

Polypropylene copolymer (PXC 2973 Propathene by ICI)

50% filler levels by weight
All figures based on talc filler = 100

Filler	Tensile Strength	Tensile Modulus	Impact Strength[a] Notched	Impact Strength[a] Unnotched
Silicates				
Talc	100	100	100	100
Coarse kaolin clay	101	90	64	122
Calcined kaolin clay	116	95	142	210
Calcium carbonates				
Ultrafine calcium carbonate	79	92	80	No break
Ultrafine calcium carbonate, stearate coated	68	78	370	No break
Ground chalk	77	78	75	No break
Ground chalk, stearate coated	71	69	218	No break

[a] Charpy impact test on 0.250 × 0.250-in. specimen machined from compression molded slabs.
*Reprinted by permission from *Plastics World*, July 1977.

Most polypropylene films are in the 2- to 3-mil range, although thicknesses up to 5 mils can be produced.

Oriented polypropylene film has several outstanding properties which make it very useful. By stretching it in the direction of extrusion, monoaxial-oriented film is produced. This film has great strength in the stretched direction, but is much weaker in the perpendicular direction. Biaxial-oriented film is stretched in both directions. The film can be stretched up to 1000% elongation, yielding increased tensile strengths of over 400%. Polypropylene film can be treated to resist shrinkage from applied heat, or can be left untreated to be useful in heat sealing or shrink packaging.

With modifications such as fillers and copolymerization, properties such as stiffness, clarity, shrinkability, and heat sealability can be tailored to specific needs. Polypropylene's high melting point, 149°C (300°F), is of great advantage, particularly in medical applications where sterilization is frequently necessary. Because of the excellent chemical and grease resistance of the polymer, food and chemical packaging industries provide a large market for the films. Polypropylene films have encroached upon the cellophane market, and polypropylene-cellophane laminations are being used for applications requiring synergistic characteristics.

The fiber and textile industries have provided a solid market base for polypropylene (see Figure 5-39). Fibers and textiles can be made almost indistinguishable from silk, cotton, or wool. They wear better and resist water and chemical attack. The largest current application is in the carpet industry. Because of polypropylene's good abrasion resistance and other physical properties, it is becoming very popular in indoor-outdoor carpeting.

Fig. 5-39. Processing flexibility of polypropylene is exemplified by a broad variety of fiber applications ranging from monofilament, through carpet backing, to woven fabrics. (*Courtesy: Amoco Chemicals Corporation, Plastics Division*)

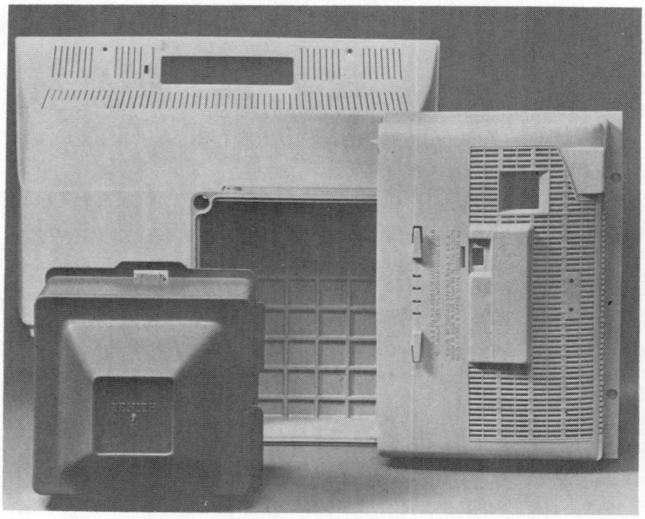

Fig. 5-40. Flame-retardant grades of polypropylene meeting standards issued by Underwriters' Laboratories and other regulatory agencies are molded into a variety of appliance parts and electrical components. (*Courtesy: Amoco Chemicals Corporation, Plastics Division*)

Polypropylene foams have enjoyed increasing popularity among the furniture and automotive manufacturers. Foams range from 0.7 to 45 lb/cu ft and are produced in five major process categories: (1) extruded, flexible packaging foam; (2) low-density, compression-molded rigid foam; (3) expansion-cast, rigid structural foam; (4) injection-molded

Fig. 5-41. Molded polypropylene chairs demonstrate the light weight, rigidity, and toughness for which the material is generally known. (*Courtesy: Exxon Chemical Co., USA*)

structural foam; and (5) rigid, extruded-foam profiles and shapes.

Being a general-purpose plastic, polypropylene has found applications in almost every field where plastics are used. All sorts of articles are made using molding techniques, and this particular group comprises the largest use of polypropylenes. The automotive industry is finding increasing uses for molded, foamed, and textile polypropylene. Appliance housings and liners are now being made from polypropylene (see Figure 5-40). The packaging industry is using an ever growing amount of polypropylene film, as well as bottles, caps, covers, and sheets. Perhaps the fastest growing use is structural foams. In the furniture industry, cabinet fascia, decorative panels, and complete furniture pieces are being made from polypropylene structural foam and molded polypropylene (see Figure 5-41).

Trade Names

Aerotuf	Extrusions	Anchor Plastics Co.
Alprodur	Chlorinated polypropylene	American Hoechst Corporation
Amcostrap	Strapping	American Manufacturing Co.
Amoco	Resin	Amoco Chemical Corporation
Arpylene	Asbestos-reinforced	Turner Brothers Asbestos Co., Ltd.
Avistrap	Strapping	FMC Corporation
Avisun		Avisun Corporation
Avi-sun	Resin	Amoco Chemical Corporation
Azdel		GRTL Co., PPG Industries
Bakelite	Resin	Union Carbide Corporation
Beamette	Monofilament	Dawbarn
Bexphane	Cast film	Bakelite Xylonite, Ltd.
Bicor	Oriented film	Mobil Chemical Co.
Carlona	Resin	Shell Chemical Co.
Celgard	Microporous film	Celanese Plastics Co.
Chevron		California Chemical Co.
Chevronfil	Resin	Chevron Chemical Co.
Copalene	Flame-retardant resin	Essex Copolymer Products
Cryovac	Biaxially-oriented film	W.R. Grace & Co.
Dapco	Resin	Davidson Products Co.
Daplen HO	High molecular weight isotactic compound	Plastic Constructions, Ltd.
Delflex	Strapping	Delta Strapping Industries, Inc.
Deloura	Expanded sheeting	Sun Chemical Corporation
Dia-Flex	Closures	Chevron Chemical Co.
Durel	Fiber	Celanese Corp. of America
Dynafilm	Film	National Distillers & Chemical Corporation
Dypro	Resins	ARCO Polymers, Inc.
Ecoro	$CaCO_3$-filled	Mitsubishi Petrochemicals Co., Inc.
Electro-film	Film and laminates	Extrudo Film Corporation
Electrolam	Resin	Extrudo Film Corporation
El Rex		Rexene Polyolefins Corporation
El Rexene		Rexene Polyolefins Corporation
Embafort	Rod and plate	M & B Plastics, Ltd.
Escon	Resin	Exxon Chemical Co.
Evolite	Rod, tube, and profile	Richard Daleman, Ltd.
Extrel	Film	Exxon Chemical Co.
Extru-Board	Corrugated sheet	Extrudyne, Inc.
Extvel	Film	Extrudo Film Corporation
E-Z FLow	Film	Extrudo Film Corporation
Fiberlene	Glass-filled compounds	Service Color Corporation
Fibretex	Spunbonded web	Crown Zellerbach Corporation
Fibrite	Fibrillated yarn	Plasticizers, Ltd.
Fillene PP	Foam	Fillite USA
Formid	Reinforced	Polymon Developments, Ltd.
Gulf		Gulf Oil Chemical Co.
Hercocel	Expandable pellets	Hercules, Inc.
Hi-Long PP Sheet	Sheet	Shin-Kobe Electric Machinery Co.
Hostalen	Resin	American Hoechst Corporation
Impolene	Tubing	Imperial Eastman Corporation
Iridon	Sheet	Commercial Plastics, Ltd.
Kastilene	Resin	Anic, S.p.A.
Kohjin-Korap	Biaxially-oriented shrinkable film	Kohjin Co.

Korap	Biaxially-oriented shrinkable film	Kohjin Co.
Lacqtene	Resins	ATO Chimie of Paris
Lorival	Tube	United Ebonite & Lorival, Ltd.
Luparen	Resins	BASF Wyandotte Corporation
Lupolen	Resins	BASF Wyandotte Corporation
Marlex	Resins	Phillips Petroleum Co.
Marvess	Fiber	Phillips Petroleum Co.
Maurylene	Film	La Cellophane, S.A.
Microfoam	Sheet foam	E.I. Du Pont de Nemours & Co., Inc.
Mitsubishi Noblem	Resin	Mitsubishi Petrochemical Co.
Moplefan	Isotactic film	Montedison, S.p.A.
Moplen	Resins	Montedison, S.p.A.
Moplen	Resins	Novamont Corporation
Mopler	Resins	Montedison, S.p.A.
Moseley	Pipe and fittings	David Moseley & Sons, Ltd.
Napryl	Resin	Rhone-Progil Direction Commerciale Matieres Plastiques
Nazarene	Resin	H & R Plastics Industries, Inc.
Noblen	Resin	Mitsui Petrochemical Industries, Ltd.
Norchem		Northern Petrochemical Corporation
Novolen	Resin	BASF Wyandotte Corporation
Olane	Resin	Amoco Chemicals Corporation
Olefane	Film	Amoco Chemicals Corporation
Olefane U	Resin	Avisun Corporation
Olefil	Resin	Amoco Chemicals Corporation
Oleflo	Resin	Amoco Chemicals Corporation
Oleflow	Pipe-grade	Amoco Chemicals Corporation
Oleform	Modified	Amoco Chemicals Corporation
Oletac	Amorphous resin	Amoco Chemicals Corporation
Oletemp	Resin	Amoco Chemicals Corporation
Olevac	Vacuum-metallized resin	Amoco Chemicals Corporation
Olex	Resin	Amoco Chemicals Corporation
Opp	Oriented film	Extrudo Film Corporation
Oyoprene	Sheet	Convert, Ets.G.
Plaskon		Allied Chemical Corporation
Plaslode	Talc- and asbestos-reinforced	Dart Industries, Inc.
Plasti Glas	Resin	Phillips Petroleum
Polycrest	Filament	Uniroyal Corporation
Poly-Em-Poly	Emulsified	Gulf Oil Chemicals Co.
Poly-Eth	Resin	Gulf Oil Chemicals Co.
Poly-Eth Hi-D	Resin	Gulf Oil Chemicals Co.
Polyflam	Compounds	A. Schulman, Inc.
Polyloom	Fiber	Chevron Chemical Co.
Polylumy	Biaxially-oriented film	Kohjin Co.
Polyorc	Pipe and fittings	Yorkshire Imperial Metals, Ltd.
Polypol	Resins	Crowley Hydrocarbon Chemicals, Inc.
Polyprene	Pipe and fittings	G & M Plastics, Ltd.
Poly-Pro	Resins	Gulf Oil Chemicals Co.
Poly Pro		Spencer Chemical Co.
Polyset	Shrinkable film	Kokoku Rayon & Pulp Co., Ltd.
Polytac	Resin	Crowley Hydrocarbon Chemicals, Inc.
Polytuf	Pipe fittings	Vistron Corporation
Procon	Fiberglass-reinforced	Dart Industries, Inc.
Proex	Monofilament	U.S. Polymeric Chemicals Co.
Pro-Fax	Resins and film	Hercules, Inc.
Profil	Reinforced	Dart Industries, Inc.
Proglas	Fiberglass-reinforced	Dart Industries, Inc.
Propafilm	Film	Imperial Chemical Industries, Ltd.
Propathene	Molding compounds	Imperial Chemical Industries, Ltd.
Prope	Monofilament	U.S. Polymeric Chemical Co.
Proplene	Film	Ludlow Corporation
Propylex	Sheet	British Celanese, Ltd.
Propylsar	Fiberglass-reinforced	Dart Industries, Inc.
Propylux	Extrusions	Westlake Plastics Co.
Prostran	Filament	Whiting Corporation
Prylene	Resin	CdF Chimie

Pryphane	Film	Rhodia, Inc.
Pyltron	Tube, Profile, and Rod	Plastic Tube & Conduit Co., Ltd.
Ribofil	Fibers	Plasticisers, Ltd.
Ripi-Prostran	Monofilament	E.B. & A.C. Whiting Co.
Sho-Allomer	Resin	Showa Yuka, K.K.
Sidathene	Film	UCB-Sidac Corporation
Silverflow	Pipe and fittings	BTR Industries, Ltd.
Simona	Fire-retardant	Plastic Constructions, Ltd.
Stabilene	Carpet backing	J. Lees & Sons
Sta-Flow		Air Products & Chemicals Corporation
Stamylon P	Resin	DSM
Sumitomo Noblen	Resin	Sumitomo Chemical Co., Ltd.
Supertac	Amorphous	Crowley Hydrocarbon Chemicals, Inc.
Tenite		Eastman Chemical Products
Torayfon	Film	Toray Industries, Inc.
Treafilm	Film	Trea Industries, Inc.
Tritor	Fibers	Plasticizers, Ltd.
Trofil	Monofilaments	Dynamit-Nobel, A.G.
Typar	Spunbonded sheet	E.I. Du Pont de Nemours & Co., Inc.
Vectra	Filament yarn	Chevron Chemical Co.
Versatex	Fabric	Mount Vernon Mills, Inc.
Vitalene	Sheet, rod, and tube	Stanley Smith & Co.
Vulkide	Sheet	I.C.I. (Hyde), Ltd.
Vypro	Film	W.R. Grace & Co.
Yupo FP	Paper	Oji-Yuka Synthetic Paper Co.

REFERENCES AND BIBLIOGRAPHY

Amoco Chemicals Corporation. *Amoco Polypropylene.* 2nd Ed., Bulletin 1975.

Avisun Corporation. *Polypropylene.* Bulletin 2a Av.

Brydson, J. A. *Plastics Materials.* 3rd Ed. Robert E. Kreiger Publishing Co., Inc., Huntington, New York, 1975.

Coffman, P. M. Stretch forming–new route to PP containers. *Plastics World*, August 1977, **35**(8):44–47.

Cowie, J. M. G. *Polymers: Chemistry and Physics of Modern Materials.* Intext Educational Publishers, New York, 1973.

Dow Chemical Co. *Dow Polypropylene.* Bulletin 171-327A, 1965.

——. *Inside Polypropylene–Its Crystallization and Fabrication.*

Dovdevani, I., Lundberg, R. D., and Alenzo, J. R. "Crystalline Polypropylene and Sulfonated EPDM Rubber Blends." Paper presentation, *SPE ANTEC*, New Orleans, May 1979.

Dubois, J. H., and John, F. W. *Plastics.* 5th Ed. Van Nostrand Reinhold Co., New York, 1974.

Dunkel, W. L., and Westlund, R. A. Jr. Polypropylene, Its Properties and Processing Characteristics. *Plastics Technical Information Sheet P-4*, Enjay Laboratories, January 1960.

Eastman Chemical Products, Inc. *General Information on Tenite Polypropylene, Properties, Processing, Applications.* Publication No. PO-2A, ECP 1167, June 1975.

——. *The Effect of Various Chemicals on Tenite Polypropylene.* Technical Report TR-19A, ECP 465, May 1968.

Exxon Chemical Co., *Exxon Chemical Polypropylene: General Technical Information.*

Funk, W. U. New developments in polypropylene film. *SPE Journal*, May 1964, **20**(5):441–444.

Gait, A. J., and Hancock, E. G. *Plastics and Synthetic Rubbers*, Pergamon Press, New York, 1970.

Hanna, R. D. "Polypropylene." In *Modern Plastics Encyclopedia*, J. Agranoff, (ed.), **52**(10A):77–79. McGraw-Hill Inc., New York 1975–76.

——. Design properties for polypropylene hinges. *Machine Design*, February 1962, **34**(3).

Hauck, J. E. Propylene plastics. *Materials in Design Engineering*, August 1966, **64**(2).

Hercules, Inc. *Pro-Fax Polypropylene Resin and Design Information*, January 1969.

——. *Pro-Fax–Information on Hercules Polypropylene.* Bulletin 500-302C, July 1960.

Houston, A. M. New types, improved processes expand use of structural foam. *Materials Engineering*, November 1975, **82**(6): 38–42.

Hug, D. P. "Polypropylene Foam." In *Modern Plastics Encyclopedia*, J. Agranoff (ed.), **52**(10A):114–115. McGraw-Hill Inc., New York 1975–76.

Kelly, P. R. PP enters container markets with high-speed solid-phase forming. *Modern Plastics*, April 1978, **55**(4):72–73.

Kresser, T. J. *Polyolefin Plastics.* Van Nostrand-Reinhold Co., New York, 1969.

Luciani, L., Monini, A., and Zaffagnini, D. "Propylene Polymerization with High-Yield Catalysts." Paper presentation, *SPE ANTEC*, New Orleans, May 1979.

Martino, R. Polypropylene, from Materials '79. *Modern Plastics*, January 1979, **56**(1):48–49.

Materials Engineering. Polypropylenes are better than ever. April 1972, **75**(4):30–34.

Materials in Design Engineering. Propylene Plastics, Materials and Processes Manual No. 238. August 1966, **64**(2):89–104.

——. Olefin copolymers–versatile plastic alloys. July 1965, **62**(2): 89–104.

Miller B. There's action in blown PP film. *Plastics World*, August 1977, **35**(8):54–55.

McCord, J. E. "Polypropylene Film." In *Modern Plastics Encyclopedia*, J. Agranoff (ed.), **52**(10A):139–141. McGraw-Hill Inc., New York 1975–76.

Modern Plastics. Big new packaging power for oriented polypropylene film. February 1979, **56**(2):50–51.

——. Polypropylene: where is it going? August 1961, **38**(12): 72–76, 142–148.

Muzzy, J. D., Hoyos, G. H., and Enuga, A. R. "Foam Fibrillated

Polypropylene Yarn." Paper presentation, *SPE ANTEC*, New Orleans, May 1979.
Plastics Design & Processing. Polypropylene enters packaging market with solid phase pressure forming. July 1979, **19**(7):39–42.
Plastics Technology. PT's Buyer's Guide to Polyolefins and Cellulosics. March 1967, **13**(3):39–52.
Plastics World. Fillers for polypropylene: silicates vs. calcium carbonate. July 1977, **35**(7):31.
Raff, R. A., and Doak, K. W. *Crystalline Olefin Polymers*, Interscience, New York, 1964.
Rodosta, J. A. "Improving Impact Resistance." Paper presentation, *SPE ANTEC*, New Orleans, May 1979.
Shearer, N. H. Jr., Guillet, J. E., and Coover, H. W. Jr. Relations between physical properties and molecular structure of polypropylenes. *SPE Journal*, January 1961, **17**(1):83–88.
Shell Chemical Co. Polypropylene, from Plastics in Packaging. *Plastics Engineering*, October 1978, **34**(10):43.
——. *Shell Polypropylene.* Bulletin SC:70–48.
Simonds, H. R., and Church, J. M. *A Concise Guide to Plastics*, 2nd Ed. Van Nostrand Reinhold Co., New York, 1963.
Simonik, J., and Drexler, J. "Atactic Polypropylene." Paper presentation, *SPE ANTEC*, New Orleans, May 1979.
Spencer Chemical Co. *Poly-Pro Polypropylene.*
Swanson, R. S. *Plastics Technology–Basic Materials and Processes*, McKnight & McKnight Publishing Co., Bloomington, Illinois, 1965.
Szulczynski, J. Z. Polypropylene is winning the bottle battle. *Plastics Engineering*, March 1978, **34**(3):47–49.

Polyolefin Copolymers

Polyethylene co Propylene—Polyallomer. Polyallomer represents the first of the propylene-ethylene copolymers to reach commercial production as a plastic. The copolymerization of propylene with ethylene by standard polymerization processes causes a pronounced loss of crystallinity, and amorphous, rubbery polymers result. A special polymerization process developed by Eastman produces ethylene-propylene block copolymers that exhibit equivalent crystallinity to that exhibited in homopolymers. "Polyallomer" is the term coined to describe the block copolymers made via this process of current commercial formulation.

Insolubility in hexane and heptane reflects the copolymer's high degree of crystallinity. Infrared spectra show polymerized segments of each of the reacted monomers. These segments exhibit crystallinity normally associated only with homopolymers of these monomers.

Polyallomer has many of the properties of high-density polyethylene and crystalline polypropylene along with the following additional characteristics: annealed density of 0.91, brittleness temperature as low as -40°C (-40°F), and impact strength of 12 ft-lb/in of notch, with less notch-sensitivity. It molds as easily as polypropylene, yet has better dynamic flexural fatigue resistance ("hingeability"), particularly at low temperatures. Typical properties of polyallomer grades are shown in Table 5-24. Table 5-25 compares the properties of the copolymer with poly(ethylene-propylene) blends.

Polyallomer flows and molds better than linear polyethylene and has superior softening point, hardness, stress-crack resistance, and mold-shrinkage properties. It is superior to rubber-modified polypropylene in color, clarity, moldability, electrical properties, and blush resistance when bent or stretched.

Polyallomer can be injection molded, extruded, and thermoformed. The initiation and growth of crystalline regions is slower in polyallomer than in highly crystalline homopolymers. Much crystalline growth occurs a considerable time after melt solidification. In practical terms, this means that the processing conditions for producing tough products are less critical for the copolymer. The very low mold temperatures needed for polypropylene and high-density polyethylene are not needed, in most cases, for polyallomer.

Some of the physical properties of this material change significantly during the 1-2-day period of polymer crystallization that occurs immediately after processing. The most noticeable change is an increase in stiffness. Consequently, mechanical properties of the copolymer should not be evaluated for 2-3 days after forming.

Polyallomer is used for blow-molded bottles; wire coating; film, sheet, and pipe extrusion; and thermoforming applications (see Figure 5-42).

Polyethylene co Vinyl Acetate. Copolymerization in polyethylene manufacture by addition of one or more dissimilar monomers reduces crystallinity. Vinyl acetate-modified resins have low crystallinity because the acetate branches interfere with crystallization. These resins are characterized by increased flexibility and resilience over a wide temperature range, and by improved clarity. Figure 5-43 shows the structure of this copolymer type.

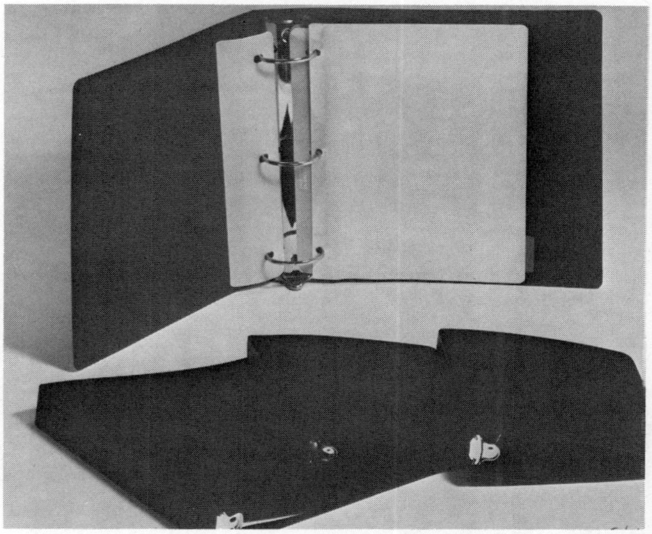

Fig. 5-42. Sheet made of polyallomer is being used in the manufacture of notebooks, stationery items, bookbacks, and in many applications as a leather substitute. (*Courtesy: Eastman Chemical Products, Inc.*)

Table 5-24. General Characteristics of Tenite Polyallomer and Properties That Do Not Vary Significantly With Formula.*

PELLET CHARACTERISTICS	
Size	1/8 inch (3.2 mm)
Shape	Cylindrical or cubical
Bulk Factor, cylindrical pellets	1.8
cubical pellets	2.2
OPTICAL PROPERTIES	
Clarity	Partially transparent in thin sections, translucent in thick sections
Refractive Index, n_D^{23}	1.492
THERMAL PROPERTIES	
Specific Heat	0.5
Coefficient of Linear Thermal Expansion	Above 50° F (10° C) 8.3-10 x 10^{-5} per °F 15-18 x 10^{-5} per °C Below 50° F (10° C) 5-5.5 x 10^{-5} per °F 9-10 x 10^{-5} per °C
Thermal Conductivity	(2-4) x 10^{-4} cal per sec per cm^2, per (°C per cm)
ELECTRICAL PROPERTIES	
Dielectric Constant (ASTM D150), 60 cps	2.3
10^6 cps	2.3
Dielectric Strength (ASTM D149), short-time, 0.050-inch-thick specimen	800-950 volts per mil 32-38 kv/mm
Dissipation Factor (ASTM D150), 60 cps	<5 x 10^{-4}
10^6 cps	<5 x 10^{-4}
Volume Resistivity (ASTM D257)	>10^{15} ohm-cm
CHEMICAL CHARACTERISTICS	Inert to most aqueous solutions except strong oxidizing acids, resistant to organic solvents and polar substances, swelled at room temperature and dissolved at high temperatures by aromatic hydrocarbons and chlorinated hydrocarbons.
MISCELLANEOUS CHARACTERISTICS	
Water Absorption	<0.01%
Molding Shrinkage	1-2%

*Courtesy: Eastman Chemical Products, Inc.

Copolymers with very low vinyl acetate content possess properties closely related to those of low-density ethylene homopolymers. The higher vinyl acetate copolymers look more like gum rubber.

For an ethylene-vinyl acetate (EVA) copolymer with a given molecular weight, it is possible to synthesize an entire series of copolymers which contain various amounts of vinyl acetate comonomer. Likewise, for a copolymer with a given vinyl acetate content, it is possible to make copolymers with many molecular weights. Table 5-26 lists a series of commercial EVA copolymers and their properties.

The presence of vinyl acetate causes copolymers to show densities consistently higher than those usually associated with low-density ethylene homopolymers. The fundamental relationship between density and physical properties associated with low-density polyethylene is not valid for the EVA copolymers. Density is more a measure of vinyl acetate content than polyethylene-type branching. Figures

Table 5-25. Comparison of Properties: Blends and Copolymers of Polyethylene and Polypropylene.*

Property, Units		ASTM Method	Tenite Polyallomer			Blends of High-Density Tenite Polyethylene and Tenite Polypropylene[a]		
			5021	5B21	5C21	5% PE	10% PE	25% PE
Flow rate, g/10 min at 230°C and 2.16 kg load		D1238	2.5	2.5	2.5	1.5	1.6	1.9
Density, g/ml		D1505	0.902	0.902	0.902	0.913	0.916	0.921
Softening point, Vicat, °C		D1525	140	135	130	143	139	139
Deflection temperature, °F	Fiber stress 264 psi	D648	140	130	120	–	–	–
	Fiber stress 66 psi	D648	210	200	180	–	–	–
Brittleness temperature, °C		D746	−10	−16	−24	+1	0	−3
Tensile strength at yield, psi		D638	4200	3800	3500	4400	4400	4150
Elongation at fracture, %		D638	400	500	500	100	200	150
Stiffness in flexure, 10^5 psi		D747	1.10	0.90	0.75	1.34	1.33	1.30
Rockwell hardness, R Scale		D785	85	75	65	91	88	83
Tensile impact strength, ft-lb/sq in.		D1822	170	200	230	145	180	180
Notched Izod impact strength, ft-lb/in. of notch		D256	1.5	2.5	5.0	0.6	0.6	0.7
Unnotched Izod impact strength, ft-lb/in. of width	At 23°C	D256	30	>80	>80	–	–	–
	At −10°C	D256	10	30	>80	–	–	–

[a]A polyethylene of 0.96 g/cc density and 0.2 melt index with a propylene of flow rate 2.5.
*Courtesy: Eastman Chemical Products, Inc.

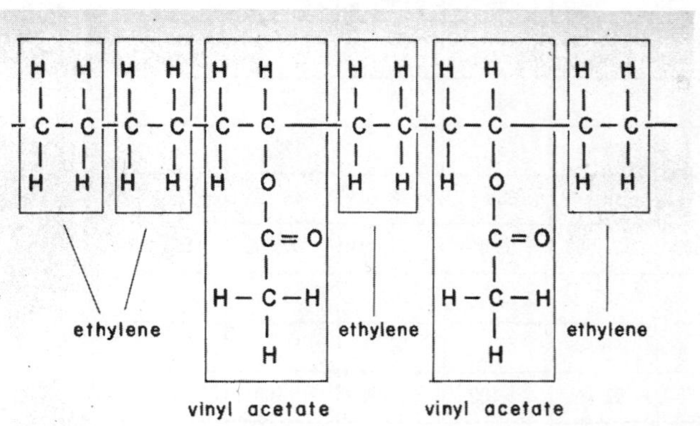

Fig. 5-43. Structure of part of an ethylene–vinyl acetate copolymer chain. (Fischer, 1967. Reprinted by permission from *Plastics Design & Processing*. Copyright 1967, Lake Publishing Corporation, Libertyville, Ill.)

5-44 through 5-46 show the relationship of selective properties to changing densities for EVA copolymer and polyethylene homopolymer. Table 5-27 shows comparative properties of EVA copolymer and two standard types of polyethylene.

As the density of the copolymer increases with increasing vinyl acetate content, the stiffness, hardness, modulus of elasticity, and softening temperature decrease. In addition, low-temperature flexibility and clarity improve as density increases. Other properties such as weatherability, chemical resistance, resistance to flexural cracking and permeability to vapors and liquids will also vary with changes in the density of the copolymer. The compatibility and ability to accept high filler loadings of the copolymer are improved as the density increases.

Accelerated tests have shown that EVA copolymers resist degradation by weathering better than conventional polyethylenes. EVA copolymers show improved weatherability as the density of the copolymer is increased. Addition of an ultraviolet stabilizer further enhances their resistance to weathering. For maximum protection against degradation by weathering, a well-dispersed carbon black compound can be used.

Chemical resistance of EVA copolymers is inferior to that of conventional polyethylene. EVA copolymers exhibit decreasing chemical resistance as copolymer density increases. Aromatic and chlorinated hydrocarbon solvents readily attack the EVA copolymer.

The barrier properties of EVA copolymers are poorer than those of low-density polyethylene. The permeability of a copolymer film to oxygen, nitrogen, and carbon dioxide increases as the density increases. Moisture vapor transmission rate (MVTR) also increases as density increases.

EVA copolymers accept high degrees of filler loading without serious degradation of their physical properties. The acceptance level for a particular filler increases as the density or vinyl acetate content of the copolymer increases. The addition of a filler to a copolymer can be ex-

Table 5-26. Typical Physical Properties: EVA Copolymer Resins.*

GENERAL PHYSICAL PROPERTIES	ASTM STD.	EVA 3130 12% Vinyl Acetate	EVA 3170 18% Vinyl Acetate	EVA 3175 28% Vinyl Acetate	EVA 3180 28% Vinyl Acetate	EVA 3185 33% Vinyl Acetate	EVA 3190 25% Vinyl Acetate
Melt Index (gram/10 min)	D 1238	2.5	2.5	6.0	15	25	2.0
Density (grams/CC)	D 792	.935	.94	.95	.95	.95	.95
Ult. Tensile Strength @ 73°F (PSI)	D 638	2800	2760	2000	1550	1440	2500
Ult. Tensile Strength @ 100°F (PSI)	D 638	1800	>1380	ND	ND	ND	ND
Ult. Tensile Strength with U.V. Stab. @ 73°F	D 638	2600	2760	2000	1300	800	2500
Ult. Elongation @ 73°F (%)	D 638	750	750	800	900	900	750
Ult. Elongation @ 100°F (%)	D 638	750	>870	ND	ND	ND	ND
Ult. Elongation with U.V. Stabilizer @ 73°F	D 638	700	740	775	850	900	750
Stiffness @ 73°F (PSI)	D 747	9500	4400	1770	1550	1000	3100
Tensile Impact @ 73°F (Ft. #/In²)	D 1822	315	315	330	228	200	280
Low Temp. Brittleness °F	D 745	<−159	<−159	<−159	<−159	<−159	<−159
Dynamic C.O.F. (To S. Steel)	D 1894	.7	1.2	2.2	2.0	Block	2.3
Index of Refraction	D 542	NC	1.494	1.485	1.432	1.480	1.491
Stress Crack Resistance (Hrs)	D 1693	>1000	>1000	>1000	>1000	>1000	>1000
Thermal Conductivity BTU/Hr Sq Ft-°F-In		2.1	2.34	2.39	2.40	2.40	2.40
Vicat Softening Point °F	D 1525	150	138	127	120	120	130
ELASTOMERIC PROPERTIES							
Hardness (Shore A)	D 1706	95	90	83	80	65	85
Hardness (Shore D)	D 1706	38	35	25	25	17	30
Bashore Resilience %		35	38	38	38	40	37
De-Mattia Flex (Cycles to ½") (Slit Cut Growth)	D 813	18,000	27,000	63,000	27,000	27,000	100,000
NBS Abrasion (% of STD.)	D 1630	380	167	23	26	18	28
Compression Set (% Set) 10 Days @ 25°C	D 395	49	46	55	56	56	56
22 Hrs @ 70°C		82	92	>100	>100	>100	>100
Compression Strength 7 Days @ 100°F 20% Set (PSI Load)		700	550	400	200	<100	300

*Courtesy: E.I. Du Pont de Nemours & Co., Inc.

pected to lower the melt index; decrease the elongation; increase the stiffness, hardness, and modulus of elasticity; and adversely affect the low-temperature flexibility. Filled EVA copolymers are generally used where opacity, self-extinguishing properties, electrical conductivity, or increased toughness are design requirements. EVA copolymers have comparatively low heat and creep resistance.

Processing techniques for the EVA copolymers are similar to those used with low-density polyethylene. Lower processing temperatures are required due to the limited thermal stability and the inherent tackiness of the EVA copolymer.

EVA resins have been crosslinked via irradiation techniques. Tensile strengths reach a maximum after exposures to dosages of 10–22 megarads, while elongation is relatively unaffected. Heat and chemical resistance are somewhat less than that of irradiated polyethylene homopolymer. As exposure is increased beyond 20 megarads, properties begin to fall off precipitously.

EVA copolymers are used in areas where flexibility, resilience, clarity, dielectric sealability, toughness, compatability, and excellent low-temperature properties are required; they can also be used in food, drug, and medical applications where FDA acceptability is required. Such

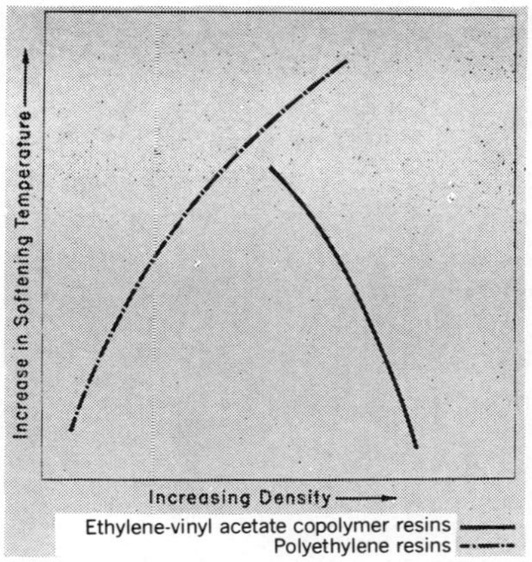

Fig. 5-44. The relation of softening temperature to densities of EVA copolymer and polyethylene resins. (Fischer, 1967. Reprinted by permission from *Plastics Design & Processing*. Copyright 1967, Lake Publishing Corporation, Libertyville, Ill.)

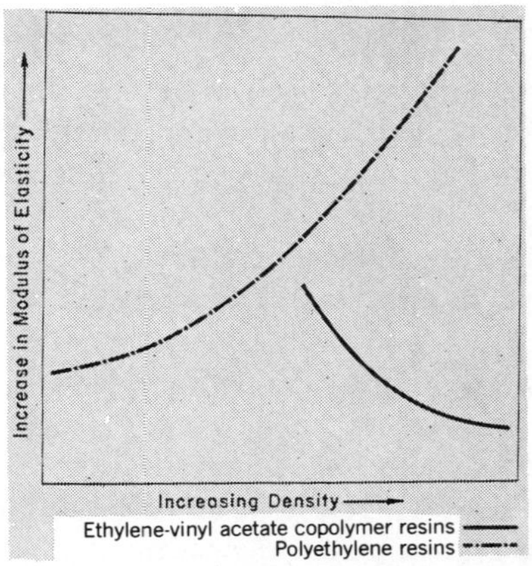

Fig. 5-45. The relation of modulus of elasticity to densities of EVA copolymer and polyethylene resins. (Fischer, 1967. Reprinted by permission from *Plastics Design & Processing*. Copyright 1967, Lake Publishing Corporation, Libertyville, Ill.)

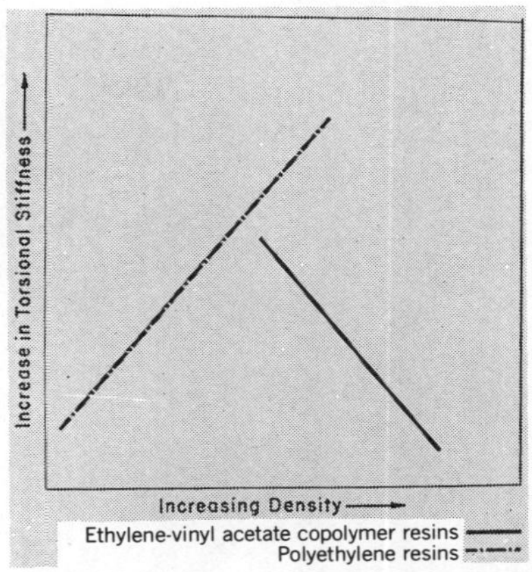

Fig. 5-46. The relation of torsional stiffness to densities of EVA copolymer and polyethylene resins. (Fischer, 1967. Reprinted by permission from *Plastics Design & Processing*. Copyright 1967, Lake Publishing Corporation, Libertyville, Ill.)

items as food, beverage, and medical tubing are being manufactured from EVA copolymers. Disposable syringes (see Figure 5-47), squeeze pumps, enema tips, and droppers are some other applications.

Film and sheeting produced from EVA copolymers find use in some types of construction film, shower curtains, disposable gloves, inflatable toys, inflatable splints, and pool liners. Items which can be injection molded, such as shoe soles, pool table bumpers, small appliance bumpers, cannister gaskets, disposable brushes, and automobile mud flaps, are applications for these EVA copolymers.

There are two basic classifications of EVA copolymers: one for copolymers designed for extrusion and molding processes, and the other for copolymers used for wax-blending and adhesive applications. EVA copolymers employed for wax blending and adhesive are, in general, of higher density and lower molecular weight than the copolymers designed

Table 5-27. Comparative Typical Properties of One EVA Copolymer and Two Polyethylene Resins.*

Physical Properties	ASTM Test Method	EVA	LDPE	HDPE
Melt index (g/10 min)	D1238	1.5	1.2	3.1
Density (g/cu cm)	D1505	0.937	0.917	0.934
Torsional stiffness (lb/sq in.)	D1043	7,500	25,000	60,000
Vicat softening temperature (°C)	D1525	69	95	114
1% Secant modulus of elasticity (lb/sq in.)	D638	7,400	23,000	57,000
Low-temperature brittleness (°C/F_{50})[a]	D746	<-76	<-76	<-76
Tensile strength (at 20 in./min)	D638			
Break (lb/sq in.)		2,100	2,000	1,700
Elongation (%)		700	550	400
Environmental stress crack resistance (hr/F_{50})[a]	D1693	>3,000	>48	<1
Hardness	D2240			
Shore D		36	45	56
Shore A		92	95	>95

[a]F_{50} = 50% failure.
*Courtesy: U.S. Industrial Chemicals Co.

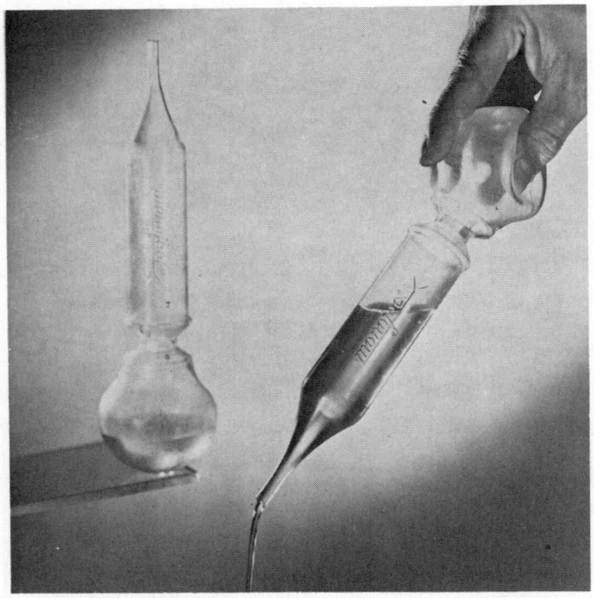

Fig. 5-47. One-piece, thin-walled, disposable syringe for surgical clean-up, with squeezable bulb and calibrated tube section, blow molded from EVA copolymer. (*Courtesy: U.S. Industrial Chemicals, Co.*)

Table 5-28. Typical Properties of EVA Resins for Wax Blending.*

	"Elvax" 150	"Elvax" 240	"Elvax" 250
Inherent viscosity at 30°C (86°F). (0.25% by wt. in toluene)	0.78	0.78	0.85
Comonomer ratio, wt. % ethylene: vinyl acetate	67:33	72:28	72:28
Melt index (ASTM D 1238-57 T)	25	25	15
Odor	Slight	Slight	Slight
Tensile strength, p.s.i.	1000	1000	2000
Elongation at break, %	700	700	750
Density, g./cc. at 30°C (86°F)	0.95	0.95	0.95
Refractive index, n_D^{25}	1.485	1.482	1.482
Softening point, ring and ball, °C (°F)	117.2 (243)	123.9 (255)	135.5 (276)

*Courtesy: E. I. Du Pont de Nemours & Co., Inc.

for use in extrusion and molding. Table 5-28 shows typical properties of these copolymer types. Table 5-29 compares the EVA's with other types of adhesive modifiers.

Combinations of waxes, EVA copolymers, and various modifiers result in upgraded hot-melt formulations. Generally, gloss, clarity, barrier and strength characteristics, and flexibility are increased in these coatings. Substrates most often coated include: paper, paperboard, foil, glassine, and corrugated cardboard. Uses for these hot-melt coatings include packaging materials for bakery and cereal products, candy, ice cream, frozen foods, and meats.

A series of copolymers are used to impart adhesiveness, flexibility, and stability to adhesive formulations. Generally, as the vinyl acetate content is increased, the EVA resins become more elastomeric, exhibit greater tackiness and adhesive properties, are more soluble in common solvents, and maintain a wide compatibility with rubbers, low-cost hydrocarbon resins, microwaxes, and synthetic resins.

Major market areas include hot-melt adhesives for bookbinding, carpet backing, and case and carton sealing.

Polyethylene co Zinc Acrylate—Ionomer. The term, "ionomer," was coined to describe a class of polymers which combine ionic forces and covalent bonds in their molecular structure. The unique properties of ionomer resins are to a great extent dependent on the presence of strong interchain forces in the long-chain molecules of the polymer structure. These interchain forces are developed between ionized carboxyl groups pendant from polymeric chains: metallic ions such as sodium, potassium, calcium, magnesium, and zinc are associated with the ionized carboxyl sites to provide a thermally reversible interchain linkage. This interchain bonding develops properties normally associated with the crosslink structure. However,

Table 5-29. Types of Hot-Melt Modifiers and Principal Functions.[a],*

Modifier Type	Viscosity	Adhesion	Tack	Flexibility	Gloss and Gloss Stability	Hardness	Cohesive and Bond Strength	Stain Resistance
EVA copolymer	X	X	X	X	X	X	X	
Hydrocarbon resins	X	X			X			
Polyterpenes Wood resins, Rosin esters	X	X	X		X		X	
Elastomers	X			X			X	
Synthetic waxes	X					X		X

[a]Boxes containing "X" indicate improved or increased function by adding modifier.
*Courtesy: U.S. Industrial Chemicals Co.

Table 5-30. Properties of Ionomers.*

		ASTM test method	
Processing	1. Molding qualities	—	Excellent
	2. Compression molding temp., °F.	—	280-350
	3. Comp. molding pressure, p.s.i.	—	100-800
	4. Injection molding temp., °F.	—	300-550
	5. Injection molding pressure, p.s.i.	—	2000-20000
	6. Compression ratio	—	2.0
	7. Mold (linear) shrinkage, in./in.	—	0.003-0.02
	8. Specific gravity (density)	D792	0.93-0.96
	9. Specific volume, cu. in./lb.	D792	30.0-29.0
	10. Machining qualities	—	Fair to good
Mechanical	11. Tensile strength, p.s.i.	D638	3500-5000
	12. Elongation, %	D638	350.0-450.0
	13. Tensile elastic modulus, 10^5 p.s.i.	D638	0.2-0.6
	14. Compressive strength, p.s.i.	D695	—
	15. Flexural yield strength, p.s.i.	D790	No yield
	16. Impact strength, ft. lb./in. of notch (½ × ½ in. notched bar, Izod test)	D256	6.0-15.0
	17. Hardness, Rockwell	D785	D50-D65 (Shore)
	18. Flexural elastic modulus, p.s.i. × 10^5	D790	0.3-0.5
	19. Compressive modulus, p.s.i. × 10^5	D695	—
Thermal	20. Thermal conductivity, 10^{-4} cal./sec./sq. cm./1 (°C./cm.)	C177	5.8
	21. Specific heat, cal./°C./gm.	—	0.55
	22. Thermal expansion, 10^{-5} in./in./°C.	D696	12.0
	23. Resistance to heat, °F. (continuous)	—	160-220
	24. Deflection temp., °F. at 264 p.s.i. fiber stress	D648	100-120
	at 66 p.s.i. fiber stress		110
Electrical	25. Volume resistivity, ohm-cm. (50% RH and 23° C.)	D257	$>10^{16}$
	26. Dielectric strength, short-time, ⅛-in. thickness, volts/mil	D149	900-1100
	27. Dielectric strength, step-by-step, ⅛-in. thickness, volts/mil	D149	2.4-2.5
	28. Dielectric constant, 60 cyc.	D150	2.4-2.5
	29. Dielectric constant, 10^3 cyc.	D150	—
	30. Dielectric constant, 10^6 cyc.	D150	—
	31. Dissipation (power) factor, 60 cyc.	D150	0.001-0.003
	32. Dissipation (power) factor, 10^3 cyc.	D150	0.0015
	33. Dissipation (power) factor, 10^6 cyc.	D150	0.0019
	34. Arc resistance, sec.	D495	<90
Optical	35. Refractive index, n_D	D542	1.51
	36. Clarity	—	Transparent
	37. Transmittance, %	—	75-85 (125 mil)
	38. Haze, %	—	3-17 (125 mil)
Resistance characteristics	39. Water absorp., 24 hr., ⅛-in. thick., %	D570	0.1-1.4
	40. Flammability[a] Burning rate, in./min.	D635	1.0
	AEB, in./ATB, sec.[b]		—
	41. Effect of sunlight	—	Requires UV stabilizer
	42. Effect of weak acids	D543	Slow attack
	43. Effect of strong acids	D543	Attacked by oxidizing acids
	44. Effect of weak alkalies	D543	Very resistant
	45. Effect of strong alkalies	D543	Very resistant
	46. Effect of organic solvents	D543	Very resistant at 75°F.

[a]This numerical flame spread rating is not intended to reflect hazards presented by this or any other material under actual fire conditions.
[b]AEB = Avg. extent of burning in inches; ATB = Avg. time of burning in seconds.
*Reprinted by permission from *Modern Plastics Encyclopedia*, McGraw-Hill, Inc.

since the ionic links are thermally reversible, these polymers may be processed at conventional temperatures in standard plastics processing equipment.

Property limitations are low stiffness and creep, low heat-distortion temperature, and poor ultraviolet resistance without stabilizers. Table 5-30 lists typical properties.

The density is low at 0.935–1.200. Light transmission of ionomers is approximately equal to 80–92% over the visible range, and has high transmittance in the ultraviolet region with certain formulations. These polymers do not crystallize; therefore, they remain transparent even in thick sections. Available commercial ionomers are nonrigid plastics, but they contain no plasticizers. Generally, good flexibility, resilience, high elongation, and excellent impact strength typify the ionomers. In general, the creep characteristics of ionomers resemble low- to medium-density polyethylene at low loadings. At high loading, the initial creep will orient the structure, and creep resistance dramatically increases in the oriented form.

Electrical properties are good. Some formulations resist stress-cracking, while others are very susceptible. Ultraviolet resistance is poor. At best, when filled with carbon black, it is as resistant as a carbon black–filled polyethylene. The base material is classified slow-burning, although self-extinguishing formulations are available.

The ionomers are resistant to a broad range of chemicals, especially water and common organic solvents. They have excellent low-temperature properties. The upper use temperature is 71°C (160°F).

Articles produced from ionomers include injection-molded housewares, toys, bottles, containers, safety shields, tool handles, closures, extruded film and sheet, electrical insulation, multiwall bags, and foil constructions for pouch and flexible packaging applications.

Both polyethylene and ethylene copolymers have been of great interest as protective coatings for metals to be used in corrosive environments. In most of such applications, interest lies in further improving thermal and chemical resistance to permit their use in even more severe environments. Crosslinking, either by chemical or irradiation techniques, is one method of upgrading these resins.

Fig. 5-48 shows the increase in tensile strength and elongation provided a compression-molded specimen of ionomer by exposure to radiation. Specimens exposed to 20 megarads did not melt at temperatures as high as 200°C (392°F), the melting point of nonradiated specimens is approximately 85°C (185°F). Chemical resistance is also substantially improved and optical properties are not significantly affected.

Polyethylene co Ethyl Acrylate. A commercial grade of polyethylene co ethyl acrylate, (PEEA) has been successfully introduced to compete with low-density polyethylene (LDPE) and plasticized PVC. At room temperature, the copolymer has about the same flexibility as plasticized PVC

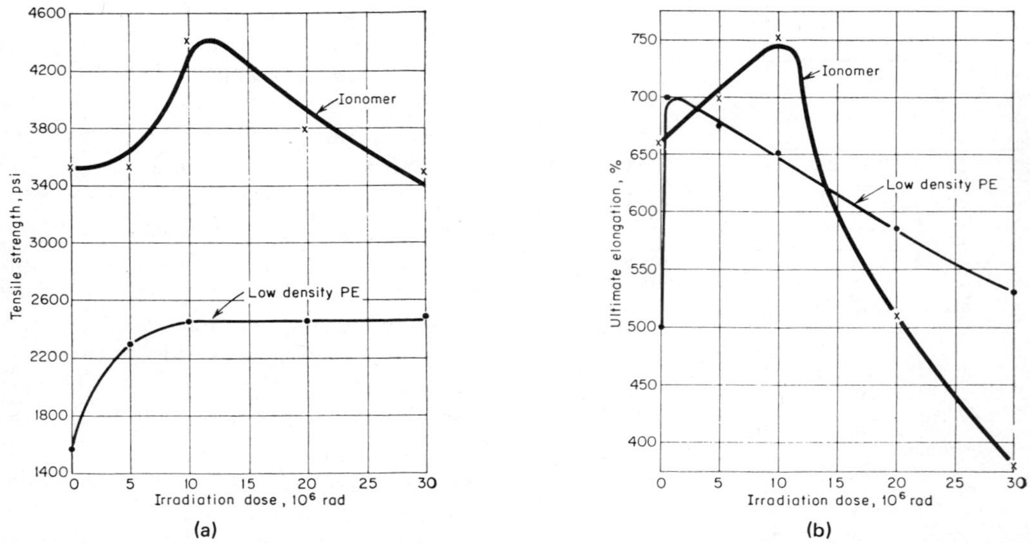

Fig. 5-48. Effects of irradiation on tensile strength (a) and elongation (b) of an ionomer compared with low-density polyethylene (PE). Comparative PE data from Lanza, *Modern Plastics*, July 1957. (Vermes, Brenner, and Pal, 1975)

and four times the flexibility of LDPE. Despite the lower rigidity, the copolymer is slightly higher in density than PE. The usual stiffness/density relationship of PE also does not apply to these copolymers, as was earlier noted in the discussion of EVA copolymers. Tensile strength of PEEA is lower than PE or PVC, as can be seen in Table 5-31 which compares physical properties of the three resin types. Tables 5-32 and 5-33 compare the electrical and chemical resis-

Table 5-31. Comparative Physical Properties of Polymers.*

					Plasticized PVC	
	Units	ASTM Test Method	PEEA Copolymer	Low-Density Polyethylene	Unfilled, Model Formulation	Typical Commercial Formulation
Melt index	Gms/10 min.	D-1238-52T	6	1.5	–	–
Density	Gms/cc	D-1505-57T	0.931	0.920	1.268	1.35
Vicat softening	°C	D-1525-58T	64	90	57.6	62
Hardness, Durometer	Shore A	D-676-55T	86	91	91	90
	Shore D		27	45	47	30
Tensile strength	PSI	D-412-51T	1,350	1,800	3,300	2,200
Elongation	%	D-412-51T	650	600	210	250
Tensile impact	ft lbs/cu. in.					
23°C		–	500	190	425	370
-30°C		–	390	70	25	40
Brittleness index	80% OK, °C	D-746-55T	-100	-80	-20	-15
Torsional stiffness at	-50°C (-58°F)	D-1043-51	155,000	180,000	450,000	400,000
	-25°C (-13°F)		23,000	70,000	280,000	200,000
	0°C (32°F)		8,000	35,000	60,000	45,000
	+25°C (77°F)		4,500	18,000	4,200	4,500
	+40°C (104°F)		1,500	10,000	1,300	1,800
	+60°C (140°F)		580	4,000	–	600
	+70°C (158°F)		350	3,000	–	450
	+80°C (176°F)		200	2,000	–	350
	+90°C (194°F)		Too Soft	1,400	–	250
	+100°C (212°F)		Too Soft	1,000	–	Too Soft
Flex cracking resistance (Ross Rubber Test Apparatus)						
@ 25°C cycles to fail			2,000,000	40,000	–	2,000,000
@ 0°C cycles to fail			2,000,000	10,000	–	135,000

*Bonotto, 1962.

Table 5-32. Comparative Electrical Properties of Polymers.*

	ASTM	PEEA	Low Density Polyethylene	Plasticized PVC
Dielectric Constant				
at 1 MC	D150	2.7	2.3	5.3
at 50 MC	D150	2.6	2.3	6.5
Power Factor				
at 1 MC	D150	0.009	0.0003	0.1
at 50 MC	D150	0.03	0.0004	0.1
Dielectric Strength				
Short Time V/mil	D149	550	550	300–400
Volume Resistivity 23°C, 50% RH, megohm-cm	D257	2.4×10^9	1×10^{11}	1×10^6

*Bonotto, 1962.

tance characteristics of the same resins. Table 5-34 shows the effect of various fillers on a PEEA copolymer.

Blends of these copolymers with PE show improved toughness over the straight homopolymers. As acrylate concentration increases, stiffness decreases and toughness increases.

The PEEA copolymer has been irradiated to induce crosslinking, with results very similar to those noted for the ionomers. Thermal and chemical stability improved, tensile strength increased, and elongation was relatively unaffected by 10–20 megarad dosages. Further exposure caused a rapid loss of properties.

The copolymer maintains its rubbery character without the use of plasticizers down to temperatures as low as -100°C (-148°F). Also, its heat stability compared to plasticized vinyls makes it easier to injection and blow mold. The resin is adaptable to extrusion, coating, vacuum-forming, and calendering techniques.

Figure 5-49 shows an example of an end use which takes advantage of the copolymer's flexibility and flex life. Other applications include coated fishing-boat anchors, injection-molded mop buckets, electrical terminal covers, and agricultural tubing, squeeze-spray bottles, flexible toys, and refrigerator gaskets.

Table 5-33. Comparative Chemical Resistance of Polymers.*

ASTM—D543 % wt. gain	PEEA	Low Density Polyethylene	Commercial* Plasticized Vinyl
Acetic Acid 5%	1.1	0.1	—
Acetone	10.4	1.3	Excessive
Carbon Tetrachloride	Excessive	39	26.8
Citric Acid 50%	0.6	0.2	0.15
Ethyl Alcohol 95%	4.2	0.14	3.2
Ethylene Dichloride	79.1	—	Excessive
Gasoline	54	16	—1.5
Heptane	Excessive	13.2	—
Kerosene	39	9.6	2.6
Lemon Oil	66	15.5	6.7
Liquid Detergent "Lux"	1.4	0.2	—
Methyl Ethyl Ketone	10.4	2.2	Partially Dissolved
Motor Oil (20-20W)	2.1	0.5	—.46
Nitric Acid 95%	9.4	0.2	—
Sodium Chloride 10%	0.2	0.1	0.2
Sodium Hydroxide 10%	0.19	0.1	3.0
Sulfuric Acid 3%	0.27	—	0.8
Tetrachloro ethylene	Excessive	42	Excessive
Toluene	Excessive	16	13.0
Turpentine	64	11.5	0.5
Water Distilled	0.1	0.015	0.5

[a]Commercial formula—90 Durometer.
*Bonotto, 1962.

Other Copolymers. New copolymers with ethylene are constantly being developed, not all of which realize a substantial degree of commercial success. Ethylene-butene co-

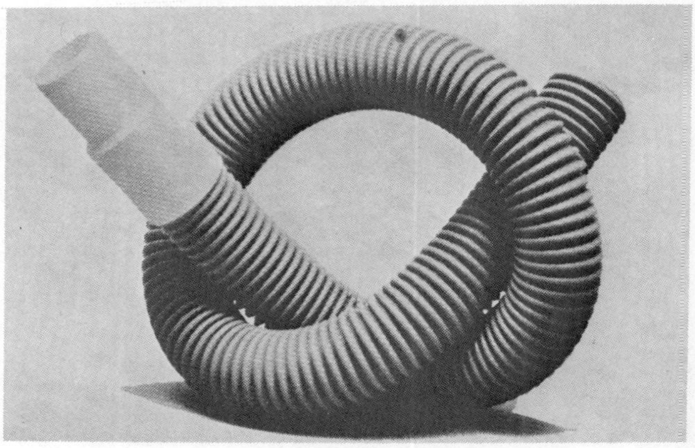

Fig. 5-49. Vacuum cleaner hose demonstrates the increased flexibility and long flex life now possible with EEA copolymers. *(Courtesy: Union Carbide Corporation)*

Table 5-34. Filled PEEA, % by Weight.*

	ASTM Test No.		PEEA 100% Control	PEEA 70% Asbestos 30%	PEEA 70% Wollastonite[b] 30%	PEEA 70% Gamaco[a] 30%
Melt index, gm/10 min	D-1238-52T		5	3	4	3
Torsional stiffness	D-1043-51-	23°C	4,500	8,600	7,900	7,000
		40°C	1,400	4,200	3,700	3,200
		70°C	350	1,000	1,000	1,000
		0°C	8,200	14,600	11,000	9,500
		-25°C	23,200	43,000	32,000	28,500
Durometer "A"						
Brittle temperature	D-676-55T		86	94	94	94
80% OK	D-746-55T		-100°C	-45°C	-45°C	-50°C
Color			Clear	Gray Black	Tan	Light Tan

[a]Calcium carbonate.
[b]Calcium metasilicate (G.L. Cabot Co.).
*Bonotto, 1962.

polymers have very good resistance to stress-cracking (see Figure 5-50 and Table 5-35), in addition to providing good rigidity and toughness. These retain the many desirable properties of linear PE, including abrasion resistance, chemical inertness, and electrical resistance. Compared to PE homopolymer they have superior load-bearing properties. Table 5-36 lists some typical properties.

Ethylene-butene copolymers can be crosslinked using sulfur-bearing agents, as in rubber vulcanization or by irradiation. They are less radiation-resistant than other materials. Physical, chemical, and thermal properties are equivalent regardless of which crosslinking method is used. This copolymer is used for wire, cable, pipe, bottle caps, and products requiring metal inserts.

Fig. 5-50. New ethylene-butene copolymer resins are approved for high-density polyethylene pipe with a hydrostatic design stress rating of 800 psi. The new material with its improved pressure rating also offers toughness, corrosion resistance, and solid fused joints which eliminate pollution. Sizes are available from $\frac{3}{4}$" through 48". (*Courtesy: Phillips Chemical Co., a division of Phillips Petroleum Co.*)

Table 5-35. Stress-Cracking Resistance of Rigid Polyolefin Copolymers.[a], *

Environment	Ethylene-Butene, hr	Polyethylene Homopolymer, hr	Propylene-Ethylene, hr	Polypropylene Homopolymer, hr
Igepal Co-630	300–400	60	>1000	>1000
Silicone DC-200	1000	43	>1000	>1000
Ethyl alcohol (100%)	1000	20	>1000	>1000
Acetic acid (glacial)	1000	9	>1000	>1000
Sodium hydroxide (5%)	670	34	>1000	>1000
Ethylene glycol	1000	40	>1000	>1000
Methyl ethyl ketone	1000	34	>1000	>1000
Diisooctyl sebacate	1000	36	>1000	>1000
Soap solution (5%)	350	20	>1000	>1000

[a]Bell Test conducted at 122 F, F_{50} values.
*Reprinted by permission from *Materials in Design Engineering*, July 1965.

Table 5-36. Properties of Ethylene-Butene Copolymers.*

Property	ASTM Test	Ethylene-Butene Copolymers
Density, g/cc	D-792	0.945–0.962
Hardness (Rockwell)	D-785	38R–67R
Stiffness in flexure, 1000 psi	D-747	100–180
Impact strength (Izod, notched), ft-lb/in.	D-256	
22.7°C (73°F)		0.7–4.0
-17°C (0°F)		–
-28.9°C (-20°F)		–
Tensile strength (2 ipm), 1000 psi	D-638	
Ultimate	(Die C)	3.1–4.3
Yield		2.9–3.0
Tensile modulus, 1000 psi	D-638 (Type I)	95–180
Tensile impact strength 22.7°C (73°F), ft-lb/in^3	D-1822	30–80
Elongation (2 ipm), %	D-638	
Ultimate	(Die C)	60–900
Yield		–
Heat deflection temperature, F	D-648	
At 66 psi		155–185
At 265 psi		–
Water absorption (24 hr), %	–	0.02–0.03
Environmental stress-cracking (F^{50}), hr	D-1693	8–350

*Reprinted by permission from *Materials in Design Engineering*, July 1965.

Little data is available for the copolymers polyethylene co methyl acrylate (EMA) and polyethylene co methyl methacrylate (EMMA). They have been proposed for applications requiring flexibility and toughness without the use of plasticizers. EMMA resins have been successfully used for extrusion coating and flame spraying. EMA films generally have inferior optical properties compared to PE. EMMA films, on the other hand, have been extruded into exceptionally clear films. EMMA materials show very good weather resistance.

Table 5-37 summarizes the various types of available copolymers, except for the ionomers which are considered by their manufacturers and many users to be a separate class of materials.

A terpolymer of ethylene, vinyl acetate, and vinyl chloride has been introduced which was designed to combine inherent flame resistance with high impact strength and easy processing on standard injection equipment. Smoke generation is purported to be lower than that of rigid PVC. Impact strength is outstanding, although other mechanical properties are considered average. Resistance to weather and UV-caused yellowing is much improved over straight PVC. Physical properties of this terpolymer are compared to other flame-resistant plastics in Table 5-38.

Table 5-37. Polyolefin Copolymers—What Kinds and Who Makes Them.*

Suppliers	Copolymers[a]	Remarks
Airco	EVA	Very high VA content (over 50%); emulsion for coatings.
Avisun	Prop-E	Introduced its propylene copolymer in May 1963. Now has many grades and types.
Dow	EEA	Additive types for waxes and adhesives and types for molding and extrusion.
Du Pont	EVA	Marketed under Elvax designation for waxes, adhesives; under Alathon name for molding, extrusion.
Eastman Chemical	Prop-E	Highly crystalline polyallomer molding and extrusion resins copolymerized by Eastman-developed process.
Enjay Chemical	Prop-E	Block copolymerized propylene copolymers for all uses.
Grace	EB, Prop-E	Wide choice of resins in both ethylene butene and propylene copolymers.
Hercules	Prop-E, EB	Entered EB competition in January 1965; propylene-ethylene in summer, 1964. Now has many types of each.
Phillips	EB	Also developing propylene copolymer and working on ethylene-butadiene type.
Rexall-El Paso	Prop-E	Jointly entered propylene copolymer field on May 1, 1965, in plant designed to produce both copolymers and homopolymers.
Spencer	EMA	Largely used for injection molding applications, some blow molding, and specialty films in which clarity is not required. Has low odor level.
Union Carbide	EEA, EVA, EB	Pioneered EEA and EVA for molding and extrusions. Has EVA for waxes and adhesives as well as molding and extrusion.
U.S. Industrial Chemicals	EVA, EB, Prop-E	First to push a broad line of EVA in molding and extrusion; broad range available.

[a]Abbreviations: EEA = ethylene ethyl acrylate; EMA = ethylene methyl acrylate; EVA = ethylene vinyl acetate; EB = ethylene butene; Prop-E = propylene ethylene.
*Reprinted by permission from *Modern Plastics*, June 1962, McGraw-Hill, Inc.

Table 5-38. Physical Properties of Injection-Molded Spirals.*

Property	Polytuf 200	Flame-Retardant PP	Flame-Retardant ABS	PP/Modified PVC
Tensile yield strength, psi	4300	4300	6100	6300
Tensile modulus, psi	220,000	410,000	310,000	360,000
Elongation @ yield, %	3	4	4	4
Elongation @ break, %	20	24	5	74
Notched Izod., ft-lb/in.				
73°F	15 HB[a]	0.3	1	15 HB[a]
0°F	1	0.3	0.3	1
HDT[b] (264 psi, annealed), F	156	185	192	140
Specific gravity	1.29	1.22	1.22	1.43
UL94 rating	V-0	V-1	V-0	V-0

[a]Hinge break.
[b]Heat-distortion temperature.
*Reprinted by permission from *Plastics Technology*, October 1975.

Trade Names

A.C.	EVA, EEA	Allied Chemical Corporation
Acralen	EVA	Mobay Chemical Co.
Airflex	VAE	Air Products & Chemicals, Inc.
Alathon	EVA	E.I. Du Pont de Nemours & Co., Inc.
Aquace	VAE	Hercules, Inc.
Aravyl	Film	UCB-Sidac Corporation
Avisun	EP	Avisun Chemical Co.
Bakelite	EA, EEA, EVA, EB	Union Carbide Corporation
Bondfast	Adhesive	Sumitomo Chemical Co.

Borinal	E-boron compounds	Allied Resinous Products, Inc.
Boronol	PE-boron	Allied Resinous Products, Inc.
Carbathene	E-*n*-vinyl Carbazole	Standard Telecommunications
Chemicat	E-imine	Alcolac, Inc.
Chemiquat	Quaternized E-imine	Alcolac, Inc.
Cobothane	EVA	Cobon Plastics Corporation
Co-mer	EVA	Union Carbide Corporation
Dacomer	EP	Diamond Shamrock Chemical Corporation
Dexon	EA, EVAA, PA, PEA	Exxon Chemical Corporation
Dipion	E-nylon 6	Soviet Origin
Dumilan	Modified EVA	Mitsui Polychemical & Du Pont Co.
Dutral-Co	EP	Montedison, S.p.A.
Dutral-Ter	EP-diene terpolymer	Montedison, S.p.A.
El Rexene	EP	Rexene Polyolefins
Elvace	EAc	E.I. Du Pont de Nemours & Co., Inc.
Elvaloy	EVA	E.I. Du Pont de Nemours & Co., Inc.
El Vax	VAE	E.I. Du Pont de Nemours & Co., Inc.
Enjay	EP	Exxon Chemical Corporation
Epsyn	EP rubbers	Copolymer Rubber & Chemical Corporation
Esprene	EP-diene terpolymer	Sumitomo Chemical Co.
EVA	EVA	E.I. Du Pont de Nemours & Co., Inc.
Evaflex	EVA	Mitsui Polychemicals Co., Ltd.
Evazote	EVA crosslinked/expanded	Bakelite Xylonite, Ltd.
Flo-Well	VCP	Air Products & Chemicals, Ltd.
Formion	Ionomer alloy	A. Schulman, Inc.
Fortiflex	EB	Celanese Plastics Co.
Get	P-copolymer	Shell Chemical Co.
Go-Mist	EP balls	Cosom Corporation
Graftmer	EVA-VC	Nippon Zeon Co.
Grex	EB, EP	Polymer Chemicals, Division of W.R. Grace & Co.
Hex One	E-copolymers	Gulf Plastics
Lacqtene	EVA	ATO Chimie of Paris
Le Rexene	P-copolymer	Rexene Polyolefins Co.
Levapren	EVA	Mobay Chemical Co.
Marlex	EB	Phillips Chemical Co.
Nipoflex	EVA	Nippon Polychemicals Co.
Nordel	EP	E.I. Du Pont de Nemours & Co., Inc.
Olemer	P-copolymer	Amoco Chemicals Corporation
Petrothene XL	Crosslinkable EVA	U.S. Industrial Chemicals Co.
Plexar	Modified EVA	Chemplex Co.
Poly Eze	E-copolymer	Gulf Oil Chemicals Co.
Polymin	E-imine	BASF Wyandotte Corporation
Poly Shrink	EVA shrink film	Chemplex Co.
Polytuf	EP, EVA-VC	Diamond Shamrock Co.
Pro-Fax	EB, EP	Hercules, Inc.
Rakoll	EVA adhesive	Sonneborn & Rieck
Royalene	EP rubber	Uniroyal, Inc.
Sho-Allomer	P-copolymer	Showa Yuka, K.K.
Spencer	EMA	Spencer Chemical Co.
Staflow	VCP	Air Products & Chemicals, Inc.
Surflex	Ionomer film	Flex-O-Glass, Inc.
Surlyn	Ionomer resin	E.I. Du Pont de Nemours & Co., Inc.
Tenite	EP (polyallomer)	Eastman Chemical Products Co.
Ultrathene	EB, EP, EVA	U.S. Industrial Chemicals Co.
Vistalon	EP	Exxon Chemicals Co.
Vulkaflex	EP cord	General Electric Co.
Vynathene	VAE	U.S. Industrial Chemicals Co.
Yukalon-EVA	EVA	Mitsubishi Petrochemical Co.
Yuka-Wrap	EVA film	Mitsubishi Petrochemical Co.
Zetafax	E-acrylic and EVA	Dow Chemical Co.
Zetafin	E-ethyl acrylate	Dow Chemical Co.

Abbreviations:

A = Acrylate (Acrylic Acid)
B = Butene
E = Ethylene
EA = Ethyl Acrylate
MA = Methacrylate
P = Propylene
VA = Vinyl Acetate
VC = Vinyl Chloride

REFERENCES AND BIBLIOGRAPHY

Allied Chemical Corporation. *A-C Polyethylene*. Bulletin 901-149.

Baumgartner, A., and Bragg, T. Here's how to run ethylene propylene copolymers profitably. *Plastics Technology*, June 1973, pp. 33–36.

Bonotto, S. A new flexible ethylene copolymer. *SPE Journal*, May 1962, **18**(5):555–560.

Bonotto, S., and Krevsky, B. H. Flexible ethylene copolymer. *Plastics World*, August 1962, **20**(8):32–37.

E.I. Du Pont de Nemours & Co., Inc. *'Elvaloy' 741 & 742 Resin Modifiers*. Bulletin E-18989, January 1978.

——. *ALATHON E/VA Ethylene-Vinyl Acetate Copolymer Resins*. Bulletin A-41133.

——. *ALATHON E/VA Ethylene-Vinyl Acetate Copolymer Resins*. Bulletin A-41134.

——. *Physical Properties–Du Pont ALATHON E/VA Ethylene-Vinyl Acetate Copolymer Resins*. Bulletin A-60150.

——. *Du Pont ALATHON Ethylene/Vinyl Acetate Copolymer Resins Injection Molding Processing Guide*. Bulletin A-48467.

——. *Extrusion of ALATHON E/VA Copolymers*. Processing Notes.

——. *Blow Molding of ALATHON E/VA Copolymers*. Processing Notes.

——. *Elvax Vinyl Resins*. Bulletin A-21382.

——. *Point . . . By Point . . . By Point . . . Clearly Exceptional for Skin Packaging Film . . . Du Pont Surlyn Ionomer Resin*. Bulletin A-71100.

——. *Du Pont Surlyn for Meat Packaging*. Bulletin A-84831.

Eastman Chemical Products, Inc. *General Information on Tenite Polyallomer, Properties, Processing, Applications*. Publication No. PO-12, March 1969.

——. *Properties and Characteristics of Tenite Polyallomer*. Materials Bulletin MB-16E, ECP 1359, January 1978.

——. *How to Utilize the Built-in Hinge Properties of Tenite Polypropylene and Tenite Polyallomer*. Technical Report TR-14C, ECP 1160, December 1973.

——. *Tenite Polyallomers–Eastman's New Class of Plastics*. Materials Bulletin MB-16A, ECP 463.

Eidman, R. A. Ethylene-vinyl acetate, from Plastics in Packaging. *Plastics Engineering*, October 1978, **34**(10):34–35.

Exxon Chemical Co. *Dexon Polymers for Glass Compounding*. Bulletin PLA-75-1629, 1975.

——. *Dexon Polymers, Adhesion to Substrates*, 1974.

Fischer, J. How to process ethylene-vinyl acetate copolymers. *Plastics Design & Processing*, August 1967, **7**(8):20–26.

Hinterwaldner, R. Graft polymers: new raw materials for hot melts? *Adhesives Age,* August 1978, **21**(8):34–38.

Krevsky, B. H., Bonotto, S., and Junghans, P. R. How to modify polyethylene with flexible E-A copolymers. *Plastics Technology*, May 1963, **9**(5):34–36.

Marziani, J. Ethylene Copolymers. Term Report, 1977, Dept. of Chem. Eng., University of Southern California.

Materials in Design Engineering. Olefin copolymers–versatile plastic 'alloys.' July 1965, **62**(1):102–103.

——. Olefin copolymers, II–flexibility combined with toughness and processability. July 1965, **62**:1.

Modern Plastics. Coextrusion/blow molding with PVA yields high-barrier polyolefin bottles. February 1979, **56**(2):20.

——. Olefin copolymers–what they can do. June 1962, **39**(10):84–88, 168–174.

——. New frontier for booming polyethylene–copolymers. May 1960, **37**(9):85–89.

Plastics Technology. New flame-retardant TP molds easily. October 1975, **2**(10):13.

——. PT's Buyer's Guide to Polyolefins and Cellulosics. March 1967, **13**(3):39–52.

Product Engineering. Plastic copolymer stays rubbery in hot or cold environment. May 1971, **42**(10):40.

Scheiner, L. L. How to process the new ionomers. *Plastics Technology*, March 1965, **11**(3):44–47.

Rosenblum, F. M. Carboxylation increases versatility of polyvinyl acetate-based adhesives. *Adhesives Age*, April 1979, **22**(4):19–22.

SPE Journal. New olefins offer excellent adhesion properties. May 1973, **29**:5.

Union Carbide Corporation. *Bakelite Ethylene Copolymer DQDA-3269 Natural 7*. Bulletin F-44760, June 1974.

——. *Discover What Flexible Bakelite Ethylene-Ethyl Acrylate Copolymers Can Do For You*. Bulletin F-43078, November 1970.

U.S. Industrial Chemicals Co. *Polyolefins for Hot Melt Formulations*. Bulletin PRTD 272-775, 1975.

——. *Ultrathene EVA Copolymers*. 4th Ed. Bulletin PTD 39-672, 1972.

——. *Twelve Design Problems. The Solution was Ultrathene EVA*. Bulletin PTD 56-568, 1968.

——. *Ultrathene Copolymers*. 3rd Ed. Bulletin PTD 39-466, 1966.

Van der Linden, R. Viscosity of EVA copolymers. *Adhesives Age*, August 1978, **21**(8):43–46.

Vermes, R., Brenner, W., and Pal, Y. Crosslinking the new ethylene copolymers. *Plastics Technology*, July 1975, **21**(7):43.

Wright, K. Y. Jr. Ionomer resins, from Plastics in Packaging. *Plastics Engineering*, October 1978, **34**(10):38–39.

Polybutenes

Trade Names

Advawax	Cincinnati Milacron Chemicals, Inc.
Indopol	Amoco Chemicals Corporation
Oppanol	BASF Wyandotte Corporation
Vistanex	Exxon Chemical Co.
Witron	Shell Chemical Co.

There are three isomeric C_4 olefins: butene-1, butene-2, and isobutylene. All are produced by petroleum refining processes, such as cracking or dehydrogenation, and are readily available. Isobutylene polymerizes much more readily than ethylene, propylene, or the other two C_4 isomers.

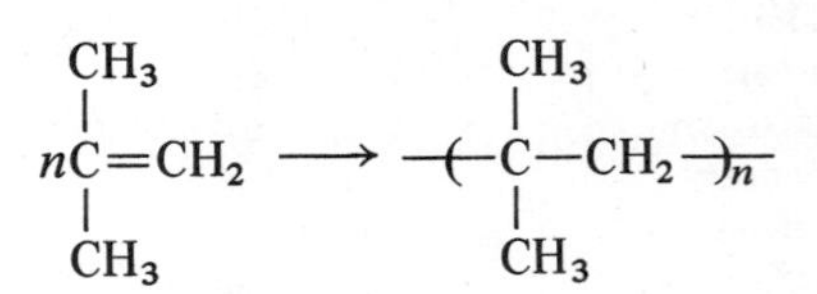

Polyisobutylene is frequently classed as a synthetic rubber since most of its applications make use of its rubberlike character, especially the grades that have been copolymerized with 1.5–4.5% polyisoprene, which gives the resultant copolymers vulcanizability. These materials have excellent resistance to ozone and, like other polymers of this class, have very low gas permeability. The saturated character of these polymers is responsible for many of these desirable properties. However, disadvantages such as poor workability, cold flow, and very poor rebound (resilience) are also a consequence of this same saturated structure.

Mixed with various fillers and other resins, polyisobutylene may be compounded into material for gaskets and caulking compounds that are oil-resistant and do not oxidize. Usually it imparts moisture resistance, low-temperature flexibility, and better adhesive properties to formulations with other resins. It is also used as a plasticizer for other waxes and resins. Table 5-39 shows a classic caulking compound formula made from polyisobutylene and the procedure used to mix it.

Until recently, polymerization of butene-1 has led to low molecular weight polymers that have been primarily useful as components of sealing and insulating compounds, adhesives, and cements. In 1974, the development of more plasticlike resins from polybutene-1 became a commercial achievement.

$$n\,\mathrm{\overset{\overset{CH_3}{|}\;CH_2}{\underset{}{CH}}{=}CH_2} \longrightarrow \mathrm{-\!\!\!\left(\overset{\overset{CH_3}{|}\;CH_2}{CH}{-}CH_2\right)\!\!\!-_{n}}$$

These polybutylene resins are high molecular weight, flexible, crystalline, isotactic polyolefins having a density of about 0.91. They have excellent flexibility, toughness, and abrasion resistance (see Figure 5-51). They accept high filler loadings (up to 75% by weight; see Figure 5-52). Other important advantages include resistance to elevated use temperatures (up to 110°C, 230°F), good barrier properties, excellent electrical insulating characteristics, and resistance to most chemical environments. They have the ability to be processed on the same equipment used for low-density polyethylene.

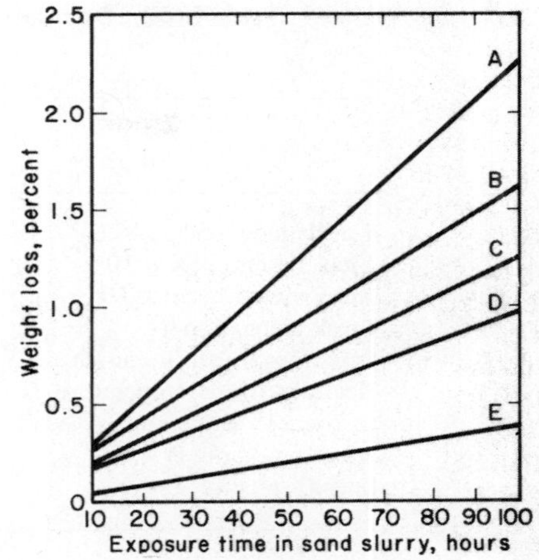

Fig. 5-51. Comparison of the abrasion resistance of various pipe materials. A: High-density polyethylene, 0.1 melt index. B: Medium-density polyethylene, 2.0 melt index. C: High-density polyethylene, zero melt index. D: Pipe-grade polybutylene, 0.4 melt index (Witron 4101); and carbon-steel pipe. E: Ultrahigh molecular weight polyethylene, zero melt index; and natural blending-grade polybutylene, 0.4 melt index (Witron 0100). (Lindegren, 1975)

Polybutylene undergoes a crystalline transformation after cooling from the melt, during which the crystallinity, density, and hardness increase, and the tensile properties attain their ultimate values. As with all crystalline polymers, the ratio of the crystalline to amorphous content, and the morphology of the crystalline regions are the main determinants of the physical properties. The change of the crystal lattice has been shown to be a function of chemical structure, molecular weight, temperature, and pressure. As a result of this crystalline transformation, polybutylene has exceptional resistance to stress-cracking and to creep under

Table 5-39. Gun-Grade Caulking Compound.*

This formulation represents the classic drying oil/polyisobutylene gun-grade caulk. Polybutene provides permanent pliability and the necessary adhesion. While relatively inexpensive this formulation has excellent properties for a variety of conventional sealant applications.

Formula

Material	Wt.%
Polyisobutylene	8.23
Blown soya oil, Z_4 viscosity	16.16
Tall oil fatty acids, 3% Rosin	0.49
Calcium carbonate	47.15
Talc	23.75
Mineral spirits	3.93
Cobalt drier, 6%	0.29
	100.00

Procedure

In the laboratory the vehicle ingredients were blended in a heavy-duty sigma blade mixer. Add fillers in the order shown and mix for 30 minutes. Thin to desired viscosity with mineral spirits. Add drier and mix to smooth consistency.

*Courtesy: Amoco Chemicals Corporation.

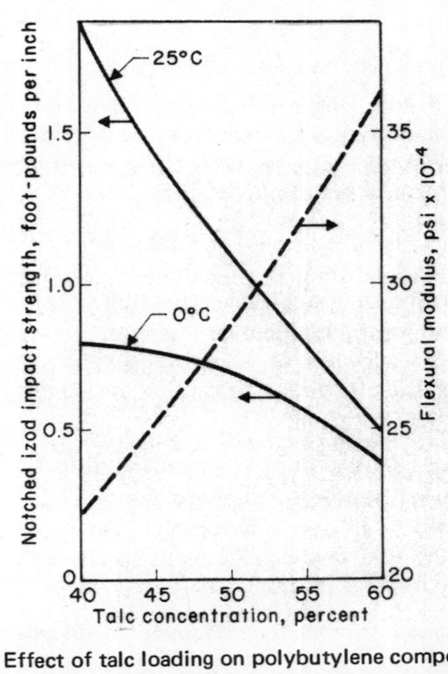

Fig. 5-52. Effect of talc loading on polybutylene compound properties. A: Flexural modulus at 25°C. B: Izod impact strength at 25°C, and C: at 0°C. (Lindegren, 1975)

Table 5-40. Comparison of Properties: Low-Density Polyethylene and Polybutylene.*

Property	Low-Density Polyethylene	Polybutylene
Tensile modulus, psi × 10^5	0.22	0.26–0.50
Yield strength, psi × 10^3	–	1.8–2.8
Break strength, psi × 10^3	1.4–2.0	4.6–5.2
Break elongation %	125–625	300–400
Izod impact ft-lb/in. notch	–	No break
Heat distortion temperature, °C(°F) (66 psi)	–	102–113 (215.6–235.4)
Brittleness temperature, °C(°F)	–30 (–22)	–(10–20) –(50–68)
Hardness, Shore D	–	55–65
Specific gravity, gm/cc	0.918–0.925	0.908–0.917

*Courtesy: Shell Chemical Co.

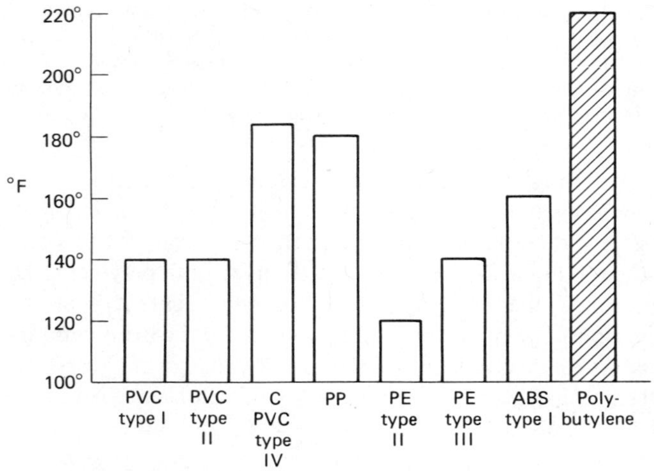

Fig. 5-53. Heat resistance of piping resins. (*Courtesy: Shell Chemical Co.*)

load. This phenomenon also allows the processor to consider postforming techniques such as cold forming. Table 5-40 compares typical properties of polybutylene with LDPE. Figure 5-53 compares the heat resistance of polybutylene with other pipe-range plastics.

Their combination of properties makes polybutylene resins adaptable to such applications as film, pressure pipe, hot water piping, gas piping, sheeting, pressure vessels, electrical insulation, protective coatings, and high-loading master batches for use with other resins. The polymers have NSF code approval for potable water use and FDA approval for food contact use. They may be used as flexibilizers for polypropylene with improved impact and heat-seal properties and are compatible in all proportions.

REFERENCES AND BIBLIOGRAPHY

Amoco Chemicals Corporation. *Indopol Polybutenes for High Quality Caulks and Sealants*. Bulletin 12-4b.

Docktor, J. I. Polybutylene, from Plastics in Packaging. *Plastics Engineering*, October 1978, **34**(10):39–40.

Foglia, A. J. Polybutylene, Its Chemistry, Properties and Applications. In *Applied Polymer Symposia*, No. 11, Interscience, New York, 1969, pp. 1–18.

Jagisch, F. C. Polyisobutylene polymers in sealants. *Adhesives Age*, November 1978, **21**(11):47–50.

Lindegren, C. R. Polybutylene bids for commodity resin markets. *Plastics Engineering*, July 1975, **31**(7):38–40.

Martino, R. Property modifiers and processing aids. *Modern Plastics*, September 1978, **55**(9):48–50.

Modern Plastics. Behind Shell's polybutylene buildup: performance plus improved economies. September 1978, **55**(9):18.

Morton, M. *Introduction to Rubber Technology*. Reinhold Publishing Corporation, New York, 1964.

Plastics Design and Processing. ASTM issues standard for polybutylene pipe. September 1974, **14**(9):6.

Product Engineering. Polybutylene, tough new plastic, is slated for pipe and film uses. December 1974, **45**(12):41.

Shannon, J. Polybutylene—a new thermoplastic for industrial piping. *Chemical Engineering*, August 30, 1976, **83**(18):121–124.

Witco Chemical, Polymer Division. *Mobil Polybutylene Blowmolding Resin*. Bulletin No. 5.

——. *Witco Polybutylene Rotational Molding Resin*. Bulletin No. 6.

——. *Polybutylene PB-131*. Bulletin 101.

——. *Witron Polybutylene Resins*. Bulletin 100.

——. *Introducing the New Polyolefin Champion—Witron Polybutylene.*

Polymethylpentene

Trade Name

TPX Mitsui Petrochemical Industries, Ltd.

As a result of the Ziegler catalysts, a polyolefin polymer with higher branching than the polybutenes has been successfully commercialized. This predominantly isotactic polymer, poly(4-methyl-pentene-1), is prepared by the dimerization of propylene.

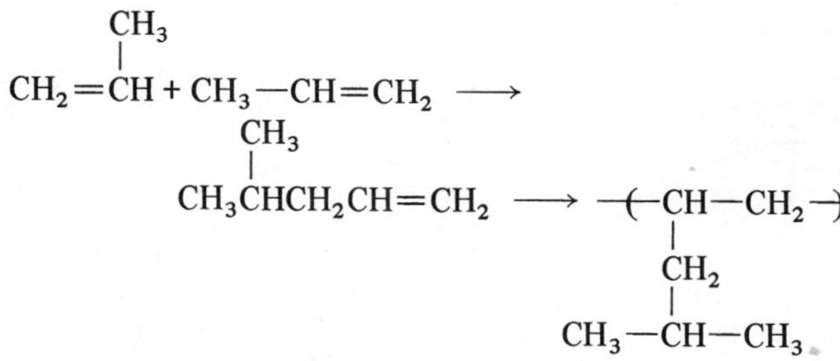

These polymers are the lowest-density plastics (0.83, compared to 0.90 for polypropylene) and have high transparency, high melting point, and excellent electrical insulation. However, they are brittle, age poorly, have high gas permeability, and are relatively expensive.

The commercially available product has a level of crystallinity of about 40%, which can be increased to about 60% with proper annealing. Like other olefinics, the electrical

Table 5-41. Test Data for Polymethylpentene at 20°C (68°F).*

Property	Test Method	Units	Typical Value
Relative density	ASTM D1505-67T		0.83
Transparency	ASTM D1746-62T	%	90
Tensile strength at yield (rate of strain 50% per min)		lb/in^2	4,000
		kg/cm^2	280
		bar	276
Elongation at break		%	15
Tensile modulus (0.2% strain, 100 sec value)		lb/in^2	210,000
		kg/cm^2	15,000
		kbar	14.7
Hardness	Rockwell: ASTM D785-65		L67–74
	Pencil		HB
Water absorption (24 hr immersion)	ASTM D570-63	%	0.01
Thermal properties			
Crystalline melting point		°C	240
		°F	465
Vicat softening point (full)	ASTM D1525-65T	°C	180
		°F	355
Coefficient of linear expansion	ASTM D696-44	in/in °F	6.5×10^{-5}
		cm/cm °C	11.7×10^{-5}
Thermal conductivity	B.S. 874A	Btu in/ft^2 hr deg F	11.6×10^{-1}
		cal cm/sec cm^2 deg C	4×10^{-4}
		mW/m deg C	168
Specific heat	ASTM C351-61	cal/g deg C	0.52
		kJ/kg deg C	2.175
Flammability	ASTM D635-63	in/min	1.0
		cm/min	2.54
		m/min	0.0254
Mould shrinkage		in/in	0.015–0.030
Electrical test data			
Permittivity (20 C) over frequency range 10^2–10^6 Hz	ASTM D150-65T		2.12
Volume resistivity	ASTM D257-66	ohm cm	$>10^{16}$
Dielectric strength on $\frac{1}{8}$ in moulded sheets,	ASTM D149-64	V/001 in	700
2 in. diameter electrodes in oil		V/m	0.018
Permeability			
Permeant			
Oxygen		cm^3 (NTP) cm/cm^2 sec (cm Hg) at 25°C	27×10^{-10}
Nitrogen		cm^3 (NTP) cm/cm^2 sec (cm Hg) at 25°C	6.5×10^{-10}
Water Vapour (for film 0.001 in. thick at 38 C & 90% r.h.)		g/m^2 24 hr	95–107

*Courtesy: Mitsui Petrochemical Industries, Ltd.

Table 5-42. Izod Impact Strengths.*

Polymer	Izod Impact Strength (ft lb./in. notched)
Acetal copolymer	1.2–1.4
Nylon 6/6	1.0
TPX grade R	0.8
Polymethylmethacrylate	0.4
Styrene-acrylonitrile copolymers	0.35–0.5
Polystyrene	0.25–0.4

*Courtesy: Mitsui Petrochemical Industries, Ltd.

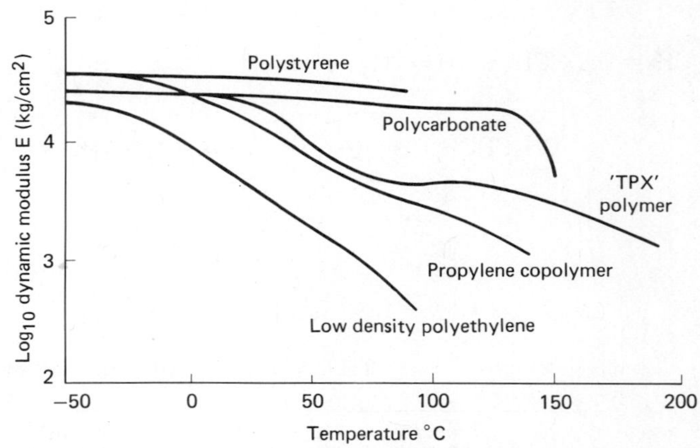

Fig. 5-54. Effect of temperature on dynamic modulus (E). (*Courtesy: Mitsui Petrochemical Industries, Ltd.*)

properties are excellent. Power factor, dielectric constant, and resistivity are equivalent to polytetrafluoroethylene without the latter's processing difficulties. Table 5-41 shows typical properties of polymethylpentene. Table 5-42 compares its impact strength with other plastics.

Its chemical resistance is typical for olefinics; that is, generally quite good (see Table 5-43). However, like polypropylene, poly(4-methyl-pentene-1) is very sensitive to oxygen. This property is aggravated by the high temperature required for processing. The use of efficient antioxidants is useful in minimizing this problem.

Table 5-43. Chemical Resistance Properties.*

		98% sulphuric acid	70% nitric acid	Concentrated hydrochloric acid	Concentrated sodium hydroxide solution	Benzene	Trichloroethylene	Heptene	Methyl ethyl ketone	Ethyl acetate	Benzaldehyde
At 20°C	"TPX" polymer	A	A	A	A	B	B	B	A	B	A
	Acrylic	B	B	B	A	C	C	C	C	C	C
	CAB	C	C	C	C	C	C	C	C	C	C
	Polycarbonate	A	A	A	C	C	C	C	C	C	C
	Polypropylene	A	A	A	A	B	B	B	A	B	A
	Polystyrene	A	B	A	A	C	C	C	C	C	C
	Polysulphone	A	B	A	A	C	C	C	C	C	C
	AS	A	B	A	A	C	C	C	C	C	C
At 60°C	"TPX" polymer	A	A	A	A	B	B	B	B	B	A
	Acrylic	C	C	C	A	C	C	C	C	C	C
	CAB	C	C	C	C	C	C	C	C	C	C
	Polycarbonate	C	A	B	C	C	C	C	C	C	C
	Polypropylene	A	A	A	A	B	B	B	B	B	A
	Polystyrene	C	C	B	A	C	C	C	C	C	C
	Polysulphone	C	B	B	A	C	C	C	C	C	C
	AS	C	C	B	A	C	C	C	C	C	C

(Column headings: Tested chemicals)

A very good
B good
C poor

*Courtesy: Mitsui Petrochemical Industries, Ltd.

The most outstanding property is the polymer's clarity. It has a light-transmission value of 90%, compared to polystyrene (88–92%) and acrylics (92%). The material is ultra-violet-sensitive, however, and thus must be protected from such radiation.

The high degree of crystallinity and the rather "knotty" side groups provide a stiffening effect on the chain. Thus, the polymer has a high crystalline melting point (245°C, 473°F) as well as a narrow melting range which permits it to retain its mechanical properties at 200°C (392°F) (see Figure 5-54). The polymer has a T_g of 50–60°C (141°F). Because of the polymer's brittleness below this temperature, fiber or rubber reinforcement is usually advised for improved toughness.

Poly(4-methyl-pentene-1) must be processed with the care required for sharp-melting polymers. It is also highly pseudo-plastic and, in the usual processing range, of low viscosity. Thus, provisions to prevent drooling in injection molding and in high-compression screws with sharp transition between feed and metering zones are recommended.

Included in the list of applications for these new resins are: laboratory and hospital ware such as syringes, beakers, burettes, and tube fittings (see Figure 5-55); thermoplastic encapsulating medium for electronic relays; central heating plant control valves; sight glasses and piping in milking machines and in laboratory plumbing traps, where its transparency and chemical resistance are desirable; replacement for stainless steel bobbins in fiber-dyeing machinery; and buttons on permanent press shirts.

REFERENCES AND BIBLIOGRAPHY

Brydson, J. A. *Plastics Materials*. 3rd Ed. Robert E. Krieger Publishing Co., Inc., Huntington, New York, 1975.

Imperial Chemical Industries, Ltd. *Adhesives for "TPX."* Information Note T205, June 1967.

——. *"TPX" Methylpentene Polymers*. Bulletin TI, February 1963.

——. *"TPX" Methylpentene Polymers*. Technical Bulletin 252, August 1966.

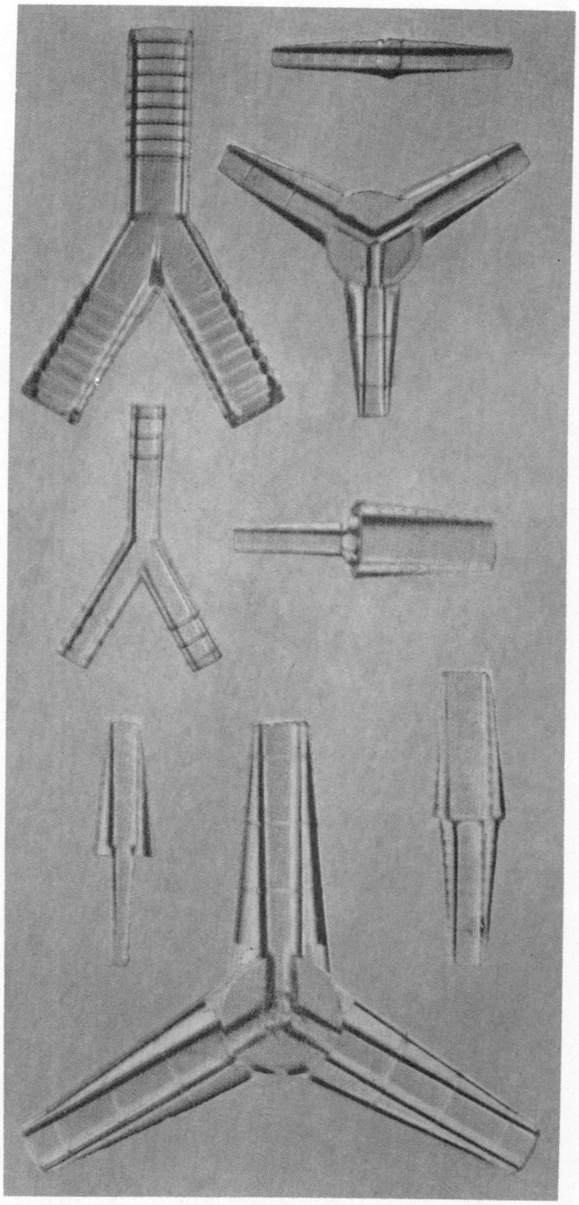

Fig. 5-55. Laboratory connector pieces. (*Courtesy: Mitsui Petrochemical Industries, Ltd.*)

——. *Injection Moulding of "TPX" Polymers*. Technological Service Note T102.

Mitsui Petrochemical Industries, Ltd. *Methylpentene Polymers "TPX."* Bulletin 2000, October 1975.

FLUOROCARBONS

A fluorocarbon is an organic chemical composed of fluorine and carbon atoms analogous to the hydrocarbons. Structures of monomers and polymers are comparable to hydrocarbons in which all the hydrogen atoms are replaced with fluorine. Two types of commercial polymers belong to this class: one is formed by the polymerization of tetrafluoroethylene, the other from trifluorochloroethylene. Many copolymers are available. The polymer of trifluorochloroethylene is not a true fluorocarbon since it contains one chlorine atom for every three fluorine atoms; but its properties are similar to those of polytetrafluoroethylene, so that it is usually considered to be in this class of polymers. Because of the long chemical terms involved, abbreviations are frequently employed. Table 5-44 shows a listing of the various polymers and copolymers, their abbreviations, and a schematic of their structure.

Homopolymers

Polytetrafluoroethylene. The fact that tetrafluoroethylene would polymerize readily was discovered accidentally. Tetrafluoroethylene (bp = -76°C, -104.8°F) had been stored at high pressure in a gas cylinder. Later it was discovered that the pressure had dropped without any evidence of leakage. Curiosity compelled a research worker to cut open the cylinder. It contained considerable quantities of a white powder, which proved to be the polymer.

The polymerization of PTFE

$$nCF_2{=}CF_2 \longrightarrow \left(CF_2CF_2 \right)_n$$

was first described by Plunkett in 1941, and semicommercial production began that same year. The monomer can be polymerized by free-radical initiators in the presence of small amounts of oxygen. Conditions of the reaction include the presence of substantial amounts of water, a pressure of approximately 1,000 psi, and temperatures ranging from 55°C to 240°C (131-464°F). If aqueous dispersions are desired as the final product, the TFE gas is allowed to react, at 0-95°C (32-203°F) and 15-500 psi, with an agitated aqueous solution containing a saturated dibasic peroxy compound, such as disuccinic acid peroxide, as the initiator. The result of this method of polymerization is a dispersion of negatively charged particles of PTFE varying in size from 0.5 to 5.0 microns. This dispersion normally contains less than 8% solids, but may be concentrated and stabilized to produce an extremely useful dispersion containing 60% solids. In addition to dispersions, PTFE is available as granules and fine powders for specially modified extrusion and molding processes.

Although PTFE polymers must be classified with the thermoplastics, they do not behave like other thermoplastic materials, since they do not become fluid even at temperatures approaching 498°C (929°F). An amorphous gel is formed at 327°C (620°F), and in this state molecular segments are mobile and the particles cohere. Most forming processes are based on this property. As temperatures increase beyond the fusion temperature of 327°C (620°F), progressively less energy is required to distort the gel; but decomposition and evaporation become appreciable above

Table 5-44. Types of Fluorocarbon Homopolymers and Copolymers.

HOMOPOLYMERS		
Polytetrafluoroethylene	PTFE	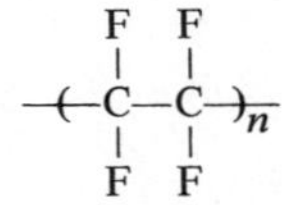
Polychlorotrifluoroethylene	PCTFE	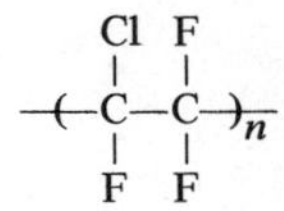
Polyvinylfluoride	PVF	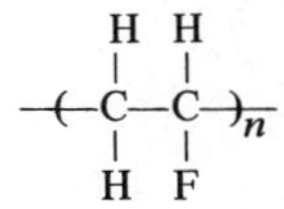
Polyvinylidene fluoride	PVDF, PVF_2	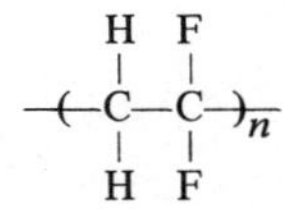
Perfluoroalkoxytetrafluoroethylene	PFA	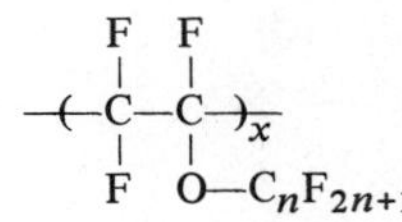
COPOLYMERS		
Fluorinated ethylene propylene	FEP	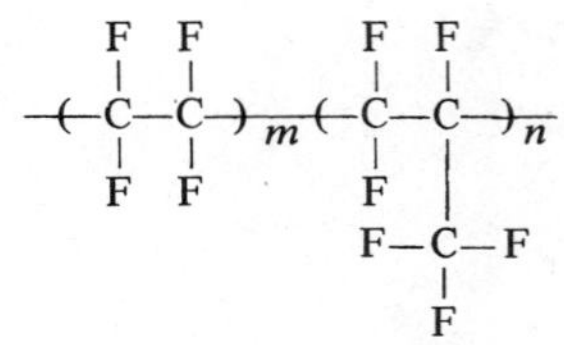
Polyvinylidene fluoride co chlorotrifluoroethylene	–	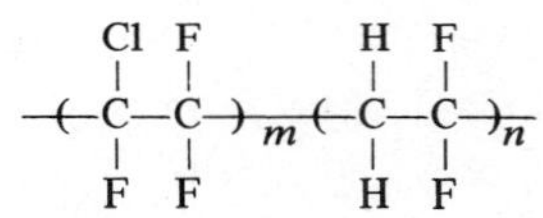
Polyethylene co chlorotrifluoroethylene	PECTFE	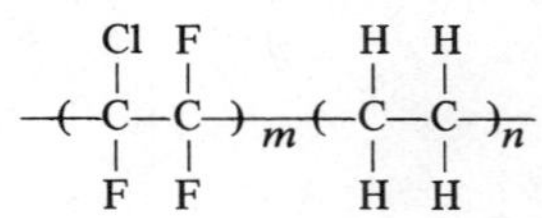
Polyethylene co tetrafluoroethylene	PETFE	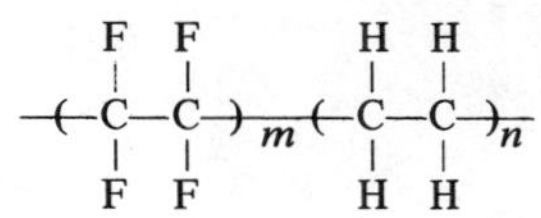
Polyvinylidene fluoride co hexafluoropropylene	–	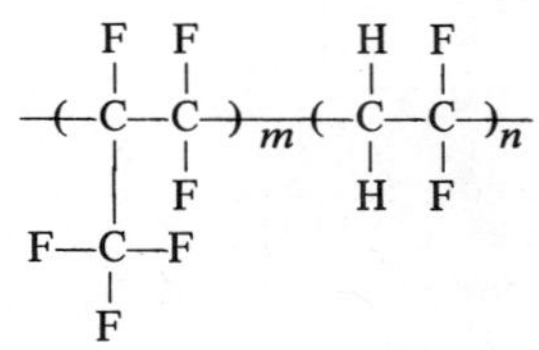
Polyvinylidene fluoride co hexafluoroisobutylene	VF_2-HFIB	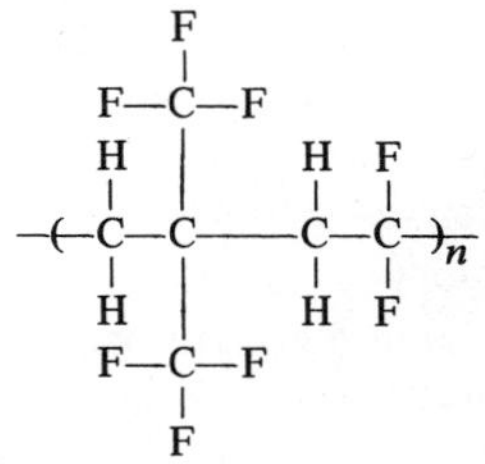

399°C (750°F), so that good ventilation must be provided whenever PTFE is exposed to higher temperatures.*

These polymers are not affected by any organic compounds, even at relatively high temperatures. Only molten alkali metals attack PTFE, forming carbon and alkali fluorides. Fluorine will react at high pressures to form CF_4. In addition to a high temperature limit of 260°C (500°F) (continuous use), temperatures as low as -73°C (-100°F) have little effect on properties. These chemical and thermal stabilities are supplemented by remarkably low dielectric constant and loss factor, which are substantially unaffected throughout the usable temperature range. Thus, PTFE is outstanding with respect to thermal and electrical properties and has the best chemical resistance of all polymeric materials. In addition, mechanical and physical properties are relatively good, particularly for many special applications. Its chief disadvantages are high cost and the relatively difficult or costly fabrication methods that must be used. It also has very poor cold-flow (creep) properties (e.g., pulleys will notch), and it has poor impermeability characteristics. PTFE also erodes under increasing rate of wear. (Note: Wear resistance should not be confused with abrasion resistance. In wear, there is a measurable erosion occurring between mating surfaces as opposed to erosion due to abrasive particle impact.) Table 5-45 shows properties of PTFE. Table 5-46 compares tensile impact properties of PTFE with other thermoplastics.

The addition of fillers to PTFE tends to rigidize the polymer, which in turn improves the wear and creep properties. The addition of fillers reduces shrinkage, compaction ratio, dielectric strength, and coefficient of expansion, while having only a slight effect on shrinkage. Fillers increase specific gravity, stiffness, compressive strength, wearability, and thermal conductivity. Since there is essentially no bonding achieved between the PTFE and the filler particles, and since the filler particles hinder PTFE coalescence, the addition of fillers lowers tensile strength, elongation, and impact strength. Table 5-47 lists four of the most popular fillers used with PTFE and the reasons for their use.

Any powdered material which will withstand the sintering temperature of PTFE may be used as a filler (up to 30% v/v). Milled glass fiber is the most widely used filler. It gives a good overall balance of properties with good chemical and electrical resistance. Bronze powders have been used for better wear and creep resistance. These formulations have better thermal conductivity than glass-filled systems, but their resistance to chemicals is poor in the presence of acids and alkalis. Carbon, graphite, and combinations of the two have better chemical resistance than bronze, and good initial wear and rubbing or sliding contact characteristics. Graphite-filled systems are relatively soft. Molybdenum disulfide is used, usually in conjunction with glass, bronze, and other fillers, to improve initial and steady-state wear, as well as to improve stiffness. Other fillers include expensive organic fillers such as poly (*p*-oxybenzoate) and polyimide powders. Tables 5-48 through 5-52 show the effect of various fillers on selected PTFE properties.

PTFE cannot be injection molded, and modifications of compression-molding techniques are employed. Forming methods are comparable with those used in powder metallurgy or ceramics. The finely divided granular polymer is cold-pressed into a preform at pressures varying from 2,000 to 10,000 psi. These preformed shapes are then removed from the mold and heated above the 327°C (620°F) gel point to fuse the particles together. Cooling in molds under pressure or repressing is frequently necessary to maintain tolerance.

Rods, tubes, and special profiles can be produced by very slow extrusion. This process involves compacting the cold powder in a screw- or ram-type extruder and forcing it out of a die having a long, heated section to provide for sintering. Production rates are very much lower than those for conventional extrusion, resulting in much higher cost. Similar methods are used to produce PTFE fiber. Table 5-53 compares fabric characteristics of PTFE with other fiber materials.

Dispersions, stabilized with dispersing agents, may be used to coat metallic or other smooth surfaces. After coating and drying, the surface must be heated above the gel point to fuse the polymer particles and produce a smooth coating. Single coats are limited to about 1.5 mils because of cracking susceptibility; but thicker coatings may be built up by multiple applications, provided each coat is fused before the next is applied. Coatings that have excellent chemical resistance may be produced in this way if the surface being coated will withstand the relatively high fusion temperatures required (see Figure 5-56).

Another unique property of PTFE surfaces permits special applications. PTFE feels waxy and slippery, and practically nothing will stick to it. Whenever it is necessary to handle extremely sticky substances such as crude rubber, adhesives, inks, bread dough, or even candy, coatings of PTFE provide surfaces from which such materials are easily removed. Dispersions may also be used to impregnate glass fabrics or mats to produce heat- and chemical-resistant laminates.

Another unusual property of PTFE is its extremely low coefficient of friction. Values as low as 0.02 have been observed, and little change occurs with either temperature or load until the gel point is approached. The static coefficient of friction is approximately equal to the dynamic value, so that starting forces are not increased when PTFE

*It is interesting to note that in the early 1960s, E.I. Du Pont de Nemours & Co. had to publish a booklet entitled "The Anatomy of a Rumor" to counteract a number of widespread misconceptions about the toxicity of PTFE. The booklet itemizes clearly the true toxicology and safe handling procedures for PTFE.

Table 5-45. Typical Properties, Forms Available, Methods of Processing and Costs for Fluoropolymers.*

	ASTM	PTFE	PTFE and 25% Glass	PCTFE	FEP	PVDF	PETFE	PETFE and 25% Glass	PECTFE	PFA	PVF	CTFE co VF_2	HFP co VF_2	HFIB co VF_2
PHYSICAL PROPERTIES														
Specific gravity	D792	2.1–2.2	2.22–2.25	2.10–2.15	2.14–2.17	1.75–1.78	1.70	1.86	1.68	2.16	1.38–1.57	1.85–2.02	–	1.88
Water absorption (24 hr), %	D570	<0.01	<0.01	<0.01	<0.01	0.03	<0.01	<0.1	<0.01	0.03	<0.5	<0.01	–	0.01
Flammability (oxygen index)	D2863	>95	–	>95	>95	43	31	–	60	–	–	>25	–	60
Hardness (Rockwell or durometer)	D1706	D50–65	D55–70	R110–115	R45	R110–115	R50	–	R93	D64	–	A45	A65	R115
Coefficient of friction (dynamic)	–	0.06	0.12	–	–	0.14–0.17	0.4[a]	0.31[a]	–	–	–	–	–	–
MECHANICAL PROPERTIES														
Strength, 1000 psi														
Tensile (yield)	D638	2.8–3.0	2.4–2.9	3.5–3.7	2.1	4.2–4.5	4	12	4.5	–	–	–	–	2
Compressive (at 0.1% offset)	D695	0.7–1.8	1.2–1.5	2.0	1.6	1.3–1.4	–	–	–	–	–	0.5-0.6	2.3	–
Elongation in ten (ultimate), %	D638	250–400	250–300	125–175	160	40–100	100–400	9	200	300	110–260	300–800	320	–
Modulus, 10^5 psi														
Tensile	D638	0.4–0.9	2.4	1.9–3.0	0.5–0.7	1.2–2.4	1.25	–	–	–	–	–	–	5.5
Compressive	D695	0.7–0.9	1.0–1.2	1.8	0.6–0.8	1.7–2.0	–	–	–	–	–	–	–	–
Flexural	D790	0.9	1.9–2.4	2.2	0.9	2.5	2.0	9.5	2.4	1.0	–	–	–	6.5
Impact strength														
Izod, notched, ft-lb/in.	D256	2.5–4.0	2.2	3.5–3.6	NB[c]	3.8	NB[c]	7	NB[c]	NB[c]	–	NB[c]	–	0.4
THERMAL PROPERTIES														
Deflection temperature, F	D648													
66 psi		250	–	196–291	158	300	220	–	240	N.A.	–	–	–	428
264 psi		130	–	151–178	124	130–195	160	410	170	N.A.	–	–	–	–
Coefficient of thermal expansion, 10^{-5} per °F	D696	5–10	4–8	4	8–11	8–9	7-10	1.6–3.2	14	12	–	–	–	3.7–4.7
Low temperature embrittlement	D746	-150	–	-150	-115	-115	-150	–	-105	N.A.	–	-(51–64)	–	–
Maximum use temp (no load)	–	550	550	390	400	300	360	410	350	–	–	–	400	536
ELECTRICAL PROPERTIES														
Dielectric strength (short term), v/mil	D149	500–550	330	530–600	550–600	260	400	–	490	–	3000–6000[d]	–	630	–

Volume resistivity, ohm-cm	D257	10^{18}	10^{15}	10^{18}	2×10^{18}	2×10^{14}– 5×10^{15}	10^{16}	–	10^{15}	–	10^{15}–10^{16}	–	2×10^{13}	–
Arc resistance, sec	D495	360	–	360	–	50–60	72	–	–	N.A.	–	–	–	–
Dielectric constant														
At 10^3 Hz		2.1	–	2.3–2.7	–	7.7–8	2.6	–	2.6	–	8.5	3.0	11.4	–
At 10^6 Hz		2.1	2.9	2.3–2.5	2.1	6.4	2.6	–	2.5	2.1	7.4	2.6	–	–
Dissipation factor	D150													
At 10^3 Hz		0.0002	–	–	0.0002	0.02	0.0008	–	0.002	–	1.6	0.029	–	–
At 10^6 Hz		0.0002	0.003	0.007–0.10	0.0007	0.17	0.005	–	0.009	0.002	–	0.032	0.0125	–
MATERIALS COST														
$/lb		3.25	3.05–3.60	4.90	4.65	3.35	6.50	–	6.50	9.00	–	–	–	–
Methods of processing														
Compression molding		X	X	X	X	X	X	X	X	–	–	X	–	–
Isostatic molding		X	X	X	–	–	–	–	–	–	–	–	–	–
Injection molding		–	–	X	X	X	X	X	X	X	–	–	–	–
Melt extrusion		–	–	X	X	X	X	–	X	X	X	–	–	–
Ram extrusion		X	–	–	–	–	–	–	–	X	X	–	–	–
Laminating		X	X	X	X	X	X	–	X	–	–	–	–	–
Coating		X	–	X	X	X	X	–	X	–	–	–	–	–
Rotational molding		–	–	–	–	–	X	–	X	–	–	–	–	–
Blow molding		–	–	–	–	–	X	–	X	–	–	–	–	–
Environmental stability														
Radiation resistance		A[b]	A	SA	A	E	E	E	E	–	–	–	–	–
Weather resistance		E	E	E	E	E	E	E	E	–	–	–	–	–
Chemical resistance (room temp)														
Weak acids		E	E	E	E	E	E	E	E	E	–	E	–	–
Strong acids		E	E	E	E	E–SA	E	E	E	E	E	E	E	–
Weak bases		E	E	E	E	E	E	E	E	E	–	–	E	–
Strong bases		E	E	E	E	E	E	E	E	E	E	–	–	–
Solvents		E	E	E	E	E–SA	E	E	E	E	E	SA	E	–
Gasoline		E	E	E	E	E	E	E	E	E	–	SA	–	–
Oils and greases		E	E	E	E	E	E	E	E	E	E	–	E	–

[a] At 10 fpm and 100 psi loading.
[b] A is attacked; E is excellent resistance; SA is slight attack.
[c] NB is no break.
[d] 0.080 in sheet
*Data extracted from: Allied Chemical Corporation, Brydson, *Materials Engineering*, *Modern Plastics Encyclopedia*.

Table 5-46. Tensile Impact Strengths of Some Thermoplastics.*

Plastic	Tensile Impact Strength at 73°F, ft-lb/in.^{2}
Acetal	19
High-density polyethylene	49
High-impact polystyrene	60
ABS	60
Impact polypropylene	68
Cellulose acetate-butyrate	130
Polycarbonate	164
Halon G-80 (PTFE)	205

*Koo et al., 1965.
Note: All the data were obtained on type S specimens per ASTM 1822-61T.

bearing surfaces are employed. For many plastics, and non-plastics as well, the static coefficient is usually greater than the dynamic coefficient. In other words, to get an object moving across the surface, one needs to apply a far greater effort than is needed to keep the object moving. This phenomenon (i.e., the ratio of static to dynamic frictional coefficients) is known as "stick-slip." Polyamides tend to have high "stick-slip"; polyacetals and polyfluorocarbons, low.

Because of the high cost of PTFE, applications are usually based on one, or a combination, of its unique properties. Its remarkable chemical resistance has led to wide use as an anticorrosive material in the process industries for tank

Table 5-48. Thermal Conductivity of Typical Filled PTFE.*

Filler	% Filler (by volume)	Thermal Conductivity Btu hr/sq ft °F/in.
None	0	1.7
Coke flour	66.7	6.6
Graphite	50	9.5
Graphite	74	19.5
Asbestos	50	4.2
Copper	50	7.5

*Courtesy: E.I. Du Pont de Nemours & Co., Inc.

Table 5-49. Deformation of Filled PTFE Systems.*

Filler	% Filler (by volume)	% Deformation Under Load 2,000 lb/sq in. 122°F, 24 hr
None	0	25
Coke flour	33.3	1–2
Coke flour	66.7	0.5–1.3
Asbestos	50	0.3
Graphite	25	1.7
Graphite	50	1.3

*Courtesy: E.I. Du Pont de Nemours & Co., Inc.

Table 5-47. Types of Fillers Used with PTFE.*

Filler	Common % Weight Concentration	Comments
Glass fiber	15% and 25%	Most widely used filling material. Greatly improves mechanical properties with least effect on electrical and chemical properties. Provides best balance of properties.
Bronze	60%	Provides for greater mechanical properties–greater hardness, lower wear, better compressive strength, thermal conductivity, etc.–than any other filler, but is less chemically resistant. Used for mechanical application and to improve electrical conductivity of TFE.
Graphite	15%	Imparts better mechanical strength and load carrying ability in rubbing applications with little effect on TFE's frictional properties, or chemical resistance. Used primarily in chemical applications.
Molybdenum –Disulfide (Moly)	5%	Generally used in small quantities in conjunction with other fillers. Greatly increases hardness and wear resistance and reduces the starting friction of TFE filled with other materials. Effect on electrical and chemical properties is negligible.

*Courtesy: Allied Chemical Corp.

Table 5-50. Wear Rate of Typical Filled Teflon Compositions.*

Filler	% Filler (by volume)	Rate of Wear mg./hr.(1)
None	0	0.74
Glass fibers plus pigment	(2)	0.0015
Molybdenum powder	58 (3)	0.0000
Graphite	(2)	0.0015
Copper	(2)	0.032
Molybdenum disulfide	(2)	0.038

(1) White, H. S., National Bureau of Standards Report No. 2882 (October 1953). Bearings ½ inch long × ¼ inch I.D.; run on Type 303 stainless steel shafts, 0.25 inch in diameter, clearance of 0.002 inch on diameter; 2,000 lb./sq.in.; room temperature; 150 r.p.m..
(2) Commercial products containing 15–25% filler by weight.
(3) 85% filler by weight.
*Courtesy: E.I. Du Pont de Nemours & Co., Inc.

Table 5-51. Electrical Properties of Filled Teflon.*

Filler	% Filler (by volume)	Power Factor, 1000 cycles	Dielectric Constant, 1000 cycles
None	0	<0.0003	2.0
Coke flour (1)	1	0.001	2.4
Powdered glass	50	0.0032	4.75
Quartz (1)	50	0.0008	3.65
Iceberg clay (2)	50	0.05	
Talc (3)	50	0.0004	4.41
Titanium dioxide (4)	50	0.05	
Glass fiber	25	0.0019	

(1) 100–200 mesh; C.H. Chrystal Co., New York, N.Y.
(2) Burgess Pigment Co., Paterson, N.J.
(3) Whittaker, Clark, and Daniels, New York, N.Y.
(4) "Ti-Pure" L., Pigments Department, Du Pont Co., Wilmington, Del.
*Courtesy: E.I. Du Pont de Nemours & Co., Inc.

linings, gaskets, tubing, pump and valve parts, packings, seals (see Figures 5-57 and 5-58), and diaphragms. The non-sticking and low friction properties of PTFE surfaces have caused it to be used for many mechanical parts, primarily bearing surfaces where lubrication is difficult or undesirable or where corrosive conditions exist. Great improvements in ease and cost of operations can frequently be made by covering critical surfaces of equipment used to process sticky materials. Its excellent electrical properties and heat resistance have found many uses for PTFE in the electrical industries, such as spacers and inserts for coaxial cables, wire insulation for high temperatures and high frequencies, and high-quality wrapping tape for insulating motors, generators, and condensers.

Table 5-52. Volume Resistivity of Filled Teflon.*

Filler	% Filler (by volume)	Volume Resistivity, ohm-cm.
None	0	$>10^{15}$
Graphite	50	87
Graphite	75	0
Coke flour	66.7	0
Asbestos	50	2×10^{12}
Copper	50	2×10^{5}
Glass fiber	25	2×10^{13}

*Courtesy: E.I. Du Pont de Nemours & Co., Inc.

Table 5-53. Typical Fabric Characteristics.*

Property	Cotton Vegetable Fiber	Nylon Polyamide Fiber	DACRON® Polyester Fiber	Glass	TEFLON® TFE-fluorocarbon Fiber	NOMEX® High Temp. Nylon Fiber
Tensile strength	Medium	Excellent	High	Very high	Medium	Excellent
Elongation	Low	Medium	Medium	Very low	Medium	Medium
Electrical insulation properties	Fair	Very good	Very good	Excellent	Excellent	Very good
Resistance to:						
Aging	Good	Good	Good	Excellent	Excellent	Good
Moisture	Fair	Very good	Excellent	Excellent	Excellent	Very good
Heat degradation	Good	Very good	Excellent	Excellent	Excellent	Excellent
Mildew	Poor to fair	Good	Good	Excellent	Excellent	Good
Strong acid	Poor	Poor	Fair to good	Poor	Excellent	Poor
Weak acid	Poor	Fair	Good	Fair	Excellent	Fair
Strong alkali	Excellent	Good	Fair	Poor	Excellent	Good
Weak alkali	Excellent	Excellent	Fair to good	Fair	Excellent	Good
Organic solvents	Excellent	Good	Good	Excellent	Excellent	Good
Flammability	Poor	Fair	Fair	Excellent	Excellent	Good
Comparative cost	Low	Moderate	Moderate	Low	High	High

*Courtesy: E.I. Du Pont de Nemours & Co., Inc.

Fig. 5-56. This open-stock cookware features DuPont's Silverstone nonstick PTFE surface. (*Courtesy: E.I. Du Pont de Nemours & Co., Inc.*)

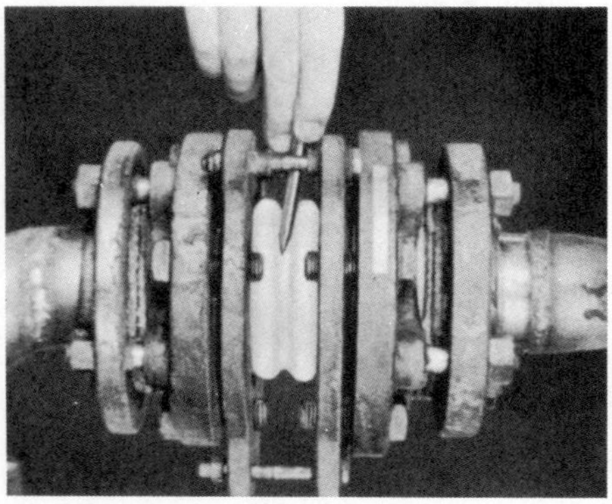

Fig. 5-57. PTFE expansion joints and seals are used in high-strength hydrogen peroxide piping systems. (*Courtesy: Allied Chemical Corporation*)

Polychlorotrifluoroethylene. The monomer, $CF_2{=}CFCl$, is a gas at atmospheric pressure with a boiling point of -26°C (-15°F). It is made from hexachloroethane and hydrogen fluoride as follows:

$$CCl_3{-}CCl_3 + HF \longrightarrow \underset{\text{trifluorotri-chloroethane}}{CClF_2{-}CCl_2F} \xrightarrow{Zn} CF_2{=}CFCl$$

At room temperature, the liquid monomer will react rapidly with oxygen and water so that it is usually handled in closed systems under nitrogen. Polymerization is carried out in the presence of chloroform using a peroxide initiator:

$$nCF_2{=}CFCl \longrightarrow \left(\begin{array}{cc} F & F \\ | & | \\ C & C \\ | & | \\ F & Cl \end{array} \right)_n$$

Fig. 5-58. Boiling acids and alkalis have no effect on PTFE laboratory equipment. (*Courtesy: Allied Chemical Corporation*)

After-treatment with cobalt trifluoride removes trace impurities of hydrogen, saturates double bonds, and replaces any loosely attached chlorine atoms with fluorine. This after-treatment improves chemical resistance and thermal stability.

A wide range of polymers can be produced by varying the conditions of polymerization. Reaction temperatures are typically 0-40°C (32-104°F) for solid polymer, higher for liquids; pressures vary from 20 to 1500 psi, with reaction times of 5-35 hours. Oils, waxes, and greases are obtained in the molecular weight range of 500-5,000, while molding powders have molecular weights in the 100,000 range. The oils may be used as a plasticizer for molding powders without appreciably altering chemical resistance.

PCTFE can be fabricated by conventional compression, injection, and transfer molding processes and by extrusion. Since a relatively wide range of properties can be obtained with different grades of molding powders, it is the most versatile of the fluorocarbons with respect to the types of products that may be produced. An important factor that affects the final physical properties of PCTFE is the ther-

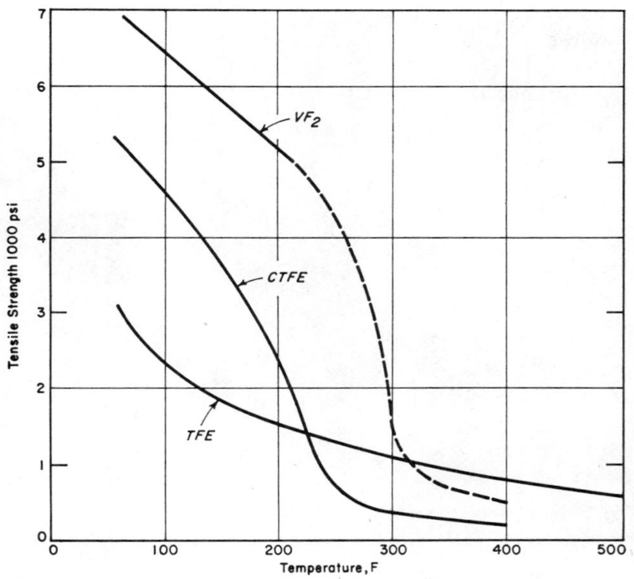

Fig. 5-59. Effect of temperature on tensile strength. (Reprinted by permission from *Materials in Design Engineering*, February 1964)

mal treatment during and after molding. It may be cooled slowly to give a highly crystalline material or quenched to produce an amorphous product. The amorphous material is more transparent, tougher, softer, and more flexible than the crystalline polymer.

PCTFE has many properties that are similar to the other fluorocarbons (see Table 5-45). In general, it has higher tensile strength, compressive strength, a wider range of flexibility and hardness, better transparency, and is easier to fabricate. Figure 5-59 shows the effect of temperature on tensile strength for PCTFE and two other polymers.

Heat resistance is not as good as PTFE, but it is much better than other thermoplastics. Chemical resistance is very good, but not as outstanding as PTFE, particularly at higher temperatures. It is swollen at room temperature by ethers, halogen derivatives, toluene, and some esters. At higher temperatures it is dissolved by these compounds. Molten alkaline metals, some metal halides, and fluorine will attack PCTFE.

Typical applications of PCTFE include gaskets, valve seats, observation windows, tubing, wire and support insulation, and coils and forms for electrical equipment. Dispersions are also available that have the same general uses as the PTFE dispersions. Table 5-54 compares the permeability of PCTFE films to other film plastics.

Polyvinylidene Fluoride. PVF_2 is polymerized from vinylidene fluoride. The monomer is made by fluorinating methyl chloroform using HF.

$$2HF + CH_3CCl_3 \longrightarrow$$

$$CH_3CF_2CL \xrightarrow{\text{pyrolysis}} CH_2{=}CF_2 + HCl$$

Table 5-54. Water Vapor Permeability Constants.*

Samples	Film thickness mils	P = constant $\times 10^{-9}$ [a]
KEL-F (dispersion)	3	3.80
KEL-F (dispersion)	8	0.09
KEL-F (extruded)	3	3.50
KEL-F (pressed)	8	0.07
Cellophane	1.5	47.00
Pliofilm (PVC)	3.2	13.00
Saran	1.0	1.35

[a]P = volume of gas in cu. cm. through 1 mm. thickness/area sq. cm./*t* (sec.)/cm. Hg differential.
*Kaufman, 1954.

Properties of special interest in the vinylidene fluoride plastics are strength, mechanical toughness, heat stability (-62-149°C, -80-300°F), and resistance to most solvents, chemicals, and ultraviolet radiation. They tend to dissolve in strongly polar solvents and discolor and embrittle when exposed to highly basic primary amines. They demonstrate extremely good weatherability, gamma radiation resistance, non-drip self-extinguishing characteristics, and very low permeability. They have exceptionally good electrical properties, and are available in colors as well as transparent forms. Noteworthy is the fact that no oxidative or thermal degradation occurs during continuous exposure to 149°C (300°F) for one year. See Table 5-45 for a comparison of properties to other fluorocarbons. Table 5-55 compares PVF_2 properties to chlorinated polyether.

PVF_2 materials may be molded by all processes, and processed by fluidized-bed coatings, spray, dip, roller, slush coating and casting.

Typical applications include high-temperature valve seals, lip seals, chemical-resistant pipe and fittings (see Figures 5-60 and 5-61), coated and lined vessels, agitators and tanks, monofilaments, coil bobbins, capacitor film, ribbon cable, protective films, and wire insulation. It is used as a base for long-life exterior coatings. It has also found wide use in heat-shrinkable tubing where its abrasion and cut-through resistance shows superiority over other polyfluorocarbons.

Polyvinyl Fluoride. PVF was first introduced as a film in the early 1960s. The manufacture of the monomer has not been disclosed, but the polymerization occurs by subjecting acetylene-free vinyl fluoride to pressures of up to 1,000 atmospheres at 80°C in the presence of water and a trace of benzoyl peroxide.

$$CH_2{=}CHF \longrightarrow \text{+}CH_2{-}\underset{\substack{|\\F}}{CH}\text{+}_n$$

To date, the polymer is mainly available as a film. PVF is a highly crystalline plastic, and it has the outstanding

Table 5-55. Comparative Properties of Polyvinylidene Fluoride.*

Properties	ASTM	RC-2525	Chlorinated Polyether[a]
Mechanical			
Ten str, 1000 psi			
At 77°F	D638	7	6
At 212°F	D638	5	3.5
Stress at yld, 100 psi			
At 77°F	D638	5.5	–
At 212°F	D638	2.5	–
Elong, %			
At 77°F	D638	300	60–160
At 212°F	D638	400	200–250
Compr str (77°F), 1000 psi	D695	10	(8.8)[b]
Mod of elast (77°F), 10^5 psi			
tensile	D638	1.2	1.6
flexural	D790	2.0	1.27
compressive	D695	1.2	1.3
Izod impact str (77°F), ft-lb/in.			
notched	D256	3	0.4
unnotched	D256	30	>33
Hardness (Shore)		D80	(R100)[c]
Physical			
Heat dist temp, °F			
66 psi	D648	300	300
264 psi	D648	195	185
Ther cond, Btu/hr/sq ft/°F/in		1.05	–
Coef of ther exp, 10^{-5} per °F	D696	8.5	4.4
Water absorp (24 hr), %	D570	0.04	0.01
Spec gr	D792	1.76	1.4
Spec vol. cu in./lb		15.7	19.8
Refrac index, n_D^{25}	D542	1.42	–
Flammability	D635	Self-ext	Self-ext
Electrical			
Vol res, ohm-cm	D257	2×10^{14}	5×10^{15}
Dielec const			
60 cps	D150	8.4	3.1
10^6 cps	D150	6.6	2.8
Dissip factor			
60 cps		0.049	0.016
10^3 cps		0.018	0.008
10^6 cps		0.17	0.01
Dielec str (short time), v/mil			
1/8-in. specimen	D149	260	400
Thin film	D149	1280[d]	4000[e]
Fabricating			
Molding temp, °F		400–550	440–465
Mold shrink, in./in.		0.020	0.005

[a]Typical properties of Hercules' Penton.
[b]Compressive stress at 1% offset.
*Riley, 1960.
[c]Rockwell R scale.
[d]0.008-in. film.
[e]0.002-in. film.

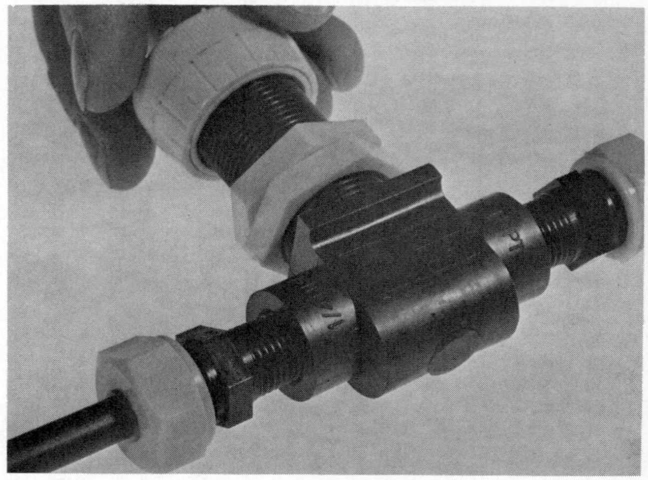

Fig. 5-60. Complete system of pipes, valves, and compression joints made of Solef PVDF from Solvay by Gielle (Italy).

Fig. 5-61. Piping and accessories in steel lined internally with Solef PVDF from Solvay by Resistoflex (Germany).

chemical resistance of the fluorocarbons, as well as excellent outdoor weatherability. It is extremely resistant to thermal degradation and maintains usable strength above 150°C (302°F), while remaining tough at -180°C (-292°F). It has low permeability to most gases and vapors and resists abrasion and staining. It resembles PVC in its low water absorption, resistance to hydrolysis, and tendency to split off hydrogen halides at high temperatures. See Table 5-45 for a list of typical properties.

The film has wide use as a protective coating in the building industry. In 0.001–0.002-inch thicknesses, bonded to wood, metal, or asphalt-based materials, it lasts many times longer than paints, enamels, or other surface coatings. It can be bonded to many substrates using many different types of adhesives. Some grades can be welded or heat-sealed.

A major application is as a surface over PVC wall coverings. These can be cleaned with soap and water or, when necessary, with strong solvents that would normally attack the PVC. They equal ceramic tile in stain resistance and cleanability, at one-fourth the installed cost.

Perfluoroalkoxytetrafluoroethylene. A recent addition to the fluoropolymer family is the fluorocarbon based on a coupling of PTFE with a perfluoroalkoxy side chain.

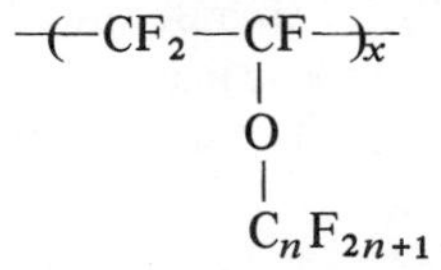

PFA resin has a melting point of just over 300°C, which is very close to PTFE (T_m = 327°C, 620.6°F) and above FEP (T_m = 270-280°C, 518-536°F). Its mechanical and chemical properties as well as its continuous operating temperature are also closer to those of PTFE (see Table 5-45 for comparative properties). However, PFA can be melt-processed by conventional molding or extrusion techniques.

The continuous-service temperature is in the 250°C (482°F) range, based on 10,000-hour exposure tests. The resin has been extruded for wire insulation, injection molded into heavy cross-section pieces, and made into thin films.

The room temperature properties are similar to both PTFE and FEP. However, at elevated temperatures, differences become apparent. Test data indicates no appreciable change in tensile strength and elongation after 5,000 hours of aging at 230°C (446°F) and 285°C (545°F). The data suggests that PFA becomes stronger as it ages at 285°C. Samples aged for 7,500 hours at 285°C show a pronounced decrease in melt-flow index, which indicates an increasing average molecular weight. This manifests itself in a 25% increase in the tensile strength and ultimate elongation. Flex life also improves several-fold with heat aging. The unusual conclusion from this data is that continued heating increases toughness.

Like other fluorocarbons, PFA is inert to inorganic acids and chemically resistant to a broad spectrum of bases, organic acids, salt solutions, and classic polymer solvents. As of this writing, no data on the effect of fluorine, fluorine-containing compounds, and alkali metals have been documented.

When exposed to fire, the polymer contributes little in fuel value, is self-extinguishing upon removal of flame, and is classified VE-O by UL-94.

It is easier to process PFA than other fluorocarbon polymers because of its high thermal stability. PFA can endure a wide range of processing temperatures, and it can tolerate some common processing variables and equipment problems.

Applications include use in heater cable, coaxial cable, hookup wire, deep oil-well cables, connector inserts or insulating bushings, and standoff insulators. PFA can also be injection molded into pump and valve parts for chemical processing equipment. Other chemical service applications include linings for pipes and chemical equipment (see Figure 5-62).

Fig. 5-62. New Teflon PFA fluorocarbon resin is used as a chemical lining for valves, pipe, pumps, and fittings, where it offers excellent mechanical characteristics at high temperatures. (*Courtesy: E.I. Du Pont de Nemours & Co., Inc.*)

Copolymers

Fluorinated Ethylene Propylene. FEP is a copolymer of TFE and hexafluoropropylene.

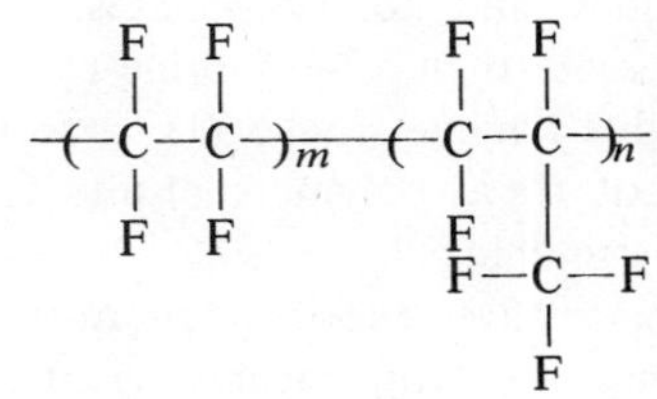

While exhibiting to a somewhat lesser degree the outstanding properties of PTFE, FEP offers processing characteristics similar to those of other thermoplastic materials. The carbon-to-fluorine ratio is the same for FEP and PTFE, and the chemical and electrical stability inherent in the C—F bond is about the same for both resins. The major difference, which is a result of the pendant CF_3 group, is a reduction in melt temperature and viscosity for FEP. Thus, in contrast to PTFE, FEP can be injection molded and extruded in conventional machinery. The cost of fabrication for many applications is therefore reduced considerably from that required to process PTFE.

The advantages of thermoplastic processability, however, have not been realized to the degree originally anticipated. Normal PTFE fabricators are equipped to work with engineering processes and can adapt to the restrictions of PTFE processing. They do not possess nor see the need for thermoplastic process equipment. Conventional fabricators, on the other hand, would have to modify their existing machines to accommodate the high melt viscosity and processing temperatures required for FEP.

FEP approaches the characteristics of PTFE in other respects, as shown in Table 5-45. It resists chemical attack almost as well, being resistant to all chemicals except the molten alkalis, fluorine, and some fluorine compounds. FEP retains toughness and strength down to temperatures approaching absolute zero. Although not as outstanding as PTFE in nonadhesiveness, it also possesses this characteristic to a degree. Films of FEP exhibit a low permeability comparable to neoprene and butyl rubber. This low permeability combined with excellent chemical resistance and relative ease of fabrication makes films of FEP a good material for lining equipment for severe corrosive service at temperatures up to 204°C (400°F).

The melt extrudability of FEP allows the fabrication of high-temperature, chemical-resistant wire insulation. Film, tubing, rods, and special profiles may also be extruded economically. Shapes that are difficult to produce by machining or by the sintering process used for PTFE may readily be produced by injection molding. Dispersions can also be made and used in a manner similar to those employed with PTFE.

Polyvinylidene Fluoride co Chlorotrifluoroethylene. Copolymers of CTFE and VF_2 range from tough, flexible thermoplastics to elastomers, depending on composition. The commercial materials are generally incapable of crystallization. They contain C—H bonds which increase chain flexibility and provide a site for crosslinking using polyamines, peroxides, and polyisocyanates. The commercial grades contain more than 50% fluorine by weight and have molecular weights ranging from 500,000 to 1,000,000. The high strength of these polymers results from crystalline orientation on stretching.

The elastomers have excellent thermal stability. Prolonged exposure to temperatures greater than 204°C (400°F) causes no scission or halogen loss. One of their outstanding properties is their chemical resistance. They will withstand the attack and penetration of powerful oxidizing agents such as propellant-grade red-fuming nitric acid and 90% hydrogen peroxide. Table 5-45 compares the properties of these resins with the other fluorocarbon polymers.

The elastomers have high tensile and tear strength with high extensibility. Other properties are comparable to other fluorocarbons, including electrical properties, gas permeability and water absorption. Their low temperature properties are adequate (brittle point of -51°C, -60°F) for many applications, but a marked increase in stiffness occurs at higher temperatures. Table 5-56 compares the water absorption properties of the copolymer (KEL-F elastomer) to that of other standard rubbers.

These copolymers find application as O-rings, chemical-resistant calendered fabrics and dipped gloves, gaskets, hose, pump impellers, aircraft ducting, extruded tube, wire, electric connectors, capacitor seals, grommets, and diaphragms. They come in latex form for such processing as casting and dipping.

Table 5-56. Water Absorption Properties of Various Rubbers.*

Rubber	Days	Mg./in.^{2}, 77°F
Smoked sheet	55	12.32
Neoprene	58	12.90
Deproteinized rubber	63	6.38
GR-S (electrical grade) (SBR rubber)	63	6.02
GR-N (nitrile rubber)	63	13.55
Chlorosulfonated polyethylene	63	5.09
KEL-F elastomer	65	3.50

*Courtesy: 3 M Co.

Polyethylene co Chlorotrifluoroethylene. PECTFE is a high molecular weight, alternating (1:1) copolymer of ethylene and chlorotrifluoroethylene. It is available in pellet and powder form and can be readily extruded as well as injection, transfer, compression, and rotationally molded. It can also be powder-coated using fluidized-bed, electrostatic-gun, and dispersion techniques.

The copolymer is a strong, highly impact-resistant material that retains useful properties over a broad temperature range. (See Table 5-45 for properties.) Its tensile strength, hardness, creep properties, impact and abrasion resistance are comparable to nylon 6. It retains its strength and impact resistance to cryogenic temperatures. The polymer can withstand elevated temperatures of 163-177°C (325-350°F). It has a lower specific gravity compared to other fluoropolymers and particularly good radiation resistance. It can be crosslinked using irradiation techniques and, in this form, can withstand 204°C (400°F) exposure for a limited time.

PECTFE is chemically resistant over a broad temperature range. It resists mineral and oxidizing acids, inorganic bases, and etchants. It is resistant to essentially all organic solvents except hot amines. It becomes slightly plasticized in a few halogenated and oxygenated solvents.

Barrier properties are good. PECTFE is non-burning and meets UL-94 VE-O standards. Electrically, PECTFE exhibits good AC loss properties with a low dissipation factor.

Except for the difference in processing temperatures (260-302°C, 500-575°F), the polymer extrudes and molds like high-density polyethylene. PECTFE rotomolds similar to high-density polyethylene, except cycle times are longer. PECTFE is not hygroscopic and does not require preprocessing.

The copolymer is particularly suitable for wire and cable insulation. Applications include computer and aircraft wire, cathodic protection wire, and nuclear control cable. As a lining or coating material, the copolymer is replacing more expensive metals and glass in chemical mixing tanks. Sheet made from PECTFE is easily machined, thermoformed, hot-air-welded, and cemented with epoxy adhesives. Valves, pumps, and pipe lined with the copolymer are used in corrosion-resistant constructions.

Drum liners, carboys, tanks, ducts, and other large components required for highly corrosive or flame-resistant service are readily rotocast from this material. Mono- and multifilament yarn from PECTFE has been woven and knit into mist eliminators, filter screens, cloth, gas-filtration bags, and fire-resistant upholstery. Film is used in laminates for aircraft interiors, flexible circuitry, flat-cable insulation, and medical packaging.

Polyethylene co Tetrafluoroethylene. PETFE is a copolymer of ethylene and tetrafluoroethylene designed to provide a synergistic balance of mechanical properties and processability. It has good thermal properties, abrasion and chemical resistance, impact strength, and electrical insulation characteristics. It is hot water and UV resistant (see Table 5-45). Dielectric constant and dissipation factor are not affected by changes in temperature and other environmental conditions. Dielectric strength and resistivity are high.

The resin exhibits good mold characteristics. Cycle times are equivalent to those of other engineering thermoplastics. Modified PETFE can be reinforced with glass fiber. This yields true reinforcement, not just filling. Because the resin will bond to the fibers, strength, stiffness, creep resistance, heat-deflection temperature, and dimensional stability are enhanced. Electrical and chemical properties approach those of the unreinforced material, while the coefficient of friction is actually lower. The use temperature of the system approaches 204°C (400°F).

Applications for injection-molded items include gears, fasteners, pump components, column packings, automotive parts, seal glands, labware, liners for valves and fittings, electrical connectors, coil bobbins, and cable ties (see Figure 5-63). Applications for wire and cable include cable for oil wells, steel plants, utility complexes, and wire for military use and computer back-panel wiring.

Polyvinylidene Fluoride co Hexafluoropropylene. This copolymer is an elastomer produced by emulsion polymerization using potassium persulphate as an initiator. It comes in various ratios with the 70:30 system having most commercial significance.

It combines high resistance to heat, fluids, and chemicals with good mechanical properties (see Table 5-45). It is cured with amine-type or peroxide coagent systems. The material can be compounded to remain serviceable for short periods at temperatures as high as 300°C (572°F), and retains useful properties indefinitely at 200°C (392°F). Resistance to the lubricants, fuels, and hydraulic fluids used in jet aircraft is unequalled by other elastomers. Mechanical properties are respectable for any elastomer, and excellent among the oil-resistant types. (Table 5-57 compares properties of various elastomers and copolymer, Viton.) It is generally superior in heat aging and oil resistance when compared to the poly(CTFE–VF_2) copolymer elastomers, but has similarly poor low-temperature properties.

Polyvinylidene Fluoride co Hexafluoroisobutylene. One of the most recent fluoropolymers to reach commercial usage is the copolymer of vinylidene fluoride and hexafluoroisobutylene. The copolymer is hard and rigid, has outstanding abrasion and scratch resistance, superior chemical resistance and electrical properties, and the low surface-energy charac-

Fig. 5-63. A variety of parts with excellent mechanical, chemical, and electrical characteristics can be easily molded from both plain and glass-reinforced Tefzel fluoropolymer resins. (*Courtesy: E.I. Du Pont de Nemours & Co., Inc.*)

Table 5-57. Typical Polymer Properties.*

Properties	Natural Rubber	Butyl	Buna N	Silicone	Neoprene	Viton® Fluoro-Elastomer	Hypalon® Synthetic Rubber	Adiprene® Urethane Rubber
Tensile strength (psi)								
Pure gum	over 3000	over 1500	below 1000	below 1500	over 3000	over 2000	over 2500	over 4000
Black loaded stocks	over 3000	over 2000	over 2000		over 3000	over 2000	over 3000	
Hardness range (shore A)	30–90	40–75	40–95	40–85	40–95	60–95	40–95	10 – 99 + (up to 80 durometer D)
Specific gravity (base material)	0.93	0.92	1.00		1.23	1.85	1.12–1.28	1.06
Vulcanizing properties	excellent	good	excellent		excellent	good	excellent	excellent
Adhesion to metals	excellent	good	excellent		excellent	good to excellent	excellent	excellent
Adhesion to fabrics	excellent	good	good		excellent	good to excellent	good	excellent
Tear resistance	good	good	fair	poor	good	fair	fair	excellent
Abrasive resistance	excellent	good	good	poor	excellent	good	excellent	outstanding
Compression set	good	fair	good	fair	fair to good	very good	fair	good
Rebound								
Cold	excellent	bad	good	excellent	very good	good	good	poor at low temp.
Hot	excellent	very good	good	excellent	very good	excellent	good	good at room temp.
Dielectric strength	excellent	excellent	poor	good	good	good	excellent	excellent
Electrical insulation	good to excellent	good to excellent	poor	excellent	fair to good	fair to good	good	fair
Permeability to gases	fair	very low	fair	fair	low	very low	low to very low	fair
Acid resistance								
Dilute	fair to good	excellent	good	excellent	excellent	excellent	excellent	fair
Concentrated	fair to good	excellent	good	fair	good	excellent	very good	poor
Solvent resistance								
Aliphatic hydrocarbons	poor	poor	excellent	poor	good	excellent	good	excellent
Aromatic hydrocarbons	poor	poor	good	poor	fair	excellent	fair	fair to good
Oxygenated (ketones, etc.)	good	good	poor	fair	poor	poor	poor	poor
Lacquer solvents	poor	poor	fair	poor	poor	poor	poor	poor
Swelling in lubricating oil	poor	poor	very good	fair	good	excellent	good to excellent	excellent
Resistance to								
Oil and gasoline	poor	poor	excellent	fair	good	excellent	good	excellent
Animal and vegetable oils	poor to good	excellent	excellent	fair	good	excellent	good	excellent
Water absorption	very good	very good	fair to good	good	good	very good	very good	good at room temp., poor at 212°F.
Oxidation	good	excellent	good	excellent	excellent	outstanding	excellent	excellent
Ozone	fair	excellent	fair	excellent	excellent	outstanding	outstanding	outstanding
Sunlight aging	poor	very good	poor	excellent	very good	very good	outstanding	good
Heat aging	good	excellent	excellent	outstanding	excellent	outstanding	excellent	good
Flame	poor	poor	poor	fair	good	good	good	fair
Heat	good	excellent	excellent	excellent	excellent	excellent	excellent	good
Cold	excellent	good	good	excellent	good	good	good	excellent

*Courtesy: E.I. Du Pont de Nemours & Co., Inc.

teristics of PTFE. Furthermore, its creep resistance at elevated temperature is quite good. The resin is melt-processable via compression and injection molding and extrusion.

The polymer can be prepared using suspension or emulsion techniques with free-radical initiators. In suspension polymerization, the liquid HFIB is added to water and initiator in a reactor at 5°C (41°F). The temperature is raised to reaction levels and VF_2 is charged, with regulation, to maintain a pressure between 150-300 psi. The resin is washed, dried, stabilized, and milled to size.

Polymerization conditions, nature and concentration of initiator, composition and concentration of emulsifier (for the emulsion process) all affect molecular weight and, thus, the properties of the copolymer. The purity of the HFIB is a critical factor in molecular weight control.

The proposed structure of the copolymer is alternating, head to tail.

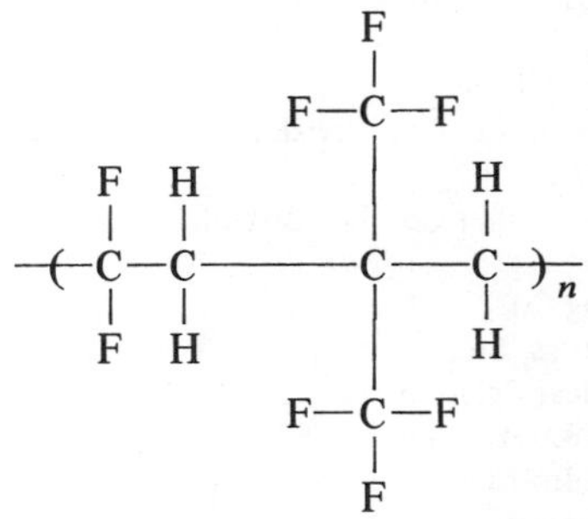

There is no evidence of blocks of $-(-CH_2-CF_2-)-$. The two trifluoromethyl groups are in a position to effectively shield the protons from thermal and oxidative attack. Optical and transmission electron microscopy has shown well-defined spherulitic structures. X-ray diffraction has confirmed a helical conformation of the backbone.

Table 5-45 lists the properties of this new resin. While its melting point is the same as for PTFE, it has superior physical properties, particularly at elevated temperatures. The surface hardness and abrasion and scratch resistance is higher than PTFE, while the impact strength at room temperature is substantially lower. The fluoropolymer exhibits superior creep resistance at elevated temperatures, equivalent to that of PTFE or FEP at lower temperatures and pressures. Figures 5-64 through 5-66 compare selected properties of the copolymer with other fluoropolymers.

Chemical resistance of the copolymer is comparable to

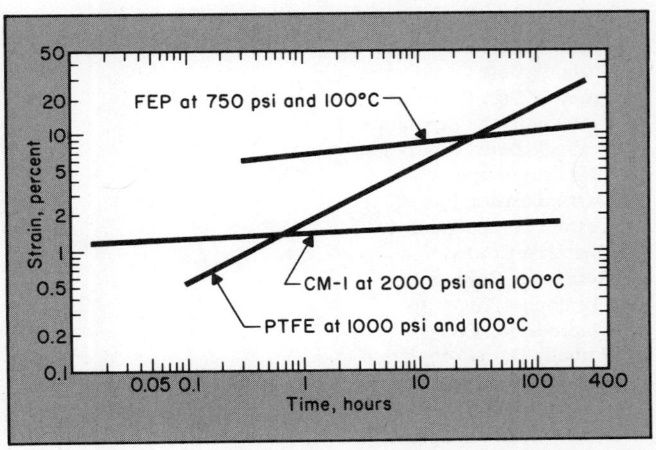

Fig. 5-64. Creep properties of two commercial fluoroplastics compared with those of the new fluorocopolymer. (Minhas and Petruccelli, 1977)

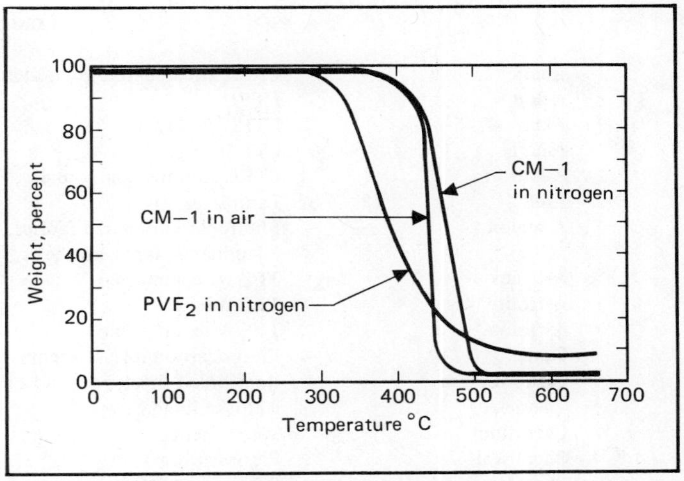

Fig. 5-65. Thermal and oxidative stability of fluorocopolymer and polyvinylidine fluoride as determined by TGA in air and nitrogen. (Minhas and Petruccelli, 1977)

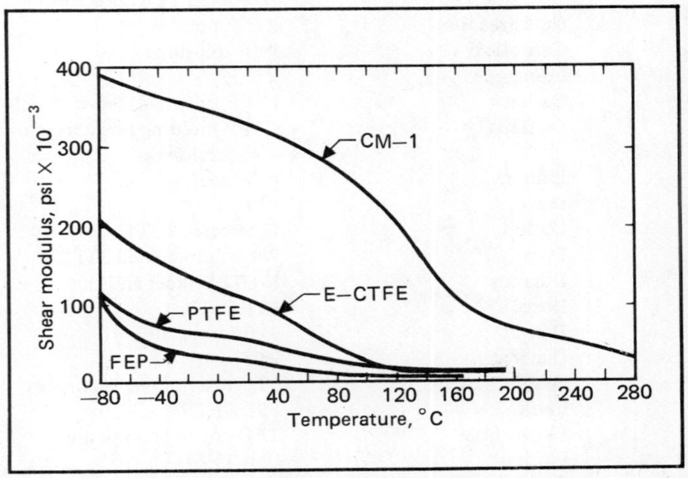

Fig. 5-66. Comparison of the torsional stiffness of the fluorocopolymer and other fluoropolymers versus temperature. (Minhas and Petruccelli, 1977)

PTFE. Strong oxidizing agents, bases, halogens, metal-salt solutions, and phosphoric acid have no effect on the polymer. Inorganic chemical resistance is equivalent to PTFE and better than PVF_2. The resin will absorb alipnatic and aromatic hydrocarbons, alcohols, amines, and chlorinated solvents, although to a lesser degree than glacial acetic acid, ketones, and esters.

Electrical-resistance properties are on the same level as for PTFE. The copolymer exhibits good adhesion to metals and excellent chemical resistance and release properties. These properties allow for use in such applications as the coating of reaction vessels, cookware, industrial and reprography rolls. Excellent adhesion is obtainable without primers and the coatings can withstand service temperatures up to 288°C (550°F).

Other end uses for the copolymer include fuel-cell separators, bearings, jet-engine fan blades, coated glass-fabric filter bags, dust precipitators, and mold-release coatings.

Trade Names

Abcite	Crosslinked coating for acrylic	E.I. Du Pont de Nemours & Co., Inc.
Aclon	CTFE	Allied Chemical Corporation
Aklar	CTFE copolymer	Soviet Origin
Alcon	CTFE	Allied Chemical Corporation
Algoflon	PTFE molding compounds	Montedison, S.p.A.
Arel	Laminates	Aremco Products, Inc.
Armalon	Fluorocarbon-coated fabrics, laminates, tapes, and felts	E.I. Du Pont de Nemours & Co., Inc.
Astrolex	TFE wire and cable	American Standard Wire & Cable Co.
Astrolite	TFE tubing	American Standard Wire & Cable Co.
Astrolon	TFE wire and cable	American Standard Wire & Cable Co.
Camie	Fluorocarbon release agents	Camie Co.
Carborundum	Reinforced sheet, rod, and tube	Carborundum Plastics, Inc.
Chemelec	PTFE with additives	United States Gasket Co., Division of Garlock, Inc.
Chemfluor	Stock shapes	Chemplast, Inc.
Chemiseal	Fluorocarbons	United States Gasket Co., Division of Garlock, Inc.
Chemlon	Fabricated PTFE products	Crane Packing Co.
Chemloy	Filled TFE	Crane Packing Co.
Chem Tite	Heat-shrinkable tubing	Fluorodynamics, Inc.
CM-1	HFIB/VF_2 copolymer	Allied Chemical Corporation
Coldlube	Powdered TFE additives	Metallurgical International, Inc.
Coldstream	PTFE powder	Metallurgical International, Inc.
Cordoflex	PVF solutions	Ferro Corporation
Crangrade	PTFE	Crane Packing, Ltd.
Daflon	PTFE tube and sleeve	Daburn Electronics & Cable Corporation
Daiflon	CTFE molding powders and dispersions	Daikin Koggo Co.
Dalbon	PVF coating	Diamond Shamrock Chemical Co.
Dalvor	PVF	Diamond Shamrock Chemical Co.
Dorlon	Fabricated PTFE	John L. Dore Co.
Durad	Polyimide-coated PTFE	Havey Super-Temperature Wires
Durasan	E-CTFE sheet and film	Allied Chemical Corporation
Durco	PTFE	Duriron Co., Inc.
Duroid	Reinforced TFE	Rogers, Inc.
Durotec	PTFE	W. S. Shamban Co.
Dyflor	VF_2 resin, tubes, laminates	Dynamit Nobel, A.G.
Ebolin	Filled PTFE	Chicago Gasket Co.
Electroplast	TFE potting compound	Toppec Manufacturing Co.
Emflon	Cast TFE	Pallflex Products Corporation
Emralon	PTFE dry film lubricants	Acheson Industries, Inc.
Enflon	Reinforced TFE	Enflo Corporation
Fiberfilm	TFE-impregnated glass paper	Pallflex Products Corporation
Fluon	Fluorocarbon resin, extrusions	Imperial Chemical Industries, Ltd.
Fluorel	HFP/VF_2 copolymer	3M Co.
Fluorex	PVF_2/acrylic film	Rexham Corporation
Fluorglas	Glass-reinforced TFE and FEP	Dodge Industries, Inc.
Fluorobestos	Asbestos/PTFE	Raybestos Manhattan, Inc.
Fluoroblack	TFE formulation	Peabody Dore Corporation
Fluoroblue	VF_2 compounds	Peabody Dore Corporation
Fluorobronz	Filled TFE	Peabody Dore Corporation
Fluorobrown	TFE formulation	Peabody Dore Corporation
Fluorocomp	Filled fluorocarbons	LNP Corporation
Fluorofil	Reinforced fluorocarbons	Dart Industries, Inc.
Fluorofill	Filled fluoropolymer	Custom Compounding
Fluorofilm	TFE tapes	Dielectrix Corporation
Fluoroflex	Fluorocarbon extrusions	Resistoflex Corporation
Fluoroglide	TFE lubricants	Chemoplast, Inc.
Fluorogold	PTFE-based	Fluorocarbon Co.
Fluorogreen	FEP compounds	Peabody Dore Corporation
Fluorolene	PTFE	LNP Corporation
Fluoroloy	Filled mold compound	Fluorocarbon Co.
Fluoron	PCTFE	Stakes Molded Products
Fluoropel	PTFE coatings	Dielectric Corporation
Fluoroplast	PTFE	United States Gasket Co.

Fluororay	Ceramic-filled PTFE	Raybestos-Manhattan, Inc.
Fluorored	TFE formulation	Peabody Dore Corporation
Fluororing	TFE molding compound	LNP Corporation
Fluoroseal	PTFE-base thread compound	Industrial Plastic Fabricators, Inc.
Fluorosint	Mica-filled PTFE	Polypenco, Ltd.
Fluorosint	Fluorocarbon base	Polymer Corporation
Fluorospar	Fluorocarbon resins	Sparta Manufacturing Co.
Fluoro-Tape	Fluorocarbon tape	Fluoro-Plastics, Inc.
Fluorotite	Fluorocarbon sealing tape	Raybestos-Manhattan, Inc.
Fluorothene	PCTFE film	Plax Corporation
Fluorulin	PTFE tapes	Joclin Manufacturing Co.
Fluran	Fluorocarbon products	U.S. Stoneware Co.
Flurene	PCTFE coating	Chemical Coatings, Ltd.
Foraflon	PVF_2	Products Chemiques Ugine Kuhlmann
Free-Flex	Metal FEP	Garlock, Inc.
Fussolon	Resin	Darkin Kogyo Co., Ltd.
Genetron	Modified molded PCTFE	Allied Chemical Corporation
Gentron	Fluorinated hydrocarbon monomers and polymers	General Chemical Division
Gore-Tex	Expanded TFE film, tubes, rods	W. L. Gore Associates, Inc.
Gylon	TFE gasket	Garlock, Inc.
Halar	PECTFE	Allied Chemical Corporation
Halcon	Fluorocarbon	Allied Chemical Corporation
Halex	Fluorocarbon molding compound	Allied Chemical Corporation
Halon	PTFE	Allied Chemical Corporation
Hiwrap	TFE tape	Simplex Wire & Cable Co.
Hostaflon	PTFE	American Hoechst Corporation
Insuron	TFE tubing	Insul-Tab, Inc.
Kalrez	Perfluoroelastomers	E.I. Du Pont de Nemours & Co., Inc.
Kel-F	PCTFE, CTFE/VF_2 copolymers	3M Co.
Koroseal	PVF/vinyl	B.F. Goodrich Industrial Products Co.
Kynar	PVF_2	Pennwalt Chemicals Co.
Leefluor	PTFE rods, sheets, tubes, molded and fabricated parts	Walt Lee Chemical Corporation
Linebacker	TFE sheet	Chemplast Inc. & Allied Chemical Corporation
Liquilon	PTFE resins	Oil Center Research, Inc.
Memorylin	TFE and FEP tubing	W. S. Shamban & Co.
Milkon	PTFE/PPS alloy	Tribol Industries, Inc.
Neoflon	TFE-copolymer	Darkin Kogyo Co., Ltd.
Penntube	TFE, FEP, VF_2	Pennsylvania Fluorocarbon Co.
Permaflon	PTFE sheet, film, rod, tube, and profile	Permali, Inc.
Plaskon	PCTFE	Allied Chemical Corporation
Plastolon	TFE powder	Plastomer Corporation
Plasto-O-Low	Fluorocarbon compounds	Plastomer Corporation
Polyflon	PTFE	Daikin Kogyo Co., Ltd.
Polyfluoron	CTFE	Acme Resin Corporation
Polylube	PTFE powder	Custom Compounding
Polymist	PTFE powders	Allied Chemical Corporation
Polytet	Filled or modified TFE	Ethylene Corporation
Raylon	TFE	Raybestos-Manhattan, Inc.
Refset	Nonflammable fluoroelastomer	Raybestos-Manhattan, Inc.
RT Duoids	Reinforced TFE	Rogers Corporation
Rulon	Modified fluorocarbon for high-temperature bearings	Dixon Corporation
Salox	Filled TFE	Allegheny Plastics, Inc.
Sealtite	TFE conduit	Anaconda Metal Hose
Sirvon	TFE	Chicago Rawhide Mfg. Co.
Solef	PVF_2	Soltex Polymer Corporation
Soreflon	PTFE	Products Chimiques Ugine Kuhlmann
T-Cote	PVF_2	Teledyne Thermatics
Te Carbolon	TFE fluorocarbons	Carbone Corporation
Tecfluorfil	Filled TFE	Thermech Engineering Corporation
Tecnoflon	Fluorinated rubbers	Montedison, S.p.A.
Tedlar	PVF film	E.I. Du Pont de Nemours & Co., Inc.
Teflon	PTFE, FEP, PFA	E.I. Du Pont de Nemours & Co., Inc.

Tefzel	PETFE	E.I. Du Pont de Nemours & Co., Inc.
Temp-R-Glass	PTFE-coated fiberglass	Connecticut Hard Rubber Co.
Temprite	PTFE magnet wire	Simplex Wire & Cable Co.
Teslar	PVF film	E.I. Du Pont de Nemours & Co., Inc.
Tetralon	Filled PTFE	Amerco, Inc.
Tetraloy	Filled PTFE	Pennwalt Corporation
Tetran	TFE	Pennwalt Corporation
Thermalon	Fluorocarbon raw stock	Dodge-Wasmund Mfg., Inc.
Thermaloy	Fluorocarbon raw stock	Dodge-Wasmund Mfg., Inc.
Tri-Flor DS	PTFE extrusion and moldings	Tri-Point Industries, Inc.
Tri-Gaspak	TFE	Tri-Point Industries, Inc.
Trikote	PTFE coating	Tri-Point Industries, Inc.
Tri-Pak	PTFE coating	Tri-Point Industries, Inc.
Triplax	Molded PCTFE	Plax Corporation
Tri-Tape	PTFE coating	Tri-Point Industries, Inc.
Trithene	CTFE film	Union Carbide Corporation
TSI	PTFE rod	Tri-Point Plastics, Inc.
Tufram	TFE/aluminum	General Magnaplate Corporation
Turbotemp	PTFE tubing	Brand-Rex Co.
Turcite	TFE	W. S. Shamban & Co.
Turcon	PTFE	W. S. Shamban & Co.
Turners	PTFE sheet, film, rod, and tube	Turner Brothers Asbestos Co., Ltd.
Tygaflor	PTFE film	Fothergill & Harvey, Ltd.
Viton	HFP/VP_2 copolymers	E.I. Du Pont de Nemours & Co., Inc.
Vottalef	CTFE	Products Chimiques Ugine Kuhlmann
Vydax	PTFE	E.I. Du Pont de Nemours & Co., Inc.
Whitcon	TFE powder	Pennwalt Corporation
Xylan	Fluorocarbon coatings	Whilford Corporation

REFERENCES AND BIBLIOGRAPHY

Albisetti, C. J. "Polyvinyl Fluoride Film." In *Modern Plastics Encyclopedia*, S. Gross (exec. ed.), **45**(14A):566–567. McGraw-Hill, Inc., New York, 1968–69.

Allen, D. B., Perkins, J. R. Jr., Reed, J. C., and Fasig, E. W. A fluorocarbon that's melt extrudable. *Plastics Engineering*, October 1974, **30**(10):48–50.

Allied Chemical Corporation. *Fluorocarbon Resins Comparison Data*. Bulletin 854-203, February 1976.

——. *Halon TFE Polytetrafluoroethylene Resin, Allied Chemical Corporation Modified PTFE Resins*. Bulletin 854-198, March 1976.

——. *Polymist F Powders*. Technical Data G-4, Form 901-146, April 1976.

——. *Polymist F Fine Particle PTFE Waxes*. Technical Data Report, Form 901-201, May 1976.

——. *Plaskon CTFE 2100 Resin (Fluorocarbon Plastic For Molding and Extrusions)*. Technical Data Report, Form 854-178, May 1975.

——. *Halar Fluoropolymer General Information Bulletin*. Form #854-147, August 1975.

——. *Halon TFE Tetrafluoroethylene Resins for Critical Chemical Processing Applications*. Bulletin 854-109, November 1974.

——. *Rotomolding Halar Fluoropolymer*. Bulletin P-1, Form #854-171, February 1973.

——. *Lubricant Uses of A-C Polyethylene*. Technical Data G-8, Form 901-151, June 1973.

——. *Chemical Resistance of Halar Fluoro-Polymer*. Bulletin C, Form #854-151, September 1972.

——. *Halon TFE for Improved Performance and Reliability in Hydraulic, Mechanical, Pneumatic Applications*. Bulletin 854-33, 1968.

——. *Halon G-700 A New TFE Resin for Heavy Stress Applications in Mechanical, Pneumatic and Hydraulic Systems*. Bulletin 854-037, 1968.

——. *Aclar Fluorohalocarbon Film–Technical Data*. Bulletin ADT 1263.

——. *Aclar Fluorohalocarbon Film–Technical Data*. Product Data Sheet A-1-660.

——. *Aclar Fluorohalocarbon Film–Lamination Procedures*. Product Data Sheet A-4-760.

——. *Halar New Generation Fluoropolymer*.

——. *Plaskon CTFE Resins (Fluorocarbon Plastics for Molding and Extrusions)*. Technical Data Report CTI-3, Form 854-03.

——. *Plaskon CTFE Resins–Injection Molding*. Technical Data Report CT2-3, Form 854-04.

Billmeyer, F. W. Jr. *Textbook of Polymer Science*. 2nd Ed. Wiley-Interscience, New York, 1971.

Brydson, J. A. *Plastics Materials*. 3rd Ed. Robert E. Krieger Publishing Co., Inc., Huntington, New York, 1975.

Coel, J. and Griskey, R. G. "Processing Properties of Vinyl Fluorides." Paper presentation, *SPE ANTEC*, New Orleans, May 1979.

DuBois, J. H. and John, F. W. *Plastics*. 5th Ed. Van Nostrand Reinhold Co., New York, 1974.

E.I. Du Pont de Nemours & Co., Inc. *Tefzel Fluoropolymer: The Tough Industrial Insulation*. Bulletin E-00481, September 1974.

——. *Teflon PFA Fluorocarbon Resin, TE9704*. Bulletin A-86400, January 1973.

——. *Du Pont Tefzel Fluoropolymer Design Handbook*. Bulletin A-84764, 1973.

——. "Tefzel" ETFE fluoropolymer: an exciting new resin. *Journal of "Teflon"*, **13**(1), 1972.

——. *Du Pont Tefzel–A New Developmental ETFE Fluoropolymer Designed for Electrical, Chemical and Mechanical Applications*. Bulletin A-73642, March 1972.

——. *Du Pont Tefzel Rugged . . . Plus–Properties and Performance, APD #4*. Bulletin A-77314, July 1971.

——. *Du Pont Tefzel Rugged . . . Plus–Glass Reinforcement of Tefzel, APD #5*. August 1971.

——. *Armalon Fluorocarbon Resin Coated Glass Fabrics and Laminates for Mechanical Applications Requiring . . . Chemical Resistance, Heat Resistance, Low Friction*. Bulletin A-32237.

——. *Armalon Fluorocarbon Resin Coated Fabrics, Laminates, Felts and Tapes*. Bulletin A-48488.

——. *Industrial Fabrics*. Bulletin A-56772.

——. *Armalon TFE-Fluorocarbon Impregnated Felt for Gasketing Under Extreme Conditions*. Bulletin A-11839, November 1966.
——. *Teflon Tetrafluoroethylene Resin–Properties, Uses, Processing Techniques*. Bulletin A-8165.
——. *The Wide World of Teflon*. Bulletin A-41200.
——. *The Facts About Teflon on Housewares*. Bulletin A-70160.
——. *Du Pont Teflon FEP-Fluorocarbon Film. Engineering Films for Industry, Technical Report*. Bulletin A-13412.
——. *Teflon PFA Fluorocarbon Resin, TE9705*. Bulletin A-80743.
——. *Du Pont Introduces New High Performance Resin*. Product Information Service PR-8.
——. *Du Pont Teslar PVF Film, Basic Property and Customer Service Information*.
Houston, A. M. New fluoroelastomers for sealing under extremes. *Materials Engineering*, October 1975, **82**(5):25–27.
Hudlicky, M. Crosslinking of Perfluoroolefin Copolymers. *Proceedings, ASTM Task Group on Geothermal Seals*, Boston, 6/78, CONF-7806141, April 1979.
Kaufman, H. S. Modified polymers containing fluorine. *Technical Association of the Pulp and Paper Industry*, July 1954, **37**(7): 140A–143A.
Koo, G. P., Jones, E. D., Riddell, M. N., and O'Toole, J. L. Engineering properties of a new polytetrafluoroethylene. *SPE Journal*, September 1965, **21**(9):1100–1105.
Larsen, H. A., DeHoff, G. R., and Todd, N. W. Injection molding FEP–fluorocarbon resin. *Modern Plastics*, August 1959, **36**(12): 89–96.
Lovell, R. T. FEP laminates cut friction, costs. *Materials Engineering*, January 1972, **75**(1):26–27.
Materials Engineering. Fluoropolymers now are tougher and easier to process. August 1972, **76**(2):30–34.
——. New moldable fluoropolymer has all the properties of PTFE. November 1972, **76**(6):39.
——. Injection molding no problem with new fluoropolymer. August 1971, **74**(2):32.
Materials in Design Engineering, Materials & Process Manual No. 215 Fluorocarbon plastics. February 1964, **59**(2):93–104.
McCane, D. I. The new melt-processable fluorocarbons. *Plastics Technology*, October 1972, **18**(10):18, 43.
Minhas, P. S. and Petruccelli, F. A new high performance fluoropolymer that can be readily melt-processed. *Plastics Engineering*, March 1977, **33**(3):60–63.
3M Co. *Kel-F Plastic Laminate, A New Structural Material*. Application Manual No. 2, Bulletin KFL-2, March 1956.
——. *Fluorel Brand Elastomer*. Bulletin Y-IFHTE.
——. *Kel-F Elastomer, Properties, Applications*. Bulletin Y-IEPB.
——. *Kel-F Elastomer, Compounding, Vulcanization, Fabrication*. Bulletin J-ECVF.
——. *Kel-F Halofluorocarbon Oils, Waxes, Greases and Alkanes*. Bulletin J-OPB.
——. *Injection Molding and Extrusion of Kel-F Halofluorocarbon Materials*. Bulletin J-HMP.
——. *Properties and Applications of Kel-F Plastic*. Bulletin Y-IKAP.
——. *Kel-F Halofluorocarbon Polymers, Physical, Electrical, Chemical, Mechanical Properties*. Bulletin H-JGIF.
——. *Kel-F 81 Plastic*. Bulletin Y-IKTDC.
——. *Properties of Kel-F Oils (and Plasticizers), Greases, and Waxes*. Technical Bulletin #5-1-52.
——. *Kel-F for Maximum Performance in Seals-Valves-Diaphragms*. Bulletin Y-ISK.
Modern Plastics. New fluoroplastic is a copolymer of ethylene and CTFE. October 1971, **48**(10):28.
Moran, L. A. Fluoroelastomers. *Proceedings, ASTM Task Group in Geothermal Seals*, Boston, 6/78, CONF-7806141, April 1979.
Osten, R. A. "TFE and FEP Resins." In *Modern Plastics Encyclopedia*, S. Gross (exec. ed.), **45**(14A):196–198. McGraw-Hill, Inc., New York, 1968–69.
Pennwalt Chemicals Co. *Kynar Vinylidene Fluoride Resin*. Product Sheet KI-65a.
——. *Kynar Vinylidene Fluoride Resin*. Data Sheet VF2R-61a.
——. *Kynar Vinylidene Fluoride Resin–Chemical Resistance Properties*. Bulletin KC-65a.
——. *Kynar Vinylidene Fluoride Resin–Properties, Applications, Fabrication*. Bulletin VF2R-62.
Plastics Design & Processing. Filled TFE fluoroplastics. July 1968, 8(7):17–19.
Plastics Technology. Here's properties rundown on new fluoropolymer. June 1970, **16**(6):19, 60.
Plastics World. Polyvinylidene fluoride resin–a study of its properties and applications in relation to commercial utilization. January 1961, **19**(1):30.
Product Engineering. Versatile fluorocarbon can be molded, extruded, and sprayed. January 1972, **43**(1):33.
——. New fluorocarbon plastic makes complex parts with thin sections. October 1971, **42**(15):32.
Riley, M. W. New fluorocarbon plastic will mold readily. *Materials in Design Engineering*, May 1960, **51**(5):9–11.
Robb, L. E. Fluorine-containing polymers. *Western Plastics*, Technical Report No. 68, December 1960, pp. 22–25.
Robertson, A. B. and Miller, W. A. Halar Fluoropolymer–A Versatile Insulation for Wire and Cable. Private Communication.
Simonds, H. R. and Church, J. M. *A Concise Guide to Plastics*. 2nd Ed. Van Nostrand Reinhold Co., New York, 1963.
SPE Journal. Allied Chemical's fluoropolymer: tough yet easy to process. November 1971, **27**(11).
Swanson, R. S. *Plastics Technology–Basic Materials and Processes*, McKnight & McKnight Publishing Co., Bloomington, Illinois, 1965.
Titterton, W. E. Fluorocarbon Resins. *Plastics World Seminar*, Los Angeles, 1975.
Titus, J. B. *Trade Designation of Plastics and Related Materials*, (revised). PLASTEC Note N9C. Plastics Technical Evaluation Center, Picatinny Arsenal, Dover, New Jersey, May 1978.
Towler, D. W. "Chlorotrifluoroethylene (CTFE)." In *Modern Plastics Encyclopedia*, S. Gross (exec. ed.), **45**(14A):198. McGraw-Hill, Inc., New York, 1968–69.
Voigt, H. B. "Modified ETFE." In *Modern Plastics Encyclopedia*, J. Agranoff (ed.), **52**(10A):22. McGraw-Hill, Inc., New York, 1975–76.
Winding, C. C. and Hiatt, G. D. *Polymeric Materials*. McGraw-Hill Book Co., New York, 1961.
Zapp, J. A. Jr. *The Anatomy of a Rumor*. Du Pont Bulletin A-26020.

VINYLS

Most versatile of the thermoplastics, vinyl polymers are also among the oldest. Laboratory polymerizations date back to 1872. First patents leading to commercial production were issued in the late 1920s. The principal members of the vinyl family are the homopolymers and copolymers of polyvinyl chloride (PVC).

The wide variety of PVC resin types available, combined with the fact that each can be modified with a multitude of compounding ingredients, results in an extremely broad spectrum of applications. Finished products are produced by every known plastics processing technique. This compounding and processing versatility is complemented by an excellent cost-performance ratio, inherent chemical resistance, and self-extinguishing properties. Table 5-58 com-

Table 5-58. Cost-Index Comparisons.*

	Cost Index–Cubic Inch Compared to PVC
ABS materials	
Low impact	86
Medium impact	92
High impact	99
High impact	98
SAN materials	
Colors	68
Styrene molding materials	
General purpose	39
Medium impact	38
Medium impact	43
Medium impact	46
Medium impact	50
High impact	49
High impact	55
High impact	69
Super high impact	75
Acrylic	141
Acetal	286
Cellulose acetate	144
Cellulose aceto-butyrate	158
Cellulose propionate	175
Nylon	278
Polyethylene, linear	54
Polycarbonate	263
Polypropylene	72
Phenolic	
G.P., black	62
Impact, black	83
Styrene-methacrylate	98
Rubber-modified acrylic	113
Vinyl, rigid	100
Metals	
Aluminum, die-cast	150
Brass, yellow	430
Magnesium, die-cast	120
Steel, CR	200
Steel, stainless	950
Zinc, die-cast	230

*Courtesy of Monsanto Co., which is no longer responsible for accuracy or content herein.

pares the cost index of vinyls to that of other plastics and metals.

Polyvinyl Chloride

Vinyl Chloride Monomer (VCM). Two commercial feedstocks are used in the production of VCM: ethylene or acetylene. Approximately 19% of the total U.S. chlorine production is for VCM manufacture.

VCM is made from acetylene in the reaction shown below.

$$HC{\equiv}CH + HCl \xrightarrow[90-140^\circ C]{HgCl_2 \text{ on charcoal}} CH_2{=}CHCl$$

The gas-phase reaction of acetylene with HCl uses mercuric chloride or other heavy metal halides as a catalyst. It is important that the gas streams be dry and free of arsine, phosphine, or sulfur. In a typical process, dry HCl is passed into a vapor blender in equimolar proportions with dry acetylene. Precautions are taken to prevent the presence of chlorine which will react explosively under these conditions. The gas mixture passes to a multitubular reactor containing the catalyst.

The product gases are cooled and stripped of excess HCl with a caustic wash; water thus produced is removed with a methanol wash. The final purification is carried out by low-temperature fractionation of aldehydes and ethylene dichloride formed from side reactions.

Because of economic considerations, most of the recent VCM production has shifted to ethylene as the base precursor material. This process utilizes an oxychlorination of ethylene. The reaction can be described by a three-step sequence.

$$(1)\ CH_2{=}CH_2 + Cl_2 \xrightarrow[30-50^\circ C]{FeCl_2} \underset{\text{ethylene dichloride (EDC)}}{ClCH_2{-}CH_2Cl}$$

$$(2)\ 2HCl \text{ (from 3 below)} + CH_2{=}CH_2 + \tfrac{1}{2}O_2 \longrightarrow ClCH_2CH_2Cl + H_2O$$

$$(3)\ ClCH_2CH_2Cl \text{ (from 1 \& 2)} \xrightarrow[\text{pumice or hot tube}]{480-510^\circ C} CH_2{=}CHCl + HCl$$

Reaction 1 is exothermic and must be carried out at low temperatures to prevent formation of tri- and tetra-chloroethanes. It is possible to use ethylene-ethane streams instead of ultrapure ethylene. Vinyl chloride is recovered from step 3 by distillation. This process is favored because, along with the lower cost of ethylene, a mixed feedstock can be used and the HCl formed in reaction 3 can be used in reaction 2. This process, with its variations involving fixed and fluidized catalyst beds and various heating and separation methods, currently dominates commercial production with 93% of the VCM made in the U.S. produced by this route.

In the early 1970s, evidence was generated that a rare form of liver cancer, as well as symptoms of "vinyl chloride disease," was linked to exposure to VCM vapors. As a consequence, plant procedures have been drastically modified to prevent such exposure. These modifications have included extensive stripping of monomer from polymer after polymerization reactions to below measurable limits to prevent possible hazard from the use of the polymer as well. The use of VCM as an aerosol propellant has also been discontinued.

Polymerization. Although vinyl chloride can be polymerized by all of the standard methods, the suspension and emulsion processes are the most important. About 78% of the U.S. and about 50% of the European processes are based on suspension techniques, with the remainder based on emulsion methods.

Suspension Polymerization. This method involves the free radical-catalyzed polymerization of discrete droplets of VCM suspended by agitation and a protective colloid such as gelatin, methyl cellulose, polyvinyl alcohol, or other synthetic gums in water. The suspending agent is dissolved in deionized and deaerated water and added to a glass-lined or stainless steel reactor. The remainder of the water, catalyst, and other additives is then added. Liquified monomer is pumped in, and the closed system is then heated to approximately 50°C (122°F) and raised to a pressure of about 125 psi.

Upon completion of the reaction, the excess monomer is removed by venting and the polymer slurry is cooled, centrifuged, and dried. Typical conversions are 92-93%. The molecular weight is controlled by the reaction temperature and the reaction rate is controlled by initiator (catalyst) type and concentration.

Emulsion Polymerization. The emulsion process involves the use of surface-active agents or soaps as emulsifiers and water-soluble catalysts (as opposed to monomer-soluble catalysts used in the suspension process.) Although agitation is necessary, it is not as important as in the suspension process, since the emulsion is stabilized by the soaps and protective colloids.

At the conclusion of the reaction, unreacted monomer is recovered by vacuum and recycled. A stabilizer like sodium carbonate is then added to the latex, and the latex is spray-dried. Particles from the emulsion process are about 1 micron in diameter, roughly $\frac{1}{100}$ the size of those produced in the suspension method.

Types of Resins. There are five basic types of PVC resins: suspension, dispersion, copolymer, blending, and solution resins.

Suspension, also called general-purpose, resins are by far the most common type of polymers. They are used in the bulk of PVC processes (i.e., extrusion, calendering, and molding).

Because of the ultrafine particle size, the dispersion (emulsion) resins are designed to be dispersed in plasticizer to form a continuous liquid phase called plastisols (see below for further information). They are considerably more expensive than general-purpose resins.

Copolymer resins, principally with vinyl acetate, are made via conventional suspension or solution processes.

Blending resins, made via the suspension technique, are intermediate in particle size between dispersion and general-purpose resins. They are utilized as partial replacements of emulsion resins in plastisol formulations in order to control

Table 5-59. Effect of Blending/Dispersion Resin Ratio on Viscosity.*

Composition					
Dispersion resin	70	60	50	40	30
Blending resin	30	40	50	60	70
Plasticizer	45	45	45	45	45
Stabilizer	4	4	4	4	4
Initial Brookfield viscosity					
2.5/5, cp	13,000	11,000	7,500	5,500	5,000
20/5, cp	11,000	9,000	6,000	4,500	4,000
Settling after 1 wk	None	None	None	Slight	Excessive

*Skiest, 1970. Reprinted by permission from *Modern Plastics*, McGraw-Hill, Inc.

such properties as viscosity (see Table 5-59), gel temperature, air release, and gloss. Figure 5-67 illustrates the procedure used to prepare dry blending resin compositions.

Solution resins are used in solution-coating applications; thus, particle properties and processing characteristics are relatively unimportant compared to the other four types of resins. Copolymers are more frequently made because of enhanced solubility in typical coating solvents. They are typically made via solution polymerization to avoid detrimental contaminants normally present as a result of the suspension and emulsion polymerization methods.

Characterization and Plasticization of PVC. Commercial PVC polymers are largely amorphous, slightly branched molecules. Individual grades of material do differ in molecular weight, molecular weight distribution, particle size, shape, and size distribution. Level of impurities also varies. X-ray studies indicate that commercial PVC has about 5% crystallinity, but laboratory techniques have increased the

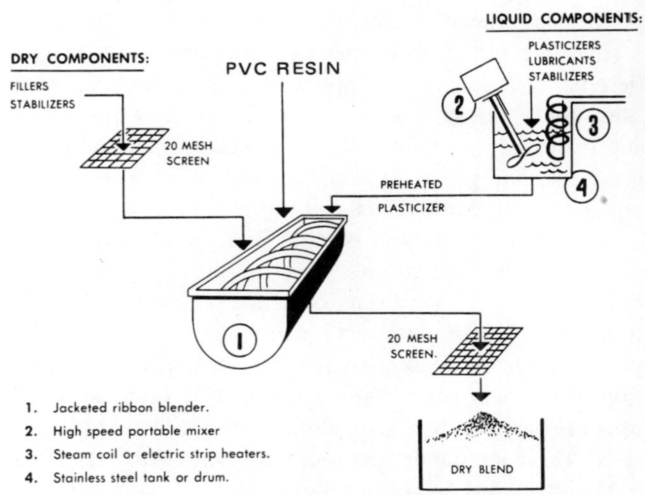

Fig. 5-67. Schematic diagram illustrating the procedure for the manufacture of dry blends in a ribbon blender. (*Courtesy: Monsanto Co.*, which is no longer responsible for accuracy or content herein.)

percentage crystallinity to as high as 50%. The presence of the chlorine renders the polymer flame-retardant. The chlorine atoms also cause an increase in the interchain attraction, which produces greater polymer hardness and stiffness compared to polyethylene. The C—Cl dipole also makes PVC more polar; thus, it has a relatively high dielectric constant and power factor.

PVC is a hard, horny, brittle polymer at room temperature and, as such, would have little utility were it not for its solubility characteristics with a class of high-boiling solvents called *plasticizers*. The polymer normally has a very limited solubility; the only solvents that are effective are those which are capable of interaction with the polymer. It has been suggested that PVC is a weak proton donor and, thus, effective solvents are weak proton acceptors. There are many materials that are suitable plasticizers for PVC. They have similar solubility parameters* and are also weak proton acceptors. These have too high a molecular weight and are too large molecularly to dissolve the polymer outright, but they are compatible and can be blended with polymers to give stable mixtures at room temperature.

Plasticized PVC products possess a broad spectrum of properties which can be varied depending on amount and types of plasticizers used. Progressive addition of plasticizers causes changes in tensile strength, elongation, and impact resistance in the direction that would be expected from a progressive softening of the composition. Actually, at a level of 10-15%, plasticization causes an increase in the degree of polymer crystallinity, which increases the cohesive forces between the chains. This is often referred to as the "antiplasticizer effect" and is dependent on the type of plasticizer used. Above this "critical" level, however, interchain forces are weakened and typical plasticization or softening action takes place. Figure 5-68 shows the effect of plasticizer content on a number of physical properties. So-called rigid PVC, the hardest of commercial types, is plasticized, generally below the critical plasticizer level.

There are other advantages to the incorporation of plasticizers, including lower processing temperatures, better flow under pressure, easier solubility in coating solvents, and increased wettability of inert fillers. Typical plasticizers include tricresyl phosphate, dibutyl phthalate, dioctyl phthalate, polymeric esters, and nitrile rubber.

PVC has a tendency to liberate HCl, particularly at temperatures approaching the upper limit of serviceability. Since no means have been found for preventing this reaction, stabilization with acid neutralizers is necessary to preclude any resultant polymer degradation (i.e., degradation of the polymer *by* the generated HCl). Standard stabilizers are based on barium, cadmium, lead, calcium, and zinc salts. Tin compounds are used for clear products where FDA restrictions do not prohibit their use.

*For an in-depth discussion of the significance of solubility paramters in polymers, refer to the references for Billmeyer, Flory, and Rodriguez.

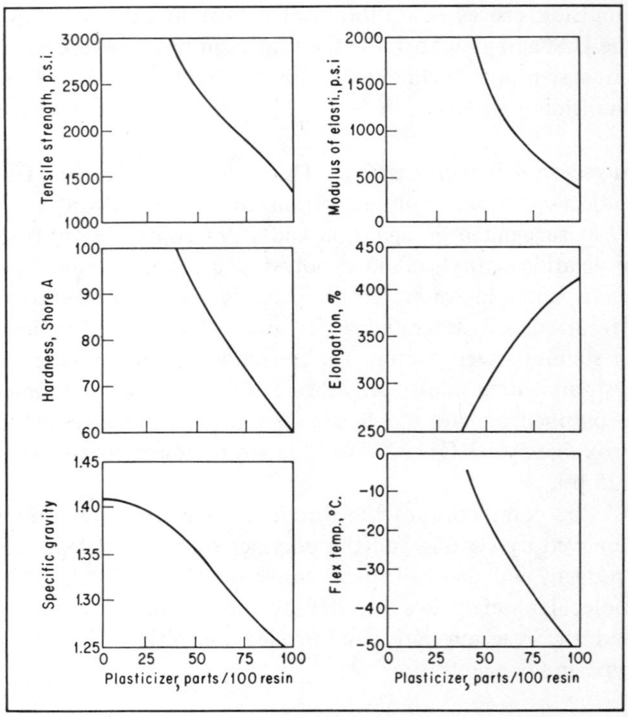

Fig. 5-68. Effect of dioctyl phthalate plasticizer content on properties of a representative PVC compound. (Buckley et al., 1968-69. Reprinted by permission from *Modern Plastics Encyclopedia*, McGraw-Hill, Inc.)

Formulation. Because of unmodified PVC's brittle nature, impact modifiers such as rubbers or copolymers are added to improve the impact strength. In fact, many other ingredients can be, and are, included in typical PVC formulations. These modifications of base PVC, with the exception of molecular weight variations, have not been fully explored because of the infinite variations available through compounding with additives. The use of additives provides a means of producing finished products with literally any desired properties.

Table 5-60 shows a list of typical ingredients in a vinyl compound as well as their function. Figure 5-69 illustrates a means for determining compound hardness and stiffness as a function of filler and plasticizer content.

Plastisols, Organosols, Plastigels. The term "vinyl paste" is a description of vinyl resin formulations in a fluid state. Pastes are highly viscous mixtures of finely divided resin *dispersed* in a liquid medium. These liquidlike pastes may be fluxed and homogenized by heat so that on cooling, a single solid phase results. When the liquid medium is plasticizer alone, the paste is called a *plastisol*. If volatile organic liquids are added as thinners, the composition is termed an *organosol*. Sometimes it is desirable to thicken the plastisol so that heavy coatings will not run off the coated article. These materials are called *plastigels*.

The plastisols and organosols are the most important of

Table 5-60. PVC Compound Ingredients.

Additive	Function	Typical Examples
1. Stabilizers	Prevent HCl- or UV-caused degradation.	See Table 5-61.
2. Plasticizers	Impart flexibility.	See Table 5-62.
3. Impact modifiers	Impart toughness.	Nitrile rubbers, ABS resins, chlorinated polyethylene.
4. Extenders and resinous modifiers	Reduce cost, viscosity, and brittleness.	See Table 5-63.
5. Lubricants and processing aids	Improve flow and processing characteristics.	SAN copolymers, PMMA, metallic stearates.
6. Fillers	Reduce cost, improve physical or thermal properties.	See Table 5-64.
7. Pigments and dyes	Impart color.	Iron oxides, Danisidine reds, ultramarines, and phthalocyanines.
8. Miscellaneous	Surfactants, thixotropic agents, bubble-release agents, fungicides, foam-blowing agents, antistatic agents, etc.	Pyrogenic silicas, silicone oils, azodicarbonamide.

the fluid types of vinyls. A plastisol is a dispersion of a special form of PVC with plasticizer as the dispersing medium. The principal distinguishing characteristic of the resin used is its particle size, which is small enough to permit it to stay in suspension without settling (approximately 1 micron as compared to perhaps 100 microns for general-purpose-grade PVC). Another key characteristic is that the resin particle resists solvation at room temperature. Thus, it is possible to suspend the resin particles in liquid plasticizer and have the mixture remain in a stable fluid condition indefinitely.

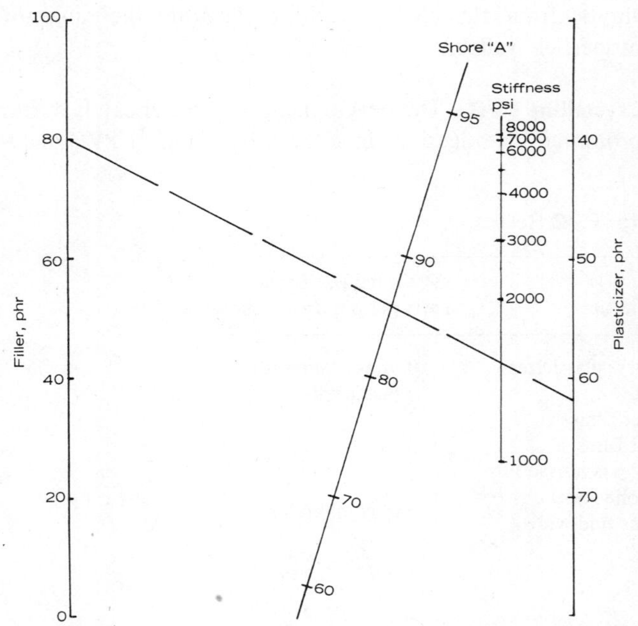

Fig. 5-69. Nomograph to determine stiffness and hardness versus filler and plasticizer content at room temperature. (Frissell, 1962)

Being in a liquid or paste condition, it is possible to shape the plastic with low pressure and, therefore, inexpensive equipment. This would also be true of solutions, but plastisols have the significant advantage of being 100% solid or nonvolatile. Therefore, they can be formed in thick sections without encountering the problems of bubbling or blistering when they are being converted to the solid state. As a result of these features, the plastisols can be cast, spread-coated, coated onto fibers, slush-molded, rotationally cast, molded, or extruded. They can also be sprayed.

The organosols are a variation involving the partial dilution of the plastisol with a mixture of solvent and non-solvent. It is a way of lowering viscosity but, since this does introduce some volatile material, applications are limited to those where the volatile matter can be evaporated conveniently. This is still better, however, than using solutions where the solids content is very low (5-30%), and toxicity, flammability, cost, and handling of the solvents may be detrimental to efficient use. Organosols typically have 75-95% solids concentration.

The formulating of plastisols involves the same principles of compounding as described for general-purpose resins, but some special problems are created. The principal problem is the fact that the liquid plasticizer is now serving two purposes. It affects fluidity as well as the mechanical properties of the fused product. Making a hard vinyl product requires a low ratio of plasticizer to resin, but this combination will lead to a very viscous, difficult-to-handle plastisol. There are several solutions to such a problem; one is the use of resin blends which, through the packing effect of different particle sizes, permits better use of the available liquid. Another solution is to replace some of the normal plasticizers with one which is capable of homopolymerizing, such as diallyl phthalate (see Chapter 6). By this means, one has a liquid component during the fluid period of processing which becomes hard but does not act as a plasticizer after fusion. A third answer, in some cases at least, is the use of an organosol. The solvent acts as a viscosity reducer before fusion, but flashes off and leaves behind a more rigid product after processing.

We have mentioned that the conversion from liquid to solid occurs through the application of heat. This conversion takes place in two steps; the first is gelation which occurs at approximately 65.5°C (150°F) and converts the plastisol to a cheesy structure which will not flow again. As the temperature goes up, the polymer molecules increase their motion and become less entangled and compact. As they spread out, a phase inversion occurs–solid resin particles dispersed in plasticizer, to plasticizer dispersed in a homogeneous resin matrix. (Figure 5-70 shows a schematic of this phenomenon.) Further heating to a temperature of 127-177°C (260-350°F), depending on the exact formulation, will lead to fusion and development of optimum strength.

An unpigmented plastisol is milky white in appearance because it is a dispersion. In the gelled state, it is still white and opaque; however, when it is properly fused it becomes transparent. This characteristic of gelling at a relatively low temperature makes possible several of the major processing methods; for example, hot-dipping and slush-molding depend upon the ability to coat a controllable amount of material onto a form. In most commercial procedures using plastisols, however, one rarely sees the gel state, as the temperature sequences it passes through usually cause it to go directly to the fused and final state.

Plastisols find a variety of end uses. They are ideally suited for spread-coating on continuous webs of fabrics or paper by conventional knife or roller-coating equipment. The subsequent fusion may be done in a tunnel dryer with zones of increasing temperature. Multiple coats may be applied with intermediate fusion. Patterns may be formed by passing the heated coating over an embossing roll.

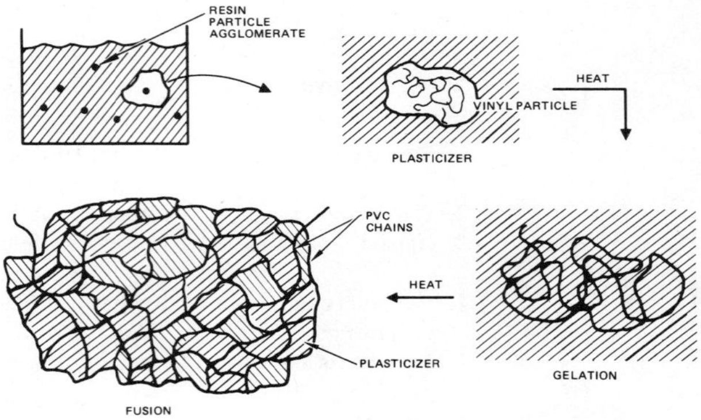

Fig. 5-70. Schematic of plastisol gelation and fusion.

Sheet metal is primed, coated, and cured to secure coatings 0.002–0.020 inch in thickness. The coatings are exceptionally resistant to scuffing, wear, and abrasion. The metal may be formed, bent, punched, or drawn without affecting adhesion. Figure 5-71 describes some applications of plastisols. Table 5-65 compares plastisol and organosol coating characteristics with other polymeric coatings.

Properties of PVC. Table 5-66 shows the properties of typical unmodified PVC. Because of the wide variety of modifications, the properties of any one given PVC formula will vary widely. The effect of plasticizers has already been discussed, and the effects of copolymerization, blending, and base polymer modification are discussed individually. Table 5-67 describes the properties of typical commercial vinyls, from flexible to rigid, to illustrate the range of properties available.

Crystalline PVC. The development of Ziegler catalysts has promoted investigations into the production of PVC mate-

Table 5-61. Stabilizers for PVC Resins.

Stabilizer	Special Properties	Normal Use Level Parts per hundred resin
Epoxy plasticizers	Heat and light stability; plasticizer; pigment dispersant	10 or as desired for plasticizing
Dibutyl tin maleate	Light stability in clear films	3
Tin mercaptide derivatives	Heat stability in clear films	3
Organic barium salts Organic cadmium salts Organic zinc salts	Provide heat stability when used in combination with one another; Ba-Cd combinations find wide general use	3 in combination
Substituted benzophenone UV screeners	Light stability	1
Inorganic and organic phosphites	Light stability	3

Table 5-62. Commonly Used Plasticizers.

- Phthalates
  - Butyl benzyl phthalate (BBP)
  - Butyl octyl phthalate (BOP)
  - Di(2-ethyl hexyl) phthalate (DOP)
  - Di-isodecyl phthalate (DIDP)
  - Di-tridecyl phthalate (DTDP)
  - Di(*n*-hexyl, *n*-octyl, *n*-decyl) phthalate (linear)
- Phosphates
  - Tricresyl phosphate (TCP)
  - Tri(2-ethyl hexyl) phosphate (TOF)
- Trimellitates
  - Tri-octyl trimellitate (TOTM)
  - Tri-iso-octyl trimellitate (TIOTM)
- Aliphatic Diesters
  - Dioctyl adipate (DOA)
  - *n*-Octyl decyl adipate
  - Dibutyl sebacate
- Benzoates
  - Diethylene glycol dibenzoate
  - Dipropylene glycol dibenzoate
- Epoxides
  - Epoxidized soybean oil
  - Octyl epoxy tallate
- Polymerics
  - Low, medium, and high molecular weights

Table 5-63. Resinous Modifiers for PVC Resins.

Resin Type	Probable Effect
Oxidizing medium oil alkyd	Increased solids; film properties essentially unchanged.
Rosin-modified alkyd	Increased solids; film properties essentially unchanged.
Styrenated alkyd	Increased solids; film properties essentially unchanged.
Silicone alkyd	Films have better adhesion; less blocking tendencies, without brittleness.
Amino resins	Films are equal or better in hardness; are non-blocking without brittleness.
Maleic rosin resins Ester gum resins Phenolic resins Coumarone-indene resins Phenol-formaldehyde rosin-modified resins	In general, all five of these resins give harder films and show better adhesion and non-blocking characteristics. However, all develop brittleness.
Petroleum resins WW rosin	Resins give harder films with better adhesion. Films are brittle with no improvement in blocking.
Acrylic resins Modified rosin resins	Selected resins are completely compatible; show some improvement in blocking, hardness or adhesion.

rials with higher crystallinity. Although the Ziegler catalysts themselves cannot be used, low-temperature free-radical polymerization using gamma radiation and alkyl boranes has been successful. Polymers produced at -50°C (-58°F) are about 65% syndiotactic and 20% crystalline. The T_g of

Table 5-64. Fillers for PVC Resins.*

Purpose	Filler
Bulk	Wood flour Sawdust Wood pulp Sisal–jute Purified cellulose Mica–rock
Reinforcement	Glass fibers and spheres Asbestos fibers Cellulosic fibers Cotton fabric Paper Synthetic fibers
Hardness	Inorganic pigments Mineral powders Metallic oxides Powdered metals Graphite Silica
Thermal Insulation	Asbestos Diatomaceous earths Ceramic oxides Silica
Chemical Resistance	Glass fibers and fabrics Synthetic fibers and fabrics Graphite Metallic oxides
Appearance	Color pigments Dyestuffs Carbon blacks Powdered metals Phosphorescent minerals Woven fabrics

*Schoengood, 1973.

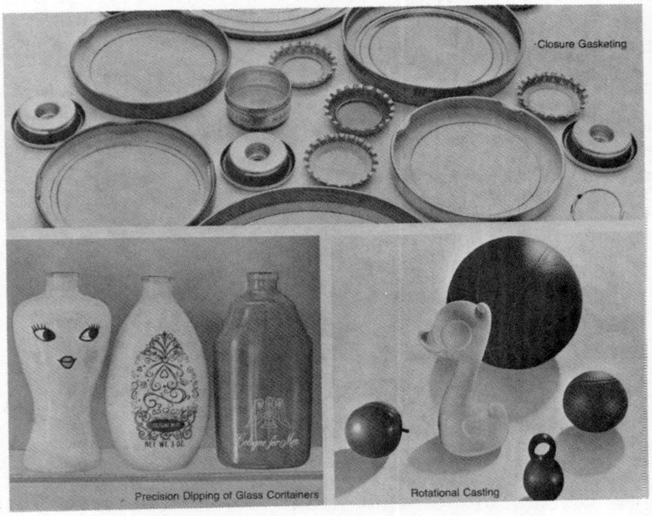

Fig. 5-71. Various applications of vinyl plastisols. (*Courtesy: Whittaker Corporation, Coatings and Chemicals Group*)

these polymers increases as the degree of polymerization decreases. These crystalline polymers do not dissolve in tetrahydrofuran or cyclohexanone at 25°C (77°F), indicating that the PVC is too weak a proton donor to overcome the increased crystallization. Crystalline PVC has greater tensile strength and creep resistance, but is more brittle than conventional PVC. Ethylene-vinyl acetate copolymers are among the more useful impact modifiers for these materials. Plasticized versions may also be prepared, although blending must occur at higher temperatures.

Polyvinyl Chloride Alloys. Alloying is tailored to optimize the hot-melt flow as well as the solid-state particle-phase relationships achieved when cooling profiles, sheet, or molded parts. The toughest, highest-temperature-service plastics are also tough in the melt stage and therefore difficult to mold. Molding temperatures may well exceed the ability of the thus thinned-out plastic to resolidify without loss of properties. Such elevated temperatures and injection pressures can cause chemical side reactions, hydrolysis, chain scission, and oxidation.

Alloys based upon PVC fill the gap between the 82°C (180°F) and 104°C (220°F) heat distortions, with toughness and service ratings from 80-105°C (176-221°F). This eases the molding problem in a large band of service requirements that encompasses flame-retardancy, electrical resistance, and moderate temperature elevations. Table 5-68 compares the properties of three alloy families, two of which are based on PVC.

Other Vinyl Polymers: Chlorinated PVC (CPVC). Post-chlorinating PVC was attempted in order to obtain polymers soluble in low-cost solvents for fiber and lacquer applications. The process involves passing chlorine gas through a PVC/tetrachloroethane solution between 50-100°C (122-212°F). The materials produced, however, have poor color and light stability, are brittle under shock, and have a comparatively low softening point.

By dropping the chlorination temperature, an improved product was developed in the mid-1960s. The reaction proceeds photochemically in an aqueous dispersion in the presence of chloroform, which acts as a swelling agent. At low temperatures and in the presence of excess chlorine, the halogen adds to carbon atoms not already having an attached chlorine atom. Chemically speaking, the material effectively resembles polyvinyl chloride co dichloroethylene.

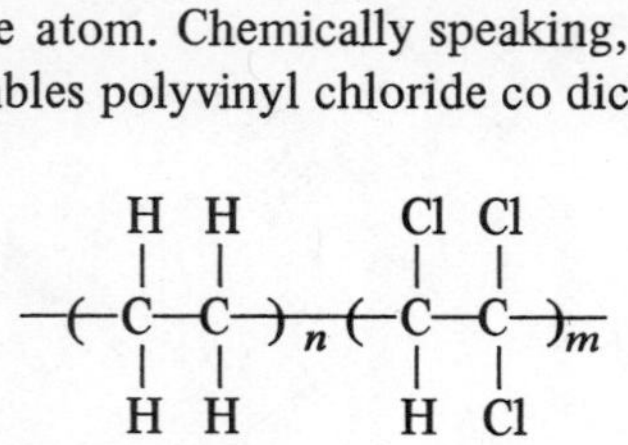

As a result of post-chlorination, melt viscosity and T_g of the polymer increase. The chlorine content is increased from 57% for typical PVC to 66-67%. The resultant polymer has a service temperature 4-16°C (40-60°F) greater than standard PVC. It retains the good chemical, self-extinguishing, and electrical characteristics of PVC. Table 5-69 lists typical properties of CPVC.

This polymer has found many uses in the plumbing and piping trade, particularly for carrying hot, aqueous-based liquids.

Polyvinyl acetate (PVAc). As a moldable plastic, PVAc has found little commercial use because of its appreciable cold flow. The monomer, vinyl acetate, is the major copolymerizing material for PVC. The polymer has a high degree of adhesion to a wide variety of substrates, so it is mainly used as an adhesive, coating base, and textile finish. It is also used for the production of polyvinyl alcohol and polyvinyl acetals.

VAc monomer is made from the reaction of acetylene and acetic acid or oxidation of ethylene. The former is now almost exclusively prepared in the vapor phase. A mixture of the reactants, including predistilled monomer, is passed through a heated tube at 210-250°C (410-482°F), using cadmium acetate, zinc acetate, and zinc silicate as catalysts. The monomer is stabilized with hydroquinone and stored.

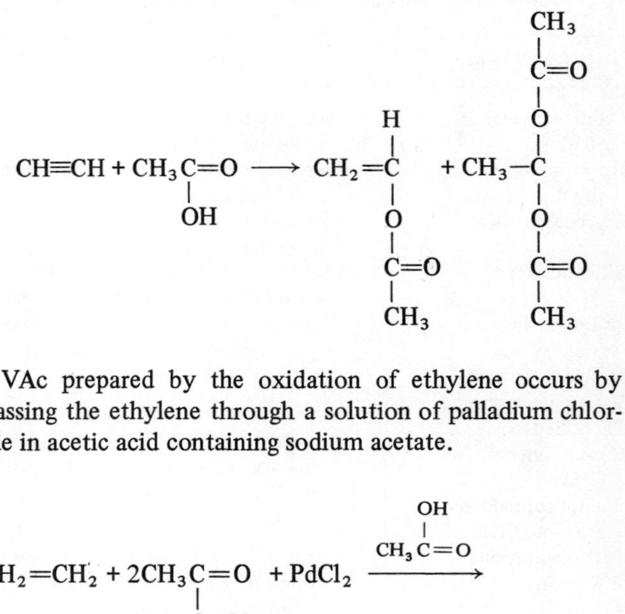

VAc prepared by the oxidation of ethylene occurs by passing the ethylene through a solution of palladium chloride in acetic acid containing sodium acetate.

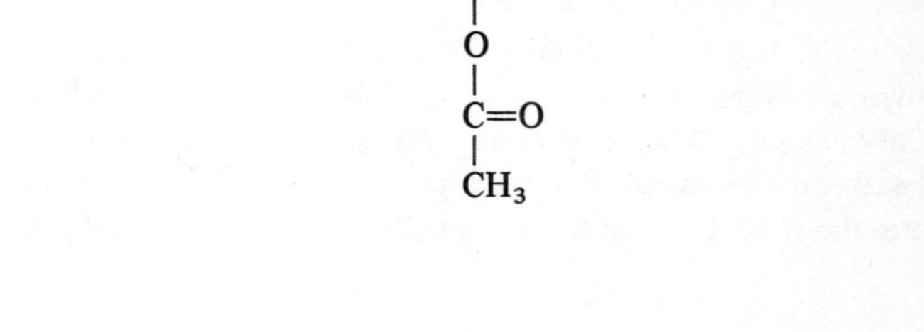

Vinyl acetate is produced, along with lesser amounts of ethylene diacetate and acetaldehyde. A vapor-phase reaction is preferred over liquid phase because it avoids problems of solvent corrosion and handling.

Although all four of the basic polymerization methods can be used, emulsion polymerization is the favored route, primarily because the bulk of polymer made is used in its latex form. In a typical system, equal quantities of VAc and water are stirred together with suitable emulsifying additives (e.g., PVA and sodium lauryl sulphate) and a water-soluble initiator (e.g., potassium persulphate).

Polymerization occurs in about 4 hours at 70°C (158°F). Cooling is critical because of the high reaction exotherm. For better reaction and particle-size control, a portion of the monomer is first polymerized and the remainder is then added gradually over 3–4 hours. Buffers such as sodium acetate are used to control pH, which allows minimization of vinyl acetate or comonomers.

Suspension polymerization is carried out in water using an oil-soluble initiator, such as benzoyl peroxide, and a dispersing agent, such as partially hydrolyzed PVAc derived from a cellulose base. Particle sizes are very large and can be varied by controlling the amount of dispersing agent, the monomer/water ratio, agitation, and polymerization rate. This process is particularly adaptable for PVAc that is to be further hydrolyzed to PVA.

In a typical solution polymerization process, 60% VAc and 40% benzene along with benzoyl peroxide is refluxed until the reaction is complete. The solvent and residual monomer are removed by distillation.

In all three processes, homopolymers and copolymers are made. Homopolymers stick well to porous or cellulosic surfaces such as wood, paper, cloth, leather, and ceramics. Homopolymer films tend to creep less than copolymer or terpolymer films. They are especially suitable in adhesives for high-speed packaging operations.

Copolymers wet and adhere well to nonporous surfaces such as plastics and metals. They form soft, flexible films, in contrast to the tough, horny films formed by homopolymers, and have a greater resistance to water. As the ratio of

Table 5-65. Coil Coating Finishes Comparative Properties and Performance Chart.

This "Comparative Properties and Performance Chart" is a composite evaluation and rating reference covering the various finishes used by the coil coating industry.

This chart has been prepared for general informational purposes only. *It is not a standard or a specification in any sense of those words.* It describes only average coatings of the generic type listed and is intended to serve only as a guide for selection of coatings systems suitable for end product requirements. Among coatings in each generic category there are, of course, varieties that are both much better and much worse than the ratings given here. This chart has been prepared to serve only as a cross index to illustrate the general comparative characteristics of the coatings listed. If used as such, it should prove to be a valuable tool to the coil coater.

As a further aid in the proper use of this chart, the instructions that were given to the committee that rated the various coatings for the properties listed are attached. It is suggested that any person using this chart first read these instructions. For example, the instructions point out that generally the ratings here are for one-coat systems on properly treated aluminum. Obviously, changing to a two-coat system or changing the substrate could conceivably substantially change the ratings given here for the various coatings. We therefore strongly urge you to consider the purpose of this chart and to use it as intelligently as possible for informational purposes.

In all cases, final recommendation of a finishing system should be obtained from the coil coater and/or the finishes supplier.

ADDENDUM: INSTRUCTIONS FOR RATINGS

The following guides should be used for rating the various coatings for the properties listed in the "Comparative Properties and Performance Chart."

1. Each type coating should be rated as a general class for each property. Example: Vinyl coatings provide flexible films, therefore they would rate high in forming properties, impact resistance, embossing, etc. Epoxy type coatings chalk rapidly and therefore would not be recommended for exterior durability, particularly in colors and tints, etc.

2. The rating for each coating should be in terms of performance as a coil finish and not in terms of maintenance finishes, appliance finishes or other uses involving the same generic type of coating.

3. A fair rating must be made for each type of coating for each property. To do this, it will be necessary to compare the performance of one type coating against another. You will find a more valid appraisal can be made if you occasionally check back and review the rating you gave to another class of coating.

4. For rating purposes, assume all the various coatings are pigmented for the best exterior durability.

5. For rating purposes, assume all the various coatings are over the same substate, namely, properly treated aluminum.

6. For rating purposes, assume all the various coatings are applied at equal film thickness (1.0 mil dry film). Exceptions to this would be the plastisols and vinyl laminates which may be applied at heavy films for embossing purposes.

7. The following rating key should be used to rate all properties:

#1 – Excellent
#2 – Good
#3 – Fair
#4 – Poor

Exceptions to the above are as follows:

Minimum Temperature: Indicate temperature. Example: 200°F., 300°F., etc.

Rating Key
#1 Excellent
#2 Good
#3 Fair
#4 Poor

Table 5-65. Coil Coating Finishes: Comparative Properties and Performance Chart.

GENERIC COATING TYPE	Ease of Application	Average Cure Rate (Based on 60 seconds)	Minimum Metal Temperature for Curing Efficiency	Color and Gloss Retention (Double time in oven)	Thermoplastic	Thermosetting	Film Hardness	Film Adhesion	Film Flexibility	Color and Gloss Retention at Elevated Temp. (on end product)	Color Retention on Aging Film (Storage)	Mar-Resistance (Fingernail Test)	Ability to achieve high gloss (above 85 units 60°)	Ability to Fabricate after Aging	Humidity Resistance
Amine-Alkyd	1	2	340	3	no	yes	2	2	3	2	2	2	1	4	2
Polyester (oil-free)[1]	2	2	400	1	no	yes	1	1	2	2	2	2	1	1	1
Vinyl – Alkyd	1	2	340	3	yes	no	2	2	2	2	2	2	2	2	1
Solution Vinyl	1	1	300	3	yes	no	2	1	1	3	2	2	2	1	1
Organosol[3, 2]	2	1	350	3	yes	no	2	1	1	3	2	3	3	1	1
Plastisol[2]	2	1	350	3	yes	no	3	2	1	3	2	3	4	1	1
Straight Epoxy	2	2	400	3	no	yes	1	1	2	3	3	1	1	3	1
Epoxy-Ester	2	2	400	3	no	yes	2	1	2	3	3	2	2	2	1
Thermoset Acrylic[1]	1	2	430	1	no	yes	2	1	2	2	1	1	1	2	1
Silicone Alkyd[1]	2	3	420	2	no	yes	2	2	3	2	1	2	1	3	2
Silicone Acrylic[1]	2	3	450	1	no	yes	2	2	3	1	1	2	1	3	2
Silicone Polyester[1]	2	3	450	1	no	yes	2	2	2	2	1	2	1	2	1
Poly-Vinyl Fluoride[2]	2	3	450	1	yes	no	2	2	1	2	1	3	4	1	1
Poly-Vinylidene Fluoride[2]	2	3	450	2	yes	no	2	2	1	2	1	3	4	1	1
Poly-Vinyl Fluoride Film Laminate	2	1	adhesive 300+	NA	yes	no	2	2	1	2	1		4	1	1
Poly-Vinyl Chloride Film Laminate	2	1	adhesive 390-405	NA	yes	no	2	1	1	3	2	2	4	1	1
Acrylic Film Laminate	2	1	adhesive 390-405	NA	yes	no	1	1	1	1	1	2	2	1	1

[1] Primers are commonly used over hot-dip galvanized steel, also possibly over cold-rolled steel. Ratings are based on primed systems when these are commonly used in commercial practice.
[2] Primer is required for adhesion.
[3] Normally applied at heavier film thicknesses.
[4] Generally applied at 6 or more mils dry film thickness.
NA = not applicable.
*Courtesy: National Coil Coaters Association.

Rating Key
#1 Excellent
#2 Good
#3 Fair
#4 Poor

(Revised by the Coatings Classification and Definitions Committee, October 1971.)*

Grease and Oil Resistance	Solvent Resistance (Aliphatic Hydrocarbon)	Solvent Resistance (Aromatic Hydrocarbon)	Solvent Resistance (Ketone or Oxygenated)	General Chemical Resistance – Acids, Alkali etc. (Spot Tests)	General Corrosion Resistance (Industrial Atmospheres)	Exterior Durability (Pigmented)	Exterior Durability (Clear Films)	Fade Resistance in Color (Properly Pigmented)	Unit Finishing Cost Range	Adaptability to Embossing (with metal)	Embossing (Film only)	Metal Marking Resistance	Resistance to Pressure Mottling in Coil	Abrasion Resistance	Stain Resistance (Household agents and foodstuffs)	Salt Spray Resistance[1]
2	2	2	3	3	2	2	3	2	L	4	NA	2	2	3	3	3
2	1	1	2	2	2	2	3	2	M	2	NA	2	2	2	2	1
2	2	3	3	3	2	3	3	3	M	3	NA	2	2	2	3	2
1	2	4	4	2	2	2	3	3	MH	1	NA	2	2	2	2	1
1	1	3	4	1	1	2	3	2	L	1	NA	3	2	1	2	1
1	1	3	4	1	1	2	3	2	L	1	1	3	3	1	2	1
1	1	1	1	1	1	4	4	4	H	4	NA	1	1	2	1	1
2	1	1	2	2	2	4	4	4	MH	3	NA	2	2	2	2	1
2	1	2	2	2	2	2	2	2	M	3	NA	2	2	2	2	1
2	2	2	2	2	2	1	2	1	H	4	NA	2	2	2	2	2
2	2	2	2	2	2	1	2	1	H	4	NA	2	2	2	2	2
2	1	1	2	2	2	1	2	1	H	3	NA	2	2	2	2	1
1	1	1	1	1	1	1	1	1	VH	1	NA	3	2	1	1	1
1	1	1	1	1	1	1	1	1	VH	1	NA	3	2	1	1	1
1	1	1	1	1	1	1	1	1	VH	1	1[4]	3	2	1	1	1
1	1	3	4	1	1	2	4	2	H	1	1[4]	3	3	1	2	1
1	1	3	4	1	1	1	1	1	H	1	1[4]	1	2	1	2	1

UNIT FINISHING COST RANGE

Coating Material Cost, ¢ per sq. ft. at 1-mil dry film,

L = Low, 0.65 or less
M = Medium, 0.65 to 0.85
MH = Medium High, 0.85 to 1.10
H = High, 1.10 to 2.50
VH = Very high, 2.20 and over

Table 5-66. Typical Properties of Polyvinyl Chloride.*

Property	Value
Tensile strength, lb/in^2	6500
Elongation at break,%	25
Tensile modulus, lb/in^2	6–12 × 10^5
Shore hardness	D75–85
Heat capacity, 25°C, cal/deg/g	0.226
Coefficient of thermal expansion, deg^{-1}	
(glass)	5.2 × 10^{-4}
(rubber)	2.1 × 10^{-4}
Glass transition temperature, °C	70–80
Refractive index	1.54

*Koleske and Wartman, 1969. Gordon & Breach Science Publishers, Inc., copyright 1969. Reprinted by permission.

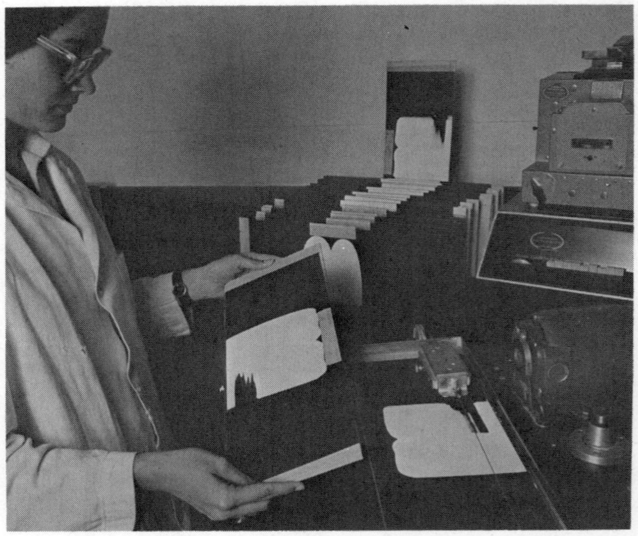

Fig. 5-72. A laboratory scrub test which is used to determine the scrubability characteristics of emulsion coatings. (*Courtesy: Air Products & Chemicals, Inc., Allentown, Pa.*)

comonomer to vinyl acetate increases, the variety of plastics to which the copolymer will adhere also increases. Comonomers containing functional groups will often adhere to specific surfaces; for example, carboxy-containing polymers adhere well to metals.

By varying the conditions for polymerization, either straight-chain or crosslinked polymer emulsions can be obtained. In order to effect crosslinking, polyvinyl alcohol must be added to the emulsion. It is the PVA that is crosslinked. Typical crosslinking agents include urea-formaldehydes, trimethylol and methylated-trimethylol melamines, glyoxal, and glutaraldehyde. These agents often need acidic catalysts such as ammonium sulphate or chloride to reduce the pH and activate the reaction.

Crosslinked polymer emulsions accept water-miscible solvents better than straight-chain emulsions. In the film state, they resist water and organic solvents better and tend to have higher heat-sealing temperatures, since they can have high molecular weights which contribute to such high heat-sealing temperatures.

The stability of PVAc emulsions allows them to accept many types of modifying additives without being damaged. Solvents, plasticizers, tackifiers, and fillers can be directly added to homopolymer and copolymer emulsions without having to be pre-emulsified, an impossible feat with elastomeric adhesive latices. Homopolymer emulsions containing partially hydrolyzed PVA as a protective colloid can accept greater amounts of these modifying additives without coagulating than any other type of emulsion.

The stability of the emulsions further permits them to be compounded in liquid-blending vessels, using simple agitators such as marine-type propellers, paddles, or turbines. Elaborate preblending techniques are unnecessary. Figure 5-72 shows a common test being performed to evaluate a PVAc emulsion adhesive.

PVAc polymers are solids which vary from gums to leather-like materials, depending on the molecular weight. Table 5-70 lists typical properties of the polymer.

The polymer is soluble in acetic acid, benzyl alcohol, most organic solvents, and swells in water. Films must be baked for complete solvent removal. The softening temperature of PVAc varies with molecular weight. The T_g of PVAc is only 29°C (84.2°F) which explains its cold-flow character and its lack of usefulness as a molding material.

The largest use of PVAc is as a dry resin for paint. Latices are low cost, light in color, have good stability, grease and oil resistance, good adhesion to a variety of substrates, and relatively high permeability to moisture vapor, so that trapped moisture can escape without creating blisters or peeling. Emulsion copolymers with dibutyl fumarate have low T_g's, are softer, have lower tensile strength and increased flexibility compared to homopolymer films.

In the textile industry, PVAc is used as a cotton sizing and a binder for nonwoven fabrics, the copolymers being superior as the fabric binder. PVAc emulsions are more colorstable and odorfree than SBR latices. PVAc is used in special inks for treating cotton and rayon and for permanent starching of clothing.

PVAc is used in binders for pigmented paper coatings, clay, wood flour, and cellulose pulp as well as an extender for chewing gum and natural rubbers.

Polyvinyl alcohol (PVA). Vinyl alcohol monomer is unstable and cannot be prepared in the free state. All efforts to prepare it result in its tautomer, acetaldehyde.

$$\underset{\displaystyle OH}{CH_2{=}\underset{|}{CH}} \rightleftharpoons \underset{\displaystyle H}{CH_3\underset{|}{C}{=}O}$$

Table 5-67. Properties of Typical Commercial Vinyls.*

		ASTM test method	Vinyl chloride and vinyl chloride-acetate molding compounds, sheets, rods, and tubes		
			Rigid [c]	Flexible unfilled	Flexible filled
Processing	1. Molding qualities	—	Fair to good	Good	Good
	2. Compression molding temp., °F.	—	285-400	285-350	285-350
	3. Comp. molding pressure, p.s.i.	—	750-2000	500-2000	500-2000
	4. Injection molding temp., °F.	—	300-415	320-385	320-385
	5. Injection molding pressure, p.s.i.	—	10000-40000	8000-25000	8000-25000
	6. Compression ratio	—	2.0-2.3	2.0-2.3	2.0-2.3
	7. Mold (linear) shrinkage, in./in.	—	0.001-0.005	0.010-0.050 (Varies with plasticizer)	0.008-0.035
	8. Specific gravity (density)	D792	1.30-1.58	1.16-1.35	1.3-1.7
	9. Specific volume, cu. in./lb.	D792	20.5-19.1	23.8-20.5	21.3-16.3
	10. Machining qualities	—	Excellent	—	—
Mechanical	11. Tensile strength, p.s.i.	D638	6000-7500	1500-3500	1000-3500
	12. Elongation, %	D638	40-80	200.0-450.0	200.0-400.0
	13. Tensile elastic modulus, 10^5 p.s.i.	D638	3.5-6.0	—	—
	14. Compressive strength, p.s.i.	D695	8000-13000	900-1700	1000-1800
	15. Flexural yield strength, p.s.i.	D790	10000-16000	—	—
	16. Impact strength, ft. lb./in. of notch (½ × ½ in. notched bar, Izod test)	D256	0.4-20.0	Varies depending on type and amount of plasticizer	
	17. Hardness, Rockwell	D785	65-85 (Shore D)	50-100 (Shore A)	50-100 (Shore A)
	18. Flexural elastic modulus, p.s.i. × 10^5	D790	3-5	—	—
	19. Compressive modulus, p.s.i. × 10^5	D695	—	—	—
Thermal	20. Thermal conductivity, 10^{-4} cal./sec./sq. cm./1 (°C./cm.)	C177	3.5-5.0	3.0-4.0	3.0-4.0
	21. Specific heat, cal./°C./gm.	—	0.25-0.35	0.3-0.5	0.3-0.5
	22. Thermal expansion, 10^{-6} in./in./°C.	D696	5.0-10.0	7.0-25.0	—
	23. Resistance to heat, °F. (continuous)	—	130-175	150-175	150-175
	24. Deflection temp., °F. at 264 p.s.i. fiber stress	D648	140-170	—	—
	at 66 p.s.i. fiber stress		135-180	—	—
Electrical	25. Volume resistivity, ohm-cm. (50% RH and 23° C.)	D257	$>10^{16}$	10^{11}-10^{15}	10^{11}-10^{4}
	26. Dielectric strength, short-time, ⅛-in. thickness, volts/mil	D149	350-500	300-400	250-300
	27. Dielectric strength, step-by-step, ⅛-in. thickness, volts/mil	D149	—	275-290	225-750
	28. Dielectric constant, 60 cyc.	D150	3.2-4.0	5.0-9.0	5.0-6.0
	29. Dielectric constant, 10^3 cyc.	D150	3.0-3.8	4.0-8.0	4.0-5.0
	30. Dielectric constant, 10^6 cyc.	D150	2.8-3.1	3.3-4.5	3.5-4.5
	31. Dissipation (power) factor, 60 cyc.	D150	0.007-0.020	0.08-0.15	0.10-0.15
	32. Dissipation (power) factor, 10^3 cyc.	D150	0.009-0.017	0.07-0.16	0.09-0.16
	33. Dissipation (power) factor, 10^6 cyc.	D150	0.006-0.019	0.04-0.14	0.09-0.10
	34. Arc resistance, sec.	D495	60-80	—	—
Optical	35. Refractive index, n_D	D542	1.52-1.55	—	—
	36. Clarity	—	Translucent, opaque	Transparent, translucent and opaque	
	37. Transmittance, %	—	76-82	—	—
	38. Haze, %	—	8-18	—	—
Resistance characteristics	39. Water absorp., 24 hr., ⅛-in. thick., %	D570	0.04-0.4	0.15-0.75	0.50-1.0
	40. Flammability [a] Burning rate, in./min.	D635	—	—	—
	AEB, in./ATB, sec. [b]		<0.16 (sheet)	—	—
	41. Effect of sunlight	—	Varies with formulation	Varies with stabilizer	Slight
	42. Effect of weak acids	D543	None	None	None
	43. Effect of strong acids	D543	None to slight	None to slight	None to slight
	44. Effect of weak alkalies	D543	None	None	None
	45. Effect of strong alkalies	D543	None	None	None
	46. Effect of organic solvents	D543	Resists alcohols, aliphatic hydrocarbons, oils; soluble or swells in ketones and esters; swells in aromatic hydrocarbons		

[a] This numerical flame spread rating is not intended to reflect hazards presented by this or any other material under actual fire conditions.
[b] AEB = Avg. extent of burning in inches; ATB = Avg. time of burning in seconds.
[c] The data for rigid vinyl sheet should be regarded as a guide only since the materials produced by the various manufacturers differ considerably from the charts and from each other.
* Reprinted by permission from *Modern Plastics Encyclopedia*, McGraw-Hill, Inc.

Table 5-68. Comparison of Three Alloy Families.*

	Heat deflect. temp (F)	Impact strength (ft-lb, notched)	Flexural modulus (psi)	Tensile strength (psi)	¢/in.^{3}
ABS/PVC					
Abtec Abson 89129 (I, E)	198	6	3.2×10^5	6000	2.93
Borg-Warner KAB (E)	158	12.5	3.0×10^5	5300	3.03
Borg-Warner KAF (I)	150	10.5	3.4×10^5	5800	2.88
Shulman Polyman 509(I)	210	2	4.3×10^5	6000	3.52
PVC/Acrylic					
Rohm & Haas Kydex (S)	165	10-15	3.3×10^5	6300	5.85
General Tire & Rubber Acrylivin (S)	151	>15	–	7000	5.10
DuPont DKE-450 (E)	189	12	3.8×10^5	6800	3.41
ABS/PC					
Borg-Warner KHS (E)	220	13	4×10^5	8000	5.37
Mobay Bayblend MC 2500 (I, E)	240	10.5	3.5×10^5	8500	3.59
Mobay Bayblend MD 6500 (I, E)	220	10.5	3.8×10^5	8500	4.29

I = Injection resin E = Extrusion resin S = Sheet stock

*Fountas, 1977.

PVA is therefore prepared by alcoholysis of PVAc.

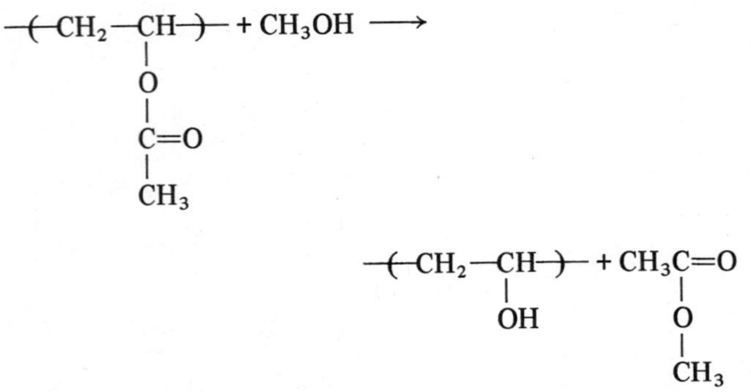

Although PVAc is the most commonly used precursor, PVA can also be produced from other vinyl esters and ethers. The esters are polymerized by Ziegler catalysts, and the ethers by polar catalysts such as boron trifluoride. PVA's formed from ethers vary widely in tacticity.

It is common, although incorrect, to use the term "hydrolysis" for the PVA alcoholysis. In fact, water does not react readily to yield PVA's and, in some situations—depending on the nature of the catalyst present—many actually retard the reaction.

Either methanol or ethanol is used as the alcoholizing agent, but methanol is generally preferred because its room-temperature miscibility with PVAc is better, and the generated product has better color. Methyl acetate is often added as a second solvent and is also formed during reaction. Acidic or basic catalysts are employed. Alkaline catalysts (e.g., caustic soda or sodium methoxide) provide more rapid reaction, and an increase of concentration will decrease residual acetate content. This is a convenient method of controlling the degree of alcoholysis. Acid-catalysis methods are less acceptable when incompletely "hydrolyzed" products are desired, because of lack of reproduceable control.

Commercial PVA is available in various grades which differ in molecular weight and residual acetate content. Alcoholysis causes scission of branched polymers at the points where branching has proceeded via the acetate group; therefore, PVA polymers have lower molecular weights than the PVAc from which they were made. Figure 5-73 shows the effect of molecular weight on tensile strength.

PVAc is atactic and amorphous. PVA is also atactic; however, it exhibits the same type of crystal lattice as polyethylene. It is believed that the hydroxyl groups are small enough to fit into the lattice without disrupting it, thus allowing the polymer to have a high degree of crystallinity.

One of the most significant effects of the hydroxyl groups on PVA is that the polymer is hydrophilic and will dissolve in water. The degree of solution is dependent on the degree of "hydrolysis" and temperature. An increase in the degree of "hydrolysis" will reduce solubility. This anomalous effect is due to the greater extent of hydrogen bonding in completely "hydrolyzed" polymers. Hydrogen bonding also causes unplasticized PVA to decompose below its flow tem-

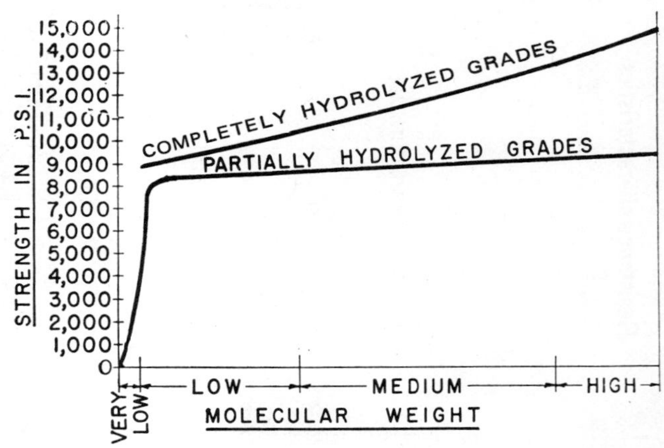

Fig. 5-73. Tensile strength of polyvinyl alcohols at 50% relative humidity. Completely hydrolyzed grades of polyvinyl alcohols show generally higher tensile strengths than partially hydrolyzed grades and also vary more with changes in molecular weight. (Grandpre, 1959)

Table 5-69. Typical Properties of CPVC.*

		ASTM test method	Chlorinated polyvinyl chloride compound
Processing	1. Molding qualities	—	Good
	2. Compression molding temp., °F.	—	350-400
	3. Comp. molding pressure, p.s.i.	—	1500-2000
	4. Injection molding temp., °F.	—	425-440
	5. Injection molding pressure, p.s.i.	—	15000-40000
	6. Compression ratio	—	2.0-2.5
	7. Mold (linear) shrinkage, in./in.	—	0.003-0.007
	8. Specific gravity (density)	D792	1.49-1.58
	9. Specific volume, cu. in./lb.	D792	18.4-17.8
	10. Machining qualities	—	Excellent
Mechanical	11. Tensile strength, p.s.i.	D638	7500-9000
	12. Elongation, %	D638	4.5-65.0
	13. Tensile elastic modulus, 10^5 p.s.i.	D638	3.60-4.75
	14. Compressive strength, p.s.i.	D695	9000-22000
	15. Flexural yield strength, p.s.i.	D790	14500-17000
	16. Impact strength, ft. lb./in. of notch (½ × ½ in. notched bar, Izod test)	D256	1.0-5.6
	17. Hardness, Rockwell	D785	R117-122
	18. Flexural elastic modulus, p.s.i. × 10^5	D790	3.8-4.5
	19. Compressive modulus, p.s.i. × 10^5	D695	3.35-6.00
Thermal	20. Thermal conductivity, 10^{-4} cal./sec./sq. cm./1 (°C./cm.)	C177	3.3
	21. Specific heat, cal./°C./gm.	—	0.33
	22. Thermal expansion, 10^{-5} in./in./°C.	D696	6.8-7.6
	23. Resistance to heat, °F. (continuous)	—	230
	24. Deflection temp., °F. at 264 p.s.i. fiber stress	D648	202-234
	at 66 p.s.i. fiber stress		215-247
Electrical	25. Volume resistivity, ohm-cm. (50% RH and 23° C.)	D257	10^{15}
	26. Dielectric strength, short-time, ⅛-in. thickness, volts/mil	D149	1220-1500
	27. Dielectric strength, step-by-step, ⅛-in. thickness, volts/mil	D149	—
	28. Dielectric constant, 60 cyc.	D150	3.08
	29. Dielectric constant, 10^3 cyc.	D150	2.8-3.6
	30. Dielectric constant, 10^6 cyc.	D150	3.2-3.6
	31. Dissipation (power) factor, 60 cyc.	D150	0.01887-0.02080
	32. Dissipation (power) factor, 10^3 cyc.	D150	0.0092-0.0108
	33. Dissipation (power) factor, 10^6 cyc.	D150	0.020
	34. Arc resistance, sec.	D495	—
Optical	35. Refractive index, n_D	D542	—
	36. Clarity	—	—
	37. Transmittance, %	—	—
	38. Haze, %	—	—
Resistance characteristics	39. Water absorp., 24 hr., ⅛-in. thick., %	D570	0.02-0.15
	40. Flammability[a] Burning rate, in./min.	D635	—
	AEB, in./ATB, sec.[b]		—
	41. Effect of sunlight	—	Slight
	42. Effect of weak acids	D543	None
	43. Effect of strong acids	D543	None
	44. Effect of weak alkalies	D543	None
	45. Effect of strong alkalies	D543	None
	46. Effect of organic solvents	D543	Resists alcohols, aliphatic hydrocarbons, oils; soluble or swells in ketones and esters; swells in aromatic hydrocarbons

[a] This numerical flame spread rating is not intended to reflect hazards presented by this or any other material under actual fire conditions.

[b] AEB = Avg. extent of burning in inches; ATB = Avg. time of burning in seconds.

*Reprinted by permission from *Modern Plastics Encyclopedia*, McGraw-Hill, Inc.

Table 5-70. Physical Properties of Polyvinyl Acetate.

Density, g/ml	
at 20°C	1.19
at 60°C	1.16
at 120°C	1.11
at 200°C	1.05
Water absorption (ASTM 48-hr test), %	2
Transition temp. (second-order), °C	30–40
Tensile strength, psi	5000
Thermal coefficient of linear expansion per °C	8.6×10^{-5}
Thermal conductivity, cal/(sec) (cm^2) (C/cm)	38×10^{-5}
Specific heat	0.389
Heat distortion point, °C	50
Burning rate	Very slow
Dielectric strength, V/mil	
at 30°C	1000
at 60°C	780
Dipole moment, Debye units	1.85
Electrical properties	

	Frequency	Dielectric constant	Power factor
at 30°C	60 cps	3.00	0.01–0.02
at 60°C	60 cps	6.00	0.07
at 20°C	1 kcps	3.15	0.025
	100 kcps	3.09	0.028
	1 Mcps	3.06	0.03
	10 Mcps	3.02	0.025

perature while providing a very tough film with a high tensile strength.

Tables 5-71 through 5-73 list the properties of PVA polymers. The most outstanding characteristics of PVA are resistance to hydrocarbons and animal and vegetable oils,

Table 5-71. Properties of Polyvinyl Alcohol.

General Properties	
Specific gravity	1.21–1.31
Specific volume (cu in./lb)	22.9–21.1
Color range	unlimited
Mechanical Properties	
Tensile strength (psi)	1000–5000
Elongation (%)	300–600
Shore hardness	10–100
Heat Properties	
Thermal coefficient of linear expansion (°C)	$(7\text{–}12)10^{-5}$
Burning rate	slow
Stability	
Water absorption (%)	30 and more
Effect of sunlight	none
Molding Conditions	
Temperature, compression	230–300°F (121–149°C)
Pressure, compression (psi)	500–1500
Chemical Resistance	
Resistant to organic solvents; swelled by or dissolves in weak acids, and weak and strong alkalis; attacked by strong acids.	

Table 5-72. Physical Properties of Extruded and Molded Polyvinyl Alcohol.

	Extruded PVA	Molded PVA
Specific gravity	1.26	1.26
Tensile strength (lb/sq. in.)	5200	2100
Elongation at break	210	440
Permanent set (%)	78	80
Cold flow (%)	63	63
Dielectric strength (volts/mil)	610	1070
Electrical conductivity (ohm/cm)	3.1×10^7	3.8×10^7

Table 5-73. Physical Properties of Hydrolyzed PVA.

Property	Fully Hydrolyzed	88% Hydrolyzed
Specific gravity	1.2–1.30	1.2–1.3
Refraction Index nD	1.50–1.53	1.50–1.53
Maximum % insoluble	0.25	0.25
pH, 4% aqueous solution	6.80	5.5–7.5
Tensile strength, dry	up to 22,000 psi	up to 22,000 psi
Heat-seal temp., dry	160–210°C	160–210°C

extremely low permeability to organic liquids and gases, abrasion resistance, toughness, and resistance to UV. These properties greatly depend on humidity; the higher the humidity, the more water absorbed. Since water acts as a plasticizer, tensile strength will decrease with a corresponding increase in elongation and tear strength.

The high polarity of PVA contributes to its chemical-resistance properties. The polymer is very resistant to hydrocarbons, will dissolve in water/ethanol mixtures, but not in pure alcohols, and will dissolve in highly polar organic solvents. The hydroxyl group is very reactive and derivatives are easily prepared. Plasticization using polar liquids capable of forming hydrogen bonds with the hydroxyl groups is possible. Glycerin is a common example.

Plasticized PVA can be compression molded or extruded. Tubing is used in lubricating equipment, for paint lacquers, dry-cleaning solvents, fire-extinguisher fluids, and refrigerants. In such applications a high degree of water resistance is not needed, but high resistance to organic solvents is essential.

PVA fibers are formed by extrusion of aqueous gels, where water acts as the plasticizer of the atactic polymer. The fiber, when drawn, retains strength and integrity at temperatures over 100°C (212°F). Fiber strength increases with extent of drawing. Fibers treated with 0.1 g/dl of $CuSO_4$ can withstand corrosion even after exposure to 180°C (356°F) for 4 hours. Special PVA formulations for fibers used in reinforced-plastics composites are not water soluble. However, the moisture pick-up of the fiber itself is 4–5%, so that a polyester premix compound using this fiber will have a similar moisture pick-up of 5%.

In general, these fibers have excellent toughness; good impact strength, abrasion resistance, and flow properties; produce smooth molded surfaces without the need for low-profile polyester resins; have one-half the density of glass fiber; good weatherability; and no hair-cracking. On the other hand, flash removal, low high-temperature modulus, broad softening range, and variable fiber lengths are inherent disadvantages.

PVA film is transparent, strong, and tear resistant and is used in bags and protective laboratory clothing. The film can be extruded using the same techniques employed for fiber, or it can be produced by allowing polymer solutions to evaporate on flat metal or glass surfaces from which the film is peeled off. Cold-water film made from highly plasticized linear PVA can be used for packaging materials. The packages can be stored at room temperature and then added to water, whereupon the film dissolves directly, dispensing its contents in premeasured portions. Typical uses include laundry detergents, paint fungicides, and medications.

Other principal markets for PVA include textile applications, as suspending agents and emulsifiers for polymerization processes, paper processing, and plastic film. PVA acts as a powerful adhesive after it has been dissolved in water and deposited on a substrate. It adheres well to hydrophilic surfaces and particularly well to cellulosic surfaces.

Either partially or completely "hydrolyzed" PVA can be used as a base for adhesives, and can be combined either with clay or with starch and clay extenders. However, PVA is primarily used as an additive for PVAc emulsion adhesives and as a base for paper substrates, after it has been modified to have high wet-track. Tackified PVA adhesives are preferred because of higher wet-tack at low viscosities, more rapid setting-speed under pressure, and lower adhesive cost per unit of glue-line measurement. A series of PVA resins with built-in tack have been produced by a controlled boration process. This process, which is analogous to that of adding borax and caustic to starch adhesives, causes aqueous solutions to develop strong wet-tack after cooking.

Polyvinyl Acetals. Reaction of aldehydes with PVA produces polyvinyl acetals. These resins are amorphous, with rigidity and softening point dependent on the nature of the aldehyde. Since the reaction between the hydroxyl groups and aldehyde is random, some hydroxyl groups become isolated and do not react. Thus, the final polymer composition will contain acetal groups, residual hydroxyl groups, and residual acetate groups from the incomplete alcoholysis of the PVAc to PVA.

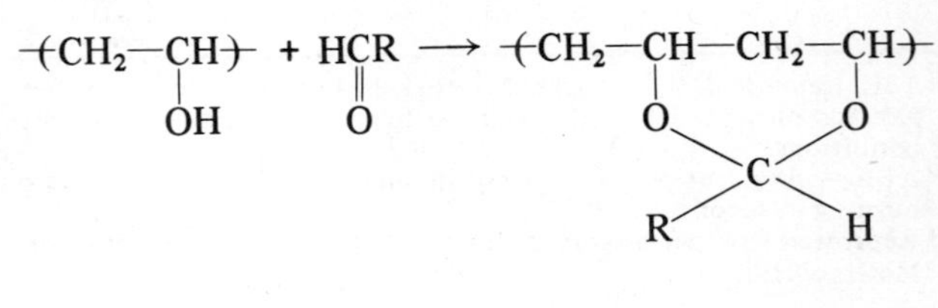

The acetals can be made from PVA or directly from PVAc without separating the alcohol.

Polyvinyl acetal from the acetaldehyde reaction has very little commercial importance. Although moldable, its properties do not warrant its use. It has been occasionally used with nitrocellulose in lacquers, wash primers, and fabric stiffeners. Table 5-74 shows properties of the two primary commercial types of polyvinyl acetal.

The manufacture of polyvinyl formal is via the direct process. PVAc is added to a mixture of acetic acid and water which has been heated to 70°C (158°F) with stirring. The catalyst (H_2SO_4) and a 40% formaldehyde solution is added and allowed to react for 24 hours at 70°C. Water is then added with rapid agitation, precipitating the granules of polymer which are then washed free of acid and dried. Grades of resin vary in hydroxyl and residual acetate content and degree of polymerization.

The hydroxyl content of commercial grades is low. The molecular weight has little effect on mechanical properties, but does influence the flow temperature. The variation in residual acetate content will affect the heat-deflection temperature, impact strength, and water absorption. The resins can be plasticized with the normal effect on properties. The polymer will be soluble in solvents of similar solubility parameters, but resists bases well and is attacked by acids.

The largest commercial use for polyvinyl formal is as a wire enamel, usually in conjunction with a phenolic resin. Polymers with low hydroxyl and acetate content are common here. Similar resins, also with phenolics, are used as structural adhesives. Some grades have been used as can coatings and wash primers. As with the acetals, molding grades have achieved no commercial significance.

By far the largest use of polyvinyl butyral is as the interlayer in safety glass, and the production of this polymer is based on its use in this application. A very pure product is required, and this is most easily prepared from PVA rather than directly from PVAc. In this process, PVA (fully "hydrolyzed") is suspended in ethanol. The catalyst (H_2SO_4) and butyraldehyde are added, and the reaction is complete after 5-6 hours at 80°C (176°F). The resulting solution is diluted with methanol, and the polymer is precipitated by the addition of water with vigorous stirring. The resin is then stabilized, washed, and dried.

For use in safety glass, the resins are plasticized (typically with dibutyl sebacate or triethylene glycol-di-2-ethyl butyrate). They are then calendered to a very thin film (approximately 15 mils) and coated to prevent blocking. To produce safety glass, the film is washed and dried and placed between two pieces of glass which are then laminated together in a heated press. Bulletproof glass is made by increasing the number of glass/resin layers in the laminate.

Polyvinyl butyral is used in safety glass because of its high adhesion to glass, toughness, light stability, clarity, and insensitivity to moisture. Other miscellaneous uses for this polymer include textile and metal coatings and in adhesive formulations.

Polyvinylidene Chloride and Copolymers. Vinylidene chloride, like vinyl chloride, has been known since the early 1800s. Successful polymers, particularly in fiber form, were introduced in the late 1930s.

There are several processes for the manufacture of vinylidene chloride monomer, the two most used commercially being the dehydrochlorination of trichloroethylene and the reaction of chlorine and acetylene.

In the former process, trichloroethylene is treated with a lime slurry or caustic soda at 90°C (194°F), the gaseous product being condensed and fractionated.

$$ClCH_2CHCl_2 \longrightarrow CH_2{=}CCl_2 + HCl$$

Pyrolysis of the trichloroethylene at 400°C (752°F) will also produce monomer.

Acetylene reacts violently with chlorine to form acetylene tetrachloride. Therefore, the reaction must take place in the presence of large amounts of inert material or must be moderated by passing the gases into antimony pentachloride. Additional acetylene and chlorine are then reacted in the presence of the tetrachloride and a ferric chloride catalyst. The product gases are condensed to recover the acetylene tetrachloride and vinylidene chloride.

$$CH{\equiv}CH + 2Cl_2 \longrightarrow Cl_2\overset{H}{C}\overset{H}{C}Cl_2$$

$$Cl_2\overset{H}{C}\overset{H}{C}Cl_2 + CH{\equiv}CH + Cl_2 \longrightarrow Cl_2\overset{H}{C}\overset{H}{C}Cl_2 + CH_2{=}CCl_2$$

The monomer must be stored under a layer of aqueous alkali or nitrogen to prevent oxidation by air which causes polymerization.

Vinylidene chloride polymerizes readily on exposure to light or free radicals, yielding a crystalline polymer with a high softening temperature which is insoluble in most organic solvents. As with PVC, any of the standard polymerization methods can be used.

Industrially, emulsion polymerization is favored, using a redox technique which takes advantage of the polymer's insolubility in the monomer. The softening temperature of the polymer can be high (185-200°C, 365-392°F), depending upon the degree of polymerization. The decomposition temperature ranges from 210°C to 225°C (410-437°F)—so close to the softening temperature that, at processing temperatures, the resins typically evolve HCl. Furthermore, since very few high-boiling liquids are compatible with the polymer, it is very difficult to plasticize.

Table 5-74. Properties of Two Commercial Types of Polyvinyl Acetal.*

		ASTM test method	Vinyl formal molding compound	Vinyl butyral molding compounds: Flexible unfilled
Processing	1. Molding qualities	—	Good	Good
	2. Compression molding temp., °F.	—	300-350	280-320
	3. Comp. molding pressure, p.s.i.	—	1000-10000	100-3000
	4. Injection molding temp., °F.	—	300-400	250-340
	5. Injection molding pressure, p.s.i.	—	10000-30000	15000-30000
	6. Compression ratio	—	—	—
	7. Mold (linear) shrinkage, in./in.	—	0.0015-0.0030	—
	8. Specific gravity (density)	D792	1.2-1.4	1.05
	9. Specific volume, cu. in./lb.	D792	23.0-19.8	26.2
	10. Machining qualities	—	Good to poor	—
Mechanical	11. Tensile strength, p.s.i.	D638	10000-12000	500-3000
	12. Elongation, %	D638	5.0-20.0	150.0-450.0
	13. Tensile elastic modulus, 10^5 p.s.i.	D638	3.5-6.0	—
	14. Compressive strength, p.s.i.	D695	—	—
	15. Flexural yield strength, p.s.i.	D790	17000-18000	Varies depending on type and amount of plasticizer
	16. Impact strength, ft. lb./in. of notch (½ × ½ in. notched bar, Izod test)	D256	0.8-1.4	
	17. Hardness, Rockwell	D785	M85	10-100 (Shore A)
	18. Flexural elastic modulus, p.s.i. × 10^5	D790	—	—
	19. Compressive modulus, p.s.i. × 10^5	D695	—	—
Thermal	20. Thermal conductivity, 10^{-4} cal./sec./sq. cm./1 (°C./cm.)	C177	3.7	—
	21. Specific heat, cal./°C./gm.	—	—	—
	22. Thermal expansion, 10^{-5} in./in./°C.	D696	6.4	—
	23. Resistance to heat, °F. (continuous)	—	120-150	—
	24. Deflection temp., °F. at 264 p.s.i. fiber stress	D648	150-170	—
	at 66 p.s.i. fiber stress		—	—
Electrical	25. Volume resistivity, ohm-cm. (50% RH and 23° C.)	D257	—	5.0×10^{10}
	26. Dielectric strength, short-time, ⅛-in. thickness, volts/mil	D149	490	350
	27. Dielectric strength, step-by-step, ⅛-in. thickness, volts/mil	D149	455	325
	28. Dielectric constant, 60 cyc.	D150	3.7	5.60
	29. Dielectric constant, 10^3 cyc.	D150	3.6	5.10
	30. Dielectric constant, 10^6 cyc.	D150	3.0	3.92
	31. Dissipation (power) factor, 60 cyc.	D150	0.013	0.115
	32. Dissipation (power) factor, 10^3 cyc.	D150	0.019	0.057
	33. Dissipation (power) factor, 10^6 cyc.	D150	0.023	0.061
	34. Arc resistance, sec.	D495	—	—
Optical	35. Refractive index, n_D	D542	1.50	1.47-1.49
	36. Clarity	—	—	—
	37. Transmittance, %	—	—	—
	38. Haze, %	—	—	—
Resistance characteristics	39. Water absorp., 24 hr., ⅛-in. thick., %	D570	0.5-3.0	1.0-2.0
	40. Flammability[a] Burning rate, in./min.	D635	—	—
	AEB, in./ATB, sec.[b]		—	—
	41. Effect of sunlight	—	Slight	Slight
	42. Effect of weak acids	D543	Attacked	Slight
	43. Effect of strong acids	D543	Attacked	Slight
	44. Effect of weak alkalies	D543	Resistant	Slight
	45. Effect of strong alkalies	D543	Resistant	Slight
	46. Effect of organic solvents	D543	Attacked by some	Resists aliphatic hydrocarbons, most oils. Swells in ketones, esters, and aromatic hydrocarbons; dissolves in alcohols.

*Reprinted by permission from *Modern Plastics Encyclopedia*, McGraw-Hill, Inc.

High molecular weight homopolymers are insoluble in almost all solvents. Tris(dimethylamido)phosphate is one of the only solvents which gives solutions that do not gel on cooling to room temperature. For the lower molecular weight resins and some of the copolymers, dioxane and tetrahydrofuran have been found suitable.

Polyvinylidene chloride is a weak, transparent, pliable plastic after fusing and quenching; however, the polymer crystallizes with aging and becomes stronger and more opaque. In copolymers, the tendency is reduced, increasingly so as the vinylidene chloride content decreases. Homopolymers and copolymers with high vinylidene chloride content possess unusually narrow softening ranges. Table 5-75 shows the properties of a typical commercial resin.

Both homo- and copolymers yellow and lose mechanical properties on heat aging. This phenomenom is believed to be caused by the photochemically catalyzed loss of HCl, similar to the process in PVC. Thus, stabilizing additives are required here also.

Copolymerization, as mentioned, reduces crystalline regularity and increases molecular flexibility. Copolymers may thus be processed at temperatures where decomposition rates are less catastrophic.

The name "saran" represents a series of vinylidene chloride copolymers with vinyl chloride or acrylonitrile. The vinyl chloride copolymers are best known. Other copolymers with vinyl acetate, allyl esters, unsaturated ethers, acrylates, methacrylates, and styrene have appeared but have not achieved significant commercial interest.

Included in the properties of saran are general resistance to organic solvents, inertness at room temperature to common acids and alkalis (with the exception of concentrated ammonium hydroxide), complete water resistance, negligible water-vapor transmission, non-burning characteristics, toughness, and durability. The copolymers are attacked by halogens and strong organic amines. Saran plastics have good stability to aging and UV attack.

Softening temperatures range from 70°C to 180°C, (158–356°F), with physical characteristics ranging from soft, flexible products to hard, rigid materials. Typical copolymer compositions have 85% vinylidene chloride, although levels as low as 73% are known.

The copolymer can be extruded and injection molded. It can be compression molded, although the resulting lack of orientation causes a loss in toughness. By stretching under controlled conditions at room temperature, orientation as well as recrystallization occurs, and much stronger products are formed. Both extruded and injection-molded articles are initially quite amorphous, but crystallization increases with aging and, thus, physical properties increase.

Saran film, with its moisture-proofing and vapor-barrier properties, plus its high degree of transparency, has become a very popular packaging material. The typical method of producing this film is through biaxially oriented, blown-film extrusion. Figure 5-74 shows such a process.

Table 5-75. Properties of Vinylidene Chloride Molding Compounds.*

		ASTM test method	Vinylidene chloride molding compounds
Processing	1. Molding qualities	—	Excellent
	2. Compression molding temp., °F.	—	220-350
	3. Comp. molding pressure, p.s.i.	—	250-5000
	4. Injection molding temp., °F.	—	300-400
	5. Injection molding pressure, p.s.i.	—	10000-30000
	6. Compression ratio	—	2.0
	7. Mold (linear) shrinkage, in./in.	—	0.005-0.025
	8. Specific gravity (density)	D792	1.65-1.72
	9. Specific volume, cu. in./lb.	D792	16.8-16.1
	10. Machining qualities	—	Good
Mechanical	11. Tensile strength, p.s.i.	D638	3000-5000
	12. Elongation, %	D638	Up to 250.0
	13. Tensile elastic modulus, 10^5 p.s.i.	D638	0.5-0.8
	14. Compressive strength, p.s.i.	D695	2000-2700
	15. Flexural yield strength, p.s.i.	D790	4200-6200
	16. Impact strength, ft. lb./in. of notch (½ × ½ in. notched bar, Izod test)	D256	0.3-1.0
	17. Hardness, Rockwell	D785	M50-M65
	18. Flexural elastic modulus, p.s.i. × 10^5	D790	—
	19. Compressive modulus, p.s.i. × 10^5	D695	—
Thermal	20. Thermal conductivity, 10^{-4} cal./sec./sq. cm./1 (°C./cm.)	C177	3.0
	21. Specific heat, cal./°C./gm.	—	0.32
	22. Thermal expansion, 10^{-5} in./in./°C.	D696	19.0
	23. Resistance to heat, °F. (continuous)	—	160-200
	24. Deflection temp., °F. at 264 p.s.i. fiber stress	D648	130-150
	at 66 p.s.i. fiber stress		—
Electrical	25. Volume resistivity, ohm-cm. (50% RH and 23° C.)	D257	10^{14}-10^{16}
	26. Dielectric strength, short-time, ⅛-in. thickness, volts/mil	D149	400-600
	27. Dielectric strength, step-by-step, ⅛-in. thickness, volts/mil	D149	400-600
	28. Dielectric constant, 60 cyc.	D150	4.5-6.0
	29. Dielectric constant, 10^3 cyc.	D150	3.5-5.0
	30. Dielectric constant, 10^6 cyc.	D150	3.0-4.0
	31. Dissipation (power) factor, 60 cyc.	D150	0.030-0.045
	32. Dissipation (power) factor, 10^3 cyc.	D150	0.060-0.075
	33. Dissipation (power) factor, 10^6 cyc.	D150	0.05-0.08
	34. Arc resistance, sec.	D495	--
Optical	35. Refractive index, n_D	D542	1.60-1.63
	36. Clarity	—	—
	37. Transmittance, %	—	—
	38. Haze, %	—	—
Resistance characteristics	39. Water absorp., 24 hr., ⅛-in. thick., %	D570	0.1
	40. Flammability[a] Burning rate, in./min.	D635	—
	AEB, in./ATB, sec.[b]		—
	41. Effect of sunlight	—	Slight
	42. Effect of weak acids	D543	None
	43. Effect of strong acids	D543	Resistant
	44. Effect of weak alkalies	D543	Resistant
	45. Effect of strong alkalies	D543	Resistant
	46. Effect of organic solvents	D543	None to slight

* Reprinted by permission from *Modern Plastics Encyclopedia*, McGraw-Hill, Inc.

Fig. 5-74. Stretch-orienting saran film by injection of compressed air into an extruded tube. The flattened tube is then slit and rolled (see upper right background). (Printed with permission of The Dow Chemical Co.)

The film may be creased or folded without appreciable effect on its low rate of moisture-vapor transmission. Its bursting strength/thickness ratio is high. Aging and UV exposure have little effect on properties. It maintains high chemical resistance, toughness, and serviceability to subzero temperatures.

Saran tubing is highly chemical resistant, noncorrosive and non-scaling. It withstands high pressures and freezing, and it is heat resistant to 80°C (176°F). It is tough, yet flexible, and possesses excellent fatigue strength. It is used in laboratories and industrial plants to transport acids and corrosive chemicals. Saran and saran-lined pipe and pipe fittings have been useful for transporting aqueous acids and alkalis normally corrosive to metal. Saran pipe is, however, brittle at low temperatures, with a low heat-deflection temperature which limits its widespread use.

A large part of saran production is for the manufacture of monofilaments. The copolymer is extruded, air-cooled, and stretched for increased orientation. Its strength, aging properties, and resistance to bacterial and insect attack make it very useful in insect screens. The principal use has been in automobile seat covers and bus and restaurant upholstery. Saran has also been made into filter cloths, cordage, and industrial brushes.

Disadvantages of saran fiber include low softening temperature, which precludes its use in ironable garments; and negligible moisture absorption, which makes it unsuitable for "breathable" fabrics such as those used in underwear. Its natural amber cast tends to discolor light hues, preventing its use in pastel and white fabrics.

Because of the difficulties in solvating the copolymers in common solvents, vinylidene chloride copolymer latices are used for coating and impregnation applications. Copolymers with acrylonitrile are more adaptable for coating purposes. The principal commercial outlets for these copolymers are as coatings for polyethylene, cellophane, paper, as well as other materials. These coatings have high moisture and gas impermeability, chemical resistance, clarity, toughness, and heat sealability. The acrylonitrile content is typically in the 5-15% range. Solubility, especially in ketones, increases, whereas impermeability decreases as acrylonitrile content increases.

PVC COPOLYMERS

The main PVC copolymers that have achieved commercial importance are those with vinyl acetate and acrylonitrile. A number of terpolymers have also appeared. Copolymerization of PVC provides polymers with a greater range of physical and mechanical properties, easier processability through lower softening temperatures, improved stability to heat, and greater solubility in a wider variety of organic solvents.

Polyvinyl Chloride co Vinyl Acetate. Blends or alloys of PVC and PVAc exhibit poor compatibility and have relatively few desirable properties. In order to optimize properties of the two resins, copolymerization is necessary.

The polymerization conditions for the copolymer are very similar to those used for pure monomer polymerization. Solution, suspension, and emulsion techniques are preferred. Aryl and acyl peroxides, tetraethyl lead, and UV radiation may be used to initiate the reactions. Variation in the reaction temperature, type and amount of initiator, and type and amount of solvent dictate the average molecular weights of the polymers obtained.

Most commercial vinyl chloride-acetate copolymers are solvent polymerized and contain 85-95% vinyl chloride. Although the polymer is sometimes precipitated with a non-solvent like water, butane or toluene are preferred; in these solvents, the solid phase will form at a desired degree of polymerization. The copolymers are then dried and masticated, and modifiers and heat-stabilizers are added.

Properties of the copolymers are a function of the molecular weight and monomer ratio. In general, the copolymers have excellent strength and good resistance to water, acids, alkalis, alcohols, and oils. They exhibit poor resistance to ammonium hydroxide, ketones, esters, aldehydes, aromatic hydrocarbons, and some organic acids. Dimensional stability is good and so is heat stability, provided stabilizers have been added.

The copolymers require less plasticization than PVC homopolymers because of the flexibility generated by the vinyl acetate present. The amount and type of plasticizer may be varied over a wide range to produce products which can be fabricated into any form, from rigid rod and tube to elastic film and flexible fiber. Compounded resins in pellet and powder form are available for molding, extrusion, and other fabricating techniques.

The lowest molecular weight resins (8,500-9,500) with vinyl chloride contents of 86% are used in surface coatings. These coatings must be baked at 125-200°C (257-392°F) in order to obtain proper adhesion and flow. They are widely used as decorative can and closure coatings and for the coating of aluminum foil, asbestos board, and paper products.

In order to improve compatability and adhesion, terpolymers are often created. For example, the addition of 1% maleic anhydride to an 86:13 VC/VAc system provides better adhesion to metal. A terpolymer obtained by partial hydrolysis of the VAc present yields improved compatibility, adhesion, higher softening temperature, and greater chemical reactivity.

For injection molding, the copolymer molecular weight averages 10,000 and has a vinyl chloride content of 85-90%. For compression molding, a molecular weight of 12,000 with a vinyl chloride content of 85-88% is preferred. Because these molding resins have minimal shrinkage during processing, relatively large moldings can be made, and accurate reproduction of such items as scales and phonograph records is possible. These copolymers have similar molding characteristics, appearance, and mechanical properties to the cellulosics.

For sheet stock used to fabricate dials, instruments, and advertising displays, copolymers in the molecular weight range of 15,000 with 90% vinyl chloride content are suitable. For elastomeric products such as tubing and wire coating, molecular weights of 20,000 with 95% vinyl chloride are typical, although compounding with plasticizers is required.

Other applications include watch crystals, drawing, calculating, and navigating instruments, and flooring tile. A typical flooring composition contains 30-40 parts of plasticizer per 100 parts of copolymer and about 400 parts of an asbestos/chalk filler.

Polyvinyl Chloride co Acrylonitrile. By the 1940s, vinyl chloride/vinyl acetate fibers had achieved some market success. In 1947, fibers based on vinyl chloride-acrylonitrile copolymers were introduced. The copolymer had 56-60% vinyl chloride and was generated via emulsion polymerization.

In 1951, a form of this copolymer, called "dynel" (a generically approved designation) was introduced. The copolymer is dissolved in acetone, filtered and deaerated, and dry-spun into tow. The tow is cut into staple lengths and crimped. The fiber has a tenacity of 0.7-0.9 grams/denier, cannot be cold drawn, but can be oriented at elevated temperatures.

The fiber is not pure white but can be bleached. Exposure to sunlight causes a mild bleaching, and prolonged exposure to elevated temperatures will cause a darkening and loss of tenacity.

Because of its outstanding chemical resistance and self-extinguishing character, the fiber is widely used in the manufacture of protective clothing and filter cloths. Flame-resistant bedding and upholstery and wash-and-wear, non-shrink wearing apparel are other major uses. One unusual application is as doll hair—the hair can be treated like human hair; that is, washed, waved, brushed, tinted, combed, and set.

Polyvinyl Chloride co Vinylidene Chloride. Mention has already been made of high vinylidene chloride-content copolymers. Copolymers with 5-20% vinylidene chloride are available. The primary effect of the vinylidene chloride addition is a reduction of the softening temperature, which falls off in a regular manner with vinylidene chloride content until a concentration of about 60% is reached. Copolymers with low vinylidene chloride contents are used where the reduction of processing and fabrication temperature is an advantage. Also, these copolymers are soluble in inexpensive solvents, allowing their use in solution-coating applications. They are used for molding, extruding, calendering, impregnating, and film casting.

Polyvinyl Chloride co Propylene. These resins were introduced to provide equivalent end-use properties of PVC, plus improvements in stability and flow characteristics. Compositions typically contain 2-10% propylene by weight. The processing ease is attributed to the introduction of stable structures as end groups and, in the polymer chain,

Table 5-76. Properties of Propylene-Vinyl Chloride Copolymers.*

		ASTM test method	Propylene vinyl chloride copolymer — Rigid
Processing	1. Molding qualities	—	Excellent
	2. Compression molding temp., °F.	—	290-360
	3. Comp. molding pressure, p.s.i.	—	100-2000
	4. Injection molding temp., °F.	—	350-400
	5. Injection molding pressure, p.s.i.	—	8000-20000
	6. Compression ratio	—	2.0-2.6
	7. Mold (linear) shrinkage, in./in.	—	0.0025-0.0035 (Inj.)
	8. Specific gravity (density)	D792	1.28-1.40
	9. Specific volume, cu. in./lb.	D792	21.4-18.05
	10. Machining qualities	—	Excellent
Mechanical	11. Tensile strength, p.s.i.	D638	5000-8000
	12. Elongation, %	D638	100.0-140.0
	13. Tensile elastic modulus, 10^5 p.s.i.	D638	3.5-4.5
	14. Compressive strength, p.s.i.	D695	7750-11700
	15. Flexural yield strength, p.s.i.	D790	10000-15400
	16. Impact strength, ft. lb./in. of notch (½ × ½ in. notched bar, Izod test)	D256	0.36-32.00
	17. Hardness, Rockwell	D785	M18-55 R107-119
	18. Flexural elastic modulus, p.s.i. × 10^5	D790	3.52-5.04
	19. Compressive modulus, p.s.i. × 10^5	D695	—
Thermal	20. Thermal conductivity, 10^{-4} cal./sec./sq. cm./1 (°C./cm.)	C177	—
	21. Specific heat, cal./°C./gm.	—	—
	22. Thermal expansion, 10^{-5} in./in./°C.	D696	10-15
	23. Resistance to heat, °F. (continuous)	—	140-175
	24. Deflection temp., °F. at 264 p.s.i. fiber stress	D648	150-170
	at 66 p.s.i. fiber stress	D648	—
Electrical	25. Volume resistivity, ohm-cm. (50% RH and 23° C.)	D257	1×10^{16}
	26. Dielectric strength, short-time, ⅛-in. thickness, volts/mil	D149	—
	27. Dielectric strength, step-by-step, ⅛-in. thickness, volts/mil	D149	335-380
	28. Dielectric constant, 60 cyc.	D150	3.1-3.7
	29. Dielectric constant, 10^3 cyc.	D150	3.1-3.6
	30. Dielectric constant, 10^6 cyc.	D150	2.8-3.3
	31. Dissipation (power) factor, 60 cyc.	D150	0.008-0.010
	32. Dissipation (power) factor, 10^3 cyc.	D150	0.01-0.02
	33. Dissipation (power) factor, 10^6 cyc.	D150	0.015-0.025
	34. Arc resistance, sec.	D495	—
Optical	35. Refractive index, n_D	D542	—
	36. Clarity	—	Transparent-opaque
	37. Transmittance, %	—	84-87
	38. Haze, %	—	6.5-11
Resistance characteristics	39. Water absorp., 24 hr., ⅛-in. thick., %	D570	0.07-0.4
	40. Flammability [a] Burning rate, in./min.	D635	—
	AEB, in./ATB, sec.[b]	D635	0.35-0.55/—
	41. Effect of sunlight	—	Slight to poor depending on grade
	42. Effect of weak acids	D543	None
	43. Effect of strong acids	D543	None to slight
	44. Effect of weak alkalies	D543	None
	45. Effect of strong alkalies	D543	None
	46. Effect of organic solvents	D543	Same as for rigid vinyl chloride

[a] This numerical flame spread rating is not intended to reflect hazards presented by this or any other material under actual fire conditions.

[b] AEB = Avg. extent of burning in inches; ATB = Avg. time of burning in seconds.

*Reprinted by permission from *Modern Plastics Encyclopedia*, McGraw-Hill, Inc.

to increased resistance to thermal degradation and reduced dehydrochlorination.

Formulations precompounded with impact modifiers, stabilizers, and lubricants are available in pellet form for injection molding and extrusion. They are available in colored, transparent, and opaque forms. Transparent grades have 85% light transmission with 5-7% haze. Maximum recommended use temperature is 60°C.

Mechanical properties such as tensile and impact strength and flexural modulus are good. Humid environments do not affect the good insulation properties because of the copolymer's low water absorption.

The compounds offer good chemical resistance to most acids, alkalis, alcohols, aliphatic hydrocarbons, greases and oils. Stress-crack resistance is also good.

An important characteristic of propylene-vinyl chloride copolymers is their ease of molding or extrusion. High thermal stability and low melt viscosity permit fabrication of large-size parts with multiple contours. Table 5-76 lists typical properties.

Applications include film and sheet, profiles, injection and blow moldings. Motor housings, connectors, toys, medical valves and filters, blow-molded bottles, and thermoformed packaging are a few of the successful end uses.

Trade Names

Abbey	Conductive vinyl	Abbey Plastics Corporation
Abson	ABS/PVC	Abtec Chemical Co.
Ace-Flex	PVC	American Hard Rubber Co.
Ace Riviclor	PVC	American Hard Rubber Co.
Acrilan	Acrylonitrile-vinyl acetate fiber	Monsanto Co.
Acrylivin	PVC/acrylic sheet	General Tire & Rubber Co.
Aero-o-buoy	PVC foam	Robinson Moulded Products, Ltd.
Agalan	PVC film	Polytherm Kassel Corporation
Agilide	PVC extrusions	American Agile Corporation
Agriplast	Vinyl linings	Agricultural Plastic Co.
Air-Bond	PVC-PVAc film	Canadian Resins & Chemicals, Ltd.
Airco	Monomers and copolymers	Air Products & Chemicals, Inc.
Alcolec	Vinyl-based resin	Allied Colloids Mfg. Co., Ltd.
Altro	PVC-coated fabrics	Welwyn Plastics, Ltd.
Ambla	PVC-coated fabrics	I.C.I. (Hyde), Ltd.
Amerifilm	Film and sheet	Ross & Roberts, Inc.
Ameriplast	Film and sheet	Ross & Roberts, Inc.
Amer-Plate	Vinyl sheet	Ameron Corrosion Control Division
Ampcoflex	Rigid PVC	Atlas Mineral & Chemicals Division, ESB Inc.
Antibak	PVC	Sington & Co., Ltd.
Anvigl	Rigid PVC extrusion	Anchor Plastics Co., Inc.
Appretan	PVAc/VAc/acrylate terpolymer	Hoechst Corporation
Aquaflex	PVC sheet and laminate	Pantasote Co.
Aquastik	PVAc adhesive	Berry Wiggins & Co., Ltd.
Arbex RY	PVC-alloy sheet	H & F Products Co.
Arkon	Vinyl foam	Arkon Products, Inc.
Arlex	PVC-coated papers	Mellowhile Products, Ltd.
Armacord	PVC-coated papers	Armoride, Ltd.
Arman	Vinyl film and sheet	Ross & Roberts, Inc.
Armodur	Rigid PVC sheet	M & B Plastics, Ltd.
Arnar	Vinyl film and sheeting	Ross & Roberts, Inc.
Astra	Flexible PVC extruded tubing	Suflex Corporation
Astradur	PVC sheet	Dynamit Nobel, A.G.
Astraglas	PVC sheet	Dynamit Nobel, A.G.
Astralit	PVC copolymer	Dynamit Nobel, A.G.
Astralon	PVC copolymer	Dynamit Nobel, A.G.
Astraprint	PVC copolymer sheet	Dynamit Nobel, A.G.
Astratherm	PVC calendered film	Dynamit Nobel, A.G.
Astrologs	PVC profiles	Minter Homes Corporation
Atlastovon	Plasticized PVC	Atlas Minerals & Chemical Division, ESB Inc.
Aurora	PVC sheet	Giovanni Crespi, S.p.A.
Bakelite	PVC, PVAc	Union Carbide Corporation
Bal-tad	PVAc adhesive	Building Adhesives, Ltd.
Barr-Hyde	Expanded vinyl	Barr Corporation
Barr-Skin	Expanded vinyl	Barr Corporation
Benvic	PVC compound	Solvay & Cie, S.p.A.
Betta Bondu	PVAc adhesive	Associated Adhesives, Ltd.

Bevaloid	PVAc adhesives	Richard Hodgson & Sons, Ltd.
Bexan	PVC monofilaments	Bakelite Xylonite, Ltd.
Blacar	PVC and compounds	Cary Chemicals, Inc.
Blavin	Vinyl extrusion compounds	Reichhold Chemicals, Inc.
Blistopex	Thermoforming PVC	Apex Coated Fabrics Co.
Boadhex Plate	PVC sheet	Nippon Zeon
Boltaflex	PVC sheet	General Tire & Rubber Co.
Boltaron	PVC sheet	General Tire & Rubber Co.
Bondall	Polyvinyl butyral	Schenectady Chemicals, Inc.
Bosilm	Adhesive-coated PVC sheet	USM Chemical Co.
Boutonniere	Vinyl asbestos flooring	American Bilbrite Rubber Co.
Brasslyfe	Vinyl lacquers	Bee Chemical Co.
Bravex	Rigid PVC	Apache Industries Corporation
Breeth-Ees	PVC sheeting	Rand Rubber Co.
Breon	PVC latices	BP Plastics, Ltd.
Bristrend	PVC/PVAc	Polymers, Inc.
Broncho	PVC composition sheet	Textileather Division, General Tire & Rubber Co.
Bubble-Pak	Saran-coated cushioning	Sealed Air Corporation
Bubble-wrap	Saran-coated cushioning	Sealed Air Corporation
Buckran	Vinyl film and sheet	Central States Products Co.
Butacite	Polyvinyl acetal	E.I. Du Pont de Nemours & Co., Inc., Polychemicals Division
Butvar	Polyvinyl butyral	Monsanto Co.
Calresin	PVC molding materials	Calresin Corporation
Calvinac	Vinyl coatings	James Lithgow Co.
Camargo	Vinyl-impregnated fabric	Breneman-Hartshorn Inc.
Capilair	Breathable PVC composite	Uniroyal, Inc.
Carena	PVC	Shell Chemical Co.
Cascorez	PVC/PVAc adhesive	Borden, Inc.
Casco Vin	PVC	Borden, Inc.
Celca-Cet	PVAc emulsions	Celanese Polymer Specialties Co.
Cellobond	PVA	BP Chemicals International
Cell Tite	Rigid vinyl foam	B.F. Goodrich Co.
Certus	Vinyl resin adhesives	Monite Waterproof Glue Co.
Chemclad	Vinyl insulation	Essex Wire Corporation
Chemcon	Conductive vinyl	Chemelec Products, Inc.
Chem-Flex	Vinyls	Eagle Picher Industries
Chemidus	PVC pipe and fittings	Chemidus Plastics, Ltd.
Chem-o-Set	Crosslinked PVC dispersion	Whittaker Corporation, Chemicals Products Division
Chem-o-Sol	Vinyl plastisols	Whittaker Corporation, Chemicals Products Division
Chemtral	PVC valves	Cabot Piping Systems
Chromalon	Metallized-PVA film	M & B Plastics, Ltd.
Chrytique	Vinyls	Chrysler Corporation
Cipco Foam	Extruded foam	Essex Wire Corporation
Cipoviol	PVC film	M & B Plastics, Ltd.
Cirrus	Vinyl floor covering	General Tire & Rubber Co.
Clearflo	Vinyl tubing	Newage Industries, Inc.
Clearseal	PVC film	Borden, Inc.
Clopane	Vinyl film	Clopay Corporation
Cobex	PVC sheet, rod, and tube	Bakelite Xylonite, Ltd.
Cobon	PVC tube	Cobon Plastics Corporation
Cobovin	Reinforced-PVC tube	Cobon Plastics Corporation
Colovin	PVC film	Borden, Inc.
Colvelta	PVC foam	Borden, Inc.
Comfortweave	Knitted vinyl	Ford Motor Co.
Conoco	PVC resins	Conoco Chemicals Division
Con-tack	PVC film	Storey Brothers & Co.
Cordo	PVC film and foam	Ferro Corporation
Corex	PVC sheet	Bakelite Xylonite, Ltd.
Corvic	PVC resins and copolymers	Imperial Chemical Industries, Ltd.
Covical	Vinyl compounds	Solcarburo, S.A.
Covinax	PVA	Franklin Chemical Co.
Covol	PVA	CPC International, Inc.
Craylene	PVC film	Commercial Plastics, Ltd.
Craypak	PVC film	Commercial Plastics, Ltd.
Craytherm	PVC sheet	Greenwich Plastics, Ltd., Giovanni, S.p.A.

Crest-Foam	Vinyl foam	Crest Foam Corporation
Cristal	PVC sheet	Crespi
Croid	PVAc	Croid, Ltd.
Crystal Clear	Press-polished vinyl	Jacobson Manufacturing Co.
Crystal Hose	PVC hose	Tigers Rubber Co., Ltd. (Japan)
Cyclon	PVC tube	Polykote, Inc.
Cyclvin	ABS/PVC alloy	Borg-Warner Corporation
Dacovin	PVC	Diamond Shamrock Chemical Corporation
Daflex	PVC tube and sleeve	Daburn Electronics & Cable Corporation
Dapco	PVC tube and sleeve	Davidson Products Co.
Daran	PVC_2 coatings	W.R. Grace & Co.
Daratak	PVAc emulsions	W.R. Grace & Co.
Darex	PVAc emulsions	W.R. Grace & Co.
Darvan	Polyvinylidene cyanide/PVAc copolymer	Celanese Corp. of America
Darvic	PVC sheet	Imperial Chemical Industries, Ltd.
D.C.	Vinyl-coated cloth	S.I. Jacobson Mfg. Co.
Decelith	PVC film, sheet, tube, and rod	VEB Eilenburger Celludoid-Werk
Deckor	PVC plastisols	Vinatex, Ltd.
Delta	PVC pipe and fittings	Foster Brothers Plastics, Ltd.
Denflex	Vinyl plastisols and primers	Dennis Chemical Co.
Denfoam	Extruded vinyl	Essex Wire Corporation
Denka	PVA	Denki Kagaku Kogyo KK, Japan
Dermide	PVC-coated fabrics	Dermide, Ltd.
Diamond	PVC and copolymers	Diamond Shamrock Chemical Co.
Diana Fyrban	Vinyl-coated and -impregnated fabrics	Breneman-Hartshorn, Inc.
Diofan	Polyvinylidene chloride	BASF, A.G.
Diorez	Polyvinylidene chloride filament	Tenneco Chemical Corporation
Domani	Vinyl flooring	American Biltrite Rubber Co., Inc.
Dorn	PVC film, sheet, and coated fabric	Gordon-Lacey Chemical Products Co., Inc.
Dow	PVC	Dow Chemical Co.
DP Hot Line	PVC pipe	Diversified Parkway Industries, Inc.
Dualam	Reinforced-PVC pipe	CPF Plastics, Ltd.
Duflex	PVC foam	Elson & Robbins, Ltd.
Duracet	PVA	Franklin Chemical Co.
Dural	Acrylic-modified vinyl	Alpha Chemical & Plastics Corporation
Duran	Vinyl upholstery	Masland Duraleather Co.
Duranyl	Rigid PVC	Mastic Corporation
Durapipe	PVC pipe	Durapipe & Fittings, Ltd.
Duraplex	PVC sheet and film	Duraplax Plastics, Ltd.
Durasol	PVC-coated materials	Masland Duraleather Co.
Duratex	Vinyl fabric	General Tire & Rubber Co.
Duratube	PVC tube	Duratube & Wire, Ltd.
Durelene	PVC tube	Plastic Warehousing Corporation
Durocel	Vinyl foam	Mastic Corporation
Durotherm	PVDC	Carlon Products Corporation
Durug	Glass-reinforced vinyl	Duracote Corporation
Duxkin	PVC film and sheeting	Harte & Co., Inc.
Dynadur	PVC tube, sheet, rod	Dynamit Nobel, A.G.
Eanplast	PVC molding compounds	East Anglia Plastic, Ltd.
Eclat	Vinyl flooring	American Biltrite Rubber Co., Inc.
Edilac	PVAc dispersions	Montedison, S.p.A.
E divil	PVAc dispersions	Montedison, S.p.A.
Ekavyl	PVC emulsions and resins	Products Chimiques, UKM
Elasti-Glass	Vinyl copolymer	S. Buchsbaum Co.
Elastofoam	Vinyl plastisols	Elastomer Chemical Corporation
Elastolex	Fabric-supported vinyl	General Tire & Rubber Co.
Elaston	PVC film	Renolite-Werke, GmbH
Elastotherm	PVC film and sheet	Renolite-Werke, GmbH
Elmer's Glue	PVC/PVAc glues	Borden, Inc.
Elvacet	PVAc emulsions	E.I. Du Pont de Nemours & Co., Inc.
Elvanol	PVA	E.I. Du Pont de Nemours & Co., Inc.
Elvax	Vinyl	E.I. Du Pont de Nemours & Co., Inc.
Emultex	PVAc/acrylic copolymers and PVAc emulsions	Revertex, Ltd.

Encor	PVAc adhesive	Standard Brands Chemical Industries
Ensocote	PVC lacquer	Uniroyal, Inc.
Encolite	Vinyl sheet foam	Uniroyal, Inc.
Envilon	PVC fibers	Toyo Kagasu, KK
Epok	PVAc	BP Chemicals International, Ltd.
Eraclene	PVAc dispersions	ANIC, S.p.A.
Eskay-Lite	Vinyl film	St. Regis Paper Co.
Eslon	PVA panels	Sekisui Chemical Co., Ltd.
Everflex	PVAc copolymer emulsions	W.R. Grace & Co.
Excelon	PVC tubing	Thermoplastic Processes, Inc.
Exon	PVC resins and copolymers	Firestone Plastics Co.
Extru-Foam	PVC foam	Extrudyne, Inc.
Extru-Glaz	PVC sheet	Extrudyne, Inc.
Extrulon	PVC	Omni Products Corporation
Extruvin	PVC	Omni Products Corporation
Fabrifoil	PVC/PVAc sheet	Canadian Industries, Ltd.
Fabrilite	Vinyl-coated fabrics	E.I. Du Pont de Nemours & Co., Inc.
Facilon	Reinforced-PVC fabric	Sun Chemical Corporation
Fashon	Vinyl film and sheet	General Tire & Rubber Co.
Fasslon	Vinyl coating	M.J. Fassler & Co., Inc.
Fasslux	Vinyl coating	M.J. Fassler & Co., Inc.
F-Bondall	Polyvinyl formal/polyvinyl butyral wire coat	Viking Wire Co., Inc.
Federan	Vinyl-coated fabric	Textron, Inc.
Federan	Vinyl-coated fabric	Airco Plastic Products
Fedlastic	Vinyl-coated fabric	Textron, Inc.
Fedwell	Vinyl-coated fabric	Textron, Inc.
Filmcelle	Embossed vinyl film	Toscany, Inc.
Flamenol	PVC	General Electric Co.
Flexac	PVAc emulsions	Airco Chemicals & Plastics
Flexbond	PVAc	Air Products & Chemicals Inc.
Flexilac	Vinyl resin solutions	Flexible Products Corporation
Flexiplast	PVAc	American Hoechst Co.
Flexmul	Vinyl emulsions	Flexible Products Corporation
Flex-o-Glas	Vinyl film	Warp Brothers
Floflex	Vinyl fluid bed powders	Flexible Products Corporation
Flotube	Vinyl tubing	Natvar Corporation
Flovic	PVC	Imperial Chemical Industries Ltd.
Flo-Well	PP/PVC copolymer	Air Products & Chemicals, Inc.
Foamex	Polyvinyl formal phenolic	General Electric Co.
Formclad	Vinyl sheet	Saturn Plastics & Engineering Co.
Formetic	Polyvinyl formal	Monsanto Co.
Formex	Polyvinyl acetal	General Electric Co.
Formtex	Polyvinyl formal	Essex Wire Corporation
Formvar	Polyvinyl formal resins	Monsanto Co.
Fortuna	Vinyl flooring	American Biltrite Rubber Co.
Fromaplas	PVC sheet	Bakelite Xylonite, Ltd.
Fromocene	PVC sheet and film	Bakelite Xylonite, Ltd.
Fromoplas	PVC sheet	Bakelite Xylonite, Ltd.
Fromotan	PVC sheet	Bakelite Xylonite, Ltd.
Fromotex	PVC sheet and film	Bakelite Xylonite, Ltd.
Frostone	Polyvinylidene chloride	President Suspender Co.
Gantrez	PVC copolymers	GAF Corporation
Garbel	Vinyl compounds	Rhone-Progil. Direction Commerciale Matieres Plastiques
Gardenia	PVC sheet	Giovanni Crespi, S.p.A.
Gelflex	Vinyl compound	Proplastic of Bognor Regis
Gelva	PVAc resins	Monsanto Co.
Gelvatex	PVAc emulsions	Monsanto Co.
Gelvatol	PVA resins	Monsanto Co.
Gem-Flex	PVC	Gemloid Corporation
Gencaseal	PVC insulation	General Cable Corporation
Genclor	PVC	ICI United States, Inc.
Genolon	PVC copolymer film and sheet	Kalle & Co., A.G.
Genotherm	PVC film and sheet	American Hoechst Corporation
Geno-Tubing	PVC tubing	American Hoechst Corporation

Genvar	Polyvinyl formal	General Cable Corporation
Geon	PVC, PVC/polyvinylidene chloride copolymers, CPVC, and other vinyl materials	B.F. Goodrich Chemical Co.
Geon/Breon	PVC	British Geon, Ltd.
Ger-Flex	Vinyl tubing	Monsanto Co.
Gold Kast	PVC film	Continental Plastic Co.
Goodallite	Vinyl-coated fabric	Goodall-Sanford, Inc.
Gorlaglass	PVC film	Montedison, S.p.A.
Gorlapack	PVC film	Montedison, S.p.A.
Gracon	PVC compounds	W.R. Grace & Co.
Graftmer	PVC graft copolymers	Nippon Zeon Co.
Gresintex	PVC	Societa del Gresling, Sala & C.
Griffco	PVA emulsions	Griffin Chemical Co.
Guard	Vinyl-coated fabric	Columbus Coated Fabrics Corporation
Gummiplast	Rubber-modified PVC	Saplast, S.A.
Guttadur	PVC	Kalle & Co., A.G.
Guttagena	PVC film and sheet	Kalle & Co., A.G.
Gu-Tubing	PVC tubing	Kalle & Co., A.G.
Hallite	PVC tubing	C. P. Hall Co.
Halvic	PVC tubing	Halvic Kunststoffwerke, GmbH
Hartex	Vinyl film and sheet	Harte & Co., Inc.
Heritage	Vinyl film and sheet	Central States Products Co.
Hi-K-80	PCV pipe and fittings	Extrudex, Ltd.
Hipact	PVC pipe and fittings	Extrudex, Ltd.
Hi Temp Geon	CPVC	B.F. Goodrich & Co.
Horco	Polyvinyl butyral	Hodgman Rubber Co.
Hostaflex	PVC copolymer	Hoechst, A.G.
Hostalit	PVC compounds	Hoechst, A.G.
Hostalit	PVC resins	Hoechst, A.G.
Hostophan	PVC emulsions	American Hoechst Corporation
Hotavinyl	PVC dispersions	American Hoechst Corporation
Hydex	PVC-coated fabric	I.C.I. (Hyde), Ltd.
Hyvin	PVC molding	Vinatex, Ltd.
Ilonca	PVC-coated fabric	Dermide, Ltd.
Indusol	Vinyl plastisols and organosols	Industrial Solvents & Chemicals, Inc.
Insular	PVC and copolymer resins	Rubber Corp. of America
Intamix	PVC compounds	Diamond Shamrock Corporation
Irrovin	Irradiated PVC	ITT Corporation
Irvinil	PVC	Great American Plastics Co.
Irv-O-Thin	Vinyl coating	3M Co.
Ixan	PVC_2 copolymers	Solvay & Cie
Jayon	Vinyl tubing	Johnston Industrial Plastics
Kanacon	PVC hose	Totaku Industrial Co., Ltd.
Kanaflex	PVC hose	Totaku Industrial Co., Ltd.
Kanevinyl	CPVC, PVC, and Copolymers	Kaneka America Corporation
Katharon	PVC sheet and film	Ross & Roberts, Inc.
Kenron	PVC resins	Sechrist Chemicals, Inc.
Key	Vinyl adhesives	Key Polymer Corporation
Khlorin	CPVC	Soviet Origin
Kleen Peel	Vinyl	Beck Equipment & Chemical Co.
Kohinor	PVC resins and compounds	Pantasote Co.
Koilset	Polyvinyl butyral coating	Red Magnet Wire Co., Inc.
Koresin	Vinyl resin on a phenolic base	Akron Chemical Co.
Korogel	Plasticized vinyl	B.F. Goodrich Chemical Co.
Korolac	Vinyl solutions	B.F. Goodrich Chemical Co.
Koroseal	PVC, PVF/PVC	B.F. Goodrich Industrial Products Co.
Korowood	Vinyl	Koro Corporation
Krene	PVC film, PVC/PVAc copolymer film and sheet	Union Carbide Corporation
Krystal	PVC sheet	Allied Chemical Corporation
Krystaltite	PVC shrink film	Allied Chemical Corporation
Ku La	PVC lining	Storeys of Lancaster, Ltd.
Kumfy Lay	PVC-coated paper	Armonde, Ltd.
Kuralon	PVC hose	Kuraray Plastics Co., Ltd.
Kurofan	PVC_2 dispersions	BASF Wyandotte Corporation

Kydene	Acrylic/PVC alloy	Rohm & Haas Co.
Kydex	Acrylic/PVC alloy sheet	Rohm & Haas Co.
Lacqvyl	PVC molding compound	ATO Chimie of Paris
Lacron	PVC molding compound	Lati, S.p.A.
Laroflex	PVC copolymer	BASF Wyandotte Corporation
Lasco Pipe	PVC pipe	Lasco Industries
Laundron	PVC film	Polyval Corporation
Leben	Thermosetting PVC	Dai Nippon Toryo Co.
Lemac	PVAc	Borden Chemical Co.
Lemol	PVA	Borden Chemical Co.
Li-Lex	PVC	P.B. Cow, Ltd.
Lintrex	Vinyl-coated fabric	Columbus Coated Fabrics
Liquid Envelope	PVC coatings	Essex Chemical Corporation
Lorival	PVC sheet, rod, tube, and profile	United Ebonite & Lorival, Ltd.
Lucalor	CPVC	Rhone-Progil, Direction Commerciale Matieres Plastiques
Lucolene	Vinyl compounds	Rhone-Progil, Direction Commerciale Matieres Plastiques
Lucolite	Clear PVC	Tenneco Chemicals, Inc.
Lucomix	Vinyl compounds	Rhone-Progil, Direction Commerciale Matieres Plastiques
Lucorex	Vinyl compounds	Rhone-Progil, Direction Commerciale Matieres Plastiques
Lucovyl	PVC and copolymers	Rhone-Progil, Direction Commerciale Matieres Plastiques
Lumite	Saran filaments	Chicopee Manufacturing
Lutanol	Polyvinyl ether	BASF Wyandotte Corporation
Lutofan	PVC emulsion	BASF Wyandotte Corporation
Lutonal	Polyvinyl ether	BASF Wyandotte Corporation
Lutrex	PVAc	American Hoechst Corporation
Luvican	Polyvinyl carbazole	BASF Wyandotte Corporation
Luvitherm	PVC	BASF Wyandotte Corporation
Maccrosal	Vinyl plastisol	Michigan Chrome & Chemical Co.
Macolite	Extruded and foam vinyl	Detroit Macoid Corporation
Macroprene	Extruded and foam vinyl	Detroit Macoid Corporation
Marimex	PVC-coated fabric	Storey Brothers & Co., Ltd.
Markren	PVC-coated fabric	Edward MacBean & Co., Ltd.
Marleyglaze	PVC sheet	Marley Extrusions, Ltd.
Marley Vinyl	PVC pipe and fittings	Marley Extrusions, Ltd.
Marshmallow	PVC foam	Sommers Plastics Products Co.
Marvinol	PVC resins and compounds	Uniroyal, Inc.
Maxproof	PVC-coated fabric	Edward Macbean & Co., Ltd.
Maxron	PVC-coated	Edward Macbean & Co., Ltd.
Maxtwill	PVC fabric	Edward Macbean & Co., Ltd.
Maxume	PVC fabric	Edward Macbean & Co., Ltd.
Mayon	Vinyl tubing	Mayon Plastics Co.
Mellocord	PVC-coated fabric	Mellowhide Products, Ltd.
Mello-Vin	PVC foam	Dr. G.L. Bonivitt Laboratories
Mellowsoc	PVC-coated fabric	Mellowhide Products, Ltd.
Mellorina	PVC-coated fabric	Mellowhide Products, Ltd.
Merithene	PVC sheet and film	Magid-Robinson Co., Inc.
Metrex	PVC sheet, pipe, and fittings	Allied Ironforenders, Ltd.
Micron	PVC coating powder	Michigan Chrome & Chemical Co.
Microsol	Vinyl plastisols	Michigan Chrome & Chemical Co.
Micro-Vent	Porous vinyl film	Canadian Resins & Chemicals, Ltd.
Micro-Vent	PVC/P(VC-Ac) film	Herculite Protective Fabrics Group
Mil-O-Sol	PVC plastisols	J.G. Milligan & Co.
Miploplast	PVC sheet	Dynamit Nobel, A.G.
Mipodur	PVC profiles	Dynamit Nobel, A.G.
Mipolam	PVC film	Dynamit Nobel, A.G.
Mipoplast	PVC film	Dynamit Nobel, A.G.
Mirrex	Rigid vinyl	Tenneco Co.
Montage	Vinyl asbestos flooring	Johns-Manville Corporation
Moviol	PVA	American Hoechst Corporation
Morvilith	PVAc	Hoechst, A.G.
Morviol	PVA	Hoechst, A.G.

Morvital	PVAc	Hoechst, A.G.
Multipruf	Vinyl-coated materials	Elm Coated Fabrics, Co., Inc.
Muntz	PVC tubing	Glymved Plastics, Ltd.
Nalgon	PVC tubing	Nalge Co.
Nappahide	PVC sheet	Armonide, Ltd.
Nappatex	PVC sheet	Commercial Plastics, Ltd.
Natvar	PVC	Natvar Corporation
Naugahyde	Vinyl-coated fabric	Uniroyal, Inc.
Nautaquilt	PVC sheet	General Tire & Rubber Co.
Nautolex	Vinyl sheet	General Tire & Rubber Co.
Nelcote	PVC emulsions	Northeastern Labs Co., Inc.
Neohide	PVC-coated fabric	Storey Brothers & Co., Ltd.
Neolux	PVC-coated fabric	Storey Brothers & Co., Ltd.
NeoVac	PVAc and copolymers	Polyvinyl Chemicals, Inc.
Nestor	PVAc	James Ferguson & Sons, Ltd.
Neufan	PVC film	Tayo Rayon Co., Ltd.
Niacet	Vinyl acetate	Union Carbide Corporation
Nieu Amsterdam	Vinyl flooring	American Biltrite Rubber Co.
Niolan	PVC-coated fabric	Storey Brothers & Co., Ltd.
Nipeon	PVC resins and copolymers	Nippon Zeon Co.
Nitrilene	PVC/butadiene/acrylonitrile copolymer	Rhein-Chemic Rheinau, GmbH
Nobilia	PVC film	Dynamit Nobel, A.G.
Nob-Lock	PVC sheet	Americoat Corporation
Non Tac	PVC sheet and molding compound	H.Z. Stauss Corporation
North	PVC-coated fabric	James North & Sons, Ltd.
Northylene	PVC-coated fabric	James North & Sons, Ltd.
Northylon	PVC-coated fabric	James North & Sons, Ltd.
Norzeen	PVC sheet	Storey Brothers & Co., Ltd.
Novon	PVC sheet	I.C.I. (Hyde), Ltd.
Nyform	Polyvinyl formal coating	Ananconda Wire & Cable Co.
Nygen-Tolex	Vinyl-coated nylon fabric	General Tire & Rubber Co.
Nylclad	Polyvinyl formal with nylon overcoat	Belden Mfg. Co.
Nyoco	PVC-coated fabric	Iow, Ltd.
Opalon	PVC resins and compounds	Monsanto Co.
Oppanol	Polyvinyl isobutyl ether	BASF Wyandotte Corporation
Orgalacqe	PVC coating powder	Aquitaine Organico
Organit	PVC sheet	Salvoy & Cie
Oriex	PVC sheet and film	Tenneco Chemicals, Inc.
Osmodrain	PVC pipe	Osma Plastics, Ltd.
Panache	PVC flooring	American Biltrite Rubber Co.
Pantasote	Vinyl sheet and coated materials	Pantasote Co.
Pantex	Vinyl film	Pantasote Co.
Paracril OZO	NBR/PVC alloy	Uniroyal, Inc.
Parcloid	PVC and P(VC-Ac) dispersions	Parcloid Chemical Co.
Pavacel	PVC sheet	British Celanese, Ltd.
Peel Kote	PVC coatings	Polykote, Inc.
Pee Vee Cee	PVC pipe	Atlas Mineral & Chemicals Division, ESB Inc.
Perbunan	PVC/nitrile butadiene	Bayer, A.G.
Perflex 66	PVC_2 film	Union Carbide Corporation
Perma-Skin	Vinyl coatings	Dennis Chemical Co.
Permafused	Vinyl coating	Anchor Post Products, Inc.
Pevalon	PVA film	May & Baker, Ltd.
Pe Ve Clair	PVC sheet	Brimar
Pevikon	PVC sheet	Kema Nord, Ab
Pexco	PVC tube	Plastic Extrusion & Engineering Co., Inc.
Pexcon	PVC tube	Plastic Extrusion & Engineering Co., Inc.
Phenolit	Phenolic/PVC blend	Soviet Origin
Pioloform	Polyvinyl acetal	Wacker-Chemie, GmbH
Plasoco	PVC-coated fabrics	Ioca, Ltd.
Plasta-Gard	Vinyl	Ashland Chemical Co.
Plasticell	PVC foam	B.T.R. Silvertown, Ltd.
Plastifilm	PVC film and sheet	Goss Plastic Film Corporation
Plastiflex	PVC compounds	General Fabricators, Inc.

Plastimayd	Vinyl film and sheet	Plastimayd Products Corporation
Plastoglass	PVC-coated fabrics	Storey Brothers & Co., Ltd.
Plastolam	PVC sheet	Storey Brothers & Co., Ltd.
Plastolene	PVC-coated fabrics	Storey Brothers & Co., Ltd.
Plastovac	PVC sheet	Storey Brothers & Co., Ltd.
Plastylite	Vinyl compounds	Plastylite Corporation
Plaxide	PVC-coated fabrics	Armoride, Ltd.
Pliaduct	PVC pipe	Dayco Corporation
Plioflex	PVC	Goodyear Tire & Rubber Co.
Plioglas	Vinyl-coated glass cloth	Flexfirm Products
Pliohyde	PVC sheet	Goodyear Tire & Rubber Co.
Pliolite	Vinyl/toluene/butadiene and vinyl/toluene copolymers	Goodyear Tire & Rubber Co.
Pliovic	PVC blending resins and copolymers	Goodyear Tire & Rubber Co.
Plyamul	PVAc adhesives	Reichhold Chemicals, Inc.
Plyhyde	PVC sheet	Plymouth Rubber Co., Inc.
Plypac	PVC film	Montedison, S.p.A.
Plyvit	PVC sheet	Montedison, S.p.A.
Polychromatic	PVC sheet and film	Ross & Roberts, Inc.
Polyclad	Vinyl	Fortin Laminating Corp., Carboline Co.
Polyclad	Vinyl coating	Carboline, Co.
Polydene	PVC alloy	A. Schulman, Inc.
Polyorc	PVC tube, pipe, and fittings	Yorkshire Imperial Metals, Ltd.
Polyplastex	Vinyl sheet & Laminates	Polyplastex United, Inc.
Polyprene	Vinyl organosol	Interchemical Corporation
Polysizer	PVA	Showa High Polymer Co.
Poly-Tex	Acrylic/PVC and PVAc/PVC copolymers	Celanese Resins
Polytrend	PVC dispersion	Tenneco Chemicals
Polyvin	PVC	A. Schulman, Inc.
Portplasto	PVC extrusions	Brimar
Prestoflex	Vinyl film and sheeting	Presto Plastic Products Co., Inc.
Prestovin	Resins	J.P. Frank Chemical & Plastic Corporation
Prima	PVC-coated fabrics	Dermide, Ltd.
Profilm	PVC	Protective Lining Corporation
Propiofan	Polyvinyl propionate	BASF Wyandotte Corporation
PVC	PVC molding compounds	Dow Chemical Co.
PVC Pearls	PVC resin	Escambia Chemical Corporation
PV-Duit	PVC conduit	Carlon Products, Inc.
Python	PVC tube, profiles, and rod	Plastic Tube & Conduit Co., Ltd.
Quelspray	PVC aerosols	Quelcor, Inc.
Quickbond	Polyvinyl butyral coatings	Bridgeport Insulated Wire Co.
Quirvil	PVC	Rumianca (USA) Chemical Corporation
Randfilm	Vinyl sheet	Rand Rubber Co.
Randflex	Vinyl sheet	Rand Rubber Co.
Ravemul	PVAc	ANIC, S.p.A.
Ravinal	PVAc dispersions	ANIC, S.p.A.
Ravinil	PVC and copolymers	ANIC, S.p.A.
Rayette	Composition/vinyl sheet	General Tire & Rubber Co.
Redo	Vinyl-coated fabric	Goodall-Sanford, Inc.
Reet	Vinyl film and sheeting	Ross & Roberts, Inc.
Regalite	PVC	Tenneco Advanced Materials, Inc.
Rescovin	PVA copolymers	Resine Sintetiche e Affini, S.p.A.
Resiavin	PVA homopolymers	Resine Sintetiche e Affini, S.p.A.
Resinite	PVC tube and shapes	Borden, Inc.
Resistoflex	PVA	Resistoflex Corporation
Respro	Vinyl sheet, film, and coated products	General Tire & Rubber Co.
Resyn	Vinyl copolymer lacquers	National Starch & Chemical Co.
Resyn 3600	Vinylidene chloride copolymer water dispersion	National Starch & Chemical Co.
Rexinite	PVC insulating materials	Borden, Inc.
Rexobond	PVC resins	Emkay Chemical Co.
Reylease	PVA	Reynolds Metals Co.
Reynofoam-V	PVC foams	Reynolds Chemical Products Co.

Reynosol	Plastisol and organosols	Reynolds Chemical Proudcts Co.
Rhenoflex	PVC	Dynamit Nobel, A.G.
Rhodopas	PVC resins	May & Baker, Ltd.
Rhoviflex	Polyvinyl butyral film	May & Baker, Ltd.
Ricon	Vinyl/butadiene resins	Richardson Co.
Rigidsol	Plastisol	Watson-Standard Co.
Rigi Flex	Vinyl extrusion	Crane Plastics, Inc.
Rigivin	PVC	Heil Process Equipment Co.
Rossella	PVC-coated fabrics	Dermicle, Ltd.
Rovana	PVC_2 and copolymers	Dow Badische Co.
Rucoam	PVC film and sheet	Hooker Chemical & Plastics Corporation
Rucoblend	PVC compounds	Hooker Chemical & Plastics Corporation
Rucon	PVC resins	Hooker Chemical & Plastics Corporation
Ruddflex	PVC sheet and film	Rudd Plastic Fabrics Corporation
Ryertex	PVC extrusions	J.T. Ryerson & Son, Inc.
Ryertex Omicron	PVC	J.T. Ryerson & Son, Inc.
Rymway	PVC pipe and fittings	Yorkshire Imperial Metals, Ltd.
Saflex	Polyvinyl butyral film	Monsanto Co.
Saraloy	Vinylidene chloride copolymer	Dow Chemical Co.
Saraloy	Vinylidene chloride sheet	Dow Chemical Co.
Saran	Vinylidene chloride copolymers, resins, coating, film, and microspheres	Dow chemical Co.
Saranex	PVC_2/PE film	Dow Chemical Co.
Saran-O-Lam	Saran casting, film, and laminates	Printon Corporation
Saran Wrap	Vinylidene chloride copolymer film	Dow Chemical Co.
Saronpac	PVC_2 film	Dow Chemical Co.
Sarovyl	PVC sheet	Saro Products, Ltd.
Satinesque	Vinyl-coated fabrics	Columbus Coated Products Corporation
Satinglo	Vinyl-coated fabrics	Columbus Coated Products Corporation
Scon	PVC/PVAc copolymers	Vinatex, Ltd.
Segrane	PVC foam	Whittaker Corporation
Seilon	Vinyl sheet	Seiberling Rubber Co.
Serfene	PVC_2 coating	Morton Chemical Co.
Shawnol	Organosols and plastisols	Adcote Chemicals, Inc.
Sheercell	PVC foam	L.B. Plastics, Ltd.
Shuthane	PVC/PU blend	Reichhold Chemicals, Inc.
Shuvin	Vinyl	Reichhold Chemicals, Inc.
Shuvinite	PVC/nitrile rubber blend	Reichhold Chemicals, Inc.
Shuvin Permabond	PVC molding compounds	Reichhold Chemicals, Inc.
Sicodur	PVC sheet and tube	Mazzucchelli Celluloide, S.p.A.
Sicoprint	PVC sheet	Mazzucchelli Celluloide, S.p.A.
Sicovinil	PVC and copolymer sheet	Mazzucchelli Celluloide, S.p.A.
Sicron	PVC resins and copolymers	Montedison, S.p.A.
Sidavine	PVC film	UCB-Sidac Corporation
Silverflow	PVC pipe and fittings	BTR Industries, Ltd.
Sintilon	PVC sheet	Imperial Chemical Industries, Ltd.
Sirpol	PVAc emulsions	Societa Italiana Resine
Sleck	PVC sheet	Vinyl Plastics, Inc.
Slovinyl	PVC sheet	Chemeiche Zavody Wilhelma Piecka, Norodni Podnik
Sobo	PVAc adhesive	Slomons Laboratories, Inc.
Solufilm	PVA films	M & B Plastics, Ltd.
Solvar	PVAc	Monsanto Co.
Solvic	PVC resins and copolymers	Solvay & Cie, S.A.
Solvitherm	CPVC	Solvay & Cie, S.A.
Solvlon	PVA film	Polyval Corporation
Sonwood	Cellulose-reinforced PVC	Sonesson Plast, Ab
Spark-L-Ite	PVC_2 monofilament	Chevron Chemical Co.
Splendex	PVC-coated fabric	I.C.I. (Hyde), Ltd.
Stabicol	PVAc/acrylate adhesive	Schotten's Chemische Fabrieken, N.V.
Sta-Flow	PVC-polypropylene copolymer	Air Products & Chemicals Co.
Sterilkote	Vinyl plastisols, organosols, and finishes	Bradley & Vrooman Co.
Stix	Vinyl adhesive	Firestone Plastics Co.
Stokes	PVC	Stokes Molded Products

Storain	PVC film	Storey Brothers & Co., Ltd.
Stor Decor	PVC film	Storey Brothers & Co., Ltd.
Storella	PVC film	Storey Brothers & Co., Ltd.
Storide	PVC-coated fabrics	Storey Brothers & Co., Ltd.
Storimo	PVC-coated fabrics	Storey Brothers & Co., Ltd.
Stormur	PVC-coated fabrics	Storey Brothers & Co., Ltd.
Storoflex	PVC-coated fabrics	Storey Brothers & Co., Ltd.
Storolan	PVC film	Storey Brothers & Co., Ltd.
Storvic	PVC film and sheet	Storey Brothers & Co., Ltd.
Sturan	PVC sheet	H.B. Strauss Corporation
Stymer	Vinyl textile sizes	Monsanto Co.
Sullvyne-Clad	Vinyl-Metal laminates	O'Sullivan Rubber Corporation
Sumikon VM	PVC compound	Sumitomo Bakelite Co.
Sumilit	PVC resin	Sumitomo Chemical Co.
Sumilite	PVC film	Sumitomo Bakelite Co.
Superlon	PVC sheet	Commercial Plastics, Ltd.
Superpipe	PVC pipe	Vinylplex, Inc.
Sur Vac	PVC film	Borden, Inc.
Swanette	Vinyl film and sheet	Harte & Co., Inc.
Sylphane	PVC film	UCB-Sidac Corporation
Synresyl	PVAc emulsions	Synres Chemical Corporation
Synskyn	Vinyl-coated glass fiber	Polyplastex United, Inc.
Synthinol	PVC insulation	Cyprus Mines Corporation
Synthite	PEs/PVC varnishes	John C. Dolph Co.
Synthovar	PVC tubing	Varflex Corporation
Talovin	PVC film and sheet	W.R. Grace & Co.
Tam Cast	Plastisols	Tamite Industries, Inc.
Tam Clad	Plastisols	Tamite Industries, Inc.
Tam Cote	Plastisols	Tamite Industries, Inc.
Tamguard	Plastisols	Tamite Industries, Inc.
Tan-O-Lite	Vinyl-coated fabric	Columbia Mills, Inc.
Tavyne	Vinyl sheet	O'Sullivan Rubber Corporation
Tegrand	PVC film	Teijin, Ltd.
Telcovin	PVC sheet, tube, and profile	Telcon Plastics, Ltd.
Tempcor	PVDC pipe	Colonial Plastics Mfg. Co.
Tenac	PVAc resins	Chicago Mastic Co.
Terezon	Vinyl-coated paper	Athol Manufacturing Co.
Terson	PVC-coated fabrics	Athol Manufacturing Co.
Texrez	PVAc solutions	Celanese Coatings & Specialties Co.
Thermasol	PVC plastisols	Lakeside Plastics Corporation
T-Lock	PVC sheet	Amercoat Corporation
Tolex	Vinyl fabrics	General Tire & Rubber Co.
Tolon	Vinyl sheet	General Tire & Rubber Co.
Touchstone	PVC sheet and film	Ross & Roberts, Inc.
Tower	Vinyl film and sheet	Canadien General Tower, Ltd.
Trocal	PVC sheet and film	Dynamit Nobel, A.G.
Tronette	PVC film	Dynamit Nobel, A.G.
Tropos	PVC leather	Dynamit Nobel, A.G.
Trosifol	Polyvinyl butyral sheets	Dynamit Nobel of America, Inc.
Trosiplast	PVC compounds	Dynamit Nobel of America, Inc.
Trovidur	PVC sheet, rod, and tube	Dynamit Nobel of America, Inc.
Trovilon	PVC film	Dynamit Nobel, A.G.
Trovipor	PVC foam	Dynamit Nobel of America, Inc.
Trovitherm	PVC film	Dynamit Nobel of America, Inc.
Trulon	PVC/PVAc copolymers	Olin Corporation
Tubolyne	PVC tubing	OFIC, Ltd.
Tuff Cote	PVC insulation	Coleman Wire & Cable Co.
Tuff-Hyde	Laminated vinyl	Barr Corporation
Tuff-Skin	Vinyl foam	Barr Corporation
Tuftarp	PVC-coated fabric	Grangate & Irwell Rubber Co.
Turbo	PVC tubing	Brand-Rex Co.
Turbogard	Crosslinked vinyl	Brand-Rex Co.
Turbolex	PVC tubing	Brand-Rex Co.
Turbotherm	PVC tubing	Brand-Rex Co.
Turbotrans	PVC tubing	Brand-Rex Co.
Turbotuf	PVC-coated glass	Brand-Rex Co.

Turbozone	PVC tubing	Brand-Rex Co.
Turon	Vinyl fabric	General Tire & Rubber Co.
Tygoflex	Vinyl plastisol	Norton Co.
Tygon	PVC and copolymer tubing	Norton Co.
Tylac	PVC/butadiene/acrylonitrile blend	Standard Brands Chemical Industries, Ltd.
Ucecrylith	PVC and copolymers	Union Chimique-Chemische Bidrijven, S.A.
Ultron	PVC film and sheet	Monsanto Co.
Ultryl	PVC and copolymers	Phillips Petroleum, Ltd.
Unichem	PVC compounds	Dart Industries, Inc.
Unichrome	Vinyl finishes	M & T Chemicals, Inc.
Union Carbide	PVC resins and compounds	Union Carbide Corporation
Uniwick	PVA foam	Unipoint Industries, Inc.
U.S. Dalex	PVC sheet	Uniroyal, Inc.
Valtex	Plastisol-based colors	Interchemical Corporation
Value	Vinyl sheet	Rand Robbin Co.
Vam	VAc monomer	U.S. Industrial Chemicals Co.
Van Cor	PVC pipe	Colonial Plastics Mfg. Co.
Vandike	PVAc	British Oxygen Chemicals, Ltd.
Varlan	PVC resins	DSM
Velbex	Vinyl	Bakelite Xylonite, Ltd.
Veloflex	Vinyl materials	Firestone Plastics Co.
Velon	PVC film and sheet, vinylidene chloride yarns and filaments, saran filament	Firestone Plastics Co.
Vent-Air	PVC foam	Barr Corporation
Ventura	PVC foam	Barr Corporation
Verel	Modacrylic fiber copolymers with VAc, VC and/or VC_2	Eastman Chemical Products, Inc.
Verilon	PVC film	Continental Plastic Co.
Veroseal	PVC film	J.P. Frank Chemical & Plastic Corporation
Versilan	Vinyl upholstery	Landers Corporation
Vestolit	PVC/PE/PVAc copolymer	Mobay Chemical Corporation
Vestolite	PVC resins	Chemische-Werke Huls, A.G.
Vestoran	PVC/PVAc copolymer	Chemische-Werke Huls, A.G.
Vichrome	Vinyl enamels	Interchemical Co.
Viclon	PVC filament and fabrics	Kureha Chemical Industries Co., Ltd.
Vicoa	PVC compounds	Vinyl Corp. of America
Vico-Vinyl	Compounds	Vinyl Corp. of America
Vicralite	Vinyl-coated products	L.E. Carpenter & Co.
Vicratex	Vinyl-coated and -impregnated products	L.E. Carpenter & Co.
Vikem	Vinyl spray	Bel-Art Products
Vinablend	PVC	Vinaflex, Ltd.
Vinac	PVAc emulsion	Air Reduction Co.
Vinacoat	PVC coating powders	Vinatex, Ltd.
Vinacron	Plastisols and organosols	Loes Enterprises
Vinalak	PVAc	Vinyl Products, Ltd.
Vinaliner	PVC sheet	Goodyear Tire & Rubber Co.
Vinaloy	PVC sheet	B.F. Goodrich Chemical Co.
Vin-Alloy	Vinyl	High Strength Plastics Corporation
Vinamul	PVAc	Vinyl Products, Ltd.
Vinapex	PVC film and sheet	Apex Coated Fabrics Co.
Vinapla-Lac	Vinyl lacquer	Schwartz Chemical Co., Inc.
Vinasol	Vinyls	Atlas Synthetics Corporation
Vinatex	PVC molding compounds	Vinatex, Ltd.
Vinavil	PVAc resins	Montedison, S.p.A.
Vinavilol	PVA resins	Montedison, S.p.A.
Vinco	PVC-coated fabrics	M. Barr & Co., Ltd.
Vin-Con	Vinyls	Norton Co.
Vine-L	PVC sheet	Nixon Nitration Works
Vinex	Vinyls	Tenneco Chemicals, Inc.
Vinidur	PVC	BASF Wyandotte Corporation
Viniku	PVC resin, compound, and film	Mitsubishi Monsanto Chemical Co.
Vinilsolum	PVA	Montedison, S.p.A.
Viniplast	PVC	Soviet Origin
Viniproz	PVC/PMMA copolymer	Soviet Origin

Vinnapas	PVAc	Wacker-Chemie GmbH
Vinnol	PVC/PVAc copolymers	Wacker-Chemie GmbH
Vinofan	PVAc copolymers	BASF, A.G.
Vinoflex	PVC/vinylisobutylether copolymer	BASF Wyandotte Corporation
Vinol	PVA	Air Reduction Co.
Vinolast	PVC insulation	American Standard Wire & Cable Co.
Vinophane	PVC film	British Cellophane, Ltd.
Vinrex	Vinyl coatings	Rexton Finishes, Inc.
Vin-So-Lite	Vinyl foam	Foam King, Inc.
Vinuran	Styrene-modified PVC	BASF Wyandotte Corporation
Vinychlon	PVC	Mitsui Chemical Industry Co., Ltd.
Vinyclair	PVC film	M & B Plastics, Ltd.
Vinylclair	PVC film	May & Baker, Ltd., Rhodia, Inc.
Vinylfilm	PVC film	Goodyear Tire & Rubber Co.
Vinylite	PVC film, sheet, molding, and extrusion compounds	Canadian Resins & Chemicals, Ltd.
Vinylite	P(VC-Ac)-coated paper	Union Carbide Corporation
Vinylite CG	Vinyl-coated fabrics	Shawinigan Chemicals Ltd.
Vinyloid	PVC coatings	Rowe Products, Inc.
Vinyl-Pane	PVC film	Warp Brothers
Vinyl Seal	Vinyl adhesive	Union Carbide Corporation
Vinyon	Vinyl copolymer fiber and yarn	Union Carbide Corporation
Vipac	PVC film	May & Baker, Ltd.
Vipla	PVC resins and compounds	Montedison, S.p.A.
Viplast	PVC resins	Montedison, S.p.A.
Viplavil	Vinyl chloroacetate copolymers	Montedison, S.p.A.
Viron	PVC resins	Industrial Vinyls, Inc.
Vista Vinyl	PVC film and sheet	Harte & Co., Inc.
Vistafilm	PVC and copolymer	Goodyear Tire & Rubber Co.
Visten	Vinyl sheet and tube	Union Carbide Corporation
Vita Tube	PVC tube	Goodyear Tire & Rubber Co.
Vitrone	PVC sheet, film, tube, and profile	Stanley Smith & Co.
Vixir	PVC suspensions and emulsions	Societa Italiana Resine
Vor Seal	PVC foam	Vorac Corporation
Vulco	PVC sheet and shapes	Vulcan Metal Products, Inc.
Vult-Acet	PVAc Latices	General Latex & Chemical Corporation
Vxgen	PVC resins	General Tire & Rubber Co.
Vyan Vinyclair	PVC film	Rhodia, Inc.
Vybak	PVAc and PVC molding, sheet, and fibers	Bakelite, Ltd.
Vybond	PVC coating powder	Plastic Coatings, Ltd.
Vycanac	Vinyl film, sheet, molding, and extrusion compounds	Canadian Resins & Chemicals Ltd.
Vycel	PVC foam	Goodyear Tire & Rubber Co.
Vydel	PVC panels and siding	Superior Plastics, Inc.
Vyflex	PVC film and sheet	Kaye-Tex Mfg. Corporation
Vyflex	PVC coating powders	Plastic Coatings, Ltd.
Vygen	PVC resins	General Tire & Rubber Co.
Vylar	PVC siding	Gotham Industries
Vylastic	PVC plastisols	Plastic Coatings, Ltd.
Vylene	Vinyl film and sheet	Elm Coated Fabric Co.
Vylon	PVC pipe	Carlon Products, Inc.
Vynaclor	PVC emulsions	National Starch & Chemical Corporation
Vynafoam	Vinyl plastisol foam	Interchemical Corporation
Vynair	PVC-coated fabrics	I.C.I. (Hyde), Ltd.
Vynaloy	PVC sheet	B.F. Goodrich Chemical Co.
Vynan	PVC film	May & Baker, Ltd.
Vynatherm	PVC film and sheet	Tenneco Chemicals, Inc.
Vynavue	PVC sheet	Fabric Leather Corporation
Vynex	PVC sheet	Nixon-Baldwin Chemicals, Inc.
Vynide	PVC-coated fabric	I.C.I. (Hyde), Ltd.
Vynlast	PVC-coated paper	I.C.I. (Hyde), Ltd.
Vynoleum	PVC-coated fabrics	Welwyn Plastics, Ltd.
Vyram	Rigid PVC	Monsanto Chemical Co.
Vyron	PVC compound	Industrial Vinyls, Inc.
Vyweld	PVC-coated fabrics	I.C.I. (Hyde), Ltd.

Wallpol	PVAc emulsions	Reichhold Chemicals, Inc.
Wascoseal	Vinyl moisture barrier	Wasco Products Inc.
Wataseal	Vinyl film and sheet	Harte & Co., Inc.
Watahyde	PVC sheet	Harte & Co., Inc.
Wautex	PVC film	Clopay Corporation
Wavelock	Nylon-reinforced PVC	Tessitura Carlo Majacchi & C.S.p.A.
Weblon	Fiber-reinforced vinyl	Weblon, Inc.
Weltreen	PVC-coated fabric	Welwyn Plastics, Ltd.
Welvic	PVC molding compounds	Imperial Chemical Industries, Ltd.
Wicanol	PVA	Story Chemical Co.
Wicaset	PVAc emulsions	Story Chemical Co.
Wilflex	Vinyl plastisols and organosols	Flexible Products Co.
Williamsburg	PVC film and sheet	Central States Products Co.
Wite Rite	PVAc adhesive	Passaic Adhesive & Chemical Co.
Wood-Lok	Vinyl copolymer adhesive	National Starch & Chemical Corporation
Wopain	PVC	Worbla, Ltd.
Wopavin	PVC sheet, tube, rod and compounds	Worbla, Ltd.
Wynene	Extruded vinyl	National Plastic Products Co.
Wynsote	Vinyl-coated fabric	Pantasote Co.
X-Link	Crosslinking vinyl/acrylic copolymer	National Starch & Chemical Corporation
X-Pand	PVC foam	Barr Corporation
Xynel	PVC sheet	Xcel Corporation
Zaflex	PVC-coated fabric	Ioco, Ltd.
Zapide	PVC-coated fabric	Ioco, Ltd.
Zerok	Vinyl coating	Atlas Mineral & Chemical Div., ESB Inc.

REFERENCES AND BIBLIOGRAPHY

Air Product and Chemicals, Inc. *Adhesives Handbook*. Bulletin AH 365, March 1965.

Anders, D. Using a planetary-gear extruder for rigid PVC. *Plastics Engineering*, May 1978, **34**(5):56–58.

Anthony, J., Quint, J., and Ward, D. Polyvinyl chloride processing technology–Part 2. *Plastics Design & Processing*, October 1967, 7(10):29–33.

Arendt, W. D. "Glycol Benzoates: Good Salvators for PVC." Paper presentation, *SPE ANTEC*, New Orleans, May 1979.

Arnold, L. K. *Introduction to Plastics*. Iowa State University Press, Ames, Iowa, 1968.

Baier, H. F. Precision for hollow PVC window frames. *Modern Plastics*, April 1978, 55(4):68–69.

Beiss, G. "Controlling Flood and Float in PVC's." Paper presentation, *SPE ANTEC*, New Orleans, May 1979.

Billmeyer, F. W. Jr. *Textbook of Polymer Science*. 2nd Ed. Wiley-Interscience, New York, 1971.

Borden Chemical Co. *What are Extruder Resins?* 1964.

Brydson, J. A. *Plastics Materials*. 3rd Ed. Robert E. Krieger Publishing Co., Inc., Huntington, New York, 1975.

Buckley, C. W., Morin, R. G., and Stockwell, A. J. "Vinyl Polymers and Copolymers." In *Modern Plastics Encyclopedia*. S. Gross (exec. ed.), **45**(14A):336–339, 347. McGraw-Hill Inc., New York, 1968–69.

Chapman, T. F., Isner, J. D., and Chapman, T. F. "Morphology of Flexible PVC." Paper presentation, *SPE ANTEC*, New Orleans, May 1979.

Chemical Products Corporation. *Designing and Producing with Chem-o-Sol*. 1964.

——. *Gasketing with Chem-o-sol PVC Dispersions*. Bulletin HCG-C7626.

DeFife, J. R. *Determining PVC Resin Porosity*. Paper presentation, *SPE ANTEC*, New Orleans, May 1979.

Dietz, G. R., and Brenis, K. L. "Vinyl Foam." In *Modern Plastics Encyclopedia*. J. Agranoff (ed.), **52**(10A):125–126. McGraw-Hill Inc., New York, 1975–76.

Durham, R. D. Jr. Designing with plastics–PVC profiles in construction. *Plastics Engineering*, June 1977, **33**(6):21–23.

Forger, G. R. Plastic alloys offer alternatives to the high cost of "overengineering." *Materials Engineering*, July 1977, **86**(1): 44–47.

Flory, P. J. *Principles of Polymer Chemistry*. Cornell, Ithaca, New York, 1952.

Friedberg, N. D., Gibson, T. A., Ambrose, A. S., and Grabenstetter, T. PVC, from Plastics in Construction: Coming on Strong in 1978. *Plastics Engineering*, July 1978, **34**(7):39–42.

Frissell, W. J. How to control properties of PVC. *Plastics Technology*, December 1962, 8(10):32–35.

Fountas, N. Plastic alloys. *Plastics World*, November 1977, **35**(11): 56ff.

Garvin, G. S. PVDC–processing, properties and applications. *Western Plastics*, Technical Report No. 98 July 1963, pp. 27–29.

Golding, B. *Polymers and Resins–Their Chemistry and Chemical Engineering*. D. Van Nostrand Co., Inc., Princeton, New Jersey, 1959.

Goldman, T. D., and Latz, J. T. Jr. "Test for Low-Temperature Impact Resistance of PVC." Paper presentation, *SPE ANTEC*, New Orleans, May 1979.

B. F. Goodrich Chemical Co. *Geon Latices*. Service Bulletin L-7, July 1958.

——. *Hi-Temp Geon CPVC*. Newsletter No. 5, Bulletin 6807-58M.

——. *Designing with Rigid Vinyl*. Bulletin 6402.

——. *Geon Solution Resins*. Service Bulletin G-15, Revised, Bulletin 6101.

——. *Geon Vinyls for Rigid Extrusions*. G-47, Bulletin 7211-AA.

——. *Injection Molding Hi-Temp Geon Materials*. Service Bulletin TSR 65-14, TF 610m 6509.

——. *Hi-Temp Geon Vinyl . . . An Engineering Thermoplastic*. G-21, Revised, Bulletin 6701-GL.

——. *Hi-Temp Geon*. Service Bulletin G-21, Bulletin 6005.

Grandpre, R. E. Polyvinyl alcohol. *Plastics World*, March 1959, 17(3):16–17.

Hollister, E. H. "Vinyl Copolymers Reexamined for Extrusion." Paper presentation, *SPE ANTEC*, New Orleans, May 1979.

Houston, A. M. With low smoke, toxicity, burning rate, can plastics fill aircraft and transit needs? *Materials Engineering*, August 1976, 84(2).

Irvine, J. L., and McGuire, S. E. Low-fogging plasticizers for automotive interiors. *Plastics Engineering*, February 1978, **34**(2): 35–38.

Kinney, G. F. *Engineering Properties and Applications of Plastics*. Wiley-Interscience, New York, 1957.

Koleske, J. V., and Wartman, L. H. *Poly(Vinyl Chloride)*. Gordon & Breach Science Publishers, Inc., New York, 1969.

Landfield, H., and Taylor, W. S. Protecting vinyl compounds from microorganisms. *Plastics Technology*, August 1962, 8(8):29–31.

Leonard, E. C. *Vinyl and Diene Monomers*. Wiley-Interscience, New York, 1970.

Malpaas, V. E., and Gaggar, S. K. "Modifying Injection-Moldable PVC." Paper presentation, *SPE ANTEC*, New Orleans, May 1979.

Martin, C. J. "Carbon Black in Vinyl Videodiscs." Paper presentation, *SPE ANTEC*, New Orleans, May 1979.

Martino, R. Polyvinyl chloride, from Materials '79. *Modern Plastics*, January 1979, **56**(1):49–50.

——. Siding boom: first wave in a surge of outdoor vinyls. *Modern Plastics*, June 1979, **56**(6):54–56.

Materials Engineering. Vinyl powder coating makes fence long-lasting. October 1977, **86**(4):54.

——. Electrical PVC grade rated for use at 175°F. October 1976, **84**(4):83.

——. Adhesives and Sealants, Manual 260. May 1976, **83**(5):73–80.

Modern Plastics. Coextrusion/blow molding with PVA yields high-barrier polyolefin bottles. February 1979, **56**(2):20.

——. PVC supply: some tightness now, but the big pinch of '81 won't happen. February 1979, **56**(2):42–43.

——. Bulk-handleable PVC dispersion resins offer new economies and efficiencies. March 1979, **56**(3):14–15.

——. EPA cuts VCM emissions 95%–and PVC output 12%. March 1979, **56**(3):49.

——. California cracks down on VCM. July 1978, **55**(7):18.

Mono-Sol Corporation. *Introducing Polyvinyl Alcohol Film.*

Monsanto Co. *Opalon 410 Stir-In Vinyl Resin . . . Formulating Quality Plastisols*. Technical Bulletin No. 1034, 6-615-0665-6.

——. *Plasticizers and Fillers: Their Effect on Properties of Vinyl Chloride Compounds*. Bulletin No. TSB 2735A 8-231-1265-92.

——. *Dry Blending of Opalon Polyvinyl Chloride Resins*. Product Information Bulletin No. 1112, 6-623-0665-92.

——. *Design Information on Vyram Rigid Vinyl*. Bulletin 5065, 6-393 0764.

——. *Powder Blending with Opalon Rigid Vinyl Resins*. Product Information Bulletin 6106, 8-321-1266-92.

Naturman, L. I. Rigid PVC. *SPE Journal*, June 1964, **20**(6):507–515.

Ott, E. E. "Propylene–Vinyl Chloride Copolymer." In *Modern Plastics Encyclopedia*, J. Agranoff (ed.), **52**(10A):89. McGraw-Hill Inc., New York, 1975–76.

Park, R. A. *Dispersion Resin Handbook*. Firestone Plastics Co.

Paul, D. H. "PVC Compounds and Plasticizers." Paper presentation *SPE ANTEC*, New Orleans, May 1979.

Paulsen, R. M. *Alloys Based Upon PVC–Polyman*. A. Schulman Inc.

Peelman, P. L., and Johnson, J. E. Coextrusion feedblocks and multimanifold dies for PVC. *Modern Plastics*, May 1978, **55**(5): 60–62.

Penn, W. S., *PVC Technology*. 2nd Ed. MacLaren & Sons, Ltd., London, 1966.

Plastics Compounding. Setting up an in-house PVC compounding operation. July/August 1979, **2**(4):31–33.

Plastics Design & Processing. New cross-linkable polyvinyl chloride vies for reinforced plastics applications. October 1975, **15**(10):8–9.

Plastics Engineering. Polyvinyl chloride, from Plastics in Packaging. October 1978, **34**(10):48–49.

——. Additives for plastics–plasticizers. January 1977, **33**(1): 32–38.

Plastics World. Joint venture to produce cellular pipe in U.S. March 1977, **35**(3):14.

Pritchard, J. G. *Polyvinyl Alcohol*. Polymer Monograph Series, 1970.

Product Engineering. Porous plastics with novel properties meet variety of design needs. February 1975, **48**(3):25–27.

Radosta, J. A. "Improving Impact Resistance." Paper presentation, *SPE ANTEC*, New Orleans, May 1979.

Rodriguez, F. *Principles of Polymer Systems*. TATA McGraw Hill Publishing Co., Ltd., New Delhi, 1974.

Schildknect, C. E. *Vinyl and Related Polymers*. John Wiley & Sons, Inc., New York, 1952.

Schoengood, A. A. Compounding, plastics engineering primer. *SPE Journal*, February 1973, **29**(2):21–36.

Shawinigan Resins Corporation *Product Reference File–Gelva . . . Gelvatol . . . Formvar . . . Butvar*.

Simonds, H. R., and Church, J. M. *A Concise Guide to Plastics*. 2nd. Ed. Van Nostrand Reinhold Co., New York, 1963.

Skiest, E. N. Blending/dispersion resin interactions *Modern Plastics*, October 1970, **47**(10):132–134.

Stanley Chemical Co. *Plastisols and Organosols in Industry*. Data Sheet No. 101.

Stoloff, A. "Acrylic Modifier for Rigid PVC." Paper presentation, *SPE ANTEC*, New Orleans, May 1979.

——. Modified-acrylic impact modifier for superior vinyl siding. *Plastics Engineering*, July 1979, **35**(7):29–31.

Summers, J. W., and Brown, R. J. Predicting surface distortion in vinyl siding–and avoiding it. *Plastics Engineering*, April 1979, **35**(4):34–36.

Swanson, R. S. *Plastics Technology–Basic Materials and Processes*, McKnight & McKnight Publishing Co., Bloomington, Illinois, 1965.

Thakkar, B. S., and Broutman, L. J. Residual Stresses and Orientation in Amorphous Polymers. Paper presentation, *SPE ANTEC*, New Orleans, May 1979.

Titus, J. B. *Trade Designations of Plastics and Related Materials*, (revised). PLASTEC Note N9C, Plastics Technical Evaluation Center, Picatinny Arsenal, Dover, New Jersey, May 1978.

Union Carbide Corporation *Plastisol Technology for Cloth Coating, Dip, Slush, and Rotational Molding*. Bulletin F-44255, February 1973.

Vaccari, J. A. One toymaker relies heavily on plastics and wood for child's play. *Materials Engineering*. May 1977, **85**(5):18–21.

Vila, G. R. 6000 Miles of Pipe. *Modern Plastics*, July 1957, **34**(11): 100–102, 203.

Walker, L. E., and Slagowski, E. L. "Mass Polymerized Impact Modified PVC." Paper presentation, *SPE ANTEC*, New Orleans, May 1979.

Winding, C. C., and Hiatt, G. D. *Polymeric Materials*, McGraw Hill Book Co., New York, 1961.

Wingrave, J. A. "Electrical Behavior of PVC." Paper presentation, *SPE ANTEC*, New Orleans, May 1979.

Wingrave, J. A., and Peden, M. C. "Rheology of PVC Dryblends." Paper presentation, *SPE ANTEC*, New Orleans, May 1979.

ACRYLICS

Acrylic resins are prepared from the esters of acrylic acid and the methyl derivative, alpha-methyl acrylic acid.

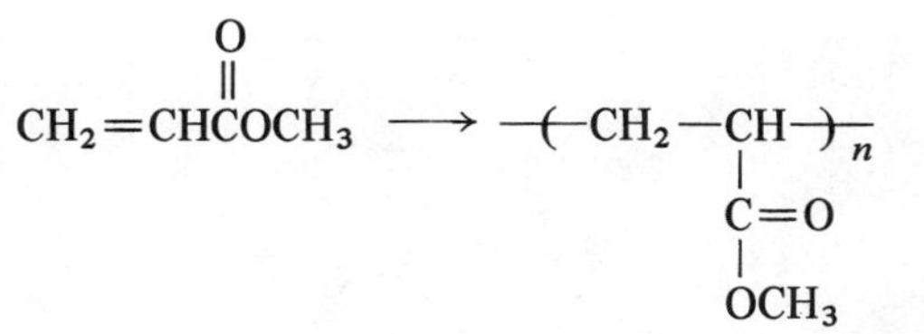

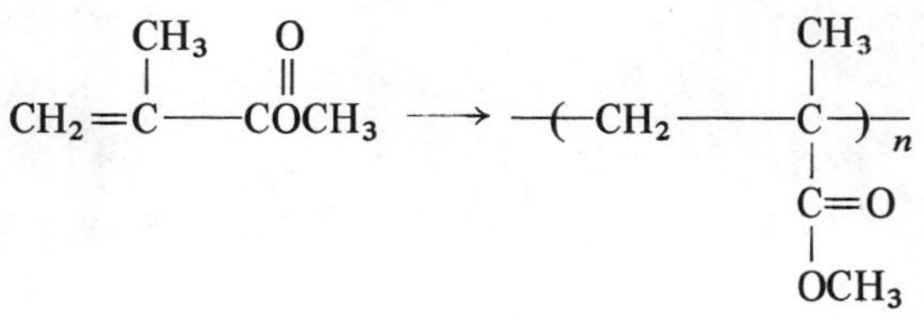

These products, especially the alpha-methyl derivative, are the molding and casting resins. Other derivatives less generally thought of as acrylics, yet related to acrylic acid, are also included in this grouping, the major one being the nitrile of acrylic acid, acrylonitrile. It is used as the major ingredient in several textile polymers as well as a basic copolymer resin. Other derivatives include acrylamide, cyanoacrylates, acrylic acid salts, copolymers, and blends.

Polyacrylates

The simple acrylic acid ester monomers are made by reacting acetylene, carbon monoxide, and an alcohol in the presence of a nickel catalyst.

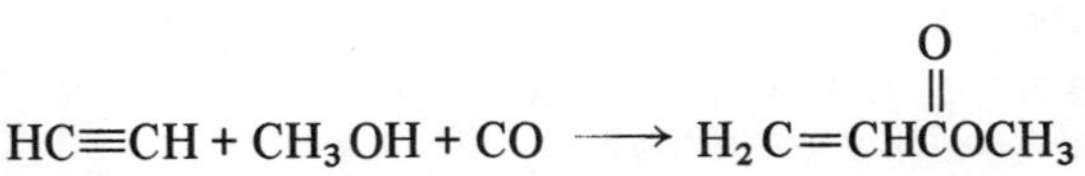

Actually, nickel carbonyl is used at the start of the reaction, and carbon monoxide is fed in to keep the reaction going. At atmospheric pressure and 30-50°C (86-122°F), an 80-90% yield of monomer is obtained. A wide range of acrylates may be prepared by this reaction. If water is used, acrylic acid is made.

These monomers react in a free-radical-type polymerization to form polymers which vary with the ester group. As a class, they do not find as wide a use as the methacrylate derivatives. Acrylic acid polymerizes violently to yield a glasslike solid which is too water susceptible for plastics use. Its salts, generally prepared by hydrolyzing the polymeric ester or the nitrile, are water soluble and find use as water-thickening agents for textile sizes. The product from hydrolyzing the nitrile has been used as a soil-conditioning agent.

These polymers all tend to be softer and more rubbery than the methacrylates. These solutions and latices are used as textile sizes and in paper making as a water-proofing agent.

The higher esters give polymers which are soft and pliable, up to waxlike solids as the molecular weight of the alcohol is increased. As the molecular weight of the monomer increases, polymerization slows considerably and low degrees of polymerization are obtained.

Polymethacrylates

In the methacrylate series the most widely used is the methyl ester, which boils at 101°C (213.8°F). The higher esters give polymers which are generally softer and have a lower softening temperature. The standard commercial polymerization is based on the reaction of acetone and hydrogen cyanide.

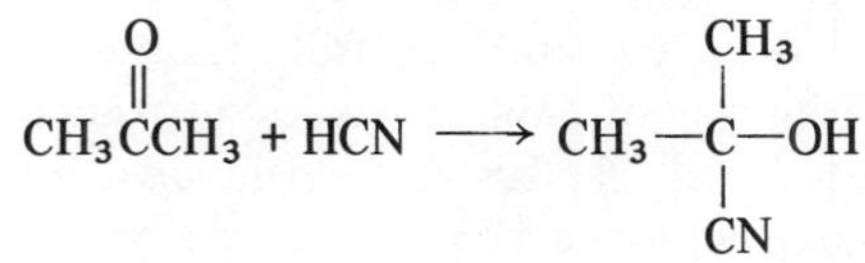

The cyanohydrin formed is treated with H_2SO_4 to yield an acrylamide sulphate which is then reacted with methanol for conversion to the appropriate ester.

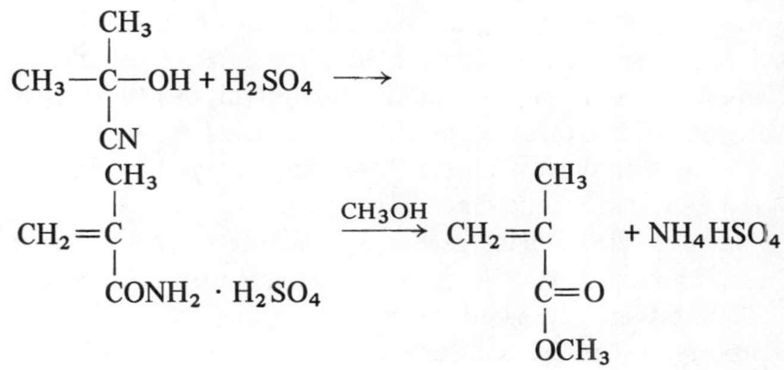

After separation, the monomer is purified by distillation.

The monomer may be polymerized in a casting process, or it may be suspension polymerized to yield a material for injection molding and extrusion. Free-radical techniques use peroxides or azobisisobutyronitrile at temperatures up to 100°C (212°F). Oxygen will affect the reaction rate and the nature of the final product because of methacrylate peroxide side reactions. Therefore, polymerizations must be carried out in a full cell or in the presence of an inert gas.

The mass polymerization of this monomer has been carefully studied, since a large volume of resin is cast into plates of glasslike clarity. This operation requires precise knowledge of temperatures and catalyst concentrations to get the most rapid polymerization consistent with bubble-free quality.

In *cell casting*, the monomer is first subjected to a partial polymerization to obtain a syruplike consistency. Either no catalyst is used, or a level of 0.02-0.05% benzoyl peroxide is used. The temperature is slowly raised to 80-100°C (176-212°F) with a minimum of agitation so that air is not whipped in and the reaction hindered. The syrup may contain 10-30% of polymer. At this time it is degassed and poured between two plates of glass (the *cell*) spaced at the edge by a flexible gasket. The use of a syrup not only shortens the reaction time in the glass cell, it also makes easier the job of containing the fluid in the cell.

The filled cell with the enclosing gasket is heated in air or water to 60–80°C (140–176°F) and then to 100–120°C (212–248°F). Rapid heating may cause the reaction to generate such heat that the monomer will boil and bubble. Shrinkage occurs during polymerization, but this loss in volume is accommodated by the flexible separating gasket.

When the reaction is complete (the total time may consume 30–40 hours, which is one of the major reasons for the relatively high cost of acrylic materials), the cells are cooled and the glass separated from the cast plastic. The piece must be free of bubbles, voids, soft centers, and surface imperfections. The surface is maintained by protecting it on both sides with a heavy paper, using a pressure-sensitive adhesive. This adhesive paper must strip cleanly from the plastic.

This type of batch-lot casting is expensive, so efforts have been made to work out a satisfactory continuous casting operation. One such method produces sheet from 3–8 feet wide in unlimited lengths, and thicknesses from 0.08–0.25 inch. It is based on a patented process using proprietary additives and methyl methacrylate (MMA) monomer. Figure 5-75 shows the machinery involved in the production of this sheet. Table 5-77 compares cast, continuous, and extruded acrylic sheet.

Casting yields clear sheets up to 100 × 120 inches in area and $\frac{1}{16}$–$\frac{3}{4}$ inch in thickness. Sheets up to 2 inches in thickness are prepared by modifications of this casting process. Cylinders and rods are also made by casting. In

(a)

(b)

Fig. 5-75. Equipment for continuous casting of PMMA sheet. (*Courtesy: Novamont Corporation*)

Table 5-77. Properties of Cast Acrylics Compared to Other Types.*

Type ⟶	Continuous	Cell-Cast	Extruded
PHYSICAL			
Specific gravity	1.19	1.19	1.19
Water Absorption, %	0.2	0.2	0.3
Heat Deflection Temperature, F	204–209	204–217	162–183
MECHANICAL			
Tensile strength, psi	10,600	10,800	9400
Tensile Mod, $\times 10^{-6}$ psi	0.47	0.47	0.45
Tensile elongation, %	5.0	5.5	4.0
Flex strength, psi	16,500	16,900	14,900
Hardness (Barcol)	50	50	47
Hardness (Rockell M)	98	98	89
Optical			
Haze, %	1	1	3
Luminous Transmission, %	92	92	91
Light reflection, %	96	97	91
Craze resistance, min[a]			
Isopropyl alcohol	60	60	Broke[b]
Lacquer thinner	10	10	Broke[b]

[a]Time for specimen to craze.
[b]Specimen broke under load.
*Reprinted by permission from *Materials in Design Engineering*, November 1965.

making these forms, it is convenient to fill a tube with the syrupy polymer and put the lower end of the tube in water at 70–80°C (158–176°F). As polymerization occurs at the lower end, the hot-water level is progressively raised. This process reduces bubble formation and keeps a fluid head to take care of shrinkage. Rod diameters of $\frac{3}{8}$–2 inches, and lengths to 46 inches are available.

The average molecular weight of most bulk-polymerized PMMA is too high to give a material which has adequate flow characteristics for extrusion and injection molding. The process is not only costly, but the additional milling and grinding required to produce pellets or powder adds to the expense and introduces a risk of contamination.

Suspension polymerization can be carried out rapidly with little exotherm. Control of particle size and agglomeration is troublesome however. Polymer beads produced are filtered after completion of polymerization, washed, and dried. Inherent contaminants in these beads are the processing additives which include suspending agents (talcs, aluminas), protective colloids (PVA), buffers (sodium hydrogen phosphates), chain transfer agents (trichloroethylene, lauryl mercaptan), and lubricants (stearic acid). The dried beads are supplied as-is, or further compounded with additives and then granulated for injection molding.

The plastic is crystal clear, lightweight, tough and durable, and flexible in thin sections. It may be cut, turned, sawed, drilled, polished, swaged, or formed. It takes lacquers and pyroxylin stains of the proper types; it is not affected by alkalis, oils, dilute acids, or dilute alcohols; and it is the most resistant of all transparent plastics to UV, moisture, and the harmful effects of sun and outdoor exposure. It has the property of transmitting light; a rod will transmit light around bends. Table 5-78 shows typical property values.

The high clarity and excellent light transmission of PMMA, coupled with its resistance to sunlight and its low density, make this resin ideally suited to production of airplane windows, cockpit closures, and the like. Cast sheets are formed by blow and vacuum molding and by a great variety of special forms and jigs. Acrylic is easily formed because it maintains its rubbery state over a broad temperature range. A high degree of optical accuracy is maintained even in curved sections, so sighting devices made from such shapes can be used with precision. Figure 5-78 shows an application that takes advantage of these characteristics.

PMMA molding powders are available in various grades which differ in hardness and ease of molding. The harder formulations show a higher heat-deflection temperature and greater compressive and tensile strength. They are used for molding automotive parts, such as tail lights, dials, and instrument panels. Their beauty in thick sections, weather-resistance, and durability make them outstanding for these end uses. The high heat-distortion formulae are satisfactory for short times at 90.5–101.7°C (195–215°F).

These resins are soluble in ketone and ester solvents as well as aromatic hydrocarbon-alcohol mixtures. One easy way of determining whether an unknown clear plastic is acrylic is to expose the surface to a ketone. If it is acrylic, it will immediately fog up, and possibly even craze. Acrylics have good water resistance, so they are unaffected by solutions of most salts, dilute alkalis, HCl, and H_2SO_4. When properly compounded, these resins make an excellent metal coating which is used in compositions as an automobile finish. This surface is resistant to weathering conditions and maintains a high luster with a minimum of care.

Fig. 5-76. Acrylite GP (general purpose) acrylic sheet is an all-purpose, high-quality acrylic sheet suitable for a variety of applications including industrial safety guards (shown here), acrylic furniture and accessories, windows, and doors. It is easily machined, weather resistant, lightweight, and never yellows. (*Courtesy: CY/RO Industries*)

The physical properties of acrylics are generally in the middle range of those in the thermoplastics family. Tensile and flexural strength, densities, and electrical properties may be presumed to be average. They are notch-sensitive, so that if parts of acrylic are to be mechanically joined, special precautions must be taken to prevent cracking in areas such as threads. The use of molded-in threaded inserts, for example, is preferred over the direct use of screws or bolts. Silicone and other polymeric-based coatings have been developed to increase the resistance of acrylic surfaces to abrasive and chemical attack. Figures 5-77 and 5-78 show the effect of one such coating on the properties of acrylic.

The outstanding properties of acrylics are transparency,

Table 5-78. Properties of Methyl Methacrylate.*

		ASTM test method	Methyl methacrylate: Cast[a]	Methyl methacrylate: Molding
Processing	1. Molding qualities	—	—	Excellent
	2. Compression molding temp., °F.	—	—	300-425
	3. Comp. molding pressure, p.s.i.	—	—	2000-10000
	4. Injection molding temp., °F.	—	—	325-500
	5. Injection molding pressure, p.s.i.	—	—	10000-20000
	6. Compression ratio	—	—	1.6-2.0
	7. Mold (linear) shrinkage, in./in.	—	—	0.001-0.004(C) 0.002-0.008(I)
	8. Specific gravity (density)	D792	1.17-1.20	1.17-1.20
	9. Specific volume, cu. in./lb.	D792	23.7-23.1	23.7-23.1
	10. Machining qualities	—	Good to excellent	Good to excellent
Mechanical	11. Tensile strength, p.s.i.	D638	8000-11000	7000-11000
	12. Elongation, %	D638	2.0-7.0	2.0-10.0
	13. Tensile elastic modulus, 10^5 p.s.i.	D638	3.5-4.5	3.8-4.5
	14. Compressive strength, p.s.i.	D695	11000-19000	12000-18000
	15. Flexural yield strength, p.s.i.	D790	12000-17000	13000-19000
	16. Impact strength, ft. lb./in. of notch (½ × ½ in. notched bar, Izod test)	D256	0.3-0.4	0.3-0.5
	17. Hardness, Rockwell	D785	M80-M100	M85-M105
	18. Flexural elastic modulus, p.s.i. × 10^5	D790	3.90-4.75	4.2-4.6
	19. Compressive modulus, p.s.i. × 10^5	D695	3.90-4.75	3.7-4.6
Thermal	20. Thermal conductivity, 10^{-4} cal./sec./sq. cm./1 (°C./cm.)	C177	4.0-6.0	4.0-6.0
	21. Specific heat, cal./°C./gm.	—	0.35	0.35
	22. Thermal expansion, 10^{-5} in./in./°C.	D696	5.0-9.0	5.0-9.0
	23. Resistance to heat, °F. (continuous)	—	140-200	140-200
	24. Deflection temp., °F. at 264 p.s.i. fiber stress	D648	160-215	165-210
	at 66 p.s.i. fiber stress		165-235	175-225
Electrical	25. Volume resistivity, ohm-cm. (50% RH and 23° C.)	D257	$>10^{15}$	$>10^{14}$
	26. Dielectric strength, short-time, ⅛-in. thickness, volts/mil	D149	450-550	400-500
	27. Dielectric strength, step-by-step, ⅛-in. thickness, volts/mil	D149	431	350-400
	28. Dielectric constant, 60 cyc.	D150	3.3-4.5	3.3-3.9
	29. Dielectric constant, 10^3 cyc.	D150	3.0-3.5	3.0-3.6
	30. Dielectric constant, 10^6 cyc.	D150	2.1-3.2	2.2-3.2
	31. Dissipation (power) factor, 60 cyc.	D150	0.05-0.06	0.04-0.06
	32. Dissipation (power) factor, 10^3 cyc.	D150	0.04-0.06	0.03-0.05
	33. Dissipation (power) factor, 10^6 cyc.	D150	0.015-0.03	0.02-0.03
	34. Arc resistance, sec.	D495	No track	No track
Optical	35. Refractive index, n_D	D542	1.48-1.50	1.49
	36. Clarity	—	Transparent to opaque	
	37. Transmittance, %	D1003	92	92
	38. Haze, %	—	1	<3
Resistance characteristics	39. Water absorp., 24 hr., ⅛-in. thick., %	D570	0.2-0.4	0.1-0.4
	40. Flammability[b] Burning rate, in./min.	D635	0.9-1.3	0.6-1.2
	AEB, in./ATB, sec.[c]		—	—
	41. Effect of sunlight	—	Nil	Nil
	42. Effect of weak acids	D543	Nil	Nil
	43. Effect of strong acids	D543	Attacked only by high concentration oxidizing acids	
	44. Effect of weak alkalies	D543	Nil	Nil
	45. Effect of strong alkalies	D543	None except by liquid ammonia (NH_3)	Attacked
	46. Effect of organic solvents	D543	Soluble in ketones, esters, aromatic and chlorinated hydrocarbons	

[a]The data for cast acrylic sheet should be regarded as a guide only since the materials produced by the various manufacturers differ considerably from the charts and from each other.

[b]This numerical flame spread rating is not intended to reflect hazards presented by this or any other material under actual fire conditions.

[c]AEB = Avg. extent of burning in inches; ATB = Avg. time of burning in seconds.

*Reprinted by permission from *Modern Plastics Encyclopedia*, McGraw-Hill, Inc.

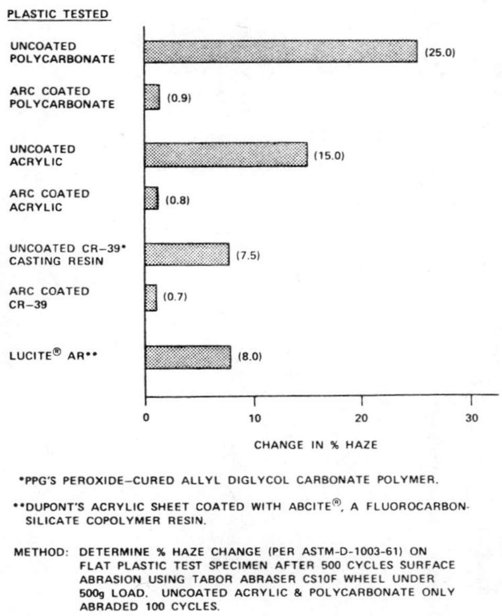

Fig. 5-77. Comparative abrasion testing. (Haluska and Galic, Dow Corning Corporation)

light piping qualities, colorability, color, and dimensional stability. These resins retain their properties in outdoor applications over very long periods of time, which may be the most outstanding characteristic of the acrylic family.

PMMA is the most widely used optical plastic (see Tables 5-79 and 5-80). It is colorless and more transparent than most types of optical glass when used in its cast form. Light transmittance of unmodified PMMA is 92% of incident light, with haze averaging less than 1%. The index of refraction is 1.49, which is the quality qualifying the acrylics for such optical applications as tail-light lenses, photographic equipment, lighting refractors, embedment casting, and the like. Very precise control of light paths is obtained by design in injection molds, and light can be turned 180° in a single reflector mold (see Figure 5-79).

Even though PMMA has 8-10 times the thermal expansion of glass, this is seldom a problem. A 1-inch-diameter lens will only change 0.003 inch when heated from 26.7°C to 65.6°C (80-150°F), and the focal length, if 2 inches, will increase only 0.008 inch over this temperature range. The optical memory of acrylics is better than glass. PMMA is the easiest of the popular optical plastics to machine and polish. However, a lubricant to reduce heat dissipation must be used. Scratch-resistant coatings have been developed for both sheet molded parts with varying degrees of success.

Color and decorative qualities of the unmodified acrylics are unmatched by those of any other plastic. The clear transparency of the resin makes it exceptionally receptive to many colorants, so that transparent, translucent, and opaque colors of almost unlimited hue, chroma, and value can be obtained. In addition, sheet and molded parts can be textured, and sheet can be embossed. Molded parts, extruded and cast sheet, and other fabrications can be decorated by hot stamping, spray painting, vacuum-metallizing, and other methods. Three-dimensional effects can be obtained by various decorating techniques.

Color stability of the unmodified acrylics is outstanding. At room temperature, at temperatures just under the resin

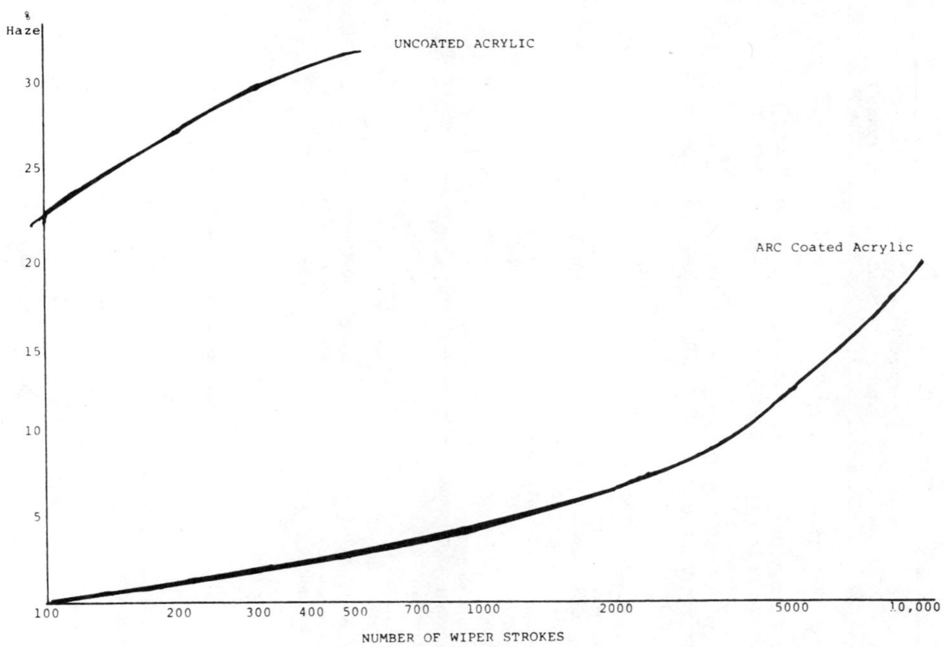

Fig. 5-78. Windshield wiper test (semilog plot). (Haluska and Galic, Dow Corning Corporation)

Table 5-79. How the Transparent Plastics Compare.*

	Acrylic			Other Thermoplastics				Thermosetting		
	Continuous cast	Cell-cast	Molded (extruded)	Polystyrene	Polycarbonate	Cellulose acetate	Vinyl chloride	Polyester	Epoxy	Urethane
Impact strength, Izod Ft-lb/in., $\frac{1}{2} \times \frac{1}{2}$	0.4–0.5	0.4–0.5	0.3–0.5	0.25–0.40 $\frac{1}{4}''$ bar	12–16 ($\frac{1}{2}'' \times \frac{1}{8}''$ bar)	1.0–3.0	0.4–2.0	0.2–0.4	0.2–1.0	Does not break
Flexural strength (psi)	16,200–17,200	12,000–17,000	13,000–17,000	8,700–14,000	13,500	6,000–10,000	10,000–16,000	8,500–23,000	13,000–21,000	–
Tensile strength (psi)	9,000–11,000	8,000–11,000	7,000–11,000	5,000–9,000	8,000–9,500	4,500–8,000	5,000–9,000	6,000–13,000	4,000–13,000	–
Elongation (%)	5	5.5	4	1.0–2.5	60–100	20–50	Up to 250	Less than 5	3–6	100–800
Modulus (psi × 10^5)	4.7	4.7	4.5	4–5	3.5	3–4	3.5–6	3–6	3.5	–
Compressive strength (psi)	11,000–19,000	11,000–19,000	12,000–18,000	11,500–16,000	12,500	18,000–25,000	8,000–13,000	13,000–36,000	15,000–25,000	20,000
Specific gravity	1.19	1.19	1.19	1.04–1.06	1.2	1.28–1.32	1.35–1.45	1.10–1.46	1.11–1.40	1.2–2.5
Hardness (Rockwell)	M95–M101	M80–M100	M85–M105	M65–M80	M70–R118	R95–R120	70–90 (Shore)	M70–M115	M80–M110	M28–R60
Thermal properties										
Heat distortion, °F	204–209	204–217	162–183	205	265–280	130–160	130–164	140–400	115–550	–
Cont. service temp, °F	140–200	140–200	140–190	150–170	250	140–220	150–175	250	250–550	190–250
Expansion (10^{-5}/°C)		5–9		6–8	6.6	10–15	5–18.5	5.5–10	4.5–6.5	10–20
Conductivity (10^{-4} cal./sec./cm^2/°C/cm.)		4–6		2.4–3.3	4.6	4–8	3.0–7.0	4	4–5	5
Electrical										
Dielectric strength (v/mil)		450–550		500–700	400	250–300	425–1300	380–500	380	450–500
Arc resistance (sec.)		no track		60–80	60–80	180–200		125	45–120	–
Optical										
Refractive index		1.49		1.59–1.60	1.59	1.49	1.52–1.55	1.523–1.57	1.61	–
Light transmission (%)		Transparent 92%		Transparent 88–92%	Transparent	Transparent	Transparent	Transparent	Transparent	Clear amber
Effect of sunlight		No visible change		Yellows	Slight color change. Slight embrittlement	No visible change	Dependent on plasticizer	Yellows	No change	No change
Chemical resistance										
Weak acids		No attack		No attack	No attack	No visible change	No attack	No attack	No attack	No attack
Strong acids	Attacked only by high concentration of oxidizing acids			Oxidizing acids	Attacked slowly	Decomposes	None to slight	None to considerable	Attacked by some	Attacked
Weak alkalies		No attack		No attack	Limited resistance	No visible change	No attack	None to slight	No attack	Slight
Strong alkalies		Attacked		No attack	Attacked	Swells	No attack	Attacked	Slight	Attacked
Organic solvents		Ketones, esters, aromatic and chlorinated hydrocarbons		Aromatic and chlorinated hydrocarbons		Ketones and esters; softened by alcohol	Resists alcohol, and oils.	Ketones, esters and chlorinated solvents	Generally resistant	None to slight
Price per lb for – $\frac{1}{4}$ in. sheet	$0.56	$0.58	$0.58	$0.20–$0.24	$1.60	$0.75	$0.65	$1.00	$1.65–4.00	$4.50

Note: From Plastics Properties Chart, *Modern Plastics Encyclopedia*, 1966, page 16, with addition of light transmission, prices, and data on continuous cast acrylic. See source for reference to ASTM test procedures.

***Courtesy: Swedlow, Inc.**

Table 5-80. Optical Properties of Transparent Plastics.*

Materials	Refractive index N_d(= 5893)	N_c(= 6563)	N_f(= 4861)	DN/DT × 10^{-5}/°C	Haze, %	Dispersion	Transmission % at 400–700 mμ
Acrylics	1.491	1.488	1.496	8.5	3[b]	57.8	92
Polystyrenes	1.590	1.585	1.604	12.0	3	30.8	89
Polycarbonates	1.586	1.581	1.598	11.8	3	30.3	88
Styrene acrylonitrile	1.567	1.563	1.578	–	3	–	88
Polymethylpentene	1.467	1.464	1.505	–	5	–	90
Allyl diglycolcarbonate	1.504	1.501	1.510	14.3	3	41.5	93
ABS	1.538	–	–	–	12	–	79–90.6[a]
Polysulfone	1.633	–	–	–	5	–	–
Cellulosics	1.46–1.51	–	–	–	–	–	–
Rigid vinyl	1.52–1.55	–	–	–	–	–	–
CTFE fluoropolymer	1.425	–	–	–	–	–	–
Ionomer	1.51	–	–	–	3–17	–	75–85
Urethane	1.50–1.60	–	–	–	–	–	–
Silicone	1.43	–	–	–	–	–	–
Epoxy	1.55–1.61	–	–	–	–	–	–
Styrene-acrylic copolymer	1.567	–	–	–	–	–	90

[a]79 at 250 mil thickness and 90.6 at 15 mil.
[b]Cast sheet has 1%.
*Almand and Byrd, 1972.

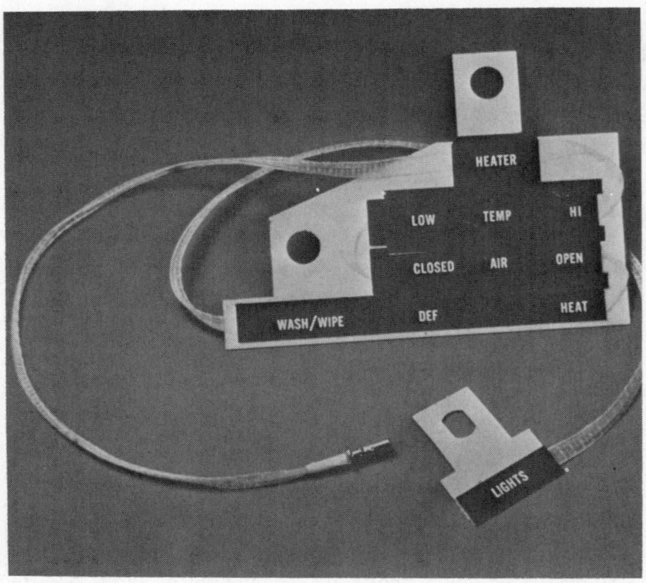

Fig. 5-79. Ribbons of Du Pont Crofon plastic fiber optics may be woven to provide bands of light in areas requiring "glow." Specific areas on the ribbon may be abraded so that light leaks from the roughened surface of the fiber in the abraded area. Here, glow areas illuminate panel graphics by spilling light through transparent lettering. (*Courtesy: E.I. Du Pont de Nemours & Co., Inc.*)

softening point, and under continuous outdoor exposure, the acrylics maintain a color integrity unmatched by other plastics regardless of their surface-temperature ratings.

Dimensional stability of acrylics is outstanding in indoor and outdoor applications where infrared, visible, and ultraviolet radiation are encountered. Moisture absorption is 0.4% or less. Only at prolonged exposure to temperatures slightly under their softening points will the acrylics lose some of their shape in finished parts.

Service temperatures of the unmodified acrylics range from 71-99°C (160-210°F), depending on the grade used and stress imposed. The highest heat resistance is offered by the cast sheet products, with extruded sheet a few degrees lower, and parts produced by injection-molding grades following closely.

The acrylics are classed as slow burning. However, they do not smoke heavily, and, in many instances when fighting fires, it is the smoke that is the greater hazard.

Acrylics are resistant to high concentrations of ultraviolet wave emissions. Used as refractors and diffusers with fluorescent lamps, they have proven satisfactory over long terms with no discernible changes in properties and color (see Figure 5-80). Acrylics have also gained use as refractors with mercury-vapor lamp sources, a further example of their stability.

The major markets of the acrylics continue to be in the

Fig. 5-80. Plexiglas acrylics can be injection molded or extruded with precision to produce hundreds of prismatic lenses which refract light, conceal lamps, and control distribution of light for efficient, low-brightness room illumination. (*Courtesy: Rohm & Haas Co.*)

fields of outdoor signs, interior and exterior lighting, and automotive lenses. Other long-established markets are appliances and the aviation industry. Spectacular growth in the architectural, building, and highway-marking areas have followed the impressive beginnings made in these applications.

Automotive exterior signal lenses are made of acrylics because they offer exceptional optical control, color integrity, toughness, and weather stability. Automobiles with original tail-light lenses installed in 1947 are still in use. Exterior and interior medallions on automobiles are also made of acrylic because of the beauty and depth obtainable. Internally lighted appliance panels perform beyond the useful lives of most appliances.

The use of acrylics has expanded with the aviation industry, where cell-cast sheet established its first commercial use in glazing. Additional applications in this industry are canopies, windows, radar plot boards, instrument panels, and landing-light covers.

Examples of the employment of acrylic panels for large architectural areas have become more prominent in recent years. The Astro-Dome in Houston, Texas is covered with white translucent acrylic panels that allow natural lighting in an enclosed area. The United States Information Agency pavilion at Expo '67 in Montreal, Quebec had 36-foot-high acrylic domes made of transparent, grey acrylic sheet. This dome structure allowed 10-foot spans between bracing.

Use of acrylic refractors in highway lane marking is a recent development of importance. Approved in several states are injection-molded reflector lenses of acrylic molded in an ABS shell and filled with a rigid epoxy formula. These are affixed to road surfaces with an epoxy adhesive at 20-foot intervals to form clearly designated lanes by day or night. At night, car headlights supply the illumination, since the beam is turned 180° back toward the driver. Thus, the highway lane is clearly shown. The abrasion resistance of the acrylics is so exceptional that the intensity of reflection is maintained at a high level, even after being passed over by literally millions of cars.

Acrylics have been in use in the dental field for many years as a substitute for conventional amalgams in filling teeth. A reaction product of bisphenol A and glycidyl methacrylate with MMA monomer in a solution of benzoin methyl ether produces a liquid which hardens to a polymeric resin on exposure to UV light. The liquid is painted on the biting surfaces of molar teeth with a brush and hardens. The coating stays in place for at least two years, and during that time reduces cavities on the protected surface by 99%.

Themosetting Acrylics

Thermosetting acrylics cure by free-radical-initiated addition polymerization, without formation of by-products, to form hard, tough, thermoset materials that are resistant to heat, solvents, and chemicals. These materials have most of the desirable properties of the epoxy and acrylic materials and have the handling characteristics of unsaturated polyester resins. In general, curing times and conditions are comparable to those of the polyesters.

The thermosetting acrylics are compatible with and copolymerize readily with most polymerizable monomers, including unsaturated silicone resins, unsaturated polyesters (with or without styrene), DAP, triallyl cyanurate, styrene, vinyl toluene, divinyl benzene, dialkyl maleates and fumarates, and acrylic and methacrylic esters. A wide range of viscosities, handling characteristics, and properties can be had through proper selection of the crosslinking monomer.

Homopolymerization can also occur but, because of initially high resin viscosity, requires either solvent reduction or heating to provide a usable viscosity. Where heat is used, it is essential to use a high-temperature catalyst such as butyl or di-t-butyl peroxide or dicumyl peroxide.

Table 5-81 shows the effect of styrene crosslinking-monomer content on the properties of one grade of thermosetting acrylic. The weight loss for unreinforced castings of this resin (Nypol) is compared to a general-purpose polyester, an anhydride-cured novolac epoxy, and a silicone resin in Figure 5-81.

Thermosetting acrylics can be cured at room temperature by using benzoyl peroxide and an amine promoter, such as dimethyl aniline. As with polyesters, air inhibition is a problem, so optimum cure is best achieved in an air-excluded environment.

Main applications include fiberglass impregnants, resins for upgrading of polyester systems, impregnating varnishes, and electrical and electronic encapsulating and potting compounds.

Modified Acrylics

Modified acrylics are offered having properties unattainable by the basic unmodified compositions. The modifications

Table 5-81. Properties of Nypol 46-4001 (a Thermosetting Acrylic) Containing Different Amounts of Styrene.*

	Percent Styrene					
	0	10	20	30	40	50
Brookfield viscosity at 77°F, cps	–	18,500	1650	195	95	23
Hardness, barcol at 77°F	43	–	45	45	–	38
Tensile strength, psi	4840	–	8548	8326	–	9545
Tensile modulus, psi × 10^6	0.528	–	0.488	0.409	–	0.463
Tensile elongation, %	1.7	–	4.3	5.1	–	5.6
Flexural strength, psi	19,165	–	19,000	19,746	–	19,390
Flexural modulus, psi × 10^6	0.495	–	0.505	0.479	–	0.489
Izod impact, unnotched, ft-lb	0.92	–	1.43	1.85	–	1.92

*Courtesy: Freeman Chemical Corporation.

are made to gain improved impact strength, higher service temperatures, toughness, and extinguishing properties, although such modifications usually cause losses in other desirable characteristics.

One method of improving the properties of cast sheeting is by multi-axial hot-stretching. The sheet transforms from a brittle, amorphous plastic to a tough, oriented laminar material. The orientation improves the impact strength, solvent and stress-crazing resistance, without disturbing the other physical properties substantially. Another method of improving impact strength is through the addition of glass-fiber reinforcement.

Impact-modified acrylics offer strength up to five times those of the unmodified resins. These modified compositions are less stiff, and lose their transparency characteristics. Weather stability and abrasion resistance are reduced also. Compared to other moderately tough thermoplastics, these modified acrylics have harder surfaces, lower water absorption, greater resistance to staining, and better color stability. Table 5-82 shows properties of impact acrylic molding compounds.

Copolymers and blends with other rubbery resins also increase toughness. Typical comonomers include butadiene and acrylonitrile as well as other rubbers.

An acrylic-PVC alloy is offered that has very high impact strength, high stiffness, exceptional thermoforming characteristics, and is classed as self-extinguishing (see Table 5-82). Corrosion resistance is superior to that of unmodified acrylics. It is opaque and has lower service temperatures than the best unmodified acrylic compositions. This alloy is used for tank liners, machine and equipment housing (see Figure 5-82), and impact-resistant automobile bumpers.

A copolymer of MMA and alpha-methyl styrene provides higher service temperature. This modified composition is clear and transparent, and has a 11–22°C (20–40°F) higher service temperature than the unmodified acrylics (see Table 5-82). The copolymer loses in processing characteristics and in stability against weather and ultraviolet emissions as a result of the modification. It is used in lens applications where conventional unmodified acrylics provide insufficient

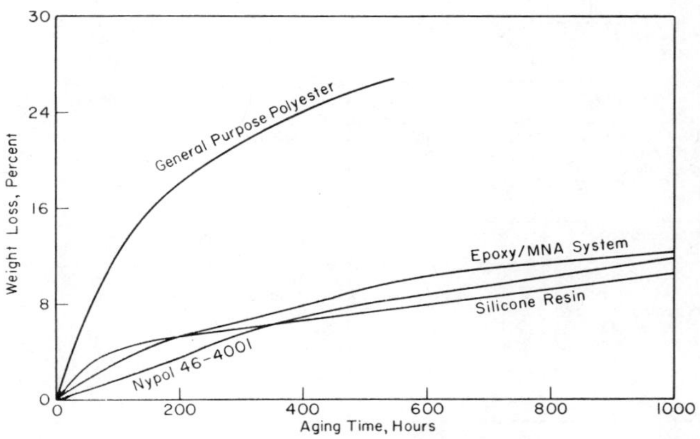

Fig. 5-81. Weight loss for various resins at 260°C (500°F). *(Courtesy: Freeman Chemical Corporation)*

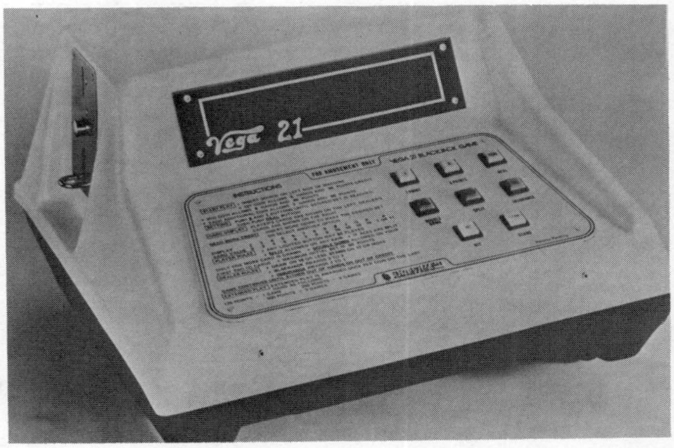

Fig. 5-82. A key component of this new, computerized, non-video blackjack machine is the ruggedly constructed cabinet of Kydex acrylic-PVC alloy sheet that resists impact, stains, and chipping, and never requires painting because its color is integral. *(Courtesy: Rohm & Haas Co.)*

Table 5-82. Properties of Impact Acrylic Molding Compounds.*

		ASTM test method	MMA styrene copolymer	Impact acrylic molding compound	MMA alpha-methylstyrene copolymer	Acrylic multi-polymer	Acrylic-PVC alloy sheet
Processing	1. Molding qualities	—	Excellent	Excellent	Excellent	Excellent	Good
	2. Compression molding temp., °F.	—	300-400	300-400	325-400	300-400	—
	3. Comp. molding pressure, p.s.i.	—	1000-8000	2000-5000	1000-10000	2000-6000	—
	4. Injection molding temp., °F.	—	330-500	400-500	350-525	400-500	—
	5. Injection molding pressure, p.s.i.	—	10000-30000	10000-20000	10000-30000	10000-20000	—
	6. Compression ratio	—	—	—	—	—	—
	7. Mold (linear) shrinkage, in./in.	—	0.002-0.006	0.004-0.008	0.002-0.006	0.004-0.010	—
	8. Specific gravity (density)	D792	1.09	1.11-1.18	1.09.1.16	1.10-1.12	1.35
	9. Specific volume, cu. in./lb.	D792	25.2	25.6-23.3	23.9-25.2	25.3-24.7	20.5
	10. Machining qualities	—	Good to excellent	Good to excellent	Good to excellent	Good to excellent	Excellent
Mechanical	11. Tensile strength, p.s.i.	D638	10000	5000-9000	10000	6000-8000	6500
	12. Elongation, %	D638	3.0	20.0-70.0	3.0	3.0-40.0	100
	13. Tensile elastic modulus, 10^5 p.s.i.	D638	4.3	2.0-4.0	4.3-4.6	3.1-4.3	3.31-3.35
	14. Compressive strength, p.s.i.	D695	11000-15000	4000-14000	11000-18000	8000-11500	8400
	15. Flexural yield strength, p.s.i.	D790	16000-19000	7000-13000	12000-19000	9500-13000	10700
	16. Impact strength, ft. lb./in. of notch (½ × ½ in. notched bar, Izod test)	D256	0.3	0.8-2.5	0.3	1.0-2.0 at 73°F 0.8-1.0 at −40°F.	15.0
	17. Hardness, Rockwell	D785	M-75	R105-R120	M75-M104	R108-119	R99-R105
	18. Flexural elastic modulus, p.s.i. × 10^5	D790	2.6-3.8	2.0-3.8	2.6-3.8	3.0-4.0	3.3-4.0
	19. Compressive modulus, p.s.i. × 10^5	D695	2.4-3.7	2.4-3.7	2.4-4.0	—	3.3-4.0
Thermal	20. Thermal conductivity, 10^{-4} cal./sec./sq. cm./1 (°C./cm.)	C177	4.0-5.0	4.0-5.0	4.0-5.0	5.3	—
	21. Specific heat, cal./°C./gm.	—	0.34	0.34	0.34	0.33	0.291-0.293
	22. Thermal expansion, 10^{-5} in./in./°C.	D696	6.0-8.0	5.0-8.0	6.0-8.0	6.0-9.0	—
	23. Resistance to heat, °F. (continuous)	—	180-200	140-195	180-200	150	—
	24. Deflection temp., °F. at 264 p.s.i. fiber stress	D648	205-212	165-194	205-220	185	160
	at 66 p.s.i. fiber stress		—	180-205	235	—	177
Electrical	25. Volume resistivity, ohm-cm. (50% RH and 23° C.)	D257	10^{15}	$>10^{16}$	10^{15}	1.7×10^{16}	10^{14}-10^{16}
	26. Dielectric strength, short-time, ⅛-in. thickness, volts/mil	D149	450	380-500	450	493	>400
	27. Dielectric strength, step-by-step, ⅛-in. thickness, volts/mil	D149	450	380-500	450	416	—
	28. Dielectric constant, 60 cyc.	D150	3.4	3.5-4.0	3.0-3.4	3.3-3.5	3.9
	29. Dielectric constant, 10^3 cyc.	D150	—	2.8-3.7	2.7	3.1-3.3	3.7
	30. Dielectric constant, 10^6 cyc.	D150	2.9	2.5-3.1	2.3-2.9	2.8-2.9	3.4
	31. Dissipation (power) factor, 60 cyc.	D150	.006	0.04-0.05	0.006-0.04	0.02-0.03	0.02
	32. Dissipation (power) factor, 10^3 cyc.	D150	—	0.02-0.04	0.03	0.02-0.03	0.02
	33. Dissipation (power) factor, 10^6 cyc.	D150	.013	0.004-0.04	0.013-0.02	0.02	0.03
	34. Arc resistance, sec.	D495	150	No track	150	—	80
Optical	35. Refractive index, n_D	D542	1.567	—	1.52-1.567	1.52	—
	36. Clarity	—	Transparent	Transparent, can be translucent opaque	Transparent	Transparent	Opaque
	37. Transmittance, %	D1003	90	55-88	90	87	—
	38. Haze, %	—	—	5 (min.)	—	6-12	—
Resistance characteristics	39. Water absorp., 24 hr., ⅛-in. thick., %	D570	.15	0.2-0.8	.15	0.3	0.06
	40. Flammability[a] Burning rate, in./min.	D635	—	1.0-1.3	1.3	1.2	—
	AEB, in./ATB, sec.[b]		—	—	—	—	0.8/5.0
	41. Effect of sunlight	—	Nil	Nil to slight strength loss	Nil	Slight color change, slight embrittlement	Some
	42. Effect of weak acids	D543	None	Practically nil	Nil	None	None
	43. Effect of strong acids	D543	—	Attacked only by high concentration oxidizing acids	Attacked only by high concentration oxidizing acids	Attacked by high concentration oxidizing acids	None
	44. Effect of weak alkalies	D543	None	Practically nil	Nil	None	None
	45. Effect of strong alkalies	D543	None	Practically nil	Nil	Very slight	None
	46. Effect of organic solvents	D543		Soluble in ketones, esters, aromatic and chlorinated hydrocarbons		Attacked by esters, ketones, aromatic and chlorinated hydrocarbons	

[a]This numerical flame spread rating is not intended to reflect hazards presented by this or any other material under actual fire conditions.

[b]AEB = Avg. extent of burning in inches; ATB = Avg. time of burning in seconds.

*Reprinted by permission from *Modern Plastics Encyclopedia*, McGraw-Hill, Inc.

heat resistance. The automotive industry uses this modified acrylic for lens interior applications.

Copolymerization with styrene yields a material with improved melt-flow characteristics (see Table 5-82). Copolymers with acrylic or methacrylic acid are superior to PMMA in hardness and abrasion resistance but are more susceptible to moisture. Copolymers with allyl methacrylate, allyl esters of dicarboxylic acids, or allyl ethers of dihydric alcohols are more scratch resistant than PMMA, but they transform the resin into a thermoset, which complicates fabrication.

A modified acrylic called a multipolymer has improved toughness, and maintains transparent characteristics, though haze is at a higher level than in the unmodified acrylics. These resins are believed to be a family of terpolymers based on MMA, acrylonitrile, and styrene in the presence of a butadiene-MMA copolymer. This family of compositions is inferior in color to the unmodified acrylics, and has relatively poor weather and UV characteristics (See Table 5-82). The compositions can be processed by blow-molding techniques, which makes it attractive to the cosmetic and food packaging industries. Other features of this resin group include FDA approval, resistance to oils and fats, low taste and odor transfer properties, excellent gas barrier, retention of strength at low temperatures, and thermoformability and heat sealability. Figure 5-83 shows an example of this type of multipolymer material.

Terpolymers based on MMA, butadiene, and styrene, which provide outstanding toughness without loss of clarity, have become of recent interest. These materials are described in some detail in the section on styrene copolymers.

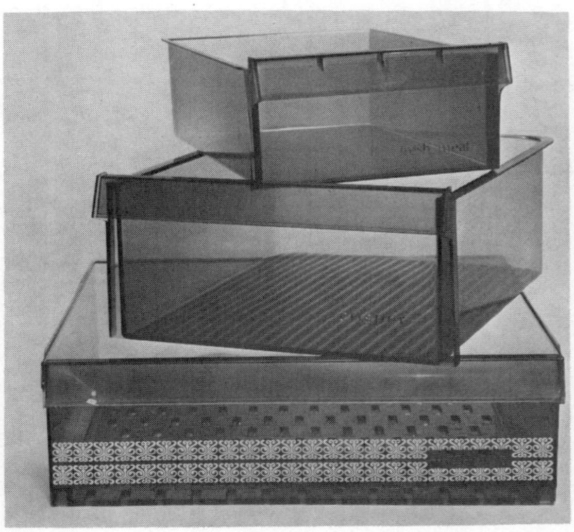

Fig. 5-83. Refrigerator crisper drawers made of XT polymer. This acrylic multipolymer has been developed for applications requiring transparency and toughness at an economical price. While it is stiff, chemical resistant, and transparent, it will not become brittle over time or with temperature extremes. *(Courtesy: CY/RO Industries)*

Acrylonitrile

Early work on the homopolymer of acrylonitrile was discouraging because of the non-uniform products obtained from emulsion polymerizations and the violence of bulk reactions. Acrylonitrile is high melting and insoluble or poorly soluble in most common solvents. However, study of both the polymerization and solubility of the polymer has developed a highly useful resin in the textile field. The monomer is made by two processes: one with ethylene oxide and hydrogen cyanide through ethylene cyanohydrin, and the other and more direct way from acetylene and hydrogen cyanide.

Ethylene oxide–hydrogen cyanide route:

$$HCN + \underset{\backslash\,O\,/}{CH_2{-}CH_2} \longrightarrow \underset{\text{ethylene cyanohydrin}}{\underset{OH\quad CN}{CH_2{-}CH_2}} \xrightarrow{-H_2O} CH_2{=}CHCN$$

Alkaline materials such as calcium hydroxide or sodium formate are used for the dehydration in the liquid phase. It is also possible to dehydrate the cyanohydrin by vaporizing it at reduced pressure and passing the vapors over activated alumina at 350°C (662°F).

Natural gas route:

$$\begin{array}{lll} CH_4 + O_2 \longrightarrow HC{\equiv}CH & & \\ \qquad\qquad\qquad\quad + & \longrightarrow & CH_2{=}CHCN \\ CH_4 + NH_3 \longrightarrow HCN & & \end{array}$$

This process can be carried out in either the liquid or gaseous phase. The gaseous process uses sodium or barium cyanide as a catalyst on a porous carrier at about 500°C (932°F). The liquid-phase process is more widely used. In one method, the reactants are passed into an aqueous solution of hydrogen chloride, cuprous chloride, and ammonium chloride at about 80°C (176°F), using a large excess of acetylene; the unreacted acetylene is recirculated. The resulting dilute solution of acrylonitrile is concentrated by continuous distillation.

Acrylonitrile is one of the most rapidly polymerizing vinyl monomers in the absence of inhibitors. Air, ammonia, ammonium carbonate, amines, and hydroquinone are effective inhibitors. Polymerization is through free-radical mechanisms.

Mass and emulsion polymerization are little used for acrylonitrile. The bulk process is difficult to control in large batches unless at very low temperatures. The solid polymer, which is insoluble in the monomer, forms a stiff slurry with the remaining monomer very early in the reaction, making the agitation for heat removal increasingly difficult. As a result, the reaction may become uncontrollable and even violent. In the emulsion process, the latex often coagulates. This effectively causes a solution-emulsion-type combination reaction, and the resultant polymer lacks molecular weight homogeneity.

The homopolymer may be prepared from an aqueous dispersion of the monomer at approximately 10–15% concentration. An oxidation-reduction catalyst system such as persulphate-bisulphite at 50°C (122°F) is effective, and, as the polymer forms, it separates from the reaction bath. The white, opaque, and crumbly polymer is separated by filtration, washed, and dried.

Unplasticized polyacrylonitrile films free of solvent are quite brittle. The homopolymer is not very thermoplastic, even upon heating, and objects can be molded only with difficulty. Thus, homopolymers are not commercially used in molding or casting applications.

The high negative polarity, the small size of the nitrile groups, and the high hydrogen-bonding capability of the polymer are responsible for its general insolubility, high softening temperature, and strength properties. It was determined very early in the polymer's history that it had all the qualities of a good fiber material. However, it could not be melt-spun, and commercial development had to await an appropriate wet-spinning process.

Dimethylformamide dissolves the polymer to give viscous spinning solutions. Dry-spinning requires cabinet temperatures of about 130°C (266°F). This temperature and higher is maintained after the solvent has evaporated, and the filaments are drawn to several times their original length. The stretching, accompanied by orientation, produces a 3-5 fold increase in tensile strength. Wet-spinning is done in concentrated salt solutions.

The acrylonitrile fiber in staple form is widely used in the manufacture of sweaters which do not shrink upon washing. It has good bulking and insulating properties. The fabric is resistant to light, mildew, moths, and smoke. Good resistance is reported to acids, alkalis, oils, and common solvents. The most outstanding property is its resistance to outdoor exposure. It has good wet and dry strength, fatigue resistance, and flex life. Filters and cloths of good chemical resistance can be manufactured.

A number of copolymers have been made for textile uses in which acrylonitrile is the major component. A principal reason for such copolymerization is that the homopolymer is not receptive to most dyes. The addition of minor quantities of vinyl pyridine and vinyl pyrrolidone greatly increases dye affinity. Vinyl chloride, vinyl acetate, and vinylidene chloride have also been used as minor constituents. In some instances, amides of acrylic acid have been used to improve dyeing properties. Vinylidene cyanide forms copolymers and also polymerizes readily to form a homopolymer.

Acrylamide, also used to copolymerize acrylonitrile, homopolymerizes to yield polyacrylamide, a white, odorless powder which is soluble in water. Its solutions are stable over a wide pH range and can be used as adhesives, thickening agents, and dispersants. The amide can be hydrolyzed to salts of polyacrylic acid. Other copolymers with styrene and styrene-butadiene will be treated in more depth in the section on styrenic copolymers.

Barrier Resins. Very recently, the development of so-called barrier resins has gained significant, though at the time of this writing, guarded commercial success. These are resins with specific properties for the food-packaging industry. Such a material must possess a glasslike luster and transparency—even at very low temperatures. It must provide an excellent barrier for oxygen and carbon dioxide, be odor- and taste-free, have good chemical resistance, be shatterproof when dropped, and be easily and safely incinerated along with household trash. One of the most promising resins meeting these requirements is a copolymer of acrylonitrile and methyl acrylate. There is even evidence that polymers of this type can be made biodegradable. The major market area of interest at this time is the soft-drink bottle market.

Acrylics in Contact Lenses and Implant Lenses. Another interesting and significant use of acrylics is in the manufacture of contact lenses. There are two kinds: standard hard lenses and recently developed soft lenses. Figure 5-84 shows a typical soft lens. Conventional hard lenses are made of PMMA, which is impermeable to oxygen and water. Oxygen is carried to the cornea in tears pumped under the lens with each blink. The soft lenses, made of hydrogels, have a remarkable ability to absorb water.

One soft lens for cosmetic use (vision correction in non-diseased eyes) is made by crosslinking poly(2-hydroxyethyl methacrylate) with ethyleneglycol dimethacrylate, then casting the hydrated material in a spinning mold. The other major hydrogel lens consists of a crossliked copolymer of hydroxyethyl methacrylate grafted onto polyvinyl

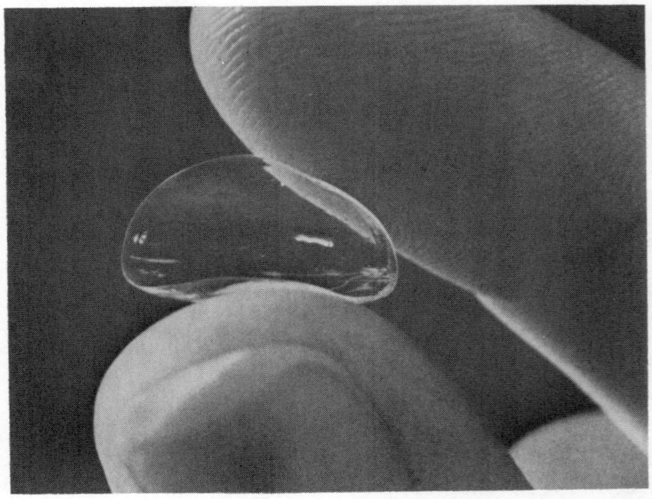

Fig. 5-84. Slight pressure of thumb and index finger demonstrates flexibility of soft lenses, which can act as bandages for many corneal disorders. (*Courtesy: Bausch & Lomb, Inc.*)

Table 5-83. Shear Strengths of Bonds Made in Eastman's Laboratories with Eastman 910 (Cyanoacrylate Type) Adhesives.*

Type of Bond	Substrate(s)	Age of Bond	Bond Strength in psi, Grade of EASTMAN 910 Adhesive: Regular	FS	EM	ET	Bond Strength in MPa, Grade of EASTMAN 910 Adhesive: Regular	FS	EM	ET
Metal to Metal	Steel	10 min	1920	1970	1670	730	13.2	13.6	11.5	5.0
		48 h	3300	3200	2890	890	22.8	22.1	19.9	6.1
	Aluminum	10 min	1480	1270	910	510	10.2	8.8	6.3	3.5
		48 h	2270	2390	1720	630	15.7	16.5	11.9	4.3
	Stainless Steel	10 min	1470	1470	850	890	10.1	10.1	5.9	6.1
		48 h	2570	2480	780	1210	17.7	17.1	5.4	8.3
	Steel/Stainless steel	10 min	140	No data	No data	No data	1.0	No data	No data	No data
		48 hr	3070				21.2			
Metal to Plastic	Aluminum/nylon	10 min	500	680	720	360	3.4	4.7	3.0	2.5
		48 h	950	900	680	450	6.6	6.2	4.7	3.1
	Aluminum/phenolic	10 min	650	850	960	370	4.5	5.9	6.6	2.6
		48 h	920	1030	910	390	6.3	7.1	6.3	2.7
Plastic to Plastic	ABS	10 min, 48 h	†	†	†	†	†	†	†	†
	Acrylic	10 min, 48 h	†	†	†	†	†	†	†	†
	Fiberglass-reinforced thermosetting polyester (FRP)	10 min	680	720	550	580	4.7	5.0	3.8	4.0
		48 h	510	320	490	910	3.5	2.2	3.4	6.3
	Nylon	10 min	330	370	320	470	2.3	2.6	2.2	3.2
		48 h	600	450	720	550	4.1	3.1	5.0	3.8
	Phenolic	10 min	†	†	†	1300	†	†	†	9.0
		48 h				1340				9.2
	Polycarbonate	10 min	790	680	710	920	5.4	4.7	4.9	6.3
		48 h	†	†	†	†	†	†	†	†
	Polypropylene, flame-treated	10 min, 48 h	†	†	†	†	†	†	†	†
	Polystyrene	10 min, 48 h	†	†	†	†	†	†	†	†
	Polysulfone	10 min	550	360	530	340	3.8	2.5	3.7	2.3
		48 h	†	†	540	†	†	†	3.7	†
	PPO	10 min	970	430	650	540	6.7	3.0	4.5	3.7
		48 h	790	520	710	710	5.4	3.6	4.9	4.9
	PTMT, unmodified	10 min	420	290	380	360	2.9	2.0	2.6	2.5
		48 h	340	430	440	530	2.3	3.0	3.0	3.7
	PTMT, glass-reinforced	10 min	890	630	690	870	6.1	4.3	4.8	6.0
		48 h	770	790	810	890	5.3	5.4	5.6	6.1
	PVC	10 min, 48 h	†	†	†	†	†	†	†	†
Rubber to Rubber	Butyl	10 min, 48 h	†	†	†	†	†	†	†	†
	EPDM	10 min	35	60	80	80	0.2	0.4	0.6	0.6
		48 h	70	50	60	85	0.5	0.3	0.4	0.6
	Natural	10 min, 48 h	†	†	†	†	†	†	†	†
	Neoprene	10 min, 48 h	†	†	†	†	†	†	†	†
	Nitrile	10 min, 48 h	†	†	†	†	†	†	†	†
	Polyurethane	10 min, 48 h	†	†	†	†	†	†	†	†
	SBR	10 min	130	140	110	140	0.9	1.0	0.8	1.0
		48 h	No data	No data	No data	155	No data	No data	No data	1.1
Rubber to Plastic	SBR rubber/phenolic	10 min, 48 h	†	†	†	†	†	†	†	†
	Neoprene rubber/FRP	10 min, 48 h	†	†	†	†	†	†	†	†
Woods and Synthetic Boards to Themselves	Balsa, 1/4 x 1 in.	24 h	†	†	†	†	†	†	†	†
	Fir, 1/4 x 1 in.	24 h	270	250	†	†	1.9	1.7	†	†
	Maple, 1/4 x 1 in.	24 h	†	570	540	960	†	3.9	3.7	6.6
	Oak, 1/4 x 1 in.	24 h	830	520	†	810	5.7	3.6	†	5.6
	Yellow pine, 13/64 x 1 in.	24 h	210	90	110	150	1.4	0.6	0.8	1.0
	Plywood, 1/4 x 1 in.	24 h	90	†	†	†	0.6	†	†	†
	Masonite board, 7/32 x 1 in.	24 h	†	†	90	†	†	†	0.6	†
	Particleboard, 3/4 x 1 in.	24 h	†	†	†	†	†	†	†	†

† Bond held, specimen broke.

*Courtesy: Eastman Chemical Products, Inc.

pyrrolidone; these "bandage" lenses are ground and polished in the dry state and rehydrated in physiologic solution.

Cyanoacrylates

Although they have been available for over 15 years, it has only been in the last few years that the cyanoacrylate adhesives have reached significant commercial success. Cyanoacrylate adhesives form strong bonds between many materials without the need for heat or catalysts.

The simplest alkyl cyanoacrylate adhesive, and the first to be offered commercially (in 1958), is based on methyl-2-cyanoacrylate. In recent years, adhesives based on higher cyanoacrylate homologs, such as ethyl, propyl, and butyl esters of cyanoacrylic acid, have been offered.

There are many features of these adhesives that distinguish them from the more common adhesives such as cellulosics, epoxies, and PVA's. Bond formation to a variety of substrates is very fast (seconds to minutes). Table 5-83 lists bonding times and strengths for various substrates. The cyanoacrylate adhesives develop very high bond strengths and the bonds are colorless. It is noted that the thinner the bond line, the higher the shear strength.

These adhesives are one-component systems that need no mixing or co-curatives. Generally, contact pressure is sufficient; and since they are 100% solids (i.e., there are no solvents present), no open time for solvent evaporation is required. These adhesives exhibit low shrinkage for good surface wetting and minimum internal stress as a result of shrinkage. The cost per weight of these adhesives is very high, but since only a few milligrams of material is required per bond in most applications, the cost per use is low.

Methyl-2-cyanoacrylate polymerizes via a base-catalyzed anionic mechanism. This polymerization is exothermic and can be initiated at low temperatures by very weak bases such as water and alcohols. Nearly all solid materials have traces of moisture adsorbed on their surfaces, enough to trigger this polymerization. Although the polymerization can also occur by free-radical mechanism, this method is of less practical importance, since nearly all industrial uses for the adhesives make use of an anionic polymerization.

The polymerization of alkyl-2-cyanoacrylates occurs without appreciable change in volume, and since no solvent is present in the adhesive to evaporate from the relatively high molecular weight bond, a bond with minimum shrinkage and maximum strength results.

Trade Names

Acco	Acrylic and methacrylate ester polymers and copolymers	Acco Polymers
Accrapak	Acrylic sheet	Chicago Plastics Sheets, Inc.
Acrifix	Acrylic adhesive	Cornelius Chemical Co., Ltd.
Acrilan	Poly (Acrylonitrile)-vinyl acetate fiber	Monsanto Co., Inc.
Acrivue	Acrylic sheet	Swedlow Inc.
Acronal	Polyacrylate	BASF Wyandotte Corporation
Acronal	Acrylic ester copolymers	BASF Wyandotte Corporation
Acropon	Acrylic finishes	DeSoto, Inc.
Acrulite	Acrylic adhesives	Rubber Latex, Ltd.
Acryester	Acrylic monomers	Mitsubishi Rayon Co.
Acryflex	Acrylic-coated fiberglass	Suflex Corporation
Acry Glass	MMA	Nitto Chemical Industry Co.
Acryl	Acrylic fiberglass	Bentley-Harris Mfg. Co.
Acrylene	Acrylic latex	Goodyear Tire & Rubber Co.
Acrylibond	Acrylic adhesive	Smooth-On, Inc.
Acrylic MF	Acrylic extrusion and molding material	Richardson Co.
Acrylite	Acrylic resins	CY/RO Industries
Acrylivin	Acrylic/PVC sheet	General Tire & Rubber Co.
Acrylux	Acrylic extrusions	Westlake Plastic Co.
Acrylweld	Acrylic adhesive	Hardman, Inc.
Acrysol	Water-soluble acrylic resins	Rohm & Haas Co.
Actomer X-80	Photocurable acrylic coating	Union Carbide Corporation
Adhesive	Cyanoacrylate adhesive	Eastman Chemical Products, Inc.
Aeress	Modacrylic fiber	Union Carbide Corporation
Alcyn	Cyanoacrylate adhesive	Evode, Ltd.
Alteco	Cyanoacrylate adhesive	Alpha Techno Corporation
Altuglos	Cast acrylic sheet, rod, and tube	Altutor, S.A.
Altulite	Acrylic resins and compounds	Altutor, S.A.
Anim	Butyl/acrylate fiber	Rohm & Haas Co.
Appretan	Acrylate copolymer	Hoechst Corporation
Aqua-Hyde	Acrylic resin solution	Lawter Chems, Inc.
Aqua-Rez	Acrylic resin solution	Lawter Chems, Inc.

Aron Alpha	Cyanoacrylate adhesive	Toagosie Chemical Industry; Vigor Co., Jadow & Sons
Aron Bright	Acrylic extruded sheet	Vigor Co., Jadow & Sons
Arontite	Anaerobic sealant	Toagosei Chemical Industry; Vigor Co., Jadow & Sons
Aroset	Thermosetting acrylics	Ashland Chemical Co.
Asterite	Acrylic sheet	Imperial Chemical Industries, Ltd.
Avdelbond A	Anaerobic adhesives and sealants	Avdel Corporation
Avdelbond B	Cyanoacrylate adhesives	Avdel Corporation
Barex	Acrylonitrile barrier resin	Vistron Corporation
Barick	Acrylic molding and extrusion compounds	J.T. Baker Chemical Co.
Bavick	Acrylic molding and extrusion	J.T. Baker Chemical Co.
Betastay	Acrylic adhesives	Essex Chemical Corporation
Bexel	SAN/acrylate sheet	Bakelite Xylonite, Ltd.
Bicor AB	Acrylic-coated PP film	Mobil Chemical Co.
Butacril	Acrylonitrile-butadiene	Ugine Kuhlmann
Butakon	Acrylonitrile-butadiene	Imperial Chemical Industries, Ltd.
Cadco	Extruded acrylic	Cadillac Plastic & Chemical Co.
Carboset 511	Acrylic resin	B.F. Goodrich Chemical Co.
Cavalon	Acrylic Adhesive	E.I. Du Pont de Nemours & Co., Inc.
Celloflex	Acrylic/melamine copolymer	Cello Chemical Co.
Centari	Acrylic enamels	E.I. Du Pont de Nemours & Co., Inc.
Chemigum	Acrylonitrile-butadiene rubbers and latices	Goodyear Tire & Rubber Co.
Chemlink	Polyfunctional acrylic and methacrylic monomers	Ware Chemical Co.
Cladlux	Acrylic/PVC cladding	Richard Daleman Group
Corian	Acrylic marble	E.I. Du Pont de Nemours & Co., Inc.
Coroc	Acrylic resins	Cook's Resins
Creslon	AN/acrylic ester copolymers	American Cyanamid Co.
Crilat	Acrylic emulsions	Montedison, S.p.A.
Crofon	Acrylic light piping	E.I. Du Pont de Nemours & Co. Inc.
Cryllex	Acrylic sheet	Bakelite Xylonite, Ltd.
Crystalex	Acrylic	Rohm & Haas Co.
Crystal-Vu-Crystic	Acrylic-silicone coating	U.S. Industries, Ltd.
Crystolex	Acrylic denture	Kerr Dental Co.
Cyanacryl	Acrylic elastomer	American Cyanamid Co.
Cyanolit	Cyanoacrylate adhesive	Denis Leader, Ltd.
Cyrolite	Impact-modified acrylic	CY/RO Industries
Dakril	PMMA casting grade	Soviet Origin
Decoloy	Acrylic-ABS composite	Borg-Warner Corporation
Degalon	PMMA	Degussa Wolgand Kunstoff-Betriebe
Delaglass	Acrylic sheet	Asahi Chemical Industry Co.
Delpet	Acrylic resins	Asahi Chemical Industry Co.
Diakon	Acrylic resins and molding compounds	Imperial Chemical Industries, Ltd.
DKE	Acrylic/PVC alloy	E.I. Du Pont de Nemours & Co., Inc.
Dralon T	Acrylic yarn	Bayer, A.G.
Duplacryl	Acrylic denture	Coralite Dental Co.
Dural	Acrylic-modified vinyl	Alpha Chemical & Plastics Corporation
Dutch Boy	Cast acrylic sheet	Landover Mfg. Co.
Dynel	Modacrylic fibers	Union Carbide Corporation
Eastman	Cyanoacrylate adhesive	Eastman Chemical Products Inc.
Eccobond KWIK	Cyanoacrylate adhesives	Emerson & Cuming, Inc.
Edimet	Methacrylic cast sheet	Montedison, S.p.A.
Elaprin	AN/butadiene and acrylic elastomers	Montedison, S.p.A.
Electro Glas	Cast acrylic rod, tube, and sheet	Glasflex, Inc.
Elvacite	Cast acrylic resin	E.I. Du Pont de Nemours & Co., Inc.
Epocryl	Epoxy acrylate	Shell Chemical Co.
Europrene	AN/butadiene copolymers	ANIC, S.p.A.
Evolite	Acrylic rod, tube, and profile	Richard Daleman, Ltd.
Evr-Kleer	PMMA	Cast Optics Corporation
Finacryl	Acrylic sheet	Croydon Plastics, Ltd.
Finnacryl	Acrylic sheet, rod, blocks, and film	Lohjan Kalk-Kitendas Oy-Lojo Kalkverk, Ab

Flair	Patterned-cast acrylic sheet	Rohm & Haas Co.
Flex-o-Crylic	Acrylic sheet	Rohm & Haas Co.
Flex-o-Glaze	Acrylic sheet	Warp Brothers
Flowcoat	Acrylic coating	Product Techniques, Inc.
Fondo	Acrylic leather	Henkel, Inc.
Gala	Acrylic rod, sheet and molding compounds	George Morrell Corporation
Gelva RA	Acrylic multipolymer solutions	Monsanto Co.
Genco	Acrylic	General Plastics Inc.
Gerlite	Acrylic sheet	Monsanto Co.
Glopaque	Acrylic sheet	U.S. Plastic & Chemical Corporation
Gravimerick	Acrylic sheet	Hermes Plastics, Inc.
Gravo Ply	Acrylic sheet	Hermes Plastics, Inc.
Habond	Acrylic cements	Hastings Plastics Co.
Hardrock	Acrylate adhesive	Denki Kagaku Kogyo, K.K.
Haze Control	Contact lenses	Tex-O-Con Optics, Inc.
Herecrol	Acrylonitrile rubber	Heresite & Chemical Co.
Hitaloid	Acrylic coating resins	Hitachi Chemical Co.
Hycar	Polyacrylic ester	B.F. Goodrich Chemical Co.
Hydron	Acrylic polymer	Woodstream-Hydron Corporation
Hydron	Loosely crosslinked hydroxyethyl methacrylate	National Patent Development Corporation
Implex	High-impact acrylic molding compounds and resins	Rohm & Haas Co.
Instant-Weld	Cyanoacrylate adhesive	Oneida Electronic Mfg., Inc.
Interpol	Acrylic resins	Freeman Chemical Corporation
Joncryl	Acrylic resins	S.C. Johnson & Son, Inc.
Kallodoc	Acrylic powder	Imperial Chemical Industries, Ltd.
Klearmount	Acrylic syrup	Vernon-Benschoff Co., Inc.
Kleer Cote	Acrylic coating	Crown Industrial Products
Koldmount	Acrylic	Vernon-Benschoff Co., Inc.
Koracryl H	Acrylic copolymer sheet	Koro Corporation
Korad A	Acrylic film	Korad, Inc.
Krilik	Acrylic coating	Guardsman Chemical Co.
Krynac	AN/butadiene	Polysar
Kydene	Acrylic/PVC alloy molding compounds	Rohm & Haas Co.
Kydex	Acrylic/PVC alloy sheet	Rohm & Haas Co.
Kyowaglas	Lead-filled acrylic sheet	Kyowa Gas Chemical Industry Co.
Kyowalite	Acrylic sheet	Kyowa Gas Chemical Industry Co.
Lancryl	Thermosetting acrylics	Lanson Chemical Corporation
Larcylic	Acrylic sheet	Larc Products, Inc.
Larodur	Acrylate coating resin	BASF Wyandotte Corporation
Latecoll	Acrylate salts	BASF Wyandotte Corporation
Lecton	Acrylic-coated fabric and laminates	E.I. Du Pont de Nemours & Co., Inc.
Loctite	Cyanoacrylate adhesives	Douglas Kane Group
Lucite	Acrylic resins and molding powders	E.I. Du Pont de Nemours & Co., Inc.
Lumasite	Glass-reinforced acrylic	American Acrylic Corporation
Luprenal	Acrylate-coating resin	BASF Wyandotte Corporation
Lustrasol	Acrylic copolymer	Reichhold Chemicals, Inc.
Lustreglow	Acrylic sheet	U.S. Plastic & Chemical Corporation
Lustrelite	Acrylic sheet	U.S. Plastic & Chemical Corporation
Lustretone	Acrylic sheet	U.S. Plastic & Chemical Corporation
Macrynal	Thermosetting acrylic resin	Cessella Farbwerke Mainkur, A.G.
Marglass	Acrylic fibers	Marglass, Ltd.
Metacel	Acrylic copolymer	Metachem Resins Corporation
Metacrylene	Methacrylate co- and terpolymer	Products Chimiques Ugine Kuhlmann
Metallex	Cast acrylic sheet	Hermes Plastics, Inc.
Methacrel	Acrylic emulsion	E.I. Du Pont de Nemours & Co., Inc.
Micro-Matte	Extruded acrylic sheet	Extrudaline, Inc.
Mirron	Vacuum-metallized acrylic rod and tube	Thermoplastic Processes, Inc.
Monocite	Methacrylate monomer	E.I. Du Pont de Nemours & Co., Inc.
Mortac	Acrylic emulsions	Asahi Chemical Industry Co.

MPL	Acrylic monomer	Borden Co.
Murcoat	Acrylic emulsions	Asahi Chemical Industry Co.
Nacrylic	Acrylic latices	National Starch & Chemical Corporation
NAS	Styrene methylmethacrylate copolymer	Richardson Co.
Nelcote	Acrylic emulsions	Northeastern Labs Co., Inc.
Nelcryl	Acrylic solutions	Northeastern Labs Co., Inc.
NeoCryl	Acrylic resins	Polyvinyl Chemicals, Inc.
Noan	PS/MMA copolymer	Richardson Co.
NuGlas	Acrylic-modified PEs	Vistron Corporation
Nypol	Thermosetting acrylic	Freeman Chemical Corporation
Orlon	Acrylic fiber	E.I. Du Pont de Nemours & Co., Inc.
Oroglas	Acrylic molding compounds and sheet	Rohm & Haas Co.
Pacrosir	Acrylic resins	Societa Italiana Resine
Pan	Polyacrylonitrile	Solvay & Cie
Paracryl	AN/butadiene copolymer	Uniroyal, Inc.
Paraglas	MMA	Kyowa Gas Chemical Industry Co.
Paraloid	Acrylic ester resin	Rohm & Haas Co.
Parcryl	Arcylic latices	Thibaut & Walkes Co.
Perlac	Acrylic sheet	Bakelite Xylonite, Ltd.
Permabond	Cyanoacrylate adhesives	Permabond International Corporation
Permacryl	Acrylic monofilaments	R. Brandt & Co., Ltd.
Perma Grain	Radiation-cured MMA/wood	ARCO Chemical Co.
Perspex	Acrylic	Imperial Chemical Industries, Ltd.
Placquer	Acrylic solutions, emulsions, and dispersions	Custom Coatings Co.
Plasi-pane	Acrylic sheet	Warp Brothers
Plex	Acrylic sheet	Rohm & Haas Co.
Plexi	Acrylic sheet and molding powders	Rohm & Haas Co.
Plexiglas	Acrylic molding powders and sheet	Rohm & Haas Co.
Plexigum	Acrylic adhesive	Rohm & Haas Co.
Plextol	Crosslinked acrylic	Rohm & Haas, Gmb. H.
Plexweld	Acrylic cement	Kidder Mfg.
Plyamul	Acrylic adhesives	Reichhold Chemicals, Inc.
Polex	Acrylic film and sheet	Southwestern Plastics, Inc.
Polycast	Acrylic sheets	Fortin Laminating Corporation
Polyglass	Cast acrylic sheet	Cast Optics Corporation
Polyglass	Acrylic	Polycast Technology Corporation
Poly-Tex	Acrylic and acrylic/PVC copolymers	Celanese Resins
Primal	Acrylic emulsions	Rohm & Haas Co.
Prime Cor X	Acrylic sheet	Primex Plastics Co.
Pronto 11	Acrylic	Vernon-Benschoff Co., Inc.
Puffee	Acrylic yarn	Pocasset
Racryl	Acrylic emulsions	R.A. Chemical Corporation
Radcrete	Radiation-cured MMA/concrete	Radiation Technology
Radstone	Radiation-cured MMA/concrete	Radiation Technology
Randacryl	Acrylic lacquer	Randolph Products Co.
Raycron	UV-cured acrylic coatings	PPG Industries
Rayoglas	Acrylic latices	California Products Corporation
Reactomer	Acrylic monomers	Thiokol Corporation
Resartglass	Cast acrylic sheet, rod, and block	Resart-IHM, A.G.
Resiacryl	Acrylic resins	Montedison, S.p.A.
Resiapat	Acrylic resins	Montedison, S.p.A.
Rexite	Acrylic adhesives	Franklin Glue
Rez-N-Bound	Acrylic cement	Schwartz Chemical Co.
Rhoplex	Acrylic-ester emulsions	Rohm & Haas Co.
Rigidite	Acrylic sheet	American Cyanamid Co.
Rohacell	Acrylic foam	Rohm & Haas Co.
Rohaglas	Acrylic sheet molding compounds	Roehm GmbH
Sadur	AN/MMA copolymer	Resart-IHM, A.G.
Scotch-Bond	Acrylic adhesives	3M Co.

Scotch-Weld	Cyanoacrylate adhesives	3M Co.
Separan	Polyacrylamide	Dow Chemical Co.
Shinkolite	Acrylic sheet	Mitsubishi Rayon Co.
Sicoglas	Cast acrylic sheet	Mazzucchelli Celluloide, S.p.A.
Skino	Methacrylate ketoxime	Mooney Chemicals, Inc.
Soflens	Hydroxyethyl methacrylate contact lens	Bausch & Lomb, Inc.
Stabicol	PVAc/acrylate adhesive	Scholten's Chemische Fabrieken, N.V.
Sta-Lok	Anaerobic sealant	Broadview Chemical Co.
Sumipex	PMMA resin and sheet	Sumitomo Chemical Co., Ltd.
Superdrop	Cyanoacrylate adhesives	Ornsteen Chemicals
Super Quick	Cyanoacrylate adhesive	IKO Co., Ltd.
Super Three Cement	Cyanoacrylate adhesive	Three Bond Co., Ltd.
Sure Kote	Acrylic lacquer	7-K Color Corporation
Swedcast 300	Continuous-cast acrylic sheet	Swedlow Inc.
Swedlow	Cast acrylic sheet	Swedlow Inc.
Swirlustre	Acrylic sheet	U.S. Plastics & Chemical Corporation
Synocure	Acrylic resins	Croy Valley Products, Ltd.
Synthacryl	Thermosetting acrylic	American Hoechst Corporation
Tame	Acrylic adhesives	B.F. Goodrich General Products Co.
Telakril	PMA/EA	Soviet Origin
Texcryl	Acrylic emulsions	R.A. Chemical Corporation
Texicryl	Acrylic emulsions	Scott Bader & Co., Ltd.
Texigel	Polyacrylate thickeners	Scott Bader & Co., Ltd.
Texilar	Acrylic	Scott Bader & Co., Ltd.
Thiacril	Polyacrylate rubber	Thiokol Corporation
Tilon	Acrylic molding compound	CMP Industries, Inc.
Torayglass	Acrylic cast sheet	Toray Industries, Inc.
Torex	Acrylic molding compounds	Toray Industries, Inc.
Transpex	Extruded acrylic sheet	Imperial Chemical Industries, Ltd.
Tuffak	Acrylic sheet and coatings	Rohm & Haas Co.
Ubabond	Acrylonitrile solvent cement	UBS Chemical Corporation
Uracryl	Acrylic resins	Unilever, Ltd.
Urifilm	Melamine-crosslinked acrylic	Advanced Coatings & Chemicals, Inc.
Valbond	Acrylic copolymer emulsions and solutions	United Merchants, Inc.
Vamac	Acrylic elastomer	E.I. Du Pont de Nemours & Co., Inc.
Vedril	PMMA molding and extrusion compounds	Montedison, S.p.A.
Vedrilcol	Acrylic adhesive	Montedison, S.p.A.
Verel	Modacrylic/AN copolymers	Eastman Chemical Products, Inc.
Vernonite	Acrylic resin	Vernon-Benshoff Co.
Versilok	Acrylic adhesives	Hughson Chemicals Div., Lord Corporation
Vicobar	AN-based plastics	Du Pont-Continental Can Co.
Vinacryl	Acrylic	Vinyl Products, Ltd.
Vinalak	Acrylic	Vinyl Products, Ltd.
Vinamul	Acrylic	Vinyl Products, Ltd.
Viniproz	PVC/PMMA copolymer	Soviet Origin
Vinlok	Acrylic emulsions	Celanese Coatings & Specialties Co.
Vitredil	Methacrylic sheet	Montedison, S.p.A.
Vult-Acryl	Acrylic latices	General Latex & Chemical Corporation
Wasco Acrylite	Acrylic structural panels	Wasco Products, Inc.
Wascolite	Acrylic	Wasco Products, Inc.
Weldmaster	Acrylic adhesive	Advance Coatings Co.
Wicacril	Acrylic emulsions	Celanese Coatings & Specialties Co.
Wresimul	Acrylic emulsions	British Paints, Ltd.
X-Link	Thermosetting vinyl-acryclic copolymer dispersions	National Starch & Chemical Corporation
XT	Acrylic multipolymer	American Cyanamid Co.
Xtrene	High-impact acrylic	Maryland Cup Corporation
Zefkrome	Acrylic fiber	Dow Badische Co.
Zefran	Acrylic fiber	Dow Badische Co.
Zerlon	MMA/styrene copolymers	Dow Chemical Co.
Zetafax	E/Ac copolymers	Dow Chemical Co.
Zetafin	E/EAc copolymers	Dow Chemical Co.
Zipbond	Cyanoacrylate adhesive	Tescom Corporation
Zip Grip	Cyanoacrylate adhesive	Devcon Corporation

REFERENCES AND BIBLIOGRAPHY

Almand, P., and Byrd, R. New developments in plastics for optical applications. *Materials Engineering*, November 1972, **76**(6):42–45.

American Cyanamid Co. *XT Polymer–Clear Plastic for Bottles.* Bulletin PKT 2C 3-2047, June 1973.

——. *XT Polymer for Thermoformed Packaging.* Bulletin PKT 5B 2/2014, April 1972.

——. *Acrylite in Architecture–Facades, Glazing, Lighting, Signs.* Form AST1, October 1968.

Brumit, T. M. Cyanoacrylate adhesives–when should you use them? *Adhesives Age*, February 1975, **18**(2):17–22.

Brydson, J. A. *Plastics Materials.* 3rd Ed. Robert E. Krieger Publishing Co., Inc., Huntington, New York, 1975.

Campbell, F., Rugg, B., Kumar, R., Arnon, J., and Brenner, W. Environmental aging studies of electron-beam-cured adhesives. *Adhesives Age,* January 1979, **22**(1):40–42.

Carlson, J. A. Nitrile resins, from Plastics in Packaging. *Plastics Engineering*, October 1978, **34**(10):40–41.

Chemical Engineering News. Polyacrylate elastomers attract new interest. November 1963, **41**(48):27–28.

Clement, L. A. A Review of acrylic resins. *Plastics Design and Processing*, February 1968, **8**(2):24–27.

CY/RO Industries. *Physical Properties of Acrylite FF Acrylic Sheet.* Bulletin AST 609, April 1977.

——. *Acrylite Acrylic Sheet Fabrication Manual.* Bulletin 6-1914, March 1976.

E. I. Du Pont de Nemours & Co., Inc. *Product Data from Du Pont, Lucite T-1000 High Impact Grade.* Bulletin E-11698, 1977.

——. *New Design Concepts With Crofon Light Guides.* Bulletin A-49415, May 1967.

——. *"Lucite" Acrylic Resin.* School Service Bulletin A-54238.

——. *Abcite AC Abrasion Resistant Sheet.* Bulletin A-77193.

——. *DKE-450 Extrusion Resin–Product Sheet.* Bulletin A-87634.

Eastman Chemical Products, Inc. *Eastman 910 Adhesives.* Publication No. PA-16B, ECP 1739, September 1977.

——. *Eastman 910 Adhesive.* Publication No. R-192.

Forger, G. R. Plastic alloys offer alternatives to the high cost of overengineering. *Materials Engineering*, July 1977, **86**(1):44–47.

Freeman Chemical Corporation. *Nypol Thermosetting Acrylic Resins–Technical Data Sheet.* May 1962.

Gitschner, H. W., Menges, G., and Putz D. "Sorption, Swelling, and Strain Behavior in Polymers." Paper presentation, *SPE ANTEC*, New Orleans, May 1979.

Golding, B. *Polymers and Resins–Their Chemistry and Chemical Engineering*, D. Van Nostrand Co., Inc., Princeton, New Jersey, 1959.

Guha, P. K., and Epel, J. N. Adhesives for bonding graphite/glass composites. *Adhesives Age*, June 1979, **22**(6):31–34.

Gush, D. P., and Kettey, A. D. "Thiol/acrylate hybrid systems may offer the best of both worlds in radiation-curable coatings. *Modern Paint & Coatings*, November 1978, **68**(1):58–66.

Hagan, J. W., Mallon, C. B., and Rifi, M. R. Recent developments in acrylic polymers for latex PSA's. *Adhesives Age*, March 1979, 22(3):29–33.

Halstead, W. G., Ringerle, J. R., and Suh, N. P. "Variable Volume Mold in Shrinkage Control." Paper presentation, *SPE ANTEC*, New Orleans, May 1979.

Haluska, L., and Galic, G. *Abrasion Resistant Coating for Transparent Plastics.* Dow Corning Corporation, Form No. 19-024.

Hayes, J. T. Water-based acrylic PSA's. *Adhesives Age*, March 1979, **22**(3):34.

Hulstein, C. Automatic application of anaerobic adhesives. *Adhesives Age*, July 1979, **22**(7):25–28.

Korad, Inc. *Korad A: Technical Notes.* October 1976.

Luskin, L. S. Acrylics, from Plastics in Construction: Coming on Strong in 1978. *Plastics Engineering*, July 1978, **34**(7):26–29.

Martino, R. Other materials–acrylics, from Materials '79. *Modern Plastics*, January 1979, **56**(1):60–61.

——. Opportunities in fiber optics: for whom, for what, for when? *Modern Plastics*, July 1978, **55**(7):38–41.

Materials Engineering. Cast acrylic sheet is easier to bond. May 1977, **85**(5):65.

——. Cyanoacrylate bonds irregular surfaces. August 1976, **84**(2):26.

——. New plastic doubles fiber optics distance. January 1975, **81**(1):31.

——. Plastics windows don't break. March 1974, **79**(3):26–27.

Materials in Design Engineering. New acrylic sheet is continuously cast. November 1965, **62**(6):5–6.

——. New thermosetting acrylic for reinforced plastics. August 1962, **56**(2):10–11.

Milazzo, B. Pressure-sensitive tapes: what's the state of the industry? *Adhesives Age*, March 1979, **22**(3):27–28.

Mock, J. A. Ultraviolet curing saves energy, speeds production and cuts costs. *Materials Engineering*, May 1976, **83**(5):40–42.

——. Acrylic coatings adhere well, have excellent weather resistance. *Materials Engineering*, August 1976, **84**(2):32–34.

Modern Paint & Coatings. Campstoves weather elements with acrylic electrodeposition coatings. November 1978, **68**(11):50.

Modern Plastics. Acrylic/glass windows for helicopters are strong, lightweight, and clear. June 1979, **56**(6):32.

——. Injection molded acrylic "cube" saves steps and doubles output. February 1978, **55**(2):28.

——. The new bottle equation: GS* + mrc† + NPt‡ = rH§. April 1975, **52**(4):41–43.

——. XT polymer–new material of design. February 1966, **43**(2):82–83, 169–171.

Parker, E. E., and St. Pierre, T. L. Formable thermosetting acrylic polymers. *Modern Plastics*, June 1961, **38**(6):133–139, 209.

Permabond International Corporation. *Perma-Lok Anaerobic Adhesives/Sealants for Controlled Bonding of Metal Parts.* Bulletin PL200, May 1977.

——. *Permabond (Alpha Cyanoacrylate) High Strength, Rapid Bonding Adhesive.* Bulletin PB1840, April 1977.

Plastics Design & Processing. High clarity thermoplastic has unusual heat, chemical and water resistance. August 1961, **1**(5):30.

Plastics Technology. The new thermosetting acrylics. May 1963, **9**(5):53–54.

Plastics World. FDA puts squeeze on nitrile beverage bottles. March 1977, **35**(3):10.

——. Monsanto mothballs its AN production. July 1977, **35**(7):8.

——. More bad news for acrylonitrile. July 1977, **35**(7):9.

——. New challenge for U.S. acrylic dollar. November 1976, **34**(11):15.

Rigbi, Z., and Goodman, S. H. Determination of optimum level of vinyl-terminated poly(butadiene co acrylonitrile) for toughening poly(methyl methacrylate). *Polymers & Plastics Materials*, Haifa, Israel, November 1977, **7**(2):28–26.

Rohm & Haas Co. *Plexiglas V & Plexiglas DR Acrylic Plastic Molding Pellets.* Bulletin PL-809a, March 1976.

——. *Kydex Acrylic PVC Alloy.* Bulletin PL-701e, August 1975.

——. *Plexiglas Sheet Physical Properties.* Bulletin PL-229L, February 1972.

——. *Physical Properties of Kydex Acrylic-PVC Sheet.* Bulletin PL-674c, September 1971.

——. *Optics–Plexiglas Molding Power Design Manual.* Bulletin PL-8976, May 1970.

——. *Glossary–Plexiglas Molding Powder Design Manual.* Bulletin PL-897e, May 1970.

——. *Korad A Surfacing of ASB.* Bulletin FM-12, October 1968.

——. *Korad A.* Bulletin FM-04, February 1967.

——. *Korad A–Acrylic Film Color Samples.* Bulletin FM-06.

——. *Solving Light Control Problems in Plexiglas.* Bulletin PL-177, November 1967.

——. *Kydex 100 Acrylic-Polyvinyl Chloride Alloy–Average Physical Properties Table.* Bulletin No. PL-6746, January 1966.

——. *Plexiglas–Handbook for Aircraft Engineers.* Manual No. PL-26, 1957.

——. *Plexiglas Acrylic Plastic Molding Pellets.*

——. *Plexiglas DR Physical Properties.* Bulletin PL-997a.

Rosato, D. V. For low weight, high performance–biaxially oriented bottles. *Plastics World*, August 1977, **35**(8):48–51.

Sanderson, T. Water-based, acrylic pressure-sensitive adhesives. *Adhesives Age*, December 1978, **21**(12):31–36.

Sherlock, J. M. Rigidized acrylics, new versatility for higher volumes. *Plastics World*, July 1977, **35**(7):56–58.

Simonds, H. R., and Church, J. M. *A Concise Guide to Plastics*, 2nd Ed. Van Nostrand Reinhold Co., New York, 1963.

Solomon, J. Eye on soft lenses. *The Sciences*, New York Academy of Sciences, November 1973, **13**(9):18–20.

SPE Journal, Acrylic activities. January 1966, **22**(1):20–21.

Stoloff, A. Modified-acrylic impact modifier for superior vinyl siding. *Plastics Engineering*, July 1979, **35**(7):29–31.

Swanson, R. S. *Plastics Technology–Basic Materials and Processes.* McKnight & McKnight Publishing Co., Bloomington, Illinois, 1965.

Swedlow, D. New casting process makes continuous acrylic sheet. *Product Engineering*, December 1965, **36**(25):89–91.

Swedlow, Inc. *Abrasion Resistant Coatings.* Engineering Bulletin 71-003, January 1972.

——. *Swedlow Aircraft Products.* Bulletin No. 202, February 1965.

——. *The Swedlow Concept.*

Thompson, R. T. "Anaerobic and Cyanoacrylate Adhesives." Paper presentation, *SPE ANTEC*, New Orleans, May 1979.

3M Co. *Scotch-Weld Brand Cyanoacrylate Adhesive.* Product Specifications CA-4 and CA-8, March 1978.

——. *Anaerobic Adhesive/Sealant Answers.* Bulletin Z-ANB (117.5) PD.

Titus, J. B. *Trade Designations of Plastics and Related Materials*, (revised). PLASTEC Note N9C, Plastics Technical Evaluation Center, Picatinny Arsenal, Dover, New Jersey, May 1978.

Western Plastics. Sierracin produces new coating material. August, 1969, p. 22.

Wilkinson, T. L. Acrylic adhesives: new way to bond aluminum auto parts. *Adhesives Age*, June 1978, **21**(6):20–23.

Winding, C. C., and Hiatt, G. D. *Polymeric Materials*, McGraw-Hill Book Co., New York, 1961.

Wood, A. S. About the new coatings-on-plastics technology. *Modern Plastics*, December 1978, **55**(12):36–39.

——. Clarity plus lots of useful extras: it's available at \$3/lb.–or 30¢. *Modern Plastics*, March 1979, **56**(3):50–53.

Woodman, J. F. "The Acrylic Resins." In *Modern Plastics Encyclopedia*. S. Gross (exec. ed), **45**(14A):136–138. McGraw-Hill, Inc., New York, 1967–68.

STYRENICS

Polystyrene

One thermoplastic which received a tremendous boost as a result of the World War II synthetic rubber program was polystyrene. Although production of this resin was already established in 1937, monomer-manufacturing capacity was not high (see Table 5-84). After the war, large quantities of monomer became available as the styrene-butadiene rubber program slowed up. Government plants were sold to private industry, which profited from the advanced know-how in monomer-making and polymerization techniques.

The monomer, which boils at 145°C (293°F), is prepared from ethyl benzene.

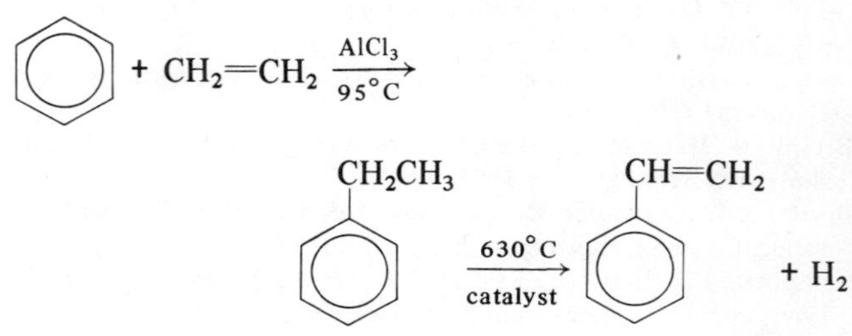

The commercial process follows a two-reaction path. Ethyl benzene is formed by the alkylation of benzene with ethylene. The ethylene and benzene are reacted in the presence of $AlCl_3$ and HCl under anhydrous conditions, forming ethyl benzene and more highly alkylated benzenes. High-purity ethylene is not required as long as the impurities are hydrogen, light hydrocarbons, or inert substances. Acetylene is not desirable because it increases catalyst consumption. The benzene used is a high-purity (99%) "styrene-grade." A small amount of ethyl chloride is present to promote high catalyst efficiency via the presence of hydrogen chloride and ethyl groups.

The purified ethyl benzene is dehydrogenated catalytically in the presence of steam to yield styrene. This is an endothermic reaction and is carried out under steam to reduce the partial pressure of the reactants. This allows for operation at reduced pressure rather than at high temperature. At higher temperatures, which allow more favorable reaction rates and equilibriums, cracking reactions occur at increasing levels. To preclude these cracking reactions at the lower temperatures, catalysts that are specific to the dehydrogenation reaction are used. Such catalysts include activated charcoal, bauxite, and zinc, magnesium, aluminum, and iron oxides.

The dehydrogenation feed must be low in ethyl benzene to preclude divinyl benzene formation. Divinyl benzene polymerizes rapidly to insoluble product during subsequent purification. The purification is the most difficult phase of styrene production. The boiling points of ethyl benzene and styrene differ by only 9°C (48°F) at atmospheric pressure. Distillation is not feasible because styrene has an increased tendency to polymerize at its boiling point. Trace amounts of impurities also increase the potential for premature polymerization. Inhibitors often used to prevent premature polymerization during storage or shipment include such materials as *p-tert*-butyl catechol and hydroquinone. Prior to use, this inhibitor is removed by washing or distillation.

The use of monomeric styrene in unsaturated polyesters will be mentioned in the section under polyester resins. The monomer serves as a solvent for the unsaturated polyester resin. Then during crosslinking, the liquid monomer reacts with the unsaturated polyester via addition across the double bonds to give an insoluble crosslinked polymer. By varying the polyester, the amount of monomer, and the catalyst, resin systems are obtainable which cure over a broad temperature range, with and without pressure, yielding a wide variety of properties and products.

The polymerization of styrene may occur via all four basic methods of polymerization, under the influence of heat alone and/or an initiator. Styrene will homopolymerize in the presence of solvents, fillers, pigments and dyes, plasticizers, and other rubbers and resins. It also coreacts with many other monomers to form a large variety of copolymers.

The choice of the polymerization method used will depend on the desired end form and on the purity and the clarity required in the final polymer. In the emulsion operation, heat-transfer problems are negligible since the reaction mass never reaches a high viscosity and agitation is easy. Water-soluble catalysts are used. The coagulated product must be washed to remove emulsifying agents, which even then contaminate the dusty, dry powder. However, if the emulsion is for a non-molding use, it may be plasticized and coated without the washing step.

Mass or bulk polymerization may be a batch or continuous process. Heat is generally the sole catalyst. Styrene of 99.5% purity is heated at 80–85°C (176–185°F) in a stainless steel or aluminum kettle. At this temperature, a 35–40% conversion is obtained in 40–60 hours. The viscous solution is then passed down a tower with zones of increasing temperature up to 200°C (392°F) to strip off the unconverted monomer and take off the polymer in a molten state. The extruded molten product is cooled and granulated. To obtain a material with the highest heat-deflection temperature, the last traces of monomer must be removed. This may be done by passing molten polymer in a thin stream through an evacuated chamber. The absence of organic catalyst and the presence of nitrogen prevent the forming of polar functional groups which would harm the good electrical properties of the polymer. Solution polymerization of styrene is not used because the problems of heat transfer and of stirring high-viscosity solutions still exist.

Polystyrene has been investigated as completely as any other plastic, at least from a theoretical standpoint. Staudinger (who, in the 1930s, proposed viscosity and other relations of macromolecules, for which he received the Nobel Prize) developed his macromolecular theory chiefly from investigations on the formation and properties of polystyrene. He carefully prepared polystyrene of different degrees of polymerization and showed that the viscosity of solutions of this material is a function of the length of the linear chain. Work on this material has supplied proof of the theory that thermoplastic and thermosetting resins are caused by linear and three-dimensional structures. If even small amounts of divinyl benzene are added to styrene before polymerization, the resulting polymer becomes insoluble and has a very much higher softening temperature. The incorporation of divinyl benzene permits occasional crosslinkages between the linear chains to give a partial three-dimensional structure.

Because polystyrene fails to show crystallinity under X-ray examination, it was thought that the structure is irregular as a result of occasional head-head, tail-tail inversions. Another suggestion was that the polymers are highly bridged. However, the newer stereospecific polymerizing catalysts have produced crystalline polystyrene. This suggests that if the benzene groups are properly oriented above the carbon chain, the molecules can assume the orderly arrangement which gives the crystal pattern structure.

Polystyrene produced by free-radical polymerization techniques is amorphous and has a syndiotactic/atactic structure. Polymerization at −30°C (−22°F) yields an isotactic polymer with a narrow molecular weight distribution. Because of the added crystallinity, the melting point is as high as 110°C (230°F) and the polymer is very brittle below its T_g.

It is very difficult to process isotactic polystyrene because of its high T_m and molecular weight. Various molding methods have been proposed and post-heat treatments at 180°C (356°F) are essential to develop a stable crystalline structure. The polymer's brittleness has hindered commercial development. Impact strengths are 20% that of the conventional amorphous materials. However, when blended with synthetic rubber or polyolefins, the resulting impact strengths are double that of the amorphous polymers.

Table 5-85 shows typical properties of various polystyrene types. The commercial general-purpose molding material is a clear, colorless pellet. A small amount of lubricant may be present to encourage molding flow and to aid in ejection from the mold. The polymer has a low density (1.05–1.08). Molded parts are rigid and water resistant, have good electrical properties, and can be produced in a wide range of colors.

Polystyrene softens just above 100°C (212°F) and becomes a viscous fluid around 185°C (365°F). The polymer is resistant to acids, alkalis, alcohols, and to vegetable oils, fats, or waxes. It dissolves easily in aromatic and chlorinated hydrocarbons, esters, and ketones. Aliphatic hydrocarbons and acetone cause swelling. The material has outstanding water-resistant properties, even though films are permeable to water vapor.

Polystyrene plastics are used unmodified for most applications, except for the occasional addition of pigments, dyes, and lubricants. A wide variety of softening temperatures and flow characteristics can be obtained through modification of the polymerization reaction to produce resins with varying molecular weights. Plasticizers are, therefore, not needed. Since the polymer is tranparent and

Table 5-84. Polystyrene Plastics

Polystyrene Product	Polystyrene General Purpose	SBR Rubber Reinforced Polystyrene	Light Stable Polystyrene	Heat-Resistant Polystyrene	Acrylonitrile Butadiene Styrene Terpolymer
Chemicals used	• Styrene	• SBR rubbers • Styrene • Stabilizers	• Styrene • Light stabilizers	• Styrene • *alpha*-Methyl styrene	• Styrene • Butadiene • Acrylonitrile • Stabilizers
Most outstanding properties	• Price • Clarity • Heat and dimensional stability • Ease of fabrication • Hardness • Rigidity	• Price • Elongation • Impact • Hardness • Toughness • Heat and dimensional stability • Ease of fabrication • Rigidity	• Price • Light stability • Dimensional stability • Hardness • Rigidity	• Price • Heat distortion • Ease of fabrication • Hardness • Rigidity	• Structural properties • Improved solvent resistance • Impact • Toughness • High gloss finish • Hardness • Rigidity
Technology of manufacturing	• Mass polymerization • Suspension polymerization • Solution polymerization	• Mass polymerization • Suspension polymerization • Solution polymerization • Mechanical mixing	• Mass polymerization • Suspension polymerization • Solution polymerization • Mechanical mixing	• Mass polymerization • Suspension polymerization • Solution polymerization	• Mass polymerization • Suspension polymerization • Solution polymerization • Emulsion polymerization • Mechanical mixing
Fabrication technology advances	• Injection molding • Compression molding • Extrusion	• Oriented films and and sheeting (1940–42) • Oriented bristles and monofilaments (1944)	• Profile extrusion	• Applying rheology to mold design • Pinpoint gating • Preplasticizing in injection molding • Weigh feeding • Mold temperature control	• Pressure and vacuum forming (1954)
Chronological order	1935–37	1948	1949	1949–52	1949–54

*Amos and Peterson.

most often used for this characteristic, fillers are rarely used.

The refractive index of polystyrene is around 1.60, which, along with its exceptional clarity, allows for the "bending" of light (comparable to PMMA). UV resistance is fair, but the polymer cannot withstand extended exposure without age degradation.

The RF electrical properties of polystyrene are excellent. Volume and surface resistivity are very high and power factor is almost zero. These properties vary little with temperature or exposure to humidity, alcohol, acids, and caustic. The material exhibits little cold flow and is thus dimensionally very stable.

The principal limitations of polystyrene are its brittleness, poor resistance to boiling water, mediocre oil resistance (softens and crazes), and low heat-deflection temperature. Because of its high rigidity, its impact strength is low. This property has been remedied by blending or alloying and through copolymerization.

A major application of polystyrene is in the manufacture of foams. The foamed material is available in a rigid, massive form which may be sawed to size. As such, it is used as a water-resistant insulating material. It has good weight-bearing properties, but has the heat limitation of all thermoplastics. It is also used in "sandwich" sheets and to produce decorative shapes and molds.

Expandable beads are also available which may be expanded in place or pre-expanded and then molded. The beads are styrene polymer compounded with a gas-releasing chemical (typically 6% of *n*-pentane). Expansion in place is

Family.*

Styrene Acrylonitrile Copolymer	Expandable Polystyrene	Methylmethacrylate Styrene Copolymer	Glass Fiber Reinforced		Stereo-specific Rubber Reinforced Polystyrenes	Flame Retardant Light Stable Polystyrene
			Polystyrene	Styrene-Acrylonitrile		
• Styrene • Acrylonitrile	• Styrene • Butadiene • Acrylonitrile • Stabilizers	• Styrene • Methylmethacrylate	• Styrene • Glass fiber	• Styrene • Acrylonitrile • Glass fiber	• Styrene • Stereo-specific rubbers • Stabilizers	• Styrene • Flame retardant agents • Light stabilizers
• Improved solvent resistance • Physical properties better than G.P. polystyrene • Hardness • Rigidity	• Low density • Insulating • Cushioning	• Light stability • Outdoor weatherability • Hardness • Rigidity • Clarity	• Low coefficient of expansion • Rigidity • Impact • Hardness	• Solvent resistance	• Low price • Rigidity • Elongation • Hardness • Impact • Structural properties • Impact properties for packaging uses • Low temperature properties	• Self-extinguishing property • Light stability • Dimensional stability • Hardness • Rigidity
• Mass polymerization • Suspension polymerization • Solution polymerization	• Mass polymerization • Suspension polymerization • Mechanical mixing • Postpressurizing of polystyrene beads	• Mass polymerization • Solution polymerization	Not known		• Mass polymerization • Solution ploymerization • Mechanical mixing	
	• Injection molding and extrusion of expandable polystyrene beads • Impact polystyrene sheet extrusion	• Blow molding • Screw injection molding machine				
1949–55	1952–58	1958	1952	1960	1961	1962

used for volumes less than 2 inches thick and when densities of over 4 lbs/ft^3 are desired. The small beads are heated in a jacketed mold or with a steam probe. At 110–135°C (230–275°F) the plastic softens, and the expanding agent decomposes and expands the beads to fill the closed cavity. Too high a temperature causes the plastic to become so soft that the released gas escapes. At too low a temperature, the gas will not expand enough to give the largest-volume bead for the lowest-density part. In pre-expansion, the beads are heated in a steam bath. This causes the beads to expand to about 40 times their previous size. The expansion is due not only to volatilization of the blowing agent, but also to an osmotic diffusion of steam into the cells formed. These beads are broken apart, packed in large volumes, slightly expanded further, and sealed together with heat. Figure 5-85 shows food containers made from expanded polystyrene.

Polystyrene and its rigid copolymers are extruded in the form of sheeting from 0.001–0.500 inch thick. Heavy gauges are preheated and vacuum-formed into large configurations such as refrigerator door liners, business machine housings, and tote boxes. Thinner stock is useful as wall tile and letters for signs. The light gauge may be oriented in two directions when it leaves the extrusion die and prior to cooling. This sheet has vastly improved flexibility and is used as a dimensionally stable photographic film base and, in still lighter gauges, as a packing material.

Oriented polystyrene is available in filament (monoaxially oriented) and film (biaxially oriented). The increase in tensile strength in the direction of stretching is offset by a

Fig. 5-85. Expanded polystyrene food containers. *(Courtesy: McDonald's Corporation)*

reduction in softening point due to the inherent instability of the oriented molecules.

Filament is produced by hot-stretching the fiber after extrusion. The film is produced by flat die extrusion at 190°C (374°F). After cooling to 120°C (248°F), the film is reheated to the optimum stretching temperature of 130°C (266°F). After this point, the film is stretched laterally by means of driven edge rollers and longitudinally by making the take-up rate far greater than the extrusion rate. Thus, stretching in both directions can be varied independently with stretch ratios of 3-4:1 commonly obtainable.

The largest application for polystyrene is in packaging. Uses include bottle caps, jars, and molded, blown, or vacuum-formed containers, as well as oriented film. The second largest use is in refrigeration and insulation for door and inner liners, lids, and trays for refrigerators and freezers. Although facing stiff competition from polyurethane foams, polystyrene still maintains a substantial hold in the insulation market as low-temperature insulation.

Moldings are widely used for housewares, toys, games, sports equipment, radio and electrical equipment, plumbing fittings, and even shoe heels.

There are several styrene polymers made from monomers with either ring or side-chain substitutions. These substituents are usually alkyl or halogen, but of course any ring or chain addition gives a new monomer polymerization possibility.

Many alkyl styrene monomers have been made. These include mono-, di-, and trimethyl styrene, mono- and diethyl styrene, and higher chains up to heptyl. These monomers yield polymers which, in general, are no great improvement over the commercial polystyrene. Vinyl toluene can be used in place of styrene in making certain rubber and finishing compositions. Divinyl benzene has already been mentioned and will be discussed again shortly.

One styrene derivative that is useful both from a price and property point of view is alpha-methyl styrene. The monomer, boiling at 165°C, is made from benzene and propylene, using aluminum and hydrogen chlorides.

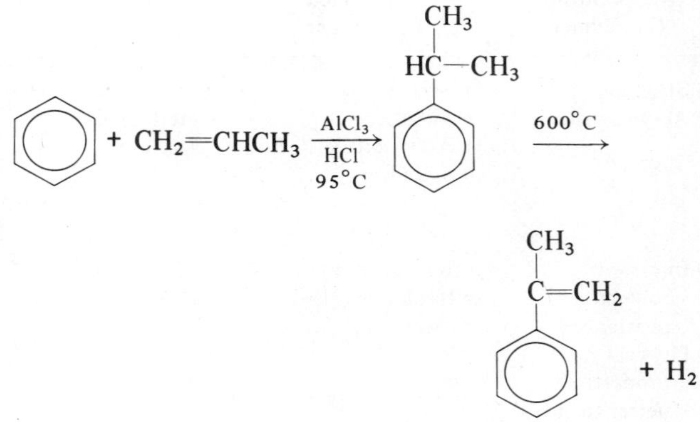

The high-purity monomer does not polymerize as easily as does styrene. It does, however, slowly convert to low molecular weight syrups and so is usually inhibited for storage and shipment.

The monomer readily polymerizes under the influence of Friedel-Crafts catalysts at low temperatures, yielding homopolymers of high molecular weight. Sodium, $AlCl_3$, $TiCl_4$, and BF_3 have been used. Interestingly, copolymerization of alpha-methyl styrene with other unsaturated monomers will occur with peroxide initiators.

The low molecular weight resins prepared from this monomer are colorless, viscous oils. These materials are useful as plasticizers and can be used as viscosity modifiers in melt-coating and calendering compositions.

The alpha-methyl groups increase the stiffness of poly-(alpha-methyl styrene). Thus, the high molecular weight modification of the polymer produces a thermoplastic with excellent heat resistance. Molded objects show no distortion in boiling water. The plastic is clear, has an unlimited coloring range, and, because of its excellent water resistance, has the good electrical properties of the styrene family. It has a higher melt viscosity, greater solvent resistance, lower elongation, and is harder than polystyrene. A copolymer with acrylonitrile offers an improvement in impact strength with better chemical and craze resistance than the simple polymer. Clarity, luster, and coloring range are maintained.

Copolymers and Alloys

A typical alloy blend is styrene polymer with a copolymer of styrene high in butadiene. This synthetic rubber is hot-milled at a 10-25% level with polystyrene to give an alloy of improved toughness which is still rigid. The materials are not highly compatible, so a somewhat laminar flow results during molding. This "plywood" effect is partly responsible for the improved toughness with minor loss of rigidity.

Numerous variations have been made in the alloying process. The alloys and their variations have allowed the exten-

Table 5-85. Properties of Polystyrene.*

		ASTM test method	Polystyrene molding compounds and sheet: Unfilled free-flowing general-purpose heat-resistant	Impact-resistant medium-impact high-impact heat-resistant	Special heat and chemical resistant type	Impact flame retardant	General purpose flame retardant	20 to 30% glass filled
Processing	1. Molding qualities	—	Excellent	Excellent	Good	Excellent	Excellent	Excellent
	2. Compression molding temp., °F.	—	265-400	250-400	300-400	250-400	265-400	—
	3. Comp. molding pressure, p.s.i.	—	1000-10000	1000-10000	1000-5000	1000-10000	1000-10000	—
	4. Injection molding temp., °F.	—	325-500	350-600	350-700	350-500	350-425	450-625
	5. Injection molding pressure, p.s.i.	—	10000-30000	10000-30000	10000-30000	10000-30000	10000-30000	15000-40000
	6. Compression ratio	—	1.6-4.0	1.6-4.0	1.6-4.0	1.6-4.0	1.6-4.0	1.6-4.0
	7. Mold (linear) shrinkage, in./in.	—	0.001-0.006(C) 0.002-0.006(I)	0.002-0.006	0.001-0.008	0.002-0.006	0.002-0.006	0.001-0.002
	8. Specific gravity (density)	D792	1.04-1.09	1.04-1.10	1.04-1.10	1.10-1.20	1.08	1.20-1.33
	9. Specific volume, cu. in./lb.	D792	26.0-25.6	28.1-25.2	26.2-24.8	25.2	25.6	22.2-20.9
	10. Machining qualities	—	Fair to good	Good	Fair to good	Good	Fair to good	Good
Mechanical	11. Tensile strength, p.s.i.	D638	5000-12000	1500-7000	6500-12000	3800-5000	7000	9000-15000
	12. Elongation, %	D638	1.0-2.5	2.0-90.0	1.4-2.5	13-25	2	0.75-1.30
	13. Tensile elastic modulus, 10^5 p.s.i.	D638	4.0-5.0	1.4-5.0	4.0-6.0	2.7-3.2	4.5	8.4-12.9
	14. Compressive strength, p.s.i.	D695	11500-16000	4000-9000	11500-16000	—	—	13500-18000
	15. Flexural yield strength, p.s.i.	D790	8000-14000	3000-12000	10000-17000	5500-6500	—	10500-20000
	16. Impact strength, ft. lb./in. of notch (½ × ½ in. notched bar, Izod test)	D256	0.25-0.40 (¼ in. bar)	0.5-8.0 (23° C.)[a] 0.4-2.5 (−25° C.) (⅛ × ½ in.)	0.35-0.60 (¼ in. bar)	1.0-1.4 (⅛ × ½")	0.25 (⅛ × ½")	0.4-4.5
	17. Hardness, Rockwell	D785	M65-M80	M10-80 R30-100	M65-M90	M10,R95	M70	M70-M95
	18. Flexural elastic modulus, p.s.i. × 10^5	D790	4.0-4.7	1.5-4.6	—	2.8-3.3	—	8.0-10.0
	19. Compressive modulus, p.s.i. × 10^5	D695	—	—	—	—	—	—
Thermal	20. Thermal conductivity, 10^{-4} cal./sec./sq. cm./1 (°C./cm.)	C177	2.4-3.3	1.0-3.0	1.9-3.0	—	2.4-3.3	—
	21. Specific heat, cal./°C./gm.	—	0.32	0.32-0.35	0.32-0.35	—	0.32	0.23-0.27
	22. Thermal expansion, 10^{-5} in./in./°C.	D696	6.0-8.0	3.4-21.0	6.0-8.0	—	6.0-8.0	1.8-4.5
	23. Resistance to heat, °F. (continuous)	—	150-170	140-175	170-220	—	—	180-200
	24. Deflection temp., °F. at 264 p.s.i. fiber stress	D648	220 max.	195 max.	180-235	175-180	195	195-220
	at 66 p.s.i. fiber stress		180-230	180-220	195-240	—	—	207-231
Electrical	25. Volume resistivity, ohm-cm. (50% RH and 23° C.)	D257	$>10^{16}$	$>10^{16}$	10^{13}-10^{17}	10^{15}	10^{15}	3.2×10^{16}
	26. Dielectric strength, short-time, ⅛-in. thickness, volts/mil	D149	500-700	300-600	400-600	500	500	350-425
	27. Dielectric strength, step-by-step, ⅛-in. thickness, volts/mil	D149	400-600	300-600	300-500	450	500	350-430
	28. Dielectric constant, 60 cyc.	D150	2.45-3.1	2.45-4.75	2.45-3.4	3.2	3.1	—
	29. Dielectric constant, 10^3 cyc.	D150	2.40-2.65	2.4-4.5	2.4-3.2	—	—	—
	30. Dielectric constant, 10^6 cyc.	D150	2.40-2.7	2.4-3.8	2.4-3.1	2.8	2.7	—
	31. Dissipation (power) factor, 60 cyc.	D150	0.0001-0.0006	0.0004-0.0020	0.0005-0.0030	0.0006	0.0006	0.004-0.014
	32. Dissipation (power) factor, 10^3 cyc.	D150	0.0001-0.0003	0.0004-0.0020	0.0005-0.0030	—	—	0.001-0.004
	33. Dissipation (power) factor, 10^6 cyc.	D150	0.0001-0.0004	0.0004-0.0020	0.0005-0.0050	0.003	0.001	0.001-0.003
	34. Arc resistance. sec.	D495	60-140	20-140	60-135	140	140	25-40
Optical	35. Refractive index, n_D	D542	1.59-1.60	—	1.57-1.60	—	—	—
	36. Clarity	—	Transparent	Translucent, opaque	Transparent	Translucent, opaque	Transparent	Translucent, opaque
	37. Transmittance, %	—	87-92	35-57	88-90	—	87	—
	38. Haze, %	—	<0.1-3.0	>77	—	—	—	—
Resistance characteristics	39. Water absorp., 24 hr., ⅛-in. thick., %	D570	0.03-0.10	0.05-0.6	0.05-0.40	—	—	0.05-0.10
	40. Flammability[b] Burning rate, in./min.	D635	<1.5(⅛")	<1.5(⅛")	<1.5(⅛")	—	—	1.0
	AEB, in./ATB, sec.[c]		—	—	—	0.5/3.0	—	—
	41. Effect of sunlight	—	Yellows slightly	Some strength loss; yellows slightly	Yellows slightly	Some strength loss; yellows slightly to rapidly	Yellows slightly	Yellows slightly
	42. Effect of weak acids	D543	None	None	None	None	None	None
	43. Effect of strong acids	D543	Attacked by oxidizing acids	Attacked by oxidizing acids	Attacked by oxidizing acids	Attacked by oxidizing acids	Attacked by oxidizing acids	Attacked by oxidizing acids
	44. Effect of weak alkalies	D543	None	None	None	None	None	Resistant
	45. Effect of strong alkalies	D543	None	None	None	None	None	Surface only
	46. Effect of organic solvents	D543	Soluble in aromatic and chlorinated hydrocarbons	Soluble in aromatic and chlorinated hydrocarbons 140-200° F.	Soluble in aromatic and chlorinated hydrocarbons 200-200° F.	Soluble in aromatic and chlorinated hydrocarbons		Soluble in aromatic and chlorinated hydrocarbons

[a]Determined on $\frac{1}{4}$-in. thick specimens.
[b]This numerical flame spread rating is not intended to reflect hazards presented by this or any other material under actual fire conditions.
[c]AEB = Avg. extent of burning in inches; ATB = Avg. time of burning in seconds.
*Reprinted by permission from *Modern Plastics Encyclopedia*, McGraw-Hill, Inc.

sion of these polymers into the molding of pipe where resistance to rough handling is necessary. Cams, gear, pipe-fittings, bobbins, and spools are also molded. Molding granules with different levels of shock resistance are available, which show increased elongation at some sacrifice in rigidity.

Styrene monomer can be copolymerized at both high and low levels with acrylonitrile. These copolymers (SAN) are an improvement over general-purpose polystyrene. They can be made in a complete range of compositions, but a balance of 15-30% of acrylonitrile is usual. Although available for a number of years, they were too high-priced initially for more than highly specialized outlets.

Because of the polar nature of the acrylonitrile, these copolymers have better resistance to hydrocarbons, oils, and greases than polystyrene. They also have a higher softening point, stress-crack and craze resistance, and enhanced impact strength without loss of transparency compared to the homopolymer. As the acrylonitrile content increases, toughness, chemical resistance, yellowing, and difficulty of molding increases. Water absorption is ten times that of polystyrene but one-tength that of cellulose acetate.

The important features of rigidity and transparency make the material competitive with polystyrene, cellulose acetate, and PMMA. In general, SAN is cheaper than PMMA and cellulose acetate, tougher than PMMA and polystyrene, and superior in most physical properties to polystyrene and cellulose acetate. Its optical and weathering properties are not as good as PMMA. Table 5-86 lists properties of unfilled and glass-filled SAN. Table 5-87 compares SAN properties with general-purpose plastics. Figures 5-86 through 5-88 compare some select properties of SAN (Tyril resin) with engineering thermoplastics.

Typical applications include dials, knobs, and covers for appliances, electrical and car equipment, picnic ware and housewares (see Figure 5-89).

A special copolymer made for a specific end use is that from styrene and divinyl benzene. This copolymer is insoluble because of its crosslinking with about 10% divinyl compound. Subsequent treatment with chlorosulfonic acid yields a polymer which has cation-capturing properties because of the free sulfonic acid groups. The product polymerized in bead form is easy to handle and remains insoluble during all phases of an ion-exchange reaction. The resin may be used as the sodium salt, which then exchanges with ions such as calcium and magnesium. Regeneration is effected by strong sodium chloride solution or to the free acid form by adding aqueous sulfuric acid.

Anion exchanges may be made by including a chloride-bearing monomer in the styrene-divinyl benzene copolymer. Reaction with a tertiary base gives a salt which yields a strong insoluble polymeric hydroxide when treated with sodium hydroxide.

The acidic and basic beads may be mixed and used to-

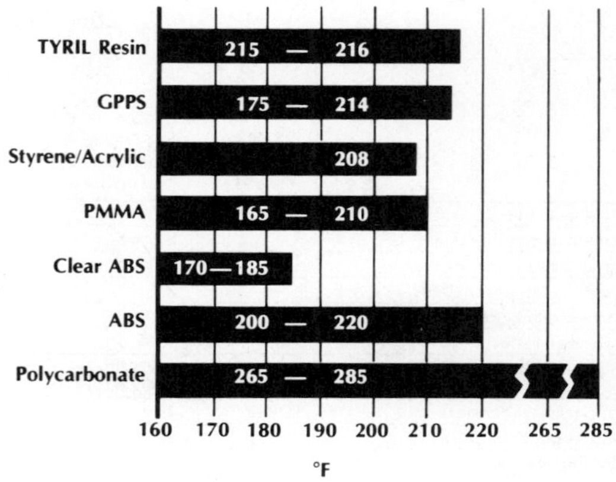

Fig. 5-86. Heat distortion[1] of Tyril resin (SAN) versus other thermoplastics at 264 psi, °F. (Reprinted with permission of The Dow Chemical Co.)

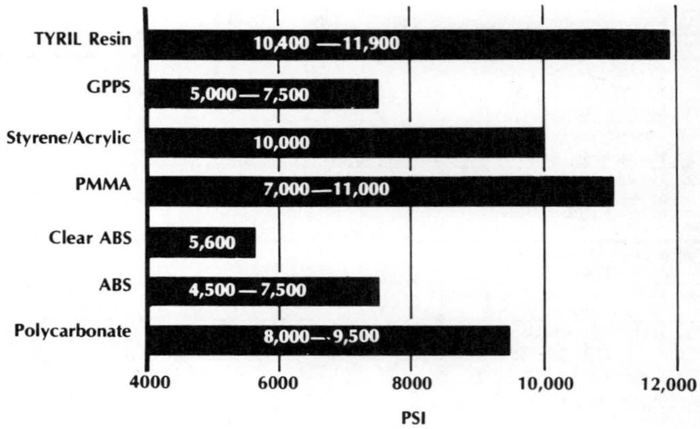

Fig. 5-87. Tensile strength[1] of Tyril resin (SAN) versus other thermoplastics. (Reprinted with permission of The Dow Chemical Co.)

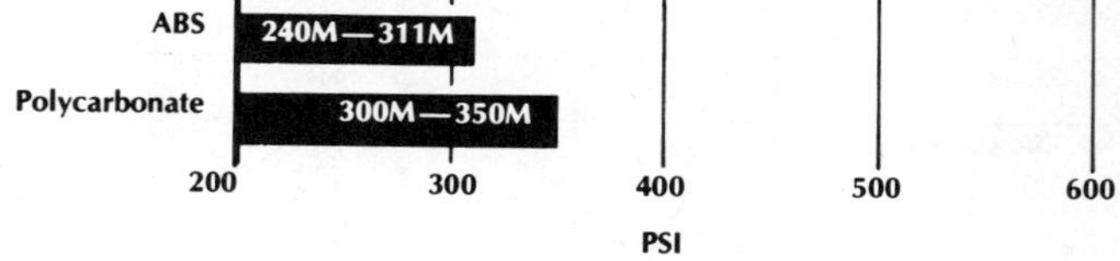

Fig. 5-88. Modulus[1] of Tyril resin (SAN) versus other thermoplastics. (Reprinted with permission of The Dow Chemical Co.)

Table 5-86. Properties of Unfilled and Glass-Filled SAN.*

		ASTM test method	Styrene-acrylonitrile copolymer Unfilled	20 to 33% glass filled
Processing	1. Molding qualities	—	Good	Excellent
	2. Compression molding temp., °F.	—	300-400	—
	3. Comp. molding pressure, p.s.i.	—	1000-10000	—
	4. Injection molding temp., °F.	—	375-575	450-575
	5. Injection molding pressure, p.s.i.	—	10000-33000	15000-40000
	6. Compression ratio	—	1.6-4.0	1.6-4.0
	7. Mold (linear) shrinkage, in./in.	—	0.002-0.007	0.001-0.002
	8. Specific gravity (density)	D792	1.075-1.100	1.20-1.46
	9. Specific volume, cu. in./lb.	D792	25.8-25.2	23.9-19.9
	10. Machining qualities	—	Good	Fair
Mechanical	11. Tensile strength, p.s.i.	D638	9000-12000	8500-20000
	12. Elongation, %	D638	1.5-3.7	1.1-3.8
	13. Tensile elastic modulus, 10^5 p.s.i.	D638	4.0-5.6	4.0-14.0
	14. Compressive strength, p.s.i.	D695	14000-17000	22000
	15. Flexural yield strength, p.s.i.	D790	14000-19000	22000-26000
	16. Impact strength, ft. lb./in. of notch (½ × ½ in. notched bar, Izod test)	D256	0.35-0.50	0.4-4.0
	17. Hardness, Rockwell	D785	M80-M90	M100, E60
	18. Flexural elastic modulus, p.s.i. × 10^5	D790	to 5.5	8.0-18.0
	19. Compressive modulus, p.s.i. × 10^5	D695	5.3	—
Thermal	20. Thermal conductivity, 10^{-4} cal./sec./sq. cm./1 (°C./cm.)	C177	2.9	—
	21. Specific heat, cal./°C./gm.	—	0.32-0.34	—
	22. Thermal expansion, 10^{-5} in./in./°C.	D696	3.6-3.8	2.7-3.8
	23. Resistance to heat, °F. (continuous)	—	140-205	200-220
	24. Deflection temp., °F. at 264 p.s.i. fiber stress	D648	190-220	190-230
	at 66 p.s.i. fiber stress	D648	—	215-240
Electrical	25. Volume resistivity, ohm-cm. (50% RH and 23° C.)	D257	$>10^{16}$	1.4-2.1 × 10^{17}
	26. Dielectric strength, short-time, ⅛-in. thickness, volts/mil	D149	400-500	510
	27. Dielectric strength, step-by-step, ⅛-in. thickness, volts/mil	D149	300-600	410
	28. Dielectric constant, 60 cyc.	D150	2.6-3.4	3.6
	29. Dielectric constant, 10^3 cyc.	D150	2.6-3.3	3.5
	30. Dielectric constant, 10^6 cyc.	D150	2.6-3.1	3.4
	31. Dissipation (power) factor, 60 cyc.	D150	0.006-0.008	0.00645
	32. Dissipation (power) factor, 10^3 cyc.	D150	0.007-0.012	0.00652
	33. Dissipation (power) factor, 10^6 cyc.	D150	0.007-0.010	0.00668
	34. Arc resistance, sec.	D495	100-150	16
Optical	35. Refractive index, n_D	D542	1.56-1.57	—
	36. Clarity	—	Transparent	Translucent, opaque
	37. Transmittance, %	—	78-88	—
	38. Haze, %	—	0.4-0.1	—
Resistance characteristics	39. Water absorp., 24 hr., ⅛-in. thick., %	D570	0.20-0.30	0.08-0.22
	40. Flammability [a] Burning rate, in./min.	D635	—	0.1
	AEB, in./ATB, sec. [b]	D635	—	—
	41. Effect of sunlight	—	Yellows slightly	Yellows slightly
	42. Effect of weak acids	D543	None	Resistant
	43. Effect of strong acids	D543	Attacked by oxidizing acids	Attacked by oxidizing acids
	44. Effect of weak alkalies	D543	None	None
	45. Effect of strong alkalies	D543	None	Surface only
	46. Effect of organic solvents	D543	Soluble in ketones, esters, some chlorinated hydrocarbons	Soluble in ketones, esters, some chlorinated hydrocarbons

This numerical flame spread rating is not intended to reflect hazards presented by this or any other material under actual fire conditions.
[b] AEB = Avg. extent of burning in inches; ATB = Avg. time of burning in seconds.
*Reprinted by permission from *Modern Plastics Encyclopedia*, McGraw-Hill, inc.

Table 5-87. Typical Properties of Styrene Acrylonitrile and Other Plastics.*

Material	SAN	Acrylic	Acetate, Type II, H6	Butyrate, Type II	Melamine, Unfilled	Polyethylene, High Density	Polystyrene, General Purpose	Polystyrene, Heat and Chemical Resistant
Mechanical Properties								
Impact Str (Izod notched), ft/lb/in.								
73 F	0.5–8.0	0.40–0.50	0.4–1.9	0.8–6.3	0.25–0.35	1.2–20	0.25–0.40	0.35–0.60
−40 F	0.6–1.8	–	–	0.4–1.7	–	–	–	–
Ten Str, 1000 psi	2.4–11	7–11	6–8.5	5.0–6.8	7–13	3.0–5.5	5–9	6.5–12
Elong, %	10–60	2–10	6–31	38–54	0.30–0.45	15–100	1.0–2.5	1.4–2.5
Heat Dist Temp (264 psi), F	170–215	150–200	–	115–227	265	120–150	150–195	180–235
Coef of Ther Exp, in./in.F $\times 10^{-3}$	3–4	27–50	–	60–95	10–25	61–72	33–34	–
Flex Str, 1000 psi	8.6–18	13–17	2–16	1.8–12	11–14	2–3	8.7–14	10–17
Mod of Elast in Ten, 10^5 psi	4–6	3.5–4.5	0.65–4.0	0.5–2.0	0.2–0.3	0.8–1.5	4–5	4–6
Compr Str, 1000 psi	2.5–11	12–18	25–36	2.1–9.4	25–30	8.5–10	11.5–16	11.5–16
Hardness (Rockwell)	R80–R118	M85–105	R112–123	R31–114	M110–125	R15	M65–M80	M65–M90
Water Absorption (24 hr), %	0.1–0.3	0.3–0.4	1.5–3.5	0.9–2.2	0.3–0.5	0.01	0.03–0.05	0.05–0.40
Flammability, ipm	0.8–1.2	slow	0.5–2.0	0.5–1.5	SE	1.0	0.32–2.5	–
Refractive Index	1.56	1.48–1.50	1.46–1.50	1.46–1.49	Opalescent	1.54	1.59–1.60	1.57–1.60
Electrical Properties								
Dielec Str, v/mil	300–400	450–550	250–600	250–400	300–400	480	500–700	300–500
Dielec Const								
60 Cps	2.4–4.75	3.5–4.5	3.5–7.5	3.5–6.4	7.9–9.5	2.3–2.35	2.45–2.65	2.45–3.4
10^6 Cps	2.4–4.75	2.2–3.2	3.2–7.0	3.2–6.2	7.2–8.2	2.3–2.35	2.45–2.65	2.4–3.1
Dissip Factor								
60 Cps	0.0004–0.034	0.05–0.06	0.03–0.38	0.01–0.04	0.03–0.08	<0.0002	0.0001–0.0003	0.0005–0.003
10^6 Cps	0.007–0.026	0.02–0.03	0.03–0.33	0.01–0.04	0.027–0.045	<0.0003	0.0001–0.0004	0.0005–0.005
Vol Res, ohm-cm	$>10^{14}$	$>10^{15}$	10^{10}–10^{13}	10^{10}–10^{12}	–	$>10^{15}$	$>10^{16}$	$>10^{15}$
Chemical Resistance[a]								
Weak Acids	E	E	G	G	G	E	E	E
Strong Acids	E	G	P	P	P	G	G	G
Weak Alkalis	E	E	G	G	G	E	G	E
Strong Alkalis	E	E	P	P	P	E	G	E
Organic Solvents	P	P	P	P	E	G	P	G
Cost								
Weight, ¢/lb.	25	46–55	40–52	40–62	42–45	28	17–18	20
Volume, ¢/cu in.	0.9	1.8–2.3	1.8–2.4	2–3	2.2–2.3	1	.6–.7	.8

[a] E–excellent, G–good, P–poor.
*Hauck, 1963.

Fig. 5-89. Tumblers, carafes, and other products benefit from the high heat resistance and improved toughness of styrene acrylonitrile resins. (Printed with permission of The Dow Chemical Co.)

gether in a column to bring about the exchange of both cations and anions in a simple liquid pass. The beads are separated by screen classification and separately regenerated.

A copolymer useful in textile applications is made by reacting styrene with maleic anhydride. This anhydride is almost unreactive in homopolymerization, but easily gives 1:1 copolymers with styrene. The copolymer swells in water and converts to the free acid, but does not dissolve. Aqueous alkali does dissolve the polymer to give a solution which is used as a textile size for acetate fibers. When the yarn is dipped into a 5% solution and dried, a thin protective coating is formed which reduces static electricity and improves abrasion resistance during handling of the loom. A simple water wash removes the coating.

Acrylonitrile-Butadiene-Styrene Terpolymers (ABS)

ABS resins were originally described as "gum plastics," a generic term for rubber-modified thermoplastic resins. High-impact styrenes were among the first "gum plastics" to emerge in the 1940s when styrene technology had improved sufficiently to make polystyrene a material of commerce. These were injection molded, extruded, and calendered into simple shapes.

ABS materials, which followed impact styrene in time, were mechanical blends of nitrile rubber and SAN copolymers. Improvements in weatherability and chemical resistance were achieved because of the acrylonitrile. Toughness also increased, but poor processability, poor appearance, and high price often offset the improvements offered.

In general, the basic building blocks of styrene, acrylonitrile, and polybutadiene are put together in a variety of ways. The two most important types of ABS resins today are blends of SAN copolymers with butadiene-acrylonitrile rubber (type I) and terpolymers of polybutadiene, styrene, and acrylonitrile (type II).

There are a number of methods to produce type I resins. Among them are: blending on a two-roll mill or in a high-intensity, internal (sigma blade) mixer, or blending of latices. The latices are then spray-dried or coagulated. At this point, the ingredients are mutually compatible and there is no marked increase in toughness. By lightly cross-linking the elastomer phase with peroxides, an incompatibility between phases occurs which significantly improves impact strength. A typical blend would consist of 64% of a 70:30 SAN polymer with 36% of a 65:35 butadiene-acrylonitrile elastomer.

In the production of type II resins, styrene and butadiene are added to polybutadiene latex and the mixture is heated to 50°C (122°F). A water-soluble initiator like potassium persulphate is added to induce polymerization. The resultant resin contains a mixture of polybutadiene, polybutadiene grafted with acrylonitrile and styrene, and SAN copolymer. The graft polymer is essential for the development of optimum properties because blends of polybutadiene and SAN lack strength and toughness.

The relationships of the various components can be con-

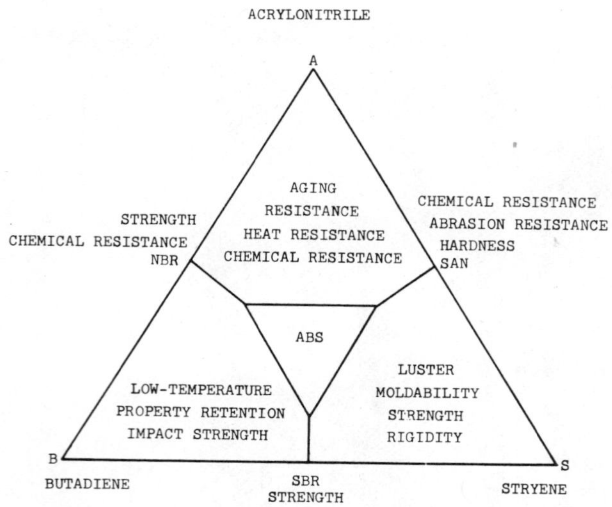

Fig. 5-90. The relative influence on properties of the monomeric constituents of ABS terpolymers. (Malpass, 1976)

trolled to take maximum advantage of a single property or a set of properties (see Figure 5-90). Controlled compatibility of the elastomers, for example, is the key to maximum impact effectiveness. For type I resins, inclusion of a nitrile rubber with a butadiene content greater than 65% increases toughness, just as high molecular weight SAN provides improved impact strength.

The variables and nonlinear effects of both the resin and elastomer phases that can be adjusted within various compositions are described in Table 5-88. The "primary" variables of those listed in Table 5-88 that affect typical properties of the final polymer are: (1) the amount of the elastomeric phase, (2) the molecular weight of the resin phase, and (3) the styrene/acrylonitrile ratio in the resin phase.

Generally, the highest-impact material goes with the lowest tensile strength and modulus. Hardness can be considered as a linear function of the elastomer over the range

Table 5-88. Variables and Nonlinear Effect of Ingredients on ABS Resins.

SAN Resin	Nitrile Elastomer
1. Molecular weight	1. Structure a. Particle size b. Particle size distribution
2. Molecular weight distribution	2. Composition
3. Styrene/acrylonitrile ratio	3. Degree and efficiency of grafting
	4. Crosslink density

which encompasses the majority of the commercial ABS materials. As a consequence, the variance in physical properties with elastomer can be plotted as a function of hardness when the other variables are held constant.

Physical properties such as tensile and impact strength and heat-deflection temperature are improved by increasing the molecular weight, although the effect is more pronounced at low to intermediate elastomer levels because of interaction effects. Increasing acrylonitrile content generally improves chemical resistance, tensile strength, and heat-distortion temperature, but impact strength is lowered. Table 5-89 shows properties of various ABS grades.

It is not necessary to restrict these resins to just acrylonitrile, butadiene, and styrene. Other monomers are used, but since they are often proprietary formulas, a complete description is difficult without reference to trade designations. In general, these materials have high impact strength, softening points equal to or greater than general-purpose polystyrene, and show excellent surface appearance on molded parts.

Processability of a resin is allied with flow. This parameter is in turn dependent on the "primary" variables listed above. Increasing graft polymer and acrylonitrile content or molecular weight decreases flow, and can result in ABS materials more difficult to form. Compromises must be made to gain maximum value in a given property. To gain impact resistance, the tensile strength, flow, and heat-deflection temperature are reduced through increases in graft content. Maximum flow will normally require sacrifices in impact and tensile strengths, and heat deflection when molecular weight is lowered. Sometimes, interactions minimize but do not eliminate these losses. Plasticizers and lubricants can also work as aids.

Being amorphous, ABS resins have low mold shrinkage and a wide melting range. The melts have a high viscosity, so large runners and gates are required in injection molding. Heat stability is not as good as polystyrene or polyethylene, and at high processing temperatures (about 250°C, 482°F) noxious vapors can be generated.

ABS can be processed on ram- or screw-injection molding machines. Structural foam parts can be molded on standard screw injection machines using an internally compounded blowing agent or on molding machines using low-pressure nitrogen for achieving the foam structure. Extrusion into pipe, sheet, and profiles as well as for blow molding is common. Extruded sheet can be thermoformed, embossed, or laminated. Packaging grades of high impermeability to oxygen and carbon dioxide can be extrusion or injection blow molded.

ABS plastics are characterized by ductile yielding down to low temperatures and up to high-abuse strain rates. The dispersed rubber particles also provide good hot strength, which is essential to thermoforming, blow molding, and profile definition. They undergo creep when loaded, and stress relaxation when deformed. Designing with ABS

Table 5-89. Properties of ABS Extrusion and Molding Compounds.*

	Property	ASTM test method	Extrusion grades	Molding grades: High-impact	High heat resistant	Medium-impact	Flame-retardant	20 to 40% glass filled
Processing	1. Molding qualities	—	—	Good to excellent			Good	Fair to good
	2. Compression molding temp., °F.	—	325-450	325-450	325-500	325-350	—	
	3. Comp. molding pressure, p.s.i.	—	—	1000 to 8000				
	4. Injection molding temp., °F.	—	—	380-525	475-550	400-525	380-500	500-550
	5. Injection molding pressure, p.s.i.	—	—	8000 to 25000				15000-40000
	6. Compression ratio	—	—	1.1 to 2.0				
	7. Mold (linear) shrinkage, in./in.	—	N.A.	0.004-0.009			0.004-0.008	0.0010-0.0020
	8. Specific gravity (density)	D792	1.02-1.06	1.01-1.04	1.05-1.08	1.03-1.06	1.16-1.21	1.23-1.36
	9. Specific volume, cu. in./lb.	D792	28.0	27.0-26.0	28.0	28.0-27.0		22.5-20.4
	10. Machining qualities	—	Good to excellent					Good
Mechanical	11. Tensile strength, p.s.i.	D638	2500-8000	4800-6300	6000-7500	6000-7500	5000-8000	8500-19000
	12. Elongation, %	D638	20.0-100.0	5.0-70.0	3.0-20.0	5.0-25.0	5.0-25.0	2.5-3.0
	13. Tensile elastic modulus, 10^5 p.s.i.	D638	1.3-3.8	2.3-3.3	3.0-3.5	3.0-4.0	3.2-4.0	5.9-10.3
	14. Compressive strength, p.s.i.	D695	5200-10200	4500-8000	7200-10000	1800-12500	6500-7500	12000-22000
	15. Flexural yield strength, p.s.i.	D790	4000-14000	8000-11000	10000-12500	11000-13000	9000-14000	16000-27000
	16. Impact strength, ft. lb./in. of notch (½ × ½ in. notched bar, Izod test)	D256	2.5-12.0 at 73° 0.6-4.0 at −40° (⅛ × ½ in. bar)	6.5-7-5 at 73° 1.5-2.5 at −40° (⅛ × ½ in. bar)	2.0-6.5 at 73° 0.7-1.6 at −40° (⅛ × ½ in. bar)	3.0-6.0 at 73° 1.0-1.5 at −40° (⅛ × ½ in. bar)	3.0-12.0 at 73° (⅛ × ½ in. bar)	1.0-2.4
	17. Hardness, Rockwell	D785	R75-115	R85-105	R100-115	R107-115	R100-120	M65-100
	18. Flexural elastic modulus, p.s.i. × 10^5	D790	1.3-4.2	2.5-3.5	3.0-4.0	3.5-4.0	3.0-4.0	9.0-13.0
	19. Compressive modulus, p.s.i. × 10^5	D695	1.5-3.9	1.4-3.0	1.9-4.4	2.0-4.5	1.3-3.1	
Thermal	20. Thermal conductivity, 10^{-4} cal./sec./sq. cm./1 (°C./cm.)	C177	4.5-8.0					
	21. Specific heat, cal./°C./gm.	—	0.3 to 0.4					
	22. Thermal expansion, 10^{-5} in./in./°C.	D696	6.0-13.0	9.5-11.0	6.0-9.3	8.0-10.0	6.5-9.5	2.9-3.6
	23. Resistance to heat, °F. (continuous)	—	140-200	140-180	190-210	160-180	130-180	200-230
	24. Deflection temp., °F. at 264 p.s.i. fiber stress	D648	(annealed) 170-225	(annealed) 205-215	(annealed) 220-240	(annealed) 200-220	(annealed) 195-225	210-240
	at 66 p.s.i. fiber stress		170-245	210-225	230-245	215-225	210-245	220-250
Electrical	25. Volume resistivity, ohm-cm. (50% RH and 23° C.)	D257	1.0-4.8 × 10^{16}	1.0-4.8 × 10^{16}	1.0-5.0 × 10^{16}	2.7 × 10^{16}	1.4-1.6 × 10^{14}	7.2-16.0 × 10^{4}
	26. Dielectric strength, short-time, ⅛-in. thickness, volts/mil	D149	350 to 500					
	27. Dielectric strength, step-by-step, ⅛-in. thickness, volts/mil	D149	350-460	400-450	360-400	370-400	390-400	
	28. Dielectric constant, 60 cyc.	D150	2.4 to 5.0					—
	29. Dielectric constant, 10^3 cyc.	D150	2.4 to 4.5					—
	30. Dielectric constant, 10^6 cyc.	D150	2.4 to 3.8					—
	31. Dissipation (power) factor, 60 cyc.	D150	0.003 to 0.008					—
	32. Dissipation (power) factor, 10^3 cyc.	D150	0.004 to 0.007					—
	33. Dissipation (power) factor, 10^6 cyc.	D150	0.007 to 0.015					—
	34. Arc resistance, sec.	D495	50 to 85				10-50	—
Optical	35. Refractive index, n_D	D542	—	—	—	—	—	—
	36. Clarity	—	Translucent in thin sections to opaque					
	37. Transmittance, %	—	33.3	28.0	33.3	33.3	—	—
	38. Haze, %	—	100	100	100	100	—	—
Resistance characteristics	39. Water absorp., 24 hr., ⅛-in. thick., %	D570	0.20 to 0.45				0.2-0.6	0.18-0.40
	40. Flammability[a] Burning rate, in./min.	D635	0.6-1.0				—	—
	AEB, in./ATB, sec.[b]		—				—	—
	41. Effect of sunlight	—	None to yellows slightly plus some embrittlement					
	42. Effect of weak acids	D543	None					
	43. Effect of strong acids	D543	Attacked by concentrated oxidizing acids					
	44. Effect of weak alkalies	D543	None					
	45. Effect of strong alkalies	D543	None					
	46. Effect of organic solvents	D543	Soluble in ketones, esters, and some chlorinated hydrocarbons					

[a] This numerical flame spread rating is not intended to reflect hazards presented by this or any other material under actual fire conditions.
[b] AEB = Avg. extent of burning in inches; ATB = Avg. time of burning in seconds.
*Reprinted by permission from *Modern Plastics Encyclopedia*, McGraw-Hill, Inc.

plastics must take these time-dependent effects into account, along with effects of temperature on mechanical properties.

ABS plastics have good chemical, wear, and stain resistance, and surface hardness. They will reproduce the mold surface whether it is polished, sandblasted, or embossed. The small and consistent mold shrinkage of ABS grades allows molding of precision parts for complex assemblies with reproducible fit between components. They are available in a variety of standard colors.

Major markets for ABS plastics include appliances, automobiles, communication devices, electronics, office equipment, furniture, luggage, pipe and fittings, recreational vehicles, toys, lawn and garden equipment, and sporting goods. Figures 5-91 through 5-94 show different products made from ABS.

Specialty grades for high heat resistance, flame retardance, electroplating, cold-forming, structural foams, and transparency are also available.

The electroplating process involves the deposition from solution of a continuous metal layer onto an etched ABS surface (see Figure 5-95). Subsequent layers of metal can then be laid down by conventional electroplating to get the required commercial electroplate composition.

In order to obtain a good bond, the polymer is first treated by an acid etching process which dissolves out some of the rubber particles at or near the polymer surface. After sensitization and activation, electroless metal deposition processes are carried out. Figure 5-96 shows a schematic of a plated ABS surface. Figure 5-97 shows a plated ABS housing for a barometer.

A recent breakthrough in ABS technology is that of the cold-forming grades. New ABS thermoplastics have been developed that can be formed at room temperature, using

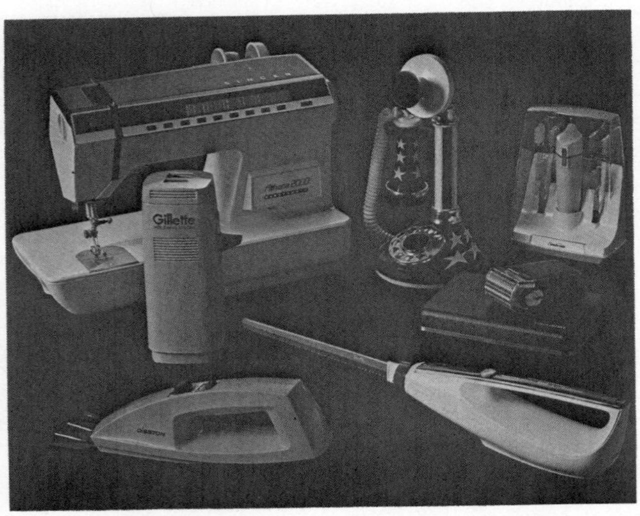

Fig. 5-91. Products made from a general-purpose ABS grade that offers good impact properties, exceptional processability, and long service life. (*Courtesy: Borg-Warner Chemicals*)

standard metal-working techniques and equipment. The main market areas of interest are in the appliance, automotive, and container industries.

In comparison with metal working, cold-forming of plastics usually requires lower tonnage. Common die materials and refined die-making techniques are used. These materials can be preprinted in flat sheet, can be stored indefinitely, and will not rust or corrode. Generally the as-formed finish is the final finish. Scrap can be reprocessed and reused.

A few other thermoplastics can be cold-drawn, but most are too brittle or lack dimensional stability to be processed in this manner. Even some metals are prone to fail if drawn to more than a 40% reduction of diameter.

A new thermoplastic, a methacrylate-butadiene-styrene terpolymer (MABS) combines good clarity with physical and chemical properties close to those of ABS. It can be produced in a number of colors or transparent. The clarity of the material is roughly comparable to other transparent plastics, with a light transmission value of 85% and a refractive index of 1.538. See Table 5-90 for a list of properties. These materials have better resistance to UV degradation; however, they have lower tensile strength and heat-deflection temperature under load.

The MABS polymers are two-phase materials, with the components only partially compatible. The refractive indices of the two phases have been closely matched via a controlled homogeneous copolymerization. This means that the compositions of the copolymers before and after reaction are equivalent. The poor aging of ABS is apparently largely due to oxidative attack of the double bond in the polybutadiene segment of the backbone. The included MMA in the MABS copolymers seems to retard this phenomenon, whereas the acrylonitrile in the ABS copolymers does not.

Processing characteristics, impact, and hardness are about the same for MABS and ABS. The new material can be decorated using conventional techniques such as painting, silk-screening, hot-stamping, and vacuum-forming. One of the new dimensions MABS brings to the decoration of ABS is second-surface finishing; the printing is applied to the reverse side of the object, thus protecting it from wear. This technique is obviously not applicable to opaque materials. Tables 5-91 and 5-92 compare physical properties of MABS with other competitive plastics.

Alloys

One of the more promising extensions of ABS's utility is through the use of alloys. The earliest use of ABS as a commercial alloy was as an impact and processing modifier for PVC. Tables 5-93 and 5-94 show comparative properties of ABS when alloyed with PVC and polycarbonate.

Because the additive efficiency of ABS is dependent on molecular weight and elastomer content, a whole family of

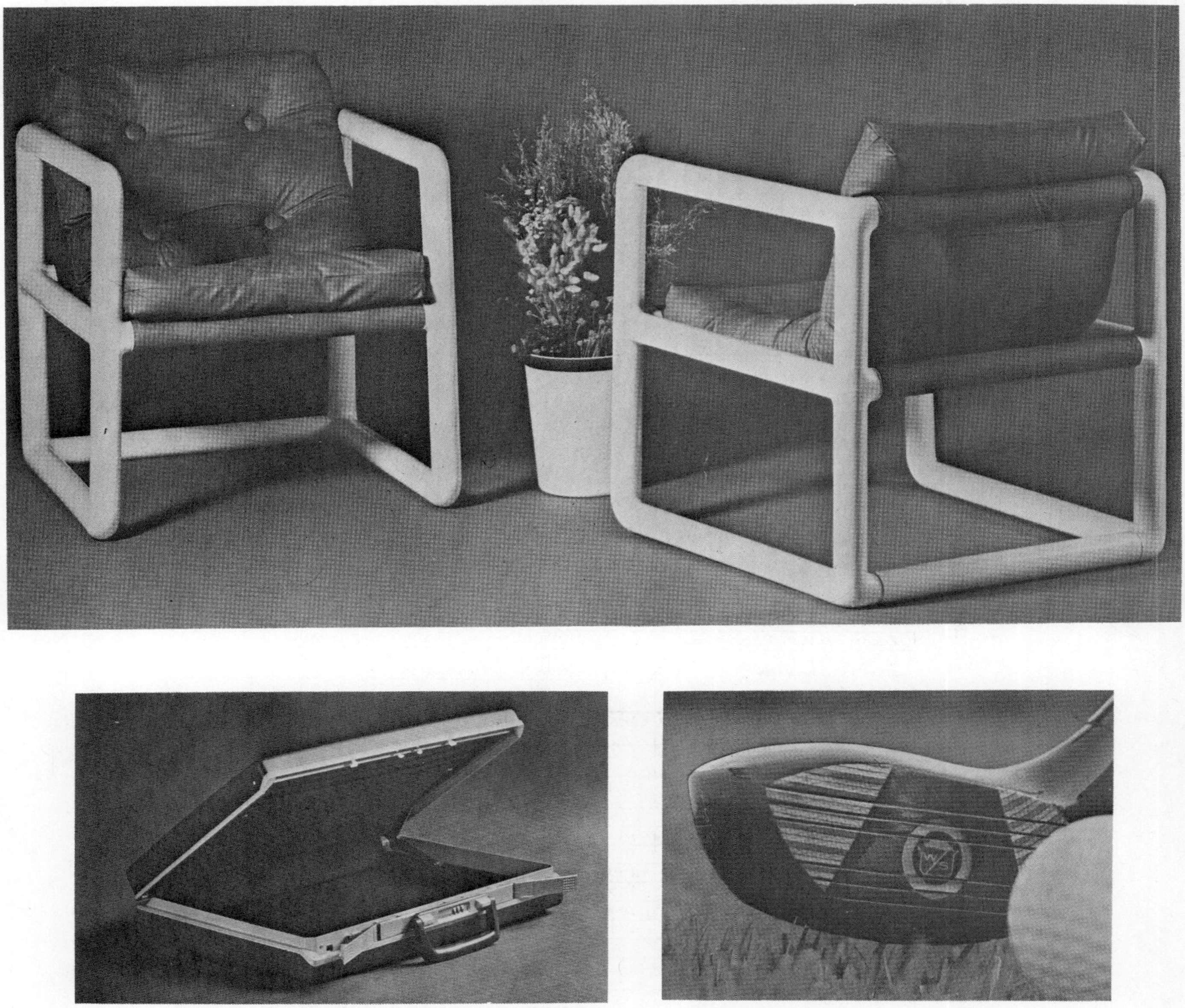

Fig. 5-92. High-impact grades of ABS offer the manufacturer excellent processability while retaining rigidity and toughness over a wide temperature range. (*Courtesy: Borg-Warner Chemicals*)

property curves could be drawn for each ABS type, depending on the type of PVC being modified. ABS-PVC combinations are also used with flexible vinyls to provide grained sheeting used in the automotive industry as crash pads and door liners. These materials can be vacuum-formed without the loss of grain which would occur with straight PVC.

The ABS also provides greater tear strength and heat-deflection temperature, improved tensile and impact strength, and lower specific gravity. These combinations can be calendered or extruded before vacuum-forming. Flame-retardancy can also be imparted to ABS compounds through alloying with PVC, but the processing stability is not good, and low stock temperatures must be used. In addition, heat-deflection temperature is sacrificed.

Polycarbonate is difficult to process and is expensive. The ABS/polycarbonate alloy has excellent processing characteristics at lower cost. The alloy has a higher heat resistance and impact strength compared to conventional ABS resins. Its resistance to flex fatigue is comparable to straight polycarbonate. Gloss and surface appearance are excellent. Its water resistance is not exceptional. It does resist immersion in cold water for 2 to 4 years, but is not recommended for extended exposure to hot water. Chemically, it resists inorganic compounds, except moist chlorine or sulfur dioxide,

Table 5-90. Properties of MABS.*

		ASTM test method	Transparent	MABS
Processing	1. Molding qualities	—	Good to excellent	Good to excellent
	2. Compression molding temp., °F.	—	320-340	400-500
	3. Comp. molding pressure, p.s.i.	—	1000-5000	4000-8000
	4. Injection molding temp., °F.	—	425-480	450-500
	5. Injection molding pressure, p.s.i.	—	10000-20000	8000-20000
	6. Compression ratio	—	1.1-1.2	1.1-1.2
	7. Mold (linear) shrinkage, in./in.	—	0.005-0.008	0.004-0.008
	8. Specific gravity (density)	D792	1.07	1.07
	9. Specific volume, cu. in./lb.	D792	26.0	27.0
	10. Machining qualities	—	Good to excellent	Good to excellent
Mechanical	11. Tensile strength, p.s.i.	D638	5600-6300	6000-8000
	12. Elongation, %	D638	25-75	20-60
	13. Tensile elastic modulus, 10^5 p.s.i.	D638	2.9-3.4	3.3-3.7
	14. Compressive strength, p.s.i.	D695	7000-10000	9500-21500
	15. Flexural yield strength, p.s.i.	D790	9600-11000	10000-120000
	16. Impact strength, ft. lb./in. of notch (½ × ½ in. notched bar, Izod test)	D256	2.5-5.0 (⅛ × ½ in. bar inj. molded)	6.0-8.0 (½ × ⅛ in. bar)
	17. Hardness, Rockwell	D785	98-107R	R102-108
	18. Flexural elastic modulus, p.s.i. × 10^5	D790	3.0-3.5	3.3-3.7
	19. Compressive modulus, p.s.i. × 10^5	D695	2.9-3.4	1.5-2.5
Thermal	20. Thermal conductivity, 10^{-4} cal./sec./sq. cm./1 (°C./cm.)	C177	—	4.5-8.0
	21. Specific heat, cal./°C./gm.	—	—	0.3-0.4
	22. Thermal expansion, 10^{-5} in./in./°C.	D696	9.0-9.5	6.0-10.0
	23. Resistance to heat, °F. (continuous)	—	122	160-200
	24. Deflection temp., °F. at 264 p.s.i. fiber stress	D648	180-186 (annealed) 168-171 (umannealed)	215-225 (annealed)
	at 66 p.s.i. fiber stress		198 (annealed) 186-190 (unannealed)	225-230 (annealed)
Electrical	25. Volume resistivity, ohm-cm. (50% RH and 23° C.)	D257	2.5×10^{15}	$1.0\text{-}5.0 \times 10^{16}$
	26. Dielectric strength, short-time, ⅛-in. thickness, volts/mil	D149	400	350-500
	27. Dielectric strength, step-by-step, ⅛-in. thickness, volts/mil	D149	400	360-400
	28. Dielectric constant, 60 cyc.	D150	3.7	3.0-4.0
	29. Dielectric constant, 10^3 cyc.	D150	—	3.0-3.8
	30. Dielectric constant, 10^6 cyc.	D150	3.2	3.0-3.5
	31. Dissipation (power) factor, 60 cyc.	D150	0.015	0.02-0.05
	32. Dissipation (power) factor, 10^3 cyc.	D150	—	0.02-0.04
	33. Dissipation (power) factor, 10^6 cyc.	D150	0.015	0.02-0.03
	34. Arc resistance, sec.	D495	125-130	—
Optical	35. Refractive index, n_D	D542	1.536	—
	36. Clarity	—	Transparent to opaque	Translucent in thin sections to opaque
	37. Transmittance, %	—	72-82	—
	38. Haze, %	—	6-12	—
Resistance characteristics	39. Water absorp., 24 hr., ⅛-in. thick., %	D570	0.4	0.5
	40. Flammability [a] Burning rate, in./min.	D635	—	—
	AEB, in./ATB, sec.[b]		—	—
	41. Effect of sunlight	—	Yellowing, some embrittlement	Good resistance to prolonged exposure
	42. Effect of weak acids	D543	Nil	None
	43. Effect of strong acids	D543	Attacked by concentrated oxidizing acids	Attacked by concentrated oxidizing acids
	44. Effect of weak alkalies	D543	Nil	None
	45. Effect of strong alkalies	D543	—	None
	46. Effect of organic solvents	D543	Soluble in ketones, esters, aromatic and chlorinated hydrocarbons	Soluble in ketones, esters, and some chlorinated hydrocarbons

*Reprinted by permission from *Modern Plastics Encyclopedia*, McGraw-Hill, Inc.

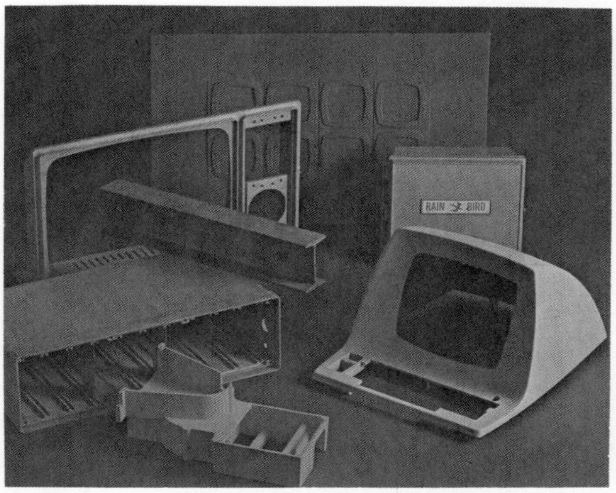

Fig. 5-93. Expandable ABS grades produce rigid cellular parts that are lightweight, attractive, and buoyant. These materials can be injection molded or expansion cast. (*Courtesy: Borg-Warner Chemicals*)

Fig. 5-94. Special-purpose ABS resins for the manufacture of pipe and fittings have high temperature resistance, excellent impact strength, and fast-setting joints. (*Courtesy: Borg-Warner Chemicals*)

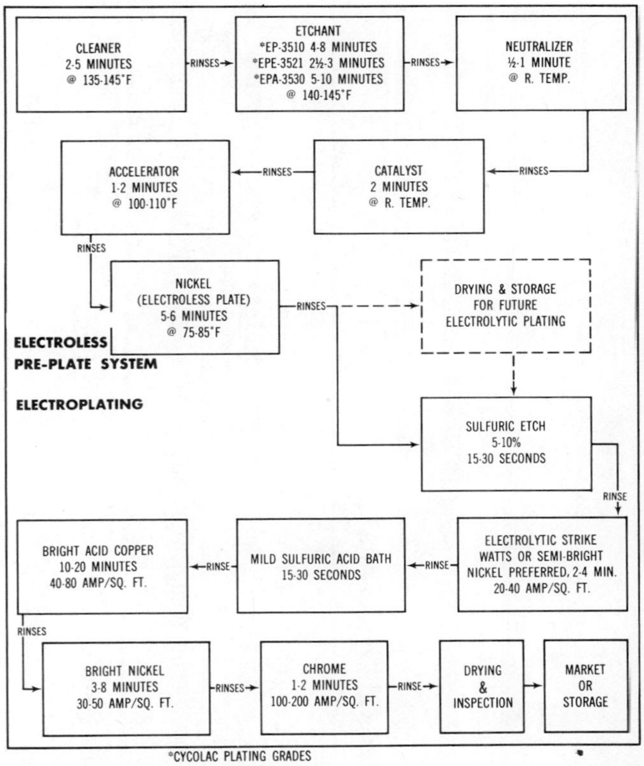

Fig. 5-95. Typical flowchart for decorative plating ABS parts. (*Courtesy: Borg-Warner Chemicals*)

Table 5-91. Comparison of Typical Impact Strengths.

	Izod Impact, ft-lb/in Notch	Falling Missile, ft-lbs
MABS	2.5	12
Acrylic	0.5	0.5
Modified acrylic	1.0	5
ABS	5.0	17
Polycarbonate	15	–

Table 5-92. Comparison of Physical Properties of MABS with Other Plastics.

	Specific Gravity	Mold Shrinkage in/in	Tensile Yield Strength 10^3 psi	Flexural Modulus 10^6 psi	Deflection Temperature @264 psi, °F	Coefficient of Thermal Expansion, 10^{-5} in/in/°F
ASTM Test	D792	D955	D638	D790	D648	D696
MABS	1.07	0.006	6.3	0.35	190	4.8
ABS	1.05	0.006	6.0	0.32	195	5.3
Polycarbonate	1.20	0.006	9.0	0.33	265	3.7
Polystyrene crystal	1.07	0.004	7.0	0.45	180	3.6
SAN	1.08	0.005	9.8	0.50	200	3.4

Table 5-93. Physical and Mechanical Properties of Types of ABS.*

MOLDING OR EXTRUSION Specs.	High Impact	Med. Impact	Low Gloss	Clear	Plateable	High Heat	Flame Retardant	Expandable*	Semi-Flexible	Extra High Imp.	ABS Polycarbonate
Tensile Strength psi	5500-6500	6000-7500	4300-5000	5600-6300	5500-6500	6000-7000	5000-8000	3600-6600	2500	4-5000	8200
Tensile Modulus ($x10^5$) psi	2.8-3.5	3.0-4.0	2.5-2.8	2.9-3.4	3.0-3.6	3.0-3.5	3.2-4.0	---	1.3	2.5-3.0	3.7
Flex. Strength psi	9000-11000	11000-13000	7200-7700	9400-11000	10000-12000	10-12500	9000-14000	4500-7500	4000	7-9000	14,300
Flex. Modulus ($x10^5$) psi	3.2-3.9	3.5-4.0	2.6-2.8	3.0-3.5	3.3-3.8	3.0-4.0	3.0-4.0	1.8-2.7	1.3	2.5-3.0	4.0
Hardness Rockwell R	100-105	107-115	81-85	98-105	103-111	100-115	100-120	75-85 (Shore D)	65 (Shore D)	80-90	118
Notched Izod Impact											
$73^{o}F$ ft.-lb./in.	4-6	2-4	4.5-8.0	2.5-5.0	4-6	2-6	4-12	--	3.6 (¼")	7-10	10.3
$-40^{o}F$ ft.-lb./in.	1-2	1-1.5	1.6-2.4	.5	1-1.5	.7-1.7	.8-1.5	--	2.5 (¼")	2-3	---
Deflection Temperature											
264 Unannealed ^{o}F	180-190	180-190	180-190	168-171	185-200	200-230	153-180	--	153	180-190	255
264 Annealed ^{o}F	200-220	200-220	205-215	180-190	205-220	220-250	160-200	--	173	200-210	--

* Properties dependent on density

*Courtesy: Borg-Warner Corporation.

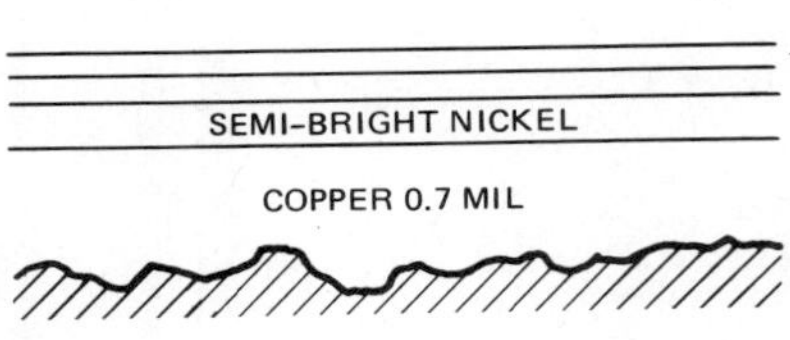

Fig. 5-96. Schematic of a plated ABS surface.

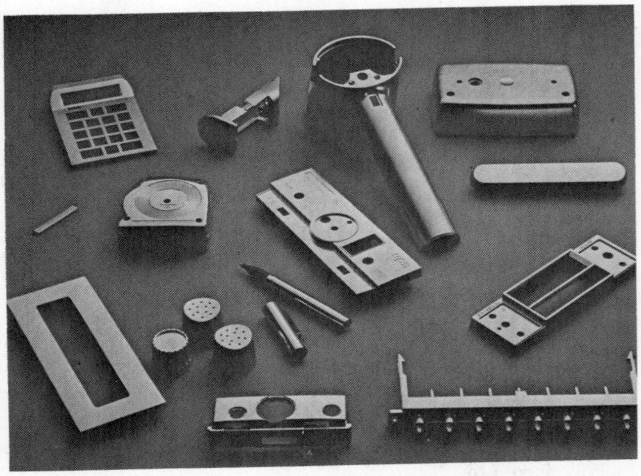

Fig. 5-97. Products take on the decorative advantages of chrome, brass, gold, or nickel finishes using plating grades of ABS. (*Courtesy: Borg-Warner Chemicals*)

Table 5-94. Effect of an ABS Modifier on Rigid PVC.*

Property	PVC	Modified PVC
Tensile strength, psi	7600	6900
Tensile modulus, psi	420,000	395,000
Tensile elongation, %	20	46
Notched Izod impact, ft-lb/in.	0.7	15.7
Specific gravity	1.40	1.36

*Courtesy: Borg-Warner Corporation.

mineral acids, except concentrated sulfuric and nitric, dilute alkali, low kauri-butanol solvents, alcohols, vegetable and mineral oils. Gasoline may cause stress-cracking, depending on the degree of stress and exposure time. The alloy swells in glacial acetic acid, high kauri-butanol solvents, and aromatic and some chlorinated hydrocarbons. It will dissolve in esters, ketones, and some chlorinated hydrocarbons.

The ABS/polycarbonate alloy can be injection molded and sheet extruded for thermoforming. Machinery used for ABS or polycarbonate can be used for the alloy. Processing temperatures range from 254°C to 278°C (490-530°F) with good flow characteristics.

Applications include sport helmets, hip pads, pump impellers, cable separators, golf-club heads, and food-serving dishes.

Trade Names

Absafil	Fiberglass-reinforced ABS	Dart Industries, Inc.
Absaglas	Fiberglass-reinforced ABS	Dart Industries, Inc.
Absasar	Fiberglass-reinforced ABS	Dart Industries, Inc.
Abselex	Extruded ABS sheet	British Celanese, Ltd.
Absinol	ABS	Allied Resinous Products, Inc.
Abson	ABS copolymer and sheet	ABTEC Chemical Co.
Abstrene	ABS and sheet	British Resin Products, Ltd.
Acrylacon	Fiberglass-reinforced SAN	Dart Industries, Inc.
Acrylafil	Fiberglass-reinforced SAN	Dart Industries, Inc.
Acrylaglas	Fiberglass-reinforced SAN	Dart Industries, Inc.
Acrylasar	Fiberglass-reinforced SAN	Dart Industries, Inc.
Acrylast	SAN monofilaments	Dawbarn Brothers, Inc.
Afcoryl	ABS	Rhodia, Inc.
Algil	Styrene copolymer monofilament	Bakelite Xylonite, Ltd.
Alta-Pak	EPS	Alta Industries, Inc.
Amoco	PS	Amoco Chemical Corporation
Amotex	EPS board, block, and sheet	American Syrup & Preserving Co.
Amphenol	PS	American Phenolic Co.
Ancorene	Impact PS	Roberts Consolidated Industries, Inc.
Ancorex	ABS extrusions	Anchor Plastics Co., Inc.
Bakelite	PS	Union Carbide Corporation
Bexan	SAN molding compounds	Bakelite Xylonite, Ltd.
Bexcoat	EPS and laminate	Bakelite Xylonite, Ltd.
Bexel	AN/S/acrylate sheet	Bakelite Xylonite, Ltd.
Bexfoam	Extruded EPS	Bakelite Xylonite, Ltd.
Bexpand	EPS	Bakelite Xylonite, Ltd.
Bi Pak	Biaxially oriented PS	A & E Plastics
Blendex	ABS	Borg-Warner Corporation
Blu Dot	EPS	Holland Plastics Co.
Bohips	Biaxially oriented PS	Sidoplax, S.A.
Boltaron	ABS sheet	General Tire & Rubber Co.
Calabro	SAN fiber	E.B. & A.C. Whiting Co.

Campcolite	ABS sheet	Campco Plastics Co.
Capolyte	ABS sheet	Chicago Plastic Sheet, Inc.
Carinex	PS molding compound	Shell Chemical Co.
Carinex	SAN	Styrene Products, Ltd.
Celatron	PS sheet	British Celanese, Ltd.
Celluliner	Resilient EPS	Gilman Brothers Co.
Cellulite	EPS	Gilman Brothers Co.
Celpak	EPS	Baxenden Chemical Co., Ltd.
Cerex	PS	Monsanto Co.
Cevian-n	SAN	Daicel, Ltd.
Chinatex	PS	Auburn Plastics, Inc.
Coberene	Thermoformable EPS sheet	Cobego, Ltd.
Cosden	PS	Cosden Oil & Chemical Co.
CP-1	PS sheet	Cast Optics Corporation
Crystal Foam-S	EPS with PS surface	Plastics Systems Corporation
Cycolac	ABS	Borg-Warner Corporation
Cycolon	ABS	Borg-Warner Corporation
Cy Con	ABS	Conneaut Rubber & Plastics
Cycopac	ABS	Borg-Warner Corporation
Cycovin	ABS/PVC alloy	Borg-Warner Corporation
Cymac	PS	American Cyanamid Co.
Dapco	ABS, PS	Davidson Products Co.
Darex	Styrene copolymer resins and latices	W.R. Grace & Co.
Decoloy	Acrylic/ABS composite	Borg-Warner Corporation
Dialux	Dimethyl itaconate/styrene	Pfizer, Inc.
Diapet	ABS	Mitsubishi Rayon Co., Ltd.
Diarex	PS molding compound	Mitsubishi Monsanto Chemical Co.
Dielectron	PS	Polypinco, Ltd.
Dikaryl	SAN	Etablissements Kuhlmann
Distrene	PS	British Resin Products, Ltd.
Dorvon	EPS	Dow Chemical Co.
Dow	PS	Dow Chemical Co.
Drave	ABS pipe	Vistron Corporation
Dri-Lite	EPS	Poly Foam Inc.
Durabass	PS filament	E.B. & A.C. Whiting Co.
Durapipe	ABS pipe and fittings	Durapipe & Fittings, Ltd.
Duratron	High-impact PS	Shell Chemical Co.
Duvalite	Reinforced PS	Duval Corporation
Dyfoam	Molded EPS	W.R. Grace & Co.
Dylark	Styrene-based polymer	ARCO Polymers, Inc.
Dylel	ABS	ARCO Polymers, Inc.
Dylene	PS	ARCO Polymers, Inc.
Dylite	PS	ARCO Polymers, Inc.
Eanplast	PS	East Anglia Plastic, Inc.
Edistir	PS	Montedison, S.p.A.
Elaprim	ABS	Montedison, S.p.A.
El Rexene	PS	Dart Industries, Inc.
Emu-Powder	PS	BASF, A.G.
Epsan	SAN/EPDM copolymer	Copolymer Rubber & Chemical Corporation
Erinold	PS molding compounds	Mobil Chemicals, Ltd.
Esbrite	PS	Sumitomo Chemical Co.
Evenglo	PS	ARCO Polymers, Inc.
Everwhite	PS	Sheffield Plastics, Inc.
Evolite	PS rod, tube, and profile	Richard Daleman, Ltd.
Expanda-Foam	PS	Expanda-Foam Inc.
Extir	EPS	Montedison, S.p.A.
E-Z Flow	PS	Polysar Resin, Inc.
Fiberlene	Glass-filled SAN and PS	Service Color Corporation
Fiber Tuff	Glass-filled PS	ARCO Polymers, Inc.
Flexafoam	EPS	Polysar Plastics, Inc.
Foam Form	EPS	Fiberez, Ltd.
Formid	ABS	Polymon Developments, Ltd.
Formos	ABS sheet	Ford Motor Co.
Fostacryl	SAN	American Hoechst Corporation
Fostafoam	EPS	American Hoechst Corporation

Fostafoam	PS	American Hoechst Corporation
Fostalite	PS	American Hoechst Corporation
Fosta-Net	EPS	American Hoechst Corporation
Fostarene	PS resins and molding compounds	American Hoechst Corporation
Fosta-Sol	Vinyl toluene	American Hoechst Corporation
Fosta Tuf-Flex	Impact PS	American Hoechst Corporation
Frost Cefoam	EPS	Expanda Foam, Inc.
Frostwhite	PS sheet and profile	Sheffield Plastics Inc.
Frostwood	Molded PS beads	Dow Chemical Co.
Fudowleon	EPS	Fudow Chemical Co.
Gedex	PS	CDF Chimie
Genco	Modified and foamed PS	General Plastics Inc.
Gilco	PS	The Gilman Brothers Co.
Gordon	PS and ABS and compounds	Carl Gordon Industries, Inc.
Grace	PS	W.R. Grace & Co.
Gravoflex	ABS sheet	Hermes Plastics, Inc.
Gresintex	ABS extrusions	Societa del Gres Ing. Sala & C.
Gulf	PS	Gulf Oil Chemicals Co.
Hi Blem	ABS	Nippon Zeon Co.
Hostapor	EPS	Hoechst, A.G.
Hostyrene	PS	Hoechst, A.G.
Hypac	High-impact PS sheet	Raritan Plastics Corporation
Iridon	PS and ABS sheet	Commercial Plastics, Ltd.
Jet Flow	PS	Polysar Resin, Inc.
Kane Ace	ABS	Kaneka America Corporation
Kane Pearl	EPS beads	Kaneka America Corporation
Kanestyrene	SAN	Kaneka America Corporation
Keltrol	Vinyl toluene copolymers	Textron, Inc.
Kostil	SAN	Montedison, S.p.A.
Kralastic	ABS	Naugatuck Chemical Div., Uniroyal, Inc.
Kralon	Rubber-modified PS and ABS	Uniroyal, Inc.
Lacqran	ABS	ATO Chimie of Paris
Lacqrene	PS	ATO Chimie of Paris
Lacqsan	SAN	ATO Chimie of Paris
Laminil	EPS core board	Lavorazione Materie Plasticke, S.p.A.
Lastilac	Reinforced ABS	Lati, S.p.A.
Lastill	Reinforced SAN	Lati, S.p.A.
Lastirol	PS film	Lati, S.p.A.
Litac	SAN	Toyo-Koatsu Industries, Inc.
Loavar	PS coating	Ashland Chemical Co.
Lorival	ABS tube and profile	United Evonite & Lorival, Ltd.
Lorkacel	PS	Ugine Kuhlmann
Lorkalene	PS	Etablissements Kuhlmann
Lorkaril	ABS	Etablissements Kuhlmann
Luran	ABS and ASA terpolymers	BASF Wyandotte Corporation
Lustran	SAN and ABS resins	Monsanto Chemical Co.
Lustrex	PS compounds	Monsanto Chemical Co.
Lustropak	ABS	Monsanto Chemicals, Ltd.
Lytron	PS emulsions and coatings	Monsanto Chemical Co.
Marbon	PS resins	Borg-Warner Corporation
Marleycel	EPS	Marley Foam, Ltd.
Micro-Por	Porous ABS tubing	International Plastics, Inc.
Monsanto	PS	Monsanto Chemical Co.
Montopore	PS molding compounds	Monsanto Chemicals, Ltd.
Myriad	EPS	American Hoechst Corporation
Nabultene	ABS Film & Sheet	Convert, Ets.G.
NAS	Styrene/methyl methacrylate copolymers	Richardson Co.
Naxolene	PS film	Convert, Ets.G.
Neo Rez	Styrene emulsion	Polyvinyl Chemicals, Inc.
Nixon	PS sheet	Nixon Nitration Works, Inc.
Noan	MMA/styrene copolymers	Richardson Co.
Novudur	ABS	Farbenfabriken Bayer, A.G.
Olo-Flex	EPS	Swedish Crucible Steel Co.
Olo-Lite	EPS	Olsonite Corporation
Pelaspan	Expandable PS	Dow Chemical Co.

Pic	PS	Cook United, Inc.
Piccoflex	Styrene copolymers	Hercules, Inc.
Piccolastic	Styrene resins	Hercules, Inc.
Piccotex	Styrene copolymers	Hercules, Inc.
Plasticor	EPS	Polysar Plastics, Inc.
Plaxacrin	SAN sheet	Monsanto Co.
Plexene	Modified PS	Rohm & Haas Co.
Pliolite	Vinyl toluene/butadiene, SAN, and vinyl toluene copolymers	Goodyear Tire & Rubber Co.
Plio-Tuf	ABS, PS resins and copolymers	Goodyear Tire & Rubber Co.
Polyblend	ABS	B.F. Goodrich Chemical Co.
Polyfiber	PS	Dow Chemical Co.
Polyflex	PS sheet	Monsanto Co., Inc.
Polyfoam	Expanded PS	Glo-Brite Foam Plastics Products
Polygrain	PS	East Anglia Plastics, Ltd.
Polylac	ABS compounds	Polychemical Co.
Polypearl	EPS	Polychemical Co.
Polypinco	Crosslinked PS	Polymer Corporation
Polyrex	PS	Poly Chemical Co.
Polysol	PS suspensions	Showa High Polymer Co.
Polystyrol	SAN	BASF Wyandotte Corporation
Polystrom-Omni	PS	Omni Products Corporation
Polytherm	EPS	Omni Products Corporation
Polytone	PS	Carl Gordon Industries, Inc.
Polyzote	EPS	Expanded Rubber & Plastics, Ltd.
Poragel	PS chromotography gel	Waters Associates, Inc.
Poron	EPS	Poron Insulation, Ltd.
Pow'r-Pak	PS	Northern Petrochemical Co.
Psv-L-N	EPS	Soviet Origin
Ravikral	ABS	ANIC, S.p.A.
Resarit	ABS	Resart-IHM, A.G.
Resiloflex	EPS	Polysar Plastics, Inc.
Resiston	PS	Allied Resin Products Co.
Resoglaz	PS	Cincinnati Milacron Chemical, Inc.
Resproid	PS-impregnated fabric	Respro, Inc.
Resticel	EPS	Societa Italiano Resine
Restil	SAN	Societa Italiano Resine
Restiran	ABS	Societa Italiano Resine
Restirolo	PS	Societa Italiano Resine
Rexolite	PS rod and sheet	Brand-Rex Co.
Rez-N-Bond	Styrene cement	Schwartz Chemical Co.
Rigi Flex	ABS extrusions	Crane Plastics, Inc.
Riji-Tuf	PS	American Hard Rubber Co.
Rimonel	PS emulsion	Cello Chemical Co.
Robex	PS sheets and film	Caleppio, S.R.L.
Ronfalin	ABS	DSM Polymer, Ltd.
Roofmate	Extruded EPS	Dow Chemical Co.
Royalite	ABS sheet	Uniroyal, Inc.
Ruddflex	PS copolymers	Rudd Plastic Fabrics Corporation
Santofoam	PS foam	Monsanto Co.
Santopearl	EPS sheet	Mitsubishi Monsanto Chemical Co.
Saroy	PS	Telcon Plastics, Ltd.
Sarrac	ABS	Saro Products, Ltd.
Satin Foam	Extruded EPS	Plastics Systems Corporation
Scorbord	Expanded PS	Dow Chemical Co.
Seamco	PS molding compound	Seamco Chemical Co.
Shalon	Styrene monofilament	Shawinigan Chemicals, Ltd.
Shalon	Styrene monofilament	Bakelite Ltd, & Polymers, Inc.
Shell	PS	Shell Chemical Co.
Shinko Lac ABS	ABS	Mitsubishi Rayon Co.
Sicoflex	ABS and sheet	Mazzucchelli Celluloide, S.p.A.
Sicostirol	PS and SAN and Sheet	Mazzucchelli Celluloide, S.p.A.
Silverflow	ABS Pipe and Fittings	BTR Industries, Ltd.
Sisal-Glaze	Styrene film	American Sisalkraft Co.
Solar	PS	Solar Chemical Corporation
Spandoplast	EPS	Baxenden Chemical Co., Ltd.

Sternite	PS and ABS molding compounds	Sterling Moulding Materials, Ltd.
Stilon	ABS/PC alloy	Soviet Origin
Stiroplasto	PS compounds	Montedison, S.p.A.
Stokofex	PS	Stokes Molded Products
Stokolite	PS	Stokes Molded Products
Structo-Foam	EPS	Stauffer Chemical Co.
Styac	SAN molding compound	Monsanto Co.
Styco	PS	Nixon-Baldwin Chemicals, Inc.
Sty Con	PS	Conneaut Rubber & Plastics
Stylafoam	Coated PS sheet	Gilman Brothers Co.
Stylour	Flocked PS sheet	Gilman Brothers Co.
Stymer	Styrene copolymer textile sizes	Monsanto Co.
Styracon	Glass-reinforced PS	Dart Industries, Inc.
Styrafil	Glass-reinforced PS	Dart Industries, Inc.
Styraflock	Flocked PS	Primex Plastics Corporation
Styraglas	Fiberglass-reinforced PS	Dart Industries, Inc.
Styrasar	Fiberglass-reinforced PS	Dart Industries, Inc.
Styrasuede	Flock-coated PS	Flock Industries, Inc.
Styrex	SAN	Dow Chemical Co.
Styrocel	S/MMA	Shell Chemical Co., Ltd.
Styrocell	EPS	Dow Chemical Co.
Styrodur	EPS	BASF Wyandotte Corporation
Styroflex	PS film	Natvar Corporation
Styrofoam	Extruded EPS	Dow Chemical Co.
Styrol	PS	Indemitsu Petrochemical Co.
Styrolux	PS rod, sheet, and tube	Westlake Plastics Co.
Styron	PS film and resin	Dow Chemical Co.
Styronal	SAN	BASF Wyandotte Corporation
Styronol	PS	Allied Resinous Products, Inc.
Styrophane	ABS film	UCB-Sidac Corporation
Styropor	Expandable PS beads	BASF Wyandotte Corporation
Styvacril	SAN	Products Chemiques, U.K.M.
Styvarene	PS molding compound	Products Chemiques, U.K.M.
Sullvac	ABS sheet	O'Sullivan Rubber Corporation
Superblow	PS	Carl Gordon Industries, Inc.
Superdense	Impact PS	Carl Gordon Industries, Inc.
Superflex	PS	Carl Gordon Industries, Inc.
Superflow	PS	Carl Gordon Industries, Inc.
Taflite	PS/EPDM	Mitsui Toatsu Chemical, Inc.
Terluran	ABS	BASF Wyandotte Corporation
Texicote	PS	Scott Bader & Co., Ltd.
Thermco	EPS	Holland Plastics Co.
Ticor	PS	Ticonderoga Chemical Corporation
Toporex	PS	Mitsui Chemical Industry Co.
Toyolac	ABS	Toray Industries, Inc.
Trilon	SAN	Triple R Industries, Inc.
Tronal	ABS sheet	Dynamit Nobel, A.G.
Trycite	PS film and sheet	Dow Chemical Co.
Tuf-Flex	PS	American Hoechst Corporation
Tufrex	ABS	Mitsubishi Monsanto Chemical Co.
Tybrene	ABS	Dow Chemical Co.
Tyril	SAN	Dow Chemical Co.
Tyrilfoam	SAN	Dow Chemical Co.
Ubatol	Modified-PS emulsion	UBS Chemical Corporation
Ugikral	ABS	Products Chemiques Ugine Kuhlmann
Uni-Crest	PS foam	United Cork Companies
Urtal	ABS molding compounds	Montedison, S.p.A.
Usalite	ABS	Uniroyal, Inc.
USS	PS	United States Steel Chemical Co.
Vacucel	EPS	Poloron Products, Inc.
Vencel	EPS	Venesta Mfg., Ltd.
Verelite	PS film and resin	Dow Chemical Co.
Vestypor	EPS	Chemische-Werke Huls, A.G.
Vestyron	PS compounds	Chemische-Werke Huls, A.G.
Vinalak	PS	Vinyl Products, Ltd.

Vinamul	PS	Vinyl Products, Ltd.
Vinuran	Styrene-modified PVC	BASF Wyandotte Corporation
Visolyte	PS film	Union Carbide Corporation
Vulkide	ABS sheet	ICI (Hyde), Ltd.
Xanfoam	Extruded EPS	W.R. Grace & Co.
Zerlon	PS/PMMA copolymers	Dow Chemical Co.
Zerok	Styrene-based coatings	Atlas Mineral & Chemical Division, ESB Inc.
Zero Lite	EPS	Johns-Manville Corporation

REFERENCES AND BIBLIOGRAPHY

Amos, J. L., and Peterson, N. R. Polystyrene plastics family.

Anis, N. The ABC's of electroplating ABS. *Plastics Engineering*, January 1977, **33**(1):14–17.

Bair, H. E., Boyle, D. J., and Kelleher, P. G. "Light and Heat Effects on ABS." Paper presentation, *SPE ANTEC*, New Orleans, May 1979.

BASF Wyandotte Corporation. Rigid plastic foams, from Plastics in Construction: Coming on Strong in 1978. *Plastics Engineering*, July 1978, **34**(7):42–44.

Bebbington, G. H. "Flame Retardant ABS Scrap." Paper presentation, *SPE ANTEC*, New Orleans, May 1979.

Beitel, J. J., Herrera, W. R., Pish, M. D., and Mitchell, D. S. "Real-life" burning behavior of polystyrene home furnishings. *Modern Plastics*, April 1979, **56**(4):88–94.

Borg-Warner Chemicals. *Marbon Plating Product News*. Vol. XIII, No. 2, May 1972.

——. *Marbon ABS Plastics Typical Property Data*. Bulletin 7179, August 1972.

——. *The Marbon Family of ABS Design Engineering Materials*. Bulletin 3654, January 1969.

——. *Marbon Chemical Introduces the First Successful Technology for the Cold Forming of ABS Plastic*. Bulletin 127-M-1494A, March 1967.

——. *Chemical Resistance of Cycolac Brand ABS Plastics*. Technical Bulletin PB-123, October 1966.

——. *Design, Converting and Handling Techniques for Plating Cycolac Brand ABS*. Technical Bulletin PB-115C.

——. *Machining of Cycolac Brand ABS*. Technical Bulletin PB-128.

——. *Cycolac Brand ABS by Marbon Division–ABS Assembly Techniques*. Technical Bulletin PB-130.

——. *Modification of PVC Compounds With Blendex ABS Resins*. Technical Bulletin B-101A.

——. *Cycovin KA Brand ABS/PVC Self-Extinguishing Resin*. Technical Report P-110A.

——. *Cycoloy Brand Alloy Grade KHP*. Technical Report P-177.

——. *Data Sheets on Grades CG, GSE, GSM, L, & LS*. Technical Publications P-538, P-533A, P-156, P-517A, & P-532A.

——. *Weatherability of Cycolac Brand ABS*. Technical Publication P-405.

——. *Cycolac ABS, An Engineering Material with Remarkable Designability and Cost Performance*.

Boundy, R. H., Boyer, R. F., and Stoesser, S. M. *Styrene–Its Polymers, Copolymers and Derivatives*, Reinhold Publishing Corporation, New York, 1952.

Brydson, J. A. *Plastics Materials*. 3rd Ed. Robert E. Krieger Publishing Co., Huntington, New York, 1975.

Corbin, J. R. "ABS Resins." In *Modern Plastics Encyclopedia*. S. Gross (exec. ed.), **45**(14A):118–123. McGraw-Hill Book Co., New York, 1968–69.

Dow Chemical Co. *Tyril Styrene-Acrylonitrile Resins*. Form No. 301-337, 1975.

——. *People Live Better With Tyril Plastics*. Form No. 301-68, 1972.

——. *Dow ABS . . . Engineered for Performance*. Form No. 301-137, 1972.

——. *Tyril Styrene-Acrylonitrile Copolymers*. Form No. 171-587, 1968.

——. *Plastics Today, The Story of Styron in the Home*. Form No. PL213A-253, 1952.

Finch, C. Polystyrene, from Plastics in Packaging. *Plastics Engineering*, October 1978, **34**(10):47–48.

Forger, G. R. Low cost polystyrene makes it as a versatile, precision material. *Materials Engineering*, October 1977, **86**(4):64–67.

——. Plastic alloys offer alternatives to the high cost of overengineering. *Materials Engineering* July 1977, **86**(1):44–47.

Fritch, L. Injection-molding ABS? Mold in quality with the machine. *Plastics Engineering*, May 1979, **35**(5):68–72.

Fritch, L. W. "The Melt Front and ABS Cavity Flow." Paper presentation, *SPE ANTEC*, New Orleans, May 1979.

Fry, W. M. *The Story of Polystyrene*. Fry Plastics International Educational Division, Los Angeles, California, 1963.

Golding, B. *Polymers and Resins–Their Chemistry and Chemical Engineering*, D. Van Nostrand Co., Inc., Princeton, New Jersey, 1959.

Han, C. D., and Yoo, H. J. "Eliminating Gas Bubbles in Structural Foam Processing." Paper presentation, *SPE ANTEC*, New Orleans, May 1979.

Hauck, J. E. Take a fresh look at low cost styrene acrylonitrile. *Materials in Design Engineering*, June 1963, **57**(6):84.

Hoa, S. V. "Fluid Mobility and Solubility in High-Impact Polystyrene." Paper presentation, *SPE ANTEC*, New Orleans, May 1979.

Helgesen, J. A., and Ferguson, L. E. ABS, from Plastics in Construction: Coming on Strong in 1978. *Plastics Engineering*, July 1978, **34**(7):23–24.

Houston, A. M. New types, improved processes expand use of structural foam. *Materials Engineering*, November 1975, **82**(6): 38–42.

Hubbauer, P., and Kern, H. E. "Molded Products from Recycled Telephones." Paper presentation, *SPE ANTEC*, New Orleans, May 1979.

Juran, R. ABS builds up its appeal. *Modern Plastics*, August 1979, **56**(8):44–45.

Klodenski, R. Coextruded sheet. *Plastics World*, August 1977, **35**(8):52–53.

Koppers Company, Inc. "Properties of Film and Sheet Extruded From Dylite Expandable Polystyrene." Chapter 46, Bulletin C-9-273, April 1962.

Kraynak, G. M. "ABS Plastics, A Versatile Thermoplastic." Paper supplied by Borg-Warner Chemicals.

Lobe, V. M., and White, J. L. "Effects of Carbon Black on a Polystyrene Melt." Paper presentation, *SPE ANTEC*, New Orleans, May 1979.

Malpass, V. E. "ABS Engineering Thermoplastics." Paper presentation, *Plastics World Seminar: Engineering Plastics II*, Los Angeles, March 1976.

Martino, R. Prize-winning structural-foam applications. *Modern Plastics*, July 1979, **56**(7):52.

——. Big buildup in EPS bead molding. *Modern Plastics*, August 1979, **56**(8):38–39.

——. Resurgence of vigor in extruded PS foam. *Modern Plastics*. August 1978, **55**(8):34–37.

Martino, R. J. Styrenics, from Materials '79. *Modern Plastics*, January 1979, **56**(1):51-53.
——. PS business finds a new growing edge: expandable bead. *Modern Plastics*, November 1978, **55**(11):50-53.
Materials in Design Engineering. Low cost, transparent plastic. July 1960, **52**(1):102-103.
Modern Plastics. New styrenics: high gloss, high impact, better processing qualities. March 1979, **56**(3):78.
——. Polystyrene in its 40th year: rejuvenated. *Modern Plastics*. March 1979, **56**(3):46-48.
——. A new idea in foam-core panels helps cut building costs 20%. March 1979, **56**(3):54-55.
——. Double-wall insulated PS shell makes a fire-protective safe. February 1978, **55**(2):28-30.
——. Toughness with a gentle touch makes ABS the choice for six molded parts. February 1978, **55**(2):30.
——. Process streamlining cuts cost of molding EPS board and shapes. November 1978, **55**(11):22.
——. The rugged polystyrenes. April 1962, **39**(4):80-89, 192-196.
Monsanto Co. *Lustran SAN Resins, Styrene Acrylonitrile Copolymer Molding Resins*. Bulletin No. 6364C, April 1979.
——. *Lustran SAN 31–New, Easy Flowing Styrene Acrylonitrile Copolymer*. Data Sheet No. 6364, September 1975.
——. *Chemical Resistance of Lustran SAN*. Product Information Bulletin 1110B, November 1974.
——. *Lustran A Styrene Acrylonitrile Copolymer* Technical Service Bulletin 3234 B, January 1967.
Moore, W. R. *An Introduction to Polymer Chemistry*, Aldine Publishing Co., Chicago, Illinois, 1963.
Mortimer, B. T. "PAMS: An Extender, Flow Promoter, and Reinforcer for Urethanes." Paper presentation, *SPE ANTEC*, New Orleans, May 1979.
Pillichody, C. T. "ABS Plastics–Yesterday, Today, and Tomorrow." Paper presentation, *Bicentennial RETEC of the Southeastern Ohio Section, SPE*, November/December 1976.
Plastics Engineering. Expanded polystyrene, from Plastics in Packaging. October 1978, **34**(10):35-37.
Plastics Technology. P.T.'s Buyer's Guide to Polystyrenes and Styrene Copolymers, December 1967, **13**(13).
Plastics World. ABS: super resin. October 1971, **29**(11):1-24.
Richmond, M. H., and Wooden, D. C. "Toxicological Compounds for Polystyrenes." Paper presentation, *SPE ANTEC,* New Orleans, May 1979.
Simonds, H. R., and Church, J. M. *A Concise Guide to Plastics*. Van Nostrand Reinhold Co., New York, 1963.
Smoluk, G. Expanded PS bead molding, from Machinery '79. *Modern Plastics*, April 1979, **56**(4):64.
Steel. Cold stamped plastics make debut. April 1967, **161**(8).
Steen, S. S. Molding ABS for electroplating. *SPE Journal*, October 1968, **24**:77-83.
Swanson, R. S. *Plastics Technology–Basic Materials and Processes*, McKnight & McKnight Publishing Co., Bloomington, Illinois, 1965.
Tallman, L. E., and Hanson, W. F. Polystyrene Course. Dow Chemical Co., November 1967.
Thakkar, B. S., and Broutman, L. J. "Residual Stresses and Orientation in Amorphous Polymers." Paper presentation, SPE ANTEC, New Orleans, May 1979.
Titus, J. B. *Trade Designations of Plastics and Related Materials*, (Revised). PLASTEC Note N9C, Plastics Technical Evaluation Center, Picatinny Arsenal, Dover, New Jersey, May 1978.
Uniroyal, Inc. *Selector Guide–Rubber, Plastics, Chemicals and Polymers*. Bulletin No. ASP-4470A.
United States Rubber, Naugatuck Chemical Division, *Kralastic for Product Engineering and Design*.
Vaccari, J. A. One toymaker relies heavily on plastics and wood for child's play. *Materials Engineering*, May 1977, **85**(5):18-21.
Walker, D. A. "Styrene Polymers and Copolymers." In *Modern Plastics Encyclopedia*. S. Gross (exec. ed.), **45**(14A):333-335, 340. McGraw-Hill, Inc., 1968-69.
West, H. J. A Methylstyrene molding compound for the electronics industry. *Western Plastics*, Technical Report No. 39, May 1958, pp. 19-21.
Winding, C. C., and Hiatt, G. D. *Polymeric Materials*, McGraw-Hill Book Co., New York, 1961.
Wood, A. S. Clarity plus lots of useful extras: it's available at $3/lb.–or 30¢. *Modern Plastics*, March 1979, **56**(3):50-53.
Zechinati, J. "Foam coextruded PS drain pipe is tough . . . and economical too. *Modern Plastics*, November 1978, **55**(11):78-80.
Zimmerman, R. L. Engineer's guide to ABS plastics. *Materials Engineering*, October 1971, **74**(5):34-39.

ENGINEERING PLASTICS

The term "engineering plastics" originally came into widespread use to denote the high-performance thermoplastic polyesters (viz., polyethylene terephthalates) that could replace metal parts in automotive, appliance, business machine, and other industrial applications where rigid engineering specifications must be met. Many other plastics materials had long been used in similar applications, yet they were never categorized as engineering plastics.

There are three key characteristics that distinguish engineering resins from other plastics materials: the plastics' mechanical and chemical properties and durability. In specific terms, the following properties are associated with "engineering" resins: (1) high strength/weight ratio, (2) retention of mechanical properties at elevated temperatures, (3) low creep, (4) low coefficients of thermal expansion, (5) good electrical properties, and (6) good chemical resistance.

Because we have chosen to group the plastics in this book by "family," a number of plastics normally categorized as "engineering" are not covered in this section. For example, ABS resins are discussed in the chapter on styrenics, phenolics in the chapter on thermosetting aldehyde-based resins, and fluorocarbons have been given a chapter of their own. Conversely, it was necessary to include some discussions of a few plastics materials not normally thought of as engineering resins for the sake of continuity (e.g., polyethers). For the most part however, the discussion that follows has as its emphasis, a description of the major engineering resins of commercial significance today.

Table 5-95 lists the design properties of leading engineering resins. Of the 14 resin types listed there, this section will cover the following seven in more detail: acetals, nylons, phenylene oxide, polycarbonate, polyphenylene sulfide, polysulfones, and thermoplastic polyester. Table 5-96 compares the material costs of engineering plastics with other plastics and metals. Table 5-97 does the same for reinforced engineering plastics and reinforced thermoplastics. Figure 5-98 compares the price change of engineering plastics with die-cast metals from 1968 onward. Figure 5-99 shows the projected growth of engineering plastics.

Table 5-95. Key Design Properties of

Polymer Class	ABS		Acetals			Epoxies	
Grade	High Impact	Reinforced	Homopolymer	Copolymer	Reinforced	Reinforced	
Filler type and concentration, percent	–	Glass fiber, 30	–	–	Glass fiber, 30	Glass fiber, 60	Mineral, 60
Molding method	Injection	Injection	Injection	Injection	Injection	Compresssion	Transfer
Molding temperature, °F	475 to 525	400 to 560	400	350 to 390	350 to 425	300 to 330	250 to 330
Molding pressure, psi × 10^{-3}	10 to 20	Varies	10 to 20	Varies	Varies	0.3 to 5	0.1 to 1
Molding time, seconds	Varies	Varies	20	Varies	Varies	Varies	Varies
Mold shrinkage, inches per inch ($\frac{1}{8}$-inch thickness)	0.007 to 0.009	0.001 to 0.0015	0.018	0.014/ 0.018[g]	0.003	0.002	0.002
Specific gravity	1.02	1.28	1.40	1.41	1.63	1.8	1.9
Tensile strength, psi	4900	14,500	10,000	8800	19,500	20,000	7500
Tensile modulus, psi × 10^{-5}	2.4	–	5.2	4.1	–	20	15
Elongation at break, percent	Varies	3 to 4	30	45	4	1.0	<1.0
Flexural strength, psi							
At room temperature	7900	18,500	14,100	13,000	29,000	30,000	12,000
At 350°F	4300[b]	–	–	–	–	–	–
Flexural modulus, psi × 10^{-5}	2.5	11.0	4.1	3.75	14.0	20	15
Compressive strength, psi	5800	14,500	–	16,000[h]	–	35,000	30,000
Compressive modulus, psi × 10^{-5}	1.7	–	6.7	–	–	25	15
Notched Izod impact strength, foot-pounds per inch of notch	7.5	1.4[c]	1.4	1.0	1.3	20	0.3
Rockwell hardness	R89	M99/R124	M94/R120	M80	M86	M100	M112
Coefficient of expansion, inches per inch per °F × 10^5	6.1	1.6	4.2 to 5.5[d]	4.7	2.4	1.0	2.0
Continuous-use temperature, °F[a]	150	155	250[e]	220	205	300	300
Dielectric constant at 1 mH	3.1	–	3.7	3.7	3.95	4.5	4.0
Dissipation factor at 1 mH	0.011	–	0.0048	0.006	0.0065	0.01	0.01
Water absorption (24-hour immersion), percent	0.20	0.14	0.25	0.22	0.60	0.12	0.05
Chemical resistance to:							
Weak acids	Good	Good	Good[f]	Good	Fair	Excellent	Excellent
Strong acids	Fair	Fair	Poor	Poor	Poor	Excellent	Good
Weak alkalies	Excellent	Excellent	Good[f]	Good	Good	Excellent	Good
Strong alkalies	Excellent	Excellent	Poor	Good	Fair	Excellent	Fair
Organic solvents	Poor	Poor	Excellent	Good	Excellent	Excellent	Good

[a]Temperature at which 50 percent of initial properties is retained after 1000 hours exposure.
[b]Measured at 160°F.
[c]Unnotched impact strength: 6 to 7 foot-pounds per inch.
[d]Varies with temperature.
[e]Retains 80 percent of initial properties.
[f]Individual tests recommended.
[g]Values are for flow direction/transverse direction.
[h]At 10 percent deflection.
[i]Value is 0.005 for $\frac{1}{4}$-inch section.
[j]Unnotched impact strength: 17 to 18 foot-pounds per inch.
[k]Value is 0.005 for $\frac{1}{4}$-inch section.
[l]Unnotched impact strength: 17 foot-pounds per inch.

Polyamides/Nylons

The polyamides are condensation products which have the amide group as an integral part of the linear chain. This family of polymers came from the endeavors of Wallace H. Carothers and co-workers to produce synthetic fibers in the early 1930s. The fibers resembled and had strength equal to natural silk, the then high-fashion material. The polyamides resemble the linear polyesters in that they are long carbon chains with recurring functional groups in the chain; however, the groups are amides instead of esters. The synthetic polyamides are also closely related to proteins and can be simply represented as an amino acid condensation product:

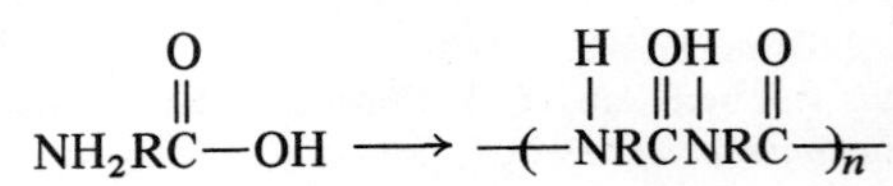

Leading Engineering Resins.*

Fluoropolymers				Nylon 66			Phenolic
PTFE	FEP	ETFE	Reinforced ETFE	Unmodified	Reinforced		General Purpose
–	–	–	Glass fiber, 30	–	Glass fiber, 30	Mineral, 40	Woodflour, 46
Compression	Injection	Injection	Injection	Injection	Injection	Injection	Compression[a]
680 to 720	–	575 to 650	560 to 650	530 to 580	500 to 575	560 to 580	310 to 380
5	–	Varies	Varies	Varies	Varies	14 to 20	2 to 5
–	–	30 to 50	Varies	Varies	Varies	Varies	10 to 90[n]
0.025/ 0.075[g]	–	0.015/ 0.020[g]	0.0025[i]	0.016	0.004[k]	0.007 to 0.010	0.005[o]
2.1 to 2.2	–	1.7	1.89	1.14	1.37	1.50	1.42
4000	–	6500	14,000	12,000	26,000	15,000	7000
–	–	1.2	–	4.1	–	–	12.0
300	300	100 to 300	5	60	3 to 4	3	0
No break	No break	No break	19,000	–	38,000	24,000	12,500
No break	No break	No break	–	–	–	–	9500[p]
0.95	0.95	20.0	10.5	4.1	13.0	10.0	10.0
–	–	–	1.150	–	24,000	–	30,000
–	–	–	–	4.1	–	–	–
–	–	>20	7.5[j]	1.0	2.0[l]	0.7	0.27
–	R25	R50	–	R118	R121/M96	R121	E92
5.5	4.6 to 5.8	7.6	1.6	4.5	1.8	3.0	1.9
500	400	300	320	300	240	355	370
2.1	2.1	2.6	3.6	3.6	3.5	5.0[m]	4.9
<0.0001	0.0005	0.005	0.005	0.02	0.018	0.2[m]	0.04
<0.01	<0.01	<0.029	0.018	1.50	0.9	0.45	0.5
Excellent	Excellent	Excellent	Excellent	Fair	Fair	Fair[f]	Excellent
Excellent	Excellent	Excellent	Excellent	Poor	Poor	Poor	Good
Excellent	Excellent	Excellent	Excellent	Good	Good	Good[f]	Good
Excellent	Excellent	Excellent	Excellent	Fair	Fair	Good[f]	Poor
Excellent	Excellent	Excellent	Excellent	Excellent	Excellent	Excellent	Good

[m]At 50 percent relative humidity.
[n]Can also be transfer and injection molded; values depend on process.
[o]Compression molded; 0.007 inch for transfer; 0.010 to 0.015 inch for injection.
[p]After 200 hours; 7000 psi after 1000 hours.
[q]Measured at 200° F.
[r]Unnotched impact strength: 9 to 10 foot-pounds per inch.
[s]Unnotched impact strength: 17 foot-pounds per inch.
[t]Unnotched impact strength: 3 foot-pounds per inch.
[u]Unnotched impact strength: 11 to 12 foot-pounds per inch.
[v]Measured at 300° F.
[w]Unnotched impact strength: 14 foot-pounds per inch.
[x]Unnotched impact strength: 9 to 10 foot-pounds per inch.
*Plastics Engineering, 1974.

Commercially, the use of amino acids would involve many difficulties. They are biologically synthesized in the human body, however, such processes are too complex and expensive to duplicate under laboratory or commercial conditions. For this reason, diamines and dibasic acids are condensed to give polymers by amide linkages. These polymers have been given the generic name "nylon."

The polyamide first used in commercial products in the United States was nylon 6/6, made from adipic acid and hexamethylene-diamine:

$$HO\overset{O}{\overset{\|}{C}}(CH_2)_4\overset{O}{\overset{\|}{C}}OH + H_2N(CH_2)_6NH_2 \longrightarrow -\!\!\left(\overset{O}{\overset{\|}{C}}(CH_2)_4\overset{OH}{\overset{\|\,|}{C N}}(CH_2)_6\overset{H}{\overset{|}{N}}\right)\!\!-_n$$

Table 5-95.

Polymer Class	Phenylene Oxide Based Resin		Polycarbonate		Polyimide	
Grade	Unmodified	Reinforced	Unmodified	Reinforced	Unmodified	Reinforced
Filler type and concentration, percent	–	Glass fiber, 30	–	Glass fiber, 30	[Poly (amide-imide)]	Glass fiber, 65
Molding method	Injection	Injection	All	Injection	Injection	Compression
Molding temperature, °F	160 to 210	525 to 600	525 to 600	540 to 650	675 to 725	450
Molding pressure, psi × 10^{-3}	12 to 18	Varies	15 to 18	Varies	18 to 30	3
Molding time, seconds	Varies	Varies	Varies	Varies	Varies	120 to 360
Mold shrinkage, inches per inch ($\frac{1}{8}$-inch thickness)	0.005 to 0.007	0.001 to 0.002	0.005 to 0.007	0.001 to 0.002	0.006	0.001
Specific gravity	1.06	1.28	1.2	1.43	1.40	1.9
Tensile strength, psi	9600	18,500	9500	18,500	26,900	27,000
Tensile modulus, psi × 10^{-5}	3.55	–	3.45	13.75	–	28.5
Elongation at break, percent	60	3 to 4	110	4 to 6	12	<1
Flexural strength, psi						
At room temperature	13,500	23,000	13,500	28,000	30,700	49,500
At 350°F	7300[q]	–	–	–	22,000	44,000
Flexural modulus, psi × 10^{-5}	3.6	11.5	3.4	12.0	6.64	32.5
Compressive strength, psi	16,400	–	12,500	22,000	32,000	32,500
Compressive modulus, psi × 10^{-5}	–	–	3.45	–	4.0	24.6
Notched Izod impact strength, foot-pounds per inch of notch	5.0	1.7[r]	16.0	3.7[s]	2.5	17.0
Rockwell hardness	R119	M93	M70/R118	M95/R118	E78/M119	M120
Coefficient of expansion, inches per inch per °F × 10^5	3.3	1.4	3.75	1.25	1.8	0.8
Continuous-use temperature, °F[a]	230	230	240	260	540	500
Dielectric constant at 1 mH	2.68	2.9	2.96	3.43	4.0	4.7
Dissipation factor at 1 mH	0.0024	0.0015	0.01	0.0075	0.009	0.007
Water absorption (24-hour immersion), percent	0.006	0.06	0.15	0.07	0.28	0.2
Chemical resistance to:						
Weak acids	Excellent	Good	Good	Fair	Excellent	Excellent
Strong acids	Excellent	Fair	Fair	Fair	Excellent	Good
Weak alkalies	Excellent	Good	Fair to poor	Fair	Fair	Good
Strong alkalies	Excellent	Fair	Poor	Fair	Poor	Poor
Organic solvents	Poor	Poor	Varies	Poor	Excellent	Excellent

[a]Temperature at which 50 percent of initial properties is retained after 1000 hours exposure.
[b]Measured at 160° F.
[c]Unnotched impact strength: 6 to 7 foot-pounds per inch.
[d]Varies with temperature.
[e]Retains 80 percent of initial properties.
[f]Individual tests recommended.
[g]Values are for flow direction/transverse direction.
[h]At 10 percent deflection.
[i]Value is 0.005 for $\frac{1}{4}$-inch section.
[j]Unnotched impact strength: 17 to 18 foot-pounds per inch.
[k]Value is 0.005 for $\frac{1}{4}$-inch section.
[l]Unnotched impact strength: 17 foot-pounds per inch.

Three types of materials are included under the heading of polyamides. The first class in this group includes nylon 6/6. The first number represents the number of carbon atoms in the diamine and the second stands for the number of carbon atoms in the diacid. Amines containing from 2 to 10 carbon atoms and acids from 2 to 18 have been used. This nomenclature is an agreed-upon form accepted throughout the industry and applied to nylons of all kinds. Thus, nylon 6/10 = hexamethylenediamine + sebacic acid.

The second class of polyamides polymerizes a single-ring molecule having an amide linkage in the ring by opening this structure during the reaction. When a 6-carbon molecule is used, the polymer is called nylon 6. Both of these polyamide types find major end uses in textiles and plastic moldings.

The third amide type is made by dimerizing methyl esters of unsaturated fatty acids, using a Diels-Alder condensation on acids from cotton, soya, and corn oils. These long-chain

(Continued)

Polyphenylene Sulfide		Polysulfone		Thermoplastic Polyester		Thermoset Polyester		Urethane
Unmodified	Reinforced	Unmodified	Reinforced	Unmodified	Reinforced	Wet Molding	SMC	Elastomer
–	Glass fiber, 40	–	Glass fiber, 30	–	Glass fiber, 30	Glass fiber, 30	Glass fiber, 60	–
Injection	Injection	Injection	Injection	Injection	Injection	Compression	Compression	Injection
600	500 to 675	650 to 750	625 to 750	450 to 500	430 to 500	200 to 260	300	350 to 450
–	Varies	15 to 20	Varies	–	Varies	0.1 to 0.2	1	6 to 15
–	Varies	20 to 70	Varies	–	Varies	45 to 90	180	–
–	0.001	0.007	0.002 to 0.003	0.017 to 0.023	0.003 to 0.0045	0.002 to 0.004	None	0.001 to 0.030
1.34	1.65	1.24	1.45	1.41	1.52	1.4 to 1.6	1.9	1.05 to 1.25
10,800	23,000	10,200	18,000	9400	19,500	12 to 16,000	14,000	4000 to 8000
4.8	–	3.6	–	–	–	–	14.0	0.1 to 3.5
3	3 to 4	50 to 100	3	15	3 to 4	0.8 to 1.8	0.8	100 to 650
20,000	32,000	15,400	24,000	14,000	28,000	16 to 24,000	33,000	–
–	–	–	15,500[v]	–	–	–	–	–
6.0	18.0	3.9	12.0	4.0	13.5	0.8 to 1.2	19.0	–
16,000	–	–	24.000	14,000	19,000	24 to 30,000	24,000	–
–	–	–	–	–	–	–	–	–
0.3[t]	1.5[u]	1.3	1.8	0.8	1.7[x]	14 to 20	18.2	No break
R123	R123	M69/R120	M92/L108	M80	M84/R119	–	M88 to M95	–
3.0	1.1	3.1	1.4[w]	5.5	1.3	–	1.11	1.0 to 4.0
450	390	300	300	260	280	–	–	200
3.22	3.80	3.07	3.49	3.1	2.0	–	–	4.6
0.0007	0.0013	0.0034	0.0049	0.02	0.02	–	–	0.075
0.02	0.01	0.3	0.2	0.09	0.06	0.1 to 0.3	0.65	0.8
Excellent	Excellent	Good	Excellent	Good	Good	Good	Fair	Poor
Excellent	Excellent	Good	Good	Fair	Fair	Good	Poor	Fair
Excellent	Excellent	Good	Excellent	Fair	Fair	Fair	Fair	Poor
Excellent	Excellent	Good	Good	Poor	Poor	Poor	Poor	Fair
Excellent	Excellent	Poor	Poor	Good	Good	Fair	Fair	Good

[m] At 50 percent relative humidity.
[n] Can also be transfer and injection molded; values depend on process.
[o] Compression molded; 0.007 inch for transfer; 0.010 to 0.015 inch for injection.
[p] After 200 hours; 7000 psi after 1000 hours.
[q] Measured at 200° F.
[r] Unnotched impact strength: 9 to 10 foot-pounds per inch.
[s] Unnotched impact strength: 17 foot-pounds per inch.
[t] Unnotched impact strength: 3 foot-pounds per inch.
[u] Unnotched impact strength: 11 to 12 foot-pounds per inch.
[v] Measured at 300° F.
[w] Unnotched impact strength: 14 foot-pounds per inch.
[x] Unnotched impact strength: 9 to 10 foot-pounds per inch.

dibasic acid esters (C_{36} and even C_{54} with trimerized acids) are then reacted with a short-chain diamine, as in preparing nylon 6/6. These amides are largely used as surface coatings and adhesives.

Preparation and Chemistry. Because of the multiplicity of precurser ingredients for nylon resins, we will bypass the manufacture of monomer and only examine the actual polymerization procedures and chemistry. (The interested reader may pursue the subject of monomer preparation in more detail by referring to the Floyd, Golding, and Kohan sources listed in the reference section.)

In the preparation of nylon 6/6*, amide-forming derivatives of the diacid can be used. These include diesters, dihalides, or diamides. However, the acid itself is normally preferred. By using the free acid, hexamethylenediammo-

*Nylon 6/6 is also referred to as nylon 66.

Table 5-96. Comparison of Material Cost (Courtesy: E.I. Du Pont de Nemours & Co., Inc.).

Before selecting an engineering material, the designer should consider other cost-reducing factors which can affect part cost in addition to the price of the plastic. These additional factors include:

- Productivity in molding
- Use of less material with no sacrifice of performance by redesigning parts with thinner walls and "cutouts."
- Reduction of the number of parts required through multifunctional design—eliminating parts, assembly and finishing operations.
- Low cost plastic assembly techniques including welding, self-tapping screws, snap and interference fits.

The material costs shown below are based on published information available on May 1, 1978 and are thought to be representative of average domestic truckload pricing for first quality materials. For a detailed cost comparison of specific materials, the latest prices should be obtained from the material suppliers.

Material Cost

Materials	Specific Gravity	lb./cu. in.	¢/lb.	¢/cu. in.
Du Pont Engineering Plastics				
Delrin® acetal resin	1.42	0.0513	100	5.13
"Delrin" 500CL acetal resin	1.42	0.0513	215	11.03
Zytel® 66 nylon resin	1.14	0.0412	116	4.78
"Zytel" 612 nylon resin	1.07	0.0387	198	7.65
"Zytel" impact modified	1.09	0.0394	126	4.96
"Zytel" ST super tough	1.09	0.0394	155	6.10
Lucite® acrylic resin	1.19	0.0430	62	2.67
"Lucite" T-1000 high impact	1.15	0.0415	79	3.28
Other Du Pont Plastics				
Alathon® polyethylene resin				
Low density	0.914–0.940	0.0330–0.0339	31.5	1.04–1.07
High density and copolymers	0.941 and up	0.0340–0.0346	30.25	1.03–1.04
Du Pont EVA copolymers	0.93–0.96	0.0336–0.0347	35–54.5	1.18–1.89
Surlyn® ionomer resin	0.94–0.96	0.0340–0.0350	68–76	2.31–2.64
Other Plastics				
Polyester (thermoplastic)	1.31	0.0473	98	4.64
Polystyrene–general purpose	1.06	0.0383	29–30	1.11–1.15
Polypropylene	0.905	0.0327	30	0.98
Polyvinyl chloride (rigid)	1.20–1.37	0.0433–0.0494	27	1.17–1.34
Styrene acrylonitrile copolymer	1.07	0.0386	45	1.74
ABS	1.04–1.06	0.0375–0.0383	48	1.80–1.84
Cellulose acetate butyrate	1.19	0.0430	113	4.86
Modified polyphenylene oxide	1.09	0.0394	113.5	4.47
Polycarbonate	1.20	0.0433	113	4.90
Polysulfone	1.24	0.0448	300	13.44
Metals				
Magnesium AZ–91B–ingot	1.81	0.0653	101	6.60
Aluminum SAE-306 (380-1% zinc)–ingot	2.77	0.100	60–61.6	6.00–6.16
Aluminum SAE-309 (360)–ingot	2.64	0.0953	60–60.5	5.72–5.77
Zinc SAE-903 ('Zamac' 3)–ingot	6.6	0.238	32.50	7.75
Brass-Yellow (#403)–ingot	8.5	0.307	62.25	19.11
Brass-85/5/5 (#115)–ingot	8.75	0.316	72.50	22.91
Steel-CR alloy–strip	7.85	0.283	30.9–34.57	8.76–9.80
Steel–hot rolled sheet	7.85	0.283	15.28	4.33
Steel-stainless 304–bar	7.92	0.286	80.00	22.89
Iron-pig, basic–pig	7.1	0.256	9.55	2.45

*Courtesy: E.I Du Pont de Nemours & Co., Inc.

Table 5-97. Comparison of Material Cost Reinforced Thermoplastic Molding Resins.*

Materials	% Reinforcing Agent[a]	Specific Gravity	lb./cu. in.	¢/lb.	¢/cu. in.
Du Pont Reinforced Engineering Plastics					
Delrin® 570	20	1.56	0.056	116	6.54
Minlon® 10B-40	40	1.50	0.0542	87	4.71
"Minlon" 11C-40	40	1.48	0.0535	87	4.65
Rynite® thermoplastic polyester	30	1.57	0.057	102	5.79
"Rynite"	40	1.70	0.061	105	6.45
Zytel® 70 G-13L	13	1.22	0.0441	124	5.47
"Zytel" 70 G-33L	33	1.38	0.0499	124	6.18
"Zytel" 70 G-43L	43	1.51	0.0546	124	6.77
"Zytel" 71 G-13L	13	1.18	0.0426	127	5.41
"Zytel" 71 G-33L	33	1.35	0.0488	127	6.19
"Zytel" 77 G-33L	33	1.32	0.0477	199	9.49
"Zytel" 77 G-43L	43	1.46	0.0527	199	10.50
Other Reinforced Plastics					
Polystyrene	20	1.20	0.0432	60	2.60
	30	1.28	0.0462	63	2.91
Polypropylene	20	1.04	0.0375	73	2.74
	30	1.13	0.0407	76	3.10
Styrene acrylonitrile	20	1.22	0.0440	73	3.22
	30	1.31	0.0472	75	3.55
ABS	20	1.22	0.0439	77	3.29
	30	1.28	0.0462	77	3.56
Polyphenylene oxide	20	1.21	0.0436	127.5	5.57
(modified)	30	1.27	0.0458	132.5	6.08
Polycarbonate	20	1.35	0.0487	172	8.39
	30	1.43	0.0516	167	8.63
Polysulfone	20	1.38	0.0498	326	16.25
	30	1.45	0.0524	304	15.92
Polyester (thermoplastic)	20	1.45	0.0524	108	5.66
	30	1.52	0.0549	105	5.77
Polyphenylene sulfide	40	1.60	0.0578	205	11.85
Polyphenylene sulfide	50	1.70	0.0615	220	13.51

[a] All compositions are glass-reinforced/filled with the exception of "Minlon" (mineral filled) and 60% polyphenylene sulfide (mineral/glass filled).
*Courtesy: E.I. Du Pont de Nemours & Co., Inc.

nium adipate is formed as a first step and, thus, by simply monitoring the pH of this salt solution, accurate control of reactant stoichiometry is obtained. The salt may be prepared by dissolving each of the reactants in a suitable solvent such as methanol, and mixing and collecting the precipitated salt which is then dissolved in water. It may also be made by adding an aqueous solution of the diamine to a stirred aqueous dispersion of the acid. This latter approach does not provide the inherent purification of a crystallization step; consequently, it is necessary to remove impurities by treating the solution with activated charcoal.

In a typical batch polymerization, the salt solution is charged to evaporators where it is concentrated to a liquor containing about 75% solids before transfer to a pressure vessel purged of oxygen. The concentrated solution is then heated to about 210°C (410°F) or until a pressure of about 17 atm is reached. Pressure is maintained while bleeding steam, and the temperature is gradually increased. Finally, the pressure is reduced to 1 atm and the temperature is increased to approximately 280°C (536°F). The T_m of dry nylon 6/6, is 269°C (576°F); this is depressed by the presence of water, and the polymerization process is designed to remove water in balance with the rate of temperature increase, to prevent melt-solidification. The polymer is forced out of the bottom of the autoclave, quenched, dried, cut into chips, blended, and packaged.

The batch manufacture of nylon 6 from caprolactam is very similar to that just described for nylon 6/6, except that caprolactam and up to 10% water are directly charged to the pressure vessel. Although the nature of the monomer precludes a concern for stoichiometric control, the nature of the reaction prevents positive assurance of exactly equal

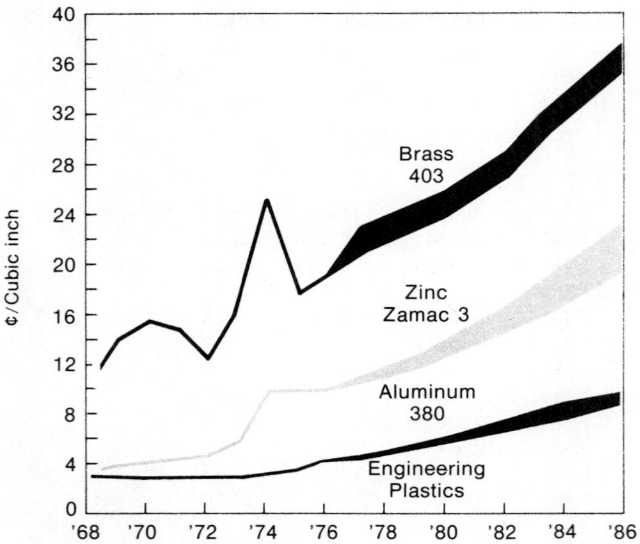

Fig. 5-98. U.S. engineering pastics growth. (*Courtesy: E.I. Du Pont de Nemours & Co., Inc.*)

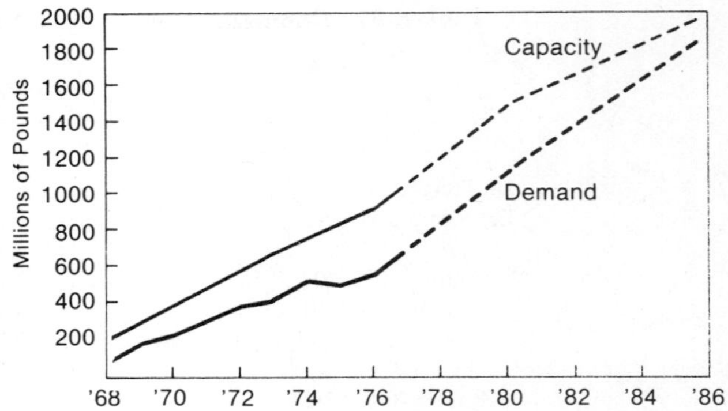

Projections By Engineering Plastics Type

Millions of Pounds

Engineering Plastic	1970	1976	Forecasts 1980	Forecasts 1986
Nylon	103	210	300	500
Acetal	62	86	180	300
Polycarbonate	40	114	250	400
Modified PPO	30	100	200	300
Thermoplastic Polyester	—	35	100	170
Polysulfone	5	15	40	70
Other	10	40	80	160
Total	250	600	1150	1900

Source: DuPont Plastic Products and Resins, Marketing Services Division

Fig. 5-99. Price comparison: engineering plastics versus die-cast metals. (*Courtesy: E.I. Du Pont de Nemours & Co., Inc.*)

balance of residual amine and acid groups in the polymer. The polymerization mechanism for nylon 6 is believed to occur in three steps.

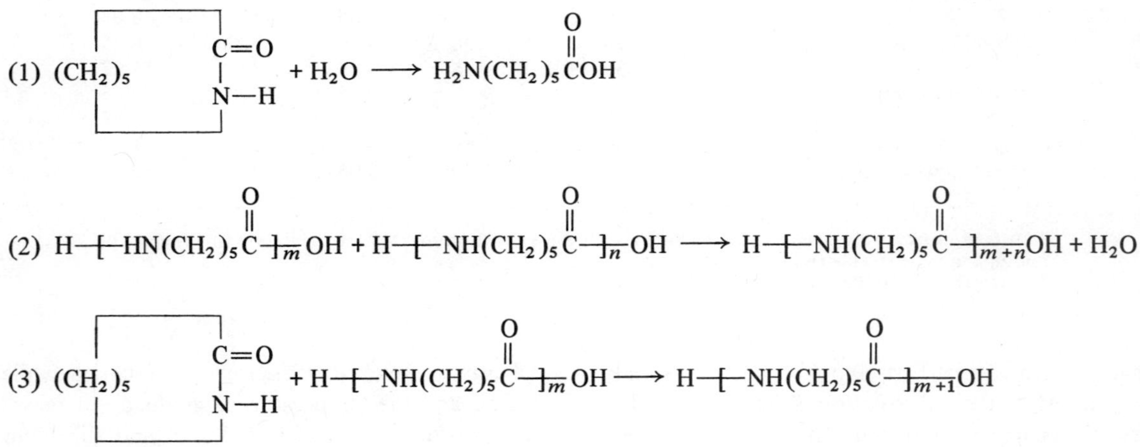

Dry caprolactam will not polymerize, but in the presence of water, the reaction proceeds well at temperatures above 200°C (392°F).

As with nylon 6/6, molecular weight can be controlled by adding amine or acid to cause an imbalance in the amine/acid ratio or by adding a monofunctional acid such as acetic acid. Nylon 6 contains about 10% extractables which plasticize the polymer. To remove them, the cut polymer is subjected to a vacuum strip or water washing and then carefully dried.

Conversion of higher-order salts into polymers such as nylon 6/10, 6/12, or 13/13 is usually accomplished in a batch process as described for nylon 6/6. As the length of the hydrocarbon chain increases, the solubility of the salt in water decreases. Therefore it is more advantageous to use, instead of water, a solvent which dissolves both the amine and acid but not the salt. For example, nylon 13/13 salt is made using ethanol. The greater hydrocarbon character of these resins implies a lower T_m, so by the time a temperature for a reasonable rate of reaction is reached, there is no need for water as a melting-point depressant.

The preparation of polycyclamides requires a specific level of steam pressure in the early stages of polymerization to assure the presence of enough water to depress the T_m, but not enough to promote diamine decomposition. Aromatic nylons are typically partially polymerized in the reaction and then discharged into an extruder where polymerization is completed.

Nylon 7 is only of commercial interest at this time in the U.S.S.R. The monomer, 7-aminoheptanoic acid, is made together with the monomer for nylon 9 via the telomerization of ethylene.

$$nCH_2{=}CH_2 + CCl_4 \xrightarrow{\text{catalyst}} Cl(CH_2CH_2)_nCCl_3 \xrightarrow[\text{acid}]{H_2O}$$

$$CH(CH_2CH_2)_n\overset{\overset{\displaystyle O}{\|}}{C}OH \xrightarrow{NH_3} H_2N(CH_2CH_2)_n\overset{\overset{\displaystyle O}{\|}}{C}OH$$

Where n = mostly 3 (7-aminoheptanoic acid) with some 4 (9-aminononanoic acid) and lesser amounts of others. Limiting the number of by-products and monomer separations are major problems of this synthesis. Polymerization is similar to that of nylon 6/6 without the stoichiometric control (as with nylon 6).

Nylon 8 has been produced, although its high cost and lack of significant property improvement has prevented successful large-scale commercialization. The monomer, capryllactam, is made from cyclooctane.

Nylon 11 is made by pyrolyzing methyl ricinoleate, from the treatment of castor oil with methanol, to an 11-carbon fragment. Through a very critical and specific bromination step, 11-aminoundecanoic acid is obtained in the form of an aqueous paste which is then charged directly into a continuous polymerization unit.

Butadiene is converted to dodecanolactam or laurolactam which is polymerized to nylon 12. Polymerization is slow because of the stability of the ring. Acidic catalysts are used, but temperatures of 300–350°C (572–662°F) are necessary for completion of the polymerization.

The exact process for the preparation of monomer for nylon 13 is not known. It has been suggested that the process is analogous to the manufacture of nylon 9 from soybean oil, but with crambe oil as the precursor.

All of the monomers with ring structures containing more than the 7 atoms of caprolactam yield polymers that contain only small amounts of extractable by-products and, thus, do not require additional extraction steps.

There is an interesting, easily performed experiment which illustrates a rather unique approach to the polycondensation of nylons. The process is termed *interfacial condensation.* Essentially the diacid (or diacid chloride) is dissolved in a chlorinated solvent and the diamine is dissolved in water. The latter solution is carefully poured over the first to form two immiscible layers. The polycondensation occurs at the interface of the two phases and the resultant polymer can be withdrawn continuously for as long as reactive monomer is available.

Modified Nylons. Modified nylons include copolymers, compositions containing additives to impart specific properties, filled and reinforced materials, and chemical-modified nylons.

Copolymers. The preparation of a copolymer requires only that the desired ingredients be added to the polymerization vessel. Any of the diamines, diacids, amino acids, and lactams mentioned, as well as many other monomers, can be used in any combination as long as the integrity of the amine/acid stoichiometry is maintained.

In general, copolymers offer lower melting points, increased flexibility and solubility and, in some cases, transparency. The most commercially significant copolymers are that of nylon 6-6/6 and 6-6/10. Copolymers of n-methyl hexamethylenediamine with adipic acid have been reported. Terpolymers, primarily of nylon 6-6/10-6/6, are available which provide a high degree of flexibility and solubility in alcohol-water systems. Their primary use is for high-impact films for coating applications.

Table 5-98 shows reported data on some copolymers.

Additives. A variety of additives may be compounded with nylons. These include heat and light stabilizers, plasticizers, lubricants, fungicides, pigments and dyes, and reinforcing and non-reinforcing fillers.

Glass-fiber-filled nylons have significantly improved creep resistance over unfilled polymers. Unfilled nylons tend to break in a tough mode when subjected to impact, and the filled materials show brittle fracture. Glass-filled polymers are, on the other hand, less notch-sensitive. Nylon grades with spherical glass beads show increased compressive and tensile strength and modulus. The glass bead components are easier to process than glass-fiber formulations because of a lower melt viscosity, uniform and predictable shrinkage, and minimum warpage.

A new series of particulate-mineral-reinforced resins has recently been introduced offering good cost/performance ratios for the replacement of die-cast metals such as zinc, bronze, and aluminum. Parts molded from these materials possess more uniform properties because of the isotropic nature of the reinforcement. Compared to conventional fiber reinforcement, such materials eliminate non-uniform dispersion and fiber orientation in parts molded with high-aspect ratio reinforcements. The basic properties of these nylons are equivalent or better than unfilled systems and maintain these properties at elevated temperatures. Some properties of one commercial grade of the mineral-filled nylons (Vydyne R) are compared to those of other engineering plastics (unfilled and glass-reinforced) in Figure 5-100.

General Properties of Nylons. Table 5-98 lists properties for a number of commercially successful nylons. Table 5-99 shows properties for the less-used types. The data reported were obtained on moldings relatively free of orientation and under carefully controlled conditions of humidity,

Table 5-98. General Properties

Properties	ASTM	Type 6 General Purpose	Type 6 30% Glass Fiber Reinf	Type 6 Flexible Copolymers	Type 6/6 General Purpose	Type 6/6 30% Glass Fiber Reinf	Type 6/6 Flexible Copolymers
Physical properties							
Specific gravity	D792	1.14	1.37	1.12–1.14	1.13–1.15	1.37	1.08
Water absorption, %	D570						
In 24 hr		1.6–2.0	1.1–1.3	0.8–1.4	1.5	0.9	2.0
Saturation		9.5	6.5–7.0	–	8.5–9.0	3.8–5.8	20
Thermal conductivity		1.2 (2.1)	1.2–1.7	–	1.7 (2.9)	1.5 (2.6)	–
			(2.1–2.9)	–	–	–	–
Btu-ft/hr-ft^2 · °F (W/m-°K)	D696						
Coef of ther exp 10^{-5}/°F		4.6–4.8	1.2–2.1	–	4.5 (8.1)	2.1 (3.8)	–
(m/m-°K)		(8.28–8.64)	(2.16–3.8)	–	–	–	–
Specific heat Btu/lb-°F		0.4	–	–	0.3	–	–
(J/kg-°K)		(1673.6)	–	–	(1255)	–	–
Brittleness temp, °F (°K)	D746	-116 (191)	–	–	-112	–	–
		–	–	–	(193)	–	–
Heat deflection temp, °F (°K)	D648						
At 66 psi (0.045 MPa)		356–378	420–425	260–350	360–470	489–507	–
		(453–465)	(489–491)	(400–450)	(455–516)	(527–537)	–
At 264 psi (0.18 MPa)		152–203	400–420	113–130	150–220	473–495	–
		(340–368)	(477–489)	(318–328)	(339–378)	(518–530)	–
Max rec serv temp, °F (°K)		250	250	175–200	250	250	250
		(394)	(394)	(353–366)	(394)	(394)	(394)
Melting point, °F (°K)	D789	410–435	410–435	410–420	482–507	480–507	300–320
		(483–497)	(483–497)	(483–489)	(523–537)	(522–537)	(422–433)
Mechanical Properties							
Ten str, 1000 psi (MPa)	D638						
At -40°F (233°K)		17.4	–	–	15.7	–	–
		(120)	–	–	(108)	–	–
At 73°F (296°K)		11.8	23.0	7.5–10.0	12.0	27.0	–
		(81)	(159)	(52–69)	(83)	(186)	–
At 170°F (350°K)		9.9	–	–	9.0	–	–
		(68)	–	–	(62)	–	–
Yld str, 1000 psi (MPa)	D638						
At -40°F (233°K)		17.4	–	–	15.7	–	–
		(120)	–	–	(108.3)	–	–
At 73°F (296°K)		11.8	–	–	12.0	–	–
		(81)	–	–	(82.7)	–	–
At 170°F (350°K)		5.9	–	–	6.5	–	–
		(41)	–	–	(44.8)	–	–
Elongation at break, %	D638						
At -40°F (233°K)		5	–	–	20	–	–
At 73°F (296°)		200	2–4	200–320	60	–	–
At 170°F (350°K)		310	–	–	340	–	–
Elongation at yield, %	D638						
At -40°F (233°K)		–	–	–	4	–	–
At 73°F (296°K)		–	–	–	5	–	–
At 170°F (350°K)		–	–	–	30	–	–
Modulus, 10^5 psi (100 MPa)							
Tensile	D638	3.8–4.4	10–12	–	4.2	12–14	–
		(26–30)	(69–83)	–	(28.9)	(83–97)	–
Flexural	D790	3.8–4.7	10–13.5	0.9–3.2	4.1–4.2	9.8–13	1.5
		(26–32)	(69–93)	(6.2–22)	(28.2–28.9)	(67–89.6)	(10.3)
Izod impact str (notched)	D256	0.8–1.2	1.5–NB	–	0.9–1.1	1.2–2.5	NB
ft-lb/in. (J/m)		(43–64)	(80)	–	(6.2–7.6)	(8.2–17)	–
Shear str, 1000 psi (MPa)	D732	8.5 (59)	10.3 (71)	–	9.6 (66)	12.5 (86)	–
Deformation under load, %	D621						
At 2000 psi (14 MPa)		1.6–1.7	–	–	1.2–1.4	–	20
at 122°F (323°K)		–	–	–	–	–	–
At 4000 psi (28 MPa)		1.8	0.1–0.9	–	–	0.1–0.6	–
at 122°F (323°K)		–	–	–	1.5	–	–
Hardness (Rockwell)	D785	R118–R120	R121	R72–R119	R118	R121	R83
Abrasion (Taber) res	D1044	–	–	–	6–8	13–25	4
mg/1000 cyc		–	–	–	–	–	–
(CS-17 wheel, 1000 g load)		5	14	3.9–16.3	–	–	–

*Reprinted by permission from *Materials Engineering*, Manual 254, March 1975.

of Nylons.*

Type 6/10		Type 6/12		Type 11		Type 12	
General Purpose	30% Glass Fiber Reinf	General Purpose	30–35% Glass Fiber Reinf	General Purpose	30% Glass Fiber Reinf	General Purpose	30% Glass Fiber Reinf
1.07–1.09	1.30–1.36	1.06–1.08	1.32–1.35	1.03–1.04	1.24–1.27	1.01–1.02	1.23
0.4	0.2–0.3	0.4	0.20	0.23–0.3	0.2	0.15–0.25	0.15
3.1–3.5	1.85–2.1	3.0	2.0	1.0–1.1	0.7	1.4	1.0
1.5 (2.6)	3.5 (6.1)	1.5 (2.6)	–	5 (2.6)	–	1.7 (2.9)	–
5.0 (9.0)	2.5 (4.5)	5.0 (9.0)	1.3 (6.5)	0.91 (1.6)	–	7.2 (12.9)	–
–	–	–	–	–	–	–	–
0.3 (1255)	–	0.3–0.5	–	0.58	–	0.28 (1171)	–
–	–	(1255–2092)	–	(2427)	–	–	–
-166	–	–	–	-94 (203)	–	–	–
(163)	–	–	–	–	–	–	–
300	420–437	330	–	154–302	–	–	–
(422)	(489–498)	(439)	–	(341–423)	–	–	–
135	400–420	180	410–415	118–131	335	115–131	230
(330)	(477–489)	(355)	(483–486)	(321–328)	(441)	(319–328)	(383)
225	250	350	–	212–250	–	175–230	–
(380)	(394)	(450)	–	(373–394)	–	(353–383)	–
405–430	405–430	406–420	406–420	367	–	347–356	347–356
(480–494)	(480–494)	(481–489)	(481–489)	(459)	–	(448–453)	(448–453)
12.0	–	1.36	–	9.9–10.0	–	–	–
(83)	–	(93.7)	–	(68–69)	–	–	–
8.5	19.0	8.8	–	8.5	–	4.8–9.2	9.0 (62)
(59)	(131)	(60.7)	–	(59)	–	(33–63)	–
5.3	–	5.9	–	6.1–6.5	–	–	–
(37)	–	(40.7)	–	(42–45)	–	–	–
12.0	–	13.6	–	–	–	11.5	–
(83)	–	(93.7)	–	–	–	(79.2)	–
8.5	17–21	8.8	–	–	–	6.0–7.8	4.3 (29.6)
(58.6)	(117–145)	(60.7)	–	–	–	(41–54)	–
4.2	–	4.3	–	–	–	3.5	–
(29)	–	(29.6)	–	–	–	(24)	–
20	–	15	–	37	–	150–300	7
20	2–4	150	–	329	–	–	–
300	–		–	400	–	–	–
10	–	8	–	–	–	5.8	–
10	–	7	–	–	–	–	–
30	–	30	–	–	–	–	–
2.8	8.0–9.5	–	–	1.78–1.85	–	1.7–2.1	–
(19.3)	(55–65)	–	–	(12.2–12.7)	–	(11.7–14.4)	–
2.8	8.0–11	–	–	1.51	–	1.7–2.2	3.0 (20.7)
(19.3)	(55–76)	–	–	(10.4)	–	(11.7–15.1)	–
0.6–1.6	1.4–3.4	0.8–1.0	2.6–3.0	0.5–0.75	2.2	0.7–1.2	1.4 (74.7)
(4–11)	(9.6–23)	(42.7–53)	(3.52)	(26.7–40.0)	(17.4)	(37–77)	–
8.4 (58)	–	8.6 (59.3)	11.0 (75.8)	–	–	–	–
4.2	–	1.6	–	–	–	–	–
–	–	–	–	–	–	–	–
–	–	–	1.0	–	–	–	–
–	–	–	–	–	–	–	–
R111	R120	R114	R118–R120	R100–R108	–	R106	–
5–6	18	5–7	–	5	–	–	–

Table 5-98a. Electrical Properties of Nylons.*

Nylon	Condition	Volume Resistivity ASTM D257 ohm-cm (ohm-m)	Dielectric Constant ASTM D150 60–100 Hz	10^3 Hz	10^6 Hz	Dissipation Factor ASTM D150 60–100 Hz	10^3 Hz	10^6 Hz
6/6	Dry	10^{15} (10^{13})	4.0	3.9	3.6	.02	.03	.02
	50% R.H.	10^{13} (10^{11})	6.0	5.1	4.1	.04–.18	.04–.16	.05–.09
	100% R.H.	10^9 (10^7)	31	29	18	.50	.23	.28
6	Dry	10^{15} (10^{13})	3.8	3.7	3.4	.01	.02	.03
	50% R.H.	10^{13} (10^{11})	13	8.3	4.2	.18	.20	.12
	100% R.H.	10^8 (10^6)	–	–	25	–	–	–
6/10	Dry	10^{15} (10^{13})	4.0	3.9	3.6	.04	.04	.03
	50% R.H.	10^{13} (10^{11})	7.4	6.1	3.7	.04	.04	–
	100% R.H.	10^{11} (10^9)	15	12	4.0	.03	–	.1
6/12	Dry	10^{15} (10^{13})	4.0	4.0	3.5	.02	.02	.02
	50% R.H.	10^{13} (10^{11})	6.0	5.3	4.0	.04	.04	.03
	100% R.H.	10^{11} (10^9)	14	–	–	–	–	–
11	Dry	10^{15} (10^{13})	–	–	–	–	–	–
	50% R.H.	10^{14} (10^{12})	–	–	–	–	–	–
	100% R.H.	10^{13} (10^{11})	–	–	–	.03	–	.06
12	Dry	10^{15} (10^{13})	4.2	3.8	3.1	.04	.05	.03
6/6 + 33%	Dry	10^{15} (10^{13})	–	4.5	3.7	–	.02	.02
glass fiber	100% R.H.	10^9 (10^7)	–	25	11	–	–	–

*Reprinted by permission from *Materials Engineering*, Manual 254, March 1975.

temperature, and rate of test. As we shall see, test conditions, particularly humidity, can drastically affect final properties. Figures 5-101, 5-102, and 5-103 show the significant plasticizing effect of moisture on various properties of nylon 6/6.

Nylons have extremely good abrasion resistance (see Figures 5-104 and 5-105), which can be improved still further through the addition of external lubricants or by processing techniques which develop a highly crystalline, hard surface (e.g., annealing). Nylons have high "stick-slip" ratios, although modifications with PTFE resins or molybdenum disulfide can affect significant reductions. Table 5-100 and Figure 5-106 compare the abrasion resistance and other properties of a molybdenum disulfide-filled nylon (Nylatron GS) with standard nylon 6/6.

The T_g's of nylons appear to be below room temperature so that the materials are somewhat flexible despite their high level of crystallinity. Their melting points are very sharp, and viscosities drop drastically above the T_m.

The nylons are reasonably good electrical insulators at low temperatures and humidities; however, these properties deteriorate rapidly as temperature and humidity increase.

Nylons have exceptionally good resistance to hydrocarbons, esters, alkyl halides, and glycols. Alcohols will swell or dissolve nylons, particularly copolymers. The resins are soluble in formic acid, glacial acetic acid, phenols, and

Table 5-99. Typical Properties of Nylons.*

Property	Nylons 7	8	9	13	13/13	6/13
Density, nominal, g/cc	1.10	1.09	1.06	1.01	1.01	1.04
Melting point, °F	405	270–307	365	356	347	410
Ultimate tensile strength, psi	–	3850–6500	NA[a]	5600	5210–5700	7680
Elongation at break, %	–	300–400	NA	110	130–290	<50
Modulus of elasticity, psi	–	0.3×10^5	NA	1.3×10^5	$1.01–1.14 \times 10^5$	1.64×10^5
Impact strength, ft-lb/in. of notch ($\frac{1}{2} \times \frac{1}{2}$ in. notched bar, Izod test)	–	>16	NA	2.1	2.6	3.7
Water absorption at 68°F, 65% RH, % (to equilibrium)	–	NA	1.5	–	–	–
Water absorption, in water at 68°F, % (to equilibrium, except as noted)	–	9.5	3.3	1.04[b]	0.75[b]	1.72[b]

a: NA = not available; b: After 5 weeks' immersion at 73.4°F.
*Kestler, 1968. Reprinted by permission from *Modern Plastics*, McGraw-Hill, Inc.

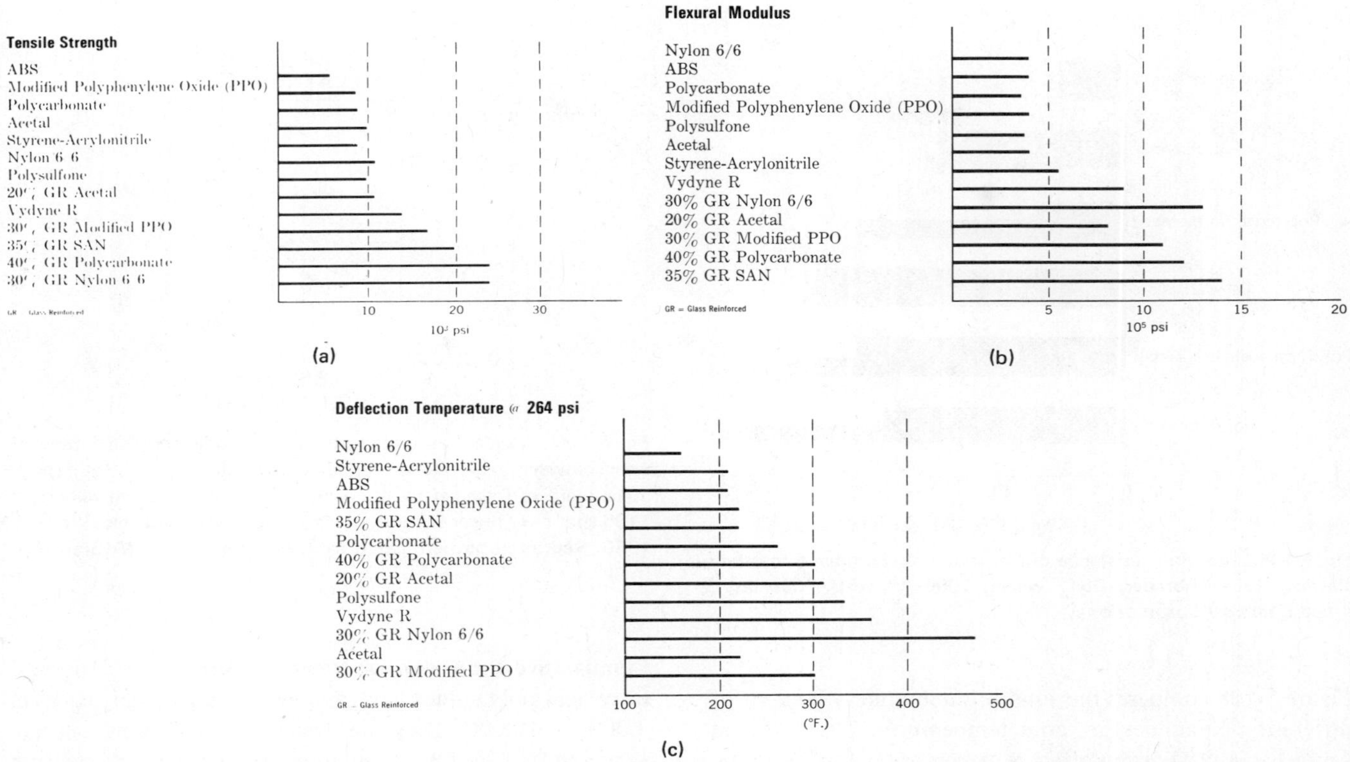

Fig. 5-100. Property comparisons: Vydyne R (mineral-filled nylon) and other thermoplastics. (*Courtesy: Monsanto Co.*, which is no longer responsible for the accuracy or content herein.)

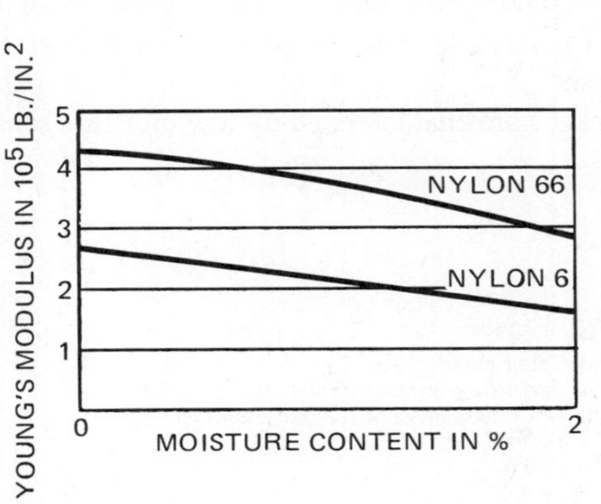

Fig. 5-101. Effect of moisture content on the Young's modulus of nylon 66 and nylon 6. (Brydson, 1975)

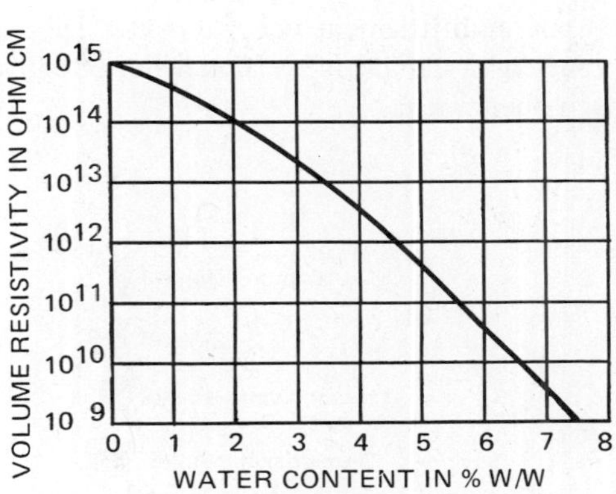

Fig. 5-103. Effect of moisture content on the volume resistivity of nylon 66. (Brydson, 1975)

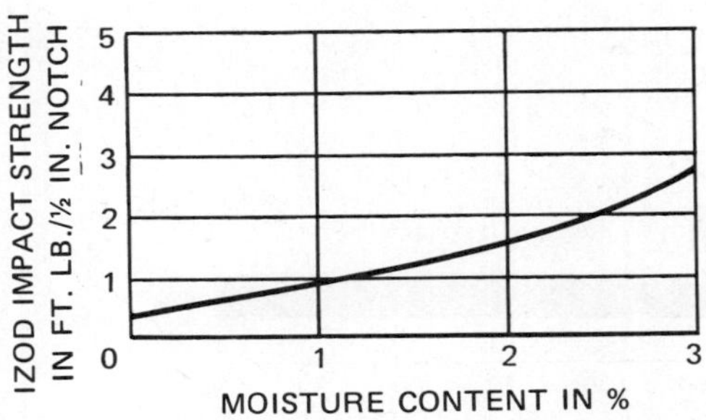

Fig. 5-102. Effect of moisture content on the impact strength of nylon 66. (Brydson, 1975. Originally published in *Engineering Materials and Design*, **1**(3):132, J. L. Riley.)

cresols. Mineral acids, particularly nitric, attack these plastics, but alkali resistance at room temperature is good. Resistance to all chemicals deteriorates with increasing exposure temperature. In general, the chemical resistance of copolymers is less than that of homopolymers.

Nylons are very hygroscopic, although the degree of absorption decreases with increase in hydrocarbon backbone length. Table 5-101 shows some comparative water absorption data of various nylons as well as some other plastics. Figure 5-107 shows the relationship between relative humidity and moisture content for three types of nylons.

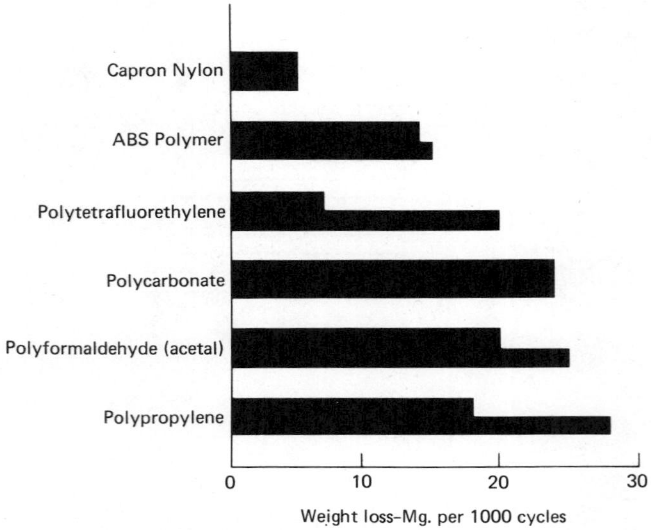

Fig. 5-104. Abrasion resistance of Capron nylon compared to other plastics: Taber Abrader, C517 wheel, 1000 gm load. (*Courtesy: Allied Chemical Corporation*)

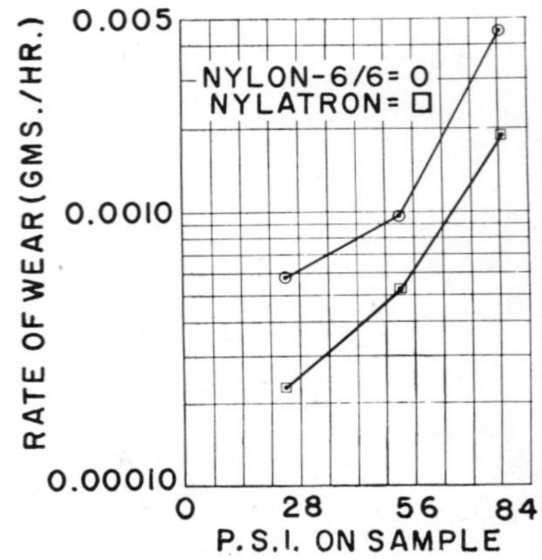

Fig. 5-106. Accelerated wear tests made to indicate relative rates of wear between molybdenum-filled and unfilled nylon. Both specimens were run against steel cleaned with CCl_4; rubbing velocity = 1165 rpm (160 linear ft/min). Surfaces were not lubricated. (Powers, 1960. Reprinted by permission from *Modern Plastics*, McGraw-Hill, Inc.)

Figure 5-108 compares the equilibrium humidity content of different polyamides at room temperature. The absorbed water has a plasticizing effect which causes a loss of tensile strength and modulus and an increase in impact strength. The effect on electrical properties has already been mentioned.

Nylons have poor UV and weathering characteristics. Discoloration and embrittlement occur on extended exposure, but heat and light stabilizers are somewhat effective in retarding such phenomena.

Comparative Properties of Newer Nylons. The dimerized fatty acid polyamides have moleuclar weights ranging from 3,000 to 10,000. They are available as solutions, suspensions, and solids for use in adhesive, coating, or extrusion applications. Variations in the acid and amine produce resins softening over a range of 40–190°C (104–374°F). These materials are mutually compatible and can thus be blended for intermediate properties in solution or melt applications.

These resins are characterized by low moisture sensitivity,

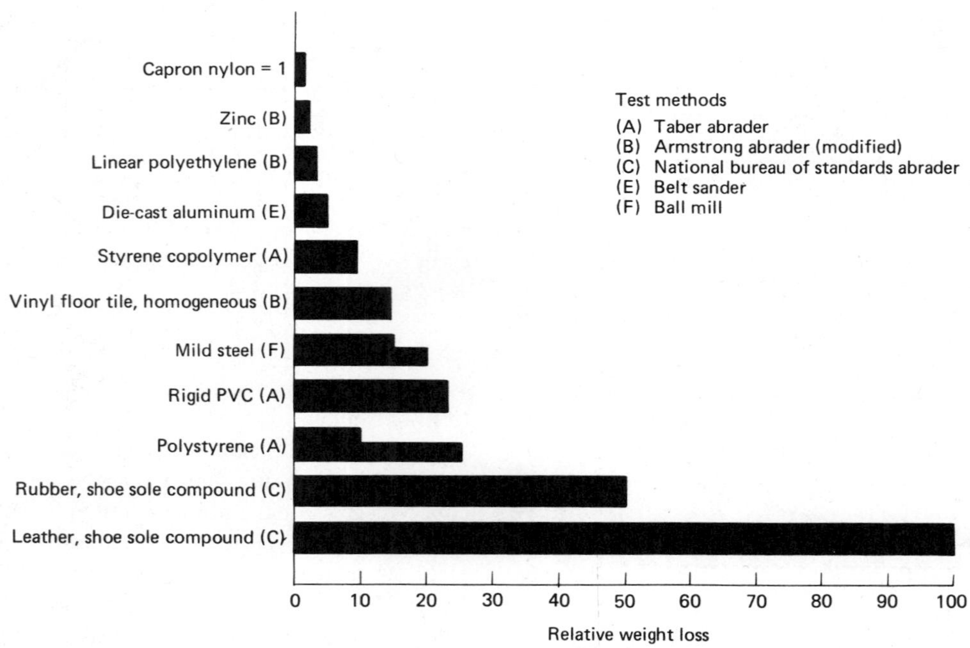

Fig. 5-105. Abrasion resistance of Capron nylon compared to miscellaneous materials. (*Courtesy: Allied Chemical Corporation*)

Table 5-100. Comparison of Physical Properties of Nylatron GS and Nylon 6/6.*

Property[a]	ASTM Method	Nylatron GS	Nylon-6/6
Tensile strength at 73°F, p.s.i.	D638-52T	12,300	11,800
Modulus of elasticity at 73°F, p.s.i.	D638-52T	575,000	400,000
Flexural strength at 73°F, p.s.i.	D790-49T	18,000	13,800
Izod impact strength at 73°F, ft.-lb./in. of notch	D256-54T	0.62	0.94
Heat distortion temp., 264 p.s.i. load, °F	D648-45T	325	200
Coefficient of linear expansion, in./in./°F	D696-44T	3.5×10^{-5}	5.5×10^{-5}
Dielectric strength (short-time), v/mil	D149-55T	356	385
Specific gravity	D797-48T	1.16	1.14
Water absorption[b]	D570-42		
24 hr, %		0.80	1.10
48 hr, %		1.14	1.48
Flammability, in./min.	D635-44	Self-extinguishing	Self-extinguishing
Color		Gray to black metallic	Buff white
Tensile impact[c], ft.-lb./sq. in.	–	118	137
Deformation under load, %	D621-51	0.8	1.0
Brittleness temperature, °C	D746-55T	–15	–30
Kinetic coefficient of friction versus steel	–	0.16 to 0.2	0.2 to 0.3

[a]The values given are average values and should not be used for specification purposes. The data for Nylatron GS were obtained using annealed test specimens.
[b]Specimens were ½ in. in diameter and ⅛ in. thick.
[c]Bragaw, C. G., *Modern Plastics 33*, 199–203, 206 (June 1956).
*Powers, 1966. Reprinted by permission from *Modern Plastics*, McGraw-Hill, Inc.

Table 5-101. Water Absorption of Various Plastics and Nylons.

Plastic	% Water Absorption	% Relative Humidity of Test
PE	<0.01	50
Polyacetals	0.25	50
PMMA	0.30	50
Nylon 6/6	1.50	25
	2.50	50
	8.00	100
Nylon 6	3.20	50
	11.00	100
Nylon 6–6/10–6/6	3.00	50
	9.80	100
Nylon 6/10	1.20	50
	3.60	100
Nylon 11	1.00	50
	1.80	100

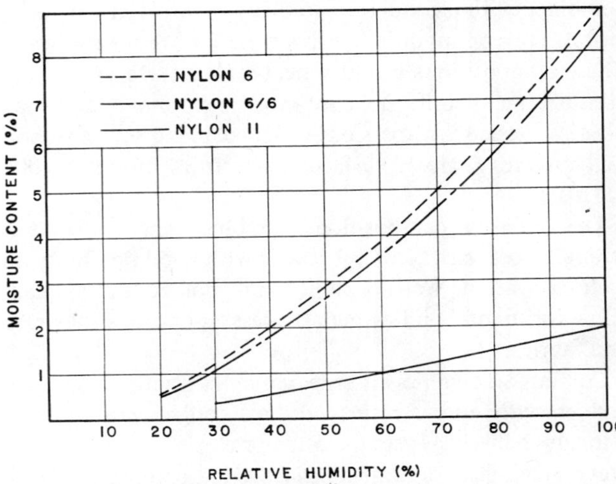

Fig. 5-107. Moisture content versus relative humidity. (Rilsan Corporation, 1976)

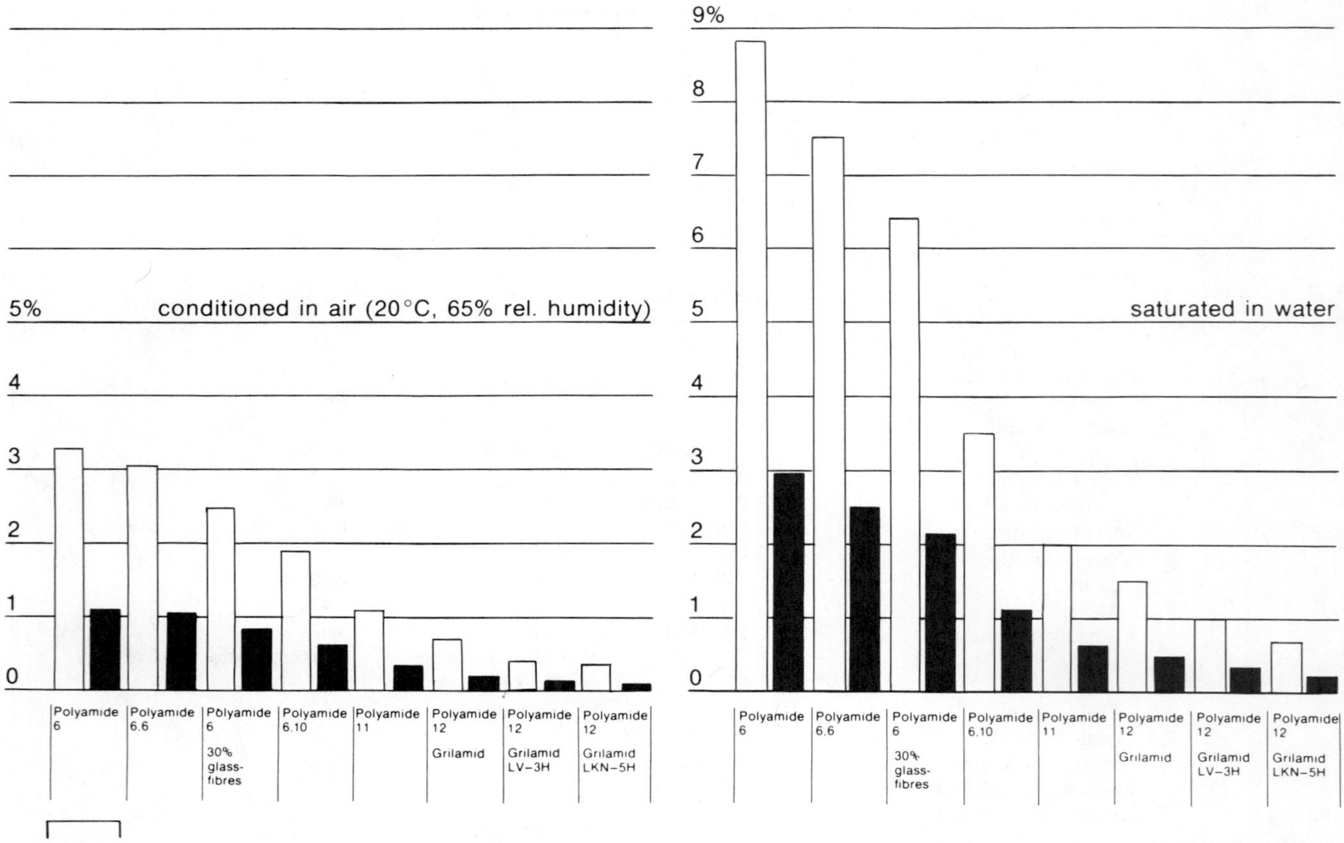

Fig. 5-108. Equilibrium humidity content of different polyamides at room temperature. (*Courtesy: Emser Werke, A. G., Zurich, Switzerland*)

flexibility with plasticizers, toughness, optical clarity, excellent adhesive properties, processing stability, and heat sealability. They find use in extrusion, solution, and powder coatings, and in flexible extruded profiles, and film. For solution purposes, aliphatic or aromatic hydrocarbons combined with alcohols are used as solvents. Coating formulations may be modified with a variety of resins and waxes. These compositions produce heat-sealing surfaces on paper, glassine, metal foil, moisture-proof cellophane, and many plastics. Aqueous dispersions when coated, dry to a smooth, dull surface. A glossy surface results from fusion at 100°C (212°F).

The adhesive compositions combine excellent adhesive strength to a variety of substrates with good flexibility and high elongation. Wetting ability and good cohesive strength allow for use in bonding metals, glass, nonporous substrates, and fabrics.

Coatings and adhesives prepared by coreacting these polyamides with epoxy resins produce tough, hard, flexible, strongly adhering systems. Substrates such as wood, concrete, steel, glass, and many plastics are compatible for use with these copolymers. Because of the excellent wetting properties of these resins, moist surfaces can be painted successfully. Even rusty surfaces can be directly overcoated to prevent further oxidation.

The castor oil-based nylons, 11 and 12, provide good chemical resistance, low water absorption, and excellent dimensional stability. Nylon 12 has lower water absorption and T_m, and better dimensional stability than nylon 11. As of this writing, the costs of both polymers are very similar, so the choice is based on application and comparative properties. The major advantage of these nylons is their very low water absorption (see Figure 5-111). In addition, nylon 11 and 12 have lower densities than nylon 6, lower cold-brittleness temperature, and better chemical and electrical resistance properties.

Nylon 11 resists stress-cracking in calcium and zinc chloride solutions, whereas nylon 6 is seriously affected. Combined with its resistance to petroleum-based chemicals, this property results in a major use in automotive molded and extruded parts. In general, these nylons show good resistance to bases, salt solutions, salt water, marine environments, oils, grease, and petroleum chemicals. Variable resistance is obtained with mineral and organic acids as well as strong oxidizing agents. Combinations of initially innocuous chemicals can combine synergistically to become destructive. Phenols and formic acid are powerful solvents for these resins.

The three major areas of use are in molded and extruded parts and as a powder coating (see Figure 5-109). The largest extruded use is for automotive hosing and tubing in truck air brakes, hydraulic and fuel lines. For powder coatings, major applications include industrial pipe and fittings, food, farm, and material-handling equipment, and institutional and outdoor furniture.

New transparent nylons have reached commerical acceptance. They are obviously amorphous materials, as opposed to the other nylons which are highly crystalline and thus opaque. Chemically, they are based on the condensation of terephthalic acid and alkyl-substituted hexamethylenediamine.

These transparent nylons process easily, have high impact resistance, and are rigid. They have high dimensional stability when exposed to temperature and pressure extremes. Scratch and abrasion resistance and electrical properties are comparable to those of other nylons. They have good stability in most aliphatic and aromatic hydrocarbons, esters, ketones, oils, and fats.

Figure 5-110 compares the moisture absorption of a transparent nylon (Trogamid T) with nylon 6 and 11. Table

Fig. 5-109. A 0.007-inch-thick Rilsan nylon 11 coating protects tubular railings, seats, and telescope stands on the World Trade Center observation decks. (*Courtesy: Rilsan Corporation*)

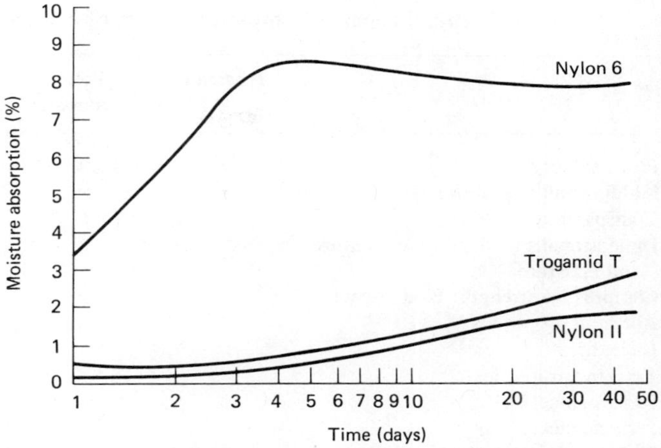

Fig. 5-110. Moisture absorption properties: nylon 6, nylon 11, and Trogamid T, a transparent nylon. (*Courtesy: Dyanamit Nobel of America*)

5-102 compares the properties of transparent nylon with other thermoplastics. Figures 5-111 and 5-112 show how tensile strength and torsional modulus of the transparent nylons vary with temperature and compare the data with other engineering thermoplastics. Figures 5-113 and 5-114 show two applications of transparent nylon.

Polycyclamides are capable of limited modification by the user. Additives such as plasticizers, dyes, fillers and reinforcing fibers can be compounded into these resins. The electrical properties are superior to nylon 6/6 at all temperatures, and this improvement, along with no substantial loss in physical properties and significant improvement in moisture resistance, dictates its replacement for nylon 6 and 6/6 in many applications, particularly where the moisture absorption of the standard nylons is marginal or unsatisfactory. The polycyclamides may be molded on equipment

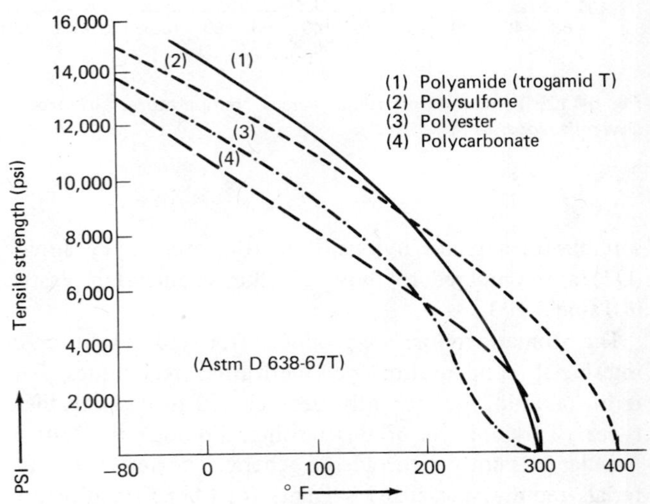

Fig. 5-111. Comparative tensile strengths of various materials as a function of temperature. (*Courtesy: Dynamit Nobel of America*)

Table 5-102. Property Comparison Chart of a Transparent Nylon (Trogamid T) and Other Thermoplastics.*

	Trogamid T	Poly Carbonate	Poly Sulfone	PPO	Methyl Methacrylate	Poly Acetal	Polyamid 6	Polyamid 12
Processability	+	o	+	+	+	o	o	o
Molding and post shrinkage	+	+	+	o	+	–	–	–
Transparency	+	+	o	o	+	–	–	–
Dimensional stability to temperature and moisture	o	+	+	+	+	+	–	–
Compressive strength: Izod impact	+	o	+	+	+	+	–	–
Abrasion properties	o	–	–	–	–	+	+	+
Dielectrics	o	o	+	+	o	+	–	–
Tracking properties	+	–	–	–	+	+	+	+
Performance: U.V.	o	+	o	–	+		Not Comparable	
Performance: X-ray	+	–	Not Comparable		+		Not Comparable	
Performance under sudden impact	+	–	o	o	–	o	+	+
Chemical resistance	+	–	–	–	o	+	+	+

Note: + = Good
o = Acceptable
– = Poor
*Courtesy: Dynamit Nobel of America.

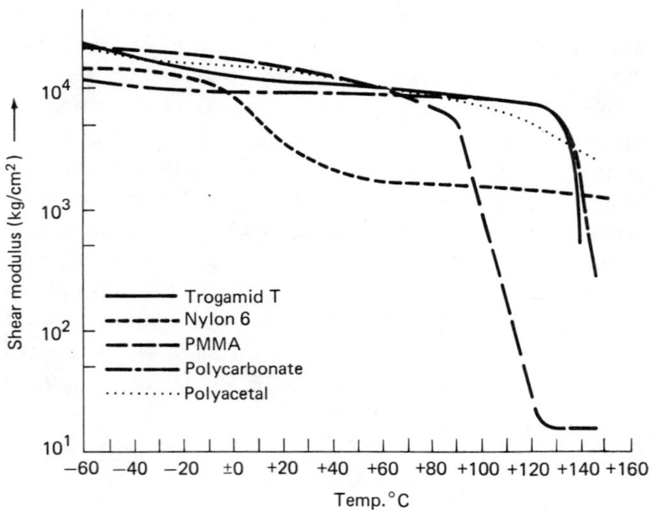

Fig. 5-112. Torsional modulus versus temperature. *(Courtesy: Dynamit Nobel of America)*

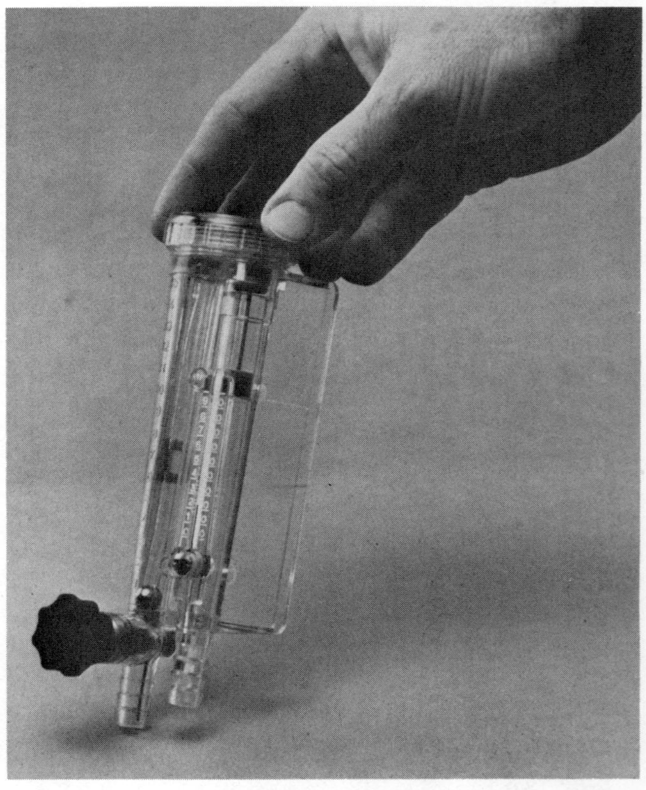

Fig. 5-113. This 176°F water flow meter is molded from Amidel transparent nylon. *(Courtesy: Union Carbide Corporation)*

suitable for aliphatic nylons. Properties of a polycyclamide (Q2) are compared to those of other engineering plastics in Table 5-103.

The aromatic nylons are produced from either solution or interfacial condensation polymerization techniques. The term "aramid" has recently been coined to describe these types of nylons. As of this writing, although the term is popular it is not yet considered generic. The best of the materials require specialized solvents for fiber formation and are produced in fibrid form by interfacial condensation. The fibers are used in yarn and cloth. They are chopped and combined with the fibrids to produce papers which have extremely high heat and chemical resistance and good electrical properties (see Figure 5-115). They are nonreceptive to dyes, and their properties in the yarn or cloth can be somewhat modified by the user.

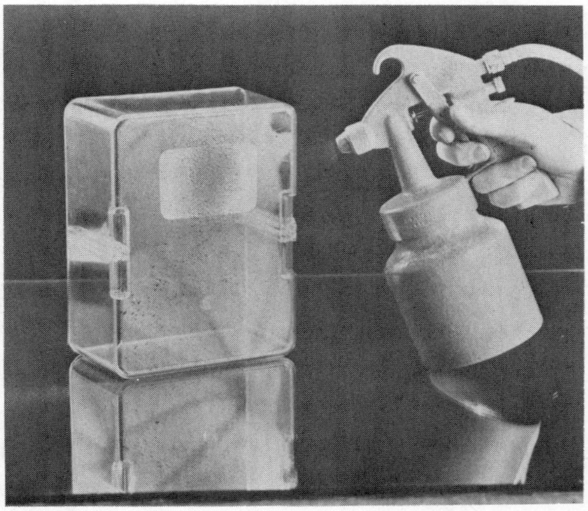

Fig. 5-114. Transparent nylon resists the solvent coating used to cover 80% of this housing, leaving a central window. (*Courtesy: Union Carbide Corporation*)

These materials provide extremely good chemical resistance against both acids and bases. Stronger acids, however, will reduce strength properties with prolonged immersion. Moisture will cause some loss of strength and elongation, and the loss is partly irreversible. Radiation resistance is outstanding. Table 5-104 describes typical properties of aromatic nylon paper. Figure 5-116 shows the effect of temperature on the tensile strength of the paper. Figure 5-117 compares the voltage endurance of aramid paper to that of other insulating papers.

The aromatic nylon yarns provide elevated temperature properties. Initial load-carrying capacity is not quite as good as with nylon fibers, but with continued aging the improved thermal stability of the aromatic polymer becomes manifest. Table 5-105 shows typical properties for aromatic nylon yarn. Aging at temperatures of 230°C (446°F) causes yellowing.

The papers may be self-bonded to make thicker laminates. The procedure is to stack the sheets, alternating machine and cross-direction of the laminate in both directions. Figure 5-118 shows the effect of temperature on laminate properties.

The yarns are used in protective fabrics exposed to high temperatures and for reinforcement of rubbers and synthetic elastomers (see the section on fillers and reinforcements for a more detailed discussion). The paper is used in electrical insulating applications (see Figure 5-119), filter bags for high-temperature gas filtration, and domestic and industrial laundry pads and covers. Honeycomb laminates have found significant use in aircraft and spacecraft construction.

An aromatic nylon has been developed recently that has a reported tensile strength value of 17,000 psi at room temperature—a heretofore unattained value for unreinforced thermoplastics. Figure 5-120 compares the tensile strength versus temperature properties of aromatic nylon with two other engineering thermoplastics. The resin exhibits exceptional wear resistance over a wide temperature range. It is available only in the form of custom-produced parts in grades containing 1–10% PTFE or graphite.

The dry powdered resin is compacted into "green" preforms on high-speed presses at room temperature, under

Table 5-103. Properties of a Polycyclamide (Q2) and Some Engineering Plastics.*

Property	Q2	Nylon 66	Acetal	Polycarbonate	ASTM method
Molding temperature, °C	325–345	290	205	275–325	
Average mold shrinkage, %	0.7	1.7	2.5	0.5	
Density, g/ml	1.12	1.14	1.43	1.20	D 1505(5)
Hardness:					
Rockwell L	110	99	104	96	D 785(5)
Rockwell M	84	52	76	54	
Tensile yield strength, psi	12,000	10,000	10,000	9,000	D 638(5)
Elongation at break, %	30	70	50	100	D 638(5)
Flexural modulus, psi $\times 10^{-5}$	3.7	3.8	3.8	3.6	D 790(5)
Notched Izod impact strength, ft-lb/in. of notch	0.7	0.9	1.2	12	D 256(5)
Deflection temperature, °C:					
Under 66 psi stress	130	124	160	150	
Under 264 psi stress	90	70	90	145	
Abrasion, sandpaper, mg/100 cycles	12	13	28	13	
Coefficient of friction, dynamic, on stainless steel	0.29	0.35	0.21	0.53	
Deformation under load at 50°C, %	0.3	2.4	1.0	0.2	D 621(5)

*Reprinted by permission from *New Linear Polymers* by Lee et al. Copyright 1967 by McGraw-Hill, Inc.

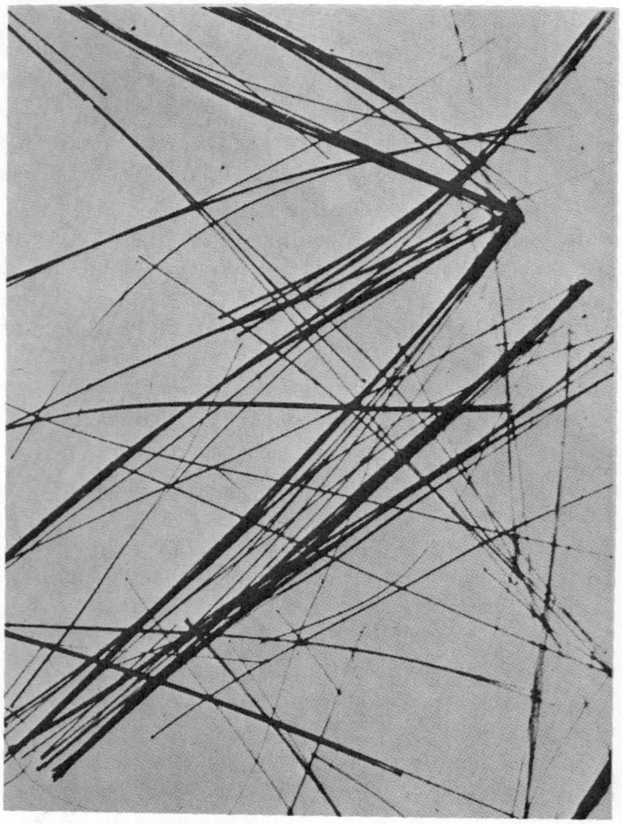

(a)

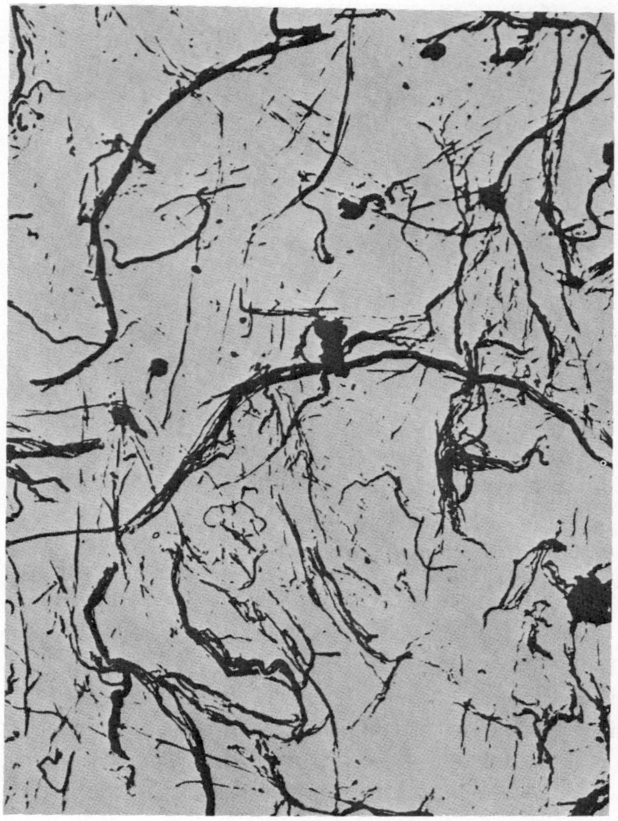

(b)

Fig. 5-115. Components of an aramid paper (Nomex Type 410). (a) Floc fibers (0.25 inch)—medium magnification. (b) Fibrid binder particles—highly magnified. (*Courtesy: E.I. Du Pont de Nemours & Co., Inc.*)

Table 5-104. Properties of Laminates of Aromatic Nylon Paper.*

Property	Value	Test (ASTM unless otherwise noted)
Specific gravity	1.3	
Tensile strength, psi	22,000	D 638
Elongation, %	5–13	D 638
Flexural strength, psi	24,600	D 790
Flexural modulus, psi	790,000	D 790
Shear strength, psi	21,300	D 752
Impact strength, ft-lb/in.	3.5	D 256
Bond strength, lb	1,300	D 229
Hardness, Rockwell F	116	D 785
Deflection temperature at 264 psi, °C	280	D 640
Abrasion resistance, mg/kc	46	Taber
Coefficient of friction	0.24	Static
Coefficient of thermal conductivity, Btu/(hr)(ft²)(°F/in.)	0.8–1.0	L 177-45
Coefficient of thermal expansion, in./(in.)(°C):		
Parallel	2.1×10^{-5}	D 696
Perpendicular	8.6×10^{-5}	D 696
Water absorption, % after 24 hours:		
Water absorption, % after twenty-four hours:		
1/32 in.	0.9–3.3	D 570
1/8 in.	0.3–1.1	D 570
1/2 in.	0.1–0.6	D 570
Compressive creep, %, 35 mil	0.68	D 352

*Reprinted by permission from *New Linear Polymers* by Lee et al. Copyright 1967 by McGraw-Hill, Inc.

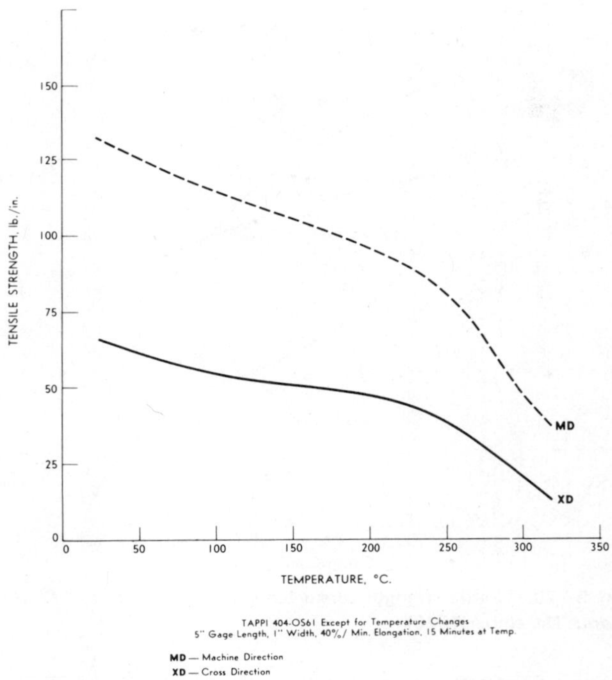

Fig. 5-116. Tensile strength versus temperature: Nomex nylon paper. (*Courtesy: E.I. Du Pont de Nemours & Co., Inc.*)

Table 5-105. Typical Properties of Nomex® High-Temperature-Resistant Nylon at 21°C (70°F).*

	Straight Test		Loop Test
Breaking strength, lb (for 200 denier yarn)			
conditioned at 65% R.H.	2.4		2.3
wet	1.8		—
Breaking tenacity, gm/denier			
conditioned at 65% R.H.	5.5		5.2
wet	4.1		—
Breaking elongation, %			
conditioned at 65% R.H.	17		14
wet	14		—
Specific gravity		1.38	
Specific heat, cal./gm./°C		0.29	
Coefficient of linear expansion (after heat-setting at 500°F) in./in./°F between 70° and 400°F		1.12×10^{-5}	
Thermal conductivity, Btu/hr/ft²/°F for 1 in. thickness		0.9	
Finish on fiber and yarn, %		1.3	

*Courtesy: E.I. Du Pont de Nemours & Co., Inc.

high pressure in precision dies. Sintering then occurs at a controlled high temperature in an inert atmosphere, and the part reaches its final density and dimensions.

Applications include bearings, thrust washers, gears, and wear surfaces that must resist high friction and hot environments.

Another recent development is the "super-tough" nylons. Their composition is unknown. The chart in Figure 5-121 describes their comparative properties qualitatively. The impact strength of super-tough nylon is dramatically higher than that of current modified and unmodified nylon 6/6 and nylon 6 under a variety of conditions. It has greater flexibility and lower tensile strength than unmodified nylon 6/6. Good chemical and heat resistance, processability, sur-

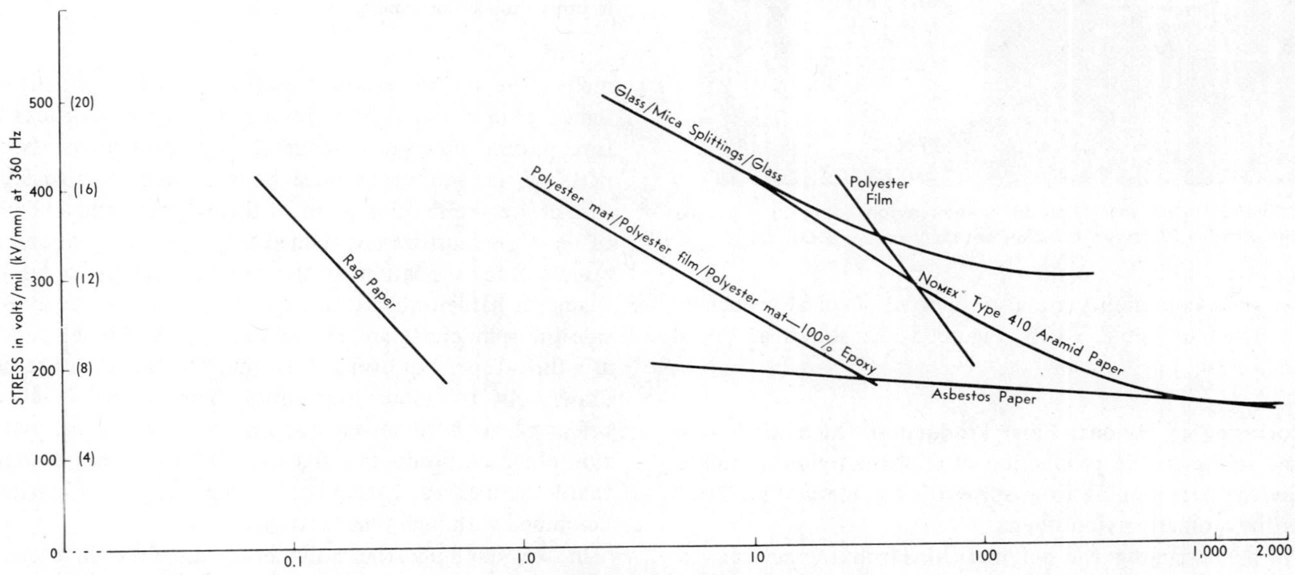

Fig. 5-117. Voltage endurance of various insulating materials (ASTM D2275). (*Courtesy: E.I. Du Pont de Nemours & Co., Inc.*)

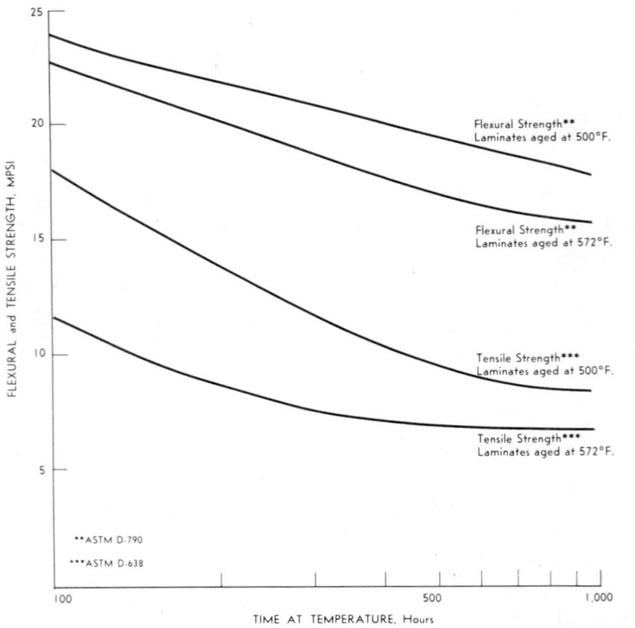

Fig. 5-118. Flexural and tensile strength of laminates made from Nomex nylon paper after aging at elevated temperatures. (*Courtesy: E.I. Du Pont de Nemours & Co., Inc.*)

Fig. 5-119. Layer insulation of Nomex nylon paper in dry tape transformer. (*Courtesy: E.I. Du Pont de Nemours & Co., Inc.*)

face, and colorability are also reported. Typical properties are listed in Table 5-106. Figure 5-122 shows a typical application.

Processing of Nylons: Fiber Production. Melt-spinning is used in the textile production of aliphatic nylon filaments, whereas wet-spinning from solvent is the method preferred for the aromatic nylon fibers.

In melt-spinning, the polymer chips from the polymerization reaction are charged into the hopper of the spinning unit and dropped onto a heated grid where the polymer

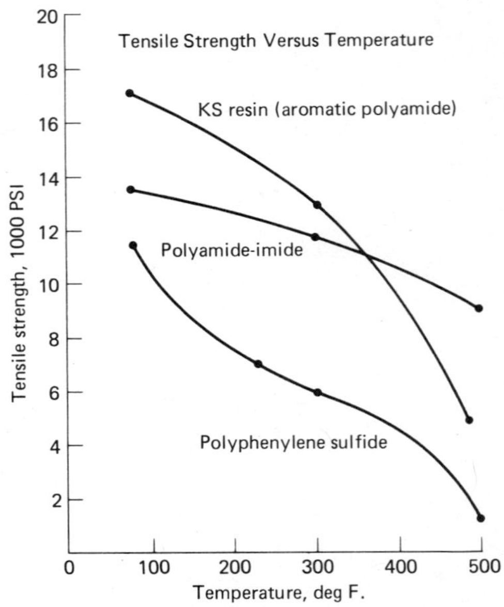

Fig. 5-120. Tensile strength versus temperature. (*Courtesy: Plastics World*, December 1975)

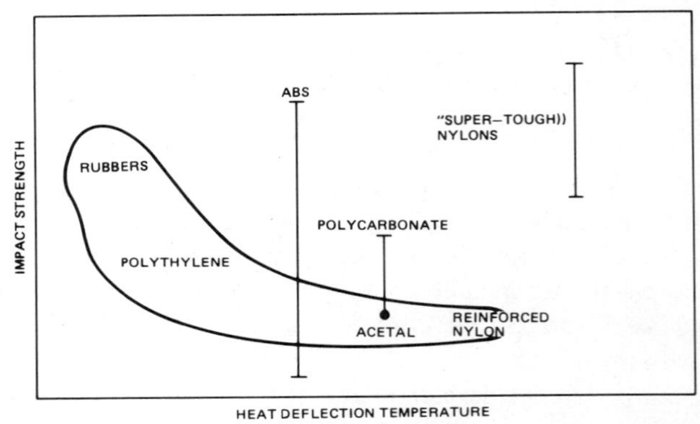

Fig. 5-121. Relative toughness of "super-tough" nylons compared to other "tough" polymers.

melts. The melt is pumped and metered by specially designed pumps and finally filtered through filter packs before passing through a spinnerette to form filaments. All parts of the equipment must be maintained at a temperature above the melting point of the polymer, and a blanket of inert gas must be maintained to prevent any undesirable effects from oxidation at the elevated temperature. The filaments harden quickly as they are exposed to the air outside the spinnerette and are wound up on bobbins operated at a thread speed of over 250 ft/min. The filaments are then drawn out to about four times their original length to orient the crystalline regions. The high degree of orientation obtained produces a filament with extremely desirable textile properties, particularly a high degree of elasticity combined with high tensile strength.

In wet-spinning, the polymer is dissolved in a strong solvent like DMF. The solution, at 128°C (262°F), is spun through a spinnerette with 0.10-mm-diameter orifices into

Table 5-106. "Super-Tough" Nylon Resin: Typical Properties.*

	UNITS	TEST	"SUPER-TOUGH" NYLON
Tensile Strength	psi	ASTM D-638	
Dry, 73°F			7,800
50% RH, 73°F			6,000
Elongation	%	ASTM D-638	
Dry, 73°F			40
50% RH, 73°F			210
Flexural Modulus	psi	ASTM D-790	
Dry, 73°F			255,000
50% RH, 73°F			125,000
Rockwell Hardness	R Scale	ASTM D-785	
Dry, 73°F			112
Izod Impact Strength	ft. lb./in.	ASTM D-256 (10 mil notch radius)	
Dry, 73°F			15
50% RH, 73°F			15-25(2)
Dry, 73°F		Special test (2 mil notch radius)	15
Heat Deflection Temperature (annealed)	°F	ASTM D-648	
66 psi			420
264 psi			160
Melting Point	°F	ASTM D-789	482-500
Specific Gravity		ASTM D-792	1.09
Mold Shrinkage	in./in.		0.013-0.018
Combustibility		UL-94	94 HB
Water Absorption 100% RH	%		6.7

(2) Samples do not break into two parts, but break part way through the sample and then bend out of way of hammer.

*Courtesy: E.I. Du Pont de Nemours & Co., Inc.

air at 225°C. The fiber is wound at 296 ft/min and drawn to 4.75 times its original length and boiled off in water. The fiber may or may not be subsequently heat-treated for even more improvement in strength properties.

Fig. 5-122. Plastic welding helmet, injection molded from Du Pont's Zytel ST super-tough nylon resin. In addition to its unusual resistance to breakage, the helmet has good heat resistance and is lighter in weight than conventional fiberglass helmets. (*Courtesy: E.I. Du Pont de Nemours & Co., Inc.*)

Molding Powders. Nylon resins can be molded in all types of screw and plunger injection-molding machines. Screw machines are preferred over plunger machines because the pressure drop is lower and the melt is more homogeneous. The general-purpose screws that are supplied with most injection-molding machines are usually suitable for molding nylons. At high output rates, a screw specifically designed for molding nylons will provide greater uniformity of melt temperature and freedom from unmelt. Nozzle temperatures are generally around 260-299°C (500-570°F) with pressures of 5,000-20,000 psi.

If the molding powder is allowed to absorb excessive amounts of water prior to molding, the mechanical properties and appearance of the molded parts are impaired. At temperatures above the melting point, hydrolysis results in a decrease in the molecular weight and toughness of the resin. At the same time, absorbed water can form steam that results in splay marks and internal bubbles in the molded part.

Other Methods. Other nylon processes include monomer casting, coating, non-melt forming, blow molding, rotational molding, and sintering. We will only discuss sintering, coating, and casting in this section, since the other topics are adequately covered elsewhere in this book.

Sintering. It is possible to prepare nylon composites using techniques similar to those in powder metallurgy. These composites have properties significantly different from standard materials. Polyamide powders with an ultimate particle size of 4-10 microns are admixed with up to 80% by volume of modifying materials, cold-pressed in cold dies, and then sintered to form the fluid part.

The powdered nylon is introduced into closed dies and pressed at room temperature at pressures of about 40,000 psi. If lower pressures are used, parts of controlled density and pore size can be prepared. Thus, porous parts saturated with lubricants, inks, and other materials for specific uses can be produced. It is important that parts are sintered in an inert environment; otherwise, excessive oxidation occurs which significantly reduces physical properties. Sintering is usually performed in a hydrocarbon-based oil medium at temperatures of 257°C (495°F) for 30 minutes to 2 hours. If the moisture content of the powder is greater than 1%, parts will frequently develop cracks.

The physical properties of pressed and sintered nylon parts are generally inferior to parts made via the standard injection-molding process. Sintered-nylon parts are used in bearings, bushings, cams, and wear plates.

Fig. 5-123. Casting from nylon 6. (*Courtesy: Cadillac Plastic & Chemical Co.*)

Casting. This process is used for large parts weighing up to 500 lb. Low-volume production is justified because mold shape and construction are simple, and costs are low. Cast nylon 6 has a higher molecular weight and crystallinity than extrusion- or injection-grade nylon 6. Thus, the final part has higher modulus, heat-deflection temperature, improved solvent resistance, lower moisture absorption, and better dimensional stability.

The process involves four steps: (1) monomer melting, (2) addition of catalyst and activator, (3) mixing the melts, and (4) casting. Caprolactam must be converted from flake to liquid form by melting under very closely monitored conditions of temperature and atmospheric humidity. Because it is very hygroscopic, exposure to high humidity will result in rapid moisture absorption, causing catalyst decomposition and incomplete polymerization. Two equally divided solutions of lactam melt are prepared; one to include alkali-metal bases as catalysts in their proper proportions, and the other containing the correct amount of co-catalyst, usually an imide or lactam. The solutions are then thoroughly mixed before being poured into suitable molds.

There are many mold materials that can be used. The basic requirement is that the material be able to contain a low-viscosity liquid at temperatures to 200°C (392°F). The molds must be designed to allow for a part shrinkage of 13–15%.

Additives such as pigments and various insoluble fillers cause problems due to inhomogeneous dispersion. All additives must be thoroughly dried before use.

For stress relief after polymerization, air or liquid annealing is used. Air annealing is performed in circulating ovens at 140–180°C (284–356°F), while liquid annealing occurs in refined mineral oil for 30 minutes to 4 hours. Figure 5-123 shows a typical part made from nylon 6 casting.

Fluidized Bed Coatings. The basic process of fluidized-bed coatings involves placing a powdered nylon in a fluidizing container and maintaining a fluidized state by the passage of gas through a porous plate. In this condition, the powder exhibits many characteristics of a fluid, including mobility, hydrostatic pressure, and observable upper-boundary layer. Individual particles are in random motion throughout the fluidized chamber. The part to be coated is heated above the T_m of the nylon powder and immersed in the fluidized bed for a short period of time (seconds). During this time the nylon particles melt on the part surface and form a coating layer. Depending on the heat capacity of the substrate, a fusion operation may or may not follow to homogenize the coating to a continuous layer.

Since the beginning of the fluid-bed technique, only a few plastic materials have achieved popular commercial success. Table 5-107 compares materials in current use in terms of various performance factors. The nylon most used for such coatings is nylon 11; however, nylon 12 has begun to show increased usage. Nylon 6 as well as a number of low-melting copolymers have also been used, although not to a significant degree. The physical and chemical properties of nylons 11 and 12 are quite similar, and the final choice is often based on economic rather than performance criteria. Both plastics exhibit similar characteristics in processing, grinding, pigmentation, and formulation.

Since these powders are 100% nonvolatile, no gaseous byproducts are evolved during coating. This is distinctly advantageous from the point of view of toxicity, solvent emission, and pollution control. Film thicknesses are 10–20 times heavier than with conventional coatings and corner and edge coverage is better. The coatings have longer service lives in corrosive and outdoor weather environments, as well as improved resistance to scratching, abrasion, and mechanical damage.

On the other hand, the large volume of material required relative to part size makes the process economics unfavorable. High inventories of nylon powder are required, and changeover because of color, formulation, or part size and configuration is more complex.

Table 5-107. Comparison of Fluidized-bed Coating Materials.*

	Nylon-11 or Nylon-12	Epoxy	Vinyl	Cellulosic	Polyethylene (Low-Density)	Chlorinated Polyether
Chemical Resistance						
Exterior durability	F	F	F	G	F	F
Water (salt or fresh)	G	G	E	VG	VG	E
Solvents						
Alcohols	E	E	E	E	E	E
Gasoline	E	E	E	G	VG	VG
Hydrocarbons	E	E	G	G	VG	E
Esters, Ketones	G	F	P	P	G	VG
Chlorinated	E	E	P	P	F	P
Salts	VG	E	E	VG	E	E
Ammonia	G	P	E	P	E	E
Alkalies	G	VG	E	F	VG	E
Mineral acids						
Dilute (10%)	F	E	E	G	E	E
Concentrated (>30%)	P	G	G	P	VG	E
Oxidizing acids						
Dilute (10%)	P	G	E	P	VG	E
Concentrated (>30%)	P	P	G	P	P	G
Organic acids:						
Acetic, formic, etc.	P	F	F	P	VG	E
Oleic, stearic, etc.	VG	E	E	F	VG	E
Mechanical and Physical Properties						
Abrasion resistance	E	VG	G	VG	F	VG
Flexibility	VG	F	E	G	VG	F
Impact resistance	VG	G	E	E	F	G
Max. service temp., °F/°C	180/82[a]	350+/177+	200+/93+	180/82	160/71	250/121
Dielectric strength	G	E	VG	VG	E	VG
Decorative Properties						
Color range	E	E	E	E	F	P
Color retention	VG	F	VG	E	VG	G
Initial gloss	G	G	VG	E	F	G
Gloss retention	E	P	G	E	F	P

[a]Up to 300°F (149°C) in nonoxidizing environment.
Key. E = excellent, VG = very good, G = good, F = fair, P = poor.
*Kohan, 1973.

Electrostatic Coating. In electrostatic powder-spray coating, a nylon powder is charged to a high potential, and the charged particles are attracted to the object to be coated (usually metallic), which is maintained at ground potential. The part is usually at room temperature, and the coating is held on the substrate by mechanical and electrostatic forces. The film is fused to a continuous coating by heating. The electrostatic powder systems are an extension of conventional electrostatic paint-spray technology.

In general, the same materials used in fluidized-bed coatings are also used for electrostatic spray coating. Nylon 11 and 12 are the only commercially used polyamides. These nylons have optimum application characteristics when the charging voltage is in the range of 30–45 KV. Optimum particle size for spraying is in the range of 10–100 microns.

This method has the same advantages as fluidized-bed techniques over conventional coating techniques. Here too, formulations are 100% nonvolatile and, with proper collection systems, almost 100% of the polymer can be used. The system is well-adapted to automation and can be run for extended periods of time. Low film thicknesses (1–3 mils) at low fusion temperatures (150–220°C, 302–428°F) can be obtained.

Applications. The nylons have found use as plastics materials for specialty purposes where their toughness, rigidity, abrasion resistance, and reasonable chemical and heat resistance are necessary. Because of their high cost, however, they have not become general-purpose materials.

The largest applications of the homopolymers have been in mechanical engineering, where they are used as cams, gears, bushings, bearings, and valve seats. Nylon hair combs and brushes (see Figure 5-124) have been widely accepted because of their flexibility, despite their high cost. In recent

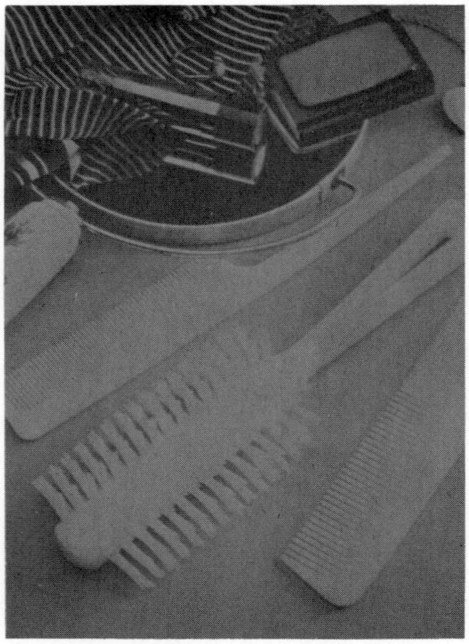

Fig. 5-124. Brushes and combs represent one of the earliest uses of nylon. (*Courtesy: American Hoechst Corporation*)

years, nylons have met increased competition from many new engineering resins, particularly polyacetals, polysulfones, poly(phenylene oxides and sulfides), and polycarbonates.

Nylon film has been used in the foodstuff and pharmaceutical packaging areas. Monofilaments have found use in brush tufting, wigs, surgical sutures, wearing apparel, parachute cloth, rope, tire cord, and sports equipment (Figure 5-125).

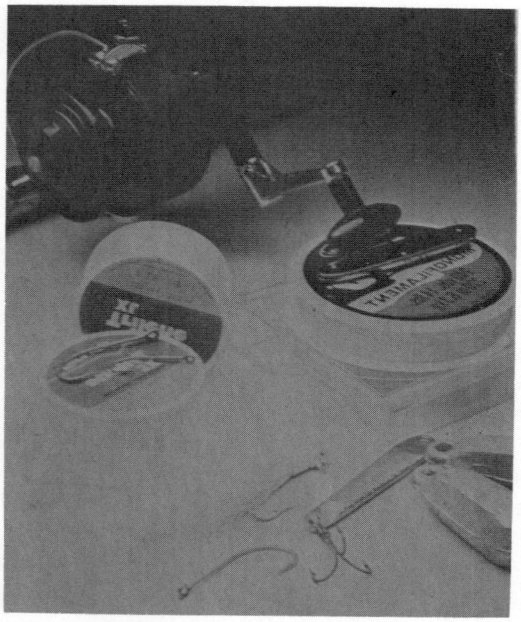

Fig. 5-125. Fishing line of oriented nylon has the tensile strength of steel, plus flexibility, low visibility, and freedom from corrosion. (*Courtesy: American Hoechst Corporation*)

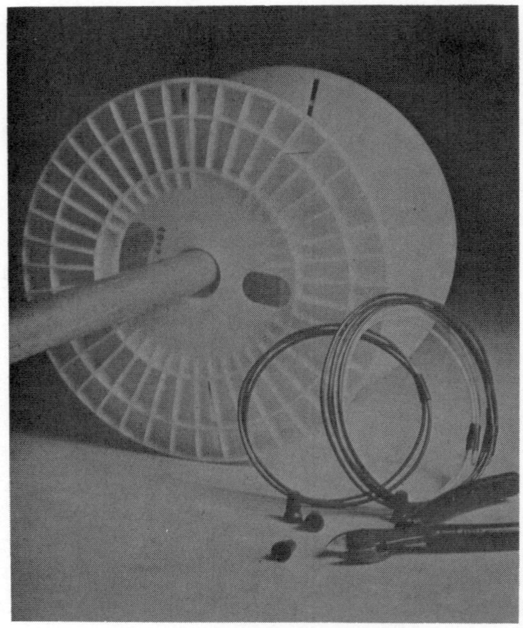

Fig. 5-126. Countless electrical coils are wound on nylon 6/6 forms. (*Courtesy: American Hoechst Corporation*)

Although the nylons are not generally considered outstanding electrically, their toughness and, to a degree, their temperature resistance, have led to applications in coil formers, bobbins, and terminal blocks (see Figure 5-126).

Glass-filled nylons are frequently used in automobile components, appliance housings, and for diverse military applications. Figures 5-127 through 5-129 illustrate a number of other end products made from nylon.

Fig. 5-127. Athletic shoe soles are made of nylon 6 because of its great toughness. (*Courtesy: American Hoechst Corporation*)

Fig. 5-128. Nylon zippers outlast traditional ones of fabric or metal. (*Courtesy: American Hoechst Corporation*)

Fig. 5-129. Spatula blades and spoons of nylon 6/6 withstand highest cooking temperatures. (*Courtesy: American Hoechst Corporation*)

Trade Names

Airloft	Acetate-nylon filament	Celanese Corporation
Akulon	Nylon 6/6	Akzo Plastics; Canada Colors & Chemicals, Ltd.
Alrac	Nylon 4 fiber	Radiation Research Corporation
Amidel	Transparent nylon	Union Carbide Corporation
Amilan	Nylon resin	Toray Industries, Inc.
Ancamide	Polyamide resin	PVO International, Inc.
Antron	Nylon fiber	E.I. Du Pont de Nemours & Co., Inc.
Avidamide	Microcrystalline nylon	FMC Corporation
Ayrlyn	Nylon filament	Rohm & Haas Co.
Azamide	Polyamide	AZ Products Co.
BASF	Nylon 6, 6/6, 6/10	BASF, A.G.
Beckalide	Polyamide	Beck, Koller & Co., Ltd.
Belding	Nylon 6/6, 8, 11, 12	Belding Chemical Industries
Blanc de Blancs	Nylon fiber	Akzona, Inc.
Blue "C"	Nylon 6/6 fiber	Monsanto Co.
Bonyl	Biaxially oriented nylon film	Kohjin Co.
Brulon	Nylon monofilament	Imperial Chemical Industries, Ltd.
Cadon	Nylon filament	Monsanto Co.
Cambrelle	Nylon fabric	ICI, Ltd.
Cantrice	Nylon fiber	E.I. Du Pont de Nemours & Co., Inc.
Caplene	Nylon 6 film	Ludlow Plastics
Capran	Nylon 6 transparent film	Allied Chemical Corporation
Caprolan	Polycaprolactam yarn, fiber, and monofilament	Allied Chemical Corporation
Capron	Nylon 6 and copolymers	Allied Chemical Corporation
Cascamid	Polyamide resins	Borden, Inc.
Catalin	Nylon 6, 6/6	Ashland Chemical Co.
Celafuse	Polyamide resins	British Celanese, Ltd.
Celanese	Nylon 6/6, 6/10	Celanese Polymer Co.
Chemiseal	Nylon 6 and glass-filled nylon	Gasket Co.
Co-Mer	Polyamide ester	Nypel, Inc.
Conaloy	Nylon blends	Continental Can Co.
Conolon	Nylon 6 film	Continental Can Co.
Coramid	Polyamides	Cooper Polymers, Inc.
Cordo Viz	Vinyl-nylon	Ferro Corporation

Cropolamid	Polyamide resins	Crosby Chems, Inc.
Cumuloft	Nylon filament	Monsanto Co.
Dartak	Nylon 6/6 film	E.I. Du Pont de Nemours & Co., Inc.
Deconyl	Nylon coating powders	Plastic Coatings, Ltd.
Deta Nylon	Polyamide rod and profile	Foster Brothers Plastics, Ltd.
Dorzan	Nylon	E.I. Du Pont de Nemours & Co., Inc.
Durethan	Nylon 6	Bayer, A.G.
Dynyl	Powdered polyamide	M & B Plastics, Ltd.
Elvamide	Nylon resin	E.I. Du Pont de Nemours & Co., Inc.
Emblem	Nylon film	Unitika, Ltd.
Emerez	Polyamide resin	Emery Industries, Inc.
Enkaloft	Nylon fiber	Akzona, Inc.
Enkalon	Polyamide 6	Akzo Plastics, N.V.
Enkalure	Nylon fiber	Akzona, Inc.
Enkamat	Nylon mats	Enak Glanstoff, A.G.
Enka Perlon	Polyamide 6	Akzo Plastics, N.V.
Enkatron	Nylon fiber	Akzona, Inc.
Ertalon	Nylon 11, 6, and 6/6	Chemplast, Inc.
Felor	Nylon filaments	E.I. Du Pont de Nemours & Co., Inc.
Fiberfil	Glass-reinforced nylon 6, 6/6, and 6/10	Dart Industries, Inc.
Fiberlene	Glass-filled nylon 6 and 6/6	Service Color Corporation
Filon	Nylon/polyester panels	Vistron Corporation
Flexten	Polyaramid tire cord	Goodyear Tire & Rubber Co.
Fosta	Nylon 6	American Hoechst Co.
Freshtuff	Ionomer/nylon co-extrusion	American Can Co.
Gental	Nylon solution	General Plastics Corporation
Genton	Nylon aqueous dispersions	General Plastics Corporation
Glasfil	Glass-reinforced polyamide	Dart Industries, Inc.
Grilamid	Nylon 12	Emser Industries, Inc.
Grilón	Nylon 6	Emser Werke, A.G.
Gril-tex	Copolyamide adhesives	Emser Werke, A.G.
Hitamide	Polyamide resins	Hitachi Chemical Co.
Hostamid	Polyamide	Hoechst, A.G.
Hytron	Nylon 11	Imperial-Eastman
Itamid	Glass-reinforced nylon	Polish Origin
Jaytron	Nylon	Johnston Industrial Plastics
Kaprolon	Nylon 6	Soviet Origin
Kapron	Nylon 6	Soviet Origin
Kelon	Nylon	W.S. Shamban & Co.
Kevlar	Aromatic polyamide fiber	E.I. Du Pont de Nemours & Co., Inc.
Latamid	Reinforced nylon 6 and 6/6	Lati, S.p.A.
Lilion	Polyamide fibers	Snia Viscosa, S.p.A.
LNP	Glass-filled or reinforced nylon 6, 6/6, and 6/10	Liquid Nitrogen Processing Corporation
Lon	Nylon 6 fiber	Polymers Corporation
Lonex	Nylon 6 monofilament	Polymers Corporation
Lowland	Nylon fibers	Akzona, Inc.
Maranyl	Polyamide compounds	Imperial Chemical Industries, Ltd.
Micropel	Nylon powders	Nypel, Inc.
Milvex	Polyamide resins	General Mills Chemical Co.
Minlon	Mineral-reinforced nylon 6/6	E.I. Du Pont de Nemours & Co., Inc.
Miramid	Polyamide 6	VEB Leuna-Werke
Moleculay	Nylon and linear polyamides	Belding Chemical Industries
Monocast	Cast nylon	Polypenco, Ltd; Polymer Corporation
Morrsylon	Nylon yarn	Reuben Zebaida
Muehlstein	Nylon 6 and 6/6	H. Muehlstein & Co., Inc.
Nailonplast	Nylon 6/6	Montedison USA, Inc.
Nichiray	Biaxially oriented nylon film	Nippon Rayon Co.
Nomex	Aromatic nylon	E.I. Du Pont de Nemours & Co., Inc.
Novamid	Nylon 6	Mitsui Petrochemical Industries, Ltd.
Ny Carbolon	Polyamide	Carbone Corporation
Nygen-Tolex	Vinyl-coated nylon	General Tire & Rubber Co.
Ny-Kon	MoS_2-filled nylon	Liquid Nitrogen Processing Co.
Nylaflow	Fiberglass-reinforced nylon 6, 6/6, and 6/10	Polymer Corporation

Nylaflow	Nylon tubing	Polypenco, Ltd.
Nylaglas	Fiberglass-reinforced polyamide	Dart Industries, Inc.
Nylam	Nylon film laminate	H & R Industries, Inc.
Nylasar	Fiberglass-reinforced nylon	Dart Industries, Inc.
Nylasint	Sintered nylon 6/6	Polymer Corporation; Polypenco, Ltd.
Nylatron	MoS_2-filled nylon 6/6	Polymer Corporation
Nylene	Nylon 6	Service Color Corporation
Nylenka	Nylon 6	Akzona, Inc.
Nylex	Nylon fiber	Polymer Corporation
Nylex	Nylon coating powders	Molytex, Ltd.
Nylode	Asbestos-reinforced nylon	Dart Industries, Inc.
Nylofil	Glass-reinforced nylon	Dart Industries, Inc.
Nyloft	Nylon fiber	Firestone
Nylonac	Nylon molding compound	ICI Australia, Ltd.
Nylrod	Nylon rods	Carmer Industries, Inc.
Nylux	Nylon rods	Westlake Plastics Co.
Nypalube	TFE-filled nylon	Nypel Corporation
Nypel	Filled nylon 6 and 6/6	Nypel Corporation
Nypelube	TFE-filled nylon	Nypel Corporation
Nyreg	Glass-reinforced nylon 6	Nypel Corporation
Nytelle	Nylon 6 fibers	Firestone Tire & Rubber Co.
Nyvel	Nylon film and laminates	Velveray Corporation
Olin	Nylon 12	Olin Chemicals Co.
Omnylon	Nylon	Omni Products Corporation
Orgamide R	Nylon 6 fibers	Aquitaine Organico
Perlon	Polyamide filaments	American Hoechst Corporation
Plaskon (See Capron)	Nylon 6-12 copolymers and molding compounds	Allied Chemical Corporation
Plaslube	Nylon	Dart Industries, Inc.
Platamid	Nylon 12 powder	Rilsan Corporation
Poliafil	Nylon 6, 6/6 filaments	Sauquoit
Polypenco	Nylon tube, rod, filament, film, and sheet	Polymer Corporation
Portex	Polyamide rod & tube	Portland Plastics, Ltd.
Poznilon	Nylon resins	Nylon Engineering, Inc.
Python	Polyamide rod, tube, and profile	Plastic Tube & Conduit Co., Ltd.
Quiana	Nylon	E.I. Du Pont de Nemours & Co., Inc.
Quill	Nylon 6/10 filament	Upjohn Co.
Reevair	Nylon fiber	Reeves Brothers, Inc.
Reside	Polyamide resins	Resine Sinteliche Affini, S.p.A.
Rilsan	Nylon 11 and 12	Rilsan Corporation
Sergene	Nylon solution	General Plastics, Inc.
Sidanyl	Polyamide film or laminate	UCB-Sidac Corporation
Siramid	Polyamide resins	Societa Italiano Resine
Skapolon	Nylon yarn	Kayser-Roth
Sniamid	Nylon 6 and 6/6	Snia Viscosa, S.p.A.
Snia Vitrid	Glass Fiber-filled nylon 6 and 6/6	Snia Viscosa, S.p.A.
Sooflex	Nylon monofilament	Soo Valley Co.
Sparkling	Nylon fiber	E.I. Du Pont de Nemours & Co., Inc.
Spectrodye	Nylon 6 filament	Akzona, Inc.
Springset	Nylon 6/6 fiber	Uniroyal, Inc.
Stren	Nylon filament	E.I. Du Pont de Nemours & Co., Inc.
Sunlon	Polyamide resin	Sun Chemical Corporation
Supranyl	Nylon 6	Mannismann-Plastic, GmbH
Supronyl	Super polyamide film	Kalle & Co., A.G.
Sutron	Nylon filament	Imperial Chemical Industries, Ltd.
Tajmir	Nylon 4 fiber	Alrac Corporation
Tarnamid	Nylon	Polish Origin
Technyl	Nylon rod and molding compounds	M & B Plastics, Ltd.
Technyl	Nylon resins, film, sheet, profiles, and dispersions	Rhone-Poulenc Polymeres
Terlon	Aromatic polyamide	Russian Research Institute
Texalon	Nylon 6 and 6/6	Texapol Corporation
Thermocomps	Filled and reinforced nylon	Liquid Nitrogen Processing Corporation
Touch	Nylon 6 filament yarn	Allied Chemical Corporation

Trogamid	Semi-aromatic transparent nylon	Dynamit Nobel Chemicals
Tynex	Nylon filaments	E.I. Du Pont de Nemours & Co., Inc.
Ultramid	Nylon 6, 6/6, and 6/10	BASF, Wyandotte Corporation
Unicast	Nylon	Allied Chemical Corporation
Vecana	Polyamide	Chevron Chemical Co.
Versalon	Polyamide resins	Henkel Resins Division
Versamid	Fatty acid polyamide resins	Henkel Resins Division
Vestamid	Nylon 12	Chemische-Werke Huls, A.G.
Vydyne	Nylon 6/6, 6/9, and 6/6–6 copolymers and compounds	Monsanto Chemical Co.
Vylor	Nylon fiber	E.I. Du Pont de Nemours & Co., Inc.
Weblox	Nylon fabric	Kendall Co.
Webnil	Nylon fabric	Kendall Co.
Well-Kor	Reinforced nylon 6 and 6/6	Wellman, Inc.
Well-A-Mold	Reinforced nylon resins	Wellman, Inc.
Wellamid	Nylon 6 and 6/6	Wellman, Inc.
Well-Blend	Blended nylon 6 and 6/6	Wellman, Inc.
Well-Fibe	Fiberglass-reinforced nylon	Wellman, Inc.
Wellon	Nylon 6/6 fiber	Wellman, Inc.
Well-Sphere	Glass sphere–reinforced nylon	Wellman, Inc.
Wellstrand	Nylon fiber	Wellman, Inc.
X-Tol	Nylon molding resin	Custom Resins, Inc.
Xylon	Nylon 6/6	Dart Industries, Inc.
Zytel	Nylon 6, 6/6, and 6/10 resins and copolymers	E.I. Du Pont de Nemours & Co., Inc.

REFERENCES AND BIBLIOGRAPHY

Allied Chemical Corporation. "Nylon-12 joins allied chemical's family of engineering plastics." *Plaskon Engineering Plastics News*, Summer, 1969.

——. *Honeycomb, Metal Bonding and Vacuum Bag Molding (Capran 80).* Industrial Information Application Note, January 1965.

——. *Capran Polyamide Film–Technical Data.* Bulletin CDT 4064.

——. *Plaskon Nylon Resins and Molding Compounds–Composite Data Sheet.*

——. *Data on Plaskon Nylon for Injection and Extrusion.*

——. *Nylon Film Made From Plaskon Nylon.* Technical Data Report 60-7.

——. *Dimensional Stability of Plaskon Nylon.* Technical Data Report 58.8.

——. *Plaskon Nylon Injection Molding Techniques.* Bulletin 5-39-18.

Baier, H. F., and Jacobi, F. "Biaxial Orientation of Film: The Future." Paper presentation, *SPE ANTEC*, New Orleans, May 1979.

Berringer, L. Engineering resins–the keys to new markets. *Plastics World*, June 1975, **33**(6):24–31.

Brydson, J. A. *Plastics Materials.* 3rd Ed. Robert E. Krieger Publishing Co., Huntington, New York, 1975.

Celanese Plastics Co. *Celanese Nylon–General Purpose Nylon 6/6.* Bulletin N1A, June 1973.

——. *Celanese Nylon 1300 Series–Crystallization Controlled Nylon 6/6.* Bulletin N13, June 1971.

Courdwelis, C. J., and Wynschenk, J. There are new things happening in the field of decorative plastics plating. *Plastics Engineering*, January 1978, **34**(1):39–40.

E.I. Du Pont de Nemours & Co., Inc. *Engineering Design with Du Pont Plastics.* Bulletins E-14974, E-20805, & E-24945, Summer 1978, Fall 1978, & Summer 1979.

——. *Properties and Performance of Nomex Type 412 Aramid Paper.* Bulletin NX-11, Form A-89150, April 1978.

——. *Concepts in Du Pont Engineering Plastics–How You Can Lower Cost, Reduce Weight and Conserve Energy.* Bulletin E-18056, May 1978.

——. *Comparison of Material Cost.* Bulletin E-21912, May 1978.

——. *Zytel ST Super Tough Nylon Resin.* Bulletin E-12230, February 1977.

——. *Properties and Performance of Nomex Type 410 Aramid Paper.* Bulletin NX-7, Form A-89145, November 1977.

——. *Minlon Data Sheets.* E-18662 & E-18659, December 1977.

——. *A New Generation of Tough Nylons.* Form E-07447, May 1976.

——. *Zytel Nylon Resins, General Guide to Products & Properties.* Bulletin E-09134, June 1976.

——. *Molding Du Pont Zytel Nylon Resins.* Bulletin A-99517, September 1975.

——. *Zytel Glass Reinforced Nylon Resins, Molding Manual.* Bulletin E-06158, December 1975.

——. *"Zytel" Nylon Molding Resins.* School Service Bulletin Form A-54238, August 1967.

——. *Properties of Nomex High Temperature Resistant Nylon Fiber.* Bulletin NP-33, Form A-16931, October 1963.

——. *Facts About Du Pont High-Temperature Resistant Nomex Aramid Paper.* Form E-21097.

——. *Nomex Aramid Fiber and Paper.* Form E-17381.

——. *Better Products Made With Du Pont Zytel Nylon Resins.* Form A-43541.

Dynamit Nobel of America, Inc. *Trogamid T–Special Polyamide.*

——. *Trogamid T Special Polyamide Clear–Transparent for Injection Molding, Extrusion and Blow Molding.* Bulletin KR 211, July 1974.

Emser Werke, A. G. *Nylon 12, Grilamid, the High Performance Polyamide.*

Floyd, D. E. *Polyamide Resins*, Reinhold Plastic Application Series, Reinhold Publishing Corporation, New York, 1958.

General Mills Chemicals, Inc. *Milvex Hot Melt Adhesive.* Bulletins III-F-1c-1, 1a-1, & 16-1.

——. *General Purpose Milvex Specialty Nylon Resins.* 1971.

——. *How to Use Versalon Resins.* Bulletin CDS 1-64, February 1964.

——. *Versalon XR-1165 Resin–Adhesive Properties on Metals.* Bulletin 4-64, April 1964.
——. *Versalon Polyamide Resins.* Bulletins CDS 4-63 & 5-63, November 1964.
——. *Versamids–The Versatile Polyamide Resins.* Technical Bulletin 11-A, 1955.
Gilbert, M. Nylons, from Plastics in Packaging. *Plastics Engineering*, October 1978, **34**(10):41–42.
Golding, B. *Polymers and Resins–Their Chemistry and Chemical Engineering.* D. Van Nostrand Co., Inc., Princeton, New Jersey, 1959.
Hauck, J. E. What's new in nylon plastics? *Materials in Design Engineering*, July 1966, **64**(1):92–95.
Hertzberg, R. W., and Manson, J. A. Fatigue testing. *Plastics World*, May 1977, **35**(5):50–53.
Houston, A. M. "New types, improved processes expand use of structural foam. *Materials Engineering*, November 1975, **82**(6): 38–42.
Kestler, J. What about the "other nylons"? *Modern Plastics*, August 1968, **45**(8):86–89.
Kjellmark, E. W. Jr., and Johannes, R. K. "Glass and Mineral Reinforced Nylons, Their Advantages, and Uses." Paper presentation, *30th Anniversary Technical Conference, Reinforced Plastics/Composites Institute, SPI, Inc.*, 1975.
Kohan, M. I. *Nylon Plastics.* John Wiley & Sons, Inc., New York, 1973.
Lee, H., Stoffey, D., and Neville, C. *New Linear Polymers.* McGraw-Hill Book Co., New York, 1967.
Materials Engineering. Glass-reinforced nylons resist high temperatures. October 1977, **86**(4):63.
——. Nylon plastics. Manual 254, March 1975, **52**(3):67–74.
——. What's new in nylons? July 1972, **49**(7):50–54.
Mannis, F. Shortcut to plastic prototypes. *Engineering Design*, E.I. Du Pont de Nemours & Co., Inc., Bulletin E-14300, Winter 1976.
Modern Plastics. New nylons: toughened, heat-stabilized, reinforced and lubricated. July 1978, **55**(7):68–70.
Monsanto Chemical Co. *Vydyne Engineering Plastics–Composite Data Sheet.* Bulletin No. 6313C, April 1975.
——. *Vydyne R Mineral Reinforced Polyamide–Engineering Data Manual.* Publication No. 6331A, November 1974.
——. *Vydyne R Engineering Plastics–A New Family of Mineral Reinforced Polyamides–Basically . . . There's No Comparison.* Bulletin No. 6299, June 1972.
Olin Chemicals. *A Report on Nylon 12.* Bulletin AD 1459, June 1968.
Nametz, R. C., Burleigh, P. H., and Moore, P. O. "New Flame Retardant For Nylon 6/6. Paper presentation, *SPE ANTEC*, New Orleans, May 1979.
Plastics Engineering. Key Design Properties of Leading Engineering Resins. December 1974, **30**(12):33–40.
——. Engineering resins primer. December 1974, **30**(12):17–32.
Plastics Technology. Look into this new mineral-reinforced nylon. August 1972, **18**(8):15–19.
Plastics World. Transparent plastics: new and improved. January 1977, **35**(1):24–26.
——. Aromatic nylon resists blistering temperatures. December 1973, **19**(12):72.
The Polymer Corporation. *Nylatron Nylon Gear Design Manual.* Bulletin BR-36A, May 1974.
——. *Nylatron GS Nylon is the Key to Better Plastic Gear Performance.* Bulletin M-GS210, January 1965.
——. *Nylatron GS Nylon and Molybdenum Disulfide Molding Compound Puts More Life in Your Products.* Bulletin GS-2a, August 1964.
Powers, T. E. Molybdenum disulfide in nylon for wear resistance. *Modern Plastics*, June 1960, **37**(6):148–154.
Rilsan Corporation. "Rilsan Nylon 11–Extrusion and Injection Molding." Paper presentation, *Plastics World Seminar, Engineering Plastics II*, Los Angeles, March 1976.
——. *The Super Nylon, Design Guide to a Remarkable Engineering Material.* 1973.
——. *Rilsan Nylon 11 Grade Designation and Physical Property Data.* Bulletin BR 2-72-1000, February 1972.
——. *Rilsan Nylon 11 Fine Powder Electrostatic Spray.* Bulletin WC, March 1972.
——. *Platamid H Powder for Fabric Bonding.* Technical Bulletin TB-7-72-3301, July 1972.
——. *The Electrostatic Deposition of Rilsan Nylon 11 Powder.* Technical Bulletin TB 11-72-1006, November 1972.
——. *Rilsan Nylon 11 Shape Extrusion, Tubing, Stock Shapes, Hose.*
——. *Processing Techniques–Rilsan Injection Moulding.*
——. *"Rilsan Nylon" Powder Coating.*
Scheiner, L. L. A molder's guide to nucleated nylon. *Plastics Technology*, April 1967, **13**(4):37–42.
Schaaf, S., and Gilbert, M. Nylon copolymer film for food packaging. *Plastics Engineering*, October 1978, **34**(10):53–55.
Simonds, H. R., and Church, J. M. *A Concise Guide to Plastics.* Van Nostrand Reinhold Co., New York, 1963.
Stott, L. L. Sintered nylon. *Modern Plastics*, September 1957, **34**(9):161–166.
Swanson, R. S. *Plastics Technology–Basic Materials and Processes*, McKnight & McKnight Publishing Co., Bloomington, Illinois, 1965.
Theberge, J., and Cloud, P. Choosing the best glass-reinforced thermoplastic. *Plastics World*, March 1977, **35**(3):44–46.
Titus, J. B. *Trade Designations of Plastics and Related Materials*, (revised). PLASTEC NOTE N9C, Plastics Technical Evaluation Center, Picatinny Arsenal, Dover, New Jersey, May 1978.
Union Carbide Corporation. *Amidel Transparent Nylon–Superior Chemical Resistance Without the Costs of Glass Fabrication.* Bulletin F-46554-A, April 1978.
Winding, C. C., and Hiatt, G. D. *Polymeric Materials*, McGraw-Hill Book Co., New York, 1961.
Wood, A. S. Engineering resins, from Materials '79. *Modern Plastics*, January 1979, **56**(1):53–55.
——. The choices get tougher in engineering thermoplastics. *Modern Plastics*, July 1978, **55**(7):42–44.
——. Action in profile extrusion shifts to engineering resins. *Modern Plastics*, November 1978, **55**(11):46–48.
——. Engineering thermoplastics get tougher. *Modern Plastics*, October 1977, **54**(10):50–52.
Young, J. D. "Engineering Thermoplastics in the Automotive Industry." Paper presentation, *174th Annual Meeting, ACS, Chemical Marketing and Economies Division*, Chicago, August 1977.
Zimmerman, H. F. "Performance of Engineering Resins in Electrical Connectors." Paper presentation, *SPE ANTEC*, New Orleans, May 1979.

Polyacetals [Poly(Oxymethylenes)]

The search for a stable polyformaldehyde polymer dates back to the 1850s, when formaldehyde was first isolated by the Russian chemist Alexander Butlerov. The first extensive laboratory studies of polymers were conducted in the 1920s by Hermann Staudinger, whose pioneering work in polystyrene we have already mentioned. It was only in the late 1950s that polymers with the requisite stability and toughness for practical use were commercialized.

Formaldehyde is a key intermediate in the manufacture of phenolic and amino resins as well as many other aldehyde-based plastics. Methanol is the main precursor for the production of formaldehyde.

$$CH_3OH + \tfrac{1}{2}O_2 \xrightarrow[300-600°C]{\text{catalyst}} CH_2O + H_2O$$

Methanol is catalytically oxidized in the vapor phase over a metal oxide catalyst. The formaldehyde produced is absorbed in water and then fractionated. A 37% solution of formaldehyde in water is separated from the excess methanol, which is then recycled to the oxidation step.

Formaldehyde is usually supplied as a stabilized aqueous solution of about 40% formaldehyde known as formalin. When formalin is the source of aldehyde for polymerization purposes, impurities present include water, methanol, formic acid, methylal, methyl formate, and carbon dioxide. The first three of these impurities retard the polymerization reaction and must be removed. For commercial polymerization, low polymers such as trioxane and paraformaldehyde are preferred because of their greater purity.

Formaldehyde will polymerize in a number of ways. Trioxane is obtained by the action of trace amounts of H_2SO_4 or hot formaldehyde vapor.

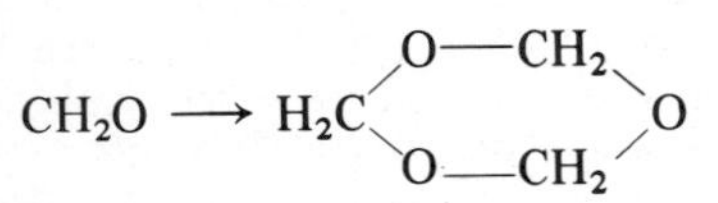

Linear polymers with $n = 50$ and having a terminal —OH group are obtained from the evaporation of formaldehyde-water solutions. Chain lengths can double if a strong acid is present.

$$CH_2O \longrightarrow HO(CH_2O)_nH$$

Evaporation of formaldehyde-methanol solutions gives a similar polymer with methoxy chain ends.

$$CH_2O \longrightarrow CH_3O(CH_2O)_nH$$

Early efforts by Staudinger produced poor polymers—brittle and thermally unstable. After an extensive research program at Du Pont, a tough, stable polymer was developed using the following procedure. First, a very pure formaldehyde is produced from an alkali-precipitated, low molecular weight polyformaldehyde which has been carefully washed with distilled water and vacuum-dried. The dried polymer is then pyrolyzed at 150–160°C((302–320°F) and the resultant formaldehyde is passed through a series of −15°C (5°F) traps. Some polymerization occurs in these traps and removes undesirable impurities from the monomer. The monomer is then charged into a reaction kettle over a rapidly stirred and carefully dried inert solvent, like heptane. Polymerization initiators include Lewis acids, amines, phosphines, arsines, and stibines, a typical one being triphenylphosphine. Stabilizers such as diphenylamine may also be added at this point. The reaction proceeds until a 20% solids content is obtained. The resultant polymer is filtered out of the solution, washed with heptane and acetone, and vacuum-dried. Molecular weight is controlled by adding trace amounts of water.

Staudinger found that diacetates of low molecular weight poly(oxymethylenes), $CH_3\overset{\overset{O}{\|}}{C}O(CH_2O)_{n<20}CH_2O\overset{\overset{O}{\|}}{C}CH_3$, were more stable than simple poly(oxymethylene glycols), $HO(CH_2O)_{n<20}CH_2OH$. However, polyformaldehydes with $n \simeq 1000$ were never esterified until the work at Du Pont occurred. At that point, the polymer's thermal stability improved substantially.

The esterification reaction is typically done with acetic anhydride, amines, and soluble alkali metal salts (like sodium acetate) as catalysts. Reaction times of 5 minutes are achieved at temperatures of 130–200°C (266–392°F) with 0.01% sodium acetate.

Beta-conidendrol or di-beta-napthyl-*p*-phenylenediamine is used as an antioxidant. Such esterified polymers degrade at a rate of 0.09%/minute at 220°C (428°F), compared to 0.8% for unesterified polymers.

Thermal stability has also been obtained through a series of copolymers developed by Celanese. The comonomer is reported to be either ethylene oxide or 1,3-dioxolane, although other cyclic ethers of the form

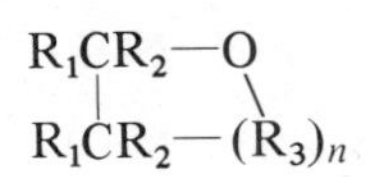

have been used. It has also been reported that the commercial copolymers have been thermally or hydrolytically treated to modify the chain terminal groups to a stable moiety, such as a hydroxyl ether.

The differences in the chemical structure of the homopolymer and copolymer account for differences in the behavior of the two polymers when they are subjected to degradative conditions. Once the end-capped chain of the homopolymer has been broken, polyformaldehyde ends are exposed and the molecular chain is free to "unzip" along its entire length. Thermal and oxidative stabilizers moderate this effect at many normal operating temperatures. Depolymerization of the copolymer discontinues at the carbon-carbon links. The copolymer end-groups are alkali-resistant; the ester end-groups of the homopolymer are not. Thus, acetal copolymer is resistant to strong bases, but the homopolymer is attacked.

The polyacetals exhibit high crystallinity. The crystalline content varies from 77% to 80%, depending on the temperature of quenching. Further annealing at 150°C (302°F) will increase the crystallinity even further. This dense crystalline structure accounts for many of the resin's properties which include stiffness, fatigue endurance, durability, high temperature resistance, good solvent resistance, low coefficient of friction and stick-slip, good appearance and creep resistance.

Polyacetals are the first plastic with strength properties approaching those of the non-ferrous metals, and in a real sense are metal-like. They are extremely rigid without being brittle and are both tough and resilient, much like spring steel. These properties are retained under adverse temperature and humidity conditions, over an extended time under stress, and during exposure to most solvents. Though denser than most other plastics, they are lighter than any of the die-casting alloys: 85% lighter than brass, 80% lighter than zinc and iron, 45% lighter than aluminum, and over 20% lighter than magnesium. Table 5-108 compares the weights of acetals and metals. Figure 5-130 compares the relative finishing costs of parts made from acetals and metals.

Although in many respects, acetal resins are similar to the nylons, they are superior in fatigue strength, creep and water resistance, and stiffness. The nylons, however, have better impact and abrasion resistance.

Table 5-109 describes the properties of polyacetal homo- and copolymers as well as a few modified compositions.

Table 5-108. Acetal Resin and Metals: Weight Comparison.*

	Specific Gravity	Weight Saving on Direct Replacement with Delrin
Delrin acetal resin	1.42	–
Magnesium die-casting alloy	1.81	22%
Aluminum die-casting alloy	2.56	45%
Zinc die-casting alloy	6.70	79%
Cast iron	7.20	80%
Steel	7.80	82%
Brass casting alloy	8.50	83%

*Courtesy: E.I. Du Pont de Nemours & Co., Inc.

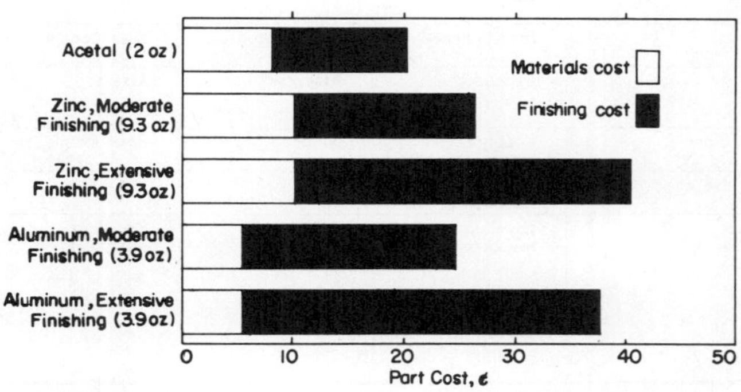

Fig. 5-130. Materials and finishing costs of acetals are quite low compared to metals having somewhat similar properties. (Hauck, 1966)

Table 5-110 compares some of these properties to those of other engineering plastics and metals. Like the nylons, the polyacetals exhibit a small but finite creep under load, and this must be given due consideration in design strategies. The creep resistance of the acetals is better than the nylons, but not as good as the polycarbonates. In terms of abrasion resistance, the acetals are superior to most plastics and die-cast aluminum, but inferior to nylon 6/6. The lowest coefficients of friction and wear are obtained with acetal against steel. With other metals, particularly aluminum, greater wear and high friction are manifest. It is not desirable to run acetal against acetal in bearing applications, because there is a tendency for high heat build-up, except with very light loads. Table 5-111 compares the weight loss of polyacetal with other materials.

The electrical insulation properties of the polyacetals are good, but not outstanding. It is the combination of toughness and temperature resistance with adequate insulation characteristics that allows for its use in electrical applications. The thermal stability of the resins are rated at 104°C (220°F) for continuous exposure in air, with short-term exposure to 132°C (270°F).

Both homo- and copolymers show outstanding resistance to organic solvents. Above 70°C (158°F), some chlorophenols have a solvating effect. Excellent resistance to aliphatic, aromatic, and chlorinated hydrocarbons is exhibited. The polyacetals have excellent resistance to hot water, although they are suggested for only intermittent use in boiling water. Resistance to inorganic reagents like strong acids, alkalis, or oxidizing agents is not good, although the copolymer is superior in hot alkali resistance. The copolymer is reported to have better thermal stability, hydrolytic stability at elevated temperatures, and easier moldability. The homopolymer has somewhat better mechanical properties and a higher melting point. Both are notch-sensitive, but have good resistance to high hydrostatic pressure. Stress-cracking has not been observed in organic solvents, but swelling will occur in solvents of equivalent solubility parameters.

Prolonged exposure to UV radiation will cause surface chalking and ultimate embrittlement. Well-dispersed carbon black reduces this tendency. The polyacetals are resistant to fungus, rodent, and insect attack, and burn slowly.

The poly(oxymethylenes) may be processed on conventional injection and blow molding and extrusion equipment. They can also be rotomolded and blown into a structural foam. Here again, standard equipment is acceptable, and for foaming, chemical blowing agents or N_2 can be used. Overheating produces formaldehyde gas, which could create the risk of damage or injury should excess gas be confined and cause an excessive pressure build-up. Although less hygroscopic than the nylons, the acetals must be stored in a dry environment. The high crystallinity which develops on cooling, results in a shrinkage of about 0.02 cm/cm.

The resins have a wide processing range. Recommended

Table 5-109. Typical Properties of Polyacetals.*

		ASTM test method	Homopolymer	Copolymer	20% glass filled homopolymer	25% glass coupled copolymer	TFE fiber-reinforced
Processing	1. Molding qualities	—	Excellent	Excellent	Good to excellent	Good to excellent	Excellent
	2. Compression molding temp., °F.	—	—	340-400	—	—	—
	3. Comp. molding pressure, p.s.i.	—	—	1000-5000	—	—	—
	4. Injection molding temp., °F.	—	380-470	360-450	350-480	380-480	370-410
	5. Injection molding pressure, p.s.i.	—	10000-20000	10000-20000	10000-20000	10000-20000	10000-20000
	6. Compression ratio	—	—	3.0-4.0	—	3.0-4.0	—
	7. Mold (linear) shrinkage, in./in.	—	0.020-0.025	0.020-(Avg.)	.009-.012	0.004 (flow) 0.018 (trans.)	0.02-0.025
	8. Specific gravity (density)	D792	1.42	1.41	1.56	1.61	1.54
	9. Specific volume, cu. in./lb.	D792	19.6	19.7	18.7	17.4	18.9
	10. Machining qualities	—	Excellent		Good to fair	Good to fair	Excellent
Mechanical	11. Tensile strength, p.s.i.	D638	10000	8800	8500-11000	18500	6900 (73°F.)
	12. Elongation, %	D638	25.0-75.0	40-75	2-7	3	12.0 (73°F.)
	13. Tensile elastic modulus, 10^5 p.s.i.	D638	5.2	4.1	10.0	12.5	—
	14. Compressive strength, p.s.i.	D695	18000 (10% defl.)	16000 (10% defl.)	18000 (10% defl.)	17000 (10% defl.)	11000 (10% defl.)
	15. Flexural yield strength, p.s.i.	D790	14100	13000	15000 (5%)	28000	—
	16. Impact strength, ft. lb./in. of notch (½ × ½ in. notched bar, Izod test)	D256	1.4 (Inj.) 2.3 (Ext.)	1.0-1.5 (1/8 × 1/2 in. bar)	0.8	1.8 (1/8 × 1/2 in. bar)	0.7
	17. Hardness, Rockwell	D785	M94,R120	M78-M80	M75-M90	M-79	M78,R118
	18. Flexural elastic modulus, p.s.i. × 10^5	D790	4.1	3.75	8.8	11.0	4.0
	19. Compressive modulus, p.s.i. × 10^5	D695	6.7	4.5	—	—	—
Thermal	20. Thermal conductivity, 10^{-4} cal./sec./sq. cm./1 (°C./cm.)	C177	5.5	5.5	—	—	—
	21. Specific heat, cal./°C./gm.	—	0.35	0.35	—	—	—
	22. Thermal expansion, 10^{-5} in./in./°C.	D696	8.1	8.5	3.6-8.1	—	7.5
	23. Resistance to heat, °F. (continuous)	—	195 (8800 hr.)	220	185-220	220	195 (8800 hr.)
	24. Deflection temp., °F. at 264 p.s.i. fiber stress	D648	255	230	315	325	212
	at 66 p.s.i. fiber stress	D648	338	316	345	331	329
Electrical	25. Volume resistivity, ohm-cm. (50% RH and 23° C.)	D257	1.0 × 10^{15}	1.0 × 10^{14}	1.2 × 10^{14}	1.2 × 10^{14}	—
	26. Dielectric strength, short-time, ⅛-in. thickness, volts/mil	D149	380	500 (90 mil)	580	580	—
	27. Dielectric strength, step-by-step, ⅛-in. thickness, volts/mil	D149	320	—	—	—	—
	28. Dielectric constant, 60 cyc.	D150	—	3.7	3.97	3.9	—
	29. Dielectric constant, 10^3 cyc.	D150	3.7	3.7	3.97	3.9	—
	30. Dielectric constant, 10^6 cyc.	D150	3.7	3.7	3.97	3.9	—
	31. Dissipation (power) factor, 60 cyc.	D150	—	—	—	—	—
	32. Dissipation (power) factor, 10^3 cyc.	D150	0.0048	0.0010 (40 mil)	0.0036	0.0025	—
	33. Dissipation (power) factor, 10^6 cyc.	D150	0.0048	0.006	0.005	0.0062	—
	34. Arc resistance, sec.	D495	129 on 15 mil spec.	240	136	136	—
Optical	35. Refractive index, n_D	D542	1.48	—	—	—	—
	36. Clarity	—	Translucent to opaque		Opaque	Opaque	Brown opaque
	37. Transmittance, %	—	—	—	—	—	—
	38. Haze, %	—	—	—	—	—	—
Resistance characteristics	39. Water absorp., 24 hr., ⅛-in. thick., %	D570	0.25	0.22	0.25-0.29	0.29	0.20
	40. Flammability[a] Burning rate, in./min.	D635	1.0-1.1		0.8-1.0	1.0	0.8
	AEB, in./ATB, sec.[b]	D635	—		—	—	—
	41. Effect of sunlight	—	Chalks slightly		Chalks slightly	Chalks slightly	Chalks slightly
	42. Effect of weak acids	D543	Resists some		Resists some	Resists some	Resists some
	43. Effect of strong acids	D543	Attacked		Attacked	Attacked	Attacked
	44. Effect of weak alkalies	D543	Resists some	None	Copolymer resists some	None	Resists some
	45. Effect of strong alkalies	D543	Not recom.	No..e	Copolymer resists some	None	Not recom.
	46. Effect of organic solvents	D543	Excellent resistance		Excellent resistance	Excellent resistance	Excellent resistance

[a]The numerical flame spread rating is not intended to reflect hazards presented by this or any other material under actual fire conditions.
[b]AEB = Avg. extent of burning in inches; ATB = Avg. time of burning in seconds.
*Reprinted by permission from *Modern Plastics Encyclopedia*, McGraw-Hill, Inc.

Table 5-110. How Properties of Acetals Compare with Competitive Materials.*

Materials	Cost, ¢/cu in.	Water Absorption[a], %	Coefficient of Thermal Expansion, per °F × 10^{-5}	Heat Deflection Temp[d], F	Deformation Under Load, %
ABS (heat resistant grades)	1.76	0.43	4.5	203	0.36[b]
Acetal	3.34	0.22–0.45	4.5–4.7	220–255	0.5[b]
20% glass fiber reinforced	4.20	0.25	1.9–2.2	315	0.5[b]
Aluminum	2.45	0	1.2	–	–
Brass	13.43	0	1.0	–	–
Nylon (type 6/6)	3.60	1.5	1.5	160	1.4[b]
20% glass fiber reinforced	6.08–6.95	0.6	0.6	470–490	0.4[c]
Polycarbonate	3.89	0.15	0.15	270	0.27[c]
20% glass fiber reinforced	7.56	0.11	0.11	295	0.22[c]
Polyphenylene oxide	4.39	0.06	0.06	375	0.1[b]
Modified	2.87	0.07	0.07	265	0.3[b]
Polysulfone	4.48	0.22	0.22	345	0.1[b]
30% glass fiber reinforced	7.40	0.20	0.20	360	0.08[c]
Zinc	4.22	0	1.5	–	–

[a] 24 hr. [b] 2000 psi, 122 F, 24 hr. [c] 4000 psi. [d] At 264 psi. *Hauck, 1966.

Table 5-111. Comparative Weight Loss of Various Materials in Different Abrasion Tests.*

	Taber	Ball Mill	Wire Drag	Sander
Zytel® nylon resin	1	1	1	1
DELRIN	2–5	4–6	5–6	3–4
Polystyrene (several types)	9–26	15–20	35	–
Terpolymer of styrene, butadiene, & acrylonitrile	9	10–20	–	–
Cellulose acetate	9–10	–	–	–
Cellulose acetate butyrate	9–15	10–20	15	–
Methyl methacrylate	2–5	10–20	20	–
Polyvinylidene chloride	9–12	–	–	–
Melamine formaldehyde (molded)	–	15–20	–	–
Phenol formaldehyde (moldings)	4–12	–	–	–
Hard rubber	–	10	–	4
Diecast aluminum	–	11	–	4–5
Mild steel	–	15–20	–	–

Test descriptions are as follows:

Taber abrasion tests made with a CS-17 wheel and a 1000-gram load; test pieces conditioned at 23°C and 50% R.H.

Ball mill abrasion tests: 2″ × ½″ × ⅛″ bars rolled in a 5″ Abbe ball mill with 25 Borundum balls and 500 ml. water.

Wire drag abrasion tests: continuous loop of fine resistance wire (spirally wrapped on a cord) pulled over a cylindrical test piece. Cord held at constant tension and pulled at about 70 feet per minute. Depth of the groove was measured after 30 minutes.

Sander abrasion test: football cleats were pushed with constant force against the belt of a wet sander.

Each number represents the weight loss in the same time based on the weight loss of "Zytel" as 1.

– Indicates no test made.

*Courtesy: E.I. Du Pont de Nemours & Co., Inc.

processing temperatures range from 165–199°C (330–390°F) for the copolymer, and about 171–182°C (340–360°F) for the homopolymer. Injection pressures are generally in the 15,000–20,000 psi range, and regrind up to 25% is acceptable.

Clear film which is stable at room temperature cannot be made by quenching. Cold-rolling can improve clarity somewhat.

Most polyacetal resins are usually supplied with additives already incorporated. These include antioxidants (phenol alkane type) and acid acceptors, to prevent attack on the acetal linkage (e.g., epoxides, nitrogen compounds, and basic salts). For further reduction in coefficient of friction, molybdenum disulfide may be added, although this usually causes problems because of inherent acidic impurities. Table 5-112 compares the characteristics of internally lubricated and fortified thermoplastics. Blends with up to 25% PTFE have coefficients of friction as low as 0.02 and can be used in more severe environments than unfilled compositions. Table 5-113 compares the properties of PTFE-filled homopolymer (Delrin AF-313) and standard homopolymer (Delrin 500).

Formulations containing up to 30% glass fiber are also available. These systems provide improved creep resistance, enhanced stiffness, and low coefficient of expansion. The main difficulties with these compounds arise from their inferior impact strength and the anisotropic nature of moldings.

The major market areas are automotive, packaging (particularly aerosol containers), hardware, appliance, industrial machinery, plumbing, oil-field pipe, and consumer. Over 80% of this business involves the replacement of metals (primarily zinc, steel, aluminum, and brass). Typical end products include conveyer belt links, molded sprockets and chains, blower wheels, cams, fan blades, check valves, pump impellers, carburetor bodies, and plumbing components

Table 5-112. Friction and Wear Characteristics of Internally Lubricated and Fortified Thermoplastics.*

Polymer	Lubricant[k]	Vol, %	Fortifier	Wt, %	Wear Fact.[a]	Wt Loss, gm[b]	Temp,[a] F	Limiting PV Short Term Svc[c] 10 fpm	100	1000	Limiting FV Sust Svc[d]	Coeff of Frict Stat[e]	Dyn[f]
Acetal copolymer	–	–	–	–	65	0.0000	100	4,000	3,500	<2,500[g]	3,000	0.14	0.21
	Mod TFE	5	–	–	40	0.0000	90	>40,000[b]	10,000	3,000	5,000	0.12	0.18
	Mod TFE	10	–	–	30	0.0000	90	>40,000	12,500	3,000	7,000	0.10	0.17
	Mod TFE	15	–	–	20	0.0000	85	>40,000	12,500	5,500	7,500	0.07	0.15
	Mod TFE	20	–	–	17	0.0000	85	>40,000	12,500	5,500	7,500	0.07	0.15
	MoS_2	10	–	–	250	0.0000	82	7,500	7,000	5,500	2,000	0.16	0.22
	Graphite	10	–	–	70	0.0000	80	10,000	5,000	3,000	2,000	0.16	0.22
	Mod TFE	15	Gl fiber	10	125	0.0073	85	>40,000	8,000	8,000	7,000	0.17	0.23
	Mod TFE	15	Gl fiber	30	200	0.0081	90	12,500	12,000	8,000	5,000	0.20	0.28
Acetal homopolymer	–	–	–	–	60	0.0000	100	4,000	3,500	<2,500	3,000	0.11	0.22
	Mod TFE	15	–	–	22	0.0000	85	>40,000	12,500	10,000	8,000	0.09	0.14
	TFE fibers	15	–	–	20	0.0004	85	>40,000	12,500	10,000	8,000	0.10	0.16
Nylon 6/6	–	–	–	–	200	0.0086	95	3,000	2,500	>2,500	2,000	0.24	0.26
	Mod TFE	15	–	–	13	0.0001	80	>40,000	27,500	8,000	7,000	0.10	0.18
	Mod TFE	15	Gl fiber	30	16	0.0014	82	17,500	20,000	13,000	10,000	0.19	0.26
	TFE fibers	15	–	–	15	0.0028	80	17,500	17,500	5,500	7,000	0.11	0.17
	–	–	MoS_2	2.5	200	0.0021	90	10,000	8,000	4,000	2,000	0.28	0.30
	Mod TFE	15	MoS_2	2.5	15	0.0016	80	15,000	17,500	5,000	7,000	0.11	0.21
Nylon 6	–	–	–	–	200	0.0081	95	3,000	2,500	>2,500	2,000	0.22	0.26
	Mod TFE	15	–	–	15	0.0001	82	>40,000	27,500	8,000	7,000	0.09	0.19
	Mod TFE	15	Gl fiber	30	17	0.0014	84	17,500	20,000	13,000	10,000	0.20	0.25
	Mod TFE	15	MoS_2	2.5	15	0.0016	85	15,000	17,500	5,000	7,000	0.13	0.22
Polycarbonate	–	–	–	–	2500[d]	0.0000	125	750	500	NR[i]	250	0.31	0.38
	Mod TFE	15	–	–	75	0.0002	95	>40,000	20,000	10,500	7,000	0.09	0.15
	Mod TFE	15	Gl fiber	30	30	0.0048	90	27,500	30,000	13,000	5,000	0.18	0.20
	Mod TFE	15	Gl fiber	40	45	0.0254	93	20,000	22,500	10,000	3,000	0.20	0.23
Chlorinated polyether	–	–	–	–	600	0.0000	100	3,000	2,000	NR	1,000	0.30	0.33
	Mod TFE	15	–	–	20	0.0000	80	20,000	17,500	3,000	10,000	0.08	0.12
	Mod TFE	15	Gl fiber	30	50	0.0062	82	15,000	12,000	10,000	5,000	0.10	0.14
Polyurethane	–	–	–	–	340	0.0013	112	2,000	1,500	NR	600	0.32	0.37
	Mod TFE	15	–	–	105	0.0011	82	<5,000	5,000	5,500	4,000	0.12	0.18
	Mod TFE	15	Gl fiber	30	35	0.0019	85	7,500	10,000	5,500	3,500	0.20	0.25
Styrene acrylonitrile	–	–	–	–	3000[j]	0.0000	110	500	300	NR	250	0.28	0.33
	Mod TFE	15	–	–	200	0.0000	80	>20,000	5,000	3,000	3,000	0.11	0.14
	Mod TFE	15	Gl fiber	30	65	0.0026	80	17,500	10,000	10,000	4,000	0.13	0.18
Oil impreg. bronze	–	–	–	–	100	0.0016	83	10,000	25,000	45,000	–	0.24	0.20
Phenolic wood flour	Mod TFE	15	–	–	15–35	0.0020	85	>40,000	45,000	58,000	17,500	0.16	0.26
Tetrafluoroethylene	–	–	Gl fiber	25	8	0.0018	81	10,000	13,000	16,000	11,500	0.08	0.16

[a]50 fpm, 40 psi, 2000 PV. [b]Metal wear ring, 54 hr. [c]Test method A. [d]Test method B. [e]40 psi. [f]50 fpm, 40 psi. [g]Low load limit of test apparatus at 1000 fpm. [h]Upper load limit of test apparatus at 10 fpm. [i]NR = Not Recommended. [j]Tested above PV limit to compare with internally lubricated compounds. [k]Mod = modified.

SOURCE: Liquid Nitrogen Processing Corp.

*Hauck, 1966.

Table 5-113. Properties Comparison: PTFE-Filled Polyacetal Homopolymer (Delrin AF-313) and Standard Homopolymer (Delrin 500).*

Property		Units	ASTM No.	Average Values[1] Delrin AF-313	Delrin 500 (for comparison)
Tensile strength and yield point[2]	−68°F	psi	D-638	10,700	14,700
	73°F			6,900	10,000
	158°F			4,700	6,900
Elongation[2]	−68°F	%	D-638	10	13
	73°F			15	40
	158°F			40	220
Flexural modulus	−68°F	psi	D-790	540,000	570,000
	73°F			350,000	410,000
	158°F			200,000	230,000
Compressive stress	1% deformation	psi	D-695	4,500	5,200
	10% deformation			13,000	18,000
Impact strength, Izod	−40°F	ft. lb./in.	D-256	0.6	1.2
	73°F			0.7	1.4
Shear strength	73°F	psi	D-732	8,000	9,500
Coefficient of linear thermal expansion	−40 to +85°F	per °F	D-696	5.8×10^{-5}	5.8×10^{-5}
	+85 to 140°F			6.8×10^{-5}	6.8×10^{-5}
Deflection temperature	264 psi	°F	D-648	244	277
	66 psi			334	342
Combustibility		in./min.	D-635	94HB	94HB
Water absorption 24 hours immersion		%	D-570	0.20	0.25
Rockwell hardness			D-785	M78, R118	M94, R120
Specific gravity			D-792	1.54	1.42
Taber abrasion (1,000 gm load, CS-17 wheel)		mg./1,000 cycles		9	20
Coefficient of friction (no lubricant) static-dynamic				0.05–0.15	0.20–0.35

[1] These values are representative of those obtained under standard ASTM conditions, and should not be used to design parts which function under different conditions. Since they are average values, they should not be used as minimums for material specifications.
[2] Determined at 0.2 in./min.
*Courtesy: E.I. Du Pont de Nemours & Co., Inc.

such as valve stems, shower heads, and complete toilet water-level assemblies. In one application in a business machine assembly, the number of total parts required was reduced from 1500 to 485, because complex molded acetal parts could be adapted to such faster assembly techniques as snap and interference fits, spin welding, and cold-heading.

Trade Names

Aldocryl	Acetal resin	Shell Chemical Co.
Alkon	Acetal copolymer	Celanese Corp. of America
Celcon	Copolymer	Celanese Plastics Co.
Delrin	Homopolymer	E.I. Du Pont de Nemours & Co., Inc.
Drelux	Acetal extrusions	Westlake Plastic Co.
Duracon	Acetal copolymer	Polyplastics Co., Inc.
Fiberlene	Glass-filled acetal	Service Color Corporation
Formaldafil	Reinforced acetal	Dart Industries, Inc.
Formaldaglas	Fiberglass-reinforced acetal	Dart Industries, Inc.
Formaldasar	Fiberglass-reinforced acetal	Dart Industries, Inc.
Hostaform	Acetal copolymer	Hoechst, A.G.
Kematal	Acetal copolymer fittings	Amcel, Ltd.
Latan	Acetal-reinforced molding compounds	Lati, S.p.A.

Plaslube	Acetal	Dart Industries, Inc.
Polypinco	Acetal shapes, extrusions, and parts	Polymer Corporation
Thermocomps	Filled or reinforced acetal	Liquid Nitrogen Processing Corporation
Ultraform	Acetal resins	BASF Wyandotte Corporation

REFERENCES AND BIBLIOGRAPHY

Baldwin, N. C. Acetal copolymer, from Plastics in Construction: Coming on Strong in 1978. *Plastics Engineering*, July 1978, 34(7):25–26.

Brydson, J. A. *Plastics Materials*. 3rd Ed. Robert E. Krieger Publishing Co., Huntington, New York, 1975.

Celanese Plastics Co. *New Applications for Acetal Copolymer in Plumbing and Automotive Applications*. Bulletin NCB-76-39, February 1976.

——. *Celcon Acetal Copolymer–Properties*. Bulletin C1A, July 1974.

——. *Celcon Acetal Copolymer–Glass-Coupled Celcon*. Bulletin C1B, July 1970.

——. *Celcon Acetal Copolymer–Injection Molding, Celcon Acetal Copolymer*. Bulletin C3A, June 1963.

E.I. Du Pont de Nemours & Co., Inc. *How You Can Lower Cost, Reduce Weight and Conserve Energy*. Bulletin E-18056, May 1978.

——. *This is Delrin AF Fiber Resin*. Bulletin E-09145, January 1977.

——. *Molding Manual Delrin Acetal Resins*. Bulletin E-09143, May 1977.

——. *Delrin Acetal Resins*. Bulletin E-20149, November 1975.

——. *Delrin AF Fiber/Resin*. Bulletin A-64528, February 1969.

——. *This is Delrin Acetal Resin*. Bulletin A-63463, November 1968.

——. *Delrin Acetal Resin*. Bulletin A-4476, June 1957.

——. *Delrin 8010, 8020 Acetal Resins*. Bulletin A-78037.

Estell, R. E. Acetal–PTFE resin blend improves properties and processing. *Plastics Design & Processing*, October 1971, **11**(10): 35–37.

Ferguson, C. R., and Kjellmark, E. W. Jr. "Acetal Homopolymers." In *Modern Plastics Encyclopedia*. S. Gross (exec. ed.), **45**(14A): 124–126. McGraw-Hill Book Co., 1968–69.

Hauck, J. E. Acetal plastics: where do they stand? *Materials in Design Engineering*, December 1966, **64**(7):86–90.

Houston, A. M. Good stability in many chemicals keeps acetals moving into new uses. *Materials Engineering*, January 1975, 81(1):36–38.

Johannes, R. K. Acetal homopolymer, from Plastics in Construction: Coming on Strong in 1978. *Plastics Engineering*, July 1978, 34(7):24–25.

Murphy, R. W. Polyacetal replaces metal in a computer print-out binder. *Plastics Engineering*, May 1977, 33(5):50–51.

Persak, K. J. Designing acetal parts. *Plastics Design & Processing*, June 1969, 9(6).

Plastics Design Forum. Acetal. January/February 1978, 3(1):85–95.

——. Putting spring into your product. September/October 1977, 2(5):85–90.

Plastics Engineering. Engineering resins primer. December 1974, 30(12):17–32.

Schar, W. C. "Acetal copolymers." In, *Modern Plastics Encyclopedia*. S. Gross (exec. ed.), **45**(14A):126–127. McGraw-Hill Book Co., 1968–69.

Theberge, J., and Cloud, P. Choosing the best glass-reinforced thermoplastic. *Plastics World*, March 1977, **35**(3):44–46.

Titus, J. B. *Trade Designations of Plastics and Related Materials* (revised). PLASTEC NOTE N9C, Plastics Technical Evaluation Center, Picatinny Arsenal, Dover, New Jersey, May 1978.

Thermoplastic Polyesters

Linear Polyesters. As the name suggests, no provision is made for converting these polymers to a three-dimensional form. Since there are only two functional groups on the precursor acid and alcohol, only linear polymers are formed. The possible variations are in the choice of alcohol, acid, and their ratios, which, along with vigor of reaction, determine the molecular weight of the polymers. In this type of polymer molecular weight is significant because it influences final properties. In the polyesters which crosslink, the original molecular weight is of small consequence, since the final crosslinking reaction carries the molecules to an insoluble product where molecular weight is of little concern.

The number of possible polymers is great, since all that is required is a new pair of acids and alcohols. If this is lacking, then a new combination using mixed acid or mixed alcohols can be tried. Since gelling is not encountered, the reaction is carried to as high a molecular weight as is consistant with mixing and with continued improvement in physical properties.

Currently there are four members of this class that have attained significant commercial success. They are:

1. polyethylene terephthalate–PET
2. poly(1,4-cyclohexylene dimethylene terephthalate)–PCDT
3. polybutylene terephthalate–PBT, also known as polytetramethylene terephthalate–PTMT
4. a copolyester of terephthalic and isophthalic acids and cyclohexane dimethanol.

Polyethylene terephthalate (PET). The oldest member of this group is the reaction product of terphthalic acid and ethylene glycol.

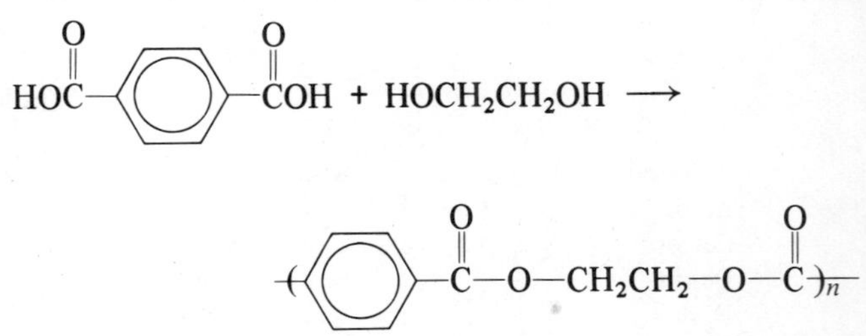

This para-substituted acid is made by oxidizing *p*-xylene. The dimethyl ester is made as a method of purifying the acid and because it reacts more readily and is more soluble in ethylene glycol. An ester interchange is first carried out in an inert atmosphere and using an alkaline-type catalyst.

After the methanol is removed, the diglycol ester is further heated under vacuum to advance the reaction by the removal of glycol. The desired degree of polymerization is obtained by viscosity control. Side reactions may occur, chief among them being the formation of some diethylene glycol which enters the polymer chain and affects polymer properties if formed in a large enough amount.

This polymer demonstrates to a high degree the property variations that come from mechanical working of resins to produce changes in orientation and crystallinity. The resin melts at approximately 255°C (491°F). If it is quickly cooled or quenched, it is almost completely amorphous and has a density of approximately 1.33. Crystallization starts at approximately 80°C (176°F) because of the slow mobilization of the molecules. At 130°C (266°F), density rises to 1.37. With proper orientation and heat setting, a density of 1.40 is obtained. These changes in state are quite significant for a tremendous change in physical properties accompanies these density modifications.

This is best demonstrated by the handling of the polymer to produce film. The previously prepared polyester is melted, filtered, extruded through a flat die, and quickly cooled to form a film. The quick melting and resulting low viscosity make handling different from many of the other thermoplastics. The clear, glossy film is of fairly good quality, but its amorphous character gives it an instability, since it will start to crystallize and become brittle and opaque above 80°C (176°F). If, however, the film is heated and stretched in two directions at right angles to each other in a special drawing machine, the molecules become oriented in the plane of the film without becoming completely oriented in either of the directions of draw. The temperature should be high enough to allow the desired degree of draw, but not so high that a high degree of molecular mobility is obtained. Should this occur, the film viscosity is too low to drag the molecules into alignment. Orientation has placed the molecules in a configuration favoring crystallization. To promote crystallization, the film is held so it cannot contract and is then heated above the draw temperature. The molecules literally "lock" into place, accounting for the final increase in film density. Subsequently, the film is stable to high temperatures and has well-balanced physical properties. Typical properties of a general-purpose grade of film are shown in Table 5-114. It is well to remember that there are many available grades, and properties may vary.

The toughness with rigidity of this polyester sheeting opens up a wide field of uses. Photographic film-base and recording tape are two big outlets (see Figure 5-131). The film is capable of long machine runs through projectors or recording systems without breaks or appreciable wear. Its high modulus of rigidity allows the use of thin gauges without creating limpness in the film. Recording tapes as thin as 0.0015 inch give long service. Figure 5-132 shows the predicted use life of 0.5-mil PET film (Mylar) versus temperature and humidity.

Fig. 5-131. Polyester film-based tapes used in audio applications. (*Courtesy: Imperial Chemical Industries, Ltd.*)

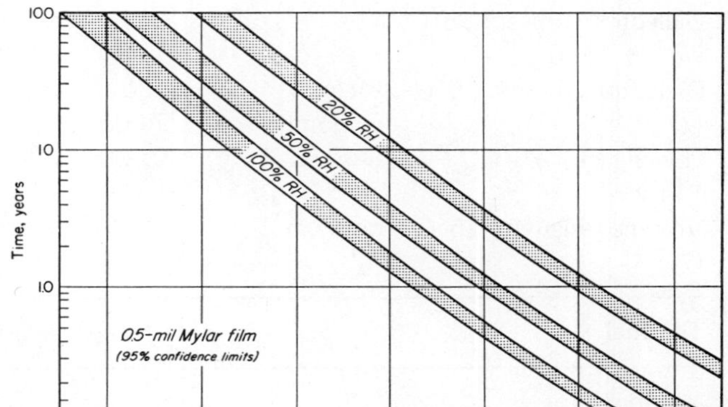

Fig. 5-132. Predicted use life of PET film. (McMahon et al., 1959)

An application which demonstrates toughness and high temperatures is the use of this polyester film as a carrier for pigment or metal leaf which is to be hot-stamped into surfaces. Under the heat and pressure of the embossing die, the supporting film conforms without tearing. Since the resin has no appreciable moisture pick-up, it does not appreciably affect the pressure in vacuum metallizers, allowing inexpensive metal effects to be produced. This resistance to moisture recommends the film for use in packaging foods and as an insulator in transformers (see Figures 5-133 and 5-134). Thinner than usual film can be used, which helps to miniaturize equipment. Toughness in these thin gauges also allows the polyester's use as a cable wrapping.

Many of the same considerations mentioned in film making also pertain to fiber forming, except here only one dimensional drawing need be used. The molten resin is extruded in fiber form at 280–295°C (536–563°F). When the fibers have been cooled sufficiently to increase viscosity, they are drawn and heat-set to give a strong, continuous fiber. If a staple fiber is desired, a high-speed cutter reduces the filament to proper lengths. A crimped-fiber effect is

Table 5-114. Properties of Polyethylene Terephthalate Films.*,**

Physical		
Density	g/cm^3	1.395-1.405
Tensile strength (M.D. and T.D.)(1)	kg/cm^2	over 1,750
	lb/in^2	over 25,000
Elongation at break (M.D. and T.D.)(1)	%	60-110
Water absorption after prolonged immersion	% by weight	0.6
Swelling due to water absorption (linear)	%	0.06
Fold endurance(2)	cycles	230,000
Electrical		
Dielectric constant 25°C (77°F) 1 kHz		2.95
Dielectric strength 25°C (77°F)(3)	kV/mm	280
	volts/mil	7,000
Power factor 25°C(77°F) 1 kHz		0.006
Volume resistivity 25°C (77°F)	ohm cm	$\sim 10^{19}$
Thermal		
Melting point	°C	~265
	°F	~510
Heat-sealing temperature (to give a surface blocking seal)	°C	~230
	°F	~445
Low temperature flexibility: −70°C(−94°F)		Excellent
−200°C(−328°F)		Some flexibility
Coefficient of expansion	per °C	27×10^{-6}
Inflammability		Burns with difficulty; passes B.S. 850
Coefficient of thermal conductivity	cal cm/cm² sec °C	4×10^{-4}
Shrinkage (linear)(4)	%	2.5-5.0

Chemical resistance to:	
dilute acids and alkalis	good
concentrated alkalis	poor
concentrated hydrochloric acid	fair
concentrated sulphuric acid	poor
greases, oils and fats	good
organic solvents, alcohols and hydrocarbons	good
ketones, esters and chlorinated compounds	fairly good
phenols, cresols and chlorinated phenols	poor
mould growth	good

M.D. = machine direction
T.D. = transverse direction

(1) ASTM D882-64T; 50% per minute strain rate;
(2) M.I.T. test; 70 kg/cm² (1,000 lb/in²) at 20°C(68°F).
(3) ASTM D149-64 or B.S. 698:1965; using 6.3 mm(¼ inch) electrodes, 50 Hz, ½ kV/sec, rise rate in air.
(4) 1 minute at 200°C(390°F).

*The values given are typical for a general-purpose-grade film, of nominal 23 microns (100 gauge) thickness and are representative of many other films.
**Courtesy: ICI Americas, Inc.

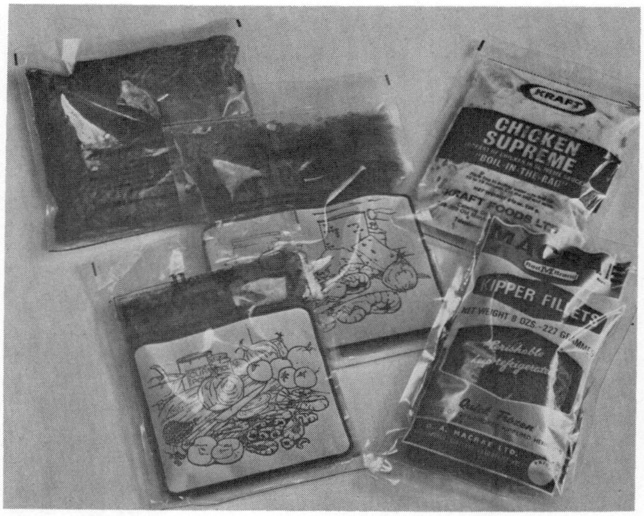

Fig. 5-133. PET withstands prolonged boiling in water and, when laminated to medium- or high-density polyethylene, provides a pack that is highly suitable for "boil-in-the-bag" applications. (*Courtesy: Imperial Chemical Industries, Ltd.*)

obtained by allowing the film to relax at a slightly elevated temperature. Fiber dyeing generally requires heat and pressure methods because of the chemical resistance of the polymer. Fabrics from this polyester dry quickly, since essentially the only water held is that retained because of the construction of the cloth.

This polyester fiber has high resilience and excellent wrinkle resistance, and, relative to most other textile fibers, properties are retained better when wet. This makes it possible to retain creases and pleats when fabrics are washed and then allowed to dry. These properties make this fiber ideally suited for "wash-and-wear" fabrics which do not need ironing after washing. However, the fabrics made from 100% polyethylene terphthalate tend to become charged with static electricity at low humidities and are uncomfortable because of lack of moisture diffusion on warm days. For these reasons, blends with hydrophilic fibers such as wool, cotton, or rayon have proved to be much more comfortable. Blends of 65% polyester fiber with cotton and wool seem to be best for most apparel applications. The percentage of polyester retains most of the desirable crease resistance and resilience, and the hydrophilic fibers greatly reduce the static electricity and moisture problems. Table 5-115 compares the properties of various plastic fibers and films.

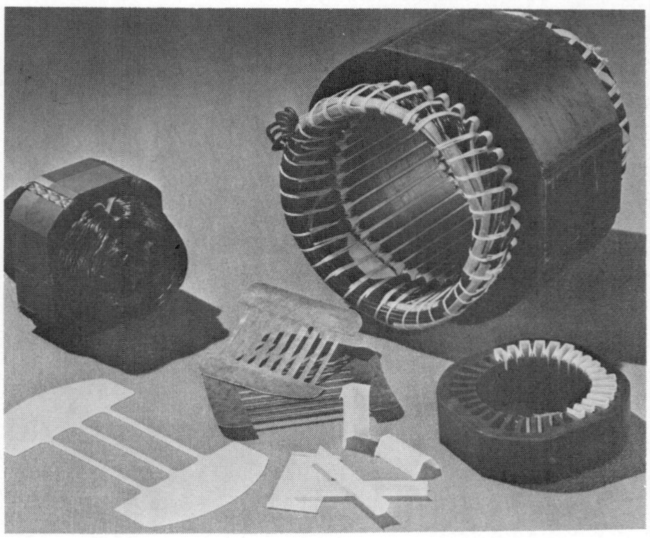

Fig. 5-134. Some applications for PET in electric motors and generators include: slot liners and closures and interphase insulation. (*Courtesy: Imperial Chemical Industries, Ltd., Plastics Division*)

In recent years the drive for the development of plastics beverage containers has promoted a strong interest in PET. PET displays excellent clarity and gloss, as well as substantial weight savings over glass. The polyester bottle has greater orientation in the hoop direction than the axial direction, such that the burst and impact strength makes the bottle shatterproof. Table 5-116 compares the drop height for various commercial bottle materials. Table 5-117 compares film impact strength of various low-cost plastics relative to the PET being used to make beverage bottles.

PET is highly impermeable to CO_2, oxygen, and water, as is seen in Table 5-118.

Acetaldehyde, a common taste component, is generated during the manufacture of the PET as well as in the production of the bottle if precautions are not taken. The aldehyde can adversely affect the taste of a soft drink. By using optimum injection-molding conditions (primarily keeping the polymer exposed to heat for a minimum amount of time), low concentrations of aldehydes can be obtained, thus protecting the taste of a beverage.

Poly(1,4-cyclohexylene dimethylene terephthalate) (PCDT). The most obvious structural difference between PCDT and

Table 5-115. Properties of Films and Fibers.*

Material	ASTM D882 Tensile Strength, psi	Elongation @ Break %
Polyester		
Oriented	25–36,000	60–100
Unoriented	8,000+	250
Polypropylene		
Oriented	25–30,000	70–100
Unoriented	3–6,000	
Vinyl	2–19,000	500
HDPE	3–10,000	10–500
LDPE	1–3,500	225–600
Polycarbonate	8–9,000	110
Glass (fiber)	700,000	5
Polyester fiber	110–170,000	12–20

*Ray et al., 1977. Reprinted by permission from *Plastics Design & Processing*. Copyright 1977, Lake Publishing Corporation, Libertyville, Ill.

Table 5-116. Drop Heights for Carbonated Beverage Bottles.*

Material	Bottle Size	Drop Height
Polyester	64-oz	90% survive 10-ft virtually all survive 6-ft
Acrylonitrile	10-oz	6½-ft
	32-oz	4-ft
Glass, plastic sleeved	10-oz	4-ft
	32-oz	2-ft
Glass, standard	assorted	less than 1½-ft

SOURCE: Polyester: Tests at Goodyear Polyester R&D. Others. "Technology Assessment in Product Development," F. D. Wharton and J. Kenneth Carver, *Chem Tech*, Sept. 1975.
*Ray et al., 1977. Reprinted by permission from *Plastics Design & Processing.* Copyright 1977, Lake Publishing Corporation, Libertyville, Ill.

Table 5-117. Film Impact Test: Pendulum Test.*

Material	kg–cm
Polyester (oriented)	25-30
Polypropylene	
(unoriented)	1-3
(oriented)	5-15
Vinyl (plasticized)	12-20
HDPE	1-3
LDPE	7-11

*Modern Packaging Encyclopedia and Packaging Guide, December, 1974.
*Ray et al., 1977. Reprinted by permission from *Plastics Design & Processing.* Copyright 1977, Lake Publishing Corporation, Libertyville, Ill.

Table 5-118. Barrier Properties of Plastics.*, **

Material	CO^2 (a)	O^2 (a)	Water Vapor Transmission (b)
Polyester	12–20	5–10	2–4
Nitrile	3	1	5
PVC	20–40	8–15	2–3
HDPE	300	110	0.5
Polypropylene	150	450	0.5
Polycarbonate		265	75

(a) = cm^3-mil/100 in.^{2}-day-1 atm at 73°F, 100% RH.
(b) = g-mil/100 in.^{2}-day at 100°F, 100% RH.

*Encyclopedia of Polymer Science and Technolgy-Supplement.
**Ray et al., 1977. Reprinted by permission from *Plastics Design & Processing.* Copyright, 1977, Lake Publishing Corporation, Libertyville, Ill.

PET is the presence of a cyclohexane ring between the two methylene groups in the backbone repeat unit.

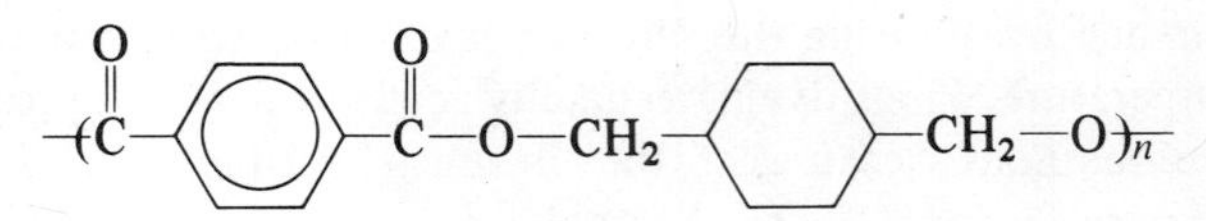

The differences in properties that occur as a result of this structural change are manifested primarily in the toughness, hydrolytic stability, density, and electrical resistance. Table 5-119 shows some selected comparative properties of PCDT and PET.

Polymers with a molecular weight of about 25,000 are processed into film in a similar manner to that described above for PET. The resultant film has a lower density than PET but is not as stiff, strong, or tough. PCDT has a higher operating temperature and is almost colorless. The film is

Table 5-119. Physical Properties of 0.5-mil Polyester Films.*

Property	PCDT	PET	Test Method
Density, g./cc.	1.226	1.390	ASTM D1505-57T (6)
Tensile			ASTM D882-56T (6)
Yield strength, psi.	10,000	14,000	
Break strength, psi.	17,000	26,000	
Elong. at break, %	45	105	
Modulus, 10^5 psi.	4.0	6.8	
Elmendorf tear strength, g./mil	6	18	ASTM D689-44 (6)
M.I.T. folding endurance, cycles	>60,000	>60,000	ASTM D643-43 Method B (6)
Mullen burst strength, psi.	28	37	ASTM D774-46 (6)
2% Heat-distortion temp. under 50 psi. initial stress, °C.	165	158	
Water absorption (24-hr. immersion, room temp.), %	0.30	0.55	ASTM D570-57T (6)
Soluble matter lost (24-hr. immersion, room temp.)	None	None	ASTM D570-57T (6)
Color	Water-white	Slightly yellow	—

*Watson, 1961.

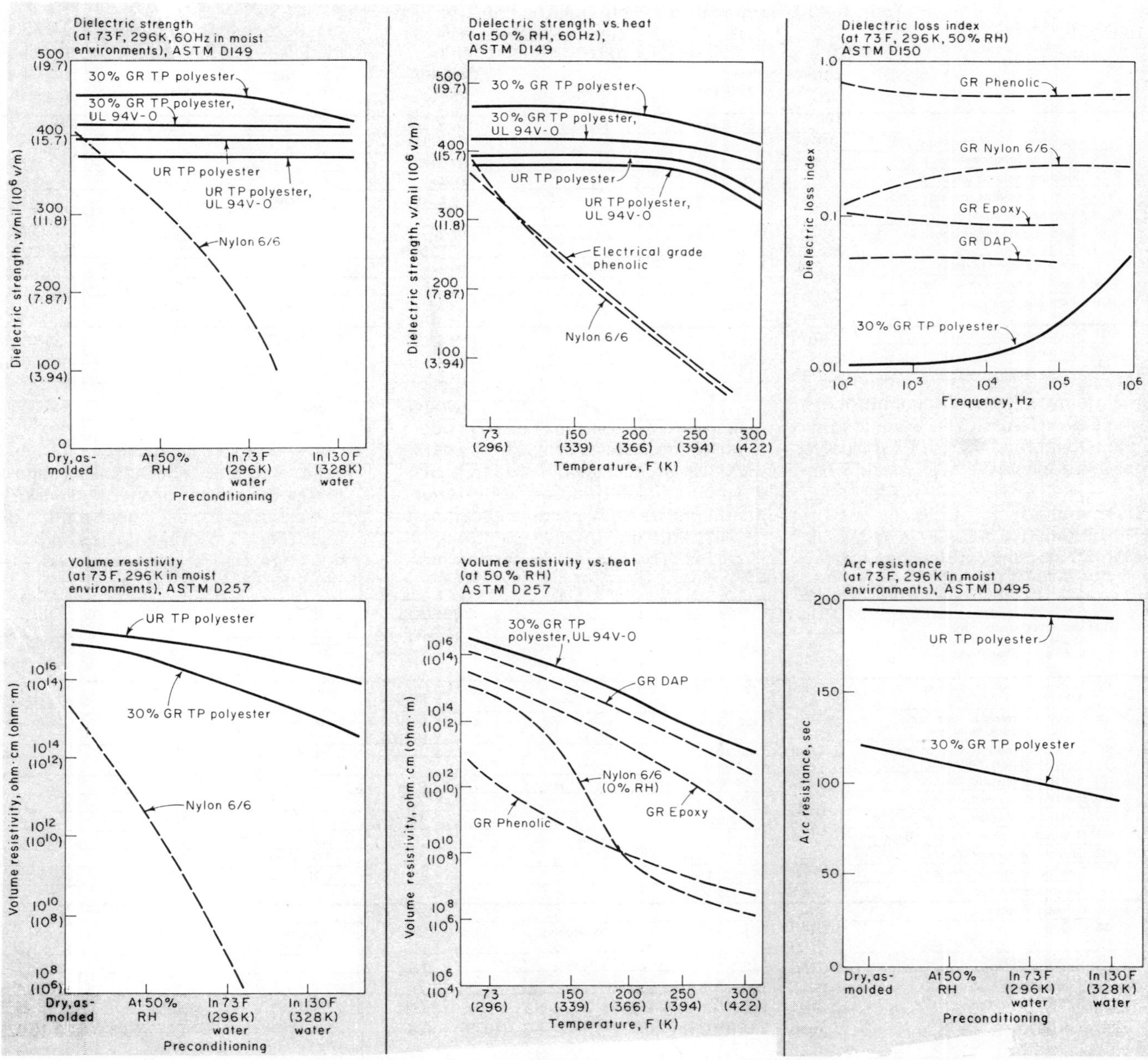

Fig. 5-135. Retention of dielectric characteristics in moist environments, after exposure to water and humidity, and at elevated temperatures is a major advantage of thermoplastic polyesters. (Vaccari, 1974)

unaffected by aromatic and aliphatic hydrocarbons, turpentine, gasoline, organic esters, mineral acids and alkalis. Although not excellent, its resistance to strong alkalis is superior to PET film. The stability of the dielectric constant and dissipation factor of PCDT is better over a broader temperature range than PET. End-use applications for PCDT are essentially the same as for PET.

Polybutylene terephthalate (PBT), polytetramethylene terephthalate (PTMT). The same basic chemistry applies to the manufacture of PBT as was described for PET. The glycol used is 1,4-butanediol, and the resulting polymer has the following structure:

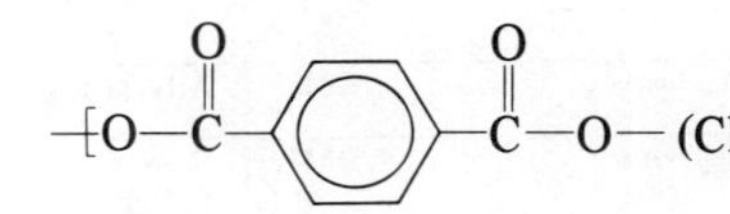

Whereas PET is used primarily for fiber, filament, and film production, PBT is designed for molding and competes with the other engineering thermoplastics. An examination of properties shows no obvious outstanding property (see Table 5-120). It appears that the essential reason for its popular acceptance is that it provides overall properties

Table 5-120. Properties of Thermoplastic Polyester (PBT-PTMT, PET).*

		ASTM test method	Non-reinforced	PCDT	18% glass filled	30% glass filled	36% glass filled	PET
Processing	1. Molding qualities	—	Excellent	Excellent	Good to excel.	Good to excel.	Good to excel.	—
	2. Compression molding temp., °F.	—	—	—	—	—	—	—
	3. Comp. molding pressure, p.s.i.	—	—	—	—	—	—	—
	4. Injection molding temp., °F.	—	437-540	440	500-540	437-530	500-550	550-660
	5. Injection molding pressure, p.s.i.	—	8000-23000	—	8000-23000	8000-23000	17000	10000-20000
	6. Compression ratio	—	—	—	—	—	—	3:1
	7. Mold (linear) shrinkage, in./in.	—	0.015-0.020	0.002	0.002-0.006	0.002-0.008	0.001-0.003	0.020-0.025
	8. Specific gravity (density)	D792	1.31-1.38	1.2	1.48-1.67	1.52	1.73	1.34-1.39
	9. Specific volume, cu. in./lb.	D792	20.2-21.1	—	18.2-18.7	18.2	17	—
	10. Machining qualities	—	Excellent	Excellent	Excellent	Excellent	Excellent	—
Mechanical	11. Tensile strength, p.s.i.	D638	8200	7290	16000-19600	17000-19000	25000	8500-10500
	12. Elongation, %	D638	50-300	210	2.0-3.0	2-4	2	3-300
	13. Tensile elastic modulus, 10^5 p.s.i.	D638	2.8	—	8	13	16	4.0-6.0
	14. Compressive strength, p.s.i.	D695	8600-14500	—	16200-18000	18000-23500	30000	11000-15000
	15. Flexural yield strength, p.s.i.	D790	12000-16700	10600	26000-26500	26000	38000	14000-18000
	16. Impact strength, ft. lb./in. of notch (½ × ½ in. notched bar, Izod test)	D256	0.8-1.0	1.0 at 73°F. 0.7 at −40°F	1.2-1.3	1.3-1.6	1.4	0.23-0.65
	17. Hardness, Rockwell	D785	M68, 78, 85, 98	R108	M90, 109	M90	M110	M94-101
	18. Flexural elastic modulus, p.s.i. × 10^5	D790	3.3-4.01	2.97	8	11.0-12.0	16	3.5-4.5
	19. Compressive modulus, p.s.i. × 10^5	D695	—	1.75	0.7 × 10^6	0.7 × 10^6	—	—
Thermal	20. Thermal conductivity, 10^{-4} cal./sec./sq. cm./1 (°C./cm.)	C177	4.2-6.9	—	7.0	7.0	7.1	3.6
	21. Specific heat, cal./°C./gm.	—	0.28-0.55	—	0.25-0.30	0.30	0.22	0.28
	22. Thermal expansion, 10^{-5} in./in./°C.	D696	6.0-9.5	4.4/°F.	1.1-2.5	1.1-6.1	2	6.5
	23. Resistance to heat, °F. (continuous)	—	122-250	—	200-300	240-350	200-300	—
	24. Deflection temp., °F. at 264 p.s.i. fiber stress	D648	122-185	154	415	428	455	—
	at 66 p.s.i. fiber stress	D648	240-374	—	465	437	480	100-106 (annealed)
Electrical	25. Volume resistivity, ohm-cm. (50% RH and 23° C.)	D257	10^{15}-2 × 10^{16}	—	20 × 10^{16}	10^{16}	>10^{16}	—
	26. Dielectric strength, short-time, ⅛-in. thickness, volts/mil	D149	420-556	420	460	460-556	—	—
	27. Dielectric strength, step-by-step, ⅛-in. thickness, volts/mil	D149	300, 400 (1.1 mm)	—	400, 550 (2.0 mm)	400, 550(2.0 mm)	—	—
	28. Dielectric constant, 60 cyc.	D150	3.29-3.3	3.1	3.9	3.7-3.81	4.1	—
	29. Dielectric constant, 10^3 cyc.	D150	3.2-3.46	—	3.8	3.6	4.0	—
	30. Dielectric constant, 10^6 cyc.	D150	3.1-3.28	—	3.7	—	3.8	—
	31. Dissipation (power) factor, 60 cyc.	D150	—	0.0004	0.003	0.002-0.0017	0.002	—
	32. Dissipation (power) factor, 10^3 cyc.	D150	0.0014-0.005	—	0.005	0.002-0.0025	0.004	—
	33. Dissipation (power) factor, 10^6 cyc.	D150	0.0022-0.03	—	0.02	0.017	0.017	—
	34. Arc resistance, sec.	D495	75-192	—	130-146	—	—	—
Optical	35. Refractive index, n_D	D542	—	—	—	—	—	1.64
	36. Clarity	—	Opaque	Transparent	Opaque	Opaque	Opaque	—
	37. Transmittance, %	—	—	70-5 mils	—	—	—	—
	38. Haze, %	—	—	0.4	—	—	—	—
Resistance characteristics	39. Water absorp., 24 hr., ⅛-in. thick., %	D570	0.08-0.09	0.2	0.02-0.07	0.07-0.08	0.05	0.1-0.2
	40. Flammability [a] Burning rate, in./min.	D635	0.4	0.8	0.5	1.1	—	—
	AEB, in./ATB, sec. [b]	D635	—	—	—	0.16 (FR gr.)	—	—
	41. Effect of sunlight	—	Discolors very slightly	—	Discolors very slightly	Colors may fade	Discolors very slightly	Embrittlement
	42. Effect of weak acids	D543	Resistant	Resistant	Resistant except hydrofluoric acid	None	Resistant except hydrofluoric acid	None
	43. Effect of strong acids	D543	Attacked by oxidizing acids	Attacked by oxidizing acids	Attacked slowly by high conc. oxidizing acids	None	Attacked slowly by high conc. oxidizing acids	Attacked by some
	44. Effect of weak alkalies	D543	Resistant	Resistant	Slight	Considerable Attack	Slight	None
	45. Effect of strong alkalies	D543	Attacked	Slight	Attacked	Considerable attack	Attacked	Attacked
	46. Effect of organic solvents	D543	Resistant to common solvents; slightly attacked by partly halogenated hydrocarbons	Attacked by ketones, hydrocarbons	Resistant to common solvents; slightly attacked by partly halogenated hydrocarbons	Swollen by some polar solvents but generally resistant	Resistant to common solvents; slightly attacked by partly halogenated hydrocarbons	—

[a]This numerical flame spread rating is not intended to reflect hazards presented by this or any other material under actual fire conditions.
[b]AEB = Avg. extent of burning in inches; ATB = Avg. time of burning in seconds.
*Reprinted by permission from *Modern Plastics Encyclopedia*, McGraw-Hill, Inc.

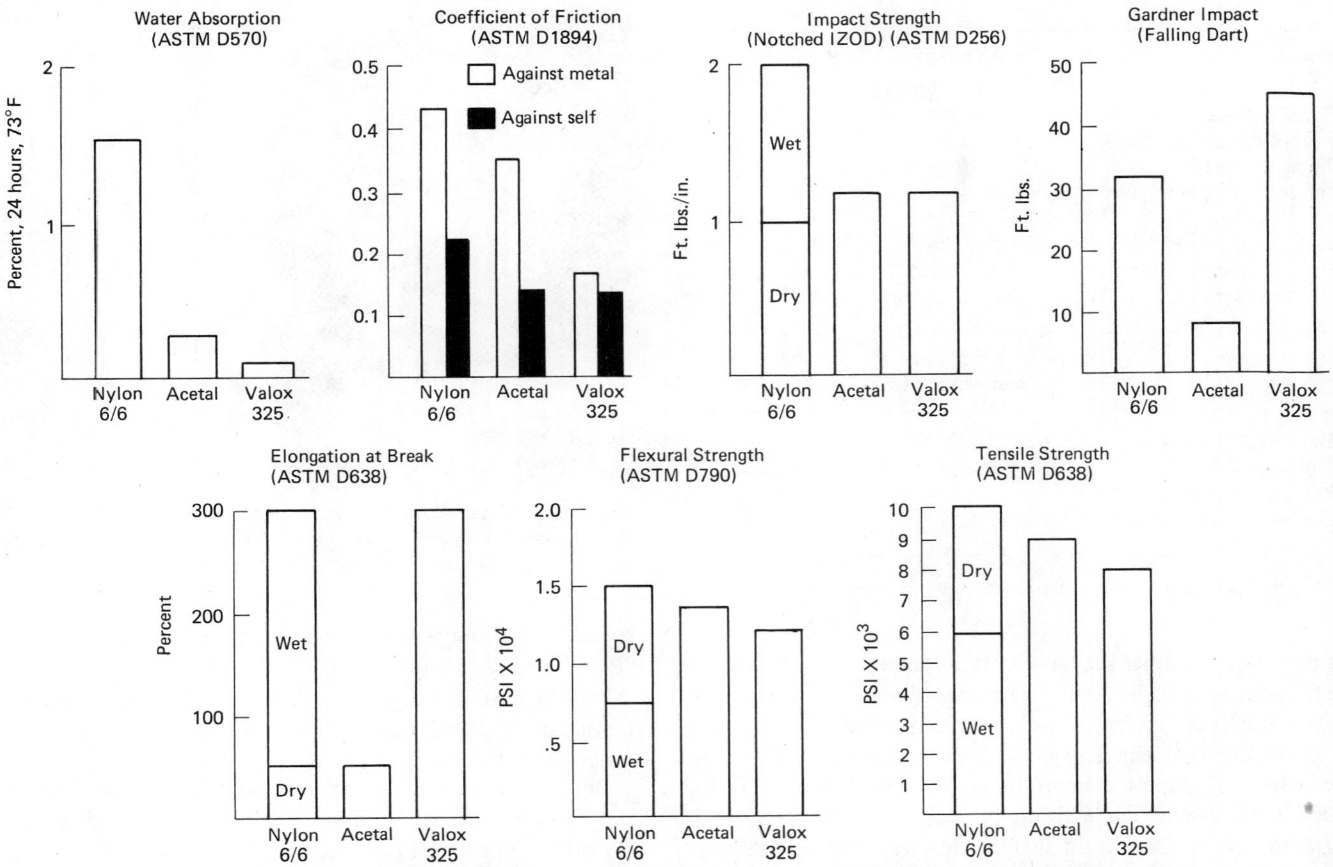

Fig. 5-136. Comparative properties: PBT (Valox 325), acetal, and nylon 6/6. (*Courtesy: General Electric Co.*)

comparable to the other engineering thermoplastics with easier moldability. Table 5-121 compares the properties of PTMT with nylon. Figure 5-135 shows that that the electrical properties of thermoplastic polyesters are somewhat better than some other engineering plastics. Figure 5-136 compares properties of PBT (Valox 325) with acetal and nylon 6/6.

The low T_g of PBT facilitates rapid crystallization at typical mold temperatures of 50°C (122°F), resulting in shorter molding cycles.

These polymers have a slight edge over nylon and polyacetal, yet all three have outstanding solvent resistance. PBT has lower water absorption, higher modulus, better dimensional stability, and lower impact strength when compared to nylon. It has lower water absorption, higher unnotched impact, lower heat-deflection temperature at high loading, and lower hydrolytic stability at high temperatures when compared to the polyacetals. It is equivalent to both nylon and polyacetals in flow and moldability properties. (Like the nylons, it must be thoroughly dried before molding.)

Glass-reinforced grades are available and have the following characteristics. Compared to reinforced nylon, they have lower water absorption, better dimensional stability, but lower heat resistance. Compared to reinforced polyacetal, they are more flame and heat resistant.

In addition, the polybutylene terephthalates have some key advantages over high-performance thermosets. Specifically, they have better processability, impact strength, design latitude, and color capability. On the other hand, the thermosets have outstanding environmental resistance, higher modulus, and heat resistance.

PBT is tough and durable for high-performance applications. It combines good abrasion resistance and a low coefficient of friction. It maintains a balance of high dielectric strength, good track resistance, and high air resistance. The resin is virtually free of the tendency to stress-crack.

Although these polyesters have low moisture absorption, they are not suggested for continuous immersion in hot water or hot aqueous solutions. They react slowly with water at elevated temperatures. Toughness and surface integrity can be reduced due to this hydrolytic degradation. PBT has excellent resistance to a wide range of chemicals including ethers, esters, hydrocarbons, and aqueous solutions of acids, bases, and salts. It is also resistant to staining from common staining agents such as grape juice, ketchup, tea, and coffee.

On the minus side, unreinforced PBT shows significant

Table 5-121. Properties Comparison: PTMT and Nylon.*

Property	Unmodified PTMT	Nylon
Tensile strength, psi	8200	8500
Elongation at fracture, %	250	220
Impact strength (−40 F)	No break	–
Impact strength notched, ft-lb/in.	1.0	1.0
Plaque Impact at 73 F	42 ft-lb	–
and at −40 F	35 ft-lb	–
Flexural strength, psi × 10^3	12	15

Property	20% glass-reinforced PTMT	20% GR nylon
Deflection temperature (264 psi), F	380	300–485
Tensile strength, psi × 10^3	14	13–30
Flexural modulus, psi × 10^5	8.7	2–19

*Reprinted by permission from *Product Engineering*, July 1971.

Fig. 5-137. Water resistance and good mechanical properties of thermoplastic polyesters have led to uses for pump and sprinkler components. (*Courtesy: General Electric Co.*)

property loss under load at elevated temperatures. It is very notch-sensitive. The resins are susceptible to attack by concentrated lyes, oxidizing and acetic acids.

Providing the resin is dry, it can be extruded and injection molded with melt temperatures ranging from 225°C to 250°C (437–482°F). With high injection speeds, glass-filled grades will produce good surface finishes. Since it is highly crystalline, it is opaque white in thick molded or extruded sections, translucent in thinner sections and, interestingly, it can be extruded into clear film in thicknesses less than 10 mils.

PBT can be structurally foamed and adapts well to printing, painting, gluing, hot-stamping, spin and sonic welding. Typical molding shrinkage values for unfilled PBT range from 18 to 24 mils/inch. Glass-reinforced grades shrink much less (2–6 mils/inch). As is typical with most glass-reinforced thermoplastics, these materials will undergo directional shrinkage.

Applications include textile machine parts, tape cassettes, gears, bearings, furniture glides, fasteners of various types, automotive parts, film, garment buttons, housings of various types, pump and sprinkler components (see Figure 5-137).

Copolyesters. This resin was introduced for the extrusion of film for formed blisters and laminated film applications. Although this polyester can be processed into excellent biaxially oriented film, the current emphasis is on an extruded thermoformable film and sheet. Extruded film is produced by chill-roll extrusion and no orientation processing is done. Complying with FDA and USDA requirements, the copolyester will withstand both ethylene oxide (ETO) and gamma ray sterilization processes. The sheet cannot be autoclaved. This copolyester has a glass transition temperature of 87°C (189°F). Forming temperatures are generally in the range of 135–177°C (275–350°F).

Only conventional hopper drying is necessary. The copolyester offers better film toughness and a higher heat-distortion point than unoriented PET. Copolyester is designed for general use in the film extrusion industry. High gloss of extruded sheet from this copolyester can create problems during wind-up and blister denesting. Antiblock and denesting concentrates are available.

Sheet from this copolyester thermoforms very rapidly with no blushing or embrittlement. Both deep and shallow draws can be made with approximately the same heating requirements. Articles formed from the copolyester are tough, even at low temperatures.

Forming cycle times in many cases are only half of those required for rigid PVC. Copolyester will seal to itself using conventional bar and impulse sealers at temperatures as low as 177°C (350°F). Both ultrasonic and radio-frequency seals have been achieved, but care should be taken when evaluating these methods since occasional brittleness has been experienced. The sheet normally seals to foils, papers, and paperboards coated with the same adhesives formulated for PVC, styrene, or cellulosics. Typical sealing conditions are 177–204°C (350–400°F), for 1 to 4 seconds.

The chemical resistance of the copolyester is somewhat better than PET, although the permeability of the film is not as good. Current cost of this resin is slightly less than nylon.

Trade Names

Arnite	PET molding powders	Akzo Plastics, N.V.
Avistar	PET film	FMC Corporation

Avlin	PET fiber	FMC Corporation
Blue "C"	PET fiber	Monsanto Co.
Celanar	PET film	Celanese Plastics Co.
Celanex	PBT (PTMT)	Celanese Plastics Co.
Chromar	Metallized PET film	Kurz-Hastings, Inc.
Cronar	PET	E.I. Du Pont de Nemours & Co., Inc.
Dacron	PET fiber	E.I. Du Pont de Nemours & Co., Inc.
Defsan	PC/PET blend	Soviet Origin
Deroton	PBT (PTMT)	Imperial Chemical Industries, Ltd.
Doray	Metallized film and laminates	Dorrie Process Co., Inc.
Dynapol	PET copolymers	Dynamit Nobel, A.G.
Ektar	PCTG copolymers	Eastman Chemical Products, Inc.
Fortrel	PET fiber	Fiber Industries, Inc.
Fybrite	PET fiber	Fiber Industries, Inc.
Gafite	PBT (PTMT)	GAF Corporation
Hostadur	PET	Hoechst Corporation
Hytrel	Polyester elastomer	E.I. Du Pont de Nemours & Co., Inc.
Jaytrel	PET	Johnston Industrial Plastics
Kodar	Copolyester	Eastman Chemical Products, Inc.
Kodel	PET fiber	Eastman Chemical Products, Inc.
Lowland	PET fibers	Akzona, Inc.
Lumirror	PET film	Toray Industries, Inc.
Martin	Dyed PET film	Martin Processing, Inc.
Melinex	PET film	Imperial Chemical Industries, Ltd.
Meltiform	Thermoformable PET sheet	ICI USA, Inc.
Montivel	PET film	Montedison, S.p.A.
Mylar	PET film	E.I. Du Pont de Nemours & Co., Inc.
Petpac	PET film	Celanese Plastics
Petra	PET film and sheet	Allied Chemical Co.
Pibiter	TP polyester	Montedison USA, Inc.
Pocan	PBT compounds	Bayer, A.G.
Polyspun	PET fibers	White & Sons, Ltd.
Quintess	PET filament	Phillips Chemical Co.
Reemay	Spunbonded polyester sheet	E.I. Du Pont de Nemours & Co., Inc.
Rynite	PET molding powder	E.I. Du Pont de Nemours & Co., Inc.
Scotchpak	PET film	3M Co.
Scotchpar	PET film	3M Co.
Snialen	TP polyester	Snia Viscoso, S.p.A.
Tenatine	TP polyester	Ciba-Geigy Corporation
Tenite	PBT (PTMT)	Eastman Chemical Products, Inc.
Terphane	PET	Rhodia, Inc.
Terylene	PET fiber	Imperial Chemical Industries, Ltd.
Ultradur	PTMT	BASF Wyandotte Corporation
Valox	PBT (PTMT)	General Electric Co.
Versel	PBT (PTMT)	Allied Chemical Co.
Vitel	TP polyesters	Goodyear Tire & Rubber Co.
Vycron	PET fiber	Beaunit Corporation
Weblox	Polyester fabric	Kendall Co.
Webril	Polyester fabric	Kendall Co.

REFERENCES AND BIBLIOGRAPHY

Baier, H. F., and Jacobi, F. "Biaxial Orientation of Film: The Future." Paper presentation, *SPE ANTEC*, New Orleans, May 1979.

Beckman, K. J., and Blitstein, M. R. Thermoplastic polyesters, from Plastics in Packaging. *Plastics Engineering*, October 1978, **34**(10):49–50.

Bonnebat, C., Roullet, G., and de Vries, A. J. Biaxially Oriented PET Bottles." Paper presentation, *SPE ANTEC*, New Orleans, May 1979.

Brydson, J. A. *Plastics Materials*. 3rd Ed. Robert E. Krieger Publishing Co., Inc., Huntington, New York, 1975.

Cabral, A. C. D., Orr, A., Stier, E. F., and Gilbert, S. G. Metallized polyester film rates high as a fatty-foods packaging material. *Modern Plastics*, March 1979, **56**:74–77.

Celanese Plastics Co. *Electrical Applications–Celanar Polyester Film*. Bulletin 088050.

——. *Celanar Polyester Film Properties–Optical*. Technical Bulletin No. D1-B.

——. *Celanar Polyester Film–Chemical Properties*. Technical Bulletin No. D1-C.

E.I. Du Pont de Nemours & Co., Inc. *"Rynite" Thermoplastic Polyester Resin*. Bulletin E-06162, December 1975.

——. *Adhesives for Mylar Polyester Film*. Industrial Finishes Technical Bulletin No. 17, 3rd Ed., January 1958.

Eastman Chemical Products, Inc. *Preliminary Information About Kodar PETG Copolyester 6763*. Publication No. PTP-10-B, February 1978.

——. *Tenite Polyterephthalate PTMT Formulas Reinforced With Glass Fibers*. Materials Bulletin MB-45A, March 1974.

——. *Unfilled Tenite Polyterephthalate PTMT Formulations*. Materials Bulletin MB-52B, March 1974.

——. *The Effect of Various Chemicals and Reagents on Tenite Polyterephthalate PTMT*. Bulletin TR-44, ECP 465, May 1972.

——. *Preliminary Information About Tenite Polyterephthalate—An Engineering-Type Thermoplastic.* Publication No. PTP-1, January 1971.

Ericson, J. E. "Packaging with Polyesters." Paper presentation, *SPE ANTEC*, New Orleans, May 1979.

GAF Corporation. *Gafite LW Thermoplastic Polyester Molding Compounds.* Bulletin 2308-018, 1979.

——. *Gafite Thermoplastic Polyester Molding Compounds.* Bulletin 2308-001R3, 1978.

Gardner, R. J., and Martin, J. R. "Humidity Effects on Thermoplastic Polyesters." Paper presentation, *SPE ANTEC*, New Orleans, May 1979.

General Electric Co. *The Big Name in Thermoplastic Polyester Introduces Valox 325.* Bulletin VAL-46, 1978.

——. "Valox Thermoplastic Polyester" Bulletin VAL-5A, February 1975.

——. *Valox Thermoplastic Polyester vs. Thermosets: A Cost Analysis.* Bulletin VAL-43, March 1975.

——. *Only One Engineering Material Combines the Best of Thermoplastics and Thermosets, Valox Thermoplastic Polyester.* Bulletin VAL-42, December 1974.

——. *Valox Resin Injection Molding.* Bulletin VAL-15B.

——. *Valox Resin Design Guide.* Bulletin VAL-50.

Houston, A. M. Thermoplastic polyesters. *Materials Engineering*, Manual 265. December 1976, **84**(7):61–68.

Imperial Chemical Industries, Ltd. *Metallizing.* Melinex Polyester Film Industry Note MX 203, 3rd Ed., 1979.

——. *"Melinex" Polyester Film for Every Need.* 3rd Ed., February 1979.

——. *Melinex" Polyester Films for Packaging.* Bulletin No. MX5, 1978.

——. *The Control of Static Charges on Film.* 'Melinex' Polyester Film Industry Note MX 1N48, 1978.

——. *An Introduction to "Melinex" Films.* Bulletin MX100, 2nd Ed.

——. *Electrical.* Melinex Polyester Film Industry Note MX 202.

Kelleher, P. G., Bebbington, G. H., Falcone, D. R., Ryan, J. R., and Wentz, R. P. "PBT Stability." Paper presentation, *SPE ANTEC*, New Orleans, May 1979.

Landfield, H. "International Update on Thermoforming Materials." Paper presentation, *SPE ANTEC*, New Orleans, May 1979.

Materials Engineering. Thermoplastic polyester cycles faster than Nylon. October 1977, **86**(4):63.

——. New copolyesters contain up to 30% glass. October 1977, **86**(4):63.

McMahon, W., Birdsall, H. A., Johnson, G. R., and Camilli, C. T. Polyester film—how long does it last? *Materials in Design Engineering*, February 1959, **49**(2):94.

Modern Plastics. One-piece molded bicycle wheel in reinforced thermoplastic polyester. April 1979, **56**(4):28.

——. PET bottle technology continues to be the subject of litigation. February 1978, **56**(2):4.

——. Beyond the soft-drink bottle: new markets in sight for PET. April 1978, **55**(4):48–49.

——. Finally, thermoplastic polyesters that mold easily. April 1972, **49**(4).

Plastics Engineering. Engineering resins primer. December 1974, **30**(12):17–32.

Plastics World. Thermoplastic polyesters. June 1977, **35**(6): 54–57.

"Thermoplastic Polyesters." Paper presentation, *Plastics World Seminar, Engineering Plastics II*, Los Angeles, March 1976.

Product Engineering. Tough thermoplastic debuts, will seek engineering product applications. July 1971, **42**(12):37.

Ray, D. D., Shriver, C. B., and Gartland, R. J. Here's why polyethylene terephthalate is the major competitor for beverage container applications. Part 1—Properties and applications. *Plastics Design & Processing*, August 1977, **17**(8):47–52.

Skelton, B. W. Fabricating polyterephthalate. *SPE Journal*, January 1973, **29**(1): 18–24.

Swanson, R. S. *Plastics Technology—Basic Materials and Processes*, McKnight & McKnight Publishing Co., Bloomington, Illinois, 1965.

Telofski, S. Thermoplastic polyesters in automotive ignition. *Plastics Engineering*, December 1977, **33**(12):21–23.

Theberge, J., and Cloud, P. Choosing the best glass-reinforced thermoplastic. *Plastics World*, March, **35**(3): 44–46.

Throne, J. L. "Cellular Structure and Foaming Agents in Resin Regions." Paper presentation, *SPE ANTEC*, New Orleans, May 1979.

Titus, J. B. *Trade Designations of Plastics and Related Materials*, (revised). PLASTEC NOTE 9C, Plastics Technical Evaluation Center, Picatinny Arsenal, Dover, New Jersey, May 1978.

Vaccari, J. A. Thermoplastic polyesters: versatile engineering materials. *Materials Engineering*, October 1974, **80**(5):48–52.

Watson, M. T. Properties of a new polyester film. *SPE Journal*, October 1961, **17**(10):1083–1087.

Wood, A. S. Engineering resins, from materials '79. *Modern Plastics*, January 1979, **56**(1):53–55.

——. The choices get tougher in engineering thermoplastics. *Modern Plastics*, July 1978, **55**(7):42–44.

——. Action in profile extrusion shifts to engineering resins. *Modern Plastics*, November 1978, **55**(11):46–48.

Zimmerman, D. Thermoplastic Polyester. In, *Modern Plastics Encyclopedia*, J. Agranoff (ed.), **52**(10A):58–61. McGraw-Hill Inc., 1975–76.

Zimmerman, D. D. "Performance of Engineering Resins in Electrical Connectors." Paper presentation, *SPE ANTEC*, New Orleans, May 1979.

Polycarbonates

Polycarbonates may be produced from a wide range of polyfunctional hydroxyl compounds. The only polycarbonate of commercial significance is derived from bisphenol A (4,4′-dihydroxy-2,2′-diphenyl-propane). This monomer is made from the coreaction of phenol and acetone under acidic conditions.

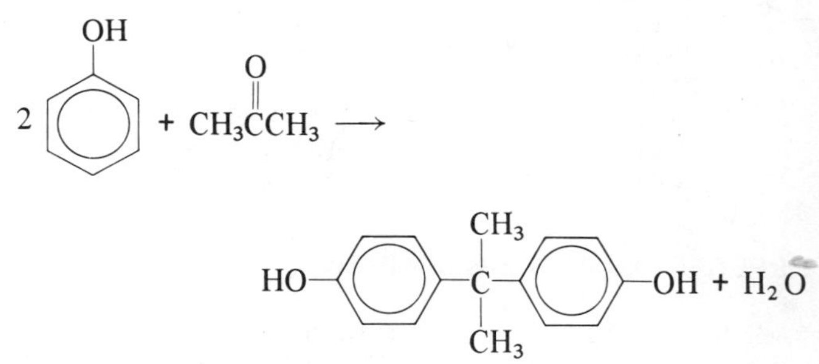

Reactant ratio, reaction conditions, and use of catalysts must be stringently controlled or else undesirable byproducts are produced. These impurities can be tolerated when the bisphenol A is used to make epoxy resins, but they do cause a loss of properties and color when used in polycarbonate resins. Removal of such impurities is normally achieved by recrystallization from chlorobenzene or aqueous alcohol.

The other monomers used are phosgene and diphenyl carbonate. The former is prepared from CO and Cl_2. The latter is the reaction product of phosgene and phenol in aqueous caustic soda solution. Tertiary amines are used as catalysts and the salt is purified by redistillation.

Polymerization. The three polymerization processes of commercial importance are (1) transesterification:

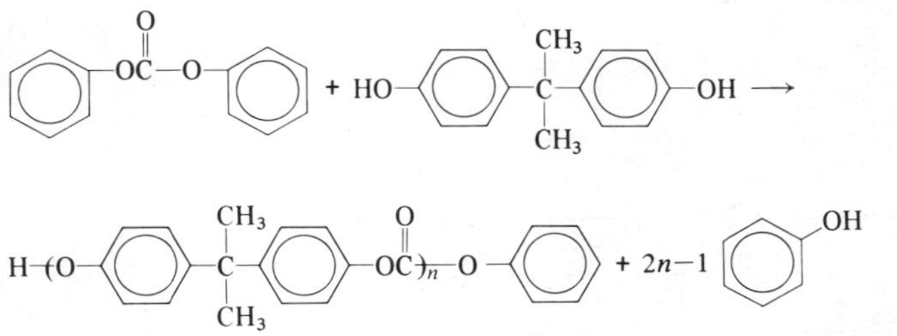

(2) Phosgenation of dihydroxy compounds in the presence of pyridine:

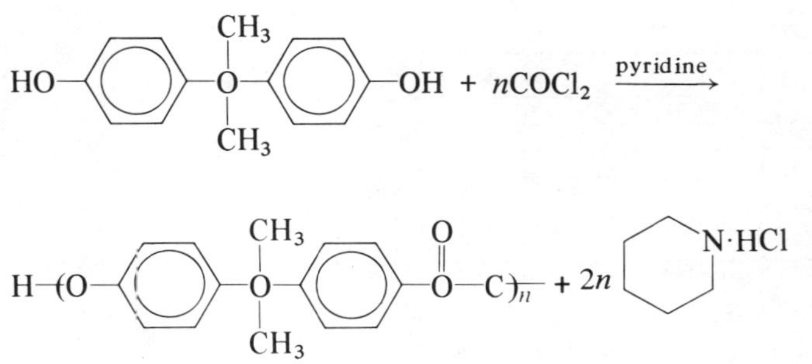

(3) Interfacial polycondensation:

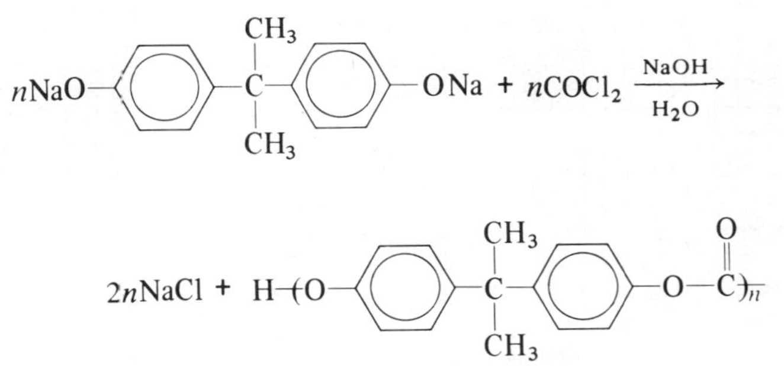

In (1), the reaction typically occurs at 180–220°C (356–428°F) at 20–30 mm Hg pressure until about 80% of the generated phenol has been removed. The temperature is then increased to 290–300°C (554–572°F) and the pressure lowered to 1 mm Hg or lower. The reaction is stopped when the melt viscosity becomes high enough to preclude pressure removal of product polymer. Excess diphenyl carbonate is added so that the bisphenol A will react before it can decompose at the elevated reaction temperatures. Hydroxides, hydrides, and amides of alkaline and alkaline earth metals, as well as zinc and antimony oxides are typical polymerization catalysts.

In (2), it is necessary to remove the HCl formed. Pyridine acts as a hydrohalide acceptor. Phosgene is bubbled through a solution of bisphenol A in pyridine at temperatures slightly above room temperature. When the solutions viscosity becomes high enough as a result of polymer formation to make proper agitation difficult, methanol is added. The polymer is insoluble in methanol and precipitates out, but the pyridine hydrochloride remains behind in solution.

In (3), the bisphenol A is dissolved in aqueous caustic soda. A polymer solvent is blended in forming a two-phase system with the caustic soda. Phosgene is introduced and polymerization occurs just inside the aqueous phase. The formed polymer dissolves in the solvent phase and is recovered by water washing, neutralization, and precipitation by evaporation.

In both (2) and (3), molecular weight is controlled by adjustment of monomer stoichiometry or use of chain stoppers such as phenol.

In the transesterification, no solvent is required, obviating the hazard problems (flammability and toxicity) normally associated with solvents. The resultant polymer does not require predensification for use in extrusion and injection molding. On the other hand, the polymer produced has a very high melt viscosity, and high temperatures and vacuum are required. This creates a need for specialized equipment and operating procedures, along with extended reaction times.

In the phosgenation processes, high molecular weight polymer is produced in fairly simple equipment under moderate reaction conditions. On the negative side, hazardous solvents and phosgene must be handled, the solvent recovered, and the polymer purified and densified.

The molecular weights produced range from 25,000 to 35,000 for molding grades. Higher molecular weight species are used for solvent casting of electrical-grade film. The polymer has a highly regular structure, yet it does not show much crystallinity. Although solution polymerization promotes some degree of crystallinity, many of the crystallites formed are destroyed during thermal fabrication. The amount of crystallization and the size of the crystallite structure decrease as the polymer's molecular weight increases.

Polycarbonate has a higher free-volume at its T_g than normally observed with other polymers. It is currently believed that the high free-volume below T_g, the damping capacity of the polymer over a wide temperature range, and the limited degree of crystallinity combine to produce polycarbonate's high level of toughness.

Polycarbonate is outstanding in toughness, transparency, and temperature resistance. The polymer also displays good rigidity up to 140°C (284°F), very good electrical insulation characteristics, and is virtually self-extinguishing. On the other hand, its cost is very high, special processing techniques are required, its UV and chemical resistance is limited, and it is notch-sensitive and susceptible to strain-crazing. Table 5-122 presents a list of typical properties for polycarbonate, both unfilled and glass filled. Table 5-123 compares these properties to other thermoplastics. Table

Table 5-122. Polycarbonate Properties.*

		ASTM test method	Unfilled	Less than 10% glass filled	10 to 40% glass filled
Processing	1. Molding qualities	—	Good to excel.	Excellent	Very good
	2. Compression molding temp., °F.	—	480-620	—	—
	3. Comp. molding pressure, p.s.i.	—	1000-2000	—	—
	4. Injection molding temp., °F.	—	480-650	520-650	575-650
	5. Injection molding pressure, p.s.i.	—	10000-20000	10000-20000	15000-40000
	6. Compression ratio	—	1.74-5.5	—	—
	7. Mold (linear) shrinkage, in./in.	—	0.005-0.007	0.002-0.003	0.001-0.003
	8. Specific gravity (density)	D792	1.2	1.25	1.24-1.52
	9. Specific volume, cu. in./lb.	D792	23.0	22.2	22.4-18.2
	10. Machining qualities	—	Excellent	Fair to good	Fair
Mechanical	11. Tensile strength, p.s.i.	D638	8000-9500	9600	12000-25000
	12. Elongation, %	D638	100.0-130.0	10.0	0.9-5.0
	13. Tensile elastic modulus, 10^5 p.s.i.	D638	3.0-3.5	4.5	5.0-17.0
	14. Compressive strength, p.s.i.	D695	12500	14000	13000-21000
	15. Flexural yield strength, p.s.i.	D790	13500	15000	17000-32000
	16. Impact strength, ft. lb./in. of notch (½ × ½ in. notched bar, Izod test)	D256	12.0-18.0 (½ × ⅛ in. bar)	4.0-5.0 (½ × ⅛ in. bar)	1.2-4.0 1.9-6.5 (¼ × ½ in.)
	17. Hardness, Rockwell	D785	M70-78 R115-125	M85	M88-95
	18. Flexural elastic modulus, p.s.i. × 10^5	D790	3.2-3.5	5.0	5.0-14.0
	19. Compressive modulus, p.s.i. × 10^5	D695	3.45	5.2	5.0-15.0
Thermal	20. Thermal conductivity, 10^{-4} cal./sec./sq. cm./1 (°C./cm.)	C177	4.6	4.8	4.9-5.2
	21. Specific heat, cal./°C./gm.	—	0.28-0.30	—	—
	22. Thermal expansion, 10^{-5} in./in./°C.	D696	6.6	3.42	1.7-4.0
	23. Resistance to heat, °F. (continuous)	—	250	275	275
	24. Deflection temp., °F. at 264 p.s.i. fiber stress at 66 p.s.i. fiber stress	D648	265-285 270-290	288 295	289-300 300-310
Electrical	25. Volume resistivity, ohm-cm. (50% RH and 23° C.)	D257	2.1×10^{16}	3×10^{16}	$4.0\text{-}5.0 \times 10^{16}$
	26. Dielectric strength, short-time, ⅛-in. thickness, volts/mil	D149	380	450	450
	27. Dielectric strength, step-by-step, ⅛-in. thickness, volts/mil	D149	380	450	400
	28. Dielectric constant, 60 cyc.	D150	2.97-3.17	3.10	3.0-3.53
	29. Dielectric constant, 10^3 cyc.	D150	3.02	—	—
	30. Dielectric constant, 10^6 cyc.	D150	2.92-2.93	3.05	3.0-3.48
	31. Dissipation (power) factor, 60 cyc.	D150	0.009	0.0008	0.0009-0.0013
	32. Dissipation (power) factor, 10^3 cyc.	D150	0.0021	—	—
	33. Dissipation (power) factor, 10^6 cyc.	D150	0.010	0.0075	0.0067-0.0075
	34. Arc resistance, sec.	D495	10-120	5-120	5-120
Optical	35. Refractive index, n_D	D542	1.586	—	—
	36. Clarity	—	Transp. to opaque	Translucent to opaque	
	37. Transmittance, %	—	85-91	—	—
	38. Haze, %	—	0.5-2	—	—
Resistance characteristics	39. Water absorp., 24 hr., ⅛-in. thick., %	D570	0.15-0.18	0.12	0.07-0.20
	40. Flammability[a] Burning rate, in./min. AEB, in./ATB, sec.[b]	D635	— —	— —	— —
	41. Effect of sunlight	—	Slight color change, slight embrittlement		
	42. Effect of weak acids	D543	None		
	43. Effect of strong acids	D543	Attacked slowly	Attacked by oxidizing acids	
	44. Effect of weak alkalies	D543	Limited resistance		
	45. Effect of strong alkalies	D543	Attacked		
	46. Effect of organic solvents	D543	Resistant to paraffinics soluble in aromatic and chlorinated hydrocarbons	Soluble in chlorinated hydrocarbons	

[a]This numerical flame spread rating is not intended to reflect hazards presented by this or any other material under actual fire conditions.
[b]AEB = Avg. extent of burning in inches; ATB = Avg. time of burning in seconds.
*Reprinted by permission from *Modern Plastics Encyclopedia*, McGraw-Hill, Inc.

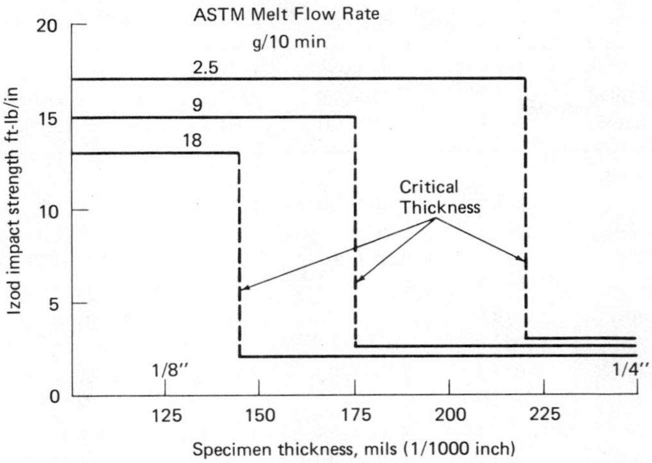

Fig. 5-138. Effect of specimen thickness on Izod impact strength. (From *Plastics World Seminar*, March 1976)

5-124 compares some selected polycarbonate properties to thermosetting polyester and some metals.

The impact strength of polycarbonate depends on its thickness. Values reported for $\frac{1}{4}$ inch-thick specimens are $\frac{1}{5}$–$\frac{1}{6}$ the value for a $\frac{1}{8}$ inch-thick specimen. Interestingly, the relationship is not linear. A discontinuity at the "critical thickness" appears for all grades (see Figure 5-138). The Izod impact test (see Chapter 3) does not correlate well with field performance. The reason is that polycarbonate is very notch-sensitive.

Polycarbonate exhibits excellent creep resistance. When there are internal molding stresses, at elevated temperatures, when the material is aged, or when exposed to a number of chemical environments under dynamic conditions, crazing may occur at much lower strain levels.

Polycarbonate has excellent optical properties. Normal moldings are clear, but yellow because of impurities in the commercial grades of bisphenol A used, and darken as a result of the chemical degradation during processing and exposure to ultraviolet light. Water-white grades can be prepared, provided ultrahigh-purity bisphenol A is used. As seen in Table 5-122, the resin exhibits a refractive index of 1.586 with only 70–80% haze. Only PMMA and polystyrene have better transparency. Table 5-125 compares polycarbonate properties to other glass substitutes.

Thermally, polycarbonate is rated for 125°C (257°F) continuous use, not subject to impact. With parts subject to impact, this temperature drops to 115°C (234°F). The heat-distortion temperature ranges from 130°C to 143°C (266–289°F), with the higher molecular weight grades giving the higher values. Low-temperature resistance is good. Electrical properties are excellent. The dielectric constant is stable over a wide temperature and frequency range.

Generally, polycarbonate has good room-temperature resistance to water, dilute organic and inorganic acids, oxidizing and reducing agents, neutral and acid salts, animal and vegetable oils and fats, and aliphatic and cyclic hydrocarbons. These resins are attacked by alkalis, amines, ketones, esters, and aromatic hydrocarbons. Good solvents are chlorinated hydrocarbons, cresol, and dioxane. The polymer has high resistance to staining by tea, coffee, fruit juices, lipstick, ink, soaps, and detergents.

Polycarbonate is available in glass-fiber-reinforced grades. As with flame-retardant grades, differences exist between producers and interchangeability must be carefully studied. The 10% glass-filled resins are a compromise which provides a small increase in modulus without a substantial loss of toughness. At higher levels of glass (20–40%), the toughness drops off markedly; however, rigidity increases, mold shrinkage decreases, and there is a significant improvement in dimensional stability. Tables 5-126 and 5-127 compare the coefficients of expansion and friction for glass-reinforced polycarbonate (Lexan) and other materials.

Processing. Polycarbonate can be injection molded, extruded, blow and rotational molded, and thermoformed. Secondary processes to which polycarbonate is amenable include: machining, vacuum-metallizing, printing, ultrasonic welding, hot-gas welding, and adhesive and solvent bonding.

Polycarbonate injection molds well if the machine and mold are suitable for the processing of high melt-temperature and high melt-viscosity plastics. The machine should be capable of achieving temperatures to 343°C (650°F). It must have good temperature control with a separate control for an open-type nozzle. The injection end should be able to handle 20,000 psi pressure.

Polycarbonate melts adhere strongly to metals and, if allowed to cool in a barrel, may, on shrinkage, pull pieces of metal away from the wall. There is little crystallization on cooling and after-crystallization has not been observed.

Polycarbonates may be cold-formed by punching and cold-formed by punching and cold rolling. The latter can improve impact resistance.

Preprocess drying is essential for polycarbonate. Any residual moisture in the resin not removed by proper drying, chemically reacts with the resin at processing temperature, reducing the molecular weight, which can result in a loss of toughness. Figure 5-139 shows the effects of resin drying on the impact strength of molded specimens.

Applications. The largest application area for polycarbonate is electronics. Figure 5-140 shows polycarbonate power connectors. Here, the good electrical insulation characteristics along with flame resistance and durability are the prime considerations. Housings for appliances consume a major share of resin production (see Figure 5-141) because of its toughness, durability, and electrical

Table 5-123. Properties Comparison Chart:

Property	Units†	LEXAN	Glass-filled LEXAN	Acetal	Polyvinyl Chloride (Rigid)	Methyl Methacrylate (Heat Resistant)	Modified Acrylic
Impact strength (per in. of notch)	ft. lbs.	12–16	2.0–2.5	1.4–2.3	0.4–15	0.3–0.4	0.5–2.3
Tensile yield strength	psi × 10^3	9–9.5	9.6–23.0	10.0	5–9	9.0–10.5	5.5–8.0
Tensile modulus	psi × 10^3	345	450–1680	410	350–600	450	230–380
Fatigue endurance limit	psi × 10^3	0.9–1.2	5.0–7.0	4.0–5.0	–	–	–
Elongation	%	110	3.5–20.0	15–75	2–40	3–5	15–30
Flexural strength	psi × 10^3	13.5	15.0–27.0	14.1	10–16	16	8.7–12
Flexural modulus	psi × 10^3	340	500–1400	410	300–400	450	280–360
Shear strength	psi × 10^3	10.0	–	7.7–9.5	–	–	7.5–9.4
Compressive strength	psi × 10^3	12.5	14–21	18	8–13	17	7.3–12
Compressive modulus	psi × 10^3	345	520–1500	–	300–400	450	250–360
Heat distortion temperature	°F (264 psi)	265–280	288–295	212	130–165	200	170–190
Heat resistance (continuous)	°F	250–270	300	185–250	120–160	190	140–195
Coefficient of thermal expansion	in/in/°F × 10^{-5}	3.75	0.93–1.79	4.5	2.78/10.28	4.5	5.0
Thermal conductivity	BTU/hr/ft²/°F/in.	1.35	1.41–1.53	1.61	0.88–2.05	1.61	1.17–1.47
Brittle temperature	°F	−215	−212	–	−50	–	–
Volume resistivity	ohm-cm	8.2 × 10^{16}	4.0–5.6 × 10^{16}	6 × 10^{14}	10^{16}	10^{14}	2 × 10^{16}
Dielectric constant	(60 cycles)	3.17	3.1–3.53	3.7	3.2–3.6	3.5	3.5–3.8
Dielectric strength (ST $\frac{1}{8}$")	volts/mil	380	456–490	465	425–1300	400–500	450–500
Power factor (60 cpm)	–	0.0009	0.008–0.0013	0.004	0.007–0.02	0.06	0.04–0.05
Arc resistance (tungsten)	sec.	120	120	129 (burns)	60–80	no tracking	no tracking
Water absorption, 24 hr	%	0.15	0.12–0.16	0.4	0.07–0.4	0.3	0.2–0.3
Rockwell hardness	–	M70–R118	85–193	M94–R118	R105–R110	M92–M103	R100–R120
Flammability	–	self-exting.	self-exting.	slow burning	self-exting.	slow burning	slow burning
Specific gravity	–	1.2	1.25–1.52	1.425	1.35–1.45	1.19	1.11–1.18
Mold shrinkage	in./in.	0.005–0.007	0.002–0.004	0.015–0.035	0.001–0.004	0.003–0.008	0.004–0.008
Color possibilities	–	unlimited	limited	limited	unlimited	unlimited	limited
Clarity	–	transparent	translucent	opaque	transparent	transparent	translucent
Light transmission $\frac{1}{8}$" section	%	86–89	–	–	89	91–93	–

*Comparative values only. Not recommended for design use. Maximum-minimum figures not given.
**Competitive data taken from technical bulletins, "Technical Data on Plastics," Manufacturing Chemists Association, February, 1957, and "Modern Plastics Encyclopedia," Breskin Publications, Inc., September, 1963. Updated Lexan data, courtesy of General Electric Co.
†Unless specified otherwise, all data derived from standard ASTM tests for plastics.

Table 5.124. A Comparison of Polycarbonate Sheet with Metals and Thermoset Materials.*

	LEXAN	Polyester Preform/Mat	Polyester Spray-Up/Mat	Polyester Premix	Aluminum Die Cast	Zinc Die Cast
Specific gravity	1.20	1.5–1.7	1.4–1.6	1.8–20	2.57–2.96	6.6
lb/in.³	0.043	0.061	0.058	0.072	0.100	0.240
Izod impact strength, lb/in. notch ($\frac{1}{8}$" TK specimen)	12–16	4–12	10–20	1.0–10	N.A.	4.3
Elongation, %	110	1.0–1.5	1.0–1.2	0.3–0.5	6–8	10
Flexural strength, psi × 10^3	13.5	45	28	26	18	–
Flexural modulus, psi × 10^6	0.34	0.8–1.8	0.8–1.8	1.6–2.0	10	–
Tensile strength, psi × 10^3	9.5	24	18	10	8	14

*Courtesy: General Electric Co.

Lexan Polycarbonate Resin and Other Plastics.*, **

Polyamide (6/6)	Cellulose Acetate Butyrate	Phenolic (Wood Flour Filled)	Phenoxy	ABS Polymer	Polypropylene (Impact)	Polyethylene (High Density)	Polyphenylene Oxide	Polysulfone
1.0–3.6	0.8–6.3	0.24–0.34	2.3	0.7–10.0	1.0–7.0	1.5–12	6.0	1.3
8.5–11.8	2.6–6.9	6.5–8.5	9.5	2.5–9.0	3.3–5.2	3.1–5.5	7.8–9.6	10.2
260–400	50–200	800–1200	380	100–410	100–170	80–160	355–380	360
1.2–3.0	–	–	–	–	–	–	–	–
60–300	40–88	–	50–100	10–140	50–550	15–100	35	50
8–14	1.8–9.3	8–12	14	3.6–13.5	5.5–10.0	1400	12.8–13.5	15.4
210–410	55–185	800–1200	410	240–450	175	75–150	360–400	370–390
2.5–9.6	–	–	–	–	3.0	–	–	–
7.2–13	2.1–22	24–36	12	2.5–11.0	4.0–8.0	2400	–	–
–	–	600–1000	–	200–370	–	50–100	–	–
150	110–200	290–340	188	165–225	210	125	220	345
270–300	140–220	300–325	170	140–250	200–320	225–250	–	–
5.55–8.33	6.11–9.44	1.67–2.5	3.2–3.8	3.33–7.22	4.0–8.5	6.11–7.22	–	–
1.52–1.70	1.17–2.34	1.17–2.05	0.1015	0.41–2.52	0.048–0.098	3.22–3.63	–	–
−112	–	–	–	–	0	−76 to −200	–	–
4.5×10^{13}	10^{10}–10^{12}	10^{11}–10^{13}	5×10^{13}	10^{13}–10^{16}	6.5×10^{16}	10^{15}–10^{16}	–	–
4.0–7.6	3.5–6.4	5.0–9.0	4.1	2.7–4.8	2.2–2.3	2.3	–	–
385	250–400	200–425	520	310–410	450–660	450–500	–	–
0.014–0.05	0.01–0.04	0.04–0.30	0.0012	0.004–0.034	0.0005	0.0002	–	–
130–140	–	5	–	71–87	185	melts	–	–
1.5	0.9–2.2	0.3–0.8	0.13	0.1–0.3	0.01–0.03	0.01	0.06	0.22
R108–R118	R31–R116	M90–M110	R123	R30–R118	R45–95	(D60–D70) (Shore)	R-119	M69–R120
self-exting.	slow burning	self-exting.	self-exting.	slow burning	slow burning	slow burning	self-exting.	self-exting.
1.14	1.12–1.22	1.32–1.42	1.18	0.99–1.10	0.90–0.92	0.941–0.965	1.09	1.3
0.012–0.030	0.002–0.005	0.005–0.009	0.003–0.004	0.003–0.008	0.010–0.025	0.020–0.050	0.005–0.007	0.0076
limited	unlimited	black & brown	–	limited	unlimited	limited	limited	limited
translucent	transparent	opaque	transparent	translucent	translucent	translucent	opaque	transparent to opaque
–	75–93	–	–	–	–	–	–	–

Table 5-125. Property Comparison: Polycarbonate and Other Glass Substitutes.

	Polished Plate Glass	Polycarbonate Sheet	Acrylic Sheet	Polyester Sheet
Optical properties	E	E	E	E
UV resistance	E	E	E	G
Oxidation resistance	E	E	E	E
Hardness	E	M	L	M
Flexural strength	G	E	G	G
Impact strength	L	E	E	G
Abrasion resistance	G	E	L	M
Thermal conductivity	H	L	L	L
Adsorption	H	L	L	L
Cost	H	H	M	L

Note: E = excellent; G = good; H = high; L = low; M = medium.

properties. Other uses include terminals, contact strips, fluorescent-lamp starter enclosures and switch plates. High molecular weight film is used in capacitor manufacture.

The use of polycarbonate in glazing is well established. The outdoor lighting industry continues to make use of polycarbonate, particularly in applications for high-vandalism areas (see Figure 5-142). Taillight lenses for cars, heretofore made from PMMA, are being made from polycarbonate because of its impact resistance. Heat-resistant sun- and safety glasses have also made commercial gains.

Where optical clarity may be dulled by repeated washings or abrasion, a number of overcoatings have been developed. One based on melamine is applied in 1-mm-thick layers to polycarbonate sheet by a proprietary process. Another is a fluorocarbon, silica-containing coating designed for acrylics, but also usable for polycarbonates. These coated sheets are

Table 5-126. Linear Thermal Coefficient of Expansion.**

Material	Coefficient of Expansion (in./in./°F x 10⁻⁵)
*Standard LEXAN	3.75
LEXAN 500	1.75
LEXAN 3412 (20% Glass)	1.49
LEXAN 3414 (40% Glass)	1.02
Acetal	4.5
Polyamide	5.5–8.3
ABS	3.3–7.22
Glass	0.4–0.5
Brass	0.93–0.97
Zinc	1.52
Aluminum	1.17–1.28
Steel	0.7–0.8

*LEXAN 101, 141, 191
**Courtesy: General Electric Co.

Table 5-127. Static Coefficient of Friction.**

Stationary Material	Movable Material	fo
*Aluminum	Aluminum	1.05
*Hard Steel	Hard Steel	0.78
*Mild Steel	Mild Steel	0.71
*Babbitt (ASTM No. 1)	Hard Steel	0.70
*Aluminum	Mild Steel	0.61
LEXAN	LEXAN	0.52
*Brass	Mild Steel	0.51
LEXAN	LEXAN (MoS_2)	0.44
*Babbitt (ASTM No. 8)	Hard Steel	0.41
LEXAN	Hard Steel	0.39
LEXAN	Brass	0.38
LEXAN (MoS_2)	LEXAN (MoS_2)	0.34
LEXAN	LEXAN (Glass Reinforced)	0.30
LEXAN (MoS_2)	Hard Steel	0.25
Polyamide	LEXAN	0.19
Polyamide	LEXAN (MoS_2)	0.17
Acetal	LEXAN	0.16
LEXAN	Hard Steel (SAE80 Gear Oil)	0.02

**Mechanical Engineers Handbook*, Lionel S. Marks, Fifth Edition, McGraw-Hill Book Company, N.Y.
(MoS_2) Coated with molybdenum disulfide.
**Courtesy: General Electric Co.

targeted for use in railroad and bus windows which undergo frequent detergent scrubbings, are subject to high breakage rates, yet still must retain high clarity.

In sporting goods, high impact strength and ability to mold thin-walled, lightweight parts have brought polycarbonate into such items as football helmets, baseball caps, and other equipment where the safety of the player is dependent on the performance of the equipment he wears.

With the advent of structural-foam techniques, polycarbonate has expanded its market potential to new areas. With densities of 0.6 g/cc and rigidity (strength to weight) twice that of metal, the moldings find extensive use in business machine housings and water-ski shoes. These moldings can also be nailed, bonded, and screwed like wood and, thus, find some use in cabinetry-type applications. Table 5-128 describes various commercial adhesives and bonding methods for foamed as well as unfoamed polycarbonate.

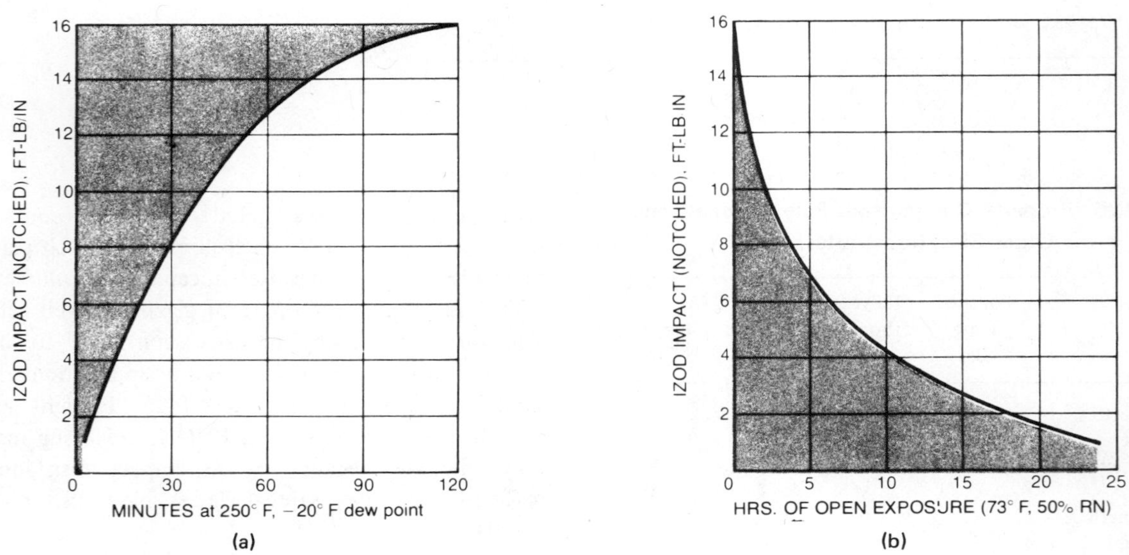

Fig. 5-139. Effects of moisture on impact strength of polycarbonate. Impact strength, a critical polycarbonate property, is significantly reduced by improper drying (a). Since moisture absorption begins immediately with exposure to room air, pellets should be removed from the dryer in small quantities so that the polycarbonate resin can be molded within 15 to 20 minutes. The loss in notched Izod impact strength due to exposure to the atmosphere (73°F, 50% RH) after drying is illustrated in (b). The data in (b) were taken on PC sample that had 16 ft-lb/in Izod strength after drying at 250°F, −20°F dew point, for 2 hrs. (*Courtesy: General Electric Co.*)

Table 5-128. Bonding Data.*

Method or Materials	For Bonds to: LEXAN	For Bonds to: Metal	For Bonds to: Rubber Wood Plastic	Hardener	Proportions (pph)	Cure Temperature (°F)	Time (hrs.)	Adhesion Quality	Tensile Shear Strength (psi)	Service Range (°F)	Color of Adhesive
Solvent Cementing											
Methylene chloride	X					Room	48	Excellent	4500–6500	–40 + 270	Clear
Ethylene dichloride	X					Room	48	Excellent	4500–6500	–40 + 270	Clear
One Part Systems											
• Room temp. curing											
Eastman 910MHT	X	X	X			Room	2–24	Excellent		70 to 212	Clear
Central Coil HB-56		X	X			Room	4–24	Good		70 to 175	Red
Vulcalock R-266-T			X			Room	Air Dry	Good			Amber
GE RTV-102	X	X	X			Room	2–24	Excellent	350	–75 to 400	White
• Hot melt adhesive											
Terrell Corp. 10 × 104	X	X	X			Melts at 350	Instantaneous	Excellent		32 to 175	Dark
Two Part Systems											
• Uralane 8089	X	X	X	Part B	100–40	Room	24–48	Very good	150		Transl. amber
Room Temp. curing											
Epon 828	X	X	X	Diethylene Triamine	100–8	Room	2–48	Excellent	2500	–40 + 250	Clear
Epon 828				Versamid 115	100–100	Room	24–48	Fair	600	32 to 160	Dark amber
Epon 828–Thiokol LP-3		X	X	DMP-30	100–100–10	Room	48	Fair		70 to 160	Dark amber
Epon Adhesive IV		X	X	Cur. Ag. A	100–6	Room	24–6 Days	Very good	1800	–60 + 190	Red-Tan
Epon Adhesive 913		X	X	Part B	100–12	Room	3–4 Days	Very good	2900	–60 + 250	Gray-Black
Bondmaster M688		X	X	CH-8	100–16	Room	4–24	Very good	2000	–40 + 212	Tan
• Elevated temp. curing											
Epon 828		X	X	Methyl Nadic Anhydride DMP-30	100–84–2	175	8 (min.)	Very good	1200	32 to 250	Light amber
Epon 828			X	Phthalic Anhydride	100–40	250	8 (min.)	Very good	1200	–40 + 250	Light amber
Epon Adhesive IV			X	Cur. Ag. A	100–6	200	45 minutes	Very good	1800	–60 + 200	Red-Tan
Epon Adhesive 913		X	X	Part B	100–12	180	3	Very good	2900	–60 + 250	Gray-Black
Bondmaster M688			X	CH-8	100–16	200	15 minutes	Very good	2000	–40 + 250	Tan
Flexicast 4002	X	X	X	Part B	100–25	212	3	Very good	600	–90 + 250	Gray-Black
Pressure- and Heat-Sensitive											
Lepage 4050			X			Room				Room	Clear
Nafco #101	X	X	X			Room		Excellent	25 psi at point of bond	–40 to 400	Clear
Kleen-Stik 295	X	X	X			120–140	<a minute	Good		–40 + 200	Amber
Kleen-Stik 298	X	X	X			120–140	<a minute	Good		–40 + 200	Amber
G-P Transmount	X	X	X			150–250	10–15 minutes	Excellent		–60 + 250	Amber
G-P Transmount	X	X	X			150–250	10–15 minutes	Excellent		–60 + 250	Amber

*Courtesy: General Electric Co.

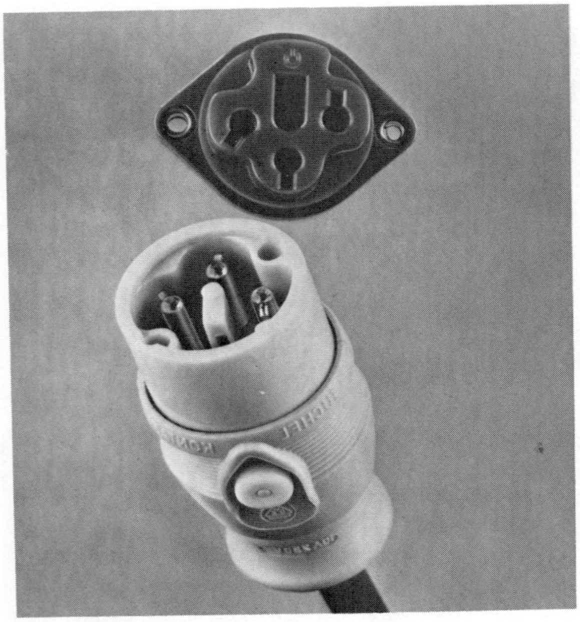

Fig. 5-140. The ability of polycarbonate to retain its electrical properties through a wide temperature and humidity range makes it a unique insulating material for such applications as heavy-duty electrical power connectors for industrial equipment and machinery, plug-and-socket components, and switch gear. (*Courtesy: Mobay Chemical Corporation*)

Fig. 5-141. A sealed beam halogen lamp with polycarbonate resin (Lexan) lens and reflector. Compared to conventional glass lamps, the new halogen lamps are almost 70% lighter, offer up to 400% greater impact strength, and improve optical efficiency by 20–30%. (*Courtesy: General Electric Co.*)

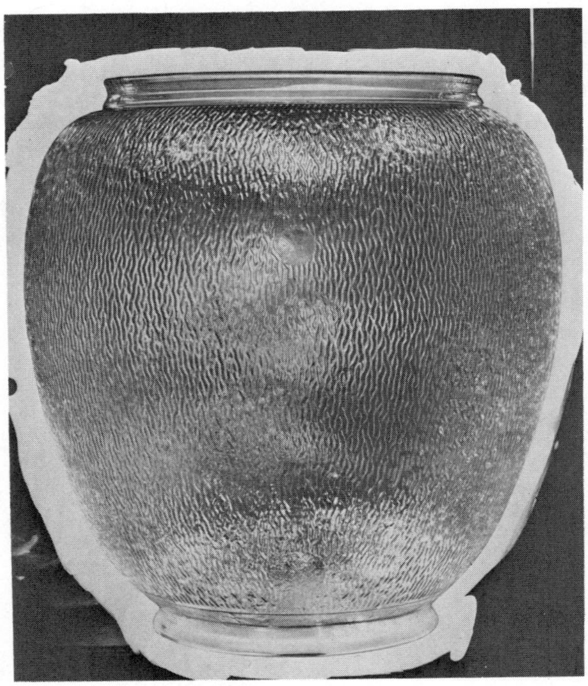

Fig. 5-142. Virtually indestructible lamp globes made of polycarbonate are gaining widespread use, particularly in high-vandalism areas. Heat-resistant polycarbonate transmits light at levels comparable to glass, repels sledgehammer blows and shotgun blasts. (*Courtesy: Mobay Chemical Corporation*)

Trade Names

Bayblend	ABS alloy	Mobay Chemical Co.
Carbaglas	Fiberglass-reinforced	Dart Industries, Inc.
Carbasur	Fiberglass-reinforced	Dart Industries, Inc.
Carlex	Light shield	Carlon Products, Inc.
Celanex	Film and sheet	Celanese Plastics Co.
Dapco		Davidson Products Co.
Defsan	PET blend	Soviet Origin
Diflon		Soviet Origin
Evolite	Rod, tube, and profile	Richard Daleman, Ltd.
Ilon		Soviet Origin
Iupilon		Mitsubishi Gas Chemical Co.
Lexan	Homopolymer	General Electric Co.
Macrofol	Film	Naftone, Inc.
Makrofol		Mobay Chemical Co.
Makrolon		M & B Plastics, Ltd.
Mem	Silicone copolymer	General Electric Co.
Merlon	Homopolymer	Mobay Chemical Co.
Mertex	PU blend	Mobay Chemical Co.
Metamarble	Meta-acrylic blend	Teijin Chemical Co., Ltd.
Novarex		Mitsui Petrochemical Industries, Ltd.
Nuclon		Beaunit Corporation
Panlite		Teijin, Ltd.
Polycarbafil	Reinforced	Dart Industries, Inc.
Polypinco	Shapes, extrusions, and parts	Polymer Corporation
Rhiamer	Block and rod	M & B Plastics, Ltd.
Rowlex	Sheet	Rowland, Inc.
Stilon	ABS alloy	Soviet Origin
Thermocomps	Filled or reinforced	Liquid Nitrogen Processing Corporation
Tuffak	Coated	Rohm & Haas Co.
Tuffak-Twinwall	Ribbed sheet	Rohm & Haas Co.
Zelux	Rod, sheet, and tube	Westlake Plastics Co.

REFERENCES AND BIBLIOGRAPHY

Baier, H. F., and Jacobi, F. "Biaxial Orientation of Film: The Future." Paper presentation, *SPE ANTEC*, New Orleans, May 1979.

Bash, N. E. Polycarbonate resins and sheets, from Plastics in Construction: Coming on Strong in 1978. *Plastics Engineering*, July 1978, **34**(7):35–37.

Brydson, J. A. *Plastics Materials.* 3rd Ed. Robert E. Krieger Publishing Co., Huntington, New York, 1975.

Christopher, W. F. Polycarbonate Resin. *SPE Journal*, June 1958, **14**(6):31–34.

Christopher, W. F., and Fox, D. W. *Polycarbonates.* Reinhold Publishing Corporation, New York, 1962.

Davis, A., and Golden, J. H. Stability of polycarbonate. *J. Chem. Soc.*, 1968, (1):45–47.

Doyle, E. N. *The Development and Uses of Polyester Products.* McGraw-Hill Book Co., New York, 1969.

General Electric Co. *Lexan Polycarbonate Resin Injection Molding.* Bulletin CDC-500F, September 1977.

——. *Lexan Resin, A Good Name to Stand On In the Appliance Industry.* Bulletin CDC 582A, October 1977.

——. *Lexan Polycarbonate Resin and Sheet–Processing Tips.* Bulletin CDC-505, March 1971.

——. *Lexan NB-155 Polycarbonate Resin–A New Non-Burning Polycarbonate With a High Heat Deflection Temperature of 155°C.* Bulletin CDC-509, October 1968.

——. *Lexan Polycarbonate Resin for Electronic Applications.* Bulletin CDC-469, June 1968.

——. *Properties Comparison Chart–Lexan Polycarbonate Resin and Other Plastics.* Bulletin CDC-363-c, January 1964.

——. *Lexan Sheet.* Bulletin CDC-470.

——. *Lexan Polycarbonate Resin.* Bulletin CDC-615.

——. *GE Plastics–World Leader in Engineering Plastics.* Bulletin CDP-74.

——. "Lexan Polycarbonate Resin–Design" Bulletin CDC-536.

Gruenwald, G. Cold-formed polycarbonate resin–properties and applications. *Modern Plastics*, September 1960, **37**(9):137–144, 225.

Haslam, J., and Willis, H. A. *Identification and Analysis of Plastics.* D. Van Nostrand Co., Inc., Princeton, New Jersey, 1965.

Hertzberg, R. W., and Manson, J. A. Fatigue testing flows make it better. *Plastics World*, May 1977, **35**(5):50–53.

Kahl, R. J. Designing with plastics–engineering polycarbonate parts. *Plastics Engineering*, January 1977, **33**(1):26–29.

Landfield, H. "International Update on Thermoforming Materials." Paper presentation, *SPE ANTEC*, New Orleans, May 1979.

Materials Engineering. Coated PC sheet resists wear and weather. July 1977, **86**(1):43.

——. New polycarbonates meet stiff FR rules. May 1976, **83**(5): 30.

Mobay Chemical Corporation. *How to Take Advantage of Merlon Polycarbonate Resin–A Profile of Application Ideas for Electrical and Electronic Designers.* 1976.

——. *A Short Course on Merlon Polycarbonate Resin.* 1975.

——. *Applications Unlimited for Merlon Polycarbonate.* 1971.

——. *An Engineering Handbook on Merlon Polycarbonate.* 1962.

Modern Plastics. Big expansion of Mobay's polycarbonates line. June 1979, **56**(6):88.
——. Polycarbonates are starting to move. July 1960, **37**(7):102–105, 178.
Plastics Engineering. Engineering resins primer. December 1974, **30**(12):17–32.
Plastics World. Transparent plastics: new and improved. January 1977, **35**(1):24–26.
——. Coating shields polycarbonate from scratches, abrasion. September 1971, **29**(9):82.
"Polycarbonate: A High Performance Plastic." Paper presentation, *Plastics World Seminar: Engineering Plastics II*, Los Angeles, March 1976.
Rammrath, H. G. "Polycarbonates." In, *Modern Plastics Encyclopedia*, S. Gross (exec. ed.), **45**(14A):260–263, McGraw-Hill, Inc., New York, 1968–69.
Sparboe, D. I. Engineers' guide to polycarbonate plastics. *Materials Engineering*, December 1970, **72**(7):26–29.
Theberge, J., and Cloud, P. Choosing the best glass-reinforced thermoplastic. *Plastics World*, March 1977, **35**(3):44–46.
Throne, J. L. "Cellular Structure and Foaming Agents in Resin Regions." Paper presentation, *SPE ANTEC*, New Orleans, May 1979.
Titus, J. B. *Trade Designations of Plastics and Related Materials*, (revised). PLASTEC NOTE N9C, Plastics Technical Evaluation Center, Picatinny Arsenal, Dover, New Jersey, May 1978.
Wallingford, J. R., & Riewe, A., "Polycarbonate" from "Plastics in Packaging" *Plastics Engineering*, October 1978, 34(10):49–50.
Wood, A. S. Engineering resins, from Materials '79. *Modern Plastics*, January 1979, **56**(1):53–55.
——. Clarity plus lots of useful extras: it's available at $3/lb—or 30¢. *Modern Plastics*, March 1979, **56**(3):50–53.
——. Action in profile extrusion shifts to engineering resins. *Modern Plastics*, November 1978, **55**(11):46–48.
——. "The Choices Get Tougher in Engineering Thermoplastics." *Modern Plastics*, July 1978, **55**(7):42–44.
Zimmerman, D. D. "Performance of Engineering Resins in Electrical Connectors." Paper presentation, *SPE ANTEC*, New Orleans, May 1979.

Polyphenylene Oxide (PPO)

PPO is made by a polymerization process known as oxidative coupling. *Oxidative coupling* is defined as a reaction of O_2 with active hydrogens from different molecules to produce water and coupled molecules. If the molecule (monomer) has two active hydrogens, oxidative coupling can continue and a polymerization results.

The PPO resins currently available are based on the copper-catalyzed oxidation of 2,6-xylenol.

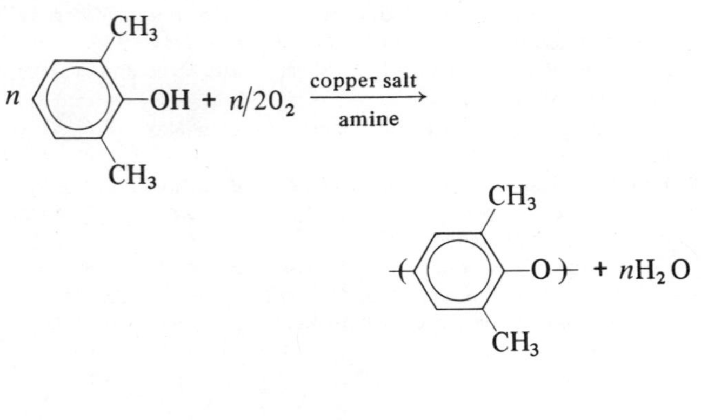

In a typical process, the 2,6-xylenol is reacted with oxygen in pyridine in the presence of copper (I) chloride for about 7 minutes at 20–46°C (68–115°F). The reaction mixture is added to methyl alcohol, filtered, and washed. A colorless solid results.

In 1965, General Electric introduced this polymer as polyphenylene oxide, even though it was in fact a substituted PPO. The unsubstituted PPO, halogenated derivatives, and PPO's with other substituent groups have not achieved any commercial importance.

The T_g of PPOs is 208°C (406°F). When polymerized, the polymer is crystalline, yet has a relatively low T_m of 257°C (527°F). The T_g/T_m ratio of 0.91 is uniquely high. Because of the small difference in T_g and T_m, there is little time for crystallization to occur on cooling, and the resulting polymer is usually amorphous. Crystallization can occur by heat treatment or solvation.

The commercial polymers have a molecular weight of 25,000–60,000 and can contain branches and even cross-links from thermal oxidation. These resins are characterized by exceptional dimensional stability. It is easy to mold to small tolerances on conventional equipment because of their low coefficient of thermal expansion, low mold shrinkage, and low water absorbance. They also maintain their mechanical properties over a wide temperature range, electrical properties over a wide temperature and frequency range, and have good resistance to aqueous environments, acids, and bases.

The price of PPOs has always been too high to permit very widespread commercial use. Consequently, in 1966, General Electric introduced Noryls—defined as "modified PPOs." Their exact chemical nature is still unknown. Initially, it was suspected that they were *o*-cresol-derived polymers, but it now appears that their key feature is that they are polymer blends, most likely with polystyrene or high-impact polystyrene.

If polystyrene ($T_g \simeq 90$°C, 194°F) is blended with poly(2,6-dimethyl phenylene oxide) in equal quantities, a transparent polymer with a single T_g at 150°C (302°F) is obtained. Both PPO and polystyrene have similar secondary transitions at approximately 116°C (241°F), and the blends also show this transition. The electrical properties of the blends and those of the individual polymers are also very similar. The viscosity of the blends is, however, lower than pure PPO, reflecting the influence of the lower-viscosity polystyrene. Thus, these Noryl resins have improved processability along with a reduced cost.

The properties of the blended resins are shown in Table 5-129. In summary, these polymers exhibit excellent mechanical properties over a temperature range from below -40°C (-40°F) to above 93°C (200°F); self-extinguishing and non-dripping characteristics; low creep; low water absorption; high modulus (excellent dimensional stability); good electrical properties; and excellent hydrolytic stability. Dimensional change is negligible in boiling water,

Table 5-129. Properties of Polyphenylene Oxides.*

		ASTM test method	Extrusion grades	Molding grades: Non-reinforced	Molding grades: 20-30% glass reinforced
Processing	1. Molding qualities	—	Excellent	Excellent	Very good
	2. Compression molding temp., °F.	—	400-460	400-460	—
	3. Comp. molding pressure, p.s.i.	—	—	500-1000	—
	4. Injection molding temp., °F.	—	—	425-600	500-650
	5. Injection molding pressure, p.s.i.	—	—	14000-20000	15000-40000
	6. Compression ratio	—	—	1.3-2.2	—
	7. Mold (linear) shrinkage, in./in.	—	0.005-0.007	0.005-0.007	0.001-0.004
	8. Specific gravity (density)	D792	1.06-1.10	1.06-1.10	1.21-1.36
	9. Specific volume, cu. in./lb.	D792	26.1-25.2	26.1-25.2	22.9-20.2
	10. Machining qualities	—	Excellent	Excellent	Excellent
Mechanical	11. Tensile strength, p.s.i.	D638	7800-9600	7800-9600	14500-17000
	12. Elongation, %	D638	50.0-60.0	50.0-60.0	4.0-6.0
	13. Tensile elastic modulus, 10^5 p.s.i.	D638	3.55-3.80	3.55-3.80	9.25-12.0
	14. Compressive strength, p.s.i.	D695	16000-16400	16000-16400	17600-17900
	15. Flexural yield strength, p.s.i.	D790	12800-13500	12800-13500	18500-20000
	16. Impact strength, ft. lb./in. of notch (½ × ½ in. notched bar, Izod test)	D256	5.0 (½ × ⅛ in. bar)	5.0 (½ × ⅛ in. bar)	2.3 (½ × ⅛ in. bar)
	17. Hardness, Rockwell	D785	R115-119	R115-119	L106-108
	18. Flexural elastic modulus, p.s.i. × 10^5	D790	3.6-4.0	3.6-4.0	7.5-11.0
	19. Compressive modulus, p.s.i. × 10^5	D695	3.7	3.7	9.0-13.0
Thermal	20. Thermal conductivity, 10^{-4} cal./sec./sq. cm./1 (°C./cm.)	C177	5.16	5.16	3.8
	21. Specific heat, cal./°C./gm.	—	.32	0.32	—
	22. Thermal expansion, 10^{-5} in./in./°C.	D696	5.2	5.2	2.2
	23. Resistance to heat, °F. (continuous)	—	175-220	175-220	240-265
	24. Deflection temp., °F. at 264 p.s.i. fiber stress	D648	212-265	212-265	270-300
	at 66 p.s.i. fiber stress		230-280	230-280	280-310
Electrical	25. Volume resistivity, ohm-cm. (50% RH and 23° C.)	D257	$>10^{16}>10^{17}$	$>10^{16}->10^{17}$	$>10^{17}$
	26. Dielectric strength, short-time, ⅛-in. thickness, volts/mil	D149	400-550	400-550	420-600
	27. Dielectric strength, step-by-step, ⅛-in. thickness, volts/mil	D149	400-550	400-550	420-600
	28. Dielectric constant, 60 cyc.	D150	2.64	2.64	2.93
	29. Dielectric constant, 10^3 cyc.	D150	2.64	2.64	2.93
	30. Dielectric constant, 10^6 cyc.	D150	2.64	2.64	2.92
	31. Dissipation (power) factor, 60 cyc.	D150	0.0004	0.0004	0.0009
	32. Dissipation (power) factor, 10^3 cyc.	D150	0.0004	0.0004	0.0010
	33. Dissipation (power) factor, 10^6 cyc.	D150	0.0009	0.0009	0.0015
	34. Arc resistance, sec.	D495	75	75	70-120
Optical	35. Refractive index, n_D	D542	—	—	—
	36. Clarity	—	Opaque	Opaque	—
	37. Transmittance, %	—	—	—	—
	38. Haze, %	—	—	—	—
Resistance characteristics	39. Water absorp., 24 hr., ⅛-in. thick., %	D570	0.066	0.066	0.06
	40. Flammability[a] Burning rate, in./min.	D635	—	—	—
	AEB, in./ATB, sec.[b]		<0.16	<0.16	<0.16
	41. Effect of sunlight	—	Colors may fade		
	42. Effect of weak acids	D543	None		
	43. Effect of strong acids	D543	None		
	44. Effect of weak alkalies	D543	None		
	45. Effect of strong alkalies	D543	None		
	46. Effect of organic solvents	D543	Soluble or swells in some aromatics and chlorinated aliphatics; resistant to alcohols		Softens in some aromatics and chlorinated aliphatics; resistant to alcohols

[a]This numerical flame spread rating is not intended to reflect hazards presented by this or any other material under actual fire conditions.
[b]AEB = Avg. extent of burning in inches; ATB = Avg. time of burning in seconds.
*Reprinted by permission from *Modern Plastics Encyclopedia*, McGraw-Hill, Inc.

Table 5-130. How Noryl Compares with Other Engineering Materials in Typical Properties.*

Property[a]	ASTM Method	Noryl	PPO	Polysulfone	Poly-carbonate	Acetal	High-Heat ABS	Nylon 6/6
Specific gravity	D-792	1.06	1.06	1.24	1.2	1.42	1.06	1.13–1.15
Water absorption, %[b]	D-570	0.066	0.06	0.22	0.15	0.25	0.36	1.5
Tensile yield strength, 10^3 psi	D-638	9.6	10–11	10.2	8.5	10	6.8	8.5
Tensile modulus, 10^5 psi	D-638	3.5	3.6–3.8	3.6	3	4	3.8	2.6–4
Elongation at fail, %	D-638	20	50–80	50–100	80	15	20–100	60
Flexural strength, 10^3 psi	D-790	13.5	14–15	15.4	13.5	14.1	11.1	8–14
Flexural modulus, 10^5 psi	D-790	3.6	3.5–3.8	3.9	3.4	4.1	4.2	4.1
Izod imp str, ft lb/in. notch	D-256	1.3	1.5–1.9	1.3	15	1.4–2.3	3.1	0.9
Hardness, Rockwell	D-785	R119	R118–120	R120	R118	R120	R112	R118
Heat deflection temperature F, 264 psi	D-648	265	375–380	345	270	255	245	150
Max rec cont use temperature		200	225–250	300	250–270	185	180	250
Flammability	D-635	Self-ext.	Self-ext.	Self-ext.	Self-ext.	1.1[c]	1.8	Self-ext.
Mold shrinkage, in./in.	D-149	0.005–0.007	0.007	0.007	0.006	0.025	0.004	0.015
Dielectric strength, v/mil[d]	D-149	550	400–550	425	400	500	415	385
						6×	>1.6×	10^{14}
Vol resistivity, ohm-cm	D-275	10^{17}	10^{17}	5×10^{16}	10^{16}	10^{14}	10^{16}	10^{15}
Dielectric constant, 60 cps[e]	D-150	2.64	2.58	3.14	3.17	3.7	2.82	4
Dissipation factor, 60 cps[c]	D-150	0.0004	0.00035	0.0008	0.0009		0.004	0.014

[a]Values shown should not be considered precise figures, but average values taken from various sources.
[b]73 F, 24 hr.
[c]Drips.
[d]Short time, $\frac{1}{8}$ in. sample.
[e]73 F, 50% RH.
*Reprinted by permission from *Plastics Technology*, June 1966.

detergents, and weak and strong acids and bases. Table 5-130 and Figures 5-143 through 5-150 compare Noryl properties with those of other engineering thermoplastics.

Glass-filled phenylene oxides provide up to double the mechanical strength, four times the modulus, an improved strength/weight ratio and good impact strength from -46°C to 121°C (-50°F-250°F).

The melt viscosity of the blended resins shows considerable shear sensitivity at the high shear ratio encountered in injection molding. A 22°C (40°F) rise in melt temperature halves the viscosity (see Figure 5-151). Unlike polycarbonate, polysulfones, and the unblended PPOs, the blended resins reputedly do not require predrying. However, for optimum appearance, a 1-hour pre-drying in a dehumidifying oven at 104-110°C (220-230°F) is recommended.

Screw machines are preferred because their rapid, uni-

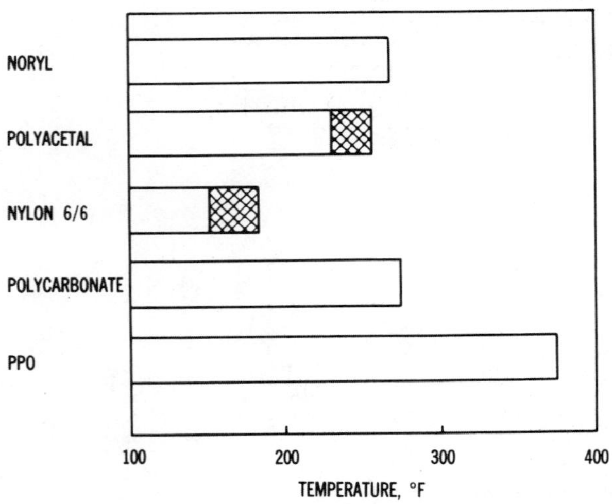

Fig. 5-143. Heat-deflection temperature comparison (ASTM D648, 264 psi). (*Courtesy: General Electric Co.*)

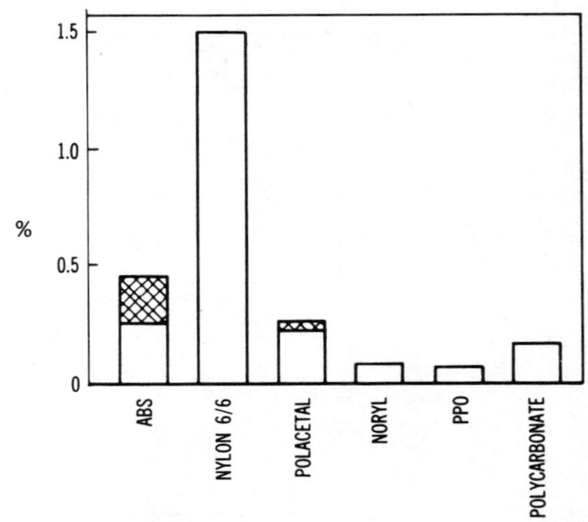

Fig. 5-144. Water absorption properties (ASTM D570, 24 hrs, 73°F). (*Courtesy: General Electric Co.*)

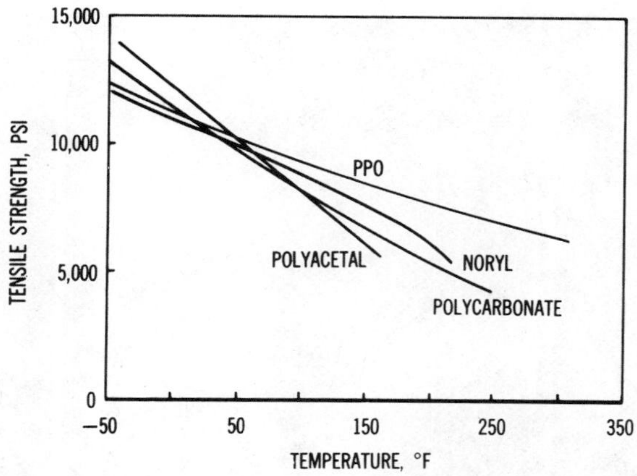

Fig. 5-145. Tensile strength versus temperature (ASTM D638). (*Courtesy: General Electric Co.*)

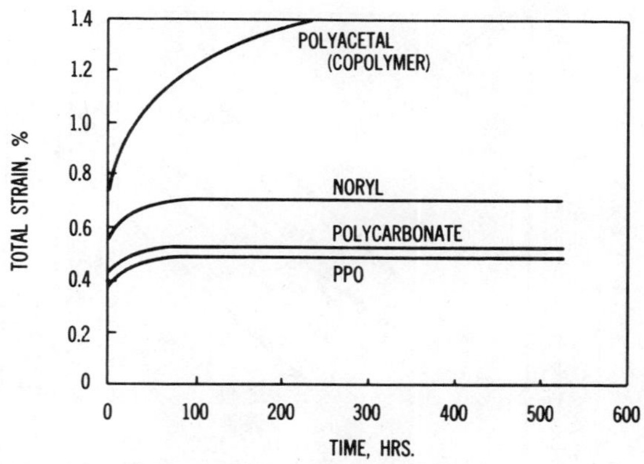

Fig. 5-148. Tensile creep properties (73°F, 2000 psi). (*Courtesy: General Electric Co.*)

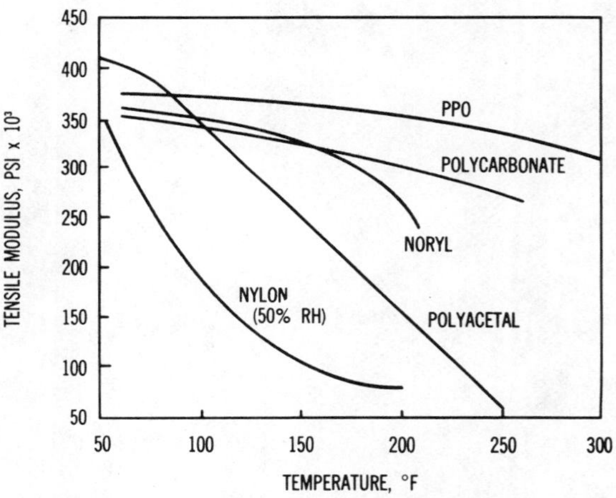

Fig. 5-146. Tensile modulus versus temperature (ASTM D638). (*Courtesy: General Electric Co.*)

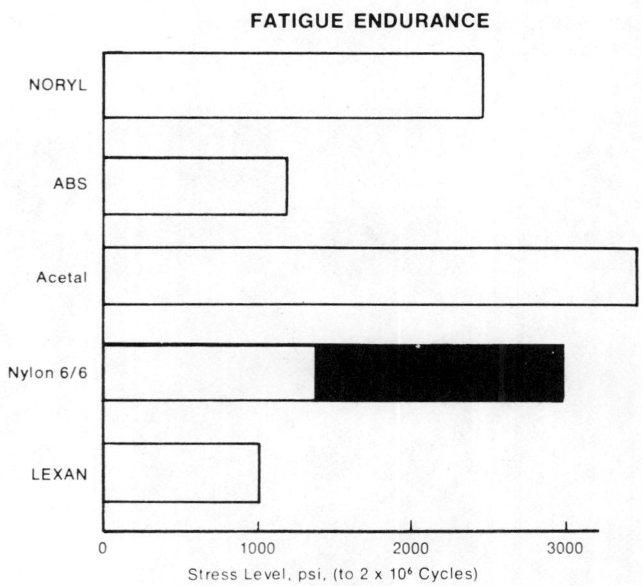

Fig. 5-149. Fatigue endurance. (*Courtesy: General Electric Co.*)

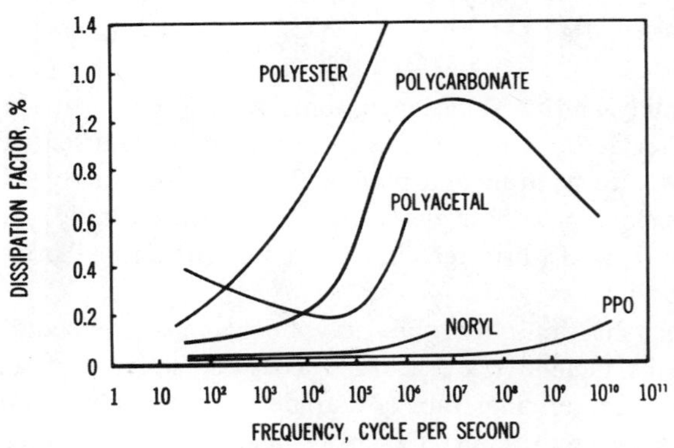

Fig. 5-147. Variation of dissipation factor with frequency (73°F). (*Courtesy: General Electric Co.*)

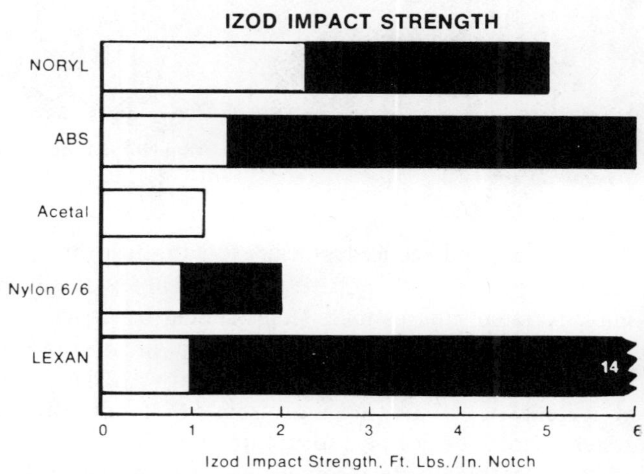

Fig. 5-150. Izod impact strength. (*Courtesy: General Electric Co.*)

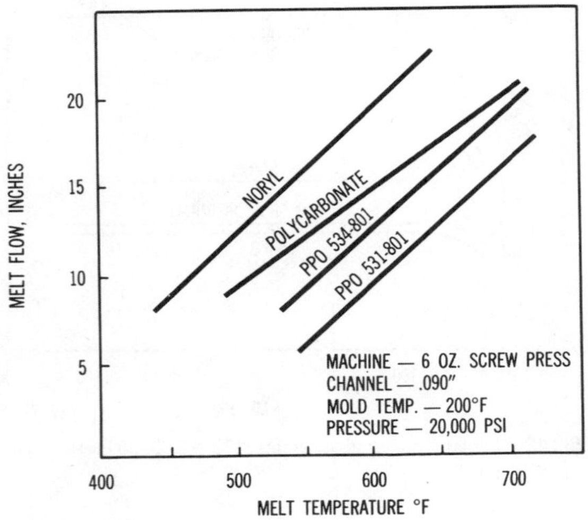

Fig. 5-151. Melt flow versus temperature. (*Courtesy: General Electric Co.*)

Fig. 5-152. Business machine housings made from Noryl. (*Courtesy: General Electric Co.*)

Fig. 5-153. Television cabinet with woodgrain finish made from Noryl. (*Courtesy: General Electric Co.*)

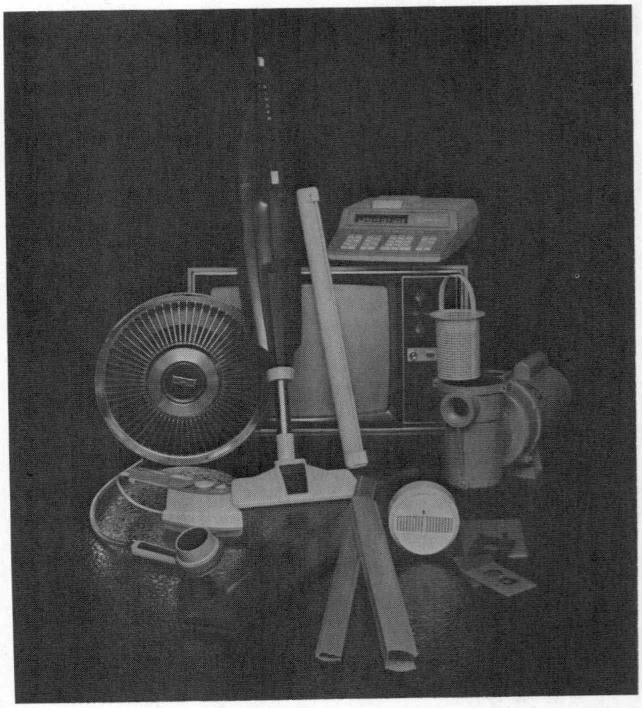

Fig. 5-154. A collage of products made from Noryl. (*Courtesy: General Electric Co.*)

form heating and reduced pressure loss through the cylinder, enable molding many parts that could not be filled in a similarly rated plunger unit. Molds should be designed to allow for material shrinkage of 0.008 in/in. A cold slug well should be provided to receive the relatively cold charge of material which first emerges from the nozzle. Runner systems should be designed so all turns are generous.

Vented extruders are recommended, which help further reduce the need for predrying. A 24:1 screw L/D (length to diameter) ratio is preferred.

The blended resins are well suited for conventional assembly and finishing operations. They are compatible with other resins and metals, do not corrode, and lend themselves to painting and plating. They can be readily solvent-cemented, adhesive-bonded and ultrasonically sealed. Table 5-131 lists recommended solvents and adhesives for Noryl resins.

Applications for the polyphenylene oxide family of resins include television and communication components such as terminal blocks and deflection yokes. Plumbing and automotive parts such as handles, vents, and brackets are a major end use. Housings, particularly those molded in the structural-foam configuration dominate the business machine and appliance markets (see Figures 5-152 through 5-154).

Table 5-131. Recommended Solvents and Adhesives for Noryl Resins.*

Adhesive	Type	Cure Time/Temp.	Tensile Shear psi	Available from
Ethylene dichloride	Solvent	48 hrs./RT	3,600	Chemical supply houses
Chloroform	Solvent	48 hrs./RT	2,900	Chemical supply houses
Methylene chloride	Solvent	48 hrs./RT	2,800	Chemical supply houses
Trichloroethylene	Solvent	48 hrs./RT	2,800	Chemical supply houses
85% methylene chloride 15% trichloroethylene	Solvent	48 hrs./RT	2,500	Chemical supply houses
Permabond (blue & white)	Cyano Acrylate Monomer	30 sec./RT	3,300	Rexco Trading Corporation New York, New York
Monobond M-2	Epoxy	30 min./250°F	5,000	Allaco Products Braintree, Massachusetts
Monobond M-5 (Electrical grade)	Epoxy	30 min./250°F	5,000	Allaco Products Braintree, Massachusetts
Flexobond	Epoxy	8 hrs./RT	3,500	Allaco Products Braintree, Massachusetts
No. 2214 (Regular)	Epoxy	40 min./250°F	5,000	3 M Corporation St. Paul, Minnesota
No. 2214 (hi-temp)	Epoxy	40 min./250°F	4,500	3 M Corporation St. Paul, Minnesota
No. 2214 (hi-flex)	Epoxy	40 min./250°F	1,900	3 M Corporation St. Paul, Minnesota
No. 1838	Epoxy	24 hrs./RT	3,500	3 M Corporation St. Paul, Minnesota
Epon 907	Epoxy	1 hr./225°F	2,100	Shell Chemical Company Pittsburgh, California
No. 2569/No. 21	Epoxy	24 hrs./RT	3,100	Epoxy Products Corporation Irvington, New Jersey
Pr 1201QG (Class 1)	Polysulfide Epoxy	24 hrs./RT	2,000	Products Research & Chemical Co. Gloucester City, New Jersey
Uralane 8089	Polyurethene	24 hrs./RT	1,200	Furane Plastics Fairfield, New Jersey
RTV 108	Silicone	–	100	General Electric Company Waterford, New York
RTV 102	Silicone	–	100	General Electric Company Waterford, New York
RTV 116	Silicone	–	100	General Electric Company Waterford, New York

*Courtesy: General Electric Co.

Trade Names

Alphalux	Rod, sheet, and tube	Westlake Plastics Co.
Ariloks		Soviet Origin
Custom Poly CR		Custom Materials, Inc.
Doryl		Westinghouse Electric Co.
Fiberlene	Glass-filled	Service Color Corporation
Noryl		General Electric Co.
PPO		General Electric Co.
Ztron		Polymer Corporation

REFERENCES AND BIBLIOGRAPHY

Borrello, R. T. Phenylene oxide-based resins, from Plastics in Construction: Coming on Strong in 1978. *Plastics Engineering*, July 1978, **34**(7):32–34.

Brydson, J. A. *Plastics Materials.* 3rd Ed. Robert E. Krieger Publishing Co., Huntington, New York, 1975.

Forger, G. R. Phenylene-oxide-based resins withstand high heat and humidity. *Materials Engineering*, September 1977, **86**(3): 76–78.

General Electric Co. *Noryl Thermoplastic Resins–FN-215 Foamable Noryl Resin.* Bulletin CDX-180, July 1972.

——. *Introducing Noryl 100 Sheet With Just the Right Performance, Properties and Economy.* Bulletin CDC-514, 1969.

——. *Glass-Filled Noryl Thermoplastic Resin–PPO Polyphenylene Oxide.* Bulletin CDX-38B, May 1968.

——. *Noryl Engineering Plastics.* Bulletin CDX-80 and Supplement.

——. *Noryl Thermoplastic Resin.* Bulletin CDX-14.

——. *Noryl Thermoplastic Resins–Decorating Techniques.* Bulletin CDX-82A.

——. *Noryl Thermoplastic Resins–Design.* Bulletin CDX-83A.
——. *Noryl Thermoplastic Resins–Engineering Plastics. . . Unmatched for Performance and Economy.* Bulletin CDX-80B.
——. *Noryl Thermoplastic Resins–Injection Molding.* Bulletin CDX-81A.
——. *PPO Polyphenylene Oxide–Technical Report, Processing.*
Gowan, A. C. "Phenylene Oxides." In *Modern Plastics Encyclopedia*, S. Gross (exec. ed.), **45**(14A):244–249. McGraw-Hill, Inc., 1968–69.
Materials in Design Engineering. New Thermoplastic stays tough at 375°F. January 1965, **61**(1):6–11.
Modern Plastics. New Phenylene oxide resin. June 1966, **43**(6):90–91, 165.
Plastics Engineering. Engineering resins primer. December 1974, **30**(12):17–32.
Plastics Technology. Noryl, PPO offshoot, complements PPO, PC and polysulfone. June 1966, **12**(6):13–17.
Theberge, J., and Cloud, P. Choosing the best glass-reinforced thermoplastic. *Plastics World*, March 1977, **35**(3):44–46.
Wood, A. S. The choices get tougher in engineering thermoplastics. *Modern Plastics,* July 1978, **55**(7):42–44.
——. Action in profile extrusion shifts to engineering resins. *Modern Plastics*, November 1978, **55**(11):46–48.

Polysulfones

The name polysulfone has been assigned to polymers with SO_2 groups in the backbone. At present there are four basic sulfone types: standard, polyaryl, polyether, and polyphenyl.

Standard Polysulfone. When a metal phenate is reacted with an "activated" aromatic halide, an aryl ether is obtained. If a dihalide is "activated" by a sulfone-containing moiety, the resultant polymer will be a polysulfone, the latter being the "activating" agent for chlorobenzene. Polysulfone is produced by the reaction between the sodium salt of 2,2-bis(4-hydroxyphenyl) propane (Bisphenol A) and 4,4′-dichlorodiphenyl sulfone.

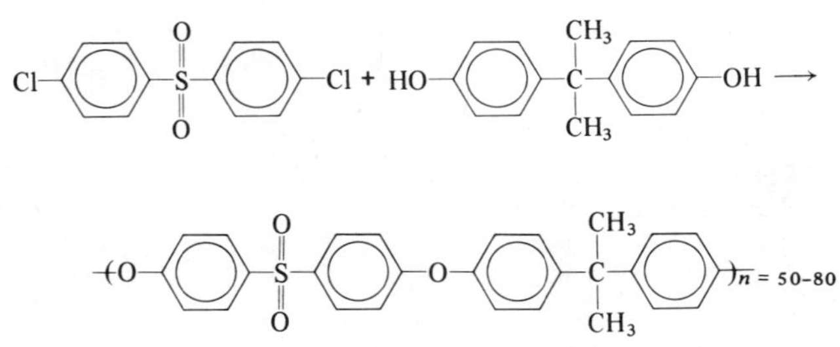

Aqueous caustic is added after heating the mixture to 60°C (140°F). A two-phase system is formed consisting of predominantly chlorobenzene and aqueous dimethyl sulfoxide containing the disodium salt. This solution is heated under reflux to 150–160°C (302–320°F). Then the 4,4′-dichlorodiphenyl sulfone is added and reacted. Methyl chloride is added to control polymerization, which causes the product to take on a light amber to orange-yellow color.

The unique feature of the chemical structure of these polymers is the diarylsulfone grouping. This is a highly resonating structure in which the sulfone group tends to draw electrons from the phenyl rings. This resonance is enhanced by having oxygen atoms para to the sulfone group. Oxidation, by definition, is a loss of electrons. Having the electrons tied up in resonance imparts excellent oxidation resistance to the polymer. Also, the sulfur atom is in its highest state of oxidation and, therefore, is not susceptible to oxidation as in other sulfur-containing polymers such as polyphenylene sulfide.

The high degree of resonance has two additional effects: (1) it increases the strength of the bonds involved, and (2) it fixes this grouping spatially into a planar configuration. This provides excellent thermal stability to the polymer and provides rigidity to the polymer chain, which is retained at high temperatures.

The excellent thermal stability of polysulfone is verified by thermal gravimetric analysis which shows that polysulfone is stable in air up to 450°C (840°F). The excellent thermal stability coupled with excellent oxidation resistance of polysulfones provides excellent melt stability for molding and extrusion.

The ether linkage imparts some flexibility to the polymer chain which gives inherent toughness to the material. In common with all tough, rigid thermoplastic polymers, polysulfone has a second, low-temperature glass transition at -100°C (-148°F). This minor glass transition is assigned to the ether linkages.

In these polysulfones, the linkages connecting the benzene rings are hydrolytically stable. Therefore, the polymers are resistant to hydrolysis and to aqueous acid and alkaline environments. This is a distinct advantage over the ester linkages present in polycarbonates and thermoplastic polyesters.

Polyethersulfone. The condensation of diphenyl ether with 4,4′-diphenyl ether disulphenylchloride yields the polyethersulfone.

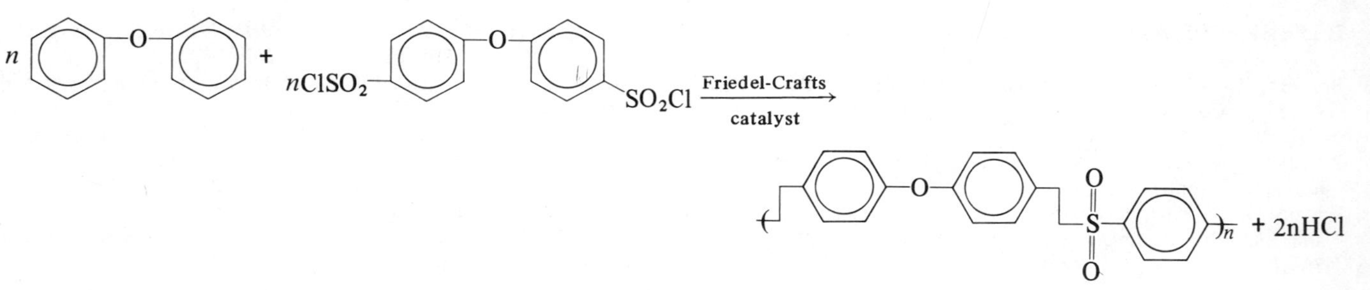

Self-condensation of 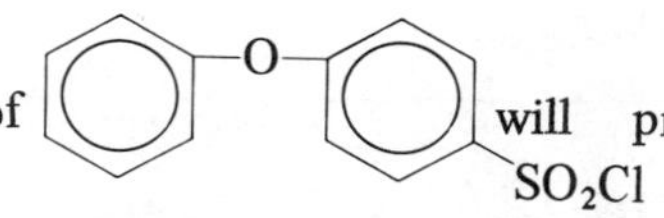 will produce the same polymer with higher molecular weights.

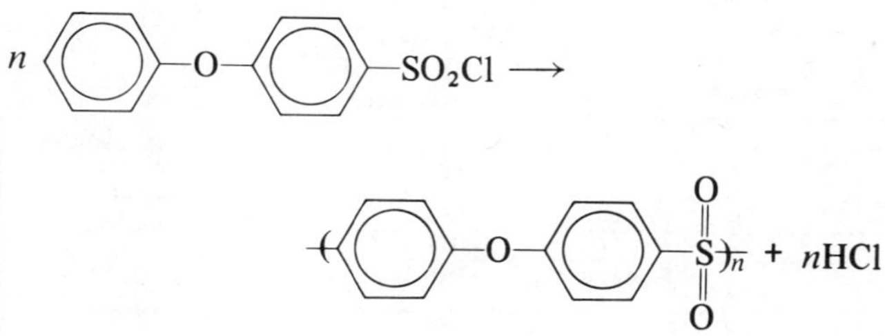

Structurally, the comments made for the standard polysulfones apply to the polyethersulfones. The replacement of the ether linkage for the isopropylidine linkage causes the T_gs to increase (see Table 5-132, second-order T_g = -70°C (-94°F).

Polyarylsulfone. If the diphenylether used to prepare the polyethersulfone is replaced with diphenyl, then the polyarylsulfone is produced.

Table 5-132. Polysulfone Temperature Properties.

	T_g, °C (°F)	Heat-Deflection Temperature °C (°F)	Continuous-Use Temperature °C (°F)
Standard polysulfone	190 (374)	174 (345)	149 (300)
Polyethersulfone	221 (430)	202 (395)	177 (350)
Polyarylsulfone	– –	204 (400)	191 (375)
Polyphenylsulfone	288 (550)	274 (525)	260 (500)

dimethyl acetamide. They will stress-crack in the presence of organic ketones, esters, aromatic hydrocarbons, and some chlorinated hydrocarbons. Filling the resin with glass fiber somewhat improves the stress-cracking resistance. Table 5-134 shows comparative stress-crack data for standard polysulfone, polyphenylsulfone, and polycarbonate.

In general, the polysulfones are most noted for their exceptional resistance to creep (see Figures 5-157 through 5-159), good high-temperature resistance (see Figures 5-160 through 5-163), rigidity, transparency, and self-extinguishing characteristics (see Figure 5-164).

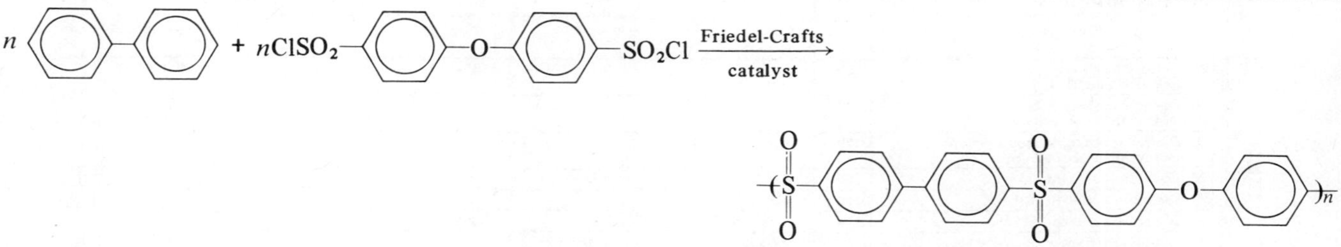

As a result of the loss of about half of the ether groups, the T_g of the polymer further increases over the values for the other two types of polysulfone.

Polyphenylsulfone. As of this writing, the manufacturer of this polymer has not disclosed the nature of the non-sulfone-containing portion of the backbone. However, reliable investigators have suggested the following structure as being most probable:

Comparing the polysulfones with the polycarbonates, which they so closely resemble, we find the former are more heat and creep resistant, whereas the latter have higher impact strengths and are cheaper. Figure 5-165 compares the effect of boiling water on properties of these two types of engineering thermoplastics.

Aside from the difference in T_g and heat-deflection temperature, the differences between the four polysulfones are not great. The other areas of difference are water absorp-

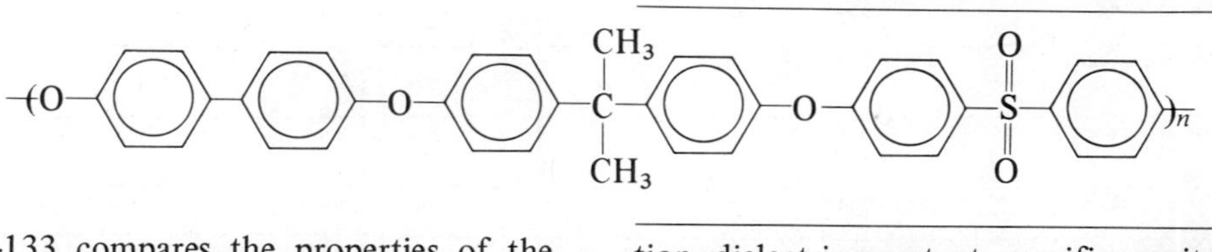

Properties. Table 5-133 compares the properties of the four types of polysulfones. The electrical insulation properties are good over a wide frequency range, although tracking resistance is not good. Figures 5-155 and 5-156 compare capacitance and loss tangent data of some polysulfones and thermoplastics.

The commercial resin grades are resistant to aqueous acids and alkalis, and aliphatic hydrocarbons. They are attacked by concentrated H_2SO_4, dimethyl formamide, and

tion, dielectric constant, specific gravity and cost. All four can be glass or carbon fiber filled for improved creep, coefficient of thermal expansion, and toughness.

Processing. The polysulfones have high softening points. The resulting melts are high in viscosity (see Figure 5-166). Thus, high processing temperatures and pressures as well as sturdy equipment are required (see Figures 5-167 and 5-168). Oriented molecular freeze-in is common, leading

Table 5-133. Properties of Polysulfones.*

		ASTM test method	Polysulfone	Polyphenyl-sulfone	Polyethersulfone		Polyaryl-sulfone
					Unfilled	40% glass filled	
Processing	1. Molding qualities	—	Excellent	—	Excellent	Excellent	Excellent
	2. Compression molding temp., °F.	—	550-600	—	590-690	600-700	700-750
	3. Comp. molding pressure, p.s.i.	—	1000	—	1000-1500	1000-2000	1000-2000
	4. Injection molding temp., °F.	—	650-750	—	590-750	600-750	700-800
	5. Injection molding pressure, p.s.i.	—	15000-20000	—	10000-20000	10000-20000	20000-50000
	6. Compression ratio	—	1.8-2.2	—	—	—	2.5
	7. Mold (linear) shrinkage, in./in.	—	0.007	0.008	0.007	0.0015	0.007-0.009
	8. Specific gravity (density)	D792	1.24	1.29	1.37	1.68	1.36
	9. Specific volume, cu. in./lb.	D792	22.3	—	19.7-20.2	16.6	20.4
	10. Machining qualities	—	Excellent	—	Excellent	Excellent	Excellent
Mechanical	11. Tensile strength, p.s.i.	D638	10200 (at yield)	10400 (at yield)	12200 (at yield)	23000	13000
	12. Elongation, %	D638	50-100 (at break)	60 (at break)	30-80 (at break)	3-4	13-20
	13. Tensile elastic modulus, 10^5 p.s.i.	D638	3.6	3.1	3.5-3.54	20	3.7
	14. Compressive strength, p.s.i.	D695	13900 (at yield)	—	—	—	17900
	15. Flexural yield strength, p.s.i.	D790	15400 (at yield)	12400	18650 (at yield)	30000	17200
	16. Impact strength, ft. lb./in. of notch (½ × ½ in. notched bar, Izod test)	D256	1.3 (¼ in. bar) 1.2 (⅛ in. bar) (−40°F.)	12 (⅛ in. bar) 8 (⅛ in. bar) (-40°C.)	1.6	1.6	1-2 at 73°F.
	17. Hardness, Rockwell	D785	M69, R120	—	M88	M98	M110
	18. Flexural elastic modulus, p.s.i. × 10^5	D790	3.9	33	3.7-3.75	16	3.95
	19. Compressive modulus, p.s.i. × 10^5	D695	3.7	—	—	—	3.4
Thermal	20. Thermal conductivity, 10^{-4} cal./sec./sq. cm./1 (°C./cm.)	C177	2.8	—	3.2-4.4	—	4.55
	21. Specific heat, cal./°C./gm.	—	0.31	—	0.26	—	—
	22. Thermal expansion, 10^{-5} in./in./°C.	D696	5.2-5.6	3.1	5.5	—	4.7
	23. Resistance to heat, °F. (continuous)	—	300-345	350-375	300-390	350-410	500
	24. Deflection temp., °F. at 264 p.s.i. fiber stress at 66 p.s.i. fiber stress	D648	345 358	400 —	397 —	420 —	525 —
Electrical	25. Volume resistivity, ohm-cm. (50% RH and 23° C.)	D257	5.0×10^{16}	3.5×10^{15}	10^{17}-10^{18}	$>10^{16}$	3.2×10^{16}
	26. Dielectric strength, short-time, ⅛-in. thickness, volts/mil	D149	425	371	400	—	350
	27. Dielectric strength, step-by-step, ⅛-in. thickness, volts/mil	D149	400	—	—	—	—
	28. Dielectric constant, 60 cyc.	D150	3.07-3.14	3.44	3.50	—	3.94
	29. Dielectric constant, 10^3 cyc.	D150	3.06-3.13	—	3.50	—	3.80
	30. Dielectric constant, 10^6 cyc.	D150	3.03-3.10	3.45	3.50	—	3.70
	31. Dissipation (power) factor, 60 cyc.	D150	0.0008	0.00058	0.001	—	0.003
	32. Dissipation (power) factor, 10^3 cyc.	D150	0.0010	—	0.0035	—	0.0035
	33. Dissipation (power) factor, 10^6 cyc.	D150	0.0034	0.00764	0.006	—	0.013
	34. Arc resistance, sec.	D495	75-122	41	65-75	—	67
Optical	35. Refractive index, n_D	D542	1.633	—	1.65	—	1.67
	36. Clarity	—	Transparent, opaque	Transparent	Transparent	Opaque	Opaque
	37. Transmittance, %	—	—	—	—	—	—
	38. Haze, %	—	5.0	—	—	—	—
Resistance characteristics	39. Water absorp., 24 hr., ⅛-in. thick., %	D570	0.22	—	0.43	0.31	1.1
	40. Flammability[a] Burning rate, in./min. AEB, in./ATB, sec.[b]	D635	— 0.4/5.0	0.28 —	— —	— —	— —
	41. Effect of sunlight	—	Strength loss, slight yellowing	Strength loss, slight yellowing	Slight yellowing, strength loss	—	Slight darkening
	42. Effect of weak acids	D543	None	—	None	None	None
	43. Effect of strong acids	D543	None	—	None	None	None
	44. Effect of weak alkalies	D543	None	—	None	None	None
	45. Effect of strong alkalies	D543	None	—	None	None	None
	46. Effect of organic solvents	D543	Aliphatic—none; aromatic—partly soluble	—	Aliphatic-none aromatic-partly soluble	Attacked by some chlorinated solvents and ketones	Soluble highly polar solvents, unaffected by most common solvents, fuel, lubricants

[a]This numerical flame spread rating is not intended to reflect hazards presented by this or any other materials under actual fire conditions.
[b]AEB = Avg. extent of burning in inches; ATB = Avg. time of burning in seconds.
*Reprinted by permission from *Modern Plastics Encyclopedia*, McGraw-Hill, Inc.

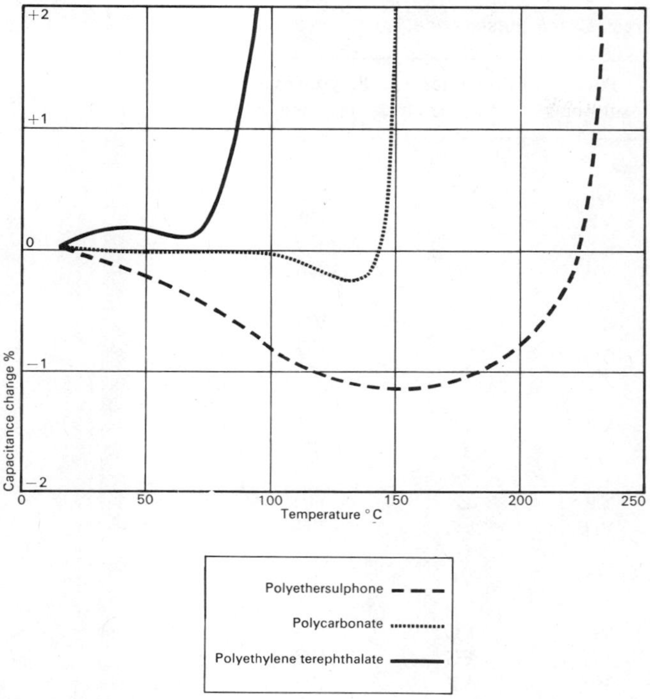

Fig. 5-155. Variation of capacitance with temperature of some thermoplastics. (Victrex® Polyethersulfone is offered by ICI Americas Inc., Wilmington, Del.)

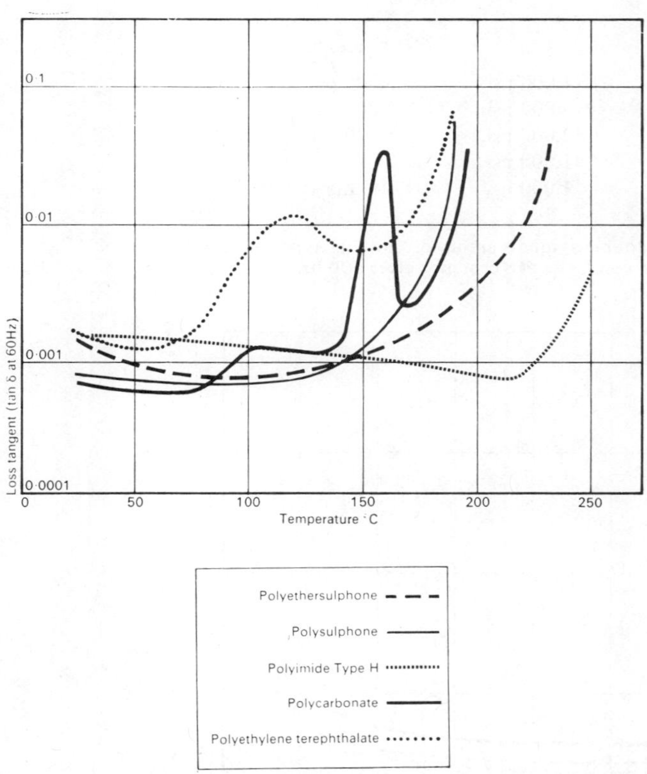

Fig. 5-156. Variation of loss tangent with temperature of some thermoplastics. (Victrex® Polyethersulfone is offered by ICI Americas Inc., Wilmington, Del.)

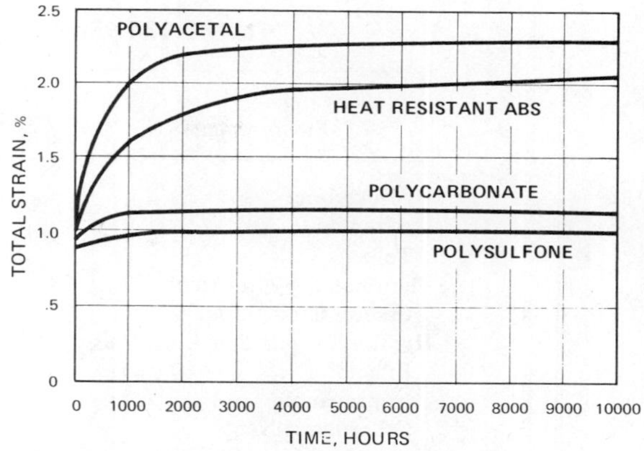

Fig. 5-157. Comparative creep behavior at 72° F; stress = 3000 psi tension. (*Courtesy: Union Carbide Corporation*)

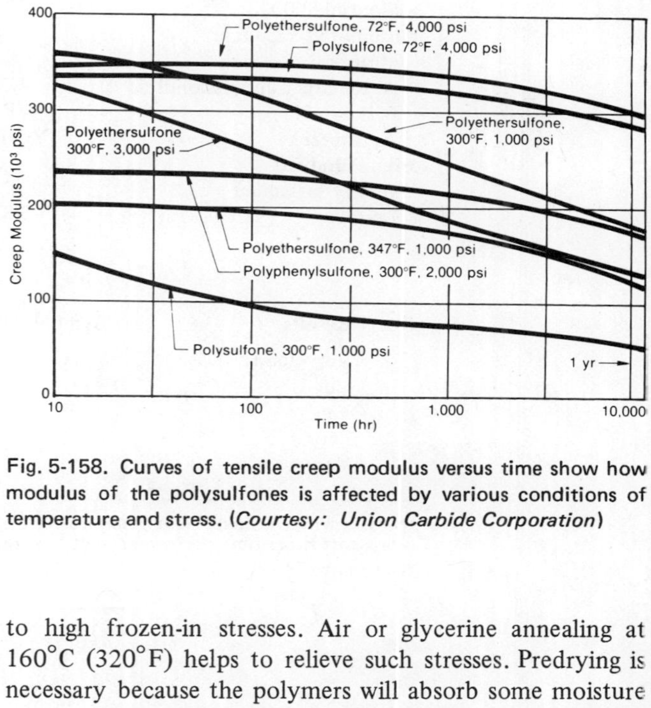

Fig. 5-158. Curves of tensile creep modulus versus time show how modulus of the polysulfones is affected by various conditions of temperature and stress. (*Courtesy: Union Carbide Corporation*)

to high frozen-in stresses. Air or glycerine annealing at 160°C (320°F) helps to relieve such stresses. Predrying is necessary because the polymers will absorb some moisture which can cause streaks and splash marks at high molding temperatures.

Screw preplasticization is preferred for injection molding. Melt temperatures range from 335°C to 400°C (635-725°F), with mold temperatures of 100-160°C (212-320°F). Generous cross-section runners and gates are needed to accommodate the high melt viscosities. Periodic purging is a necessity because degraded polymer tends to build up on the cylinder wall. This build-up is black and pulls away from the wall into the melt, discoloring the molded part.

Higher molecular weight grades of polysulfones are used for extrusion and blow molding. Melt temperatures range from 300°C to 360°C (572-680°F) with mold temperatures of 70-95°C (158-203°F).

Table 5-134. Environmental Stress-Crack Resistance.*

Environment	Poly-carbonate	Poly-sulfone	Polyether-sulfone	Polyphenyl-sulfone
Hydrocarbons				
n-Heptane	F	G	VG	VG
Toluene	VP	VP	P	VP
Chlorinated Hydrocarbon				
Carbon tetrachloride	VP	VP	VG	VG/G
Hydroxyl-Containing Chemicals				
Ethanol	G	F	F	VG
Isopropanol	G	G	VG/G	VG
Cellosolve solvent	P	P	P	F
Ketone and Ester Chemicals				
Acetone	VP	VP	VP	VP
Ethyl acetate	VP	VP	P	P
Fuels				
Unleaded gasoline	VP	VP	G	G/F
Jet fuel	G	VG	VG	VG
Hydraulic Fluid				
Skydrol 500A	VP	VP	VP	P
Automotive Products				
Motor oil	VG	VG	VG	VG
Automatic transmission fluid	VG	VG	VG	VG
Brake fluid	VP	VP	VP	P
Antifreeze	VG/F	VG/G	VG	VG
Home products				
Jet dry surfactant	F/P	P	VG/F	VG/F
Lestoil	VG/G	VG	VG	VG

Key: Results	Symbol	Criteria 1 hr	Criteria 500 hr
Very good	VG	>4,000 psi	>3,000 psi
Good	G	>3,000 psi	>2,000 psi
Fair	F	>2,000 psi	>1,000 psi
Poor	P	>1,000 psi	>0 psi
Very poor	VP	<1,000 psi	<500 hrs at 0 psi

In cases where a single term could not be applied, double symbols are used. The first symbol represents short-time performance (1 hr); and the second, the performance after 500 hr.
*Dreger, 1978.

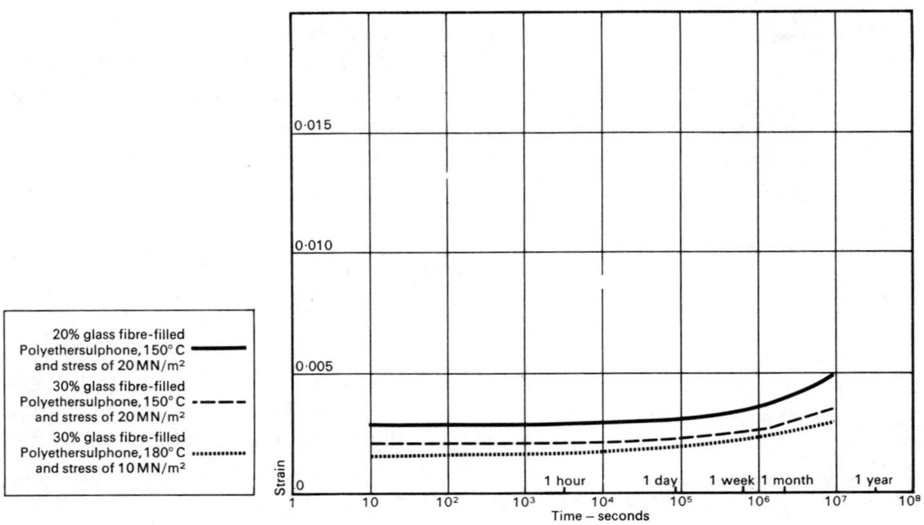

Fig. 5-159. Tensile creep of polyethersulfone at elevated temperatures. (Victrex® Polyethersulfone is offered by ICI Americas Inc., Wilmington, Del.)

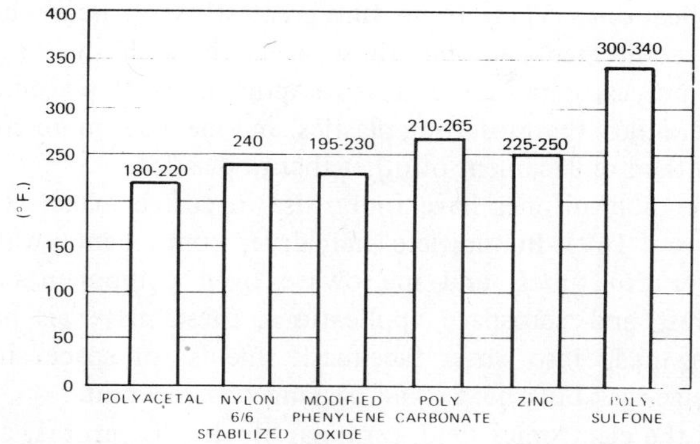

Fig. 5-160. Continuous-service temperature in air. (*Courtesy: Union Carbide Corporation*)

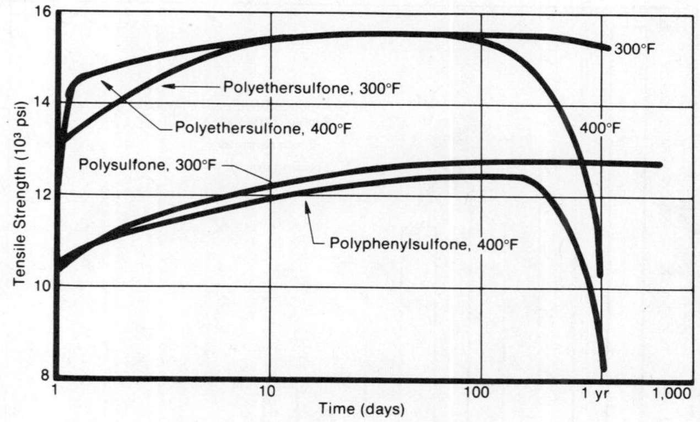

Fig. 5-161. Heat aging appreciably increases tensile strength of all three polysulfone polymers. A similar increase occurs in modulus of elasticity and deflection temperature. (Dreger, 1978)

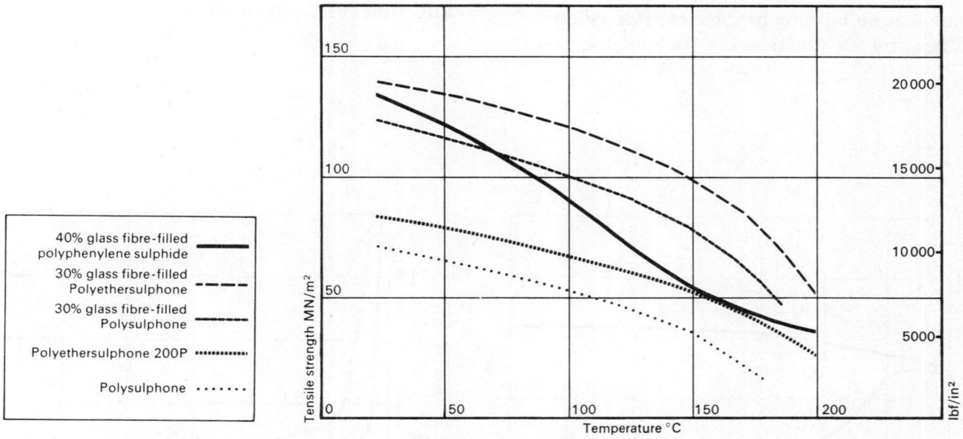

Fig. 5-162. Effect of temperature on tensile strength of some high-temperature thermoplastics. (Victrex® Polyethersulfone is offered by ICI Americas Inc., Wilmington, Del.)

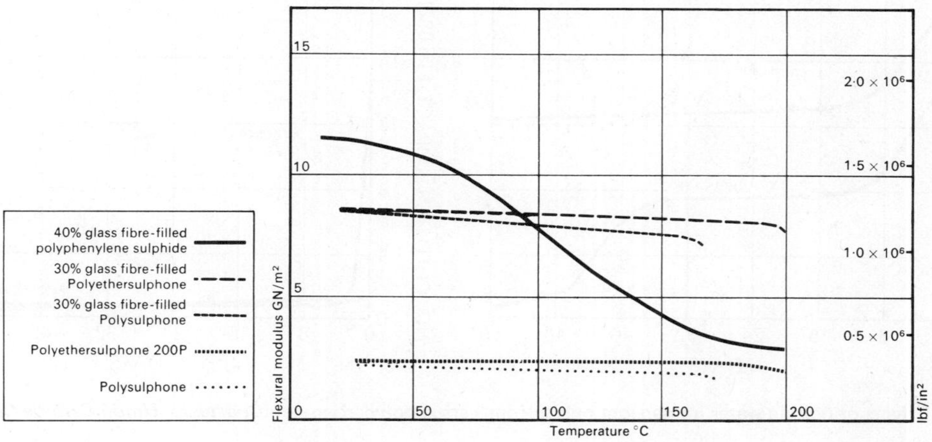

Fig. 5-163. Effect of flexural modulus of some high-temperature thermoplastics. (Victrex® Polyethersulfone is offered by ICI Americas Inc., Wilmington, Del.)

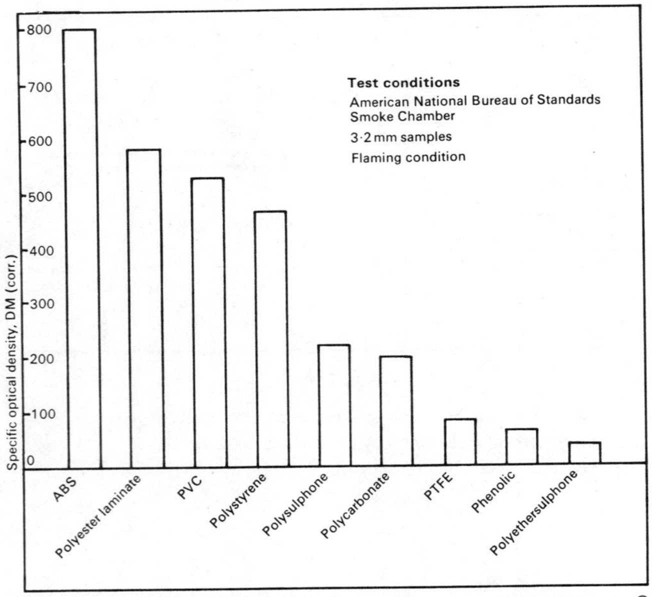

Fig. 5-164. Smoke emission on burning of some plastics. (Victrex® Polyethersulfone is offered by ICI Americas Inc., Wilmington, Del.)

Applications. These resins find greatest use in areas where the requirements exceed those available with the lower-cost polycarbonates and aromatic polyethers. Replacement of ceramics, thermosetting plastics, and metals is more common than replacement of other thermoplastics.

The polysulfones have found use in coffee makers (see Figure 5-169), humidifiers, hair-dryer components, water-heater dip tubes, and microwave oven components. In aircraft and aerospace applications, these materials have been made into outer face-mask shields for space suits, luggage-rack bulk heads, and passenger service units.

In the electronics field, terminal blocks, IC carriers, connectors, capacitor film, printed circuit boards, and battery cases have been made out of polysulfones.

Dialysis components, surgical instruments, and hospital feeding trays are examples of the medical usage of these materials. Other miscellaneous uses include camera and digital watch casings, electroplated print-out wheels, overhead projector transparencies and projector components, and fuel cell components.

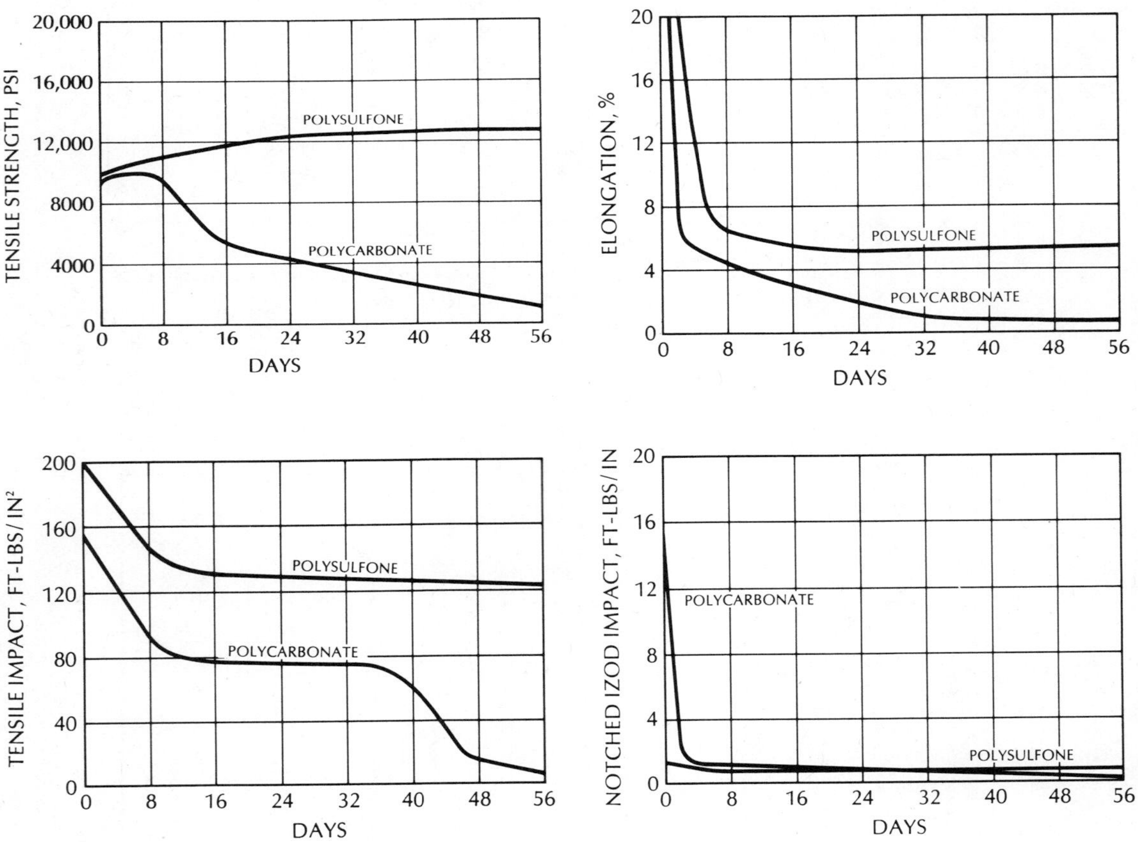

Fig. 5-165. Effects of boiling-water immersion; polysulfone versus polycarbonate. (*Courtesy: Union Carbide Corporation*)

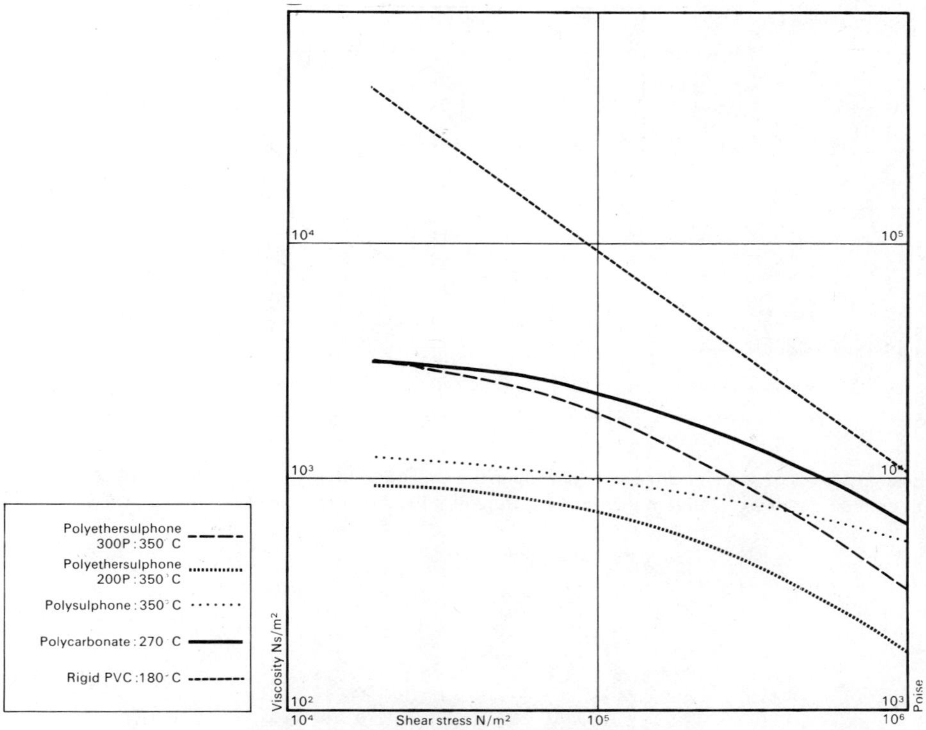

Fig. 5-166. Viscosity/shear-stress curves for some thermoplastics. (Victrex® Polyethersulfone is offered by ICI Americas Inc., Wilmington, Del.)

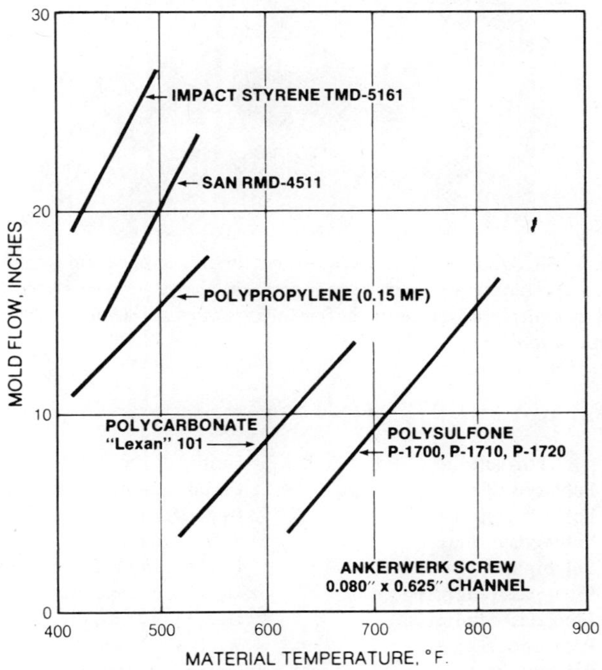

Fig. 5-167. Effect of material temperature on mold flow (0.08-inch wall thickness). (*Courtesy: Union Carbide Corporation*)

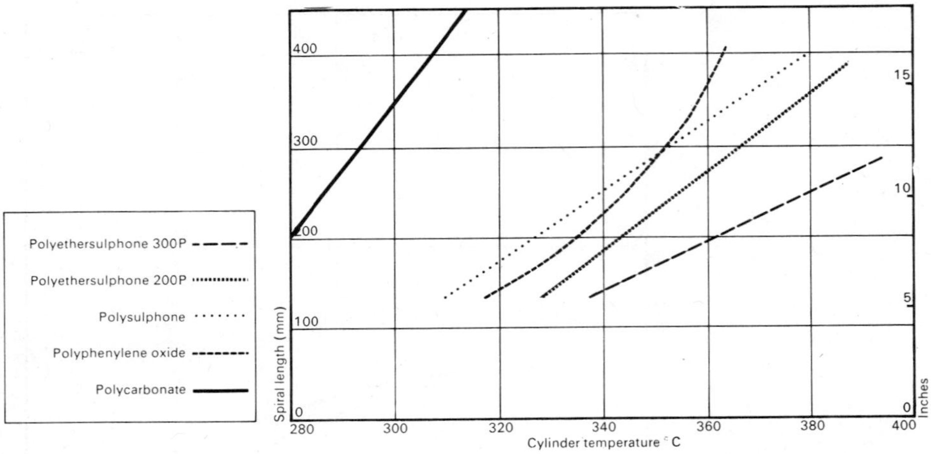

Fig. 5-168. Spiral flow characteristics of some engineering thermoplastics (injection pressure = 96 MN/m², 14,000 ft-lb/in²; mold temperature = 120°C; 6 mm = section thickness). (Victrex® Polyethersulfone is offered by ICI Americas Inc., Wilmington, Del.)

Fig. 5-169. Udel polysulfone was used for this application because of its combination of high heat resistance, clarity, impact strength, and chemical resistance to coffee oils. (*Courtesy: Union Carbide Corporation*)

Trade Names

Arylon T	ABS/polysulfone	Uniroyal, Inc.
Astrel 360	Polyarylsulfone	Carborundum Corporation
Polyethersulfone	Polyethersulfone	Imperial Chemical Industries, Ltd.
Polysul'son	Polyarylsulfone	Soviet Origin
Radel	Polybiphenoylsulfone	Union Carbide Corporation
Sulfasar	Fiberglass-reinforced	Dart Industries, Inc.
Sulfil	Fiberglass-reinforced	Dart Industries, Inc.
Thermalux	Polysulfone	Westlake Plastics Co.
Ucardel	Polysulfone	Union Carbide Corporation
Udel	Polysulfone	Union Carbide Corporation
Victrex	Polyethersulfone	Imperial Chemical Industries, Ltd.

REFERENCES AND BIBLIOGRAPHY

Askins, D. R. "Residual Stress Levels in Rotating Bands." Paper presentation, *SPE ANTEC*, New Orleans, May 1979.

Brydson, J. A. *Plastics Materials.* 3rd. Ed. Robert E. Krieger Publishing Co., Inc., Huntington, New York, 1975.

Chasin, D. G., and Rigby, R. B. "Polyethersulfone." In, *Modern Plastics Encyclopedia*, J. Agranoff (ed.), **52**(10A):75-76. McGraw-Hill, Inc., New York, 1975-76.

Danyliuk, I. M., and Woodhams, R. T. "Mica Reinforced Polyethersulfone." Paper presentation, *SPE ANTEC*, New Orleans, May 1979.

Dearing, R. K. "The Time Modified Modulus For Design With Polysulfone in Steam or Hot Water." Paper presentation, *SPE 34th Annual Technical Conference*, Atlantic City, New Jersey, April 1976.

Dreger, D. R. Polysulfone coming on strong for electrical applications. *Plastics Engineering*, February 1977, **33**(2):53-56.

ICI United States, Inc. *Polyethersulfone the High Temperature Engineering Thermoplastic*. Bulletin PES 11, April, 1977.

——. *Polyethersulfone: Flammability and Smoke Properties.* Bulletin P-3, July 1976.

——. *Polyethersulfone the High Temperature Engineering Thermoplastic*. Bulletin P-4, December 1976.

——. *Polyethersulfone Data for Design*. Technical Note PES 101, 2nd Ed.

Penney, W. H. "Polyaryl Sulfone." In *Modern Plastics Encyclopedia*, J. Agranoff (ed.), **52**(10A):52. McGraw-Hill, Inc., New York, 1975-76.

3M Co. *Astrel Brand 360 Plastics*. Bulletin Y-1APF.

Union Carbide Corporation. Technology Letter Engineering Polymers:

No. 101. *The Radiation Response of Udel Polysulfone*, September 22, 1976.

No. 102. *Radel Polyphenylsulfone–Comparative Chemical Resistance Under Stress*, January 24, 1977.

No. 103. *Environmental Stress Rupture–'Udel' Polysulfone.*

No. 104. *Fumes Generated During Injection Molding "Udel" Polysulfone and "Radel" Polyphenylsulfone*, March 4, 1977.

No. 105. *Udel Polysulfone Usage in Microwave Heating Applications*, February 14, 1977.

No. 106. *Polysulfones as Printed Circuit Substrates*, May 4, 1977.

No. 107. *Udel Polysulfone Combustion Characteristics,* May 17, 1977.

No. 108. *Polysulfone Chemical Resistance*, June 13, 1977.

No. 110. *Chemical Resistance of Glass-Filled Udel Polysulfone*, June 9, 1977.

No. 111. *Food Resistance of Udel Polysulfone in Microwave Cooking*, July 18, 1977.

No. 113. *Environmental Stress Crack Resistance of Amidel, Radel, and Udel Polymers.*

——. Designing for Performance. For Value:

No. 1. *Introducing Radel Polyphenylsulfone*, Fall 1976.

No. 2. various articles, 1977.

No. 3. *Reinforced Polysulfone and Heat Aging Polyphenylsulfone*, 1977.

No. 4. *Polysulfone in Medical Hardware and Coloring Polysulfone*, 1977.

——. *Plating Udel Polysulfone*. Product Data Bulletin F-46064A, June 1977.

——. *Radel Polyphenylsulfone–The "Tough" High Temperature Thermoplastic*. Bulletin 46264A, June 1977.

——. *Udel Medical Grade Polysulfone*. Bulletin F-46022A, June 1977.

——. *Resistance of Udel Polysulfone to Boiling Water and Steam*. Bulletin F-46622, June 1977.

——. *Guide to Chemical Resistance of Udel Polysulfone*. Bulletin F-46706, August 1977.

——. *Udel Polysulfone for Injection Molding and Extrusion*. Product Data Bulletin F-42072D, July 1976.

——. *Techniques for Molding Polysulfone*. Bulletin F-41782B, August 1970.

Walton, R. K. "Polysulfone." In *Modern Plastics Encyclopedia*, J. Agranoff (ed.), **52**(10A):82-84. McGraw-Hill, Inc., New York, 1975-76.

——. "Polysulfone." Paper presentation, *Plastics World Seminar, Engineering Plastics II*, Los Angeles, March 1976.

Wood, A. S. New day dawns for circuit boards. *Modern Plastics*, June 1977, **54**(6):52-54.

Polyarylates

Trade Names

Ardel	Union Carbide Corporation
Durel	Hooker Chemicals & Plastics Corporation
RTP (glass-reinforced)	Fiberite Corporation

A new family of engineering thermoplastics has recently been introduced. Neat and glass-reinforced versions are available. Although the chemistry of these new resins is still unpublished as of this writing, the polymer is meant to compete with the polycarbonates, thermoplastic polyesters, phenylene oxides, and polysulfones.

The key characteristics of these new materials are high heat-deflection temperatures, excellent UV resistance, toughness, transparency, excellent flexural recovery, high elastic limits, good electrical and mechanical properties, including outstanding creep resistance.

Table 5-135 shows properties of one of the two commercially available neat resins, Ardel. Table 5-136 compares some basic physical properties of the other commercial polymer, Durel, with typical polycarbonate and polysulfone properties.

The material is amorphous and thus subject to environmental stress-cracking, particularly in the presence of aromatic or aliphatic hydrocarbons.

Molding pellets must be predried before molding. Typical molding conditions include barrel temperatures of 327-349°C (620-660°F), with mold temperatures of 93-149°C (200-300°F). Injection pressures of 16,000-19,000 psi administered very rapidly through round gates are recommended.

REFERENCES AND BIBLIOGRAPHY

Domine, J. D. "Polyarylate: A Tough New Polyester." Paper presentation, *SPE ANTEC*, New Orleans, May 1979.

Fiberite Corporation. *RTP 2009 Series Polyarylate Glass Reinforced Product Data*.

Union Carbide Corporation. *Designing for Performance–for Value.* Bulletin F-47197, No. 7, 1979.

——. *Ardel Polyarylate–The Tough Weathereable Thermoplastic.* Bulletin F-47141, November 1978.

Table 5-135. Typical Properties of Ardel Polyarylate D-100.

Property	ASTM Method	Value English	SI
GENERAL (72°F unless otherwise noted)			
Density	D1505	75.5 lb/ft³	1.21 Mg/m³
Mold Shrinkage	D995	0.009 in./in.	0.009 mm/mm
Water Absorption	D570		
24 hours at 72°F		0.27%	0.27%
30 days at 72°F		0.71%	0.71%
MECHANICAL (72°F unless otherwise noted)			
Tensile Strength at Yield	D638	9,500 psi	65.5 MPa
Tensile Elongation at Yield	D638	8.0%	8.0%
at Rupture	D638	50.0%	50.0%
Tensile Modulus	D638	290,000 psi	2,000.0 MPa
Flexural Strength	D790	11,000 psi	75.9 MPa
Flexural Modulus at 5% Strain	D790	310,000 psi	2,140.0 MPa
Notched Izod Impact			
1/8 in. specimen	D256	4.2 ft-lb/in.	220 J/m
Direct Impact,			
1/8 in. plaque, 3/4 in. radius dart		>68 ft-lb	>92 J/mm
THERMAL			
Heat Deflection Temp at 264 psi, 1/4 in. specimen	D648	345°F	174°C
ELECTRICAL (Data at 72°F, 50% R.H.)			
Dielectric Strength 1/8 in. specimen	D149	400 V/mil	16.8 kV/mm
Arc Resistance	D495	125 Sec	125 Sec
Dielectric Constant	D150		
60 Hz		2.73	2.73
10^3 Hz		2.71	2.71
10^6 Hz		2.62	2.62
Dissipation Factor	D150		
60 Hz		0.0008	0.0008
10^3 Hz		0.005	0.005
10^6 Hz		0.02	0.02
COMBUSTION			
Oxygen Index, 1/4" specimen	D2863	34	34
Flammability	UL94	V-0 (0.062 in.)	V-0 (1.5 mm)

These Numerical Flame Spread Ratings or flammability ratings are not intended to reflect hazards presented by these or any other materials under actual fire conditions.

*Courtesy: Union Carbide Corporation.

Table 5-136. Comparison of Basic Physical Properties of Clear High-Temperature Plastics.*

	Durel 400	Polycarbonate	Polysulfone
Specific Gravity	1.2	1.2	1.26
Mold Shrinkage, in/in	.008	.006	.007
Tensile Strength, psi	10,000	9,000	10,000
Tensile Modulus, psi	3×10^5	3.3×10^5	3.5×10^5
Notched Izod ⅛", 73°F, ft-lb/in	5.5	17	1.2
Notched Izod ⅛", −40°F, ft-lb/in	4.0	3-15	1.2
Notched Izod ⅛" after 3 months at 212°F (100°C), ft-lb/in	4.0	1.6	0.9
Tensile Impact, ft-lb/in²	177	181	163
Endurance Limit in Flexural Fatigue (10^7 cycles), psi	2400	2400	1000
HDT 264 psi Annealed, °F (°C)	356 (180)	284 (140)	352 (178)
HDT 264 psi, °F (°C)	320 (160)	258 (126)	352 (178)
HDT 66 psi, °F (°C)	345 (174)	279 (137)	351 (177)
Taber Weight Loss, mg/1000 cycles	4-8	4-16	15
Oxygen Index, %	38	26	32
UL 94-⅛"	V-0	V-2	V-2
UL 94-1/16"	V-0	V-2	Unlisted

*Courtesy: Hooker Chemical & Plastics Corporation.

——. *Ardel Polyarylate D-100, Effect of Temperature on Physical Properties.* Technology Letter No. 122, Issue 1.

Hooker Chemicals & Plastics Corporation. *Durel Polyarylate: Versatile Engineering Polymer*, various data sheets.

Polyphenylene Sulfides (PPS)

Trade Names

Milkon	PFFE alloy	Tribol Industries, Inc.
Ryton	Resins	Phillips Petroleum Co.
Tribolon	Polyimide alloy	Tribol Industries, Inc.

The commercial polymers are produced by the reaction of *p*-dichlorobenzene with sodium sulfide in a polar solvent.

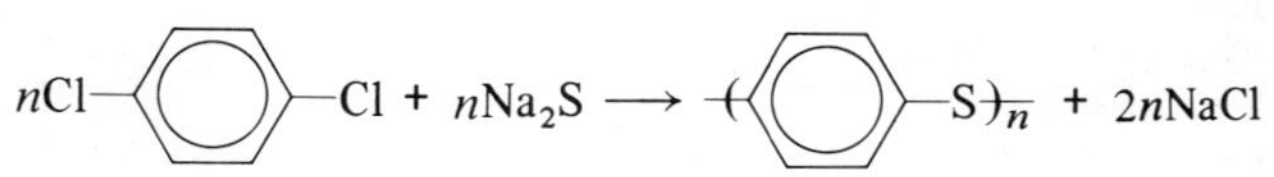

The polymers have a T_g of 150°C (302°F) and a T_m of 285°C (545°F). They are stable in air for short exposures to around 350°C (662°F). Extended exposure to oxygen above this temperature causes crosslinking to occur.

There are three basic types of polyphenylene sulfide currently available. The first to be developed was a branched polymer. Processing of this resin is based on similar methods to those used for PTFE (sintering, ram compression molding, etc.). It has no crystalline melting point, exhibits a very high melt viscosity, and can be reprocessed. Chemical resistance, resilience, and memory are somewhat better than the linear resin discussed below.

The second resin developed was a linear polymer which can be processed by compression molding, including laminating with fiberglass, and which is then oxidatively crosslinked at temperatures of 343°C (650°F) or more. The resin has unusually good wetting action on most substrates and all common fillers. Filler loadings of 70% by weight or greater are possible. Typically used fillers include iron oxide, asbestos floats, and chopped glass.

Moderate molecular weight grades constitute the third type of resin developed. These materials can be injection molded at 340–370°C (644–698°F) and then further crosslinked via air aging, as is done with the linear polymers.

Table 5-137 shows available data for the moldable resins. Table 5-138 shows properties of the injection-moldable resins.

Although there are differences in properties between the three resins, it is possible to list some generalizations. The materials have a low coefficient of linear expansion, particularly in the direction of flow. Predrying is recommended, although moisture pick-up is gradual. All the resins have excellent chemical resistance. There is no known solvent for these resins at temperatures below 191–204°C (375–400°F).

PPS's not only have excellent mechanical properties, but they retain these properties over a temperature range from

Table 5-137. Properties of PPS Compression-Molding Compounds.*

	(unfilled)	(50% Asbestos)
Density, gms/cc	1.34	2.67
Tensile, psi	10,800	10,800
Elongation, %	3	2
Flexural Modulus, psi	600,000	1,160,000
Flexural Strength, psi	20,000	14,000
Tensile Modulus, psi x 10^{-5}	4.8	7.0
Hardness, Shore D	86	91
Rockwell Hardness	R124	R123
Compressive Strength, psi	16,000	24,000
Notched Izod Impact, ft-lbs/in.		
at 75°F	0.2	–
at 300°F	1.0	–
Unnotched Izod Impact, ft-lbs/in.		
at 75°F	3	–
Heat Deflection Temperature, °F at 264 psi	278	425+
Maximum Recommended Service Temperature	500	500
Thermal Expansion, 10^{-5}, in/in/°F	3.0	2.0
Thermal Conductivity, btu/hr/sq. ft/°F/in.	2.0	–
Burning Rate, ASTM D635	Non-burning, non-drip	Non-burning, non-drip
Mold Shrinkage, in/in.	.010	–
Chemical Resistance	Excellent	Excellent
Hydrolytic Stability	Excellent	Excellent

*Courtesy: Phillips Chemical Co.

cryogenic levels to 260°C (500°F). The crosslinkable grade has the highest Underwriters' Laboratories temperature index of any thermoplastic and the highest values for heat-deflection temperature.

The resins are rated UL-94 VO, based on an inherent inflammability (i.e., no flame-retardents need be added).

Electrically the PPS's are very stable over a wide range of temperature, frequency, and exposure to moisture. Volume resistivity and insulation resistance, both dry and wet, are exceptionally good. The arc-resistance and arc-tracking properties are variable depending on grade.

These polymers have been extensively tested for nuclear radiation resistance. Excellent resistance to both neutron and gamma radiation has been recorded.

PPS's do not stress-crack but are notch-sensitive. A comparison of the thermal conductivity of PPS with other thermoplastics is shown in Table 5-139.

Table 5-138. Nominal Properties for PPS Injection-Molding Resins.*

Properties	Unfilled	40% Glass Filled
Specific gravity	1.34–1.80	1.64
Water absorption, %, ASTM D-570	0.020–0.030	0.010
Mold shrinkage, in/in.	0.002–0.006	0.002
Injection molding temperature, °F	600	625
Tensile strength, psi	10,800–14,000	21,400
Elongation, %	0.7–3.0	3
Flexural strength, psi	20,000–19,200	34,000
Flexural modulus, psi × 10^5	6.0–18.07	22.0
Compressive strength, psi	16,000	21,000
Rockwell Hardness	R-121–124	R-123
Shore D Hardness	86	91
Izod Impact, ft. lb/in.		
Notched 75°F	0.3–0.5	1.5
300°F	1.0	1.8
Unnotched 75°F	3.0	6.0
Deflection temperature at 264 psi, °F	278–>500	425+
Maximum recommended service temperature, °F	500	500
Thermal expansion, 10^{-5} in/in/°F	2.8-3.0	2.2
Thermal conductivity, btu/hr/sq. ft/°F/inch	2.0	–
Flammability, in/min.	Non-burning	Non-burning
UL94 flammability classification	"SE-O"	"SE-O"
Dielectric strength, volts/mil	340–595	490
Dielectric constant, 1 kc	3.11	3.80
Dissipation factor, 1 kc	0.0004	0.0037
Volume resistivity, ohm/cm	1.5-3.0 × 10^{16}	2.2 × 10^{16}
Arc resistance, seconds	185	160
Chemical resistance	Excellent	Excellent
U.V. resistance	Good	Good

*Courtesy: Phillips Chemical Co.

Table 5-139. Thermal Conductivity.*

Polymer	Temperature (°F)	Thermal Conductivity cal/sec/cm², (°C/cm)
Linear PPS	100	6.84 × 10^{-4}
	200	7.07 × 10^{-4}
	300	7.35 × 10^{-4}
Nylon 6/6	–	5.85 × 10^{-4}
Polyacetal	–	5.5 × 10^{-4}
High-density polyethylene	–	11-12.4 × 10^{-4}
Polypropylene	–	2.8 × 10^{-4}
Polystyrene	–	2.4-3.3 × 10^{-4}
Penton	–	3.1 × 10^{-4}
PPO	–	4.5 × 10^{-4}
Polysulfone	–	6.2 × 10^{-4}
Polytetrafluoroethylene	–	6.0 × 10^{-4}
Polycarbonate	–	4.6 × 10^{-4}

*Courtesy: Phillips Chemical Co.

The PPS's have found significant use in coating applications such as cookware and metal tire-mold surfaces (see Figure 5-170). Table 5-139 shows some selected properties of formulated PPS coatings.

Other end-use areas are in fluid transport systems. Pumps and components such as vanes, impellers, housings, and diffusers are made from or coated with PPS.

The largest end use at this time is in electric-electronic components. Included in this area are such items as interconnection devices, coil forms, bobbins, yokes, and terminal boards (see Figures 5-171 and 5-172).

Fig. 5-170. A liquid slurry of polyphenylene sulfide is spray-coated onto a skillet. The part will then be heated for 45 minutes at 700°F to cure the coating. (*Courtesy: Phillips Chemical Co.*)

Table 5-140. Properties of Polyphenylene Sulfide Coatings.*

Property	PPS/TiO_2 (3/1)	$PPS/TiO_2/PTFE$ (3/1/0.3)
Hardness, pencil	2H	2H
Mandrel bend, 180°, 3/16″	Pass	Pass
Elongation (ASTM D522), %	>32	>32
Reverse impact, in.-lb	>160	>160
Abrasion resistance, Taber mg loss/1000 rev, CS-17 wheel	50	57
Chemical resistance	Excellent	Excellent
Thermal stability	Excellent	Excellent
Color	Light Tan	Light Tan

*Courtesy: Phillips Chemical Co.

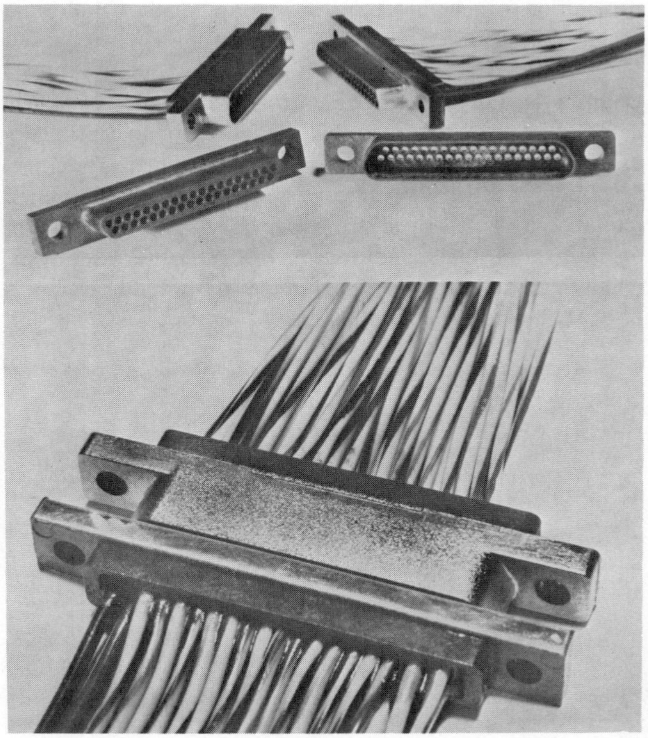

Fig. 5-171. Ryton PPS connectors shown with wires are molded within a die-cast aluminum band shell. The Ryton parts, shown with holes, separate the circuits, holding them in position. (*Courtesy: Phillips Petroleum Co.*)

REFERENCES AND BIBLIOGRAPHY

Brydson, J. A. *Plastics Materials*. 3rd Ed. Robert E. Krieger Publishing Co., Inc., Huntington, New York, 1975.

Boeke, P. J. "Polyphenylene Sulfide–Phase II." Paper presentation, *Plastics World Seminar, Engineering Plastics II*, Los Angeles, March 1976.

Osberg, E. New engineering material for the rubber industry. *Rubber & Plastics News*, April 1973.

Phillips Petroleum Co. *Properties and Uses of Anti-Friction Compounds*. Bulletin TSM-262, October 1974.

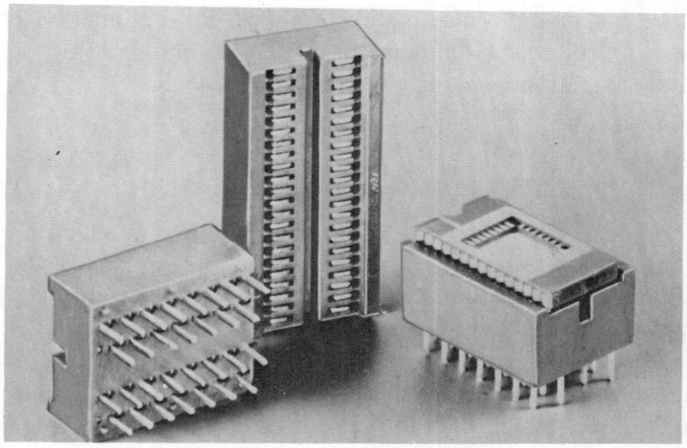

Fig. 5-172. A new series of Robinson-Nugent burn-in and test sockets, molded of glass-filled Ryton PPS. The socket contacts are gold-plated beryllium copper. (*Courtesy: Phillips Chemical Co.*)

——. *Ryton Polyphenylene Sulfide Resins*. 2nd Ed., April 1969.

——. *Ryton Polyphenylene Sulfide–Corrosion Resistance Guide*. Bulletin 5017 DWO 1694.

——. *Technical Information on Ryton Polyphenylene Sulfide Resins–Properties . . . Processing*. Bulletin 100.

——. *Comparing Ryton Molding Grade Resins With Other Engineering Plastics*. Bulletin 12733.

——. *Corrosion Resistance Guide*. Bulletin 13685.

Plastics Design & Processing. "Why polyphenylene sulfide is IN television. July 1973, **13**(7).

Powell, W. D. "Ryton Polyphenylene Sulfide for Corrosion Protection." Paper presentation, *31st Annual Conference & Exhibition of the Society of the Plastics Industry of Canada*, Vancouver, 1973.

Schooler, G. N. "Polyphenylene Sulfide." In *Modern Plastics Encyclopedia*, J. Agranoff (ed.), **52**(10A):76. McGraw-Hill, Inc., New York, 1975–76.

Polyethers

Several polymers containing ether linkages in the chain are becoming increasingly important. Polyglycols are the major class in this group. Although they are not considered to be engineering polymers, they provide the background necessary for a discussion of the other two resin members of this group—the chlorinated polyethers and the polyaryl ethers—both of which are considered to be in the engineering plastics family.

Polyglycols. Although polyglycols can be considered the condensation products of difunctional glycols, they are actually produced from the corresponding oxides as follows:

$$HOCH_2\overset{R}{\overset{|}{C}}HOH + \underbrace{CH_2-\overset{R}{\overset{|}{C}}H}_{O} \xrightarrow{NaOH}$$

$$HOCH_2\underset{H}{\underset{|}{\overset{R}{\overset{|}{C}}}}(OCH_2\overset{R}{\overset{|}{C}}H)_n CH_2\overset{R}{\overset{|}{C}}HOH$$

The glycol molecule required to start chain formation is easily produced by the presence of a small amount of water in the reaction medium.

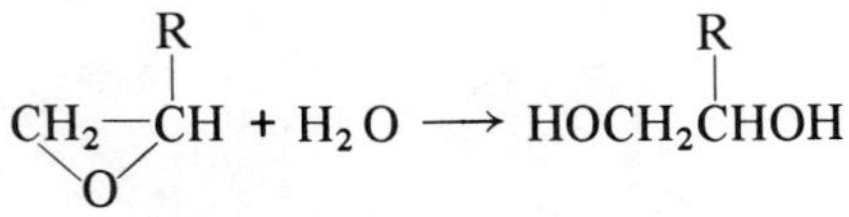

When R is an H atom, the product is polyethylene glycol or polyethylene oxide. This polyether is unique in that it is water soluble even up to very high molecular weights. Materials ranging in molecular weight from 200 to 600 are liquids. Up to about 10,000, they are waxy solids; at very high molecular weights, they are resinous plastics, resembling polyethylene in appearance and bulk properties. These polyethers have many uses, depending on their water solubility, lubricating properties, waxy feel, and the fact that the ends of the chains have two-OH groups which can readily react to produce derivatives or form higher polymers.

They are used in cosmetics and pharmaceuticals because of their blandness, water solubility, and lubricity. Their water solubility and thickening properties make them useful in water-based paints, paper coatings, adhesives, sizes, inks, polishes, and mold-release agents. They may be converted by reactions involving the end hydroxyl groups to esters, amines, and other chemical intermediates. The fatty acid esters have gained prominence as detergents and as wetting, penetrating, and dispersing agents. Of particular interest is the production of other polymers by reaction with diisocyanates to produce polyurethanes (see Chapter 6).

Even the very high polymers of ethylene oxide, having molecular weights that may range up to several million, retain the water-solubility characteristics of the lower liquids and waxes. Some properties of a medium-high molecular weight ethylene oxide polymer are shown in Table 5-141. Water solutions of these polyethers have very high viscosity and are highly pseudoplastic. The enormous change of viscosity with shear rate at very low rates of shear give these materials very unusual rheological properties. The highest molecular weight materials have interesting plastic properties. They are highly crystalline and yet have very good ductility. Films are heat sealable and water soluble, suggesting use in water-soluble packages and capsules.

Polyethers produced from propylene oxide are next in importance. They are available commercially in a range of molecular weights. Up to a molecular weight of 750, these materials are also water soluble, but only slightly so at higher values. However, they are much more soluble in various solvents.

Polypropylene glycols have some of the same uses as the polyethylene glycols and are important intermediates for polyurethanes via reaction with diisocyanates (see Chapter 6).

Table 5-141. Some Properties of a Medium-High Molecular Weight Ethylene Oxide Polymer.*

Tensile strength	1,800–2,400 lbf in^{-2} (13–22 MN/m^2)
Yield strength	1,000–1,500 lbf in^{-2} (7–11 MN/m^2)
Ultimate elongation	700–1,200%
Shore A hardness	99
Melting point	66°C
Brittle temperature	–50°C

*Brydson, 1975.

Many mixed or copolymerized polyethers of a wide range of molecular weights are possible. Since they all have primary hydroxyl groups on the ends of the chains, they are joining the well-established polyglycols as intermediates in the production of chemical derivatives and higher molecular weight polymers.

Chlorinated Polyether. The corrosion resistance of chlorinated polyether has made this thermoplastic exceptionally useful in the design and production of equipment for chemical and processing systems. Its monomer, a chlorinated oxetane, is synthesized from pentaerythritol and, under carefully controlled conditions involving the use of catalysts, is polymerized into a very high molecular weight (average = 250,000–350,000) polyether.

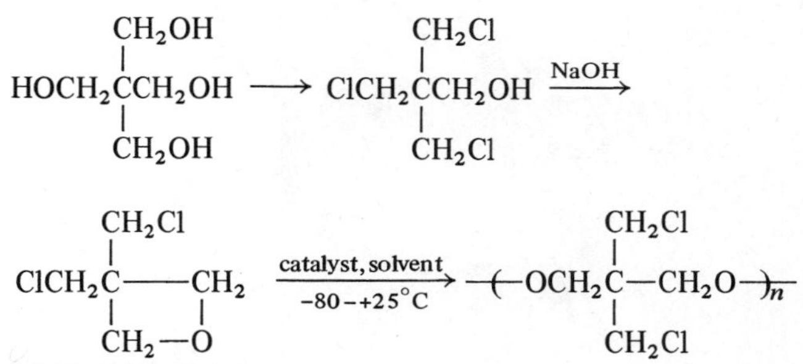

3,3-bis(chloromethyl)-
1-oxacyclobutane

Typical catalysts include BF_3 or BF_3-etherate. Solvents include hexane, benzene, liquid SO_2, or chlorinated solvents.

Chlorinated polyether is characterized by high chemical and thermal stability resulting from its crystallized structure. Chlorine constitutes approximately 46% of the polymer's weight. Chlorinated polyether is linear in nature and crystalline in character. It differs from most halogenated polymers in that it is extremely resistant to thermal degradation at molding and extrusion temperatures. Chlorinated polyether crystallizes at a very fast rate, resulting in the production of essentially strain-free components, stable over a wide range of chemically active environments. Properties of the polymer are shown in Table 5-142. Although

Table 5-142. Properties of Chlorinated Polyether (Brydson).*

Property	ASTM Test	Value	Units
Tensile strength (23°C)	D638	6,000	lbf in^{-2}
		41	MN/m^2
Modulus of elasticity (23°C)	D638	150,000	lbf in^{-2}
		1,100	MN/m^2
E_B	D638	130	%
Flexural strength (23°C)	D790	5,000	lbf in^{-2}
		35	MN/m^2
Flexural modulus (23°C)	D790	130,000	lbf in^{-2}
		900	MN/m^2
Izod impact strength (23°C)	D758	0.4	ft-lb per in notch
Deflection temperature	D648		
264 lbf in^{-2}		99	°C
66 lbf in^{-2}		140	°C
Thermal conductivity	D696	0.13	W/mK
Volume resistivity	D257	1.5×10^{18}	Ωm
Power factor			
60–10^6 Hz (23°C)		~0.011	
60 Hz (100°C)		0.045	
10^4–10^6 Hz (100°C)		~0.003	
Dielectric constant (varying according to temperature and frequency)		2.8–3.5	
Dielectric strength		155	Kv/cm
Flammability	D635		self extinguishing

*First published in *Plastic Institute Transactions*, vol. 30, no. 127 (Plastics Institute, London, 1962).

not outstanding, these properties are not extremely limiting, when one considers its exceptional chemical resistance.

The corrosion resistance of chlorinated polyether to more than 300 chemicals and reagents is highly documented. It possesses resistance to both organic and inorganic agents (with the exception of strong oxidizing agents) at temperatures up to 121°C (250°F) or higher. The thermal conductivity of chlorinated polyether was measured at 0.91 Btu/hr/ft^2/°F/in, or considerably lower than most thermoplastics with corrosion resistance. As a heat-insulating material, chlorinated polyether is excellent, and lined components often eliminate the need for additional insulation.

In processing equipment, there are injection-molded valves and pump heads, extruded pipe in standard lengths and diameters with threaded or socket connections, and precision machine gears, pins, bushing, and wobble plates. Low-cost metal components such as pipe, valves, pumps, and metal parts offer improved performance when lined with chlorinated polyether. Continuous, pinhole-free coatings of chlorinated polyether can be applied to pipe and fittings, valves, pumps, vessels, and other chemical-processing components. A number of coating techniques are available and are especially useful where difficult configuration or custom-built items require corrosion protection. These techniques are: (1) fluidized bed, (2) water-suspension coating (spray, drive off water), and (3) flock coating (part heated, chlorinated polyether powder sprayed on and fused, two coats used).

Polyaryl Ether. A new high-heat engineering thermoplastic, chemically described as a high-impact modified polyaryl ether and believed to be a copolymer of ABS and a polyethersulfone is now available. As of this writing, information about the chemistry and polymerization of this material has not been made available. The outstanding feature of this material is that it has a unique combination of three important, desirable properties: (1) a heat-deflection temperature that is exceeded only by two other engineering plastics; (2) the highest impact strength of all engineering plastics, except polycarbonate in thin sections; and (3) excellent flow characteristics that permit easy filling of intricate sections and thin walls.

Fabricating advantages include a low shrinkage factor that permits precision molding of void-free parts. Modified polyaryl ether resists the tendency toward sinks, and even thick sections can be molded without voids or sinks. It is not moisture sensitive and requires only normal drying before molding.

The chemical resistance of modified polyaryl ether to environments, including boiling water, is excellent. It can be solvent-welded and can be painted with commercial

Table 5-143. Properties of Polyarylether.*

		ASTM test method	Polyaryl ether
Processing	1. Molding qualities	—	Excellent
	2. Compression molding temp., °F.	—	—
	3. Comp. molding pressure, p.s.i.	—	—
	4. Injection molding temp., °F.	—	540-590
	5. Injection molding pressure, p.s.i.	—	10000-20000
	6. Compression ratio	—	2.6-1.8
	7. Mold (linear) shrinkage, in./in.	—	0.007
	8. Specific gravity (density)	D792	1.14
	9. Specific volume, cu. in./lb.	D792	25.2
	10. Machining qualities	—	Excellent
Mechanical	11. Tensile strength, p.s.i.	D638	7500
	12. Elongation, %	D638	25.0-90.0
	13. Tensile elastic modulus, 10^5 p.s.i.	D638	3.2
	14. Compressive strength, p.s.i.	D695	—
	15. Flexural yield strength, p.s.i.	D790	11000
	16. Impact strength, ft. lb./in. of notch (1/2 × 1/2 in. notched bar, Izod test)	D256	8.0 (1/2 × 1/4 in. bar)
	17. Hardness, Rockwell	D785	R117
	18. Flexural elastic modulus, p.s.i. × 10^5	D790	3.0
	19. Compressive modulus, p.s.i. × 10^5	D695	—
Thermal	20. Thermal conductivity, 10^{-4} cal./sec./sq. cm./1 (°C./cm.)	C177	7.13
	21. Specific heat, cal./°C./gm.	—	0.35
	22. Thermal expansion, 10^{-5} in./in./°C.	D696	3.6
	23. Resistance to heat, °F. (continuous)	—	250
	24. Deflection temp., °F. at 264 p.s.i. fiber stress	D648	300
	at 66 p.s.i. fiber stress		320
Electrical	25. Volume resistivity, ohm-cm. (50% RH and 23° C.)	D257	1.5×10^{16}
	26. Dielectric strength, short-time, 1/8-in. thickness, volts/mil	D149	430
	27. Dielectric strength, step-by-step, 1/8-in. thickness, volts/mil	D149	—
	28. Dielectric constant, 60 cyc.	D150	3.14
	29. Dielectric constant, 10^3 cyc.	D150	2.85
	30. Dielectric constant, 10^6 cyc.	D150	3.10
	31. Dissipation (power) factor, 60 cyc.	D150	0.006
	32. Dissipation (power) factor, 10^3 cyc.	D150	0.004
	33. Dissipation (power) factor, 10^6 cyc.	D150	0.007
	34. Arc resistance, sec.	D495	180+
Optical	35. Refractive index, n_D	D542	—
	36. Clarity	—	Transl. to opaque
	37. Transmittance, %	—	—
	38. Haze, %	—	—
Resistance characteristics	39. Water absorp., 24 hr., 1/8-in. thick., %	D570	0.25
	40. Flammability[a] Burning rate, in./min.	D635	—
	AEB, in./ATB, sec.[b]		—
	41. Effect of sunlight	—	None to yellows slight embrittlement
	42. Effect of weak acids	D543	None
	43. Effect of strong acids	D543	Resists some
	44. Effect of weak alkalies	D543	None
	45. Effect of strong alkalies	D543	None
	46. Effect of organic solvents	D543	Soluble in chlorinated aromatics, esters, ketones

*Reprinted by permission from *Modern Plastics Encyclopedia*, McGraw-Hill, Inc.

industrial finishes without pretreatment or priming. Table 5-143 shows its properties.

Modified polyaryl ether has a heat-deflection temperature of 149°C (300°F). This, combined with fine processability, is one of the most important and unusual characteristics of the material. This temperature is exceeded only by PPO and polysulfone (see Table 5-144). Impact strength is not thickness-dependent, as is the case with polycarbonate, but is relatively constant (see Table 5-145). Modified polyaryl ether flows 40% farther than modified PPO and has a 19.5°C (35°F) higher heat-deflection temperature. It flows

Fig. 5-173. Spiral flow of Arylon T polyaryl ether. (*Courtesy: Uniroyal Chemical Co.*)

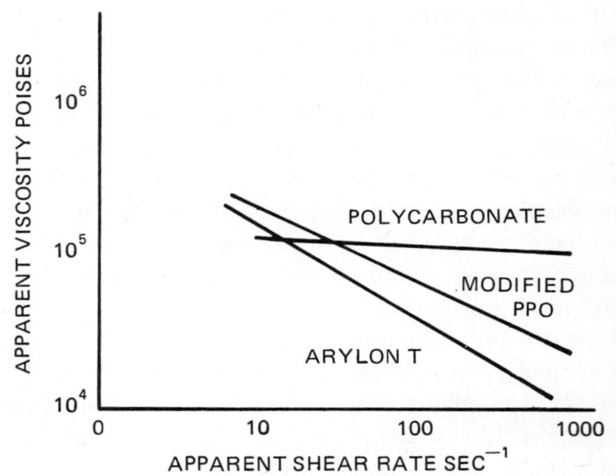

Fig. 5-174. Apparent viscosity versus shear rate of Arylon T polyaryl ether. (*Courtesy: Uniroyal Co.*)

Table 5-144. Heat-Distortion Values for Arylon T, a Polyaryl Ether, and Other Engineering Thermoplastics.*

	Arylon T	Acetal	Nylon 6/6	PPO	Modified PPO	Poly-sulfone	Poly-carbonate
264 psi	300°F	230°F	150°F	375°F	265°F	345°F	265°F
66 psi	320°F	315°F	360°F	–	280°F	360°F	285°F

*Courtesy: Uniroyal Chemical Co.

Table 5-145. Notched Izod Impact Values for Arylon T, a Polyaryl Ether, and Other Engineering Thermoplastics.*

	Arylon T	Acetal	Nylon 6/6	PPO	Modified PPO	Poly-sulfone	Poly-carbonate
$\frac{1}{4}$ inch bar	8.0 ft-lbs	–	–	1.5 ft-lbs	1.3 ft-lbs	1.3 ft-lbs	2.0 ft-lbs
$\frac{1}{8}$ inch bar	8.0 ft-lbs	1.2 ft-lbs	1.0 ft-lbs	–	–	1.3 ft-lbs	12.0 ft-lbs

*Courtesy: Uniroyal Chemical Co.

120% farther than polycarbonate and has a 16.6°C (30°F) higher heat-deflection temperature. It flows 250% farther than polysulfone. Figures 5-173 and 5-174 show differences in mold flow and apparent viscosity.

Modified polyaryl ether has excellent resistance to organic solvents, except chlorinated aromatics, esters, and ketones; it appears to resist hydrolysis, withstanding prolonged immersion in boiling water. The effect of seven days' immersion in common chemicals produces only negligible weight changes.

Within its modified property ranges, polyaryl ether has a truly broad potential in automotive components, appliances, instruments, radio and television components, business machine parts, recreation helmets, snowmobile parts, plumbing valves and fixtures, and special machinery such as vending machines, power-tool housings, and fluidic controls.

Trade Names

Actol	Polyethers	Allied Chemical Corporation
Arylon	Polyaryl ether	Uniroyal, Inc.
Caradol	Polyether polyols	Shell Chemical Co.
Carbowax	Polyethylene glycols	Union Carbide Corporation
Cord	Polyethers	CPC International, Inc.
Dithane	Difunctional polypropylene glycols	Pelcon Corporation
Isonol	Polyfunctional polyol	Upjohn Co.
Lipol	Polyoxyethylene ethers	PVO International
Niax	Polyol polyethers	Union Carbide Corporation
Penton	Chlorinated polyether	Hercules, Inc.
Pluronic	Polypropylene and polyethylene glycols	BASF Wyandotte Corporation
Polymeg	Polytetramethylene ether glycols	Quaker Oats Co.
Polyox	Polyol polyethers	Union Carbide Corporation
Suprether	Polyether foam	Wm. T. Burnett & Co.
Tetronic	Polyether tetrols	BASF Wyandotte Corporation
Thanol	Polyether polyols	Texaco, Inc.
Voranol	Polyol polyethers	Dow Chemical Co.

REFERENCES AND BIBLIOGRAPHY

Brydson, J. A. *Plastics Materials.* 3rd Ed. Robert E. Krieger Publishing Co., Inc., Huntington, New York, 1975.

King, N. E. "Chlorinated Polyether." In *Modern Plastics Encyclopedia*, S. Gross (exec. ed.), 45(14A):194–199. McGraw-Hill Book Co., 1968–69.

Loy, R. K. "Polyaryl Ether." In *Modern Plastics Encyclopedia*,

J. Agranoff (ed.), 52(10A):51. McGraw-Hill, Inc., New York, 1975–76.

Plastics Technology, Water-soluble PE oxide film: latest entrant in packaging field. November 1964, 10(11):67.

Simonds, H. R., and Church, J. M. *A Concise Guide to Plastics*, 2nd Ed. Van Nostrand Reinhold Co., New York, 1963.

Uniroyal, Inc. *Arylon T-Polyaryl Ether.* Bulletin No. 820-B30B.

Winding, C. C., and Hiatt, G. D. *Polymeric Materials.* McGraw-Hill Book Co., New York, 1961.

Miscellaneous Specialty Thermoplastics

Phenoxy. Phenoxy resins are an outgrowth of epoxy resin technology. They bridge the gap between the thermoplastic and thermosetting polymer fields.

The synthesis of phenoxy resins occurs via the reaction of a dihydric phenol and epichlorohydrin in the presence of alkali. Any dihydric phenol or combination of dihydric phenols may be used. However, commercial resins are primarily based on bisphenol A.

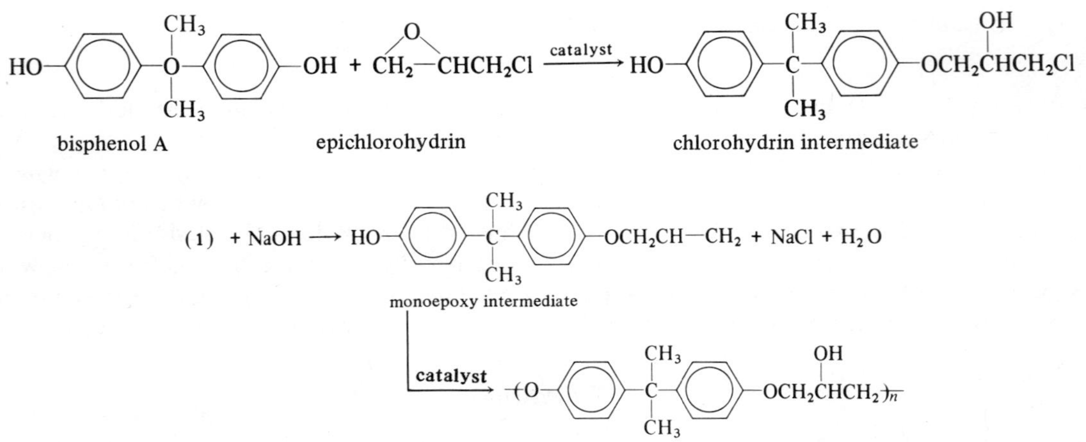

(1)

(2)

Typical catalysts include lithium hydroxide and benzyl trimethyl ammonium hydroxides. The reaction can be carried out in solution using such solvents as methyl ethyl ketone. Reaction temperatures vary between 80–160°C (176–320°F), depending on the molecular weight of the final polymer desired.

When producing low molecular weight phenoxies, an excess of epichlorohydrin over the stoichiometric requirement is used. Mild reaction conditions can then be used which protect against side reactions (caused by the ring being opened by water) and a subsequent epoxy-aliphatic hydroxyl reaction.

For higher molecular weight species, there is a greater likelihood of reaction with the secondary alcohols on the polymer backbone. This side effect is minimized by the use of a monohydric primary alcohol, which reacts preferentially over the secondary alcohols and thereby prevents branch-chain formation and crosslinking.

When the final phenoxy resin is to be supplied as a solution, stripping of the reaction solvents is not necessary. After washing, additional solvents may even be added to adjust flow and evaporation characteristics. If the resin is produced from solventless procedures, the cooled, brittle material may be broken into small pieces and dissolved in appropriate solvents.

If solvent-free polymer is required, the synthesis may take place in water-soluble solvents. After completion of the reaction, the mass is dissolved into water, the phenoxy being insoluble, and the resulting solid shreds and fibers of phenoxy are filtered out and dried. They may then be powdered or pelletized if desired.

Films may be produced from the solvent phase by pouring the resin solution onto a two-roll mill. The solvent evaporates as the solution passes between the rolls, and the remaining solid resin is sheeted off.

Aside from the use of various dihydric phenols and alteration of molecular weight, there are two other main methods of modifying phenoxies. The first is through esterification of the secondary alcohols along the chain with free acids or acid halides and/or anhydrides. The use of monofunctional reactants preserves the soluble, thermoplastic nature of the resultant polymer.

The second method of modification is to create a thermosetting system. This may be achieved through the addition of drying oils (see Chapter 6, alkyds) which cure through the influence of oxygen and catalysts. Alternatively, polyfunc-

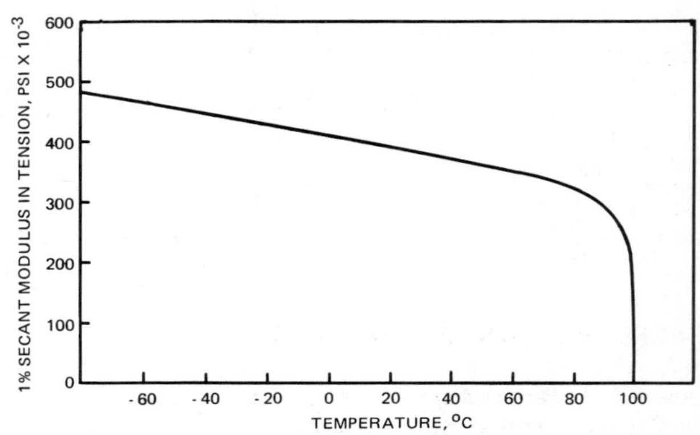

Fig. 5-175. Modulus of elasticity of phenoxy resin versus temperature. (From *New Linear Polymers* by Lee et al. Copyright 1967 by McGraw-Hill, Inc. Used with permission of McGraw-Hill Book Co.)

Table 5-146. Properties of Phenoxy Molding and Extrusion Materials.*

Property	Phenoxy PRDA-8060 and -8030	Phenoxy PRDA-8100	Phenoxy PRDA-8160
Price (6/15/65), tankloads, \$/lb	0.75	0.78	1.00
Specific gravity	1.18	1.17	1.32
Pound volume cost, ¢/in.^{3}	3.2	3.3	4.8
A. Mechanical Properties at 25°C			
Tensile yield strength, psi	9,000	8,000	9,000
Flexural strength, psi	14,000	12,000	14,500
Ultimate elongation, %	50–100	100	50
Flexural modulus, psi	400,000	340,000	400,000
Izod impact, ⅛-in. bar, ft-lb/in.	2.3	12	1.5
Charpy impact at -40°, ft-lb/in	No break	No break	No break
Rockwell hardness, R scale	123	118	123
B. Thermal Properties			
Deflection temperature, °C:			
At 264 psi	87	86	80
At 66 psi	91	92	86
Coefficient of thermal expansion in./(in.)(°F) × 10^5	3.5	3.4	3
Flammability	Burns slowly	Burns slowly	Self-ext. nondrip
C. Electrical Properties			
Dielectric constant			
At 60 cycles	4.1	4.1	4.1
At 1 Mc	3.8	3.8	3.7
Dissipation factor at 60 cycles	0.0012	0.0012	0.0009
Dielectric strength, volts/mil ⅛-in. plaque, short time	520	400	505
D. Chemical Resistance			
To acids	Excellent	Excellent	Excellent
To alkalies	Excellent	Excellent	Excellent
To solvents	Poor	Poor	Poor
E. Miscellaneous			
Mold shrinkage, in./in.	0.004	0.004	0.003
Weatherability	Poor-fair	Poor	Under test
Color	Transparent light straw	Transparent straw	Opaque white
Water absorption in 24 hours, %	0.13	0.13	–
Bearing characteristics	Poor	Poor	Poor
Creep[a]	8	7	7
O_2 permeability, cc/(100 in.^{2})(day)(mil)	5.8	–	–
MVT, g/(100 in.^{2})(day)(mil)	3.5	–	–

[a]10 = best; 1 = very poor.

*From *New Linear Polymers* by Lee et al. Copyright 1967 by McGraw-Hill, Inc. Used with permission of McGraw-Hill Book Co.

tional coreactants, such as urethanes, methylol-containing resins, epoxy resins, and anhydrides may be added, which will react with the alcoholic hydroxyls of the phenoxy.

Typical properties of phenoxies are shown in Table 5-146. The data show that phenoxies have moderately good impact resistance, relatively high strength, good elongation and creep resistance. The mechanical properties of an uncross-linked phenoxy will be a function of its molecular weight. Crosslinking will reduce flexibility and improve thermal and chemical resistance. Crosslinking with methylol-containing materials requires heat. Properties of room temperature-cured phenoxy-polyurethane systems show improved temperature, weatherability, and abrasion resistance properties, better resistance to staining and blushing, and higher hardness. Chemical resistance is somewhat better than straight phenoxy. As with most aromatic polymers, color retention, UV resistance, and weatherability are generally poor.

Adhesives based on phenoxies have demonstrated room-temperature lap-shear strengths of over 3,000 psi. The electrical properties are equivalent to those of crosslinked epoxys, thus, they make excellent insulators.

Strength retention during thermal exposure is inferior to that obtained with many linear polymers. However, the properties present a rather flat profile with temperature over the recommended operating range of -60–80°C (-76–176°F) (see Figure 5-175).

Phenoxy is resistant to acids and alkalis, has poor solvent resistance, especially in ketones, has good resistance to

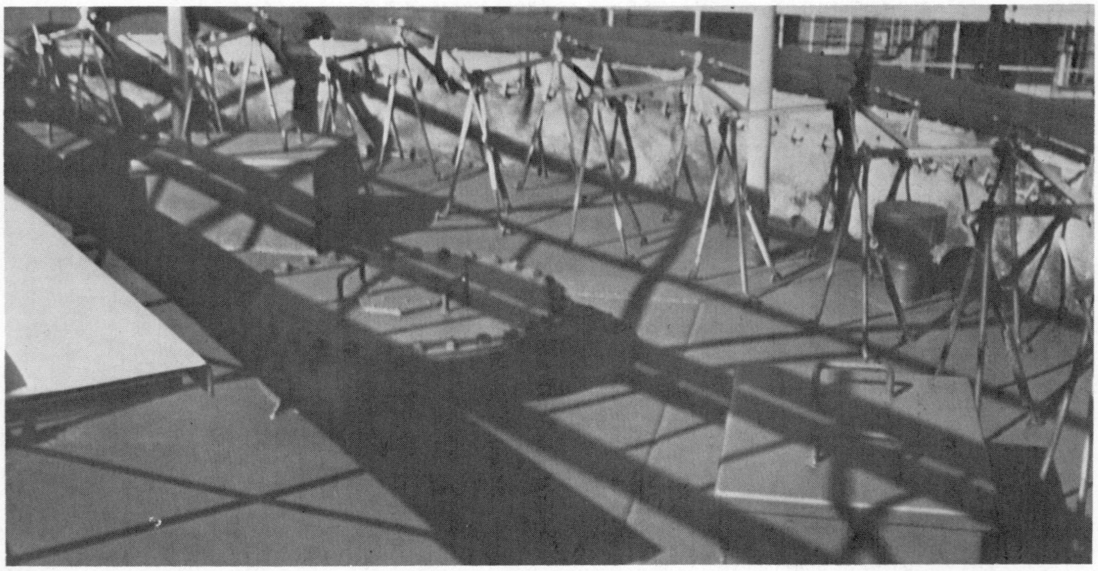

Fig. 5-176. Roof of fuel storage tank coated against corrosion with Eponol resin/red iron oxide primer. (*Courtesy: Shell Chemical Co.*)

aliphatic hydrocarbons, and resists staining from common household agents rather well. The permeability of phenoxy, particularly to O_2 and other gases, is lowest of any melt-processable polymer. Oxygen permeability is $\frac{1}{20}$ that of HDPE, and its moisture vapor transmission is about the same as that of rigid vinyl, higher than PE, and much lower than styrene and cellulosics.

The phenoxy resins suitable for molding and extrusion have molecular weights that range from 25,000 to 40,000. They will pick up moisture, so predrying is required.

Above 175°C (347°F), these resins display nonelastic behavior. At higher processing temperatures, the melt viscosity is very highly temperature-dependent. Typical molding temperatures are 230–275°C (446–527°F); thus, minor adjustments in temperature can cause substantial change in viscosity, thereby simplifying adaptations to equipment and application demands. Machinery suitable for impact polystyrene is adequate for phenoxy. Operating pressures of 14,000 psi are typical. Mold design follows general practice for most general-purpose plastics, except for slightly more draft than usual in deep draw sections because of the low shrinkage of the polymer.

Extrusion proceeds easily at about 230°C (446°F), using a high-compression-ratio polyethylene- or nylon-type screw.

Table 5-147. Comparative Performance: Eponol 55, a Phenoxy Lacquer, Versus Other Lacquer-Type Resins and Converted Epon Resin Systems.*

	Lacquer Resins (All systems force dried and aged seven days at room temp.)				Converted EPON Resin Systems			
Test Conditions	EPONOL 55 Resin	Vinyl Resin	Acrylic Resin	Nylon Resin	EPON Resin 1001 Amine-cured (room temp.)	EPON Resin 1007 Phenolic-cured (400°F)	EPON Resin Ester Melamine-cured (400°F)	EPON Resin 1007 Urea-cured (375°F)
Chemical Resistance:								
20% NaOH (11 days–200°F)	8	6	1	1	8	9	1	5
30% H_2SO_4 (14 days–room temp.)	10	10	7	1	10	10	8	3
10% HCl (14 days–room temp.)	9	9	6	1	8	5	5	2
Distilled H_2O (20 days–room temp.)	10	7	5	1	8	10	10	10
Distilled H_2O (11 days–200°F)	6	7	1	1	4	10	8	4
Environmental Resistance								
Humidity–1000 hrs. (100°F and 100% r.h.)	8	$7\frac{1}{2}$	$8\frac{1}{4}$	1	–	–	–	–
Humidity–2500 hrs. (100°F and 100% r.h.)	8	7	6	1	–	–	–	–
Heat Resistance	7	1	9	5	7	6	4	3

Key to Ratings: 10 = unaffected
0 = complete failure
*Courtesy: Shell Chemical Co.

The elasticity of the polymer at the lower processing temperature facilitates blow molding through minimization of parison sag and runaway.

Machining of phenoxies is easily accomplished through the use of sharp tools operated at slow surface speeds. Phenoxies may be hot-gas-welded or solvent-cemented without difficulty.

The biggest use for phenoxies is as a vehicle for coatings formulations (see Figure 5-176). Table 5-147 compares the performance of phenoxy lacquers and other lacquer resins. They are exceptionally useful in primer applications where drying speed, compatability with various top coats, and high adhesive strength are required. They are used in automotive and marine primers as well as in heavy-duty maintainance primers.

Molding compounds are limited to applications not requiring service temperatures in excess of 80°C (176°F). The combination of good impact strength, clarity, and impermeability makes the resins attractive for blow molding of cosmetics, foodstuff, and household chemical bottles.

As adhesives, the biggest use for these resins is in wood and plastic sandwich constructions and as production adhesives for wood surfaces.

Trade Names

Araldite	Ciba Products Corporation
Bakelite PXXX	Union Carbide Chemicals Corporation
D.E.R.	Dow Chemical Co.
Epi-Rez	Celanese Resins Co.
Eponol	Shell Chemical Co.

REFERENCES AND BIBLIOGRAPHY

Lee, H., Stoffey, D., and Neville, K. *New Linear Polymers.* McGraw-Hill Book Co., New York, 1967.

Shell Chemical Co. *Eponol Resins for Surface Coatings.* Bulletin SC: 73-29.

Parylene (Poly-Para-Xylylene). Para-xylene is pyrolyzed in steam at 950°C (1742°F) to produce the dimer di-*p*-xylylene.

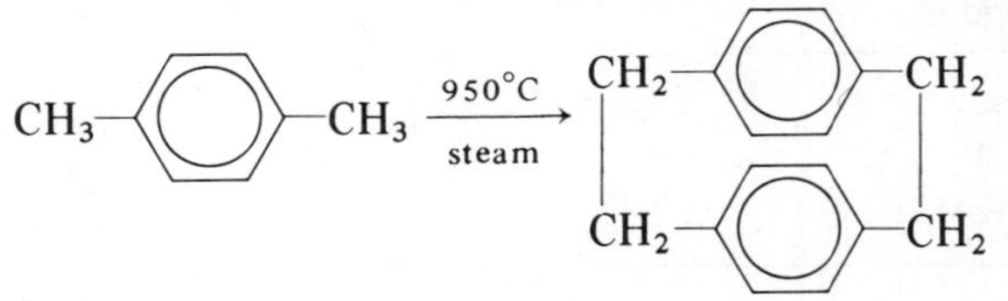

p-xylene di-*p*-xylylene

This dimer, when purified, is the monomeric precursor for an unusual specialty polymer, parylene. The polymer is unusual in that it is polymerized *in situ* in the end-use application.

Polymerization occurs via a vacuum, vapor-phase pyrolysis of the dimer to a reactive intermediate.

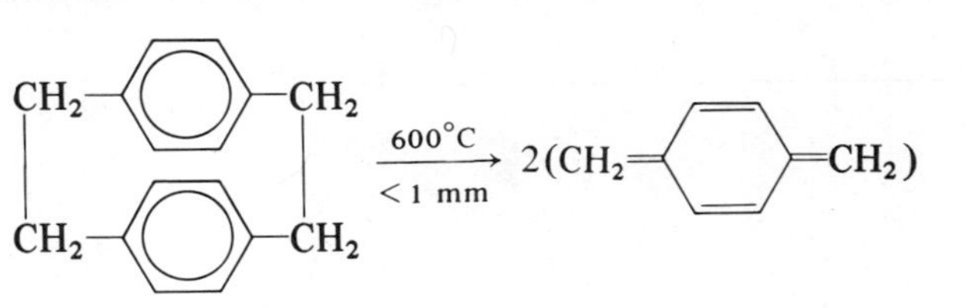

This intermediate then spontaneously polymerizes via a free-radical mechanism upon cooling.

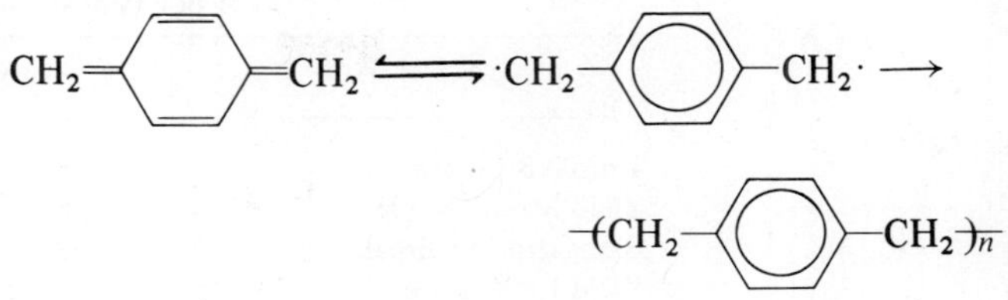

(a)

A quantitative yield of a linear polymer is obtained having a molecular weight of about 500,000.

Three types of parylene have reached commercial levels of acceptance, (a) parylene N (poly-*p*-xylylene), (b) parylene C (poly-monochloro-para-xylylene), and (c) parylene D (poly-dichloro-para-xylylene).

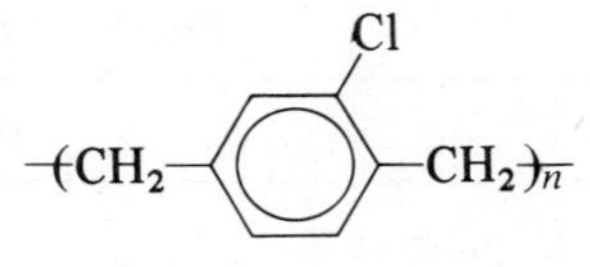

(b)

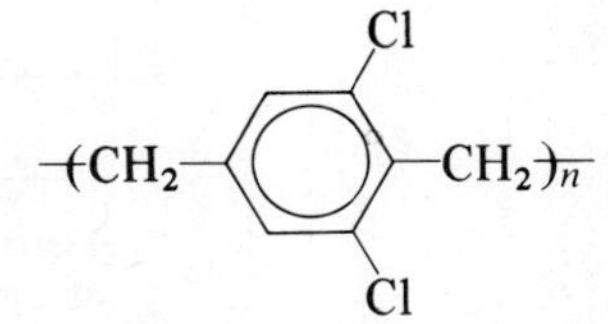

(c)

The application of parylene requires special, although not complex or bulky, equipment. It consists essentially of a vaporizer, a pyrolysis unit, and a deposition chamber. A schematic of such a unit is shown in Figure 5-177. Figure 5-178 shows an actual production coater.

Objects to be coated are placed in the deposition chamber, where the vapor coats them with polymer. Condensation coating does not run off or sag like conventional coating methods. It is not "line-of-sight" as in vacuum-metallizing. The vapor evenly coats edges, points, and internal areas. Figure 5-179 shows 2-mil copper foil coated to a depth of 0.5 mil with parylene. Although the vapor is pervasive, holes can be coated without bridging. Masking easily prevents areas from being coated where desired. Objects to be coated remain at or near room temperature, thus preventing possible thermal damage. Because of the quantitative nature of the reaction, coating thickness is accurately and simply controlled by manipulation of the dimer charge to the vaporizer.

Current available deposition chambers can accomodate large parts (28,000 in^3, 5 ft long and 1.5 ft high). Single coat thicknesses range from 0.25 to 1.5 mils (0.00025-0.00015 inch).

Table 5-148 lists the properties of films prepared from the three types of parylene. Additives are not generally used, and property changes occur through selection of precursor dimer. However, fluorescence can be imparted to poly-*p*-xylylene coatings through the addition of about 0.25 weight percent of anthracene. This fluorescence provides a means of checking the uniformity of coating and is essential in microelectronic applications.

Melting points of these resins range from 290°C to 400°C (554-752°F). Glass-transition temperatures range from

Table 5-148. Properties of Parylene Films.*

Typical Physical and Mechanical Properties

	Parylene N	Parylene C	Parylene D
Tensile Strength, psi	6,500	10,000	11,000
Yield Strength, psi	6,100	8,000	9,000
Elongation to Break, %	30	200	10
Yield Elongation, %	2.5	2.9	3
Density, g/cm^3	1.11	1.289	1.418
Coefficient of Friction			
Static	0.25	0.29	0.33
Dynamic	0.25	0.29	0.31
Water Absorption, 24 hours	0.06 (0.029")	0.01 (0.019")	–
Index of Refraction, n_D23°C	1.661	1.639	1.669

Typical Electrical Properties

	Parylene N	Parylene C	Parylene D
Dielectric Strength, Short Time, volts/mil at 1 mil	7,000	5,600	5,500
Volume Resistivity, 23°C, 50% RH, ohm-cm	1×10^{17}	6×10^{16}	2×10^{16}
Surface Resistivity, 23°C, 50% RH, ohms	10^{13}	10^{14}	5×10^{16}
Dielectric Constant			
60 Hz	2.65	3.15	2.84
10^3 Hz	2.65	3.10	2.82
10^6 Hz	2.65	2.95	2.80
Dissipation Factor			
60 Hz	0.0002	0.020	0.004
10^3 Hz	0.0002	0.019	0.003
10^6 Hz	0.0006	0.013	0.002

Typical Barrier Properties

	Gas Permeability cm^3-mil/100 in.^{2}-24 hours-atm (23°C)						Moisture Vapor Transmission, g-mil/100 in.^{2}-24 hours, 37°C-90% RH
Polymer	N_2	O_2	CO_2	H_2S	SO_2	Cl_2	
Parylene N	7.7	39.2	214	795	1,890	74	1.6
Parylene C	1.0	7.2	7.7	13	11	0.35	0.5
Parylene D	4.5	32	13	1.45	4.75	0.55	0.25

Typical Thermal Properties

	Parylene N	Parylene C	Parylene D
Melting or Heat Distortion Temperature, °C	405	280	>350
Linear Coefficient of Expansion, (10^{-5}/°C)	3.5	6.9	–
Thermal Conductivity, (10^{-4} cal/sec/cm^2-°C/cm)	~3	–	–

Data recorded following appropriate ASTM method.
*Courtesy: Union Carbide Corporation.

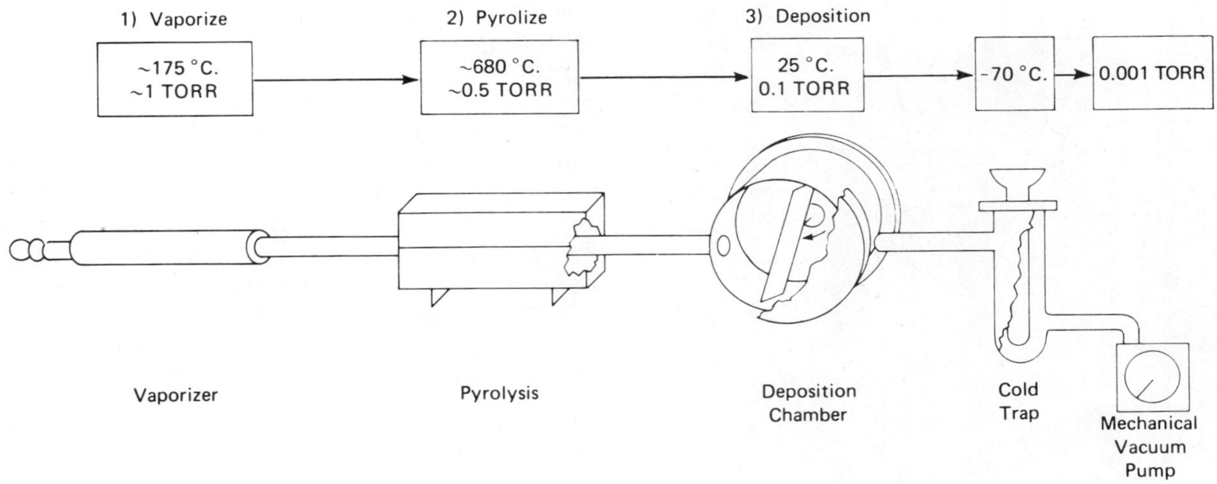

Fig. 5-177. Diagram of the parylene process. (*Courtesy: Union Carbide Corporation*)

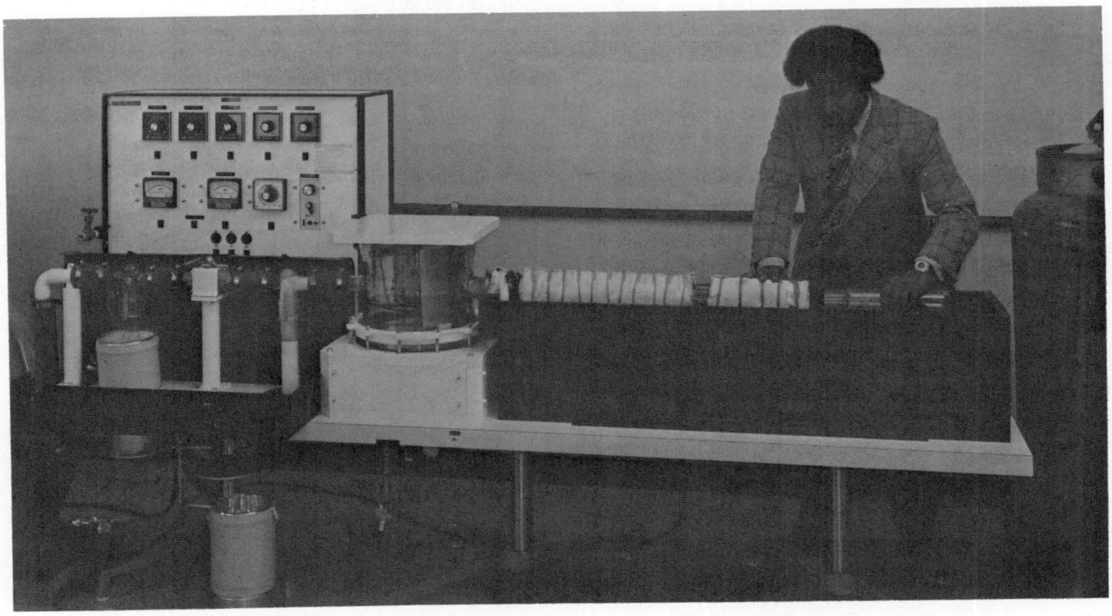

Fig. 5-178. Production coater for parylene. (*Courtesy: Hughes Aircraft Co.*)

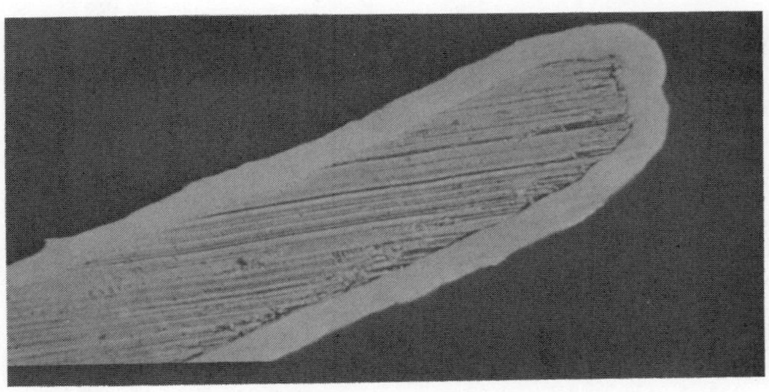

Fig. 5-179. Cross-section of a 2-mil copper foil with $\frac{1}{2}$-mil parylene coating. (*Courtesy: Union Carbide Corporation*)

60°C to 100°C (140–212°F). The cryogenic properties of the film are excellent. Physical properties are unaffected by thermal cycles from 2°K to room temperature. Thermal endurance in air is poor. Short-term (1000 hr) exposure is limited to 93–129°C (200–265°F) and long-term (10 years) exposure to 60–100°C (140–212°F). In inert atmospheres or in the absence of air, properties are maintained up to 216–279°C (420–535°F).

These polymers are generally insoluble up to 150°C (302°F). At 270°C (518°F) they will dissolve in chlorinated biphenyls, but the solution gels upon cooling below 160°C (320°F). Weather resistance of the resins is poor. Embrittlement is the primary consequence of exposure to UV.

Fig. 5-180. Circuit module coated with parylene. (*Courtesy: Hughes Aircraft Co.*)

The first significant commercial application of parylenes was as a dielectric film in high-performance, precision electrical capacitors. Since then, their use as circuit board and electronic module coatings has come to dominate. The coatings protect the units from airborne contaminants, moisture, salt spray, and corrosive vapors, while maintaining excellent insulator protection (see Figure 5-180). The coatings are also extensively used in the protection of hybrid circuits.

Ferrite toroids are conveniently encapsulated with parylene. The polymer sheath provides environmental protection without affecting part dimensions, shapes, or magnetic properties.

One of the most recent end uses involves the fabrication of free-standing films of parylene. These films are ultra-thin (250Å–3 microns) and are called pellicles. They are used as beam splitters in optical instruments, windows for nuclear radiation measuring devices, dielectric supports for planar capacitors, and for extremely fast-responding, low-mass thermistors and thermocouples.

Trade Names

Parylene N, –C, –D	Union Carbide Corporation

REFERENCES AND BIBLIOGRAPHY

Baker, T. E., Bagdasarian, S. L., Fix, G. L., and Judge, J. S. Characterization of vapor-deposited paraxylyene coatings. *Journal of the Electrochemical Society*, June 1977, 124 (6): 879–900.

Lee, H., Stoffey, D., and Neville, K. *New Linear Polymers*. McGraw-Hill Book Co., New York, 1967.

Gorham, W. F. "Parylene Polymers." In *Modern Plastics Encyclopedia*, S. Gross (exec. ed.), 45(14A):230–234. McGraw-Hill, Inc., New York, 1968–69.

Union Carbide Chemical Co. *Parylene Conformal Coatings*. Bulletin No. F-43427A, July 1974.

6
Thermosets

INTRODUCTION

Before we begin our discussion of the various plastics which comprise the family of thermosets, it is important that an understanding of the concepts of *gelation* and *curing* are clear. Typically, the crosslinking mechanism is one of low molecular weight moieties combining chemically to form the three-dimensional network structures described in Chapter 2. As with all chemical reactions, the kinetics are time, temperature, and mass dependent. It is irrelevant whether the initial reactants are solids, liquids, or gases—the reaction principles are the same. However, the explanation of the practical facets of crosslinking is considerably simplified if it is assumed that the initial reactants are low-viscosity liquids.

As the molecular weight of the forming polymer increases, the viscosity increases. The crosslink density concurrently increases, and a point in time is reached when the conversion from liquid to solid becomes irreversible. This point is known as the *gel time* of the system.

Figure 6-1 is a sketch of the viscosity/time relationship of a hypothetical polymer undergoing crosslinking. Point A represents the initial viscosity of the reactants. As the reaction proceeds to point B, the exothermic heat of reaction causes a typical reduction in viscosity. By the time point C is reached, the growing molecular mass is causing the viscosity to increase at a faster rate than the exothermic heat can drive it down. At point D, the gel point (time), the conversion to solid is macroscopically manifest and the viscosity is rapidly going to infinity. With some thermoset systems (epoxies, polyesters, and polyurethanes), this point can be seen quite accurately as a small gel-like lump in the curing liquid. With other systems, this point is very difficult to narrow down, as the viscosity is increasing very slowly and the entire mass gets uniformly thicker. Still other systems are solids to begin with, and the gel point can only be observed by consecutively measuring its flow under pressure with time.

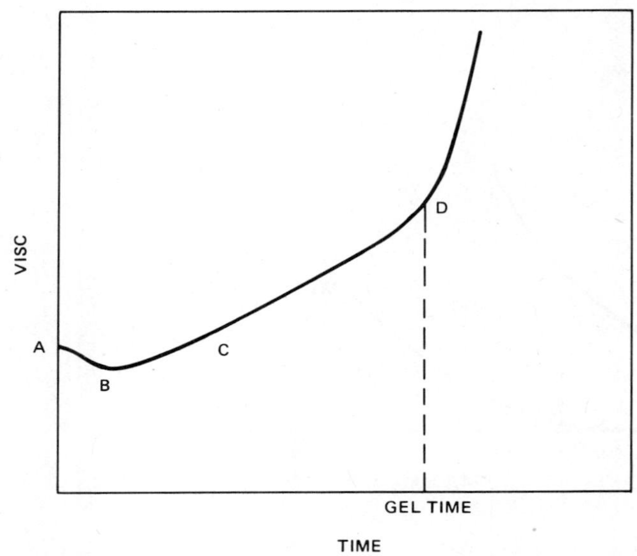

Fig. 6-1. Viscosity versus time for a hypothetical thermosetting polymer at constant temperature.

Figure 6-2 shows the effect on gel time (GT) of increasing the temperature. At T_1 the reaction kinetics are very slow, and, in the time span shown, gelation has yet to occur. At T_4, the reaction is very rapid, and gelation occurs very fast (usually accompanied by large amounts of exothermic heat).

At T_0, the system is essentially dormant. This is the basis of the so-called one-can or one-component systems. The coreactants will remain stable at T_0 because the reaction has been effectively stopped. For crosslinking to occur, the temperature must be increased to a point appropriate for the anticipated processing methods.

Figure 6-3 shows the effect of mass on gel time. The curves look very similar to those shown in Figure 6-2. As the mass of reactants increases, the time to gelation decreases. The reason for this phenomenon is that polymeric materials are inherently thermal resistant. Thus, as more exothermic heat is generated by the larger quantity of reactants, more heat is retained in the polymerizing mass. The heat cannot conduct away from the system to the environment, and the temperature of the system increases. The kinetics then change, as described in Figure 6-2. A gallon bucket of a curing epoxy system may gel in 30 minutes at 25°C (73°F). If the same system is poured into a wide shallow metal pan sitting on a cold concrete floor, it may take 8 to 16 hours to gel.

Shelf life and *working life* are terms related to and dependent upon the gel time of a thermosetting system. Working life is the time available for processing the system from the point the reactants are first brought together. Only under very unusual conditions does the working life ever equal or exceed the gel time.

Shelf life represents the amount of storage time during

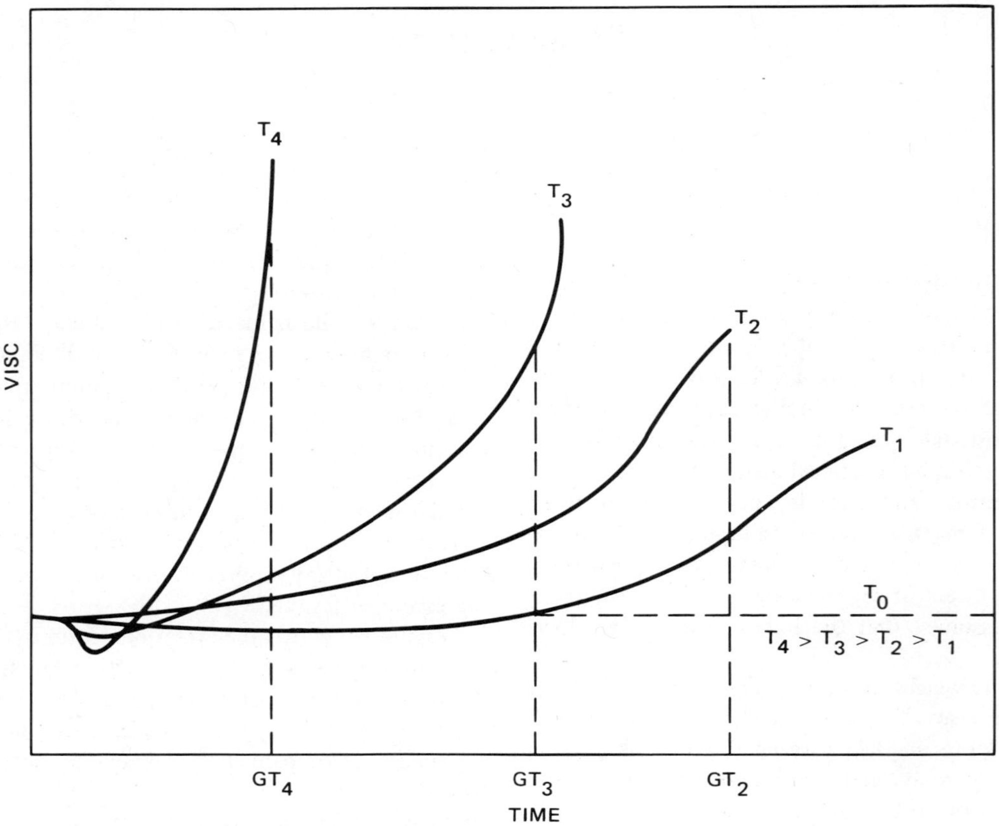

Fig. 6-2. Influence of temperature on gel time of thermosetting resins.

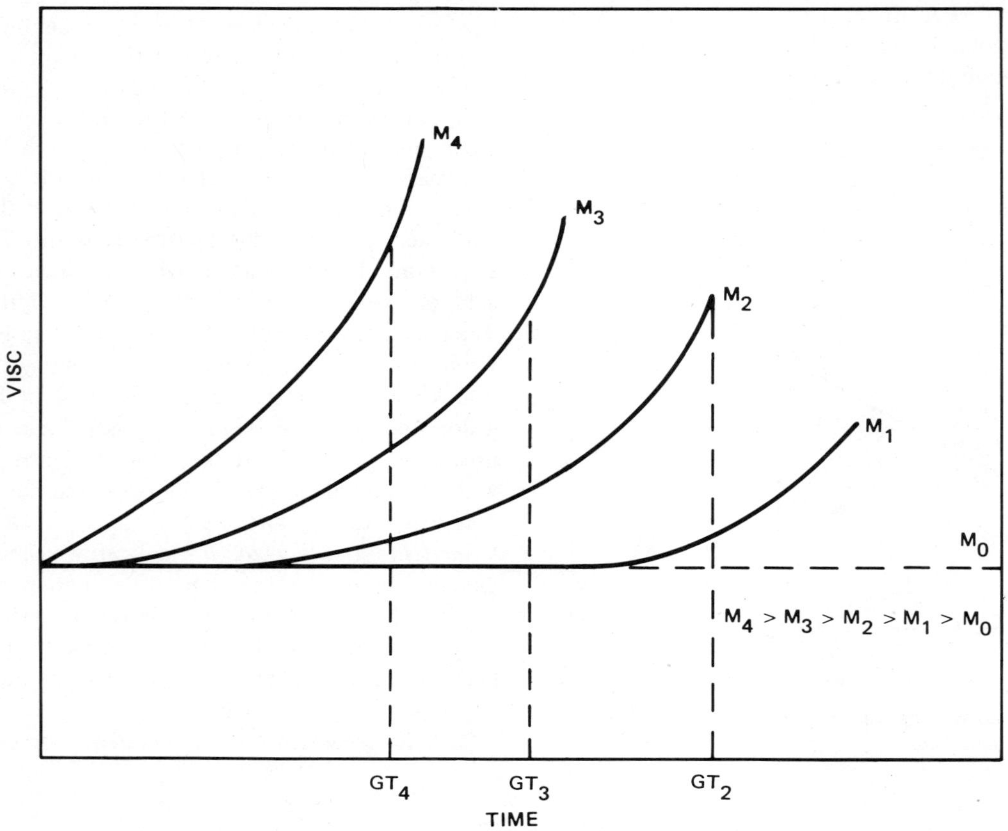

Fig. 6-3. Influence of resin mass on gel time of thermosetting resins.

which a thermosetting system will remain stable (i.e., the time available for storing the material before processing begins). One-can systems are designed to provide very long shelf lives at room temperature or below. Shelf life can be influenced by other factors besides reaction kinetics. For example, contamination of the resins by moisture, carbon dioxide, or other airborne particulates can seriously affect the gelation characteristics of epoxies, polyurethanes, and silicones.

Curing schedules reflect the programmed control of the crosslinking mechanism after gelation. Once the irreversible conversion point to crosslinked solid has been passed, the system continues to react until it begins to take on the form of the final polymeric structure. This structure dictates the properties of the polymer.

If the mechanical properties of a polymer are measured as a function of reaction time, a curve such as is schematically outlined in Figure 6-4 results. The gelation period shown represents the period before gelation, wherein the system has essentially no plastics properties, merely the properties of the unreacted monomers. As the polymer structure begins to develop, so do the properties, until the reaction approaches completion. Most curves are asymptotic at this point. Typically 80–90% of the ultimate properties will develop during the very early part of the asymptotic plateau. Further curing is often not value-effective, since the time span needed to generate the additional 5–10% of properties is disproportionately long.

Optimization of a cure schedule (program of temperature, pressure, and other processing variables) is usually achieved by superimposing a series of property/time curves derived from variations in cure schedule. The time which provides the optimum combination of properties is then selected from this configuration (see Figure 6-5).

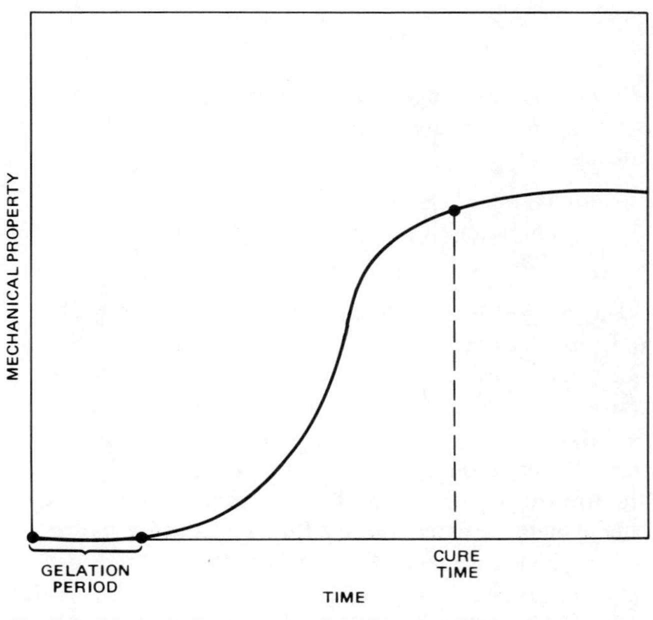

Fig. 6-4. Mechanical property of a thermosetting polymer versus time.

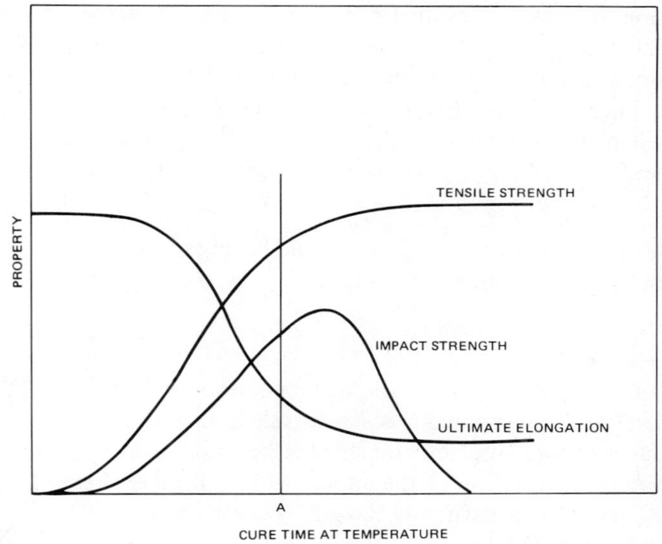

Fig. 6-5. Optimization of cure schedule for thermosetting plastics. Optimum properties occur at point A.

Since curing is still the phenomenon of the polymerization reaction, it is also dependent on time, temperature, and concentration of reactants. Alteration of the first two variables provides the basic mechanism for controlling the processing of curable resins. Thus, by allowing the cure to progress to an intermediate level (a degree of chain elongation and crosslinking is allowed to occur) and then dropping the temperature, a shelf-stable monomer/polymer system may be obtained. This condition is often referred to as *B-staging.* B-staging will be discussed in greater depth under the sections on phenolics.

Other methods of influencing gelation and cure characteristics include *prepolymerization* and *adducting*. Prepolymerization is a method of increasing the molecular weight (usually by chain extension only) to an intermediate value. The resultant "prepolymer" has sufficient functionality left to allow for a subsequent reaction which will complete the crosslinking.

In adducting, one of the monomers is "capped" chemically with the coreacting monomers. Figure 6-6 shows such a technique in schematic form. Reaction (1) is a representa-

(1) Standard Polymerization

$$nF_1AF_1 + nF_2BF_2 \longrightarrow -\!\!(AF_1F_2B)\!\!-_{2n}$$

(2) Adduction

$$1F_1AF_1 + 2F_2BF_2 \longrightarrow F_2BF_2F_1AF_1F_2BF_2 \text{ (adduct)}$$

(3) Completion of Polymerization

$$nF_2BF_2F_1AF_1F_2BF_2 + nF_1AF_1 \longrightarrow -\!\!(AF_1F_2B)\!\!-_{2n}$$

F_1AF_1 = Monomer A, with two functional groups F_1

F_2BF_2 = Monomer B, with two functional groups F_2

Fig. 6-6. The process of monomeric adduction.

tion of a standard linear polymerization. Monomer A's functional group (F_1) coreacts with monomer B's functional group (F_2). When reacted in equimolar quantities, a linear polymer results of the form: $-(\!-AF_1F_2B-\!)-$, where F_1F_2 represents the result of the functional-group reaction (e.g., F_1F_2 would be a single bond if F_1 and F_2 were double bonds, or F_1F_2 would be an ester, $-\overset{\overset{O}{\|}}{C}-O-$, if F_1 was a hydroxyl, —OH, and F_2 was a carboxylic acid, $-\overset{\overset{O}{\|}}{C}-OH$).

In reaction (2) one mole of monomer A is reacted with two moles of monomer B. The result is an adduct with F_2 functionality available for further reaction. Reaction (3) shows the reaction of the adduct with additional monomer A, resulting in essentially the same polymer as was obtained in reaction (1).

Gelation and cure are both dependent on the stoichiometric relationship between coreactant monomers. Part of the "art" of thermoset formulation is the ability to correctly balance the stoichiometric equivalents of each reacting species, such that the resultant polymer yields the desired properties. (See additional discussions of stoichiometry considerations in the sections on unsaturated polyesters and epoxies.)

With these fundamental ideas clarified, we can now proceed to the discussions of the various classes of plastics in the thermoset family.

ALDEHYDE-BASED RESINS

Traditionally, the discussion of thermosetting plastics begins with phenol-formaldehyde resins. They were not only the first of their kind, but their history encompasses all the facets of terminology, chemistry, manufacture, and processing that has since characterized the entire thermosetting family of plastics.

Subsequent development of other resins based on the copolymerization of formaldehyde led to the group of materials called aminoplasts (urea- and melamine-formaldehydes). It is thus not inconsistent to consider all three of these polymer groups as part of a larger family, namely aldehyde-based thermosetting resins. We have included the furan resins in this group as well, although traditionally they are dealt with separately. However, furan resins are based on an aldehyde, furfuryl aldehyde, and many of the commercial products are based on furan-urea and furan-phenolic copolymers.

Phenolics

Adolf Bayer first described reactions between phenols and aldehydes in 1872. Between 1905 and 1914, Leo H. Baekeland produced the first commercially exploitable product based on these reactions. As stated earlier, these were the first truly synthetic resins, and were called phenolics or phenoplasts. In order to make them work adequately for widespread industrial use, however, Baekeland had to solve two key problems: (1) the unmodified resins produced were very brittle, and (2) steam voids, formed as a result of the hot water vapor generated from the condensation reaction (from the reaction exotherm) caused porosity in the molded part. Baekeland's technique was to (1) add a filler, such as wood flour, which improved the resin's toughness, and (2) mold under heat and pressure greater than the vapor pressure of water. The latter prevented bubbling and speeded molding time because higher temperatures could be used.

Phenols will react with aldehydes to form condensation products if the ortho and para positions on the benzene ring are unoccupied. For reasons to be discussed shortly, formaldehyde is the preferred aldehyde.

Phenol is obtained from the fractional distillation of coal tar as well as from synthetic processes. There are at least six known commercial processes, and their use is most often a function of current economics. The cumene process is presently among the more important.

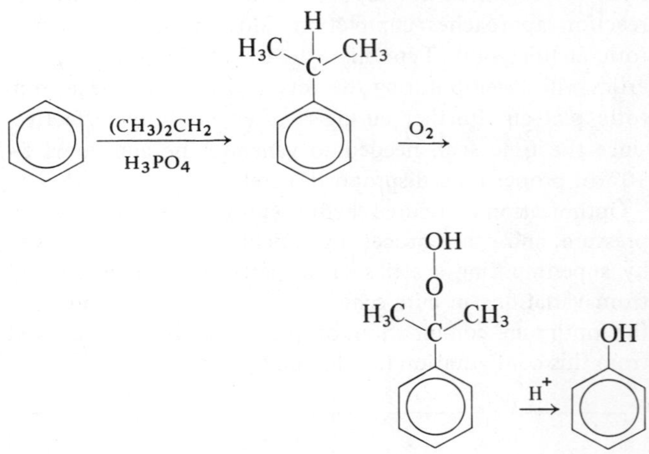

Formaldehyde is made from methanol. The details of its preparation have been described previously (Chapter 5, Polyacetals).

$$CH_3OH + O_2 \xrightarrow[350\text{–}450°C]{\text{Fe-Mo Oxide}} HCHO$$

The precise mechanism of the reaction between phenol and formaldehyde is still the subject of study. Whatever the exact mechanism may be, the phenolic hydroxyl group activates the benzene ring, such that a methylol group ($-CH_2OH$) joins the nucleus at the ortho and para positions. When alkaline catalysts are used, phenol alcohols are formed—called *resols.* If acid catalysts are used, diphenyl methane derivatives are formed—called *novolacs.*

The phenol alcohol condensation to resinous form is very complex. A classification first established by Baekeland, divides the process into three stages. Each stage is not clearly exclusive, as the mechanism passes gradually from one to the next.

The first is the *A-stage* or resol. This represents the initial condensation product consisting principally of phenol alcohols. The resin at this point is soluble in organic solvents and has a thermoplastic nature.

The second stage is the *B-stage* or resitol. This represents a degree of condensation wherein the molecular weight has increased to such a point that the solubility has decreased. Some crosslinking has occurred, the resin is soft and fusible when hot, but hard and brittle when cold.

The last stage is the *C-stage* or resite. This represents the final stage of polymerization. The bulk of crosslinking has occurred and the resultant resin is completely insoluble and infusible.

The phenol alcohols condense rapidly to a three-dimensional macromolecule crosslinked by methylene bridges. Water is the simple molecule that is eliminated. With alkaline catalysts, the reactions that occur will depend on the structure of phenol, reaction time, and temperature. The most important reactions that occur are the formation of methylene (a) and ether (b) bridges.

(a)

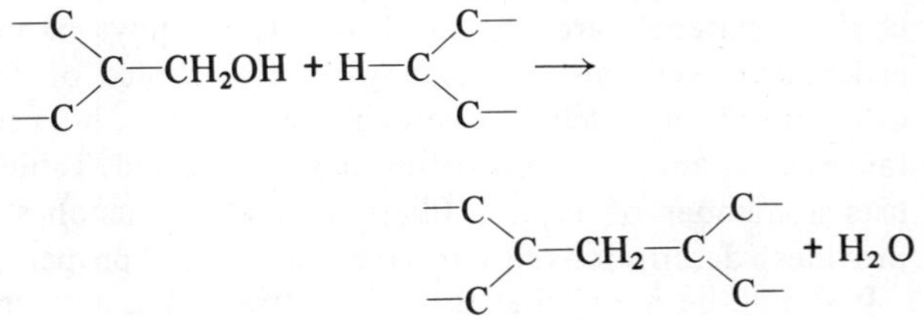

(b)

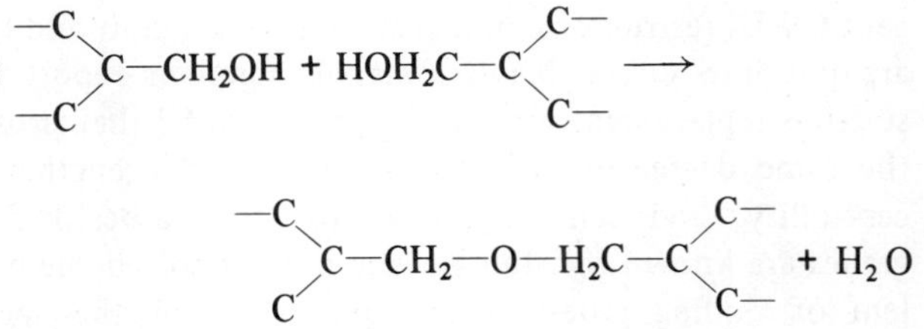

When phenol is condensed with formaldehyde under acidic conditions, the novolacs formed are permanently fusible and soluble. The novolacs are long-chain polymers (n = 8-12) in which the benzene rings are connected only by methylene bridges. The structure for the simplest type of novolac, poly(dihydroxy-diphenylmethane) is:

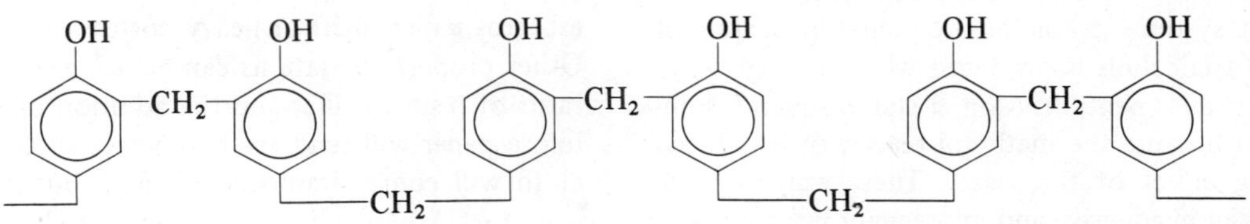

This is only one of many possible structures however.

The novolacs are thermoplastic in this form; that is, fusible and soluble. To convert them to a thermoset, further formaldehyde is added in an alkaline environment. The commonest way to achieve this crosslinking is through the addition of hexamethylenetetramine ("hexa"). Upon exposures to heat, the "hexa" decomposes into methylene groups for crosslinking and ammonia to provide an alkaline catalyst.

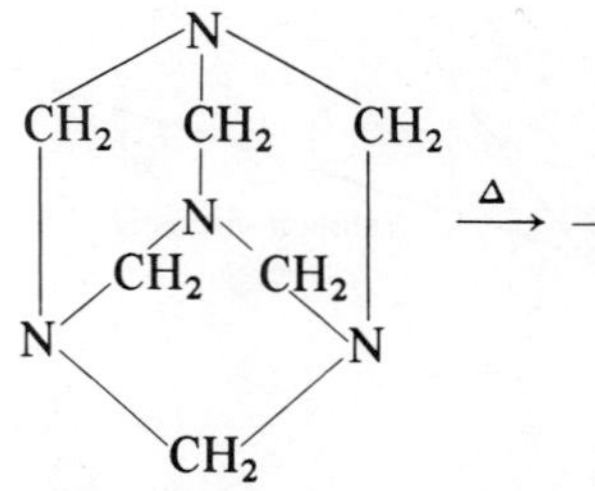

It is apparent that the basic difference between a resol and a novolac is the presence of one or more free methylol groups on the resol. The crosslinking occurs through the reaction of these methylol groups.

Formaldehyde is the preferred aldehyde for three important reasons. First, resins made from formaldehyde have shorter curing times with concomitant increased production rates compared to other aldehydes. Second, formaldehyde participates in fewer side reactions, allowing for more predictable end-product properties. Third, formaldehyde is the cheapest aldehyde.

Other aldehydes are used when specialty resins or properties are required. Included in this grouping are acetaldehyde, butyraldehyde, acrolein, and furfuraldehyde (fufural). The latter has achieved the largest commercial utility of those materials listed.

The most important phenols are the monohydric type. They are readily available and lowest in cost. Included in this grouping are resols and xylenols. Resorcinol is used for "cold-curing" (temperature from -4°C to 93°C, 25-200°F) resins.

Phenolic resins may be manufactured in two different ways. One-step resins are prepared by charging all the ingredients into the reaction kettle (phenol, formaldehyde, catalyst). An alkaline catalyst such as sodium carbonate or ammonia is used, and the molar ratio of phenol to formaldehyde is about 1:1.25. Condensation is interrupted while the resin is still thermoplastic and soluble. However, it is heat reactive and requires only further heating to complete the crosslinking reaction to an infusible, insoluble state.

Two-step resins are prepared by charging only part of the necessary formaldehyde and using an acid catalyst. The ratio of phenol to formaldehyde in this first step is

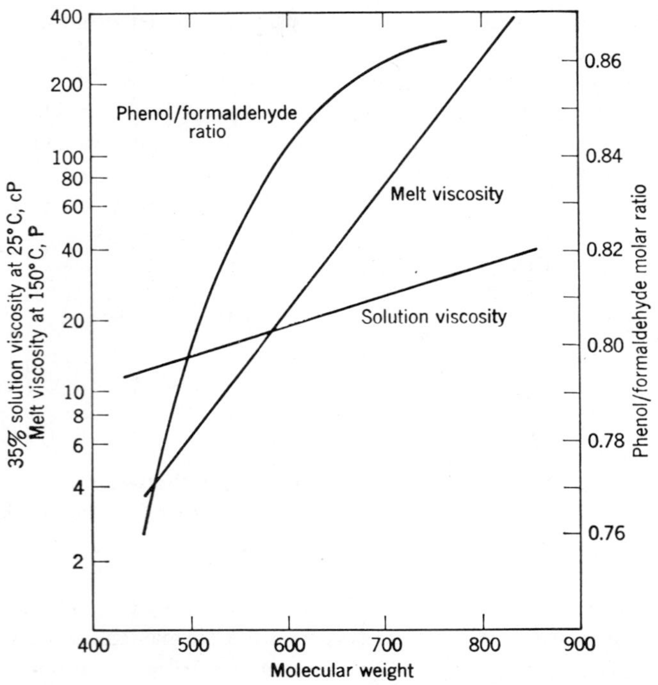

Fig. 6-7. Relationships between molecular weight of acid-catalyzed novolac resins and phenol/formaldehyde ratio, melt viscosity, and viscosity of ethanol solutions of 35% resin content. (Keutgen, 1968)

about 1:0.08. Typical catalysts include hydrochloric, sulfuric, formic, and aromatic sulfonic acids. If the amount of formaldehyde is the same, then resins using an acid catalyst are formed about three times as fast as resins made using an alkaline catalyst. This first step produces a stable novolac. (Figure 6-7 shows the relationship between molecular weight and phenol/formaldehyde ratio of acid-catalyzed novolacs.) Molding powders may then be prepared at this point by mixing fillers, pigments, and other compounding ingredients, together with the additional "hexa" to complete the cure. A final phenol/formaldehyde ratio of approximately 1:1.5 is obtained after the "hexa" is added.

Industrial production typically proceeds as follows. The phenol is charged to the resin kettle first, followed by the formaldehyde, catalyst, and modifiers. Because of the resin viscosity, adequate agitation and side scraping is required for thorough mixing, particle contact, and heat transfer from the kettle jacket. The heat of reaction is high and reaction rate is fast, so reflux techniques are used for temperature control rather than jacket or coil cooling.

With one-step systems, a condensate consisting largely of phenol di- and trialcohols is produced which is very water soluble. With catalyst neutralization, a stable aqueous solution is obtained because the methylol reactivity is reduced by the diluting effect of the water. These water-soluble phenolics are used in adhesive and impregnation formulas.

If a solid resol is required, dehydration takes place under vacuum. The heating must be done very carefully, since overheating can initiate crosslinking. The reaction is highly exothermic and can quickly run away. As soon as dehydration is completed, the resin is dumped into shallow cooling pans and the resulting brittle solid is broken up into small pieces. These chunks can then be ground or milled to a powder.

In the two-step systems, the reaction mass will separate into two phases at the end of the first step. The lighter phase consists of the water of reaction plus the water from the formalin. The heavier phase is the novolac. The water is either decanted or distilled off. Then the resin is further dried by vacuum heating.

For molding powders, the powdered resin is blended with lubricants and "hexa." Then, fillers and colors are blended in a ribbon blender. For laminates, alcoholic solutions of resins are prepared and the webs impregnated with the solutions. Often the solvents are then driven off and the B-stage "advanced" by passing the impregnated webs through forced-air ovens. The process is known as "prepregging" and results in a tack-free or slightly tacky sheet that is then cut and plied to form the final laminate through curing in a hot press.

Fillers greatly improve the properties of most thermosets. They are always incorporated in phenolic molding compounds; consequently, all plastics articles made from this class of materials are opaque. Not only are physical properties improved, but, depending on the choice of filler, other properties such as electrical conductivity, heat resistance, cost, and flow properties may be affected. Table 6-1 lists a number of typical fillers used with phenolics and provides a description of their effects on selected properties.

It should be noted that the use of asbestos as a common phenolic filler will become increasingly limited as the stringent OSHA restrictions on asbestos use throughout industry are put into effect. Major materials suppliers report that suitable replacement grades have been found that provide the same degree of heat resistance, impact strength, processability, and relatively low cost that asbestos-filled grades are known for. But no one has yet solved the problem of finding asbestos-free equivalents for the special high-impact, high heat-resistant grades which, though only a small percentage of total sales of asbestos-filled phenolics, are still vital in some electrical and automotive applications.

Table 6-2 lists the commercial asbestos-free phenolic molding compounds available as of this writing. Figure 6-8 shows an application for such compounds.

Laminates represent a special category of fillers. Webs are impregnated with phenolic and may be cellulose paper, asbestos paper, light or heavy cotton cloth, nylon, or glass. Other property variations can be achieved by adjusting the ratio of resin to filler and the addition of plasticizers. Cellulose paper will tend to show water susceptibility. Cotton cloth will confer improved strength but still not be water resistant. Asbestos paper will give products useful in electrical applications (low moisture uptake), but these laminates will be relatively heavy. Asbestos and glass fillers will be extra heat resistant.

The National Electrical Manufacturers Association

Table 6-1. Fillers Used with Phenolics.

1. Organic fillers—all have relatively poor heat resistance.
   a. Wood flour—cheapest and commonest; good appearance and flow; poor to fair electrical properties; fair impact and good tensile strength; low heat conductivity.
   b. Walnut-shell flour—smooth surface with medium gloss.
   c. Cotton flock—improves impact but reduces flow.
   d. Chopped textile fabric—improves impact even more, but flow becomes poorer.
   e. Twisted cord—highest impact with good flow; poor water absorption.
   f. Rubber—questionable whether this addition should be listed as a filler since, when the rubber and phenolics are used together, a conventional filler such as wood flour, cotton flock, or asbestos is also added; improvements in impact and flexibility.
2. Inorganic fillers—have some advantages over organic fillers, but are not as widely used except for special-purpose molding powders; better dimensional stability, water absorption, heat resistance, and chemical resistance compared with wood flour and other cellulose fillers.
   a. Asbestos—good heat resistance, chemical resistance, and water resistance; lowers flexibility and tensile strength and increases specific gravity.
   b. Diatomaceous silica—good appearance, heat resistance, and electrical properties; very abrasive.
   c. Mica—excellent electrical properties, heat resistance, and dimensional stability; does not wet well; causes sticking and poor mold release.
   d. Graphite—improved frictional properties; electrical resistance decreases.
   e. Fibrous glass—good impact and electrical properties; good high-temperature resistance.

Fig. 6-8. The upper housing and grease drip-tray of this burger cooker are made from asbestos-free Genal phenolic molding compound. (*Courtesy: General Electric Co.*)

(NEMA) has graded standard phenolic laminates used for electrical and electronic applications. A brief list and description of these grades is shown in Table 6-3. Descriptions of melamine, epoxy, and silicone laminates are provided as well.

Phenolics are capable of exhibiting a large degree of property variance through chemical as well as physical modification; thus, it is difficult to become very specific when discussing properties. In general, phenolics are characterized by dimensional stability, good chemical resistance (except to strong alkalis and oxidizing acids), and excellent thermal and electrical resistance properties. Most tables of data are published for molding powder grades. Table 6-4 shows such a table.

If we examine the properties based upon the end-use form we can become somewhat more specific. Phenolic moldings are made using numerous types of compounds containing fillers and reinforcements to obtain specific engineering properties. These compounds include grades designed for general-purpose service, heat resistance, impact strength, high electrical insulating characteristics, dimensional stability, chemical resistance, and for contact with foods, cosmetics, and drugs.

Phenolic moldings are regarded as only moderate in strength relative to other plastics; however, they are more rigid than most thermoplastics and many other thermosets. They are relatively heavy (density), yet their stiffness-to-weight ratio is very high. They are not particularly tough or impact resistant unless modifications are made which also result in reduced moldability. Table 6-5 compares some properties of phenolics with those of other thermosets. Figure 6-9 compares some properties of injection-moldable phenolics with those of standard engineering thermoplastics.

Phenolic moldings retain their excellent electrical insulating properties at elevated temperatures. The heat-resistant grades provide the best arc resistance. At high electrical loads, some glass-reinforced thermoplastics outperform phenolics.

Creep resistance (see Figure 6-10), mold shrinkage, and dimensional stability of phenolics are very good. Organic fillers provide the least stability, while glass-fiber fillers provide the best stability.

Grades of phenolic are available which can withstand extended exposure to 232°C (450°F) and limited exposure to 343°C (650°F). They are very resistant to burning (UL ratings of VE- and HE-0).

Phenolic resins are dark brown to black, but further darkening and loss of gloss will occur upon exposure to weathering and UV. Moldings may even crack, craze, and/or lose mechanical and electrical properties.

Phenolic moldings are resistant to weak acids, organic solvents, hydrocarbons, detergents, and cleaning fluids. Stress-crack resistance is high. Phenolics are attacked by strong alkalis and oxidizing acids; however, there are available grades with resistance to strong non-oxidizing acids.

Table 6-2. Available Asbestos-Replacement Molding Phenolics.*

Material Type/ Supplier	Grade	Molding Type	Distinguishing Characteristics
General-purpose			
Durez	111	Comp., Trans., Inj.	Two-stage; general-purpose.
	118	C, Tr., Inj.	Two-stage; general-purpose.
Medium-impact			
Durez	123	Comp., Trans., Inj.	Two-stage; shock resist.; for elec. circuit breakers, auto terminal blocks, appliance motor bases, covers; UL temp. index, 150 C (302 F).
Electrical			
Durez	145	Comp., Trans., Inj.	
Plenco	509	C, Tr., Inj.	Two-stage; fast cure; for wiring devices and electrical control applications.
	586	Comp., Trans.	One-stage; brown; UL flammability rating of 94V-O @ 0.02 in.
Reichhold	25344	C, Tr., Inj.	Two-stage; dimensionally stable under hot, humid conditions.
Medium-heat-resistant/electrical			
Durez	152	Comp., Trans., Inj.	Two-stage; good molded appearance; for electrical, telecommunication, appliance applications.
	153	Comp., Trans., Inj.	Two-stage; good molded appearance and shock resistance; for electrical and appliance applications.
	156	Comp., Trans., Inj.	Two-stage; good molded appearance; for automotive, appliance and electrical applications; UL temp. index, 150 C (302 F); UL flammability rating of 94V-1 @ 0.02 in., 94V-0 @ 0.04 in.
G.E.	Genal 4300E	Inj.	Fast cycle; good surface finish and rigidity; granular or pellet; stiffest flow recommended for granular form (pellet only for cold-manifold); IEC track (CTI), 210 v.
	Genal 4301E	Inj.	Exc. surface finish; rigid set; resists tumbling; good moisture resistance.
	12553E	Comp., Trans.	Two-stage; excellent surface finish; rigid set; ejection distortion minimal; undercuts to be avoided; resistance to tumbling.
	12983E	Comp., Trans.	Two-stage; good surface finish; rigid set; IEC track (CTI), 200 v.
Plenco	400	Comp., Trans.	Two-stage; excellent surface finish; fast rate of cure.
	414	C, Tr., Inj.	Two-stage; for electrical applications.
	485	Comp., Trans., Inj.	Two-stage; rigid set; low-shrinkage; excellent surface finish; physical strengths similar to those of g-p grades.
	466	C, Tr., Inj.	Two-stage; excellent surface finish; good dimensional stability.
	527	Comp., Trans., Inj.	One-stage; improved mechanical strength; resistance to cracking in wet/dry exposures; minimum of free ammonia.
Reichhold	25171	Comp., Trans., Inj.	Two-stage; good surface; fast cure; IEC track (CTI), 220v; 94V-0 @ 1/32 in.
	25346	C, Tr., Inj.	Two-stage; excellent surface appearance.
	25378	C, Tr., Inj.	94V-0 @ 1/32 in.; good molding in inj.
Glass-reinforced grades for high-impact, high-heat resistance			
Acme resin	P-1000 P-1050 P-1100 P-1200	Comp., Trans., Inj.	Two-stage; short-glass-fiber-reinforced; for electronic/electrical applications.
Fiberite	4000 Series	C, Tr., Inj.	For electronic/electrical applications.
	FM-5064 FM-8130 FM-4030-190	Comp., Trans.	Long-glass-fiber reinforced; high-impact.
Cold-manifold molding grades			
G.E.	Genal P4000	–	General-purpose; in pellet form.
	Genal P4202	–	Medium-impact; in pellet form.
Plenco	600	–	General-purpose.
Reichhold	25000	–	Two-stage; fast cure; good appearance.
	25002	–	General-purpose.
	25398	–	Two-stage; electrical; asbestos-filled; 94V-0 @ 1/32 in.
Rogers	RX 611	–	High-impact, short-glass-filled; good electrical properties and heat resistance.
	RX 625	–	Med-impact; short-glass-filled; gd elec. properties and heat resistance.
	RX 660	–	Med-impact; short-glass-filled; gd heat resistance and elec. properties.

*Murray, 1977.

Table 6-3. NEMA Grades of Phenolic Laminates.*

Major Application Requirements	Typical Use	Micarta Grade	NEMA Grade	Military Specifications Type and Number	Base Material	Color	Resin
Paper Base Materials							
High tensile, flexural and compressive strength (uses where machining or humidity characteristics are not important)	Insulating washers	213	X	–	Paper	Tan	Phenolic
		423	X	–	Paper	Black	Phenolic
Electrical insulation in dry or humid conditions where fair mechanical strength and good machining are needed	Switchboard panels; relay and switch bases	219	XX	LP 513a-PBG	Paper	Tan	Phenolic
		429	XX	–	Paper	Black	Phenolic
High humidity resistance; good dimensional stability, excellent resistance to splitting	Panelboards	254	XXX	LP 513a-PBE	Paper	Tan	Phenolic
Electrical insulation in applications requiring low dielectric losses in severe humidity conditions	Punched parts for electronic equipment	H-17900	XXP	–	Paper	Tan	Phenolic
Low 60-cycle power factor	Tap-changer boards	H-18720	–	–	Paper	Natural	Phenolic
Economical plate with fair electrical strength	Panelboards	H-2898	X	–	Paper	Natural	Phenolic
Cold shearing and cold punching under normal conditions; low loss at high humidity	Radio and TV terminal boards	H-5639	XXXPC	LP 513a-PBEP	Paper	Tan	Phenolic
Self-extinguishing; punchable, good dimensional stability	Commercial computer terminal boards	H-9454	FR-3	Mil-P-22324-PEE	Paper	Cream	Epoxy
Excellent cold punching and cold shearing characteristics under normal conditions	Intricate punched parts requiring little or no preheating	H-18331	XPC	–	Paper	Chocolate	Phenolic
		H-18332	XPC	–	Paper	Tan	Phenolic
		H-18333	XPC	–	Paper	Black	Phenolic
Good cold shearing and cold punching under normal conditions	Intricate punched parts requiring little or no preheating	H-19128	XP	–	Paper	Tan	Phenolic
		H-19242	XP	–	Paper	Black	Phenolic
		H-19243	XP	–	Paper	Chocolate	Phenolic
Ultra-soft punch plate	Thick punched parts	H-23655	XPC	–	Paper	Tan	Phenolic

Table 6-3. *(Continued)*

Major Application Requirements	Typical Use	Micarta Grade	NEMA Grade	Military Specifications Type and Number	Base Material	Color	Resin
Cotton Cloth Base Materials							
Good machining qualities; mechanical and electrical strength; moisture resistance; toughness; good appearance	Marine relay bases; terminal boards; radio parts	221	LE	Mil-P-15035-FBE	Fine weave fabric	Tan	Phenolic
		431	LE	–	Fine weave fabric	Black	
Low moisture absorption; toughness; fair dimensional stability	Paper mill doctor blades; pump valves; marine bearings; piston and packing rings	223	–	Mil-P-18324	Medium weave fabric	Tan	
High mechanical strength; good appearance, strength in punching	Fine tooth gears; radio parts; terminal boards	238	L	Mil-P-15035-FBI	Fine weave fabric		
Good mechanical properties with especially high impact strength	Gears; pinions; spacers	262	C	Mil-P-15035-FBM	Medium weave fabric	Tan	
Good mechanical properties; good machining qualities	Gears; pinions	H-25126	–	–	Fine weave fabric		
Low moisture absorption; good dimensional stability; good resistance to acids and alkalis	Valves and plating tanks	281	–	–	Medium weave fabric		
Low-voltage low-frequency electrical performance; mechanical performance in moderately humid conditions	Marine switchboard panels; small gears and pinions; radio parts	286	CE	Mil-P-15035-FBG			
		H-24721	CE	Mil-P-15035-FBG			
		496	CE	–	Medium weave fabric	Black	
Low coefficient of friction; self-lubricating (graphite filled)	Textile mill bearings; pistons; packing rings	400	–	Mil-P-5431			
Good mechanical heat resistance and low abrasion	Bearings; slate wedges and gears	H-24907	-	–	Synthetic fiber	Tan	

Table 6-3. *(Continued)*

Major Application Requirements	Typical Use	Micarta Grade	NEMA Grade	Military Specifications Type and Number	Base Material	Color	Resin
Glass Cloth Base Materials							
High mechanical strength and hardness; high arc and heat resistance; excellent electrical properties under dry conditions; self-extinguishing	Switchboard panels; arc barriers; circuit-breaker parts; structural electrical parts	259-2	G5	–	Glass cloth	Gray	Melamine
Retention of high mechanical strength at temperatures up to 150°C; self-extinguishing; retention of 50% of initial flexural strength at 150°C	Rotor-slot insulation, structural members at elevated temperature	H-23842	G-11	Mil-P-18177-GEB	Glass cloth	Green	Epoxy
		H-2497	G-11	Mil-P-18177-GEB	Glass cloth	Green	Epoxy
Good machining qualities; high flexural, impact and bond strength at room temperatures; good electrical properties under dry and humid conditions; should not be used under mechanical load at elevated temperatures	Terminal boards	H-22033	G-10	Mil-P-18177-GEE	Glass cloth	Green	Epoxy
High mechanical strength; high arc and heat resistance; excellent electrical properties under dry and humid conditions; self-extinguishing	Marine switchboard panels and structural parts	H-12426	G-9	Mil-P-15037-GME	Glass cloth	Gray	Melamine
Self-extinguishing; good machining and punching qualities; high flexural, impact and bond strength at room temperatures; should not be used under mechanical load at elevated temperatures	Terminal boards	H-16010	FR-4	Mil-P-18177-GEE	Glass cloth	Green	Epoxy
Excellent heat and arc resistance; good electrical properties under humid conditions; good mechanical properties; self-extinguishing	Class H insulation; heating appliance insulation	H-17825	G-7	Mil-P-997D	Glass cloth	White	Silicone
Excellent heat and arc resistance; good electrical properties under humid conditions; self-extinguishing	Radio transmitter parts; Class H transformers, motors and generators low-loss HF radio and radar insulators	20201-1	G-7	Mil-P-997-GSG	Glass cloth	White	Silicone
Excellent heat and arc resistance; good electrical properties under humid conditions; stability at high temperature; self-extinguishing		20201-2	G-7	Mil-P-997-GSG	Glass cloth	White	Silicone

*Courtesy: Westinghouse Electric Corporation, Micarta Division.

Table 6-4. Properties of Phenolics.[a,*]

Type →	Test[b]	Injection Molding Grades	Compression and Transfer Molding Grades							Foam
Property		General Purpose	Improved Impact	High Impact	Heat Resistant	General Purpose	Improved Impact	Very High Impact	Heat Resistant	Rigid, No Surface Skin
Physical properties										
Specific gravity	D792	1.37–1.42	1.36–1.38	1.37–1.39	1.45–1.54	1.36–1.42	1.36–1.40	1.50–1.78	1.50–2.0	2–4 (32–64)[i]
Water absorption (24 hr), %	D570[c]	0.28–0.5	0.8–1.1	0.8	0.2–0.3	0.2–0.7	0.4–1.0	0.2–0.4	0.15–0.4	<3
Electrical properties										
Dielectric strength[d], v/mil	D149	234–340	325–370	300	327–360	300–340	260–400	320–400	210–400	–
(10^6 v/m)		(9.2–13.4)	(12.8–14.6)	(11.8)	(12.9–14.2)	(11.8–13.4)	(10.2–15.7)	(12.6–15.7)	(8.3–15.7)	–
Dielectric constant (60 cyc, dry)	D150	7.8–18.8	12.4–13.0	4.2[g]	10.6–22.5	7.1–13.1	7.9–2.1	5[g]	8.6–17	–
Dissipation factor (60 cyc, dry)	D150	0.13–0.81	0.50–0.53	0.05[g]	0.22–0.50	0.04–0.34	0.13–0.64	0.02–0.04[g]	0.13–0.22	–
Volume resistivity, ohm-m	D257	0.9×10^8–	0.3×10^{10}–	2×10^{11}	2.4×10^8–	0.15×10^{10}–	0.18×10^{10}–	7×10^{10}–	0.5×10^{10}–	–
		6.1×10^{10}	0.18×10^{10}		5×10^{11}	2×10^{10}	6×10^{10}	2×10^{11}	2×10^{10}	–
Arc resistance, sec	D495	–	–	60	180	–	–	–	180–184	–
Thermal properties										
Mold shrinkage, in./in.	D955	0.005–0.012	0.009–0.011	0.002–0.006	0.003–0.0075	0.0035–0.009	0.002–0.007	0.004–0.008	0.004–0.006	–
Coef of ther exp 10^{-5}/F	D696	1.5–1.6	1.78–1.83	1.1[h]	1.2–1.4	1.2–2.3	0.94–1.8	0.67–0.72	0.78–1.5	0.5
(10^{-5}/K)		(2.7–2.9)	(3.2–3.3)	(2)[h]	(2.2–2.5)	(2.2–4.1)	(1.7–3.3)	(1.2–1.3)	(1.4–2.7)	(0.9)
Ther cond, Btu · ft/hr · ft^2 · F	Special	0.131–0.135	0.116–0.121	–	0.145–0.174	0.131–0.184	0.135–0.179	–	0.140–0.309	0.017–0.018
(W/m · K)		(0.226–0.234)	(0.201–0.209)	–	(0.251–0.301)	(0.226–0.318)	(0.234–0.310)	–	(0.243–0.536)	(0.029–0.031)
Heat deflection temp[e], F	D648	330–375	325–335	320–330	360–380	300–340	320–335	400–500	340–390	–
(K)		(439–464)	(436–441)	(433–439)	(455–466)	(422–444)	(433–441)	(478–533)	(444–472)	–
Max service temp, F	Special	300–375	315–325	–	450	300–375	300–325	–	400–450	270
(K)		(422–464)	(430–436)	–	(505)	(422–464)	(422–436)	–	(478–505)	(405)
Flammability	D635	Nonburning	Nonburning	–	Nonburning	Nonburning	Nonburning	–	Nonburning	–
	UL 94	94 HB	94 HB	–	V-0	V-1, HB	HB	–	V-0, V-1, HB	–
Mechanical propreties										
Tensile strength, 1000 psi	D651	6–7.4	6–7	6.1–6.5	7–7.7	5.5–7	5–7	7.2–9	3.8–9	0.02–0.055
(MPa)		(41–51)	(41–48)	(42–45)	(48–53)	(38–48)	(34–48)	(50–62)	(26–62)	(0.14–0.38)
Compressive strength, 1000 psi	D695	27–33.4	27–30	24–29	25–32.9	30–35	25–30	27–28	17–32	0.02–0.09
(MPa)		(186–230)	(186–207)	(165–200)	(172–227)	(207–241)	(172–207)	(186–193)	(117–221)	(0.14–0.62)
Flexural strength, 1000 psi	D790	9–14	9–10	10	11–15.6	8.5–11.5	7–11	15.8–17	7–12	0.025–0.065
(MPa)		(62–97)	(62–69)	(69)	(76–108)	(58.6–79)	(48–76)	(109–117)	(48–83)	(0.17–0.45)
Flexural modulus, 10^6 psi	D790	0.9–1.07	1	1.1–1.3	1–1.5	0.95–1.3	0.8–1	1.5–2.3	1–1.5	–
(MPa)		(6.2–7.4)	(6.9)	(7.6–9)	(6.9–10.3)	(6.6–9)	(5.5–6.9)	(10–16)	(6.9–10)	–
Impact strength[f], ft · lb/in.	D256	0.29–0.30	0.42–0.50	0.6–1.9	0.28–0.38	0.27–0.36	0.4–0.5	4.5–12	0.28–0.36	–
(J/m)		(15–16)	(22–27)	(32–101)	(15–20)	(14–19)	(21–27)	(240–640)	(15–19)	–
Hardness, Rockwell M	D785	110	114	105–110	102–105	100–115	95–105	110–115	110–115	–

[a]Sources: General Electric and/or Union Carbide except for high- and very-high impact grades (Fiberite).
[b]ASTM unless otherwise indicated.
[c]And/or special tests.
[d]Short time; 60 cyc/sec.
[e]At 264 psi.
[f]Notched Izod.
[g]10^6 cycles.
[h]1.9 ft · lb/in.-impact grade.
[i]Density lb/ft^3 (kg/m^3).
[j]At 10%.
*Courtesy: *Materials Engineering*, July 1974.

Table 6-5. Comparative Properties of Phenolics and Other Thermosets.*

	Alkyd		Allylic (DAP)		Epoxy		Melamine		Phenolic		Polyester Premix		Silicone		Urea	
Density, lb/cu in.	0.08	–	0.05	–	0.07	–	0.05	–	0.06	–	0.06	–	0.07	–	0.05	–
Flexural strength, 1000 psi	8	15	10	20	12	60	10	23	10	45	6	25	7	18	10	18
Flexural modulus of elasticity, 10^5 psi	20	28	7	13	22	30	15	19	10	33	7	20	14	25	14	–
Tensile strength, 1000 psi	4	10	5	7	9	25	6	10	7	21	–	–	4	6	7	–
Impact strength, ft-lb/in. notch	0.4	12	1	6	0.4	30	0.3	12	0.3	27	3	24	0.3	10	0.3	–
Hardness (Rockwell)	M116	–	M108	–	M110	–	M120	–	M112	M120	M45	M100	M89	–	M118	–
Volume resistivity, ohm-cm	10^{14}	–	10^{14}	10^{16}	9×10^{15}	–	10^{13}	–	10^{11}	10^{13}	10^{12}	10^{14}	10^{14}	–	10^{11}	–
Dielectric strength (short-time), v/mil	375	–	375	400	375	–	320	430	300	425	345	420	350	400	350	–
Dielectric constant (10^6 cps)	5	4.2	4	3.3	4.75	–	5.7	–	5.5	4	5.2	6.4	3.8	–	6.6	–
Dissipation factor (10^6 cps)	0.017	0.014	0.02	0.01	0.013	–	0.03	0.02	0.05	0.005	0.008	0.022	0.006	0.005	0.03	–
Arc resistance, sec	140	180	100	160	123	–	120	186	35	180	140	180	300	420	130	–
Max cont service temperature, F	275	300	325	500	400	500	210	400	300	550	250	300	500	700	170	–
Flammability	Yes	Self-ex	Yes	Self-ex	Yes	Self-ex	Self-ex	–	Yes	Self-ex	Yes	Self-ex	Yes	Self-ex	Self-ex	–
Colorability	Yes	–	Poor	–	Yes	–	Exc	–	Poor	–	Yes	–	Yes	–	Exc	–

*Riley, 1962. Value in left-hand column = mean average for general-purpose grades. Value in right-hand column = generally accepted high value in compounds formulated to optimize that particular property. Data are only typical and approximate, and not representative of specifications.

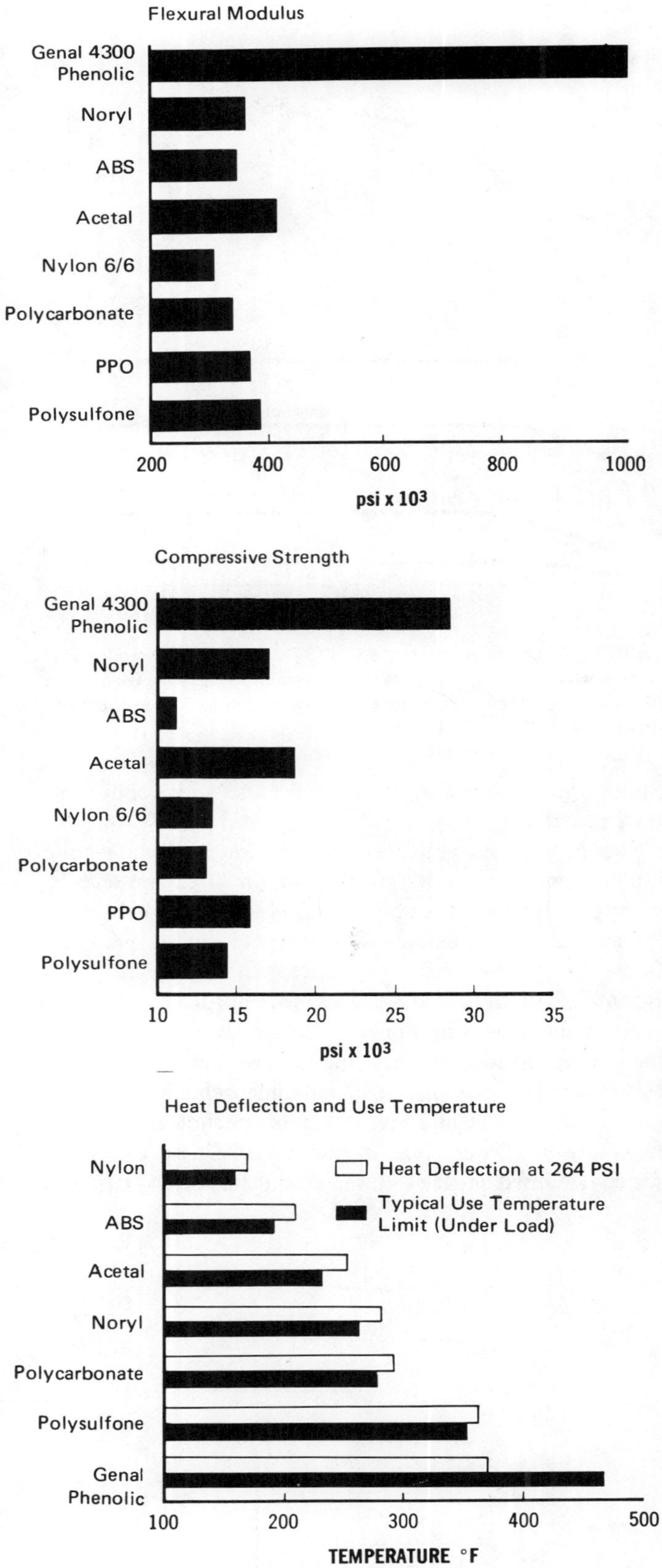

Fig. 6-9. Comparative properties of injection-moldable phenolics and other engineering thermoplastics. (*Courtesy: General Electric Co.*)

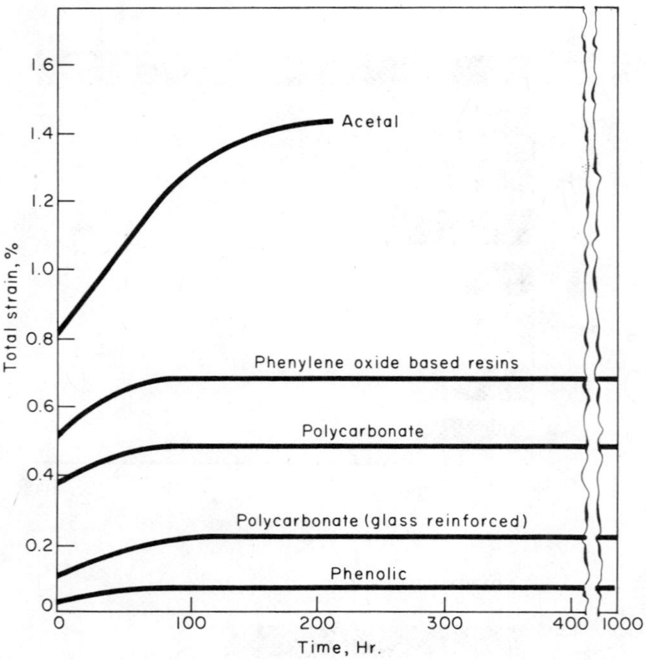

Fig. 6-10. High creep resistance (for 2000 psi at 73°F) of phenolics is indicative of their good dimensional stability in service. (Reprinted by permission from ***Materials Engineering,*** Manual 249, September 1974)

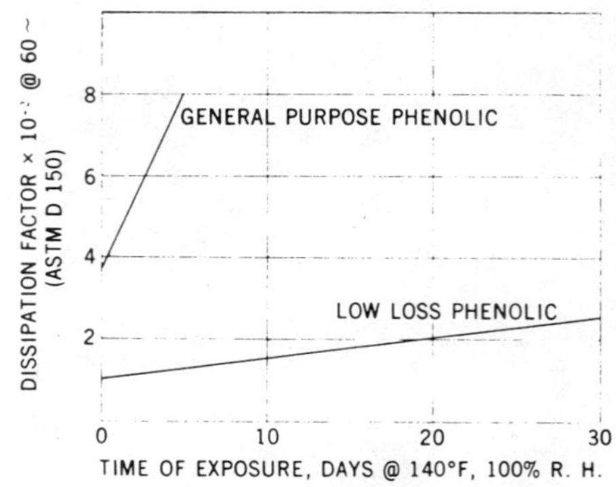

Fig. 6-12. Effect of humidity on dissipation factor of phenolics. (Martino, 1967)

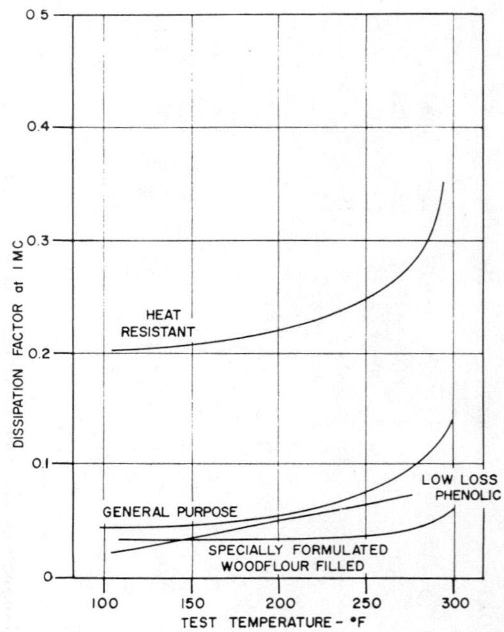

Fig. 6-13. Effect of temperature on dissipation factor. (Martino, 1967)

Special grades are also available for use with foods, cosmetics, and drugs.

The effects of moisture and temperature on phenolic molding-powder properties are shown in Figures 6-11 through 6-16 and Tables 6-6 through 6-8.

Phenolics can tolerate the incorporation of graphitic fillers to improve wear resistance and reduce the coefficient of friction. Glass-fiber cloth-reinforced moldings, such as those used as firewalls and ablative shields, resist high temperatures, do not melt, have low thermal diffusivity, and are strong, tough, and dimensionally stable.

In adhesives, major advantages of phenolics are their resistance to water, fungus, and creep. These adhesives, which are used in metal-to-metal and metal-to-plastic designs, also have excellent oil and grease resistance and thermal stability.

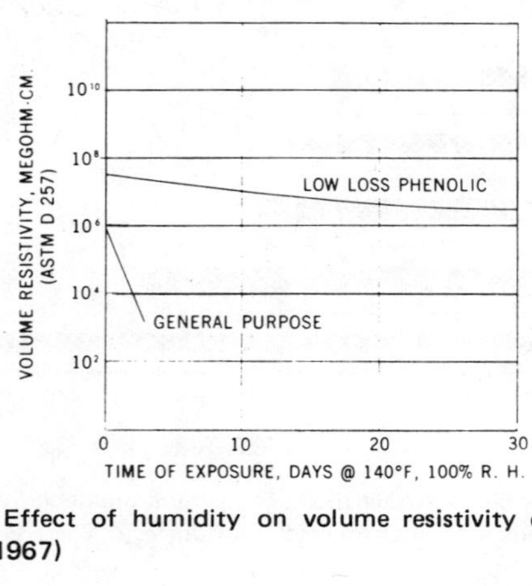

Fig. 6-11. Effect of humidity on volume resistivity of phenolics. (Martino, 1967)

In fibrous insulation, including mineral-wool batts, and glass-fiber insulations, phenolics impart: water resistance, thermal stability, mechanical strength, and resistance to solvents and other chemicals.

Phenolic foams have a multicellular, semipermeable structure. They are produced in densities of 0.5 to 20 lb/ft^3 (8-320 kg/m^3). They are rigid foams, which have relatively high heat resistance and good moisture, chemical, and solvent resistance. However, the phenolic foams have low mechanical properties, and they are friable.

Syntactic phenolic foams, which are composed of hollow microspheres in a phenolic matrix, also are available. These foams are strong, nonfriable, and are produced in densities of 47-81 lb/ft^3 (750-1300 kg/m^3).

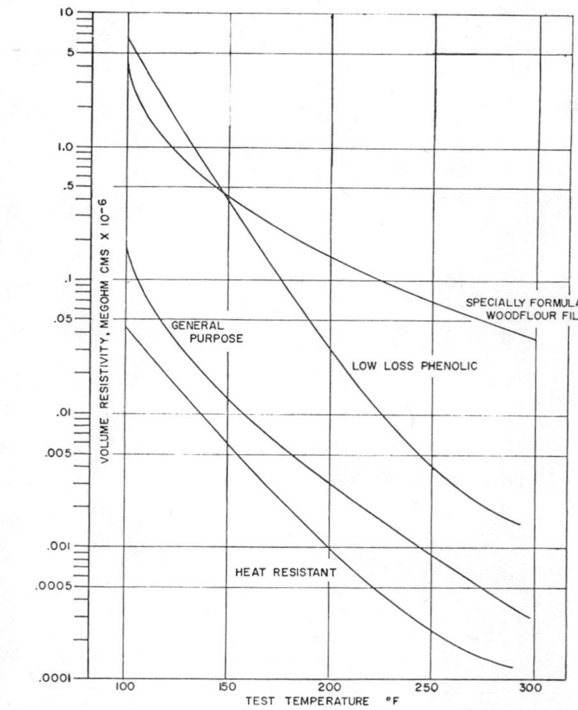

Fig. 6-14. Effect of temperature on volume resistivity. (Martino, 1967)

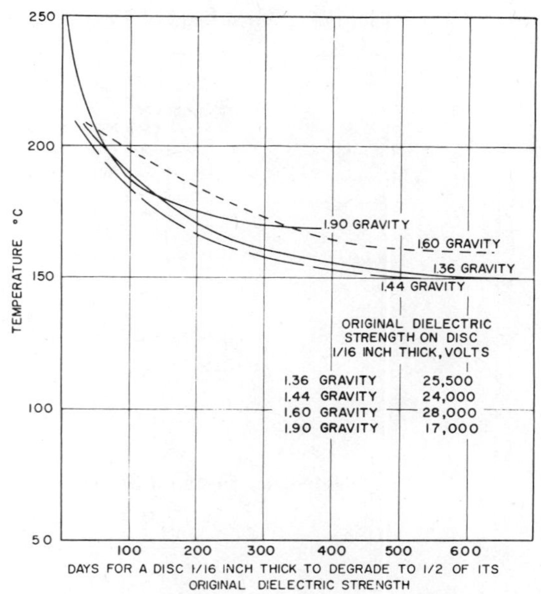

Fig. 6-15. Effects of time and temperature on dielectric strength of several typical phenolic compounds. (Martino, 1967)

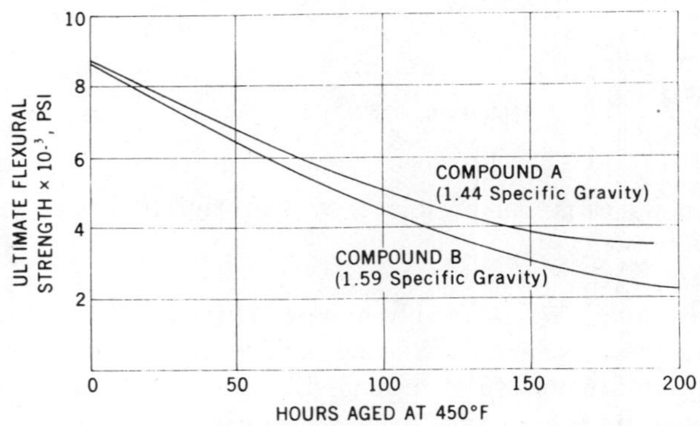

Fig. 6-16. Effect of 450°F aging on flexural strength of phenolics. (Martino, 1967)

Processing: Molding. Phenolic parts can be made by compression, transfer, and injection molding. Compression molding is best for parts requiring extremely close dimensional control. Transfer molding is recommended for parts with small, delicate, molded-in inserts. Screw injection is the preferred method for injection molding these resins. Barrel temperatures range between 93-110°C (200-230°F), with mold temperatures of 171-199°C (340-390°F).

Coloring or decorating can be readily accomplished by molding textured surfaces and/or by subsequent painting, chrome plating, vacuum-metallizing, or hot stamping.

Table 6-6. Effect of Water on the Mechanical Properties of Two Phenolic Compounds.*

	Flock-Filled One-Step		Wood-Flour-Filled Two-Step	
	Control	25 Days in Water @ 84°C	Control	10 Days in Water @ 84°C
% change in weight (tensile specimens)		2.78		7.7
Tensile strength, psi	7900	6700	7600	3000
% change in weight (flex specimens)		2.56		6.7
Flex. strength, unnotched, psi	9700	7900	10300	4800
Flex. modulus, psi × 10^{-6}	0.75	0.62	0.95	0.45
Flex. work to bk., unnotched, ft-lb/in.^{3}	0.69	0.50	0.50	0.23

*Martino, 1967.

Table 6-7. Shrinkage of Two Intermediate-Gravity Phenolic Compounds after Aging.*

Specimen Thickness	Shrinkage in in./in. After 126 Days at 150°C (305°F) 1.44 Sp Gr	After 126 Days at 150°C (305°F) 1.60 Sp Gr	After 60 Days at 190°C (374°F) 1.44 Sp Gr	After 60 Days at 190°C (374°F) 1.60 Sp Gr
1/2 in.	.0094	.0046	.0084	.003
1/4 in.	.0100	.0046	.0090 (40 days)	.003
1/8 in.	.0094	.0055	.0122	.005
1/16 in.	.0082		.0150	

*Martino, 1967.

Table 6-8. Flexural Strength of Two Intermediate-Gravity Phenolic Compounds after Aging.*

Specimen Thickness	Flexural Strength, psi After 168 Days @ 150°C (305°F) 1.44 Sp Gr	After 168 Days @ 150°C (305°F) 1.60 Sp Gr	After 60 Days @ 190°C (374°F) 1.44 Sp Gr	After 60 Days @ 190°C (374°F) 1.60 Sp Gr
1/2 in.	9400	9000	5500	6150
1/8 in.	9500	8900	4500	5400
1/16 in.	8000		3700	

*Martino, 1967.

Phenolic parts can be joined by mechanical fasteners or adhesives.

Laminating. Water- or solvent-based resins can be used to form prepregged sheet. Staging is done in large air-blown ovens at temperatures around 127-163°C (260-325°F). Staging times may vary from 30 seconds for impregnated thin paper to 10 minutes for impregnated canvas. Hydraulic presses are used for laminating the sheet. They can be of various sizes, some large enough to handle a bank of 4′ X 10′ laminates. Curing pressure may vary from 50 to 1200 psi.

Production of rock wool insulation is similar to laminating. The inorganic oxide fibers are coated with a water-based phenolic formulation. The coated rock wool is compressed and fed through a drying oven at temperatures of about 232°C (450°F), which drives off residual moisture and cures the resin.

Fig. 6-17. Phenolic transmission reactor. (*Courtesy: Durez Division, Hooker Chemicals and Plastics Corporation*)

Applications: Molding compounds. Some of the typical applications for phenolic moldings in the automotive, appliance, and electrical industries include:

Automotive—rotors, fuse blocks, coil towers, solenoid covers and housings, ignition parts, distributor caps, connectors, power brake and transmission components (Figure 6-17), carburetor spacers, ash trays, and instrument housings.

Appliances—handles, knobs, timing motor housings, blower wheels, mixer bases, can-opener fronts, toaster end-panels, iron handles, heater bases for electric cookware, lamp housings, and cases for personal-care products.

Electrical—circuit breakers, switch and relay housings, thermostat components, receptacles, terminal blocks, stiffener boards, light sockets and baffles, meter bases, coil bobbins, control housings, and tube sockets.

Laminates. Two general groups are available: industrial and decorative. Industrial applications include propellers, pulleys, and antenna masts in the aviation industry; suction box covers and doctor blades in paper mill equipment; coil forms and tube sockets in radios; gears, bobbins, and bearings in textile machinery; and switchboard panels, terminal blocks, and armature wedges in telephones and power transmission equipment.

Decorative laminates are used as automobile interior panels, interior wall tiles in building construction, and table and desk tops in the furniture trade. They are also used as a facing for plywood. This composite structure consists primarily of plywood, but has a surface with the wear and scratch resistance of a laminated plastic.

Adhesives. Phenolic resins are the principal bonding agents for the asbestos used in friction materials such as brake linings, clutch facings, and automatic-transmission discs. Phenolics are used to bond abrasives and grits for grinding wheels and abrasive papers.

Impregnation of wood and wood dust, particles, and shavings has enabled the extended use of wood and wood by-products that used to go to waste. The most well-known application is the manufacture of particle-board. Molded articles can be made from these wood-filled compositions such as toilet seats, croquet balls, and shuffleboard discs. Exterior-grade plywood is bonded with phenolic resins for added resistance to moisture.

Metal Casting. For making molds, phenolics have been used since before World War II. The earliest systems were composed of one-step resins and sand that were cold-pressed into shapes for cores. The shell-molding procedure has since become the dominant technique. The phenolic is coated on the sand grains which are then dumped on a metal pattern which has been heated to 232°C (450°F).

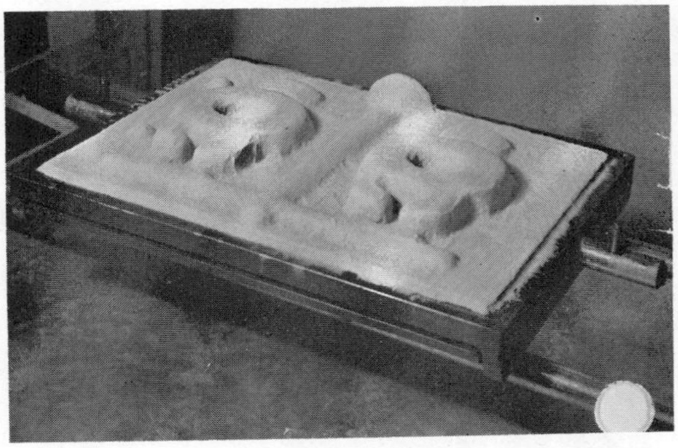

Fig. 6-18. Pattern with adhering shell, following removal from the dump box. The shell as shown here is "green" and must be cured. It is about to be charged to the oven for curing. (*Courtesy: Monsanto Co.*, which is no longer responsible for accuracy or content)

Fig. 6-19. Cured shell following its ejection from the pattern. A second matching shell-half will complete the mold unit. (*Courtesy: Monsanto Co.*, which is no longer responsible for accuracy or content)

A shell of sand, having the shape of the pattern and bound together by the fused resin, is formed very quickly. Curing at 427–538°C (800–1000°F) for a minute or two, results in a finished mold (see Figures 6-18 and 6-19). Molten metal is poured into the mating mold halves made by this process, then cooled, and a finished part is produced.

Coatings. Phenolics are used as the sole film-former or to fortify drying oils in various coating applications. They are used to line drums and cans, milk and beer processing equipment, water tanks, and air conditioning equipment. They have been applied as dip coatings for electrical capacitors and condensers. These resins enhance the properties of emulsion floor waxes. Ablative coatings achieved great success on reentry shields of spacecraft.

AMINO RESINS

Urea-Formaldehydes

In 1923 a product called "organic glass" was introduced, which was based on a reaction product of urea-formaldehyde. It was a cast resin, softer than and lacking the weather resistance of glass. By 1926 molding powders based on a combination of urea and thiourea were produced. However, these materials were short-lived because they were slow curing, lacked heat stability, and could not yield the properties of the forthcoming urea- and melamine-based polymers. In 1928 the first successful, pure urea-formaldehyde polymer was produced as a replacement for the heavy enamel cast-iron frames of scales then in use. Although the phenolics existed at that time, they were all dark colored, and the ureas represented the first thermosetting, colorless materials available.

Urea is made by several methods, all dependent on the reaction of ammonia and CO_2 under high pressure and temperature. The product formed is ammonium carbonate which splits off water to form urea.

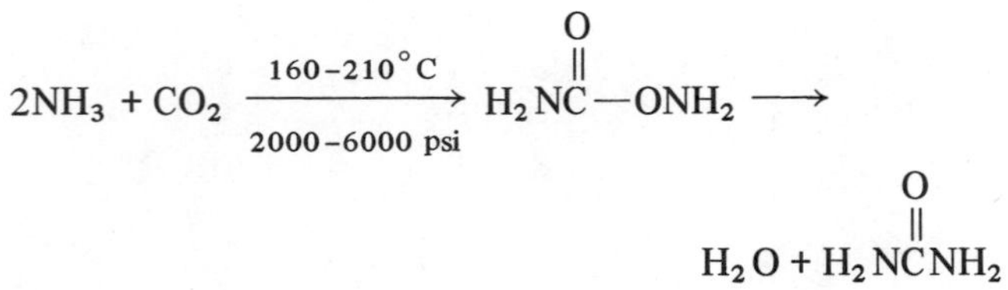

As with phenolics, the polymerization chemistry of urea with formaldehyde is very complex, and a comprehensive understanding is still lacking. The urea is known to react with formaldehyde in a neutral or alkaline environment to yield mono- and dimethylol ureas.

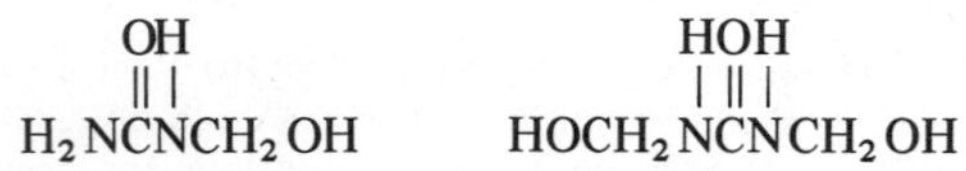

Tri- and tetramethylol ureas are not known.

These monomers will upon heating, form water-soluble condensation products, which roughly correspond to B-stage phenolics, although they are more soluble. It is believed that with acidification and heating the reaction proceeds as follows:

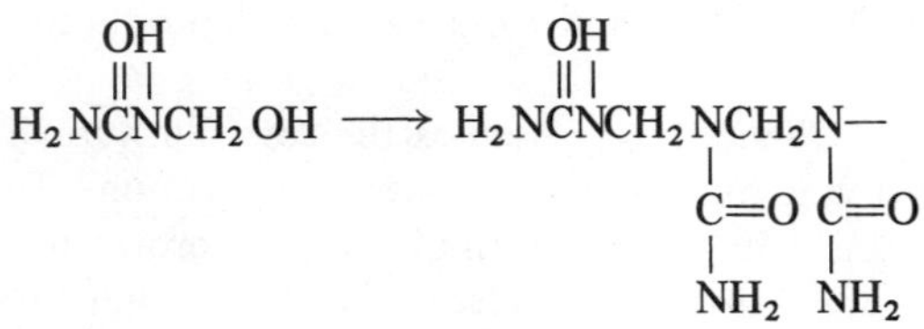

An insoluble, infusible product results when dimethylol urea is used, indicating that crosslinking has occurred.

The industrial manufacture of the resins consists of the addition of aqueous solutions of urea and formaldehyde in a ratio of about 1 : 1.3–1 : 1.8 to a reaction kettle. Because the clarity of the final resin is critical, ultra-clean glass-lined or stainless steel reactors are used. A weakly attractive catalyst such as dilute caustic, ammonia, or an alkanolamine is added.

The initial reaction takes place at room temperature and results in the water-soluble mono- and dimethylol ureas. The solution is then heated to boiling and held until a viscous mass develops, which is subsequently partially vacuum-dehydrated.

There are two procedures used to prepare molding powders. In the first method, the filtered liquid resin is mixed with fillers, and the damp mixture is screened and then oven dried. The resulting coarse powder is blended with any other compounding ingredients, and the blend is ball-milled to fine powder.

The second method consists of spray-drying the liquid resin in hot furnace gases. The dried dust is separated using cyclone separators and dry-blended with other compounding ingredients.

In the unfilled form, ureas are clear, transparent solids resembling glass. However, this form is comparatively soft and too easily scratched. In addition, the polymerization reaction apparently continues extremely slowly, even after molding. This reaction causes strains that result in surface cracks or crazing, spoiling the surface polish of the clear resin. They are not completely water resistant, so that continued exposure to water also dulls the surface. The use of proper fillers moderates both of these problems. High-grade cellulose furnishes compositions which are colorless and translucent. Lower-cost cellulose fillers yield less expensive compositions which are darker in color. A high percentage of filler is used to prevent surface cracking.

The effect of adding alpha-cellulose to urea resins is different from the effect of adding fillers to phenolic resins. Close examination of molded parts shows no evidence of individual particles of cellulose physically admixed with the resin. The consensus seems to be that there exists either a solid solution of cellulose in resin or possibly the chemical formation of cellulose semi-acetals. Wood flour is used in lower-grade powders because the resultant molded parts are more opaque than those with alpha-cellulose. Other fillers that have been used to impart special properties include cotton, nylon, walnut-shell flour, and some mineral fillers. Compounding ingredients include plasticizers, colorants (pigments and dyes), lubricants, and stabilizers.

Acids are used as curing agents for completion of the crosslinking process. Two types, direct and latent, are employed. The direct acids are organic and inorganic acids, acid salts, and acid esters, which are effective at all temperatures. The latent catalysts include salts and esters which develop acidity upon heating, often in conjunction with the moisture released as a result of the condensation polymerization. In fact, in the absence of moisture, the room temperature effectiveness is nil. Examples of direct catalysts include sodium acid sulfate and acid alkyl phosphates. Typical latent catalysts are amine hydrochlorides, benzoyl or phthaloyl mercaptobenzothiazole, and sodium sulfite.

Urea itself, thioureas, or substituted ureas may be added to molding powder formulas to prevent excessive shrinkage. They combine with excess formaldehyde and free methylol groups which normally impair the resin's water resistance. They also improve flow and cure by acting as plasticizers.

For use in adhesives, the resins are supplied in solution, although dry powders are provided which will convert to liquid systems upon the addition of water. These liquid systems are prepared by condensing the urea and formaldehyde under slightly acidic conditions to a predetermined level of B-stage. The solution is then neutralized and the water removed by vacuum distillation until the desired solids level or viscosity is obtained. Conversion to dry powder is achieved by evaporation over hot rolls and subsequent scraping or by spray-drying.

Urea adhesives are almost always supplied as two-component systems; the resin being one component and the acid catalyst, the curing agent component (hardener). The curing agent is blended into the resin just prior to use;

the mix is applied to the substrate and cured until optimum properties are obtained.

If the urea-formaldehyde condensation is carried out in the presence of an alcohol which has limited compatability with water, resins can be produced which are compatible with organic coatings, solvents, and resins. Typical of these types of resins are propylated and butylated urea-formaldehydes. When plasticized with oil or phenolic-modified alkyd resins and properly dried, they produce hard, solvent-resistant, colorless films that have found wide application in the paint and coatings industry.

The importance of the urea-formaldehyde resin depends partly on the fact that the resin is water-white and clear. Thus, light-colored or pastel shades can be produced. However, the water-white color of this resin is not its only advantage, since its tensile strength and hardness are better than the phenolics. The water resistance and impact strength, on the other hand, are considerably lower. The resin in the final cured state is thermosetting and resembles the phenolics in heat resistance, although it is inferior in this respect.

Table 6-9 shows the properties of a general-purpose alpha-cellulose-filled urea.

Urea-formaldehyde thermosetting molding powders are supplied in a partly polymerized state so that final curing under pressure takes place in the hot mold. The molded product is insoluble, infusible, tasteless, odorless, hard, and tough. It has good light-diffusing qualities and its electrical properties, including non-tracking characteristics, are good. It is completely resistant to oils, greases, and organic solvents, but is attacked by concentrated acids and alkalis.

Heat-resistant ureas cannot be compounded with inorganic fillers because of the unique effect of alpha-cellulose in generating the resin's strength and molding properties.

Ureas do not carbonize like the phenolics when subjected to electric arcing. Their dielectric strength is high. The ureas have unusually high surface hardness which can make machining difficult.

Applications include: buttons, bottle caps, lighting fixtures, food equipment, toilet seats, stove and refrigerator parts, cosmetic and jewelry containers, electrical wiring devices, radio cabinets, bathroom accessories, cabinet hardware, electrical appliance components, and many types of housings.

One of the more unusual applications for ureas is as a typing element. The choice of urea for the job was dictated by the material's impact strength, its resistance to acids, its dimensional stability, and its adaptability to mass-production precision molding.

Urea adhesives are used as plywood bonding agents and veneering. Decorative laminates using light-colored urea adhesives, backed with plywood, are used for table, desk, and counter tops.

Alkyd-urea coating formulations are in widespread use for coating metal kitchen and bathroom fixtures, hospital equipment, and refrigerators. Urea-phenolic-alkyd systems have found much acceptance as automotive primers and metal primers for washing machines and laundry equipment.

Melamine-Formaldehydes

Melamine reached significant commercial production in 1939. It is prepared from calcium carbide which is made from lime and coke in an electric furnace.

$$3C + CaO \xrightarrow{\Delta} CaC_2 + CO$$

The calcium carbide is converted to calcium cyanamide via the exothermic reaction with nitrogen.

$$CaC_2 + N_2 \longrightarrow CaCN_2 + C$$

The $CaCN_2$ is acidified with dilute sulfuric acid or aqueous CO_2 to the free cyanamide which rapidly dimerizes to dicyandiamide.

$$CaCN_2 \xrightarrow{H^+} H_2CN_2 \longrightarrow \begin{array}{c} H_2NC{=}NH \\ | \\ HN{-}C{\equiv}N \end{array}$$

Further reaction at high temperature and moderate pressure, often in the presence of ammonia, produces the trimer, melamine.

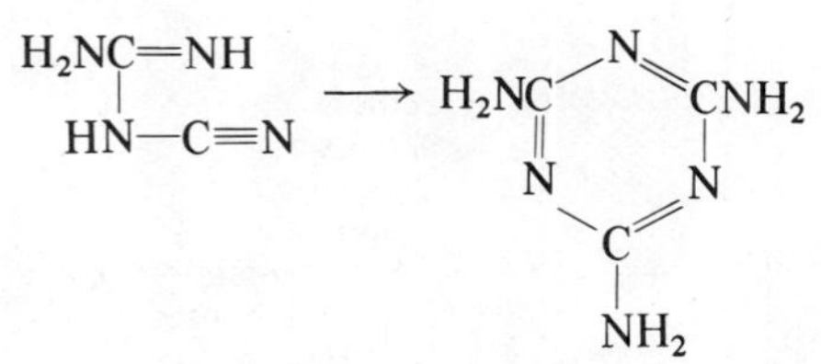

Methylol melamines are formed when melamine reacts with formaldehyde under neutral or slightly alkaline conditions. All six hydrogen atoms can be replaced with methylol groups. However, for commercial grades of resins, only two to three groups are introduced. The condensation is carried out in corrosion-resistant equipment at temperatures at or above 90°C (194°F). Aqueous formaldehyde is used and water is removed by vacuum evaporation. This technique operates at low temperatures which prevent the resin reaction from running away and causing crosslinking in the reactor. Control of pH is not as critical for these reactions as with urea resins.

The resins used in molding powders usually have a melamine-formaldehyde ratio of 1:2–1:3. The conversion to C-stage is more rapid than ureas at molding temperatures. Also, alkaline fillers may be added which will not interfere with the curing mechanism. Other alkaline-curing resins, like phenolics, may also be added.

For adhesive and coating resins, mechanisms similar to those employed with ureas are used. Adhesive systems may

Table 6-9. Properties of Urea Molding Compounds.*

PROPERTY		ASTM test method
Raw Material		
Bulk factor (granular form)	2.2 — 2.5	
Preformability	excellent	
Molded		
Physical		
Specific gravity	1.5	D792
Water absorption, % [24 hours at 77°F (25°C)]	0.6	D570
Flammability,(1) ⅛ in.(2)	V-O	UL 94
Mechanical		
Tensile strength, psi	7,000	D638
Compressive strength, psi	30,000	D695
Flexural strength, psi	12,000	D790
Flexural modulus, psi	1.3×10^6	D790
Impact strength, Izod, ft-lb/in. of notch	0.3	D256 method A
Rockwell hardness, M-scale	118	D785
Thermal		
Coefficient of linear thermal expansion/°C, max	2.7×10^{-5}	D696
Thermal conductivity, g-cal/sec/cm²/°C/cm	8×10^{-4}	
Heat resistance(3) °F, max	170	
°C, max	77	
Heat distortion temperature, (HDT), °F	275	D648
°C	135	
Electrical		
Arc resistance, sec	120	D495
Dielectric strength v/mil		D149
Short time, ⅛ in.	350	
Step-by-step, ⅛ in.	275	
Dielectric constant		D150
60 Hz	7.9	
1 KHz	7.0	
1 MHz	6.6	
Dissipation factor		D150
60 Hz	0.04	
1 KHz	0.03	
1 MHz	0.03	
Tracking index	600+	IEC-112
Chemical		
Effect of visible light(4)	negligible	
Resistance to organic solvents, oil and grease	excellent	
Resistance to weak acids and alkalies	poor	
Resistance to strong acids and alkalies	poor	
Molding Parameters (compression molding)		
Mold temperature, °F	275 — 325	
°C	135 — 163	
Mold pressure, psi	2,000 — 8,000	
Mold shrinkage, in./in.	0.006 — 0.008	
Post mold shrinkage in./in. [170°F (77°C), 96 hours]	0.006 — 0.008	

(1) Many grades listed with Underwriters Laboratories.
(2) Injection grades measured at .150."
(3) Based on no color change.
(4) Pastels gray in sunlight.

*Courtesy: Allied Chemical Corporation.

be cured via hot-pressing without catalysts, or acid catalysts may be added for more rapid hardening. They cannot, however, be cold-cured, even with strong acid catalysts because their reaction kinetics are just too slow, particularly below 50°C (122°F). Formulation ingredients are the same as those used in urea systems. For coating systems, pH and temperature control is critical because of the tendency for more rapid methylol condensation at the expense of methylol etherification.

Table 6-10 shows properties of three standard filled-melamine molding powders. Since the pure resins are colorless, molding powders made from these resins can be produced in light, pastel colors. These plastics are tasteless and odorless.

Molding powder properties are usually compared with those of urea molding powders. Although the melamines are more expensive, their properties are generally superior. For example, alpha-cellulose-filled systems produce translucent moldings that are harder and more heat and chemical resistant. In fact, alpha-cellulose is not essential for good part strength; thus, asbestos can be used to increase heat resistance even more. Melamines have 50–75% less water absorption than ureas. They can withstand 100°C (212°F) exposure continuously, which is some 15°C (27°F) higher than for ureas. Consequently they can be washed repeatedly in dishwashers or sterilizers without damage. These resins are also glossier and less readily stained.

The scratch and mar resistance of melamines is not particularly good. Improvements in surface-finish hardness are achieved through overcoating with special resin-impregnated-rayon/paper compositions.

The superior properties of melamines over ureas carry over to coating systems. They are less viscous, have a greater compatibility with hydrocarbon solvents, are more rapidly cured, and yield films having considerably better heat, water, and chemical resistance. These resins, like ureas, must be plasticized, usually with an oil-modified alkyd. They require twice as much alkyd as ureas in order to provide films of comparable hardness, adhesion, and flexibility. Also, like melamine adhesives they cannot be cold-cured (below 80°C, 176°F) and are seldom used in air-drying finishes.

The most well known use of melamine molding powders is in the manufacture of institutional and low-cost household dinnerware. Ash trays, desk sets, knobs, buttons, dials, lavatory units (see Figure 6-20), electrical appliance housings, aircraft and automotive ignition parts, circuit breakers, terminal blocks, switch gears, and connectors have also been fabricated from melamine systems.

Decorative laminates are based on a composite construction of phenolic-impregnated kraft paper core with an alpha-cellulose-melamine overlayer. Further overprinting or decoration may be applied with clear or pigmented melamines or melamine-impregnated foils.

Melamine adhesives have been primarily used for wood bonding. Lower alkyl ethers of methylol melamines cured

Fig. 6-20. Melamine molding compound shows excellent physical properties as a lavatory material. (*Courtesy: American Cyanamid Co.*)

with diammonium acid phosphate have been used in shrinkproofing wool. They have also found use as textile finishing agents because they provide water repellancy and crease resistance.

Varnishes made from melamines have been used in electrical equipment because of their combination of excellent insulating, non-tracking, heat- and oil-resistance properties.

Furan Resins

Furan resin is a generic term for many resinous products that contain the heterocyclic unsaturated furan ring in their molecular structures. Commercially, the furfuryl alcohol polymer is most important.

Although the first furan compound, furoic acid, was discovered in 1780, the first patent on furan polymers was not issued until 1928. The basic material for production of furan resins, furfural (furfuraldehyde), was discovered by Doebereiner in 1832. Large quantities of furfural were not available until 90 years later.

Pentosans from corn cobs and rice hulls are the main sources of furfural. Considering their curing mechanisms, furan resins are divided into two groups: (1) alkaline cured and (2) acid cured. Table 6-11 shows the properties of an asbestos-filled furan system.

In 1953, a patent was issued for a furan resin capable of being converted to the final, infusible state in the presence of an alkaline material. This resin possesses a number of unique properties, including the ability to (a) cure in the presence of, or under, water; (b) mix with alkaline fillers or powdered metal; and (c) be applied over acid-reactive surfaces such as steel or concrete without primer. Unfortunately, the alkaline-cured resin has some disadvantages, such as slow cure, and limited storage stability.

There are many methods of preparing acid cured furan resin, but, in general, a furfural alcohol condensation product can be prepared with neutral or acid catalyst or by thermal catalysis. Resinification is dependent on hydrogen ion concentration and is always accelerated by heating. The resulting polymer is composed of furan nuclei linked to-

Table 6-10. Properties of Melamine-Formaldehyde Molding Compounds.*

		ASTM test method	No filler	Alpha cellulose filled	Cellulose filled	Flock filled	Asbestos filled	Macerated fabric filled
Processing	1. Molding qualities	—	Good	Excellent	Good	Good	Good	Good
	2. Compression molding temp., °F.	—	300-330	280-370	290-360	275-300	280-340	275-330
	3. Comp. molding pressure, p.s.i.	—	2000-5000	1500-8000	1500-6000	4000-8000	1000-7000	4000-8000
	4. Injection molding temp., °F.	—	—	290-340	—	—	—	—
	5. Injection molding pressure, p.s.i.	—	—	15000-20000	—	—	—	—
	6. Compression ratio	—	2.0	2.1-3.1	2.2-2.5	4.0-7.0	2.1-2.5	5.0-10.0
	7. Mold (linear) shrinkage, in./in.	—	0.011-0.012	0.005-0.015	0.006-0.008	0.006-0.007	0.005-0.007	0.003-0.004
	8. Specific gravity (density)	D792	1.48	1.47-1.52	1.45-1.52	1.50-1.55	1.70-2.0	1.5
	9. Specific volume, cu. in./lb.	D792	18.7	18.8-18.2	20.0-19.1	18.5-17.8	16.3-13.8	18.5
	10. Machining qualities	—	—	Fair to good	Good	Good	Fair	Good
Mechanical	11. Tensile strength, p.s.i.	D638	—	7000-13000	5000-9000	7000-9000	5500-7000	8000-10500
	12. Elongation, %	D638	—	0.6-0.9	0.6	—	0.30-0.45	0.6-0.8
	13. Tensile elastic modulus, 10^5 p.s.i.	D638	—	12.0-14.0	11.0	—	19.5	14.0-16.0
	14. Compressive strength, p.s.i.	D695	40000-45000	40000-45000	33500	30000-35000	30000	30000-35000
	15. Flexural yield strength, p.s.i.	D790	11000-14000	10000-16000	9000-11500	13000	9000-11000	12000-15000
	16. Impact strength, ft. lb./in. of notch (½ × ½ in. notched bar, Izod test)	D256	—	0.24-0.35	0.27-0.36	0.40-0.45	0.28-0.40	0.6-1.0
	17. Hardness, Rockwell	D785	—	M115-M125	M115-M125	—	M110	M120
	18. Flexural elastic modulus, p.s.i. × 10^5	D790	—	1.1 × 10^4	—	—	—	—
	19. Compressive modulus, p.s.i. × 10^5	D695	—	—	—	—	—	—
Thermal	20. Thermal conductivity, 10^{-4} cal./sec./sq. cm./1 (°C./cm.)	C177	—	7.0-10.0	6.5-8.5	—	13.0-17.0	10.6
	21. Specific heat, cal./°C./gm.	—	—	0.4	—	—	—	—
	22. Thermal expansion, 10^{-5} in./in./°C.	D696	—	4.0	4.5	—	2.0-4.5	2.5-2.8
	23. Resistance to heat, °F. (continuous)	—	210	210	250	250	250-400	250
	24. Deflection temp., °F. at 264 p.s.i. fiber stress	D648	298	350-370	266	—	265	310
	at 66 p.s.i. fiber stress		—	—	—	—	—	—
Electrical	25. Volume resistivity, ohm-cm. (50% RH and 23°C.)	D257	—	0.8-2.0 × 10^{1}	—	—	1.22 × 10^{12}	10^9-10^{10}
	26. Dielectric strength, short-time, ⅛-in. thickness, volts/mil	D149	—	270-300	350-400	300-330	410-430	250-350
	27. Dielectric strength, step-by-step, ⅛-in. thickness, volts/mil	D149	—	240-270	250-350	—	280-320	200-300
	28. Dielectric constant, 60 cyc.	D150	—	7.9-9.5	6.2-7.6	—	6.4-10.2	7.6-12.6
	29. Dielectric constant, 10^3 cyc.	D150	—	7.8-9.2	6.0-7.5	—	9.0	7.1-7.8
	30. Dielectric constant, 10^6 cyc.	D150	—	7.2-8.4	4.7-7.0	—	6.1-6.7	6.5-6.9
	31. Dissipation (power) factor, 60 cyc.	D150	—	0.030-0.083	0.019-0.033	—	0.07-0.17	0.07-0.34
	32. Dissipation (power) factor, 10^3 cyc.	D150	—	0.015-0.036	0.013-0.034	—	0.07	0.03-0.05
	33. Dissipation (power) factor, 10^6 cyc.	D150	—	0.026-0.045	0.032-0.060	—	0.041-0.050	0.036-0.041
	34. Arc resistance, sec.	D495	100-145	110-140	95-135	110-150	120-180	115-125
Optical	35. Refractive index, n_D	D542	—	—	—	—	—	—
	36. Clarity	—	Opalescent	Translucent	Opaque	Opaque	Opaque	Opaque
	37. Transmittance, %	—	—	29.9 (0.50 in.)	—	—	—	—
	38. Haze, %	—	—	—	—	—	—	—
Resistance characteristics	39. Water absorp., 24 hr., ⅛-in. thick., %	D570	0.3-0.5	0.1-0.6	0.34-0.80	0.16-0.30	0.08-0.14	0.3-0.6
	40. Flammability Burning rate, in./min.	D635	—	—	—	—	—	—
	AEB, in./ATB, sec.		—	—	—	—	—	—
	41. Effect of sunlight	—	Colors may fade	Pastels yellow	Slight color change	Slight color change	Slight color change	Slight discoloration
	42. Effect of weak acids	D543	None	None	None to slight	None to slight	None to slight	None
	43. Effect of strong acids	D543	—	Decomposes	Decomposes	Decomposes	Decomposes	Decomposes
	44. Effect of weak alkalies	D543	None	None	None	None	Very slight attack	None
	45. Effect of strong alkalies	D543	—	Attacked	Decomposes	Decomposes	Slight attack	Attacked
	46. Effect of organic solvents	D543	None	None	None	None on bleed-proof colors	None	None on bleed-proof colors

*Reprinted by permission from *Modern Plastics Encyclopedia*, McGraw-Hill, Inc.

gether by methylene bridges, and possesses a terminal hydroxyl group.

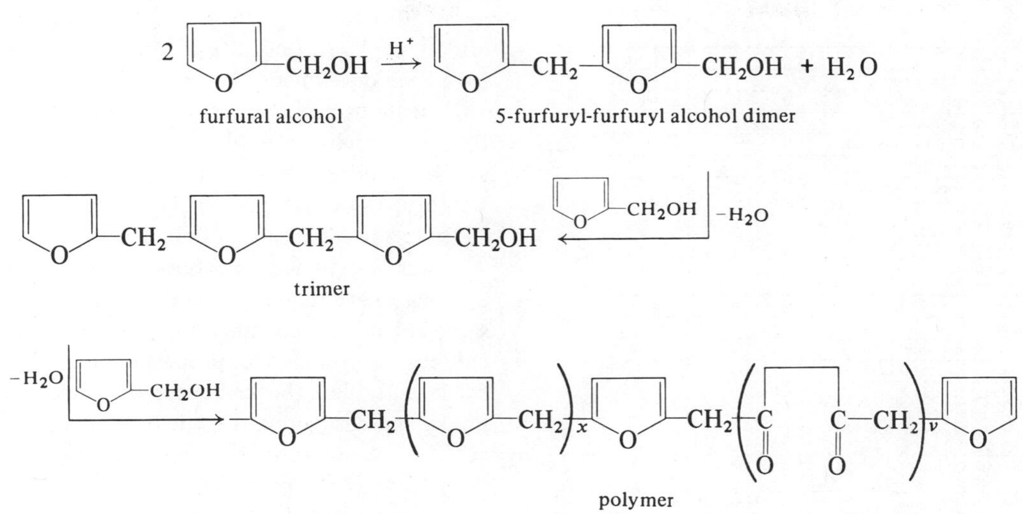

Resins prepared by a neutral catalyst such as activated alumina or thermal catalysis contain a large percentage of nonreactive by-products that remain in the system and have a negative effect upon the resin's properties. The by-products also prevent thick layers of resin from curing properly, even at a baking temperature of 149°C (300°F). As a result, attempts to use the resin for casting or molding purposes have been unsuccessful. Therefore, acid-catalyzed furfuryl alcohol resins have primarily been used for fabrication of corrosion-resistant processing equipment, castings, resinous chemical-resistant cements, as battery sealers, carbon bonding, temporary binders for various products, and sometimes as coatings.

Generally, the furan polymers of greatest commercial interest can be divided into three broad classes: (1) unmodified furfuryl alcohol polymers, (2) aldehyde-modified furfuryl alcohol polymers, and (3) polymers made from furfuryl alcohol, urea, and formaldehyde.

The first two types of polymers are noted for their outstanding chemical resistance, high carbon content and ease of cure at room temperatures; they are used in corrosion-resistant equipment, in carbon and graphite products, and as foundry binders of the no-bake type. For applications such as carbon brushes, electrodes, and various vitrified products, these resins are unsurpassed. Upon firing, these liquids yield the theoretical amount of carbon to produce a carbon-carbon bond necessary to achieve a homogeneous, hard, fused mass. The furfuryl alcohol-urea-formaldehyde polymers are used in adhesive applications like wood gluing and foundry binders.

All furan resins are dark in color and have a reddish-black appearance; when catalyzed to a set, they are black. The acid-catalyzed resins are extremely sensitive to strong acids and resinify readily in their presence. Addition of excess catalyst will result in a violent reaction. Therefore, only recommended acids or acid dilutions must be used to provide controllable polymerization. When these acid-catalyzed resins are heated under controlled conditions, they trigger themselves and, before exothermic heat is dissipated, develop such high temperatures that *vesiculation* (the formation of small spherical cavities produced by bubbles of gas formed while the mass is in a molten state) of the setting mass occurs. Polymer mass is an important factor in the control of curing speed. For curing thin layers of resin, more curing agent is required because the heat of reaction is readily dissipated; whereas thick layers of resin require less, since the heat is contained in the mass a longer time.

The use of furan resins for coating purposes is quite limited, due to the low adhesion to nonporous surfaces (a fact we shall return to shortly), brittleness, and shrinkage. Figure 6-21 shows a plaster of paris object whose surface has been impregnated with the resin.

The biggest use of furan resins in the corrosion-resistance field is the manufacture of chemical-resistant cements and equipment. Furan resin cements have been used for many years to bond acid-proof brick. Surfaces of the brick may be saturated with alkaline substances, mineral and organic acids, as well as many solvents, singularly or in combination. Furan cements can be used in those areas where it would be impossible to use other construction materials. They offer corrosion protection to concrete and steel structures, lengthening their life. As pump base cements, they can be used in chemical processing plants, metal finishing plants, petroleum refineries, fertilizer plants, and pulp and paper mills. Floors and walls of structures such as process tanks, continuous-strip pickle lines, process towers, collecting sumps, neutralizing tanks, pits, manholes, and tank cappings can be protected with the aid of furan cements.

Table 6-11. Properties of Asbestos-Filled Furan Resin.*

		ASTM test method	Asbestos filled
Processing	1. Molding qualities	—	Good
	2. Compression molding temp., °F.	—	275-300
	3. Comp. molding pressure, p.s.i.	—	100-500
	4. Injection molding temp., °F.	—	—
	5. Injection molding pressure, p.s.i.	—	—
	6. Compression ratio	—	—
	7. Mold (linear) shrinkage, in./in.	—	—
	8. Specific gravity (density)	D792	1.75
	9. Specific volume, cu. in./lb.	D792	15.8
	10. Machining qualities	—	Fair to good
Mechanical	11. Tensile strength, p.s.i.	D638	3000-4500
	12. Elongation, %	D638	—
	13. Tensile elastic modulus, 10^5 p.s.i.	D638	15.8
	14. Compressive strength, p.s.i.	D695	10000-13000
	15. Flexural yield strength, p.s.i.	D790	600-9000
	16. Impact strength, ft. lb./in. of notch (½ × ½ in. notched bar, Izod test)	D256	—
	17. Hardness, Rockwell	D785	R110
	18. Flexural elastic modulus, p.s.i. × 10^5	D790	—
	19. Compressive modulus, p.s.i. × 10^5	D695	—
Thermal	20. Thermal conductivity, 10^{-4} cal./sec./sq. cm./1 (°C./cm.)	C177	—
	21. Specific heat, cal./°C./gm.	—	—
	22. Thermal expansion, 10^{-5} in./in./°C.	D696	—
	23. Resistance to heat, °F. (continuous)	—	265-330
	24. Deflection temp., °F. at 264 p.s.i. fiber stress	D648	—
	at 66 p.s.i. fiber stress		—
Electrical	25. Volume resistivity, ohm-cm. (50% RH and 23° C.)	D257	—
	26. Dielectric strength, short-time, ⅛-in. thickness, volts/mil	D149	—
	27. Dielectric strength, step-by-step, ⅛-in. thickness, volts/mil	D149	—
	28. Dielectric constant, 60 cyc.	D150	—
	29. Dielectric constant, 10^3 cyc.	D150	—
	30. Dielectric constant, 10^6 cyc.	D150	—
	31. Dissipation (power) factor, 60 cyc.	D150	—
	32. Dissipation (power) factor, 10^3 cyc.	D150	—
	33. Dissipation (power) factor, 10^6 cyc.	D150	—
	34. Arc resistance, sec.	D495	—
Optical	35. Refractive index, n_D	D542	—
	36. Clarity	—	Opaque
	37. Transmittance, %	D1003	—
	38. Haze, %	—	—
Resistance characteristics	39. Water absorp., 24 hr., ⅛-in. thick., %	D570	0.01-0.20
	40. Flammability Burning rate, in./min.	D635	—
	AEB, in./ATB, sec.		—
	41. Effect of sunlight	—	None
	42. Effect of weak acids	D543	None to slight
	43. Effect of strong acids	D543	Attacked by oxidizing acids
	44. Effect of weak alkalies	D543	None
	45. Effect of strong alkalies	D543	None to slight
	46. Effect of organic solvents	D543	Resistant

*Reprinted by permission from *Modern Plastics Encyclopedia*, McGraw-Hill, Inc.

These cements are widely used in the food industry for dairies, ice cream plants, kitchens, breweries, meat packing plants, bakeries, soft drink plants, and food processing plants. Specialized cements that even risk the corrosive effects of hydrofluoric acid and strong caustic solutions have been developed.

Cements using furan resin have numerous advantages over those in which a phenolic, epoxy, or polyester resin is used as the binder. Primarily, they (1) have better storage stability, (2) cause practically no dermatitis problem, and (3) have a broader range of chemical resistance.

It was mentioned previously that furan resins, such as coatings, have poor adhesion to nonporous surfaces. The same is true for cements using acid-cured furan resins. They exhibit poor adhesive properties to red-shale acid-proof brick and tile. The cement binds mechanically to these bricks and tiles along the cut or serrated edges, but, where surfaces are smooth, the mechanical adhesion is inadequate to bond materials properly.

Resins have been developed to overcome this deficiency. They are similar to the crosslinked-type furfuryl alcohol resins and exhibit, in addition to other excellent properties, adhesion to nonporous surfaces. Unlike the rest of the acid-cured furan resinous cements, their adhesion to smooth-surfaced red-shale brick tile and glass is excellent.

Furans have also found wide use in the manufacture of grinding wheels and foundry molds. Their use in the form of fiber-reinforced composites is still comparatively uncommon. A typical furan-based composite possesses good heat and chemical resistance, excellent surface hardness, and is inherently nonflammable.

These low- to medium-cost materials are mixed with a suitable acidic hardener just prior to use, and the fiber is then quickly impregnated and cured at room temperature. A cured furan resin is black in color and tends to be brittle. The latter characteristic has often limited the use of furan/glass-fiber laminates to the area of chemical process vessels and associated plant equipment. The cure behavior is somewhat unpredictable. The very low viscosity of furan resins also makes impregnation difficult, and this has contributed to their restricted role in composite systems.

A recent development of a furan-based glass-fabric prepreg makes an alternate system available for use in molding applications where phenolic and other relatively nonburning, chemically resistant polymer systems previously have dominated. In addition to the good thermal and fire resistance of laminates produced from furan prepregs, very little smoke is generated under conditions of forced combustion. The mechanical characteristics of the furan laminates described here compare well with those of many types of phenolics and polyester systems. This combination of properties makes molded furan prepregs attractive for numerous applications in the chemical processing industry as well as for interior aircraft parts, where the pos-

Fig. 6-21. Impregnating plaster of paris object with liquid furane resin creates artistic textured effect. (*Courtesy: Furane Plastics, Inc.*)

sible limitations of low flexibility and color are not of critical importance.

For industrial applications a low-viscosity prepolymer is formed by treating furfuryl alcohol with suitable acidic materials. Unless care is taken, the strongly exothermic reaction may become uncontrollable and too high a level of condensation may result. However, with precautions, a satisfactory liquid polymer can be produced. This prepolymer is hardened by chain extension and crosslinking, with the evolution of water, by further acidic treatment. The acid for this final step in the polymerization generally is *p*-toluenesulfonic acid or phosphoric acid. With this approach, cure may be effected at room temperature.

Because they bring about an almost immediate cure, the usual acidic catalysts cannot produce a furan-preimpregnated composite that has acceptable shelf-life. Under storage conditions, stability of at least several days is usually required. Hence, it is necessary to use latent curing catalysts which become chemically active only upon exposure to heat. After suitable stabilization, Lewis acids, such as aluminum trichloride or boron trifluoride offer the best latency properties.

Trade Names

Ablaphene	P/F Resins	Products Chemiques, UKM
Adlock	P/F molding compounds	M & T Chemicals, Inc.
Admino	M/F	Ashland Chemical Co.
Aerolite	U/F	Ciba-Geigy Corporation
Aerophen	P/F adhesive	Ciba-Geigy Corporation
Air-in-Oval	Furans	Air Products & Chemicals, Inc.
Alberit	M/F, M/P, and P/F molding compounds	Canadian Hoechst, Ltd.
Alcylite	P/F resins and molding compounds	Alcylite Plastics & Chemical Corporation
Alkor	Furan resin	Atlas Minerals & Chemicals Div., ESB, Inc.
Alnoval	Novolac resins	American Hoechst Corporation
Alresen	P/F resins	American Hoechst Corporation
Amberol	P/F resins	Rohm & Haas Co.
Aminocel	U/F resins	Montedison S.p.A.
Aminolac	M/F resins	Products Chimiques Ugine Kuhlmann
Amres	P/F, U/F, and M/F	Pacific Resins & Chemicals, Inc.
Aqualite	P/F laminate	NVF Co.
Aquapearl	P/F	Ashland Chemical Co.
Arboneeld	Dimethylolurea	E.I. Du Pont de Nemours & Co., Inc.
Arcolite	P/F	Consolidated Molded Products

Arochems	Modified P/F	Ashland Chemical Co.
Arodures	U/F	Ashland Chemical Co.
Arofene	P/F	Ashland Chemical Co.
Ashlam	M/F and P/F sheet, rod, and profile	Mica & Micanite Supplies, Ltd.
Avisco	U/F	FMC Corporation
Bakelite	P/F	Union Carbide Corporation
Beckacite	Modified P/F	Reichhold Chemicals, Inc.
Beckadrin	P/F	Beck, Koller & Co., Ltd.
Beckamine	U/F	Reichhold Chemicals, Inc.
Beckophen	P/F	Beck, Koller & Co., Ltd.
Beckopol	Modified P/F	Reichhold Chemicals, Inc.
Beetle	U/F	American Cyanamid Co.
Biplos	U/F and M/F molding powders	British Industrial Plastics, Ltd.
Calcerite	Amino casting compound	M & T Chemicals, Inc.
Cascomel	M/F adhesives	Borden, Inc.
Cascote	Furan resin	Borden, Inc.
Cassopal	U/F resin	Cassella Farbwerke Mainkur, A.G.
Castastock	P/F resin	Ashland Chemical Co.
Catabond	P/F	Ashland Chemical Co.
Catabox	Furan P/F, U/F	Ashland Chemical Co.
Catacol	P/F	Ashland Chemical Co.
Catacore	P/F	Ashland Chemical Co.
Catacote	P/F	Ashland Chemical Co.
Catadec	M/F prepreg	Ashland Chemical Co.
Catafilm	P/F	Ashland Chemical Co.
Cataform	P/F	Ashland Chemical Co.
Catalac	P/F	Ashland Chemical Co.
Catalay	M/F prepreg	Ashland Chemical Co.
Catalin	Cast P/F	Ashland Chemical Co.
Cataplas	P/F resin	Catalin, Ltd.
Catavar	P/F laminates	Ashland Chemical Co.
Celeron	P/F	Budd Co./Polychem
Cellobond	P/F	BP Chemicals International
Celloflex	Acrylic/M/F copolymer	Cello Chemical Co.
Celoron	P/F	Budd Co.
Chase-glasterro	P/F asbestos fiberglass	Chase & Sons, Inc.
Colfoam	U/F microballoons	Colton Chemical Co.
Coloron	Molded P/F	Budd Co.
Coltrock	P/F	Colt's Plastics Co.
Coltwood	High-impact P/F	Colt's Plastics Co.
Compregnite	P/F solution	Borden Co.
Consoweld	M/F laminates	Consolidated Water Power & Paper Co.
Corfoam	P/F	Rezolin, Inc.
Coroc	U/M, P/F	Cook's Resins
Cosmolite	P/F	Colton Chemical Co.
Croid	P/F, U/F	Croid, Ltd.
Crystle	Clear P/F	Marblette Corporation
Cymel	M/F molding compounds	American Cyanamid Co.
Daka-Ware	U/F	Harry Davies Molding Co.
Decola	M/F	Sumitomo Bakelite Co.
Deligno	P/F molding compounds	Kreidl, Rutter & Co., Verainigte Chemische Fabriken
Deresit	P/F resins and glues	Kreidl, Rutter & Co., Verainigte Chemische Fabriken
Desamin	M/F resins	Kreidl, Rutter & Co., Verainigte Chemische Fabriken
Desurit	U/F adhesives	Kreidl, Rutter & Co., Verainigte Chemische Fabriken
Diaron	M/F resins	Reichhold Chemical Co.
Duraglas	P/F prepreg	Natvar Corporation
Duralon	Furan resins	Norton Co.
Durestos	P/F prepreg and molding	Turner Brothers Asbestos Co., Ltd.
Durez	P/F resins and molding compounds	Durez Plastics Div., Hooker Chemical Corporation
Durite	P/F molding compounds	Borden Co.
Duron	P/F resins	Firestone Rubber & Latex Products Co.
Durophen	Plasticized P/F	American Hoechst Corporation
Dynapor	P/F foam	Dynamit Nobel, A.G.
Dyphene	P/F	Sherwin-Williams Co.
Dyphenite	Modified P/F	Sherwin-Williams Co.

E-Case	Filled and glass-reinforced P/F	Epoxy Products Co.
Edolite	P/F moldings, laminates, and varnishes	Mitsubishi Edogawa Chemical Co.
Edolyte	P/F	Mitsubishi Edogawa Chemical Co.
Elegon	M/F prepreg	Micanite & Insulators Co., Ltd.
Elo	P/F molding	Birkbys, Ltd.
Encap	P/F molding compounds	Hooker Chemical Co.
Epok	P/F, U/F	BP Chemicals International, Ltd.
Ervamine	M/F molding powder	Products Chimiques Ugine Kuhlmann
Ervaphene	P/F resins	Products Chimiques Ugine Kuhlmann
Fablonite	M/F laminates	Commercial Plastics, Ltd.
Fabrez	U/F	Reichhold Chemicals, Inc.
Fabrolite	P/F molding	Associated Electrical Industries, Ltd.
Fiberite	Reinforced P/F and M/F	Fiberite Corporation
Fiberloy	Reinforced P/F sheet	Rogers Corporation
Fiberplast	P/F and P/furfuryl molding compounds	International Textile Co.
Fluosite	P/F resins and molding compounds	Montedison, S.p.A.
Foamex	Polyvinyl formal-phenolic wire coating	General Electric Co.
Foramine	U/F adhesives	Reichhold Chemicals, Inc.
Forasite	P/F adhesives	Reichhold Chemicals, Inc.
Formica	M/F laminates	Formica Corporation
Foundrez	U/F and R/F resins	Reichhold Chemicals, Inc.
Fudowlite	P/F molding compounds	Fudow Chemical Co.
Furathane	Epoxy/furan	Atlas Mineral & Chemical Div., ESB, Inc.
Furatone	Furan resins	3M Co.
Furesir	Furan resins	Societa Italiana Resine
Furnane	Furan cement	Atlas Minerals & Chemical Div., ESB, Inc.
Fursatil	U/furan, P/furan, and U/F resins	Fordath Engineering Co., Ltd.
Gabbrocel	U/F resins	Montedison, S.p.A.
Gabrite	U/F resins	Montedison, S.p.A.
Gedelite	P/F resins	CDF Chimie
Gemstone	P/F casting resin	Knoedler Chemical Co.
Genal	P/F compounds	General Electric Co.
Glasfloss	P/F-impregnated glass	Hooker Chemical Co.
Gusco	Epoxy-phenolic coatings	A. Gushmer, Inc.
Heresite	P/F resins and coatings	Heresite & Chemical Co.
Hitafran	Furan resins	Hitachi Chemical Co.
Hitanol	P/F resins	Hitachi Chemical Co.
Hitterlite	M/F laminates	Hitachi Chemical Co.
Hostaset	M/F and P/F molding compounds	Hoechst, A.G.
Hycar	P/F, P/furfuryl molding compounds	B. F. Goodrich Chemical Co.
Indur	P/F molding compounds	Reilly Tar & Chemical Corporation
Insulspray	U/F insulation	Borden, Inc.
Insurok	P/F and P/furfuryl molding compounds	Spaulding Fibre Corporation
Iporka	U/F resin and foam	BASF Wyandotte Corporation
Isomin	M/F molding compounds	James Ferguson & Sons, Ltd.
Jet-Kote	Furan coat	M & T Chemicals, Inc.
Kaston	P/F coating resins	Poly Resins
Kastor Acrylon	P/F and P/furfuryl molding compounds	Borden, Inc.
Kauramin	M/F resin	BASF Wyandotte Corporation
Kauresin	P/F	BASF Wyandotte Corporation
Kaurit	U/F	BASF Wyandotte Corporation
Kollerbond	Epoxy/P/F	Beck, Koller & Co., Ltd.
Kombiplast	U/F adhesives and resins	Kreidl, Rutter & Co.
Krumbhaar	P/F resins	Lawter Chemicals, Inc.
Kys-ite	P/F and P/furfuryl molding compounds	Keys Fiber Co.
Lamalac	M/F, P/F prepreg, sheet, and rod	Mica & Micanite Supplies, Ltd.
Lamelite	M/F resin	Montedison, S.p.A.
Lamicoid	U/F and P/F laminates	3M Co.

Lamin-Art	M/F laminates	Eagle-Picher Co.
Lamophene	P/F and P/furfuryl molding compounds	Poly Resins
Lamorok	P/F laminates	Rogers Corporation
Lauxite	P/F, U/F, and M/F adhesives	Monsanto Chemical, Ltd.
Luphen	P/F resins	BASF Wyandotte Corporation
Luvipal	M/F resin	BASF Wyandotte Corporation
Madurit	M/F	Cassella Farbwerke Mainkur, A.G.
Maprenal	M/F	American Hoechst Corporation
Melacel	M/F-impregnated cellulose fiber	Pall Trinity Micro Corporation
Melaglaze	M/F/guanamine	Allied Chemical Corporation
Melamine	M/F molding compounds	Fiberite Corporation
Melamit	M/F molding compounds	A/S Syntesia
Melan	M/F and U/F resins	Hitachi Chemical Co.
Melantine	M/F	Ciba-Geigy Corporation
Melbrite	M/F resins and molding compounds	Montedison, S.p.A.
Melit	M/F resins	Societa Italiana Resine
Melmac	M/F resins and molding compounds	American Cyanamid Co.
Melmex	M/F molding compound	B.I.P. Chemicals, Ltd.
Melolam	M/F	CIBA (A.R.L.), Ltd.
Melsir	M/F molding powders	Societa Italiana Resine
Melsprea	M/F molding compound	S.P.R.E.A.
Meltron	M/F resins	Bostik South, Inc.
Melurac	M/F, U/F adhesives	American Cyanamid Co.
Meno	M/F prepreg	Micanite & Insulators Co., Ltd.
Methylon	P/F coating resins	General Electric Co.
Micarta	P/F laminates	Westinghouse Electric Co.
Micoid	P/F laminates	Mical Insulator Co.
Moldarta	Molded P/F	Westinghouse Electric Co.
Moldesite	P/F molding compound	S.P.R.E.A.
Monolite	P/F	Monowatt Electric
Mouldrite	M/F, P/F, and U/F molding compounds	Imperial Chemical Industries, Ltd.
Naxomine	U/F	Convert, Ets. G.
Neillite	P/F and P/furfuryl molding compounds	Watertown Manufacturing Co.
Neopress	P/F molding, laminating, and fabric systems	Loven Chemical of California
Nestor	P/F	James Ferguson & Sons, Ltd.
Nestorite	P/F and U/F molding compounds	James Ferguson & Sons, Ltd.
Norsophen	P/F	CDF Chimie
Ohmoid	P/F laminate	Wilmington Fibre Specialty Co.
Paradura	P/F	Paramet Chemical Co.
Paxolin	P/F prepreg, sheet, rod, and profile	Micanite & Insulators Co., Ltd.
Penacolite	P/F resins and adhesives	ARCO Polymers, Inc.
Pentaphen	P/F compounds	Pennwalt Corporation
Permalite	M/F molding compounds	Fiberite Corporation
Permanite	Furan	Maurice A. Knight Co.
Permelite	Reinforced M/F molding compounds	Fiberite Corporation
Permelite	M/F and P/F molding compounds	Melamine Plastics, Inc.
Phenall	P/F molding compounds	Allied Chemical Corporation
Phen-Alloy	P/F	High Strength Plastics Corporation
Phenoboard	P/F laminates	Robert Farkas
Phenocel	P/F laminates	Plastilight, Inc.
Phenocel	P/F foam	Robert Farkas
Phenodur	P/F	American Hoechst Corporation
Phenoline	P/F	Carboline Co.
Phenolit	P/F/PVC blend	Soviet Origin
Phenolite	P/F laminates	NVF Co.
Phenolls	P/F	General Electric Co.
Phenoweld	P/F adhesives	Hardman Inc.
Phenuren	P/F resins	BASF Wyandotte Corporation
Pionite	M/F laminates	Pioneer Plastics Corporation

Plaskon	U/F and M/F resins	Allied Chemical Corporation
Plaspreg	Furan resin	M & T Chemicals, Inc.
Plastitool	P/F casting resin	Calresin Corporation
Plastopal	U/F	BASF Wyandotte Corporation
Plenco	P/F and M/F resins and molding powders	Plastics Engineering Co.
Pliofoam	U/F foam	Goodyear Tire & Rubber Co.
Plyamine	U/F adhesives	Reichhold Chemicals, Inc.
Plyophen	P/F resins, laminates, adhesives, and varnishes	Reichhold Chemicals, Inc.
Poilopas	U/F molding powders	Products Chimiques Ugine Kuhlmann
Pollopas	U/F compounds	Dynamit Nobel Corporation
Polychem	P/F and M/F molding compounds	Budd Chemical Co.
Polymerin	M/F-alkyd enamels	Interchemical Corporation
Poly-phen	Cast P/F, P/F, and P/furfuryl	Stepan Chemical Co.
Poly-ria	U/F molding compound	Stepan Chemical Co.
Polytool	P/F tooling compound	Reichhold Chemicals, Inc.
Progilite	P/F molding powders	Products Chimiques, UKM
Protan	P/F prepreg	Micanite & Insulators Co., Ltd.
Protectophene	P/F and P/furfuryl molding compounds	Poly Resins
Pyroloy	P/F resins	Raybestos-Manhattan, Inc.
Qua Corr	Furan resins	Quaker Oats Co.
Rapco Foam	U/F foam	Rapco, Inc.
Raycolite	P/F molding compound	Rayon Processing Co. of Rhode Island
Realite	M/F laminates	Reiss Assoc., Inc.
Resalfen	P/R resins	Resine Sintetiche e Affini, S.p.A.
Resamin	U/F and M/F resins	American Hoechst Corporation
Resart	M/F molding compound	Resart-IHM, A.G.
Resfurin	U/F	Resine Sintetiche e Affini, S.p.A.
Resiafen	P/F resins	Resine Sintetiche e Affini, S.p.A.
Resi Chem	U/F and M/F resins	Brown Co.
Resimene	U/F and M/F resins	Monsanto Co.
Resinox	P/F resins and compounds	Monsanto Co.
Resloom	M/F and ethylene-urea resins	Monsanto Co.
Resmelin	M/F resins	Resine Sintetiche e Affini, S.p.A.
Resnit	P/F prepreg	Leicester, Lovell & Co., Ltd.
Resophen	P/F resins	Plastrigil
Resophene	P/F resins	Products Chimiques, UKM
Resurfene	P/F resins	Resine Sintetiche e Affini, S.p.A.
Rhonite	U/F resins	Rohm & Haas Co.
Richelain	P/F, U/F	Richardson Co.
Richlite	P/F laminates	Rainier Plywood Co.
Rockite	P/F molding compounds	British Resin Products, Ltd.
Ryercite	P/F	Joseph T. Ryerson & Son, Inc.
Ryertex	P/F laminate	Joseph T. Ryerson & Son, Inc.
Saduren	M/F resins	BASF Wyandotte Corporation
Scarab	U/F molding compound	B.I.P. Chemicals, Ltd.
Sierra	P/F and U/F materials	Sierra Electric Corporation
Sirfen	P/F resins and molding compounds	Societa Italiana Resine
Sirfenol	Modified P/F	Societa Italiana Resine
Sirit	U/F resins	Societa Italiana Resine
Siritle	U/F molding powders	Societa Italiana Resine
SNAP	P/F laminate	Fiberite Corporation
Splyset	U/F adhesive	National Casein Co.
Spraymask	P/F resin	Adhere, Inc.
Spreacol	U/F adhesive	S.P.R.E.A.
Sta-Form	U/F	Georgia-Pacific Corporation
Stalpic	P/F coatings	A. Gusmer, Inc.
Standlite	P/F resins	Hitachi Chemical Co.
Stello	M/F laminate	Shin-Etsu Chemical Co.
Sternite	P/F molding compounds	Sterling Moulding Materials, Ltd.
Stratoclad	Glass-reinforced P/F	Spaulding Fibre Co.
Stylplast	U/F molding compounds	FMC Corporation
Sumikon PM	P/F	Sumitomo Bakelite Co.
Sumilit	P/F laminate	Sumitomo Bakelite Co.

Sumilitersin PR	P/F resin	Sumitomo Bakelite Co.
Super-Beckacite	P/F resins	Reichhold Chemicals, Inc.
Super-Beckamine	M/F resins	Reichhold Chemicals, Inc.
Superstone	U/F molding compounds	Lacrinoid Products, Ltd.
Sylplast	U/F molding compounds	Sylvan Plastics, Inc.
Synsize	U/F resins	Synco Resins, Inc.
Synthane	M/F and P/F laminates	Synthane Corporation
Synthite	P/F varnish	John C. Dolph Co.
Syn-U-Tex	M/F and U/F resins	Celanese Coating Co.
Synvanite	U/F and P/F resins	Synvar Corporation
Synvar	P/F and P/furfuryl molding compounds	Synvar Corporation
Synvaren	P/F resins	Synvar Corporation
Synvarite	P/F resins	Synvar Corporation
Synvar-Ite U	U/F resins	Synvar Corporation
Synvarol	U/F resins	Synvar Corporation
Tamanol	Terpene phenolic resin	Arakawa Chemical Co.
Taylor	P/F laminates, sheet, rod, and tube	Taylor Fiber Co.
Tecolite	P/F molding compounds	Tokyo Shibauro Electric Co.
Tegit	P/F molding compounds	Garfield Mfg. Co.
Tego	P/F	Rohm & Haas Co.
Tessilite	P/F resins and compounds	Montedison, S.p.A.
Tetra-Foam	P/F foam	Stepan Chemical Co.
Tetra-Phen	P/F resins	Georgia-Pacific Corporation
Tetraria	Amino resins	Georgia-Pacific Corporation
Thermoflex	P/F	Evercoat Chemical, Inc.
Thermomix	Filled P/F	Johns-Mansville Aerospace Products
Thermorez	U/F resin	USM Corporation
Thor	P/F, U/F	Leicester, Lovell & Co., Ltd.
Toolplastik	P/F casting	Rezolin, Inc.
Trevarno	P/F impregnations	Hexcel Corporation
Troliton	P/F molding compounds	Dynamit-Nobel of America, Inc.
Troporit	P/F foam	Dynamit-Nobel, A.G.
Tube-Cote	P/F	Tuboscope, Inc.
Tuffite	P/F	S.R. Browne
Tyglas	P/F prepreg	Fothergill & Harvey, Ltd.
Tygon	Furan resin	Norton Co.
Uformite	U/F and M/F resins	Rohm & Haas Co.
Ultrapas	M/F resin	Dynamit-Nobel of America, Inc.
Uniplast	P/F	Universal Plastics Co.
Unirez	P/F	Union-Camp Corporation
Urac	U/F adhesives	American Cyanamid Co.
Uramol	U/F molding compounds	Gordon Chemicals, Inc.
Uravar	P/F foam	Synthetic Resins of Speke
Urecoll	U/F adhesives	BASF Wyandotte Corporation
Urifilm	M/F crosslinked acrylic coatings	Advanced Coating & Chemicals, Inc.
Uropal	U/F/butyl resins	Products Chimiques Ugine Kuhlmann
Uroplas	U/F molding compound	S.P.R.E.A.
Valite	P/F molding compounds	Valentine Sugars, Inc.
Valmel	Methylated methylol melamines	United Merchants, Inc.
Valrez	U/F derivatives	United Merchants, Inc.
Varcum	P/F	Reichhold Chemicals, Inc.
Vulkadur	P/F resin	Mobay Chemical Co.
Xylok	P/F/aralkyl ethers	Ciba-Geigy Corporation
Xylox	Phenol/aralkyl resin	Albright & Wilson, Ltd.
Zol	P/F foams	Plastrigil

REFERENCES AND BIBLIOGRAPHY

Allied Chemical Plastics Division. *Molding and Handling of Plaskon Urea and Melamine Molding Compounds*. Bulletin 80-113, 1965.

——. *Handbook on the Molding and Handling of Plaskon Urea and Melamine Molding Compounds*. 1953.

——. *New Design Opportunities with Plaskon General Purpose Urea–A Low-Cost Cellulose-Filled Molding Compound*. Bulletin 843-09.

——. *Plaskon® Thermosets Urea Molding Compounds*. 612-300, October 1978.

American Cyanamid Co. *Molding Compounds · Resins*. Bulletin 040-3000, July 1958.

——. *Cymel Molding Compounds*. September 1956.

——. *Cymel Melamine Laminating Resins*. May 1956.

Ashland Chemical Co. *Specialty Phenolic Resins*. Bulletin 1356.

Barton, H. D. Phenolic and furan resins in chemical equipment. *Industrial and Engineering Chemistry*, July 1964, **56**(7):66–69.

BASF Wyandotte Corporation. Rigid plastic foams, from Plastics in Construction: Coming on Strong in 1978. *Plastics Engineering*, July 1978, **34**(7):42–44.

Borro, E. F. Sr. Phenolics–more versatile than ever. *Modern Plastics*, October 1964, **42**(2):102–104.

Colclough, W. G., Harding, J., and Meyers, C. Y. New heat resistant phenolic molding materials. *SPE Journal*, March 1961, **17**(3):225–256.

Florentine, F. P. Asbestos free phenolics stand up to heat, impact and arc tracking. *Materials Engineering*, September 1976, **84**(3):76–77.

Furane Plastics Inc. *Furane Resins–What They Are and Where They Are Used*. Product Review Bulletin EP-58-3 (Mod. 1).

General Electric Co. *Introducing Genal Pellets–Look What You Don't Get*. Bulletin CDC-482, February 1972.

——. *New Thermoset Compound for Injection Molding*. Bulletin CDC-474B, May 1970.

——. *Genal Injection Phenolics and Genal Pellets for Injection Molding*. Bulletin CDC-484.

Golding, B. *Polymers and Resins–Their Chemistry and Chemical Engineering*, D. Van Nostrand Co., Inc., Princeton, New Jersey, 1959.

Hauck, J. E. Don't overlook new, improved phenolics. *Materials in Design Engineering*, February 1966, **63**(2):72–75.

Haywood, K. W., Windover, W. R., and Gauronski, J. "Recycling Thermosets." Paper presentation, *SPE ANTEC*, New Orleans, May 1979.

Hooker Chemical Co. Introducing Durez SI injection grade phenolic molding materials. *The Durez Molder*, Issue One, 1973.

——. *Durez–Facts on Thermosets–Phenolics, Alkyds, Diallyl Phthalates*. Bulletin D204, 1967.

——. *Durez–Thermosetting Molding Compounds*. Bulletin D400.

Keegan, J. F. "Phenolics." Paper presentation, *Plastics World Seminar–Engineering Plastics II*, Los Angeles, March 1976.

Keutgen, W. A. "Phenolic Resins." In *Kirk-Othmer Encyclopedia of Chemical Technology*. Vol. 15, 2nd Ed. John Wiley & Sons, 1968, pp. 176–208.

Laurie, W. A. Dimensional stability improvement of melamine impact molding compounds. *SPE Journal*, December 1961, **17**(2):1310–1313.

Lidstone, J. B. and Jost, H. D. Designing phenolic parts for heat resistance. *Western Plastics*, Technical Report No. 113, November 1964, pp. 27–29.

Martino, C. F. Phenolics, too good to be outmoded. *SPE Journal*, March 1967, **23**(3):96–101.

McGrady, C. Thermosets, from Materials '79. *Modern Plastics*, January 1979, **56**(1):58–59.

Materials Engineering. Adhesives and sealants, Manual 260. May 1976, **83**(5):73–80.

——. Phenolics make headway in automotive parts. May 1976, **83**(5):31.

——. Phenolic plastics, Manual 249. September 1974, **80**(3):89–96.

Materials in Design Engineering. Thermosets are primarily used for . . . . July 1962, **56**(1):107.

Modern Plastics. New asbestos-free phenolics for metal replacement. April 1979, **56**(4):98.

——. Live-sprue injection molding is more than just an interim step. November 1978, **55**(11):49.

——. New ways to decorate melamine. March 1967, **44**(3):98–100.

——. Hard surface for melamine. November 1961, **39**(11):103, 216.

——. Precision-molded urea sphere. November 1961, **38**(11):101.

——. Plastics in the product revolution: decorated thermosets. June 1960, **37**(6):96–100.

——. The specialty phenolics. December 1959, **36**(12):81–84, 204–205.

Monsanto Chemical Co. *The Shell Molding–Dry Blend Process*. Product Information Bulletin No. 1073.

——. *Manufacturing Information on Decorative Melamine Laminates–Resins, Techniques*. Product Information Bulletin No. 1073.

Murray, T. A. Asbestos-free and cold-manifold phenolics, it's a changing scene for an old workhorse. *Plastics Technology*, August 1977, **23**(8):51–56.

Plastics Engineering. Engineering resins primer. December 1974, **30**(12):17–32.

Power, G. E. High pressure laminates, from Plastics in Construction: Coming on Strong in 1978. *Plastics Engineering*, July 1978, **34**(7):31–32.

Quaker Oats Co. *Quacor Furan Resins for Fiberglass Reinforced Equipment to Handle Solvent, Acids and Bases*. Bulletin 400.

Reichhold Chemicals, Inc. Phenolics, from Plastics in Packaging. *Plastics Engineering*, October 1978, **34**(10):42.

Reichhold Chemicals, Inc. *RCI Plyophens in Laminates and Related Fields*. Technical Bulletin PLY-1.

Riley, M. R. Engineer's guide to thermosetting molding compounds. *Materials in Design Engineering*, Manual No. 197, July 1962, **56**(1):101–107.

Rutan, R. M. Urea formaldehyde, from Plastics in Packaging. *Plastics Engineering*, October 1978, **34**(10):50.

Simonds, H. R. and Church, J. M. *A Concise Guide to Plastics*. 2nd Ed. Van Nostrand Reinhold Co., New York, 1963.

Sloats, M. A. Eliminating glue-line failure in bonding hardwood. *Adhesives Age*, June 1979, **22**(6):18–20.

Sneller, J. Thermoset injection molding becomes more reliable and more economical. *Modern Plastics*, June 1979, **56**(6):44–47.

Swanson, R. S. *Plastics Technology–Basic Materials and Processes*. McKnight & McKnight Publishing Co., Bloomington, Illinois, 1965.

Titus, J. B. *Trade Designations of Plastics and Related Materials*, (revised). PLASTEC NOTE N9C, Plastics Technical Evaluation Center, Picatinny Arsenal, Dover, New Jersey, May 1978.

Union Carbide Corporation. *Bakelite S Phenolics for Injection Molding*. Bulletin F-43455, February 1972.

Westinghouse Electric Corporation. *Micarta Convolute Tubing, Filament-Wound Tubing*. Bulletin 47-WIM-T.

——. *Micarta Industrial Plate, Polyester Laminates, Rods, Angles, Channels, Zees*. Bulletin 47-WIM-P.

Wilson, J. B., Jay, G. L., and Krahmer, R. L. Using resin properties to predict bond strength of oak particle board. *Adhesives Age*, June 1979, **22**(6), 26–30.

Wood, A. S. More muscle, higher heat: they power phenolic molding compound "Revival." *Modern Plastics*, July 1979, **56**(7):50–51.

Young, R. H. "Aromatic Polymers," from *Proceedings of Technical Seminar*, ASTM Task Group on Geothermal Seals, CONF-7806141, Boston, June 1978.

Zimmerman, D. D. "Performance of Engineering Resins in Electrical Connectors." Paper presentation, *SPE ANTEC*, New Orleans, May 1979.

POLYESTERS

The polyester class of polymers is formed by the condensation of polyfunctional acids and alcohols. The reactants must be polyfunctional, or only low molecular weight compounds will be formed. Once this requirement is met, the number of functional groups reacting must be known, for this will determine the configuration of the molecules; that is, whether they are linear or three-dimensional. The feature of unsaturation is also an important factor. This allows an acid such as maleic, which is bifunctional with reference to condensation, to become tetrafunctional when its unsaturation reacts by addition.

For simplicity of presentation, this section will be divided into alkyds, unsaturated polyesters, and allyls, even though commercial polyesters are not always so separated. Linear polyesters were treated separately in Chapter 5, Thermoplastic Polyesters.

Alkyds

The world "alkyd" was coined by R. H. Kienle in 1929 to designate the "al" of alcohol and "cid" (changed to "kyd") from acid, the two reactants used. In actual practice, this polyester is so generally modified that the term alkyd is understood to include such modifications.

If a dibasic acid such as phthalic is reacted with a monohydric alcohol, no polymer is formed. Simple dibasic acid esters are produced. Millions of pounds of these simple esters are manufactured each year as plasticizers. The polyfunctional acid used to produce plasticizers is generally phthalic, though adipic, azelaic, sebacic, citric, and tartaric are also widely used. The alcohols vary from methanol, to normal and branched molecules of eight to ten carbons. If difunctional alcohols are employed, then linear polymers are possible, and, when functionalities of three (or higher) with either acids or alcohols are attained, a thermosetting structure is possible. These resins are the most versatile of all the coating resins because of the large variety of raw materials and modifying agents that may be used in their preparation. Most of the various acids and alcohols that might be used are given in Table 6-12.

The various possible combinations from this list are almost unlimited, even if the possibility of modification of these esters with other materials were not present. Not all the acids and alcohols listed are commercially important at present, but they have been used at least in experimental quantities, and some of them would be used more generally if their cost was not quite so high.

Among the acids or anhydrides, phthalic is most commonly used because of its low-cost preparation from the oxidizing of naphthalene or *o*-xylene, and because of its excellent heat and ultraviolet stability. Heating with glycerol produces materials at first soluble in water and alcohol, but later becoming an infusible gel.

Table 6-12. Typical Acids and Alcohols Used to Make Alkyds.

ACIDS	
Adipic	Malic
Azelaic	Pimelic
Camphoric	Rosin and other terpenes
Citric	Sebacic
Cyclopentadiene	Styrene
Glutaric	Suberic
Isophthalic	Tartaric
Maleic and addition products	Terephthalic
	Vinyl acetate
ALCOHOLS	
Diethylene glycol	Pentaerythritol
Ethylene glycol	Polyethylene glycol
Glycerine	Propylene glycol
Mannitol	Sorbitol

Unmodified glyceryl phthalate resins have poor solubility characteristics and have not, therefore, been widely used. However, the modification of alkyds with oils and resins, as well as internal modification through the use of more than one alcohol or acid, will upgrade the final resin's properties to desirable levels.

For example, one purpose of modification is to alter the gel time/cure rate characteristics of a resin. This can be achieved by incorporating reactants with different functionalities and relative reaction rates. Plasticization and flexibility may be introduced via long-chain fatty acid modifiers. Inclusion of unsaturation in either the base reactants and/or modifying oils and resins enables further crosslinking to occur.

Since these polymers are used almost exclusively as resins for the preparation of varnishes, paints, and other coatings, their compatibility with oils and other resins is an extremely important factor. To achieve this compatibility, the anhydrides and glycerol are reacted to produce a complex ester-interchange product. Direct fusion of the reactants may be used, or an inert solvent may be added. In a typical fusion reaction, the oil (e.g., linseed oil) is heated

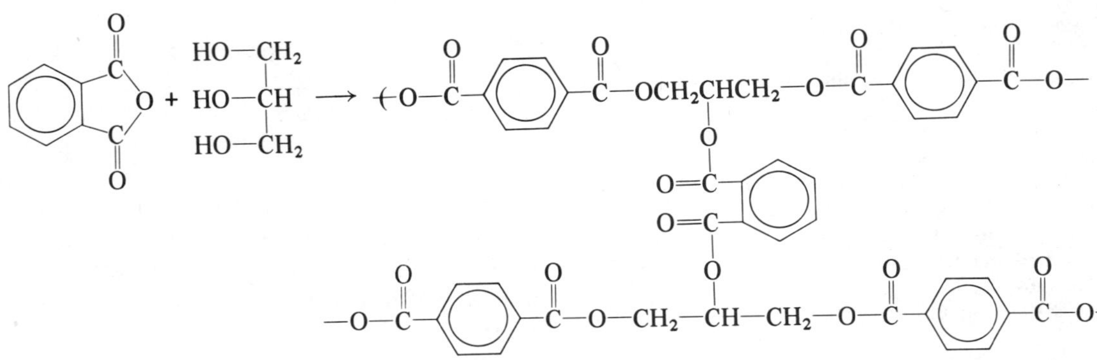

with a major part of the glycerol and with lime as catalyst to bring about ester-interchange and produce monoglycerides. This intermediate is more receptive to subsequent additions than is the oil alone.

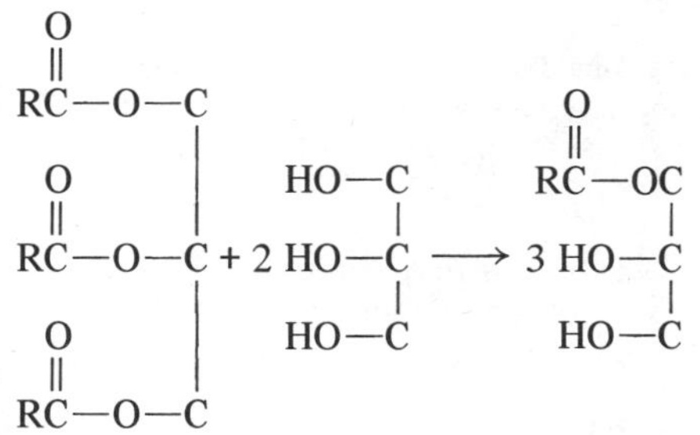

The phthalic anhydride and remaining glycerol are stirred in and further heated to the desired viscosity and low acid number. The total number of acid and alcohol functional groups must be balanced stoichiometrically in order to determine the proper amounts of anhydride, glycerol, and oil.

The actual manufacturing process is comparatively simple. Stainless steel or aluminum kettles are generally used. They are either electrically heated or direct-fired and are equipped with agitators, special reflux condensers, means of applying an inert atmosphere, and vacuum. The raw materials are heated to 200-250°C (392-482°F), which is the reaction temperature. The time required to complete the reaction varies from 5 to 16 hours. At the end of this period, the resins are usually incorporated in solvents such as toluene, xylene, or coal-tar naphthas, and sold as solutions. Some of the paint companies add the molten resin directly to varnish oils. In other cases, the liquid is poured into drying pans and allowed to cool to a hard, solid resin.

The use of these modified resins depends on the ability to arrest the polymerization reaction before gelation. At that point the resin can be dissolved, and, if required, other ingredients such as drying oils, pigments, or thixotropes may be added. When coated onto a substrate, the solvent evaporates, and the residual resin completes its curing by thermal or oxidative crosslinking to an infusible, insoluble form.

Depending on the kinds of oils used, the alkyd may have drying, semi-drying, or non-drying properties; and depending on the amount of oils used, the resin may be classed as short, medium, or long oil-modified. Oils, such as linseed and dehydrated castor, give air-drying resins, while more saturated oils, such as soya, dry slowly and generally require baking. However, these less reactive compositions tend to show less yellowing on aging.

The amount of oils used is termed *oil-length*. The percent phthalic anhydride used in resins of short, medium, and long "length" are 40-50, 30-40, and 20-30, respectively. The ratio of resin to oil naturally affects water resistance, weathering, rate of drying, flexibility, and many other of the characteristics of the surface finish. Long oil alkyds are generally favored for exterior finishes, whereas medium and short oil resins are used principally in interior applications.

Other modifiers are also used. Resins and ester gums may be added. Phenolic resin addition improves water and alkali resistance. Urea and melamine resins shorten the baking time and improve color retention and surface hardness. These resins are especially useful in non-drying alkyds for improving film strength.

The alkyds are used for every conceivable type of coating from printing inks to coated fabrics. They are the most widely used resins in durable air-drying outdoor paints and, together with the phenolics, epoxies, and ureas, supply the baking finishes and high-grade enamels and varnishes for homes, appliances, furniture, automative, and many other industries. They also have some use as adhesives.

Alkyd Molding Compounds. The ingredients used in alkyd molding compounds are similar to those in alkyd coating resins; however, they are rarely oil-modified. Instead, they are primarily modified through choice of acid or glycol. Typical reactants include maleic, fumaric, adipic, phthalic, and diallyl phthalic acids or anhydrides, and ethylene or diethylene glycol. The resin binder used in molding compounds is tailored to give the desired end properties to suit the requirements of the application of the molded part.

Alkyd molding compounds are categorized in four groups, based on the physical form in which they are supplied. These forms include granular, putty, rope, and bulk. Table 6-13 describes these forms and how they are used, and provides a generalized listing of properties. (Table 6-19 provides a more comprehensive listing of alkyd properties and compares them to unsaturated polyester systems.) Figure 6-22 shows some of these forms.

The polymerization exotherm is quite high with these resins and causes a substantial amount of volumetric shrinkage. Consequently, unfilled systems are not suitable for accurate, reproducible moldings. Bulk fillers are added to moderate the reaction and thereby control shrinkage and internal stresses and strains. These fillers also reduce cost, control flow properties, impart heat resistance, provide an improved surface appearance, increase electrical resistance, reduce moisture absorption, or impart other more specific properties. They are, as a rule inexpensive, uniform in size, and chemically inert. They must not interfere with the curing mechanism nor disturb the resin stability. A list of commonly used bulk fillers along with the resin properties they influence is shown in Table 6-14.

Fibrous fillers may also be added to molding compounds to impart additional mechanical strength and stability. Asbestos fibers impart heat resistance, enhance surface smoothness, and improve tensile strength, but they do not provide appreciable impact strength. Sisal and cellulose

Table 6-13. Forms and Properties of Alkyd Molding Compounds*

A. FORMS

Putty. The putty form is very free flowing and is suited for encapsulation of resistors, capacitors, coils, and transformers. The putty is usually extruded into ribbon form by the user and molded by compression or plunger molding.

Granular. The granular form is readily handled by automatic bulk feeding devices. Typical applications for this form are electrical sockets, vacuum tube bases, and automotive ignition parts. They are usually produced by compression molding. Softer flowing granular alkyds for plunger molding now are available, as well as short glass fiber reinforced types.

Rope. Extruded rope is a convenient way of handling medium length glass fiber reinforced alkyd molding compounds. Mold charges are prepared by cutting lengths of the rope. Compression and hot or cold plunger molding are suitable methods for processing this form.

Bulk. Long glass fiber reinforced alkyd molding compounds for the highest strength and impact resistance are available in bulk form. The bulk material can be preformed by machine into uniform size pills or formed by hand into a loose ball. Compression, pot-transfer, or hot-plunger molding are all suitable for handling bulk-form compounds.

B. PROPERTIES

Property	Granular	Putty	Rope	Bulk
Arc resistance (sec)	175	175	160	130
Compressive strength (psi)	15,000	15,000	20,000	18,000
Dielectric strength (v/mil)				
Short time	375	375	300	150
Step-by-step	325	325	250	125
Flame resistance	Self-Exting.	Non-burn.	Self-Exting.	Self-Exting.
Flexural strength (psi)	7,500	7,500	14,000	12,000
Impact strength (ft-lb)	0.30	0.33	3.0	6.0
Tensile strength (psi)	3,500	3,500	6,000	3,500
Water absorption, 48 hr @ 50°C(%)	0.5	0.5	0.5	1.5
Max continuous operating temp (°C)	–	–	130°C	130°C

*Howald, 1968. Reprinted by permission from *Plastics Design & Processing*. Copyright 1968, Lake Publishing Corp., Libertyville, IL.

Fig. 6-22. Types of alkyd molding compounds. (*Courtesy: American Cyanamid Co.*)

fibers lower the specific gravity of the molding compounds, but a loss of properties is noticed on exposure to moisture or heat. Synthetic fibers, such as polyester, nylon, or acrylic, give good moisture resistance and electrical properties, and also freedom from internal voids in the molded part.

The most outstanding fibrous reinforcement is fiberglass. The strength-to-weight ratio of fiberglass is several times higher than steel. The high tensile strength, low elongation, and inertness of fiberglass result in glass-reinforced alkyd molding compounds with excellent rigidity, dimensional stability, and flexural and impact strengths.

Alkyds are characterized by rapid curing and low molding pressures. These resins have heat resistance, dimensional stability, and electrical properties over a wide temperature range.

Applications include appliance connectors, pump impellers, relays, switches, thermostats, and timers. The automotive industry uses alkyd molding compounds in brush holders, coils, distributors, housings, solenoids, and switches (see Figures 6-23 and 6-24). In electronics they are used in

Table 6-14. Common Fillers for Alkyd Molding Compounds.*

- Antimony Oxide—used to impart flame resistance, usually in conjunction with a chlorinated or brominated resin.
- Asbestos—fibrous in nature, forms a strong bond with the resin to impart toughness and stiffness, and is a good flow control agent.
- Barium Sulfate—high specific gravity and opaque to X-rays.
- Calcium Carbonate—near white appearance, high filler loadings possible with good flow, imparts excellent high frequency electrical properties.
- Clay—lowers moisture absorption, raises hardness and flexural strength, excellent in high voltage applications.
- Hydrated Alumina—imparts heat, arc-tracking, and flame resistance.
- Mica—good flow control agent, imparts high dielectric strength.
- Silica and Talc—inert, inexpensive fillers.
- Titanium Dioxide—very white appearance, raises dielectric constant.
- Wollastonite—imparts hardness.

*Howald, 1968. Reprinted by permission from *Plastics Design & Processing*. Copyright 1968, Lake Publishing Corporation, Libertyville, Ill.

bobbins, capacitors, coils, relays, encapsulated components, molded circuitry, resistors, and transformers.

Circuit breakers, contactors, and overload protectors for motors as well as starters, interlock switches and switch covers are another major application area. A large nu-nber of television parts are made from alkyd moldings, such as coils, corona caps, sockets, transformers, terminal boards, and tuner strips.

Unsaturated Polyesters

The oil-modified alkyd resins already discussed are quite evidently also unsaturated polyesters, since the modifying oil is generally unsaturated. The resins discussed in this section, however, are those in which the polyfunctional acid, or alcohol, or both, contain the unsaturation which is thus introduced into the polymer chain. Once the polyester is formed by the condensation reaction, the subsequent conversion to insolubility is an addition polymerization step, generally induced by the use of a monomer such as styrene, which acts as a crosslinking agent. The latter reaction, of course, proceeds without the evolution of by-products. Typical reactions are:

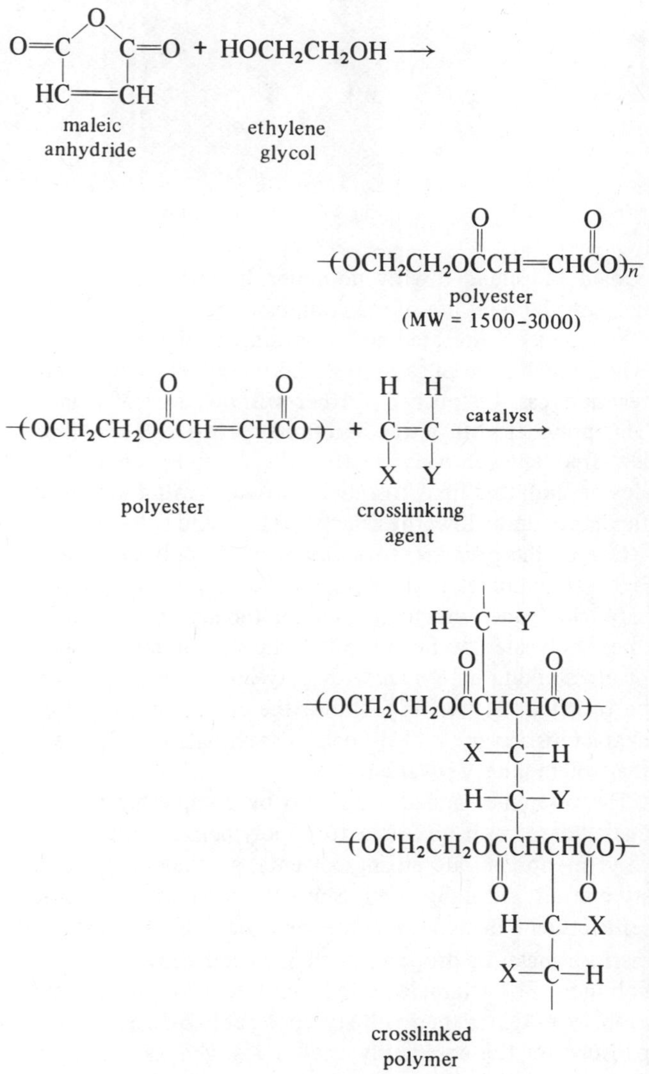

Typical crosslinking agents are styrene, diallyl phthalate, vinyl acetate, and methyl methacrylate.

In a typical preparation, maleic anhydride is heated with ethylene or diethylene glycol to a temperature range of 165-220°C (329-428°F) under an inert gas. The water of reaction is removed, while glycol is returned by reflux. Care must be exercised since the unsaturated acid may react by addition to produce a gelled mass. When the acid number or viscosity indicates a desired degree of reaction, the resin is

Fig. 6-23. Slip ring of reinforced polyester molding compound, used in alternator rotor assembly. (*Courtesy: American Cyanamid Co.*)

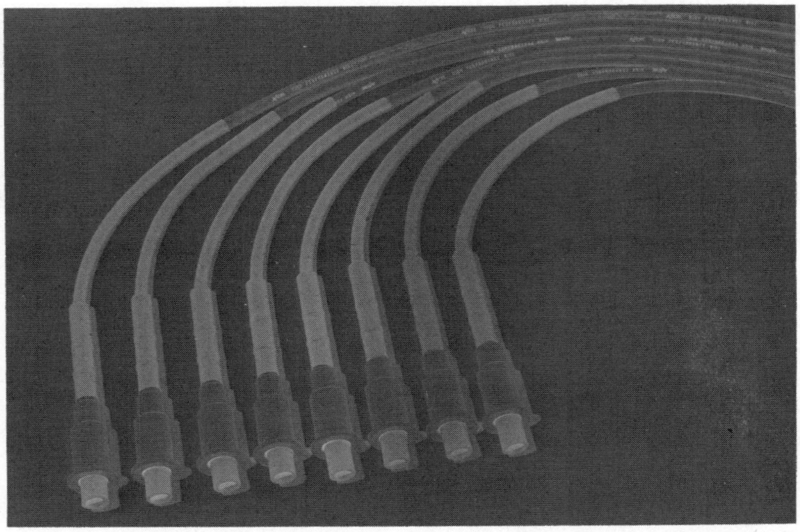

Fig. 6-24. Spark plug wire set, alkyd connectors. (*Courtesy: American Cyanamid Co.*)

cooled and blended with monomer. Inhibitors are added to increase the shelf life of the composition.

Mixtures of saturated and unsaturated acids may be used. When a high ratio of saturated acid is used, the condensation reaction can be pushed further without fear of gelation. This polymer with fewer functional points will crosslink at less frequent intervals, so that the composition will cure slower and the final insoluble product will have greater flexibility and a lower resistance to heat and solvents.

One of the prime responsibilities of the polyester chemist is to stoichiometrically balance the monomeric reactants through proper matching of their molar ratios. Once the linear polyester has been formed, the stoichiometric balance of crosslinking monomer with polyester unsaturation must be satisfied. Figure 6-25 shows the effect on gel and cure characteristics of a DAP/maleic/isophthalate polyester as the stoichiometry is varied.

The monomer added must also be compatible with the unsaturated resin and able to copolymerize with it. Most vinyl monomers are strong solvents, so that acrylates and styrene are generally used. Since the monomer added as a resin diluent becomes a component part of the final polymerized mass, its properties will influence those of the final polymer. For example, ethyl acrylate will yield a more rubbery polymer than will styrene. Table 6-15 shows comparisons for the commonly used acids, glycols, and monomeric crosslinking agents, and the characteristics they impart to the final resin system. Table 6-16 shows the effect various monomers have on the system shrinkage and exotherm. Figure 6-26 shows the effect of styrene content on physical properties. Table 6-17 shows the effect of the addition of methyl acrylate monomer on the flexural properties of polyester laminates.

Most of the monomers which are added to polyester systems are somewhat unstable at room temperature. The

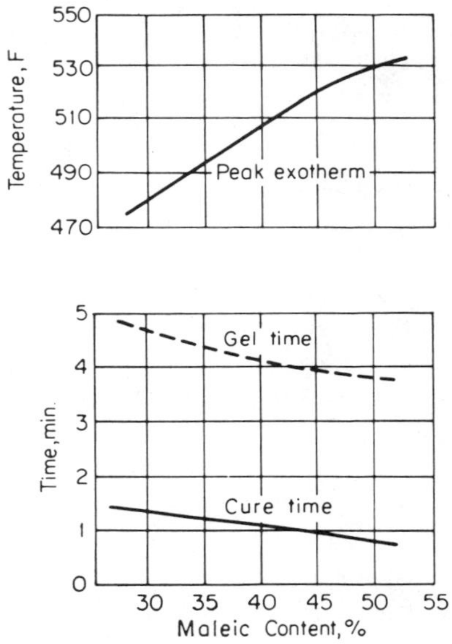

Fig. 6-25. Effect of maleic functionality on gel and cure time. (Beacham et al., 1963)

addition reaction between the unsaturated groups will progress (although slowly), without the intervention of catalysts or promotors. Consequently, in order to control shelf-life and viscosity, inhibitors are incorporated into the polyester system. An ideal inhibitor must impart stability, provide an adequate pot-life, and not adversely affect the cure characteristics of the resin. Once the inhibitor is expended, polymerization can proceed normally. The most widely used inhibitors are hydroquinone and *p*-tert-butylcatechol.

The control of gelation and cure is the most critical part of polyester resin technology. Although polyester resins gel under varying conditions, it is extremely difficult to arrest the crosslinkage reaction at the gel stage. Gelation is fol-

Table 6-15. Unsaturated Polyester Components.*

Components	Ingredients	Characteristics
Unsaturated anhydrides and dibasic acids	a–Maleic anhydride	a–Lowest cost, moderately high heat deflection temperature (HDT)
	b–Fumaric acid	b–Highest reactivity (crosslinking), higher HDT, more rigidity
Saturated anhydrides and dibasic acids	a–Phthalic (orthophthalic) anhydride	a–Lowest cost, moderately high HDT, provides stiffness, high flex, and tensile strength
	b–Isophthalic acid	b–Higher tensile and flex strength, better chemical and water resistance
	c–Adipic acid, azelaic acid, and sebacic acid	c–flexibility (toughness, resilience, impact strength); adipic acid is lowest in cost of flexibilizing acids
	d–Chlorendic anhydride	d–Flame retardance
	e–Nadic methyl anhydride	e–Very high HDT
	f–Tetrachlorophthalic anhydride	f–Flame retardance
Glycols	a–Propylene glycol	a–Lowest cost, good water resistance and flexibility, compatibility with styrene
	b–Dipropylene glycol	b–Flexibility and toughness
	c–Ethylene glycol	c–High heat resistance, tensile strength, low cost
	d–Diethylene glycol	d–Greater toughness, impact strength and flexibility
	e–Bisphenol-A adduct	e–Corrosion resistance, high HDT, high flex and tensile strength
	f–Hydrogenated bisphenol-A adduct	f–Corrosion resistance, high HDT, high flex and tensile strength
Monomers	a–Styrene	a–Lowest cost, high reactivity, fairly good HDT, high flex strength
	b–Diallyl phthalate	b–High heat resistance, long shelf life, low volatility
	c–Methyl methacrylate	c–Light stability, good weatherability, fairly high HDT
	d–Vinyl toluene	d–Low volatility, more flexibility, high reactivity
	e–Triallyl cyanurate	e–Very high HDT, high reactivity, high flex and tensile strength
	f–Methyl acrylate	f–Light stability, good weatherability, moderate strength

*Reprinted by permission from *Modern Plastics*, February 1964.

lowed immediately by a self-propagated cure with considerable evolution of heat, which in thick sections and with a highly reactive resin can cause a temperature rise to over 204°C (400°F). Molded or laminated structures can be made with low or zero pressure. In closed systems, there is no loss of volatiles during cure.

Polyester resins can be cured from room temperatures up to 149-163°C (300-325°F). The addition of nothing but an organic peroxide is satisfactory for cure above 82°C (180°F), but cure at lower temperatures requires the use of both a peroxide catalyst and a promotor.

Most of the chemical catalysts or initiators used for curing unsaturated polyesters are peroxides. Ninety percent of the peroxides used consist of four types: benzoyl peroxide, cumene hydroperoxide, di-tert-butyl peroxide (most commonly used for heat-initiated polymerization), and methyl ethyl ketone peroxide (MEKP), (normally used for room-temperature curing). Of these four, MEKP is the most used, representing better than 50% of sales. The promoters most often used are organic cobalt salts and dimethyl- or diethylaniline. The peroxide-promoter systems generally are used as follows:

Table 6-16. How Monomer Affects Shrinkage and Exotherm.*

Monomer	Polymer Volume Shrink, %	Peak Exotherm, F
Styrene	17	300
Vinyl toluene	12.6	–
Vinyl acetate	27	290
n-vinyl carbazol	7.5	–
Acrylonitrile	26	280
Methyl methacrylate	21	245
n-Butyl methacrylate	14.3	–
Diallyl phthalate	11.8	180
Allyl diglycol carbonate	14.0	–
Triallyl cyanurate	–	305

*Maudru, 1968.

Benzoyl peroxide: (1) No promoter, for cure at 82°C (180°F) or higher. (2) Diethylaniline or dimethylaniline promoter for cure at room temperature.

Cumene Hydroperoxide: Used to a limited extent to initiate gel in the 60-71°C (140-160°F) range, particularly on continuous sheeting equipment. Cobalt can be used as the promotor, but a special promoter is generally preferred. An advantage of cumene hydroperoxide is a minimum tendency to bleach colored pigments used in colored, corrugated exterior sheeting. It is, however, common practice to use a mixed catalyst; for example, 0.5% benzoyl peroxide, and 0.5% cumene hydroperoxide, the latter for gel at 60-71°C (140-160°F) and the benzoyl peroxide for completion of cure at a final temperature, usually well above 82°C (180°F).

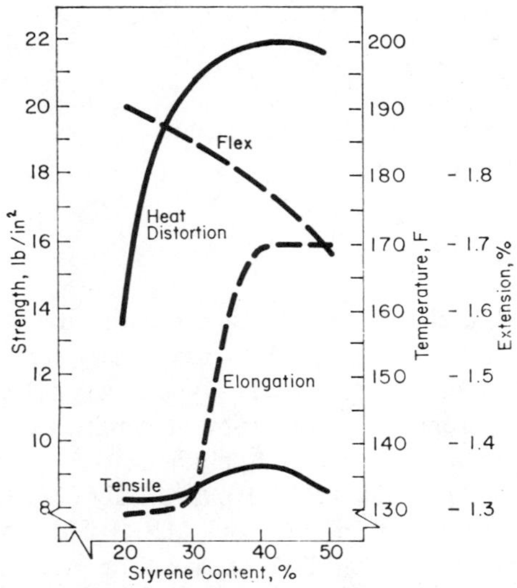

Fig. 6-26. Styrene affects polyester system's tensile, elongation, and heat-distortion properties. Note abrupt transition in elongation for styrene contents between 30-40% in 40/60 maleate/phthalate resin. (Maudru, 1968)

Table 6-17. Flexural Strength of Polyester Laminates Using Methyl Methacrylate Monomer.*

Resin[a]	Resin %	Styrene %	MMA %	Thickness Inches	Flexural Strength, psi	Flexural Modulus 10^6 psi
A	80	20	0	0.125	23,040	–
A	80	15	5	0.110	24,640	–
A	80	10	10	0.115	30,260	–
A	80	5	15	0.106	30,300	–
A	80	0	20	0.122	28,680	–
B	80	20	0	0.111	20,100	0.98
B	80	10	10	0.117	27,260	1.25
A	80	10	10	0.134	26,320	1.13

All laminates contained nominal four ounce glass fiber mat, 28% glass by weight.
[a]A Polylite 8100A, Reichhold Chemicals Company.
B Selectron 5155, Pittsburgh Plate Glass Company.
*Slone, 1960.

Methyl ethyl ketone peroxide: Used almost exclusively for cure at room temperature with cobalt promoter. By far the most widely used catalyst-promoter system.

Typical catalyst and promotor concentrations, based on resin, are:

Heat Cure: Benzoyl peroxide, 0.25-1.0%.

Intermediate Cure: Cumene hydroperoxide 0.5-1.0% with special promoter.

Room-Temperature Cure: (1) Methyl ethyl ketone peroxide 0.5-2.0%, with cobalt naphthenate, 6% solution, 0.10-0.24%. (2) Benzoyl peroxide 0.5-1.5% with diethylaniline 0.1-0.4%.

These concentrations are varied depending on the mass being cured: a larger mass requires less catalyst than a smaller quantity of the same resin.

Benzoyl peroxide is available as dry granules and also as a pasty dispersion in tricresyl phosphate containing 50% peroxide. The paste form dissolves in the polyester resin more readily than does the dry form. Cumene hydroperoxide is a thin liquid and the lowest-cost polyester catalyst available. Methyl ethyl ketone peroxide is generally supplied as a 60% solution in dimethyl phthalate.

NOTE: *It is important that special care be taken to avoid the contamination of organic peroxides with accelerators or promoters used in polymerization reactions. These materials should never be added directly to one another or consecutively to a resin unless one ingredient is thoroughly mixed in before adding the other. In some cases, vigorous or explosive decompositions may result if direct contamination occurs. Information on the proper handling, storage, and use of peroxides and additives should be obtained from suppliers.*

Many other materials have a retarding or accelerating effect on the gel and/or cure of polyester resins. Some additives are capable of stabilizing the resin before the peroxide is added and then speeding up the gelation of the catalyzed resin. Other additives can speed up gel and then slow down the cure. This is sometimes advantageous, since it allows more time for dissipation of the heat of cure, thus lowering the peak exotherm and reducing cracking tendencies. In most cases, however, these additives are incorporated by the resin manufacturer. Fully promoted systems are available, containing a carefully balanced system of promoters and stabilizers and requiring only the addition of catalyst.

Figures 6-27 and 6-28 show typical exotherm curves for polyester systems using common peroxides and accelerators. Table 6-18 shows the properties of various catalyst/accelerator systems used in polyesters.

Curing of specially formulated systems can also be achieved through irradiation techniques, particularly exposure to UV wavelengths. This is achieved through the incorporation of benzoin or other UV-activated catalysts into the system. The main advantage to such procedures is very rapid, low-temperature curing without chemical contamination. Irradiation techniques have proven very successful in the processing of coatings, adhesives, and laminated structures. Figure 6-29 shows that very thick laminates can be cured very rapidly using UV techniques.

Because such wide design latitude is available in the construction of polyester systems, a discussion of resin properties per se is impossible. Table 6-19 shows the spectrum of properties of the two main categories of polyesters, namely casting and glass-reinforced systems and compares them to those of alkyd systems. In general, they are resistant to chemicals, moisture, and heat cycling. They can be made rigid or flexible, and their density and mechanical properties can be substantially modified through judicious choice of compounding ingredients.

Compounding ingredients include the bulk and reinforcing fillers discussed above with phenolics and alkyds. Other ingredients include colorants (pigments and dyes), rheologic agents (thixotropes, viscosity depressants), and other miscel-

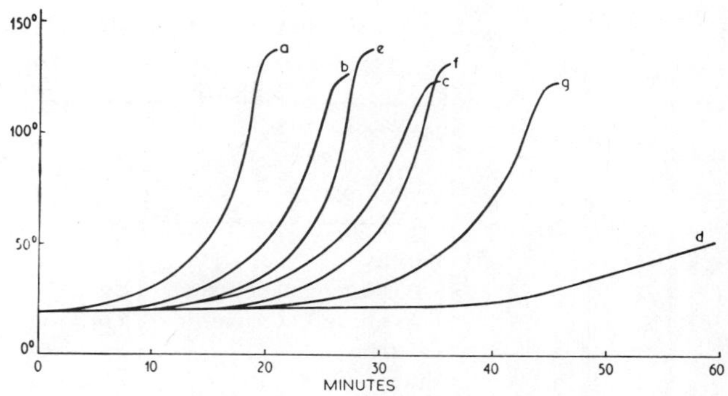

Fig. 6-27. Gel times and exotherms obtained at 20°C with MEK peroxide (50% solution) as catalyst and cobalt naphthenate (1% metal) as accelerator in varying proportions as follows:

Curve	Catalyst	Accelerator
a	4%	2%
b	3%	2%
c	2%	2%
d	1%	2%
e	4%	1%
f	3%	1%
g	2%	1%

(Cywinski, 1957).

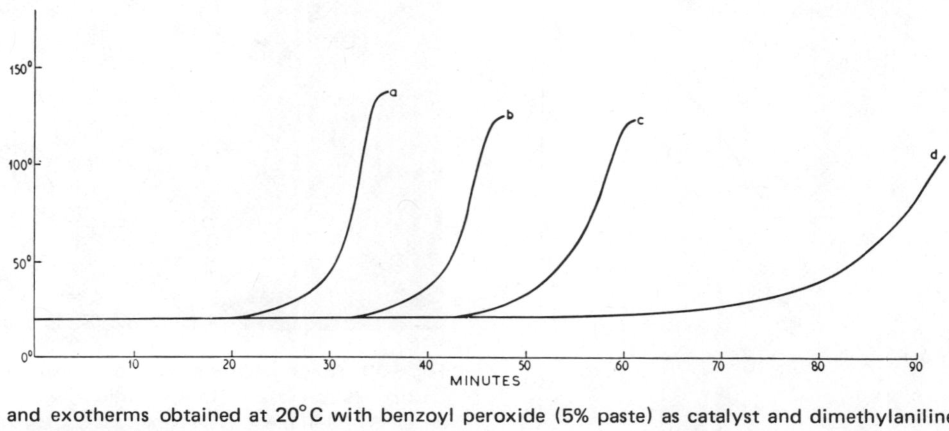

Fig. 6-28. Gel times and exotherms obtained at 20°C with benzoyl peroxide (5% paste) as catalyst and dimethylaniline as accelerator (50% solution) in varying proportions as follows:

Curve	Catalyst	Accelerator
a	4%	2%
b	3%	2%
c	4%	1%
d	3%	1%

(Cywinski, 1957).

Table 6-18. Properties of Various Peroxide-Accelerator Systems for Polyesters.*

Property	Benzoyl peroxide + tertiary amine	Ketone peroxide + cobalt	Ketone peroxide or hydroperoxide + vanadium
Potlife			
Polyester and peroxide	Long	Short	Short
As above + accelerator	Long	Long	Short
Gel time	Variable	Variable	Short
Cure	Fast	Moderate	Very fast
Sensitivity to:			
Air inhibition	Strong	Moderate	Strong
Moisture	None	Strong	Strong
Glass finish	None	Faint	Moderate
Filler and pigment	Seldom	Mostly strong	Reacts with oxides, hydroxides, carbonates, etc.
Color after curing	Faint	Faint	Moderate
Discoloration			
By light	Strong	None	None
By heat	Strong	Faint	Faint

*Brinkman et al., 1968.

laneous additives such as UV absorbers, fungicides, and reodorants.

Casting resins are used to produce buttons, synthetic marble, decorative castings (Figure 6-30), and clear or pigmented surface coatings. For example, to make synthetic marble, calcium carbonate is blended into a suitable resin formulation. Very near the end of the mixing cycle, a coloring agent is added and the mixer allowed to turn over once or twice more which produces the swirled effect found in natural marble (Figure 6-31).

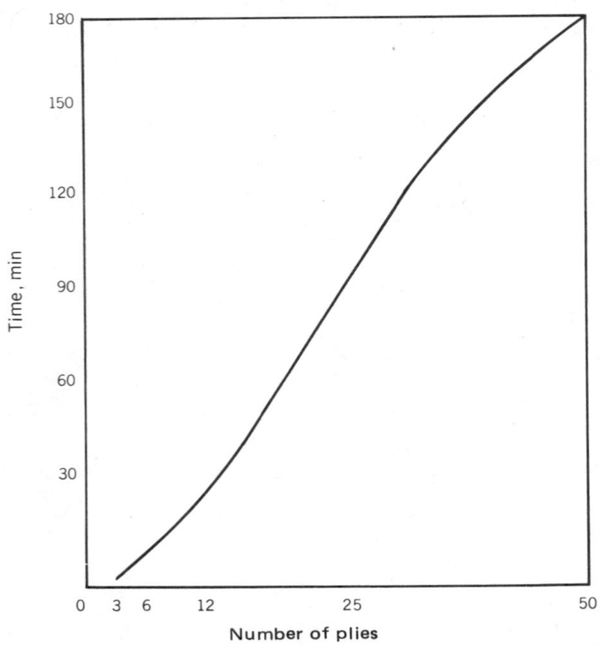

Fig. 6-29. Exposure time necessary to cure laminates with 3 to 50 plies, Cordopreg UVFR/181-MPC-1. (Anderson, 1968)

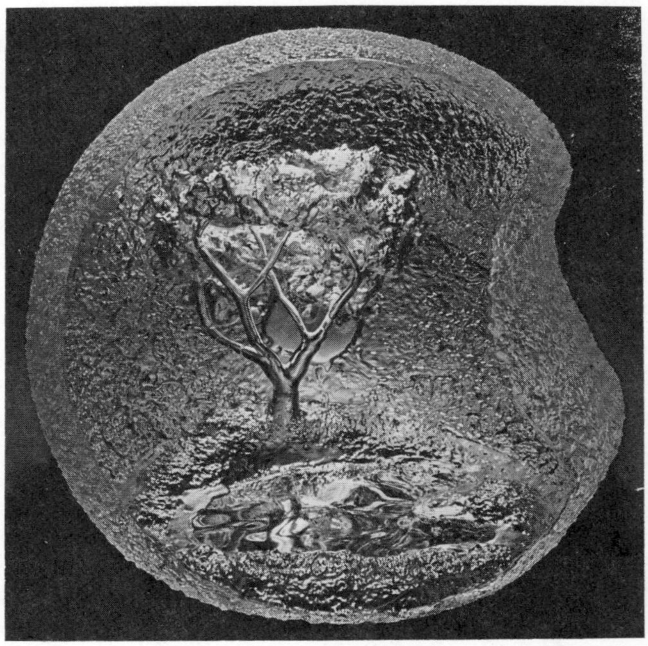

Fig. 6-30. Resin sculpture by Los Angeles sculptor Caryl Craig. Casting resins are used in applications where clarity, good appearance, or color blends are paramount. (*Courtesy: Caryl Craig and Silmar Division, Vistron Corporation*)

In recent years the manufacture of "synthetic wood" has

Fig. 6-31. Ortega Onyx (synthetic or cultured marble) has now proven to be a major component for bathroom units, unilavs, pullmans, wall panels, bathtubs, and shower pans, as well as a major item in furniture. *(Courtesy: Oretga Onyx, Inc., subsidiary of Gruber Systems, Inc., Valencia, Ca.)*

increased the usage of polyesters significantly (Figure 6-32). The "wood" is based on polyester casting resins that are filled with lightweight fillers such as glass or ceramic microspheres, vermiculite, or pecan-shell flour. The systems are cast in flexible molds which accurately reproduce the veneer of wood.

A newer procedure is to use water-fillable polyesters. The resins are formulated to accept up to more than 50% by weight of water as a filler. The water becomes trapped in the cured resin in very tiny dispersed particles. Parts made from these resins have a higher density than parts from standard polyesters and low-density fillers. While they can reproduce wood grain and parts can be nailed or stapled, they are not intended for use in structural furniture parts but in decorative applications such as fascia and lamp bases.

Fig. 6-32. Duplication of moldings and patterns is achieved with flexible polyester resin systems and low-cost fillers. Staples or screws can be used as structural parts around metal or wooden cores; staining or painting matches the wood. *(Courtesy: Cal-Style Furniture)*

Premix molding compounds are made from unsaturated polyester systems in much the same way as described earlier for alkyd molding compounds, and they are used in the same applications. Additionally, these formulations have been modified to accommodate fabrication into essentially B-staged sheets that are then molded using matched-die molding techniques. Such systems are called SMC (for sheet molding compounds), BMC (for bulk molding compounds), TMC (for thick molding compounds), and SPMC (for solid polyester molding compounds).

General-purpose SMC is used principally for structural applications in which a smooth appearance is not required. A high glass content (30-35% by weight) is used to achieve high tensile strength. Low-profile SMC is more suitable for applications that must have a smooth surface and generally good appearance. The strength of low-profile SMC is less than that of general-purpose compounds because less glass-fiber reinforcement is used, and the individual fiber lengths are shorter. Typically, low-profile compounds contain 26% glass fibers cut from $\frac{1}{2}$-1-inch lengths; general-purpose compounds contain up to 35% glass cut from 1-3-inch lengths. Both products are available in continuous sheet form, approximately 24-60 inches wide. Typical properties are shown in Table 6-20.

TMC production is in some ways similar to SMC production. But while SMC now has about a $\frac{1}{4}$-inch-thickness limit, TMC can be produced up to 2 inches thick, and even in billet form for compression and injection molding. Table 6-21 shows a comparison between TMC, BMC, and SMC.

SPMC's differs from the other molding compounds in its physical form. It is a convenient and free-flowing dry pellet containing up to 20% glass. The extrusion processing technique used to produce SPMC, however, permits the addition and wetting of glass fibers with minimal fiber-bundle degradation. The resulting pellets can be easily fed to an in-line screw injection molding machine to mold parts with impact strengths approaching 2 ft-lb/in.

SPMC was developed for use in automatic molding systems such as injection, transfer, and compression molding. It contains no asbestos and is currently available in various colors, flow grades, and shrinkage factors. Table 6-22 compares performance characteristics of SPMC with other thermosetting systems and some engineering thermoplastics. Typical molded properties are shown in Table 6-23.

Recently, a series of improved structural molding compounds have been commercialized. Trademarked XMC and HMC by PPG Industries, they are distinguished from other BMCs by their very high percentage of fiberglass (up to 80% by weight). The XMC uses filament winding techniques to incorporate continuous fiberglass strands. The polyethyl-

Table 6-19. Properties of Casting and Glass-Reinforced

		ASTM test method	Cast polyester		Glass reinforced polyester				
			Rigid	Flexible	Preformed chopped roving	Premix chopped glass	Woven cloth	Sheet molding compound	Low shrink compound
Processing	1. Molding qualities	—	—	—	Excellent	Excellent	Hand lay up	Good	Good
	2. Compression molding temp., °F.	—	—	—	170-320	280-350	Rm. temp.-250	270-350	270-330
	3. Comp. molding pressure, p.s.i.	—	—	—	250-2000	500-2000	0.300	300-1200	>400
	4. Injection molding temp., °F.	—	—	—	—	—	—	—	270-330
	5. Injection molding pressure, p.s.i.	—	—	—	—	—	—	—	2000
	6. Compression ratio	—	—	—	—	—	—	—	1.0
	7. Mold (linear) shrinkage, in./in.	—	—	—	0.000-0.002	0.001-0.012	0.000-0.002	0.001-0.004	0.0001
	8. Specific gravity (density)	D792	1.10-1.46	1.01-1.20	1.35-2.30	1.65-2.30	1.50-2.10	1.65-2.60	1.60-2.40
	9. Specific volume, cu. in./lb.	D792	25.2-19.0	27.4-23.0	20.5-13.9	—	18.5-13.2	—	—
	10. Machining qualities	—	Good	Fair	Good	Good	Good	Good	Good to excel.
Mechanical	11. Tensile strength, p.s.i.	D638	6000-13000	500-3000	15000-30000	3000-10000	30000-50000	8000-20000	4500-20000
	12. Elongation, %	D638	<5.0	40.0-310.0	0.5-5.0	0.5	0.5-2.0	—	—
	13. Tensile elastic modulus, 10^5 p.s.i.	D638	3.0-6.4	—	8.0-20.0	10.0-25.0	15.0-45.0	—	10.0-25.0
	14. Compressive strength, p.s.i.	D695	13000-30000	—	15000-30000	20000-30000	25000-50000	15000-30000	15000-30000
	15. Flexural yield strength, p.s.i.	D790	8500-23000	—	10000-40000	7000-20000	40000-80000	10000-36000	9000-35000
	16. Impact strength, ft. lb./in. of notch (½ × ½ in. notched bar, Izod test)	D256	0.2-0.4	>7.0	2.0-20.0	1.5-16.0	5.0-30.0	7.0-22.0	2.5-15.0
	17. Hardness, Rockwell	D785	M70-M115 50-75 (Barcol)	84-94 (Shore D)	50-80 (Barcol)	50-80 (Barcol)	60-80 (Barcol)	50-70 (Barcol)	40-70 (Barcol)
	18. Flexural elastic modulus, p.s.i. × 10^5	D790	—	—	10-30	10-20	—	10.0-22.0	10.0-25.0
	19. Compressive modulus, p.s.i. × 10^5	D695	—	—	—	—	—	—	—
Thermal	20. Thermal conductivity, 10^{-4} cal./sec./sq. cm./1 (°C./cm.)	C177	4.0	—	—	10.0-16.0	—	—	—
	21. Specific heat, cal./°C./gm.	—	—	—	—	0.25	—	—	—
	22. Thermal expansion, 10^{-5} in./in./°C.	D696	5.5-10.0	—	2.0-5.0	2.0-3.3	1.5-3.0	2.0	0.6-3.0
	23. Resistance to heat, °F. (continuous)	—	250	250	300-350	300-350	300-350	300-400	300
	24. Deflection temp., °F. at 264 p.s.i. fiber stress	D648	140-400	—	>400	>400	>400	375-500	375-500
	at 66 p.s.i. fiber stress		—	—	—	—	—	—	—
Electrical	25. Volume resistivity, ohm-cm. (50% RH and 23° C.)	D257	10^{15}	—	10^{14}	10^{12}-10^{15}	10^{14}	10^{14}-10^{15}	$10^{12} \times 10^{14}$
	26. Dielectric strength, short-time, ⅛-in. thickness, volts/mil	D149	380-500	250-400	350-500	345-420	350-500	380-450	380-450
	27. Dielectric strength, step-by-step, ⅛-in. thickness, volts/mil	D149	280-240	277	300-450	275-390	300-400	350-400	300-400
	28. Dielectric constant, 60 cyc.	D150	3.00-4.36	4.4-8.1	3.8-6.0	5.3-7.3	4.1-5.5	4.4-6.3	4.5-6.0
	29. Dielectric constant, 10^3 cyc.	D150	2.8-5.2	4.5-7.1	4.0-6.0	4.68	4.2-6.0	4.4-6.1	4.5-6.0
	30. Dielectric constant, 10^6 cyc.	D150	2.8-4.1	4.1-5.9	3.5-5.5	5.2-6.4	4.0-5.5	4.2-5.8	4.2-6.0
	31. Dissipation (power) factor, 60 cyc.	D150	0.003-0.028	0.026-0.310	0.01-0.04	0.011-0.041	0.01-0.04	0.007-0.021	0.1-0.2
	32. Dissipation (power) factor, 10^3 cyc.	D150	0.005-0.025	0.016-0.050	0.01-0.05	—	0.01-0.06	0.007-0.015	0.1-0.2
	33. Dissipation (power) factor, 10^6 cyc.	D150	0.006-0.026	0.023-0.060	0.01-0.03	0.008-0.002	0.01-0.03	0.016-0.024	0.01-0.02
	34. Arc resistance, sec.	D495	125	135	120-180	120-240	60-120	121-202	—
Optical	35. Refractive index, n_D	D542	1.523-1.570	1.537-1.550	—	—	—	—	—
	36. Clarity	—	Transparent to opaque		Transl. to opaq.	Opaque	Transl. to opaq.	Opaque	Opaque
	37. Transmittance, %	—	—		—	—	—	—	—
	38. Haze, %	—	—		—	—	—	—	—
Resistance characteristics	39. Water absorp., 24 hr., ⅛-in. thick., %	D570	0.15-0.60	0.50-2.50	0.01-1.00	0.06-0.28	0.05-0.50	0.10-0.15	0.01-0.25
	40. Flammability[a] Burning rate, in./min.	D635			0.7-2.0			—	—
	AEB, in./ATB, sec.[b]		—	—		<0.16 (FR gr.)		—	—
	41. Effect of sunlight	—	Yellows slightly	—	Slight	Depends on pigmentation	Slight	Slight	Slight
	42. Effect of weak acids	D543	Ortho & gen. purp. iso resis. some at RT:chem. resis. iso resis. to 170° F.; Bis. & HBPA to 210° F.; chlor. anhyd. resis. most at RT, some to 350° F.					None	None
	43. Effect of strong acids	D543	Ortho & gen. purp. not resis. strong or oxidizing; chem. resis. iso resis. strong to 170° F., not resis. low-concen. oxidizing at RT; Bis & HBPA resis. strong to 210° F., oxidizing to 170° F., chlor. anhyd. resis. most at RT, some to 250° F.					—	Attacked by high conc.
	44. Effect of weak alkalies	D543	Ortho & gen. purp. resis. some at RT; chem. resis. iso to 170° F. and Bis. & HBPA to 210° F.; chlor. anhyd. resis. some to 210° F.					Slight	Slight
	45. Effect of strong alkalies	D543	Ortho & gen. purp. iso & chem. resis. iso not resis.; chem. resis. Bis. resis. to 170° F.; chlor. anhyd. resis. some to 170° F.					Attacked	Attacked
	46. Effect of organic solvents	D543	Ortho & gen. purp. iso not resis.; chem. resis. iso resis. some & Bis & HBPA most at RT; chlor. anhyd. resis. most at RT, some to 200° F. Effect of salts: Ortho & gen. purp. iso resis. some at RT; chem. resis. iso resis. to 170° F., Bis & HBPA to 120° F.; chlor. anhyd. resis. most to 250° F.					Slight	Slight to considerable

a—This numerical flame spread rating is not intended to reflect hazards presented by this or any other material under actual fire conditions.
b—AEB = Avg. extent of burning in inches; ATB = Avg. time of burning in seconds.
*Reprinted by permission from *Modern Plastics Encyclopedia*, McGraw-Hill, Inc.

Polyesters, and Alkyds.*

Alkyd				
Granular and putty types, mineral filled	Asbestos filled	Glass filled	Synthetic fiber filled	Vinyl ester
Excellent	Excellent	Excellent	Excellent	Excellent
270-350	280-320	290-350	280-350	270-300
100-4000	500-2000	1000-6000	500-3000	100-200
280-390	330-380	280-380	280-380	—
2000-20000	10000-20000	2500-20000	2000-20000	—
1.8-2.5	2.0-2.5	>1.0-11.0	>1.0-5.0	—
0.003-0.010	0.004-0.007	0.001-0.010	0.005-0.0010	—
1.60-2.30	1.65-2.20	2.03-2.33	1.24-2.10	1.03-1.083
17.3-12.0	16.8	13.2	22.4-13.2	—
Poor to fair	Poor to fair	Poor to fair	Fair to excellent	—
3000-9000	4500-9000	4000-9500	4500-8000	10500
—	—	—	—	3.5-5.5
5.0-30.0	20-30	20.0-28.0	20.0	—
12000-38000	22500	15000-36000	20000-30000	—
6000-17000	8000-10000	8500-26000	10000-20000	17800-18400
0.30-0.50	0.45-0.50	0.50-16.0	0.50-4.50	—
60-70 (Barcol) 98 (E scale)	M99	55-80 (Barcol) 95 (E scale)	95 (M scale) 76 (E scale)	—
20	20-30	20.0	17.0	4.7-4.9
20-30	—	—	—	—
12.2-25.0	—	15.0-25.0	7.0-16.5	—
0.25	—	0.25	—	—
2.0-5.0	—	1.5-3.3	3.0-5.5	—
300-450	450	450	300-430	200-220
350-500	315	400-500	245-430	—
—	—	—	—	185-215
10^{13}-10^{15}	6.6×10^{8}	6×10^{12} 1.5×10^{15}	10^{8}-10^{16}	2.6×10^{15}
350-450	380	250-530	365-500	445-541
300-375	290	200-400	325-450	—
5.1-7.5	—	5.9-7.3	3.8-5.0	4.38-4.88
5.0-6.4	5.2	5.4-7.4	3.7-4.9	4.35-4.72
4.6-6.1	4.5	5.2-6.8	3.6-4.7	4.15-4.27
0.009-0.06	—	0.007-0.041	0.012-0.026	0.0097-0.0210
0.007-0.040	0.11	0.007-0.04	0.001-0.03	0.0055-0.0210
0.006-0.040	0.04-0.06	0.008-0.023	0.010-0.016	0.013-0.019
75-240	138	130-420	85-185	—
—	—	—	—	—
Opaque	Opaque	Opaque	Opaque	—
—	—	—	—	—
—	—	—	—	—
0.05-0.50	0.14	0.03-0.5	0.05-0.20	0.2
—	—	—	—	—
<0.16 (FR gr.)		—	—	—
None	None	None	None	—
None	None	Poor to fair	None	Good
Attacked	Slight	Poor to fair	Slight	Good
Attacked	None	Poor to fair	None	Good
Decomposes	Slight	Poor to fair	Slight	Good
Fair to good	None	Fair to good	Fair to good	Fair to good

Table 6-20. Typical SMC Properties.*

Property	Low-Profile Grade	Structural Grade
Tensile strength (psi)	14,000	16,000
Flexural strength (psi)	25,000	30,000
Flexural modulus (psi)	1.4×10^6	1.4×10^6
Impact strength		
Izod, notched (ft-lb/in.)	10–12	12–15
Surface finish (μin.)	200	600
Shrinkage (in./in.)	0–0.001	0.002–0.0025

*Keown, 1971.

Table 6-21. TMC: Comparison with SMC and BMC.*

Property	TMC	SMC	BMC
Thickness	up to 2 in.	up to $\frac{1}{4}$ in.	Bulk form
Form	Continuous	Continuous	Batch process
Filler concentration	High	Limited	High
Mechanical properties	High	High	Low
Fiberglass degradation	Low	Low	High
Molding methods	Compression, injection	Compression	Compression, injection

*Reprinted by permission from *Plastics World*, July 1977.

ene film-wrapped, uncured composite is removed from the winding mandrel for further molding to shape.

The HMC utilizes chopped glass strands and is further processed like the XMC. Major applications for these materials are in automotive structures where high strength-to-weight ratios are in greatest demand. Examples include doorbeams, transmission and radiator supports and wheels.

SMC and BMC parts usually have smooth, glossy surfaces with improved flexural strength, modulus, and compressive strength, both at room and elevated temperatures. Tensile strength is normally unaffected at these temperatures.

Most polyester resin is used in reinforced plastics. Glass fiber is by far the most important reinforcement. The glass takes the form of woven cloth, yarn, surfacing mat, continuous roving, and chopped rovings and strand. The techniques for fabricating fiberglass-reinforced polyester parts are described in more detail in Chapters 8 and 17.

The most significant feature of glass-reinforced polyesters is the high tensile and flexural strength obtainable, which in many constructions and on an equal weight basis can far exceed the values for steel. Glass filaments have tensile strengths ranging from 180,000 psi to 300,000 psi. Rigid-cast polyester resin usually has a tensile strength in the order of 10,000 psi. In a laminate, the resin serves mainly to transmit the load from fiber to fiber and to protect the fiber from damage through abrasion and shear, thereby preserving the inherent strength of the filament. Laminated, unidirectional, continuous composite when tested parallel to the filament direction shows tensile strengths as high

Table 6-22. Price/Performance Characteristics of SPMC Compared to Other Selected Materials.*

	Pre-Molding Properties				Post-Molding Properties					
Materials[a]	Cost	Material Handling	General Moldability	Shrinkage Control	Mechanical Strength	Electrical Resistance	Heat Resistance	H_2O Absorption	Specific Gravity	Pigmentation
Phenolic, heat resistant	++	+	–	–	–	+	++	+	0	=
Polyester, SPMC	+	+	0	+	+	++	++	+	–	+
Polyester, granular alkyd	+	+	–	–	=	++	+	+	–	+
Polyester, BMC	+	=	=	++	++	++	++	=	–	–
Melamine	+	+	–	=	–	++	+	–	0	=
PPO, modified-glass filled	–	+	0	0	+	–	–	++	++	+
Nylon, 6-6 type glass filled	–	+	+	0	+	–	0	–	+	+
Polycarbonate, glass filled	=	+	0	0	++	–	–	++	+	++
PBT, glass filled	–	+	+	0	+	+	0	++	0	+

++ excellent + very good 0 average – below average = poor
[a] All systems are UL 94 V-O rated grades.
*Shreve, 1977.
Reprinted by permission from *Plastics Design & Processing*. Copyright 1977, Lake Publishing Corp., Libertyville, IL.

Table 6-23. SPMC After-Molded Properties.*

Property	Typical Values
Specific gravity	1.8–2.0
Mold shrinkage, mil/in.	1–6
Tensile strength, 10^3 psi	5–8
Flexural strength, 10^3 psi	10–12
Flexural modulus, 10^3 psi	1.3–2.0
Izod Impact strength, ft-lb/in. notched	1.0–2.5
Deflection temperature, 264 psi, °F	450–500+
Maximum continuous use temp, (Est.), °F	266–356
U.L. 94 flammability rating	HB–VO
Water absorption, 24 hr., %	0.1–0.3
Dielectric strength, S.T., v/mil	350–600
Arc resistance, sec.	185–195
Tracking index, IEC, volts	600+

*Shreve, 1977.
Reprinted by permission from *Plastics Design & Processing*. Copyright 1977, Lake Publishing Corp., Libertyville, IL.

Fig. 6-33. This fire boat calls for resins that cure at room temperature in either wet lay-up or spray-up production of large parts. Resins must meet rigid Coast Guard and Navy specifications and strict fire ratings. (*Courtesy: Silmar Division, Vistron Corporation*)

as 150,000 psi, approaching the strength of the reinforcing filament itself. Tested at right angles, the tensile strength will be as low as that of the resin. Intermediate values (40,000–60,000 psi) are obtained with laminated woven strand or roving.

The phenomenal strength of bonded unidirectional filament is strikingly demonstrated in filament-wound structures. Pressure tanks made by this technique weigh far less than steel tanks of equal bursting pressure.

Laminated cloth structures are used for weight saving with no loss in strength in many applications such as boat hulls (Figure 6-33), car and truck bodies, radomes, sheeting and roofing, ducting, furniture, swimming pools, trays, and sports equipment.

Other types of reinforcing fibers such as sisal, asbestos, or synthetic fibers are seldom used in open-mold fabrication but are frequently used together with glass fiber in formulating BMCs and SMCs.

Vinyl Esters

The class of corrosion-resistant polyester resins commonly referred to as "vinyl esters" are, chemically, the reaction products of epoxy resins and ethylenically unsaturated monocarboxylic acids with styrene added as a coreactant (see Figure 6-34). A wide variety of vinyl ester resins can

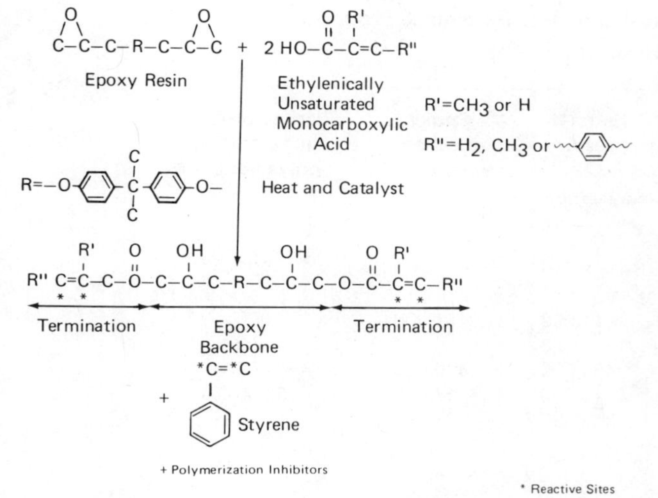

Fig. 6-34. Vinyl ester resins—chemical description. (*Courtesy: Shell Chemical Co.*, manufacturers of Epocryl vinyl ester resins)

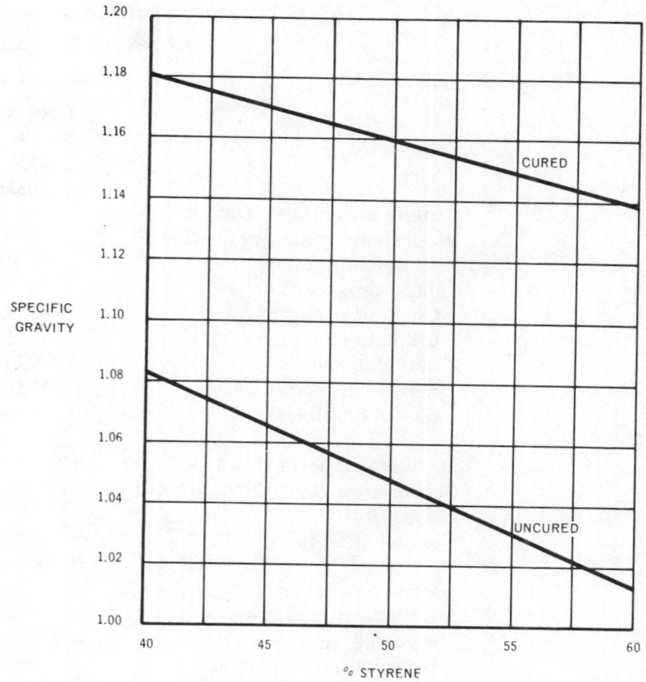

Fig. 6-35. Specific gravity of vinyl ester resins at various styrene dilution levels—temperature = 72°F. (*Courtesy: Dow Chemical USA*)

be prepared by simply altering the epoxy resin and/or the monocarboxylic acid. The typical crosslinking monomer is styrene. Figure 6-35 shows the effect of styrene content on the specific gravity of a vinyl ester resin.

Three types of vinyl ester resins are commercially available: (1) "conventional" or "resilient" grades based on bisphenol-A type epoxies; (2) "fire-retardant" grades based on tetrabromobisphenol-A epoxies; and (3) "high heat" grades based on novolac epoxies. The need for maximum corrosion resistance has limited the unsaturated monocarboxylic acid to methacrylic acid. The methacrylate termination of the epoxy backbone is particularly desirable over other vinyl ester linkages because of the shielding effect that the methyl group has on the adjacent ester group.

A new development in this family of resins is based on a bisphenol/fumarate backbone with the acrylate termination obtained through a urethane linkage. This resin combines the better characteristics of a bisphenol/fumarate polyester with those of the vinyl ester family.

Table 6-24 compares some selected properties of a conventional, a high heat, and an epoxy novolac vinyl ester with a bisphenol A/fumarate polyester. The vinyl esters perform well in acidic or alkaline environments at temperatures up to their heat-deflection temperature (approximately 121°C, 250°F). Their curing is more predictable and complete compared to other polyesters. They wet and adhere extremely well to glass and provide resiliency without significant loss of modulus in fibrous glass composite systems.

The biggest use for vinyl ester resins has been in reinforced plastics equipment such as scrubbers, pipe and tanks, and chemically resistant coatings such as tank and flue stack linings (see Figure 6-36).

Allyl Resins and Monomers

The allyl resins are thermosetting polyester materials which retain their desirable physical and electrical properties on prolonged exposure to severe environmental conditions, and can be molded into complex and intricate components.

The resins are used not only in molding compounds, but in pre-impregnated glass cloth for reinforced plastics applications, and as sealants, insulating coatings, and decorative laminates. Allyl monomers are employed widely as crosslinking agents in other polyester systems.

The principal allylics are the diallyl esters of phthalic acid (DAP) and isophthalic acids (DAIP).

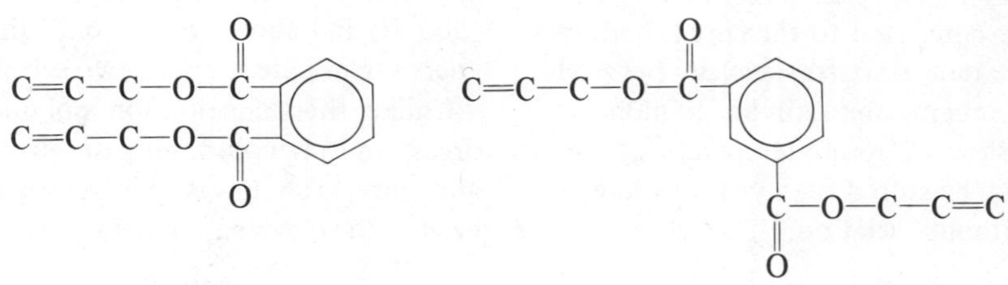

Table 6-24. A Comparison of Mechanical and Heat Resistance Properties of Several Corrosion-Resistant Resins.*

	Epocryl resin 480 (high heat)	Epocryl resin 322 (conventional)	Epoxy novolac vinyl ester	Bisphenol-A fumarate polyester
Casting properties[1], 1/8" (Room temp. gel. + 2 hrs. at 300° F)				
Heat deflection temp., °F	250	210	250	225
Tensile properties at 77° F				
Ultimate strength, psi	12,200	11,000	11,300	7,800
Ultimate elongation, %	5.3	5.8	3.4	2.5
Modulus, psi	470,000	450,000	470,000	440,000
Flexural strength, psi	17,000	17,000	17,000	15,000
Barcol hardness	30-35	30-35	45-50	
Laminate properties[1] (ASTM C-581) (Room temp. cure + 2 hrs. at 300° F)				
Glass content, %	25	25	25	25
Flexural strength, psi, 77° F	17,800	18,000	16,800	15,000
Flexural modulus, psi, 77° F	1,000,000	900,000	850,000	750,000
Tensile properties, 77° F				
Ultimate strength, psi	15,800	15,100	14,800	11,200
Modulus, psi	1,300,000	1,100,000	1,250,000	950,000
Ultimate elongation, %	1.8	2.1	1.2	1.0
Castings 1/8"				
Flexural strength, psi, 77° F				
Initial strength	17,000	17,000	16,800	15,000
After 500 hrs at 300° F	17,800	16,000	17,500	—
After 500 hrs at 400° F	14,500	14,000	14,000	—
Modulus, psi, 77° F				
Initial strength	475,000	450,000	475,000	440,000
After 500 hrs at 300° F	480,000	425,000	480,000	—
After 500 hrs at 400° F	465,000	445,000	460,000	—
Wt. loss, %				
After 500 hrs at 300° F	0.5	2.4	0.6	—
After 500 hrs at 400° F	2.0	6.0	2.5	—
Laminates (ASTM C-581 type)				
Flexural strength, psi				
at 77° F	17,800	18,200	16,500	15,000
at 250° F	15,900	5,000	14,200	6,000
Flexural modulus, psi				
at 77° F	910,000	800,000	850,000	750,000
at 250° F	550,000	190,000	540,000	230,000

[1] Cure conditions: 100 parts by weight resin, 2.0% MEKP (60%), 0.2 CoN, 2 days at 77°F + 2 hours at 300°F.
*Courtesy: Shell Chemical Co., manufacturers of Epocryl resins.

They are available commercially as monomers or as partially polymerized prepolymers. Both monomers and polymers—or mixtures—are converted to the cured thermoset stage by heat and/or free-radical sources, usually benzoyl peroxide or tert-butyl perbenzoate. Allylic homopolymerization is extremely slow at room temperature, and catalyzed compositions may be stored for a year or more at ambient temperatures without gelation. The cure rate increases rapidly with temperature, however, and fast cures are obtained at temperatures of approximately 149°C (300°F) and above. Figure 6-37 shows the change in hardeners versus cure time for two select diallyl systems.

Unlike the condensation polymerization of melamines, ureas, and phenolics, no gases evolve in allylic crosslinking and cure. The low vapor pressures of allylic monomers, even at the elevated temperatures required for cure, coupled

Fig. 6-36. Epocryl-vinyl-ester-lined scrubber stack illustrates corrosion resistance of the polymer. (*Courtesy: Shell Chemical Co.*, manufacturers of Epocryl resins)

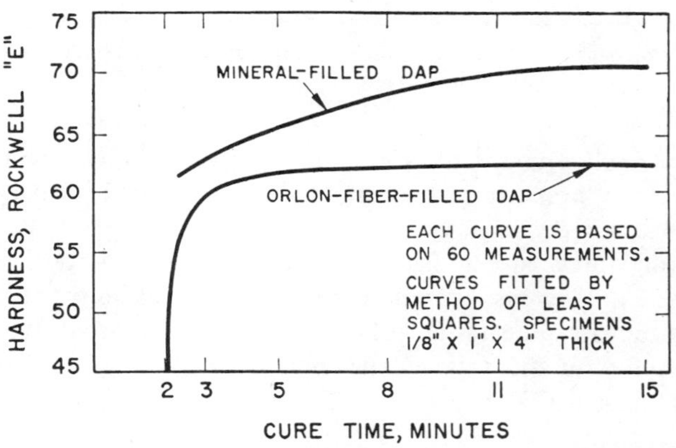

Fig. 6-37. Effect of cure time on hardness of Orlon-fiber-filled and glass-titanium-dioxide-filled diallyl phthalates. (Keller et al., 1959)

with excellent wetting of reinforcing fillers and fibers, low mold shrinkage, and lack of low heat distortion permit rapid cures and defect-free parts. Table 6-25 shows typical properties of allyl prepolymers. In general, parts made from these resins yield excellent dimensional stability with low mold and negligible post-mold shrinkage. They have excellent physical and electrical properties, arc resistance, corrosion resistance, and withstand heat, cold, moisture, sunlight, acids, alkalis, and fungi to a high degree.

Thermosetting molding compounds comprise the largest area of utility for allylic resins. Molding performance is excellent, yielding rigid pieces at mold temperatures. Diallylorthophthalate compounds are designed for continuous use at temperatures up to approximately 177°C (350°F), while the isophthalates are useful to approximately 232°C (450°F). Electrical properties are excellent—low loss factor, high surface and volume resistivity, and high arc resistance—and are retained under conditions of high temperature and humidity.

Table 6-25. Properties of Allyl Prepolymers.*

Property	Diallyl Phthalate	Diallyl Isophthalate
Specific Gravity	1.27	1.26
Rockwell Hardness, (M)	115	120
Heat Distort. (°C.)	155	240
Izod Impact	0.3	0.3
Compressive Str. (PSI)	24,000	24,000
Flex. Str. (psi)	9,000	8,000
Flex. Modulus $\times 10^{-6}$ (psi)	0.5	0.5
Tensile Str. (psi)	4,000	4,000
Volume Resist. (ohm-cm)	10^{16}	10^{17}
Surface Resist. (ohms)	10^{15}	10^{13}
Diel. Str. (v/mil)	450	420
Diel. Constant (10^3 cycles)	3.4	3.2
Dissip. Factor (10^3 cycles)	.011	.009
Arc Resistance	118	125
Water Absorp. 24 hr (%)	0.1	0.1
Thermal Coef. of Expan. /°C$\times 10^{-5}$	2.96	2.00

*Beacham, 1967. Reprinted by permission from *Plastics Design & Processing*. Copyright 1967, Lake Publishing Corporation, Libertyville, Ill.

Molding compounds are prepared by blending together prepolymer resin, additional allyl monomer, catalyst, filler and pigments, and other compounding ingredients. Fibrous fillers such as glass, acrylic, polyester, asbestos, and cellulosic fibers are commonly used. Glass fiber produces moldings with the best all-around properties, with longer fibers yielding greater impact strength. Particulate mineral fillers such as calcium carbonate, clay, calcium silicate, and barytes are used for flow control and cost reduction. Table 6-26 shows typical properties of various types of allylic molding compounds.

These molding compounds can be compression, transfer, or screw-injection molded. Molding cycle times range from $\frac{1}{2}$ to 4 minutes at temperatures of 160–168°C (320–335°F). The high heat-deflection temperature allows parts to be removed without deformation using standard knockout pins. Pressures vary between 500–10,000 psi, depending on the molding technique and compound being run. Because no volatile by-products are generated, mold cycling or breathing is not required.

Typical applications for DAP moldings include insulators, connectors, circuit boards and terminals. Other end uses include detonators and caps, tool and appliance handles, pump impellers, parts for medical and biological sterilizers, missile components, camera and equipment cases, and corrosion-resistant parts for chemical and food processing equipment.

The use of glass cloth and roving pre-impregnated with allylic resins has increased markedly. Storage stability, ease

Table 6-26. Typical Properties of Allylic Molding Compounds with Various Fillers.*

Property	Diallyl Phthalate					Diallyl Isophthalate		
	Acrylic	Polyester	Mineral	Short Glass	Long Glass	Mineral	Short Glass	Long Glass
Flex. Str. (psi)	10,000	11,500	11,400	12,300	18,000	13,000	14,000	19,000
Flex. Mod. $\times 10^{-6}$ (psi)	0.71	0.64	1.2	1.2	1.2	1.2	1.3	1.4
Compres. Str. (psi)	30,000	30,000	23,300	20,300	30,000	25,000	30,000	29,000
Tensile Str. (psi)	6,000	5,000	5,500	6,000	7,000	8,000	8,000	9,000
Izod Impact (ft-lb)	1.2	4.5	0.33	0.42	6-16	0.5	1	6
Heat Distort. Temp. (°F)	265	290	335	335	400	450	600	600
Mold Shrink. (in./in.)	.009	.010	.006	.006	.002	.006	.002	.002
Post Mold Shrink. (in./in.)	.001	.0006	.001	.001	.0007	.0007	.0002	.0002
Specific Gravity	1.31-1.45	1.39-1.62	1.55-1.65	1.62	1.62	1.55-1.65	1.68	1.63
Rockwell Hardness (M)	108	105	100	103	103	106	108	105
Diel. Constant (10^3 cycles)	3.7	3.8	5.2	4.6	4.3	5.1	4.4	4.0
(10^6 cycles)	3.3	3.4	4.5	4.4	4.2	4.5	4.1	3.9
Dissip. Factor (10^3 cycles)	.020	.006	.04	.009	.004	.04	.004	.004
(10^6 cycles)	.015	.012	.04	.007	.010	.04	.010	.010
Surface Resist. (ohm)	10^{15}	10^{15}	10^{15}	10^{16}	10^{13}	10^{13}	10^{13}	10^{16}
Vol. Resist. (ohm-cm)	10^{14}	10^{14}	10^{14}	10^{15}	10^{14}	10^{14}	10^{14}	10^{13}
Heat Resist., Contin. (°F)	300	300	400	400	400	500	500	500

*Beacham, 1967. Reprinted by permission from *Plastics Design & Processing*. Copyright 1967, Lake Publishing Corporation, Libertyville, Ill.

Table 6-27. Properties of Allylic Glass Cloth Laminates.*

Property	Diallyl Phthalate Prepolymer[1]	Diallyl Isophthalate Prepolymer[2]
Flex. Str. (psi)		
Ambient	70,000	68,000
Wet, 30 days	61,000	63,000
300°F.	45,000	46,000
400°F.	19,500	34,000
Flex. Mod. (psi) $\times 10^{-6}$		
Ambient	3.0	3.1
Wet, 30 days	3.0	3.3
300°F.	2.6	3.8
400°F.	—	2.7
Tensile Str. (psi)	47,000	50,500
Compres. Str. (psi)	41,000	49,000
Diel. Constant		
10^3 cycles	4.9	4.8
10^6 cycles	4.5	4.5
10^9 cycles	4.2	4.3
Loss Factor		
10^3 cycles	.040	.058
10^6 cycles	.060	.038
10^9 cycles	.024	.018

Notes: [1]Dapon 35, FMC Corporation, N.Y.C.
[2]Dapon M, FMC Corporation, N.Y.C.

*Beacham, 1967. Reprinted by permission from *Plastics Design & Processing*. Copyright 1967, Lake Publishing Corporation, Libertyville Ill.

of handling and fabrication, combined with outstanding electrical properties which are retained at elevated temperatures, are the incentives for prepreg use. Flame-retardant formulations are available with equivalent properties, as well as newer systems with exceptional chemical resistance.

Laminate preparation using allyls can be achieved using vacuum-bag and matched-die molding methods (described in Chapter 17). Typical cure schedules are 20–35 minutes at temperatures of 93–177°C (200–350°F). Some selected properties of glass-reinforced DAP laminates are shown in Table 6-27. Figure 6-38 shows the heat aging resistance of DAP laminates as retention of flexural modulus.

The specific characteristics of these prepregs have led to applications in the manufacture of tubing, ducting, radomes, junction boxes, and aircraft and missile parts where resistance to heat, weather, and chemicals and/or excellent electrical characteristics are needed.

DAP prepolymer is a versatile surfacing medium for a wide range of decorative laminate constructions. The resin imparts a durable and attractive surface to such materials as hard board, particle-board, and plywood. These laminates are used by the furniture and paneling industries.

Allyl sealants are gaining increased acceptance in the vacuum impregnation of porous, nonferrous metal castings to replace sodium silicate, tung oil, and polyesters. Pores of aluminum, magnesium, copper, and copper-bearing alloys as well as ceramic and other compositions are currently being sealed with these resins. They offer advantages over polyester types in their freedom from styrene odor, excellent pot-life, low toxicity, low evaporation losses dur-

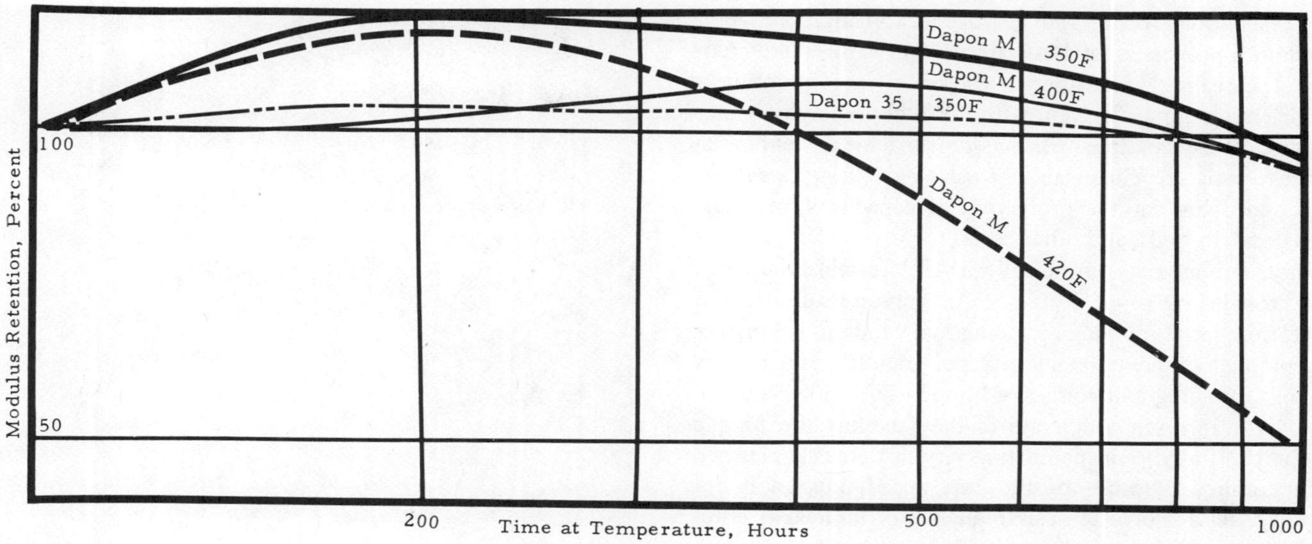

Fig. 6-38. Heat aging flexural modulus properties of DAP laminates tested at room temperature—181 E glass cloth, Volan A finish. (*Courtesy: FMC Corporation*)

Table 6-28. Properties of Polymers Produced from Allyl Diglycol Carbonate Monomer.*

(All tests in accordance with ASTM Standards and Procedures, if applicable.)

Property	Value
Specific Gravity, 25°C	1.32[1]
Refractive Index, $\eta^D/_{20}$	1.50[2]
Dispersion Factor	57.8
Light Transmission, ¼" Thickness, %	89-92[3]
Hardness, Rockwell	M95-M100
Modulus of Elasticity in Flexure, 10^5 psi 25°C. (77°F.)	2.5-3.3
Tensile Strength, psi	5000-6000
Modulus of Elasticity in Tension, 10^5 psi	3.0
Compressive Strength, Ultimate, psi	22,500
Modulus of Elasticity in Compression, 10^5 psi	3.0
Impact Strength, 25°C., ft. lbs. per inch Izod, notched	0.2-0.4
Izod, unnotched	2.3
Maximum Recommended Operating Temperature Under No Load Continuous Service	100°C. (212°F.)
Intermittent (1 hr. duration)	150°C. (302°F.)
Specific Heat, cal./g.°C	0.55
Thermal Expansion, Linear Coeff. per °C., −40 to 25°C. (−40 to 77°F.)	8.1 x 10^{-5}
25 to 75°C. (77 to 167°F.)	11.4 x 10^{-5}
75 to 125°C. (167 to 257°F.)	14.3 x 10^{-5}
Thermal Conductivity Btu/hr./ft.²/in./°F	1.45
Burning Rate, in. per min.	0.35
Abrasion Resistance Modified Taber, times methyl methacrylate	30-40
Modified Falling Emery, times methyl methacrylate	9-12
times glass	0.9-1.2
Warpage	0.00
Water Absorption, 24 hrs./25°C	0.2[4]

(1) Glass: 2.5
(2) Glass: 1.52
(3) Glass: 92%; Acrylics: 90-92%
(4) Cast phenolics: 0.3 to 0.4.

SPECIFICATIONS

Plastics made from monomer meet the following specifications:
JAN-P-77, type G-3 (allyl-based resin);
MIL-P-77A, type G-3 (allyl-based resin);
MIL-P-8257A; USAF-12040.
L-P-516a (Type G-3)

*Courtesy: PPG Industries, Inc.

ing evacuation cycles, and absence of oozing. The good insulating properties of DAP are used in varnish-like coatings for sealing and other purposes by incorporating a sufficient amount of solvent with the allyl prepolymer compound. These coatings, when cured, offer excellent resistance to moisture, chemicals, corrosive gases, fungi, weathering, and aging for the protection of capacitors, resistors, transformers, coils, and other parts.

Allylic monomers, principally DAP, are widely used in the crosslinking of polyesters. DAP polyesters are used in preform or mat molding including mat binding, laminating by prepreg or lay-up, and in rope, granular, and premix "gunk" molding compounds. Although DAP polyesters do not cure at room temperature, they exhibit the unique property of allylics in permitting very fast cures at elevated temperatures. Because of the cure speed plus good hot rigidity, allylic polyesters can frequently be molded at faster rates than those crosslinked with styrene. Better moldability, including lower mold shrinkage, allows greater freedom of formulation. In addition, the monomer is used as an extender/carrier for the catalysts used with conventional polyesters containing styrene and as a vehicle for pigments and other additives.

A number of other allyl esters have seen some commercial utility, primarily because of the readiness with which they form fluid, soluble prepolymers. These prepolymers can then be cured with heat and peroxide catalysts and can copolymerize with styrene and other vinyl monomers. One of the allyl esters to be produced on a relatively large scale is diethylene glycol bis(allyl carbonate) (CR-39).

$$HOCH_2CH_2OCH_2CH_2OH + COCl_2 \longrightarrow ClC(=O)OCH_2CH_2OCH_2CH_2OC(=O)Cl$$

$$ClC(=O)OCH_2CH_2OCH_2CH_2OC(=O)Cl + CH_2{=}CHCH_2OH \longrightarrow \begin{matrix} CH_2{=}CHCH_2OC(=O){-}OCH_2CH_2 \\ | \\ O \\ | \\ CH_2{=}CHCH_2OC(=O){-}OCH_2CH_2 \end{matrix}$$

This monomer has the ability to form clear castings having high scratch resistance, good optical properties, and a high heat-deflection temperature. Like the other allyls, its cure is relatively slow. It is primarily used as a casting resin where greatest advantage can be taken of its good optical properties and only infrequently used in composite structures. Table 6-28 shows typical properties of the polymer, and Figure 6-39 shows some common applications.

Another allylic monomer then has been found useful is triallyl cyanurate.

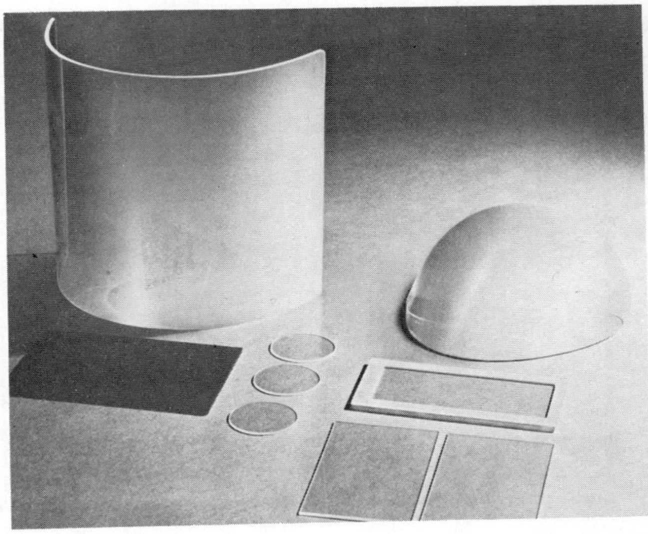

Fig. 6-39. The versatility of sheets cast from allyl diglycol carbonate monomer is shown by these typical products: clockwise starting at upper left, full-face lens for fire-fighter's hood, lens for gas mask, instrument gauge cover (in panel), two cover plates for welding hoods, three lenses for welder's goggles, and gold-coated cover plate for welding hood. (*Courtesy: PPG Industries, Inc.*)

$$HCN + Cl_2 \longrightarrow ClCN + HCl$$

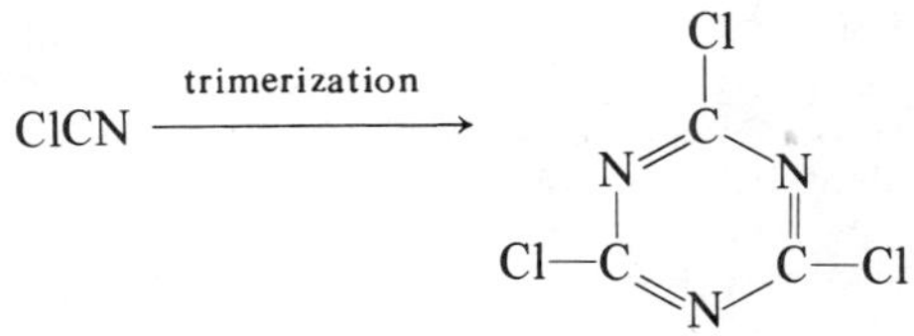

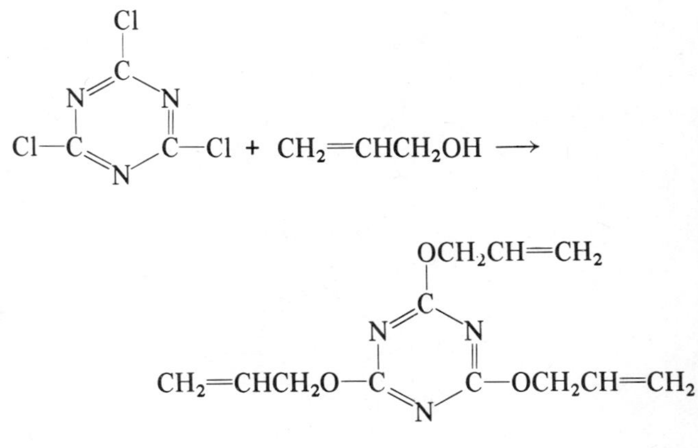

When used as a crosslinking agent in unsaturated polyester resin systems, triallyl cyanurate causes an increase in the overall system's heat-deflection temperature. Figure 6-40 shows the effect of heat aging on the flexural strength of a triallyl cyanurate laminate.

Recent use of this monomer has dropped so significantly that production is reported to have been terminated.

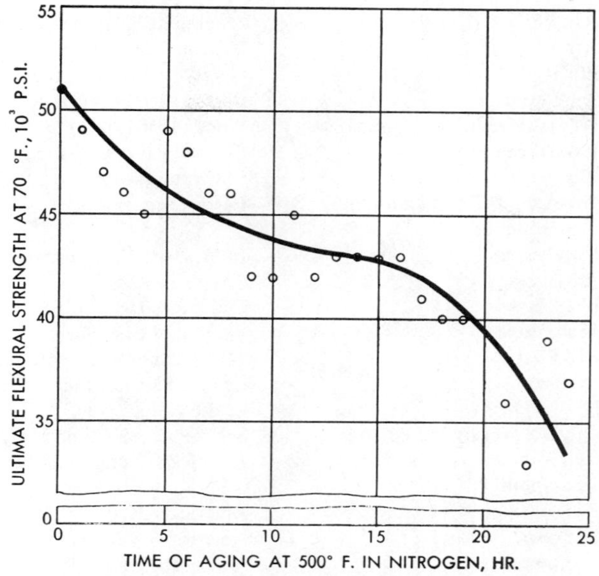

Fig. 6-40. Effect of aging on flexural strength at 75°F of typical triallyl cyanurate alkyd glass-reinforced laminate. (Wahl and Preston, 1957. Reprinted by permission from *Modern Plastics*, McGraw-Hill, Inc.)

Trade Names

Acpal	PEs resin	Midwest Manufacturing Co.
Actoglas	Glass-reinforced PEs	Natvar Corporation
Advaco	Isophthalate-modified PEs	Advance Coatings Co.
Alberit	PEs Molding Compounds	Canadian Hoechst, Ltd.
Alkanex	Polyester Coatings	General Electric Co.
Alkonex	PEs Materials	General Electric Co.
Alkydal	Alkyd resins	Mobay Chemical Co.
Almanex	PEs coatings	General Electric Co.
Alpolit	PEs resins	American Hoechst Corporation
Alsynite	Reinforced polyester	Reichhold Chemicals, Inc.
Altek	PEs resins	Alpha Chemical Corporation
Amberlac	Alkyd resins	Rohm & Haas Co.
Amester	Polyester and alkyd resins	American Alkyd Industries
Ampreg	Polyester coatings	Atlas Mineral & Chemicals Div., ESB Inc.
Apex	Reinforced polyester	Apex Electrical Manufacturing Co.
Aquaplex	Alkyd emulsions	Rohm & Haas Co.
Aquasize	Waterborne PEs coating	Whittaker Corporation
Aquaspray	Waterborne PEs coating	Whittaker Corporation
Amorex	Reinforced polyester	Valtronic Corporation
Aroflat	Alkyd resins	Ashland Chemical Co.
Arolon	Alkyd resins	Ashland Chemical Co.
Aroplaz	Alkyd resins and coatings	Ashland Chemical Co.
Aropol	Polyester and alkyd resins	Ashland Chemical Co.
Astorloy	PEs resin	Reichhold Chemicals, Inc.
Atlac	Polyester resins	ICI Americas, Inc.
AZ Rez	PEs	AZ Products Co.
Baygal	PEs resins	Mobay Chemical Company
Beckolester	PEs resins	Beck, Koller & Company, Ltd.
Beckosol	Alkyd resins	Reichhold Chemicals, Inc.
Bec-Qua-Sols	Alkyd	Beck, Koller & Company, Ltd.
Benthal	Alkyd resin intermediate	Monsanto Chemical Co.
Bont-Tite	PEs	Oatey Co.
Casthane	Molded PEs	Chicago-Allis Manufacturing Corporation
Chempol	Alkyd resins	Freeman Chemical Corporation
Chemreinforced	Rfd PEs	Chemold Co.

Cimglas	Rfd PEs	Cincinnati Milacron Chemical Co.
Clearlite	Rfd PEs	Vistron Corporation
Cleartuf	PEs resin	Goodyear Tire & Rubber Co.
Cocor	Allyl ester	Cast Optics Corporation
Conolite	Polyester resins and laminates	Woodall Division, LOF Plastics, Inc.
Copalam	Reinforced PEs	Cottrell Paper Co., Inc.
Cord	PEs	CPC International, Inc.
Cordopreg	Polyester and alkyd molding compounds	Ferro Corporation
Corezyn	Vinyl esters	Interplastic Corporation
Coroc	Alkyd resins	Cook's Resins
Corocrete	PEs	Ceilcote Co.
Corrolite	Vinyl esters	Reichhold Chemicals, Inc.
Corrulux	Rfd PEs	Johns-Manville Corporation
Cosmic	DAP compounds	Cosmic Plastics, Inc.
Croid	PEs	Croid, Ltd.
Crystic	Polyester resins and coatings	Scott Bader & Company, Ltd.
Cyglas	Reinforced polyester molding compounds	American Cyanamid Co.
Cyplex	PEs coatings	American Cyanamid Co.
Cypor	PEs resin	American Cyanamid Co.
Cyzac	Laminating resins	American Cyanamid Co.
Daiso Dap	DAP resin	Osaka Soda Co.
DAP	DAP monomer	FMC Corporation
Dapolyl	Allyl resins and adhesives	Ciba-Geigy, Ltd.
Dapex	DAP	Acme Resin Co.
Dapmix	DAP monomer	FMC Corporation
Dapon	DAP prepolymer	FMC Corporation
Dapol	DAP molding compounds	Fudow Chemical Co.
Daponite	DAP laminates	FMC Corporation
Derakane	Vinyl esters	Dow Chemical Co.
Desmophen	PEs	Mobay Chemical Corporation
Diall	DAP molding compounds	Allied Chemical Corporation
Diolen	PEs	Akzo Plastics, N.V.
Dion	Polyester Resins	Diamond Shamrock Chemical Co.
Dion Cor-Res	PEs	Diamond Shamrock Chemical Co.
Dion-Iso	Isophthalic PEs	Diamond Shamrock Chemical Co.
Diorez	Isophthalic PEs	Briggs & Townsend, Ltd.
Dow	Vinyl ester resins	Dow Chemical Co.
Dry-Ply	PEs and alkyd molding material	Flexfilm Products
Duraglas	PEs Prepreg	Natvar Corporation
Dura-Lite	PEs sheet stock	Homalite Corporation
Duraplex	Alkyd resins	Rohm & Haas Co.
Durez	Polyester and DAP molding compounds	Hooker Chemical Corporation
Duroflex	PEs resins	Allied Chemical Corporation
Dyal	Alkyd and styrenated alkyd coatings	Sherwin-Williams Co.
Dypol	Polyester resins	Sherwin-Williams Co.
E-Case	Filled and reinforced DAP	Epoxy Products Co.
Elastonol	PEs polyols	BASF Wyandotte Corporation
Epocryl	Vinyl esters	Shell Chemical Co.
Epoester	PEs polyols	Commercial Chemical Co.
Epok	PEs and alkyd	BP Chemicals International, Ltd.
Ervadiol	PEs resins	Products Chimiques Ugine Kuhlmann
Ervamix	PEs molding compound	Products Chimiques Ugine Kuhlmann
Ervapon	PEs molding compound	Products Chimiques Ugine Kuhlmann
Ervapreg	PEs Prepreg	Products Chimiques Ugine Kuhlmann
Estolan	PEs Prepreg	Lankro Chemicals, Ltd.
Estafoam	Polyester foam	Vanguard Products, Inc.
Esterol	Alkyd resin	Paramet Chemical Co.
Estoglas	Reinforced polyester	Plastilight, Inc.
Estral	PEs resin	Indpol
Everlite	Reinforced polyester	Everlite Corporation
Falkaloid	Modified alkyd	Falk Co.
Fibercore	Reinforced PEs and alkyd molding compounds	Plumb Chemical Corporation

Fiberpane	Reinforced PEs Laminates	Vistron Corporation
Fibrez	Reinforced PEs molding compounds	General Plastics Corporation
Filabond	Reinforced PEs resins	Synthetic Resins, Ltd.
Filon	Reinforced polyester	Vistron Corporation
Filopane	Reinforced PEs sheet	Vistron Corporation
Fire-Snuf	Reinforced polyester	Resolite Corporation
Flomat	Reinforced PEs	Fothergill & Harvey, Ltd.
Formadall	Reinforced polyester	Woodall Industries, Inc.
Formrez	Polyester resins	Witco Chemical Company, Inc.
Fostercast	PEs resin	Westinghouse Electric Corporation
Fosterite	PEs resin	Westinghouse Electric Corporation
Fudowlite	PEs molding compounds	Fudow Chemical Co.
Fybrazine	PEs molding compounds	A. Zollinger, Inc.
Gabraster	PEs resins	Montedison, S.p.A.
Gedeglyps	Alkyd resins	CDF Chimie
Gel-Kate	PEs resins	Glidden Co.
Genite	Rfd PEs	General Tire & Rubber Co.
Genpol	PEs resin	General Tire & Rubber Co.
Gillfloor	Rfd PEs	M.C. Gill Corporation
Gilliner	Rfd PEs laminate	M.C. Gill Corporation
Glasbord	Rfd PEs laminate	Kemlite Corporation
Glaskyd	Reinforced DAP-alkyd molding compounds	American Cyanamid Co.
Glasrod	Pultruded PEs	Glastic Corporation
Glastrusions	Pultruded polyester	Glastrusions, Inc.
Glidpol	PEs resin	Glidden Co.
Glykon	Polyester resin	General Tire & Rubber Co.
Glyptal	Alkyd resins	General Electric Co.
Hapol	PEs resin	Hastings Plastics, Inc.
Hasko-Struct	Reinforced polyester	Haskelite Manufacturing Corporation
Hatco	Polyester resins	W.R. Grace & Co.
Havamold	Sheet molding compound	Haveg Industries, Inc.
Hawley-Glass	Reinforced PEs SMC	Hawley Products Co.
Haysite	Reinforced polyester	Synthane Taylor Corporation
Heirloom	Alkyd coatings	McCloskey Varnish Co.
Hetrolac	Polyester lacquer	Hooker Chemical Corporation
Hetron	Vinyl esters	Ashland Chemical Co.
Hi-Therm	Reinforced PEs/P/F	John C. Dolph Co.
Homalite	Cast polyester	SGL Industries, Inc.
Hostaset	PEs molding compounds	Hoechst, A.G.
Icdal	Alkyds	Dynamit Nobel, A.G.
Impolex	PEs resins	ICI, Ltd.
Insulfoam	PEs foam	Robinson Foams, Ltd.
Insulstrice	Reinforced polyester	Cincinnati Development Mfg. Co.
Insultruc	Reinforced polyester	Cincinnati Development Mfg. Co.
Insultrue	Rfd PEs laminate	Cincinnati Development Mfg. Co.
Interchem	Rfd PEs laminate	Interchemical Corporation
Interchemical	Polyester resin	Interchemical Corporation
Isolite	Polyester resin	Schenectady Chemicals, Inc.
Isonel	Polyester enamel	Schenectady Chemicals, Inc.
Jek-O-Mer	Alkyd resins	Alkydol Laboratories
Jel-O-Mer	Alkyd resins	Alkydol Laboratories
Keeplus	Filled PEs	APV-Kestner, Ltd.
Kelkyd	Alkyd resin	Textron, Inc.
Koplac	PEs resin	ARCO Polymers, Inc.
Kriston	Allyl ester casting resins	B.F. Goodrich Chemical Co.
Lamellon	PEs	Scado, B.V.
Lamidal	Reinforced polyester	Woodall Industries, Inc.
Laminac	Polyester resins	USS Chemicals Co.
Lankyd	Alkyd resin	Lanson Chemical Corporation
Lanpol	Silicone/PEs	Lanson Chemical Corporation
Legwal	PEs	Bayer, A.G.
Lexorez	PEs polyols	Inolex Corporation
Lipal	PEs polyols	PVO International, Inc.
Ludopal	PEs coatings	BASF Wynadotte Corporation
Macamoll	PEs resins	Cossella Farbwerke Mainkur, A.G.

Marco	Polyester and alkyd resins	W.R. Grace & Co.
Marco Resin	PEs resins	Scott Bader & Co., Ltd.
Marcothix	Polyester resins	W.R. Grace & Co.
Micro-Mix	Filled PEs	Micro-Materials, Inc.
Mirasol	Alkyd resins	C.J. Osborn Co.
Modulite	Reinforced PEs	SGL Industries, Inc.
Modulite	Reinforced polyester	Modular Molding Co.
Morester	Polyester resins	Morton Withers Chemical Co.
Multanil	Polyester resins	Mobay Chemical Co.
Multron	Polyester resins	Mobay Chemical Co.
Natro-Curm	DAP monomers	Natrochem, Inc.
Naugatile	PEs coatings	Uniroyal, Inc.
Naxol	Alkyd resins	Convert, Ets. G.
Nestorite	Alkyd molding compounds	James Ferguson & Sons, Ltd.
Nestrapol	Rfd PEs	Polycolor (Rumania)
Norpal	PEs molding compounds	Jotun
Norsodyne	PEs resins	CDF Chimie
Norsomix	PEs compounds	CDF Chimie
NuGlas	Acrylic-modified PEs	Vistron Corporation
Nureco	PEs and alkyd molding compounds	Nureco, Inc.
Palatal	Polyester cast resin	BASF Wyandotte Corporation
Paraplex	Polyester and alkyd resins	Rohm & Haas Co.
Pedigree Products	PEs	P.D. George Co., Inc.
Peogen	PEs resins	American Petrochemical Corporation
Perma-Flexit	PEs resins	Baird Dynamic Co.
Peropal	Filled alkyd compound	Perstorp Ferguson, Inc.
Petrex	Alkyd resins	Hercules, Inc.
Phenoline	PEs	Carboline Co.
Phthalkyd	Alkyd resins	Hitachi Chemical Co.
Plaskon	Alkyd, DAP, and polyester resins	Allied Chemical Corporation
Plast	PEs	Fibre-Glass Developments
Plastaloy	Reinforced polyester	Atlas Mineral & Chemicals Division, ESB Inc.
Pla-Tank	Reinforced polyester	Haveg Industries, Inc.
Pleogen	Polyester resins	Whittaker Corporation
Plexton	Reinforced polyester	G.B. Lewis Co.
Pneumacel	PEs foam	E.I. Du Pont de Nemours & Co., Inc.
Politen	Rfd PEs	Theiten Fibre Co.
Polnac	PEs resin	S.P.R.E.A.
Polyall	Alkyd molding compounds	Mesa Plastics Co.
Polycast	MMA/allyl resin	Fortin Laminating Corporation
Poly-Dap	DAP molding compounds	Hitco Co.
Poly-Fiber	Reinforced polyester	Poly-Fiber, Inc., Division, Calresin Corporation
Polyglas	Pultruded PEs	Westinghouse Electric Corporation
Polylite	Polyester resins	Reichhold Chemicals, Inc.
Polymaster	PEs	W.A. Mitchell & Smith, Ltd.
Polymerin	Melamine-alkyd enamels	Interchemical Corporation
Polyset	PEs	Hitachi Chemical Co.
Polysol	PEs solutions	Showa High Polymer Co.
Polytool	Polyester tooling compounds	Reichhold Chemicals, Inc.
Premi-Glas	Reinforced polyesters	Premix, Inc.
Pre-Imp	Reinforced PEs molding compound	Flexfirm Products
Protectoplas	Reinforced PEs	Protective Plastic Co.
Relcote	PEs resins	Reliance Universal, Inc.
Resaflex	Alkyd resin	Resine Sintetiche e Affine, S.p.A.
Resalol	Alkyd resin	Resine Sintetiche e Affine, S.p.A.
Resalpar	Alkyd resin	Resine Sintetiche e Affine, S.p.A.
Resaltix	Thixotropic alkyds	Resine Sintetiche e Affine, S.p.A.
Resibond	PEs	Maurice A. Knight Co.
Resichem	PEs resins	Brown Co.
Resiglas	PEs resins	Kristal Kraft, Inc.
Respol	PEs resins	Montedison, S.p.A.
Restiroid	Styrenated alkyds	Societa Italiana Resine
Restral	Styrenated alkyds	Resine Sintetiche e Affini, S.p.A.
Rezimac	Modified alkyds	IMC Chemical Group, Inc.

Rezyl	Alkyd varnishes and coatings	American Cyanamid Co.
Rhodester	PEs resin	May & Baker, Ltd.
Rigidon	Reinforced polyester	Heil Process Equipment Co.
Rigiline	PEs resin	Heil Process Equipment Co.
Rigolac	PEs resin	Showa High Polymer Co.
Robi-Foam	PEs foam	Robinson Foams, Ltd.
Rockite	PEs molding compounds	British Resin Products, Ltd.
Rosite	PEs and alkyd molding compounds	Rostone Corporation
Roskydal	UV crosslinked PEs	Bayer, A.G.
Ruco	PEs	Hooker Chemical & Plastics Corporation
Ruggid	Reinforced PEs parts	H.P. Ruggles Company, Ltd.
Satusply	Reinforced polyester	Uniroyal Corporation
Selectron	Polyester resin	PPG Industries, Inc.
Silmar	Polyester resins	Vistron Corporation
Siralkyd	Alkyd resins	Societa Italiana Resine
Sirester	PEs resins	Societa Italiana Resine
Sniatron	PEs resins	Snia Viscosa, S.p.A.
Sterpon	PEs resins	Convert, Ets. G.
Stratyl	PEs resins	Rhone-Progil Direction Commerciale Matieres Plastiques
Structoform	SMC	Fiberite
Structoform	PEs sheet	Standard Oil Co.
Structoglas	Reinforced polyester	Reichhold Chemicals, Inc.
Stypol	Polyester resin	Freeman Chemical Corporation
Styresol	Styrenated alkyds	Reichhold Chemicals, Inc.
Styretex	Styrenated alkyds	Celanese Coating Co.
Super Alkyd	Alkyd resins	Thibaut & Walker Co.
Super Astroloy	PEs sheet	Reichhold Chemicals, Ltd.
Super-Beckosol	Isophthalic acid alkyds	Reichhold Chemicals, Inc.
Syloy	Reinforced polyester	The General Tire & Rubber Co.
Synolite	PEs resins	Synres Chemical Corporation
Synresate	Alkyd resins	Synres Chemical Corporation
Syntex	Alkyd resins	Celanese Coatings Co.
Synthe-Mul	Alkyd emulsions	Reichhold Chemicals Co.
Synthite	PEs/vinyl varnish	John C. Dolph Co.
Synvar	PEs resins	Synvar Corporation
Synvarex	Alkyds resins	Synvar Corporation
Syretex	Styrenated alkyds	Celanese Coatings Co.
Tacon	PEs sheet	Walker & Zanger, Inc.
Teglac	Alkyd coatings	American Cyanamid Co.
Tere-Cast	PEs compounds	Reichhold Chemicals, Inc.
Tere-Gard	PEs compounds	Reichhold Chemicals, Inc.
Terlenka	PEs compounds	Akzo Plastics, N.V.
Thermaflow	Reinforced polyester	Atlas Powder Co.
Thermoflow	PEs compounds	Atlas Powder Co.
Toshiba Premix	Alkyd molding compounds	Toshiba Chemical Products Co., Ltd.
Trevarno	Reinforced polyester	Hexcel Corporation
Trak-Less	Alkyd molding compounds	American Cyanamid Co.
Tube-Cote	PEs	Tuboscope, Inc.
Tyglas	PEs prepreg	Fothergill & Harvey, Ltd.
Ucelflex ME	PEs	Union Chemique Belge
Ukadiol	PEs resins	Union Camp Chemical Co.
Ukapon	PEs resins	Union Camp Chemical Co.
Ukapreg	PEs prepreg	Union Camp Chemical Co.
Unirez	Alkyd resin	Union Camp Chemical Co.
Valco	DAP resins	United Merchants, Inc.
Varex	PEs resins	McClosky Varnish Co.
Varkyd	Alkyd solutions	McClosky Varnish Co.
Veemo	PEs	Belding Heminway Co., Inc.
Verdoc	PEs coating	Ferro Corporation
Vestopol	PEs resins	Chemische-Werke Huls, A.G.
Vibramix	Polyester molding compounds	Uniroyal Corporation
Vibrin	Polyester resins	Uniroyal Corporation
Vibrin-Mat	Reinforced PEs	W.R. Grace & Co.
Vibro-Flo	PEs coating	Armstrong Products Co.

Vitof	Polyester resins	Goodyear Tire & Rubber Co.
Vitralon	PEs coating	Pratt & Lambert, Inc.
Vitroplast	Polyester cement	Atlas Mineral & Chemicals, Division, ESB Inc.
Vituf	PEs resin	Goodyear Tire & Rubber Co.
Vycron	Tere- and isophthalic PEs	Goodyear Tire & Rubber Co.
White Bond Tite	PEs	Oatey Co.
White Diamond	PEs	Oatey Co.
White Tiger	PEs	Oatey Co.

REFERENCES AND BIBLIOGRAPHY

Allied Chemical Corporation. *Diall (Diallyl Phthalate) Molding Compounds, Molding Techniques.* Bulletin 841-85, 1966.

——. *Plaskon Alkyd Molding Compounds, Molding Techniques.* Bulletin 841-63, 1965.

——. *Plaskon Polyester Resins, Premix Molding.* Bulletin 851-36.

American Cyanamid Co. *Cyanamid Glaskyd and Cyglass Molding Compounds, Typical Properties & Trouble Shooters Guide.* Bulletin GLT14, 2-20141, November, 1972.

——. *Laminac Polyester Resin Selector.* Bulletin LM 400, 1-1217, December, 1971.

——. *Glaskyd Glass Reinforced Alkyd Molding Compounds.* Bulletin GLT-12A, May, 1967.

——. *Glaskyd Glass Reinforced Alkyd Molding Compounds.* Bulletin GLT-11A, June, 1967.

Amoco Chemicals Corporation. *How to Improve Long-Term Performance of Gel Coated Laminates.* Bulletin IP-69, January, 1979.

——. *AMOCO Terephthalic Acid/Applications Unlimited.* Bulletin TA-11, November, 1978.

Anderson, A. Reaction injection molding: process control, reinforcements, new materials build up markets. *Modern Plastics*, April 1978, **55**(4):66–67.

Anderson, N. D. Ultraviolet curing of polyester-preimpregnated materials. *SPE Journal*, April, 1968, **24**(4):81–84.

Ashland Chemical Co. *Hetron 902 and 903 Vinyl Ester Resins.* Technical Bulletin 1343-1.

Beacham, H. H. Diallyl phthalate resin and monomer. *Plastics Design and Processing*, April 1967, 7(4):20–23.

Beacham, H. H., Litwin, J., and Johnston, C. W. How to formulate heat resistant DAP polyesters. *Plastics Technology*, May 1963, 9(5):44–46.

Brinkman, W. H., Damen, L. W. J., and Maira, S. Accelerators for peroxide curing of polyesters. *Modern Plastics*, October 1968, **45**(10):167–172.

Brueggemann, W. H. Blistering of gel-coat laminates. *Modern Plastics*, June 1979, **56**(6):80–84.

Cassoni, J. P., Gallagher, R. B., and Kamath, V. R. Effect of compounding variables on peroxyketal-cured polyesters. *Modern Plastics*, April 1979, **56**(4):78–82.

Cassoni, J. P., Harpell, G. A., Wong, P. C., and Zupa, A. H. Room-temperature curing-minus the russian roulette. *Plastics World*, February 1977, **35**(2):55–57.

Crites, S. M., and Ozbun, L. Anatomy of a glass-reinforced-polyester motor housing. *Plastics Engineering*, January 1979, **35**(1):39–42.

Cywinski, J. W. A survey of some curing methods of polyester resins. *British Plastics* (incorporated into *European Plastics News*), October 1957, **30**:2–7.

Dow Chemical Co. *Derakane 51ON Resin, A New Brominated, Novolac-Base Vinyl Ester Resin.* Form No. 296-001, 1979.

——. *Derakane Vinyl Ester Resins for Matched Die Molding.* Bulletin 170-197A, 1968.

FMC Corporation. *Dapon Resin Diallyl Phthalate.* Technical Bulletin No. 20.

Foster, L. P., and Miller, S. B. Inhibitors for unsaturated polyesters. *Modern Plastics*, March 1967, **44**(3):163–167, 224.

Funke, W., and Hamann, K. Structure and composition of cross-linked polyester. *Modern Plastics*, May 1962, **39**(5):147–150, 226.

Golding, G. *Polymers and Resins—Their Chemistry and Chemical Engineering.* D. Van Nostrand Co., Inc., Princeton, New Jersey, 1959.

Haveg Industries, Inc. *Havamold Sheet Molding Compound Data Sheets, Series HM 9100, 9200, 9300, & 9310.*

Hillman, S. L., Palmer, M., and Davis, J. H. A new generation of corrosion-resistant polyesters based on TMPD Glycol .... *Plastics Engineering*, June 1979, **35**(6):32–35.

Howald, J. M. A report of what's available in ... alkyd molding compounds. *Plastics Design & Processing*, January 1968, 8(1):18–19.

Interplastic Corporation. *CoRezyn VE 8100 & VE 8300 Vinyl Ester Resins.* Bulletin VE-1-76, January 1976.

Kalek, R. L., and Hammill, J. L. "Visible Light Cured Polymers." Paper presentation, *SPE ANTEC*, New Orleans, May 1979.

Keller, L. B., McClone, W. R., and Woodin, D. H. Molding of diallyl phthalate compounds. *Plastics Technology*, May 1959, 5(5):38–42, 56.

Keown, J. A. Designing with sheet-molding compounds. *Machine Design*, February 1971, **43**(3):111–113.

Kerle, E. J., and Fisher, J. J. Hydrogenated bisphenol-A in chemical resistant polyester. *Western Plastics*, Technical Report No. 111, August 1964, pp. 25–27.

Koppers Co., Inc. *Koppers Polyester Technology.*

Litwin, J., and Beacham, H. H. Catalysis and inhibition of diallyl phthalate polyesters. *Modern Plastics*, July 1965, **42**(7):133–136, 140–145, 185.

Longenecker, D. M., Greth, G. G., and Dixon, B. G. Processing plastics—visible-light curing of polyesters. *Plastics Engineering*, May 1977, **33**(5):52–55.

Mallick, P. K. "Fatigue Characteristics of High Glass Content Sheet Molding Compounds." Paper presentation, *SPE ANTEC*, New Orleans, May 1979.

Materials Engineering. Unsaturated polyester is dielectric to near 400°F. May 1977, **85**(5):64.

Maudru, W. F. What you should know about working with polyester resins. *Plastics Technology*, July 1968, **14**(7):45–47.

McGrady, C. Thermosets, from Materials '79. *Modern Plastics*, January 1979, **56**(1):58–60.

Mettes, D. G. RP and PUR offer critical values in designing pipe for tough service. *Modern Plastics*, November 1978, **55**(11):54–55.

Mock, J. A. Ultraviolet curing saves energy, speeds production and cuts costs. *Materials Engineering*, May 1976, **83**(5):40–42.

Modern Plastics. Clarity, cure control, fast action: all from new additive for polyester. September 1978, **55**(9):22.

——. A polyester for every purpose. February 1964, **41**(2):82–85, 148–155.

Monsanto Chemical Co. *The Chemistry and Processing of Alkyd Resins*, 1952.

Morrison Molded Fiber Glass Co. *MMFG EXTREN & Special Products.* Form No. MMFG 876-2.

Nelson, D. L. Introducing low-cost dicyclopentadiene into unsaturated polyester resins. *Modern Plastics*, September 1978, **55**(9).

Peterson, C. R., and Dalluge, M. What furniture manufacturers should know about unsaturated polyesters. *SPE Journal*, November 1970, **26**(11):33–38.

Plastics Design & Processing. A key to processing components successfully–casting polyester resins. August 1970, **10**(8):23–26.
Plastics Engineering. Engineering resins primer. December 1974, **30**(12):17–32.
Plastics World. Best of SMC and BMC–and then some. July 1977, **35**(7):32.
Rostone Corporation. *Rosite Data Sheet.* April 1979.
Rowe, E. H., and Howard, F. H. Toughening additive for BMC & SMC. *Modern Plastics*, August 1978, **55**(8):58–59.
Selley, J. Polyesters, from Plastics in Construction: Coming on Strong in 1978. *Plastics Engineering*, July 1978, **34**(7):37–39.
Selley, J. E., and Vaccarella, P. W. Controlling flammability and smoke emissions in reinforced polyesters. *Plastics Engineering*, February 1979, **35**(2):43–47.
Shell Chemical Co. *Epocryl 320 Resins.* Technical Bulletin SC: 323-79, January 1979.
——. *Epocryl Resin 480.* Technical Bulletin SC: 122-78R, January 1978.
——. *Epocryl Resins Corrosion Control Guide.* Bulletin SC: 161-77, January 1977.
——. *Heat Resistant Vinyl Ester Resin.* Technical Bulletin SC: 116-76, June 1976.
——. *Epocryl Resins 321 and 322.* Technical Bulletin SC: 16-76, August 1976.
Shreve, T. Solid polyester molding compound: a thermoset/thermoplastic hybrid. *Plastics Design & Processing.* November 1977, **17**(11):38–41.
Silver, W. C. "Encapsulating with Polyesters." Paper presentation, *SPE ANTEC*, New Orleans, May 1979.
Simonds, H. R., and Church J. M. *A Concise Guide to Plastics.* 2nd Ed. Van Nostrand Reinhold Co., New York, 1963.
Slone, M. C. Methyl acrylate as a monomer in glass fiber reinforced polyester laminates. *SPE Journal*, October 1960, **16**(10):1123–1127.
Sundstrom, G. A., and Davis, T. "BMC Can Replace Phenolic." Paper presentation, *SPE ANTEC*, New Orleans, May 1979.
Swanson, R. S. *Plastics Technology–Basic Materials and Processes.* McKnight & McKnight Publishing Co., Bloomington, Illinois, 1965.
Sweitzer, C. W., Lyon, F., and Grabowski, T. S. Cure of carbon black-unsaturated polyester mixtures. *Industrial & Engineering Chemistry*, November 1965, **47**(11):2380–2385.
Titus, J. B. *Trade Designations of Plastics and Related Materials*, (revised). PLASTEC NOTE N9C, Plastics Technical Evaluation Center, Picatinny Arsenal, Dover, New Jersey, May 1978.
Vistron Corporation. *Silmar Polyester Resins.* Bulletin 775, 1975.
——. *How to Design With Daylight–A Guide to the Facts and Figures You Need to Know About Translucent Fiberglass Panels–Filon.* Bulletin No. 160, October 1971.
——. *General Brochure on Polyester Resins–Silmar.*
——. *Fire Retardant–Information on FR Polyester Resins From Silmar.*
Wahl, N. E., and Preston, H. M. Properties of polyester–triallylcyanurate glass-reinforced laminates at elevated temperatures. *Modern Plastics*, October 1957, **34**(10):153–166.
Western Plastics. Problem and remedies in polyester molding. March 1960, pp. 48–51.
Wood, S. Polyesters: furniture sleeper? *Modern Plastics*, August 1969, **46**(8):50–54.
Zimmerman, A., and Wright, V. What's happening with SMC's? *Modern Plastics*, February 1969, **46**(2):36–40.
Zimmerman, D. D. "Performance of Engineering Resins in Electrical Connectors." Paper presentation, *SPE ANTEC*, New Orleans, May 1979.

EPOXIES

The term epoxy refers to a chemical group consisting of an oxygen atom bonded with two carbon atoms already united in some other way. The simplest epoxy is a three-membered ring, to which the term alpha-epoxy or 1,2-epoxy is applied. Ethylene oxide, $\overset{O}{\widehat{CH_2{-}CH_2}}$, is an example of this type. There are others–1,3- and 1,4-epoxies–but they are not important for this discussion. The ethylene oxide ring is termed the oxirane ring and the glycidyl group is $\overset{O}{\widehat{CH_2{-}CH}}{-}CH_2{-}$. Both of these terms are frequently used in the epoxy resin trade.

An epoxy resin, epoxide (Europe), is defined as any molecule containing more than one alpha-epoxy group (whether situated internally, terminally, or on cyclic structures) capable of being converted to a useful thermoset form. The term is used to indicate the resins both in the thermoplastic (uncured) and thermosetting (cured) state.

Credit for the synthesis of the materials first designated as epoxy resins–those derived specifically from epichlorohydrin and bisphenol A–is shared by Dr. Pierre Castan of Switzerland and Dr. S. O. Greenlee of the United States.

The most common commercial epoxy resin is the diglycidyl ether of bisphenol A (DGEBA). It is made by reacting bisphenol A and epichlorohydrin as follows:

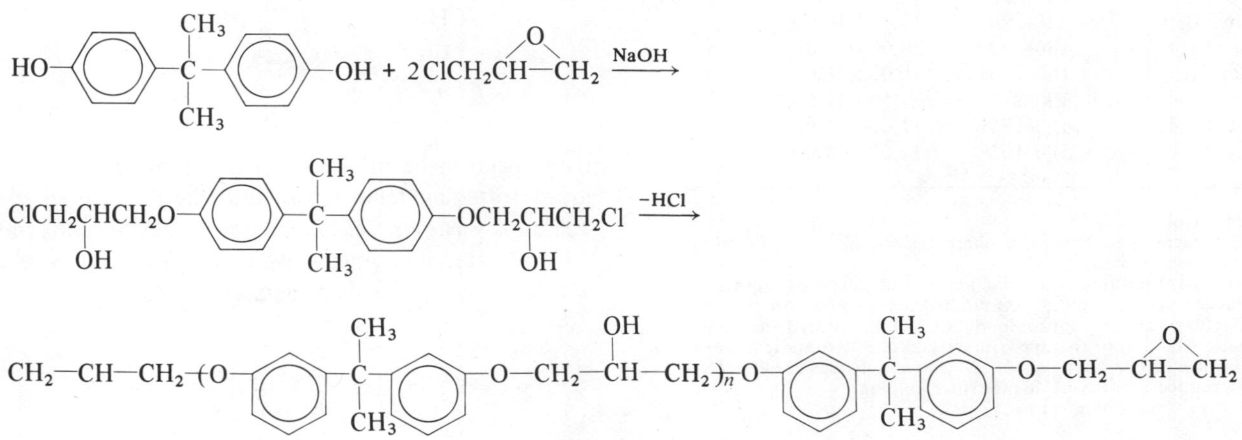

The caustic catalyzes the reaction to produce the chlorohydrin intermediate, acts as the dehydrohalogenating agent, and neutralizes the formed HCl. Reaction typically occurs over 16 hours at 110°C (230°F), with 30% aqueous caustic solution being slowly added. The resulting organic layer is separated, dried with sodium sulfate, and fractionally distilled under vacuum. The DGEBA having a molecular weight of approximatley 380 is the basic liquid epoxy resin in commercial use, although purified versions of molecular weights as low as 344 (n = essentially 0) are being produced. Higher molecular weight products (n = 1–10) may be obtained by reducing the amount of excess epichlorohydrin and reacting under more strongly alkaline conditions. Tables 6-29 through 6-33 list commercial grades of standard DGEBA resins.

We mentioned in our discussion of phenolics, the manufacture of phenol-formaldehyde materials called novolacs. These can be epoxidized with epichlorohydrin to form glycidyl ethers with the following general structure:

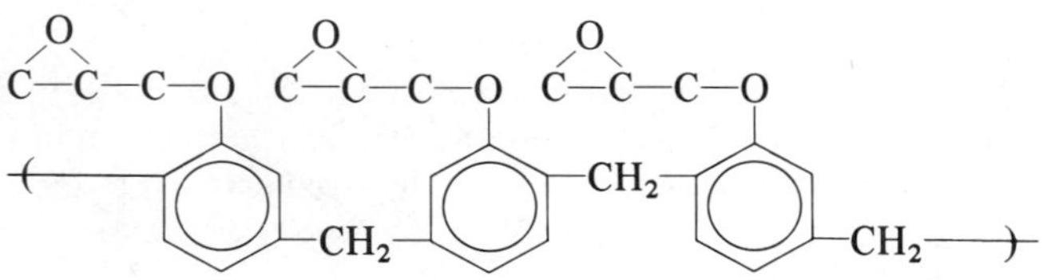

The number of glycidyl groups per molecule per resin is dependent upon the number of phenolic hydroxyls in the starting novolac, the extent to which they are reacted, and the extent to which the lowest molecular species are polymerized during synthesis. A wide variety of novolacs may be used, based on different starting phenols, molecular weights, and available phenolic hydroxy groups. The epoxies generated have functionalities that vary between 2.5 and 6.0, which allows for a high degree of crosslinking during cure for the development of higher heat and chemical resistance. Table 6-34 lists some commercial grades of epoxidized novolacs.

Table 6-29. Standard Undiluted BIS Resins.*

	EEW	Viscosity (cps @ 25°C)	Color[1]
D.E.R. 331	182–190	11,000–14,000	3
Epon 8280[2]	185–192	11,000–15,000	3 max.
Epon 828	185–192	11,000–15,000	3 max.
Araldite 6010	185–196	12,000–16,000	2
Epi-Rez 5101	180–200	13,000–15,000	3
Epi-Rez 510	180–200	10,000–16,000	3
Epi-Rez WD-510	190–216	7,500–12,500	3
Epotuf 37-140	180–195	11,000–14,000	3
Gen Epoxy 190	180–195	11,000–14,000	3

[1] Gardner-Holdt.
[2] Special vacuum-casting resin characterized by rapid foam breakdown under vacuum.

Standard undiluted resins for all general purposes requiring performance up to 400°F. Aliphatic polyamines or polyamides satisfactory up to approximately 230°F. Anhydrides such as phthalic satisfactory to approximately 200°F. Aromatic amines and anhydrides satisfactory to 400°F. The anhydrides are effective viscosity reducers to permit higher filler loading.
*Courtesy: Dow Chemical Co.

Table 6-30. Lowest-Viscosity Resins.*

Products	EEW	Viscosity (cps @25°C)	Color[1]	Chlorides[2]
D.E.R. 332	172–176	4,000–6,000	75 APHA Max	
Epon 825	172–178	4,000–6,000	1 Max	
Epi-Rez 508	171–177	3,600–5,500	1	0.20% Max
Araldite 6004	187 Max	5,000–6,400	2 Max	

The low equivalent weight resins are virtually pure diglycidyl ethers of Bisphenol A. They are the lowest viscosity undiluted Bisphenol A resins available. They are so pure, however, that they crystallize on storage. The crystals melt on warming above 125°F. They fit all the general uses of D.E.R. 330 or D.E.R. 331 resins with the following advantages:

a. Increased HDT
b. Lower viscosity
c. More chemical uniformity
d. Longer pot life with most curing agents
e. Better wetting of glass reinforcements
f. Very pale color
g. Better electrical properties

[1] Gardner
[2] Parr Bomb
*Courtesy: Dow Chemical Co.

Glycidyl ether resins based on bisphenol F

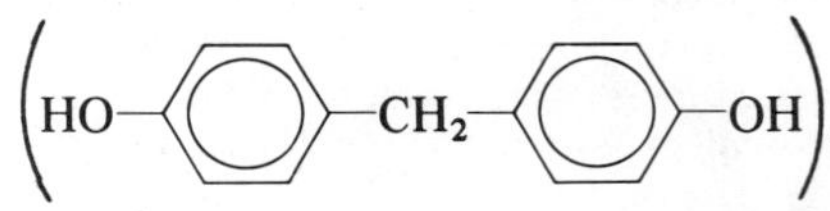

glycerol 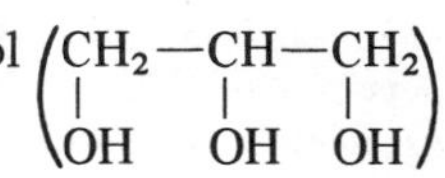 , and cashew nut oils

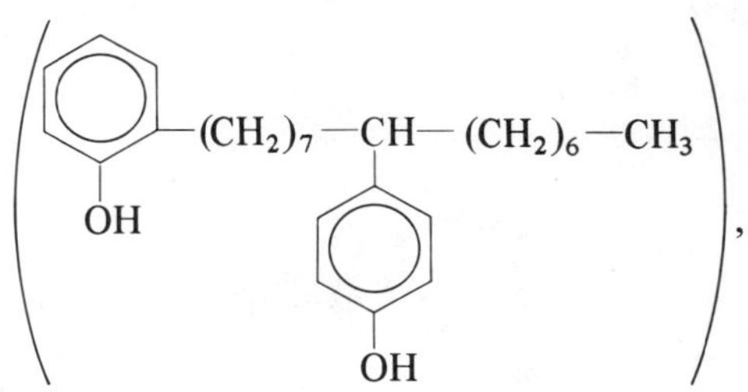

have been made in order to build in more flexibility and lower softening temperature into the final cured system. Table 6-35 lists a series of commercial flexible epoxy resins.

Unsaturated monomers have been epoxidized by reaction with hydrogen peroxide or peracetic acid.

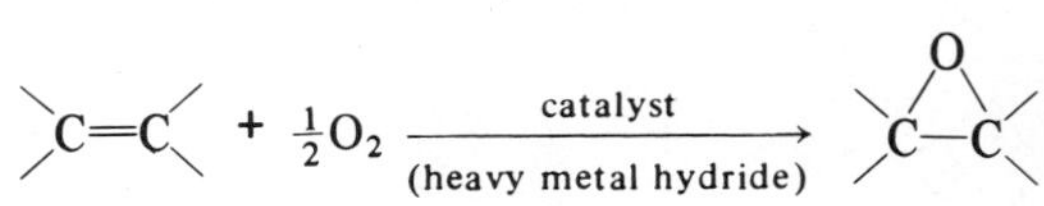

Table 6-31. High-Viscosity Resins.*

Products	EEW	Viscosity (cps @25°C)	Color[1]
D.E.R. 317	192–203	16,000–25,000	5
D.E.R. 337	230–250	400–800[2]	3[2]
Araldite 6020	196–208	16,000–20,000	3
Epon 830	190–198	17,000–22,500	9
Epon 834	230–280	410–970[2]	5[2]

The lower EEW resins in this series have the same general properties as D.E.R. 331 except viscosity.

The higher EEW resins are very viscous liquids finding main use in coatings or adhesive systems where solvents may be used to reduce viscosity.

As EEW increases, pot life is shorter, HDT decreases, exotherms decrease; impact, elongation, and adhesion improve.

[1] Gardner
[2] @70% NV in DOWANOL* DB glycol ether solvent
*Courtesy: Dow Chemical Co.

Two classes of epoxy resins are formed from such reactions:
(a) cyclic aliphatic resins such as:

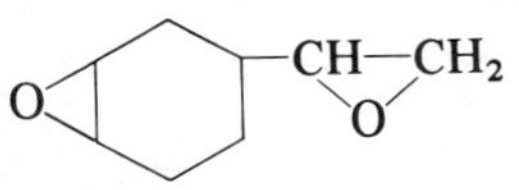

vinylcyclohexane dioxide

(b) acyclic aliphatic resins such as epoxidized polybutadiene, epoxidized soya or linseed oils, and epoxidized polyglycols.

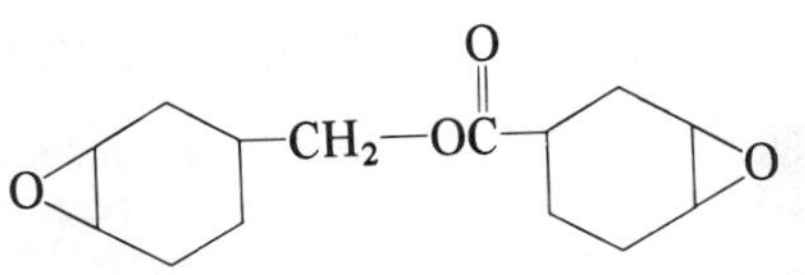

3,4-epoxycyclohexylmethyl-3,4-epoxycyclohexane carboxylate

The resins in the first grouping provide systems with properties such as lower viscosity and color, yet develop higher crosslink densities on cure. The resins in the second grouping are almost always used in conjunction with other standard resins for modification of such properties as cure rate, flexibility, and heat-deflection temperature. Table 6-36 shows the commercial grades of peracid resins. Some of the resins are considered to be reactive diluents.

Resins have recently been introduced that are based on hydantoin,

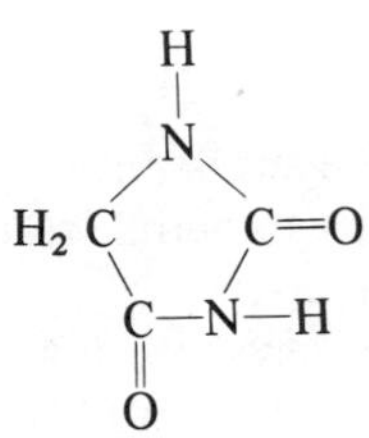

Table 6-32. Low-Melting Solid Resins.*

Products	EEW	Viscosity[1]	Color[2]	Durran's SP °C
D.E.R. 661	475–575	G–J	1	70–80
Gen Epoxy 525	475–525	G–J	1	70–80
Epon 1001	450–550	D–G	4 Max	
Epi-Rez 520 C	450–525	C–G	3	65–75
Epotuf 37-301	450–525	E–J	3	65–75
Araldite 7065	455–500	G–J	2	68–78
Araldite 7071	450–550	D–G	2	65–75

Primary uses in amine cured protective coatings and for pre-preg glass cloth for electrical laminates.

D.E.R. 661 resin modified with polyamines or polyamides is used where high chemically resistant performance is required coupled with a room temperature or low bake application.

Blends of ketone solvents (MEK or MIBK) with aromatics (Xylene or Toluene) are generally suitable for thinning these systems. Higher boiling solvents such as glycol ethers can be used in amounts of 5–15% to improve flow and film surface properties.

Systems of D.E.R. 661 resin can be used on all substrates–metal, wood, glass, masonry by all applications–brushing, spraying, dipping, etc.

Coatings end uses include pipe and drum linings, maintenance finishes, marine finishes.

[1] Gardner-Holdt at 40% NV in DOWANOL* DB @25°C
[2] Gardner @40% NV in DOWANOL DB @25°C
*Courtesy: Dow Chemical Co.

Table 6-33. High MW Solid Resins.*

Products	EEW	Viscosity[1]	Color[2]	Durran's SP °C
D.E.R. 667	1,600–2,000	Y–Z_1	3	115–130
Gen Epoxy 1800	1,600–2,000	Y–Z_1	1	113–123
Epon 1007	2,000–2,500	Y–Z_1	5 Max	
Araldite 6097	2,000–2,500	Z–Z_2	3	125–135
Araldite 7097	1,650–2,000	X–Z	3	113–123
Epi-Rez 540-C	1,400–1,800	X–Z	3	120–130
Epi-Rez 540-F	1,600–2,000	X–Z_1	4	127–133
Epotuf 37-307	1,550–2,000	V–Z_1	3	116–124
Epotuf 37-306	2,000–2,500	Y–Z	3	NS

Optimum epoxy coating can be obtained by modifying D.E.R. 667 resin with urea or melamine formaldehyde or phenolic resins. These systems in the blended solution form have excellent pot life and can be stored for several months without noticeable viscosity change. To cure the coating, high bakes of 300°F to 400°F from 15 to 30 minutes are required Phenolic modified systems require the maximum bake schedule for complete cure. The addition of 1% phosphoric acid will catalyze the cure at somewhat lower temperatures. Ketone and aromatic solvents are used to thin D.E.R. 667.

End-use applications include tank and drum linings, wire enamels, collapsible tube coatings, metal furniture finishes.

[1] Gardner-Holdt @40% NV in DOWANOL* DB @25°C
[2] Gardner @40% NV in DOWANOL DB @25°C
*Courtesy: Dow Chemical Co.

Table 6-34. Epoxy Novolac Resins.*

Products	EEW	Viscosity (cps @25°C)	Color[1]	Durran's SP °C	Solvent/% NV
D.E.N. 431	172-179[4]	1,100-1,700[2]	3		100
D.E.N. 438-EK85	176-181[4]	600-1,600	2		MEK/84-86
D.E.N. 438	176-181	20,000-50,000	2		100
D.E.N. 438-A85	176-181	500-1,200	2		Acetone/84-86
D.E.N. 438-MK75	176-181[4]	200-600	2		MIBK 74-76
D.E.N. 439	191-210	4,000-10,000[3]	3	45-58	
D.E.N. 439-EK85	191-210[3,4]	4,000-10,000	3		MEK/84-86
Ciba EPN 1138	176-181	20,000-50,000[2]	2		
Ciba EPN 1138 A-85	176-181				Acetone/85%
Ciba EPN 1139	172-179	1,400-2,000[2]	3		100
Epi-Rez 5155	175-190	30,000-90,000[2]	10		100
Epi-Rez 5109	180-200	6,000-16,000	10		100

The multi-functional epoxy novolacs have greater heat and chemical resistance than Bisphenol A-derived resins when cured with appropriate hardeners.

[1] Gardner
[2] At 125°F
[3] 85% in MEK
[4] On solids
*Courtesy: Dow Chemical Co.

Table 6-35. Flexible Epoxy Resins.*

Products	EEW	Viscosity (cps @25°C)	Color[1]	Type
D.E.R. 736	175-205	30-60	1	Polyglycol
D.E.R. 732	305-335	55-100	1	Polyglycol
D.E.R. 741	364-380	2,500-4,500	125 APHA	
Epon 812	150-170	120-200	3 Max	Glycerine
Epon 872	650-750	1,500-3,000[2]	10 Max	
Epon 872 X-75	650-750	1,500-2,500	10	75% NV Xylene
Epon 871	390-470	400-900	12 Max	
Araldite 508	400-455	2,000-5,000	5	Polyglycol
Epotuf 37-151	450-550	30,000-70,000	4	
Epi-Rez 505	550-650	300-500	8	
Epi-Rez 5060	270-320	220-520	8	
Epi-Rez 5132	400-450	40,000-60,000	Dark	
Epi-Rez 50821	245-270	900-1,300	4	
Epi-Rez 50833	220-240	1,750-2,250	5	

[1] Gardner
[2] 75% NV in Xylene
*Courtesy: Dow Chemical Co.

Table 6-36. Peracid Epoxies.*

Products	EEW	Viscosity (cps @25°C)	Color	Composition
ERL 4206	70-74	<5	1[1]	Vinyl cyclohexane dioxide
ERL 4221	131-143	350-450	1[1]	Epoxy cyclohexyl methyl epoxy cyclohexane carboxylate
ERL 4289	205-216	500-1,000	1[1]	Cycloaliphatic epoxide
Araldite				
CY-178	213	900	1	Cycloaliphatic epoxies
CY-179	140	350	1	Cycloaliphatic epoxies
RD-4	77	20	1	Cycloaliphatic epoxies
CY-182	160	P-U[2]	2-3	Cycloaliphatic epoxies
CY-183	154	L-Q[2]	1-2	Cycloaliphatic epoxies
ERL 4234	133-154	7,000-17,000[5]	2[1]	Epoxycyclohexyl-spiroepoxycyclohexane dioxide
ERL 4090	390-430	450[3]	2[1]	Modified cycloaliphatic epoxies
ERL 4205	91-102	<100[4]	2[1]	Cycloaliphatic epoxide

[1] Gardner, Max.
[2] Gardner-Holdt
[3] 40°C
[4] 45°C
[5] 100°F
*Courtesy: Dow Chemical Co.

which can be cured with most conventional epoxy curing agents.

The most valuable single property of epoxy resins is their ability to transform readily from the liquid (or thermoplastic) state, to tough, hard, thermosetting solids. The conversion normally occurs without the evolution of by-products. Conversion is accomplished by the addition of a chemically active compound known as a curing agent (hardener, activator, or catalyst). Depending on the particular agent, curing may be accomplished at room temperature with heat produced by exothermic reaction, or may require application of external heat.

The cured structure may be a homopolymer (one composed essentially of epoxy resin molecules linked together through their own reactive sites), a heteropolymer (one composed essentially of epoxy resin molecules linked together through the reactive sites of the curing agent), or a mixture of both types. Curing agents are commonly categorized as alkaline or acidic; they may be catalytic or polyfunctional crosslinking agents which become chemically bound in the final three-dimensional structure.

Catalytic curing agents are added at low levels (0-5 phr*) because their action is truly catalytic. The similarity between the use of epoxy catalysts and polyester catalysts often causes confusion for the user. If the epoxy catalyst is truly catalytic, then the application principles are the same as with polyesters. However, most epoxy curing agents are polyfunctional coreactants, as will be seen below. Consequently, the user must now take into account the stoichiometric balance between reactive species. The polyester chemist has already established the correct ratios of polyester and crosslinking monomer and it only remains for the user to determine the appropriate peroxide catalyst level.

With epoxies, a formulating supplier will often establish the correct stoichiometric ratio between reactants and supply the final system as a two-component package. However, many users develop their own formulas and consequently must be prepared to establish the correct reactive proportions.

The calculation of the amount of amine curing agent to be used is basically quite simple. For example, in determining the amount of diethylenetriamine (DETA) to be used with DGEBA of 380 molecular weight, draw the structure of the DETA:

$$N(H_2)—CH_2—CH_2—N(H)—CH_2—CH_2—N(H_2),$$

and compute the molecular weight as follows:

4 carbons	= 4 × 12 =	48
3 nitrogens	= 3 × 14 =	42
13 hydrogens	= 13 × 1 =	13
	molecular weight =	103

There are five amine hydrogens functionally reactive with an epoxy group (circled). Therefore:

$$\frac{103 \text{ grams/mole}}{5 \text{ equivalents/mole}} = 20.6 \text{ grams/equivalent}$$

Thus, 20.6 grams of DETA is used per equivalent of epoxy. If the DGEBA has an equivalence of 190, 20.6 grams of DETA is used with 190 grams of resin or

$$\frac{20.6}{190} = 10.8 \text{ grams of DETA per 100 grams of DGEBA.}$$

*phr = parts per hundred of epoxy-containing resin.

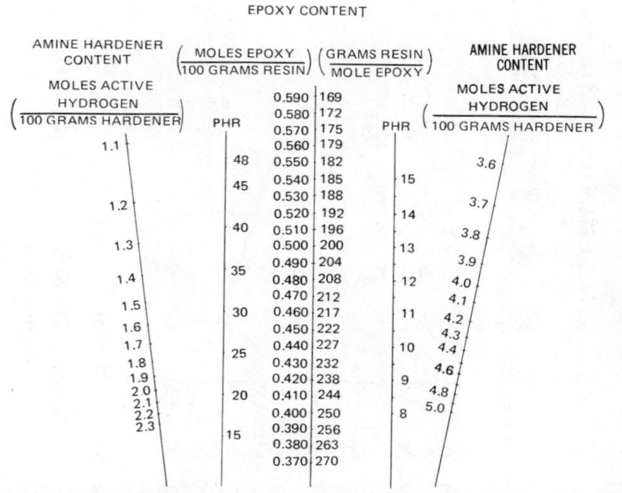

Fig. 6-41. Nomograph calculator for epoxy resin/hardener ratios. (Redrawn from *TSN 99*, Ciba Products Co.)

When the chemical structure is unknown (as with proprietary curing agents), the amine equivalent (or active hydrogen equivalent) is provided from which the appropriate mix ratio can be calculated. Often, nomographs such as the one shown in Figure 6-41 are provided to simplify the calculations.

As will be seen with acid anhydride curing agents, one anhydride group provides two carboxylic acid groups, and it is these carboxylic acid groups that react with the epoxy. Thus it is important to remember that one anhydride group has a functionality of two in calculating the stoichiometry of the epoxy-anhydride reaction.

It has also been determined that the actual stoichiometric equivalency does not always produce a cured resin with optimum properties. It is therefore not uncommon to see data sheets suggesting the use of a particular curing agent at a level of 80-110% of its true stoichiometry. This most often occurs with aromatic amine and acid anhydride curing agents. Figure 6-42 shows the effect of diethylenetriamine (DETA) and triethylenetetramine (TETA) level on two properties of the resultant cured resin. Figure 6-43 illustrates the effect of stoichiometric ratio variance on deflection temperature (DT) and weight loss of phthalic anhydride (PA) systems. Table 6-37 shows the recommended concentrations for various amine curing agents. Note the wide spectrum of pot-life values. Also note that the recommended level of DETA is 8 phr, 80% of the theoretical value calculated above.

Certain curing agents, particularly the polyamides and amidopolyamines, are sufficiently complex that a precise determination of the functional equivalency is not possible. Mix ratios are chosen, therefore, on the basis of the cured resin properties. Table 6-38 shows the wide variance of hardener level available with an amidopolyamine and the resulting properties obtainable.

The alkaline curing agents employed in epoxy resin tech-

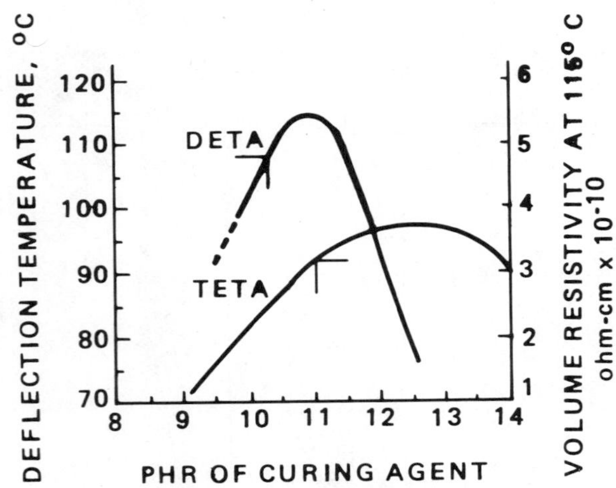

Fig. 6-42. Effect of concentration of DETA and TETA on deflection temperature of DGEBA, (From *Handbook of Epoxy Resins* by Lee and Neville. Copyright 1967 by McGraw-Hill, Inc. Used with permission of McGraw-Hill Book Co.)

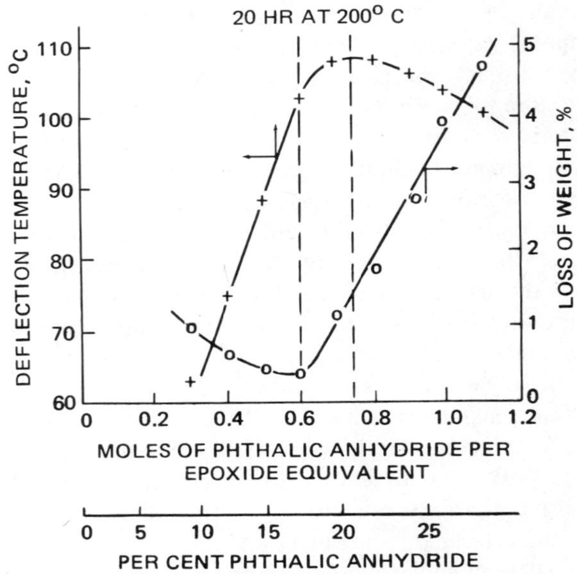

Fig. 6-43. Weight loss and deflection temperature versus curing-agent concentration for DGEBA (MW = 400) cured with PA. (From *Handbook of Epoxy Resins* by Lee and Neville. Copyright 1967 by McGraw-Hill, Inc. Used with permission of McGraw-Hill Book Co.)

nology are Lewis bases, inorganic bases, primary and secondary amines and amides.

Lewis bases are those compounds containing an atom with an unshared electron pair in its outer orbital. Typical examples are tertiary amines such as benzyldimethylamine, trimethylamine, pyridine, diethylethanolamine, and 2,4,6-tris(dimethylaminomethyl) phenol.

The polymerization through the epoxy groups is thought to take place as follows:

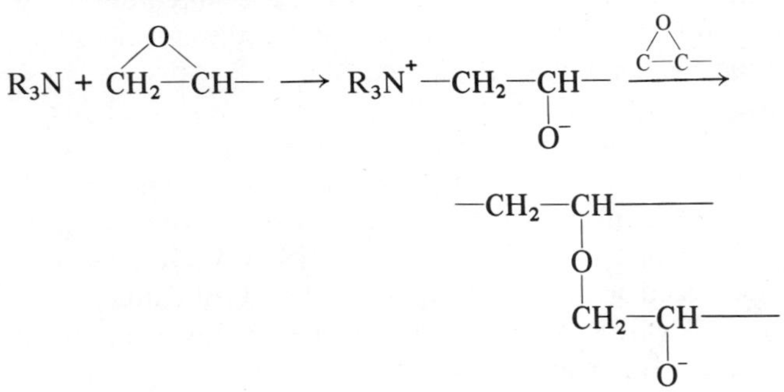

In actual practice, very few epoxy systems are cured with tertiary amines alone, as the resulting polymers do not yield particularly good properties. These amines, however, are widely used as accelerators for acid anhydride curing agents.

Although curing with inorganic bases such as alkali metal hydroxides is possible, it is not done commercially, primarily because reaction control is imprecise and final properties do not warrant the extra care needed to handle such systems.

Primary and secondary amines form the largest class of epoxy curing agents, especially for "cold-curing" systems. The epoxy-primary amine reaction is:

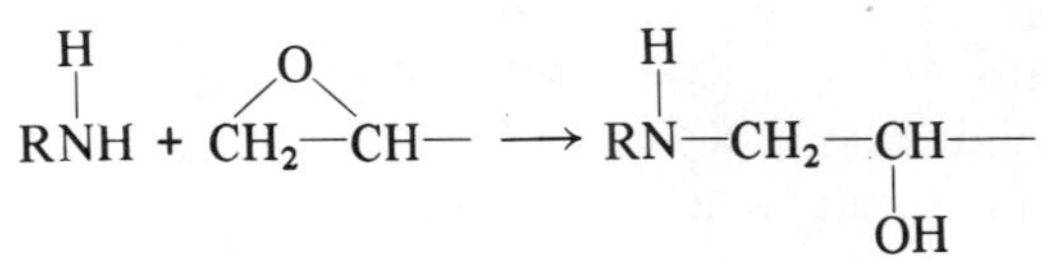

Table 6-37. Data for Cure of Epon 828 Liquid Resin.*

Agent	Recommended Concentration (phr)	Pot Life (Room Temp.) (min.)	Optimum Cure Schedule (min.)	Optimum Cure Schedule (°C)
Diethylentriamine	8	53	30	115
Dimethylaminopropylamine	6	55	60	80
Diethylaminopropylamine	6	210	60	115
Benzyldimethylamine	6	229	60	80
Piperidine	6	360	180	80
Diethylamine	12	360	150	115
Pyridine	15	25	120	80
Dicyandiamide	6	1440	30	165

*Shell Chemical Corp.

*From *Handbook of Epoxy Resins* by Lee and Neville. Copyright 1967 by McGraw-Hill, Inc. Used with permission of McGraw-Hill Book Co.

Table 6-38. Properties of DGEBA Cured with an Amidopolyamine from Tall-Oil Fatty Acid and TETA.*

Property	Amidopolyamine, phr			
	50	75	100	150
Pot life, minutes	75	63	57	60
Gel time, minutes	78	66	61	135
Exotherm, °C	209	190	169	133
Tensile strength, psi, cured 7 days at 25°C	8,400	6,000	1,780	200
Elongation, %	9	12	70	93
Tensile strength, psi, cured 7 days at 25°C + 70 hr at 100°C	8,620	6,160	2,670	620
Elongation, %	8	10	55	130
Hardness, Shore D, after 24-hr cure	80	74	56	10
Ultimate hardness/time elapsed days	80/1	76/2	60/2	15/2
Volume resistivity, ohm-cm	1.2×10^{14}	3.4×10^{13}	4.5×10^{12}	1.1×10^{11}
Surface resistivity, ohms	1.4×10^{11}	4.7×10^{9}	1.2×10^{9}	1.9×10^{9}
Dielectric constant	3.4	3.5	4.4	7.2
Dissipation factor	0.039	0.062	0.126	0.238

*From *Handbook of Epoxy Resins* by Lee and Neville. Copyright 1967 by McGraw-Hill, Inc. Used with permission of McGraw-Hill Book Co.

The epoxy–secondary amine reaction is:

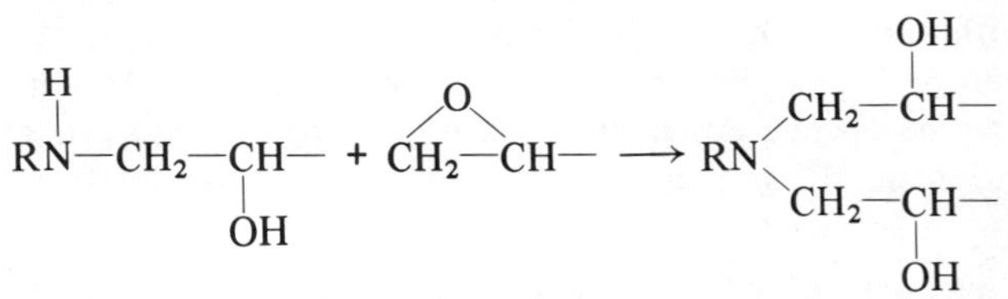

The hydroxyls formed are also capable of further reaction via the mechanisms to be described further on in this discussion.

Generally, the tertiary amines formed by the reaction of two epoxy groups with a primary amine are ineffective catalysts. Examples of typical primary amines are diethylenetriamine, triethylenetetramine, diethylaminopropylamine, *m*-phenylenediamine, diaminodiphenylsulfone, and aminoethylpiperazine.

Modified amines have also been used to improve overall cured resin performance. Modification most frequently occurs through adducting. Adducts of primary amines and alkyl mono- and diepoxides, ethylene and propylene oxide, acrylonitrile and aldehydes as well as Schiff's bases and amine-acid salt complexes have found wide use.

In general, aliphatic amines provide roughly equivalent properties up to the heat-deflection temperature of the systems. They will, as a rule, effect crosslinking at temperatures below 100°C (212°F). Aromatic amines, on the other hand, generally require exposure to temperatures over 100°C (212°F) for cure. The resulting resins have higher heat and chemical resistance along with improved physical properties.

Provided the stoichiometric ratio of amine to epoxy functionality is maintained, amines may be blended to generate "copolymerized" resins which exhibit desirable intermediate properties.

Piperidine and diethyanolamine are representative of secondary amines. Secondary amines, when used alone, may be considered a special class of tertiary amines in that after the secondary hydrogens have reacted, the resulting crosslinking is believed to occur via the tertiary amine mechanism described earlier. These curing agents generally have limited temperature resistance, poorer chemical resistance, and equivalent physical and electrical properties compared to primary aliphatic polyamines.

Many so-called amides used in epoxy resin formulations are actually amidopolyamines containing sufficient free amine hydrogens to react by the procedures described above. The reaction mechanisms have not been developed in the literature on the epoxy resins, although presumably the mechanisms are similar to those encountered with the primary and secondary amines.

These amidopolyamines generally introduce reduced volatility and skin irritation potential, more convenient epoxy/curing agent mix-ratios, and increased flexibility and impact strength to formulated systems. Other amides used are based on amine-terminated fatty polyamides such as those described in the chapter on nylons. Here again they provide improved resistance to skin irritation and increased flexibility. Urea- and melamine-formaldehydes, guanamines, as well as many other nitrogen-containing compounds have found specialized use as epoxy resin hardeners.

Imidazoles, such as:

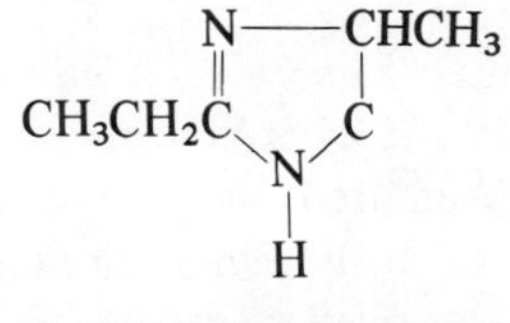

2-ethyl-4-methylimidazole

and dicyandiamide,

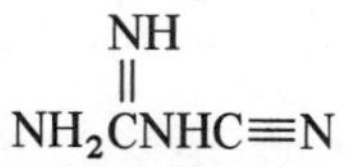

dicyandiamide

have long been used in so-called one-can or one-component epoxy systems because of their reactive latency at room temperature. Cure is heat activated and the resultant polymers generally exhibit high temperature, chemical, and electrical resistance.

The acid curing agents normally used in epoxy resins technology are Lewis acids, phenols, dibasic organic acids, and carboxylic acid anhydrides.

Lewis acids are those compounds containing empty orbital positions in the outer shell in one of the atoms. They are attracted to areas of increased electron density. Typical examples are the BF_3 adducts, the most common being BF_3-monoethylamine, and metal halides such as aluminum and ferric chloride.

It is believed that the cure mechanism for the adducts follows one of two paths. In the first, the complex dissociates to yield a proton and

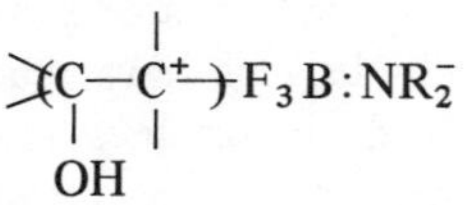

which is capable of chain propagation. In the second mechanism, the dissociation may be BF_3 and a secondary amine, which will then cause crosslinking independently.

The Lewis acids are more effective catalysts for glycidyl ether resins and epoxidized olefins. They are highly reactive and often are "capped" through adducting to control reaction rate and exotherm, the amine adducts being the most common. Once blocked, many Lewis acids provide the reactive latency required for one-component systems.

Phenols will coreact with epoxies at elevated temperatures. They will bring about independent etherification through reaction of epoxy groups with available hydroxyl groups.

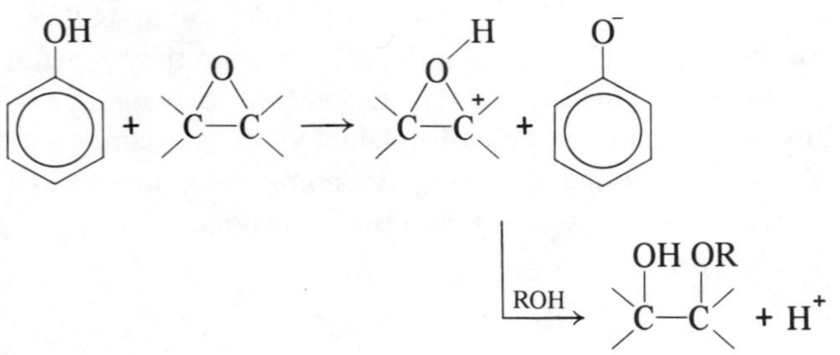

Phenols are often used to accelerate other curing reactions. Substituted phenols and phenol formaldehyde resoles and novolacs have also been used.

Organic acids react in a fashion similar to the phenols. First, an addition esterification occurs.

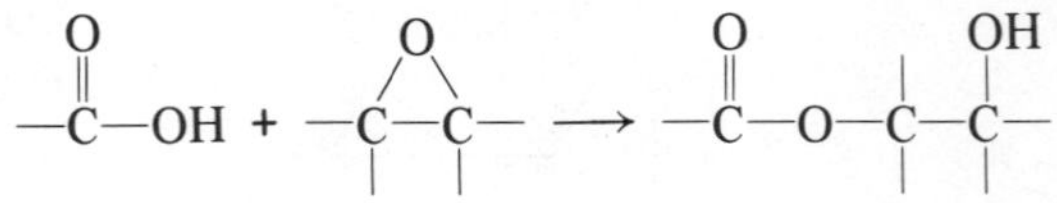

The alcoholic hydroxyl would then react preferentially by the mechanism described for phenol cures. Both reactions, addition esterification and etherification, are temperature dependent, with higher temperatures favoring the former, and lower temperatures the latter. The extent to which etherification will occur also depends on the nature of the carboxyl group and the location of the epoxy group in the molecule.

Tertiary amines are used as catalysts in order to enhance the esterification, exclude the etherification, and accelerate the rate of reaction. Polyhydric phenols and organometallic catalysts have also been used for these purposes.

Typical acids employed as accelerators include carboxy-terminated polyesters, dimerized and trimerized fatty acids, and maleic, oxalic, and phthalic acids.

In epoxy chemistry, only cyclic anhydrides are used. Typical examples are phthalic, dodecylsuccinic, maleic, hexa- hydro- and tetrahydrophthalic, Nadic methyl, and chlorendic. These anhydride groups will not react directly with epoxy groups. In order for the reaction to occur, it is first necessary that the anhydride ring be opened. The ring may be opened either by active hydrogens present as water or hydroxyls, or by a Lewis base.

Ring opening by H_2O:

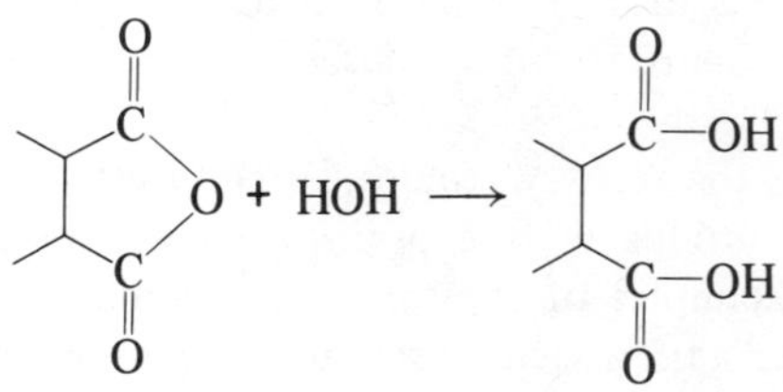

Ring opening by OH:

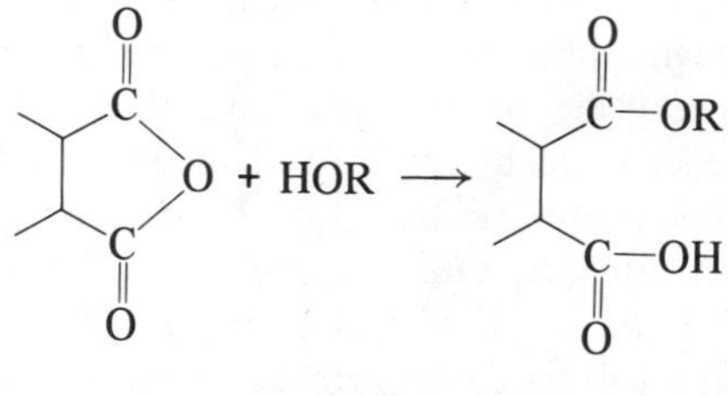

The final step would be an addition esterification as described for organic acids.

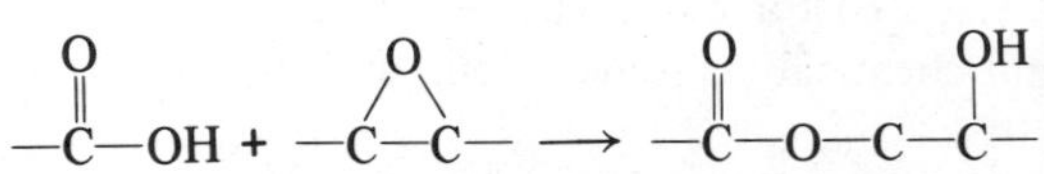

Anhydride systems generally provide low viscosity and long pot-life. They exhibit improved high-temperature stability over the amines and better physical and electrical properties above their heat-deflection temperature. They generally require elevated temperature cures and postcuring at temperatures above 150°C (302°F) is often needed for maximization of properties.

Reactions of the anhydride with the resin are complex and depend on a number of factors: gel time and temperature, postcure time and temperature, presence or absence of accelerator, type of accelerator, amount of hydroxyl present, ratio of anhydride to epoxy (A/E), and amount of free acid.

In almost all commercial epoxy resin/anhydride formulations, some form of accelerator is employed. Acid accelerators favor etherification, and basic accelerators favor esterification. The optimum A/E ratio, and thus the cured properties, are determined to a large degree by the accelerator used.

The variety of accelerators suggested for anhydride cures is considerable. Tertiary amines are more conventionally used accelerators. Preferred tertiary amines for catalyzing the epoxy-anhydride reaction are benzyldimethylamine (BDMA) and tris(dimethylaminomethyl)-phenol (DMP-30). Optimum amounts of tertiary amine accelerators in terms of cured properties fall in the 1-5 phr range, with the value depending on the specific system.

A variety of mercaptan-containing (-SH) compounds have been investigated as epoxy resin coreactants. Commercially, polysulfides having the following general structure have found the most acceptance.

$$HS{-}\!(\!-C_2H_4{-}O{-}CH_2{-}O{-}C_2H_4{-}S{-}S\!-\!)_n{-}C_2H_4{-}OCH_2OC_2H_4{-}SH$$

These compounds contain occasional mercaptans randomly distributed along the chain, but for the most part are considered bifunctional. These curing agents generally are used to develop more flexible epoxy systems.

In recent years polymercaptans have been marketed as extremely fast (seconds to minutes) room-temperature curing agents. They also enable curing to occur at very low temperatures (-40-0°C, -40-32°F). There are BF_3 adducts which will cure in seconds at room temperature, but they are not active at lower temperatures. These polymercaptans are the basis of the so-called five-minute epoxy repair kits.

Specialized epoxy systems, primarily surface coatings and inks, have been developed recently which cure under the influence of UV radiation. Such systems cure very rapidly at room temperature with little change in ultimate properties compared to standard curing formulas. Equipment costs however are high, and short-run production is not economic.

In the past, many techniques were tried to develop a usable, one-component, shelf-stable epoxy system. Most of these techniques were based on latent catalysts selection and formulation methods. As a rule, these systems were slow curing, had poor shelf-life and minimum pot-life, and required sensitive processing.

With the development of reaction injection molding (RIM) processing, pultrusion, and thermoset injection molding, the need for more adaptable epoxy formulas became more urgent. Consequently, a variety of new "moldable" epoxies have been introduced. These new systems can be adapted for traditional processes such as compression and transfer molding and filament winding, but also for reaction and liquid injection molding, pultrusion, and SMC and BMC processing.

Two of these new systems bear noting. The first is a series of epoxies designed for SMC and BMC processing. They are exceptionally moldable. Fiberglass-reinforced parts from 0.050 to 1.000 inch can be made without loss of fiber/matrix adhesion or generation of interior or surface voids. Typical fiberglass-reinforced epoxy properties such as strength, heat resistance, fatigue, deformation, dimensional stability, corrosion resistance, and electrical properties are maintained.

The second is a series of formulations in powder, pellet, and liquid form. The mineral-filled granular compound is very shelf-stable with an ultrafast cure at temperatures around 149°C (300°F). It offers a 600-volt IEC Tracking Index, 220 seconds arc-track resistance, and a V-O flammability rating. No refrigeration is required for storage. This resin is designed primarily for the electronic and electrical industry.

A glass-reinforced pelletized resin is designed for metal replacement applications and is moldable in injection molding machinery. The resin has unlimited colorability, high gloss, high temperature modulus with notched Izod impact, excellent chemical resistance, and requires only low molding pressures.

The third resin is available as a one-component liquid for pultrusion, filament winding, and composite fabrication (SMC and BMC), as well as liquid injection processing. It is volatile-free, which simplifies fabrication conditions.

Not all epoxy groups will react with all curing agents. When reaction does occur, the kinetics may vary considerably. In general, basic types of curing agents react with the terminal carbon atom and the acid-catalyzed reactions proceed via attack on the oxygen atom.

Some further generalizations about the various curing agents can be made. The primary amines provide well-cured resins, even at room temperature, that are rigid, adhesive, and chemically resistant. The aromatic diamines have improved heat and chemical resistance, but require a heat cure. The cure is often a two-step cure, one at a lower temperature to minimize exotherm, and one at a higher temperature to maximize properties. Most amines are skin sensitizers, unless adducted or otherwise modified, and fumes are frequently objectionable.

Amides require heat cure and can provide the latency required for one-component systems. Fatty amides can be cured at room temperature. They increase impact strength

Table 6-39. Curing Agents for Solvent-Free Epoxy Coatings.*

CURING AGENT	ADVANTAGES	DISADVANTAGES
POLYAMIDES	GOOD WATER RESISTANCE; LOND POT LIFE; GOOD FLEXIBILITY	POOR CHEMICAL RESISTANCE; SLOW CURING; POOR SURFACE FINISH
AMIDO-AMINES	LOW VISCOSITY; GOOD ADHESION TO CONCRETE; GOOD WATER RESISTANCE	POOR CHEMICAL RESISTANCE; SLOW CURING
ALIPHATIC AMINES	GOOD CHEMICAL RESISTANCE	TOXIC; GREASY SURFACE FINISH
ALIPHATIC AMINE ADDUCTS	LESS TOXIC THAN STRAIGHT AMINES	POOR SURFACE FINISH
MODIFIED ALIPHATIC AMINES	RAPID CURE IN THIN FILMS; CURE AT LOW TEMPERATURES	SHORT POT-LIFE
AROMATIC AMINES	EXCELLENT CHEMICAL RESISTANCE; CURE UNDER ADVERSE CONDITIONS; BLOOM-FREE FINISH	POOR COLOUR AND COLOUR STABILITY
CYCLOALIPHATIC AMINES	LOW VISCOSITY; GOOD COLOUR AND COLOUR STABILITY	EXPENSIVE

*Courtesy: Pacific Anchor Chemical Co.

and flexibility, and reduce the skin-sensitization potential. Table 6-39 shows a generalized comparison of various amine and amide curing agents used in solvent-free epoxy coatings.

Acid anhydrides provide improved thermal stability over the amines. Cured resins are usually brittle and require flexibilization. The anhydrides are relatively free of harmful dermatitic effects although some, when heated, release noxious fumes.

As with most thermosetting plastics and many thermoplastics, particularly vinyls, epoxy resins can be further modified through formulation and compounding. We have often referred to the effect of reinforcing and non-reinforcing fillers, lubricants, plasticizers, and other compounding ingredients. In the case of two-component systems, both resin and curing agent may be compounded. The properties of epoxy systems are thus susceptible to a very wide spectrum of modifications and adjustments.

Table 6-40 shows the composition of a typical epoxy formulation. By varying the ingredients in Table 6-40, one can generate systems which range in character from low-viscosity clear liquids, to highly filled viscous pastes or free-flowing molding powders. The systems can be adapted to a wide variety of processing techniques.

We have already discussed the characteristics of the various resins and curing agents; the following section examines in more detail the other items listed in Table 6-40.

Table 6-40. Typical Epoxy Formulation.

Epoxy resin
Resinous modifier
Epoxy-containing reactive diluent
Non-reactive diluents
Fillers (reinforcing and/or non-reinforcing)
Colorants (pigments and dyes)
Rheological additives (thixotropes, viscosity suppressants)
Property promoters (adhesion promoters, wetting agents)
Processing aids (deaerating agents, mold release agents)
Curing agent(s)
Catalysts and accelerators

Resinous modifiers—A variety of heteropolymers and alloys can be developed when epoxies are combined or coreacted with suitably compatible resins. Included in this group are resins such as polyvinyl chloride, unsaturated and saturated polyesters, isocyanates and polyurethane resins, silicone resins, furfural resins, coal tar and other phenolic modifiers, acrylic resins, and butadiene-acrylonitrile resins. Most are used as low-cost extenders; others are used to impart specialized properties which are, in large part, contributed by the modifying resins. Such specialized properties include improved peel strength, anti-cratering and chalk resistance of coatings, wetting and adhesion to substrates, chemical resistance, and flexibility.

Epoxy-containing reactive diluents—These are either monoepoxy molecules or low-viscosity epoxy resins (shown in Tables 6-41 and 6-42). The former includes butyl, phenyl, cresyl, or allyl glycidyl ether (BGE, PGE, CGE, AGE), styrene oxide, and glycidyl methacrylate. These diluents are generally highly skin sensitizing, provide significant viscosity reduction at relatively low concentration, and adversely affect cured properties. Table 6-43 shows the viscosity reduction efficiency of a number of monofunctional diluents.

The low-viscosity epoxy resins include such materials as butadiene dioxide, vinyl cyclohexane dioxide, and the diglycidyl ether of resorcinol. These diluents may be used at high concentrations to achieve viscosity reduction with little effect upon cured properties and, in some instances, to improve selected properties.

Non-reactive diluents—These materials generally degrade cured properties to a degree that is a direct function of the concentration used. They modify not only the system viscosity but will affect pot-life and exotherm as well. Examples of such diluents include dibutyl and dioctyl phthalate (DBP, DOP), monomeric styrene, bisphenols, and hydrocarbon oils.

Many commercial epoxy resins come preblended with reactive and non-reactive diluents. Tables 6-44 and 6-45 present a selected listing of available types.

The effect of diluent concentration on the viscosity of DGEBA blends is shown in Figure 6-44. The effect of

Table 6-41. Typical Epoxy Resin Reactive Diluents.*

Butadiene dioxide (mol. wt. 86)

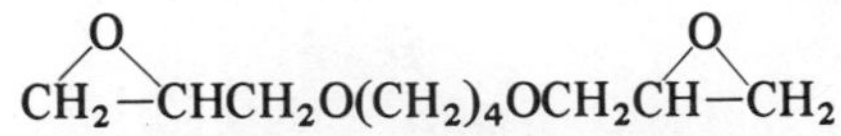

Butanediol diglycidyl ether (mol. wt. 202)

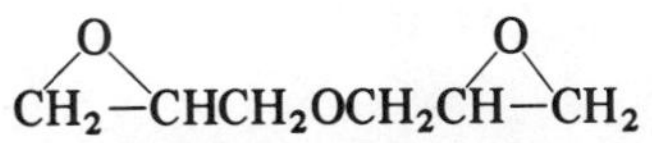

Diglycidyl ether (mol. wt. 130)

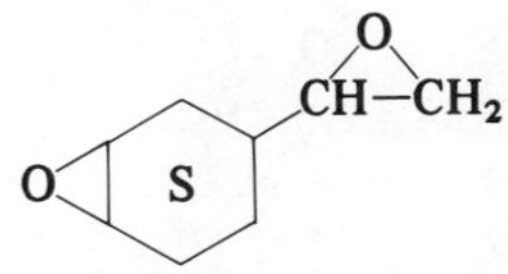

Vinylcyclohexane dioxide (mol. wt. 140)

Diethylene glycol diglycidyl ether (mol. wt. 218)

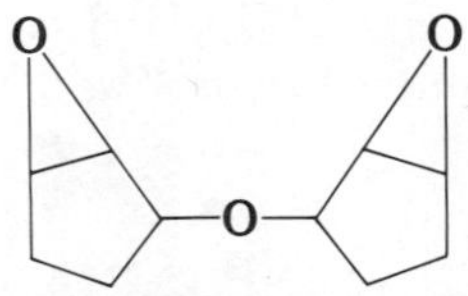

Bis(2,3-epoxycyclopentyl) Ether (mol. wt. 182)

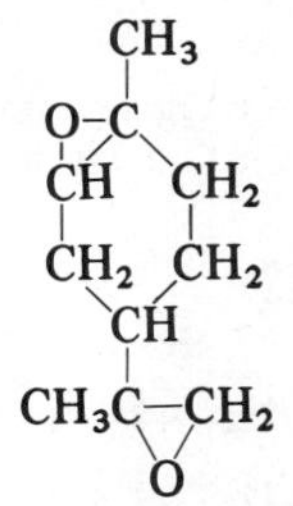

Limonene dioxide (mol. wt. 168)

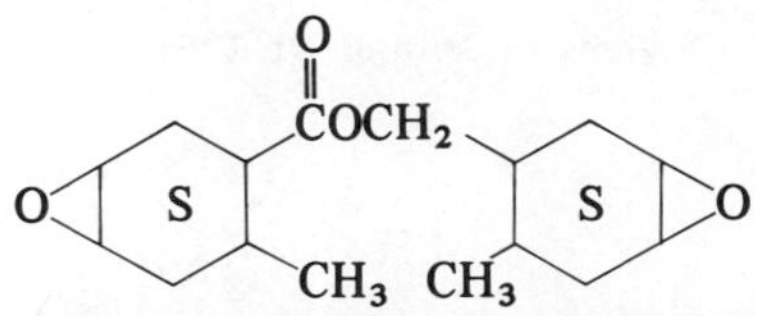

3,4-Epoxy-6-methylcyclohexylmethyl 3,4-epoxy-6-methylcyclohexane carboxylate (mol. wt. 280)

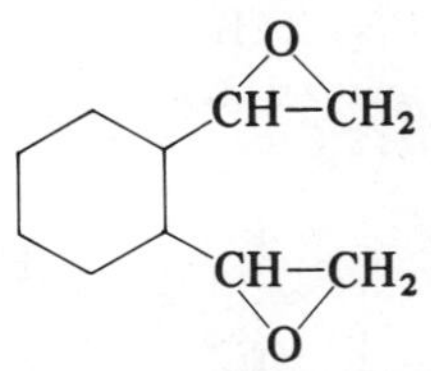

Divinylbenzene dioxide (mol. wt. 162)

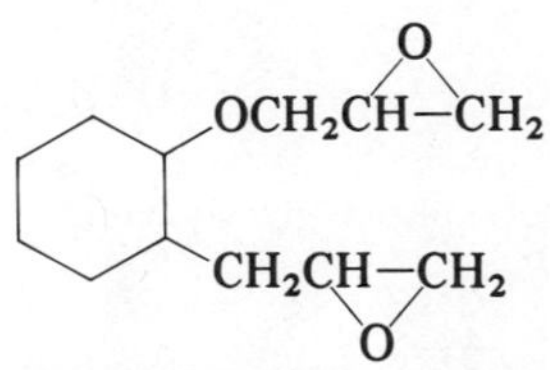

2-Glycidyl phenyl glycidyl ether (mol. wt. 206)

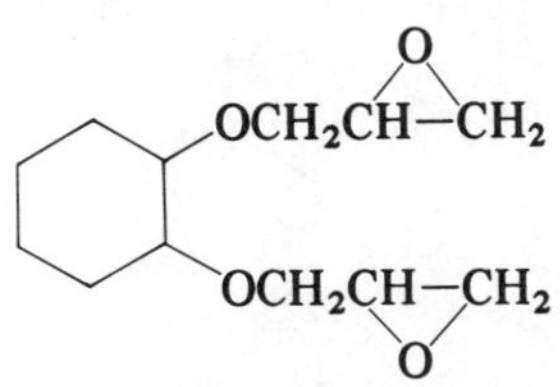

Diglycidyl ether of resorcinol (mol. wt. 222)

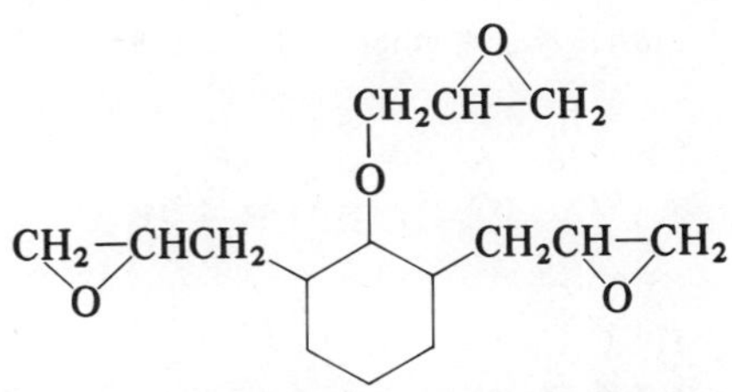

2,6-Diglycidyl phenyl glycidyl ether (mol. wt. 262)

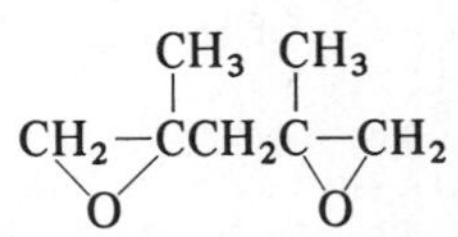

Dimethylpentane dioxide (mol. wt. 128)

*From *Handbook of Epoxy Resins* by Lee and Neville. Copyright 1967 by McGraw-Hill, Inc. Used with permission of McGraw-Hill Book Co.

diluents on properties of an epoxy-amine system is shown in Table 6-46.

Fillers—As in other plastics materials, fillers are employed for a variety of reasons. They include materials like powdered metals, alumina, silica, mica, talc, calcium carbonate, barium sulfate, wollastonite, carbon, and graphite. The addition of progressively more filler increases system viscosity and specific gravity. Specialized fillers impart thixotropy and some, like hollow glass or phenolic microballoons, create a syntactic foam of reduced density. Generally, with the addition of fillers, pot life is extended and reaction exotherm is reduced. Particulate fillers only mar-

Table 6-42. Commercial Epoxy-Containing Monofunctional Reactive Diluents.*

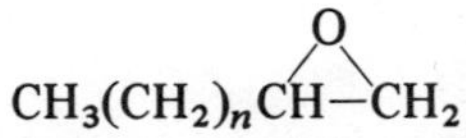

Olefin oxides, b.p. 168°C at 37 mm

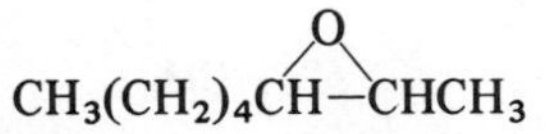

Octylene oxide (mol. wt. 128) 85%, 15% 1,2-isomer b.p. 77°C at 45 mm

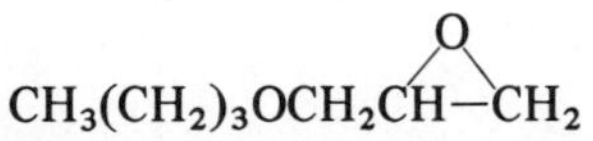

Butyl glycidyl ether (mol. wt. 130)

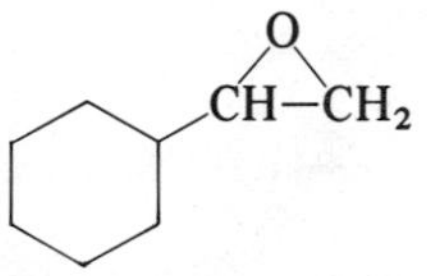

Styrene oxide (mol. wt. 120)

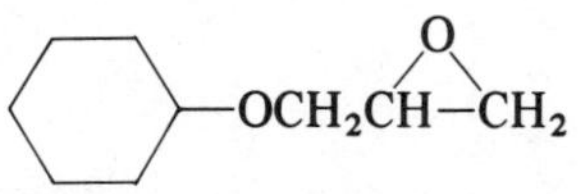

Phenyl glycidyl ether (mol. wt. 150)

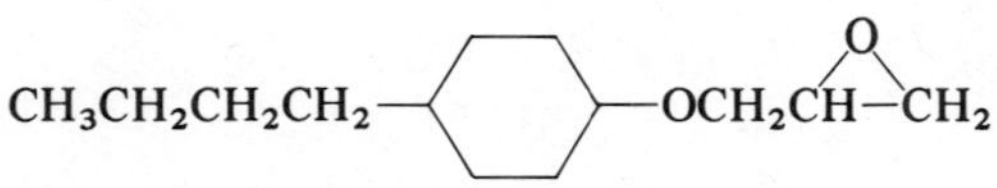

p-Butyl phenol glycidyl ether

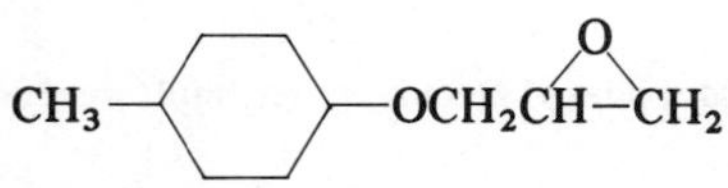

Cresyl glycidyl ether (mol. wt. 165)

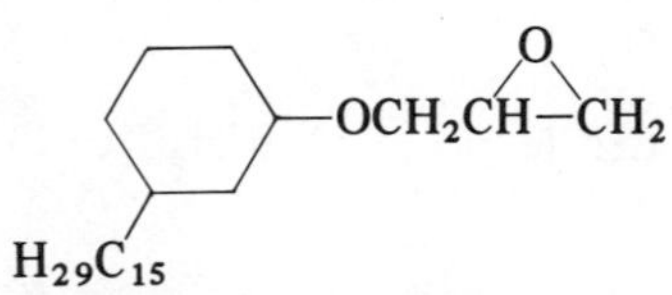

3 (Pentadecyl) phenol glycidyl ether

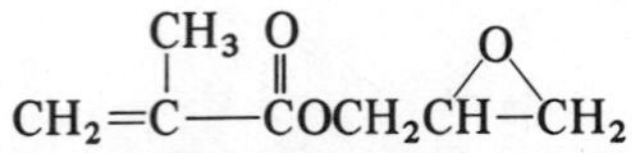

Glycidyl methacrylate (mol. wt. 142) b.p. 75°C at 10 mm

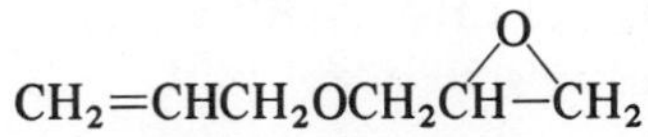

Allyl glycidyl ether (mol. wt. 114)

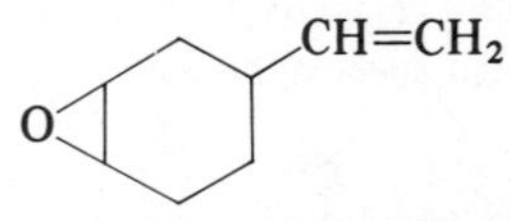

Cyclohexene vinyl monoxide (mol. wt. 124), b.p. 169°C at 760 mm

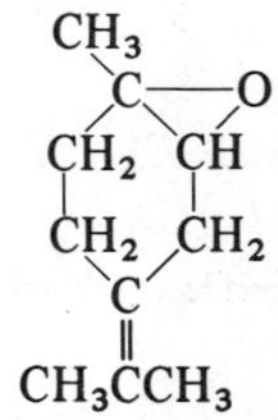

Dipentene monoxide (mol. wt. 152), b.p. 75°C at 10 mm

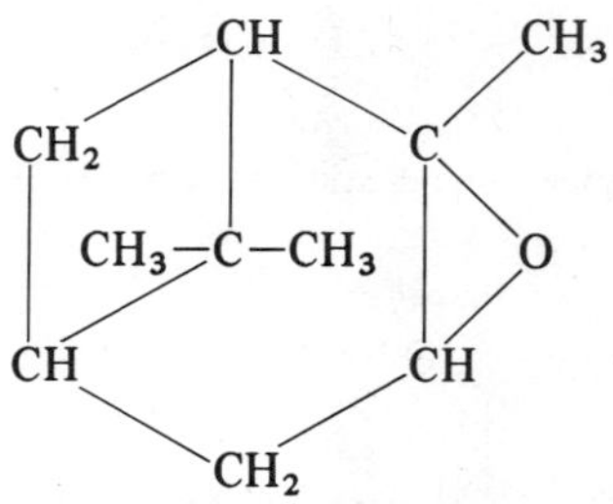

α-Pinene oxide (mol. wt. 152), b.p. 62°C at 10 mm

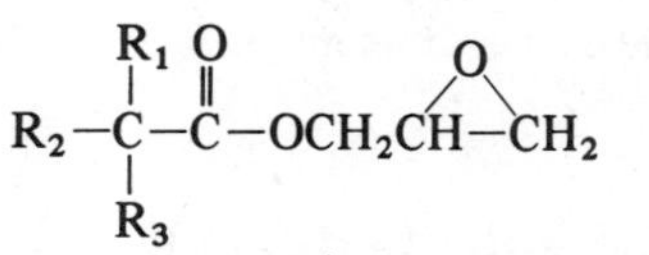

($C_{12-14}H_{22-26}O_3$) Glycidyl ester of *tert*-carboxylic acid (mol. wt. 240–250), b.p. 135°C at 760 mm

*From *Handbook of Epoxy Resins* by Lee and Neville. Copyright 1967 by McGraw-Hill, Inc. Used with permission of McGraw-Hill Book Co.

ginally affect physical properties, whereas fibrous reinforcing fillers impart significant increases in tensile, compressive and flexural strengths and moduli. Table 6-47 shows the effect of fillers on the properties of an epoxy-amine system. Thermal and electrical conductivity can be varied widely—in some cases of metallic fillers, an epoxy can be transformed from an insulator to a conductor. Chemical and weather resistance is only marginally affected by the inclusion of fillers.

It becomes readily apparent that a listing of epoxy prop-

Table 6-43. Viscosity Reduction Efficiency of Various Monofunctional Diluents Used with DGEBA.*

Diluent	Phr Diluent Required to Obtain Listed Viscosity: 1,500 Centipoises	500 Centipoises	200 Centipoises
High-efficiency diluents:			
Allyl glycidyl ether	6.8	13.0	20.0
Octylene oxide	7.2	13.2	20.5
Butyl glycidyl ether	7.2	13.5	21.0
Styrene oxide	5.3	19.3	30.0
Medium-efficiency diluents:			
Glycidyl methacrylate monomer	10.0	19.0	30.0
Phenyl glycidyl ether	15.0	29.0	49.0
Low-efficiency diluent:			
Cresyl glycidyl ether	19.0	38.0	65.0
3-(Pentadecyl) phenol glycidyl ether	27.0	64.0	–

*From *Handbook of Epoxy Resins* by Lee and Neville. Copyright 1967 by McGraw-Hill, Inc. Used with permission of McGraw-Hill Book Co.

Table 6-44. Lowest-Viscosity Diluted Resins.*

Products	EEW	Viscosity (cps @ 25°C)	Color[1]	Diluent
Araldite 506	172–185	500–700	3	BGE–12%
Araldite 507	185–192	500–700	7	CGE
Araldite 509	189–200	500–700	NS	Epoxide 7
Epi-Rez 504	170–180	150–210	3	BGE–20%
Epi-Rez 5071	180–195	500–900	3	BGE–12%
Epi-Rez 5077	185–200	500–700	7	CGE
Epon 815	175–195	500–700	5	BGE–12%
Epotuf 37-130	175–195	500–900	3	BGE–12%
Epotuf 37-137	185-200	500–700	3	CGE

These resins are similar to D.E.R. 331 resin diluted with a reactive diluent to reduce viscosity. This permits easier laminating and/or higher filler loading for cost reduction. The resins are suitable for most general purposes where use temperatures will not exceed 200°F and are generally used with polyamine or polyamide hardeners.

*Courtesy: Dow Chemical Co.
[1] Gardner

Table 6-45. Medium-Viscosity Diluted Resins.*

Products	EEW	Viscosity (cps @ 25°C)	Color[1]	Diluent
Araldite 502	232–250	2,100–3,600	3	DPB
Epi-Rez 50840	190–210	2,000–5,000	5	Water Dispersible
Epon 820	180–194	4,000–10,000	5	PGE

General comments on utility of PGE-containing resins same as for D.E.R. 330 and D.E.R. 331 resins, but the PGE introduces greater toxicity and dermatitis problems.

The DBP modified resins are generally suitable for uses only at ambient temperatures, are cured with polyamines and polyamides, and are generally softer, less brittle, and less solvent resistant than 100% reactive resins.

*Courtesy: Dow Chemical Co.
[1] Gardner

erties is highly dependent on the nature of the formulation. As with vinyls and phenolics, a broad tabulation can be made and is shown in Table 6-48.

To summarize epoxy resins as a class we can list the following generalizations. Epoxy resins are resins ranging from low-viscosity liquids to high molecular weight solids which readily convert to the thermosetting phase upon the admixture of the proper curing agent. There are other resins—phenolics, polyesters, acrylics, etc.—which cure in a similar fashion, but the epoxy resins possess a rather unique combination of properties. The liquid resins and their curing agents can form low-viscosity, easy-to-process (or modify) systems. Epoxy resins cure quickly and easily at practically any temperature from 5°C to 150°C (41–302°F), depending on the selection of the curing agent. One of the most advantageous properties of the epoxy resins is their low shrinkage during cure. Phenolic casting resins which evolve water reveal high shrinkage, as do the acrylic and polyester resins which must rearrange and reorient considerably in the liquid and semi-gelled phase. Epoxy resins react with very little rearrangement and with no volatile by-products being evolved.

Table 6-46. Effect of Diluent on Properties of TETA-Cured DGEBA.*

Diluent and Mixed Viscosity, Centipoises	Pot Life at 23°C	Exo-therm, °C	Water Absorption, Weight Increase, %	Flexural Strength, psi	DT, °C	Weight Loss in 48 hr at 200°C, %
None	40	200	0.73	20,300	120	1.89
Octylene oxide:						
1,500	55	205	1.10	14,185	74	5.20
500	72	180	1.18	11,440	65	Unstable
200	100	164	1.04	7,797	58	Unstable
Styrene oxide:						
1,500	48	200	0.71	19,403	93	2.39
500	69	193	0.80	18,854	82	4.29
200	84	178	1.06	16,303	70	6.20
3(Pentadecyl) phenol glycidyl ether:						
1,500	54	164	1.19	14,010	76	2.84
500	68	148	1.48	6,443	55	4.15
Butyl glycidyl ether:						
1,500	55	204	1.01	18,433	101	2.12
500	63	207	1.32	17,237	85	2.72
200	76	193	1.63	16,400	75	3.77
Allyl glycidyl ether:						
1,500	48	204	1.11	18,817	95	1.86
500	59	194	1.51	18,047	76	2.37
200	66	189	1.72	15,827	74	3.14
Phenyl glycidyl ether:						
1,500	47	198	1.04	19,717	–	2.37
500	49	209	1.26	19,523	–	3.28
200	54	198	1.35	19,600	–	4.93
Cresyl glycidyl ether:						
1,500	46	202	0.98	22,067	92	2.98
500	46	204	1.24	20,000	78	4.40
200	51	201	1.35	15,230	63	6.58
Glycidyl methacrylate:						
1,500	41	208	1.30	18,507	105	2.93
500	43	224	1.64	17,510	95	3.78
200	46	224	2.02	18,400	87	4.43

*From *Handbook of Epoxy Resins* by Lee and Neville. Copyright 1967 by McGraw-Hill, Inc. Used with permission of McGraw-Hill Book Co.

Although both phenolics and polyesters are cheaper than epoxies, they have other inherent disadvantages aside from shrinkage that make epoxy a preferred material. Cure times for phenolics can be inordinately long and the water of condensation must be removed for optimization of properties. In critical applications, the peroxy catalysts commonly employed with polyesters may cause corrosion of embedded or contacted metals such as copper. Similar objections apply to resins based on the polymerization of styrene or acrylate esters. In addition, they are thermoplastics and soften with heat.

The silicones have excellent electrical properties combined with usefulness over a wide temperature range, but they are very expensive, require high curing temperatures, and do not adhere as well to metals, glass, or other resins.

Fluorocarbons, also expensive but outstanding in electrical properties and wide temperature range, are not adaptable to no-pressure casting techniques.

Figure 6-45 compares the weight loss at 175°C of various polymers with epoxy.

Because of their chemical make-up, chiefly the presence of polar hydroxyl and ether groups, the epoxy resins are excellent adhesives. The resins cure with low shrinkage, so that the various surface contacts set up between the liquid epoxy resin formulation and the adherends are not disturbed during cure. Adhesive strength, without the need for either open time or high pressures, is perhaps the best obtainable in contemporary plastics technology. The strength of properly formulated epoxy resins usually surpasses that of other types of casting resins. This is probably in part a result of their low shrinkage, which minimizes stresses that otherwise would weaken the mechanical structure.

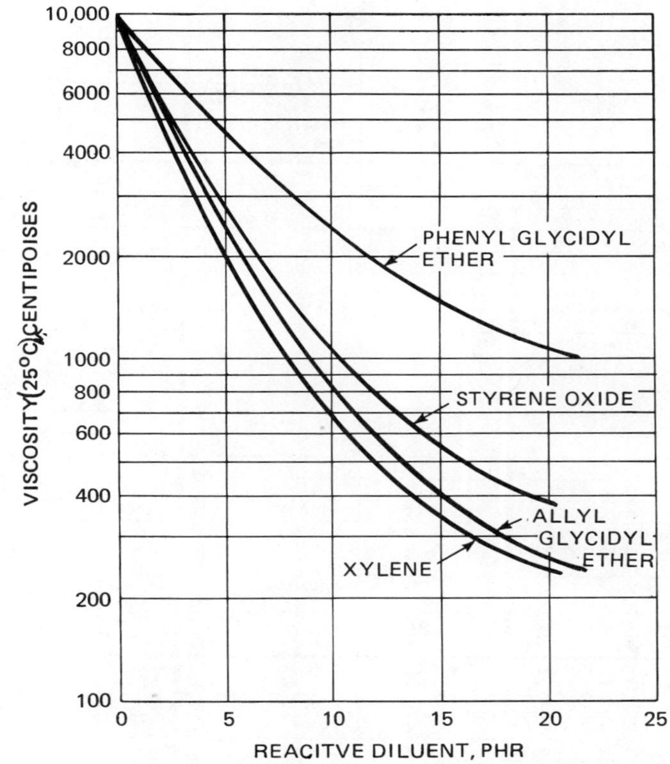

Fig. 6-44. Diluent concentration versus viscosity for DGEBA. (From *Handbook of Epoxy Resins* by Lee and Neville. Copyright 1967 by McGraw-Hill, Inc. Used with permission of McGraw-Hill Book Co.)

Table 6-47. Effect of Fillers on Viscosity, Exotherm, Modulus of Rupture, and Shrinkage of DGEBA Cured with Primary Amine Curing Agent.*

	Viscosity at 25°C, centipoises				Peak exothermic temperature rise, °C				Working life, minutes				Modulus of rupture psi	Shrinkage, %
					% filler				% filler					
Filler	20% filler	35% filler	50% filler	65% filler	20	35	50	65	20	35	50	65	20% filler	20% filler
Unfilled	1,100				223°C				48 min				18,174	0.91
Asbestine	3,850	23,650	100,000	–	158	122	55	–	53	60	73	–	10,920	0.48
Albalith	2,700	9,600	95,000	–	169	133	76	–	47	66	85	–	6,896	0.59
Silica	2,750	6,800	51,500	–	170	115	53	–	52	72	95	–	16,875	0.77
Mica	2,100	4,800	54,500	–	159	111	51	–	50	71	94	–	12,358	0.66
Calcium carbonate	2,200	3,500	22,600	100,000	186	125	55	33	49	57	70	80	11,970	0.64
Flint powder	2,200	4,300	39,500	–	178	113	53	–	52	75	96	–	17,566	0.41
Quartz	2,100	3,750	13,150	100,000	178	113	51	34	49	74	107	130	14,100	0.67
Cryolite	1,900	2,750	6,150	20,900	166	113	56	36	54	71	107	120	15,483	0.54
Portland cement	1,950	2,400	4,100	19,300	183	147	77	41	52	62	89	102	14,370	0.69
Limestone	1,700	3,250	10,000	31,700	179	128	59	37	49	71	90	95	12,433	0.47
Atomized aluminum 101	2,100	3,100	4,600	34,500	174	127	47	37	56	68	100	105	13,710	0.80
Barytes	1,700	2,450	4,200	17,200	179	152	83	38	49	60	84	116	17,417	0.71
Atomized aluminum 100	1,900	2,950	5,500	33,500	176	115	52	35	53	70	99	105	–	–
Atomized aluminum 120	1,800	3,000	5,900	35,000	178	128	53	33	52	67	100	105	14,790	–

*From *Handbook of Epoxy Resins* by Lee and Neville. Copyright 1967 by McGraw-Hill, Inc. Used with permission of McGraw-Hill Book Co.

Table 6-48. Properties

		ASTM test method	Encapsulating grades: Mineral filled	Encapsulating grades: Glass filled	Phenol novolac epoxy resins	Cycloaliphatic epoxy resins	Ortho cresol novolac epoxy resins
Processing	1. Molding qualities	—	Good to excel.	Good	Good	—	Good
	2. Compression molding temp., °F.	—	250-330 (Transf.)	280-330 (Transf.)	250-330	—	250-330
	3. Comp. molding pressure, p.s.i.	—	50-1000 (Transf.)	50-1000 (Transf.)	100-800	—	100-800
	4. Injection molding temp., °F.	—	—	—	—	—	—
	5. Injection molding pressure, p.s.i.	—	—	—	—	—	—
	6. Compression ratio	—	2.0	3.0	—	—	—
	7. Mold (linear) shrinkage, in./in.	—	0.004-0.010	0.004-0.008	—	—	—
	8. Specific gravity (density)	D792	1.7-2.1	1.7-2.0	1.16-1.21	1.16-1.21	1.16-1.21
	9. Specific volume, cu. in./lb.	D792	14.0	14.0	—	—	—
	10. Machining qualities	—	Fair	Fair	Good	Good	Good
Mechanical	11. Tensile strength, p.s.i.	D638	4000-10000	5000-15000	6000-12000	8000-12000	—
	12. Elongation, %	D638	—	—	2.0-6.0	2.0-10.0	—
	13. Tensile elastic modulus, 10^5 p.s.i.	D638	—	—	—	—	—
	14. Compressive strength, p.s.i.	D695	18000-30000	18000-30000	—	15000-20000	—
	15. Flexural yield strength, p.s.i.	D790	6000-15000	8000-20000	—	10000-13000	—
	16. Impact strength, ft. lb./in. of notch (½ × ½ in. notched bar, Izod test)	D256	0.3-0.45	0.5-2.0	—	—	—
	17. Hardness, Rockwell	D785	M100-M112	M100-M112	—	—	—
	18. Flexural elastic modulus, p.s.i. × 10^5	D790	—	—	—	—	—
	19. Compressive modulus, p.s.i. × 10^5	D695	—	—	—	—	—
Thermal	20. Thermal conductivity, 10^{-4} cal./sec./sq. cm./1 (°C./cm.)	C177	4.0-10.0	4.0-10.0	—	—	—
	21. Specific heat, cal./°C./gm.	—	—	—	—	—	—
	22. Thermal expansion, 10^{-5} in./in./°C.	D696	3.0-6.0	3.0-5.0	—	—	—
	23. Resistance to heat, °F. (continuous)	—	300-450	300-450	—	—	—
	24. Deflection temp., °F. at 264 p.s.i. fiber stress	D648	225-450	225-450	300-500	200-450	300-500
	at 66 p.s.i. fiber stress		—	—	—	—	—
Electrical	25. Volume resistivity, ohm-cm. (50% RH and 23° C.)	D257	$<10^{14}$	$>10^{14}$	10^{15}	10^{15}	10^{15}
	26. Dielectric strength, short-time, ⅛-in. thickness, volts/mil	D149	250-400	250-400	—	—	—
	27. Dielectric strength, step-by-step, ⅛-in. thickness, volts/mil	D149	250-400	250-400	—	—	—
	28. Dielectric constant, 60 cyc.	D150	3.5-5.0	3.5-5.0	3.7	3.2	2.98
	29. Dielectric constant, 10^3 cyc.	D150	3.5-5.0	3.5-5.0	3.7	—	2.95
	30. Dielectric constant, 10^6 cyc.	D150	3.5-5.0	3.5-5.0	3.4	3.2	2.88
	31. Dissipation (power) factor, 60 cyc.	D150	0.01	0.01	0.003	0.005	0.005
	32. Dissipation (power) factor, 10^3 cyc.	D150	0.01	0.01	0.012	—	0.007
	33. Dissipation (power) factor, 10^6 cyc.	D150	0.01	0.01	0.024	0.0016	0.010
	34. Arc resistance, sec.	D495	120-180+	120-180+	—	Excellent	Excellent
Optical	35. Refractive index, n_D	D542	—	—	—	—	—
	36. Clarity	—	Opaque		Transparent	Transparent	Transparent
	37. Transmittance, %	—	—		—	—	—
	38. Haze, %	—	—		—	—	—
Resistance characteristics	39. Water absorp., 24 hr., ⅛-in. thick., %	D570	0.03-0.20	0.04-0.20	Excellent	—	Good
	40. Flammability[a] Burning rate, in./min.	D635	—	—	—	—	—
	AEB, in./ATB, sec.[b]		—	—	—	—	—
	41. Effect of sunlight	—	Slight	Slight	Darkens	None	Darkens
	42. Effect of weak acids	D543	None	None	None	None	None
	43. Effect of strong acids	D543	Slight	Slight	—	Attacked by some	Attacked by some
	44. Effect of weak alkalies	D543	Slight		None	None	None
	45. Effect of strong alkalies	D543	Slight to attacked		None	Attacked by some	Attacked by some
	46. Effect of organic solvents	D543	Slight	Slight	None	None	None

a—This numerical flame spread rating is not intended to reflect hazards presented by this or any other material under actual fire conditions.
b—AEB = Avg. extent of burning in inches; ATB = Avg. time of burning in seconds.
c—Filled with microballoons or other hollow spheres.
*Reprinted by permission from *Modern Plastics Encyclopedia*, McGraw-Hill, Inc.

of Epoxies.*

Cast resins		Cast resins		Molding compounds		
No filler	Silica filled	Aluminum filled	Flexibilized	Glass fiber filled	Mineral filled	Low density[c]
—	—	—	—	Excellent	Excellent	Good
—	—	—	—	300-330	250-330	250-320
—	—	—	—	300-5000	100-3000	100-2000
—	—	—	—	—	—	250-300
—	—	—	—	—	—	100-1500
—	—	—	—	3.0-7.0	2.0-3.0	3.0-7.0
0.001-0.01	0.0005-0.008	0.001-0.005	0.001-0.01	0.001-0.005	0.002-0.008	0.006-0.010
1.11-1.40	1.6-2.0	1.4-1.8	1.05-1.35	1.6-2.0	1.6-2.0	0.75-1.00
24.9-20.0	17.3-13.9	—	20.5	15.4-14.0	14.2-13.4	
Good	Poor	Excellent	Good	Fair	Fair	Good
4000-13000	7000-13000	7000-12000	2000-10000	10000-20000	5000-10000	2500-4000
3.0-6.0	1.0-3.0	0.5-3.0	20.0-70.0	4		
3.5	—	—	0.01-3.5	30.4		
15000-25000	15000-35000	15000-33000	1000-14000	25000-40000	18000-40000	10000-15000
13300-21000	8000-14000	8500-24000	1000-13000	10000-60000	8000-15000	5000-7000
0.2-1.0	0.3-0.45	0.4-1.6	3.5-5.0	2.0-30.0	0.3-0.4	0.15-0.25
M80-M110	M85-M120	M55-M85	—	M100-M110	M100-M110	
—	—	—	—	$2.5\text{-}4.5 \times 10^6$	$1.4\text{-}2.2 \times 10^6$	$5.0\text{-}7.5 \times 10^5$
—	—	—	—	—	—	—
4.0-5.0	10.0-20.0	15-25	—	4.0-10.0	4.0-30.0	4.0-6.0
0.25	0.20-0.27	—	—	0.19	—	—
4.5-6.5	2.0-4.0	5.5×10^{-6}	2.0-10.0	1.1-3.5	2.0-5.0	—
250-550	250-550	200-500	250-300	300-500	300-500	—
115-550	160-550	190-600	RT-250	250-500	250-500	200-250
—	—	—	—	—	—	—
$10^{12}\text{-}10^{17}$	$10^{13}\text{-}10^{16}$	<10	$1.3\text{-}15.0 \times 10^{14}$	$>10^{14}$	$>10^{14}$	$10^{13}\text{-}10^{14}$
300-500	300-550	—	235-400	300-400	300-400	380-420
380	—	10-200	235-400	300-400	300-400	375-400
3.5-5.0	3.2-4.5	>8	3.0-6.0	3.5-5.0	3.5-5.0	—
3.5-4.5	3.2-4.0	—	3.0-5.0	3.5-5.0	3.5-5.0	—
3.3-4.0	3.0-3.8	—	3.0-6.0	3.5-5.0	3.5-5.0	2.0-3.0
0.002-0.010	0.008-0.03	>0.1	0.010-0.040	0.01	0.01	—
0.002-0.02	0.008-0.03	—	0.012-0.050	0.01	0.01	—
0.030-0.050	0.02-0.04	—	0.018-0.090	0.01	0.01	0.005-0.012
45-120	150-300	—	50-180	120-180	150-190	120-150
1.55-1.61	—	—	—	—	—	—
Transparent	Opaque	Opaque	Transparent	Opaque	Opaque	Opaque
—	—	—	—	—	—	—
—	—	—	—	—	—	—
0.08-0.15	0.04-0.10	0.1-4.0	0.27-0.50	0.05-0.20	0.04	0.2-1.0
—	—	—	—	—	—	—
—	—	—	—	—	<0.16 (FR gr.)	—
None	None	None	None	Slight	Slight	None
None	None	None	None	None	None	None
Attacked by some	Attacked by some	Attacked by some	Attacked by some	Negligible	None	Slight
None	None	None	None	None	None	Slight
Slight	Slight	Some	Slight	None	Slight	Slight to attacked
Generally resistant	Generally resistant	Chlorinated affects	Generally resistant	None	None	Slight

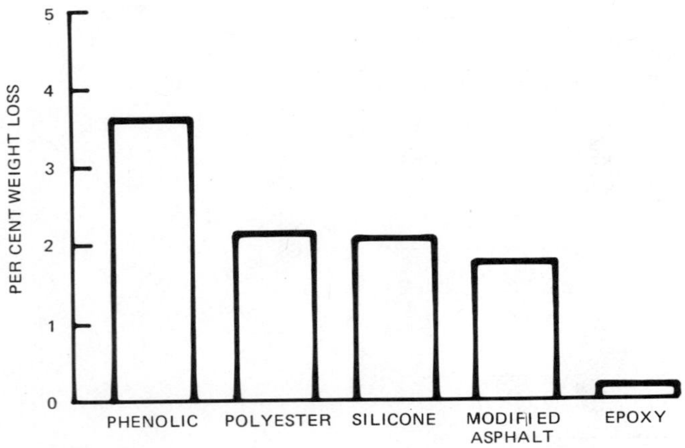

Fig. 6-45. Comparison of cured DGEBA with various other polymers in terms of weight loss after 15 hours at 175°C. (From *Handbook of Epoxy Resins* by Lee and Neville. Copyright 1967 by McGraw-Hill, Inc. Used with permission of McGraw-Hill Book Co.)

Many materials require treatment at the surface prior to bonding so that the adhesive will actually bond to a fresh surface rather than to a contaminant on the surface. Some examples of these contaminants are metal oxides, oils, moisture, and dirt. The effect of surface treatment on adhesive bond strength is illustrated in Table 6-49, where adhesive properties are compared for treated and non-treated aluminum and polystyrene bonded with an epoxy-polyamide adhesive. Some common surface treatment methods for various substrates are listed in Table 6-50.

Acid etching generally gives best results when bonding metals with epoxy resin adhesives. Typical methods for acid etching stainless steel and aluminum are given in Table 6-51.

Epoxy resins are excellent electrical insulators. The chemical resistance of the cured epoxy resin depends considerably on the curing agent used. Selectively outstanding chemical resistance can be obtained by specification of the proper material. Overall, most epoxy resins possess extremely high resistance to caustics and fairly good acid resistance. The epoxy resins are probably the most versatile of the contemporary plastics. The basic properties may be modified in many ways: by blending of resin types, by selection of curing agents, and by use of modifiers and fillers.

Because of their versatility, the epoxy resins are used in thousands of industrial applications. Indicative of the range of applications are the following: adhesives for aircraft honeycomb structures, for paint brush bristles, and for concrete topping compounds; body solders and caulking compounds for the repair of plastics and metal boats and automobiles; casting compounds for the fabrication of short-run and prototype molds, stamping dies, patterns, and tooling; caulking and sealant compounds in building and highway construction applications and in applications where high orders of chemical resistance are required.

Other applications include potting and encapsulating compounds and impregnating resins and varnishes for electric and electronic equipment (see Figures 6-46 and 6-47); laminating resins for airframe and missile applications, for filament wound structures, (see Figure 6-48), and for tool-

Table 6-49. Effect of Surface Treatment on Bond Strength.*

Substrate	Surface Treatment	Tensile Shear Strength, psi
Aluminum	Solvent wipe	1500
	Solvent wipe and surface abrasion	2000
	Alodine coating	2000
	Chromate etch	3500
Polystyrene	None	469 adhesive failure
	Flame-sprayed surface	>556 polystyrene broke before glue lines failed

*Courtesy: Dow Chemical Co.

Table 6-50. Surface Treatment Methods for Adhesive Bonding.*

Substrate	Surface Treatment
Steel	Solvent degrease, abrasion or acid etch
Aluminum	Solvent degrease, abrasion or acid etch
Rubber	Abrasion or acid etch
Glass	Solvent degrease
Thermoset plastics	Degrease and abrade
Thermoplastics	Degrease and acid etch or flame treatment

*Courtesy: Dow Chemical Co.

Table 6-51. Acid Etch Surface Treatments for Stainless Steel and Aluminum.*

Aluminum

1. Immerse metal for 10 minutes in a 180°F bath of:
   - 27% sulfuric acid (96%)
   - 70% water
   - 3% sodium dichromate
2. Rinse with distilled water.
3. Air dry.

Stainless Steel

1. Acetone wipe.
2. Immerse 15 minutes in a solution of:
   - 35 parts sodium dichromate
   - 100 parts sulfuric acid.
3. Rinse with distilled water and air dry.

*Courtesy: Dow Chemical Co.

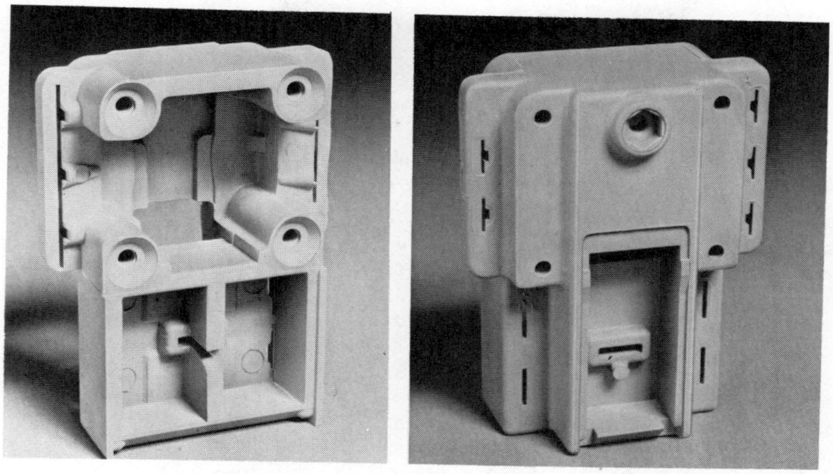

Fig. 6-46. This electronic thermostat cover was injection molded from Arnox one-component resin because of its excellent electrical properties and dimensional stability. (*Courtesy: General Electric Co.*)

Fig. 6-47. A good example of Arnox one-component molding compound's excellent moldability and encapsulation effectiveness is this coil and bobbin application. The bobbin, weighing 5 ounces, was transfer molded on a 60-second cycle. (*Courtesy: General Electric Co.*)

ing fixtures. Table 6-52 compares the specific strength of fiberglass-reinforced epoxy with various metals.

Epoxy-based solution coatings are used as maintenance and product finishes, marine finishes, masonry finishes, structural steel coatings, tank coatings, aircraft finishes, appliance primers, automotive primers, can and drum linings, furniture finishes, and collapsible tube coatings. Table 6-53 lists application properties of epoxy resin finishes. They are used for concrete-floor paints, gym and floor varnishes, spar varnishes, etc. Epoxy resins are also used in decorative floor applications; as chemically resistant mortars and floor topping compounds; in printing inks, in fabric treating applications; in dental, surgical, and prosthetic applications; for breaking petroleum emulsions; for lightweight, chemically resistant foams. The epoxy resins are used as additives for a variety of other plastics materials, such as vinyl and acrylic resins and natural and synthetic rubbers.

Fig. 6-48. Pipe being fabricated from epoxy-impregnated fiberglass using the filament-winding technique. (*Courtesy: General Electric Co.*)

Dermatitis is a serious problem with epoxy resins and especially with some of the amine curing agents, as many are potential skin irritants. This potential varies with the individual working with the materials, the environment he works in, and the formulation itself. In general, good formulating can reduce the hazard but not eliminate it. On the other hand, personal and work-area cleanliness is unquestionable the best preventative. A serious case has rarely developed where good housekeeping was standard practice.

Table 6-52. Comparative Specific Strengths.*

Property	Glass-Epoxy Filament-Wound	Glass-Epoxy Fabric	Aluminum 7075-T6	Titanium Ti-155A	Tool Steel 5GR-MOA
Density, pci	0.070	0.070	0.010	0.163	0.281
Tensile strength, ksi	250	100	82	145	280
Specific tensile strength[1]	3.6	1.4	0.81	0.89	1.00
Tensile modulus, psi × 10^{-6}	9.0	3.5	10.3	16.0	30.0
Compressive strength, ksi	180	90	72	140	260
Specific compressive strength[1]	2.6	1.3	0.71	0.86	0.93
Compressive modulus psi × 10^{-6}	9.0	3.0	10.5	18.0	30.0
Flexure strength, ksi	270	100	–	–	–
Flexure modulus, psi × 10^{-6}	7.0	5.0	3.9	6.2	11.0
Shear strength, ksi	25	30	49	85	170
Specific shear strength[1]	0.36	0.43	0.49	0.52	0.61
Bearing strength, ksi	65	60	147	273	465
Specific bearing strength[1]	0.93	0.86	1.46	1.67	1.66

[1] Strength divided by density, psi × 10^6.
*From *Handbook of Epoxy Resins* by Lee and Neville. Copyright 1967 by McGraw-Hill, Inc. Used with permission of McGraw-Hill Book Co.

Table 6-53. Application Properties of Epoxy Resin Finishes.*

Type of Finish	Method of Application	Approximate % Nonvolatile at Spray Viscosity: Clear Finishes	Approximate % Nonvolatile at Spray Viscosity: Pigmented White[1]	Film Build Per Normal Spray Coat	Typical Curing Schedules	Typical End Uses
Epoxy-phenolic	Spray, roller coat, dip coat, flow coat	30	45	0.7–1.1 mils	20 min. at 375°F	Can, drum and tank linings; pipe linings; wire coatings and impregnating varnishes.
Epoxy-urea	Spray, roller coat, dip coat, flow coat	30	45	0.7–1.1 mils	20 min. at 300°F[2] 20 min. at 385°F[3]	Can linings; appliance primers: clear coatings for brass hardware, jewelry and vacuum metalized plastics; foil coatings, and coatings for laboratory and hospital furniture.
Epoxy-polyamine	Spray, brush	40	55	1.0–2.0 mils	Air dry or bake 10 min. at 250°F	Linings for transmission pipelines; coating for machinery, process equipment and structural steel for use in chemical plants.
Epoxy-polyamide	Spray, brush	40	55	1.0–2.0 mils	Air dry or bake 10 min. at 250°F	Same general uses as epoxy-polyamine finishes.
Epoxy resin esters (unmodified)	Spray, roller coat, dip coat, flow coat, brush	30–35	50	0.7–1.1 mils	Air dry or bake 20 min. at 325°F	Overprint varnishes; primers for autos, appliances and other equipment; foil coatings; original and maintenance finishes for structural steel.
Melamine modified epoxy resin esters	Spray, roller coat, dip coat, flow coat	30–35	50	0.7–1.1 mils	30 min. at 300°F	Same general uses as unmodified epoxy resin ester finishes.

*Skiest, 1958.
[1] Titanium dioxide, P/B = 1/1.
[2] Clear finishes.
[3] White finishes.

Trade Names

Able-	Formulations	Ablestik Laboratories
Abo-	Formulations and curing agents	Abatron, Inc.
Abrac	Epoxidized oils and esters	A. Boake Roberts & Co., Ltd.
Actoflex	Glass-reinforced	Natvar Corporation
Actoglas	Glass-reinforced	Natvar Corporation
Acu-pak	Adhesive	Allied Resin Corporation
Adlock	Prepreg	M & T Chemicals, Inc.
Aerobond	Adhesive tape	Dexter Corporation
Aggre Bond	Formulations	Techform Laboratories, Inc.
Aggrepox	Formulations	Biwax Corporation
Agipon	Fabrications and reinforced epoxy	American Agile Corporation
Agomet	Adhesives	Bush Beack & Segner Bayley, Ltd.
Airtemp	Compounds	Aries Laboratories, Inc.
Ajicure	Curing agents	Ajinomoto Co., Inc.
Alfane	Cement	Atlas Minerals & Chemicals Div., ESB Inc.
All-	Formulations	Bacon Industries
Altocast	Compounds	Alto Development Corporation
Ancamide	Curing agents	Pacific Anchor Chemical Co.
Ancamine	Curing agents	Pacific Anchor Chemical Co.
Ankpoxy	Adhesives	Pacific Anchor Chemical Co.
Apco	Formulations and curing agents	Applied Plastics Co., Inc.
Apogen	Resins	Shaefer Chemicals
Aquapaint	Adhesives and coatings	W.F. Rees, Ltd.
Aquapon	Coatings	PPG Industries
Aracast	Resins	Ciba-Geigy Corporation
Araldite	Resins and curing agents	Ciba-Geigy Corporation
Armobond	Coatings	Deerfield Coatings, Inc.
Arnox	Compounds	General Electric Co.
Aroflint	Epoxy/polyester resins	Ashland Chemical Co.
Ashlam	Prepreg, sheet, rod, and profile	Mica & Micanite Supplies, Ltd.
Atlas	Resins	Degussa, Inc.
Autoplax	Adhesives	Automobile Plastics, Ltd.
Bakelite	Cycloaliphatic resins	Union Carbide Corporation
Beckipox EP	Resins	American Hoechst Corporation
Berlon	Formulation	National Beryllise Corporation
Beta-S	Curing agents	Tenneco Chemical, Inc.
Betastay	Adhesives	Essex Chemical Corporation
Bi Cast	Compounds	Bishop Manufacturing Corporation
Bilaseal	Formulation	B.I.C.C., Ltd.
Blac-Tac	Adhesives	Dexter Corporation
Bondstrand	Reinforced pipe	Amercoat Corporation
Borester	Curing agents	U.S. Borax Research Corporation
Calb-	Formulations	Calbar, Inc.
Calthane	Formulations	Cal Polymers, Inc.
Carbo-	Formulations	Carboline Co.
Cardolite	Resins	3M Co.
Casamid	Curing agents	T. Swan & Co., Ltd.
Cat-A-Lac	Coating	Bostik-Finch, Inc.
C-Cure	Curing agents	R.H. Carlson Co., Inc.
Cepox	Resins	Chemical Development Corporation
Chemfast	Resins and paints	Truscon Laboratories
Chemlene	Reinforced pipe	A.O. Smith Inland Corporation
Chemline	Reinforced pipe	Dow Smith, Inc.
Chem-pon	Resins	Eagle Picher Industries, Inc.
Chockfast	Resins	Philadelphia Resins Corporation
Coast	Formulation	Hexcel, Inc.
Colma-	Adhesive	Sika Chemical Corporation
Compox	Formulations	Halliday-Armstrong
Cona-	Resins and curing agents	Conap, Inc.
Concord	Formulations	Morton Chemical Co.
Concresive	Adhesives	Adhesive Engineering Co.
Copon	Compounds and curing agents	Reliance Universal, Inc.
Copy Cast	Molding compounds	Schramm Fiberglass Products, Inc.
Corlar	Enamels and prepregs	E.I. du Pont de Nemours & Co., Inc.

Coroc	Resins	Cook's Resins
Corocrete	Compounds	Ceilcote Co.
Corrocote	Resins	Chemical Coating & Engineering Co., Inc.
Crystal-clear	Coatings	Epoxy Coatings Co.
Custom-	Formulations	Custom Materials, Inc.
Delta Bond	Adhesives	Wakefield Engineering, Inc.
DEN, DER	Resins and curing agents	Dow Chemical Co.
Design Cast	Resins	MacNaughton-Brooks, Inc.
Devcon	Resins	Chemical Development Corporation
Devran	Resins	Devoe & Raynolds Co.
Diad	Curing agent	Henkel Corporation
Dolphon	Resin and formulations	John C. Dolph Co.
Double Bubble	Adhesive	Hardman, Inc.
Dow	Resins	Dow Chemical Co.
Dri-Kote	Powders	Dexter Corporation
Durachem	Coatings	Maas & Waldstein Co.
Durafoam	Foam	D & R Pilot Plants, Inc.
Duraglas	Prepreg	Natvar Corporation
Duromeen	Curing agents	Armak Co.
E-	Compounds	Allied Products Co.
Ecco-	Formulations	Emerson & Cuming, Inc.
Enal	Novolac resins	Soviet Origin
Enkp	Novolac resins	Soviet Origin
Eno Enos	Novolac resins	Soviet Origin
Ent	Novolac resins	Soviet Origin
Enta	Novolac resins	Soviet Origin
Epacron	Compounds	Loes Enterprises
Epiall	Compounds	Allied Chemical Corp.
Epibond	Adhesives	M & T Chemicals, Inc.
Epiclon	Resins	Celanese Coatings Co.
Epi-cure	Curing agents	Celanese Coatings Co.
Epikote	Resins	Shell Chemical Co., Ltd.
Epikure	Curing agents	Shell Chemicals, U.K.
Epilink	Curing agents	Interstab, Ltd.
Epilok	Curing agents	Interstab, Ltd.
Epiphen	Resins	Monomer-Polymer Chemical Co.
Epi-Rez	Resins	Celanese Coatings Co.
Epi-Tex	Epoxy ester resins	Celanese Coatings Co.
Epi-Var	Epoxy ester resins	Celanese Coatings Co.
Epo-Cap	Formulations	Hardman Co., Inc.
Epocast	Formulations	M & T Chemicals, Inc.
Epochlor	Chlorinated resins	Shell Chemical Co.
Epocrete	Formulations	Hardman Co., Inc.
Epocryl	Epoxy/acrylates	Shell Chemical Co.
Epocure	Curing agents	Hardman Co., Inc.
Epodite	Resins	Showa High Polymer Co.
Epofoam	Foam	M & T Chemicals Inc.
Epoglas	Reinforced epoxy	Plastilight, Inc.
Epolac	Resin	Lawrence Adhesive & Chemical Co.
Epolast	Compound	Hardman Co., Inc.
Epolite	Formulations	Rezolin, Inc.
Epoloid	Coatings	Rowe Products, Inc.
Epolux	Resins and coatings	Steelcote Manufacturing Co.
Epomin	Amine adducts	Societa Italiana Resine
Epon	Resins and curing agents	Shell Chemical Corp.
Epophen	Resins	Leicester, Lovell & Co., Ltd.
Eposet	Formulations	Hardman Co., Inc.
Eposir	Resins	Societa Italiana Resine
Epo-Tek	Formulations	Epoxy Technology
Epotherm	Formulations	Transene Co., Inc.
Epotuf	Resins and curing agents	Reichhold Chemicals, Inc.
Epoweld	Compound	Hardman Co., Inc.
Epoxical	Tooling resins	U.S. Gypsum Co.
Epoxin	Resins	BASF Wyandotte Corporation
Epoxi-patch	Adhesives	Dexter Corporation
Epoxiplate	Powders	Armstrong Products Co.

Epoxybond	Adhesive	Atlas Mineral & Chemical Division, ESB Inc.
Epoxy-Coat	Coating	Devcon Corporation
Epoxylab	Adhesives	Tra Con, Inc.
Epoxylite	Formulations	Epoxylite Corporation
Epoxyn	Resins, adhesives, and coatings	Co-Polymer Chemicals, Inc.
E-Solder	Adhesive	Allied Products Corporation
Estabex	Oils and esters	Akyo Chemic Nederland, B.V.
Estoxy	Coatings	Halliday-Armstrong, Ltd.
Estynox	Epoxy plasticizers	NL Industries, Inc.
Farbaloy	Powders	Farboil Co.
Fel-poxy	Adhesives	Fel-Pro, Inc.
Fiberloy	Reinforced sheet	Rogers Corporation
Flaymbar	Coatings	Ocean Chemicals, Inc.
Flexipon	Resins, coatings, and adhesives	Flexible Products Corporation
Flintflex	Powders	E.I. du Pont de Nemours & Co.
Flomat	Reinforced	Fothergill & Harvey, Ltd.
Furnane	Cements	Atlas Minerals & Chemicals Division, ESB Inc.
Fybrglas	Reinforced	Schramm Fiberglass Products, Inc.
Genepoxy	Resins	Henkel Corporation
Gepol	Reinforced insulators	General Electric Co.
Glasply	Reinforced tubing	SGL Industries, Inc.
Glasshesive	Adhesive	Adhesive Engineering Co.
Grilonite	Resins	Emser Werke, A.G.
Gusco	Coatings	A. Gushmer, Inc.
Hapex	Formulations	Hastings Plastics, Inc.
Harcure	Curing agents	Pennwalt Corporation
Helmicol	Resins	Canadian Helmitin Co.
High Strength	Compounds	High Strength Plastics Co.
Hightemp	Formulations	Hightemp Resins, Inc.
Hi-Peel	Resins	Baker Industries, Inc.
Humble-Weld	Adhesives	Exxon Corporation
Hycon	Coatings	Shell Chemical Co.
Hydantoin	Resins	Ciba-Geigy Corporation
Hydrepoxy	Water-based formulations	Allied Products Corporation
Hyflo	Molding powders	Dexter Corporation
Hyseal	Formulations	Dexter Corporation
Hysol	Formulations	Dexter Corporation
Isochem-	Formulations	Isochem Resins Co.
Kenbond	Formulations	Kenics Corporation
Kencast	Formulations	Kenics Corporation
Kenseal	Formulations	Kenics Corporation
Kenthane	Formulations	Kenics Corporation
Koller	Formulations	Beck, Koller & Co., Ltd.
Kopox	Resins	ARCO Polymers, Inc.
Kopoxite	Resins	ARCO Polymers, Inc.
Koropon	Coatings	DeSoto Chemical Coatings, Inc.
Krelite	Formulations	Key Polymer Corporation
Kwik-Patch	Adhesives	Bacon Industries, Inc.
Lamalac	Prepreg, sheet, and rod	Mica & Micanite Supplies, Ltd.
Lancast A	Curing agent	AZ Products, Inc.
Lancure	Curing agent	Lanson Chemical Corporation
Lankroflex	Plasticizers	Lankro Chemicals, Ltd.
Laromid	Curing agents	BASF Wyandotte Corporation
Laromin	Curing agents	BASF Wyandotte Corporation
Lay-up-	Formulations	Bacon Industries, Inc.
Lee-	Formulations and curing agents	Leepoxy Plastics
Lefkoweld	Adhesive	Leffingwell Chemical Co.
Lekoweld	Adhesives	Techform Laboratories, Inc.
Lekutherm	Resins	Mobay Chemical Corporation
Lopox	Resins	CdF Chimie
Lytex	SMC and BMC	Morton Chemical Division
Magna-	Formulations	Magnolia Plastics, Inc.
Mara-	Formulations	Allied Products Corporation
Marglass	Prepreg	Marglass, Ltd.
Mereco	Adhesives	Metachem Resins Corporation
Meta-Cote	Coatings	Wallace Clark & Co., Inc.

Metalset	Formulations	Smooth-On Mfg. Co.
Metlhesive	Adhesives	Dexter Corporation
Metre-	Formulations and curing agents	Metachem Resins Corporation
Micabond	Formulations	Dexter Corporation
Micaply	Laminates	Mica Corporation
Miccron	Coatings	Michigan Chrome & Chemical Co.
Minit-Cure	Adhesive	Bacon Industries, Inc.
Minit-Grip	Adhesive	High Strength Plastics Corporation
Minit-Man	Adhesive	Kristel-Kraft, Inc.
Minit-Tronic	Adhesive	Bacon Industries, Inc.
Mistapox	Compounds	M.R. Plastics & Coatings, Inc.
Moldcast	Formulations	Bacon Industries, Inc.
Molycor	Tubing	A.O. Smith Corporation
Monomid	Coatings	Steelcote Mfg. Co.
Monopoxy	Adhesives	Hardman Co., Inc.
Nepoxide	Coatings	Atlas Minerals & Chemicals Division, ESB Inc.
Norbak	Formulations	Rexnord Co.
Novoloy	Powders	Amicon Corporation
Orgalacqe	Powders	Aquitaine Organico
Oxiron	Resins	FMC Corporation
Paxolin	Prepreg	Micanite & Insulators Co., Inc.
Pedigree Products	Resins	P.D. George Co., Inc.
Permachem	Castings	Permali, Ltd.
Perma Rez	Castings	Permali, Ltd.
Perma-Skin	Coatings	Dennis Chemical Co.
Phillyclad	Resins	Philadelphia Resin Corporation
Placquer	Resins	Custom Coatings Co.
Plast	Resins	Fibre-Glass Developments
Plastic Steel	Formulations	Devcon Corporation
Plenco	Molding compounds	Plastics Engineering Co.
Polycast	Tooling resins	Poly Resins
Polychem	Molding compounds	Budd Chemical Co.
Polydride	Curing agents	R.T. Vanderbilt Co.
Poly-Gard	Compounds	Sterling Varnish Co.
Poly-Link	Curing agents	Celanese Coatings Co.
Polymer Systems	Molding compounds	Adams & Associates, Inc.
Polyox	Resins	Reichhold Chemicals, Inc.
Polyset	Molding compounds	Morton Chemical Co.
Polysol	Suspensions and dispersions	Showa High Polymer Co.
Polythread	Reinforced	A.O. Smith-Inland Corporation
Polytool	Tooling compound	Reichhold Chemicals, Inc.
Poxy-Finish	Coatings	Polyastics Corporation
Poxy-Gard	Compounds	Reichhold Chemicals, Inc.
Pre-Lam	Laminates	Precision Laminates Corporation
Propol Betabond	Formulations	Protective Polymers of Shepreth
Quik-I-Poxy	Adhesive	Leech Products, Inc.
Ramalacq	Coatings	Whittaker Corporation
Ramaset	Finishes	Whittaker Corporation
Randac	Molding compound	Mitchell Rand Mfg. Corporation
Randalite	Reinforced rods, tubes and sheets	Randalite
Reapoxy	Casting Resins	Read Plastics, Inc.
Red Thread	Reinforced tubes	Dow Smith, Inc.
Reliabond	Adhesives	Reliable Mfg. Co.
Ren-	Formulations	Ren Plastics
Re-New-Coat	Coatings	Truscon Laboratories
Repox	Resins	Resine Sintetiche e Affini, S.p.A.
Rezklad	Formulations	Atlas Mineral & Chemicals Division, ESB Inc.
Ripoxy	Resins	Showa High Polymer Co.
Rolox	Compounds	Hardman Co., Inc.
Rotabond	Compounds	Isochem Resins Co.
Rox'N Pox	Adhesives	Biwax Corporation
Safe Grip	Adhesives	United Gilsonite Laboratories
Samicalok	Reinforced	Micanite & Insulators Co., Ltd.
Sandbind	Formulations	Bacon Industries, Inc.
Scotch-	Formulations	3M Co.
See Cure	Formulations	Epic Resins Corporation

Shurbond	Compounds	Anchor Alloys, Inc.
Silpox	Adhesive	Starnetics Co.
Smooth-On	Adhesive	Smooth-On Mfg. Co.
Sonite	Formulations	Smooth-On Mfg. Co.
Speed-Rex	Coatings	Truscon Laboratories
Spiralloy	Reinforced epoxy	Hercules Co.
Sterilkote	Coatings	Bradley & Vrooman Co.
Stycast	Formulations	Emerson & Cuming, Inc.
Super Coilife	Compound	Westinghouse Electric Corporation
Super Poxec	Adhesives	Watsco, Inc.
Sym-	Formulations	Symplastics, Inc.
Synprep	Foam prepreg	Whittaker Corporation
Tayloron	Reinforced	Synthane-Taylor Corporation
Telcoset	Powders	Telcon Plastics, Ltd.
Terpox	Resins	Tenneco Chemical, Inc.
Thermoset	Resins	Thermoset Plastics, Inc.
Thirty-Thirty	Adhesives	Bacon Industries, Inc.
Tonox	Curing agents	Uniroyal Corporation
TRA-	Formulations	Tra-Con, Inc.
Trav-Mar	Formulations	X-Pando Corporation
Trevarno	Reinforced epoxy	Hexcel Corporation
Tru-	Formulations	Walter Kidde & Co., Inc.
Tube-Cote	Formulations	Tuboscope, Inc.
Tuf Film	Formulations	Products Techniques, Inc.
Tuf-Lite	Formulations	H.B. Fuller Co.
Tuff Tube	Reinforced	Koppers Co.
Twenty-Twenty	Adhesive	Bacon Industries, Inc.
Twinweld	Adhesives	High Strength Plastic Corporation
Tyglas	Prepreg	Fothergill & Harvey, Ltd.
Unalloy	Adhesive	Dexter Corporation
Unipoxy	Formulations	Kristal Kraft, Inc.
Uniset	Formulations	Amicon Corporation
Unox	Cycloaliphatic epoxies	Union Carbide Corporation
Upox	Formulations	Isochem Resins Co.
Vanamid	Curing agents	R.T. Vanderbilt Co., Inc.
Vanclad	Powders	Vanguard Powder Coatings, Ltd.
Vandride	Curing agents	R.T. Vanderbilt Co., Inc.
Vanoxy	Resins	R.T. Vanderbilt Co., Inc.
Verdoc	Coatings	Ferro Corporation
Verifilm	Reinforced	American Cyanamid Co.
Versamid	Curing agents	Henkel Corporation
Versapox	Coatings	Union Tank Car Co.
Vibro-Flo	Powders	Armstrong Products Co.
Vinacoat	Powders	Vinatex, Ltd.
Vi Pox	Adhesives	Adhesives Products Corporation
Vitralon	Powders	Pratt & Lambert, Inc.
Vorpon	Formulations	Vorac Corporation
Weldfast	Formulations	Fibercast Corporation

REFERENCES AND BIBLIOGRAPHY

Aluminum Company of America. *Adhesive Bonding Alcoa Aluminum.* Form 50-12039, 1967.

Anderson, A. Reaction injection molding; process control, reinforcements, new materials build up markets. *Modern Plastics*, April 1978, **55**(4):66–67.

Bruins, P. F., ed. *Epoxy Resin Technology*. John Wiley & Sons, Inc. New York, 1968.

Brydson, J. A. *Plastics Materials.* 3rd Ed. Robert E. Krieger Publishing Co., Inc., Huntington, New York, 1975.

Buchi, G. "New Epoxy Resins for High Voltage Applications." Paper presentation, *SPE ANTEC*, New Orleans, May 1979.

Cagle, C. V. *Adhesive Bonding Techniques and Applications.* McGraw-Hill Book Co., New York, 1968.

Catsiff, E. H., Dee, H. B., and Seltzer, R. Hydantoin epoxy resins. *Modern Plastics*, July 1978, **55**(7):54–58.

Celanese Resins. *Epoxy Adhesives.* Technical Bulletin 0867.

Ciba Products Co. *A Glimpse at the Remarkable World of Ciba Epoxies.*

——. *Araldite Epoxy Resins in the Electrical and Electronics Industries.* Bulletin 650,517/R469.3M.

——. *Ciba Epoxies, Coatings, Electrical, Electronic, Reinforced Plastics, Adhesives, Building Materials.* Bulletin CR170, Supp. CC6.

Davis, W. Flame retardants for thermosets, part II/epoxies. *Plastics Compounding*, July/August 1979, **2**(4):53–58.

DeGisi, S. L., and Smith, C. H. Improved bondability of molded rigid urethane foam. *Adhesives Age*, July 1978, **21**(7):35–37.

Dow Chemical Co. *Dow Liquid Epoxy Resins.* Form No. 190-224, 1976.

——. *Dow Flexible Epoxy Resins.* Bulletin 190-105, 1972.
——. *Dow Epoxy Resins–Adhesives.* Technical Data Report No. 35-J, 190-39, April 1971.
——. *Introducing Dow Epoxy Resins.* Bulletin 190-17, 1969.
——. *Epoxy Resin Manual, Comparative Data.*
Epoxylite Corporation. *Epoxylite High Temperature Compounds.* Technical Bulletin L-800, March 1979.
Gaschke, M. M. Epoxy resins, from Plastics in Packaging. *Plastics Engineering*, October 1978, **34**(10):33–34.
General Electric Co. *Arnox Resin, The Processable Epoxide–Aerospace Performance With Total Processability.* Bulletin A-2.
——. *Arnox Resin, The Processable Epoxide For Filament Winding.* Bulletin A-1.
——. *Arnox Resin Product Data Sheets 1000, 2000, 3110 & 3210.* Bulletins A-4, 5, 6, & 7.
Goodman, S. H., and Tunick, S. A. "Problems in Testing Electrically Conductive Structural Adhesives." Paper presentation, *Conference on Structural Adhesives and Bonding*, El Segundo, Ca., 1979.
Graham, J., and O'Conner, J. Low-temperature-cure epoxy with high temperature properties. *Adhesives Age*, July 1978, **21**(7): 20–23.
Guha, P. K., and Epel, J. N. Adhesives for bonding graphite/glass composites. *Adhesives Age*, June 1979, **22**(6):31–34.
Gulf Oil Chemicals Co. *Lytex, The Shapes of Tomorrow From Gulf Today.* Bulletin SP14800.
——. *Lytex High Performance Epoxy-Based Sheet Molding Compound–Technical Data.* Bulletin 20035-B.
Hunter, W. L., and Zettek, R. J. "Water in Plastic-Encapsulated Semi-Conductors." Paper presentation, *SPE ANTEC*, New Orleans, May 1979.
Jackson, L. C. Improving adhesion by gas-plasma contaminant removal. *Adhesives Age*, September 1978, **21**(9):34–38.
Kubee, F. "Plastic Patterns New Tools and Materials." Paper presentation, *64th Casting Congress, American Foundrymen's Society*, Philadelphia, May 1960.
Lee, H., and Neville, K. *Handbook of Epoxy Resins*, McGraw-Hill Book Co., New York, 1967.
——. *Epoxy Resins in Their Application and Technology.* McGraw-Hill Book Co., New York, 1957.
Lewis, A. F., and Gounder, R. N. Advanced composite/elastomer structural laminates. *SAMPE Journal*, November/December 1978, **14**(6):13–20.
May, C. A., and Tanaka, Y., eds. *Epoxy Resins: Chemistry and Technology*, Marcel DeKiler, New York, 1973.
Materials Engineering. Graphite-epoxy cuts weight of tool. September 1976, **84**(3):68.
——. Epoxy cuts cost of liquid rubber parts. January 1975, **81**(1):31.
McGrady, C. Thermosets, from Materials '79. *Modern Plastics*, January 1979, **56**(1):58–59.
Misra, S. C., Manson, J. A., and Vanderhoff, J. W. Coatings from epoxy latexes may be used as effective substitutes to solvent-based coatings. *Modern Paint & Coatings*, December 1979, **68**(12):27–31.
Modern Plastics. Big news in thermosets–including a revolutionary "one-component" epoxy. July 1979, **56**(7):14.
Pacific Anchor Chemical Corporation. *Comparison Chart of Competitive Epoxide Curing Agents.*
Parker, B. G., and Smith, C. H. "Shelf-Life of Epoxy Prepregs & Film Adhesives." Paper presentation, *SPE ANTEC*, New Orleans, May 1979.
Plastics Engineering. Engineering resins primer. December 1974, **30**(12):17–32.
Product Engineering. Two new families of epoxy resin vie for electrical mechanical uses. July 1975, **46**(7):38.
Pschorr, F. E. Epoxies, from Plastics in Construction: Coming on Strong in 1978. *Plastics Engineering*, July 1978, **34**(7):29–31.
Shell Chemical Co. *Epon Resins for Electrical & Electronic Embedment.* Technical Bulletin SC:226-78, December 1978.
——. *Epon Resins for Fiber Glass Reinforced Plastics*, Bulletin SC:72-25, January 1972.
Simonds, H. R., and Church, J. M. *A Concise Guide to Plastics.* 2nd Ed. Van Nostrand Reinhold Co., New York, 1963.
Sinclair, J. "Recent Developments in Epoxy Resin-Based Coatings." Paper presentation, 1975 Meeting, Federation of Societies for Coatings Technology in Los Angeles, October 1975.
Skeist, L. *Epoxy Resins.* Reinhold Publishing Corporation, New York, 1958.
Titus, J. B. *Trade Designations of Plastics & Related Materials*, (revised), PLASTEC NOTE N9C, Plastics Technical Evaluation Center, Picatinny Arsenal, Dover, New Jersey, May 1978.
Union Carbide Corporation. *Union Carbide Cycloaliphatic Epoxide Systems.* Bulletin F-42953A, June 1976.
——. *Bakelite Liquid Epoxy Resins and Hardeners.* Bulletin F-42461, October 1969.
——. *Epoxide Reactions.* Bulletin F-41346A, 1966.
United States Sypsum Co. *Tooling and Pattern Making Materials Guide.* IGL Bulletin No. 420, August 1965.
——. *Industrial Tooling and Pattern Making With Epoxical Plastic Tooling Resins.* IGL Bulletin No. 418, September 1965.
——. *Epoxical Plastic Tooling Resin-Laminating Procedure.* IGL Bulletin No. 400.
——. *Epoxical Plastic Tooling Techniques.* IGL Bulletin No. 402.
——. *Epoxical Die Surface Resin.* IGL Bulletin No. 406.
——. *Epoxical Core Fill.* IGL Bulletin No. 407.
——. *Epoxical Tooling Plastic for Checking Die Casting Dies.* IGL Bulletin No. 425.
Vandover, L., Berg, J. C., and Foshay, R. W. "UV Curable Epoxy Resins." Paper presentation, *Nepcon West*, Anaheim, Ca., February 1976.
Wittenwyler, C. V. Achieving performance balance in fire-retardant epoxy systems. *Modern Plastics*, December 1978, **55**(12):62–66.
Wolff, R. V. Effects of moisture on mean strength of bonded composite-to-metal joints. *Adhesives Age*, June 1978, **21**(6):33–36.

POLYURETHANES

The term "polyurethane" is more one of convenience than of accuracy, since these polymers are not derived by polymerizing urethane-containing monomers, nor are they usually polymers containing primarily urethane groups. The polyurethanes include those polymers which contain a significant number of urethane groups, regardless of what the rest of the molecule may be. Usually these polymers are obtained by the combination of a polyisocyanate with reactants which have at least some —OH groups; for example, polyethers, castor oil, and simple glycols. Other reactive groups such as amino and carboxyl may also be present in the monomers. Thus, a typical "polyurethane" may contain, in addition to urethane groups, aliphatic and aromatic hydrocarbon, ester, ether, amide, and urea groups. These polymers are also sometimes called simply "urethane" and sometimes "isocyanate polymers."

Urethanes can be considered esters of the unstable carbamic acid, or amide esters of carbonic acid: $-\mathrm{N}-\overset{\overset{\displaystyle \mathrm{O}}{\|}}{\mathrm{C}}-\mathrm{OR}$, and the urethane group has the characteristic configuration:

$-\mathrm{N}-\overset{\mathrm{O}}{\overset{\|}{\mathrm{C}}}-\mathrm{O}-$. Thus, a polyurethane has a significant number of these urethane groups, although not necessarily repeating in a regular order.

Polyurethane can be formed by a variety of methods, although the most widely used production method is the reaction of di- or polyfunctional hydroxyl compounds with di- or polyfunctional isocyanates. The general structure of a linear polyurethane derived from a dihydroxy compound, HOROH and a diisocyanate, OCNR′NCO, can be represented by the following formula:

$$\left(\mathrm{RO}\overset{\mathrm{O}}{\overset{\|}{\mathrm{C}}}\overset{\mathrm{H}}{\overset{|}{\mathrm{N}}}\mathrm{R'}\overset{\mathrm{H}}{\overset{|}{\mathrm{N}}}\overset{\mathrm{O}}{\overset{\|}{\mathrm{C}}}\mathrm{O}\right)_n$$

The functionality of the —OH-containing component as well as the isocyanate can be increased to three or more to form branched or crosslinked polymers. Other structural changes can be made at will. For example, the nature of R may be changed drastically, such as in molecular weight and type (polyether, polyester, simple glycol), and mixtures of these polyhydroxyl compounds can be used. Similarly, the nature of the R′ may be altered, as in a change from an aliphatic isocyanate to an aromatic isocyanate. For these reasons the polyurethanes are almost unique in that crosslinking, chain flexibility, and intermolecular forces can be varied widely and almost independently. It is to be expected, then, that the polyurethanes include fibers, soft and hard elastomers, flexible and rigid foams, coatings for many purposes, and highly crosslinked plastics.

While the chemistry and technology of polyurethanes are of relatively recent origin, the chemistry of organic isocyanates dates back over a hundred years. In 1937, Professor Otto Bayer and co-workers discovered the diisocyanate addition polymerization that resulted in the preparation of many different types of polyurethanes and polyureas. Initial work consisted of reactions of diisocyanates with diamines, yielding polyureas which were, however, infusible and strangely hydrophilic, hence less interesting for plastic or fiber applications. However, the reaction of diisocyanates with glycols such as 1,4-butylene glycol, led to the formation of polyurethanes possessing properties that made them of interest in the production of plastics and fibers, especially of bristles. Prior to 1952, the polyurethanes were characterized by good properties and high prices. It was not until the advent of flexible urethane foams, with their high strengths and very low densities that large-scale commercial use of polyurethanes seemed assured.

The flexible foam industry was started largely on Bayer know-how, utilizing machinery imported from Germany. Other Bayer developments in the coatings, elastomer, and adhesive fields were introduced that also employed polyesters as the principal resin component. In order to reach much broader markets, American companies soon developed cheaper resins, such as polyethers derived from propylene oxide, and castor oil, to partially replace the more expensive polyesters in the urethane systems. For technical and economic reasons the use of these polyethers now exceeds that of all other resins.

Preparation of Diisocyanates

Current diisocyanate production methods are based on the phosgenation of amines or amine salts.

$$\underset{\text{solution}}{\mathrm{R(NH_2)_2}} + \underset{\text{phosgene}}{\mathrm{COCl_2}} \xrightarrow[\text{temperatures}]{\text{low or moderate}}$$

$$\mathrm{Cl}\overset{\mathrm{O}}{\overset{\|}{\mathrm{C}}}\overset{\mathrm{H}}{\overset{|}{\mathrm{N}}}\mathrm{RNH_2 \cdot HCl} \xrightarrow[120\text{–}150^\circ\mathrm{C}]{\mathrm{COCl_2}} \mathrm{R(NCO)_2} + 4\mathrm{HCl}$$

The most important diisocyanates produced are tolylene diisocyanates (TDI)

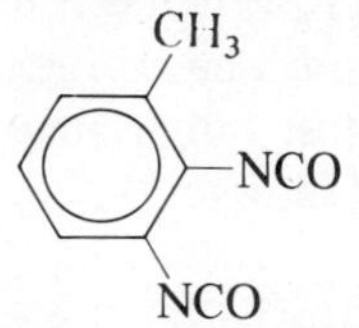

(an 80/20 and 65/35 isomeric mixture of 2,4 and 2,6 are most commonly used), naphthalene diisocyanate, hexamethylene diisocyanate and derivatives, triphenyl methane-pp′p″-triisocyanate and diphenylmethane 4,4′-diisocyanate (MDI).

The TDI mixtures are made from nitrated toluene. Further nitration of the products (2- and 4-nitrotoluene in approximately a 50:50 ratio) produces the dinitrotoluenes, the resulting ratio of isomers being a function of the degree of separation of the 2-nitrotoluene.

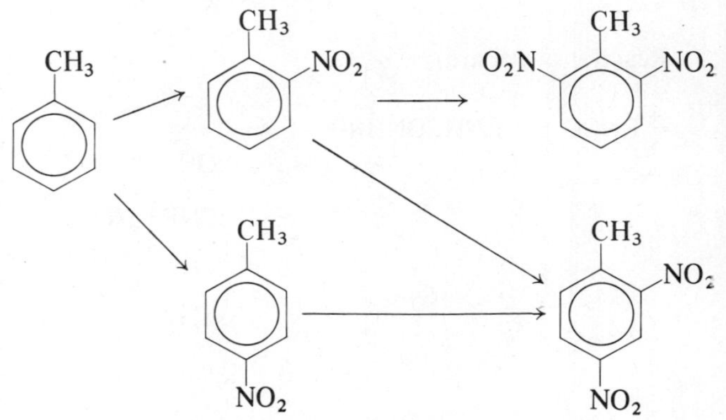

The nitro compounds are reduced to amines, using iron powder and water which are then reacted with phosgene. Although there are many phosgenation reactions, most commonly phosgene is bubbled through a solvent slurry of the amine hydrochloride at 140°C (284°F). Typical solvents include nitrobenzene, toluene, and *o*-dichlorobenzene.

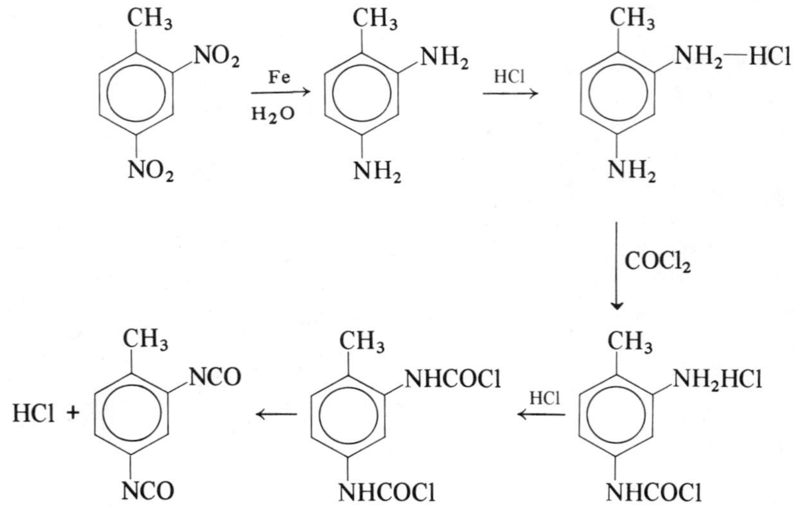

Naphthalene and hexamethylenediamine are the precursers for their respective diisocyanates.

Isocyanates are toxic and can cause respiratory reactions. Although toxicity can be reduced through prepolymerization techniques, care must be exercised during use.

The reactions of most importance for the generation of polyurethanes are as follows:

1. Reactions with amines:

$$-RNCO + -R'NH_2 \longrightarrow -RNH\overset{O}{\overset{\|}{C}}NHR'- \text{ (disubstituted ureas)}$$

2. Reactions with amides:

$$-RNCO + -R'CONH_2 \longrightarrow -RNHCONHCOR'- \text{ (acyl ureas)}$$

3. Reactions with ureas:

$$-RNCO + -RNHCONHR'- \longrightarrow -R\underset{\substack{|\\ C=O\\ |\\ N-H\\ |\\ R'\\ |}}{N}\overset{O}{\overset{\|}{C}}\overset{H}{\overset{|}{N}}R- \text{ (biurets)}$$

4. Reactions with alcohols:

$$-RNCO + -R'OH \longrightarrow -R\overset{H}{\overset{|}{N}}\overset{O}{\overset{\|}{C}}OR'- \text{ (urethane or carbamate)}$$

5. Reactions with urethanes:

$$-RNCO + -R\overset{H}{\overset{|}{N}}\overset{O}{\overset{\|}{C}}OR'- \longrightarrow -R\underset{\substack{|\\ C=O\\ |\\ N-H\\ |\\ R\\ |}}{N}\overset{O}{\overset{\|}{C}}OR'- \text{ (allophanate)}$$

6. Reactions with carboxylic acids:

$$-RNCO + -R'\overset{O}{\overset{\|}{C}}OH \longrightarrow \left(-R\overset{H}{\overset{|}{N}}\overset{O}{\overset{\|}{C}}O\overset{O}{\overset{\|}{C}}R'-\right) \longrightarrow -R\overset{H}{\overset{|}{N}}\overset{O}{\overset{\|}{C}}OR'- + CO_2$$

7. Reaction with water:

$$-RNCO + H_2O \longrightarrow \left(-R\overset{H}{\overset{|}{N}}\overset{O}{\overset{\|}{C}}OH-\right) \longrightarrow -RNH_2 + CO_2$$

It is obvious that reactions 1, 2, and 7 produce the precursors to a crosslinking reaction via reaction 3. Similarly, reaction 6 leads to potential crosslinking via reaction 5. The CO_2 liberated from reactions 6 and 7 can, under controlled conditions, be trapped within the crosslinking polymer matrix, such that a foam structure is formed. Table 6-54 shows the relative reactivity of the various isocyanate coreactants discussed.

Isocyanate coreactants of commercial significance usually

Table 6-54. Order of Descending Reactivity with Isocyanates.*

Primary Amines
Primary Alcohols
Water
Secondary and Tertiary Urethane Hydrogens
Carboxylic Acid Groups
Amides

*Courtesy: Dow Chemical Co.

belong to one of the following groups: polyesters, amines, polycaprolactones, polyglycols, natural and hydroxyl-containing oils. The polyesters are most often made from adipic and phthalic acids, ethylene or propylene glycol, and triols like glycerol. Although amines are highly reactive, certain sterically hindered species such as 3,3-dichloro-4,4-diaminodiphenyl methane (MOCA) have been developed for controlled polymerization. Polyhydroxy glycols are produced from corresponding oxides such as ethylene, propylene, and butylene oxide, from epichlorohydrin, tetrahydrofuran, glycerol, sorbitol, and sucrose. Polycaprolactones are derived from ε-caprolactone and methyl-ε-caprolactones, using diol and triol initiators and titanium catalysts.

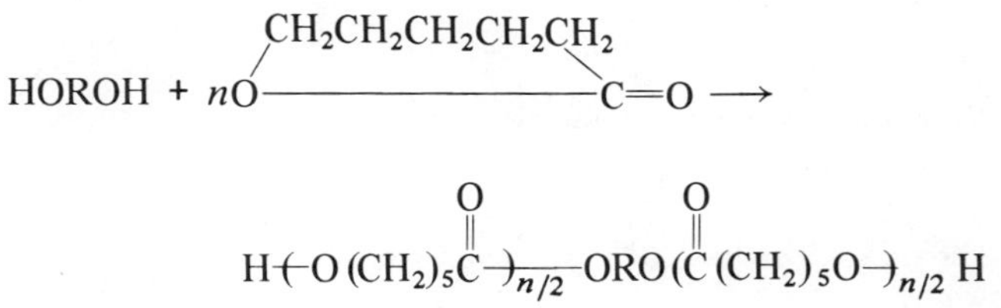

Castor oil and oil-based alkyds represent commonly used hydroxy-based oils.

The structure of the isocyanate group is such that it will have the greatest electron density on the oxygen atom and least on the carbon atom, with the nitrogen atom having an intermediate net negative charge. Any substance which can increase the electron acceptance of the carbon atom in the NCO group will catalyze the isocyanate-active hydrogen reaction. Thus, nucleophilic or basic substances have found use as catalysts for urethane reactions. Many compounds of this type have been used, including weak base salts of inorganic acids, tertiary amines, and organometallic compounds.

Polymerization and/or crosslinking will occur depending on the amount and nature of the diisocyanates, stoichiometric amount and type of comonomer, type of catalyst used, and conditions of curing. As with other plastics, a variety of formulating ingredients can be used to further modify the systems in order to generate specific end-use products. For example, with blowing agents and polyethers, flexible foams can be produced; with blowing agents and polyesters, rigid foams. Without blowing agents and with suitable precautions to exclude moisture, cast elastomers can be manufactured; with minimal crosslinking or extremely long chain length, polyurethane rubber (elastomers). Adhesives can be made with solvents and adhesion promoters, and with solvents and driers, coatings of all sorts can be produced.

Formation of Prepolymers

To produce a prepolymer, the hydroxy-bearing coreactant is reacted with an excess of isocyanate which gives a chain-extended, isocyanate-terminated molecule.

Isocyanate (excess) + Polyol or Polyester ⟶
IPUPUPI (U = Urethane)

Typical formulas used in establishing the proper stoichiometric balance of ingredients are shown in Table 6-55. The effect of stoichiometry on a representative adhesive property (peel strength) and on the chemical changes that result are shown in Figures 6-49 and 6-50.

These prepolymers (when protected from moisture, contaminants, and heat) are reasonably stable and have reduced toxicity. They can then be further reacted with additional coreactants to form linear or network polymers. A quasi-prepolymer can be made by using a much larger excess of isocyanate which yields a resin with a much lower viscosity and a higher percentage of available reactive isocyanate sites.

Properties and Applications

We will discuss properties and applications by examining each broad group of urethane materials end uses: flexible and rigid foams, elastomers, coatings, fibers, and other miscellaneous applications.

Flexible Foams. Flexible urethane foams can be defined as those urethane foams having a high tensile-to-compressive-strength ratio (15–70:1), high elongation, a fast recovery rate, and a high elastic limit. The presence of water and/or auxiliary blowing agents such as monofluorotrichloromethane or methylene chloride will cause the polymeric mass of isocyanate and comonomer to expand into a cellular foam.

Polyols with hydroxyl equivalent weights of less than 200 are usually used for rigid foam. Flexible foams usually use polyols having equivalent weights between 500 and 2000. The polyol functionality required for flexible foam is 2 and 3. Triols and blends of diols and triols are used.

Rigid foams require a much more tightly crosslinked structure and therefore use higher-functionality polyols. Functionalities of 3, 4, 5, 6, and 8 are typical of polyols used for rigid foam. The octols initiated with sucrose and the hexols initiated with sorbitol are the most common

Table 6-55. Formulas for Urethane Calculation.*

1. Equivalent weight of any product $= \dfrac{\text{molecular weight}}{\text{number of reactive sites}} = \dfrac{\text{M.W.}}{\text{X}}$

2. Equivalent weight of a polyol $= \dfrac{\text{atomic wt. of hydroxyl}}{\text{\% hydroxyl in polyol}} = \dfrac{17}{\text{\% OH as a fraction}}$

3. Equivalent weight of a polyol (from phthalic anhydride esterification test) $= \dfrac{\overset{\text{mol. wt. of KOH}}{(56.1)(1000)}}{\text{OH number}}$

4. Hydroxyl number, (OH no.) = (33) (% OH)

5. Percent hydroxyl, (% OH) $= \dfrac{\text{(hydroxyl no.) (atomic wt. of OH)}}{56.1} = \dfrac{\text{(OH number) (17) (100)}}{56.1\ (1000)} = \dfrac{\text{OH number}}{33}$

6. Molecular weight of any polyol $= \dfrac{\text{(functionality of polyol) (56.1) (1000)}}{\text{OH number}}$

therefore, diol m.w. $= \dfrac{(2)\ (56.1)\ (1000)}{\text{OH number}} = \dfrac{112{,}200}{\text{OH number}}$

triol m.w. $= \dfrac{(3)\ (56.1)\ (1000)}{\text{OH number}} = \dfrac{168{,}300}{\text{OH number}}$

7. Stoichiometric weight TDI required for 100 parts polyol = (equivalent weight TDI) $\left(\dfrac{100}{\text{equivalent weight polyol}}\right)$

Note: equivalent weight of TDI = 87

8. Stoichiometric weight TDI required for X parts water = (equivalent weight TDI) $\left(\dfrac{\text{X parts water}}{\text{equivalent weight water}}\right)$

Note: Molecular wt. H_2O = 18 but this reacts with 2 NCO groups so water is difunctional for TDI, therefore, its equivalent is 9.

9. The total weight of TDI required for reaction.
The total amount of TDI used in the formulation is the product of the TDI index and the sum of the TDI necessary for reaction with the polyol and water components:

$$\text{Total wt. of TDI} = \left(\frac{\text{TDI Index}}{100}\right)(\text{eq. wt. TDI})\left[\frac{100}{\text{eq. wt. polyol}} + \frac{\text{parts } H_2O}{\text{eq. wt. } H_2O}\right]$$

The following example illustrates how this equation may be used to calculate the weight of TDI to be used in the formulation shown.

Conditions: VORANOL CP 3000, (OH No. = 54.0) – 100.0 parts
Water added to formulation – 3.6 parts
TDI Index–105 – X parts

$$X = \text{Total weight TDI} = \left(\frac{\text{TDI Index}}{100}\right)(\text{eq. wt. TDI})\left[\frac{100}{\text{eq. wt. polyol}} + \frac{\text{parts } H_2O}{\text{eq. wt. } H_2O}\right]$$

$$X = \left(\frac{105}{100}\right)(87)\left[\frac{100}{\dfrac{(56.1)\ (1000)}{54}} + \frac{(3.6)}{9}\right]$$

X = 45.3 parts by weight

10. Equivalent weight of adduct or prepolymer $= \dfrac{\text{atomic wt. NCO}}{\text{\% access NCO}} = \dfrac{42}{\text{\% excess NCO as a fraction}}$

11. Parts adduct required for 100 parts polyol $= \left(\dfrac{\text{eq. wt. adduct}}{\text{eq. wt. polyol}}\right)(100)$

$$= \frac{\left(\dfrac{42}{\text{\% excess NCO}}\right)}{\left(\dfrac{17}{\text{\% OH}}\right)} = (2.47)\left(\frac{\text{\% OH}}{\text{\% NCO}}\right)$$

Table 6-55. *(Continued)*

12. $\text{TDI Index} = \dfrac{\text{no. of equivalents TDI used}}{\text{no. of equivalents polyol + water used}}$

13. *Calculations for Formulating a Prepolymer to a Specified Percentage NCO

111.7 = eq. wt. VORANOL RN 490, 15.2% OH

87.0 = eq. wt. TDI

198.7 = mixture of 0% NCO

Let X = amt. of TDI needed to bring contents up from 0% NCO to 31.3% NCO

48.27% X = 31.3%

198.7 + X

Where 48.27 = % NCO in TDI

48.27% X = 31.3% (198.7 + X)

4827 X = .313 (198.7 + X)

X = 366.2

366.2 + 87 = 453.2 gms TDI per 111.7 gms VORANOL RN 490

*This is a calculation only, not a procedure for preparing a prepolymer.

*Courtesy: Dow Chemical Co.

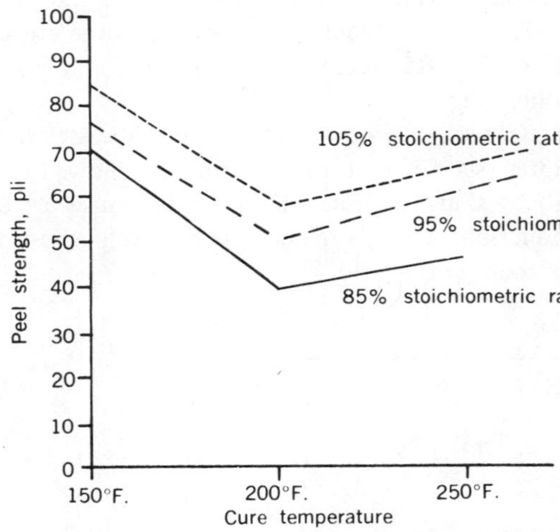

Fig. 6-49. Effect of urethane structure on adhesion. (Swanson and Price, 1973)

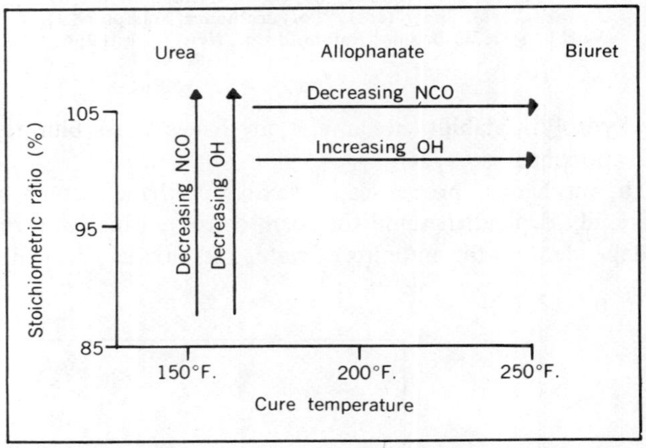

Fig. 6-50. Chemical changes in urethane adhesives. (Swanson and Price, 1973)

types of polyols used for rigid foam production. Polyesters are normally reacted directly with monomeric isocyanates ("one-shot" technique), whereas polyethers may be reacted with either monomeric isocyanates or prepolymers.

Catalysts are included to force the crosslinking and chain-extension reactions to completion, and to control the rate of CO_2 evolution. A silicone or surfactant is added to control cell size, green strength, and impact moldability. Fillers such as barytes, clay, and phosphorous-containing fire retardants can be added to improve flame resistance, give the product specific physical properties and reduce costs. Temperature, water content, auxiliary blowing agent, catalyst and surfactant also play major roles in determining the physical properties of the finished product. Table 6-56 shows a typical flexible polyether foam formulation.

The blowing reaction is that of isocyanate with water. This reaction is accompanied by evolution of carbon dioxide as a by-product, which acts as a blowing agent for the polymer. Water adds to isocyanates to form a carbamic acid. The latter, being unstable, decomposes to the corresponding amine and carbon dioxide. The amine reacts with another isocyanate group to form a symmetrical urea. The disubstituted urea may react further with more isocyanate to form biurets.

Many tertiary amine compounds act as catalysts for the blowing reaction. Among these are triethylenediamine, N-ethylmorpholine, and dimethylaminoethanol.

The preceding gelling and blowing reactions occur competitively during the urethane polymerization. This is completely true with the "one-shot" technique. However, the use of prepolymers allows the various reactions to also be carried out in stages to some degree.

The reactions above produce several different bond types or linkages. These bonds vary in thermal stability and resistance to hydrolysis. In order of decreasing thermal stability we find alkyl amines, alkyl urethanes, aryl urethanes, allophanates, biurets, and diaryl ureas. In sequence of decreas-

Table 6-56. A Typical Formulation for One-Shot Flexible Polyether Foam.*

Component	Parts by Weight
Polyether triol, 3000 m.w.	100
80:20 TDI	38
Stannous octoate	0.5
N-ethylmorpholine	0.5-1.0
Tetramethyl-1, 3-butane diamine	0.1
Silicone copolymer X-520	1.0
Water	2.9

The above components are combined so that they are pumped into the foam machine mix-head in four streams as follows:

1. Polyether triol
2. 80:20 TDI
3. Stannous octoate and ethyl morpholine
4. Diamine, water and X-520

*Landrock, 1969. Originally appeared in *Manufacture of Plastics*, W. M. Smith (ed.), "Polyurethanes," chapter 12, Vol. 1, by J. W. Britain, Reinhold Co., New York, 1964.

ing hydrolytic stability are amides, urethanes, ureas, biurets, and allophanates.

The mechanical properties of flexible urethane foams are markedly dependent upon the foam density, which in turn is dependent on the amounts of water and auxiliary blowing agent (fluorocarbon), and the amount of isocyanate used in the formulation. The type and concentration of catalyst and the cell size of the foam are other variables affecting the density. All flexible urethane foams stiffen at low temperatures, with most polyester-based foams becoming rigid more rapidly than polyether-based foams. Generally, the solvent resistance of flexible urethane foams is good. Water and aqueous solutions do not affect polyester foams, but solvents such as hydrocarbons cause some swelling. Aromatics, chlorinated solvents, esters, ketones, and alcohols cause severe swelling of the polyether foams. With most solvents the strength properties of urethane foams are markedly affected when tested wet; however, upon drying the original properties are regained. Most commercial flexible foams burn with a rate somewhat less than latex rubber. Flexible urethane foams vary from self-extinguishing to burning rates of 1-10 in/min. All cellular materials tend to fatigue; that is, lose height and load-bearing capacity with use. Polyether urethane foams are as good as or better than polyester urethane foams, as can be seen from the data presented in Table 6-57. Figure 6-51 compares the stress-strain curves for 2 lb/ft^3 flexible foams made from a polyether and a polyester.

The techniques used in producing foams are discussed in a separate chapter (see Chapter 15). Figure 6-52 shows three basic techniques that are used. They lend themselves to in-house production methods of slab stock as well as *in-situ* generation of foam in the field.

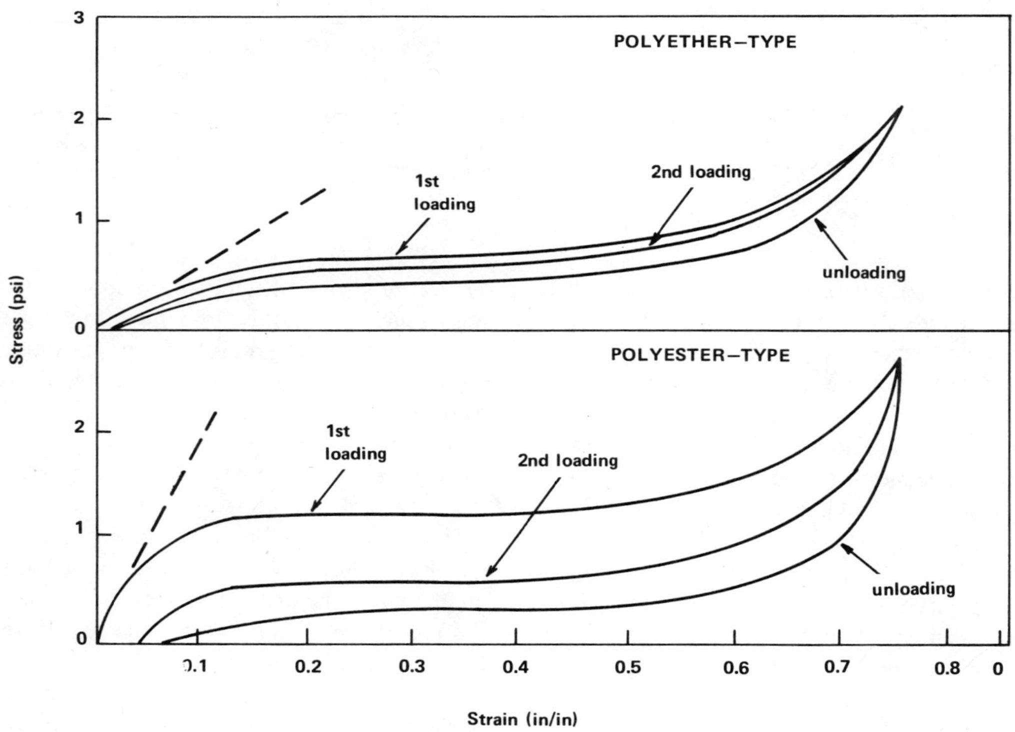

Fig. 6-51. Stress-strain curves for 2 lb/ft^3 flexible urethane foams—note the greater resilience of the polyether foams. (Landrock, 1969. Originally appeared in *Polyurethanes*, 2nd ed., by B. A. Dombrow, Reinhold Publishing Corporation, New York, 1965.)

Table 6-57. Properties of Flexible Polyurethane Foams Based on Polyether and Polyester Polyols.*

	Test Method	Polyether	Polyester
Mechanical Properties			
Density (lb/cu ft)		0.94–2.8	0.94–2.8
Compression set (%)	ASTM 1564-64T		
@ 50% strain		1–5	3–7
@ 75% strain		2–8	5–10
@ 90% strain		2–10	6–15
Indentation hardness characteristics (kg)	BS 3667, Pt. II, 1963		
@ 25% indentation		3–21	13–21
@ 50% indentation		5–26	16–30
@ 65% indentation		7–38	26–49
Tensile strength (kg/sq cm)	BS 3369-61	0.55–1.60	1.5–2.4
Elongation at break (%)	BS 3379-61	100–280	200–450
Tear strength (kg/cm)	ASTM 1564-64T	0.20–0.55	0.50–0.90
Resilience (%)	ASTM 1564-64T	20–55 (usually 45–55)	20–35
Cell count (cells/linear inch)		15–70	15–70
Aging characteristics	BS 3379-61		
(a) Heat aging 16 hrs @ 140°C % retained tensile strength		>70	>70
(b) Humid aging 3 hrs @ 105°C in steam % retained tensile strength	BS 3379-61	>70	>70
Thermal Properties			
Thermal conductivity (BTU/in/sq ft/hr/°F)	ASTM C177-45	0.22–0.25	0.22–0.45
Flammability	ASTM 1692-59T		
(a) unprotected		burns	burns
(b) protected by adding flame retardant to formulation		self-extinguishing	self-extinguishing

*Landrock, 1969.

Molded flexible urethane foams are also extensively used. An approach to foam molding that has seen rapid acceptance is integral skin molding. In this process, a single foaming composition (such as is shown in Table 6-58) is charged into a mold and forms a tough, continuous outer skin while the core is expanded to a cellular structure. Thus, the foam's outer skin has a higher density than the core (see Figure 6-53).

Table 6-58. One-Shot Integral-Skin Polyurethane Foam Formulation.*

Ingredient	Parts by Weight
Polyether triol	100.0
Aromatic diamine	20.0
Silicone surfactant	0.8
Tertiary amine catalyst	0.3
Organo tin catalyst	0.02
Tolylene diisocyanate	21.3
Monofluorotrichloromethane	12–16

*Zwolinski, 1969.

Table 6-59 compares the properties of flexible polyurethane foams with other commercial flexible foam materials.

Applications of Flexible Foams. Comfort cushioning in furniture is currently the biggest outlet for flexible urethane foam. Providing light weight, great strength, and ease of fabrication, urethane foam can also be sewn, stretched, stapled, and hog-ringed (metal-clipped to hook onto seat frames). The urethane foam has great skid resistance which keeps fabric webbing in place. Compared to the latex foam traditionally used in furniture, urethane is more economical. The fabrication and application of slab-stock foam is easier and faster than using animal hair or other filling materials, and the end product provides greater confort and retains its shape better (see Figure 6-54).

Urethane foam mattresses have found acceptance because of their superior durability, freedom from odor, nonallergenic and self-extinguishing properties, ease of cleaning, and resistance to fungus growth.

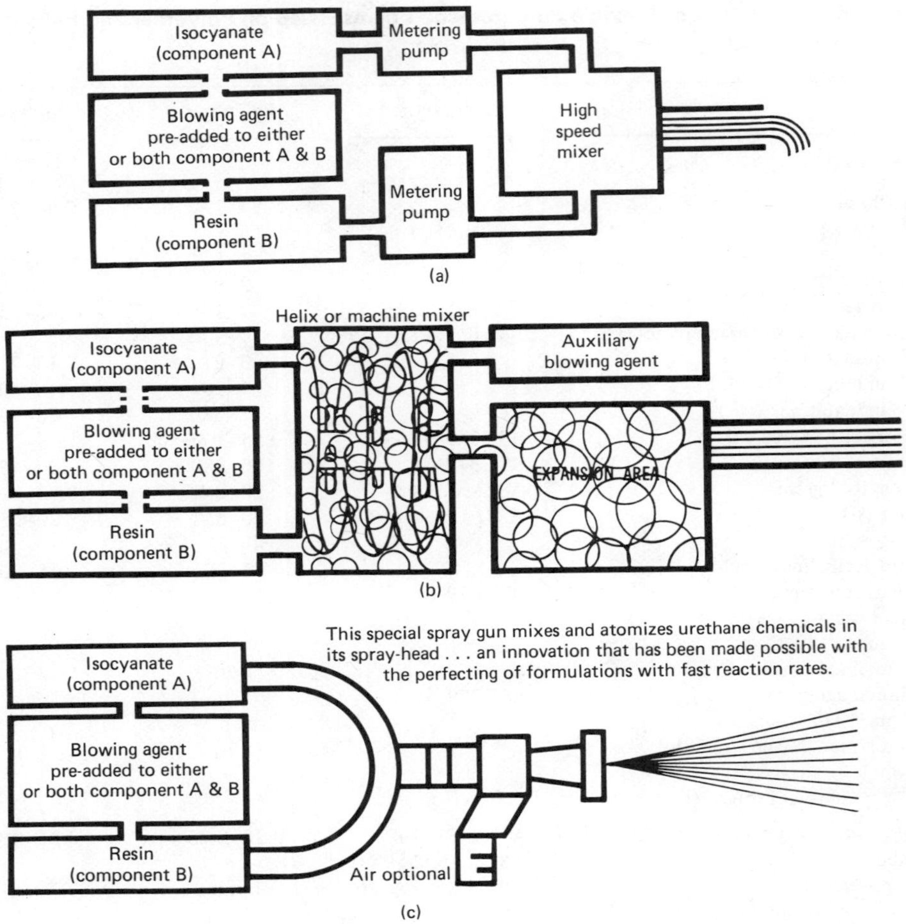

Fig. 6-52. Basic urethane foaming techniques. (*Courtesy: CPR Division, Upjohn Co.*)

(a) In the pour-in place method (which can now be done on the job), two liquids—an isocyanate and a resin—are mixed with a gas-producing blowing agent and poured directly into a cavity or mold. In a matter of minutes, this mixture foams to 30 times its volume and sets into a strong crosslinked cellular plastic composed of uniform gas-filled, closed cells (more than 214 million/ft^3).

(b) Frothing is a modification of the pour-in-place method in which the mixture is dispensed partially pre-expanded—like aerosol cream. Frothing requires special equipment and an extra blowing agent for immediate pre-expansion; final expansion then occurs as the chemical reaction goes to completion. Since frothing exerts less pressure on the forms, fewer forms or jigs are needed. It allows easier and more uniform filling of narrow-walled molds and cavities.

(c) Specially designed spray guns mix and atomize urethane foam so that insulation can be sprayed into place right on the job. Because of its low K-factor, as little as half as much of this extremely light foam is necessary to provide the insulation that is needed.

Fig. 6-53. Automobile arm rest illustrates the used of molded integral-skin foam (*Courtesy: Upjohn Co.*)

Rigid and Semirigid Foams. Rigid urethane foams may be characterized as having a low tensile/compressive ratio (1:0.5), low elongation (<10%), a low recovery rate from distortion, and a low elastic limit. The glass-transition temperature of a rigid foam is above room temperature. Some of the properties common to both rigid polyether and polyester foams are: (1) excellent thermal insulation properties, especially when blown with fluorocarbon, (2) combination of high strength with light weight, (3) good heat-resistance properties, (4) good energy-absorption properties for sound deadening or vibration dampening, and (5) excellent adhesion to metal, wood, glass, ceramics, and fabrics.

Typical mechanical properties of rigid urethane foams are shown in Table 6-60. Comparative data for rigid urethane and other rigid cellular plastics are shown in Table 6-61.

Table 6-59. Physical Properties of Commercial Flexible Cellular Plastics.*

Property	Polyurethane				Polyvinyl Chloride			Latex Foam Rubber			Expanded Neoprene		Expanded Acrylo-nitrile-Butadiene Elastomer	Expanded Natural Rubber		Expanded Butyl Rubber		Expanded SBR[3]	Silicone		
	Super-Soft	Polyether Polyurethane			Closed-Cell		Open Cell														
Density (lb/cu ft)	1.5	1.1	1.8	2.4	3.5	7.0	6.0	5.0	8.1	10.0	7	12	10–25	3.5	20	8–9	14–19	4.5	7–9	10–15	20–24
Mechanical Properties																					
I.L.D., lb	16	14	36	50.0	75	175	25	18	39	64	–	375	–	–	–	80–100	180–275	–	50	100	500
Tensile Str., psi	12	9	20	16.9	–	–	32	15	–	–	–	110	40	5	30	–	–	80	5	15	60
Tensile Elong., %	190	220	258	163	–	–	–	310	–	–	–	500	–	–	–	–	–	–	40	45	200
Compr. Set,[1] %	15	3.5	–	–	15	20	7.8	10	–	–	–	–	–	–	–	–	–	–	0	0	1.0
Rebound Resilience, %	31	–	49	55	–	–	–	73	–	–	–	–	–	–	–	39–36	36–16	–	–	–	–
Tear Strength, lb/in	–	–	3.3	1.6	–	–	–	–	–	–	–	–	–	–	–	–	–	–	1.0	3.0	5.0
Thermal Properties																					
Max. Service Temp., °F	–	–	257	–	–	–	–	–	–	–	221	158	212	158	158	–	–	158	293	293	392
Thermal Cond.[2]	–	–	0.28	–	0.24	0.28	–	–	0.35	–	0.28	0.45	0.25–0.30	0.25	0.30	–	–	0.21	0.31	0.36	0.60
Flammability	–	–	–	–	–	–	–	–	–	–	Self	Ext.	–	–	–	–	–	–	–	–	–

*Landrock, 1969. Originally appeared in *Encyclopedia of Polymer Science and Technology*, 3:80–130, "Cellular Materials," by R. E. Skochdopole, Wiley Interscience, New York, 1965.
[1] 50% deflection
[2] BTU/(hr) (sq ft) (°F/inch)
[3] Styrene-butadiene rubber

Fig. 6-54. Removing flexible foam cushioning from the mold. (*Courtesy: Upjohn Co.*)

These foams can be processed by the same techniques described for flexible foams. In most systems, but particularly with polyester-based systems, the reactions are slower and manufacturing conditions are somewhat less critical. Both prepolymer and quasi-prepolymer systems are used to reduce the hazards involved in handling one-shot systems.

Products intermediate to the flexible and rigid foams (so called semirigid foam) are obtained from systems derived from castor oil, moderate molecular weight triols, or polyesters that have some trifunctional hydroxy compound in the structure. Figure 6-55 compares the tensile properties of the three basic foam types.

Applications of Rigid Foams. Typical applications of rigid foams include: refrigeration insulation; refrigerated trucks and trailers insulation; insulation of pipes and tanks (see Figure 6-56; Tables 6-62 and 6-63 present thermal property data of rigid foams, Figure 6-57 compares the thermal conductivity of thermal insulation materials); structural uses: wall panels and roof tops (see Figure 6-58; Table 6-64 compares acoustic characteristics of urethane foams), pour-in-place wall insulation; aircraft: radios, wing tips, fuselage insulation; military uses: foam-in-place foxholes; void filling and ship insulation, for buoyancy properties (the buoyancy of urethane foam is compared with other buoyant materials in Table 6-65); packaging of engine parts and electronic tubes; electronics: transformer potting; and aerospace applications in space capsules.

One of the major uses of rigid polyurethanes is in the manufacture of furniture—particularly as a substitute for wood. Casting and molding techniques have been developed that not only replicate the appearance and grain of wood, but also the feel, handling and workability properties of wood. Highly intricate shapes can be produced for aesthetic purposes, yet the foam is easily adaptable to normal assembly methods (see Figure 6-59).

Elastomers. Several chemical steps are usually involved in preparing urethane elastomers. These steps may be isolated

Table 6-60. Typical Mechanical Properties of Rigid Urethane Foams.*

Density lb/cu ft	*Tensile Modulus* ↑↑ psi	*Tensile Modulus* ┼→ psi	*Tensile Strength* ↑↑ psi	*Tensile Strength* ┼→ psi	*Compression Modulus* ↑↑ psi	*Compression Modulus* ┼→ psi	*Compression Strength* ↑↑ psi	*Compression Strength* ┼→ psi
2	1 250	500	65	36	780	430	25	20
4	4 100	1 600	140	80	2 400	1 400	80	70
6	8 300	3 400	200	130	4 700	2 800	160	140
8	13 600	6 000	300	170	7 500	4 500	250	230
10	20 000	9 500	375	210	10 700	6 600	460	350
12	27 300	13 700	450	260	14 400	9 000	580	480
14	35 600	18 200	530	300	18 500	11 700	620	600
16	44 800	23 000	620	350	23 000	14 600	770	750

↑↑ measured parallel to grain. ┼→ measured across grain.

The foams were prepared in boxes 7 in x 7 in cross section and they were about 12 in high. These figures give some indication of how the strength depends on density.

*Landrock, 1969.

Table 6-61. Physical Properties of Commercial Rigid Cellular Plastics.*

Property	Polyether Polyurethane FIP[4]	Polyether Polyurethane Board	Polystyrene Extruded	Polystyrene Molded	Epoxy	Phenol-Formaldehyde	Polyethylene	Polyethylene	Polyvinyl Chloride	Silicone	Cellulose Acetate	Urea-Formaldehyde	Ebonite
Density (lb/cu ft)	2.1	2.3	1.9	2.0	2.3	2.0	2.0	29	2.2	3.5	6-7	1.8	4.0
Mechanical Properties													
Compr. Str., psi	37	50[3]	35	35	25	25	—	—	40	6.2	125	8	40
Tensile Str., psi	47	50	70	45	40	15	25	1800	50	—	170	—	40
Flexural Str., psi	60	60	70	60	—	45	—	—	—	—	147	17	—
Shear Str., psi	16	30	40	—	—	11	—	—	45	—	140	—	—
Compr. Mod., psi x 10^3	0.30	1.0[3]	1.0	0.75	0.37	—	—	—	1.5	—	—	—	2.5
Flexural Mod., psi x 10^3	—	1.0	2.5	2.4	—	—	—	—	—	—	—	0.7	—
Shear Mod., psi x 10^3	0.176	0.5	0.9	—	—	—	—	—	1.5	—	—	—	—
Thermal Properties													
Thermal Cond.[1] (initial)	0.11	0.11	0.26	0.16	0.11	—	—	—	0.19	0.28	—	—	—
Thermal Cond.[1] (equilibrium)	0.16	0.16	0.26	0.24	0.15	0.20	0.35	—	—	—	—	—	—
Coeff. of Therm. Exp.[2]	—	2.7	3.5	3.5	—	1-3	—	—	—	—	2.5	—	2.8
Flammability	A	A	A	A	A	FR	B	B	FR	FR	B	FR	FR
Max. Service Temp., °F	—	250	175	180	200	250	160	—	—	650	350	120	122
Specific Heat, Btu/lb	~.22	~.22	0.29	0.29	—	0.38	—	—	—	—	—	0.40	0.33
Electrical Properties													
Dielectric Constant @ 10^9 cps	—	1.04	<1.05	1.03	—	—	1.05	1.55	—	1.09	1.12	—	—
Dissipation Factor @ 10^9 cps x 10^{-4}	—	13	<4.0	7.0	—	—	2.0	40	—	—	20	—	—
Chemical Properties													
Water Absorp., lb/sq ft	0.04	<0.04	0.8	nil	—	—	0.04	—	—	0.284	—	—	—
(10 ft head)											4.5		1.5
Water Absorp., Vol. %	—	<2.0	—	<1.0	—	100	4.0	—	—	—	—	—	—
M.V.T., Perm-In.	1.0	<2.5	1.5	2.0	1	21	<0.4	—	—	41.2	—	—	—

1-BTU/(hr) (sq ft) (°F/inch)
2-Inches/inch/°F x 10^{-4}
3-Load applied parallel to thickness direction
4-Foam-in-place
A-Burns, can be made flame retardant
B-Burns
FR-Flame retardant

*Landrock, 1969.

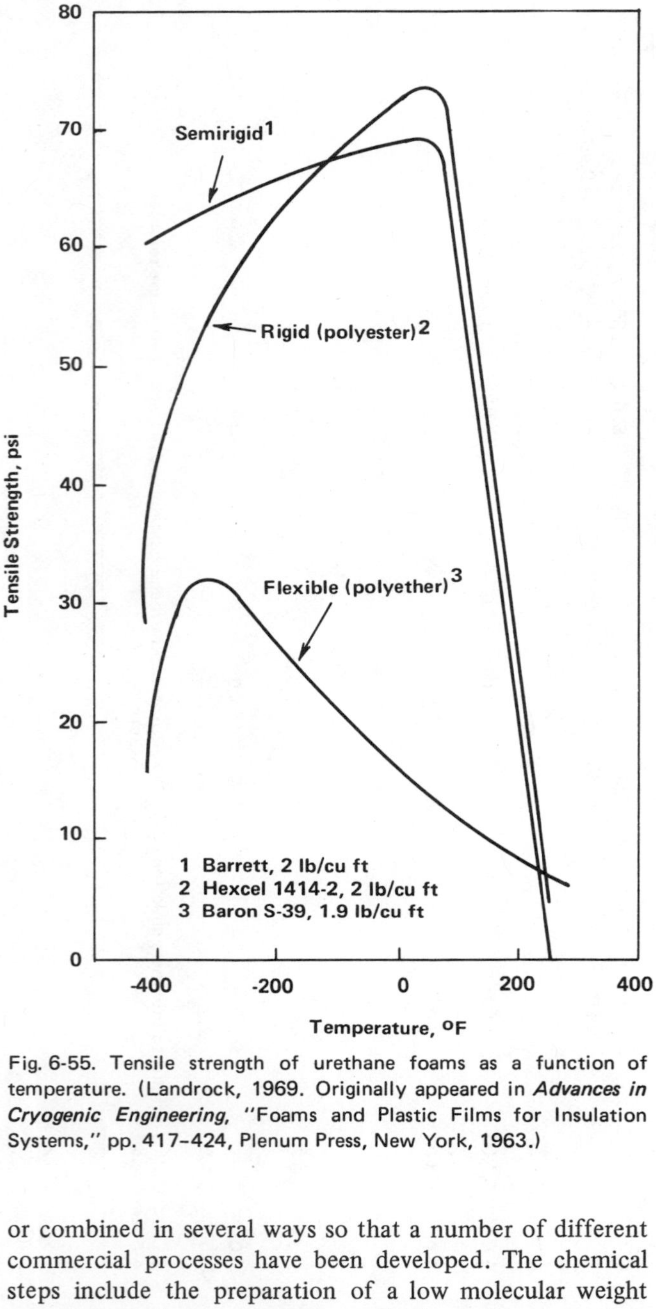

Fig. 6-55. Tensile strength of urethane foams as a function of temperature. (Landrock, 1969. Originally appeared in ***Advances in Cryogenic Engineering,*** "Foams and Plastic Films for Insulation Systems," pp. 417–424, Plenum Press, New York, 1963.)

Fig. 6-56. CPR Isonate urethane spray foam is applied to a tank with a special high-volume impingement spray gun. (*Courtesy: Upjohn Co.*)

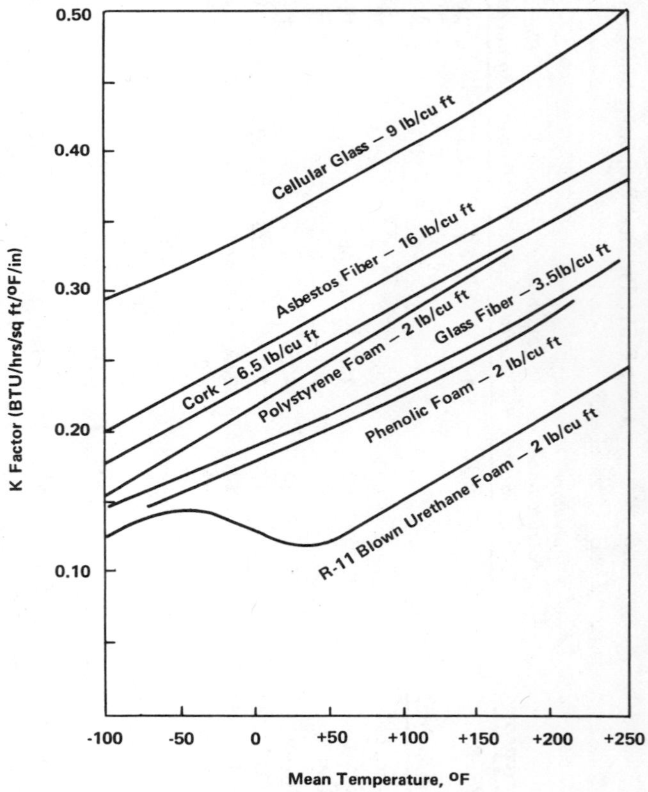

Fig. 6-57. Thermal conductivity versus temperature for typical thermal insulation materials. (Landrock, 1969. Originally appeared in ***Product Engineering,*** How to evaluate the rigid plastic foams, June 21, 1965, **36**(13):59–68.)

or combined in several ways so that a number of different commercial processes have been developed. The chemical steps include the preparation of a low molecular weight prepolymer, chain extension to a higher molecular weight soluble polymer, and crosslinking. These chemical steps, generally characteristic of all urethane preparations, may be combined in several ways. Commercial systems most frequently employed fall into the categories of "casting," "millable gum," and "thermoplastic."

The "casting" system of preparing elastomers utilizes a technique in which the reactants are mixed in the liquid state, poured into a mold, and permitted to cure to a solid, rubbery state in the mold. Additional curing may be used after demolding. The "millable gum" technique is designed to be closely analogous to the processing of more conven-

Table 6-62. Typical Thermal Properties of Plastics Foams Used in Construction.†

Property	Polystyrene*	Urethane*	Polyethylene	Wood
Density, lb/cu ft	1.0–3.0	1.5–2.5	2.0	
Insulation Value (K-factor, BTU-in/hr °F ft^2)	0.24–0.30	0.14–0.16	0.35	
Linear Coefficient of Thermal Expansion, in./in.-°F	4×10^{-5}	5×10^{-5}	8×10^{-5}	
Maximum Temperature for Continuous Use, °F	170-180	250	160	
Heat of Combustion				
BTU/lb	16,000	11,000	16,000	8,000
BTU/cu ft	32,000	22,000	32,000	320,000
BTU/board foot	2,660	1,840	2,660	26,600
Ignition Temperature (ASTM D 1929-62T)				
Flash Ignition Temp., °F	650–700	600	650	500
Self Ignition Temp., °F	735–915	975	660	500
Surface Flame Spread (ASTM E 84–61 "Tunnel Test")	10–25**	40–80**	Non FR	100 (Red Oak)

* Based on a density of 2.0 lb/cu ft ** Flame-retardant grades

†Landrock, 1969. Originally appeared in *Journal of Cellular Plastics*, 1(3):355–362, July 1965, Cellular plastics in construction applications and fire protection by Society of the Plastics Industry, Inc., Cellular Plastics Division.

Fig. 6-58. Applying preshaped urethane foam insulation on a roof top. (*Courtesy: Upjohn Co.*)

Fig. 6-59. Intricately carved furniture pieces, like this china cabinet, have the appearance, weight, feel, and screw-retention of wood, and yet are molded of polyurethane in single units. (Cabinet is by Jasper Stylemasters Plastics, Jasper, Ind. *Courtesy: Upjohn Co.*)

Table 6-63. Thermal Conductivities of Typical Thermal Insulation Materials Showing Efficiencies Relative to Urethane Foam.*

	K-Factor @ 75°F	*Weight Equivalent*	*Thickness Equivalent*
Cellular Glass (9.0 lb/cu ft)	0.39	12.6	2.8
Polystyrene Foam (2.0 lb/cu ft)	0.26	1.9	1.9
Glass Fiber (3.6 lb/cu ft)	0.23	2.9	1.6
R-11 Blown Urethane Foam (2.0 lb/cu ft)	0.14	1.0	1.0
Cork (6.5 lb/cu ft)	0.27	6.2	1.9
Asbestos Fiber (16.0 lb/cu ft)	0.30	17.6	2.2

*Landrock, 1969. Originally appeared in *Technical Information—CPR 9000 Series Rigid Slab Stock*, bulletin CPR-2605, CPR Division, Upjohn Co.

Table 6-64. Sound Absorption Characteristics of Urethane Foams and Architectural Acoustical Materials.*

	Weight lbs/sq ft	*NRC Specification Range*
Urethane Foams		
Flexible Polyether Urethane Foam (2.7 lbs/cu ft).	0.22	0.60-0.70
Flexible Polyester Urethane Foam (3.2 lbs/cu ft).	0.26	0.75-0.85
Flexible Castor Oil Based Urethane Foam (2.0 lbs/cu ft).	0.17	0.70-0.80
Architectural Acoustical Materials		
Regularly Perforated Cellulose Fiber Tile.	1-1.5	0.70-0.80
Random Perforated Cellulose Fiber Tile.	0.96-1.5	0.70-0.80
Fissured Cellulose Tile (3/4 inch thickness).	0.77	0.60-0.70
Perforated Mineral Fiber Tile.	1.7	0.65-0.75
Fissured Mineral Fiber Tile (15/16 inch thickness).	1.45	0.65-0.75
Textured Mineral Fiber Tile.	0.96	0.65-0.75

Sample thickness is one inch unless otherwise specified.

*Landrock, 1969. Originally appeared in *Sound Absorption Properties of Urethane Foams* by R. A. Stengard, E.I. Du Pont de Nemours & Co., Inc., bulletin A-7906, October 1958.

Table 6-65. Buoyancy of Various Substances.*

Material	Density lb/ft^3	Buoyancy Factor
Rigid urethane foam	2	60
Polystyrene foam	1	61
Balsa wood	6	56
Cork	7	55
Pine wood	38	24
Oak wood	48	14

*Landrock, 1969. Originally appeared in *Rigid Urethane Foam* by Union Carbide Corporation, 1964.

tional rubber. A storage-stable, soluble polymer is first prepared, usually with the —OH component in slight excess. This polymer may be prepared using a prepolymer as an intermediate, or in a "one-shot" process, eliminating the prepolymer step. This gum is then cured by milling or otherwise mixing with a polyisocyanate and heating. If the gum contains groups susceptible to free-radical attacks, such as unsaturated or methylene groups, it may also be cured by peroxide or sulfur curing agents.

The "thermoplastic" technique employs a reaction sequence (one-shot or prepolymer) leading to a chain-extended, high molecular weight, largely soluble polymer. This stage may be achieved using a slight excess of active H

Table 6-66. Properties of Polyurethanes.*

		ASTM test method	Cast urethane		Urethane elastomers thermoplastic
			Liquid	Unsaturated	
Processing	1. Molding qualities	—	Good to excellent	Excellent	Good to excellent
	2. Compression molding temp., °F.	—	185-250	—	285-350
	3. Comp. molding pressure, p.s.i.	—	100-5000	—	100-5000
	4. Injection molding temp., °F.	—	—	—	350-450
	5. Injection molding pressure, p.s.i.	—	—	—	1000-20000
	6. Compression ratio	—	—	—	3.0-1
	7. Mold (linear) shrinkage, in./in.	—	0-0.020	—	0.001-0.030
	8. Specific gravity (density)	D792	1.10-1.50	1.05	1.05-1.25
	9. Specific volume, cu. in./lb.	D792	27.0-22.0	—	26.5-22.0
	10. Machining qualities	—	Fair to excellent	Good	Fair to excellent
Mechanical	11. Tensile strength, p.s.i.	D638	175-10000	10000-11000	2000-8400
	12. Elongation, %	D638	100-1000	3-6	100-650
	13. Tensile elastic modulus, 10^5 p.s.i.	D638	1.0-10.0	—	0.1-3.5
	14. Compressive strength, p.s.i.	D695	20000	—	20000
	15. Flexural yield strength, p.s.i.	D790	700-4500	19000	700-9000
	16. Impact strength, ft. lb./in. of notch (½ × ½ in. notched bar, Izod test)	D256	25.0 to flexible	0.4	Does not break
	17. Hardness, Rockwell	D785	10A-90D (Shore)	30-35 (Barcol)	65A-80D (Shore) M28,R60
	18. Flexural elastic modulus, p.s.i. × 10^5	D790	0.1-1.0	6.1	0.1-1.3
	19. Compressive modulus, p.s.i. × 10^5	D695	0.1-1.0	—	0.04-0.09
Thermal	20. Thermal conductivity, 10^{-4} cal./sec./sq. cm./1 (°C./cm.)	C177	5.0	—	1.7-7.4
	21. Specific heat, cal./°C./gm.	—	0.42-0.44	—	0.40-0.45
	22. Thermal expansion, 10^{-6} in./in./°C.	D696	10.0-20.0	—	10.0-20.0
	23. Resistance to heat, °F. (continuous)	—	190-225	200	To 190
	24. Deflection temp., °F. at 264 p.s.i. fiber stress	D648	Varies over wide range	190-200	Varies over wide range
	at 66 p.s.i. fiber stress			—	
Electrical	25. Volume resistivity, ohm-cm. (50% RH and 23° C.)	D257	2.0×10^{11}-10^{15}	6.9-10^{16}	2.0×10^{11} -0.11×10^{14}
	26. Dielectric strength, short-time, ⅛-in. thickness, volts/mil	D149	300-500	—	330-630
	27. Dielectric strength, step-by-step, ⅛-in. thickness, volts/mil	D149	400-500	830-840	530
	28. Dielectric constant, 60 cyc.	D150	4.0-7.5	3.4-3.5	5.4-7.6
	29. Dielectric constant, 10^3 cyc.	D150	4.0-7.5	3.4-3.5	5.59-7.60
	30. Dielectric constant, 10^6 cyc.	D150	6.5-7.1	—	4.21-5.10
	31. Dissipation (power) factor, 60 cyc.	D150	0.015-0.017	0.008	0.015-0.048
	32. Dissipation (power) factor, 10^3 cyc.	D150	0.050-0.060	0.005	0.043-0.060
	33. Dissipation (power) factor, 10^6 cyc.	D150	—	—	0.050-0.100
	34. Arc resistance, sec.	D495	0.1-0.6	—	122
Optical	35. Refractive index, n_D	D542	1.50-1.60	—	1.50-1.60
	36. Clarity	—	Clear to opaque	Clear	Clear to opaque
	37. Transmittance, %	—	—	—	—
	38. Haze, %	—	—	—	—
Resistance characteristics	39. Water absorp., 24 hr., ⅛-in. thick., %	D570	0.02-1.5	0.1-0.2	0.7-0.9
	40. Flammability Burning rate, in./min.	D635	—	—	—
	AEB, in./ATB, sec.		—	—	—
	41. Effect of sunlight	—	None to yellow	—	Strength loss, slight yellowing
	42. Effect of weak acids	D543	Slight	None	Slight to dissolves
	43. Effect of strong acids	D543	Attacked (moderate)	Slight to moderate	Moderate to dissolves
	44. Effect of weak alkalies	D543	Slight	None	Slight to dissolves
	45. Effect of strong alkalies	D543	Slight to attacked	Slight	Moderate to dissolves
	46. Effect of organic solvents	D543	None to moderate	Aliphatic—none; aromatic—moderate	Resists most solvents; most grades are insoluble but swell in polar solvents; specialized soluble grades available for coatings and adhesives

*Reprinted by permission from *Modern Plastics Encyclopedia*, McGraw-Hill, Inc.

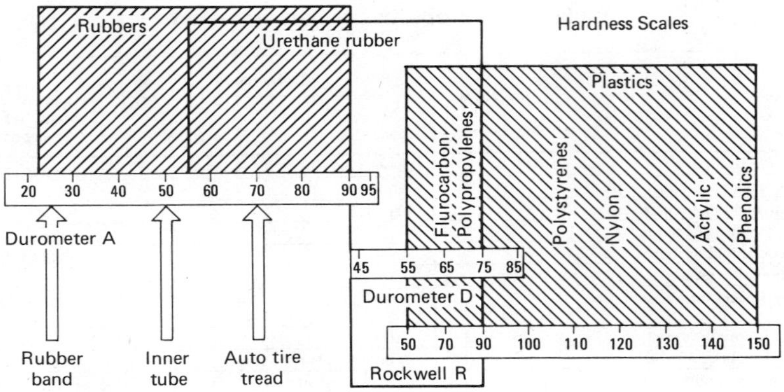

Fig. 6-60. Hardness of urethane elastomers. (*Courtesy: Du Pont Elastomer Chemicals Department*)

component, giving a polymer essentially the same as "millable gum," although a lower molecular weight may be preferred for the gum.

With conventional elastomers, the engineer or designer must usually choose between hardness and elasticity. This desirable combination of properties can sometimes be achieved in a limited hardness range by blending with plasticizers, waxes, reinforcing agents, and fillers. However, it is generally not possible to obtain conventional rubbers with high hardness and resilience, possessing at the same time good abrasion resistance or shock-absorbing properties. Urethane elastomers provide such a combination of properties. They can be formulated over a wide range of mechanical and physical properties (see Table 6-66). Hardness can be varied over a wide range, extending from Shore A of approximately 10 to Shore D of 75 (see Figure 6-60). The tensile strength of most urethane elastomers is excellent, even without the use of fillers. The impact strength of urethane elastomers is compared to three engineering thermoplastics in Table 6-67. With those urethanes having strong intermolecular attractions, fillers usually serve to increase the modulus somewhat, but do not improve the ultimate tensile strength. Elongation is generally high, although it is usually reduced as the hardness is increased. Urethane elastomers have greater load-bearing capacity than other elastomers of comparable hardness and elasticity. This allows design of smaller parts with a resulting saving in weight and possibly in material costs.

Urethane elastomers, especially those based on polyesters and on polyether-diamine combinations, have unusually high tear strength. The abrasion resistance of urethane rubbers is excellent. Parts made from urethane rubber having a Shore A hardness of 78–83 have worn ten times longer than parts made from natural or synthetic rubber. This property, coupled with a high coefficient of friction, has led to their extensive use in friction drives. For applications requiring low friction, such as oscillating journals and knuckle-joint bearings, additives can be added to urethane elastomers to make them essentially self-lubricating and more abrasion resistant. Depending upon the application, urethane rubbers can be formulated to exhibit low or high resilience. The electrical properties of urethane elastomers are good and are of about the same order of magnitude as those of phenolic resins. The harder grades exhibit generally better electrical properties than the softer grades.

Urethane elastomers have outstanding resistance to ozone and oxygen and aliphatic hydrocarbons. Reversible swelling occurs with aromatic hydrocarbons. Hydrolytic decomposition can occur with exposure to acids, alkalis, and the prolonged action of water and steam.

Because of the high heat buildup due to hysteresis effects, the flex fatigue resistance of polyurethane is not good in large masses. Thus, large truck and auto tires cannot be made from these rubbers even though it would be highly desirable to take advantage of the exceptionally good abrasion resistance in such an application.

Table 6-67. Impact Resistance Izod, ft-lb/in (ASTM D-256).*

	Nylon	Acetal	Acrylic	Urethane Elastomer	
Hardness–durometer type	Rockwell R 108	Rockwell R 120	Rockwell M 103	Shore D 75	(Rockwell R 90)
Notched, 75°F	2.0	1.4	0.3	15.0	–
Notched, –40°F	0.5	1.2	0.3	1.1	–

*Courtesy: Du Pont Elastomers Chemicals Department.

The thermoplastic grades lend themselves to processing via normal thermoplastic processing techniques such as injection molding and extrusion. Most are based on polyester or polycaprolactone systems. Table 6-68 compares the properties of two commercial types of elastoplastic urethanes.

Urethane elastomers are used in small tires and wheels, shoe heels and soles (see Figure 6-61), friction drives and belts, pump-housing linings, gaskets, mallet heads, shock-absorbing pads, cable jacketing, potting compounds (see Figure 6-62), solid propellant binder, caulking and sealing compounds, extruded wire and cable insulation (see Figure 6-63), and fuel hose. Injection-molded parts include gears, bushings, O-rings (see Figure 6-64), bottle grippers, ski boots, and sports shoes.

Coatings. Urethane coatings possess a combination of a wide range of flexibility coupled with toughness, excellent electrical properties, and good abrasion and chemical resistance. They can be used either as clear varnishes or in pigmented form, and can be cured either at room temperature or at elevated temperatures. A disadvantage of these coatings has been their tendency to yellow on exposure to light. This yellowing tendency may be reduced by employing aliphatic rather than aromatic isocyanates. However, the toxicity of the aliphatic isocyanates, significant vapor pressures, and cost are serious deterrents for their commercial use. In an effort to extend the pot life of the coatings and to employ a one-component system, "blocked" isocyanates have been developed. The ability of these "blocked" urethanes at elevated temperatures to react like free isocyanates has led to their successful use as coatings for the electrical industries, especially for wire enamels. Another type of one-can urethane coatings is the "urethane oils," reaction products of isocyanate with drying oils or their derivatives. With the successful development of coatings for clear wood varnishes, leather finishes, maintenance paints, and other exterior finishes, urethane coatings are tending to move

Fig. 6-62. Cast urethane encapsulant. (*Courtesy: Du Pont Elastomer Chemicals Department*)

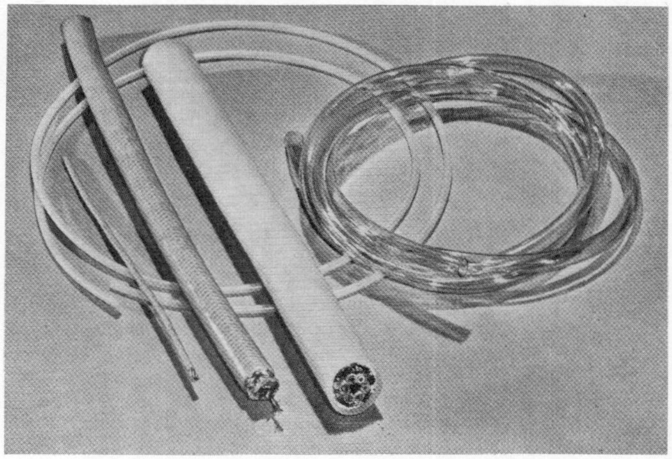

Fig. 6-63. Extruded cable insulation. (*Courtesy: CPR Division, Upjohn Co.*)

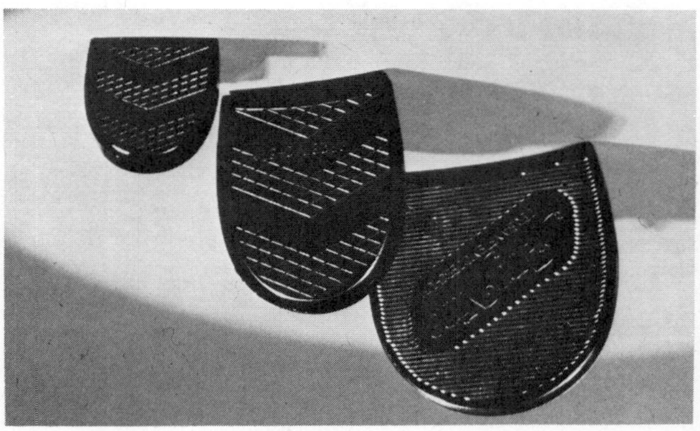

Fig. 6-61. Shoe heels are among the many injection-molded goods which can be made using thermoplastic urethanes. Those shown in the picture were made with PLASTOTHANE™ TPUs from Thiokol Corporation. (*Courtesy: Thiokol Corporation, Chemical Division*)

Fig. 6-64. Injection-molded gears, bushings, shock mounts, and O-rings. (*Courtesy: Mobay Chemical Corporation*)

Table 6-68. Detailed Physical, Electrical, Thermal and Chemical Properties of Polyurethane Elastoplastics.*

Properties	ASTM Test Method	Polyester Base	Polyether Base
Shore A hardness	D2240	–	78–92
Shore D hardness	D2240	37–63	–
Tensile modulus: 100% elongation (psi)	D412	1000–2500	500–1200
300% elongation (psi)	D412	2500–5000	1000–3000
Tensile strength (psi)	(2 in./min.)	5500–7500	5000–7500
Elongation at break (%)	–	350–600	500–700
Elongation set, after break (%)	–	50–90	40–70
Tear: die C (pli)	D624	750–950	500–700
trouser (pli)	D1938	350–450	300–450
split (pli)	D470	150–200	150–200
Compression set, method B, 22 hours:			
at room temperature (%)	D395	20–25	20–25
at 158°F (%)	D395	25–35	35–45
at 212°F (%)	D395	35–45	50–70
Resilience, Bashore rebound (%)	–	25–30	30–35
Bell brittle point (°F)	D746	–40 to –50	–80
Thiokol bend test (–50°F)	D736	Pass	Pass
Taber abrasion: H-18 wheel; 1000 gram load (grams loss in 5000 revolutions)	–	250–650	200–300
Specific gravity	–	1.25	1.12–1.15
Specific volume (cu in./lb)	–	22.2	23.5–24.5
Processing temperatures:			
Injection molding; melt temperature (°F)	–	415–445	390–425
Extrusion; melt temperature (°F)	–	–	370–420
Compression molding; platen temperature (°F)	–	390–415	330–360
Volume resistivity; 500 vdc, 1 minute:			
at room temperature (ohm/cm × 10^{12})	D257	–	10–12
at 50°C (ohm/cm × 10^{12})	D257	–	0.6–0.8
at 70°C (ohm/cm × 10^{12})	D257	–	0.2–0.3
Dielectric constant: at 60 cycles	D150	–	5.5–6.5
at 10^3 cycles	D150	–	5.0–6.0
at 10^6 cycles	D150	–	4.0–5.0
Dissipation factor: at 60 cycles	D150	–	0.045–0.050
at 10^3 cycles	D150	–	0.040–0.045
at 10^6 cycles	D150	–	0.070–0.075
Dielectric strength (v/mil)	D149	–	450–550
Arc resistance (sec)	D495	–	100–150
Immersion in Fuel A; 7 days; room temperature:			
Hardness	D471	40–60 D	80–90 A
Maximum tensile strength (psi)	D471	5500–7500	5000–7500
Volume change (%)	D471	+0.2 to +0.4	+4 to +6
Immersion in Fuel B; 7 days; room temperature:			
Hardness	D471	35–55 D	75–85 A
Maximum tensile strength (psi)	D471	5000–7000	3000–4000
Volume change (%)	D471	+3 to +12	+20 to +30
Immersion in ASTM oil #1; 5 days; 250°F:			
Hardness	D471	43–63 D	75–90 A
Maximum tensile strength (psi)	D471	5500–7500	5000–7000
Volume change (%)	D471	+0.5 to +1.0	+1.0 to +2.0
Immersion in ASTM oil #3; 5 days; 250°F:			
Hardness	D471	40–60 D	70–85 A
Maximum tensile strength (psi)	D471	5500–7500	4000–6000
Volume change (%)	D471	+2.5 to +3.5	+20 to +30
Physical properties at 212°F:			
Hardness	D2240	30–45 D	60–80 A
Tensile strength (psi)	D412	2000–4500	2000–4000
Mold linear shrinkage (in./in.)	–	0.010 to 0.020	0.010 to 0.020
Ozone resistance: 1000 hrs; 50 pphm; 100°F	D518	No effect	No effect
Weatherometer; 1000 hrs	All methods D740	Excellent retention of physical properties	
Preferred method of processing	–	Injection molding	Extrusion and injection molding

*Courtesy: Uniroyal Chemical.

gradually from the specialty finish field into broader markets.

Fibers. Polyurethane fibers were first developed in Germany in an attempt to obtain fibers similar to nylon. Under the Textile Fiber Products Identification Act, the generic name of "spandex" fibers was selected to designate elastomeric fibers in which the fiber-forming substance is a long-chain polymer consisting of at least 85% segmented polyurethane.

The elastic behavior of the spandex fibers is due to the alternate arrangements of "soft" segments consisting of either polyester- or polyether-urethane blocks, and "hard" segments, containing generally aromatic, urea, and sometimes urethane groups as the rigid components. The soft sections impart the ability to elongate, while the hard sections tie the elastic chains together by hydrogen bonding or crystallite formation, or, in some cases, primary cross-linking points, thus preventing excessive flow of the polymer (stress decay) when kept under tension and allowing it to snap back when the tension is released. The combination of properties such as light weight, strength, and sheerness accounts for the success of spandex in the elastomeric fiber market, particularly for foundation garments and swim wear.

Compared to nylon 6/6 fibers, the spandex fibers have a higher tensile strength, discolor less on exposure to air, have higher acid resistance and lower moisture absorption. On the other hand, they are harder to dye and handle.

Other Applications. Isocyanates and several types of isocyanate derivatives have been used to render fabrics water repellent, to improve the wet strength and reduce the shrinkage of fabrics. Urethanes have been used in combination with vinyl compounds to make coatings. Some thermoplastic polyurethane compounds have been made into adhesives. These materials are hydroxy-terminated polymers supplied in granular form. They are soluble in ketones, esters, as well as other solvents. They provide fast crystallization rates, giving rapid development of green strength, good heat and hydrolysis resistance, good low-temperature performance, resistance to plasticizer migration, and exceptional adhesion, particularly to flexible substrates such as vinyl, leather, rubber, and fabric. Primary uses are to bond shoe uppers to soles, and for laminating plastic films, foils, and fabrics.

POLYISOCYANURATES

Although rigid polyurethanes are excellent thermal insulators, they have a limited level of fire retardance, even when containing phosphorous- or halogen-based fire retardants. Polyisocyanurates not only have good resistance to burning and flame spread, but can also withstand service temperatures up to 150°C (302°F) (see Table 6-69). Polyisocyanurate foams retain the good hydrolytic stability and low thermal conductivity of rigid polyurethane foams.

Table 6-69. Typical Physical Properties of Polyisocyanurates.*

TRYMER CPR 9545	ASTM Method	Typical Values
Nominal Density, LBS/FT³	D1622	1.8-2.3
k Factor, BTU/FT²-HR-°F/IN	C177	0.15 @ 75°F. Mean Temp.
Closed Cell Content, %	D1940	93
Coefficient of Linear Expansion, IN/IN-°F.	—	5×10^{-5} @ −20°F. to 160°F.
Water Vapor Permeability, PERM-IN	C355	2-3
Compressive Strength		
Parallel, psi	D1621	25
Perpendicular, psi	D1621	17
Compressive Modulus		
Parallel, psi	D1621	700
Perpendicular, psi	D1621	450
Tensile Strength		
Parallel, psi	D1623	27
Perpendicular, psi	D1623	24
Shear Strength		
Perpendicular, psi	C273	24

Underwriters' Laboratories, Inc. Fire Hazard Classification (UL 723)

CAUTION: This numerical flame spread rating is not intended to reflect hazards presented by this or any other material under actual fire conditions.

(Based on 100 for untreated red oak)				
Thickness, inches	0.65	1	2	3-4
Flame Spread	25	25	25	25
Fuel Contributed	N.D.	N.D.	5	5
Smoke Developed	55	70	135	135-200

*Courtesy: CPR Division, Upjohn Co.

Polyisocyanurates are formed via the trimerization of an isocyanate. The most commonly used isocyanate is a modified form of MDI.

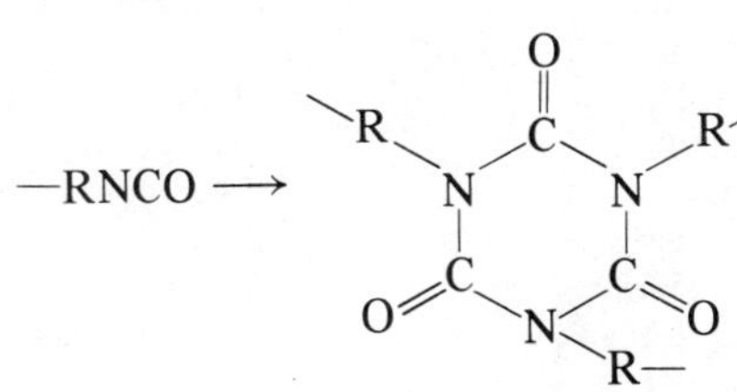

Typical catalysts used are alkali metal phenolates, alcoholates, carboxylates, and compounds containing *o*-(dimethylaminomethyl)phenol subgroups. Fluorocarbons are the only blowing agents used in the absence of any isocyanate-water reaction. The resulting foams have high crosslink density and are consequently very brittle. Combinations of polyisocyanurate and polyurethane have been developed. Trimerization in the presence of low molecular weight polyols (<300) have produced foams which are adaptable to both one-shot and prepolymer foaming techniques.

Trade Names

Acoustazip	Foam	Accessible Products Co.
Actol	Polyether polyols	Allied Chemical Corporation
Adaphah	Resins	American Cyanamid Co.
Adiprene	Elastomers	E.I. Du Pont de Nemours & Co., Inc.
Afonic	Foam	Scott Paper Co.
Aggre Bond	Formulation	Techform Laboratories, Inc.
Airborne	Foam	Albert Trostel & Sons Co.
Airfoam	Foam	Reeves Hardifoam, Ltd.
Airfoam	Foam	Goodyear Tire & Rubber Co.
Air-thane	Foam	Perlon Corporation
Alfa	Elastomer	N L Industries, Inc.
Allfoam	Polyether and polyester form	Robinson Foams Ltd.
Alpco	Resin	Olin Corporation
Amxco Foam	Foam	American Excelsior
Analac	Magnet wire	Anaconda Wire & Cable Co.
Ancrol	Resins	Newton Chambers & Co., Ltd.
Andur	Prepolymers	Anderson Development Co.
Arcon	Resins	Allied Resin Corporation
Arma	Elastomer	N L Industries, Inc.
Arolast	Prepolymer	Ashland Chemical Co.
Arothane	Prepolymers and foam resins	Ashland Chemical Co.
Artfoam	Foam	Strux Corporation
Arvetone	Adhesive	Arveta, S.A.
Astrocoat	Coatings	Olin Corporation
Astro Lam	Reinforced	Astronautic Industries, Inc.
Autofroth	Foam	Olin Corporation
Baxpak	Foam	Baxenden Chemical Co., Ltd.
Baydur	Plastics	Mobay Chemical Co.
Bayflex	Foam	Mobay Chemical Co.
Baymidur	Isocyanates	Mobay Chemical Co.
Beckurane	Resins	Beck, Koller & Co., Ltd.
Beethane	Coatings	Bee Chemical Co.
Beldure	Coatings	Belden Corporation
Cab-Curalac	Coatings	Whittaker Corporation
Calthane	Formulations	Cal Polymers, Inc.
Caprolast	Elastomer	Inter-Polymer Research Corporation
Caradate	Isocyanates	Shell Chemical Co.
Caradol	Polyether polyols	Shell Chemical Co.
Carbamac	Oil-modified	Commercial Solvents Corporation
Carolux	Foam	North Carolina Foam, Inc.
Carstan	Foam	Cincinnati Milacron Chemicals, Inc.
Carthane	Formulation	Upjohn Co.
Castethane	Elastomers	Upjohn Co.
Casthane	Molded	Chicago-Allis Mfg. Co.
Castocure	Curing agents	Witco Chemical Corporation
Castomer	Resins	Witco Chemical Corporation
Cat-A-Last	Coatings	Bostik-Finch, Inc.
Catapol	Elastomers	Arnco
Celpak	Foam	Dacar Chemical Products Co.
Celthane	Foam	Dacar Chemical Products Co.
Chemglaze	Coatings	Lord Mfg. Co.
Chem-o-thane	Formulations	Whittaker Corporation
Chempol	Resins	Freeman Chemical Corporation
Chem-Thane	Resins	Eagle Picher Industries, Inc.
Cipfoam	Foam	Essex Wire Corporation
Clocel	Foam	Baxenden Chemical Co., Inc.
Cloudfoam	Foam	Holiday Inns of America, Inc.
Conacure	Curing agents	Conap, Inc.
Conathane	Formulations	Conap, Inc.

Conefoam	Foam	Cone Mills Corporation
Cord	Polyethers	CPC International, Inc.
Coustex	Foam	Scott Paper Co.
CPR	Foam	CPR Division, Upjohn Co.
Curalon	Curing agent	Uniroyal, Inc.
Curathane	Coatings	Whittaker Corporation
Cureprene	Polyols	Iharo Chemical Industry Co., Inc.
Curon	Foam	Reeves Brothers, Inc.
Cush-N-Foam	Foam	Hudson Cush-N-Foam Corporation
Cuvertin	Coatings	Henkel & Cie, GmbH
Cyanaprene	Elastomer	American Cyanamid Co.
Cyanaset	Curing Agents	American Cyanamid Co.
Cybond	Adhesive	American Cyanamid Co.
Cynaprene	Rubber	American Cyanamid Co.
Cyno-Glass	Reinforced	Sun Chemical Corporation
Cytor	Thermoplastic	American Cyanamid Co.
Daltocel	Resins	Imperial Chemical Industries, Ltd.
Daltoflex	Solutions	Imperial Chemical Industries, Ltd.
Daltolac	Resins	Imperial Chemical Industries, Ltd.
Daltomold	Thermoplastic	Imperial Chemical Industries, Ltd.
Daycollan	Elastomer	Dayco Corporation
Desmocoll	Adhesive	Mobay Chemical Corporation
Desmodur	Isocyanates	Mobay Chemical Co.
Desmoflex	Resin	Mobay Chemical Corporation
Desmopan	Thermoplastic	Mobay Chemical Corporation
Desmophen	Polyesters and polyethers	Mobay Chemical Corporation
Dielectric foam	Foam	Scott Paper Co.
Diorez	Polyesters	Briggs & Townsend, Ltd.
Disogrin	Elastomer	Disogrin Industries, Inc.
Dithane	Polyols	Pelcon Corporation
Dow	Resins	Dow Chemical Co.
Dulon	Foam	Dunlop Co., Ltd.
Duralux	Foam	Cone Mills Corporation
Durane	Coatings	Raffi & Swanson, Inc.
Dura-Soft	Fiber	Reeves Bros., Canada, Ltd.
Duraspan	Spandex fiber	International Latex
Durathane	Coatings	PPG Industries
Durelast	Thermoplastic	Briggs & Townsend, Ltd.
Durethane	Coatings	PPG Industries
Dyalon	Elastomer	Thombert, Inc.
Elastacast	Thermoplastic and thermoset elastomers	Acushnet Process Co.
Elastogram	Elastomers	BASF, A.G.
Elastolit	Thermoplastic	North American Urethanes, Inc.
Elastollyx	Thermoplastic	North American Urethanes, Inc.
Elastolur	Coatings	North American Urethanes, Inc.; BASF Wyandotte Corporation
Elastonate	Prepolymers	North American Urethanes, Inc.; BASF Wyandotte Corporation
Elastonel	Curing agents	North American Urethanes, Inc.
Elastonol	Polyester polyols	North American Urethanes, Inc.; BASF Wyandotte Corporation
Elastopel	Thermoplastics	North American Urethanes, Inc.
Elastothane	Rubber	Thiokol Chemical Corporation
Elfoam	Foam	Elliott Co. of Indianapolis, Inc.
Endurethane	Resin	Blair Process Co., Inc.
Epibond	Adhesives	M & T Chemicals, Inc.
Epithane	Formulation	Daro Products
Epofoam	Foam	M & T Chemicals, Inc.
Epoxiglas	Reinforced foam	A. B. Chance Co.
Ertcopal	Resin	Easthampton Rubber Thread Co.
Estane	Thermoplastic	B. F. Goodrich Co.
Estolan	Polyesters	Lankro Chemicals, Ltd.
Everlon	Foam	Stauffer Chemical Co.
Extone	Tubing and extrusions	Extron Corporation
Fastcast	Resin	Arnco

Fawthane	Coatings	Pan Chemical Corporation
Flam-less	Foam	Bulstrode Plastics & Chemical Co., Ltd.
Flepol	Resin	Flexible Products Corporation
Flexi-Cast	Resin	Urethane Products Canada, Ltd.
Flexifoam	Resin	Foamade Industries
Flexipol	Foam	Flexible Products Corporation
Flexocel	Foam	Baxenden Chamical Co., Ltd.
Foamall	Foam	Woodall Industries, Inc.
Foamart	Foam	Adhesive Products Corporation
Foamedge	Foam	Sterling Aldefer Co.
Foampac	Foam	Donald Macpherson Group, Ltd.
Foamthane	Foam	Pittsburgh Corning Corporation
Foamwrap	Foam	Sealed Air Corporation
Fomofill	Foam	Fomo Products
Fomol	Foam	Atlas Chemical, Ltd.
Fomosoft	Foam	Fomo Products
Fomospray	Foam	Fomo Products
Formrez	Polyester polyols	Witco Chemical Co., Inc.
Fortified Foam	Foam	Phillips-Fosure Corporation
Froth-Pak	Foam	Insta Foam
Ful Fram	Foam	Fuller-O'Brien Corporation
Genthane	Elastomers	General Tire & Rubber Co.
Glas Pan	Spandex fiber	Globe Mfg. Co.
Gold Bond Zer-O-Cel	Foam	National Gypsum Co.
Gurane	Rubber	Michelin & Cie
Handi-Foam	Foam	United Foam Corporation
Hartmoltropen	Foam	Mobay Chemical Co.
Hathane	Resin	Hastings Plastics, Inc.
Hetrofoam	Foam	Hooker Chemical Co.
Hightemp	Formulations	Hightemp Resins, Inc.
Honey Foam	Foam	Dimensional Plastics Corporation
Hylene	Isocyanates	E.I. Du Pont de Nemours & Co., Inc.
Hylon	Foam	E.I. Du Pont de Nemours & Co., Inc.
Hy-Thane	Resins	Permali, Inc.
Impranil	Prepolymer and adhesives	Mobay Chemical Corporation
Imron	Coating	E.I. Du Pont de Nemours & Co., Inc.
Indpol Monothane	Formulations	Synair Corp.
Instant-Set Polymer	Resin	Dow Chemical Co.
Insul	Varnish	Easton R.S. Corporation
Insulcote	Coatings	Easton R.S. Corporation
Insulfoam	Foams	Robinson Foams, Ltd.
Iporane	Elastomers	IPRC Corporation
Isobestos	Coated asbestos	Natvar Corporation
Isochem	Foams	H-S Enterprises, Inc.
Isochem-	Foams and formulations	Isochem Resins Co.
Isocon	Isocyanates	Lankro Chemicals, Ltd.
Isoderm	Foam	Upjohn Co.
Isofil	Resin	Synair Co.
Isofoam	Foam resins	Witco Chemical Corporation
Isogen	Isocyanates	General Tire & Rubber Co.
Isoglas	Coated cloth	Natvar Corporation
Isolastane	Coated cloth	Natvar Corporation
Isolastube	Coated fiberglass	Suflex Corporation
Isolon	Foam	Englander Co., Inc.
Isomelt	Enamel	Schenectady Chemicals, Inc.
Isonate	Foam	Upjohn Co.
Isoplex	Coated cloth	Natvar Corporation
Isothane	Foam	Bernel Foam Products Co., Inc.
Jet Foam	Foam	Holiday Inns of America, Inc.
J. Thane	Formulation	Texaco, Inc.
Kalex	Formulations	Hardman, Inc.
Kalspray	Foam	Baxenden Chemical Co.
Kayfoam	Foam	Kay-Metzeler, Ltd.
Ken-U-Thane	Formulations	Kenrich Petrochemicals, Inc.
Korel	Film	Foamade Industries
Korofoam	Foam	Koro Corporation

K-Prene	Formulation	Di Acro Kaufman
Kryptane	Resin	Kryptonics Sales Co.
Lankrothane	Coatings and adhesives	Lankro Chemicals, Ltd.
Last-A-Foam	Foam	General Plastics Mfg. Co.
Leoperm	Resin	Scott Paper Co.
Lexorez	Polyester polyols	Inolex Corp.
Lock Foam	Foam	Nopco Chemical Co.
Lumitol	Curing Agents	BASF Wyandotte Corporation
Lux-Foam	Foam	Morristown Foam Corporation
Marafoam	Foam	Allied Products Corporation
Marathane	Resin	Allied Products Corporation
Mearthane	Elastomers	United Shoe Machinery Corporation
Medwood	Resin	Adhesive Products Corporation
Megapren	Rubbers	Rhein-Chemic Rheinau, GmbH
Melofoam	Formulations	Guardian Chemical & Equipment Co., Inc.
Mertex	Polycarbonate blend	Mobay Chemical Co.
Microthane	Foams	Cal Polymers, Inc.
Millathane	Elastomers	Technical Sales & Engineering, Inc.
Minit Foam	Resin	Kerr Chemicals, Inc.
Mira-Dip	Coatings	Kristal Kraft, Inc.
Miraglas	Coatings	Polymer Industries, Inc.
Mista Flex	Elastomers	M. R. Plastics & Coatings, Inc.
Mista Foam	Foams	M. R. Plastics & Coatings, Inc.
Mista Froth	Foams	M. R. Plastics & Coatings, Inc.
Moldflo	Resin	Kornylak, Ltd.
Moltopren	Foam	Bayer, A.G.
Mondur	Isocyanates	Mobay Chemical Co.
Monopol	Elastomers	Indol
Monothane	Elastomers	Synair Corporation
Multanil	Resins	Mobay Chemical Co.
Multithane	Resins	J. A. Richards Co.
Multi-Tube	Tubing	Molded Products Co.
Multrathane	Elastomers	Mobay Chemical Co.
Multron	Polyester polyols	Mobay Chemical Co.
Nacconate	Isocyanates	Allied Chemical Corporation
Nafil	Isocyanates	Chase Chemical Corporation
Nelthane	Emulsions	Northeastern Labs Co., Inc.
Neo Rez	Coatings	Polyvinyl Chemicals, Inc.
Neothane	Coatings	Goodyear Tire & Rubber Co.
Niax	Polyester polyols	Union Carbide Corporation
Nimbus	Foam	General Tire & Rubber Co.
Non-Dis V	Rubber	Goodyear Tire & Rubber Co.
Nopcocell	Foam	Diamond Shamrock Co.
Nopcofoam	Foam	Diamond Shamrock Co.
Nopcolast	Elastomers	Diamond Shamrock Co.
Nopco Lock Foam	Foam	Diamond Shamrock Co.
Nopcothane	Coatings	Diamond Shamrock Co.
Norwind	Coated fabric	Seton Co.
Novitane	Compounds	B. F. Goodrich Chemical Co.
Nyglathane	Formulations	Nypel, Inc.
Orthane	Resin	Ohio Rubber Co.
Ostamer	Foam	BASF Wyandotte Corporation
Oxirane	Polyols and isocyanates	Oxirane Chemical Co.
PAPI	Polyisocyanates	Upjohn Co.
Pedigree Products	Resins	P. D. George Co., Inc.
Pellethane	Thermolastic	Upjohn Co.
Perflex E	Film	Union Carbide Corporation
Pinpoly	Foam	Holiday Inns of American, Inc.
Plasti-Foam	Foam	Polystructures, Inc.
Plasti Shell	Resins	Commercial Chemical Co.
Plastone	Formulations	Upjohn Co.
Plastothane	Thermoplastic	Thiokol Co.
Platilon	Film	R. H. Cole, Ltd.
Pleogen	Formulations	Whittaker Corporation
Pluracol	Polyether polyols	BASF Wyandotte Corporation
Pluragard	Foam	BASF Wyandotte Corporation

Pluronic	Resins	BASF Wyandotte Corporation
Polane	Coatings	Sherwin-Williams Co.
Polyaire	Foams	American Rubber & Plastics Corporation
Polycel One	Foams	Coplanar Corporation
Polycin	Prepolymers	N L Industries, Inc.
Polycoat	Resin	Showa High Polymer Co.
Polyfoam	Foam	Copolymer Rubber & Chemical Corporation
Polyfoam	Foam	General Tire & Rubber Co.
Polyfoam-Hull Fill	Foam	Copolymer Rubber & Chemical Corporation
Poly-G	Polyether polyols	Olin Corporation
Polyisocyanurat	Foam	Bayer, A.G.
Polymeg	Polyols	Quaker Oats Co.
Polymercast	Elastomers	U.S. Polymeric
Poly-Q	Elastomers	Arnco
Poly-Thane	Foams	Pelron Corporation
Polythane	Foams	Chemical Components, Inc.
Poly Tile	Foams	Polyfilm Corporation
Polytool	Tooling Compounds	Reichhold Chemicals, Inc.
Polyur-Alloy	Formulations	High Strength Plastics Corporation
Poron	Foams	Rogers Corporation
Powerthane	Formulations	Tool Mate Corporation
Pressure Bottle	Foam	Midwest Manufacturing Corporation
Propylan	Polyether polyols	Diamond Shamrock, Inc.
Purelast	Adhesives and coatings	Polymer Systems Corporation
Purethane	Elastomer	Easton R.S. Corporation
Pyrell	Foam	Scott Paper Co.
Q-Pak	Foam	Scott Paper Co.
Q-Thane	Thermoplastic adhesives	K. J. Quinn & Co.
Quasilan	Prepolymers	Lankro Chemicals, Ltd.
Quelfam	Formulations	Baxenden Chemical Co.
Ranthane	Resins	Randolph Products Co.
Redcote	Formulations	Indpol, Inc.
Reevane	Coated fabric	Reeves Brothers, Inc.
Ren-	Formulations	Ren Plastics
Rest Foam	Foam	Stauffer Chemical Co.
Reyhosol	Formulations	Hoover Ball & Bearing Co.
Reyno-Foam	Foams	Hoover Ball & Bearing Co.
Reynufoam-P	Foams	Reynolds Chemical Products Co.
Richlux	Foams	E. R. Carpenter Co.
Rim	Formulations	Union Carbide Corporation
Rimthane	Formulations	Upjohn Co.
Robi-Foam	Foams	Robinson Foams, Ltd.
Royalar	Molding compounds	Uniroyal, Inc.
Roylar	Thermoplastics	Uniroyal Chemical Co.
Rubical	Polyether Polyols	Uniroyal, Inc.
Rubicast	Foams	Uniroyal, Inc.
Rubigard	Foams	Uniroyal, Inc.
Rubinate	Isocyanates	Rubicon Chemical, Inc.
Ruco	Polyester polyols	Hooker Chemical & Plastics Corporation
Rucothane	Resins and coatings	Hooker Chemical & Plastics Corporation
Safecore	Foam	Kaiser Aluminum & Chemical Corporation
Safecrest	Foam	Crest-Foam Corporation
Safe-T-Foam	Foam	Chicago Foam Products
Saf-T-Foam	Foam	Holiday Inns of America, Inc.
Sanfoam	Foam	Sanyo Chemical Industries, Ltd.
Sannix	Polyether polyols	Sanyo Chemical Industries, Ltd.
Sanprene	Resins	Sanyo Chemical Industries, Ltd.
Scotch-	Adhesives	3M Co.
Scott-	Foam	Scott Paper Co.
Seal Coat	Varnish	C. J. Webb, Inc.
See Cure	Resins	Epic Resins Corporation
Sharnet	Adhesive	B. F. Goodrich Chemical Co.
Shellertex	Foam	Sheller Manufacturing Corporation
Shock-Sorb	Foam	Leewood Corporation
Shuthane	PVC blend	Reichhold Chemicals, Inc.
Silicane	Elastomers	Bacon Industries, Inc.

Skinn	Coated fabrics	Collegen Corporation
Solithane	Elastomer	Thiokol Co.
Superthane	Resin	Thibout & Walker Co.
Suprasec	Resin	Imperial Chemical Industries, Ltd.
Suprathene	Foam	U.S. Mineral Products Co.
Supresac	Resins	ICI United States, Inc.
Syntex	Ester Resins	Celanese Coatings Co.
Synthite	Varnishes	John C. Dolph Co.
Techniform	Laminates	Union Carbide Corporation
Tedimon	Isocyanates	Montedison, S.p.A.
Tenn-Foam	Foam	Morristown Foam Corporation
Tenpelle	Coated Fabric	Tenneco Chemicals, Inc.
Tetronic	Polyether tetrols	BASF Wyandotte Corporation
Texfoam	Formulations	Texas Urethanes
Texin	Thermoplastic	Mobay Chemical Co.
Texthane	Foam	Texas Urethanes
Thanate	Isocyanates	Texaco, Inc.
Thanol	Polyether polyols	Texaco, Inc.
Therma Thane	Foam	National Gypsum Co.
Thermazote	Foam	Expanded Rubber & Plastics, Ltd.
Thixane	Formulation	Whittaker Corporation
Thorane	Foam	Dow Chemical Co.
Thurane	Foam	Dow Chemical Co.
Timbura	Foam	Ben Chair, Ltd.
Tiprene	Rubber	Tigers Rubber Co., Ltd.
Tranco	Formulations	Trancoa Chemical Corporation
Trostel	Elastomer	Albert Trostel & Sons Co.
Trymer	Isocyanurate resins and foams	Upjohn Co.
Tube-cota	Formulations	Tuboscope, Inc.
Tufcote	Foam	Specialty Composites Corporation
Tuf Film	Formulations	Products Techniques, Inc.
Tuftane	Film and sheet	B. F. Goodrich Chemical Co.
Unicote	Coatings	Unipoint Industries, Inc.
Unifoam	Foam	Wm. T. Burnett & Co., Inc.
Unithane	Resins	Thiokol Corporation
Upox	Epoxy Blends	Isochem Resins Co.
Urabond	Coatings	Poly Resins, Inc.
Uradur	Resins	Synthetic Resins, Ltd.
Urafil	Reinforced	Dart Industries, Inc.
Uraglas	Reinforced	Dart Industries, Inc.
Uralane	Formulations	M & T Chemicals, Inc.
Uralast	Coatings	Unilever, Ltd.
Uralite	Formulations	Hexcel Corporation
Uraloid	Coatings	Rowe Products, Inc.
Urapol	Elastomers	Poly Resins, Inc.
Urecon	Formulations	Easton R.S. Corporation
Urefoam	Foams	Atlas Mineral & Chemicals Division, ESB Inc.
Urefome	Foams	Textile Adjuncts Corporation
Urefroth	Foam	Polytron Sales Corporation
Urelast	Elastomer	Easton R.S. Corporation
Urepan	Elastomer	Bayer, A.G.
Urepot	Formulations	Easton R.S. Corporation
Ureputty	Formulations	Easton R.S. Corporation
Urifilm	Coatings	Advanced Coatings & Chemicals, Inc.
Uroflex	Formulations	North American Urethane, Inc.; BASF Wyandotte Corporation
Urolast	Formulations	North American Urethane, Inc.; BASF Wyandotte Corporation
Uropac	Formulations	North American Urethane, Inc.; BASF Wyandotte Corporation
Urotuf	Resins	Reichhold Chemicals, Inc.
Urutuf	Resins	Reichhold Chemicals, Inc.
USM Suprathane	Foam	United States Mineral Products Co.
U-Thane	Foam	Upjohn Co.
Uvithane	Oligomers	Thiokol Corporation
Varkydane	Vehicles	McClosky Varnish Co.

Vibrafoam	Foam	Uniroyal, Inc.
Vibrathane	Intermediates	Uniroyal, Inc.
Vithane	Resins	Goodyear Tire & Rubber Co.
Voranate	Resins	Dow Chemical Co.
Voranol	Polyether polyols	Dow Chemical Co.
Vordnol	Resins	Dow Chemical Co.
Vorite	Formulations	N L Industries, Inc.
Vulafoam	Prepolymers	General Latex & Chemical Corporation
Vulathane	Coatings	General Latex & Chemical Corporation
Vullkollan	Elastomer	Mobay Chemical
Vultathane	Coatings	General Latex & Chemical Corporation
Vyrene	Spandex fiber	Uniroyal, Inc.
Witcobond	Coatings and adhesives	Witco Chemical Co.
Zer-O-Cel	Foam	National Gypsum Co.
Z-Foam	Foam	Norton Co.

REFERENCES AND BIBLIOGRAPHY

American Cyanamid Co. *CYTOR Thermoplastic Polyurethane*. Bulletin 7-2621, February 1979.

——. *Hydrolytic Stability With Cyanaprene Polyester Urethane Elastomers*. Bulletin 4-2726, December 1974.

——. *Cyanaprene Thermoplastic Polyurethane Elastomers*. Bulletin 4-2611, May 1974.

——. *Engineering Properties of Cyanaprene Urethane Elastomers*. Bulletin 3-2682, June 1973.

Anderson, A. Reaction injection molding; process control, reinforcements, new materials build up markets. *Modern Plastics*, April 1978, **55**(4):66–67.

BASF Wyandotte Corporation. Rigid plastic foams, from Plastics in construction: Coming on strong in 1978. *Plastics Engineering*, July 1978, **34**(7):42–44.

Brydson, J. A. *Plastics Materials*. 3rd Ed. Robert E. Krieger Publishing Co., Inc., Huntington, New York, 1975.

Cox, H. W., and Iobst, S. A. Molding polyurethane parts on standard injection machines. *Plastics Engineering*, May 1978, **34**(5):49–52.

DeGisi, S. L., and Smith, C. H. Improved bondability of molded rigid urethane foam. *Adhesives Age*, July 1978, **21**(7):35–37.

Dow Chemical Co. *For the Urethane Industry . . . Voranol Polyols From Dow*. Bulletin No. 111-1006, 1969.

Eakin, J. L. Urethane structural foams: key properties give them a competitive edge. *Plastics Engineering*. July 1978, **34**(7):56–60.

E.I. Du Pont de Nemours & Co., Inc. *Engineering Properties of Du Pont Adiprene Urethane Rubber*. Bulletin A-88436.

——. *Facts About Adiprene Urethane Rubber*. Bulletin A-45945.

Galbreath, D. Urethane foam, from Plastics in Packaging. *Plastics Engineering*, October 1978, **34**(10):50–51.

Guha, P. K., and Epel, J. N. Adhesives for bonding graphite/glass composites. *Adhesives Age*, June 1979, **22**(6):31–34.

Harper, R. C., and Reber, D. H. Glass fiber filled polyurethane monomers in RIM equipment. Paper presentation, *SPE ANTEC*, New Orleans, May 1979.

Hemann, J. H., and Fang, S. J. Glassy waves in urethane rubber. Paper presentation, *SPE ANTEC*, May 1979.

Hooker Ruco Division. *Rucoflex Polyesters as Replacements for Polycaprolactones*. Laboratory Report 79-7, 1977.

Houston, A. M. Filled urethanes fight noise and abrasion in gears. *Materials Engineering*, October 1976, **84**(4):84–85.

Jackson, L. C. Improving adhesion by gas-plasma contaminant removal. *Adhesives Age*, September 1978, **21**(9):34–38.

Kresta, J. E., Shen, C. S., and Lin, S. I. "Catalysis in Urethanes, Steric and Solvation Effects." Paper presentation, *SPE ANTEC*, New Orleans, May 1979.

Landrock, A. H. *Polyurethane Foams: Technology, Properties and Applications*. Plastec Report 37, Plastics Technical Evaluation Center, U.S. Army ARRADCOM, Dover, New Jersey, January 1969.

Lin, C. "Polyurethanes from Polyester." Paper presentation, *SPE ANTEC*, New Orleans, May 1979.

Mettes, D. G. RP and PUR offer critical values in designing pipe for tough service. *Modern Plastics*, November 1978, **55**(11):54–55.

Miller, B. The "high mods" RIM urethanes gain new muscle. *Plastics World*, June 1977, **35**(6):36–40.

Mobay Chemical Co. *Texin–The Elastomer That Molds Like Plastic, Works Like Rubber and Wears Like Metal!* 1964.

Modern Paint & Coatings. Urethane coatings protect Gulf acid plant. September 1978, **68**(9):56.

Modern Plastics. PUR is the medium for plant growth. March 1978, **56**(3):30.

——. New urethane system makes soft-to-hard parts without switching polymers. April 1979, **56**(4):20–22.

——. More usable material, less waste in new system for flat-top urethane buns. August 1978, **55**(8):20.

Nieske, F. W., and Gahimer, F. H. *Hydrolytic Stability of Polyurethane and Polyacrylate Elastomers in Humid Environments*. Naval Avionics Facility Publication TR 1372, February 1969.

Plastics Engineering. Engineering resins primer. December 1974, **30**(12):17–32.

Rybny, C. B., Faust, E. E., and Purgason, R. R. Polyester-based urethanes. . . . *Paint and Varnish Production*, April 1973, **63**(4): 23–26.

Saunders, J. H., and Frisch, K. C. *Polyurethanes–Chemistry and Technology*, (2 volumes), Wiley-Interscience, New York, 1962.

Simonds, H. R., and Church, J. M. *A Concise Guide to Plastics*. 2nd Ed. Van Nostrand Reinhold Co., New York, 1963.

Skowronski, M. J., and DeLeon, A. "Polyol in Isocyanurate Foam." Paper presentation, *SPE ANTEC*, New Orleans, May 1979.

Swanson, F. D., and Price, S. J. Chemistry of urethane adhesives with silane coupling agents. *Adhesives Age*, June 1973, **16**(6): 23–28.

Szycher, M., Poirier, V., and Dempsey, D. "Urethane Elastomers for Cardiac Assist Devices." Paper presentation, *SPE ANTEC*, May 1979.

Tarin, P. M., Prokoi, B., and Schwarz, E. G. "Particle Size of Reinforced High-Modulus RIM Urethane." Paper presentation, *SPE ANTEC*, New Orleans, May 1979.

Thiokol Chemical Division. *Solithane LFTD1 Urethane Prepolymer*. Bulletin TD 990, March 1977.

——. *Plastothane Thermoplastic Urethanes for Adhesives*. Bulletin TD-1780, September 1977.

——. *Plastothane Thermoplastic Urethanes for Injection Molding and Extruding*. Bulletin TD-1781, September 1977.

Titus, J. F. *Trade Designations of Plastics and Related Materials*, (revised). PLASTEC NOTE N9C, Plastics Technical Evaluation Center, Picatinny Arsenal, Dover, New Jersey, May 1978.
Uniroyal Chemical Co. *Selection Guide: Roylar Elasto-Plastics*. Bulletin 740-B32, ASP-6.
Upjohn Co., CPR Division. *Urethane and Isocyanate Building Insulation*. Bulletin 7.14/Cp.
——. *CPR Urethane Foam: Pour-Spray-Froth*. Bulletin CPR 38J6.
——. *Urethane Elastomers*. Bulletin NMT-1.
——. *Castethane CPR Urethane Elastomers*. Bulletin CPR 1367.
——. *Pellethane–Processing Information for Urethane Elastoplastic Compounds*. Bulletin CPR PELL K9 HLI.
——. *Castethane-Pellethane Urethane Elastomers*. Bulletin BA774.
Wood, A. S. Polyurethane foams, from Materials '79. *Modern Plastics*, January 1979, **56**(1):57.
Zwolinski, L. M. Molded integral-skin urethane foam. *SPE Journal*, September 1969, **25**(9):24–27.

SILICONES

The word "silicones" was first used around the turn of the twentieth century when F.S. Kipping and his students coined the term as the name for the kinds of materials we now associate with the word. Kipping called them "silicoketones" or "silicaones." The silicoketones, however, differed in a very important respect from ordinary ketones. The oxygen in a ketone is bonded to a single carbon atom and the ketones are monomers. The corresponding oxygen in silicoketones, though, cannot be attached to a single silicon atom by a double bond; instead, the oxygen atoms form a "bridge" between two silicon atoms. Thus, the silicones are polymers and have none of the usual properties of ketones.

The name silicones, therefore, is something of a misnomer. In the strictest sense, it should be applied only to silicon polymers containing Si—O—Si linkages and a significant proportion of organic groups which are attached directly to the silicon atoms.

There is no common or popular term to describe the various kinds of monomers from which silicones are made: silanes (SiH_4), halosilanes (SiH_3X) and alkoxysilanes (ROSi). The word silicone is also sometimes stretched to cover these. This use of the word, however, is incorrect. The word silicone is also used rather loosely to cover polymers and copolymers of any type containing silicon. This, again, is incorrect. Silicones are those polymers and copolymers based on Si—O chains.

Silicones differ in structure from all other classes of organic polymers in that the chain backbone is composed of recurring silicon and oxygen atoms and does not contain carbon atoms. This chain is modified and given special properties by the introduction of various organic radicals on the silicon atoms.

In one sense, the basic structure of the silicones resembles that existing in quartz. In the crystal of beta-quartz, each silicon atom is linked to four separate oxygen atoms, and each oxygen atom is in turn linked to two different silicon atoms. This structure could be considered to be made up of siloxane chains bound in a rigid assembly completely crosslinked. A single chain has the following structure, where each silicon atom is linked through oxygen to similar chains:

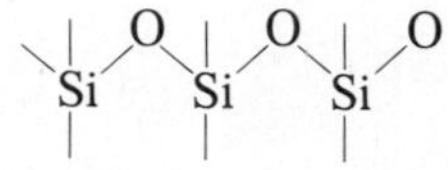

If a chain is produced with two substituted organic groups on each silicon atom, crosslinking would be prevented and a linear or chain polymer would result.

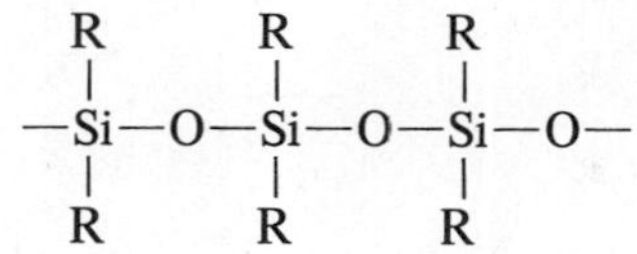

This is a polyorganosiloxane and is the most common silicone polymer. If the organic substitution were left incomplete, such organosiloxane chains could be cross linked by oxygen bridges:

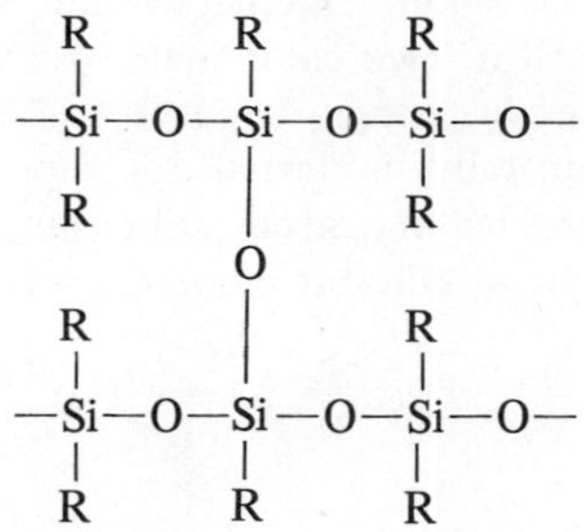

Occasional tetrafunctional silicon atoms would give a still higher degree of crosslinking.

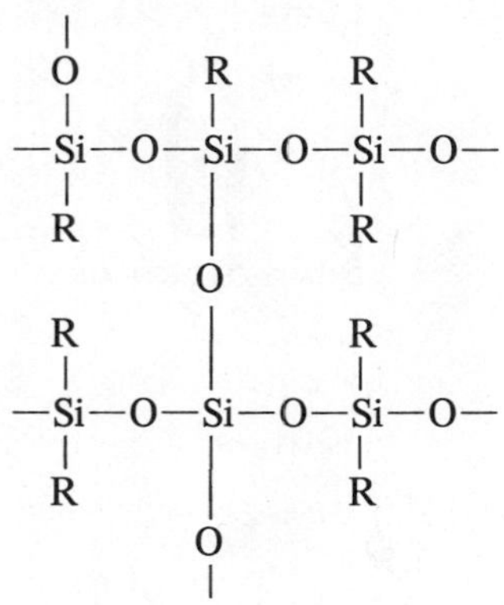

The most common way of crosslinking to a three-dimensional structure, whether to produce either elastomers or thermosetting resins, is to use a peroxide. Most common is benzoyl peroxide and the mechanism is as follows:

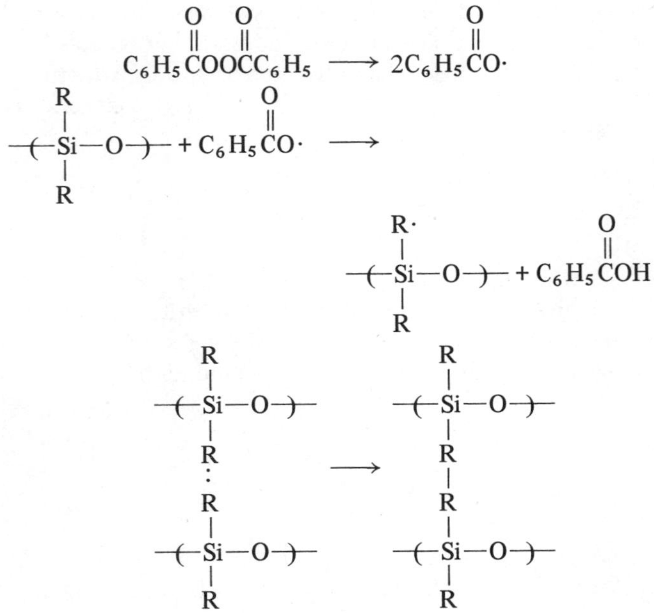

So-called room-temperature vulcanizing (RTV) systems are also available. They come in one- and two-component systems and do not require external heating for cure.

Curing agents for two-component RTV compounds normally are metallic soaps. Dibutyl tin dilaurate is the curing agent generally preferred for most applications. Most frequently added in concentrations ranging from 0.1% to 0.5% by weight, it cures at moderate speeds and is considered the most reliable RTV catalyst. Tin octoate is the fastest curing agent available (cure within 30 minutes at room temperature). Levels up to 0.5% by weight are used. Platinum catalysts are available for addition polymerized systems.

The one-component systems give off acetic acid or methyl alcohol during cure. These systems require air-borne moisture to cure, and the by-product evolution can cause shrinkage, offensive odor, and/or corrosion of contacting metals like copper.

A complete series of products is theoretically possible, ranging from linear polymers which are relatively soft and low melting (in the case of the silicones, they are fluids at room temperature), to the excessively hard, high-melting quartz itself. The properties of the resins should then extend from those of true liquids, through the thermoplastic stage, then a rubbery stage, and finally to a complex three-dimensional structure characterized by extreme heat resistance, inertness, and hardness. This vast range of properties is as great as the entire range of all the other polymers and is the reason why the silicones are presented as a separate class. It should be realized, however, that the actual grades of silicone resins now in production are only relatively isolated members of this much larger class, and production, as well as technical difficulties, prevent development of silicones having the complete range of possible properties. Figure 6-65 shows the types of commercial silicone products that derive from organo chlorosilanes.

The starting monomers for silicone production are the

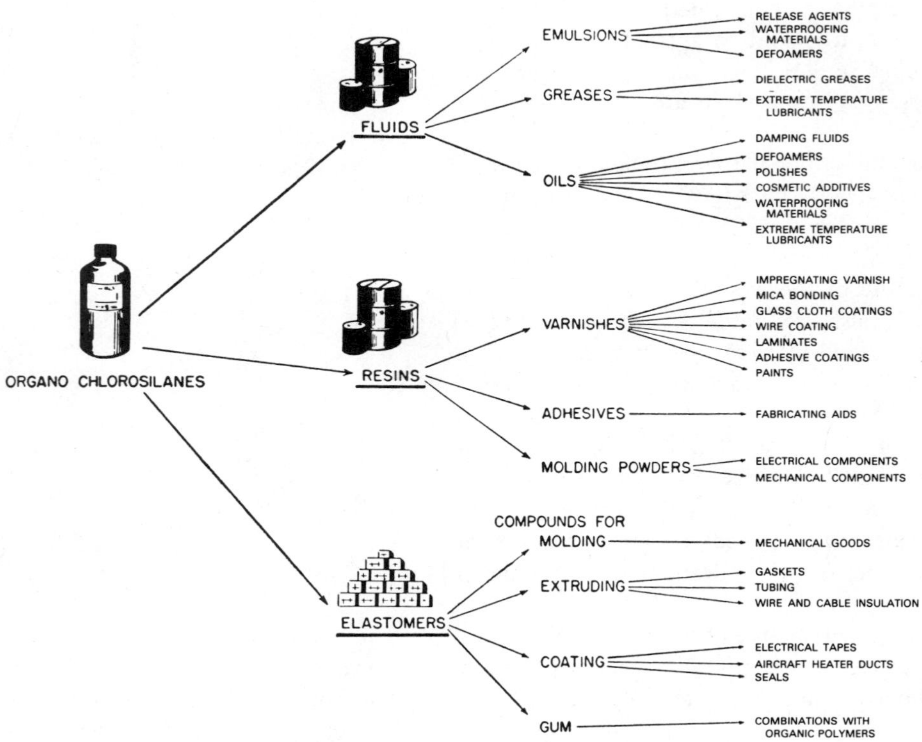

Fig. 6-65. Types of silicone products. (Lewis, 1962)

alkyl- or arylhalosilanes of the type R_aSiX_b where $a + b = 4$. The condensation polymerization reaction requires a preliminary hydrolysis of the halogen atoms.

$$R_aSiX_b + bHOH \longrightarrow R_aSi(OH)_b + bHCl.$$

The condensation reaction which follows frequently takes place so much more rapidly than the hydrolysis reaction that the siloxanes can be considered to be transitory intermediates, although this is not true of the aryl and some of the higher alkyl compounds.

The functionality of the silane monomer is given by b. R_3SiX is monofunctional, R_2SiX_2 is difunctional, and so on. Thus, the dihalo compounds are the linear chain formers and, theoretically, at least, the tri- and tetrahalo compounds might be considered to be polyfunctional monomers that would form three-dimensional networks. The trihalo compounds alone do form a few useful products, but the tetrahalo compounds hydrolyze to silicic acid.

From the foregoing it may be concluded that the degree of organic substitution determines the extent of crosslinking, which in turn governs the properties of the polymer. Solubility, fusibility, hardness, etc., may be expected to depend primarily on the degree of crosslinking, but the mechanical and chemical properties will change with the kinds of radicals which are attached to the silicon. Thus, the properties will vary with the type of substituent group, the R/Si ratio, and the degree of polymerization. This might be compared with previous classes of ethenic compounds, in which polyethylene forms the parent chain and all the other polymers are formed by substituting other atoms or radicals for hydrogen. Figure 6-66 shows the relation between viscosity and molecular weight for polydimethylsiloxanes.

As an example of the above, methyl silicone is in reality a general name for products of varying properties, from oils to elastomers. The oils are composed of linear chains of relatively low molecular weights; resins are formed by crosslinked structures; and elastomers are high molecular weight products. Although dimethyldichlorosilane, $(CH_3)_2SiCl_2$, is the principal raw material for all these products, modifications produced by varying reaction conditions and adding controlled amounts of other silanes have a great influence on the product produced. If a mixture of di- and monochlorosilanes is hydrolyzed, the chief product is a linear polymer in which the chains of dimethylsiloxane units are terminated by monofunctional trimethyl units as follows:

$$CH_3-\underset{CH_3}{\overset{CH_3}{\underset{|}{\overset{|}{Si}}}}-O-\!\!\left(\underset{CH_3}{\overset{CH_3}{\underset{|}{\overset{|}{Si}}}}-O\right)_{\!n}\!-\underset{CH_3}{\overset{CH_3}{\underset{|}{\overset{|}{Si}}}}-CH_3$$

The trimethylsiloxane units "cap" the ends of the chains, preventing further condensation and imparting a high degree of chemical stability. The resulting products are highly stable oils that are not affected by high temperatures and most chemicals. These linear methylpolysiloxanes are the bases of the "silicone oils."

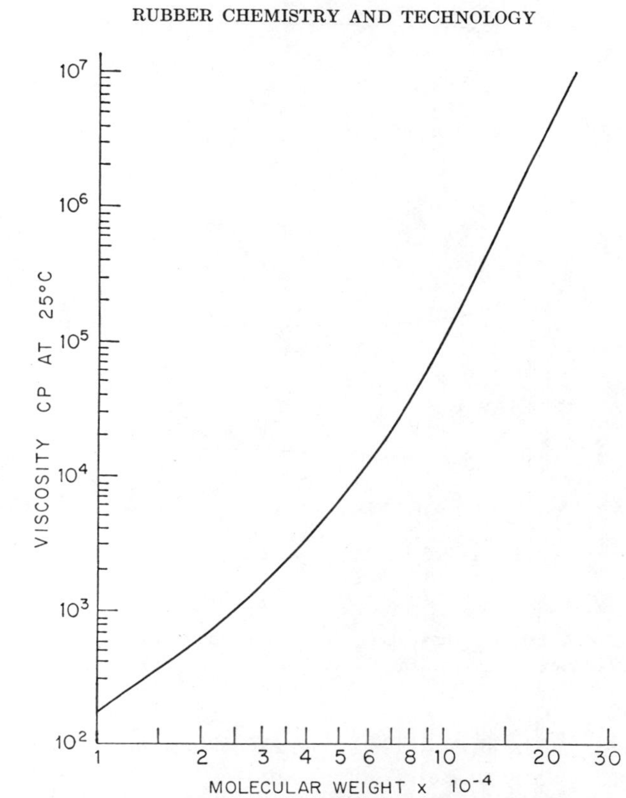

Fig. 6-66. The relation between viscosity and molecular weight for polydimethylsiloxanes. (Lewis, 1962)

The rubbers are made by what is in effect a polymerization of previously prepared ring polymers. The dimethyl polymer is treated with a reagent that opens silicon-oxygen bonds. Since no trifunctional groups are present, no crosslinking occurs; and the opened bonds continue to unite with one another until aggregates of 4,000 or more $(CH_3)_2SiO$ units are formed. The elastomer is milled with inorganic fillers on mixing rolls, and a curing catalyst (usually a peroxide) is added. Although the mechanical properties are not outstanding, it is well suited to applications where thermal stability and resistance to chemicals are more important than tensile or tear resistance. Figures 6-67 and 6-68 show the thermal stability of silicone rubber.

The methyl silicone resins require crosslinking to obtain their resinous state. Their properties are dependent on their molecular weight and the degree of crosslinking. The ratio of methyl groups to silicon is a measure of the amount of crosslinking and serves as a convenient means of indexing various products. CH_3/Si ratios of 1.2–1.5 give colorless solids with densities decreasing from 1.2 to 1.06. The time required to harden them at 100°C (212°F) increases from 2 to 24 hours. Below 1.2 methyl groups per silicon

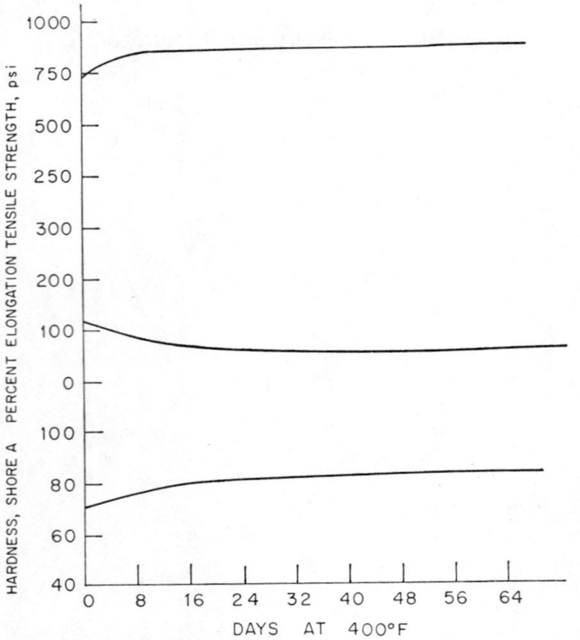

Fig. 6-67. Effect of heat aging on room-temperature properties of silicone rubber. (Lewis, 1962)

atom, the resins are sticky syrups which cure to hard, brittle, glossy masses. Above a ratio of 1.7, the products are oily and relatively volatile.

The ethyl group attached to silicon in place of the methyl group in a siloxane chain produces a softer, more soluble, and slower curing product. A ratio of two ethyl groups per silicon atom is a linear polymer which is an oil regardless of the molecular weight. A R/Si ratio of one produces materials which are suitable for coating compositions. Propyl, butyl, and amyl silicones of the $(R_2SiO)_n$ type are colorless, oily liquids which resemble ethyl silicone in appearance. In general, the larger alkyl groups cause the resin to cure more slowly, to remain softer, and to oxidize more readily.

The alkyl silicones of high R/Si ratios are oily liquids. Resinous properties are produced by a considerable amount of crosslinkage resulting in a much lower R/Si ratio. Aryl silicones on the other hand, are brittle, weak solids, even with high R/Si ratios. By attaching both alkyl and phenyl groups to the same silicon atom or by cocondensing alkyl and phenyl silicons to make a copolymer, resins have been produced with advantageous intermediate properties. Marked improvements are produced in strength and toughness. Thus the most important resins now in production are mixed methyl-phenyl and ethyl-phenyl siloxanes plus the previously described crosslinked methyl silicones.

On a commercial scale, resins are prepared by the batch hydrolysis of chlorosilane blends. For crosslinking, some trichlorosilanes must be added. The chlorosilanes are dissolved in a suitable solvent system and then an aqueous additive solution is added for reaction control. The reaction is normally highly exothermic, although if phenylchlorosilanes are present, additional heat may be required to effect the desired degree of hydrolysis.

Upon completion of the reaction, the polymer-solvent layer is separated from the aqueous acid layer and neutralized. Solvent is removed by distillation until the required solids content is obtained. The resin at this point consists of a mixture of cyclic, linear, branched, and crosslinked polymers of low molecular weight but with a substantial amount of hydroxyl end groups. The molecular weight is increased by heating with a catalyst such as zinc octoate at 100°C (212°F). The viscosity, which is a measure of the molecular weight, is maintained until the desired values are reached. The resins are then cooled and stored in containers that will not advance the polymerization further.

The crosslinking occurs *in situ* during manufacture of the final product. Heating the resin with catalysts such as triethanolamine or metal octoates completes the crosslinking process.

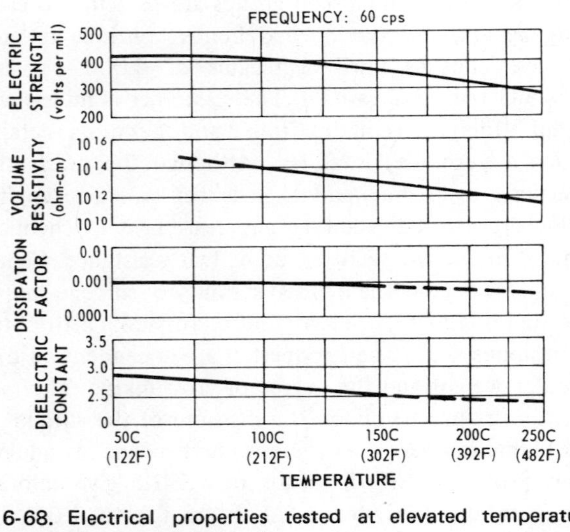

Fig. 6-68. Electrical properties tested at elevated temperatures. (*Courtesy: General Electric Co.*)

Production Methods

Grignard Method. The monomers are prepared from silicon tetrachloride by the stepwise substitution for chlorine by the Grignard reaction (see Figure 6-69).

$$CH_3MgCl + SiCl_4 \longrightarrow (CH_3)SiCl_3 + MgCl_2$$
$$CH_3MgCl + CH_3SiCl_3 \longrightarrow (CH_3)_2SiCl_2 + MgCl_2$$

Tri- and tetraalkyl silanes are also produced, and excess silicon tetrachloride remains in the reaction products. This mixture is first separated from the $MgCl_2$ and then distilled to remove the solvents and to separate the various components. The trichloride may be recycled to add one or more alkyl groups to produce the desired dichloride, or it can be

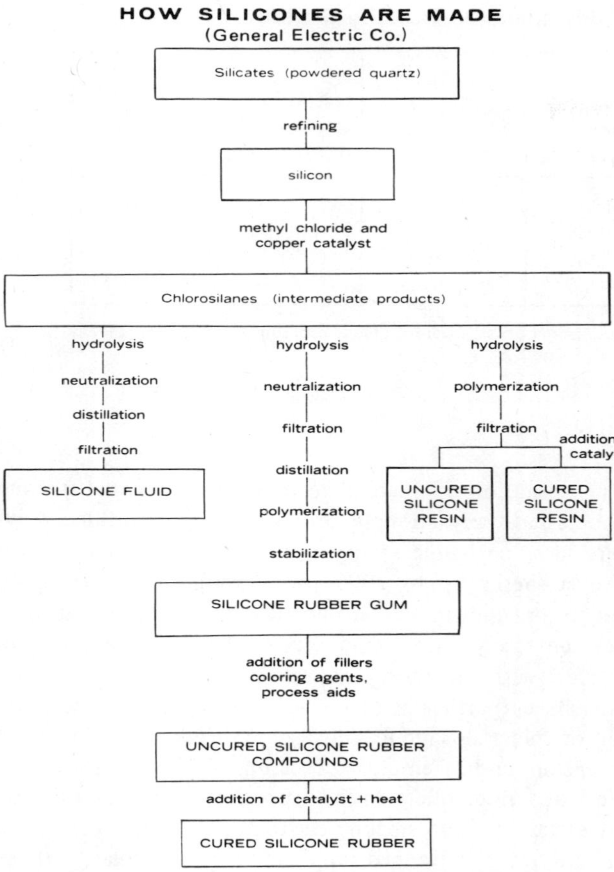

Fig. 6-69. How silicones are made. (*Courtesy: General Electric Co.*)

used to produce crosslinked resins. The monochloride may be used to cap the ends of the polysiloxane chains.

The Grignard method is a substitution process that results in a mixture of all five possible products ($SiCl_4$, $RSiCl_3$, R_2SiCl_2, R_3SiCl, and R_4Si), the proportion of each in the mixture being a function of the molar ratio of Grignard reagent to silicon tetrachloride.

The Grignard process is extremely flexible. It can be used for the preparation of alkyl- and aryltrichlorosilanes as readily as for the preparation of the dialkyl- or diaryl-dichlorosilanes. A wide variety of organic groups may be attached to silicon by the one general procedure. Disadvantages involve the handling of large quantities of anhydrous ether and reactive reagents, the separation of precipitates under anhydrous conditions, and a high weight ratio of raw materials to products. On economic grounds, the chief drawbacks are the multiplicity of steps and the dependence upon $SiCl_4$ (which is only 13.5% silicon) as a source of silicon.

Direct Method. The other process that is used in the production of dialkyldichlorosilanes is the direct method. This process utilizes the action of alkylchlorides or bromides upon elementary silicon to produce a mixture of alkyl- or arylhalosilanes of the type R_aSiX_b. A copper-copper oxide catalyst is employed.

$$2CH_3Cl + Si \longrightarrow (CH_3)_2SiCl_2$$

The operations for converting silane intermediates into polymers are the same regardless of the method used for producing the intermediates.

Properties and Applications

The qualities that characterize silicone fluids, elastomers, and resins as outstanding engineering materials may be considered in the broad categories of (1) thermal stability, (2) surface properties, (3) inertness, and (4) electrical properties. They are, however, extremely expensive, have poor physical properties, and have been known to revert to a pseudomonomeric form upon exposure to certain high temperatures and humid environments.

Methylsilicone polymers are commercially available in the form of oils and fluids useful at temperatures above those permissible for their carbon-chain counterparts. The oils have flow points as low as -84°C (-120°F), may be used at 149°C (300°F) without oxidation or sludging, and have a relatively small change of viscosity with temperature. Typical properties of a number of grades used as lubricating additives in thermoplastics are shown in Table 6-70. The effect of these fluids on the lubricity of a typical thermoplastic (nylon 6/6) is shown in Table 6-71.

Silicone fluids are available which are useful wherever there is a need for a liquid with a lower rate of change of viscosity at low temperatures or extremely low volatility at elevated temperatures. These properties, together with the liquid's inertness towards metals, coatings, and gasketing materials, indicate uses as damping fluids, gauge fluids, and dash-pot liquids. Because of their exceptional resistance to chemicals, they have also proved useful as impregnants for asbestos packing and gaskets in chemical pumps.

The silicone fluids are also used as liquid dielectrics, because of their extremely low power loss, low water absorption, and stability to heat. Their solutions in chlorinated solvents may be used to make surfaces water repellent and prevent loss of resistivity due to moisture.

Emulsions are made from these fluids that have a wide range of uses based on water repellency and anti-foaming and anti-sticking properties. A major outlet is in the production of polishes and waxes containing silicones. Other applications involve mold-release compounds, defoaming agents, (see Figure 6-70), and water repellents.

When the fluids are combined with thickening agents, various types of greases are produced for special lubrication problems. Different grades are available to permit a total range of working temperatures from -57°C (-70°F) to as high as 260°C (500°F), a much wider range than can be covered with conventional greases. Special plug-cock grease is available for use in contact with chemicals that dissolve

Table 6-70. Typical Properties of Silicone Fluids—Standard Viscosity Grades.*

Viscosity at 77F (25C), centistokes	Flash Point	Pour Point†	Specific Gravity at 77F (25C)	Refractive Index at 77F (25C)	Surface Tension at 77F (25C) dyne/cm	Thermal Conductivity at 77F‡ (25C)	Volatility,** percent
12500	610F (321C)	−51F (−46C)	0.975	1.4035	21.5	0.00038	2.0
30000	610F (321C)	−47F (−43C)	0.975	1.4035	21.5	0.00038	2.0
60000	610F (321C)	−42F (−41C)	0.976	1.4035	21.5	0.00038	2.0
100000	610F (321C)	−28F (−33C)	0.977	1.4035	−−−	−−−	2.0

† ASTM D 97-39 Sec. 5-7. Due to the effects of supercooling this test method yields pour points lower than the temperatures at which these fluids solidify when held at such temperature for a longer period.
‡ gm-cal/sec/cm²/C differential/1-cm thickness.
** Determined by heating a 2 gram sample in a 50 milliliter beaker for 24 hours at 302 F (150 C). The heating is carried out in an air circulating oven.
*Courtesy: Dow Corning Corporation.

other greases for severe temperature conditions. A variety of this grease is marketed as a stop-cock grease for laboratory glassware.

Elastomer grades may be compounded to produce silicone rubbers that can withstand surface temperatures of 260°C (500°F) and as low as -73°C (-100°F). They are inert, resist oxidation, and have excellent resistance to weathering and water. Figure 6-71 shows the effect of cold on properties of typical silicone rubbers.

Figures 6-72 and 6-73 compare a few select properties of silicone rubber versus standard organic rubbers after aging at 204°C (400°F) and exposure to cold temperatures. Table 6-72 shows a summary of physical and electrical properties of silicone rubber.

Incorporation of fillers is necessary in order that silicone rubbers have any strength properties at all. Unfilled polymers have negligible strength, whereas reinforced rubbers have strengths up to 2000 psi. Carbon blacks, typically used in organic rubbers, do not reinforce silicones; instead they adversely affect the electrical resistance and can interfere with the curing mechanism. Therefore, pyrogenic-type silicas (particle sizes = 30–300 Å) are used. Incorporation of fillers also reduces cost, cure shrinkage, and thermal expansion, and often aids processing. Table 6-73 shows the effect of various fillers on the properties of silicone rubbers.

A series of fluorosilicone elastomers has been developed primarily for advanced applications, particularly those requiring additional chemical resistance combined with the

Table 6-71. Effect of Silicones on the Lubricity of Nylon 6/6.*

Unreinforced Nylon

Property	No Lubricant	3% Silicone	10% Silicone	15% PTFE	5% MoS_2
Coefficient of friction					
100 psi, 100 ft/min					
Static	.35	.20	.18	.20	.30
Dynamic	.38	.12	.14	.22	.35
Wear rates–(100 psi, 50 ft/min)					
(in./hr. × 10^{-3})	28	.4	.10	1.8	20
PV limits @ 100 ft/min	7,200	30,000	60,000	14,500	10,000

Glass-Reinforced Nylon

Property	No Lubricant	3% Silicone	15% Silicone	5% MoS_2
Coefficient of friction				
Static	.33	.11	.20	.20
Dynamic				
(100 psi, 100 ft/min)	.40	.15	.22	.29
Wear rate–(100 psi, 50 ft/min)				
(in./hr. × 10^{-3})	.14	0.8	1.2	9
PV limit @ 100 ft/min	10,000	29,800	21,000	20,000

*Reprinted by permission from *Plastics World*, January 1978.

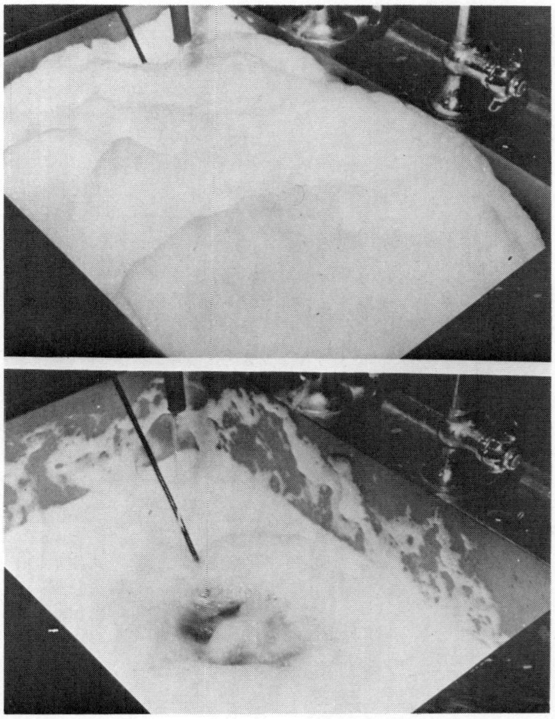

Fig. 6-70. A few drops of silicone anti-foam eliminates a large volume of foam and suds from industrial processes. (*Courtesy: General Electric Co.*)

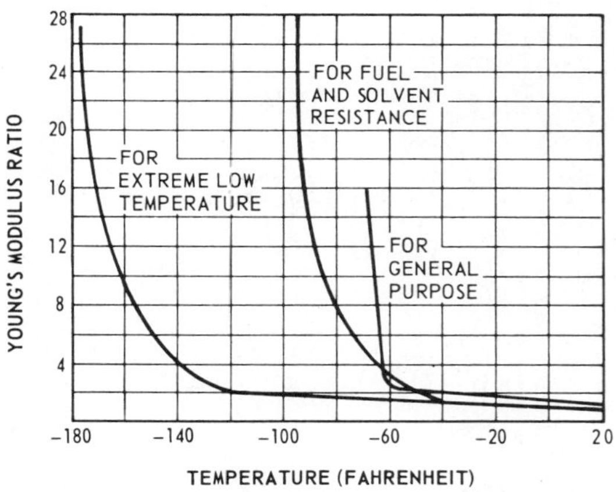

Fig. 6-71. Effect of cold on hardness and flexibility of typical silicone rubbers. (*Courtesy: General Electric Co.*)

environmental stability of standard silicones. They are based on methyltrifluoropropylsiloxane:

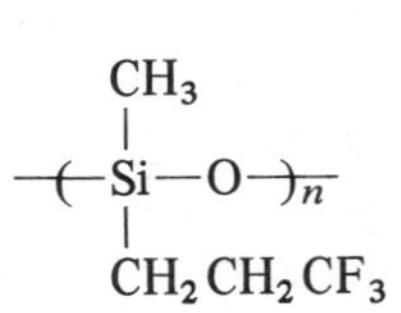

Typical properties are shown in Table 6-74. Figure 6-74 shows one of the major end uses of this type of fluorosilicone elastomer.

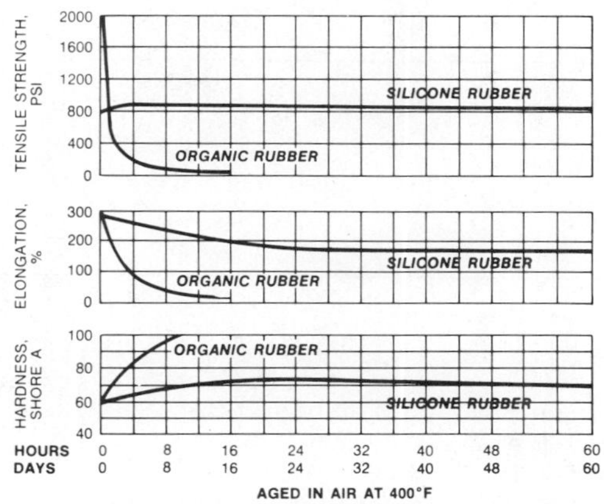

Fig. 6-72. Stability of silicones to heat. (*Courtesy: General Electric Co.*)

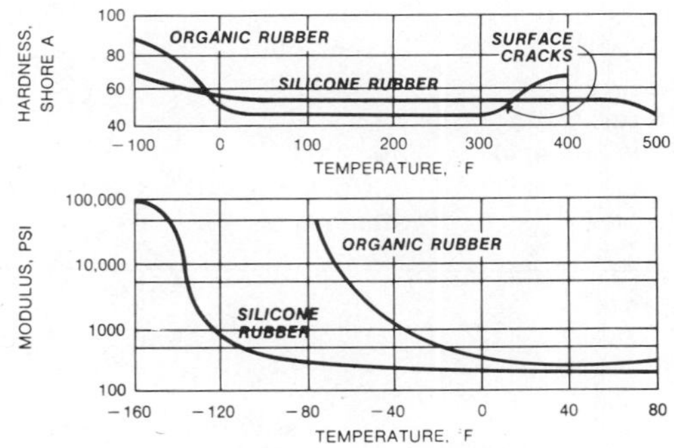

Fig. 6-73. Resistance of silicones to low temperatures. (*Courtesy: General Electric Co.*)

Typical components made of silicone rubbers include molded mechanical parts, gaskets, caulking and sealing compounds, O-rings, rollers, and foams. Rod, tubing, special profiles, and wire and cable insulation are fabricated by extrusion. Various coated fabrics also have many applications. Of interest for its surprising behavior is the methyl silicone polymer which is known as "Silly Putty," a material which is plastic to slowly applied forces. Under static load, even its own weight, this composition deforms and flows like a viscous liquid, but when struck or thrown it rebounds like a high-grade rubber.

Silicone resins have very good heat resistance but have

Table 6-72. Properties of Silicone Rubber.*

Volume resistivity, ohm-cm	1×10^{14}–1×10^{16}
Dielectric strength, volts/mil	400–700
Dielectric constant, 60 Hz	2.95–4.00
Power factor, 60 Hz	0.001–0.010
Durometer hardness, Shore A	25–90
Tensile strength, psi	up to 1600
Elongation, %	up to 700
Tear resistance, lb/in	up to 250
Compression set, %	down to 5

*Courtesy: General Electric Co.

Table 6-73. Fillers Suitable for Use with Silicon Gums.*

	Type	Particle size mean diameter (millimicrons)	Surface area, sq m/g	Reinforcement produced in silicone gums: Tensile strength range, psi	Reinforcement produced in silicone gums: Elongation range, %
Reinforcing fillers					
Santocel	Silica aerogel	30	110–150	600–1000	200–350
Cab-O-Sil	Fumed silica	15–20	175–200	600–1800	200–600
Shawinigan black	Acetylene black	45	78–85	600–900	200–350
Semi-reinforcing and non-reinforcing fillers					
Celite Super Floss	Flux calcined diatomaceous silica	1000–5000	<5	400–800	75–200
Celite 350	Calcined diatomaceous silica	1000–5000	<5	400–800	75–200
Whitetex clay	Calcined kaolin	1000–5000	<5	400–800	75–200
Witcarb R	Precipitated calcium carbonate	30–50	32	400–600	100–300
Isco 1240 silica	Ground silica	5000–10000		100–400	200–300
Neo Novacite	Ground silica	1000–10000		100–400	200–300
Minusil	Ground silica	5000		100–400	200–300
XX-78 zinc oxide	Zinc oxide	300	3.0	200–500	100–300
RY-2196 iron oxide	Iron oxide	<1000		200–500	100–300
Superpax	Zirconium silicate			400–600	100–300
Titanox RA	Titanium dioxide	300		200–500	300–400

*Lewis, 1962.

Table 6-74. Properties of a Fluorocarbon Silicone Rubber.*, **

Properties after 24 hr/300° F	
Sp. g.	1.4
Shore A	55
Tensile (psi)	1,000
Elongation	170%
Compression set 22 hrs/300° F	22%
Stiffening temp.	−78° F
Brittle point	−90° F

% Volume swell (72 hr immersion)	Fluorocarbon rubber	Me_2SiO rubber
Iso-octane	+15	>+150
Xylene	+19	>+150
CCl_4	+21	>+150
Ethanol	+5	+2
Acetone	>+150	+15
Butyl acetate	>+150	>+150

* Based on
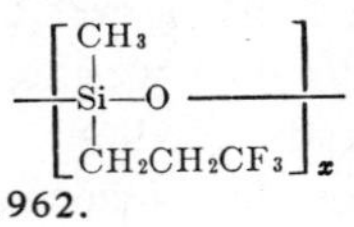

**Lewis, 1962.

very poor mechanical properties. It is believed that the source of the poor physical properties is the tendency of the polymers to form ring structures with consequent low crosslinking efficiency as well as low intermolecular forces. Table 6-75 shows properties of glass-fiber-filled molding compositions.

Resins with high phenyl content are compatible with just about all thermoset resins except non-modified alkyds. Silicones are very water resistant and excellent electrical insulators. Although most resins maintain their properties under conditions of high humidity and elevated temperatures, instances of polymeric reversion have been recorded.

One of the most important applications of the silicone resins has been for electrical insulation. Since they are not affected by temperatures that cause burning or charring of organic resins and varnishes, they are combined with glass fibers or cloth, asbestos, or mica to produce reliable and safe insulating materials for electric motors and coils. Since motors can be permitted to run hotter with this type of insulation, they can be made much smaller because more heat can be dissipated per unit weight of surface. Motors can also be installed in hotter zones, such as those in hot-air ducts, if the insulation will withstand the temperatures (see Figure 6-75).

Table 6-75. Silicone Properties.*

		ASTM test method	Cast resin: Flexible (including RTV)	Molding compounds: Glass fiber filled
Processing	1. Molding qualities	—	Excellent	Good
	2. Compression molding temp., °F.	—	—	310-360
	3. Comp. molding pressure, p.s.i.	—	—	1000-5000
	4. Injection molding temp., °F.	—	—	—
	5. Injection molding pressure, p.s.i.	—	—	—
	6. Compression ratio	—	—	2.0-6.0
	7. Mold (linear) shrinkage, in./in.	—	0-0.006	0-0.005
	8. Specific gravity (density)	D792	0.99-1.50	1.80-1.90
	9. Specific volume, cu. in./lb.	D792	—	15.4-14.6
	10. Machining qualities	—	—	Fair
Mechanical	11. Tensile strength, p.s.i.	D638	350-1000	4000-6500
	12. Elongation, %	D638	100-10000	—
	13. Tensile elastic modulus, 10^5 p.s.i.	D638	900.0	—
	14. Compressive strength, p.s.i.	D695	100	10000-15000
	15. Flexural yield strength, p.s.i.	D790	—	10000-14000
	16. Impact strength, ft. lb./in. of notch (½ × ½ in. notched bar, Izod test)	D256	—	0.3-8.0
	17. Hardness, Rockwell	D785	15-65 (Shore A)	M80-M90
	18. Flexural elastic modulus, p.s.i. × 10^5	D790	—	10-25
	19. Compressive modulus, p.s.i. × 10^5	D695	—	—
Thermal	20. Thermal conductivity, 10^{-4} cal./sec./sq. cm./1 (°C./cm.)	C177	3.5-7.5	7.0-9.0
	21. Specific heat, cal./°C./gm.	—	—	0.19-0.22
	22. Thermal expansion, 10^{-6} in./in./°C.	D696	8.0-30.0	2.0-5.0
	23. Resistance to heat, °F. (continuous)	—	500	>600
	24. Deflection temp., °F. at 264 p.s.i. fiber stress	D648	—	>900
	at 66 p.s.i. fiber stress		—	—
Electrical	25. Volume resistivity, ohm-cm. (50% RH and 23° C.)	D257	10^{14}-10^{15}	10^{14}
	26. Dielectric strength, short-time, ⅛-in. thickness, volts/mil	D149	550	200-400
	27. Dielectric strength, step-by-step, ⅛-in. thickness, volts/mil	D149	550	125-300
	28. Dielectric constant, 60 cyc.	D150	2.75-4.20	3.3-5.0
	29. Dielectric constant, 10^3 cyc.	D150	—	3.2-4.5
	30. Dielectric constant, 10^6 cyc.	D150	2.6-2.7	3.2-4.3
	31. Dissipation (power) factor, 60 cyc.	D150	0.001-0.025	0.004-0.030
	32. Dissipation (power) factor, 10^3 cyc.	D150	—	0.0035-0.0200
	33. Dissipation (power) factor, 10^6 cyc.	D150	0.001-0.002	0.002-0.020
	34. Arc resistance, sec.	D495	115-130	200-250
Optical	35. Refractive index, n_D	D542	1.43	—
	36. Clarity	—	Clear grades available	Opaque
	37. Transmittance, %	—	—	—
	38. Haze, %	—	—	—
Resistance characteristics	39. Water absorp., 24 hr., ⅛-in. thick., %	D570	0.12 (7 days at 77°F.)	0.2
	40. Flammability Burning rate, in./min.	D635	—	—
	AEB, in./ATB, sec.		—	—
	41. Effect of sunlight	—	None	None
	42. Effect of weak acids	D543	Little or none	None to slight
	43. Effect of strong acids	D543	Slight to severe	Slight
	44. Effect of weak alkalies	D543	Little or none	None to slight
	45. Effect of strong alkalies	D543	Moderate to severe	Slight to marked
	46. Effect of organic solvents	D543	Insoluble but swells in some	Attacked by some

*Reprinted by permission from *Modern Plastics Encyclopedia*, McGraw-Hill, Inc.

Fig. 6-74. Fluorosilicone rubber O-rings in the flexible fuel-line couplings of a jetliner. (*Courtesy: Dow Corning Corporation*)

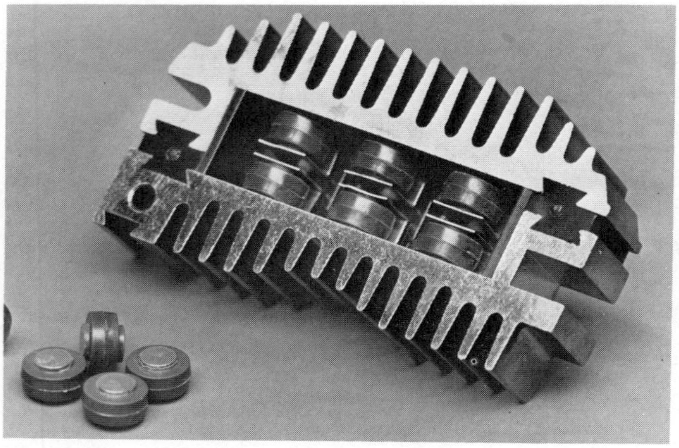

Fig. 6-75. Diodes for automobile alternators molded of silicone plastic have stable electronic insulating properties over long periods of time. (*Courtesy: Dow Corning Corporation*)

Fig. 6-76. This 40-lead dual-in-line device, encapsulated in a thermosetting silicon-epoxy molding compound, is the "brain" of Fairchild's digital clock/calendar circuit. It is designed for high reliability in both industrial and household equipment. (*Courtesy: Dow Corning Corporation*)

Table 6-76. Typical Properties of Silicone-Epoxy Hybrid.*

As Supplied	
Color	black
Form	granular
Filler type	fused silica and short glass fibers
Molding Properties	
Spiral flow, 177°C (350°F) and 1000 psi, inches	25
Flow duration at 177°C (350°F), seconds	12
Hot hardness, Shore D at 177°C (350°F), 2-in.-diameter disc at 1.5 min.	65
Physical Properties:[1]	
Flexural strength,[2]	
As molded, psi	17,000
Postcured, psi	17,000
Flexural modulus, psi[2]	2.2
Shrinkage	
As molded, in/in	0.005
Postcured	no change
Coefficient of thermal expansion, (TMA)[2]	
(-70–150°C) in/in/°C	3.0×10^{-5}
(150–250°C) in/in/°C	5.9×10^{-5}
Tensile strength, psi[3]	7,960
Izod impact, notched[3]	0.3
Compressive strength, psi[3]	28,700
Thermal conductivity $cal/sec/cm^2/cm/°C$ (Colora)	16×10^{-4}
Flammability, per UL 94V-0	pass[4]
Flammability, limited oxygen index	27
Water absorption, percent[3,6]	0.1
Specific gravity[3]	1.84
Electrical Properties:	
Dielectric strength, volts/mil[3]	
in air	246
in oil	338
Volume resistivity, ohm–cm[3]	
Condition A[5]	1.0×10^{15}
Condition D[6]	1.0×10^{15}
Dielectric constant	
Condition A at 10^2 Hz	3.61
at 10^6 Hz	3.43
Condition D at 10^2 Hz	3.67
at 10^6 Hz	3.51
Dissipation factor	
Condition A at 10^2 Hz	0.004
at 10^6 Hz	0.010
Condition D at 10^2 Hz	0.005
at 10^6 Hz	0.01

[1] All specimens transfer molded and postcured 4 hours at 177°C (350°F), unless otherwise stated.
[2] $\frac{1}{4}'' \times \frac{1}{10}'' \times 1\frac{1}{2}''$ specimen.
[3] Standard ASTM specimen.
[4] As tested by Dow Corning Corporation, does not imply UL recognition.
[5] As received postcured 4 hrs at 177°C (355°F).
[6] Immersed in water 24 hrs at 23°C (75°F) before testing.
*Courtesy: Dow Corning Corporation.

A recent development is an epoxy-silicone-hybrid molding compound. It is used to encapsulate sensitive electronic components. The material is reputed to offer the strength, salt-spray resistance, and electrical stability of epoxies,

Fig. 6-77. Water beads up on silicone-treated brick. Masonry water repellents made with silicone resins allow moisture vapor to pass through, but stop water from entering. (*Courtesy: Dow Corning Corporation*)

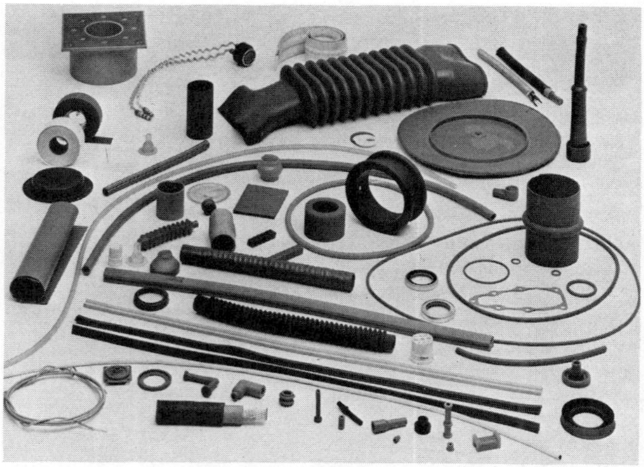

Fig. 6-78. Fabricators use silicone rubber to make limitless varieties of parts for thousands of applications, ranging from hot-air starter ducts for jet airplanes that withstand superheated air hotter than molten solder, to home freezer door seals that do not embrittle at extremely low temperatures. (*Courtesy: General Electric Co.*)

along with the processability and mold release of standard silicones (see Figure 6-76). Properties of this new material are shown in Table 6-76.

The silicones are also used alone or in combination with other resins in protective coatings for extreme surface conditions. They provide good surface life even when exposed to temperatures in the range of 260–649°C (500–1200°F) and are used for exhaust ducts or stacks for hot gases.

Exposure of various materials to the vapors of organosilicon halides can also be used to develop a water-repellent film of silicone on the surface. Vapors of chlorosilane react with adsorbed moisture on the surface to be treated, forming a silicone layer. Ceramic articles to be treated are preconditioned at 50–90% relative humidity, exposed a few minutes to the vapors of the chlorosilane, and then aired to allow the byproduct HCl to volatilize. Treated surfaces show a high angle of contact for water droplets. Moisture does not condense on such a surface as a continuous film but as tiny isolated droplets. The electric resistance of surfaces so treated remains high even under humid conditions.

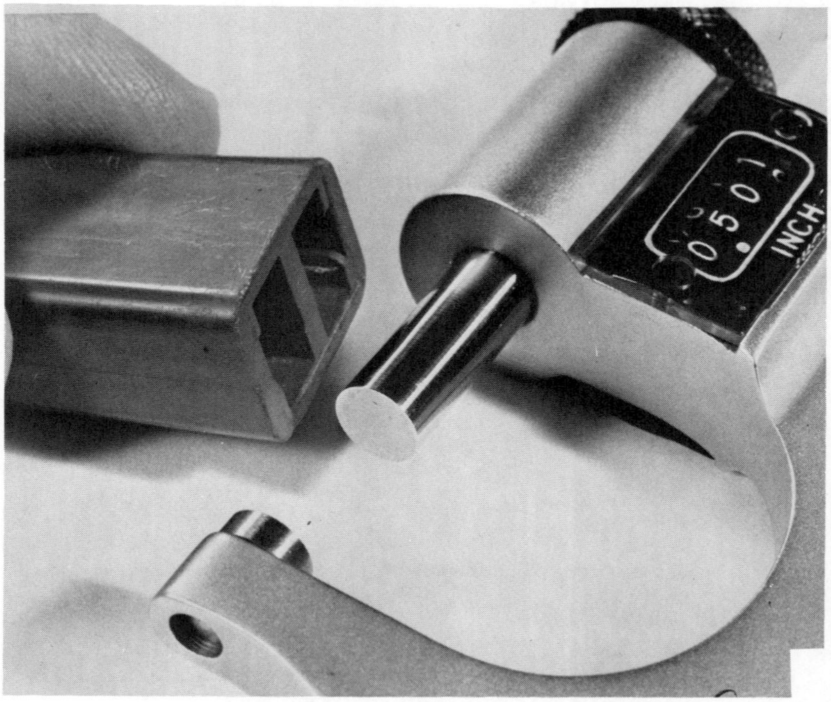

Fig. 6-79. Silicone plastic electrical connector was developed for use with exhaust system sensors by a major U.S. auto manufacturer. It withstood thousands of hours of exposure to temperatures up to 370°C, and thermal shocks from 370°C to −65°C. (*Courtesy: Dow Corning Corporation*)

Other areas of use for silicone polymers are as follows:

1. Construction: one-part (single package) silicone rubber as glazing compounds, curtain wall sealants, and joint sealers; silicone primers and silane adhesion promotors for bonding polysulfide, polyurethane, epoxy, and other non-silicone-type sealants; silanes as coupling agents in reinforced plastics, and as anti-matting agents on glass wool insulations; silicone resins as water repellents for the protection of above-ground masonry (see Figure 6-77).

2. Automotive: elastomers as molded boots for spark plugs and wire connectors (Figures 6-78 and 6-79), and as sealants and caulks; fluid as adhesion promoters for facilitating the bonding of non-silicone-type sealants.

3. Appliances: elastomers for 260–316°C (500–600°F) extruded and molded seals, oven door seals, wire insulation, sealants and adhesives; fluids as lubricants for clothes dryers; silicone resins for use as binders for various high-temperature paints.

4. Aerospace: elastomers for oxygen hoses, space suits (reinforced Dacron) and cabin seals; compositions for ablation shields; matrices for solid fuels, and binders for fuel; weatherproof coatings for electrical connectors; resin laminates for supersonic plane parts; molded oxygen and emergency masks; in-flight feeding tubes; insulation for electrical wiring; encapsulants for semiconductors, windshield laminating material; surface water-repellent after-landings.

5. Medical: elastomers for synthetic human parts; and for medical equipment to be implanted or used near the body. Special medical-grade types of silicone materials are necessary for these applications.

Trade Names

Aerospace	Adhesives and sealants	Dow Chemical Co.
Bakelite	Resins	Union Carbide Corporation
Baysilone	Resins	Mobay Chemical Co.
Berlon	Rubber compound	National Beryllia Corporation
Blu-Sil	Molding compound	Perma-Flex Mold Co.
Calrite	Molding compound	Chemore Corporation
Camie	Release agents	Camie Co.
Chase-Glasterra	Reinforced	Chase & Sons, Inc.
Cho-Bond	Adhesive	Chomerics, Inc.
Cohrlastic	Rubber	Connecticut Hard Rubber Co.
Conolon	Compounds	Whittaker Corporation
Coppercone	Dyes and sheet	Gladen Enterprises, Inc.
Cru-Rez	Resin	Crucible Chemical Co., Inc.
Crystal-Vu-Crystic	Silicone-acrylic	U.S. Industries, Inc.
Custom-Load	Resin	Custom Materials, Inc.
Densil	Adhesives	Dennison Manufacturing Co.
Dow Corning	Resins	Dow Corning Corporation
Drimix	Silanes	Kenrich Petrochemicals, Inc.
Drival	Emulsions	United Merchants & Manufacturers
Durasyl	Rubber	Bentley-Harris Mfg. Co.
Ecco-	Formulations	Emerson & Cuming, Inc.
Econo Pak	Mold-release	Stoners Ink Co.
Elastosil	Sealants	Wacker-Chemie, GmbH
Electrisil	Rubber	General Electric Co.
GE	Resins	General Electric Co.
Hadbar	Elastomer	Purolator Products, Inc.
Insulgrease	Compound	General Electric Co.
Insuljel	Dispersion	General Electric Co.
Irrosil	Rubber	General Electric Co.
Isochem-	Formulations	Isochem Resins
Lanpol	Silicone polyesters	Lanson Chemical Corporation
Level-Wrap	Rubber	Continental-Diamond Fibre Corporation
Lubricone	Lubricant	Kano Labs, Inc.
Megasil	Compound	Midsil Corporation
Megapren	Rubbers	Rhein-Chemic Reinau, GmbH
Mem	Silicone-polycarbonate copolymer	General Electric Co.
Metacote	Coating	Metachem Resins Corporation
Mettcone	Rubbers	Gladen Enterprises, Inc.
Mica-Seal	Reinforced	McMillan International Co.
Moxfoam	Foam	Moxness Products, Inc.
Omni	Sealants	Omni Technik, GmbH
Pio-Con	Mold-release	Piochem Industries, Inc.
Real Ease	Release	Borco Chemicals, Inc.
Releaso	Mold-release	Bel-Ray Co.
Repel	Water repellent	Reliance Chemical Products Co.

Rhino Bib	Sealants	Devcon Corp.
Rhodorsil	Resins	Rhodia, Inc.; May & Baker, Ltd.
RTV	Room-temperature vulcanizing silicones	General Electric Co.
Sarcon	Rubber	Fuji Polymer Industries Co., Ltd.
Scurasil	Elastomer	Rhodia Corporation
Shure-Stick	Formulation	Pallflex Products, Inc.
Silastic	Rubber	Dow Corning Corporation
Silastomer	Rubber	Midsil Corporation
Silastoseal	Adhesive	Midsil Corporation
Silcast	Rubber	General Electric Co.
Silcocell	Surfactants	Uniroyal, Inc.
Silcolapse	Fluids and emulsions	Imperial Chemical Industries, Ltd.
Silflex	Reinforced resin	Verflex Corporation
Silgan	Resins	SWS Silicones, Inc.
Silglaze	Sealant	General Electric Co.
Silgrip	Adhesives	General Electric Co.
Silicol	Rubber sheet	Colonial Rubber Co.
Silicure	Curing agents	Tenneco Chemical, Inc.
Silite	Sealants and adhesives	Devcon Corporation
Silopren	Rubber	Bayer, A.G.
Silpruf	Sealant	General Electric Co.
Silvercone	Rubber	Gladen Enterprises, Inc.
Simconex	Formulations	Simplex Wire & Cable Co.
Si-O-	Resins, formulations, and fluids	Stauffer Chemical Co.
Slipicone	Release agent	Dow Corning Corporation
Solucote	Coatings	Soluol Chemical Co., Inc.
Soundfoam	Foam	Soundcoat Co.
Spandofoam	Foam	Baxenden Chemical Co., Ltd.
Spenkel	Resins	Textron, Inc.
Spenlite	Aliphatic resins	Textron, Inc.
Spraythane	Resin	Thiokol Corporation
Sproth	Foam	Baxenden Chemical Co., Ltd.
Stafoam	Foam	Olin Corporation
Stanfoam	Foam	Cryo-Therm, Inc.
Stepanfoam	Foam	Stepan Chemical Co.
Stepanpak	Foam	Stepan Chemical Co.
Stycast	Formulations	Emerson & Cuming, Inc.
Super-Porseal	Resin	Truscon Laboratories
Sylgard	Resins	Dow Corning Corporation
Sylkyd	Silicone-alkyds	Dow Corning Corporation
Syl-Off	Release coatings	Dow Corning Corporation
Tac-Fre	Release agents	Taconic Plastics, Inc.
Texturecone	Rubber	Gladen Enterprises, Inc.
Transpello	Coatings	Ranco Industrial Products Corporation
Trevarno	Reinforced	Hexcel Corp.
Ucarsil	Silicones and silanes	Union Carbide Corporation
Union Carbide	Resins	Union Carbide Corporation
Varsil	Resin or rubber-coated glass	New Jersey Wood Finishing Co.
Velvasil	Fluids	General Electric Co.
Versilube	Lubricants	General Electric Co.
Viscasil	Fluids	General Electric Co.
Vol-Sil	Coated tubing	Voltube Corporation

REFERENCES AND BIBLIOGRAPHY

Antonen, R. C., Michael, K. W., and Engleman, J. H. "New Hybrid Electronic Molding Compounds, Long Term Life Testing." Paper presentation, *18th Annual Conference of the Electronic Materials Committee of the Metallurgical Society of AIME*, Salt Lake City, June 1976.

Bobear, W. J. Silicone rubber reactions during high temperature testing. *Rubber Age*, September 1966, General Electric Bulletin No. CDS-867.

Brydson, J. A. *Plastics Materials*. 3rd Ed. Robert E. Krieger Publishing Co., Inc., Huntington, New York, 1975.

Copper, J. From liquid to thermoset elastomer: efficient new silicone technology. *Modern Plastics*, February 1979, **56**(2):69–72.

DeGisi, S. L., and Smith, C. H. Improved bondability of molded rigid urethane foam. *Adhesives Age*, July 1978, **21**(7):35–37.

Dow Corning Corporation. *Information About Semiconductor-Grade Molding Compounds*. Form No. 23-262A, 1977.

——. *Now, A New Generation of Semiconductor-Grade Molding Compounds*. Bulletin No. 23-251, 1977.

——. *Silicone Fluids: Lubricating Additives for Thermoplastics*. Bulletin 22-222a, January 1976.

——. Silicone Seminar. Presentation to Hughes Aircraft Co., 1976.

——. *Information About Silicone Fluids*. Bulletin No. 22-290, September 1974.
——. *From Ideas Come Creative Solutions By Dow Corning*. Bulletin No. 01-206, 1970.
——. *Aerospace Materials Engineering By Dow Corning*. Bulletin No. 01-133, 1968.
——. *Working with Dow Corning Aerospace Sealants and Coatings*. Bulletin No. 08-317, 1968.
Electronic Engineer's Design Magazine. Silicone rubber: space-age insulating material. GE Bulletin No. CDS-341, September 1966.
General Electric Co. *Silicone Rubber Technical Information for Designers and Specifiers of Rubber Parts*. Technical Data Book S-1E.
——. *RTV Silicone Rubber*. Technical Data Book S-35.
——. *GE World of Silicones*. Bulletin CDS-580.
Hyde, J. F. Chemical background of silicones. *Science*, February 19, 1965, **147**:829–836.
Lake, N. E. Give the slip to your hangups with silicone fluids. *Plastics Engineering*, Dow Corning Reprint Bulletin No. 22-352, November 1975, **31**(11).
Lewis, F. M. The science and technology of silicone rubber. *Rubber Chemistry and Technology*, December 1962, **35**(5):1222–1275.
Lockhart, F. J. *Silicone Plastics For High Temperature Service in Automotive Applications*. Dow Corning Corporation.
Lynch, W. *Handbook of Silicone Rubber Fabrication*, Van Nostrand Reinhold Co., New York, 1978.
Materials Engineering. Silicone tape bonds many materials. August 1976, **84**(2):26.
Merrill, D. F. Silicone PSA's: types, properties and uses. *Adhesives Age*, March 1979, **22**(3):39–41.
Michael, K. W., Bank, H. M., and Antonen, R. C. "New Hybrid Electronic Molding Compounds." Paper presentation, *26th Annual Electronic Components Conference, The Institute of Electrical and Electronic Engineers*, San Francisco, April 1976.
Modern Plastics. Liquid silicone eases parts release, cuts molding costs and cycle time. August 1978, **55**(8):26.
Plastics World. Silicone additives reduce resin friction. January 1978, **36**(1):32.
——. Internal lubricant eliminates rejects in molding grained surface. Dow Corning Reprint Bulletin No. 22-213, July 1974, **32**(7).
Poole, D. R. Building conductive silicone elastomers into electrical devices. *Plastics Engineering*, June 1978, **34**(6):25–28.
Rhodia Inc. *Rhodia Silicones*. Bulletin 326, 1977.
Simonds, H. R., and Church, J. M. *A Concise Guide to Plastics*, 2nd Ed. Van Nostrand Reinhold Co., New York, 1963.
Smith, R. F. Silicone additives permit decorating plastic parts right out of the mold. *Plastics Design & Processing*, August 1977, **17**(8):53–56.
Titus, J. B. *Trade Designations of Plastics and Related Materials*, (revised). PLASTEC NOTE N9C, Plastics Technical Evaluation Center, Picatinny Arsenal, Dover, New Jersey, May 1978.
Union Carbide Corporation. *Silane Coupling Agents in Mineral-Filled Composites*. Bulletin F-43578A, April 1979.
Winding, C. C., and Hiatt, G. D. *Polymeric Materials*, McGraw-Hill Book Co., New York, 1961.

HIGH-TEMPERATURE POLYMERS

Most commercial polymeric materials do not possess the physical, chemical, or mechanical properties required for extended use in the extreme operating environments engendered by modern technology. Extensive research and development has been conducted and is currently being conducted throughout the world to find polymeric materials having useful thermal, oxidative, and hydrolytic stability plus physical and mechanical properties at or above 538°C (1000°F).

The fact remains that most polymeric materials available are not suited for prolonged service in air above 350°C (662°F). Even the words "thermal stability" have some inherent ambiguity. Thermal stability must be defined in terms of environment and exposure time. For example, thermogravimetric analysis indicates that polyimides are stable in air to about 400°C (752°F). However, in practice, they cannot be used continuously above about 300°C (572°F). Tables 6-77 and 6-78 show some comparative heat-resistance data for a variety of polymers. Table 6-79 shows thermal properties of the more heat-resistant thermoplastics. Tables 6-80 and 6-81 show the effect of heat exposure on tensile strength of heat-resistant thermoplastics. Figure 6-80 shows the heat capabilities of high-temperature plastics as reflected by flexural modulus. As pointed out in Chapter 3, a limiting oxygen index (LOI) greater than 25 represents a flame-resistant plastic. Table 6-82 shows that most heat-resistant thermoplastics are inherently non-burning materials. Table 6-83 shows that almost all high-temperature polymers are costly and relatively dense, so that cost per fabricated part is also high.

Most of the new high-temperature polymers are thermoplastic organic polymers having phenylene, heterocyclic, or chelate groups in the polymer chain. Table 6-84 lists the many types of polymers that have been examined for high-temperature use. Those underlined are materials we have discussed earlier. Those starred are to be dealt with in more detail in this section.

There are two basic approaches to improving the thermal stability of a polymer, both based on increasing the material's rigidity—crystallinity and crosslinking. To promote formation of crystalline structures, the common procedures used include incorporation of mer units that exert strong interchain forces, inclusion of aromatic rings and/or bulky side chains, or improvement of polymer regularity to create a tighter interlock with adjacent molecules. Obviously these techniques are primarily used with thermoplastic polymers.

For thermosetting materials, denser crosslinking is the approach most often taken. The objective is to generate aromatic or heterocyclic structures as well as chain interconnection through the crosslinking mechanisms (see Figure 6-81). The most stable form is the so-called ladder polymer. It has been speculated that the ladder configuration probably represents the upper limit of temperatures for a completely organic polymer structure and, chemically, 649°C (1200°F) is the ultimate.

Although ladder polymers provide higher molecular order than regular polymers, they create tremendous processing problems. It appears that the more complex the molecule, the more difficult it is to fabricate. As the molecule becomes more stable, it increases its resistance to melting and flow during fabrication.

Table 6-77. Maximum Service Temperature of Plastics and Elastomers.*

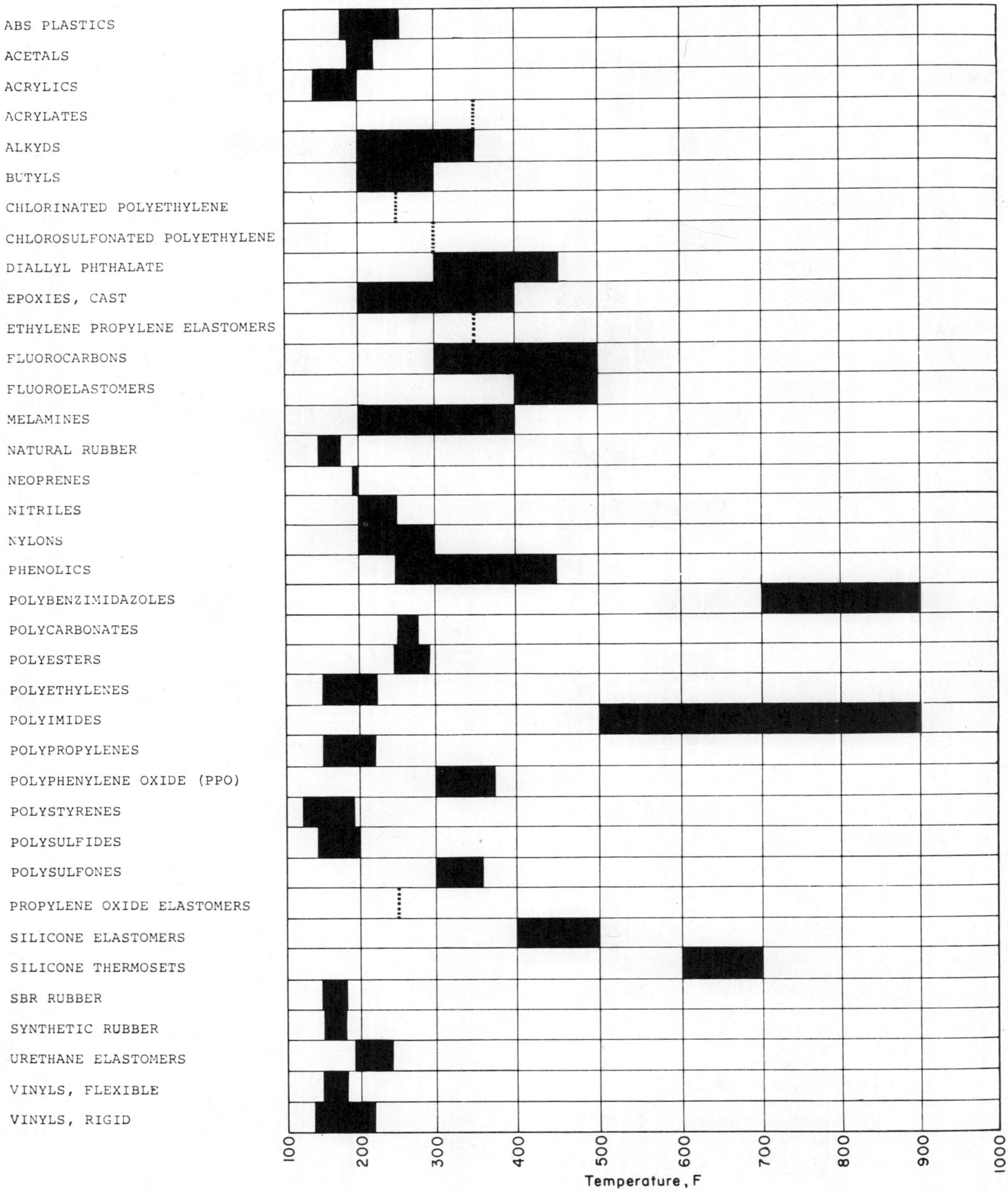

*Howeth et al., 1965.

Table 6-78. Heat-Deflection Temperature of Plastics.*

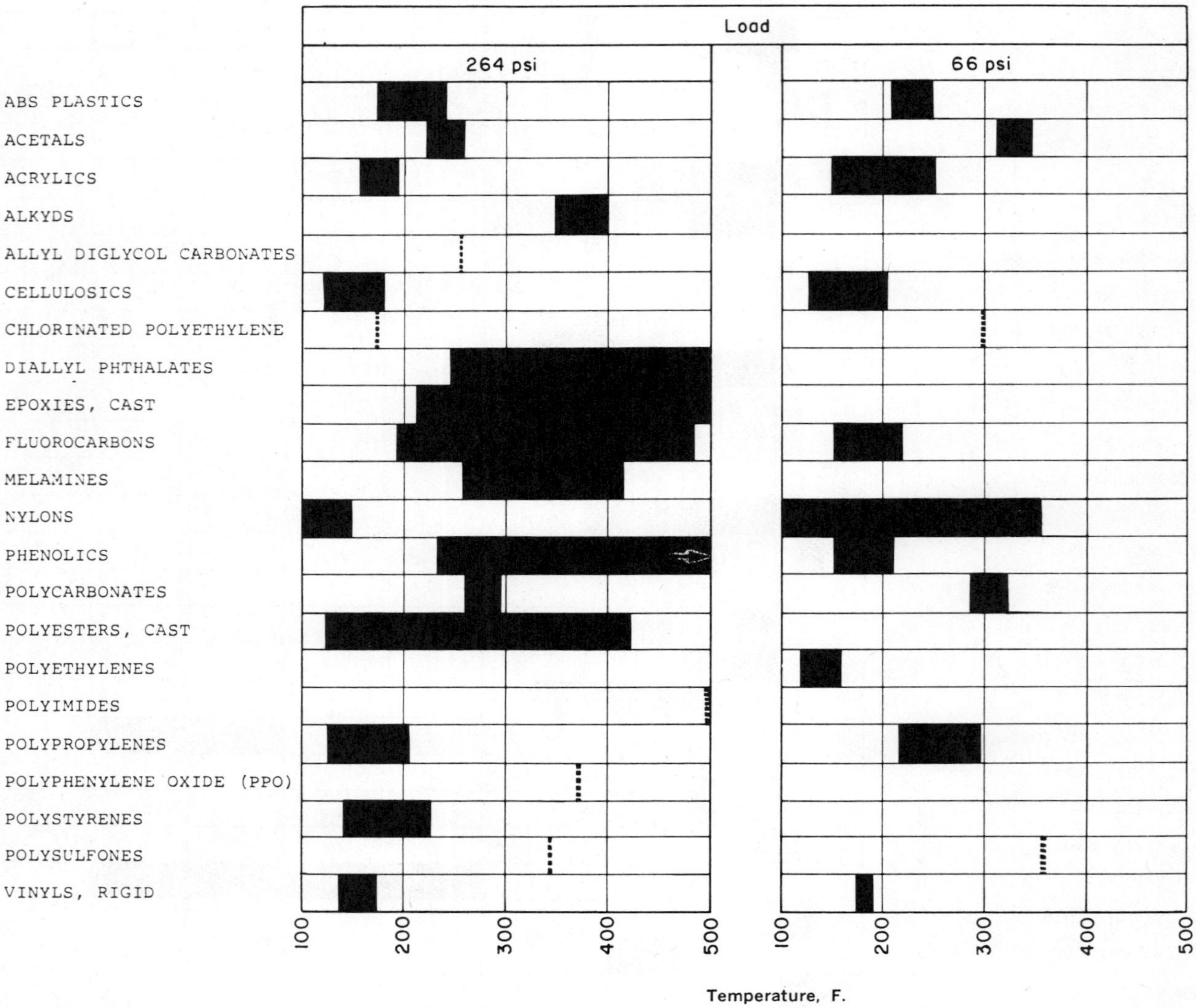

*Howeth et al., 1965.

Table 6-79. Typical Thermal Properties of Heat-Resistant Thermoplastics*

Thermoplastic	Continuous use temp, F (K)	Heat deflection temp[a], F (K)	Glass transition temp, F (K)	Melting point, F (K)
Chlorotrifluoro-ethylene (CTFE)	350 (450)	150 (339)	—	428 (493)
Ethylene-chlorotri-fluoroethylene (E-CTFE)	300 (422)	170 (350)	—	464 (513)
Ethylene-tetrafluoro-ethylene (ETFE)	302 (423)	165 (347)	—	522 (545)
Fluorinated ethylene propylene (FEP)	400 (478)	124 (324)	—	518 (543)
Perfluoroalkoxy (PFA)	500 (533)	118 (321)	—	585 (580)
Polyamide-imide	410 (483)	525 (547)	532 (551)	—
Polyarylsulfone	365 (458)	525 (547)	547 (559)	—
Polyethersulfone	347 (448)	397 (476)	442 (501)	—
Polyimide 2080	500 (533)	536 (533)	590 (583)	—
Polyphenylene sulfide, 40% glass reinf	356 (453)	>500 (>533)	190 (361)	540 (555)
Poly-p-oxybenzoate	392 (473)	560 (566)	—	745 (669)
Polysulfone	302 (423)	345 (447)	374 (463)	—

[a]—at 264 psi (1.82 MPa)
*Houston, 1975.

Table 6-80. Effects of Thermal Aging on Tensile Strength.*

Thermoplastic	% glass reinf	Initial ten str, 1000 psi (MPa)	Ten str, 1000 psi (MPa) after 1500 hr at 400 F (478 K)	Ten str, 1000 psi (MPa) after 1500 hr at 500 F (533 K)
Ethylene-tetra fluoroethylene (ETFE)	20	11.3 (77.9)	11.2 (77.2)	2.3 (15.9)
Fluorinated ethylene propylene (FEP)	20	5.0 (34.5)	4.7 (32.4)	4.5 (31.03)
Polyamide-imide	—	27.4 (188.9)	22.5 (155.1)	22.0 (151.7)
Polyarylsulfone	—	13.1 (90.3)	12.75 (87.9)	7.6 (52.4)
Polyethersulfone	40	22.7 (156.5)	11.3 (77.9)	10.5 (72.4)
Polyphenylene sulfide	40	23.2 (159.9)	16.5 (113.8)	13.8 (95.2)
Poly-p-oxybenzoate	—	13.9 (95.8)	13.75 (94.8)	9.0 (62.1)
Polysulfone	40	20.3 (139.9)	4.3[a] (29.6)	melted[b]

[a] After 750 hr. [b] After 100 hr. (Source, LNP Corp.).
*Houston, 1975.

Table 6-81. Tensile Strength of Heat-Resistant Thermoplastics at Elevated Temperatures.*

Thermoplastic	Ten str at 68 F (293 K), 1000 psi (MPa)	Ten str at 300 F (422 K), 1000 psi (MPa)	Ten str at 500 F (533 K), 1000 psi (MPa)
Chlorotrifluoro-ethylene (CTFE)	4.0 (27.6)	0.7 (4.8)[a]	—
Ethylene-chlorotri-fluoroethylene (E-CTFE)	7.0 (48.3)	0.8 (5.5)[b]	—
Ethylene-tetrafluoro-ethylene (ETFE)	6.5 (44.8)	—	2.0 (13.8)[c]
Fluorinated ethylene propylene (FEP)	4.0 (27.6)	0.5 (3.4)	—
Perfluoroalkoxy (PFA)	4.5 (31.0)	—	1.8 (12.4)[d]
Polyamide-imide	26.9 (185.5)	15.2 (104.8)	7.5 (51.7)
Polyarylsulfone	13.1 (90.3)	8.7 (59.9)	3.23 (22.3)
Polyethersulfone	22.7 (156.5)	4.9 (33.8)[e]	—
Polyimide	17.1 (117.9)	8.3 (57.2)	5.5 (37.9)[f]
Polyphenylene sulfide, 40% glass reinf	23.2 (159.9)	6.9 (47.6)	3.0 (20.7)
Poly-p-oxybenzoate	13.9 (95.8)	9.3 (64.1)	3.9 (26.9)
Polysulfone	17.3 (119.3)	2.3 (15.6)	—

[a]—at 250 F (394 K) [b]—at 212 F (373 K)
[c]—at 450 F (505 K) [d]—at 480 F (522 K)
[e]—at 350 F (450 F) [f]—at 550 F (561 K)
*Houston, 1975.

Perhaps the final evolutionary stage for high-temperature plastics will be the replacement of carbon and/or silicon as the main element. For example, a polymer based on boron nitride has been fabricated which withstands temperatures up to 2200°C (4000°F).

Inorganic polymers have greater stability primarily because they rely on ionic bonding instead of organic covalent bonding. These polymers are generally insoluble, except for strong inorganic acids, and are sensitive to moisture and oxidative effects.

Only a few of the advanced organic polymers have reached commercial status. The remainder of this section will be devoted to a close examination of three of them—the polyimides, the polybenzimidazoles, and the poly-*p*-oxybenzoates.

Table 6-82. Limiting Oxygen Index for Heat-Resistant Thermoplastics.*

Thermoplastic	LOI
Ethylene-chlorotrifluoroethylene (E-CTFE)	60
Ethylene-tetrafluoroethylene (ETFE)	30
Fluorinated ethylene propylene (FEP)	>95
Perfluoroalkoxy (PFA)	>95
Polyamide-imide	42
Polyethersulfone	40
Polyimide	44
Polyphenylene sulfide, 40% glass rein	46.5
Poly-p-oxybenzoate	37
Polysulfone	30

*Houston, 1975.

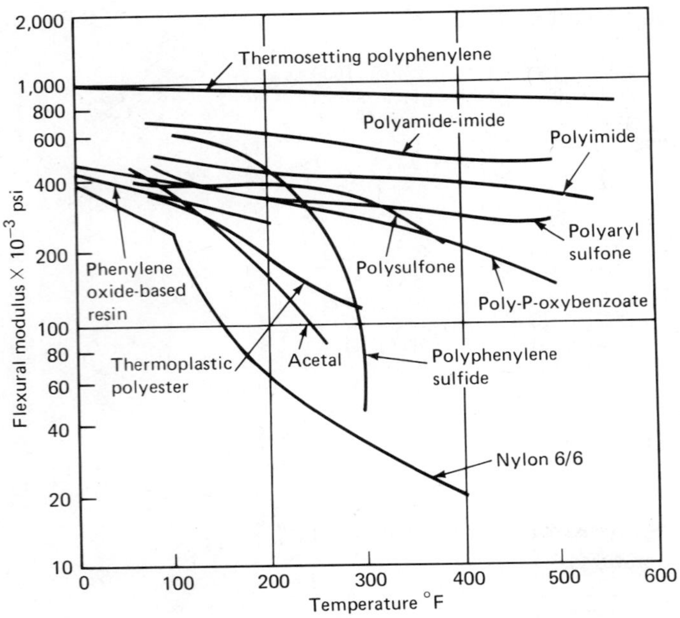

Fig. 6-80. Heat capabilities are reflected by flexural modulus. (Reprinted by permission from ***Product Engineering,*** January 1975)

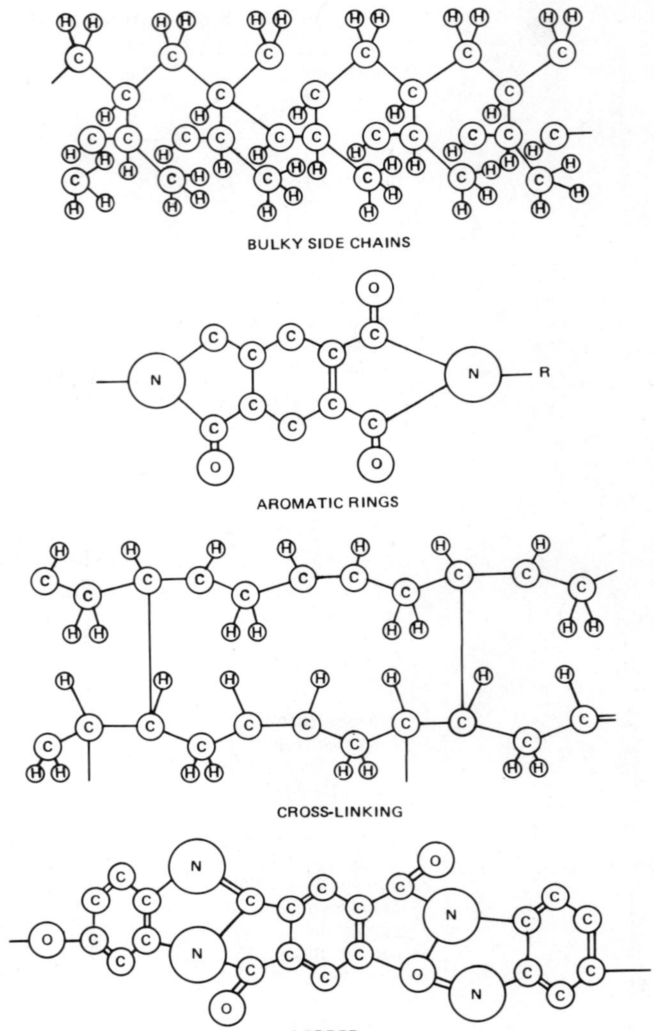

Fig. 6-81. Types of polymeric structures which provide heat resistance (West, 1969)

Polyimides

Polyamide-Imides and Polyester-Imides. Polyamide- and polyester-imides bridge the property gap between the aromatic and cyclic polyamides (see Chapter 5) and the aromatic polyimides. Early materials, based on aliphatic reactants, were not of commercial importance. With the introduction of aromatic dianhydrides the reaction assumed significance. Much of the early work was under government sponsorship and was aimed at the development of high molecular weight heat-resistant polymers for aerospace adhesive and laminating applications in the

Table 6-83. Cost Comparison of Heat-Resistant Thermoplastics.*

Thermoplastic	Specific gravity	$/lb	$/kg	¢/cu in.	¢/cu cm
Chlorotrifluoro-ethylene (CTFE)[a]	2.13	7.45	16.39	57.34	3.49
Ethylene-chlorotri-fluoroethylene (ECTFE)[b]	1.68	6.50	14.30	39.46	2.41
Ethylene-tetrafluoro-ethylene (ETFE)[c]	1.70	6.00	13.20	36.84	2.25
Fluorinated ethylene propylene (FEP)[d]	2.13	5.20	11.44	39.99	2.44
Perfluoro alkoxy (PFA)[e]	2.13	9.00	19.80	69.23	4.23
Polyamide-imide[f]	1.4	8.00	17.60	40.47	2.47
Polyether sulfone[g]	1.37	5.00	11.00	24.75	1.51
Polyimide[h]	1.4	10.00	22.00	50.58	3.09
Polyphenylene sulfide[i]	1.64	3.00	6.60	17.34	1.06
Poly-p-oxybenzoate[j]	1.4	14.50	31.90	73.35	4.48
Polysulfone[k]	1.24	2.00	4.40	8.96	0.546

[a] Plaskon CTFE 2100 (Allied). [b] Halar ECTFE 300 (Allied). [c] Tefzel 200 (Du Pont). [d] Teflon FEP 100 (DuPont). [e] Teflon PFA 9704 (Du Pont). [f] Torlon 4203 (Amoco). [g] 200 P (ICI). [h] Polyimide 2080 (Upjohn). [i] Ryton R-4 (Phillips). [j] Ekkcel I-2000 (Carborundum). [k] Udel (Union Carbide).

*Houston, 1975.

Table 6-84. Polymeric Materials for Extreme-Temperature Applications.*

Polymer	Comments
Polyphenyls	Decompose at 530°C (986°F); infusible, insoluble polymers.
Polyphenylene oxide	Decomposes close to 500°C (932°F); heat cures above 150°C (302°F) to elastomer; usable heat range -135-185°C (-211-365°F).
Polyphenylene sulfide	Melts at 270-315°C (578-599°F); crosslinked polymer stable to 450°C (842°F) in air; adhesive and laminating applications.
Polybenzyls; polyphenethyls	Fusible, soluble, and stable at 400°C (752°F); low molecular weight.
Parylenes (poly-*p*-xylylene)	Melt above 520°C (968°F); insoluble; capable of forming films; poor thermal stability in air; stable to 400-525°C (752-977°F) in inert atmosphere.
Polyterephthalamides	Melting points up to 455°C (851°F); fibers have good tenacity, elongation, modulus.
Polysulfanyldibenzamides	Melting points up to 330°C (626°F); soluble; good fiber properties.
Polyhydrazides	Dehydrate at 200°C (392°F) to over 400°C (752°F) to form polyoxadiazoles; good fiber properties.
Polyoxamides	Some melting points above 400°C (752°F); give clear, flexible films.
Phenolphthalein polymers	Melting points of 300°C (572°F) to over 400°C (752°F); formable into fiber and film.
Hydroquinone polyesters	Soluble polymers with melting points of 335°C (635°F) to over 400°C (752°F).
Polyhydroxybenzoic acids	Films melt at 380-450°C (716-842°F); stable to oxidation but not to hydrolysis; tough, flexible films; good thermal stability.
Polyimides*	Commercial film, coating, and resin stable up to 600°C (1112°F); continuous use up to 300°C (572°F).
Polyarylsiloxanes*	Good thermal stability 400-500°C (752-932°F); coatings, adhesives.
Carboranes	Stable in air and nitrogen at 400-450°C (752-842°F); elastomeric properties for silane derivatives up to 538°C (1000°F); adhesives.
Polybenzimidazoles*	Developmental laminating resin, fiber, film; stable 24 hours at 300°C (572°F) in air.
Polybenzothiazoles	Stable in air at 600°C (1112°F); cured polymer soluble in concentrated sulfuric acid.
Polyquinoxalines	Stable in air at 500°C (932°F); tough, somewhat flexible resins; make film, adhesive.
Polyphenylenetriazoles	Thermally stable to 400-500°C (752-932°F); make film, fiber, coatings.
Polydithiazoles	Decompose at 525°C (977°F); soluble in concentrated sulfuric acid.
Polyoxadiazoles	Decompose at 450-500°C (842-932°F); can be made into fiber or film.
Polyamidines	Stable to oxidation up to 500°C (932°F); can make flexible elastomer.
Pyrolyzed polyacrylonitrile**	Stable above 900°C (1625°F); fiber resists abrasion with low tenacity.
Polyvinyl isocyanate ladder polymer	Soluble polymer that decomposes at 385°C (725°F); prepolymer melts above 405°C (761°F).
Polyamide-imide*	Service temperatures up to 288°C (550°F); amenable to fabrication.
Polysulfone	Thermoplastic; use temperature -102°C (-152°F) to greater than 150°C (302°F); acid and base resistant.
Polybenzaylene benzimidazoles (pyrrones)	Thermally stable to 600°C (1112°F); insoluble in common solvents; good mechanical properties.
Polybenzoxazoles	Stable in air to 500°C (932°F); insoluble in common solvents except sulfuric acid; nonflammable; chemical resistant; film.
Ionomer	High melt and tensile strength; tough; resilient; oil and solvent resistant; adhesives, coatings.
Diazadiphosphetidine	Thermoplastic up to 350°C (662°F); thermosetting at 357°C (707°F); cured material has good thermal stability to 500°C (932°F); amenable to fabrication.
Phosphorous amide epoxy	Soluble B-staged material; amenable to fabrication; good thermal stability.
Phosphonitrilic	Retention of properties in air up to 399°C (750°F).
Metal polyphosphinates	Polymers stable to better than 400°C (752°F).
Phenylsilesesquioxanes (phenyl-T ladder polymers)	Soluble; high molecular weight; infusible; improved tensile strength; high thermal stability to 525°C (977°F) in air; film forming.

Note: Underlined materials are discussed in Chap. 5. Starred materials are discussed in Chap. 6. Double starred materials are discussed in Chap. 8.
*Washburn and Karle. Courtesy: McDonnell Douglas Astronautics Co.

300°C (572°F) range. The general finding was that the aromatic polyamide-imide polymers were less thermally stable than the aromatic polyimides.

Despite the less than ideal thermal resistance, interest in polyamide-imides as electrical insulation continued. A special resin from methylenedianiline and trimellitic anhydride was developed, followed by molding and extrusion grades, some based on terephthalic acid.

Polyamide-imides are easier to prepare than high molecular weight aromatic polyamides or polyimides. Three methods of synthesis have been employed. Using interfacial condensation, an amine-terminated polymer can be formed from the reaction of a diacid chloride with excess primary amine. This polymer is then reacted in solution with a dianhydride and the resultant polymer ring is closed using techniques described in the section on aromatic polyimides.

A polymer can result from the reaction of a diprimary amine and trimellitic anhydride. This resin will contain alternating amide-imide links. Finally, a dianhydride may be reacted with excess diamine to obtain a low molecular weight prepolymer which is further reacted with a diacid chloride, utilizing either solution or interfacial condensation techniques.

By varying the reaction ratios of the acid chloride and the amine, amide-containing precursors can be developed with virtually any molecular weight, which act as a control for the final amide-imide ratio. The overall reaction proceeds as follows:

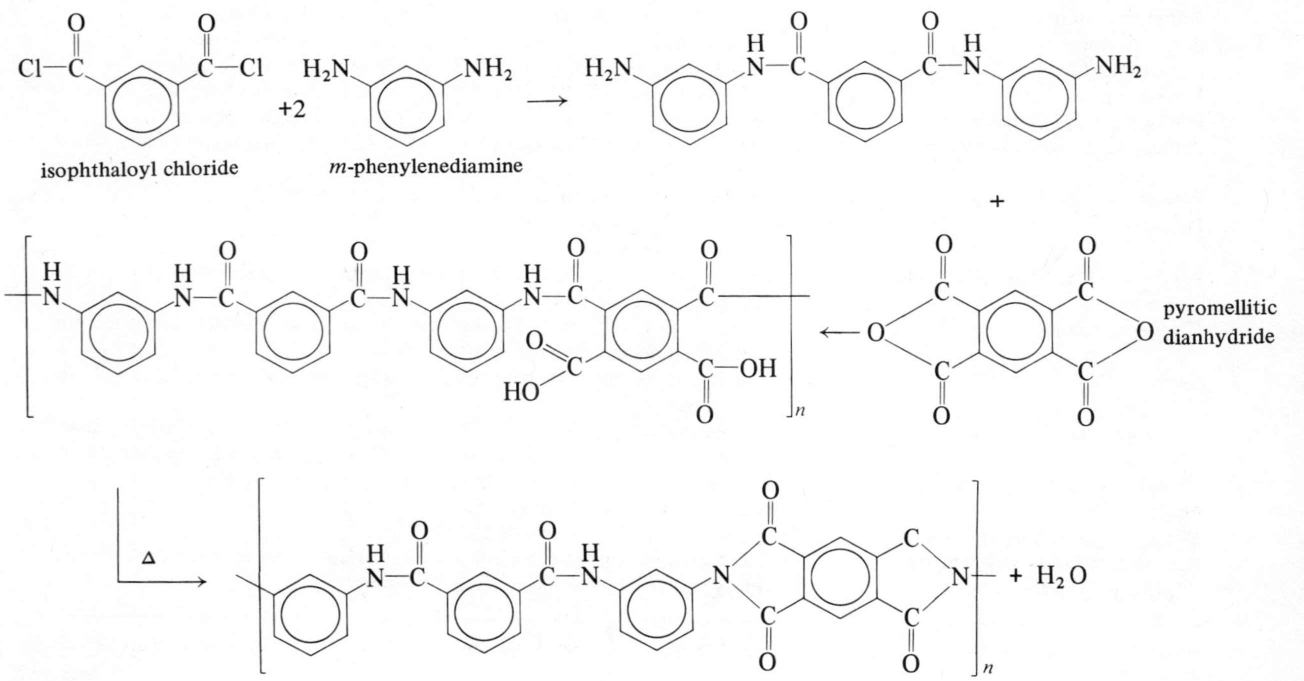

The polyamide-imides from trimellitic anhydride are produced from the 4-acetyl chloride of trimellitic anhydride and a diamine. Further heating produces a polyimide with the following generalized structure:

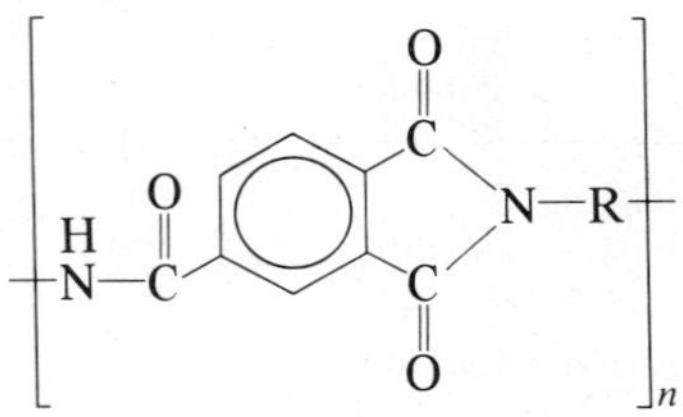

As the amide-imide link ratio increases, the flexibility of the final polymer increases; but this increase is not preserved as the temperature approaches its upper limit. However, the higher-ratio resins are easier to process because the water eliminated in the ring-closing step is less.

The polyester-imides are produced by essentially the same technique as the polyamide-imides. In this case, the precursor is a dianhydride containing aromatic ester links. Such dianhydrides are prepared from trimellitic anhydride and a diacetoxy derivative of an aromatic diol. The dianhydride from hydroquinone is reputed to offer the best properties.

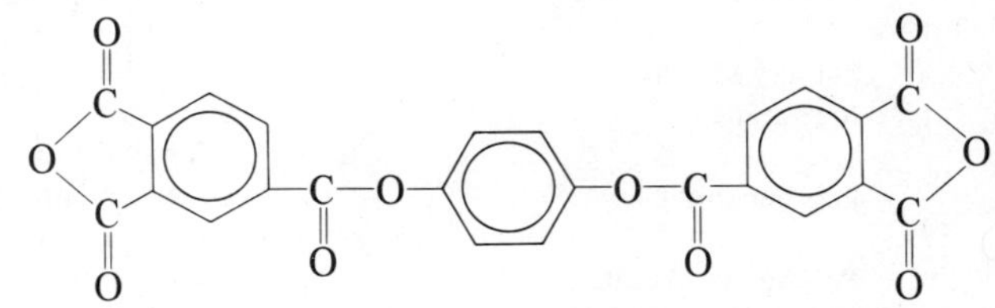

dianhydride from hydroquinone and trimellitic anhydride

There are a large number of modifications that can be made in the backbone chains of these resins. Choice of specific amine-terminated amides can be varied as well as their chain length. Combinations of acid chlorides can be used for property modification. The amide linkage seems to control the level of thermal stability obtainable.

Solvents commonly used for polyamide-imides include dimethylformamide, N-methyl pyrrolidone, dimethyl sulfoxide, and pyridine. Commercial solutions can be further thinned with solvents like aromatic and ester hydrocarbons and aliphatic or cycloaliphatic ketones. When diluted to low solids levels, the polymers are more stable than comparative aromatic polyimide solutions.

Polyamide-imide resins are highly stable and perform exceptionally well over a temperature range of -51-260°C (-60°F-500°F). They exhibit high mechanical and impact strength as is shown in Tables 6-85 and 6-86. At 260°C (500°F), properties of polyamide-imides almost equal those found in other engineering plastics at room temperature. Only 4% of the tensile strength is lost after a 2000-hour exposure to 260°C (500°F). Retention of properties at -51°C (-60°F) is excellent. Tensile and flexural properties increase, while impact strength decreases.

Polyamide-imide resins resist many acids and most organic and inorganic materials including fluorinated and chlorinated solvents. They are attacked by strong bases. The resins are fire resistant, exhibiting a UL-94 V-O (nonflammable) rating and a high limiting oxygen index.

Rheologically, these polymers show sharp melting around 329°C (625°F). They are highly shear sensitive as Figure 6-82 shows. A ten-thousandfold reduction in viscosity occurs with a thousandfold increase in shear rate.

Table 6-85. Polyamide-Imide Physical Properties at 73° F, 300° F, 500° F.*

Property	Units	Typical Value
Density	Gm/Cm^3	1.04
Tensile strength @ break	PSI	
@ 73°F		26,900
@ 300°F		15,200
@ 500°F (@ yield)		7,500
Elongation @ break	%	
@ 73°F		12
Flexural strength @ break	PSI	
@ 73°F		30,700
@ 300°F		22,600
@ 500°F		10,800
Modulus in flexure	PSI	
@ 73°F		664,000
@ 300°F		524,000
@ 500°F		431,000
Compressive modulus	PSI	
@ 73°F		402,000
Izod impact strength	Ft-lb/in of notch	2.5
Heat deflection temperature	°F	525
Flammability		94 V-O
Limiting oxygen index	%	42

*Courtesy: Amoco Chemicals Corporation.

Table 6-86. Polyamide-Imide Physical Properties at 73° F, -60° F.*

Property	Units	Value
Tensile strength @ break	PSI	
@ 73°F		26,900
@ -60°F		31,700
Elongation @ break	%	
@ 73°F		12
@ -60°F		16
Flexural strength @ break	PSI	
@ 73°F		30,700
@ -60°F		40,200
Flexural modulus	PSI	
@ 73°F		664,000
@ -60°F		847,000
Izod impact strength	Ft-lb/in of notch	
@ 73°F		2.5
@ -60°F		0.9

*Courtesy: Amoco Chemicals Corporation.

Polyamide-imides are readily injection molded or extruded. Some modifications of normal processing units are necessary to accommodate the unusually high strength and high melt viscosity of the resins. Any modern, typical screw-injection molding machine with a low compression-ratio screw but without a nonreturn check valve can be used. Pressures may vary from 20,000 psi to 50,000 psi, depending on the complexity of the parts. Mold shrinkage is fairly low at 0.007 in/in. Sprues and runners are generally designed oversized. As a result there is a somewhat greater amount of regrind. However, the regrind can be reused up to as much as 50% without substantial loss in mechanical properties—except for elongation which will drop substantially.

Because of the high melt viscosity at the typically low shear rates encountered during extrusion, the profiles

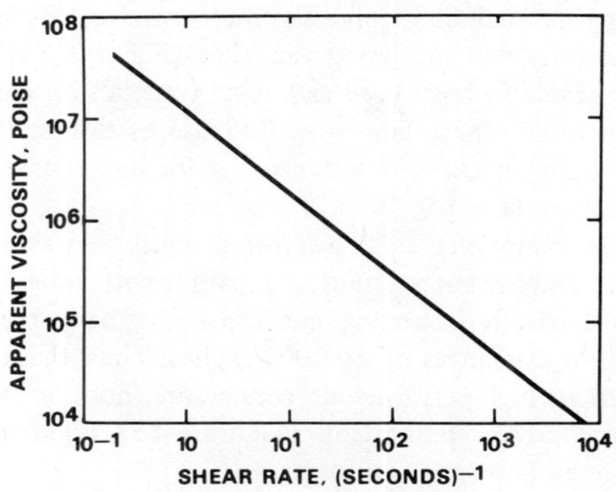

Fig. 6-82. Apparent viscosity as a function of shear rate for polyamide-imide at 340°C. *(Courtesy: Amoco Chemicals Corporation)*

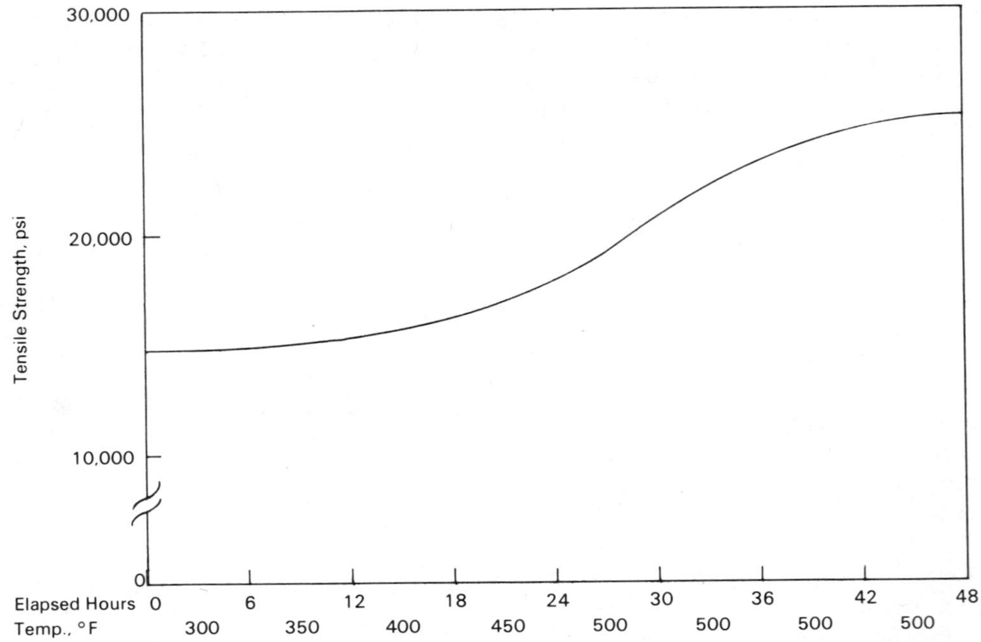

Fig. 6-83. Tensile strength change during typical annealing cycle. (*Courtesy: Amoco Chemicals Corporation*)

produced are generally simple in shape (e.g., circular or square). Single- or twin-screw extruders are used. Low-compression or low-working screw designs should be used. Air cooling is usually satisfactory.

In order to develop optimum properties, heat treatment (annealing) is strongly recommended (see Figure 6-83). Typically, the temperature is increased gradually from 104°C to 249°C (220°F–480°F) and held at the latter temperature for at least 24 hours. For large parts, holding the temperature at 104°C (220°F) for many hours may be required to allow all moisture and other volatile materials to diffuse from the parts. In general, a slow rate of temperature increase is preferred.

There are two main areas of use for polyamide-imides; (1) electrical and electronic components, and (2) friction- and wear-resistant parts. In the former area, the solvent-based resins (including polyester-imides) are used as magnet wire coatings and insulating varnishes (Figure 6-84). The liquids have also been used as binder materials for composite structures. These laminates find use as slot liners and sticks, phase insulation, and sleeving for high-temperature electrical insulation.

Molded components for electronics must be stable, temperature and solvent resistant, compatible with adhesive systems, electrically insulating, and capable of being precision molded to tolerances of ±0.0002 inches. Thus, these resins find use in high-performance connectors, burn-in sockets, microwave oven components (Figure 6-85), solenoid bobbins, arc barriers, and coil forms.

Graphite- and PTFE-filled systems are being used in high-quality, friction- and wear-resistant appliance bearings, light- or medium-duty machinery, valve plates, piston rings, poppet valves, and automobile door hinge bushings.

Fig. 6-84. Polyamide-imide varnishes are used for electrical insulation in high-temperature applications. (*Courtesy: Rhone-Poulenc, Inc.*)

Aromatic Polyimides. In theory, high molecular weight polyimides may be synthesized from any dianhydride and any diprimary amine. When both precursors are aromatic, commercially useful polymers with high heat resistance are obtained.

The general reaction takes place in two steps: (1) formation of a polyamic acid, followed by (2) imidization.

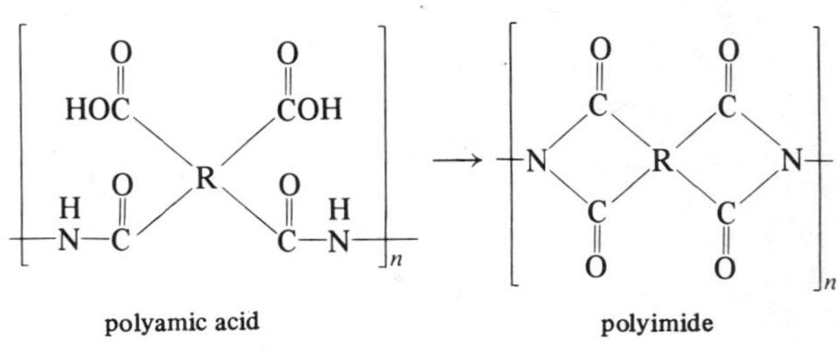

The polyamic acid is formed in solution utilizing a solvent which is active enough to permit reaction between the monomers. Such solvents include dimethylformamide, dimethyl sulfoxide, and N-methyl-2-pyrrolidone. Formation occurs under anhydrous conditions around 50°C (122°F) or lower. Higher temperatures can cause premature imidization with release of water.

The sequence of ingredient addition, method of combination, and purity will strongly influence the resultant polymer's molecular weight. Molecular weight control is managed through the use of chain stoppers such as phthalic anhydride.

After formation of the polyamic acid, the imide is obtained by heating to temperatures of approximately 300°C (572°F) or, at lower temperatures (⩽100°C, 212°F), through the use of cyclizing agents such as acetic anhydride and pyridine.

Common monomers currently used include diamines such as *m*-phenylenediamine, 4,4′-methylenedianiline, and 4,4′-diaminodiphenyl ether. Pyromellitic dianhydride is the predominant anhydride used, although bis(3,4-dicarboxyphenyl) ether dianhydride and 3,3′,4,4′-benzophenone tetracarboxylic dianhydride have found commercial acceptance for specific resin systems.

The overall course of the reaction is:

Fig. 6-85. Microwave transparency, physical strength, and heat resistance were the properties required in these parts designed for applications in the microwave industry and made from polyamide-imide. (*Courtesy: Amoco Chemicals Corporation*)

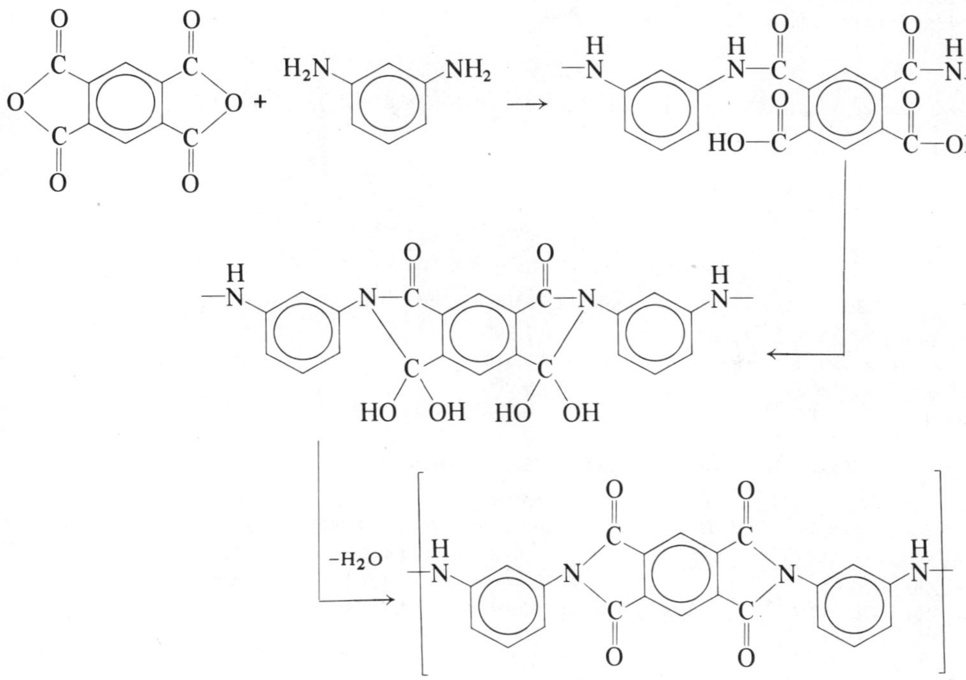

The reaction is beset by two major difficulties. First, the regenerated ingredients have a propensity for transimidification which results in low molecular weight polymers. Second, the polyamic acids exhibit poor storage stability as a result of the solution viscosity depression that occurs as water is liberated from the imidization reaction. The water may split polymeric chains and cause a reversion to acid and amine monomeric species. Unliberated water may create porosity and void stresses in the final polymer. Control of these problems is effected through the inclusion of water acceptors like acetic anhydride and catalysts like pyridine.

Polycarbodiimides have received increased attention in recent years because of the ease of their preparation. The overall reaction consists of an isocyanate function reacting with another isocyanate to yield a carbodiimide and carbon dioxide.

$$-RNCO + OCNR- \xrightarrow{\text{phospholene oxide}} -(R-N=C=N-R)- + CO_2$$

The reaction is quite rapid and produces high yields. High molecular weight polymer is easily obtainable under mild conditions. These polymers show useful high-temperature properties and have been found to improve the performance of polyurethane elastomers and rigid foams. Typical properties of compression-molded polycarbodiimide are shown in Table 6-87.

Modification of polyimide systems is limited to the selection of the anhydride and amine monomers. Crosslinking agents have been introduced through the use of more highly functional amines. Adhesive formulations incorporating fillers and stabilizers have been produced and supplied, generally supported on glass cloth carriers in a B-staged form.

Before imidization, the polyamic acid may be converted to a silver salt. As the polyimide forms, free metal particles are uniformly distributed throughout the polymer matrix. The dispersed particles are very small in size ($<0.8\mu$) and thus do not interfere with transparency, yet they will provide a conductive path through the polymer. Similar techniques have been used with copper and iron.

A class of thermosetting polyimides, which cure by an addition polymerization process, has been developed. The prepolymers are acetylene-terminated oligomers with molecular weights of less than 2000. When heated to 200°C (392°F) or above, polymerization and cure occurs via a process which may involve the trimerization of the terminal acetylenic groups into aromatic rings.

The oligomers are readily soluble in dimethylformamide or N-methyl pyrrolidone, but the latter is preferred for both the synthesis of the resins and for varnishes because it is more completely removed from the resin upon drying than are other solvents. The varnishes are useful in the production of fiber-filled molding materials or laminates. Several prepolymers of this class are soluble in chloroform, a rather unusual solubility characteristic for polyimides.

Data, obtained by thermomechanical analysis, show that

Table 6-87. Typical Properties of Compression-Molded Materials.*

Molded at 356°F, 2000 psi	
Density, g/cc	1.2
Tensile strength, psi	10,800
Tensile modulus, psi	265,000
Elongation, %	7.3
Flexural strength, psi	16,000
Flexural modulus, psi	342,000

IZOD IMPACT

Heat Treated hr at 392°F	Unnotched ft-lb/in	Notched ft-lb/in (1/4" thickness)
0	11.7	1.9
1	8.7	1.4
50	2.8	1.2

PROPERTIES OF FILLED RESIN
Molded 375–390°F, 4000–8000 psi

Filler	Flexural Strength, psi	Flexural Modulus, psi
40% (wt) milled glass fibers	17,500	809,000
40% (wt) 1/8" glass fibers	15,800	950,000
70% (wt) A24 grit	6,000	1,015,000
50% (wt) Ferro PF10 glass beads	8,900	690,000

EFFECT OF TEMPERATURE ON MECHANICAL PROPERTIES

Curing Schedule	T_g, °C 10⁹ Dynes/cm²	Tensile Strength, psi	Tensile Modulus, psi	Elongation, %
As Molded (175°C)	190	10,820	264,000	7.3
1 hr at 200°C	201	10,700	191,000	11.3
4 hr at 200°C	208	10,000	201,000	8.4
24 hr at 200°C	215	7,800	212,000	5.4
500 hr at 200°C	246	7,200	255,000	3.1

ELECTRICAL PROPERTIES

Dielectric constant	
600 Hz	3.50
1000 Hz	3.47
Dissipation Factor	
1000 Hz	0.0009

*Alberino et al., 1974,

these resins have T_g's as high as 400°C (752°F). Weight loss of graphite composites show values of only 0.9% after 500 hours in air at 260°C (500°F) and 2.8% after 2 hours in air at 427°C (800°F). When formulated into adhesives, values of 4100 psi on titanium adherends were obtained, with high strengths continuing to be observed after aging at 288°C (550°F).

These resins have been used in composites, adhesives, molding powders, solid lubricants, and high-temperature coatings.

Thermoplastic Polyimides. Polyimides are usually supplied in a low molecular weight prepolymer state; but thermoplastic polyimides are supplied in a completely polymerized and imidized form, and the only molding requirement is that the temperature be above the glass-transition level.

Since the resin—supplied in either a dry molding powder or as a solution in N-methyl-2-pyrrolidone for use in laminating, adhesion, coating, or for solution casting of film—is delivered unfilled, a processor can blend the exact amount of reinforcement necessary for a specific application. Scrap materials can be reprocessed because the polyimide is thermoplastic.

Polyimide compounds exhibit high tensile, flexural, and impact strengths; low wear rates and coefficients of friction; excellent retention of mechanical properties at 250°C (480°F); several thousand hours of thermal stability at 200–250°C (390–480°F); easy moldability and machineability (thermoplastic grades); exceptional creep resistance; and insensitivity to fire, chemical reagents, solvents, oils, cryogenic temperatures, radiation, and moisture (see Tables 6-88 and 6-89).

Polyimides are resistant to most organic solvents and dilute acids. However, they are not recommended for long-term exposure to caustic solutions, including ammonia and hydrazine.

Reinforcements include graphite, graphite fibers, molybdenum disulfide, and PTFE in self-lubricating grades; and glass, graphite, or boron fibers in high-strength structural molding grades. Table 6-90 shows the effect on polyimide properties caused by the addition of these fillers.

Thermoplastic polyimides can be compression molded in standard equipment at 350°C (660°F) at pressures between 4,000 and 5,000 psi.

A number of versions of the polyimides have been made available as films, varnishes, laminating resins, adhesives, and molding compounds. Commercial polyimide compounds can be categorized in terms of anticipated end use. There are those systems directed at applications where mechanical properties are of prime importance, and these compounds are fiberglass reinforced. Other grades are directed to applications where low wear rates, low coefficients of friction, and good dimensional stability under load are required. For these product areas, graphite-, molybdenum disulfide-, and PTFE-filled polyimides have been developed.

A special grade of polyimide is specifically formulated to be free-sintered. Typical free-sintered parts exhibit approximately 20% lower physical properties and 5% lower densities than corresponding parts prepared by hot compression molding.

One of the interesting characteristics of thermoplastic polyimides is their convertability by wet and dry spinning methods into synthetic textiles. Flame-resistant applications include aircraft furnishings, protective clothing, special household furnishings (drapes, upholstery), and toys. High-temperature applications include industrial filters, hot-gas filtration, corrosive chemical filtration, and electrical insulation for cables and wires. Typical laminate applications include radomes, circuit boards (see Figure 6-86), turbine blades, and other parts requiring high-temperature performance, fire resistance, and good electrical properties.

Other applications include the use of polyimides in film, laminating varnish, high-temperature adhesives (circuitry and aerospace structure bonding), high-temperature coatings (dip for electrical components), and as a molding compound for self-lubricating, structural, or friction precision parts.

Polybenzimidazoles (PBI)

The early polybenzimidazoles (PBI) were synthesized from aromatic bis-*o*-diamines and aliphatic dicarboxylic acids or their derivatives. A class of linear polymers from 3,3′-diaminobenzidine and diphenyl isophthalate was produced in the late 1950s which had molecular weights in the range of 55,000–90,000, had T_m's greater than 770°C (1418°F), were stable in nitrogen to 600°C (1112°F), and did not de-

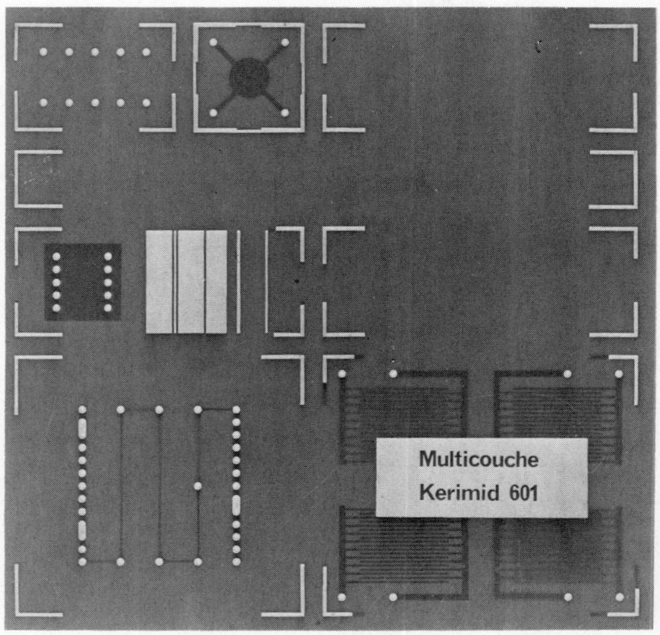

Fig. 6-86. Polyimide multilayer circuit boards have found numerous applications in computers and in aerospace, aircraft, and industrial electronics. (*Courtesy: Rhone-Poulenc, Inc.*)

Table 6-88. Typical Properties of Thermosetting Polyimide.*

Properties	ASTM	Base Resin	15% Graphite Filled	50% Glass Fiber Reinforced
PHYSICAL AND MECHANICAL PROPERTIES				
Specific gravity	D792	1.43	1.51	1.75
Hardness (Rockwell)	D785	45-58E	32-44E	44-56E
Ten str, ksi (MPa)	D1708			
73 F		11-14.5	6.3-11.5	3.7-7.1
		(75.8-100)	(43.4-79.2)	(25.5-48.9)
Elev temp, F (C)	@ 572	4.8-6.3	4.4-6.0	–
	(300)	(33.1-43.4)	(30.3-41.3)	–
Flex str, ksi (MPa)	D790			
73 F		15.5-18.5	11.4-18.0	6-12
		(106.8-127.5)	(78.5-124)	(41.3-82.7)
Elev temp, F (C)	@ 572	8.4-10.5	6.2-9.9	–
	(300)	(57.9-72.3)	(42.7-68.2)	–
Flex mod, ksi (MPa)	D790			
73 F		440-460	505-570	620-1000
		(3031.6-3169.4)	(3479.5-3927)	(4271.8-6890)
Elev temp, F (C)	@ 572	260-275	308-360	–
	(300)	(1791.4-1894.8)	(2122.1-2480.4)	–
Compr str, ksi (MPa)	D695			
73 F		36.7-45	29-34	–
		(252.9-310.1)	(199.8-234.3)	–
Elev temp, F (C)*	@ 572	18.5-19.3	9.3-9.8	–
	(300)	(127.5-132.9)	(64.1-67.5)	–
Shear str, ksi (MPa)	D732	12.3-13.7	10.9-11.6	–
		(84.7-94.4)	(75.1-79.9)	–
Impact str (Izod)	D256			
ft-lb/in. (notched) (l/cm)		0.7-1.4	0.5-0.6	–
		(2.42-4.84)	(1.73-2.07)	–
Poisson's ratio		0.41	0.41	–
Flex fatigue limit (MPa)		6.5	6.5	–
ksi at 10^7 cycles		(44.8)	(44.8)	–
Deformation under load, %	D621			
2000 psi (13.78 MPa) at 122 F (60 C)		0.14	0.10	–
Wear rate, in./1000 hr				
In nitrogen (PV=25,000)		0.010-0.015	0.004	–
In air (PV=25,000)		0.25-1.2	0.09	–
Coef of friction				
Static in air		0.35	0.30	–
Steady state				
In air (PV=25,000)		0.29	0.24	–
In nitrogen (PV=25,000)		0.04-0.09	0.06-0.08	–
THERMAL PROPERTIES				
Coef of linear exp, in./in./°F × 10^{-5}	D696	2.8-3.4	2.3-3.6	1.0-2.1
Thermal conductivity, Btu/hr/sq ft/°F/in.	C177	2.3-2.6	4.7-7.1	2.8-3.3
Specific heat, Btu/lb/°F (J/KgK)		0.27	–	–
		(1.13)	–	–
Deflection temp, 264 psi (1.82 MPa), F (C)	D648	680	680	–
		(357)	(357)	–
ELECTRICAL PROPERTIES				
Dielec constant at 10^5 Hz	D150			
73 F		3.41	7.60	–
Elev temp, F	@ 506	3.01	–	–
Dissipation factor at 10^5 Hz	D150			
73 F		0.0052	0.0040	–
Elev temp, F	@ 506	0.0052	–	–
Dielec str, v/mil	D149			
Short time (80 mils)		560	250	–
Vol resistivity, ohm-cm	D257	10^{16}-10^{17}	1.5 × 10^{15}	–
Surface resistivity, ohm-cm	D257	10^{15}-10^{16}	–	–
Arc resistance, sec	D495	230	–	–
Water absorption, %	D570			
24 hr		0.32	–	–
Saturation		3	–	–

*Reprinted by permission from *Materials Engineering*, Manual 245, February 1974.

compose completely at 900°C (1652°F). The general synthesis can be described as follows, although the actual reaction is considerably more complex.

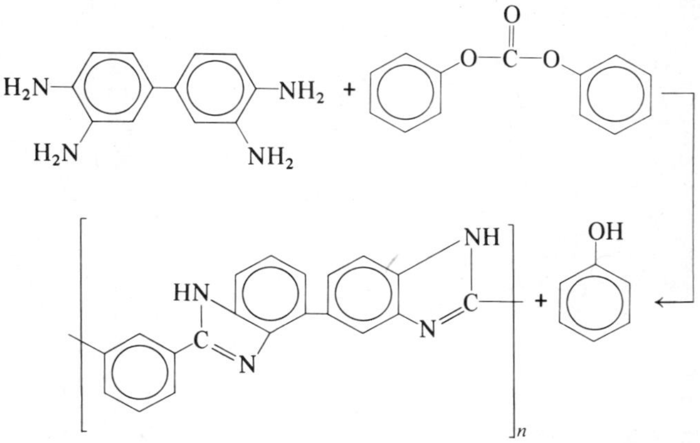

To obtain optimum thermal stability, blends of isophthalates and terephthalates have been most promising. It has been suggested that in the formation of prepolymer via a melt polymerization, the polymer contains both benzimidazole and hydroxybenzimidazolene structures.

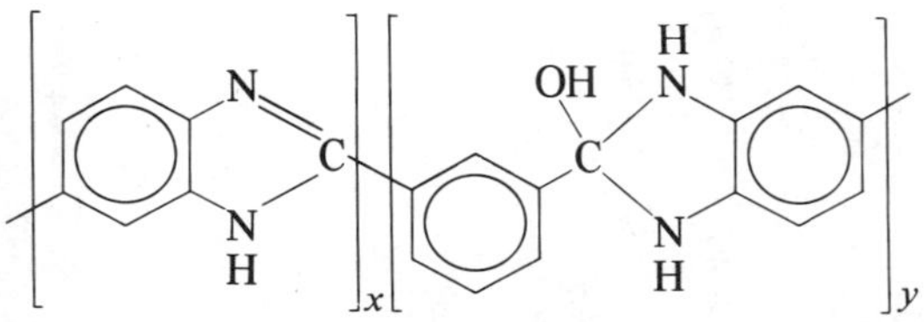

$x:y = 0.25:1.0$

Further heating to 400°C (752°F) results in final conversion to polymer. If equimolar quantities of monomers are heated, high molecular weight polymer is formed which is soluble only in formic or sulfuric acids.

These polymers, although thermoplastic in structure, in fact, cannot be further processed. Thus, the reaction is typically interrupted at an intermediate prepolymer stage to produce soluble, lower-melting compounds which can be processed in the liquid phase. An alternative approach is to use excess amine to generate an amine-terminated prepolymer which is coreacted with an acid-terminated prepolymer. The final polymerization can occur in a hot-melt or solution.

Large amounts of volatiles are generated from the polymerization reaction. These volatiles are often entrapped, causing the formation of low molecular weight polymer and porous parts. Although in practice, porous parts show better strength properties than more dense structures, the porosity does decrease humidity, water, and oxidation resistance. Techniques to minimize volatile release during final processing include use of highest practical molecular weight prepolymers, use of coreactants that release a smaller molecular species of volatile (e.g., isophthalamide instead of diphenyl ester to give off NH_3 instead of phenol), and combinations of the two.

Modification of the base polymer is primarily achieved through variation of precursor reactants or blending of the acid esters. The optimization desired usually involves a trade-off between polymer solubility and ease of processing with improved thermal stability.

Commercial PBIs are primarily used in laminating and adhesive formulations. Although hot-melt systems are available, solution processing is more convenient. Common prepolymers have limited solubility in relatively strong solvents; consequently they are only available in low-solids solutions. Typical solvents include dimethylacetamide, dimethylformamide, and N-methylpyrrolidone.

Adhesive systems, particularly those in contact with ferrous metals, are sensitive to oxidation during high-temperature aging. Arsenic-based antioxidants such as arsenic thioarsenate are used to retard this phenomenon. Adhesive systems also contain fillers (typically aluminum powder) to reduce shrinkage and coefficient of thermal expansion. Carrier films such as glass cloth are used to facilitate the application of the adhesives.

Property data for PBI's is usually reported for fully formulated adhesive and laminate systems.

Laminates. Laminate reinforcements include glass, graphite, carbon, or organic fibers. Cloth pattern, finish, processing methods, polymer wettability, and adhesion all influence final laminate properties. Strength properties are preserved to quite high temperatures after short-time aging. With longer aging, thermal stability weakens, beginning at temperatures around 260°C (500°F). Compressive strengths break rather sharply after 100 hours of aging at 260°C (500°F), tensile strengths are relatively flat, and flexural strengths decline slowly. Figure 6-87 shows comparative

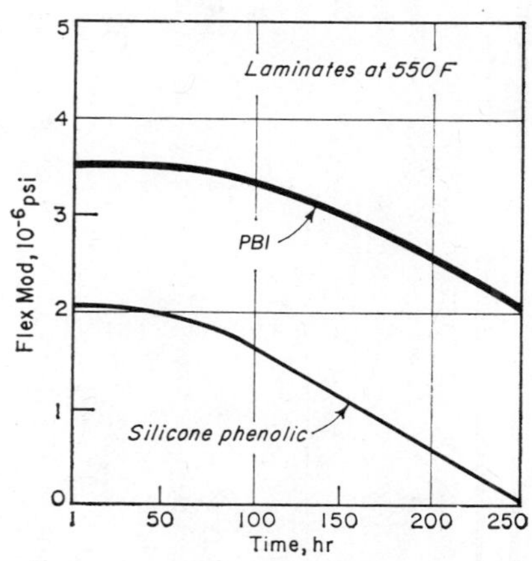

Fig. 6-87. Laminate based on polybenzimidazole (PBI) has better flexural strength than a high-temperature silicone-phenolic laminate. (Reprinted by permission from *Materials in Design Engineering,* May 1963)

Table 6.89. Typical Properties of Addition and Thermoplastic Polyimide Moldings.*

		Poly(amide-imide)s			Polyimides					
	ASTM	(Torlon) Unfilled	PTFE Filled	Graphite Filled	(Kinel) Glass Fiber Reinf	Graphite Filled	(Upjohn 2080) Unfilled	Graphite Filled	(P13N) Unfilled	Glass Fiber Reinforced
PHYSICAL AND MECHANICAL PROPERTIES										
Specific gravity	D792	1.41	1.44	1.45	1.9	1.45	1.4	–	1.33	1.83
Hardness (Rockwell)		104E	95E	98E	120M	110M	99E	–	91D	95D
Ten str, ksi (MPa)	D638									
73 F		13.3 (91.6)	10.4 (72.3)	13.0 (89.6)	27 (186)	5.7 (39.3)	17.1 (117.8)	10.5 (72.3)	7.3 (50.3)	14.3 (98.5)
Elev temp, F	500 F (260C)	8.9 (61.3)	4.6 (31.7)	5.2 (156.4)	22.7 (35.8)	3.6 (24.8)	4.4 (30.3)[a]	– –	5.7 (39.3)	9.3 (64.1)
Flex str, ksi (MPa)	D790									
73 F		23.4 (161.2)	20.8 (143.3)	22.1 (152.3)	49.5 (341)	12.8 (88.2)	28.8 (198.4)	13.7 (84.4)	10–12 (68.9–82.7)	30–40 (206.7–275.6)
Elev temp, F	500 F (260C)	14.2 (97.8)	9.6 (66.1)	7.7 (53.1)	35.5 (244.6)	9.2 (63.4)	5 (34.5)[a]	4 (27.6)	6–8 (41.3–55.1)	25–30 (172.3–206.7)
Flex mod, ksi (MPa)	D790									
73 F		700 (4823)	660 (4547.4)	710 (4892)	32.5 (223.9)	9 (62)	480 (3307)	530 (3652)	460–490 (3169–3376)	2000–2400 (13,780–16,536)
Elev temp, F	500 F (260C)	450 (3100.5)	460 (3169.4)	500 (3445)	24.2 (166.7)	7.5 (51.7)	160 (1102)[a]	210 (1447)	300–330 (2067–2274)	1700–1900 (11,713–13,091)
Compr str, ksi (MPa)	D695									
73 F		35 (241.1)	33.9 (233.6)	33.9 (233.6)	32.5 (166.7)	20 (137.8)	29.9 (206)	20.8 (143.3)	37.4 (257.7)	32.6 (224.6)
Elev temp, F	500 F (260C)	– –	– –	– –	18.4 (126.8)	– –	– –	– –	– –	21.8 (150.2)
Shear str, ksi (MPa)	D732	18.8 (129.5)	5.3 (36.5)	14.6 (100.6)	– –	– –	– –	– –	– –	– –
Impact str (Izod) ft-lb/in. (notched) (J/cm)	D256	0.7 (2.42)	0.6 (2.07)	0.8 (2.76)	17 (58.7)	0.25 (0.86)	0.7 (2.42)	– –	– –	– –
Water absorption, % 24 hr	D570	0.28	0.27	0.27	0.75	0.6	0.6	–	0.4	1.8
THERMAL PROPERTIES										
Coef of linear exp, in./in./°F	D696	19	–	–	8×10^{-6}	10.5×10^{-6}	28×10^{-6}	–	–	–
Thermal conductivity, Btu/hr/sq ft/°F/in.	C177	1.1	–	–	(3.48)	11.3	–	–	–	–
Deflection temp, 264 psi (1.82 Mpa), F (C)	D648	565 (296)	575 (302)	565 (296)	– 660	– –	– 270–280	– –	– –	– –
Oxygen index, %	D2863	–	–	–	–	–	44	–	31.6	–

ELECTRICAL PROPERTIES

Dielec constant at 10^5 Hz										
73 F	D150	3.7	—	—	—	—	—	—	4.33	6.00
Dissipation factor at 10^5 Hz										
73 F	D150	0.001	—	—	—	—	—	—	0.006	0.006
Dielec str, v/mil	D149								—	—
Short time (80 mils)		430	—	—	—	—	—	—	372	357
Vol resistivity, ohm-cm	D257	0.7×10^{15}	—	—	—	—	—	—	1.56×10^{16}	4.56×10^{15}
Surface resistivity, ohm-cm	D257	$>10^{15}$	—	—	—	—	—	—	3.9×10^{14}	1.7×10^{14}
Arc resistance, sec	D495	—	—	—	—	—	—	—	83	182

[a] 550 F (288 C).
*Reprinted by permission from *Materials Engineering*, Manual 245, February 1974.

data of flexural modulus of PBI and a silicone-phenolic laminate as aging progresses at 288°C (550°F). Laminate performance at cryogenic temperatures is somewhat better than at room temperature.

Humidity resistance of laminates is lower than would be predicted, with water absorption values quite high. This phenomenon is more a function of the laminate porosity than of some inherent characteristic of the polymer matrix.

The electrical properties of PBI laminates are in the range of typical thermosetting plastics but are quite flat with temperature (see Figure 6-88).

Adhesives. Optimum properties of PBI adhesives are obtained when they are used with appropriate primers and are glass cloth supported. Stainless steel lap-shear strengths of 3000 psi can be obtained at room temperature if the primer used is based on a solution of the same base PBI. If a polyimide primer is used, improvements in adhesion up to 35% can be obtained. Peel resistance of PBI adhesives is quite good.

Short-time strength retention of PBI adhesive is good. At 260°C (500°F), the adhesive strength begins to decrease after about 250-hour exposure (see Figures 6-89 and 6-90).

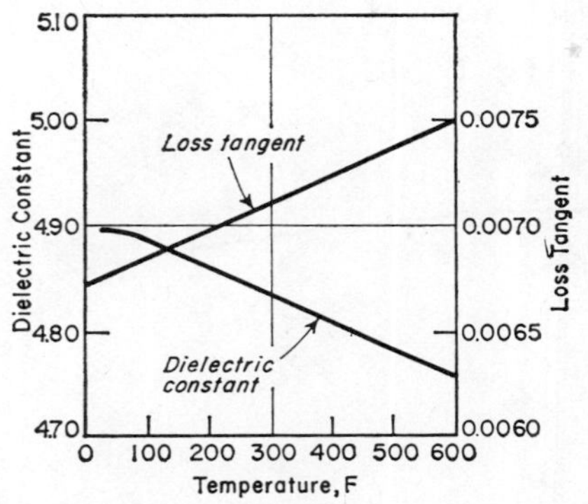

Fig. 6-88. Electrical properties of PBI laminates are quite stable over a broad range of temperatures. (Reprinted by permission from *Materials in Design Engineering*, May 1963)

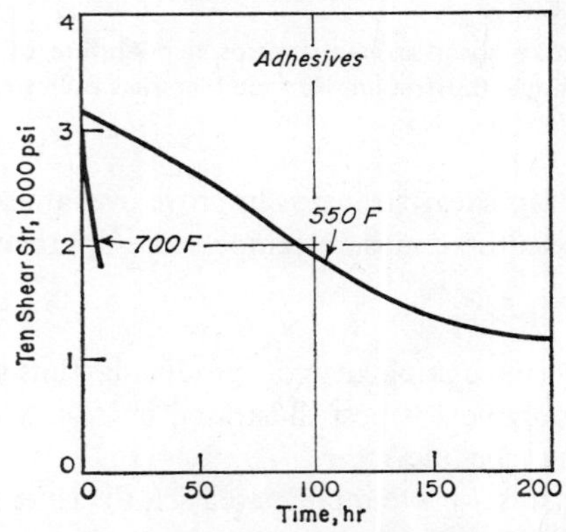

Fig. 6-89. Adhesives based on PBI have better long-term aging strengths than previously available high-temperature adhesives. (Reprinted by permission from *Materials in Design Engineering*, May 1963)

Table 6-90. Typical Properties of Filled Polyimide.*

Filler Type		None,	Molybdenum Disulfide	PTFE	Graphite	Graphite Fiber[b]
Filler concentration, volume percent		Unfilled	15	28	30	40
Property	ASTM Method	Values[a]				
Specific gravity						
Tensile strength, psi × 10^{-3}	D638	1.4	1.51	1.52	1.51	NR[c]
Tensile modulus, psi × 10^{-3}	D638	17.1	11.5	5.9	7.2	11.3
Elongation at break, percent	D638	188	970	720	1140	2010
Compressive strength, psi × 10^{-3}	D695	10.0	3.0	2.0	1.6	1.3
Compressive modulus, psi × 10^{-3}	D695	29.9	37.3	17.8	22.0	35.4
Strain at rupture, percent	D695	295	480	340	470	1200
Flexural strength, psi × 10^{-3}	D790	0.7	30.9	9.6	6.8	7.2
Flexural modulus, psi × 10^{-3}	D790	28.8	21.5	10.1	17.9	20.3
Coefficient of thermal expansion		481	780	512	1020	2157
(50 to 200°C), inches per inch per °F × 10^6	D696	28	27	30	8	2.7
Heat-deflection temperature, °F	D648	550	550	550	550	500
Hardness, Rockwell M scale	D785	99	111	93	103	NR

[a]Values vary with sample orientation. All values given were measured perpendicular to the direction of ram travel.
[b]Random fibers; base resin different from that used in other grades in the table.
[c]Not reported.
*Hensel, 1977.

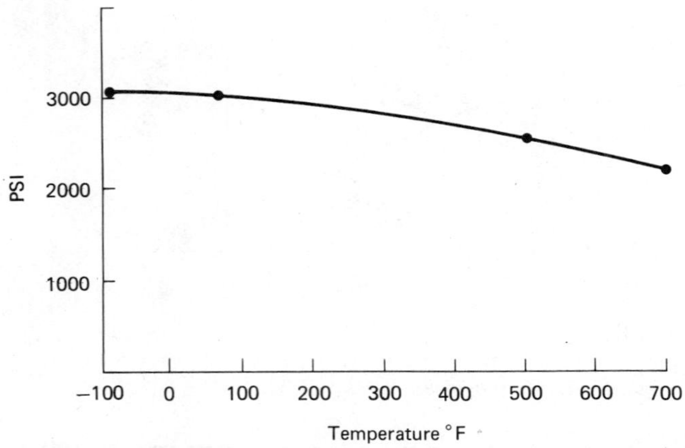

Fig. 6-90. Tensile shear strength versus temperature of polybenzimidazole adhesive. (Suffredini, Narmco Materials Bulletin)

Fatigue and lap-shear strengths improve over those at room temperature as the exposure temperature approaches cryogenic levels.

Processing. The complexity of producing finished goods from these polymers is best illustrated by following a typical processing schedule.

For laminates, a preimpregnated cloth is exposed to 120°C (248°F) with pressure gradually increasing from contact to 200 psi. The temperature is then increased to 370°C (618°F) and held for 3 hours. The laminate is then cooled to 100°C (212°F), removed from the press, and then postcured under dry nitrogen or vacuum according to the following schedule:

24 hours at 315°C (600°F)
24 hours at 345°C (650°F)
24 hours at 370°C (700°F)
24 hours at 400°C (750°F), and
8 hours at 425°C (800°F).

Following this cycle, an additional cure of 3 hours at 370°C (700°F) in air optimizes final properties.

Supported tape adhesive processing involves a cure of 1 hour at 220°C (428°F) plus 1 hour at 315°C (600°F), at a pressure of 200 psi. A dry nitrogen postcure follows this operation.

It becomes fairly obvious that the processing requirements for these polymers severely restrict their areas of application. The aerospace industry is about the only field that has the needed equipment and facilities. Consequently, end uses for the PBI's have been limited to the high-temperature-resistant adhesives and secondary structural members (e.g., radomes) used in this high-cost, high-technology area.

Poly-p-oxybenzoates and Copolymers

The first of this family to be developed was poly-*p*-oxybenzoate.

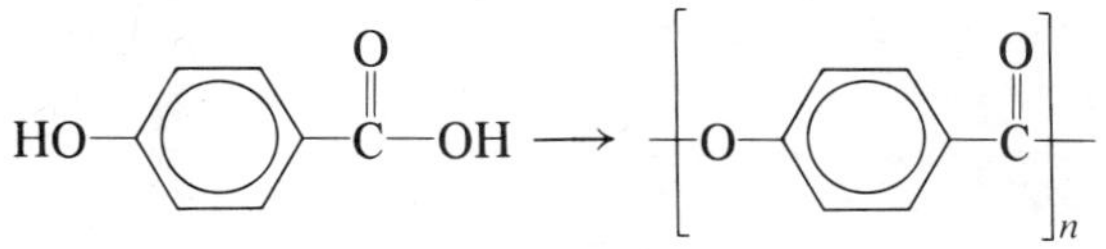

p-hydroxybenzoic acid

It has a unique combination of polymeric properties. The process used to make this polymer permits for the first time actual formation of a thermally stable, "tractable" aromatic polyester. At 538°C (1,000°F), the polymer will not melt; however, at temperatures of 427°C (800°F), it begins to display a nonviscous flow behavior similar to metals. This behavior permits fabrication of filled or unfilled forms by metallurgical techniques such as sintering, high-energy-rate forming, and plasma spraying. It can be fabricated at unusually rapid rates because of the absence of volatile by-products. Properties can be further tailored for specific application requirements by the use of a wide variety of modifiers, fillers, and forming techniques.

This polymer shows negligible loss of weight or physical properties, even when exposed to high ambient temperatures for extended periods of time. In air at 260°C (500°F) for 1,500 hours there is only 1.5% weight loss, while mechanical levels are retained at 95% of the rated values.

The poly-*p*-oxybenzoate has the highest known thermal conductivity of any known commercial resin. This high thermal conductivity provides a mechanism for dissipating heat from localized hot spots which may arise during extensive use. It has a low coefficient of thermal expansion.

The polymer has outstanding load-carrying capacity combined with excellent wear characteristics, and it is also self-lubricating. It displays high mechanical rigidity over a broad range of temperatures. Its wear resistance is greatly enhanced by its inherent self-lubricating properties. Its ability to flow under shear and to form a self-healing film protects both the wear and the abrading surfaces. This behavior also permits conformity of countermoving surfaces to an extremely close tolerance, such as in abradable seals.

The electrical insulating characteristics are quite exceptional. The dielectric strength is 660 volts/mil with a dissipation factor of 0.000198 (this is fantastically low for a material with high thermal conductivity). Resistivity is about average at around 10^{15} ohms-cm.

Unlike many plastics, under static or dynamic load, this polymer does not tend to deform because of its unusually high compressive strength and rigidity.

It has outstanding resistance to almost all aliphatic and aromatic solvents at temperatures as high as 100°C (212°F). It resists swelling in organic solvents and has very low water absorption at room and elevated temperatures.

Not only can parts be made rapidly using compression molding but, with plasma spray, pinholes can be eliminated by merely applying a shear force to the surface and letting it self-heal.

End uses include bearings and seals in pumps and fittings in chemical handling systems, abradable seals in moving parts like jet engines, in electrical applications that require insulation or evenly controlled conduction such as substrates, circuit boards, and switch parts.

Polyphenylenesulfide, blended with poly-*p*-oxybenzoate, gives an injection-moldable engineering plastic. Some properties are sacrificed as a consequence of alloying, though the blend is said to resist continuous 316°C (600°F) temperatures and is high in flexural and impact strength at high temperatures. The material still retains high thermal conductivity, load-carrying capacity, wear resistance, mechanical rigidity, electrical resistance, and resistance to chemical attack.

Two copolymers have also been offered commercially—one an injection molding resin and one for compression molding. The first is derived from terephthalic acid, *p*-hydroxybenzoic acid, and dihydroxy bisphenol.

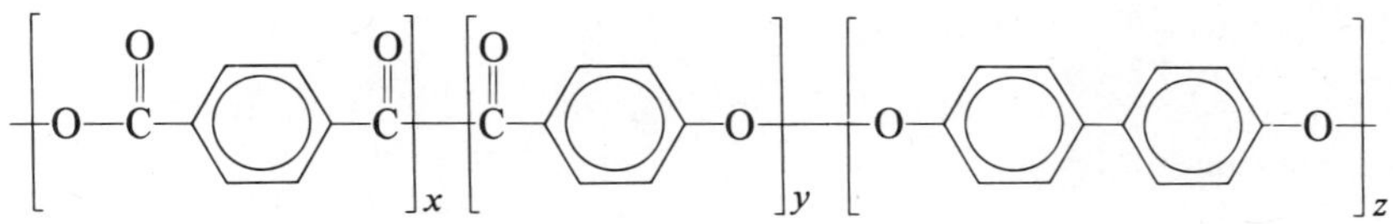

Table 6-91 shows typical properties of this resin. The resin is characterized by long-term thermal resistance, low moisture absorption, retention of properties to 288°C (550°F) (see Figures 6-91 and 6-92), radiation resistance, no corrosive outgassing, outstanding electrical properties, resistance to flammability, and thin-wall moldability. The resin is highly crystalline and mold shrinkage is very isotropic.

This copolymer is very shear sensitive, and too much working or high heat during processing can cause fibrolyzation (excessive oriention) leading to over crystallinity. As a consequence, regrind cannot be used nor can molds with more than six cavities for thin walled parts because of the polymer working that occurs. On the other hand, the shear thinning of the virgin polymer is the key feature that provides the flow characteristics necessary for molding thin sections with adequate green strength as well as the molding of thick wall sections without voids or surface sink-marks. A glass-bead-filled version for improved dimensional stability and reduced shrinkage is also available.

Table 6-91. Physical Properties of Injection Molding Grade of Poly-*p*-oxybenzoate Copolyester.*

Property		ASTM	Units	Typical Values
Molding				
Molding Temperature			°F	750
Molding Pressure			psi	5-10,000
Shrinkage (1)			in/in	.005
(2)			in/in	.018
Mechanical				
Tensile Strength	73°F	D-638	psi	12,000
	500°F		psi	2,500
Tensile Modulus	73°F		psi	300,000
Elongation	73°F		%	7-10
Flexural Strength	73°F	D-790	psi	12,000
	500°F		psi	3,500
Flexural Modulus	73°F		psi	700,000
	500°F		psi	250,000
Compression Strength	73°F	D-695	psi	10,000
Compression Modulus	73°F		psi	150,000
Impact Strength	Notched	D-256	ft-lb/in	1.6
	Unnotched		ft-lb/in	6
Deformation Under Load	(2000 psi)	D-621	%	3.5
Rockwell Hardness		D-785		R-88
Thermal				
Heat Distortion Temperature	(264 psi)	D-648	°F	540
Coefficient of Thermal Expansion		D-696	10^{-5} in/in/°F	1.6
Thermal Conductivity		Cenco Fitch	10^{-4} cal/(sec)(cm)(°C)	6.81
			BTU/(hr)(ft²)(°F/in)	1.98
Electrical				
Dielectric Constant	10^3 Hz 73°F	D-150		3.16
	10^6 Hz 73°F			2.87
Dissipation Factor	10^3 Hz 73°F			.010
	10^6 Hz 73°F			.025
Dielectric Strength	0.125 in	D-149	volts/mil	350
Volume Resistivity	(Observed)	D-257	ohm-cm	10^{15}
Surface Resistivity			ohm	10^{15}
Arc Resistance	Tungsten Rod Method		sec	100
Other				
Density		D-792	g/cm³	1.39
Water Absorption		D-570	%	.01
Flammability		UL-94		V-O
		LOI		40

(¹) Flow direction
(²) Cross Flow direction
*Courtesy: Dart Industries.

Table 6-92. Physical Properties of Compression Molding Grade of Poly-*p*-oxybenzoate Copolyester.*

Property		ASTM	Units	Typical Values
Mechanical				
Tensile Strength	73°F	D-638	psi	10,000
	500°F		psi	3,000
Tensile Modulus	73°F		psi	190,000
Elongation	73°F		%	10
Flexural Strength	73°F	D-790	psi	15,000
	500°F		psi	5,000
Flexural Modulus	73°F		psi	450,000
	500°F		psi	125,000
Compressive Strength	73°F (1)	D-695	psi	20,000
	73°F (2)		psi	19,000
Compressive Modulus	73°F (1)		psi	300,000
	73°F (2)		psi	209,000
Impact Strength	Notched Izod	D-256	ft-lb/in	0.4
	Unnotched		ft-lb/in	2.0
Deformation Under Load	(2000 psi)	D-621	%	1.02
Thermal				
Coefficient of Thermal Expansion	73-500°F (1)	D-696	10^{-5} in/in/°F	3.7
	(2)			4.3
Thermal Conductivity	100°F	Cenco Fitch	10^{-4} cal/(sec)(cm)(°C)	9
			BTU/(hr)(ft^2)(°F/in)	2.61
Heat Distortion Temperature	(264 psi)	D-648	°F	572
Flammability		UL-94		V-O
Limiting Oxygen Index			%	32
Other				
Density		D-792	g/cm^3	1.35
Rockwell Hardness		D-785		E-57
Water Absorption		D-570	%	0.04
Radiation Resistance			rads	10^{10}
Electrical				
Dielectric Constant	100 Hz 73°F	D-150		3.7
	1 KHz 73°F			3.7
	1 MHz 73°F			3.5
	2000 MHz 73°F			2.8
Dissipation Factor	100 Hz 73°F	D-150		0.005
	1 KHz 73°F			0.008
	1 MHz 73°F			0.0151
	2000 MHz 73°F			0.004
Dielectric Strength	0.125 in	D-149	volts/mil	450
Volume Resistivity	(Observed)	D-257	ohm-cm	10^{15}
Surface Resistivity			ohm	10^{15}
Arc Resistance	Tungsten Rod Method		sec	127

(1) In Mold direction
(2) Cross Mold direction
*Courtesy: Dart Industries.

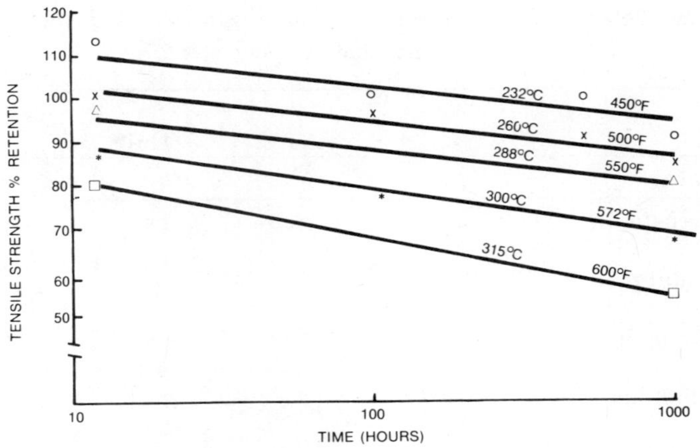

Fig. 6-91. Effect of heat aging on tensile properties of Ekkcel I-2000 poly-*p*-oxybenzoate copolyester. (*Courtesy: Dart Industries*)

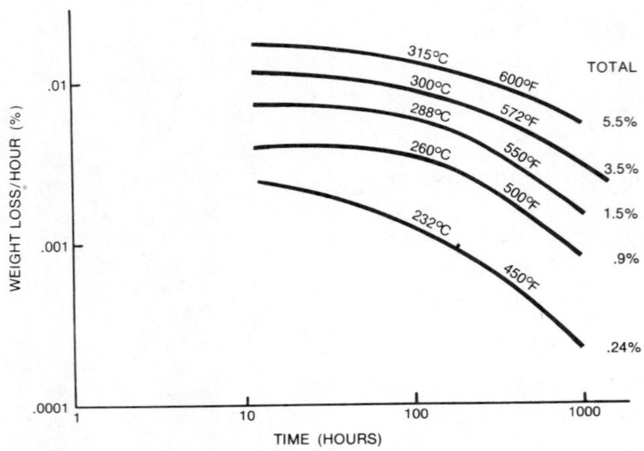

Fig. 6-92. Rate of weight loss versus time for Ekkcel I-2000 poly-*p*-oxybenzoate copolyester. (*Courtesy: Dart Industries*)

Table 6-93. Comparative Heat-Deflection Temperatures of Poly-*p*-oxybenzoate Copolyesters and Other High-Temperature Polymers.*

	Heat Deflection @ 264 psi °C (°F)
Ekkcel I-2000	282 (540)
Ekkcel I-2006 (30% glass)	310 (590)
Ekkcel C-1000	300 (572)
Astrel 360 (polyaryl sulfone)	274 (525)
Polyimide	243 (470)
Polyamide-Imide	274 (525)
Polyphenylene Sulfide	204 (400)

*Cook, 1976.

Among the end uses that have incorporated this polymer successfully are dishwasher parts, arc cages, and electronic contacters resistant to 300°C (572°F).

The second copolymer was developed for compression molding applications. It is somewhat different from other compression molding materials in that the proper processing technique is more like a confined sintering method. Typical mold conditions are 371°C (700°F) and 10,000 psi. The chemical structure is:

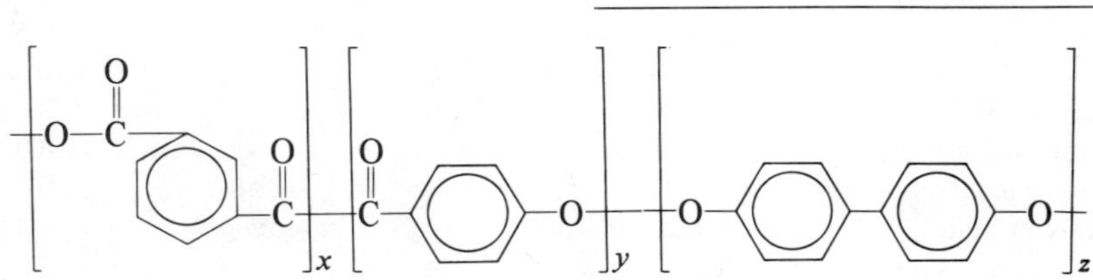

and is derived from isophthalic acid instead of terephthalic acid. The material never goes through a melt phase and behaves similarly to a sintering compound. Some crosslinking is believed to occur in the confined heated mold.

Table 6-92 shows typical properties of this resin. The resin is characterized by excellent friction and wear properties, long-term thermal stability, radiation resistance, low moisture absorption, strength at temperatures in excess of 288° (550°F), and relative molding ease. Table 6-93 compares the heat-deflection temperatures of injection and compression moldable grades with other high-temperature polymers.

Grades of resin are also supplied with 30% glass for improved dimensional stability, 25% PTFE for improved lubricity, and 20% graphite for better heat dissipation.

Typical applications for this group of resins include hull valve seats, packings, piston rings, thrust washers, bearings, ferrules, insulators, and machine parts.

As of this writing, complete chemical-resistance data has not been published. In general however, both injection and compression molding materials resist most aromatic and chlorinated hydrocarbon solvents.

Trade Names

Allex	Polyimide magnet wire	Essex Wire Co.
Amoco	Poly(amide-imide)	Amoco Chemical Co.
Anamid	Poly(ester-imide) and poly(amide-imide magnet wire	Anaconda Wire & Cable Co.
Aramidyl	Poly(amide-imide) resin	Westinghouse Electric Corporation
Armored Polythermaleze	Poly(amide-imide) magnet wire	Belden Corporation
Baymid	Polycarbodiimide foam	Bayer of Lever Kusen
Ben-Har	Polyimide varnish	Bentley-Harris Mfg. Co.
Dalite	Polyimide film and wire	Daburn Electronics & Cable Co.
De Beron	Polyimide	DeBeers Laboratories, Inc.
DSC-40	Polycarbodiimide	Upjohn Co.
Durad	Polyimide-coated PTFE	Haveg Super-Temperature Wires
Ekkcel	Poly(*p*-oxybenzoate) copolyester	Dart Industries
Ekonol	Poly(*p*-oxybenzoate)	Dart Industries
Envex	Thermoplastic polyimide	Rogers Corporation
Ferro	Reinforced polyimide	Ferro Corporation
Feuralon	Polyimide parts	Bermol Co.
H Film	Polyimide film	E.I. Du Pont de Nemours & Co. Inc.
Hexcel	Reinforced polyimide	Hexcel Corporation
Imidalloy	Polyimide products	Toshiba Chemical Products Co.
Imidite	Polybenzimidazole	Whittaker Corporation
Isomid	Poly(ester-imide) film and magnet wire	Schenectady Chemicals, Inc.
Kapton	Polyimide film	E.I. Du Pont de Nemours & Co., Inc.
Kemid	Polyimide rod and sheet	Chemplast, Inc.
Kerimid	Polyimide	Rhodia Inc.
Kimel	Polyimide	Rhodia Inc.
Lashclad	Polyimide adhesive	Lash Laboratories
Meldin	Polyimide parts	Dixon Corporation
Meltbond	Polyimide adhesives	Whittaker Corporation
Omega	Poly(ester-amide-imide) wire enamel	Westinghouse Electric Corporation
P13N	Polyimide	Ciba-Geigy Corporation
Polyimidal	Polyimide	Raychem, Inc.
Polyimide 2080	Thermoplastic polyimide	Upjohn Co.
Pyralin	Reinforced polyimide	E.I. Du Pont de Nemours & Co., Inc.
Pyralux	Polyimide film laminate	E.I. Du Pont de Nemours & Co., Inc.
Pyre-ML	Polyimide coating	E.I. Du Pont de Nemours & Co., Inc.
Rhodeftal	Poly(amide-imide)	Rhodia Inc.
Rigimid	Polyimide laminates	Riggs Engineering Corporation
Simatherm	Poly(ester-imide) magnet wire	Siemens Corporation
Skybond	Polyimide	Monsanto Co.
Spacewrap	Polyimide film	Hercules, Inc.
Sparmon	Polyimide parts	Sparta Mfg.
Tetra Temp	Polyimide film	Royal Industries
Thermid	Polyimide	Gulf Oil Chemical Corporation
Torlon	Poly(amide-imide)	Amoco Chemical Co.
Tribolon	Reincorced polyimide	Tribol Industries
Tritherm	Polyimide	P.D. George Co.
Vespel	Polyimide parts	E.I. Du Pont de Nemours & Co., Inc.
XWE-80	Poly(amide-imide)	Schenectady Chemicals Inc.
YF	Reinforced polyimide	LNP Corporation

REFERENCES AND BIBLIOGRAPHY

Alberino, L. M., Farrissey, W. J., and Sayigh, A. A. R. Preparation and properties of carbodiimide oligomers. *ACS Polymer Preprints*, September 1974, 15(2):77ff.

Amoco Chemicals Corporation. *Injection Molding Torlon Resins*. Technical Information Bulletin TAT-3, March 1975.

——. *Chemical Resistance of Torlon Resins*. Technical Information Bulletin TAT-4, March 1975.

——. *Annealing Torlon Parts to Obtain Optimum Properties*. Technical Information Bulletin TAT-5, June 1975.

——. *Extrusion of Torlon Resins*. Technical Information Bulletin TAT-6, October 1975.

——. *Machining Operations for Amoco Torlon Engineering Resins*. Technical Information Bulletin TAT-1(REV), December 1974.

——. *Torlon Molding Resins.*

——. Amoco AI Precision Molded Components" Bulletin HT-9a.

——. *Guidelines for Compression Molding of Torlon Powders*. Technical Information Bulletin TAT-2, October 1974.

Bateman, J. H., Geresy, W. Jr., and Neiditch, D. S. Soluble polyimides derived from phenylindane diamine: a new approach to heat resistant protective coatings. Ciba-Geigy Corporation.

Bilow, N., and Landis, A. L. "Recent Advances in Thermosetting Acetylene-Substituted Polyimides." Paper presentation, *SAMPE Symposium* Seattle, October 1976.

Bilow, N., Keller, L. B., Landis, A. L., Boschan, R. H., and Castillo, A. A. "New Developments in Acetylene-Terminated Polyimides." Paper presentation, *SAMPE Symposium*, Anaheim, California, May 1978.

Bilow, N. "Acetylene-Substituted Polyimides as Potential High Temperature Coatings." Paper presentation, *ACS/JCS Joint Meeting*, Honolulu, April 1979.

Blatz, P. S. NR-150 polyimide precursor adhesive solution developed. *Adhesives Age*, September 1978, **21**(9):39–44.

Business Week. Industrial plastics that will take the heat. June 1 1974, pp. 36D–37F.

Campbell, F., Rugg, B., Kumar, R., Arnon, J., and Brenner, W. Environmental aging studies of electron-beam-cured adhesives. *Adhesives Age*, January 1979, **22**(1):40–42.

Carborundum Co. Various bulletins on Ekkcel.

Chemical and Engineering News. Polyimide polymers under development. August 1964, **44**(35):24–25.

Cook, R. E. "Ekkcel Presentation." *Plastics World Seminar–Engineering Plastics II*, Los Angeles, March 1976.

Cottis, S. G. Aromatic polyester plastics. *Modern Plastics*, July 1975, **52**(7):62–63.

Darmory, F. P. High performance thermosets. *Plastics Design and Processing*, September 1974, **14**(9):18–21.

Emery, D. B. "Torlon Resins–Applications and Processing Technology." Paper presentation, *Plastics World Seminar–Engineering Plastics II*, Los Angeles, March 1976.

General Electric Co. *Gemon Thermoset Polyimide, Glass Reinforced Molding Compound*.

Hensel, J. D. When, where, and how to use polyimides. *Plastics Engineering*, October 1977, **33**(10):20–23.

Houston, A. M. How heat resistant thermoplastics compare at elevated temperatures. *Materials Engineering*, June 1975, **81**(6): 2ff.

Howeth, M. S., Thornton, H. R., and Burroughs, J. E. Materials vs. heat. *Materials in Design Engineering*, December 1965, **62**(7): 99–134.

Lee, H., Stoffey, D., and Neville, K. *New Linear Polymers*, McGraw-Hill Book Co., New York, 1967.

Materials Engineering. Polyimide stable in air at 700°F. October 1975, **84**(4):20.

——. Polyimide plastics. Manual 245, February 1974, **79**(2): 69–76.

Materials in Design Engineering. New polymer looks good for 1000°F. May 1963, **57**(5):92.

Miska, K. H. Adhesives and sealants take the heat up to 600°F. *Materials Engineering*, May 1977, **85**(5):57–58.

Monsanto Chemical Co. *Skybond 700 High Heat Resistant Resin*. Technical Bulletin No. 5042B.

Plastics Design & Processing. More information on molded polyimide–Du Pont's new high temperature thermoplastic. February 1963, **3**(2):20–21.

Plastics Engineering. Engineering resins primer. December 1974, **30**(12):17–32.

Plastics Technology. More data reported on new PBI resins. May 1963, 9(5):55.

Product Engineering. New high temperature plastics feature easier processing, lower costs. January 1975, **46**(1).

Rausch, K. W. Jr. "Polyimide Polymers and Composites." Paper presentation, *ASTM Task Group on Geothermal Seals*, CONF-7806141, April 1979.

Rhodia, Inc. *Kimel Polyimide Molding Powders Self-Lubricating Materials*. Bulletin No. 6453, February 1974.

——. *Kerimid 601 High Temperature Composite Polyimide Resin*. Bulletin No. 6273, May 1974.

——. *Kimel Fiberglass Reinforced Polyimide Molding Compounds*. Bulletin No. 6433, May 1974.

——. *Rhone-Poulenc Polyimide Materials*. Bulletin No. 6485, May 1974.

Rohm & Haas Co. *Kamax Brand 201 Engineering Thermoplastic Molding Resin*. Bulletin No. EP-3, March 1979.

——. "Kamax Brand Engineering Thermoplastic Molding Resin–Processing of Kamax Molding Resins" Bulletin No. EP-4, April 1979.

St. Clair, A. K., Slemp, W. S., and St. Clair, T. L. High-temperature adhesives for bonding polyimide film. *Adhesives Age*, January 1979, **22**(1):35–39.

Spaulding, L. D. "Tensile vs. Shrink Properties in Molding Amide-Imide Polymer." Paper presentation, *SPE ANTEC*, New Orleans, May 1979.

Suffredini, L. P. *Imidite: A Commercial Realization of a Recent Laboratory Curiosity*. Narmco Materials Bulletin.

Theberge, J. E. Arkles, B., and Cloud, P. Comparing high temperature plastics. *Machine Design*, February 6, 1975, **47**(3):73–77.

Titus, J. B. "Trade Designations of Plastics and Related Materials (revised). PLASTEC NOTE N9C, Plastics Technical Evaluation Center, Picatinny Arsenal, Dover, New Jersey, May 1978.

Upjohn Co. *Polyimide 2080 Molding Power*. Bulletin PC, No. 18, October 1974.

——. *DSC-40 Polycarbodiimide–Processing Information*.

——. *Highlights: Polyimide Fibers*. Report No. 3, Bulletin A-3473-R1.

——. *Highlights: Polyimide 2080*. Report No. 5, Bulletin A-2327.

——. *Highlights: Laminates From Polyimide 2080*. Report No. 10, Bulletin A-2998.

Van Boskirk, R. L. New polymers described at 20th SPE ANTEC. *Modern Plastics*, March 1964, **41**(3):222.

Washburn, R. M., and Karle, D. W. *New Polymers for Thermal Extremes*.

Watson, R. W., and Chen, Y. T. "Poly(Amide-Imide) Resins in Injection-Molded Parts." Paper presentation, *SPE ANTEC*, New Orleans, May 1979.

West, P. "Polymer chemists seek plastics that beat heat. *Product Engineering*, November 1969, **40**(22):84–88.

Wood, A. S. New day dawns for circuit boards. *Modern Plastics*, June 1977, **54**(6):52–54.

7
Elastomers

An elastomer may be defined as a natural or synthetic material which exhibits rubberlike properties of high extensibility and flexibility, is available in varying hardnesses, and, most important, has the ability to snap back quickly after being extended to at least twice its original length. This last requirement thus precludes plasticized PVCs and flexible low-density polyethylenes from being considered elastomers.

Until the middle sixties the only materials then considered to be elastomers were natural rubber or the various types of synthetic rubbers. Parts made from these materials were formed to shape in the time-consuming heat-and-pressure process commonly called vulcanization in the rubber industry. The analogous process in the plastics industry is the curing cycle used for thermosetting plastics such as phenolics, melamines, and epoxies. Despite the fact that the processes were similar in both fields and that the final product in each case was a thermoset polymer, in general, rubber or elastomeric parts were only made in "rubber" shops and thermoset plastics parts were processed mainly in "plastics" shops.

In the latter part of the sixties a new class of elastomers were introduced. These were thermoplastic elastomers (TPEs) which met the conventional elastomer requirements of extensibility and rapid retraction, but which also had a number of advantages (and some disadvantages) when compared to the older heat-curing types of rubber. (Figures 7-1 through 7-7 show some of the many parts made from TPEs—the new "rubbers.")

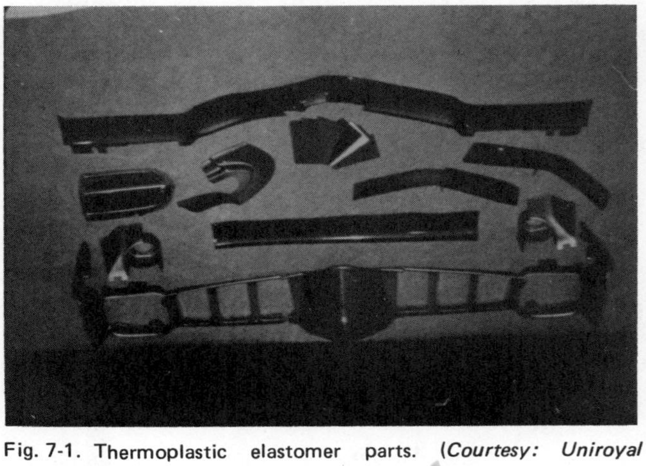

Fig. 7-1. Thermoplastic elastomer parts. *(Courtesy: Uniroyal Chemical Co.)*

The major advantage of the new materials was that, since they were thermoplastic, they could be rapidly processed by all the conventional thermoplastic processes (e.g., injection molding, extrusion, blow molding, thermoforming). Other advantages of using the TPEs included:

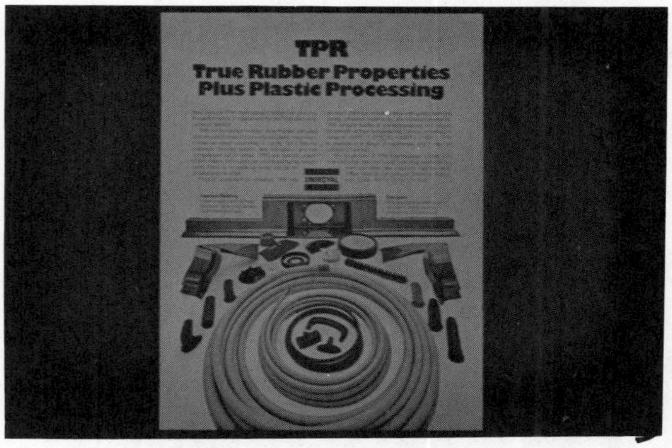

Fig. 7-2. Thermoplastic elastomer parts. *(Courtesy: Uniroyal Chemical Co.)*

Fig. 7-3. Light-duty cart wheels—injection molded. *(Courtesy: Uniroyal Chemical Co.)*

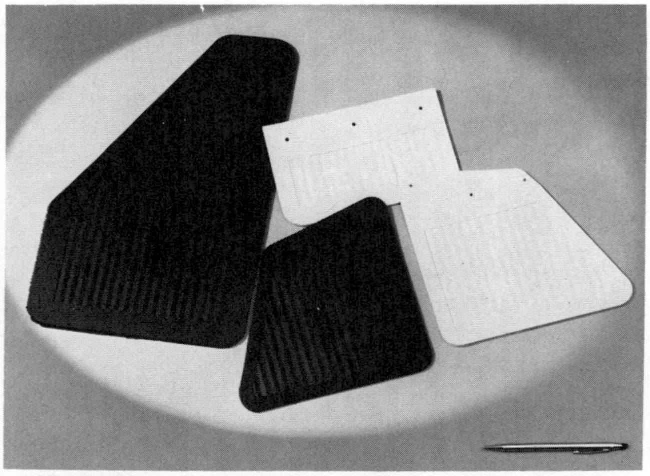

Fig. 7-4. Automotive studguards—injection molded. (*Courtesy: Uniroyal Chemical Co.*)

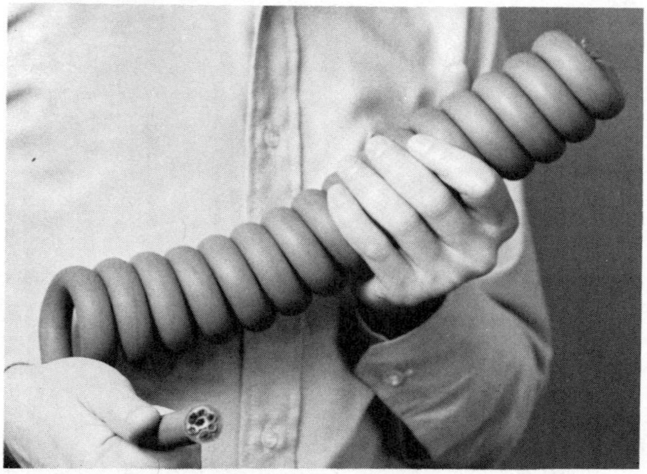

Fig. 7-5. Extruded cable covers. (*Courtesy: Uniroyal Chemical Co.*)

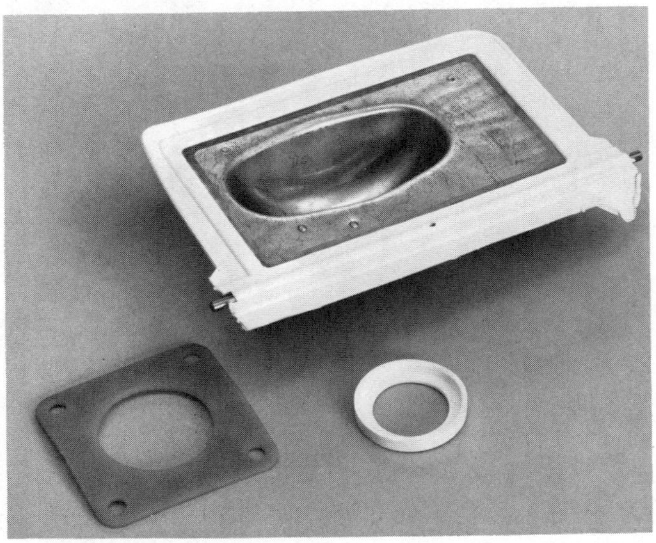

Fig. 7-6. Injection molded gaskets. (*Courtesy: Uniroyal Chemical Co.*)

Fig. 7-7. Blow molded automotive duct parts. (*Courtesy: Uniroyal Chemical Co.*)

1. Elimination of the vulcanization cycle, which also meant that curing agents and, in most cases, fillers were not needed, and the mixing and compounding operation could be left out.
2. Scrap and trimmings could be reused.
3. Most of the TPEs had very good low-temperature flexibility.
4. High coefficients of friction were characteristic of some of these materials.
5. Some TPEs were available with high tensile strengths at normal temperatures.
6. Since carbon blacks were not required for reinforcements, a wide range of colors or clear materials could be obtained.
7. LDPE, HDPE, or styrenes could be added to increase the impact resistance or tear strength.
8. Because plasticizers were not used, the use of TPEs in medical applications was more acceptable.

There were also disadvantages of the new thermoplastic elastomers. Among these were:

1. Some of the materials had poor high-temperature resistance properties; others had poor low-temperature properties.
2. Some TPEs had poor solvent resistance or poor weather resistance.
3. High oxygen permeability (a disadvantage or an advantage, depending on the application) characterized some of these materials.
4. In some cases the costs were quite high compared to conventional rubber stocks.

In spite of the disadvantages, however, a number of "plastics" shops are now utilizing the thermoplastic elastomers to turn out parts formerly thought of as being exclusively in the realm of rubber fabricators. The rubber industry has also improved its product and processes, so that today conventional rubber is being molded in injection presses at speeds considerably above those of a few years ago.

Before describing the characteristics of the various rubber or elastomer types, it is necessary to define the characteristics of elastomers in general so that the differences between plastics and elastomers will be clearly understood. In addition, there are words and phrases peculiar to the world of elastomers which should be explained.

ELASTOMER CHARACTERISTICS

Hardness. Possibly the most distinguishing characteristic of elastomeric materials is their wide range of hardnesses. As applied to elastomers, hardness refers to the resistance of the material to indentation by means of an instrument known as a Shore durometer–ASTM D2240 (see Figures 7-8 through 7-10). The Shore durometers are available with two scales: the A scale for common rubber stocks and the D scale for harder rubbers and soft plastics.

Most rubber stocks are rated on the A scale from a hardness of approximately 30 to as high as 100 for hard rubber. While the 30–100 range is usual for elastomers, there is one material, Norsorex polynorborene from American Cyanamid, which can be obtained as low as 10 on the A scale or up as high as 40 with appropriate fillers and processing. Table 7-1 shows the approximate Shore A hardnesses of a number of familiar rubber articles.

Another durometer scale overlapping the A scale at approximately 85 and extending it to higher values of hardness is the Shore D. Figure 7-11 illustrates the Shore A and D scale relationships and also shows where the thermoplastic elastomers fit in–between the harder thermoset rubbers and the softer, flexible thermoplastic polymers. As shown, there is considerable overlap among the different materials, thus accounting in part for the occasional con-

Fig. 7-8. Quadrant style durometer. (*Courtesy: Shore Instrument Mfg. Co., Inc.*)

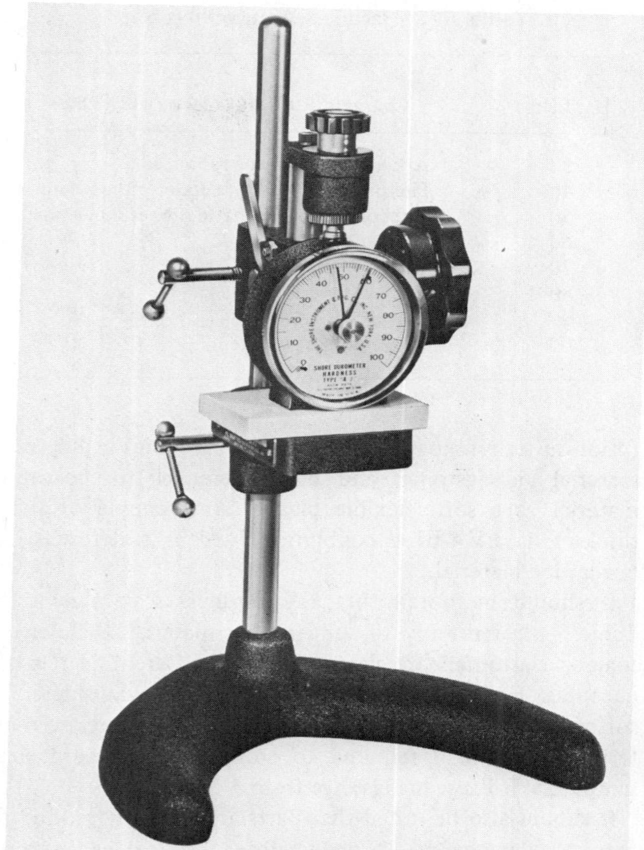

Fig. 7-9. Stand mounted Shore "A" durometer. (*Courtesy: Shore Instrument Mfg. Co., Inc.*)

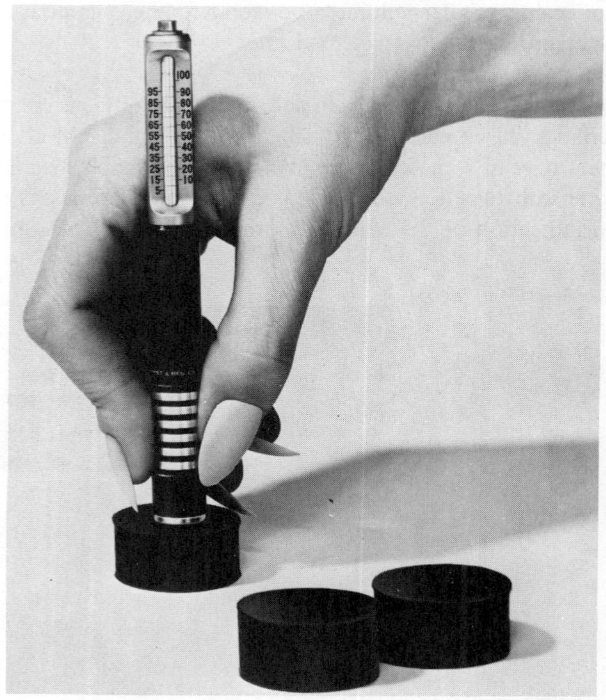

Fig. 7-10. Pencil type durometer–for use in confined spaces. (*Courtesy: Shore Instrument Mfg. Co., Inc.*)

Table 7-1. Hardness Approximations.

Shore A Hardness	Common Rubber Articles (when new)
30	Artgum eraser, crepe rubber sole
40	Pinkpearl or Parawhite eraser, rubber band
50	Rubber stamp, synthetic rubber inner tube
60	Rubber heel, tire tread
70	O-ring, Ruby eraser
80	Typewriter eraser
90	Typewriter roll
100	Battery case, pipe stem

fusion in terminology; that is, some authorities calling one material an elastomer and others referring to the same material as a soft, flexible plastic. An example of such thinking is EVA-EEA copolymer, which is definitely a borderline material.

It should be noted that the hardness scale shown in Table 7-1 refers only to solid rubber material. Cellular or foamed elastomers are also available in different degrees of hardness. However, since these materials are considerably softer than solid rubber stocks, their hardnesses are given in terms of the load required to compress a standard size sample 25%. These loads range from 5 to 150 psi.

It should also be noted that parts are normally produced with a tolerance of ±5 on whatever nominal hardness is requested. Another characteristic of rubber parts is that solids are incompressible. Under load, solid rubber parts will deform if unconfined, but will not compress to a noticeably smaller volume. Sponge or foamed material can be compressed to a smaller volume.

Damping and Resilience. In addition to the obvious characteristics of extensibility, resilience, and snapback, elastomers may in some cases act as dampers or shock absorbers to absorb the energy of shocks or vibration. Hardness is *not* an indication of the energy-absorbing ability of the rubber stock. The damping ability of a rubber material is expressed in terms of percent Yerzley resilience (ASTM D945). The lower the Yerzley number, the better the damping properties. Rubber stocks used for damping usually run from 50 to 60% Yerzley resilience.

Compression Set. Since many rubber parts are used in a compression mode, the compression set properties are particularly important. *Compression set* refers to the permanent deformation which takes place in a gasket, seal, or O-ring after prolonged use. The figure which is given for compression set is the amount of permanent set in the rubber compared to the original deformation. For example, an 0.250-inch-thick gasket is compressed to 0.200-inch thickness, or 20%. (The gasket is unconfined.) After some period of use it is removed, and the final thickness is 0.230 inch. The compression set is then:

$$\frac{0.250 - 0.230}{0.250 - 0.200} \times 100 = 40\%$$

Most rubber materials show compression sets ranging from 25 to 45%. In stating the compression set, the data must include the exposure temperature, time, and amount of compression as well as the final set; otherwise the compression set figure is meaningless. In stating rubber formulation characteristics, the compression set is usually given as performed in accordance with ASTM Test D-395. Method B. In this test, a special compression set test "button" is compressed in a fixture a prescribed amount (usually 25%) and held in an oven for a period of time at an elevated temperature, as agreed upon (see Figure 7-12). By running the test at an elevated temperature (which varies for different rubber stocks) the test is, in effect, an accelerated aging test, providing some idea of the material's long-term aging properties. For gaskets, seals, and similar applications, a compression set test is one of the most important tests which can be made on the material. (Method A, infre-

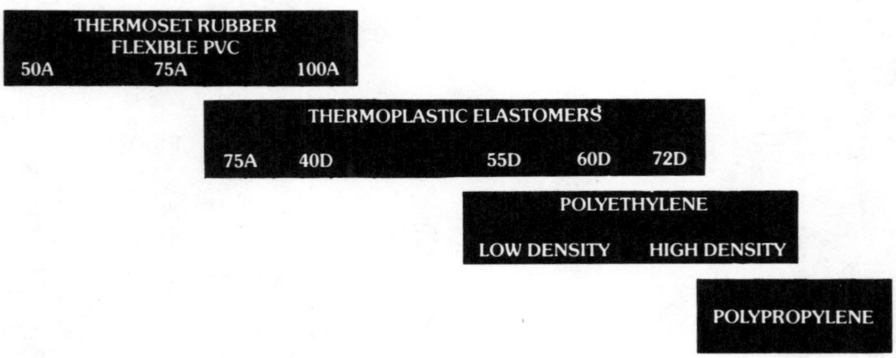

Note: Letter designations refer to Shore hardness tests.

Fig. 7-11. Durometer scale relationships and hardness ranges of the thermoplastic elastomers.

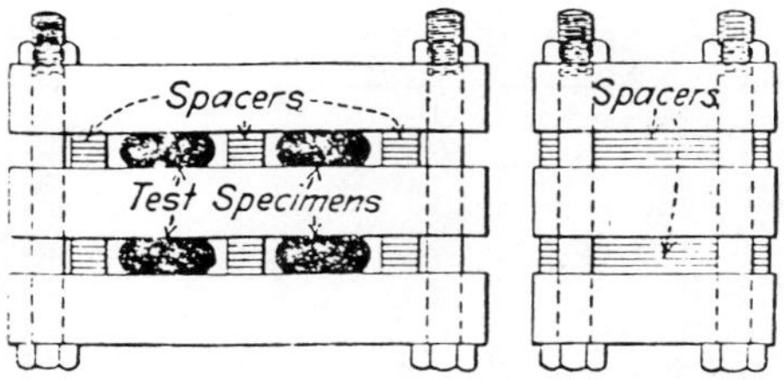

Fig. 7-12. Device for compression set test under constant deflection, Method B. (*Courtesy: ASTM D395*)

quently used, subjects the button to a constant *load* by a spring or weights, rather than a constant *deflection* as in Method B. The latter test is considered to be "more true to life" for seals and gaskets.)

Tensile Strength, Modulus, and Elongation. The tensile strength and elongation of rubber are determined in the conventional manner, using small dumbbell samples which are pulled to failure. The cross-sectional area measurements, made before stretching, are used to determine the tensile strength in a manner similar to that used in determining the strength of metals or plastics.

The meaning of tensile modulus when applied to rubber is different from that term's meaning when applied to plastics. The tensile modulus of a plastic indicates a ratio of stress to strain. Tensile modulus of rubber, however, is always expressed as the load required to stretch the rubber a certain percentage. For example, a 100% modulus of 200 psi is the load required to stretch the rubber 100%. A material with a 300% modulus of 500 psi requires 500 psi to stretch it 300%.

Elongation is expressed in percent similar to plastics; however; the figures are always much higher. Elongations range from 200 to 700% as a rule. It should be noted that the elongation figure always represents an additional amount which is added to the original length; 300% elongation means that the specimen has been stretched to four times its original length.

Glass Transition Point and TR10. The glass transition point, T_g, also known as the glass transition temperature, is that temperature at which the rubber becomes a glassy, brittle solid. Since rubber parts should not be operated or subjected to flexure at such a low temperature (although they may be stored at or below the T_g), a more meaningful characteristic, for a designer, is the TR10 temperature.

The TR10 temperature is obtained by stretching a sample of rubber 250% (or some other agreed-upon amount), freezing it at some temperature below the T_g, and then uniformly warming the sample until it retracts 10% of the original elongation. This temperature is a realistic figure of what the minimum operating temperature might be for an actual part. The actual operating temperature might be higher or lower depending on the part geometry and the flexibility required during operation.

MAJOR ELASTOMER TYPES

The usual practice when classifying *thermoset* elastomers is to classify them into five types depending on the major characteristics of the base polymer. The classifications are: non-oil-resistant materials, oil-resistant materials, heat-resistant materials, weather-resistant materials, and specialty elastomers. The major characteristic of the polymer also determines its type of application. Most of the thermoset elastomers are available in hardnesses from 40 to 80 Shore A; some are available considerably softer and/or harder.

Thermoplastic elastomers are divided into four different types (at the time of writing). These divisions, however, are based on the chemistry of the polymer, and not particularly on its physical or chemical characteristics. Thus, the thermoplastic elastomers are divided into the following families: styrene-butadiene block copolymers, olefinic rubbers, thermoplastic urethanes, and copolyesters. (Some ethylene–vinyl acetate copolymers and ionomers show "rubbery" characteristics and are considered borderline elastomers. These are discussed in Chapter 5.) While the basic chemistry of each family is different, they are similar in that they can all be injection molded, and are all softened appreciably at temperatures which the thermoset elastomers can easily withstand. The olefinics, the urethanes, and the copolyesters are mainly available in hardnesses at the high end of the Shore A or low to high Shore D scale (at the time of writing). The styrene block copolymers can be obtained in low Shore hardnesses down to 30 Shore A.

Styrene-Butadiene Block Copolymers (SBS)

The first of the thermoplastic elastomers was produced by the Shell Chemical Company in the early 1960s. The term block copolymer refers to the fact that the polymer is made from two or more blocks of dissimilar molecules. In the first TPE, the copolymer consisted of a butadiene center with a styrene block at each end. The result was an orderly three-block molecule—the butadiene acting as an elastomeric center with a glass transition temperature well below room temperature, and the styrene acting as a rigid plastic with a transition point of approximately 100°C (200°F). Thus, with the correct proportion of butadiene to styrene (≈70:30), a material was produced which was an elastomer at room temperature, but on heating above 100°C (200°F) would melt and flow. Figure 7-13 shows the simple three-block SBS molecule.

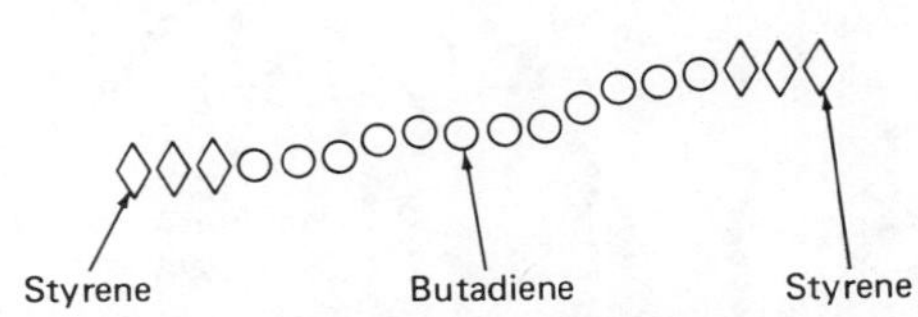

Fig. 7-13. Simple three-block styrene-butadiene-styrene molecule. (*Courtesy: Shell Chemical Co.*)

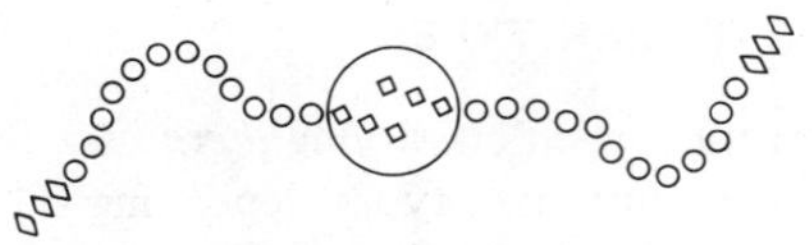

Fig. 7-14. (*Courtesy: Shell Chemical Co.*)

Because of the thermodynamic incompatibility of the butadiene and styrene end-blocks, the styrene blocks in differing molecules associate in submicroscopic particles, either in small groups (Figures 7-14 and 7-15) or in larger groups to form three-dimensional structures (Figure 7-16). The styrene groups are called *domains*, and the continuous phase occupied by the butadiene moiety is called the *rubber network*. The rubber phases, or chains, are thus chemically bound to the styrene end-blocks and are, at the same time, intertwined and entangled with other rubber chains and immobilized by the domain formation. However, since the crosslinking is physical rather than chemical, simple heating of the polymer above the styrene T_g results in melting, allowing injection molding, extrusion, blow molding, etc.

The first block copolymers were made, as described, using butadiene as the rubber phase. Shortly afterward, copolymers were also made using isoprene as the elastic phase. Both of these blocks, it should be noted, are unsaturated, resulting in materials having a tendency toward further reaction; thus, ultraviolet, oxygen, and ozone resistance is poor. The newer block copolymers are now made with a saturated center of ethylene-butylene, resulting in a material with superior weathering properties.

The SBS copolymers, in addition to being processable as thermoplastic elastomers, can also be used very satisfactorily as modifiers for low- and high-density polyethylene, polypropylene, and polystyrene. In most cases, the SBS is used as a toughening agent for the parent polymer. The

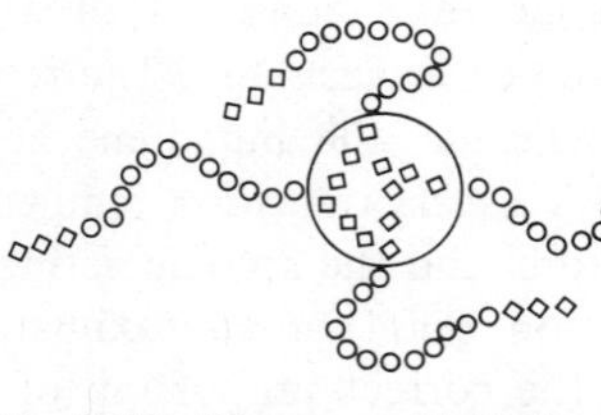

Fig. 7-15. (*Courtesy: Shell Chemical Co.*)

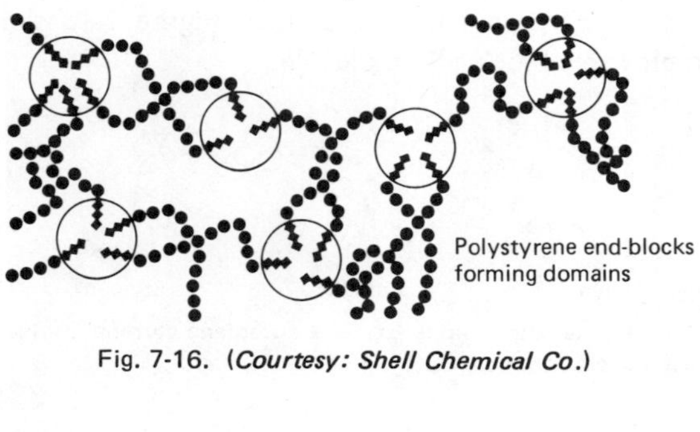

Fig. 7-16. (*Courtesy: Shell Chemical Co.*)

Table 7-2. Typical SBS Properties.

PROPERTY	RANGE
Hardness, Shore A	25 to 95
Flexural modulus, psi	500 to 20,000
300% modulus, psi	100 to 1800
Elongation at break, %	100 to 1200
Tensile strength, 80°F, psi	500 to 5000
Resilience, %	20 to 85
Specific gravity	0.9 to 1.2[a]
Deformation temperature, °F	100 to 270
Tear strength, lb/in.	Up to 500
Brittleness temperature	Down to <-100F
Volume resistivity, ohm-cm	Up to 10^{16}
Dielectric constant, 60 Hz to 1 MHz	2.4 to 3.4
Dissipation (power) factor, 60 Hz to 1 MHz	0.001 to 0.009

[a]Most common range; can be wider for special compounds.

SBS materials (there are a number of types and hardnesses) can also be modified by the incorporation of various fillers, plasticizing oils, other resins, notably styrenes, and, to a lesser extent, olefinics.

In summary, the SBS block copolymers are used as thermoplastic elastomers having a range of hardnesses, good low-temperature performance -50°C (-60°F), rubbery feel, high coefficient of friction, resistance to cutting and cracking, and good abrasion resistance. They are also among the lowest priced of the TPEs. Their major disadvantage—poor elevated-temperature resistance—is said to be improved in the latest, unsaturated polymers. Another disadvantage, high compression set, is also claimed to be improved in the newer materials. Resistance to water, humidity, most dilute acids and bases, and some polar solvents is good. Aliphatic and especially aromatic hydrocarbons attack the block copolymers. Principal uses, at the time of writing, are for shoe soles; however, a number of automotive applications are being sought. These include sight shields, bumper guards, and kick plates, where low costs and good paintability make the use of these copolymers particularly attractive. Table 7-2 summarizes the typical property values (except chemical and environmental) for the block copolymers.

Olefinic Rubbers

The olefinic elastomers were first introduced by the Uniroyal Chemical Company in 1972 as mechanical blends of ethylene-propylene rubber (EPR) or ethylene-propylene-diene (EPDM) and polypropylene. Since the introduction of the first materials, several variations on the original theme have been developed by a number of producers. Table 7-3 summarizes the physical and mechanical properties of one group of materials, the Telcars produced by the B.F. Goodrich Company.

Like the SBS polymers, all the olefinics are available as pellets or granules which are usually molded in the as-

Table 7-3. Typical Properties of Telcar Olefinics.*

	ASTM METHOD	TELCAR TR 100	TELCAR TR 101	TELCAR TR 301	TELCAR TR 302	TELCAR TR 400	TELCAR TR 402	TELCAR TR 405
Specific gravity	D-792	0.89	0.89	0.84	0.84	0.91	0.93	0.88
Hardness, Shore A	D-2240	93	95	83	71	90	78	54
Shore D		45	55	–	–	–	–	–
Original properties	D-412							
Stress strain								
Yield tensile strength, psi.								
72°F (22°C)		1780	2560	400-550	300-400	585	320	200
100°F (38°C)		–	–	450	250	440	250	100
150°F (66°C)		–	–	175	60	260	130	–
Ultimate tensile strength, psi.								
72°F (22°C)		1800	2150	900-1150	800-1000	2575	1250-1500	650
100°F (38°C)		–	–	500	300	820	365	175
150°F (66°C)		–	–	175	75	305	157	–
250°F (121°C)		>390	>750					
Ultimate elongation, %.								
72°F (22°C)		290	70	750-850	750-850	870	900	1000 NB
100°F (38°C)		–	–	900	900	900	900	NB
150°F (66°C)		–	–	900	–	NB	NB	–
250°F (121°C)		>620	>620	–	–	–	–	–
Modulus at 100% elongation, psi.								
72°F (22°C)		1780	–	450	310	600	360	200
Modulus at 300% elongation, psi.								
72°F (22°C)		–	–	550	400	700	375	200
Flexural modulus, psi.								
-20°F (-29°C)		76,000	162,000	22,000	12,500	40,000	10,150	3750
0°F (18°C)		–	–	16,000	8000	15,000	7200	2300
72°F (22°C)		23,000	53,600	4500	5000	4000	2450	900
100°F (38°C)		–	–	2750	1225	2500	1080	216
160°F (71°C)		5800	22,600	–	–	–	–	–
Stiffness, psi.								
-20°F (-29°C)	D-747	93,500	155,000	–	–	–	–	–
72°F (22°C)		33,700	64,000	–	–	–	–	–
160°F (71°C)		12,700	22,600	–	–	–	–	–
Tear strength, Die C, ppi.								
72°F (22°C)	D-624	450	635	290	175	310	185	110
250°F (121°C)		105	240	–	–	–	–	–
Vicat softening point, °C	D-1525	59	91	47	40	53	43	30
Compression set Method B, %	D-395							
22 hours at 72°F (22°C)		45	48	37	32	24	28	30
22 hours at 158°F (70°C)		84	52	92	90	70	58	67
Brittleness temperature, °C	D-746	<-80	-58	-65	-65	-60	-60	-65
Clash-Berg stiffness, psi	D-1043							
Modulus of rigidity at								
-40°C (-40°F)		–	–	18,000	8000	14,000	4500	1450
-35°C (-31°F)		27,000	67,000	–		–	–	
-10°C (14°F)		–		3400	1400	–	–	–
T_f°C		-41	-15	–	–	–	–	–

*Courtesy B. M. Walker, *Handbook of Thermoplastic Elastomers*, Van Nostrand Reinhold, 1979.

received condition. The molded or extruded materials have good tensile strengths and elongations, and excellent low-temperature flexibility in most grades. They have the lowest specific gravities of the TPEs, 0.9 to 1.02. All have good to excellent weathering properties and oxygen and ozone resistance, and, with the addition of UV absorbers, good UV resistance. The olefinics exhibit good chemical resistance to water, dilute acids and bases. Some aliphatics attack the olefinics, and all aromatics and halogenated solvents attack the materials, causing bad swelling and loss of strength.

Applications for the olefinics include the same automotive applications as the SBS materials. Other typical uses include wire and cable insulation and mechanical applications, where a broad temperature range from -50° to 120°C (-60° to 250°F) make these polymers particularly desirable. Flexibility, feel, and good colorability are reasons for their use in many types of sporting goods.

Thermoplastic Polyurethanes (TPUs)

Polyurethane elastomers were an outgrowth of the urethane foam chemistry which was developed in Germany in World War II and later in the United States. The typical reaction to produce a urethane elastomer is similar to that used to produce foams; that is, an isocyanate is reacted with a polyol. The isocyanate usually used is one known as MDI (diphenylmethane-4, 4′-diisocyanate). The polyol is either an aliphatic polyester or a polyether. At a similar hardness, the polyester urethane will exhibit better hardness, toughness, and oil resistance. The polyether type will always have better hydrolytic stability and fungus resistance and will generally show better low-temperature flexibility and higher resilience.

Because of the tremendous number of polyester and polyether materials available and the numerous combinations which can be made, a great many formulations are possible. One characteristic shared by all the TPUs is the requirement for *thorough drying* of the pellets before molding or extruding. The many formulations of the urethanes result in elastomers ranging in hardnesses from 65 Shore A (definitely elastomeric) to as high as 80 Shore D (in the plastics region). Elongations can be as high as 750% for the softer materials. These properties are accompanied by tensile strengths ranging from 5000 to as high as 8000 psi—by far the highest of any elastomeric material, thermoset or thermoplastic. The urethanes also exhibit excellent abrasion resistance, again the highest of the TPEs. Compression set values, up to 100°C (212°F), are also very good.

The outstanding abrasion resistance of the urethanes is demonstrated in Table 7-4 which shows the comparative Taber resistance, on a number of different polymeric materials, run with a CS-17 wheel, 1000-gram load at 23°C (73°F).

In addition to the outstanding abrasion resistance of all the urethanes, mention should be made of the good cut resistance and load-bearing capacity of these very versatile polymers. It was this combination of properties which helped the revival of skate boards and roller skates, the wheels of which are a major end use for the urethanes.

Urethanes in general show good resistance to ozone and

Table 7-4. Taber Abrasion Tests.*

MATERIAL	LOSS (mg)
Urethane	0.5-3.5
Nylon 6/10	16
Polyester film	18
Nylon 11	24
High-density polyethylene	29
Polytetrafluoroethylene	42
Nitrile rubber	44
Nylon 6/6	49
Low-density polyethylene	70
High-impact PVC	122
Natural rubber	146
Normal-impact PVC	160
Styrene butadiene rubber	177
Plasticized PVC	187
Butyl rubber	205
ABS	275
Polychloroprene	280
Polystyrene	324
Nylon 6	366

*Courtesy B. M. Walker, *Handbook of Thermoplastic Elastomers*, Van Nostrand Reinhold, 1979.

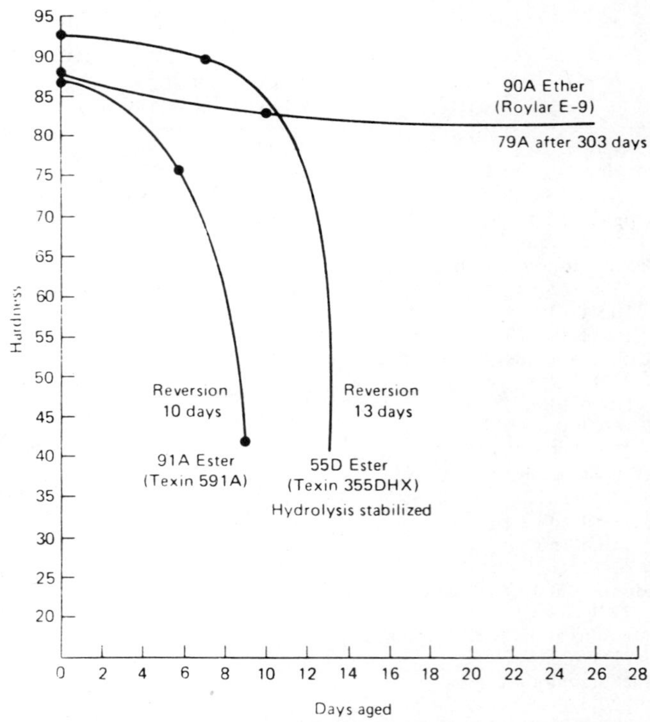

Fig. 7-17. Hydrolysis comparison of ether and ester-based TPU versus days ages at 95% RH and 200°F. (*Courtesy: B. M. Walker, Handbook of Thermoplastic Elastomers, Van Nostrand Reinhold, 1979*)

oxidation, with the polyester types being somewhat the better for oxidative resistance, particularly at the higher temperatures. Both types must be protected by UV absorbers—or better yet, by opaque pigments—if maximum resistance is to be secured. The definitely superior hydrolytic resistance of the polyether type over polyesters is shown in Figure 7-17.

Both urethane types show good resistance to weak acids and bases, oils, and aliphatic solvents. Most aromatics and halogenated solvents attack both polymers. The general chemical resistance of the urethanes is shown in Table 7-5.

The very wide spread of hardnesses available in the urethanes is illustrated in Figure 7-18. Typical physical properties are shown in Table 7-6.

Because of the wide spread of hardnesses and the many other good properties of these versatile polymers, it would be expected that their applications range would be extremely wide. Claims have been made that urethanes have replaced metals and a number of rigid or semi-rigid plastics in a variety of applications, as well as being used in many cases where only urethane's combination of properties made for success. Figure 7-19 shows some products made of urethane and Table 7-7 lists additional uses.

Polyester Thermoplastic Elastomers

Polyester elastomers, at the time of writing, were being produced solely by the Du Pont Company in hardnesses from 40 Shore D to 72 Shore D. They, like the harder urethanes, fill the gap between the "really rubbery" ma-

Table 7-5. Chemical Resistances of the Urethanes.*

CHEMICAL RESISTANCE	POLYESTER TYPES	POLYETHER TYPES
Acids		
Formic, 20%	Poor	Poor
Sulfuric 20%	Fair-Good	Fair
Sulfuric, 30%	-	Fair
Alcohols		
Isopropyl	Poor	Fair-Poor
Methyl	Fair	-
BASE		
Sodium hydroxide, 20%	Fair-Poor	Fair
Detergent		
Mr. Clean	Good	-
Fuels		
ASTM Fuel A	Good	Good
ASTM Fuel B	Fair-Good	Fair
ASTM Fuel C	Fair	Fair-Poor
Gasoline, 100 Octane	Good	Fair
Kerosene	Good	-
Glycols		
Ethylene	Good	Good
Ethylene/water 50/50	Good	Good
Propylene	Good	Good
Propylene/water 50/50	Good	Good
Oils		
ASTM #1	Excellent	Good
ASTM #2	Excellent	Good
ASTM #3	Excellent-Good	Good-Fair
Brake Fluid Type A	Fair-Poor	Poor
Detergent 20W	-	Good
Non-Detergent 20W	-	Good
Skydrol Type B	-	Dissolves
Transmission Type A	Excellent	Good
Plasticizer		
Didecyl adipate	Good	-
Dioctyl adipate	Good	-
Dioctyl phthalate	Good	Fair
Tricresyl phosphate	-	Poor
Salt solutions		
Calcium chloride, Saturated	Good	-
Sodium chloride, Saturated	Good	Good
Synthetic perspiration	Good	Good
Solvents		
Benzene	Fair-Poor	Poor
Carbon tetrachloride	Fair	Poor
Cyclohexanone	Dissolves	Dissolves
Dimethyl formamide	Dissolves	Dissolves
Dimethyl sulfoxide	Dissolves	Dissolves
1,4 Dioxane	Dissolves	Dissolves
Ethylene dichloride	Poor	-
Methyl ethyl ketone	Fair-Poor	Poor
N-Methyl-2 Pyrrolidone	Dissolves	Dissolves
Perchloroethylene	Fair	Poor
Pyridine	Dissolves	Dissolves
Tetrahydrofuran	Dissolves	Dissolves
Toluene	Fair	-
Trichloroethylene	Poor	Poor
HYDROLYSIS RESISTANCE		
Water immersion		
23°C	Poor-Fair	Good
70°C	Poor	Fair-Good
DRY HEAT RESISTANCE		
Properties:		
At elevated temperature	Fair-Good to 130°C	Fair-Good to 80°C
After aging at elevated temp.	Fair-Good to 100°C	Fair-Good to 100°C

WEATHERING	UNPIGMENTED	PIGMENTED	UNPIGMENTED	PIGMENTED
Outdoor				
Arizona	Poor	Fair-Good	Poor	Good
Florida	Poor	Fair-Good	Poor	Good
Ohio	Poor	Fair-Good	Poor	Good
Accelerated				
Fade-Ometer	Poor	Fair-Good	Poor	Good
Weatherometer	Poor	Fair-Poor	Poor	Good

*Courtesy B. M. Walker, *Handbook of Thermoplastic Elastomers*, Van Nostrand Reinhold, 1979.

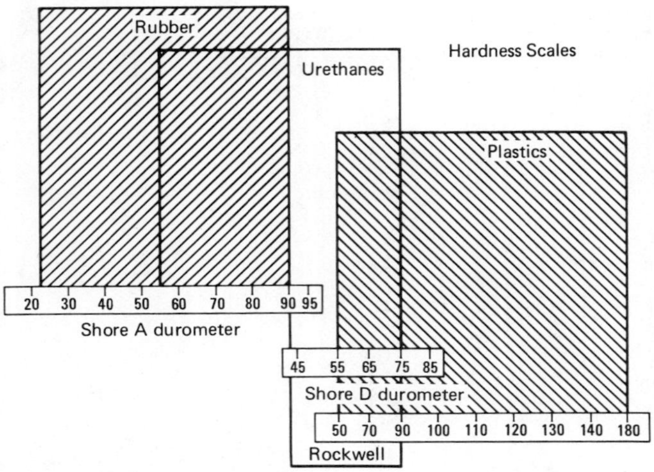

Fig. 7-18. Hardness range of the urethanes. (*Courtesy: B. M. Walker, Handbook of Thermoplastic Elastomers, Van Nostrand Reinhold, 1979*)

Fig. 7-19. The uses of urethane. (*Courtesy: The Upjohn Company*)

terials and flexible plastics shown in Figure 7-11. The materials, designated as Hytrel by Du Pont, consist of random block copolymers in which amorphous elastomeric polyalkylene ether terephthalate soft segments are joined to crystalline tetramethylene terephthalate units. Like the styrenic block copolymers, the crystalline phase acts as the reversible crosslinker, and the proportion of the crystalline phase determines the hardness, softening point, modulus, and so on. The higher the crystalline portion, the higher will be the hardness and the other properties dependent on hardness.

The mechanical properties of the polyester elastomers are in some respects quite unique: under certain tests the material acts very "rubbery," yet when tested for another property it acts more like a flexible plastic. In attempting to assess the comparative properties of the polyesters, the materials of reference are usually the urethanes, in similar hardnesses. Thus, as compared to a urethane of the same

Table 7-6. Physical Properties of Cyanaprene Thermoplastic Polyurethanes.*

Property	Test Method	1880	1890	1850	1857
Hardness, Shore A or D	ASTM D2240	80A	90A	50D	57D
Ultimate tensile strength, psi	ASTM D412	6000	5500	5500	8000
Tensile modulus, psi	ASTM D412				
at 100% elongation		700	1100	1800	2250
at 200% elongation		900	1600	2300	2850
at 300% elongation		1400	2200	3000	3850
Ultimate elongation, %	ASTM D412	750	700	650	600
Elongation set at failure, %	ASTM D412	15	35	45	37
Compression modulus, psi	ASTM D575				
at 5% deflection		100	200	400	410
at 10% deflection		250	450	800	990
at 25% deflection		750	1250	2250	2960
Compression set, %	ASTM D395-B				
22 hours at 158°F, 25% deflection		28	27	30	29
22 hours at 212°F, 25% deflection		42	41	40	41
22 hours at 250°F, 25% deflection		88	59	69	86
Abrasion resistance, taber	ASTM D1044				
H-18 wheel, mg per 1000 cycles		10	35	46	156
Tear strength, Die C, pli	ASTM D624	550	720	820	800
Brittle point, °F	ASTM D-746	-90	-90	-90	-90
Impact resilience (percent rebound)	ASTM D-2632	37	29	31	21
Abrasion, NBS	ASTM D-1630	413	840	780	340
Coefficient of linear thermal expansion, 10^{-5} °F	ASTM D696	11.2	11.2	11.2	6.4[1] 8.3[2]
Specific gravity	ASTM D792	1.25	1.25	1.27	1.27

*Courtesy American Cyanamid Co.
[1] At 15°C [2] At 100°C

Table 7-7. Typical Urethane Applications.*

Toplifts	Tractor parts
Football shoe cleats	Sewing machine parts
Football shoe soles	Textile equipment parts–loom spindle gaskets, loom picker
Automotive front end seals	Covers for grease fitting
Automotive steering gear parts	Flexible couplings
Automotive gaskets	Pump impellers
Automotive assembly line equipment	Fastening devices
Industrial O-rings, diaphragms, and seals	Shock absorbing bumper (automotive)
Military and space agency cable jacket	Aircraft fuel connections
Atomic Energy Commission cable jacket, wire insulation and transparent tubing	Protective caps for connecting rods for automotive engines
Jacket for commercial geophysical cable	Drive wheels for band saw
Industrial extruded tubing	Molded treads on belt sander drive assembly
Jackets for offshore oil well hydraulic bundles	Cattle name tags
Solution coating for synthetic leather	Artificial hip bones for dogs
Hammer head tip	Endcaps for fluorescent lights
Drive wheels for snowmobiles	Electrical and electronic connectors
Industrial and consumer appliances using TEXIN gears	Bearings
Use of MERLON®/TEXIN composites	Computer parts
Camera and projector parts–mostly gears	Scraper rings for hydraulic equipment
Snow chains for automobiles	Jacket for wire rope
Molded drive belts	Sheet and film
Conveyor belt facings	Compounded with resins such as ABS and vinyl for vacuum cleaner hose, etc.
Vacuum formed parts	Windshield wipers
Jackets for flexible hydraulic hose	Custom profile extrusions
Tubing for handling abrasive solids and liquids	Jacket for glass rope
Back-up pads for disc sanders	Skid blocks for gas pipelines
Material handling rollers	
Horseshoes	

*Courtesy The Upjohn Company.

hardness, the polyesters, at room temperature, exhibit lower tensile strengths and lower elongations. However, the polyesters show considerably higher tensile strengths at temperatures much above the 121°C (250°F) maximum allowed for the urethanes (see Figure 7-20).

As Figure 7-20 shows when the polyesters are strained over approximately 50% they show a definite yield point (as do the urethanes). When the strain does exceed the yield point the polyesters exhibit varying amounts of irreversible strain; that is, a tensile set condition is found. As an example, at 100% elongation the tensile set, depending on the particular compound, can vary from 18% for the softest material to as high as 50% for the hardest polymer. Despite the possibility of tensile set, which would suggest the material is becoming brittle, the polyesters show outstanding flexural resistance, as shown in Table 7-8. In both types of flexing tests listed in Table 7-8, the samples are pierced. The high number of cycles shown for the polyesters indicates a low degree of notch sensitivity.

In most other mechanical properties, the polyester elastomers are comparable to the urethanes (except in ultimate tensile strength) as shown in Table 7-9.

The values shown for the compression set of the varying types in Table 7-9 are from tests conducted in accordance with ASTM D-395, Methods A and B. Method B, in which a test button is compressed 25% for an agreed-upon time and temperature, is the method most in use for materials employed as gaskets or seals. The figures shown for Method B are considerably higher than those shown for the Method A tests. (Method A is a constant load test). The reason for the higher Method B test results is that the polyesters display a hysteresis effect when compressed at such a high (for the polyesters) deflection. Under a constant and a fairly high load, the polyesters show very satisfactory set properties compared to urethanes and plasticized nylon 11, as shown in Table 7-10. Table 7-11 further demonstrates the higher load-carrying abilities (and greater stiffness) of the polyesters as compared to the urethanes.

The low-temperature properties of the polyester elastomers are also somewhat of an anomaly when compared to the urethanes of the same hardness, since they show a higher apparent modulus of rigidity at room temperature and yet at temperatures below -18°C (0°F) they are lower in rigidity, as shown in Figures 7-21 and 7-22.

The environmental and chemical resistance of the polyester elastomers is, in general, quite similar to that of the urethanes. The polyesters, however, because of their higher temperature resistance are also more resistant to long-term heat aging and to the effects of oxidation. Heat aging characteristics can also be improved by the addition of antioxidants. The use of the antioxidants is, however, at the expense of color stability, since the antioxidant will cause

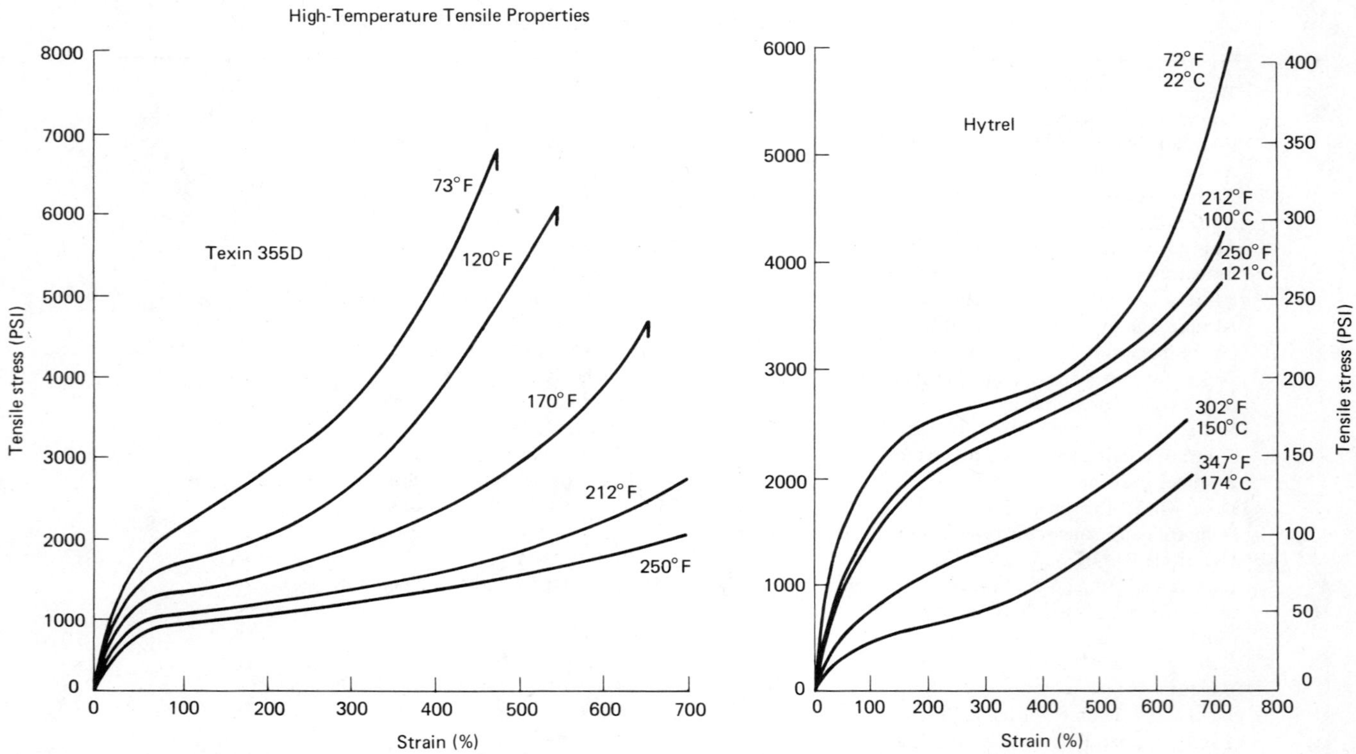

Fig. 7-20. High-temperature properties—Hytrel versus Texin. (*Courtesy: B. M. Walker, Handbook of Thermoplastic Elastomers, Van Nostrand Reinhold, 1979*)

Table 7-8. Flex Resistance of Polyesters and TPUs.*

	Hardness (ASTM D2240)	De Mattia Flex, Pierced (ASTM D813) Kilocycles to Failure 73°F [23°C]	250°F [121°C]	Ross Flex, Pierced (ASTM D1052) Kilocycles to 5X Cut Growth 73°F [23°C]	-40°F [-40°C]
Hytrel	40D (92A)	216	108	>300[a]	>12[b]
TPU-Ester	80A	1.5	1.5	30	Immediate
TPU-Ester	91A	216	18	>300[a]	Immediate
TPU-Ether	90A	3	Immediate	144	Immediate
Hytrel	55D	72	18	>300[a]	>12[c]
TPU-Ester	55D	1.5	Immediate	84	Immediate
Hytrel	63D	–	–	280	Immediate

Flex Fatigue Resistance

[a] No change in the length of the pierced area after 300,000 cycles.
[b] Test terminated after 3X cut growth.
[c] Test terminated after 4X cut growth.

Note: Properties were measured on injection-molded test specimens.
*Courtesy B. M. Walker, *Handbook of Thermoplastic Elastomers*, Van Nostrand Reinhold, 1979.

yellowing of the polymer. For maximum resistance to the effects of ultraviolet radiation, the use of UV absorbers or pigmentation with an opaque colorant is recommended. Hytrel compounds have been found to be quite resistant to the newest environmental factor—nuclear radiation. Neat (uncompounded) formulations of all types withstand 10^7 rads with little apparent damage, as shown in Table 7-12.

As previously mentioned, the chemical resistance of the polyester elastomers is similar to that of the urethanes. Their excellent resistance to oils and greases at elevated temperatures and good flexibility at subzero temperatures make them very attractive for use in automobiles and snowmobiles. For applications such as fuel lines and hoses, where softness and a low modulus of rigidity are not re-

Table 7-9. Typical Mechanical Properties of Polyester Elastomers.*

Property[b]	Units	ASTM Method	Hytrel Polyester Elastomer Hardness (Durometer) Grades 40D	55D	63D	72D
Tensile strength	psi [MPa]	D638	3700 [25.5]	5500 [37.9]	5700 [39.3]	5700 [39.3]
Ultimate elongation	%	D638	450	450	350	350
25% modulus or yield point	psi [MPa]	D638	1100 (M25) [7.6]	2000 (M25) [13.8]	2500 (Yield) [17.2]	3800 (Yield) [26.2]
Stress at 25% compression	psi [MPa]		1800 [12.4]	3850 [26.5]	No data –	No data –
Flexural modulus	psi [MPa]	D790	7000 [48.3]	30,000 [207]	50,000 [345]	75,000 [517]
Resilience, Bashore	%	–	62	53	43	Not applicable
Compression set resistance, 22 hours at 158°F [70°C]						
25% deflection	%	D395B	60[a]	56[a]	Not applicable	Not applicable
Constant load (1350 psi) [9.3 MPa]	%	D395A	27	4	2	2
Tear strength						
Die B	lb/in. [kN/m]	D624	631 [110]	935 [164]	1055 [185]	Not applicable
Die C	lb/in. [kN/m]	D624	700 [122]	900 [158]	850 [149]	Not applicable
Resistance to flex cut growth						
Ross (pierced)	cycles to failure	D1052	$>3 \times 10^5$	$>3 \times 10^5$	2.8×10^5	Not applicable
DeMattia (pierced)	cycles to failure	D813	$>2 \times 10^5$	$>7 \times 10^4$	Not applicable	Not applicable
Notched impact, Izod						
at 75°F [24°C]	ft-lb/in. [J/cm]	D256A	>20 (No [10.6] break)	>20 (No [10.6] break)	>20 (No [10.6] break)	3.9 [2.1]
at −40°F [−40°C]	ft-lb/in. [J/cm]	D256A	>20 (No [10.6] break)	>20 (No [10.6] break)	0.5 [0.3]	0.8 [0.4]
Taber abrasion						
CS-18 wheel, 1000 g load	mg/1000 cycles	D1044	3	5	8	13
Softening point, vicat	°F [°C]	D1525	234 [112]	356 [180]	363 [184]	397 [203]
Heat distortion temperature						
66 psi [0.5 MPa]	°F [°C]	D648	No data	315 [157]	No data	330 [166]
264 psi [1.8 MPa]	°F [°C]	D648	No data	110 [43]	No data	155 [69]
Brittleness temperature	°F [°C]	D746	<−94 [<−70]	<−94 [<−70]	<−94 [<−70]	<−94 [<−70]
Coefficient of linear expansion	in./in./°C [°C^{-1}]	D696	20×10^{-5} [20×10^{-5}]	18×10^{-5} [18×10^{-5}]	17×10^{-5} [17×10^{-5}]	21×10^{-5} [21×10^{-5}]
Water absorption, 24 hours	%	D570	0.6	0.5	0.3	0.3
Specific gravity	–	D720	1.17	1.20	1.22	1.25

[a] Can be improved by annealing.
[b] Properties were measured on injection-molded test specimens.
*Courtesy Dupont Co.

Table 7-10. Compression set Resistance (Measured by ASTM D-395, Method A, 1350 psi [9.3 MPa] load)*

	Hardness (ASTM D2240)	Compression Set, % after 22 hours at: 73°F [23°C]	158°F [70°C]	212°F [100°C]
Hytrel	40D (92A)	11	27	33
TPU-Ester	91A	15	27	44
TPU-Ether	90A	8	24	33
Hytrel	55D	<1	4	8
TPU-Ester	55D	3	9	28
Hytrel	63D	<1	2	4
Plasticized Nylon 11	65D	1	3	5
Hytrel	72D	<1	2	5

*Courtesy B. M. Walker, *Handbook of Thermoplastic Elastomers*, Van Nostrand Reinhold 1979.

Table 7-11. Modulus in Compression or Tension*

	Hardness (ASTM D2240)	Compression Modulus, psi [MPa] (ASTM D695)	Tensile Modulus, psi [MPa] (ASTM D638)
Hytrel	40D (92A)	7400 [51.0]	7800 [53.8]
TPU-Ester	91A	4000 [27.6]	5600 [38.6]
TPU-Ether	90A	4800 [33.1]	5600 [38.6]
Hytrel	55D	12,500 [86.2]	20,000 [138]
TPU-Ester	55D	8800 [60.7]	–
TPU-Ether	55D	–	–
Hytrel	63D	20,000 [138]	50,000 [345]
Hytrel	72D	–	75,000 [517]

Note: Properties were measured on injection-molded test specimens.

quired, the low permeability to fuels, as compared to fuel-resistant thermosetting rubbers, places the polyesters at an advantage. Table 7-13 shows the results of fuel permeability tests of Hytrels and thermoset rubbers.

Table 7-14 is a summarized list of the effects of a number of chemicals on the different Hytrel materials. In general, the chemical resistance of the elastomers increases as the polymer hardness increases.

The electrical properties of none of the TPEs have been discussed, since all but the urethanes have good electrical properties for power frequencies and direct current. None of the TPEs should be considered for high-frequency insulation, except as an outer, abrasion-resistant jacket on a high-frequency cable. Urethanes are somewhat poorer in electrical characteristics, since their tendency to absorb water lowers their dielectric strength and insulation resistance.

Manufacturers of all the polymers discussed are continually bringing out new formulations which presumably will be improvements on the original materials. Furthermore, improvements in any given polymer can be made by the addition of one or more additives or reinforcing fillers. The improvement in any one property, however, should be balanced against the fact that improving one property may cause a deterioration of another. An example might be the

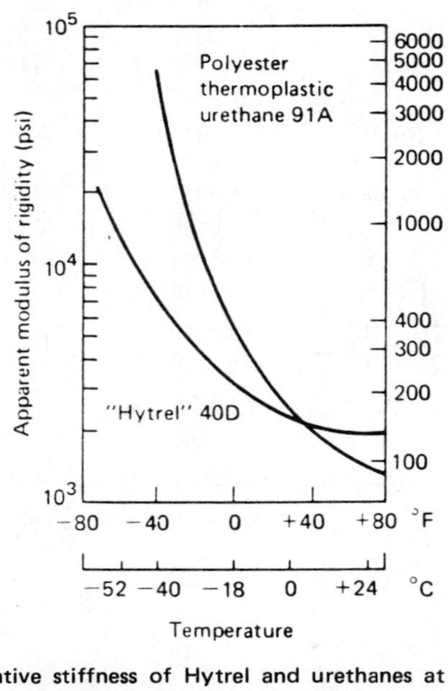

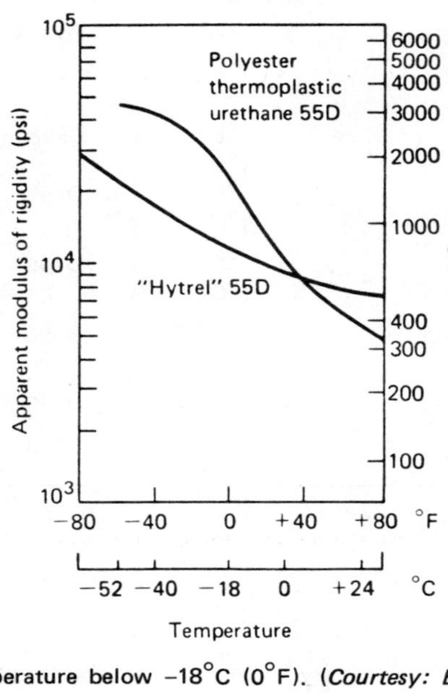

Fig. 7-21 and 7-22. Relative stiffness of Hytrel and urethanes at temperature below −18°C (0°F). (*Courtesy: B. M. Walker, Handbook of Thermoplastic Elastomers, Van Nostrand Reinhold, 1979*)

Table 7-12. Effect of Radiation on Hytrel Formulations.*

Stability of Hytrel Polyester Elastomers to Radiation Electron Beam, 2 Mev., 30°C (86°F)/50% RH, Radiation Dosage in Rads

	Hytrel 40D	Hytrel 55D	Hytrel 63D
Original			
Tensile strength, psi (kgf/cm^2)	3500 (246.0)	5090 (357.8)	5225 (367.3)
Elongation at break, %	610	570	450
100% modulus, psi (kgf/cm^2)	1275 (89.6)	2025 (142.4)	2790 (196.1)
300% modulus, psi (kgf/cm^2)	1620 (113.9)	2660 (187.0)	3750 (263.6)
10^5 Rads			
Tensile strength, psi (kgf/cm^2)	3315 (233.0)	4415 (310.4)	5270 (370.5)
Elongation at break, %	595	525	425
100% modulus, psi (kgf/cm^2)	1185 (83.3)	1980 (139.2)	2750 (193.3)
300% modulus, psi (kgf/cm^2)	1620 (113.9)	2525 (117.5)	3830 (269.2)
10^6 Rads			
Tensile strength, psi (kgf/cm^2)	3500 (246.0)	4280 (300.9)	4910 (345.2)
Elongation at break, %	600	530	430
100% modulus, psi (kgf/cm^2)	1285 (90.3)	1980 (139.2)	2680 (188.4)
300% modulus, psi (kgf/cm^2)	1700 (119.5)	2500 (175.8)	3900 (274.2)
10^7 Rads			
Tensile strength, psi (kgf/cm^2)	3380 (237.6)	4220 (296.7)	5120 (359.9)
Elongation at break, %	535	450	415
100% modulus, psi (kgf/cm^2)	1230 (86.5)	2150 (151.1)	2840 (199.7)
300% modulus, psi (kgf/cm^2)	1760 (123.7)	2680 (188.4)	4000 (281.2)

*Courtesy B. M. Walker, *Handbook of Thermoplastic Elastomers*, Van Nostrand Reinhold, 1979.

use of a reinforcing filler to raise the tensile strength, with a corresponding increase in stiffness.

No particular mention has been made in the foregoing discussion of the processing methods for the various polymers. In general, all the elastomers are designed to be injection molded, extruded, thermoformed, and so on; however, these will be carried out at different temperatures and pressures, depending on the polymer and type.

In order to facilitate comparisons of the thermoset rubbers and the thermoplastic elastomers, Table 7-15 lists the various types by the thermoset designations of non-oil resistant, oil resistant, and so on. Also listed are the operating ranges and the major characteristics of each polymer. When the particular type of elastomer is made by only one supplier, the source is also shown. Following the thermoset elastomers, the thermoplastic elastomers are listed, along with their representative trade names and sources. Each material is entered in the table roughly in the order of usefulness, based on its classification; that is, the best—from the standpoint of performance—is at the top, the poorest (in that ranking) is the lowest. It should be understood that the same material may be shown in two classifications, not necessarily in the same order. The order of rankings does *not*, however, apply to the thermoplastic elastomers, except insofar as they fit into the large classifications.

All the elastomer types can be produced as either solid

Table 7-13. Permeability[a] of Fuel-Resistant Elastomers (ASTM D-814)*

Fuel	55D Hytrel	63D Hytrel	Neoprene[b] WHV	Hypalon[b] 48	Nitrile Rubber Med. High Nitrile
Reference fuel B	0.43 [1.6]	0.024 [0.088]	6.9 [25]	2.7 [9.9]	1.2 [4.4]
Reference fuel C	0.74 [2.7]	0.060 [0.22]	17.4 [64]	8.2 [30]	2.9 [11]

*Courtesy Du Pont Co.
[a] Units of permeability: Fluid oz/24 hr · ft^2 at 73° F, 25 mil film thickness [mm^3/s · m^2 at 23° C, 0.64 mm film thickness]
[b] Reg U.S. Trademark of the Du Pont Company.

Table 7-14. Chemical Resistance of Hytrels.*

Chemical	Rating*
Acetic acid, 20%	A
Acetic acid, 30%	A
Acetic acid, glacial	A
Acetic acid, glacial (100°F −38°C)	B
Acetic anhydride	T
Acetone	B
Acetylene	A
Aluminum chloride solutions	T
Aluminum sulfate solutions	T
Ammonium chloride solutions	A
Ammonium hydroxide solutions	T
Ammonium sulfate solutions	A
Amyl acetate	B
Amyl alcohol	A
Aniline	C
ASTM oil #1 (300°F −149°C)	A
ASTM oil #3 (300°F −149°C)	A
ASTM reference fuel A (158°F −70°C)	A
ASTM reference fuel B (158°F −70°C)	A
ASTM reference fuel C	A
ASTM reference fuel C (158°F −70°C)	B
Asphalt	T
Barium hydroxide solutions	T
Beer	A
Benzene	B
Borax solutions	A
Boric acid solutions	A
Bromine, anhydrous liquid	X
Butane	A
Butyl acetate	B
Butyric acid	T
Calcium chloride solutions	A
Calcium hydroxide solutions	T
Calcium hypochlorite, 5%	A
Carbon bisulfide	B
Carbon dioxide	A
Carbon monoxide	A
Carbon tetrachloride	C
Castor oil	B
Chlorine gas, dry	X
Chlorine gas, wet	X
Chloroacetic acid	X
Chlorobenzene	X
Chloroform	C
Chlorosulfonic acid	C
Citric acid solutions	A
Copper chloride solutions	A
Copper sulfate solutions	A
Cottonseed oil	A
Cyclohexane	A
Dibutyl phthalate	A
Diethyl sebacate	A
Dioctyl phthalate	A
Epichlorohydrin	X
Ethyl acetate	B
Ethyl alcohol	A
Ethyl chloride	C
Ethylene dichloride	C
Ethylene glycol	A
Ethylene oxide	A
Ferric chloride solutions	T
Fluosilicic acid	T
Formaldehyde, 40%	B
Formic acid	B
FREON−11**	T
FREON−12**	T
FREON−113** (130°F −55°C)	A
FREON−114**	T
Gasoline	A
Glue	A
Glycerin	A
n-Hexane	A
Hydrazine	C
Hydrochloric acid, 20%	B
Hydrochloric acid, 37%	C
Hydrocyanic acid	T
Hydrofluoric acid, 48%	X
Hydrofluoric acid, 75%	X
Hydrofluoric acid, anhydrous	X
Hydrogen	A
Hydrogen sulfide	A
Isooctane	A
Isopropyl alcohol	A
JP−4	A
Kerosene	T
Lacquer solvents	B
Lactic acid	T
Linseed oil	T
Lubricating oils	A
Magnesium chloride solutions	T
Magnesium hydroxide solutions	T
Mercuric chloride solutions	T
Mercury	A
Methyl alcohol	A
Methyl ethyl ketone	B
Methylene chloride	C
Mineral oil	A
Naphtha	A
Naphthalene	B
Nitric acid, 10%	B
Nitric acid, 30%	C
Nitric acid, 60%	C
Nitric acid, 70%	C
Nitric acid, red fuming	C
Nitrobenzene	C
Oleic acid	A
Oleum, 20-25%	C
Palmitic acid	A
Perchloroethylene	C
Phenol	C
Pickling solution (20% nitric acid, 4% HF)	X
Pickling solution (17% nitric acid, 4% HF)	X
Potassium dichromate solutions	T
Potassium hydroxide solutions	A
Pydraul 312C	A
Pyridine	X
SAE #10 oil	A
Sea water	A
Silicone grease	A
SKYDROL 500	A
Soap solutions	A
Sodium chloride solutions	A
Sodium dichromate, 20%	T
Sodium hydroxide, 20%	A
Sodium hydroxide, 46½%	B
Sodium hypochlorite, 5%	A
Soybean oil	T
Stannous chloride, 15%	T
Steam (212°F −100°C)	B
Steam (230°F −110°C)	C
Stearic acid	T
Styrene	X
Sulfur, molten	T
Sulfur dioxide, liquid	T
Sulfur dioxide, gas	T
Sulfuric acid, up to 50%	A
Sulfuric acid, 50-80%	C
Sulfuric acid, 60%	C
Sulfuric acid, 90%	C
Sulfuric acid, 95%	C
Sulfuric acid, fuming (20% oleum)	C
Sulfurous acid	B
Tannic acid, 10%	A
Tartaric acid	T
Tetrahydrofuran	B
Toluene	B
Trichloroethylene	C
Triethanolamine	C
Trisodium phosphate solutions	A
Tung oil	T
Water (158°F −70°C)	A
Water (212°F −100°C)	B
Xylene	B
Zinc chloride solutions	A

*Rating Key
A—Fluid has little or no effect.
B—Fluid has minor to moderate effect.
C—Fluid has severe effect.
T—No data—likely to be compatible.
X—No data—not likely to be compatible.

Unless otherwise noted, concentrations of aqueous solutions are saturated.
All ratings are at room temperature unless specified.
**FREON is a registered trademark of E. I. Du Pont de Nemours & Co. (Inc.).

*Courtesy Du Pont Co.

Table 7-15. Major Elastomer Families.

Type	Polymer and Source	Typical Applications	Useful Temperature Range, °F	Major Characteristic
Non-oil Resistant (swelling or decomposition in presence of oil, fuel, etc.)	Natural rubber	Elastic cord, tires	-65 to 220	High resilience, abrasion resistance, tear strength.
	Styrene-Butadiene GRS (thermoset)–various	Tires	-65 to 220	Good abrasion resistance.
	Butyl–Exxon	Inner tubes	-65 to 220	Very low gas permeability.
	Ethylene-Propylene (EPR and EPDM)–various	Automobile weather stripping, seals	-60 to 300	Excellent weather resistance.
	TPEs			
	Styrene-Butadienes: Kraton–Shell Solprene–Phillips	Footware, toys, household items	-80 to 150	Tough, rubbery materials; 35 to 95 Shore A; inexpensive; can be blended with polyethylene or polypropylene to increase toughness.
	Ethylene-Ethyl Acetate (EEA), Ethylene-Vinyl Acetate (EVA)–U.S. Ind. Chem.	Toys, household items, wire insulation, flexible hose, toys, household items, medical devices	-60 to 150	High flexibility; can be clear; can be filled; moderate price.
Oil Resistant (slight swelling in oils, fuels, etc.)	Perfluoroelastomer: Kalrez–Du Pont	Seals, O-rings for special applications	-20 to 550	Excellent high-temperature, oil, and chemical resistance; very expensive; 65–80 Shore A.
	Fluorocarbon: Viton-Du Pont Fluorel–3M Co.	Seals, O-rings for high-temperature applications	-10 to 400	Very good resistance to chemical deterioration; poor low-temperature properties; expensive; 60–80 Shore A.
	Acrylonitrile (Buna N)–various	Hydraulic O-rings, aromatic fuel seals	-65 to 250	Low swell and good strength in oil, fuel, etc.; good abrasion resistance; inexpensive; 40–80 Shore A.
	Neoprene (Chloroprene)–Du Pont	Gasoline hose, cable covers	-20 to 300	High strength, good abrasion resistance, excellent weathering, best general-purpose rubber, 40–95 Shore A, moderate price.
	Polysulfide–Thiokol Corp.	Sealing material	-65 to 225	Good resistance to many solvents, low mechanical and thermal properties, 40–85 Shore A, inexpensive.
	Fluorosilicone–Dow Corning	Seals, O-rings	-120 to 350	Very good flexibility at low temperatures, good resistance to high temperatures, oils and gasolines, expensive, 50–80 Shore A.
	Chlorinated Polyethylene (CPE)–Dow Chemical		-10 to 300	High temperature and combustion resistant, good weathering, 50–95 Shore A, moderate price.
	Acrylic, Ethylene	Automotive transmission seals	0 to 400	Good chemical resistance, poor low-temperature properties, 55 Shore A, moderate price.
	Acrylic: Cyanacryl–American Cyanamid Vamac–Du Pont	Seals	-30 to 250	50 Shore A, moderate price.
	TPEs			
	Polyurethane: Estane–Goodrich Pellethane–Upjohn Plastothane–Thiokol Corp. Roylar–Uniroyal Texin–Mobay Cyanaprene–American Cyanamid	Seals, cable jacketing, abrasion-resistant parts	-65 to 200	High strength and excellent abrasion resistance, 75 Shore A to 60 Shore D, moderate price.

Table 7-15. *(Continued)*

Type	Polymer and Source	Typical Applications	Useful Temperature Range, °F	Major Characteristic
	Polyester: Hytrel–Du Pont	Couplings, gears, mechanical parts	−65 to 300	Very tough, high-performance material; moderate to high-priced; 40–50 Shore D.
	Olefinic Rubbers: Somel–Du Pont, Telcar–Goodrich, TPR–Uniroyal, Vistaflex–Exxon	Household items, mechanical parts	−60 to 250	Medium hardness and performance materials, 50 Shore A to 50 Shore D, inexpensive.
Heat resistant	Perfluoroelastomer: Kalrez–Du Pont	Seals, O-rings	−20 to 550	See above.
	Silicones: Dow Corning, General Electric Co.	Seals, O-rings, etc., not requiring oil resistance	−100 to 450	Available as curable gum stock, RTV materials, and one-part air-curing; excellent weathering and electrical properties; 30–90 Shore A; moderately expensive.
	Fluorocarbon: Viton–Du Pont, Fluorel–3M Co.	Seals, O-rings, etc., not requiring low-temperature flexibility below −20°F	−20 to 400	Poor low-temperature flexibility, high heat resistance, good weathering and electrical properties, good oil resistance.
	Fluorosilicone: Dow Corning	Seals, O-rings, etc., in contact with oil.	−120 to 350	See above, except good low-temperature properties, fair mechanical properties.
	Acrylic: Cyanacryle–American Cyanamid	Auto transmission seals	0 to 450	See above.
	TPEs			
	Polyester: Hytrel–Du Pont	Mechanical parts	−65 to 300	See above.
Weather, light, UV, and ozone resistant	Ethylene-Propylene: Nordel–Du Pont, Epcar–Goodrich, Vistalon–Exxon	Dust seals, fairings exposed to weather	−65 to 300	Excellent weather resistance, poor oil resistance, good thermal and mechenical properties, resistant to phosphate ester fluids, inexpensive.
	Chlorosulfonated Polyethylene: Hypalon–Du Pont	Flexible coatings, tarps, etc.	−50 to 300	Excellent weather resistance, need not be black, relatively inexpensive.
	Silicones and Fluorosilicons–Dow Corning and G.E. Co.	Fairings, seals, etc. requiring flexibility at low temperatures	−100 to 450	See above, expensive.
	Fluorocarbons: Viton–Du Pont	Fairings, seals, O-rings, etc., requiring good weather resistance and chemical resistance	−20 to 400	See above, expensive.
	Neoprene–Du Pont	Cable covers, etc., where good all-around properties are required	−20 to 300	See above, moderately expensive.
	TPEs			
	Polyester: Hytrel–Du Pont	Mechanical parts, seals	−65 to 300	See above.
	Olefinics: Somel–Du Pont, Telcar–Goodrich, TPR–Uniroyal	Seals, noncritical items	−60 to 250	See above.
Specialty Rubbers: Low gas permeability	Butyl	Seals, O-rings, not in contact with oil	−65 to 250	Best for gas retention, medium physical properties, medium cost.
	Fluorocarbons: Viton–Du Pont	Seals, O-rings, in contact with oil and solvents	−20 to 400	See above.
High gas permeability	Silicones and Fluorosilicones	"Breathable" diaphragms and seals	−100 to 450	See above.

Table 7-15. *(Continued)*

Type	Polymer and Source	Typical Applications	Useful Temperature Range, °F	Major Characteristic
Damping and energy absorbing	Silicone, Fluorosilicone, Chlorobutyl Polyurethane	Shock absorbers, mounts, and vibration dampers	-65 to 300	Very low resilience, "dead" materials, moderate physicals. Silicones and fluorosilicones can be used over wider temperature range than chlorbutyl or polyurethane.
Electrically conductive	Silicone filled with electrically conductive particles	RFI shielding	-100 to 400	Conductive materials with fair physical properties.
Thermally conductive	Silicone and other polymers filled with thermally conductive material	Conforming gaskets between hot parts and heat sink	-100 to 450	High thermal condùction and conformance to rough surfaces.
Room-temperature vulcanizable (RTV)	Silicones	Form-in-place seals, coatings, prototype parts	-120 to 450	Many types differing mainly in hardness and mechanical properties.
Thermoplastic elastomers TPEs	Styrene-Butadiene, Ethylene Polymers, Polyesters Urethanes, Olefinic Elastomers	Various parts not to be operated at temperatures over 120-140°F (except polyester, 300°F)	–	Thermoplastics, injection moldable at production rates. Used for high production of light-colored or transparent parts. Polyesters for high-strength harder parts, urethanes for abrasion resistance.
Extremely low Shore	Polynorborene: Norsorex–American Cyanamid	Very soft dampers, etc.	-65 to 160	Very soft to hard, good mechanical properties.

molded parts, raw stock slabs, or as cellular materials. However, not all the polymers were, at the time of writing, being produced as cellular materials. The cellular materials are made in the same manner as in the manufacture of cellular plastics; that is, a blowing agent is incorporated into the polymer before final processing with heat. During the molding (or extruding) operation, the polymer softens, the blowing agent decomposes to liberate a gas which then foams the polymer, and the polymer is then cured (vulcanized) or allowed to cool.

Cellular materials are made as open-celled or closed-celled stock. The open-celled material consists of foamed material whose cells are predominantly interconnected, so the material can readily breathe or act as a sponge to absorb water. Closed-cell material has predominantly nonconnected cells which are essentially gas tight. Closed-cell materials would, thus, tend to change volume with rapid changes in external pressure. Material to specification MIL-C-3133, (MIL-STD-670) Grade TO-7 indicates an open-cell material; Grade TE-7 indicates an expanded, closed-cell stock.

The surface of thermoset cellular parts is usually either mica-covered or rough, with a screenlike or similar pattern on the "skin" of each surface. Such a finish is necessary since, in the expansion of the foam in the mold, it is necessary to have a surface which allows the rubber to "breathe" and get rid of the excess gas. A mica-coated surface or a screen in the mold allows this "breathing." The "skin" can be sliced or ground off, if required, thus exposing the cellular surface. Thermoplastic foams can be made with smooth skins.

FABRICATION TECHNIQUES

In addition to injection, compression, transfer molding, extrusion, and foaming, two types of elastomers (available in liquid form) can be cast to produce useful articles.* The two families are the urethanes and silicones. In both cases, compounds can be formulated to be cured at an elevated temperature or at room temperature. In the case of the silicones, the room-temperature curing materials are called RTVs, (room-temperature vulcanizing materials). The silicones, in addition, are available either as two-part materials which are mixed together just before casting, or as single-component materials which are cured when in contact with atmospheric moisture. Both the urethanes and the silicones can also be formulated to produce room-temperature-cured foams.

Dipping is an inexpensive method of making parts with thin, fairly regular wall thicknesses. It is used for production of parts such as boots, gloves, and fairings. Since only

*A liquid rubber solution based on cis-1,4-polyisoprene DPR (Hardman, Inc., Belleville, N.J.) can be cast into parts with very low Shore A hardnesses (25–40) and tensile strengths of 100 to 150 psi.

Table 7-16. Comprehensive Compilation

Ratings are in this decreasing order: Outstanding, Excellent, Good, Fair, Poor			Non-Oil-Resistant Rubbers			
Generic Name			Natural rubber	Styrene-Butadiene (GRS) (SBR)	Butyl	EP rubber
Weight of Base Elastomer		Lb/in³	0.033	0.034	0.033	0.031
		Specific gravity	0.93	0.94	0.92	0.86
Physical Properties for Elastomer Compounds	Durometer range		30–100	40–100	30–100	30–90
	Resilience		O	G	F	G
	Tensile strength, psi (reinforced)		4000	2000	2000+	2000–3000
	Elongation % (reinforced)		500	450	300–800	500
	Creep, room temperature		E	E	F	F
	Compression set		G	G	F	F
	Electrical resistivity		E	E	E	E
	Impermeability, gas		G	F	O	G
Resistance Properties — Mechanical	Resistance to	Impact	E	E	G	G
		Abrasion	E	E	G	G
		Tear	E	F	G	P
		Cut growth	E	G	E	G
Resistance Properties — Temperature	Tensile Strength; Psi, at	250°F	1800	1200	1000	2000
		400°F	125	170	350	400
	Elongation, %, at	250°F	500	250	250	300–500
		400°F	80	60	80	0–120
	Creep at 212°F		G	G	F	F
	Heat aging at 212°F		G	G	E	E
	Flame resistance		P	P	P	P
	Low Temperature	Stiffening, °F	-20–-50	0–-50	-10–-40	-20–-50
		Brittle pt, °F	-80	-80	-80	-90
Resistance Properties — Environmental	Weather		F	F	E	E
	Oxidation		G	G	E	G
	Ozone		P	P	E	E
	Radiation		F–G	G	P	P
	Water		E	E	E	G–E
	Acid		F–G	F–G	E	G–E
	Alkali		F–G	F–G	E	G–E
	Gasoline, kerosene, etc. (aliphatic HCs)		P	P	P	P
	Benzol, toluol, etc. (aromatic HCs)		P	P	F–G	F
	Degreaser solvents (halogenated HCs)		P	P	P	P
	Alcohol		G	F	E	P
	Synthetic lubricants (diester)		P–F	P	F	P–F
	Hydraulic fluids	Silicates	P	P–F	F	F–G
		Phosphates	P–F	P	G	G–E
Subjective Properties	Taste		F–G	F–G	F–G	G
	Odor		F–G	G	G	G
	Nonstaining		P–G	P–G	G	G
Bonding to rigid materials			E	E	F–E	P
Typical applications			Elastic cord, tires	Tires	Air and gas containers	Weather-resistant parts

Note: O = outstanding; E = excellent; G = good; F = fair; P = poor.

of Rubber Properties.

Oil-Resistant Rubbers				Heat-Resistant Rubbers			
Neoprene	Acrylonitrile (BUNA-N)	Polysulfide	Polyurethane	Silicone	Fluorosilicone	Fluoroelastomers	Acrylics
0.044	0.036	0.048	0.039	0.036	0.049–0.06	0.05–0.07	0.040
1.23	1.00	1.34	1.05	0.95	1.35–1.65	1.4–1.95	1.10
40–95	20–90	20–80	55–100	25–80	40–80	65–90	40–90
E	G	F	G–E	P–E	P–G	F	G
3000+	1000–3500	500–1500	4000–8000	600–1500	800–1350	1500–3000	1700
650–85	400–600	200–550	250–800	90–800	100–480	100–450	450
F–G	G	P	G–E	F–E	F–G	G	F
F–G	G	P–F	E	G–E	G–E	G–E	F–G
F	P	F	G	E	E	G	F
G	G	E	G	P	P	E	G
G	F	P	E	P–F	P–F	P–G	P
G–E	E	P–F	E	P–G	P–G	G	F–G
G	G	P–F	E	P–G	P–G	P–G	F
G	G	P	F–E	P–G	P–G	P–G	G
1500	700	700	1800	850	850	300–800	1300
180	130	Under 25	200	400	350	150–300	225
350	120	140	300	350	160	100–350	400
0–100	20	Undero 25	140	200	150	50–160	150
F–G	E	P	E	E	E	G–E	F
G	G	G	F–G	E	E	O	E
G	P–F	P	P–F	F–G	F–G	E	P–F
+10– –20	+30– –20	–10– –45	–10– –30	–60– –120	–70	+20–+30	+35–+10
–45	–65	–60	–60	–90– –140	–90– –120	+10–+20	–10
E	G	E	E	E	E	E	E
G	F–G	E	E	E	E	O	E
E	P	E	E	E	E	O	E
F–G	F–G	F–G	G	F–E	F	F–G	P–G
G	E	G	G	G	G	G	F
G	G	F	P–F	P–G	G	G–E	F
G	F–G	G	P–F	P–F	G	P–G	P
G	E	E	E	P–F	G	E	E
P	G	E	P–F	P	F	E	P
P	P	F–G	F–G	P–F	F–G	G	P
F	E	G	G	G	G	E	P
P	F–G	G	P	P–F	G	F–G	G
G	F	P–F	–	P	G	G	G
P	P	P–F	P	G	G	P	P
F–G	F–G	P–F	G	G	G	F–G	F–G
F–G	G	P	G	G	G	G	F–G
G–E	P–G	P–F	G	O	O	P–G	G
G–E	G–E	F–G	F–G	F–E	G	P–G	G
Weather- and oil-resistant parts	Gasoline- and oil-resistant parts	Solvent-resistant parts	Abrasion-resistant parts	High- and low-temperature, non-oil-resistant parts	High- and low-temperature, oil-resistant parts	High-temperature solvent-resistant parts	High-temperature, oil-resistant parts

Table 7-16.

Ratings are in this decreasing order: Outstanding, Excellent, Good, Fair, Poor				Weather-Resistant Rubbers	Specialty Rubbers
Generic Name				Chlorosulfonated polyethylene (Hypalon)	Various types
Weight of Base Elastomer			Lb/in³	0.040	Electrically conductive, Thermoplastic, Thermally conductive, Damping, High and low permeability For specific details consult manufacturer
			Specific gravity	1.10	
Physical Properties for Elastomer Compounds		Durometer range		50–95	
		Resilience		G	
		Tensile strength, psi (reinforced)		1500–2500	
		Elongation % (reinforced)		250–500	
		Creep, room temperature		F	
		Compression set		F–G	
		Electrical resistivity		G	
		Impermeability, gas		E	
Resistance Properties	Mechanical	Resistance to	Impact	F–G	
			Abrasion	G	
			Tear	F–G	
			Cut growth	F–G	
	Temperature	Tensile Strength; Psi, at	250°F	500	
			400°F	200	
		Elongation, %, at	250°F	60	
			400°F	20	
		Creep at 212°F		F	
		Heat aging at 212°F		E	
		Flame resistance		G	
		Low Temperature	Stiffening, °F	-30--50	
			Brittle pt, °F	-60	
	Environmental	Weather		E	
		Oxidation		E	
		Ozone		O	
		Radiation		F–G	
		Water		G	
		Acid		E	
		Alkali		E	
		Gasoline, kerosene, etc. (aliphatic HCs)		F	
		Benzol, toluol, etc. (aromatic HCs)		P–F	
		Degreaser solvents (halogenated HCs)		P–F	
		Alcohol		G	
		Synthetic lubricants (diester)		P	
		Hydraulic fluids	Silicates	G	
			Phosphates	P–F	
Subjective Properties		Taste		F–G	
		Odor		G	
		Nonstaining		E	
Bonding to rigid materials				F–G	
Typical applications				Weather-resistant coatings	

(Continued)

Thermoplastic Elastomers					Polynor-Bornene
Styrene-butadiene	Olefinic	Urethanes	Polyester	EVA and EEA	
0.033–0.038	0.031–0.037	0.042–0.045	0.043	0.033–0.034	–
0.93–1.113	0.88–1.02	1.15–1.25	1.20	0.91–0.95	–
35–95A	65–90A, 50D	75A–65D	92A–63D	80A–60D	10A–80A
High	Low	High-Damp	Low	Medium	Medium
500–3000	650–2000	3000–7000	5800–6400	800–4000	2500
400–1350	180–250	200–500	500–800	50–1200	–
–	–	–	–	–	–
P	P–G	F–G	G	P	–
G	High	P	G	F	G
–	–	–	–	–	–
–	–	–	–	–	–
F–G	F–G	O	G	F–G	G
G	G	E	E	F–G	F
–	–	–	–	–	–
–	–	–	–	–	–
–	–	–	–	–	–
–	–	–	–	–	–
–	–	–	–	–	–
Maximum use	–	–	–	–	–
temp, °F	250	200	250–300	140	158
150 P	P	P	P	P	P
–40	–60	–40	–70	–20	–65
–	–76	–	–94	–	–
F–P	E	G–F	G	G–F	G–P
F	E	G–F	G	G–F	G–P
F	E	G–F	G	G–F	G–P
–	–	–	–	–	–
G	E	P	G	E	E
F	G–F	F	G	G	E
F	–	–	G	–	–
Very poor	G–F	E	G	P	F–G
Very poor	P	G	G–F	P	P
Very poor	P	F	P	P	P
Very poor	G–F	F–P	G	F	G
–	–	–	–	–	–
–	–	–	–	–	–
–	–	–	–	–	–
–	–	–	–	–	–
–	–	–	–	–	–
–	–	–	–	–	–
G	P	G	G	F	–
Gaskets, seals, footware	Gaskets, seals, toys	Household items, toys	Gears, industrial parts	Household items, toys	Gaskets, seals, toys

a master (usually a male) is required, inexpensive aluminum tooling can be used. This technique is used in many cases for prototype parts, since neither the physical properties nor the outer surface finish are comparable to parts made in a two-part matched metal mold under high temperature and high pressure conditions. Dipped parts are usually made only of natural rubber, neoprene, or silicone RTV. However, since rubber prototype parts are usually made to check the flexing or other dynamic action of the part, any one of these can be used to simulate the performance of the final part—provided the hardness of the prototype is the same as the final part.

Properties Charts. A comprehensive compilation of the properties of the various polymers is given in Table 7-16. Because thermoset materials are more frequently used, the data in Table 7-16 is somewhat more complete for those than for the thermoplastic materials. Tables 7-17 to 7-21 give additional quantitative information on the thermoplastic elastomers (except for the environmental data). Table 7-21 gives the major processing characteristics of the thermoplastic elastomers.

Specifications. Contrary to the format used for metals, specifications for elastomer compounds are based on performance requirements and normally do not specify the chemical composition of the compound. In some cases, however, the base polymer is specified, along with the specific requirements for tensile strength, elongation, heat resistance, oil resistance, and so on. This system of not

Table 7-17. Physical and Mechanical Properties.

Property	ASTM	SBS Copolymers	Olefins	Urethanes	Polyesters
Hardness, Shore A	D-2240	25-95	54-95	65	–
Hardness, Shore D	D-2240	–	–	80	40-72
Flexural modulus, psi	D-790B	500-20,000	900-56,000	–	7000-75,000
Tensile strength, psi 20 in/min	D-412	500-5000	650-2800	3000-8000	3700-5700
Elongation, %	D-412	100-1200	70-1000	200-700	350-450
100% modulus, psi	D-412	–	200-2035	–	925-4100
Tear strength, pli, Die C	D-624	Up to 500	110-310	550-820	700-900
Compression set, %, 72°F	D-395B	–	20-48	10-50	–
Compression set, %, 158°F	D-395B	–	52-92	–	56-60 (38 annealed)
Oven aging, 250-275°F	–	P	E-G	F-P	G-E
Resilience, Bashore, %	D-2632	20-75	–	–	43-62
Flex cut resistance Demattia, cycles to fail	D-813	–	–	–	7×10^4-2×10^5
Ross, cycles to 5X growth	D-1052	3.5×10^3	–	–	3×10^5
Abrasion resistance, 1000 gm, CS-17 wheel, mg/M revolutions	D-1044	P	3-12	0.08-0.6	3-8
Adhesion properties	–	G-E	P	G	G
Rubbery feel/traction	–	E	G	G	F
Specific gravity	D-720	0.9-1.2	0.84-1.02	1.10-1.25	1.17-1.25

Note: E = Excellent; G = Good; F = Fair; P = Poor.

Table 7-18. Thermal Properties.

Property	ASTM	SBS Copolymers	Olefins	Urethanes	Polyesters
Heat-deflection temperature, °F	D-548	–	–	–	–
@66 psi fiber stress	–	110-130	66-250	176	315-330
@264 psi fiber stress	–	–	–	158	110-155
Vicat softening point, °F	D-1525	140-167	–	196-340	234-397
Coefficient of thermal expansion, in/in/°C	D-696	1.3×10^{-4}	1.2-3.8×10^{-4}	1.2-1.7×10^{-4}	2×10^{-4}
Maximum operating temperature, °F	–	150	250	200	300
Brittleness Temperature, °F	D-746	-60	-60	-80	-94

Table 7-19. Electrical Properties.

Property	ASTM	SBS Copolymers	Olefins	Urethanes	Polyesters
Volume resistivity, ohm-cm	D-257	10^{16}	$1.1\text{-}5 \times 10^{16}$	1×10^{12}	$1\text{-}5 \times 10^{13}$
Dielectric strength, volts/mil	D-149	–	600-800	–	645-900
Dissipation factor	D-150	.001-.004	.0003-.003	.015-.100	–
Dielectric constant, 60 Hz to 1 MHz	D-150	2.4-3.4	2.2-2.3	4-7	–

Table 7-20. Environmental and Chemical Resistance.

Resistance to:	SBS Copolymers	Olefins	Urethanes	Polyesters
Water	G	E	F-P	G
Polar solvents	G	G	F	F-G
Aliphatic hydrocarbons	VP	F	E	F-G
Aromatic hydrocarbons	VP	P	G	F-G
Halogenated solvents	VP	P	F	P
Weathering	P-G	E	P-F	G
Ultraviolet radiation	P-F	F-G	F	F-G
Oxygen	P-F	G	F-G	G
Ozone	P-F	G	F-G	G

Note: VP = very poor; P = poor; F = fair; G = good.

specifying the base polymer is used since, in many cases, a compound may be made of two different polymers—or different formulations of the same polymer—and still meet the requirements of the specification. However, the two compounds may differ significantly in some characteristic not cited in the specification. In selecting an elastomer specification, the designer should then consider, as a minimum, the following performance requirements:

1. Low Temperature—What is the lowest operating temperature at which the compound must exhibit flexibility? (Parts are not flexed in storage.)
2. High Temperature—Will the elastomer be exposed to high temperatures and for how long?
3. Age and Environment Resistance—Will the elastomer be exposed to the weather for long periods?
4. Solvent Resistance—Must the elastomer withstand continuous fluid immersion or just occasional spillage?

Table 7-21. Processing Characteristics.

Process	SBS Copolymers	Olefinics	Urethanes	Polyesters
INJECTION MOLDING				
Melt temperature, °F	400-450	325-450	375-425	350-420
Screws, L/D ratio	16:1 or more	16:1-24:1	15:1 or more	18:1-30:1
Screws, RPM	–	–	20-40	60
Screws, compression ratio	2:1-3:1	2:1-3:1	$2\frac{1}{2}$:1-3:1	$2\frac{1}{2}$:1-4:1
Molding pressure	–	–	–	Maximum without flash
Mold temperature, °F	80-150	80-120	60-150	70
Shrinkage, in/in	0.01-0.02	0.001-0.05	0.005-0.03	0.001-0.05
EXTRUSION				
Screw, L/D ratio	20:1 or more	24:1 or more	24:1	18:1 or more
Screw, compression ratio	$2\frac{1}{2}$-$3\frac{1}{2}$:1	$3\frac{1}{2}$:1	$2\frac{1}{2}$-$3\frac{1}{2}$:1	$2\frac{1}{2}$-4:1
Barrel temperature, °F	360-440	300-395	360-440	310-440
Die temperature, °F	410-460	300-395	390	340-450
Purging materials	HDPE or PP	–	Polystyrene or acrylic	HDPE or acrylic
Drying requirements	Not usually	No	Yes	Not usually
Runnerless molds	Yes—thin sections	Yes	Yes—with caution	Yes
Mold-release requirements	Yes	No	Yes	Yes

5. Miscellaneous Requirements—Are there special requirements, such as abrasion resistance, low gas permeability, or electrical properties?
6. Costs—Where costs are particularly significant, such as in high-volume consumer items, then thermoplastic elastomers should certainly be considered, with the material called out by chemical type.

A thorough analysis of the part by the designer, using the above requirements as a guide, should then help in the selection of the optimum material for the application. While the various compounds are not normally designated by chemical composition, as stated above, as a matter of practicality the compounds are grouped together by the base polymer; that is, a silicone would be the only base for one class, ethylene propylenes would be a logical base for another group, and so on. Table 7-16 then lists typical major requirements for parts and recommended elastomer types and applicable specifications, and Table 7-17 gives specific, quantitative characteristics.

Trade Names

Thermoplastics

Olefinics	
ET Polymer	Allied Chemical Co.
Polytrope	A. Schulman, Inc.
Ren-Flex	Ren Plastics
Somel	E.I. Du Pont de Nemours & Co., Inc.
Telcar	B.F. Goodrich Chemical Co.
TPR	Uniroyal Chemical Co.
Block Polymers (Styrene-butadiene)	
Kraton	Shell Chemical Co.
Solprene	Phillips Petroleum Co.
TPE	Elexar Chemical Co.
Polyester	
Hytrel	E.I. Du Pont de Nemours & Co., Inc.
Polyurethanes	
Adiprene	E.I. Du Pont de Nemours & Co., Inc.
Castathane	CPR Division, Upjohn Co.
Conathane	Conap Corporation
Cyanaprene	American Cyanamid
Estane	B.F. Goodrich Chemical Co.
Multrathane	Mobay Chemical Co.
Pellethane	CPR Division, Upjohn Co.
Roylar	Uniroyal Chemical Co.
Solithane	Thiokol Chemical Co.
Texin	Mobay Chemical Co.
Vibrathane	Uniroyal Chemical Co.
Ethylene–Vinyl Acetate	
Elvax	E.I. Du Pont de Nemours & Co., Inc.
Vynathene	USI Chemicals

Thermoset Elastomers

Acrylic	
Cyanacryl	American Cyanamid Co.
Hycar 4051, 4054, etc.	B.F. Goodrich Chemical Co.
Thiacril	Thiokol Chemical Co.
Chlorinated Polyethylene	
CPE	Dow Chemical Co.
Chlorosulfonated Polyethylene	
Hypalon	E.I. Du Pont de Nemours & Co., Inc.
Chloroprene	
Neoprene	E.I. Du Pont de Nemours & Co., Inc.
Epichlorhydrin	
Hydrin	B.F. Goodrich Chemical Co.
Ethylene-Acrylic	
Vamac	E.I. Du Pont de Nemours & Co., Inc.
Ethylene-Propylene	
Epcar	B.F. Goodrich Chemical Co.
Nordel	E.I. Du Pont de Nemours & Co., Inc.
Royalene	Uniroyal Chemical Co.
Vistalon	Enjay Corporation
Fluorinated Polymers	
Fluorel	3M Co.
Kalrez	E.I. Du Pont de Nemours & Co., Inc.
Kel-F	3M Co.
Viton	E.I. Du Pont de Nemours & Co., Inc.
Nitrile	
Hycar 1092C50, etc.	B.F. Goodrich Chemical Co.
Chemigum	Goodyear Tire & Rubber Co.
FR-N	Firestone
Nitrex	Uniroyal Chemical Co.
Paracril	Uniroyal Chemical Co.
Polyisoprene	
Ameripol	B.F. Goodrich Chemical Co.
Natsyn	Goodyear Tire & Rubber Co.
Shell Isoprene	Shell Chemical Co.
Polynorbornene	
Norsorex	American Cyanamid Co.
Polysiloxane	
Silicone	General Electric Co.
Silicone	Union Carbide Corporation
Silastic	Dow Corning Corporation

REFERENCES AND BIBLIOGRAPHY

Abrams, E. J. Two-part liquid rubbers. *Rubber Age*, March 1974, **106**(3):53.

American Cyanamid Co. *Norsorex Polynorbornene, A New Era in the Performance and Processing of Elastomers*. Bulletin 7-281210M9, 1977.

——. *Cyanaprene Thermoplastic Polyurethane Elastomers*. Bulletin 5-26082500, January 1975.

——. *Cyanacryl C Acrylic Elastomer*. Bulletin 3-2619-500, January, 1973.

Barney, A. L., Keller, W. J., and van Gulick, N. M. A high-performance fluorocarbon elastomer. *Journal of Polymer Science*, John Wiley & Sons, Inc., 1970, **8**(A-1):1091–1098.

Bonotto, S. Thermoplastic ethylene copolymers compared to other flexible materials. *Rubber Age*, November 1966, **98**(11):79.

Boyce, H. L. Designing molded rubber parts. *Mechanical Engineering*, September 1970, **92**(9):30.

Coleman, J. F. "Castable Reinforced Elastomers II, New Processing Methods." Paper presentation, American Chemical Society, Rubber Division, Paper No. 57, 107th Meeting, May 1975. *J. Elastomers Plast.*, 1976, **8**(2).

Cooper, J. From liquid to thermoset elastomer: efficient new silicone technology. *Modern Plastics*, February 1979, **56**(6):69.

Corbett, D. P., and DeMello, J. G. The CPE's take on the specialty elastomers. *Machine Design*, November 14, 1974, **46**(27):162.

Deanin, R. D. Butadiene/styrene thermoplastic elastomers. *SPE Journal*, January 1967, **23**(1):45.

Dow Chemical Co. *Dow CPE Elastomers for Oil and Heat Resistant Rubber Goods*. Form no. 305-334-74, 1974.

Dow Corning Corporation. *The Tough Differences Between Fluoroelastomers*. Form no. 09-269, 1969.

E.I. Du Pont de Nemours & Co., Inc. *An Engineering Profile, Nordel Hydrocarbon Rubber*. Bulletin A-64117, January 1969.

——. *The Engineering Properties of Viton Fluoroelastomer*. Bulletin A-65053, April 1969.

——. *Engineering Properties of Adiprene Urethane Rubber*. Bulletin A-42331, May 1969.

——. *Engineering Guide to the Du Pont Synthetic Rubbers*. Bulletin A-67246, November 1969.

——. *Engineering Properties of Hypalon Synthetic Rubber*. Bulletin A-52005, February 1967.

——. *Du Pont Neoprene*. Bulletin A-33448, November 1963.

——. *Chemical Resistance Guide for Du Pont Kalrez*. Bulletin E-07251.

Du Pont Kalrez Perfluoro Elastomer. Bulletin E19829, May 1980.

——. The Many Handiworks of Hytrel. Bulletin E37901, October, 1980.

Enjay Chemical Co. *Vistalon/The Design Rubber that Stays Young*. Bulletin SYN-67, December 1973.

Erich, F. R. (ed.) *Science and Technology of Rubber*. Academic Press, New York, 1978.

Flanders S. Miniguide to elastomers. *Materials Engineering*, June 1970, **71**(6):38.

——. Energy-absorbing elastomers soak up shock and vibration. *Materials Engineering*, June 1968, **67**(6):48.

Garshman, A. Thermoplastic elastomers: sizing up a crowded field *Plastics Technology*, June 1978, **24**(6):59.

B.F. Goodrich Chemical Co. *An Introduction to Epcar 5465 EPDM Elastomer.* Bulletin 7106.

——. *Injection Molding Elastomers*. Bulletin 7502-AK.

——. *Numbering System for Epcar Elastomers*. Bulletin EPB-1.001.

——. *Selector Chart*. Bulletin EPB-1.002-R.

——. *Processing Epcar EPDM Polymers*. Bulletin EPB-2.001-R.

——. *Epcar 545–Black/Oil Study*. Bulletin EPB-3.001.

——. *Heat Under the Hood*. Bulletin E-1.

——. *Properties of Hycar, Epcar, Hydrin, Ameripol*. Bulleti E-2.

——. *Estane Thermoplastic Polyurethanes, Versatile Above All*. Bulletin G-45.

——. *Hycar Reactive Liquid Polymers*. Bulletin RLP-1.

——. *Telcar Thermoplastic Elastomers*. Bulletin T-1.

Goodyear Tire & Rubber Co. *Natsyn 450–Oil Extended Polyisoprene.* Bulletin NS-25, February 1967.

Hall, A. There are profits for processors in thermoplastic rubber. *Plastics World*, November 18, 1974, **32**(11).

Hall, D. Rubber that doesn't act natural. *Machine Design*, July 12, 1973, **45**(13):109.

Hall, D., and Allen, E. The many faces of rubber. *Materials & Process Technology*, January/February 1972.

Hallman, R. W., Shah, A. K., and Moore, K. C. Compounding EPDM. *Rubber Age*, November 1971, **103**(11):57.

Hartman, P. F., Eddy, C. L., and Koo, G. P. A report on the new elastomeric thermoplastics. *SPE Journal*, May 1970, **26**(5):62.

Hauck, J. E. Blended elastomers resist elements. *Materials Engineering,* March 1968, **67**(3):50.

——. New high temperature elastomers last longer, cut maintenance. *Materials Engineering*, April 1967, **65**(4):108.

——. Low temperatures no problem with new elastomers. *Materials in Design Engineering*, April 1966, **63**(4):78.

——. EPT elastomer–the "crackless" rubber. *Materials in Design Engineering*, August 1964, **50**(2):79.

Hollis, E. W. Injection molding of rubber. *Rubber Age*, November 1961, **90**(2):261.

Houston, A. M. Natural rubber: dynamic in many ways. *Materials Engineering*, May 1977, **85**(5):66.

——. How chlorinated elastomers compare in overall performance. *Materials Engineering*, November 1976, **84**(6):26.

International Packings Corporation. How to protect elastomer parts from ozone attack. *Materials in Design Engineering*, August 1966, **64**(2):82.

Kehrer, G. P., and Hays, W. R. "Fabricating with Liquid Silicone Rubber." Paper presentation, *16th Annual TLARGI Foundation Conference*, Los Angeles, California, June 1977.

King, W. H. A fresh look at elastomers. *Machine Design*, January 25, 1973, **45**(2):106.

Materials Engineering. Rubbers and plastics fight heat under the hood of automobiles. February 1973, **77**(2):48.

——. An engineer's guide to natural rubber. April 1970, **71**(4):22.

——. *Rubber Parts*. Materials and Processes Manual No. 241, May 1967, **65**:99.

Minter, H. F., and Meir, J. F. Guide to fluid resistance of rubber and elastomers. *Materials Engineering*, January 1975, **81**(1).

Mobay Chemical Co. *Applications for Texin Urethane Elastomers*. 1970.

Nagel, W. B. Designing with rubber. *Machine Design*, Part 1: June 23, 1977, **49**(14):84. Part 2: **49**(16):78. Part 3: **49**(17):100.

Ogintz, S. M. "Perfluoroelastomer O-ring Seals for Severe Chemical and High Temperature Applications." American Society of Lubrication Engineers, *32nd ASLE Annual Meeting*, Montreal, Quebec, May 9–12, 1977.

Palmer, M. Try this route to non-MOCA PU elastomers. *Plastics Technology*, October 1977, **23**(10):95.

Panek, J. R., and Perrine, V. H. Engineer's guide to polysulfide rubber. *Materials Engineering*, April 1971, **73**(4):38.

Phillips Petroleum Co. *Compounding and Processing Solprene Radial Block Copolymers*. Report 804-P-TR.

——. *Typical Applications and Processing Characteristics of Solprene Radial Teleblock Copolymers*. Bulletin 4000.

——. *Modification of Polyethylene Film with Solprene Thermoplastic Rubber*.

Plastics Design Forum. Thermoplastic elastomers. March/April 1979.

Plastics World. Elastomers: more versatile than ever. December 1976, **34**(12):38.

Product Engineering. New elastomers take cars and abuse. March 1978, **49**(3):41.

——. Rubbery copolymer forms base for molded products and adhesives. April 1977, **23**(4):41.

——. Rubbery polymers match plastics and rubbers on cost/performance. March 1975, **46**(3):25.

——. Low-cost rubbery thermoplastic is processed by injection molding. October 1975, **46**(10):29.

——. New low-cost thermoplastic behaves like natural rubber. September 12, 1966, **37**(18).

Samuels, H., and Joslin, C. A. Process/production report: EEA copolymers. *Plastics Design & Processing*, March 1975, **15**(3).

Scheiner, L. L. Those TP elastomers are stirring up a frenzy. *Plastics Technology*, May 1973, **19**(5):36.

Shell Chemical Co. *Formulating Kraton Thermoplastic Rubber Crumb*. Bulletin SC: 72-3.

——. *Melt Processing with Kraton Thermoplastic Rubber Crumb*. Bulletin SC: 72-4.

——. *Solution Processing with Kraton Thermoplastic Rubber Crumb*. Bulletin SC: 72-5.

Siebert, A. R. "Castable Reinforced Elastomers I, Material Properties." Paper presentation, *American Chemical Society, Rubber Division*, Paper no. 56, 107th Meeting, May 1975. *J. Elastomers Plast*., 1976, 8(2).

Spenadel, L. EPT for Automotive sponge. *Rubber World*, August 1964, **150**(5):69.

Stoeck, P. F. The rubber jungle–and how to find your way out. *Materials Engineering*, January 1970, **71**(1):24.

Theberge, J. E., and Arkles, B. Stronger thermoplastic elastomers. *Machine Design*. February 12, 1976, **48**(3):113.

Thiokol Chemical Corporation. *Thiacril 55 Polyacrylate Rubber*. Bulletin AR-3, September 1964.

3M Co., Chemical Division. *Fluorel Brand Elastomer*. Bulletin Y-IFHTE (34.2) JO, 1959.

3M Co., Chemical Products Group. *Kel-F Elastomer*. Bulletin J-EPB (68.5) MP.

Uniroyal Chemical Co. *Uniroyal Urethane Elastomers*. Form no. 740-B31.

——. *Uniroyal TPR Thermoplastic Rubber*. Bulletin ASP–3495.

United States Rubber. *Royalene, The Crackless Rubber*. Form 570-B4A.

U.S. Industrial Chemicals Co. *Vynathene VAE Elastomers*. Bulletin PTD-68-975, 1975.

Vaccari, J. A. New heat resistant elastomers are more versatile. *Materials Engineering*, May 1977, **85**(5):45.

——. Relative performance of rubbers and elastomers. *Materials Engineering*, July 1974, **80**(1):26.

——. Be cautious in selecting rubbers for weathering resistance. *Materials Engineering*, June 1973, **77**(6):60.

Walker, B. M. *Handbook of Thermoplastic Elastomers*. Van Nostrand Reinhold Co., New York, 1979.

Wallis, T. E. Try hard rubber for low cost parts. *Materials in Design Engineering*, March 1959, **49**(3).

Weaver, E. P. New TPR grades move into entrenched rubber markets. *Elastomerics*, May 1978, **110**(5):21.

Wells, S. C. Thermoplastic-type elastomers. *Machine Design*, December 28, 1972, **44**(31):52.

Zaiser, G. L. Opportunities for thermoplastic elastomers. *Plastics Engineering*, May 1978, **34**(5):61.

8
Fillers and Reinforcements

Fillers are defined as relatively inert substances added to plastics compounds to reduce their cost and/or to improve physical properties, particularly hardness, stiffness, and impact strength. A filler differs from a reinforcement in two respects. Fillers are generally small particles, and they do not markedly improve the tensile strength of a product; reinforcements are fibrous and do markedly improve the tensile strength.

The most commonly used general-purpose fillers are clays, silicates, talcs, carbonates, asbestos fines, and paper. Some fillers act as pigments (e.g., chalk, titanium dioxide, and carbon black), some impart lubricity (e.g., molybdenum disulfide, PTFE, and graphite), and some increase electrical and thermal conductivity and magnetic properties (e.g., barium sulfate, powdered metals, and metal oxides). Most fillers increase the specific gravity of the matrix polymer.

Reinforcements are strong, inert, fibrous materials incorporated in polymeric masses to improve physical properties. To be effective, a reinforcement must form a strong adhesive bond with the resin for which purpose adhesion-promoting substances known as *coupling agents* are often applied to the fibers. Typical reinforcements include asbestos, boron fibers, carbon and graphite fibers, ceramics, glass fiber, chopped paper, macerated and flock fabrics, and synthetic fabrics. The term *reinforcing pigments* is used to describe pigments, such as carbon black, which improve physical properties, particularly of elastomers, as well as impart color.

There is some confusion between the terms "filler" and "extender." They serve different purposes, yet some fillers do act as extenders. Extenders are used as diluents for the more expensive resins in order to reduce cost. They modify physical properties and have even provided reinforcement. Typically thought of as being extenders are wheat, rye, and natural protein flours, soluble lignin, and reground resins.

A symbiotic relationship between resin and included ingredients such as fillers and reinforcements is essential to optimize composite properties. The designer may select from a wide variety of resins and fillers and is not limited to any special ratio of ingredients.

Some fillers are inert, but there is considerable evidence that a strong attraction exists between fillers having chemically active sites and the functional groups in the polymer backbone. The term "filler" is used to describe the nonresinous component of a composite system regardless of the degree of interaction between the fiber and the polymer chain.

The forces between active sites on filler surfaces and functional groups on polymer chains are weak. However, these forces are additive and become more significant with long polymer chains because of the great numbers of these weak bonds. Fillers reduce crystallization tendencies and solubility, and increase the T_g, hardness, and modulus of elasticity. Changes in impact resistance and coefficients of expansion of filled polymers are associated with changes in polymer chain mobility of these composites. The interaction between fillers and polymers in solution often results in an increased viscosity.

The principal types of fillers are listed in Table 8-1. Table 8-2 summarizes typical characteristics of various fillers. These materials vary widely in availability, utility, cost, composition, and physical and chemical properties. The optimum amount of resin is the minimum required to produce a monomolecular layer on the filler surface.

Most of the reinforcing fillers are high molecular weight compounds. Finely divided fillers have little effect on thermoplastic composites, but their effectiveness is increased when a small amount of crosslinking in the polymer matrix occurs. Fibrous fillers are more useful than nonfibrous fillers. A fiber length of at least 100 microns is essential for significant reinforcement of thermoplastics.

ORGANIC MATERIALS

Wood Products

Kraft Paper. Kraft paper may be impregnated with resin and compression molded as laminate sheet. The impregnated paper may also be dried and chopped to yield a product that can be compression molded.

Wood Flour. Wood flour is obtained by attrition mill or hammer mill grinding of fibrous softwoods and hardwoods. Dr. Baekeland added wood flour to phenolic resins at the turn of the century in order to reduce the brittleness of his new phenol-formaldehyde materials. This inexpensive filler reduces mold shrinkage and provides a filled thermosetting resin with good electrical properties and impact resistance. Wood flour is used to reinforce urea resins when dark-colored products are acceptable.

Shell Flour. Shell flour is obtained by grinding walnut, pecan, or peanut shells. It contains lignin and furfural, which

Table 8-1. Types of Fillers for Plastics.*

Organic Materials
- A. Cellulosic products
  - 1. Wood products
    - a) Kraft paper
    - b) Chips
    - c) Coarse flour
    - d) Ground flour
      - 1. Softwood flour
      - 2. Hardwood flour
      - 3. Shell flour
  - 2. Comminuted cellulose products
    - a) Chopped paper
    - b) Diced resin board
    - c) Crepe paper
    - d) Pulp preforms
  - 3. Fibers
    - a) Alpha cellulose
    - b) Pulp preforms
    - c) Cotton flock
    - d) Textile byproducts
    - e) Jute
    - f) Sisal
    - g) Rayon
- B. Lignin-type products
  - 1. Ground bark
  - 2. Processed lignin
- C. Synthetic fibers
  - 1. Polyamides (nylon)
  - 2. Polyesters (Dacron)
  - 3. Polyacrylonitrile (Orlon)
- D. Carbon
  - 1. Carbon black
    - a) Channel black
    - b) Furnace black
  - 2. Ground petroleum coke
  - 3. Graphite filaments
  - 4. Graphite "whiskers"

Inorganic Materials
- A. Silica products
  - 1. Minerals
    - a) Sand
    - b) Quartz
    - c) Tripoli
    - d) Diatomaceous earth
  - 2. Synthetic materials
    - a) Wet process silica
    - b) Pyrogenic silica
    - c) Silica aerogel
- B. Silicates
  - 1. Minerals
    - a) Asbestos
      - 1. Chrysotile
      - 2. Amosite
      - 3. Anthophyllite
      - 4. Crocidolite
      - 5. Tremolile
      - 6. Actinolite
    - b) Kaolinite (China clay)
    - c) Mica
    - d) Talc
    - e) Wollastonite
  - 2. Synthetic products
    - a) Calcium silicate
    - b) Aluminum silicate
- C. Glass
  - 1. Glass flakes
  - 2. Hollow glass spheres
  - 3. Solid glass spheres
  - 4. Milled fibers
  - 5. Fibrous glass
    - a) Filament
    - b) Rovings
    - c) Woven roving
    - d) Yarn
    - e) Mat
    - f) Fabric
- D. Metals
  - 1. Steel
  - 2. Tungsten
  - 3. Titanium
  - 4. Beryllium filaments
  - 5. Molybdenum filaments
- E. Boron filaments
- F. Metallic oxides
  - 1. Ground material
    - a) Zinc oxide
    - b) Aluminum
    - c) Magnesia
    - d) Titania
  - 2. "Whiskers"
    - a) Aluminum oxide
    - b) Beryllium oxide
    - c) Magnesium oxide
    - d) Thorium oxide
    - e) Zirconium oxide
- G. Calcium carbonate
  - 1. Chalk
  - 2. Limestone
  - 3. Precipitated calcium carbonate
- H. Polymeric fillers
- I. Other fillers
  - 1. "Whiskers" (non-oxide)
    - a) Aluminum nitride
    - b) Beryllium carbide
    - c) Boron carbide
    - d) Silicon carbide
    - e) Silicon nitride
    - f) Tungsten carbide
  - 2. Barium ferrite
  - 3. Barium sulfate

*Seymour, 1968–69. Reprinted by permission from *Modern Plastics Encyclopedia*, McGraw-Hill, Inc.

Table 8-2. Some Fillers and Reinforcements and Their Contributions to Plastics.*

Filler or reinforcement	Properties Improved*													
	Chemical resistance	*Heat resistance*	*Electrical insulation*	*Impact strength*	*Tensile strength*	*Dimensional stability*	*Stiffness*	*Hardness*	*Lubricity*	*Electrical conductivity*	*Thermal conductivity*	*Moisture resistance*	*Processability*	*Recommended for use in*[a]
Alumina, tabular	•	•				•								S/P
Alumina trihydrate, fine particle			•				•					•	•	P
Aluminum powder										•	•			S
Asbestos	•	•	•	•		•	•	•						S/P
Bronze							•	•		•	•			S
Calcium carbonate[b]		•				•	•	•					•	S/P
Calcium metasilicate	•	•	•	•		•	•	•				•	•	S/P
Calcium silicate		•				•	•	•						S
Carbon black[c]		•				•	•			•	•		•	S/P
Carbon fiber										•	•			S
Cellulose				•	•	•	•	•						S/P
Alpha cellulose			•		•	•								S
Coal, powdered	•											•		S
Cotton (macerated/chopped fibers)			•	•	•	•	•	•						S
Fibrous glass	•	•	•	•	•	•	•	•				•		S/P
Fir bark													•	S
Graphite	•				•	•	•	•	•	•	•			S/P
Jute				•			•							S
Kaolin	•	•				•	•	•	•			•	•	S/P
Kaolin (calcined)	•	•	•			•	•	•				•	•	S/P
Mica	•	•	•			•	•	•	•			•		S/P
Molybdenum disulphide							•	•	•			•	•	P
Nylon (macerated/chopped fibers)	•	•	•	•	•	•	•	•	•				•	S/P
Orlon	•	•	•	•	•	•	•	•				•	•	S/P
Rayon			•	•	•	•	•	•						S
Silica, amorphous			•									•	•	S/P
Sisal fibers	•			•	•	•	•	•				•		S/P
TFE-fluorocarbon						•	•	•	•					S/P
Talc	•	•	•			•	•	•	•			•	•	S/P
Wood flour			•		•	•								S
Altered novaculite	•	•	•	•		•	•	•			•		•	S/P

†The chart does not show differences in degrees of improvement: calcined kaolin, for example, generally gives much higher electrical resistance than kaolin. Similarly, differences in characteristics of products under one heading, such as talc (which varies greatly from one grade to another and from one type to another), also are not distinguished.

a–Symbols: P–in thermoplastics only; S–in thermosets only; S/P–in both thermoplastics and thermosets. b–In thermosets, calcium carbonate's prime function is to improve molded appearance. c–Prime functions are imparting of UV resistance and coloring; also is used in crosslinked thermoplastics.

*Seymour, 1968–69. Reprinted by permission from *Modern Plastics Encyclopedia*, McGraw-Hill, Inc.

contribute to good flow properties, and cutin, a wax which provides gloss, luster, and moisture resistance. Molded products which contain shell flour have low shear strength and lower resistance to impact than those containing wood flour.

Cellulose Fibers. The smallest coherent unit of cellulose retaining characteristics of the bulk fiber is called a *microcrystal.* When added to liquid resins, these microcrystals produce thixotropic systems and, thus, may be used for viscosity control. When alpha-cellulose is used as a filler, light-colored moldings with fair resistance to impact are produced. This filler—obtained by the alkaline treatment of wood pulp—yields readily processed powders characterized by low shrinkage and low resistance to moisture and finds wide use in amino molding compounds.

Phenolic molding compounds of medium resistance to impact are obtained by using processed cotton flock or caustic-purified cotton linters. Cotton flock-filled molding powders are characterized by high bulk factor and low moisture resistance. High-impact molding resins incorporate chopped fabric fillers.

Jute fibers, yarn, and fabric have been used for reinforcements in phenolic composites, but these are characterized by low strength. Moderate-impact molding powders are produced from chopped sisal fibers.

Rayon may be used both as flock and as yarn. The impact resistance is a function of its fiber length. Lignin is the natural binder for cellulose in wood. A product containing this type filler may be obtained by grinding the bark of Douglas fir trees or by extending wood flour with a derivative of lignosulfonic acid. All of these dark-colored fillers have been used with resins to produce molding powders. Since the lignin contributes to flow, less resin is required.

Synthetic Organic Fibers

Nylon (Caprolan and others, see Chapter 5) is available as flock, filament, and chopped fabric. Being thermoplastic, it flows at many molding temperatures, causing considerable mold shrinkage. Nylon-filled thermosetting resins are used as ablative materials. Wear-resistant parts have been molded from nylon fiber–reinforced phenolics. Nylon-reinforced polyesters have been molded to produce shower stall bases.

Aromatic polyamide fibers or aramids (e.g., Kevlar) form composites with excellent strength and heat resistance. Composites have low densities because the specific gravity of the aramid fiber is relatively low (1.45). The composites thus show high specific strength values.

Du Pont refers to all of its aromatic nylons as "aramids," including the materials discussed earlier in Chapter 5 (Nomex). The material discussed here has not, as of this writing, been sufficiently characterized chemically and thus can only be identified as an aramid fiber.

Three types of Kevlar aramids are available: a high tensile strength, high modulus type for reinforcement of plastics; a similar fiber for use in ropes, cables, coated fabrics, and protective clothing; and a type designed to reinforce rubber products such as tires, belts, and hose. The first type is of interest here because of its applications in plastics technology.

Figure 8-1 compares the tensile strengths and moduli of the aramid fibers (Kevlar 29 and 49) with other common plastics reinforcements. The tensile strength is better than twice that of nylon or polyester and 15% better than E-type fiberglass. Modulus is twenty times greater than nylon, ten times greater than polyester, and twice that of E-glass. Elongation is low and density of the fiber is 40% lower than glass. Aramids are as strong as steel at $\frac{1}{5}$ the weight.

Yarns and rovings are available, comprised of multiple, continuous, round cross-section filaments with 0.0005-inch diameters. Unlike glass, the aramid fiber is relatively insensitive to fiber surface defects.

The thermal stability of aramid fibers is good. Even after exposure for 7 days to temperatures as high as 180°C (356°F), the physical property losses are less than half the initial values. Essentially no embrittlement or degradation occurs during cryogenic exposure. Because of its slightly negative coefficient of thermal expansion in the longitudinal direction, composites made with polyaramid fibers exhibit excellent dimensional stability.

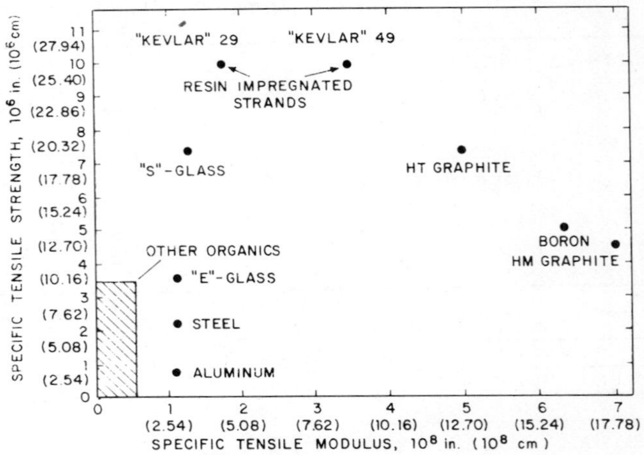

Fig. 8-1. Specific tensile strength and specific tensile modulus of reinforcing fibers. (Miner, 1977)

Aramid fibers have good chemical resistance, except to strong acids and bases. They are not dyeable and have an equilibrium moisture level of 3.5–4.5%. The presence of such surface moisture can adversely affect interlaminar shear properties in some resin systems. Predrying is thus strongly recommended. The fibers are supplied unfinished or sized. The effect of sizings and coupling agents is still undetermined.

Aramid fibers have high stress-rupture life and fracture toughness, good cyclic tensile fatigue resistance, and fair damping characteristics. Environmental stability of composites is generally good, except in extremely hostile environments. The main disadvantages of the aramid fibers include: high moisture absorptivity (mentioned above), low compressive strength, and difficulty of machining.

Polyester fibers (Dacron, Fortrel, Kodel, etc.) reinforce both thermosetting and thermoplastic resins. Because of the ability of these fibers to serve as nucleating agents for crystalline thermoplastics, the fiber-thermoplastics composites have unusually high specific strength. Abrasion and impact resistance of polyester composites increases when reinforced with woven roving of polyethylene terephthalate (as opposed to fibrous glass). These polyester fibers are solvent resistant and high melting and can reinforce thermoplastic, phenolic, and epoxy resin systems.

Low-density composites with improved resistance to impact and weathering have been obtained using PVA fibers as a reinforcement for polyesters. Most resins adhere to these fibers and no surface treatment is required. Polyurethane properties have been improved through reinforcement with PVA fibers.

Carbon-Based Fillers and Reinforcements

Carbon black is the principal reinforcing filler for all types of elastomers. It is also used as a pigment and filler for various thermoplastics and thermosets. Carbon black reduces the sizes of crystallites and improves resistance to stress-cracking and ultraviolet degradation of polyolefins.

Reinforcing carbon black is made up of crystallites consisting of layers of carbon atoms. *Channel blacks*, with particle sizes of 200–300 Å, are produced by burning natural gas in a limited supply of air and allowing the products to deposit on cold metal surfaces. *Furnace blacks*, with particle sizes of 300–800 Å, are produced by burning gas or oil in a limited supply of air and removing the products by centrifugation or electrostatic precipitation. Thermal, acetylene, and lamp black and ground petroleum coke are coarser than the more widely used reinforcing blacks.

Filaments of carbon were first made commercially in 1880 when Thomas Edison used them in electric light bulbs. It was not until 1954 that true textile forms of carbon fibers were produced commercially. Carbon and graphite fibers are made from any carbonaceous fibrous raw material that pyrolyzes to char, does not melt, and leaves a high carbon residue. The most commonly used source materials are rayon and polyacrylonitrile (PAN-based). Figure 8-2 shows a schematic representation of the production process for various forms of carbon reinforcement.

To produce carbon fiber, rayon is reduction-heated to 371°C (700°F) and then flash-heated to 1371°C (2500°F). A 45–55% weight loss occurs. Graphite fiber is derived from carbon fiber as the result of extra or post-thermal treatment which converts the fiber to the crystalline phase—bulk-carburized cloths are passed through the graphitization stage by reheating to 1093°C (2000°F) for a period of two weeks, or by flash-heating to 2760°C (5000°F). Direct graphitization is achieved on polyacrylonitrile yarns.

A recent development is pitch-based fibers. These have physical properties comparable to PAN-based fibers, but their cost is more like the rayon-based type.

General-purpose carbon and graphite yarns exhibit a tenacity (strength/size of yarn) in the range of 2.3–3.8 gram/denier (weight/9000 m of fiber). Typical room-temperature properties are listed in Table 8-3. Figure 8-3 shows various forms of fibrous carbon.

High-modulus graphite yarns were introduced in 1965. These continuous filament yarns demonstrate high tensile strength with the customary graphite characteristic of retaining fiber strength at high temperature. The extremely fine filaments making up these high-modulus yarns give them flexibility similar to other graphite fiber forms. These yarns can be handled in much the same manner as glass fibers and conventional graphite yarns.

Carbon yarns are characterized by high surface area, higher electrical volume resistivity, and a lower carbon assay than graphite yarns. Carbon yarns have a tendency to pick up moisture, will wet with resins easily, and provide low thermal conductivity yarn-resin composites.

The graphite yarns have low surface area, high carbon

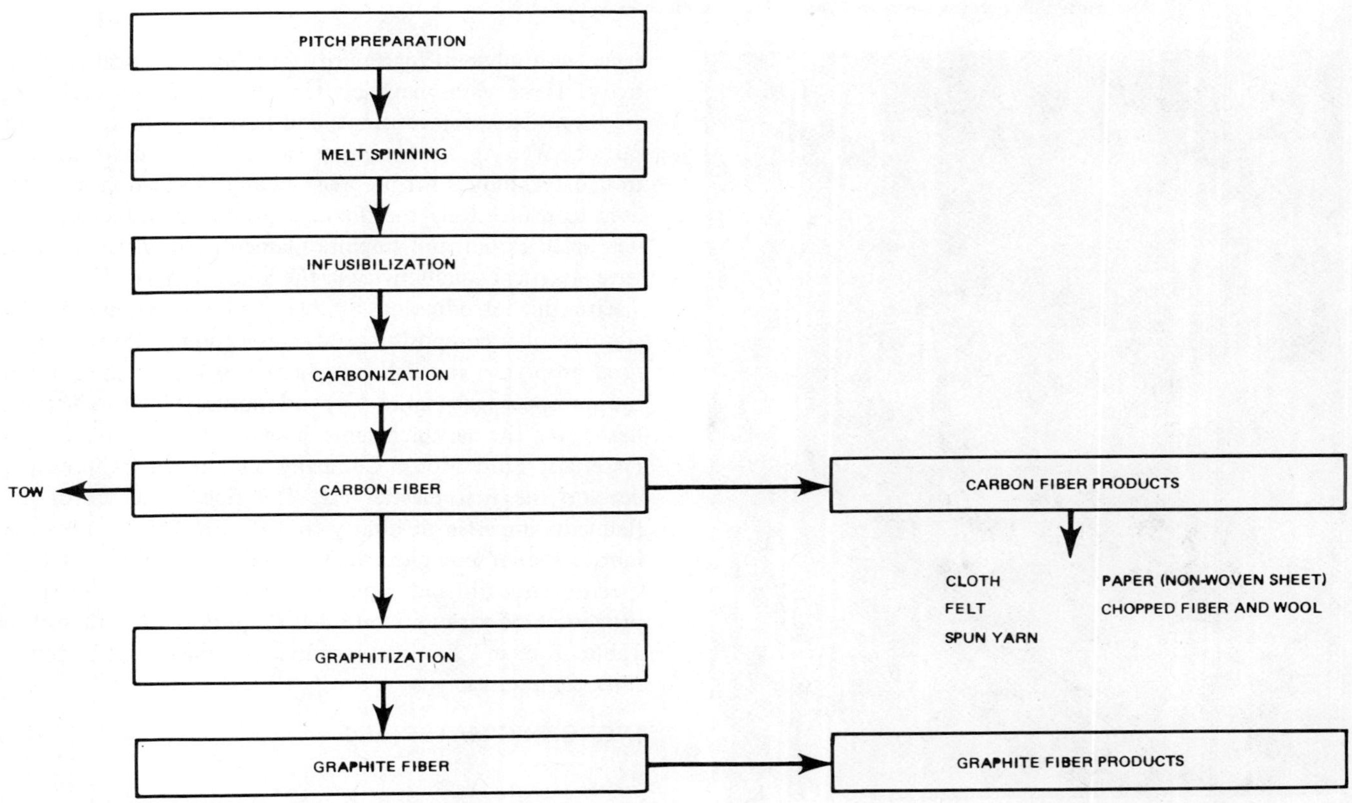

Fig. 8-2. Schematic representation of carbon and graphite fiber manufacture. (*Courtesy: Kureha Chemical Industries Co., Ltd.*)

Table 8-3. Typical Carbon and Graphite Filament and Yarn Properties at 20°C.*

		General Purpose		High Modulus
Filament		Carbon	Graphite	Graphite[2]
Tensile strength, psi	1-in. ga.	120,000	90,000	285,000
Tenacity, g/denier	1-in. ga.	6.2	5.3	13.7
Elongation at break, %		2.0	1.5	0.6
Elastic recovery, %		100	100	100
Modulus of elasticity, psi	1-in. ga.	6,000,000	6,000,000	50,000,000
Stiffness, g/denier	1-in. ga.	310	350	2300
Density, Hg minimum, g/cc		1.53	1.32	1.63
Equivalent diameter, microns		9.5	8.9	6.6
Carbon assay, w/o		90	98.8	99.9
Volume electrical resistivity, ohm-cm × 10^6		6000	3500	–
Surface area, m^2/g		130	4	1
Thermal conductivity, Btu/ft./(hr.) (sq. ft.) (°F.)		13	22	68
Specific heat, 70°F., Btu/(lb.) (°F.)		0.17	0.17	0.17
Mean 70°F. to 2700°F.		0.40	0.40	0.40
Two-ply Yarn[1]				
Yield, yards/lb.		3100	4100	6355
Breaking strength, lb. (average)	5-in. ga.	12	6	6.4
Tenacity, g/denier	5-in. ga.	3.8	2.3	4.7
Elongation at break, %		2.4	1.3	0.53
Yarn diameter	in., approx.	0.03	0.03	0.015
pH, ASTM D-1512-60		7.0	6.0	7.5

1-Single ply, 5 ply, 10 ply, and 30 ply yarn constructions have also been made commercially.
2-25 million and 40 million p.s.i. elastic modulus fibers are also commercially available.
*Reprinted by permission from *Modern Plastics*, McGraw-Hill, Inc.

Fig. 8-3. Forms of fibrous carbon. (*Courtesy: Kureha Chemical Industries Co., Ltd.*)

assay with inherent high purity, and high electrical conductivity. These yarns have less tendency to absorb moisture, are more oxidation resistant, and have more lubricity than carbon. Wetting with resins is not difficult, although controlled techniques for prepregging and fabrication must be used to realize fully the filament properties in the composites. Both carbon and graphite filaments are available which have electrical conductivity in the longitudinal direction.

With the introduction of high-modulus graphite yarns, graphite fiber composites now offer a combination of structural properties and light weight that is being put to use in the aerospace (see Figure 8-4) and sporting goods industries. Based on the advancements in graphite fiber-reinforced materials, Ford Motor Company announced its intent to develop the first plastics car. The fine diameters of the filaments impart a flexibility to the yarns that permits the fabrication of complex shapes and the retention of high specific strength and modulus. Tables 8-4 and 8-5 show properties of various reinforced thermoplastic composites. Tables 8-6 and 8-7 present data for thermoset graphite-fiber composites.

INORGANIC MATERIALS

Glass

Glass flakes used in polymeric protective coatings improve plastics corrosion resistance and form an impermeable

Fig. 8-4. Antenna dish made from aluminum honeycomb with graphite/epoxy face sheets. (*Courtesy: Hughes Aircraft Co.*)

barrier to vapors. In epoxy, polyester, and urethane flooring and wall surfacing, glass flakes give a tough surface with excellent chemical and abrasion resistance. Glass in granular form adds excellent non-skid, non-slip properties to epoxy, polyester, or urethane coating systems for shipboard deck and industrial floor use.

Soda-lime glass in solid spherical form—smooth, transparent, hard, lightweight—has proved to be an effective reinforcement/filler in both thermoplastics and thermosets. Available in diameters of 2-500 microns, it reinforces the mechanical, electrical, chemical, and thermal properties of base resins, often with cost savings. Diameters of 6-50 microns are most popular with plastics.

The perfect spherical shape and smooth surface keep resin viscosity low, permitting high loading of spheres and good mold flow. Spheres set up more predictable and uniform stress patterns in moldings than do irregularly shaped particles or short fibers, with resultant gains in physical properties. The spheres improve flexural strength and modulus, and yield smooth, reinforced surfaces, corners,

Table 8-4. Properties of Nylon 6/6 Composites.*

Property	ASTM Method	Unreinforced Resin	Reinforcing Fiber (%): 20 Carbon	30 Carbon	40 Carbon	20 Carbon, 20 Glass	40 Glass
Specific gravity	D792	1.14	1.23	1.28	1.34	1.40	1.46
Water absorption (%)	D570	–	–	–	–	–	–
24 hr	–	1.6	0.6	0.5	0.4	0.5	0.6
Equilibrium after continuous immersion	–	8.0	2.7	2.4	2.1	–	3.0
Mold shrinkage, $\frac{1}{8}$ in.	D955	–	–	–	–	–	–
Av section (10^{-3} in./in.)	–	15.0	2.0–3.0	1.5–2.5	1.5–2.5	2.5–3.5	3.5–4.0
Tensile strength (psi)	D638	11,800	28,000	35,000	40,000	34,000	31,000
Tensile elongation (%)	D638	10	3–4	3–4	3–4	3–4	2–3
Flexural strength (psi)	D790	15,000	42,000	51,000	60,000	49,000	42,000
Flexural modulus (10^6 psi)	D790	0.4	2.4	2.9	3.4	2.8	1.6
Shear strength (psi)	D732	9,600	12,000	13,000	14,000	13,000	12,000
Impact strength, Izod (ft-lb/in.)	D256	–	–	–	–	–	–
Notched, $\frac{1}{4}$ in.	–	0.9	1.1	1.5	1.6	1.8	2.6
Unnotched, $\frac{1}{4}$ in.	–	–	8.0	12.0	13.0	16.0	19.0
Heat-distortion temperature at 264 psi (F)	D648	150	495	495	500	500	500
Coefficient of thermal expansion (10^{-5} in./in.-deg. F)	D696	4.5	1.40	1.05	0.80	1.15	1.4
Thermal conductivity (Btu/hr/ft^2/°F/in.)	C177	1.7	5.5	7.0	8.5	6.4	3.6
Surface resistivity (ohms/sq)	–	10^{15}	25–30	3–5	1–3	–	10^{15}

*Theberge et al., 1974.

Table 8-5. Properties of Various Thermoplastic Composites.*

Property	ASTM Method	Polysulfone			Polyester			Polyphenylene Sulfide			Ethylene Tetrafluoroethylene			Vinylidene Fluoride-PTFE	
		Unrein-forced	30% Glass	30% Carbon	Unrein-forced	30% Glass	30% Carbon	Unrein-forced	30% Glass	30% Carbon	Unrein-forced	30% Glass	30% Carbon	Unrein-forced	20% Carbon
Specific gravity	D792	1.24	1.45	1.37	1.32	1.52	1.47	1.34	1.56	1.45	1.70	1.89	1.73	1.76	1.77
Water Absorption (%)	D570														
24 hr		0.20	0.20	0.15	0.08	0.06	0.04	0.20	0.04	0.04	0.020	0.018	0.015	0.04	0.03
Equilibrium after continuous immersion		0.60	0.58	0.38	–	0.45	0.23	–	–	0.10	–	–	0.24	–	–
Mold shrinkage, $\frac{1}{8}$-in. Av. section (10^{-3} in./in.)	D955	7–8	2–3	1–2	17–23	3–4	1–2	10	2	1	15–20	2–3	1.5–2.5	–	2.5–3.5
Tensile strength (psi)	D638	10,200	18,000	23,000	8,000	19,500	20,000	10,800	20,000	27,000	6,500	14,000	15,000	5,500	12,300
Tensile elongation (%)	D638	50–100	3–4	2–3	10	3–4	2–3	3–4	3–4	2–3	150	4–5	2–3	100	3–4
Flexural strength (psi)	D790	15,400	24,000	32,000	13,000	28,000	29,000	20,000	29,000	34,000	10,000	19,000	20,000	7,000	17,200
Flexural modulus (10^6 psi)	D790	0.39	1.20	2.05	0.35	1.35	2.00	0.60	1.60	2.45	0.20	1.05	1.65	0.20	1.10
Shear Strength (psi)	D732	9,000	9,500	9,500	7,000	8,000	8,000	–	–	–	6,000	7,000	7,000	6,000	7,500
Impact strength, Izod (ft-lb/in.)	D256														
Notched, $\frac{1}{4}$ in.		1.2	1.8	1.2	0.3	1.6	1.2	0.3	1.4	1.1	N.B.	7–8	4–5	2.3	2.6
Unnotched, $\frac{1}{4}$ in.		60.0	14.0	6–7	25	9–10	4–5	3–4	8–9	5–6	N.B.	17–18	10.0	–	7–8
Heat-distortion temperature at 264 psi (F)	D648	345	365	365	155	430	430	278	500	500	165	460	465	190	295
Coefficient of thermal expansion (10^5 in./in.)	D696	3.1	1.4	0.6	5.3	1.2	0.5	3.0	1.3	0.6	4.2	1.7	0.8	8.5	3.0
Thermal conductivity Btu-in./hr-ft^2-deg (F)	C177	1.8	2.2	5.5	1.1	3.2	6.5	2.0	2.8	5.2	1.65	2.8	5.6	2.1	3.0
Surface resistivity (ohms/sq)	–	10^{16}	10^{16}	1–3	10^{15}	5×10^{15}	2–4	10^{16}	10^{16}	1–3	5×10^{14}	10^{15}	3–5	5×10^{14}	3–5
Flammability	UL Subj 94	V-O	V-O	V-O	–	–	–	V-O	V-O	V-O	V-O	V-O	V-O	V-O	V-O

*Theberge et al., 1974.

Table 8-6. Properties of Graphite Fiber/Polyester Resin Composites.

50% v/v Type I Fibre (not surface treated)	
Flexural strength	120,000 lb/in^2
Flexural modulus	20 x 10^6 lb/in^2
Tensile strength	130,000 lb/in^2
Tensile modulus	28 x 10^6 lb/in^2
Coefficient of thermal expansion per ^oC :–	
along fibre	– 0.73 x 10^{-6}
across fibre	+29 x 10^{-6}
Thermal conductivity	0.04 cal/sec/cm/^oC
Tensile fatigue @ 30,000 ± 20,000 lb/in^2	more than 20 x 10^6 cycles without failure
Tensile/compressive fatigue @ 0 ± 40,000 lb/in^2	more than 20 x 10^6 cycles without failure
Impact strength (Izod)	19.8 (unnotched)
	18.7 (notched)

Table 8-7. Properties of Graphite Fiber/Epoxy Resin Composites.

60% v/v Type I Fibre (surface treated)	
Flexural strength	120,000 – 125,000 lb/in^2
Flexural modulus	25 – 28 x 10^6 lb/in^2
Tensile modulus	30 – 40 x 10^6 lb/in^2
Shear strength	7,000 – 10,000 lb/in^2

and edges in molded parts. Addition of glass spheres to a resin mix decreases the burning rate of a molded part.

In thermoplastics, solid glass spheres have been used successfully as reinforcement/filler in ABS, nylon, and vinyl. The spheres improve tensile strength, reduce elongation, and sharply increase flexural modulus and compressive strength. They are also used in epoxy, polyester, and phenolic resins. In polyester, they have made possible a low-viscosity, highly loaded, hand lay-up system, having physical properties equal to those of high-pressure moldings. Addition of glass spheres to glass fiber–epoxy laminates has proved to increase flexural strength and modulus without significant loss of other properties. Replacing quartz as a filler, glass spheres increase modulus and flexural, tensile, and impact strengths. Figure 8-5 shows some examples of plastics products modified with glass microbeads.

Cellular glass modules with multiple closed cells have also been used as resin fillers. Low-density syntactic foams with controlled specific gravity may be obtained when hollow glass spheres are added to resinous composites. Tables 8-8a and 8-8b present comparative properties of epoxy and polyester filled systems. Figure 8-6 shows an example of a high-technology application of epoxy syntactic foam.

Fibrous glass, which was the first reinforcing fiber used in structural composites, continues to be the principal reinforcement for plastics. It increases stiffness, strength, impact resistance, heat-deflection temperature, hardness and dimensional stability. It also decreases the coefficient of thermal expansion and can improve the electrical properties of the composite.

The strength of a reinforced plastic is related to its ability to transfer external stress on the resin matrix to the reinforcement. When a spherical filler is used, there is little interaction between its surface and the resin, so the energy generated by the applied stress is dissipated through filler particle movement (slippage of the particle past the matrix). Since the movement may pull the particle away from the matrix, the interface becomes a source of microcracks. Further energy exchange will cause crack propagation and eventual failure. Crack propagation can be reduced in three ways. One is to improve interfacial adhesion through the use of coupling agents. Another way is to use softer matrices since they resist crack formation and are more resilient to

Fig. 8-5. Glass beads (25-micron diameter) were used in these parts to increase abrasion resistance, rigidity, and compressive stress. (***Courtesy: Potters Industries, Inc.***)

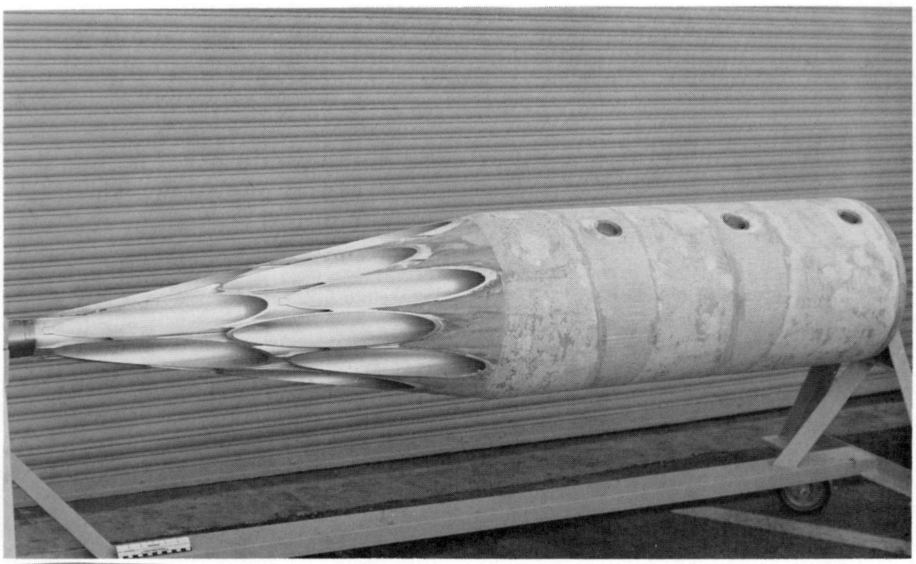

Fig. 8-6. Prototype rocket launcher constructed from an epoxy syntactic foam, glass-reinforced epoxy, and aluminum. (***Courtesy: Hughes Aircraft Co.***)

Table 8-8a. Properties of Polyester-Glass Mat Laminates Containing Hollow Glass and Mineral Fillers.*

	Glass Microballoon particles	Mineral filler
Filler, wt-%	2.1	25.2
Reinforcement, wt-%	15	40
Density, g/cc	1.44	1.82
Tensile strength, psi	14,400	15,000
Flexural strength		
Dry, psi	26,900	25,000
Wet, psi	19,700	17,500
Modulus in flexure, 10^6 psi	1.74	1.3
Water absorption, wt-%	0.6	0.5

*Alford and Veatch, 1961. Reprinted by permission from *Modern Plastics*, McGraw-Hill, Inc.

Table 8-8b. Properties of Filled Epoxy Resin at Equivalent Volume-% Loading.*

	Glass Microballoon filler	Silica (200 mesh) filler
Density, g/cc	0.6	1.7
Flexural strength, psi	4400	4200
Linear expansion, in./in./°F	17×10^{-6}	16×10^{-6}
Dielectric constant, 10^{10} cy./sec	1.9	3.6
Dissipation factor, 10^{10} cy./sec	0.015	0.028
Volume resistivity, ohm-cm	1×10^{12}	4×10^{12}

*Alford and Veatch, 1961. Reprinted by permission from *Modern Plastics*, McGraw-Hill, Inc.

crack propagation. The third method is to use fibrous reinforcements in which the stress becomes concentrated at the fiber ends. This allows the far larger area along the fiber length to continue to function as a stress-transfer surface and prevent interfacial crack formation.

Fibrous glass filaments are comprised of lime-aluminum-borosilicate glass that is relatively soda free. This glass has great durability and is known as "E" glass because of its good electrical characteristics. Glass surfacing and overlay mats are formed from a low soda, more chemically resistant, glass called "C" glass, which is highly durable and resists strong acids.

While E-glass fibers have greater mechanical strength, their chemical resistance is lower than C-glass fibers. Consequently E-glass is most frequently used for reinforcements. When first made, E-glass fibers manifest tensile strengths of about 500,000 psi. Damage to the fibers during normal handling reduces this strength so that it seldom exceeds 280,000 psi in commercial form.

For a considerable length of time, E-glass was the industry standard. Indeed, it is still used in many applications and for general-purpose use provides an excellent structural material at minimum cost. However, to realize even greater composite strength-to-weight ratios further improvements were needed.

Chemical compounds called *sizings* are applied to fiber or film surfaces to fill pores and irregularities. The sizes improve the fiber surface uniformity and the surface compatibility with adhesives, encapsulating or coating resins. This in turn improves interfacial stress transfer between resin and filler/reinforcement.

Improved sizings for glass fibers were developed through the use of nonaqueous application methods and upgraded compositions. The resulting increase in strength of composites fabricated with these sized fibers was not the result of any change in glass formulation, but reflected the increased utilization of the virgin strength of the glass through improved adhesion and stress transfer between the glass and polymeric matrix.

The next level of improvement was achieved through the development of "S" glass. This glass is produced via new fiber-production methods involving operation of fiber-drawing equipment at significantly higher temperatures. S-glass is inherently strong, and when combined with optimized sizings, its composite strength also dramatically increases.

S-glass composites are used in the majority of aerospace applications. The glass is used to the greatest extent in combination with epoxy resins. The resultant composites maintain excellent structural characteristics up to the degradation temperature of the epoxy systems.

Glass fibers of controlled diameter are manufactured by steam or air blowing, flame blowing, and mechanical pulling processes (see Figure 8-7). Most fibers used for plastics reinforcement are produced by mechanical pulling. In all three processes, molten glass flows through a small electric crucible, or "bushing," with a number of metering orifices in its bottom. In mechanical pulling, continuous fibers are produced by drawing filaments from each stream of molten glass as it drops from an orifice. Streams are pulled into desired diameters by a high-speed winder. As the winder speed increases, the filament diameter decreases. A number of filaments are formed simultaneously, one to an orifice. A gathering device collects the filaments into a bundle, called a "strand," and applies a protective binder surface treatment.

Because glass filaments are easily abraded by each other, a protective surface treatment (size) is applied after the filaments are formed and before they are wound. This sizing treatment also bonds the filaments together into a strand so they can be handled, and may include a coupling agent which subsequently improves adhesion of resin to the glass.

Several surface treatments (finishes) have been developed to improve the wetting and adhesion characteristics of glass fibers. Common to all these treatments are properties that allow one part of the chemical compound to adhere to the

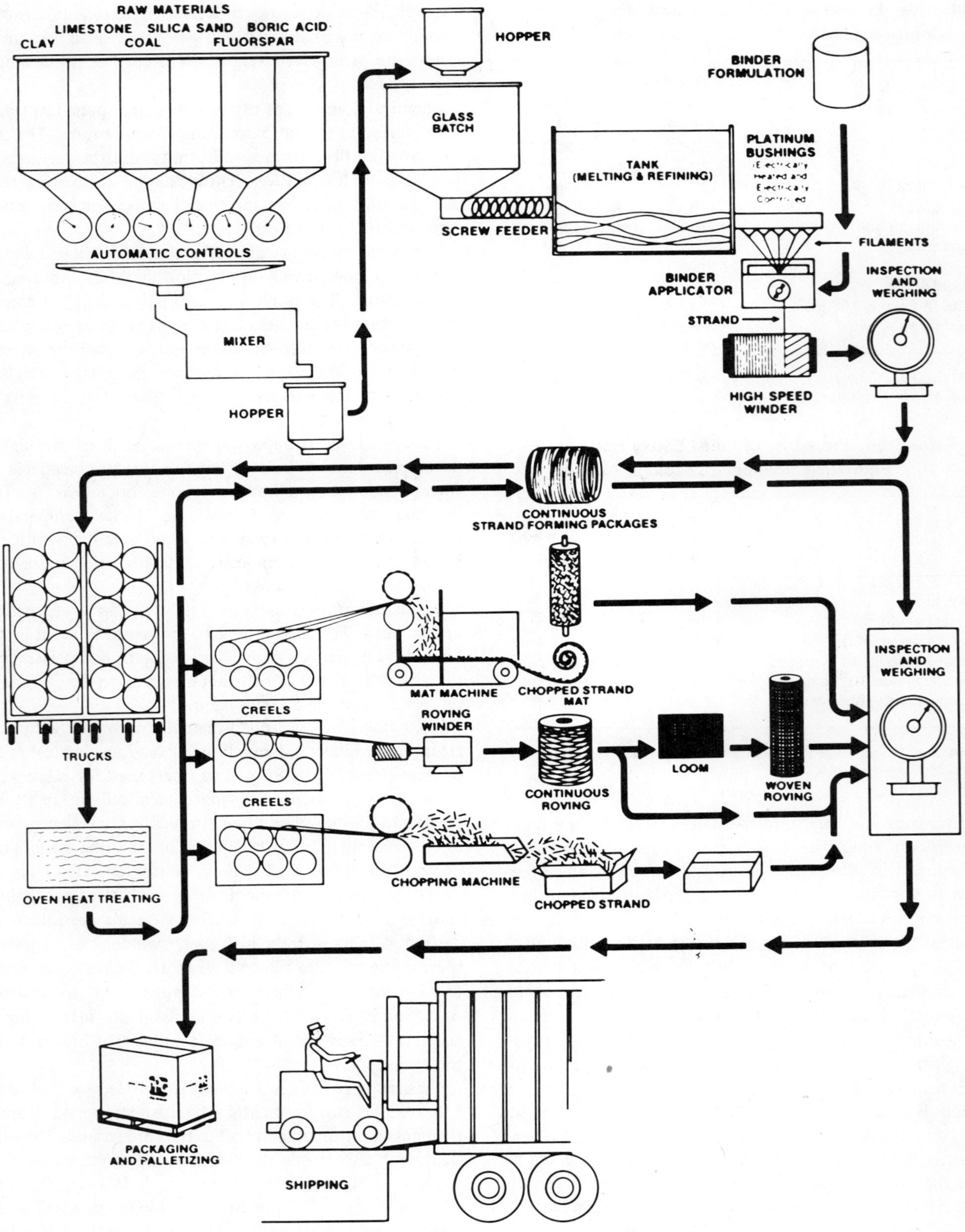

Fig. 8-7. A capsule view of the fiber glass manufacturing process for yarn, roving, and chopped strand takes the fiber from raw materials batch stage to finished products. (*Courtesy: PPG Chemicals*)

fiber surface while another part of the same molecule reacts with the resin matrix, thus effecting a chemical-mechanical molecular bridge. Table 8-9 lists a number of commercial surface treatments for glass fiber and the types of resin systems for which they were designed.

As with finishes, coupling agents are ambifunctional molecules with the ability to chemically bridge the interface between fiber and resin. The coupling agents most used are those based on silane chemistry. Table 8-10 shows the chemical structure of the more common kinds.

For thermosetting matrices, a description of the exact bonding mechanism is somewhat controversial. What is clear is that the organofunctional portion of the silane reacts with the functional groups in the thermoset system and becomes covalently bonded to the matrix. Silanes must therefore be chosen on the basis of coreactivity with the matrix resin. Other considerations are water solubility, resin wettability of the fiber surface, and functional-group catalytic activity.

In most cases, performance benefits are particularly noted in the retention of physical properties after exposure to a wet environment. In addition, moderate to excellent improvement is noted in the initial dry strengths of the composite. Table 8-11 shows performance data for various silane-treated thermoset systems.

It was originally believed that the improvement in wet strength is caused by the hydrophobicity of the silanes. The silanes prevent water molecules from migrating through the

Table 8-9. Commercial Glass Fiber Surface Treatments and Finishes.*

Greige goods:	Fabric as woven containing 2 percent binder or sizing (starch and oil) yields poor moisture resistance and poor physical properties.
111 Finish:	Fabric is heat treated to 600°F, thereby removing some of the sizing. Used with melamines.
112 Finish:	Fabric is heat treated continuously for 65 hours at 600°F, thereby removing all the sizing. Used with silicones.
114 Finish:	Methacrylate chromic chloride is applied to the 112 fabric and allows copolymerization with the resin binder, primarily polyesters, thus improving the glass-to-resin bond.
Volan A:	Similar to the 114 finish except that the chlorine is hydrolized off, yielding a fabric with good wet strength, wettability, and drapability. Used with polyesters and epoxies.
A 172 Finish:	A vinyl silane-type finish yielding excellent physical properties with polyester binders.
Finishes: (Garan and 136)	Unsaturated trichlorosilane finishes allowing for the silicone atoms to adhere to the glass and areas of unsaturation to copolymerize with the resin binder (polyesters and silicones).
A 1100 Finish:	Amino silane type yielding excellent physical properties with phenolic, epoxy, and melamine resins (not recommended for polyester binders).

*Davis, 1962.

Table 8-10. Chemical Structure of Silane Coupling Agents.*

Organic Functional Group	Chemical Structure
Vinyl	$CH_2{=}CH_2Si(O\overset{\overset{O}{\|}}{C}\,CH_3)_3$
Amine	$H_2N\,CH_2CH_2NH(CH_2)_3Si(OCH_3)_3$
Chloroalkyl	$Cl(CH_2)_3Si(OCH_3)_3$
Epoxy	$\underbrace{CH_2CH_2}_{O}CH_2O(CH_2)_3Si(OCH_3)_3$
Methacrylate	$CH_2{=}\overset{\overset{CH_3}{\vert}}{C}{-}\overset{\overset{O}{\|}}{C}O\,CH_2CH_2\overset{\overset{(CH_3)_2}{\vert}}{N^+}\,Cl^-\,(CH_2)_3Si(OCH_3)_3$
Mercapto	$HS(CH_2)_3Si(OCH_3)_3$
Styrylamine	$CH_2{=}CH{-}C_6H_4{-}CH_2\overset{+}{N}H_2CH_2CH_2NH(CH_2)_3Si(OCH_3)_3\ Cl^-$
Isothiouronium chloride	$(H_2N)_2C\overset{+}{S}\,(CH_2)_3Si(OH)_3\ Cl^-$

*Collins, 1975–76. Reprinted by permission from *Modern Plastics Encyclopedia*, McGraw-Hill, Inc.

Table 8-11. Performance Data with Silane Coupling Agents.*

	Flexural Strength,[a] 1000 psi		% Improvement[b]	
	Dry	Wet	Dry	Wet
LAMINATES				
Polyester				
181–112 Glass Finished	87	79	43	240
Silane Loading	72	67	18	190
Washed	80	66	30	185
Washed, Aged, Uncured	79	77	34	165
Carmalized	74	52	6	145
Sized	76	69	35	200
Epoxy				
181–112 Glass	92	67	24	107
Melamine				
181–112 Glass	95	94	125	450
CASTINGS				
Polyester				
Quartz	20	15	75	130
Kaolin Clay	12	9	25	22
Calcium Silicate	18	16	38	100
Aluminum Silicate	15	14	25	100
Calcined Kaolin Clay	15	14	15	27
Calcium Carbonate	12	9	20	nil
Epoxy				
Quartz Filler	21	20	nil	122
MOLDINGS				
Polyester Premix				
Clay, Chopped Glass	23	17	29	25–69
Phenolic				
Glass Beads	390[c]	280[c]	nil	5,500
Sand	777[c]		32	[d]
Aluminum Oxide	5.9	4.6	15	114

[a]Laminates prepared according to MIL R-7375 B (polyester), MIL R-9299 (phenolic), MIL R-9300 (epoxy), and MIL R-15037 C (melamine). Testing procedures conform to Federal Specification LP-406 b.
[b]% Improvement based on control containing no silane.
[c]Tensile strength, psi.[d] No control available.
*Sterman and Marsden, 1963.

resin/fiber interface which would destroy adhesion and reduce composite strength. Recently it has been proposed that the silane/fiber bond is hydrolyzable, but reversibly so. Under such conditions water may actually aid composite performance, since breaking and reforming of siloxane bonds provide a means of stress relaxation at the organic/inorganic interfaces. Such stresses as those caused by the difference in expansion coefficients of the fiber and matrix are thus reduced, and overall composite properties are improved.

With thermoplastic resins, the interaction of the silane with the matrix is not as clear as with thermosets. Since improved adhesion does occur as a consequence of the incorporation of silanes, it is suggested that there is some sort of chain scission or condensation reaction and/or hydrogen bonding between the silane and the thermoplastic.

Tables 8-12 and 8-13 show the effect of silane coupling agents on thermoplastic properties. Figures 8-8 and 8-9 show two examples of fiberglass-reinforced thermoplastics. Table 8-14 presents a selected listing of fibrous glass forms with their characteristics and uses.

Rovings are rope-like bundles of continuous, untwisted strands used in such fabrication processes as preform, press molding, filament winding, spray-up, pultrusion, and centrifugal casting (see Figure 8-16). Several forms of roving are available. Spun roving is among the least expensive, consisting of a single continuous strand that is looped and wound into roving-like form. Rovings produced for use in gun spray-up processes can be easily cut. Figures 8-10 and 8-11 show applications which take advantage of the strength and chemical resistance of roving-reinforced thermoset plastics.

Another low-cost reinforcement is *chopped strand*, available as bulk fibers (see Figure 8-12). Continuous strands are cut into lengths of $\frac{1}{8}$ inch to 2 inches. Chopped strand is used in wet slurry preforming, and special types of strand

Table 8-12. Effects of Silanes on Properties of Filled Thermoplastics.*

Resin	Filler[a]	Silane	Flex. Str. Room Temp.	Flex. Str. 150 F	Tens. Str. Room Temp. 100 F	Tens. Str. % Elong.	Wet Electricals[d] Diel. Const.	Wet Electricals[d] Dissip. Factor
Polystyrene	Unfilled	–	9-14	–	–	–	–	–
	Calcium silicate[b]	–	5.9	5.7	–	–	6.9	0.098
	Calcium silicate	A-174	7.0	5.4	–	–	3.5	0.037
	Calcined clay[c]	–	6.2	6.0	–	–	3.4	0.041
	Calcined clay	A-186	6.3	5.0	–	–	3.1	0.010
Polyvinylchloride Flexible	Unfilled	–	–	–	1-3.5	200–400	–	–
	Calcined clay	–	–	–	2.6	240	–	–
	Calcined clay	A-1100	–	–	3.6	270	–	–
Polycarbonate	Unfilled	–	12.5	6.3[e]	–	–	2.8	0.001
	Calcium silicate	–	10.2	6.8[e]	–	–	3.9	0.035
	Calcium silicate	A-1100	13.8	7.9[e]	–	–	3.7	0.013
Polyethylene	Calcined clay	–	–	–	–	–	3.0	0.082
	Calcined clay	A-174	–	–	–	–	2.7	0.003
	Calcium silicate	–	–	–	–	–	4.2	0.147
	Calcium silicate	A-174	–	–	–	–	2.9	0.014

[a]All systems contain 100 phr of filler except plasticized PVC, which contains 43 phr. [b]Wollastonite, (Cab-O-Lite, Cabot Corp.). [c]Whitex, (Southern Clays Co.). [d]Samples conditioned 16 hr at 100% R.H. at 120 F. [e]At 250 F. SOURCE: Silicones Div., Union Carbide Corp.
*Hosford, 1965.

have been developed for incorporation into premix compounds. Use of chopped strand in reinforced thermoplastic molding greatly increases strength, stability, and tolerance levels in the molded product. In general, the physical properties of a reinforced thermoplastic improve as the percentage of glass content goes up. Figure 8-13 shows an application for chopped glass in epoxy.

Fig. 8-8. Award-winning ventilating fan made from glass-reinforced nylon. *(Courtesy: Fiberfil Division, Dart Industries)*

The two basic types of reinforcing mats are chopped strand and continuous filament (see Figure 8-14). Mats provide nondirectional reinforcement in hand lay-up, matched metal die molding, and panel laminating processes. Chopped strand mats are available with either chemical binders or mechanical stitching (needled). Continuous filament and chopped strand mats provide essentially the same degree of reinforcement. The major difference is in handling characteristics. Needled mats are softer and more drapable, and are available in heavy weights. Bonded mats are somewhat stiffer, but stronger. Continuous strand mats can be molded into more complex shapes without risk of tearing.

Milled fibers are short nodules of filamented glass produced by hammer milling continuous strands. Lengths range from $\frac{1}{16}$ to $\frac{1}{32}$ inch (see Figure 8-15). They are used in premix, as anticrazing filler reinforcement in casting resins, and in resin additives that provide greater body and dimensional stability.

An extremely smooth surface can be obtained on a reinforced plastic molding by placing a thin mat of staple monofilaments over the reinforcement. Such a mat is called a *surfacing* or *overlay mat* or *veil.* This type of mat is from 10 to 30 mils thick and provides little actual reinforcement. It does allow, however, a resin-rich surface layer to be incorporated into the molded part. Surfacing mats are relatively strong and stiff and are applied when molding flat areas or parts of simple contour. Overlay mats are fluffy,

Table 8-13. Flexural Properties of Glass-Reinforced Thermoplastic Laminates.*

	Room Temperature		Aged[c]		H. Temp.	
Silane[a]	Strength, 1000 psi	Modulus, 10^6 psi	Strength, 1000 psi	Modulus, 10^6 psi	Strength, 1000 psi	Modulus, 10^6 psi
POLYSTYRENE[b]						
Control						
No Glass	9.5	–	–	–	–	–
Glass: 181–112[m]	24.4	2.1	15.8	1.8	16.8[p]	1.8[p]
A-174	48.4	2.6	31	2.4	28.9[p]	2.1[p]
A-186	42.8	2.4	33.5	2.3	26.3[p]	1.8[p]
A-1100	30	1.8	23.5	1.6	18.4[p]	1.3[p]
NYLON[d]						
Control						
No Glass	8-14	–	–	–	–	–
Glass: 181–112[m]	26.2	1.6	18.4	1.3	13.5[l]	1.0[l]
A-1100	54.1	2.3	40.9	2.0	26.1[l]	1.6[l]
A-186	49.9	2.2	41.3	2.1	25.1[l]	1.5[l]
A-187	47	2.0	41.5	2.1	27.9[l]	1.6[l]
ACRYLIC[e]						
Control						
No Glass	13-17	–	–	–	–	–
Glass: 181–112[m]	42.6	2.3	27.6	2.1	26.8[p]	2.1[p]
A-1100	62.1	2.4	53.3	2.2	34.0[p]	2.1[p]
A-174	47.4	2.7	42.1	2.4	33.8[p]	2.4[p]
A-186	43.8	2.4	37.3	2.2	29.3[p]	2.1[p]
A-187	33.9	2	31.6	1.9	22.2[p]	1.7[p]
PVC[f]						
Control						
No Glass	10-16	–	–	–	–	–
Glass: 181-112[m]	23.8	–	20	–	–	–
A-1100	36.2	–	32.9	–	–	–
A-186	32	–	30.6	–	–	–
A-187	33.2	–	24.4	–	–	–
Y-2967[g]	44.1	–	40.3	–	–	–
POLYCARBONATE[h]						
Control						
No Glass	11-13	–	–	–	–	–
Glass: 181-112[m]	38.6	–	21.1	–	29.6[i]	–
A-1100	51.1	–	37.9	–	35.7[i]	–
A-186	44.8	–	39.8	–	38.1[i]	–
A-187	45.2	–	37.4	–	35.6[i]	–
ABS[j]						
Control						
No Glass	10-16	–	–	–	–	–
Glass: 4-14[m]	18.9	1.6	14.0[p]	1.6[p]	3.8[k]	1.0[k]
A-186	41	2.3	29.3[p]	2.2[p]	3.3[k]	–
A-187	46.5	2.3	34.2[p]	2.1[p]	17.0[k]	1.6[k]
A-174	44.6	2.4	35.9[p]	2.4[p]	14.8[k]	1.7[k]
A-1100	38.8	2.5	9.3[p]	1.4[p]	<1[k]	–

[a]Silane loading 0.5 weight per cent based on the glass fabric. [b]Resin content 40%. [c]Aged 16 hr at 100% RH/120 F, except where noted; tested at room temperature. [d]Resin content 42%. [e]Resin content 43%. [f]Resin content 5.5%. [g]Y-2967 is chemically modified A-1100 and contains 62% silane in ethanol; loading given is based on contained silane. [h]Resin content 47%. [i]At 250 F. [j]Resin content 38%. [k]At 150 F. [l]At 400 F. [m]Heat-cleaned. [n]All laminates were 11 plies 181 cloth. [p]At 200 F. SOURCE: Silicones Div., Union Carbide Corp.

*Hosford, 1965.

Table 8-14. Characteristics and Uses of Different Forms of Fibrous Glass.*

Type	General Description	Process	Glass Content	Typical Applications
Chopped strands	Strands cut to lengths of $\frac{1}{8}$ to 2 inches, sold in bulk	Premix molding, wet slurry preforming, thermoplastic screw injection molding	15–45	Electrical and appliance parts, ordinance components, industrial parts
Rovings	Rope-like bundle of strands of glass fibers	Filament winding, continuous panels, matched die molding, sprayup, pultrusion, centrifugal casting	25–80	Pipe, automotive bodies, rod stock, rocket motor cases, ordinance
Reinforcing mats	Chopped or continuous strands in non-woven random matting	Matched die molding, hand layup, centrifugal casting	20–45	Marine applications, translucent sheets, truck and auto body panels
Yarns	Twisted yarns with afterfinish	Fabric weaving, filament winding, unidirectional reinforcement	50–70	Fishing Rods
Surfacing and overlay mats	Non-reinforcing monofilaments in random mat	Matched die molding, hand layup, filament winding	5–15	Automobile bodies, housings, etc., where smooth surfaces are required
Woven fabrics	Woven cloths from fibrous glass yarns with afterfinish	Hand layup, vacuum bag, autoclave high pressure laminating	45–65	Aircraft structures, marine, ordinance, electrical flat sheet and tubing, radomes, aircraft skin
Woven roving	Woven fibrous glass strands—coarser and heavier than fabrics	Hand layup	40–70	Large containers, marine
Non-woven fabrics	Unidirectional and parallel rovings in sheet form	Hand layup, filament winding	60–80	Aircraft structures
Sheet molding compound (SMC)	Mat or chopped fibers, semi-solid filler resin mix, and catalyst in relatively dry flowable sheet	Matched die molding	20–35	Automobile bodies and components, housings

*Paulus, 1968–69. Reprinted by permission from *Modern Plastics Encyclopedia*, McGraw-Hill, Inc.

Fig. 8-9. A design award was given to this wheel balancer cabinet made from glass-reinforced polypropylene. (*Courtesy: Fiberfil Division, Dart Industries*)

Fig. 8-10. Bondstrand® reinforced thermosetting resin pipe and fittings in sewage treatment plant handles large volume of domestic and industrial wastes. Note compatible mating of steel and reinforced thermosetting resin flanges. (*Courtesy: Ameron, Inc.*)

Fig. 8-11. Extensive configuration of Bondstrand® pipelines is part of water flood system in a California oil field. (*Courtesy: Ameron, Inc.*)

Fig. 8-12. Chopped fiberglass strands. (*Courtesy: Owens-Corning Fiberglas Corporation*)

more drapable, and harder to handle, but can be used on parts of more complex shape.

Yarns are higher-cost reinforcement produced on special textile equipment by twisting fibrous glass strands from the basic forming package (see Figure 8-16). Section beams, produced by wrapping many parallel yarns on a mandrel, are used primarily in fabric weaving. They permit a large number of yarns to be handled conveniently as a unidirectional reinforcement.

The first type of fibrous glass reinforcement developed was woven fabric (see Figure 8-16), a high-cost reinforcement widely specified because of its high strength characteristics. Breaking strength, fabric thickness, fabric weight, pliability, and finish are factors to be considered when

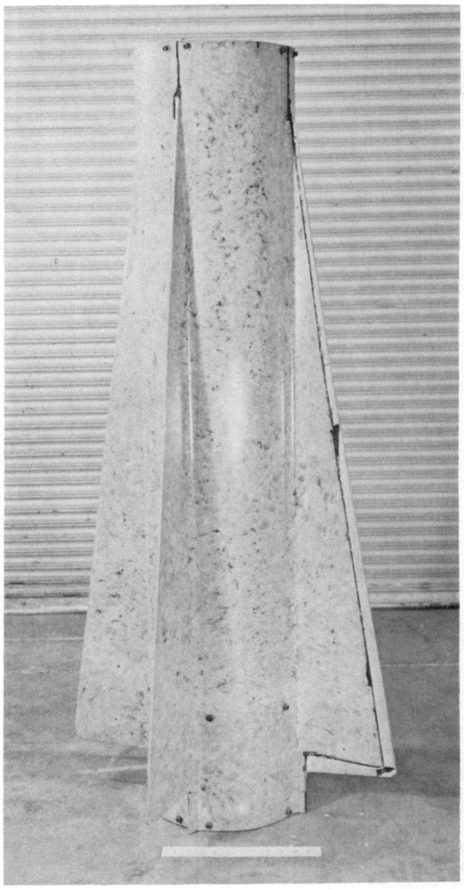

Fig. 8-13. Compression-molded rocket body quadrant made of epoxy and chopped glass. (*Courtesy: Hughes Aircraft Co.*)

Fig. 8-14. Reinforcing mats of fiberglass. (*Courtesy: Owens-Corning Fiberglas Corporation*)

Fig. 8-15. Milled fiberglass. (*Courtesy: Owens-Corning Fiberglas Corporation*)

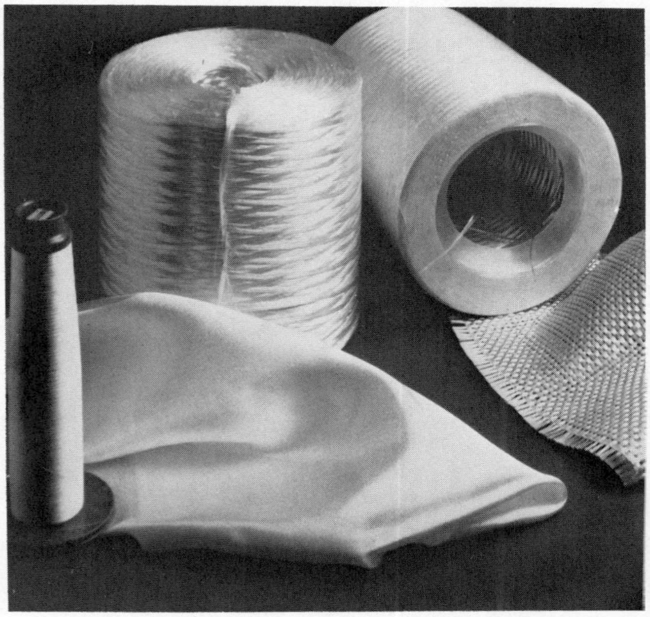

Fig. 8-16. Various forms of fiberglass yarn, fabric, and roving. (*Courtesy: Owens-Corning Fiberglas Corporation*)

fabric selection is made. The most common weaves used in fabrics include plain or taffeta weave and satin weave (see Figure 8-17). Taffeta is a repeated diagonal or bias weave. Satin weave is more pliable and more easily conformed to three-dimensional contours.

High strength can be attained with woven roving, a thick reinforcement made with either conventional or spun roving (see Figure 8-16). Woven roving is available in a variety of weaves, costs less than conventional fabrics, and is sometimes called roving cloth. It is produced by weaving rovings into coarse, heavy, drapable fabrics. Tables 8-15 through 8-19 show typical properties of a cross-section of glass-

Plain taffeta weave
The plain weave is the most simple and commonly used. In this type of weave, the warp and filling yarns cross alternately.

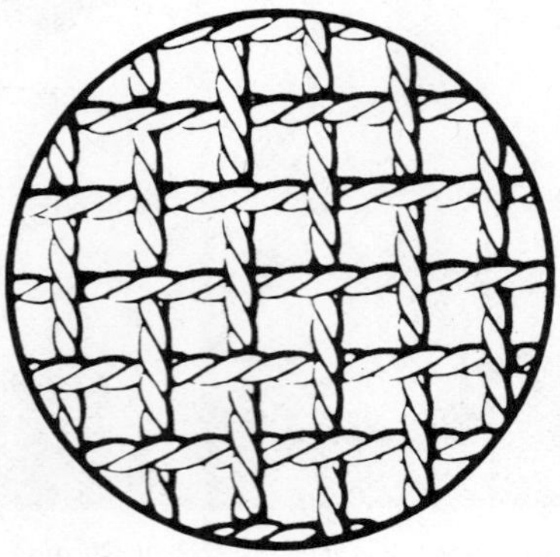

Twill weave
A basic weave characterized by a diagonal rib, or twill line. Each warp end floats over at least two consecutive picks.

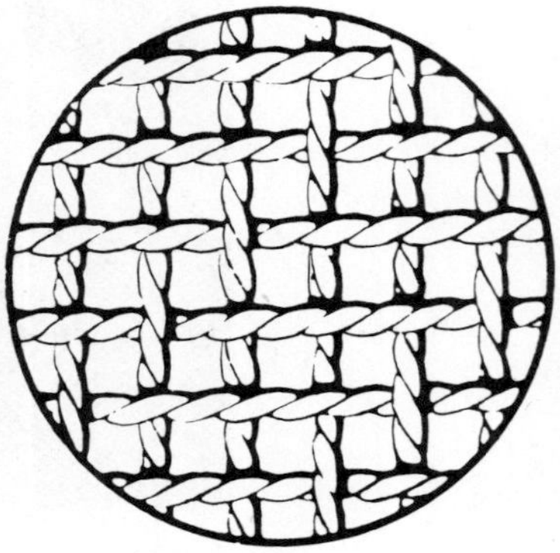

Leno Weave
A leno weave is a locking type weave in which two or more warp yarns cross over each other and interlace with one or more filling yarns. It is used primarily to prevent shifting of yarns in open fabrics.

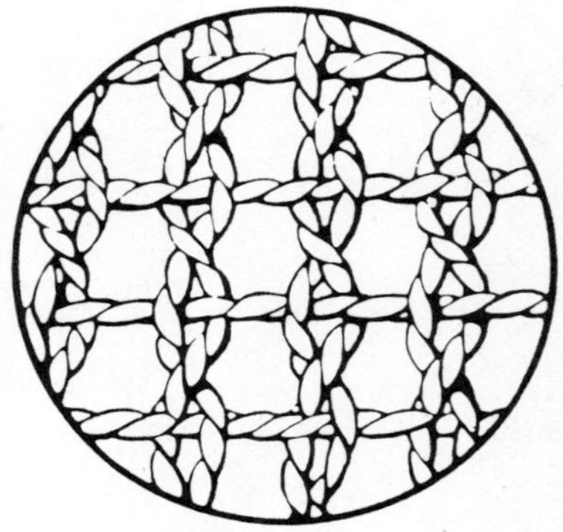

Basket weave
In this type of weave, two or more warp yarns cross alternately with two or more filling yarns.

Long Shaft Satin
This construction has one warp end weaving over three or more and under one filling pick.

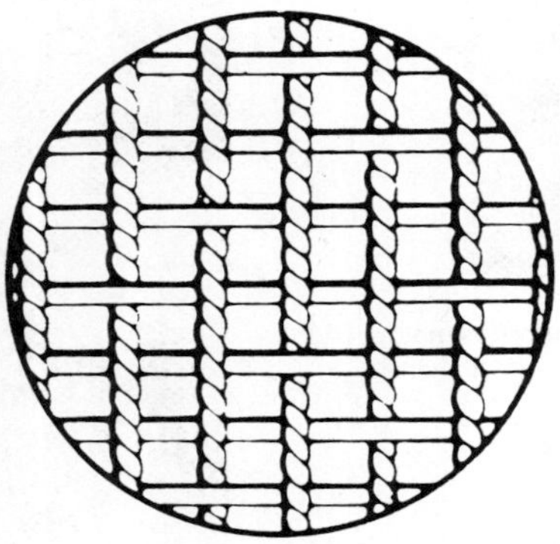

Mock Leno Weave
A mock leno weave is exactly what the name implies. It is an open type weave that resembles a leno and is accomplished by a system of interlacing that draws a group of yarns together and leaves a space between groups. The warp yarns do not actually cross each other as in a true leno and, therefore, no special attachments are required for the loom.

Fig. 8-17. Types of fiberglass weaves. (*Courtesy: PPG Chemicals*)

reinforced thermoplastics. Table 8-20 shows typical properties of epoxy/glass composites.

A special manufacturing process produces extremely fine, continuously hollow glass filaments that are gathered into multifilament strands. These hollow fibers permit production of a much thicker reinforced plastics part, with an equal weight of glass, than is possible using solid fibers. Hollow fiber filaments are treated as drawn to assure abrasion resistance, high tensile strength, and good adhesion between glass and matrix. Buckling failures of all types are averted because of the increase in rigidity. In addition, hollow fibers allow weight savings of up to 40% with only moderate loss in strength. Hollow fibers are of particular value in high-performance filament winding. Table 8-21 compares properties of this hollow glass with other epoxy/glass composites.

Other Silica-Based Products

Sand, quartz, tripoli, and diatomaceous earth are naturally occurring forms of silica which differ in particle size, degree of crystallinity, and hardness. Large volumes of graded silica sand are used in shell molding processes, resin cements, and chemically resistant epoxy flooring. Sharp silica sand is often spread over uncured surfaces of resin cements to provide abrasive qualities on the surface. Naturally occurring quartz has been used as a filler with phenolic and epoxy resins to provide ablative insulators for nose cones, space capsules, and rocket motors.

Diatomaceous earth, or fossil flour, is obtained from deposits of diatom skeletons. The commercial products are obtained by heat treatment which removes much of the water from the naturally occurring substance. The use of

Typical values of 30% dispersed glass fiber fortification.

For additional aid in material comparisons, these resins have been grouped as follows: Styrenics, Olefins, Nylons, Misc. Crystallines, Arylates, and Miscellaneous.

For additional fortification levels and other fortifiers such as TFE, MoS_2, see specific product data sheets.

Table 8-15. Design Properties Profile of LNP Thermocomp® Fortified Thermoplastics.*

Base Polymer	LNP Thermocomp Series Code	Property	Physical			Mechanical				Thermal				Electrical		
			Specific Gravity	Water Absorption, 24 Hrs.	Mold Shrinkage	Tensile Yield Strength	Flexural Modulus	Izod Impact Strength Notched/ Unnotched	Rockwell Hardness	Deflection Temperature @264 psi	Long Term(4) Service Temperature	Coefficient of Linear Thermal Expansion	Flammability	Dielectric Strength (ST)	Dielectric Constant 60 Hz– 10^6 Hz	Dissipation Factor 60 Hz– 10^6 Hz
		Units	–	%	in./in.	psi	10^3 psi	ft.-lbs./in.	–	°F	°F	10^{-5} in./in./°F	–	volts/mil.	–	–
		ASTM	D792	D570	D955	D638	D790	D256	D785	D648	–	D696	U.L.	D149	D150	D150
Acrylonitrile-Butadiene-Styrene (ABS)	AF		1.28	0.14	0.001	14,500	1,100	1.4/6-7	M99,R124	220	155	1.6	94HB*	–	–	–
Styrene-Acrylonitrile (SAN)	BF		1.31	0.10	0.0005-0.001	17,400	1,500	1.0/3-4	M94,R123	215	140	1.8	94HB†*	–	–	–
Polystyrene	CF		1.28	0.05	0.0005-0.001	13,500	1,300	1.0/2-3	M92	215	120	1.9	94HB	550	2.81-2.81	0.0007-0.0008
Styrenic Terpolymer	NF		1.32	0.10	0.0015	16,000	925	2.5/8-9	M74	300	185	1.35	94HB	–	–	–
Polyethylene (HDPE)	FF		1.17	0.015	0.0030	10,000	900	1.1/8-9	R85	260	185	2.7	94HB	–	–	–
Polypropylene	MF		1.13	0.03	0.004	9,800	800	1.6/5-6	M57,R111	295	220	2.0	94HB†*	475	2.30-2.20	0.001-0.003
Polypropylene	MFX(1)		1.13	0.02	0.0035	13,500	800	1.9/10-12	M57,R111	310	220	2.1	94HB	–	–	–
Nylon–Type 6	PF		1.37	1.1	0.0035	23,000	1,200	2.3/20	M92,R121	420	215**	1.7	94HB†*	450	4.20-3.60	0.009-0.018
Nylon–Type 6/12	IF		1.30	0.20	0.0035	22,000	1,100	2.4/20	M93,R120	415	210**	1.5	94HB	440	4.20-3.50	0.013-0.015
Nylon–Type 6/10	QF		1.30	0.20	0.0035	21,000	1,100	2.4/20	M93,R120	420	210**	1.5	94HB†*	440	4.20-3.50	0.013-0.015
Nylon–Type 6/6	RF		1.37	0.9	0.004	26,000	1,300	2.0/17	M96,R121	490	240**	1.8	94HB†*	440	4.20-3.50	0.009-0.0180
Amorphous Nylon	XF		1.35	0.19	0.003	21,500	1,150	1.2/6-8	R119	285	185**	1.8	94HB	–	–	–
Acetal	KFX(1)		1.63	0.30	0.003	19,500	1,400	1.8/8-10	M86	330	220	2.2	94HB	525	3.95-3.95	0.0035-0.0065
Polyphenylene Sulfide	OF		1.56	0.04	0.002	20,000	1,600	1.4/8-9	R121	500	355	1.3	94VO†	510	3.88-3.78	0.003-0.007
Thermoplastic Polyester (PBT)	WF		1.52	0.06	0.003	20,000	1,200	2.6/16-18	M84,R119	430	280	1.2	94HB*	510	3.60-2.00	0.002-0.020
Polycarbonate	DF		1.43	0.07	0.001	18,500	1,200	3.7/17-18	M95,R118	300	260	1.3	94V1†*	480	3.50-3.43	0.0010-0.0075
Polysulfone	GF		1.45	0.20	0.002-0.003	18,000	1,200	1.8/14	M92,L108	365	300	1.4	94VO†	480	3.55-3.49	0.0019-0.0049
Polyether Sulfone	JF		1.60	0.34	0.002-0.003	20,000	1,225	1.5/10	M98	415	350	1.8	94VO	460	3.80-3.76	–
Modified Polyphenylene Oxide	ZF		1.28	0.06	0.001-0.002	18,500	1,150	1.7/9-10	M93	310	195	1.4	94HB	550	2.90-2.90	0.0010-0.0015
Thermoplastic Polyurethane	TF		1.46	0.25	0.004	8,200	190	9.5/28-29	D65††	170	140	2.5	94HB	–	–	–
Polyester Elastomer	YF		1.42	0.17	0.003	8,500	320	5.0/16	D70††	340	150	2.5	94HB	–	–	–
FEP	LF(2)		2.21	0.01	0.002-0.004	6,000	800	8.0/17	D63††	350	390	2.4	94VO	475	2.55-2.52	0.0020-0.0002
ETFE(3)	UF		1.89	0.02	0.004-0.006	14,000	1,500	7.5/17-18	R74	460	350	1.6	94VO	410	3.6-3.6	0.004-0.005

(1) Chemically Coupled (2) 20% Glass Fiber Fortified (3) Based on DuPont's Tefzel (4) Definition - Air temperature which would cause the 30% glass reinforced resin to lose 50% of its mechanical properties at 100,000 hours. † U.L. Recognized †† Shore D Hardness *U.L. Recognized Flame Retardant Versions Available. **Values on Heat Stabilized (H.S.) Compounds

*Courtesy: LNP Corporation, Malvern, Pa.

Table 8-16. Guide for Glass Fiber Reinforced

	Glass Fiber Content (wt. %)	Specific Gravity	Tensile Strength (psi x 10^3)	Tensile Elongation (%)	Tensile Modulus (psi x 10^5)	Flexural Strength (psi x 10^3)	Flexural Modulus (psi x 10^5)	Compressive Strength (psi x 10^3)	Heat Deflection Temperature @ 264 psi (°F)	Impact Strength, Izod Notched (ft-lb/in.)	Coefficient of Linear Thermal Expansion (10^{-5} in./in./°F)
Test Method (ASTM)		D792	D638	D638	D638	D790	D790	D695	D648	D256	D696
ABS (acrylonitrile-butadiene-styrene)	10	1.10	9.5	3.0	6.7	14.8	6.5	12.0	208	1.2	2.3
	20	1.22	11.0	2.0	7.4	15.5	7.1	14.0	210	1.1	2.1
	30	1.28	13.0	1.4	9.1	16.8	9.3	15.0	212	1.0	1.7
Acetal	10	1.54	10.5	2.4	9.6	15.5	8.8	10.0	255	1.0	2.9
	30	1.63	12.0	2.0	11.2	16.5	10.5	11.8	325	0.8	2.4
Nylon 6	15	1.25	15.0	4.0	8.5	23.0	7.8	14.0	385	1.5	1.7
	30	1.37	24.0	3.0	10.5	29.0	10.0	24.0	400	2.2	1.5
Nylon 6/6	13	1.23	14.0	4.0	9.0	25.0	6.5	13.5	475	1.0	1.5
	30	1.37	25.0	3.0	13.0	34.0	13.0	27.0	485	2.0	1.3
Nylon 6/12	30	1.30	19.6	4.0	12.0	28.0	11.0	20.0	390	2.2	2.2
Polycarbonate	10	1.26	12.0	9.0	7.5	16.0	6.0	14.0	280	2.0	1.8
	30	1.43	17.5	2.0	12.5	20.5	10.0	17.0	290	2.4	1.3
Polyester (thermoplastic)	30	1.52	19.0	4.0	12.0	28.0	11.5	18.0	415	1.8	1.4
Polyethylene	10	1.04	5.2	4.0	3.7	6.6	3.6	5.0	230	1.4	3.0
	30	1.18	8.6	3.0	7.3	12.9	7.1	6.0	255	1.7	2.1
Polyphenylene Oxide (modified)	20	1.21	14.5	5.0	9.25	18.5	7.5	17.6	290	1.8	2.0
Polyphenylene Sulfide	40	1.64	22.0	3.0	20.5	37.0	19.0	21.0	510	1.5	1.1
Polypropylene	10	0.98	6.3	4.0	3.6	7.8	3.5	6.0	260	0.8	2.6
	20	1.04	6.5	3.0	5.4	8.3	5.2	6.5	270	1.1	2.4
	30	1.12	6.8	2.0	6.4	9.2	6.2	6.8	280	1.3	2.1
Polypropylene (chemically coupled)	10	0.98	7.2 to 8.6	4.0	5.3	10.4 to 13.6	5.1	6.2 to 6.4	275 to 285	1.2 to 1.4	2.5
	20	1.04	8.2 to 9.8	3.0	5.6	11.8 to 15.4	5.4	6.4 to 6.8	288 to 295	1.3 to 1.5	2.3
	30	1.12	9.8 to 12.0	2.0	6.7	13.1 to 19.0	6.6	6.5 to 7.0	294 to 300	1.3 to 1.7	2.0
Polystyrene	20	1.20	11.0	1.5	11.0	15.5	9.9	15.0	210	1.0	2.3
	30	1.29	13.5	1.2	13.1	17.5	12.4	18.0	212	1.0	1.8
Polystyrene High Heat Copolymer	20	1.22	13.0	1.2	12.0	19.0	11.5	16.0	240	1.1	2.2
Polystyrene High Heat Terpolymer	30	1.35	12.0	1.8	9.4	17.8	8.3	11.0	300	1.5	2.0
Polysulfone	20	1.38	14.0	2.5	8.7	20.0	8.5	18.0	355	1.2	1.5
	40	1.55	18.0	1.5	16.8	25.0	15.5	20.0	365	1.5	1.3
Polyurethane	10	1.22	4.8	48.0	0.95	6.2	0.90	5.0	130	14.0	3.4
PVC (polyvinyl chloride)	20	1.58	14.0	3.0	1.1	21.0	10.0	12.0	180	1.5	2.3
SAN (styrene acrylonitrile)	20	1.22	14.5	1.8	12.5	19.0	11.0	17.5	215	1.2	2.1
	35	1.35	16.0	1.4	15.0	22.5	13.5	21.0	220	1.0	1.6

Materials
The glass fiber contents shown represent the most common ratios used with that kind of thermoplastic resin; many other glass fiber formulations are available.

Flammability
Most of these resins can be compounded for a V-0 flammability rating (UL94), however, desirable properties tend to be reduced by the flame retardant additives and cost is increased. PVC and polyphenylene sulfide have a V-0 rating due to their natural characteristics. PC has a V-2 rating unreinforced, V-1 with glass added, and when brominated and glass reinforced it is rated V-0. Polysulfone has a V-1 rating when unreinforced; the addition of glass fiber reinforcement reduces this to V-0. Nylon 6/6 unreinforced has a V-2 rating; the addition of glass improves this to HB.

High Temperature Applications
The maximum temperature for continuous use is listed for general reference. The actual value for each material must be based on the specific load, duration, frequency, and environment of that particular application.

High Performance Plastics.*

Maximum Temperature Continuous use (°F)	Water Absorption 24 Hrs. (%)	Volume Resistivity (Ohm-Cm)	Dielectric Strength Dry (V/mil)	Mold Shrinkage (in./in.)	Cost Per Cu. In. ($)
	D570	D257	D149	D955	
170	0.3	10^{15}	450	0.003	.0276
180	0.3	10^{16}	465	0.002	.0317
180	0.2	10^{16}	480	0.002	.0345
230	0.22	10^{14}	510	0.006	.0557
260	0.2	10^{14}	480	0.005	.0506
200	1.8	10^{15}	420	0.007	.0550
230	1.3	10^{15}	410	0.004	.0558
225	1.0	10^{16}	530	0.007	.0542
260	0.9	10^{16}	500	0.004	.0558
230	0.2	10^{13}	500	0.004	.0939
260	0.14	10^{15}	440	0.005	.0646
270	0.12	10^{15}	480	0.003	.0779
250	0.06	10^{16}	600	0.003	.0532
180	0.08	10^{16}	680	0.005	.0201
200	0.06	10^{16}	610	0.003	.0218
240	0.06	10^{17}	420	0.003	.0489
450	0.01	10^{16}	510	0.002	.1450
180	0.05	10^{16}	440	0.007	.0203
190	0.05	10^{16}	440	0.006	.0222
210	0.04	10^{16}	420	0.006	.0243
200	0.05	10^{16}	430	0.006	.0221
230	0.04	10^{16}	425	0.006	.0244
250	0.04	10^{16}	425	0.006	.0271
170	0.1	10^{16}	410	0.002	.0238
180	0.1	10^{16}	415	0.001	.0279
220	0.28	10^{16}	400	0.003	.0303
240	0.10	10^{16}	400	0.002	.0542
325	0.2	10^{16}	500	0.004	.1181
340	0.18	10^{16}	480	0.002	.1192
110	0.4	10^{12}	380	0.007	.1000
150	0.09	10^{13}	420	0.002	.0445
180	0.24	10^{16}	490	0.002	.0286
190	0.21	10^{16}	500	0.001	.0336

*Reprinted by permission from *Plastics World*, March 1977, **35**(3): 65.

this type filler in polyurethane foams provides improved compressive strength and reduces shrinkage.

Amorphous silica is obtained by precipitating silica from a solution of sodium silicate. In one commercial process, $CaCl_2$ is added and the calcium silicate produced is treated with HCl. In other processes, the Na_2SiO_3 solution is acidified by gaseous CO_2 or HCl. Much of the reinforcing properties of these light-colored fillers is associated with active silanol surface groups which are destroyed at elevated temperatures.

Pyrogenic silica is obtained by flame or fume processing of $SiCl_4$ in the presence of H_2 and O_2. Pyrogenic silica, also known as fumed silica, has an average particle size of 0.015 micron, low free-moisture content, and a very large surface area (about 200 m^2/gram). Functions of the silica in either thermosetting or thermoplastic systems are to thicken the resin and control thixotropy, reinforce a resin, suspend pigments and fillers, prohibit film blocking, improve flow properties of dry powders, and/or increase electrical properties of cable compounds. Dispersion methods, system polarity, and other parameters will influence the degree of efficient usage, but only small concentrations are usually required to achieve performance levels.

Asbestos

The six fibrous minerals normally classified as asbestos occur in two major groups: (1) fibrous serpentine, or chrysotile; and (2) the amphiboles which include crocidolite, amosite, anthophyllite, actinolite, and tremolite. Actinolite and tremolite have generated no commercial plastics significance.

Chrysotile comprises better than 90% of the total asbestos production. It is a hydrated magnesium silicate with double layers of silica tetrahedrons and magnesium hydroxide. The amphiboles are various combinations of magnesium, iron, calcium, and sodium hydrated silicates with a basic structure of Si_4O_{11} groups in a double-chain configuration. Table 8-22 compares the properties of the various asbestos types.

The mining of asbestos is a difficult process. It results in a product that generally has a wide spectrum of fiber lengths and included impurities. Various classification schemes based on fiber length, color, or purity are used.

Uses of asbestos include the manufacture of friction materials such as brake linings and blocks, transmission bands, and clutch facings, where the asbestos is bonded with phenolic and other resins. In the production of flooring tile, asbestos is used with asphalt and vinyl.

The premix technique is a common production method. The asbestos is mixed with the resin and other ingredients, and the mix is then molded to its final shape. Another method involves putting the asbestos into the form of a paper, felt, or mat which is then impregnated with the resin, formed, and cured to the final part. Other techniques include spray in place and spray and mold.

Table 8-17. Effect of Different Concentrations of Glass Fibers on Physical/Mechanical Properties (As Exemplified by Nylon 6/6).*

	% of glass fiber, by wt.						
Property	0	10	20	30	40	50	60
Specific gravity	1.14	1.21	1.28	1.37	1.46	1.57	1.70
Specific volume (cu. in./lb.)	24.3	22.9	21.6	20.1	19.0	17.6	16.3
Tensile strength (psi x 10^3)	12	13	19	25	31	32	33
Tensile elongation (%)	60	3.5	3.5	3.0	2.5	2.5	1.5
Flexural strength (psi x 10^3)	15	20	29	34	42	46	50
Flexural modulus (psi x 10^5)	4.0	6.0	9.0	13	16	22	28
Compressive strength (psi x 10^3)	4.9	13	23	27	28	29	30
Heat deflection temp. @ 264 psi (°F)	150	470	475	485	500	500	500
Thermal expansion (10^{-5} in./in./°F)	4.5	1.6	1.4	1.3	1.2	1.0	0.9
Water absorption, 24 hr. (%)	1.6	1.1	0.9	0.9	0.6	0.5	0.4
Mold shrinkage (10^{-3} in./in.)	15	6.5	5	4.0	3.5	3.0	2.0

Data from commercially available materials.

*Reprinted by permission from *Plastics World*, March 1977, 35(3):65.

Table 8-18. Effect of Various Fillers/Reinforcements on Physical/Mechanical Properties (as Exemplified by Nylon 6/6).*

Property	Unrein-forced	Glass fiber	Carbon (graphite)	Mineral	Carbon/glass fiber	Mineral/glass fiber	Glass bead
Reinforcement content (% by wt.)	0	40	40	40	20C/20G	20M/16.5G	40
Specific gravity	1.14	1.46	1.34	1.50	1.40	1.42	1.44
Tensile strength (psi x 10^3)	12	31	40	15	34	17.5	13
Flexural modulus (psi x 10^5)	4.0	16	34	11	28	9.5	8
Impact strength, notched/unnotched (ft-lb/in.)	0.9/6	2.6/19	1.6/13	0.7/8	1.8/16	1/13	1/5.5
Heat deflection temperature @ 264 psi (°F)	150	500	500	440	500	470	190
Thermal expansion (10^{-5} in./in./°F)	4.5	1.4	0.8	3.0	1.15	2.5	2.0
Mold shrinkage (10^{-3} in./in.)	15	4	—	9	—	7	—
Water absorption, 24 hr. (%)	1.6	0.6	0.4	0.45	0.5	0.5	0.65

*Reprinted by permission from *Plastics World*, March 1977, 35(3):65.

Table 8-19. Thermal Aging Effect on Tensile Strength at 73°F of High-Temperature Thermoplastics (psi).*

	Aging time (hrs.) at 500F						
Resin (% glass fiber by weight)	0	100	250	500	750	1000	1500
Ethylene-tetrafluoro ethylene (ETFE) (20)	11,300	11,500	10,000	7000	5000	3800	2300
Fluorinated ethylene propylene (FEP) (20)	5000	5100	4800	4700	4700	4600	4500
Polyamide-imide (0)	27,400	27,200	26,600	26,000	24,500	23,500	22,000
Polyarylsulfone (0)	13,100	11,500	10,500	10,000	9500	8400	7600
Polyethersulfone (40)	22,700	15,600	14,800	14,300	13,700	12,200	10,500
Polyimide (30)	13,000	15,000	14,300	13,400	12,800	12,000	11,200
Poly-p-oxybenzoate (0)	23,000	18,000	16,300	16,000	15,500	15,100	13,000
Polyphenylene sulfide (40)	23,200	16,400	16,000	15,500	15,000	14,500	13,800

*Reprinted by permission from *Plastics World*, March 1977, 35(3):65.

Table 8-20. Typical Properties of Glass Roving and Epoxy Composites.*

Properties	*E/801*	*E/HTS*	*S/HTS*
Rovings			
Ignition loss, %	0.65	1.3	1.4
Weight per lineal yard, g.	0.647	0.647	0.604
Ultimate tensile (strand test), p.s.i.	280,000	350,000	450,000
End count	20	20	20
Epoxy 787 composites			
NOL tensile, p.s.i.	180,000	215,000	290,000
NOL shear,			
Room temp.	10,500	11,000	11,500
250° F.	2,000	2,200	3,000
Room temp. after 2-hr. boil	9,500	10,500	11,000

*Martenson, 1965. Reprinted by permission from *Modern Plastics*, McGraw-Hill, Inc.

Table 8-21. Performance Data on New Versus Older Types of Glass Fibers.[a,*]

Material	Bulk Glass		Epoxy/Glass Composites				Outstanding Characteristics	Typical Applications
	Density, lb/cu in.	Dielec Const[b]	Spec Gr	Dielec Const[b]	Loss Tan	Ultimate Ten. Str. 1000 psi[c]		
S(994)-HTS	0.090	5.6	2.0	4.7	0.016	413	Exceptionally high tensile strength and strength-to-weight ratio	Internal pressure vessels, rocket motor cases
D-556	0.078	4.0	1.8	3.5	0.005	191	Excellent electrical stability over wide range of frequency and temperature.	Electrical radomes
Hollow[d] Glass	≈0.055	≈4.1	1.6	3.6	0.015	275	St/wt Ratio equiv to E glass, improved electrical performance, low density, greater stiffness	Lightweight structures, radomes, hydrospace vehicles
E-HTS	0.092	6.2	2.2	5.0	0.018	326	Improved structural properties at slight increase in cost	Internal pressure vessels, tanks
E Glass	0.092	6.2	2.2	5.0	0.018	260	Available in wide variety of finishes, and end counts at low cost	Commercial tubes, tanks, cylinders. Military radomes, structural parts

[a]Compiled from data developed by Brunswick Corp. [b]At 9.375 kmc.
[c]Computed for all-glass strength. [d]Still in developmental stages.
*Carter and Moorefield, 1963.

Short fibers of asbestos impart higher modulus and greater hardness to the finished product compared to unfilled systems, but do not substantially improve other properties. Table 8-23 compares advantages and disadvantages of asbestos as a reinforcing fiber and filler.

Unlike glass, asbestos can be wetted by almost any thermoplastic or thermosetting resin. Problems do arise when agglomeration of longer fibers prevents development of a homogeneous mix.

Another consideration in the use of asbestos is the health problem. Since asbestos workers have developed statistically high incidences of mesothelioma and lung cancer, it is essential that appropriate precautions be implemented when processing the material. Exposure levels must be below OSHA-proscribed limits.

Polymeric Fillers

Surface characteristics of reinforced thermoset polyesters have been improved by the addition of finely divided polyethylene. These materials are referred to as low-profile composites. Polystyrene and phenolics have been blended into styrene-butadiene copolymers in order to improve elevated-temperature storage modulus.

Table 8-22. Properties of the Six Forms of Asbestos.*

Property	Chrysotile	Crocidolite	Amosite	Anthophyllite	Tremolite	Actinolite
Chemical formula	$Mg_6[(OH)_4 Si_2O_5]_2$	$Na_2MgFe_5[(OH) Si_4O_{11}]_2$	$MgFe_6[(OH) Si_4O_{11}]_2$	$(Mg, Fe)_7[(OH) Si_4O_{11}]_2$	$Ca_2Mg_5[(OH) Si_4O_{11}]_2$	$Ca_2(MgFe)_5[(OH) Si_4O_{11}]_2$
Color	white, gray, green, yellowish	blue	gray, yellow brown	gray, yellow brown	gray-white, greenish, yellowish, bluish	greenish
Hardness, mohs	2.5–4	4	5.5–6.0	5.5–6.0	5.5	6
Tensile strength, psi	824,000 max 281,000–436,000	876,000 max 469,000–605,000	287,000 max 148,000–203,000	4000 –	1000–8000 –	1000 –
Modulus of elasticity, psi	23.2×10^6	27.1×10^6	23.6×10^6	–	–	–
Fusion point, °F	2770	2180	2550	2675	2400	2540
Specific heat, Btu/lb, °F	0.266	0.201	0.193	0.210	0.212	0.217
Specific gravity	2.4–2.6	3.2–3.3	3.1–3.25	2.85–3.1	2.9–3.2	3.0–3.2
pH	10.3	9.1	9.1	9.4	9.5	9.5
Refractive index	1.50–1.55	1.7	1.64	1.61	1.61	1.63
Texture	soft-semibrittle	soft-semibrittle	coarse, brittle	brittle	brittle	brittle
Fiber diameter, A	180–300	600–900	600–900	600–900	600–900	600–900
Surface area, BET $m_2$gm.	17–60	9–10.5	8–9	6–7	6–9	–
Coefficient of thermal cubical expansion in.^{3}/in.^{3}	5×10^{-5}	–	–	–	–	–
Electric charge in water slurry	plus	minus	minus	minus	minus	minus
Isoelectric point	11.3–11.8	–	–	–	–	–

*Axelson, 1968–69. Reprinted by permission from *Modern Plastics Encyclopedia*, McGraw-Hill, Inc.

Table 8-23. Asbestos in Plastics.*

Advantages	Disadvantages
As a Reinforcing Filler	
Higher modulus	increase in viscosity of resin
Some improvement in tensile	fair electrical resistance
Flow control in mold	somewhat abrasive
Good surface finish	possible polymer degradation
Heat stability	high density
Dimensional stability	
Reduced creep	
Chemical resistance	
Higher hardness	
Arc resistance	
Low water absorption	
As a Reinforcing Fiber	
Low cost	fair impact strength
Easy flash removal	dark color
Flow control in mold	possible mixing difficulties
Minimum fiber degradation	
Good flexural strength	
Low water absorption	
Heat resistance	
Good electrical properties	
Chemical resistance	

*Axelson, 1968–69. Reprinted by permission from *Modern Plastics Encyclopedia*, McGraw-Hill, Inc.

The lubricity of nylon and acetal is improved through the addition of PTFE powders. PTFE powder has also been added to polyphenylene sulfide for upgrading of the latter's non-stick character. Table 8-24 shows wear and frictional properties data of some PTFE-filled resins.

Other polymer-polymer composites include polystyrene in silicone and polyethylacrylate in PVC.

Just as glass microballoons have been added to polymeric matrices to develop the low-density, high-strength syntactic foams, there are a number of hollow polymeric microspheres that perform the same function. These include phenolic, epoxy, and polyvinylidene-acrylonitrile copolymers. These microballoons often provide the composite with a woodlike character (nailable, sandable, etc.). Table 8-25 shows a comparison of syntactic foams used in deep water floats made from epoxy or glass microballoons.

Cellulose acetate butyrate has been used in polyester prepolymer formulations to reduce the shrinkage of sheet molding and bulk molding compounds (see Chapter 6 section on polyesters for detailed description).

Metals

Bronze and aluminum powders have been added to acetals and nylons to produce conductive moldings suitable for plating. Zinc has been added to coatings for corrosion

Table 8-24. Plastic-on-Plastic Wear and Frictional Properties.*

Stationary Surface: Base Resin	Stationary Surface: Filler System	Moving Surface: Base Resin	Moving Surface: Filler System	Wear Factor Stationary Surface 10^{-10}in^{3}min./ft./lb./hr.	Coefficient of Friction: Static (40 psi)	Coefficient of Friction: Dynamic (40 psi, 50 FPM)
Polycarbonate	10% PTFE	Nylon 6/10	15% PTFE 30% Glass Fiber	35	0.04	0.06
Polycarbonate	15% PTFE 20% Glass Fiber	Nylon 6/10	15% PTFE 30% Glass Fiber	15	0.05	0.07
Acetal Copolymer	20% PTFE	Nylon 6/6	20% PTFE	30	0.03	0.04
Nylon 6/10	15% PTFE 30% Glass Fiber	Polycarbonate	10% PTFE	15	0.04	0.06
Nylon 6/10	15% PTFE 30% Glass Fiber	Polycarbonate	15% PTFE 20% Glass Fiber	18	0.05	0.07
Nylon 6/6	20% PTFE	Acetal Copolymer	20% PTFE	12	0.03	0.04

*Courtesy: LNP Corporation, Malvern, Pa.

Table 8-25. Syntactic Foams—Deepwater Float Materials.*

Foam Density lb./ft.^{3}	Glass Microballoons?	Epoxy Macroballoons?	Uniaxial Compressive Yield Strength psi	Hydrostatic Compressive Strength psi	Method of Preparation	Resin System
24	Yes	Yes	1,700	1,800	Pack-in-Place	Polyester
27	Yes	Yes	2,800	2,300	Pack-in-Place	Epoxy
30	Yes	Yes	2,500	3,000	Infiltration	Polyester
32	Yes	No	7,300	7,000	Pack-in-Place	Polyester
34	Binary	No	-----	16,000	Infiltration	Epoxy
35	Yes	No	10,500	14,000	Vacuum Cast	Epoxy
38	Yes	No.	10,000	14,000	Cast	Polyester
42	Yes	No	10,000	15,000	Cast	Polyester

*Rowe, 1973.

resistance. Lead-filled polymers are used as neutron and gamma ray shields.

Epoxy resin–aluminum formulations have been used to cast form tools and dies, and for producing metal cements and coatings. Silver-, gold-, platinum-, and copper-filled epoxies and silicones have been used as conductive adhesives.

Heat-resistant composites have been made from resins with beryllium, chrome, molybdenum, niobium, stainless steel alloys, tantalum, titanium, and tungsten.

Mineral Fillers

Wollastonite. Wollastonite is a calcium metasilicate which has been used in thermosetting resin molding compounds, paints, and ceramics. Wollastonite is a high-melting acicular mineral available in very fibrous to finely divided grades.

Barytes. Barytes are a form of barium sulfate. This filler has an exceptionally high specific gravity (4.4) along with chemical inertness and low water solubility. Plastics filled with barytes are opaque to x-rays.

Natural barium sulfates having selected particle-size distributions impart increased density and load-bearing properties and improved processing, drape and handling properties to polyurethane foam. Plastisol compositions containing barytes, particularly for sealing rings on food jar covers, require less resin/plasticizer, yet yield optimum properties.

Clay. Commercial kaolins are available in a variety of forms and particle sizes. The properties of air-floated clays are controlled by the deposit from which they are mined, since the air-flotation equipment is not capable of close

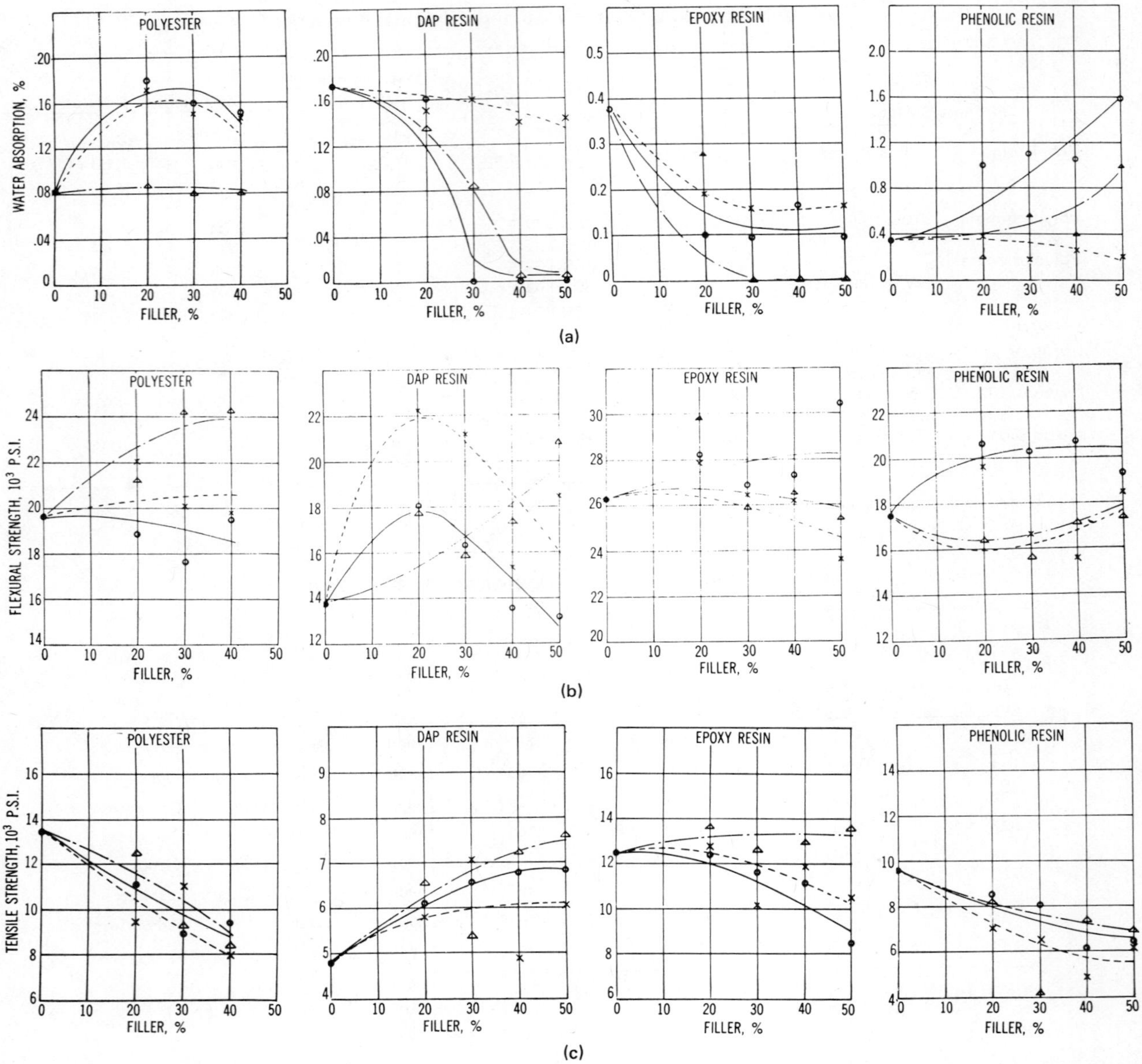

Fig. 8-18. (a) Effect of kaolinite fillers on water absorption of reinforced plastics: ◉ = plates; X = stacks; △ = 50:50 mixture. (b) Effect of kaolinite fillers on flexural strength of reinforced plastics: ◉ = plates; X = stacks; △ = 50:50 mixture. (c) Effect of kaolinite fillers on tensile strength of reinforced plastics: ◉ = plates; X = stacks; △ = 50:50 mixture. (Larson, 1958. Reprinted by permission from *Modern Plastics*, McGraw-Hill, Inc.)

particle-size fractionating. Water-classified clays offer close particle-size control as well as control of color, brightness, pH, and impurity content. Calcined clays are thermally treated to remove the water of hydration. This treatment alters the crystal structure of the product and changes many of the other characteristics such as brightness, particle size, hardness, and oil absorption.

Kaolin clays are used both as fillers and extenders in both thermosetting and thermoplastic resins. In polyester premix, preform, and mat moldings for matched die techniques, the clay aids in consistency and moldability control. The clay contributes to chemical resistance, low water absorption, and good electrical properties (Figure 8-18 shows the effect of clays on selected properties of some thermosetting plastics). High and low oil-absorption fillers* are often combined with kaolin clays for synergistic properties. Bentonite clays are often used as thickening agents and also provide chemical resistance and better reverse impact values in sheet molding compounds. Calcined clay improves the insulation resistance of plasticized vinyl.

Calcium Carbonate. There are two major types of calcium carbonate: natural ground limestone and precipitated

*Oil absorption represents a measure of the absorbable surface area of a filler particle. The higher the value, the more resin, plasticizer, processing oil, etc. that will be absorbed. This factor is significant in plastics since the percentage of filler in a formula influences final properties.

calcium carbonate. Dry- or wet-ground limestones range in size from a minimum of 0.5 to 44 microns. Precipitated $CaCO_3$ is manufactured by several processes, the most common being soda ash–calcium chloride, ammonium carbonate–calcium chloride, and carbonated milk of lime reactions.

Both precipitated and natural ground limestones are higher in whiteness and brightness than other fillers. Plastics formulations most often have precipitated grades, uncoated large-particle grades, or wet-ground grades treated for low oil-plasticizer absorption. Because of their low oil-plasticizer absorption and particle-size uniformity, the calcium carbonates provide easy processing, little or no abrasion of processing equipment, and can be used in high filler levels. They do not affect thermoset resin physical properties except at very high loadings. They do reduce physical properties in thermoplastics proportionate to the quantity incorporated, especially when dry-ground grades are used.

Altered Novaculite. Altered novaculite is typically 99.5% quartz (SiO_2). It is mined from an Arkansas mountain range having three strata of novaculite. The microcrystalline version used in plastics is found in the upper stratum of the deposits. Altered novaculite has the basic hardness of quartz, but lacks the abrasive character. The grinding energy requirements are comparatively low since the mineral is so friable. Grades of various particle-size distribution are usually accomplished by air separation. Novaculite is extremely wettable and mixes easily. It is regarded as hydrophilic, but not as hygroscopic.

Altered novaculite is used primarily in epoxy systems where good rheological flow, heavy loading, and low abrasive coefficient are called for. Other areas of use include polyurethane foam, PVC, phenolics, and polyesters.

Boron Fibers

Boron filaments are made by chemically vapor-depositing the boron from a boron trichloride and H_2 gas mixture onto a 0.0005-inch-diameter tungsten wire substrate. The product of this operation is a 4-mil-diameter continuous filament having a density around 2.65 g/cc, a modulus of $55\text{–}57 \times 10^6$ psi, and tensile strengths around 4.5×10^5 psi.

Boron filaments are usually used in the form of epoxy-prepreg tapes in which the filaments are carefully collimated and tensioned to provide a one-filament-thick tape, varying in width from $\frac{1}{8}$ to 3 inches. Generally, these tapes are composed of 65–70% filaments in the epoxy resin matrix. Some work with epoxy novolacs, phenolics, and polyimides has been done. Figure 8-19 shows the relative position, strengthwise, of boron/epoxy composites with other structural materials. Table 8-26 shows some selected properties of a composite.

Table 8-26. Boron-Epoxy Composite Properties.*

Typical Acceptance Criteria (psi)		Typical Values (psi)
4.0-mil Boron-Epoxy		
Longitudinal Flexural Strength		
RT	225–245,000	280,000
350–375°F	180–200,000	235,000
420°F	130,000	–
Longitudinal Flexural Modulus		
RT	27×10^6	27×10^6
350–375°F	22×10^6	23×10^6
420°F	21×10^6	–
Transverse Flexural Strength		
RT	11,000	12,500
350–375°F	8,000	10,500
420°F	7,000	–
Transverse Flexural Modulus		
RT	2.0×10^6	2.2×10^6
350–375°F	0.8×10^6	0.9×10^6
Horizontal Shear		
RT	13,000	16,000
350–375°F	6,000	7,000
420°F	4,000	–
Longitudinal Tensile Strength		
RT	200,000	230,000
350–375°F	160,000	200,000
Longitudinal Tensile Modulus		
RT	30×10^6	30×10^6
350–375°F	27×10^6	27×10^6
Transverse Tensile Strength		
RT	7,000	11,000
350–375°F	3,500	6,000
Transverse Tensile Modulus		
RT	3×10^6	3×10^6
350–375°F	1×10^6	1×10^6
Longitudinal Compressive Strength		
RT	375,000	375,000
350–375°F	140,000	140,000
5.6-Mil Boron-Epoxy		
Longitudinal Flexural Strength		
RT	250,000	300,000
350°F	200,000	250,000
Longitudinal Flexural Modulus		
RT	27×10^6	30×10^6
350°F	22×10^6	23×10^6
Transverse Flexural Strength		
RT	14,000	15,000
350°F	10,000	12,000
Longitudinal Tensile Strength		
RT	200,000	250,000
350°F	160,000	200,000
Horizontal Shear		
RT	14,500	15,500
350°F	7,000	9,000

*Courtesy: Composite Technology, Inc., Broad Brook, Conn.

Miscellaneous Fibers

Single-Crystal Fibers. Fantastically high fiber strengths, equivalent to those forces which bond atoms together, are possible with reinforcing materials known generically as "whiskers." Whiskers are extremely thin single-crystal fibers produced under controlled growth conditions of such materials as Al_2O_3, AlN, BeO, B_4C, C, and SiC.

Whiskers have polygonal cross-sections and have a high

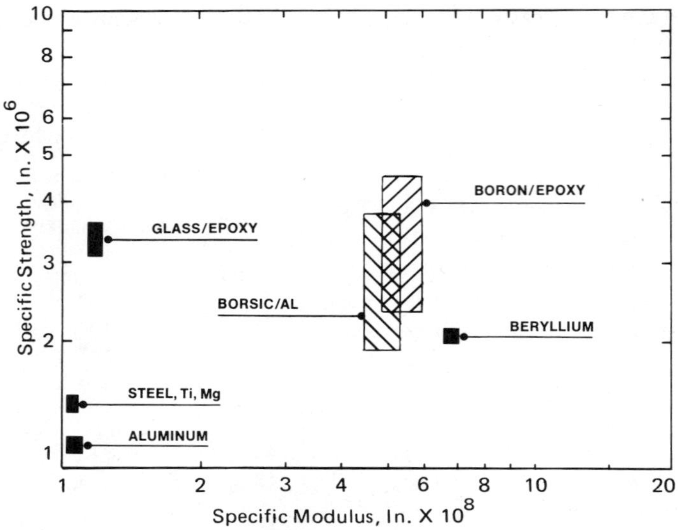

Fig. 8-19. Comparative strengths of composites and structural materials. *(Courtesy: Composite Technology, Inc., Broad Brook, Conn.)*

level of structural perfection. They are produced in diameters of 1–30 microns with aspect ratios (length/diameter) of 500–10,000:1. Their physical properties differ from bulk counterparts because crystal defects are absent at such small diameters. Conventional materials contain many grain boundaries, voids, dislocations, and other imperfections which are stress foci. Because of their internal perfection, whiskers have strengths that are not strictly related to surface perfection. This gives whiskers an unusual toughness and nonfriability when compared to polycrystalline filaments such as fibrous glass.

In comparing the tensile strengths of whiskers with that of other reinforcing fibers such as glass and boron, a five- to tenfold increase in basic fiber strength is observed. Whiskers have the elongation of glass fibers (3–4%) and the modulus of boron (60–100 $\times 10^6$ psi). The major disadvantage of whiskers is control of orientation in a matrix. This leads to anisotropic composites with batch to batch variation.

Physical properties of the more prominent whisker materials are shown in Table 8-27. Pilot plant supplies of sapphire (Al_2O_3), SiC, silicon nitride, and AlN are available.

The high aspect ratio of whiskers is related to the amount of surface available for bonding by the composite. No fatigue effects have been observed in whiskers. They can be handled roughly and can be milled and chopped or otherwise worked on without having their strengths impaired.

Quantity of whiskers used is a function of desired end properties. Up to 5% will retain grain refinement in the matrix and produce a sizeable increase in mechanical properties. From 5% to 25% will improve mechanical properties further and increase performance at elevated temperatures. Eighty to ninety percent will provide the greatest gain in mechanical, electrical, and thermal properties. Space applications are the prime areas of whisker usage.

Ceramic Fibers. Ceramic fibers are made from the oxides of aluminum, beryllium, magnesium, thorium, and zirco-

Table 8-27. Physical Properties of Whiskers.*

Material	Density gm/cm^3	Melting point °F.	Tensile strength psi $\times 10^6$	Young's modulus psi $\times 10^6$	Strength density in. $\times 10^6$
Aluminum oxide	3.9	3780	2–4	100–350	14–28
Aluminum nitride	3.3	3990[1]	2–3	50	13–21
Beryllium oxide	1.8	4620	2.0–2.8	100	31–43
Boron carbide	2.5	4440	1	65	11
Graphite	2.25	6500	3	142	37
Magnesium oxide	3.6	5070	3.5	45	27
Silicon carbide (alpha)	3.15	4200[1]	3–5	70	26–44
Silicon carbide (beta)	3.15	4200[1]	1–3	100–150	0.9–26
Silicon nitride	3.2	3450	0.5–1.5	55	4.2–13

[1] Disassociates.
*Shyne and Milewski, 1968–69. Reprinted by permission of *Modern Plastics Encyclopedia*, McGraw-Hill, Inc.

nium. They are produced via chemical vapor deposition, melt drawing, fiber spinning, and extrusion techniques.

Plastics composites with ceramic fiber reinforcements exhibit good physical properties, high strength/weight ratios, and resistance to high temperatures and environmental attack. The reinforcement is selected for retention of high tensile strength and elastic modulus. The ceramics have low thermal expansion which minimizes residual stresses in the composite. The longer the fiber, the more effective it is as reinforcement.

Trade Names

"A" Glass	Glass	Sun Chemical Corporation
Acetylene Black	Carbon black	Gulf Oil & Chemical Co.
Aero Rove	Fiberglass	Glass Fibers Products, Inc.
Aerosil	Fumed silica	Degussa Corporation
Afton	Aluminum silicate	Cyprus Industrial Mining
Agro	Rayon fiber	Beaunit Corporation
Albacar	Calcium carbonate	Pfizer, Inc.
Albex	Calcium carbonate	Pfizer, Inc.
Alcoa	Alumina and aluminum powders	Alcoa Corporation
Allied Whiting	Calcium carbonate	Akron Chemical Co.
Almeg	Aluminum powders, flakes, etc.	Aluminum Metallurgical Granules
Alsibronz	Ground mica	Franklin Mineral Products Co.
Al-Sil-Ate	Clays	Freeport Kaolin Co.
Anklorlok	Asbestos	Anchor Packaging Co., Inc.
Antron	Nylon fiber	E.I. Du Pont de Nemours & Co., Inc.
Apex	Clay	Thompson, Weinman & Co.
Aplex	Asbestos	Asbestos Corp. of America
Araton	Glass rovings, strands, mats	Owens-Corning Fiberglass Corporation
Arbonier	Cellulose	ITT Rayonier, Inc.
Aristocrat	Rayon fiber	Beaunit Corporation
ARO	Carbon black	J. M. Huber Corporation
Arogen	Silica	J. M. Huber Corporation
Arrow	Carbon black	J. M. Huber Corporation
ASP	Mineral filler	Minerals & Chemicals Corp. of America
Astra	Papers	Brown Co.
Astrocarb	Carbon fabrics	J. P. Stevens & Co.
Astroquartz	Quartz-filament fabric	J. P. Stevens & Co.
Astrosil	High-silica fabric	J. P. Stevens & Co.
Atomite	Calcium carbonate	Thomson, Weinman & Co.
Avceram	Ceramic filaments	FMC Corporation
Avicron	Rayon fiber	FMC Corporation
Avlin	Polyester fiber	FMC Corporation
Avril	Rayon fiber	FMC Corporation
Avron	Rayon fiber	FMC Corporation
Ballotini	Glass beads	Potters Industries, Inc.
Barden	Kaolin clay	J. M. Huber Corporation
Baymal	Colloidal alumina	E.I. Du Pont de Nemours & Co., Inc.
Beaverwhite	Talc	Cyprus Industrial Mining
Belastraw	Rayon fiber	Akzona, Inc.
Belcobalsa	Balsa core	Balsa Ecuador Lumber Co.
Beta	Superfine glass filament, yarn, etc.	Owens-Corning Fiberglas Corporation
Black Knight	Lampblack	Pfizer, Inc.
Black Pearls	Carbon black	Cabot Corporation
Blanc de Blancs	Nylon fiber	Akzona, Inc.
Borsic	Boron monofilament	Composite Technology, Inc.
Briglo	Rayon fiber	Akzona, Inc.
Brite Sorb	Precipitated magnesium silicate	Philadelphia Quartz Co.
Buca	Clay	J. M. Huber Co.
Cab-O-Sil	Fumed silica	Cabot Corporation
Calidria	Asbestos	Union Carbide Corporation
Caloforts	Stearate-coated calcium carbonate	Berkshire Chemicals, Inc.
Calwhite	Calcium carbonate	Georgia Marble Co.
Camel-Carb	Calcium carbonate	Harry T. Campbell Sons Corporation
Camel-Kote	Calcium carbonate	Harry T. Campbell Sons Corporation
Camel-Tex	Calcium carbonate	Harry T. Campbell Sons Corporation
Camel-WITE	Calcium carbonate	Harry T. Campbell Sons Corporation
Cantrice	Nylon fiber	E.I. Du Pont de Nemours & Co., Inc.
Caprolan	Nylon fiber	Allied Chemical Corporation

Carbium	Calcium carbonate	Diamond Shamrock Chemical Co.
Carbo-Spheres	Carbon microballoons	Vetrotex International
Carbovar	Carbon black	Cabot Corporation
Carthage	Calcium carbonate	Calcium Carbonate Co.
Celion	Carbon fibers	Celanese Corporation
Celite	Diatomaceous fillers	Johns-Manville Corporation
Celkate	Magnesium silicates	Johns-Manville Corporation
Celramic	Glass modules	Pittsburgh Corning Corporation
Certimesh	Woven wire mesh	Pall Corporation
Chalkelle	Rayon fiber	FMC Corporation
Chem Carb	Calcium carbonate	Engelhard Minerals & Chemicals Corporation
Chemeglass	Fiberglass	Chemicovens, Ltd.
Chemiglass	Fiberglass	Chemicovens, Ltd.
Civona	Rayon fiber	Beaunit Corporation
Colfoam	Urea microballoons	Colton Chemical Co.
Coloray	Rayon fiber	Courtaulds
Columbia	Fiberglass	Mutual Industries, Inc.
Comiso	Rayon fiber	Beaunit Corporation
Conductex	Carbon black	Continental Carbon Co.
Continental	Carbon black	Continental Carbon Co.
Continex	Carbon black	Continental Carbon Co.
Corcel	Volcanic ash	Interpace Corporation
Corval	Rayon fiber	Courtaulds
Cosmic Black	Bone black	Ebonex Corporation
Cosmostrand	Fiberglass	Owens-Corning Fiberglass Corporation
Costyreneblak	Carbon black	Cities Service Co.
Covinylblak	Carbon black	Cities Service Co.
Cowaxblak	Carbon black	Cities Service Co.
Crinofil	Rayon fiber	Beaunit Corporation
Crystalon	Rayon fiber	Akzona, Inc.
Cupersil	Rayon fiber	Beaunit Corporation
Cupioni	Rayon fiber	Beaunit Corporation
Cupracolor	Rayon fiber	Beaunit Corporation
Dacote	Calcium carbonate	Diamond Shamrock Chemical Corporation
Dacron	Polyester fiber	E.I. Du Pont de Nemours & Co., Inc.
Davenite	Muscovite mica	Akron Chemical
Deeglas	Chopped glass, strand, mats, etc.	Deeglas Fibers, Ltd.
Densite	Gypsum	Bestwall Gypsum Co.
Delovisca	Rayon fiber	Delaware Rayon
Delray	Rayon fiber	Delaware Rayon
Desertalc	Talc or magnesium silicate	Desert Minerals, Inc.
Dixie	Carbon black	Ashland Chemical Co.
Dream Slub	Rayon fiber	Beaunit Corporation
Drikalite	Calcium carbonate	Thompson, Weinman & Co.
Dryca-Flo	Calcium carbonate	Sylacauga Calcium Products Co.
Dulesco	Rayon fiber	FMC Corporation
Duraglas	Fiberglass	Natvar Corporation
Duragold	Bronze powders	Claremont Polychemical Corporation
Duramite	Calcium carbonate	Thompson, Weinman & Co.
Durmite	Calcium carbonate	Thompson, Weinman & Co.
Dylok	Rayon fiber	IRC
Eccofloat	Epoxy microspheres	Emerson & Cuming, Inc.
Eccospheres	Glass, metal, epoxy, and ceramic microspheres	Emerson & Cuming, Inc.
Economat	Fiberglass mat	Euroglas Sales, Ltd.
Electro	Calcium carbonate	Calcium Carbonate Co.
Elf	Carbon black	Cabot Corporation
Elftex	Carbon black	Cabot Corporation
Emtal	Talc	Engelhard Minerals & Chemical Corporation
Encron	Polyester fiber	Akzona, Inc.
Englo	Rayon fiber	Akzona, Inc.
Enkaloft	Nylon fiber	Akzona, Inc.
Enkalure	Nylon fiber	Akzona, Inc.
Enkatron	Nylon fiber	Akzona, Inc.
Esellar	Silica	J. M. Huber Corporation
Eternagold	Bronze powder	Atlantic Powdered Metals, Inc.
Evergold	Bronze powder	Atlantic Powdered Metals, Inc.
Excello White	Calcium carbonate	Tamms Industries Co.

Excelsion	Carbon black	Cities Service Co.
Exeterglas	Glass fabric	Exeter Manufacturing Co.
Extendospheres	Aluminum silicate hollow spheres	Industrial Minerals & Chemicals
FP	Alumina fibers	E.I. Du Pont de Nemours & Co., Inc.
Fabmat	Chopped and woven glass	Fiber Glass Industries
Fabran	Glass fabric	Paper Corporation of United States
Fabricfil	Chopped rayon	Rayon Processing Co. of Rhode Island
Fabrifil	Chopped rag	Microfibres, Inc.
Famco-Fiber	Fiberglass and fabrics	Famco Machine Co.
Featheray	Rayon fiber	Akzona, Inc.
Feather Brand	Balsa core	International Balsa Co.
Fiberglas	Fiberglass	Owens-Corning Fiberglas Corporation
Fiberglass	Chopped fiberglass	Fiberglass, Ltd.
Fibro	Rayon fiber	Courtaulds
Filflox	Cut rayon	Rayon Processing Co. of Rhode Island
Filfrac	Cut-cotton fiber	Rayon Processing Co. of Rhode Island
Filomat	Fiberglass mat	Fiber Glass Industries, Inc.
Filospun	Fiberglass roving	Vetreria Italiana Balzaretti Modigliani, S.p.A.
Firmex	Carbon black	Cities Service Co.
Flaikona	Rayon fiber	Beaunit Corporation
Flakeglas	Glass flakes	Owens-Corning Fiberglas Corporation
Floform Thermax	Carbon black	R. T. Vanderbilt
Florence	Zinc oxides	New Jersey Zinc Co.
Format	Chopped glass	Fiber Glass Industries, Inc.
Fortafil	Graphite fiber	Great Lakes Carbon Co.
Fortrel	Polyester fiber	Fiber Industries, Inc.
Furnex	Carbon black	Cities Service Co.
Fybex	Titanate whiskers	E.I. Du Pont de Neumours & Co., Inc.
Fybrglass	Fiberglass	Eraser Company, Inc.
Fybrite	Polyester fiber	Fiber Industries, Inc.
Gamaco	Calcium carbonate	Georgia Marble Co.
Gama-sperse	Calcium carbonate	Georgia Marble Co.
Garalum	Alumina paper	Armak Co.
Garan	Fiberglass roving	Johns-Manville Corporation
Garanmat	Fiberglass mat	Johns-Manville Corporation
Garolin	Clay	Groxton & Garry, Ltd.
Glasterra	Glass asbestos	Chase-Foster, Inc.
Glasweb	Glass tape and cloth	Atlas Asbestos Co.
Glitter	Rayon fiber	Beaunit Corporation
Glomax	Calcined aluminum silicate	Georgia Kaolin Co.
Gold Bond	Asbestos	National Gypsum Co.
Glufill	Nut shell flour	Agrashell, Inc.
Grafil A	Carbon fiber	Courtaulds, Ltd.
G-White	Calcium carbonate	Calcium Carbonate Co.
Hakwenka	Calcium carbonate	Royal Industries
Hesgon	Glass tape	Burlington Glass Fabrics Co.
Highflex	Silane-coated alumina trihydrates	Salem Industries
Hi-Pflex	Calcium carbonate	Pfizer, Inc.
Hi Sil	Precipitated silica	PPG Industries
Hitco	Carbon fiber and cloth	Hitco Co.
Hitron	Graphite yarn	Hitco Co.
Hi-White	Kaolin clay	J. M. Huber Corporation
Hybon	Fiber glass	PPG Industries
Hydral	Hydrated alumina	Alcoa
Hydrite	Kaolin clays	Georgia Kaolin Co.
Hydrocal	Gypsum	United States Gypsum Co.
Iceberg	Aluminum silicate	Burgess Pigment Co.
Icecap K	Aluminum silicate	Burgess Pigment Co.
Imsil	Amorphous silica	Illinois Minerals Co.
Insulbatte	Glass mat	J. P. Stevens & Co., Inc.
Insulglass	Glassine	3M Co.
Intersperse	Carbon black	International Colloids, Ltd.
Isomica	Mica sheet, rod, and tube	3M Co.
Iso Sil-Cote	Silver-coated copper	Isochem Resins Co.
Ivo	Bone black	Cities Service Co.
Kadox	Zinc oxides	New Jersey Zinc Co.

Kalite	Calcium carbonate	Diamond Shamrock Co.
Kalmac	Calcium carbonate	Georgia Marble Co.
Kanamite	Glass microscpheres	Ferro Corporation
Kaobrite	Clay	Thiele Kaolin Co.
Kaofill	Clay	Thiele Kaolin Co.
Kaogan	Aluminum silicate	Georgia Kaolin Co.
Kaoglass	Clay	Thiele Kaolin Co.
Kaolloid	Aluminum silicate	Thompson, Weinman & Co.
Kaopaque	Aluminum silicate	Georgia Kaolin Co.
Kaophile	Aluminum silicate	Georgia Kaolin Co.
Kaophobe	Aluminum silicate	Georgia Kaolin Co.
Kaowhite	Clay	Thiele Kaolin Co.
Kemat	Chopped glass	Fiberglass Industries, Inc.
Ketjenblack	Carbon black	Armak Polymer Additives
Kevlar	Aromatic polyamide fiber	E.I. Du Pont de Nemours & Co., Inc.
Kingcote	Cut-cotton fiber	Rayon Processing Co. of Rhode Island
Kingcote	Cotton flock	Microfibers, Inc.
Kodel	Polyester fiber	Eastman Chemical Co.
Kolorbon	Rayon fiber	Akzona, Inc.
Kotamite	Calcium carbonate	Thompson, Weinman & Co.
Krecobeads	Solid graphite spheres	Kureha Chemical Industry
Kroma-	Iron oxides	Pfizer, Inc.
Kryolith	Mineral filler	Pennwalt Corporation
Kureha	Graphite and carbon fiber	Kureha Chemical Industries
Kynol	Phenolic fiber	Carborundum Co.
Lastex	Rayon fiber	Uniroyal, Inc.
Lektroset	Rayon fiber	IRC
Leno-Web	Nowoven fabric	Kendall Co., Inc.
Lexon	Nowoven fabric	Kendall Co., Inc.
Liasil	Glass roving	Fiber Glass Industries, Inc.
Lilion	Nylon fibers	Snia Viscosa, S.p.A.
Lithafrax	Lithium aluminum silicate	Carborundum Co.
Lon	Nylon fiber	Polymer Corporation
LoVel	Silica	PPG Industries
Lowland	Nylon, polyester, and rayon fibers	Akzona, Inc.
Magcarb	Magnesium carbonate	Merck & Co.
Maglite	Magnesium oxide	Merck & Co.
Magnamite	Graphite fibers	Hercules, Inc.
Mapico	Iron oxides	Cities Service Co.
Marble-Mix	Calcium carbonate	Sylacauga Calcium Products Co.
Marglass	Glass fabric and rovings	Marglass, Ltd.
Marincate	Magnesium trisilicate	Merck & Co.
Matesa	Rayon fiber	Beaunit Mills Corporation
McNamee Kaolin	Kaolin clay	R. T. Vanderbilt Co., Inc.
Metafil	Metallized glass fibers	MB Associates
Metso	Sodium sesquisilicate pentahydrate	Philadelphis Quartz Co.
Micabond	Mica	Budd Co.
Michigan	Wire cloth	AMBAC Industries, Inc.
Microballoon	Glass microspheres	Sohio Chemical Co.
Microbestos	Asbestos	Johns-Manville Corporation
Micro-Fil	Limestone	Calcium Carbonate Co.
Micro-Mica	Mica	Norwegian Talc A/S
Micromite	Calcium carbonate	Tamms Industries Co.
Micronex	Carbon black	Cities Service Co.
Microtalc	Talc	Pfizer, Inc.; Norwegian Talc A/S
Micro-White	Calcium carbonate	Sylacauga Calcium Products
Millical	Calcium carbonate	Diamond Shamrock Co.
Minex	Aluminum silicate	American Nepheline Corporation
3M	Glass microballoons	3M Co.
Mirrorgold	Bronze powder	Atlantic Powdered Metals, Inc.
Mistron	Talc	Cyrpus Industrial Minerals Co.
Mobaloy	Glass roving	Ferro Corporation
Modiglass	Fibrous glass	Reichhold Chemicals, Inc.
Modmor	Carbon fibers	Morganite Research and Development, Ltd.
Mogul	Carbon black	Cabot Corporation
Molylube	Molybdenum disulfide	Bel-Ray Company, Inc.

Molysulfide	Molybdenum sulfide	Climax Molybdenum Co.
Monarch	Carbon black	Cabot Corporation
Mono-Glass	Glass filaments	Fram Corporation
Mosher	Calcium carbonate	Calcium Carbonate Co.
Multemat	Glass mat	Fiberglass, Ltd.
Multiflex	Calcium carbonate	Diamond Shamrock Co.
Multi-Strate	Rayon fiber	Beaunit Corporation
Muscovite	Mica	English Mica Co.
Narcon	Rayon fiber	Beaunit Corporation
Narene	Rayon fiber	Beaunit Corporation
National	Carbon, graphite, and boron nitride	Union Carbide Corporation
Neosil	Silica	Tamms Industries Co.
Neo-Spectra	Carbon black	Cities Service Co.
Neotex	Carbon black	Cities Service Co.
Newbray	Rayon fiber	New Bedford Rayon
New-Dull	Rayon fiber	New Bedford Rayon
Newlow	Rayon fiber	New Bedford Rayon
Non-Fer-Al	Calcium carbonate	Diamond Shamrock Co.
Noramite	Asbestos	North American Asbestos Corporation
Novacite	Novaculite	Malvern Minerals Co.
Novakup	Silane-treated novaculite	Malvern Minerals Co.
Novonette	Nonwoven fabric	Kendall Company, Inc.
Nuf	Glass fabric	Ferro Corporation
Nulok	Silane-treated clay	J. M. Huber Corporation
Nuploglas	Fiberglass	Nupla Manufacturing Co.
Nupron	Rayon fiber	IRC
Nytal	Talc	R. T. Vanderbilt Co., Inc.
Omya	Calcium carbonate	Pluess-Stoufer, Inc.
Opal Black	Silica/carbon black	H. M. Royal Co.
Optiwhite	Aluminum silicate	Burgess Pigment Co.
Orcon	Graphite fabric	Orcon Corporation
Ozide	Zinc oxide	Sherwin-Williams Co.
Paragon	Clay	J. M. Huber Co.
Paramat	Glass roving and mats	Fiberglass, Ltd.
Paritex	Nonwoven fabrics	G. Harrison, Ltd.
Philblack	Carbon black	Phillips Petroleum Co.
Plasfill	Mica	Hancock International
Plastibest	Asbestos	Johns-Manville Corporation
Ply-Mat	Chopped glass and roving	Johns-Manville Fiber Glass Division
Polyclay	Aluminum silicate	Burgess Pigment Co.
Polyfil	Kaolin clay	J. M. Huber Corporation
Polyspun	Polyester fiber	White & Sons, Ltd.
Powminca	Asbestos	Powhatan Mining Co.
PPG	Fiber glass	PPG Industries
Protex	Zinc oxides	New Jersey Zinc Co.
Purecal	Calcium carbonate	BASF Wyandotte Corporation
Pyrocides	Alumina trihydrates	Industrial Minerals & Chemicals, Inc.
Pyrosleeve	Fiberglass	Bentley-Harris Manufacturing Co.
Pyrotex	Asbestos felt	Raybestos Manhattan, Inc.
Q Cel	Inorganic microspheres	Philadelphia Quartz Co.
Quincy	Calcium carbonate	Calcium Carbonate Co.
Quso	Silica	Philadelphia Quartz Co.
Raco	Carbon black	Ebonex Co.
Raybrite	Alpha-cellulose reinforcement	ITT Rayonier, Inc.
Raven	Carbon black	Cities Service Co.
Reevair	Nylon fiber	Reeves Brothers, Inc.
Refrasil	High-silica glass	Hitco Co.
Regal	Carbon black	Cabot Corporation
Rigilor	Carbon and graphite fibers	LeCarbone-Lovraine
Rigimesh	Wire cloth	Pall Corporation
Rovcloth	Glass roving	Fiber Glass Industries, Inc.
Rovlok	Fibrous glass	Fiber Glass Industries, Inc.
Rovmat	Chopped-glass mat	Fiber Glass Industries, Inc.
Royal Spectra	Carbon black	Cities Service Co.
Santocel	Silica aerogels	Monsanto Co.
Saran Microsphere	Saran hollow microsphere	Dow Chemical Co.

Sevacarb	Carbon black	Severn Valley Chemical Industries, Ltd.
Shasta White	Calcium carbonate	Pfizer, Inc.
Silanox	Fumed silica	Cabot Corporation
Silbon	Rayon paper	Kohjin Co.
Silcron	Silica	SCM Corporation
Silene	Precipitated silica	PPG Industries
Silvex	Aluminum pigments	Silberline Manufacturing Co., Inc.
Sivco	Nut shell flour	Bernard Sirotta Co.
Snobrite	Hydrated aluminum silicate	Thompson, Weinman & Co.
Snowflake White	Calcium carbonate	Thompson, Weinman & Co.
Snow White	Calcium sulfate	United States Gypsum Co.
Solka-Floc	Alpha-cellulose flock	Brown Co.
Sparkling	Nylon fiber	E.I. Du Pont de Nemours & Co., Inc.
Sparmite	Barium sulfate (barytes)	Pfizer, Inc.
Spherefil	Alumina/silicate	Penn Virginia Materials Corporation
Spheron	Carbon black	Cabot Corporation
Stan-White	Calcium carbonate	Harwick Standard Chemical Co.
Statex	Carbon black	Cities Service Co.
Stay/Steel	Stainless steel flake	Pfizer, Inc.
Sterling	Carbon black	Cabot Corporation
Stonelite	Pulverized limestone	Ohio Lime Co.
Stranglas	Glass yarn	Dart Industries, Inc.
Superba	Carbon black	Cities Service Co.
Super Carbovar	Carbon black	Cabot Corporation
Supercoat	Calcium carbonate	Thompson, Weinman & Co.
Superjet	Lampblack	Pfizer, Inc.
Superlith	Lithopone, zinc sulfide, clay	C. J. Osborn, Co.
Supermite	Calcium carbonate	Thompson, Weinman & Co.
Super-Multiflex	Calcium carbonate	Diamond Shamrock Chemical Co.
Super P Flex	Calcium carbonate	Pfizer, Inc.
Supreave	Fiberglass	Fiberglass, Ltd.
Supremat	Glass mat	Fiberglass, Ltd.
Supreme	Hydrated aluminum silicate	Hammell Gillespie, Inc.
Surfex	Calcium carbonate	Diamond Shamrock Co.
Suspenso	Calcium carbonate	Diamond Shamrock Co.
Suzorite	Mica	Marietta Resources International, Ltd.
Syloid	Micronized silica	W.R. Grace & Co.
Synthamica	Mica	Mycalex Corporation of America
Talcron	Talc	Pfizer, Inc.
Tammesilite	Silica	Tammesco, Inc.
T-Carb	Calcium carbonate	Thompson-Haywood Chemical Co.
Tempgard	Fiberglass	J. P. Stevens & Co., Inc.
Terom	Polyester fiber	Fibrex of Savinesti
Terylene	Polyester fiber	Imperial Chemical Industries
Texas	Carbon black	Sid Richardson Carbon Co.
Textilmat	Glass mat	Fiber Glass Industries, Inc.
Thermax	Carbon black	R. T. Vanderbilt Co., Inc.
Thermica	Mica	Mykroy Ceramics Corporation
Thornel	Carbon and graphite fiber	Union Carbide Corporation
Tissuglas	Glass paper	Pall Flex Products Corporation
Tissuquartz	Quartz fiber	Pall Flex Products Corporation
Topel	Rayon fiber	Courtaulds, Inc.
Trans-Oxide	Iron oxides	Hilton-Davis Chemical Co.
Trevarno	Glass fabrics	Hexcel Corporation
Tricon	Glass fabric	J. P. Stevens & Co.
Troyfelt	Fiber felts	Troy Mills, Inc.
Troytuf	Fiber felts	Troy Mills, Inc.
Tuflam	Fiber glass	Plastics Design & Engineering, Ltd.
Tygamat	Glass mat	Fothergill & Harvey, Ltd.
Tyglas	Glass fabric and rovings	Fothergill & Harvey, Ltd.
Ty-Pak	Glass rovings	Johns-Manville Fiber Glass Division
Ucar	Carbon and graphite, boron nitride, silicon nitride	Union Carbide Corporation
Ultrafine	Antimony trioxide	Chemetron Corporation
Ultrastrand	Glass mat and roving	Gustin-Baker Manufacturing Co.
Unifab	Glass fabric	Ferro Corporation
Unifil	Glass mats	Certain Teed Corporation

Uniflo	Glass mat and roving	Vetreria Italiano Balgoretti Modigliani, S.p.A.
Uniform	Fiberglass	Ferro Corporation
Uniformat	Chopped-glass mat	Ferro Corporation
Uniglass	Glass fabric	United Merchants Industrial Fabrics
Unirove	Woven-glass roving	Ferro Corporation
Uniroving	Fiberglass rovings	Ferro Corporation
Unistrand	Glass roving	Ferro Corporation
Unitape	Woven-glass tape	Ferro Corporation
United	Carbon black	Ashland Chemical Co.
Vansil	Wollastonite	R. T. Vanderbilt Co., Inc.
Varon	Clays	Groxton & Garry, Ltd.
Vertol	Talc	Vermont Talc, Inc.
Vetrotex	Fiberglass	Vertexil S.A. Saint Gobain
Vi-Cal	Ground limestone	Pfizer, Inc.
Vicron	Calcium carbonate	Pfizer, Inc.
Vitro-Flex	Fiberglass mat	Johns-Manville Corporation
Vitron	Fiberglass rovings	Johns-Manville Corporation
Vitro-Strand	Fiberglass strands	Johns-Manville Corporation
Vulcan	Carbon black	Cabot Corporation
Vylor	Nylon fiber	E.I. Du Pont de Nemours & Co., Inc.
Waterground Mica	Mica	English Mica Co.
Weblene	Nonwoven fabrics	Kendall Co., Inc.
Weblox	Nonwoven fabrics	Kendall Co., Inc.
Webloy	Nonwoven fabrics	Kendall Co., Inc.
Webril	Nonwoven fabrics	Kendall Co., Inc.
Wey Chem	Lignin fillers	Weyerhaeuser Co.
Whitetex	Aluminum silicate	Southern Clays, Inc.
Witcarb	Calcium carbonate	Witco Chemical Co.
Witcolblak	Carbon black	Witco Chemical Co.
Witcolith	Carbon black	Witco Chemical Co.
Wollastonite	Wollastonite	Interpace Co.
Wyex	Carbon black	J. M. Huber Corporation
Zeolex	Silicate	J. M. Huber Corporation

REFERENCES AND BIBLIOGRAPHY

Alford, H. E., and Veatch, F. Glass microballoon particles, a low-density filler. *Modern Plastics*, November 1961, **38**(11):141–150, 223.

Ameron, Inc. *Bondstrand Fiberglass-Reinforced Plastic Pipe & Fittings.* Bulletin BB-1, February 1977.

Axelson, J. W. "Asbestos Fillers and Reinforcements." In *Modern Plastics Encyclopedia*, S. Gross (exec. ed.), **45**(14A):586–588. McGraw-Hill, Inc., New York, 1968–1969.

Banks, M. T. Plastics for the tough jobs. *Machine Design*, September 1979, **51**(17):88–90.

Berg, K. R., and Filippi, F. J. Advanced fiber-resin composites. *Machine Design*, April 1971, **43**(15):160–168.

Bolen, G. N. Selecting glass reinforcements. *Plastics World*, February 1975, **33**(2):32–35.

Bortz, S. A. "Ceramic Fiber Reinforcements." In *Modern Plastics Encyclopedia*, S. Gross (exec. ed.), **45**(14A):606–607. McGraw-Hill, Inc., New York, 1968–69.

Cabot Corporation. *Cab-O-Sil, The Dry Process Pyrogenic Silica.* Form XM926.

Carborundum Co. Data sheets on Boron Nitride Fibers, 1973.

Carter, J. S., and Moorefield, S. A. A new look at the new glass fibers. *Plastics Technology*, September 1963, **9**(9):42–44.

Celanese Corporation. *Celion Carbon Fibers-Material Properties Composites.* April 1978.

Chambers, W. E. Low-cost, high-performance carbon fibers. *Mechanical Engineering*, December 1975, **97**(12):37–41.

Chamis, C. C., Lark, R. F., and Sullivan, T. L. "Boron/Aluminum-Graphite/Resin Advanced Fiber Composite Hybrids." NASA Technical Note TN D-7879, February 1975.

Cloud, P. J. The role of fillers and reinforcements in plastics. *Machine Design*, September 1975, **47**(17):94–97.

Collins, W. "Silane Coupling Agents." In, *Modern Plastics Encyclopedia*, J. Agranoff (ed.), **52**(10A):234–236. McGraw-Hill, Inc., New York, 1975–76.

Composite Technology, Inc. *Composite Materials.* Aerospace Information Sheets, 1978.

Copeland, J. R., and Rush, O. W. Reinforcing with wollastonite filler makes for a tougher propylene. *Modern Plastics*, March 1979, **56**(3):68–70.

Davis, R. Fibrous reinforcement for RP. *Reinforced Plastics*, September/October 1962, pp. 10–12.

Dow Corning Corporation. *Dow Corning Organosilicon Chemicals.* Bulletin 23-181, July 1974.

Dowell, M. B. Using carbon fibers to reinforce plastics. *Plastics Engineering*, April 1977, **33**(4):31–35.

Du Mond, T. C. Now fiber composites also reduce costs. *Iron Age*, June 16, 1975, **76**(12).

Economy, J. Present status and future for high strength fibers. *SAMPE Journal*, November/December 1976, **12**(6):5–9.

——. Boron nitride fibers. *Research/Development*, June 1967, **18**(6).

E.I. Du Pont de Nemours & Co., Inc. *Characteristics and Uses of Kevlar 49 Aramid High Modulus Organic Fiber.* Bulletin K-2, February 1978.

——. *Kevlar 49 Aramid in Aircraft Applications.* Preliminary Information Memo 398, March 1978.

——. *Properties and Performance of NOMEX Type 412 Aramid Paper.* Bulletin NX-11, April 1978.

——. *Dyeing and Finishing NOMEX Type 450 Aramid.* Bulletin NX-1, May 1976.

——. *Dust Collection With Filter Bags of NOMEX Aramid Fiber.* Bulletin NX-2, June 1976.

——. *Making Garments From Fabrics of NOMEX Aramid Fiber.* Bulletin N-274, August 1974.

——. *Yarns and Fabrics of NOMEX Aramid Staple and Tow.* Bulletin N-275, November 1974.

——. *Laundering and Drycleaning Garments of NOMEX High Temperature-Resistant Nylon.* Bulletin N-257, March 1972.

——. *Properties and Characteristics of "Kevlar" 49 Yarn and Composites.* Bulletin A-97958.

——. *NOMEX Aramid.* Collection of Bulletins E-02079.

Emerson & Cuming, Inc. *Eccospheres Microballons–Hollow Glass and Ceramic Microspheres.* Bulletin 206-037, February 1977.

——. *Eccospheres EP & HB–Lightweight Epoxy Spheres.* Bulletins 14-2-10 and 14-2-11A, July and September 1976.

Ferro Corporation. *Ferro S-1014 Ultra-High Strength Glass Fiber Products and Zero Twist Fabrics.* Bulletins FG-101 and 102.

Fiberfil Corporation. *Guide to Fiberglass Reinforced Thermoplastic Resins.* Form No. 475-7854, 1975.

Fiberite Corporation. *Graphite/S-2 Glass Hybrid Material Comparison.* April 1978.

——. *The Graphite Fabric Concept.*

Forest, J. D. Advanced composite materials in spacecraft. *Modern Plastics*, September 1970, **47**(9):136–140.

Forsyth, R. A. "Carbon and Graphite Fibers for Plastics." In *Modern Plastics Encyclopedia*, S. Gross (exec. ed.), **45**(14A): 604–606. McGraw-Hill, Inc., New York, 1968–69.

Forsyth, R. B. "New Low Cost 'Thornel' Fibers From Pitch Precursor." Paper presentation, *19th National SAMPE Symposium and Exhibition*, Buena Park, California, April 1974.

Fothergil & Harvey, Ltd. *Carboform Preimpregnated High Modulus Carbon Fibre.* Publication NO. 32, 1970.

——. *Resin Impregnated High Modulus Carbon Fibre.* Publication No. 23 (2nd Ed.), April 1969.

General Electric Co. and AFML, Wright-Patterson AFB. "Boron Reinforcing Filaments." In *Modern Plastics Encyclopedia*, S. Gross (exec. ed.), **45**(14A):601–602. McGraw-Hill Inc., New York, 1968–69.

Great Lakes Carbon Corporation, *Fortafil.* Technical Data Sheet (revised), June 1977.

——. *GLC Graphite For Diversified Industrial Applications.* Bulletin CP, May 1973.

——. *Fortafil High Performance Graphite Fiber Products.* 1972.

Hand, A. J. Super-strong graphite. *Popular Science Monthly*, April 1975, **210**(4):2–4.

Hercules, Inc. *Magnamite Graphite Fibers.* Bulletin 200-269B, April 1978.

——. *Background Information, Magnamite Graphite Fiber Composites.* February 1974.

——. *Advanced Composites Products and Capabilities Summary.* Bulletin 51198H, March 1972.

Hexcel Corporation. *Total Graphite.* Bulletin 1M107.

Hiler, D. C. Carbon fiber composites. *Plastics World*, July 1977, **35**(7):63–65.

Hiler, D. C., Atkins, K. E., and Gentry, R. R. *Conductive Fiber Reinforced Plastics* Union Carbide Corporation reprint.

——. Mold electrically conductive FRP parts with carbon fiber reinforcement. *Plastics Design & Processing*, January 1978, **18**(1): 2–6.

Hosford, R. C. A guide to fillers and reinforcements for thermosets. *Plastics Technology*, February 1965, **11**(2):34–44.

Houston, A. M. Microspheres cut costs, improve properties of plastics. *Materials Engineering*, April 1975, **81**(4):58–59.

Johns-Manville Fiber Glass. *Fiber Glass Textile Yarn.* Bulletin FTX-5A, September 1960.

Katz, H. S. Carbon/graphite reinforcements are coming of age. *Plastics Compounding*, March/April 1979, **2**(2):18–25.

Katz, H. S., and Milewski, J. V. (eds.) *Handbook of Fillers and Reinforcements for Plastics*, Van Nostrand Reinhold Co., New York, 1978.

Kendall Co. *An Introduction to Kendall Nonwoven Fabrics.* Bulletin AS-802.

Kliger, H. S. Carbon fibers–designer's tool for high-modulus parts. *Plastics Design Forum*, May/June 1977, **2**(3):36–40.

Kolek, R. L. New reinforcing fiber fills void between glass and carbon fiber. *Materials Engineering*, September 1973.

Kureha Chemical Industries. *Kureha Carbon Fiber.* December 1973.

Larson, G. P. Kaolinite fractions, their effect on physical properties of reinforced plastics. *Modern Plastics*, May 1958, **35**(9):157–166, 236–237.

Liquid Nitrogen Processing Corporation. *LNP Internally Lubricated Reinforced Thermoplastics.* Bulletin 254-278, 1978.

Loud, S. N. Jr. S-2 glass. *Plastics World*, March 1979, **37**(3):47–49.

Martenson, J. A. Materials for filament-wound, high temperature structural uses. *Modern Plastics*, May 1965, **42**(5).

Materials Engineering. High performance fibers stabilize thermoplastics. January 1974, **79**(1):54–55.

——. SMC reduce FRP costs and raise strengths. January 1973, **77**(1):38–41.

——. Graphite and carbon composites move closer to less exotic uses. July 1973, **78**(1): 59–61.

Miner, L. H. "Kevlar Aramid Fibers for High Performance Composites." Paper presentation, *Society of Automotive Engineers International Automotive Engineering Congress and Exposition*, Detroit, Michigan, March 1, 1977.

Milewski, J. V. The field is wide open for reinforcements. *Plastics Engineering*, November 1978, **34**(11):23–31.

Miska, K. H. Detroit looks to advanced composites to shed weight, improve fuel economy. *Materials Engineering*, September 1977, **86**(3):66–68.

Mitsubishi International Corporation. *Krecosphere.* August 1975.

Modern Plastics. Fillers and reinforcements. July 1979, **56**(7):41–48.

——. New developments in reinforcements and fillers. July 1977, **54**(7):49–58.

Mohr, J. G. (ed.). *SPI Handbook of Technology and Engineering of Reinforced Plastics/Composites*, 2nd Ed. Van Nostrand Reinhold Co., New York, 1973.

Monsanto Co. *A Guide to Corrosion Control Through Reinforced Plastics.*

Morganite Research and Development Limited. *The New World of Modmor High Modulus Carbon Fibres.* Bulletin R25, October 1968.

Owens-Corning Fiberglas Corporation. *An Introduction to Fiberglass-Reinforced Plastics/Composites.* Publication No. 1-PL-6305-A, December 1976.

——. *Textile Fiber for Industry.* Publication No. 5-GT-5-442-A, March 1972.

——. *S2 Glass Fiber High Performance/Low Cost Reinforcement.* Collection of bulletins, Publication No. 8-ASP-7900.

PPG Industries. *PPG Fiber Glass Textile Products.* Form F 375A 744.

Paulus, H. J. "Fibrous Glass Reinforcements." In *Modern Plastics Encyclopedia*, S. Gross (exec. ed.), **45**(14A):592–594. McGraw-Hill, Inc., New York, 1968–69.

Plastics Design & Processing. Injection molding performance of polypropylene can be improved with limestone filler. June 1979, **19**(6):42–45.

Plastics World. Glass bubbles fill composite core. 1977, **35**(1):19.

——. Commercial graphite fibers debut. November 1973, **31**(11): 62–65.

Potters Industries, Inc. *Solid Glass Spheres–New Dimension in Reinforced Plastics.*

Powers, W. M., and Shave, R. G. Hollow carbon microspheres. *SAMPE Journal*, April/May 1970, **6**(2):69–73.

Product Engineering. Tiny bubbles beef up plastics, cut costs and reduce weight. October 1977, **48**(10):39–41.

——. Strongest synthetic fiber yet fills a host of design needs. September 1974, **45**(9):49–51.

——. New fiber and fillers boost performance of plastics and rubber. January 1973, **44**(1):49–51.

——. Downward price trend widens markets for graphite composites. December 1973, **44**(12):35–37.

Raia, D. C. Glass microspheres–a new era in fillers. *Plastics World*, April 1974, **32**(4):54–55.

Ritter, J. "Sphere-Filled Composites Theory, Testing, Applications." Paper presentation, *25th Annual Technical Conference, Reinforced Plastics/Composites Division, The Society of the Plastics Industry, Inc.*, 1970.

Rosato, D. V., and Grove, C. S. Jr. *Filament Winding, Its Development, Manufacture, Applications, and Design.* Wiley-Interscience, New York, 1964.

Rowe, P. E. "Glass Microballoon Spheres as a Filler for Plastics." Paper presentation, *Symposium on Fillers and Reinforcements for Plastics, Division of Organic Coatings and Plastics Chemistry, American Chemical Society and Plastics Institute of America*, Chicago, August 1973.

Seymour, R. B. Additives for plastics–fillers and reinforcements. *Plastics Engineering*, August 1976, **32**(8):29–34.

——. "Fillers for Polymers." In *Modern Plastics Encyclopedia*, S. Gross (exec. ed.), **45**(14A):570–581. McGraw-Hill, Inc., New York, 1968–69.

——. "Fibrous Reinforcements for Plastics." In *Modern Plastics Encyclopedia*, J. Agranoff (ed.), **52**(10A):187–191. McGraw-Hill, Inc., New York, 1975–76.

——. "Fillers for Plastics." In *Modern Plastics Encyclopedia*, J. Agranoff (ed.), **52**(10A):194–200. McGraw-Hill, Inc., New York, 1975–76.

Shepler, R. E., Towne, M. K., and Saylor, D. K. Carbon fiber composites. *Machine Design*, May 1979, **51**(12):89–95.

Shyne, J. J., and Milewski, J. U. "Whiskers (Single-Crystal Fibers.)" In *Modern Plastics Encyclopedia*, S. Gross (exec. ed.), **45**(14A): 602–604. McGraw-Hill, Inc., New York, 1968–69.

Society of the Plastics Industry, Inc. *Plastics Engineering Handbook.* 3rd Ed. Reinhold Publishing Corporation, 1966.

Soldatos, A. C., Burhans, A. S., Eckstein, B. H., and Spence, G. B. High-performance epoxy/graphite fiber composites. *Modern Plastics*, December 1971, **48**(4):62–66.

Sterman, S., and Marsden, J. G. Improved thermosetting systems. *Plastics Technology*, May 1963, **9**(5):38–41.

Sturgeon, D. L. G., Riewald, P. G., and Lacy, R. I. *"Kevlar" 49 Composites for Sporting Goods and Recreational Equipment.* E.I. du Pont de Nemours & Co., Inc., January 1974.

Theberge, J. E., Arkles, B., and Robinson, R. Carbon fibers add muscle to plastics. *Machine Design*, February 7, 1974, **46**(3): 102–107.

Theberge, J., and Cloud, P. Choosing the best glass-reinforced thermoplastic. *Plastics World*, March 1977, **35**(3):44–46.

Thompson, T. D. Applications for kaolin fillers. *Plastics Compounding*, May/June 1979, **2**(3):58–68.

3M Co. *How to Choose and Use "3M" Brand Glass Bubbles As A Resin Extender.* Bulletin L-GB1B500.

——. *When More is Needed Than Conventional Materials Can Give–Scotchply Reinforced Plastics.* Bulletin SP-RNB.

Union Carbide Corporation. Thornel Bulletins: 465-204AH, 215ij, 216cj, 218bb, 219, 222, 223, 224, 225, 226, 227, 228, 229, 230, 232, 233, 234, 235, and 236.

Vaughan, D. J. Coupling agents for use in aramid-reinforced composites. *Modern Plastics*, October 1977, **54**(2):66–67.

Whittington, L. R. *Whittington's Dictionary of Plastics.* Technomic Publishing Co., Inc., Stamford, Connecticut, 1968.

Wood, A. S. Coming: cut-rate super fibers. *Modern Plastics*, December 1973, **50**(12):56–59.

Woodhams, R. T. For low cost, high strength performance try mica reinforced thermoplastics. *Plastics Design & Processing*, August 1978, **18**(8):17–23.

9
Films

By general agreement in the plastics industry, plastics *films* are any plastics materials made in flat form with a thickness of 10 mils or less. Flat stock with a thickness greater than 10 mils is referred to as *sheet*. This convention is not always strictly adhered to, however. Some sources will refer to materials up to 20 mils in thickness as film, while others consider a thickness greater than 3 mils to be sheet. For our purposes, the 10-mil measurement will be maintained as the cutoff point.

Films are made from any of the commonly used plastics materials; however, the majority of films are thermoplastics and can thus be easily produced by the common processes of solvent casting, extruding, and/or calendering.

The base materials for the films are (with possible slight variations) identical to the materials used to make moldings or extrusions. However, the films themselves are, in many cases, subjected to special treatments or processing or are combined in ways to produce a final material with considerably improved properties over the original material. The three most important of the special treatments (described more fully below) are orientation, coating, and lamination. In each case, the special processing is used to obtain an improvement in some particular characteristic of the original material or in several related characteristics. In the case of lamination, the combination of two or more materials produces a synergistic effect, with the laminate properties superior to the properties of the original base films.

FILM PRODUCTION TECHNIQUES

Extrusion

The most common method for producing film, and sheet, is by extrusion from a melt. Virtually any thermoplastic can be extruded into film and sheet using the set-up illustrated in Figure 9-1. When used with an appropriate film die, materials up to 6 feet wide can be extruded at thicknesses as low as 0.001 inch. The material can then be stretched to as low as 0.00025 inch in thickness.

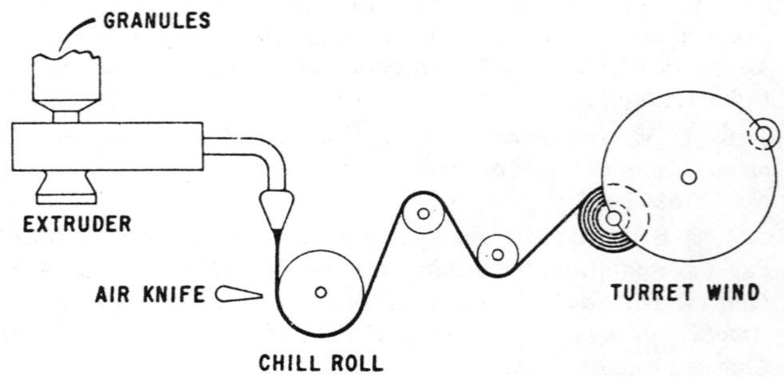

Fig. 9-1. Film extrusion using an air knife to insure good contact of the film to the cooling roll. (Park, 1969)

One of the major problems in the production of film by extrusion is the maintenance of uniform thickness throughout the width of the film. To allow for variance in the die opening, the die lips are equipped with adjustment bolts along the width of the die. While the measurements and adjustments can be made manually, the latest technique is the use of a beta ray nuclear gauge scanner across the film width. The scanner continually monitors the film thickness and instantly indicates the need for adjustments in the die lips. Automatic controllers can also be tied in with the scanners to allow the fastest response. Other types of scanners use x-ray and infrared sensors to accomplish the same purpose.

If a thick spot is allowed to develop in the extruded film for any length of time, there will be a tightly wound area, or bulge, flanked by more loosely wound areas on the roll. This unevenness can create problems in the original winding process and later in the unwinding of the roll.

One way to minimize the formation of a regularly occurring thick or thin spot is to cause randomization of the defect(s). Figure 9-2 shows how this can be accomplished by means of stationary deckle rods and internal deckle bars which control the width and position of the film. An oscillating choke bar (shown in Figure 9-3) can be used to produce the same randomization in thicker films and sheets.

Another extrusion system, used very extensively for thin films of polyethylene and polypropylene, is the "bubble" process. In this procedure, the polymer is extruded as a

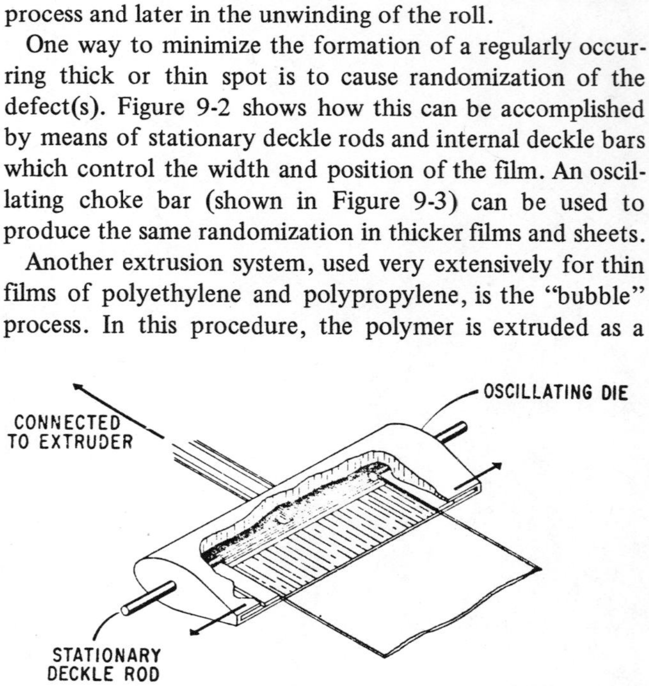

Fig. 9-2. Oscillating die and stationary deckles. (Park, 1969)

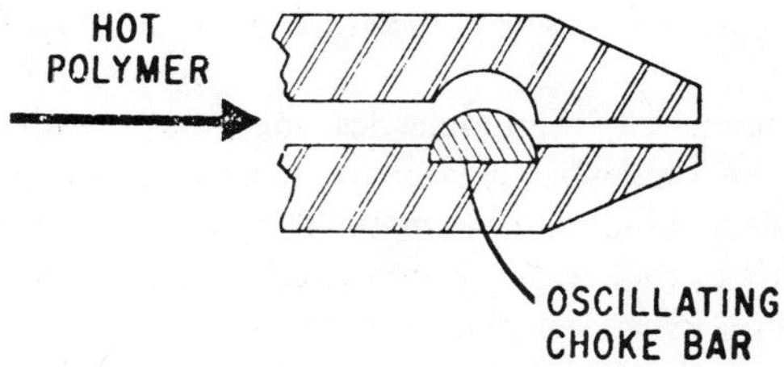

Fig. 9-3. Choke bar oscillator. (Park, 1969)

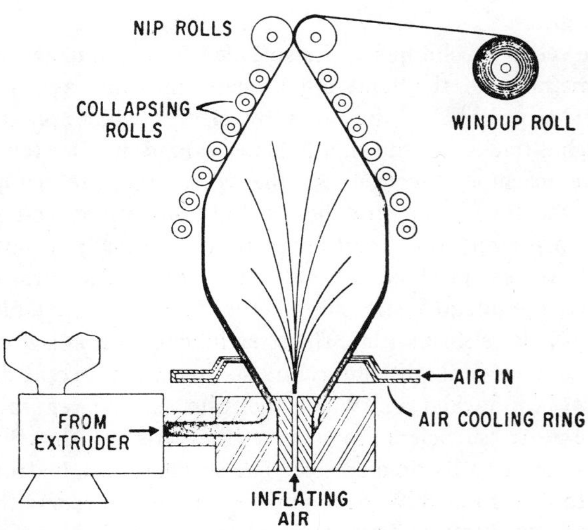

Fig. 9-4. Bubble process, used mainly with polyethylene. (Park, 1969)

relatively thick tubular extrusion and is then blown up by internal air pressure to form a thin-walled tube, as shown in Figure 9-4. The tube is then slit to form a sheet of film up to 10 feet in width, or is left as a flattened tube known commercially as "lay-flat tubing." To improve gauge uniformity in this process, the circular extrusion die, the air cooling ring, or both are frequently rotated (see also Chapter 11 on extrusion). While low-density polyethylene and polypropylene are the materials most used for this process, polyvinyl chloride, polyamides (nylons), and some polystyrenes have also been used.

Cellophane (although not strictly considered to be a plastics material) is made by an extrusion process in which a solution of dissolved cellulose (viscose) is extruded through a slot into an acid bath; this insolubilizes the solution and regenerates the cellulose.

Solution Casting

Another method of obtaining thin films is to cast a solution of the polymer onto a belt which is sent through a heating tunnel. By precise adjustment of the temperature in the tunnel and the belt speed, the solvent is removed and a finished film is recovered at the end of the tunnel. Cellulosics, such as cellulose acetate, have been extensively made by this process. Films made by this technique have exceptionally fine gloss and freedom from defects such as die lines which can never be completely eliminated from extruded film. Figure 9-5 illustrates the process.

Miscellaneous

A process used almost exclusively for polyvinyl chloride resins is calendering. A formulation of resin, plasticizer, colorants, stabilizers, etc. is heated and passed through polished and heated rollers which are successively placed closer and closer together as shown in Figure 9-6. This same process is also used in production of rubber sheeting or fabric-reinforced plastic or rubber sheeting.

Skiving (see Figure 9-7) is a method of continuously cutting thin sheets or films from a round section of material which is not normally amenable to extruding or solution casting. In practice, it has been used almost exclusively for production of films of Teflon TFE (polytetrafluoroethylene).

SPECIAL FILM PROCESSES

As previously mentioned, there are three processes used to enhance certain properties of films: (1) orientation, (2) coating, and (3) lamination.

Orientation. Orientation consists of a controlled stretching of the polymer chains of the film in such a way as to

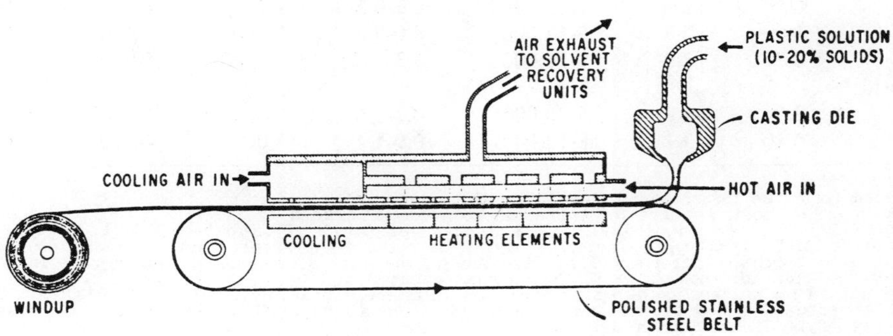

Fig. 9-5. Solvent belt casting process. (Park, 1969)

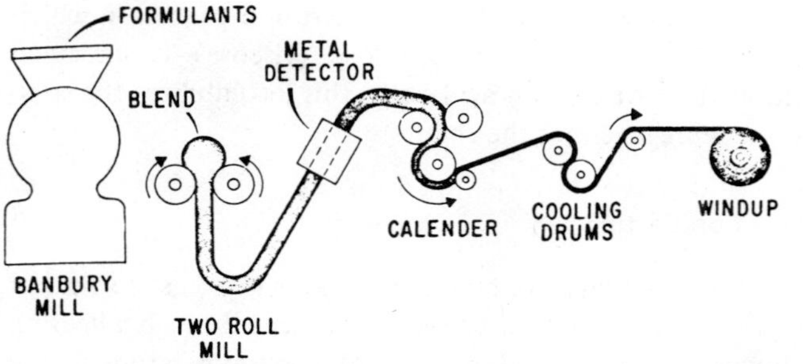

Fig. 9-6. Calendering process, mainly for PVC film. (Park, 1969)

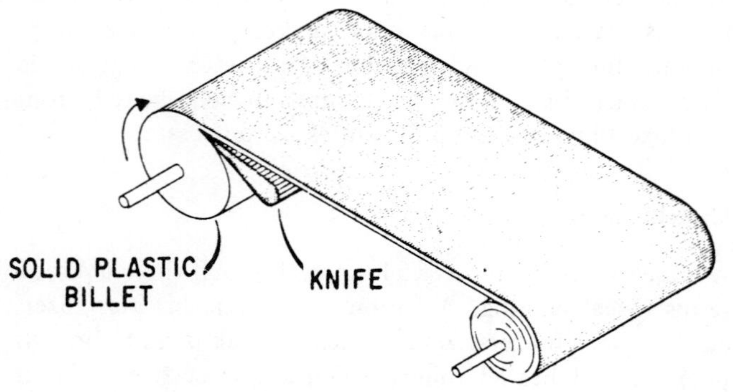

Fig. 9-7. Skiving process, mainly for Teflon TFE. (Park, 1969)

orient the chains in a direction parallel to the plane of the film. The orientation may be monoaxial or biaxial, with the latter being preferred. Biaxial orientation, for most films, improves tensile strength, toughness, flexibility, and shrinkability. It may also improve the clarity of transparent films and decrease the gas and moisture permeability of some films.

In general, all thermoplastics may be oriented; the benefits are somewhat greater for some materials than for others, depending on the material and/or the particular characteristic that is to be improved. There are two main methods by which orientation is accomplished. The first is by use of the bubble technique of film manufacture. During bubble formation, the biaxial stretch given the polymer not only thins the material, but also strengthens and orients the film.

The second technique is the so-called flat sheet or tentering method. In the tentering process, the film, as a flat sheet, is grasped at the sides by clamps which run on diverging tracks. As the film is being wound, it is stretched in the machine direction. At the same time, the clamps cause the film to be stretched in the transverse direction. It is only rarely that the film is stretched equally in both directions to produce a "balanced" film. The stresses created by unequal stretching result in an oriented film which is also a shrink film. The stretching operation is done in most cases while the film is hot, so that when it cools the stresses are locked in. When the film is reheated to a temperature sufficient to release the stresses, the film shrinks back to its original size. The shrinkage can be from 30% to as high as 80%, depending on the polymer and the original orientation treatment. A correspondingly high shrink stress can also be developed in the film. Table 9-1 shows some of the properties of common shrink films.

Coatings. A variety of coatings are used to impart special characteristics or improve some property of a film. In

Table 9-1. Properties of Shrink Films.[a,*]

Types of shrinkable films	*Most popular gage*	*Cost–most common ga., 1,000 sq. in.*	*Tensile strength*	*WVTR*	*Oxygen perm.*	*Maximum shrink*	*Shrink tension range*	*Film shrink temp. range[d]*
	mils	*cents*	*p.s.i.*	*g.m./mil[b]*	*cc./mil[c]*	*%*	*lb./sq. in.[e]*	*° F.*
Polyester	0.65	6.0	17-25,000	2.7-3.4	20-40	40-55	700-1500	160-250
Polyethylene[f], Type I	1.50	2.0	1600-1800	0.8-0.9	10,000	15-40[g]	Below 50	220-250
Crosslinked, Type I	1.0	3.5	8-13,000	0.6-0.7	7,500	70-80	150-500	160-240
Crosslinked, Type II	0.75	3.0	14-19,000	0.3-0.4	3,500	70-80	250-1000	200-280
Polypropylene	0.50	2.5	15-27,000	0.2-0.3	1,700	70-80	300-600	220-330
Polystyrene	1.0	2.3	9-12,000	4.0-9.0	3,500	40-60	100-600[h]	210-270
Polyvinyl chloride	0.75	2.9	8-16,000[h]	3.3-140[h]	20-180[h]	50-70[h]	150-300[h]	150-300
Polyvinylidene chloride copolymer	0.60	4.0	6-20,000[h]	0.2-1.4[h]	15-300[h]	30-60[h]	50-150[h]	150-210
Rubber hydrochloride	0.40	3.1	8-12,000	0.9-1.4	3,000	40-50	150-350	150-230

This table is reproduced from the 1963 **Modern Packaging Encyclopedia.**
a—The reported data for gage, cost, etç. are for the usual commercial grades.
b—ASTM—Method #E96-53T. Units = gm./100 in. sq./24 hr./per mil at 100° F. and 90% R.H.
c—ASTM—Method #D1434-58. Units = cc./sq. meter/24 hr/mil at room temperature and 1 atmosphere.
d—% Shrink (change from original) determined by five-second immersion of marked film sample in water for temperatures below 212° F. or in silicone oil above 212° F. (Tests run in Cryovac laboratory.)
e—Shrink tension determined by immersing film strip clamped in frame containing BLH Type SR4 Strain Gauge and reading maximum value. Hot water used below 212° F. and silicone oil used above 212° F. (Test run in Cryovac Laboratory.)
f—Type I is low density. Type II is medium density.
g—Shrink is directional, varying from 15% maximum in the transverse direction to 40% in the longitudinal direction.
h—Varies with film grade and method of manufacture.
*Reprinted by permission from *Modern Plastics*, November 1962, McGraw-Hill, Inc.

general, the coatings are thin (<0.001 inch) and are applied as lacquers, solvent dispersions, aqueous dispersions, etc. They are applied to one or both sides of the film, depending on the requirements. The reasons for use of coatings are:

1. To act as heat-sealing adhesives for materials which do not heat seal (e.g., cellophane) or for films which might show bad shrinkage effects if heated to a sealing temperature (e.g., polyester film). In both cases the coating seals at a moderate temperature which precludes danger to the film.
2. To improve the barrier properties to passage of gases or moisture.
3. To change the handling characteristics (decrease static buildup, fogging, sticking; increase slip, abrasion resistance). In many cases the addition of the coating will increase the "machinability" of the film. In this case the term machinability refers to the fact that the film can be handled by automatic machines—usually for packaging purposes.

There are other types of coatings applied to films which determine the entire character and use of the film. Such coatings are photographic emulsions, magnetic tape coatings, decorative films used in hot-stamping operations, etc.

Heat-Sealable Coatings. Heat-sealable coatings are made of materials with relatively low melting temperatures such as nitrocellulose, polyvinylidene chloride, PVC, LDPE, and Teflon FEP.

Barrier Coats. A heat-sealing coating can often function as a barrier coat to enhance the substrate in some particular way. An example is the use of heat-sealable polyvinylidene chloride coatings with cellophane to provide gas and moisture resistance. A number of proprietary coatings are used with various films for special purposes, and the film vendor should be contacted for specific information about them.

Machinability Coatings. The type of coating used to improve the machine handling of any particular film depends on the film and on its intended application. The film vendor should be consulted regarding any problem with the film—after the type has been selected for specific characteristics such as gas or moisture transmission, and electrical or chemical properties. There are a number of different coatings which can be easily applied to many types of films to improve their external surface properties, and, in some cases, these coatings will improve other properties as well.

Lamination. Laminations are combinations of two or more plastics films or a film and a paper, aluminum foil, or other material. The reason for the laminate is obvious—to produce a composite with properties which are superior to those which might be obtained in the same thickness of a single layer of only one of the plies. Usually the results are that the laminate will pick up the best points of each of the constituents. For example, a polyethylene-cellophane combination will have the stiffness, low elongation, and good printability of cellophane, and at the same time be a good moisture barrier and be heat sealable because of the polyethylene. These properties could be expected by combining 1 mil of polyethylene with 1 mil of cellophane, whereas a 5-mil film of either material alone might not give as good results in all respects.

Normally when combining an extrudable material such as polyethylene with a solid substrate such as paper, the polymer is simply extruded directly onto the paper creating a resin-coated paper film. The resin mechanically locks into the paper. When the plastic is extruded onto a material such as aluminum foil, it is usually necessary to either treat the foil by etching the surface or to use a primer which is specially formulated for the particular resin. Some of the primers which have been used include: polyalkyleneimine, organic titanates, polyurethanes, polyesters, and colloidal silica. The choice of the best primer for any particular polymer depends on the polymer, the processing conditions, and the final usage of the completed film. Figure 9-8 illustrates treatment of the substrate and the extrusion process.

Multilayer lamination of film to film is rarely done because of the difficulty in controlling the process, air entrapment, and so on. However, multilayer films are very com-

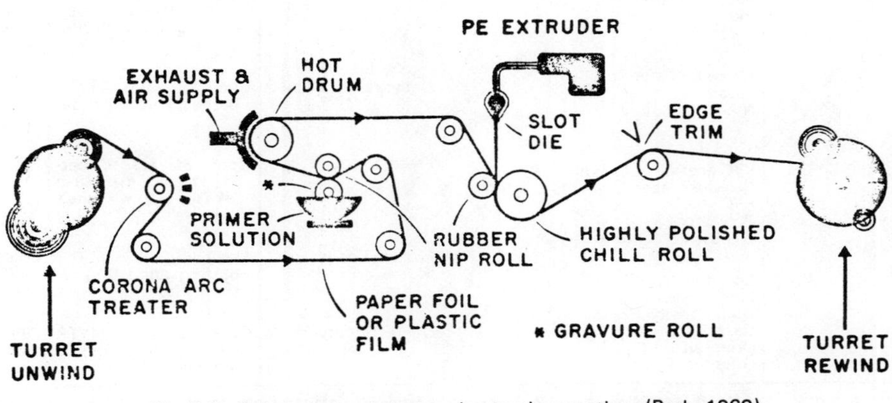

Fig. 9-8. Substrate treatment and extrusion coating. (Park, 1969)

monly made by a process known as *co-extrusion*, in which two or more polymers are simultaneously extruded through one die. In this process, the various melts are introduced into the die lips under conditions of laminar flow such that there is no intermixing, but bonding occurs at the layer interfaces. Chapter 11 illustrates the types of equipment used to make such films. The multilayer films produced by co-extrusion include such combinations as LDPE/HDPE/LDPE, LDPE/PVDC/LDPE, LDPE/ionomer, nylon/ionomer, EVA/PP/EVA, and many others, with as many as five, six, or seven layers.

APPLICATIONS OF PLASTICS FILMS

By far the most common uses for plastics films are for various packaging applications: heat-sealed bags for fresh produce and meat, thermoformed skin packages for meat, and thermoformed blister packages for dry goods, hardware items, and similar parts. The obvious advantages of visibility and reduction of pilferage as well as the tight fit of the skin packages and the well-formed blisters, all add up to excellent acceptance by both commercial interests and the consuming public.

Films are also used for such diverse applications as electronic capacitors (in the form of metallized foils), high-temperature wire insulation, thermal insulation of spacecraft, and in fabrication of high-altitude balloons for research purposes. The diversity in the applications is accompanied by a similar diversity in requirements for the various applications. For food packaging, the moisture vapor transmission rate (MVTR) can be all-important, since most foods require that the moisture be retained. However, some foods require fresh oxygen to be admitted, while others must minimize contact with atmospheric oxygen.

Table 9-2. Properties and Costs of Basic Packaging Films.*

Property		Cost Range ¢/1000 sq in.[a]	Film
A. Moisture Vapor Transmission Rate[b]			
1. 0.2-1.5	1.4	1.1-6.6	Low-density PE
	0.8		Medium-density PE
	0.5		High-density PE
	0.4-1.0		Cast PP
	0.2-1.2		Moisture-proof cellophane
	0.2-0.4		Oriented PP
	0.2		Saran
2. 1.5-10.0	4.4	2.4-8.8	Oriented PS
	9.11		PVC
	4.0		PVC-acetate copolymer
	4.0		Cellulose acetate
	8.1		Nylon 6
	2.5		Polyester
3. >10.0	90-100	2.5-9.7	Nonmoisture-proof cellophane
	40-50		Semimoisture-proof cellophane
	80		PVA
B. Oxygen permeability[c]			
1. nil-1.0	0.1-0.2	2.5-9.7	All cellophane (dry)
	0.2		Saran
	1.0		Nylon
	nil		PVA
2. 1.0-75.0	15	8.0-50.0	PVC-acetate copolymer
	4.5		Cellulose acetate
	40-45		PVC
	10		Polyester
	65-75		CTFE fluorocarbon
3. 75.0	550	1.1-2.4	Low-density PE
	280		Medium-density PE
	125		High-density PE
	97-250		Cast PP
	213-300		Oriented PS
	150		Oriented PP

[a] All films are listed in increasing price order. [b] MVTR—cc/100 sq in./mil/24 hr at 70-80 F and 1 atmosphere. [c] gm/100 sq in./mil/24 hr at 100 F, 90-100% RH.
*Sacharow, 1968.

Table 9-3. Permeability Data for Selected Laminates.[a],*

Lamination	Oxygen[b]	Water (MVTR)[c]
Nylon 6/PE	25-40	.33
Polyester/ Saran/PE	3-5	.30
Oriented PP/ Cellophane/PE	5	.31
Saran-Coated PP/PE	7	.25
Cellophane/PE	3	.20
Saran-Coated PVA/PE	5	—
Polycarbonate/ PE	1,400	—

[a] These laminates are in the 10.0-20.0/1000 sq in price range.
[b] ml./sq.m./24 hr at 73 F., 50% RH.
[c] gm./100 sq.in./mil./24 hr at 100 F., 90-100% RH.
*Sacharow, 1968.

Table 9-4. How Film and Film Combinations are Used.[a],*

Product	Properties Required	Films and Laminates Used
Meat		
Fresh	2 wk shelf life; good O_2 rate of permeation; low MVTR.	PVC, PE, cellophane.
Cured	Good O_2 barrier; low MVTR; 8-10 wk shelf life.	Polyester/saran/PE Nylon/saran/PE
Cheese	Good O_2 barrier; low MVTR; flexibility; machinability; 10 wk shelf life.	Cellophane/saran/PE; polyester/PE; polyester/PE; polyester / Saran / PE; Nylon / Saran/PE; oriented PP/PE/cellophane/PE; oriented PP/Saran/PE.
Boil-in-Bags	Flexibility; low MVTR; heat resistance; stress resistance; 1 yr shelf life	Polyester/PE; nylon/PE; oriented PP/PE; nylon 11; polycarbonate/PE.
Coffee	Good O_2 barrier; flexibility; stress resistance; 6 mo shelf life.	Polyester/saran/PE; nylon/saran / PE; Clysar/saran; oriented PP/saran/PE; oriented PP/aluminum foil/PE.
Dehydrated Foods	Good O_2 and MVTR barrier; 2 yr shelf life.	Paper/aluminum foil/PE; cellophane / aluminum foil / PE; polyester/aluminum foil/PE.
Fresh Produce	Good O_2 rate of permeation; medium MVTR; 1-2 wk shelf life.	PE; cellophane; polystyrene; PVC; saran.
Frozen Foods	Flexibility; low MVTR barrier; good O_2 1 yr shelf life.	Coated paperboard containers; PVC; Pliofilm, PE; nylon; saran.
Confectionery	Good O_2 and MVTR barrier; flexibility; 6 mo shelf life.	Cellophane; oriented PP; foil laminates.

[a] Many film laminates may be used for different food products. The choice is dictated by economics and overall feasibility.
*Sacharow, 1968.

Data is thus supplied for all food packaging films on the MVTR characteristics and the oxygen permeability. Table 9-2 lists the MVTRs and the oxygen permeability of the commonly used food packaging films. Table 9-3 shows similar information for selected laminates.

The same food item may be packaged in a number of different films or film laminates. The actual choice will most often be dictated by availability and overall economic and usage feasibility. As an example, Table 9-4 shows a number of different food items, their requirements, and the possible films which might be used.

Packaging of food items is one of the largest uses of films in the packaging field, but is is not the only use for films in packaging. There are a number of other uses for films in which properties other than MVTR and oxygen permeability are of importance. Table 9-5 is a brief compilation of the important characteristics of the most commonly used film-forming materials, with emphasis on their non-food packaging characteristics. All these materials are transparent for optimum packaging usage.

A more complete listing of all the pertinent properties of the various films is given in Table 9-6, and a list of trade names is given in Table 9-7.

In addition to the single films listed above, there are many "standard" co-extruded laminates available. The most popular of these are listed in Table 9-8.

All the single layer films and the coextruded laminates are available in many thicknesses.

TESTING

Tests for film materials include all the usual physical property tests common to other plastics; tensile strength, elongation, modulus, dielectric strength, etc. However, because they are used as very thin materials, and more importantly because the majority of the usages are as packaging materials, there are a number of tests made which relate directly to their use as packaging materials. The test methods, in almost all cases, are standard tests developed by various testing associations or organizations whose primary interest is in packaging materials. Thus, the test methods cited should be able to be performed by any well-equipped test laboratory.

Optical Tests. One of the major reasons for the use of plastics films in packaging is the transparency of almost all the materials. Because not all the materials are equally transparent, nor are all shipments of the same material equally transparent, there are several tests to determine the various aspects of transparency. These include measure-

Table 9-5. Packaging Films—Major Characteristics.

Material	Characteristics
Acrylics	Weather and UV resistance.
Cellulosics: CA, CAB, CAP, CA_3	Good scuff, oil, and grease resistance. High gas permeability, medium MVTR, medium clarity. CAB: good outdoor weathering; major uses: skin and blister packaging.
Fluoroplastics	Excellent electrical, chemical, and thermal properties. $CFCl_3$: lowest MVTR of any transparent film, LOX (liquid oxygen) compatible and flexible at cryogenic temperatures; uses: packaging of hygroscopic pharmaceuticals.
Ionomers	Very tough, oil and grease resistant, heat sealable through oil or grease.
Polyamides	Good general-purpose, tough, strong material. Useful from −100°F to +300°F.
Polycarbonates	Strong, flexible material. Attacked by many chemicals. Must be dried thoroughly before processing or bubbles result.
Polyesters	Tensile strengths up to 40,000 psi in tensilized (oriented) condition. Good high-temperature properties (original "boil-in-bag" material). Very good electrical properties. Metallized film used for thermal insulation on spacecraft.
Polyethylenes	Tough, low cost, versatile. Used for all types of packaging: trash bags, bread wraps, shrink film, antistatic film, etc.
Polyimides	Highest temperature resistance of all films. Excellent chemical and electrical properties. Heat sealable using FEP coating.
Polypropylenes	Strong, tough, low gas transmission, good chemical and grease resistance, best when oriented. Poor low-temperature properties.
Polystyrenes	High gas permeability, good for produce packaging, not recommended for outdoor usage. Many grades available. Good thermoforming capabilities for cups, blister packs, etc.
Polyurethanes	Very tough, good tear and abrasion resistance. Can be extruded and calendered onto fabric.
Polyvinyl chlorides	Very versatile, strong, tough. Used for fresh meat and produce packaging.
Polyvinylidene chlorides	Extremely low moisture and gas transmission rates. Good chemical resistance. Good shrink characteristics.
Polyvinyl fluoride	Excellent weather and UV resistance, good chemical resistance.
Polyvinylidene fluoride	Good UV, nuclear, weather, and chemical resistance; good electrical properties.

Table 9-6. Film and Sheeting Property Chart.

Generic Name	Specific Gravity	Sq In Per Lb One Mil Thickness	Thickness Range	Manufacturing Method (extrusion, biaxial, orientation, casting, etc.)	Availability (in sheets, rolls, tapes)	Maximum Width, in.	Colorability
ASTM No.	D-1505	–	–	–	–	–	–
1. Acrylic	1.14–1.26	22,000–24,000	0.002–0.006	extrusion	rolls	108	good
2. Acrylonitrile butadiene styrene	1.01–1.30	25,600–27,400	0.0075–2.000	calendering, extrusion, press laminated	sheets, rolls	60	good
3. Acrylonitrile-styrene copolymer	1.08–1.09	25,400	0.001–0.020	biaxially oriented	rolls	105	excellent
4. Cellophane	1.40–1.50	11,600–25,000	0.0008–0.0017	extrusion	rolls, sheets, ribbons	47–52	good
5. Cellulose acetate	1.28–1.43	21,000–22,000	0.0005–0.250	casting, extrusion	sheets, rolls, tapes	60	excellent
6. Cellulose acetate butyrate	1.19–1.22	23,800	0.0011–0.150	casting, extrusion	sheets, rolls	40–48	good to excellent
7. Cellulose propionate	1.20	23,000	0.003–0.250	extrusion	sheets, rolls	48	good
8. Cellulose triacetate	1.28–1.31	21,200–22,000	0.0005–0.200	casting	sheets, rolls	47	–
9. Ethyl cellulose	1.15	23,800	0.002–0.015	extrusion	rolls, sheets	42	excellent
10. Ethylene chlorotrifluoroethylene copolymer (E-CTFE)	1.66–1.68	16,600	0.005–0.090	extrusion	rolls, sheets, tubes	54	good
11. Ethylene methacrylate	0.942	29,280	0.04–0.250	extrusion, blown, casting	rolls, tubing, sheets	60	excellent
12. Ethylene-tetrafluoroethylene copolymer (ETFE)	2.15	16,300	–	extrusion	rolls, sheets, tubes	48	good
13. Ethylene-vinylacetate copolymer	0.924–0.920	29,000	0.00075 and up	extrusion	rolls, sheets, tape, tubing	480	–
14. Fluorinated ethylene propylene copolymer	2.15	12,900	0.0005–0.060	extrusion	tapes, rolls	48	excellent
15. Ionomer	0.94	29,400	0.001–0.01	extrusion	rolls, sheets, tapes, tubes	72	good
16. Polyamide (nylon 6/6)	1.14	24,350–25,700	0.0005–0.030	extrusion	rolls	84	good
17. Polyamide (nylon 6) (biaxially oriented)	1.16	23,870	0.0005–0.001	extrusion, orientation	rolls, tubing	86	good
18. Polyamide (nylon 11 and 12)	1.03	26,630–27,200	0.0005–0.030	extrusion	rolls, sheets, tubes	28	good
19. Polybutylene							
20. Polycarbonate	1.20	23,000–23,100	0.001–0.250	extrusion	sheets, rolls	45–54	excellent
21. Polyester	1.380–1.399	20,000–21,000	0.00015–0.014	biaxially oriented, extrusion	rolls, tapes, sheets	67–120	–
22. Polyether sulfone	1.37	20,000	0.001–0.250	casting, extrusion	rolls, sheets	52	brown
23. Polyethylene (low density)	0.910–0.925	30,000	0.0004–0.250	calendering, casting, extrusion, biaxially oriented	sheets, rolls, tapes, tubes	480	excellent
24. Polyethylene (medium density)	0.926–0.940	29,500	0.001–0.250	calendering, casting, extrusion, biaxially oriented	sheets, rolls, tapes, tubes	240	excellent
25. Polyethylene (high density)	0.941–0.965	29,000	0.0004–1.000	calendering, extrusion, stress relieving	sheets, rolls, tapes, tubes	60	excellent

Table 9-6. (*Continued*)

Clarity	Shelf Life	Flammability	Maximum-minimum Use Temperature Range °F	U.V. Resistance	Heat Sealing Range °F	Dimensional Change At High R.H.%	Generic Name No.	Gas Permeability cc/100 sq in 24 hour/mil (O_2, N_2, CO_2, SO_2)	Water Vapor Permeability Gm/24 hr/100 sq in/mil @ 95°F. 90% R.H.
–	–	CS-192	D-759	D-1435	–	–	–	D-1434	E-96 (B)
excellent	indefinite	slow burning	−65 to 215	excellent	yes	–	1	–	8 (3 mil)
poor	good	slow burning to self-extinguishing	−65 to 215	poor to good	–	–	2	–	–
excellent	good	slow burning	−80 to 185	fair	–	–	3	CO, 210, $N_2$9, O_2 36	12.0–14.0
good	indefinite	slow burning	0 to 350	good	250–300	2.5	4	CO_2 0.4–0.6, H_2 1.2–2.2 N_2 0.5–0.6, O_2 0.5–0.8	0.4–1.34
excellent	indefinite	slow burning	−15 to 200	good	350–450	1.5	5	O_2 117, N_2 40, CO_2 1000	100–200
good to excellent	indefinite	slow burning	−30 to 180	good	–	–	6	CO_2 6,000, N_2 250, O_2 950	$\frac{4.5\text{–}6.5}{730^{x}}$
good	indefinite	self-extinguishing	−30 to 200	good	350–450	0.5	7	High	100–200
excellent	indefinite	self-extinguishing	0 to 400	fair	non-sealing	0.5	8	CO_2 880, N_2 15,O_2 30, 0	50–150
excellent	indefinite	slow burning	−75 to 250	good	non-sealing	–	9	CO_2 7,800, N_2 100 O_2 3,500	4.8–14.2
good	indefinite	self-extinguishing	−80 to 400	good	475–500	2.5	10	CO_2 110, N_2 10, O_2 25	0.6
fair	indefinite	slow burning	−80 to 150	fair	175–350	nil	11	O_2 335, CO_2 4,780	4.1
good	indefinite	self-extinguishing	–	good	525	–	12	CO_2 250, N_2 30, O2 10	0.04
–	–	burns	120 to 180	fair	150–300	–	13	CO_2 6,000, N_2 400, O_2 840	2–3
good	indefinite	non-flammable	−415 to 400	good	575–700	none	14	O_2 3,250, N_2 1,500, CO_2 3,150	–
good	indefinite	–	–	–	–	–	15	O_2 600	–
fair to good	indefinite	self-extinguishing	−100 to 300	poor to fair	400–450	3.4	16	CO_2 160, O_2 25, (50% R.H.)	19.0
good	indefinite	slow burning	−100 to 300	good	400–450	3.4	17	O_2 1.2–1.4	10–11
good	indefinite	slow burning	–	–	320–390	–	18	CO_2 153–336, N_2 3.4–18, O_2 34–92, H_2 323	4.0
							19		
excellent	excellent	self-extinguishing	−150 to 275	fair to good	350–440	none	20	CO_2 1,075, O_2 300, N_2 50	11.0
excellent	excellent	slow burning to self-extinguishing	−75 to 300	fair to good	325–450	11×10^{-6}	21	CO_2 16, O_2 6, N_2 1	1.7
fair	indefinite	slow burning to self-extinguishing	−100 to 400–450	poor	500–550	–	22	CO_2 950, N_2 40, O_2 230	–
good to excellent	indefinite	slow burning	−70 to 180	fair	230–400	none	23	CO_2 2,900, O_2 550, N_2 180	1.2–1.5
excellent	indefinite	slow burning	−70 to 200	fair	250–400	none	24	CO_2 990, O_2 280	0.7–0.9
good	indefinite	slow burning	−70 to 250	fair	260–425	none	25	CO_2 6,200–3,300, O_2 800–450, N_2 100–50	0.3–0.4

Table 9-6. (*Continued*)

Generic Name	Water Absorption 24 hr Immersion %	Resistance to Acids	Resistance to Alkalies	Resistance to Greases and Oils	Resistance to Organic Solvents	Resistance to Water and/or Detergents	Resistance to Oxidizing Agents
ASTM No.	D-570	D-543	D-543	D-722	D-546	–	–
1. Acrylic	1.4–16	good	good	good	paramatrics, (aliphatics)	good	–
2. Acrylonitrile butadiene styrene	0.2–0.3	fair to good	good	excellent	fair to excellent	–	fair
3. Acrylonitrile–styrene copolymer	0.5	good	excellent	good	excellent to good	–	good
4. Cellophane	45–115	poor	poor	excellent	poor	good	fair
5. Cellulose acetate	4–9	poor to good	poor	good	poor	good	poor
6. Cellulose acetate butyrate	0.1–3.4	poor to good	poor	good	poor	excellent	poor
7. Cellulose propionate	1.6	poor	poor	good	poor	excellent	poor
8. Cellulose triacetate	2.0–4.5	good to poor	poor	good	fair	–	poor
9. Ethyl cellulose	2.5–7.5	fair to poor	excellent	good	poor	–	poor
10. Ethylene chlorotrifluoro-ethylene copolymer (E-CTFE)	<0.02	good	good	good	good	good	good
11. Ethylene methacrylate	nil	fair to good	good	poor	poor	excellent	poor
12. Ethylene–tetrafluoroethylene copolymer (ETFE)	<0.02	good	good	good	good	good	good
13. Ethylene–vinylacetate copolymer	<0.01	fair	fair	poor	fair	good	–
14. Fluorinated ethylene propylene copolymer	0.01	excellent	excellent	excellent	excellent	excellent	good
15. Ionomer	0.4	good	good	good	good	good	–
16. Polyamide (nylon 6/6)	9.5	poor	excellent	excellent	excellent	good	poor to fair
17. Polyamide (nylon 6) (biaxially oriented)	–	poor	fair	excellent	good	good to poor	poor to fair
18. Polyamide (nylon 11 and 12)	0.27–0.25	poor	good	good	good	good to poor	poor to fair
19. Polybutylene							
20. Polycarbonate	0.15	excellent	fair	good	poor	fair	good
21. Polyester	<0.8	fair to excellent	fair to good	excellent	good to excellent	excellent	good
22. Polyether sulfone	2.1	good	good	good	good to poor	good	–
23. Polyethylene (low density)	0–0/8	excellent	excellent	poor	good (60°C)	–	good
24. Polyethylene (medium density)	nil	excellent	excellent	good	good	fair	fair
25. Polyethylene (high density)	nil	excellent	excellent	good	good (80°)	–	poor

Table 9-6. (*Continued*)

Tensile Strength lbs sq in @ RT	Elongation %	Tearing Strength (Elmendorf GM/mil)	Folding Endurance	Dielectric Constant 10^3 cps	Dielectric Constant 10^6 cps	Dielectric Constant 10^9 cps	Dissipation Factor 10^3 cps	Dissipation Factor 10^6 cps	Dissipation Factor 10^9 cps	Dielectric Strength, V/mil	Volume Resistivity ohm-cm
D-882	D-882	D-1004	D-2172	D-150	D-150	D-150	D-150	D-150	D-150	D-149	D-257
4,400	130	25	6,000+	4.4	3.5	–	0.046	0.040	–	2,400	10^{16}
1,700–7,800	–	–	–	2.5–3.0	2.5–3.0	–	–	–	–	350–400	10^{16}
10,000–15,000	20–60	–	–	–	–	–	–	–	–	–	–
7,000–18,000	10–50	2–20	500	3.2 (un-coated)	–	–	0.015 (un-coated)	–	–	2,000–2,500 (uncoated)	10^{11}
8,000–16,400	25–35	2–25	250–2,000	3.6	3.2	3.2	0.016	0.33	–	3,200–5,000	10^{10}–10^{15}
4,100–9,700	40–100	3–10	1,200–1,500	2.9	2.5	2.8	0.013	0.03–0.04	–	3,400	10^{11}–10^{15}
5,000	80	25	1,000–2,000	–	3.2	–	–	0.015	–	–	–
9,000–16,000	10–40	4–10	3,500–4,000	3.2–4.5	3.3–3.8	3.2	0.016	0.033	–	3,700	10^{13}–10^{15}
8,000–10,000	20–30	7–36	250,000	3.1	3.0	–	0.002–0.20	0.010–0.06	0.046	1,500 (10 mil)	10^{15}
8,000–10,000	150–250	900–1,300	250,000	2.6	2.6	–	0.002	0.013	–	5,000–6,000	10^{16}
1,500–1,800	600–1,000	60 MD 150 TD	6,000+	–	3.1	–	–	0.15	--	300–500	10^{15}
7,000–8,000	300	600–900	–	2.6	2.6	–	0.008	0.015	–	3,500	10^{16}
1,000–3,500	400–800	–	very high	2.7–2.9	2.7–2.9	–	0.01–0.02	0.01–0.02	–	5,000	10^{9}
3,000	300	125	4,000	2.0–2.50	2.0–2.05	2.05	<0.002	0.0004	0.0015	7,000	10^{19}
5,000	250–450	30	very high	–	–	–	–	–	–	–	–
9,000–12,000	300–400	50–90	>250,000	4.0	3.4	–	–	–	–	850	4.5×10^{13}
32,600–35,000	100–120	–	–	–	–	–	–	–	–	–	–
7,000–12,000	250–400	–	good	3.8	3.1	–	0.05	0.03	–	1,800	10^{13}
8,000–10,500	85–105	20–25	250–400	2.99	2.93	2.89	0.0015	0.010	0.012	1,500	10^{16}
25,000–40,000	50–120	12–40	230,000	3.2	3.0	2.8	0.005	0.016	0.003–0.008	7,500	10^{18}
10,000–12,000	20–150	7–16	–	3.5	3.5	–	0.0035	0.006	–	2,400 (4 mil)	10^{17}–10^{18}
up to 12,000	200–800	50–300	very high	2.2	2.2	2.2	0.0003	0.0003	0.0003	470	10^{16}
up to 17,000	50–650	50–300	very high	2.2	2.2	2.2	0.0003	0.0003	0.0003	500	10^{16}
2,400–6,100	10–650	15-300	very high	2.3	2.3	2.3	0.0005	0.0005	0.0005	500	10^{16}

Table 9-6. (*Continued*)

Generic Name	Specific Gravity	Sq In Per Lb One Mil Thickness	Thickness Range	Manufacturing Method (extrusion, biaxial, orientation, casting, etc.)	Availability (in sheets, rolls, tapes)	Maximum Width, in.	Colorability
26. Polyimide	1.42	19,400	0.0005–0.005	casting	rolls, tapes	54	–
27. Polymethyl methacrylate	1.18–1.19	23,400	0.005–5.000	extrusion, biaxially oriented	sheets, rolls	43	excellent
28. Polymethyl pentene	0.83	33,360	0.003–0.030	extrusion	rolls, sheets	48	good
29. Polyparabanic acid	1.3	20,800	0.001–0.005	casting	rolls	50	amber
30. Polypropylene, cast	0.9	30,700	0.0003–1.000	casting, extrusion	sheets, rolls, tapes	60	excellent
31. Polypropylene, balanced, oriented	0.91	30,500–30,600	0.0005–0.0015	orientation	rolls	72–80	–
32. Polypropylene, balanced, oriented, coated	0.94	29,500	0.0009–0.0011	orientation	rolls	72–80	–
33. Polystyrene (casting, extrusion)	1.05–1.06	26,100	0.0004 and up	casting, extrusion	sheets, rolls	76	excellent
34. Polystyrene (oriented)	1.05–1.06	26,100–26,300	0.0003–0.020	oriented, biaxially oriented	sheets, rolls	76	excellent
35. Polysulfone	1.24–1.25	22,355	0.0001–0.250	extrusion	sheets, rolls	52	good
36. Polytetrafluoroethylene (Teflon TFE)	2.14–2.17	12,700–13,000	up to 0.250	skiving, casting, extrusion	sheets, rolls, tapes	–	good
37. Poly(perfluoroalkoxytetrafluorethylene) (PFA)	2.13–2.16	–	–	casting, extrusion, molding	rolls, sheets	–	good
38. Polytrifluorochloroethylene (Kel-F)	2.1–2.2	13,000	0.0005–0.020	casting, extrusion,	sheets, rolls, tapes, tubes	–	good
39. Polyurethane	1.11–1.24	23,000	0.0005 and up	calendering, casting extrusion	sheets, rolls, tubes	68	good
40. Polyvinyl alcohol	1.21–1.31	21,600	0.0005–0.012	casting, extrusion	sheets, rolls, tapes	–	excellent
41. Polyvinyl chloride (PVC) (rigid)	1.35–1.48	20,000–23,000	0.001–2.000	calendering, press, casting, extrusion	sheets, rolls, tapes	54	excellent
42. Polyvinyl chloride (PVC) (flexible)	1.24–1.45	20,000–24,000	0.0005–0.375	calendering, press, casting, extrusion	sheets, rolls, tapes	80	good to excellent
43. Polyvinyl fluoride (Tedlar)	1.38–1.57	20,000	0.0005–0.002	extrusion	rolls	–	good
44. Polyvinylidene fluoride (Kynar)	1.75–1.78	16,000	0.002–0.020	extrusion	sheets, rolls	50	good
45. Rubber hydrochloride	1.11	25,000	0.0004–0.0025	casting	ribbons, sheets, rolls, tapes	–	–
46. Vinyl chloride–acetate copolymers (rigid)	1.30–1.39	21,000	0.001–0.030	calendering, extrusion, casting	sheets, rolls	–	excellent
47. Vinyl chloride–acetate copolymers (flexible)	1.20–1.35	21,000	0.001–0.030	calendering, extrusion, casting	sheets, rolls, tapes, tubes	–	excellent
48. Vinylidene chloride–vinyl chloride copolymer	1.57–1.68	16,300–23,000	0.004–0.010	calendering, biaxially oriented, extrusion, casting	rolls	–	excellent
49. Vinylnitrile rubber	1.18–1.21	22,800–23,700	0.001–0.020	casting, extrusion, calendering	sheets, rolls, tapes, tubes	–	–

Table 9-6. (*Continued*)

Clarity	Shelf Life	Flammability	Maximum-minimum Use Temperature Range °F	U.V. Resistance	Heat Sealing Range °F	Dimensional Change At High R.H.%	Generic Name No.	Gas Permeability cc/100 sq in 24 hour/mil (O_2, N_2, CO_2, SO_2)	Water Vapor Permeability Gm/24 hr/100 sq in/mil @ 95°F. 90% R.H.
good	excellent	self-extinguishing	–450 to 600	good	does not heat seal	22×10^{-6}	26	CO_2 450, O_2 25, N_2 6	5.4
excellent	excellent	slow burning	– to 190	good to excellent	375–450	0.35	27	–	0.5
fair	indefinite	burns	–	poor	285–400	–	28	–	–
good	indefinite	slow burning	–320 to 500	–	250–350	–	29	–	–
excellent	indefinite	slow burning	0 to 275	fair	260–375	none	30	CO_2 800, O_2 240, N_2 60	0.7
excellent	indefinite	slow burning	–60 to 240	fair	narrow	none	31	CO_2 540, O_2 160, N_2 15	0.25–0.45
excellent	indefinite	slow burning	–60 to 225	fair	200–275	none	32	CO_2 520, O_2 130	0.40
excellent	–	slow burning	–55 to 200	–	–	–	33	–	0.5–0.7, 6.0
excellent	indefinite	slow burning	–80 to 175	fair	280–330	none	34	AIR 105, CO_2 1,080, H_2 2,420, N_2 49.5, O_2 310	7.0–10.0
good	indefinite	–	–100 to 350	poor	–	–	35	CO_2 405, N_2 40, O_2 90	18
poor	indefinite	non-flammable	–450 to 500	excellent	does not heat seal	–	36	–	–
good	indefinite	self-extinguishing	–450 to 500	excellent	600–700	–	37	–	–
good	indefinite	non-flammable	–423 to 300	excellent	325–375	–	38	He 600–1,150, O_2 1.3–58, N_2 9.3–19, CO_2 11.7–225	0.025
good	indefinite	slow burning	–100 to 190	poor to good	275–375	–	39	CO_2 450–1650, N_2 41–119, O_2 75–327	40–75
excellent	–	slow burning	–40 to 420	good	240–380	none	40	He 0.2, H_2 nil, CO_2 nil, O_2, nil, (0% R.H.)	10.0 est
good to excellent	excellent	slow burning to self-extinguishing	–50 to 200	fair to good	340–400	negligible	41	CO_2 970, O_2 150, (50% R.H.)	4.0
good to excellent	indefinite	slow burning to self-extinguishing	–50 to 200	poor to good	275–350	very slight	42	CO_2 5,000, O_2 500 (50% R.H.)	0.8–20.0
good	indefinite	slow burning to self-extinguishing	–100 to 225	excellent	400–425	none	43	CO_2 15, O_2 3, N_2 0.25 H_2 58.1	5.2
fair	indefinite	self-extinguishing	–80 to 275	good	375	–	44	CO_2 5.5, N_2 9, O_2 14	2.6
good	poor	self-extinguishing	–20 to 205	poor	225–250	–	45	CO_2 6–13,000, O_2 3.2–3.250	0.008–0.03 (MP)
excellent	–	slow burning to self-extinguishing	–50 to 200	poor to good	300–350	none	46	CO_2 970, O_2 150, (50% R.H.)	4.4
excellent	–	slow burning to self-extinguishing	–60 to 200	poor to good	280–350	none	47	CO_2 970, O_2 150, (50% R.H.)	0.28
excellent	indefinite	self-extinguishing	0 to 200 (dry) 300 (wet)	good	230–315	none	48	CO_2 3.8–150, O_2 8–24, NE. 12–10	0.02–0.6
–	–	slow burning	32 to 200	–	–	–	49	low	–

Table 9-6. *(Continued)*

Generic Name	Water Absorption 24 hr Immersion %	Resistance to Acids	Resistance to Alkalies	Resistance to Greases and Oils	Resistance to Organic Solvents	Resistance to Water and/or Detergents	Resistance to Oxidizing Agents
26. Polyimide	2.9	good to excellent	poor	excellent	excellent	–	good
27. Polymethyl methacrylate	negligible	good	good	good	fair	–	poor
28. Polymethyl pentene	0.005	good	good	fair	fair	good	–
29. Polyparabanic acid	2.8	good	poor	good	good	–	poor
30. Polypropylene, cast	0.005 or less	excellent	excellent	good	good	good	fair
31. Polypropylene, balanced, oriented	nil	excellent	excellent	good to excellent	good	good	fair
32. Polypropylene, balanced, oriented, coated	nil	excellent	coating attacked	excellent	good	good	fair
33. Polystyrene (casting, extrusion)	–	good	excellent	good	poor to excellent	–	good
34. Polystyrene (oriented)	0.04–0.06	good	excellent	good	poor to excellent	good	good
35. Polysulfone	0.3	good	good	good	poor	good	–
36. Polytetrafluoroethylene (Teflon TFE)	<0.01	excellent	excellent	excellent	excellent	excellent	good
37. Poly(perfluoroalkoxytetra-fluorethylene) (PFA)	0.03	excellent	excellent	excellent	excellent	excellent	excellent
38. Polytrifluorochlorethylene (Kel-F)	nil	excellent	excellent	excellent	excellent	excellent	excellent
39. Polyurethane	0.55–0.77	poor to good	poor to good	good	good	fair to good	–
40. Polyvinyl alcohol	55	poor	poor	excellent	excellent	excellent	good
41. Polyvinyl chloride (PVC) (rigid)	negligible	excellent	excellent	good	poor to excellent	excellent	very good
42. Polyvinyl chloride (PVC) (flexible)	negligible	good to excellent	good to excellent	good to excellent	poor to good	good to very good	good to very good
43. Polyvinyl fluoride (Tedlar)	0.5	excellent	excellent	excellent	excellent	excellent	excellent
44. Polyvinylidene fluoride (Kynar)	0.04	excellent	fair to poor	excellent	fair to good	good	excellent
45. Rubber hydrochloride	5	good	good	excellent	good	–	–
46. Vinyl chloride–acetate copolymers (rigid)	negligible	very good to excellent	excellent	excellent	poor to excellent	excellent depends on plasticizer	good to very good
47. Vinyl chloride–acetate copolymers (flexible)	negligible	excellent	excellent	fair to excellent	poor to excellent	very good	very good
48. Vinylidene chloride–vinyl chloride copolymer	negligible	excellent	good to excellent	good to excellent	good to excellent	excellent	excellent
49. Vinylnitrile rubber	negligible	good	good	excellent	good	–	–

Table 9-6. *(Continued)*

Tensile Strength lbs sq in @ RT	Elongation %	Tearing Strength (Elmendorf GM/mil)	Folding Endurance	Dielectric Constant 10^3 cps	Dielectric Constant 10^6 cps	Dielectric Constant 10^9 cps	Dissipation Factor 10^3 cps	Dissipation Factor 10^6 cps	Dissipation Factor 10^9 cps	Dielectric Strength, V/mil	Volume Resistivity ohm-cm
25,000	70	8	10,000	3.5	3.4	–	0.003	0.010	–	7,000	10^{18}
8,000–12,000	4–12	–	–	3.5–4.0	3.0–3.5	2.58	0.040	0.030	0.009	400	10^{15}
2,700–4,000	10	–	–	2.1	2.1	–	0.003	–	–	–	10^{16}
16,000	10	500	25,000	3.4	3.4	–	0.004	–	–	5,700	5×10^{16}
3,200–10,000	300–500	5–1,800	very high	2.2	2.2	2.2	0.0003	0.0003	0.0003	3,000–4,500	3×10^{15}
25,000–30,000	45–85	5–7	very high	2.2	2.2	2.2	<0.0002	<0.0002	<0.0002	7,000–10,000	3×10^{18}
30,000	70	8	very high	2.2	2.2	2.2	<0.0002	<0.0002	<0.0002	7,000–10,000	3×10^{18}
7,100–12,100	3–40	25	–	2.4–2.7	2.4–2.7	2.4–2.7	0.0005	0.0005	0.0005	5,000	10^7–10^{16}
7,100–12,100	3–40	–	–	2.4–2.7	2.4–2.7	2.4–2.7	0.0005	0.0005	0.0005	5,000	10^7–10^{16}
8,400–10,600	64–110	9–12	–	3.05	3.05	3.00	0.0008	0.0034	0.0041	7,500	5×10^{16}
1,500–4,000	100–350	10–100	very high	2.0–2.1	2.0–2.1	2.0–2.1	0.0002	0.0002	0.0002	≈2,500 (1 mil)	10^{18}
4,000–4,500	300–480	–	–	2.06	2.06	2.06	0.000035	0.000080	0.00115	2,000	10^{18}
6,300–10,000	50–400	8–26	–	2.5–2.7	2.3–2.4	2.3	0.022–0.24	0.009–0.017	0.004	1,000–3,700	10^{18}
5,000–12,000	200–700	220–710	–	5.2–7.5	5.5–7.1	–	0.040–0.060	–	–	600–1,300	1-3.0 $\times 10^{11}$
6,500–12,000	200–600	250–800	very high	–	–	–	–	–	–	–	–
7,000–14,000	10–175	10–700	very high	3.0–3.3	2.8–3.1	2.8	0.009–0.017	0.006–0.017	0.019	425–1,300	10^{16}
1,400–17,000	100–510	60–1,400	very high	4.0–8.0	3.3–4.5	–	0.07–0.16	0.04–0.14	–	250–1,000	10^{11}–10^{14}
7,000–19,000	110–260	12–48	excellent	–	–	–	–	–	–	–	–
5,200–7,400	25–500	–	–	7.46	6.1	2.98	0.019	0.059	0.110	260–1,280	2×10^{14}
3,500–5,000	stretches	60–1,600	250,000	3.0	3.0	–	–	0.006	–	–	10^{13}
3,000–8,000	depends on plasticizer	10–30	very high	3.0–3.3	2.8–3.1	2.8	0.0009–0.017	0.006–0.019	0.019	425–1,300	10^{16}
2,000–4,500	250–300	30–1,400	–	4.0–8.0	3.3–4.5	–	0.070–0.160	0.04–0.140	–	250–1,000	10^{11}–10^{14}
8,000–20,000	35–110	10–7,100	>500,000	3.9–4.5	3.0–4.0	2.7	0.052–0.063	0.050–0.080	0.016	3,000–5,000	10^{12}–10^{15}
2,500–4,000	–	200–960	very high	–	–	–	–	–	–	–	–

Table 9-6a. Film Materials.

No.	Generic Type	No.	Generic Type
1.	Acrylic	26.	Polyimide
2.	Acrylonitrile butadiene styrene	27.	Polymethyl methacrylate
3.	Acrylonitrile–butadiene copolymer	28.	Polymethylpentene
4.	Cellophane	29.	Polyparabanic acid
5.	Cellulose acetate	30.	Polypropylene (cast)
6.	Cellulose acetate butyrate	31.	Polypropylene (balanced, oriented)
7.	Cellulose propionate	32.	Polypropylene (balanced, coated)
8.	Cellulose triacetate	33.	Polystyrene (cast extruded)
9.	Ethyl cellulose	34.	Polystyrene (oriented)
10.	Ethylene chlorotrifluoro copolymer	35.	Polysulfone
11.	Ethylene methacrylate	36.	Polytetrafluoroethylene (Teflon TFE)
12.	Ethylene–tetrafluoroethylene copolymer	37.	Poly(perfluoroalkoxytetrafluorethylene) (PFA)
13.	Ethylene–vinyl acetate copolymer	38.	Poly(trifluoromonochloroethylene) (Kel-F)
14.	Fluorinated ethylene–propylene copolymer	39.	Polyurethane
15.	Ionomer	40.	Polyvinyl alcohol
16.	Polyamide (nylon 6/6)	41.	Polyvinyl chloride (PVC) (rigid)
17.	Polyamide (nylon 6, biaxially oriented)	42.	Polyvinyl chloride (PVC) (flexible)
18.	Polyamide (nylon 11 and 12)	43.	Polyvinyl fluoride (Tedlar)
19.	Polybutylene	44.	Polyvinylidene fluoride (Kynar)
20.	Polycarbonate	45.	Rubber hydrochloride
21.	Polyester	46.	Polyvinyl chloride–acetate copolymer (rigid)
22.	Polyether sulfone	47.	Polyvinyl chloride–acetate copolymer (flexible)
23.	Polyethylene (LD)	48.	Vinylidene chloride–PVC copolymer
24.	Polyethylene (MD)	49.	Vinylnitrile rubber
25.	Polyethylene (HD)		

Table 9-6b. Film Sources.

Company	Generic Type
Adam Spence Corporation	14, 36
Alamo Industries, Inc., Film Division	23, 24, 25, 30, 31, 32
Allied Chemical Corporation	10, 14, 16, 17, 36, 38, 41, 42
Allied Resinous Products, Inc.	16, 17, 18, 23, 24, 25, 30, 31, 32
American Hoechst Corporation, Film Division	16
Auburn Plastics, Inc., Extrusion Division	2, 5, 6, 23, 24, 25, 30, 31, 32
Avisun Corporation	30, 31, 33
BXL Industrial Products Group, Ltd.	2, 5, 23, 24, 25, 30, 31, 32, 33, 41, 42
Baldwin-Montrose Chemical Co., Inc.	42
Blank, Arthur & Co.	5, 7, 14, 20, 21, 23, 33, 36, 46, 47, 49
Borden Chemical Co., Columbus Coated Fabrics Division	41, 42, 47
Budd Co., Polychem Division	36
Cadillac Plastic & Chemical Co.	2, 5, 6, 16, 17, 18, 20, 33, 41, 42
Cee Jay Extruders, Inc.	23
Celanese Plastics Co.	5, 21
Cellu-Craft Inc.	4
Central Bag & Supply Co.	23, 24, 25, 30, 31, 46, 47
Chemplast, Inc.	14, 36
Chevron Chemical Co.	30, 31, 32
Chicago Gasket Co.	14, 36
Chicago Molded Products Corporation	2, 6, 23, 24, 25, 33
Clopay Corporation, Plastic Film Division R.H. Cole, Ltd. Consolidated Thermoplastics, Inc.	16, 17, 23, 24, 25, 30, 31, 32, 43, 47, 49
Continental Can Co., Inc.	23, 24, 25
Continental Group, The Inc./Continental Diversified Industries, Teapak, Inc.	5, 16, 20, 21, 30, 31, 39, 46, 47
Crane Packing Co., Teflon Products Div.	36
Crown Zellerbach, Flexible Packaging Div.	4, 23, 24, 25, 30, 31
Cryovac Div. of W.R. Grace	30, 31, 48

Table 9-6b. *(Continued)*

Company	Generic Type
Joseph Davis Plastics Co.	23, 24, 25, 34, 41, 46
Deerfield Plastics Co., Inc.	23, 24, 25
Deluxe Packages, Inc.	4
Dixico, Inc.	30, 31
Dow Chemical Co.	9, 23, 24, 25, 34, 48
E.I. Du Pont de Nemours & Co., Inc.	4, 10, 12, 14, 15, 21, 23, 24, 25, 26, 36, 37, 43
Eastman Chemical Products, Inc.	5, 7, 8, 21
Ethyl Corporation Visqueen Division	23, 24, 25, 42
Excello Filmpak, Inc.	23, 24, 25, 30, 31
Extrudo Film Corporation	23, 24, 25
Exxon Chemical Co.	29
Favorite Plastic Co.	13, 15, 16, 19, 23, 24, 25
FMC Corporation	4, 41, 42
Firestone Plastics Co.	23, 24, 25, 39, 41, 42, 46
Flex-O-Glass, Inc.	6, 7, 13, 15, 19, 23
Foamade Industries	–
Foster Grant Co.	16
Garlock, Inc.	16, 17, 18, 33
General Electric Co.	20
General Films, Inc.	16, 23, 24, 25, 30, 31
The General Tire & Rubber Co.	2, 23, 24, 25, 41, 42
Gering Plastics Co.	23, 24, 25, 42
B.F. Goodrich Co.	39, 41, 42
Goodyear Tire & Rubber Co.	42, 45
Gordon-Lacey Chemical Products Co., Inc.	42, 47
W.R. Grace & Co., Cryovac Division	30, 31, 32, 48
W.R. Grace & Co., Elm Coated Fabrics Division	2, 34, 41, 42, 46, 47
Gulf Oil Chemicals Co.	11
H & R Plastics Industries, Inc.	2, 3, 23, 24, 25, 33, 41, 42
Harte & Co., Inc.	2, 23, 24, 25, 41, 42
Hercules, Inc.	21
Imperial Chemical Industries, Ltd., Plastics Division	21, 22, 30, 31, 32, 41, 42, 46
Island Extrusion Corporation	5, 6, 7, 21
Kaye-Tex Manufacturing Corporation	42
Kleerpak Mfg. Co.	15, 16, 25, 30, 31
Korad, Inc.	1
Lariat Packaging Co.	5, 6, 7, 13, 15, 19, 23, 24, 25, 30, 31
Lustro Plastics Co.	2, 3, 5, 6, 7, 21, 30, 31, 33, 34
3M Co.	21, 38
Martin Processing, Inc.	21
Midwest Plastic Products Co.	2, 3, 5, 6, 23, 24, 25, 33
Mitsubishi Petrochemical Co.	13
Mitsui Petrochemical Industries, Ltd.	28
Mobay Chemical Corporation	21
Mobil Chemical Co., Film Dept.	30, 31, 32
Mono-Sol	40
Monsanto Co.	3, 34
Meyercord Co.	46, 47
Natvar Corporation	34
Nixon-Baldwin Division, Tenneco Chemicals, Inc.	2, 5, 6, 41, 46, 47
Northern Petrochemical Co.	13, 15, 23
Olin Mathieson Chemical Corporation	4
Pantasote Co.	41, 42, 46, 47
Paramount Packaging Corporation	4, 23, 30, 31
Pervel Industries	42, 47
Pennwalt Corporation	44
Plastex Process Co.	16, 17, 18, 21
Polymer Corporation	16, 17, 18, 20, 21, 36
Poly Plastic Products, Inc.	23, 24, 25
RJR Archer, Inc.	46, 47
Rand Rubber Co.	42, 48, 49

Table 9-6b. *(Continued)*

Company	Generic Type
Reliance Plastic & Chemical Corporation	2, 3, 23, 24, 25, 30, 31, 32, 33
Reynolds Metals Co.	23, 24, 25, 42, 47
Rohm and Haas Co.	1, 27
Ross & Roberts, Inc.	2, 23, 24, 25, 41, 42, 46, 47
Rotuba Extruders, Inc.	33
Rubber Corp. of America	41, 42
St. Regis Paper Co.	4, 23, 30, 31
Seilon, Inc.	2, 23, 24, 25, 41, 46
W. S. Shamban & Co.	14, 36
Shawinigan Chemicals, Ltd.	36, 41, 42, 46, 47
Sinclair-Koppers Co.	23, 24, 25
Southern Plastics Co.	2, 3, 5, 6, 16, 17, 18, 20, 23, 24, 25, 38, 41, 42, 46, 47
Standard Packaging Corporation	2, 3, 4, 5, 6, 7, 13, 14, 15, 16, 17, 20, 21, 23, 24, 25, 30, 31, 36, 38, 40, 42, 46, 47
Tenneco Chemicals, Inc.	46, 47, 49
Union Carbide Corporation	23, 24, 25, 34, 35, 39, 41, 42, 46, 47
U.S.I. Film Products	13, 15, 19, 20, 23, 25, 30, 31
Waljohn Plastics, Inc.	33
Weldotron	23, 24, 25, 30, 31, 48
Westlake Plastics Co.	2, 3, 5, 6, 16, 17, 18, 20, 22, 23, 24, 25, 33
XCEL Corporation	5, 6, 8, 42, 46, 47

ments of gloss, haze measurements, and see-through or clarity tests.

Gloss Measurements. The surface gloss of a film is measured by determining the amount of light which is reflected from the film's surface at a given angle (usually 45 degrees). The reflected light value (gloss) is expressed as the fraction of the original light. A gloss meter, shown in Figure 9-9, is used. The higher the reading shown by the photocell, the higher the gloss. As a rule, materials with the highest readings are considered to have the most consumer appeal because of their inherent sparkle.

Haze Measurements. A material may have a highly reflective surface but still scatter incident light because of a multiplicity of internal reflections. By using a specially designed haze meter, a measurement may be obtained of the percent of light scattering. The haze meter, shown in

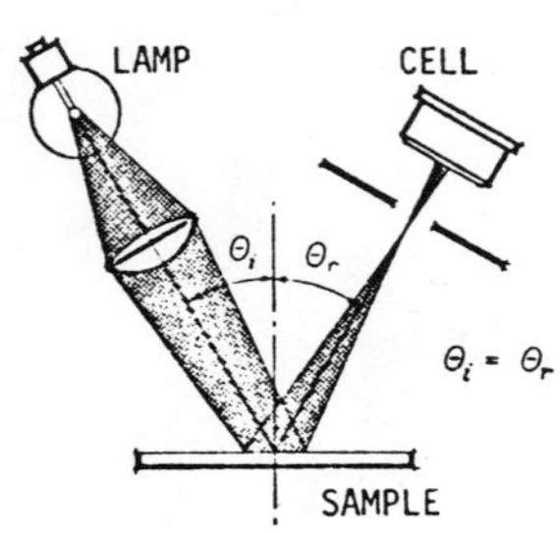

Fig. 9-9. Gloss meter. *(Courtesy: Plastics Design & Processing,* June 1967)

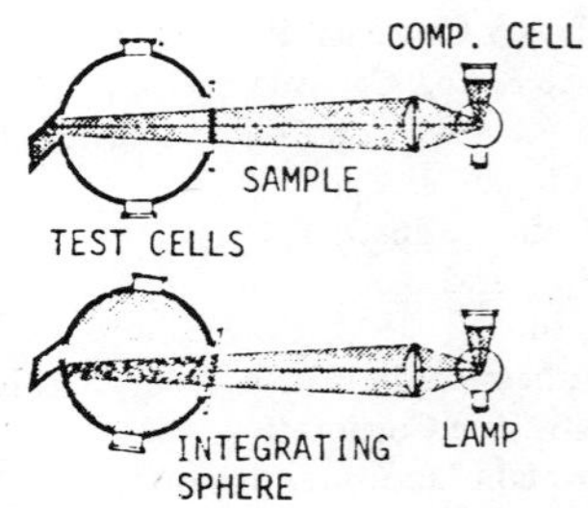

Fig. 9-10. Integrating sphere haze meter. *(Courtesy: Plastics Design & Processing,* June 1968)

Figure 9-10, is used to detect the transmitted light which has passed through the sample. By rotating the integrating sphere, the measure of the scattered light is determined. Low readings indicate less scattering and a clear film.

See-Through (Clarity) Measurements. Measures of clarity can be made by using the meter shown in Figure 9-11, which determines the ratio of the light intensity with and without the film sample in the light path. The same results can be obtained, subjectively, by a trained observer looking

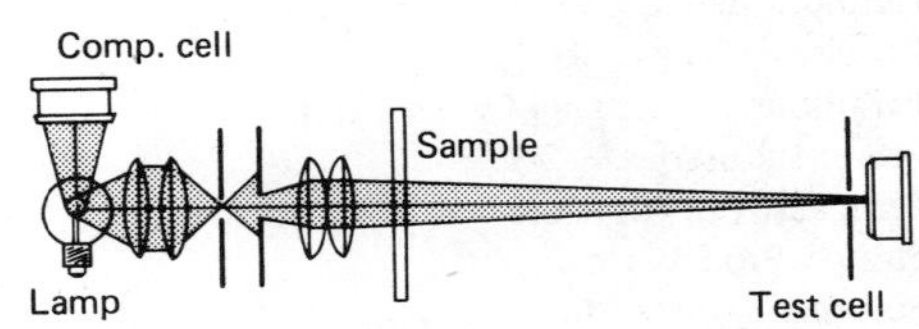

Fig. 9-11. Clarity meter. *(Courtesy: Plastics Design & Processing,* June 1968)

Table 9-7. Trade Names—Films.

Material	Trade name	Source
Acrylic	Korad	Korad
Cellulosics	Tenite I and II	Eastman Chemical
EVA	Polyshrink	Chemplex
Fluoroplastics	Teflon TFE and FEP	Du Pont
	Aclar, Halon	Allied Chemical
	Kel-F	3M
Ionomer	Surlyn	Du Pont
Polyamide	Zytel	Du Pont
	Capran	Allied Chemical
Polycarbonate	Lexan	General Electric
Polyester	Mylar, Cronaflex	Du Pont
	Melinex	ICI Americas
	Scotchpak	3M
	Videne	Goodyear Tire & Rubber
	Celanar	Celanese
Polyether sulfone	Victrex	Westlake Plastics
Polyethylene (oriented)	Clysar	Du Pont
Polyimide	Kapton	Du Pont
Polypropylene	Cryovac	W.R. Grace
	Celgard (microporous)	Celanese
Polystyrene	–	–
Polyurethane	Perflex E	Union Carbide
	Tuftane	B.F. Goodrich
	Korel	Foamade Industries
Polyvinyl alcohol	PVA	Monosol
	Solvlon	Polyvol
Polyvinyl chloride	–	Firestone Plastics
Polyvinylidene chloride	Saran	Dow Chemical
Polyvinyl fluoride	Tedlar	Du Pont
Polyvinylidene fluoride	Kynar	Pennwalt

Table 9-8. Trade Names—Film Laminates.

Material	Trade name	Source
LDPE/HDPE	Multifilm M-5	U.S.I. Chemicals
HDPE/EVA	Co/Ex	St. Regis Paper
Nylon/Ionomer	TLS-68	National Poly Products
LDPE/Saran/LDPE	Saranex-14	Dow Chemical
LDPE/Saran/EVA	Saranex	Dow Chemical
HDPE/Ionomer/Nylon/EVA	Zeelon-400	Crown Zellerbach
PP-PE/OPP/PP-PE*	Bicor OP-400S	Mobil Chemical

*PP-PE = polypropylene-ethylene copolymer; OPP = oriented polypropylene.

at samples of the film and comparing known acceptable materials with new shipments. By looking at newsprint or other similar objects through the film at varying distances, it will be found that the film clarity has a considerable effect on the distance at which the newsprint can be easily read. With some practice, one can acquire the ability to make fairly accurate judgments of film clarity.

Permeability Tests. Other tests of great importance to the use of films in food packaging are those which measure permeability to moisture and to various gases, mainly oxygen. In most cases, the film is to act as a barrier material; however, for produce packaging a "breathable" material is required. While the passage of water vapor or a gas is essentially the same, in actual testing it is somewhat easier to perform two types of tests to determine the two different values. In testing for MVTR, the film is clamped over a dish containing a known amount of dehydrating agent. The assembly is then placed in a controlled humidity cabinet for a specified time at a specified temperature. After the time period is up, the weight gain of the dehydrating agent is measured which allows for the calculation of the film's MVTR.

The gas permeability is determined by using the film as a diaphragm between two chambers of a diffusion cell. One side of the cell is pressurized with the test gas, and the other side of the cell is evacuated, so the gas is driven through the film under pressure, usually one atmosphere. The permeability value is usually reported in cc of gas/m^2/24 hours at STP.

Machinability Tests. Machinability of films, as previously discussed, refers to the capability of the film to be processed on high-speed packaging machinery. There are a number of characteristics which are required for efficient film usage in a packaging machine, but which can not be easily evaluated by "normal" plastics material testing techniques. Some of the most important of these characteristics are: slip, curl, stiffness, blocking, running quality, and dimensional stability.

Slip. Slip refers to the coefficient of friction of the film. There are several methods used to determine this quality, all based on measurement of the force required to move the material under a given load either over a machine mandrel or over itself. Inclined planes, spring testers, and Instron testing machines are used to determine the forces.

Curl. Curl is a phenomenon associated with composite films which may be composed of two or more films having different moisture absorption rates and possibly differing thermal coefficients of expansion. Films may be checked for curl by simply cutting a piece the size of the proposed package and suspending it in air for a day or so. It is then compared with similar or previous shipments which have been found to run satisfactorily in the machine. Placing the film in a humidity chamber, to thoroughly saturate it, may remove the major portion of the curl. The highly humidified material should then be checked to be sure that it will meet all other requirements.

Stiffness. Stiffness is a measure of what is called "hand" for fabrics. It roughly defines the drapability of the film. There are several methods of measurement; one of the best is the Taber Stiffness Meter, which simply bends a small sample through a certain angle and measures the required force.

Blocking. Blocking refers to the adhesion between layers of film when wound on a roll, or when stacked up for some period of time. The tendency for films to block after being stored can interfere seriously with smooth machine operation. One method of quantitatively determining the extent of block is to test two sheets as shown in Figure 9-12. By holding the sheets together for a given length of time under a specified load and then pulling at a given speed, a definite figure can be obtained for the block.

Running Quality. The running quality refers to the ability of the material to be satisfactorily run on high-speed packaging machines. In most cases, the measurement must be made on the equipment with which the film will be used. Besides the obvious characteristics of low slip, minimum blocking, minimum curl, etc., such intangibles as static electricity effects must be ascertained, the effect of too much moisture absorption on sealability must be tested, etc. Again, all this must be subject to trial in the equipment to be used. If static generation is problem, this can be cured by ionization bars and other static eliminators.

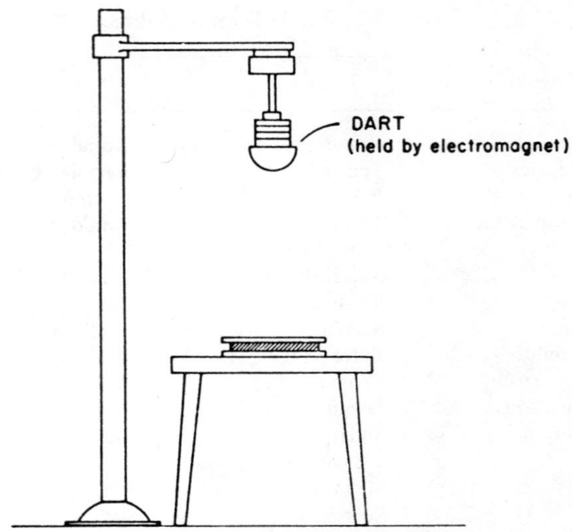

Fig. 9-13. Falling dart test. (Park, 1969)

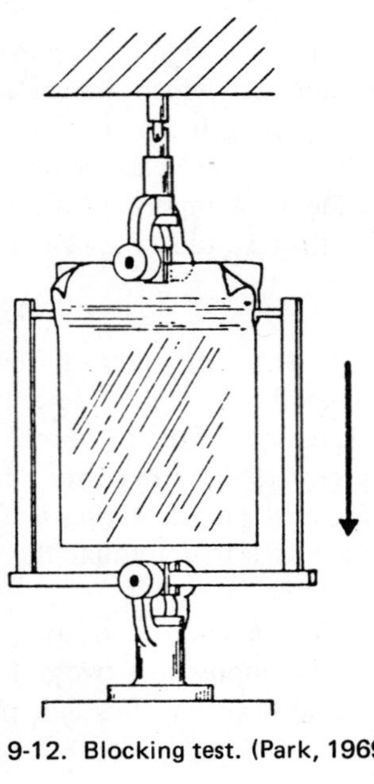
Fig. 9-12. Blocking test. (Park, 1969)

Dimensional Stability. In most cases, dimensional stability can, like running quality, be tested by a practical test. This can consist of subjecting a finished package to the most severe thermal or moisture conditions which might be encountered in service. If it is found that instability is due to excessive moisture pickup, then a coated film of the same type might prove to be the most expedient solution.

Mechanical Tests. Standard mechanical tests used for other types of plastics are, in general, also used for testing of films. These include tensile strength, elongation, and modulus determinations, using standard dumbbell-type specimens. Tear strength tests are made using rubber-type specimens. There are a number of impact tests which can be used, including tensile impact and falling dart; however, standard Izod tests could probably not be used because of the sample thickness. The falling dart test, where the film acts as a diaphragm, is probably the easiest and the most reproducible of all the impact tests. Figure 9-13 illustrates the falling dart test.

Probably the only two mechanical tests which are used extensively for films (and paper) are a stress-flex test and burst test. The stress-flex test measures the ability of a film to withstand a number of backward and forward folds at a given rate. As general rule, the temperature of testing as well as the speed is specified. The burst test is simply a test to determine the pressure required to burst the film when it is positioned as a diaphragm. It is frequently used in place of a tensile test. Both tests find their greatest utility as quality-control measures.

Film Tests Summary. Table 9-9 lists the various test methods for films by the applicable test designations as

Table 9-9. Flexible Film Testing Procedures.*

Property	ASTM Method	PI Method	TAPPI Method	BS Method
Gloss	D-523-53T	T-3633	T-480	—
Haze	D-1003-61	—	—	—
See-through	D-1746-60T	—	—	—
Slip	D-1894-63	—	—	2782, 311A
Curl	D-826-47 (Paper)	—	—	—
Rigidity	D-747-63	T-3610	T-451, 489	—
Running quality	Machine trials must be conducted.			
Caliper	Obvious Measurement			
Dimensional stability	—	—	—	2782, 106F & G
Tensile strength	D-882-61T	T-3608, 3615 (for wet and dry paper)	T-404, 456 (for wet, and dry paper)	2782, 301F
Elongation	D-882-61T	—	—	2782, 301E
Tear strength	D-1922-61T	—	—	—
Impact strength	D-1709-62T	—	—	2782, 306F
Stress-flex	D-790-63	—	—	—
Burst strength	D-774 (paper)	T-3606 (paper)	T-403 (paper)	1133 (paper)
Water vapor permeability	E-96-63T	T-3623	T-464, 448	1133, 2782, 513A-D
Gas permeability	D-1434-63	—	—	2782, 514A

Note: This table lists references to most of the test methods discussed. In several methods, the test was originally designed for paper and paperboard. However, they can be adapted to flexible plastic films. It also is noted that some tests for films have not been standardized by several organizations. The preferable method to follow in all testing is the ASTM procedure. Other procedures may be consulted for comparison purposes.
*Courtesy: *Plastics Design & Processing*, June 1967.

Table 9-10. List of Standards Organizations.*

Organization	Address	Comments
American Society for Testing and Materials (ASTM)	1916 Race St., Phila., Pa. 19103	Widely used and constantly revised
The Packaging Institute (PI)	342 Madison Ave., New York, N. Y. 10017	Often similar to other test methods
The Technical Association of the Pulp and Paper Industry (TAPPI)	360 Lexington Ave., New York, N. Y. 10017	Relates to composites of paper and film
The British Plastics Federation (BS)	47 48 Piccadilly, London, W. I., England	Widely used in Europe and useful for comparison studies

*Courtesy: *Plastics Design & Processing*, June 1967.

developed by the different organizations. Table 9-10 lists the standards organizations which have developed these tests.

REFERENCES AND BIBLIOGRAPHY

Anderson, D. C., and Arlidge, D. B. Vacuum deposited films up appearance, performance. *Materials Engineering*, August 1968, 68(2):77.

Bird, V. How to identify plastic films. *Plastics Technology*, September 1963, 9(9):40.

Cabral, A. C. D., Orr, A., Stier, E. F., and Gilbert, S. G. Metallized polyester film rates high as a fatty-foods packaging material. *Modern Plastics*, March 1979, 56(3):74.

Celanese Plastics Co. *Celgard Microporous Polypropylene Film, An Innovation in Porous Technology*. Bulletin 05M053.

E.I. Du Pont de Nemours & Co., Inc. *Du Pont Tedlar PVF Film*. Bulletin A-38809, Technical Bulletin TD-1B.

——. *Polyimides*. Bulletin A-31715.

Hathaway, C. T. Orientation characterization in plastic film and sheet. *SPE Journal*, June 1961, **17**(6):567.

Hauck, J. E. Plastic films as engineering materials. *Materials in Design Engineering*, October 1965, **62**(4):102.

Meyer, F. J. How to make roll-chilled polyolefin films commercially. *Modern Plastics*, August 1959, **36**(12):97.

Modern Plastics. Pretreated polyester film adds performance values and cuts costs. July 1978, **55**(7):50.

——. In films for packaging, the future belongs to composite structures. March 1976, **53**(3):36.

——. The promise of polyester film. December 1966, **43**(4):110.

——. Fabulous inflatables. October 1962, **40**(2):99.

——. Polypropylene: what should film users look for? December 1961, **38**(4):87.

——. For versatility unlimited: film laminations. August 1960, **37**(12):79.

Paper, Film and Foil Converter. Anti-static films reduce potential packaging hazard. March 1967.

Park, W. R. R. ed., *Plastics Film Technology*. Van Nostrand Reinhold Co., New York, 1969.

Plastics Engineering. Plastics in packaging. October 1978, **34**-(10):29.

Product Engineering. Foils, films and paper combine to solve tricky design problems. March 1975, **46**(3):31.

Riley, M. W. Where to use polyester film. *Materials in Design Engineering*, January 1961, **17**(1):104.

Sacharow, S. Polystyrene film in packaging. *Plastics Design & Processing*, January 1972, **12**(1):22.

——. Polyester films in packaging. *Plastics Design & Processing*, February 1972, **12**(2):21.

——. Polyamide film in packaging. *Plastics Design & Processing*, March 1972, **12**(13):17.

——. Polyvinylidene chloride film in packaging. *Plastics Design & Processing*, April 1972, **12**(4):18.

——. Edible and water soluble films in packaging. *Plastics Design & Processing*, September 1972, **12**(9):18.

——. New plastics for packaging. *Plastics Design & Processing*, January 1969, 9(1):14.

——. Films for food packaging. *Plastics Technology,* March 1968, **14**(3):27.

——. Plastics for packaging and the FDA. *Plastics Design & Processing*, July 1968, 8(7):14.

——. Testing flexible plastic films. *Plastics Design & Processing.* 7(6):28.

——. Surface-treating flexible plastic films. *Plastics Design & Processing*, March 1967, 7(3):23.

——. Heat-shrinkable flexible plastic films. *Plastics Design & Processing*, June 1966, **6**(6):24.

——. Flexible plastic film laminations. *Plastics Design & Processing*, September 1966, **6**(9):17.

——. Flexible plastic films—chemistry and uses. *Plastics Design & Processing*, June 1965, **5**(6):24.

Sawyer, J. W., and Stuart, R. E. Ionomer laminates—new metal plastic composites. *Modern Plastics*, June 1967, **44**(10).

Schaff, S. Nylon copolymer film for food packaging. *Plastics Engineering*, October 1978, **34**(10):53.

Schaar, C. H. Flame process for film perforation. *Modern Plastics*, September 1962, **40**(1):145.

Schroeder, G. O. Recent advances in structured films. *SPE Journal*, May 1969, **25**(5):42.

Thor, C. J. B. Polyethylene films. *SPE Journal*, February 1958, **14**(2):44.

Vermillion, J. L. Improving transparency in polypropylene film. *Plastics Technology*, June 1962, 8(6):32.

Ward, E. J. Film slitting and winding methods. *Plastics Design & Processing*, September 1978, **18**(9).

Zukor, L. J. Biaxially oriented polypropylene. *Plastics Technology*, October 1961, 7(10):44.

10
Injection Molding

Injection molding is a process of heating thermoplastic granules or pellets in a cylinder until they are liquid, and then injecting the hot liquid into a relatively cold mold where it freezes and takes the shape of the mold cavity. No chemical change takes place in the material, and consequently, the process may be repeated over and over again with the same material, provided no degradation takes place. This process is thus distinguished from the injection molding of thermoset materials, since with thermosets, the mold is hot and the liquid undergoes a chemical reaction causing it to solidify.

A variation of the basic injection molding process is reaction molding (RIM) or liquid injection molding (LIM), where cold or warm, highly reactive thermoplastic or thermosetting liquids are first injected into a mixing head, and then into a warm mold where the reaction continues and the mixture solidifies.

The major advantages of injection molding, of any type, are the speed with which parts can be made, the possibility of obtaining finished, flash-free parts, and the fact that many parts can be made simultaneously. A further advantage, when molding *thermoplastic* granules, is that any scrap from flash, sprues, and runners, may be reused by simply regrinding and mixing with virgin material.

In all three processes the main variables are: (1) the resin(s), (2) the machine type, and (3) the process variables of time, temperature, and pressure. Proper choice of the first two and correct adjustment of the last three allow the satisfactory fabrication of many types of injection-molded parts, ranging from fractions of an ounce in weight to many pounds. However, in addition to the correct choice of the optimum material for the particular application, the best machine, and the right adjustments, there are a number of other factors which must be considered so that parts may be made competitively. These have to do with such things as the type of molds used, the use of automation and ancillary equipment.

MOLDING MACHINES

Credit for the original injection molding machine goes to I. S. and J. W. Hyatt, two brothers, who in 1872 obtained a patent for a machine to inject molten cellulose nitrate into a closed mold held in a vertical hydraulic press. Unfortunately, the thermal instability of cellulose nitrate made the use of the new machine impractical. However, in the early 1920s the development of an improved injection molding machine paralleled the development of moldable cellulose acetate. This material, in the form of molding pellets, could be melted in safety, and the melt could be readily injected into a cold mold. It can be said that these two developments truly ushered in the Plastics Age.

The first injection molding machines (see Figure 10-1) were manually powered, but they were soon improved with electric motors and hydraulic systems to actuate the movable components of the press. The first injection molding presses were actually quite similar to older, metal die-casting machines, in that plastics pellets were fed into one end of the machine, heated in a cylinder to a liquid state, and then injected by a ram into a mold, where the plastic was solidified. After solidification, the mold was opened and the part was ejected. Figure 10-2 illustrates such an early molding press, commonly called a ram- or plunger-type injection molding press. This type of press, while now obsolete, illustrates well the basic principles of injection molding.

The injection molding machine shown in Figure 10-2 basically consists of three systems: heating, melting, and injection (on the right), clamp portion (on the left), and controls (in the center). In the injection unit, the hopper is used to supply the pellets to the volumetric feeder, which in turn feeds a controlled, intermittent volume of pellets to the heating cylinder. The ram, actuated by the small hydraulic plunger, then forces the pellets past the spreader, or torpedo, into the nozzle and subsequently into the mold. The function of the spreader is to force the granules into a thin stream which can be more efficiently heated and homogenized by the cylinder before entering the nozzle. Figure 10-3 illustrates the specific action of the spreader or torpedo. In the clamp section of the machine (Figure 10-2), the large hydraulic cylinder is used to actuate the clamp plunger which opens and closes the mold. Since the injection pressures exerted by the ram can range from 3000 psi to as high as 30,000 psi, the need for the much larger clamp cylinder is obvious. Figure 10-2 shows the time, pressure, and heater controls which enable the press operator to accurately control those variables. Not shown is the mold, which can also have its own heating or cooling controls. Another item not shown is a weigh feeder, used in place of the less expensive volumetric feeder. With the weigh feeder, a more accurate charge of pellets can be delivered to the cylinder, which results in more accurate injection pressure

Fig. 10-1. First U.S. injection molding machine, made in 1931. (*Courtesy: HPM Corporation*)

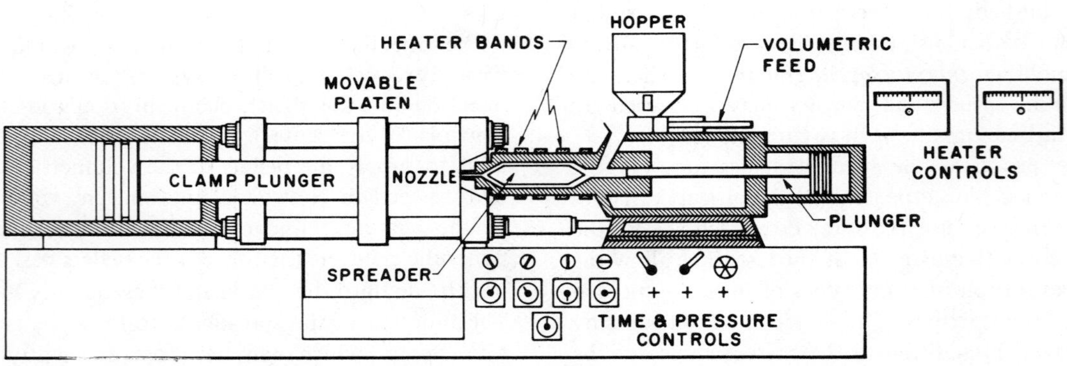

Fig. 10-2. Ram (plunger) injection molding machine (1940s and 1950s).

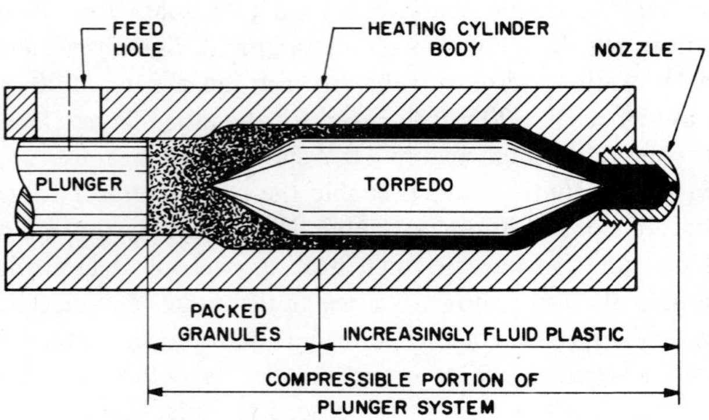

Fig. 10-3. Torpedo or spreader action.

control, thus allowing full injection pressure to be used for each shot without the danger of flashing* the material in the mold.

Ram injection molding machines, as described above, were initially rated in terms of the number of ounces of cellulose acetate which could be injected in each shot. Thus, a machine would be described as a "4-ounce machine," "a 12-ounce machine," and so on. Today the machines are described in terms of ounces of polystyrene per shot or in terms of pounds of plasticating capability per

*Flashing = thin sections produced at mold parting lines due to injection of excess material or excess pressure or both.

Table 10-1. Clamp Pressure versus Capacity in Ounces.[a], *

Tons of Clamp	Ounces of Capacity
Plunger (Ram) Machine	
50-75	1.5-4.0
100-200	2-9
225-350	6-20
375-450	20-30
475-600	40-48
650-1000	32-80
Reciprocating Screw Machine	
50-75	2.5-5.0
100-200	1.5-17.0
225-350	6-44
375-450	18-60
475-600	48-88
650-1000	43-165
1000-1500	106-220
1800-2500	140-420
2700-4000	265-720

[a] Ounces of polystyrene per shot.
* *Western Plastics*, October 1965.

hour; for example, 100 pounds per hour of polystyrene, to as high as over a 1000 pounds per hour. Another way in which machines are currently rated is in terms of the maximum clamping pressure which can be exerted. Such a rating then might be a 75-ton or 250-ton machine, meaning the maximum clamping pressure on the mold would be 75 or 250 tons. Table 10-1 shows the approximate relationship of clamp tons to ounces of polystyrene for ram machines

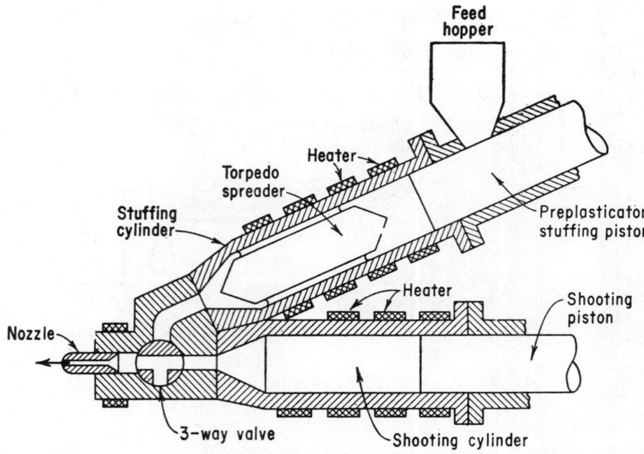

Fig. 10-5. A ram-type preplastication system including a three-way valve that can be set for filling the injection (shooting) cylinder; for injecting plastic melt into the mold with the stuffing cylinder (preplasticator) shut off as here; or for purging the preplasticator. (Reprinted with permission from *Modern Plastics Encyclopedia* (1968–69). Copyright 1968 McGraw-Hill, Inc. All rights reserved.)

and the currently used reciprocating screw machines. The ram or plunger machines were used as described up through the 1950s with few basic changes other than size increases. Figure 10-4 illustrates a typical 12- to 16-ounce ram machine, complete with controls, timers, heat indicators for four cylinder zones, and the necessary mechanical safety interlocks.

With the advent of newer polymers which were harder to plasticate and the need for larger machines which could produce more parts per hour, new designs were developed. One of the first of the improved machines was known as a ram preplasticator or two-stage plunger (see Figure 10-5).

Fig. 10-4. Ram- or plunger-type injection molding machine of 1952. (*Courtesy: HPM Corporation*)

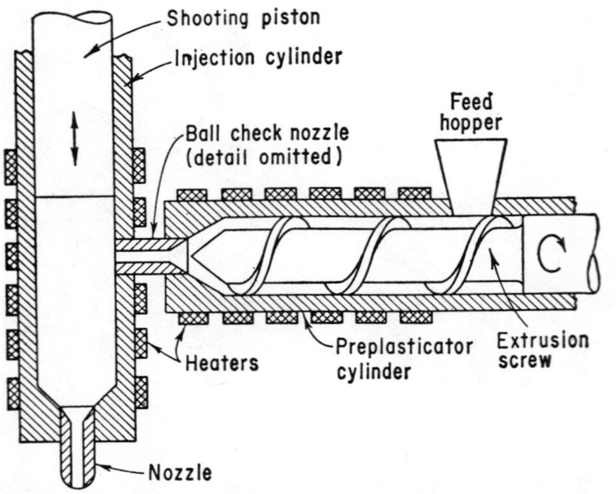

Fig. 10-6. A screw preplastication system in which an extrusion screw in the preplasticator cylinder (above right) feeds plasticated melt to the injection (shooting) cylinder (above left). (Reprinted with permission from *Modern Plastics Encyclopedia* (1968–69). Copyright 1968 McGraw-Hill, Inc. All rights reserved.)

In this system, the first stage is similar to a conventional plunger machine, in that the material is heated and liquefied in the cylinder. However, instead of delivering the melt directly to the mold, the material is injected into the second cylinder which in turn injects it into the mold. The two-stage machine met the requirements of producing a larger volume of somewhat superior plastified melt per hour, but it did not meet all the requirements for plastification quality.

An improvement in the two-stage plunger machine (shown in Figure 10-6) replaced the preplasticating ram with a rotating screw. The screw extrudes the plastic, which is melted by the combination of mechanical work by the screw and the heat from the heater bands. The use of the screw resulted in more thorough plastification, combined with better control of the melt temperature through the mixing action of the screw. Like the two-stage ram machine, this type was capable of producing shots which were larger and of better quality than could be produced on a single-stage ram machine.

The two types of two-stage machines, while capable of meeting the requirements of increased plastification and large shot capacity, had the disadvantages of increased complexity and cost. Thus, in the early 1960s, the reciprocating screw machine was developed, and is currently the type most generally in use. The reciprocating screw machine (the injection section of which is schematically illustrated in Figure 10-7) has the advantage of screw plastification and ram injection in the same unit. In this design, as the screw rotates it is forced backward by the buildup of material in the front portion of the cylinder. At a predetermined point, rotation stops and the nozzle moves forward, acting as a ram to force the melt into the mold. It remains forward until the mold gates (where the molten plastic enters the mold cavity) freeze, then pumps back to repeat the cycle while the mold opens (see Figure 10-8).

It was originally believed that the reciprocating screw mechanism would be mainly applicable only to the larger machines; however, both very large machines and small machines are currently being used (Figures 10-9 and 10-10 illustrate both sizes).

The greater efficiency of the reciprocating screw machine is shown in Table 10-1 which allows a comparison of the ram versus the screw machine. Above 350 tons clamping capacity, the screw is clearly superior in capacity because of its higher plastification efficiency. Another instance of the greater efficiency of the screw machine is its capability to mold parts whose weight is above the actual capacity of the shooting cylinder. In this process, known as "flow molding," the machine acts first as an extruder, pumping the mold partially full; it then injects the remainder by the ram under full pressure. To successfully accomplish flow molding, the walls of the part have to be sufficiently thick and the mold must be hot enough so the part will not freeze off before the entire sequence is completed. By use of maximum extrusion rate (i.e., screw speed) and rapid injection just after the extrusion, cold-flow surface blemishes can be avoided.

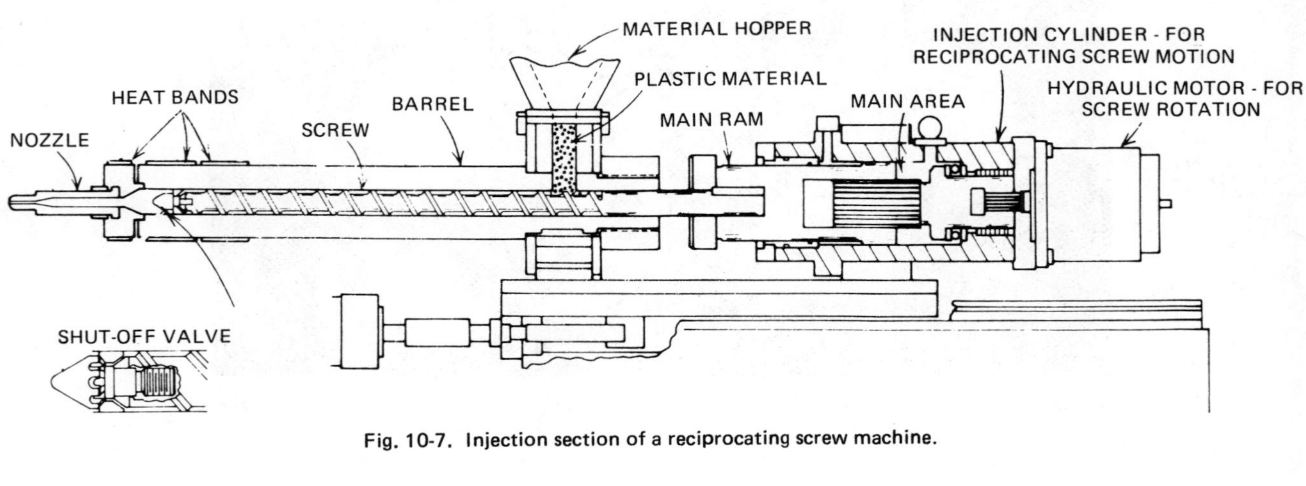

Fig. 10-7. Injection section of a reciprocating screw machine.

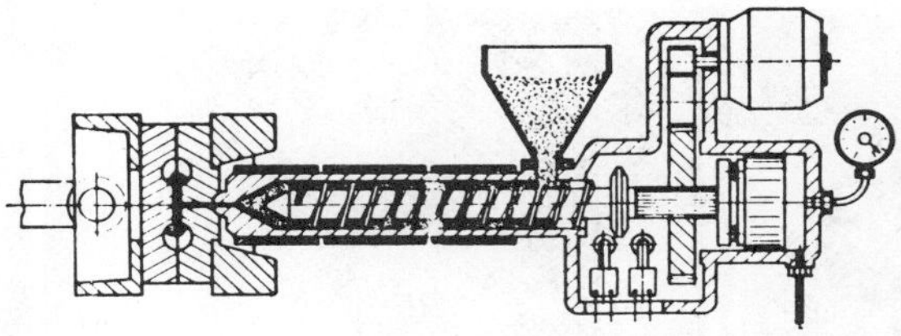

SCREW ADVANCES, MOLD FILLS

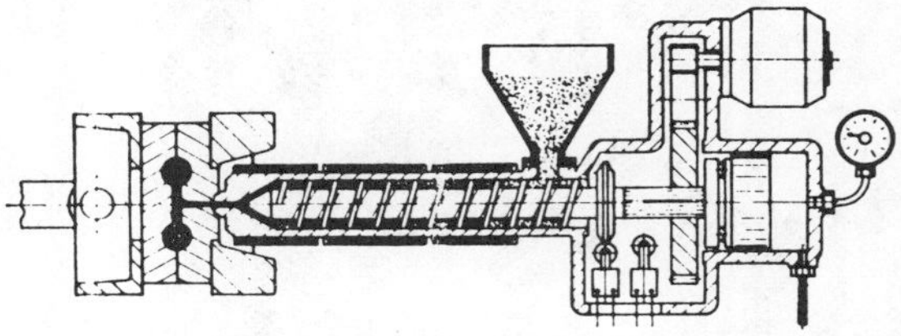

SCREW HOLDS PRESSURE

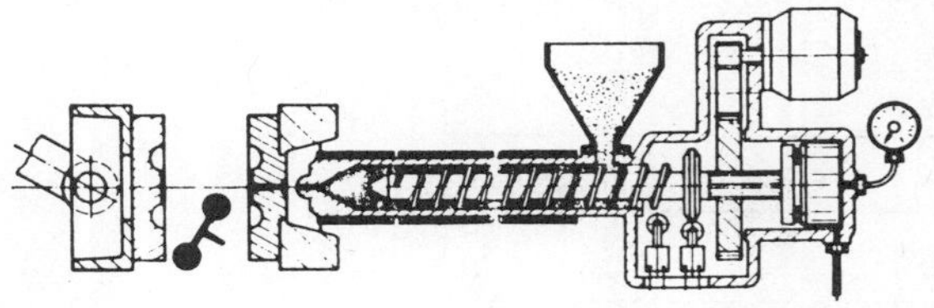

Fig. 10-8. Process cycle of a reciprocating screw machine. (From *Rheology of Polymers* by Severs.)

Fig. 10-9a. One-ounce reciprocating screw machine in horizontal (conventional) position. (*Courtesy: Boy Machines, Inc.*)

Fig. 10-9b. One-ounce reciprocating screw machine in vertical position. Allows injection at the parting line. (*Courtesy: Boy Machines, Inc.*)

Components of the Reciprocating Screw Machine

As mentioned previously, two major features of the injection molding machine are the injection section and the clamping mechanism. The subsidiary units which make up these major components, however, can result in considerable differences in what appears to be the same type of machine. These units–including the injection screw, valves, nozzles, drive motors, clamping mechanisms, and controls for screw speed, heater bands, and injection timers–are described in the next sections.

The Injection Screw. As shown in Figures 10-7 and 10-11, the injection screw is generally a screw of constant pitch and of varying depth of flights. It is the variation in the flight depths which determines the compression ratio (flight depth in feed zone/flight depth in metering zone) of the molten plastic and which assures that the plastic will be driven forward under a high positive pressure. The first section of the screw, where the pellets enter, is known as the feed zone; it has the greatest flight depth so as to assure easy entry of the pellets and their conveyance forward to the center compression or transition zone. It is in the compression zone that the greatest amount of heat is added by the shearing action of the screw. Here, the plastic is turned to liquid and most of the air is eliminated. The liquid plastic is then conveyed to the forward or metering zone, where it is further plasticated to change any unmelted pellets to a liquid. Eventually the material is forced past the nozzle into the mold.

Injection screws are rated by their length/diameter (L/D) ratios. The longer the length, the better the plastification; and the greater the diameter, the more capacity the screw will have at any given rpm. Current practice is to use screws with L/D ratios of 20:1 to as high as 36:1. In general, screws are made of two types of material: a tough steel for the body of the screw and an overlay of a wear-resistant alloy or ceramic to give the longest life without degradation.

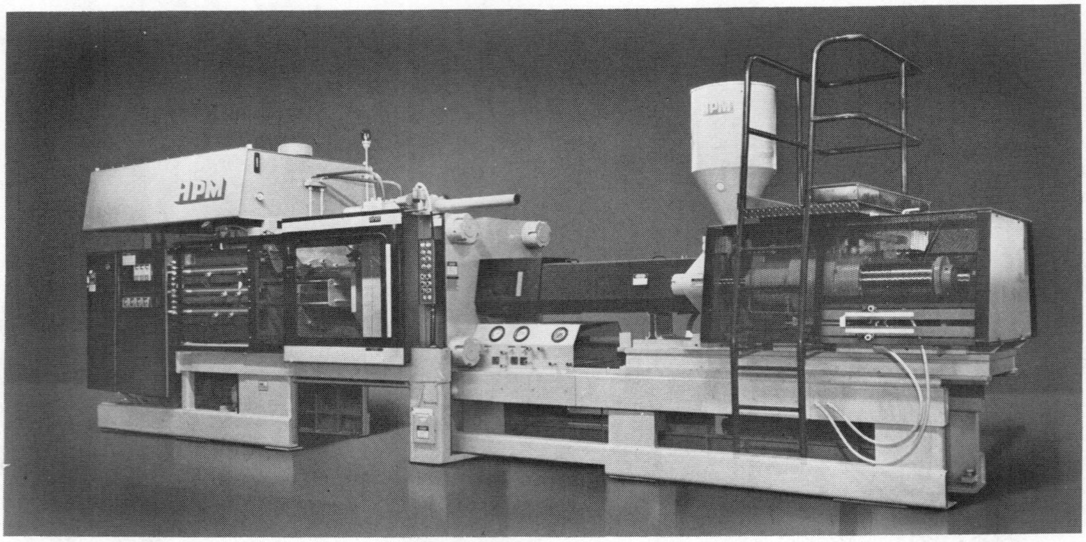

Fig. 10-10. 500-ton reciprocating screw injection molding machine (1978). (*Courtesy: HPM Corporation*)

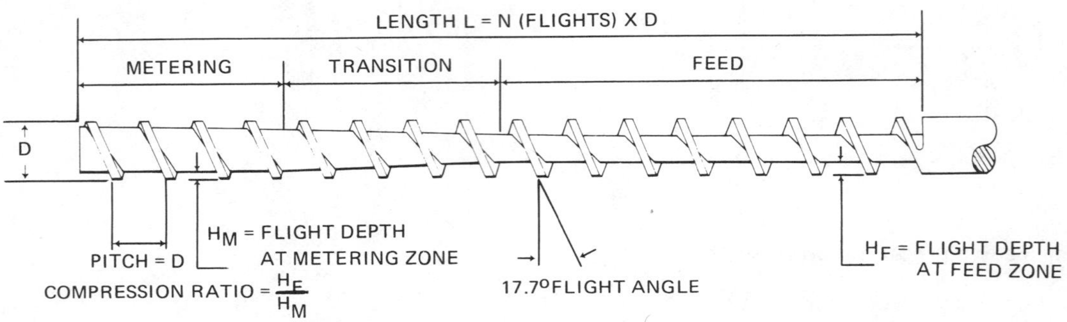

Fig. 10-11. Profile of a conventional one-stage injection molding screw. (Reprinted by permission from *Modern Plastics*, McGraw-Hill, Inc.)

A promising new development in injection screw design is the use of a fluted mixing section at the nozzle end of the screw (see Figure 10-12). This type of screw, which has been used for extrusion for some time, has been found to improve the uniformity of melt temperatures and the dispersion of additives when used in an injection molding machine. Figures 10-13a and b illustrate the mixing action. Table 10-2 shows the temperature variation over ten con-

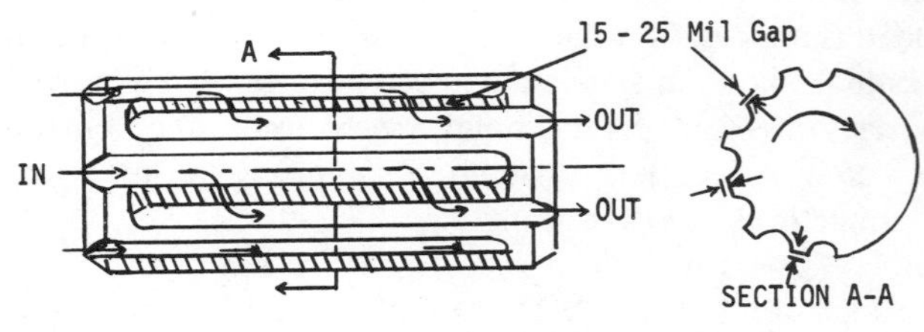

Fig. 10-13a. Union Carbide fluted mixing section. (*Courtesy: Feed Screws, Inc.*)

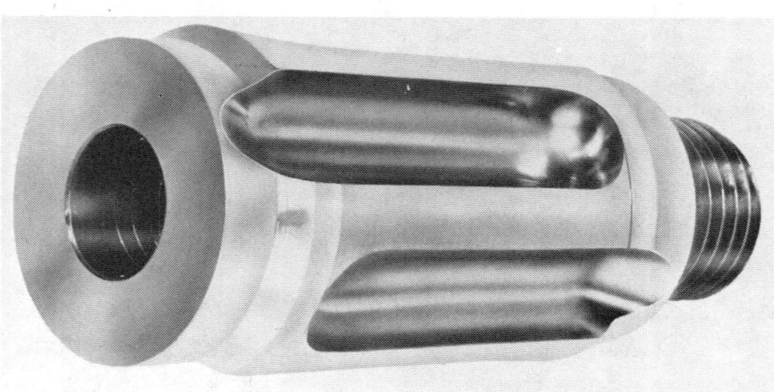

Fig. 10-12. Fluted mixing section used at end of screw. (*Courtesy: Feed Screws, Inc.*)

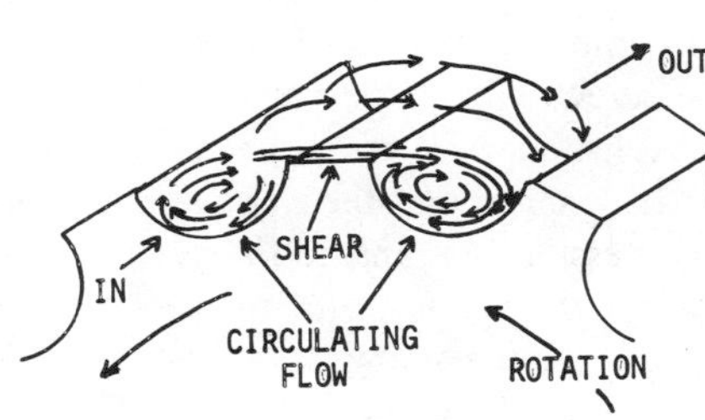

Fig. 10-13b. Mixing and shear mechanisms in fluted mixing section. (*Courtesy: Feed Screws, Inc.*)

Table 10-2. Temperature Variation Over 10 Consecutive Shots (Fluted Mixing Section).*

	Material temperature, °F	
Shot number	Conventional screw	Modified screw
1	453	476
2	450	476
3	453	474
4	440	473
5	440	474
6	444	472
7	448	475
8	450	475
9	453	477
10	450	477
Range	440 to 453	472 to 477
Variation	13	5

*Screw speed, 125 rpm; back pressure, 375 psi; cycle 80 seconds; shot size, 250 grams.
Courtesy: *Plastics Engineering*, April 1976.

secutive shots with the modified screw and a conventional screw. A slight loss in plasticating capacity per hour is, however, a result of using the modified screw.

Valves. Most screws incorporate some kind of check valve at the metering end to prevent backflow along the flights during the pumping and injection period. Figure 10-14 illustrates such a valve. Similarly, a nozzle shutoff valve is usually used at the nozzle to prevent drooling from the nozzle during the mold-open period. When heat-sensitive materials such as PVC are being molded, the possibility of degraded particles hanging up in the nozzle or at the screw tip makes these kinds of valves impractical. For such materials, a smooth self-cleaning tip is recommended. This type of nozzle tip, shown in Figure 10-15, minimizes material hangup and contamination of the molded part with particles of degraded material. A backpressure valve in the

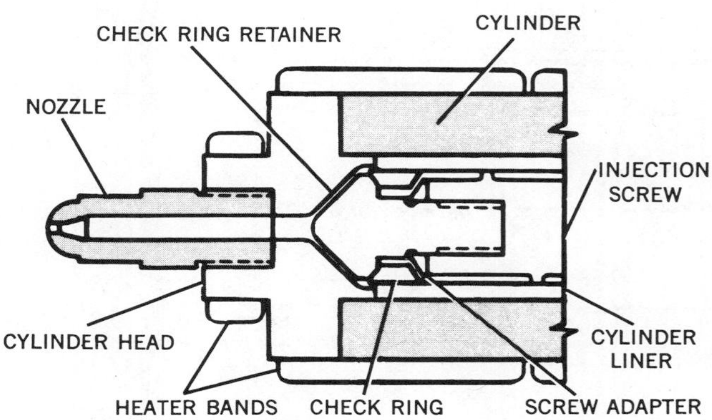

Fig. 10-14. Check valve at end of screw, used to prevent feedback during pumping and injection. (*Courtesy: USI Clearing, division of U.S. Industries*)

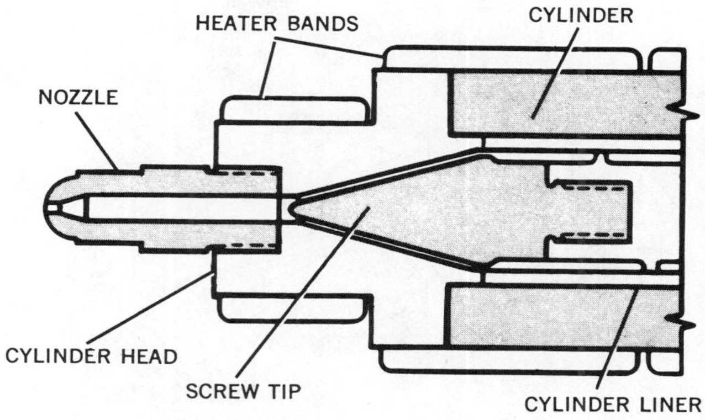

Fig. 10-15. PVC-type ("smearhead") screw tip for heat-sensitive materials. A self-cleaning screw tip minimizes material hangup and contamination of the molded part. (*Courtesy: USI Clearing, division of U.S. Industries*)

screw injection cylinder mechanism is used to regulate the oil pressure during the return pumping action of the screw. By increasing the backpressure, the amount of work undergone by the plastic can be regulated to cause more shear, higher temperatures, and a greater degree of plastification, usually accompanied, however, by a slower cycle. Each material thus requires a separate adjustment of the backpressure screw rpm, and barrel heater temperature in order to achieve maximum production speed, consistent with absence of material degradation. These adjustments are particularly important for heat-sensitive plastics.

Nozzles. A nozzle is used between the end of the cylinder and the sprue bushing to act as a simple "push-to-seal" connection between the cylinder and sprue. The primary function of the nozzle is to allow injection of the molten resin into the mold with a minimum of leakage. A secondary function, with some nozzles, is to prevent resin leakage from the mold when the pressure is relieved during screw retraction and/or prevent hot-melt "drool" from the cylinder during the open-mold portion of the cycle. Figure 10-16 shows the relationship of the nozzle to the cylinder and mold with regard to position and orifice sizes.

General-purpose or "standard" nozzles are made in varying diameters, to accommodate the cylinder bores and sprue diameters, and in varying lengths dependent on the heating requirements. Generally, the standard nozzles are made with a drilled straight bore, restricted at the outlet end and tapered as required to match the cylinder bore at the inlet end (see Figure 10-17). Other bore configurations, as required by the resin type, are also used. In addition to the correct type of internal construction of the nozzle, it is extremely important that the melt be injected at the optimum temperature in order to minimize pressure drop and/or nozzle drool. In almost all nozzle installations, therefore, either external heating bands are used, or internal cartridges or heating coils are installed, as shown in Figure 10-18. Along with the heating assemblies, a separate ther-

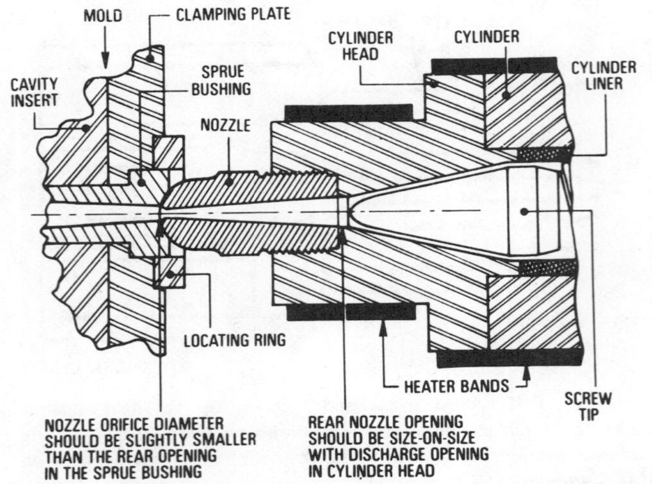

Fig. 10-16. Nozzle placement. (*Courtesy: B.F. Goodrich Chemical Co.*)

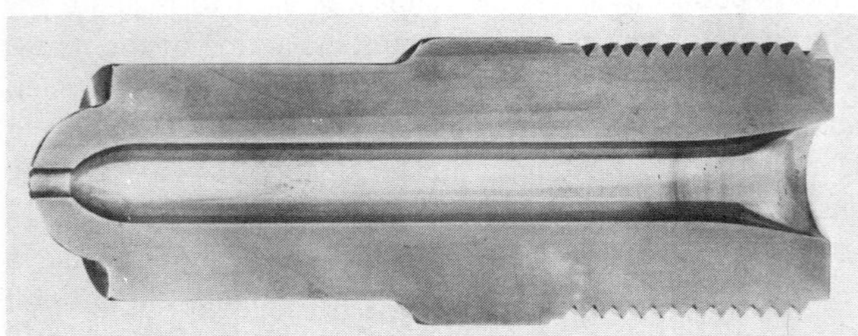

Fig. 10-17. General-purpose or standard nozzle. (*Courtesy: IMS Co.*)

mocouple and controller are always recommended. Nozzle heaters should not be connected to or be controlled by the barrel heating and controlling system if consistently uniform production is expected.

Materials such as nylon which show a very low melt viscosity can be molded best with a nozzle known as a reverse taper or melt valve nozzle (see Figure 10-19). In this case, the reverse taper is used to freeze off the melt or prevent drooling between shots (hence the term melt valve). Its use with other polymers such as cellulosics and ABS is justified because of better flow, ease of cleaning, and improved sprue breaks without long strings. Many of the nozzles, either general-purpose or special, can be obtained with removable tips, as illustrated in Figure 10-20. Since the major wear and damage to the nozzle occurs at the tip, savings through the use of replaceable tips can be appreciable. For nozzles without removable tips, radius refinishing tools are available. Most nozzles are made of steel, heat treated to the proper hardness at the tip and face seal. However, for use with certain corrosive polymers, such as PVC, nozzles can be obtained made of stainless steel or Hastelloy. Chrome-plated nozzles have not proved to be satisfactory because of the tendency of the plating to flake off from crucial corners and surfaces.

Resins are no longer purchased as "ready-to-mold" materials in every case. The inclusion of a colorant with the resin and mixing by the screw have become commonplace with most processors. Very often, other ingredients are added to be mized in—these could include antioxidants, plasticizers, lubricants, flame retardants, and light and heat stabilizers. To aid the screw in its mixing operation, it is possible to include a static mixing element in the nozzle as shown in Figure 10-21. Other static or "motionless" mixers are made which can be installed between the nozzle and the cylinder output to accomplish the same thing. Such devices have been used more extensively in the past on extruders, where they are installed between the extruder outlet and the die. Their use is claimed to allow a reduction in the amount of colorant or other additive required because of the more intensive mixing. (See Chapter 11 for illustrations and additional details.)

The usage of nozzles so far described has been in terms of the nozzle leading into a sprue bushing of a mold as shown in Figure 10-16. Other uses for temperature-controlled nozzles are as components in "runnerless molding," where

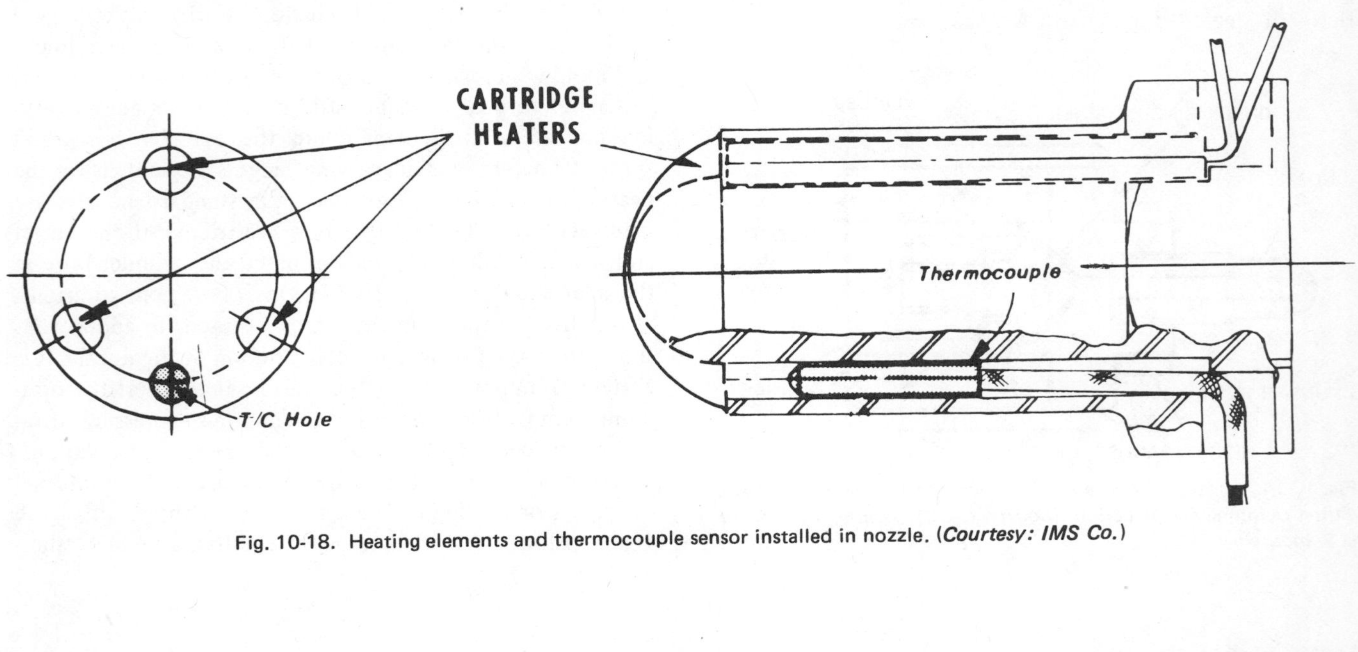

Fig. 10-18. Heating elements and thermocouple sensor installed in nozzle. (*Courtesy: IMS Co.*)

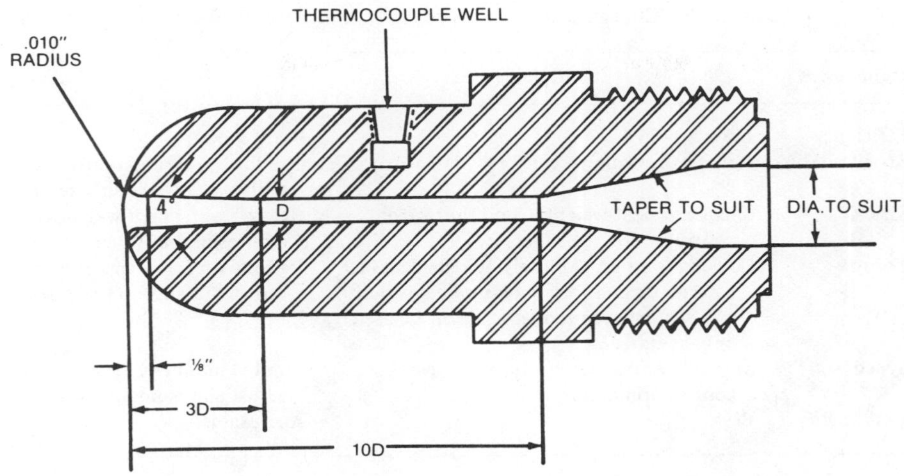

D MINIMUM = ⅛", D TYPICAL = 3/16" - ¼"

Fig. 10-19. Reverse taper or melt valve nozzle used for nylon, ABS, and certain other polymers. (*Courtesy: E.I. Du Pont de Nemours & Co., Inc.*)

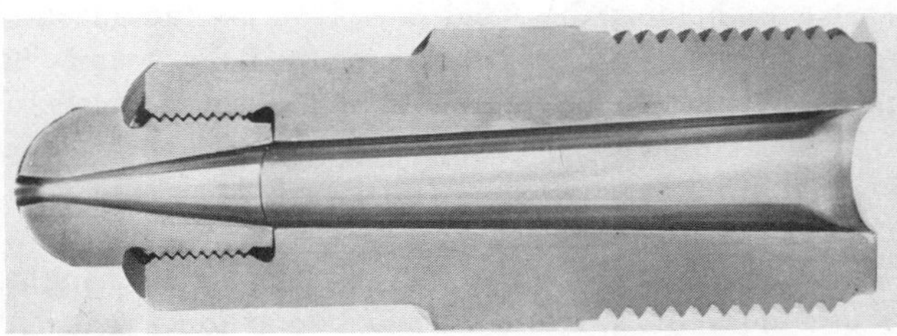

Fig. 10-20. Removable tip nozzle. (*Courtesy: IMS Co.*)

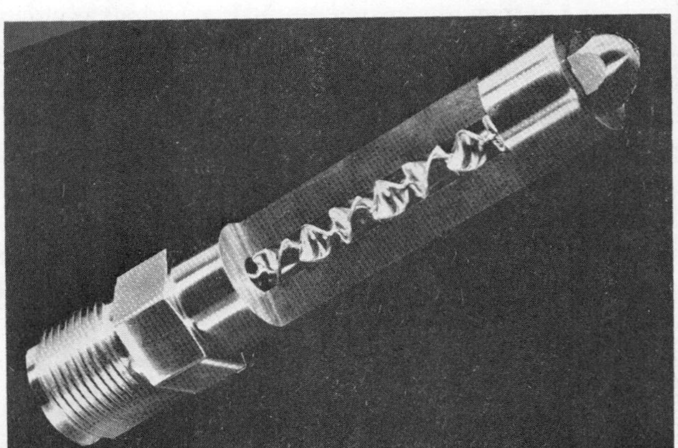

Fig. 10-21. Mixing nozzle. (*Courtesy: Kenics Co.*)

the nozzle leads directly into the mold cavity, thus eliminating the sprue and runner system altogether. (See the section on molds and special techniques in this chapter for a discussion on sprues and runners and "runnerless molding.")

Drive Motors. The screw is usually driven by either an electric motor and a gear train, or a hydraulic motor. In general, electric motors are more efficient and have higher torque and lower costs than the hydraulic motors. However, the high starting torques may prove to be a liability when starting a cold machine, since the possibility exists that the screw may be broken by the high torque. Thus, electric motor drives must incorporate some type of overload protection such as thermal switches or friction clutches to prevent excess screw loads. Speed change in the electric motor drive is usually accomplished by changing the gear ratios, thus limiting the range of speeds available.

Hydraulic drives consist of an electric motor driving a hydraulic pump, which in turn furnishes the power for the hydraulic motor. The use of the hydraulic motor results in a stepless speed-variation capability, even when a gearbox is used. Particularly desirable are direct drives which attach to the screw with no gearing, speed reducers, thrust bearings, etc. Such a system provides minimum noise and vibration and requires minimum maintenance. Table 10-3 compares the important characteristics of hydraulic and electric drives for screws.

Heating Cylinders or Barrels. The heating cylinder or barrel contains the screw and is made to fit the screw with very close tolerances. Since the barrel is also subject to considerable wear and/or corrosion during use (from decomposition products, abrasive fillers, etc.), it, like the screw, is usually made of two different materials. The barrel may be made of a special steel which is then nitrided on the inner surface to produce a very hard, wear-resistant surface. Alternatively, bimetallic barrels are used. The outer portion of the barrel may be a low-carbon steel or, for greater strength, an alloy steel. An inner liner of a wear-resistant alloy may be shrunk into the barrel, or it may be centrifugally cast in place and then machined to dimension. Several types of inner liners are available, depending on whether abrasion resistance or corrosion resistance is required.

Table 10-3. Characteristics of Hydraulic and Electrical Drives for Screws.

Characteristic	Hydraulic	Electrical
Efficiency	Low: 60–75%.	High: 95%.
Screw safety	Relief valve will prevent screw damage.	Overload protection difficult, particularly with small-diameter screws.
Size	Small and light-weight. Good shot-weight control.	Heavy. Poorer shot-weight control. More stresses in the system.
Torque	Constant output. Infinitely adjustable. Must have maximum torque requirements built into system even though it might be used only at startup.	Varies with screw speed. Excellent characteristics at startup low screw speeds.
Speed	Stepless control easily adjusted. Gives best control of molding conditions.	Limited number of speeds. More difficult to adjust and maintain.
Melt quality	Best.	Acceptable.

Vented barrels are a new development, said to be a considerable improvement for some materials over the conventional nonvented type. Figure 10-22 shows a vented barrel and a two-stage screw which is used with it. The major advantage of the vented barrel is in running materials which are naturally hygroscopic—such as ABS, polycarbonate, cellulosics, or nylon. The use of a vented barrel is claimed to eliminate predrying, with the saving of time and money in drying and capital equipment. Other advantages claimed are absence of splay marks, improved physical properties for some moisture-sensitive polymers, and removal of nonaqueous volatiles. This last factor may be of considerable use to molders of ABS which is subsequently to be plated.

The Clamp Section. The clamping mechanism is the second major portion of the injection press. The function of the clamp mechanism is to make sure the following sequence is carried out:

1. Bring the mold together quickly
2. Close the mold gently
3. Part it slowly
4. Move it rapidly to full open
5. Strip the mold slowly

While carrying out the above actions, the clamp must be capable of holding the mold closed during the injection stroke, at which time pressures of up to 30,000 psi may be applied. At the same time, the clamp mechanism must also be capable of opening the mold in a minimum time, holding the mold open as required to strip the parts from the mold, and then closing it for the next shot.

There are essentially two types of clamping mechanisms in use today—fully hydraulic and hydraulically actuated toggle mechanisms. The fully hydraulic is, as the name implies, a hydraulic cylinder or cylinders which advance, clamp, and retract the platen through a hydraulic circuit consisting of high- and low-pressure pumps, valves, and limit switches. The toggle mechanism is a mechanical assembly used to amplify the force exerted by a small hydraulic cylinder. This mechanical advantage may be as high as 20:1 or, in some cases, 50:1. Thus, the toggle hydraulic system can be a good deal smaller, and less expensive, than the system used for the fully hydraulic clamp. There are, however, a number of other significant differences which

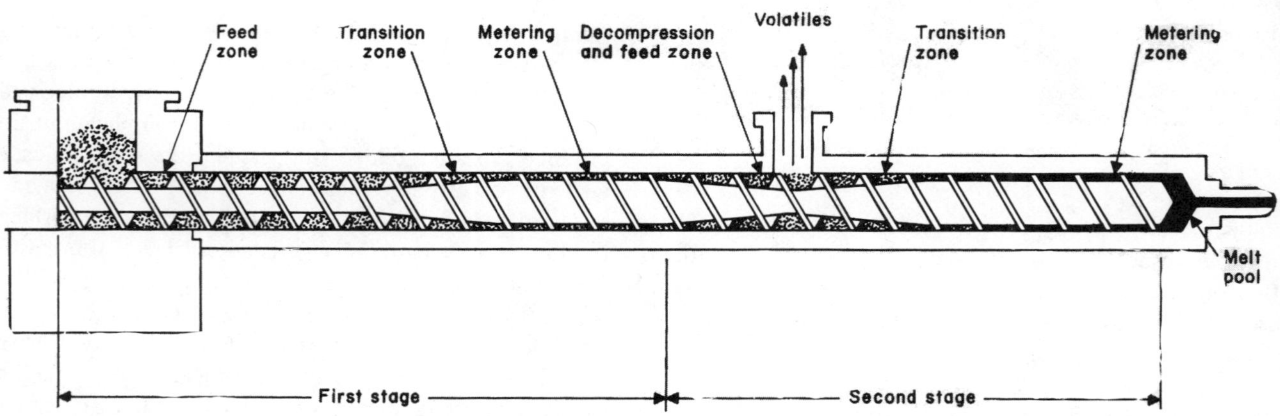

Fig. 10-22. A typical two-stage, vented reciprocating-screw and barrel assembly. (R. DeCapite and C. S. Gudermuth, *Plastics Engineering*, January 1976)

influence the choice of one or the other system for a particular application. In general, toggle clamps have been used for smaller machines (less than 450 tons of clamping force), while most larger presses use the fully hydraulic system.

Hydraulic Clamping. In a fully hydraulic clamp mechanism, a main cylinder and ram are used to exert the full clamp force. In addition to the main ram, there is a smaller ram (frequently located inside the main ram) which is used for rapid advance, mold opening, and retraction. The small ram is called the jack ram and is actuated to provide high-speed movement until the platen is within a fraction of an inch of closing. The main ram then takes over and applies the full clamping force for the mold lockup. In addition, a sensing circuit is usually incorporated so the full pressure will not be applied if there is any obstruction in the mold, such as an unejected sprue or runner, or a hungup part. Figure 10-23 shows the action of the jack ram and the main ram in opening and closing the mold. In some of the newer machines, the main ram is made so the travel is extremely small, not exceeding $\frac{1}{4}$ inch; this minimum movement of oil saves oil and energy.

The clamping force exerted by the cylinder is a direct function of the pressure available from the pump and the area of the ram, in accordance with the formula:

$$F = PA$$

where

F = force (lb)
P = pressure (lb/in^2, psi)
A = area (in^2)

Thus, a press with a 12-inch-diameter ram operating at a pressure of 3000 psi would have a clamping force of:

$$F = 3000 \times \pi \times 6^2 = 339{,}293 \text{ lb}$$

$$= \frac{339{,}293}{2000} = 170 \text{ tons}$$

Note that the pressure of the oil in the ram is expressed in psi (lb/in^2), while the force is always in pounds or tons. In the example shown, the machine would probably be rated as a 150-ton press.

The hydraulic clamp system has a number of special features which are unique to this mechanism:

1. By the use of simple limit switches, the length of stroke can be easily adjusted to fit various sizes of molds.
2. Positive, low-pressure mold protection can be assured by use of a controllable, low-pressure relief valve in the jack ram circuit. During the slow-close stroke, the valve can be adjusted to just overcome the friction loads, thus stopping if anything else impedes the movement.
3. The tonnage being applied is readily discernible by simply reading a pressure gauge.
4. The machine is self-protecting. If the injection pressure times the projected area of the part exceeds the rated tonnage of the clamp, then oil is forced through the relief valve, thus allowing the mold to flash. No abnormal stresses to the clamp mechanism, tie rods, or other parts of the press are applied.
5. The mechanism requires minimal maintenance. Since all the moving parts are running in oil, the mechanism is self-lubricating. Normal maintenance consists of an infrequent sealing-gland adjustment to tighten the

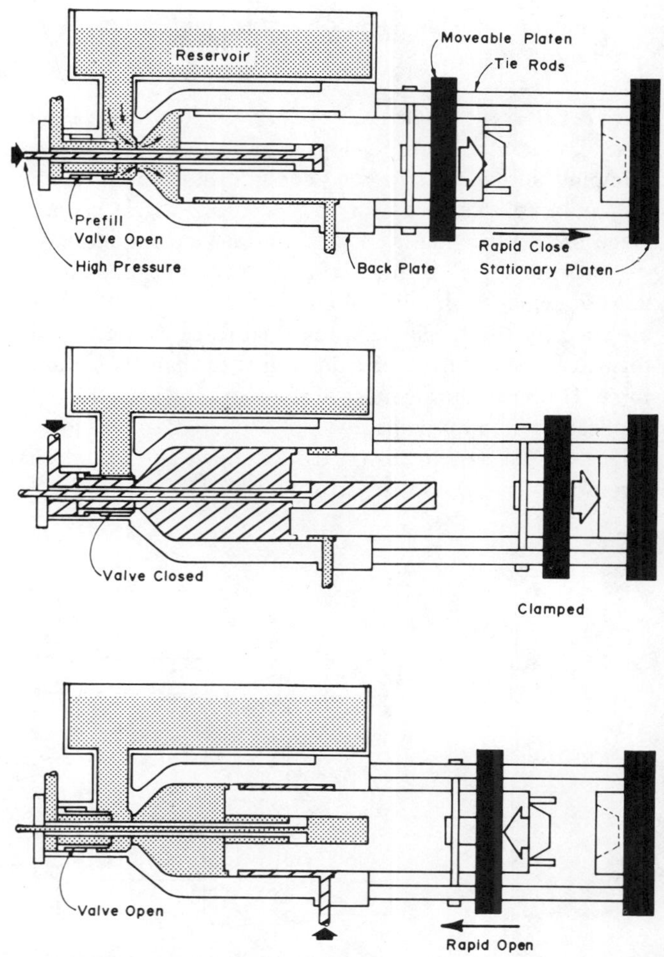

Fig. 10-23. Diagram of high-speed hydraulic clamping system using a "jack" ram and main ram (Van Dorn system). On rapid forward travel, the high-pressure oil is pumped into the jack ram. Final closing and lockup is accomplished by diverting the high-pressure oil to the main clamp piston. To open the mold, oil is pumped to the face of the main ram, driving it back to the set open position. (Reprinted by permission from ***Plastics Design & Processing:*** September 1970. Copyright 1970, Lake Publishing Corporation, Libertyville, Ill.)

chevron packing, and occasional lubrication of the tie-rod bushings.

Toggle Clamps. Toggle clamps are used for the same purposes as are hydraulic clamps—closing the mold halves, clamping during injection and cooling, and then retracting the mold so the parts can be ejected. A toggle clamp is made of two bars jointed together at one end, and arranged so that the "knee" formed by the joint can be straightened out by a small hydraulic cylinder. Because of amplification of force by the toggle bars, the final force exerted by the toggle can be as high as 50 times greater than the cylinder force. However, during most of the toggle's travel, the force application is approximately 1:1. Figure 10-24 shows a typical toggle system. Figure 10-25 shows another mechanism for obtaining a longer stroke.

There are a number of specific advantages to the use of a toggle clamp, including:

1. Automatic deceleration of the moving platen. The speed decreases as the leverage increases in force.
2. Inherent high speed. The automatic speed decrease allows the highest speed to be used for the greater part of the stroke, without danger to the platen as it closes.
3. Automatic development of the highest clamping force and the maximum breakaway force. Figure 10-24 shows how the maximum forces are developed in the last half-inch of movement.
4. The toggle is inherently a self-locking system. The hydraulic pump need not be kept on during the injection and cooling portion of the stroke, thus resulting in an energy savings.

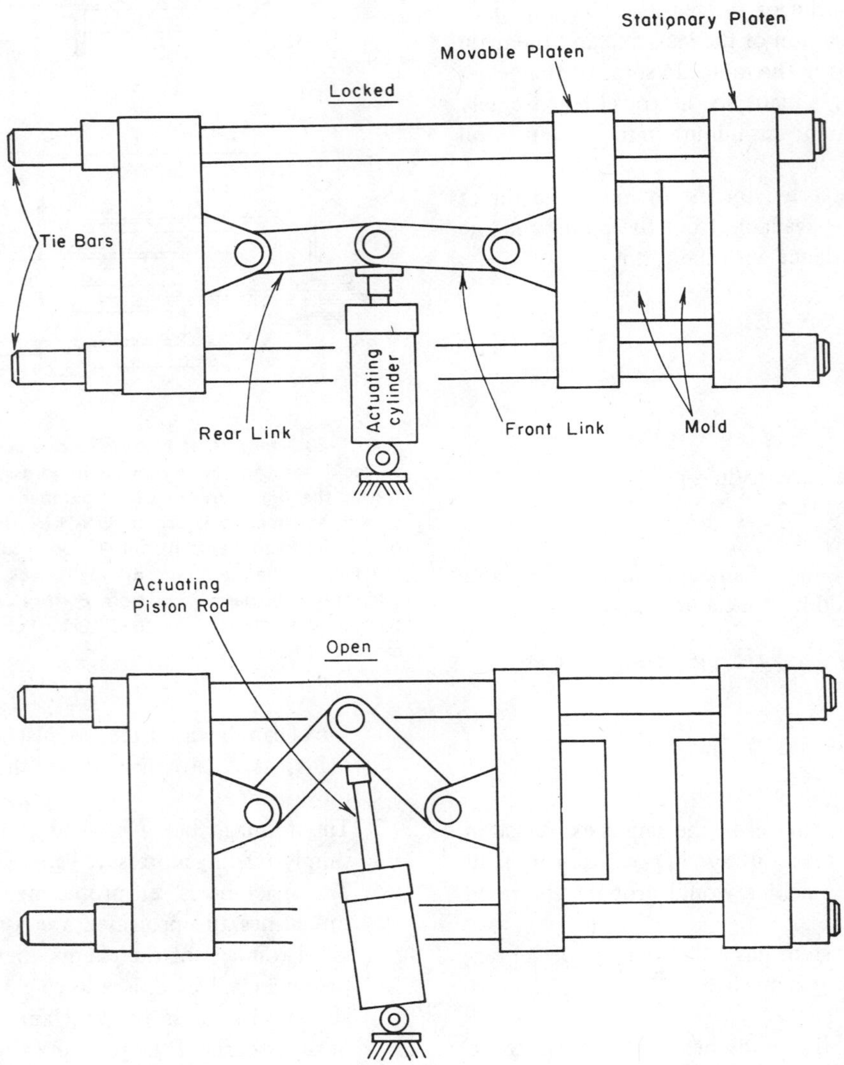

Fig. 10-24. Typical toggle clamp system. Observe that the single toggle is generally mounted in a vertical plane along the center line of the platen. Hydraulic cylinder extends to straighten toggle links and pivots to maintain alignment as links extend. (Debreceni, 1968. Reprinted by permission from *Plastics Technology*)

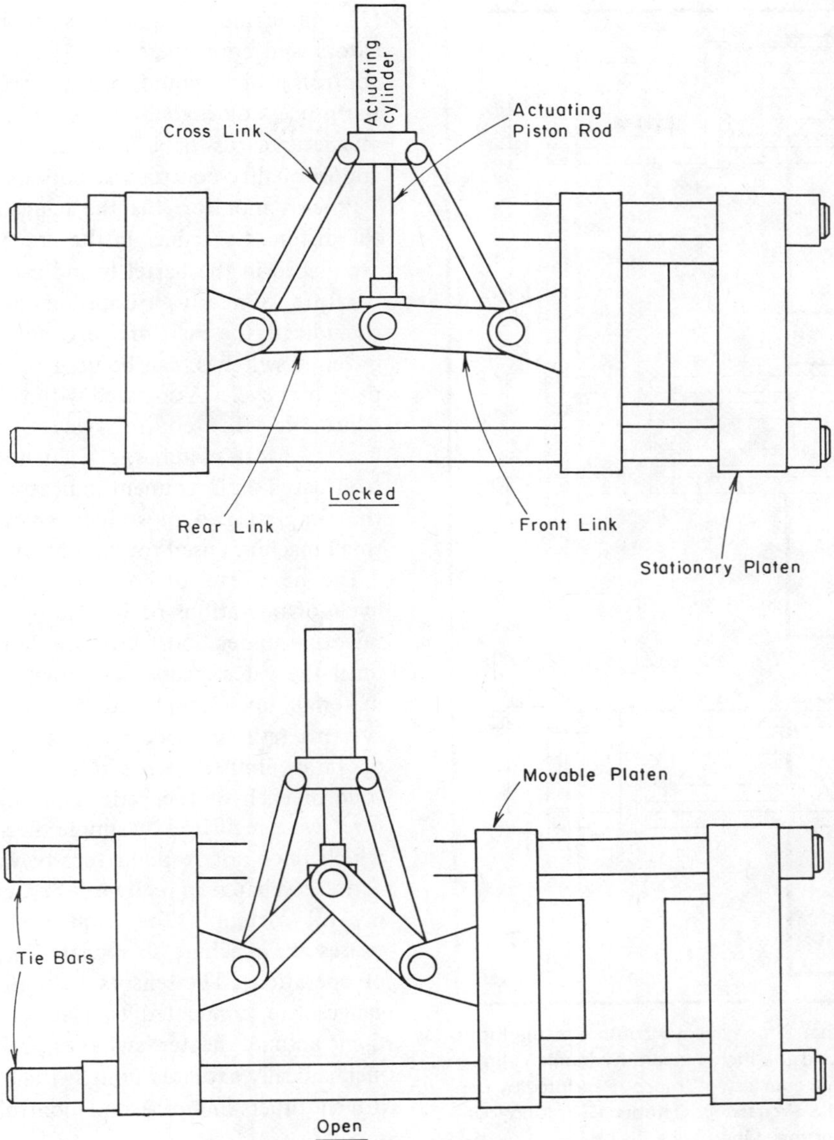

Fig. 10-25. Long-stroke toggle system. In monotoggle, operating cylinder is not fixed to the base and rides with the toggle arms as they extend. Result: longer stroke without loss of leverage. (Debreceni, 1968. Reprinted by permission from *Plastics Technology*)

5. Less expensive. Because a smaller cylinder can be used and less power is needed, there can be significant cost savings in the original machine cost and further savings during its use.
6. Single-point mold-height adjustment. Many toggle systems are now available with a means of easily adjusting the toggle to accommodate varying mold thicknesses. By means of a hand-operated or motor-driven chain or gear system, all four tie-bar nuts can be simultaneously rotated to give quick, easy, precise movements of the toggle as required for any given mold height. This can be of considerable advantage to a "job" shop where frequent mold changes are the rule.

Some disadvantages of the toggle clamp are:

1. The toggle does not easily allow for measurement of the clamp force.
2. Because of unequal rates of heating, the toggle bars may expand differently and thus affect the clamp force. The hydraulic clamp automatically compensates for this by action of the relief valve or addition of more oil in the cylinder.
3. The toggle system is more susceptible to wear at the joints. Thus, more attention must be paid to lubrication and general maintenance. Automatic greasing and oiling systems are available, however.

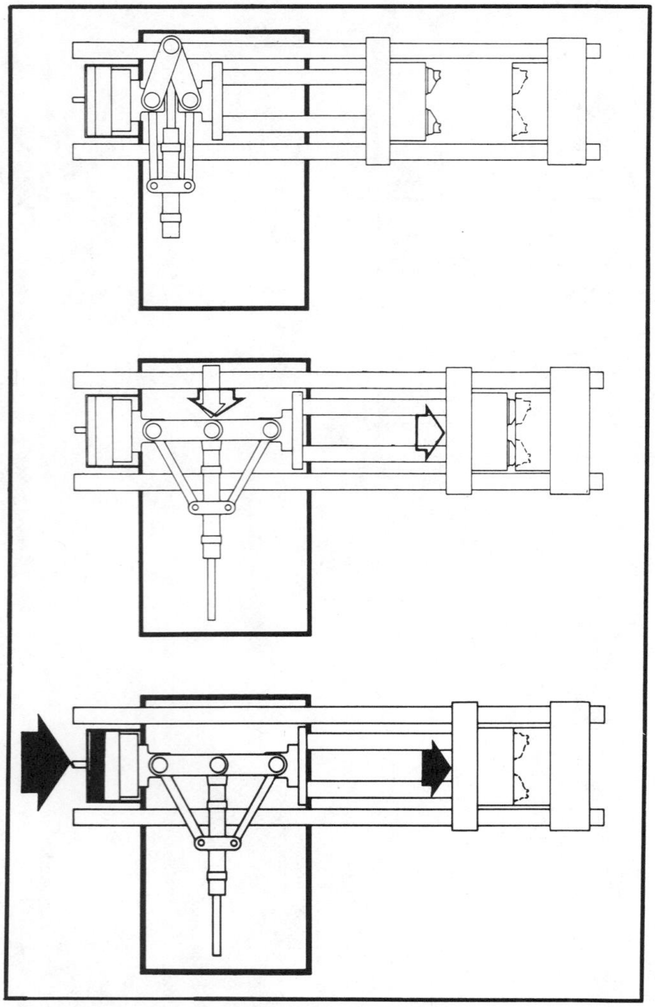

Fig. 10-26. Toggle-ram system. This system incorporates the toggle for fast platen movement and the high-pressure hydraulic cylinder for clamping force. (Drawing: courtesy of Impco. Reprinted by permission from ***Plastics Design & Processing:*** October 1970. Copyright 1970, Lake Publishing Corportion, Libertyville, Ill.)

Because both clamping systems have definite advantages and disadvantages, there have been mechanisms developed which attempt to use the best features of both. An example of one such system is shown in Figure 10-26. This system combines the high-speed movement and automatic slowdown and buildup of force of the toggle, with the precision control of tonnage and the reliability and power saving of a short-stroke hydraulic cylinder.

Control Systems

The third important component of the injection molding machine is the control system. The controls regulate the functioning of the injection system and clamping mechanism so that the process can be accomplished at the right temperature and pressure, and in the proper sequence.

Control systems in injection molding machines can be: (1) simple indicating process monitors, (2) process indicators and controllers, or (3) the ultimate–computerized controls which would indicate process conditions, control the process during variations, and, by means of punched or magnetic cards or a tape cassette, could direct startup and/or quality-control instructions to the machine.

Process monitors, as the name implies, simply monitor the different variables in the process. Thus, thermocouples are placed in the barrel to indicate the barrel or melt temperature. Similarly, a Bourdon tube pressure gauge is used to indicate the melt pressure, and electric timers, actuated by limit switches, can be used to show the cycle length or parts of a cycle. Associated with the indicators are manually adjustable flow-control valves, pressure-control valves, heater on-off switches, etc. Such a system of manual control, based on instrument indicators, would only be used on the simplest and most inexpensive machines, or possibly small machines used for laboratory tests.

The next level of control involves regulating the actual cycle of operations performed by the machine–mold close, injection under boost pressure, hold under a lower pressure until the gates freeze, withdraw screw and plasticize new material, open mold and eject parts. The earliest control systems (still in wide use) used timers and cam- or rod-operating limit switches to control the duration or magnitude of each of the various phases of the cycle (temperatures were regulated by simple thermostats). Such a system, which does not respond to any variations in the machine action or change in melt viscosity, etc., is called a "sequence control system." The sequence control system simply causes the machine to repeatedly perform the same cycle of operations. The sensors used–thermocouples or pressure gauges–are connected to electromechanical relays which then actuate heater switches, solenoid valves, etc. The mechanically actuated limit switches and the electric motor-driven timers indicate and control the movement of the screw or ram.

Temperature measurement and control were accomplished at first by simple on-off regulation; that is, the heaters were kept on until the required set-point temperature was reached, then the heaters were turned off and remained off until a lower set point was reached. Wide variations led to the development of "proportion controllers." In this type of control system, the heat is supplied in proportion to the amount of heat required; that is, the proportion of time-on to time-off periods would change in proportion to the deviation of the sensed temperature from the desired set point. This type of control of heating and cooling results in smoother thermal cycling with less under- and overshoots.

Pressure continued to be sensed by the use of Bourdon tube gauges (a flattened tube bent into an incomplete circle, which when pressurized tends to straighten out). The slight movement causes actuation of electromechanical relays to control solenoid valves in the hydraulic system.

The great majority of injection molding machines made in the late sixties or early seventies use such indicating and controlling systems.

In the latter part of the sixties, the era of solid-state controls was initiated. The first solid-state applications were the elimination of relays. Relays, with their moving parts, are subject to small delays in actuation and also to deterioration from wear during operation. Solid-state relays, silicon-controlled rectifiers (SCRs) with no moving parts, respond very rapidly to the minute voltages produced by a thermocouple or the slight changes in resistance produced by a wire resistance sensor or a thermistor (a solid-state device which changes resistance with temperature changes). Associated with the use of SCRs are transistors and diodes used in amplifier circuits. The SCRs, in addition to their applications as relays, were also used as contactors to handle the large amounts of current needed to actuate the heaters.

Pressure sensors were also developed: from the Bourdon tube sensor, to the strain gauge (a wire network which stretches under pressure and changes resistance) then to a completely solid-state piezoelectric crystal which generates a small electric current under pressure. Such pressure indicators (strain gauge, and piezoelectric), called transducers, could now be installed anywhere in the machine to indicate hydraulic pressures, or, in the mold cavity to indicate melt pressures. By the use of a transducer installed under an ejection pin or at a separate location in a mold, the part itself could be used to control the process. If the transducer indicates that the mold is filling too slowly, then additional heat can be called for, or additional pressure can be applied if the temperature is already at the upper limit.

With the advent of solid-state devices in the sequence-control systems, the next improvement was the development of true process controls in which the sensor detects a variation in the process itself; for example, a change in viscosity of the hydraulic oil as the machine heats up would alert the operator to make changes in the timer and pressure settings. Such a system, where the operator makes the changes as indicated by the machine controls, is called an "open-loop" system.

To have even better control of machine or melt variations, it is necessary that the loop be closed, so that the control system will detect the process anomaly and apply the needed corrective action. In this "closed-loop" process-control system, emphasis, at the time of writing, was being placed on the molding process itself as the controlling agent. Thus, pressure transducers in the mold, placed under an ejector pin or separately in the mold cavity or cavities, could signal when primary or boost pressure should be changed to secondary or hold pressure.

"Adaptive control" is a higher degree of control sophistication in which the system is used to select the optimum processing conditions in a process which may have several simultaneous variables. A cavity pressure sensor can be a

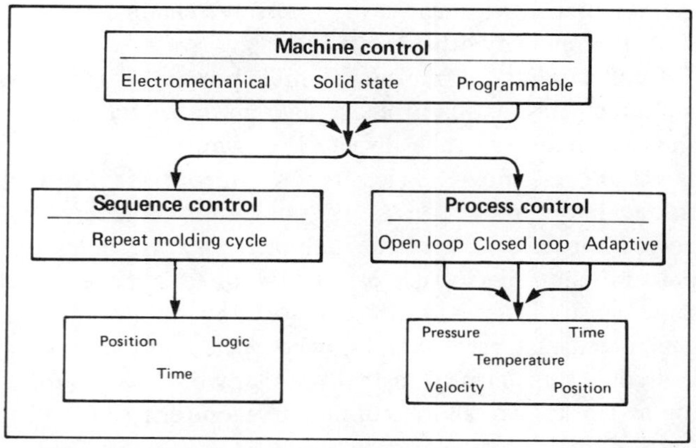

Fig. 10-27. Injection machine control systems. (Courtesy: *Plastics World*, December 1978)

simple adaptive control when it must determine the optimum cavity pressure to produce uniform part densities despite changes in oil temperature or melt viscosities and temperatures during repetitive cycles.

The sequence-control systems, with or without solid-state components, and the more elaborate process control systems can all be programmed for any given cycle, depending on how the system components are adjusted and interconnected. Figure 10-27 shows the various options and capabilities.

The control systems described above have the system's logic determined by the type of interconnections made between the components. Thus, the systems are referred to as "hard wired," in contrast to the ultimate in control systems in which microprocessors (integrated circuit chips) are used. A microprocessor control system is essentially a computer. In this system, data from the various sensors is compared with instructions which have been stored in its memory. In case of any discrepancy, the computer takes the required remedial action and, at the same time, notifies the operator of the malfunction. Alternatively, the computer may make a printed sheet or magnetically recorded tape, so that a record may be kept and, simultaneously, the process conditions may be shown on a cathode ray tube.

Computer control as described above is not necessarily confined to one machine or even to one production facility. The computer may be used by a small factory on a time-sharing basis with a number of users or, in a larger facility, to control up to twenty machines. Depending on the type of parts, automatic inspection equipment can also be designed to service a number of machines through the computer.

One very promising use for the computer memory is as an automatic setup operator. In this case, the computer is fed all the information needed to produce a part when a particular mold is placed in the press. After activation, the computer then automatically adjusts the machine for such variables as barrel temperature, injection speed, hold time, mold temperature, melt-cavity pressure, clamp positions, shot sizes, and any other desired variables.

An illustration of the use of mold-pressure controllers, mold-temperature controllers, ram-position transducers, and a cathode ray tube display is shown in Figure 10-28. In the process shown, a closed-loop control system is used to facilitate production of a double-shot molding using polycarbonate and urethane polymers. By use of cavity-pressure controllers, each part in the mold is made at the same density. The use of the cavity-temperature controllers guaranteed that parts would not be ejected until each resin had cooled sufficiently so that warpage would be at a minimum. The display allows continuous monitoring of the entire system.

While computer control of the injection machine may represent the ultimate in control efficiency for the process, it is possible to incorporate many of the benefits of solid-state controls and, to some extent, computer control, at a much lower cost by using a separate unit to control just one element of the process. At the time of writing, cavity-pressure control was considered to be one of the best techniques to be utilized in controlling the process, since in effect the part itself, during its production, controls the process. Figure 10-29 shows one such device—a mold "pressure transfer controller." The maker, the Barber-Colman Company, states that it:

Fig. 10-29. Mold pressure transfer controller. (*Courtesy: Barber-Colman Co., Industrial Instruments Division*)

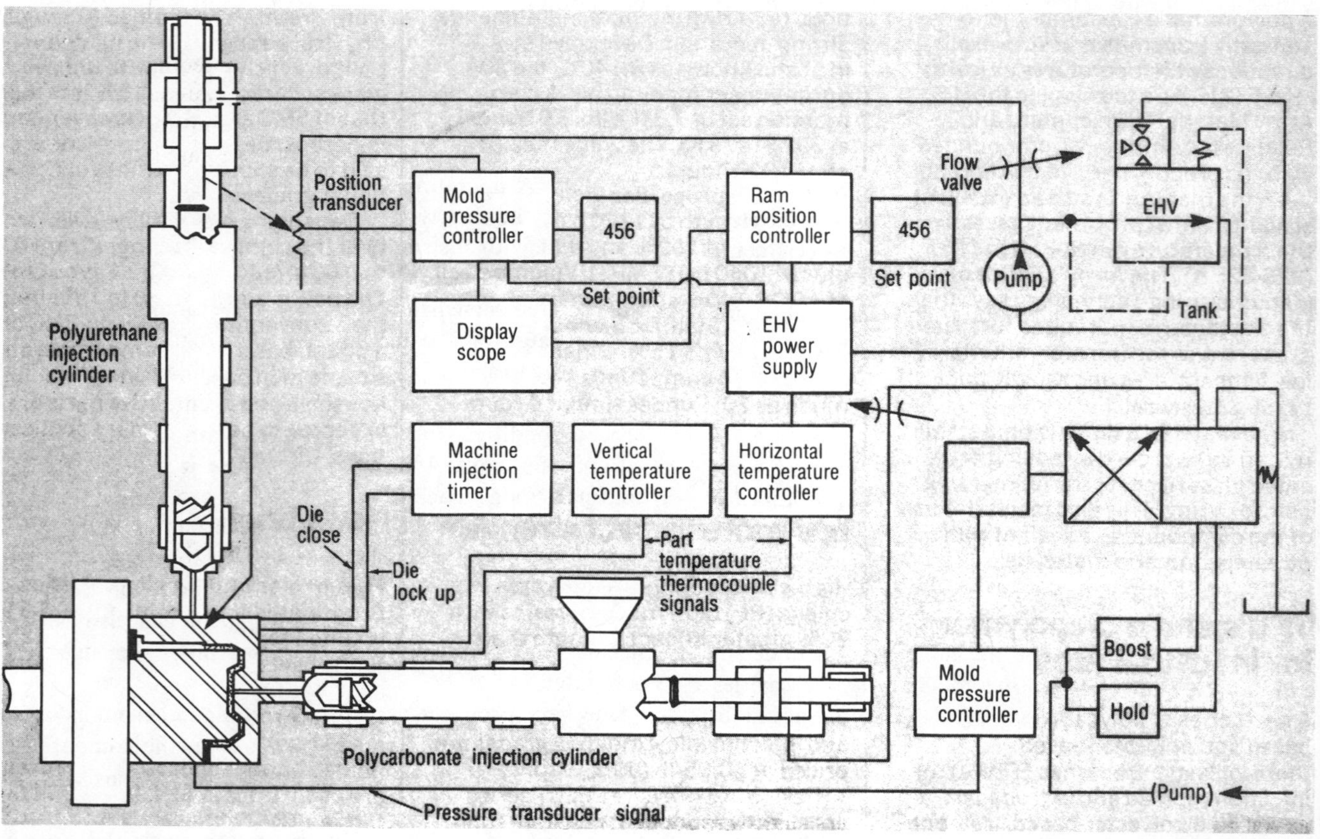

Fig. 10-28. Flow diagram of a closed-loop-controlled, double-shot injection molding system used to mold a precision part in polycarbonate and urethane. (Courtesy: *Materials Engineering*, December 1977)

Controls melt pressure in the mold
Protects molds from excessive pressures
Reduces startup time
Eliminates short shots
Allows in-cycle control
Requires no hydraulic system modifications
Optimizes boost time to molding conditions
Reduces scrap rate and inspection time
Shows peak pressure retention on a digital display

The manner in which the pressure controller accomplishes these tasks is shown in Figures 10-30, 10-31, and 10-32.

The description of one type of process controller is given in detail to show what a modern solid-state controller can do to adjust and accommodate the machine to variations in one aspect of the molding cycle. Similar controllers can be used to compensate for changes in other variables in the cycle (e.g., melt-temperature changes, changes in ram speed, shot size, etc.). For additional information on controls, see Chapter 20.

Molds

All injection molding machines (ram type, two-stage preplasticization, reciprocating screw, etc., whether having hydraulic clamps or toggles) have one item in common—a mold. The purpose of the mold is to shape the injected liquid plastic to the required dimensions, and to hold the material until it cools and can be ejected as a solid part. In its simplest form, the mold consists of two plates: the stationary plate which usually contains the cavity into which the plastic is injected, and the moving plate which is attached to the clamp. The movable portion usually contains the ejection mechanism—pins or a plate which is used to eject the parts. Both plates are also usually cored so that water, or another fluid, can be circulated to maintain a constant temperature. Because of the need for the ejection mechanism, the coring, and the cavities, the mold is gen-

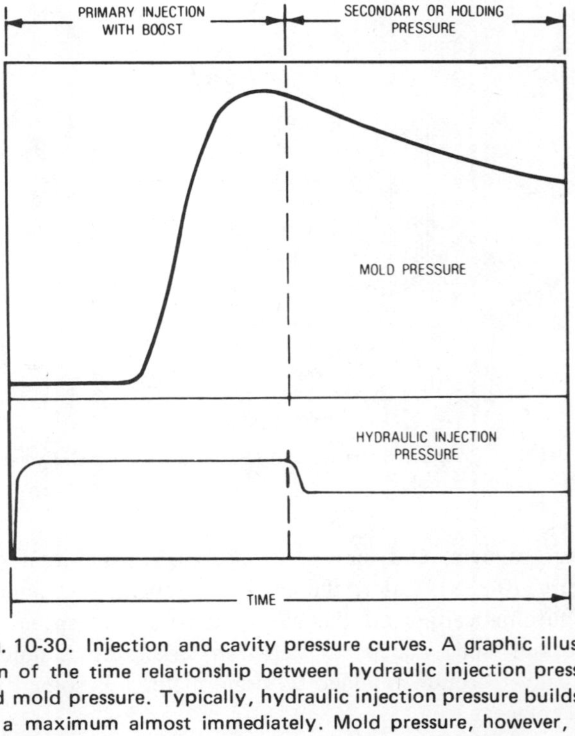

Fig. 10-30. Injection and cavity pressure curves. A graphic illustration of the time relationship between hydraulic injection pressure and mold pressure. Typically, hydraulic injection pressure builds up to a maximum almost immediately. Mold pressure, however, lags hydraulic pressure due to the fact that the mold must fill before force is exerted on the measuring transducer. Assuming that hydraulic pressure and flow are maintained constant, mold pressure from shot to shot should build up at the same rate and reach the same peak value. However, variations in the plastic itself, in the barrel temperature controllers, or any other parameter that might affect viscosity, will change the mold pressure curve.

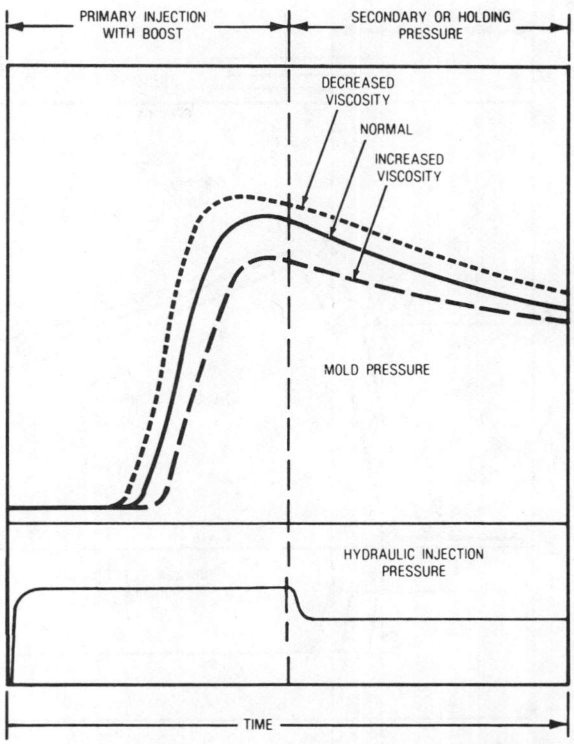

Fig. 10-31. Effect of viscosity on mold pressure. Different curves are obtained because of changes in material viscosity. Note that hydraulic injection pressure is the same in all cases and the differences are the result of material changes only. The figure shows three different curves—one for normal viscosity, one for increased viscosity, and one for decreased viscosity. With a decrease in viscosity, mold pressure builds up sooner and also reaches a higher peak value. The result on the molded part, in this case, will be increased part weight and flashing if pressure builds up significantly. With an increase in viscosity, it takes longer to fill the mold and, consequently, longer for mold pressure to build up. In this case, because it does take longer to fill the mold, the part does not completely pack and results in a lighter part and possibly even a short shot.

From the above description, it follows that the control system must be capable of compensating for both increased and decreased viscosity conditions. With a decrease in viscosity, injection pressure should be cut off early in the cycle to prevent overpacking and flashing. With an increase in viscosity, injection pressure must be applied longer and at the same time be high enough in value to insure that the mold fills and packs.

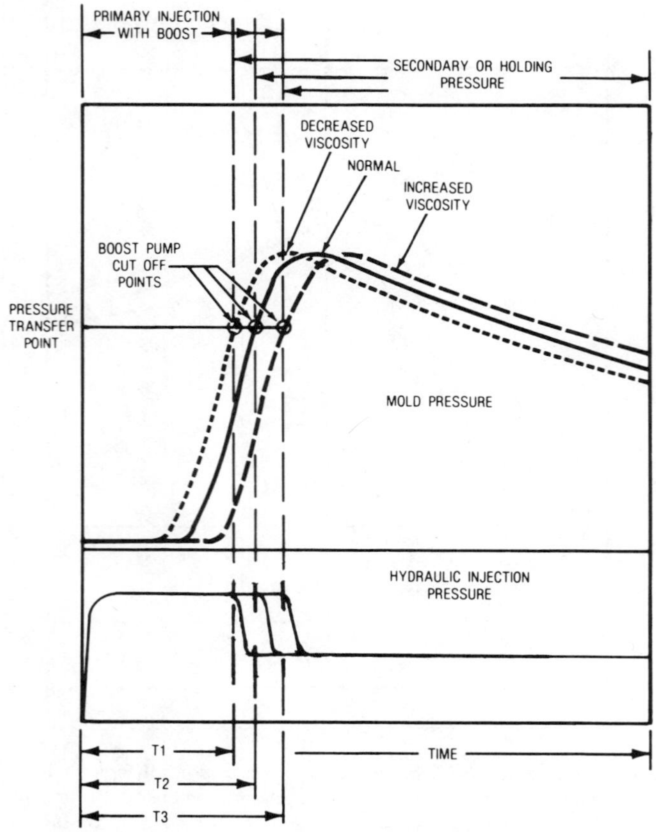

Fig. 10-32. Effect of 916 PTC on mold pressure with different apparent viscosities. The curves show the results of applying the Pressure Transfer Controller (PTC) to the different conditions described above. Note that with the 916, primary injection pressure can be increased thereby causing a faster rate of fill. This is possible because the controller monitors mold pressure during fill and cuts off primary injection pressure when mold pressure reaches a predetermined value. By being able to fill the mold faster, boost time can be reduced thus reducing the overall cycle time.

As shown in the figure, due to increased injection pressure, the mold fills faster and consequently mold pressure builds faster. When mold pressure reaches the Pressure Transfer Point, primary injection pressure is cut off and secondary or holding pressure is applied. Note that even though primary injection pressure is cut off, mold pressure continues to increase due to the dynamics and response of the hydraulic system. For all three conditions, injection pressure is the same, but the length of time that it is applied, varies, dependent on the apparent viscosity of the material. In this manner, injection pressure time is adjusted as a function of mold pressure and results in more consistent molded parts despite changes in material viscosity.

erally considerably more complicated than the simple two-plate mold described. Figure 10-33a shows a schematic cross-section drawing of a typical mold in which the melt is fed to the part cavity from one side. Such a mold would be known as a two-plate mold, despite the presence of the extra plates shown. In use, the molten plastic enters the mold from a nozzle through the sprue bushing; from the sprue, the melt is conveyed through the runners to a restriction, the gate, and then into the cavity to form the part. On cooling, the entire molded piece, including parts, gates, runners, and sprues, is ejected as one item (see Figure 10-33b). Later, the parts are separated, manually or automatically, and the gates, runners, and sprues are reground and, added to virgin material, are remolded. (Usually reground is added to virgin material in some fixed percentage; e.g., 1:5.)

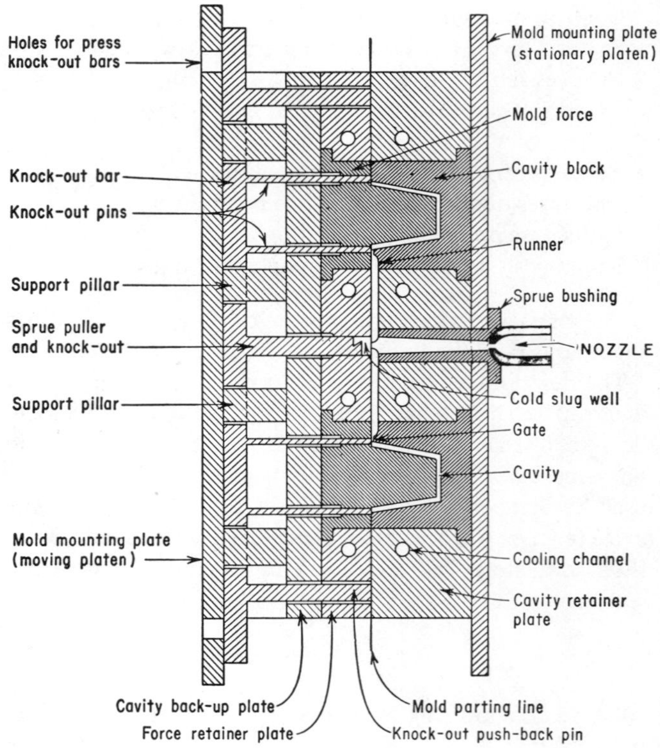

Fig. 10-33a. Two-plate, two-cavity, edge-gated mold.

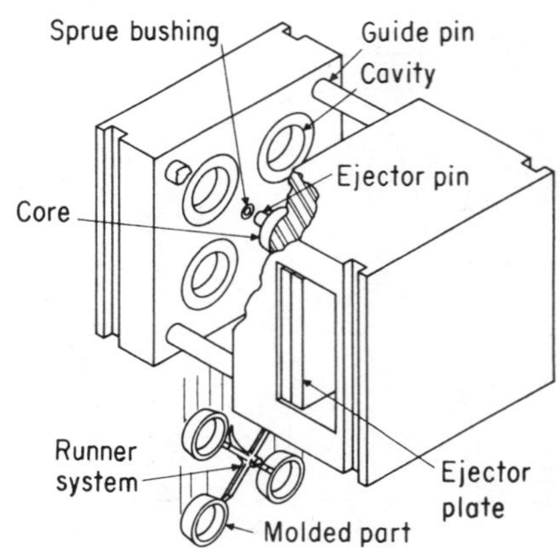

Fig. 10-33b. Connected parts, runners, and sprue being ejected from a two-plate mold.

A method of automatic removal of gates, runners, and sprues at the time of molding is to use a three-plate mold. In this type of mold, the third plate is known as the runner plate and is used to house the runner and gate system. Figure 10-34a illustrates the three-plate mold and Figure 10-34b shows the ejection of the separated parts from the

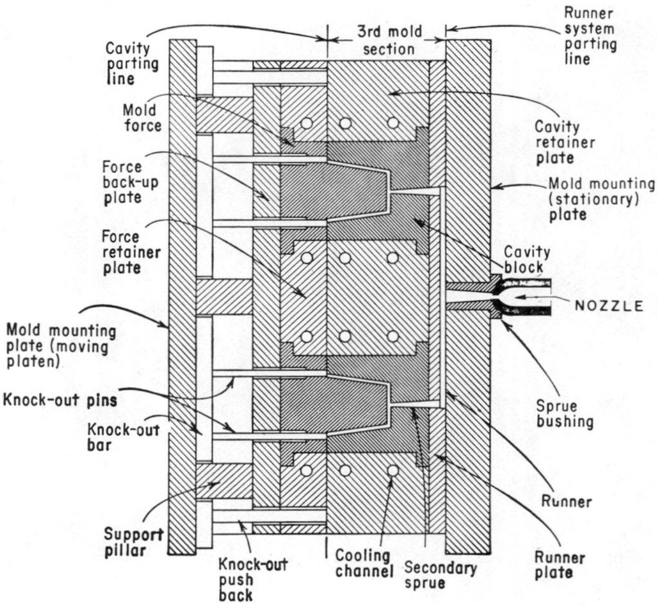

Fig. 10-34a. Three-plate, two-cavity, center-gated mold.

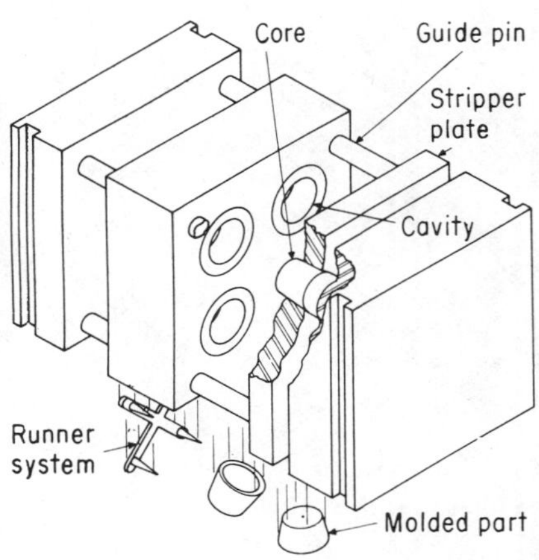

Fig. 10-34b. Separated parts and runner system being ejected from mold.

runner system. In a third type of mold, known as a "runnerless mold," no runners, gates, or sprues are ejected at all, since they are not allowed to solidify during the mold cycle. (See section on special techniques in this chapter and also Chapter 20.) For certain parts, separation from the gate can be done at the time of molding by use of a special gate, known as a "tunnel gate" or "submarine gate" (see Figure 10-35).

Figure 10-36a, b, c and d shows, somewhat more clearly than the previous three figures, the relationship between the sprue, runners, gates, and cold slugs, not mentioned previously. Also shown in Figure 10-36a, b and c are the cavities arranged in a balanced configuration; that is, all cavities are equidistant from the sprue and all runners are the same length. Such a multicavity configuration is considered optimum, since theoretically all the cavities are filled at the

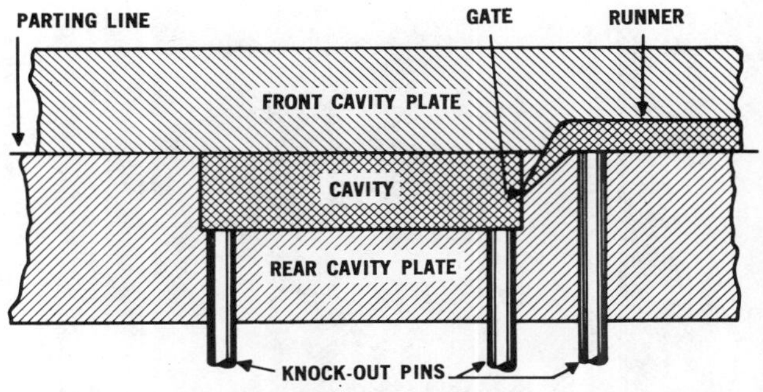

Fig. 10-35. Tunnel or submarine gate for automatic part-gate separation on opening of the mold.

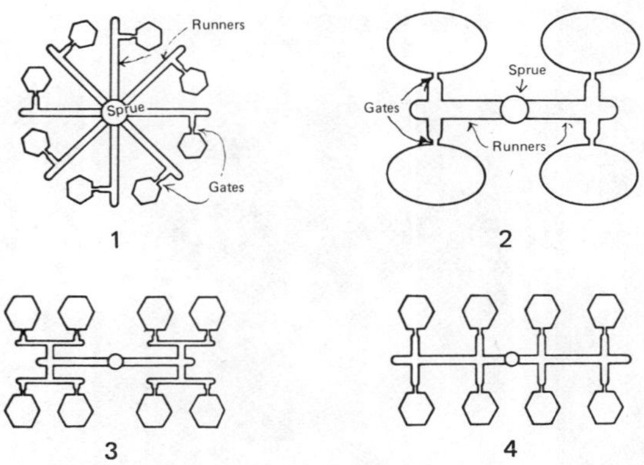

Fig. 10-36. Relationships among sprue, runner, and gates. Regardless of cavity configuration, relationships remain the same. (1) Sprue delivers material from nozzle to runner. (2) Runners distribute molten plastic to gates. (3) Gates (small restricted entrance to cavities) allow parts to be easily broken off, leaving only a small mark on part. The sprue, runners, and gates are usually ground up and, mixed with virgin material, are remolded. (4) Cold slug extension allows the first, cooled material to enter the runner, not the cavity. Later, hot material fills the cavity.

same time and at the same pressure. Contrast this with Figure 10-36d, which is unbalanced, since the two outermost cavities would be filled after the inner cavities, with consequent possibilities of warpage or density differences.

The molds shown in Figure 10-36 are used to make several parts of the same type at each shot and so are known as balanced multicavity molds. Another type of multicavity configuration, known as a "family mold" (also unbalanced), is used when several different parts are to be molded at the same time. The advantage of such a mold is that all the parts are identical in color, and, if the parts are to be assembled together as is the usual case, they can be kept together until ready for assembly. The disadvantage of such a configuration is that, since the parts are not all the same size or shape, the cycle is determined by the time to mold and cool the largest part. Typical applications for family molds are for toy model kits such as airplanes, racing cars, or ships. In such cases, having all the parts attached to each other is an advantage for inspection and packaging (see Figure 10-37).

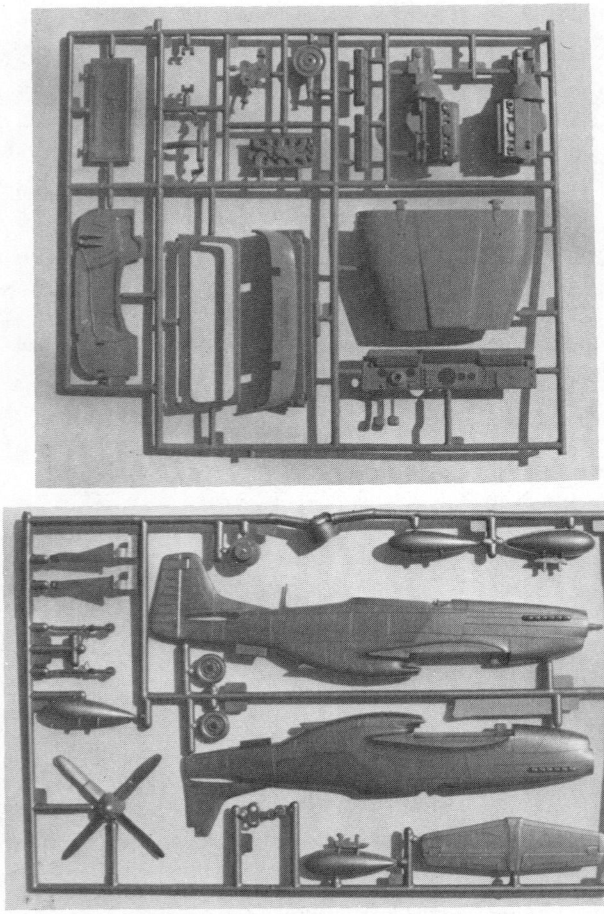

Fig. 10-37. Parts molded in "family" molds.

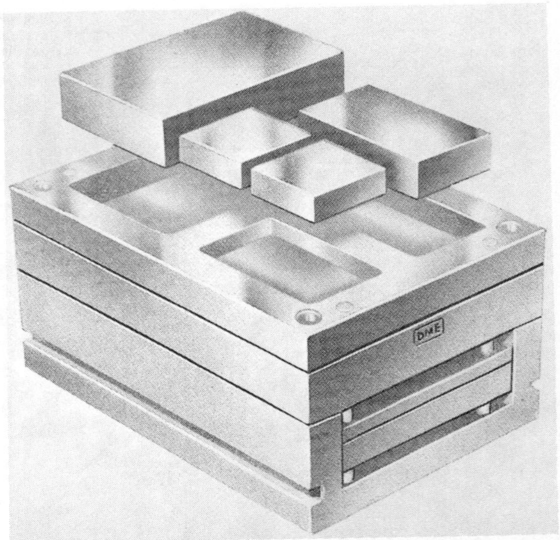

Fig. 10-38. Mold base with cavity inserts. (*Courtesy: D-M-E* Co.)

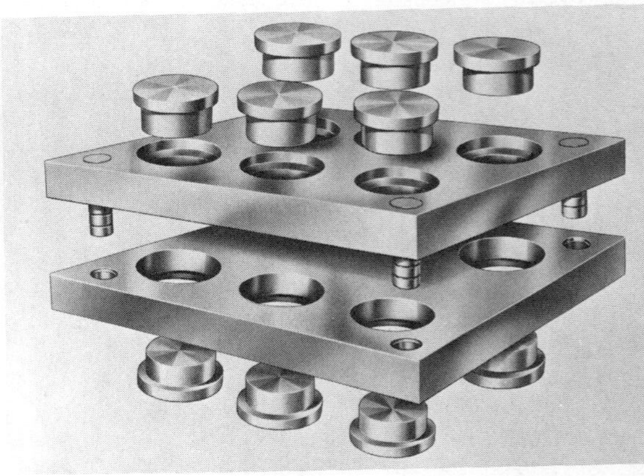

Fig. 10-39. Mold base with cavity inserts and leader pins. (*Courtesy: D-M-E Co.*)

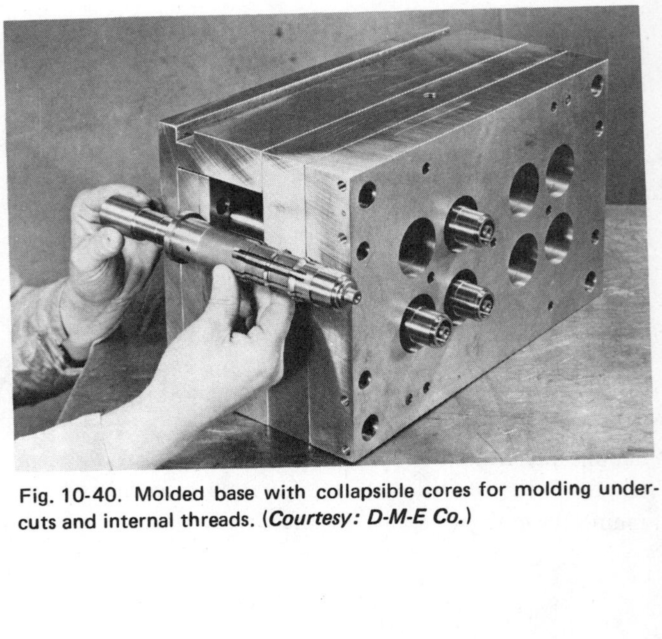

Fig. 10-40. Molded base with collapsible cores for molding undercuts and internal threads. (*Courtesy: D-M-E Co.*)

Injection molds, depending on the quantity of parts required and the method of construction, are made mainly of steel. Alloy steels of various types are used for cavities which are to be machined and later heat treated to a specified hardness. Auxiliary plates are usually of low-carbon steel. Low-carbon steel and nickel-alloy/low-carbon steels are used for cavities which are to be hobbed. Various nonferrous alloys, principally aluminum, are used for molds which are experimental or are to be used for very short runs.

Experimental or short-run molds are commonly machined from large blocks of metal; however, production molds are usually made using commercially available standard mold bases and other mold components such as leader pins and leader-pin bushings, ejector pins, and sprue bushings. The use of such parts, which are available in a variety of sizes to fit most machines, results in considerable savings of time and labor for the average tool fabricator, since the components are furnished with finely ground surface finishes and machined to close tolerances. Thus, the mold maker need only add the cavities and the runners and gates. Coring is also provided, as specified. Figures 10-38 and 10-39 show typical mold bases fitted with blocks or rounds which can be separately machined. Figure 10-40 shows a base fitted with collapsible cores which are also commercially available. For use with corrosive plastics, these bases and inserts are available in stainless steel alloys.

The important considerations in the design of a mold are as follows:

Sprue design	should be as short as possible.
Runner design	large, generous runners, to deliver sufficient molten plastic without too much waste.
Gate design	as small as possible, consistent with good cavity fill.
Cavity design	highly variable, but should allow quick filling, tapered if required for easy removal, oversize for material shrinkage, and a good surface finish.
Mold coring	should maintain an even temperature throughout the cavity, unless otherwise required.
Ejector system	must operate automatically on opening the mold. Pins, rings, stripper plates, and/or compressed air all used, whichever results in minimum part markoff.

Additional information on specific details of mold design and construction is given in the section on mold design in Chapter 20.

Auxiliary Equipment and Materials for Injection Molding

The injection machines, controls, and molds described previously are all required to do injection molding. However, there are a number of auxiliary pieces of equipment whose use is very necessary if successful, competitive, and profitable molding is to be carried out. In some cases, the equipment described below is used primarily for convenience or to speed the job; however, for some materials or parts, the special apparatus or auxiliary material is a must if the job is to be done at all.

Mold-Temperature Regulation. Circulating liquid/heat-transfer systems are used to maintain a constant temperature in the mold and, thus, improve the quality of the molded product. They bring the mold to a certain temperature when first starting up and remove heat from the mold as the hot plastic is continually injected, thus maintaining the set temperature. This is done by circulating thermostatically controlled water, or other fluids, through the cored passages in the mold. The mold-temperature regulator is usually a self-contained system equipped with a pump, thermostatic controls, heaters, reservoirs, etc. Such a unit (see Figure 10-41), which is usually portable, can be used to control one machine or as many as three zones at once. The unit shown in Figure 10-41 uses water at normal inlet temperatures. Other units, less common, are available with refrigerators for below-normal-temperature use.

Fig. 10-41. Mold-temperature regulator. *(Courtesy: Mokon Division, Protective Closures, Inc.)*

Another method of obtaining cold water for mold or machine cooling is by the use of a cooling tower. In this technique, the warm water is pumped to a tower on the roof and, as it descends down a slatted tower, is cooled by evaporation. Such a system is well suited to be a central water-cooling system for a bank of machines, although it does not offer quite the precise control or convenience of an individual regulator. Instead of depending solely on the wind to cause evaporation, some units are fitted with motor-driven fans installed in the tower. In either case, the water must be treated and filtered to minimize scale and salt buildup in the mold and cooling towers, since the water which evaporates is distilled water and the residual water is thus higher in dissolved chemicals. Because of this, in many cases, the tower water is not used to cool the mold but simply to cool a chiller condenser. The mold coolant then is a recirculating fluid which contains rust inhibitors and other chemical additives to reduce buildup in the mold.

Drying Equipment. Almost all plastic granules either absorb moisture or carry some moisture on their surface. In either case, the pellets must be dried prior to molding because the absorbed moisture can cause streaks and bubbles on the molded part. Materials which readily absorb moisture from the air are known as hygroscopic. These materials include nylon, ABS, polycarbonate, cellulosics, and methyl methacrylate. Other materials such as polyethylenes and styrenes, which do not normally absorb much moisture, can carry appreciable amounts on their surface, due to improper temperature and humidity cycling during storage or transit. An example would be removal of styrene from a cold warehouse into a warm mold room, which could cause moisture condensation on the normally very dry styrene pellets.

Hopper Driers. The most efficient method of drying the plastic pellets is in the hopper, using hot, dry air. Modern

Fig. 10-42. Hopper dryer-vacuum loader on individual machine. (*Courtesy: Conair, Inc.*)

hopper driers combine the use of hot air with desiccants and vacuum to dehumidify the air, thus removing the moisture much more completely than by use of hot air alone. In addition, the hopper drier can act as a preheater, so that the cycle time can be reduced. Hopper driers are available either as individual machine units (Figure 10-42), or a number of machines can be hooked up to a central dehumidifier, as shown in Figures 10-43 and 10-44. In such a system, each machine is furnished only with cool, dehumidified air from the central source. An individual preheater at each machine is then used to heat the dry air to the particular temperature required for the material that is being run. In that manner, one machine might be running nylon and next to it another machine might be running polycarbonate at a considerably higher temperature.

Ovens. Ovens were the original method of drying plastic pellets, and, for some plastics such as polyethylenes or vinyls which only require surface moisture removal, they can still be used satisfactorily. The plastic granules to be used in an oven should be dried in shallow pans, not over $1\frac{1}{2}$ inches deep. Furthermore, the materials should be used

Fig. 10-43. Central dehumidifier combined with individual hopper dryer–vacuum loaders for each machine. (*Courtesy: Conair, Inc.*)

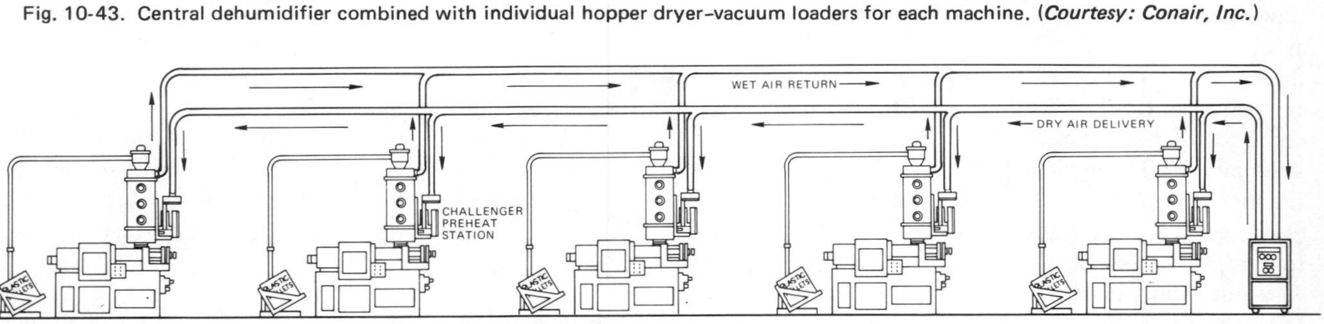

Fig. 10-44. Central dehumidifying system combined with individual hopper dryer–vacuum loaders for each machine. (*Courtesy: Conair, Inc.*)

as soon as possible after being removed from the oven, instead of being allowed to stand in the normal shop atmosphere.

Vacuum Hopper Loaders. In conjunction with the use of a hopper drier, vacuum can also be used as a simple system to insure continuous feeding to the machine hopper. By the use of the vacuum hopper loader, manual feeding is eliminated and the first step toward automation is taken. Various types of units are made so that a simple loader can be used or one that automatically feeds two materials in a predetermined ratio. Commonly, one material is virgin and the other is regrind. These units, like the others discussed, can be individual units for each machine, or the vacuum can be from a central source and the material fed from one or several drums or silos.

Granulators. Plastic granulators, formerly called scrap grinders, are used to regrind runners, sprues, and damaged or off-specification parts so the resultant newly made pellets can be blended with virgin material and remolded. Like the hopper-loading equipment, granulators are available either as central units or as individual beside-the-press machines. In general, the central units are used for larger installations (over fifteen machines), or perhaps where fewer presses are used to mold the same items for long runs. The beside-the-press units have the advantage that the regrind can be used immediately, while the job is running, without, in most cases, the necessity of drying. Other advantages of using individual units include: scrap does not have to be stored, thus saving floor space; possibility of contamination is minimized; and a number of different materials may be granulated simultaneously. Figure 10-45 shows an automated drying and feeding process for an injection molding machine.

A large central granulator has the advantage of taking up less floor space than does an equivalent number of individual units. It is also easier to isolate the noise of a central grinder than a number of beside-the-press units which, of course, are on the production floor. The lower maintenance cost for the one large unit and its somewhat lower initial cost are additional advantages enjoyed by the user of the large granulator. The choice of a granulator type should thus be made only after a careful consideration of all these relevant factors.

Regardless of the type of granulator chosen, either one may be incorporated into a vacuum hopper loader system, so that a completely automatic, "closed-loop" system (not to be confused with "closed-loop" control) can be achieved. In order to have a completely automatic system, the injection press is usually fitted with some type of parts separator—runners and sprues go to the grinder and parts are deposited in a bin or a conveyor belt. Granulators are available with various types of knives and sizes so scrap as thin as film and small parts may be reground, or the machine may be of a size to regrind large pipes, refrigerator doors, or similar massive parts. One requirement common to all types and sizes of granulators is that they may not exceed the OSHA sound-level requirements. At the time of writing, the maximum was 90 dB. In judging sound levels it must be noted that two 90 dB machines operating 6 feet apart would expose the operator to 93 dB. In planning an installation, this factor should be taken into consideration, either by obtaining machines with an 85 dB maximum, or fitting the units with sound-absorbing housings.

Color Mixing. To obtain various colors in the molded plastics parts, the molder has the choice of buying colored pellets or coloring clear material. The latter method is obviously more economical, since stocking a small number of color concentrates and a clear material is much cheaper than attempting to maintain a large stock of various colored

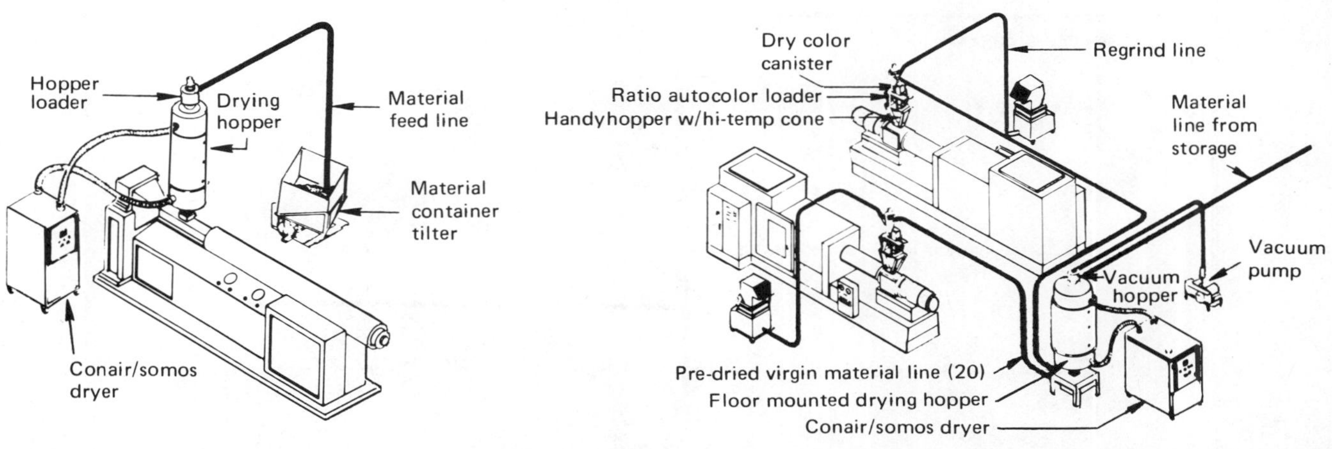

Fig. 10-45. Automation at the injection molding machine. Drying hoppers and hopper loaders work together to provide a fully automated material delivery and preconditioning system. Dry colorants can be used on hygroscopic materials without blowing off the color and creating dusting problems by simply pre-heating and removing moisture before the color is added at the machine hopper. Colored regrind can be processed as it becomes available by adding it to virgin material using automatic proportioning and mixing equipment, as illustrated. (*Courtesy: Conair, Inc.*)

materials. The original method of coloring clear materials, when ram injection presses were used, was the use of mixing drums. With the introduction of the screw machine, it became possible to add the color concentrate in with the bulk of the material in the hopper, and then depend on the screw to intensively mix the two. A newer technique is the use of liquid colorant which is automatically metered to the machine as the screw rotates. As shown in Figure 10-46, the colorant is introduced at the screw, and thus does not require contamination of a whole hopper full of material when only a small amount of colored plastic is required. (See Chapter 19 for a fuller discussion of coloring techniques.)

When the mixed colors are required in very large amounts (700 lb to 4000 lb at a time), other types of mixers are available which operate apart from the molding machine. In this case, the two materials are mixed in large mixing chambers and the mixed stock is then stored in drums or silos.

Part Extractors and Conveyors. In line with the trend toward automation in larger mold shops, and some small shops, part extractors are available which insure the complete removal of all parts and runners from the mold. The use of the extractors then minimizes the need for operators. In addition, the part extractors can reduce the overall cycle time by elimination of the time to open and close the press safety gate. Assuming the machine operates with a 15-second cycle and that it takes 3 seconds for each opening and closing of the gate, a saving of 1200 hours per year can be realized if the machine operates 24 hours a day for 250 days a year. This saving, combined with the saving in labor cost, can be appreciable. The extractors can be air, hydraulic, or electro-mechanically operated. The extractors either use jaws and various types of grippers, or simply a stiff bristle brush which sweeps the parts and runners out. The extractors are usually specially "tailored" for each machine or mold.

In order to further the automation concept, if the operator has been eliminated, then it is natural to also install a conveyor system on the press. The conveyor will then remove parts from the machine and deposit them directly in the inspection or shipping area. Available conveyor systems range from simple portable belt conveyors which are placed under a horizontal machine's part-drop area, to elaborate conveyors which accumulate parts from several machines and deposit them in one work area. Other types of conveyors are available which use plant air or vacuum to transport parts through tubes. These latter systems are similar to the units which are used with the vacuum hopper loaders, only operating over somewhat longer distances.

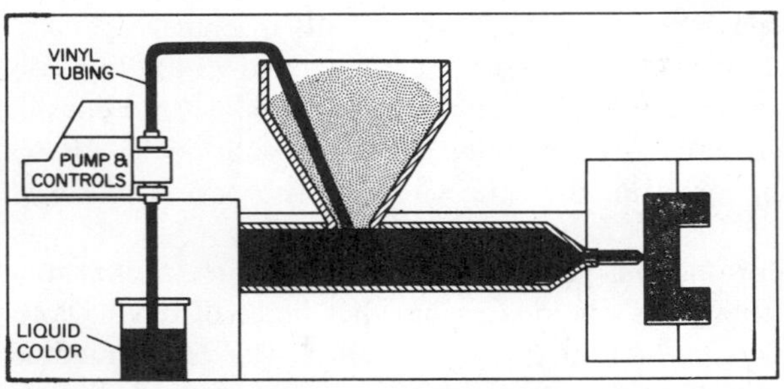

Fig. 10-46. Resin coloring by liquid colorant introduction. (Reprinted by permission from *Plastics Technology*, November 1971)

The Mold Cycle

The mold cycle for a typical screw injection molding machine is shown in Figure 10-47. As the figure reveals, the mold cycle is made up of a number of elements, any

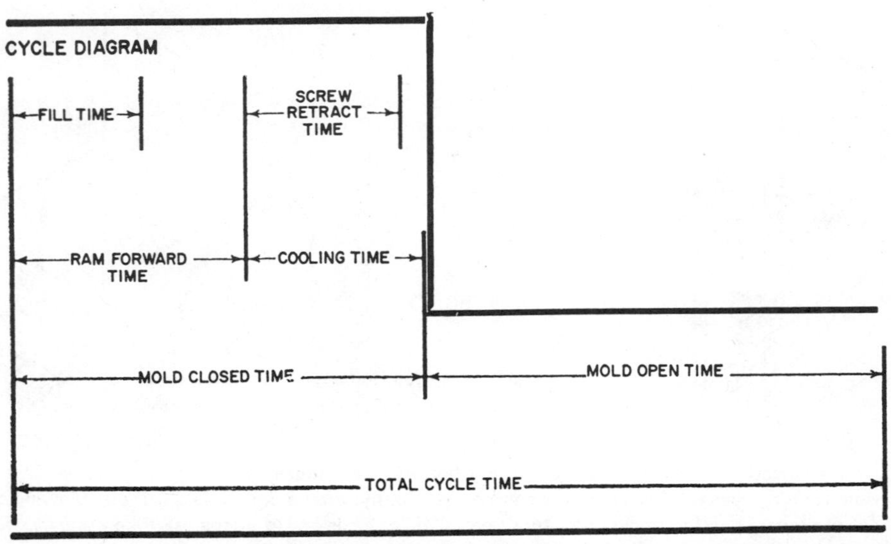

Fig. 10-47. Reciprocating screw injection mold cycle. (Sherlock, 1970. Reprinted by permission from *Plastics Technology*)

one of which may possibly be shortened, or accelerated, thus condensing the entire cycle. Some of the steps which can be taken to compress the cycle are described below.

Preparation of a Mold Diagram. To prepare a mold diagram, a map is made showing the machine conditions needed for the production of satisfactory parts in a minimum of time. Maximum and minimum pressures for various cylinder temperatures are determined. The mold temperature and the other elements of the mold cycle are kept constant. As shown in Figure 10-48, starting at a low temperature and a low pressure, a series of short shots will be obtained. By increasing the pressure, several satisfactory shots will be produced. At some point, as the pressure continues to be increased for each shot, the mold will begin to flash or stick. This then becomes one point on the flash or stick line.

A second series of points is generated at a higher temperature. This procedure is continued with successive temperature increases until a temperature is reached at which obvious polymer deterioration is taking place. By connecting all the last short-shot points, a curve is drawn which indicates the minimum pressure-temperature combination at which satisfactory parts can be obtained. Connecting all the flash-stick lines also indicates the maximum pressure-temperature combinations which can be used to produce satisfactory parts. The area between the curves is then the mold area within which, theoretically, any combination of pressure or temperature should produce good parts. The larger the area, the more latitude there is for any given material. Note the specification that "theoretically" all good parts will be made if the molding conditions are kept within the molding area. In actuality, some of the full parts may show streaking, bubbles, or other defects. If a record is kept of the conditions of molding each piece, it may be found that some defects are consistently present when parts are molded in one particular area of the diagram. Thus, this portion of the mold diagram may be established

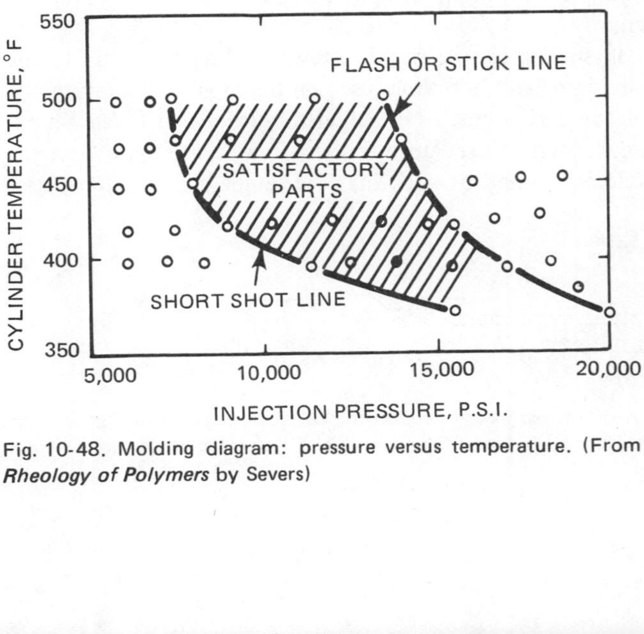

Fig. 10-48. Molding diagram: pressure versus temperature. (From *Rheology of Polymers* by Severs)

as a "forbidden" area. Other areas may be found which consistently produce good parts. The optimum area would be the one characterized by the highest temperatures and pressures necessary to produce parts which are completely satisfactory. The use of the highest temperatures and pressures would then result in a minimum fill time, shown in Figure 10-47.

The object of producing a minimum cycle time is, of course, to produce parts with maximum efficiency. The development of the molding diagram is only part of the process by which the molder can determine the conditions needed to produce the shortest overall cycle time for any given part. The mold diagram, once developed would only establish the optimum pressure-temperature combination. Other tests would then be made to determine the shortest times for the other elements shown in Figure 10-47.

In order to discover the conditions for a minimum cycle, one would start, at a given mold temperature, with the highest pressure and cylinder temperature which would result in satisfactory parts. The next step is to make a series of shots, at a constant pressure-temperature combination, but decreasing the ram-forward time. At some point, a time will be found below which no satisfactory parts can be made. During the tests, the cooling times should remain constant. After establishing the minimum ram-forward time, the same procedure should be repeated by decreasing the cooling time, with the ram-forward time remaining at the constant minimum. This would establish a minimum molding cycle for any given cylinder and mold temperature. The procedure should then be repeated with either higher or lower mold temperatures, as indicated, until an optimum is reached.

The establishment of an optimum fill time, ram-forward time, and a cooling time affects the total time the resin remains in the cylinder during transit. The net effect should be a lowering of the time in the cylinder, so that possibly a higher cylinder temperature may be feasible. Another series of test shots should be made at various increased cylinder temperatures, establishing a minimum cycle for each cylinder temperature. The criteria in each case will be production of satisfactory parts, and the limits would be either short shots, from below the minimum cycle times, or deteriorated parts which were shot at excessive temperatures and/or excessive times. A plot of the minimum cycle times versus the cyclinder temperatures would look like Figure 10-49.

Like the pressure-temperature molding diagram, the molding area in the minimum cycle diagram may also have some "forbidden" areas because of production of consistent defects. In general, creating a minimum cycle diagram, while seemingly quite tedious, should prove very valuable, particularly when establishing conditions for a new mold or a new resin. With familiarity, shortcuts will be used by the experienced molder or setup man who, in any case, has been using such a procedure instinctively.

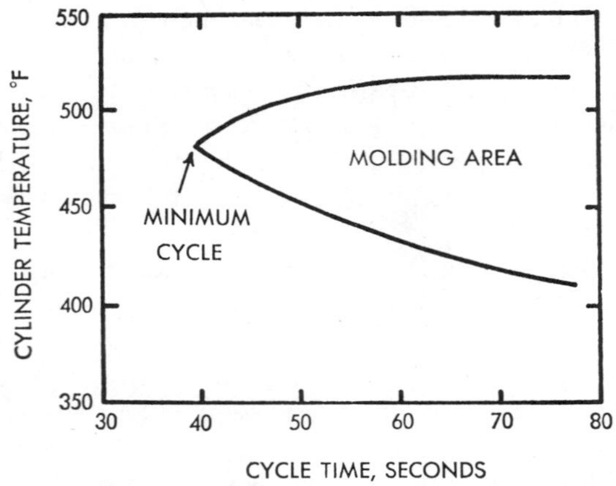

Fig. 10-49. Minimum cycle diagram. (From *Rheology of Polymers* by Severs)

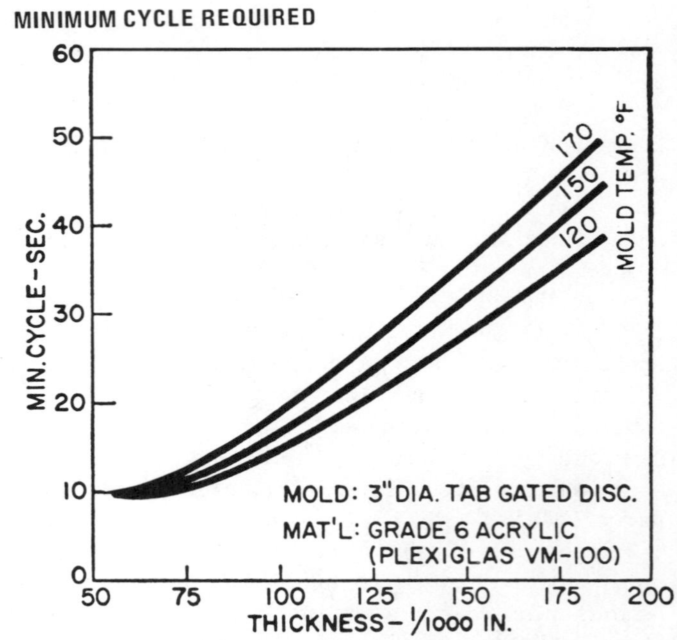

Fig. 10-50. Minimum cycle required to mold warp-free parts of various thicknesses using different mold temperatures. (Sherlock, 1970. Reprinted by permission from *Plastics Technology*)

In addition to developing minimum cycles in the machine-process conditions, it is possible to further reduce the molding cycle by the proper design or redesign of a part. By designing for a minimum cycle time at the outset, it is possible to produce parts at the lowest price or the highest profits. There are a number of specific factors in the design of a part which have an influence on the time required to mold it.

The first, and most important, consideration is the thickness of the part; the thicker the part, the longer the cycle time. Figure 10-50 shows the relationship between the thickness and the molding cycle time. It should be noted that a reduction in part thickness from $\frac{3}{16}$ inch to $\frac{1}{8}$ inch results in a cycle-time reduction of from 40 seconds to 24 seconds at a mold temperature of 124°F. The diagram also shows the importance of using a minimum mold temperature—note the time difference between 120°F and 170°F.

Another consideration relating to thickness is the importance of uniformity of thickness in the molded part. Since the part is usually molded at the cycle time required for the thickest section, it is obvious that a few thick sections in a part can be responsible for an increase in the total cycle which is not commensurate with the amount of thick sections in the molding. Figure 10-51 shows how, by proper redesign, a part may be made at a lower cycle time without appreciably weakening the part. (See also Chapter 21.)

Other important factors in obtaining a shorter cycle include sidewall draft and generous radii. By using a sidewall draft of not less than $\frac{1}{2}°$ per side and rounding the corners, the part can be ejected while still somewhat hot, with minimum sticking and corner or edge breakage.

In the mold design, a number of steps can be taken to assure the fastest cycle without any change in the part design. Cooling lines in the mold should be overly generous so that not only the part, but also the runners will be cooled and thus assure easy ejection. At the same time, a generous number of ejection pins should be provided so the part can be removed with minimum sticking or warpage when the molding is removed while still fairly hot. Since the plastic will be injected as rapidly as possible, venting of the cavity must be provided so that the air will be removed quickly to prevent burning of the plastic. Ribs, sidewalls, and the body of the part should be polished in the direction of draw to minimize sticking during ejection and possible breakage or cracking.

In setting up the mold, attention should be paid to the circulator and hose lines used on the mold-temperature regulator. The circulator or chiller should have sufficient pumping capacity to set up turbulent flow in the mold, and the velocity should be such that the temperature of the outlet

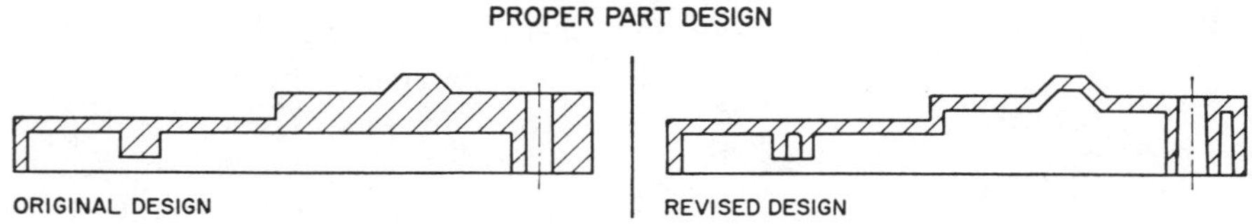

Fig. 10-51. In this cross-sectional view of a thermoplastic housing, note how in revised design, section thicknesses are smoothed out and the massive rib is reduced to two smaller ones, enabling far easier molding at shorter cycle times. (Sherlock, 1970. Reprinted by permission from *Plastics Technology*)

water will be within 3°C (5°F) of the inlet water or coolant temperature. Hoses should be as short as possible to obtain maximum efficiency from the circulator. Long hose lengths cause pressure drops in the system and absorb heat from the air rather than the mold.

Referring back to the mold-cycle elements shown in Figure 10-47, the first element, fill time, can be reduced to a minimum by correct choice of gate size and location. Gates which are too small can give rise to jet marks, splay, or degradation when injection speeds are high. Gate location can sometimes be chosen to minimize the effect of the gate markings, by having the jet impinge on a noncritical appearance area.

The second element, ram-forward time, depends on the time required for the gate to solidify enough to prevent any backflow from the cavity. For any material, this time depends on the part thickness, the gate size, and, to some extent, the runner size. Here, it is important to use the minimum size gate which will still adequately fill the mold, since this is the governing factor. The mold should be built with the smallest gate initially; after trials, it can be enlarged if required.

The screw-retract time, the third key element, should ideally be slightly less than the cooling time, as shown in Figure 10-47. In the event that the screw-retract time is too long, it may be possible to reduce the backpressure slightly without affecting the melt, or it may be possible to increase the screw speed. A third alternative is to place the mold in a molding machine with a larger shot capacity, thus the screw would have less distance to travel, with a consequent shorter travel time. Associated with the lower screw-retract time is a mold temperature which is kept at a minimum in order to result in the shortest cooling time.

Last, the mold-open time can be reduced by using a mold setup which assures a minimum distance for the mold to open—just enough so the part can be ejected or removed easily. Furthermore, the clamp-slowdown switches should be activated at the last safe point, not several inches from mold closure.

One technique used to reduce the molding cycle time is the inclusion of 1% to $2\frac{1}{2}$% of #200 Dow Corning silicone fluid mechanically mixed in with the molding pellets. With polystyrene pellets, it was possible to boost the mold-fill rate and inject at a slightly lower temperature, which produced a reduction in molding time of up to 25%.

Reducing the mold cycle to its minimum is not without some drawbacks. First, the total shrinkage of the part will be somewhat higher than if a longer cycle is used. This may cause serious problems if the part is one component of an assembly, or if extremely tight tolerances are required on dimensions. The second factor which should be considered is that a molding made at a minimum cycle may have much higher molding stresses locked in than the same part made at a longer cycle. Thus, to assure adequate and satisfactory service life, such parts should be tested to determine that the part's usefulness has not been jeopardized. The tests may consist of simple heating, which may reveal warpage, or, depending on the material, coating the part with a moderate solvent for the polymer, which may show up cracks or crazing. In each case, the part made at the minimum mold cycle should be compared with a similar part made with a longer cycle.

Special Techniques

The descriptions of the molding operations, presses, and molds have been concerned up to this point with so-called conventional injection molding. In this process, dry pellets are heated, liquefied, and then injected into a mold containing a sprue, runners, and gates. After the molten plastic has cooled, the part is ejected with the sprue, runners, and gates which are subsequently granulated and remolded or, in some cases, discarded. There are, however, a number of variations on this process which are used to increase production or quality or both.

Runnerless Molding. Runnerless molding is not, as the name appears to indicate, molding without runners. Rather it is the use of special mold designs in which the runners and sprues are kept in a fluid state and are never actually ejected with the molded parts. There are a number of these different mold configurations with names such as "hot runner mold," "hot manifold mold," "insulated runner mold," "internally heated hot runner," etc. They are all similar in that in each case the mold contains runners and/or a sprue to feed molten plastic to the part cavity; however, they differ in the manner employed to keep the runners and sprue molten.

Hot Runner Molds. The concept of runnerless molding originated in the late 1950s. Tests were made using heaters in the runner plate of a three-plate mold so the runner-sprue system would remian molten and the mold need not be opened at this plate. At the same time, an air-gap insulated nozzle at the cavity gate was supposed to be kept cool enough so the plastic would freeze at this point after injection and thus break off and allow ejection of only the part. Figure 10-52 illustrates this original design.

It was soon found that temperature control of the insulated nozzle was extremely critical. If the nozzle became too hot, resin would drool when the mold was open, and if it was allowed to become too cool, the gate would freeze. Either condition would cause problems with the next shot. Production, particularly with multicavity molds, was very nonuniform. Despite the poor production record with the early designs, mold makers continued their research—a satisfactory runnerless mold could result in a number of cost savings and other advantages to an injection molder.

Insulated Runner Molds. Soon after work began on the hot runner molds, it was found that, when working with

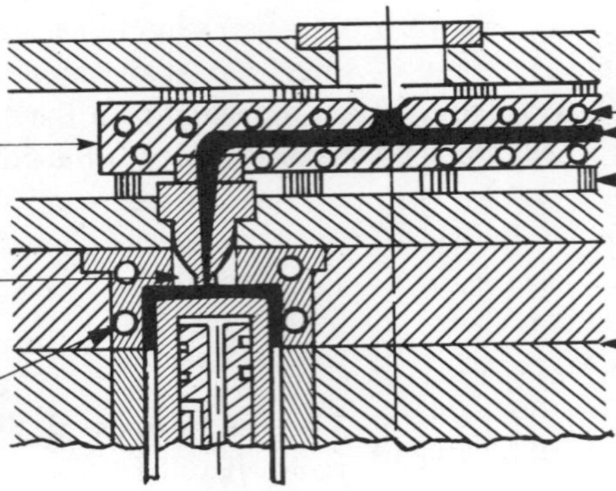

Fig. 10-52. Hot runner (hot manifold) mold.

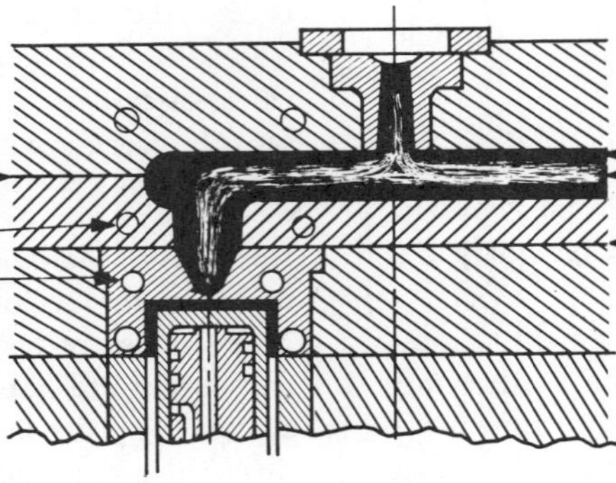

Fig. 10-53. Insulated runner mold. A runnerless mold in which no additional heat is supplied to the runner. Wide-diameter insulated runner allows the outer circumference of the melt to solidify against runner walls while the central core remains molten in preparation for the next shot.

polyethylene, the high insulating qualities of the resin could be used to produce a new type of runnerless mold. Runner systems were increased from $\frac{1}{4}$-inch diameters to as high as 1 inch in diameter. After a few shots with such large-diameter runners, an outer shell of the runner freezes to the mold wall, allowing an inner, molten liquid portion to pass through freely. Figure 10-53 shows just how flow occurs and clearly illustrates why this mold configuration is called the "insulated runner" design. This system, while working fairly well with most olefinic resins and some polystyrenes, still had the serious disadvantage of requiring very accurate control of the gate temperatures in order to have satisfactory production rates. Another disadvantage of the insulated runner system is that in the event of prolonged delay in the molding cycle, the runners may freeze up completely, necessitating a separation of the runner plate so the runners can be removed. Nevertheless, since such molds are relatively inexpensive, they are still in use, particularly if modified as described below.

Heated Gate Feed System. The next logical development was a means for introducing the melt into the gate under controlled temperature conditions. A number of slightly differing components called "hot tip bushings," "heated probes," "mold probes," are available from several manufacturers. In each case, the probes are inserted into the mold at the end of the heated runner (see Figure 10-54).

In some of the bushing or probe designs, the probe is hollow and the resin passes through it to the cavity gate; in so doing, it is heated by an internal heater. In some designs, the probe or bushing is simply a heater which is inserted into the end of the runner and is surrounded by molten resin. In almost all bushings or probes, in addition to the heater, the component will have a thermocouple either built in or the mold will incorporate a thermocouple very close to the gate. Figure 10-55 illustrates a typical "hot tip bushing" which contains an internal heating element and a thermocouple, and features resin passage through the bushing.

Other bushings or probes are made which, in addition to maintaining the desired gate temperature (always considerably lower than the hot runner), will also stop flow at the gate to effect a positive resin shutoff. This is accomplished mechanically by one design where a spring pushes the conical end of the probe into the gate opening when the melt pressure is reduced by the screw retraction.

Figure 10-56 shows one such mechanically actuated flow-stopping probe. Also shown is the newer technique of heating the runner passages using a heater tube containing a cartridge heater and a thermocouple. Such a system (known as an internally heated hot runner), contrasted with car-

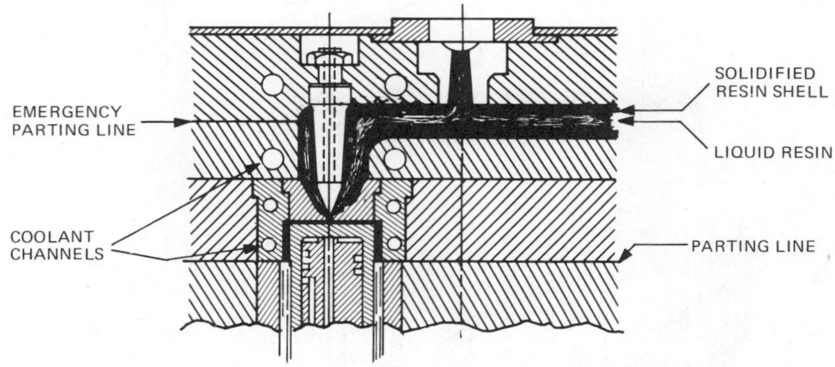

Fig. 10-54. Heated probe installation used in conjunction with an insulated runner.

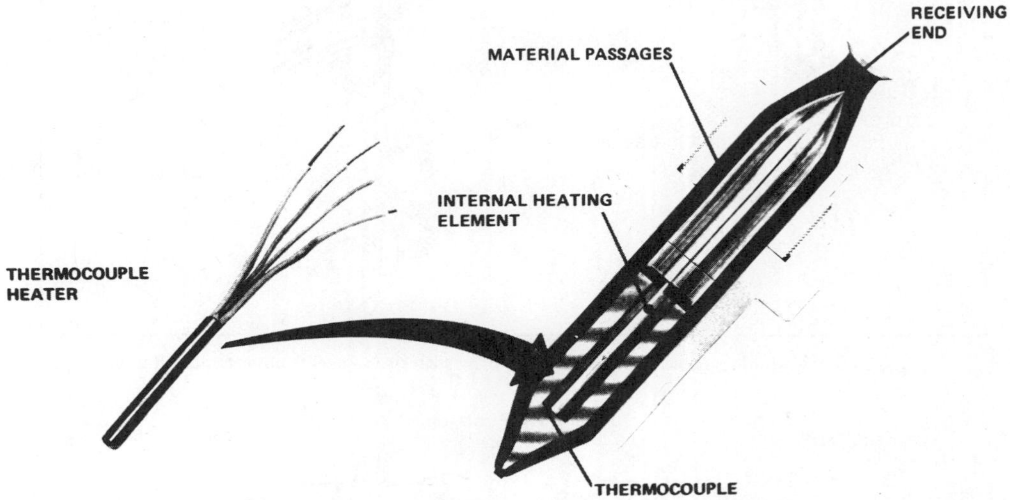

Fig. 10-55. Internally heated "hot tip" bushing.

tridges installed in the runner manifold, is said to result in considerable energy savings. Only the resin is heated, while the manifold, insulated from the heater by the resin, remains at a relatively low temperature.

The newest (at the time of writing) method of gate-flow control is by means of a "spear" probe (Spear System, Inc., Chatsworth, California). This type of probe fits into the exit end of a runner, similar to other hot bushings or probes, and is designed with two independently controlled heating elements. One element, in the body of the probe, maintains the melt at the hot manifold temperature. The other heater, at the probe tip (the spear), controls the flow of melt into the cavity.

At the start of each cycle, the solid resin is melted, thus allowing flow through the gate into the cavity. At the conclusion of injection, the spear tip heat is turned off, and, due to the cooling channels in the mold near the gate, the melt is quickly frozen solid again. During the clamp-open time in the mold cycle, the tip preheats the melt to just below the melting point, thus reducing the cycle time for remelting. Figure 10-57 illustrates the action.

In order to obtain automatic cycling with the "spear" probe, it is necessary to use electronic controls. Thus, the complete system consists of the probe or probes and an electronic control system that consists of timers for the tip on-off cycle and high-heat and preheat cycles. Current regulators for the tip high-heat and preheat cycles and for the body heater must also be provided. In addition, the control system ties in the manifold heater to the body heater so a uniform runner temperature is maintained at all times. Figure 10-58 shows a portable control system connected to a six-probe runnerless mold. (For additional details see Chapter 20.)

Regardless of the type of probe, or whether a heated manifold or insulated runners are used, it is important that the gate-feed nozzles, probes, or bushings be controlled by some type of closed-loop thermal control system if satisfactory results are to be obtained. Systems similar to the "spear" control system, but possibly not as elaborate, are available for all other probe types.

When a mold is used containing a single cavity, heated and temperature-controlled sprue bushings are available which can be used to feed directly from the nozzle to the cavity, with complete elimination of the sprue. For simple,

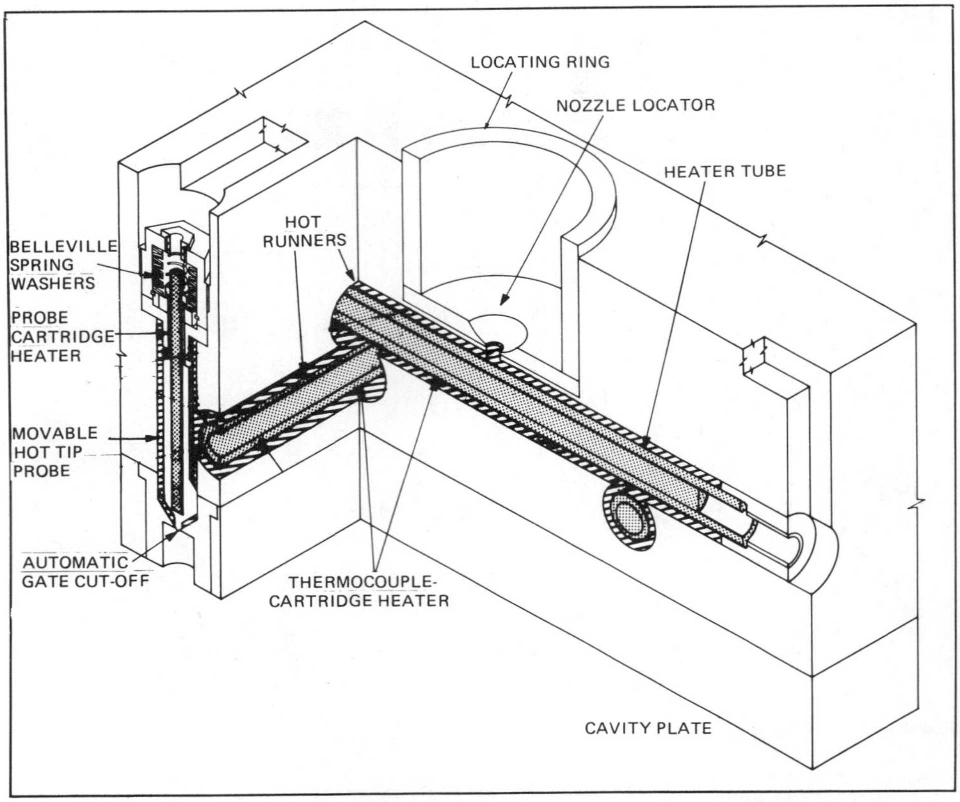

Fig. 10-56. Runnerless mold using automatic flow-stop probe and tubular runner heaters.

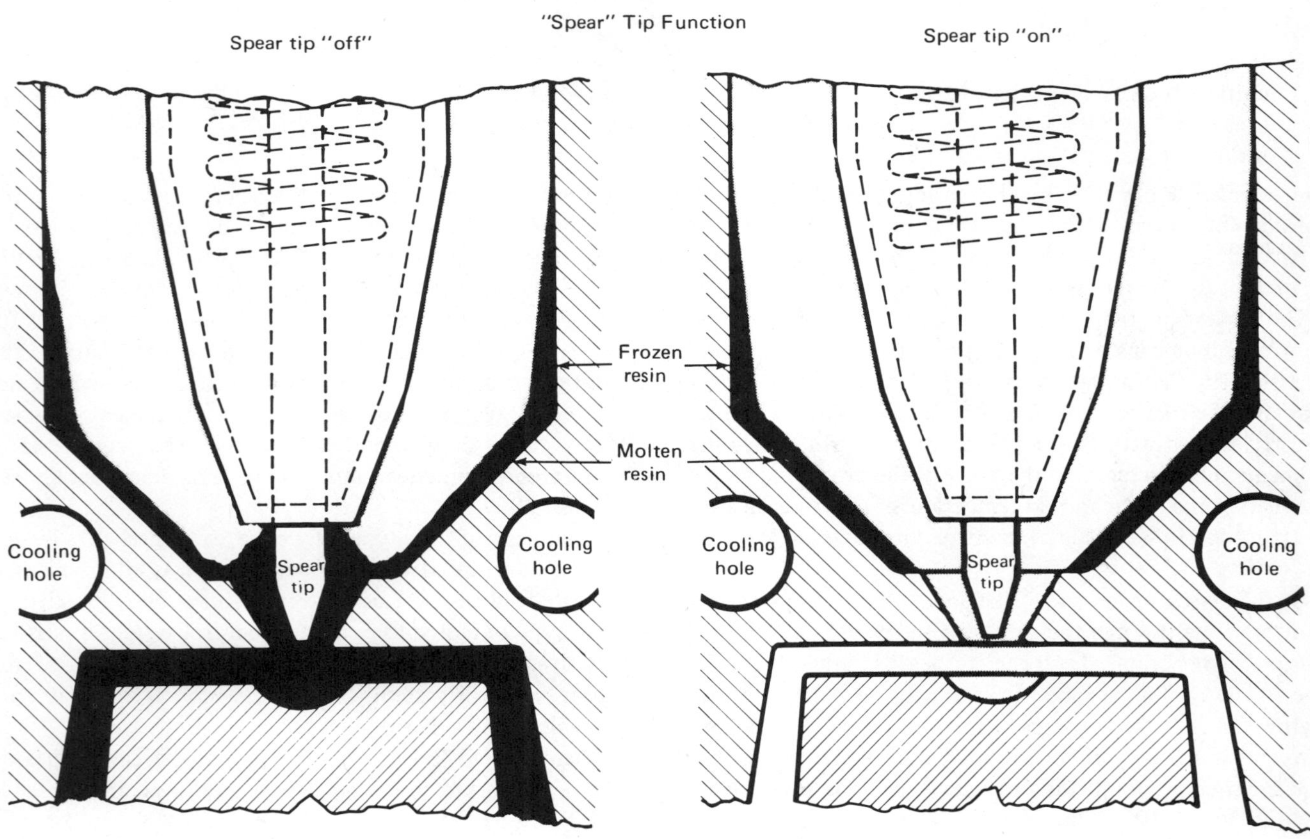

Fig. 10-57. "Spear" tip action.

Fig. 10-58. "Spear" tip system connected to six-probe mold.

noncritical molded items, it is possible to control the bushings manually by variable voltage transformers. The control of hot manifolds is also, in many cases, done manually, since the need for critical control is not as great as with the gate-feed systems. Temperatures should, of course, be indicated by means of thermocouple probes and pyrometers.

Advantages of Runnerless Molding. There are a number of distinct advantages in the use of runnerless molds, particularly if equipped with hot-tip gate feeds and adequate temperature-control equipment. These advantages include the following.

1. Hot runners cut costs. There are savings in raw materials not used for production of runners. Other significant savings result from elimination of scrap regrinding, with the possibility of mixing in dirt, burned resin, etc.
2. Cycles can be shorter by elimination of the runner cooling time (often the heaviest section molded).
3. More cavities may be possible in a mold since full plasticating capacity of the machine may be used for parts by elimination of the runner system.
4. Runnerless molds can operate at lower pressures and temperatures than a similar cold runner mold. With lower pressures the parts show less residual stress, thus resulting in better dimensional stability over long periods of time. Lower plastics temperatures allow shorter cooling periods, with increased productivity.
5. With full controls, the system is particularly well suited for automatic cycling.
6. Parts made with minimum variations in pressures and temperatures result in better part quality and a reduction in sink marks and flow lines.
7. Because of the precise control of temperature possible at the gate, the system is claimed to be particularly well suited to molding of engineering materials and heat-sensitive resins such as polycarbonate, nylon, acetal, and PVC.

Stacked Molds. Another technique for increasing production is the use of stacked molds, also called two-level or double-deck molds. Stacked molds, as the name indicates, are essentially two molds fed by a single sprue. Figure 10-59 illustrates the concept. The illustration shows two two-plate molds; however, the more modern molds are built as runnerless molds using controlled-temperature hot-tip bushings at each cavity, since such a construction results in much more uniform parts production. It should be noted that doubled production from a stacked mold is not actually achieved, since opening and closing the mold is somewhat longer than a single-level mold. However, increases of 80% are routinely being achieved. It should also be noted that even though approximately double the number of parts are being made at each shot, the projected pressurization load in the mold is the same as with a single-level mold. Thus, the same machine could be used for a 12- or a 24-cavity mold. On the other hand, a smaller high-speed machine could be used with a stacked mold to produce 16 parts per shot on a faster schedule than the larger machine producing 24 parts per shot. The savings in cycle time, machine costs, and a reduced mold cost might make its use preferable. The concept has been found to be most suitable for production of small flat objects such as lids, tape recorder cassettes, and shallow boxes.

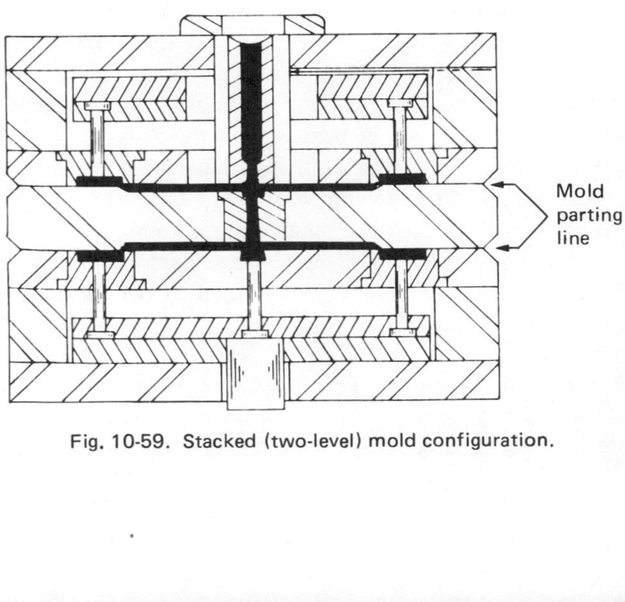

Fig. 10-59. Stacked (two-level) mold configuration.

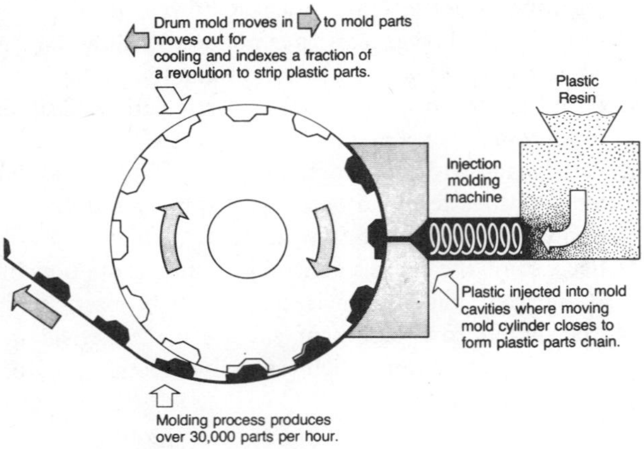

Fig. 10-60. Continuous chain injection molding.

Continuous Chain Injection Molding. Considerably increased production rates (up to 30,000 parts per hour) are claimed for a proprietary process patented by the American Can Company. In this technique (see Figure 10-60), a rotary mold is used and parts are ejected as a continuous chain, which is later broken up or fed to machines for further processing. The entire process can be completely automated for small precision parts and, with the use of hot runner molds, is said to result in considerable savings over conventional methods.

Multicolor Molding. Multicolor molding is used for applications such as typewriter keys and automobile taillight clusters. The original processes used to produce such objects first molded those whose color had the highest melting point. Then, after transferring the mold to another machine, the second color would be molded where the cavity insert had been removed. Current practice is to use a rotatable mold in the machine, which is now equipped with two injection screws, but one nozzle. By the proper valving, mold manipulation, and temperature control, it is possible to mold parts with two and three colors.

In multicolor molding, it is necessary that the two materials be mutually compatible. It has been found that if two colors or materials are butt-welded together the resulting bond is not very strong. Thus, in the design stage it is always well to consider a design in which the final material completely surrounds the first material so that shrinkage will help to give a mechanical lock to the assembly. In the event that such a design is not feasible then consideration should be given to molding the two colors first, with the relatively weak butt-joint, then molding an overlay of a transparent material over both parts. Such a system has been used satisfactorily for production of two- and three-color automobile taillight assemblies.

Reaction Injection Molding. Reaction injection molding (RIM) is a process in which two low molecular weight, highly reactive, low-viscosity liquids are injected at a high pressure into a small mixing chamber and then into a mold cavity. In the mold, the polymerization reaction takes place as the part is formed. At the time of this writing, the liquid components are almost exclusively the polyols and isocyanates (or prepolymers) used to form polyurethanes. It is expected, however, that in the future the process will be used in the molding of parts made of other liquid monomers such as polyesters, epoxies, and nylons. This high-pressure metering-mixing and injection process has also been called "Liquid Injection Molding" (LIM), and "Liquid Reaction Molding" (LRM), and, since the majority of parts made so far are elastomeric, it has also been labeled "Liquid Injection of Molded Elastomers" (LIME). A variation of the RIM process is known as RRIM, "Reinforced Reaction Injection Molding." The object is to produce materials with a higher modulus (stiffer) than can be attained in normal RIM formulations. The reinforcing materials include milled glass, mica, wollastonite, and a small amount of glass fibers. With the small-particle reinforcements, modulus increases up to three times the unreinforced materals have been obtained.

The RIM process offers a number of advantages over conventional injection molding. The primary one is the ability to mold large, complex polyurethane parts quickly and with relatively low-cost tooling, as compared to "conventional" injection mold tooling. The parts may be solid elastomers or microcellular in structure. The usual polyurethane characterictics of high abrasion resistance, light weight, structural strength, corrosion resistance, sound deadening, and colorability are present. Parts as large as 100 pounds in weight can be molded in a single shot in 2 to 3 minutes. The actual throughput ranges from 100 pounds per minute to as high as 660 pounds per minute. Figure 10-61 shows typical RIM parts.

The process for RIM (or LRM, etc.) is shown diagrammatically in Figure 10-62. The five major equipment components are:

1. Storage tanks. Capable of being maintained at a con-

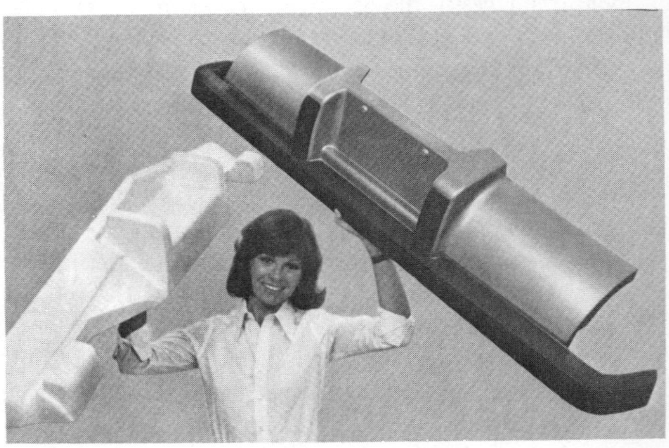

Fig. 10-61. Typical RIM parts. Maximum weight saving in soft-face systems is achieved with thin-walled RIM urethane fascia masking an energy-managing urethane foam bumper.

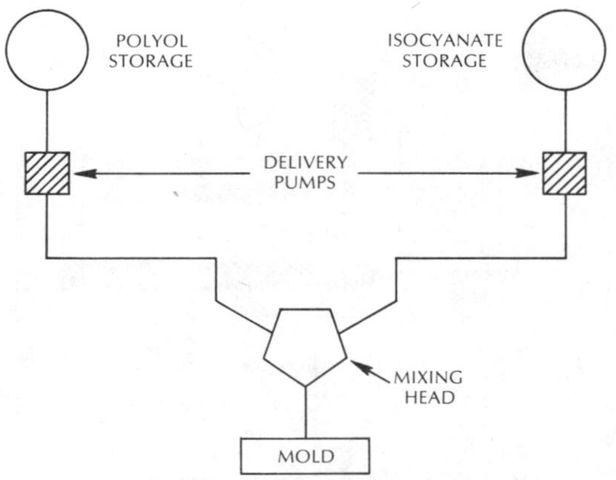

Fig. 10-62. The RIM process. (*Courtesy: Union Carbide Corporation*)

stant temperature of 27–38°C (80–100°F) and fitted for recirculation of fluids.

2. Metering units. High-pressure pumps capable of delivering the liquid materials to the mixing head at pressures of 1500 to 3000 psi. The pumps must also be capable of completely filling the mold within 2 to 3 seconds.
3. The mixing head. The mixing head is considered the most crucial piece of equipment, since in its small mixing chamber (1–6 cubic centimeters) the two or more liquids, at up to 3000 psi, must impinge and mix thoroughly before being transferred to the mold cavity. The mixing heads are usually self-cleaning, between injections, by means of the recirculating action of the pumps. Thus, no solvent flush is required to keep the mixing head clean.
4. The mold. Since the mold pressures are relatively very low (50–70 psi) molds can be made of aluminum, Kirksite, or even epoxy. For high production, the molds are usually of polished or nickel-plated steel. Because the mold pressures are low, the clamping pressures are correspondingly low. One departure from regular injection mold technology is that mold orientation is very important. The mold must be positioned so that the flow enters with minimum turbulence and so that rapid gas venting takes place. Depending on the compound, the mold temperatures range from ambient to a maximum of approximately 66°C (150°F).
5. The press. Since the process is designed for rapid production, it is ideal for automation. Thus, most presses are fully automatic in opening and closing of the mold, feeding resin, demolding, etc. Since the clamp pressures are low, 300-ton presses are usually sufficient for such large parts as automobile bumpers, furniture, business machines housings, and similar items.

The RIM process is still in its infancy compared to other plastics processes. Nevertheless, certain process generalizations can be made which can serve as guides in setting up or in troubleshooting. Some of these rules from *Advances in RIM Technology* by V. L. Johnson of the Admiral Equipment Company, are as follows:

> Higher temperature components will give lower viscosities, hence better mixing. The low viscosities, however, may also create turbulent flow conditions. Turbulent flow can be eliminated through sprue design or by lowering the total output of the metering unit which will give a longer injection time. Higher temperature components will also give a shorter cream time. Component temperatures should be as low as practical to obtain good metering and mixing. If the time for foam formation is too long for the output rate and the size of the part being molded, then the catalyst level in the formulation should be raised, keeping the catalysts in the same ratio. The maximum amount of catalyst in the formulation will give the fastest cure and shortest demold time. The maximum catalyst level will be determined by the injection time for the largest part. This injection time should be at least $\frac{1}{2}$ to $\frac{3}{4}$ of the cream time. Mold temperature has the greatest effect on skin definition and surface defects. Too high a mold temperature will give poor skin definition, too low a mold temperature will give increased cycle times, thick skins, and occasionally surface defects.

While the RIM process, as previously stated, is still relatively new, certain specific design rules have been developed by some processors. These are summarized as follows:

1. Wall sections should be as thin as possible ($\frac{1}{4}$ to $\frac{1}{2}$ in.) to minimize cycle time. Cross-sections can be varied within the part, but should not be less than $\frac{1}{4}$ inch. The thickest section of the part will control cycle time.
2. Rounded corners, rather than sharp edges, should be used when possible. Inside corners should have a minimum radius of $\frac{1}{8}$ inch to assure adequate skin formation.
3. Draft angles of 1 to $1\frac{1}{2}$ degrees should be used. Avoid undercuts unless side pulls can be incorporated into the mold. Cutouts or holes perpendicular to the ejection direction should be considered as undercuts.
4. Slots must be oriented parallel to the flow of the material and should be designed to facilitate flash removal. Grooves should be tapered at least $1\frac{1}{2}$ degrees to facilitate demolding.
5. Place the mold parting line in a location that permits maximum venting.
6. Dimensional control in RIM is mostly related to the quality of the mold and its dimensional accuracy. With careful process control, part dimensions can be maintained within ±0.1 to 0.2%, allowing for the characteristic urethane shrinkage of 0.4 to 0.6%.

Structural Foam Molding. Structural foam molding, as the name implies, is a process in which foamed structural parts are produced by injection molding. The process differs from the RIM process in that virtually any granular or pelletized thermoplastic resin may be used, instead of only liquid polyurethane components. It is similar to the RIM process in that the method is ideally suited for the fabrication of large, structural parts with high strength-to-weight ratios. Like the RIM tooling, structural foam tooling can be lightweight and fairly inexpensive, and is frequently made of aluminum. (For further information, see Chapter 15.)

Two processes are used to produce structural foam parts: the low-pressure process and the high-pressure process. The pressure differences refer to the pressures used on the injected melt to feed the mold cavity, and not to the final pressure in the in the mold, as is the usual case in referring to injection mold pressures.

Low-Pressure Process. In this process, a special press may be used, consisting of an extruder, one or more accumulators, and a means of clamping and unclamping the mold platens. An extruder is used to feed the accumulator(s), since in many cases the mold platens may range in size up to 10 × 12 feet or larger, and take a shot size of 100 pounds of resin or more. The plastic melt in the low-pressure process consists of granules to which have been added either a chemical blowing agent, or nitrogen gas which is fed directly into the barrel of the extruder. The extruder then mixes the blowing agent (which decomposes to liberate gas) or the nitrogen gas with the granules to produce a foamed melt. The melt is then injected into an accumulator which expands to absorb the full shot size, but at the same time maintains the plastic under enough pressure to prevent foaming. When sufficient molten material accumulates to form a part, it is injected into the mold cavity at a pressure of 2000 to 3000 psi, but this only partially fills the mold. The resin mass then rapidly expands, completely filling the mold at a pressure of 200 to 400 psi. The part then cools, and, while cooling, the accumulator is again being filled with a new shot of resin. After cooling, the part is ejected, and the process is repeated. Since this process is used mainly for production of very large parts, structural foam presses are commonly made with multiple nozzles—as many as 45 have been used for one mold—and as many as four or more accumulators are used on one press. The use of many nozzles insures complete fill of one large, intricate part, or many small parts may be molded at one shot. The parts may be of different sizes since the nozzles can be individually controlled. Because the final pressure in the mold is relatively low, compared to conventional injection molding, the total clamp pressure may be may be as low as 300 to 500 tons, even for the large platen sizes described. The molds may be fairly inexpensive aluminum or even plastic molds, where low production quantities are anticipated. Figure 10-63 illustrates the low-pressure process, using nitrogen gas as the blowing agent.

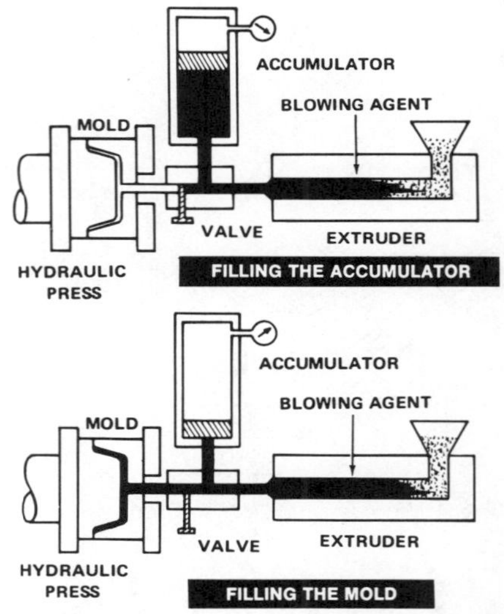

Fig. 10-63. Low-pressure structural foam injection molding. (*Courtesy: Union Carbide Corporation*)

While the description given above (and the diagram) applies to very large machines for injection molding structural foam parts, a conventional injection molding press may also be used for production of small versions of the same type of parts. In such a case, a suitable dry-powder blowing agent (chemical blowing agent or CBA) is simply mixed with the resin prior to placing it in the machine hopper. A short shot (experimentally determined) is then injected into the mold cavity, and the foaming melt is allowed to expand to fill the mold cavity. (For more complete information on structural foam injection molding, see Chapter 15.)

One of the major drawbacks to the low-pressure process is the difficulty in producing parts with a satisfactory surface finish, free of swirl marks from the injected and expanded foam. At the time of writing, a number of factors were being examined in order to determine the optimum techniques required to produce parts without the characteristic swirl marks. Some of the methods being recommended were: longer cycles, molds specifically designed for heating and cooling, fast mold fill, and venting the mold at least 30% around the sealing edge.

High-Pressure Process. In the high-pressure process, a chemical blowing agent is mixed with the resin prior to its being placed in the injection press hopper. The molten resin–gas mixture is then injected directly into a special type mold cavity at a pressure of approximately 15,000 psi. Enough melt is injected to completely fill the mold cavity. The mold then expands, allowing the center of the resin to foam, thus producing a part with a firm, solid skin and a foam core. Other high-pressure processes use an accumulator. Another process, also using either fixed-volume or expandable molds and two extruders, uses two resins to produce sandwich structure parts. Figure 10-64 illustrates the use of two resins, injected sequentially to produce

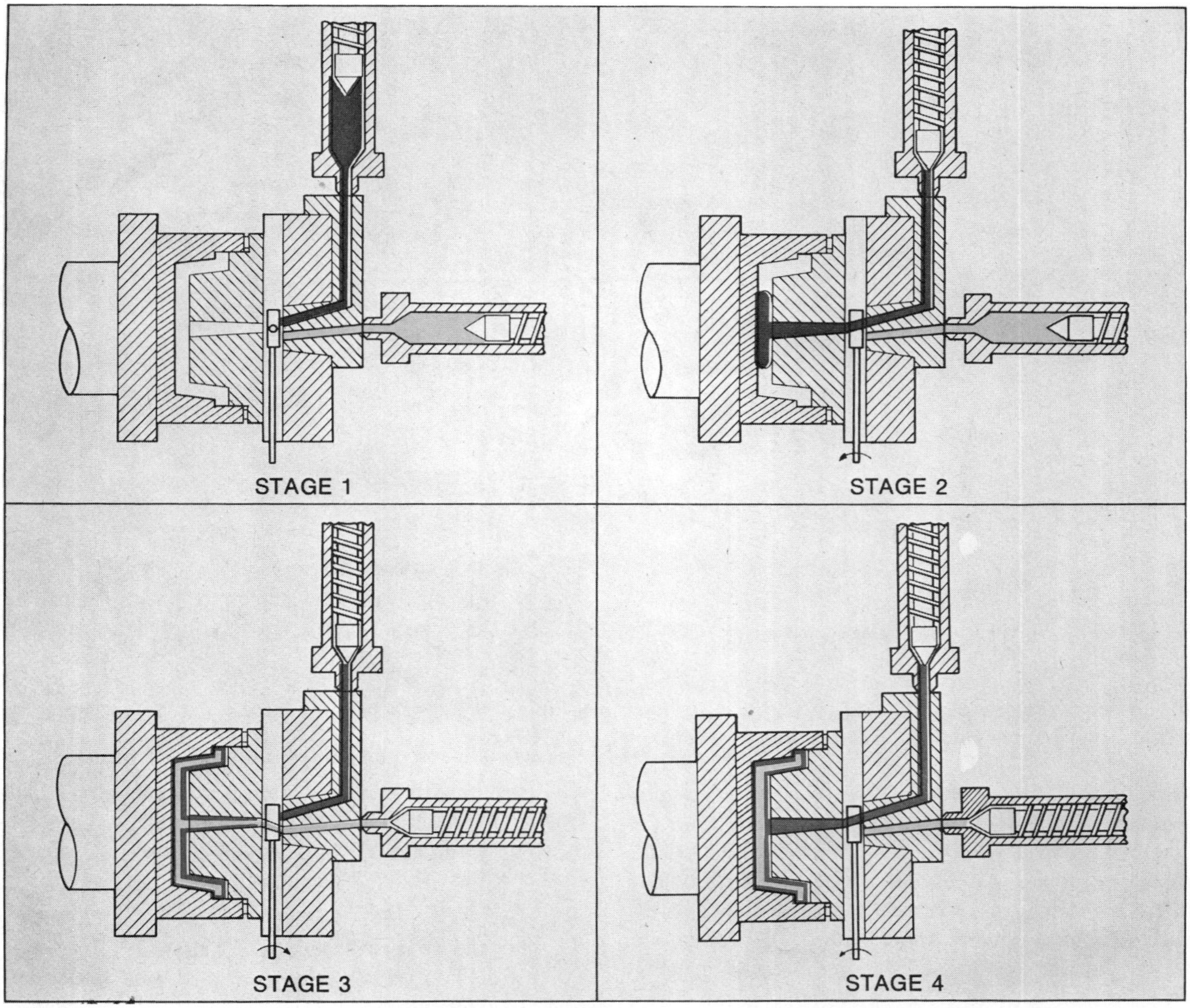

Fig. 10-64. Shown in four stages is the sequence involved in filling the mold in the ICI (Imperial Chemical Industries, Great Britain) sandwich-molding process. First the skin polymer is injected, then the core polymer is injected. Finally, the skin polymer is injected again to clear the sprue and seal the skin. (Courtesy: *Plastics Machinery & Equipment*, January 1978)

sandwich structure parts. A fixed-volume mold is shown, but the same process can also be used with an expanding mold. In either case, the major advantage of the high-pressure process, freedom from swirl marks, is achieved.

In general, the use of expanding molds for bicomponent molding will raise the mold cost 10 to 20% over a conventional mold, and the bicomponent injection press will be from 20% to 40% higher in price than a conventional press. However, substantial savings in materials and finishing costs may be realized compared to other methods of making the same type of parts. Material savings can come about by using an inexpensive resin, or even regrind, for the foamed interior. The savings in finishing costs can be realized in the absence of swirl marks, permanent colors, etc.

Another technique for the production of sandwich structures, utilizing a low-pressure process and conventional non-expanding molds, is known as the "co-injection process." In this method, two or three channel nozzles are used to inject two resins simultaneously. By injecting the two resins at very high rates, laminar flow is obtained and the core and skin resins remain separate. The result is a structural foam sandwich part with a solid skin, free of blemishes, which has been produced at a rate comparable to regular injection molding. Figure 10-65 shows a three-channel co-injection nozzle assembly. This is used to produce the same type of parts which might be made by the valving and sequential injection shown in Figure 10-64. However, there is a special advantage in the use of the three resins in the structure. The outer skin could be a hard, stiff material, not too compatible with the inner foam core. The intermediate resin could then be one which is compatible with both, thus resulting in a structural foam part which, because

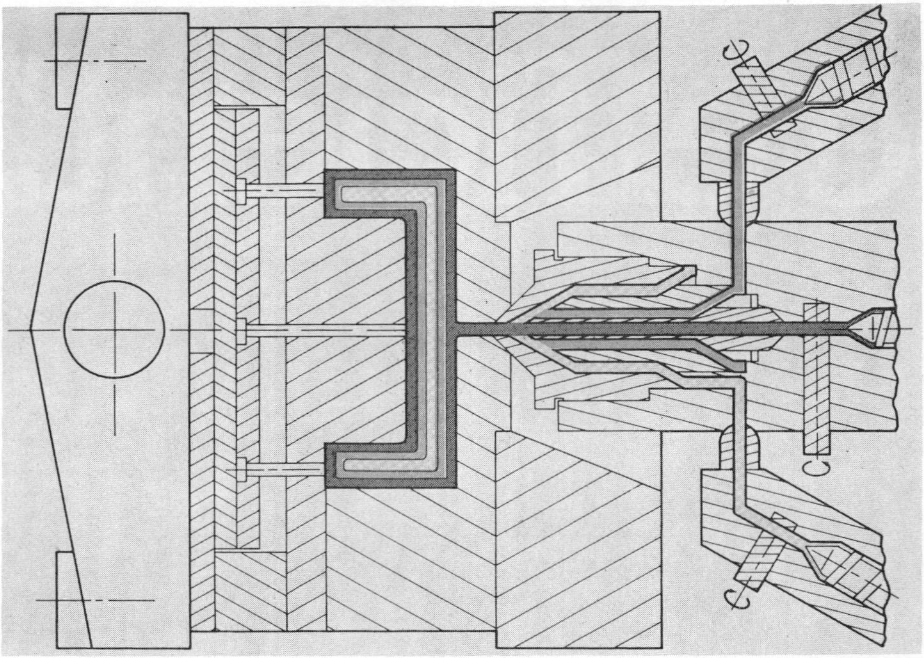

Fig. 10-65. Three-component co-injection nozzle used in production of structural foam parts.

of the better bond between the skin and core, is stronger than if just two components had been employed

Molding Press Selection Criteria. The selection of an injection molding press for any given job should be governed, to a large extent, by the economics of the job; that is, the press chosen should be the smallest machine which will do a satisfactory job. Such a press will probably be the cheapest to run in terms of energy costs, amortization costs, etc. However, the choice of a press for any given mold should be made on the basis of some logical conclusions, rather than simply trial and error. The following guidelines (supplied by Farrel Machine Group, USM Corporation), while somewhat "rule of thumb," can nevertheless be used as a basis for machine selection. Consideration, of course must also be given to the mold and part design, the material, and the cycle requirements.

RULE ONE

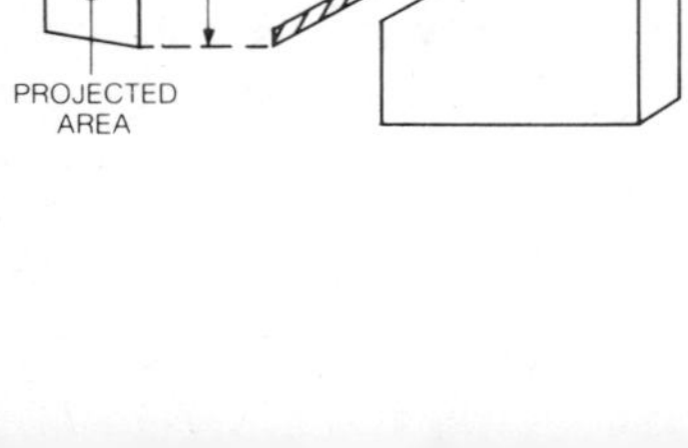

Projected area vs. clamping force

The machine clamping force (in *tons*) must be at least:

3 times the total projected area (in *sq. in.*) of all the molded parts and runners for a complete shot. This allows 3 tons per sq. in. of projected area for parts having thin walls, deep draw or thick bosses.

or

2 times the total projected area to allow 2 tons per sq. in. of projected area for parts with thick walls, shallow draw and no thick bosses.

RULE TWO

Part weight vs. machine injection capacity

The total weight of parts, sprue and runner system for a complete shot should not exceed 90% of the machine's rated maximum injection capacity for the particular material molded. The requirement for a high degree of melt uniformity may further reduce the usable shot size to 75% of its rated maximum.

Since machines are rated in ounces of styrene, the injection capacity of the machine should be adjusted for other materials according to approximate values.

To Convert oz. of Styrene	
To	**Multi. By**
H.D.P.E.	.90
POLYPROPYLENE	.85
ABS	.99
NYLON	1.08
PVC	1.23

EQUIPMENT NEEDED CONSISTS OF: 1) hot plate capable of maintaining surface temperature of 525 F ± 25° F, 2) 75 × 25 mm glass microscope slides, 3) tweezers capable of handling ⅛ in. pellets and 4) conventional wooden tongue depressors. Total cost for this equipment should not exceed $25. Following are the steps to follow in running the test:

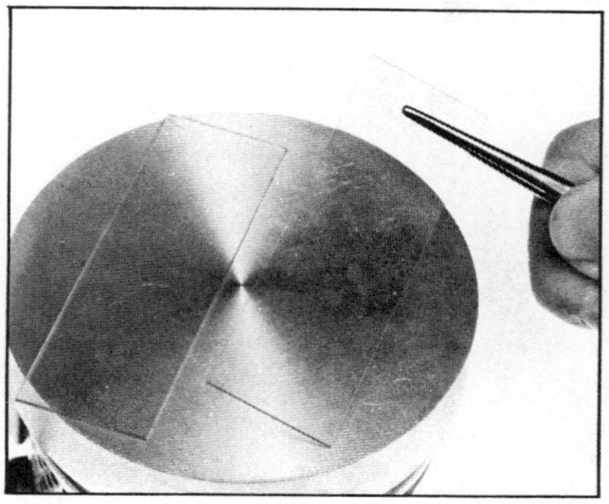

1. Plug in hot plate (be sure surface is clean) and calibrate it to a surface temperature of 525 ± 25° F. Place two glass slides on surface for 1-2 min.

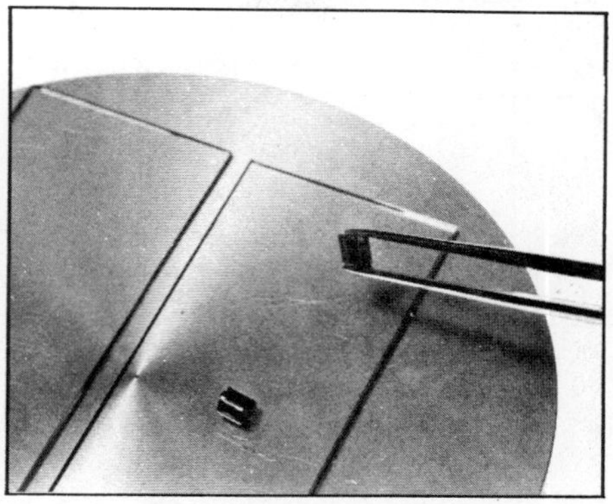

2. By this time the glass surface temperature should have reached 450-500° F. Use your tweezers to place four or five pellets on one of the glass slides.

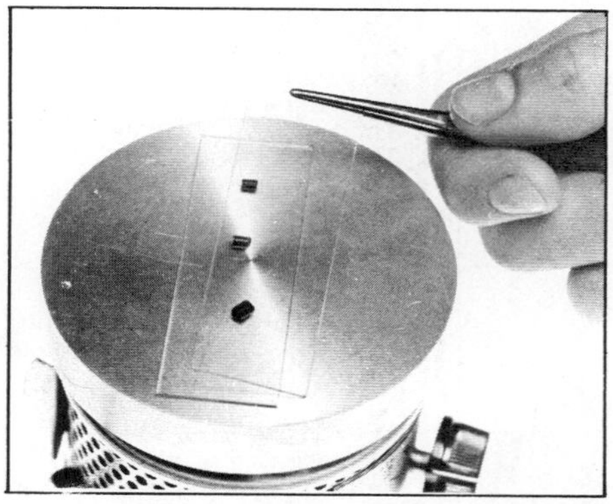

3. Now place a second hot slide over the first one to sandwich the pellets between them.

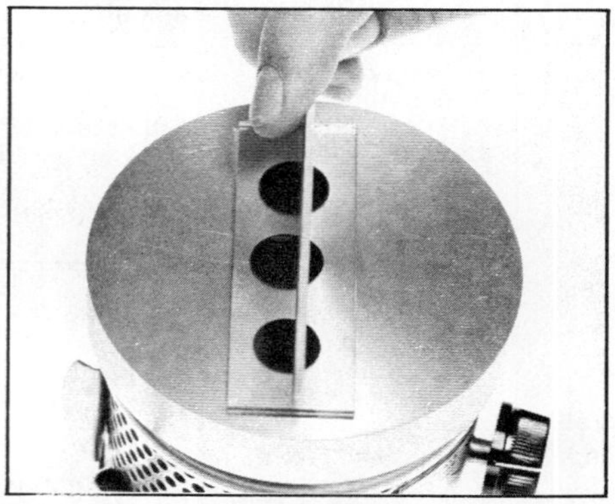

4. Press a tongue depressor on the top of the sandwich until the pellets flatten out to about ½ in. dia.

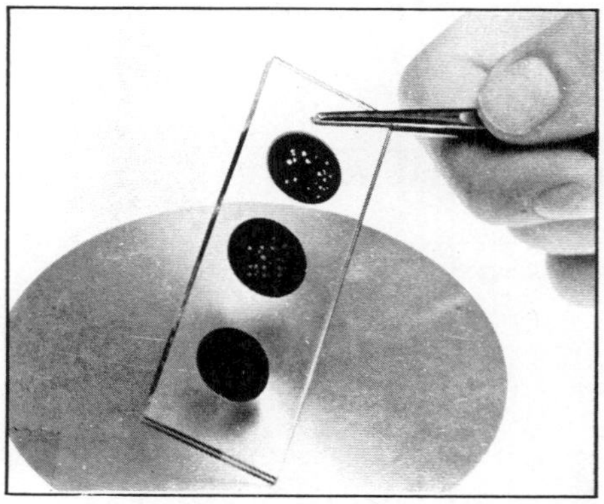

5. Remove sandwich and allow to cool. Amount and size of bubbles indicate percentage of moisture, correlating bubbles with moisture.

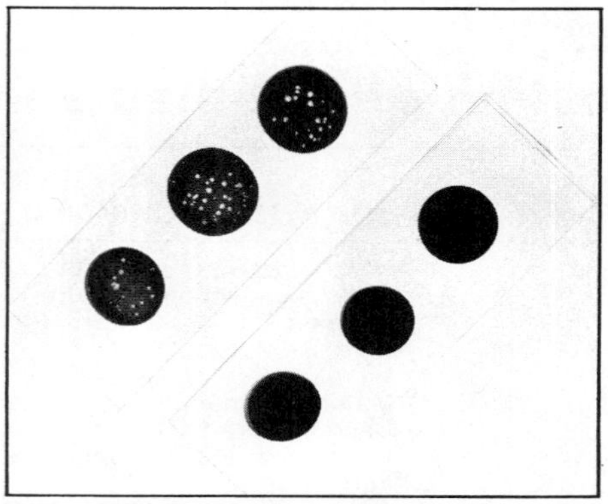

6. Here are typical results. Slide at right indicates dry material; slide at left indicates moisture-laden material. One or two bubbles may be only trapped air.

Fig. 10-66. TVI test. (*Courtesy: General Electric Company's Plastics Division, LEXAN Products Department*)

Table 10-4. Troubleshooting Guide for Injection Molding.*

1. The Problem of Brittleness

POSSIBLE CAUSE	SUGGESTED SOLUTION: INJECTION MOLDING	SUGGESTED SOLUTION: SCREW INJECTION MOLDING
1. Material too cold	a. Increase rear cylinder temperature b. Increase nozzle temperature so material keeps moving	a. Increase rear cylinder temperature b. Increase nozzle temperature so material keeps moving c. Decrease screw speed
2. Material overheated resulting in break-down of molecular structure	a. Decrease temperatures across the cylinder b. Residence time of material in cylinder too long for size of shot — change to smaller capacity machine. For maximum performance, shot size should be no more than 2/3 the rated capacity of the machine c. Decrease overall molding cycle	a. Decrease temperatures across the cylinder b. Residence time of material in cylinder too long for size of shot — change to smaller capacity machine. For maximum performance, shot size should be no more than 2/3rds the rated capacity of the machine c. Decrease overall molding cycle d. Reduce back pressure
3. Mold surface too cold	a. Throttle flow of coolant through mold b. Increase mold temperature. Fosta Nylons can	a. Throttle flow of coolant through mold b. Increase mold temperature. c. Reduce injection pressure
4. Too much reground material	a. Reduce amount of reground material blended with virgin material. The product application usually dictates the amount of reground	a. Reduce amount of reground material blended with virgin material. The product application usually dictates the amount of reground
5. Contamination by foreign material	a. Inspect molding powder for evidence of contamination b. Thoroughly purge heating cylinder c. Thoroughly clean feed system and keep hopper covered d. Insure reground material being properly handled	a. Inspect molding powder for evidence of contamination b. Thoroughly purge heating cylinder c. Thoroughly clean feed system and keep hopper covered d. Insure reground material being properly handled
6. Rate of injection too slow	a. Increase injection pressure b. Increase booster time setting c. Pre-position plunger d. Maintain a minimum feed cushion e. Use weigh feeder	a. Increase injection pressure b. Increase booster time setting c. Maintain a minimum feed cushion
7. Gates, runners and sprue too small	a. Increase size as necessary b. Relocate gates to obtain optimum filling	a. Increase size as necessary b. Relocate gates to obtain optimum filling
8. Original design of part too thin	a. Redesign, if possible, to strengthen weak section	a. Redesign, if possible, to strengthen weak section
9. Parts not properly conditioned	a. Condition parts in warm water	a. Condition parts in warm water

2. The Problem of Bubbles and Voids

POSSIBLE CAUSE	SUGGESTED SOLUTION: INJECTION MOLDING	SUGGESTED SOLUTION: SCREW INJECTION MOLDING
1. Internal shrinkage after outer surfaces have set-up	a. Increase effective pressure in cavities as follows: (1) Increase injection pressure (2) Increase plunger forward time (3) Increase booster time (4) Increase rate of injection (5) Maintain minimum feed cushion (6) Decrease pre-pack or stuffing time (7) Increase size of gates, nozzle, runners and sprue (8) Apply external lubricant to granules b. Slowly cool part in warm water after ejection.	a. Increase effective pressure in cavities as follows: (1) Increase injection pressure (2) Increase plunger forward time (3) Increase booster time (4) Increase rate of injection (5) Maintain minimum feed cushion (6) Decrease pre-pack or stuffing time (7) Increase size of gates, nozzle, runners and sprue (8) Apply external lubricant to granules b. Slowly cool part in warm water after ejection. c. Decrease screw speed for last 10% of fill
2. Material too hot	a. Decrease cylinder temperatures b. Decrease overall molding cycle	a. Decrease cylinder temperatures b. Decrease overall molding cycle
3. Inadequate venting causing entrapment of air	a. Vent cavities at point of final fill b. Relocate gate to obtain optimum filling c. Decrease mold clamping force	a. Vent cavities at point of final fill b. Relocate gate to obtain optimum filling c. Decrease mold clamping force
4. Condensation on surfaces of cavities	a. Wipe surfaces thoroughly with solvent such as acetone b. Throttle flow of coolant through mold	a. Wipe surfaces thoroughly with solvent such as acetone b. Throttle flow of coolant through mold
5. Moisture in material	a. Preheat material b. Use hopper dryer	a. Preheat material b. Use hopper dryer

Table 10-4. (*Continued*)

3. The Problem of Crazing

POSSIBLE CAUSE	SUGGESTED SOLUTION: INJECTION MOLDING	SUGGESTED SOLUTION: SCREW INJECTION MOLDING
1. Injection pressure too high	a. Decrease injection pressure b. Decrease plunger forward time c. Decrease booster time	a. Decrease injection pressure b. Decrease plunger forward time c. Decrease booster time
2. Melt temperature of material injected into cavities not uniform	a. Increase cylinder temperatures b. Increase overall molding cycle c. Use machine with greater cylinder capacity	a. Increase cylinder temperatures b. Increase overall molding cycle c. Use machine with greater cylinder capacity
3. Knockout pins working unevenly	a. Adjust or repair knockout pins	a. Adjust or repair knockout pins
4. Part improperly designed	a. Redesign part to avoid large thickness variations, sharp corners, threads, and other areas of high stress concentration.	a. Redesign part to avoid large thickness variations, sharp corners, threads, and other areas of high stress concentration.
5. Chemical attack by paints, lacquers, solvent vapors, oil, and grease	a. Avoid contact with such materials	a. Avoid contact with such materials
6. Metal inserts too cold	a. Clean and warm inserts before using	a. Clean and warm inserts before using

4. The Problem of Discoloration

POSSIBLE CAUSE	SUGGESTED SOLUTION: INJECTION MOLDING	SUGGESTED SOLUTION: SCREW INJECTION MOLDING
1. Degradation of material including pigment	a. Decrease cylinder temperatures b. Decrease overall molding cycle c. Residence time of material in cylinder too long for size of shot — change to smaller capacity machine. For maximum performance, the shot size should be no more than ⅔ the rated capacity of the machine	a. Decrease cylinder temperatures b. Decrease overall molding cycle c. Residence time of material in cylinder too long for size of shot — change to smaller capacity machine. For maximum performance, the shot size should be no more than **2/3 the rated** capacity of the machine d. Reduce screw speed e. Check for improper seating of non-return valve
2. Contamination by foreign material	a. Inspect molding powder for evidence of contamination b. Thoroughly purge heating cylinder c. Thoroughly clean feed system and hopper d. Insure reground material being properly handled	a. Inspect molding powder for evidence of contamination b. Thoroughly purge heating cylinder c. Thoroughly clean feed system and hopper d. Insure reground material being properly handled e. Increase feed
3. Charred material hanging up in cylinder or nozzle	a. Thoroughly purge heating cylinder b. Remove and clean nozzle c. Check alignment of nozzle tip with sprue bushing d. Inspect heating cylinder for evidence of cracks	a. Thoroughly purge heating cylinder b. Remove and clean nozzle c. Check alignment of nozzle tip with sprue bushing d. Inspect heating cylinder for evidence of cracks e. Reduce injection speed
4. Granules scoring plunger	a. Provide proper clearance between cylinder and plunger b. Apply additional external lubricant to granules	a. Provide proper clearance between cylinder and plunger b. Apply additional external lubricant to granules
5. Gas entrapment	a. Vent cavities at point of final fill b. Relocate gate c. Decrease mold clamping force	a. Vent cavities at point of final fill b. Relocate gate c. Decrease mold clamping force

Table 10-4. (*Continued*)

5. The Problem of Flashing

POSSIBLE CAUSE	SUGGESTED SOLUTION: INJECTION MOLDING	SUGGESTED SOLUTION: SCREW INJECTION MOLDING
1. Material too hot	a. Decrease cylinder temperatures b. Decrease overall molding cycle c. Decrease injection pressure until desired cylinder temperature is attained	a. Decrease cylinder temperatures b. Decrease overall molding cycle c. Decrease injection pressure until desired cylinder temperature is attained
2. Injection holding pressure too high	a. Decrease injection dwell pressure b. Decrease booster time c. Decrease plunger forward time	a. Decrease injection dwell pressure b. Decrease booster time c. Decrease plunger forward time
3. Projected area too large for available clamping force	a. Reduce number of cavities b. Install mold in larger machine	a. Reduce number of cavities b. Install mold in larger machine
4. Inadequate clamp setting	a. Adjust clamping mechanism to obtain maximum holding pressure	a. Adjust clamping mechanism to obtain maximum holding pressure
5. Foreign matter and flash acting as high spot on mold land	a. Remove and thoroughly clean mold	a. Remove and thoroughly clean mold
6. Cores, cavity inserts and mold surfaces out of register	a. Remove mold, carefully inspect, repair and correct register to obtain balanced filling	a. Remove mold, carefully inspect, repair and correct register to obtain balanced filling
7. Injection pressure unevenly distributed in cavities	a. Properly balance cavity layout to obtain uniform filling. The uniform filling of cavities results in parts having greater strength b. Check cross-section of molded part for uniformity	a. Properly balance cavity layout to obtain uniform filling. The uniform filling of cavities results in parts having greater strength b. Check cross-section of molded part for uniformity
8. Packing of material in cavities	a. Decrease pre-pack or stuffing time b. Reduce feed c. Starve feed with weigh feeder d. Use valve type nozzle	a. Decrease pre-pack or stuffing time b. Reduce feed c. Starve feed with weigh feeder d. Use valve type nozzle
9. Inadequate venting forcing material from cavity areas	a. Inspect and clean vents b. Enlarge vents to not more than 0.0005″ in depth	a. Inspect and clean vents b. Enlarge vents to not more than 0.0005″ in depth

6. The Problem of Flow Marks

POSSIBLE CAUSE	SUGGESTED SOLUTION: INJECTION MOLDING	SUGGESTED SOLUTION: SCREW INJECTION MOLDING
1. Material too cold	a. Increase cylinder temperatures b. Increase overall molding cycle	a. Increase cylinder temperatures b. Increase overall molding cycle
2. Mold surface too cold	a. Increase mold temperature	a. Increase mold temperature
3. Gates and runners too restrictive	a. Increase size of gates and runners. For optimum filling, the use of full round gates and runners is recommended. Naturally, the geometry of the part will determine the dimensions of the gates and runners. For thin section, best results are achieved with rectangular cross-section gates	a. Increase size of gates and runners. For optimum filling, the use of full round gates and runners is recommended. Naturally, the geometry of the part will determine the dimensions of the gates and runners. For thin section, best results are achieved with rectangular cross-section gates b. Increase front temperature
4. Jetting during injection	a. Decrease rate of injection b. Increase size of gates, runners and sprue	a. Decrease rate of injection b. Increase size of gates, runners and sprue
5. Cold slug	a. Increase size of cold slug well b. Increase size of runner extensions c. Use internally heated nozzle and sprue d. Retract injection carriage from mold every shot as temporary measure	a. Increase size of cold slug well b. Increase size of runner extensions c. Use internally heated nozzle and sprue d. Retract injection carriage from mold every shot as temporary measure

Table 10-4. (*Continued*)

7. The Problem of Mica Specks (Silver Streaks)

POSSIBLE CAUSE	SUGGESTED SOLUTION	
	INJECT ON MOLDING	SCREW INJECTION MOLDING
1. Material too cold resulting in slow filling	a. Increase cylinder temperatures b. Increase overall molding cycle	a. Increase cylinder temperatures b. Increase overall molding cycle
2. Mold surface too cold resulting in slow filling	a. Throttle flow of coolant through mold	a. Throttle flow of coolant through mold
3. Moisture and volatiles in material	a. Preheat material b. Use hopper dryer. Nylon moldings possessing both superior appearance and strength can best be obtained by keeping the moisture content of the pellets below 0.28%. Never use a can of nylon which has been opened for a period of time without drying it.	a. Preheat material b. Use hopper dryer. Nylon moldings possessing both superior appearance and strength can best be obtained by keeping the moisture content of the pellets below 0.28%. Never use a can of nylon which has been opened for a period of time without drying it.
4. Material too hot	a. Decrease nozzle temperature b. Decrease cylinder temperatures c. Decrease overall molding cycle	a. Decrease nozzle temperature b. Decrease cylinder temperatures c. Decrease overall molding cycle
5. Cold slug	a. Increase size of cold slug well b. Increase size of runner extensions c. Use internally heated nozzle and sprue d. Retract injection carriage from mold every shot as temporary measure	a. Increase size of cold slug well b. Increase size of runner extensions c. Use internally heated nozzle and sprue d. Retract injection carriage from mold every shot as temporary measure
6. Contamination by foreign material	a. Inspect molding powder for evidence of contamination b. Thoroughly purge heating cylinder c. Thoroughly clean feed system and hopper d. Insure reground material being properly handled	a. Inspect molding powder for evidence of contamination b. Thoroughly purge heating cylinder c. Thoroughly clean feed system and hopper d. Insure reground material being properly handled
7. Reground runners and sprue	a. Grind runners and sprue to same size as granules b. Feed back correct weight of granules with each shot c. Apply additional external lubricant as necessary	a. Grind runners and sprue to same size as granules b. Feed back correct weight of granules with each shot c. Apply additional external lubricant as necessary
8. Excessive condensate and lubricant on mold surface	a. Thoroughly clean mold surface b. Avoid excessive lubrication of granules and mold	a. Thoroughly clean mold surface b. Avoid excessive lubrication of granules and mold
9. Gates and runners too restrictive	a. Increase size of gates and runners	a. Increase size of gates and runners

8. The Problem of Nozzle Drooling and Foaming

POSSIBLE CAUSE	SUGGESTED SOLUTION	
	INJECTION MOLDING	SCREW INJECTION MOLDING
1. Nozzle temperature too high	a. Decrease nozzle temperature.	a. Decrease nozzle temperature. b. Relieve internal pressure before valve shuts off
2. Material too hot	a. Decrease cylinder temperatures b. Decrease overall molding cycle c. Insert screen in nozzle	a. Decrease cylinder temperatures b. Decrease overall molding cycle c. Insert screen in nozzle
3. Nozzle orifice too large	a. Use nozzle with smaller orifice and reverse taper at tip b. Use valve type nozzle	a. Use nozzle with smaller orifice and reverse taper at tip b. Use valve type nozzle
4. Moisture in material	a. Preheat material b. Use hopper dryer	a. Preheat material b. Use hopper dryer
5. Residual pressure in cylinder too high	a. Decrease injection pressure b. Decrease plunger forward time c. Decrease booster time	a. Decrease injection pressure b. Decrease plunger forward time c. Decrease booster time

Table 10-4. (*Continued*)

9. The Problem of Scorching

POSSIBLE CAUSE	SUGGESTED SOLUTION INJECTION MOLDING	SCREW INJECTION MOLDING
1. Material too hot	a. Decrease cylinder temperatures b. Decrease overall molding cycle. Seldom is scorching a problem with Foster Grant Nylons since they can be molded with lower cylinder temperatures than the 6/6 type nylons	a. Decrease cylinder temperatures b. Decrease overall molding cycle. Seldom is scorching a problem with Foster Grant Nylons since they can be molded with lower cylinder temperatures than the 6/6 type nylons
2. Moisture in material	a. Preheat material b. Use hopper dryer	a. Preheat material b. Use hopper dryer
3. Gas entrapment	a. Vent cavities at point of final fill b. Relocate gate c. Decrease mold clamping force	a. Vent cavities at point of final fill b. Relocate gate c. Decrease mold clamping force
4. Rate of injection too fast	a. Decrease plunger speed b. Decrease booster time c. Decrease injection pressure d. Increase size of gates	a. Decrease plunger speed b. Decrease booster time c. Decrease injection pressure d. Increase size of gates

10. The Problem of Short Shots

POSSIBLE CAUSE	SUGGESTED SOLUTION INJECTION MOLDING	SCREW INJECTION MOLDING
1. Injection pressure too low	a. Increase injection pressure b. Increase plunger forward time c. Increase booster time	a. Increase injection pressure b. Increase plunger forward time c. Increase booster time
2. Material too cold	a. Increase overall molding cycle b. Increase cylinder temperature. To insure quality in the molded part, complete and uniform melting in the shot zone of the heating cylinder is necessary	a. Increase overall molding cycle b. Increase cylinder temperature. To insure quality in the molded part, complete and uniform melting in the shot zone of the heating cylinder is necessary c. Increase heat at front zone
3. Mold surface too cold	a. Throttle flow of coolant through mold b. Increase mold temperature	a. Throttle flow of coolant through mold b. Increase mold temperature
4. Insufficient feeding	a. Increase feed until plunger consistently stops ⅛" or less from its maximum stroke b. Check feed ports for bridging of granules c. Use weigh feeder	a. Increase feed until plunger consistently stops ⅛" or less from its maximum stroke b. Check feed ports for bridging of granules c. Check for improper seating of non-return valve d. Increase back pressure
5. Excessive feeding	a. Reduce feed cushion thereby increasing effective pressure of melt b. Control overfeeding of lubricated granules by use of feed dam	a. Reduce feed cushion thereby increasing effective pressure of melt b. Control overfeeding of lubricated granules by use of feed dam
6. Melt flow too restricted	a. Increase size of gates, runners and sprue as necessary b. Use nozzle with larger orifice c. Check nozzle for contamination and tramp metal	a. Increase size of gates, runners and sprue as necessary b. Use nozzle with larger orifice c. Check nozzle for contamination and tramp metal
7. Back pressure of entrapped air causing resistance	a. Vent cavities at point of final fill b. Reduce rate of injection c. Reduce clamping force d. Lower mold surface temperature	a. Vent cavities at point of final fill b. Reduce rate of injection c. Reduce clamping force d. Lower mold surface temperature
8. Cold slug clogging gates, runners and sprue	a. Increase size of cold slug well b. Increase size of runner extensions c. Use internally heated nozzle and sprue d. Retract injection carriage from mold every shot as temporary measure	a. Increase size of cold slug well b. Increase size of runner extensions c. Use internally heated nozzle and sprue d. Retract injection carriage from mold every shot as temporary measure

Table 10-4. *(Continued)*

11. The Problem of Sink Marks

POSSIBLE CAUSE	SUGGESTED SOLUTION: INJECTION MOLDING	SUGGESTED SOLUTION: SCREW INJECTION MOLDING
1. Thermal contraction of material during set-up	a. Increase effective pressure in cavities as follows: (1) Increase injection pressure (2) Increase plunger forward time (3) Increase booster time (4) Increase rate of injection (5) Maintain a minimum feed cushion (6) Decrease pre-pack or stuffing time (7) Increase size of gates, nozzle, runners and sprue (8) Apply external lubricant to granules b. Redesign part with thinner section c. Reduce mold surface temperature	a. Increase effective pressure in cavities as follows: (1) Increase injection pressure (2) Increase plunger forward time (3) Increase booster time (4) Increase rate of injection (5) Maintain a minimum feed cushion (6) Decrease pre-pack or stuffing time (7) Increase size of gates, nozzle, runners and sprue (8) Apply external lubricant to granules b. Redesign part with thinner section c. Reduce mold surface temperature d. Reduce temperature at front and center sections
2. Parts too hot when ejected	a. Decrease cylinder temperature b. Increase mold cooling time c. Reduce mold surface temperature d. Slowly cool part in warm water (100-110°F.) immediately after injection	a. Decrease cylinder temperature b. Increase mold cooling time c. Reduce mold surface temperature d. Slowly cool part in warm water (100-110°F.) immediately after injection
3. Insufficient material sealed in-cavities	a. Increase feed b. Increase dwell time of plunger c. Redesign gates to obtain proper sealing. The land of the gate, or the distance between the end of the runner and the entrance to the cavity, should be kept as short as possible	a. Increase feed b. Increase dwell time of plunger c. Redesign gates to obtain proper sealing. The land of the gate, or the distance between the end of the runner and the entrance to the cavity, should be kept as short as possible
4. Material too hot or too cold	a. Adjust cylinder temperatures b. Adjust mold surface temperature	a. Adjust cylinder temperatures b. Adjust mold surface temperature
5. Part section uneven	a. Gate at heavy section b. Increase dwell time of plunger c. If possible, redesign part with more uniform section	a. Gate at heavy section b. Increase dwell time of plunger c. If possible, redesign part with more uniform section
6. Moisture in material	a. Preheat material b. Use hopper dryer	a. Preheat material b. Use hopper dryer

12. The Problem of Sprue Sticking

POSSIBLE CAUSE	SUGGESTED SOLUTION: INJECTION MOLDING	SUGGESTED SOLUTION: SCREW INJECTION MOLDING
1. Nozzle seated improperly	a. Realign nozzle with sprue bushing	a. Realign nozzle with sprue bushing
2. Orifice of nozzle larger than orifice of sprue bushing	a. Replace with nozzle having smaller orifice b. Replace with sprue bushing having larger orifice. The diameter of the sprue bushing should be at least 1/16" greater than both the main runner and the nozzle bore. The recommended taper of the sprue bushing is from 3.5° to 5° included angle	a. Replace with nozzle having smaller orifice b. Replace with sprue bushing having larger orifice. The diameter of the sprue bushing should be at least 1/16" greater than both the main runner and the nozzle bore. The recommended taper of the sprue bushing is from 3.5° to 5° included angle c. Relieve internal pressure before shifting valve
3. Sprue puller too small	a. Enlarge sprue puller b. Increase undercut	a. Enlarge sprue puller b. Increase undercut
4. Material too cold — sprue unable to break away from nozzle	a. Increase nozzle temperature until sprue breaks away from nozzle every shot	a. Increase nozzle temperature until sprue breaks away from nozzle every shot
5. Material too hot — sprue breaking at base	a. Decrease nozzle temperature b. Lower cylinder temperature c. Increase mold cooling time d. Enlarge sprue at base	a. Decrease nozzle temperature b. Lower cylinder temperature c. Increase mold cooling time d. Enlarge sprue at base e. Increase nozzle temperature

Table 10-4. (*Continued*)

POSSIBLE CAUSE	SUGGESTED SOLUTION INJECTION MOLDING	SCREW INJECTION MOLDING
6. Inadequate taper or sprue diameter too small	a. Increase taper b. Replace with sprue having larger diameter	a. Increase taper b. Replace with sprue having larger diameter
7. Too much material packed into sprue	a. Decrease injection pressure b. Decrease plunger forward time c. Increase mold cooling time to allow proper shrinkage of sprue d. Increase mold temperature in sprue bushing area e. Increase feed cushion	a. Decrease injection pressure b. Decrease plunger forward time c. Increase mold cooling time to allow proper shrinkage of sprue d. Increase mold temperature in sprue bushing area e. Increase feed cushion
8. Material too brittle — sprue breaking inside sprue bushing	a. See Section I on Brittleness	a. See Section I on Brittleness

13. The Problem of Sticking in Cavities

POSSIBLE CAUSE	SUGGESTED SOLUTION INJECTION MOLDING	SCREW INJECTION MOLDING
1. Mold defective as follows:		
(a) Surface poorly polished	a. Polish in same direction as part is ejected	a. Polish in same direction as part is ejected
(b) Burrs, dents, undercuts or damaged surface	b. Repair and polish	b. Repair and polish
(c) Insufficient draft or taper on cavities and sprue	c. Provide maximum allowable draft	c. Provide maximum allowable draft
(d) Knockout system, including ejector pins, stripper rings and sprue pullers, not ejecting properly	d. Make necessary adjustments to obtain desired ejection	d. Make necessary adjustments to obtain desired ejection
(e) Cavities made of soft porous metal	e. Replace with hardened steel cavities	e. Replace with hardened steel cavities
(f) Core shifting and cavity misalign-ment	f. Realign cores with cavities	f. Realign cores with cavities
2. Overpacking material in cavities	a. Decrease injection pressure b. Decrease plunger forward time c. Decrease booster time d. Reduce feed e. Decrease pre-pack or stuffing time f. Remove undercuts	a. Decrease injection pressure b. Decrease plunger forward time c. Decrease booster time d. Reduce feed e. Decrease pre-pack or stuffing time f. Remove undercuts g. Reduce fill speed for last 5% of filling
3. Injection pressure unevenly distributed in cavities	a. Maintain constant feed cushion b. Balance gates and runners for uniform filling of cavities	a. Maintain constant feed cushion b. Balance gates and runners for uniform filling of cavities
4. Parts too soft when ejected	a. Increase mold cooling time b. Decrease cylinder temperatures and nozzle temperature c. Reduce mold surface temperature	a. Increase mold cooling time b. Decrease cylinder temperatures and nozzle temperature c. Reduce mold surface temperature
5. Film on cavity surfaces	a. Wipe surfaces thoroughly with solvent such as acetone	a. Wipe surfaces thoroughly with solvent such as acetone
6. Parts remaining on injection half of mold	a. Use mold temperature differential to keep parts on ejection half of mold b. Use undercut runner as puller to keep parts on ejection half of mold c. Apply light spray of mold release lubricant on mold d. Increase nozzle temperature sufficiently to keep material flowing	a. Use mold temperature differential to keep parts on ejection half of mold b. Use undercut runner as puller to keep parts on ejection half of mold c. Apply light spray of mold release lubricant on mold d. Increase nozzle temperature sufficiently to keep material flowing

14. The Problem of Surface Lamination

POSSIBLE CAUSE	SUGGESTED SOLUTION: INJECTION MOLDING	SUGGESTED SOLUTION: SCREW INJECTION MOLDING
1. Contamination by foreign material	a. Inspect molding powder for evidence of contamination b. Thoroughly purge heating cylinder c. Thoroughly clean feed system and keep hopper covered d. Insure reground material being properly handled. The end-use application will usually determine the ratio of reground material to virgin material. Precautions should be taken to insure that the reground material has been properly dried; otherwise poor moldings will be the result	a. Inspect molding powder for evidence of contamination b. Thoroughly purge heating cylinder c. Thoroughly clean feed system and keep hopper covered d. Insure reground material being properly handled. The end-use application will usually determine the ratio of reground material to virgin material. Precautions should be taken to insure that the reground material has been properly dried; otherwise poor moldings will be the result
2. Moisture in material	a. Preheat material b. Use hopper dryer	a. Preheat material b. Use hopper dryer
3. Material too cold	a. Increase cylinder temperature b. Increase overall molding cycle	a. Increase cylinder temperature b. Increase overall molding cycle
4. Mold surface too cold	a. Increase mold surface temperature	a. Increase mold surface temperature
5. Rate of injection too slow	a. Increase plunger speed	a. Increase plunger speed
6. Gates and runners too restrictive	a. Increase size of gates and runners	a. Increase size of gates and runners

15. The Problem of Surface Sticking

POSSIBLE CAUSE	SUGGESTED SOLUTION: INJECTION MOLDING	SUGGESTED SOLUTION: SCREW INJECTION MOLDING
1. Poor mold surface	a. Polish mold	a. Polish mold
2. Material too cold	a. Increase cylinder temperatures b. Increase overall molding cycle	a. Increase cylinder temperatures b. Increase overall molding cycle
3. Mold surface too cold	a. Throttle flow of coolant through mold b. Increase mold temperature	a. Throttle flow of coolant through mold b. Increase mold temperature
4. Excessive condensation, grease lubricant, etc. on mold surface	a. Thoroughly clean mold surface b. Avoid excessive lubrication of granules and mold. Careful judgement should be exercised when applying external lubricant to nylon.	a. Thoroughly clean mold surface b. Avoid excessive lubrication of granules and mold. Careful judgement should be exercised when applying external lubricant to nylon.
5. Injection pressure too high causing uneven filling of cavities	a. Decrease injection pressure	a. Decrease injection pressure

16. The Problem of Surface Waviness

POSSIBLE CAUSE	SUGGESTED SOLUTION: INJECTION MOLDING	SUGGESTED SOLUTION: SCREW INJECTION MOLDING
1. Material too cold	a. Increase cylinder temperatures b. Increase overall molding cycle	a. Increase cylinder temperatures b. Increase overall molding cycle
2. Mold surface too cold	a. Throttle flow of coolant through mold b. Increase mold temperature	a. Throttle flow of coolant through mold b. Increase mold temperature
3. Injection pressure too low	a. Increase injection pressure b. Increase ram speed. The use of a fast ram speed is preferred when molding nylon. Furthermore, it is of the utmost importance that the ram speed be maintained during the filling portion of the injection cycle. Fast ram speeds tend to eliminate surface waviness and other surface defects such as weld lines, sink marks, etc.	a. Increase injection pressure b. Increase ram speed. The use of a fast ram speed is preferred when molding nylon. Furthermore, it is of the utmost importance that the ram speed be maintained during the filling portion of the injection cycle. Fast ram speeds tend to eliminate surface waviness and other surface defects such as weld lines, sink marks, etc.
4. Rate of injection too slow	a. Increase plunger speed b. Increase booster time c. Increase size of gates, nozzle, runners, and sprue d. Apply external lubricant to granules	a. Increase plunger speed b. Increase booster time c. Increase size of gates, nozzle, runners, and sprue d. Apply external lubricant to granules

Table 10-4. *(Continued)*

17. The Problem of Warping

POSSIBLE CAUSE	SUGGESTED SOLUTION: INJECTION MOLDING	SUGGESTED SOLUTION: SCREW INJECTION MOLDING
1. Parts too hot when ejected	a. Decrease cylinder temperatures b. Increase mold cooling time c. Reduce mold surface temperature d. Slowly cool part in warm water (100-110°F.) immediately after ejection e. Insert part in shrink jig immediately after ejection	a. Decrease cylinder temperatures b. Increase mold cooling time c. Reduce mold surface temperature d. Slowly cool part in warm water (100-110°F.) immediately after ejection e. Insert part in shrink jig immediately after ejection
2. Distortion due to internal strains	a. Decrease injection pressure b. Decrease plunger forward time c. Decrease booster time d. Increase mold cooling time e. Increase mold surface temperature f. Maintain a constant feed cushion	a. Decrease injection pressure b. Decrease plunger forward time c. Decrease booster time d. Increase mold cooling time e. Increase mold surface temperature f. Maintain a constant feed cushion
3. Distortion of heavy section only	a. Increase mold cooling time b. Increase cylinder temperatures to obtain uniform melt temperature. It is essential that the melt temperature in the shot zone remains uniform. This practice results in lower residual stresses, improved mechanical properties, and better dimensional stability in the molded part. c. Slowly cool part in warm water (100-110°F.) immediately after ejection d. Insert part in shrink jig immediately after injection	a. Increase mold cooling time b. Increase cylinder temperatures to obtain uniform melt temperature. It is essential that the melt temperature in the shot zone remains uniform. This practice results in lower residual stresses, improved mechanical properties, and better dimensional stability in the molded part. c. Slowly cool part in warm water (100-110°F.) immediately after ejection d. Insert part in shrink jig immediately after injection
4. Differential shrinking of part having irregular cross-section	a. Redesign part with a more uniform cross-section to obtain balanced filling	a. Redesign part with a more uniform cross-section to obtain balanced filling
5. Knockout pins working unevenly	a. Adjust or repair knockout pins	a. Adjust or repair knockout pins
6. Improper gate design	a. Relocate gate so melt flows from heavy to thin section	a. Relocate gate so melt flows from heavy to thin section
7. Parts bending and sticking	a. See Section 13 on Sticking in Cavities	a. See Section 13 on Sticking in Cavities

18. The Problem of Weld Lines

POSSIBLE CAUSE	SUGGESTED SOLUTION: INJECTION MOLDING	SUGGESTED SOLUTION: SCREW INJECTION MOLDING
1. Material too cold	a. Increase cylinder temperatures b. Increase overall molding cycle	a. Increase cylinder temperatures b. Increase overall molding cycle
2. Insufficient pressure at weld	a. Increase effective pressure at weld as follows: (1) Increase injection pressure (2) Increase plunger forward time (3) Increase booster time (4) Increase rate of injection (5) Maintain minimum feed cushion (6) Decrease pre-pack or stuffing time (7) Increase size of gates, nozzle, runners and sprue (8) Apply external lubricant to granules	a. Increase effective pressure at weld as follows: (1) Increase injection pressure (2) Increase plunger forward time (3) Increase booster time (4) Increase rate of injection (5) Maintain minimum feed cushion (6) Decrease pre-pack or stuffing time (7) Increase size of gates, nozzle, runners and sprue (8) Apply external lubricant to granules b. Decrease injection speed
3. Mold surface too cold	a. Throttle flow of coolant through mold b. Increase mold temperature	a. Throttle flow of coolant through mold b. Increase mold temperature
4. Gas entrapment	a. Vent cavities at point of final fill b. Relocate gate c. Decrease mold clamping force	a. Vent cavities at point of final fill b. Relocate gate c. Decrease mold clamping force
5. Distance from gate to weld area too long	a. Use multiple gates and runners.	a. Use multiple gates and runners.
6. Excessive condensate and lubricant on mold surface near weld areas	a. Thoroughly clean mold surface b. Avoid excessive lubrication of granules and mold	a. Thoroughly clean mold surface b. Avoid excessive lubrication of granules and mold

*Courtesy: Foster Grant Co.

RULE THREE

Mold size vs. platen size

The mold must fit within the available mold mounting surface between the tie rods.

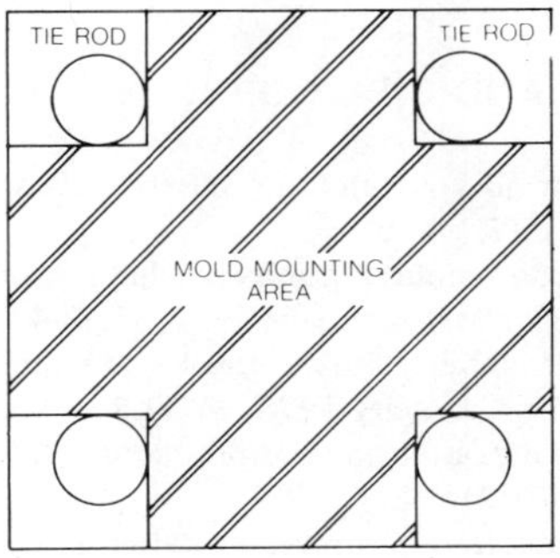

RULE FOUR

Mold thickness vs. closed daylight

Mold thickness when shut must be greater than the minimum closed daylight of the machine. For mechanical clamp machines only the mold thickness must be less than the maximum closed daylight in order to build clamp tonnage.

RULE FIVE

Part depth vs. open daylight

For part removal, open daylight must be greater than twice the depth of the part plus the thickness of the closed mold.

Equation:

Open daylight > 2X (part depth) + (mold thickness)

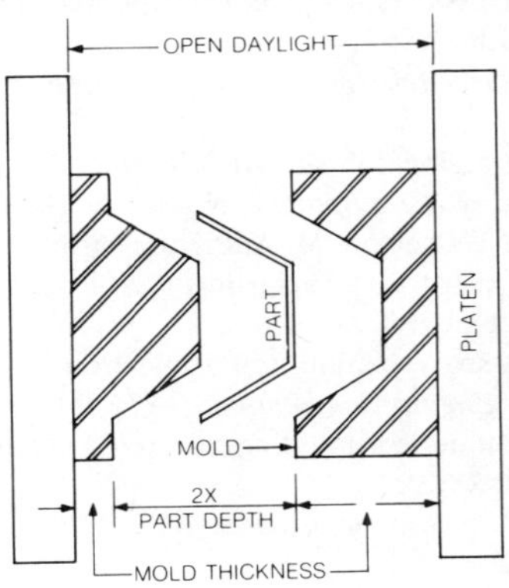

RULE SIX

Part depth vs. clamp stroke

On hydraulic clamp machines the clamp stroke must be equal to or greater than the open daylight minus the mold thickness and greater than two times the depth of the molded part.

Equations:

a) Clamp stroke ⩾ (open daylight) - (mold thickness)

b) Clamp stroke > 2X (part depth)

On mechanical clamp machines the clamp stroke must be equal to the open daylight minus the mold thickness and greater than 2 times the depth of the molded part.

Equations:

a) Clamp stroke = (open daylight) - (mold thickness)

b) Clamp stroke > 2X (part depth)

RULE SEVEN

Cycle time

The cycle time equals machine time plus cooling time. Typical examples listed give approximate machine times based on minimum injection and hold time, full clamp stroke and hydraulic ejection as the mold opens. Approximate cooling times are based on general purpose plastic with the mold having adequate cooling capability and a part with uniform wall thickness.

Machine Size (tons)	Machine Time	Wall Thickness (inches)	Cooling Time
75-200	4 sec.	.030	3 sec.
300-500	8 sec.	.045	5 sec.
700-1000	10 sec.	.095	12 sec.
1500-2500	15 sec.	.1875	20 sec.

Equation:

Cycle time* = (machine time) + (cooling time)

*For manual part removal add 5–10 seconds.

Example:

The part to be molded has a uniform wall thickness of .045 inches and the machine selected is a Farrel 375 ton unit.

Cycle time:

8 sec. machine time + 5 sec. cooling time

13 sec. total cycle for automatic operation

RULE EIGHT

Screw recovery vs. cycle time

For minimum cycle time, screw recovery time must be equal to or less than cooling time.

$$\frac{\text{Screw recovery} = \text{Shot Wt. (oz.)}}{\text{Recovery Rate (oz./sec.)}} \leqslant \text{Cooling time}$$

Process Control and Troubleshooting

Process control, as differentiated from the controls used on the machine, means all the steps taken to assure that a satisfactory part will be made, and that such parts will be made consistently.

Process control would then include the following:

1. Premolding operations. Storage and handling of the resin to insure that clean, uncontaminated materials are delivered to the molding area.
2. Dry resin. Most resins must be thoroughly dried before

being introduced into the press cylinder. This drying could take place either in the press hopper or outside. In either case, unless moisture, splay marks, or bubbles on the surface are of no consequence, simple tests should be made periodically to be sure the drying technique is performing satisfactorily. One such test is to weigh a small amount of resin on an analytical balance before and after heating for the normal drying period, and then after heating the same resin for double and triple the normal time. If the same weight is found after the normal, double, and triple heating times, then it can be concluded that the normal heating is removing all the moisture. Another simple method of testing for moisture, the TVI (Tomasetti Volatile Indicator) test, developed by the General Electric Company, is shown in Figure 10-66.

3. A sound design. The part should contain a minimum of thick and thin sections which could cause sink marks (indentations) and warpage.
4. A molding machine adequate for the task. The part weight should utilize from 50% to 75% of the machine's capacity for optimum results. The three machine variables are time, temperature, and pressure. The machine should be capable of varying any of these conditions within the limits recommended by the resin manufacturer. After being adjusted, the machine must be capable of operating consistently for long periods.
5. A properly designed mold. The mold must be properly gated so the material will flow in at the correct rate. Temperature regulation should be such that proper cooling will take place to minimize shrink marks, warpage, etc. The mold should be designed and gated so that even filling will take place, etc.
6. Post-molding operations. The post-molding operations of trimming and cooling should be done to leave minimum mark-off, warpage, or variable shrinkage.

The above list outlines only some of the major factors which must be taken into account in production of satisfactory parts. If, in spite of having taken account of all the factors described, problems are still encountered, then a more extensive investigation should be made of the possible causes and remedies for the difficulties. Fortunately, all the suppliers of molding materials furnish guides for the proper molding of their resins. Usually these guides include a section on troubleshooting, in which typical malfunctions are listed, along with probable causes and possible solutions. One such troubleshooter's guide, prepared by the Foster Grant Company, Inc., is shown in Table 10-4. Although prepared specifically for the molding of nylon, the information given is applicable to all types of thermoplastic injection molding. This guide is considered to be particularly valuable because: (1) the information is divided into 18 different types of common problems, each with a number of possible causes and solutions, and (2) the solutions are given in terms of modern screw injection machines and in terms of older ram machines which are still being used in many shops. In the guide, the ram data is simply called injection molding, and the screw data is called screw injection molding.

REFERENCES AND BIBLIOGRAPHY

Allen, E. O. Know all your molding variables. *Plastics Engineering*, July 1974, **30**(7):37.

Allen, P. E. Injection-mold venting: the hidden processing parameter. *Plastics Engineering*, January 1979, **35**(1):43.

Avery, J. A. Melt stability–that's the key to molding the PBT's. *Plastics Technology*, January 1978, **24**(1):87.

Blake, W. T. Flow measurement and product quality. *Plastics World*, March 1963, **21**(3):21.

Bonner, R. M. Evaluation of screws for nylon. *SPE Journal*, October 1963, **19**(10):1069.

Bringer, R. P., and Morneau, G. A. Processing of CTFE plastics. *Plastics Design & Processing*, August 1968, **8**(8):17.

Broderick, W. M. Part extractors keep injection machines humming. *Plastics Engineering*, March 1976, **32**(3):32.

Brown, P. Runnerless molding systems. *Plastics Technology*, Part 1: September 1977, **23**(9):5. Part 2: October 1977, **23**(10).

Bulter, R. A. Injection machine screw and barrel wear. *British Plastics*, June 1970, **43**(6):139.

Burns, R. L., and Spoo, B. H. How to injection mold polyarylsulfone. *Plastics Design & Processing*, April 1972, **12**(4):26.

Carleton, P. S., Ewen, J. H. Jr., Shah, T. M., Reymore, H. E. Jr., and Sayigh, A. A. R. *A Thermoplastic RIM System for Automotive Fascia*. Society of Automotive Engineers, Inc., Warrendale, Pa., 1976.

Carvalho, T. B. Plan your plant right. *Plastics World*, October 1977, **35**(10):68.

——. How to improve cost efficiency in injection. *Plastics World*, February 17, 1975, **33**(2):42.

Celanese Plastics Co. *Celcon Acetal Copolymer Injection Molding*. Bulletin C3A, June 1967.

——. *Fortiflex Polyethylene Injection Molding*. Bulletin P3A, June 1971.

Clark, D. C. Increasing control over the injection molding process. *Plastics Machinery & Equipment*, November 1977, **7**(11):29.

Cloud, P. J., and Wolverton, M. A. Predict shrinkage and warpage of reinforced and filled thermoplastics. *Plastics Technology*, November 1978, **24**(11):107.

Conway, F. The story behind liquid injection molding of polyurethanes. *Plastics Engineering*, January 1975, **31**(1):25.

Csaszar, E. J. Runnerless molding without hangups. *SPE Journal*, February 1972, **28**(2):20.

Danien, W. R. A processor looks at RIM. *Plastics World*, June 21, 1976, **34**(6):32.

Davis, M. A. Processing plastics–servocontrolled injection molding. *Plastics Engineering*, April 1977, **33**(4):26.

Debreceni, T. Why choose a toggle clamp? *Plastics Technology*, January 1968, **14**(1):39.

De Capite, R., and Gudermuth, C. S. The low-down on vented screws for injection molding. *Plastics Engineering*, January 1976, **32**(1):37.

——. *The Vented Reciprocating Screw Plasticator*. HPM Division, Koehring Co., May 1976.

Dow Chemical Co. *Injection Molding Technology Fundamentals*. Form 171-289B, December 1966.

Dreger, D. R. Lower cost reinforced-plastics parts with resin injection molding. *Machine Design*, May 20, 1976, **48**(10):103.

——. Reaction-injection molding. *Machine Design*, April 3, 1975, **47**(7):148.
DuBois, J. H. Reaction injection molding: a new horizon in plastics. *Plastics World*, April 1975, **33**(4).
——. Purging injection cylinders. *Plastics World*, August 1964, **22**(8):26.
DuBois, J. H., and John, F. W. *Plastics*. Van Nostrand Reinhold, New York, 1974.
E.I. Du Pont de Nemours & Co., Inc. *Molding Manual for Delrin Acetal Resins*. Bulletin A-34149, February 1964.
——. *Du Pont Suryln Ionomer Resin Injection Molding Processing Guide*. 2nd Ed. Bulletin A-60070.
Eakin, J. L. Obtaining optimum quality for RIM parts with good mold design. *Plastics Machinery & Equipment*, June 1977, 7(6).
Eastman Chemical Products, Inc. *Injection Molding Tenite Cellulosic Plastics*. Publication Number PC-2A, ECP 2549, November 1973.
——. *Injection Molding Tenite Polypropylene and Tenite Polyallomer*. Bulletin TR-8A, ECP 463, June 1970.
——. *Molding Very Thin Sections From Tenite Polypropylene and Tenite Polyallomer*. Bulletin TR-23, ECP 463, February 1966.
——. *Injection Molding of Tenite Polyethylene*. Bulletin 58-2342, 1956.
Enjay Laboratories. *Escon Polypropylene Injection Molding*. Bulletin PLA-64-803.
Erwin, T. Why choose a hydraulic clamp? *Plastics Technology*, January 1968, **14**(1):35.
Evans, W. B. Two experts view computer control for injection molding: in injection molding, a computer is simply overkill. *Plastics World*, February 1973, **31**(2):50.
——. How to avoid damage to injection barrels and screws when molding RP. *Modern Plastics*, March 1972, **49**(3):68.
——. What about flow molding? *Plastics Technology*, October 1964, **10**(10):42.
Farris, J. Your guide to choosing hot-runner temperature controls. *Plastics Technology*, April 1974, **20**(4):51.
Ferrari, R. J., Silverwood, H. A., and Salisbury, W. C. What's available in RIM materials. *Plastics Technology*, May 1976, **22**(5):39.
Fiberite Corporation. *General Molding Conditions for Fiberite Reinforced Thermoplastics*.
Fishberg, L. D., Longstreet, D. C., and Olivo, A. R. Determining shrinkage in injection molding phenolics. *SPE Journal*, March 1970, **26**(3):34.
Frimberger, H. H., and Evans, W. B. Flow molding of ABS plastics. *SPE Journal*, August 1964, **20**(8):744.
Fritch, L. W. Jr. Is there a cure for splay? *Plastics Engineering*, April 1975, **31**(4):61.
Gellert, J. U. Valve-gate system conveys melt at ideal molding conditions. *Plastics Engineering*, February 1976, **32**(2):16.
General Electric Plastics. *Injection Molding Valox*. Bulletin VAL-15A, May 1974.
——. *Noryl Thermoplastic Resins Injection Molding*. Bulletin CDX-81, August 1974.
——. *New Thermoset Compound for Injection Molding*. Bulletin CDC-474B, May 1970.
——. *Injection Molding Noryl Thermoplastic Resin*. Bulletin CDX-47, November 1967.
Glatt, J. E. Combined injection molding and secondary operations create new products. *Plastics Design & Processing*, April 1961, **1**(1):14.
B. F. Goodrich Chemical Co. *Injection Molding General Purpose Abson ABS Materials*. Bulletin A-2.
——. *Injection Molding Guide for Abson 89129 ABS Compound*. Bulletin A-3.
——. *Injection Molding Elastomers*. Bulletin E-4.
W.R. Grace & Co. *Molder's Guide to Injection Molding Grex High Density Polyethylene*. Bulletin ET 5397, December 1957.
Gregory, R. B. Vented-barrel injection molding: drawbacks and advantages. *Plastics Machinery & Equipment*, April 1977, **7**(4):20.
Gudermuth, C. S. Vented injection machines: time for a good look. *Plastics Technology*, April 1976, **22**(4):61.
Gulf Oil Co., Plastics Division. *A Short Course in Polyethylene Molding*. Bulletin SP 12101 CD, April 1969.
Hadden, C. W., and Talbot, H. M. Top performance polycarbonate molding. *Plastics Technology*, May 1974, **20**(5):33.
Halstead, W. G., Rinderle, J. R., and Suh, N. P. Vary the volume–control the shrinkage. *Plastics Engineering*, June 1979, **35**(6):41.
Harris, H. E. Second generation digital control system for total control of processing plants. *Plastics Design & Processing*, November 1975, **15**(11).
Harry, D. J. Temperature effects on part shrinkage during pressure-controlled injection molding. *Plastics Design & Processing*, June 1973, **13**(6):16.
Hauck, J. E. Injection machine of tomorrow? *Modern Plastics*, July 1970, **47**(11):71.
Haycock, W. M. Processing the new thermoplastics PPO and polybutylene. Part 1: How to injection mold PPO. *Plastics Technology*, February 1966, **12**(10):35.
Hoffman, D. B., and Sarasohn, N. Pressure transducers–which, why and specmanship. *Plastics Design & Processing*, July 1972, **12**(7):20.
Hold, P. Two experts view computer control for injection molding: the time to investigate computer control is now. *Plastics World*, February 1973, **31**(2):50.
Holt, H. New techniques in shrinkage control. *SPE Journal*, June 1964, **20**(6):519.
Houston, A. M. What you should know about plastics for fatigue applications. *Materials Engineering*, March 1977, **85**(3):31.
Hubbauer, P. Injection molding for quality. *Plastics Engineering*, August 1973, **29**(8):37.
Hunkar, D. Process control reduces problems encountered in injection molding thermosets. *Plastics Design & Processing*, October 1976, **16**(10):8.
Husky Manufacturing & Tool Works, Ltd. *Technical Planning of Injection Molding Operations*. Ontario, Canada, 1972.
Imig, C. S. Mold temperature effects in polyethylene molding. *Modern Plastics*, December 1956, **33**(4):149.
Imperial Chemical Industries, Ltd. *Injection Moulding of "TPX" Polymers*. Technical Service Note T102, April 1971.
Inslee, T. Hydraulic vs. toggle clamps–which one is right for you? *Plastics Technology*, April 1974, **20**(4):47.
Johnson, L. I. Strain free injection molding. *Modern Plastics*, June 1963, **40**(10):111.
Johnson, V. L. *Advances in RIM Technology*. Admiral Equipment Co., subsidiary of Upjohn Co.
Jones, D. C. Automating molded parts removal increases injection molded productivity. *Plastics Design & Processing*, May 1979, **19**(1):34.
Kavanagh, J. A. The new mold, its problems and some solutions. *SPE Journal*, July 1961, **17**(7):653.
Keiser, N., Cirker, L. H., and Kohl, P. D. The rotating spreader–a device for superior injection molding. *SPE Journal*, April 1961, **17**(4):340.
Kircher, K., and Menges, G., Combining polyurethanes and polystyrenes . . . . *Plastics Engineering*, October 1976, **32**(10):37.
Kuo, Y. Predicting part moldability–improved techniques cut guesswork. *Plastics World*, October 1978, **36**(10):64.
Lachowecki, W. Molding RTP's? Here's how to prevent machine wear. *Plastics Technology*, April 1972, **18**(4).
Lake, N. E. Give the slip to your hangups with silicone fluids. *Plastics Engineering*, November 1975, **31**(11):37.

Laufer, B. Machine replacement: when? *Plastics Technology*, April 1976, **22**(4):47.

Levy, S., and Du Bois, J. H. The limitations of injection molding on product design. *Plastics Design Forum*, January/February 1977.

Lord, H. A. Eliminate splay in injection molding by cavity pressurization. *Plastics Engineering*, March 1978, **34**(3):51.

Martin, M. C. Don't "blow" hot runner molding savings with second rate temperature control. *Plastics World*, February 18, 1974, **32**(2):44.

Marx, E. The graphics route to figuring costs. *Plastics Technology*, April 1976, **22**(4):67.

Miller, B. In-plant scrap recovery—a good idea gets better. *Plastics World*, January 19, 1976, **34**(1):22.

——. Runnerless molds—an idea to cut cycle time, save power. *Plastics World*, December 22, 1975, **33**(12):38.

Misko, J. C. Souping up injection screws with fluted mixing sections. *Plastics Engineering*, April 1976, **32**(4):48.

Mobay Chemical Co. *Injection Molding Merlon Polycarbonate*. Thermoplastics Processing 83-M7.

Modern Plastics. Thermoset injection molding becomes more reliable and economical. June 1979, **56**(6):44.

——. Something extra in resin drying. November 1978, **55**(11):54.

——. Coinjection: new molding technology. July 1976, **53**(7):40.

——. Molding filled materials is a lot easier now. July 1975, **52**(7):44.

——. RIM new high-speed processing option. June 1974, **51**(6):54.

——. Injection machine of tomorrow. July 1970, **47**(7):71.

——. Cooling it, but how. April 1969, **46**(4):142.

——. New standard spiral flow test. February 1968, **45**(2):104.

——. Time-saving nomograms for injection and blow molding. May 1968, **45**(5).

——. Guide to moldability tests. August 1965, **42**(12):111.

Monsanto Co. *Injection Molding of Lustrex and Lustran Styrene Polymers*. Product information bulletin no. 1104B, April 1976.

Morse, A. R. Selecting the right injection molding machine nozzle. *Plastics Design & Processing*. Part 1: June 1968, 8(6). Part 2: July 1968, 8(7).

——. What's happening in injection plant accessory equipment. *Plastics Design & Processing*, December 1963, **3**(12):16.

Neely, J. K. Automatic injection molding. *SPE Journal*, May 1961, **17**(5):475.

New Britain Plastics Machine, Division of Litton Industrial Products, Inc. *Planning Guide for Profitable Injection Molding*. Bulletin NB PG, September 1974.

Noren, D. W. A new cure for mold cooling headaches. *Plastics Technology*, April 1978, **24**(4):122.

Norfleet, J. S. Molding "maps" guidelines to productivity. *Plastics World*, May 1977, **35**(5):53.

Olson, R. Structural-foam tooling. *Plastics Technology*, June 1976, **22**(6):38.

Paggi, L. Molding cycles. *SPE Journal*, April 1959, **15**(4).

Parker, F. J. Injection molding of BMC with minimum strength loss. *Modern Plastics*, December 1977, **54**(4):54.

Parr, T. W. Injection molders—understand those specs. *Plastics Technology*, April 1975, **21**(4):40.

Parr, T. W., and Jacks, J. H. Injection machine specs—a look at accessories. *Plastics Technology*, May 1975, **21**(5):41.

Paulson, D. C. Guide to injection machine control. *SPE Journal*, January 1971, **27**(1):37.

Peterson, D. Injection molders speak out on electronic process control. *Plastics Design & Processing*, March 1975, **15**(3).

Phillips Chemical Co. *Technical Information on Ryton Polyphenylene Sulfide Resins.* Bulletin 9415.

Plastics Design & Processing. Static mixers increase part quality, decrease colorant requirements by 20%. June 1976, **16**(6):20.

——. Coolers and chillers for faster injection molding cycles. June 1972, **12**(6):12.

——. Developments in hydraulic clamping systems. September 1970, **10**(9):26.

——. Developments in mechanical clamping system. October 1970, **10**(10).

Plastics Engineering. The low-down on vented screws for injection molding. January 1976, **32**(1):137.

Plastics Machinery & Equipment. New screw design improves melt-temperature control, March 1978, 8(3).

Plastics Technology. Gate temperature control speeds hot-runner molding. May 1979, **25**(5):30.

——. Heat pipes: new tool for process heating/cooling. March 1977, **23**(3):15.

——. Flame tests for structural foam parts. November, 1977, **23**(11):88.

——. Machine replacement: when. April 1976, **22**(4):47.

——. Here's more on liquid resin molding. July 1976, **22**(7):23.

——. A processor's guide to chilling and cooling needs. November 1973, **19**(11):39.

——. Injection molders: now you can liquid-color plastics. November 1971, **17**(11).

Plastics World. Motionless mixers. May 1978, **36**(5):40.

——. What processors should know about process controls. December 1978, **36**(12):32.

——. Everything you wanted to know about electronic process control, but . . . . February 1973, **31**(2):46.

——. Mixer nozzle cuts injection molding cycle by 10–15%. November 19, 1973, **31**(11):39.

Prasad, A. Injection mold coolant line design can determine mold cycling efficiency. *Plastics Design & Processing*, August 1976, **16**(8).

Product Engineering. Urethane foam process molds large rubbery parts in a hurry. October 1975, **46**(10):23.

Reinbacher, W. R. Injection molding for growth and profit. *SPE Journal*, August 1972, **28**(8):40.

Richardson, P. N., and Houston, J. C. Metering screws in a screw injection molding machine. *SPE Journal*, January 1965, **21**(1):44.

Robinson, A. H. Reap more profit from your injection shop. *Plastics Technology*, September 1975, **21**(9):35.

Rockenbaugh, R. E. Convert any mold to hot-probe runnerless molding. *Plastics Technology*, April 1977, **23**(4):89.

Roobol, N. R., Wenzel, U. R., and Molloy, J. R. Analyzing the processing variables in liquid injection molding of urethanes. *Modern Plastics*, March 1976, **53**(3):58.

Rubin, I. I. *Injection Molding, Theory and Practice*. John Wiley & Sons, Inc., New York, 1972.

Scheiner, L. L. Now you can injection mold rigid PVC. *Plastics Technology*, September 1963, **9**(9):32.

Schott, N. R. Cutting energy use—checking the forces at play. *Plastics World*, November 1978, **36**(11):71.

Scott, J. N., Peters, D. L., and Boeke, P. J. Insulated runner systems for injection molding. *SPE Journal*, September 1959, **15**(9):785.

Seres, A. How to set up and run a runnerless molding system. *Plastics Technology*, March 1970, **16**(3):44.

Severs, E. T. *Rheology of Polymers*. Reinhold Publishing Corporation, New York, 1962.

Sherlock, J. M. Injection molders: here's how you can cut that cycle. *Plastics Technology*, June 1970, **16**(6):52.

Silverwood, H. A. Gauging the future of reaction injection molding. *Modern Plastics*, August 1975, **52**(8):56.

Skross, R. E., and Walker, R. H. Processing tips for molding poly(amide-imide) resins. *Plastics Engineering*, January 1976, **32**(1):35.

Slyh, R. E. Handling finished parts. *SPE Journal*, March 1970, **26**(3):43.

Smith, J. V., and Scott, J. N. Injection molding rigid polyethylene. *Modern Plastics*, September 1957, **35**(1).

Sokolow, N. N. Which injection screw? . . . and why? *Modern Plastics*, March 1978, **55**(3):60.

SPE Journal. New injection system slashes cycle times. July 1972, **28**(7):35.

Staub, R. B. Effects of basic polymer properties on injection molding behavior. *SPE Journal,* April 1961, **17**(4):345.

Stinson, S. C. Computers in injection molding–the key is the interface. *Plastics Technology*, April 1972, **18**(4):44.

Tenney, J. H. Fine-tune your process with today's controls. *Plastics Engineering*, December 1975, **31**(12):15.

Thayer, W. J. Is there an optimum control for injection molding? *Plastics World*, March 17, 1975, **33**(3):45.

Titlebaum, R. P. Liquid injection molding: low-cost process for high-performance parts. *Plastics Engineering*, May 1975, **31**(5):30.

Todd, J. L. How high speed injection can increase your productivity. *Plastics Design & Processing*, October 1978, **18**(10):18.

Tsutsumi, S. Probes for runnerless injection molds enhance viscosity control. *Plastics Engineering*, August 1978, **34**(8):22.

Uglione, H. L. Survey of a new concept–liquid resin molding. *Polymer Engineering and Science*, November 1970, **10**(6):345.

Union Carbide Corporation. *Injection Molding of Bakelite Polyethylene*. Bulletin J-2229 036-1, 1961.

——. *Bakelite Polysulfone for Injection Molding and Extrusion*. Bulletin F-42072.

Usab, M., and Moritz, D. Injection molding of UPVC, Part 1. *Western Plastics*, January 1967, p. 17.

U.S. Industrial Chemicals Co. *Polyolefin Injection Molding: An Operating Manual*. Bulletin P2-1277.

VanSickle, R. W. Mold design technology for polyethylene. *Plastics Technology*, January 1958, **4**(1):35.

Waters, C. E. Sizing chiller to mold can save a lot. *Modern Plastics*, April 1969, **46**(4):147.

Weir, C. I. Introduction to injection molding. *SPE Processing Series*, 1975, Library of Congress Catalog No. 75-10034.

Western Plastics, Technical information on Marlex high density polyethylene. September 1958.

Williams, R. F. Jr. Effect of process variables on mold shrinkage. *Modern Plastics*, September 1967, **45**(1):185.

Wood, R. Multicomponent injection molding for sandwich construction. *Plastics Machinery & Equipment,* January 1978, 8(1).

——. Stack molds increase output in injection molding machine. *Plastics Machinery & Equipment*, April 1978, 8(4):35.

——. Multicolor injection molding in Europe. *Plastics Machinery & Equipment*, October 1978, 8(10).

——. Clamping systems for injection molding: a review of the options. *Plastics Machinery & Equipment*, October 1977, 7(10):21.

——. Nozzle designs for injection molding of thermoplastics. *Plastics Machinery & Equipment*, July 1976, **6**(7).

Wroten, C. D. A guide to estimating the cost of plastic parts. *Machine Design*, May 18, 1972, **44**(10):123.

11
Extrusion

Extrusion is a process in which plastics materials, usually in the form of pellets, are continuously formed into tubing, rods, sheet, pipe, wire and paper coating, films, and a myriad of profile shapes. "Continuously" is the key word, since each of the above items could be made by other techniques—but only in relatively short lengths. The extrusion process is used whenever extended lengths of a product are required, or when a large number of short pieces are required which can be cut from such extended lengths. An extruder can be used to compound materials by mixing two or more components together. It can also blend virgin with regrind stock and, in fact, is the machine used to form long filaments which, when chopped up, are the pellets used to injection mold or extrude special shapes.

In addition to the basic extruding operation, the processor may be required to perform a number of post-extruding operations such as forming round to oval shapes, cutting sections to length, blowing the part to a larger or different shape, machining the extrusion, punching, etc. When combined with the various post-extruding operations, the extrusion process is a most useful and versatile technique. Besides being highly versatile, the extrusion process is one which can result in the highest rate of output per hour of any plastics process. Some large extruders have a throughput of as high as 2000 pounds per hour of pipe or a similar part. Most extruders contain a single screw and function basically in the same manner as a common meat grinder, except that the product is a continuous length resembling the material obtained from a toothpaste tube. There are some extruders containing two or more screws, and some extruders have been made which have no screws. The great majority of the extruders in use today, however, are of the type known as single-screw extruders.

The single-screw extruder consists essentially of the following seven machine elements:

1. A feed throat into which the pellets are placed.
2. The barrel, which contains the screw and is equipped, on its exterior, with elements for heating and cooling.
3. The screw, which acts to plasticate the pellets, heat them to a liquid state, and convey them to the die.
4. A motor and gear reduction system used to turn the screw.
5. Screen packs and breaker plates used to "filter" the melt and create backpressure.
6. The die, which determines the shape of the extrusion.
7. Thermocouples, pressure gauge valves, and other instrumentation used to indicate and control the machine variables of temperature, pressure, and screw-rotation speed.

Figures 11-1a and b illustrate a single-screw extruder and show the various heating zones on the barrel. Contrast this machine with the one shown in Figure 11-2, one of the first extruders (rubber "tuber") built in the United States around 1880.

The extrusion process begins with the pellets being fed into the machine through the hopper and feed throat. The screw, forcing the pellets through the tightly fitting barrel, then plasticates and compresses them until a liquid melt is obtained. The melt is then pumped through the die, which shapes the extrusion. On leaving the die, the extrusion is cooled, sized, solidified, and drawn off by some type of take-off equipment. The final product is then cut to length, reeled up, coiled, etc.

The sequence of feeding, plasticating, and extruding is common to all extrusion processes: tubing production, film extrusion, wire and paper coating, etc. The further sequences of cooling, sizing, and taking-off differ from process to process. However, in almost all the techniques, when the machine adjustments have been correctly made and the machine is fed continuously, the process can be completely automatic. The operator then merely acts as an inspector, making periodic adjustments to assure a product of uniform quality.

In order for the operation to be completely automatic, however, for any given type of extrusion, the right machine must be used: the right screw, the right barrel, pull-off equipment of the type best suited for that extrusion, etc. All of this adds up to a precision machine and precision operation with regard to time, temperature, pressure, speed of screw rotation, speed of pull-off, and cooling time. How these factors combine in a process that produces thousands of pounds per hour in a continuous operation over long periods of time is described in the following sections.

The Screw

The most important element in the extruder is the screw. While any extruder can be used to produce almost any type of product in almost any material, in order to produce the product efficiently and competitively, the usual practice

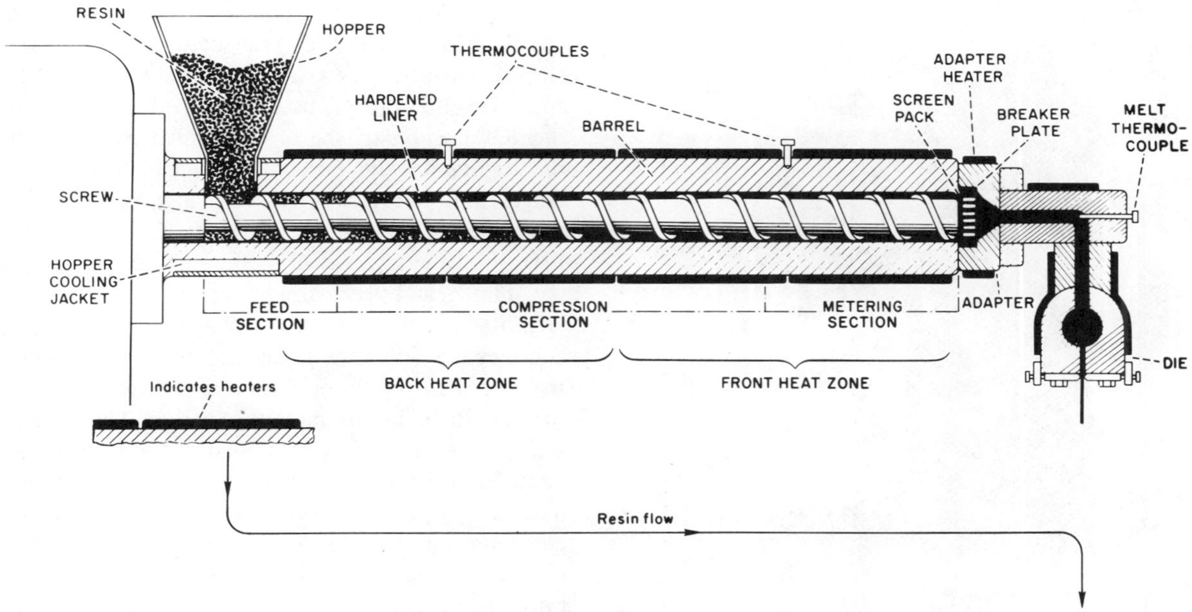

Fig. 11-1a. Cross-section of an extruder. The die is turned at a 90° angle from its normal position so that an end view can be shown. (*Courtesy: U.S. Industrial Chemicals Co.*)

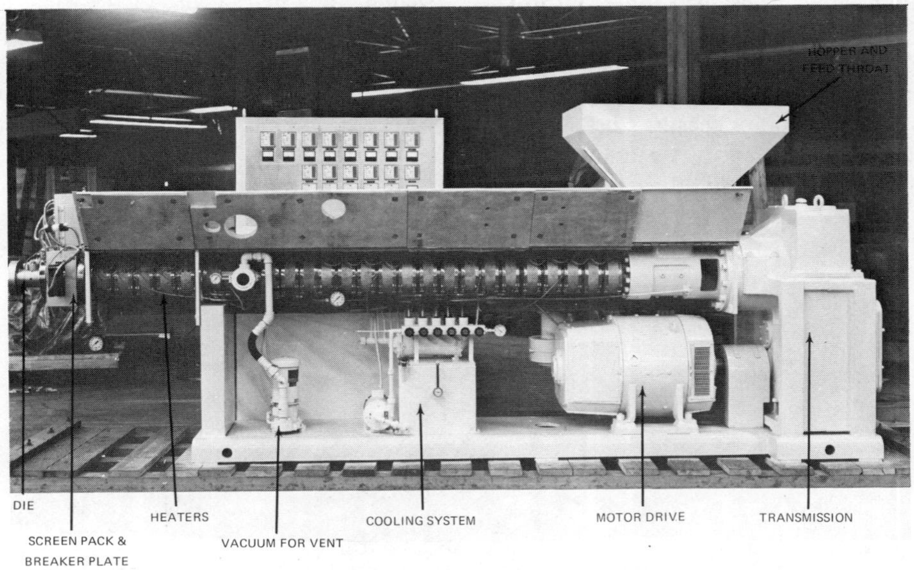

Fig. 11-1b. Modern extruder. (*Courtesy: Welex, Inc.*)

is to use a screw specially made for that material. Less frequently, screws are made for production of a particular product (e.g., extrusion of film).

As shown in Figure 11-3, the screw is usually divided into three sections: the feed section, the transition section, and the metering section. The function of the feed section is to feed material smoothly into the transition section and later into the metering section. The feed section is always designed with the channel (flight) depth greater than either of the other two sections. This assures a conveying capacity well in excess of the other two sections, so there will always be material forced into the transition zone. The helix angle and the lead or pitch of the flights are constant throughout the feed section, and indeed are usually kept constant through the entire screw length, mainly for reasons of economy in machining.

In the transition section, the channel depth is gradually decreased, thus increasing pressure on the granules and causing an increase in the frictional heat generated. In this section of the screw, the pellets are plasticated into a

Fig. 11-2. One of the first extruders built in the United States. (*Courtesy: John Royle & Sons*)

smooth, homogeneous melt which is then forced into the metering section.

The metering section of the screw is usually made with a constant channel depth, which, however, is less than either the feed or the transition section. Because the channel depth is constant, the melt is pumped at a uniform rate through the metering section to the die. Figure 11-4 illustrates these sections (in a somewhat exaggerated manner compared to Figure 11-3).

Extruder screws are classified by their length-to-diameter (L/D) ratio, which is also the designation used to classify barrels, since the screw fits the barrel with very little clearance. The L/D ratio is usually defined as the length of the flighted portion of the screw divided by the diameter of the flights (or the inside barrel diameter). Ratios in use range from as low as 12:1 to a high of 30:1. The most common L/D ratios are 20:1 and 24:1. The higher the L/D ratio, the higher will be the surface available for shearing, mixing, and plasticating the granules.

Another method of describing a screw is in terms of compression ratio: the comparison of the channel depth in the first flight of the feed section and the channel depth of the last flight in the metering section. Thus, a screw with a flight-channel depth of $\frac{1}{4}$ inch at the feed section and a depth of $\frac{1}{8}$ inch at the metering section would be said to have a compression ratio of 2. Such a screw would be characterized as a low-compression-ratio screw. A screw with a compression ratio of 4 would be a high-compression-ratio screw.

Most screws in the United States are made of low-alloy SAE 4140 steel. In order to keep wear of the flights at a minimum, the tips of the flights are usually hardened by welding on a hard-facing alloy such as Stellite #6, by flame hardening, or, less frequently, by nitriding. The first treatment is used where the highest degree of abrasion resistance is required; flame hardening is used for "regular" wear. Chrome plating is also used when corrosion resistance is required, such as when PVC is being extruded. Chrome-plated screws can also be cleaned of plastic somewhat easier than nonplated screws. Stainless steel screws are also used when corrosion resistance is required, but these screws

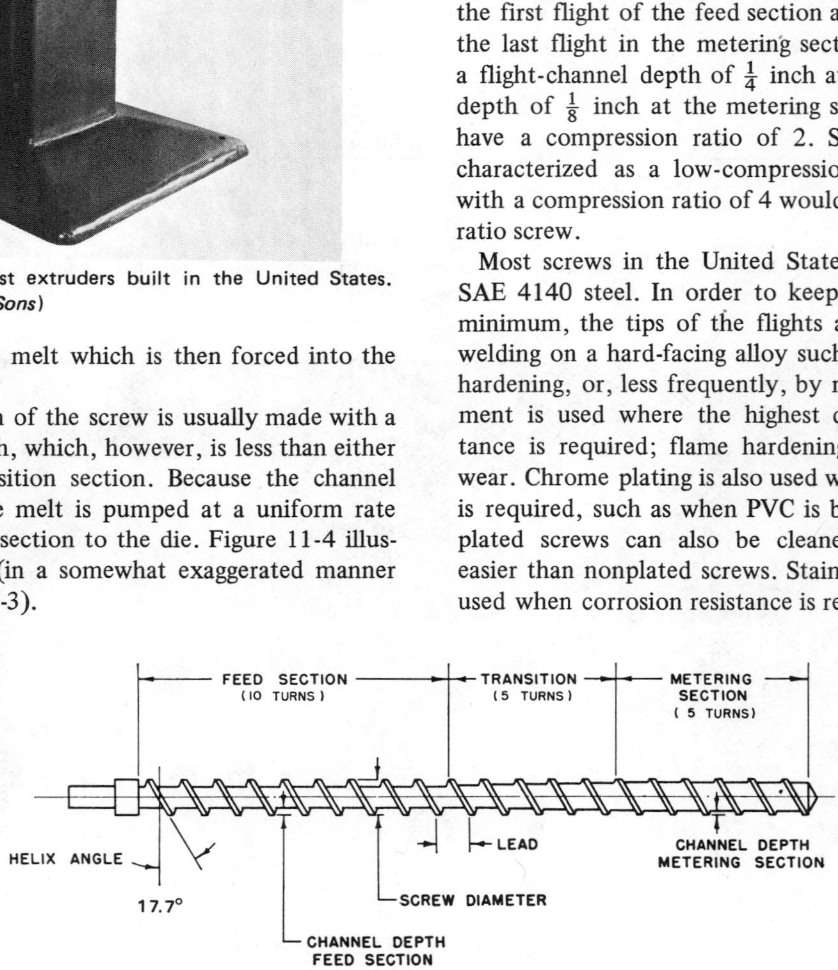

Fig. 11-3. Screw terminology.

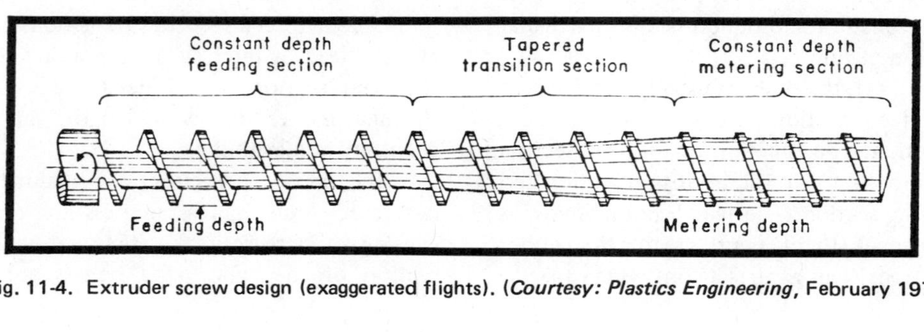

Fig. 11-4. Extruder screw design (exaggerated flights). (*Courtesy: Plastics Engineering*, February 1977)

are said to be unsuitable for use with Saran (polyvinylidene chloride). In addition to special treatment of the flights to minimize wear, most screws are cored out for a portion of their length for heating and cooling, either by steam, water, or oil. The latter is the preferred means since it is noncorrosive.

The end of the screw is usually machined to a 120° point, although some screws have points as sharp as 60° (for heat-sensitive materials such as PVC), and some screws have points which are almost blunt. Since the purpose of the screw point is to prevent formation of a stagnant area just forward of the screw end where material might decompose, screws are also made with removable tips so the point can be changed to fit the material and/or die.

The description of the screws given above applies to the great majority of the screws in use. However, in the quest for increased quality and higher production rates, a number of improved screw designs have been developed. Most of the new screws have had changes incorporated into the metering section. These changes include slots in the end portion of the metering section, pins or slotted mixing rings on the end, overlapping flights, etc. In each case, the intention is to improve homogeneity of the melt and smooth out the pulsations in the flow so more material can be put through in a given time, with higher quality and a minimum increase in temperature.

Figure 11-5 illustrates some of the types of mixing sections used on new screws. Figure 11-6 shows a patented screw design which moves unmelted material from the discharge end of the screw back upstream. It is reported to increase production considerably without a corresponding increase in power consumption.

In addition to making changes in the traditional metering section of the screw, a number of designs have been made in which the change is in the center portion of the screw. Again, the object is to increase productivity by better mixing. One such change is the incorporation of a flight of left-hand threads among the right-hand threads, thus causing a backward flow toward the feed section which results in more intensive mixing. Another design is the incorporation of a spirally restricted section in the center portion of the screw which causes more working of the higher-viscosity materials and allows the lower-viscosity materials to pass through. This also results in more homogeneity of the melt.

In both types of screw improvements, either at the metering end of the screw or in the center, in addition to more intensive mixing, the melt is forced to flow through extremely narrow passageways, narrower than the normal flight depths. Thus, the thinner layer of plastic is heated more uniformly and is melted more thoroughly, resulting in better mixing, compounding, or elimination of small unmelted particles. These "special" types of screws are usually designed for use with one particular material and are by no means general-purpose screws.

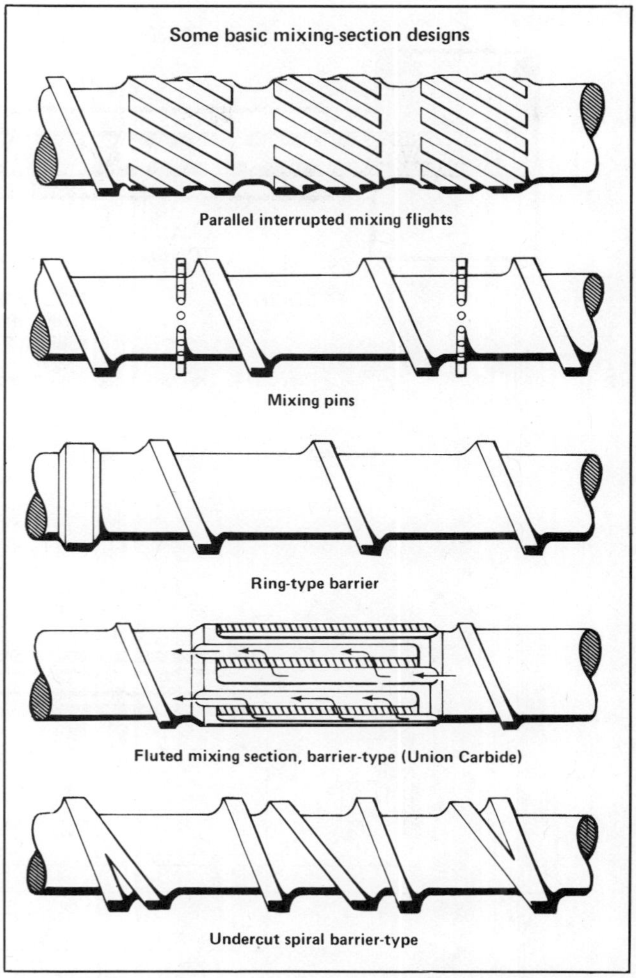

Fig. 11-5. Special mixing sections in extruder screws. The fluted mixing section (also called a Maddox mixer) can also be installed at the screw tip. (*Courtesy: Plastics World*, July 1978)

Possibly the most useful of the improved screws is the type used in conjunction with a vented barrel. In this design, a deep-flighted decompression or extraction zone is included just forward of the metering (transition) zone, as shown in Figure 11-7. When the highly compressed melt arrives at the low-pressure extraction section, it immedi-

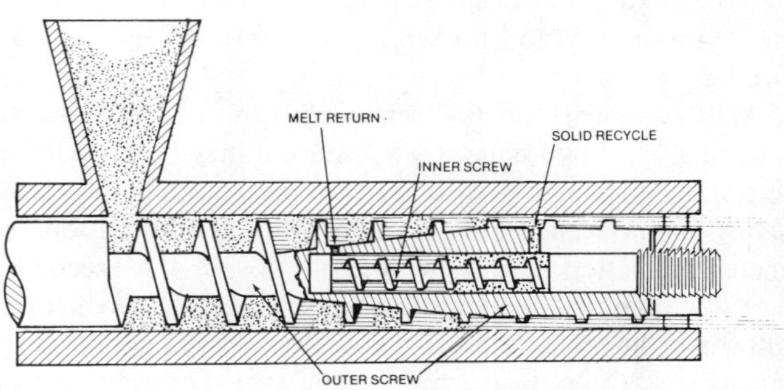

Fig. 11-6. Reverse-flow screw using hollow screw as a secondary barrel to heat further, melt, and convey material back to upstream end. (*Courtesy: Scientific Process and Research, Inc.*)

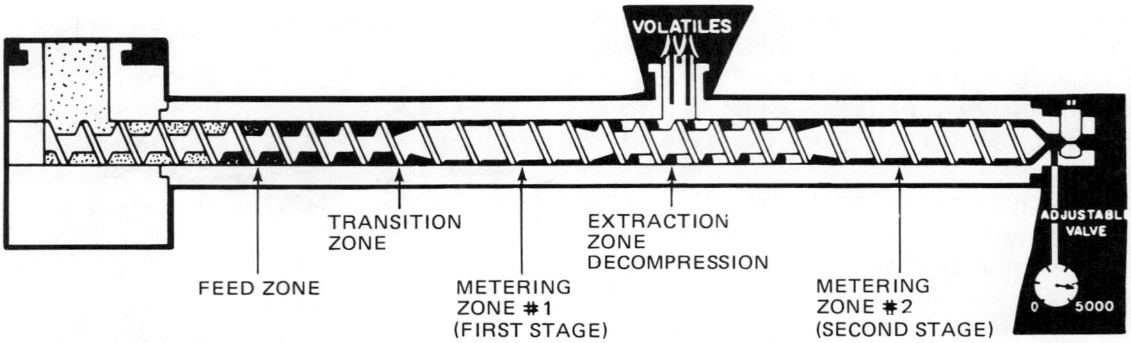

Fig. 11-7. Double-metering-zone screw. (*Courtesy: Prodex Corporation*)

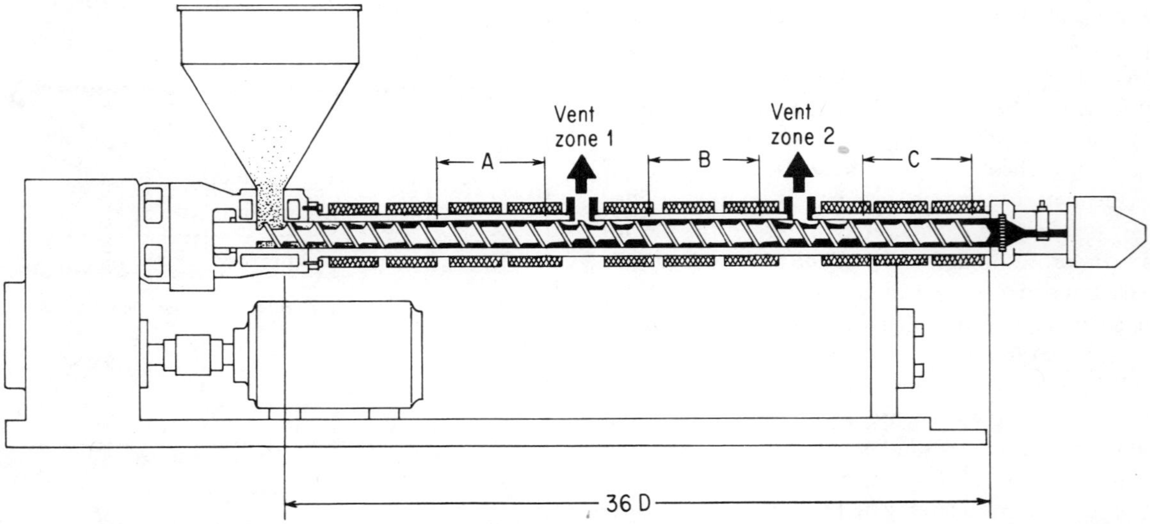

Fig. 11-8. Cross-section of double-vented extruder, showing triple-metering state, 36:1 L/D screw. Dual vent zones exhaust moisture and volatiles, eliminating need to predry hygroscopic ABS. A, B, and C are the three metering stages. (Reprinted by permission from *Modern Plastics*, McGraw-Hill, Inc.)

ately foams up due to the release of the compressed air and liberated moisture. The air and moisture is then allowed to escape or is drawn up through the vent in the barrel and the resin is recompressed by the second metering zone. By correct design of the vent and optimum placement, only the volatiles are lost through the vent. In the event that volatiles extraction is not required, all vented-barrel extruders have a provision for sealing the vent and operating as ordinary extruders.

Another name for the screw used in a vented-barrel extruder is a "two-stage screw," since it may be considered as acting as two screws, one ahead of the other. The first screw has a feed section, followed by the transition or metering section. The new feed section in the "second" screw is then deep enough so the pressure is lowered, thus allowing the escape of the volatiles. Carrying this even further, a "three-stage screw" using two vent sections has been made and is said to be very successful in removing air and moisture from non-predried ABS which is normally quite hygroscopic. Figure 11-8 shows such an extruder which has an extraordinary L/D ratio of 36:1.

Two interesting new developments in screw and barrel configurations are shown in Figure 11-9. In these variations of the "normal" configurations, one machine uses a square screw inside a hexagonal barrel and the other uses a hexagonal screw inside a hexagonal barrel. The developer, Mitsubishi, claims increases of throughput as high as 100% for the "HM" (high-melt) screw. In addition to increasing the output, Mitsubishi asserts that melt homogeneity is improved and that the melt temperature can be reduced. At the time of writing, these claims have not been proved by long-term usage (see Table 11-1).

The Barrel

The barrel, also called the cylinder, contains the screw on an extruding machine. It is thus similar to the screw in dimensions; the inside diameter of the barrel is very close to

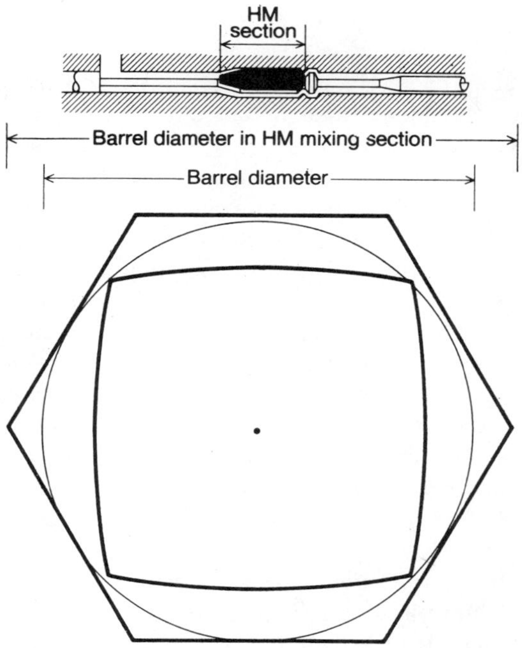

Fig. 11-9a. Profile of two-stage Mitsubishi HM screw shows diameter increase in mixing section; end view shows relationship. (Reprinted by permission from *Modern Plastics*, McGraw-Hill, Inc.)

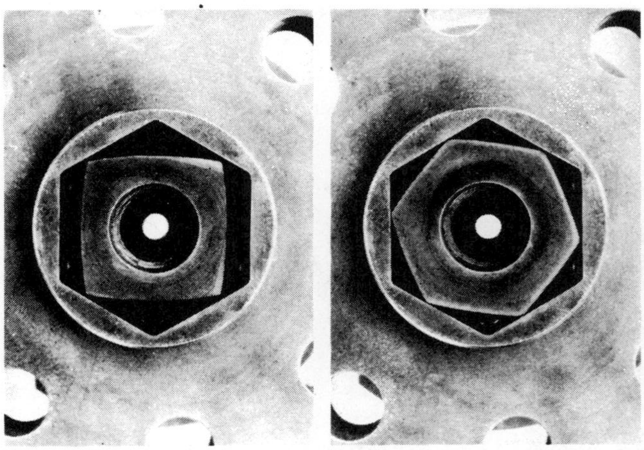

Fig. 11-9b. Two versions of Mitsubishi HM mixing section: square screw in hexagonal barrel and hexagonal screw and barrel. Shallow, interrupted flights are not visible. (Reprinted by permission from *Modern Plastics*, McGraw-Hill, Inc. Photo: Mitsubishi)

the outer diameter of the metering section of the screw flights, and the length is almost the same as the screw. Therefore, many extruders are designated by the L/D ratio of the barrel. In addition to containing the screw, the barrel is also a source of heat and cooling for the plastics material, since it is always surrounded by heaters and cooling coils.

Most barrels are currently of bimetallic construction. A tough outer section, frequently of 4140 steel alloy, is fitted with an inner lining of a centrifugally cast, very hard

Table 11-1. Comparing HM Screw Performance.[a],*

Material (melt index)	Screw type	Screw speed, r.p.m.	Melt temp., °C.	Output, kg./hr.
PP (1.3)	HM	153	263	398
	Regular	140	263	166
PS (g.-p.)	HM	150	249	483
	Regular	121	251	250
HDPE (0.2)	HM	100	237	390
	Regular	100	256	251

[a]Using 90-mm, 27D HM screw versus 90-mm, 29D normal screw.
*Reprinted by permission from *Modern Plastics*, McGraw-Hill, Inc.

and abrasion-resistant alloy. The alloy most commonly used in the United States is Xaloy, which is available in three variations. Xaloy 101 is a nickel-chromium-boron-iron base alloy used to provide wear resistance at a reasonable cost. Xaloy 306, a cobalt-nickel-chromium-boron alloy, is used for maximum corrosion resistance and good wear resistance. Xaloy 800 is a composite of tungsten-carbide particles dispersed in a hard, corrosion-resistant nickel-base alloy matrix, and is used to provide the ultimate in abrasion resistance combined with excellent corrosion resistance. The Xaloy cylinder liners are commonly used at thicknesses of $\frac{1}{16}$ inch; however, thicknesses of up to $\frac{1}{4}$ inch have been made. The standard bore surface finish of a honed Xaloy bimetallic cylinder is 8–12 rms. Less frequently used in the United States, although common in Europe, are barrels with nitrided inner surfaces, which provide fair abrasion resistance with minimum cost.

Since the extruding pressures can be very high inside the barrel, most are made to withstand pressures up to 10,000 psi. Some barrels have been made to withstand pressures as high as 15,000 psi. Such a barrel pressure might be encountered in an extruder with an L/D ratio of 40:1.

The barrel, which is rigidly attached at the drive portion of the extruder, is supported along its length by sliding contacts which allow for longitudinal expansion and contraction as the barrel heats and cools. In addition, auxiliary supports may be provided at the die end to prevent deformation when heavy dies are used.

Heating of the barrel is required for startup of almost all materials and is definitely required for running of a few materials such as a low-viscosity polyethylene, in which not enough heat is generated by friction to cause the plastic to stick to the barrel surface. Most heaters are electrical, the least expensive of which are simple metal bands containing resistance wires insulated by mica or cement. Somewhat more expensive are tubular heaters imbedded in half-round aluminum castings. The same casting in some cases is also cored or contains cast-in stainless steel cooling coils. The most expensive heating system is induction heating, in which case the heating takes place in the cylinder wall itself, with no additional or external heaters required. This system offers the advantage of quick and uniform heating, and very easy cooling, since there are no heater bands to cool off. The heaters, regardless of the type, should be capable of heating the barrel to temperatures as high as approximately 290°C (550°F). To obtain these temperatures within a reasonable length of time, watt densities of from 25 to 50 watts per square inch are used.

Since there are three or more distinct zones in the screw length, there are usually three or more heating zones on the barrel. Small extruders of approximately 2-inch diameter commonly have three heating zones, while larger extruders will have four or five. Each zone will have its own temperature sensors and controller. Regardless of the type of resistance-heater element installed on the barrel, it is important that it makes good contact with the barrel over its entire surface to insure maximum heat transfer, and to minimize the chance of premature burnout of the heater. Normal expansion and contraction of the barrel and the heater bands will gradually cause loose contact between the heaters and the cylinder. Periodic inspection should therefore be made to insure good contact at all times.

Cooling of the barrel is sometimes more important than heating, particularly with the trends toward higher outputs, which mean higher melt temperatures, and greater possibility of resin degradation. Cooling may be done with air, water or, in some cases, a special liquid. Air cooling is usually done by means of fans which blow room-temperature air, through sheet-metal ducts, over the heaters and barrel. This system is simple, inexpensive and has the advantage of minimum thermal shock to the barrel. On the other hand, it has the disadvantage of being lower in efficiency than a liquid cooling system. Most liquid systems in use today employ water as the coolant, either circulated in separate cooling coils or in coils cast into the heater blocks. The simpler systems use water directly from the plant supply. A superior system is a closed-loop circulating system using distilled or specially treated water to minimize fouling of the cooling lines. Such a system would have its own pump, a distribution manifold, and solenoid-operated valves for each zone. In addition, there would be a heat exchanger to cool the hot water and steam before returning it to the reservoir. The major disadvantage of the circulating liquid system is the thermal shock which may be imparted to the barrel and heaters by the sudden circulation of cold water. Figure 11-10 shows a cast-in tubular heater system and a closed-circuit cooling coil system with the accompanying pump, valves, and gauges.

In addition to containing the screw, and heating and cooling the plastics granules, many barrels also contain the vent outlet for volatiles removal. Although the vent tube is usually shown in drawings on top of the cylinder, in most modern installations, the vent tube is positioned on the side of the cylinder. The side installation has the following advantages over a vertical vent:

1. If volatiles condense, the condensate will not drip back into the barrel.
2. In the case of vent flow during startup, with an open vent, venting material will not flow over the barrel and heaters.

Fig. 11-10. Closed-circuit cooling system. (*Courtesy: Welex, Inc.*)

3. Side vents can be easily observed through the glass port.
4. Side vents are easy to clean and there is less possibility of dropping objects into the screw flights.

A typical side-vent setup, including vent, vacuum pump, and sight glass, is shown on the left of Figure 11-10.

The barrel has two other functions in addition to those previously described. At one end, the plastics granules are fed in; at the other end, the molten material is extruded. The end where the granules are introduced into the extruder is bolted to what is appropriately known as the feed section. A large hopper tops the feed throat where the plastic is actually fed into the barrel. The hopper is usually equipped with a sight glass to show the amount of material inside, a side discharge, and a shutoff valve. Optionally, the hopper can be equipped with a crammer feeder which, by forcing the resin into the throat, maintains positive feeding at all times, particularly for very fluid or low-density materials (see Figure 11-11).

Below the hopper is the feed throat proper. This is usually rectangular to accommodate the hopper, except when a crammer feeder (a screwlike device) is used, in which case the feed throat will have a circular opening. The feed throat is also usually water jacketed so that cooling water may be circulated. This is done to prevent melting of the granules at the throat, and subsequent "bridging" across the throat, which could then result in a loss of feed and a consequent suspension of production. Feed throats are commonly chrome plated to minimize corrosion by the possibly moist pellets, and to minimize sticking of the granules as they enter the barrel. Figure 11-12 shows a typical feed throat, with the water line and an opening for inspection and cleaning.

The exit end of the barrel is attached to an end gate which commonly contains a screen pack, a breaker plate, a valve, a pressure gauge, and some means of attaching a die to the end gate. Usually, the end gate is made so it can be easily swung away from the barrel while supporting the die, so the breaker plate and screen can be serviced. Figure 11-13 shows an end gate and a die mounted on the machine in the closed position. The open position, showing access to the breaker plate and screen pack, can be seen in Figure 11-14.

The screen pack and breaker plate mentioned above are placed in the end gate just at the end of the screw. The screen pack consists of a number of discs of wire cloth of varying mesh sizes, whose function is to screen out contaminants, metal hardware, or unmelted resin particles from the melt stream. The breaker plate is a heavy disc of metal, perforated with a number of holes or slots, ranging from $\frac{1}{8}$ to $\frac{3}{16}$ inch in size. Its function is to support the screen pack against the pressure of the melt. A secondary function of the screen pack is to help develop pressure in the melt, to cause better mixing and/or higher melt temperature.

Fig. 11-11a. Crammer feeder mounted on an extruder. (*Courtesy: Welex, Inc.*)

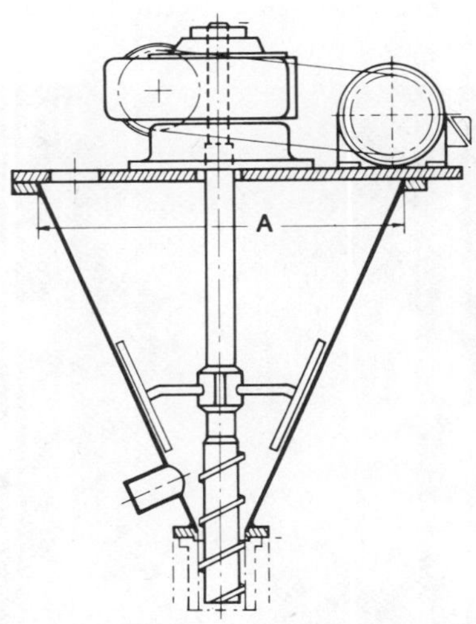

Fig. 11-11b. Cross-sectional view of crammer feeder. (*Courtesy: Welex, Inc.*)

One of the problems connected with the use of the screen pack and breaker plate to develop pressure is that as the screens became clogged, the pressure could, in some cases, rise very suddenly, leading to an increase in melt temperature and subsequent melt degradation. In addition, the necessity to replace the screen causes downtime on the machine, with a loss of production. One method of reducing the machine downtime is to use a screen changer, illustrated

Fig. 11-12. Extruder feed throat. Note water-cooling line to prevent jamming by molten resin. (*Courtesy: Welex, Inc.*)

in Figure 11-15. The unit operates in a manner similar to a slide projector. When a screen change is indicated, the unit, hydraulically or mechanically, slides a new screen pack into place and removes the old screen. The entire operation can occur with no interruption in the extruder operation, but marking on the extrudate is possible. Other screen changers are available which operate continuously, resulting in little or no marking on the product.

To minimize the effect of screen clogging on pressure variations, many modern extruders rely on valves which

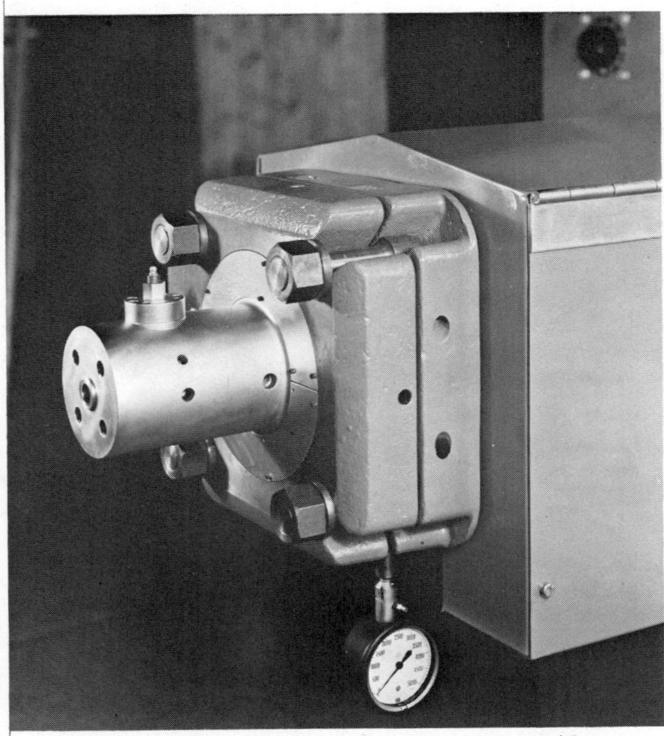

Fig. 11-13. End gate and die in operating position. (*Courtesy: Welex, Inc.*)

Fig. 11-14. End gate open for servicing screen pack and breaker plate. (*Courtesy: Welex, Inc.*)

Fig. 11-15. Hydraulically actuated screen changer. (*Courtesy: The Berlyn Corporation*)

are placed in or near the gate section. In general, two types of valves are used. One type uses the screw tip as the valve by arranging for the screw to be moved forward or back, thus resulting in a variable clearance at the gate. The second valve type is a variation of the needle-valve principle. Both types are shown in Figure 11-16.

The pressure gauge shown on the bottom section of the end gate in Figure 11-13 is usually a simple Bourdon tube pressure gauge, the stem of which is full of silicone grease. The function of the grease is to transmit the pressure to the tube. Since it is silicone, it will withstand the high temperatures without degradation. The tube is mounted on the bottom of the barrel so the grease will not run into the melt. Electronic transducers can also be used, particularly if remote indication or a graphic record is required.

Extruder screw speeds range from 20 to 200 rpm. In general, electric motors are used which run at speeds of 1750 or 1800 rpm. In order to turn the screw at the desired lower speed, a gear-reducer unit is used which also has the function of increasing the torque needed to turn the screw. The power requirements generally run to approximately 1 horsepower for 5 to 10 pounds of material extruded per hour. Thus, a 1000-pound-per-hour machine would require a 100 to 200-horsepower motor. The differences arise from the fact that some resins, such as PVC, require considerably more power to cause adequate shearing and melting than do other resins, such as polyethylene.

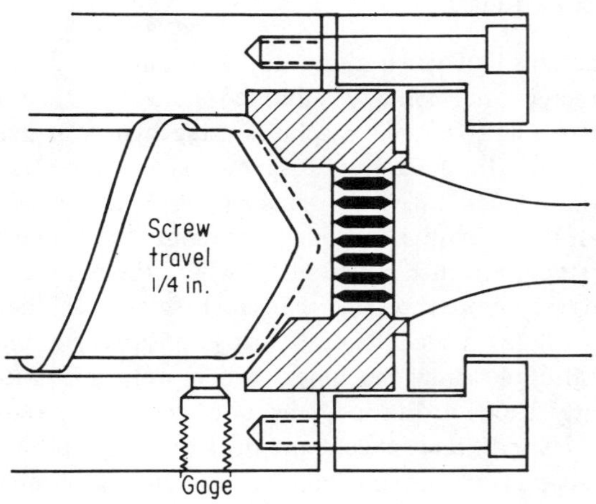

Fig. 11-16a. Valving by change of tip distance from die. (Reprinted with permission from *Modern Plastics Encyclopedia*, 1968. Copyright 1968, McGraw-Hill, Inc. All rights reserved.)

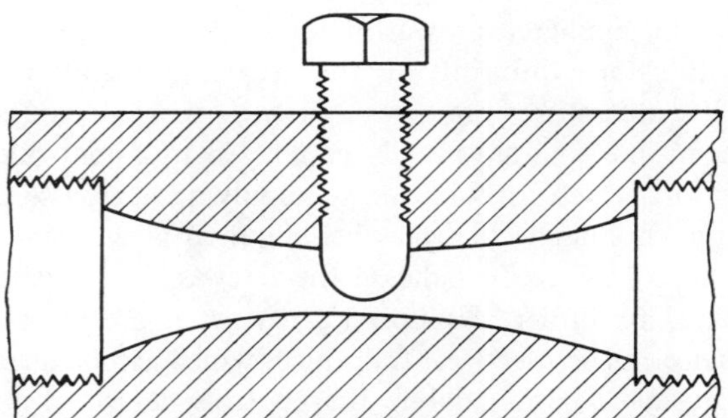

Fig. 11-16b. Needle valve placed in flow constriction near die. (Reprinted with permission from *Modern Plastics Encyclopedia*, 1968. Copyright 1968, McGraw-Hill, Inc. All rights reserved.)

The gear-reducer unit is commonly made with interchangeable gears so a range of speeds is available. Other speed reducers, particularly on smaller extruders, use a simple system of belts and pulleys. In addition to speed reduction by the gear or belt system, continuous variation in speed is normally available by means of a magnetic clutch drive, or the motor can be a DC type, running on AC through a solid-state rectifier system. The DC motor is easily regulated to 1% accuracy in speed control. The DC motor also has good torque at low speeds. A magnetic clutch drive (eddy current drive) can also be made to give good regulation at relatively low cost; however, low-power units require air cooling and high-power units require water cooling. A higher noise level than other drives and a somewhat reduced efficiency at low speeds are also disadvantages of magnetic drive regulation.

In generating the considerable pressures in the barrel, the screw also generates a very high backpressure against the thrust bearing which supports the screw. Thrust bearings are rated in terms of their B-10 life in hours, at a given pressure and rpm. The B-10 life is a statistical figure indicating the number of hours at which 10% of the thrust bearings are expected to fail, at the given pressure and rpm. The B-10 life is inversely proportional to the screw rpm and is approximately inversely proportional to the cube of the pressure. Thus, exceeding the normal rating of screw speed and/or pressure for long periods can seriously reduce the bearing life.

Fig. 11-17. Remote-control panel for a large extruder. (*Courtesy: Welex, Inc.*)

Controls

All extruders are equipped with several basic indicating and controlling instruments. At the minimum, controls are used to indicate screw speed, power consumption, and barrel temperature and pressure. Of all the instruments, one of the most important is an ammeter or a wattmeter, used to indicate the power being used by the motor. An increase in power can mean an increase in friction in the barrel, excessive working of the pellets, or a sudden cooling of the melt. A decrease could mean a loss of material in the feed hopper, bridging in the hopper, or excessive heat in the melt. The ammeter or wattmeter is thus most useful in indicating the uniformity of the extruding operation. The barrel temperatures are measured with thermocouples set in each zone, and are usually connected to a controller of some sort which, in addition to supplying heat, also actuates the cooling mechanism when required. A tachometer is used to continuously indicate the screw speed. Pressure is indicated by either a Bourdon gauge set at the die end of the barrel, or an electrical transducer which can be installed at the same location. While the controls described are all spoken of as indicating instruments, they may also be in the form of indicating and recording instruments, so graphical or other records can be kept of each machine's production.

In addition to the basic instruments and controls described, the machine may have auxiliary meters and instruments: ammeters or pilot lights on the heaters to show the power consumption or possible burnout, timers to show the hours of usage, preheat time switches, special motor-start packages, etc. The control panel, containing all the indicating instruments, may be mounted on the machine in the case of smaller extruders, or it may be in a separate panel away from the machine, as is shown in Figures 11-17 and 11-18. Almost all control units today are solid-state models.

Special Extruders

The descriptions given above apply in almost all cases to single-screw, horizontally mounted machines. There are, however, multiple-screw machines using two or as many as four screws. In general, multi-screw machines are used for materials which are heat sensitive, such as PVC. Other special uses for these models, predominantly the twin-screw type, are in compounding, where the mixing action is said to be more thorough than single-screw machines, and with materials which require steady, positive pumping action. Multiple-screw machines, which were introduced in the early 1960s in Europe, have not been used extensively in the United States except for rigid PVC pipe production and specialized mixing and compounding operations. Figure 11-19 illustrates one such extruder.

Another type of extruder, once used exclusively for plasticating materials for the record industry, is now being

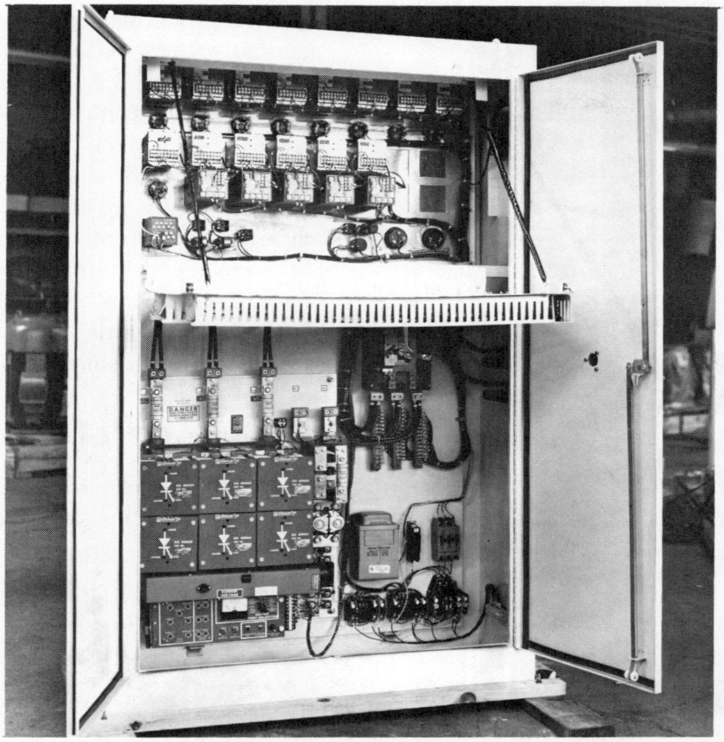

Fig. 11-18. Solid-state controls used in modern extruder. (*Courtesy: Welex, Inc.*)

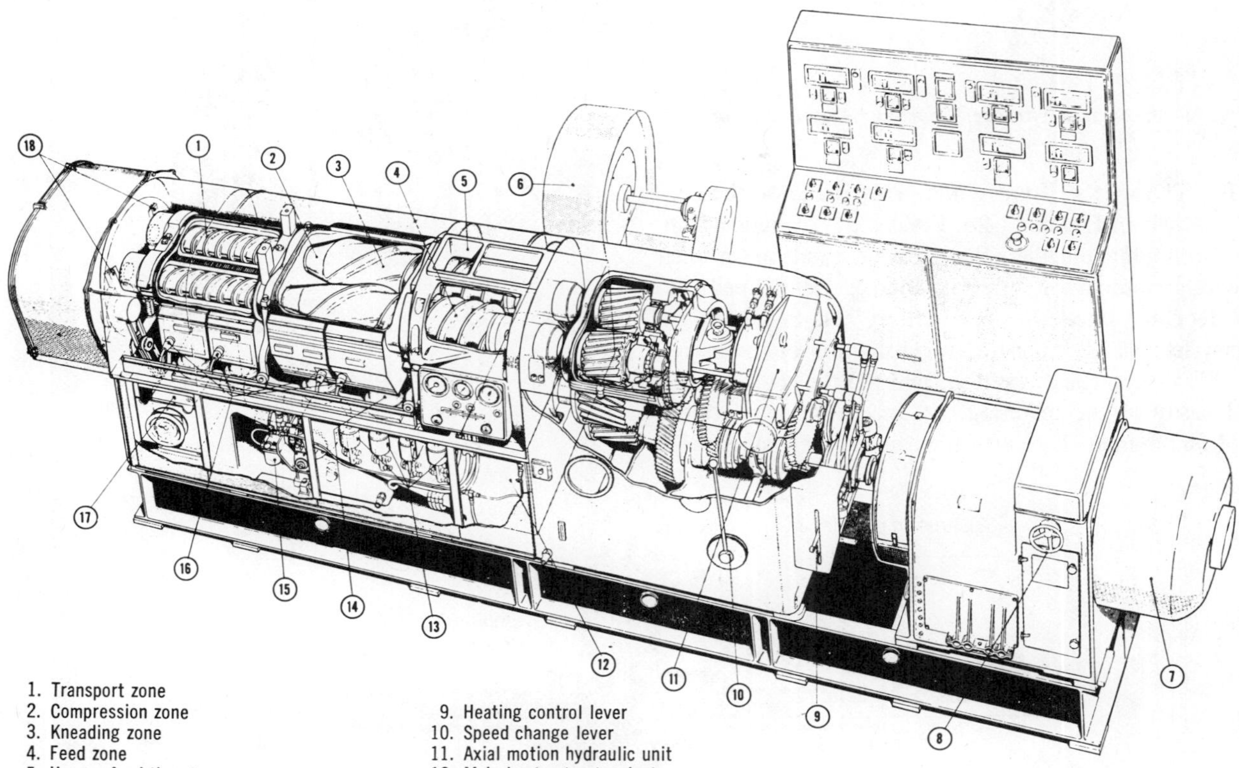

1. Transport zone
2. Compression zone
3. Kneading zone
4. Feed zone
5. Hopper feed throat
6. Heat extractor fan
7. Main drive motor
8. Drive motor speed regulator
9. Heating control lever
10. Speed change lever
11. Axial motion hydraulic unit
12. Main heat extractor duct
13. Heat duct valve control units
14. Heater ducts
15. Axial motion control unit
16. Heater plugs
17. P.I.V. gearbox driving granulating cutters
18. Belt-driven granulating cutters

Fig. 11.19. Twin-screw extruder. (*Courtesy: SPE Journal*, September 1962)

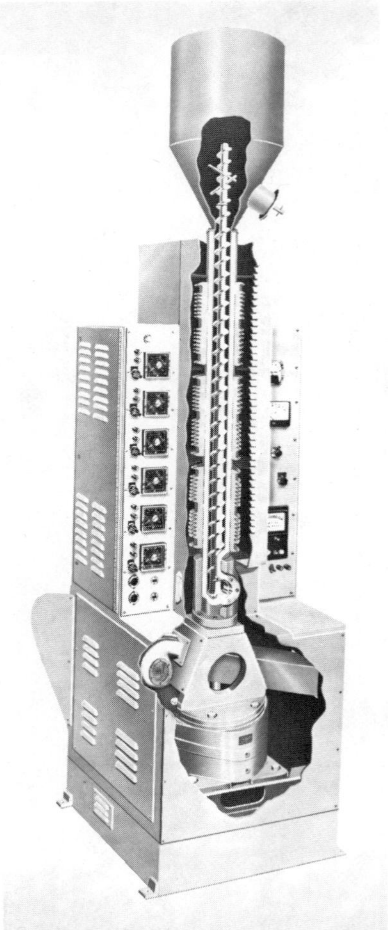

Fig. 11-20. Vertical extruder. (*Courtesy: Aragon Division,*

used for all aspects of extruding. This is the vertical extruder, shown in Figure 11-20. Besides the obvious advantage of occupying much less floor space than a conventional extruder, a number of other advantages are claimed: (1) better feeding, since the screw itself acts much like a crammer feeder; (2) no cantilever distortion of the barrel, since it is in a vertical position; and (3) the screw can be cooled easily at the metering end for better temperature control of the melt. The vertical extruders are mainly made in 1 to 2-inch diameters with L/D ratios of 31.5 to 1.0. Currently, these machines are being used for rigid and flexible profiles, wire coating, pipe and tubing, blow molding, and general compounding.

Other types of extruders (see Figure 11-21a) have been made which work without a screw and depend on friction between a rotating and a stationary plate to develop the shear action. Still other extruders (see Figrue 11-21b) have been made which use a ram, or two rams, feeding a manifold, to result in continuous extrusion. None of these "non-conventional" extruders, however, have been adopted to any great extent commercially.

Dies and Take-Off Equipment

The extrusion operation covers a large variety of specialized operations: profile shapes of many kinds, wire coating,

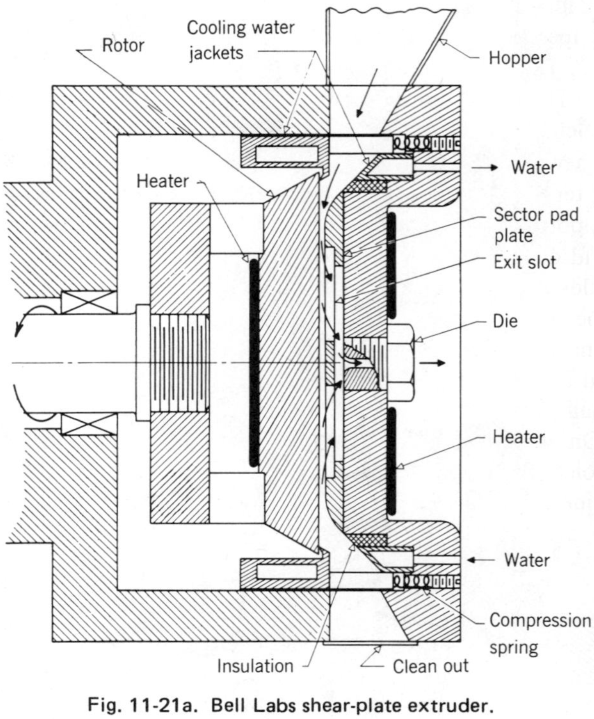

Fig. 11-21a. Bell Labs shear-plate extruder.

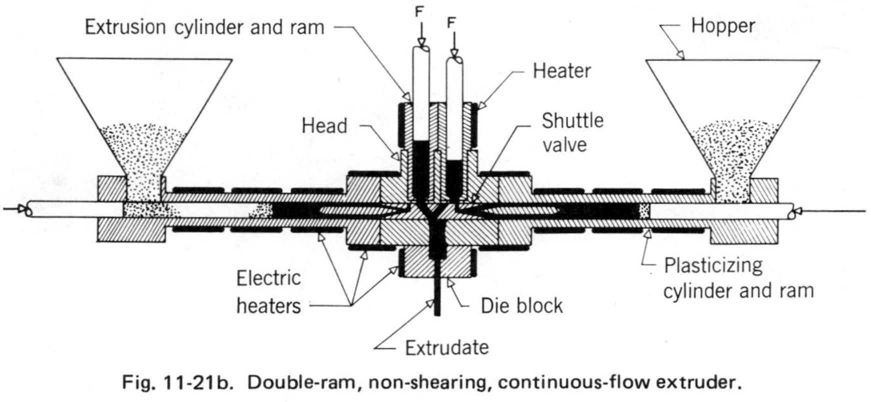

Fig. 11-21b. Double-ram, non-shearing, continuous-flow extruder.

production of pipe and tubing, coating of paper and cloth, film and sheet products, etc. All of these operations have one common element—the extruder. Up to the die, the machines are more or less similar. The dies for each type of product, however, are different, as are the kinds of take-off, sizing, cutting, coiling, and stacking equipment. A description of these various extruding operations is thus in order.

Pipe and Tubing. The extrusion of pipe and tubing is one of the major uses of the extruder, and is one of the simpler operations. Pipe is generally defined as material over $\frac{1}{2}$ inch in diameter with a rigid wall. Common materials for pipe include polyethylene, rigid PVC, ABS, and cellulose acetate. Tubing is usually defined as being flexible and less than $\frac{1}{2}$ inch in diameter. Tubing is commonly made of polyethylene, flexible PVC, ABS, and nylon.

The die used for the extrusion of either pipe or tubing is cylindrical in shape, since such a configuration is the cheapest to machine, and a heater band fits it easily and well. The die can be either direct-center fed or can be side fed. The side-fed or cross-head die, while somewhat cheaper, is being phased out by the center-fed die because of the higher quality produced by the latter. Figure 11-22 shows a center-fed die (also called in-line) in which the mandrel is supported by a spider, and Figure 11-23 illustrates a cross-head die. In both cases, the mandrel, also called the pin or male member, is used to form the inner diameter of the pipe or tube, and is rigidly supported. The die bushing, or female part of the die, is used to form the outer diameter and is attached to the die body by adjustable screws. Thus, small changes can be made in the wall thickness if required.

On leaving the die, the extruded pipe or tube must be cooled and sized. In the case of tubing, the simplest technique is what is known as free extrusion, where the tubing

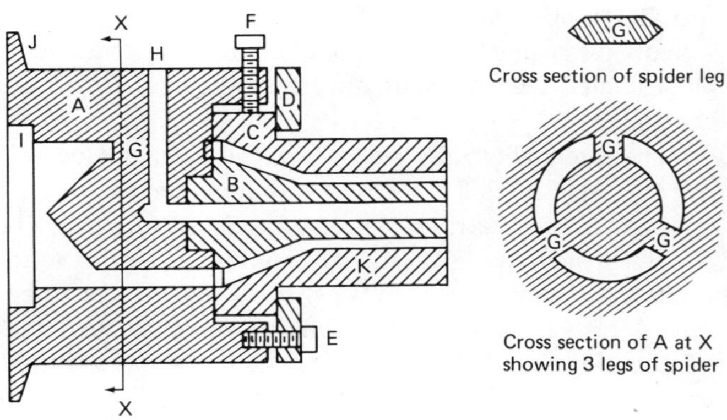

A. Die body
B. Mandrel, pin, male die part
C. Die, die bushing, female die part
D. Die retaining ring
E. Die retaining bolt
F. Die centering bolt
G. Spider leg
H. Air hole
I. Seat for breaker plate
J. Ring for attachment to extruder
K. Die land

Fig. 11-22. Pipe or tubing die for in-line extrusion. (*Courtesy: Society of the Plastics Industry, Inc.*)

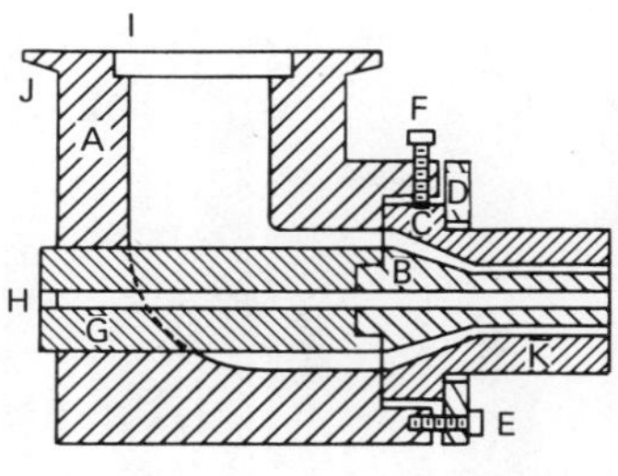

A. Die body, crosshead
B. Mandrel, pin, male die part
C. Die, die bushing, female die part
D. Die retaining ring
E. Die retaining bolt
F. Die centering bolt
G. Mandrel holder
H. Air hole
I. Seat for breaker plate
J. Ring for attachment extruder
K. Die land

Fig. 11-23. Pipe or tubing die for cross-head extrusion. (*Courtesy: Society of the Plastics Industry, Inc.*)

is extruded into a water bath, cooled immediately, and the size is established. The tubing is then drawn off and coiled. For larger diameter, thicker walled and rigid pipe, a more elaborate procedure is required. The pipe, as it is extruded, may pass directly into a water trough and then through a series of sizing plates, or female mandrels, which gradually decrease in size to compensate for the material shrinkage. Air pressure in the pipe, or a vacuum in the water trough, forces the pipe against the sizing member to insure good contact while cooling is taking place. In addition to the sizing which results from contact with the sizing plates or mandrels, the pipe or tubing is drawn down by the pulling mechanism to decrease the diameter. Drawdown of tubing is usually much greater than the drawdown for pipe. The drawdown ratio is expressed as the ratio of the cross-sectional area of the die opening to the tube cross-sectional area. Thus, tubing emerging from a die opening with an outer diameter of 0.4 inch and an inner diameter of 0.3 inch is commonly drawn down to an outer diameter of 0.3 inch and an inner diameter of 0.25 inch. The cross-section area of the die would be $\pi(0.2^2 - 0.15^2)$. The cross section area of the tube would likewise be $\pi(0.15^2 - 0.125^2)$. A comparison of the two areas indicates a ratio of 3.1 to 1. Tubing is commonly drawn down in the range of 2 to 6 times the original diameter. The drawdown of pipe, on the other hand, only ranges from 10% to 20%.

Take-off equipment for tubing can consist of simple coiling equipment. Pipe take-off machines, however, are usually made to grip the cooled pipe firmly, but with not enough pressure to cause distortion in the pipe. Pullers can be either caterpillar-type tread or rubber tire types (some makes of pullers actually use tires). Figure 11-24 illustrates one type of take-off equipment. A traveling saw which cuts the pipe to length or coiling machines complete the pipe extrusion line.

Operation of the pipe or tubing line is largely automatic once the proper conditions have been established. Pipe is extruded fairly slowly, from 20 to 50 feet per minute, depending on the pipe size and the cooling capacity of the equipment. Because of the fairly thick walls on most pipe, the cooling rate is usually the limiting factor in production. For this reason, the material is extruded at as low a melt temperature as possible. This is also advantageous in mini-

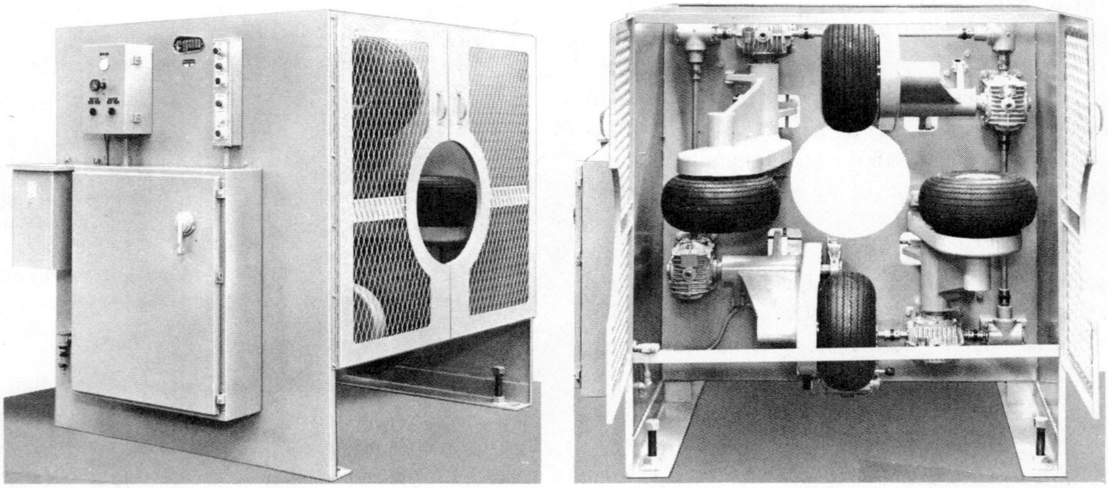

Fig. 11-24. Pipe pulling equipment using pneumatic tires for large pipe diameters. (*Courtesy: Egan Machinery Co.*)

mizing degradation in PVC pipe. Fortunately, since the die openings are quite large, obtaining good flow at the low melt temperature is usually no problem. Tubing is produced at rates of 150 to 300 feet per minute, and is fed directly from the cooling bath to the coiling machine.

Pipe, and to a lesser extent tubing, is usually made to definite specifications. Therefore, a number of standard tests are used to determine the quality of the material. These include burst-strength, impact, weathering, creep- and chemical-resistance tests. The exact test methods are published in ASTM specifications; industry, government, and sometimes proprietary specifications are other sources.

Film. Production of film is another very large application in the extruding world. Two methods of making film are in use. One method is known as the blown film technique, and the other produces what is called cast film (the extruded film is "cast" onto a chilled roller). Blown film production is considerably greater than film made by casting, since the blown film has a number of advantages over the cast film product. In general, the blown film can be made at a faster rate than the cast film; it is stronger and tougher; and the equipment to produce the same amount of film by the blown film process takes up considerably less space than does a line for production of cast film.

Film is defined by convention as material less than 0.01 inch thick. Material over 0.01 inch in thickness is called sheet. By far the greatest amount of film is made of polyethylene, with polypropylene, PVC, polystyrene, and nylon also being used.

Blown Film. The blown film process consists of extruding the molten plastic through a circular die to form a thin-walled tube. The tube is almost always extruded vertically, and, as it leaves the die, internal air pressure, introduced through the die mandrel, expands the tube anywhere from 1.5 to 2.5 times the die diameter. Figures 11-25 and 11-26 illustrate the blown film extrusion process. Because the film is extruded vertically, buildings housing such a process must have ceilings 35 to 50 feet high.

The die used in making the blown film is similar to a pipe or tubing die in that it is circular and may be either center fed or side fed. There are, however, a number of important differences. The die is made, in many cases, so that it can be rotated or oscillated around the inner mandrel during the extrusion operation; it is also possible for the mandrel itself to rotate. The rotation minimizes the effect of any small irregularities which might be present on one part of the die or mandrel. Such irregularities can show up as a thick spot on a wound roll, uneven thickness of one edge, etc. Other extrusion dies are made with inner mandrels which are adjustable axially so that the opening (land length) may be changed slightly to allow tubing of various thicknesses to be extruded. Another die design uses a spiral mandrel to eliminate the spiders which are usually used to support the mandrel. This is done to produce a continuous annular melt feed, without the faint weld lines caused by the spiders.

Immediately above the die, an external air ring is located which is used to chill the tubing with room-temperature or refrigerated air. Refrigerated air, introduced internally, is also used on some installations. This forced cooling (also rotated to increase uniformity) enables the extruder to be run at higher speeds, thus producing more material per minute. Figure 11-27 shows a rotating spiral-manifold co-extrusion die. Figure 11-28 shows the components of such a die.

In the blowing operation, the tubing diameter is increased considerably above the original die diameter. The ratio of the final diameter to the original diameter is known as the blowup ratio, and is obtained by simply dividing the diameter of the bubble by the die diameter. Blown tubing is also

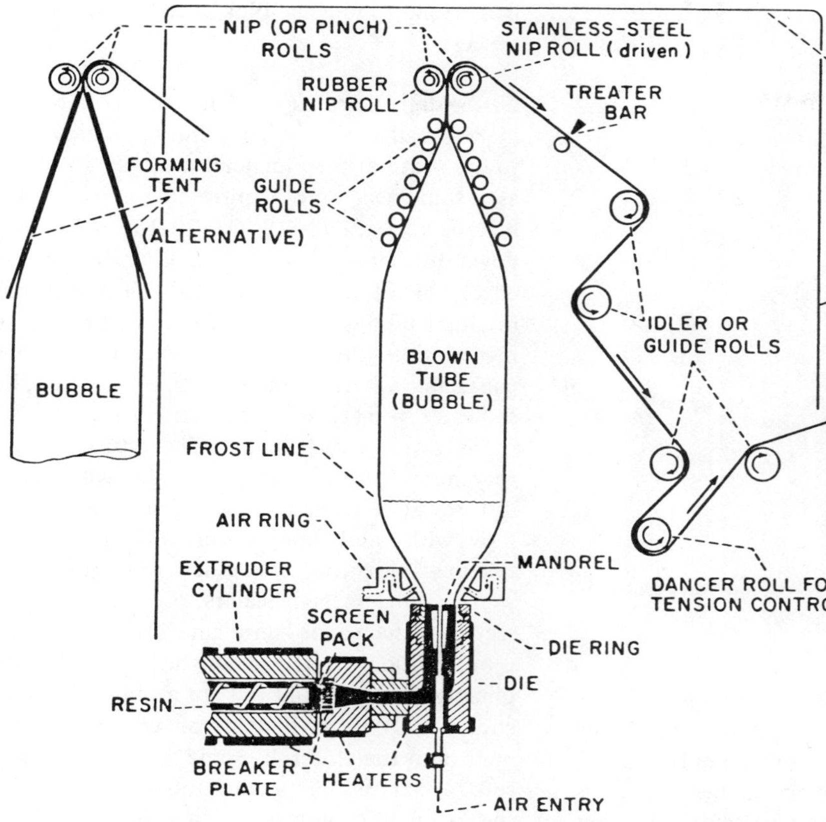

Fig. 11-25. Blown film process. (*Courtesy: U.S. Industrial Chemicals Co.*)

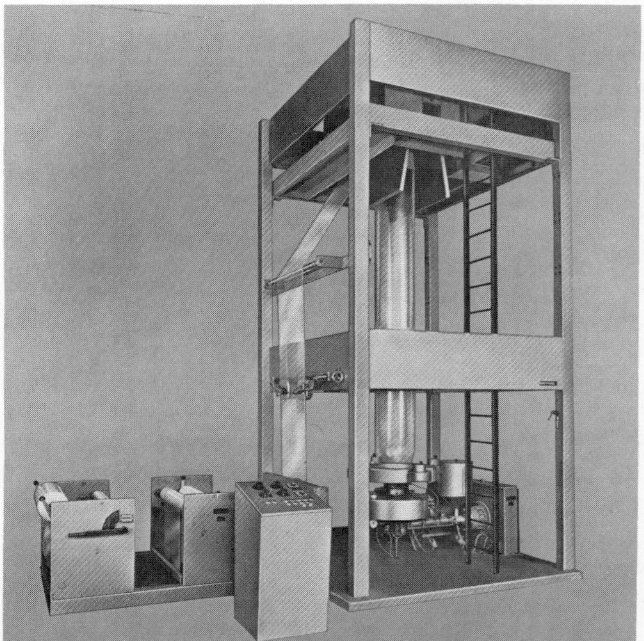

Fig. 11-26. Blown film extrusion. (*Courtesy: American Barmag Corporation*)

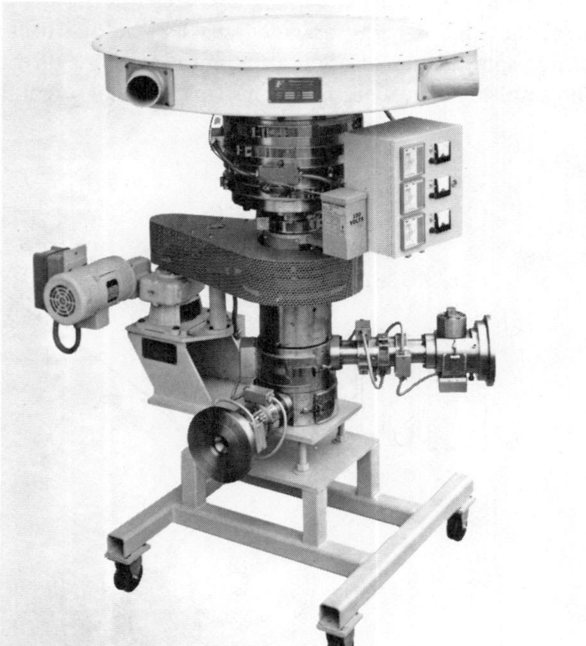

Fig. 11-27. Rotating spiral-manifold co-extrusion die with air ring on top. (*Courtesy: Filmaster Design, Inc.*)

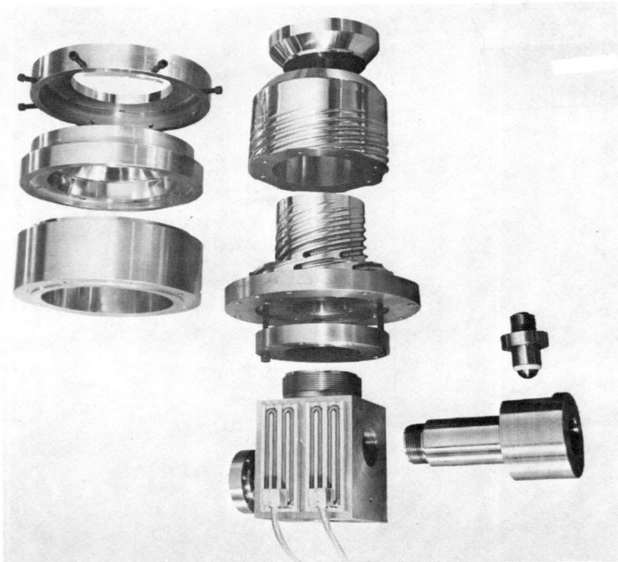

Fig. 11-28. Disassembled co-extrusion blown film die: bottom-fed spiral design. (*Courtesy: Filmaster Design Inc.*)

known as lay-flat tubing, since, after being blown, it is flattened by the collapsing frame and the nip rolls at the top of the apparatus. Lay-flat tubing is commonly made in widths of 12 to 40 inches, but tubing has been made as wide as 120 inches, which is split to give film 240 inches wide.

The blowing of the film not only increases the width of the product but also orients the film slightly in the transverse direction, thus increasing its strength. At the same time, the film is being drawn down by the pull rolls, imparting some degree of orientation in the machine direction. The resulting film, because of this small biaxial orientation, is a stronger, tougher film than film made by the casting process.

Cast Film. So-called cast film is made by extruding the melt (usually polyethylene) through a large die equal in size to the width of the film being cast. The material is extruded as a thin sheet onto a mirrored surface, chill roll, (which may be kept at 248–302°C (120–150°F), and is then drawn down further by other rolls. Prior to final winding, the edge, which is usually somewhat rough, is trimmed off. The resultant product is similar to blown film but usually has a higher gloss, more suitable than most blown film for packaging where attractiveness is an asset. Figure 11-29 shows a typical cast film extrusion setup.

The slot die used in the film-casting operation is made very massive so its size and weight will enable it to resist distortion. It is made with a generous manifold for resin flow with minimum pressure drop, adjustable jaws for thickness variation, deckle bars for adjustment of the film width, and attached heaters. Figure 11-30 shows a cross-section of such a die and Figure 11-31, a complete die.

In addition to the die, which deposits the hot melt directly onto the chill roll, an air blast is often used at the chill roll to help eliminate air entrapment and force the melt onto the chill roll. Some installations use several chill rolls, at varying temperatures, before delivering the film to the windup equipment. Other installations use another roller directly above, or below, the chill roll, which acts as a calender roll to form film of a definite thickness. Alternatively, the calender roll may be embossed or engraved so that the film is given a pattern as it passes through.

Additives. Both blown and cast film operations use special resin formulations which include a number of additives to

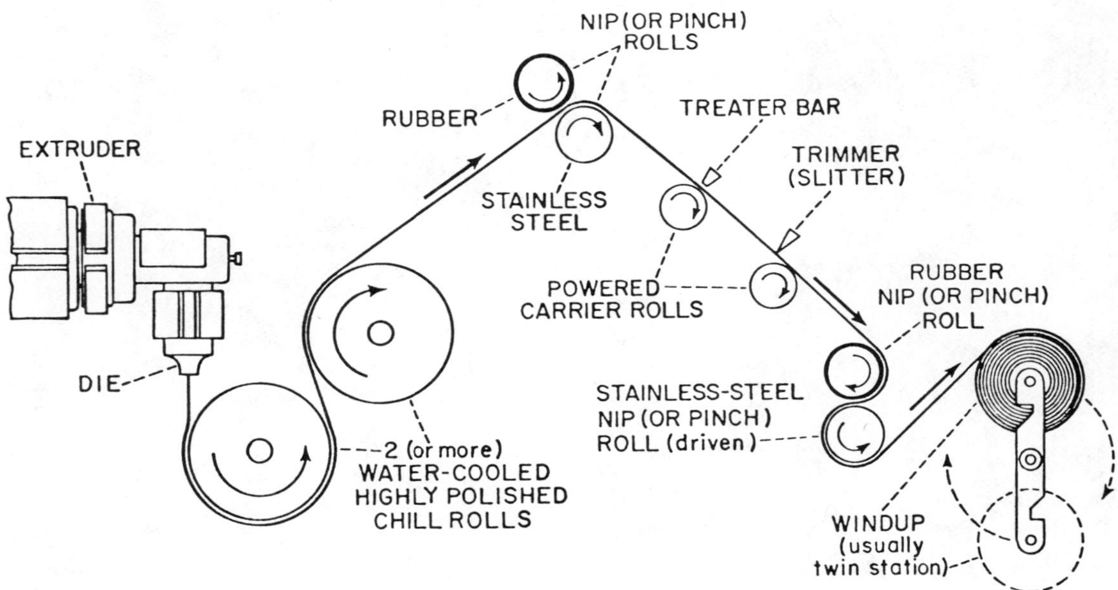

Fig. 11-29. Schematic drawing of chill-roll flat-film equipment. (*Courtesy: U.S. Industrial Chemicals Co.*)

impart various characteristics to the film. Among the most important additives are materials to control slip and decrease blocking tendencies. Slip is defined as the tendency of the film to slide on itself and is related to the material's coefficient of friction. This property is important in the behavior of a stack of bags in an automatic bagging machine. Blocking is the tendency of a tightly wound roll of film to stick to itself and form a solid block of plastic which cannot be easily unwound.

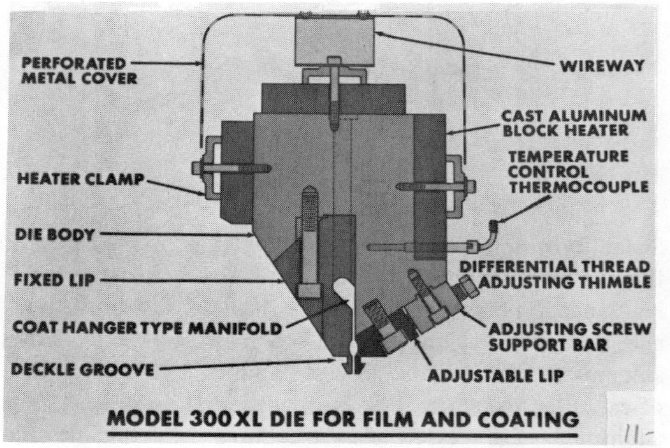

Fig. 11-30a. Cross-section of cast film die. (*Courtesy: Black Clawson Co.*)

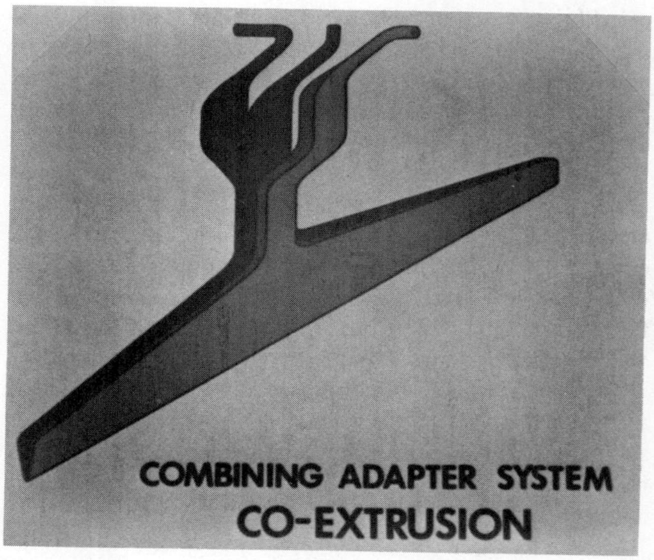

Fig. 11-30b. "Coat hanger" manifold configuration for three-ply co-extrusion. (*Courtesy: Black Clawson Co.*)

Other additives used are colorants, antioxidants, lubricants, light stabilizers, etc. In addition to additives in the resin formulation, both types of film may be treated to accept ink just before being wound up. In some cases, the material is treated by flame or high-voltage electric discharge and then printed and heat-sealed into bags which are then wound into a roll, or individually cut and stacked for shipment. The treatment for ink retention is mainly used for polyethylene and polypropylene, which normally will not accept ink. However, after the treatment, which causes a surface oxidation to hold ink or other marking material, it is important that the film be printed immediately. Any handling, rolling, etc. could rub off the oxidized surface and impair the quality of the marking; therefore, treatment and marking are done prior to winding whenever possible.

Winding. Whether tubing is blown or cast, winding can be done by two methods: surface winding or center winding. In both cases, it is important that the roll of film be wound with an even tension and a minimum of irregularities across the width. In surface winding, the roll of film rotates itself by friction against another rotating roll. Figure 11-32 illustrates the setup used for blown film. This technique has the advantage of economy and simplicity; as the roll enlarges, it simply moves up the roller track. However, there are disadvantages to this method. Tension buildup can cause the roll to be very tightly wound. Die irregularities can result in bands or bumps which stretch the film excessively, producing wrinkles and puckers. This can interfere with printing and marking.

In center winding, the roll rotates because of the turning of the center shaft upon which it is mounted. The shaft can be turned in three ways: (1) constant tension, (2) constant torque, or (3) tapered tension winding. The result of any of these methods is a roll with an even tension throughout, less stretching over bumps, and less tendency toward the blocking which can occur when excessive tension is applied

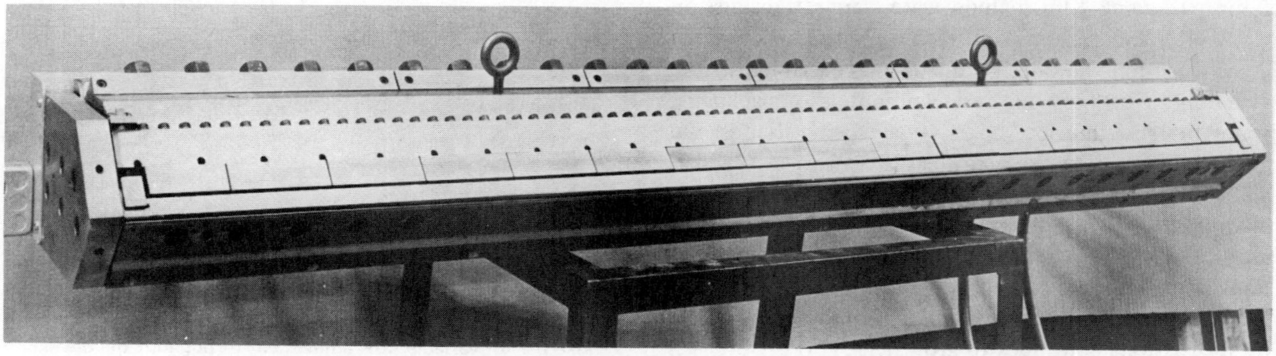

Fig. 11-31. Cast film die assembly. (*Courtesy: Johnson Plastics Machinery, Leesona Corporation*)

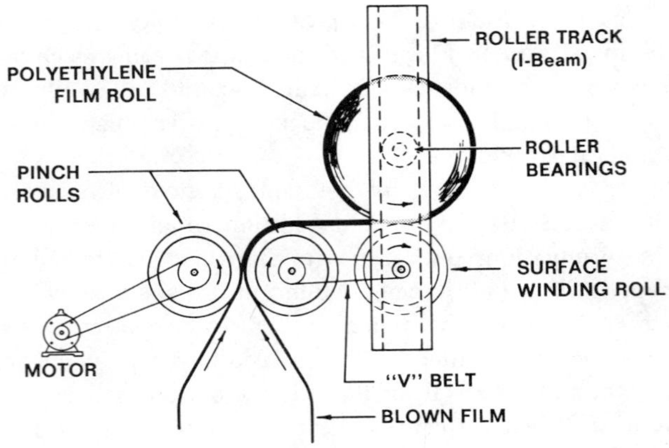

Fig. 11.32. Surface winding of blow film (*Courtesy: E. I. DuPont de Nemours & Co., Inc.*)

to a roll. Another advantage of the center winding system is that this method may be more easily adapted to the use of an automatic cutoff and/or a quick-change turret roll system than the surface winder. (See Figure 11-40 for diagram of turret roll assembly.) The disadvantages of the center winding method are the higher initial cost and maintenance costs. A minor disadvantage is that the take-up rolls must be more accurately synchronized with the extruder output than is necessary with the surface winding system.

Testing. Like pipe and tubing, film is made to a number of specifications and is tested for uniformity and quality by standard and special tests. Tensile and elongation tests are used to determine the strength of the film in both directions. Orientation or film imbalance can be determined by measuring the shrinkage when the film is heated in hot oil, glycerine, or an oven. Uneven shrinkage will indicate the direction of maximum stress or orientation; that is, the higher the shrinkage in any direction, the higher the orientation in that direction. Other tests for strength are falling dart tests or burst tests which can be used to measure the allover strength. Transparent packaging film, which must have high "see-through" properties, is checked optically for total light transmission, haze, and gloss. Color determinations are made with special instruments. Other important tests for packaging film include water-vapor transmission rate (WVTR), and transmission rate for various gases such as oxygen, carbon dioxide, or nitrogen. Toxicity tests are run on film to be used for food containers or packaging to insure that noninjurious additives are being used in the formulation. The FDA, in many cases, will specify which additives may be used, thus avoiding the necessity of testing.

Plastics films are used to make bags in various sizes and thicknesses; heavy film is needed in outdoor construction work; stiff film becomes "skin" packaging material; and the list of applications continues to grow. (See also Chapter 9.)

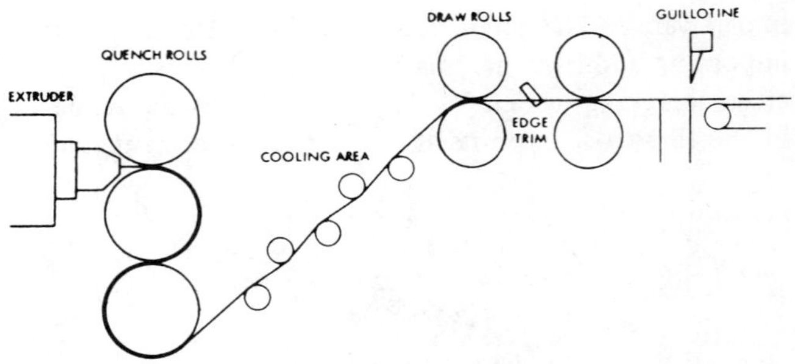

Fig. 11-33. Sheet extrusion line.

Sheet Extrusion. By convention, material over 0.01 inch thick is called sheet or sheeting. The most used material is high-impact polystyrene, although high-density polyethylene, PVC, ABS, cellulosics, and nylon are also used in large amounts. In general, the process is similar to the extrusion of cast film, except that a slightly different die is used, and additional rollers assure that a definite sheet thickness is produced. Figure 11-33 illustrates the sheet extrusion process which is used to make sheet as wide as 60 inches.

A massive center-fed die is used for sheet extrusion, as in cast film production; however, the manifold is usually considerably larger in the sheet die to accommodate the larger melt being extruded. In order to extrude this greater amount of melt, the die gap is considerably larger than is used to cast film. The hot melt might normally be expected to run out of the center portion of the large die gap, unless extremely high pressures were used in the die. Instead, a choker bar is used to regulate the thickness of the melt emerging from the die. Figure 11-34 shows a cross-section of a sheet die and illustrates the choker bar (which is flexible and adjustable by a row of screws) and the adjustable die blades or jaws.

The quench rolls shown in Figure 11-33 are maintained at a precise temperature by means of circulating water; oil is sometimes used if higher temperatures are required, as in the case of polyolefins. Typical temperatures for ABS sheet are: top roll, 347–437°C (175–225°F); middle roll, 302–

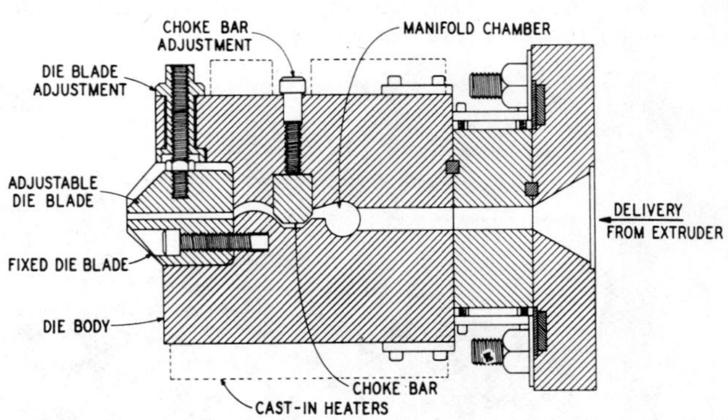

Fig. 11-34. Sheet extrusion die for high-impact stytrene. Note choke bar for variable flow adjustment throughout the die length.

437°C (150–175°F); bottom roll 347–437°C (175–225°F). Once the exact temperature has been determined, it is usually maintained to a precision of 2°C (±5°F) for optimum quality. In addition to cooling the hot extrudate, the rolls, especially the middle roll, are usually chrome plated and polished to a high finish, since this will to a large extent determine the surface gloss of the sheet. To a lesser extent, the rolls are also used to smooth out minor variations in the thickness of the sheet.

On leaving the cooling rolls, the sheet travels to the pulling rolls, a distance of approximately 10 feet; during this time, it is usually supported on small rollers and may also be cooled by fans. The pulling rolls are usually adjusted to run a little faster than the cooling rolls so that the drawdown amounts to about 10%, considerably less than the drawdown used to make film. From the pull rolls, the sheet is usually cut automatically and stacked as flat sheets, cut to a definite length. Sometimes thin sheet is, however, stored in large-diameter rolls.

Embossed patterns can also be made in the sheet by engraving a design in the middle roll. This design is frequently a leatherlike grain pattern. For lighting fixtures, a transparent sheet is frequently embossed with a prism pattern to act as a diffuser. Laminated sheets can also be made with the three-roll setup by simply feeding a film into the nip along with the extrudate. Two colored sheets can be made in this manner, or a very high-gloss surface can be obtained by using a highly polished transparent film which bonds to the hot sheet.

Another technique to produce a sheet or film having two or more different materials laminated together is known as co-extrusion. In this process, which requires a good deal of "know-how" to produce satisfactory material, two or more extruders are used simultaneously to feed a special die which has an adapter and feed block between it and the extruders. Figures 11-35 and 11-36 reveal the cross-sections of some of these multiple-manifold dies. Figure 11-37 shows a machine which uses two extruders to produce two- or three-layer sheet or film. The complexity of a feed block used to extrude a five-layer film (ABCBA coatings) is illustrated in Figure 11-38. The resultant material may be a sheet which has the best properties of all three resins. A four-extruder combination which could be used to produce a seven-layer composite is shown in Figure 11-39. Thus, materials can be made with improved surface appearance; fire retardance can be built in; chemical resistance can be improved; ultraviolet resistance can be strengthened, etc.

Extrusion Coating. Coating paper, metal foil, fabric, or other materials is another large application of the extrusion process. Commercial packaging is one of the many product lines. Polyethylene-coated paper to make milk cartons represents the largest consumption of coated materials. The water-barrier properties and the heat-sealing characteristics make for an optimum container material.

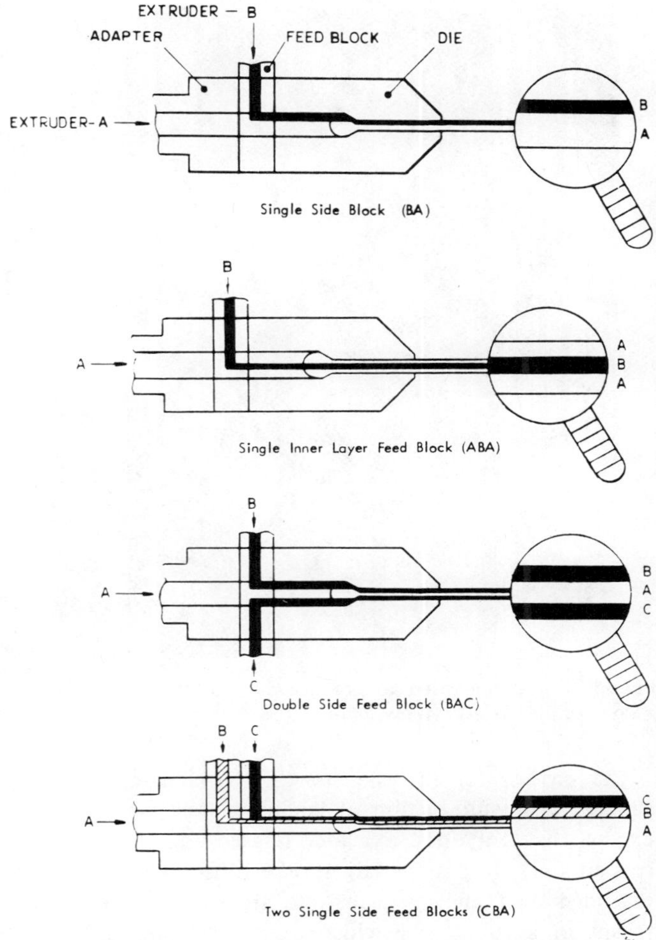

Fig. 11-35. Multiple-manifold dies used in co-extrusion. (*Courtesy: Welex, Inc.*)

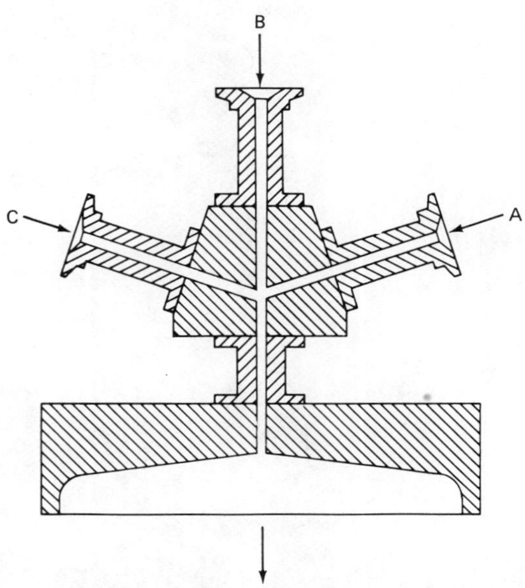

Fig. 11-36. Cross-section of a three-component co-extrusion adapter and coat hanger die.

Fig. 11-37. Machine used to produce two- or three-layer AB or ABA sheet. (*Courtesy: Welex, Inc.*)

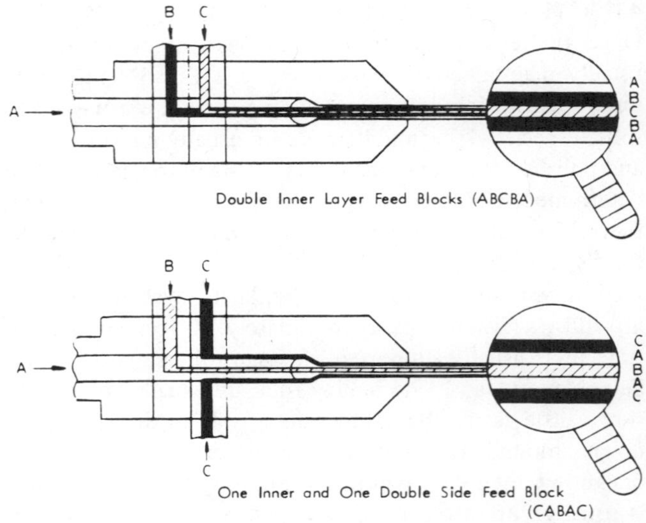

Fig. 11-38. Multiple feed blocks used to produce five-layer co-extrusions. (*Courtesy: Welex, Inc.*)

The process for extrusion coating is, in many respects, similar to the production of laminated cast film or sheet, except that only one extruder is used and the substrate (paper, foil, etc.) is heated. Another difference is that the extruded melt and the substrate are forced together by means of a rubber roll which exerts pressures as high as 100 lb/linear inch. This forces the resin into the fibers of the paper, or produces maximum adhesion to a metal substrate. In order to increase adhesion, a number of techniques are employed: primers are used on the substrate; vacuum is applied through the substrate; flame treatment or electrical discharge changes the surface of the substrate.

The die used for the coating operation is similar to the one used for film or sheet, but no choker bar is used since there is no need for thickness adjustment of the melt. The

Fig. 11-39. Four-extruder co-extrusion line with seven-layer feed block. Total capacity = 1000 kg per hour. (*Courtesy: Welex, Inc.*)

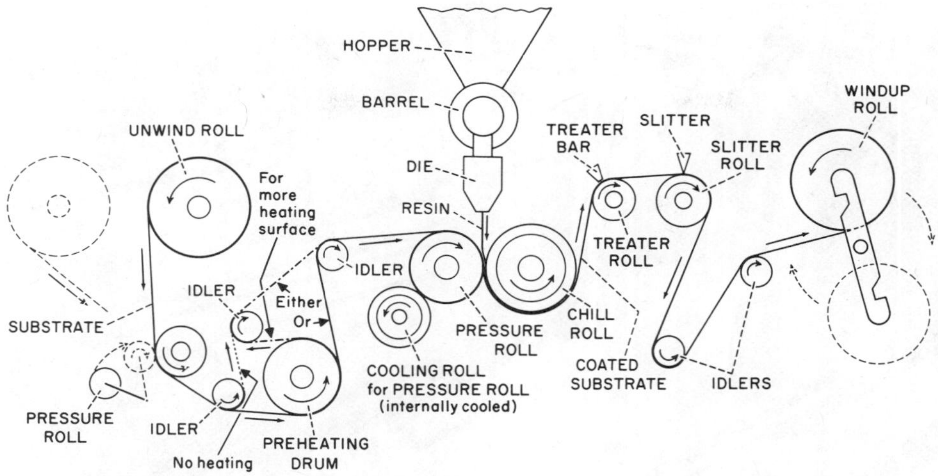

Fig. 11-40. Schematic cross-section through the entire extrusion coating setup, with unwind and rewind equipment. (*Courtesy: U.S. Industrial Chemicals Co.*)

Fig. 11-41. Overall view of the extrusion coating equipment in the USI Polymer Service Laboratories. Right: unwind roll. Center: extruder with hopper, barrel, adapter, and die. From the die, the melt is drawn into the nip between the pressure roll and water-cooled chill roll, whose pressure forces the hot film onto the paper. Left, between adapter and die: slitter roll. The picture shows a kraft paper coating run. One operator (right) is checking coated-substrate tension at the rewind while the other operator is recording run conditions in the log book. (*Courtesy: U.S. Industrial Chemicals Co.*)

Fig. 11-42. Overall view of a very large extrusion coating outfit. Lower left: unwind station. Left of operator: extrusion coater with pressure and chill rolls. Above nip of these rolls: die with adapter above its center, at the head of the extruder barrel (upper center). Upper right: rewind station. Barrel diameter = 8 inches. Barrel L/D ratio = 24:1. Die width = 104 inches. (*Courtesy: U.S. Industrial Chemicals Co.*)

die, however, is made with adjustable jaws and deckle bars so that many widths can be made with one die. Figure 11-40 shows a typical extrusion coating setup with a heating roll for the substrate, pressure roll, an electrical-discharge treater roll for increased adhesion, a slitter roll for trimming, and a turret-type center wind for a non-interrupted take-off. Figures 11-41 and 11-42 show the actual equipment used for a laboratory setup and a high-quantity production assembly.

Wire and Cable Coating. Possibly the earliest use of an extruder was for the purpose of coating wire with rubber insulation. The machine used was called a "tuber" (see Figure 11-2). The covering of wire with a plastic for either the primary electrical insulation or for mechanical protection, where it is called jacketing or sheathing, is still one of the major uses for the extruder. Polyethylene and vinyls are the most commonly used resins for primary insulation. The polyethylene is usually in the solid or foamed, low- or medium-density form, but crosslinked types are also being used. Nylon is used frequently for jacketing because of its strength, toughness, abrasion resistance, and good heat resistance.

The coating of wire and cable is similar to extrusion coating in that the substrate is preheated before coating to improve adhesion and reduce shrinkage stresses in the insulation. Preheating of thin wire is done by passing a current through the wire just prior to applying the plastic. Thicker wires and cables are heated by a gas flame or hot gas.

Two types of dies are used for coating wire and cables. One is known as a pressure-coating die since, as shown in Figure 11-43, the wire is coated inside the die at pressures often as high as 5000 psi. The second type of die is known as a tubing die and, as shown in Figure 11-44, the wire is

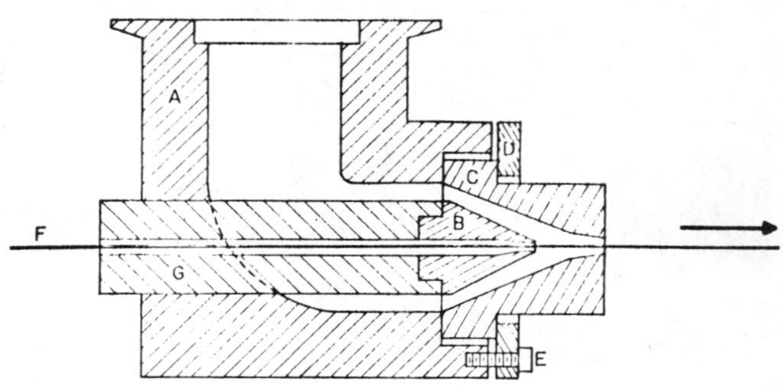

A. Die body, crosshead
B. Guider tip
C. Die
D. Die retaining ring
E. Die retaining bolt
F. Wire
G. Core tube

Fig. 11-43. Pressure-coating die.

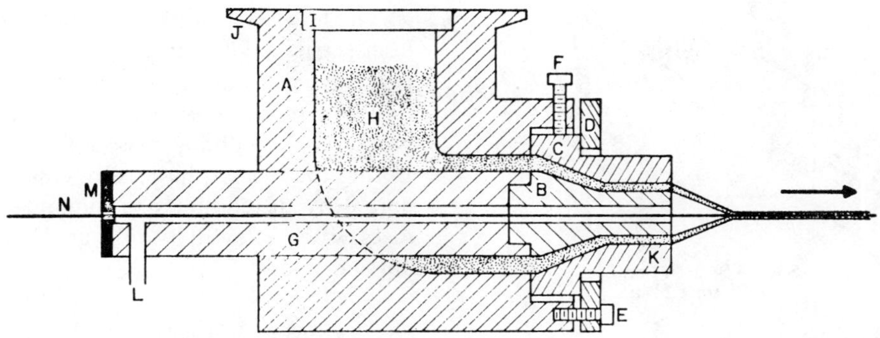

A. Die body, crosshead
B. Guider, male die part
C. Die, die bushing, female die part
D. Die retaining ring
E. Die retaining bolt
F. Die centering bolt
G. Core tube
H. Molten plastic
I. Seat for breaker plate
J. Ring for attachment to extruder
K. Die land
L. Vacuum connection
M. Wire guide and vacuum seal
N. Wire

Fig. 11-44. Tubing die for wire insulation.

coated outside of the die with a vacuum used to aid in pulling the resin onto the wire. In both cases, the dies are known as cross-head dies because the wire usually enters at a 90° angle to the extruder flow, although entrance angles of 45° and 30° have been used to affect a more streamlined flow. The choice of what type of die to use depends on a number of factors involving the wire size, whether the insulation is primary or secondary, the available equipment, etc. Tubing dies are usually used with small wires and thin coverings, since the strain on the wire is less with this kind of die. Pressure dies are used when multiple wires must be covered, because the melt can be forced into the wire interstices. Jacketing is usually run through tubing dies.

Small-diameter wire can be coated at a rate of up to 4000 feet per minute. In order to be able to pay-off wire at such a high rate, without excessive strain on the wire or the reel, the wire is taken off from a special reel which is similar to a spin-casting fishing reel. Large wires and cables are coated at a considerably slower rate and so are paid-off from a conventional wire reel.

After leaving the die, the coated wire is pulled through a water-filled cooling trough which may be 30 to 100 feet or more in length. In cooling the wire, it is important to sustain a gradual rate to avoid thermally shocking the insulation which could lead to excessive strains. Thus, it is not uncommon to have several zones in the trough, each at a different temperature. The wire is pulled through the die and the cooling trough by means of a capstan-like device, which exerts a considerable pull without excessive strain. From the capstan, the wire goes to the windup equipment, which is usually a center winding device on a turret.

The coating of wire and cable is one of the few extruding processes in which the product is tested while it is being made. After the wire emerges from the cooling trough, it is tested for diameter and concentricity mechanically, electrically, or by photoelectric devices. The insulation is also tested for pinholes, breaks, or thin spots by means of a high-voltage spark tester. Foamed insulation, used for TV transmission lines, is tested by capacitance measurements to indicate the degree of foaming in the coating. Since the coated wires and cables are used mainly for electrical applications, as direct current (DC) conductors and as carriers of alternating current (AC) at low and high frequencies, there are a great many commercial and government standards and specifications to which the products must conform. Thus, there are a large number of tests, in addition to those outlined above, which are also applied to the coated wires and cables.

Profile Extrusion. The extruding of special shapes—table edges, window channels, automobile trim, tapes, rods, filaments, and all types of contours—is referred to as profile extrusion. In this type of production, in many cases handled by small specialty shops, extrusion becomes more an art than a science. The equipment used for the many types of products ranges from standard to unique, it must often be made in the extruder's shop, since only he can make the particular die, cooling trough, winder, etc. The same is true of the formulations used. Many of the shapes and profiles use vinyl compounds which are specially formulated by the processor for a particular job.

One of the important facts of life for the profile extruder is that many of the jobs are comparatively short-run. Consequently, dies are made as cheaply as possible. A shape die might consist of an adapter fastened to the end of the extruder which leads the melt to a simple die which is bolted to the adapter. The die could be merely a steel plate with a hole of the desired shape placed over the adapter.

Shrinkage is the greatest problem in designing successful dies for use in profile extrusion. During cooling, a part will

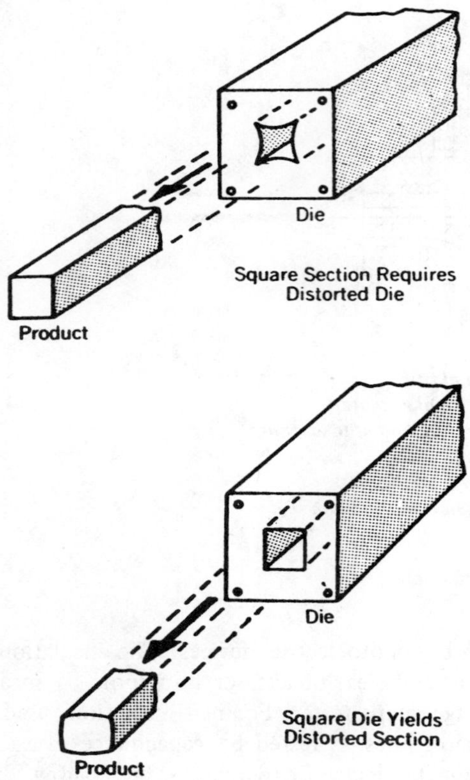

Fig. 11-45. Compensating die. (From *Plastics Engineering Handbook of the SPI* edited by Frados)

shrink irregularly if it is made up of thick and thin sections. Flat areas may bulge or become concave, depending on the total size and shape of the extrusion. Figure 11-45 illustrates the use of a compensating die for distortion. Another method for minimizing shrinkage is to use amorphous resins which show less shrinkage, such as vinyls, styrenes, cellulosics, and acrylics, instead of crystallizing resins such as polyethylene, polypropylene, nylon, or acetal. A third technique to insure minimum shrinkage and warpage is proper cooling during passage through the cooling trough. During this phase, the part may be cooled by air and supported by metal guides to minimize warpage. In extreme cases, water followed by air is used, and the air may be both hot and cold.

Post-Extrusion Processing

Post-extrusion processing refers to special processing which may be done to the extrudate, usually just after it emerges from the die and before the material has a chance to cool. When the material is worked in such a state it is known as in-line processing, as opposed to cutting, forming, or other processing which might be done on the cold extrusion. In-line processing is usually done very close to the extruder, and is done automatically, with little or no extra labor on the part of the machine operator. The extra processing, which may involve shaping, cutting, reforming, or a surface modification of the extrudate, can considerably increase the value of the extrusion without materially increasing the cost, it may also be done to enable the use of a lower-cost die; for example, flattening a tubular extrusion into an oval, so that a much cheaper circular die can be used.

One of the most common post-extrusion operations, other than a simple reshaping, is embossing the surface of an extrudate. Figure 11-46 shows several methods of embossing one or both sides of an extrusion, either with a shallow pattern or a deeper embossed pattern. The technique using a single roll, backed up by a roll underneath or a backup surface, is the one most used. The embossing roll

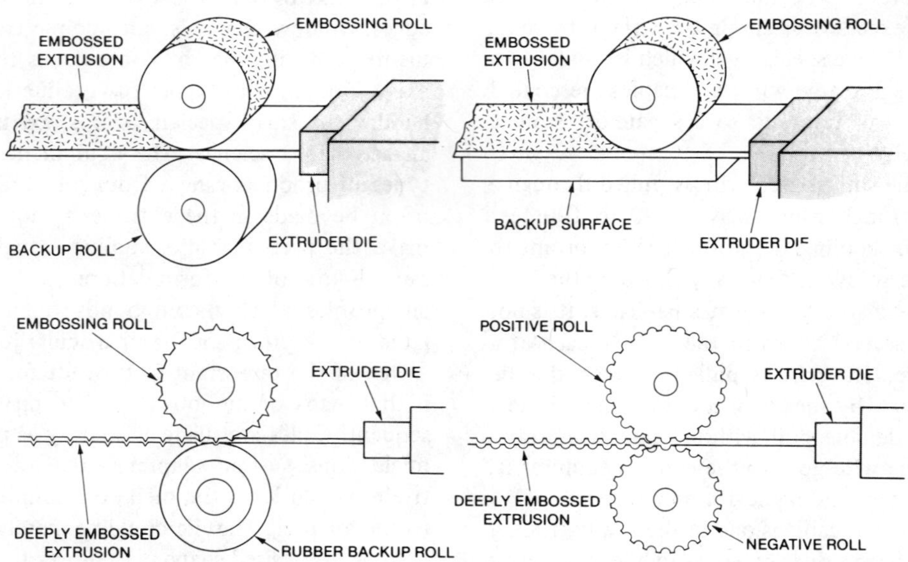

Fig. 11-46. Embossing techniques for extrusions. (Reprinted, with permission, from *Plastics Machinery & Equipment*, August 1978. Copyright 1978, Industry Media, Inc.)

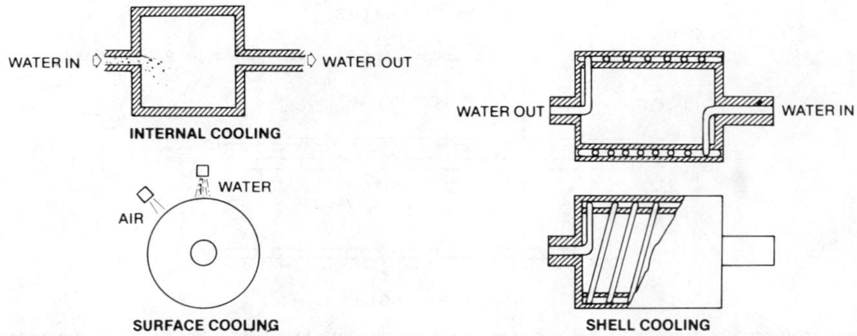

Fig. 11-47. Several alternative methods for cooling the embossing roll are shown schematically here. (Reprinted, by permission, from *Plastics Machinery & Equipment*, August 1978. Copyright 1978, Industry Media, Inc.)

can be a metal roll with the pattern etched into the surface, engraved on the surface, or, in some cases, the roll is an electroformed part made from a master. For very soft materials, the roll can even be a silicone roll which has been cast against a master pattern.

The use of two rolls to obtain deep patterns has the disadvantage that the rolls must always be kept in perfect registration if the pattern is not to suffer. The extrudate must also be carefully controlled with regard to temperature so no tearing occurs, while at the same time it must be kept soft enough to obtain a deep pattern.

One of the problems in using rolls for forming patterns is that the rolls gradually pick up heat from the extrusion, which causes the quality of the pattern to suffer. One way to assure uniformity is to cool the rolls, by either surface cooling with air or water, or—even better—by using hollow rolls which can be water cooled as illustrated in Figure 11-47.

Other patterns can be obtained by the use of vacuum drums, or vacuum-forming dies as shown in Figure 11-48. With the drum it is again important that adequate cooling be provided since so much of the drum is in contact with the extrudate. By the use of an elongated pressure-forming die, shown in Figure 11-48, cooling can be accomplished by

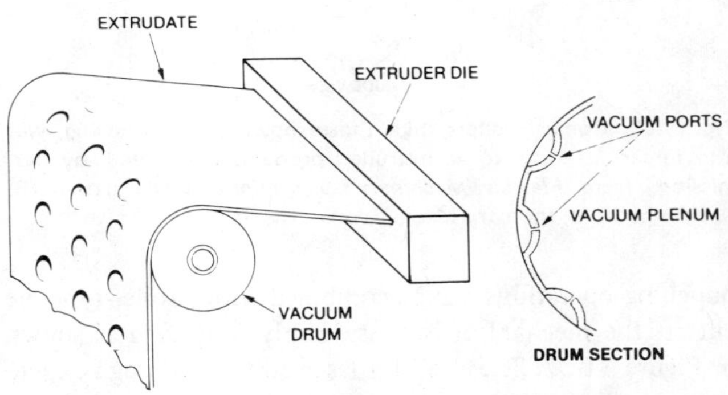

Fig. 11-48. An embosser that uses a vacuum drum to form the pattern (like the one shown) is a process that resembles a vacuum-forming technique. (Reprinted, by permission, from *Plastics Machinery & Equipment*, August 1978. Copyright 1978, Industry Media, Inc.)

simply blowing air over the die surface not contacting the plastic. Also, because the die can be made long, as shown, a cutter bar can be attached to one of the dies so parts of a given length can be easily obtained.

Figure 11-49 shows two methods of forming corrugated or wave-formed sheets. These dies could also be equipped with cutters, either at the sides or at the ends, to form parts of a definite size. The dies shown in Figure 11-49 are called transverse forming dies, in contrast to the former dies which are considered in the main to be axial forming dies, or simple embossing dies.

When the extrusion is to be made in the form of a coil or spring, a rotating mandrel can be used, as shown in Figure 11-50. In this case, it is necessary that the mandrel be driven in perfect synchronism with the rate of extrusion so the extrusion will be free of any drag or mismatch which could cause shape alteration. A counter-roll is usually incorporated into the assembly so the coil can be made in definite lengths. Typical parts made by this process are coiled telephone cords, fluid lines, springs, decorative items, etc.

Continuous or very long curved sections can be formed without a mandrel by using puller rolls operating at different speeds, as shown in Figure 11-51. Similar parts, with curvatures going in a different plane, can also be made using conical rollers as illustrated in Figure 11-52. In this case, because of the differing surface speed on the cone, the extrudate curves towards the small end of the cone. It can be seen that by running the cones at different speeds the stock can be caused to curve in two planes simultaneously.

Other forming techniques include the production of twisted material used for either decorative or mechanical purposes. One method of obtaining twisted stock is the use of a rotating ring and a stationary holder ring as shown in Figure 11-53. A more sophisticated technique is the use of crossed belts as shown in Figure 11-54. By changing the angle of the belts, it is possible to vary the pitch of the spiral.

Using synchronized drive mechanisms, it is also possible to perform machining and punching operations on the extrusion, as illustrated in Figures 11-55 and 11-56. When the

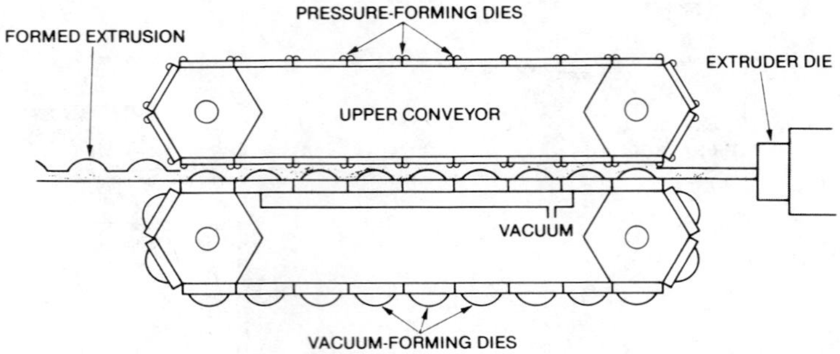

Fig. 11-49a. This in-line vacuum/pressure former uses a series of forming dies mounted on a continuous conveyor device. (Reprinted, by permission, from *Plastics Machinery & Equipment*, August 1978. Copyright 1978, Industry Media, Inc.)

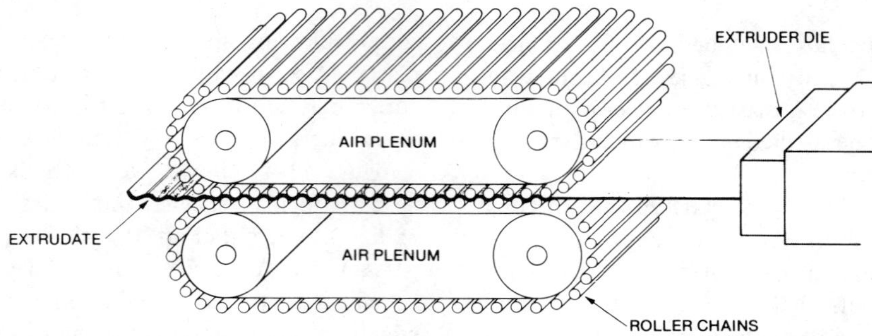

Fig. 11-49b. This simple corrugator also uses a conveyor arrangement to make corrugated sheet. (Reprinted, by permission, from *Plastics Machinery & Equipment*, August 1978. Copyright 1978, Industry Media, Inc.)

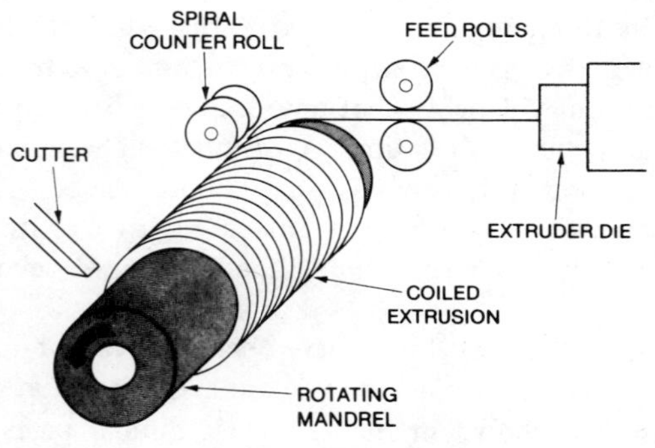

Fig. 11-50. Coiled products like telephone cord can be formed using a rotating mandrel that is driven in synchronization with the puller. (Reprinted, by permission, from *Plastics Machinery & Equipment*, August 1978. Copyright 1978, Industry Media, Inc.)

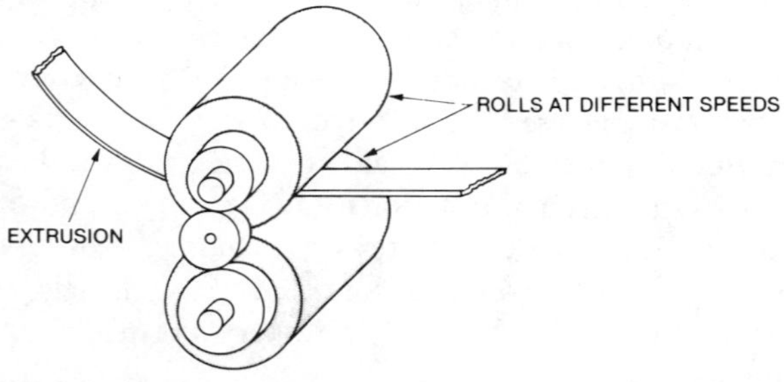

Fig. 11-51. Two puller rolls driven at different speeds will result in the production of a curved extrusion. A typical layout is seen here. (Reprinted, by permission, from *Plastics Machinery & Equipment*, August 1978. Copyright 1978, Industry Media, Inc.)

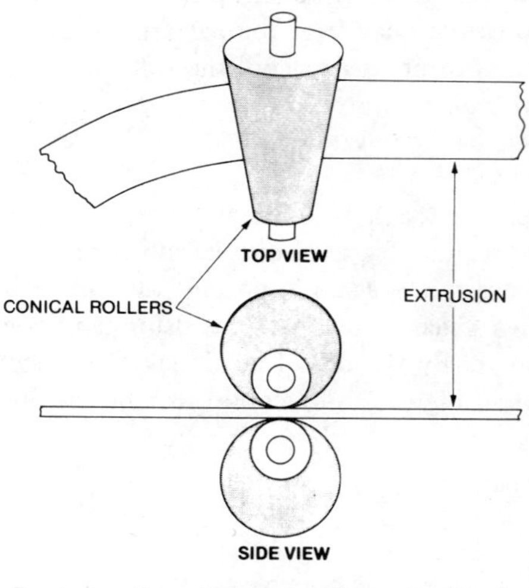

Fig. 11-52. Conical rollers, like those shown in the drawing, will also impart a curve to an extruded product. (Reprinted, by permission, from *Plastics Machinery & Equipment*, August 1978. Copyright 1978, Industry Media, Inc.)

punching operations can be combined with a roller-type die cutter, the mechanism is considerably simplified as shown in Figure 11-57. The use of a laser cutter, although expensive for short runs, can be used for special cutting operations as shown in Figure 11-58. Because of the speed with which the laser cuts can be made, this technique can be very useful in cutting many small holes simultaneously, particularly if the cuts are made in thin stock.

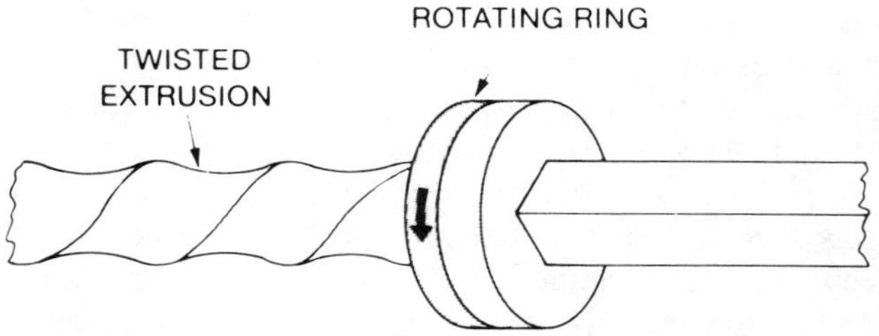

Fig. 11-53. An internal-twist die uses a rotating ring to twist an extruded square rod. (Reprinted, by permission, by *Plastics Machinery & Equipment*, August 1978. Copyright 1978, Industry Media, Inc.)

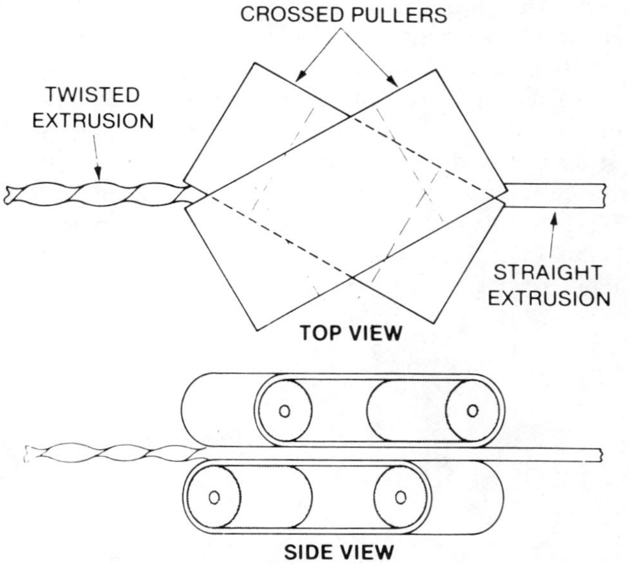

Fig. 11-54. Crossed puller belts can be used to twist just about any shape, and the system has the added advantage of pulling the extrusion along the line. (Reprinted, by permission, from *Plastics Machinery & Equipment*, August 1978. Copyright 1978, Industry Media, Inc.)

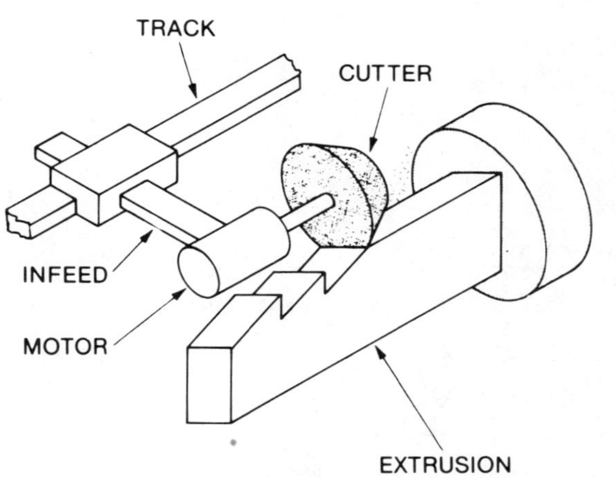

Fig. 11-55. A typical cutter head used for in-line machining of heavy extrusions is pictured. This configuration is used to produce a saw-tooth edge on a profile extrusion. (Reprinted, by permission, from *Plastics Machinery & Equipment*, August 1978. Copyright 1978, Industry Media, Inc.)

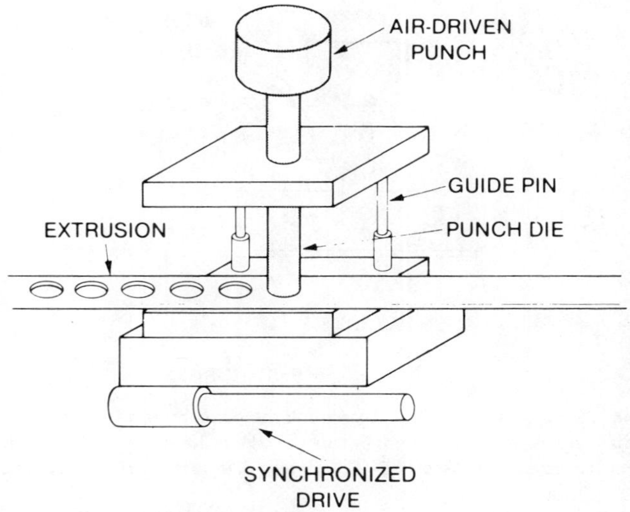

Fig. 11-56. An in-line punching unit like this must be mounted on a movable track and must be synchronized with the movement of the extrusion. (Reprinted, by permission, from *Plastics Machinery & Equipment*, August 1978. Copyright 1978, Industry Media, Inc.)

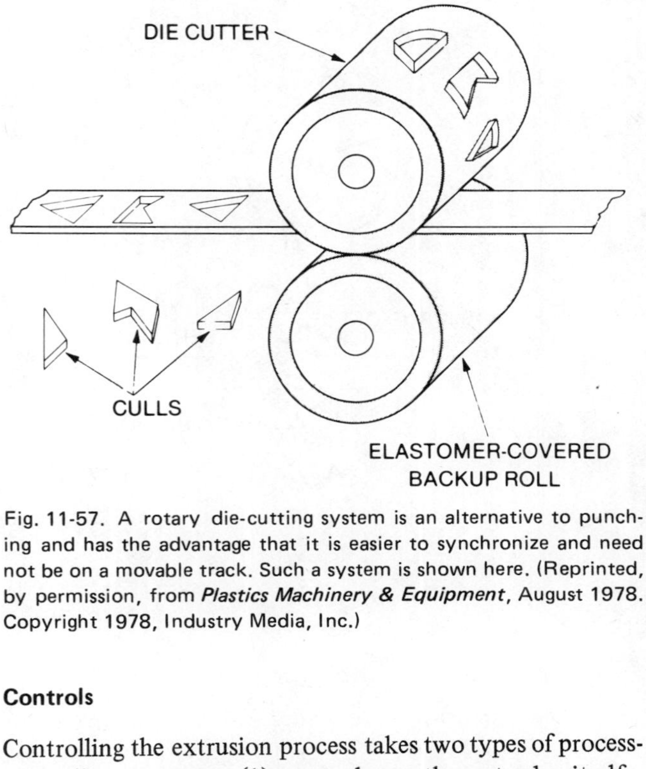

Fig. 11-57. A rotary die-cutting system is an alternative to punching and has the advantage that it is easier to synchronize and need not be on a movable track. Such a system is shown here. (Reprinted, by permission, from *Plastics Machinery & Equipment*, August 1978. Copyright 1978, Industry Media, Inc.)

Controls

Controlling the extrusion process takes two types of process-controlling systems: (1) controls on the extruder itself—temperature and pressure indicators and controllers on the barrel, ammeters, or wattmeters to show the power being used, and a tachometer to show the screw-rotation rate, and (2) controls on the finished product—film-thickness indicators, pipe-wall thickness indicators, width indicators, etc. The second type, controls on the extruded product, are not merely indicators, but are usually interconnected with the process controllers so that any variation in the product is

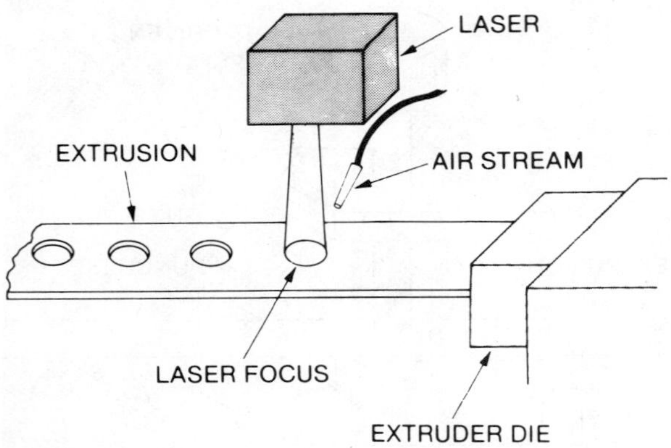

Fig. 11-58. Laser cutting, a particularly good method for cutting small holes in thin-walled extrusion. (Reprinted, by permission, from *Plastics Machinery & Equipment*, August 1978. Copyright 1978, Industry Media, Inc.)

immediately reflected as a corrective change in the appropriate process control.

The process-control systems which work directly with the extruder conditions are very similar to those used in injection molding. As such, they can be closed loop, microprocessor controlled, so that in addition to minimizing process variations they can also incorporate memory systems and/or cathode ray tube or paper or magnetic tape displays or printouts. However, since extrusion is a continuous process, it is probably of greater importance that the materials be gauged immediately on extrusion and cooling in order to keep waste to a minimum. In the early days, unfortunately, "on-line gauging" was only applicable in a few instances, and checking consisted of cutting samples at intervals and measuring the sheet or wall thickness or other pertinent dimensions. Currently there are a number of techniques for very satisfactory gauging of sheet, film, pipe, and tubing. Since extrusion is mainly used for the above type of products, it is not surprising that most of the methods are for some type of thickness measurement. Figure 11-59 shows ten different on-line gauging techniques ranging from simple direct dial indicators, to air pressure, magnetic, and sonic instruments, and complex optical and electrical methods with correspondingly higher sensitivities and costs. Table 11-2 lists some of the restrictions of different gauging techniques. Table 11-3 shows some of the applications for the various gauging methods and the limitations of each.

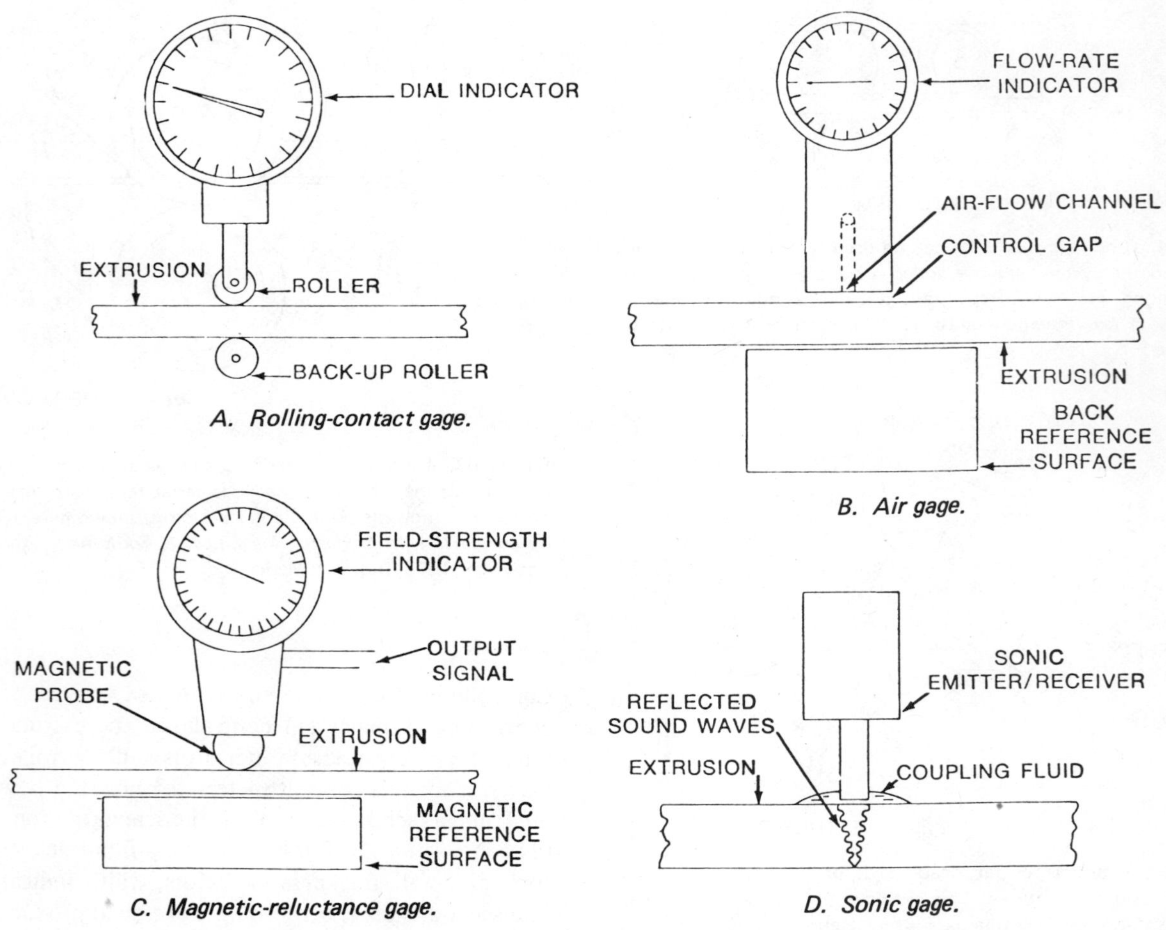

Fig. 11-59. On-line extrusion thickness gauging techniques. (Reprinted, by permission, from *Plastics Machinery & Equipment*, January 1978.) Copyright 1978, Industry Media, Inc.)

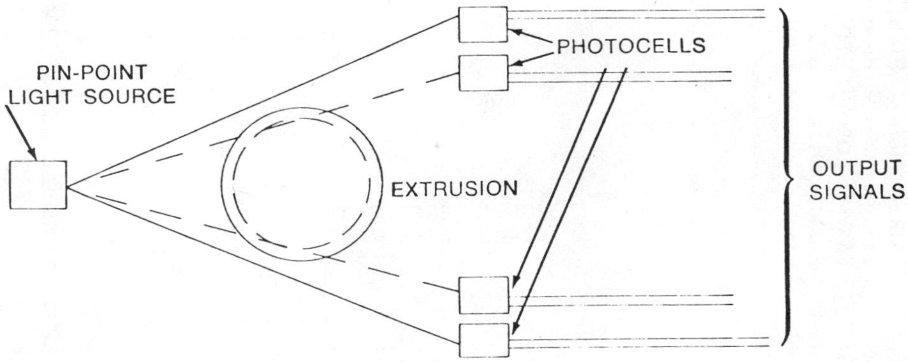

E. Optical-intercept gage.

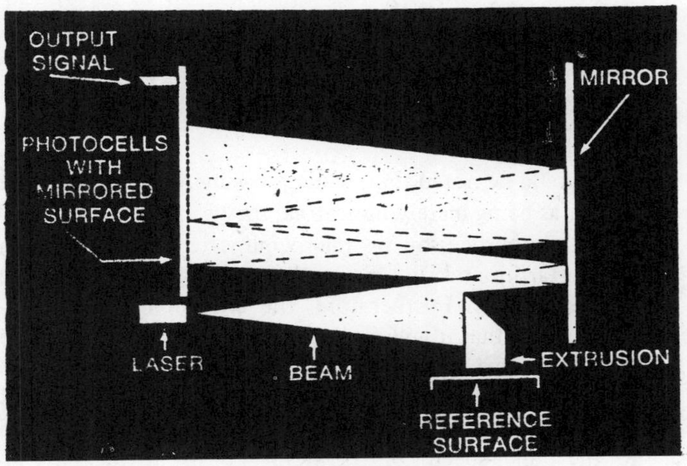

F. Laser-intercept gage.

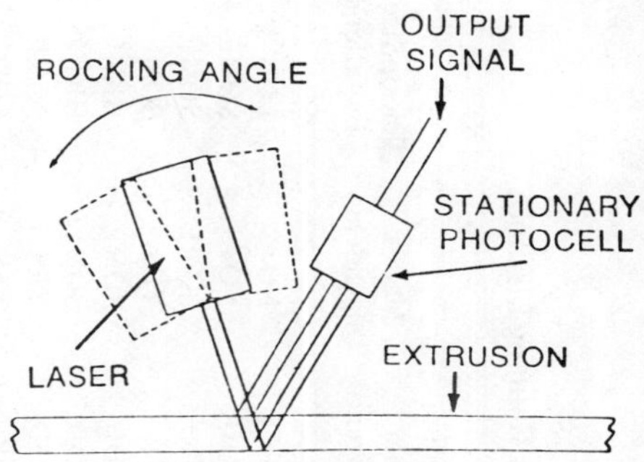

G. Laser-interferometry gage.

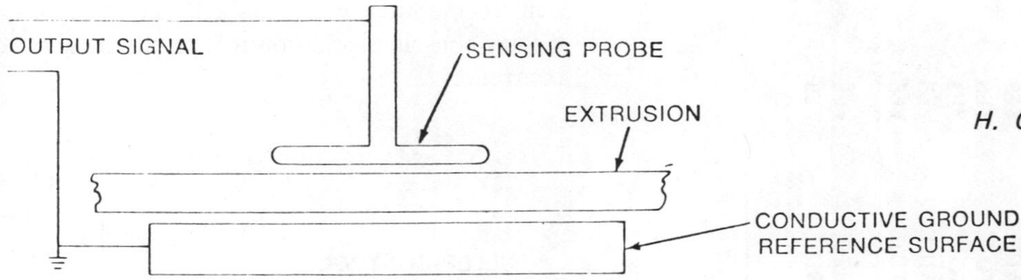

H. Capacitance gage.

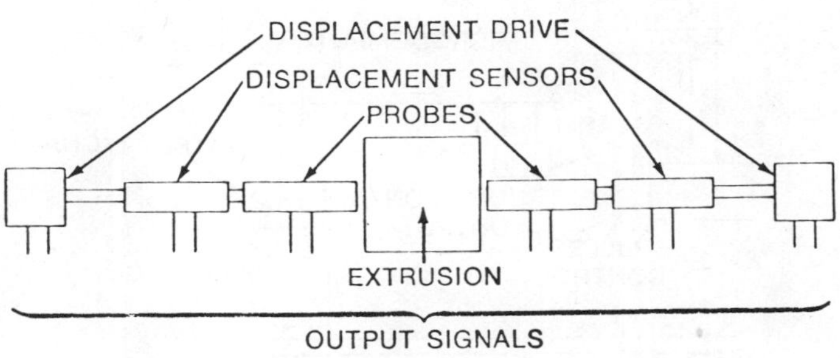

I. Proximity gage.

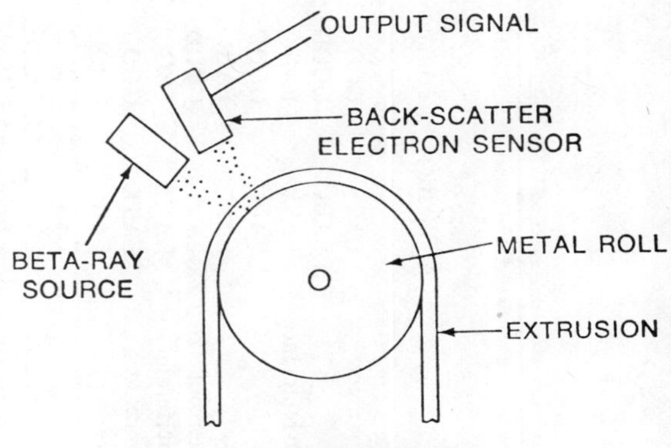

J. Beta-ray gage.

Fig. 11-59. (*Continued*)

Table 11-2. Restrictions on Various Gauging Methods.*

	Ease of Application	Accuracy	Dimensional Range	Transverse Measurement	Reflex Measurement	Freedom from Interference	Material Dependency	Sensor-Size Limitations	Complexity of Equipment	Ease of Calibration	Cost	Sensitivity
Rolling-contact	EASY	GOOD	WIDE	YES	NO	GOOD	NO	LOW	LOW	EASY	LOW	MED.
Air	FAIR	GOOD	WIDE	YES	NO	GOOD	NO	SOME	MED.	EASY	MED.	HIGH
Magnetic-reluctance	EASY	FAIR	TO $\frac{1}{4}$"	YES	POSS.	FAIR	SOME	SOME	HIGH	EASY	MED.	MED.
Sonic	FAIR	GOOD	TO 1"	YES	YES	FAIR	YES	SOME	HIGH	FAIR	MED.	MED.
Optical	FAIR	FAIR	WIDE	YES	YES	GOOD	SOME	HIGH	MED.	FAIR	MED.	MED.
Laser-intercept	EASY	GOOD	WIDE	YES	YES	GOOD	SOME	HIGH	MED.	FAIR	MED.	HIGH
Laser-interferometry	HARD	EXC.	LTD.	NO	YES	GOOD	YES	SOME	HIGH	EASY	HIGH	HIGH
Capacitance	EASY	GOOD	MED.	YES	NO	FAIR	YES	LOW	HIGH	EASY	LOW	MED.
Proximity	FAIR	GOOD	WIDE	YES	NO	FAIR	SOME	LOW	HIGH	EASY	MED.	HIGH
Beta-ray	FAIR	GOOD	LTD.	NO	YES	GOOD	SOME	LOW	HIGH	EASY	HIGH	HIGH

*Reprinted, by permission, from *Plastics Machinery & Equipment*, January 1978. Copyright 1978, Industry Media, Inc.

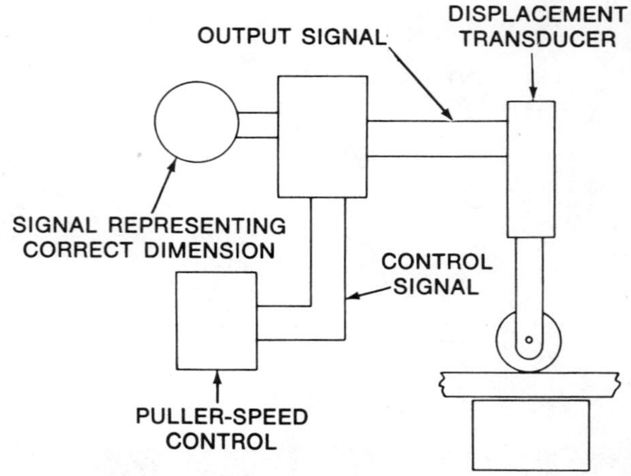

Fig. 11-60. Feedback control used to vary puller speed for control of extrusion dimensions.

It should be noted that, while the gauging instruments are shown as being merely indicators, they can all be connected to a microprocessor or microcomputer, so information about startup and running conditions can be inserted in the memory banks. Furthermore, the computer can be programmed so that it will "learn" by using its memory to store results of previous runs and act to restore equilibrium when conditions warrant. Figure 11-60 shows a simple feedback control system using a displacement transducer connected to the puller-speed control; if the extrusion thickness increases, the puller is speeded up to restore the correct dimension. Figure 11-61 shows a more elaborate system using a microcomputer which can activate an alarm system and a display console in addition to its functioning as a closed-loop controller.

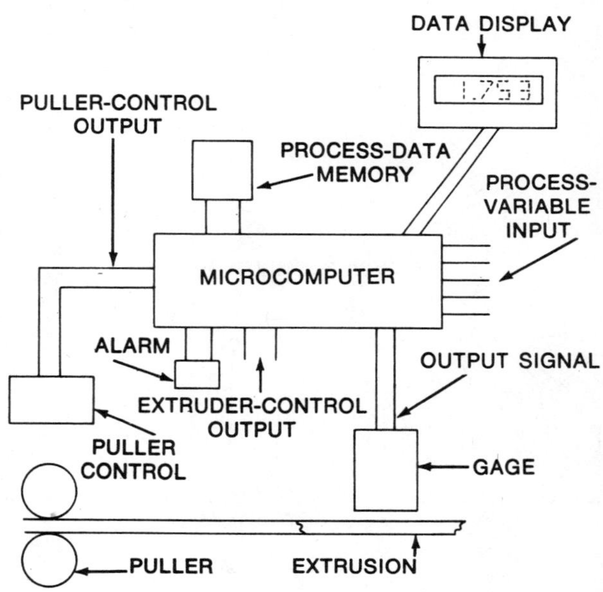

Fig. 11-61. Control loop using microcomputer.

Table 11-3. Applications and Limitations of Gauging Methods.*

Gauging Method	Material Limitations	Applications Limitations	Type of Extrusion
Roller - contact	None	Requires rigid material	All
Air	None	Requires reasonably flat surface	Sheet, profile
Magnetic-reluctance	Nonmagnetic materials with thin walls	Position control important	Sheet, tube
Sonic	Material with good sound transmission	Material with good sound transmission	Pipe, sheet
Optical	None	For dimensional control	All
Laser-intercept	None	For dimensional control	All
Laser-interferometry	May be transparent in infrared	For dimensional control	Thin tube, sheet
Capacitance	Materials with higher dielectric constant	For thickness control	Sheet, tube
Proximity	Materials with higher dielectric constant	Requires nonconductive materials	All
Beta-ray	None	Requires relatively smooth surface	Thin sheet

*Reprinted, by permission, from *Plastics Machinery & Equipment*, January 1978. Copyright 1978, Industry Media, Inc.

Auxiliary Extrusion Equipment

In order to extrude any of the products described previously, the minimum equipment necessary is obviously an extruder and some type of cooling and/or pull-off equipment. However, in order to produce parts or raw stock in a truly cost-effective manner, it is essential for the rate of production to be as rapid as possible and the quality of the products to be as high as possible. To assist production plants in reaching that goal, a number of auxiliary pieces of equipment have been developed which are used in one or more extrusion processes.

Crammer Feeders. A crammer feeder is used at the feed end of the extruder when a very fluid or low-density material is being extruded. The use of a crammer feeder provides positive feed to the throat of the extruder to insure an adequate supply of materials which tend not to be caught by a screw unless forced in (such as light, powdery polyethylene, rigid PVC, or polypropylene). The crammer feeder can also be used as a vacuum hopper when air is to be excluded. (Figure 11-11 shows a crammer feeder.)

Mixing Sections. Most extruders today, striving for improved mixing and higher output per hour, use a mixing section in the metering portion or on the end of the screw. The mixing section may be in the form of pins, reverse flights, barriers, etc. However, if one screw must be used for many materials, one of the most successful types is the Maddox mixing section, originally patented by Union Carbide. The section, previously described under Injection Screws, is approximately two screw flights in length. On extruders, it used at the screw tip and consists of alternating inlet and outlet flutes, as shown in Figure 11-62. (Figure 10-13 illustrates how the polymer enters the inlet flute and is forced over the narrow barrier in which it is sheared and intensively mixed.) The severe working of the resin at this point very effectively mixes any colorant, hot or cold particles, or additives which may be present. Since the gap is very shallow, no unmelted particles can get through, and only homogeneous melt is delivered to the die.

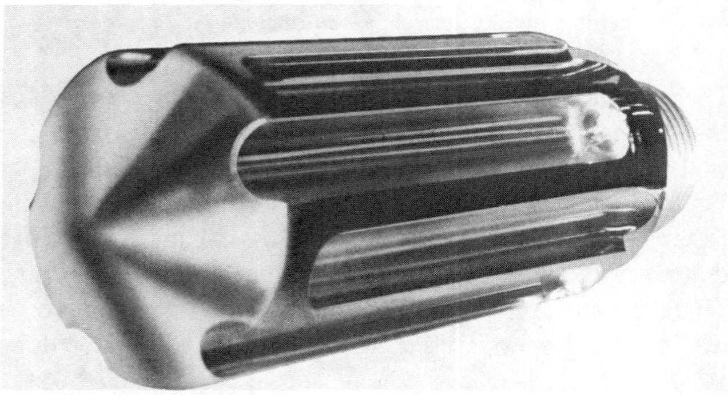

Fig. 11-62. Maddox fluted extruder screw tip used for increased shear and plastication. (*Courtesy: Feed Screws, Inc.*)

Post-Extrusion Mixing. Another method for improving the mixing of the extrudate and, in addition, for improving the thermal homogenization of the melt is the installation of an electrically heated mixing element located in line between the screw and the die. The device, known as a motionless mixer, Melt Blender (Koch Engineering Co., Inc.), or

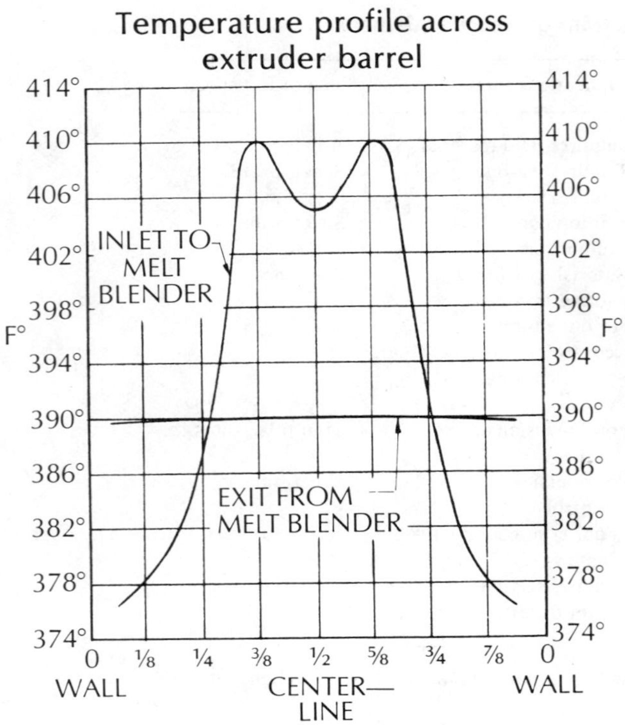

Fig. 11-63. Reduction in extruder barrel temperature profile by use of a melt blender. (*Courtesy: Koch Engineering Co., Inc.*)

Fig. 11-64. Koch Melt Blender used for extrusion. Melt homogenization is said to produce 32,768 subdivisions with four, non-moving elements. (*Courtesy: Koch Engineering Co., Inc.*)

Thermogenizer (Kenics Corporation), is claimed to reduce the temperature differential of 32°F in a melt, from the wall to the center line, to less than ±1°F, as shown in Figure 11-63. Other advantages claimed for the devices are faster extrusion rates and, because of the more extensive mixing, a reduction in colorant and additive requirements. There is, however, some pressure drop in the mixer, so its use in a machine operating at full capacity should be done with some caution. Figure 11-64 shows both the external and internal views of the Melt Blender. Figure 11-65 shows a typical installation on a sheet line.

Screen Changers. Equipment was previously described which could change a screen very quickly using a mechanism resembling a photo slide changer (Figure 11-15). Another concept in screen changing is one which involves a roll of woven screen. By means of automatic timing equipment, the melt flow pushes the screen across the breaker plate continuously. This system is said to eliminate the momentary blockages and pressure surges encountered with the slide-plate screen changer. This type of changing is claimed to be particularly beneficial to blown film, wire coating, sheet, and film. Savings in screen material as well as the benefits

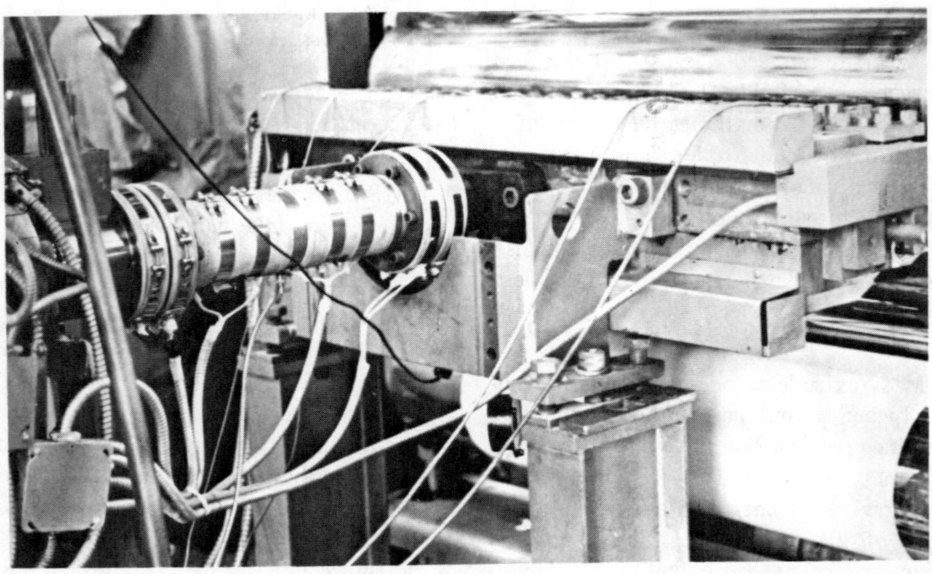

Fig. 11-65. Attachment of melt blender between extrusion nozzle and die on extruder. (*Courtesy: Kenics Corporation*)

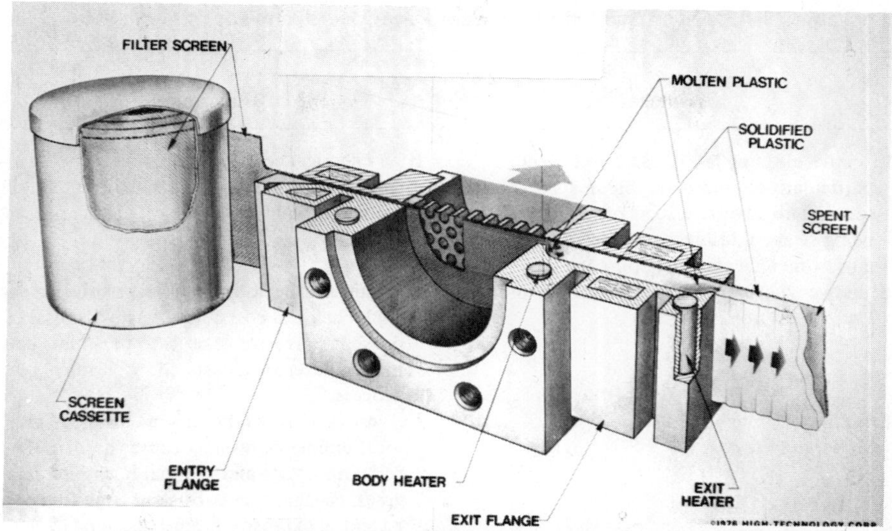

Fig. 11-66. Automatic continuous screen changer using woven wire belt screen. (*Courtesy: High-Technology Corporation, Hackensack, New Jersey.* AutoScreen® is a registered trademark.)

of completely automatic operation are other benefits. Figure 11-66 illustrates one type of such equipment. Another type of continuous screen changer is one somewhat like the "slide changer" mechanism, but with the advantage of a number of screens which can be pushed continuously across the breaker plate (see Figure 11-67). In both continuous changers, a novel technique, using the melt characteristics of the resin itself, prevents leakage.

Breaker Plate and Die Cleaning. As production continues, the screen, breaker plate, and die surfaces can accumulate burned and decomposed plastic residues which can seriously interfere with achievement of a quality product. When this happens, one or both pieces of equipment must be disassembled from the extruder and cleaned. Manual techniques, such as torching, wire brushing, scraping, and sanding have been used, but generally, because of damage to the die and high labor costs, these methods are not considered to be satisfactory. Other methods which are in use include chemical solvents, hot caustic baths, molten salt baths, burn-off furnaces, vacuum pyrolysis, and fluidized-bed systems. Table 11-4 lists the various methods used to clean extruder parts and their advantages and disadvantages.

Because of OSHA regulations for safety, ecological requirements for non-pollution, and possible damage to hardware by scratching, thermal distortion, or possible corrosion, the techniques in greatest use are hot salt baths or fluidized beds. The latter has the advantage of being somewhat safer for personnel and equipment, and is claimed to require very little maintenance and be noncorrosive to parts. It is also

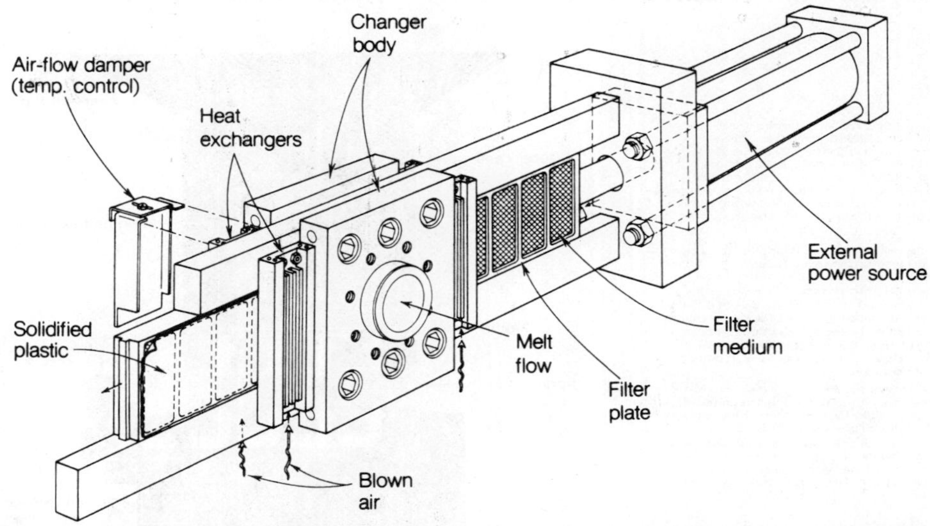

Fig. 11-67. Continuous screen changer, using separate elements and hydraulic cylinder power. (Reprinted, by permission, from *Modern Plastics*, McGraw-Hill, Inc.)

Table 11-4. Advantages and Disadvantages of Hardware Cleaning Methods.*

Cleaning Method	Advantages	Disadvantages	Relative Cost of Capital Equipment
Chemical solvents	Effective cleaning job in cases where the particular polymer is soluble in a suitable solvent; no adverse effects on the parts being cleaned. Mildest treatment of parts of all commercial cleaning approaches.	High chemical costs, high solvent recovery and/or waste disposal costs, and flammability and/or toxic hazards. Slow cleaning process.	High
Hot caustic solution	Effective where the polymer is soluble in hot caustic solutions.	Hot caustic solutions may be corrosive to parts. High chemical costs, high waste-disposal cost, possible corrosive hazards to personnel and high maintenance costs. Slow cleaning process.	Low to moderate
Burn-off furnace	Effective cleaning job in some cases. Moderately fast.	Leaves residues behind in some cases. High local temperatures may cause thermal distortion in parts and thermal release of acid gases, causing part corrosion. High fuel costs, potential explosive hazard and long cycles. Frequently requires steam inertizing that may cause part corrosion.	Moderate
Hand/mechanical	Effective cleaning job in some cases. No chemical disposal or energy costs.	Part damage may occur due to physical abuse—dents, scratches, elongation of holes. High labor costs and high part-replacement costs. Slowest cleaning process.	Very low
Salt-bath	Most effective cleaning job. Only residue is salt, which may be cleaned with water wash. Fastest cleaning process.	High chemical costs, high waste-disposal cost, corrosive burn hazards to operating personnel, and high maintenance cost due to equipment corrosion. High installation cost.	Moderate to high
Fluidized-bed	Effective cleaning job. Generally noncorrosive; little or no thermal degradation. Low chemical costs; no waste disposal; no corrosive burn hazards; low installation and maintenance costs. Relatively fast.	Occasional residue may require secondary cleaning operations, necessitating additional handling step.	Moderate to high
Vacuum pyrolysis	Usually an effective cleaning job. Generally noncorrosive to the part; little or no thermal distortion. No chemical costs, no waste disposal cost, no corrosive burn hazards; low installation costs; low to medium maintenance costs. Moderately fast.	Occasional residue may require secondary cleaning operations, necessitating additional handling step.	High

*Reprinted, by permission, from *Plastics Machinery & Equipment*, May 1977. Copyright 1977, Industry Media, Inc.

Air is supplied to the bath either from an air line or directly from a small air compressor. It enters a chamber in the fluidized bath through an air flow control valve. The air under pressure is forced through a diffuser plate (porous plate) made of either ceramic or stainless steel containing pores or small holes. The action of the porous plate is to distribute the air flow over the whole area of the bath.

When the velocity of the air is low, the air just passes through the spaces between particles. If the air flow is further increased, a state is reached when the weight of the particles is counterbalanced by the drag force exerted on them by the air. This is termed incipient fluidization. This will be seen as an increase in the height of the bed. If the air flow is further increased beyond this point, bubbles will break on the top surface and the particles will start moving inside the enclosure.

Heaters immersed in the fluidized medium raise the temperature and by cycling the heaters on and off with the controller the temperature of the medium can be controlled over a wide range.

A pressure switch or fail safe unit is installed which operates from the air supply to the bath and turns off the heaters if the main air supply drops and fluidization is lost. This protects the heaters against burn out.

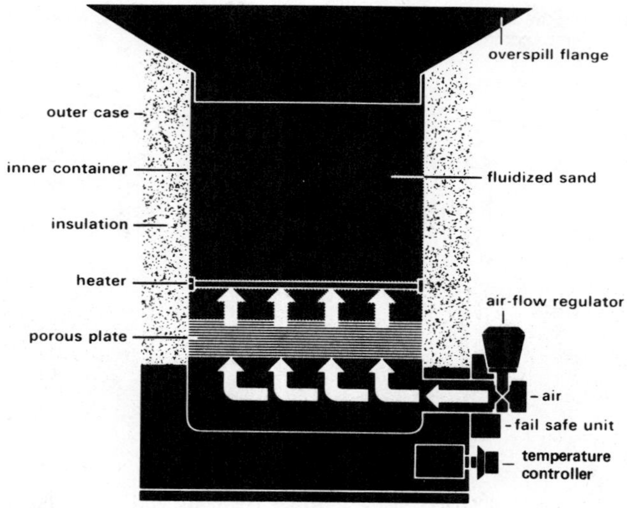

Fig. 11-68. Principle of fluidized bed.

Fig. 11-69. Typical "dirty" screens and breaker plate. (*Courtesy: Techne Incorporated, Princeton, New Jersey*)

Fig. 11-70. Screens and breaker plate afted fluidized-bed cleaning. (*Courtesy: Techne Incorporated, Princeton, New Jersey*)

Fig. 11-71. Various sizes of fluidized beds for screen and breaker plate cleaning. (*Courtesy: Techne Incorporated, Princeton, New Jersey*)

quickly started up since the cleaning material (aluminum oxide particles) does not solidify on cooling. Figure 11-68 shows the principle of a fluidized bed in which the aluminum oxide is suspended in air to act as a hot abrasive in removing the burned-off resin. Figures 11-69 and 11-70 show screens and breaker plates before and after cleaning. Figure 11-71 shows various sizes of units.

Take-Off and Cutting Equipment for Pipe and Tubing. In order to produce pipe and tubing in definite lengths, or even very short pieces of similar extrudate, special take-off and cutting machines are available. The take-offs feed the extrudate to the cutters at a very precisely controlled speed and the cutters, either knives or saws, travel with the extrudate as they cut, or cut through extremely rapidly, stopping the material for a few milliseconds while making the cut. The cutters measure the extrudate as it passes through, so parts as small as washers can be cut at rates as high as 8000 per minute, or lengths of pipe can be cut 10 or 15 feet long.

REFERENCES AND BIBLIOGRAPHY

Adams, R. L. Producing pigmented thermoplastics in volume. *SPE Journal*, April 1971, **27**(4):954.

Baier, H. F. HDPE "paper." *Plastics Technology*, February 1976, **22**(2):41.

Barr, R. A. New screws drive extruder performance up and up. *Plastics Technology*, October 1972, **18**(10):67.

Barr, R. A., and Chung, C. I. Effects of radial screw clearance on extruder performance. *SPE Journal*, June 1966, **22**(6):71.

Berylyn, G. E. The fundamentals of screen changers. *Plastics Design & Processing*, April 1977, **17**(4):55.

——. Screen changer is "truly" continuous. *Modern Plastics*, December 1976, **53**(12):63.

Bernhardt, E. C. Calculating extruder performance. *Modern Plastics*, June 19, 1972, **49**(6).

——. *Processing of Thermoplastic Materials*. Reinhold Publishing Corporation, New York, 1959.

Borg-Warner Chemicals. *Tips for Solving Extrusion Problems*. Bulletin PB-180, 10915.

Burkhardt, K., Hermann, H., and Jakopin, S. Plasticizing and mixing in twin-screw extruders. *Plastics Compounding*, November/December 1978, p. 73.

——. Twin screw extruders. *Plastics World*, November 1977, **35**(11):64.

Campbell, W. B. Single-screw versus twin-screw extruders. *Plastics World*, June 1968, **26**(6):70.

Celanese Plastics Co. *Fortiflex Polyethylene Extrusion*. Bulletin P3B, January 1967.

——. *Extrusion of Celcon Acetal Copolymer*. Bulletin C3B, February 1966.

Celanese Polymer Co. *Extrusion of Celanese Acetate*. Bulletin A3C, December 8, 1960.

Chung, C. I. A guide to better extruder screw design. *Plastics Engineering*, February 1977, **33**(2):34.

Comey, J. Take a look at pultrusion for high-performance, complex shapes. *Plastics Design & Processing*, May 1976, **16**(5).

——. Designing thermoset shapes for pultrusion. *Machine Design*, December 23, 1971, **43**(24):45.

Doherty, R., and Cotter, P. J. External vacuum sizing of thermoplastic pipe and tubing. *SPE Journal*, October 1964, **20**(10):1093.

Dominghaus, H. Recent developments in processing plastics on single and multiple screw extruders. *SPE Journal*, July 1964, **20**(7):593.

Dowd, L. E. Troubleshooting HDPE processing problems. *Plastics Design & Processing*, May 1979, **19**(5):40.

Doyle, R. Extrusion of sheet from high density polyolefins. *SPE Journal*, February 1961, **17**(2):165.

Dreger, D. R. Now: profile extrusions from high-performance plastics. *Machine Design*, May 29, 1975, **47**(10):42.

E. I. Du Pont de Nemours & Co., Inc. *The Extrusion of Polyethylene Film*. Bulletin A-52858.

——. *Extrusion Coating*. Bulletin A-28451, March 1968.

Eastman Chemical Products, Inc. *Extrusion, Tenite Polyethylene, an Eastman Plastic*. Bulletin T 56 2283, 1956.

Fisher, E. G. *Extrusion of Plastics*. Interscience Publishers, New York, 1964.

Fitch, A. L. Vertical extruders with a difference. *Plastics Engineering*, May 1974, **30**(5):20.

Frados, J. (ed.) *Plastics Engineering Handbook of the SPI*. Van Nostrand Reinhold Co., New York, 1976.

Frizelle, W. G. An extrusion primer: how to calculate extruder output. *Plastics Technology*, September 1970, **16**(9):46.

Gardiner, R. A. *Thermal Requirements of Extruders*. David-Standard Division, Crompton & Knowles Corporation.

Garshman, A. Extruding TPU blown film: the how's and the why's. *Plastics Technology*, February 1979, **25**(2).

Gatto, C. Methods of sizing extruded plastics pipe and tubing. *Plastics World*, December 1964, **22**(12):16.

General Electric Co., Chemical Materials Department. *Lexan Polycarbonate Resins Extrusion*. Technical Report CDC-501, August 1961.

Gregory, R. B. How plastics melt in an extruder. *SPE Journal*, June 1971, **27**(6):49.

——. How friction affects extruder operation. *SPE Journal*, October 1969, **25**(10):55.

——. Advances in extrusion die design. *Modern Plastics*, March 1967, **44**(7).

Griff, A. L. *Plastics Extrusion Technology*. Reinhold Publishing Corporation, New York, 1968.

Gulf Oil Co., Plastics Division. *A Technical View of Extrusion Coating*.

Haine, W. A. Improving tubular-film extrusion. *Plastics Technology*, August 1962, **8**(8):25.

Heilmayr, P. F. PVC pipe keeps rolling along. *Plastics Engineering*, January 1976, **32**(1).

Hoberg, C. E. Extrusion of ionomer resins. *SPE Journal*, September 1966, **22**(9):29.

Hollander, E. R. Jr. TP polyesters. *Plastics Technology*, February 1976, **22**(2):29.

Johnson, J. E. Co-extrusion. *Plastics Technology*, February 1976, **22**(2):45.

Johnson, R. W. Paste extrusion of filled TFE-fluorocarbon resin for wire insulations. *SPE Journal*, February 1961, **17**(2):151.

Kent, J. R., and Walker, G. E. Trouble-free extrusion of ABS sheet. *Plastics Technology*, November 1964, **10**(11):46.

Klein, I. Extrusion—an update on melt technology. *Plastics Design & Processing*, Part 1: August 1973, **13**(8). Part 2: September 1973, **13**(9). Part 3: October 1973, **13**(10):20.

——. Computer design of plasticating extruder screws. *Modern Plastics*, September 1969, **46**(9).

Klein, I., and Klein, R. Plasticating extrusion, a new look. *Modern Plastics*, June 1979, **56**(6):65.

Koehring Co., HPM Division. *Co-Extrusion by Prodex*.

Kruder, G. A., and Ridenour, R. E. New concepts spur development of ultra-high-speed extrusion. *Plastics Engineering*, April 1979, **35**(4):29.

Ladd, C. F. A checklist for safe extruder operation. *Plastics Engineering*, January 1978, **34**(1):47.

Levy, S. Gaging and control of extrusion processes. *Plastics Machinery & Equipment*, June 1979, **9**(6):31.

——. On-line gaging of extrusion dimensions. *Plastics Machinery & Equipment*, January 1978, **8**(1).

——. In-line coating of rigid substrates. *Plastics Machinery & Equipment*, February 1978, **8**(2).

——. Maximizing flow through profile extrusion dies. *Plastics Machinery & Equipment*, April 1978, **8**(4):24.

——. Processes and equipment for in-line post extrusion forming. *Plastics Machinery & Equipment*, August 1978, **8**(8):26.

——. Melt rheology: its effect on profile-extrusion dies. *Plastics Machinery & Equipment*, September 1978, **8**(9):28.

——. Diamond-wire saw cuts precision profile dies. *Plastics Machinery & Equipment*, August 1977, **7**(8):22.

——. Cooling techniques for profile extrusion. *Plastics Machinery & Equipment*, September 1977, **7**(9):29.

LFE Corporation. *Extruded Plastic Pipe Thickness Control Systems.* Bulletin 5531-SB1, March 1976.

Lindau, J. W. III. Extrusion then and now. *Plastics World*, September 1968, **26**(10):50.

Lounsbury, D. C. Basic downstream equipment for extruded blown film handling. *Plastics Design & Processing*, June 1977, **17**(6).

Mac Dermott, C. P. Future possibilities for extruded plastics. *SPE Journal*, June 1964, **20**(6):523.

Maddock, B. H. The status of extrusion science as related to polymer properties. *SPE Journal*, August 1963, **19**(8):707.

——. Measurement and analysis of extruder stability. *SPE Journal*, December 1964, **20**(12):1277.

Maddock, B. H., and Smith, D. J. Extruder design by computer printout. *SPE Journal*, May 1972, **28**(5):12.

Maddock, B. H., and Matzuk, P. P. Theory and operation of the two-stage, single-screw extruder. *SPE Journal*, April 1962, **18**(4):405.

Martin, J. D. Pultrusion: the other process. *Plastics Engineering*, March 1979, **35**(3):53.

Maxwell, B., and Scalora, A. J. The elastic melt extruder–works *without screw. Modern Plastics*, October 1959, **36**(10):107.

McCormick, W. S. If your extruder throughput is falling you may have feed problems. *Plastics World*, October 22, 1975, **33**(10):36.

Meadows, D. Dual extrusion–the technique of joining two plastics into one part through extrusion. *Plastics Design & Processing*, December 1968, **8**(12):18.

Merz, E. H. The rheology of thermoplastics, a primer. *SPE Journal*, August 1963, **19**(8):735.

Miller, B. S. Extrusion process saves days in finishing extrusion dies. *Plastics World*, December 17, 1973, **31**(12):54.

Miller, R. L. Feeding the single screw extruder. *SPE Journal*, November 1964, **20**(11):1183.

Mobay Products Co. *Processing Merlon Polycarbonate Resin by Extrusion.* TIB No. 47-M3, May 1, 1960.

Modern Plastics. Now extrusion control is affordable by everybody. February 1979, **56**(2):38.

——. Controls for molding: which one's for me? March 1977, **54**(3):46.

——. Toward perfect screw design. November 1977, **54**(11):41.

——. Extrusion-casting of composite films. April 1968, **45**(4):130.

——. New approach to screwless extruders. November 1962, **40**(3):119.

Mohr, W. D., Squires, P. H., and Starr, F. C. Flow patterns in a single screw extruder. *SPE Journal*, September 1960, **16**(9).

Monica, R. E. What a plastics engineer should look for when planning to buy auxiliary equipment for extrusion. *Plastics Technology*, September 1959, **5**(9):6.

Monsanto Co. *Sheet Extrusion of Lustran I.* Extrusion Technical Report No. 5, April, 1964.

——. *Fundamentals of Thermoplastic Extrusion*. Product Information Bulletin No. 1029, 1962.

Murray, T. A. Here's your guide to die design. *Plastics Technology*, February 1978, **24**(2):99.

——. Extruder screen changers: the choices widen. *Plastics Technology*, October 1977, **23**(10):65.

Naitove, M. H. Co-extrusion. *Plastics Technology*, February 1977, **23**(2):61.

——. Now co-extrude foam-core ABS pipe. *Plastics Technology*, December 1976, **22**(12):50.

Nissel, F. R. Recent developments in sheet extrusion systems. *SPE Journal*, January 1970, **26**(1):25.

Nissel, F. R., and Shaver, L. D. Finished product extrusion from rigid PVC dryblend. *SPE Journal*, January 1966, **22**(1):29.

Peek, W. Tackle rigid PVC with twin screws. *Plastics Engineering*, July 1974, **30**(7):23.

Peelman, P. L. Proper maintenance procedures can minimize your extruder downtime. *Plastics Design & Processing*, May 1977, **17**(5).

Pettit, G. A. and Eckman, R. L. Electronic die-pressure control holds key to quality extrusion. *Plastics Technology*, May 1964, **10**(5):39.

Pickens, S. H. Pultrusion–the accent is on the long pull. *Plastics Engineering*, July 1975, **31**(7):16.

Plastics Design & Processing. Startup and troubleshooting tips to increase blown film line efficiency. January 1978, **18**(1):20.

——. Static mixers increase part quality, decrease colorant requirements by 20%. June 1976, **16**(6).

——. A survey of trends in extrusion. February 1968, **8**(2):14.

Plastics Design Forum. Design considerations for extrudable products. November/December 1976, p. 44.

Plastics Engineering. A new twist in mixing extruders. May 1974, **30**(5):40.

Plastics Machinery & Equipment. Troubleshooting sheet extrusion problems. Part 1: June 1977, **7**(6):46. Part 2: July 1977, **7**(7):29.

Plastics Technology. Wear-resistant coating protects screws longer. February 1978, **24**(2):11.

——. Fluidized-bed cleaning cuts extruder die maintenance costs. February 1977, **23**(2):85.

——. Melt, mix better with these new screws. February 1977.

——. Screwless mixer/extruder gives new look to compounding. October 1975, **21**(10):14.

——. Blown film system offers new approaches to bubble cooling, width and gauge control. December 1973, **19**(12):11.

——. Look into this foam process for extruding profiles. February 1969, **15**(2).

——. New developments in extrusion coating. September 1966, **12**(9).

——. Two unique extruders. January 1963, **9**(1).

Plastics World. Extrusion productivity: tuning up for the 80s. May 1979, **37**(5).

——. How to stretch your additive dollar: motionless mixers. May 1978, **36**(5):40.

——. Extruder productivity: big gains in 5 years. July 1978, **36**(7):48.

——. Extruder control: not "if," but "when." January 1977, **35**(1):32.

——. New dimensions in sheet extrusion. January 1977, **35**(1):28.

——. Extruding the "tough" ones. August 1976, **34**(8):34.

——. Now you can extrude irregular shapes from PVC powder. November 19, 1973, **31**(11):68.

——. Hard-surfaced screws protect productivity in extrusion and injection. December 1972, **30**(12):58.

Port, C. O. Dual-vent extruder shortcuts the need to predry ABS. *Modern Plastics*, October 1973, **50**(10):148.

Prall, G. How to build a blown polyethylene film line. *Plastics Technology*, July/August 1968, **14**(7,8):48, 43.

Prause, J. J. How to care for the screw and barrel. *Plastics Technology*, February 1969, **15**(2):49.

——. Twin-screw extruders–when to use and why. *Plastics Technology*, November 1967, **13**(11):41.

Prodex Corporation. *How to Choose an Extruder*. Bulletin E-3, 2nd Ed.

Richardson, P. N. *Introduction to Extrusion*. Society of Plastics Engineers, Greenwich, Connecticut, 1974.

Ryder, L. B. Some comments on vented extruder design. *SPE Journal*, August 1961, **17**(8):731.

Saint-Amour, J. D. Diagnose pipe extrusion problems with on-line thickness gauging. *Plastics Technology*, February 1979, **25**(2):57.

Saltzman, G. A. Wear data aid selection of barrel-screw combinations. *Plastics Design & Processing*, October 1975, **15**(10).

Schard, M. P., and Vera, O. E. Extrusion of small diameter polybutylene pipe. *Plastics Engineering*, October 1977, **33**(10):45.

Scheiner, L. L. Slit film/monofilament extrusion. *Plastic Technology*, February 1977, **23**(2):53.

——. Processors' guide to mixing, blending and compounding equipment. *Plastics Technology*, June 1967, **13**(6):39.

——. Extruding and forming PS foam sheet. *Plastics Technology*, Part 1: August 1966, **12**(8):41. Part 2: September 1966, **12**(9). Part 3: October 1966, **12**(10). Part 4: November 1966, **12**(11).

——. Extruder screws–where we are . . . where we're headed. *Plastics Technology*, November 1965, **11**(11):50.

Schroeder, G. O. Coextrusion: when is it competitive? *Plastics World*, December 1978, **36**(12):50.

Schutz, F. C. Dual screw extruders. *SPE Journal*, September 1962, **18**(9):1147.

Sisco, P. E. Jr. Design and construction of dies for blown and cast film. *Plastics Machinery & Equipment*, March 1978, 8(3).

Smits, L. G. P. Vented extruders–a review. *SPE Journal*, September 1964, **20**(9):999.

SPE Journal. Nomenclature of the screw extruder. March 1962, **18**(3):295.

——. Paste extrusion of filled TFE-fluorocarbon resin for wire insulations. February 1961, **17**(2):151.

Squires, P. H. Extruder heating and cooling systems. *SPE Journal*, August 1962, **18**(8):863.

Staffin, H. K. Cleaning plastics processing hardware. *Plastics Machinery & Equipment*, May 1977, 7(5).

Summers, P. H. Fluidized bed–safe, efficient method for cleaning dies, molds and molding equipment. *Plastics Design & Processing*, December 1974, **14**(12):20.

Summerville, A. B. Static elimination equipment and dust particle removal systems ease processing of extruded film. *Plastics Design & Processing*, February 1976, **16**(2):198.

Tadmor, Z., and Klein, I. *Principles of Plasticating Extrusion*. Reinhold Publishing Corporation, New York, 1970.

——. The effect of design and operating conditions on melting in plasticating extruders. *Polymer Engineering and Science*, January 1969, **9**(1):1, 11.

U.S. Industrial Chemicals Co. *Polyethylene Film Extrusion . . . An Operating Manual*. Bulletin PL 10-170, 2nd ed., 1970.

——. *Polyethylene Extrusion Coating . . . An Operating Manual*. Bulletin PL 11-569, 2nd Ed., 1969.

Van Ness, R. T. *Screw Design and Mixing*. E. I. Du Pont de Nemours & Co., Inc.

Von Hassell, A. Systems for internal bubble cooling. *Plastics Technology*, May 1977, **23**(5):55.

Young, J. F. Sheet extrusion. *Plastics Design & Processing*, February 1968, **8**(2):19.

Westover, R. F. Design your own extruder. *Plastics World*, June 1968.

——. Progress in . . . extrusion. *Plastics Technology*, May 1966, **12**(5).

——. New views of extrusion. *SPE Journal*, May 1965, **21**(5):455.

——. A hydrodynamic screwless extruder. *SPE Journal*, December 1962, **18**(12):1473.

Western Plastics. Coextrusion unlocks. February 1972.

Willer, A. M. Temperature and pressure–important factors in extrusion and injection molding. *Plastics Design & Processing*, August 1968, **8**(8):29.

Windish, C. J. Polypropylene extrusion–a guide to procedure and processing conditions. *Plastics Design & Processing*, March 1962, **2**(3):22.

12
Blow Molding

Blow molding is a centuries old technique of forming hollow articles from a molten material. The process was used in ancient times by Egyptians and Babylonians to blow glass and molten amber into small containers and decorative figurines. Basically, the process consists of forming a tube of the melt and introducing air into the tube to expand it, either to a free-blown object or against a mold to form it to a definite size and shape.

Blow molding of plastics is essentially the same process as outlined above, except that the materials and the equipment have been changed and made considerably more complex. The original technique developed for blow molding of plastics may be considered an outgrowth of the extrusion process, since an extruder was the most vital piece of equipment in the procedure. In the first techniques developed, the extruder was used to produce a tube which was, and still is, called a parison. The hot parison, immediately on emerging from the die, is clamped in a mold; air is introduced and, after a cooling period, a hollow container is ejected from the mold. This basic process is illustrated in Figure 12-1.

After some years of extrusion blow molding, and after the introduction of the screw-injection blow molder, a technique was developed whereby the parison could be injection molded, with some distinct advantages and some distinct limitations.

Materials

Theoretically, any thermoplastic resin might be used for a blow molding operation. However, only those materials which show high hot strength and good stretch properties at the extrusion temperature are suitable for formation of a parison and the subsequent blowing. The materials which show these properties and which are most commonly used include: high-density polyethylene, made into rigid bottles and similar items; low-density polyethylene used for more flexible items; polypropylene, used for high gloss, heat resistance, stiffness, and good strength; medium-impact polystyrene and PVC, used for general-purpose articles where transparency is required, at moderate cost; and polycarbonate, where transparency and high strength are required at a somewhat higher cost. Acetal, nylon, ionomers, and acrylonitrile-styrene are other materials which might be used, but either because of cost and/or sharp melt points or other properties, they have not been proved to be as suitable as the materials first mentioned.

Extrusion Blow Molding Processes

There are a number of different methods of extrusion blow molding which have been developed; such factors as the size of the part, the number of parts to be made, and the type

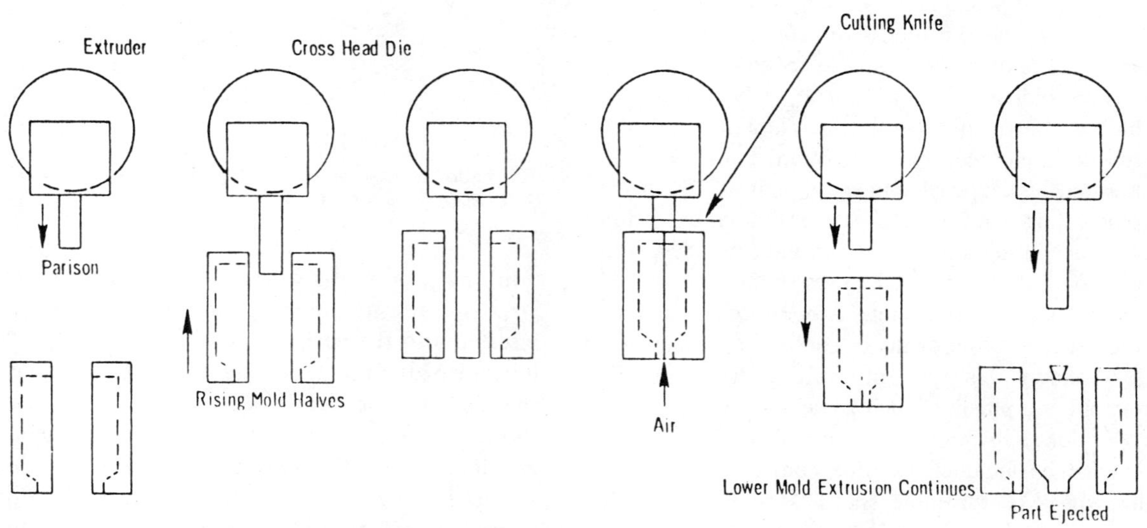

Fig. 12-1. Basic extrusion blow molding process. (*Courtesy: Gulf Oil Chemicals Co.*)

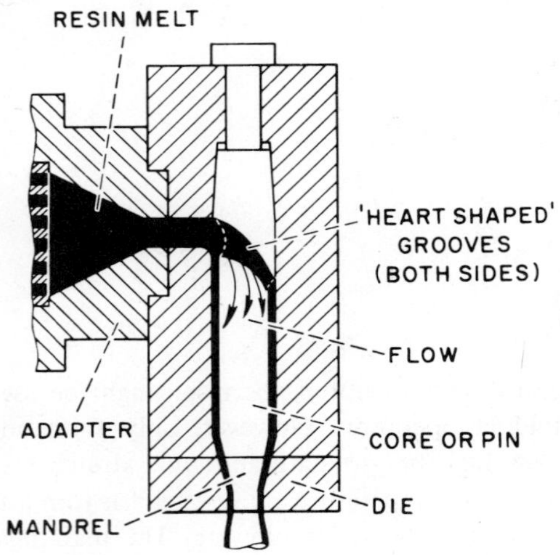

Fig. 12-2. Side-fed die with "heart-shaped" grooves. (*Courtesy: U.S. Industrial Chemicals Co.*)

of part will influence the decision about which technique will be employed. In all cases, however, the extruder will probably have an L/D ratio of at least 20:1 to insure complete and thorough plasticization of the melt. After leaving the extruder, the melt is directed downward through a cross-head die which then forms the tubular parison. It is the function of the die to direct the melt flow smoothly around the core pin with no hangups or stagnant areas which could show as striations in the parison, or subsequent marks on the blown container. Figure 12-2 shows a schematic of a side-fed die with streamlined "heart-shaped" grooves on each side of the core pin to insure even flow around the core pin and minimum markoff at the point of convergence. In actual production, the extruder is run either continuously or intermittently, as outlined below.

Continuous Extrusion Blow Molding. The continuous extrusion method is used extensively for the production of small bottles from approximately 6 ounces to 1 gallon. As the name indicates, the extruder is run continuously, using a multiplicity of molds to pick off the parison, blow, cool, and eject. The molds may be on a vertical wheel or a horizontal table, as shown in Figure 12-3. These methods are well suited for large-quantity production. Another technique, somewhat less expensive in equipment, is the "rising mold" method shown in Figure 12-4. Continuous extrusion can also be done using a manifold and multiple dies, controlled by automatically actuated valves so the dies are fed in sequence. Thus, only one mold is being fed at a time, while in the other molds blowing or cooling is taking place. Figure 12-5 shows this type of multiple-die, multiple-stationary-mold equipment, which is well adapted for high-speed production. Obviously, in any of these methods, provisions must be included for fine control of the screw speed, the wheel or turntable speed, sequencing of the valves, etc.

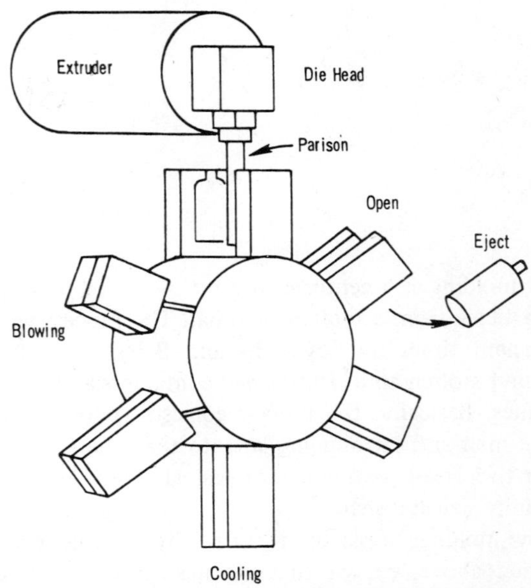

Fig. 12-3a. Vertical-wheel-type, continuous extrusion blow molding machine. (*Courtesy: Gulf Oil Chemicals Co.*)

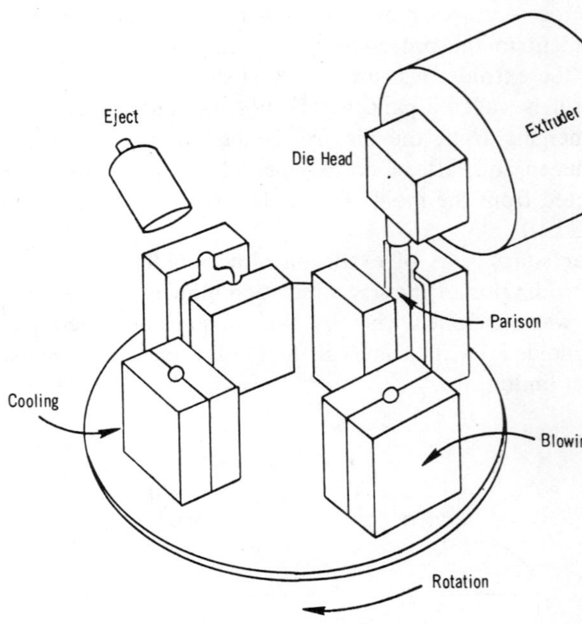

Fig. 12-3b. Horizontal rotating, blow molding machine. (*Courtesy: Gulf Oil Chemicals Co.*)

In any process, blowing is accomplished by introducing air, CO_2, or a mixture of gases through a blow pin which is inserted into the neck of the container, either at the top or bottom of the mold, or by a needle which is inserted into the parison. The needle is inserted into the parison above the neck area so the markoff can be trimmed away after ejection, or into the body of the part, where the hole will seal itself when the needle is withdrawn. Blowing pressures range from 50 to 150 psi, depending on the material and

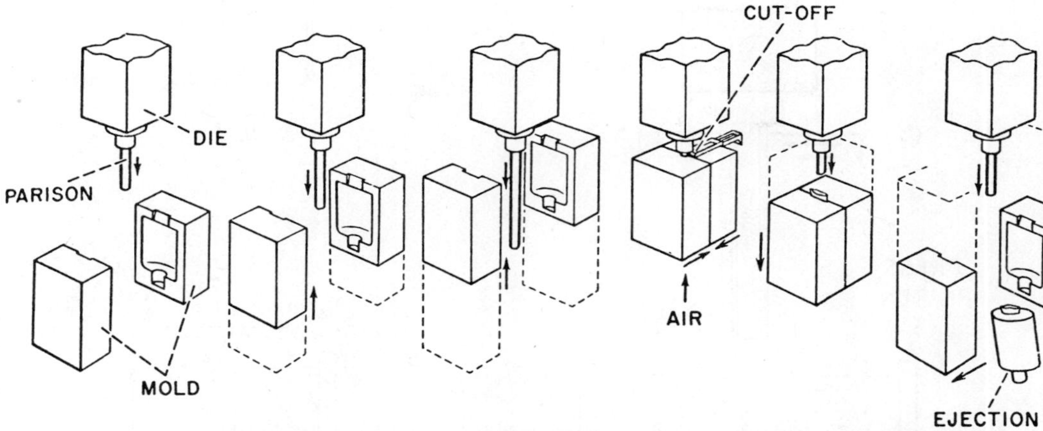

Fig. 12-4. "Rising mold" method of blow molding. (*Courtesy: U.S. Industrial Chemicals Co.*)

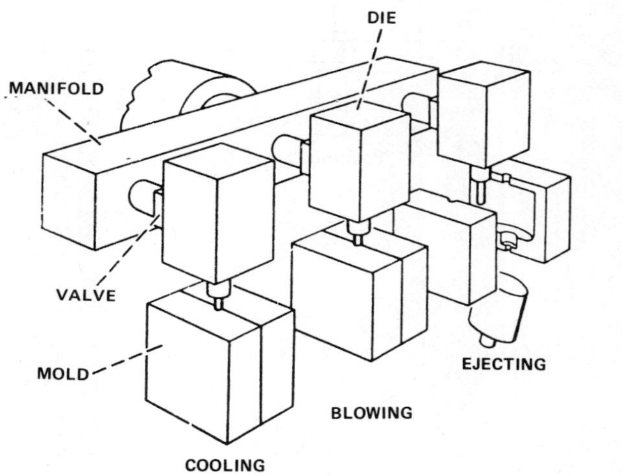

Fig. 12-5. Multiple-die, multiple-stationary-mold equipment. (*Courtesy: U.S. Industrial Chemicals Co.*)

part. The higher pressures have the advantage of better pickup of detail from the mold, but have the disadvantage of requiring stronger molds. In Figure 12-6, a process known as a parison-transfer technique which employs a bottom blow pin is illustrated. The parison-transfer process is used in fabrication of large bottles such as a 5-gallon carboy, where moving a heavy mold, such as in the "rising mold" process, might be impractical.

As might be expected from the multiplicity of processes which can be used to produce the same type of parts, a large variety of machine designs are available. Figures 12-7 and 12-8 show a relatively small machine which can be used to produce bottles up to $\frac{1}{2}$ gallon in capacity at a rate of approximately 220 per hour. In Figure 12-9, a larger machine is shown which incorporates four die heads, four molds, and an integral trim fixture to produce as many as 720 gallon-containers per hour. A large continuous extrusion, rotary-table machine which can produce up to 6000 units per hour is presented in Figure 12-10.

Intermittent Extrusion Blow Molding. When making very large parts, the parison becomes so large and heavy that it sags because of its own weight, and becomes thinner and thinner as it is extruded. Figure 12-11 illustrates the phenomenon. One method of overcoming the thinning is to use a programmed parison (described below). Another technique is to "squirt" the melt through the die so quickly that it doesn't have time to sag. This is accomplished by intermittent extrusion, in which the melt is allowed to accumulate until a large charge forms which is then pushed through the die very rapidly.

Two methods are used to deliver material to the die intermittently. In one method, a reciprocating screw is used, as in injection molding, to inject a specific amount of material through the die to form the parison. In this manner, a parison can be delivered in 1 or 2 seconds which might take 5 to 10 seconds if delivered by the normal extrusion technique. During the subsequent blowing and cooling, the screw is retracting and accumulating another charge. The blowing and cooling times remain the same; however, the overall reduction in cycle time can be significant. The use of the reciprocating screw machine thus allows the production of larger objects than could be made with a normal extruder of the same barrel size, and can do so at an increased production rate.

The second method of intermittently extruding the parison utilizes an external accumulator, which can be many times larger than the volume in a reciprocating screw barrel. This method is used for the production of very large barrels and containers (55 to hundreds of gallons). In this technique, one or as many as four extruders are used to continually fill the accumulator. After the correct charge has been accumulated, a hydraulic piston drives out the melt at a very rapid rate to form a relatively uniform parison, which is then placed in the mold and blown. Figure 12-12 shows a large blow molding machine with a die-head accumulator which can produce 55-gallon barrels. Figure 12-13 shows a machine built to use four extruders simul-

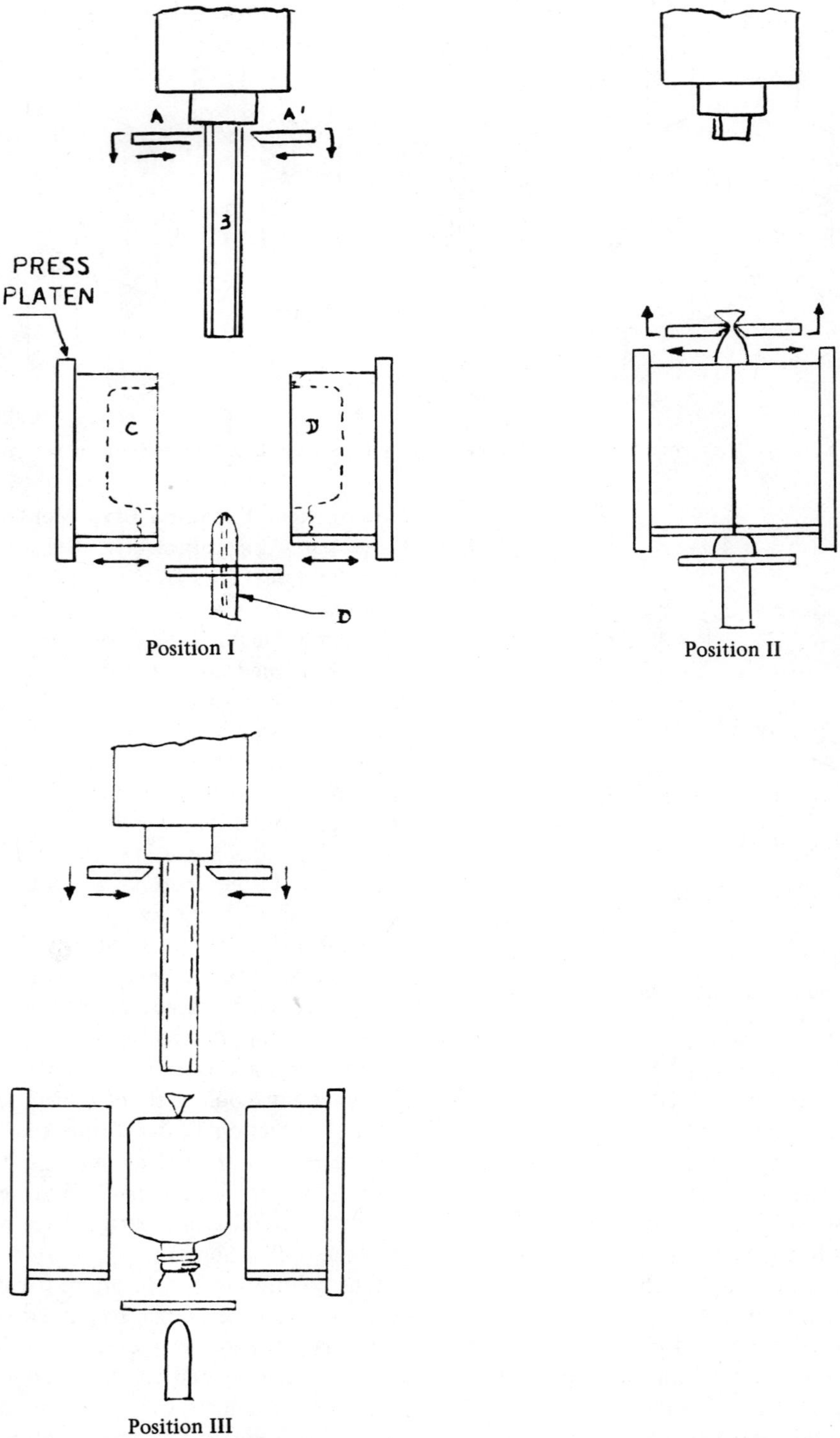

Fig. 12-6. In the parison-transfer blow molding process, the parison is extruded and cut off by knives A and A′ which then transport the parison down between the mold halves. Inflation air enters through bottom blow pin D. (From *Plastics* by DuBois and John)

Fig. 12-7. Small and compact blow molding machine, showing extruder, extruding head, and mold-holding mechanism. (*Courtesy: Rocheleau Tool & Die Co., Inc.*)

Fig. 12-8. Small blow molding machine showing pumps for hydraulic motor on extruder and control panel. (*Courtesy: Rocheleau Tool & Die Co., Inc.*)

taneously feeding the accumulator head. Figure 12-14 shows typical parts made by such large extruders.

Injection Blow Molding

Injection blow molding, as the name implies, uses an injection molding press to produce the parison; however, this is not accomplished in the same manner as previously described for intermittent extrusion. In injection blow molding, the parison is injected into a preform cavity, and around a core pin, in the exact quantity required to form the container. The preform mold is kept at a precisely controlled tempera-

Fig. 12-9. Four-die machine able to produce gallon containers at a rate of 720 per hour. (*Courtesy: Hoover Universal, Inc., Blowmolding Machinery Division*)

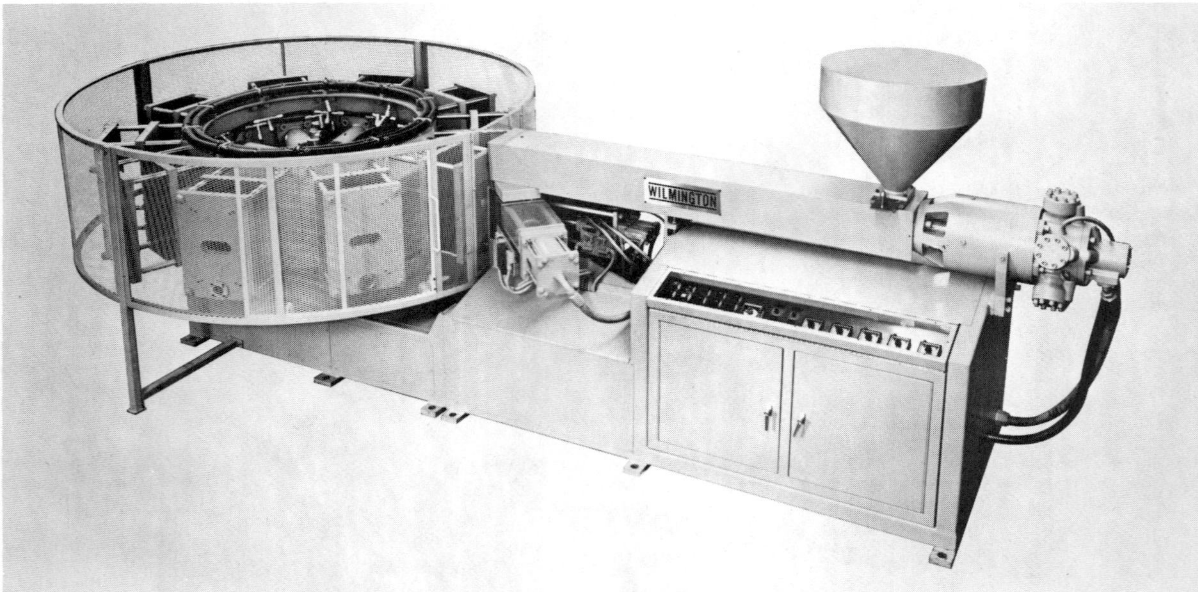

Fig. 12-10. Continuous extrusion blow molder with rotary die table. (*Courtesy: Wilmington Plastics Machinery*)

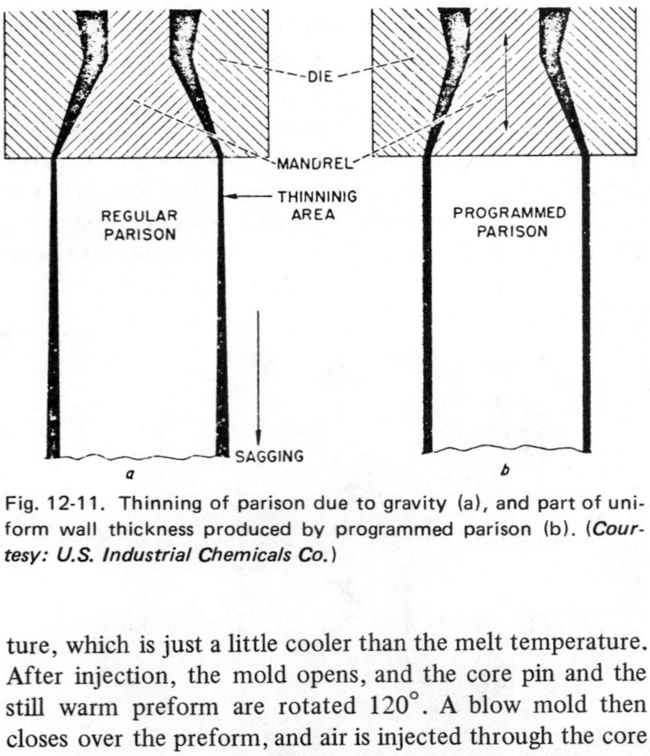

Fig. 12-11. Thinning of parison due to gravity (a), and part of uniform wall thickness produced by programmed parison (b). (*Courtesy: U.S. Industrial Chemicals Co.*)

Fig. 12-12. Blow molder using an accumulator to produce parts greater than the extruder capacity. (*Courtesy: Kautex Maschinebau GmbH, Bonn, West Germany*)

ture, which is just a little cooler than the melt temperature. After injection, the mold opens, and the core pin and the still warm preform are rotated 120°. A blow mold then closes over the preform, and air is injected through the core pin. After the bottle is blown, it is rapidly cooled by contact with the walls of the blow mold, which are kept at 104–122°C (40–50°F) by cold air or fluid circulating through the mold passageways. The mold then opens, a second 120° rotation occurs, and the part is stripped from the core pin. Then a third 120° rotation of the transfer head returns the core pin to the preform injection mold, and the cycle is repeated. Figure 12-15 illustrates such a three-station injection blow molding system. Figure 12-16 shows the actual appearance of an injection blow molder.

The injection blow molding process has a number of advantages.

1. It produces parts which are completely finished, with no scrap to be removed.
2. The parts can be made with closer-tolerance neck finishes, since they are actually injection molded at a

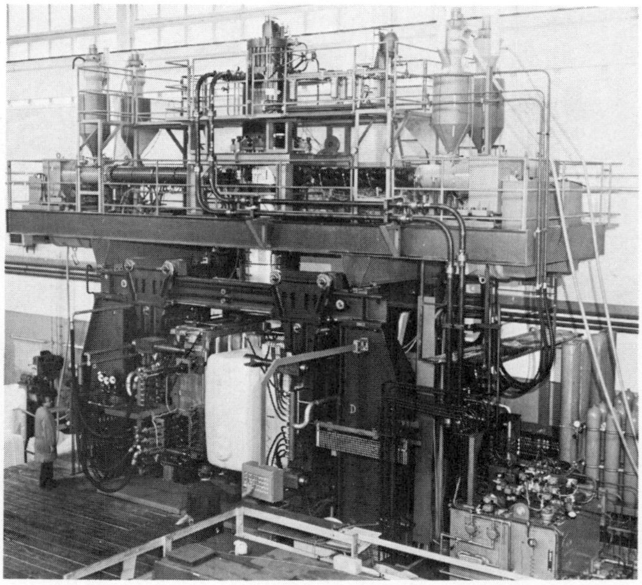

Fig. 12-13. Large blow molding machine using four extruders feeding an accumulator. (*Courtesy: Kautex Machinebau GmbH, Bonn, West Germany*)

considerably higher pressure than when formed by extrusion blow molding.

3. Uniform wall thicknesses can be easily obtained and held during long production runs.
4. Preforms for unsymmetrical jars or bottles are practical.
5. There is no pinch-off seam at the bottom, which might become a weak area.
6. Since the process can be fully automatic, with up to eight or ten cavities molded at once, it is well suited for high-volume production of bottles and jars.

The major disadvantage of the injection blow molding process is the limitation, at the time of writing, to a maximum of approximately 32 ounces in volume. Also, the technique is not suited for production of handles such as those used on detergent bottles. The fact that two molds are required for each part makes the process more expensive than extrusion molding, except when large-volume production is done. In this case, the saving in secondary operations (scrap regrinding, etc.) may offset the extra mold costs.

Cold Preform Blow Molding

A third production process, not widely used, consists of extruding a number of parisons, or tubing which is then cut to length to serve as parisons, which are allowed to cool. They can then be conveyed to the blow molding machine, which may be at another facility. The parisons are then heated either selectively, so particular shapes may be obtained, or to just the thermoelastic state, so that on stretching, a small amount of cold working occurs, resulting in a material which is oriented and strengthened. Because the

Fig. 12-14. Typical large parts made by accumulator-type blow molding machines. (*Courtesy: Kautex Machinebau GmbH, Bonn, West Germany*)

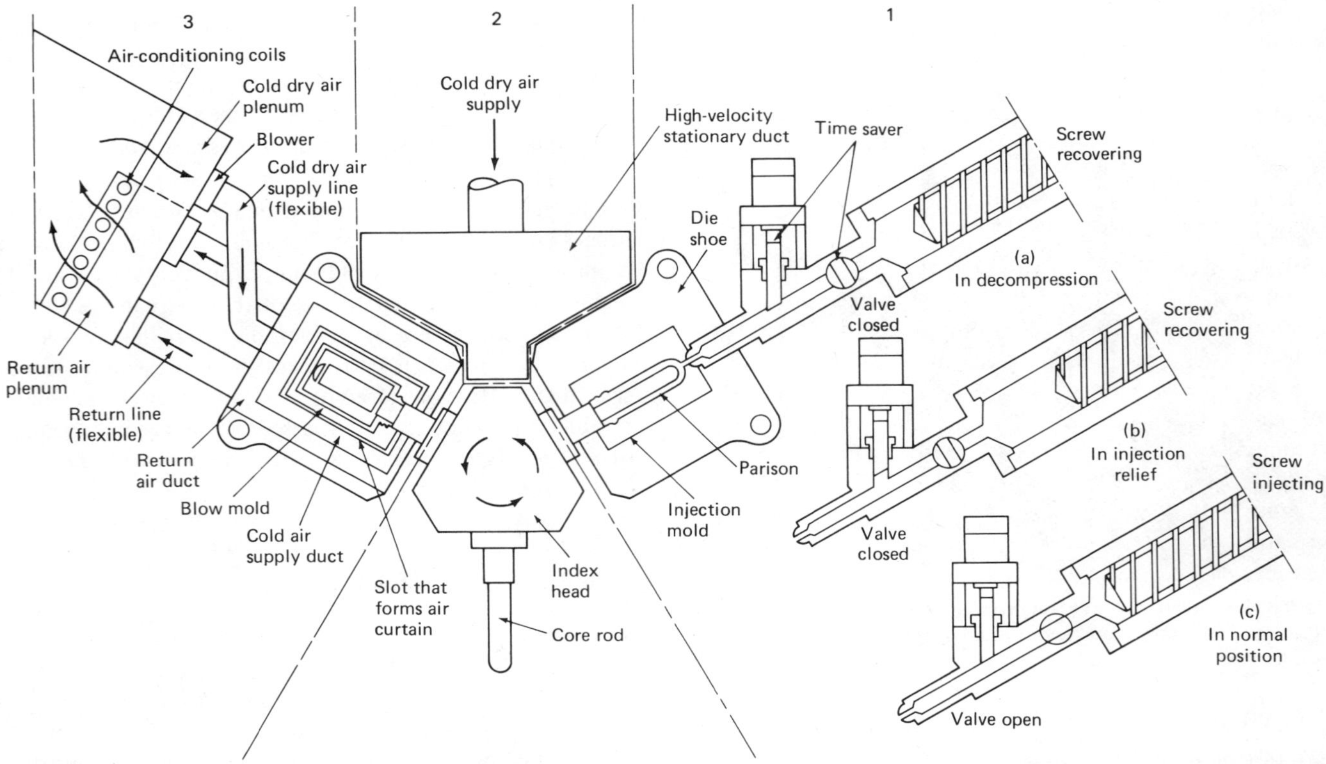

Fig. 12-15. Major savings can be made in injection blow molding cycles: 1. Time Saver—Gives machine additional screw recovery time: (a) When mold is filled, time-saver valve is closed and screw recovery begins. (b) With valve still closed, time-saver plunger retracts, sucking material back. Nozzle now retracts, without material drool, and open/close/rotate cycle starts. (c) Valve is open and machine is ready to inject next shot. 2. Parison cooling—Eliminates need for blow delay. Ducts mounted above and below parison path (between injection and blow stations) force dry, cool air onto injected parison. 3. Blow mold cooling—Air-dry system forces cool, dry air around outside of blow mold, preventing moisture buildup, thus permitting lower mold-cooling temp. (Not shown: internal cooling via air/water mist, which also markedly cuts cooling time.) 4. Core rod conditioning—Air/water missing device is mounted on stripper mechanism. (Reprinted by permission from ***Plastics Technology***, December 1973)

Fig. 12-16. Three-station injection blow molding machine. (*Courtesy: Rainville Co., Inc.*)

material is stronger, the container can be made lighter, and thus effect a saving in the total amount of material used.

Special Techniques

Two special techniques are worthy of note. One is called orientation or stretch-blow molding. In this process, the parison is extruded and, while hot, is stretched biaxially—lengthwise by mechanical stressing, and diametrically by blowing to result in hoop orientation. The process starts by the continuous extrusion of the parison, which is first clamped in a preform mold. In the preform mold, the parison is blown into a preform shape, the neck is accurately sized, and most important, the preform is cooled to the optimum temperature for orientation (tempered). When the preform has been correctly "conditioned" to the specified temperature, shuttle clamps remove the preform and place it in a stretch-blow mold, where it is mechanically stretched by an internal rod, and at the same time blown to the final dimensions. Figure 12-17 illustrates the four steps of: (1) extruding, (2) preforming, neck calibrating, deflashing, and conditioning, (3) stretch-blowing, and (4) discharging the finished containers.

The process is particularly useful for heat-sensitive materials such as PVC and nitriles, since the parison is only subjected to one heat cycle to obtain the orientation and blowing, rather than two as in the cold preform method. In addition to the tensile strength, the process is also claimed to improve impact strength, stiffness, clarity, and gloss, and to lower vapor transmission. In order to obtain all these advantages in material properties, the preform mold must precisely control the temperature during the "conditioning" period. To accomplish this, the preform mold is provided with a number of cooling passages divided into as many as four or five zones, with temperatures ranging from 65°C to 80°C (150-175°F). Figure 12-18 shows a preform mold with four cooling zones. Also illustrated is a typical stretch-blow mold and the stretch-blow pin. Figure 12-19 shows a heat history, time-temperature plot of the entire stretch-blowing cycle.

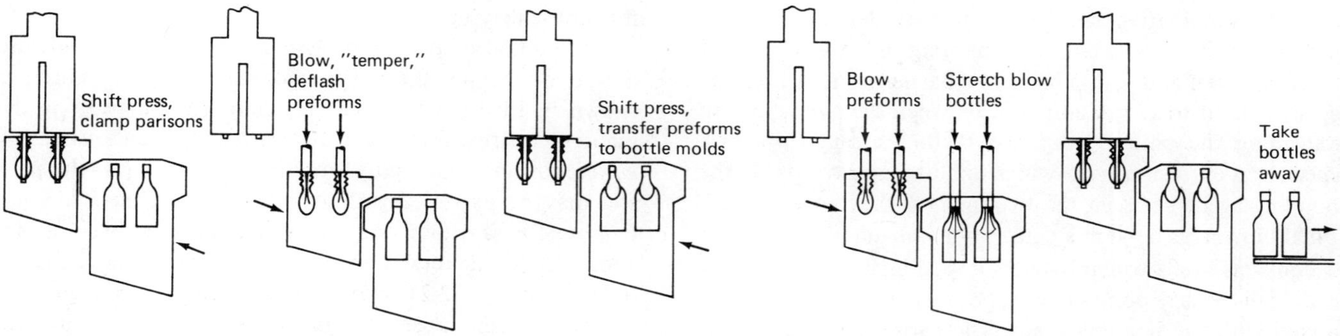

Fig. 12-17. Four steps in stretch-blow molding. Operational sequence traces process steps from parison extrusion through "hot" preform and finished biaxially oriented bottle. Note how shuttling press transfers tempered preform to stretch-blowing station. (Reprinted by permission from *Modern Plastics*, McGraw-Hill, Inc.)

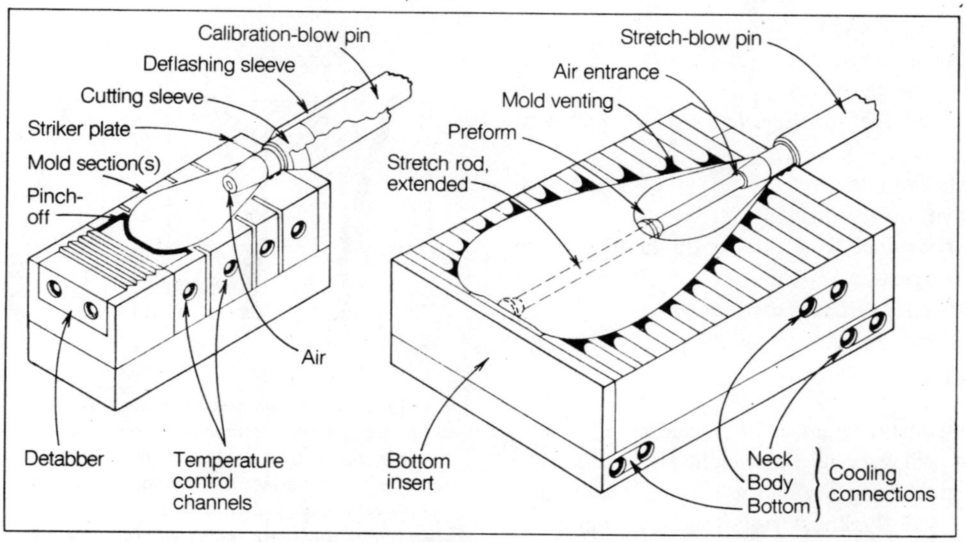

Fig. 12-18. Typical preform mold with four cooling zones (left). Blow mold with stretch-blow pin (right). (Reprinted by permission from *Modern Plastics*, McGraw-Hill, Inc.)

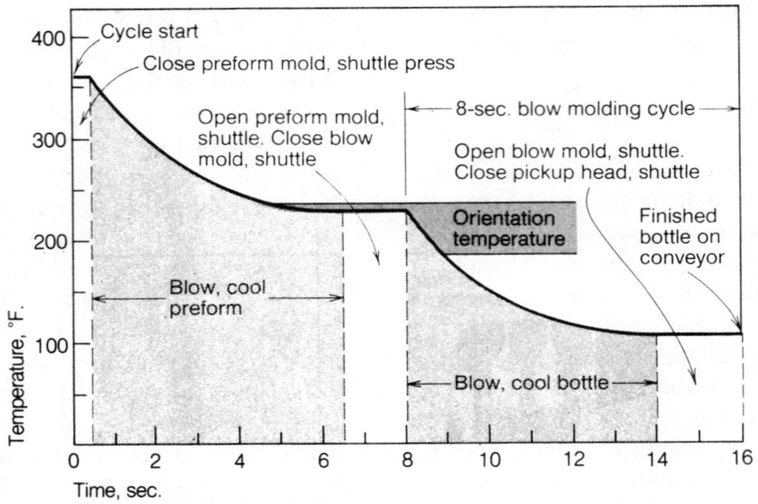

Fig. 12-19. Time-temperature plot for complete cycle shows single heat history to which plastic is subjected. Plot is typical cycle for 22-ounce PVC container.

A second unique technique is known as co-extrusion blow molding. In this method, a second, smaller extruder is used in combination with the main extruder to produce a parison which is a laminate, consisting of a thick layer of body material and a thin layer of a barrier resin. A bonding agent is used to insure adhesion between the two layers. In extruding the parison, the two melts are sent through a special co-extrusion head which simultaneously applies the bonding agent between the two melts as they are extruded. Final blowing is done in a conventional manner.

Body materials which have been used include polypropylene, high-density polyethylene, or low-density polyethylene mixed with the bonding agent. The barrier material used is nylon, which may be either on the inside of the container or on the outside. Some of the advantages claimed for the co-extrusion process are:

1. With nylon on the outside:
   a. Excellent aroma barrier
   b. Brilliant surface gloss
   c. Surface ready for printing (no flame treatment required)
   d. Good scratch resistance
2. With nylon layer on the inside:
   a. Excellent resistance to a number of chemicals with corrosive properties
   b. Excellent barrier to hydrocarbons

Parison Programming

As mentioned previously, when extruding a large parison, the wall thickness will vary as the weight of the material increases and it begins to sag (see Figure 12-11). One method of minimizing the wall-thickness variation is by increasing the pressure of the melt in the die either by regulation of the hydraulic pressure of a reciprocating screw or, possibly, by pressure variations of an accumulator ram. Not too much variation can be obtained by this latter technique, so it is not widely used.

The method of parison programming which is in wide use is that of varying the position of the internal mandrel, as shown in Figure 12-20. The actual method of varying the mandrel position is by use of an electrical-hydraulic assembly which automatically programs the parison. Thus, the parison may be extruded with an even wall thickness to compensate for sag, or it may be programmed to have several variations in thickness along its length for irregularly shaped objects. Figure 12-21 shows a typical parison programming mechanism which can be cam, timer, or electronically programmed. Figure 12-22 shows a programmed parison for an irregularly shaped bottle. Programming systems are available which offer over a hundred variations in one length of parison. In addition to allowing the production of more uni-

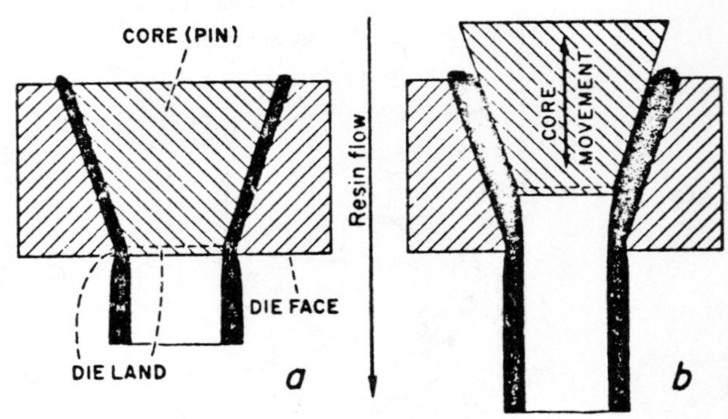

Fig. 12-20. Automatic parison programming. One means to overcome sagging to a degree is automatically raising the core inside a variable die (shown in *a* as flush with the die face) by means of a timer (or automatic programming device). Such gradual increase of the die clearance, shown in *b*, takes place during the parison-forming stage. Wall thickness increases gradually and thus compensates for sagging at the upper end of the parison. (*Courtesy: U.S. Industrial Chemicals Co.*)

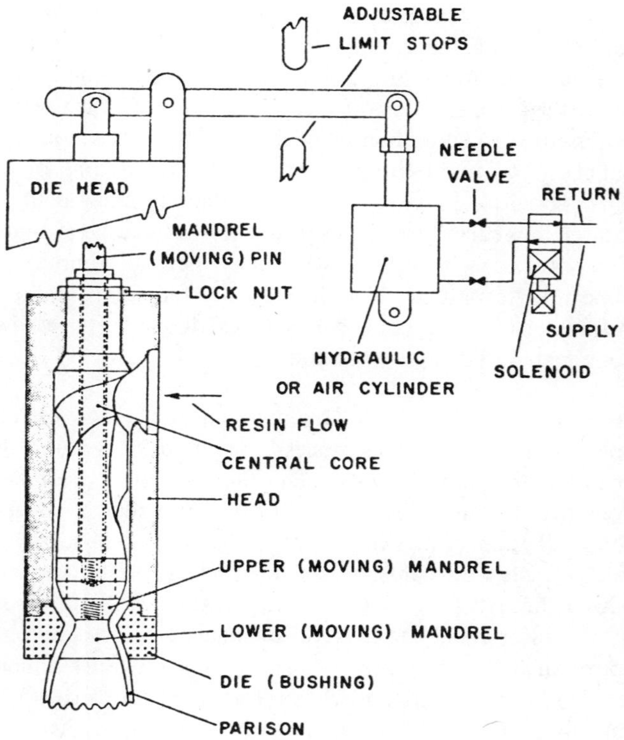

Fig. 12-21. Automatic parison programming system. (*Courtesy: U.S. Industrial Chemicals Co.*)

form wall thickness in parts, a programmed parison will also save material and cooling time by not having thicker walls in one area simply because an adjacent area must be blown to a greater extent. More advanced systems of process and parison control utilize microprocessor chips as the basis of the control system. Such systems will control parison thickness and length, barrel and mold temperatures, mold sequencing and coolant flow, screw speeds, pre-blow air pressure, blow-pin indexing, and pinch-off sequencing. Like injection and extrusion control systems, the microprocessors can incorporate memory systems, visual displays, and magnetic tape or printout recordings.

Blow Molds

Unlike molds used in injection molding, blow molds may be made of aluminum, since they do not have to withstand the high pressures encountered in injection molding. Aluminum molds are also largely used in blow molding since they have excellent heat-transfer qualities, are lightweight, relatively low in cost, and, in certain aircraft alloys, can be used for long production runs with negligible wear. To reduce the wear in the pinch-off section, a beryllium-copper or steel insert will frequently be used in the pinch-off and/or the neck area. Where long production runs are required, cast steel molds are often used, particularly if a long pinch-off

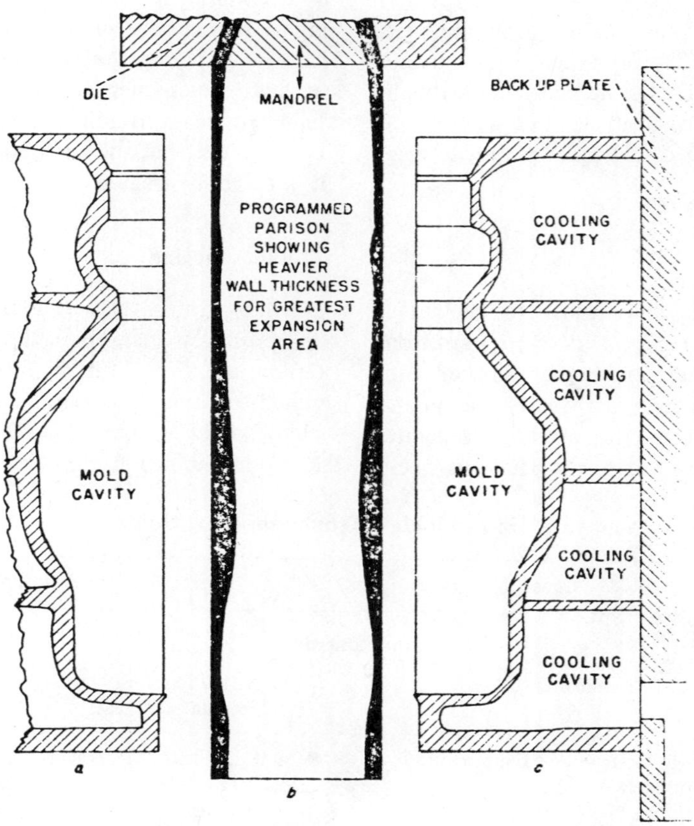

Fig. 12-22. A programmed parison designed to fit a particular mold configuration. (*Courtesy: U.S. Industrial Chemicals Co.*)

line is required. With thin walls and properly designed cooling passages, the lower rate of heat transfer of steel can be compensated for to result in cooling times equal to aluminum or beryllium-copper molds. Zinc alloys are also used for molds since they can be cast easily and are relatively inexpensive. For high production runs, they have not proved to be as durable as steel or aluminum, even when the pinch-off and neck areas have steel or beryllium-copper inserts. Another disadvantage of the zinc alloys is their relatively high density as compared to aluminum, which makes their use for large dies somewhat impractical.

Depending on the material, the molds are made by casting, machining, or, in some cases, electroforming. Casting is the most popular technique for aluminum molds and is used extensively. Regardless of the production technique used to make the mold, all the molds must be provided with some type of cooling passages. In small, nonintricate molds, simple drilled holes are used. The majority of molds, however, are made with three or more distinct channels or compartments through which the cooling medium is circulated.

To insure minimum cooling time, it is important that the mold be adequately vented. Entrapped air or air which is not removed rapidly enough from the mold can hold the plastic away from the mold wall and thus prevent or delay efficient cooling. The vents can be in the form of parting-line vents, or small drilled or machined passageways (approximately 0.005 to 0.009 inch deep) placed in strategic locations.

In the search for decreased cycle time, other methods of cooling are also used. Liquid CO_2 or chilled or moistened air is introduced into the interior of the container through the neck. Substantial reductions in cycle time can be accomplished, and the cooling air can also act as the blowing medium.

Postmold Operations

By proper design of the mold, blow pin, and associated parts, it is possible to produce containers on an extrusion blow molding machine that are completely finished and ready for filling. However, in a great many cases, the pinch-off and the flash at the neck must be removed by a separate trimming operation. In addition, in many parts the neck must be reamed and sized in order to produce a finished part. Some of the equipment used for these operations includes: guillotines, for fast trimming where tolerances are large; fly cutters, used to square up a guillotine cut; reamers, used to size the interior of the neck; trim dies, routers, punches, etc., used to remove flash, excess material, or to separate parts. In most cases the trimming equipment is located immediately adjacent to the blow molding machine, if it is not a part of the blow molder. When the trimming is done in a second machine, it is usually fed the containers automatically via a conveyor belt, which also removes the containers to the next operation.

Printing and Decorating. Included in the post-molding operations are the various processes required to mark and/or decorate the part. In particular, the most used material, polyethylene, must be surface treated in order to assure good adhesion of the ink or other marking fluid. Usually this consists of a flame treatment in which the bottles are passed through a circular flame ring as the bottles drop out of the blow mold, or the bottles may be rotated on the conveyor belt as they pass by a flame, or some other similar scheme is used. Corona discharge treatments are also occasionally used, and, somewhat more infrequently, a chemical treatment, with chromic acid, is used whenever the inside surface of the bottle is to be treated.

In addition to printing on the container, other methods of decorating include: hot stamping, silk screening, spraying through masks, and offset printing. This latter technique involves applying the marking on a silicone rubber roller, or belt, which then contacts the part to print the mark. Label applications adhesively bond paper to the container. In the event that the label completely encircles the bottle, it need not be treated.

Troubleshooting

Inevitably things will go wrong. Temperatures drop or go up, pressures change, parts wear and upset alignment or timing, etc. When this happens, the cause must be determined and the remedy must be applied as promptly as possible. To aid in doing this the troubleshooting guide (Table 12-1) should be consulted.

Table 12-1. Blow Molding Troubleshooting Guide.

Problem	Cause(s)	Solution(s)
	In Parison Formation	
1. Parison curl	1. Non-uniform walls	1. Align die. Provide even heat.
2. Rough parison	2a. Too low melt temperature 2b. Melt fracture	2a. Build heat progressively throughout system. 2b. Go outside of melt-fracture/shear-rate region. i. raise temperature ii. streamline die approach iii. raise or lower pressure

Table 12-1. *(Continued)*

Problem	Cause(s)	Solution(s)
3. "Doughnut" formation	3a. Mandrel not up to temperature 3b. Mandrel does not extend below die 3c. Die face dirty 3d. Mold scraping die	3a. Allow mandrel to reach equilibrium with rest of system. 3b. Adjust mandrel for .015–.020 inch (minimum) protrusion. 3c. Clean die. 3d. Reposition mold.
4. Contamination	4a. Degraded material in system 4b. Hangup areas in system 4c. Poor housekeeping	4a. Thoroughly purge to clean system. Avoid prolonged shutdowns with material held at high temperatures. 4b. Improve streamlining within the head. Check welds and joints, disassemble and clean. 4c. Keep area clean; containers, etc. protected. Enclose resin-conveying system. Inspect regrind to be sure it is clean.
5. Neck down	5a. Improper material selection 5b. Melt temperature too high 5c. Cycle too slow	5a. Use higher-density or lower-melt-index material. 5b. Lower melt temperature. 5c. Raise screw speed.
6. Die lines (including weld lines)	6a. Blend of materials (differing density and melt index) 6b. Nicks or scratches in die/mandrel 6c. Impeded flow in die and head-stagnation and hangup	6a. Use maximum 3:1 spread of melt index in blends. (Purge one material completely before introducing another.) 6b. Polish orifice surfaces. 6c. Streamline flow. Raise mold temperature. Increase blowing pressure.
	During Blow Molding	
1. Insufficient blowing	1a. Air supply restricted or off 1b. Pinch-off too sharp 1c. Pinch-off too hot, no weld	1a. Provide adequate air (hose may be pinched-off-valve off, etc.) 1b. Provide sufficiently broad pinch-off area (0.02 to 0.12 inch land). 1c. Cool mold uniformly.
2. Part blowout	2a. Mold separation 2b. Part too hot 2c. Parison blowout 2d. Too large a blowout ratio	2a. May need machine of larger clamping capacity. 2b. Longer cooling. 2c. Use higher-melt-strength material. 2d. Use larger die.
3. Webbing handle (in jugs)	3. Low flare	3. Lower shear rate on material: heat head and die, raise melt temperature, decrease parison drop speed.
4. Excessive cycle	4a. Part too hot 4b. Inadequate cooling on mold 4c. Improper material choice	4a. Decrease melt temperature. Decrease part wall thickness. Increase blowing air pressure. 4b. Decrease mold coolant temperature. 4c. Increase material density. Increase melt index (with decrease in stock temperature).
	After Molding	
1. Poor surface	1a. Melt temperature too low 1b. Die area too cold 1c. Mold temperature too low (water condensate on mold) 1d. Mold finish 1e. Insufficient vent 1f. Insufficient air 1g. Low melt index material	1a. Raise melt temperature. 1b. Increase die/mandrel temperature. 1c. Raise mold temperature. Air condition shop. 1d. May be too rough, or too smooth (vapor honing usually good). 1e. Vent molds. 1f. Increase blowing pressure. Increase size of air orifice. 1g. Increase melt index.
2. Indented parting line	2a. Mold temperature too low 2b. Blowing pressure too low 2c. Air entrapment	2a. Raise mold temperature. 2b. Increase blowing pressure. Increase size of air orifice. 2c. Improve mold venting.

Table 12-1. *(Continued)*

Problem	Cause(s)	Solution(s)
3. Warpage (top and bottom)	3. Insufficient cooling	3. Increase cooling time. Provide more cooling in pinch-off areas of mold. Reduce mold temperature.
4. Parison tail sticking to part	4a. Parison too long 4b. Mold base pinch off improperly designed	4a. Reduce cycle time for shorter parison tail. 4b. Design pinch-off to "compression-cool" tail.
5. Flashing	5a. Melt too hot 5b. Blowing pressure too high 5c. Mold parting 5d. Vents too large	5a. Reduce melt temperature. 5b. Reduce blowing pressure. 5c. Increase clamp on mold. Decrease blowing pressure. 5d. Reduce vents to 0.003 inch deep × 0.04 to 0.15 inch long stepped deeper to edge of mold.
6. Variable wall thickness	6a. Melt too hot (parison sag) 6b. Inadequate melt strength 6c. Die out of alignment 6d. Excessive blowup ratio 6e. Non-uniform temperature on die	6a. Decrease melt temperature. 6b. Lower melt index. Increase material density. 6c. Adjust and center die. 6d. Change die dimensions. 6e. Put uniform heat on die and head area.
7. Variable bottle weight	7. Extruder sagging	7. Check type of screw used. Check screen pack. Decrease extrusion rate. Increase screw cooling.
8. Excessive shrinkage	8a. Wall thickness 8b. Mold temperature too high 8c. Melt temperature too high 8d. Inadequate blowing pressure 8e. High shrinkage material	8a. Decrease wall thickness. Make walls more uniform. 8b. Decrease mold coolant temperature. Increase mold coring. 8c. Decrease melt temperature. 8d. Increase blowing pressure. 8e. Decrease material density.
9. Post-molding distortion	9a. Part removed too hot 9b. Cooling differential in thick and thin areas	9a. Cool longer in mold. 9b. Consider cooling jigs, water quench.
10. Low impact or environmental stress-crack resistance	10a. Part design 10b. Material sheared cold 10c. Too high melt index 10d. Too high density	10a. Eliminate sharp corners. Make walls uniform. 10b. Raise cylinder heats to peak near feed zone. (Increase material temperature). 10c. Lower melt index. 10d. Lower material density.
11. Sticking to blow pin	11a. Excessive heat build-up around blow pin	11a. Reduce stock temperature. 11b. Add cooling to blow pin by drilling 4 or 8 evenly spaced perimeter holes 0.010–0.015 inch diameter.
12. Gels	12a. Cold pellets 12b. Contamination	12a. Allow resin to reach equilibrium temperature before introducing to machine. Increase stock temperature or use heated hopper. 12b. Check feed stock for regrind cleanliness.

REFERENCES AND BIBLIOGRAPHY

Antonopoulos, J. A., and Hohman, F. G. Stretch-blow molding PVC and nitriles. *Modern Plastics*, September 1976, **53**(9):82.

Bailey, F. Z., and Bailey, M. E. The what and why of injection-blow molding. *Modern Plastics*, April 1968, **45**(4):136.

Berringer, L. D. Blow-molding controls–without them you're out of the running. *Plastics World*, March 18, 1974, **32**(3):46.

Cancio, L. V., and Joyner, R. S. Why resin selection is crucial to blow molding of lighter bottles. *Modern Plastics*, January 1977, **54**(1):72.

Celanese Plastics Co. *Fortiflex Polyethylene Blow Molding*. Bulletin P3C, August 1971.

——. *Blow Molding Celcon Acetal Copolymer*. Bulletin C3C 126040, December 1966.

Cyanamid Industrial Chemicals and Plastics Division. *Blow Molding of XT Polymer*. Bulletin PKT-1A 1-1905, November 1971.

DuBois, J. H., and John, F. W. *Plastics*. 5th Ed. Van Nostrand Reinhold Co., New York, 1974.

E.I. Du Pont de Nemours & Co., Inc. *Blow Molding of Alathon Polyolefin Resins*. Bulletin A-24964, October 1962.

Farrell, J. Processor's guide to injection blow molding. *Plastics World*, April 19, 1976, **34**(4):34.

Goldsberry, H. H., and Fox, A. J. Blow molding expandable polystyrene. *SPE Journal*, April 1962, **18**(4):448.

Gulf Oil Co., Plastics Division. *A Short Course in Blow Molding Linear Polyethylene*. Bulletin SP-12161.

Heldrich, G. C. The history of blow molding. *SPE Journal*, May 1961, **17**(5):435.

Holzmann, R. Big-tank blow molding is now big business and growing. *Plastics Technology*, September 1971, **17**(9):41.

Hunkar, D., and Keyes, D. Should you switch to electronic parison programming? *Plastics Technology*, November 1970, **16**(11):50.

Iwawaki, A., Kojima, K., and Shitara, Y. Coextrusion–blow molding for gas tanks and industrial parts. *Modern Plastics,* March 1977, **54**(3):70.

John, F. W. Blow molding fundamentals. *Plastics World*, July 1966, **24**(7):54.

Johnson, W. C. Cost comparison of rotational casting with injection molding and blow molding. *SPE Journal*, January 1964, **20**(1):56.

Joyner, R. S. Troubleshooting HDPE processing problems. Part 2: Blow molding. *Plastics Design & Processing*, April 1979, **19**(4).

Kaplan, T., and Gordon, I. B. Designing molds for blow molding. *Plastics Technology*, February 1962, **8**(2):42.

Lippens, A. M., and Johnson, E. S. How to blow mold rigid PVC. Part 2: Parison forming and blowing techniques. *Plastics Technology*, June 1966, **12**(6):52.

McCubbin, C. G., and Berwick, R. L. Blow molding polycarbonate. *Modern Plastics*, June 1964, **41**(4):130.

Modern Plastics. What's exciting about stretch-blow? February 1976, **53**(2):34.

——. What's exciting about stretch-blow? It adds muscle to beverage bottles. March, 1976, **53**(3):44.

Morgan, B. T., Peters, D. L., and Branscum, T. E. Controlling wall distribution in blow molded parts. *Modern Plastics*, August 1963, **40**(8):114.

Naturman, L. I. Industrial blow molding. *SPE Journal*, December 1963, **19**(12):1233.

Peters, D. L., Morgan, B. P., and Cooper, W. E. Why it pays to build low cost prototype blow molds. *Plastics Technology*, September 1968, **14**(9):34.

Pickering, G. E. Orientation blow molding: (1) Why? (2) Where to now? *Plastics Technology*, October 1976, **22**(10):23.

——. Molds for blow molding. *SPE Journal*, October 1959, **15**(10):891.

——. Progress in blow molding containers. *Plastics Design & Processing*, May 1970.

Plastics World. What's causing those blow-molding problems? February 1979, **37**(2):50.

——. Injection blow molding–a process that's found its place. November 18, 1974, **32**(11):59.

——. Parison programmer traces new profit profile for blow molders. April 1971, **29**(4):48.

Sackett, R. D. An introduction to extrusion for blow molders. *SPE Journal*, May 1961, **17**(5):443.

Scheiner, L. L. Close-up: blow molding, the equipment scene. *Plastics Technology*, October 1976, **22**(10):49.

Schmidt, P. H. Custom-select your own blow molding system. *Plastics Technology*, July 1976, **22**(7):81.

Schoeller, R. E. Blow mold designs change to meet new demands. *Plastics Technology*, May 1971, **17**(5):48.

——. Recent advances in blow mold design. *SPE Journal*, July 1964, **20**(7):609.

Trieter, R. A. Selecting injection blow machines–the in-depth approach. *Plastics Technology*, May 1974, **20**(5):27.

Tusch, R. L. Blow molding die design. *SPE Journal*, July 1963, **19**(7):632.

Union Carbide Corporation. *Blow Molding of Bakelite Polyethylenes*. Bulletin F-41785 036-9.

U.S. Industrial Chemicals Co. *Polyethylene Blow Molding . . . An Operating Manual*. Bulletin PL 12-1272, 1970.

Waechter, C. J. The techniques of blow molding. *Plastics Design & Processing*, October 1967, **7**(10):10.

Waechter, C. J., and Pomper, A. W. Ram accumulators for blow molding. *SPE Journal*, October 1961, **17**(10):1065.

Wechsler, R. L., and Baylis, T. H. Blow molding polyethylene. *Modern Plastics*. Part 1: May 1959, **36**(5):107. Part 2: June 1959, **36**(6):115.

13
Thermoforming

Thermoforming is one of the oldest plastics processes. Ancient Egyptians discovered that tortoise shell, when heated in hot water, could be formed into various pleasing shapes. Cellulose nitrate, one of the first synthetic plastics, was shaped by heat into celluloid collars, dolls, and other articles. Although the thermoforming technique was available, the process was not used to any great extent for a number of years. Shortly before and during World War II, it was revived. During World War II it was used to form contour maps for troop commanders in the field. A more spectacular use for the technique was the production of "bubble canopies" for fighter aircraft.

Thermoforming, as the name implies, is the forming of a plastics part by the application of heat. The forming actually involves heat and pressure. The pressure can be that brought about by a vacuum on one side of the sheet, by applied air pressure, or by clamping the hot sheet between a male and female mold. The process is also called vacuum forming since much of the early work was done using vacuum to furnish the forming pressure.

There are a number of advantages to the thermoforming process. One of the most significant is that the tooling, in comparison to other methods of making plastics parts, is relatively inexpensive. Since the technique is particularly suited to the production of very large parts, cost savings can be impressive. Many parts with very thin walls can be made only by this method; for example the familiar plastic cold drink cups with wall thicknesses of a few thousandths of an inch which could probably not be made by injection molding. The process is also ideal for the production of small numbers of parts—samples, prototypes, etc.—again because of the relatively low cost of tooling.

There are also some distinct limitations to the technique. First, the process is confined to the use of sheet material. Therefore, all parts must have fairly uniform wall thicknesses. Thick ribs or mounting bosses cannot be obtained, although formed ribs can be designed in, or rolled edges, as found on drinking cups, can be made to give edge strength.

From the original production of contour maps and fighter plane canopies, the process now is used for such items as refrigerator door liners, automobile dashboard panels, cheese containers, soft drink cups, signs, packaging, and many similar items of uniform wall thickness. In addition to the above, many unusual shapes can be made, as shown in Figure 13-1. Thermoforming and adhesive bonding or "double-sheet forming" can be used to make hollow or tubular shapes in many configurations.

The thermoforming process, as practiced by industry, is composed of five separate steps, with at times a sixth step added. The five steps are:

1. Clamping. The material is usually held in frame of some kind.
2. Heating. The material is heated to the desired temperature for forming or orientation.
3. Forming. Done by vacuum, pressure, and/or a mechanical assist of some kind.
4. Cooling. The part is cooled sufficiently to be removed without distortion.
5. Trimming. Excess material, usually the areas in the clamps, is removed.

A sixth step, decorating, is sometimes also included to add trademarks, designs, etc. Usually, however, the decorations are applied to the sheet prior to forming. When this is the case, the decorations may be applied in a distorted shape so that after forming they are in the correct shape.

Materials

Virtually any thermoplastic sheet material can be thermoformed. Most sheet material used in thermoforming is made by the extrusion process; however, calendered and cast sheet is also used. There are certain special characteristics which a good thermoforming material must possess so that successful forming can be done. Successful thermoforming entails the ability of the material to be deep drawn into a cavity over a female mold without tearing or excessively thinning in one area. Unfortunately, not all thermoplastic sheet materials have these properties to the same degree.

Plastic Memory. One of the most important desirable properties of a material is plastic memory; its ability to become "rubbery" when heated, so the formed part will fit tightly around a mold. Because of this property, parts which have been incorrectly formed can be reheated outside of the mold, and, if the plastic memory is strong enough, they will revert to a flat sheet again, which can be subsequently re-thermoformed. Acrylics are good examples of materials with this property.

Hot Strength and Hot Elongation. Somewhat allied to plastic memory are the two properties of hot strength and hot elongation. Hot strength refers to the characteristic of

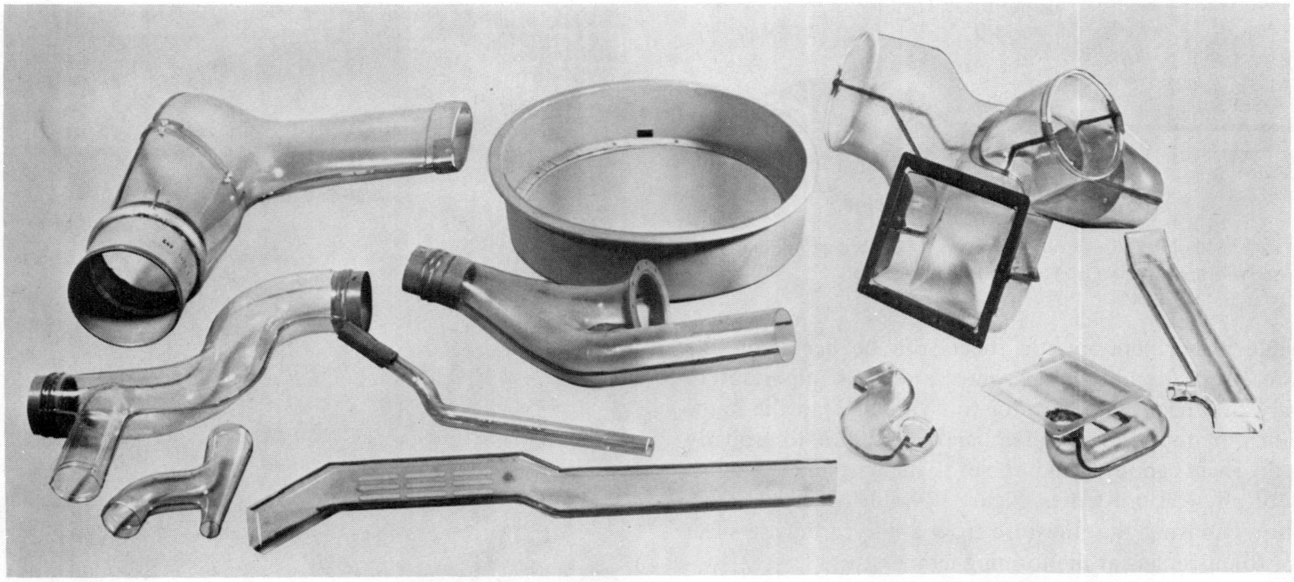

Fig. 13-1. Complex shapes made by thermoforming. (*Courtesy: Mobay Chemical Corporation*)

the material to remain strong when heated. Some materials lose all their strength when heated and become soft and putty-like, but others stay strong and elastic. These latter materials can be stretched around or over a die, whereas the low hot-strength materials would tend to tear quickly. In fact, strong hot-strength materials are also referred to as high hot-tear-strength materials. A material with good hot elongation can be stretched as much as 500% to 600% when hot, while materials with low hot elongations might only stretch 15% to 20% before tearing.

Forming-Temperature Range. A very important characteristic is the forming range of the material. It is desirable for a good thermoforming material to become soft and rubbery over a wide range, rather than have a sharp melting point. In this respect, some materials such as acrylics may be formed over a range of from 125-175°C (260-350°F), while other materials may melt just a few degrees above their softening range.

The selection of the correct type and grade of material is extremely important. When choosing the optimum material, it is important to determine if an oriented grade is required. Orientation may be in the lengthwise, or machine, direction of the sheet and/or in the transverse direction. Material which is oriented in both directions is said to be biaxially oriented, and is usually considered to be the best grade of material because of higher properties in all directions. Sheet which is oriented in only one direction is termed uniaxially oriented.

Oriented thermoplastic sheet has higher tensile and impact strength than non-oriented sheet and, in some types of material, higher resistance to crazing and better weather resistance. However, it is important to remember that in the orientation process the sheet is necessarily stretched in one or both directions which imparts a certain amount of stress of built-in "elastic memory." These stresses tend to pull the sheet back to its original size if the temperature becomes high enough. Thus, there are definite temperature limitations placed on oriented sheet which will be thermoformed. Since the act of thermoforming also imparts a certain amount of orientation to the material, the selection of oriented sheet will be influenced by the degree of forming to be done and the temperatures which will be used. The forming of the oriented sheet must also be done with a great deal more care to maintain the narrow range of time and temperature which is optimum for the particular grade of sheet material being used.

Forming Processes

Heating. Material to be formed can be heated in a variety of ways, ranging from simple oven heating, oil baths, and hot platens, up to complicated infrared heaters which heat one side or both sides of the sheet at the same time. Most modern thermoforming machines use infrared radiant heaters since they are fast, easily turned on and off, and allow the use of selective heat; that is, simple masks or shades can be used to prevent heat from reaching any particular area, as shown in Figure 13-2. In general, the heating units should be capable of reaching a temperature of 538°C (1000°F), and should have a capacity of from 3 to 6 kilowatts per square foot. The higher heat ranges would, of course, be used for thicker material. The most important characteristic of the heating system is that the heat supplied be uniform, unless specifically required otherwise. In addition to being uniform, the heating equipment should be capable of heating the film or sheet in 15 to 40 seconds in order to be commercially useful.

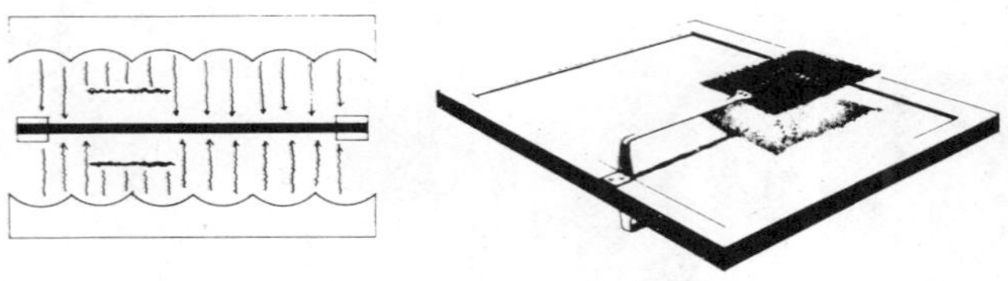

Fig. 13-2. Shading or preferential heating by means of metal masks. (Printed with permission of The Dow Chemical Co.)

While most thermoplastic sheets can be heated in the "as-received" condition, in some cases it is important to predry the sheet material prior to clamping it in the frame and heating to the forming temperature. Failure to properly dry the sheet can result in bubbles in the part and a possible loss of physical properties. Figure 13-3 illustrates the effect of not removing the moisture from a polycarbonate sheet prior to processing at the forming temperature.

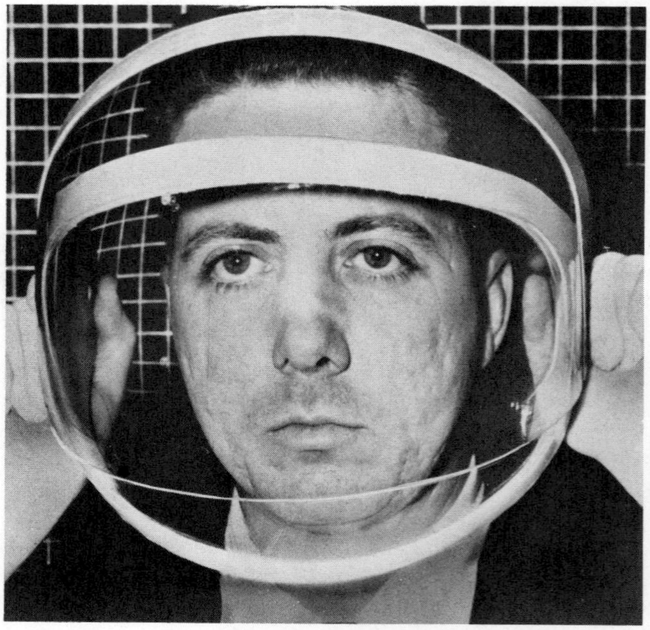

Fig. 13-4. The free-blowing method is used to produce items with maximum optical qualities such as this visor. (*Courtesy: Mobay Chemical Corporation*)

Forming. There are a number of forming techniques in use, but all are basically variations of two simple processes in which the heated sheet is moved by (1) air in the form of an applied vacuum and/or pressurized air, or (2) mechanical draw assists which force the sheet into the desired contour. In most cases, the two processes are combined to result in a wide variety of procedures for making many types of parts. In all the processes except one, the hot sheet is drawn into a female mold, or drawn over a male mold, or the two molds are used together to form the part. The one exception is known as free-forming or free-blowing, where the hot sheet is blown into the contour desired without contacting a mold. Usually this is done by drawing the sheet into a female cavity, where the extent of draw is determined by electric-eye devices. Free-blowing is used when the highest optical quality is required, such as in aircraft canopies, lenses, and helmet visors (as shown in Figure 13-4).

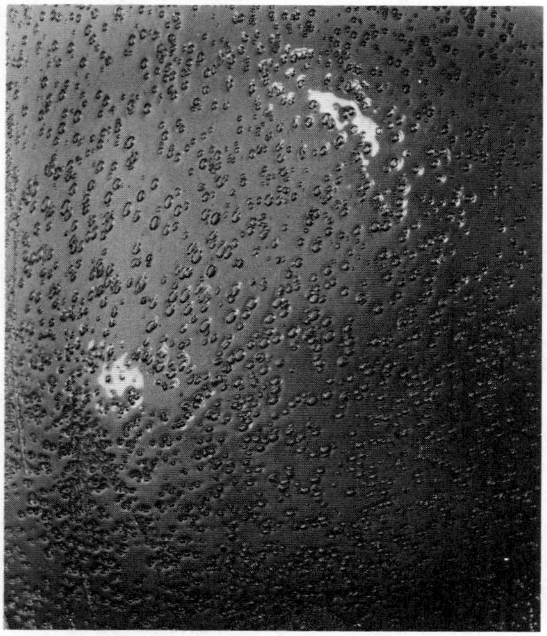

Fig. 13-3. Effect of heating polycarbonate sheet which has not been predried. (*Courtesy: Mobay Chemical Corporation*)

In forming processes involving a mold, whenever the hot sheet comes into contact with the mold, chilling takes place, and the sheet will no longer stretch at that point. In some processes, an effort is made to prevent such contact so that maximum stretch, or draw, is achieved. In other processes, the sheet is deliberately chilled by being touched to the mold or a plug to create a thicker section of the part at that spot.

A technique to assure more uniform distribution of the drawn material is the procedure known as prestretch. In this procedure, the heated material is blown into a bubble or billow prior to any contact with a mold or a plug. The forming of the billow results in a uniform prestretching of the material, which can be used either to make a part with more uniform wall thicknesses, or to make deeper draws than would normally be possible by simply using the mold alone.

Straight Forming. Straight forming, by vacuum and by air pressure, is shown in Figures 13-5 and 13-6. Vacuum forming is the least complex process and was the first of the thermoforming processes used in high volume. The process is well suited for production of low-profile parts not requiring a deep draw. As shown, the process consists of heating a sheet until it softens and starts to sag, after

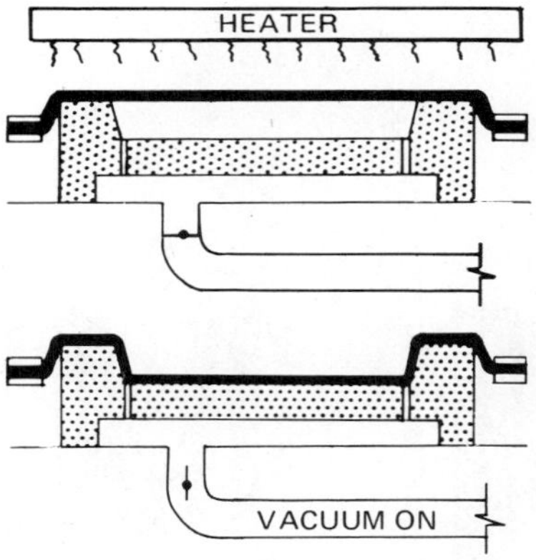

Fig. 13-5. Straight forming: vacuum.

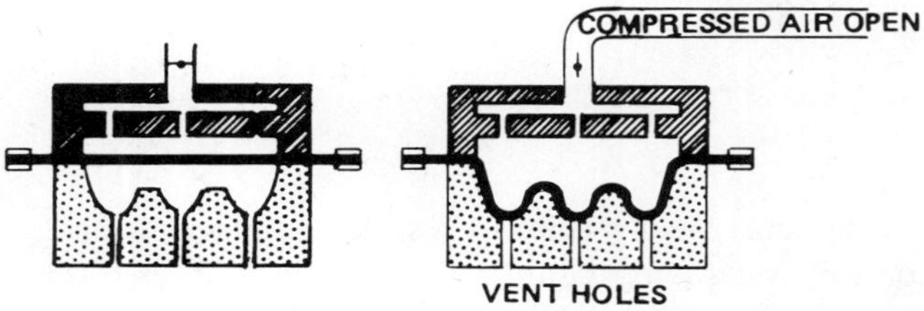

Fig. 13-6. Straight forming: pressure.

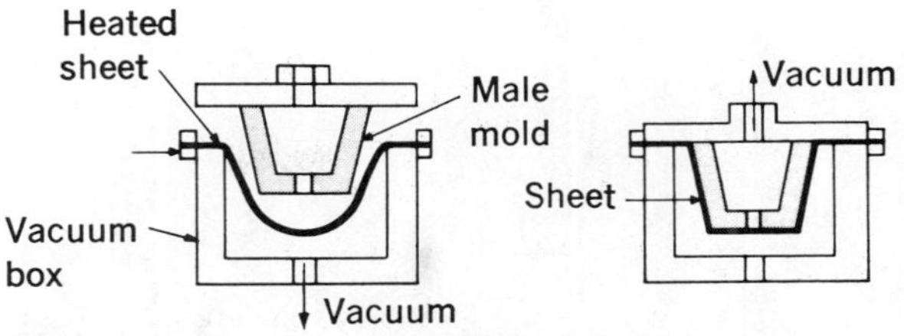

Fig. 13-8. Snapback forming. (*Courtesy: Mobay Chemical Corporation*)

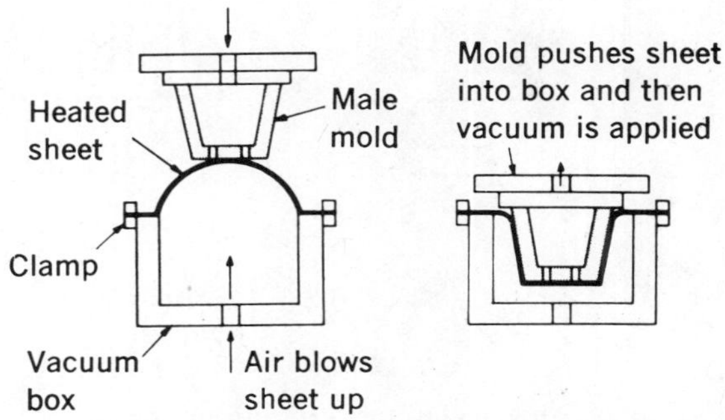

Fig. 13-9. Forming with a billow snapback is recommended for any parts requiring a uniform, controllable wall thickness. (*Courtesy: Mobay Chemical Corporation*)

which the vacuum is turned on or pressure is applied and the sheet is forced against the female die. It should be noted that the sharpest detail is on the outside of the sheet, as well as the most markoff. Thinning will be greatest in the corners. In using a female mold, as illustrated, no draft allowance is needed since, in almost all cases, the part will release from the die as it cools and shrinks. Thus, parts can be made with vertical sides or slight undercuts.

Drape Forming. When deeper drawing is required than can be obtained by straight forming, the hot sheet is often draped over a male mold, mechanically prestretched, and then formed to the final contour by vacuum or pressure, as shown in Figure 13-7. In this type of part, the greatest detail is on the inner surface which contacts the mold. Molds used in drape forming must have some draft so the part will release, since shrinkage of the sheet will cause it to lock onto the mold if draft is not provided.

Snapback Forming. In snapback forming, the "elastic memory" of the sheet material is used to advantage in forming the part. The heated sheet is drawn into a vacuum chamber to a predetermined depth, the mold is inserted into the bubble and firmly pressed onto the sheet. The vacuum is released and the sheet material is then allowed to snap back onto the mold. The mold must be vented to allow the escape of air, and sometimes a vacuum is also used to obtain better detail in the formed part. Figure 13-8 shows this type of process.

Reverse-Draw Forming. This technique, also known as billow snapback, is very similar to snapback forming except that, as shown in Figure 13-9, the bubble is blown upward into the air, or toward a heater, if required. The mold is lowered into the bubble, and internal air pressure forces the sheet against the mold. Vacuum may also be used to obtain maximum adherence to the mold contour.

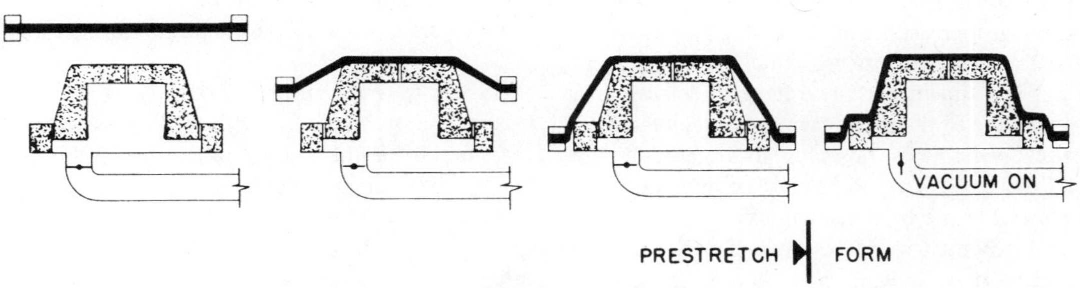

Fig. 13-7. Drape forming. (Printed with permission of The Dow Chemical Co.)

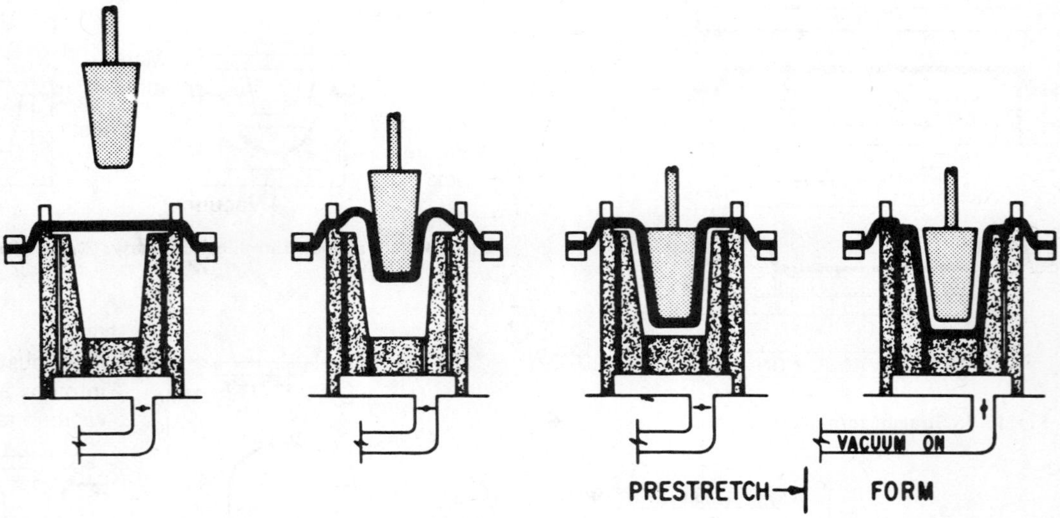

Fig. 13-10. Plug-assist forming. (Printed with permission of The Dow Chemical Co.)

Parts made by this technique can differ considerably in material distribution compared to parts made by snapback forming.

Plug-Assist Forming. This procedure combines straight vacuum forming and drape forming, except that the drape mold is somewhat smaller than the final vacuum mold and is referred to as "the plug." Figure 13-10 illustrates the procedure. The plug forces the hot sheet down into the cavity, while at the same time the air is compressed, causing the sheet to billow up around the plug. The sheet is therefore prevented from contacting the cavity edges as it is stretched, thus giving more even material distribution. The plug stops before reaching the bottom of the female cavity and vacuum is then employed to obtain the final adherence to the mold contour. This technique is used largely when deep draws are required with minimum thinning at the edges of the part.

The plug-assist procedure has a number of variables which, when correctly controlled, can be used to produce optimum parts with complex shapes and draw depths up to the maximum for that particular material. One very important variable is the ratio of the clamp area to the projected area of the part. Large-area, deep-drawn parts will have clamp areas considerably larger than the cavity area, thus allowing for an increased amount of billowing, which in turn results in more uniform wall thicknesses in deep-drawn parts. Shallow-drawn parts will have the clamp area correspondingly closer to the mold cavity's projected area.

Another critical variable is the plug itself. Usually the plug occupies 70% to 90% of the cavity volume and conforms to the general shape of the cavity, with smooth radii on the edges and no detail work on the surfaces. The plug temperature during the drawing operation is important. For this reason, plugs are often made of aluminum and have heating and cooling coils to insure accurate temperature control. When temperature-controlled plugs are used, they are usually run at the highest temperature at which sticking or excessive thinning does not occur. Too low a temperature will cause the part bottom to become thicker, and it may harden or freeze to the plug. For moderate draws, nonconductive plugs of wood, thermosetting plastics, and plaster may be used.

The speed of plug entry is a third significant variable. For most parts, the fastest entry possible will usually make the best parts, not only from the standpoint of higher production rates, but also in reduction of "drag marks."

The compression of air in the cavity as the plug enters is what causes the material to billow up around the plug. The billowing is what prevents the hot sheet from touching the edges of the mold as it is being stretched. Fast plug entry is thus one way to insure adequate billowing. However, too fast a plug entry can cause excessive billowing, thinning, or both. The mold, therefore, should be equipped with vents which may be valved to allow for variation in the air flow. Another technique to minimize excessive billowing with a fast plug entry is to stop the plug halfway down, or wherever is found to be satisfactory. Usually this type of procedure must be determined experimentally. In extreme cases, vacuum may be used to aid venting, using valves which are controlled by the plug entering the billow.

Plug-Assist, Reverse-Draw Forming. Figure 13-11 illustrates reverse-draw forming with a plug assist. This procedure differs from the regular plug-assist technique in that hot pressurized air is used to form an upward billow in the sheet prior to the plug entry. Large-area, deep-drawn parts are often made this way in order to minimize the thick or heavy section which is usually in the center of the plug when this procedure is not used.

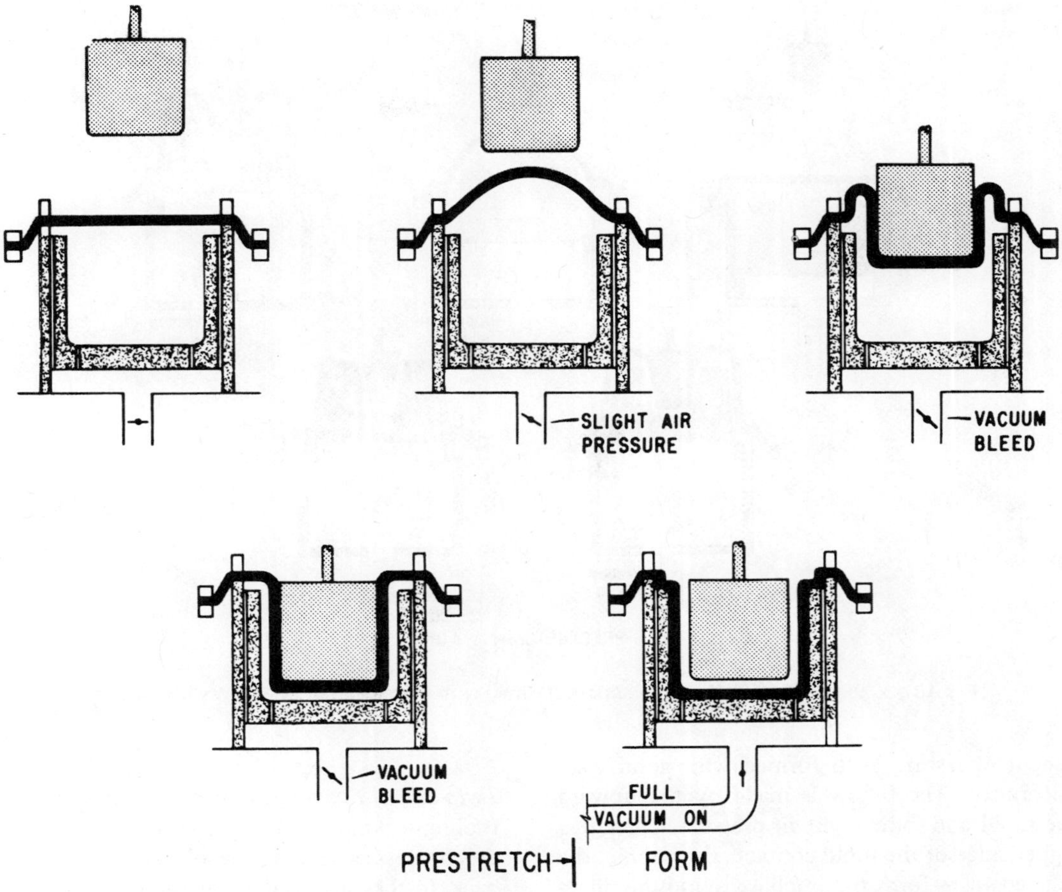

Fig. 13-11. Plug-assist, reverse-draw forming. (Printed with permission of The Dow Chemical Co.)

Air-Slip Forming. This technique is similar to snapback forming, but the billow is formed differently and considerably different tooling is used. Figure 13-12 illustrates the procedure in which the billow is pushed upward by the rising mold compressing the air ahead of it. At the proper point, the mold stops and the vacuum is turned on, thus tightly contouring the sheet to the mold. This process is used where improved material distribution is required.

Plug-Assist, Air-Slip Forming. Complex parts are often made by a combination of air-slip and plug-assist forming, as shown in Figure 13-13. In this method, high ridge sec-

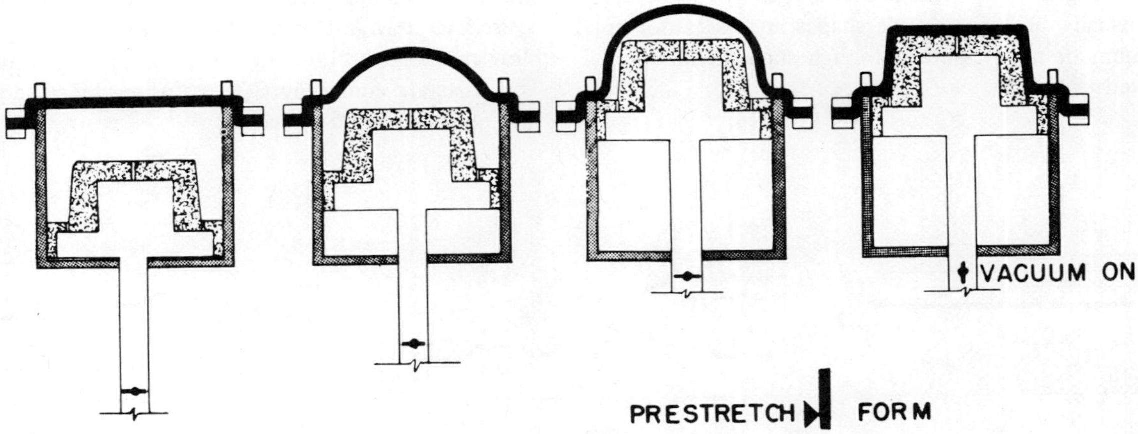

Fig. 13-12. Air-slip forming. (Printed with permission of The Dow Chemical Co.)

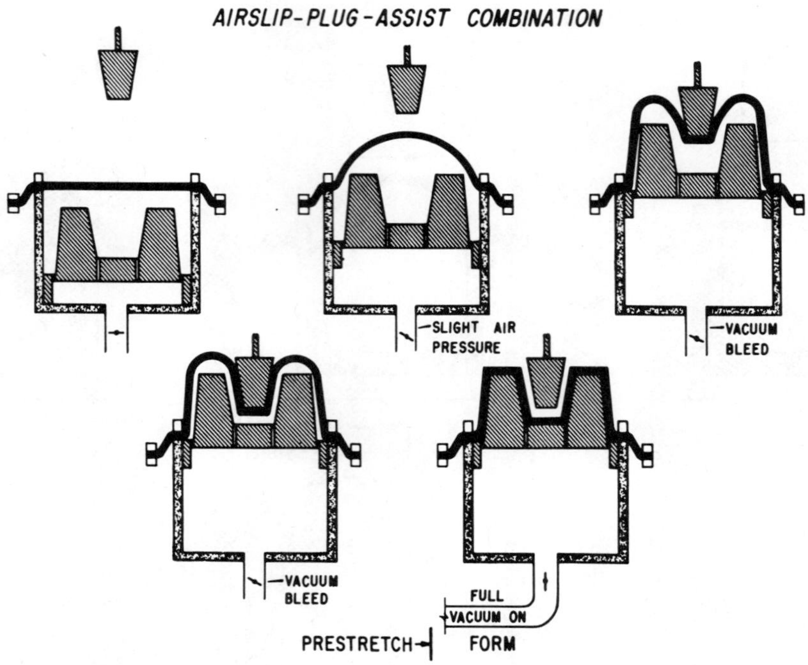

Fig. 13-13. Plug-assist, air-slip combination. (Printed with permission of The Dow Chemical Co.)

tions and deep cavities are both formed with good wall-thickness uniformity. The billow is made by the upward motion of the mold and some slight air pressure. Then, just before the upper ridge of the mold contacts the bubble, the plug is introduced to re-form the bubble around the ridges and into the cavity.

Matched-Tool Forming. In this procedure, the heated sheet is introduced between two tools and then is pressed to the desired shape, as shown in Figure 13-14. This process produces excellent reproductions of mold detail and also gives the highest dimensional accuracy. It is also the most expensive technique, since the tooling must be very accurately made.

A variation of the matched-tooling technique is known as plug and ring forming. This procedure, shown in Figure 13-15, is usually used for simple shapes and does not involve vacuum or air pressure, although such assists could be employed.

Twin-Sheet Forming. Another variation of the matched-tooling technique is known as twin-sheet forming. Actually, the process is one of the oldest plastics processes, dating back to the earliest days of cellulose nitrate processing. In this procedure, two hot sheets are clamped between two tools, and hot pressurized air is injected between the sheets. A hollow part is immediately produced which accurately reflects the contours of the top and bottom tool. This process was used for many years to produce hollow dolls, among other types of articles. Figures 13-16, 13-17, and 13-18 illustrate the process, which may be done as a single operation using two sheets at a time (Figure 13-16), or as a continuous operation using two rolls of material (Figure 13-17) or using two rolls of film and a series of molds on endless belts (Figure 13-18). In each case, either a vacuum is used to draw the sheets to the mold surfaces or, as in blow molding, a blow pin is used to pressurize the sheets. The parts are cooled by CO_2 gas using the same blow pin, thus achieving rapid cycles.

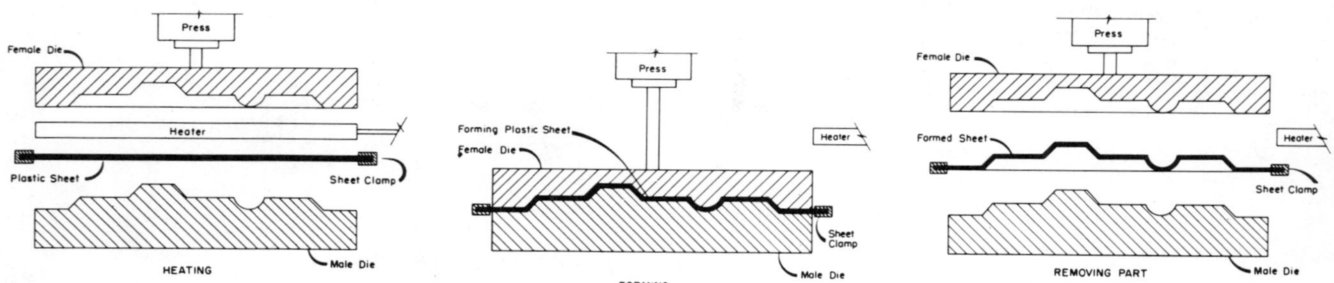

Fig. 13-14. Matched-die forming. (Printed with permission of the Dow Chemical Co.)

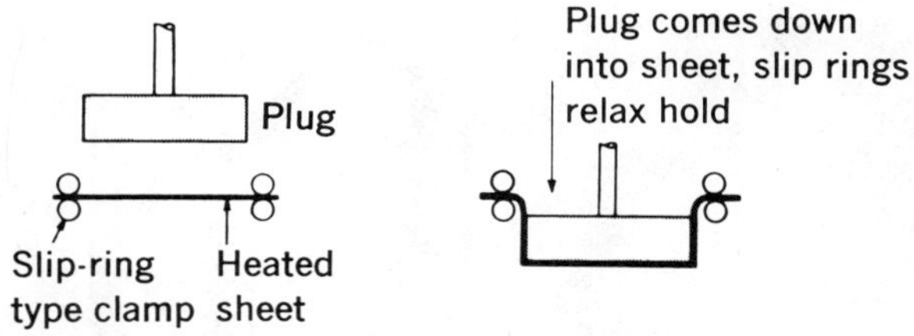

Fig. 13-15. Vacuum forming with plug and slip-ring produces stronger parts with heavier wall sections in deep-draw setups. *(Courtesy: Mobay Chemical Corporation)*

Cooling. After forming, the part must be cooled sufficiently to be removed from the mold without distortion, sticking, or other damage. This stage in some operations can be the most time-consuming phase of the entire process. To cool the part in the minimum time, the molds are usually made of materials having high thermal conductivity, such as cast or hollow aluminum. Cooling coils are usually incorporated into cast aluminum molds used for large production quantities. However, since the part is only cooled on the side which contacts the mold, fans are also frequently used to cool the part on the side away from the mold. Cooling the part too rapidly can cause large thermal stresses with resultant distortion or reduction of physical properties. For this reason, many molds are maintained at temperatures of 35–65°C (100–150°F), or even higher if thermal stresses are a problem. Because some molds are run warm, they are often made of poor thermal conductors such as wood, plastic, or plaster, in which case fan cooling is usually the only cooling technique. Water sprays can be used but rarely are, since they cause spotting.

After cooling, the parts are sometimes removed from the mold by building up air pressure inside the mold which forces the part off. In other cases, the part will shrink away

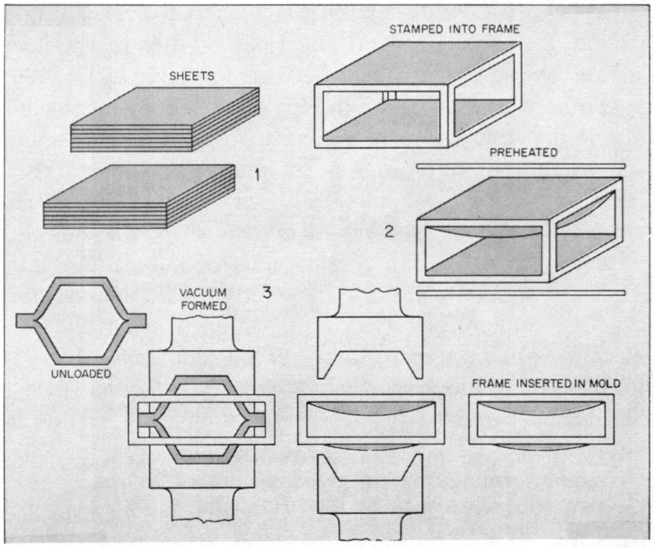

Fig. 13-16. Single-station, twin-sheet forming. (Jones, 1969. Reprinted by permission from *Plastics Technology*)

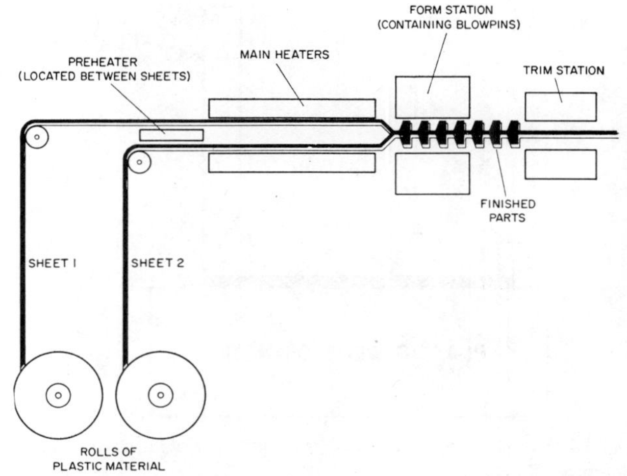

Fig. 13-17. Continuous twin-sheet thermoforming. (Jones, 1969. Reprinted by permission from *Plastics Technology*)

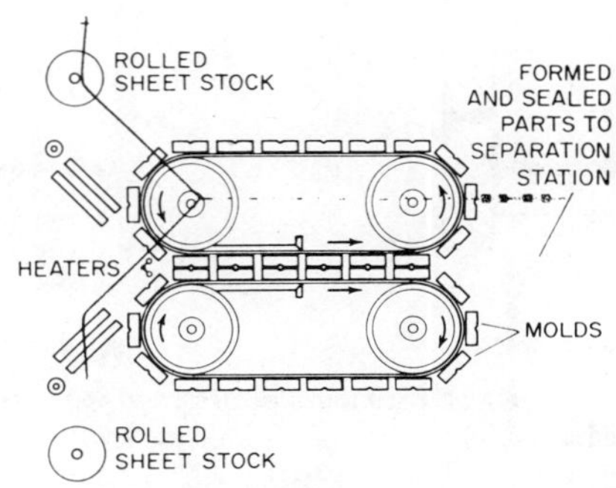

Fig. 13-18. Moving molds clamp onto the sheets in twin-sheet forming. (Jones, 1962. Reprinted by permission from *Plastics Technology*)

from the mold, which is then simply removed, or the clamp frame will be raised from the mold.

Trimming. The last operation on the part is trimming. Before deciding on the type of trimming operation, the extent of part shrinkage should be determined. As it cools, the part will shrink in all directions. Most of the shrinkage will take place by the time the part has reached room temperature; however, some shrinkage may continue to be noticed for 15 to 30 minutes after cooling.

Trimming may be done in the mold by using somewhat complicated equipment attached to the mold or the clamp. This is usually done where high production is required, coupled with noncritical tolerances, as for example in forming soft drink cups. For larger parts, with lower production rates or where close tolerances are required, special trimming equipment is used. For small quantities, trimming devices such as shears, band or circular saws, or routers can be used. Where higher production is required, cutting dies such as steel-rule dies are employed for stock which is below 0.050 inch in thickness. Heavier, machined cutting dies known as clicker dies are used in a clicker press for heavier

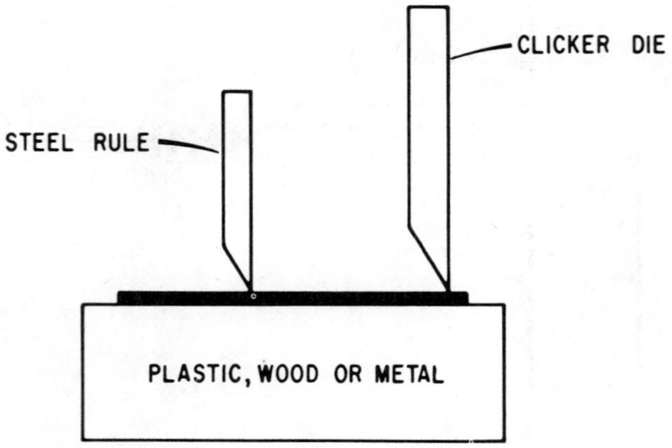

Fig. 13-19a. Cutting dies. (Printed with permission of The Dow Chemical Co.)

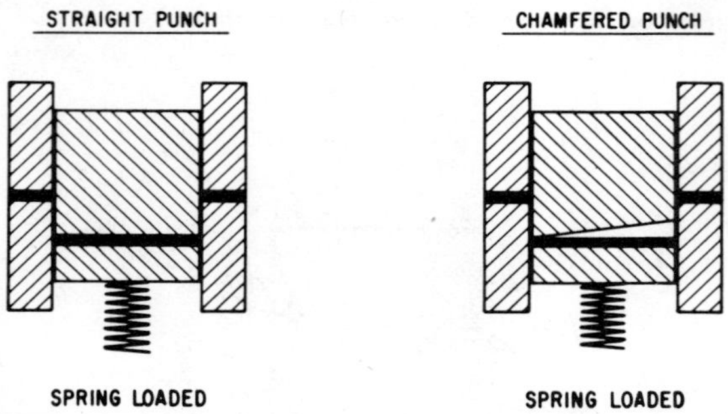

Fig. 13-19b. Shearing or banking dies. (Printed with permission of The Dow Chemical Co.)

Fig. 13-20. Small, laboratory-type thermoforming machine. (*Courtesy: Comet Industries, Inc.*)

material. Matched-shearing dies are also used for material with a thickness range from 5 to over 200 mils. Figure 13-19 shows the dies discussed.

Machine Types

There are many different types of thermoforming machines; however, each type should have the following characteristics to perform efficiently:

1. A method of heating the sheet quickly and evenly, preferably from both sides simultaneously
2. A clamping arrangement
3. A way to raise and lower the mold
4. A vacuum and/or an air pressure system
5. Controls for timing the heating, vacuum, mold or plug travel and speed, cooling, etc., preferably automatically actuated once the process is started
6. Trimming equipment, either as part of the machine or closely associated with it

The machine types range from fairly simple, laboratory-size machines such as the one shown in Figure 13-20, to more elaborate machines which take in rolls of sheet stock at one end and put out finished, trimmed parts at the other end. Figure 13-21 illustrates a machine of that type used to produce 42,000 meat trays an hour. For the production of very large parts such as boat hulls or outdoor signs, single-station machines are used. Other large machines are constructed with a rotary bed on which sheet is heated at one station, then rotated to a forming station, then to a cooling station where the part is removed (see Figure 13-22).

Another machine, shown in Figure 13-23, combines blending of the resin and colorant, extruding, a thermoforming station, a four-color, dry-offset printing station, a trim press, and a scrap granulator, all in one machine. The parts produced are mainly shallow-drain items such as jar lids.

Items such as thermoformed cups, food containers, and medical disposables are manufactured in a unique machine called a Monoformer (Hayssen Manufacturing Co.). The machine combines extrusion, compression, and thermoforming to produce completed parts (see Figure 13-24). Since there is no waste material, a trimming operation is not necessary. The three steps in the process: preforming, container forming and ejection, performed by means of a rotary indexing mechanism, are illustrated in Figure 13-25. Parts may be formed and ejected as shown in Figure 13-26, or the machine may be set to deliver counted stacks of parts automatically, as shown in Figure 13-24.

Fig. 13-21. Continuous thermoformer using a roll of polystyrene foam stock. (*Courtesy: Brown Machine Co.*)

A somewhat similar process, called SFP for "Scrapless Forming Process" (Dow Chemical Co.), is shown in Figure 13-27. In this process, blanks cut from co-extruded, multi-layered sheet are used instead of molten slugs, and the blanks are heated to a temperature *above* the softening point, but *below* the melting point of the material. The softened blanks are formed into preforms by a high-pressure forging process, followed by plug-assist pressure forming. This type of forming, at a temperature below the melting point, is also called solid-phase forming. Other solid-phase-forming (SPF) techniques are described later in this chapter in the section on cold forming.

Fig. 13-22. Rotary thermoforming machine. (*Courtesy: Brown Machine Co.*)

Fig. 13-23. Machine for processing shallow-drawn parts. It blends, extrudes, thermoforms, prints, trims, and reclaims scrap. (*Courtesy: Brown Machine Co.*)

Fig. 13-24. Hayssen #140-30 Monoformer automatic cup-forming machine. (*Courtesy: Hayssen Manufacturing Co., Sheboygen, Wis.*)

Like the Monoformer, the SFP process has the advantage of eliminating scrap, which results in a savings in materials and regrinding. Other advantages include:

1. Multilayer materials can be used which combine desirable properties of two plastics. For example, containers of high-impact styrene coated with a layer of Saran have low permeability to gases and water vapor and the excellent physical properties of the styrene.
2. In the disc-forging operation, the material is biaxially oriented, resulting in improved physical properties.
3. Because of the improvement in physical properties, lower-cost resins may be used, or the parts may be made lighter in weight.

Special machines are being developed to perform the SFP operations. Figure 13-28 shows one of these machines which, like the Monoformer, will be used mainly for production of cups, cheese container tubs, and similar parts as shown in Figure 13-29.

Packaging

One of the most common applications of thermoformed materials is in the field of packaging. Two types of packag-

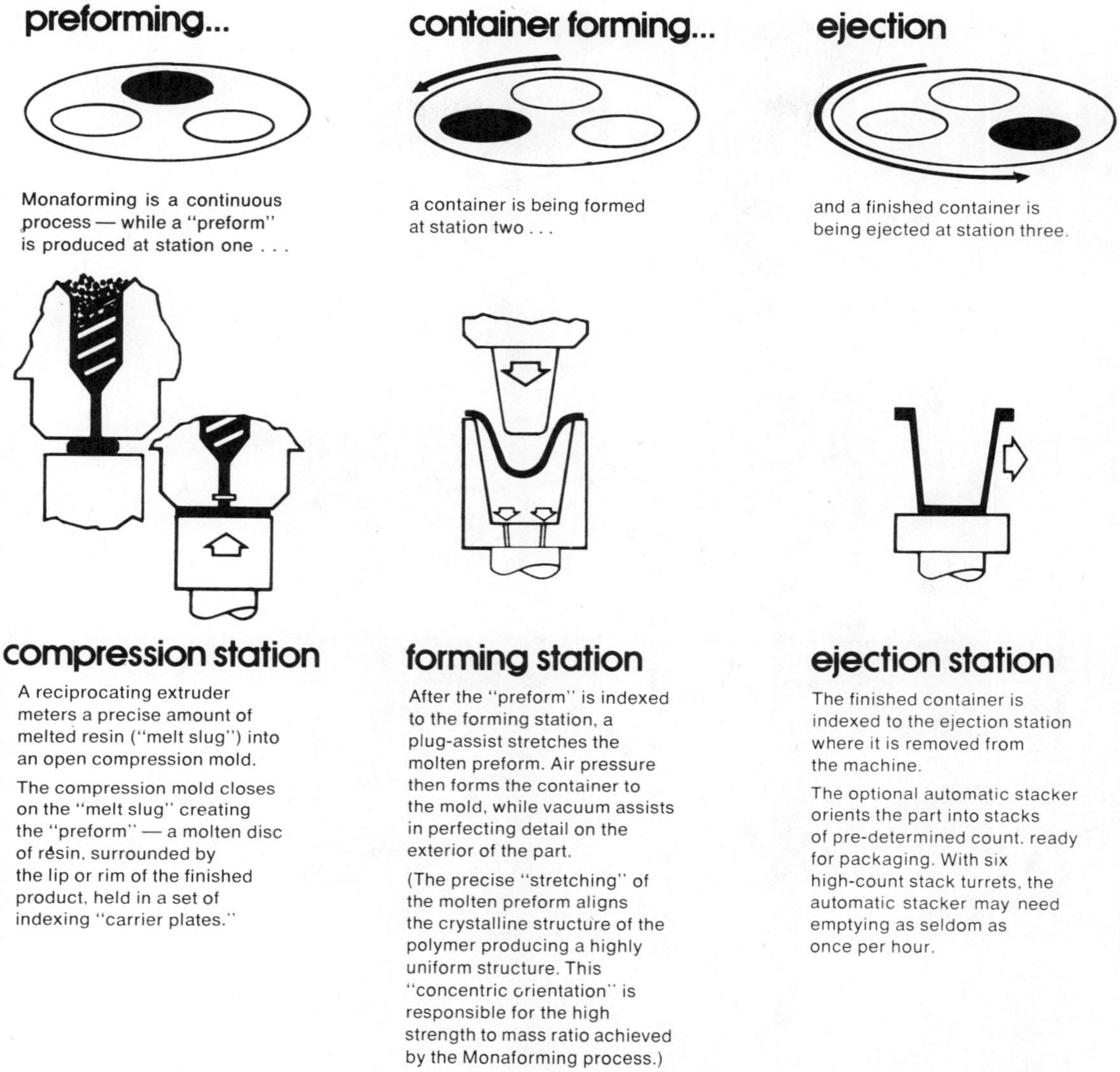

Fig. 13-25. Three steps in the Monoformer process. (*Courtesy: Hayssen Manufacturing Co., Sheboygen, Wis.*)

Fig. 13-26. Forming and ejection section of the Monoformer. (*Courtesy: Hayssen Manufacturing Co., Sheboygen, Wis.*)

ing are in use: skin packaging and blister packages. In both instances, the packages are of clear material so the merchandise can be readily viewed and, at the same time, be protected from possible damage while it is being handled. Clear acetate film is often used for both types of packaging, since it is readily formed and can be bonded easily to a cardboard substrate.

For skin packaging, the item or items are placed on pieces of perforated cardboard. The heated sheet, or film, is then placed over the parts and drawn down by vacuum to form a tight "skin" over the parts. The film, where it contacts the cardboard, adheres to the treated surface. After trimming, the packages are ready for display.

Blister packaging is done in the same manner except that a mold is used to form a "blister" or "bubble." The part is placed in the "blister," which is then bonded or heat sealed to a previously prepared cardboard or other surface.

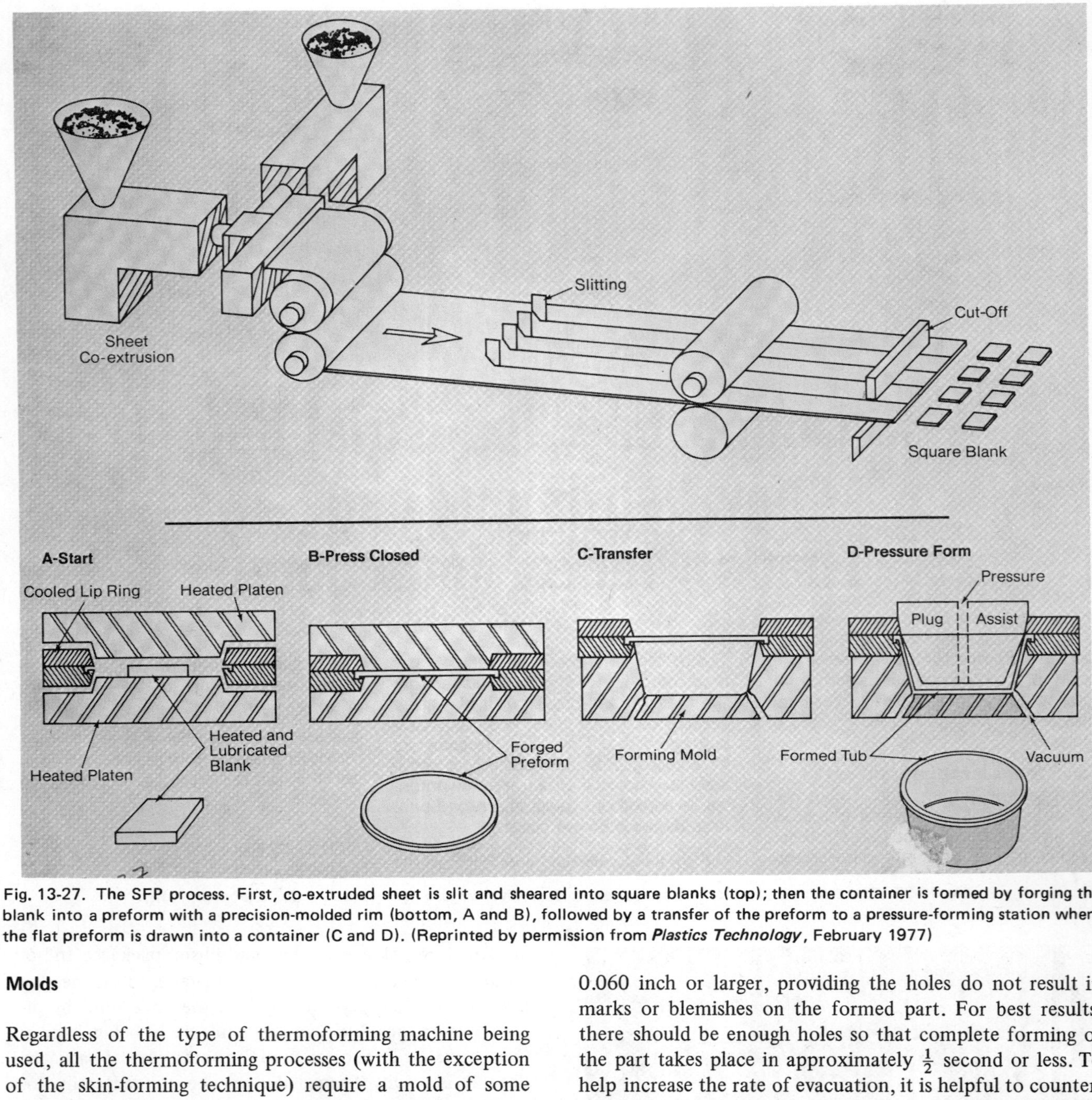

Fig. 13-27. The SFP process. First, co-extruded sheet is slit and sheared into square blanks (top); then the container is formed by forging the blank into a preform with a precision-molded rim (bottom, A and B), followed by a transfer of the preform to a pressure-forming station where the flat preform is drawn into a container (C and D). (Reprinted by permission from *Plastics Technology*, February 1977)

Molds

Regardless of the type of thermoforming machine being used, all the thermoforming processes (with the exception of the skin-forming technique) require a mold of some kind. Prototype molds are frequently made of wood or plaster. For production quantities, however, material with good thermal conductivity should be used, since the mold should be kept at a constant temperature. Aluminum is generally chosen since it is easily cast or machined, and cooling coils can be readily incorporated in the mold.

Vacuum holes should be located in the mold at the point where the sheet touches last, to minimize the possibility of trapping air and not completely forming the part. For thin sheets, the holes should have diameters of 0.010–0.015 inch. For heavier sheets, the holes may be 0.040–0.060 inch or larger, providing the holes do not result in marks or blemishes on the formed part. For best results, there should be enough holes so that complete forming of the part takes place in approximately $\frac{1}{2}$ second or less. To help increase the rate of evacuation, it is helpful to counterbore the holes to 0.250-inch diameters on the underside of the mold.

In general, male molds are cheaper to make and allow deeper draws than female molds. Female molds, however, allow easier release and give better detail on the outer surface of the part. Some female molds may not require any draft angle on the mold to affect easy release, since the part may shrink away. Most female molds should incorporate a 2–3° draft angle; male molds should have 5–7° angles for easy release. The molds should, of course, always be designed oversize to allow for the final shrinkage

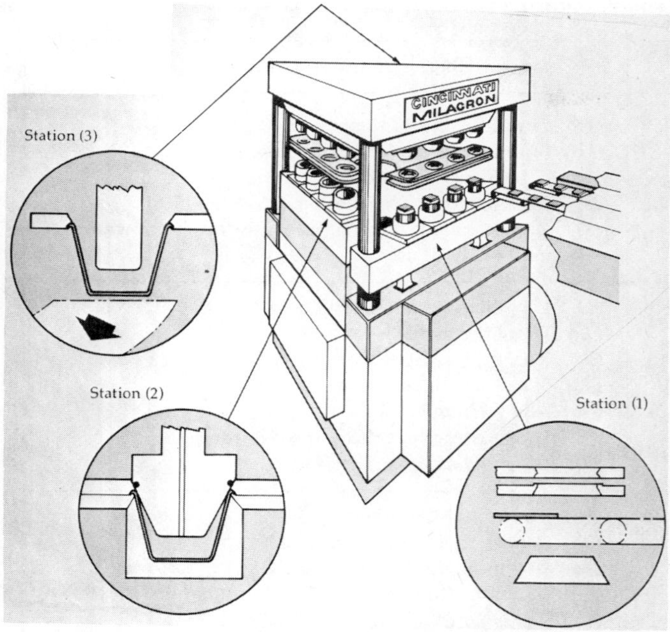

Fig. 13-28. Proposed SFP machine. (Printed with permission of The Dow Chemical Co.)

Fig. 13-29. SFP-formed multilayer containers. (*Courtesy: Dow Chemical USA*)

of the part after complete cooling. Mold surfaces need not, in most cases, be highly polished, since (except for vacuum-formed parts drawn into a female mold) the finish of the part is determined by the original sheet finish. In fact, too high a polished finish on a mold can result in air entrapment; this can, be remedied, however, by simply sanding with a fine grit paper.

Process Variables

Material Variables. There are a number of variables which can arise from variations in the sheet properties, from differing lots of the same material, or from different materials, with apparently the same specifications. Some of the variables which can affect the sheet materials are:

1. Sheet thickness. Variations in sheet thickness can cause uneven heating, uneven drawing, or both. Thickness variations should not be over 4-8% for high-quality production.
2. Sheet viscosity or melt index variations. This is mainly applicable to olefins. If the sheet varies in melt index from sheet to sheet or from lot to lot, it can cause problems in maintaining uniformity in the forming process.
3. Sheet orientation. For maximum uniformity, the sheet used should be unoriented. If an oriented sheet must be used, it is important that the degree of orientation be the same in each sheet, and that the sheet be placed in the clamping frame in exactly the same manner each time; that is, the orientation must be the same each time.
4. Sheet density. Variations in sheet density (especially of olefins) can cause variations in tensile strength, shrinkage, hardness, stiffness, and softening temperature or time.

Process Variables. During the heating, forming, and cooling phases of the process, a number of variables can affect the uniformity of the final part. Some of these are:

1. Heating temperature and final sheet temperature. If either the heating or the absorption of heat is, for any reason, nonuniform, then the part will not be uniform.
2. Prestretching. Nonuniformity in the prestretching process will affect the final part. If the prestretch is done by blowing a billow upward, drafts can cause localized cooling of the billow, with consequent nonuniform forming later.
3. Air or plug temperatures. Air and/or plugs used to prestretch or form the part should be kept at a controlled temperature to insure maximum uniformity in the part. In general, to avoid part chilling, the air and plug should be kept at a temperature approximately 10% lower than the sheet temperature.

Mold Variables. The final forming and cooling of the part depend on the mold, either male or female. Some possible mold variables include:

1. Vacuum holes. These should be as small as possible, so that no marks will be on the final part. Slots are better than holes, but are more expensive.
2. Speed of evacuation. In general, the faster the evacua-

Table 13-1. Thermoforming Troubleshooting Guide.*

Situation	Probable Causes	Suggested Courses of Action
Blisters or bubbles	(1) Overheating (2) Wrong sheet type or formulation (3) Poor storage conditions	(1a) Lower the heater temperature (1b) Increase distance of heater from the clamped sheet (1c) Attach masks or baffles (2) Obtain correct formulation (3) Do not remove material from moistureproof wrap until ready to use
Blush or change in color intensity	(1) Insufficient heating (2) Mold is too cool (3) Assist is too cool (4) Sheet cools before it is completely formed (5) Too deep a draw (6) Poor mold design	(1) Lengthen heating cycle (2) Warm the mold (3) Warm the assist (4a) Speed the drape action (4b) Add vacuum holes (5) Use heavier-gauge sheet or softer formula (6) Use mold of proper design
Sticking to the mold	(1) Rough or improperly designed mold	(1a) Make mold smoother (1b) Increase the taper on male plugs (1c) Use mechanical release assists (1d) Use air pressure to blow piece from mold (1e) Use mold-release agents
Incompletely or improperly formed pieces	(1) Sheet is cold (2) Insufficient vacuum (3) Vacuum holes are plugged up	(1a) Lengthen the heating cycle (1b) Bring heater closer to sheet (2) Check vacuum system (3) Clean, relocate, or add vacuum holes
Warped or distorted pieces	(1) Poor mold design (2) Sheet removed while too hot	(1) Redesign mold using proper tapers and ribs (2a) Increase cooling cycle (2b) Use water-cooled molds
Webbing or bridging	(1) Insufficient vacuum (2) Sheet is overheated (3) Long parallel molds with extrusion direction parallel (4) Poor mold layout or design (5) Sharp corners on deep draw	(1a) Check vacuum system (1b) Add more vacuum holes (2a) Shorten heating cycle (2b) Increase heater distance from the sheet (3) Move sheet 90° in relation to mold (4) Use mechanical drape or plug assists (5) Increase radius
Bad surface markings	(1) Markoff (due to trapped air) (2) Markoff (due to accumulation of plasticizer on mold) (3) Mold is cool (4) Mold is too hot (5) Improper mold composition (6) Mold surface too highly polished (7) Mold surface too rough	(1a) Slow draping action (1b) Add more vacuum holes (2a) Use a temperature-controlled mold (2b) Have mold as far away from the sheet as possible during the heating cycle (2c) Shorten the heating cycle (if too long) (2d) Wipe the mold (3a) Warm the mold (3b) Bring the heater closer to the sheet (4) Provide cooling for mold (5) Avoid phenolic molds with clear transparent sheet (6) Remove high surface gloss from mold (7) Smooth surface
Excessive post shrinkage	(1) Sheet removed from mold while still hot	(1) Increase cooling time
Pinholing or rupturing	(1) Vacuum holes too large (2) Uneven heating	(1) Partially plug up holes with wood or solder or completely plug and redrill (2) Attach baffles to the top clamping frame

*Courtesy: Celanese Plastics Co.

tion, the better the part will form. In order to speed up the removal of air it is necessary, in many cases, to counterbore the vacuum hole on the inside of the mold. A $\frac{1}{4}$-inch counterbore to within $\frac{1}{8}$ inch of the surface will generally suffice. In other cases, vacuum holes will have to be placed very close to each other in order to obtain fast enough evacuation.

3. Mold temperatures. The mold temperature can affect the forming of the part and the final size by affecting the shrinkage. Thus, the temperature which has been found to be optimum should be maintained as closely as possible if size variations cannot be tolerated.
4. Mold surface. For some materials, such as low-density polyethylene, a highly polished mold surface will trap air and interfere with swift evacuation. A lightly sandblasted surface will alleviate this problem.

Troubleshooting

Table 13-1 lists some general problems occasionally encountered in thermoforming with their usual causes and their remedies.

Cold Forming (Solid-Phase Forming)

Another method of forming plastics sheet stocks is known as "cold forming" or "solid-phase forming," since the plastics materials are usually not really cold when formed. Before being formed, the sheet stock is heated to a temperature below the melting point, contrary to normal thermoforming, thus the term "cold forming" came into use. Solid-phase forming is somewhat more descriptive, since the sheet is in the solid phase when formed, rather than in a molten state.

The process of solid-phase forming of plastics is analogous to the metal-forming operations of forging, stamping, drawing, etc., in that the plastics material is formed at a relatively high pressure between two dies. Just as there are various techniques of forming metal, so there are several different ways of forming plastics in the solid phase, including (1) forging, (2) rubber-pad forming, (3) fluid forming, and (4) drawing and stamping.

Forging. Plastics forging is done in a manner very similar to metal forging, and the same type of equipment is used. Figure 13-30 illustrates the process in which the billet, preheated to approximately 572°C (300°F) for polypropylene, is placed in the die and forged to shape. Mechanical presses can be used, although hydraulic presses are favored because they are easier to control for the slight dwell time usually allowed after the full pressure is applied.

The pressure required is determined by the part thickness, and is inversely proportional to the thickness. Thus, a thin part requires the highest pressure, while a thicker part requires less pressure and usually has better quality. The system is therefore complementary to injection molding; that is, injection molding of a part with a wall thickness of 0.080 inch is entirely feasible, whereas injection molding of a part with a wall thickness 0.50 inch becomes somewhat impractical because of the long cooling time. On the other hand, forging a part 0.080 inch thick would require a pressure of approximately 30 tons per square inch, and the quality would be poor. Forging a part with a wall 0.50 inch thick would require a pressure of less than 4 tons per square inch, would be made in considerably less time than if injection molded, and would be of good quality.

Rubber-Pad Forming. In rubber-pad forming, a thick pad of solid rubber is used as one-half of the die set, and it acts as a hydraulic force under pressure (see Figure 13-31). The mold side of the part reproduces the mold-surface detail very well in this technique, but it is not as sharp on the pad side. Rubber-pad forming requires a dwell time of 10 to 15 seconds on the preheated sheet to limit springback caused by plastic memory. The method is particularly well suited for shallow draws on materials from 0.10 to 1.0 inch in thickness. Interestingly, where vacuum-drawn parts would show thinning at the corners, in rubber-pad forming, the corners are the toughest areas because of the cold working of the material.

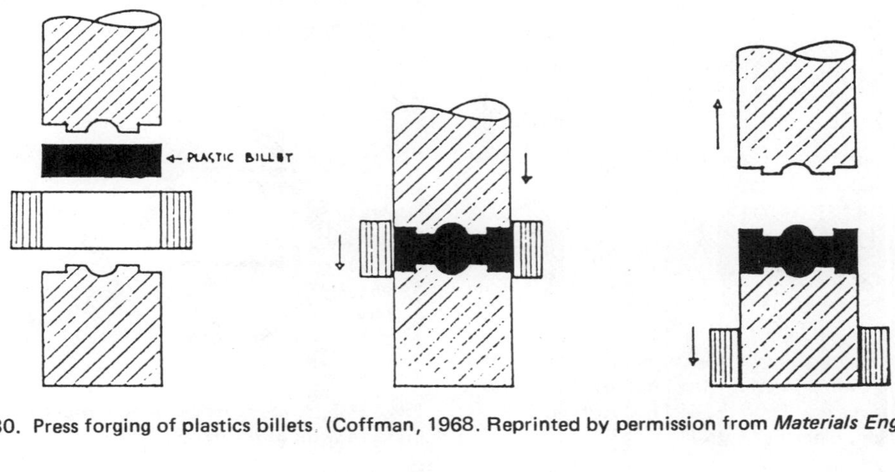

Fig. 13-30. Press forging of plastics billets. (Coffman, 1968. Reprinted by permission from *Materials Engineering*)

RUBBER PAD FORMING

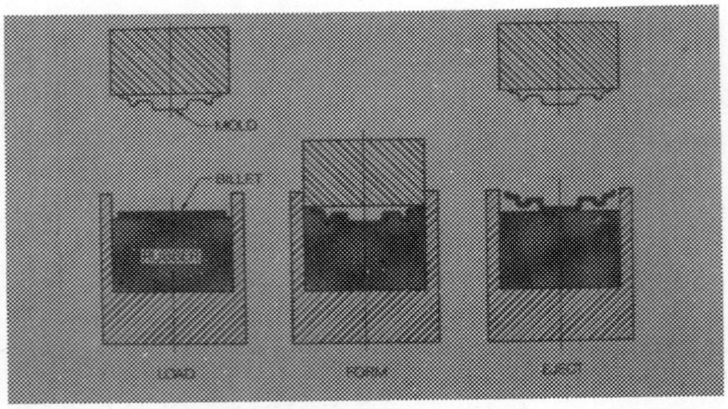

Fig. 13-31. Rubber-pad forming.

Fluid Forming. In fluid forming, a fluid is retained behind a flexible diaphragm (usually rubber), and the diaphragm is forced against the heated sheet causing it to conform tightly to the mold contour. There are several variations on the fluid-forming technique which are similar to rubber-pad forming but which can produce deeper parts with lower pressure (see Figures 13-32 and 13-33). Like rubber-pad forming, the billet is forced into position and then held for a short dwell period.

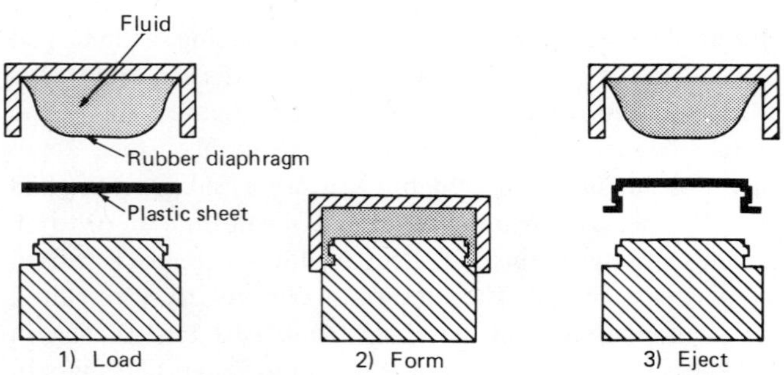

Fig. 13-32. Diaphragm forming with the fluid remaining behind the diaphragm. (Reprinted by permission from *Plastics Design & Processing*, February 1971. Copyright 1971, Lake Publishing Corporation, Libertyville, Ill.)

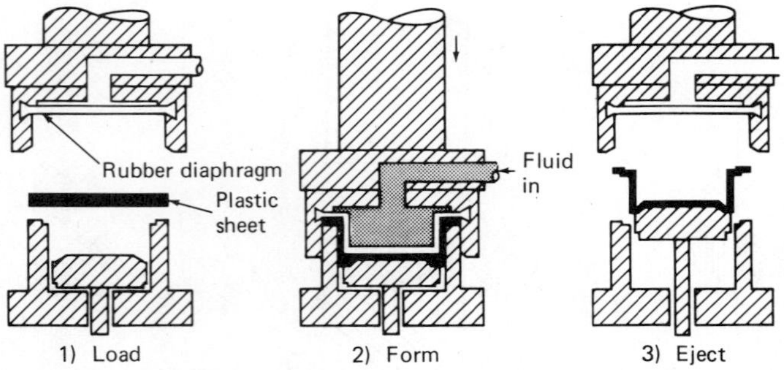

Fig. 13-33. Diaphragm forming using a diaphragm with the fluid introduced as part of the forming cycle. (Reprinted by permission from *Plastics Design & Processing*, February 1971. Copyright 1971, Lake Publishing Corporation, Libertyville, Ill.)

HYDROSTATIC FORMING

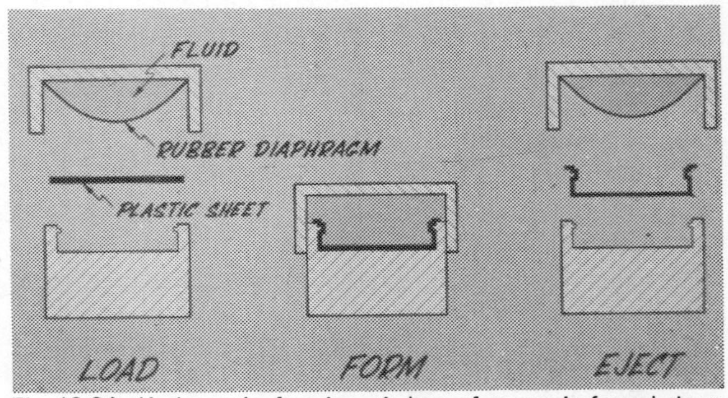

Fig. 13-34. Hydrostatic forming. A bag of water is forced down onto plastics sheet that has been placed over a plug in the shape of the desired part.

In one version of fluid forming, called hydrostatic forming, the fluid is retained by the diaphragm in a loose baglike manner, as shown in Figure 13-34. A second version of this process uses fluid which is pumped into the space behind the diaphragm. This method has the advantage of accommodating more sizes and shapes of parts than does the fixed diaphragm-fluid combination.

In a third technique, the use of a press is dispensed with entirely. The diaphragm and fluid are contained in one-half of a large clam-shell mold, the other half of which contains the die. On closing the mold, fluid is introduced into the top half of the die, thus forcing the part to conform to the die as shown in Figure 13-35. Because no press is required, this method is particularly suited for production of very large parts.

Drawing and Stamping. Drawing and stamping are done in a manner very similar to that used for the same type of metal parts. Tooling and presses can be the same as those used for metal. When drawing or stamping very shallow parts, in many cases, the material is not heated, particularly if the sheet is of the type specially made for cold forming. In some parts, distorted decorative designs are placed on the parts prior to drawing; after being formed, they assume

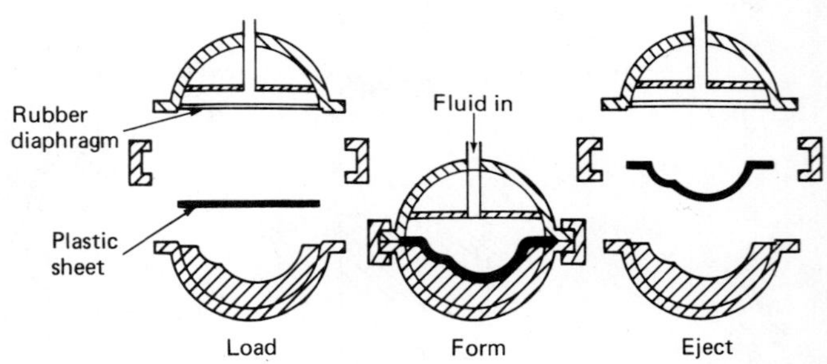

Fig. 13-35. This spherical shape provides excellent retention of pressure without the use of a press.

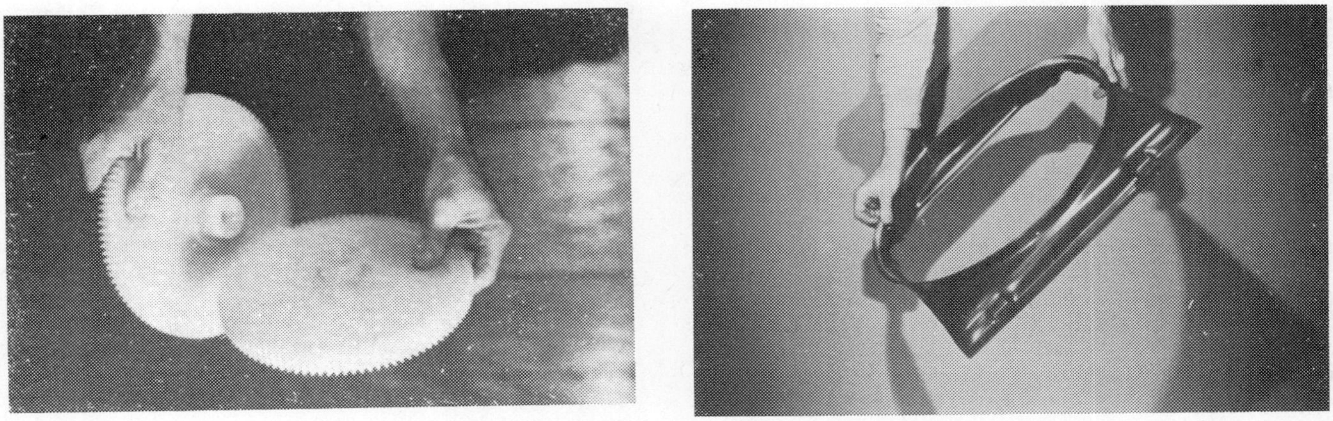

Fig. 13-36. Solid-phase formed parts.

the correct shape. This is the only practical method of decorating some parts.

Somewhat related to cold drawing are cold heading and thread-rolling operations. Plastic rod is used in forming rivets and screws in the same type of equipment used to form similar metal parts. Usually the processes are carried out at room temperature. Coining, a stamping operation, is used to form integral hinges in parts such as briefcases and containers of various types which require attached lids (see Figure 13-36).

The advantages and disadvantages of the cold-forming (solid-phase forming) techniques may be summarized as follows:

Advantages:

1. The processes are particularly well suited for production of thick parts.
2. The processes can be used with ultrahigh molecular weight materials which are difficult to form by other methods.
3. Thick parts can be made quicker and cheaper than by injection molding.
4. Tooling costs can be considerably cheaper than for injection molding.
5. Because of cold working, the impact resistance of the material may be improved.
6. Parts can be made with no trimming required.
7. No sprue marks or weld lines are present.

Disadvantages:

1. There is a definite cost to preparing and heating the sheet material or billets.
2. The process is not suited for production of thin-walled parts.
3. Parts, because of built-in stresses, have low creep resistance in tension.
4. Because of the built-in stresses, the parts usually have fairly low, maximum operating temperatures.

REFERENCES AND BIBLIOGRAPHY

Allison, D. E., and Doyle, R. Thermoforming techniques for high density polyolefin sheet. *Modern Plastics*, June 1961, **38**(6):123.

American Cyanamid Co. *XT Polymers for Thermoforming.* Bulletin PKT 12 2-2057, February 1973.

Biery, J. Molds and plugs for production thermoforming. *Plastics Technology*, September 1965, **11**(9):41.

Boden, H., and Gronemeir, U. Thermoformable rigid urethane-foam sheeting and sandwich constructions. *Modern Plastics*, August 1976, **53**(8):52.

Bosche, E. R. Thermoformed plastics parts. *Machine Design*, January 7, 1971, **43**(1):94.

Charles, P. F. Vacuum systems for thermoforming. *Plastics World*, April 1963, **21**(4):34.

Celanese Corporation of America. *Thermoforming Fortiflex A.* Product Data Bulletin P2B, September 15, 1958.

Celanese Plastics Co. *Forming Cellulose Acetate.* Product Data Bulletin B-13, Form P-8, January 1, 1960.

——. *Celanese Plastics Thermoforming.* Booklet 056125, 4th Ed.

Coffman, P. M. Thick thermoplastic slabs via rubber ring forming. *Plastics Engineering*, June 1978, **34**(6):37.

——. Stretch-forming: a simple process with a lot of potential. *Plastics Engineering*, August 1977, **33**(8):14.

——. Forming plastics by metalworking techniques. *SPE Journal*, January 1969, **25**(1):50.

——. Cold form thermoplastics to shape thick parts fast. *Materials Engineering*, April 1968, **67**(4):74.

Curtis, C. J. Jr. How to thermoform polycarbonates. *Plastics Technology*, November 1962, **8**(11):41.

Deeds, D. C. Deep draw bubble forming techniques for ABS polymers. *SPE Journal*, May 1963, **19**(5):471.

Delmonte, J. Porous, high temperature plastic tools for vacuum forming. *Plastics Design & Processing*, November 1965, **5**(11):12.

Dow Chemical Co. *Forming of Large Area, Deep Drawn Parts.* Bulletin 171-168-85C-561, May 1961.

——. *Fundamentals of Sheet Forming.* Bulletin Form No. 302-41-69.

——. *SFP Scrapless Forming Process.* Bulletin Form No. 5-303-74-76 and Form No. 0-307-119-76.

Doyle, R., and Allison, D. E. Designing Thermoforming Molds for High Density Polyethylene. Reprint from *Technical Papers*, **VI** and *SPE Journal*, March 1960, **16**(3).

Dreger, D. R. Stampable plastics challenge sheet steel. *Machine Design*, December 9, 1976, **48**(23):134.

Eastman Chemical Products, Inc. *Vacuum Forming Sheet Made of Tenite Polyallomer and Other Tenite Polyolefin Plastics.* Bulletin TR-13 ECP 465, September 1963.

Eastman Kodak Co. *Guide to Thermoforming.* Bulletin 4-U-02, 1964.

Falkin, S. How to choose molds for thermoforming. *Modern Plastics*, March 1964, **44**(7).

Fortner, C. P. Oriented thermoplastic sheet and film. *Materials in Design Engineering*, December 1959, **50**(7):94.

Fox, A. J. Thermoforming PS foam sheet and foam-polystyrene laminates. *Modern Plastics*, April 1962, **39**(8):117.

General Electric Co., Chemical Materials Department. *Lexan Polycarbonate Film and Sheet, Thermoforming.* Technical Report CDC-503.

Griff, A. L. Selecting sheet for thermoforming. *Modern Plastics*, July 1961, **38**(7):97.

Haddad, E. A. When and how to use pressure forming. *Modern Plastics*, August 1960, **37**(8):85.

Hauck, J. E. Diaphragm forming: a metalworking technique applied to plastics. *Modern Plastics*, June 1969, **45**(6):82.

Jones, E. Twin-sheet forming: a big future lies ahead. *Plastics Technology*, April 1969, **15**(4):947.

Kostur, R. E. Flexibility is key to thermoforming progress. *Plastics Design & Processing*, November 1963, **3**(11):34.

Limbert, F. J. Polycarbonate thermoforming. *SPE Journal*, October 1968, **24**(10):87.

Lynch, J. Guidelines for selecting web fed thermoforming machinery. *Plastics Design & Processing*, March 1979, **19**(3):29.

Margolis, J. M. Basic thermoforming technology. *SPE Journal*, July 1963, **14**(7):600.

McDonald, E. J. Jr. Vacuum forming ABS. *Plastics Technology*, December 1965, **11**(12):35.

McTier, R. F. Distortion printing and vacuum forming of thermoplastic sheet. *SPE Journal*, July 1962, **18**(7):741.

Mobay Chemical Co. *Thermoforming Merlon Polycarbonate.* Thermoplastics Processing 89-M9.

Modern Plastics. Cold stamping ABS. April 1968, **45**(4):124.

——. Warm forging thermoplastics. May 1968, **45**(5):116.

——. "Giant" thermoformed parts. September 1960, **38**(1):91.

McConnell, W. K. "Thermoforming." In *Modern Plastics Encyclopedia*, McGraw-Hill Book Co., New York, 1969–70.

Monsanto Chemical Co., Plastics Division. *Production Techniques for Vacuum Forming Thermoplastic Sheets.* Production Information Bulletin No. 1030.

——. *Production Techniques for Vacuum Forming Thermoplastic Sheets.* Production Information Bulletin No. 1092.

Naitove, M. H. New vistas for thermoforming, *Plastics Technology*, November 1975, **21**(11):27.

Naturman, L. I. Cold forming: where to next? *Plastics Technology*, April 1969, **15**(4):39.

Plastics Design & Processing. Up-date on rigidized thermoformed parts—new technology expands applications into diverse fields. January 1974, **14**(1):19.

——. Solid phase forming of plastics. February 1971, **11**(2):24.

——. Thermoforming—an overview of an advancing industry. July 1971, **11**(7).

Plastics Technology. Scrapless forming process makes multilayer containers. February 1977, **23**(2).

Product Engineering. Forged plastics promise complex parts that are strong, economical. April 26, 1971, **42**(8):50.

——. Special process forges thick plastic parts from billets. April 8, 1968, **39**(8):102.

Ryba, M. A. Your guide to thermoforming phenylene oxide based resins. *Plastics Technology*, June 1979, **25**(6):77.

Scheiner, L. L. How to buy thermoformers. *Plastics Technology*, June 1965, **11**(6):87.

Smith, R. W. Sophistication comes to thermoform tooling. *Plastics Technology*, May 1971, **17**(5):52.

United States Gypsum Co. *Vacuum Forming with Hydrocal and Ultracel.* IGL Bulletin No. ITT-55.

Ward, L. G. Thermoplastics stamping seeking stamp of approval. *Plastics Engineering*, March 1979, **35**(3):47.

Warshavsky, M., and Tokita, N. Cold drawability of thermoplastic sheets. *SPE Journal*, August 1970, **26**(8):55.

Wehrenberg, R. H. II. New materials and processes strengthen thermoformed plastics. *Materials Engineering*, April 1979, **89**(4):50.

Wise, J. R. *Automation in Thermoforming.* AAA Plastic Equipment, Inc., Fort Worth, Texas.

Zelnick, D. Thermoforming—the process and the machinery. *Plastics Design & Processing*, May 1968, **8**(5):12.

Zukor, L. J. Innovations in thermoforming techniques. *SPE Journal*, July 1963, **19**(7):607.

14
Rotational Molding

Rotational molding, also known as rotoforming or rotomolding, is a process in which finely ground powders, usually thermoplastics, are heated in a rotating mold until melting or fusion occurs. (The process was originally done with liquid PVC plastisols and was called slush molding.)

The melting or fusion of the resin results in a uniform coating lining the inner surface of the mold. After the resin cools, it is removed as a finished plastics part. Rotoforming thus offers another technique for producing large, complex plastics parts by a method which is quite different from all other methods used to make similar parts. The advantages and disadvantages of the process are listed below.

Advantages of Rotational Molding:

1. The tooling required is usually very simple and relatively inexpensive.
2. The process is well suited for making very large and/ or very complex parts with single or double walls.
3. Parts produced are strain free.
4. Parts with square corners are thicker at the corners rather than thinner, as is the case with thermoformed parts.
5. Parts are usually made with very uniform wall thicknesses, except as indicated above.
6. Parts are usually made with no weld lines, sprue marks, ejection marks, etc.
7. No scrap, or very little, is produced.
8. Hollow parts are easily produced with simple tooling.
9. Usually no secondary operations are required.
10. Wall-thickness changes can be easily made with no need for new tooling or modification of the starting raw material.
11. The process can be operated with relatively unskilled labor.

Disadvantages of Rotational Molding:

1. Material costs are relatively high since most materials are produced as pellets and then reduced to a fine powder.
2. The process is not suited for production of parts with wall thicknesses of less than 0.030 inch.
3. It is not suited, generally, for large production runs of small parts.

Rotational molding is well suited for prototype production of large parts such as boats (up to approximately 12 feet long), camper tops, missile containers, gasoline tanks, toys, and similar items. The technique is also used for production quantities of such parts. Figures 14-1, 14-2, and 14-3 show typical rotational-molded items.

Fig. 14-1. Rotoformed polyethylene campers. (*Courtesy: Hollowform, Inc.*)

Materials

The first materials used to make parts in rotating molds were vinyl plastisols. These vinyl dispersions were used mainly to make toys, beach balls, artifical flowers, and similar objects. The process consisted then, as now, of placing the liquid in the closed mold, then rotating and heating the mold until the plastisol was fused. Because the plastisols used were thick, viscous liquids the process acquired the name "slush molding."

Today, most rotoformed parts are made using finely ground powder, and the process is usually referred to by the more descriptive term rotational molding. With the introduction of powdered polyethylene, the rotoforming industry began producing the wide variety of products available now. Almost any thermoplastic can be used for rotoforming, and parts are being made from cellulosics, nylon, polycarbonates, acetal, styrenes, etc. However, the greatest number of parts are still being made from polyethylene of both high and low density. The latest type of polyethylene is a crosslinkable material which has considerably improved strength, environmental resistance, and toughness compared to conventional polyethylenes.

Fig. 14-2. Tanks and various containers rotoformed from polyethylene and polypropylene. (*Courtesy: Hollowform, Inc.*)

Another new technique is production of composite moldings in which two resins are used to form the part. One resin melts first and forms the outer skin of the part. The second resin then melts, fuses to the first, and forms the inner skin. A variation of this process is the production of foamed parts in which two or three resins are used. First one resin melts and forms the outer skin. Then the foaming resin melts and foams. The third resin then melts and fuses to the foam to form the inner skin. This process is being used in the production of small boats, camper tops, and similar parts. The exact resins and the forming techniques are still considered proprietary.

The use of thermoset resins in rotoforming processes is still in an early stage. However, should satisfactory resins be

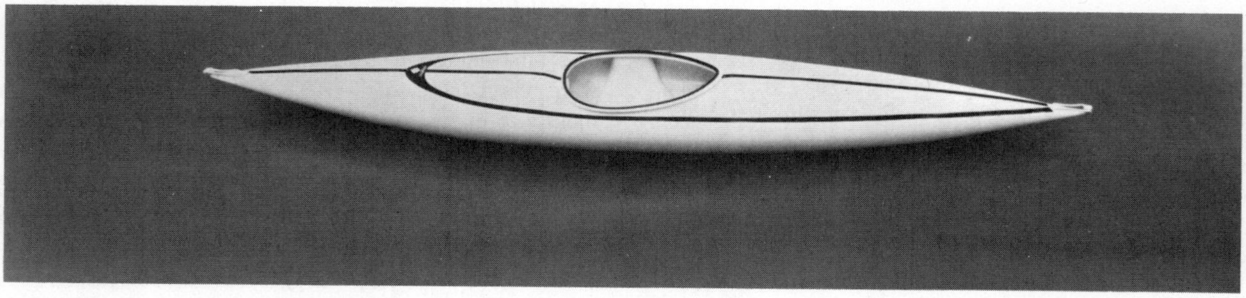

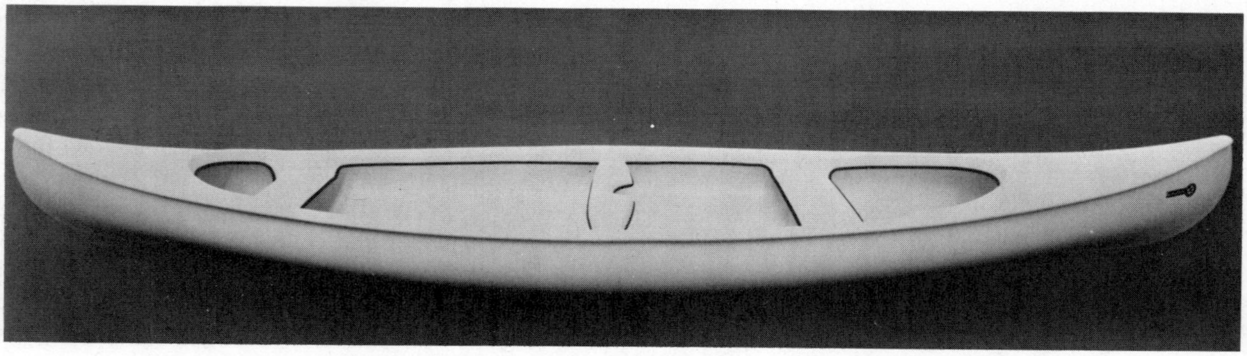

Fig. 14-3. Rotoformed kayak and canoe of crosslinked polyethylene. (*Courtesy: Hollowform, Inc.*)

developed which would not be brittle, or if techniques are developed in which fibrous reinforcements can be added uniformly, then rotational molding could well become a large-volume application. The advantage here is that it might be possible to eliminate the heating stage entirely, thus removing this step from the process. Or, alternatively, the parts might be removed from the mold and just be given a post-cure, which would still be a considerable advantage since the molds would not be tied up during cure.

Process Steps

Rotoforming is usually described as having four distinct steps—loading, molding, cooling, and unloading. In the loading step, the mold is loaded, or charged, with the requisite amount of powder or liquid. (To produce uniform parts this step should be done by weighing the raw material.) The mold is then closed and the part is formed by rotating the mold in two axes, while at the same time heating it. Heating is done either by hot air, by placing the mold assembly in an oven, or, in some cases, by a hot liquid—either oil or a molten salt. Biaxial rotation, as shown in Figure 14-4, can be done using a mold on the end of an offset arm or using a series of molds on the end of a center swing arm, which in effect is a radius arm powered from a central hub. In either case, a ratio of approximately 1:4 is held between the speed of rotation of the major axis and the minor axis. Typical speeds for the major axis are 3 to 10 rpm, and the

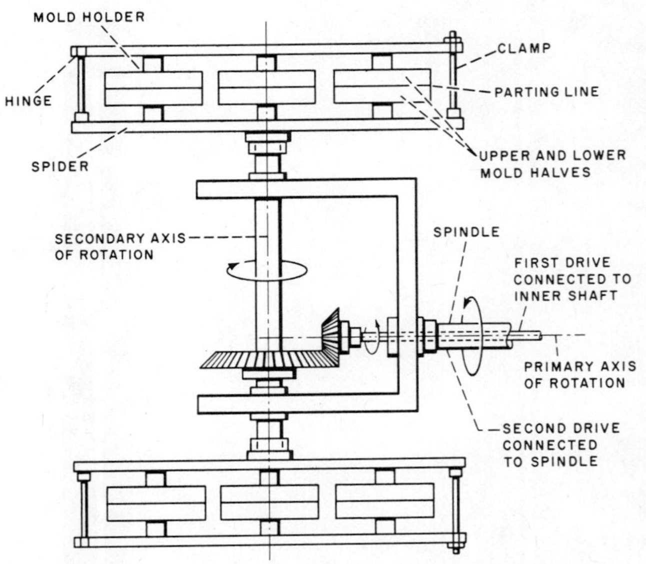

Fig. 14-4. Biaxial rotation used for rotary molding. This schematic shows a typical system used to obtain mold rotation in two planes perpendicular to each other. The spindle is turned on the primary axis (right) while the molds are rotated on the secondary axis (center). Usually, the ratio of the two simultaneous rotations is determined by the gear (center), which may be exchangeable, or, occasionally, by two motor drives being available. A ratio of 4:1 usually gives good results. However, in some cases, other ratios have proven to be superior. (*Courtesy: U.S. Industrial Chemicals Co.*)

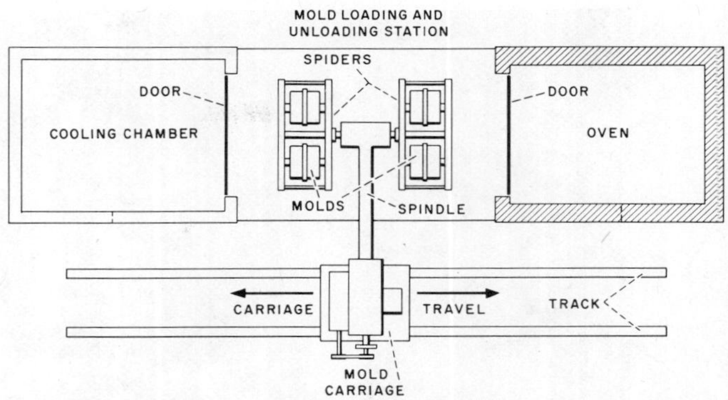

Fig. 14-5. Schematic of a single-spindle rotational molding machine. The mold carriage moves spiders and molds back and forth on a track through the mold loading and unloading station, the oven, and the cooling chamber. (*Courtesy: U.S. Industrial Chemicals Co.*)

minor axis speed would be 12 to 40 rpm. The actual speeds used are usually determined experimentally and are dependent on the part size, geometry, material, heating rate, etc.

Heating of the molds causes the resin to melt, and gravitational forces, not centrifugal forces, distribute the material uniformly over the entire inside surface of the mold. The heating period may be anywhere from 1 to 10 minutes, again depending on the part, the wall thickness desired, and the type of resin being used. At the end of the heating period, the mold assembly is moved to the cooling station, at which time it is cooled by a water spray and/or cold air. After cooling, the mold assembly is moved to the loading station where unloading and loading take place. For low-volume production, a simple, single-spindle, rotating mold on a reciprocating unit is used for loading, molding, and cooling, as shown in Figure 14-5.

Where high production is required, a multiple-spindle unit is used. In this type of machine, one or more spindles can be undergoing a different stage of the process, as illustrated in Figure 14-6. With such an assembly, there are no inter-

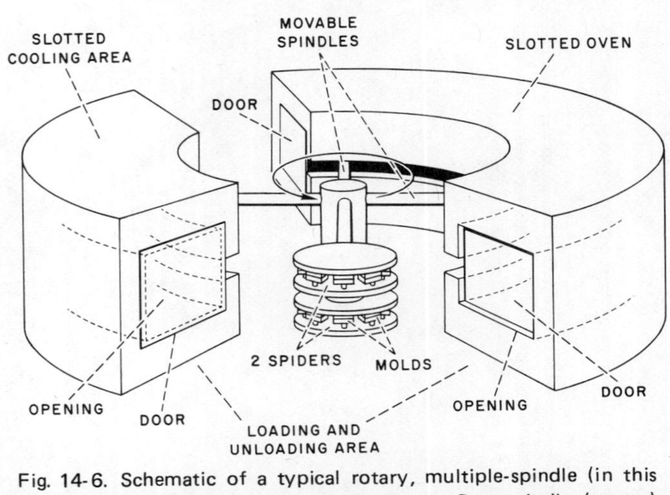

Fig. 14-6. Schematic of a typical rotary, multiple-spindle (in this case a four-spindle) mold-handing system. One spindle (center) is in the loading-unloading area, two are in the oven, and one is in the cooling area. (*Courtesy: U.S. Industrial Chemicals Co.*)

Fig. 14-7. Three-stage rotoforming. The loading station is in the foreground, the heating chamber is on the right, and the cooling chamber is on the left. (*Courtesy: McNeil Akron Division*)

Fig. 14-8. Rotary molder used to make camper tops. Cooling chamber is shown on the right. (*Courtesy: Hollowform, Inc.*)

ruptions at any time during the cycle, but the times must be coordinated so each step is completed in the same time. This can be done by adjusting the size of the oven or cooling chamber, the rate of heat input, etc. Figure 14-7 shows an actual three-stage production unit for relatively small parts. Figure 14-8 shows one of the largest rotary molders made. This unit was used to make the camper tops shown in Figure 14-1.

Molds

The molds used in rotoforming are generally of very simple, light construction since no pressure is developed during the molding operation, nor are any water cooling passages needed. The major requirement is that the mold be capable of closing tightly so no powder or liquid will run out of the parting line, or form flash.

Cast aluminum is one of the most popular materials for molds. Other mold-making materials are steel and electroformed copper or nickel. Steel is used for large molds, and electroformed molds are used where small, complex shapes are required, or if many molds of the same type are desired. The wall thickness for aluminum molds is usually $\frac{1}{4}$ inch, while steel molds may be as thin as 0.040 to 0.075 inch, providing that the shape is one that can resist warpage.

Figure 14-9 shows a measured amount of polyethylene being placed into one-half of a mold, and Figure 14-10 shows both halves of the mold. The completed part, a giant chess piece, is shown in Figure 14-11.

The inner surfaces of the molds are usually coated with a semi-permanent release agent, or, for long production, the surface can be coated with a baked-on coating of Teflon TFE. With the use of the release agents and the natural tendency of the material to shrink away from the mold, release is usually not a problem, even where slight undercuts are incorporated.

In some cases, molds must be vented to decrease the pressures required to keep the mold closed, and to minimize the possibility of mold distortion or part blowout due to internal pressure. With a vent a thinner mold can often be used. The vent is placed in an inconspicuous area of the mold, and usually consists of a tube filled with glass wool which allows for breathing, but does not allow passage of the powder during rotation. Figure 14-12 shows the use of a vent in a bottle mold. When open sections are required in a part, the mold in that area will be covered with an insulating material which acts to prevent resin melting at that spot.

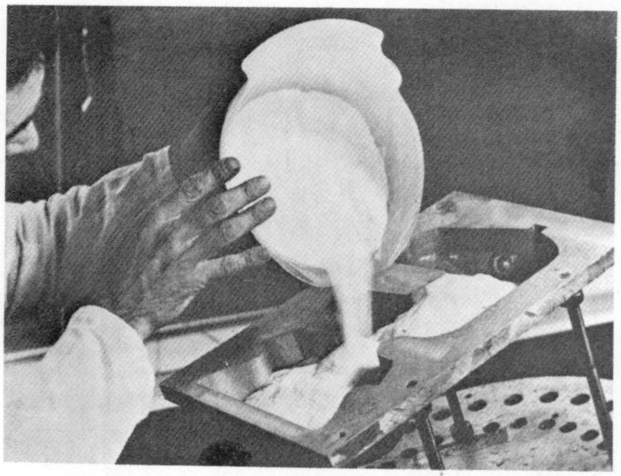

Fig. 14-9. A measured amount of powdered polyethylene is being poured into the bottom half of this large mold designed for one of six different giant chess pieces. (*Courtesy: U.S. Industrial Chemicals Co.*)

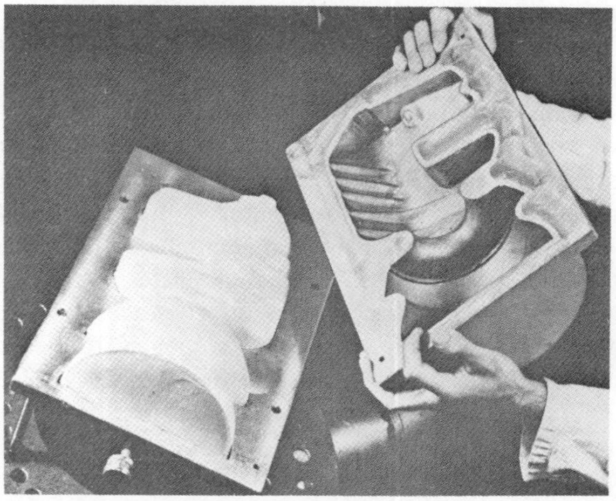

Fig. 14-10. The chess-piece mold is opened. The cooled molding rests in the lower half of the mold. (*Courtesy: U.S. Industrial Chemicals Co.*)

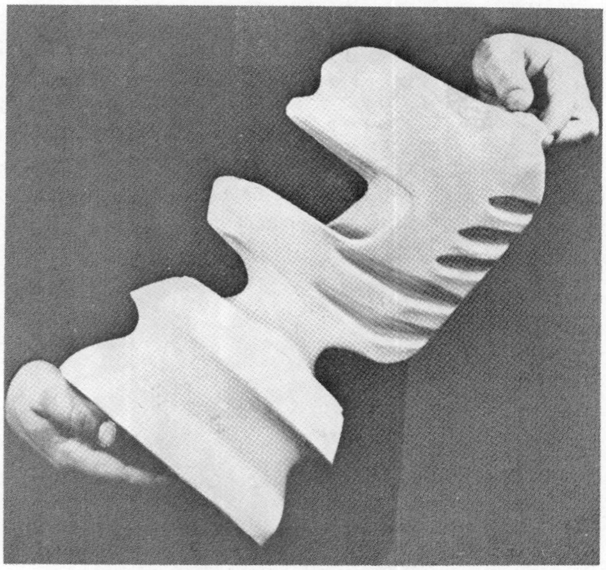

Fig. 14-11. The molding, one of the giant chess pieces molded from polyethylene powder, has been removed from its mold. (*Courtesy: U.S. Industrial Chemicals Co.*)

Heating and Cooling

Heating the molds, as previously described, is mostly done by means of a hot-air oven. Oven temperatures can be

200°C (400°F) to as high as 370°C (700°F), or even higher for some materials. The major requirement of the oven, other than the ability to furnish the hot air in large quantities and at the right temperature, is that the air flow be even, with a minimum of hot or cold spots. Liquid heating, although considerably faster, has the disadvantage of requiring thicker molds and more perfect parting-line closures. The liquid is usually a molten salt, which may lead to corrosion problems with some mold materials unless precautions are taken to wash the molds after each use. A third heating technique is the use of high frequency (RF heaters), which is efficient but quite costly. Infrared heat can also be used, but there are distinct limitations. While the infrared heat can be very fast and efficient, its use is confined to simple shapes and single molds, since the "shadowing effect" of one mold on another can cause very uneven heating.

Cooling is done by means of a cold-water spray in the majority of applications. The spray is sometimes combined with cold air for greater flexibility. The molds must be rotated during the cooling cycle to prevent distortion in the molded part. Too rapid cooling can, in some cases, cause warpage in the finished part. However, since the heating stage is usually the pacing item (heating time determines times for all other operations), cooling can almost always be adjusted satisfactorily with little difficulty.

Special Equipment

Mold filling for small production runs is usually done manually. However, for long production runs automatic fill-

Table 14-1. Rotational Molding Troubleshooting Guide.*

Problems	Possible Solutions
If the production rate is too low (or the heating cycle too long)	Raise the oven temperature, Use a higher-melt index powder, Use a lower-density powder, Use a thinner-walled mold.
If the powder fuses across deep, narrow draws in the mold, resulting in incomplete mold fill	Increase the speed of rotation, Use a lower-density powder, Use a higher-melt index powder, Modify the mold, if possible.
If the piece sticks in the mold	Clean the mold surface and apply a suitable mold release agent, Decrease the heating cycle or oven temperature. Increase the cooling cycle, Vent the mold Modify the mold so that the part tapers to a slightly larger dimension toward the end of the mold through which it is removed.
If the molded piece contains bubbles or if the piece has a rough inside surface	Raise the oven temperature, Increase the heating cycle, Use a higher-melt index powder, Use a lower-density powder, Improve mold wall uniformity, Decrease the wall thickness, if possible, by decreasing the amount of powder in every charge, Make sure the interior surface of the mold is dry.
If the piece is yellow, brown, or otherwise discolored	Decrease the heating cycle, Lower the oven temperature, Check the mold wall for contamination such as rust.
If the piece is warped	Decrease the cooling rate, Lower the oven temperature, Rotate the mold during cooling, Vent the mold, Improve the wall uniformity, Allow short air cooling before water cooling.
If the piece is brittle	Use a low-density powder, Use a lower-melt index powder, Redesign the part to eliminate sharp corners,
(and is incompletely fused or has a rough inside surface)	Raise the oven temperature, Increase the heating cycle.

*Courtesy: U.S. Industrial Chemicals Co.

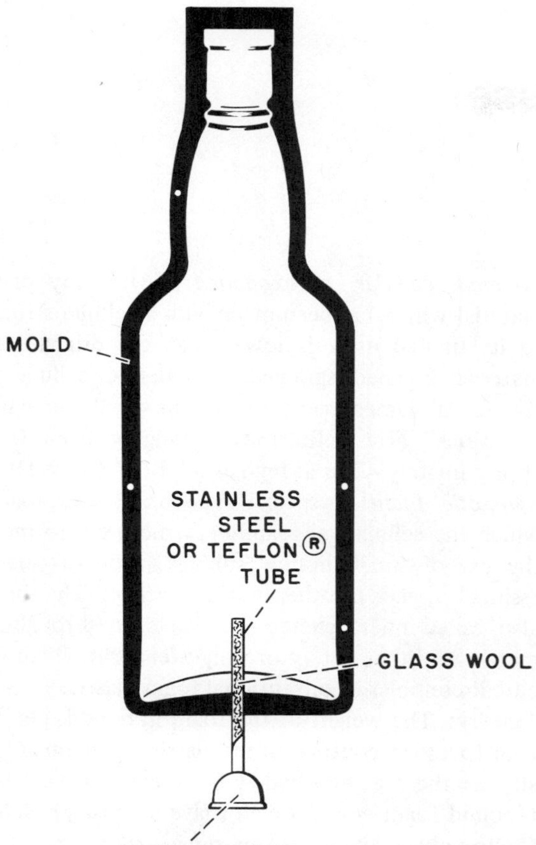

Fig. 14-12. Schematic of a vented rotational mold for a display bottle. (*Courtesy: U.S. Industrial Chemicals Co.*)

ing equipment is used with either liquid or powders. Since the apparent density of a powder can vary depending on the mesh, the fillings are usually done on a weight basis.

Molds can be designed to allow introduction of an inert gas, such as nitrogen or carbon dioxide, to replace air. With such a system, the resin can be heated to the highest temperatures, while at the same time minimizing the danger of oxidation, discoloration, odors, etc. The physical properties of the final part can even be improved, compared to parts made in an air atmosphere, since thermal and oxidative degradation are reduced.

In addition to the introduction of an inert gas, the mold can also be manifolded to allow introduction of a coolant such as cool air followed by a water spray. Internal cooling by such a technique can reduce the cooling time and, in the case of crystalline materials such as linear polyethylene, can result in improved physical properties by lowering the crystalline content, thus improving the toughness of the final product.

Finishing

Most rotationally molded items are made complete, with perhaps some slight finishing at the mold line. For olefin materials, the surface can be treated by flame or electric discharge methods to promote ink and paint adhesion. The use of pigment in resins should be kept to a minimum (0.1-1.0%) in order not to affect the mechanical properties of the finished part.

Table 14-1 presents a troubleshooting guide for rotational molding.

REFERENCES AND BIBLIOGRAPHY

Bakelite Review. Rotational molding. October, 1965.

Brenner, W., and Vermes, R. Rotational molding today: new materials, equipment. *Materials Engineering*, January 1968, **67**(1):64.

Brezinski, J. J., Cavender, K. D., and Olhoft, G. V. Improve your rotomolding operation. *Plastics Technology*, October 1973, **19**(10):75.

Carrow, G. E. Here's what you should know about rotomolding crosslinkable HD polyethylene. Part 2: Part design and mold construction. *Plastics Design & Processing*, March 1974, **14**(3):21.

E.I. Du Pont de Nemours & Co., Inc. *Rotational Molding Process Variables*. Bulletin A-50788.

Ebert, R. O. Progress in powder molding. *Plastics Technology*, October 1968, **14**(10):69.

Emmerich, A. What's new in rotomolding? *Plastics Technology*, October 1978, **14**(10):87.

Faribo Manufacturing Co. *The Wonderful World of Rotational Molding*.

Johnson, D. M. Hollow parts molded of polyethylene. *Materials in Design Engineering*, February 1963, 57(2):70.

MacAdams, J. L. Predict properties of rotomolded parts. *Plastics Technology*, June 1976, **22**(6):45.

Mobay Chemical Co. *Merlon Polycarbonate–Rotomolding*. 1973.

Modern Plastics. Rotomolding: on the brink of something big. September 1976, **53**(9):69.

———. Big push starts in powder rotomolding. March 1968, **45**(3):64.

———. Rotational molding machine has continuous four-spindle operation. December 1968, **45**(12):44.

———. New dimensions in rotomolding. April 1966, **43**(8):86.

———. Powder molding is starting to move. July 1961, **38**(11):80.

Nickerson, J. A. Rotomolding comes of age. *Machine Design*, November 12, 1970, **42**(21):164.

Plastics Design & Processing. Recent advances in rotational molding. May 1971, **11**(5):14.

Plastics World. Will Whittaker spawn a new generation of rotomolders? January 1971, **29**(1):33.

———. Rotational molding. January 1968, **26**(1):55.

Product Engineering. Rotation molding gets a boost from new approach. September 9, 1968, **39**(9):83.

Ramazzotti, D. J. Rotational molding–a quick update. *Plastics Engineering*, December 1975, **31**(12):32.

———. How to plan a rotational molding facility. *Plastics Technology*, January 1972, **18**(1):19.

Rosato, N. New machinery: vertical indexing rotational molders. *Plastics World*, September 1968, **26**(9):66.

Sawyer, A. T. Jr. For rotomolding: after tanks, then what? *Plastics World*, April 1973, **31**(4):36.

U.S. Industrial Chemicals Co. *Symposium II on Rotational Molding with Linear Powdered Polyethylene*. Bulletin PL-25, September 1966.

———. *Symposium I on Rotational Molding*. Bulletin PL-13, 1964.

———. *Molding Macrothene Polyethylene Powder*. Bulletin PTD-36, April 1963.

USM Corporation. *USM Rotary Matic Rotational Molding Machines*.

Wittman, C. A. Rotational molding: bigger parts, faster production. *Materials Engineering*, October 1969, **70**(4):40.

15
Foam Processes

PLASTICS FOAMS

Foamed plastics materials have been assuming more and more importance in the plastics industry, and there are a number of reasons to account for this. First is probably the extreme versatility of the foamed materials. Plastics foams may be found in the form of very soft cushioning materials used in upholstery, clothing interlayers, automobile seats, vibration absorbers, etc. Semirigid foams are used for flotation devices, marine bumpers, special electrical insulation on television cables, packaging, and a host of other applications. Rigid foams of various materials are used in the production of airplane parts, boats, electronic encapsulation, and in many furniture applications where wood was formerly used. In addition to the use of plastics foams as special electrical insulation, the foams are also used as thermal insulation and as acoustical insulation, either as sound-absorbing material or as barriers to the transmission of sound.

The versatility of plastics foams or cellular materials is not only expressed in the many forms in which the materials are made, but also in the many processes which are used to produce the foams. These include the production of various shapes by casting, extruding, injection molding, thermoforming, and reaction injection molding (RIM). The shapes may be foamed in place to produce slabs, sheet stock, or profiles, or, in the case of a special type of foam known as syntactic foam, may be formed in place by a mixing and troweling action. Foam versatility also includes the fact that any plastics material, thermoplastic or thermoset, may be produced in a cellular form.

Another extremely important factor contributing to the increasing use of rigid foams for all sorts of applications is that frequently a part may be made lighter, stiffer, or both, when a foam is used in place of a solid plastics part. By substituting a foamed material for a solid part made either of plastic or in some cases metal, the part may be improved and/or made more cheaply, simply because of the savings in material quantity and weight.

Definitions

There are a number of terms used in the foam industry which are either peculiar to foams or are used in a sense somewhat different from that of normal usage. Some terms have also been adapted from another industry or product. This section provides definitions of these terms.

1. *Foamed, cellular, or expanded plastic:* any plastics material which has been made with a cellular structure so its final density is lower than the original, solid material. Foamed, sponged, expanded, or cellular plastics are all names used for materials which are made in "densities" (see definition 3) ranging from 0.5 to approximately 40 or as high as 60 lb/ft^3 (60 pcf).
2. *Syntactic foam:* a special type of plastics foam in which the cellular, or foamlike, structure is formed by the use of small, hollow spheres made of phenolic resin, clay, glass or other hollow particles. The spheres, also called microspheres, are dispersed throughout a matrix of either epoxy or polyester resin which when cured resembles a conventional foam plastic.
3. *Density*: The weight of the foam in pounds per cubic foot (pcf) (in contrast to the classical meaning of density as the weight divided by the volume). Thus a 6-pound foam would weigh (have a density) of 6 pcf. Commonly available foams range from 2 to 20 pcf. Syntactic foams, however, usually range from 10 to as high as 60 pcf, depending on the type and the percentage of microspheres used in the resin.
4. *Closed-cell foam* (unicellular): a foam in which the greatest proportion of the cells are *not* connected together by open passageways. Such a material (in a flexible grade) when squeezed under water absorbs water only in the surface cells and a few cells under the surface. Closed-cell foams are used for flotation devices and similar items. They would not be used satisfactorily for sound absorption.
5. *Open-cell foam* (interconnecting cells): a foam in which the greatest proportion of the cells are connected by open passageways to other cells. Such a foam will absorb water throughout its mass when squeezed under water. It would also be a good sound-absorbing material.
6. *Compression resistance*: another method classifying foams is in terms of the force needed to compress one square inch of flexible foam 25%. In the case of rigid foam, the force in psi needed to cause permanent crushing is usually called the compressive strength.
7. *Compression set*: a term which is exclusively applied to flexible foams and/or solid rubberlike materials. It refers to the percent of permanent compression in a sample after a definite sized specimen has been compressed for a given period of time at a given tempera-

ture. Common test procedures utilize a one square inch sample compressed 25% for 22 hours at 70°C (158°F), and then measured 20 minutes after removal from the oven and the compression test fixture.

8. *K factor*: refers to the heat insulating properties of the material. It indicates the thermal conductivity of the foam in BTUs per hour, per square foot, per inch of thickness, under a thermal difference of 1 degree Fahrenheit. In general, plastics foams have K factors ranging from 0.15 to 0.35 at room temperature. As the temperatures increase, the K factors increase. When test temperatures are not stated it is assumed the K factor refers to room temperature tests. It should be understood that the lower the K factor, the better the insulating qualities are. Thus, a foam with a K of 0.2 is better than one of 0.35. (Another method of classifying foams is in accordance with the R factor. "R" indicates the resistance of the material to the transmission of heat. $R = \frac{1}{k}$. Thus, the higher the R factor, the better the insulating properties.)
9. *Blowing agent*: the particular material which causes any given plastic to foam. Two types of blowing agents are in common use; (1) gas which is introduced into the molten or liquid plastics material, or (2) chemicals which are incorporated into the plastic and which, at a given temperature, decompose to liberate gas. In either case, the gas, if evenly dispersed, expands to form the cells in the plastic. There are a number of different ways to bring about the cell formation, depending on the gas being used, the chemical blowing agent (CBA), the type of plastic resin, and/or the particular process being used.

General Production Methods

Plastics foams are principally made by seven different methods:

1. Incorporating a chemical blowing agent (CBA) into the polymer to form a gas by decomposition at a given elevated termperature. Usually these CBAs are in the form of fine powders which can be evenly dispersed in either a liquid resin or mixed with molding pellets. The usual gas is nitrogen liberated from organic materials known as azo compounds. A typical CBA is azobisformamide (ABFA), also called azodicarbonamide. CBAs are available which decompose at temperatures from 110°C (230°F) to as high as 280°C (537°F). Thus, a CBA may be selected to match any polymer melting point or processing temperature desired.
2. Injecting a gas, usually nitrogen, into a molten or partially cured resin. The gas may be injected into the resin either in the barrel of an extruder or injection press, or into a large mass in an autoclave. In either case, when the pressure is decreased, the gas expands and forms the cellular structure.
3. A bifunctional material, such as an isocyanate, may be combined with a polyester or other liquid polymer. During the polymerization reaction to form a solid polymer, the isocyanate also reacts to liberate a gas which forms the cells. This is the basis of some polyurethane foam techniques.
4. Volatilization of a low-boiling liquid either by the heat liberated by an exothermic reaction, or by externally applied heat. Commonly used liquids are fluorocarbons (Freons). This is the usual technique in production of polyurethane foams.
5. Whipping air into a colloidal resin suspension and then gelling the porous mass. Foamed latex rubber is made in this manner.
6. Incorporating a nonchemical, gas-liberating agent into the resin mix, which when heated releases a gas. Such a material could be a gas adsorbed onto the surface of finely divided carbon.
7. Expansion of small beads of a thermoplastic resin by heating an internally contained blowing agent. This technique is employed for expansion of polystyrene beads used to make cups, packaging, mannequin heads, etc.

Specific Foam Materials: Types, Production Methods, and Usage

FLEXIBLE FOAMS

Polyurethanes

Urethane foams demonstrate the extreme versatility of the foam materials better than any other plastics family. Among the urethanes are found the very soft, cushioning stock, the semirigid foams, and the very hard, rigid materials. The densities of these materials can range from as low as 1 pcf, or lower, to as high as 60 pcf, corresponding to a completely solid material.

Polyurethane foams, in general, are produced by the reaaction of a diisocyanate and a polyol or a mixture of polyols which are hydroxy terminated. Such materials are polyesters or polyethers, or even castor oil. The reaction between the isocyanate and the polyol produces the urethane resin. A further reaction between a slight excess of isocyanate and water, added to the polyol, then produces carbon dioxide gas to form the cellular structure of the foam. This was the original method of producing polyurethane foams developed by Germany shortly before World War II. The simplified formulas shown below illustrate the reactions.

$$\underset{\text{isocyanate}}{R{-}N{=}C{=}O} + \underset{\text{hydroxy compound}}{R'{-}OH} \longrightarrow \underset{\text{urethane}}{RNHCOOR'}$$

The second reaction to produce the cell-forming CO_2 gas is:

$$\underset{\text{isocyanate}}{R{-}N{=}C{=}O} + \underset{\text{water}}{H_2O} \longrightarrow \underset{\text{substituted carbamic acid}}{RNHCOOH} \longrightarrow \underset{\text{amine}}{RNH_2} + \underset{\text{carbon dioxide gas}}{CO_2\uparrow}$$

It should be noted that the above reaction, in which the isocyanate is reacted with a small amount of water to form the gas, is now considered obsolete. In its place a low-boiling, inert liquid fluorocarbon, is mixed in with the polyol. When the polymerization reaction takes place, the exotherm is sufficient to cause the fluorocarbon to volatilize and act as a blowing agent to form the cells. It should also be noted that a typical urethane formulation, in addition to the isocyanate, polyol, and blowing agent, will also have incorporated, usually in the polyol, a catalyst and a surface-active agent. The catalyst is, in most cases, an organo-tin compound, and is used to accelerate the reaction so gelation will take place before the gas is dissipated. The surface-acting agent, called a surfactant, is usually a silicone liquid and is used to obtain a uniform, fine cell structure. These are the basic ingredients of a polyurethane foamant, which is usually a two-part mixture. There may also be included certain modifiers to impart special characteristics such as fire retardance, high strength, colors, etc.

The isocyanates which are used today include: toluene-2,4-diisocyanate, known as TDI; diphenylmethane-4,4′-diisocyanate, MDI; and polymethylene polyphenylisocyanate, Papi. In what is known as an 80/20 mixture,* TDI is used to make both flexible and rigid foams. MDI is used mainly for rigid foams and in applications where lower toxicity is desired than can be obtained by the use of TDI. It is also frequently mixed with TDI. Papi is used for production of high-temperature-resistant foams, fast-gelation foams which can be sprayed against vertical walls, and for rigid foams with the highest dimensional stability.

While the isocyanates are mainly limited to the three types cited (other isocyanates are used very infrequently), the range of polyols is almost limitless. The polyols can be short- or long-chain materials with few or many crosslinking sites. By the appropriate choice of the polyol, the foam can be "tailored" to produce a flexible foam, a semirigid material, or a rigid material of many possible densities, depending on the amount of blowing agent, the catalyst used, the type of surfactant, etc. Polyols with low molecular weights and many sites for crosslinking will produce rigid foams, while long-chain resins with few crosslinking sites produce flexible foams. Table 15-1 illustrates two foam formulations. In one case, the polyol-to-isocyanate ratio is 1:1 by weight; in the other, the ratio is over 2:1. The flexible foam is made with a polyol having fewer reactive sites per mole—thus, the greater weight required to fully react with the isocyanate.

---

*80/20 mixture = 80%, 2,4-TDI

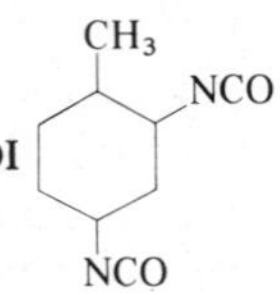

and

20%, 2,6-TDI

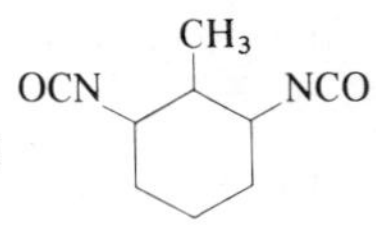

Table 15-1. Representative Polyurethane Foam Formulations.*

Rigid Foam			
Component A	100.0 parts	Isocyanate	
Component B	100.0 parts	Polyol	65.0 parts
		Surfactant	1.0 parts
		Catalyst	1.5 parts
		Blowing Agent	32.5 parts
Flexible Foam			
Component A	46.5 parts	Isocyanate	
Component B	102.6 parts	Polyol	98.0 parts
		Surfactant	1.0 parts
		Blowing Agent (water)	3.5 parts
		Amine Catalyst	0.1 parts
Component C	2.2 parts	Polyol	2.0 parts
		Organotin Catalyst	0.2 parts

*Sweets, 1967. Reprinted by permission from *Plastics Design & Processing*. Copyright 1967, Lake Publishing Corporation, Libertyville, Ill.

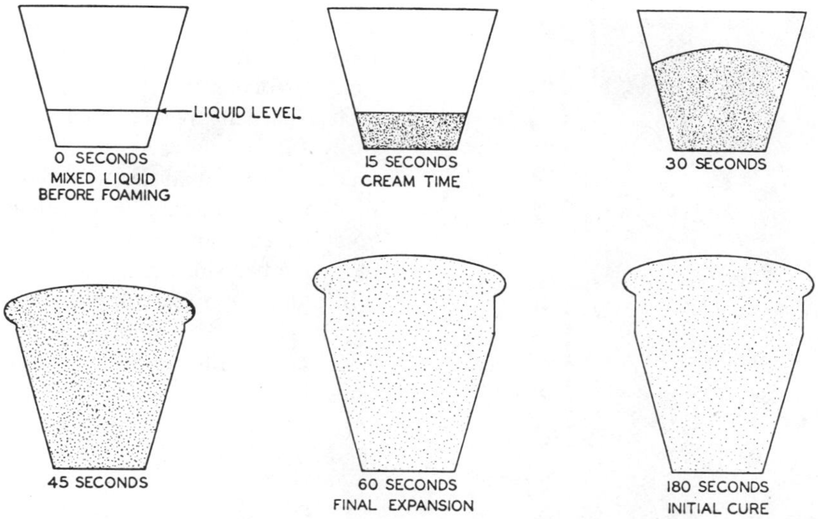

Fig. 15-1. Illustration of the unrestrained rise of urethane foam from a clear mixed liquid in the bottom of a cup to a full cure. (Reprinted by permission from *Plastics Design & Processing*, December 1967. Copyright 1967, Lake Publishing Corporation, Libertyville, Ill.)

Foams made from polyethers are currently finding more acceptance than polyester foams because of better molding qualities, greater softness and resiliency, and better resistance to hydrolysis conditions.

The foamant can be supplied as "one-shot" material, in which case all the ingredients are mixed at one time. This was the original foam process. However, because of the toxicity of the TDI (lung and eye irritation), methods were developed to tie up the TDI and so reduce the volatility. Such formulations are known as prepolymer formulations; part of the polyol is reacted with an excess of the TDI to form the prepolymer. The prepolymer is still a liquid and is later reacted with the balance of the required polyol, catalyst, and other components to form the foam.

Depending on the particular formulation, urethane foams may be compounded to gel or "cream" in very short periods of time (see Figure 15-1), or may take several minutes to react, so the materials can be mixed and then poured into a container such as an airplane trim tab, boat rudder, etc. Equipment for foaming can be nothing more elaborate than a bucket, scales, and a drill motor for mixing. On the other hand, for large-scale production, automatic metering, mixing, and conveying equipment is required to produce the large slabs or buns which are made for slicing into cushions, packaging, filters, etc. Other types of equipment are used where the formulations are made so the foam can be sprayed on walls for thermal insulation. In any case, it is extremely important that the materials be weighed very accurately, that the mixtures be mixed for the same time and at the same temperature each time, and that the curing conditions be uniform so that homogeneous material will be made continuously, or batch to batch. Figure 15-2 shows a typical metering and mixing unit and an accompanying heated hose and spray gun. Figures 15-3 and 15-4 illustrate the unit being used for packaging and for remedial roofing.

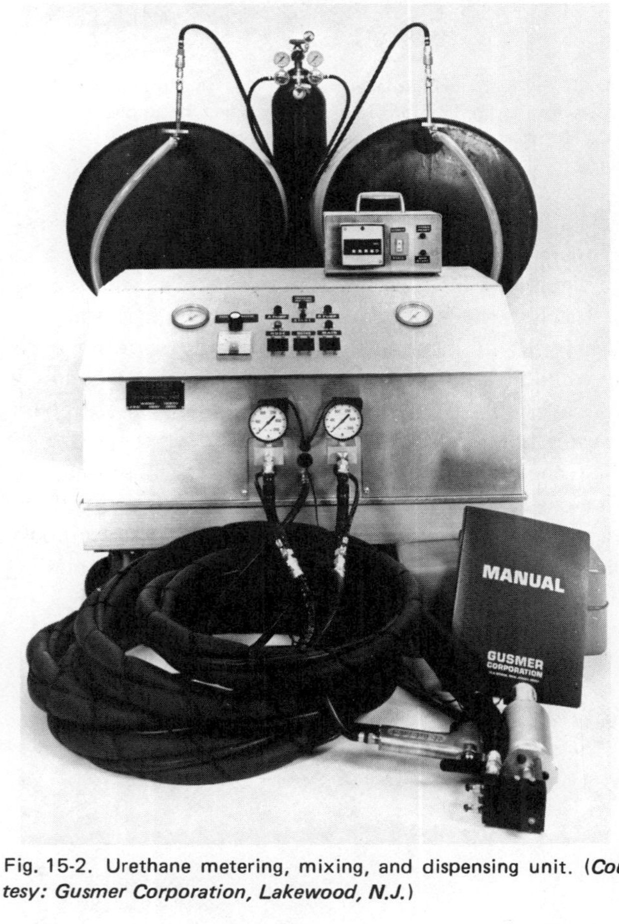

Fig. 15-2. Urethane metering, mixing, and dispensing unit. (*Courtesy: Gusmer Corporation, Lakewood, N.J.*)

The metering and mixing unit shown in Figure 15-2 would normally be called a "pour-in-place" unit, since in operation the material is simply mixed and poured (or sprayed) into a cavity or onto a surface. For most applications, including

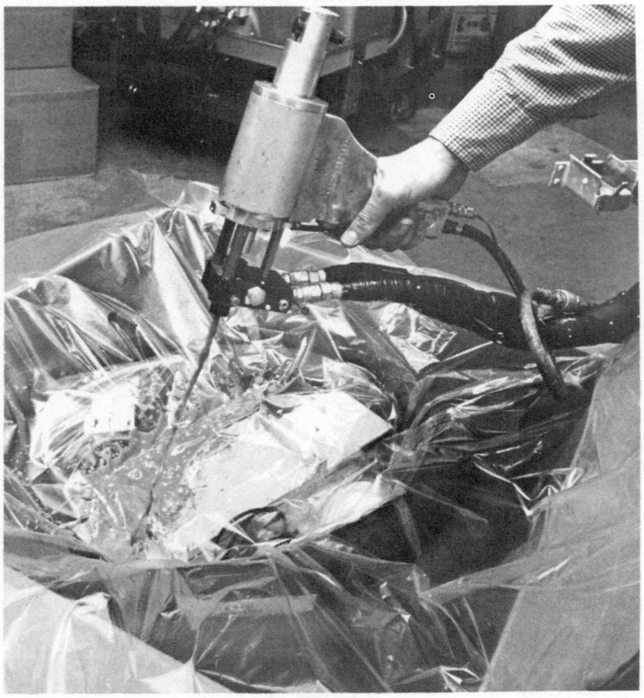

Fig. 15-3. Foamant being placed in container to form 1/2 pcf foam for a packaging application. Note how perfect fit and cushioning results. (*Courtesy: Gusmer Corporation, Lakewood, N.J.*)

those illustrated, such a unit is entirely satisfactory. However, there are applications where the foam is required to rise a significant distance between two vertical walls spaced close together. Such an application might be insulating the vertical walls of a refrigerated truck body. However, the foam will not rise the 6 to 8-foot distance to give a homogeneous layer in one shot. Even with two or more pours homogeneity may still not be achieved.

The technique used to obtain a low-density, homogeneous foam (1.5–2.0 pcf) is known as "frothing." In the frothing process the equipment is very similar to that used for pour-in-place, except that Freon-12 is introduced into the mixing

Fig. 15-4. Sprayed polyurethane foam being-applied as remedial roofing. (*Courtesy: Gusmer Corporation, Lakewood, N.J.*)

chamber as a liquid, and, because of the reaction exotherm, the mixed material is ejected from the gun as a frothy foam, similar in consistency to an aerosol shaving cream. With such a partially reacted material it is possible to satisfactorily fill narrow cavities with a lightweight foam that has high insulating qualities. It should be noted that operation of the froth dispensing equipment is considerably more complicated than the pour-in-place machine and requires a more highly trained and skilled operator.

In preparing a foam-in-place polyurethane, whether made by an automatic proportionating machine or by means of scales and a drill motor, there are a number of factors, besides the choice of isocyanate and polyol, which can influence the final density and uniformity of the foam. Some of these factors are:

1. The size of the pour. Generally, the larger the pour, the lower the final density—using the same material. This is because the exothermic heat can not be dissipated as rapidly from a large pour as from a small one. Thus, in the large pour, the gas expands more and produces a lower-density foam.
2. Mold construction and geometry. When a mold is made of an insulating material, such as wood, plaster, plastic, etc., the density of the foam will be somewhat lower than if the mold were made of metal. The mold made of the insulating material tends to retain the exothermic heat of the reaction, thus causing greater expansion of the gas, with a consequent lower-density foam.

   On the other hand, if the mold is made of a metal, which is a good heat conductor, or if the mold is cooled when the foamant is placed in it, the reaction will be inhibited. The effect will be to cause a denser foam to be formed, at least on the surface of the foam contacting the mold. By the correct choice of mold temperature, foam ingredients, special additives, etc., the inhibition can be enough to completely block foaming for an appreciable thickness of polymer. Thus, a part can be made with an integrally molded skin on the outer surface. Such parts are used for automobile upholstery armrests, dashboard insulation, etc.

   The same type of effect can be obtained by some mold geometries; that is, thin-cavity molds, because of faster heat loss, result in higher-density parts than the same foamant placed in a thicker-cavity mold.
3. Preheating of molds and foam components. Preheating either the molds or the materials tends to help produce lower-density foams. Conversely, cooling the molds or the components tends to produce higher-density foams.
4. Mixing techniques. In general, the better or more intensively the formulation can be blended, the lower the foam density which is obtained.
5. Restraint pattern. The amount of restraint in a given

mold helps determine the final density; that is, more restraint usually results in a higher-density foam.

Urethane foams, both flexible and rigid, may be fabricated into various parts from slab stock by all the usual techniques. These include band saws, serrated knives, contour cutters, die cutters, and hot wire cutters. Foams may be shaped by heat forming, using a die heated to approximately 260°C (500°F), and may be welded using dielectric or conventional heat-sealing equipment. Of course, a number of cements, including rubber, epoxy, or urethanes, may be used to adhere the foams to themselves or to other surfaces. It should also be remembered that the foam, during formation, is also a good adhesive and will stick to walls of the mold or container unless a parting agent is used.

Foamed Vinyls

Open-Cell Vinyl Foams. Foamed or cellular vinyls are usually made from plastisols. Two methods are used to form predominantly open-cell foams: (1) a chemical blowing agent technique, and (2) a mechanical frothing process, in which a gas is also used as part of the blowing mechanism. Closed-cell, predominantly rigid, foam is also made using chemical blowing agents.

In preparation of a soft, open-cell vinyl foam, using a chemical blowing agent, the plastisol is first chosen for the characteristics desired; that is, the resin and plasticizer ratio would be one in which a sufficient quantity of plasticizer assures production of a soft, flexible product. To the plastisol is added a paste made of a powdered blowing agent dispersed in a plasticizer. The chemical blowing agent (CBA) could be any one of a number of organic materials, which when heated decompose to liberate a gas, usually nitrogen. One class of materials used for a large majority of foams is the azocarbonamides, and other azo compounds, which can be obtained to decompose at temperatures of from approximately 120–220°C (250–425°F). Thus, a CBA can be picked to accommodate practically any required resin fusion temperature. It is extremely important that, at the gas evolution temperature, the viscosity characteristics of the plastisol allow complete and even expansion, and that a uniform cell structure is formed. Factors which influence these properties are the particle size of the blowing agent, the type and quantity of the resin and plasticizer, and the specific type of stabilizer and/or activator used to initiate decomposition of the blowing agent. Soft, very flexible vinyl foams used for garment insulation, upholstery, and similar applications are made with this process. Somewhat harder foams used as underlays for rugs and flooring can also be made with this technique, but require different resins and lower plasticizer contents.

The second technique used to produce a large amount of vinyl foams, particularly sheets such as flooring underlay, wall coverings, and other applications requiring relatively close thickness tolerances, is the mechanical frothing process. In this method, the plastisol is mixed with a given amount of air in a high-shear, temperature-controlled mixing head. The resulting product, resembling shaving cream, is cast onto a belt or fabric and knifed to a definite thickness. Passage through an oven or heating tunnel then causes fusion of the plastisol. Fusion temperatures for all the plastisol foams range from 125–160°C (257–325°F). The times may be from 1 to 3 minutes for thin sheets of from 10 to 20 mils, to as high as 7 to 10 minutes for materials as thick as $\frac{3}{4}$ inch. Thicknesses over 1 inch present a problem, because the insulating properties of the foam prevent the heat from reaching the center of the foam in a reasonable length of time. Dielectric heating can, however, be used very successfully to produce foam as thick as 2 inches in a very short period of time.

Closed-Cell Vinyl Foams. Closed-cell foam is made in a manner somewhat similar to the technique used to produce open-cell foam with chemical blowing agents, except that much higher pressures are used, and the process is accomplished in two steps. First, the vinyl plastisol, containing the blowing agent, is placed in a mold in which very little space is left for expansion. The mold is then heated, causing decomposition of the blowing agent and, at the same time, fusion of the foam. This step raises the internal pressure in the mold to anywhere from 2000 to as high as 16,000 psi. At these high pressures, the gas is dissolved in the resin in the form of microscopically small bubbles. The part is then allowed to cool and harden in the mold before being removed.

The second step consists of reheating the molded part at approximately 93–120°C (200–250°F), at which time the resin softens and the gas expands to form a closed-cell foam. With this technique it is possible to produce foams with a density as low as 2 pcf. Because of the two-step procedure, the process is considerably slower than foaming open-cell materials. Consequently, it is used mainly for specialty items such as marine flotation parts, special athletic mats, and similar items. Since the parts made by the process are all made at high pressures, another name for these materials is "pressure sponge."

Cellular Polyethylene

Polyethylene is the only other high-production foam material which is truly flexible. However, polyethylene foams are considerably less flexible and resilient than the most flexible of the urethanes. Consequently, they are not used for upholstery applications, but are found in packaging, marine bumpers, sporting goods padding, etc. Densities commonly range from 2 to 10 pcf.

Cellular polyethylene is usually produced by extrusion. In this technique, a blowing agent is mixed in with the polyethylene pellets, and, during the mixing and shearing in the barrel, the pellets melt and the blowing agent decomposes. The liberated gas dissolves in the molten resin, but since the mass is under high pressure it remains as a solid. On extrusion, the gas expands and forms the cellular material. To date, extrusion is the process most used to produce bulk or raw stocks of polyethylene foam, mainly in the form of planks, rods, or billets with a cross-section up to approximately 50 square inches (see Figure 15-5).

Crosslinked Polyethylene Foams. Another type of polyethylene foam is crosslinked foam, made by either using a chemical crosslinking agent or using radiation. In the chemical process, which is not amenable to extrusion, polyethylene pellets are first blended with the crosslinking agent, usually a peroxide, and then blended with the blowing agent. Blending can be done on a two-roll rubber mill, or using a screw-type blender. After blending, the material is placed in a mold and then heated to a temperature first high enough to cause crosslinking to take place and then, at a higher temperature, to cause decomposition of the blowing agent.

Radiation crosslinked foam is made by first blending the component materials and then extruding the mixture as a thin, ($\frac{3}{8}$-inch maximum) sheet. The sheet is then subjected to electron beam radiation, after which it is heated to cause foaming of the sheet. The thickness of the sheet is limited by the inability of the radiation to penetrate too deeply, and the width of the sheet is determined by the die size and the radiation equipment. Sheets up to 48 inches wide are commonly made.

The characteristic chemical, thermal, and mechanical properties of the extruded and crosslinked foams are in general very similar. The crosslinked sheets, however, have

Fig. 15-5. Extruded polyethylene foam products. (Printed with permission of The Dow Chemical Co.)

a somewhat smaller and more uniform cell size and have a slightly softer feel than does the extruded material.

Cellular polyethylene has been used, mainly in the form of extruded rods, discs sliced from rods, sheets, etc., for applications such as spacers, lightweight bumpers on boats, and bottle caps. The major application, however, is primary insulation on communication cable such as coaxial cables, television lead-ins, and high-frequency applications. Foamed polyethylene has all the inherently good properties of solid polyethylene, such as chemical, moisture, and abrasion resistance, toughness, and flexibility. There is a considerable advantage in the use of the cellular material for high-frequency applications. The foam has a dielectric constant of approximately 1.5, compared to 2.3 for the solid stock, which results in lower losses.

Polyethylene-Related Foams. Related to the polyethylene foams are foamable ethylene copolymers. These materials are furnished as already compounded foamable materials which can be processed either by extrusion or by injection molding. These copolymers offer the advantages of low rigidity and good toughness at normal and low temperatures. The other physical properties are, in general, similar to those of comparable-density polyethylenes.

Ionomer foams are also related to the polyethylene foams, both in chemical structure and in properties. The ionomer foams, however, are only available in the form of sheet stock. The stock is distinguished by an extraordinarily tough integral skin, as well as high tensile strength, high tear strength, and excellent low-temperature flexibility.

Silicone Foam

Flexible silicone foams are unique in that they expand and cure at room temperature. For this reason they are known as RTV (room temperature vulcanization) foams. They are supplied as two-part mixtures, a resin and a curing agent, usually used in the ratio of approximately 5 parts of curing agent to 100 parts of the liquid silicone. After thorough mixing, of the resin and curing agent, the foam starts to expand and cure within about 5 to 10 minutes. Cure is complete in 24 hours at 25°C (77°F).

The foam is usually elastomeric, with a typical density of 12 pcf if free blown and about 16.5 pcf if blown in a mold. The material has a tensile strength of about 20 psi and elongation of about 75%. The compression-deflection load for 25% compression is approximately 1-2 psi at room temperature. Like all silicones, the foam has very good resistance to both high and low temperatures. One unique use for these foams is as medical prostheses; for example, as breast implants.

RIGID FOAMS

Rigid foams are made of the same types of polymers used to form flexible and semirigid foams, except that these materials are substantially nonreturnable when compressed; that is, compression of more than a few percent usually causes catastrophic failure of the foam. Rigid foams are usually made in the lighter densities of from 2 to 6 pcf. They are also usually closed-cell materials, good for heat insulation, but not too satisfactory as sound absorbers or insulators.

Rigid foams, like the flexibles, can be made by a variety of processes. These include cast-in-place, injection, extrusion, reaction injection molding (RIM), and spray-up. The spraying process is almost exclusively used for rigids in such applications as wall insulation, roofing, and even house construction. Of particular interest in the production of rigid foams is the fact that such foams can be made with integral skins ranging from very thin to, relatively, very thick. The thick-skinned foams are mainly used for structural foam applications, in which case they are also usually made in the higher densities.

Materials are available as slab stock; cast, injection molded, or extruded into special shapes; or as sprayed-on coatings.

Polyurethanes

Polyurethane rigid foams are made using basically the same general chemistry employed in formulating the flexibles or the semirigids. However, a higher percentage of crosslinks per molecule (compared to molecules of flexible foam) is used, thus creating greater rigidity. Polyester polyols are very frequently used for the rigid foams, which therefore have good bonding characteristics. The low-density materials, frequently used for thermal insulation, are made with fluorocarbon blowing agents, resulting in foams with densities of approximately 1 to 4 pcf and K factors as low as 0.12. The thermal efficiency of these materials, compared to other common insulations, is shown in Figure 15-6. (The maximum temperature shown for the urethanes is 120°C (250°F), for the most common materials.)

One of the major characteristics of the rigid foams is the relatively high compressive strength developed by the

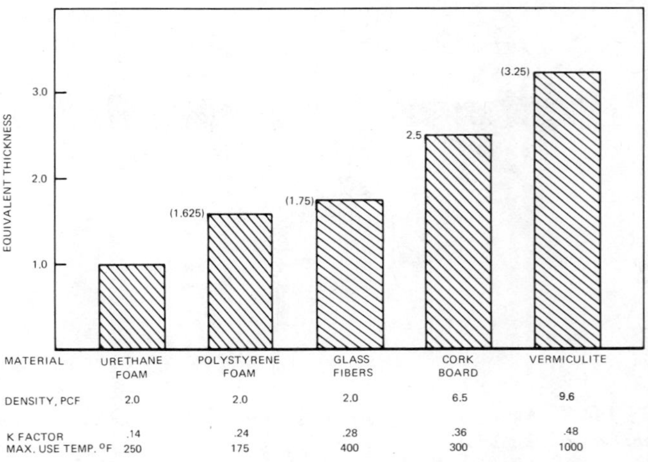

Fig. 15-6. Thermal efficiency of various insulation materials.

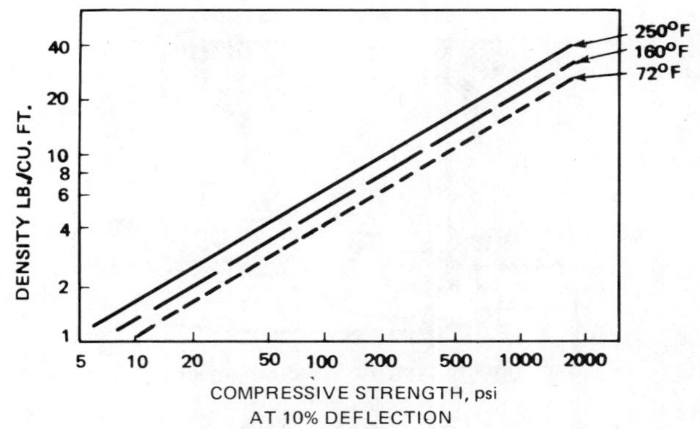

Fig. 15-7. Load-bearing characteristics of typical rigid PU foams.

higher-density foams. Figure 15-7 shows typical compressive strengths available from various densities. It is these high strengths, coupled with either an integral skin or a skin of wood or metal, which make such composites particularly attractive to aircraft designers or designers of other transportation media.

Typical liquid foam formulations, using water as the blowing agent and morpholine as the catalyst, are shown in Table 15-2. Note the increase in density as the amount of water is decreased.

Possibly one of the fastest growing processes for the production of parts of rigid or semirigid urethane foam is the reaction injection molding process, also called RIM. In this process, the two ingredients, polyol (containing the catalyst and blowing agent) and isocyanate, are metered and injected, at a high pressure, into an impingement mixing chamber and, later at a lower pressure, into a mold. Figure 15-8 shows a simplified representation of the process.

Between shots, the metering and dispensing unit continuously circulates both components at a thermostatically controlled temperature through the mixing chamber. The circulation can be at a low or a high pressure. Figure 15-9 shows a typical metering unit. From the metering-dispensing unit, the liquids are circulated through the impingement mixing head, shown in Figure 15-10. In the head, the only moving parts are a valve and a hydraulic actuator, as shown in Figure 15-11. In the circulation mode (Figure 15-11a), materials continually circulate, maintaining the correct

Table 15-2. Rigid Polyurethane Foam Formulations.

	Parts by Weight				
Polyester polyol	100	100	100	100	100
Isocyanate	170	140	117	103	95
Water	4	2	1	0.35	–
Tween 21	1	1	1	1	1
N-methyl morpholine	1	1	–	–	–
Density, pcf	2.0	3.3	5.0	10	15
Compressive strength, lbs/in² at yield point	25	55	100	275	575

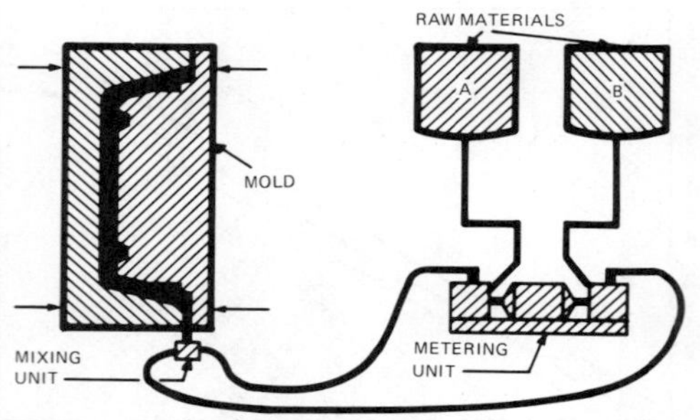

Fig. 15-8. Diagrammatic sketch of reaction injection molding (RIM).

temperature, pressure, and the head at instant readiness for the foam shot. On actuation (Figure 15-11b), the sliding valve cuts the flow to the return ports and the swirling liquids mix in the impingement chamber and are then pumped into the mold.

This process can be used to very rapidly produce diverse parts with an integral skin such as automobile bumers, dashboards, business machines housings, furniture, and similar parts (see Figures 15-12, 15-13, and 15-14). Since low pressure (25–50 psi) is used in the mold, large-area parts can be made using inexpensive aluminum, Kirksite, or even epoxy molds. The clamping forces required are also relatively low.

Differing from conventional injection molding presses, the mold clamp is usually a separate unit from the metering and pumping assembly, connected, however, by high-pressure hoses which feed the mixing head. In molding very large parts it is extremely important that gas entrapment be minimized. To insure this, the clamping press is usually

Fig. 15-9. Metering and dispensing unit to supply materials to the mixing head. (*Courtesy: Battenfeld Corporation of America*)

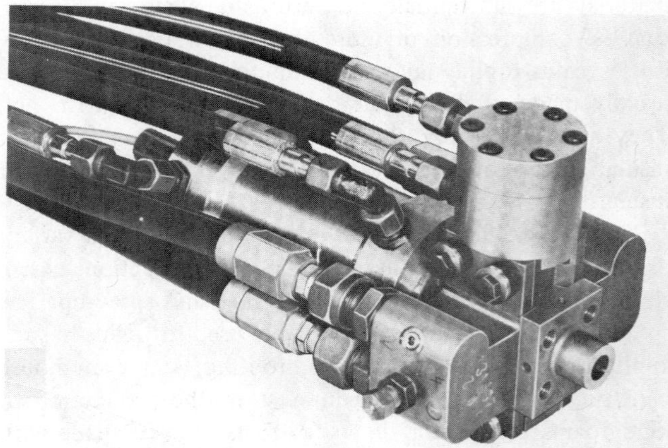

Fig. 15-10. Impingement mixing head. (*Courtesy: Battenfeld Corporation of America*)

made so the mold can be tilted and rotated, after clamping and injection, to allow the gas to escape from the highest point in the mold. Figure 15-15 illustrates a typical mold-clamping unit which allows rotation in two planes.

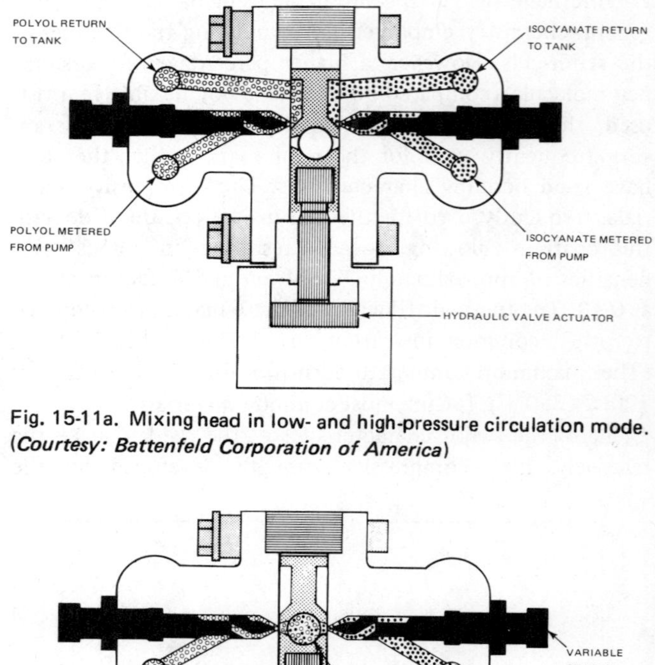

Fig. 15-11a. Mixing head in low- and high-pressure circulation mode. (*Courtesy: Battenfeld Corporation of America*)

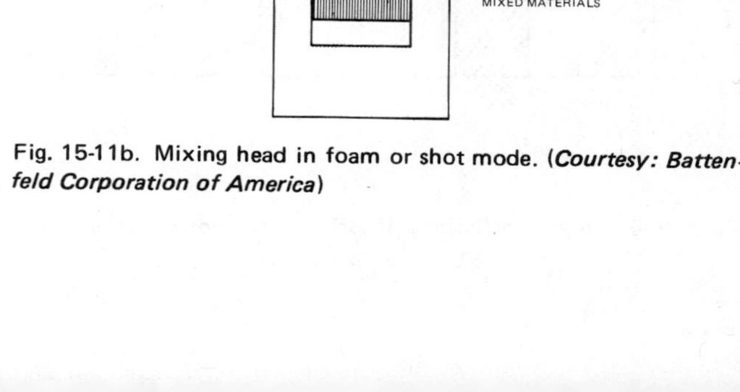

Fig. 15-11b. Mixing head in foam or shot mode. (*Courtesy: Battenfeld Corporation of America*)

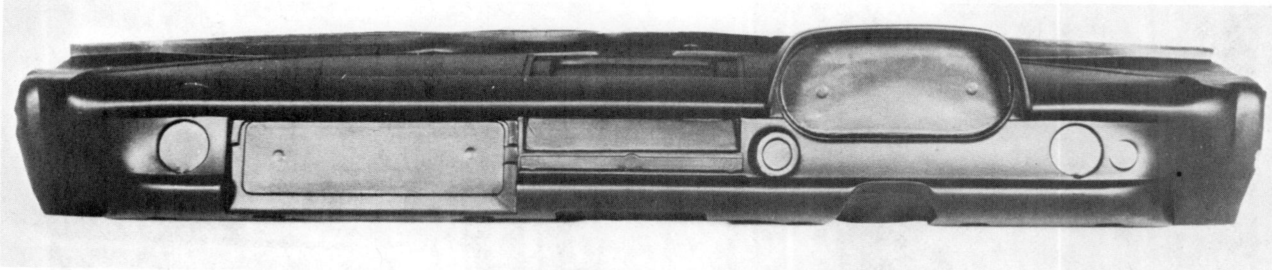

Fig. 15-12. Padded dashboard for automobile made by RIM process. (*Courtesy: Battenfeld Corporation of America*)

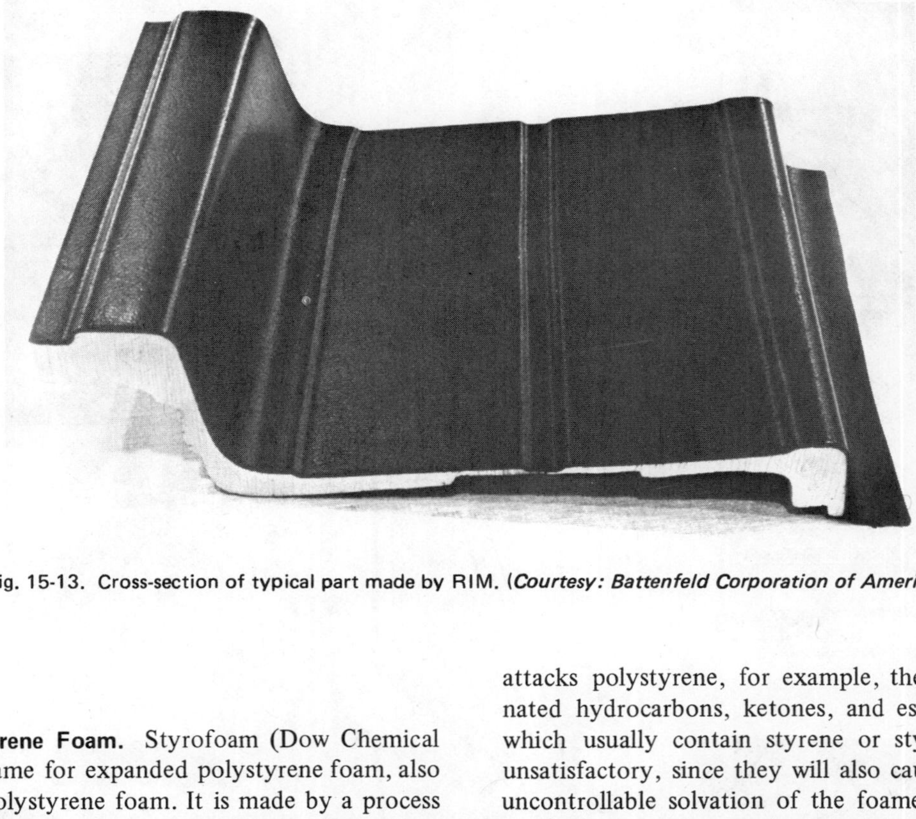

Fig. 15-13. Cross-section of typical part made by RIM. (*Courtesy: Battenfeld Corporation of America*)

Polystyrenes

Expanded Polystyrene Foam. Styrofoam (Dow Chemical Co.) is the tradename for expanded polystyrene foam, also called extruded polystyrene foam. It is made by a process similar to that used for production of foamed polyethylene; a blowing agent is mixed with the polystyrene pellets and the materials are extruded through a die in the form of rods, sheets, or boards. The materials have a closed-cell structure with very low thermal conductivity and low moisture absorption. Boards, sheets, and rods are made in densities of 2.0, 3.0, and 4.5 pcf. The cross-sectional areas of the boards are limited by the size of the die or the capacity of the machine. Table 15-3 lists properties of the extruded foam, and Figure 15-16 shows Styrofoam boards as extruded and trimmed.

Like cellular polyethylene, foamed polystyrene can be fabricated with all common woodworking tools, hot wire cutters, etc. Tools must be kept very sharp for best results. Bonding of styrene foam can be satisfactorily done provided the adhesive does not contain a solvent which readily attacks polystyrene, for example, the aromatic or chlorinated hydrocarbons, ketones, and esters. Polyester resins which usually contain styrene or styrene derivatives are unsatisfactory, since they will also cause rapid and usually uncontrollable solvation of the foamed surface. Adhesives containing water or similar non-solvents are satisfactory if used in applications where the solvent can evaporate. Adhesives which set by chemical action, such as phenolics, epoxies, resorcinols, ureas, etc., are very satisfactory for this application.

One of the serious limitations of polystyrene foam is its rather low maximum operating temperature of approximately 80°C (175°F). It is, however, used very satisfactorily for thermal insulation in buildings, particularly for refrigerated warehouses. A self-extinguishing type is available. Major uses for this foam include insulation, flotation apparatus, decorations, and packaging (see Figures 15-17 and 15-18).

Thin, extruded polystyrene foam is made into thermoformed trays and cartons for packaging meats, eggs, and produce. Foam sheet of this type is made by injecting a

Fig. 15-14. Assortment of items made by RIM process. (*Courtesy: Battenfeld Corporation of America*)

liquid blowing agent directly into the plastic melt in the barrel of the extruder. Since the sheet is extruded "to size," it has a smooth skin which makes it particularly attractive for meat trays and egg cartons. Its heat-insulating properties also make it desirable as a container for hot sandwiches in fast-food restaurants.

Expandable Polystyrene. Expandable polystyrene is the usual name for the beads used to make molded polystyrene foam parts, and the parts themselves are frequently called molded foam polystyrene. Instead of the extruded and chopped particles or pellets usually used as plastic molding granules, tiny spheres of polystyrene are used. The beads or spheres are approximately $\frac{1}{32}$–$\frac{1}{16}$ inch in diameter, and each contains a small quantity of a hydrocarbon liquid, usually propane, which is used as the blowing agent.

Cellular structures produced with these beads are usually made by a two-step process. The first step consists of a

Fig. 15-15. RIM clamp. Note capability for rotation in two planes. (*Courtesy: Battenfeld Corporation of America*)

Fig. 15-16. Trimmed Styrofoam boards. (Printed with permission of The Dow Chemical Co.)

Fig. 15-17. Styrofoam used as a flotation device. (Printed with permission of The Dow Chemical Co.)

pre-expansion of the virgin beads by heat. The heat may be by steam, hot-air oven, radiant heat, or even hot water. Steam, however, is the most used medium. In the pre-expansion step, the beads are brought to almost the density required in the molded part. After the pre-expansion step, the beads are stored for 6 to 12 hours to allow them to reach equilibrium. The beads are then conveyed to the mold, usually by air, where they expand to the final dimensions. Steam is the preferred mode of heating, either by introducing live steam through perforations in the mold

Table 15-3. Properties of Extruded Styrofoam.*

Average Properties (1)	ASTM TEST METHOD	STYROFOAM BRAND PLASTIC FOAMS			DEVELOPMENTAL PRODUCTS		MOLDED POLYSTYRENE FOAM**	POLYURETHANE*** FOAM (Unfaced)
		IB	SM/TG	RM	XFS-43001	XFS-4249		
Physical Properties								
Density (lb/cu ft)	C303-56	1.7	2.0	2.3	2.4	2.3	1.0	2.0
Compressive Strength (2)	D1621-73	40	40 (3)	45 (3)	60 (3) (min)	45	12	23
Water Resistance Properties								
Water Absorption (% by volume)	D2842-69	1	1	1	1	1	3	1-5
Test Precision, ± 1%								
Water Vapor Permeability (perm-inch)	C355-64	0.9	0.6	0.4	0.4	.4	3.0	2.0
Thermal Properties								
Thermal Conductivity "K" (Btu/hr/sq ft/°F/in)								
1. Aged (Design) Values (5)	C177-76E (4) and/or 76E (4)							
at 25°F mean	C518	0.22	0.175	0.175	0.175	0.175	0.24	0.17 (7)
at 40°F mean		0.23	0.185	0.185	0.185	0.185	0.25	0.16
at 75°F mean		0.25	0.20	0.20	0.20	0.20	0.27	0.16***
2. Fresh (as manufactured) Values (6)								
at 40°F mean		0.16	0.11	0.11	0.11	0.11	0.25	0.11

*Courtesy: Dow Chemical Co.

Fig. 15-18. Tongue-and-groove Styrofoam for construction purposes. (Printed with permission of the copyright owner, The Dow Chemical Co.)

itself, or by means of steam probes which are withdrawn as the beads are expanding. The final expansion of the beads fills the voids originally made by the probes. During the expansion, the beads melt together slightly, adhering and, at the same time, forming a smooth outer skin exactly mirroring the mold contours. Figure 15-19 illustrates the two-step process.

Cooling of the expanded part is the last step in the process. Because the material is an excellent thermal insulator, it takes a relatively long time to remove the heat from the part sufficiently so it can be extracted from the mold without distortion. Cooling is usually done by a water spray on the mold. In order to assure easy removal of the part, mold-release agents are frequently used.

Expandable polystyrene foam can be easily distinguished from expanded polystyrene foam by observing the outer surface of the part. The round bead outlines can be readily observed, usually even on the smooth outer surface. If the part is cross-sectioned the round beads can be clearly seen. The expanded foam, on the other hand, shows a smooth skin on the outer surface, if it has not been trimmed. If the surface is a trimmed edge, a true foam is seen; that is, thin walls and open cells.

Expandable foam has the same elevated-temperature limitations as the expanded product; thus, it is also widely used where low-temperature insulation is required, such as in refrigerated structures and coolers. Since the expandable foam can be molded into discrete shapes, it is a common material in molded products like portable ice chests and beer carriers. The expandable foam is also used for flotation devices in the form of inexpensive molded surf boards, small boats, and miscellaneous floats. Packaging devices such as molded containers for acid bottles and fragile electronics parts, bulk packaging beads or chips are other

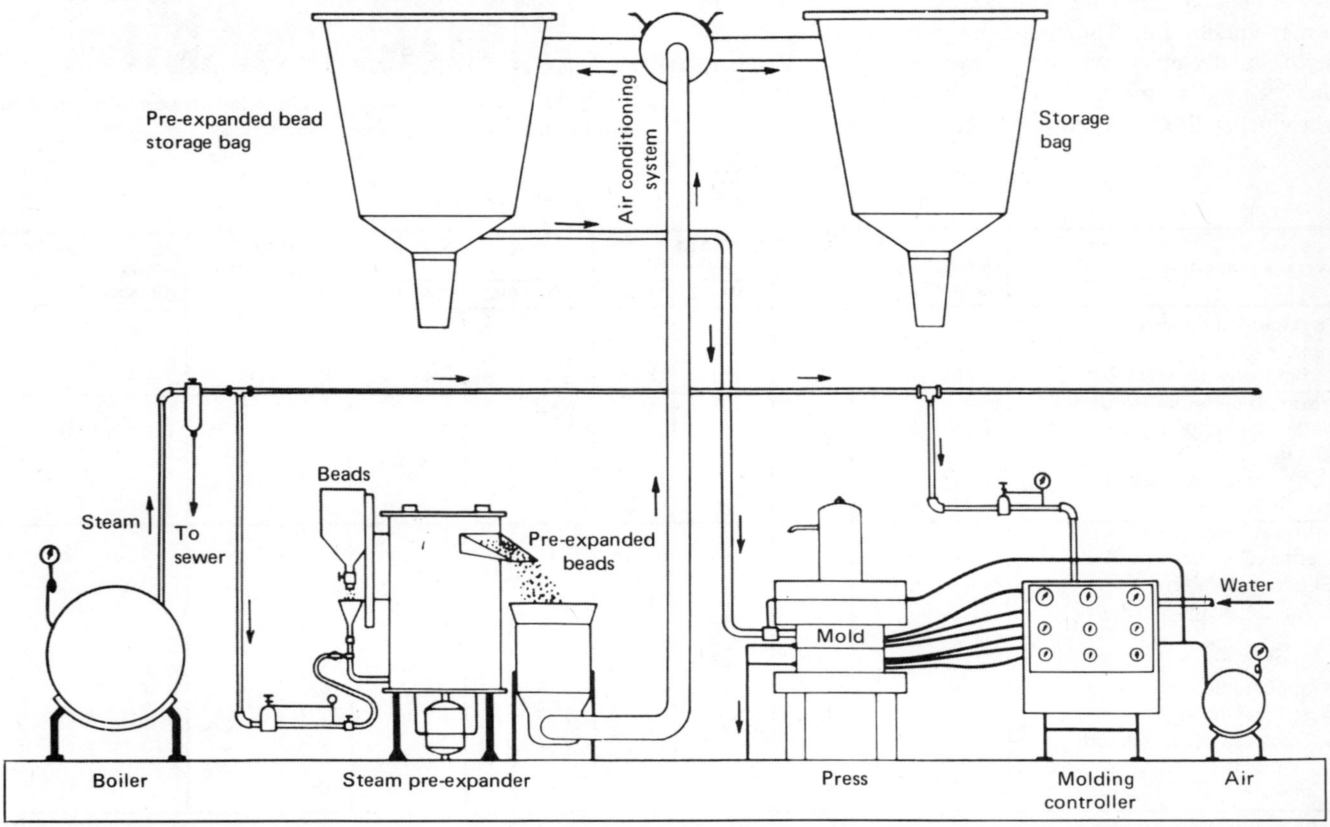

Fig. 15-19. Flowchart for expandable PS foam molding first shows pre-expansion of beads to required density. Next, beads are conveyed by air to storage bags for aging and later loaded into molds for final expansion. (Reprinted by permission from *Plastics Technology*, July 1964)

common applications. Since the material can be molded to virtually any size, it is also possible to have boards considerably wider and thicker than would be possible with the expanded foam.

Cellular Cellulose Acetate

Cellular cellulose acetate is also known by the abbreviation CCA or the tradename Strux (Aircraft Specialties Corporation). It is made in a manner somewhat similar to the one used in the production of expanded polystyrene. A mixture of molten cellulose acetate, dissolved in a volatile solvent, is extruded through a die to produce boards ($1'' \times 4''$, $\frac{3}{4}'' \times 6''$, and $\frac{1}{2}'' \times 4''$) and $2\frac{1}{4}$-inch diameter rods. The density of the foam ranges from 6 to 8 pcf. The material, in physical properties, is similar to the higher-density expanded polystyrene, except that it has a considerably higher temperature operating range, being useful up to 175°C (350°F). It has somewhat better chemical resistance than Styrofoam, except that care must be taken with adhesives containing esters or ketones. Although higher in cost than polystyrene foam, it is used in many of the same applications; for example, flotation devices, aircraft sandwich structure cores, and similar applications where its special properties are considered necessary.

Thermosetting Foams

Foams made from thermosetting resins are similar to thermoplastic foams in appearance and general physical properties. Heat resistance is usually better than most thermoplastic foams, but the materials are somewhat more brittle. Thermoset foams are made using the same general processes as those for thermoplastics; that is, at some point in the procedure the resin is liquid, heat is used to activate the blowing agent, and the foam is then allowed to cool and harden. The major difference is that the heat may be generated from the heat of reaction, and the "hardening" may be the result of the polymerization of the resin. (In these respects, the reactions are the same as for urethane foams—which may or may not be thermosets depending on the molecular chain length or the degree of crosslinking in the molecule.)

Epoxy Foams

Epoxy foams are made in two types. One is a powder foamant which contains powdered epoxy resin, such as Shell Epon 1031, and powdered diaminodiphenyl-sulfone, the hardener, and a powder CBA. The mixture is heated and liquefied, and the CBA causes foaming after which the curing agent causes hardening. The other is a liquid foamant which uses liquid resins, curing agents, blowing agents, and surfactants. The curing agents chosen are usually ones which develop a highly exothermic reaction so little or, in some cases, no external heat is required. The foams range from 2 pcf to as high as 38 pcf. Like the urethanes, the foams, may be handled a short time after cooling, perhaps as little as 15 minutes; to achieve full strength, however, the materials should be postcured for several hours as recommended by the manufacturer. The low-density foams (2 to approximately 7 pcf) are usually serviceable up to temperatures as high as 82°C (180°F), while the higher-density materials have been used up to 200°C (400°F). Epoxy foams have been made for use as pour-in-place materials, either as powders or liquids, and as sprayable materials. Their excellent electrical characteristics and adhesive properties have led to their use in many electrical and electronic applications as potting and encapsulating agents.

Phenolic Foams

Phenolic foams are typically made by catalysis of a liquid phenolic resin to which has been added a small amount of a blowing agent and a silicone surfactant. The condensation reaction of the phenolic resole during polymerization liberates a small amount of water which, along with the blowing agent, is volatilized by the exothermic reaction, thus causing the polymer to foam. The catalyst is usually an acid such as sulfuric, hydrochloric, or phosphoric for ambient temperature foams, and a milder organic acid such as benzenesulfonic or toluenesulfonic where it is possible to add heat to the reactants. Formulations containing strong mineral acids such as sulfuric must be used with caution because they will severely corrode any metals they may contact.

The foams can vary in density from as low as 0.3 pcf to as high as 50 pcf, although the average is 2–4 pcf. Density variations, using the same resin, can be made by changing the amount or type of catalyst, blowing agent, or surfactant. The resin type may also be changed. The reaction can be carried out in a cream time of as fast as 1-2 seconds and a foam rise time of 50–60 seconds. A full cure, however, is usually not realized for at least 24 hours at ambient temperature and possibly an additional 8 to 10 hours at 60–95°C (140-200°F) for a postcure, depending on the resin.

Low-density phenolic resins are distinguished by their extreme brittleness. Other physical properties, however, are in line with foams of similar densities. The low-density foams are also characterized by a low K factor, a property which makes them particularly useful as thermal insulation, especially for "foam-in-place" insulation. Ordinary phenolic foams are combustible and have a tendency to "punk" (continue glowing and burning without a visible flame) after the original source of combustion is removed. However, blends of boric/oxalic acid catalysts are said to considerably increase the time before combustion takes place, and to eliminate the "punking" effect after the ignition source is removed.

Another characteristic peculiar to phenolic foams is the high degree of cell interconnections. Thus, a low-density foam will absorb water throughout its structure in a short time after being immersed. One application that takes advantage of this property is a self-watering mount for floral displays. One method of inhibiting the water absorption and/or furnishing the foam with a hard, protective skin is to cast the foam in a mold whose temperature is lowered to the point where the resin, in immediate contact with the mold, does not exotherm. An alternative technique is to cast the foam between plywood, metal, plastic, or paper skins.

Urea-Formaldehyde Foams

Urea-formaldehyde foam (UFC) is another type of thermosetting foam made by a condensation reaction in which the urea-formaldehyde resin is mixed with air, an aqueous detergent, and an acid catalyst. The foam is commonly generated by mixing the air in with the resin and the catalyst in an applicator machine. Depending on the strength and amount of catalyst, the foam, which resembles a froth on emerging from the applicator gun, will set up, with no exothermic heat, in 10 to 120 seconds. Full hardness and complete drying, however, will take from several days to a month or more, depending on the temperature, humidity, and permeability of the surrounding walls. Prior to setting up, the foam can be made to flow into crevices, around pipes, or can be hand troweled to the desired shape or surface condition.

The finished foam is a material with an approximate weight of 0.7 pcf, with very low compressive strength. The dried foam has a nominal thermal conductivity (K factor) of 0.2 at 21°C (70°F) and 0.18 at 2°C (35°F). When used at thicknesses of $1\frac{1}{4}$ to 2 inches thick or more, it is said to be a good sound insulator. One of its most interesting properties is its relative noncombustibility, since a temperature in excess of 650°C (1200°F) is required to cause the material to decompose. On removal of the source of ignition, the material immediately ceases burning or decomposition. The gases released act as a fire extinguisher during decomposition. The foam has low water vapor permeability, but, because of its partially interconnected cells, it will hold considerable quantities of water for slow release.

Because of its fragility, the material is mainly in use as home thermal and sound insulation where it is installed *in situ*, and in mines as a fire-stop material. It has also been used as packaging for lightweight parts, and for such diverse applications as raising sunken ships. In the United States the material is known as Rapco-Foam (Rapperswill Corporation).

Syntactic Foams

Syntactic foams are made using a resin matrix to which has been added hollow spheres of various materials. The resulting product is a foamlike material which is made without the use of a blowing agent. The most common matrix resins are epoxies and polyesters, although urethanes, PVC plastisols, and phenolic resins have also been used. Indeed, any polymer which can be made liquid, either before final polymerization or by heat, can be used as the binding matrix. The "foaming agents" can be glass microspheres, with diameters from 20 to 200 microns, or other spheres with diameters ranging from 0.05 mm to 2 cm, which are classed as macrospheres. The materials used to form the spherical particles are glass, phenolic resins, silica, and naturally occurring materials such as perlite and coal dust. The most used materials are glass and phenolic spheres.

The syntactic foams are made by simply mixing the micro- or macrospheres into the catalyzed resin until the desired consistency is obtained. In most cases the materials are mixed to a puttylike state or, if a casting material is desired, to a state in which the material can just be cast. The usual ratio of filler to resin is approximately 60% filler by volume.

A summary of the physical and environmental properties of the various foam-in-place, extruded, expanded, and syntactic rigid foams is given in Tables 15-4 and 15-5. Table 15-6 shows the relative costs (1965) and processing factors for the various foams.

STRUCTURAL FOAM

Structural foam is the term usually used for foam produced in an injection molding press and made of almost any thermoplastic resin. The major difference between structural foam and the other foams discussed is that the structural foam is produced with a hard integral skin on the outer surfaces and a cellular core in the interior. A second difference is that structural foam is used almost exclusively for production of molded parts, and rarely for production of raw stock; that is, slabs, bars and billets.* With some polymers, parts made by the RIM process are also considered to be of structural foam.

Structural foam parts made by injection molding employ the same principles of foaming used in all other foam processes discussed; that is, the resin is liquified by heat, and then the blowing agent, either a gas or a CBA, causes cell formation when pressure is decreased as the resin mixture is injected into the mold. On cooling, the finished part is ejected. Depending on the conditions of foaming, and mainly on the amount and type of blowing agent used, the foam can range from an almost solid resin to a material with a density approximately 75% of the solid resin. Figure 15-20 illustrates density variations with changing amounts of blowing agent in an injection-molded foam.

In Figure 15-20 it should be noted that increasing the amount of blowing agent reduces the density of foams **a** through **d**. However, foam **e**, with a thickness of $\frac{3}{8}$ inch, has a density of 0.8 with the same amount of blowing agent as

*Profile shapes made by extrusion are the exceptions.

Table 15-4. Properties of Rigid Foams.*

Property	Density, lb/cu ft		Structure % closed	Thermal conductivity BTU/in/hr/sq ft/F		Expansion coeff, Fx10⁻⁵	Compressive modulus, psi		Compressive strength, psi		Shear strength, psi		Tensile strength, psi		Flexural strength, psi		Dielectric constant		
ASTM standard	D1622			C177			D1621		D1621		C273		D1623		C203		D150		
Density	min	max		min	max		min	max	min	max	min	max	min	max	min	max	min	max	cps
Urethane																			
CO_2 blown	1.0-1.5	65	40-100	0.21	1.0		300	350,000	10-20	18,000	10-15	2,100	15-20	4,000	30	13,000	1.1	4.5	10^7
Freon blown	1.5	15-20	80-95	0.11	0.37	3-6	300	25,000	15-25	1,500	10-20	200	10-20	900	40	1,500	NA	NA	
Froth	1.5	4	60-95	0.11	0.16		300	3,000	15	100	11	65	10	125	30	50	NA	NA	
Polystyrene																			
Extruded	1.3	4.5	85	0.24	0.33	4.0	500	5,000	10	140	15	95	35	185	32	170	1.05	—	10^3
Molded	0.5	10	100	—	0.77	3.0	280	—	8	200	13	90	31	220	50	330	1.02	1.16	10^3
Self-expanded	5.0	10	—	0.24	—				45	120									
Epoxy																			
Pre-formed	5.0	38	90 (fine)	0.65	—	0.85			50	6,000					200	4,500	2.0	7.0	16^6
Pack-in-place	15	25	90 (fine)	0.24	0.80	0.85			600	3,000					500	2,500	2.5	7.0	10^6
Foam-in-place	2.0	8	93 (fine)	0.11	0.50	2.3	570	14,300	13	110	38	360	—	—			1.55	—	10^6
Silicone																			
Heat activated powder	12	18	10-60 (fine)	0.3	—	—			25-70	325							1.26		10^9
Room-temp liquid	3.0	20	10-60 (fine)	0.16	0.3	—			8	115							1.08	1.20	10^5
Phenolic																			
Lowest density	0.3	1.0	—	0.22		1.3			1	5	5		10		20				
Low density		2.0	10	0.20		—			5	25	14		20		40		1.5		10^{10}
Medium density		4.0	—	0.20		—			40	100	30		42		90				
High density	15	22	50	0.30		2.5			300	1,100	90		920		260				
Syntactic																			
Phenolic spheres	10	68	closed	0.2	1.2														
Epoxy filler							100,000	500,000	200	25,000			100	13,000	150	21,000			
Polyester filler							20,000	50,000	200	15,000			30	2,400	200	3,500			
Glass spheres	15	68	closed	0.3	1.2														
Epoxy filler						0.5-2.0	200,000	500,000	500	25,000			150	13,000	400	21,000	1.45	3.9	10^{10}
Polyester filler						1.0-10	40,000	500,000	300	15,000			45	2,400	400	3,500			
Epoxy spheres, epoxy filler	10	25	closed	0.6-0.8		2.2-3.6	—	—	400	1,200									

*Gerstin, 1965. Courtesy: *Product Engineering*, Morgan Grampion Publishing Co.

Table 15-5. Environmental Characteristics.*

Property	Maximum temperature, F			Water resistance				Vibration resistance	Fungus and vermin attack	Chemical resistance, degree of attack							Effect on materials in contact
	Short-term	Continuous	Flame resistance	Vapor trans, perms/in.		Water absorption, lb/sq ft, 2 in. hd				Chlorinated solvents	Ketones, esters	Aromatic hydrocarb.	Alcohol	Oxygen	Acids and bases		
															Dilute	Conc.	
ASTM Standard			D1692	C355		D2127 for 24 hrs.			D1924								
Density				min	max	min	max										
Urethane																	
CO_2 blown	600	450	Burns; self-	0.04	6.0°	0.015	0.10°	Varies									
Freon blown	350	250	extinguishing	0.3	6.0-8.0	(48 hr immersion		No effect	None	Degrades on prolonged attack		None	None	None	None	Degrades except for nonoxidizing	None
Froth	350	250	with additives	1.0	6.0	under 10 ft head)		No effect									
Polystyrene																	
Extruded		175	Burns; self-	0.04	1.7	0.04	0.20	Degrades									
Molded		175	extinguishing	1.1	4.5	0.02	0.15	Damps	None	Destroys		Degrades	None	Dis-colors	None, to dis-color	Degrades	None
Self-expanded		—	with additives	—	—	0.01	—	—									
Epoxy																	
Pre-foamed	—	500	Self-extinguishing			0.1% weight									None	Degrades	
Pack-in-place	—	600	Self-extinguishing			gain per		No effect	None	Degrades			None		None	Degrades	None, except on flavors
Foam-in-place	450	350	Burns			MIL-P-12420									Degrades	Destroys	
Silicone																	
Heat-activated powder	650		Does not burn	—		2.3-3.2		Degrades under severe vibration	None	Degrades				None	Acids discolor, bases degrade	Acids degrade, bases destroy	None
Room-temp liquid	650		Self-extinguishing	41.2		0.28											
Phenolic																	
Density to 2.0		250	Self-extinguishing,	71		—		Degrades,			Swells		Swells				
4.0		250	but burns under	63		—		very brittle	Heavy	None	Swells	None	Swells	None		Destroys	Attacks ferrous, stains brass and copper
20		200	certain conditions	NA		0.1% weight gain		Degrades			None		None				
Syntactic																	
Phenolic spheres			Self-extinguishing with additives														
Epoxy filler										None		None	None	—	Dis-colors	Degrades	
Polyester filler									None	Degrades							
Glass Spheres	400		Self-extinguishing														No effect on metal
Epoxy filler										None		None	None	—	None		
Polyester filler										Degrades					Discolors		
Epoxy spheres, epoxy foam	450	350	Burns					No effect		Degrades	None	None	None	None	Degrades	Destroys	

°Except open-cell polyester

*Gerstin, 1965. Courtesy: *Product Engineering*, Morgan Grampion Publishing Co.

Table 15-6. Cost and Production Characteristics.*

	Material cost $/lb	Number of systems available	Number of components	Maximum dimensions of part	Skin	Blowing agent	Reaction rate	Exotherm F	Curing temp, F	Production methods: Foam-in-place	Spray	Mold	Mold material	Workability and fabrication methods
Urethane	0.45-3.00	Unlimited	2-3	Board stock 2 ft x 4 ft 1000 lb in high densities, 40 cu ft										
CO2 blown					Heavy	Chemical	2 sec to 5 min	350	260-300	yes	yes	yes	Paper to metal	Outstanding with hand or power wood-working tools
Freon blown					Heavy	Physical, chemical		350	Room temp					
Froth					Light		15-60 sec	250						
Polystyrene Extruded or molded	0.45-1.00	8 or 9	Boards, billets, beads	20 in x 4 ft; extrusions, 12 ft long	0.01 to 1.0 in	Chemical			230-250 (molded)	no	no	yes	Metal	Outstanding; hot wire, thermoforming
Epoxy	Medium to high	Unlimited												
Pre-foamed			Sheet, rod	1 ft x 2 ft, 6 ft long	Slight	None	None	High	None	no	no	no**	None	Outstanding workability; wood-working tools, good machinability
Pack-in-place			Putty	not determined	None	None	None		None	no	yes	yes	Carboard	
Foam-in-place			1 or 2		None	Chemical	½-3 min		70-350	yes	yes*	yes	or metal	
Silicone														
Heat-activated powder	High	2	1	not determined	Slight, soft	Chemical	4 min	Slight	480	yes	no	yes	Paper	Friable; wood-working tools
Room-temp liquid	5 to 10	1	2			Physical	15 min		70	yes	yes	yes	or metal	
Phenolic	Low	5	1-3		None	Chemical	½-30 min	yes	200-250	yes	no	Powder	metal	Brittle at low density
Syntactic	0.30-6.00	Unlimited	2		None	None	varies	None	RT to 100	no	some	yes	any	Metal working tools

*Except powder **Post form 250-300 F

*Gerstin, 1965. Courtesy: *Product Engineering*, Morgan Grampion Publishing Co.

in **b**, which has a density of 0.93 and is $\frac{1}{8}$ inch in thickness. Thus, it may be seen that, as a rule, thin-wall parts require a higher percentage of blowing agent to obtain the same density as a thicker part. This is because of the cooling effect of the metal mold walls on the resin which inhibits, to some extent, bubble formation in material contacting the metal surfaces.

As previously stated, almost any thermoplastic resin, combined with a blowing agent, may be used to produce a structural foam. In all cases, the resin's original properties of heat resistance, chemical resistance, and most electrical properties remain the same as for the solid resin. (The dielectric constant of foamed material is improved over the solid resin—the reason for the use of foamed polyethylene for television cables.) Tables 15-7, 15-8, 15-9, and 15-10 show typical properties of commodity-type resins when foamed, and foamed with glass-fiber reinforcements. Table 15-11 shows typical properties obtainable with engineering-type thermoplastics. For a more extensive list of the properties, the resin manufacturer's literature should be consulted. Table 15-12 lists the common chemical blowing agents (CBAs) used with the various resins. The choice of whether a CBA or nitrogen gas is used depends on the type of equipment available.

Most structural foam is produced in the form of complete parts by injection molding, rather than as bulk raw stock as in many other foams. Since a complete part is made, with a skin on each outer surface, the process is ideally suited for fabrication of parts such as business machine housings, furniture sections, tote boxes, automobile dashboards, and similar parts or components in which light weight and stiffness are required.

Some of the distinct advantages in using structural foam are:

Table 15-7. Physical Properties of High-Density Polyethylene Structural Foam.*

Resin Density — 0.962 gm/cm³
Melt Index — 8.0 gm/10 min.

Sample Thickness	*1/4"*	*3/8"*	*1/2"*	*ASTM Test No.*
Density g/cc	0.6	0.6	0.6	
Flexural Modulus psi in 25°C air	120,000	112,000	118,000	D—790—66
Flexural Strength psi — 25°C — 5% Strain	2,730	2,450	2,510	D—790—66
Tensile Str/Elongation psi — 25° C/%	1310	1260	1390	D—638—64T
Compressive Strength psi — 10% Deflection				
Perpendicular to Skin	1,090	978	1,090	D—695—63T
Parallel to Skin	1,840	1,640	1,860	
Coefficient Linear Exp. x 10⁻⁵ in./in. °C	12.0	11.9	11.8	D—696—44 (61)
Heat Distortion Temp.				
°F @ 66 psi	129.6	138.2	131.8	
°F @ 264 psi	93.5°F	99°F	103°F	
Vicat Softening Point °F	254°F	244°F	244°F	D—1525—65T
Dielectric Strength Volts/mil.	309	217	194	D—149—64
Durometer Hardness "D" Scale	54	51	55	D—2240—64T
Un-notched Izod, ft.-lbs.	2.50	2.23	2.31	
Rod Drop Impact, ft.-lbs.	3.25	3.5	7.75	
Shrinkage Average — 0.0175 in./in.				

*Courtesy: Structural Foam Division, The Society of the Plastics Industry, Inc.

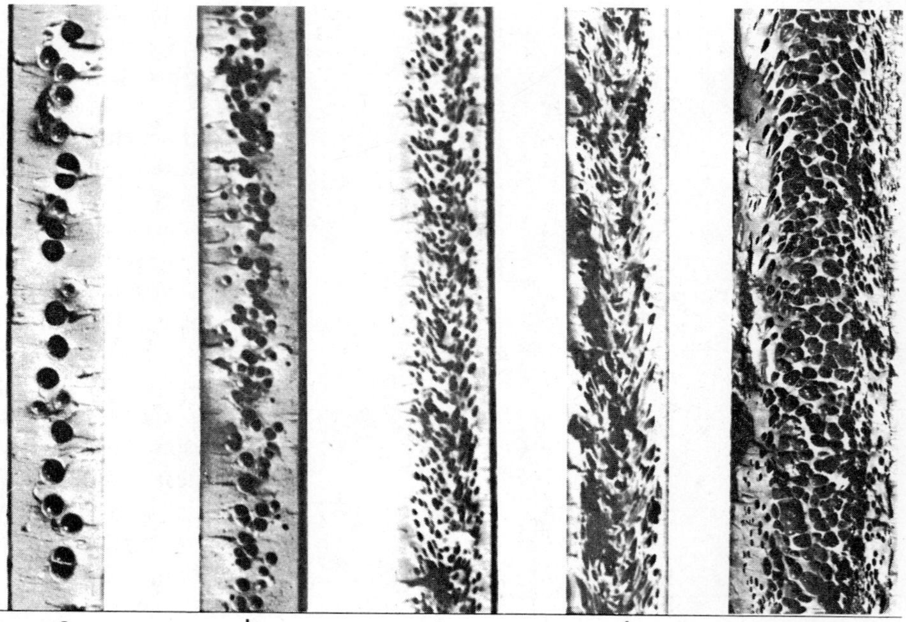

	a	b	c	d	e
Thickness, inches	⅛	⅛	⅛	⅛	⅜
Density	1.0	0.93	0.8	0.72	0.8
% Blowing Agent	¼	½	1	1½	½

Fig. 15-20. Blowing agent versus density. (*Courtesy: General Electric Co.*)

1. A weight reduction of approximately 25% is achieved with the same or greater stiffness in the part.
2. A savings in the amount of resin used is possible.
3. There is more rigidity in the part with the same weight.
4. Larger parts can be made using the same equipment, since lower molding pressures are used.
5. The parts produced are largely stress-free, with consequently less warpage.
6. Sink marks are minimized or completely eliminated.
7. Tooling can be considerably cheaper than if the parts were made using solid resin.
8. Existing injection molding presses can be used, although special foam molding equipment will be more efficient for larger parts.

Disadvantages associated with the production of structural foam include:

1. The surface finish, particularly in the material made by the low-pressure process, may show swirl marks.
2. Cycle times can be considerably longer than molding of comparable solid parts, because of the longer cooling time required.

Table 15-8. Physical Properties of Polypropylene Structural Foam.*

Sample Thickness	*1/4"*	*3/8"*	*1/2"*	*ASTM test No.*
Density g/cc	.6	.6	.6	
Flexural Modulus psi	120,000	110,000	105,000	D—790—66
Flexural Strength psi	3,200	3,000	2,500	D—790—66
Tensile Strength psi	2,000	2,300	1,500	D—638—65T
Compressive Strength				
Perpendicular to skin	1,600	2,300	3,200	D—695—63T
psi — 10% Deflection				
Parallel to skin	3,400	2,800	4,000	D—695—63T
Coefficient Linear Exp.				
x 10^{-5} in./in. °C	9.4	9.2	9.3	D—696—44 (61)
Heat Distortion Temp.				
°F @ 66 psi	167	162	184	D—648—56
°F @ 264 psi	132	115	118	D—648—56
Vicat Softening Point °F	301	286	302	D—1525—65T
Durometer Hardness				
"D" Scale	63	54	64	D—2240—64T
Notched Izod Impact				
ft.-lbs./in. of Notch	•	•	•	D—256—56
Un-notched Izod, ft.-lbs.	1.26	.99	.95	
Rod Drop Impact, ft.-lbs.	.75	2.25	3.75	

•Values obtained were too low to be meaningful in this test.

*Courtesy: Structural Foam Division, The Society of the Plastics Industry, Inc.

Table 15-9. Physical Properties of High-Impact Polystyrene Structural Foam.*

Sample Thickness	*1/4"*	*3/8"*	*1/2"*	*ASTM Test No.*
Density g/cc	.7	.7	.7	
Flexural Modulus psi	210,000	190,000	195,000	D—790—66
Flexural Strength psi	4,500	4,000	4,000	D—790—66
Tensile Strength psi	1,800	1,500	1,200	D—638—64T
Compressive Strength				
Perpendicular to skin	3,300	2,300	2,000	D—695—63T
psi — 10% Deflection				
Parallel to skin	5,200	3,700	3,600	D—695—63T
Coefficient Linear Exp.				
x 10^{-5} in./in. °C	9.0	8.4	8.6	D—696—44 (61)
Heat Distortion Temp.				
°F @ 66 psi	189	187	193	D—648—56
°F @ 264 psi	176	161	165	D—648—56
Vicat Softening Point °F	204	204	204	D—1525—65T
Durometer Hardness				
"D" Scale	74	70	70	D—2240—64T
Notched Izod Impact				
ft.-lbs./in. of Notch	1.05	1.10	1.00	D—256—56
Un-notched Izod, ft.-lbs./in.	2.3	1.25	1.14	
Rod Drop Impact, ft.-lbs.	1.25	2.25	6.75	

*Courtesy: Structural Foam Division, The Society of the Plastics Industry, Inc.

Table 15-10. Properties of Fiberglass-Reinforced Structural Foams.*

Property Unit	Tensile Strength PSI	Flexural Modulus PSI x 10	Flexural Strength PSI	Izod Impact Strength ft.lb./in.	Un-notched Impact Strength ft.lb./in.	Deflection Temperature Under Load @264 PSI F	Density gm/cc	Mold Shrinkage in./in
30% glass filled nylon	15,500	11.4	24,700	1.2	6.2	380	.992	.005-.011
20% glass filled styrene	5,000	7.5	8,500	0.6	1.5	190	.84	.002-.005
20% glass filled SAN	6,100	7.8	9,500	0.6	2.1	202	.85	.002-.005
20% polycarbonate	14,000	5.8	14,200	2.1	7.1	276	.94	.005-.009
40% polycarbonate	19,300	10.0	17,800	1.8	6.1	272	1.06	.003-.006
20% polypropylene	4,000	4.2	6,000	1.4	3.5	162	.67	.004-.010
20% ABS	7,000	7.5	12,000		3.5	210	0.84	.003-.008

Compliments of Dart Industries Fiberfill Division

*Courtesy: Structural Foam Division, The Society of the Plastics Industry, Inc.

Table 15-11. Typical Properties of Structural Foams Produced From Engineering Thermoplastics.*

Property (Metric)	Method of Testing	Modified Phenylene Oxide	Modified Polycarbonate	Thermoplastic Polyester
Specific Gravity, 73 F (22.8 C)	—	0.8	0.0	1.2
Heat Deflection Temperature	ASTM-D648			
F @ 66 psi (C @ 46,404.6 Kgs./sq. meter)		205 (96.1)	280 (137.8)	405 (207.2)
F @ 264 psi (C @ 185,618.4 Kgs./sq. meter)		180 (82.2)	260 (126.7)	340 (170.9)
Thermal Conductivity, BTU/hr./ft./F/inc. (Cal./hr./cm./C/gm.-cm.)	ASTM-C177	0.86 (1.06)	1.05 (1.30)	1.10 (1.36)
UL Listing**	UL Standard 94	94 V-O 94-5V	94 V-O 94-5V	94 V-O
Tensile Strength, psi (Kgs./sq. meter x 10)				
@ 73 F (22.8 C)	ASTM-D638	3,300 (2,320)	5,370 (3,775)	9,910 (6,967)
@ 200 F (93.2 C)	—	—	3,400 (2,390)	6,350 (4,465)
Elongation at Break, % @ 73 F (22.8 C)	ASTM-D638	15	3.9	1.3
Tensile Modulus, psi (Kgs./sq. meter x 10) @ 73 F (22.8 C) (Tangent)	ASTM-D638	226,000 (159,000)	275,000 (193,350)	1,028,000 (722,790)
Flexural Strength, psi (Kgs./sq. meter x 10)				
@ 73 F (22.8 C)	ASTM-D790	6,500 (4,570)	10,300 (7,241)	17,900 (12,585)
@ 200 F (93.2 C)	—	—	5,820 (4,092)	8,400 (5906)
Flexural Modulus, psi (Kgs./sq./meter x 10)				
@ 73 F (22.8 C)	ASTM-D790	260,000 (18,280)	325,000 (228,507)	1,000,000 (703,100)
@ 200 F (93.2 C)	—	130,000 (9,140)	270,000 (18,983)	390,000 (27,420)
Flexural Stiffness-to-weight ratio, @ 73 F (22.8 C)	—	100	124	160
Compressive Strength (10% deformation) psi (Kgs./sq./meter x 10)	ASTM-D695	5,000 (3,515)	5,000 (3,515)	11,300 (7,945)
Creep (300 hrs. 73 F (22.8) @ 1,000 psi) (703,100 Kgs./sq./meter) % strain	ASTM-D674	0.6	0.41	0.12
Charpy Impact, un-notched ft.lbs./in. (Kg./meter/cm.)	ASTM-D256	7.6 (1.05)	10.2 (1.41)	5.3 (.733)
Gardner Impact, ⅝" (1.58 cm.) Nose: 1" (2.54 cm.) Base lbs./in. (grams/cm.)	—	23 (4107.8)	56 (10001.6)	11 (1964.6)
Dielectric Strength volts/mil.				
Perpendicular	ASTM-D149	216	220	152
Parallel	—	192	204	200
Mold shrinkage, in./in. x 10- (cm. x 10-)	ASTM-D1299	6-8	5-7	—
Flow direction		—	—	4-6
Perpendicular Flow Direction		—	—	6-10

Compliments General Electric Company

*Courtesy: Structural Foam Division, The Society of the Plastics Industry, Inc.

Table 15-12. Commercial Organic CBAs for Structural Foam.*

Chemical	Decomposition Temperature °F (°C)	Use Temperature Range In S.F. °F (°C) [a]	Typical Usage Level In S.F. % [b]	Trademark•	Manu-facturer [c]
Azodicarbonamide	400 (204)	350-450 (177-232)	.5	Azocel	F
				Celogen AZ	UC
				Ficel AC	FIC
				Kempore	NPD
				Porofor ADC	B
				Vinyfor AC	E
Non-Plate-out Modified Azodicarbonamide	400 (204)	350-450 (177-232)	.5	Celogen AZNP	UC
				Ficel EP-A	FIC
				EP-B, EP-C	
				EP-D	
				Kempore MC	
P,P'-Oxybis (benzene sulfonyl hydrazide)	315 (157)	300-350 (149-177)	.75	Celogen OT	UC
				Ficel OB	FIC
				Neocellborn	E
				Nitropore OBSH	NPD
Proprietary Hydrazide	390 (199)	350-450 (177-232)	.75	Celogen CB	UC
P-Toluene Sulfonyl Semicarbazide	440 (227)	400-500 (204-260)	.75	Celogen RA	UC
5-Phenyl Tetrazole	480 (249)	480-600 (249-316)	.3	Expandex 5 PT	NPD
Trihydrazine Triazine Type	500 (265)	500-600 (260-316)	.4	Ficel THT	FIC
HT 550 (developmental proprietary)	500 (288)	500-650 (260-343)	.3	Celogen HT 550	UC

•Reg. U.S. Patent Office.
a. At the lower end of range activation may be required.
b. Actual use level will depend on part thickness, part design, desired density, molding conditions and other factors.
c. Code: B-Farbenfabriken Bayer, Leverkusen, Germany.
F=Fairmont Chemical Co., Inc., Newark, N.J. through Sobin Chemicals, Inc., Boston, Mass.
UC=Uniroyal Chemical, Div. of Uniroyal, Inc., Naugatuck, Ct.
E=Ewia Chemical Ind., Co. Ltd. Japan through A&S Corp., Verona, N.J.
FIC=Fisons Industrial Chemical Div., Fisons, Ltd. Loughborough, England.
NPD=National Polychemical Div., Stepan Chemical Co., Wilmington, Mass.

[1]The products listed herein are proprietary to the corresponding manufacturers and are presented here simply for illustrative purposes. SPI makes no warranty as to the merchantability or fitness for particular purpose for any of the listed products.

***Courtesy: Structural Foam Division, The Society of the Plastics Industry, Inc.**

Injection molding is the largest application of structural foam resins. Foamed, extruded profiles are also made into raw stock, in many cases as wood substitutes, for furniture trim and household moldings, picture frames, retainers for glass partitions, etc. Figure 15-21 shows some of the many injection-molded parts which can be made and the various materials used in the production of the parts.

Structural Foam Molding Techniques

Injection molding of structural foam is commonly done by two techniques: low-pressure molding and high-pressure molding. In each case, the pressures refer to the pressure in the mold and not to the molding barrel pressures, as is usually the case when referring to injection molding. In Chapter 10, a brief outline was given of the two processes; it should be reviewed in conjunction with this more detailed description of the processing techniques.

The Low-Pressure Process. The low-pressure process makes use of either specially made injection-molded presses, modified "regular" injection presses, or, even for small parts, standard injection-molded presses. In general, the process consists of extruding molten resin containing a chemical blowing agent or nitrogen gas under pressure into an accumulator(s) until sufficient resin for the shot has been collected. The liquified material (it is not foaming because it is under relatively high pressure) is then injected into the mold as a "short shot." Figure 10-63 shows the type of equipment used. Not shown are the large number of

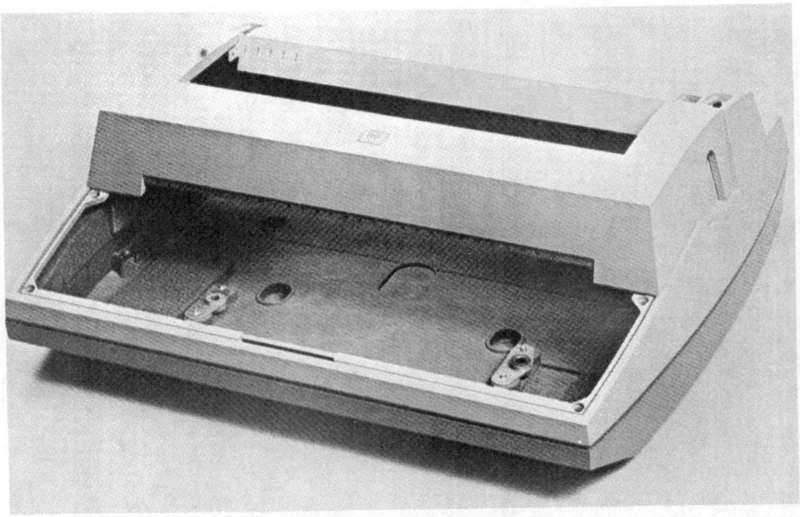

Fig. 15-21a. Business machine cabinetry in structural foam provides an economic alternative to metal fabrication, and insulates for sound, heat, and vibration. Bosses and other supporting members are molded in. Consoles shown are Redactron's Editing Typewriter (left) and Control Data's Terminal. Molder: Cahiers Plastics. Plastic: Modified PPO. (*Courtesy: Union Carbide Corporation*)

Fig. 15-21c. Cabinet front for Capehart stereo shows the ability of structural foam to reproduce fine detail and graining right in the mold. Surface finishing follows wood techniques. Molder: Capehart Corporation. Plastic: Modified PPO. (*Courtesy: Union Carbide Corporation*)

Fig. 15-21b. Display racks are tough, attractive, and stand up to outdoor exposure. Two shelves, one upright, and eight clips (no hardware needed) are all molded in one 25-pound shot. Molder: Dinaco, Inc. Plastic: Polystyrene. (*Courtesy: Union Carbide Corporation*)

Fig. 15-21d. 140-gallon tub shows large-part capability of Union Carbide's process. One-piece, 75-pound tub is molded on a 200 metric ton press; 27 nozzles deliver melt. Molder: Funabashi Kasei (Japan). Plastic: HDPE. (*Courtesy: Union Carbide Corporation*)

Fig. 15-21e. Door and side lights from Ottawa, in a variety of shapes, don't warp, rot, or absorb moisture in outdoor use. Units are molded in two mating pieces that assemble like sandwiches. Molder: Michigan Plastic Products. Plastic: Polystyrene. (*Courtesy: Union Carbid Corporation*)

Fig. 15-21f. Waste receptacles are seamless one-piece moldings from a 26-pound shot; the door is molded separately. Wall thickness is 5/16 inch. Molder: Contico. Plastic: HDPE. (*Courtesy: Union Carbide Corporation*)

Fig. 15-21g. Forklift pallets are easily molded despite the complexity of their designs. Although weighing only 25–35 pounds, they will support dynamic loads of several thousand pounds. Molder: Polycel (top); Menasha Corporation (bottom). Plastic: HDPE (top); HIPS (bottom). (*Courtesy: Union Carbide Corporation*)

Fig. 15-22. Accumulator-type high-speed structural foam injection molding machine. (*Courtesy: Beloit Corporation*)

nozzles or the manifold system required to insure an even flow of resin to all areas of a very large part. Figure 15-22 shows the appearance of a specially made structural foam injection molding press containing the screw injection unit and the accumulator. Figure 15-23 illustrates the "working" parts of the press.

The specific steps in the process are as follows:

1. Resin, in pellet form, is fed into the extruder hopper. If a chemical blowing agent is to be used, the resin is preblended with the blowing agent. If gas is used as the blowing agent, nothing is added to the resin.
2. The resin melts, and the nitrogen gas, if used, is added to the melt in the forward end of the barrel.
3. The resin melt is fed into the thermostatically controlled accumulator, under pressure high enough to prevent expansion (2000-3000 psi). Several shots may be required to fill the accumulator to the desired shot weight.
4. When the required shot size has been attained, the shutoff valve is opened and the accumulator plunger forces the "short shot" into the mold cavity, partially filling it, at a pressure of 200-400 psi.
5. The shutoff valve then closes, and the gases, either from the chemical blowing agent or the injected nitrogen gas, cause the melt to expand completely. By correct temperature control of the mold, the resin forms a skin where it contacts the mold surfaces and, at the same time, expands in the interior of the part.
6. The part then cools, during which time the accumulator is again being filled up.
7. The mold opens, the part is ejected, and the cycle is then repeated.

The above process, because the resin is injected into the mold at a relatively low pressure (200-400 psi), is ideally suited for production of very large parts, since the declamping force on the mold is relatively very low. At the same time, because the pressures are low, the parts are molded in a stress-free and warp-free condition, with no sink marks, even opposite heavy ribs or bosses.

To obtain parts with the optimum characteristics outlined above requires stringent processing controls. These include very accurate and uniform control of the temperatures in the press, accumulator(s), and the mold. All fill pipes and transfer and accumulating devices should be insulated to minimize temperature losses and fluctuations. The mold should be designed to allow smooth flow, with a minimum of 90° bends or other abrupt changes in resin flow. Insulation of the nozzles, or the use of Teflon tips, will help prevent nozzle "freeze off."

Additional information to aid in optimum processing of structural foam is shown in Table 15-13.

Tooling. Because of the low pressures in the mold, tooling can be relatively inexpensive cast aluminum, forged aluminum plate, Kirksite, or similar low-cost, easy machining material. This is particularly important when very large parts are to be made, since the savings in material and machining costs can be impressive when compared to conventional steel injection molding tooling. Also, the use of aluminum tooling results in shorter cycles, since the parts can cool faster because of the better heat conductivity, compared to steel.

The mold should be designed to require minimum resin travel from the nozzle to the mold cavity. Multiple nozzles are usually required if any length is to be traversed; and

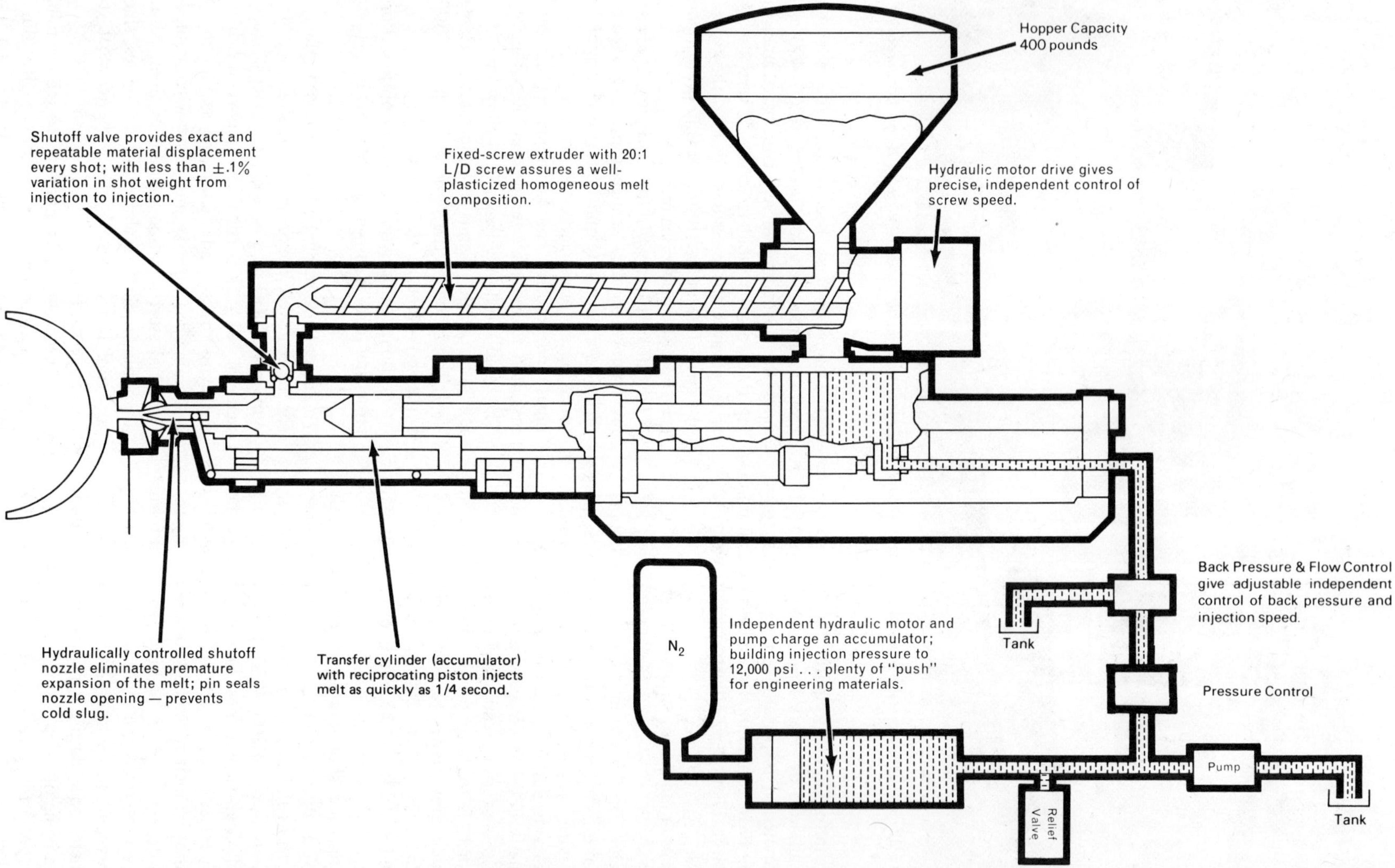

Fig. 15-23. Cross-section of machine shown in Figure 15-22. Note that the accumulator is below the screw. Machine can be used with nitrogen gas or CBAs to blow the foam. It can also be used as a conventional injection molding press. (*Courtesy: Beloit Corporation*)

Table 15-13. Guidelines for Optimum Structural Foam.*

FOR THESE RESULTS →	Maximum Flow; Smooth Surface	Fine Cell Structure; No Open Cells	Uniform Density; Low Part Weight	Uniform Density; Stress-Free; Warp-Free	Uniform Density; Short Cycle; Low Mfg. Cost	Strong Knit Lines
FOLLOW THESE GUIDELINES ↓						
Melt viscosity	Low	High	Low	Low	Low	N.A.
Blowing agent concentration	High	Low	High	N.A.†	Low	N.A.
Material density	High	High	Low	N.A.	Low	High
Wall thickness, (especially near end of flow path)	Don't reduce	Don't increase	Don't reduce	Don't reduce	Keep low and uniform	As thick as practical
Mold temperature	High	N.A.	High	High	Low	N.A.
Injection speed (with proper venting)	Fast	Fast	Fast	Fast	Fast	Fast

†N.A. = Not applicable.
*Courtesy: *Plastics World*, September 1976.

sprues, runners, and gates should be 1 to 2 times larger than those used for solid materials. Venting is very important and can be generous since the low pressures will not force material up the vents. End or tab gating is recommended with larger knockout pins for injection (see Chapter 20).

Multiple-cavity foam molding usually refers to several cavities on a rotary table, rather than several cavities shot at once, as is the case when speaking of conventional injection molding. Since the cooling time of foam moldings is appreciably longer than solid parts, of the same weight, the use of a rotary table allows extension of the cooling time, while at the same time several parts can be molded. Figure 15-24 shows such an arrangement for a four-cavity table, utilizing a main clamp for the injection portion of the cycle and auxiliary clamps for the cooling portion. With such a system, production can be increased considerably over a conventional machine, particularly for small parts. If the molding press is equipped with a programmed injection system, then the molds can be of different sizes, shapes, and weights with equally good results.

One of the major problems encountered in the low-pressure molding process is the production of an unwanted "swirl" pattern on the surface of the molded part. New techniques of molding, such as slower injection rates, precise temperature control of the mold and the resin melt, and selection of the optimum type and concentration of the blowing agent for the particular resin, are said to aid considerably in alleviating the problem.

High-Pressure Foam Molding. In the high-pressure foam molding process, resin is preblended with the chemical blowing agent prior to being placed into the hopper of the injection molding press. A full shot of molten resin is injected into the mold, completely filling it under relatively

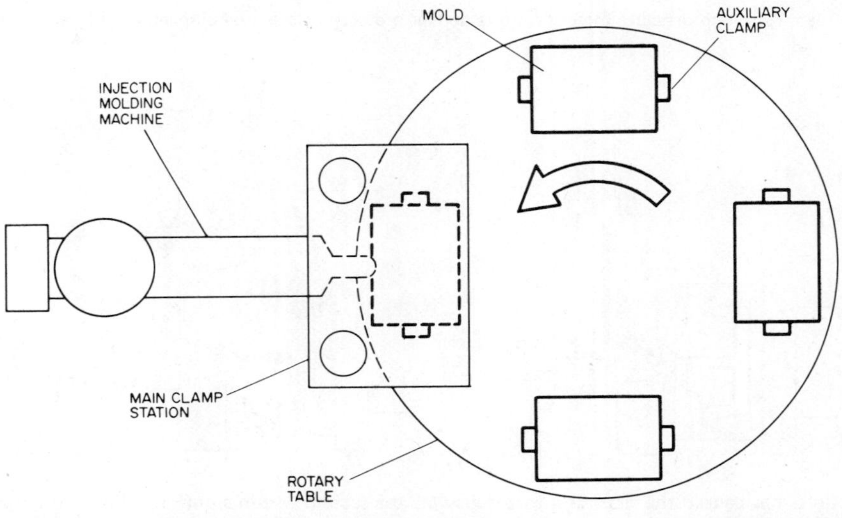

Fig. 15-24. This four-station rotary-table injection molding machine operates off a single clamp. (Reprinted by permission from *Plastics Technology*, April 1972)

high pressure. The resin is allowed to contact the cooler mold surfaces for a precisely controlled interval, long enough to form a skin of the desired thickness. The mold cavity is then enlarged by movement of the mold elements so that the still molten resin in the interior will foam. The result is a structural foam part with integral skins and a foamed core, which is similar in appearance and physical properties to a part made by the low-pressure process, except that the surface is free of "swirl" marks and, in fact, very accurately reproduces any textured mold surface. The major advantage of the high-pressure process is very good surface quality on the finished part. This advantage is somewhat offset by the necessity of using relatively complicated and expensive tooling, compared to the low-pressure process.

Sandwich Foam Molding. Sandwich foam molding also called co-injection molding may be considered a special form of high-pressure foam molding, in that two injection units are used to produce the skin and core as separate operations. First, the skin material is injected; then, through the same sprue, the core foamant is injected. Laminar flow in the core resin assures that the inner, unfoamed resin core will be completely surrounded by the skin material. The mold cavity then expands, allowing the core to foam. Figure 15-25 illustrates the process and the equipment.

One of the major advantages in the sandwich foaming process is that two different resins may be used for the skin and core. Thus, a hard, relatively expensive material can be used for the outer skin and an inexpensive resin, or even regrind and a blowing agent, may be used for the core. If two different resins are used, they should be compatible; for example, polyethylene and polypropylene, acrylic and ABS, PVC and SAN, etc. Figure 15-13 shows a cross-section of a part which could also have been made by co-injected sandwich foam.

The process described above, using the expanding mold and high pressure, is a proprietary one patented by the ICI Company of Great Britain. In addition, the same organization has developed a low-pressure process for sandwich molding. In this technique, which is also proprietary, the skin polymer is injected first and then a short shot of the foaming component follows. The foamable core then expands, forcing the skin resin into the extremities and filling the mold, which in this case does not expand. The

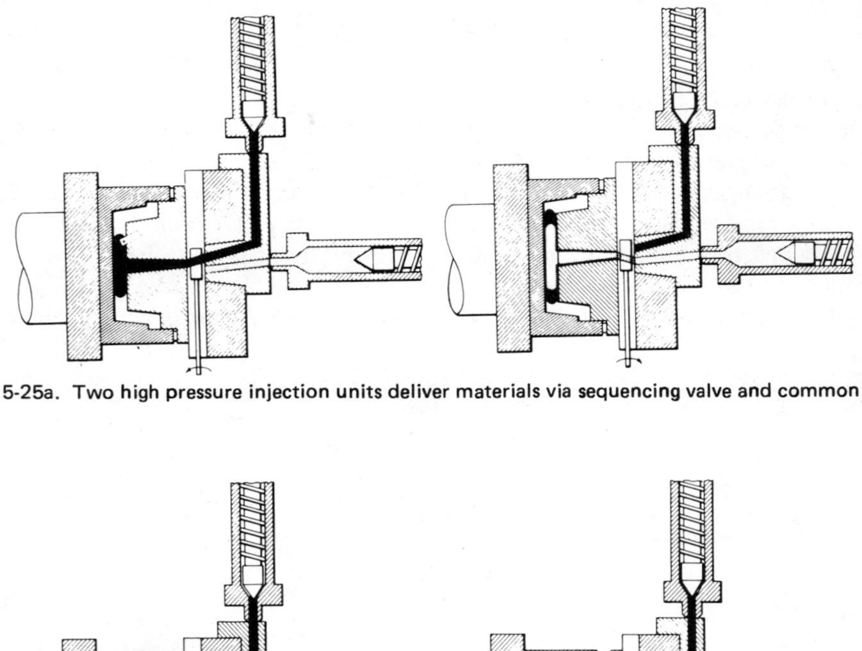

Fig. 15-25a. Two high pressure injection units deliver materials via sequencing valve and common sprue.

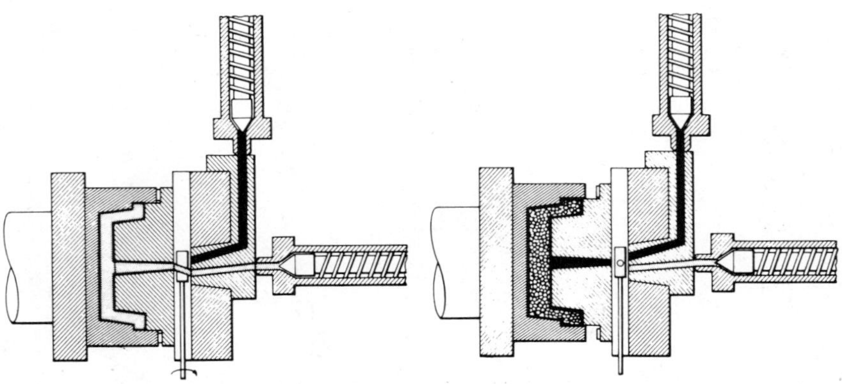

Fig. 15-25b. Laminar flow (left) has caused the first resin to encapsulate the second within a uniform skin. Total time is from 3 to 10 seconds. After shutting the valve (right), the positive "vertical Flash" mold is enlarged, and the core material, which alone contains blowing agent, foams. Expansion ratio (final/initial cavity thickness) is preset to the desired value.

obvious advantages are the use of lower pressure and less expensive tooling, since it need not resist high pressures or expand.

Troubleshooting. The information given in Table 15-14, while applicable mainly to low-pressure foam molding, can

Table 15-14. Troubleshooting Guide for Foam Molding.*

Melt temperature: Portable pyrometers with sensitive thermocouples are invaluable tools and a necessity for good production control through the entire system. Melt-temperature checks at the dump valve and nozzles help the operator to balance the heat profile for proper processing. The desired melt-temperature profile should be established in the mixing extruder zone of the equipment. Throughout the accumulator, pipe, and manifold areas only enough heat should be employed to maintain the desired melt temperature.

Heat: Overheating resin produces brittle parts; inadequate heat produces heavy or unfilled parts.

Venting: Masking tape placed at several locations around the open mold can be used as a test to check for adequate venting. Approximately 5-10 mils of venting $\frac{1}{2}$ inch wide every 2 inches is recommended.

Blowing agents: Polycarbonate and polyester resins may be attacked by many commercial blowing agents. Only recommended blowing agents should be used. Contamination by a blowing agent can occur when resins are mixed in the same container where non-recommended agents were previously used.

Runners and gates: Generous sized runners and gates must be employed to get the material into the mold cavity as quickly as possible. Restrictions in the runner system will be the limiting factor for filling the cavity.

Processing Considerations:

Condition	Check Point
Heavy part	–Shot size too great –Stock temperature too low –Fill time to slow –Gas pressure too low –Not enough blowing agent (chemical system)
Flash	–Clamp pressure too low –Shot size too great
Short fill	–Stock temperature too low –Nozzle not fully open –Shot size too small –Mold not properly vented –Heater band not functioning
Post blow	–Cooling cycle too short –Overpacked shot –High gas pressure –Non-uniform manifold temperature –Mold temperature too high –Too much blowing agent (chemical system)
Material discoloration	–Stock temperature too high and/or residence time too long –Non-uniform heating resulting in hot spots –Nozzle burn –Dead spots where material flow is restricted
Poor weld line strength	–Stock temperature too low –Inadequate venting –Flow length too long –Injection speed too slow
Slow injection speed	–Stock temperature too low –Nozzle not fully open –Poor temperature balance –Injection gas pressure too low
High extruder amperage	–Stock temperature too low
Nozzle freezing	–Mold hung on nozzle openings –Nozzle temperature too low –Improper nozzle shutoff –Tips not insulated
Poor surface appearance	–Gas pressure too high –Mold not properly vented –Stock temperature too low –Injection speed too slow –Flow length too long
Large "glossy" void	–Mold not properly vented –Stock temperature too high –Gas pressure too high –Flow length too long –Shot size too small –Non-uniform melt temperature
Brittle parts	–Wet material –Stock temperature too high –Blowing agent contamination –Contamination by grease, oils and degraded regrind material

Table 15-14. (*Continued*)

Quality-Control Considerations:

In establishing finished-part, quality-control procedures, the following categories should be taken into consideration as well as the recommended quality-control procedures.

Considerations	QC Procedures
Part weight and density	Construct part weight vs. density chart
Surface finish	Visual inspection
Discoloration	Visual inspection
Incomplete fill	Visual inspection
Impact strength	Drop ball testing procedure
Skin thickness	Break part; visual inspection (Do not cut or saw)
Cell structure	Break part; visual inspection (Do not cut or saw)
Voids	X-ray; high-intensity lighting, break part
Post blowing	Visual inspection
Sink marks	Visual inspection
Warpage	Visual inspection
Tolerances	Part measurement
Poor paint adhesion	Scratch surface (cross-hatch); adhesive tape test (put on/pull off)

*Courtesy: General Electric Co.

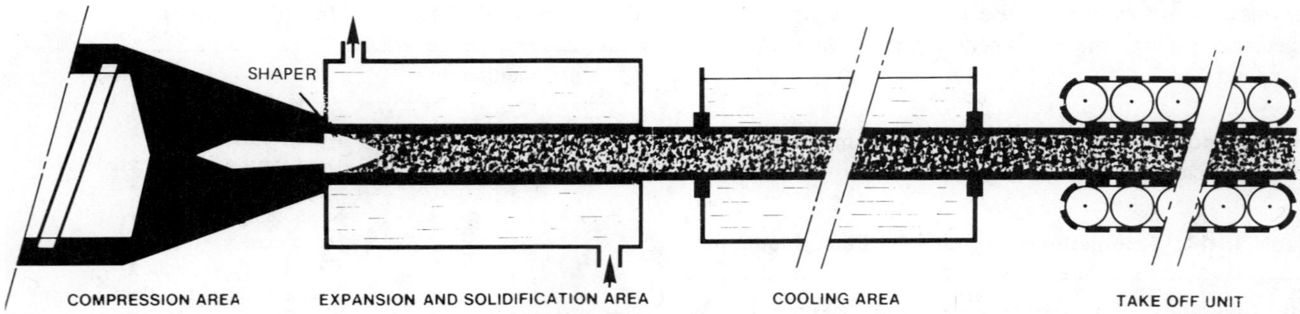

Fig. 15-26. The "Celuka" process.

also be used in the diagnosis of problems in high-pressure and sandwich molding. In general, most of the problems are caused by poor temperature control, inadequate venting, wet or contaminated material, and poor or inadequate gating.

Extruded Structural Foam. Molten resin containing a blowing agent, or mixed with an injected gas, can be extruded in a normal manner from a "conventional" extruder. However, the product usually does not have the thick, solid skin one expects from structural foam. In fact, as previously discussed, polystyrene or polyethylene foams are made as large extruded blocks with little or no appreciable skin. Polystyrene and PVC structural foams can be extruded with good skins, however, by the use of special formulations and a number of proprietary methods. These extrusions are largely confined to wood substitute shapes such as picture frame stock, moldings, glass partition moldings, etc. The densities obtained are in the range of 0.5 to 0.7 gram/cm^3, considerably below the density usually found in extruded polystyrene or polyethylene.

At the time of writing, the "Celuka" process, a French development, is said to be the process which gives a thicker skin than any other extruding technique and at the same time, results in a lower overall density, 0.4 to 0.5. This then allows a considerable weight saving in material. The unique feature of the "Celuka" process is the use of a special extruder die-head and a cooling shaper immediately adjacent to the die. In operation, the resin is forced past the die to form a hollow extruded profile of the desired shape. The profile then goes into the shaper which immediately cools the outer surface, forming the skin. The still hot resin then expands *inward*, thus forming the "solid" foamed extrusion. Figure 15-26 illustrates the role of the die and shaper.

The foamed extrusions, made by either the proprietary or the "Celuka" process, are all strong, stiff woodlike materials which can be nailed, screwed, or adhesive bonded. The finish of the skins can be either dull or glossy and can be made with a wood grain finish from the extruder, eliminating further finishing. Other profiles are finished by being painted, hot stamped, film wrapped, or rotogravure printed in a variety of patterns.

Rotational Molded Foam. Rotational molding can also be used to produce structural foam by two different techniques. In both techniques conventional rotomolding equipment is used. The resin is simply mixed with blowing agents to form the foamant material, when using polyolefins. When using ABS, a specially compounded resin, containing the blowing agent, is employed.

One technique for the production of the rotomolded part consists of first molding a skin of high-density polyethylene on the inner surface of the mold. Then a charge of lower-melting polyethylene resin and blowing agent is introduced into the mold and fused and blown into foam. In the second system, known as the sinter technique, proprietary methods are used. These are believed to consist of simply placing the requisite amount of material in the mold and heating or sintering until expansion and fusing take place.

REFERENCES AND BIBLIOGRAPHY

Abbott, R. A., and Robertson, E. J. Spray application of rigid urethane foams. *SPE Journal*, May 1962, **18**(5):561.

Alexander, P. D. Foam–expanding into all markets. *Plastics World*, July 1968, **26**(7):38.

Angell, R. G. Jr., Hartsing, T. F. Jr., and Rocky, J. F. Jr. Plan your line for the payoff in molding structural foam. *Modern Plastics*, April 1973, **50**(4):57.

Allied Chemical Corporation, General Chemical Division. *Questions and Answers About Rigid Urethane Foam.*

Allied Chemical Corporation, National Aniline Division. *Test Methods for Evaluation of Flexible Urethane Foam.* Technical Bulletin TS-9, 1961.

——. *Calculations in Urethane Formulation.* Bulletin No. 120158, December 1958.

——. *Polyester Rigid Urethane Foams.* Bulletin No. 71558.

Aylesworth, R. D., Boehringer, R. H., Moore, D. T., and Smith, M. H. Polyesters of dimer acids as intermediates for urethane foams. *Modern Plastics*, May 1958, **35**(7):145.

Bakelite Co., Division of Union Carbide Corporation. *Syntactic Foam.* Technical Release No. 4 J-868, March 1958.

——. *Packaging in Foam Based on "Bakelite" Phenolic Resin.* Technical Release No. 4 J-888, December 1955.

Baruth, E. J. The ramifications of urethane foam molding, what designs are possible, what production techniques are used. *Plastics Design & Processing*, September 1961, **1**(9):10.

Benning, C. J. Progress in thermoplastic foams. *Plastics Technology*, October 1968, **14**(10):67.

Berger, S. E., Cataldo, F. L., Dwyer, F. J., and Stone, H. Toluene diisocyanate 2, 6-isomer–its effect on urethane polymers. *SPE Journal*, February 1961, **17**(2):170.

Borg-Warner Chemicals. *Expandable ABS–Molding Grade Cycolac JP.* Technical Report P-156, 91124.

Boudreau, R. J. How silicone surfactants affect polyurethane foams. *Modern Plastics*, January 1967, **44**(5):133.

Bryant, R. M., and Stewart, H. F. The chemistry of silicone surfactants for urethane foams. *Cellular Plastics*, March/April 1973, **9**(2).

Bryson, F. E. Plastic foams–the flexibles. *Machine Design*, November 11, 1971, **43**(21):145.

——. Plastic foams–the rigids. *Machine Design*, November 25, 1971, **43**(22):92.

Carwin Co. *New Information on the Preparation of Rigid Urethane Foams.* Technical Bulletin No. 101, May 1963.

Chambers, R. E. *Structural Properties of Rigid Foams.* Massachusetts Institute of Technology, Bulletin R63-13, February 1963.

Cochran, L. D., and Osborn, C. W. Injection molding thermoplastic foams. *SPE Journal*, September 1969, **25**(9):20.

Crane Plastics, Inc. *Foamextrusions.* 1972.

Currier, V. How to compound high-density urethane foams. *Plastics Technology*, August 1966, **12**(9):35.

D'Ancicco, V. V. Heat resistant urethane polymers. *SPE Journal*, February 1958, **14**(2):34.

Davis, R. Foams and reinforced plastics. *Reinforced Plastics*, January/February 1964, p. 8.

Dickert, E. A., and Toone, G. C. Chemistry and flame retardancy of rigid urethane foams. *Modern Plastics*, January 1965, **42**(5):197.

Dietz, A. G. H. Rigid plastics foams in building. *Modern Plastics*, December 1958, **36**(4):91.

Divis, R. R. Foam boards duplicate wood down the line. *Plastics Engineering*, February 1976, **32**(2):19.

Dougan, T. P. Where to use urethane foams. *Materials in Design Engineering*, January 1959, **49**(1):36.

Dow Chemical Co. *Styrofoam.* Bulletin 157-50, January 1959.

——. *Styrofoam, a Superior Insulation for Low Temperatures.* Form No. 157-3-57.

——. *Pelaspan Expandable Polystyrene.* Form No. 171-414-68.

——. *Foamable Thermoplastic Spheres.* Form No. 190-101-69.

Dow Corning Corporation. *Silicone Additives for Prepolymer-Type Urethane Foams.* Bulletin 05-027, October 1962.

E.I. Du Pont de Nemours & Co., Inc. *Structural Foam Injection Molding Zytel FE 5082 and FE 5083 Nylon Resins.* Bulletin E-06160, October 1975.

——. *Rigid and Semi-Rigid Urethane Foam.* Bulletin A-3705.

Dworkin, D. Urethane foam processing techniques. *SPE Journal*, December 1961, **17**(12):1269.

Dwyer, F. J., Kaplan, M., Piver, E. L. Jr., and Stone, H. Technology of super soft flexible urethane foams. *Modern Plastics*, May 1964, **41**(9):139.

Eakin, J. L. Urethane structural foams: key properties give them a competitive edge. *Plastics Engineering*, July 1978, **34**(7):58.

Eastman Chemical Products, Inc. *Foamable Tenite Polypropylene P2635-08AA.* Materials Bulletin MB-55B.

Ehrlich, A., and Patton, T. C. Urethane foams based on castor polyols. *Modern Plastics*, March 1964, **41**(7):154.

Emerson & Cuming, Inc. *Lightweight, Cast Synthetic Wood.* Bulletin 826, April 1969.

Frados, J. (ed.) *Plastics Engineering Handbook.* 4th Ed. Van Nostrand Reinhold Co., New York, 1976.

Ferrigno, T. H. *Rigid Plastic Foams.* Reinhold Publishing Corporation, New York, 1963.

——. The use of extenders in flexible urethane foams. *SPE Journal*, June 1960, **16**(6):638.

Fiberfil Division, Dart Industries, Inc. *Fiberfil Glass Reinforced Thermoplastic Foam Resins.* Form No. 71-9871.

Finchke, F. K., and Wilson, G. R. Cellular aromatic polyimides. *Modern Plastics*, April 1969, **46**(4):108.

Frisch, K. C., and Robertson, E. J. Advances in technology and uses of rigid urethane foams. *Modern Plastics*, October 1962, **40**(2): 165.

Gallegher, R. B., Kamath, V. R., Kamens, E. R., and Hurpell, G. A. AZO foaming agents for unsaturated polyesters. *Modern Plastics*, June 1976, **53**(6):58.

Gariano, R. Tooling up for molding engineering structural foam. *Plastics Design & Processing*, November 1973, **13**(11).

Gemeinhardt, P. G., Reid, G. E., Bingham, S. A., Eakin, J. L., and Szabat, J. F. One-shot flexible polyether urethane foam. *SPE Journal*, April 1964, **20**(4):357.

Gemeinhardt, P. G., Szabot, J. F., Morecroft, A. S., and Saunders, J. H. Molding flexible polyether urethane foams. *SPE Journal*, October 1960, **16**(10):1117.

General Electric Co. *Structural Foam Resins.* Bulletin SFR-1, April 1973.

——. *Structural Foam Resins, Noryl FN-215 Foamable Resin.* Bulletin ESF-001, September 1972.

——. *The Handbook of Engineering Structural Foam.* Bulletin SFR-2.

Gerstin, H. Rigid plastic foams. *Product Engineering*, June 21, 1965, **36**(6):59.

Gmitter, G. One-shot urethane foam production. *Plastics Technology*, July 1963, **9**(7):28.

W. R. Grace & Co., Dewey and Almy Chemical Division. *HYPOL Laboratory Procedures and Foam Formulations.*

——. *HYPOL Flame Retardant Properties.*

Gross, L. H., and Angell, R. G. Swirl-free foam parts. *Plastics Technology*, May 1976, **22**(5):13.

Gusmer Corporation. *Model HII Hydraulically Powered Proportioning Unit.* Bulletin 575-7.

Hansen, R. H. Production of fine cells in the extrusion of foams. *SPE Journal*, January 1962, **18**(1):77.

Harding, R. H., and James, B. F. Thermal conductivity and structure in rigid urethane foams. *Modern Plastics*, March 1962, **39**(7):133.

Harper, C. A. Foams and low density compounds. *Insulation*, May 1961, p. 34.

Harris, W. D. Finding the way to a smooth surface in structural foam. *Plastics Engineering*, May 1976, **32**(5):26.

Heck, R. L. III. Selecting chemical blowing agents. *Plastics Compounding*, November/December 1978, p. 52.

Heiss, H. L., Saunders, J. H., Morris, M. R., Davis, B. R., and Hardy, E. E. Preparation of polymers from diisocyanates and polyols. *Industrial and Engineering Chemistry*, July 1954, **46.**

Hersch, P. What about the newer expanded thermoplastic foam technology. *Plastics Technology*, February 1967, **13**(2):40.

——. Choosing and using the right silicone surfactant for urethane foams. *Plastics Technology*, November 1967, **13**(11):44.

——. Thermosetting foams. *Plastics Technology*, May 1966, **12**(5):66.

Holdredge, E. C. Plan, analyze, scrutinize before buying a low-pressure system. *Plastics Technology*, August 1975, **21**(8):33.

Hosford, R. C. New machinery for PS foam molding. *Plastics Technology*, July 1964, **10**(7):28.

Johnson, R. W., Montgomery, P. W., O'Brien, D. L. Plastic foams hold up 5000 fathoms down. *Materials Engineering*, September 1967, **66**(3):74.

——. Here's what's happening in high performance syntactic foams. *Plastics Design & Processing*, April 1973, **13**(4):14.

Jones, R. E. Rigid urethane foam–process variables and test procedures. *Plastics Technology*, Part 1: October 1961, **7**(10):27. Part 2: November 1961, **7**(11):43.

Joslyn, W. G. PVC plastisol foam technology. *Rubber Age*, February 1974, p. 49.

Kaplan, M. One-shot rigid urethane foam molding reduces costs, improves properties. *Plastics Technology*, August 1964, **10**(8):38.

Kiessling, G. C. Extruded structural foam. *Cellular Plastics*, November/December 1976, p. 337.

Kraus, R. J., and Straight, R. J. Update on rigid cellular PVC. *Plastics World*, July 1978, **36**(7):58.

Laclair, R. C. Blowing agents for structural foam molding. *Plastics Technology*, March 1976, **22**(3):31.

Lambert, F. H. Foam molding goes automatic. *Modern Plastics*, February 1961, **38**(6):107.

Lasman, H. R. Foaming agents for polyolefins. *SPE Journal*, September 1962, **18**(9):1184.

LeBras, L. R. Recent developments in rigid urethane foams. *SPE Journal*, April 1960, **16**(4):420.

Levy, S. Foam profiles: new processes, new tools, new products. *Plastics Machinery & Equipment*, November 1977, **7**(11):11.

Lindstrom, R. Thermoplastic foams. *Plastics Technology*, May 1966, **12**(5):68.

Mack, G. P. Effect of tin catalysts on physical properties of cellular urethane. *Modern Plastics*, December 1964, **42**(4):148.

Marchetti, R. Selecting resins for thermoplastic structural foam. *Plastics Design & Processing*, April 1979, **14**(4):20.

Materials Engineering. Coated plastic foams live a lot longer than uncoated. September 1973, **78**(3):56.

——. Foamed plastics show the way to big, lightweight, structural parts. March 1973, 77(3):48.

Materials in Design Engineering. Plastic Foams. Materials and Processes Manual No. 236, May 1966.

——. Quick spot test for polyurethane foams. June 1959, **49**(6).

——. Self-extinguishing urethane foams. July 1961, **54**(1):10.

Matonic, D. E. Molding PC structural foam on standard injection machines. *Plastics Engineering*, September 1976, **32**(9):31.

Matonic, D. E., and Kaminski, A. Standard machines mold PC structural foam. *Plastics Technology*, September 1976, **22**(9):57.

McElroy, W. R. *The Techniques of Prepolymer Preparation.* Mobay Chemical Co.

Miller, B. S. What's putting that fizz in the structural foam biz? *Plastics World*, July 23, 1973, **31**(7):38.

Minnesota Mining & Manufacturing Co. *"Scotchply" XP–241 Syntactic Foam.* Bulletin P-RSL (961.5)R2, Technical Data Sheet No. 11, January 1969.

Mobay Chemical Co. *Urethane Foam Technical Information Bulletins.* 1958–1960.

——. *Flexible Urethane Foam.* Bulletin 4a/Mo.

——. *Urethane Foam in Marine Uses.*

Modern Plastics. Through-the-nozzle gas injection produces class A foam parts. March 1980, **57**(3).

——. Structural foam surface finishes are practically perfect now. October 1979, **56**(10).

——. The action starts in polyester foam. June 1978, **55**(6):37.

——. Surprising structural foams . . . engineering types. March 1975, **52**(3):50.

——. Structural foam: it's looking more solid every day. March 1974, **51**(3):38.

——. Engineering foams set to invade metal markets. April 1973, **50**(3):78.

——. Expandable filler catches on. August 1969, **46**(8):55.

——. Now–foam with concentrates. May 1968, **45**(5):88.

——. New foam for new markets. December 1968, **45**(12):68.

——. The urethanes grow up. Part 1: March 1959, **36**(7):91. Part 2: April 1959, **36**(8):109.

Monsanto Corporation. *Injection Molding of Styrene and ABS Structural Foam Parts.* Form 183 4852, 1970.

Murray, T. A. Foam reservoir molding: the processing story. *Plastics Technology*, October 1978, **24**(10):93.

Naitove, M. H. Foam profile extrusion excites industry-wide scramble. *Plastics Technology*, February 1973, **19**(2):42.

Nass, L. I. Compounding of low density vinyl foams. *Modern Plastics* March 1963, **40**(7):151.

Neary, W. J. What to look for in urethane foam machinery. *Plastics Technology*, March 1965, **21**(3):50.

Nopco Chemical Co. *Nopcofoam Pour-In-Place Urethane for Insulation.* Bulletin PSP-103-16, 1961.

——. *Chemical and Physical Properties of Nopcofoam.* Bulletin 459.

Olson, R. Structural-foam tooling. *Plastics Technology*, June 1976, **32**(6):38.

Peterson, D. Structural foams create new trends in furniture. *Plastics Design & Processing*, August 1974, **14**(8):12.

Plastics Design & Processing. For smooth structural foam finish co-injection has the answer. June 1979, **14**(6):26.

——. Casting polyurethane foam. August 1970, **10**(8):19.

Plastics Design Forum. Prototyping for structural foam. July/August 1978, p. 72.

——. Advances in SF technology open new doors for designers. July/August 1980, p. 60.

Plastics Engineering. GE aims at computer market with new polycarbonate foam. March, 1978, **34**(3):11.

Plastics Technology. Delayed-action catalysts aid PU foam processing. July 1977, **23**(7):15.

——. Now TS polyesters are foamable. January 1976, **22**(1):47.

——. Injection and structural-foam molding. November 1976, **22**(11):59.

——. Thermoplastic microspheres are on scene for syntactic foam. April 1969, **15**(4):67.

——. Simple tests uncover PU processing faults. January 1967, **13**(1):49.

——. Load-bearing properties of flexible urethane foam. April 1962, 8(4):26.

Plastics World. Guidelines for better structural foam. September 1976, **34**(9):28.

——. Foam extrusion technique narrows the gap between plastics and wood. February 1972, **30**(2):36.

——. Structural foam in your future? It's closer than you think! April 1972, **30**(4):40.

——. Structural-foam process. October 1970, **28**(10):51.

Product Engineering. Structural foams: rigidity at low cost. March 1979, **50**(3):29.

——. Plastic foams in sheets and planks form new lightweight products. March 1977, **48**(3):21.

——. Variety of cushioning materials absorb impact, energy and shock. May 1977, **48**(5):25.

——. Smooth finishes for structural foam are the promise of new technology. November 1976, **47**(11):23.

——. Weight-saving foam plastic sandwiches form auto chassis. April 1973, **44**(4):29.

——. Economical foam process makes composites without adhesives. September 1972, **43**(9):38.

——. Plastic foam composites make tough, lightweight products. September 28, 1970, **41**(9):40.

Randa, S. K. Foaming FEP for wire insulation. *Plastics Technology*, April 1964, **10**(4):42.

Recktenwald, G. W. Urethane foam in the appliance industry. *SPE Journal*, June 1964, **20**(2):528.

Renshaw, J. T., Cannon, J. A., and Hinchen, J. D. How formulations and process variables affect vinyl foams. *SPE Journal*, August 1970, **26**(8):47.

Resnick, I. Performance of glass spheres/epoxy syntactic foam. *Modern Plastics*, September 1965, **43**(1):144.

Roe, R. R. *Plastic Foams in Furniture.* Technical Report No. 79, *Western Plastics*, December 1961, p. 19.

Ryder, L. B. When to go to foam. *Plastics Engineering*, November 1975, **31**(11):19.

Sanger, M. H., Gmitter, G. T., and Maxey, E. M. How polyether foams compare. *Materials in Design Engineering*, March 1958, **48**(3):101.

Saunders, J. H. The formation of urethane foams. *Rubber Chemistry and Technology*, December 1960, **23**(5).

Saunders, J. H., Steingiser, S., Gemeinhardt, P. G., Morecroft, A. S., and Hardy, E. E. Properties of flexible urethane foams. *Chemical & Engineering Data Series*, American Chemical Society, April 1958, **3**(1):153.

Sayad, R. S., and Williams, K. W. There is an alternative to fluorocarbon blowing agents. *Plastics Technology*, June 1979, **25**(6):71.

Scheiner, L. L. Structural foam machines: what the buyer needs to know. *Plastics Techology*, September 1974, **20**(9):54.

——. Structural thermoplastic foams: an emerging technology. *Plastics Technology*, December 1968, **14**(12):35.

Seele, M. How to cut costs with one-shot urethane systems. *Plastics Technology*, January 1963, **9**(1):42.

Serlin, I., Markhart, A. H., and Lavin, E. Polyimide foams. *Modern Plastics*, July 1970, **47**(7):120.

Shea, F. How to extrude foam film. *Plastics Technology*, February 1978, **24**(2):83.

Smith, H. A. Here's a guide to making flexible molds for foam furniture. *Plastics Technology*, September 1968, **24**(9):38.

Smith, J. B. Frothing rigid urethane foams. *Plastics Technology*, January 1964, **10**(1):35.

SPI Cellular Plastics Conference (9th). Paper presentations: (1) Manufacture and Use of Polystyrene Foams, H. Muller-Tamm. (2) Novel Foams from Styrene-Acrylonitrile Copolymers, A. R. Ingram. (3) Physical Property Modifications of Low Density Polyethylene Foams, R. E. Skochdopole. (4) The Production Chemist and Flexible Urethane Foam Technology, G. D. Walmsley. (5) Improved Semi-Flexible Urethane Foams, P. G. Gemeinhardt, J. F. Szabatm, and J. K. Backus. (6) Flame-Retardant Characteristics of Rigid Urethane Foam, H. Piechota. (7) Air Flow Measurement and Its Relation to Properties of Flexible Urethane Foams, R. E. Jones and G. Fesman. *Modern Plastics*, November 1964, **42**(3):139.

Stepan National Polychemicals Division. *Chemical Blowing Agents for Injection Molding Thermoplastic Structural Foam.* Technical Bulletin 00R-04-0972.

Stinson, S. C. Developments are breaking fast on the red-hot foams front. *Plastics Technology*, September 1972, **8**(9):33.

Society of the Plastics Industry, Inc., Structural Foam Group, Cellular Plastics Division. *Structural Foam.*

Supnik, R. H., and Silberberg, M. Properties of foams and laminates under shock loading. *SPE Journal*, January 1959, **15**(1):43.

Sweets, M. Foam processing—a technology on the move. *Plastics Design & Processing*, December 1967, 7(12):12.

Swenson, S. B., and MacClarence, J. Versatile partnership: reinforced plastics-urethane foam. *Modern Plastics*, August 1961, **38**(12):78.

Thiokol Chemical Corporation. *Mold Release Agents for Use with Rigithane 112.* Bulletin R-112-2, January 1959.

——. *Catalysts for Use with Rigithane 112.* Bulletin 112-3-A, January 1959.

——. *Thiokol's Rigithane 112 Urethane Foaming Resin for Use in Preparing Semi-Rigid and Rigid Foams.* Bulletin R-112-1, February 1959.

——. *Molding with Rigithane 112 Foaming Resin.* Bulletin R-112-5, February 1959.

——. *Self-Extinguishing Properties of Rigithane 112 Foams.* Bulletin R-112-6, April 1959.

——. *Techniques and Equipment for Mixing Rigid Foams.* Bulletin R-112-7, April 1959.

Thompson, E. J., Reymore, H. E., Grieve, R. L., and Sayigh, A. A. R. To process structural urethane foams, get a handle on your design factors, molding parameters, tooling needs. *Plastics Technology*, September 1973, **19**(9):30.

Thorpe, L. R. A new look at some foamed plastics. *Plastics Design & Processing*, November 1963, **3**(11):39.

Ugine Industries Inc. *Celuka Process.*

Union Carbide Corporation. *Structural Foam Molding with UDEL Polysulfone.* Bulletin F-46063 A, June 1977.

——. *Structural Foam.* Bulletin F-44669 B, November 1974.

——. *Structural Foam.* Bulletin F-42919, February 1971.

——. *Bakelite Foamable Ethylene Copolymers.* Bulletin 7-1526 046-8 BA-326-290 B.

Universal Lumber Co., Aircraft Division. *Strux.*

Upjohn Co., Polymer Chemicals Division. *The Preparation of Rigid Urethane Foams Based on PAPI.* Technical Bulletin 101.

——. *Analysis of PAPI—Procedures.* Technical Bulletin 103.

——. *One-Shot Sprayable Foams Based on PAPI.* Technical Bulletin 104.

Wallace, C. W. High-speed injection molding of structural foam. *Plastics Engineering*, November 1976, **32**(11):53.

Warren, D. "Intaglio"—a new concept in foam plastic parts. *SPE Journal*, November 1970, **26**(11):31.

Webb, J. W. A guide to making PVC foam plastisols without blowing agents. *SPE Journal*, February 1971, **27**(2):43.

Weir, C. L. The action is swinging to structural foams. *Plastics Technology*, April 1972, **18**(4):37.

Welex, Inc. *A Few Words about Welex Foam Sheet Lines.*

Wendle, B. C. Structural foam molding. *Plastics Design & Processing*, September, 1968, **8**(9):20.

Whitman, R. D. How to make integral-skin urethane foams. *Plastics Technology*, February 1970, **16**(2):44.

Wilson, M. G. Glass reinforced foam for injection molding. *Plastics Design & Processing*, January 1972, **12**(1):12.

——. Guide to working with reinforced thermoplastic foams. *SPE Journal*, June 1971, **27**(6):35.

Winkler, J. Effects of foaming catalysts on aging of urethane foams. *SPE Journal*, November 1956.

Woodland, J. H. Rigid urethane foam—a substitute for styrene-polyester casting compound. *SPE Journal*, December 1964, **20**(12):1309.

Yeates, T. E., and Mehltretter, C. L. Urethane foams from methyl glucoside polyethers. *Modern Plastics*, November 1961, **39**(3):151.

Zuckerman, J. L. The art of molding flexible urethane foam. *SPE Journal*, May 1966, **22**(5):56.

16
Compression, Transfer, and Injection Molding of Thermosets

Compression molding is one of the oldest molding techniques still in use. It probably originated when Egyptian or Roman brickmakers placed a ball of clay into a wooden form and then forced it to assume the contours of the form by pushing it with a tight-fitting tamp. The heat of the sun then baked the brick to the final "cure." In essence, this is the same procedure used today, except that thermosetting resins are used instead of clay, pressures and temperatures are considerably higher, and the time is less.

Modern compression molding started with the onset of the rubber industry in the 1840s, when Thomas Hancock patented a technique for molding rubber. Figure 16-1 illustrates the process and the simple mold used to produce a cuplike object.

In addition to the mold, which today is usually made of steel, some method must be provided to force the mold together, and thus form the resin (either rubber or plastic) to the contours of the mold. This pressure is usually brought about by the use of a hydraulic press which puts the material and the mold under a compressive load—hence the name compression molding. Compression molding presses range from small, laboratory types which exert a few tons of force through hand-operated hydraulic cylinders, similar to an automobile jack, to large, motor-driven, pump-equipped presses which can exert thousands of tons of force on the mold. Figures 16-2, 16-3, and 16-4 show the appearance of some of these presses from the very simple, hand-operated press to the more complicated, motor-driven presses. In all types, the principles of operation are the same: the mold is opened, a measured amount of powdered molding material—pellet or chopped resin-impregnated fabric—is placed in the cavity, the mold is gradually closed while the resin is heated and liquified. When the the material has softened enough so it will completely fill the mold cavity, the mold is then closed, as shown in Figure 16-1, and the material is held in the closed mold under pressure until hardening or "cure" takes place, after which the mold is opened and the part is removed while still hot.

The compression molding sequence for the fabrication of a phenolic-fiberglass missile nose cone is illustrated in Figures 16-5 through 16-9. The type of tooling and equipment shown is representative of that used for low-volume, large-part production.

From the brief description of compression molding given above, one can readily see some of the fundamental problems of this method of molding; for example, the necessity of measuring the resin charge fairly accurately, closing the mold slowly while the plastic is heating, and softening, and then holding the hot material under pressure until cure has taken place.

The method of compression molding as outlined above was used until the early 1920s when the need for large quantities of automotive and electrical parts spurred the development of a new technique for molding thermosetting plastics—transfer molding. Transfer molding consists of simply melting the plastic in a separate chamber and then injecting it into the closed mold. In principle, it is similar to ram injection molding of thermoplastic materials. The name transfer was adopted to distinguish it from the original process—compression molding. In addition to being somewhat faster than compression molding, transfer molding has other advantages which will be described in more detail later on.

With the development of screw machine injection mold-

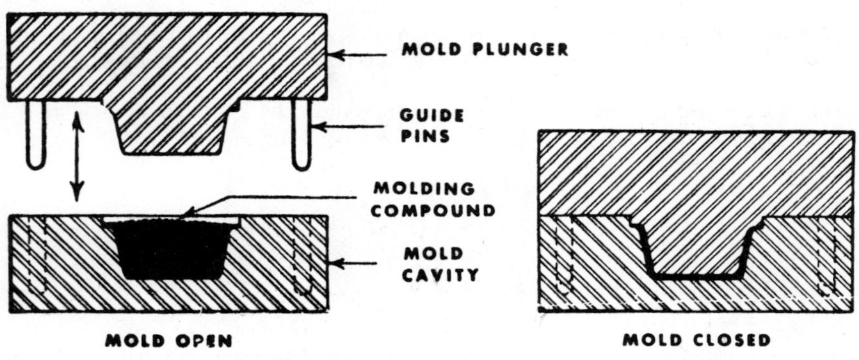

Fig. 16-1. Compression molding. (From *Plastics Engineering Handbook of the SPI* edited by Frados)

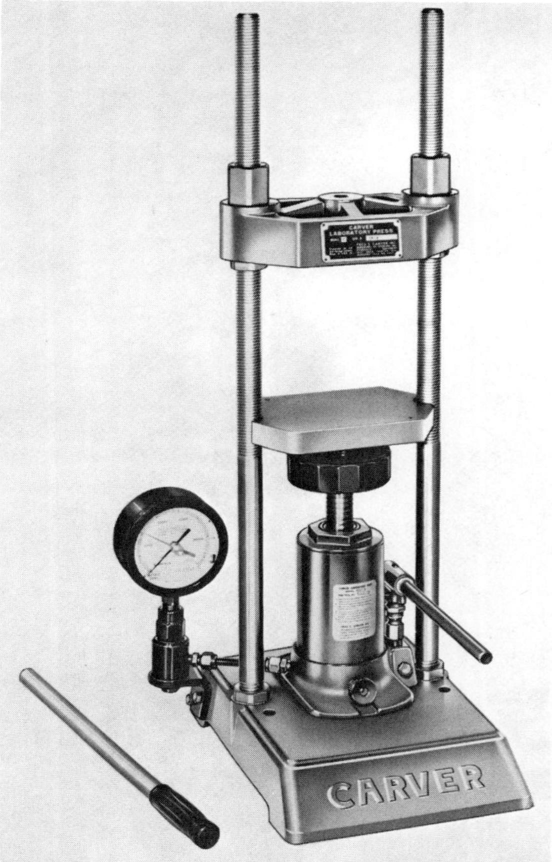

Fig. 16-2. Compression molding press. (*Courtesy: Fred S. Carver, Inc.*)

Fig. 16-4. A 400-ton compression molding press for SMC and BMC. (*Courtesy: EEMCO Division, Zurn Industries, Inc.*)

Fig. 16-3. A 75-ton press. (*Courtesy: Clifton Hydraulic Press Co.*)

ing, it was inevitable that this type of equipment would be adapted to thermoset materials. This then resulted in another increase in the production rate of thermoset parts which, with the development of fast-curing materials, enabled thermoset molders to compete very satisfactorily with many types of thermoplastic molding. Thus, today, all three methods of molding are still in use, each with certain advantages and disadvantages which are more fully described below.

MATERIALS

The great majority of all materials used in compression molding are thermosets. While compression molding could be done using thermoplastic resins, this is only rarely done. The time needed to heat up the mold, soften the resin, and then cool the mold again, so the part can be removed, makes this process completely impractical and unprofitable. A single exception might be the molding of an acrylic lens, where compression molding might give a transparent part with no sign of flow marks, or other inhomogeneities. Thus, for practical purposes, compression molding materials are all thermosets, as are transfer materials.

While there are a great many different types of thermoset molding compounds, all types contain a low molecular

Fig. 16-5. Missile cone being molded in male and female compression mold. (*Courtesy: Hughes Aircraft Co.*)

weight resin which is cured to a crosslinked, high molecular weight resin. In addition, with but a few exceptions, all the resins are combined with a filler system to form the complete molding compound. A typical molding compound might have as many as twelve ingredients divided into two main constituents: (1) the resin portion, consisting of the resin, a curing agent, a plasticizer, and an internal lubricant which acts as a mold-release agent; and (2) the reinforcing filler, which might consist of some type of fibers, a clay filler, pigments, and UV inhibitors. Not all compounds contain all the ingredients listed, and some might contain more. The resin portion creates the general character of the compound: the chemical resistance, the cure schedule, the flammability, electrical properties, dimensional stability, etc. The filler and reinforcement materials affect the mechanical properties such as tensile strength, impact strength, and, to a lesser extent, the heat resistance, dimensional stability, and electrical properties of the compound. To emphasize any particular quality, the appropriate resin or filler is chosen. As an example, a melamine resin would be used if a very hard compound was required. If, in addition to being hard, the resin was also to have superior electrical properties, then it would probably have a mineral filler such as mica, and if high strength was required, then glass fibers would probably be included. In naming the compound, the resin is usually given first followed by the filler. Thus, the hypothetical compound described above would probably be called a melamine-glass compound, if glass was the principal reinforcement.

Short descriptions of the major thermosetting materials are given below:

Phenolics. The phenolic compounds have rightly been called the "workhorse" materials of the plastics industry, since they, by far, represent the greatest tonnage of any thermosetting resin used. The reasons for their extensive use are low cost, good mechanical properties, good heat and chemical resistance, excellent moldability, and many varieties, ranging from general-purpose materials filled with wood flour or cotton flock, to strong, impact-resistant compounds using chopped-glass fabric as the reinforcing filler. In general, phenolic compounds contain approximately 40% phenolic resin and approximately 50% filler, dyes, pigments, plasticizer, and lubricants. They are usually molded at pressures of 2000–5000 psi and temperatures ranging from 130–195°C (270–380°F). The principal limitations of the phenolics are their color—usually dark brown or black—and their electrical properties, which are not of the highest caliber.

Ureas. Urea-formaldehyde resins, originally developed as transparent "organic glass," were found to have poor water

Fig. 16-6. Part removed from mold, still on male or punch section of mold. (*Courtesy: Hughes Aircraft Co.*)

Fig. 16-8. One method of part removal. Part is being manually unscrewed from the male portion of the mold. (*Courtesy: Hughes Aircraft Co.*)

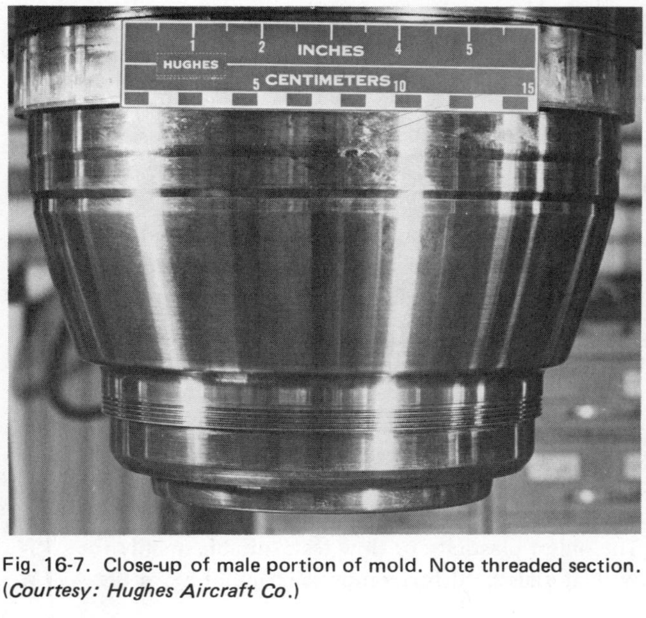

Fig. 16-7. Close-up of male portion of mold. Note threaded section. (*Courtesy: Hughes Aircraft Co.*)

Fig. 16-9. The completed part made from chopped phenolic resin-impregnated fiberglass fabric. (*Courtesy: Hughes Aircraft Co.*)

and weather resistance. However, when filled with pure cellulose fibers (alpha-cellulose), the resulting molding compounds have many of the same processing and physical characteristics as phenolics. In addition they have the significant advantage of a light color and the ability to be blended in shades ranging from white to pastels. Unfortunately, the molding materials have poor resistance to outdoor exposure, and so are used almost exclusively for indoor applications such as light switches, buttons, and bottle closures. Because they do not carbonize when an electric arc is passed upon their surface, as do the phenolics, they are particularly well suited for electrical applications. Ureas are also distinguished by exhibiting extremely hard surface hardness, which somewhat limits their machinability.

Melamines. Melamine, like urea and phenol, can be reacted with formaldehyde to produce a resin which is combined with a number of fillers to produce a series of molding powders. The processing temperatures and pressures are similar to those used with phenolics. The melamine resins are similar to ureas in that they are colorless and can be compounded into any number of light or dark colors. Unlike the ureas, however, they can be compounded with many types of fillers, rather than just mainly cellulose fibers. The cured resins show the highest surface hardness of any plastics. Like the ureas, they do not carbonize when an electric arc strikes the surface, and their water and weather resistance is superior to the ureas. They are used extensively for applications where their combination of hardness, colorability, electrical properties, and heat resistance make them particularly desirable. Their major use, however, is for dishes. When used for dishwear, benzoguanamine is also included in the reaction to produce materials with improved stain resistance.

The three resins described above, the phenolics, ureas, and melamines, are all *condensation* reaction products; that is, they all give off water during the initial resin formation reaction and during the final curing reaction. Thus, to produce a cured material with best physical properties, they must all be molded at substantially high pressures (2000 psi or higher) to counteract the "blowing" effect of the steam produced. The series of resins described below are all *addition* reaction products; thus, they do not give off water during the final cure. Consequently, they may be molded at considerably lower pressures than the condensation resins. The significance of this is considerable. Less costly equipment and tooling can be used, and size limitations, imposed by the necessity for high pressures, can be minimized or eliminated.

Alkyds. Alkyd resins are polyester materials combined with catalysts, crosslinking monomers, fillers, reinforcements, and pigments. Since they are colorless, they can be obtained in many colors, although they are usually furnished in a light tan or beige because of the filler or reinforcement. They have the advantage of a fast cure at a relatively low pressure, and a temperature approximately the same as required for phenolics; for example 165°C (330°F). They are mainly used for electrical applications in such parts as distributor caps and rotors for automobiles and switch parts.

Allylics. Allylic resins are also polyesters, usually diallyl phthalate or diallyl isophthalate. These materials are characterized by excellent dimensional stability, excellent electrical properties, and heat resistance. They are consequently used for high-quality electrical parts.

Epoxies. Epoxy resins may be considered the "glamour" resins in the thermosets, in that they can be cured at very low pressures, at low temperatures, and in very short periods of time. They are, however, among the highest priced of the thermosets. They are used mainly where very thin-walled electrical parts are required, since they can be compounded to show very good flow at the relatively low pressures sometimes required for parts which are epoxy encapsulated by compression molding.

Silicones. Silicone molding materials are usually compounded of silicone resins, an inert silica filler, a small amount of catalyst, a release agent, and possibly a pigment. These resins are characterized by having excellent long-term, high-temperature resistance and electrical properties. They are also relatively high priced compared to the other resins.

Polyimides. Polyimide resins are distinguished by being the most heat resistant of any of the polymers. They can be kept for thousands of hours at temperatures as high as 315°C (600°F), and for short periods of time at up to 500°C (930°F). Because of their heat-resistant properties, they are always combined with similar heat-resistant reinforcements; for example, fiberglass, quartz fibers, and graphite. Condensation-type polyimide resins are molded at high temperatures and pressures similar to phenolic resins. Addition-type polyimides may be molded at lower pressures. They are among the highest-priced plastics materials available.

MATERIALS TESTING

Any of the materials listed above can be obtained in a variety of plasticities, or flows, depending on the conditions of time, temperature, and pressure required to fill varying types of compression or transfer molds. However, since all the materials are thermosets, and are therefore catalyzed, their original plasticity can slowly change after aging, particularly if they are stored at elevated temperatures. It thus becomes important to have a means of specifying the original flow properties, and of testing the material after an aging period.

The oldest plasticity or flow test, suitable mainly for stiff-flow materials used for compression molding, is the ASTM

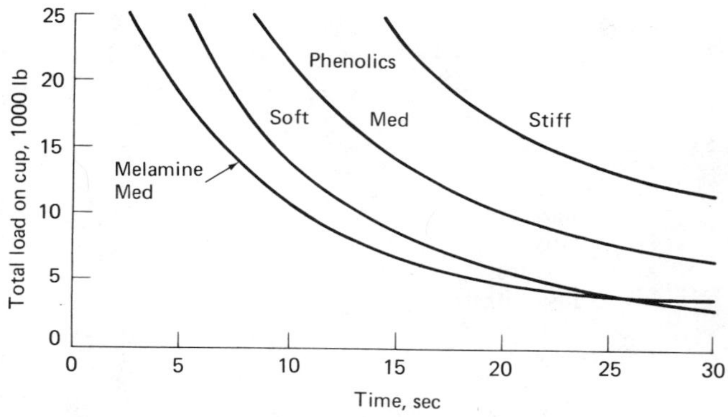

Fig. 16-10. Pressure-closing time curves. Note that curves for different hardnesses of the same material are parallel; a different material curve is not. (Reprinted by permission from *Plastics Technology*, February 1966)

D-731 Cup Flow Test. In this test, a given quantity of material is formed into a small cup at a standard, or specified, temperature and pressure, and the time to close the mold is an indication of the plasticity of the material. The results are reported in seconds required to close the mold at a given pressure. This test does not show whether a material would be suitable for use in any particular mold, but rather indicates if one batch of material is like a previous batch, or if a material has changed significantly after a storage period. Not all material suppliers rate their compression molding materials in exactly the same way, nor are the tests performed in exactly the same way. In many cases, the results on any given material are shown as a pressure-closing time curve, as shown in Figure 16-10. These curves may be furnished by the resin supplier using a standard mold, or, better yet, by the molder using a production mold.

The cup test described above is mainly applicable to stiff-flow, compression molding materials. With the growth of transfer and injection molding thermosets, it was found that the cup test was inadequate to satisfactorily indicate the flow properties of the, softer, high-flow compounds. As a consequence, a number of different tests were devised to measure flow under transfer or injection molding conditions. The most commonly used of these tests are two types of spiral-flow tests in which the resin is injected into the center of a mold, consisting of an Archimedean spiral. The spiral is marked in inches, starting at the center and continuing out for five, ten, or more turns, depending on the mold type. The five-turn mold, known as the Mesa spiral test mold, has a length of 48 inches and is used for transfer molding of materials in the 1000–10,000-psi range. At 2920 psi, the usual flow is 3 inches for a short-flow material, and up to approximately 12 inches for a long-flow compound. The ten-turn mold, known as the IBM spiral, has a total length of 102 inches and is used for testing very soft, long-flow materials used for encapsulation, where the pressures range from 25 to 1000 psi. Figure 16-11 shows the cup sample and the two types of spiral moldings. Unfortunately, even though the cup test is an ASTM standard, it is not done in the same way by all users. In the case of the spiral molds, there is no agreed-upon standard for the industry, at the time of writing, so the method of testing is usually decided upon by the resin manufacturer, or the mold and resin suppliers.

MATERIAL FORMS

In general, all of the above molding materials are available in the form of molding powders which range from fine granules to distinct pellets. In addition, any of the materials may be compressed into small discs, rods, or other shapes, as desired, in order to obtain small amounts of a definite shape or weight for easier handling. The phenolic resins are

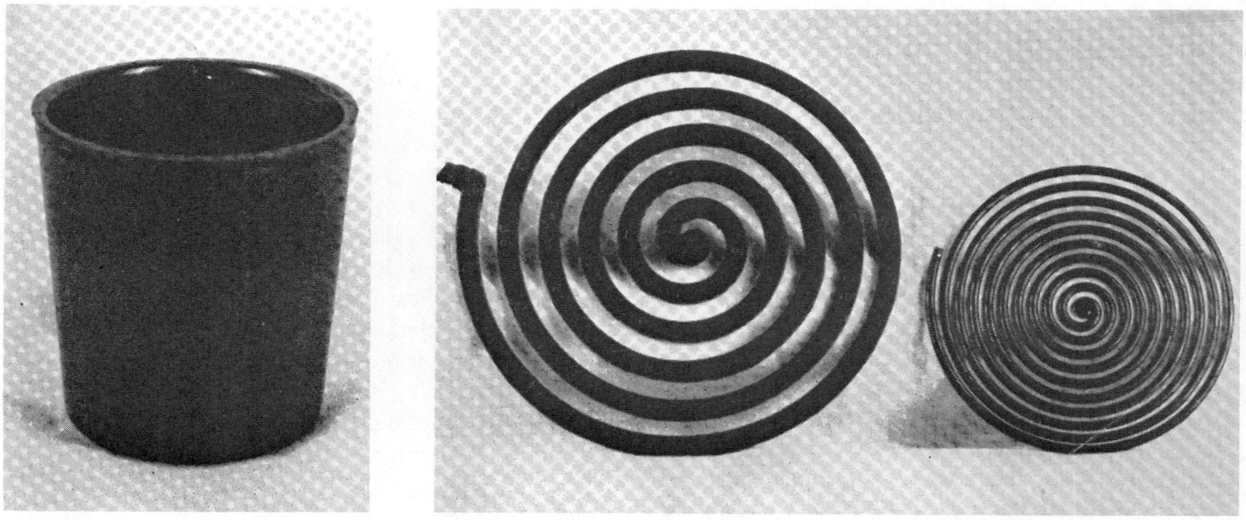

Fig. 16-11. Cup and spiral test pieces. Mesa spiral (center) and IBM spiral (right). (Spiwak, 1965. Reprinted by permission from *Plastics Technology*)

probably the ones available in the greatest variety of forms. In addition to being easily obtainable in the form of powders or pellets, the phenolic resins are also furnished in the form of resin-impregnated, chopped fibers or fabrics. Polyester molding materials are also furnished in the form of a catalyzed, puttylike material which is extruded in the form of rope. The molder can then simply cut off the desired length of material for a part instead of weighing a small quantity. Polyesters are also compounded as large masses of resin mixed with the reinforcement, catalyst, pigment, lubricants, etc. In this form they are called, appropriately, bulk molding compound or BMC. Another term for BMC is premix.

Polyesters are also available in the form of a catalyzed sheet, usually resin-impregnated, chopped-glass fibers or a glass fiber mat, held between two polyethylene films. Such material is called sheet molding compound or SMC. BMC and SMC materials are usually molded at pressures of 100-3000 psi. Regardless of the physical form, all of the materials are molded in compression presses; these differ, however, in size, shape, pressure available, and type of mold used.

Bulk Factor and Preforms

As mentioned above, the type of mold used will differ depending on the material used and the type of press. One of the factors influencing the type of mold chosen is a characteristic of the molding compound known as the bulk factor. The bulk factor is defined as the ratio of the volume of the molded part to the volume of the loose powder, or:

$$\text{Bulk factor} = \frac{\text{volume of loose powder}}{\text{volume of molded part}}$$

Thus, a fluffy material such as flock-filled phenolic would occupy much more space in a mold cavity prior to molding than would a mineral-filled phenolic. After being molded, of course, parts of either material will occupy the same volume. Bulk factors can vary from as low as 1.1 to as high as 18, as shown in Table 16-1. Obviously then if a light, fluffy material is being molded, the cavity would have to be much deeper to contain the material than if a dense molding compound is being molded. In preparing a compression mold, therefore, it is necessary that a cavity be made deep enough to accommodate the highest bulk factor material which might ever be used. In experimental work, or if several different materials are to be processed in the same mold, this could lead to a considerable increase in machining time and cost of the mold.

One method of minimizing the effects of varying bulk factors, with respect to the size of the mold cavity, is by the use of preforms. Preforms are simply small pellets or tablets of molding material that are formed to shape in a special preform mold at room temperature, so no curing takes place, only densification. In addition to densification, or bulk factor reduction, there are a number of other advantages to the use of preforms. These are:

1. The mold charges are of uniform weight and size.
2. Mold cavities are filled more easily, without the waste of loose powder.
3. Preforms heat faster, thus reducing the mold cycle.
4. Because the mold cavity can be made smaller, the mold cost is reduced.

Preforms are usually made in automatic presses at a very rapid rate. The most usual shape is a round, flat-face disc; however, rectangular shapes, balls, and odd shapes are frequently used. The object, of course, is to make a small, dense mass of molding powder in a shape that can be easily handled, will heat up rapidly in the mold, and most impor-

Table 16-1. Bulk Factor of Thermosetting Molding Compounds.[a,*]

Material	Bulk Factor
Alkyd	
Granular	1.8-2.4
Putty	1.1-2.0
Asbestos	1.8-2.5
Diallyl phthalate (DAP)	
Asbestos	2.2-2.3
Orlon	4.2
Dacron	4.2-7.0
Glass fiber	2.2-7.0
Epoxy	
Unfilled	2.0
Mineral	1.8-2.8
Glass fiber	2.5-3.0
Melamine	
Cellulose	4.0-7.0
Asbestos	2.1-2.5
Macerated fabric	5.0-10.0
Glass fiber	5.0-10.0
Phenolic	
General purpose	2.1-4.4
Asbestos	2.0-14.0
Mica	2.1-2.7
Macerated fabric	3.5-18.0
Glass fiber	2.0-10.0
Powdered metal	2.0-4.0
Polyester (premix)	
Mineral-filled	2.0-2.8
Glass fiber	4.0-18.0
Silicone	
Asbestos	6.0-8.0
Mineral	1.7-2.0
Glass fiber	6.0-9.0
Urea	
Cellulose	2.2-3.0

[a]Bulk factor is also known as compression ratio.
*Milby, 1973. Courtesy: McGraw-Hill, Inc.

tantly, will flow satisfactorily to all parts of the mold cavity. There are some times, however, when preforms are not most desirable. Some reasons for not using preforms are:

1. When the run is too small to justify the cost.
2. If the mold is of such an intricate shape that only a powder will satisfactorily fill it.
3. If the material has such a high bulk factor (e.g., chopped-glass phenolic) that it can not be made into good preforms.

While preforms can be made by the individual molding shop using the existing presses run cold, it is usually more economical to have them made by companies that specialize in preforming. Such shops will have the fast-acting presses, similar to drug-tableting machines, and will also have a stock of standard molds for various sizes and shapes of preforms.

As indicated above, preforms are used extensively for long runs when the molds are filled manually, or when the preforms are automatically placed in the mold cavity. In the newest types of molding presses, described below, however, the preforms are made in the press, or preforms are not used at all. Instead, the powder, or pellets, are accurately weighed and placed in the mold cavity, where rapid cure takes place.

Preconditioning

Preconditioning of molding powder includes the forming of the preform and a preheating process, which is used for two purposes. The molding powder or preforms, usually phenolic, are heated at a specific temperature to: (1) remove any moisture from the material which might have been picked up during storage or as the result of removing the material from a low-temperature storage area and opening the package before it was warmed to room temperature, and (2) to reduce the cycle time in the mold by presenting prewarmed powder or pellets to the mold cavity. Three main methods of preheating the molding thermosets are:

1. *A circulating hot-air oven*, heated by gas or electrical heaters. Good air circulation is a must in order to minimize possibility of hot spots. If the molding material is slightly set up, a pan of water placed in the oven will sometimes provide a plasticizing effect so the material will flow more easily. This technique should never be used for parts where maximum electrical properties are required.
2. *Infrared heaters* are also used where faster heating is desired. Such heaters can be incorporated directly in the press hopper. There should be an agitator in the hopper, however, to prevent hot spots. This technique is used largely with automatic presses where molding powders or small pellets are mainly used.
3. *Radio frequency or dielectric heating* is the fastest method of heating and is particularly effective with preforms. The preforms are usually placed in a nonmetallic paper or plastic tray which is positioned between two flat-plate electrodes in the preheater. By action of the radio frequency current (27-80 megahertz to as high as 140 megahertz) the preforms can be brought to a temperature of 130-150°C (270-290°F) in 7 to 10 seconds. The preforms are heated from the inside out, so that maximum removal of volatiles occurs, with minimum stresses developing in the preform. The process is similar to the cooking action in the familiar home microwave oven which, in fact, was developed from the plastics preheater. In placing a batch of preforms in the preheater, it is important that all be of the same thickness so that all may be heated at the same rate. It is also important that an automatic timer be used, rather than manual timing, again to insure maximum uniformity. By adjusting the cure cycle to take advantage of the preheated material, it is possible to reduce the overall cure cycle 20-40%. Figure 16-12 shows a typical radio frequency preheater.

In addition to reduction in the overall cycle time, it has been found that preheated material results in parts with better dimensional stability and more freedom from internal stresses and strains. This is due to the fact that the

Fig. 16-12. Radio frequency preheater for thermosets. (*Courtesy: Reeve Electronics, Inc.*)

material flows easier in the mold than does non-preheated material. Because of this easier flow, larger parts can be made in a given press than would normally be possible.

One special method of heating, applicable only to premix materials, is the heating of the premix ingredients during the actual mixing operation. In this case, the materials are mixed in a heated pot or container and then are immediately molded. This process is particularly well suited for automation.

THE MOLDING PROCEDURE AND MOLDS

Compression Molding

As previously described, the molding procedure for thermosetting materials consists basically of placing a weighed or measured charge of material into a heated mold and then holding the resin in the closed mold until cure has taken place. In the case of a compression mold, the resin is placed into the open cavity, or female section of the mold, and then the male (also called force or punch) section of the mold is lowered into the cavity to compress the resin and force it to assume the mold contours.

The description given above details only the essentials of the process. For optimum results with any given resin, the following additional factors must be considered:

1. Mold Temperature. The mold must be heated to the correct temperature for that particular resin and mold. These temperatures range from 130-180°C (270-360°F) for phenolics, ureas, melamines, epoxies, and polyesters. Polyamide-imides can range up to 350°C (650°F).
2. Mold Pressure. The mold pressure to be used must be adjusted for each resin and mold. This will vary roughly from 1000 to 6000 psi and will also vary depending on whether the resin is preheated or not. If preheating is used, the pressure can be reduced considerably. In addition, the pressure will vary with the cavity depth. A rule of thumb is to add 700 psi for each additional inch of depth when molding conventional high-pressure phenolics. For low-pressure materials, the rule is to add 250 psi for each inch of depth.

   For depths over 4 inches, preheating is recommended, and for cavities over 12 inches it is a must.
3. Press Closing Speed. Depending on the particular resin and/or mold configuration, the press closing speed should be adjusted to result in the fastest closing commensurate with good resin flow and no mold damage.
4. Degassing or Breathing. Before final closing of the mold, it is usually degassed or allowed to breath. This procedure consists of bringing the mold almost to full closure and then opening it slightly one or more times to allow the entrapped air, moisture, and liberated volatiles to escape. This action helps to insure a sound part, free of internal voids, blisters, or entrapped air pockets.
5. Final Cure Time. The final cure time may be as low as 20 seconds for a very small, thin-walled part, 1-3 minutes for larger parts, and as long as 24 hours for some massive, thick-walled aerospace rocket nozzles. The objective here is to produce a completely cured part, free of overcured or undercured sections. In the case of light-colored epoxies, polyesters, ureas, and melamines, overcure can usually be detected by a change in the color. Phenolics, because of their predominantly dark colors, can not be so easily checked. Undercured parts, of any type, are difficult to detect unless the undercure is so severe that the parts are noticeably soft. Where undercure is suspected, there are various solvent-extraction tests which show whether a part is cured satisfactorily or not.

MOLDS

There are basically five different types of compression molds. They differ only in the amounts of compression which can be applied to the molding powder, and in the methods used to "land" the male portion of the mold to obtain an exact size.

Flash Mold

The simplest mold type is the flash mold. Figure 16-13 shows a typical flash mold, the ejector pins used to remove the part from the mold, and the mold cores used to contain electric heaters or steam lines. In Figure 16-13a, the preform is shown in place, prior to molding. Figure 16-13b shows the cured part with the flash forced out just as the mold closes. A flash mold is usually fairly inexpensive to make and produces uniformly dimensioned parts. It does, how-

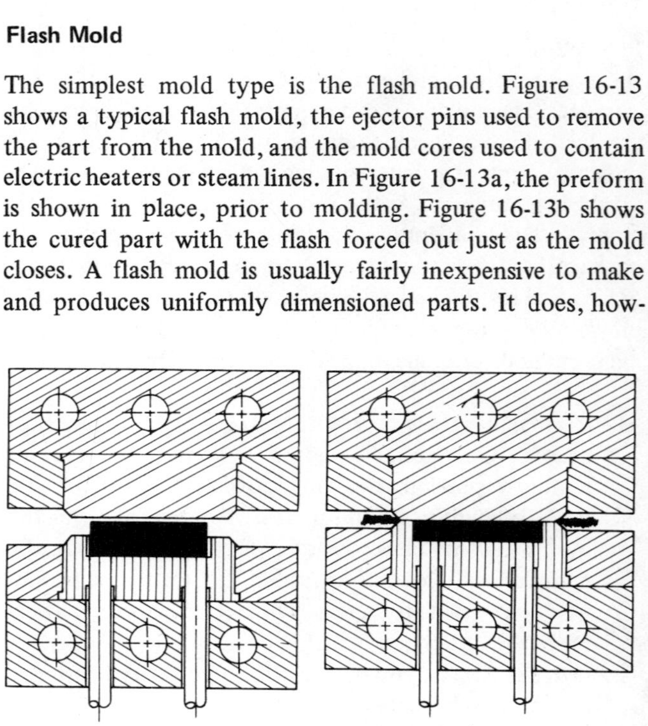

a. b.

Fig.,16-13. Open flash mold with preform in place, at left, and closed, at right, showing excess material forced out as flash. (Reprinted with permission from *Modern Plastics Encyclopedia* 1968. Copyright 1968 McGraw-Hill, Inc. All rights reserved.)

ever, have some drawbacks. Probably the most important is the fact that the parts produced are not of the highest density, since the molding material is never put under very high compression. Also, in order to assure complete mold filling and some compression on the part, the preform is usually 10–20% heavier than is required to make the part. Thus, this type of mold is wasteful of material. Nevertheless, flash molds are used wherever large quantities of small, thin parts, such as poker chips, are not needed in the highest quality.

Fully Positive Mold

The fully positive mold, shown in Figure 16-14, is the opposite of the flash in that it will produce parts of the highest density, since the full compression load can be applied on the molding and not on the mold. However, it does have a disadvantage in that, unless the amount of molding material is weighed very accurately, the parts will not be the same size each time in the direction of the applied pressure. Consequently, other mold types are used in an attempt to combine the dimensional accuracy of the flash mold and the good density obtained in a fully positive mold.

Semi-Positive Mold

The semi-positive mold, illustrated in Figure 16-15 combines some of the compression loading possible with the fully positive mold with, at the last moment, a flash cut-off and a land to assure good dimensional accuracy. A minor disadvantage of this type is the fact that, if the land is made very narrow, so as to cause a quick, thin cut-off of the flash, the lands will wear very quickly, thus resulting in dimensional instability.

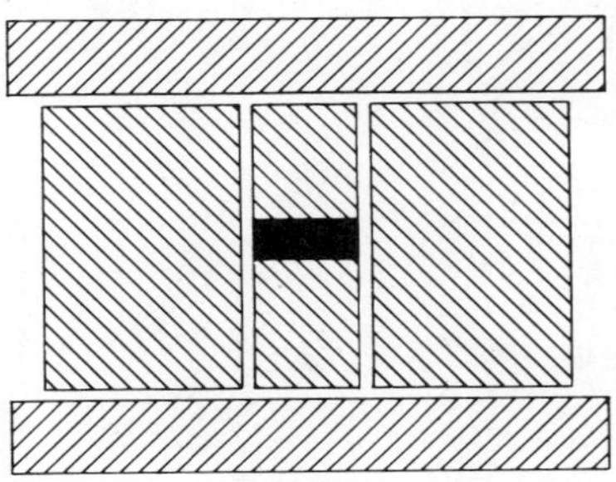

Fig. 16-14. Fully positive mold.

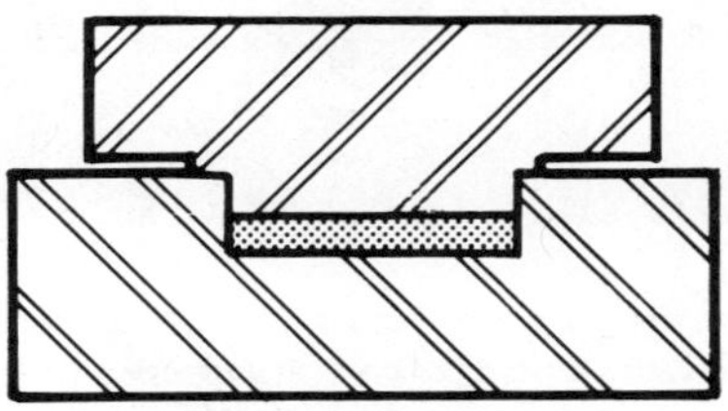

Fig. 16-15. Semi-Positive mold.

Fig. 16-16a. Internally landed positive mold.

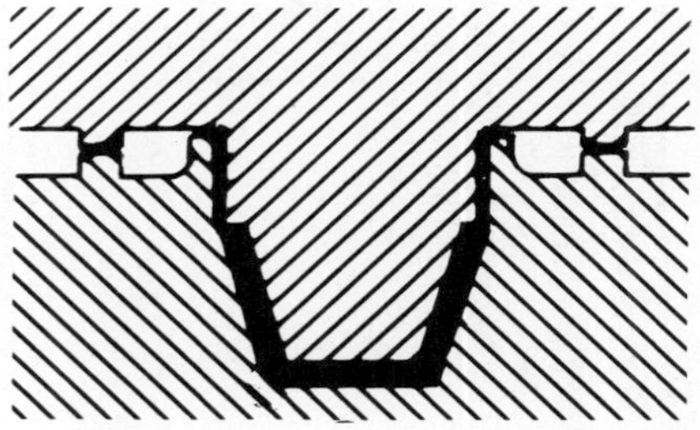

Fig. 16-16b. Externally landed positive mold.

Landed Positive Mold

Landed positive molds are similar to semi-positive molds except that the telescoping portion is deeper to allow more compression on the molding prior to landing. Usually this type of mold incorporates substantial lands which will not wear down quickly. Figure 16-16a shows a landed positive mold with internal lands and Figure 16-16b shows a landed positive mold with external lands. Despite the deeper telescoping of the male into the female mold portion, the final compression load on the part is not as great as in the fully positive mold. Dimensional accuracy is better however.

TRANSFER MOLDING

The previous discussion of the molding procedure applies to either compression molding or transfer molding; however, the description of the molds applies only to compression molding. In the transfer molding process, the molds are always used in the closed position, thus assuring good dimensional accuracy and uniformity in all parts. The male and female components of the mold are then made in whatever form or shape is best for molding and removal of the part.

As described earlier, transfer molding consists of heating and liquifying a thermosetting resin in one chamber and then transferring the liquid into the mold cavity under sufficient pressure to fill the cavity and produce a dense uniform part. There are two methods of transfer molding; both are essentially a transfer process but slightly different equipment is used for each. One method is called "pot" transfer and employs an ordinary compression molding

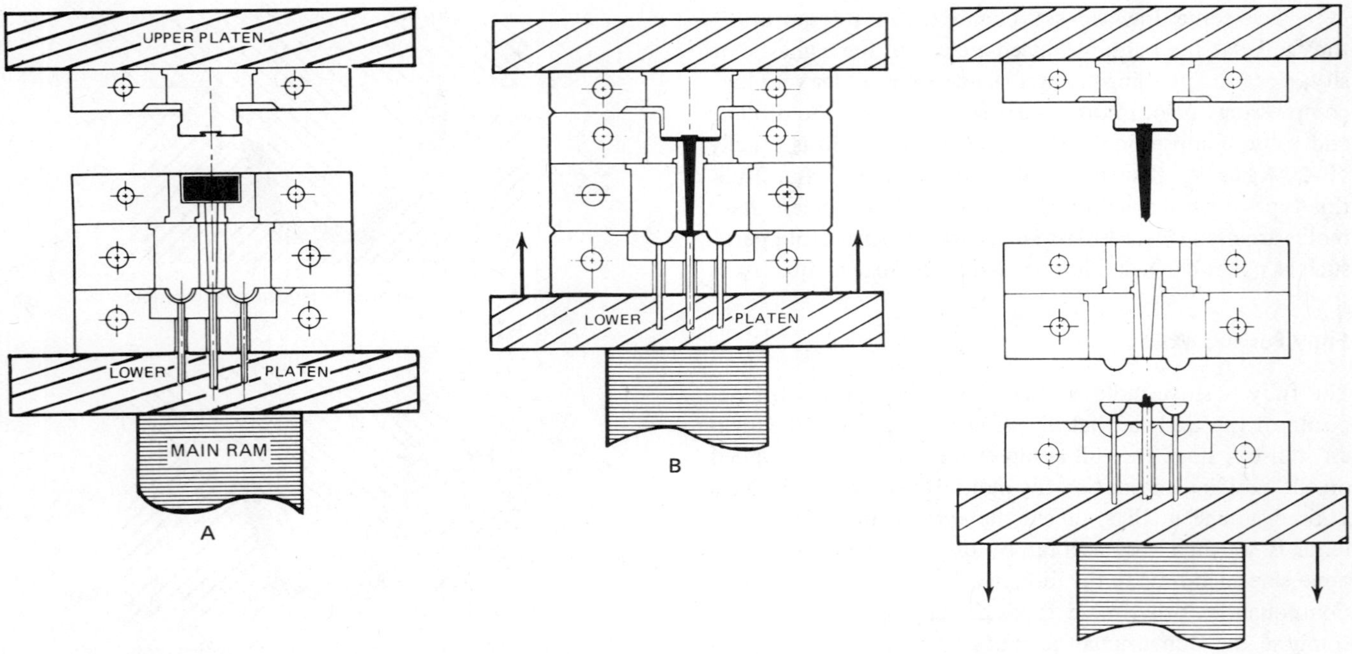

Fig. 16-17. "Pot" transfer molding. A: the preform is placed in the "pot" of the partially opened mold. B: the mold is closed, forcing the resin into the cavities. C: the mold is completely opened, allowing removal of the parts and the sprue as cull. (Ejection pin mechanism not shown.)

press and a mold with a built-in "pot," or transfer chamber, where the resin is heated and liquified. The plunger which forces the liquid resin into the cavity is part of the upper section of the mold, and it is actuated by the press platen, as shown in Figure 16-17. This process requires no special equipment, other than the "pot" mold. An important design consideration is that the area of the pot must always exceed the area of all cavities, sprues, and runners by at least 15%, since the total force exerted by the pot is required to keep the mold from opening and flashing when the liquid resin is injected into the cavities.

The second type of transfer process is called "plunger" or auxiliary ram molding. This method employs almost the same type of mold used in the "pot" process, except that the material is injected by use of an auxiliary ram which is part of the transfer press (see Figure 16-18). The press platens here are used only to hold the mold sections together at the required clamping force. This clamping force must always be 20–25% above the total force which might be exerted by all the cavities, sprues, and runners when operating under full auxiliary ram pressure. If this is not so, then the mold will tend to open during injection. Figure 16-19 illustrates a "plunger" mold.

Fig. 16-18. Transfer molding press. Note auxiliary ram which extends through the upper platen. (*Courtesy: Clifton Hydraulic Press Co.*)

Transfer Mold Design

In both types of transfer molds, the liquid resin is injected into the cavity or cavities by means of a sprue, runners, and gates. For best results, the mold design should take into account the type and number of parts to be molded and the

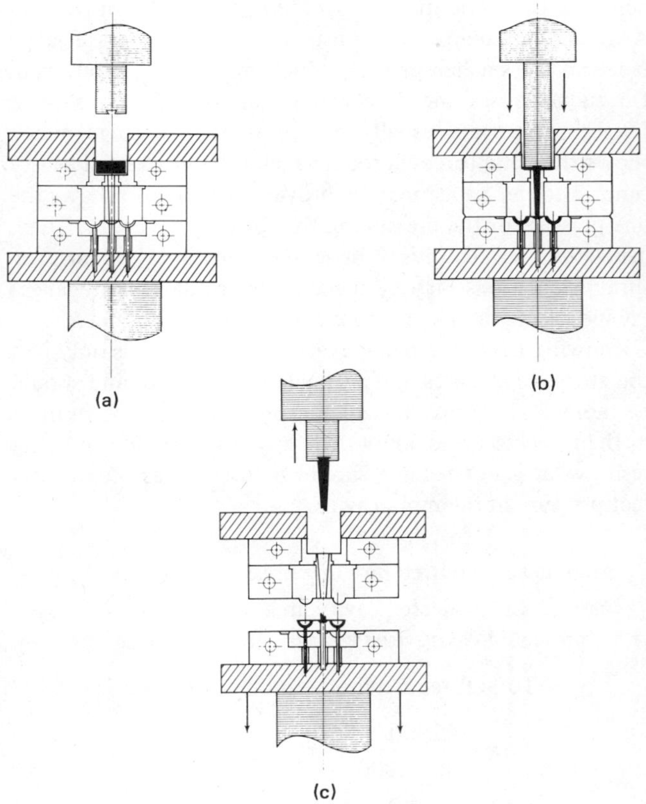

Fig. 16-19. "Plunger" type transfer molding.

material to be used. Other important considerations are discussed in the following sections.

Gates. The gates should always be as small as possible so the part can be broken off with a minimum of marking. However, they should not be so small that they cause a severe restriction in flow. Thus, an initial design should have the smallest gate possible, which can then be enlarged if necessary. (See Chapter 20 for an extended discussion of sprues, runners, and gates.)

Runners. Runners should be as short as possible while maintaining minimum flow restriction. Runner shape is important. The best and most expensive runners are full-round. The most popular are half-round or trapezoidal in cross-section, since they are easily cut into one section of the mold.

Vents. Vents for air removal are very important. They most often are in the form of small grooves in the parting line of the mold from the edge of the cavity to the outer mold edge. In most cases, the vent is placed just opposite the gate entrance. The vents are usually 1-3 mils deep and 25-125 mils wide. They allow hot air and gases to escape which might cause porosity and/or a burned spot on the part if these were to remain in the cavity.

Advantages of Transfer Molding

Molding Cycles. The loading time can be reduced since one large, preheated preform can be used. The preform itself can be of a simple shape. In going through the runners and gates, the flow restriction can be large enough to add a sensible quantity of heat to the liquid resin, thus saving time in the chamber.

Tighter Dimensional Tolerances. Since the material is injected into a closed mold, there is little or no flash, and dimensions across the parting line are held closely. Also, since there is little flash, finishing costs are lower.

Insert Molding. Because the material enters the cavity as a liquid, the pressure against fragile inserts is less, and there is less tendency for pins or thin mold sections to bend.

Encapsulation. Transfer molding is the only feasible method to encapsulate delicate electronic parts such as capacitors and transistors, where maximum density is required with low molding pressures.

"Family" Molds. Transfer molding is well suited for producing parts of different sizes which must be assembled together. Here, one "shot" allows production of several parts all connected together by the runners, and all cured at the same time. In compression molding, each part would probably be molded separately and cured for different times, which might result in slight color differences.

Mold Costs. The mold costs in some cases may be less than for a comparable compression mold because the cavity can be made smaller, since there is no necessity for a deep well, and mold sections can be made thinner, since the stresses on the mold are lower in a transfer mold. The wear and tear on a transfer mold is usually less than on a compression mold so that dimensional accuracy is kept longer and refinishing is not required as often.

Disadvantages of Transfer Molding

Mold Costs. While some transfer molds might cost less than a comparable compression mold, in general "pot" type molds cost more than plunger molds because of their additional complexity. These costs may be, in some cases, more than a compression mold used to make the same article.

Material Costs. Per part, the material cost is higher for transfer molding since the cull (material left in the well, sprue, runners, and gates) cannot be reused. For small parts, this extra material may be a sizable factor. Automatic compression molding may be cheaper when very large quantities of simple, uniform cross-section parts such as bottle caps, buttons, etc., are required.

Mechanical Properties. Under certain circumstances, the physical properties of transfer-molded parts can be less in particular directions. An example might be a tendency to crack around inserts or at weld lines where the reinforcement materials might line up in one direction, instead of being homogeneously scattered in all directions.

PRESS CAPACITIES AND PRESSURE CALCULATIONS

Whether a part is compression molded or transfer molded, there is usually a specific pressure which should be used to get the optimum part. In most cases, this particular pressure must be calculated using the hydraulic line pressure, ram size, and the cavity sizes. When auxiliary rams are used, this further complicates the calculations.

Molding presses, particularly compression and transfer presses, are rated in terms of the press capacity; that is, the maximum force which the press can exert. Thus, a 20-ton press would be capable of exerting a total *force* of 40,000 pounds on a mold. It should be noted at this point the term *force* is used to indicate the total *load* which could be placed on a mold. The force is derived from the hydraulic fluid pressure, in terms of pounds per square inch (psi, in the United States), times the area of the ram which is being pushed by the fluid. The basic formula is then:

$$\text{Force} = \text{Line Pressure} \times \text{Ram Area}$$

As an example, a compression press has a ram with a diameter of 8 inches and operates at a line pressure of 3000 psi. What is the maximum press capacity?

$$\text{Maximum Force} = \frac{3000\ (\text{psi}) \times 4^2 \times \pi\ (\text{in}^2)}{2000\ \text{lb/ton}}$$

$$= 75\text{-ton capacity}$$

Another frequently needed calculation is the force, in tons, exerted by the press at any given pressure. To simplify such calculations, it is well to use a constant, "K," which is the ram area divided by 2000 (area/2000). For the 8-inch-diameter ram mentioned above:

$$K = \frac{50\ (\text{rounded-off})}{2000} = 0.025$$

Thus, 0.025 times the indicated pressure at any time gives the force in tons. Obviously, K for a different ram diameter would be different; for example, $K = 0.014$ for a 6-inch ram and for a 12-inch ram, $K = 0.057$.

Some presses are equipped with gauges to show the line pressure at any given time. Thus, to know the total force being exerted, one must know the ram diameter or the ram area. The ram diameter will usually be marked on the gauge face on the smaller presses, which may be hand pumped. On such presses the K constant can be used as shown. Larger presses are usually equipped with a gauge which shows the total force, in tons, which is being exerted at any time. Another gauge may be provided which also shows the line pressure or the pressure in the hydraulic cylinder. Then, using the formula given above and knowing the ram pressure, calculations may be made to determine if both gauges are indicating the same amount of force at any time.

Knowing the force being exerted on the mold is only half the story. In all cases, the particular resin compound should be molded at a pressure which is optimum for that resin. It is then necessary to know, for any given mold, and that resin, what gauge reading should be used to assure the correct pressure in the mold cavity.

Assume the required molding pressure to be 2500 psi
Assume the projected cavity area to be 8.5 in^2
(projected cavity area = cavity area at parting line)

$$\text{Total force required} = \text{pressure} \times \text{area}$$

$$\text{Force} = \frac{2500 \times 8.5}{2000} = 10.6\ \text{tons}$$

If the press gauge can be read directly in tons, then the pressure is simply increased until the gauge reads 10.6. However, if the press has a pressure gauge calibrated in psi, then further calculations are necessary.

Assume the press has an 8-inch diameter ram.
To determine the ram pressure needed, in psi, use the equation:

$$10.6 = \text{pressure} \times K$$

$$K = 0.025\ (\text{previously determined})$$

$$\frac{10.6}{0.025} = 424\ \text{psi}$$

Similarly, a 6-inch-diameter ram would require 757 psi to exert the same force (10.6 tons), and a 12-inch ram would require 188 psi.

Another technique for determining the hydraulic pressure to be used for any given cavity is to simply compare the areas of the ram and cavity, and multiply the ratio times the required cavity pressure.

In the example used above, the area of the 8-inch ram is approximately 50 in^2 and the cavity area is 8.5 in^2.

$$\frac{8.5}{50} \times 2500 = 425\ \text{psi}$$

For a 12-inch ram, the area is 113 in^2. Thus:

$$\frac{8.5}{113} \times 2500 = 188 \text{ psi}$$

If a press is equipped with an auxiliary ram for transfer molding, it is vital that the total force exerted by the cavities, runners, and sprue(s) (which comes from the auxiliary ram) does not exceed the force exerted by the main ram to keep the mold closed. Should this happen, then the mold would tend to open during injection and produce excessive flash. To guard against this, the force of the main ram should always exceed the force exerted by the auxiliary ram by a ratio of approximately 5:1. Thus, even when the total area of the cavities, sprue(s), and gates is considerably higher than the auxiliary ram area, the main ram force will always keep the mold closed.

AUTOMATIC MOLDING

With the increased demand for thermosetting products and the competition of the rapid molding thermoplastics, it was inevitable that some form of automatic molding for thermosets would be developed.

The first "automatic" presses were simply compression presses. Molding powder was automatically weighed out for each cavity and was then dropped in. After a short period of cure, the parts were ejected from the opened mold and an arm or a "comb" would sweep the parts into a bin and the cycle would be repeated. These first presses were soon combined with preformers and with preheaters, so the cycle time was reduced. Somewhat the same cycle took place for the first transfer molding presses.

With the development of the screw injection molding press, it also seemed inevitable that the same technology would be applied to thermoset molding. There are, however, some basic differences in thermoplastic and thermoset injection molding machines. In both processes the screw is used to heat and convey the resin to the mold. In the thermoplastic press, the screw is made with a definite compression ratio so it both heats and conveys the resin. In the thermoset machine, the screw is used mainly to convey the resin to the nozzle. The barrel temperature on a thermoplastic press can run as high as 315°C (600°F), but on a thermoset press the temperature is normally maintained in the 65–150°C (150–300°F) range in order to prevent premature cure in the barrel. The mold temperatures are reversed in a similar way. In the thermoplastic press, the hot resin is injected into a cold mold where it "sets up"; in the thermoset process, the warm resin is injected into a hotter mold in order to cause the material to cure.

The barrel in a thermoset press is generally shorter than in a thermoplastic press of a corresponding size. Also, temperature control of the barrel, and the mold, must be very accurately maintained so there is sufficient softening of the resin to allow good flow, but, at the same time, it must prevent cure in the barrel. The mold must also be hot

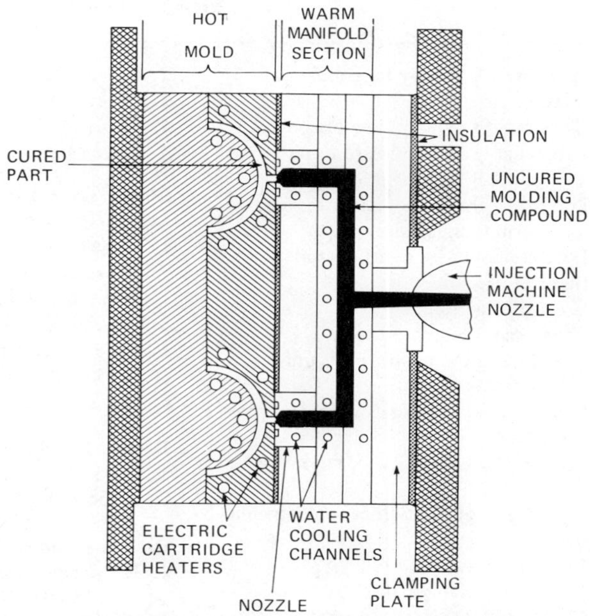

Fig. 16-20. Typical warm runner mold for thermosets. (*Courtesy: Union Carbide Corporation*)

enough to cure the resin rapidly, but, at the same time, it must allow the resin to completely fill the mold prior to onset of cure. Thus, good temperature regulation is a must.

Today, thermosetting resins are successfully molded in screw machines and in screw machines which also use rams for the actual injection of the resin in the mold. In both cases, neither preforms nor preheating is used.

Keeping pace with machinery developments, the resin manufacturers are able to adapt for automatic machines materials which cure very rapidly, so production is able to compete effectively with thermoplastic parts on a time-cycle basis.

Automatic equipment costs considerably more than manual or semiautomatic equipment. However, the automatic machines may be as much as two to five times as fast as the hand-operated compression or transfer presses. One device being used with some automatic transfer equipment to save time and material is a "warm runner" mold, in which the sprue and runners do not set up and the formerly discarded cull is no longer wasted. Figure 16-20 shows the type of mold used where the sprue and runner are kept at 90–110°C (200–230°F). For best results with this type of mold, a special resin is used, one with a long liquid "life" at an intermediate temperature, but which cures rapidly at the higher curing temperature. Warm runner molding is also called "cold runner," "hot manifold," or "runnerless" molding.

TROUBLESHOOTING

Table 16-2 is a troubleshooting guide for thermoset molding. Using one or more of the suggested remedies listed below each of the defects can often effect dramatic improvements.

Table 16-2. Troubleshooting Checklist for Molded Thermosets.*

Defect: Blistering, Heavy Section
1. Increase cure.
2. Close mold slowly on low pressure, then apply high pressure to purge gas.
3. Vent mold if plunger molding.
4. Preheat to eliminate moisture. Increase preform temperature.
5. Breathe mold as soon as possible.
6. Increase pressure.
7. Use stiffer material.
8. Lower mold temperature.
9. Check mold temperature for uniformity.
10. Vent mold at dead ends with vent pins.

Defect: Porosity (not filled out in every section)
1. Try a range of plasticities to determine best grade.
2. Increase pressure.
3. Lower mold temperature.
4. Close mold faster if external porosity, apply high pressure sooner; close mold slower if internal porosity, apply low pressure longer.
5. Preheat material.
6. Adjust charge weight.
7. Try a higher density material to obtain greater "back pressure" in the mold.
8. Breathe mold earlier.
9. Use semi-positive molds to minimize porosity.

Defect: Mold Sticking.
1. Raise mold temperature.
2. Preheat to eliminate moisture.
3. Clean mold.
4. Use stiffer material, particularly with one-steps.
5. Remove dents and undercuts in mold cavity and force.
6. Polish mold.
7. Increase cure.
8. Reduce pressure; use softer material if plunger molding.
9. Determine whetner knockout pins are properly adjusted.
10. Be sure cavities have enough draft–¼° recommended.

Defect: Dull Surface
1. Adjust mold temperature–usually reduce temperature for better appearance.
2. Polish mold.
3. Chrome plate.
4. Close mold more slowly, longer on low pressure.

Defect: Orange Peel (rough surface with extremely pimply surface)
1. Close slowly, on low pressure, delay 2 or 3 sec. before applying high pressure.
2. Preheat (radio frequency).
3. Use stiffer material.
4. Use finer ground material.
5. Use large preforms and few as possible.
6. Use lower mold temperature.

Defect: Pitted Surface
1. Use stiffer material.
2. Preheat.

Defect: Clouded or Segregated Surface
1. Reduce molding temperature (for colors, 300 to 310°F).
2. Preheat preforms.
3. Adjust plasticity.
4. Adjust pressure and closing speed.

Defect: Dimpled Surface
1. Close slowly, longer on low pressure, no breathe, then high pressure.
2. Use stiffer material.
3. Lower mold temperature.
4. Increase charge weight.
5. Shape preforms more closely to shape of piece.
6. Increase pressure.

Defect: Warped Piece
1. Heat mold more uniformly.
2. Use stiffer material.
3. Preheat if soft plasticity.
4. Increase cure.
5. Use lower shrinkage material and/or material more rigid at discharge.
6. Redesign mold; use larger diameter knockout pins.
7. Adjust temperature of force and cavity–lower temperature may reduce warpage.
8. Preheat for irregular shaped pieces.
9. Cool uniformly and use shrink fixtures–a one-, perhaps 2-dimensional fixture.

Defect: Cracking after Molding
1. Increase wall thickness around inserts.
2. Check knockout pins.
3. Correct size of shrink fixture.
4. Use more flexible material.
5. Increase radii and ribs.
6. Oven anneal 2 hr. at 200°F.

Defect: Poor Electrical Properties
1. Close slowly.
2. Preheat ½ hr. at 185°F.
3. Increase cure.
4. Reduce temperature.
5. Use stiffer grade.

Defect: Weak Mechanically
1. Increase temperature.
2. Increase cure.
3. Increase pressure.
4. High frequency preheat.
5. Increase charge.
6. Afterbake to increase strength, if economical. Try 2 hr. at 250°F.

Defect: Pimpling
1. Breathe mold.
2. Use higher fines material.
3. Use softer material.
4. R. F. preheat.
5. Lower mold temperature.

Defect: Burned Marks
1. Reduce preheat temperature.
2. Reduce mold temperature.
3. Plunger molding: open gates; be sure vents are clear; check plunger travel time to see if it is between 5 and 12 sec. (faster close will entrap gas; slower close will cause burn marks).
4. Avoid or minimize breathing.

Defect: Thick Flash
1. Reduce mold charge–particularly important in semi-positive molds.
2. Reduce mold temperature.
3. Increase high pressure.
4. Close on low pressure, apply high pressure sooner.
5. Eliminate breathe; follow (4).
6. Use softer grade of material if more pressure not available.
7. Be sure of at least 0.002 in. between force and cavity.
8. Be sure overflow grooves in cavity or force are clean.
9. In plunger molds, reduce injection pressure and/or increase clamping pressure.

*Reprinted by permission from *Modern Plastics Encyclopedia*, McGraw-Hill, Inc.

HIGH-PRESSURE LAMINATING

A process somewhat similar to compression molding is the production of laminated sheet stock, rods, and tubes. Most laminates consist of phenolic resin-impregnated paper, woven cotton fabrics, and, to a lesser extent, asbestos and glass fabrics. Other laminates are available in which the resin is either melamine, silicone, epoxy, or polyester. These laminates are made up as raw stock, similar to metal sheets and shapes, and are sold as raw materials to be

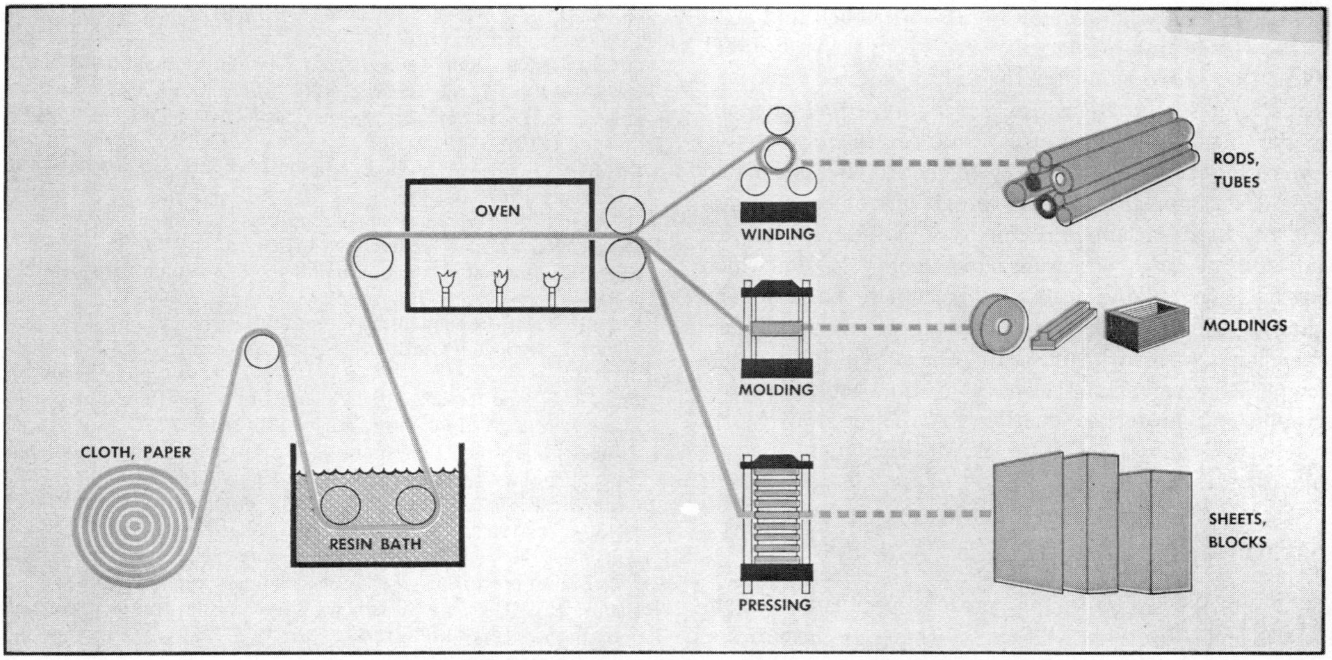

Fig. 16-21. High-pressure laminating processes. (*Western Plastics*, July 1965)

fabricated into components of electrical assemblies, timing gears, and a variety of mechanical devices.

The general process for production of such raw stock (see Figure 16-21) is to lay up sheets or rolls of resin-impregnated paper or fabric until the desired thickness is obtained. The laminated stack or roll is then placed between highly polished chrome-plated platens or rollers, and subjected to heat and pressure until the material is thoroughly cured. The usual temperature range is 135–195°C (280–380°F), and the pressure may range from 2000 psi to as high as 5000 psi. In the production of laminates, the press usually contains several heating platens so a number of laminate

Fig. 16-22. Typical laminating press used to make sheet stock. (Press manufactured by The French Oil Mill Machinery Company, Pigua, Ohio)

stacks are cured simultaneously. Tubular stock and solid rods are made in a similar manner, except using different molds. Decorative laminates for use as counter tops, wall boards, and similar items are made using paper phenolic for the majority of the plies, and for the next-to-the-topmost ply, a one-sheet layer of melamine-impregnated fabric or an impregnated-paper rotogravure of a wood grain pattern is used. Melamine is employed as the resin impregnant since it cures as a transparent, colorless film which allows the underlying decorative pattern to be seen. The topmost ply is usually a thin sheet of alpha-cellulose impregnated with melamine resin. On curing, the top ply becomes transparent and serves as a thin, hard, abrasion-resistant and protective coating for the decorative ply. Figure 16-22 shows the type of wide-bed press used to make the laminates.

POSTFORMING TECHNIQUES

A particular thermoforming process peculiar to high-pressure laminates is known as postforming. In this process, laminates, either fabric- or paper-based phenolics, with or without a decorative melamine top surface, can be thermally formed *once* into simple curves or shallow draws. The laminates are commercially available from certain of the high-pressure laminates suppliers as postforming grade. Essentially, the materials are laminates which have not been completely cured. The forming operation completes the cure and changes the material to a fully cured thermoset. Postforming grades are used in applications such as coved or curved sink tops and industrial parts.

REFERENCES AND BIBLIOGRAPHY

Allied Chemical Corporation. *Mesa Spiral Test*. Quality Control Procedure No. 113, November 8, 1967.

——. *Spiral Reliability Check*. Quality Control Procedure No. 114, November 8, 1967.

——. *Diall Molding Compounds Molding Techniques*. Bulletin 841-85, 1966.

——. *Molding and Handling of Plaskon Urea and Melamine Molding Compounds*. Bulletin 80-113, 1965.

American National Standards Institute. *Measuring the Spiral Flow of Low-Pressure Thermosetting Molding Compounds*. Bulletin D 3123, 1972.

——. *Molding Index of Thermosetting Molding Powder*. Bulletin D 731.

Bainbridge, R. W. Phenolic molding compounds. *Plastics Design & Processing*, December 1967, **7**(12):26.

Bauer, S. H. Processing plastics–recycle thermoset scrap effectively. *Plastics Engineering*, March 1977, **33**(3):44.

——. Thermoset molders can recycle scrap into filler and profit. *Plastics Design & Processing*, April 1977, **17**(4):25.

——. Now–Injection mold all types of thermosets. *Plastics Design & Processing*, October 1972, **12**(10):11.

Borro, E. F. Sr. New processes in molding thermosets. *Western Plastics*, June 1962, p. 29.

——. Compression and transfer molding. *SPE Journal*, July 1962, **18**(7):785.

Durez Division, Hooker Chemical Co. *Durez Molder*. Issue No. 3, 1976.

Deeley, C. W., and Terenzi, J. F. Guide to moldability tests. *Modern Plastics*, August 1965, **42**(12):111.

Dreger, D. R. Smooth, ripple-free plastic panels. *Machine Design*, June 21, 1979, **51**(12):108.

Du Bois, J. H. Advances in thermoset molding. *Plastics World*, February 1967, **25**(2):46.

Duncan, G. G. Injection mold reinforced thermosets–with fewer headaches. *Plastics Technology*, August 1973, **19**(8):35.

Erley, W. Thermoset injection molding–for the uninitiated. *Plastics Design & Processing*, January 1973, **13**(1).

Fina, P. E. Consistency in thermoset molding requires updated compound testing. *Plastics Design & Processing*, October 1977, **17**(10).

Fina, P. E., and Hersch, P. Make sense out of flow tests for thermosets. *Plastics Technology*, February 1966, **6**(2):27.

Fishbert, L. D., and Longstreet, D. C. New data on injection molding phenolics. *Modern Plastics*, June 1969, **46**(6):92.

General Electric Co. *Phenolic Injection Molding*. Bulletin CDC-465, November 1967.

Grigor, J. M. Jr. The reciprocating screw injection machine for thermoset processing. *SPE Journal*, January 1968, **24**(1):63.

Hauck, J. E. Thermoset plastic parts mold faster, cost less. *Materials Engineering*, May 1978, p. 78.

——. Transfer molded epoxies upgrade electronic parts. *Materials in Design Engineering*, March 1966, p. 76.

Hersch, P. Thermoset molding: a state of the art report. *Plastics Technology*, August 1968, **14**(8):33.

——. Reciprocating screw. *Plastics Technology*, May 1967, **13**(5):35.

Hoffman, K. R. How to evaluate moldability of thermosets. *Modern Plastics*, April 1967, **44**(8):131.

Hoffman, K. R., and Vellturo, J. W. What processing does to reinforced thermoset molding compounds. *Plastics Technology*, July 1970.

Hosford, R. C. News in thermosetting processing. *Plastics Technology*, May 1964, **10**(5):43.

Hull, J. L. Transfer molding–past, present and future. *SPE Journal*, January 1962, **18**(1):87.

——. Important considerations in the compression and transfer molding of epoxies. *Plastics Design & Processing*, December 1961, **1**(12):18.

Keegan, J. F. Trim costs of thermoset molding. *Plastics Technology*, May 1974, **20**(5):40.

Lipphardt, D. R. A molder's report on thermoset screw plastication. *Modern Plastics*, July 1967, **44**(11):114.

McGlone, W. R., and Keller, L. B. Molding shrinkage. *Modern Plastics*, Part 1: February 1958, **35**(6):125. Part 2: March 1958, **35**(7):117.

——. Volume and pressure effects in compression molding. *Modern Plastics*, Part 1: March 1957, **34**(7):173, Part 2: April 1957, **34**(8):137.

McNamara, J. B. How uniformity in molding powder influences processing of phenolics. *Modern Plastics*, November 1963, **41**(3):135.

Milby, R. *Plastics Technology*. McGraw-Hill Book Co., New York, 1973.

Miller, H. L. The effects of humidity on dimensional stability of post-baked phenolic parts. *Plastics Design & Processing*, October 1975, **15**(10):19.

Modern Plastics. Transfer molding gets a lot faster–and a lot more efficient. January 1979, **56**(1):82.

——. Thermoset injection molding becomes more reliable and more economical. June 1979, **56**(6):44.

——. Compression mold in seconds. April 1971, **48**(4):98.

——. New standard spiral flow test. February 1968, **45**(2).

——. Injection molding–a new day for thermosets. December 1966, **44**(4):96.

Naitove, M. H. "Dry" pelletized polyesters bid for phenolic applications. *Plastics Technology*, April 1977, **23**(4):95.

Paci, R. Screw transfer vs. screw in-line thermoset molding. *SPE Journal*, March 1968, **24**(3):56.

——. Molding thermosetting materials. *Plastics Design & Processing*, November 1967, **7**(11):12.

Paci, R., and Helliwell, S. S. High-speed thermoset molding. *SPE Journal*, March 1971, **27**(3):49.

Papoojian, R. S. Experience on thermoset screw injection molding equipment. *SPE Journal*, August 1968, **24**(8):67.

Plastics Design & Processing. Transfer molding tips for epoxy parts. October 1979, **9**(10):18.

——. Injection molding of thermosets. August 1968, **8**(8):25.

Plastics Technology. Cold-sprue molding: halfway to the cold-manifold process. January 1978, **24**(1):11.

——. The new face of thermoset molding. June 1972, **18**(6):36.

——. Cycles speeded in compression molding, but not in transfer. October 1972, **18**(10):51.

——. Simple, low-cost method checks moisture content of your resin. March 1969, **15**(3):12.

——. Two new modifications of thermoset molding methods. February 1963, p. 42.

Plastics World. Dry thermoset polyesters–where they fit in the thermoset picture. January 1979, **37**(1):68.

Raech, H. Jr. Jet molding short run parts. *Western Plastics*, Technical Report 104, January 1964, p. 19.

Raghupathi, N., and Mallick, P. K. Preheating to decrease cure cycles in compression molded FRP. *Plastics Design & Processing*, January 1979, **19**(1):39.

Rheinfrank, G. B. You can injection-mold alkyd and polyester compounds at high speeds. *SPE Journal*, September 1971, **27**(9):30.

Riddell, R. M. Tailoring the properties of compression molded plastics. *Machine Design*, July 10, 1980, **52**(13):106.

Scherer, W. R. The production of laminated plastic tubing. *Western Plastics*, Technical Report No. 23, September 1956, p. 16.

Smith, D. F. Two tests for phenolic molding compound. *SPE Journal*, September 1968, **24**(9):61.

Society of the Plastics Industry, Inc. *Compression and Transfer Molding*. New York, 1966.

——. *Method of Test for Spiral Flow*. Bulletin EMMI 1-66.

Spiwak, L. Thermoset flow tests. *Plastics Technology*, May 1965, **11**(5):41.

Stricklin, J. E. Injection molding thermosets through the cold runner method–part 2. *Plastics Design & Processing*, April 1971, **11**(14):18.

Sundstrom, D. W., and Walters, L. A. Molding conditions versus strength of thermosets. *Modern Plastics*, November 1969, **46**(11):106.

Vaill, E. W. Technical advances in automatic molding. *SPE Journal*, March 1967, **23**(3):60.

——. Practical phenolic molding techniques. *SPE Journal*, May 1961, **17**(5):463.

Woebecken, W. Effects of processing techniques on structure of molded parts. *Modern Plastics*, December 1962, **40**(4):146.

Zecher, R. F. What does transfer molding offer today? *Insulation*, January 1967, p. 37.

——. Transfer molding of the new thermoset compounds. *SPE Journal*, January 1964, **20**(1):61.

Zecher, R. F., and Trembley, J. F. The future of electronic preheating and automatic thermosetting molding. *SPE Journal*, September 1967, **45**(1):58.

17
Reinforced Plastics/Composites Processing

In the plastics industry, the term "reinforced plastics" is generally understood to refer to products made from polymers that are reinforced with fibrous reinforcements in the form of discrete fibers, fiber strands, or woven or nonwoven fabrics. The term does not generally include sheet, rod, and tubing made by the high-pressure laminating process. The reinforced plastics industry as such, and the use of the term, probably started with the almost simultaneous development of long, thin, nonbrittle fibers and polyester resins which became available in the late 1930s and early 1940s. During and shortly after World War II, the term "fiberglass-reinforced plastics" referred almost exclusively to parts made with polyester resins and glass fibers. Frequently, the term "FRP" or "GRP" was also used to describe this new branch of the plastics world.

In the early 1960s, thermoplastic resins reinforced with glass fibers became available for injection molding. These materials were also described as FRP, so that term took on another meaning. Since these new, injectable materials are processed in substantially the same manner as other, nonreinforced thermoplastics, their processing will not be dealt with in this chapter.

As discussed above, the terms reinforced plastics, FRP, or GRP originally meant polyester reinforced with glass fibers in almost all cases. Because polyester resins can be cured at no or very low pressures (0–50 psi, since volatiles are not given off), still another term for parts made with these materials was "low-pressure reinforced plastic." Glass-reinforced phenolic, melamine, and urea parts were also made, but were simply called glass-reinforced thermosets. The later development of the epoxy resins in the early 1950s made another low-pressure process available, in contrast to the high pressures required by the phenolics and the amino resins.

In the late 1960s, a number of new "exotic" fibers were developed for use primarily with epoxy resins. These new reinforcement fibers included boron, graphite, silicon carbide, as well as improved glass fibers. In the early 1970s, an aromatic polyamide fiber, Kevlar, was developed by Du Pont. The purpose of all these new fibers was to allow production of reinforced parts having higher tensile strengths, moduli, and flexural properties, with the same, or lighter, weight as conventional parts. For specific information on these reinforcement fibers see Chapter 8.

With the development of the new "exotic" fibers, which cost as much as $500 or more a pound, a new term was adopted to distinguish them from the older glass-reinforced materials. This term was "composite," which signified that the material was made from two constituents, each of which contributed a quality that made the whole greater than the sum of the parts. The term composite can, of course, also be applied to the older glass-reinforced plastics or to any filled plastics part. However, at the time of writing, composites refer mainly to plastics parts made using epoxy or polyimide resins as a matrix, glass, boron, graphite, or Kevlar as the reinforcement, and processed at pressures below 200 psi.

Perhaps the major reason for the ready acceptance and wide use of the low-pressure process in the 1940s and 1950s was the fact that for the first time, parts could be made in any size without the limitations imposed by the capabilities of the press. Thus, large objects that formerly required massive presses and tremendous tools and molds to withstand the enormous pressures could now be made, and were made, in many garages and backyard operations. Tooling was very simple, ranging from plaster and wooden molds to simple aluminum castings or sheet metal molds. Another distinct advantage of this new method of processing was that it was possible to make parts with intricate sections, undercuts, three-dimensional curves, etc., simply because the tooling was inexpensive. It was possible and economical to make a plaster tool for every part and then destroy the tool in removing the part.

These parts were in most respects equal to or superior to similar parts made with high-pressure technology. The parts had the good strength, low density, and good electrical and chemical resistance of other thermoset materials. Large structures could be made with relatively inexpensive tooling and, in many cases, the parts could be erected at the building site. The additional advantage of being able to make repairs very easily, quickly led to the use of these materials by a number of industries besides the aircraft industry, which was the first large-volume user of FRP and the advanced composites. These industries now include automotive, sporting goods, marine construction, home appliances, chemical processing, and many others. The various industries using these new materials employ a number of processing techniques, ranging from simple hand lay-up work to highly sophisticated automatic molding machines, which can turn out parts as large as truck cabs on a schedule similar to high-pressure compression or transfer molding. In many cases, the same part can be made by any one of

Fig. 17-1. Fiberglass-reinforced plastic can deliver more strength per unit than any unreinforced plastic, and most metals. Regardless of the processing method selected, tooling usually represents a very small part of the product unit cost. Fiberglass-reinforced plastic can be designed to provide a wide range of tensile, flexural, and impact strength properties. (*Courtesy: Owens-Corning Fiberglas Corporation*)

Fig. 17-3. Fiberglass-reinforced plastics parts hold their forms and shapes under severe mechanical and environmental stresses and can be formed into any shape the designer may desire—simple or complex, large or small. (*Courtesy: Owens-Corning Fiberglas Corporation*)

several procedures, but the final choice may be dictated by the number of parts required, the cost of tooling, or the type of labor available. Conversely, the same part may be made of several types of resins or differing reinforcements, depending on the function of the part or the environment, etc. Selection of the optimum material and process then becomes a matter of knowing all the factors and parameters which limit these very versatile materials and processes. Figures 17-1, 17-2, 17-3, and 17-4 show some of the many diverse types of parts made and the advantages supplied by the materials used in their production.

Fig. 17-2. Fiberglass-reinforced plastic will not rust or corrode. There are many resin systems from which to choose (depending on particular application requirements) that provide resistance to almost every chemical environment. (*Courtesy: Owens-Corning Fiberglas Corporation*)

Fig. 17-4. It is possible for a single fiberglass-reinforced plastics structure to replace an assembly of many parts and associated fasteners. Assembly handling time is reduced. Color can be molded into the laminate for long-lasting good appearance. (*Courtesy: Owens-Corning Fiberglas Corporation*)

MATERIALS

Resins

The purpose of the resin in reinforced plastics is mainly to hold the fibers together and to transmit the loads from fiber to fiber. In order to do this, the uncured resin should wet the reinforcement well; should be cured at room temperature or by heat; and, when cured, should have good adherence to the fibers, good heat resistance, and good electrical and chemical properties. In addition, the resin should be available in a wide range of viscosities, should be compatible with pigments, filler materials, and, if possible, be cured at temperatures from ambient to a maximum of approximately 150°C (300°F).

The first resins to be used in the low-pressure process were the polyester resins, which continue to be the mainstay polymers used in this process. In almost all respects, the polyesters met the requirements listed above:

1. Polyester resins are available in liquid forms in many viscosities, so good fiber impregnation is relatively easy.
2. Using different catalysts, the resins can be cured at room temperature or at an elevated temperature or by ultraviolet light. This allows a part to be worked on all day, without its setting up, and then be cured overnight in an oven, or by using UV radiation from a lamp (sunlight could also be used).
3. Cured materials have good strength-to-weight ratios.
4. Cured materials have excellent electrical and chemical properties.
5. Costs are relatively low.

The major disadvantages of the polyester resins were the relatively high shrinkage which occurred when curing took place. Also, particularly under high humidity or moisture conditions, less than ideal adherence to the fibers necessitated special treatments for the fibers in order to upgrade the performance of the laminates. A minor disadvantage, particularly for thin fiberglass-reinforced parts, was a relatively high brittleness. Special formulations have been produced which show increased toughness. Vinyl ester resins, introduced in the late 1970s, are said to be tougher and have higher temperature resistance than conventional polyester resins.

Following the success of the polyester resins, epoxies were developed in the late 1940s. Fabricators now had a material with all the good properties of the polyesters—many viscosities; the ability to be cured at room temperature or at an elevated temperature, and, in some cases, with UV light; good electrical and chemical properties. The epoxy resins also had some special properties of their own which made them superior to the polyesters for many applications: their low shrinkage and—most important—excellent adhesion to most materials. This last gave laminates with higher strength-to-weight ratios.

Both the polyesters and the epoxies were cured at pressures ranging from 0 to 15 psi, at first; later, pressures as high as 50-100 psi were used. It was soon found that, while high pressures were not required, since there were no volatiles given off in the polymerization reaction, increased pressures would result in a more highly densified part. This densification took place by compression of the dissolved air, with the result that the parts were stronger.

Soon after the introduction of the epoxies, some of the older resins were modified so they too could be used with this new low-pressure technology and new resin types were also developed. Today there are a number of thermosetting resins, each with its unique characteristic which makes it specially useful when reinforced with fiberglass or other fibrous reinforcements. Table 17-1 summarizes the various resins currently in use, their major characteristics, and the processes in which they are used.

Table 17-2 shows in a more quantitative manner the differences in the various unmodified (neat) resin systems.

Reinforcements

The original material used, in any large quantity, for low-pressure moldings and the material still most often used is glass fiber in the form of woven fabrics, nonwoven sheet material, rovings, chopped fibers, milled glass, and various other forms. Table 17-3 lists the major glass-fiber reinforcements, characteristics, typical applications, and so on.

The various forms of fiberglass reinforcements shown in Table 17-3 are all made from Type E (electrical grade) glass. Some of the forms are also made in Types S and S2 glass. Type S is the highest-strength glass fiber available and is generally used only for aircraft or military applications. Type S2 is a high-strength, moderate-cost commercial version of the Type S fiber. Table 17-4 shows the properties of the three fibers. Figure 17-5 shows tensile strengths of S glass and E glass at various temperatures.

Regardless of the particular form of the glass fibers, in virtually every application, the fibers are treated and finished in a special manner to promote maximum adhesion of the resin to fiber. First, the original lubricant, also called a sizing, used to aid in drawing the filament, is removed. This process is known as heat cleaning, and is done on all fibers. For some resins this is the only treatment. For other resins a finish known as a coupling agent is applied to the fiber, to result in increased adhesion, particularly under high humidity or moisture conditions. Table 17-5 lists some of the finishes employed and the resins for which they are best suited.

The most widely used finishes are the chrome complexes and the silanes. Before using any glass fibers as reinforcements, the type of finish should be determined.

In addition to the glass fibers as reinforcement materials, other common fibers have been and are being used. These include asbestos, which has more strength as an unwoven mat or felt than does glass. Asbestos is also used for its heat resistance and good chemical resistance when combined with an appropriate resin. Nylon and polyester woven fabrics have been used, particularly when special surfaces are required in the completed part.

Table 17.1 Resins Used in Reinforced Plastics.*

Resin	Properties	Processes
Polyesters (Properties listed also apply to some polyesters formulated for thermoplastic processing by injection molding.)	Simplest, most versatile, economical, and most widely used family of resins, having good chemical resistance, especially to acids. Many forms available for many processing techniques; room-temperature cure, elevated-temperature cure, or UV cure. Usually used with gel coats for improved surface finish. Best finishes are obtained in modified low-shrink/low-profile resins. Can be colored easily.	Compression molding Filament winding Hand lay-up Mat molding Pressure bag molding Continuous pultrusion Injection molding Spray-up Centrifugal casting Cold molding Encapsulation
Epoxies	Excellent mechanical properties, dimensional stability, chemical resistance (especially to alkalis), low water absorption, self-extinguishing (when halogenated), low shrinkage, good abrasion resistance, very good adhesion properties. Many forms available for varied processing techniques; room-temperature cure, elevated-temperature cure, UV exposure cure.	Compression molding Filament winding Hand lay-up Continuous pultrusion Pressure bag molding Encapsulation Centrifugal casting
Phenolics	Good acid resistance, good elec-electrical properties (except arc resistance), high heat resistance, limited colorability, low to moderate in cost.	Compression molding Continuous laminating
Silicones	High heat resistance low water absorption, excellent dielectric properties, high arc resistance, good weatherability fire retardant, high costs.	Compression molding Injection molding Encapsulation
Melamines	Good heat resistance, high impact strength, very hard, self-extinguishing.	Compression molding
Diallyl phthalate	Good electrical insulation, low water absorption.	Compression molding
Polyimides	Highest temperature resistance, good chemical resistance, high cost.	Compression molding

*Courtesy: Owens-Corning Fiberglas Corporation.

In preparing the composites using the "exotic" fibers, only boron, graphite, and Kevlar were, and are, being used to any great extent. Boron was mainly used where the fibers could be laid in relatively straight configurations, since these fibers are very stiff and brittle. Graphite fibers* are now used for the majority of high-modulus applications, since they are considerably cheaper than boron and will get more so, because they are lower in specific gravity and they are capable of being formed easily into three-dimensional curves and bent into small radii. A special characteristic of the graphite-reinforced parts is a very low coefficient of expansion, which makes such materials very useful for high-precision parts such as optical element parts, space antennas, etc. To the public, graphite's main use is as a new material for golf club shafts, tennis racquet frames, sailboat masts, and other articles requiring high stiffness and light weight.

*Graphite fibers are sometimes called carbon fibers; however, carbon fibers may not always be graphite.

Table 17-2. Comparison Table of Unmodified Resin Systems.*

Property	Polyesters	Epoxies	Phenolics	Melamines	Silicones	Polyimides
Specific gravity	1.10–1.46	1.10–1.4	1.30–1.36	1.40–1.48	1.30–1.34	1.43
Color possibilities	Very good	Good	Limited	Very good	Good	Limited
By-products of cure	None	None	H_2O	H_2O	H_2O, RCOOH	H_2O, ROH
Molding pressures	0–high	0–high	Low-high	Medium-high	Low-high	High
Molding temp., °F	<70–320	<70–330	270–320	270–320	290–350	400–660
Shrinkage, %	2–8	0.1–0.4	1–1.2	1–1.5	1–1.5	3–4
Tensile strength, 10^3 psi	6-10	4–13	6–9	7–9	4–5	12.5–17.0
Elongation, %	5	2–6	1.5–2.0	–	–	7.5–10.0
Modulus of elasticity tension, 10^5 psi	3–6.5	4–6	4–5	7.5–10	–	2.0
Compressive strength, 10^3 psi	13–36.5	15–21.5	12–15	40–45	9–15	30–40
Flexural strength 10^3 psi	8.5–18.5	13–21	11–17	11–14	9–14	19–29
Impact strength (IZOD)	0.2–0.4	0.2–1.0	0.25–0.40	0.24–0.40	0.25–6.0	0.7–1.5
Heat resistance, °F (continuous)	250–500	250–550	160–250	210	>600	400–600
Heat distortion, °F	140–400	115–550	165–260	298	>900	270–680
Water absorption, %, (24 hrs.)	0.15–0.60	0.08–0.15	0.03–0.04	0.3–0.5	0.1–0.5	0.24–0.3
Dielectric strength V per mil	280–420	400–500	350–400	300–400	200–500	560
Dielectric constant (60-10^6 cps)	2.8–5.2	3.3–5.0	4.0–7.5	4.3–7.6	3.2–5.2	3.43–3.6
Dissipation (power) factor (60-10^6 cps)	0.003–0.028	0.002–0.050	0.01–0.15	0.015–0.080	0.0008–0.01	0.0018–0.0034
Arc resistance, sec	125	45–120	Tracks	100–145	250–350	230
Burning rate	Slow to self extinguishing	Slow to self extinguishing	Very low	Self-extinguishing	None to slow	None to slow
Price Range, $/lb	0.30–0.50	0.65–1.30	0.12–0.40	0.40–0.60	5.00–7.50	5.00–50.00
Effect of sunlight	Yellows	None	Darkens	Slight color change	None to slight	None to slight
Effect of weak acids	None	None	None	None	None to slight	None
Effect of strong acids	None to considerable	Attacked by some	Decomposed by OX. acids none to slight reg. org. acids	Decomposes	Slight	Slowly attacked to none
Effect of weak alkalis	None to slight	None	Slight to marked function of PH	None	None to slight	Slowly attacked
Effect of strong alkalis	Attacked	Slight	Decomposes	Attacked	Slight	Attacked
Effect of organic solvents	Attacked (ketones and chlorinated solvents)	Generally resistant	Attacked by some	None	Attacked by some	Very resistant
Machining qualities	Good	Good	Excellent	Fair	Fair to good	Good
Major limitations	Cure shrinkage	Dermatitis, difficult to mold release	Color	High cost	Very high cost	High cost, color
Major advantages	Ease of fabrication, clarity with flame retardancy, moderate dissipation (power) factor	Low shrinkage, excellent adhesion	Good general properties, low cost	Arc resistance, scratch resistance, colorability	Heat resistance, low dissipation (power) factor	Heat resistance

*Davis, 1963.

Table 17-3. Commonly Used Glass-Fiber Reinforcements.*

GLASS FORM	GENERAL DESCRIPTION	LAMINATE NORMAL GLASS CONTENT, WT. %	PRINCIPAL PROCESS	TYPICAL APPLICATIONS	RELATIVE GLASS COST
Continuous Strand	Gathered filaments supplied on a cylindrical forming package	50-80*	Filament winding	Pipe	Lowest
		70-80	XMC® composite/matched die molding	Manhole steps, structural auto parts	
Chopped Strand	Bulk cut strands, 1/8 to 2 inches	10-30	Premix and bulk molding compound/matched die and injection molding	Auto, appliance, business machine, and electrical components & housings	Low
		10-45	Thermoplastic extrusion compounding/injection molding		
Continuous Roving	Gathered bundle of one or more continuous strands wound into a cylindrical package	50-80*	Filament winding	Pipes, tanks, booms	Low
		25-40	Sheet mldg cpd, preform,	Chairs, auto front ends,	
		50-70	HMC® composite/matched die molding	ski-mobile housings, structural parts	
		25-35	Spray-up, unfilled	Tanks, fume hoods, truck parts	
		15-25	filled	Tub/shower units	
		20-30	Continuous laminating	Translucent greenhouse & patio roof panels, garage doors & industrial skylights	
		25-80*	Continuous pultrusion	Electrical & corrosion resistant stock; fishing rods	
Reinforcing Mat	Chopped or continuous strands in non-woven random matting, lightly bonded with resin	20-45	Hand lay-up	Boats; corrosion resistant ducts, fume hoods; pools	Medium
		25-35	Press, cold & matched die molding	Electrical laminates Housings, shrouds, trays	
		20-30*	Centrifugal casting	Low pressure water & chemical tanks	
Surfacing Mat (Veil)	Non-reinforcing monofilaments in random matting	0.5-10*	Hand lay-up Matched die molding	Used with reinforcing mat & fabrics for improved surface finish and as inside resin rich barrier surface in corrosion resistant products.	High
Woven Roving	Heavy, drapeable fabrics made from continuous roving	up to 45* in combination	Hand lay-up	Boats, tanks, large containers & housings	Medium
Woven Roving Mat	Chemically bonded woven roving & chopped strand mat	with other types	Hand lay-up	Same as above	High
Woven Fabrics	Woven textile yarn with after-finish. Available in many styles.	up to 65*	Hand lay-up, bag molding, press lamination	Boats, electrical laminates, aircraft & aerospace applications	High

***These ranges reflect the reinforcement content of that specific portion of a multi-type reinforcement structure.**
***Courtesy: PPG Industries, Inc.**

Polyaramid fibers (Kevlar), introduced in the early 1970s, gave materials with twice the modulus or stiffness of glass parts, 25% higher tensile strength, and approximately $\frac{1}{3}$ less weight. For a full description of the various reinforcement materials consult Chapter 8.

Release Agents

In molding any of the reinforced materials, regardless of the type of reinforcement or the resin used, one characteristic is common–the tendency of the part to stick to the mold surface. Thus, some type of release agent (also called parting agent or mold release) is always used to facilitate removal of the part from the mold. The release agent functions by forming a barrier film between the mold surface and the part. There are a great many materials used as release agents; however, all can be classified as either

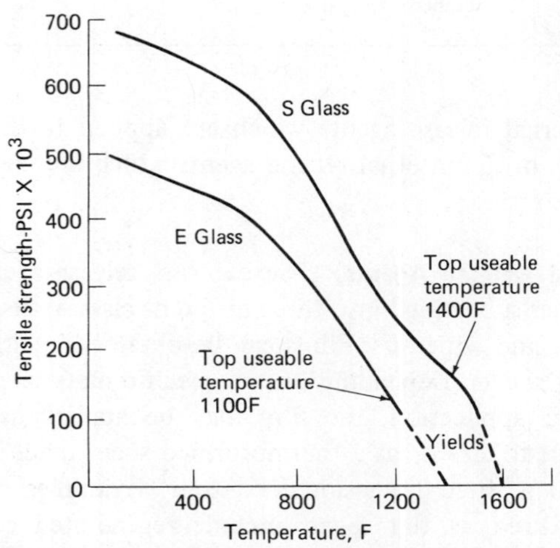

Fig. 17-5. Virgin glass filament tensile strengths at various temperatures. (*Courtesy: Owens-Corning Fiberglas Corporation*)

Table 17-4. Properties of Glass Fibers.†

PROPERTIES OF GLASS FIBERS		E Glass	S,S2 Glass
Physical Properties*			
Specific Gravity—Fibers Gms/cc		2.60	2.49
Density—Lbs./cu. in.		.094	.090
Mechanical Properties*			
Virgin Tensile Strength 70°F, PSI x 10^3		500	665
Modulus of Elasticity			
72F, PSI x 10^6		10.5	12.6
72F, PSI x 10^6 (After heat compaction)		12.4	13.5
1000F, PSI x 10^6 (After heat compaction)		11.8	12.9
Elongation at 72F, %		4.8	5.4
Thermal Properties**			
Coefficient of Expansion in./in./°F x 10^{-6}		2.8	3.1
Specific Heat at 75F		0.192	0.176
Softening Point F		1555	1778
Strain Point F		1140	1400
Annealing Point F		1215	1490
Electrical Properties**			
Dielectric Constant at 72F	1 MHz	6.33	5.34
	10 KHz	6.13	5.21
Power Factor at 72F	1 MHz	0.001	0.002
(Loss Tangent)	10 KHz	0.0039	0.0068
Acoustical Properties*			
Velocity of Sound Calculated Ft./Sec.		17,500	19,200
Measured Ft./Sec.		18,000	—
Optical Properties**			
Index of Refraction		1.547	1.523

*Properties were determined on glass fibers.
**Properties were determined on bulk glass.
†Courtesy: Owens-Corning Fiberglas Corporation.

Table 17-5. Fiberglass Coupling Agent Finishes.

Finish	Resin	Base
111	Melamine	Partly desized
112	Silicones	Heat cleaned, desized
114	Polyester	Chrome complex
Volan-A	Polyester	Chrome complex
Y-1100	Epoxies and silicones	Silane
136	Polyester and silicones	Silane
NOL-24	General purpose	Silane
A-172	Polyester	Silicone
301	Modified polyesters	—
T-31	General purpose	—

(1) external release agents which are applied to the mold surfaces, or (2) internal release agents which are combined with the resin.

External Release Agents. The external release agents can be further subdivided into three different classes; films, film formers, and wipe-ons. Films can be obtained in either roll form or sheets. Depending on the specific material and the intended application, the film may be applied as a flat sheet cut to size or as a thermoformed sheet made to conform to a three-dimensional contour. Examples of films usually used as flat sheets include regenerated cellulose (cellophane), polyvinyl alcohol (PVA), polyester (Mylar), polyamide (nylon), and Teflon TFE. Easily thermoformed films are polyethylene, cellulose acetate, polyvinyl chloride (PVC), and various fluorocarbon films (e.g., Teflon FEP, Kel-F). However, all the fluorocarbons are considerably more expensive than the other types of films.

In all cases, the films will act as a satisfactory barrier to prevent mold sticking; in almost all cases, the film will stick to the part after cure, when it can be easily stripped off. In picking the particular film to be used, the first consideration is the temperature at which the part is to be cured. If the temperature is above approximately 70°C (160°F), then polyethylene and cellulose acetate are automatically eliminated or should be used only after a small test part is made satisfactorily. Depending on the type of PVC, it may be useful up to approximately 105°C (220°F). Cellophane, PVA, and nylon films can be used up to 150°C (300°F) or somewhat higher, depending on the particular film and the application. Fluorocarbon films can be used at temperatures from 150°C to 260°C (300–500°F), depending on the film.

The second category of external release agents are the film-forming materials applied either by brush or spray gun onto the mold surface. Included in this category are such materials as cellulose acetate lacquers, PVA and nylon solutions, heat-cured fluorocarbon coatings, and possibly varnishes and/or automobile-type paint fillers which can be sanded to a high-gloss finish. These film-forming materials are generally used over porous surfaces such as wood or plaster-of-paris molds. They are also used where the surface must be filled and smoothed so as to remove scratches, pits, or the grain pattern of a wooden form. The film formers are also frequently used in conjunction with a wax, silicone, or fluorocarbon type of wipe-on material.

Wipe-on release agents constitute the third category of release agents. These materials are applied with a wiping rag, in the case of certain waxes; a brush, for oily materials; or a spray, usually in the form of an aerosol can. In most cases, after the initial application, the mold is wiped smooth with a rag, hence the term "wipe-on."

There are a number of wax preparations which have been used, ranging from pure carnauba wax, which is generally applied only to hot molds, to various soft paste wax formulations similar to automobile waxes (Simonize has been used). The major advantage of the waxes is the high gloss which can be obtained if the material is correctly applied and polished. Another advantage is that the mold can be used immediately, with no drying or curing period required.

A number of silicone materials are available in the form of solutions which are wiped or brushed on. Some are air dried and others must be heat cured. Silicones are also available in the form of aerosol solutions, which are either air dry or heat curing. A third type of widely used mold-release agents are the aerosol spray cans of fluorocarbon particles dispersed in a volatile solvent-carrier. The major advantages of this type of release agent are the excellent release properties and the high-temperature resistance of the fluorocarbon (polytetrafluoroethylene–Teflon TFE), which means they can be used with polymers curing at the highest temperatures. The final gloss, however, may not be as high as that obtained when using a paste wax or its equivalent.

In the use of any mold-release agent, it is important to first follow the resin manufacturer's recommendations for his particular resin. The use of the wrong mold release can result in vastly increased time schedules for part removal, or can result in damaged surfaces or parts because of mold sticking. It is also important that the mold release be chosen with regard to any postmold processing on the part; for example, if the part is to be bonded or painted, the mold release must either by completely removed or be one which is compatible with the bonding agent or the finish. (This is particularly important when silicone release agents are used.) In case of any doubt, it is usually best to lightly sand the surface of the part prior to applying the adhesive or the finish. It should also be borne in mind that mold surfaces used for the first time must be "broken in." Thus, the first few parts made may not release too easily; however, after several parts have been made, release becomes progressively easier as the pores and grain boundaries are filled in.

Internal Release Agents. The chemical compounds used as internal release agents are added to the resin by the manufacturer to minimize mold sticking after cure. Zinc and calcium stearates and similar metallic soaps are used for polyester and phenolic resins, silicone oils and high-melting-point waxes are used in epoxies and organo-phosphates, and waxes are used in polyester resins. The internal release agents all work by being squeezed or exuded onto the surface during cure of the resin.

Gel Coats

When polyester resins were first being introduced for use with fiberglass or other fiber reinforcements, it was noticed that, when the parts were molded at 0–15 psi, the surfaces showed a distinct three-dimensional fiber pattern. This was found to be caused by the shrinkage of the resin away from the glass fibers during curing. Since these early parts were almost exclusively for aircraft, this condition could not be tolerated for aerodynamic and aesthetic reasons. A remedy was soon developed in the use of the gel coat.

The usual gel coat is a polyester resin heavily filled with a mineral filler and a pigment that, because of its very high filler-to-resin ratio, shows very little shrinkage. In addition, the gel coat, when properly applied and cured, results in a hard, smooth coating 10–30 mils thick, which is colored throughout, thus minimizing the effect of scratches. In some instances, a clear gel coat is used to improve the stain resistance, gloss, and depth of coating when applied over cultured marble, or a chemical-resistant, clear gel coat is used over chemical-resistant laminates to improve the resistance to corrosive environments. While most gel coats are used with polyester resins, epoxy gel coats are also used, primarily for improvement in the surface quality of the cured reinforced plastics part.

A typical polyester gel coat formulation might contain the following component material:

Resin: A low-viscosity (500–1000 cps) resin is used so it can be easily filled and will allow entrapped air to escape. Usually a high-impact grade is preferred to insure freedom from chipping caused by impact or thermal stresses. Concentration in the complete formulation is 25–95% by weight, with the lower concentration predominating.

Fillers: The most widely used fillers are calcium carbonate (marble) and hydrated aluminum silicate. Other fillers include metal powders, carbides, oxides, silicates, sulfides, carbon, and graphite, depending on the particular physical, chemical, or electrical properties desired. The fillers are used primarily to reduce the resin shrinkage, lower the exotherm, increase the hardness, increase the thermal conductivity and dimensional stability, and change the density and opacity of the resin. The filler concentrations range from 5% to 75% by weight.

Thixotropic Agents: Colloidal silica (Cab-O-Sil) and magnesium aluminum silicate clays (bentonite) are used as thixotropic agents to prevent sagging and running of the gel coat when applied to vertical surfaces. They are also used to minimize filler settling and increase pigment efficiency. They are used in concentrations of 2-15% by weight.

Pigments: Pigments are dispersed into the resin to act as coloring agents. They are used in concentrations of 2-10% by weight.

Solvents: Solvents such as acetone are added to the formulation to thin the material to spraying consistency. The minimum amount of solvent is used since its use can result in attack on the mold release, resulting in sticking parts, poor surfaces, etc.

Application. Catalyzed gel coats are applied to the release-coated mold surface by brush, roller coat, or spray. This latter method is the most used process by far. The coat is allowed to gel, and, while it is still tacky, the reinforcing fiber and resin are applied. If done correctly, the bond between the gel coat and the fiber-reinforced laminate will be an excellent one. Application of the reinforcing material before gelation of the gel coat may result in fiber exposure or the impression of the fiber or fabric. However, most gel coats are made using air-inhibited resin so that the gel coat automatically cures with a tacky surface.

Gel coats currently are used extensively for such items as shower stalls and bath tubs, outer surfaces of boats, campers, truck bodies, swimming pools, radomes, and a host of other parts where a smooth, hard, tough, and colored surface is a necessity (see Figure 17-6).

Fig. 17-6. Gel coat being applied to the mold surface to form the outer, colored surface of the laminate. (*Courtesy: Owens-Corning Fiberglas Corporation*)

Storage and Handling of Materials

Resins. In general, all resins used in reinforced plastics should be stored in a cool, dry place. Storage below 18°C (65°F) will materially lengthen the life of polyester and phenolic resins. Epoxy resins are not as sensitive to the effects of storage temperatures as are polyesters and phenolics, and so need not be stored below normal room temperatures. (Resins which have been catalyzed, and/or advanced to the B-stage should be stored in accordance with the manufacturer's recommendations—which may call for refrigerated storage.)

Resins in storage should be kept in tightly sealed containers to prevent the evaporation of volatile monomers and solvents. Again, this precaution applies mainly to polyester and phenolic resins, but it may also apply to low-viscosity grades of epoxies which may contain reactive diluents. A further precaution, which applies mainly to polyester resins, is that these materials should be stored in containers that exclude light such as steel drums, tin cans, or brown light-proof bottles.

Associated with the resins are various monomers and solvents such as styrene, alcohol, and diglycidal ether. These materials should be stored under the same conditions as the resins they are used with. Since these liquid materials all have relatively high vapor pressures, adequate ventilation should be provided to prevent the accumulation of fumes which could become a fire hazard. Likewise, adequate grounding should be provided on all dispensing equipment to dissipate any static charges.

When using any material that has been stored under refrigeration, it is important that the container *not* be opened until the resin has reached room temperature. In this way, condensation of moisture on the resin will be avoided, and the resin will also have reached normal room-temperature viscosity conditions.

Catalysts, Hardening Agents, and Accelerators. All catalyst, hardening agents, or accelerators should be stored in a manner similar to their corresponding resin. Catalysts for polyesters, however, should be treated with an additional precaution. As powerful oxidizing materials, they should be stored so that they can be easily isolated in the event of fire. In the storage of these materials under refrigeration conditions, a crystalline precipitate will be occasionally found. Simply warming the material to room temperature or to 50-60°C (120°-140°F) will usually restore the normal condition.

Reinforcements. There are no toxic or fire hazards in the storage or handling of glass-fiber reinforcements. Organic reinforcements could present the possibility of smoke or fumes in a fire, as is common with most organics. The only special precautions for the reinforcement materials is storage in a dry atmosphere to reduce moisture pickup to a minimum to insure high-quality moldings.

Catalyzed Resins and B-Staged Fabrics. Refrigerated storage is necessary for all catalyzed resins and B-staged fabrics. The lower the temperature, the longer will be the usable life. Catalyzed resins should be kept in sealed containers and B-staged fabrics should be wrapped in several layers of cellophane or polyethylene film when in storage. When the containers or the wrapped materials are removed from the refrigerator for use, they should be allowed to reach room temperature before being opened or unwrapped to avoid possible moisture condensation on the catalyzed material.

Ancillary Materials. Ancillary materials such as parting agents, sealants, films, fillers and pigments may all be stored under normal room-temperature storage conditions.

Receiving Inspection Tests

Associated with the problem of correctly storing and handling various reinforced plastics materials is the problem of insuring that the correct material has been received before storing. While there are a number of laboratory tests which can be used to characterize a resin or an impregnated fabric, there are also a number of comparatively simple tests which can be run by receiving personnel to help identify material or to detect mislabeled stock. Some of these tests are:

Resins:

Specific gravity	Requires a hydrometer and a thermometer.
Viscosity	Requires a Zahn cup, a Brookfield viscometer, or similar instrument and a thermometer.
Odor	Useful to identify styrene, alcohol, etc.
Color	Can be used to indicate the presence or absence of accelerators in polyester resins, or some types of hardeners for epoxy resins.

B-Staged Materials:

Weight	Requires a scale and ruler to determine the ounces or grams per square yard. Can indicate the resin content and/or the presence of the correct reinforcement material.
Flow test	Requires a hot plate. With some experience, this test can be used to indicate the state of cure of a B-staged material.
Gel time	Requires a hot plate. Can indicate the working life of a given material.

It should be noted that the above tests are not recommended as substitutes for standard laboratory tests, but rather as helpful adjuncts to these tests. In general then, these tests would be used as comparison tests; that is, the new batch is compared with the values found on the old batch or with a standard sample. Another use for these tests is for checking materials which have been in storage for some time, in which case the original values can be used as standards.

Material Preparation Techniques (Raw Stock)

Reinforced plastics consist essentially of two components: the fibrous material either in the form of woven fibers, mat, or as simple cut fibers or rovings, and a liquid resin, which is later converted to a solid. In any case, it is necessary that the fibers be thoroughly coated with resin and that all the air in the fiber texture be replaced with resin. The resulting product should then be a limp and easily manipulated, fiber-resin composite which can be made to conform to a mold contour. In some cases, this flexibility in the composite material is not realized until the material is heated for the molding operation. In all cases, however, the proper resin should be chosen so its gel time and viscosity is such as to prevent air from reentering the laminate due to gravitational flow of the resin while it is still liquid.

Wet Lay-Up. The simplest method of impregnating continuous lengths of mat, fabrics, or rovings with resin is to pass the material through a bath and then through squeeze rolls. Such a procedure can be used to produce material with resin contents which vary approximately ±3% and can be produced with resin contents of approximately 30-50%. This procedure is used where materials are required for immediate use or only in small or moderate amounts. The final product, depending on the particular resin, is usually quite tacky and somewhat difficult to handle, particularly if stored rolled up.

Prepregs or B-Staged Materials. A process producing a more easily handled and stored raw stock is one in which the final product is known as a preimpregnated fabric, mat, or roving. In this process, the fibers are drawn through a bath, squeeze rolls, and then through one or more heating zones. In the heaters, the laminate is first stripped of solvents and other volatiles and then the resin is advanced to the B-stage to result in a material which is relatively dry and only slightly tacky. The degree of B-staging achieved is a function of the temperature, speed of travel, resin content, and resin type. This process is used for laminates to be cured at high temperatures or by ultraviolet radiation, and usually results in materials having relatively long shelf lives.

Premixes. Other names for premixes are dough molding compounds or, when special low-shrink additives are incorporated, bulk molding compounds (BMC). Premixes are mixtures of polyester resin, a mineral filler (usually calcium carbonate, clay, silica clay, etc.), reinforcing fibers (glass, sisal, nylon, etc.), appropriate catalysts, pigments, and possibly lubricants. A typical formula is:

Polyester resin:	20-35%
Mineral filler:	45-55%

Chopped glass fibers:	10-25%
Catalyst, pigment:	2-3%
Lubricant:	0.5-2%

The premix formulation is usually made by first mixing the resin and filler to a substance which may vary from a thick pourable liquid to a soft putty. To this material, the fibers are carefully added. Mixing of these compounds is very critical. Filamentizing or breaking of the glass-fiber reinforcement due to excessive or improper mixing causes very rapid degradation of the physical properties of the resultant moldings (see Figure 17-7).

The glass-fiber length has a considerable effect on the final properties of the molding compound. The optimum length has been found to be $\frac{1}{2}$-$\frac{5}{8}$ inch, although fiber lengths up to $1\frac{1}{4}$ inches have been used. Shorter fibers result in low material strength while longer fibers do not disperse well in the mixing operation, are apt to break up, and may interfere with flow in the molding operation.

Usually the premixed material is wrapped in polyethylene film to minimize loss of the volatile styrene, unless the BMC is to be used immediately. Depending on the catalyst used, the material usually may be stored for 3-4 weeks at room temperature.

In many cases, the premix will be extruded or rolled into long lengths or cords so that preforms may be easily obtained by simply chopping off the desired length. In this case, the preforms are defined by length rather than weight as is the usual case. Because of its high filler content, polyester premix or BMC is among the cheapest of the plastic molding compounds. Its excellent surface finish, good dimensional stability, (due to its heavy loading), moderate strength, and easy, rapid processing make for increasing use in a wide variety of high-volume applications. Epoxy resin-based BMC has also been made; however, costs are considerably higher.

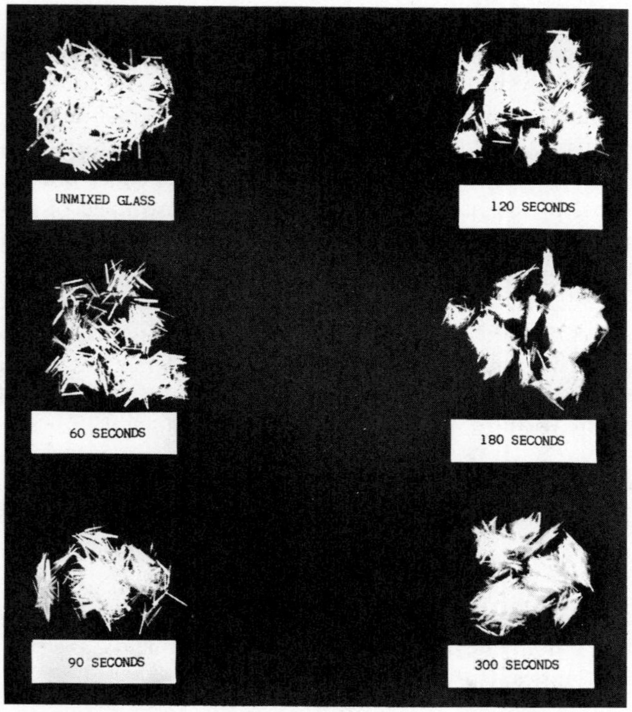

Fig. 17-7. Effect of excessive mixing on glass-fiber reinforcement. (*Courtesy: PPG Industries, Inc.*)

Premix materials are molded by the traditional methods of compression and transfer molding. The main advantage of the use of premixes is that these may be molded at fairly low pressures with the fast cure cycles and high strengths associated with glass-reinforced parts.

Sheet Molding Compounds (SMC). A significant improvement over ordinary prepreg or B-staged fabrics or mats was the development, in the 1950s and early 1960s, of sheet molding compound, commonly known as SMC. This class of materials, made in sheets up to approximately $\frac{1}{4}$ inch in thickness, consists basically of 1-2 inch long chopped strand glass-fiber mat or chopped glass-fiber roving as the reinforcement material, combined with a polyester resin, and a number of additives. (Vinyl esters are used in some cases where maximum toughness and/or high-temperature flexural strength is required.) The additives are responsible for the considerable differences between B-staged materials and SMC. The principal additives in SMC, besides the catalyst, are thickeners, fillers, thermoplastic polymers, and pigments. (Figure 17-8 and Table 17-7 show the processing equipment and a typical formulation.)

The function of the thickeners in the SMC (principally magnesium oxide) is to cause a controlled increase in viscosity of the resin impregnant, after thorough fiber wetting has taken place. The thickening of the resin then insures a uniform flow of reinforcement and resin to all parts of the mold when heat and pressures are applied. This greatly minimizes the problems of "resin-rich" and crazed areas, or "resin-starved" areas found in B-staged material when heated and pressed.

The addition of up to approximately 30% by weight of powdered thermoplastic resins results in a considerable decrease in shrinkage of the polyester impregnant. Such resin formulations are called low-shrinkage or low-profile resins. SMC containing thermoplastic resins (acrylics, PVC, styrene copolymers, polyethylene, and others) can be compression molded to extremely smooth, high-gloss surfaces that very accurately reproduce the polished mold surface. Gel coats are not used, but the molded parts are commonly finished by painting, using automotive primers and finish coats. The major uses for SMC were originally for automobile body, hood, and fender parts, but since then many applications in home appliances, sporting goods, electronics, etc., have been developed.

Thick Molding Compounds (TMC). In 1976 a new method of preparing a sheet molding compound was developed.

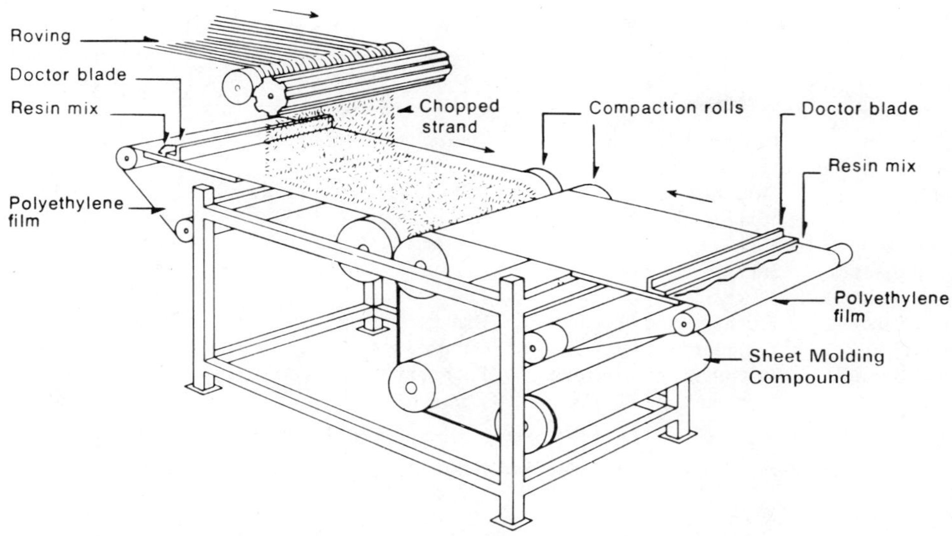

Fig. 17-8. Manufacture of SMC. (*Courtesy: PPG Industries, Inc.*)

Since the new materials could be made in sheets, similar to SMC but considerably thicker (up to 2 inches), the material was named TMC. Because it could be made in substantial thicknesses, it had many of the characteristics of BMC, with, however, improved physical properties and the advantage of being a continuous process instead of the batch process usually employed in making BMC.

Figures 17-8 and 17-9 show the types of machines used in the manufacture of SMC and TMC. As illustrated, the SMC fibers are saturated with the resin by means of two "doctor" blades which meter the polymer onto the polyethylene films. The resin must be very accurately controlled to obtain a uniform impregnation and, because of the method of saturation, the fiber layer cannot be too thick. In contrast, impregnation in the TMC machine takes place in the nip of two rapidly counter-rotating rolls, whose spacing can be varied to provide controlled, thorough fiber saturation. The formulations used for SMC and TMC can be the same, and are almost the same as in a BMC formulation. Table 17-6 compares the three different processes. Tables 17-7 and 17-8 show the comparisons in physical properties between SMC and TMC, and BMC and TMC. Despite the apparent superiority of TMC at the time of writing, SMC and BMC are the materials most widely used.

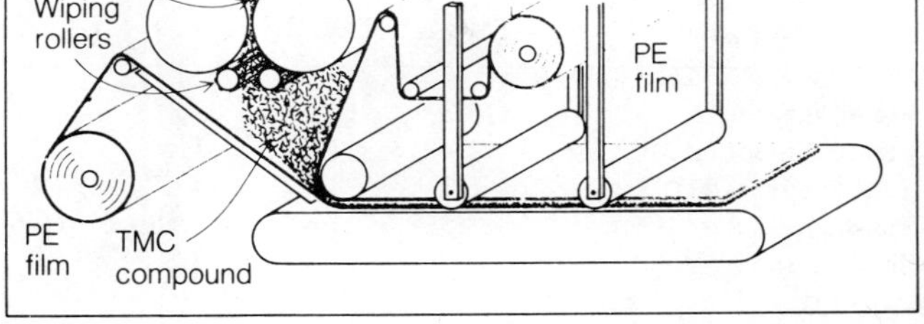

Fig. 17-9. Schematic of TMC process. (Reprinted by permission from *Modern Plastics*, McGraw-Hill, Inc.)

Equipment for General Use

Resin Mixers. Resin mixing equipment for the industry is as diverse as its processing requirements and runs the gamut from the bent rod in the hand drill motor to the continuous, vacuum-deaerated, proportional pump, in-line mixer.

The type of mixing equipment used is normally dictated by the type of manufacturing operation. For example, the average job shop, because of relatively short runs of any particular resin mix, employs propeller-type mixers. The Lightning portable drum mixer, converted bakery dough breaks, and other types of rotary and planetary paddle mixers are typical.

For high-production molding and encapsulation, metered, shot-dispensing mixing equipment is used to eliminate pot life problems, provide shot weight control, allow multi-station dispensing from a single system, and improve housekeeping.

The continuous laminator or extruder will use an in-line, proportional metering pump system to get higher running speeds with hotter catalytic systems while providing a steady, controllable flow of resin mix.

The boat manufacturer and producers of other large-area hand lay-up or spray components use pressure pot, spray mixing units because of their compatibility with the rest of the processing equipment and the ability to deposit a rela-

Table 17-6. Comparison of TMC, SMC, and BMC Processes.*

Factors	TMC	SMC	BMC
Compound form, thickness and weight	Unlimited; continuous	Limited; continuous	Batch; bulk form usually
Productivity	Over 3000 lb./hr. per ft. width of machine	2000 lb./hr. per ft. width of machine	Batch; process usually
Filler concentration	High	Limited	High
Mechanical properties	High	High	Low
Glass fiber degradation	Low	Low	High
Scrap, waste, handling	Minimum; less PE film	Higher; multilayers	Higher
Molding versatility	Compression; injection	Compression	Compression: injection
Glass impregnation	Instantaneous; controlled; no kneading area	Poor control; long kneading areas	Poor; blade mixing causes fiber degradation

*Reprinted by permission from *Modern Plastics*, McGraw-Hill, Inc.

tively large volume of resin mix over extensive surfaces rapidly.

Whatever system is used, it should be accompanied by deaeration of the mix and/or the base resin if high-quality end product is to be attained.

Impregnation Equipment. Although there is some wet impregnation equipment in custom lay-up plants around the country, companies specializing in custom impregnating for the reinforced plastics fabricator operate the major portion of impregnation equipment. The equipment employed by these impregnators, although basically the same in operating principle, varies widely in specific detail. The four commonly preimpregnated forms of reinforcement are fabrics, rovings, mat, and chopped fibers. Impregnation of the sheet materials (fabric, woven rovings, and SMC mat) is done in the same general manner and usually the product is a B-staged material or SMC. Impregnation of rovings as continuous fibers or as chopped fibers to produce BMC is done by separate processes and may or may not result in B-staged stock.

The equipment to do fabric, mat, or roving impregnation basically consists of unwind stations for polyethylene support films (top and bottom), an unwind station for the reinforcement, an impregnating area where the resin is applied, a compacting area, an oven curing tunnel or tower for B-staging, and a rewind station where the laminate assembly is wound. Production of chopped fiber sheet

Table 17-7. Comparison of Properties: TMC and SMC.[a, *]

Formulation (parts by wt.)	TMC-1	SMC
Resin	70	70
Low shrink additive	30	30
Calcium carbonate	120	120
t-Butyl perbenzoate	1	1
Zinc stearate	4	4
Thickener	3	3
Glass fiber (1 in.), %	28	28
Weight, oz./sq. ft.	48	10
Properties	**TMC-1**	**SMC**
Flexural strength, 10^3 psi	26.5	28.6
Flexural modulus, 10^6 psi	1.43	1.44
Tensile strength, 10^3 psi	14.0	13.2
Tensile modulus, 10^6 psi	1.58	1.58
Izod impact, ft.-lb./in. of notch	10.1	8.8

[a] Specimens cut from disks molded at 1400 psi for 2 min. at 145°C., using 12-inch-diameter loading charge covering 60% of the mold area.
*Reprinted by permission from *Modern Plastics*, McGraw-Hill, Inc.

Table 17-8. Comparison of Properties: TMC and BMC.[a, *]

Formulation (parts by wt.)	TMC-2	TMC-3	BMC-1	BMC-2
Resin	70	70	70	70
Low shrink additive	30	30	30	30
Calcium carbonate	180	180	180	180
t-Butyl Perbenzoate	1	1	1	1
Inhibitor	0.03	0.03	–	–
Zinc stearate	5	5	7	7
Black pigment dispersion	10	10	10	10
Magnesium oxide	0.8	0.8	0.4	0.4
Glass fiber (1 in.), %	15	20	–	–
Glass ($\frac{1}{4}$ in.), %	–	–	15	20
Properties	**TMC-2**	**TMC-3**	**BMC-1**	**BMC-2**
Flexural strength, 10^3 psi	15.3	19.9	9.8	14.0
Flexural modulus, 10^6 psi	1.43	1.54	1.44	1.72
Tensile strength, 10^3 psi	6.7	8.0	4.0	4.1
Tensile modulus, 10^6 psi	1.51	1.65	1.58	1.76
Shear strength, 10^3 psi	8.3	9.1	7.2	7.7

[a] See footnote of Table 17-7 for molding conditions.
*Reprinted by permission from *Modern Plastics*, McGraw-Hill, Inc.

molding compound (SMC) is substantially the same except that the sheet is formed at the machine, and the resin impregnation is somewhat more complicated.

Methods of applying and controlling the resin content vary widely from straight immersion and squeeze rolls to doctor blade techniques and many forms of roll coating. Probably the most critical part of an impregnating unit is the B-staging area. Proper temperature zoning and control is extremly critical if quality and reproducibility are to be achieved. This dictates the use of complex control and read-out systems in this area. Many types of heating are used, such as gas, electricity, steam, etc. The heating medium is usually determined by local economics.

Impregnation of continuous fiber roving, primarily used for filament winding, consists of simply running the fibers through a resin bath, then through squeeze rolls, or a compaction die, and then applying the fiber directly to the part. Equipment requirements are relatively simple.

Preparation of premix bulk molding compound (BMC) consists of mechanically mixing chopped fibers, resin, catalyst, fillers, and other additives all together in a variety of methods. One method employs a water-jacketed chamber (50°C, (120°F) maximum operating temperature) and spiral or sigma mixing blades and is used for batch preparations. Such a system featuring automatic weighing and feeding of the component materials is shown in Figures 17-10 and 17-11.

Continuous mixing and extruding equipment is also used. One method employs a rotating screw similar to conventional extrusion equipment except that the screw L/D ratio is less to minimize glass-fiber degradation. Automatic metering and feed mechanisms are used to assure a steady, proper flow of ingredients to the mixer. Figure 17-12 illustrates this type of equipment. The product from such a mixer can be a puttylike rope which is then simply cut to length for molding, thus eliminating the weighing operation.

Other continuous systems for premix include chopping impregnated roving and then pelletizing, running resin and reinforcements through meshing gears, and chopping roving and impregnating the fibers on a moving belt similar to SMC production.

Fig. 17-11. Tilt-type batch mixer used for BMC. (*Courtesy: Baker-Perkins, Inc.*)

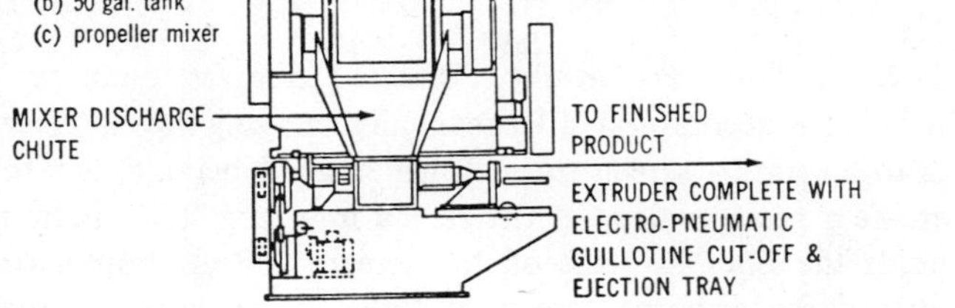

Fig. 17-10. Automated glass, resin, and filler feed system built around a spiral or sigma blade mixer for compounding batch-type premix and BMC mixing equipment. (From ***SPI Handbook of Technology and Engineering of Reinforced Plastics/Composites***, edited by Mohr.)

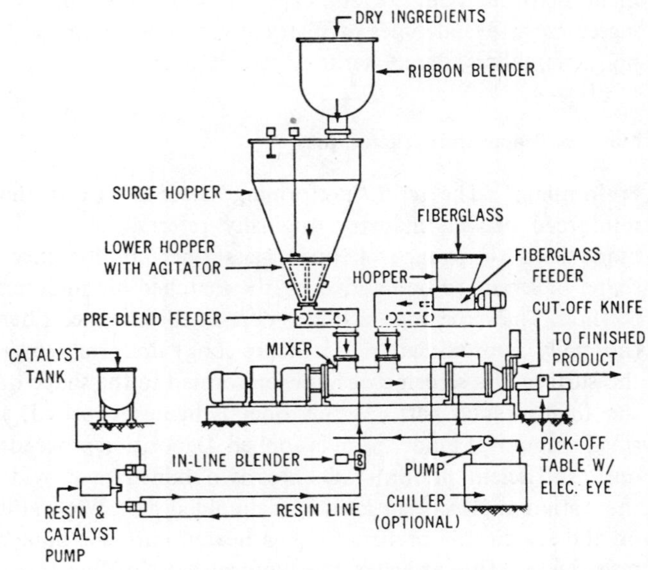

Fig. 17-12. Continuous-feed automated system for feeding glass, resin, and filler to a continuous, in-line mixer for preparation of premix and BMC compounds while keeping glass degradation to a minimum. (From ***SPI Handbook of Technology and Engineering of Reinforced Plastics/Composites***, edited by Mohr.)

Vacuum Equipment. Choosing the type of vacuum pump to be used in a reinforced plastics plant is determined by the specific operation. In plants where there is a preponderance of custom vacuum-bag laminating production, small individual pumps are preferred as opposed to a central system. The ability to move the pump with the job has definite advantages. Conversely, in a plant where the system used is primarily deaeration or vacuum impregnation, a central system with a large receiver works best. In either case, precautions should be taken to insure that there is adequate trap capacity between every job and the pump, or the system will not remain functional for long. Even with the best trap and filter system, some of the resin manages to reach the pump in vapor form. The floating, eccentric vane-type of cylindrical pump seems to tolerate this type of service better than most.

Presses. Specialized presses for the reinforced plastics industry have become standard items with almost all press manufacturers. They differ from most other types of presses in that they have relatively low ram forces compared to their platen areas (see Figure 17-13).

This image is shifting in some parts of the industry, however. The premix molders are looking for increased tonnages and faster closure speeds. The military job shops who are molding ablative components for rocketry are using extremely high molding pressures on outside components. The result is a rebirth of the demand for presses in the 2000-ton category.

Miscellaneous Equipment. Depending on the particular process being used, all or only part of the following equipment may be used: various capacity scales for weighing ingredients, several types of fiber cutters, spray equipment, polishers, rollers, Barcol hardness testers, etc.

Material Preparation Techniques

Preforming.* The term "preforming" as it is used in the reinforced plastics industry originally referred to a technique used to arrange chopped glass fibers in the exact shape of a part to be molded by the matched-die molding method. The basic approach is to deposit the chopped fiber (normally approximately 1.5 inches long) uniformly over the surface of a screen that has been formed to the shape of the final molded part. As this fiber is being deposited, a small amount of binder resin is applied. Deposition proceeds until sufffficient preform thickness is obtained to provide the optimum glass-resin ratio in the molded part. While still on the screen, the preform is then heated until the binder resin cures. This provides the preform with sufficient integrity to be stripped from the screen and transported to the press for final molding.

*Preforms in reinforced plastics molding should not be confused with preforms as used in compression and transfer molding.

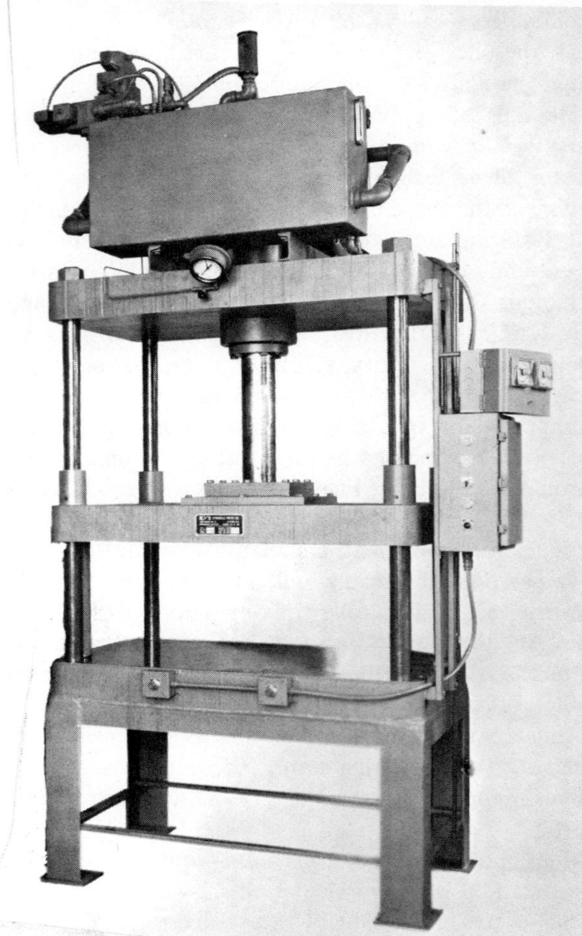

Fig. 17-13. 75-ton reinforced fiberglass molding press. (*Courtesy: Modern Hydraulic Corporation, Clifton, N.J.*)

The term preform is also currently applied to the use of fully impregnated SMC material which is precut or cut and stapled to fit a mold prior to press curing. The use of SMC, with low-profile resins, for smaller parts or for parts requiring a high-gloss finish to automobile smoothness qualities is quite general. The older methods of preform are still in use for larger parts such as truck bodies, boats, camper tops, etc.

There are two different methods commonly used for producing the non-SMC preforms: directed fiber preforming and plenum chamber preforming.

Directed Fiber Preforming. Directed fiber or open preforming is accomplished by manually spraying the cut fiber onto a rotating screen through which air is being drawn to create a pressure drop at the screen interface. This suction holds the fiber in place on the screen while the operator directs the airborne screen of fibers to all areas of the screen. The operator also intermittently sprays on the proper amount of an aqueous emulsion polyester or acrylic liquid binder. When the fiber buildup is complete, the preform is oven cured and removed from the screen (see Figures 17-14 and 17-15).

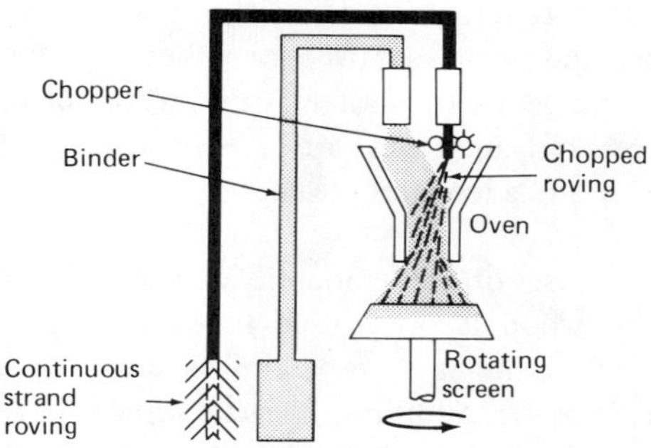

Fig. 17-14. Preform by directed fiber method. (*Courtesy: Owens-Corning Fiberglas Corporation*)

Aqueous emulsion binders used in the preforms are usually based upon about 5% concentrations of polyester or acrylic resins, although other types of resin are sometimes used for special purposes. The percentage of binder applied to the preform varies between 2% and 10% of the finished weight.

Several types of machines are available, and machine configuration varies with size and desired production rate. On average-size preforms, a common type is a four-station rotary mount for the screens, where while one screen is being sprayed, the second screen is in the oven-curing position, the third screen is being stripped, and the fourth screen is in the ready position. Extremely large preform machines have only one screen. After preforming, a hood lowers over the screen to form an oven, and the air is redirected through a heat exchanger to provide curing temperatures.

Fig. 17-15. Preform being produced by the directed fiber method on a rotating preform screen to be used in a compression molding process. (*Courtesy: Owens-Corning Fiberglas Corporation*)

This method has the disadvantage of depending almost entirely on operator skill for uniformity of product. Its advantages are the ability to produce preforms with varying wall thicknesses and built-up areas, and the wide range of possible machine sizes.

Plenum Chamber Preforming. Plenum chamber preforming is quite similar to the directed fiber method and varies primarily in the method of applying the fiber to the screen. This was the first type of preforming developed and is an adaptation of the old felt-hat forming machines. The rotating screen is positioned in a plenum chamber and air is drawn through the screen as in the directed fiber method. The roving cutter is mounted at the top of the plenum and the cut fiber falls onto a rotating spinner which distributes the fiber uniformly in the plenum chamber. The fibers then "snow" down into the screen. Normally, liquid resin binder is applied by spray gun, but in some machines, powder binder is sifted on from a vibrator feeder. Binder is cured by methods previously described (see Figure 17-16).

Types of plenum preforming equipment available range from single-screen units with separate curing ovens to multiple-station, fully automatic units.

The advantages of this process are uniformity of product after initial setup and capability for complete automation with resultant reduction in labor costs. Disadvantages are restriction to uniform wall thickness, difficulty of initial setup on some types of parts, and the compounding of the fiber distribution problem as size increases.

In both of these air forming methods, care must be exercised in the selection of the proper binder, the amount used, and its distribution in the preform, if specific molding requirements are to be met. Most molding problems can be solved or created in the preform machine.

Derivation of preforming parameters is usually empirical and varies widely from part to part. The texture of the preform, as determined by binder percentage and type, must be determined for each molding. For example, a shallow molding with fine detail requires a soft, fluffy preform, where a molding with long vertical walls will wrinkle or wash unless the preform is quite stiff. Color and/or translucency requirements may also dictate the type and amount of binder. For the finest smooth-finish SMC, impregnation should be done with a low-profile resin.

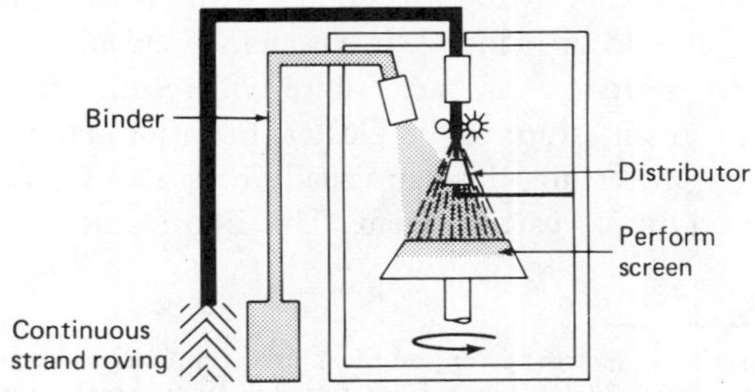

Fig. 17-16. Preform by plenum chamber method. (*Courtesy: Owens-Corning Fiberglas Corporation*)

Fabric and Material Tailoring. When only small numbers of parts are required, tailored sections may be cut from impregnated raw stock by the use of hand shears or knives. For production of large numbers of parts, inexpensive steel rule dies may be used, or the material may be piled up in a number of layers and cut, from a pattern, using a special textile-cutting machine. If very simple shapes are required, such as small discs, then commercially available dies may be procured at nominal costs from machine shop supply houses and similar sources.

Tooling

The unique ability of reinforced plastics to be hand fabricated on a variety of inexpensive tooling materials is a mixed blessing. Although it has allowed the rapid exploration of a multitude of potential FRP markets, it has conversely been a substantial deterrent in the evolution from a hand art to a stable, well-capitalized industry.

The apparent simplicity and low cost of tooling have caused more product and business failures than any other factor. Just as it is possible to form a piece of sheet metal with a rawhide mallet and a sand bag, so it is possible to make an FRP* part with only a plaster mold. But it is a rare case where this should be considered a production technique.

In the final analysis, as with every raw material, the quality, economics, and salability of an FRP end product are solely dependent upon the quality of the tooling on which it is produced.

Of the many materials that will be described in the following paragraphs, each has a specific area of usefulness. Be certain, however, that in contemplating the manufacture of an FRP item, the tooling chosen is in every way compatible with all facets of the product requirements.

Soft Tools

Plaster. Because of the relative simplicity of producing complex shapes directly from the design, plaster is very frequently used to produce master tools from which other types of tools are produced, or, in some cases, it actually becomes the mold or mandrel.

Its excellent dimensional stability and ease of working make it a highly desirable medium for producing masters from which to evolve other tooling. If, however, it is to be used as a mold or mandrel, it has an inherent problem that must be overcome or part failure will result. This is the presence of moisture in the plaster. For room-temperature types of processing, this can usually be sealed with a film-forming type of parting agent. This film former must not be of the water-soluble type or the results will be a one-piece part and mold. If oven or other high-temperature cures are to be used, plaster is not recommended. Even where breakout mandrels are a necessity, plaster cores must be dry and adequately sealed.

Wood. Because of its workability in initial pattern development, wood is often used for prototyping tools. In general, it has the same problems as plaster as an actual mold material. In addition, it lacks dimensional stability and has a relatively soft working face. It has limited usefulness in the room-temperature types of processing and even here is usually faced with some harder material. Figure 17-17 shows a wood tool used as a master to form epoxy-fiberglass tools.

Reinforced Plastics. Industries producing large-area, complex-shaped parts with the room-temperature fabricating processes have found that FRP itself makes excellent production tools. Such tools are capable of almost unlimited size and shape complexity at nominal cost. Ease of tool duplication, once the master is established, is another outstanding advantage. Development of special tooling gel coats of both the polyester and epoxy types has resulted in very fine surface finishes on parts molded from these tools and has decreased their production maintenance (see Figure 17-18).

The development of high-temperature epoxy resins, metallic fillers, etc., has greatly increased the effectiveness of FRP tools in heat-cure processing. Even with these advances, very careful analysis of the specific product problems should be made before deciding on an FRP tool for a heat cure.

Miscellaneous Tooling Materials. Because of the numerous types of processing, and more particularly the ability to

Fig. 17-17. Wood tool used as a master for fabrication of epoxy-fiberglass tools. (*Courtesy: Hughes Aircraft Co.*)

*Because fiberglass is by far the most used reinforcement material, the term FRP is used in this section for fiberglass or any other reinforced thermoset plastics part or composite.

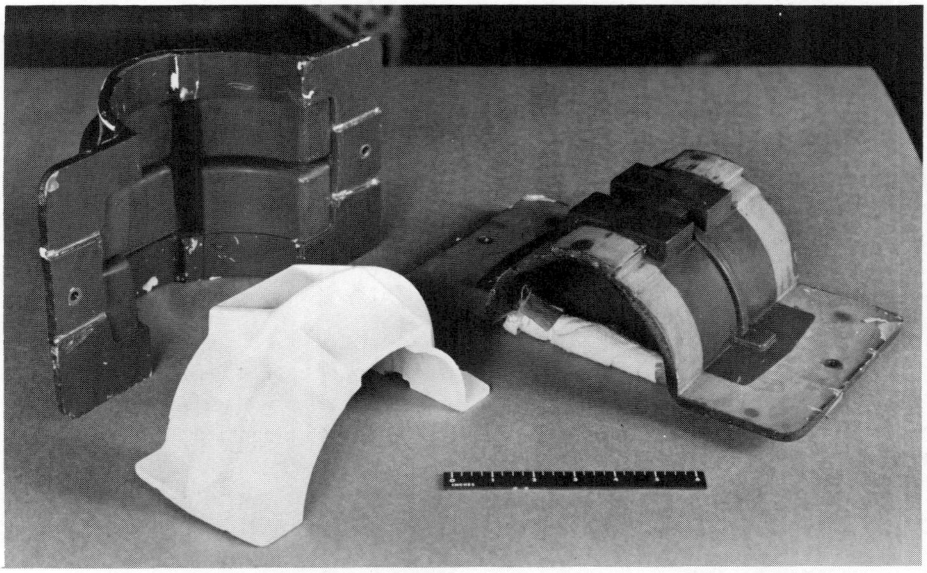

Fig. 17-18. Epoxy-fiberglass tools and part made on plastics tools. (*Courtesy: Hughes Aircraft Co.*)

form very complex shapes in FRP materials, a demand is created for diverse and unusual tools and tooling aids. For example, in vacuum bag processing, fine detail or small, deep-drawn areas are hard to produce because of bag bridging. This problem has been nicely solved by casting local male plugs from room-temperature-vulcanizing silicone rubber. This material is uniquely suitable for this job for many reasons. It is readily castable and cures at room temperature. Its high-temperature resistance allows repeated reuse, and its chemical inertness permits direct contact with the resins with neither inhibition nor part-sticking occurring. For these same reasons, it is often used as the male die in the rubber plunger molding technique. Urethane rubbers can also be used in the same way with, however, some temperature limitations. Advantage can be taken of the very high coefficient of expansion of silicone rubber to cause pressure to be developed in a closed two-part mold, by simple heating.

Another group of materials in this same miscellaneous category are those used for expendable or "breakaway" mandrels and molds. These fall into three general categories:

1. Mechanical Breakouts: The materials used here are plaster and other gypsum products, cast resins, ceramic, and glass.
2. Meltouts: Low-melting alloys, salts, and a variety of high-temperature wax compounds perform this function.
3. Soluble Mandrels: There are a number of commercially available, castable salts with relatively low shrinkage that are designed for this type of application. All these materials are water soluble. In addition, there are some gypsum products that are compounded to have high solubility.

Hard Tools

The types of hard tools used in the industry are, of course, dictated by the particular fabricating technique involved. One effective method of categorizing is to divide them into single-mold processes covering contact, vacuum bag, pressure bag, autoclave, and hydroclave processing, and double-mold tooling, covering the various matched-die, press molding techniques.

Single-Mold Materials. The low pressures used in single-mold processes allow the use of sheet metal, fabricated plate, castings, etc., in a wide variety of metals. The particular choice of material and fabricating technique depends on size, contour complexity, tolerance requirements, and other specifics of the job at hand and is usually relatively straightforward, particularly in small molds. Reinforced plastics components, however, find many of their most attractive market areas in the large, complex-shaped piece. Such molds are extremely expensive to produce in metal by normal methods, so two approaches have been developed that might be noted as they are somewhat unique to this industry.

The first of these is the production of large molds by metal spraying over a release-agent-treated plaster or cast phenolic master. A variety of metals may be sprayed onto these masters, and common practice is to use a fairly dense metal against the master to provide mold surface hardness and then to make the major buildup with a low-melting material like aluminum, or an epoxy-aluminum pellet mixture. One of the advantages of this approach is the ability to include steam or electrical heating coils directly onto the mold.

The second method of making large-area, complex molds is that of electroforming. A cast phenolic matrix is normally used. This matrix, or master, is treated with a conductive coating over a release agent, and standard electrolytic plating procedures are followed. Most of these tools are plated from nickel and are left in the tank until wall thicknesses as high as 0.375 inch are achieved.

The two techniques described above are sometimes combined, with the electroforming providing the surface and the metal spray contributing bulk and/or heating coil attachment.

Double-Mold Materials (Male and Female). The comparatively low pressures used in most matched-metal die molding of reinforced plastics permits the use of die materials not feasible in other phases of the plastics industry. But, although cast aluminum and Kirksite dies may produce small quantities of acceptable parts, steel dies are the best answer for high production. With the lower pressure requirements, these need not necessarily be heat-treated tool steel. Cast steels, such as Meehanite, are adequate even for production runs. These may be flame hardened on the shear edges or other critical wear points.

MOLDING PROCESSES

In the various molding processes described below, the techniques run from simple, no-pressure methods with low-cost tooling, to increasingly high-pressure processes, with consequently higher-cost tooling. It should be borne in mind that as the molding pressure is increased, density is also increased, and there is a corresponding increase in mechanical properties.

Hand Lay-Up Techniques

Contact Molding. Also called open-mold fabricating, contact molding is one of the oldest forming techniques and is the simplest of the reinforced plastics processes.

Contact molding has certain limitations. First, only one side of the part has a good, smooth surface, the side contacting the mold. Therefore, the choice of a male or female mold is dictated by which surface must be the good one, not by fabrication conditions or the part contour. The process is essentially a slow one, and the quality of the part made will depend on the skill of the operator to a large extent. It is well suited for fabrication of very large articles and/or small production runs. It is, therefore, frequently used for production of prototypes, regardless of the eventual high-production process. It is important to note that the physical properties of parts made by this process will be lower than those made by almost any other process.

The contact molding process basically consists of applying a coat of resin to the release-coated mold (see Figure 17-19). Usually this first resin coat is a pigmented gel coat to insure an optimum surface. After the gel coat has become tacky, the reinforcement layer is laid on the resin-

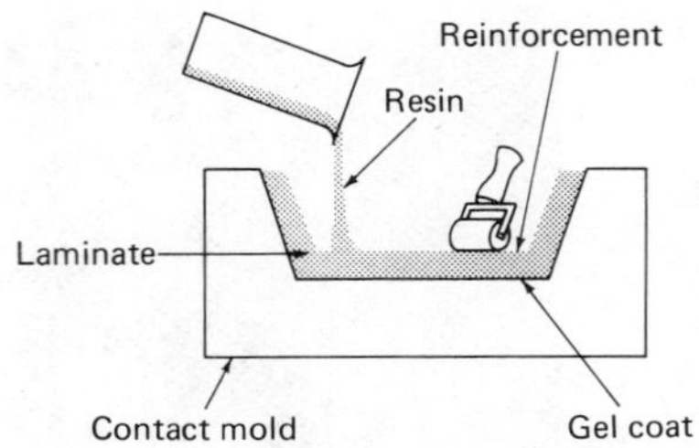

Fig. 17-19. Hand lay-up. *(Courtesy: Owens-Corning Fiberglas Corporation)*

coated gel coat. This layer might be a sheet of surfacing mat if a very fine surface is required, or it could be a layer of dry or impregnated fabric or mat. The use of dry fabric laid onto a resin layer is preferred since this results in the least air entrapment. Additional catalyzed resin is then brushed or roller coated on. Impregnation of the reinforcement is assisted by the use of a roller or squeegee, which is used to promote wetting and to push air bubbles out. Time should be allowed for complete wetting, after which more resin may be added for the next layer. Figure 17-20 shows the sequence of steps in the process. Figure 17-21 shows a mold being made with thin mat for optimum surface characteristics.

After addition of all the required fabric or mat layers, the part is allowed to remain undisturbed until gelation takes place. After gelation, the assembly may be heated to accelerate the final curing process. If possible, the exposed surface should be covered with cellophane or PVA film to inhibit the evaporation of any volatile monomers, and minimize the possibility of air inhibition for some polyester resins.

The choice of materials to be used in this process is dictated by the requirements of the part and the contours to be met. Nevertheless, certain guidelines can be given. First, the resin must be thin enough to be easily impregnated into the reinforcement, at the same time it must be thixotropic enough so no drainage occurs on vertical sections before gelation takes place. The reinforcement material must be one suited for the contours of the part. The catalyst system must be one which allows sufficient time for fabrication of the part; for example, in production shops, a heat-activated catalyst is used so the part can be worked on all day and oven cured at night. Do-it-yourselfers commonly use room-temperature cure catalyst and, thus, work on small sections at a time. Alternatively, a "sunshine cure" ultraviolet catalyst might be used so the part can be worked on all day, in the shade, then placed in the sun for cure.

The equipment required for this process includes the following: resin mixers, including special stirrers which introduce a minimum of air; scales and graduated cylinders for addition of catalyst to the resin; containers for mixing resin—preferably expendable cardboard; squeegees and rollers for rubbing out the resin and brushes for application; table and scissors for cutting fabric and mat. Highly desir-

Fig. 17-20a. Applying parting agent to the mold. (*Courtesy: Owens-Corning Fiberglas Corporation*)

Fig. 17-20b. Gel coat application. (*Courtesy: Owens-Corning Fiberglas Corporation*)

Fig. 17-20c. Fiberglass mat application. (*Courtesy: Owens-Corning Fiberglas Corporation*)

Fig. 17-20d. Resin application to mat. (*Courtesy: Owens-Corning Fiberglas Corporation*)

able, but not absolutely necessary, are a viscometer to check the resin viscosity and a hot bath and gel timer. A Barcol hardness checker is also very useful to indicate the degree of cure. Figure 17-22 shows the hardness tester and the way it is used: the instrument is simply pressed onto the surface of the cured part and the dial is read.

Vacuum Bag Molding. Curing a laminate under pressure will almost always result in a higher-quality hand lay-up laminate than one made by contact or open-mold fabrication. The simplest method of pressurizing the lay-up is the use of a vacuum bag (see Figure 17-23). Vacuum bag molding is very similar to contact molding in that the tooling can be the same, the lay-up is made in the same manner, and the same materials are used. The difference, however, is that a form-fitting bag is used on the outer surface of the part, and a vacuum is drawn so the lay-up is pressurized to atmospheric pressure (approximately 15 psi). The result is that higher reinforcement loading is obtained in the finished part with higher density and a consequent increase

Fig. 17-20e. Use of a roller to promote thorough wetting of a glass mat. (*Courtesy: Owens-Corning Fiberglas Corporation*)

Fig. 17-21. Application of resin to a glass mat during a hand lay-up process. A thin mat is used with many laminations so a minimum of air entrapment will result. (*Courtesy: Owens-Corning Fiberglas Corporation*)

Fig. 17-20f. Finished part. Note smooth inner surface due to gel coat application. (*Courtesy: Owens-Corning Fiberglas Corporation*)

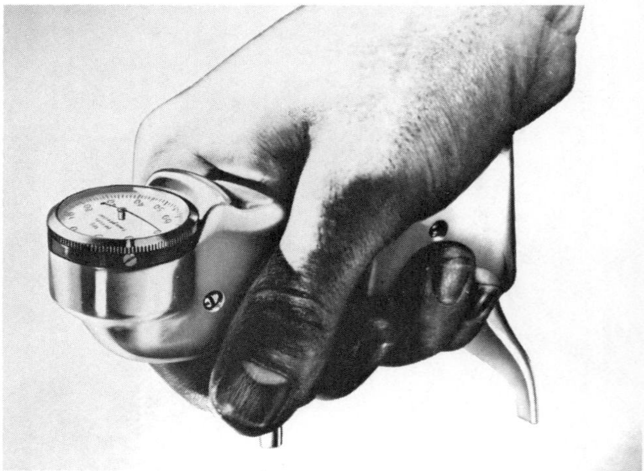

Fig. 17-22. Barcol hardness tester. (*Courtesy: Barber-Colman Co., Industrial Instruments Division*)

in the physical properties. Also, a better surface condition results on the side away from the mold.

In the vacuum molding process, heat curing can be more readily done since the resin content is lower and the pressure can be used to inhibit the flow of resin. Because of the heat, however, the bag must be of a material which is heat resistant. It should also not be affected by the resin and should not affect resin cure. Film materials suitable for bag manufacture include cellophane, polypropylene, nylon, and fluorocarbons. Neoprene, silicone, and butyl rubbers are also used to make bags. In order to obtain the

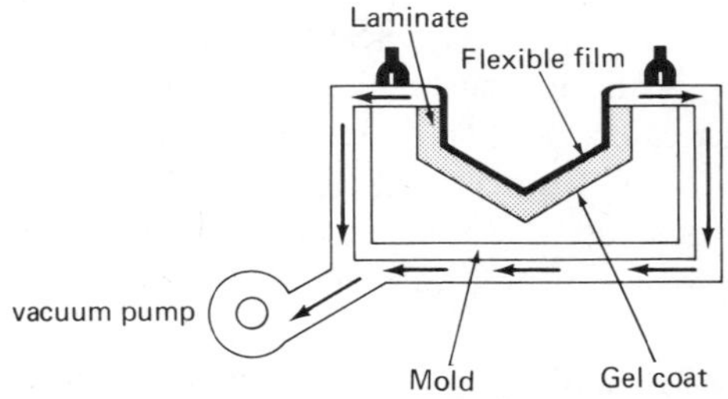

Fig. 17-23. Vacuum bag molding. (*Courtesy: Owens-Corning Fiberglas Corporation*)

Fig. 17-24. Form-fitting vacuum bag. Note rubber tubing and connections. (*Courtesy: Hughes Aircraft Co.*)

best parts, the bag should be tailored to the part so that wrinkles will be at a minimum (see Figure 17-24).

When the laminate is ready for bagging, it is important that a bleeder (breather) sheet be incorporated in the lay-up so the part can be rubbed out to eliminate air bubbles and excess resin. Figure 17-25 illustrates the vacuum-bag assembly configuration and Figure 17-26 shows a process sequence. Figure 17-27 shows a typical hand lay-up shop and a flat sheet being vacuum formed. Figure 17-28 shows a coated plaster tool used for hand lay-up and vacuum bagging. A prototype part made by hand lay-up and vacuum bagging is shown in Figure 17-29. The only equipment required for the vacuum molding process, in addition to that needed for contact molding, is the vacuum pump and associated resin traps to prevent pump damage, and the special bags, breather sheets, and vacuum lines.

Pressure Bag Molding. Using a tailored rubber bag is essentially the same as contact and vacuum molding except that the tooling is considerably stronger and is made to resist pressures from 5 psi to as high as 50 psi or higher.

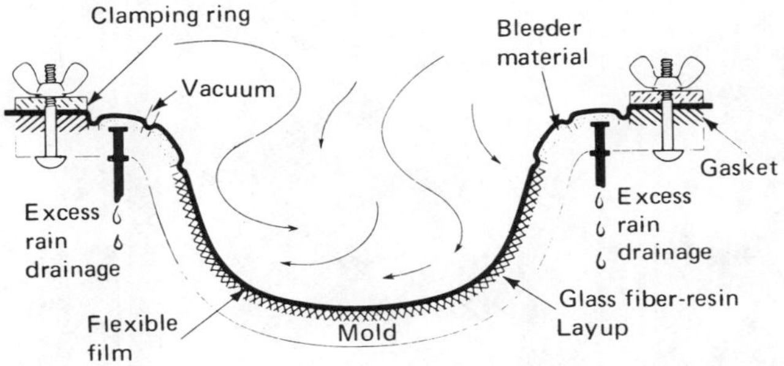

Fig. 17-25. Vacuum bag mold. (*American Cyanamid Co., Plastics and Resins Division*)

The resulting parts, when made at higher pressures, are considerably more dense and of a higher reinforcement content than parts made by the methods discussed so far. This technique, however, is mainly applicable to female molds. Bags used on male molds have a tendency to pucker and develop folds. In actual practice, the pressure bag technique is combined with the use of vacuum to aid in removing most of the entrapped air (see Figures 17-30 and 17-31).

Autoclave Molding. Autoclave molding is similar in most respects to pressure bag molding except that, since the cure is carried out in the autoclave, the tooling need not be pressure resistant in the same manner as the pressure bag tooling is required to be. A simple cast Kirksite or aluminum tool is ideal since the tool is only in compression. With the use of the autoclave, it is possible to use pressures as high as 100 psi. Parts made by this process are dense and exhibit excellent physical properties when the lay-up is expertly made (see Figures 17-32 and 17-33).

The principal disadvantage of the use of the autoclave technique is the cost of installation and the limitations imposed by the size of the autoclave on part sizes (see Figure 17-34).

In the three processes described above, the initial laminate must be laid up by hand and, therefore, the skill of the operator is of utmost importance. The improvement brought about by the additional pressure cannot undo poor resin distribution or bad placement of the fibrous reinforcement, etc.

Spray-Up Technique

Spray-up is perhaps the only FRP molding process that is a genuine Hollywood production. This process was first used

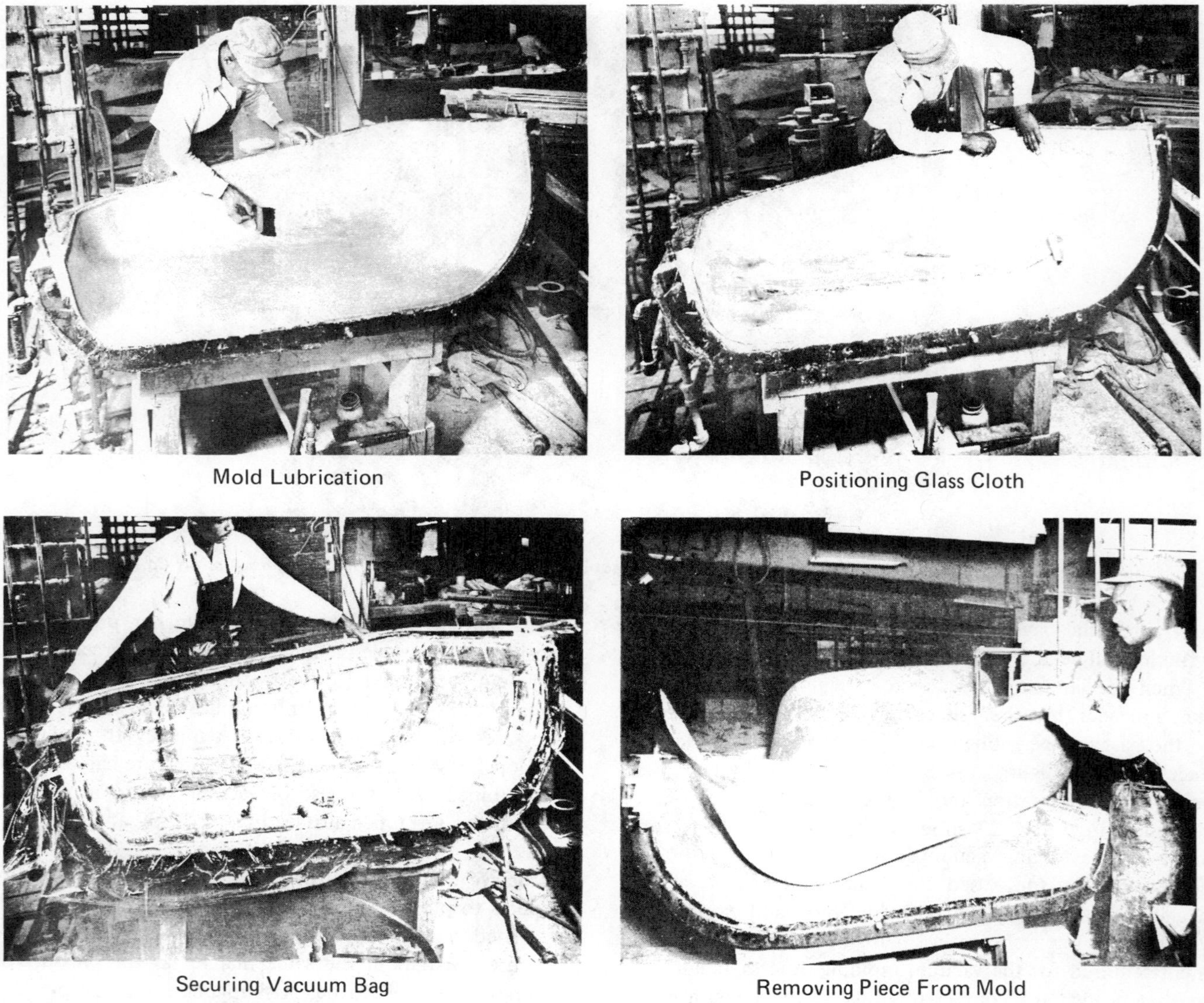

Fig. 17-26. Process sequence for vacuum bag lamination. (*Courtesy: American Cyanamid Co., Plastics and Resins Division*)

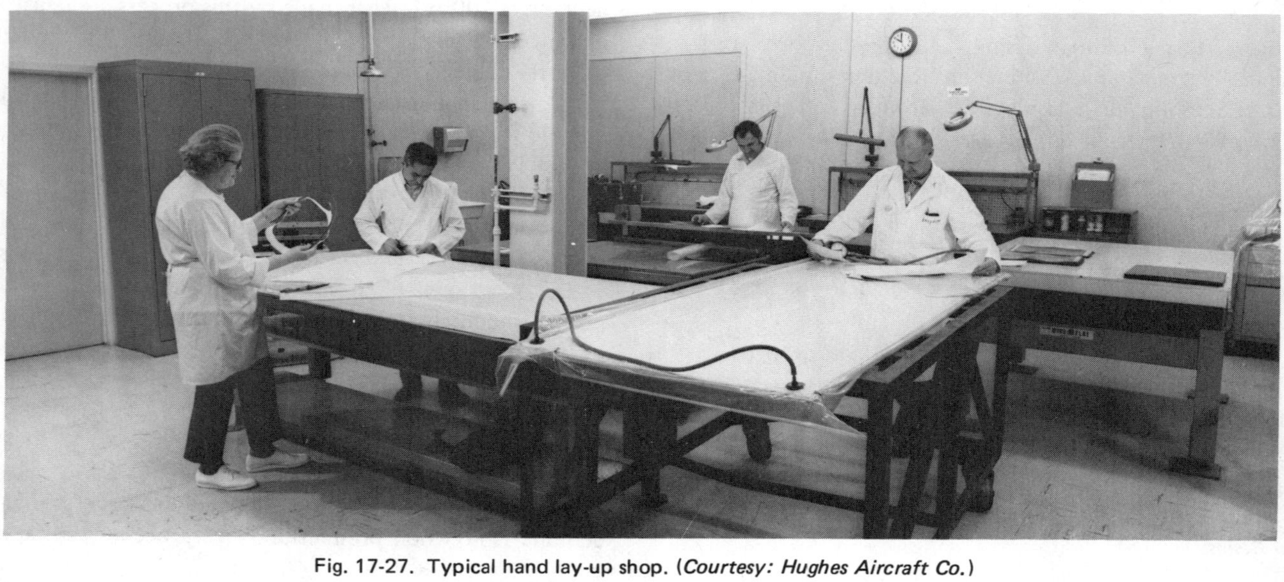

Fig. 17-27. Typical hand lay-up shop. (*Courtesy: Hughes Aircraft Co.*)

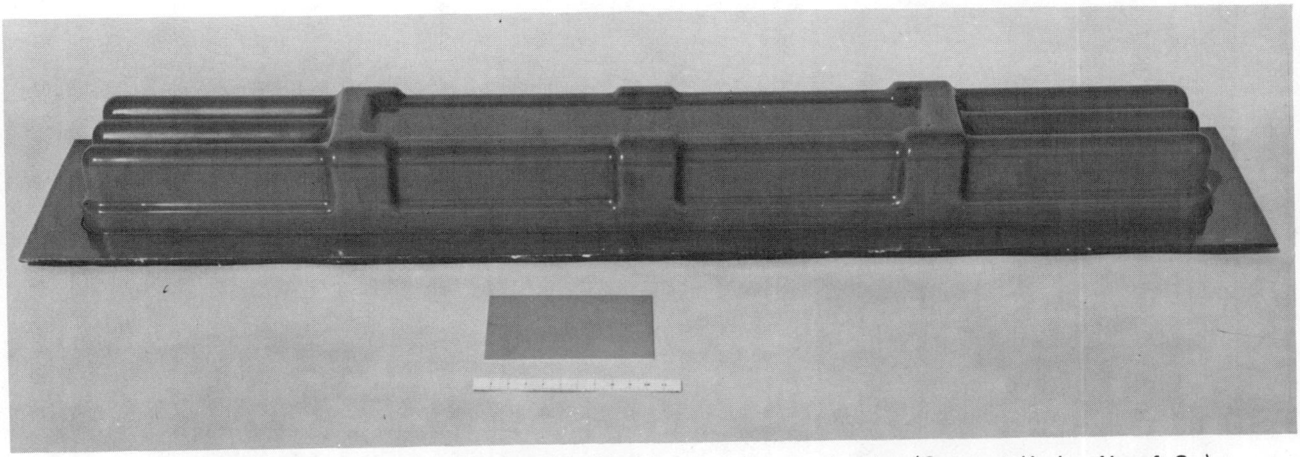

Fig. 17-28. Release-agent-coated plaster tool used for hand lay-up and vacuum bagging. (*Courtesy: Hughes Aircraft Co.*)

Fig. 17-29. Prototype part made by hand lay-up and vacuum bagging, using tool shown in Figure 17-28. (*Courtesy: Hughes Aircraft Co.*)

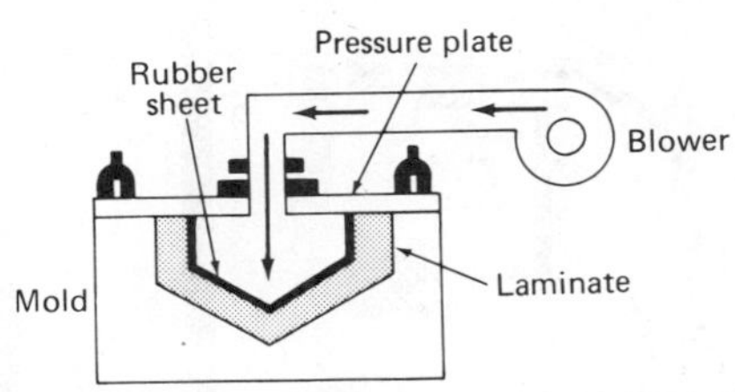

Fig. 17-30. Pressure bag molding. (*Courtesy: Owens-Corning Fiberglas Corporation*)

in the early 1940s as a method of making studio sets. The equipment was very crude originally, but in a period of 10 years, good units were developed and a commercial manufacturing process evolved.

A spray-up machine consists basically of a roving cutter (which may be mounted remotely or directly on the spray head), and a two-component resin-catalyst delivery system (see Figure 17-35). This resin delivery system has taken a number of forms: pressure pot systems and airless systems, double-nozzle guns which mix catalyst and resin in the air, and single mixing head guns. The glass delivery system also varies: with the remote system, the chopped roving is fed to the gun by hose rather than chopping right at the point of delivery, and some guns do not chop at all, but deliver the roving as a continuous strand which is laid in a swirl pattern.

Regardless of configuration, all the units function in the same manner. They combine a catalyzed resin with an accelerated resin and wet the glass fibers with this mixture as it is sprayed on the mold surface. Some guns can also be used to spray gel coats; others can spray resin, catalyst, and a powdered filler simultaneously.

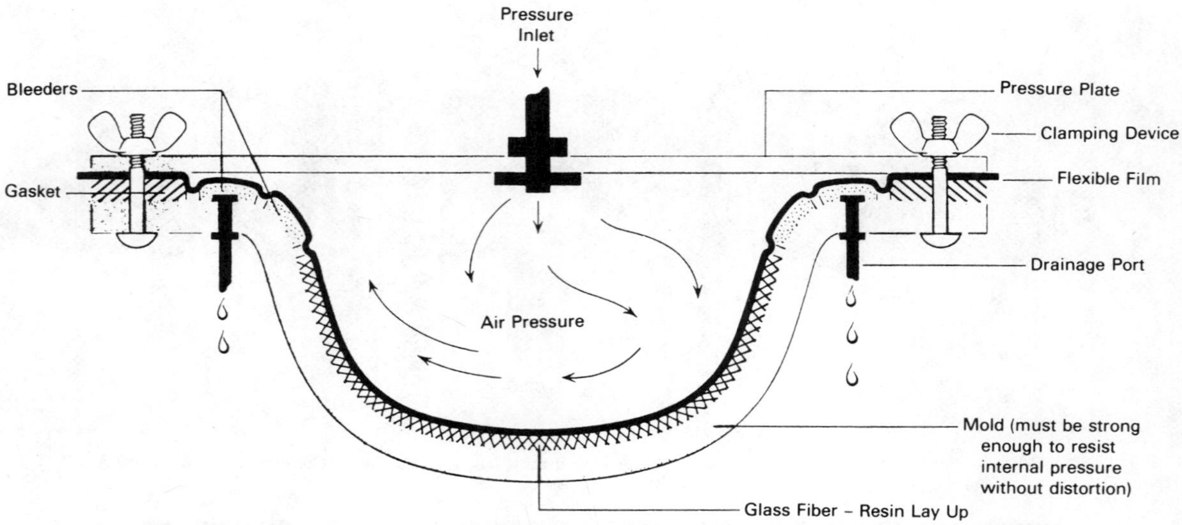

Fig. 17-31. Pressure bag molding process. (*Courtesy: American Cyanamid Co., Plastics and Resins Division*)

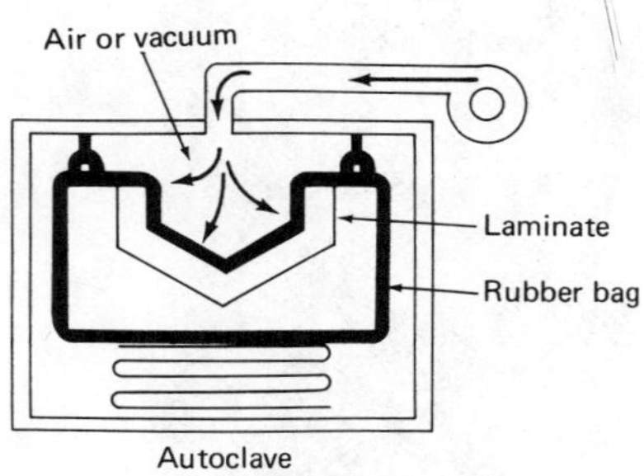

Fig. 17-32. Autoclave molding. (*Courtesy: Owens-Corning Fiberglas Corporation*)

This would seem to be a very simple and rapid technique for building almost any large reinforced plastics structure. Unfortunately, however, the sprayed composite must be rolled by hand to remove the air and achieve reasonable density. This brings the labor cost up to the equivalent of hand lay-up on structures requiring good physical properties, and the process will not consistently yield properties as high as mat lay-up.

This process is finding increasingly greater acceptance in areas where physical properties are not of paramount importance. Large outdoor display signs, tank linings,

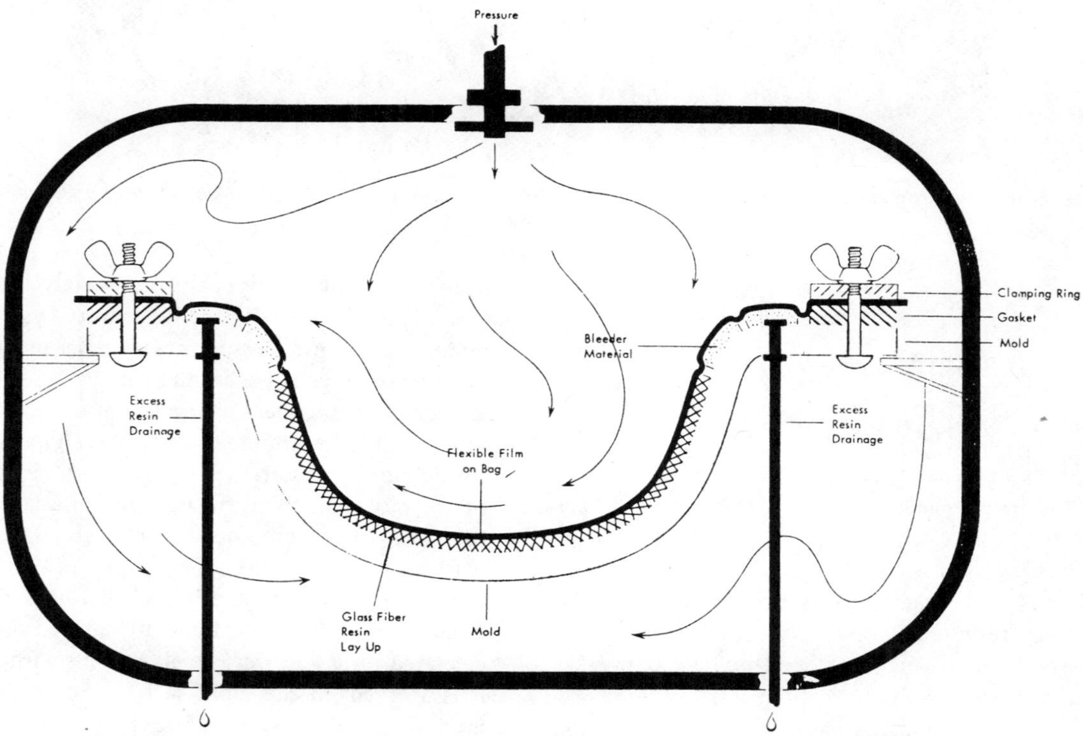

Fig. 17-33. Autoclave molding. (*Courtesy: American Cyanamid Co., Plastics and Resins Division*)

roofing, shower and tub enclosures, and other large-area, detailed moldings are typical applications. Gel coating and spray-up are very compatible, and their combination produces moldings with very fine single-surface finishes. Figure 17-36 shows spray-up equipment. Figure 17-37 illustrates use of rollers to compact the deposited material. This process is ideally suited for use of "robot" operators as shown in Figure 17-38.

Matched-Die Molding

Matched-die compression molding is the technique developed to facilitate mass production of reinforced plastics parts (see Figure 17-39). Two methods are used: a low-pressure technique where pressures run from 50 psi to 300 psi, and a high-pressure molding where pressures may run up to 3000 psi or higher. Either technique is easily adapted

Fig. 17-34. Vacuum-bagged part being placed in an autoclave for 100-psi cure. (*Courtesy: Hughes Aircraft Co.*)

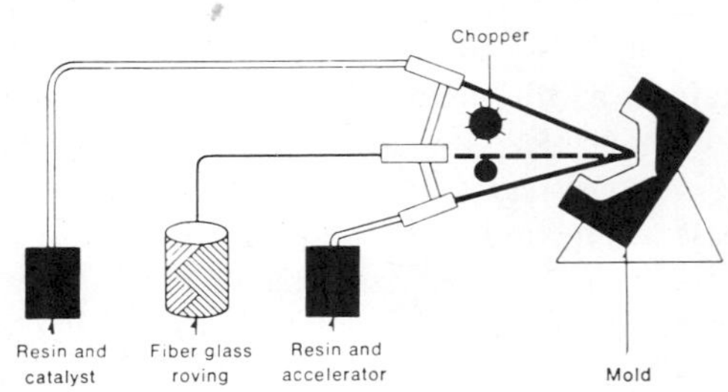

Fig. 17-35. Spray-up technique.

Fig. 17-36. Spray-up process using fiber-chopping gun and resin-catalyst two-head spray gun. (*Courtesy: Owens-Corning Fiberglas Corporation*)

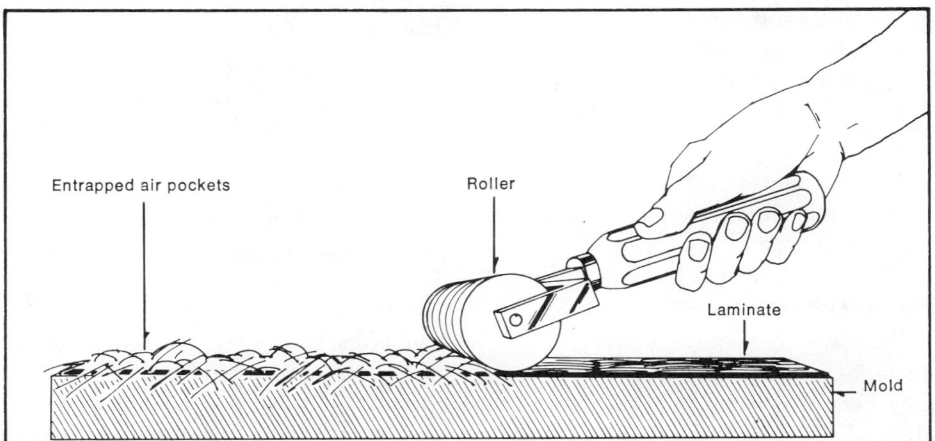

Compacting operation

Fig. 17-37. Use of rollers to compact spray-up. (*Courtesy: PPG Industries, Inc.*)

Fig. 17-38. Robot fabrication of boat hull by spray-up process. (*Courtesy: Owens-Corning Fiberglas Corporation*)

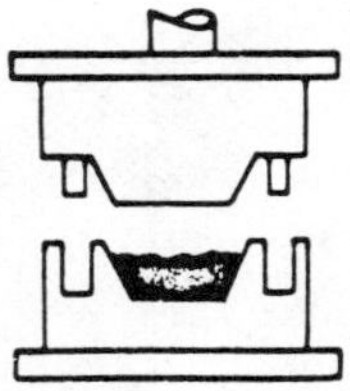

Fig. 17-39. Matched-die molding.

for automation. Usually the preform is heated and compressed, but matched tools can also be used for "cold forming," where only the resin exotherm is used for cure.

In addition to the obvious advantage of relatively high speed, matched-die molding has a number of other advantages:

1. Good finish is obtained on all surfaces.
2. Parts can be made with very little trimming needed.
3. The parts are very reproducible.
4. The process is relatively independent of operator skill.
5. Lower-cost reinforcements may be used.

In the low-pressure process, preforms–made by any of the methods previously described–are combined with the resin at the mold just prior to molding, or SMC is used. Molding pressure is used: (1) to distribute the resin throughout the fibers and (2) to trim the preform to the required dimensions. In order to perform the operation most efficiently, the press should have a dual closing rate–fast, to bring the two parts of the mold together rapidly, and an infinitely variable slow rate, to effect final closing. While manual control can be used for such work, more consistent quality can be obtained in presses that have automatically controlled closing, as well as automatically monitored temperatures, pressures, and time.

Tooling for the matched-die process consists of metal molds. These molds may be made of aluminum, zinc, or Kirksite where relatively short runs are anticipated, or steel for long runs. In any case, the molds should be designed to trap resin and shear the fibers during the last $\frac{1}{16}$ to $\frac{1}{8}$ inch of movement. The mating, shearing surfaces must be accurately machined and hardened with clearances of 0.002–0.004 inch. Heating is usually done by hot oil, steam, or electric heaters which are thermostatically controlled.

Pinch-off should be used only as a means of cutting fibers, not for location, or rapid tool wear will result. Guide pins should be used for locations. The pins should be well out of the resin area to prevent fouling. For the same reason, ejector pins are usually not used in such molds. Removal of parts is done with compressed air, and, in some cases, an air passage will be included in the mold, which is closed with tape during the molding operation.

The molding operation is usually done with the male or punch on the bottom of the mold and the preform placed over the punch. This tends to minimize dragging or creasing of the preform as the mold is closed. To obtain the optimum in reproducible parts, the preform and resin loading should be as consistent as possible. Molds are coated with mold-release compound just as in all other processes. In selecting the resin, a combination of temperature, catalyst concentration, and resin formulation is used which causes gelation in a time long enough for complete resin flow and short enough to be commercially feasible. A typical time for a $\frac{1}{8}$-inch-thick part is approximately 3 minutes.

High-pressure molding commonly uses preimpregnated SMC mat or fabrics, either chopped to small pieces or cut to a specific configuration. Molding is then done at pressures up to 3000 psi and is very similar to normal compression molding.

A variation of matched-die molding is known as rubber plunger molding. In this process, a heated female cavity is used into which the preform (or lay-up) is placed. The rubber male plunger is then forced into the cavity, touching first at the bottom, then deforming so as to put pressure all over the part. This process has the advantage of lowered tool costs, since only one machined tool is required. Slight undercuts can be produced by this method, depending on the part. It, of course, has the disadvantage of having only one good surface–the side contacting the punch is usually similar to a pressure-bag surface. Definite limits exist for the size of parts that can be produced by this method, since large rubber punches are impractical. Small punches made with RTV silicone, however, work very satisfactorily.

Premix Molding

One of the fastest and largest growing segments of the plastics industry is premix or BMC molding. The process is actually a form of compression or transfer molding using the dough or puttylike type of molding compound called premix. (The generic term is "gunk," and for many years the process was known as "gunk molding," until the industry decided the term lacked sophistication.)

In many cases, the molder is also the compounder, as shown in Figure 17-40. As previously described, BMC or premix can be made by careful mixing of the ingredients to form the puttylike materials. One of the methods of quantity production is to feed directly from the mixer to an extruder which then prepares long lengths of material that are cut up to be fed to the presses.

Molding may be done at pressures as high as several tons, or as low as several hundred pounds, such is the versatility of the possible compound preparations. The major problems in processing premix compounds come about in the possibility of nonhomogeneous compound being prepared or a nonisotropic part being made. Both of these problems can be dealt with by careful attention to mixing procedures, and by attention to part design details which would tend to accentuate resin segregation or trap filler during compound flow in the mold. The definite advantages in the use of BMC are rapid cycles at lower pressures with lower-cost materials.

The marketing areas for premix moldings are somewhat different from those for other forms of reinforced plastics. Moldings of this type are generally most competitive in areas where die castings and sand castings were formerly used. They are finding wide acceptance in the electrical and electronic fields, business machine housings, home appliances, etc., because of their unique combination of good physical and electrical properties. The ability to mold in delicate inserts to precise tolerances also contributes to their popularity in these fields. Figures 17-41 and 17-42 show typical parts made of BMC material. The type of press used to mold such parts is shown in Figure 17-43.

Injection Molding

Injection molding of polyester- and epoxy-reinforced materials is done in a manner similar to injection molding of thermoplastics, except that the barrel and screw, or plunger, is kept at a relatively low temperature and the material is

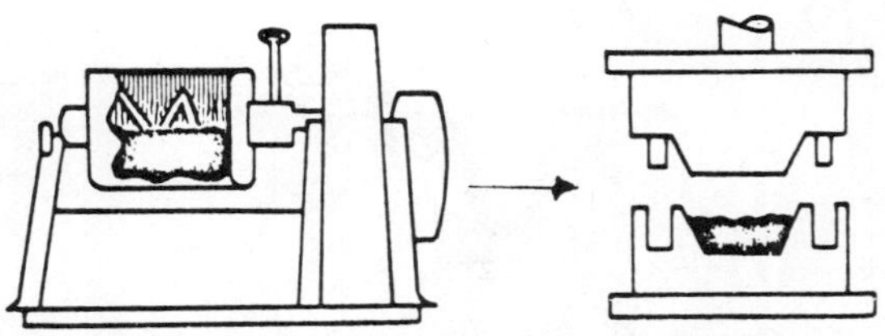

Fig. 17-40. Premix preparation and molding.

Fig. 17-41. Typical BMC part. Note cutaway section to show uniformity of wall thickness. (*Courtesy: Owens-Corning Fiberglas Corporation*)

injected into a mold kept at 150–175°C (300–350°F) (see Figure 17-44). Compounds with a maximum reinforcement fiber length of approximately $\frac{1}{2}$ inch are used. The process offers the possibilities of high production rates for very complex parts, with almost complete automation. Since parts are injected into a closed mold, flash is minimal and postmold trimming is also kept to a minimum.

Continuous Lamination

Continuous laminating is accomplished using a wide variety of reinforcing media such as fiberglass mats, fabrics, and papers. The plies of reinforcement are drawn through a resin-impregnating system, which may take the form of an impregnating tank, doctor blade, roll coater, or other type

Fig. 17-42. BMC molded part. Note thick and thin sections—both satisfactorily cured. (*Courtesy: Owens-Corning Fiberglas Corporation*)

Fig. 17-43. Molding press used for large BMC parts. (*Courtesy: Owens-Corning Fiberglas Corporation*)

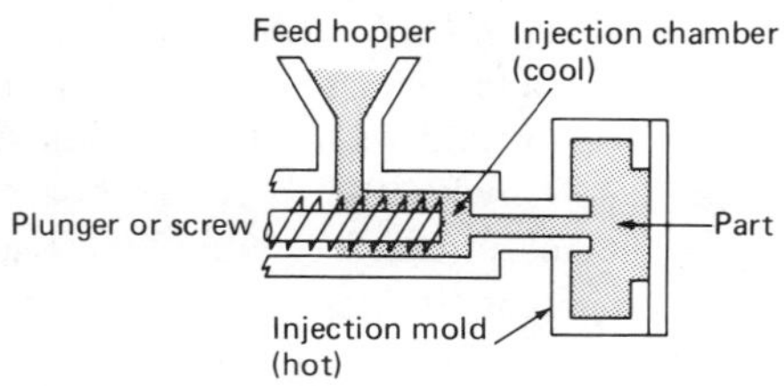

Fig. 17-44. Injection molding of thermosets. (*Courtesy: Owens-Corning Fiberglas Corporation*)

of resin-metering device. After impregnation, a carrier sheet of cellophane or polyethylene film is applied to each face of the web. This sandwich then passes through a set of squeeze rolls for removal of excess resin and air, and final thickness determination. The web next traverses a long tunnel oven with accurately controlled temperature zones. The cure being completed, the web emerges from the oven as a rigid sheet (see Figure 17-45). After passing through the puller rolls, the sheet is edge trimmed by abrasive or diamond saws, the cellophane is stripped, and the sheet is cut to length by a flying cutoff saw.

There is some quantity of laminated decorative flat sheet produced by the continuous method, as well as translucent glazing panels and structural sheet for military and electronic applications. The largest tonnage produced by this process, however, is used in corrugated sheet for the building industry. Although there is still some corrugated sheet produced by hand laminating and press molding tech-

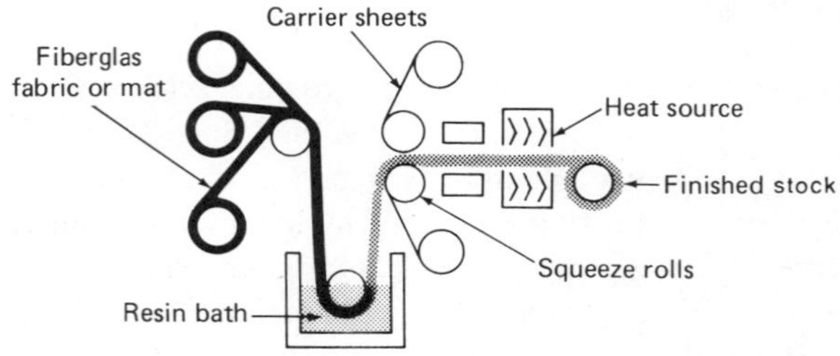

Fig. 17-45. Continuous laminating. (*Courtesy: Owens-Corning Fiberglas Corporation*)

niques, the vast majority of the production is on continuous laminators.

On these laminating lines, most producers start with roving which they chop and blow down on a moving belt—effectively producing their own reinforcing mat in an in-line operation. This is similar to SMC production except that the material is not B-staged.

The process is identical to the flat sheet process described, with the addition of forming shoes in the oven stage to impart the desired corrugation configuration.

Pultrusion

Pultrusion is a word coined to describe the process for making structural profiles continuously from reinforced plastics. In this process, the required number of roving packages are creeled up to produce the desired profile. (Approximately 400 strands of 60 end roving are required for each square inch of profile cross-sectional area.) The roving strands are drawn through an impregnating tank and a squeeze bushing for removal of excess resin and air (see Figure 17-46).

There are two techniques in use for the production of pultrusions; these differ only in the method of curing the extruded shape. In one method, the tunnel oven process, the pultrusion passes through a long oven for cure. The second method uses radio frequency energy to affect a cure in a much shorter period of time.

In the tunnel oven process, the wet roving bundle to be made into rod stock passes first through a shaping bushing, which gives it the general desired circular shape. It is then circumferentially and spirally wound with cellophane tape. Passing through the heated curing zones in the oven accom-

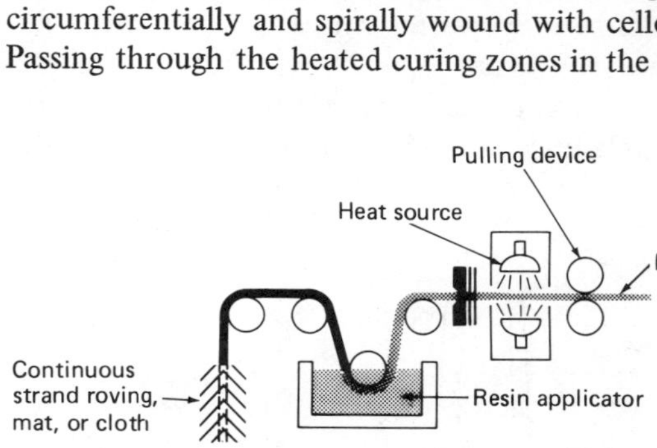

Fig. 17-46. Continuous pultrusion. (*Courtesy: Owens-Corning Fiberglas Corporation*)

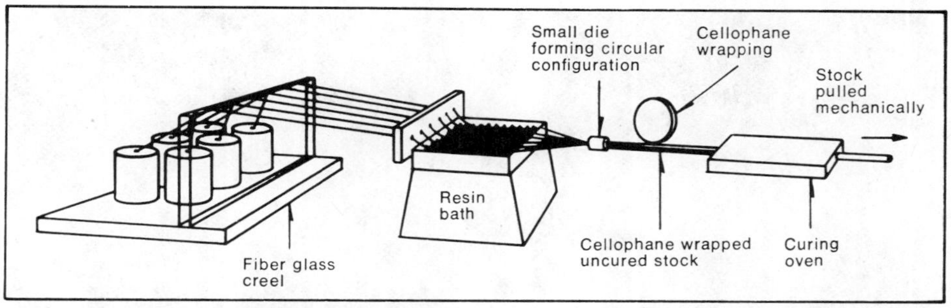

Pultrusion process using cellophane

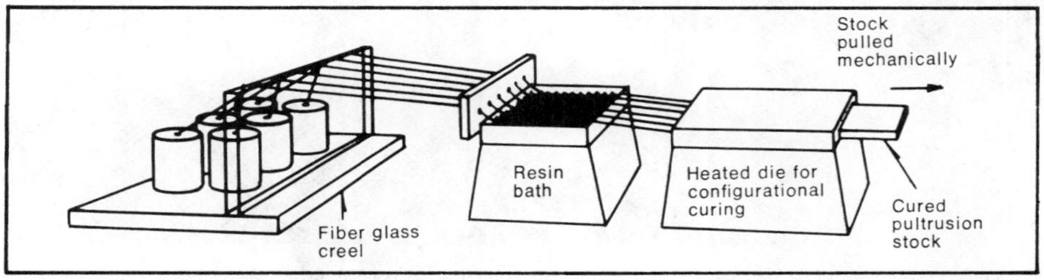

Pultrusion process using a die

Fig. 17-47. Pultrusion processes. (*Courtesy: PPG Industries*, Inc.)

plishes the cure, and the rod stock exits through the pulling device where it is cut to length (see Figure 17-47).

Square and rectangular bars and other simple shapes can also be produced satisfactorily by this technique, particularly if they are finished in a secondary operation. However, because there is no shape restraint during cure, it is difficult to produce complex profiles or to retain dimensional tolerances or high finishes without secondary finishing.

In the second process, the wet roving bundle is drawn into a shaping die where it is subjected to radio frequency cure. Because of the extremely fast cure attainable by this dielectric heating system, the final cure takes place entirely within the short shaping tube. This means that any extrudable profile may be produced by this method and high surface finishes are inherent without secondary finishing (see Figures 17-48 and 17-49).

The field of reinforced plastics structural shapes is relatively new, but markets are developing rapidly. Items such

Fig. 17-48. Glastruder pultrusion machine: standard complete machine with roving creels, wet-out tank, RF cabinet, die station, controls, stock gripper-pullers, and flying cutoff saw. (*Courtesy: Goldsworthy Engineering*)

Fig. 17-49. Glastruder fitted with both radio frequency and induction heating for advanced composites. (*Courtesy: Goldsworthy Engineering*)

Fig. 17-50. Large-tank construction by on-site pultrusion of an epoxy-fiberglass, 1/2-inch thick "tape," which has tongue-and-groove edge configurations. The "spider" is then used to coat the edges with epoxy cement and lay the tape in the desired shape. (*Courtesy: Goldsworthy Engineering*)

as fishing rods and archery bows are well established in the sporting goods field. Various other items include hammer and ax handles, construction I beams and T beams, and even on-site tank construction (see Figure 17-50).

Resin Transfer Molding (RTM)

Resin transfer molding is an adaptation of an old technique known as the Marco process which was developed shortly before World War II. In this process, the reinforcing material, commonly continuous-strand, glass-fiber mat is

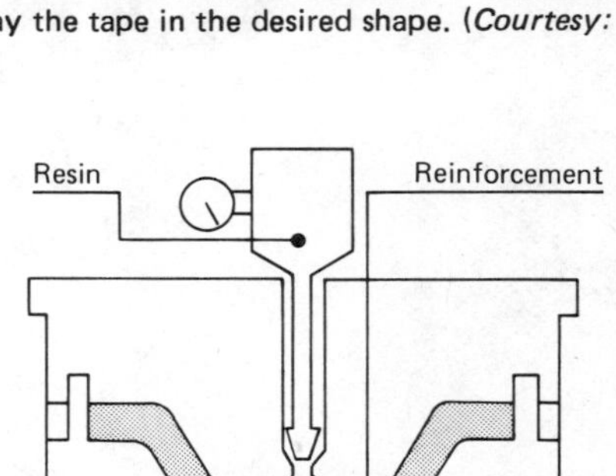

Fig. 17-51. Resin transfer molding. (*Courtesy: Owens-Corning Fiberglas Corporation*)

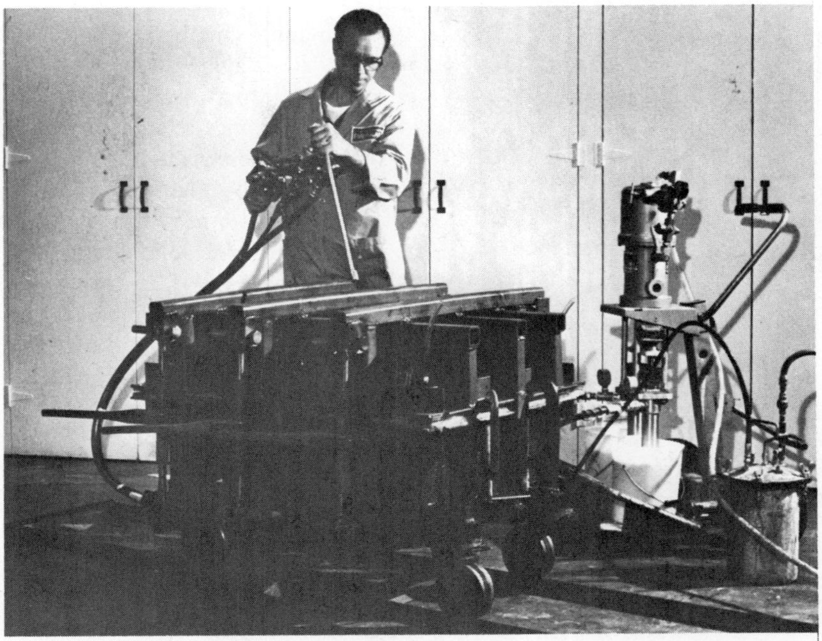

Fig. 17-52. Resin injection in RTM process. Note closed mold. (*Courtesy: Owens-Corning Fiberglas Corporation*)

clamped between the two halves of a male and female mold. Through injection ports, room-temperature-cured, catalyzed resin is injected, and air is displaced through other ports. Usually the resin is injected at the bottom, and air exits through top ports which have transparent tubing inserts. When the resin is seen in the tubing, injection is stopped (see Figures 17-51 and 17-52).

A variation of the RTM technique is known as cold press molding. In this technique, a tightly fitted male and female mold is also used. In this case, the reinforcement, continuous-strand mat or similar material, is placed in the female half of the mold and a catalyzed resin is poured into the reinforcement. Closing the mold distributes the resin. Some experimentation is usually required to determine the most uniform resin distribution, but once this has been done, consistency is claimed to be easily accomplished (see Figure 17-53).

Fig. 17-53. Parts made by RTM process. (*Courtesy: Owens-Corning Fiberglas Corporation*)

The advantages of both RTM and cold press molding are:

1. Smooth surfaces on both sides of the part.
2. Fast cycle times.
3. Lower part costs than hand lay-up or spray-up.
4. Uniform wall thickness.
5. Good dimensional control.
6. Low-cost plastic molds can be used since high pressures are not involved.
7. The processes are particularly adaptable for moderate quantities of large parts—up to hundreds of pounds or square feet in area.

Filament Winding

Filament winding is a technique in which the part is literally wound to shape (see Figure 17-54). In this process, continuous strands (rovings) are fed through a catalyzed resin bath and then are wound over a mandrel to form the part (see Figures 17-55 and 17-56). Alternatively, woven or unidirectional prepreg tapes are used with the advantage of higher production rates and certain wind patterns, etc. The prepregs can cost more but have the possibility of a more uniform resin content. On the other hand, the use of a woven

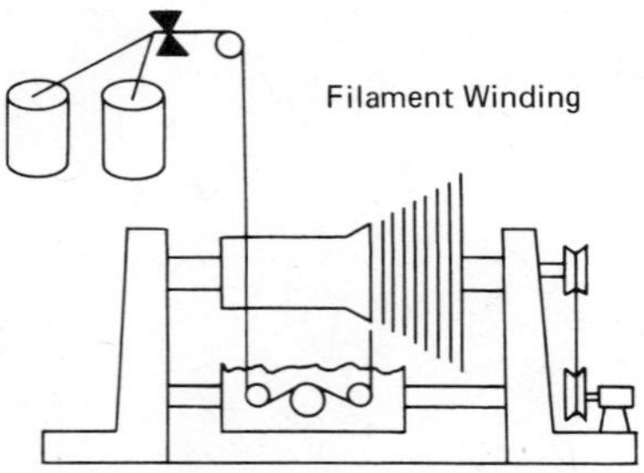

Fig. 17-54. Filament winding.

tape does not result in a part with the maximum content of unidirectional fibers, so for some applications this does not result in the highest strengths.

Depending on the angle of wind, extremely high strengths can be obtained, particularly if the winding is all done in the circumferential direction. The winding machine may act like a lathe, laying the strands in a pattern that is 90° to the axis of the part, or the machine may be programmed to wind the strands at some helix angle which results in more or less reinforcement in the axial direction as well as circumferentially. Figure 17-57 (pp. 742-743) shows some of the wind patterns which may be used, the equipment required, and some consideration for each pattern.

Parts made by filament winding can have the highest strengths of any plastic parts in one direction or in several directions, since the ratio of reinforcement to resin can be very high. In most cases, parts are made using automated equipment so very little operator skill is required. As a consequence, parts can be very uniform, and, since the reinforcement material is low in cost, high-quality parts can be made very economically.

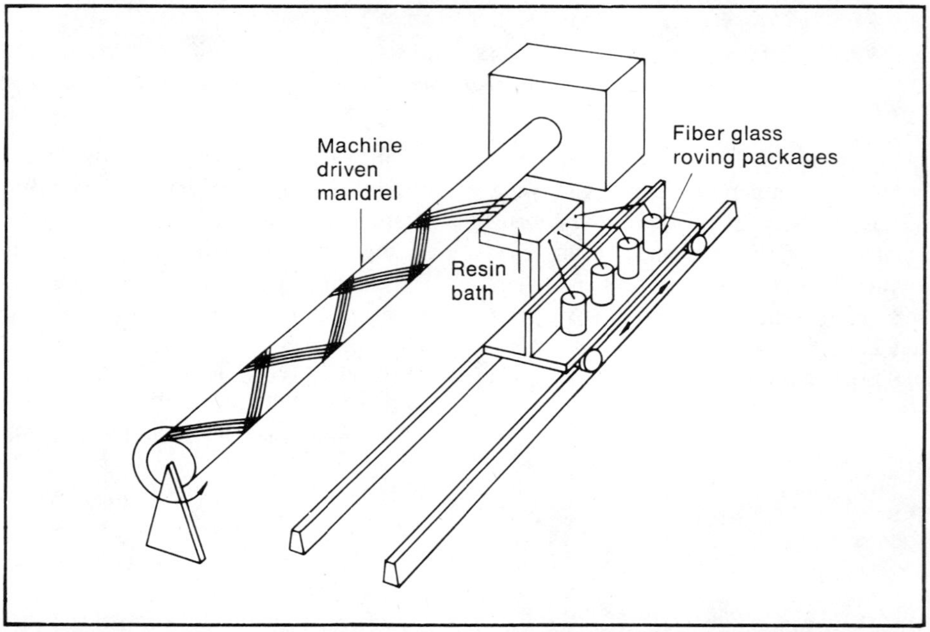

Filament winding system with combination traversing resin bath and traversing creel unit.

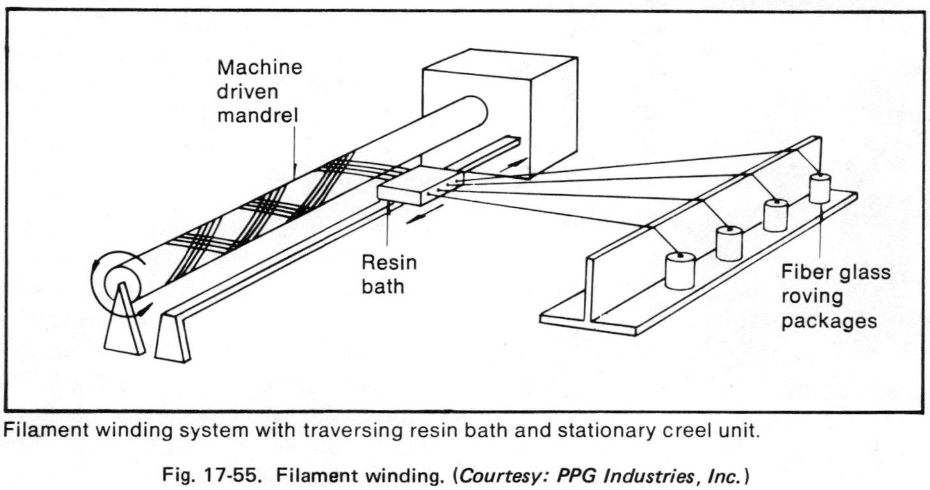

Filament winding system with traversing resin bath and stationary creel unit.

Fig. 17-55. Filament winding. (*Courtesy: PPG Industries, Inc.*)

Fig. 17-56. Filament winding process used to make a large cylinder. (*Courtesy: Owens-Corning Fiberglas Corporation*)

Tooling for filament winding can be any type of body of revolution which can be withdrawn from the cured part. Mandrels are made of steel, aluminum, and plaster. Some tools must be made collapsible to facilitate their removal. Other tools are made of breakaway plasters, or of materials which can be dissolved out. Inflatable mandrels have also been used.

Because of the high strengths obtainable with the process, epoxy resins are usually used with fiberglass, graphite, or Kevlar roving reinforcements. When graphite is used, a prepregged fiber or tape is frequently chosen, since this can be made with closely controlled resin contents from companies that specialize in preparing such raw materials. The use of prepregged tapes also allows certain angles of winding which could not be done with "wet" fiber winding because of slippage.

Applications for filament winding include such items as corrosion-resistant piping, various types of high-pressure-resistant tanks and containers, rocket motor nozzles, and similar parts. Tank closures, metal fittings, metal threaded sections, etc., can be easily incorporated into the part as it is being wound. Since the epoxy resins used to impregnate the reinforcement are also excellent adhesives, they are almost always used when metal inserts are required.

In the 1960s there were a number of large-size small-volume projects, primarily connected with the space program or for the military, that required fabrication of huge bodies of rotation. These included very large filament-wound tanks, rocket motor nozzles, entry cones, etc. Because only a few of each type were to be made, a number of ingenious techniques were developed to accelerate the normal filament winding process and/or to eliminate the need for very large-sized processing equipments chiefly used for application of pressure to the part. Figures 17-58 and 17-59 (page 744) show a method of fabricating flat laminates using unidirectional prepreg tape.

One system for speeding up the filament winding process was the use of the "snap cure" method of tape winding. In this process, the preimpregnated tape, just prior to being laid down on the mandrel, is heated with a quartz lamp, which quickly tackifies the resin and starts the cure. The tape is then pressed down on the mandrel with a hydraulically actuated roller at a pressure of 400–800 psi, instead of the usual 200–400 psi. Several rollers are used at once, assuring good densification. This system, which was developed for use with phenolic-impregnated tape, was said to result in fast cures, the elimination of volatiles, and the production of very dense and uniform parts.

Another system for pressurization of very large parts, without the use of an autoclave was known as the "cable-clave" process. In this process a large part such as a rocket nozzle was wound, either with filament rovings or tape. Two form-fitting rubber bags were then placed over the wrapped part, and then a number of loose, thin metal strips were placed over the bags to cover the total bag area. A steel cable was then wrapped completely around the entire structure. Fluid at pressures up to 2000 psi was then pumped into the space between the two rubber bags. The steel strips prevented bag extrusion between the cables, and the cables restrained the bags so that very highly densified structures could be made. Heating was done by either heating the structure in an oven or heating the pressurizing fluid.

A second system for obtaining very high pressures for large structures was known as the "hydroclave" process. In this method, it was proposed that filament-wound parts would be placed in a rubber bag, and the entire assembly would then be placed in a pressure vessel, similar to an autoclave except that instead of using steam for pressurization and heat, the vessel is filled with a liquid as the pressurizing agent. By the right choice of liquids, pressures up to 1000 psi at reasonable temperatures were expected. [Steam pressure at 1000 psi will require a temperature of 285°C (546°F). Conversely, at 162°C (325°F)–a common cure temperature–steam pressure is only 82 psi.] Figure 17-60 (page 744) illustrates the two techniques.

An obvious disadvantage of the hydroclave process is that an enormously costly pressure vessel would be required if the parts were very large. One method proposed to eliminate requirement for the costly hydroclave was known as the "deep submergence technique." In this method, the parts would again be enclosed in a rubber bag and surrounded by electric heating coils, which in turn would be surrounded by a bag completely sealing the assembly. The structure would then be submerged in the ocean to whatever depth was required to obtain satisfactory pressurization; for example, at 2000 feet the pressure would be approximately 1000 psi. While this system appears feasible for fabrication of extremely large parts, the logistics of transporting the assemblies out to sea, supplying the enormous power re-

TYPE OF WINDING	CONSIDERATIONS	MACHINERY REQUIRED
Hoop or Circumferential 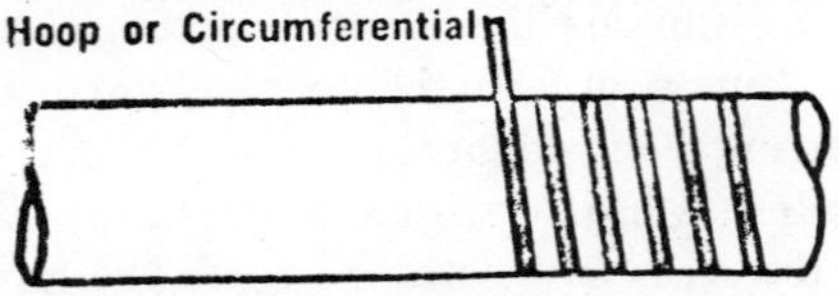	High winding angle. Complete coverage of mandrel each pass of carriage. Reversal of carriage can be made at any time without affecting pattern.	Simple equipment. Even a lathe will suffice.
Helix With Wide Ribbon 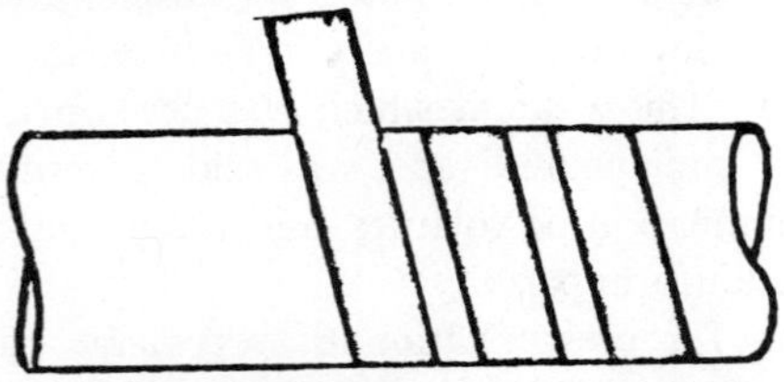	Complete coverage of mandrel each pass of carriage. Reversal of carriage can be made at any time without affecting pattern	Simple equipment with provision for wide selection of accurate ratios of carriage-to-mandrel speeds. Powerful machine and many spools of fiber required for large mandrel.
Helix With Narrow Ribbon and Medium or High Angle 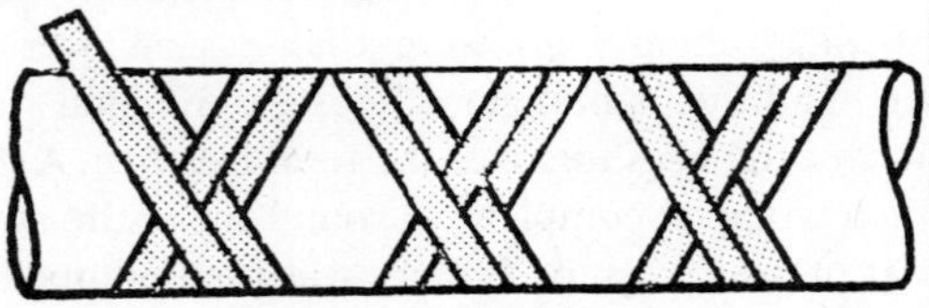	Multiple passes of carriage necessary to cover mandrel. Programmed relationship between carriage motion and mandrel rotation necessary. Reversal of carriage must be timed precisely with mandrel rotation. Dwell at each end of carriage stroke may be necessary to correctly position fibers and prevent slippage.	Precise helical winding machine required. Ratio of carriage motion to mandrel rotation must be adjustable in very small increments. Relationship of carriage to mandrel positions must be held in selected program without error through carriage reversals and dwells. Relationship between carriage position and mandrel rotation must be progressive so that pattern will progress.
Helix with Low Winding Angle 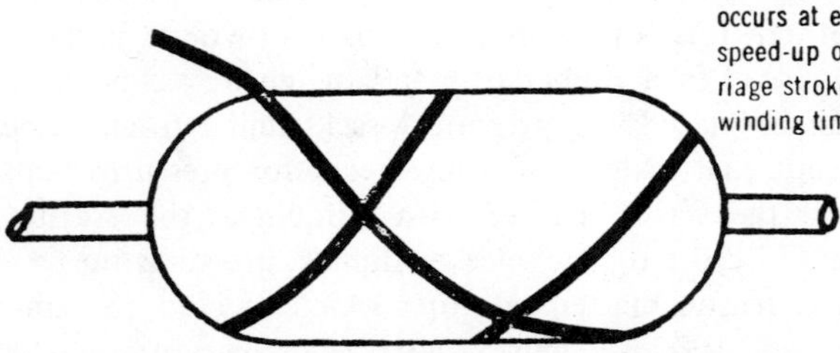	Fibers positioned around end of mandrel close to support shaft. Characteristics of "helix with narrow ribbon" apply. Fibers tend to go slack and loop on reversal of carriage. Fibers tend to group from ribbon into rope during carriage reversal. Mandrel turns so slowly that extremely long delay occurs at each end of carriage stroke and speed-up of mandrel at each end of carriage stroke is highly desirable to shorten winding time.	Similar machinery required as for "helix with narrow ribbon." Take-up device for slack fibers is necessary if cross-feed on carriage is not used. Cross-feed on carriage is required for very low winding angles. Programmed rotating eye can be used to keep ribbon in flat band at carriage reversal. Mandrel speed-up device must be programmed exactly with carriage motion or pattern will be lost. Polar wrap machine can be used for narrow ribbons with winding angle below about 15° without take-up device or mandrel speed-up being required.
Zero or Longitudinal 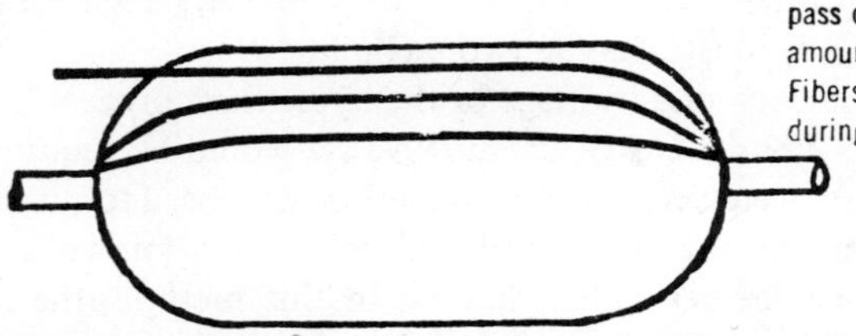	Mandrel must remain motionless during pass of carriage and then rotate a precise amount near 180° while carriage dwells. Fibers must be held close to support shaft during mandrel motion or fibers will slip.	Precise mandrel indexing required. Simple two-position cross feed on carriage sufficient. Vertical mandrel machine and pressure follower for ribbon sometimes required to preserve ribbon integrity.
Polar Wrap 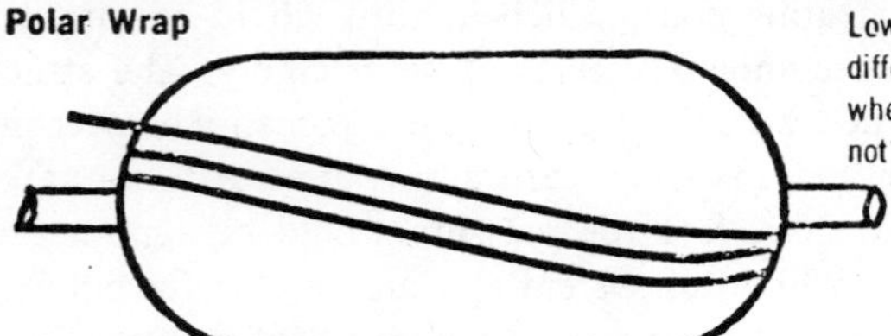	Low angle wrap. Fibers may be placed at different distances from centers at each end when geodesic (non-slipping) path does not have to be followed.	Polar wrap machine with swinging fiber delivery arm desirable for high-speed winding. Helical machine with programmed cross-feed will wind polar wraps more slowly.

Fig. 17-57. Machinery required for different types of filament winding. (Reprinted by permission from *Plastics Design & Processing:* January 1965. Copyright 1965, Lake Publishing Corporation, Libertyville, Ill.)

Cone

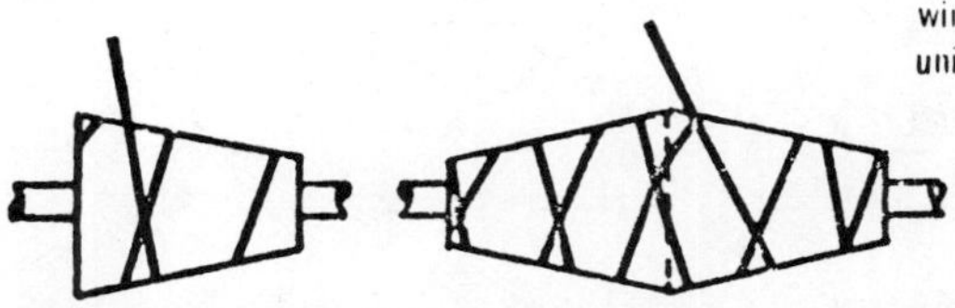

General considerations same as for helical winding except that carriage motion is not uniform.

Programmed non-linear carriage motion required. Other machine requirements same as for helical winding.

Simple Spherical

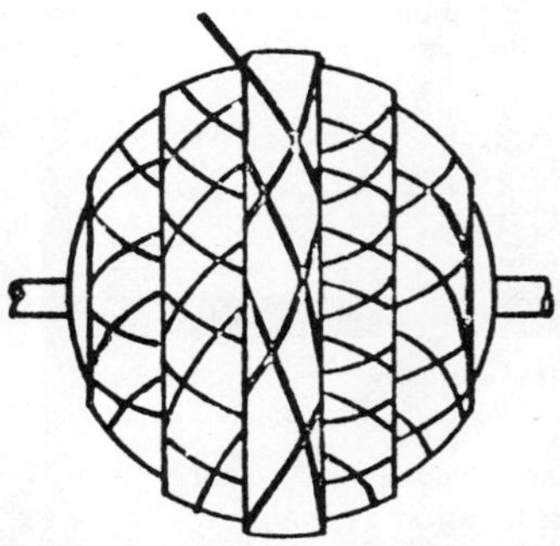

Planer windings at a particular angle result in a heavy build-up of fibers at ends of wrap. For more unifrom strength, successive windings at higher angles are required.

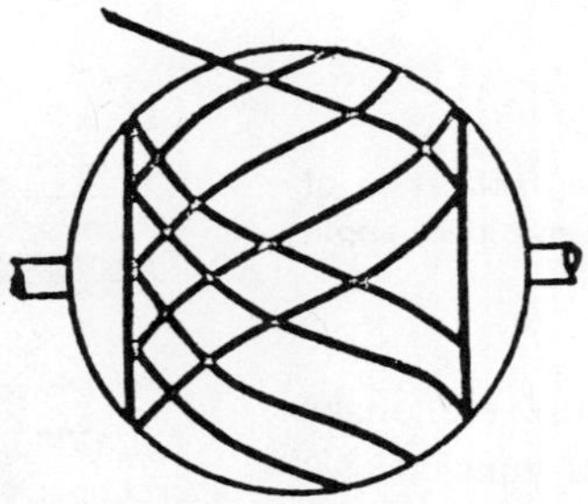

Sine wave motion of carriage is required for carriage with no cross-feed. At low angles of wind, cross-feed is necessary because carriage travel becomes excessive. Polar wrap machine may be used if range of axis inclination is large enough.

Simple Ovaloid

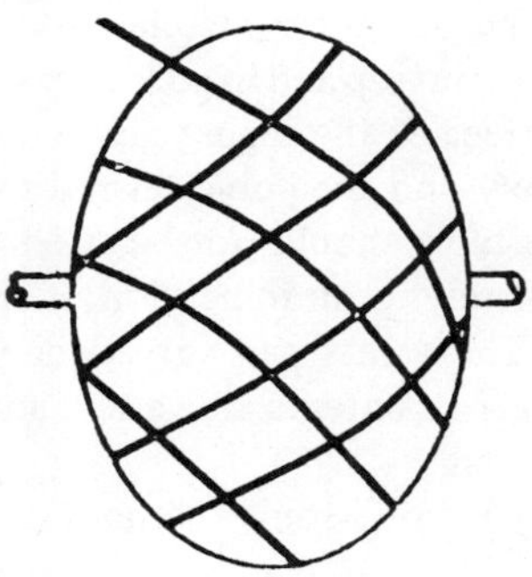

Similar to simple spherical winding but with different carriage or cross-feed motion.

Helical machine with programmed carriage or cross-feed. Polar wrap machine can be used where geodesic (non-slipping) path is in a plane.

True Spherical

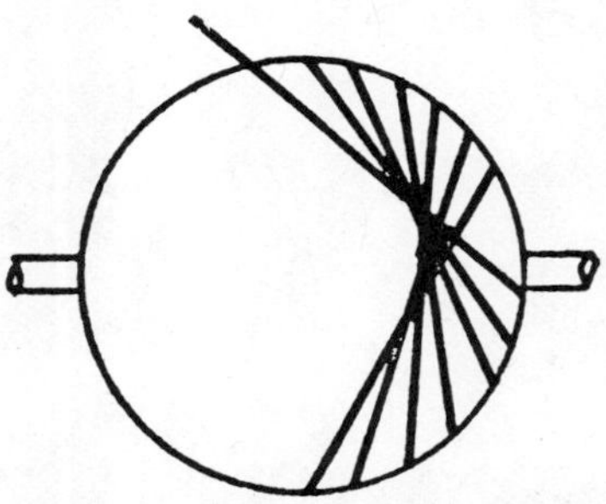

Path of fibers programmed to give uniform wall thickness and strength to all areas on sphere.

Special machine best approach. Otherwise complex programming of all motions of helical machine required.

Miscellaneous

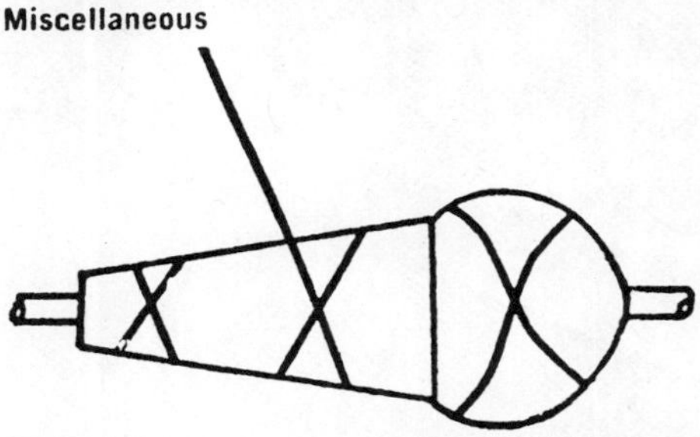

For successful filament winding, it must be possible to hand-wind with no sideways slipping of fibers on mandrel surface.

Machine to reproduce motions of hand winding. Programmed motions in several axes may be required.

Fig. 17-57. *(Continued)*

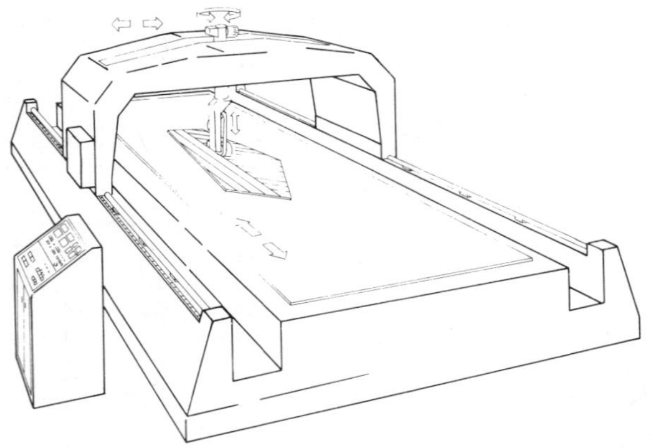

Fig. 17-58. Atlas tape-placement machine used for fabrication of high-strength composite laminates for aircraft and aerospace applications. (*Courtesy: Goldsworthy Engineering*)

quirements, holding the parts steady during cure, etc., make implementation in the near future difficult to imagine. No parts are known to have been made by this process. Descriptions of this and the other novel processes have been included to show the range of thought and creativity which have gone into developing the various techniques. It is hoped that these descriptions may stimulate further research in other techniques and processes.

Design Guide

Table 17-9 compares the mechanical properties of plastics products obtained from the various processes. All values for the reinforced parts in that table are for glass reinforce-

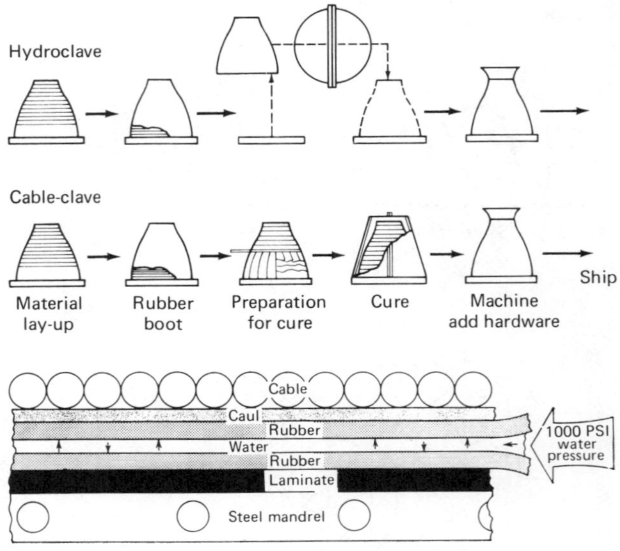

Fig. 17-60. Hydroclave and cable-clave techniques. The cross-sectional view shows pressurization of the part. (Reprinted by permission from *Plastics Technology*, February 1963)

ments. The superior properties of the matched-tool molded parts, as compared to the spray-up and hand lay-up materials, should be noted. The effect of the higher glass contents of the filament-wound epoxy and the polyester rod stocks are also clearly shown. This table should not be taken as completely definitive, but merely as an indication of the range of possible properties. The values can vary widely as the resins are changed, the glass contents are varied, and different cure techniques are used.

Table 17-10 (page 746) shows some of the design possi-

Fig. 17-59. Atlas tape-placement machine. (*Courtesy: Goldsworthy Engineering*)

Table 17-9. Mechanical Properties of Reinforced and Non-Reinforced Plastics.*

Material	% Glass Fiber By Weight	Flexural Strength psi × 10^3	Flexural Modulus psi × 10^5	Tensile Strength at Yield psi × 10^3	Tensile Modulus psi × 10^5	Compressive Strength psi × 10^3	Ultimate Tensile Elongation %	Impact Strength Izod Ft-lb/in of notch
Glass Fiber-Reinforced Thermosets								
Sheet molding compound (SMC)	15–30	18–30	14–20	8–20	16–25	15–30	0.3–1.5	8–22
Bulk molding compound (BMC)	15–35	10–20	14–20	4–10	16–25	20–30	0.3–0.5	2–10
Preform/mat (compression molded)	25–50	10–40	13–18	25–30	9–20	15–30	1–2	10–20
Cold press molding-polyester	20–30	22–37	13–19	12–20	–	–	1–2	9–12
Spray-up–polyester	30–50	16–28	10–12	9–18	8–18	15–25	1.0–1.2	4–12
Filament wound–epoxy	30–80	100–270	50–70	80–250	40–90	45–70	1.6–2.8	40–60
Rod stock–polyester	40–80	100–180	40–60	60–180	40–60	30–70	1.6–2.5	45–60
Molding compound–phenolic	5–25	18–24	30	7–17	26–29	14–35	0.25–0.6	1–6
Hand lay-up (cloth)	30–40	26–30	12–20	14–24	20–25	20–35	1–2	25
Glass Fiber-Reinforced Thermoplastics								
Acetal	20–40	15–28	8–13	9–18	8–15	11–17	2	0.8–2.8
Nylon	6–60	7–50	2–26	13–33	2–20	13–24	2–10	0.8–4.5
Polycarbonate	20–40	17–30	7.5–15	12–25	7.5–17	14–24	2	1.5–3.5
Polyethylene	10–40	7–12	2.1–6	6.5–11	4–9	4–8	1.5–3.5	1.2–4.0
Polypropylene	20–40	7–11	3.5–8.2	5.5–10.5	4.5–9	6–8	1–3	1–4
Polystyrene	20–35	10–17	8–12	10–15	8.4–12.1	13.5–19	1.0–1.4	0.4–4.5
Polysulfone	20–40	21–27	8–15	13–20	15	21–26	2–3	1.3–2.5
ABS (acrylonitrile butadiene styrene)	20–40	23–26	9.2–15	11–16	6–10	12–22	3–3.4	1–2.4
PVC (polyvinyl chloride)	15–35	20–25	9–16	14–18	10–18	13.4–16.8	2–4	0.8–1.6
Polyphenylene oxide (modified)	20–40	17–31	8–15	15–22	9.5–15	18–20	1.7–5	1.6–2.2
SAN (styrene acrylonitrile)	20–40	15–21	8.0–18	13–18	9–18.5	12–23	1.1–1.6	0.4–2.4
Thermoplastic polyester	20–35	19–29	8.7–15	14–19	13–15.5	16–18	1–5	1.0–2.7
Unreinforced Thermoplastics								
Acetal	–	13–14	4	8–10	4–5	5	25–60	1.2–2.3
Nylon	–	5–18	2–4	9	2–5	7–10	29	1–4
Polycarbonate	–	13	3	9–11	3.5	12	100–130	16
Polyethylene (high density)	–	–	0.7–2.6	4	0.6–1.5	2.7–3.6	30–900	0.6–20.0
Polypropylene	–	5–8	1.2–2.7	3–5	1.2	3.7–8	200–700	0.5–20.0
Polystyrene (high impact)	–	3–10	1–5	3–5	2–4	4–9	15–30	0.7–3.6
Polysulfone	–	1.5	4	10	3.6	14	50–100	1.3
ABS (high heat)	–	9	3–4	6.6	2.8–4.1	6.8–12.5	10–20	2.5
PVC	–	13–16	4	6–7	4	–	–	2–20
Polyphenylene oxide (modified)	–	15	4	10	3.7	15	50–100	1.5–1.9
SAN	–	9.7–17.5	5	9–11	5	14–17	2.5–3.7	0.4
Metals								
Gray cast iron	–	10	N.Av.	15–30	120	25	1	4–4.4
Low carbon steel (cold rolled)	–	28	300	29–33	300	28	38–39	N.Ap.
Stainless steel	–	30–35	280	30–35	280	30	50–60	8.5–11.0
Aluminum, wrought	–	20	100	6–27	100	N.Av.	30–40	N.Ap.
Aluminum, die cast	–	8–26	100	8–26	100	9	6–8	N.Ap.
Magnesium, die cast	–	14	65	8–30	65	10–14	4–6	3
Zinc, die cast	–	N.Av.	N.Av.	10–25	N.Av.	N.Av.	10	4.3
Brass, plain yellow wrought	–	14	150	14	150	N.Av.	60–65	N.Ap.

*N.Av. = not available.
N.Ap. = not applicable.

bilities of the various processes. Again, the table is not intended as a complete guide to all possible design features, but can be used as a basis for comparing what can be done with each technique.

Sandwich Structure Processing

The use of sandwich structures is said to have originated in World War II with the development of birch-faced, balsa

Table 7-10. Design Guide.*, **, †

		COMPRESSION MOLDING			INJECTION MOLDING (THERMO-PLASTICS)	COLD PRESS MOLDING	SPRAY-UP AND HAND LAY-UP
		SHEET MOLDING COMPOUND	BULK MOLDING COMPOUND	PRE-FORM MOLDING			
Minimum inside radius, inches	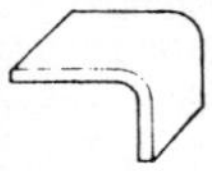	1/16″	1/16″	1/8″	1/16″	1/4″	1/4″
Molded-in holes	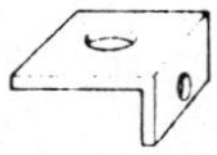	yes*	yes*	yes*	yes*	no	large
Trimmed in mold		yes	yes	yes	no	yes	no
Core pull & slides		yes	yes	no	yes	no	no
Undercuts	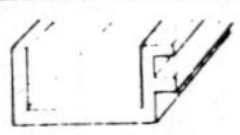	yes	yes**	no	yes	no	yes**
Minimum recommended draft, in./°	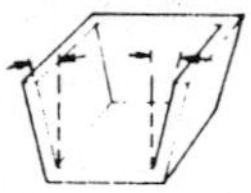	1/4″ to 6″ depth: 1° to 3° 6″ depth and over: 3° or as required				2° 3°	0°
Minimum practical thickness, in.	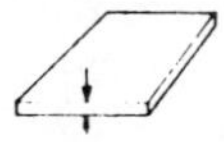	.050″	.060″	.030″	.035″	.080″	.060″
Maximum practical thickness, in.	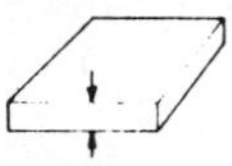	1″	1″	.250″	.500″	.500″	no limit
Normal thickness variation, inches	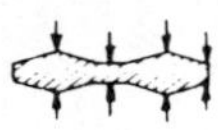	±.005	±.005	±.008	±.005	±.010″	±.020″
Maximum thickness build-up, heavy build-up & increased cycle	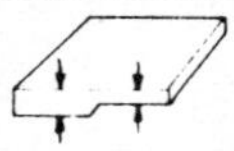	as req'd.	as req'd.	2-to-1 max.	as req'd.	2-to-1 max.	as req'd.
Corrugated sections	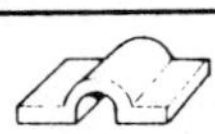	yes	yes	yes	yes	yes	yes
Metal inserts		yes	yes	not recommended	yes	no	yes
Bosses	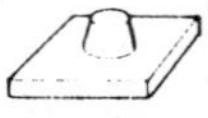	yes	yes	yes	yes	not recommended	yes
Ribs	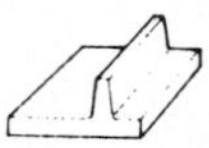	as req'd	yes	not recommended	yes	not recommended	yes
Molded-in labels		yes	yes	yes	no	yes	yes
Raised numbers		yes	yes	yes	yes	yes	yes
Finished surfaces (reproduces mold surface)	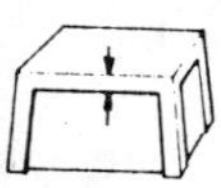	two	two	two	two	two	one

*Parallel or perpendicular to ram action only.
**With slides in tooling, or split mold.
†Courtesy: Owens-Corning Fiberglas Corporation.

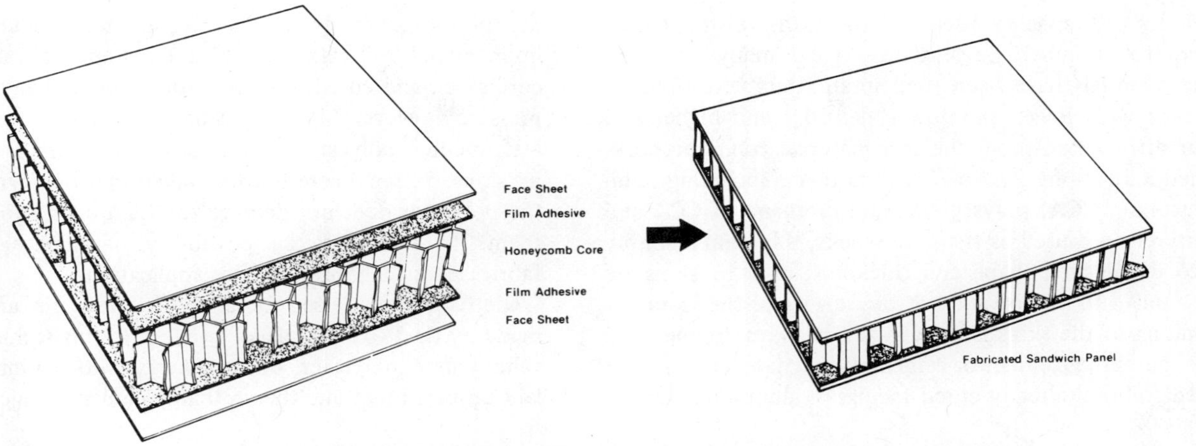

Fig. 17-61. Honeycomb sandwich construction. (*Courtesy: Hexcel Corporation*)

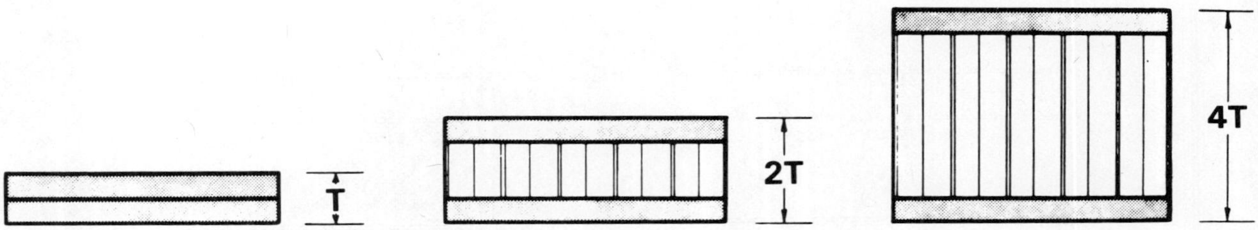

Stiffness = 1
Relative Deflection = 1.000
Approximate Weight = 1

Stiffness = 7
Relative Deflection = 0.140
Approximate Weight = 1

Stiffness = 37
Relative Deflection = 0.027
Approximate Weight = 1

Fig. 17-62. The honeycomb core spreads the facings to increase stiffness while adding an insignificant amount of weight. (*Courtesy: Hexcel Corporation*)

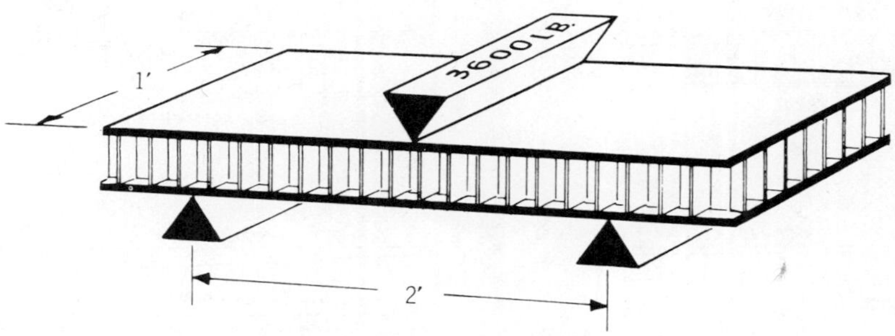

Honeycomb Sandwich—Lightest of Materials.

MATERIAL	Deflection (In.)	Weight (Lb).
Honeycomb Sandwich	.058	7.79
Nested "I" Beams	.058	10.86
Steel Angles	.058	25.9
Magnesium Plate	.058	26.0
Aluminum Plate	.058	34.2
Steel Plate	.058	68.6
Glass Reinforced Plastic Laminate	.058	83.4

Fig. 17-63. Deflection of honeycomb panels versus monolithic panels. (*Courtesy: Hexcel Corporation*)

wood core sandwiches used for the skins of the famed "Mosquito" bomber. Since then a great many core and facing materials have been used in an attempt to produce materials with lower and lower densities and higher and higher stiffnesses. Among the core materials tried were cork, foamed hard rubber, various plastics foams including cellulose acetate, CCA, polystyrene, and urethane. If CCA and polystyrene are used as the core stocks, the foam is simply cut to the desired shape and thickness, and the skins are then bonded to the top and bottom surface of the foam.

Bonding of the skins is done in several ways. In one technique, an epoxy cement or cement-impregnated scrim cloth is used to bond already cured facings or aluminum skins to the plastics core. In another technique, noncured resin-impregnated and catalyzed skins are simply laid up on the core stock and cured, usually with a vacuum bag to furnish pressure; however, higher pressures have also been employed.

If uncured polyester-impregnated fabrics are used as lay-up skins, extreme care must be taken that the styrene from the polyester does not damage the CCA or the polystyrene foam. To minimize the possibility of damage, B-staged fabrics are usually used for this application.

Urethane foams used in sandwich structures are usually made by the "foam-in-place" process. In this technique, the skins (cured polyester, epoxy fiberglass, or aluminum) are laid up in a mold and the urethane foamant is injected into

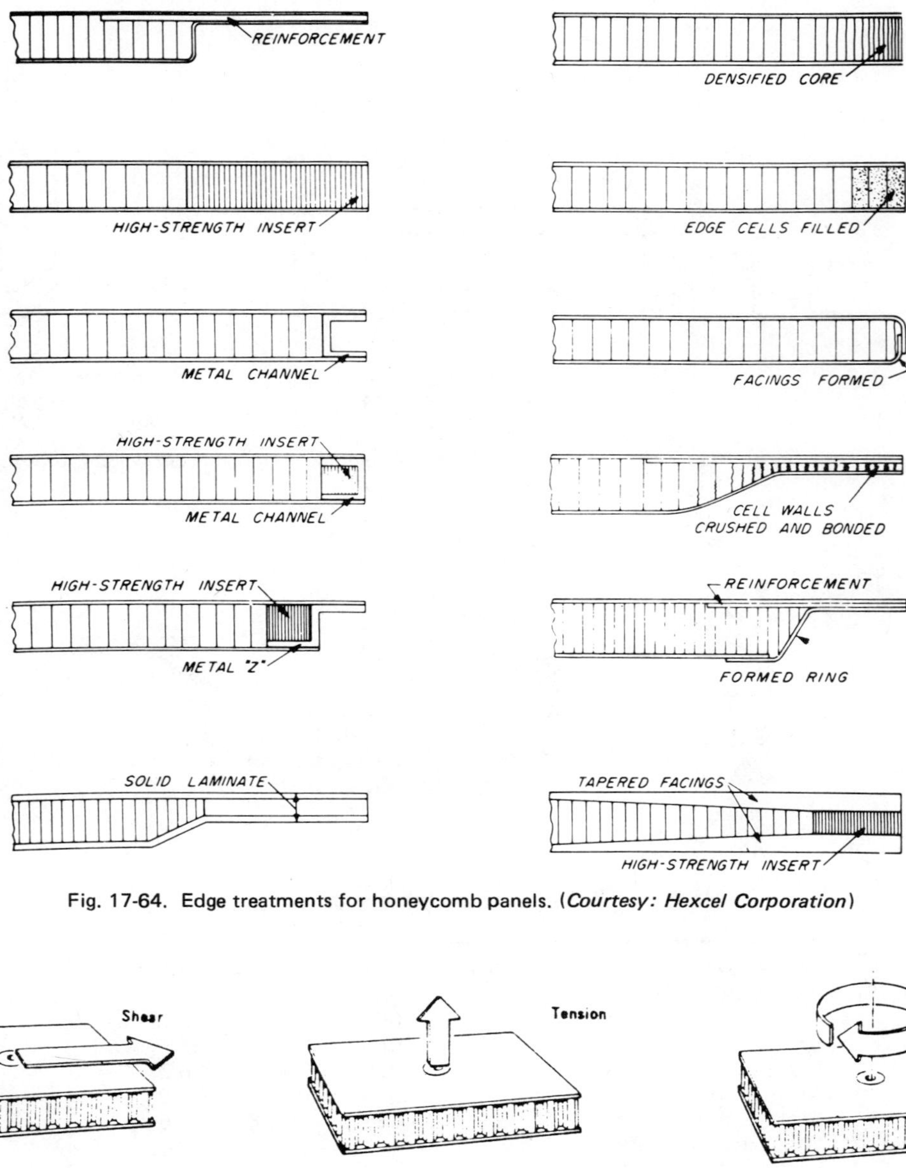

Fig. 17-64. Edge treatments for honeycomb panels. (*Courtesy: Hexcel Corporation*)

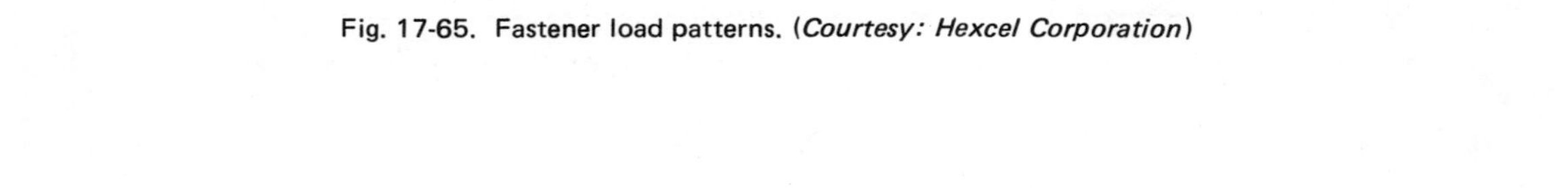

Fig. 17-65. Fastener load patterns. (*Courtesy: Hexcel Corporation*)

the space between the skins. The adhesive qualities of the urethane resin provide adequate bond strength to the core-to-skin interface. In preparing this type of foamed sandwich structure, a mold must be provided that has adequate structural integrity to resist the considerable gas pressure developed by the urethane when it foams. Alternatively, the skins, and the mold, may be made with a number of pressure-relief holes, if the design permits.

The most successful sandwich in terms of strength-to-weight ratio and the one most used today is honeycomb. Honeycomb sandwiches are made of a variety of materials, including polyester and epoxy-impregnated fiberglass, paper impregnated with phenolic resin, and several types of metallic honeycomb structures. Figures 17-61 and 17-62 (page 747) show a honeycomb core and one method of fabricating a sandwich panel. Also shown is the increase in stiffness of two facing sheets when separated "2T" and "4T" by means of a honeycomb core.

As shown, the honeycomb is sandwiched between two facing sheets which may be impregnated fiberglass or metallic skins; for the lowest costs, the entire structure may be of phenolic resin-impregnated paper. Regardless of the material the processing is similar; that is, an adhesive is used to bond the skins to the core. In the case of an aluminum core and aluminum face skins, the adhesive could very well be an adhesive-impregnated scrim cloth as shown. Alternatively, the core could be roller coated with the adhesive and the coated facings could be bonded to the core by a vacuum bag, or bagged and autoclaved, or even press cured. For the lowest possible weight, such as used on spacecraft panels, the face sheets could be prepreg fabric which is then simply bonded to the core by pressure. The final product can be a panel with many times the stiffness of an equivalent weight of a solid metal or plastics part. The exact stiffness-to-weight ratio, of course, depends on the core and facing thicknesses, the type of core used, and the cell size of the core. Design data for determining all the parameters to obtain any desired characteristics are available from the honeycomb manufacturers' and aircraft design manuals. Figure 17-63 (page 747) shows a comparison of weights for the same deflection under load for honeycomb and a number of monolithic materials in various shapes.

Honeycomb panels are inherently very light, since they are composed of two thin facings and a lightweight core. Therefore, making structural attachments to the panel is a matter of special design. In order to transmit a load into the panel it must be terminated at the ends or edges in such a manner that the load can be put into the panel without crushing the core. A number of end or edge teminations have been developed that transmit the structural loads

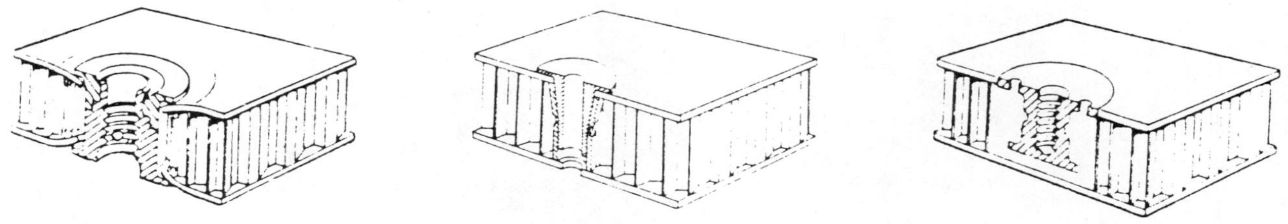

Molded-in fasteners will provide the best structural strength since the adhesive bonds the fastener to both facings and core. They add more weight to the panel (approximately 6 grams) and are more time-consuming to install. Total cost is about $4.00 installed.

Grommet-type fasteners consist of a plug and sleeve which are installed from opposite sides of the panel. They may be the simple force-fitted type or mechanically locked to one or both facings. They are not as strong as the molded-in type since the load is transferred to a smaller portion of one or both facings. Typical weight per fastener is approximately 2 grams. Cost per fastener installed is approximately $2.00.

Sleeve-type fasteners consist of two parts, both of which are installed from one side of the panel. They lock mechanically to the top facing only. Hence, this type will not carry high loads in shear or torque. Tension loads should be applied primarily in one direction. Weight is about 1.5 grams and prices may range from 15¢ in large volume to about $1.00 per fastener.

Variations in each category include:

a) Flush, raised, or countersunk heads
b) Clearance hole or threaded through
c) Blind, threaded–with or without self-locking device
d) Floating nuts which simplify critical alignments
e) Materials include aluminum, stainless steel, carbon steel, and special alloys

A wide range of thread types, sizes and fastener lengths are available in each category. Installation of fasteners is relatively simple. However, careful layout and locating the holes is important. Hole sizes and installation techniques vary with fastener types.

Fig. 17-66. Sandwich panel fastening techniques. (*Courtesy: Hexcel Corporation*)

through a relatively large area of the panel and at the same time prevent edge crushing. Some of these close-out designs are shown in Figure 17-64 (page 748). It should be noted that where rigid foam is shown as a supporting component of the core, this foam is usually a syntactic foam of approximately 20 pcf density. The foam is used to fill the cell walls and encapsulate the insert at the same time.

In addition to attachment of the honeycomb panels to another structure, it is often necessary to attach other assemblies to the panel by means of threaded connections or through holes. The problem remains the same as that found in attaching the panel to any other structure—a means must be found for transferring the load to the relatively fragile panel structure. There are a number of special fasteners designed for such sandwich panels. In all cases they are designed so they can be installed in a finished sandwich, either on the assembly line or in the field. The loads imposed on a fastener can be one or a combination of several as shown in Figure 17-65 (page 748).

Because of the thin facings, in almost all cases, the fasteners are designed to go through the entire sandwich so the loads are placed on both facings equally. The fasteners are

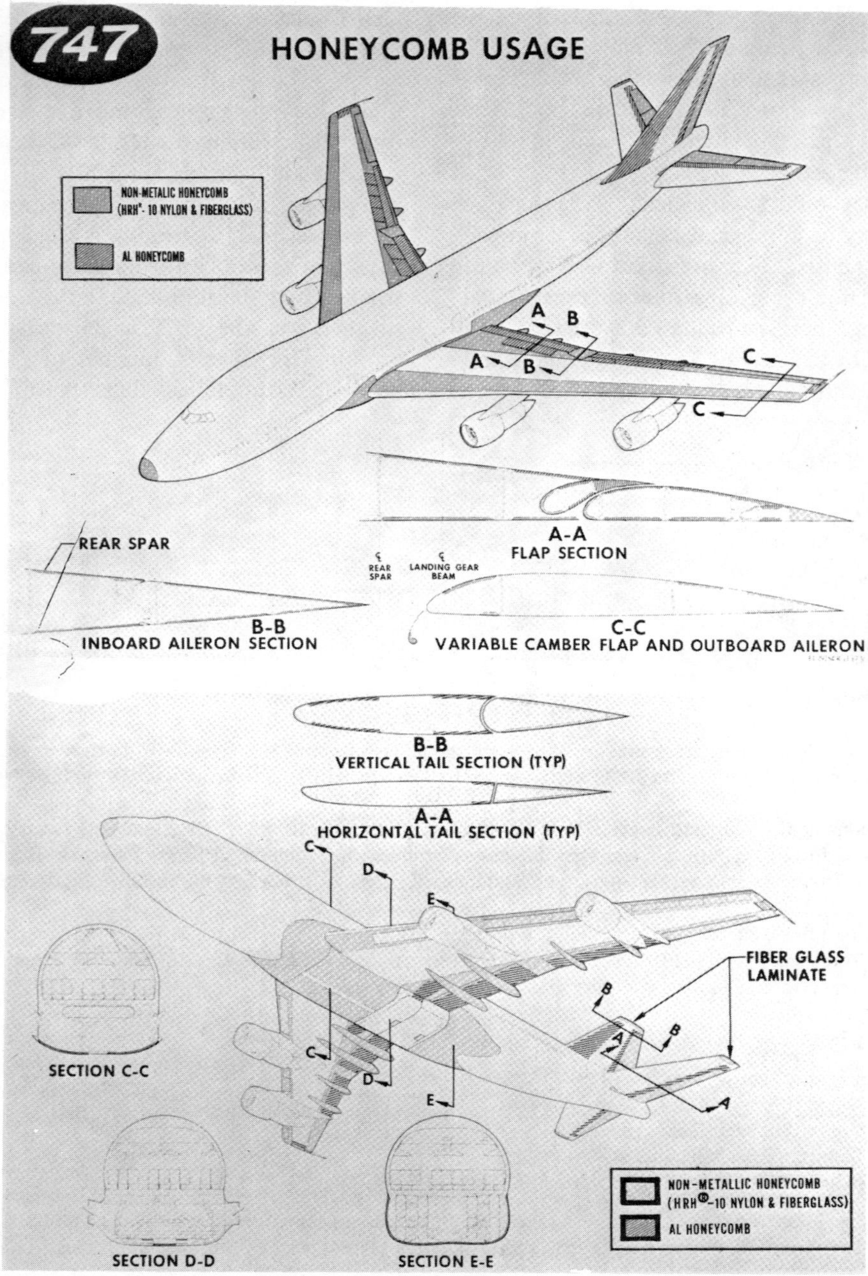

Fig. 17-67. Aircraft applications of honeycomb panels. (*Courtesy: Hexcel Corporation*)

also made, in almost all cases, as through-fittings so the fastener will resist the local crushing load, rather than the panel. Figure 17-66 (page 749) illustrates the three major categories of fasteners for sandwich panels.

Because honeycomb panels are relatively high priced, except for the paper-based material, almost all applications are in the aircraft and aerospace industries. Figures 17-67 (page 750) and 17-68 show applications in 747 and A-10 aircraft.

Foam Reservoir Molding (FRM). Another method of preparing a sandwich structure is known as foam reservoir molding, also called elastic reservoir molding (ERM). Essentially the process consists of impregnating an open-cell urethane foam with a controlled amount of epoxy, or other structural resin. Dry glass fabric or mat facing sheets are then placed on either side of the foam and the assembly is placed in a mold as shown in Figure 17-69. In closing the mold, the resin escapes from the foam and saturates the facing sheets. When cured, the resulting laminate has a density considerably lower than other FRP laminates. The physical properties are claimed to be superior to an SMC part of comparable weight, as shown in Table 17-11.

The process is said to be very competitive to SMC molding in that lower pressures are used (15–100 psi versus 88–1000 psi for SMC molding). Thus, the capital investment for large presses is less. Since this process is mainly suited for large parts, the savings can be considerable. The process cannot be used, however, where sharp curves, ribs, or bosses

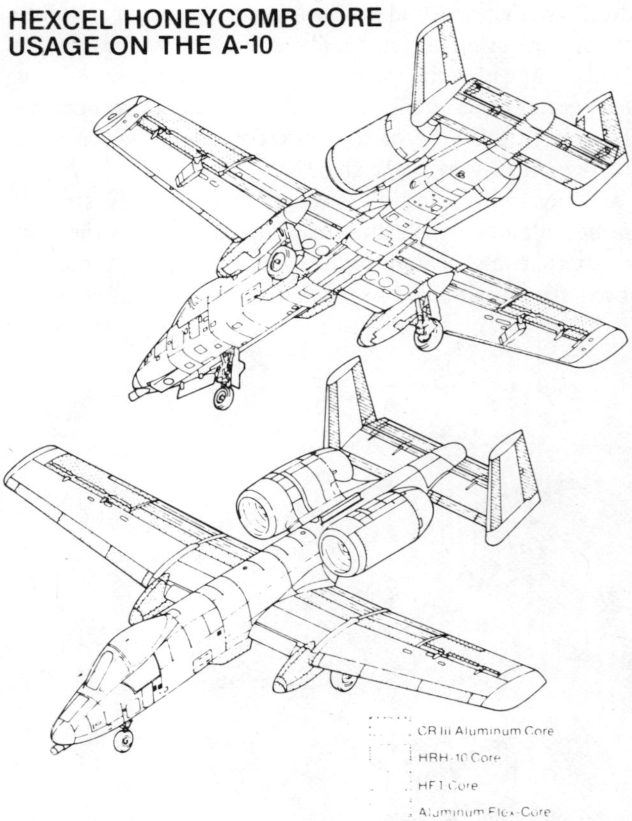

Fig. 17-68. Honeycomb panels in A-10 aircraft. (*Courtesy: Hexcel Corporation*)

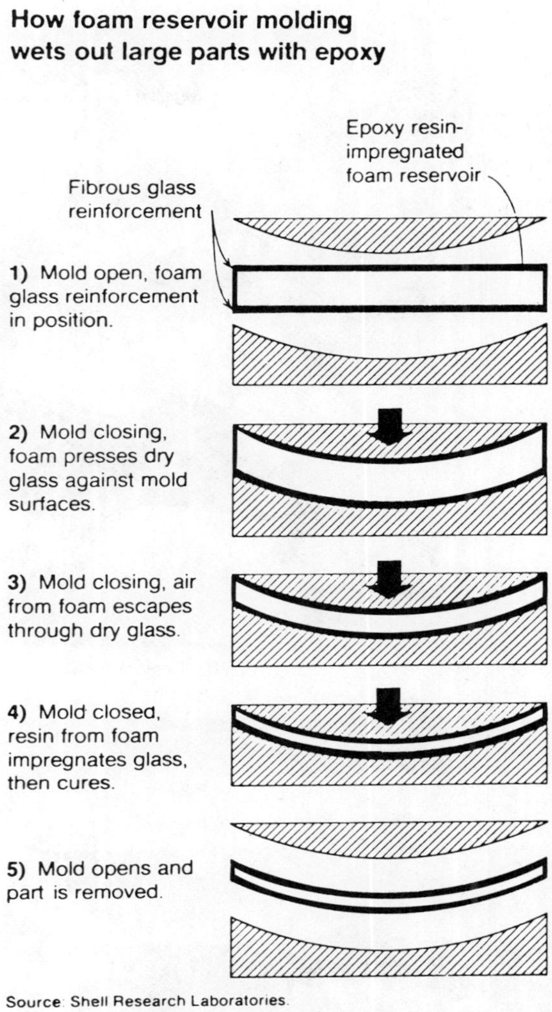

Fig. 17-69. Foam reservoir molding. (Reprinted by permission from *Modern Plastics*, McGraw-Hill, Inc.)

Table 17-11. Comparative Properties: FRM and SMC.[a,*]

Resin System	FRM Epoxy	FRM Vinyl/Ester	FRM Polyester	SMC Polyester
Glass Content, oz/sq ft	7.0	7.0	7.0	7.0
Density, lb/sq ft	1.0	1.0	1.0	1.6
Tensile Str., 10^4 psi	1.62	1.35	1.30	1.00
Flexural Str., 10^4 psi	4.5	3.0	3.8	2.4
Flexural Mod., 10^6 psi	2.0	1.7	1.6	1.6

[a] Laminate thickness in all cases is 0.157 inch. Data supplied by Composite Technology Corporation.
*Murray, 1978.

Fig. 17-70. Typical FRM part. (*Courtesy: Composite Technology Corporation*)

Fig. 17-71. Part made of FRM. (*Courtesy: Composite Technology Corporation*)

are required. Proposed applications for the FRM process include fabrication of automobile body parts (see Figures 17-70 and 17-71).

Still another method of forming a sandwich structure consists of using an integrally woven three-dimensional structure in which the facings are attached to the core during the weaving operation, rather than by adhesive bonding. The core in this type of structure consists of woven webs, as shown in Figures 17-72 and 17-73.

Additional compression strength and bending strength can be obtained by placing foam inserts between the webs or, alternatively, polypropylene inserts can be used. The woven structure can be impregnated before or after the in-

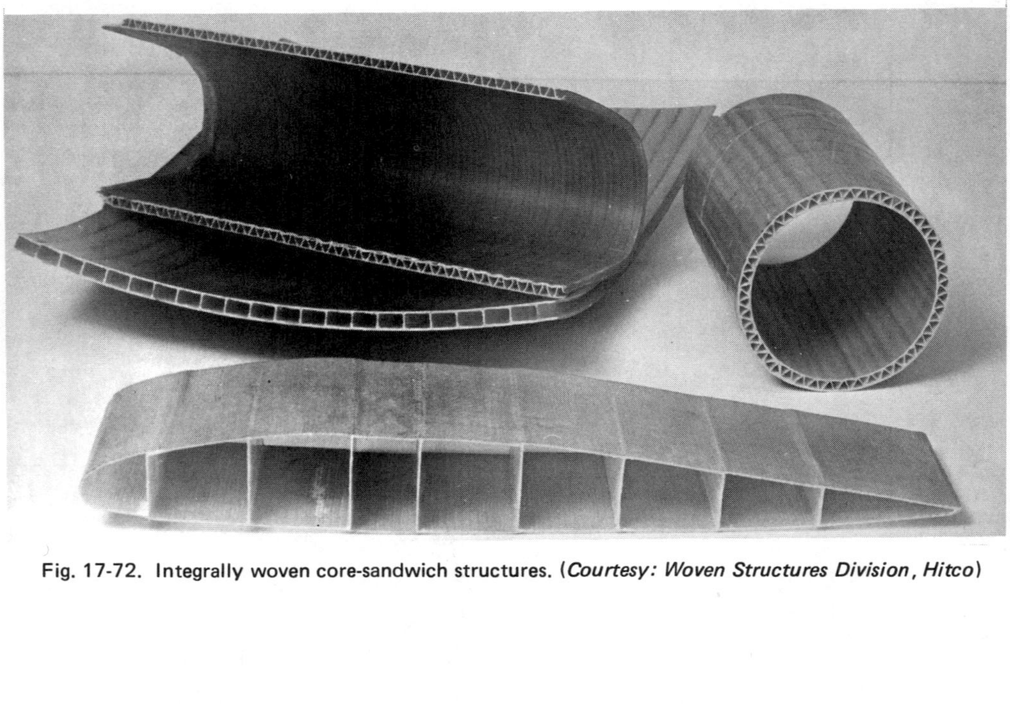

Fig. 17-72. Integrally woven core-sandwich structures. (*Courtesy: Woven Structures Division, Hitco*)

Fig. 17-73. Three-dimensional structure being woven. (*Courtesy: Woven Structures Division, Hitco*)

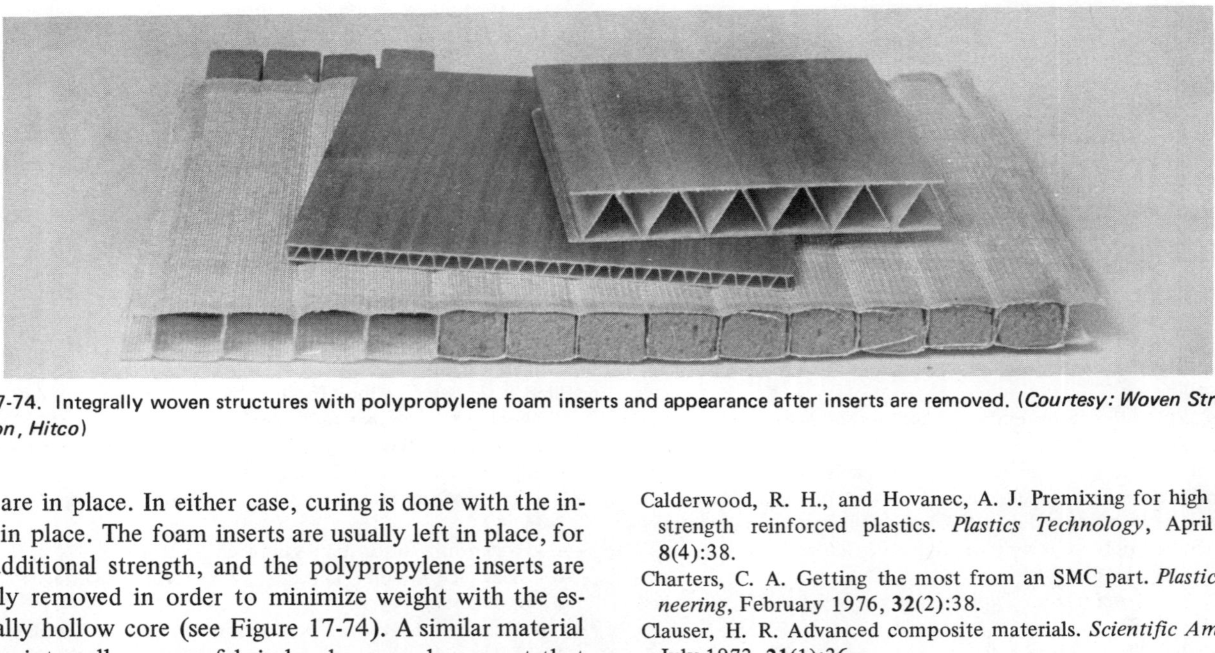

Fig. 17-74. Integrally woven structures with polypropylene foam inserts and appearance after inserts are removed. (*Courtesy: Woven Structures Division, Hitco*)

serts are in place. In either case, curing is done with the inserts in place. The foam inserts are usually left in place, for the additional strength, and the polypropylene inserts are usually removed in order to minimize weight with the essentially hollow core (see Figure 17-74). A similar material to the integrally woven fabric has been made, except that the webs are formed by simply sewing. The advantage of this material is its greater versatility, since webs which taper, or change width or height, can be easily fabricated.

REFERENCES AND BIBLIOGRAPHY

Ackley, R. H. "XMC and HMC structural molding compounds." SAE Paper 760053, February 1976.

Anderson, B. C. Reinforced plastics for deep submersible vehicles. *SPE Journal*, September 1965, **21**(9):1090.

Bastone, A. L. Structural SMC–new design horizon for FRP. *Plastics World*, February 1978, **36**(2):64.

Beck, E. W. Silicone rubber puts on squeeze for high quality composite layups. *Plastics World*, June 16, 1975, **33**(6):45.

Bishop, R. B. Mold Release. *SPE Journal*, May 1969, **25**(5):51.

Calderwood, R. H., and Hovanec, A. J. Premixing for high impact strength reinforced plastics. *Plastics Technology*, April 1962, **8**(4):38.

Charters, C. A. Getting the most from an SMC part. *Plastics Engineering*, February 1976, **32**(2):38.

Clauser, H. R. Advanced composite materials. *Scientific American*, July 1973, **21**(1):36.

Clavadetscher, D. J., and Bradish, F. W. Jr. Competition for new applications pushes improvements in SMC and BMC. *Plastics Design & Processing*, March 1977, **17**(3):43.

Chamis, C. C. Important factors in fiber composite design: part 1. *Modern Plastics*, September 1969.

Charters, C. A. Getting the most from an SMC part. *Plastics Engineering*, February 1976.

Cobb, A. L. Elastic reservoir molding: economical processing alternative. *Modern Plastics*, July 1979, **54**(7):64.

Conference on Advanced Composite Technology. "Advanced Composites Technology." Paper presentation, March 14–16, 1977, El Segundo, California.

Cryer, R. E. Asbestos reinforced plastics resist heat and chemicals. *Materials in Design Engineering*, April 1966,

Davis, J. W. Standards for filament wound plastics. *SPE Journal*, July 1964, **20**(7):601.

Davis, J. W., and Modig, G. R. Preimpregnated epoxy nonwoven filament materials. *SPE Journal*, June 1966, **22**(6):79.

Davis, R. Gel Coats. *Reinforced Plastics*, May/June 1963,

——. Resin systems for reinforced plastics. *Plastics World*, February 1963,

Davis, R. L. Fibrous reinforcement run-down. *Plastics World*, November 1963, **21**(11):60.

DeDani, A. *Glass Fibre Reinforced Plastics*. Interscience Publishers, New York, 1960.

Della Rocca, R. J. Reinforced plastics in boat hull construction. *Plastics Technology*, Part I: November 1957, **3**(11):904. Part II: December 1957.

Dismukes, G. S. Guidelines for selecting FRP Reinforcement. *Plastics Design & Processing*, January 1978, **18**(1):34.

Dreger, D. R. Smooth, ripple-free plastic panels. *Machine Design*, June 21, 1979, **89**(6):108.

——. Structural SMC aims at automotive market. *Machine Design*, December 8, 1977, **49**(23):44.

——. Resin injection molding. *Machine Design*, May 20, 1976, **48**(10):103.

——. Which process for FRP parts? *Machine Design*, September 19, 1974, **46**(18):128.

Eakins, W. J. Shear failure mechanism in parallel filament glass-resin composites. *SPE Journal*, April 1963, **19**(4):379.

Eig, M. Recent RP progress: a perspective. *Plastics Technology*, August 1967,

——. Commercial machines for filament winding. *Plastics Technology*, May 1964, **10**(4).

Epstein, G., and Bandaruk, W. Crazing of filament-wound composites. *Plastics Technology*, August 1963, **21**(8):38.

Feuer, S. S., and Torres, A. F. The effects of laminating techniques and laminate compositions on corrosion resistance of reinforced polyesters. *SPE Journal*, October 1961, **17**(10):1071.

Fischer, L. How to predict structural behavior of RP laminates. *Modern Plastics*, June 1960, **37**(10):120.

Forester, R. H., Akins, D., and Holtby, F. Reinforcement of epoxy resins with metal fibers. *Modern Plastics*, April 1963, **40**(8):117.

Frados, J. (ed.) *Plastics Engineering Handbook of the Society of Plastics Engineers*. 4th Ed. Van Nostrand Reinhold Co., New York, 1976.

Fried, N. Role of resin in reinforced plastics. *Modern Plastics*, September 1963, **41**(1):155.

Gallagher, R. B., and Kamath, V. R. Organic peroxides review. *Plastics Design & Processing*, Part 1: Room temperature cure of polyester resin systems. July 1978, **18**(7):38. Part 2: High temperature cure of polyesters with organic peroxide and azo initiators. August 1978, **18**(8):62.

Gaylord, M. V. *Reinforced Plastics–Theory and Practice*. Koppers Co., Inc., 1963.

Goldsworthy, W. B. Process trends and equipment for advanced composites. *Plastics Machinery & Equipment*, October 1978, 8(10):22.

Griffith, R. M. Reducing blistering in SMC molding. *Plastics Design & Processing*, February 1977, **17**(2):10.

Greunewald, R. G., and Walter, O. 15 year performance with SMC. *Modern Plastics*, May 1975, **52**(5):66.

Hartung, M. Resin transfer molding: how it's done on a large scale. *Plastics Technology*, March 1979, **25**(3):73.

Hauck, J. E. FRP parts take to the air to prove structural strength. *Materials Engineering*, December 1967, **66**(7):54.

——. High production of reinforced plastic parts. *Materials in Design Engineering*, September 1966, **64**(3):87.

Hayes, C. E. The spray-up process–promising fabrication technique. *SPE Journal*, June 1961, **17**(6):571.

Hersch, P. What you should know about filament commercial winding. *Plastics Technology*, September 1967, **13**(9):35.

Hexcel. *Mechanical Properties of Hexcel Honeycomb Materials*. TSB 120, 1974.

Hiler, D. C., Atkins, K. E., and Gentry, R. R. Mold electrically conductive FRP parts with carbon fiber reinforcement. *Plastics Design & Processing*, January 1978, **18**(1):28.

Hosford, R. C. Guide to spray-up equipment for RP. *Plastics Technology*, April 1964,

Howard, J. S. When to design for filament winding. *Product Engineering*, October 26, 1964, **30**(20):102.

Jackson, R. S. Premixes. *Reinforced Plastics*, September/October, 1962, p. 20.

Kline, G. M. 15-year performance experience with SMC. *Modern Plastics*, May 1975, **52**(5).

Kuebeler, G. C., and Jordan, C. E. *Advanced Composites . . . The New "Diet" Material for Structural Applications*. Hercules, Inc.

Longnecker, D. M., Greth, G. G., and Dixon, B. G. Processing plastics–visible-light curing of polyesters. *Plastics Engineering*, May 1977, **33**(5):52.

Loud, S. N. Fiber glass for composite flywheels, *SAMPE Journal*, May/June 1977, **13**(3):20.

Machine Design. Carbon fibers add muscle to plastics. February 7, 1974,

Marker, L. F., and Ford, B. Flow and curing behavior of SMC during molding. *Modern Plastics*, May 1977, **54**(5):64.

Martin, H. A. How to use glass fabrics for designing desired-strength tubular products. *Plastics Design & Processing*, February 1963, 3(2):22.

Materials in Design Engineering. Premix or preform moldings? December 1963, **58**(7):77.

——. Boron fibers for stronger composites. October 1962, **56**(4):11.

——. Filament wound reinforced plastics: state of the art. Special Report No. 174, August 1960.

McAbee, E., and Chmura, M. Effects of rate on the mechanical properties of glass reinforced polyester. *SPE Journal*, April 1963, **19**(4):373.

McClean, W. G. How to select a filament winding machine. *Plastics Design & Processing*, January 1965, **5**(1):18.

McGlone, W. R., and Damico, J. Porous reinforced plastic laminates. *SPE Journal*, April 1964, **20**(4):369.

Miller, E. R. What to look for in SMC production machines. *Plastics Technology*, December 1970, **16**(12):23.

Modern Plastics. New technologies mean better molded RP corrosion parts. February 1979, **56**(2):44.

——. It's modern times for reinforced plastics spray-up. June 1977, **54**(6):42.

——. New route to RP/foam sandwich. April 1975, **52**(4):52.

——. New RP molding material: low profile polyester. February 1968, **45**(2):84.

——. The day of RP mass production has arrived. February 1966, **43**(6):72.

——. The new prepregs . . . new laminate construction . . . and what they mean. December 1965.

——. Filament winding advances. February 1963, **40**(6).

——. Guide to RP molding presses. October 1962, **40**(2):126.

——. Progress in spray-up. September 1961, **39**(1).

——. Spray-up–exciting economies for reinforced plastics. May 1959, **36**(9):85.

Mohr, J. G. (ed.) *SPI Handbook of Technology and Engineering of Reinforced Plastics/Composites.* 2nd Ed. Van Nostrand Reinhold Co., New York, 1973.

Molyneaux, M. *A Guide to Carbon Fiber Composite Production*. Publication No. 51. Fothergill & Harvey, Ltd.

Moore, W. Bag molding with UV curing prepreg. *Plastic Technology*, November 1968, **14**(11):46.

Morgan, P. *Glass Reinforced Plastics*. Interscience Publishers, New York, 1961.

Morris, E. E., Sanger, M. J., and Darms, F. J. Filament-wound structures. *Space/Aeronautics*, February 1967, p. 86.
Morrison, R. S. Matched metal die molding of high-strength reinforced plastics. *Plastics Design & Processing*, January 1968, **8**(1):8.
Murray, T. A. Foam reservoir molding–the processing story. *Plastics Technology*, October 1978, **24**(11):93.
Naitove, M. H. New technology aplenty stimulates the entire RP field. *Plastics Technology*, March 1979, **25**(3):61.
——. New technologies spur RP growth. *Plastics Technology*, March 1976, **22**(3):39.
National Electrical Manufacturers Association. *Filament Wound Structures.*
Okuto, H., Hirano, H., and Yotsuzuka, M. Having trouble evaluating processability of SMC? *Plastics Technology*, August 1973, **19**(8):43.
Oleesky, S. S. Advanced techniques for rocket engine case and nozzle fabrications. *Reinforced Plastics*, January/February 1963,
Outwater, J. W. Filament wound internal pressure vessels. *Modern Plastics*, March 1963.
Owens-Corning Fiberglas Corporation. *Structural SMC: Material, Progress and Performance Review*. Publication No. 5-TM-8364-A, April 1979.
——. *The Hand Lay-Up Spray-Up Guide*. Publication No. 5-PL-9342, July 1979.
——. *FRP–An Introduction to Fiberglas-Reinforced Plastics/Composites*. Publication No. 1-PL-6305A, December 1976.
——. *Resin Transfer Molding, An Intermediate FRP Molding Process*. Publication No. 5-CCR-7032, October 1975.
——. *Fiberglass/Plastics Design Guide*. Publication No. 5-EA-6758, December 1974.
——. *Glossary of Terms for Filament Winding*. Publication No. 2-ASP-2417, January 1963.
——. *Designing Matched-Die Molded Fiberglass-Reinforced Plastic Parts*. Publication No. 1-PL-2383.
Parker, L. F., and Ford, B. Flow and curing behavior of SMC during molding. *Modern Plastics*, May 1977, **54**(5).
Pastorino, R. L. A guide to sheet molding catalysts. *SPE Journal*, May 1972, **28**(5):18.
Paul, J. T., Jr., and Thomson, J. B. Importance of voids in the filament wound structure. *Modern Plastics*, December 1965, **43**(4):145.
Petersen, R. S., and Yarborough, F. B. *Graphite/Epoxy Structures–Design and the Manufacturing Interface*. Bulletin 77-DE-15. American Society of Mechanical Engineers, Design Engineering Conference and Show, May 9–12, 1977.
Petker, I., Segimoto, M., and Kertesz, J. J. Processability of prepreg material: Part 1. *Modern Plastics*, September 1965, **43**(1):129.
Petker, I., Norton, J. H., and Feltzin, J. Processability of prepreg material: Part 2. *Modern Plastics*, October 1965, **43**(2):132.
Plastics Design & Processing. Quick curing with visible light speeds up polyester pipe processing. April 1977,
——. Resin transfer molding cuts labor, tooling costs for short run production. May 1977, **17**(5):47.
——. Practical guidelines to designing parts in reinforced thermosets. January 1971, **11**(1).
——. Quality control pays off in reinforced plastics. November 1966, **6**(11):25.
Plastics Design Forum. Exploring the frontiers of RP design. January/February 1978, p. 18.
Plastics Technology. Resin transfer molding: news in equipment, materials, tooling. May 1979, **25**(5):34.
——. Make molds with built-in release surface. May 1978, **24**(5).
——. "TMC"–new look in polyester molding compounds. July 1977, **23**(6):11.
——. Hydroclave and cable-clave techniques. February 1963.
——. SMC's improving in pigmentation, flame retardance. March 1973, **19**(3).
——. Integral weaving strengthens structural panels. August, 1963, **9**(8).
Plastics World. Current trends in electricals, the case for FRT. April 1979, **37**(4):69.
——. For better damage resistance–toughened SMC. June 1978, **36**(6):62.
——. Reinforced plastics: making a stronger case against metals. June 1978, **36**(6):51.
——. What's new in reinforced plastics. March 1977, **35**(3):57.
——. Choosing the right boss for SMC. May 1977, **35**(5):34.
——. Three intermediate-volume FRP processes. April 21, 1975, **33**(4):58.
Plessis, H. D. How good are GRP boats? *SPE Journal*, July 1970, **26**(7):20.
PPG Industries, Inc. *Fiber Glass Reinforced Plastics–By Design*. Publication F352-R-785, and Form No. F352-R-969.
Product Engineering. Filament winding materials and techniques build complex shapes and structures. June 1977, **48**(6):25.
——. Reinforced plastics bid for metal replacement jobs. August 1975, **46**(8):23.
——. Reinforced plastics guide shows engineers how to improve designs. August 1971, **42**(8):22.
——. When to design for filament winding. October 26, 1964.
Raghupathi, M., and Mallick, P. K. Preheating to decrease cure cycles in compression molded FRP. *Plastics Design & Processing*, January 1979, **14**(1).
Raia, D. C. Organic fiber puts new muscle in reinforced plastics. *Plastics World*, December 17, 1973, **31**(12):46.
Rawlinson, C. C., and Kinley, J. R. Economics of low-pressure RP molding. *Modern Plastics*, July 1976, **53**(7):52.
Reinforced Plastics. The future of composites. May/June 1965, p. 18.
——. RP is the superior material for fatigue applications. May/June 1965, p. 8.
——. Design procedure for reinforced plastics products. May/June 1964, p. 8.
——. Extruding glass fiber reinforced resin. September/October 1962.
Riley, M. W. Low pressure reinforced plastics. *Materials in Design Engineering*, February 1960, **51**(2):103.
——. Reinforced plastics for high production. *Materials in Design Engineering*, October 1962, **56**(4):118.
Rizzi, M. A. Here are details of the new . . . total injection system for polyester premix. *Plastics Technology*, May 1971, **17**(5):11.
Rolston, J. A. Reinforced plastics for rotating structures. *SPE Journal*, April 1963,
Rosato, D. V. An introduction to filament winding. *Reinforced Plastics*, May/June 1964.
Rosato, D. V., and Grive, C. S. *Filament Winding*. Interscience Publishers, New York, 1964.
Schulte, E. L. Improved autoclave method gives low-void RP parts. *Modern Plastics*, May 1969, **46**(5):112.
Schwartz, H. S. Principles of fiber reinforced structural plastic components. *SAMPE Journal*, June/July 1966, **1**(4):19.
Scott Bader and Co., Ltd. *Polyester Handbook*. 1961.
Sidwell, D. R., and Yarborough, F. B. "Graphite/Epoxy Fabric Utilized on the C4 Trident Missile Program." Paper presentation, 20th National SAMPE Symposium and Exhibition, San Diego, California, April 29–May 1, 1975.
Skinner, P. N., and Brelant, S. Quality control for filament winding. *Plastics Technology*, July 1963, **9**(7):28.
Stevens, P. J. B., and Polentz, L. M. Honeycomb: part 1. *Materials in Design Engineering*, March 1960.
Strauss, E. L. Effects of stress concentration on the strength of

reinforced plastics laminates. *SPE Journal*, October 1959, **15**(10):894.

Suffredini, L. P. Stiffer reinforced plastics. *Materials in Design Engineering*, October 1961.

Technical Conferences Associates. *Advanced Composites Technology*. Papers presented at The Conference of Advanced Composite Technology, March 14–16, 1977, El Segundo, California.

Theberge, J. E., Arkles, B., and Robinson, R. Carbon fibers add muscle to plastics. *Machine Design*, February 7, 1974, **47**(3):102.

Throckmorton, P. E., Hickman, H. M., and Browne, M. R. Origin of stress failure in glass reinforced plastics. *Modern Plastics*, November 1963, **41**(3):140.

Tickle, J. D. Pultrusions step up the challenge to structural steels. *Machine Design*, October 20, 1977, **49**(20):163.

——. New materials and systems simplify thick pultrusion processing. *Plastics Design & Processing*, October 1977, **17**(10):29.

——. Lower cost tooling carves out a big market for pultrusion. *Plastics Engineering*, October 1975, **31**(10):36.

Tortolano, F. W. Medium-volume FRP processes–are they for you? *Plastics World*, March 17, 1975, **33**(3):55.

——. High-strength FRP outperforms steel. *Plastics World*, November 17, 1975, **33**(11):46.

Torres, A. F., and Feuer, S. S. How to measure uniformity and predict molding characteristics of polyester premixes. *Modern Plastics*, April 1959, **36**(8):124.

Turner, K. B. Filament winding–some basic principles. *Modern Plastics*, December 1959, **37**(4):116.

Union Carbide Corporation. *Bakelite Low Profile Additives*. Bulletin F46567.

Waldeck, J. W. Selecting mold-release agents for FRP. *Plastics World*, January 19, 1976, **34**(1):40.

Wehrenberg, R. M. II, and Stedfeld, R. More muscle in reinforced plastics. *Materials Engineering*, March 1979, **89**(3):23.

Welhart, E. K. How to fabricate advanced composites. *Materials Engineering*, August 1977, **86**(2):50.

West, P. Finishes for glass fiber reinforced plastics. *Materials in Design Engineering*, October 1964, **60**(10):106.

Western Plastics. Finished surface for boats imparted directly by molds. August 1959,

Whitaker, R. A. *Reinforced Plastic Fabrication Processes*. Bulletin 684C, Automotive Engineers–American Society of Naval Engineers National Aero-Nautical Meeting, Washington, D. C., April 8–11, 1963.

White, R. B., and Jackson, R. S. Problems with premix moldings. *Modern Plastics*, Part 1: March 1959, **36**(7):117. Part 2: May 1959, **36**(9):115. Part 3: July 1959, **36**(11). Part 4: September 1959, **37**(1):112.

Williams, G. L. Lower molding pressures brighten outlook for shorter-run SMC parts. *Plastics World*, May 17, 1976, **34**(5):47.

Wilson, J. L. Should you go to filament winding? *Modern Plastics*, November 1962,

Yamada, M., Iwai, T., Matsumoto, K., and Walton, J. P. TMC–new system for RP molding. *Modern Plastics*, November 1978, **55**(11).

Yates, D. N., Torczyner, R. D., and Sidwell, D. R. "Development and Application of Woven Graphite Products for Aerospace Structures." AIAA Paper No. 75-774. AIAA/ASME/SAE 16th Structures, Structures Dynamics, and Materials Conference. Denver, Colorado, May 27–29, 1975.

18
Fastening and Joining Techniques

Fastening and joining of plastics parts to each other or to metal can be done by four general techniques:

1. Mechanical fasteners such as screws, rivets, etc.
2. Mechanical means such as snap-in fits and press fits.
3. Welding is applicable only to thermoplastics. A number of welding processes are used, but all are similar in that the materials or parts to be joined are heated until melting takes place, and the molten surfaces are held together until the joint is cooled, after which pressure is released. The processes include hot gas welding, heated tool welding or heat sealing, spin welding, ultrasonic welding, and dielectric or induction welding.
4. Adhesive bonding is employed with both thermosets and thermoplastic materials, but is used mostly for the latter.

There are a number of subdivisions within each category, and since there are many new techniques developed each year, the examples given should be taken as merely representative of the many materials, devices, and processes which are in use. Table 18-1 shows a comparison of the various techniques as they apply to different materials.

MECHANICAL FASTENERS

Screws

Screws that are inserted directly into the plastic represent the type of mechanical fastener most often used. There are two basic types of screws: thread-forming and thread-cutting. In both cases, the screws are forced into holes, either molded or drilled, that are usually 1-pitch diameter in size. Figure 18-1 shows the pitch diameter of a screw. This dimension is always available from the screw manufacturer, who should also be consulted before a decision is made concerning the screw-hole combination and the type of screw to be used for a given material.

Thread-Forming Screws. Thread-forming screws are used for thermoplastics; they function by displacing and compressing material as they are forced into the hole. The compression and displacement of the thermoplastic material, combined with the inherent memory of the plastic, insures a tight fit with this type of screw in a thermoplastic. The same type of screw used in a thermoset material, however, may result in cracking around the hole because the thermoset will not flow around the screw threads. The four USA Standard thread-forming types most often used are shown in Figure 18-2.

Types B and BP are mainly used for general-assembly operations with most thermoplastics. Type BP is somewhat easier to insert than type B because of the pointed end which speeds screw alignment during assembly operations. Type U is a drive screw used where disassembly is not intended. Type L is a special combination thread-forming and thread-cutting screw used for nylon. In general, thread-forming and thread-cutting screws are used for assembly when the screws are not intended to be removed for disassembly of the part. (It is possible to remove types B, BP, and L up to five times, but type U should never be removed.)

There are also screws with proprietary thread designs. Figure 18-3 illustrates one type which can be pushed or screwed into a hole and will then strongly resist pull-out or disengagement.

Thread-Cutting Screws. Thermoset materials are fastened using thread-cutting screws. The six most common types are those shown in Figure 18-4; they differ mainly in the torque needed to drive the screw. Because of their greater pitch, the BF and BT screws require lower driving torques and consequently show lower boss stresses and somewhat lower holding power than the other screws.

In using either thread-forming or thread-cutting screws, the molded or drilled hole should have a chamfer at the top to guide the screw in and minimize the chance for misalignment. For maximum efficiency the hole should be tapered to match the screw taper. The wall thickness of a mounting boss should be at least equal to the screw diameter. Holes for thread-cutting screws should always be longer than the screw to allow room for the chips, as shown in Figure 18-5.

Inserts. In the event that the parts will require repeated assembly and disassembly during their lifetime, then internally threaded metal inserts should be considered. In addition to taking the wear and tear of frequent screw insertion and removal, the inserts can carry mechanical stresses above the limits of threaded plastic; they can also serve as electrical contacts and even be employed for their decorative value. Inserts should always be used sparingly since they are placed in the mold or part by hand, unless

Table 18-1. How Joining Methods Compare.*

Plastic	Method: Mechanical Fasteners	Ultrasonic Welding	Thermal Welding	Spin Welding	Solvent Welding	Adhesives	Remarks
THERMOPLASTICS							
ABS	G	E–G	G	G	G	G	–
Acetals	E	G	G	G	NR	F	Surface treatment needed
Acrylics	G	G	G	F–G	F–G	G	Molded difficult to solvent weld
Cellulosics	G	G–P	G	E	G	E	Don't solvent weld CA to CAB or CAP
Chlorinated polyether	G	P	G	NR	G	G	–
Fluorocarbons	G	NR	NR	NR	NR	G	PFEP, PVF_2 can be spin, thermal welded; PVF_2 can be ultrasonic welded
Nylons	G	G	G	G	NR	F	–
Phenoxies	F	E	G	G	G	G	–
Polycarbonates	G	E	G	G	G	G	–
Polyethylenes	P	G–P	G	G	NR	P–F	Surface treatment needed
Polyimides	G	NR	NR	NR	NR	F–G	High temp adhesives needed
Polyphenylene oxides	G	G	G	E	E–G	G	–
Polypropylenes	P–F	G–P	G	E	NR	P–F	Surface treatment desirable
Polystyrenes	F–G	E–P	G	E	G	G	Impact grades difficult to solvent weld
Polysulfones	G	E	E	VG	G	G	–
Polyurethanes	NR	NR	NR	NR	G	G	–
Vinyls	F	F–P	G	F	G	F–G	–
THERMOSETS							
Alkyds	G	NR	NR	NR	NR	E	–
DAP	G	NR	NR	NR	NR	E	–
Epoxies	G	NR	NR	NR	NR	E	–
Melamines	F	NR	NR	NR	NR	NR	Material is very notch sensitive
Phenolics	G	NR	NR	NR	NR	E	–
Polyesters	G	NR	NR	NR	NR	E	–
Silicones	F–G	NR	NR	NR	NR	G	–
Ureas	F	NR	NR	NR	NR	G	Material is very notch sensitive

E = excellent, G = good, F = fair, P = poor, NR = not recommended.
*Reprinted by permission from *Materials Engineering*, March 1972.

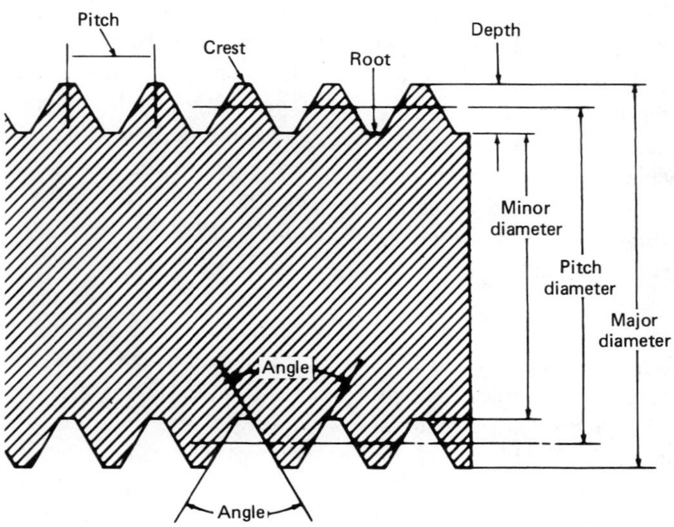

Fig. 18-1. Terminology for screw threads.

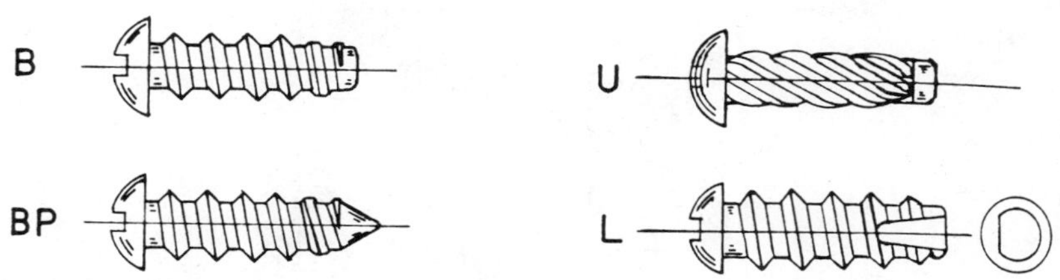

Fig. 18-2. Thread-forming screws most often used in thermoplastic materials. (From *Plastic Product Design* by Beck)

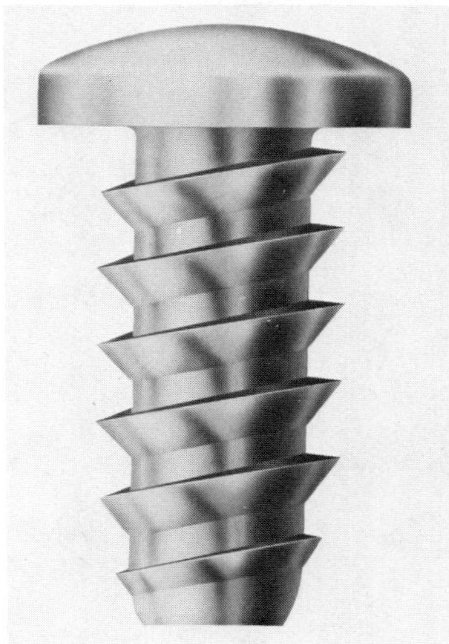

Fig. 18-3. Pushtite proprietary screw for thermoplastics with very high pull-out strength. (Pushtite is a registered trademark of Camcar Division of Textron Inc.)

large quantities justify automatic equipment. In either case, their use can be expensive.

There are two general types of inserts: molded-in inserts and postmolded inserts. For both types, the insert is locked in by the plastic. Molded-in inserts are made with medium or coarse diamond knurl to provide adequate gripping power. The inserts shown in Figure 18-6 represent only some of the most common types in use: **A** is a blind-hole insert with internal threads; **B** is an open-ended female insert with internal threads; **C** is a blind-hold female insert with a counter-bore on the threads. In each case, the slight variation is used to gain some particular advantage, either to improve the molding operation (e.g., to prevent resin from flowing into the insert during injection

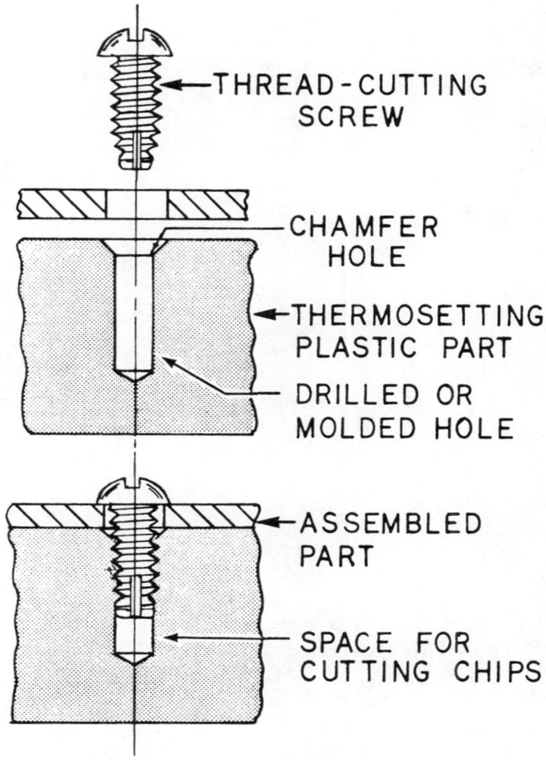

Fig. 18-5. This demonstrates a metal thread-cutting screw assembly in a thermosetting plastic material. (From *Plastic Product Design* by Beck)

as might be the case with insert **I**) or to act as an electrical plug as shown in **M**.

Postmolded inserts, as the name implies, are installed after the part has been molded. Like screws, postmolded inserts are installed in holes drilled or molded in the part;

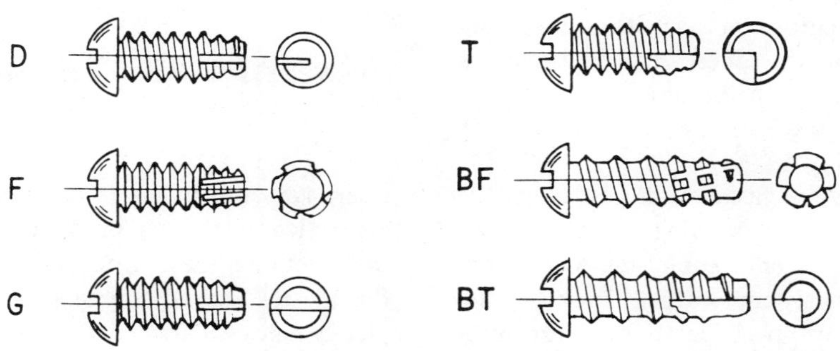

Fig. 18-4. Thread-cutting screws most used for thermosets. (From *Plastic Product Design* by Beck)

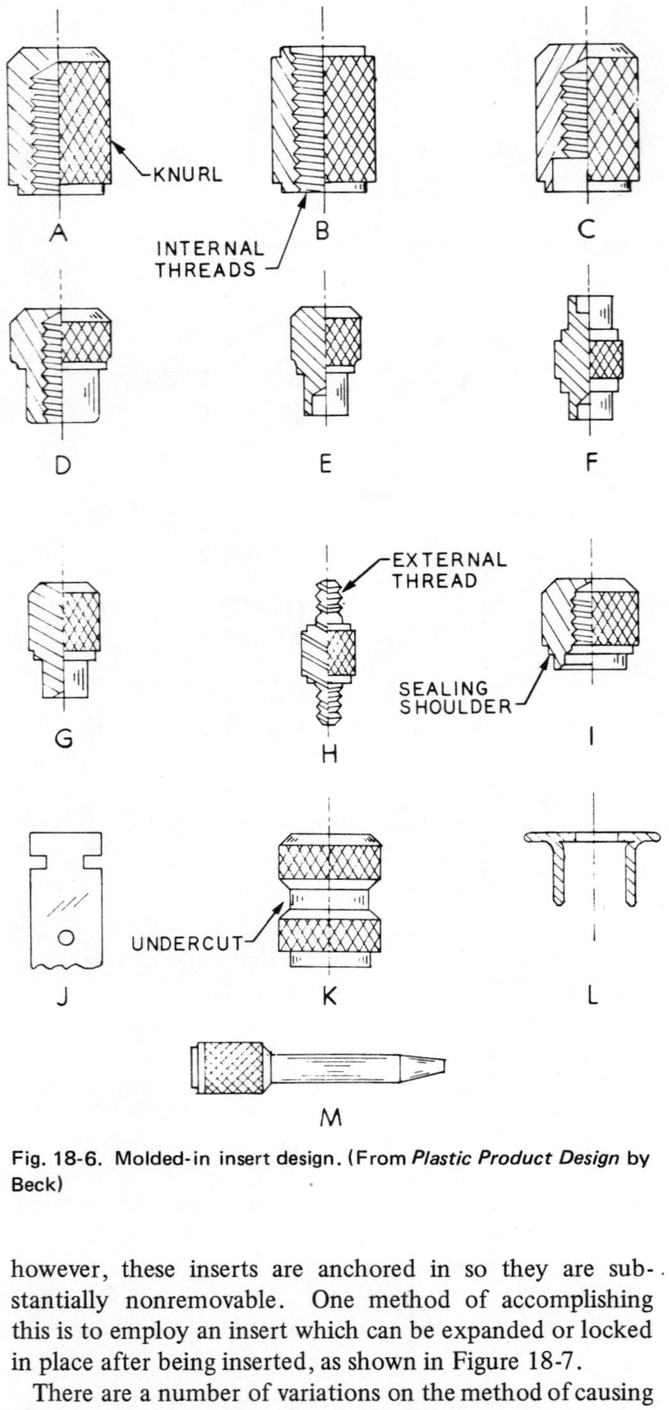

Fig. 18-6. Molded-in insert design. (From *Plastic Product Design* by Beck)

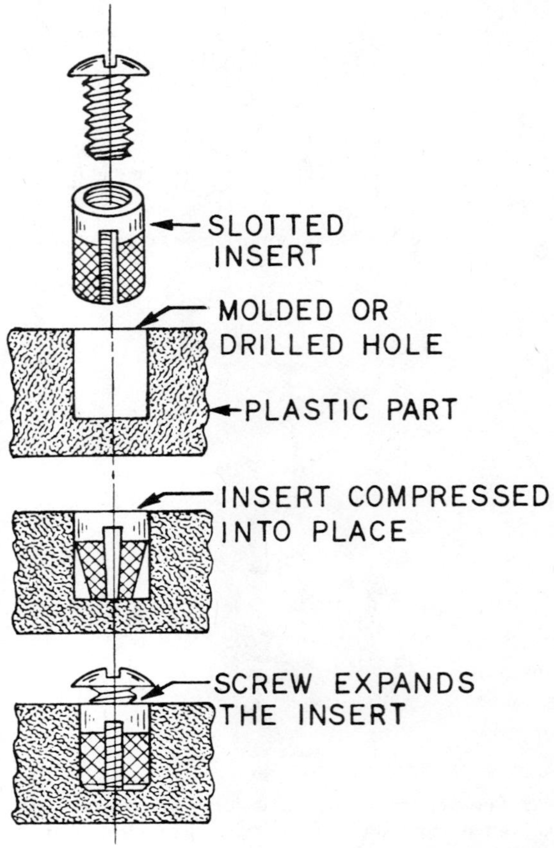

Fig. 18-7. A metal expansion-type insert. The double-slotted insert is expanded when the screw is installed. (From *Plastic Product Design* by Beck)

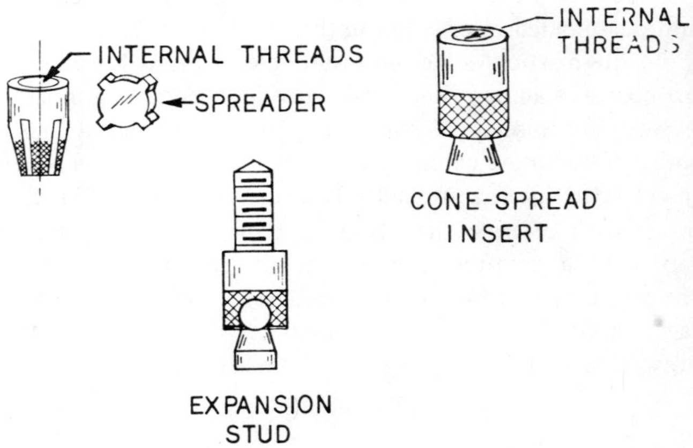

Fig. 18-8. Expansion designs for postmolded inserts. (From *Plastic Product Design* by Beck)

however, these inserts are anchored in so they are substantially nonremovable. One method of accomplishing this is to employ an insert which can be expanded or locked in place after being inserted, as shown in Figure 18-7.

There are a number of variations on the method of causing the insert to expand internally. These include spreaders forced into the insert cavity from the top, cone spreaders forced into the bottom of the insert, etc. Some of these designs are shown in Figure 18-8.

Another technique used for installation of permanent postmolded inserts is ultrasonic bonding to lock in the insert. This method, which is only applicable to thermoplastics, does require special equipment. However, the process is very rapid and can well be adopted for automation or semi-automation. Figure 18-9 shows how the insert is placed in the recommended hole, a pneumatically controlled ultrasonic probe presses down the insert, and the frictional heat from the ultrasonic vibrations melts the plastic around the insert. The resulting part is completely encapsulated, with a minimum of stresses applied to the surrounding plastic. Cooling is complete in a frac-

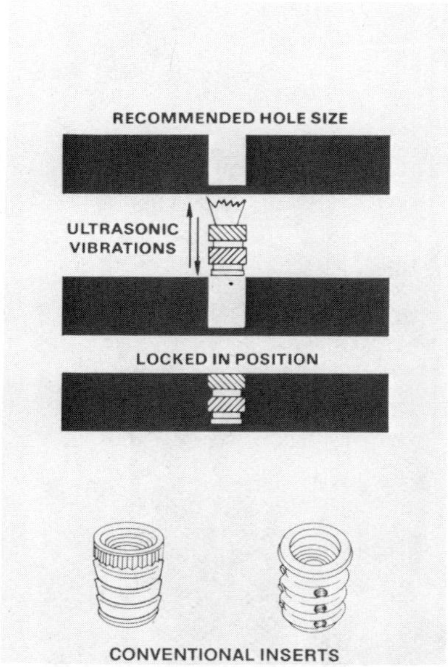

Fig. 18-9. Ultrasonic bonding with conventional inserts. (*Courtesy: IN-X Fastener Corporation, Newark, N.J.*)

tion of a second as soon as the probe is withdrawn. Figure 18-10 illustrates one type of postmold insert for ultrasonic bonding, said to have higher pull-out than conventional inserts.

Inserts can also be installed in thermoplastic parts by means of induction heat or by conduction, using the same principles as installation by ultrasonic vibration; that is, the thermoplastic resin is melted and the insert is forced in the hole, with the molten plastic encapsulating the insert.

Thermal insertion by induction requires a special induction coil which heats up the metal part just prior to and during insertion. Conduction heating requires somewhat simpler equipment, such as a soldering iron-type of heater whose operating temperature can be accurately controlled.

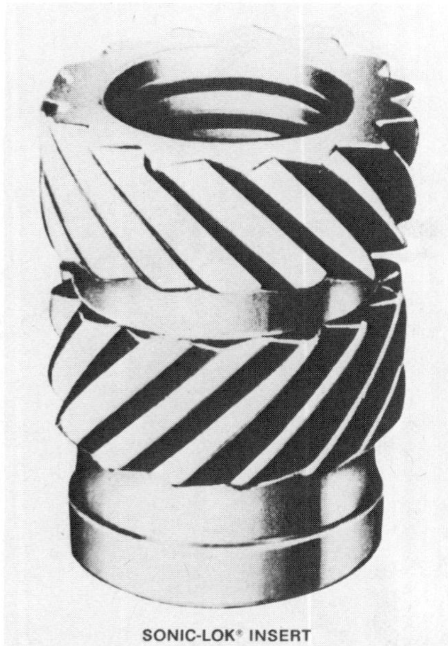

Fig. 18-10. Sonic-Lok® insert. (*Courtesy: IN-X Fastener Corporation, Newark, N.J.*)

Figures 18-11 and 18-12 show commercially available equipment, hand held and on a stand, both with thermostatically controlled heat probes.

Thermal insertion of postmolded inserts by conduction, while fairly inexpensive, has a number of drawbacks. First, it is usually difficult to control the heating process perfectly so that there is always a possibility of material degradation due to the plastics' being so close to the source of heat. Second, the plastic may draw away at the top of the hole from the hot metal insert, thus resulting in lowered reten-

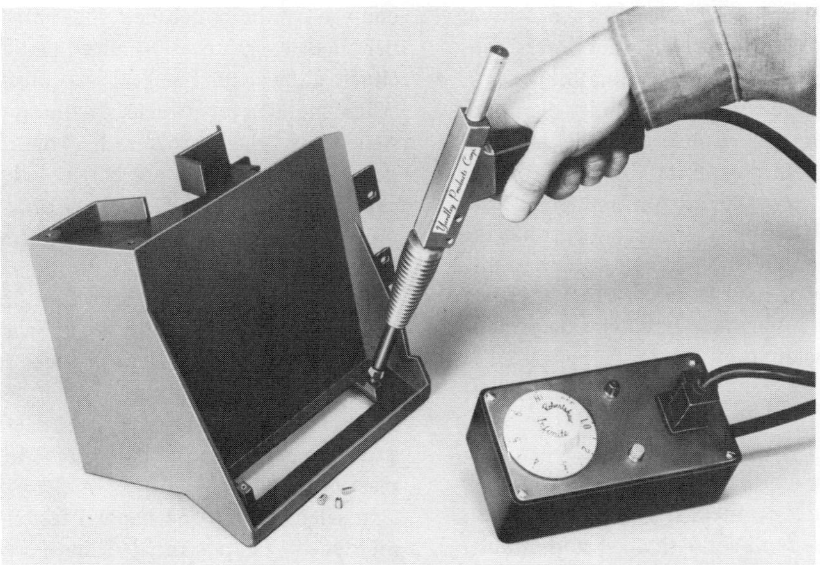

Fig. 18-11. Hand-held thermal postmold insert tool. (Yardley Thermal Inserter manufactured by Yardley Products Corporation, Yardley, Pa.)

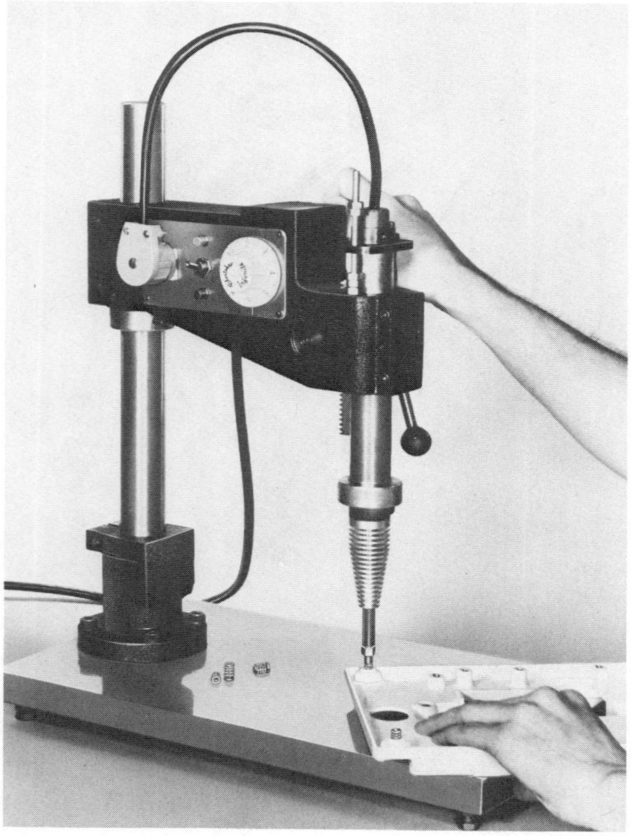

Fig. 18-12. Stand-mounted thermal insert tool. (Yardley Thermal Inserting Press manufactured by Yardley Products Corporation, Yardley, Pa.)

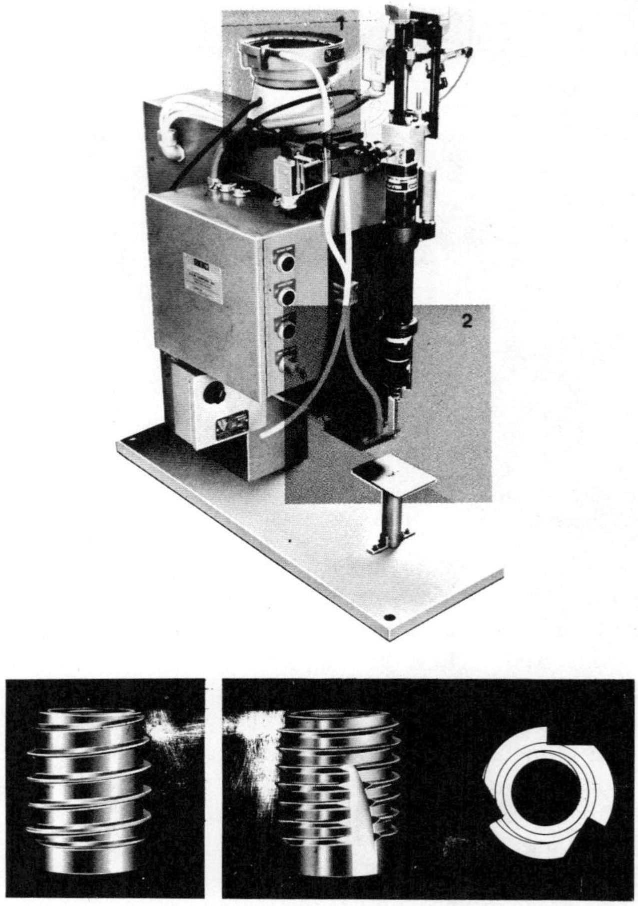

Fig. 18-13. Self-tapping postmold inserts and installation equipment. (*Courtesy: C.E.M. Co., Inc.*)

tion strength. In cooling, the plastic, which may have been overheated, may develop excessive hoop stresses which will at some later date show up as cracks. Another drawback is that, because of the time to heat the insert before insertion, the process in considerably slower than other postmold insert operations. Thermal insertion by conduction is thus best suited for low-volume or prototype production.

Other types of postmolded inserts include diamond-shaped wire coils which are pushed into the hole and lock in place, forming internal and external threads. Another similar insert is available as a thread-forming or thread-cutting type, similar to self-tapping screws, except that the insert has internal threads and also is claimed to have higher pull-out torques because of the external thread shapes. Figure 18-13 shows these inserts and the type of production equipment used for installation.

Rivets

Riveting of plastics is yet another process used to fasten plastics to plastics or plastics to metals. The types of rivets used include solid shanks, semi-solid shanks, and tubular rivets. In general, when riveting plastics to metal, the head of the rivet is placed next to the plastics surface to avoid damage by the clinch end. The important point to remember in using rivets is to allow sufficient edge distance and clinch allowance. Figure 18-14 illustrates the terminology. When metal rivets are used, the application should be designed for shear loads rather than tensile loads if at all possible. Blind rivets are particularly useful for such applications as fastening metal trim onto plastics parts, holding molded escutcheons and nameplates to metal panels, etc. Figure 18-15 shows the use of blind rivets.

An adaptation from metal assembly fasteners is the use of speed nuts and clips, as shown in Figure 18-16. The speed nut is used where it is necessary to have an assembly which will resist vibration, since when fully tightened the nut acts as a locking nut. The speed clip is designed to be snapped over the molded projecting boss or stud on the plastics part.

In selecting any mechanical fastener—self-tapping screws, molded-in or postmolded inserts, even speed nuts and clips—one of the most important factors to consider is the possibility of excessive stresses being set up by the

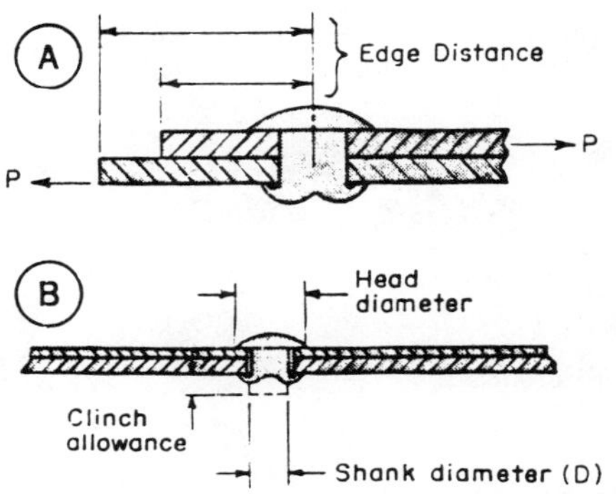

Fig. 18-14. Edge distance and clinch allowance. Depending on the thickness and structural strength of the plastic, the distance between the rivet hole centerline and the part edge should be as high as three rivet shank diameters (A). Clinch allowance for rivets used in plastics assemblies ranges from 50% to 70% of the rivet shank diameter (B). However, prototype riveting testing will indicate the optimum clinch allowance for a particular assembly. (Reprinted by permission from *Plastics Design & Processing*: March 1967. Copyright 1967, Lake Publishing Corporation, Libertyville, Ill.)

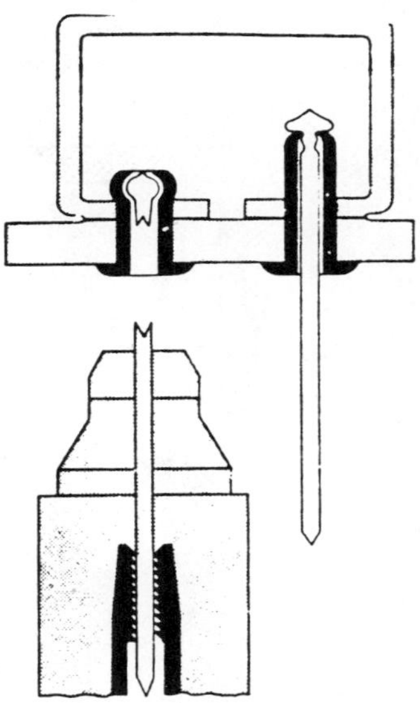

Fig. 18-15. Blind rivet usage. To set rivet, pull mandrel through hollow cylinder. (Reprinted by permission from *Materials in Design Engineering*, September 1959)

differences in the thermal coefficients of expansion of the plastics and the metal part. While this problem is not too significant with filled materials such as phenolics, materials such as polystyrene exhibit a coefficient of thermal expansion which is 6 to 7 times that of steel. Under certain environmental conditions, this could lead to cracking of the plastic boss in which the screw or insert is molded. Preheating of the inserts will help reduce the molding stresses. Another technique to minimize stresses is the use of brass or aluminum inserts, since they will more nearly match the plastics' coefficients of expansion. The use of filled thermoplastics will also result in lower coefficients of expansion as well as higher pull-out strengths.

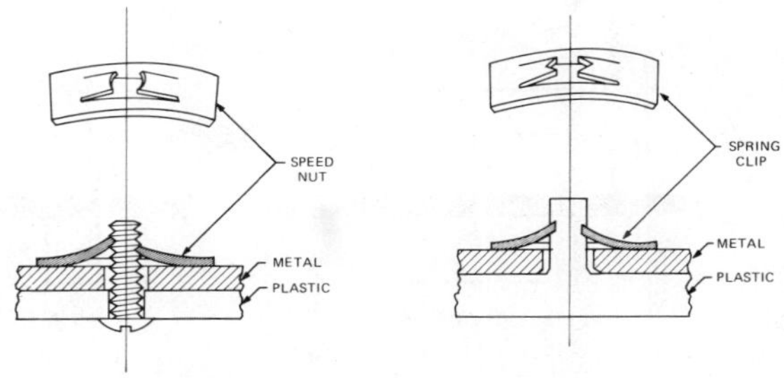

Fig. 18-16. Use of speed nuts and speed clips.

Another important consideration in using mechanical fasteners is the question of minimum wall thickness in the boss or stud supporting the insert. In general, the fastener manufacturer should make a recommendation, which will, of course, depend on the type of material used and the loads anticipated. However, a rule of thumb is that *the wall thickness should in any case not be less than the diameter of the insert*.

For thermoplastics, another rule of thumb is to avoid inserts with sharp threads or knurls since they may cause excessive stress concentrations leading to cracks after a short period of time, or after exposure to low temperatures.

MECHANICAL MEANS OF FASTENING

Mechanical means of fastening include various methods of making snap fits, press fits, hinges, etc., using the molded plastics parts only. The major advantages of such fastening techniques are that thermal expansion problems are usually eliminated; the fastener, or possibly one-half, is made at the time the part is molded; assembly is usually very fast; and, once the tooling is amortized, the cost is negligible. Some disadvantages are that the joint may not be too strong or be able to withstand repeated usage. Figure 18-17 shows some of the many designs for mechanical fastening and hinging, and Table 18-2 lists the optimum applications for various polymers.

WELDING

Welding is defined as the process of uniting, fusing, or bringing (metal or plastics parts) into intimate contact. With regard to plastics, only thermoplastic materials can be welded, by either thermal techniques or by the use of solvents. In both cases the processes are similar; the materials are softened by heat or the solvent, then are joined together, and held, under pressure, until the joint has cooled or the solvent has evaporated. The major differences between the thermal processes–hot gas welding, hot plate welding, spin, ultrasonic, and induction or dielectric welding–are the methods of generating and/or applying the heat to the parts. The differences in solvent welding are in the solvents used. (Solvent processes are discussed under adhesive bonding.)

With the exception of polytetrafluoroethylene (Teflon

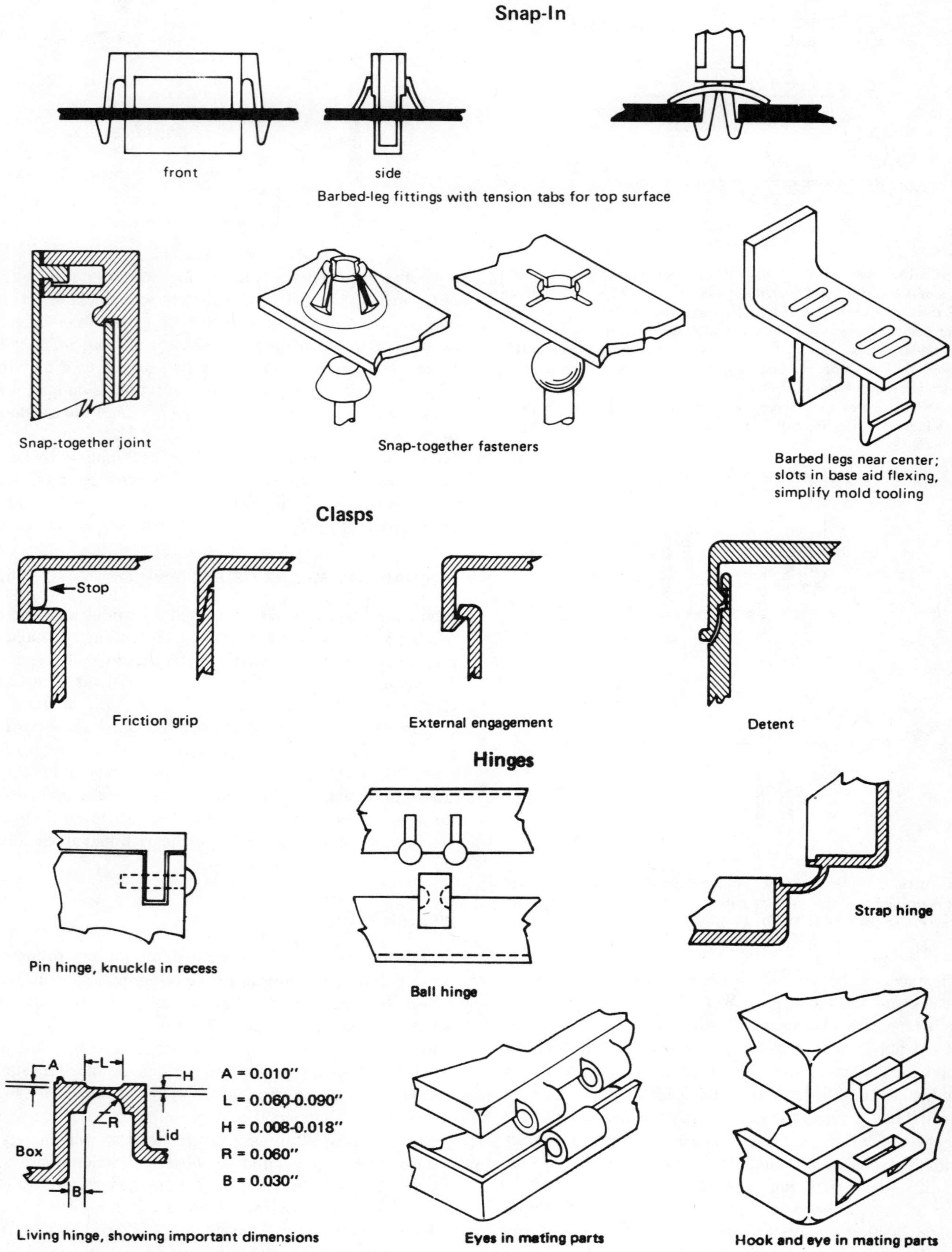

Fig. 18-17. Design ideas for molded mechanical fasteners. (*Courtesy: Plastics World*, September 1972)

Table 18-2. Materials for Plastics "Hardware."*

Material	Application						
					Hinges		
	Snap-In	Snap-On	Clasp	Drive-Pin	Knuckle and Pin	Ball-Grip	Integral
ABS	✓	✓	✓		✓	✓	
Acetal	✓	✓	✓	✓	✓		
Acrylic			✓		✓		
Cellulosic		✓					
Fluorocarbon	✓		✓				
Polycarbonate	✓	✓		✓	✓	✓	
Polyethylene	✓	✓	✓				✓
Polyamide	✓	✓		✓	✓	✓	
Polypropylene		✓	✓				✓
Polystyrene			✓			✓	
Polyurethane				✓			
Vinyl	✓	✓	✓		✓		

*Courtesy: *Plastics World*, September 1972.

TFE), and possibly some of the other fluoroplastics, all thermoplastics can be joined by one or more welding processes. In almost all cases, only the same plastics materials are welded together–the major exception is the bonding of PVC to acrylic which can be welded together with some success. With some exceptions, the bond strengths obtained by the various welding processes for any given material are approximately the same, as shown in Table 18-3. Thus, the choice of the process to be used for any material usually depends on the type of the part, the type of joint desired, the number of parts, and the available equipment.

Hot Gas Welding

Hot gas welding is very similar to metals welding with a torch except that a direct flame is not used. Compressed air, or nitrogen, heated by passage through a heating element impinges on a plastic welding rod and the work by means of a welder, as shown in Figure 18-18. The weld is made by fanning the gun from side to side while heating the rod and the beveled joint. As the materials soften, the rod is forced into the joint, applying pressure as the joint cools. The quality of the welded joint is largely dependent on the skill of the operator, although in any case the strength of the bond will rarely exceed 80% of the parent material. This process is inherently slow, mainly suited for low-volume production of parts such as tanks, pipes, fume hoods, chemical equipment, and similar items. The thickness of materials welded by this process is usually from

Table 18-3. Percent Tensile Strength Retention with Different Welding Techniques.*

	Original Tensile Strength, Psi	Hot Air Welding	Friction Welding	Hot Plate Welding	Dielectric Welding	Solvent Welding	Adhesive Bonding	Polymerization Welding
Thermosetting plastics								
Epoxy	7,000–13,000	–	10–15	10–15	–	–	50–80	60–100
Melamine	7,000–13,000	–	–	–	–	–	50–80	60–100
Phenolic	6,000–9,000	–	–	–	–	–	50–80	60–100
Polyester	6,000–13,000	–	–	–	–	–	50–80	60–100
Thermoplastics								
Acrylonitrile butadiene styrene	2,400–9,000	50–70	50–70	50–70	50–80	30–60	40–60	–
Acetal	8,000–10,000	20–30	50–70	20–30	–	–	–	–
Cellulose acetate	2,000–8,500	60–75	65–80	65–80	–	90–100	50–80	–
Cellulose acetate butyrate	3,000–7,000	60–75	65–80	65–80	–	90–100	50–80	–
Ethyl cellulose	2,000–8,000	50–70	50–70	50–70	–	80–90	50–80	–
Methyl methacrylate	8,000–11,000	30–70	30–50	20–50	–	40–60	40–60	60–90
Nylon	7,000–12,000	50–70	50–70	50–70	–	–	20–40	–
Polycarbonate	8,000–9,500	35–50	40–50	40–50	–	40–60	5–15	–
Polyethylene	800–6,000	60–80	70–90	60–80	–	–	10–30	–
Polypropylene	3,000–6,000	60–80	70–90	60–80	–	–	20–40	–
Polystyrene	3,500–8,000	20–50	30–60	20–50	–	25–50	20–50	–
Polystyrene acrylonitrile	8,000–11,000	20–60	20–50	20–50	30–50	25–60	20–50	–
Polyvinyl chloride	5,000–9,000	60–70	50–70	60–70	60–70	50–70	50–70	–
Saran	3,000–5,000	60–70	50–70	60–70	60–70	50–70	50–70	–

*Harris, 1970.

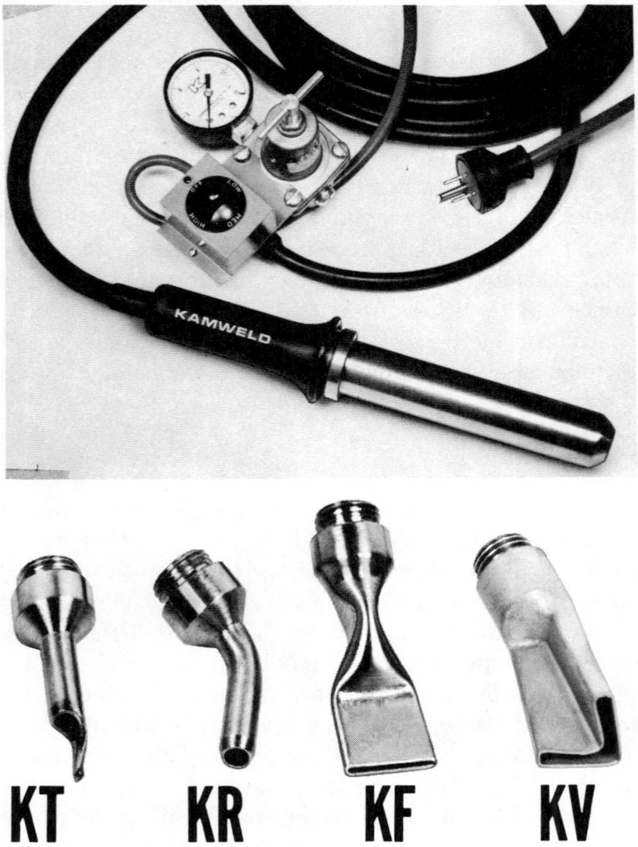

Fig. 18-18. Hot gas hand-welding gun and various types of tips. (*Courtesy: Kamweld Products Co., Inc., Norwood, Mass.*)

$\frac{1}{16}$ inch to as thick as $\frac{3}{8}$ inch. Figures 18-19 and 18-20 show the hand welding technique.

In an effort to produce faster welds, a "speed" tip is sometimes used on the gun. This is basically a tip which feeds the welding rod automatically while simultaneously heating the rod end and the joint through two orifices in the tip. With the use of the "speed" gun, considerable increases in the speed of welding can be made since the rod

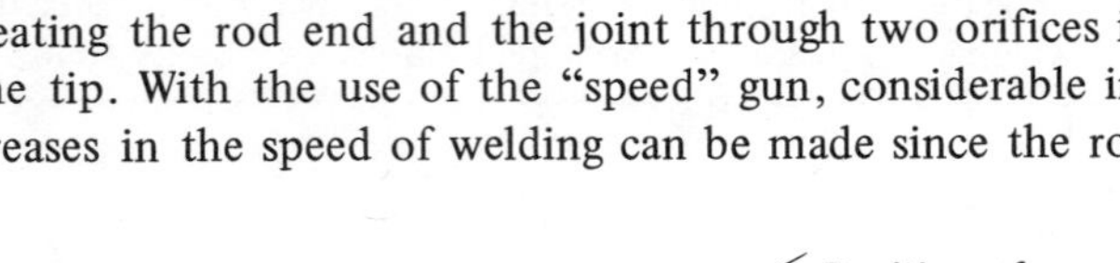

Fig. 18-19. Hand welding process. (*Courtesy: Kamweld Products Co., Inc., Norwood, Mass.*)

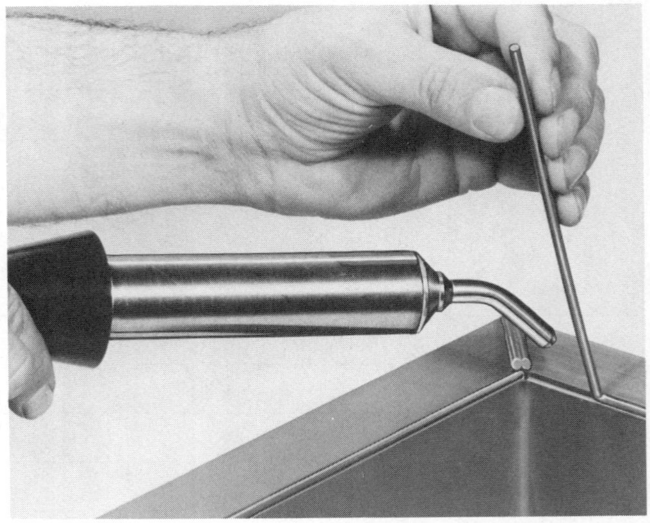

Fig. 18-20. Use of standard tip in repair of ductwork. (*Courtesy: Seeyle, Inc.*)

is held in the correct position and is easily applied. Figures 18-21 and 18-22 show the use of the speed gun. Figure 18-23 shows some representative parts made by hand welding and speed welding.

In either hand or speed welding, it is necessary that the materials be clean, free of grease, and that the joints be beveled to 60° or 45° so a correct bead can be laid down. Welding rods are either round or triangular, with the latter rod being available in large sizes so joints can be filled in one pass rather than several, as illustrated in Figure 18-24. The most commonly welded materials are polyvinyl chloride (PVC), polyethylene, and polypropylene. Other thermoplastics which have been welded include ABS (some blends), cellulosics, acetals, acrylics, and polycarbonates.

Hot Plate Welding

Hot plate welding, or heated tool welding, as the process is known, can be used as a simple manual method of welding sheet, pipes, tubes, and similar parts which have a wall thickness of $\frac{1}{16}$ inch or greater. The process, illustrated in Figures 18-25 and 18-26 (p. 769), consists of heating the edges of the sheet or pipe on a hot plate or special tool until a molten bead is formed on each edge. The sheet should be held with just enough pressure to form the bead, but not enough to force the melt away from the sheet. As soon as the two beads are formed, the sheets should quickly be brought into intimate contact, again using just enough pressure to bring the parts into good contact. Too much pressure will cause a starved weld bead resulting in lowered strength.

Equipment required for heated tool welding can be as simple as the thermostatically controlled hot plate or tools shown in Figure 18-25, preferably with aluminum or nickel

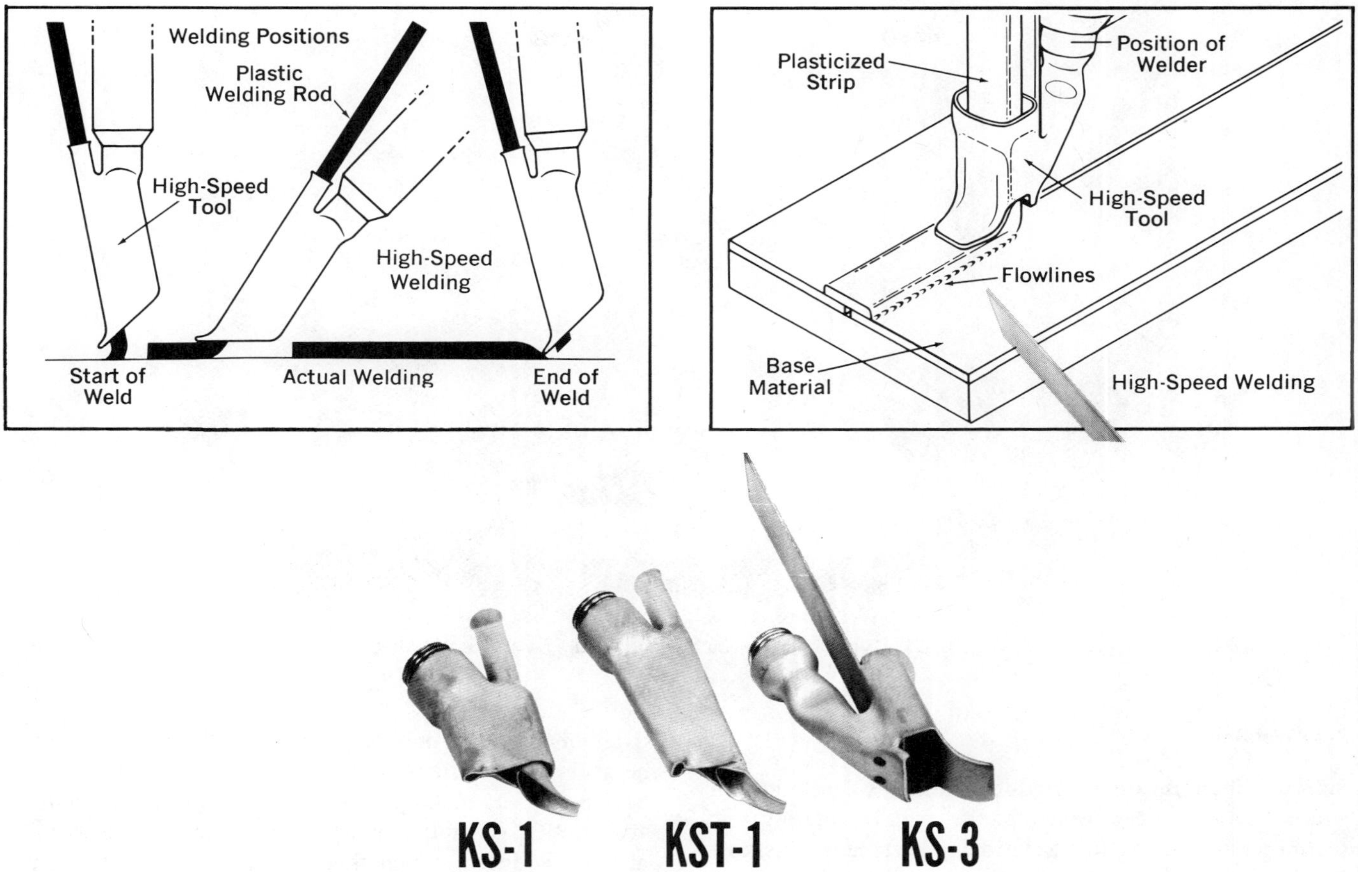

Fig. 18-21. High-speed welding with automatic rod feed. Shown also are various types of high-speed tips. (*Courtesy: Kamweld Products, Co., Inc., Norwood, Mass.*)

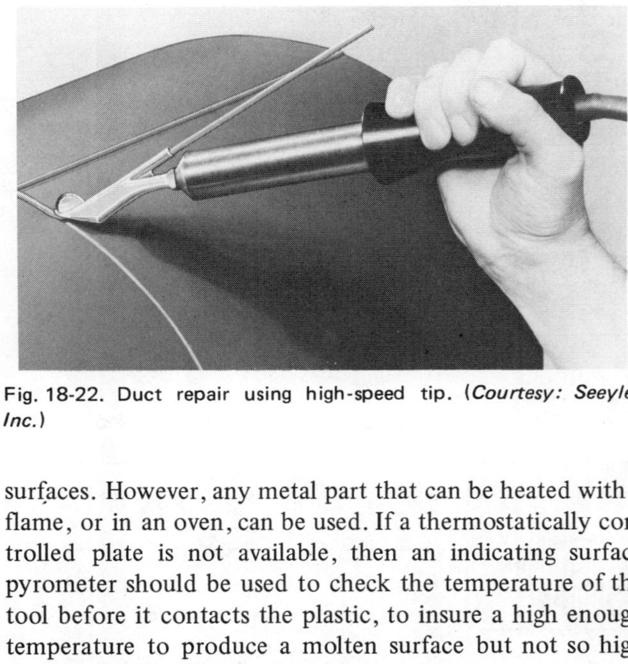

Fig. 18-22. Duct repair using high-speed tip. (*Courtesy: Seeyle, Inc.*)

surfaces. However, any metal part that can be heated with a flame, or in an oven, can be used. If a thermostatically controlled plate is not available, then an indicating surface pyrometer should be used to check the temperature of the tool before it contacts the plastic, to insure a high enough temperature to produce a molten surface but not so high as to cause degradation of the plastic.

A variation of the heated tool process is shown in Figures 18-27 and 18-28, where an electrically heated wire is used to cause melting of the plastic. In actual use, which is mainly for sample or prototype production, the wire is clamped between the two plastics parts and the correct time and current requirements are developed experimentally, while pressure is continuously maintained. The shape and amount of the wire insert is, of course, determined by the parts to be joined. Grooves for the wire help to maintain its position during the melting and pressurization phase; this does not necessarily result in stronger joints, but may help in making the joints more uniform. The same technique can be used to cure a thermosetting adhesive in a joint, when bonding thermoplastics or thermoset parts.

The processes illustrated in Figures 18-25, 18-26, and 18-27 are mainly for experimental and prototype production, since they are inherently slow and depend to a great extent on operator skill. Heated tool welding can be mechanized, however, to produce fast, uniform welds on many types of parts. Figures 18-29 and 18-30 illustrate an actual machine used for and the principle behind the mechanized Bielomatik process. Joint configurations to hide the weld bead are shown in Figure 18-31. Parts made using the process are shown in Figure 18-32.

Fig. 18-23. Typical parts made by hand welding and speed welding.

Spin Welding

Spin welding (also called friction welding) is a technique in which frictional heat, generated by rapidly rotating one thermoplastic part against a stationary part, causes melting. By stopping the rotating part as soon as the melt is produced and then applying pressure, an excellent weld results within 1 to 25 seconds. The equipment required is simply a drill press or a lathe which has a brake for quick stopping, or a clutch to release the motor torque. A more elaborate setup would also include an air cylinder to put pressure from 20 to 200 psi on the joint. Figures 18-33 and 18-34 illustrate the equipment and the process.

The spin welding process can be used for attaching studs and bosses to sheet stock, for bonding hollow members together, etc. The only requirement is that the joint must be a circular one, although the parts need not necessarily be circular. In making a spin welded joint, the spinning and friction generation must be continued until a definite melt is produced, which then produces flash in the joint. Should this be objectionable, then there are joint designs which confine the flash to the interior of the joint, if the part is hollow. Figure 18-35 shows typical joint configurations for hollow parts, some of which are designed to confine the flash to the interior of the part.

Ultrasonic Bonding

Another technique for using frictional heat to cause fusion and subsequent bonding is by means of ultrasonic vibration.

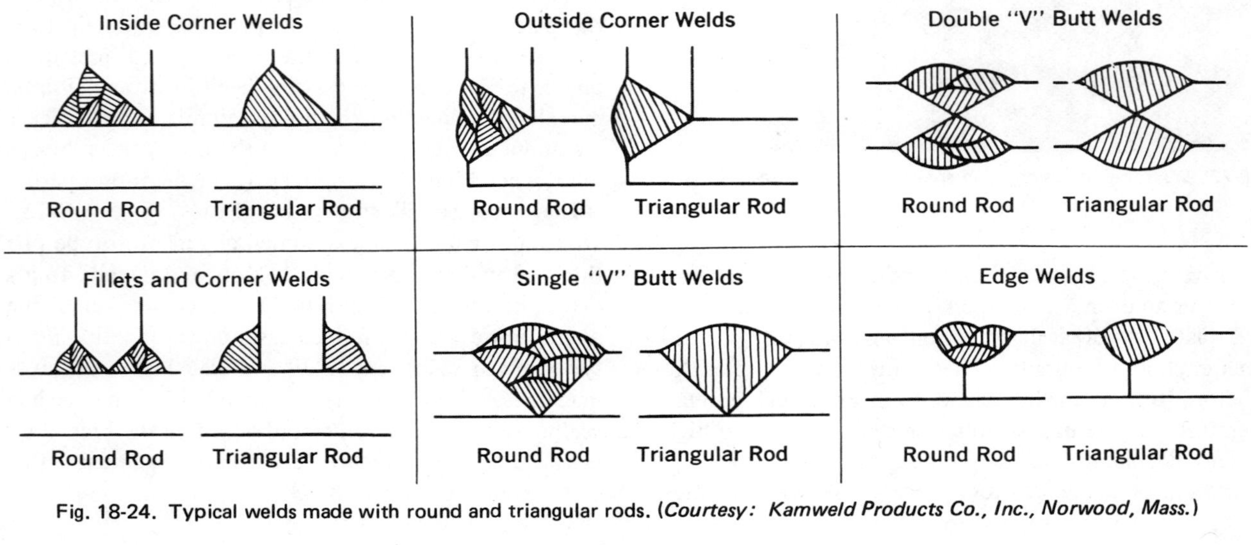

Fig. 18-24. Typical welds made with round and triangular rods. (*Courtesy: Kamweld Products Co., Inc., Norwood, Mass.*)

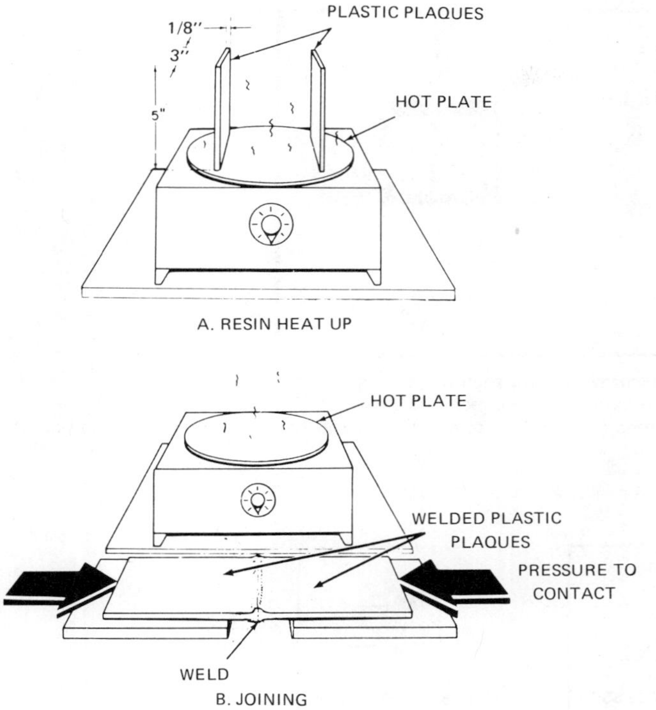

Fig. 18-25. Hot plate welding of sheet stock.

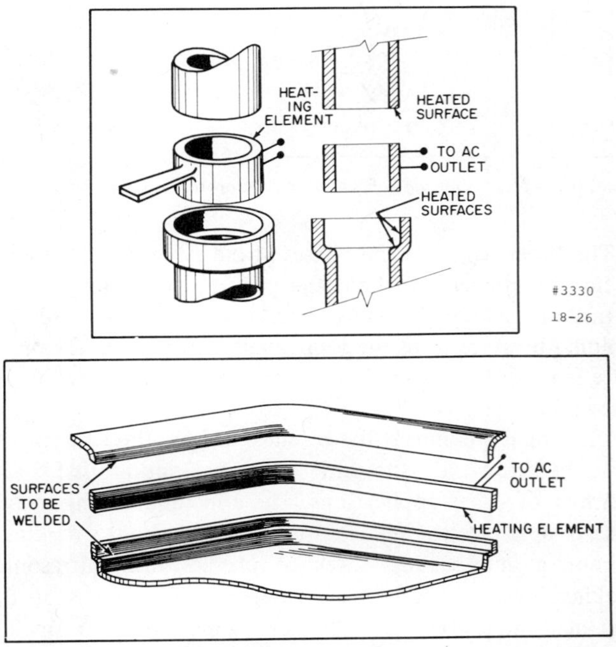

Fig. 18-26. Hot plate welding of circular or irregular joints. (*Courtesy: Celanese Plastics Co.*)

The ultrasonic energy is generated and transmitted to the plastics part by means of special equipment consisting essentially of five components: a high-frequency power supply, a transducer or converter, a tool to transmit the vibrations to the parts, a fixture to hold the parts, and a cycle timer to program the process (see Figure 18-36, p. 772).

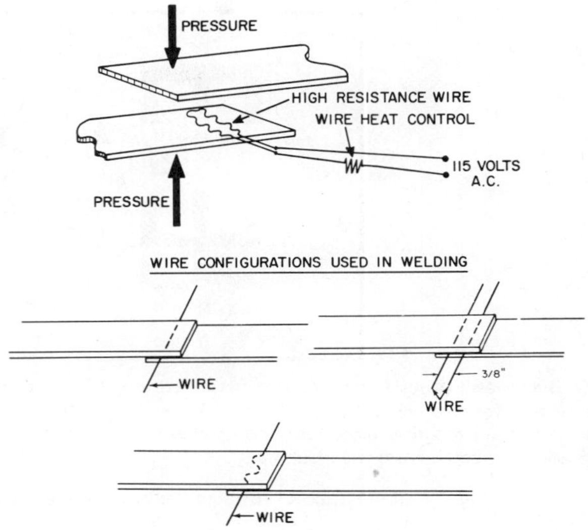

Fig. 18-27. Hot wire welding. (*Courtesy: Celanese Plastics Co.*)

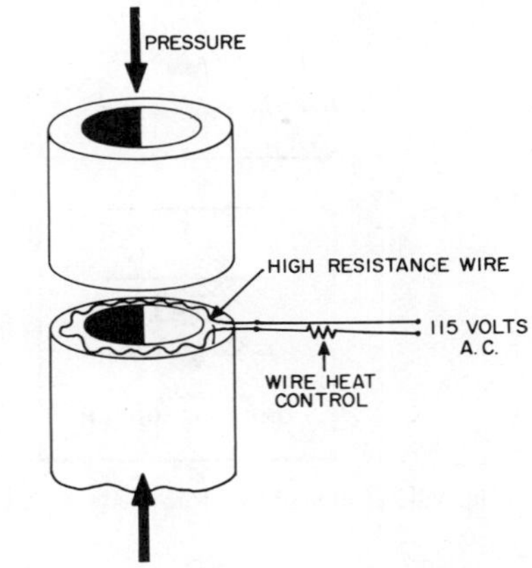

Fig. 18-28. Hot wire welding for circular joints. (*Courtesy: Celanese Plastics Co.*)

Fig. 18-29. Bielomatik hot tool machine used to produce welded polypropylene battery boxes. (*Courtesy: Werner & Pfleiderer Corporation*)

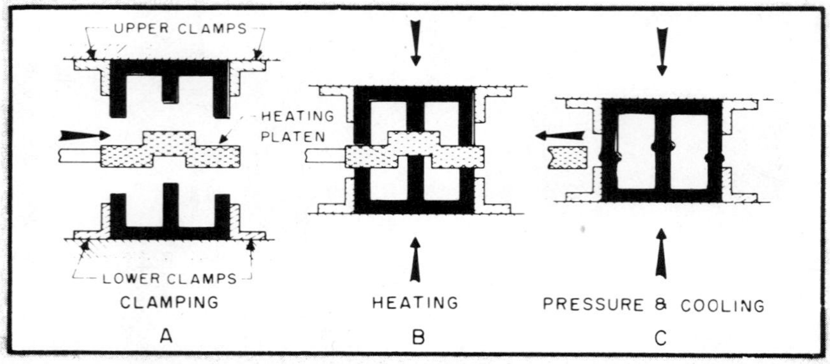

In the Bielomatik process, the parts to be joined are aligned in holding fixtures (A) and then pressed against a heating platen with a precise pressure and for a precise length of time (B). After the proper plasticized depth has been achieved, the heating platen retracts and the two parts are pressed together under very controlled pressure and time conditions (C). After cooling, the holding fixtures retract, and the part is automatically ejected from the machine.

Fig. 18-30. Method of forming the bond in the Bielomatik process. (*Courtesy: Werner & Pfleiderer Corporation*)

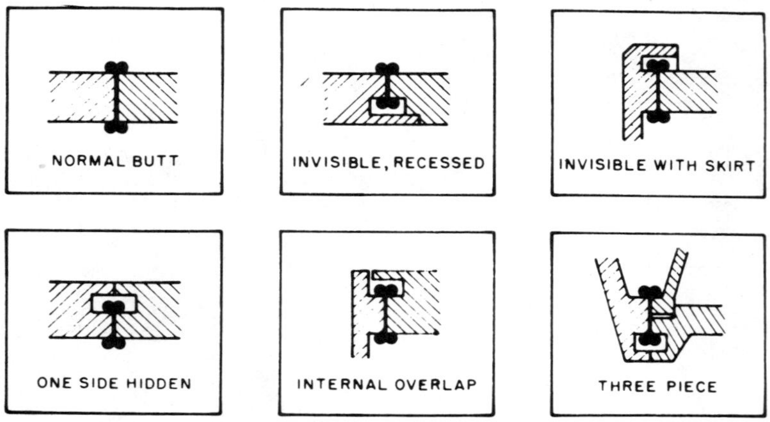

Fig. 18-31. Invisible joints made with the hot tool technique. (*Courtesy: Werner & Pfleiderer Corporation*)

In operation, the power supply is used to change the low-frequency, 60 cycle line current to high-frequency 20 kHz (20,000 cycles per second) current.* The power supply is also called a high-frequency generator, or sometimes an oscillator. The transducer receives the high-frequency electrical oscillations and converts the electrical energy to mechanical vibrations, by means of a piezoelectric ceramic, lead zirconium titanate, which converts 90% of the electrical energy into mechanical energy. (Older units used magnetostrictive units made of nickel laminations which only had approximately 30% efficiency.) The mechanical energy is then transferred from the transducer to the tool, a titanium horn, which contacts the plastic. Titanium is used because of its good strength-to-weight ratio and good acoustical properties which enable it to transmit the high frequency well. The horn serves two functions:

1. It directs the ultrasonic vibrations directly to the plastics parts being welded.
2. It applies pressure to the molten surfaces to form the weld and holds the pressure during the dwell time.

*Frequencies of 10, 15 and 40 kHz are also used.

The timer regulates the phases of the cycle: (1) the application of pressure to hold the part, (2) initiation of the ultrasonic energy for the correct length of time, and (3) holding pressure until the joint cools, after which the pressure is released and the horn retracts. A typical machine is shown in Figure 18-37.

Ultrasonic welding is the primary use for the equipment described; however, the same apparatus can be used for a number of other applications. These include driving metal inserts into plastics, metal to plastics assembly by staking, trimming and sealing edges of plastics film, ultrasonic staking, degating parts from runners, and reactivation of hot-melt adhesives, or other dry adhesives to form an adhesive bond (see Figure 18-38).

To secure optimum performance in any of the ultrasonic applications, it is necessary that the horn used be specially designed for the particular part and the material to be bonded. It is thus recommended that each application be discussed with the manufacturer of the ultrasonic equipment before attempting to make the bond. Figure 18-39 shows some of the horn designs which have been used.

Ultrasonic bonding, while applicable only to thermo-

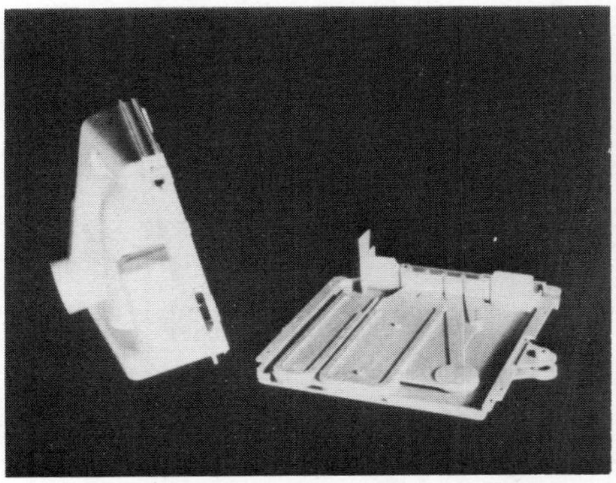

Glass reinforced polypropylene detergent dispenser made from three pieces

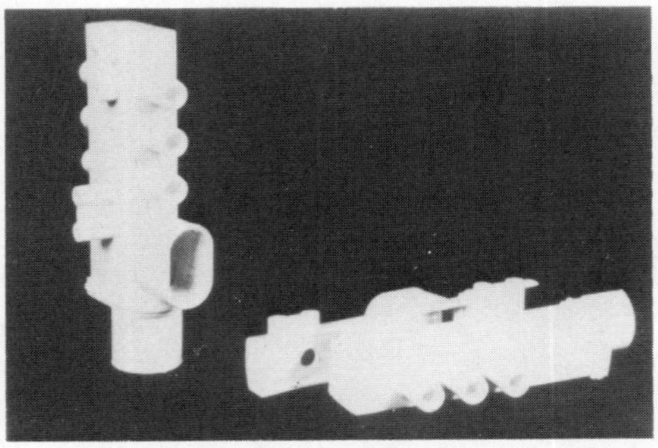

Talc filled polypropylene dishwasher housing

Three piece acrylic tail light assembly two acrylic lenses welded to an ABS frame

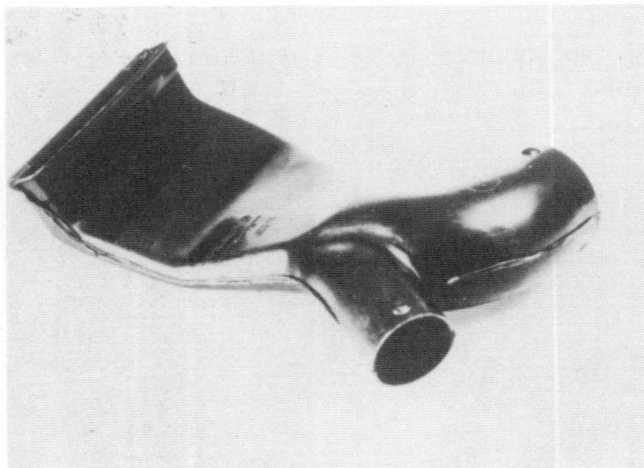

ABS air conditioning ducts

Appliance cannister made from highly filled polyethylene

Fig. 18-32. Various parts made using the Bielomatik hot tool welding process. (*Courtesy: Werner & Pfleiderer Corporation*)

plastics, cannot be used to bond all thermoplastics with equal results. The harder, more rigid thermoplastic materials such as general-purpose polystyrene, acrylics, polycarbonate, and ABS are, in general, good transmitters of acoustical energy, so can be depended on to bond well ultrasonically. The softer materials such as polyethylene, polypropylene, plasticized vinyls, ethylene vinyl acetates, and similar materials effectively absorb acoustical vibrations instead of transmitting them, so, in general, do not form such good welds. A good rule of thumb for judging whether a material will weld ultrasonically is the sound which results when the material is dropped on a hard surface. A sharp, metallic ring indicates good acoustical transmittance and good weldability. A dull thud indicates absorption of the impact vibrations, so the material will not ultrasonically bond as well.

More rigid materials can be welded with the joints farther away from the welding horn than can materials which ab-

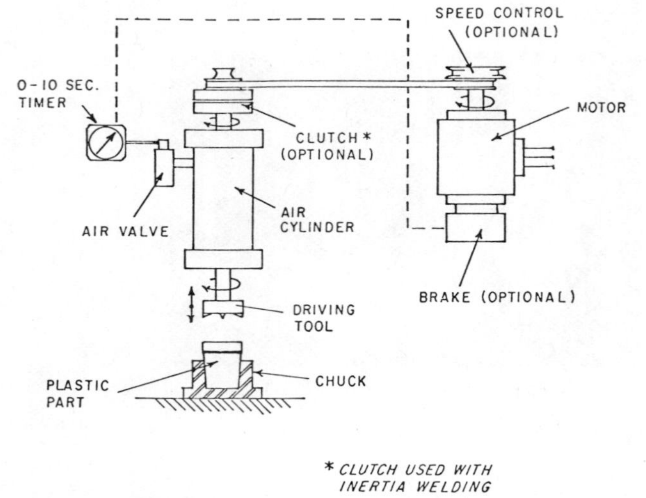

Fig. 18-33. Spin welding equipment. (*Courtesy: E.I. Du Pont de Nemours & Co., Inc.*)

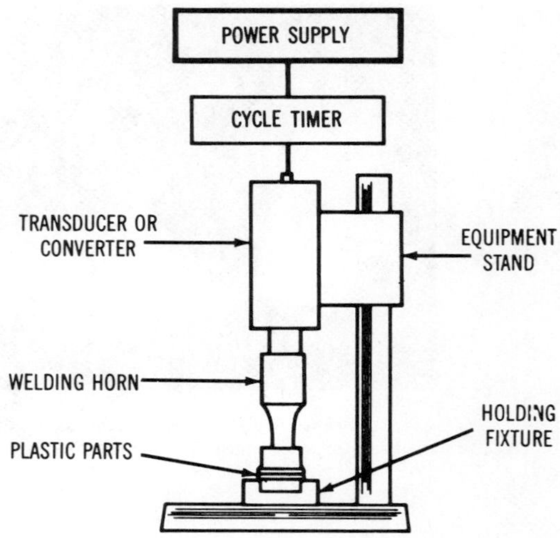

Fig. 18-36. Basic schematic of an ultrasonic welding machine. (*Courtesy: E.I. Du Pont de Nemours & Co., Inc.*)

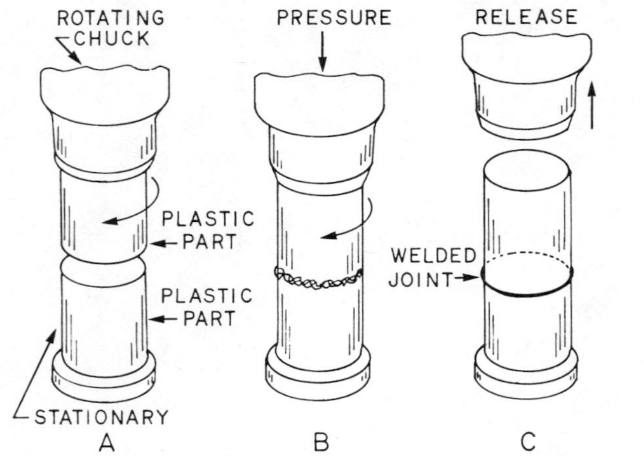

Fig. 18-34. The principle of friction or spin welding. (From *Plastic Product Design* by Beck)

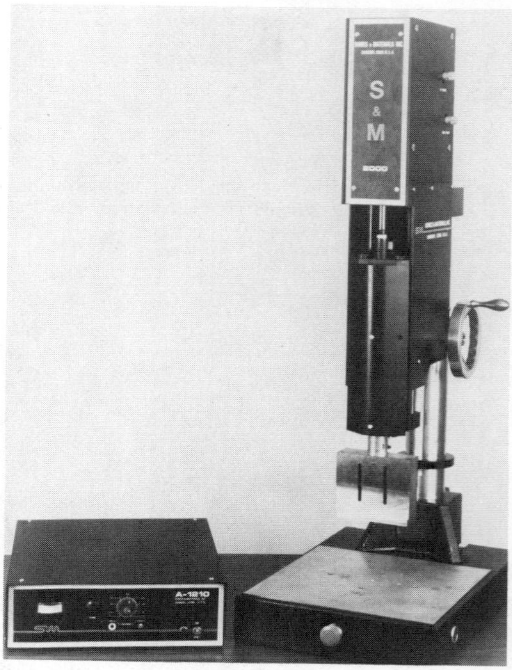

Fig. 18-37. Typical ultrasonic welder. (*Courtesy: Sonics & Materials, Inc., Danbury, Conn.*)

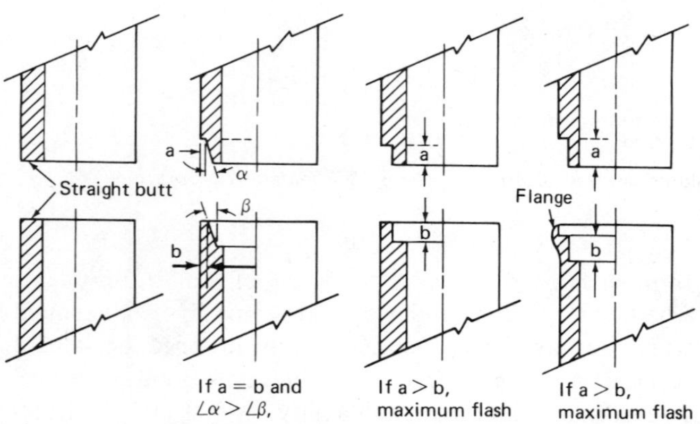

Fig. 18-35. Typical joint configurations for spin welding hollow members, with flash direction. (*Courtesy: Celanese Plastics Co.*)

sorb the vibrational energy. Thus, materials which can only be welded with the horn within $\frac{1}{4}$ inch or less are said to require "near field" welding. Joints which can be made more than $\frac{1}{4}$ inch from the horn are termed "far field" joints. Table 18-4 lists the types of ultrasonic applications for various materials, near and far field weldability, bond quality, etc.

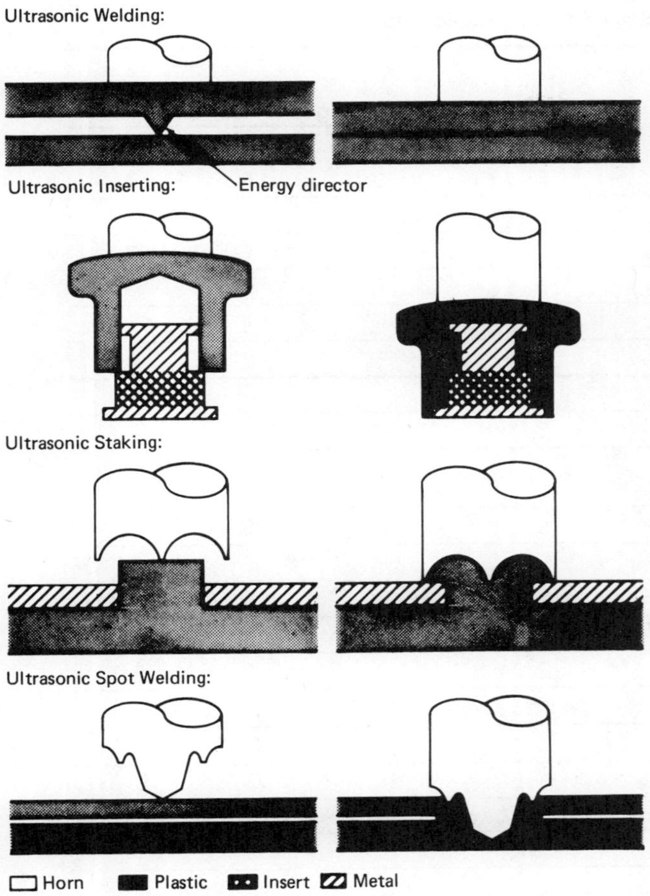

Fig. 18-38. Ultrasonic machine functions. (*Courtesy: Branson Sonic Power Co.*)

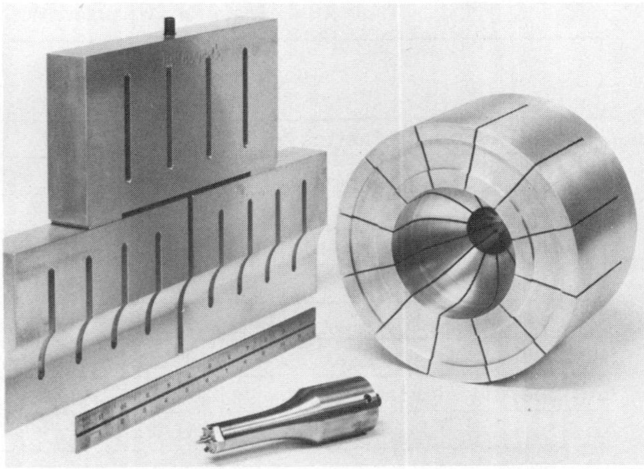

Fig. 18-39. Horn designs used for various parts. (*Courtesy: Sonics & Materials, Inc., Danbury, Conn.*)

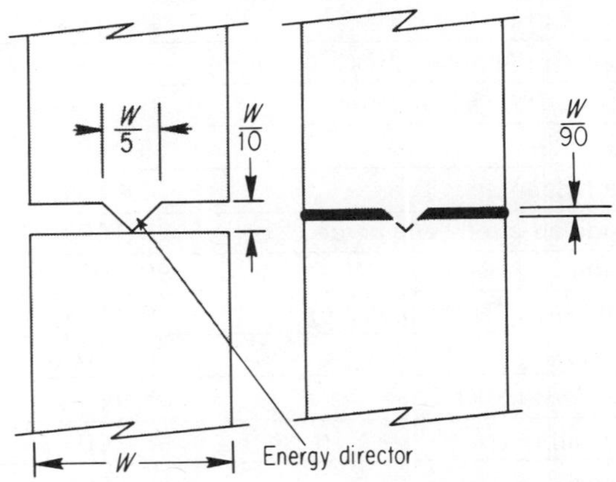

Fig. 18-40. Recommended proportions of energy director for welded plastics joints. Height of director should be at least 0.005 inch. When walls are thicker and height $W/10$ is greater than 0.020 inch, two or more directors should be used; the sum of their heights should equal $W/10$. (Kolb, 1967)

Materials to be welded should have similar melt temperatures. A difference of 16°C (30°F) between two materials (of the same type resin) can interfere with the formation of a good weld, since one material will melt before the other does, thus preventing a true weld. In general, similar molecular structures are also required to result in satisfactory welds. Thus, polystyrene and polycarbonate would probably not weld together well even if the melt temperatures could be matched.

Some of the advantages of ultrasonic bonding include:

1. It is a clean process; there are no fumes, solvents, or adhesives normally required.
2. No heat is required, either for the welding operation or subsequently, as for adhesive drying. Thus, ovens are not required.
3. The operation is very fast: 1–2 seconds.
4. It is an excellent method of making hermetic seals.

To form the highest-quality welds, and particularly if a hermetic seal is required, the joint should be correctly designed. Each joint must incorporate an energy director dimensioned as shown in Figure 18-40. The energy director acts to focus the vibrational energy on a restricted volume of plastic so it is brought to its melting temperature. The energy director should always be kept as small as possible at the point of contact. As shown, the height should not be over one-tenth the wall thickness, and the recommended width is one-fifth the wall thickness. If correctly designed, the volume of plastic in the energy director will be just sufficient to completely flow over all surfaces of the joint. Figure 18-41 shows a number of recommended types of joints, each incorporating an energy director to facilitate the melting action and fill the joint space. For optimum welds tongue-and-groove joints are used.

Any joints, but particularly tongue-and-groove or a pin-and-socket joint, should not be fitted so tightly that vibratory motion is restricted. However, if appearance is a major

Table 18-4. Material Characteristics with Regard to Ultrasonic Applications.†

Material	% Weld strength*	Spot weld	Staking and inserting	Swaging	Welding Near field†	Welding Far field†
General-purpose plastics						
ABS	95-100+	E	E	G	E	G
Polystyrene unfilled	95-100+	E	E	F	E	E
Structural foam (Styrene)	90-100[a]	E	E	F	G	P
Rubber modified	95-100	E	E	G	E	G-P
Glass filled (up to 30%)	95-100+	E	E	F	E	E
SAN	95-100+	E	E	F	E	E
Engineering plastics						
ABS	95-100+	E	E	G	E	G
ABS/polycarbonate alloy (Cycoloy 800)	95-100+[b]	E	E	G	E	G
ABS/PVC alloy (Cycovin)	95-100+	E	E	G	G	F
Acetal	65-70[c]	G	E	P	G	G
Acrylics	95-100+[d]	G	E	P	E	G
Acrylic multipolymer (XT-polymer)	95-100	E	E	G	E	G
Acrylic/PVC alloy (Kydex)	95-100+	E	E	G	G	F
ASA	95-100+	E	E	G	E	G
Methylpentene	90-100+	E	E	G	G	F
Modified phenylene oxide (Noryl)	95-100+	E	E	F-P	G	E-G
Nylon	90-100+[b]	E	E	F-P	G	F
Polyesters (Thermoplastic)	90-100+	G	G	F	G	F
Phenoxy	90-100	G	E	G	G	G-F
Polyarylsulfone	95-100+	G	E	G	E	G
Polycarbonate	95-100+[b]	E	E	G-F	E	E
Polyimide	80-90	F	G	P	G	F
Polyphenylene oxide	95-100+	E	G	F-P	G	G-F
Polysulfone	95-100+[b]	E	E	F	G	G-F
High-volume, low-cost applications						
Butyrates	90-100	G	G-F	G	P	P
Cellulosics	90-100	G	G-F	G	P	P
Polyethylene	90-100	E	E	G	G-P	F-P
Polypropylene	90-100	E	E	G	G-P	F-P
Structural foam (Polyolefin)	85-100	E	E	F	G	F-P
Vinyls	40-100	G	G-F	G	F-P	F-P

Code. E=Excellent. G=Good. F=Fair. P=Poor.
*Weld strengths are based on destructive testing. 100+% results indicate that parent material of plastic part gave way while weld remained intact.
†Near field welding refers to joint ¼ in. or less from area of horn contact; far field welding to joint more than ¼ in. from contact area.
[a] High-density foams weld best.
[b] Moisture will inhibit welds.
[c] Requires high energy and long ultrasonic exposure because of low coefficient of friction.
[d] Cast grades are more difficult to weld due to high molecular weight.

†Courtesy: Branson Sonic Power Co.

factor, then the joint can be designed with tightness in certain areas so as to confine the flash to the interior of the part, as shown in Figure 18-41.

Location of the joint relative to where the horn contacts the part is extremely important, second only to the joint design and the type of material. The joint should be as close to the contact point of the ultrasonic tool as possible, consistent with good overall design and molding practice. Shortening the acoustical path will minimize the loss of energy, resulting in a faster cycle and higher strength. Minimum path

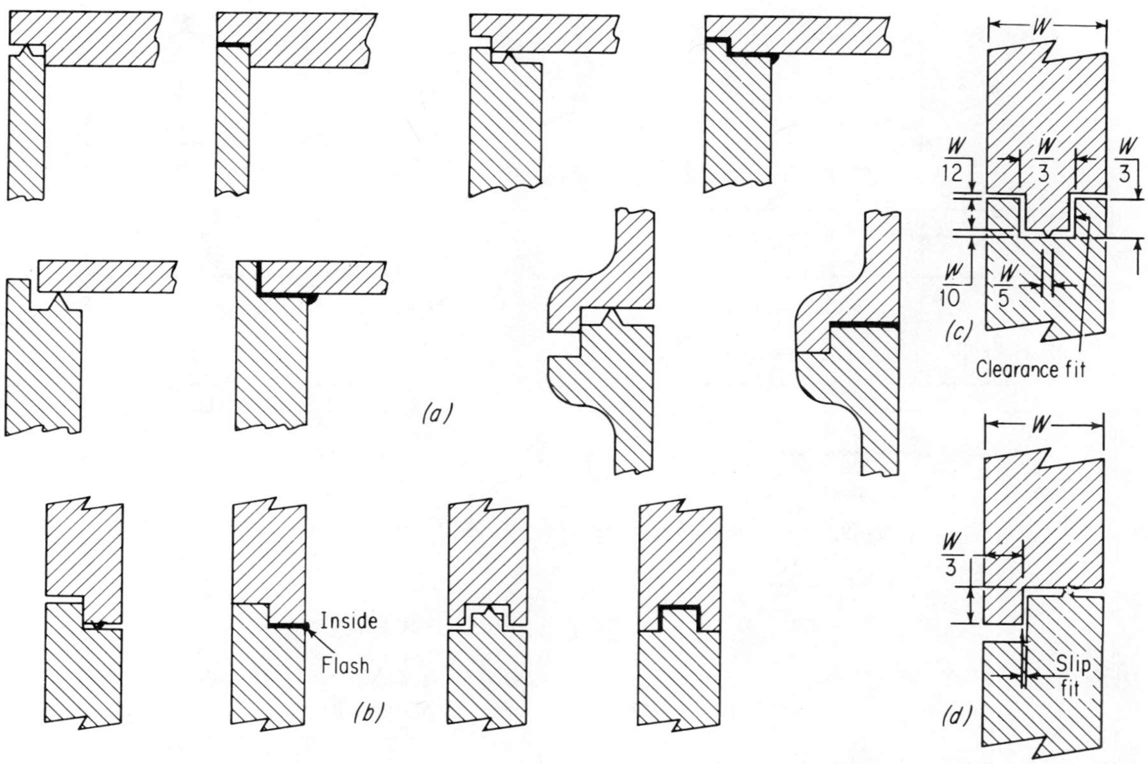

Fig. 18-41. Recommended joint configurations for ultrasonic welding. Variations of basic joints are shown in *a*; configurations shown in *b* control the flow of material for appearance reasons; recommended tongue-and-groove joint proportions and a simpler modification are indicated in *c* and *d*. (Kolb, 1967)

lengths, less than $\frac{1}{4}$ inch, are known as "near field" welds and path lengths longer than $\frac{1}{4}$ inch are called "far field" welds. A change in design from a "far field" weld to a "near field" weld path could make the difference between a fair weld and an excellent weld exhibiting maximum tensile strength (see Figure 18-42).

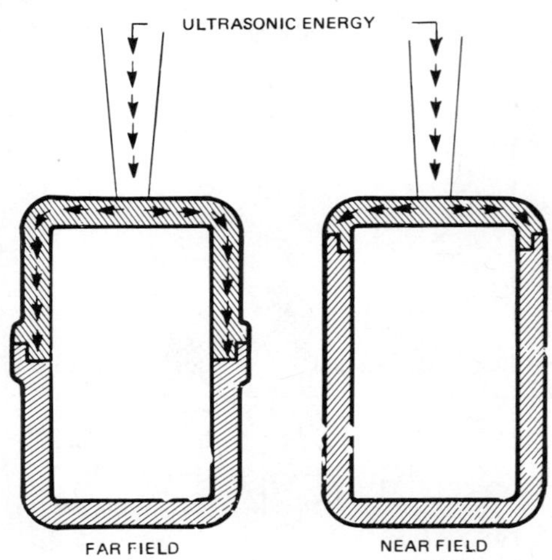

Fig. 18-42. Near and far field joints.

Ultrasonic Staking

Ultrasonic staking is a technique particularly well suited for assembly of metal parts to plastics. In general, the metal part will have a hole in it through which the plastics stud protrudes. By a controlled, light pressure during the vibratory phase, the horn forms a head on the stud, which may be either standard or low profile. The standard head is usually twice the stud diameter and the height is one-half the diameter. The low-profile head, which is used because of its somewhat better appearance, is only $1\frac{1}{2}$ times the stud diameter, and has a height of one-fourth the diameter. Because the low-profile head has the lesser dimensions, its holding power is only approximately one-third that of the standard head. Figure 18-43 illustrates staking.

Ultrasonic Spot Welding

Ultrasonic spot welding is a technique in which extruded or cast sheets of any size can be spot welded together in a manner analogous to spot welding of metal sheets. In this process, using high-powered equipment and a portable, pistol grip tool with a pointed probe, sheets up to $\frac{1}{4}$ inch thick can be joined to each other.

To make the spot weld, the tip of the probe, while vibrat-

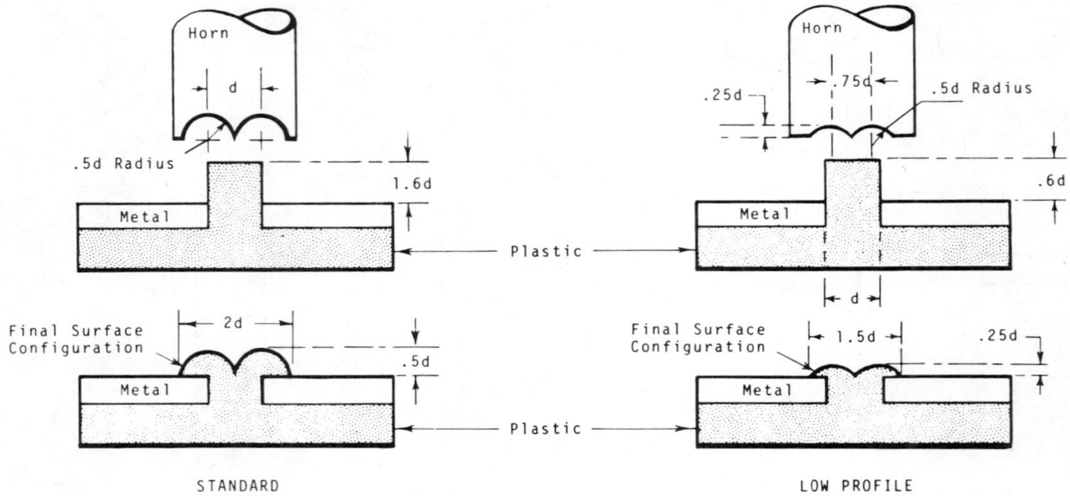

Fig. 18-43. Ultrasonic staking proportions. (*Courtesy: Branson Sonic Power Co.*)

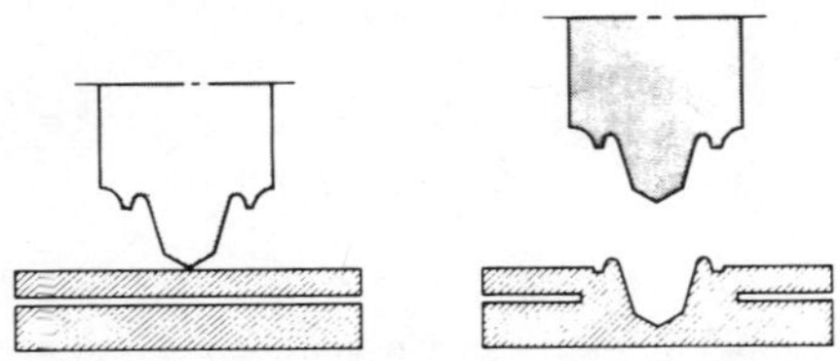

Fig. 18-44. Ultrasonic spot welding. (Reprinted by permission from *Modern Plastics Encyclopedia*, 1977-78, McGraw-Hill, Inc.)

ing, is forced through the top sheet, and the displaced molten plastic rises to the top and forms a ring around the hole, due to the shape of the tip. With further penetration of the tip, the frictional heat causes more molten plastic to flow into the cavity and join the two sheets together, not however, filling the cavity. In this process, the softer plastics such as polyethylene, polypropylene, PVC, high-impact styrene, etc., give the best results. Figure 18-44 illustrates the process.

Ultrasonic Swaging

Ultrasonic swaging, illustrated in Figure 18-45, is a technique in which ultrasonics are used to melt and re-form the plastic to lock another part, usually metal, in place.

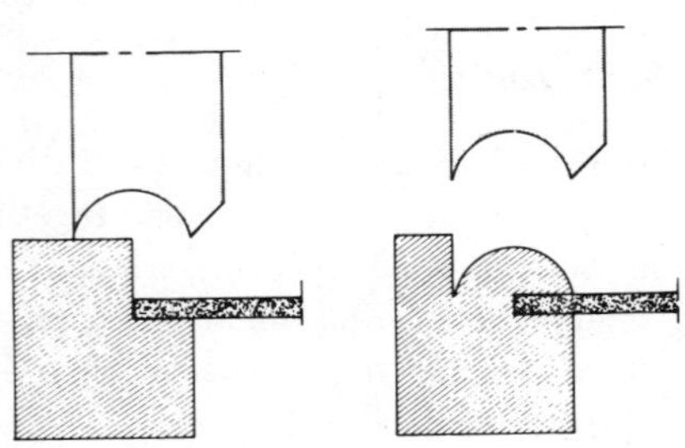

Fig. 18-45. Ultrasonic swaging. (Reprinted by permission from *Modern Plastics Encyclopedia*, 1977-78, McGraw-Hill, Inc.)

Vibration Welding

Vibration welding is a technique which employs frictional heat, produced by *low-frequency* vibrations, to effect a weld in thermoplastic parts. It differs from ultrasonic welding in that the vibration frequency is 90 to 120 Hz, corresponding to normal AC line frequencies, although frequencies as high as 300 Hz are also possible. Another difference between vibration welding and ultrasonic welding is that the joint does not require an energy director.

The usual joint in vibration welding is a butt joint, and it can be considerably larger than that capable of being made by ultrasonic welding. Parts as large as 22 inches long and 12 inches wide have been made successfully.

For optimum joints, careful design is essential, as in ultrasonic welding. Figure 18-46 illustrates joint designs. By using flash traps, nearly invisible joints can be made.

Clamping pressures of 200 psi are commonly used in making the welds. To assure that the pressure will be applied

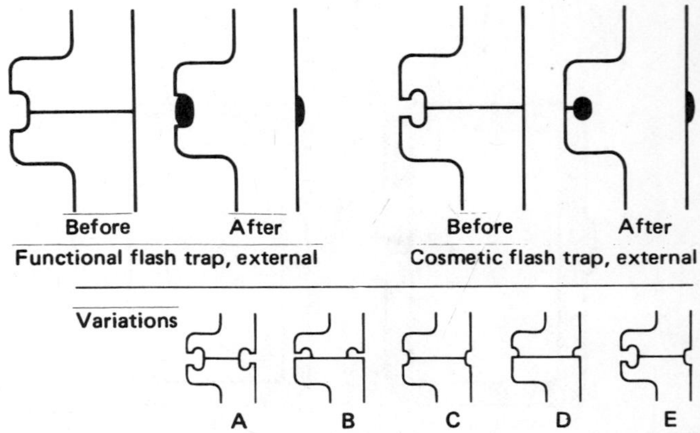

Fig. 18-46. Flash-trap designs. Top row: operation of two basic designs. Bottom row: variations of flash-trap designs. (*Courtesy: Branson Sonic Power Co.*)

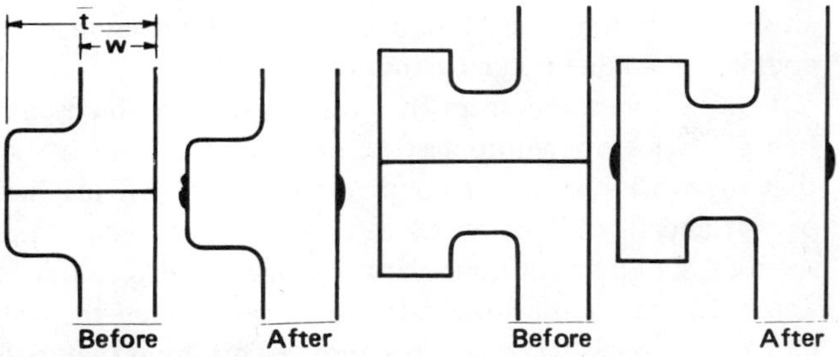

1. w = wall thickness
t = weld surface, 2 to 2.5 x w (for maximum strength
t minimum = .6 x M where M = total relative motion

Fig. 18-47. Typical vibration welding joints. Plain butt weld and joint with added groove to aid in clamping. (*Courtesy: Branson Sonic Power Co.*)

satisfactorily, the butt joints should be at least 2–2$\frac{1}{2}$ times the wall thickness. Frequently, grooves are incorporated in the flange for additional rigidity and to assist in the establishment of the high clamp pressure directly on the joint area (see Figure 18-47).

Electromagnetic Induction Bonding

Electromagnetic induction bonding uses the principle of induction heating of magnetic materials by a high-frequency field to produce heat in any thermoplastic which contains magnetic particles or a metal screen. In practice, the magnetic particles or the metal screen are placed only at the interface of the two materials to be welded, so that heating and fusion of the plastic are confined only to the joint (see Figure 18-48).

Equipment for this process consists of three basic components:

1. A high-frequency induction generator.
2. Specially formed work coils.
3. Fixturing equipment for holding and pressurizing the work. In some cases the work coils can be part of the fixturing equipment.

The induction generator usually has a power output of 1-5 kw at a frequency 4-30 mHz. The most used frequencies, however, are in the range of 4-10 mHz. Included in the generator is a water pump and heat exchanger so that the work coils and the generator itself can be cooled during operation.

The work coil or inductor generally consists of square or round, hollow copper tubing, or machined copper bar stock. The function of the work coil is to surround the joint and form the magnetic field in the metallic materials. The work coil then acts as the primary of a transformer, and the magnetic particles are the secondary. Heating then takes place by a combination of hysteresis and eddy currents, with the latter supplying most of the heat. Heating is very rapid, in approximately 0.1 second for the average joint. To keep the coils from gradually heating up with continuous production, the hollow center is used for circulation of the cooling water.

Work coil design is possibly the most important factor in obtaining a satisfactory joint in a reasonable length of time. The shape of the work coil is determined by the particular joint; however, regardless of the shape, it is most important that the coil be coupled as closely as possible to the joint for highest efficiency. Since a square coil lends itself to being closer to the joint throughout one surface, it is the preferred cross-section. The square coil also can be used most efficiently as a pressure source over the entire area of the joint. Figure 18-49 illustrates loose and close coupling and various work coils, all designed for different types of joints.

Fixtures to hold the work during the welding operation are required so that pressure can be applied to the joint dur-

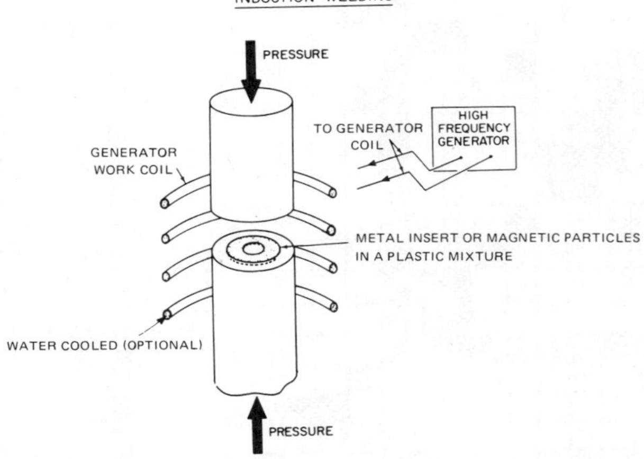

Fig. 18-48. Basic principles of electromagnetic induction bonding. (*Courtesy: Celanese Corporation*)

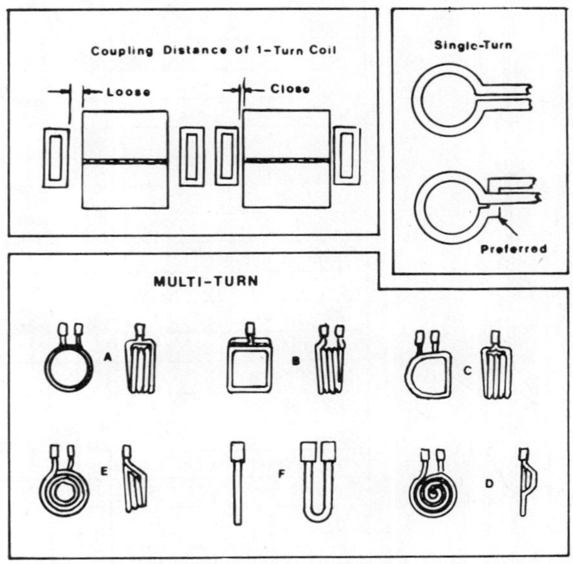

Fig. 18-49. Coil placement and designs. (*Courtesy: EMABond, Inc.*)

ing, and just after, heat generation. No metallic parts should be in the immediate vicinity of the work coil during application of the generator current. The use of the coil itself to apply pressure is therefore very desirable, if the coil and joint design is such that it can be used. If the coil cannot be used for pressurization, then nonmetallics such as phenolic blocks, epoxy-fiberglass, etc., should be employed. Extremely high pressures are not needed; for example, 10–25 psi for polyolefins, 50–75 psi for PVC, up to 125 psi for polymers such as Noryl, polycarbonate, nylon, etc. The pressure can be applied either hydraulically or pneumatically, with the latter used predominantly, because of its simplicity. Appropriate electronic controls are used to program the required heat, pressure, and dwell time or to hold cycles automatically.

As previously stated, the source of heat is the magnetic susceptor material placed in the joint area. This can be either a metallic screen, which remains in the part, a special liquid containing submicron magnetic particles, or a molded plastics gasket, extruded rod, or a stamped plastics sheet which contains the minute magnetic particles. These particles—usually iron oxide or stainless steel (special ferrites or graphite powders have also been used)—are either suspended in the liquid carrier or mixed in with various forms of plastics carriers. In the case of the plastics tape, extruded shape, or molded part, the carrier is usually the same material as the material to be joined. These welding fillers are available from companies specializing in producing such materials in all the commonly used resins. Special mixtures can, of course, be obtained on a custom basis.

It should be noted again that this process can be used to join all types of thermoplastics, provided the joint is such that the work coil can be coupled satisfactorily. Thus, hard or soft and flexible materials can be joined in thick or thin sections, and polar or nonpolar materials can also be joined. Table 18-5 lists the materials which can be satisfactorily joined, including some combinations of different materials.

Joint design is also an extremely important factor in producing satisfactory welds, in particular in producing hermetic seals in containers of different types. The principle to be followed in producing such seals is to place, in the cavity to be sealed, a volume of filler which will just fill the void when the parts are brought together. The filler in this case would be an extruded rod or other profile or a molded gasket preform (see Figure 18-50). On heating the two parts and applying pressure to the joint, the preform melts and effectively seals the cavity void. Figure 18-51 illustrates the formation of inside and outside sealing welds for containers, and Figures 18-52 and 18-53 illustrate welding polyethylene pipe thread fittings to 15-, 30-, and 55-gallon drums.

Another application of electromagnetic bonding, somewhat unique to this process, is the reactivation of hot-melt cements. Normally, parts to be bonded with hot melts must be joined immediately on application of the cement. However, by using a hot melt containing magnetic particles, the cement can be applied to one of the components and al-

Table 18-5. Compatibility of Various Polymers to Electromagnetic Bonding.*

■ Compatible Combinations	ABS	Acetal	Acrylic	Nylon	PC	PE	PP	PS	PVC	PU	CAB	SAN	PSUL	X-Link PE	UHMWPE	Paper	PPO	TFE	FEP	Ionomer
ABS	■		■			■	■	■	■			■				■				
Acetal		■														■				
Acrylic	■		■		■			■	■			■				■				
Nylon				■												■				
Polycarbonate			■		■								■			■				
Polyethylene	■					■	■							■	■	■				
Polypropylene	■					■	■	■				■				■				
Polystyrene	■		■				■	■	■							■	■			
Polyvinylchloride	■		■					■	■	■						■				
Polyurethane									■	■	■					■				
Cellulose Acetate Butyrate										■	■					■				
SAN	■		■				■					■				■				
Polysulfone					■								■			■				
Cross-Linked PE						■								■	■	■				
UHMWPE						■								■	■	■				
Paper	■	■	■	■	■	■	■	■	■	■	■	■	■	■	■	■				
Polyphenylene Oxide								■									■			
TFE Teflon																		■	■	
FEP																		■	■	
Ionomer																				■

*Courtesy: *Iron Age*, April 19, 1976.

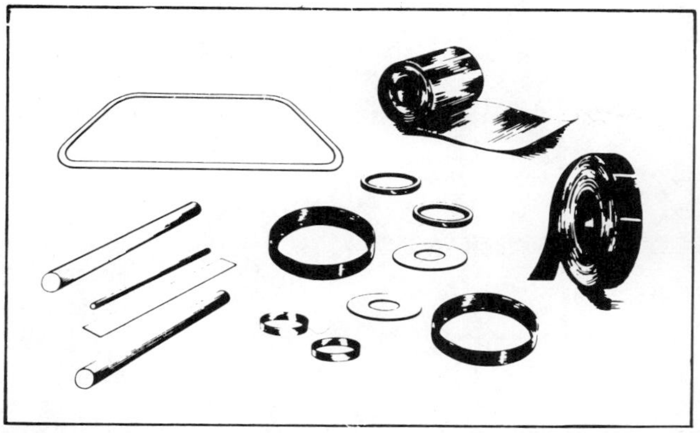

Fig. 18-50. Preforms used in electromagnetic bonding. (*Courtesy: EMABond, Inc.*)

lowed to harden. Later, the two component parts can be fixtured, exposed to the magnetic field, and pressurized. The cement will be reactivated and a bond will be made. While hot-melt cements are not usually used for structural bonds, particularly on thermoset materials, this technique is one method by which electromagnetic induction could be used to bond thermoset materials.

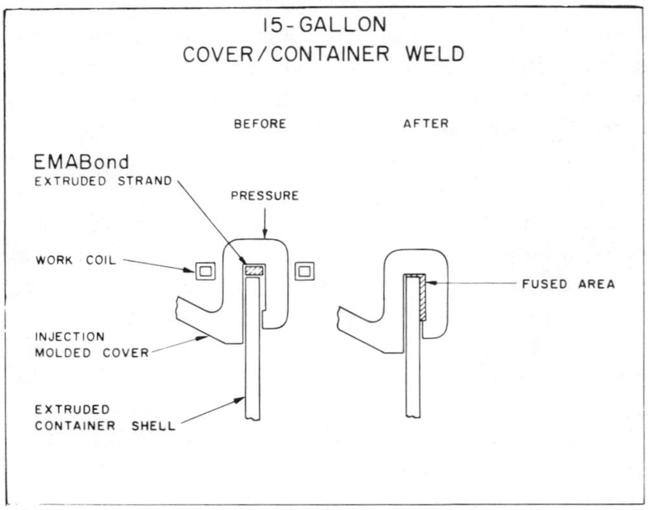

Fig. 18-51. Extruded-molded containers assembled by induction bonding. (*Courtesy: EMABond, Inc.*)

Fig. 18-52. Blow-molded polyethylene 15-, 30-, and 55-gallon containers with induction-welded polyethylene pipe threads. (*Courtesy: EMABond, Inc.*)

Sealing of film or sheeting, commonly called heat sealing, while mainly done by other techniques not requiring a filler material, can also be done electromagnetically either on an intermittent basis, using a hair pin work coil up to 15 feet long, or on a continuous basis as shown in Figure 18-54. In either intermittent or continuous sealing, a liquid filler can be applied before sealing, or an extruded tape can be fed in between the sheets or films. The tape can be pre-extruded or the sealing operation can be made in-line with

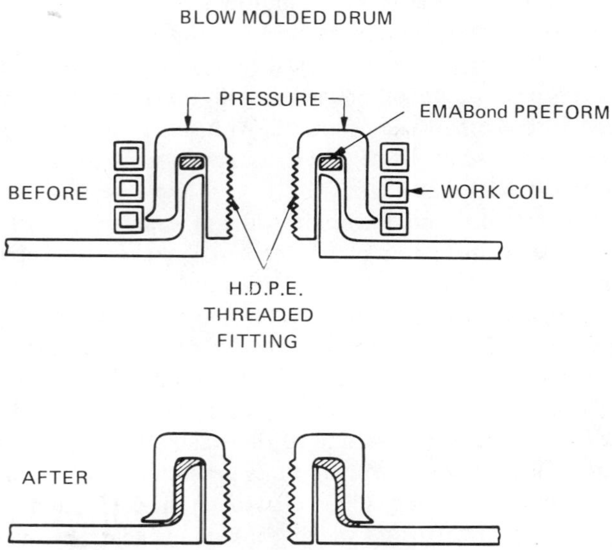

Fig. 18-53. Electromagnetic welding of polyethylene pipe-fitting to blow-molded drum. (*Courtesy: EMABond, Inc.*)

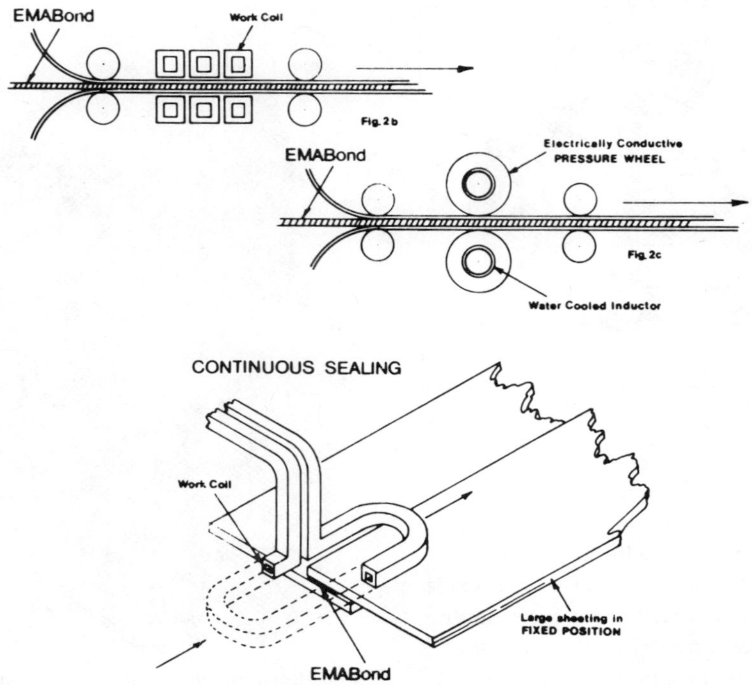

Fig. 18-54. Continuous sealing of films and sheets by electromagnetic induction bonding. (*Courtesy: EMABond, Inc.*)

the extruder. Bonding speeds up to 100 feet per minute are claimed. Further information on bonding of films and sheets is given in the next section on Heat Sealing.

Heat Sealing

Heat sealing is a term used almost exclusively for the joining of homogeneous film laminates, thin sheets of flexible plastics, and, occasionally, rigid films and sheets. As the name suggests, the joining is done by heat, and, since the materials welded together are quite thin, particular attention must be paid to the process conditions if the welds are to be satisfactory. The material must be brought to the fusion temperature, then pressurized just enough to form the weld, and then held, under pressure, until the joint is strong enough so the part can be moved. With too high a temperature, the films may be degraded; too much pressure and the joint will be thinned excessively; too low a temperature and/or pressure and the joint will not weld sufficiently. The time is also important. If the part is moved too soon, the joint may tear, too long a dwell time is simply a waste of time, and also raises the possibility of the joint sticking to the tool. Thus, whatever thermal process is used, it must be one which is capable of precise process control.

There are a number of heating techniques which can be used to perform heat sealing of films and thin sheets, and some which can not. Among those processes which can be used are heated tool welding, ultrasonic, dielectric, and induction welding. Hot gas welding and spin welding, because of the difficulty of accurate control of the three parameters of welding—temperature, pressure, and time—are usually not suitable for welding of materials less than $\frac{1}{16}$ inch thick.

Heated Tool Welding. Heated tool welding is the simplest method of welding plastics films together. It is usually employed for polyethylene or polypropylene films of 1–6 mils in thickness. Films over 6 mils are difficult to heat quickly by conduction, and other processes such as ultrasonic or induction heating will produce the thicker seals more rapidly.

It its basic form, the process consists of using some type of heated bar, or a hand-held sealing iron, similar to a small domestic iron. For best results the bar or sealing iron should be thermostatically controlled, and the sealing surface should be covered with a resilient material such as silicone rubber over which a thin, Teflon-coated, glass fabric is used to prevent sticking. Alternatively, the Teflon-coated sheet could be put over the area of film to be sealed, and then removed after the joint is cool. The operation can be done manually, but the results are greatly dependent on the skill of the operator, particularly if the two films are nonlaminated films; for example, pure polyethylene or polypropylene, nylon, etc.

Laminated or composite films, such as polyethylene-coated cellophane, polypropylene-coated Mylar, or Teflon FEP-coated Kapton, etc., can be easily sealed by hand, since the low-melting adhesive coating fuses at a much lower temperature than does the supporting film. However, for high productivity most heat sealing, using a heated bar, is done with automatically controlled equipment incorporating the bar as part of a stand which also includes pneumatic pressurization, controlled electronically.

Included in the category of heated tool welders are rotary sealers which, as the name indicates, are used to form long, continuous seals, or intermittent seals as the film is passed through the wheels. Figure 18-55 shows various types of

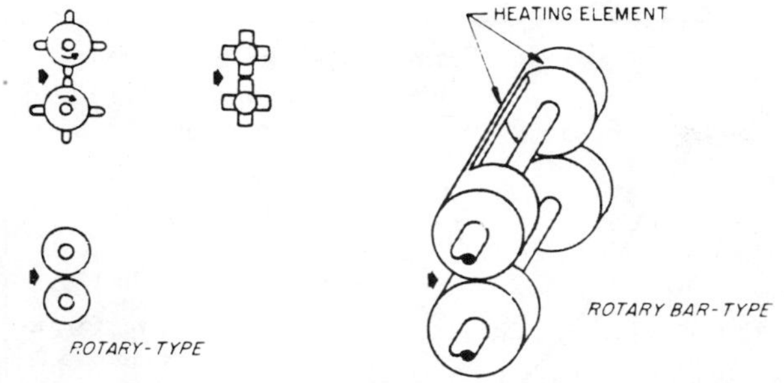

Fig. 18-55. Rotary heat sealer designs. (Reprinted by permission from *Plastics Design & Processing*: December, 1966. Copyright 1966, Lake Publishing Corporation, Libertyville, Ill.)

designs, including one which forms the side seams of a bag and intermittently seals across the end of the bag. Rotary sealers suffer from the disadvantage that the joint seal is not kept under heat or pressure for an appreciable length of time.

An improved type of rotary sealer is a band sealer. In this machine, the film is carried between two endless metal bands, which first pass through a heating section and then through a cooling section; at the same time, the film is kept under pressure. Continuous band sealers are usually incorporated into assembly-line packaging machines.

Impulse bar sealers are another type of bar sealer in which the bar is kept cool, except for a brief electronically controlled interval in which an impulse of electric current causes a Teflon-coated nichrome ribbon to heat up very rapidly, usually less than a second, thus fusing the film which is between the jaws. The film is kept under pressure until it cools, after which the jaws open (see Figure 18-56). The cool jaw of the impulse sealer can be air or water cooled. However, because of the intermittent heating and cooling, the rate of seal formation is usually not as fast as with band sealers. Impulse sealers are used extensively for polyethylene and nylon bags and similar packaging products. Figure 18-57 shows a production sealer which may be controlled manually or automatically cycled.

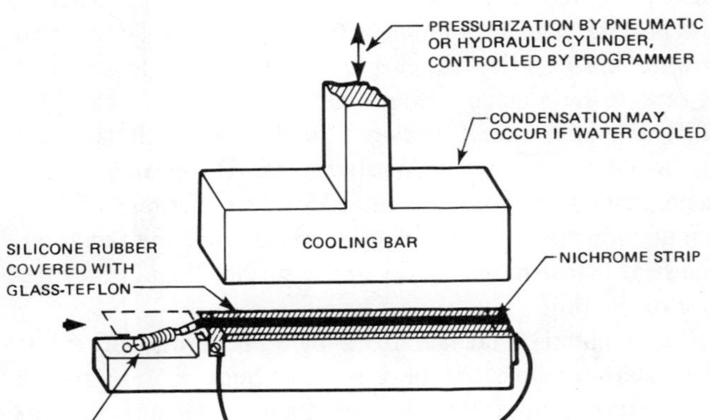

Fig. 18-56. Impulse bar sealer. For thicker films, both jaws may be equipped with heater ribbons.

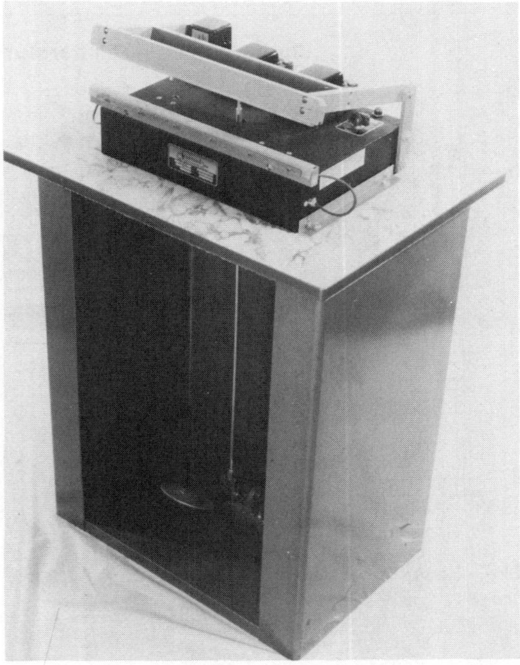

Fig. 18-57. Typical foot-operated impulse bar sealer. Seals and trims edge of bag. Can make up to 15 seals per minute in 15-mil polyethylene. Available up to 96-inch sealing width. (*Courtesy: Vertrod Corporation*)

The majority of seals made by hot bars and impulse sealers are for thin-gauge polyethylene used for bags and similar packaging uses. Because of the great number of such articles made and used annually, there have been some definite relationships worked out between the time, temperatures, and pressures or clamp times required to produce satisfactory seals. Some of these rules are:

1. The greatest temperature flexibility can be achieved by working with short clamp times at the highest temperatures which do not produce degradation. Figure 18-58 illustrates this principle.
2. With increasing pressure, the time and temperature can be reduced, but at the expense of flexibility. Thus, machine conditions would have to be more closely controlled, as shown in Figure 18-59.
3. For a given clamp time, the higher the resin density, the higher the minimum temperature required for sealing.
4. The lower the melt index of the resin, the larger will be the operating range in terms of clamp times and temperatures.
5. A gusseted bag, because of more layers of film to be sealed, takes appreciably more time for sealing than does an ungusseted bag. There is also less flexibility in the machine conditions, since in sealing the gusseted bag at times four layers of film must be sealed, but in other areas only two layers of film are sealed. The machine conditions must be chosen so that without making readjustments in the temperature, clamp time,

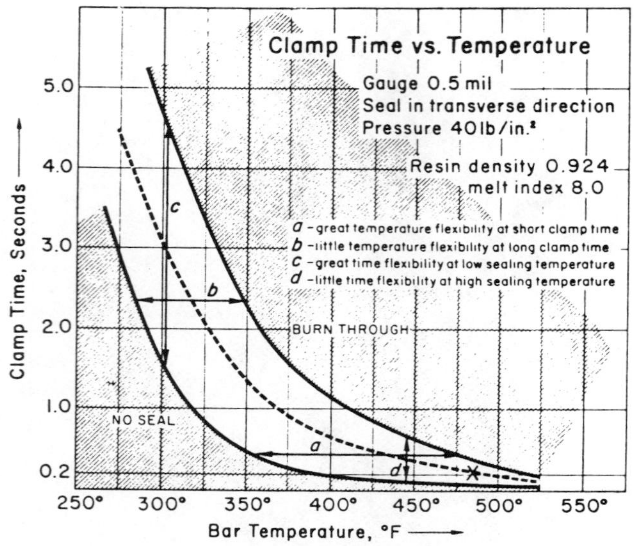

Fig. 18-58. Clamp time versus temperature. (*Courtesy: U.S. Industrial Chemicals Co.*)

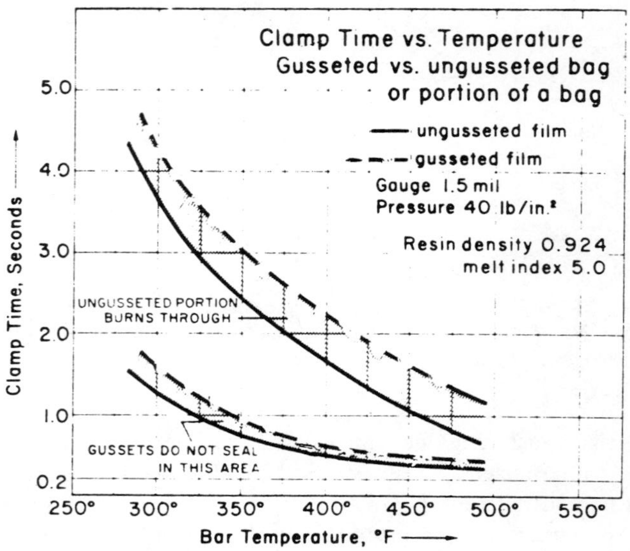

Fig. 18-60. Gusseted bag processing. (*Courtesy: USI Film Products*)

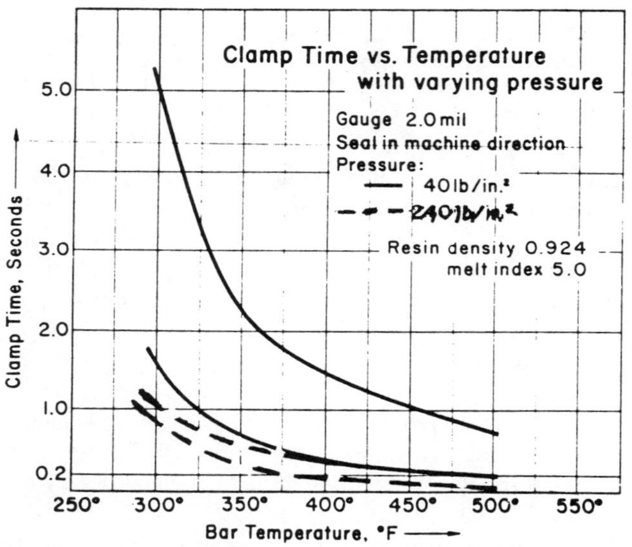

Fig. 18-59. Clamp time versus temperature with varying pressure. (*Courtesy: U.S. Industrial Chemical Co.*)

or pressure, satisfactory seals will be obtained in both areas. The sealer must then be adjusted so that seals will be made using conditions shown within the shaded area of Figure 18-60; that is, the machine must seal below the upper limit of the two-layer sealing range, and above the lower limit of the four-layer range shown. Impulse heater sealing machines are made with solid-state controls for the clamp time, pressure, and temperature so that reliable production seals can be made within the stringent process requirements.

Three major problems in hot tool heat sealing have been found, in many cases, to be related to improper design or use of the heat sealing equipment. These problems are: non-uniform seals, weak seals, and film sticking to the sealing jaws. Table 18-6 shows possible causes for these seal defects and suggests corrective action to be taken.

Dielectric Heat Sealing. Dielectric heat sealing, also known as high-frequency or radio-frequency welding or sealing, is a technique in which electricity is used to generate heat inside the body of the material, instead of introducing it to the outer surfaces, as in heated tool sealing. Thus, in dielectric sealing, the outer surfaces of the films are usually cooler than the inside surfaces. It is therefore possible to have better control of the weld with regard to thinning out and surface marking on the film.

The principle of dielectric welding is that the films are placed between two plates between which a high-frequency alternating current is generated, thus setting up a rapidly oscillating field inside the films. As the polarity of the plates continually reverses, the positive and negatively charged parts of the plastics molecules attempt to realign themselves in accordance with the changing field. That is, the positive pole of the molecule tries always to be oriented toward the negative plate, and the negative pole attempts always to be aligned toward the positive plate. This rapid shifting of the molecules causes "molecular frictional" heat to be developed in the plastics film. The heat is evenly generated throughout the film. However, because of heat conduction to the metal platens, convection, and radiation, the heat is lost more rapidly at the surface than in the interior of the film.

Polar materials characterized by ionic bonds are most amenable to dielectric heat sealing. Such materials have high dipole moments; that is, their molecules are not symmetrical. Another way to describe these materials is to refer to them as having high dielectric losses or being "lossy." Polyvinyl chloride, a very asymmetrical molecule, is a good

Table 18-6. Problems, Causes, and Corrective Measures in Hot Tool Heat Sealing.*

Problem	Possible Causes	Suggested Corrective Action
A. Nonuniform seals (e.g., burn-through on one portion of seal)	1. Excessive temperature range across sealing bar	Modification or replacement of sealing jaw or wire so that a temperature range of 15°F or less is maintained across its length
	2. Nonuniform pressure at time of seal	Alignment of sealing jaws and use of a resilient, high-temperature-resistant backing pad
	3. Poor control of pressure in the lower pressure ranges	Installation of pressure-regulating equipment that is accurate in the 20 to 60 lb/in.^{2} pressure range
B. Weak seals	1. Seal not cooled sufficiently before stress is applied to it	a. Installation of a water-cooled backing jaw or external air jets to promote faster cooling of the sealed area b. Use of impulse sealers (wire) usually recommended for in-line filling and sealing operations, since less heat is present, permitting more rapid cooling
	2. Extrusion of film between sealer jaws forming a weak, brittle line of polymer	a. Installation of serrated sealing jaws which will permit polymer flow within the seal area b. Installation of pressure-regulating equipment that is accurate in the 20 to 60 lb/in.^{2} pressure range
C. Film sticking to the sealing jaws	Direct contact between metal sealing jaws and film	a. Installation of a nonsticking membrane between sealing jaw and film b. Application of a nonsticking sealing jaw

*Courtesy: U. S. Industrial Film Products.

example of a "lossy" compound. It is one of the easiest materials to seal dielectrically and one of the ones most frequently used. Other molecules easily sealed with this technique are nylons and urethanes. Molecules such as polyethylene, polypropylene, and fluorocarbons are symmetrical molecules and consequently are not "lossy" enough to respond to the alternating field by generating heat. Such materials cannot be sealed dielectrically.

Dielectric sealers are commonly built to operate at an FCC assigned frequency of 27.12 mHz and do not, at this frequency, require a shielded room. Higher-frequency equipment can be used, but usually requires special shielding for the machine, or installation in a shielded room to insure against unwanted signals being radiated. The higher frequencies are used for the more difficult to seal materials, or thicker pieces. The RF generators are made in powers from 1 to 30 kw, for normal production work. New machines, used in the automotive industry for sealing urethane-PVC-cardboard composites and similar parts are made with powers of 50 kw and up. Because of the high powers required, dielectric sealing is usually confined to sealing or welding of films.

In addition to the radio-frequency generator, the dielectric welding machine, like the heated tool welder, has provisions for applying pressure for a preset time, by means of a pneumatic cylinder; less frequently, hydraulic or mechanical actuators are also used. Solid-state timers are provided for preheat pressurization, high-frequency on, and pressurization during cooling. Temperature control is provided by varying the generator output, or the time of application. Usually the lower platen is stationary and the upper platen contains the die, or tool, which forms the shape of the weld. Since pressures are not excessive, normally under 100 psi, neither the platens nor the supporting structure is made as massive as are corresponding components used in compression or transfer presses (see Figure 18-61).

The tool in the dielectric sealer is used to generate heat in the films and to apply pressure, a function similar to that of the bar in hot tool welding. However, by suitable design the die can also be made to form simple tear-off seals, or can cut the seal at the time of welding. The heat sealing dies are usually made of brass or aluminum bar stock bent to the desired shape and mounted on an aluminum plate. The bar can be anywhere from $\frac{1}{2}$ to 2 inches high. For high-speed production some dies are made with water-cooling provisions. For normal sealing applications, the bottom surface of the die should be flat; however, if special designs or embossing effects are required, the die can be engraved with various patterns.

Compared to heat sealing using hot tools, the initial cost of dielectric sealing is usually considerably higher. However, operating and labor costs are usually much lower, and effi-

Fig. 18-61. Dielectric sealer and rotary turntable for load and unload stations. Used for making billfolds, book covers, inflatables, etc. (*Courtesy: Cosmos Electronic Machine Corporation*)

ciency is usually higher with relatively nontechnically trained labor. For sealing of thin films, 1–2 mils thick, dielectric sealing may be the only method to use, since conductive sealing may deteriorate the surface before a satisfactory weld can be made. In some cases there may be an energy savings using dielectrics, since the heat can be applied so selectively, or over a local area, whereas other methods of heating may make heating a larger area necessary.

The use of dielectric sealing has made possible the very large array of all types of low-cost items made of film and sheet, particularly using vinyl plastics. These items include many types of household goods, advertising and novelty items, rainware, inflatable toys and air mattresses, checkbook covers, baby pants, etc. The dielectric sealer can also be combined in a production line with other equipment to cut and decorate the sealed articles, as shown in Figure 18-62.

Continuous sealing can be done with a dielectric sealer using a drum that is engraved or embossed with the seal pattern. A metal belt, partially encircling the drum, conveys the film stock past the die. Double metal bands have also been used where high-speed production is required for simple seals on light films. Sealers using these high-speed components usually have higher-frequency generators to obtain the temperature rise in the shortest periods of time.

The power requirements to produce satisfactory seals vary with the square inches of seal in short, compact seals. The requirements increase in direct proportion to the linear inches of seal in narrow, long seals, and also depend on the type and thickness of material to be sealed. As a general guide the following rules apply:

1. Seals in thin films generally require more power than the same seals in thick films because of heat losses.
2. Compact dies require approximately 1 kw for each 4 square inches of seal.
3. Narrow seals require approximately 1 kw for each 24 inches.

Table 18-7 gives specific power requirements for seals of various areas and thicknesses.

Ultrasonic Heat Sealing. While not used as extensively as dielectric or hot tool sealing for films, ultrasonics can be used to seal certain "hard-to-seal" materials. The best known of these are polyester films (e.g., Mylar), used in making belts for business machines, etc., and polypropylene film used in making "seamless" tubes, etc. The major advantage in the use of ultrasonic welding is the absence of heat during the seal formation, thus minimizing the shrinkage, puckering, or warpage which might occur in a heat-affected film. This elimination of heat effects is particularly important in fabrication of parts using an oriented film, such as might be

Fig. 18-62. Integrated sealer, cutter, and silk screen decorator machine. (*Courtesy: Cosmos Electronic Machine Corporation*)

Table 18-7. Heat Sealing Guide.*

Sealing Area (Square Inches)							
Material	Generator Output						
Total Thickness	$\frac{1}{2}$-kw	1-kw	2-kw	3-kw	6-kw	8-kw	10-kw
8 gauge	1.2	2.0	4.2	6.0	13.2	16.0	20.0
12 gauge	1.6	3.0	5.8	8.6	17.5	24.0	30.0
16 gauge	1.8	3.5	6.8	10.2	20.2	27.0	35.0
20 gauge	2.0	4.0	7.6	11.4	22.6	32.0	40.0
24 gauge	2.2	4.2	8.2	12.4	24.4	33.6	42.0
30 gauge	2.4	4.6	9.0	13.6	26.8	36.8	46.0
36 gauge	2.5	4.9	9.6	14.5	28.8	39.2	49.0
40 gauge	2.6	5.1	10.0	15.1	30.0	40.8	51.0
60 gauge	2.8	5.6	11.2	16.8	33.6	44.8	56.0

The sealing guide is based on the following assumptions:

1. Material is good-quality vinyl.
2. Brass electrodes of good design and construction at production temperatures (no pre-heaters).
3. Presses and generators are industry-approved.
4. The sealing time is approximately 4 seconds.

Heating units add approximately 20% to capacity.

To determine the sealing area:

1. Measure running inches of seal
2. Multiply by the width of the seal in inches

For example: A bar seal 32″ long and $\frac{1}{16}$″ wide has an area of 2 square inches (32 × $\frac{1}{16}$).

Tear seals are actually a combination of a bar sealing section and a tear sealing section. The area of each section must be computed separately. The procedure outlined above is used to determine the bar sealing area. For the tear sealing section, measure the running inches of tear seal and divide by 8. Eight is a constant for the tear seal section regardless of profile or thickness of tear seal edge.

For example: A $\frac{1}{32}$″ straight tear seal that is 32″ long would have a sealing area of 5 square inches. The bar sealing section is $\frac{1}{32}$″ wide. Area of bar seal is 32″ × $\frac{1}{32}$″ or 1 square inch. Tear seal area equals 32″ divided by 8 or 4 square inches. Total area equals 1 square inch plus 4 square inches or 5 square inches.

*Courtesy: Cosmos Electronic Machine Corporation.

used in making a precision belt for scientific tape recorders, etc.

Figure 18-63 illustrates the technique for fabricating a butt-welded belt using a polyester film. This type of seal is reported to result in joints having 85–100% of the parent material's strength.

In addition to sealing of thin films by ultrasonics, synthetic fabrics such as Dacron, Orlon, and nylon can be bonded with the same technique. Other types of materials which can be ultrasonically sealed are various plastics-coated metal foils and papers. The process is particularly amenable to papers, since there is no heat to scorch the paper.

Ultrasonics has a unique advantage over other types of sealing in that the vibratory energy, in addition to heating

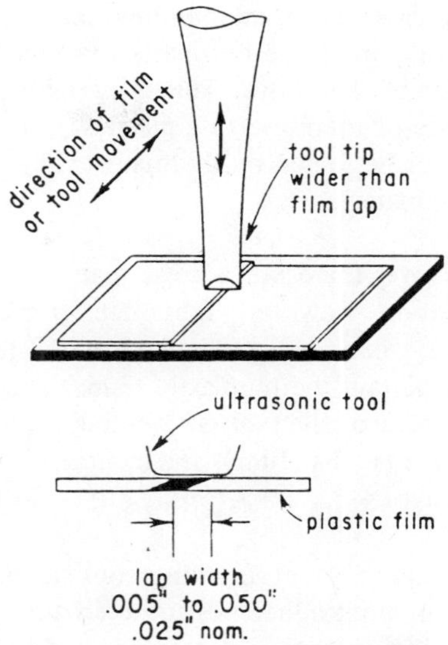

Fig. 18-63. Schematic drawing showing how a simulated butt weld is made between two pieces of plastics film. Overlapping edges are smoothed into diagonal weld across cross-section of the film (lower sketch) so that total thickness at the joint does not exceed thickness of single film. (Deans, 1961)

the material, also has a natural cleansing action. Seals can be made very satisfactorily through films coated with grease, oil, various powders, dirt, ink, etc. This process can then be used to seal the ends of plastic or foil pouches filled with oil, cosmetics, food products, etc.

ADHESIVE BONDING

Adhesive bonding is the process in which an adhesive is used to attach two like or unlike plastics together, or plastics to non-plastics materials. An adhesive, based on a plastics resin, may also be used to bond two metal sheets together, or a nonmetallic, nonplastic sheet may be joined to metal. In any case, there are thousands of combinations of materials which might be joined to each other, and there are a corresponding number of adhesives and adhesive modifications which can be used for the varying situations.

There are a number of reasons why adhesives are in use for all types of parts and materials. Some of these are:

1. An adhesive bond can be over an entire surface, so there is no local stress concentration as might result from a rivet or screw.
2. The joining method is inherently inexpensive. No expensive equipment may be needed.
3. The joint may be made completely sealed against air or water.
4. The joint is noncorrosive.
5. The joint may be used to take up stresses which arise from different coefficients of expansion of two differing materials.
6. The joint may be used as a damping medium.

7. In some cases adhesive bonding may be the only method of joining the two parts together to result in any or all of the above. This is particularly true for parts made of thermosetting plastics, since such materials cannot be joined by thermal welding, as can thermoplastic materials.

As stated above, there are a great many adhesive types and formulations; however, they can conveniently be grouped into six basic categories, with one category applicable only to certain thermoplastic resins. The classes are: solvent bonding, also called solvent welding; liquid adhesives; mastics; hot melts; dry films; and contact- and pressure-sensitive adhesives.

Solvent Bonding. Solvent bonding consists essentially of softening one or both adherend surfaces wtih a solvent and then pressing the two surfaces together and holding under pressure until all the solvent is absorbed and eventually evaporated away. The resemblance to thermal welding is apparent; that is, softening, contact, hold under pressure, etc. The method is only applicable to amorphous plastics such as cellulosics, acrylics, polystyrene, etc. Crystalline materials such as polyethylene, polypropylene, fluorocarbons, acetals, etc., cannot, except under special circumstances, be solvent bonded.

There are a number of solvents used to soften plastics which are amenable to this technique, and, in most cases, more than one solvent can be used for a given material.

Since the solvent is used only to soften the plastic, and does not remain in the joint, any good solvent for the resin may be used, with equal bond strength expectations. The choice of the solvent depends more on such factors as drying time, odor, toxicity, and flammability. Table 18-8 lists a number of common solvents and some of their physical properties. It should be noted that the drying time of a solvent-bonded joint is directly related to the boiling point of the liquid. Those boiling at the highest temperatures naturally take the longest time to dry completely. The drying time of a solvent-bonded joint is also influenced by how

Table 18-8. Properties of Solvents Used in Solvent Bonding.

SOLVENT	BOILING POINT, °C.	FLAM-MABLE	MAX. ALLOW-ABLE BREATH-CONC. IN USE, PPM	ODOR
ACETONE	56	YES	1000	PENETRATING, STRONG, SWEET
BUTYL ACETATE	126	YES	500	MILD
"CELLOSOLVE"	135	YES	100	MILD
CYCLOHEXANONE	155	YES	100	PEPPERMINT, STRONG
DIOXANE	101	YES	100	MILD
ETHYL ACETATE	77	YES	400	MILD
ETHYLENE DICHLORIDE	84	YES	100	MODERATE
FORMIC ACID	100	YES	10	STRONG, PUNG. PENETRATING
GLACIAL ACETIC ACID	118	YES	10	STRONG, PENETRATING
METHYL ETHYL KETONE	80	YES	80	MODERATE
METHYL CELLUSOLVE	125	YES	25	MILD
METHYL ISO-BUTYL KETONE	127	YES	?	MILD
METHYLENE CHLORIDE	40	NO	500	MODERATE, PENETRATING
N, N-DIMETHYL-FORMAMIDE	307	YES	?	STRONG
N-METHYL-PYRROLIDONE		YES	?	STRONG
O-DICHLOROBENZOL	180	YES	50	STRONG
PERCHLORETHYLENE	121	NO	200	MODERATE
PROPYLENE OXIDE	35	YES	21	STRONG
TETRAHYDROFURAN	6.5	YES	100	MODERATE, PENETRATING
TOLUENE	111	YES	200	MODERATE
XYLENE	144	YES	200	MODERATE

much solvent is used, how thick the joint is, the surrounding temperature, etc. In many cases a joint will appear dry on the outside, due to rapid loss of solvent, but it may still be soft inside.

The use of very rapid-drying solvents such as acetone or ethyl acetate for bonding cellulosics is very common; however, when such solvents evaporate too quickly there is a danger that the joint will "blush," acquire a white discoloration, brought about by lowering of the surface temperature and consequent precipitation of water vapor from the air, which cause resin precipitation on the joint surface. In many cases solvent mixtures are used which include, with the rapid-drying solvent, 10–20% of a slow-drying solvent such as methyl Cellosolve. This is used to retard the rapid drying, or fast evaporation, which would lower the surface temperature. Appropriately enough such solvents are called "retarders." In some cases they are only mixed in with the fast solvents on days when the humidity is particularly high and the danger of "blushing" is increased.

Regardless of the solvent used for bonding operations, good ventilation and exhaust hoods should be used at all times. People have varying tolerance levels for solvents, so it should not be assumed that all workers can work with the same solvents. Anyone showing allergic symptoms to the solvent or the vapor should be transferred to another area.

In addition to the use of the pure solvent, or a mixture of solvents for use in softening the plastic surface, solvent "dopes" are also made. These doped solvents consist of a pure solvent in which there has been dissolved a small amount of the resin, powdered or shaved for easy solution. The doped solvents are used since, containing as much as 50% resin, they dry much slower, allowing more working time, and being much more viscous or "bodied up," they can be used to fill small gaps in the joint, which normally might not be closed by the pure solvent alone, or would require a much longer softening period. Doped cements are very commonly made up for such applications as bonding of cellulosics of all kinds, acrylics, and styrenes. They can be easily made by the user or are readily available from commercial sources. Table 18-9 lists various plastics and recommended solvents (or doped solvents) which have been used to make bonds.

Alternative Adhesive Bonding Methods. Other than solvent bonding, all the other classes of adhesives–liquids, mastics, hot melts, etc.–may be used for bonding all types of plastics, with, of course, some formulations giving much superior bonds for certain plastics than for others. The five other categories of adhesives which are available are shown in Table 18-10. Also included in this table are the advantages, limitations, and processing considerations of the various types, so comparisons can be made with regard to applica-

Table 18-9. Recommended Solvent-Plastics Combinations.

	SOLVENTS																		
PLASTIC	ACETONE	CYCLOHEXANONE	N, N-DIMETHYL-FORMAMIDE	DIOXANE	ETHYL ACETATE	ETHYLENE DICHLORIDE	FORMIC ACID	GLACIAL ACETIC ACID	METHYL ETHYL KETONE	METHYL CELLUSOLVE	METHYLENE CHLORIDE	N-METHYL-PYRROLIDONE	O-DICHLOROBENZOL	PERCHLOR-ETHYLENE	TETRAHYDRO-FURAN	TOLUENE	80 TOLUOL-20 ETHANOL	TRICHLOR-ETHYLENE	XYLENE
	X					X			X		X			X	X			X	
ACRYLIC						X		X			X								
CELLULOSE ACETATE	X				X				X	X									
CELLULOSE ACETATE BUTYRATE	X	X			X	X			X	X							X		
ETHYL CELLULOSE					X	X													
NYLON							X												
POLYARYL ETHER	X				X														
POLYARYL SULFONE			X									X	X						
POLYCARBONATE					X			X		X									
POLYSTYRENE		X		X	X			X		X	X	X	X	X		X			X
POLYSULFONE		X				X			X		X		X	X		X			X
POLYVINYL CHLORIDE	X	X								X			X						
PPO BASE (NORYL)						X					X			X					
STYRENE-ACRYLONITRILE	X							X				X	X						
VINYLIDENE CHLORIDE		X	X								X		X						
VINYLCHLORIDE-VINYLACETATE	X	X						X		X			X						

Table 18-10. Adhesive Categories for Bonding of Plastics.*

Technique	Description	Advantages	Limitations	Processing Considerations
ADHESIVES				
1. Liquids Solvent, Water Base, Anaerobics	Solvent-and-water-based liquid adhesives, available in a wide number of bases–e.g., polyester, vinyl–in one or two part form fill bonding needs ranging from high-speed lamination to one-of-a-kind joining of dissimilar plastics parts. Solvents provide more bite, but cost much more than similar base water-type adhesive. Anaerobics are a group of adhesives that cure in the absence of air, with a minimum amount of pressure required to effect the initial bond. Adhesives are used for almost every type of plastic. Includes epoxies and two-part "no-mix" acrylics; i.e. each surface is coated with one component.	Easy to apply; adhesives available to fit most applications.	Shelf and pot life often limited. Solvents may cause pollution problems; water-base not as strong; anaerobics toxic.	Application techniques range from simply brushing on to spraying and roller coating-lamination for very high production. Adhesive application techniques, often similar to decorating equipment, from hundreds (e.g., a glue pump for $400) to thousands of dollars with sophisticated laminating equipment costing in the tens of thousands of dollars. Anaerobics are generally applied a drop at a time from a special bottle or dispenser.
2. Mastics	Highly viscous single- or two-component materials which cure to a very hard or flexible joint depending on adhesive type. Includes epoxies, urethanes, rubber base, silicones, etc.	Does not run when applied.	Shelf and pot life often limited. Can be used for structural bonds.	Often applied via a trowel, knife or gun-type dispenser; one-component systems can be applied directly from a tube. Various types of roller coaters are also used. Metering-type dispensing equipment in the $2500 range has been used to some extent.
3. Hot Melts	100% solid adhesives that become flowable when heat is applied. Often used to bond continuous flat surfaces. Made from polyethylene, saturated polyester, or polyamides, as granules, pellets, rope, or slugs.	Fast application; clean operation.	Virtually no structural hot melts for plastics.	Hot melts are applied at high-speeds via heating the adhesive, then extruding (actually squirting) it onto a substrate, roller coating, using a special dispenser or roll to apply dots or simply dipping.
4. Film	Available in several forms including hot melts, these are sheets of solid adhesive. Mostly used to bond film or sheet to a substrate. Includes epoxies, phenolics, polyamides, and elastomer base.	Clean, efficient.	High cost. Can be used for precision structural bonds.	Film adhesive is reactivated by a heat source; production costs are in the medium-high range depending on heat source used.
5. Pressure Sensitive	Tacky adhesives used in a variety of commercial applications (e.g., cellophane). Often used with polyolefins. Usually rubber base materials.	Flexible	Bonds not very strong. Not used for structural bonds.	Generally applied by spray with bonding effected by light pressure.

[a]Because of the thousands of formulations available within the various adhesive categories, it is not practical to attempt an analysis here of which adhesives will satisfy a particular application's need. However, typical adhesives in each class are:
Liquids: 1. Solvent–Polyester, vinyl, phenolics, acrylics, rubbers, epoxies, polyamide; 2. Water–acrylics, rubber, casein; 3. Anaerobics–cyanacrylate; Mastics–rubbers, epoxies; Hot Melts–polyamides, PE, PS, PVA; Film–epoxies, polyamide, phenolics; Pressure Sensitive–rubbers.
*Travernicht, 1970.

bility to any given part or situation. In using the information in Table 18-10, it should be noted that a given adhesive type (e.g., an epoxy) may be available in several different categories, such as a liquid of fairly low viscosity, or a thick, nonrunning mastic, or even as a dry film which can be cut to a definite shape and placed exactly where required. In each case the final bond strength could be substantially the same.

Before bonding, the surfaces of both adherends should be free from all traces of dirt, oil, mold release, etc. The following surface treatments are being used:

1. Wiping or dipping the surface with a chemical cleaner such as acetone, isopropyl alcohol, or any other solvent which will remove contaminants and not attack the surface appreciably

2. Abrasive cleaning with light sandpaper, sandblast, vapor honing, or a wire brush
3. Degreasing in a vapor which will not attack the plastic

In addition to the above treatments, polyethylene, polypropylene, fluorocarbons, and acetals require special treatments to assure a surface which will accept the adhesive. In the case of the olefins, the surfaces can be prepared by corona (high voltage) discharge treatment, a flame treatment, or a chemical dip utilizing chromic acid (concentrated sulfuric acid 150 parts by weight, water 12 parts, and potassium dichromate 7.5 parts), for approximately 30 seconds at 70°C (158°F). Regardless of the method used, the result is an oxidized surface which will accept the adhesive and allow a bond to be made. For the fluorocarbons and acetals there are various commercially available proprietary solutions which are used.

A relatively new method of surface preparation, applicable to any plastics surface, is known as a plasma etch treatment. The parts to be bonded are placed in a vacuum chamber and are exposed to gas particles which are electrically charged or exist as free radicals, due to excitation by a radio-frequency current. In this condition, the gas particles are extremely reactive and promote crosslinking of the polymer's surface molecules, which forms a skin that will accept an adhesive. Or the reactive gas particles will create free radicals on the polymer surface which will then enter into strong chemical bonds with adhesives, inks, or coatings.

Currently, the process is a batch system, with cycles ranging from 2 minutes to as long as 15 minutes. It is thus not a high-speed, high-production process. However, for certain applications or materials it can be very useful. First, it will treat all types of materials and shapes, thus assuring equal treatment to all parts of a difficult-to-treat part. The treatment is safe, requiring no special disposal equipment or safety precautions. Since the treatment is done at room temperature there are no thermal problems of warpage, shrinkage, etc. The major disadvantage is the initial cost of the equipment which, at the time of writing, was high. Figure 18-64 illustrates the equipment for a small plasma etch reactor.

Selection of the Bonding Material and Process.

The success of any adhesive-bonded joint or structure depends mainly on the particular adhesive selected; however, the design of the joint and the method of implementing the process also enter into the ultimate success or failure of the bond. It is impossible to cover here every aspect of the suitability of various adhesives for various applications, or all the factors that influence the choice of one material or one design over another. However, it is possible to describe some of the factors governing a great majority of applications.

First, every bonding application should be subjected to a "problem analysis." This will consist of a series of questions, the answers to which will often indicate exactly the type of adhesive and the joint design to be used. Some typical questions are as follows:

1. What materials are to be joined?
2. What are the surface conditions? Or what conditions can be obtained?
3. What are the thicknesses of the adherends?
4. What is the joint design? Or what joint design can be used?
5. What is the end use of the part? Will it be subjected to high stresses, vibration, continual load?
6. What are the highest and lowest temperatures to which the part will be subjected in service or in storage?

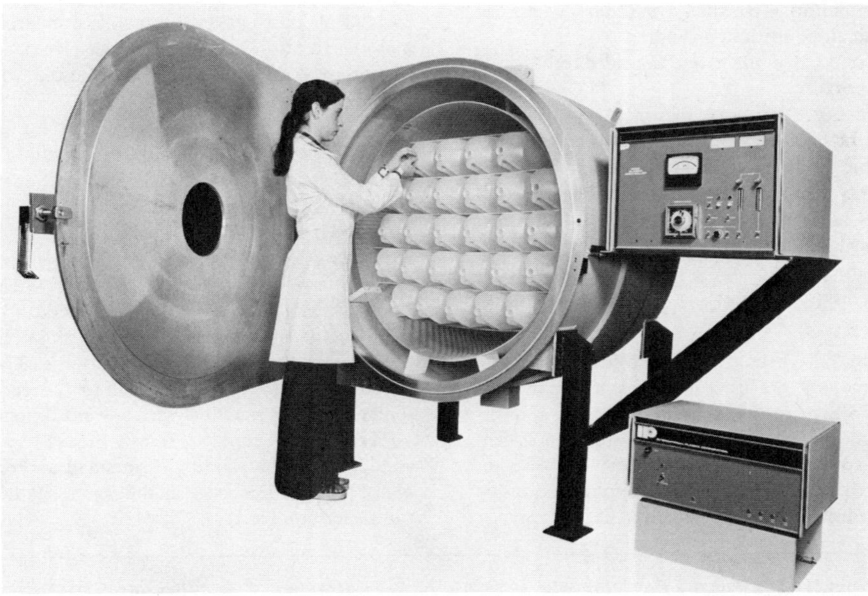

Fig. 18-64. Plasma etch reactor. (*Courtesy: Dionex, Inc.*)

7. What strengths are required in the joint? Tensile, shear, peel, compression, impact, etc.?
8. What are the environmental stresses? Humidity, chemical exposure–oils, solvents, fuel, etc.? For example, the gasoline resistance required in a plastics part for a minimum of 5000 hours at an under-hood temperature of 200°F is vastly different from that required for a toy to be used in a home.
9. Is precise placement of the adhesive required? This would dictate the use of a dry film adhesive.
10. Are there any conditions present which might limit the type of bonding tools or fixtures used?

Table 18-11 presents the various types of adhesives and the substrates for which they are suited. By a thorough consideration of the ten questions, and any others which might arise, and perusal of the table, selection of the optimum adhesive system for any situation should be simplified.

Joint Design

The design of the joint is almost as important as the selection of the adhesive system. The shape of the parts is of course largely determined by the end use, but there is almost always some latitude within mechanical or artistic limits for designing to achieve maximum bond strength. Joint design requirements are relatively simple and can be summed up in the following four statements. The joint should be designed to:

Table 18-11. Assembly of Fabricated Parts.*

Adhesive Bonding

The variety of adhesives available for plastic bonding is constantly growing and because of this, adhesive bonding is often the most durable, efficient, and economical method for plastic assembly.

Tables A through D are guides for choosing an adhesive for a particular type of plastic. To use the charts:

1) Locate the substrates on a particular chart.
2) Determine the adhesives suggested for bonding these substrates.
3) Note the advantages/disadvantages of the listed adhesives to narrow the choice of candidates.
4) Contact adhesive suppliers for additional information, testing laboratory facilities, and recommendations for adhesive application.

Design and Cost Considerations

The most important factor in choosing an assembly method is to insure that the strength and the durability of the adhesive will be adequate for the final product. It is important to eliminate candidates which might develop a high initial strength but would not provide adequate durability under the required environmental conditions. For example, the trim on the exterior of an automobile requires a much more durable adhesive than a part bonded to the exterior of a computer in an air-conditioned room.

Proper joint design can maximize the strength and durability of the adhesive and minimize costs.

Assuming equal performance, the next factor in choosing an assembly system is to find the most economical method of assembly which will exhibit acceptable durability. The entire assembly system must be considered, not only from the standpoint of adhesive costs but including such factors as labor, extra operations for cleaning parts or installing mechanical fasteners, and equipment for drilling, tapping, or heat-curing of adhesives.

Properties of Adhesive Classes

In Table C the adhesive types have been divided into four classes.

Elastomeric adhesives generally are used in the assembly of elastomers and in areas where a high degree of joint flexibility is desirable. The majority of them are solvent-based adhesives which generally are spray- or brush-applied and allowed to dry to a certain tack before assembly of the parts. They often require extensive venting to eliminate the hazards of the solvent materials.

Thermoplastic resins have a wide variety of applications both in the area of bonding plastics to themselves and bonding to other substrates. Generally, they are solvent-based or applied to parts as melts as elevated temperatures. They are used widely for assembly of thermoplastics and where speed is required on a production line.

Thermosetting resins are by far the most versatile, durable, and environmentally resistant adhesives. Cure methods generally involve the use of heat or mixing two components, but single-component, room-temperature curing systems also are available. Application methods include automated glue guns or automated pressure-time applicators. Single-component thermosetting adhesives can be dispensed through all methods.

Assembly Methods

The various techniques that are available for applying adhesives offer advantages for particular applications.

Brushing is a low-cost method which permits application of adhesive to controlled areas. For high production, flow brushes can be used. Uniform application is difficult to achieve with the method, and it usually is labor-intensive.

Dipping is an excellent large-area method for completely coating parts. Roll application is possible for large sheets and panels. The method is not easily used with solvent cements, however, and a limitation is that masking is required for partial application of adhesive.

Roll coating is extremely efficient for uniform coating of large areas. A variety of roll coating methods is available, with automated systems the most efficient. Roll coating is of limited use for solvent-based adhesives: solvent loss causes inconsistency in coating thickness.

Spraying is a rapid method for large-area bonding. Small spray systems can coat precise areas and reach inaccessible areas. All spray systems require venting and operator protection.

Stencil/screening permits precise amounts of adhesive to be dispensed rapidly over complex patterns. The method easily can be automated. Often it is limited to flat surfaces, however. Stencil/screening is not practical with solvent cements.

Pressure-time applicators can be used to place beads or drops of adhesive precisely, suiting it for small- rather than large-area bonding. Many types of tips are available–gun, needle, spray, etc. The method can be automated.

*Reprinted by permission from *Modern Plastics Encyclopedia*, 1978–79, McGraw-Hill, Inc. V. R. Allies and B. D. Murray, authors.

Table 18-11A. Adhesive Types for Bonding Nonplastics to Thermoplastics.[a]

	Nonplastics						
Thermoplastics	**Ceramic**	**Fabric**	**Leather**	**Metal**	**Paper**	**Rubber**	**Wood**
ABS	6,16	3,5	16,23	23,25,26	4,42	5,16,21,25	4,26,42
Acetal	16,23	4,23	4,23	4,23	4,16,23	4,16	16,23
Cellulosic[b]	4,16	4,5,42	4,5,42	3,4	16,42	1-5,16	4,16
Ethyl cellulose	14,16	14	14	14	14,16	14,16	14,16
Nylon	4,23,26	3,4	3,4,26	3,23,25,26	4,41	2,3,25	3,4,26
Phenylene oxide-based resin	5,24	4	16,31,36	23,25,31,37	4,42	3,4	16,26,42
Polycarbonate	16,23,26,36	23,36	16,23,36	23,25,26	16,36	5,16,25,36	16,23,26,36
Polyethersulfone	16,23,25	3,4	4,31,36	23,25	5,41	2,6,16,25	16,26,36
Polyethylene	3,16,41	3,41	3,41	3,31,41	16,41	3,16,41	3,16,41
Polyethylene terephthalate	26,26	5,36	5,26,36	26,36	5,36	13,36	26,36
Polymethylmethacrylate	3,4,26	4	3,4,16,42	3,4,25	42	1-5	3,4,26,42
Polyphenylene sulfide	16,23	3,5,42	16,31	4,23,25,26	4,16,41	5,16,25	16,26
Polypropylene[c]	1,16,41	1,41	1,41	1,2	1,2,16,41	1,2,16,41	1,2,16,41
Polystyrene	16,41,42	3,5	5,31,36	25,31	5,16,31,36	2,16,16,25	16,31,36
Polyurethane	4	5,36	4,5	4,5	5,36	5,36	36
Polyvinyl chloride	4,5,26	4,5,42	4,5,16,41,42	3,4,15,26,36	42	4,5,15	4,26,36,42
Tetrafluoroethylene[c]	23	22	22	22,23	22,23	23	23

a—Adhesive number code is shown in Table III.
b—Cellulose acetate, cellulose acetate butyrate, cellulose nitrate.
c—Special surface treatment recommended.
Table developed by Bostik Div., USM Corp., and updated by the present authors.

Table 18-11B. Adhesive Types for Bonding Nonplastics to Thermosets.[a]

	Nonplastics						
Thermosets	**Ceramic**	**Fabric**	**Leather**	**Metal**	**Paper**	**Rubber**	**Wood**
Diallyl phthalate	5,24	36	31,36	31	31,36	31	31,36
Epoxy	23,31	4	4	23,31	4	4	23,31
Melamine	3,16	4	3,4	4	16,41,42	2,3,4,16	3,16
Phenolic	3,16	4	3,4	3,25	16,42	3,4,16,25	3,16,42
Polyester	3,26	4	5,26	5,26	41	1-5	3,26
Urea	4,16	4,42	3,4	3,4	16,42	1-5,16	3,16,42

a—Adhesive number code is shown in Table 18-11C.
Table developed by Bostik Div., USM Corp., and updated by the present authors.

1. Make the bonded area as large as possible
2. Make the maximum proportion of bonded area contribute to bond strength
3. Stress the adhesive in the direction of its maximum strength
4. Minimize stress in the direction in which the adhesive is weakest.

Generally, good design avoids types of loads and joints which concentrate stresses in a small area or on an edge. Because adhesives are usually strong in shear, joints which put the adhesive in shear are preferable. By concentrating on joints which place the loads in shear, the effects of peel or cleavage stresses are minimized. Of course, joints which place the adhesive under a compressive load are always acceptable. Figure 18-65 illustrates various types of stresses encountered in adhesive-bonded joints.

In designing a joint, a compromise must usually be made between use of the best possible joint and the practicalities of the expense of making the joint, from the standpoint of machining the joint, application of the adhesive and/or auxiliary pieces, etc. Figure 18-66 shows various lap-joint

Table 18-11C. Adhesive Number Code for Tables 18-11A,-B, and -D.

Elastomeric
1) Natural rubber
2) Reclaim
3) Neoprene
4) Nitrile
5) Urethane (also thermosetting)
6) Styrene butadiene

Thermoplastic resin
11) Polyvinyl acetate
12) Polyvinyl alcohol
13) Acrylic
14) Cellulose nitrate
15) Polyamide
16) Hot-melt copolymer blends (EVA, EEA, etc.)
25) Cyanoacrylate

Thermosetting resin
21) Phenol formaldehyde (phenolic)
22) Resorcinol phenolic
23) Epoxy
24) Urea formaldehyde
26) Reactive acrylate monomer systems
31) Butyral phenolic
33) Phenolic nylon
36) Polyester (also thermoplastic)
37) Anaerobic
38) Silicone

Miscellaneous
41) Rubber latices (natural or synthetic-water based)
42) Resin emulsions (water-based)

Table 18-11D. Adhesive Types for Bonding Plastics to Plastics.[a]

Surfaces	ABS	Acetal	Cellulosic[b]	Ethyl cellulose	Nylon	Phenylene oxide-based resin	Poly-carbonate	Polyether sulfone
ABS	4,5,23,25,26	4,16	4,16,36	5,16	5,23,25,26	5,23,25,26,36	5,25,26	5,25,26,36
Acetal	4,16	4,5,16,23,31	4,5,16	4,5,16	5,23	4,5,16,23	4,16	5,16,23
Cellulosic[b]	4,16,36	4,5,16	4,5,14,16,36	4,5,14,16,36	5,36	4,36	4,16	5,36
Ethyl cellulose	5,16	5,16	5,14,16	4,5,14,15,16, 36	5,36	4,5,36	5,16	5,36
Nylon	5,23,25,26	5,23	5,36	5,36	3,5,22,23,25, 26,36	3,5,23,26,37	25,26	5,25,26,36
Phenylene oxide-based resin	5,23,25,26,36	4,5,16,23	4,36	4,5,36	3,5,23,26,37	5,13,23,25, 26,36	4,16,25	23,25,36
Polycarbonate	5,25,26	4,16	4,16	5,16	25,26	4,16,25	4,15,16,25,26	25,26
Polyether sulfone	5,25,26,36	5,16,23	5,36	5,36	5,25,26,36	23,25,36	25,26	25,26
Polyethylene[c]	5,16	5,23,31	5,36	5,15,16,36	5,23,36	23,31,36	15,16	5,31,36
Polyethylene terephthalate	5,25,26,36	4,5	4,5,36	4,5,36	5,25,26,36	4,5,36	4,25,26	5,13,25,26
Polymethyl-methacrylate	25,26	5,31	5,36	5,36	5,25,26,36	5,31,36	25,26	5,13,25,31
Polyphenylene sulfide	23,25	4,5,16,23	4,14,16	4,16	5,22,23,25	23,25,26	4,15,16,25	23,25
Polypropylene[c]	5,16	5,23,31	5,36	5,36	5,23,36	23,31,36	15,16	5,31,36
Polystyrene	16,25	5,16,23,31	5,16,36	5,16,36	5,23,25,36	23,31,36	16,25	5,13,25,31
Polyurethane	4,25,36	4,5,23	4,5,36	4,5,36	5,23,36	4,23,36	4	5,25,36
Polyvinyl chloride	4,25,26	4,5	4,5,36	4,5,36	5,26,36	4,36	4,26	5,13,25,26
Tetrafluoro-ethylene	15,23	23	5	5,15	5,22,23	5,23	15	5
Diallyl phthalate	4,31	4,23,31	4,36	4,5,36	3,5,23,36	3,5,23,31,36	4	5,31,36
Epoxy	4,25	4,23,31	4,36	4,36	3,23,36	4,23,25,31,37	4	25,31,36
Melamine	4,16	4,16,23,31	4,16,36	4,16,36	3,23,36	4,23,31	4,16	23,31,36
Phenolic	5,16,25	4,5,16,23,31	4,5,16,36	4,5,16,36	3,5,23,25,36	3,23,31,37	4,16,25	5,25,31
Polyester	23,26,36	4,23,31	4,36	4,36	23,26,36	4,23,26,36,37	4,26	26,31,36
Urea	4,16	4,16,23,31	4,16	4,16	3,23	4,23,31	4,16	31

a—Adhesive number code is shown in Table III.
b—Cellulose acetate, cellulose acetate butyrate, cellulose nitrate.
c—Will require surface treatment.
Table developed by Bostik Div., USM Corp., and updated by the present authors.

Table 18-11D. (*Continued*)

Surfaces	Poly-ethylene[c]	Polyethylene terephthalate	Polymethyl-methacrylate	Polyphenylene sulfide	Polypro-pylene	Polystyrene	Polyurethane	Polyvinyl chloride
ABS	5,16	5,25,26,36	25,26	23,25	5,16	16,25	4,25,36	4,25,26
Acetal	5,23,31	4,5	5,31	4,5,16,23	5,23,31	5,16,23,31	4,5,23	4,5
Cellulosic[b]	5,36	4,5,36	5,36	4,14,16	5,36	5,16,36	4,5,36	4,5,36
Ethyl cellulose	5,15,16,36	4,5,36	5,36	4,16	5,36	5,16,36	4,5,36	4,5,36
Nylon	5,23,36	5,25,26,36	5,25,26,36	5,22,23,25	5,23,36	5,23,25,36	5,23,36	5,23,36
Phenylene oxide-based resin	23,31,36	4,5,36	5,31,36	23,25,26	23,31,36	23,31,36	4,23,36	4,36
Polycarbonate	15,16	4,25,26	25,26	4,15,16,25	15,16	16,25	4	4,26
Polyether sulfone	5,31,36	5,13,25,26	5,13,25,31	25,26	5,31,36	5,13,25,31	5,25,36	5,13,25,26
Polyethylene[c]	5,15,16, 23,31,36,41	5,36	5,31,36	5,23	5,23,31,36,41	5,23,31,36	5,23,36	5,36
Polyethylene terephthalate	5,36	4,5,13,25,26, 36	5,13,25,26,36	5,25,26	5,36	5,13,25,36	4,5,36	4,5,13,26,36
Polymethyl-methacrylate	5,31,36	5,13,25,26,36	2,5,6,13,25, 26,31,36	5,25,26,36	5,26,31	2,5,6,13,25, 31,36	5,36	5,13,26,36
Polyphenylene sulfide	5,23	5,25,26	5,25,26,36	23,25,26	5,23,26	5,23,25,36	5,23,36	5,25,26,36
Polypropylene[c]	5,23,31,36,41	5,36	5,26,31	5,23,26	5,15,16,23, 31,36,41	5,23,31,36	5,23,36	5,36
Polystyrene	5,23,31,36	5,13,25,26	2,5,6,13,25, 31,36	5,23,25,36	5,23,31,36	2,5,6,13,16, 23,25,36	5,23,36	5,13,36
Polyurethane	5,23,36	4,5,36	5,36	5,23,36	5,23,36	5,23,36	4,5,23,36	4,5,36
Polyvinyl chloride	5,36	4,5,13,26,36	5,13,26,36	5,25,26,36	5,36	5,13,36	4,5,36	4,5,11,13,26, 36,42
Tetrafluoro-ethylene	5,15,23	5	5	5,22,23	5,15,23	5,23	5,23	5
Diallyl phthalate	23,31,36	4,5,36	5,31,36	25,31,36	23,31,36	23,31,36	4,23,36	4,36
Epoxy	23,31,36	4,36	31,36	4,23,36	23,31,36	23,31,36	4,23,36	4,36
Melamine	16,23,31,36	4,36	31,36	4,23,36	23,31,36	16,23,31,36	4,23,36	4,36
Phenolic	5,16,23,31,36	4,5,25,36	5,25,31,36	4,16,25,26	5,23,31,36	5,16,23,25, 31,36	4,5,23,36	4,5,36
Polyester	23,31,36	4,26,36	26,31,36	4,26,36	23,31,36	23,31,36	4,23,36	4,26,36
Urea	16,23,31	4	31	3,23	23,31	16,23,31	4,23	4

a—Adhesive number code is shown in Table III.
b—Cellulose acetate, cellulose acetate butyrate, cellulose nitrate.
c—Will require surface treatment.

Surfaces	Tetrafluoro-ethylene[c]	Diallyl phthalate	Epoxy	Melamine	Phenolic	Polyester	Urea
ABS	15,23	4,31	4,25	4,16	5,16,25	23,26,36	4,16
Acetal	23	4,23,31	4,23,31	4,16,23,31	4,5,16,23,31	4,23,31	4,16,23,31
Cellulosic[b]	5	4,36	4,36	4,16,36	4,5,16,36	4,36	4,16
Ethyl cellulose	5,15	4,5,36	4,36	4,16,36	4,5,16,36	4,36	4,16
Nylon	5,22,23	3,5,23,36	3,23,36	3,23,36	3,5,23,25,36	23,26,36	3,23
Phenylene oxide-based resin	5,23	3,5,23,31,36	4,23,25,31,37	4,23,31	3,23,31,37	4,23,26,36,37	4,23,31
Polycarbonate	15	4	4	4,16	4,16,25	4,26	4,16
Polyether sulfone	5	5,31,36	25,31,36	23,31,36	5,25,31	26,31,36	31
Polyethylene[c]	5,15,23	23,31,36	23,31,36	16,23,31,36	5,16,23,31,36	23,31,36	16,23,31
Polyethylene terephthalate	5	4,5,36	4,36	4,36	4,5,25,36	4,26,36	4
Polymethyl-methacrylate	5	5,31,36	31,36	31,36	5,25,31,36	23,31,36	31
Polyphenylene sulfide	5,22,23	23,31,36	4,23,36	4,23,36	4,16,25,26	4,26,36	3,23
Polypropylene[c]	5,15,23	23,31,36	23,31,36	23,31,36	5,23,31,36	23,31,36	23,31
Polystyrene	5,23	23,31,36	23,31,36	16,23,31,36	5,16,23,25, 31,36	23,31,36	16,23,31
Polyurethane	5,23	4,23,36	4,23,36	4,23,36	4,5,23,36	4,23,36	4,23
Polyvinyl chloride	5	4,36	4,36	4,36	4,5,36	4,26,36	4
Tetrafluoro-ethylene[c]	5,15,22,23	5,23	23	23	5,23	23	23
Diallyl phthalate	5,23	3,4,5,23,31, 36	3,4,23,31,36	3,4,23,31,36	3,4,23,31,36	4,23,31,36	3,4,23,31
Epoxy	23	3,4,23,31,36	3,4,23,31,36	3,4,23,31,36	3,4,23,31,36	4,23,31,36	3,4,23,31
Melamine	23	3,4,23,31,36	3,4,23,31,36	3,4,16,23,31, 36	3,4,16,23,31, 36	4,23,31,36	3,4,16,23,31
Phenolic	5,23	3,4,23,31,36	3,4,23,31,36	3,4,16,23,31, 36	3,4,5,16,23, 25,31,36	4,23,31,36	3,4,16,23,31
Polyester	23	4,23,31,36	4,23,31,36	4,23,31,36	4,23,31,36	4,23,26,31,36	4,23,31
Urea	23	3,4,23,31	3,4,23,31	3,4,16,23,31	3,4,16,23,31	4,23,31	3,4,16,23,31

a—Adhesive number code is shown in Table III.
b—Cellulose acetate, cellulose acetate butyrate, cellulose nitrate.
c—Will require surface treatment.

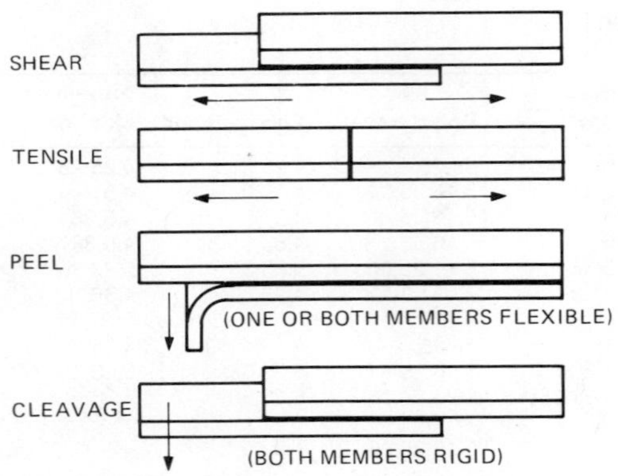

Fig. 18-65. Basic stresses in adhesive bonds.

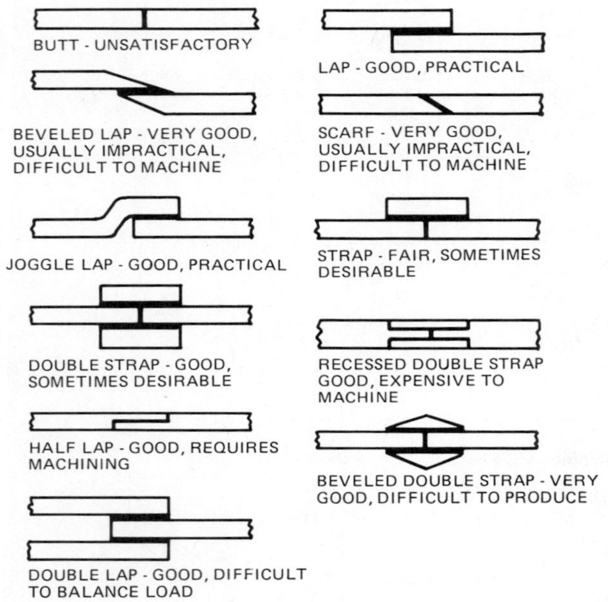

Fig. 18-66. Lap joint configurations. (*Courtesy: Plastics World*, June 17, 1974)

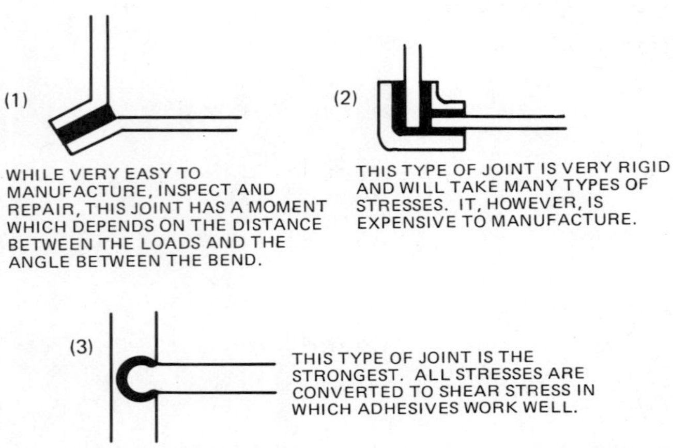

Fig. 18-67. Perpendicular joint configurations. (*Courtesy: Plastics World*, June 17, 1974)

configurations and Figure 18-67 shows perpendicular joints. Both figures include information on expense of production.

REFERENCES AND BIBLIOGRAPHY

Mechanical Fasteners

Aronson, R. B. How fasteners get a good grip in plastics. *Machine Design*, September 21, 1978, **50**(18):126.

Beck, R. D. *Plastic Product Design*. Van Nostrand Reinhold Co., New York, 1970.

Baechtle, C. R. With simple, integral designs, "hardware" joins the plastics revolution. *Plastics World*, September 1972, **30**(9):45.

Boots Aircraft Nut Division, Townsend Co. *Banc-Lok, The Modern Fastener*. Beaver Falls, Pa., May 1965.

Buchanan, T. C. Riveting plastics–a guide to good assembly practice. *Plastics Design & Processing*, March 1967, **7**(3):14.

Camcar Screw & Manufacturing. *Plastite by Camcar*. Form TT-10-475.

C. E. M. Co., Inc. *CEM, Designing with Self-Tapping Inserts*. Bulletin No. T-478.

Cheney, A. J., and Ebeling, W. E. Joining plastics parts. *SPE Journal*, March 1958, **14**(3):31.

Dodge, G. E. A look at post molding inserts. *Plastics World*, January 1964, **22**(1).

Du Bois, J. H. Inserts and fasteners. *Plastics World*, July 1960, **18**(7):20.

Duffy, W. B. Mechanical fasteners for plastics. *Plastics World*, October 21, 1974, **32**(10):46.

Edwards, C., and Smith, E. Seven mechanical fasteners speed assembly of thermoplastics. *Materials Engineering*, December 1974, **80**(7):24.

Ehner, W. J. Techniques for joining metal inserts and molded plastics. *Plastics World*, November 1964, **22**(3):28.

Elco Industries, Inc. *Plasform Threadforming Screws for Plastics*. Publication No. E-1045/85.

Harris, W. D. *Joining of Plastics*. Interpretive Reports Committee of the Welding Research Council, 1970.

Heli-Coil Corporation, Phelps Manufacturing Division. *Standard Dodge Expansion Inserts*. Bulletin P-101-116, July 1, 1963.

IN-X Fastener Corporation, Newark, N.J. *Self-Clinching Nuts*.

——. *Sonic-Lok Threaded Ultrasonic Inserts*. Catalog TI-1272, December 1971.

Linebarrier, H. L. Adhesives and fasteners join in rivet-bonded assemblies. *Materials Engineering*, December 1975, **82**(7):35.

Machine Design. Plastic joining. Fastening and Joining Reference Issue, November 18, 1976, **48**(22).

Materials Engineering. Seven mechanical fasteners speed assembly of thermoplastics. December 1974, **80**(7):24.

——. Basic guide to fasteners. April 1973, **77**(4):48.

——. How to fasten and join plastics. March 1971, **73**(3):32.

Materials in Design Engineering. Riveting plastics is fast, economical. September 1959, **50**(3):12.

Miska, K. H. Plastic fasteners are light, insulating, nonmagnetic and cut corrosion. *Materials Engineering*, February 1976, **83**(2):34.

Oest, L. Tapping screws for plastics assemblies. *Plastics World*, August 1963, **21**(8):44.

Phelan, C. S. The forgotten corner of bonding–how sandwich panels are attached. *Adhesives Age*, September 1974, **17**(9):23.

Plastics Design & Processing. Stapling meets assembly requirements for all-terrain vehicles. February 1971, **11**(2):17.

Plastics World. Fasteners for plastics–a wealth of choices. June 1979, **37**(6):50.

——. Plastic fastener system puts it all together. October 1971, **29**(10):31.

Price, G. W. Developments in insert design for thermoset fastening. *Plastics Design & Processing*, March 1979, **19**(3):13.

Product Engineering. Fastener study shows what's best for fiberglass sheet molding compounds. June 1977, **48**(6):36.

Sprow, E. E. The cost of fastening plastics. *Machine Design*, July 27, 1972, p. 72.

Strasser, F. Inserts that stay put. *Machine Design*, November 11, 1976, **48**(21):142.

Strauss, E. L. Fastening reinforced plastics: how to design mechanical joints. *Materials in Design Engineering*, **57**(2).

Weiss, M. D. How to use mechanical fasteners with reinforced plastics. *Materials in Design Engineering*, June 1959, **49**(6):84.

Yardley Products Corporation. *Yardley Fiber-Sert*. Bulletin No. 4300.

——. *Yardley Zap-Serts*. Bulletin No. 4500, June 1976.

——. *Yardley Trisert Inserts*. Bulletin No. 4600, August 1976.

Hot Gas and Miscellaneous Welding Processes

Celanese Plastics Co. *Welding Celcon Acetal Copolymer*. Bulletin C3F, October 1964.

Harris, W. D. *Joining of Plastics*. Interpretive Reports Committee of the Welding Research Council, 1970.

Hurley, D. Design potentials of heat-welding plastic parts. *Plastics Engineering*, **34**(4):29.

——. "Welding Thermoplastics." In *Modern Plastics Encyclopedia*, McGraw-Hill, Inc., New York, 1971–72.

Iron Age. Precise control is key to welding of plastics. May 3, 1976, **27**(18):45.

Kaminsky, S. J. "Welding Thermoplastics." In *Modern Plastics Encyclopedia*, McGraw-Hill, Inc., New York, 1976–77.

Kaminsky, S. J., and Williams, J. A. *Handbook for Welding and Fabricating Thermoplastic Materials*. Kamweld Products Co., Norwood, Mass.

Kamweld Products Co. *Guide to Plastics Welding*.

Nicholson, C. K. Use heat to join heat-resistant plastic films. *Materials Engineering*, February 1963, **77**(2):71.

Plastics Design & Processing. Using heat guns and welding torches in plastics fabrication. November 1970, **10**(11):24.

Prost, H. W. Heat welding–a good way to make complex assemblies. *Plastics Design & Processing*, November 1973, **13**(11):18.

Shyman, M. L. Welding clear plastics–pre-engineering spots the glitches. *Plastics World*, June 1979, **37**(6):86.

Seeyle Plastics Co., Minneapolis, Minn. *All About Welding Plastics*.

Stemmler, R. M. How to spin-weld acetals. *Plastics Technology*, May 1963, **9**(5):42.

Werner & Pfleiderer Corporation. *A Designer's Guide to Welding, Joining and Assembling Plastics Components*. Bulletin WP-W-PTG-78-1, January 1978.

Ultrasonic and Vibration Welding

Bongo, R. Weld joints for plastics. *Machine Design*. April 21, 1977, **49**(8):275.

Branson Instruments, Inc. *A Guide to Ultrasonic Plastics Assembly*. 1973.

Branson Sonic Power Co., Danbury, Conn. *Joint Design for Ultrasonic Welding*. Bulletin 56a, 1967.

Deans, H. *Molded Part Design Considerations for Ultrasonic Fabrication of Molded Plastic Parts*. Ultra Sonic Seal Division, Kleer Vu Industries, Inc. 1965.

Dreger, D. R. Welding plastics by vibration welding. *Machine Design*, December 26, 1974, **46**(24):42.

Kish, F. A. How to ultrasonically weld the new engineering thermoplastics. *Plastics Technology*, Part 1: Part design and welding techniques. September 1972, **18**(9):40. Part 2: Horn design and fabrication. December 1972, **18**(12)41.

Kolb, D. J. Designing plastic parts for ultrasonic assembly. *Machine Design*, March 16, 1967, **39**(3).

——. Assembling thermoplastics by ultrasonic vibration. *SPE Journal*, November 1966, **22**(11):21.

Materials in Design Engineering. Ultrasonics joins plastics quickly and simply. September 1966, **64**(3):112.

Mengason, J. "New Designs through Vibration Welding." Paper presentation, *SAE International Automotive Engineering Conference*, Detroit, Michigan, 1977.

——. Vibration welding–a more versatile way to joint plastics. *Plastics Engineering*, July 1976, **32**(7):33.

Miller, B. Ultrasonics–the sound approach to plastics assembly. *Plastics World*, December 16, 1974, **32**(12):64.

Miller, E. M., and Craft, D. I. Ultrasonically weld the TP polyesters. *Plastics Technology*, October 1974, **20**(10):44.

Modern Plastics. New advances in efficient welding from ultrasonic's horns of plenty. October 1976, **54**(10):50.

——. In films, sealing the unsealable. October 1976, **54**(10):42.

——. Ultrasonics can join BIG parts, too. October 1968, **45**(10).

Obeda, E. How to get good ultrasonic welds. *Modern Plastics*, November 1964, **41**(11):130.

Plastics Design & Processing. Ultrasonics for ultra fast assembly. October 1970, **10**(10):16.

——. Ultrasonics simplify attachment of metal fasteners and parts to plastic moldings. May 1965, **5**(5):14.

Plastics Technology. Three new ways to fabricate plastics–ultrasonically. August 1978, **24**(8):37.

——. Welding rigid spheres: good application for ultrasonic bonding. May 1964, **10**(5):11.

Plastics World. Vibration welding could make that "impossible" design possible. September 1976, **34**(9):23.

——. Way-out sound makes long, clean welds, cuts out post-assembly operations. October 1972, **30**(10):60.

——. Designing for ultrasonic bonding demands special know-how. June 1970, **28**(6):50.

Port, C. O. Inserting metal parts ultrasonically. *Modern Plastics*, February 1974, **51**(6):62.

Product Engineering. Ultrasonic joining gains favor with better equipment and know-how. January 1977, **48**(1):37.

——. Vibration technique joins large, complex plastic components. May 1977, **48**(5):64.

——. Ultrasonics is newest way to join plastic to metal. May 23, 1966, **37**(5):101.

Scarpa, T. J. Joining plastics with ultrasonics. *Plastics Technology*, January 1962, **8**(1):22.

Sherry, J. R. Thermoplastic compatibility: the key to successful ultrasonic assembly. *Plastics Design & Processing*, April 1974, **14**(4):10.

——. Plastic welding with ultrasonics: new technique enhances value of plastics in industry. *Engineer/Scientist*, May 1969, p. 35.

Soloff, R. S. New concepts in ultrasonic sealing. *Modern Plastics*, March 1964, **41**(3):125.

Sonics & Materials, Inc. *Horns for Ultrasonic Plastic Assembly Applications*. 1972.

——. *Horn Tips for Ultrasonic Plastic Assembly Applications*. Technical Bulletin 7401.

Electromagnetic Induction Welding

Capolongo, P. Plastics and induction heating. *Plastics World*, November 1963, **3**(11):48.

Chookazian, M. Electromagnetic bonding takes its place among conventional methods for bonding and sealing rigid thermoplastics. *Plastics Design & Processing*, May 1978, **18**(5):43.

——. "Design Criteria for Electromagnetic Bonding of Thermoplastics." Paper presentation, *Society of Plastics Engineers National Technical Conference*, Cleveland, Ohio, October 6, 1976.

——. Electromagnetic bonding of thermoplastics: another method of joining like and unlike plastics. *Plastics Design & Processing*, September 1975, **15**(9):18.

——. Bonding plastics by induction heating. *SPE Journal*, October 1970, **26**(10):49.

EMABond, Inc. Englewood, N.J. *Electromagnetic Bonding of Thermoplastics*.

Hellerbond, Columbus, Ohio. *Hellerbond, New Technology for Joining Materials*.

Plastics Design & Processing. An overview of techniques for heat joining plastic parts. September 1979, **19**(9):38.

Plastics World. Electromagnetic bonding–it's fast, clean and simple. July 1970, **28**(7):50.

Product Engineering. Heat generated by magnetic field welds plastics together. September 1971, **42**(9):34.

RF and Hot Tool Heat Sealing

Arutt, K. "High Frequency Heat Sealing/Embossing." In *Modern Plastics Encyclopedia*, McGraw-Hill, Inc., New York, 1975–76.

Black, P. Dielectric heating and sealing today. *Plastics Technology*, August 1967, **13**(8):33.

Campbell, G. Improved cycle times, reliable seals show bright future for fusion bonding. *Plastics Design & Processing*, February 1978, **18**(2):37.

Castagna, E. G. High frequency bonding: Part 1. *Plastics Technology*, March 1964, **10**(3):32.

Farkas, R. D. Electronic heat-sealing. Modern Plastics, March 1958, **35**(7):109.

Knight, R. M., and Funk, W. U. Heat sealing of polyethylene film. *Modern Plastics*, December 1957, **35**(4):133.

Landvatter, G. Heat-sealing PE films. *Plastics Engineering*, November 1973, **29**(11):42.

McKelvey, J. M., and Strome, T. H. Sealability of polyethylene films. *Modern Plastics*, June 1959, **36**(10):107.

Modern Plastics. Happy birthday, dielectric sealing: you're $750 million old today. *Modern Plastics*, June 1969, **46**(10):64.

Nicholson, C. K. Use heat to join heat resistant plastic films. *Materials Engineering*, February 1973, **77**(2):71.

Sacharow, S. Heat sealing of plastic films. *Adhesives Age*, November 1970, **13**(11):8.

——. Sealing flexible plastic films. *Plastics Design & Processing*, December 1966, **6**(12):18.

Scheiner, L. L. High frequency bonding: Part 2. *Plastics Technology*, April 1964, **10**(4):45.

Sealomatic Electronics Corporation, Brooklyn, N.Y. *Electronic Sealing*.

U.S. Industrial Chemicals Co. *Polyethylene Processing Tips*. Series V, No. 5, September 1960.

U.S. Industrial Film Products. *Heat Sealing Polyolefin Films*. Bulletin PL 19-265, 1965.

Adhesives and Adhesive Bonding

Adhesives Age. Second-generation acrylic adhesives. September 1976, **19**(9):21.

Aronson, R. B. Adhesives are getting stronger in many ways. *Machine Design*, February 8, 1979, **51**(3):54.

——. PABST: new reliability for adhesive bonding. *Machine Design*, April 6, 1978, **50**(7):20.

Barie, W. P. Jr., and Franke, N. W. High temperature epoxy adhesives based on BTDA. *Adhesives Age*, October 1971, **14**(10):36.

Bittence, J. C. Engineering adhesives. *Machine Design*, June 10, 1976, **48**(11):92.

Blatz, P. S. NR-150 polyimide precursor adhesive solution developed. *Adhesives Age*, September 1978, **21**(9):39.

Bragole, R. A. "The Adhesive Bonding of Polyolefins." Paper presentation, *103rd Meeting of the Rubber Division of the American Chemical Society*, Detroit, Michigan, May 1–4, 1973.

——. *Factors Affecting the Adhesion of Urethanes to Plastic Substrates*. USM Corporation.

Brumit, T. M. Cyanoacrylate adhesives–when should you use them? *Adhesives Age*, February 1975, **18**(2):17.

Bryson, F. E. Heat resistant adhesives. *Machine Design*, June 15, 1972, **44**(14):124.

Carroll, K. W. How to apply adhesives. *Product Engineering*, November 22, 1965, **36**(11):93.

——. Adhesives for bonding foams. *Plastics World*, May 1961, **19**(5):40.

Carter, J. S. Using automatic dispensers to apply discrete amounts of adhesive. *Adhesives Age*, May 1979, **22**(5):27.

Coover, H. W. Jr. Advances in polymer chemistry provide key to new adhesives. *Adhesives Age*, February 1968, **11**(2):20.

Coover, H. W. Jr., Joyner, F. B., Shearer, N. H. Jr., and Wicker, T. H. Jr. Chemistry and performance of cyanoacrylate adhesives. *SPE Journal*, May 1959, **15**(5):413.

Crane, L. W. Hammermesh, I., and Maus, L. Surface treatment of cured epoxy composites to improve adhesive bonding. *SAMPE Journal*, March/April 1976, **12**(2):6.

Degisi, S. L., and Smith, C. H. Improved bondability of rigid urethane foam by plasma treatment. *Adhesives Age*, July 1978, **21**(7):35.

Delmonte, J. Bonding thermoplastics with improved urethanes. *Adhesives Age*, April 1968, **11**(4):27.

De Lollis, N. J. Durability of structural adhesive bonds: a review. *Adhesives Age*, September 1977, **20**(9):41.

——. Theory of adhesion, mechanism of bond failure and mechanism of bond improvement. Part 1: Evolution and present status of the theories of adhesion. *Adhesives Age*, December 1968, **11**(12):21.

Dreger, D. R. Joining advanced composites. *Machine Design*, May 8, 1980, **52**(9):89.

——. Hot melt adhesives put it all together. *Machine Design*, January 9, 1975, **47**(1):88.

Eastman Chemical Products, Inc. *Eastobond Hot Melt Adhesives*. Publication No. PA-5A, ECP 1741, October 1974.

Epstein, G. Adhesives for space systems. Part 1: Requirements and applications. *Adhesives Age*, June 1972, **15**(6):18.

Epstein, G., Smallen, H., and Carter, R. L. "Bonding of Dissimilar Materials for Space Systems Applications." Air Force Report No. SAMSO-TR-72-180, Aerospace Report No. TR-0073 (3250-20)-1, July 31, 1972.

Fabian, R. J. 41 new adhesives for bonding parts. *Materials Engineering*, July 1970, **72**(1):34.

Gould, B. Guide to adhesives selection with a pictorial checklist. *Adhesives Age*, October 1973, **16**(10):19.

——. How to get lower costs and higher quality in the bonding of reinforced plastic/honeycomb sandwich panels. *Plastics Design & Processing*, January 1963, **3**(1):26.

Graham, J. A. Structural adhesive bonding. *Machine Design*, October 7, 1976, **48**(19):118.

——. Adhesives that can live with oil. *Machine Design*, December 8, 1977, **49**(23):183.

Graham, J. A., and O'Connor, J. E. Epoxy with low-temperature cure and high-temperature properties developed. *Adhesives Age*, July 1978, **21**(7):20.

Guha, P. K., and Epel, J. N. Adhesives for bonding graphite/glass composites. *Adhesives Age*, June 1979, **22**(6):31.

Hart-Smith, L. J. Designing adhesive bonds. *Adhesives Age*, October 1978, **21**(10):32.

Hauck, J. E. New high temperature adhesives: easier to apply, stand more heat. *Materials Engineering*, April 1967, **65**(4):84.

Hauser, R. L. Ultra adhesives for ultrasonic bonding. *Adhesives Age*, March 1969, **12**(3):26.

Herot, F. C. Engineered dispensing systems for adhesives. *Adhesives Age*, May 1980, **23**(5):23.

Hoehn, R. Non-woven textiles as adhesive carriers. *Adhesives Age*, December 1970, **13**(12):28.

Huber, H. F., and Stroble, H. G. Applications and future uses of capsular adhesives. *Adhesives Age*, November 1967, **10**(11):28.

Ishai, O., and Girshengorn, T. Strength of bonded aluminum-CFRP single lap joints. *Adhesives Age*, July 1978, **21**(7):25.

Jackson, L. C. Improving adhesion by gas-plasma contaminant removal. *Adhesives Age*, September 1978, **21**(9):34.

Kapsh, J. P. Anaerobics: novel chemical locks. *Product Engineering*, November 1978, **49**(11):60.

Kuhbander, R. J., and Aponyi, T. J. New concepts in fatigue-resistant adhesives. *Adhesives Age*, September 1976, **19**(9):27.

Kutscha, D. Mechanisms of bonding and joint performance. *Adhesives Age*, November 1968, **11**(11):37.

MacDonald, N. C. Standard test methods for adhesives. *Adhesives Age*, September 1972, **15**(9):21.

MacGrandle, T. D. Film adhesives. *Machine Design*, July 8, 1971, **43**(13):77.

Machine Design. Fastening and Joining Reference Issue—Adhesives. November 17, 1977, **49**(22):158.

Martin, R. A. How to apply adhesives. *Materials Engineering*, July 1970, **72**(1):38.

Materials Engineering. Adhesives and sealants. May 1976, **83**(5):73.

——. High temperature adhesives are easy to apply, usable up to 700°F. May 1973, **77**(5):60.

——. Hot melt adhesives take seconds. November 1972, **76**(6):74.

Miska, K. H. Adhesives and sealants take the heat up to 600°F. *Materials Engineering*, May 1977, **85**(5):57.

——. Hot melt adhesives are easy to apply, set rapidly, bond many materials. *Materials Engineering*, June 1977, **85**(6):52.

——. Second generation acrylic adhesives speed processing at lower cost. *Materials Engineering*, November 1976, **84**(6):40.

——. Which low temperature adhesive is best for you? *Materials Engineering*, May 1975, **81**(5):52.

——. Try solventless adhesives to protect environment and worker. *Materials Engineering*, November 1975, **82**(6):50.

Modern Plastics. Now: adhesives for large-part assembly. October 1978, **55**(10):40.

Monsanto Chemical Co. *Cementing Monsanto Lustrex Styrene*. Product Information Bulletin 1001.

O'Malley, W. J. Silicone pressure-sensitive adhesives for flexible printed circuits. *Adhesive Age*, June 1975, **18**(6):17.

Palermo, F. C., and Korpman, R. Extrusion coating with pressure-sensitive adhesives. *Adhesives Age*, September 1978, **21**(9):25.

Pascuzzi, B. Bonding supersonic aircraft with polyimide resin systems—a study of thermal aging. *Adhesives Age*, February 1971, **14**(2):26.

Pearce, M. B. Jr. An introduction to anaerobics. *Materials Engineering*, July 1973, **78**(1):50.

Peters, R. A., and Logan, T. J. Microvoid epoxy adhesives for high peel and shear strength. *Adhesives Age*. April 1975, **18**(4):17.

Petrie, E. M. Joining the "engineering" plastics—how adhesives compare with other bonding methods. *Adhesives Age*, August 1980, **23**(8):14.

Phelan, C. S. The forgotten corner of bonding, how sandwich panels are attached. *Adhesives Age*, September 1974, **17**(9):23.

Phillips Petroleum Co. *Solprene . . . Radial Teleblock Copolymers for Adhesives*. Bulletin 4950.

——. *Contact Adhesives Based on Radial Block Copolymers of Butadiene and Styrene*. Bulletin 302.

Plastics Design & Processing. Three prime factors in adhesive bonding of plastics, (1) adhesive selection, (2) surface preparation, (3) applying the adhesive. June 1968, **8**(6):10.

Plastics Technology. New way to void-free bond lines. March 1965, **11**(3).

Plastics World. Adhesive joint design for plastics applications. June 17, 1974, **32**(6):62.

Product Engineering. Key design factors make adhesive bonds more reliable. February 1977, **48**(2):38.

——. Simple acrylic adhesive systems bond oily metals at room temperature. July 1977, **48**(7):23.

——. Adhesive performance is important but not everything, says expert. June 1976, **47**(6):39.

——. Heat-resistant meltable adhesives solve processing dilemmas. December 1976, **47**(12):21.

——. Wide range of adhesives meet all plastic and metal assembly needs. April 1975, **46**(4).

——. Should your design be joined with a structural adhesive? July 1973, **44**(7):19.

——. Instant adhesives gain ground on other fastening methods. September 1971, **42**(9):40.

Raia, D. C. Adhesives—the king of fasteners. *Plastics World*, June 17, 1974, **32**(6):59.

Rauhaut, H. W. Assembling plastic parts with adhesives. *Plastics Design & Processing*. Part 1: April 1971, **11**(4):22. Part 2: May 1971, **11**(5):24.

Reinhart, T. J. Jr. Engineering properties of adhesives. *Adhesives Age*, July 1973, **16**(7):35.

Reinhart, T. J. Jr., and Scardino, W. M. Composite to metal bonding using structural thermosetting adhesives. *Adhesives Age*, February 1975, **18**(2):23.

Ross, M. C., McAbee, A., and Bodner, M. J. Effect of surface exposure time on bonding of engineering thermoplastics. *SAMPE Journal*, May/June 1976, **12**(3):15.

Schneberger, G. L. Basic concepts of adhesive bonding. *Adhesives Age*, January 1980, **23**(1):42.

——. Chemical aspects of adhesive bonding. *Adhesives Age*. Part I: Chemical concepts. February 1970, **13**(2):21. Part II: Physical principles. March 1970, **13**(3):28. Part III: Chemistry in the bonding shop. April 1970, **13**(4):41.

——. Polymer structure and adhesive behavior. *Adhesives Age*, April 1974, **17**(4):17.

Skiest, I. (ed.) *Handbook of Adhesives*. 2nd Ed. Van Nostrand Reinhold Co., New York, 1977.

Smith, D. R. How to prepare the surface of metals and non-metals for adhesive bonding. *Adhesives Age*, March 1967, **10**(3):25.

Smith, J. C. Silicone adhesives for joining plastics. *Adhesives Age*, June 1974, **17**(6):27.

Snogren, R. C. *Handbook of Surface Preparation*. Palmerton Publishing Co., Inc., New York, 1974.

——. Selection of surface preparation processes. *Adhesives Age*. Part 1: Process selection guide, July 1969, **12**(7):26. Part 2: Appendix of surface preparation processes. August 1969, **12**(8):36.

St. Clair, A. K., Slemp, W. S., and St. Clair, T. L. High temperature adhesives for bonding polyimide film. *Adhesives Age*, January 1979, **22**(1):35.

Sterman, S., and Toogood, J. B. How to promote adhesion with silicones and silanes. *Adhesives Age*, July 1965, **8**(7):34.

Stokes, E. P. Adhesives for the electrical/electronic industry. *Insulation/Circuits*, September 1970, **16**(9):51.

3M Co. *Jet Melt Bonding System Application Equipment and Accessories*. Bulletin Z-JEF (92.25)MP.

Toy, L. E. Plastics/metals: can they be united? *Adhesives Age*, October 1974, **17**(10):19.

Thrall, E. W. Jr. PABST program test results. *Adhesives Age*, October 1979, **22**(10):22.

Trauernicht, J. O. Bonding and joining: weigh the alternatives. *Plastics Technology*. Part 1: Solvent cements, thermal welding. August 1970, **16**(8):43. Bonding and joining still an art. Part 2: Adhesives and mechanical fasteners. September 1970, **16**(9):49.

Tremain, H. E. Stearic acid for PE-metal bonds. *Plastics Technology*, February 1963, **9**(2):11.

Union Carbide Corporation. *Adhesives Intermediates*. Bulletin F-42325, April 1970.

Vaughn, R. E. Bonding failure mechanisms and their cure. *Adhesives Age*, September 1976, **19**(9):37.

Vaughn, R. W., Sheppard, C. H., and Baucom, R. Polyimide adhesives for weld-bonding titanium. *Adhesives Age*, July 1977, **20**(7):19.

Weber, C. D., and Gross, M. E. Modified epoxy adhesives speed honeycomb bonding. *Materials Engineering*, May 1974, **79**(5):92.

Weggemans, D. M. Adhesives application charts. *Adhesives Age*, October 1973, **16**(10):31.

Wehrenberg, R. H. II. Adhesives that conduct electricity. *Materials Engineering*, January 1979, **89**(1):38.

Zalucha, D. New acrylics structurally bond unprepared metals. *Adhesives Age*, February 1979, **22**(2):21.

19
Miscellaneous Processes

CASTING

Casting is one of the oldest and simplest processes for producing parts from liquid or molten materials. Essentially it is a low-pressure process requiring little in the way of tooling except a container in the shape of the desired part, and possibly ovens, if the resin is a heat-curing material, or one which has to be molten prior to casting. An important reason for the use of the casting process is that extremely large parts can be made with relatively inexpensive tooling.

Materials

While not all plastics materials can be cast, there are both thermoplastic polymers and thermosets which are castable. Of the thermoplastic materials, acrylics, nylons, urethanes, and PVC plastisols are the materials in most general use. In the thermoset field the materials in the greatest use (at the time of writing) are phenolics, polyesters, epoxies, silicones, and urethanes. (Urethanes are included as both thermoplastics and thermosets because the percentage of cross-linking agent employed will determine its characteristics.)

Thermoplastics

Acrylics. Methyl methacrylate liquid monomer, or a monomer-polymer mixture catalyzed with an organic peroxide, is employed in the production of acrylic sheet stock used for signs, aircraft canopies, display cases, and items such as rods, tubes, etc. While sheet stock can be made by extrusion, or even by injection molding, such processes inevitably result in products showing stress marks, striations, etc. Thus, for production of sheets of the highest optical quality, casting is the production process most often used.

One method of sheet production involves casting the monomer-polymer mix between two plates of polished glass, separated by a resilient gasketing material. The resilient gasket and a variable ratio of polymer in the mixture are used to compensate for the high shrinkage (up to 20% for some thicknesses) during the polymerization reaction. Sheets made by this process are necessarily limited in size to the area of the largest glass plates available—currently up to 120 inches in width and 168 inches long. Thicknesses of the acrylic sheets made by this process range from 0.030 to 4.25 inches. Not all thicknesses are available in all the lengths and widths given.

Sheet materials are also produced in a continuous process by several manufacturers. In this technique the catalyzed casting material, a monomer-polymer slurry, is introduced into the space between two highly polished, moving stainless steel belts. The thickness of the sheet is determined by the thickness of a resilient gasket used to space the belts apart, and the sheet width is determined also by the spacing between the gaskets on each side of the belts.

In both production processes, polymerization of the cast liquid is done under carefully controlled conditions of heating. The cell-cast materials are either oven, autoclave, or liquid heated at temperatures up to approximately 100°C (200°F). The temperature cycles vary with the size and thickness of the sheet, to keep the exotherm under control. A runaway reaction could, of course, produce bubbles and other nonhomogeneities in the sheet. After the heating cycle, the sheet is cooled between the plates and then removed as a finished product, which usually only requires trimming. Sheets made by the continuous process are polymerized in a similar manner by running the belts through several heat zones and ultimately cooling the material so it can be removed. These sheets can be as long as 600 feet and have been made as wide as 108 inches. For materials of the highest optical quality, such as for military aircraft, final processing may consist of flat bed polishing in a manner similar to that used in production of plate glass.

Rod and tube stocks are also made by casting. Rods are cast into metal molds or into tubular plastic bags made of film, such as polyethylene lay-flat tubing. Tubes are usually made by centrifugally casting. In this process, a given amount of acrylic mix is placed into a tube or pipe which is rotated during the heating cycle. After removal from the molds, both the rods and tubes are usually centerless ground and polished to produce the highest optical quality.

Embedments of biological specimens or small objects for decorative purposes are made in a manner similar to cell casting of sheet, except for a few special steps required because of the embedded part. If a hard part such as a metal paperweight or a display item is desired, then methyl methacrylate casting mix is used. Butyl methacrylate, which results in a soft embedment, is used if the finished part is to be sliced in a microtome for microscopic examination.

To embed a specimen, a small amount of the monomer-polymer mix is placed in the holding receptable which has the depth required to hold the specimen. This amount is then degassed, under vacuum; heated, at atmospheric pressure, to approximately 35-50°C (100-120°F); and held at this temperature until gelation occurs. The specimen, which has been thoroughly dried, is then placed on the gelled base. Additional casting material is then added, degassed, and placed into an oven or autoclave for the final cure. After cure, the casting should be removed from the mold and annealed at 110°C (225°F) for 2 hours for each $\frac{1}{2}$ inch of thickness.

In general acrylic castings are made without fillers which would interfere with the transparency of the final casting. However, all types of raw stock, sheet, rods, tubes and special shapes can be obtained in a great variety of dyed transparent or translucent colors.

Nylon. Nylon 6, caprolactam, can be cast as a catalyzed liquid monomer to form very large parts, with weights up to 1500 pounds. The process, which is proprietary, consists of *in-situ* polymerization, and results in a finished part with tolerances of ±1% or ±0.030 inch, after shrink allowances are made. To prevent excessive shrinkage or warping, the length-to-thickness ratio is generally kept to 20:1 or less. A refinement on the conventional gravity casting technique is the use of low-pressure closed molds. Parts made by the low-pressure process show less shrinkage effects and allow greater length-to-thickness ratios and somewhat more intricate designs than can be made at atmospheric pressure. Because of the relatively low costs of the tooling, of either type, short runs to produce massive parts have been economically feasible. *In-situ* rotational molding to produce hollow parts such as tubing, fuel tanks, etc., can also be done. Relatively little heat is required to initiate polymerization, and since the exotherm is slight, control is very simple.

Materials used for nylon casting include both unfilled monomer and filled monomer. Unfilled castings are used for applications such as heavy-duty bearings, gears, propellors, etc. Mineral-filled mixtures have also been cast where a stiffer, less resilient material is required. Filling of the monomer also results in lower shrinkage, which allows parts to be made to tolerances approaching those of injection molding. Figures 19-1, 19-2, and 19-3 show some of the types of parts made by casting of nylon.

PVC Plastisols. Plastisols, described more fully in Chapter 5, are medium- to high-viscosity dispersions of polyvinyl chloride in liquid plasticizers. Since only the application of heat is required to change the liquids to a solid, they are very simple to use in casting small parts, or in related operations of dip coating, rotational molding, slush molding, spread coating, and pouring. In each of these processes, liquid in a mold or spread onto a fabric is heated to the fusion temperature, at which time the dispersion gels and after a short time solidifies. The fusion temperature ranges from 135°C (275°F) to as high as 175°C (350°F). Since reaching this temperature takes an appreciable time, molds containing solid castings are not too large or the time can become excessive enough to make such castings uneconomical. However, for fabric spread coating the process is ideal, since the cast liquid on the fabric is first lowered in viscosity by going through a heating zone set at about 35-65°C (100-150°F). This allows the plastisol to firmly impregnate the fabric pores. Further heating at the fusion temperature then completely solidifies the coating to form artificial leather, plastic-coated tablecloths, tarpaulins, etc.

Dip coating is done in a slightly different manner, although the same principles apply. In dip coating, a metal male tool such as a glove form or a rain boot form is heated to 200-230°C (400-450°F) and is then dipped in the cold liquid plastisol. The resin immediately fuses around the form, with the material on the actual surface of the tool being the most completely fused and the resin on the outside only partially gelled. The form is then withdrawn and placed in an oven, or, if it is on a conveyor belt, run through a heating tunnel where the plastisol is completely gelled. The form is then cooled and the part is stripped from the tool. By having a large tank of plastisol, kept at room tempera-

Fig. 19-1. Cast nylon bearings require no lubrication. (*Courtesy: The Polymer Corporation*)

Fig. 19-2. Cast nylon conveyor buckets. *(Courtesy: The Polymer Corporation)*

ture, and having a number of tools on a conveyor belt this process is well adapted to automation. It is indeed the process used to form doll heads, arms and legs, and to make gloves, rain boots, spark plug protectors, drain pipe seals, and many other items requiring the flexible, elastomeric properties of the vinyl plastisols.

Rotational molding, described in Chapter 14, is somewhat akin to the casting process in that the plastisol is placed

Fig. 19-3. Cast nylon gears run quieter and longer than steel gears in paper-making equipment. *(Courtesy: The Polymer Corporation)*

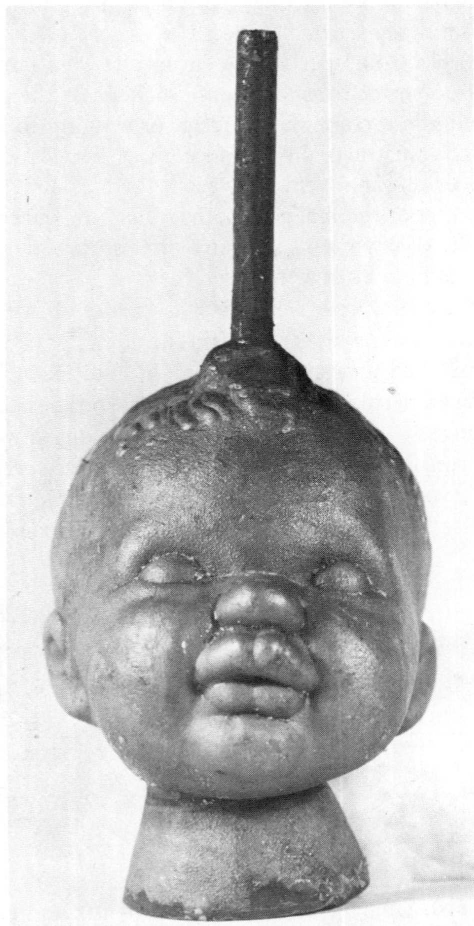

Fig. 19-4. Nickel electroformed molds used in PVC slush molding. *(Courtesy: Electroforms)*

inside a mold which is then heated to the fusion temperature, while being rotated, after which it is cooled and the part is removed. Slush molding, however, is even more like dip coating in that a hot mold is filled with plastisol and the material is immediately poured out. The temperature of the mold and its heat capacity determine the final thickness of the molded part, as in the dip coating process.

Thin and lightweight tooling for plastisol casting can be very inexpensive, since high pressure resistance is not required. Frequently such tools are electroformed over a plaster mold to make surfaced female tools of good quality at an economical price (see Figure 19-4).

ENCAPSULATION

Encapsulation (also called potting) is a process in which parts, mainly electronic units or assemblies, are coated with any one of a variety of plastics materials in order to provide electrical insulation, resistance to mechanical shock, thermal shock, and protection of the part from the effects of moisture and chemical attack. There are a number of techniques for encapsulation of parts, and since many of them are somewhat akin to casting, the process can be considered an offshoot of the casting process.

Initially electronic parts were encapsulated in pure resin, usually a polyester or an epoxy. In such a case the part was simply hung in a container of resin, or was dipped into the resin, which was then allowed to cure, either at room temperature or in an oven. However, failures due to resin brittleness, poor thermal conductivity, and unequal thermal coefficients of expansion indicated that more sophisticated materials and methods were called for.

One of the first improvements was the use of filled resins which gave materials with coefficients of expansion more closely matched to that of the part. In addition, the filled materials had better thermal conductivity so thermal shocks were reduced, and increased heat dissipation allowed the units to run cooler. Currently encapsulation is performed using resins containing fillers such as mica, aluminum oxide, milled glass fibers, and many others. While epoxies are the resins most generally used, polyesters, filled and unfilled, silicones, urethanes, and polysulfides are also employed.

The encapsulation operation may be as simple as pouring the filled, catalyzed resin over the part, or it may be improved by use of a vacuum to remove all the air from the component and the resin (see Figure 19-5). In many cases the electronic part is placed in a temporary container such as one made of Cerrobend, or polyethylene and a filler such as aluminum oxide is packed around the part. The packing may be done by a vibration table to insure maximum compaction. The resin is then introduced, under vacuum conditions and the entire assembly is cured. After curing, the container may or may not be removed. Such a process is usually used for large parts such as transformers and similar units. Figures 19-6 and 19-7 illustrate a simple technique for encapsulation of small parts using a Cerrobend mold as a nonpermanent container for the resin encapsulant. After cure, the container is simply broken or stripped from the encapsulated part. When the container is kept around the part permanently, the process is known as potting.

By the proper choice of fillers it is possible to match expansion rates of the electronic part and the encapsulant, increase the thermal conductivity of the part, and considerably upgrade the electrical and mechanical properties of the assembly. It should be noted that some assemblies are encapsulated in an unfilled elastomeric silicone which may or may not be transparent. When such a room-temperature-curing resin is used, the encapsulant, a gel-like material, absorbs the thermal stresses by simply deforming, thus not stressing the electronic components. When transparent gels are used, the job of locating and replacing a damaged or defective component is simplified (see Figure 19-8).

The casting techniques described above are mainly used for small quantities of parts and assemblies, or very large parts where automatic equipment might not be practical. The majority of individual parts, such as resistors, capacitors, transistors, etc., are currently encapsulated by transfer molding. Initially, phenolic resins were used to encapsulate parts such as resistors which could withstand the relatively high pressures in the transfer mold. Later, as low-pressure

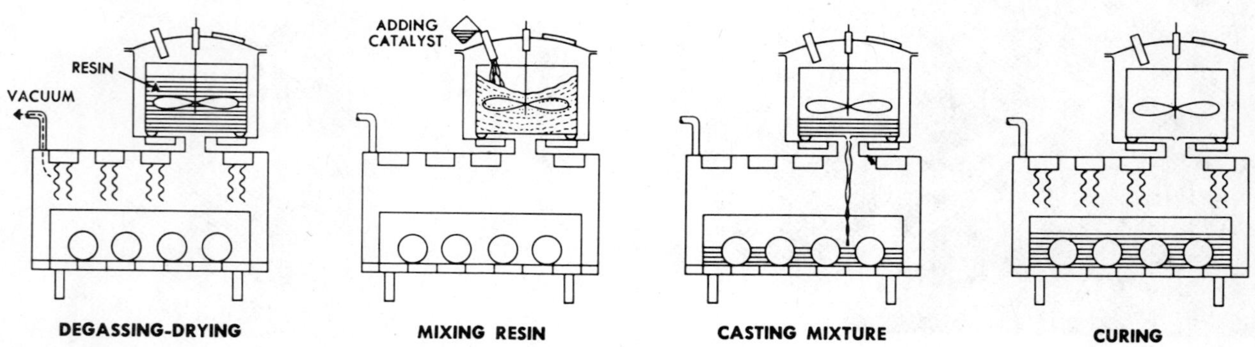

Fig. 19-5. Schematic diagrams showing the four steps in a vacuum potting process. Operations consist of, from left to right: degassing and drying of component to be potted, mixing resin under vacuum, casting resin, and curing the potted components. (Harper, 1961)

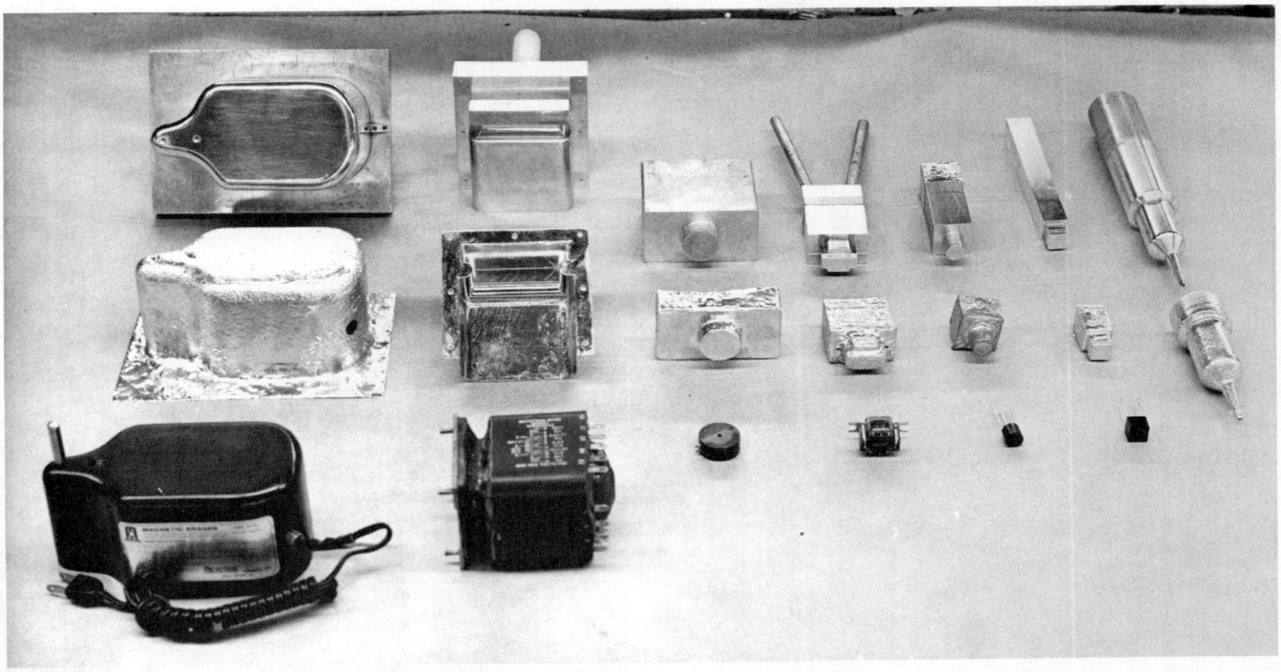

Fig. 19-6. Nonpermanent molds made by dipping the solid forms into molten Cerrobend which "freezes" around the form. (*Courtesy: Cerro Metal Products, a division of the Marmon Group, Inc.*)

epoxy resin molding powders and polyester molding materials became available, these became the preferred materials for high-volume encapsulation. Figure 19-9 shows a typical transfer encapsulation press and Figure 19-10 illustrates the molding process. The latest technique, at the time of writing is still transfer molding except that the encapsulants are liquid resins. These resins can be injected at considerably lower pressures than the low-pressure epoxies and polyesters formerly in use. The liquid materials are injected at pressures as low as 50 psi, which in many cases is less than a tenth of the pressure used with the low-pressure molding powders. The first liquid molding systems used two components, catalyst and resin, which were brought together in a mixing chamber and then injected into the mold. The latest systems utilize a one-component, precatalyzed polyester which is simply injected into the hot mold. The system is claimed to be simpler, and faster and use the lowest pressures of any of the transfer encapsulation systems. The advantages of the lower pressures are that part deformation is reduced, lower clamping pressures are required, and somewhat cheaper tooling may be used. While polyester resins are the materials in widest use for the one-component process; epoxies, silicones, and urethanes can also be used. It is interesting to note that, while special transfer encapsulating presses are available, it is also possible to install a liquid resin conversion unit onto a standard transfer or compression molding press.

Fig. 19-7. Encapsulation by pouring resin around the part. The mold is later broken away and remelted. (*Courtesy: Cerro Metal Products, a division of the Marmon Group, Inc.*)

MACHINING

Even though it is the object of molding operations to turn out complete parts whenever possible, there are a number of reasons why plastics are machined or require special finishing operations. First, most injection-molded thermoplastic parts have gates and flash to be removed. Other parts formed by rotational molding, thermoforming, blow molding, etc., rarely can be made without areas where the part is gripped, or where material is introduced. However, removal of excess material is only one reason for machining of plas-

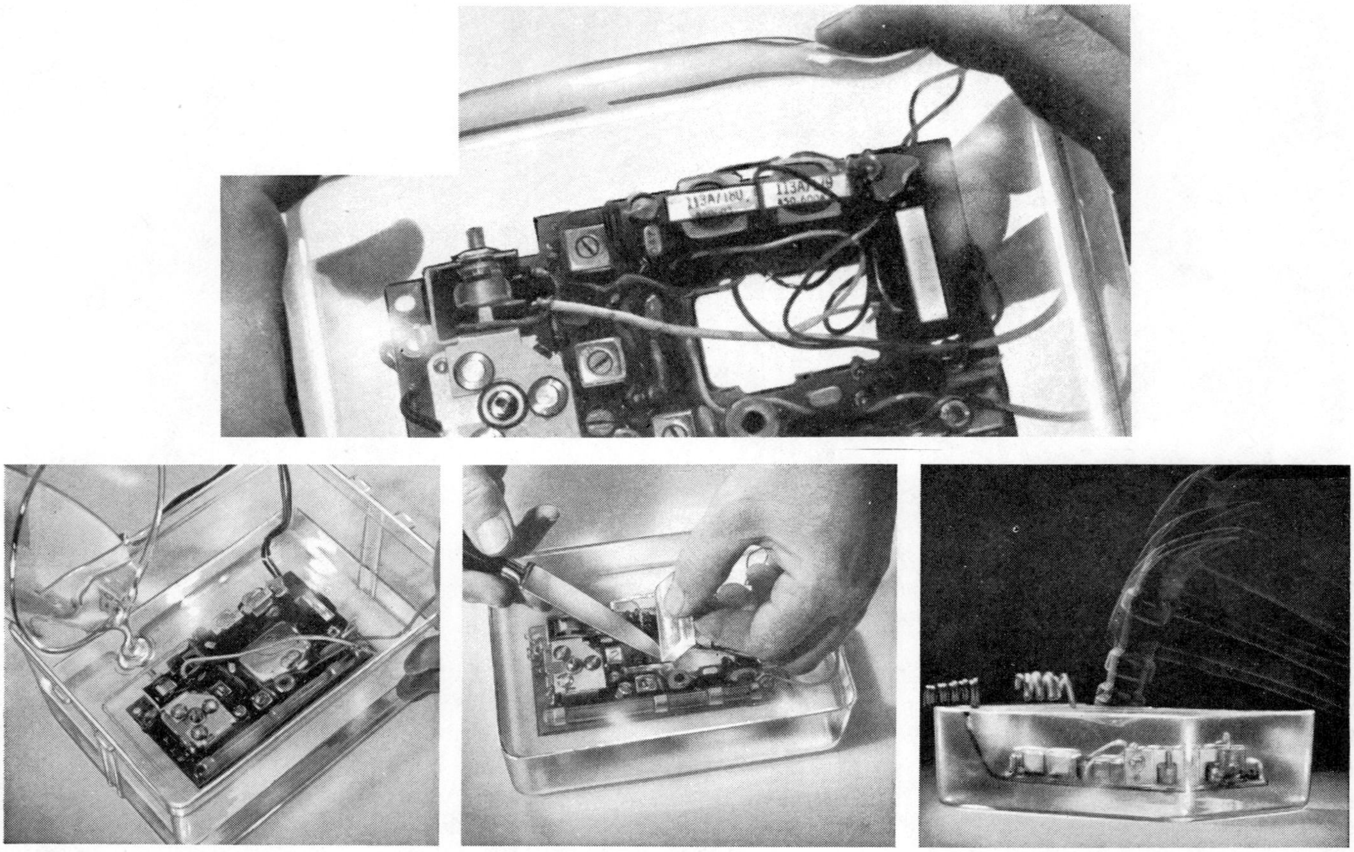

Fig. 19-8. Use of clear silicone gel encapsulant. (*Courtesy: General Electric Co.*)

tics. In many cases it is considerably more economical to produce a few prototype parts by machining raw stock than to invest in molds which later might have to be extensively revised. It should also be noted that in many cases, machining, can supplement the molding operation. An example might be the drilling or tapping of holes in molded parts where inclusion of removable core pins in the mold could make the tooling costs considerably more expensive. A similar saving might be in the inclusion of a machined undercut in a part which, if molded in, could make the tooling prohibitively expensive. Last, for many parts, production by high-speed automatic screw machines has proved to be cheaper than conventional molding methods, particularly when the quantities run into thousands or tens of thousands instead of hundreds of thousands or millions.

Machining of plastics, in general, is similar to machining of yellow brass and other relatively soft metals. There are, however, a number of differences, which if ignored by the machinist, can result in out-of-tolerance parts, lowered production, and, at the least, rapid tool wear. Each class of plastics is unique and requires its own set of machining rules; however, all plastics to some extent share the following characteristics:

1. *Heat sensitivity*. All plastics materials show a greater sensitivity to heat than do most metals. This sensitivity is most noticeable in unfilled thermoplastics, somewhat less in filled thermoplastics, and thermoset materials show the least effect of heat during the machining operation. The specific effect of the heat buildup is softening of the plastic, gumming and discoloration, accompanied by rough machining and/or distortion.

2. *Poor heat conductivity*. Because all plastics, filled or unfilled, are poor heat conductors, the frictional heat developed by turning, drilling, sawing or any other operation is largely not conducted away from the surface. Thus the heat buildup is considerably more rapid than in a similar metal operation, and is concentrated on the surface of the plastic, where it will have the most damaging effect. An accompanying effect is the fact that the heat buildup is also considerably higher at the tool, leading to more rapid tool deterioration. The use of air jets, water, or specific coolants

Fig. 19-9. Transfer and encapsulation molding press. (*Courtesy: Rucker-PHI*)

for each plastic is a necessity if high-volume production rates are required.

3. *High coefficient of expansion*. Plastics have thermal coefficients of expansion roughly ten times that of metals. The initial heating from the machining causes expansion in the plastic, which in turn causes more friction against the tool, resulting in additional heat. Adequate cutting tool clearances must be used to minimize the frictional heat of rubbing. Because of the high thermal coefficient of expansion, in production of close tolerance parts it is customary to allow the part to cool prior to making the final cut, which is usually considerably less in depth than the previous cuts. For this reason, particularly for unfilled thermoplastic parts, measurements are made prior to the final cut and after the final cut, when the part has cooled down to room temperature. For the highest precision, the measurements should be made at a given temperature, which should be stated on the blueprint, specification, etc.

4. *Low modulus*. Plastics materials, because of their lower modulus as compared to metals, can be easily distorted by the cutting tool, or even the holder or chuck in which the part is kept during the operation. To minimize the possibility of this distortion, the part should be adequately supported at all times. This is particularly true during operations such as lathe cutting where the part is not supported on a solid bed such as in drilling or milling.

5. *Plastic memory*. Adequate tool relief (clearance) should be provided so that friction is kept to a minimum. Increased friction can, of course, lead to additional heat, which in turn causes expansion. After the tool is removed, the expansion, which may have been prevented by the tool, shows up as a change in dimensions. This effect, however, may not show up until several days have passed. This is sometimes shown in the appearance of tapered or smaller holes in a part which had initially shown the proper dimen-

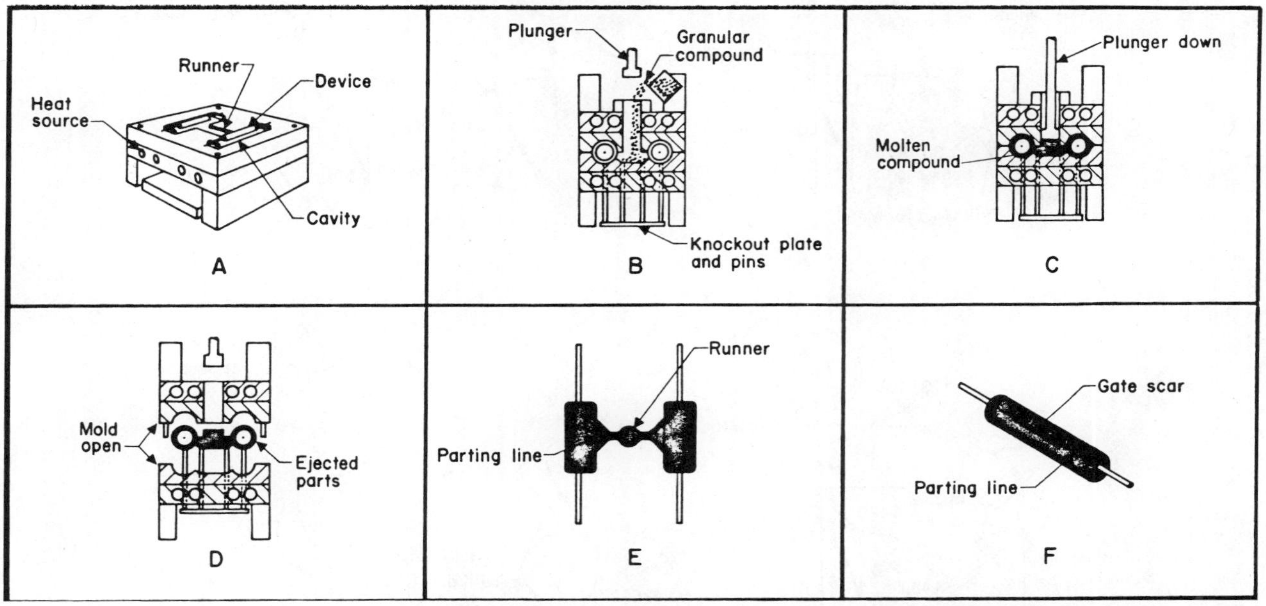

Fig. 19-10. Sequence diagram of transfer mold. (A) Perspective view of transfer mold. (B) Mold closed with devices in position for encapsulations, plunger retracted, and granular compound fed into heating chamber. (C) Plunger moves downward, forcing molten compound around devices in cavities. (D) Mold opens and knockout pins eject encapsulated devices. (E) Encapsulated devices as molded. (F) Encapsulated device showing parting line and gate scar. (*Courtesy: Plastics Engineering*, April 1978)

sions. It can also manifest itself in turned parts which have larger diameters than when initially checked just after cutting.

6. *Annealing*. Some materials, particularly acrylics and polystyrenes, can show severe crazing weeks or months after being machined due to locked-in machining stresses. To minimize these stresses such parts should be annealed. Usually the annealing temperature is approximately 5–6°C (10°F) below the highest temperature the part can sustain without distortion. The time required for the annealing operation is usually a function of the part thickness, and geometry and the amount of stress relief required. The annealing time period must be sufficiently long to allow the part to reach equilibrium temperature throughout the entire piece, and then long enough for all the stresses to be relieved. Heating media used include water baths, oil baths, circulating air ovens, and infrared lamp ovens. Water baths are probably the most satisfactory, from the standpoint of providing uniform heat which is easily kept at a given temperature.

7. *Tool wear by glass-reinforced plastics*. Cutting-tool wear by glass-reinforced plastics is considerably higher than when doing the same type of machining on non-glass-reinforced materials. Another factor frequently overlooked is that of machine wear by the dust produced by the machining. This dust can be introduced into bearings, machine ways, and other precision parts, with consequent inordinate rates of wear. The remedy is an adequate suction hose positioned as close as possible to the cutting tool at all times.

Based on the particular characteristics of plastics that make their machining somewhat different than the same operations on metals, five simple general rules can be given for the machining of plastics:

1. Use only sharp tools with honed and polished cutting edges.
2. Use tools which are designed to have adequate clearance.
3. Use appropriate work supports whenever possible.
4. Use tools at high speeds and low rates of feed, except when drilling. (Too low a feed when drilling can cause excessive heat.)
5. Provide adequate and appropriate coolants.

Specific recommendations for tool geometry and various machining operations are shown in Tables 19-1 through 19-10. These recommendations represent starting suggestions; individual cases can be run at higher or lower speeds, feeds, etc.

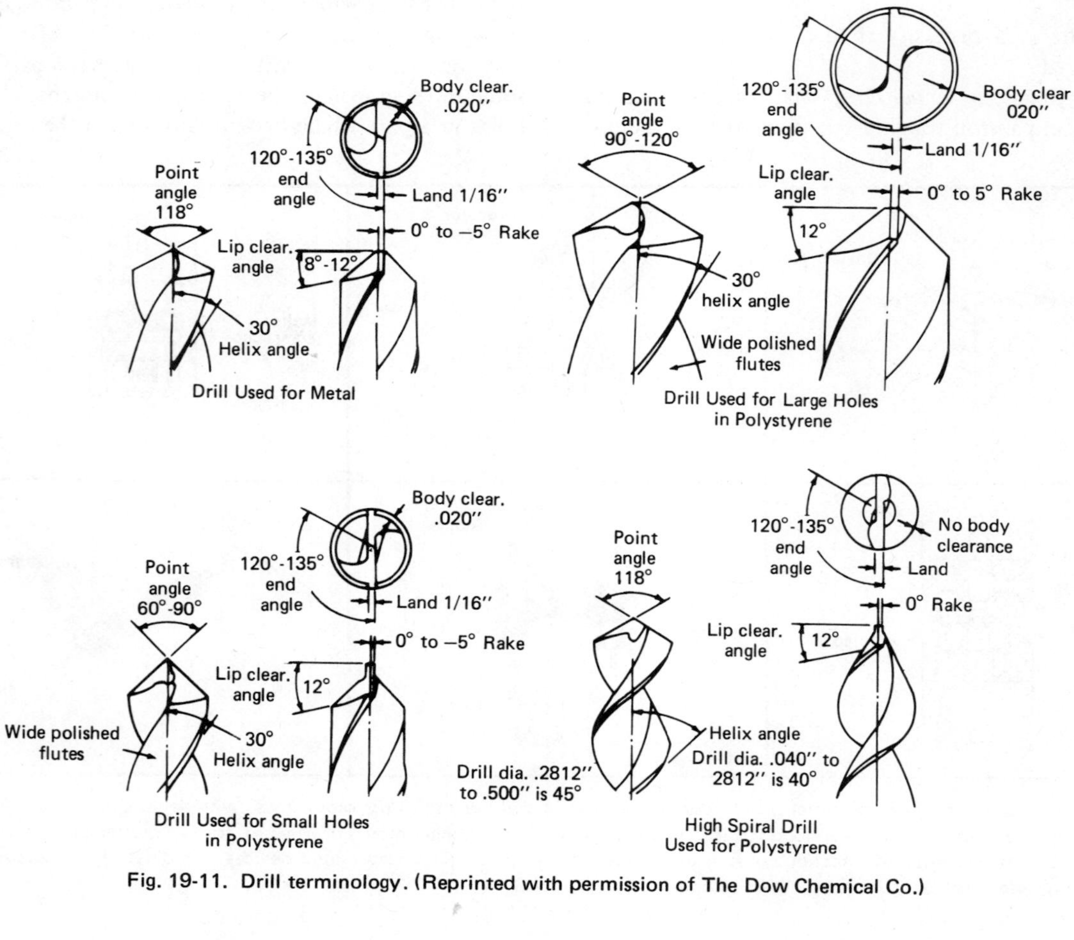

Fig. 19-11. Drill terminology. (Reprinted with permission of The Dow Chemical Co.)

The information given in Figures 19-11, 19-12, and 19-13 and Tables 19-1 through 19-10 applies, in general, to all types of plastics and plastics composites—with one exception. The exception includes all types of composite materials using Kevlar aramid fiber as the reinforcement. Because of the extreme toughness of this fiber, special cutting and machining techniques and equipment are required. Information about the procedures required for the Kevlar-reinforced materials is available from Du Pont.

Drilling

Drilling is a particularly severe operation because of restricted chip flow, inherently poor rake angles, and variable speed across the cutting edge. While drills designed for wood or metal can be used, special drills for plastics will give the best results. These drills are made with a point angle of 60° to 90° (as contrasted with a metal drill point angle of 118°), and lip clearance angles of 10° to 15°. The helix angle is usually larger than in a metal drill, since this aids in chip removal. See Figure 19-11 for drill terminology and Table 19-1 for specific drill recommendations.

Friction created by chips trapped at the drill point and in the flutes is the major cause of generated heat during the drilling operation. During drilling of deep holes, therefore, the drill should be withdrawn frequently to remove the chips. The drill speed, in general, will depend on the type of material and the size of the hole. Speeds are usually decreased with increasing hole diameters; with harder materials, speeds are increased. Too slow a feed will cause increased frictional heat, rather than less heat. The feed should be the greatest that will not cause gumming or burning. Water-based or other types of appropriate coolants, or air jets should be used whenever possible. Wood or scrap plastics should be used under the part being drilled to minimize chipping as the drill breaks through. For this same reason, the drill feed should be slowed down somewhat on breaking through, to minimize chipping.

Satisfactory and consistent drilling of thermoplastics is a skill that demends considerable practice. A skilled operator mindful of the principles discussed in this section and equipped with quality tools can do the job with good results.

Generalized drill speeds recommended for thermoplastics are listed in Table 19-2. More specific recommendations are given in Table 19-4.

Drilling of thermosetting plastics and heavily filled thermoplastics is somewhat easier than drilling unfilled thermoplastics because of the better heat dissipation and lower gumming tendencies, particularly with thermosets. General recommendations for drill speeds are shown in Table 19-3. Slower speeds should always be used if overheating is encountered. (Table 19-4 contains specific recommendations.)

Table 19-1. Drill Geometry.*

Material	Helix Angle, Deg.	Point Angle, Deg.	Clearance, Deg.	Rake, Deg.
Polyethylene	10–20	70–90	9–15	0
Rigid polyvinyl chloride	27	120	9–15	–
Acrylic (polymethyl methacrylate)	27	120	12–20	–
Polystyrene	40–50	60–90	12–15	0
Polyamide resin	17	70–90	9–15	–
Polycarbonate	27	80	9–15	–
Acetal resin	10–20	60–90	10–15	–
Metal	30	118	8–12	0–5

*Reprinted by permission from *Modern Plastics Encyclopedia*, McGraw-Hill, Inc. Original material appeared in *Machining of Plastics* by A. Kabayashi, published by McGraw-Hill, Inc.

Turning

Tables 19-5 and 19-6 should be taken as starting recommendations for turning of plastics materials. The general rules concerning heat buildup, low modulus, and high expansion should not be ignored. The work should be supported as much as possible to minimize chatter and prevent stock deflection from the cutting tool. The tools

Table 19-2. Recommended Drill Speeds for Thermoplastics.

Drill Size	RPM	Drill Size	RPM
No. 1 through No. 16	2500	$\frac{5}{16}$ in.	1700
No. 17 through No. 32	3000	$\frac{3}{8}$ in.	1300
No. 33 and up	5000	$\frac{7}{16}$ in.	1000
$\frac{1}{16}$ in.	5000	$\frac{1}{2}$ in.	1000
$\frac{1}{8}$ in.	3000	A through D	2500
$\frac{3}{16}$ in.	2500	E through M	1700
$\frac{1}{4}$ in.	1700	N through Z	1300

Table 19-3. Recommended Drill Speeds for Thermosets.

Drill Size	RPM	Drill Size	RPM
No. 1 through No. 10	1700	$\frac{5}{16}$ in.	1300
No. 11 through No. 27	2500	$\frac{3}{8}$ in.	1000
No. 28 through No. 41	3000	$\frac{7}{16}$ in.	600
No. 42 and up	5000	$\frac{1}{2}$ in.	600
$\frac{1}{16}$ in.	5000	A and B	1700
$\frac{1}{8}$ in.	3000	C through O	1300
$\frac{3}{16}$ in.	2500	P through Z	1000
$\frac{1}{4}$ in.	1700		

Table 19-4. Drilling.*

Material	Hardness	Condition	Speed ft./min.	Feed, in./revolution — Nominal hole diameter, in. 1/16	1/8	1/4	1/2	3/4	1	1½	2	HSS tool material
Thermoplastics												
Polyethylene, polypropylene, fluoroplastic, butyrate	31 to 116R_R	Extruded, molded, or cast	150–200	0.002	0.003	0.005	0.010	0.015	0.020	0.025	0.030	M10, M7, M1
High-impact styrene, acrylonitrile-butadiene-styrene, modified acrylic	83 to 107R_R	Extruded, molded, or cast	150–200	0.002	0.004	0.005	0.006	0.006	0.008	0.008	0.010	M10, M7, M1
Nylon, acetal, polycarbonate	79 to 100R_M	Molded	150–200	0.002	0.003	0.005	0.008	0.010	0.012	0.015	0.015	M10, M7, M1
Acrylic	80 to 103R_M	Extruded, molded, or cast	150–200	0.001	0.002	0.004	0.008	0.010	0.012	0.015	0.015	M10, M7, M1
Polystyrene, medium- and low-impact	70 to 95R_M	Molded or extruded	150–200	0.001	0.002	0.003	0.004	0.005	0.006	0.007	0.008	M10, M7, M1
Thermosets												
Paper or cotton base	50 to 125R_M	Cast, molded, or filled	200–400	0.002	0.003	0.005	0.006	0.010	0.012	0.015	0.015	M10, M7, M1
Glass fiber, graphitized, and asbestos base	50 to 125R_M	Cast, molded, or filled	200–250	0.002	0.003	0.005	0.008	0.010	0.012	0.015	0.015	M10, M7, M1

*Reprinted from *Machining Data Handbook*.

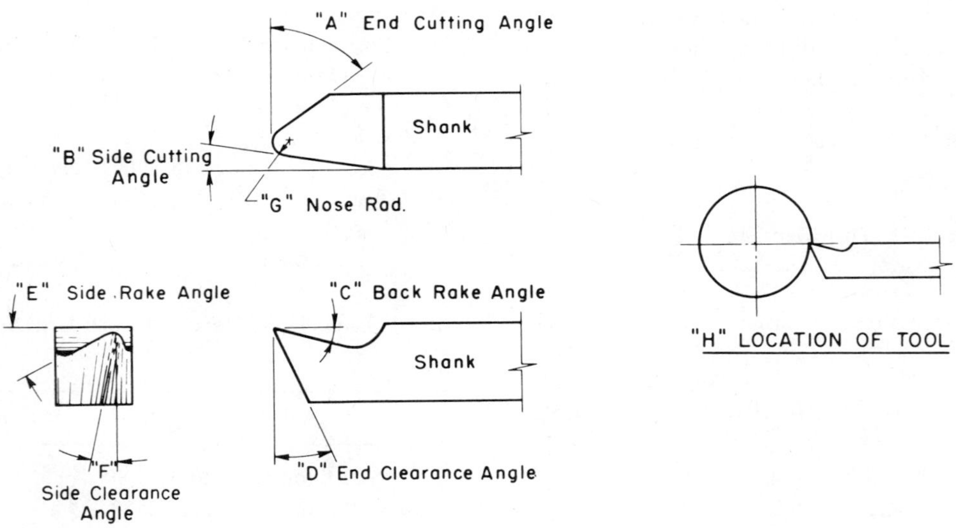

Mat'l.	A	B	C	D	E	F	G	H
Med. Steel	20°	10°	+8°	6°	14°	4°	3/16	At ℄
Polystyrene	20°	10°	−2°	12°	10°	10°	3/16	2° above ℄

Fig. 19-12. General-purpose lathe tool design. (Reprinted with permission of The Dow Chemical Co.)

Table 19-5. Turning, Single-Point, and Box Tools.[a]

Material	Hardness	Condition	Depth of cut, in.	High-speed steel tool: Speed, ft./min.	High-speed steel tool: Feed, in./rev.	High-speed steel tool: Tool material	Carbide tool: Speed, ft./min.	Carbide tool: Feed, in./rev.	Carbide tool: Tool material
Thermoplastics									
Polyethylene, polypropylene, fluoroplastic, butyrate	31 to 116R_R	Extruded, molded, or cast	0.150 0.025	250–350 300–400	0.010 0.002	M2, T5 M2, T5	400–450 450–500	0.010 0.002	C-2 C-2
High-impact styrene, acrylonitrile-butadiene-styrene, modified acrylic	83 to 107R_R	Extruded, molded, or cast	0.150 0.025	250–350 300–400	0.015 0.005	M2, T5 M2, T5	400–450 450–500	0.015 0.005	C-2 C-2
Nylon, acetals, polycarbonate	79 to 100R_M	Molded	0.150 0.025	300–400 400–500	0.010 0.002	M2, T5 M2, T5	500–600 600–700	0.015 0.005	C-2 C-2
Acrylics	80 to 103R_M	Extruded, molded, or cast	0.150 0.025	250–300 300–400	0.008 0.005	M2, T5 M2, T5	450–500 500–600	0.010 0.005	C-2 C-2
Polystyrenes, low- and medium-impact	70 to 95R_M	Molded or extruded	0.150 0.025	75–100 150–200	0.005 0.001	M2, T5 M2, T5	200–300 350–400	0.010 0.002	C-2 C-2
Thermosets									
Paper and cotton base	50 to 125R_M	Cast, molded, or filled	0.150 0.025	500–1000 1000–2000	0.012 0.005	M2, T5 M2, T5	750–2000 1000–3000	0.012 0.005	C-2 C-2
Glass fiber and graphite base	50 to 125R_M	Cast, molded, or filled	0.150 0.025	400–500 500–1000	0.012 0.005	M2, T5 M2, T5	500–1000 750–1500	0.012 0.005	C-2 C-2
Asbestos base	50 to 125R_M	Molded	0.150 0.025	650–750 750–1000	0.012 0.005	M2, T5 M2, T5	700–1000 750–2000	0.010 0.005	C-2 C-2
Silica fiber-reinforced phenolic resin (Refrasil)	55 to 75[b]	Molded	0.050	—	—	—	200	0.010	C-2

[a] Reprinted from *Machining Data Handbook.*
[b] Barcol hardness.

Table 19-6. Turning Tools, Single Point and Box.[a]

	High-Speed Steel and Cast Alloy Tool					Carbide Tools: Brazed		Carbide Tools: Throwaway				
Material[b]	Back Rake Angle, Deg.	Side Rake Angle, Deg.	End Relief Angle, Deg.	Side Relief Angle, Deg.	Side and End Cutting Edge Angle, Deg.	Back Rake Angle, Deg.	Side Rake Angle, Deg.	Back Rake Angle, Deg.	Side Rake Angle, Deg.	End Relief Angle, Deg.	Side Relief Angle, Deg.	Side and End Cutting Edge Angle, Deg.
Thermoplastics	0	0	20–30	15–20	10	0	0	0	0	20–30	15–20	10
Thermosets	0	0	20–30	15–20	10	0	15	0	15	5	5	15

[a] Reprinted from *Machining Data Handbook.*
[b] Material of all hardnesses.

Table 19-7. End Milling—Slotting.[a]

Material	Hardness	Condition	Depth of Cut, in.[b]	Speed, ft./min.	Feed, in./tooth — Width of Slot, in. ¼	½	¾	1–2	Tool Material
Thermoplastics	31R to 125R_R	Extruded, molded, or cast	0.250	270–450	0.002	0.003	0.005	0.008	M2, M7
			0.050	300–500	0.001	0.002	0.004	0.006	M2, M7
Thermosets	50 to 125R_M	Cast, molded, or filled	0.250	125–180	0.002	0.003	0.005	0.008	M2, M7
			0.050	140–200	0.001	0.002	0.004	0.006	M2, M7
Reinforced plastics	55 to 75[c]	Molded	0.250	760	–	–	0.004	0.005	C-2 Carbide
Silica fiber-reinforced phenolic resin (Refrasil)			0.050	850	–	–	0.003	0.004	C-2 Carbide

[a]Reprinted from *Machining Data Handbook*.
[b]Measured parallel to axis of cutter.
[c]Barcol hardness.

Table 19-8. End Milling—Peripheral.[a]

Material	Hardness	Condition	Depth of Cut, in.[b]	Speed, ft./min.	High-Speed Steel Tool — Feed, in./tooth — Cutter Diameter, in. ¼	½	¾	Tool Matl.	Speed ft./min.	Carbide Tool — Feed, in./tooth — Cutter Diameter, in. ¼	½	¾	Tool Material
Thermoplastics	31 to 125R_R	Extruded, molded, or cast	0.050	500–750	0.004	0.006	0.010	M2, M7	1300–1500	0.004	0.006	0.010	C2
			0.015	750–1000	0.003	0.005	0.008	M2, M7	1500–2000	0.003	0.005	0.008	C2
Thermosets	50 to 125R_M	Cast, molded, or filled	0.050	200–300	0.005	0.006	0.010	M2, M7	1300–1500	0.005	0.006	0.010	C2
			0.015	400–500	0.004	0.005	0.008	M2, M7	1500–2000	0.004	0.005	0.008	C2
Reinforced plastics	55 to 75[c]	Molded	–	–	–	–	–	–	–	–	–	–	–
Silica fiber-reinforced phenolic resin (Refrasil)			0.015	–	–	–	–	–	1300	0.004	0.005	0.008	C2

[a]Reprinted from *Machining Data Handbook*.
[b]Measured parallel to axis of cutter.
[c]Barcol hardness.

should be sharp, honed, or polished to minimize friction and should have the general dimensions shown in Figure 19-12. A water-soluble coolant, or another appropriate type, should be applied whenever possible.

Face Milling

Table 19-9 gives starting recommendations for face milling of thermoplastics and thermosets. The feed shown is given in inch/tooth. Since milling cutters usually have multiple cutting edges this feed should be converted into inch/minute of table travel. Too high a table feed will result in a rough surface; too low a feed travel will result in excessive heat that can cause melting, surface cracks due to high temperature, loss of dimensions, and a poor surface finish. Use of a compatible cutting fluid is recommended. Tools should be honed or polished to reduce frictional drag which can increase the heat buildup.

Sawing

It is possible to saw all types of plastics with circular, jig, and band saws. In some cases an abrasive cutting wheel can also be used. However, whenever possible, band saws should be the type chosen because of their better dissipation of heat and their greater versatility. An air jet coolant or a lubricant such as zinc stearate can be used.

The preferred saw blade is a buttress (skip tooth) type

Table 19-9. Face Milling.[a]

Material	Hardness	Condition	Depth of Cut, in.[b]	High-Speed Steel Tool: Speed, ft./min.	High-Speed Steel Tool: Feed, in./tooth	Tool Matl.	Carbide Tool: Speed, ft./min.	Carbide Tool: Feed, in./tooth	Tool Material
Thermoplastics	31 to 125R_R	Extruded, molded, or cast	0.150	500–750	0.016	M2, M7	1300–1500	0.020	C-2
			0.060	750–1000	0.004	M2, M7	1500–2000	0.005	C-2
Thermosets	50 to 125R_M	Extruded, molded, or filled	0.150	200–300	0.015	M2, M7	1300–1500	0.015	C-2
			0.060	400–500	0.005	M2, M7	1500–2000	0.005	C-2
Reinforced plastics Silica fiber-reinforced phenolic resin (Refrasil)	55 to 75[c]	Molded	0.060	–	–	–	1300	0.009	C-2

[a] Reprinted from *Machining Data Handbook.*
[b] Measured parallel to axis of cutter.
[c] Barcol hardness.

with three or four teeth per inch for thick stock and 10 to 14 for thin stock (see Figure 19-13). The blade should have zero front rake. Narrow blades result in less frictional heat generation because of the lower contact area. A high tooth set will reduce friction but will remove more material and possibly give a rough cut. A buttress blade, instead of the usual full tooth band, will also reduce friction, permit better coolant application, and allow better chip removal.

Circular saws should be hollow ground and chrome plated for minimum friction. The set should be a little higher than normal. Jig saws should also have a set a little greater than normal to assure adequate blade clearance. Silicon carbide wheels are also used for cutting both thermosets and thermoplastics. In all cases an air blast or water should be used to prevent overheating.

The proper rate of feed is best learned by experience. However, too high a rate of feed will lead to overheating of the blade, gumming of the material, and loading of the saw teeth. At the end of the cut the pressure should be relieved in order to minimize chipping. Tables 19-10, 19-11, and 19-12 give specific recommendations for sawing various plastics with different types of blades. For more complete information of machining consult *Machining Data Handbook* or the *Plastics Engineering Handbook*.

FINISHING

Finishing of plastics generally entails the removal of flash, gate projections, and overall cleanup of parts just prior to shipping. It includes removal of burrs from a machining operation, possible removal of small defects in the part due to nicks and scratches in the mold, which are visible as small protrusions, and the application of waxes or polishes to improve the appearance of the part.

Removing flash and defects from small quantities of parts is usually done manually using files, scrapers, or specially made tools. Examples of such tools are shown in Figure 19-14. By the use of the correct tool, flat, concave, and

Standard or Precision Tooth

Buttress or Skip Tooth

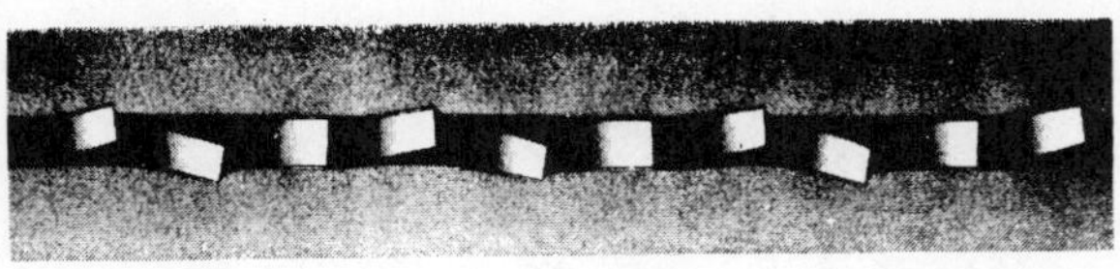

Regular Tooth Set

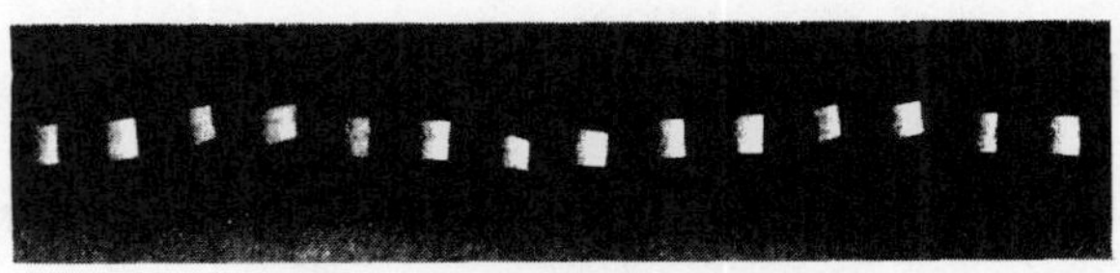

Wavy Tooth Set

Fig. 19-13. Typical band saws—standard or precision tooth; buttress or skip tooth; regular tooth set and wavy tooth set. (*Courtesy: Borg-Warner Chemicals*)

Table 19-10. Power Band Sawing, Vertical, High-Speed Steel Blades.*

Material	Hardness BHN	Condition	Material Thickness (in.)	Tooth Form†	Pitch (teeth per in.)	Band Speed (fpm)
Thermoplastics						
Polyethylene	$31R_R$	Extruded,	$<\frac{1}{2}$	P	10–14	4000‡
Polypropylene	to	molded,	$\frac{1}{2}$–1	P	6	3000‡
Fluorocarbons	$116R_R$	or cast	1–3	B	3	2500‡
Butyrates			>3	B	3	1500‡
High-impact styrene	$83R_R$	Extruded	$<\frac{1}{2}$	P	10–14	2300‡
Acrylonitrile-butadiene-	to	molded,	$\frac{1}{2}$–1	P	8	2000‡
styrene	$107R_R$	or cast	1–3	P	6	1500‡
Modified acrylic			>3	B	3	1000‡
Nylon	$79R_M$	Molded	$<\frac{1}{2}$	P	10–14	3000‡
Acetals	to		$\frac{1}{2}$–1	P	6	2500‡
Polycarbonates	$100R_M$		1–3	B	3	2000‡
			>3	B	3	1500‡
Acrylics	$80R_M$	Extruded,	$<\frac{1}{2}$	P	10–14	4000‡
	to	molded,	$\frac{1}{2}$–1	P	6	3000‡
	$103R_M$	or cast	1–3	B	3	2500‡
			>3	B	3	1500‡
Polystyrenes,	$70R_M$	Molded	$<\frac{1}{2}$	P	8–10	2300‡
medium and low impact	to	or	$\frac{1}{2}$–1	P	6	2000‡
	$95R_M$	extruded	1–3	B	3	1500‡
			>3	B	3	1000‡
Fluorocarbons	$31R_R$	Extruded,	$<\frac{1}{2}$	P	8–10	2300‡
	to	molded,	$\frac{1}{2}$–1	P	6	2000‡
	$116R_R$	or cast	1–3	B	3	1500‡
			>3	B	3	1000‡
Cellulose acetate	$49R_R$	Molded	$<\frac{1}{2}$	P	10–14	4000‡
	to		$\frac{1}{2}$–1	P	6	3000‡
	$112R_R$		1–3	B	3	2500‡
			>3	B	3	1500‡
Nylon	$79R_M$	Molded	$<\frac{1}{2}$	P	10–14	3000‡
	to		$\frac{1}{2}$–1	P	6	2500‡
	$100R_M$		1–3	B	3	2000‡
			>3	B	3	1500‡
Thermosetting Plastics						
Melamine-formaldehyde	$114R_M$	Unfilled,	$<\frac{1}{2}$	P	10–14	5500‡
	to	molded	$\frac{1}{2}$–1	P	8	5000‡
	$119R_M$		1–3	B	3	4000‡
			>3	B	3	3000‡
	$50R_M$	Cast, molded,	$<\frac{1}{2}$	P	10–14	5500‡
	to	laminated-	$\frac{1}{2}$–1	P	6	5000‡
	$93R_M$	asbestos or	1–3	B	3	4000‡
		fabric filler	>3	B	3	3000‡
	$100R_M$	Cast or	$<\frac{1}{2}$	P	10–14	5500‡
	to	molded,	$\frac{1}{2}$–1	P	6	5000‡
	$105R_R$	no filler	1–3	B	3	4000‡
			>3	B	3	3000‡
Phenolics	$108R_M$	Cast, molded	$<\frac{1}{2}$	P	10–18	75‡
	to	or laminated	$\frac{1}{2}$–1	P	8	50‡
	$115R_M$	with glass	1–3	P	6	50‡
		or mineral	>3	–	–	–
		filler				
Urea-formaldehyde	$94R_E$	Molded	$<\frac{1}{2}$	P	10–14	5000‡
			$\frac{1}{2}$–1	P	6	4000‡
			1–3	B	3	3500‡
			>3	B	3	2200‡

* Reprinted from *Machining Data Handbook.*
† P–Precision; C–Claw; B-Buttress.
‡ Recommended speed for high carbon blades.

Table 19-11. Power Band Sawing, High Carbon Saw Blades.*

Material	Hardness	Condition	Pitch, teeth/in. Minimum Thickness of Material, in.: $\frac{1}{4}$ and Under	$\frac{1}{4}$–$1\frac{1}{2}$	$1\frac{1}{2}$ and Over	Speed (fpm)
Thermoplastics						
Cellulose acetate	$49R_R$ to $112R_R$	Molded	10	6–10	3–6	3000
Acrylics	$80R_M$ to $103R_M$	Extruded, molded, or cast	14	6–10	3–6	3000
Nylon	$79R_M$ to $100R_M$	Molded	14	4–10	3–4	2500
Polystyrenes	$70R_M$ to $95R_M$	Molded or extruded	10	4–6	3–4	2000
Thermosets						
Melamine-formaldehyde	$114R_M$ to $119R_M$	Unfilled molded	14	6–10	3–4	4500
Phenolics	$50R_M$ to $93R_M$	Cast, molded laminated asbestos or fabric filler	14	6–10	3–6	4500
	$100R_M$ to $105R_M$	Cast or molded, no filler	14	6–10	3–6	4500
	$108R_M$ to $115R_M$	Cast, molded or laminated with glass fiber or graphitized cloth	14–18	8–10	6–8	2000
Urea-formaldehyde	$94R_E$	Molded	14	6–10	3–6	–

*Reprinted from *Machining Data Handbook.*

Table 19-12. Saw Blade Data.

Saw Type	Teeth/In.	Thickness of Cut (in.)	Speed (ft./min)
Band	10 to 18	Up to $\frac{1}{4}$	2500–3000
		$\frac{1}{4}$ up to 1	2000–2500
		1 up to $1\frac{1}{2}$	1500–2000
		$1\frac{1}{2}$ and greater	800–1200
Circular	10 to 12	$\frac{1}{4}$ to $\frac{1}{8}$	6000–8000
	8	$\frac{1}{4}$ to 4	6000–8000

convex surfaces can be scraped smooth. In many cases the use of a scraper is preferred to that of a file, since the file may leave marks where the scraper will not.

Removal of flash from large quantities of parts can be done by a variety of techniques, depending on the type of part, its shape, material, and number of parts and the type of finish required. Circular parts can be lathe cut, or deflashed by an automatic screw machine if the part shape is amenable to automation. Another technique, particularly for thermoformed parts or parts with flat surfaces at the flash line, is to die cut the flash off.

One of the simplest and most inexpensive of all flash removal techniques is the use of tumbling barrels. In this process the parts are placed in flat-sided barrels, usually octagonal shaped, which contain partial shelves inside to insure tumbling rather than rolling. The barrels are usually made of steel outer shells, lined on the inside with either hard rock maple wood or with neoprene rubber. The latter lining is preferred since it lasts longer and resists any liquids or chemicals which might be used (see Figure 19-15).

Before the tumbling process begins, the barrel is charged with a mixture of parts and various tumbling media to fill approximately 50–60% of the barrel's volume. The media used will depend on the type of parts and the action desired. For thermosetting parts where only flash removal is required, lignum vitae balls $\frac{1}{2}$–1 inch in diameter are used in a roughly 1 : 1 ratio, or in whatever combination is determined best by experiment. The barrel is usually rotated from 5 rpm to 30 rpm, with the optimum speed again determined experimentally (see Figure 19-16). In such an operation there may be several openings in the barrel, covered by coarse screen so the broken flash particles are automatically ejected (see Figure 19-17).

Barrel tumbling can also be used to do grinding and polishing, in which case the barrel is charged with an abrasive or a polishing wax, in addition to tumbling media. For thermoplastics a number of materials are available. These include wood, felt, nut shells, and metal parts ranging in size from $\frac{1}{16}$ inch to as large as $\frac{3}{4}$ inch in diameter. It is important that the correct size be used, particularly if the inside of holes or corners have to be finished. Media too large to penetrate the opening will, of course, be ineffective.

Frequently, to speed up the flash and protrusion removal, the barrel will be charged with pellets of dry ice in addition to the tumbling media. The dry ice freezes the flash and makes it more brittle so a cleaner break will result. In this case, however, the barrel does not contain the screened openings, but is sealed completely, except for a few small holes to allow the carbon dioxide gas to escape. By the use

Scraping is the operation of smoothing a plastic surface by the use of a sharp-bladed tool, which is held so that the blade rides over the surface and has no tendency to dig under the surface. A knife blade may be used for scraping, if the blade is slanted so that the sharp edge is at the back of the blade instead of being the forward edge.

TRIANGULAR SCRAPER

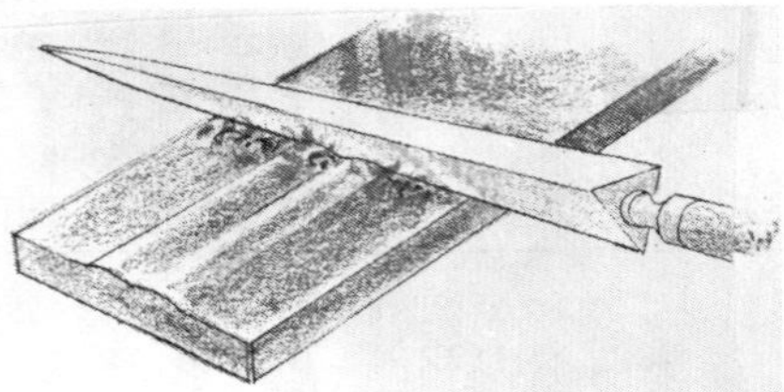

A standard scraper is a three-edged tool of triangular cross-section, the sides making up a scraping edge being at an angle of sixty degrees. The blade is tapered so that the scraping edge runs from a fairly long straight section at the handle end to a curved section near the point. This variation in shape allows the careful scraping of a great variety of straight, concave and convex surfaces. A small scraper will reach every part of many uneven complicated surfaces which might otherwise require the use of files of three or four different varieties.

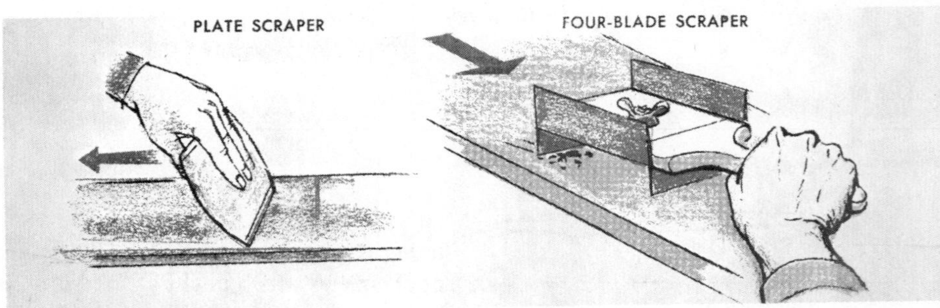

Some scrapers are made in the form of an oblong metal plate, sized to be conveniently held between the thumb and three fingers. The ends of the plate are straight while the side edges are slightly convex to allow scraping along narrow bands. Very delicate work may be done on flat surfaces with these tools. The 4 bladed scraper is a common variation of the plate scraper.

Holding Scrapers

Scrapers should be grasped firmly and moved across the plastic surface at a slight angle to give a shear cut and avoid chatter. The side of the scraping blade should be held at an acute angle with the plastic surface to avoid cutting into the surface. Either drawing or pushing strokes are satisfactory if the edge is turned so as to avoid digging under the surface. In working on delicate surfaces, let your thumb rest on the smooth surface near the scraping line as a gauge and guide.

Scraper Edges

The edges of the scraper must be free from burrs, dents or scratches which might mar the surface of the plastic.

Fig. 19-14. Special finishing tools. (*Courtesy: Structural Foam Division, The Society of the Plastics Industry, Inc.*)

of the right combination of tumbling media and abrasive it is possible to obtain a fair polish on parts at a fraction of the cost of doing it manually. If several grades of abrasives are used, first coarse and then fine, followed by a polishing compound, it is possible to reduce the size of parts, round corners and edges, as well as polish the part.

Tables 19-13, 19-14, and 19-15 detail the types of media used, the compounds for cutting, deburring, and polishing, and specific guides for tumbling of thermoplastics. Figure 19-18 shows typical parts deflashed, ground, and polished by tumbling.

For parts that are too large to be successfully tumbled, or too fragile to be tumbled, there are other types of equipment where the items are kept on a rotating table and blasted with crushed fruit pits, nut shells, or nylon pellets. Where large quantities of parts are required, or continuous

Fig. 19-15. Tumbling barrel for flash removal. (*Courtesy: Tumb-L-Matic Co.*)

production is done of the same items, the blasting can be arranged to be done on a conveyor belt. Figures 19-19 and 19-20 show parts being loaded into rotating table and drum blasting machines. Figure 19-21 shows a conveyor belt deflashing system. Figures 19-22, 19-23, and 19-24 show typical parts before and after deflashing. (See pp. 818 and 819.)

Polishing

Molded parts normally have a polished surface, reproducing the mold surface, since such a surface releases most easily. There are times, however, when the part may require additional polishing to cover small defects, or accentuate the surface. The simplest technique is to buff the part with muslin wheels to which a cutting compound, such as tripoli wax, is applied. Stitched wheels are used where fast-cutting action is desired, and loose wheels are used

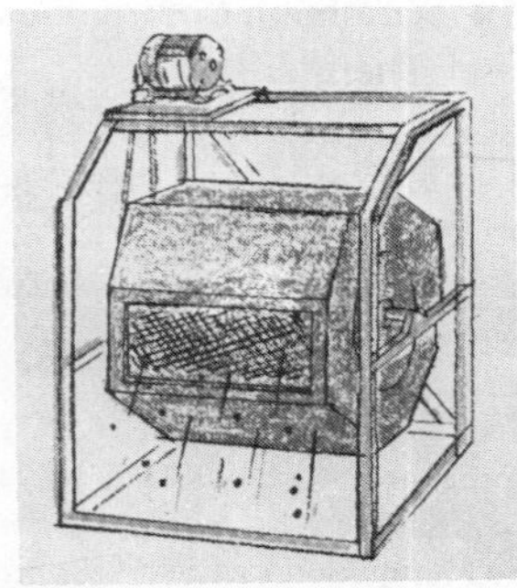

Fig. 19-17. Tumbling barrel with an open side for flash particle ejection. (*Courtesy: Structural Foam Division, The Society of the Plastics Industry, Inc.*)

Table 19-13. Tumbling Media Used for Thermoplastics.*

Medium	Type	Size Range in.	Use
A	Maple wood pegs	$\frac{1}{16}$ by $\frac{5}{16}$ to $\frac{1}{4}$ by $1\frac{1}{4}$	Cutting, deburring, polishing
B	Maple wood balls	$\frac{3}{8}$ diam. to $\frac{7}{8}$ diam.	Cutting, deburring, deflashing, polishing
C	Felt blocks	$\frac{1}{2}$ in. sq. to $1\frac{1}{4}$ in. sq.	Cutting, deburring
D	Lignum vitae	$\frac{1}{2}$ in. sq. to 1 in. sq.	Cutting, polishing
E	Ivory nut chips	$\frac{1}{8}$ to $\frac{7}{8}$	Cutting, deburring, deflashing, polishing
F	Steel shot and shapes	$\frac{1}{16}$ to $\frac{1}{2}$	Deflashing

*Ashmore, 1960. Reprinted by permission from *Modern Plastics*, McGraw-Hill, Inc.

where less cutting is required or where the wheel must be pliable enough to adapt to irregular contours on the part. The loose wheel also is less likely to cause scorching of the part than the stiffer wheel.

The usual procedure when cleaning up surface irregularities is to first use the stiffer wheel with the recommended

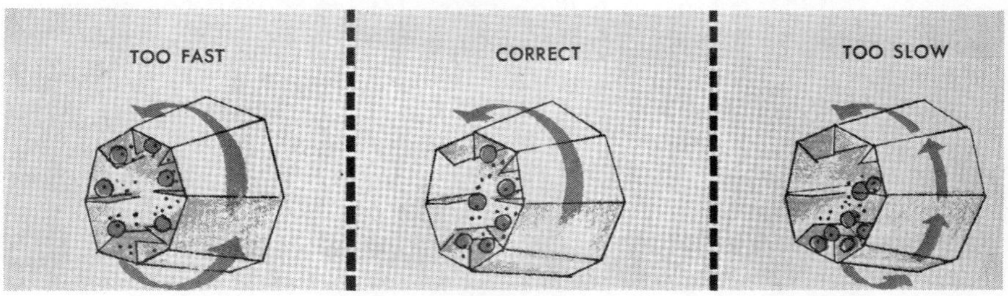

Fig. 19-16. The correct speed of rotation depends on the load size, the media, and the parts, and must be found experimentally. (*Courtesy: Structural Foam Division, The Society of the Plastics Industry, Inc.*)

Table 19-14. Tumbling Compounds Used for Thermoplastics.*

Compound	Composition	Use
A	Graded hardwood sawdust impregnated with oils and coarse abrasives.	Cutting and deburring
B	Graded hardwood sawdust impregnated with oils and medium abrasives.	Cutting, deburring, and smoothing
C	Graded hardwood sawdust impregnated with oils and fine abrasives.	Smoothing
D	Graded hardwood sawdust impregnated with oils and polishing powders.	Smoothing, polishing, & lustering
E	Cream base with extra fine abrasives and polishing powders.	Smoothing, very light cutting, & polishing
F	Cream base with polishing powders.	Lustering

*Ashmore, 1960. Reprinted by permission from *Modern Plastics*, McGraw-Hill, Inc.

cutting compound. Then, after the surface has been polished so the irregularities can no longer be seen, the part is cleaned, with soap and water or an appropriate solvent. After all traces of the cutting compound have been removed, the part is given a final buff with a dry, loose muslin wheel. At this time the part must be moved continually, or the wheel is moved continually, so no streak marks develop because of excessive heat.

A special procedure, known as ashing, is employed when relatively large amounts of material must be cut away. Various compounds of pumice or rouge are used on the wheel, which is kept wet. If done correctly this procedure will correspond to wet sanding and, because it is kept wet, has a cooling effect which aids in preventing ripples or flow marks from appearing.

The polishing, buffing, and ashing procedures described can all be done manually, and for small quantities are commonly done by skilled operators. There are also buffing machines which can be set up to do one or more of the operations completely automatically. Their use, however, must be justified by large-volume production for long periods of time on the same item.

COLORING, COATING, AND DECORATING

Coloring

One of the basic appeals of plastics parts is the capacity for through and through coloration, for all thermoplastics and for most thermoset materials. Thermoplastics, since they are used far more extensively in a colored state than are thermosets, have naturally received the major share of attention with regard to methods of inducing coloration, and the colorants which are available.

The first methods of obtaining colored parts, in cellulosics, consisted of procuring colored resins from the polymer producers. In this case the manufacturers mixed the colorants—organic pigments, inorganic pigments, and, to a lesser extent, transparent dyestuffs—with the resins during fabrication. The colorants were mixed in on hot rolls, in Banbury mixers, and in some cases while the resins were in a liquid or pasty state. With the thorough mixing obtained in these processes, which was similar to grinding in paint pigments in a vehicle, it was possible to mold uniform, well-colored parts with the ram-type injection machines in use.

Dry Powder Coloring. With the growth of the plastics industry after World II, the necessity to stock large inven-

Table 19-15. Guide to Tumbling Some Thermoplastics.*

Plastic to be Finished	Approx. Size of Parts oz.	Finish Desired	Typical Media Used[a]	Approx. Ratio Parts to Media	Typical Tumbling Compound[b]	Tumbling Cycle, 2 cu. ft. hr.
Acrylic	1 to 4	Rough[c]	A, B, C, E	1:1	A, B	4 to 24
		Smooth	A, B, E	1:1	D, E	8 to 60
		Polish	A, B	1:1	F	4 to 8
Acetate	1 to 8	Rough	A, B, C, E	1:2	A, B	8 to 48
		Smooth	A, B, E	1:2	D, E	16 to 48
		Polish	A, B	1:2	F	8 to 16
Butyrate	1 to 6	Rough	A, B, C, E	1:2	A, B, C	8 to 48
		Smooth	A, B, E	1:2	D, E	16 to 48
		Polish	A, B	1:2	F	8 to 24

[a] See Table 19-13 for identification.
[b] See Table 19-14 for identification.
[c] "Rough" means cutting operation such as deflashing, cleaning holes, etc.
*Ashmore, 1960. Reprinted by permission from *Modern Plastics*, McGraw-Hill, Inc.

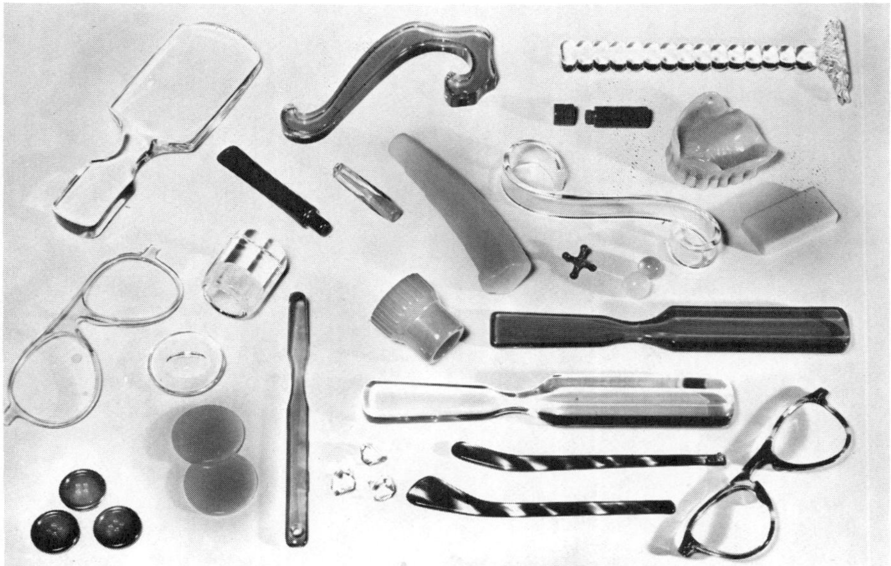

Fig. 19-18. Typical parts finished in a tumbling barrel. (*Courtesy: Tumb-L-Matic Co.*)

tories of many colored resins led to the development of another technique for obtaining colored materials. This was called in-house coloring and consisted of mixing uncolored "natural" materials with powdered colorants, using in most cases, 55-gallon tumbling barrels. This system, which allowed a molder to stock a large amount of so-called crystal pellets only, and then use a small stock of many colors, still had a number of drawbacks. Probably the worst drawback was that the parts, which initially were produced in ram injection machines or were extruded, did not come out in the highest quality at all times. Streaks, and nonuniform coloration were all too common. Other drawbacks were the dust which necessitated separated mixing rooms, and the relatively high amount of labor required for the operation. Nevertheless this system is still in use for production of some low-cost parts.

Fig. 19-19. Rotating drum deflasher using grit blasting. (*Courtesy: Wheelabrator-Frye Inc., Materials Cleaning Systems Division, Mishawaka, Ind.*)

Fig. 19-20. Rotary table deflashing. (*Courtesy: Wheelabrator-Frye Inc., Materials Cleaning Systems Division, Mishawaka, Ind.*)

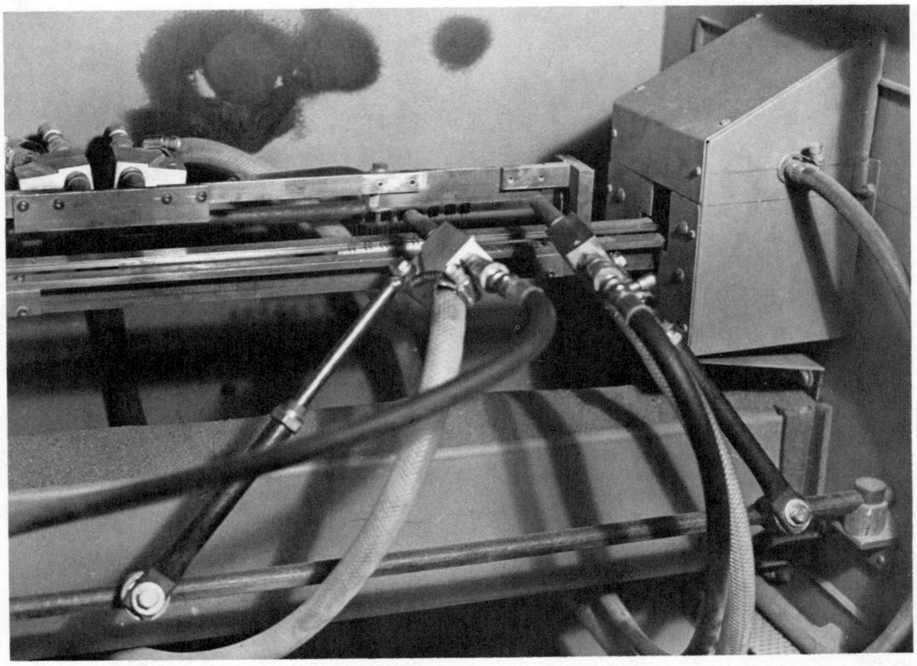

Fig. 19-21. Conveyor belt deflashing. (*Courtesy: Advanced Plastics Machinery Corporation, Ivyland, Pa.*)

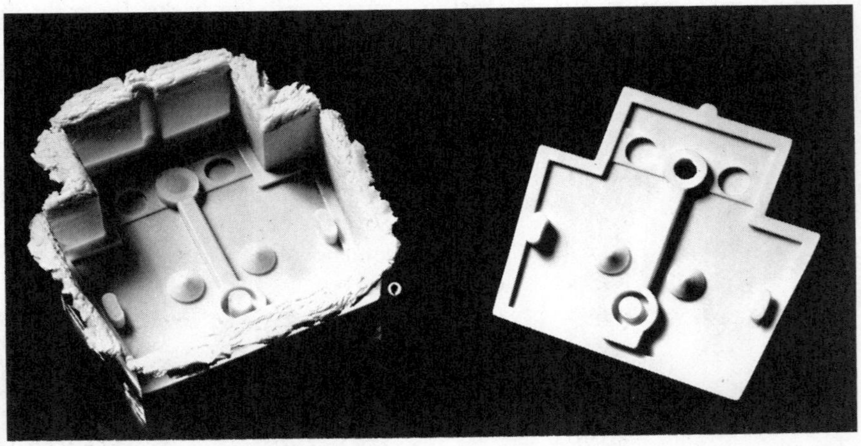

Fig. 19-22. Molded part before and after flash removal by blasting. (*Courtesy: Advanced Plastics Machinery Corporation, Ivyland, Pa.*)

Color Concentrates. With the advent of the screw injection machine in the 1960s and larger L/D ratio extruders, significant improvements were made in resin-pigment mixing in the barrel of the machine. A further improvement was the development of color concentrates. Color concentrates are small pellets of from 50 to 80% pigment dispersed in a polymer carrier. The carrier can be any of the thermoplastic resins and is usually made up as a stock concentrate by the concentrate manufacturer, or it will be produced as a custom concentrate for the molder. In any case, the pigment is thoroughly dispersed in the resin using hot rollers, Banbury mixers, etc., just as the resin manufacturer did previously. The molder has the same advantage of having to stock only the "natural" material in large quantities, and small quantities of the color concentrate.

Another advantage to the use of color concentrates is that the dust problem associated with the powder colorants is eliminated. First, because the pellets are inherently dust free. Second, their introduction brought about the development of automatic color blenders which could be attached to directly feed into the hopper of the molding press or the extruder. Figure 19-25 illustrates blenders used to mix virgin material and colorant, or two additional materials—regrind and colorant.

In the selection and use of colorant systems, either the dry powder or the color concentrate, two terms are of par-

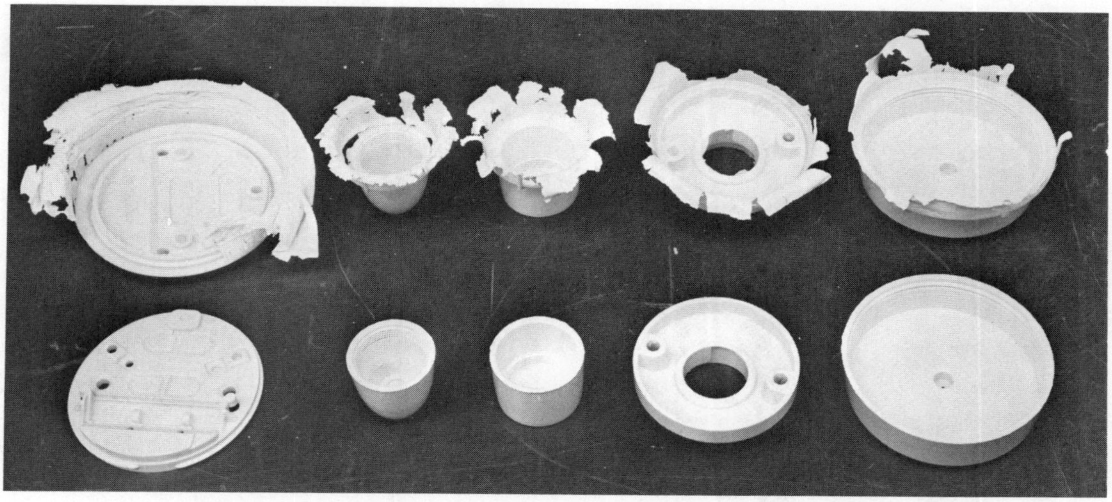

Fig. 19-23. Compression-molded parts before and after air-impact deflashing. (*Courtesy: Advanced Plastics Machinery Corporation, Ivyland, Pa.*)

Fig. 19-24. Typical parts deflashed by blasting. (*Courtesy: Advanced Plastics Machinery Corporation, Ivyland, Pa.*)

ticular importance. One is the "let-down ratio" of the colorant which refers to the number of pounds of natural resin which can be satisfactorily colored by 1 pound of colorant. The usual let-down ratios are 10:1 to as high as 30:1. The second term is "metamerism" which refers to the phenomenon exhibited by two surfaces that appear to be of the same color when viewed under one light source, but which are not matched in color when viewed under a different light; for example, sunlight the first time and incandescent light the second. To prevent this, when specifying colors which must match other colors, the viewing light source should also be specified.

In general, the L/D ratio of the press or extruder will determine how much let-down is possible with any given concentrate. The higher the L/D ratio, the greater will be the let-down. Other factors which have an influence are the speed of the screw, particularly in an extruder (too fast a screw speed will lower the let-down ratio), and the temperature profile of the barrel. Increasing the temperature at the feed section slightly, 5–10°C, (10–15°F) is said sometimes to result in a higher let-down ratio.

When using a color concentrate, it should be noted that opacity is not a function of the pigment as a general rule. Opacity in many cases can be increased at less expense by

Fig. 19-25. Powder concentrate blender. (*Courtesy: Whitlock, Inc.*, manufacturer of 680 Series Blender)

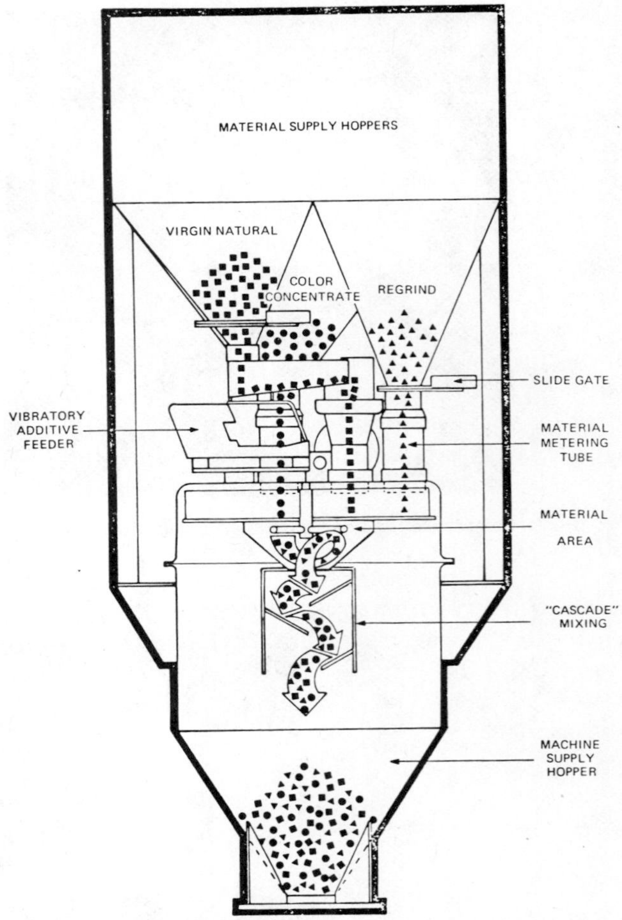

Fig. 19-25a. Schematic of powder concentrate blender. (*Courtesy: Whitlock, Inc.*)

increasing the amount of carbon black or titanium dioxide in a colorant, rather than increasing the more expensive pigment. Many molders stock titanium or carbon black concentrates simply to increase the opacity of items such as semi-transparent films used for trash bags and similar applications.

Color concentrates were originally developed to be a source of color for the resins, but they have also taken on several other functions. Concentrates are now available which, in addition to the colorant, can also contain fire-retardant chemicals, blowing agents, UV absorbers, antioxidants, and other additives. Prior to use of these MFCs (multifunctional concentrates), they should be thoroughly checked out. This precaution should also be taken to be sure the carrier resin in the concentrate is compatible with the matrix resin to be colored. One of the most common carriers is low-density polyethylene (LDPE), which is usually satisfactory for use with other olefinics; for example, HDPE or PP. It is in fact advantageous for use in these matrices because its lower melting range assures thorough melting and dispersing in the matrix. The use of an LDPE carrier would probably not be satisfactory in a polystyrene or an acrylic matrix, just to name a few. There are so-called universal color concentrates available which are said to blend in with most resins. They should also be checked out carefully before being used in production.

Liquid Color Concentrates. There are also commercially available liquid color concentrates which are metered directly into the throat of the molding press or extruder. These materials are said to offer let-down ratios as high as 100:1. The liquids are materials with a viscosity approximating that of a heavy oil. In use, the metering pump, which is usually mounted on the liquid reservoir, is tied in directly with the screw rotation; that is, it only pumps on the return stroke of the injection screw and when used on an extruder may be driven by the extruder screw to compensate for screw speed fluctuations.

The use of the liquid concentrate can cause a decrease in power requirements due to the lubricating properties of the liquid carrier. At the same time, because of the small quantity of carrier, which must be absorbed by the resin, the change in physical properties of the matrix resin is said to be negligible. Like the color concentrates, the liquid colorants can also incorporate blowing agents, UV absorbers, etc. The cost per pound of the liquid colorants is considerably higher than pellet concentrates, but with the higher let-down ratio is said to be lower in ultimate cost.

Coloring Thermosets. The coloring of thermosetting plastics has some special problems not encountered in the coloring of thermoplastic materials. These have to do largely with the fact that thermosets undergo a chemical reaction when they are molded. They are also, in many cases, used at elevated temperatures, and a number of them yellow with age or from exposure to ultraviolet radiation from the sun.

In almost all cases, molding powders, such as phenolics, uras, melamines, epoxies, and alkyds, are procured from the manufacturer in a colored state, if desired. With the exception of the phenolics, all the others can be made in many colors, light and dark, and most are available in stock colors from the manufacturers. The phenolics, because of the natural tan to brown color of the polymer, and subsequent yellowing with age, are usually available only in brown, black, or dark red, green or blue.

Thermosetting polymers available in liquid form, such as epoxies, polyesters, and urethanes, are usually colored by means of paste colorants which are added directly to the liquid resin. The paste colorants are made by grinding the pigments into the appropriate carrier by companies which specialize in such work. (Simple mixing of the powdered pigment in the resin by the molder usually does not result in a uniform dispersion.) In selecting a colorant system to be used with a liquid resin system, of any type, the following points should be considered:

1. The colorant must be stable at the molding temperature or at the use temperature of the part.
2. The colorant must not affect the polymerization reaction. As an example, certain pigments change the pH of a urethane reaction which can drastically change the reaction time.
3. The colorant should not affect the physical properties of the final part; for example, over 1% of colorant can reduce the tensile and tear strength of urethane foam.
4. For outdoor use, colorants should usually be inorganic pigments. This is particularly important with polyesters, which are exposed to the weather in many cases.

COATING

Coating of a plastics part implies either the application of a paint coating, or a metallizing coating, or possibly the application of a film coating over the surface. Of the various coating methods, the application of paint is the simplest, and is the most used when either a large or a small portion of the surface is to be coated. There are two main reasons why plastics parts are coated after molding: (1) to produce a new color or several colors on the part, and (2) to obtain additional protection from the effects of UV and weathering. Some other reasons are to increase the abrasion resistance of the surface, to produce greater chemical resistance, to obtain electrical shielding, and possibly to cover defects such as mold marks, swirl patterns, etc. All of these can be obtained with a paint coating, to one degree or another.

Paint Coatings

There are paint coatings available for all types of plastics; however, not all types of paints can be used on all plastics materials. The great majority of plastics paint coatings, at the time of writing, fall into two classes—lacquers and enamels. Lacquers essentially consist of resins dissolved in a solvent. They are usually air dried, by the evaporation of the solvent. Enamels are also solutions of resins in solvents; however, they dry or cure by polymerization induced by heat and/or oxygen. Other types of paints are coming into favor because of either air pollution restrictions, or the need to conserve power used for baking or drying. These include water-base paints, mainly acrylics, and radiation-cured paints, which include polyester, acrylic, epoxy, and epoxy-acrylic based materials. The majority of the radiation-cured paints are cured by ultraviolet radiation; however, there are paints commercially available which are electron beam cured. The main reason for the use of the radiation-cured materials is that the paints are dried in seconds, with no heat and with very little or no solvents used. The major disadvantage currently with the UV-curing paints is the fact that clear coatings can be cured only up to a thickness of approximately 1 mil, and pigmented coatings are limited to a maximum thickness of approximately 0.5 mil. Electron beam coatings can be cured up to a thickness of 2-3 mils, but at a considerably higher cost for the paint, and for the equipment needed to cure the paint. In both cases, the radiation-cured coatings are best suited for flat or nearly flat surfaces since the radiation is a straight line energy source, and would not be effective on round parts, internal surfaces, etc.

Powder coatings are available which require no solvents. These materials are used in electrostatic systems where the coatings can be applied to all surfaces of a part, without rotation, and with virtually complete elimination of overspray. Since the powder coatings must be fused after application, their use is therefore almost always confined to thermosets that can withstand the fusing temperatures which range from 120°C to 175°C (250-350°F). For a more complete discussion of powder coatings, see the section on Powdered Coatings in this chapter.

The selection of a paint system to be used for any given plastic part depends, to the greatest extent, on the type of plastic material to be painted and to the processing history of the part. The heat-distortion temperature of the plastic will determine whether or not a baking enamel can be used. The HDT should be at least 10°-15°C(20-30°F) above the baking temperature of the paint. The tendency of the plastic to be attacked by the paint solvent must be considered. In this respect the possibility of residual stresses in the part should also be taken into consideration, since the paint solvent could cause crazing at the stress points. Acrylics and styrenes particularly should be annealed prior to coating, if this has not already been done. The possibility of mold-release residues and/or plasticizer or other additives migrating into the coating at a later date, and causing bubbles or loss of adhesion, must also be determined before deciding on a paint system.

Before application of any coating to the plastics parts, they should be thoroughly cleaned of all traces of mold release, dirt, oil, or finger marks. For certain materials, this could consist of a simple wash in a detergent solution. For other plastics, a solvent wipe is also required in order to prepare the surface for the paint coating. Olefins, fluorocarbons, and acetals all require a special surface treatment to insure that the coating, or a primer, will show satisfactory adhesion. For the olefins, flame treatments are the most used (see Figure 19-26), but chemical oxidizing solutions and plasma etch are also employed (see Figure 19-27). Acetals can be treated by a combination of a chemical dip (satinizing) and heat treatment as shown in Figure 19-28. Table 19-16 is a suggested selection guide for a number of the common plastics.

In addition to the conventional paint coatings and powder coatings, there are a number of less conventional coatings which are applied to different plastics for specific purposes, usually not having to do with colorability. Some of these special coatings are as follows:

1. Intumescent coatings: Fire-resistant coatings applied to a part in a thickness of 50 to 60 mils which, during a fire, foam up to a thickness of $\frac{1}{2}$ to 1 inch, and protect the part from the flame and high temperature.
2. Anti-static coatings: Applied to minimize the tendency of the part to acquire or hold a static charge, which might attract dust or dirt. Usually function by absorbing moisture from the air. Not too effective in very dry atmospheres.

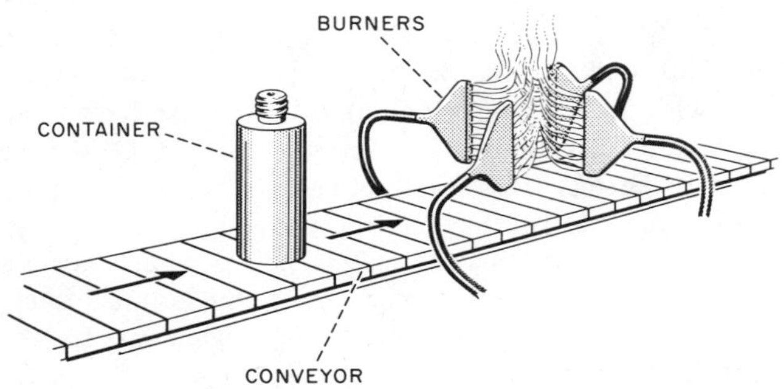

Fig. 19-26a. A conveyor carries the molded item to be treated through the flame. If the area to be treated is small, impingement by only one flame may be adequate. If a larger, ring-shaped area of a surface requires treatment, several flames may be provided, such as the four flames shown in this drawing. (*Courtesy: U.S. Industrial Chemicals Co.*)

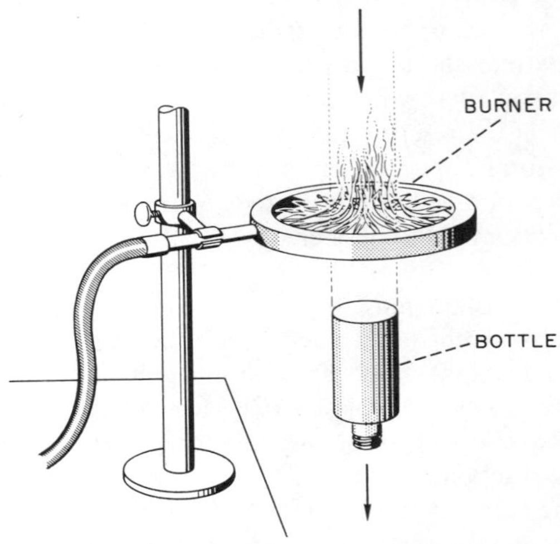

Fig. 19-26b. A cylindrical item is treated while falling through a ring-shaped burner. (*Courtesy: U.S. Industrial Chemicals Co.*)

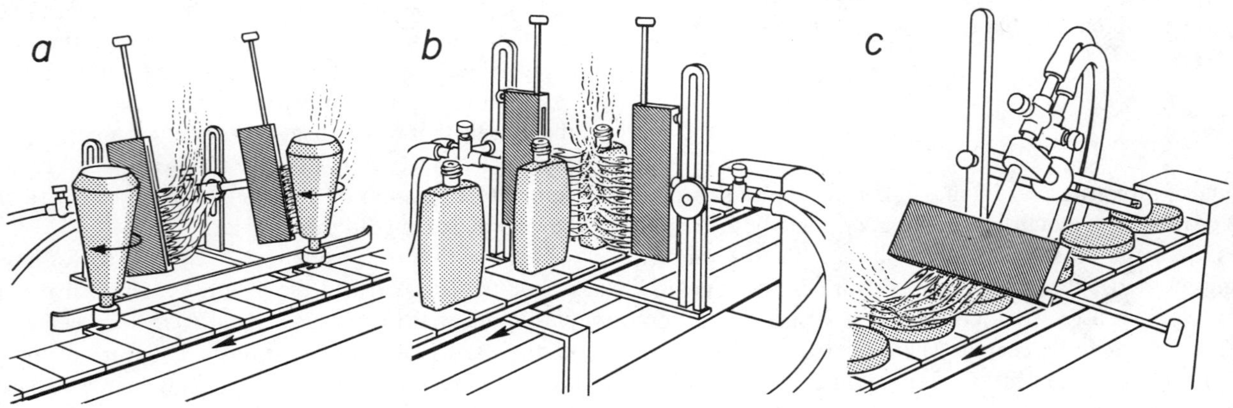

Fig. 19-26c. These drawings show how the burners on the machine in Fig. 19-26b can be adapted to a great variety of occurring shapes and sizes of moldings to be flame treated for printability and adhesion. (A) The burners are tilted at an angle to treat cone-shaped bottles. (B) Flat bottles are treated on both sides simultaneously by passing burners mounted on opposite sides of the conveyor belt. (C) Only one burner is required to treat the top surface of flat moldings. (*Courtesy: U.S. Industrial Chemicals Co.*)

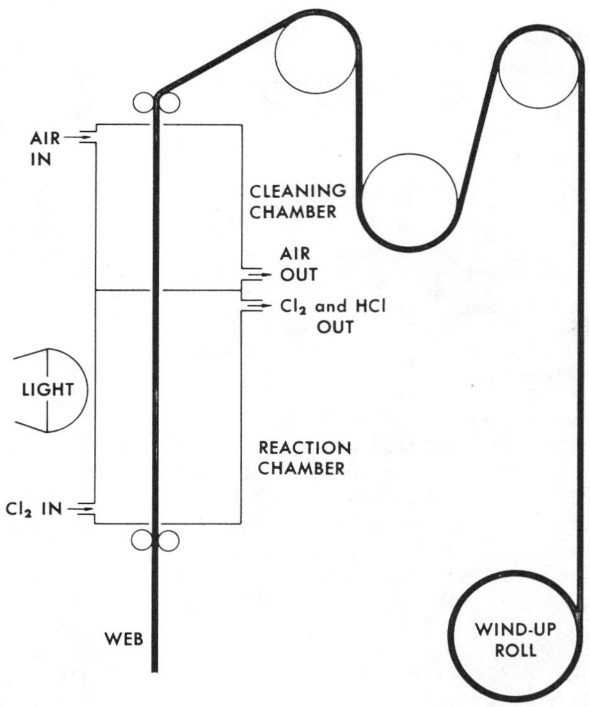

Fig. 19-27. Chlorination treatment process for olefins. (Printed with permission of The Dow Chemical Co.)

3. Abrasion-resistant (AR) coatings: Transparent coatings used on acrylic and polycarbonate to increase the abrasion resistance of the surface.
4. Electrically conductive coatings: Coatings usually containing metal particles (mostly silver or copper), or graphite to render the coating electrically conductive. Some with fairly high resistance are used only as static charge dissipators, others with much lower electrical resistance are used to conduct current so they can be used as heating films, or as radio-frequency shields in electronic applications. Transparent electrically conductive coatings are also available, but require special application techniques.
5. Organosol coatings: Used where a thick coating, 50 to 60 mils, is required in spray or dip application, and where the coating is required to exhibit a slight amount of resilience. An example might be the coating used on a metal steering wheel on a car or an airplane.

Paint Application Techniques. There are a number of different methods which can be used for application of the paint coating; the choice usually depends on the number of parts to be painted and the type of part. In general, any of the paints can be applied by any of the methods listed below. Probably the most used is conventional spraying, which is either done manually or automatically. Automatic conventional spraying, however, requires elaborate equipment to be successful and is usually confined to large-volume production.

Electrostatic spraying is done by establishing a high-voltage electrostatic charge between the paint and the part, which is usually grounded. Because of this potential difference the paint is attracted to the part and will wrap around it, thus eliminating a large portion of the overspray which is usually wasted in conventional spraying. Electrostatic spraying is therefore well adapted to automation, because of the paint savings and the possibility of eliminating rotating fixtures required for many conventional spray applications.

Other methods of paint application include roller coating, wiping, and dipping. Flow coating is another technique which, as the name indicates, is simply allowing the paint coating to flow down the surface of the part. Usually this surface is flat. Silk screen coating is a technique in which the coating is wiped through a pattern on a screen. The term silk screen is used since the original screens were of silk; currently, screens are made of nylon fabric, polyester

Table 19-16. Selection Guide to Paints for Plastics.*

Plastic	Coatability	Pretreatment	Recommended Coating Types
ABS	Good	Solvent wipe	Vinyl, modified vinyl, polyurethane
Acrylic	Good	Solvent wipe	Acrylic, modified acrylic, nitrocellulose, polyurethane, epoxy
Cellulose acetate and butyrate	Good	Solvent wipe	Acrylic, vinyl, nitrocellulose
Nylon	Fair	Detergent wash	Acrylic, vinyl, polyurethane
Phenolic and melamine	Difficult	Solvent wipe	Alkyds, polyurethane, epoxy, acrylic
Polycarbonate	Fair	Primer	Polyurethane, acrylic, modified acrylic
Polyester	Good	Solvent wipe	Polyurethane, epoxy
Polyolefins	Difficult	Flame solvent or primer	Polyurethane, nitrocellulose, modified acrylic
Polyphenylene oxide-based	Fair	Primer	Polyurethane
Polyurethane	Good	Primer	Polyurethane
Polyvinyl chloride	Fair	Solvent wipe or primer	Polyurethane, nitrocellulose

*Reprinted from *Plastics Engineering Handbook of the SPI* edited by Frados.

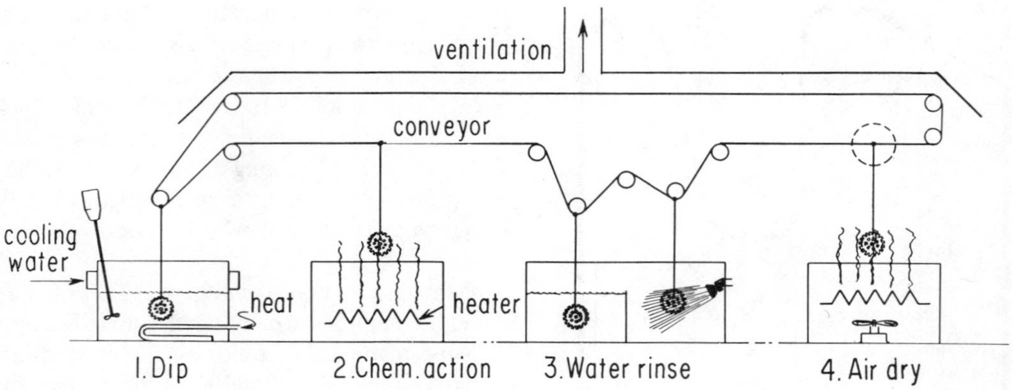

"SATINIZING" FORMULATION*

Ingredients	Concentration (% by weight)
Kieselguhr	0.5
Dioxane	3.0
Para-Toluene Sulfonic Acid (p-TSA)	0.3
"PERCLENE" perchlorethylene	96.2

No preferential method is required for mixing the ingredients. The kieselguhr, dioxane, and p-TSA may be combined into a premix to which PERCLENE may be added whenever "Satinizing" is to be performed.

Treatment Process

Step One: Dip The article is immersed in the bath for 10 to 30 seconds at 175° to 250°F. Cleansing of the part is simultaneously accomplished by the degreasing action of the PERCLENE. This immersion step initiates the "Satinizing" action. Parts should not touch each other or other objects except the hooks by which they are held.

*Courtesy: E.I. Du Pont de Nemours & Co., Inc.

Step Two: Chemical Action Transfer parts immediately from the treatment bath to an air oven adjusted to a temperature of 100°–250°F. This step completes the "Satinizing" treatment. Degree and quality of treatment are easily controlled by the exposure time and the air temperature. The maximum holdup time at 250°F should be 1 minute. At lower temperatures, the time in the oven must be extended to give results comparable to 250°F for 1 minute.

Step Three: Water Rinse Water rinsing "kills" chemical reaction by flushing chemicals from the surface. Cold water may be used, although water at 160° to 175°F will be more efficient. One convenient procedure is to hold the treated article momentarily in water and then rinse with a hot water spray. Thorough washing is essential for achieving uniformly smooth, satin-like surfaces which are essential for producing highly reflective finishes such as in vacuum metallizing. For routine painting, however, traces of the treating composition need not be removed.

Step Four: Air Dry Adjust rate of drying to suit the production schedule. The rate depends on air temperature and humidity. After drying, the article is ready for immediate use, or for packaging and shipment.

Fig. 19-28. Schematic drawing of the chemical surface treatment process for acetal resins showing how it could be adapted for the continuous production of molded parts. (Reprinted by permission from *Modern Plastics*, McGraw-Hill, Inc.)

fabric, or stainless steel mesh. The process is well adapted for application of two or more colors, and/or patterns. It is applicable to curved surfaces as well as flat surfaces (see Figure 19-29).

Silk screening offers one method of obtaining two colors or patterns on a painted part. Another method, particularly

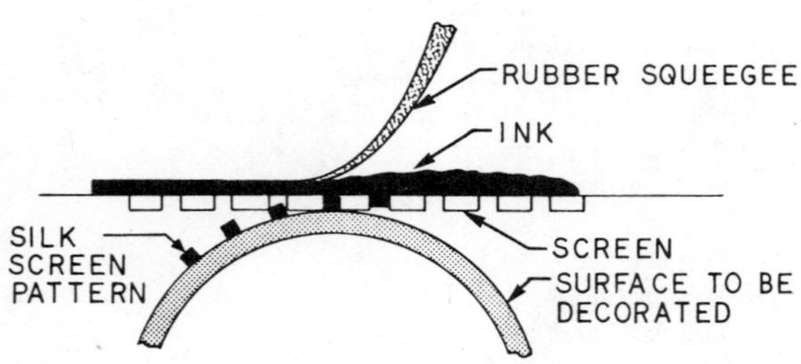

Fig. 19-29. Silk screening is a process of transferring paint through an open screen onto a flat or gently rounded surface. (From *Plastic Product Design* by Beck)

for spray painted parts, is the use of spray masks which, as the name indicates, are used to block or prevent paint from reaching certain areas of the part. The mask may be of paper, vacuum-formed plastic, cast rubber, fiberglass plastic, or electroformed metal. The major requirements are that the mask fit tightly to the part, and not be affected by the paint so it can be reused a number of times. Possibly the most satisfactory, for prolonged use, are various types of electroformed masks, which can be made to fit the part perfectly, with cutouts where the paint is to appear. Figure 19-30 shows how such masks are used.

Hot Stamping

Hot stamping is another process where a decorative coating (a colored film, a rotogravure, or other printed or wood-grained film or a metallized film) is transferred to the surface of the part by heat and pressure, applied for a precisely

PLUG TYPE MASKS

For Protecting Depressed Areas

Plugs suspended from bridging fill depressed sections, preventing paint leakage while spraying backgrounds. Also used for protecting areas to be vacuum plated after spraying other surfaces.

BLOCK CUT-OUT

Plane Surface Masks

A cut-out area outlines dial calibration lines or lettering too small for practical individual masking. It confines paint to a small area, reducing wiping passes.

LIP TYPE MASKS

For Painting Sunken Areas

A lip extending down depressed sidewalls protects background and prevents fogging. Arched bridging of free sections (loose insides of letters and numerals, such as "8") eliminate ties.

LIP TYPE MASKS

For Capping Raised Areas

The lip protects the embossed areas and desired areas of sidewalls from paint. Bridging holds the capping in place.

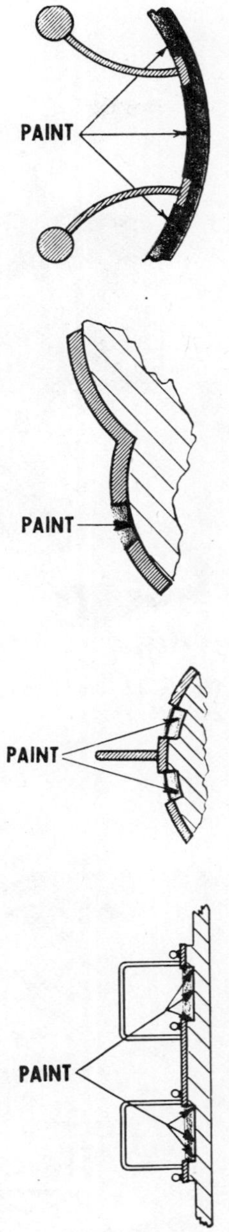

Fig. 19-30. Electroformed paint masks. (*Courtesy: Plastics Design & Processing*)

timed interval. It is a completely dry process, requiring no pretreatment of the plastic, even olefins, and requires a special machine and a heated die to apply the film, in roll form, to any thermoplastic and to some thermosets.

The films are normally applied to flat surfaces or to surfaces with only two-dimensional curvatures, although some films are available which can be applied to spherical or three-dimensional surfaces. The decorative coating is mounted on a polyester, cellophane, or acetate carrier film and may contain as many as six or eight layers in its makeup. The first, or topmost, layer is a wax or low-melting adhesive so it will part easily from the carrier base. The other layers are first a transparent, abrasion-resistant coating, then the decorative coating which may have three or four layers to give it a "deep" three-dimensional or multicolored appearance, and finally a thermally activated adhesive layer, sometimes called a "size" (see Figure 19-31, p. 826).

In hot stamping, the decorative film, also called a roll leaf, is placed between the die and the plastic, with the size facing the plastic. The hot die presses the film into the plastic which softens and takes the form of the engraving of the die. Simultaneously, the colored coating is released from the carrier film and bonds itself to the depression in the plastic. When the die retracts from the plastic, the plastic resolidifies leaving a dry, colored impression of the die. If done at the right temperature, pressure, and with the right dwell time, the process can be automated to give rapid, excellent markings time after time.

To insure good adhesion of the roll film to the plastic, it is important that the die be pressed against the plastic with a good deal of force, and that the correct temperature be maintained at all times. Dies are made of silicone rubber or a soft metal such as zinc, magnesium, copper, or brass. Because of the ability to compensate for small thickness variations in the plastic, silicone rubber is used very frequently, particularly if a roller is required in order to coat large surfaces such as a television cabinet or a table top, box, etc.

Raised or embossed effects are easily achieved by having the three-dimensional effect in the plastic and forcing the film into the contours, or using a three-dimensional or embossed film. In either case, using a soft silicone roller, which will withstand the temperature of from 120°C to 190°C (250-375°F), the embossed pattern is firmly established.

Machines used to apply the hot stamping materials range from small machines which stamp simple patterns such as trade marks or identification codes or decorations, (see Figure 19-32), to large presses which use rollers to completely cover a television set cabinet or an automobile dashboard with a wood grain pattern (see Figure 19-33, p. 826). Many of the small machines are either manually or automatically operated, while most of the large machines are fully automatic, so that the three parameters of temperature, pressure, and time can be controlled as uniformly as possible. In-mold decorating of thermoformed or injection-molded items can also be done using hot stamping foil as shown in Figure 19-34 (p. 826).

Parts to be decorated by hot stamping should be designed with this process in mind from the beginning of the design phase. Some of the factors to be considered, in order to obtain maximum efficiency in the process, are:

1. Use a uniform wall thickness; avoid stamping over ribs and bosses, as they tend to cause shrinks in part.
2. Raised patterns, numbers, or letters should be at least 0.020 inch high, and preferably 0.030 inch or more. This will minimize overstamping in adjacent areas.

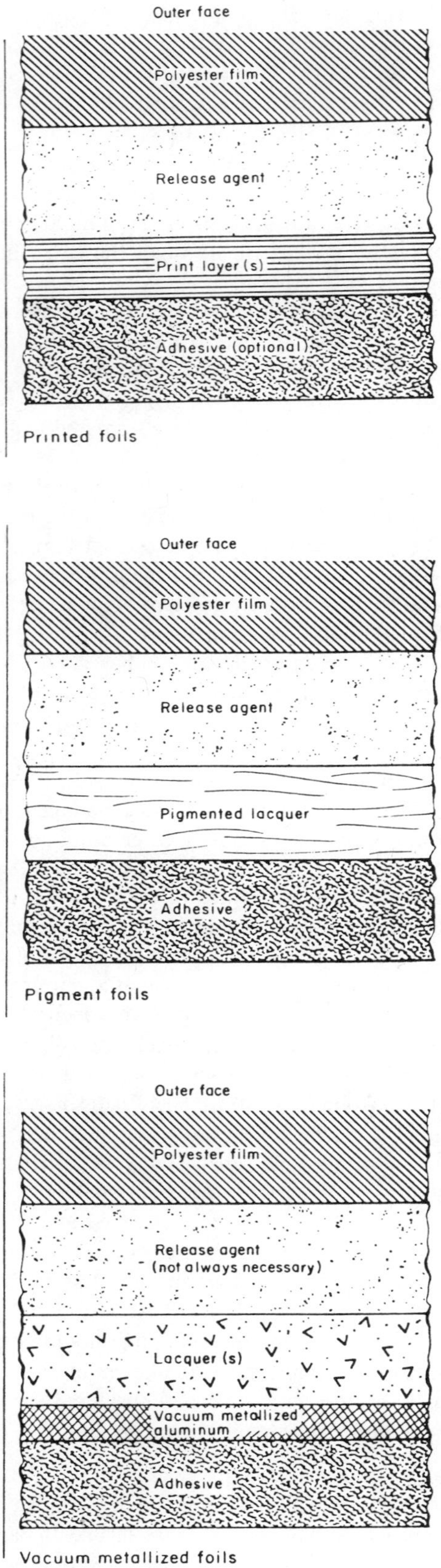

Fig. 19-31. Three types of hot stamping foils. (Reprinted by permission from *Materials Engineering*, January 1975)

Fig. 19-32. Small foil marking machine. (*Courtesy: Kingsley Machine Co.*)

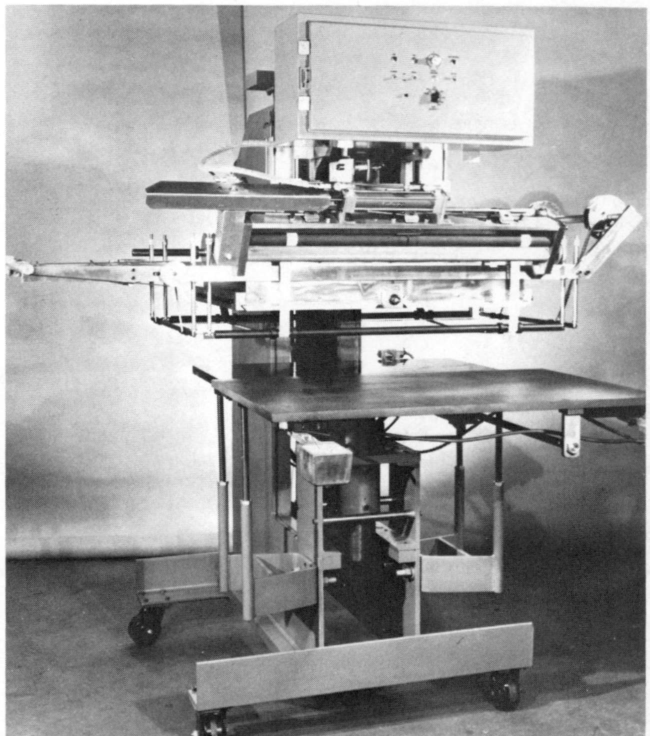

Fig. 19-33. Large hot stamping machine. (*Courtesy: Franklin Manufacturing Corporation, Norwood, Mass.*)

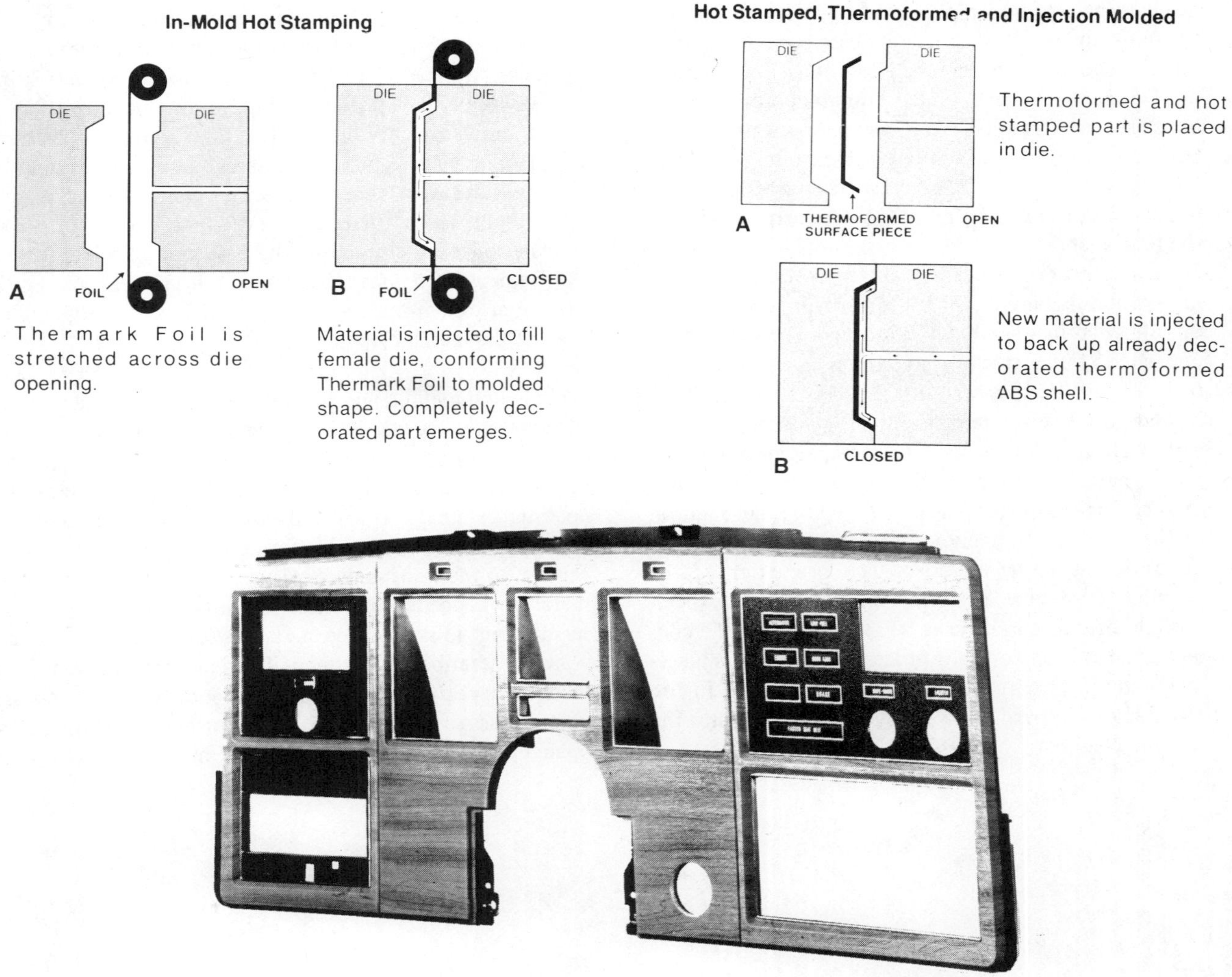

Fig. 19-34. In-mold decorating techniques. (*Courtesy: Thermak, an Avery International Co.*)

3. Edges of patterns and letters should have a slight radius, rather than being sharp. This will minimize cutting of of the silicone pad, thus insuring a longer life to the die.
4. Areas to be stamped should be as flat and parallel as possible.
5. Dimensions must not vary from cavity to cavity in multicavity molds. Variations as small as 0.001 inch can cause difficulty in stamping.

Metallic Finishes

Metallic finishes are applied to plastics parts for decorative purposes, to furnish electrical conductivity, to minimize electromagnetive interference, and to result in a surface with improved solvent resistance and/or abrasion resistance. These finishes are truly metallic and should not be confused with paints containing metal particles. There are, in general use, two types of metallic coatings; vacuum-applied metallic coatings, usually very thin aluminum; and electroplated coatings which can be relatively quite thick and may be of many different metals, including copper, nickel, chrome, gold, silver, etc. The choice of which coating or process is to be employed is usually determined by the type of part and the environment in which it is used.

Vacuum Metallization. The use of a vacuum chamber to deposit a metal coating on a part or on one or both surfaces of a plastics film or sheet is called vacuum metallization. The process consists of several steps which include the following:

1. The base coat application. This consists of applying a clear, high-gloss material to the substrate which is then used as a base for the subsequently applied metal. Each base coat is specially formulated to have high adherence to the particular substrate, to smooth over any small surface irregularities in the substrate, and to act as a

mechanical and molecular bonding surface for the vacuum-applied metal. Improper base coats or improperly applied coatings can cause low adhesion or failure of the metal coating under humid conditions. For best base coat results it is imperative that the parts, before coating, be absolutely clean, and preferably molded without mold-release agents. Areas to be free of metallization can be masked prior to application of the base coat.

2. Metal deposition. Deposition of the metal takes place in a vacuum chamber, which is normally evacuated to 0.2–14 microns ($2\text{–}4 \times 10^{-4}$ mm Hg). The metal to be deposited, in most cases aluminum, is placed in the form of small wires on tungsten filaments which are heated to whatever temperature is required to cause vaporization of the wires. The metal vapor condenses on the parts which are on rotating racks inside the vacuum chamber. Correct placement, racking, and rotation of the parts is extremely important if a uniform metal deposition is to be obtained. The evacuation of the chamber may take 8 to 15 minutes, but the metal flashing off takes as little as 30 seconds. Since the condensation is a line-of-sight process, correct parts placement and rotation are of major importance. Figure 19-35 shows typical racks ready for insertion into the vacuum chamber.
3. Top-coat application. The application of a top coat over the metal surface is the last step in the process. The top coat can be a clear, abrasion-resistant coating, or it may be a tinted coating, so the final color may be a simulation of gold, copper, brass, or any other transparent shade. The top coat is also a specially formulated coating with maximum adhesion to the deposited metal, which at the same time must not affect the adhesion of the metal to the base coat. The top coat is expected to protect the metal from wear, oxidation, or any environmental effects which could cause dulling of the metallic coating.

While any metal could theoretically be vacuum deposited, in actual practice aluminum is the choice in the great majority of the cases, partly because of its relatively low cost, and also because of its highly reflective, brilliant surface. The coating, in most cases, is only one to several millionths of an inch thick which is why the top coat is required. An exception to the transparent top-coat requirement is when the metal is deposited in what is known as second-surface metallizing. (First-surface metallization is the process previously described.) In second-surface metallization, the metal is deposited on the rear or second surface of a transparent plastics or glass part. A final protective coating of an opaque, heavily pigmented lacquer may be used. Second-

Fig. 19-35. Vacuum metallization chamber. (*Courtesy: Galileo Corporation of America, U.S. subsidiary of Officine Galileo, Florence, Italy*)

surface metallization is extensively used for ornamental items such as automotive horn or hub cap emblems, while first-surface metallization is used for parts such as knobs, bezels, etc.

Design Guidelines. The design of parts to be metallized should take into account the characteristics and limitations of the process. Some of the guidelines are as follows:

1. Sharp corners, edges, and serrations should be avoided. The top coat will fail at these areas first when exposed to abrasion.
2. Molds should be highly polished.
3. Gates should be adequate to reduce stress which might be evident as craze marks on application of the base coat.
4. Mold releases should not be used if at all possible.
5. Internal lubricants, added by the resin manufacturer to aid in molding, should be minimized, since these can interfere with good adhesion of the base coat or the metal.
6. Low-density areas in the part, caused by too rapid mold filling and insufficient dwell, may cause irregularities in the base coat such as dullness and loss of adhesion.

Electroplating. Until the early 1960s, electroplating of plastic materials was essentially an encapsulating operation; that is, the part was completely surrounded by the metal coating. This was necessary since there was practically no adhesion of the metal deposit to the plastics surface. With the advent of "platable" grades of ABS, and after that other types of plastics, plating of plastics became a recognized method of plastics finishing. Adhesion of up to 15 pounds per inch of width for the plate is commonly obtained by subjecting the part to several special treatments prior to the actual plating step, which is identical to normal electroplating of metal substrates.

The preplating treatments used for most plastics parts consist of cleaning, etching, neutralizing, catalyzing, accelerating, and electroless nickel or copper deposition. After these treatments, the part is ready for deposition of whatever metal or metals are to be used for the final coating. Figure 19-36 illustrates the process for ABS polymers which, with slight variations, is used for all plastics plating.

The preplating cleaning solutions consist of mild alkaline cleaners, or detergent solutions, depending on the type of plastic. The purpose, of course, is to remove all oils, greases, fingerprints and residual mold-release agents from the surface of the parts. Since most plastics are hydrophobic, the cleaning process may not be too effective. Therefore, parts should be handled as little as possible and should be as clean as possible from the moment of demolding. The cleaning process should not be relied upon to insure satisfactory parts for further processing.

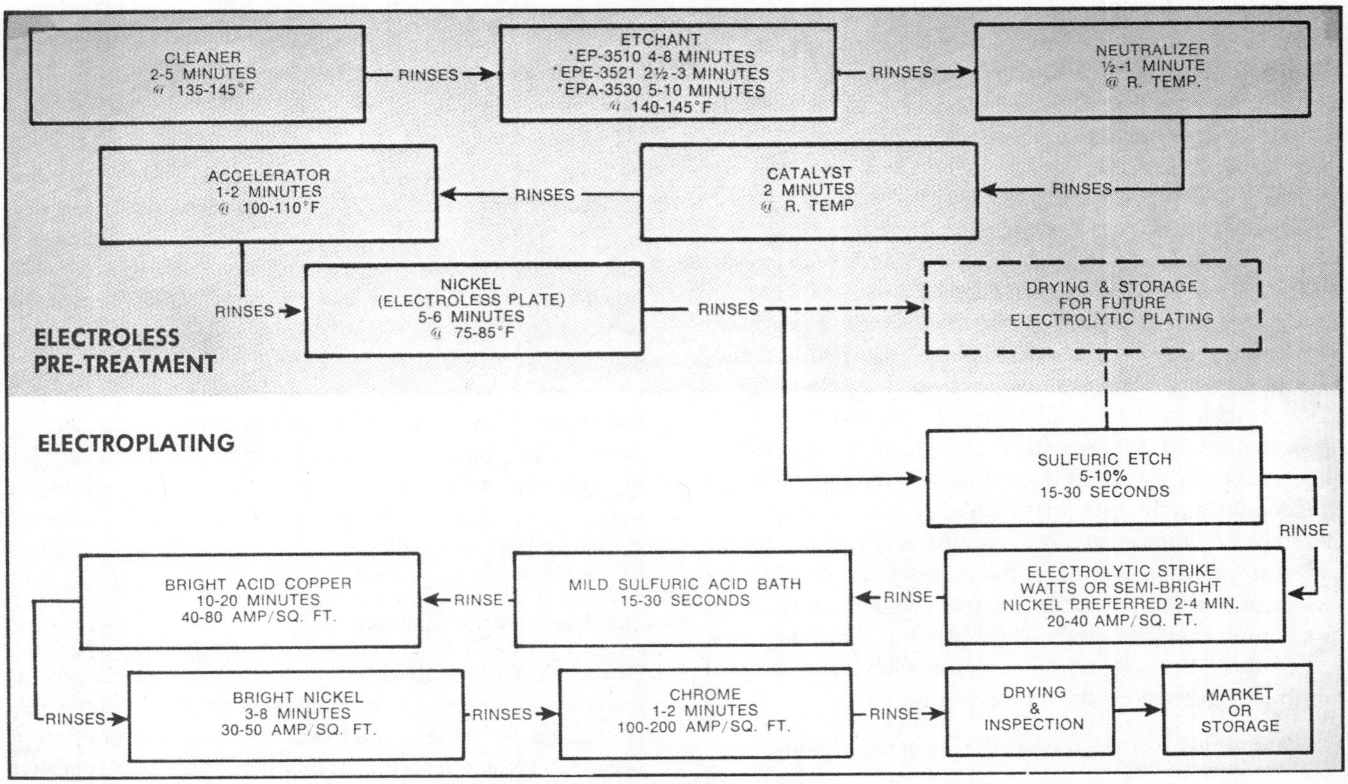

Fig. 19-36. Electroless preplating and electroplating sequence for ABS.

The etchant solutions are materials which act on the surface of the material, particularly ABS, to soften it and then allow the etchant to selectively attack the butadiene domains on the surface. In addition to selectively attacking the surface to produce a porosity or a microetch, the etchant renders the surface hydrophilic, and also removes any remaining foreign matter on the surface. The microetch is used to provide the mechanical bond between the surface and the electroless nickel or copper which is the first conductive coating. These solutions are usually strong oxidizing agents, such as acidic chromium oxides.

Neutralizers are used to prevent any carryover of the oxidizer solutions, which would interfere with the successive steps in the process. The catalysts (also called activators) are used to introduce acidic solutions of tin and palladium salts. The tin compounds are absorbed into the surface of the plastic and also carry with them the palladium ions which act as "seeds" to catalyze the electroless deposition of the nickel or copper. The accelerator acts to remove the excess tin salts, which might interfere with the electroless deposition. The nickel or copper is then deposited to result in a metal thickness of 0.001–0.0025 mils. Because of the extreme thinness and fragility of these conductive coatings, the parts at this point must be handled with great care so they will not become degraded which would be reflected in the final electroplated coating.

The first electrodeposited coating is usually a soft, ductile copper coating deposited from an acidic copper solution. The soft copper is used to obtain a smooth coat for the succeeding coats, and, most important, it acts as a stress-absorbing layer for the next coatings which are usually nickel and chrome. The soft copper acts as a stress absorber for the stresses induced by the nickel and chrome deposits, but also acts to absorb thermal stresses induced by differing thermal coefficients of expansion of the plastic and the metals.

The majority of parts being plated, at the time of writing, were ABS plated with bright chrome finishes. However, while ABS was being used for the largest number of applications, there were a significant number of parts being made in polypropylene, polyphenylene oxide–based material, and, to a lesser extent, polysulfone and polyaryl ether. These parts are being used for all types of applications such as plumbing, appliances, electrical and electronic parts. The greatest user by far, however, will be the automotive industry. Because automobile parts such as radiator grills are subject to possibly the most severe conditions of thermal cycling and environmental exposure, the industry has developed a special test which combines thermal cycling and exposure to environmental chemicals. This test is known as the CASS test because it makes use of copper-accelerated acetic acid salt spray. The steps in the testing process are:

1. Three cycles of the following temperature conditions:
   180°F for 2 hours
   room temperature for 1 hour
   -30°F for 1 hour

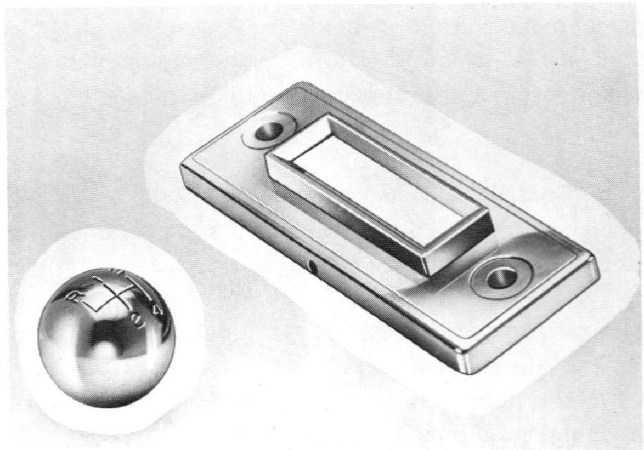

Fig. 19-37. Electroplated plastics parts.

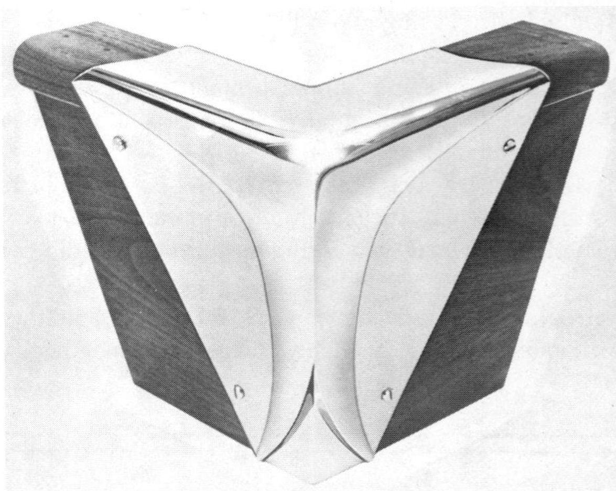

Fig. 19-38. Electroplated plastics part.

2. Exposure to copper-accelerated acetic acid salt spray for 16 hours.

Satisfactory results after exposure to the CASS test, while not a guarantee that the material and/or the process will work in the desired application, does indicate a high-quality coating (see Figures 19-37 and 19-38).

Design Guidelines. To obtain the optimum utility and appearance in a plated plastics part, the requirements for plating should be taken into account when the part design, the mold design, and the molding process are chosen. Relevant design parameters are illustrated in Figure 19-39.

Powdered Plastics Coatings

Powdered plastics of many types are also used as coating materials; however, the powders are used almost exclusively as coatings for metals, and possibly for some thermosetting plastics. The reason for this is that the powders are deposited on the part and then are fused, either during the deposition step or immediately afterward. Because the substrate parts must of necessity be heated above the melting or fusion

Basic design elements

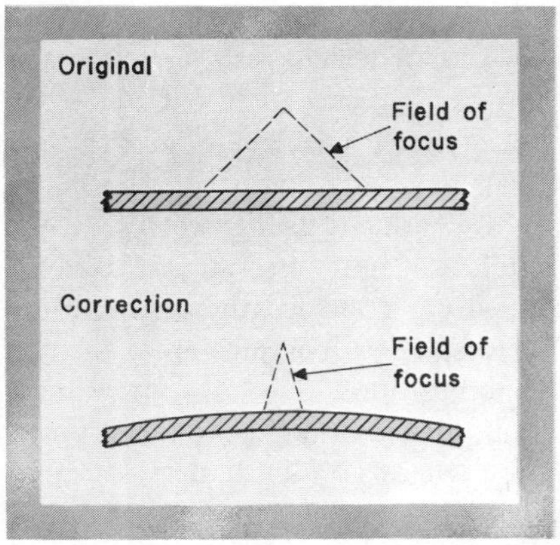

Fig. 19-39a. Crowning large planes camouflages surface irregularities.

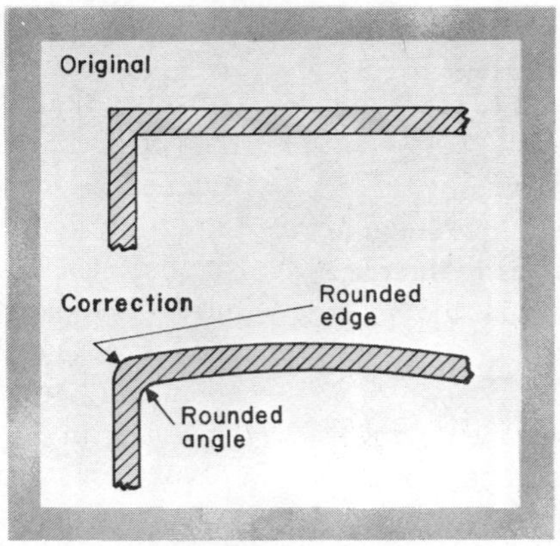

Fig. 19-39b. Angles should be as large as possible and sharp edges avoided.

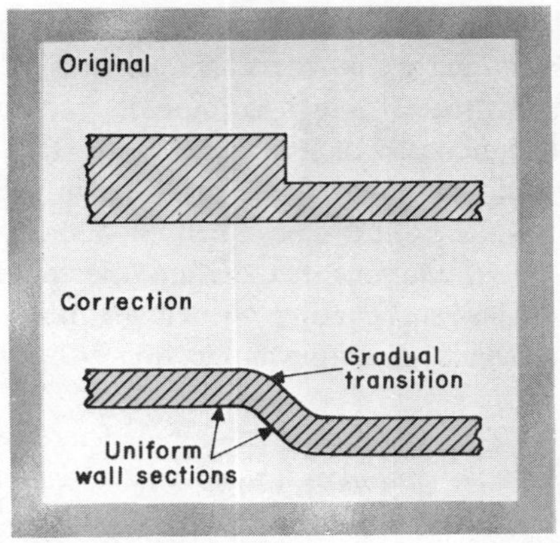

Fig. 19-39c. Uniform wall thickness and gradual transitions are desirable.

Recesses

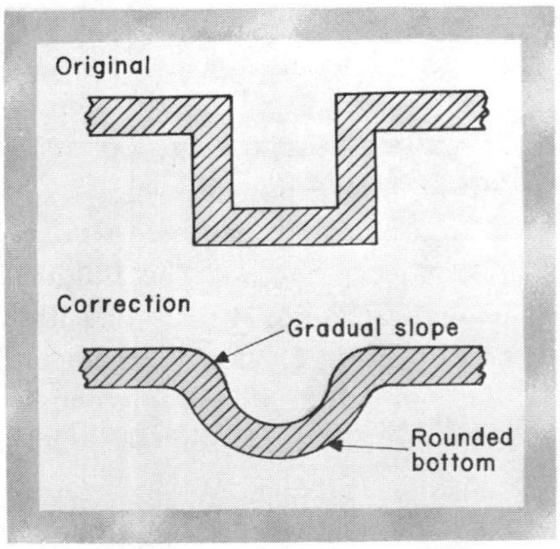

Fig. 19-39d. Flat-bottom grooves should be rounded and their depth limited to 50% of their width. Both internal and external edges should be chamfered or rounded.

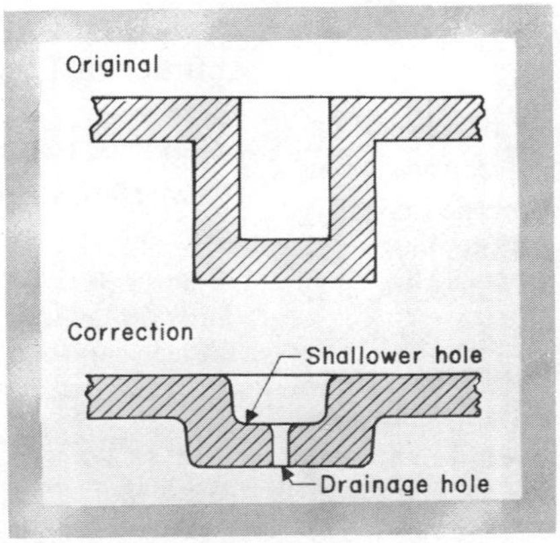

Fig. 19-39e. Blind holes should not be any more than 50% deeper than they are wide. Deep, small holes (less than $\frac{7}{32}$ inch in diameter) trap air that prevents electrodeposition.

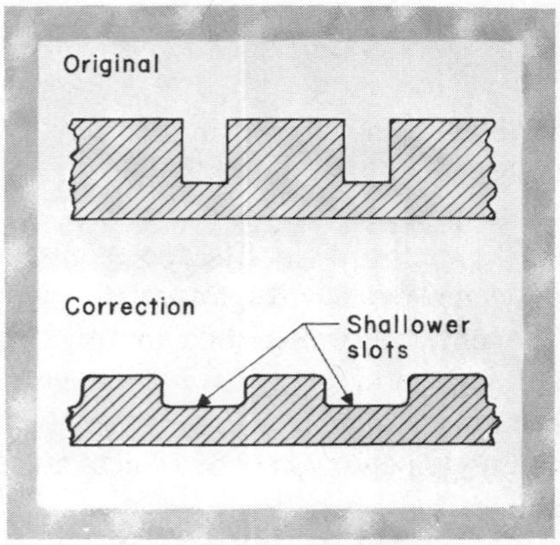

Fig. 19-39f. Slots and indentations should be twice as wide as they are deep. If corners are rounded, plate buildup will be lower in high-current density spots.

Protuberances

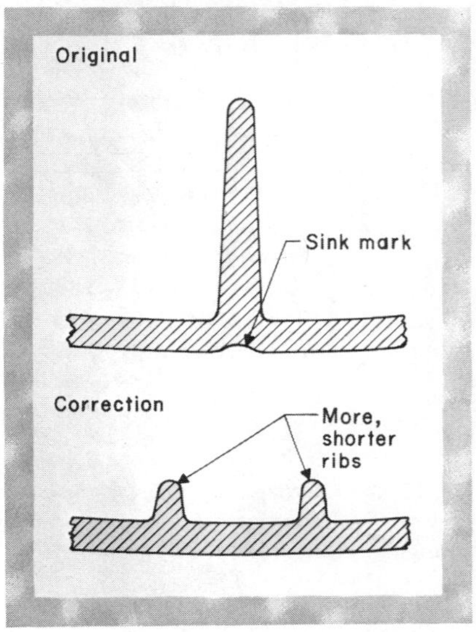

Fig. 19-39g. Ribs are electroplated most easily when they are no more than half as thick and 1-$\frac{1}{2}$ times as tall as the adjacent part wall thickness.

Fig. 19-39h. Bosses and studs should be as short as possible. Plating will be best if their height is less than twice their diameter. (Figs. 19-39a through 19-39h: Courtesy of *Plastics Engineering*)

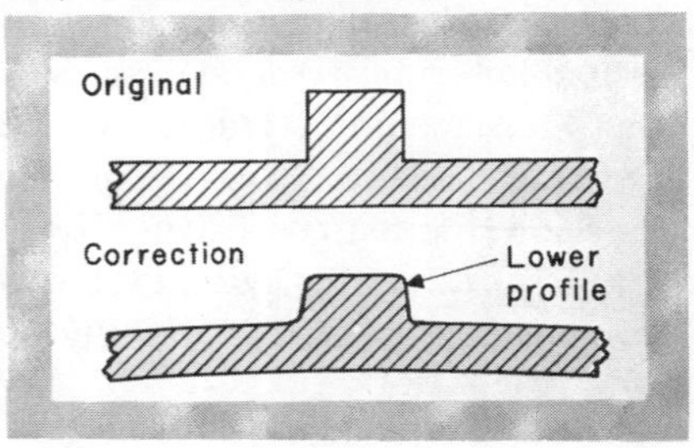

point of the powder, this automatically confines the process to materials which will not be distorted or otherwise affected by the high temperature. There are basically two processes by which the powders are applied to parts: (1) the fluidized-bed process (which is the older method), and (2) spraying techniques, which include electrostatic spraying, (the most used technique), flame spraying, and plasma spraying.

While both fluidized-bed and spraying techniques require special equipment, there are certain advantages in their use of powder coatings which are not evident in the older methods of coating application utilizing liquids.

1. No solvents are required. Therefore there are no pollution problems, no fire hazards, and less chance of pinholes in the coating.
2. Certain polymers, not usable as coatings by other means, because of insolubility or high molecular weight, can be applied only as powders.
3. Heavy coats can be obtained in one application.
4. Both processes are well suited for automation with minimum material losses by overspray or dragout.

Materials. In theory, any thermoplastic polymer should be usable for either fluidized-bed or spraying treatments. In actual practice, only a few resins are used for the majority of the applications. The thermoplastics used include polyethylene, PVC, nylon, chlorinated polyether, and cellulosics. The only thermoset used in any large quantity is epoxy, although polyesters have been used. Mixtures have also been made of PVC with epoxy, polyester and polyethylene, and nylon-epoxy, all of which are employed for special applications.

Polyethylene is used primarily for its low cost and ease of application. The coatings, applied by either technique, have low water absorption, excellent chemical resistance, and good electrical insulation. Polyethylene coatings are used for food handling equipment, chemical apparatus, and for automotive applications such as battery hold-downs.

PVC coatings are used for their excellent impact strength, resilience, and salt spray and weathering resistance. They are also used for general-purpose electrical insulation. Special primers are required for optimum adhesion. The coatings are used for chain-link fence wire coatings, on outdoor furniture, and similar applications. Two types of vinyl powders are used—dry blends and melt mixed. The dry blends are cheaper, but the melt-mixed materials are superior in weather resistance and should be used for most outdoor applications.

Nylon is usually chosen if a low-friction coating is required which also must exhibit good wear and abrasion resistance. Nylon coatings also require a primer for best adhesion. Nylon 11 is generally used for its good abrasion resistance, toughness, and low friction coefficient. Nylon 6/10 is used sometimes as a substitute for nylon 11, but at an increase in cost. Nylon coatings are used extensively for marine applications where they are used to completely encapsulate fittings, bolts, and other hardware items.

Chlorinated polyether coatings are used mainly on chemical equipment where they show excellent resistance to many corrosive liquids at temperatures up to 120°C (250°F). They also show good electrical insulation properties, low water absorption, and can be steam sterilized. The major limitation to the use of this coating is its high cost.

Cellulosics, mainly cellulose acetate butyrate, are used for outdoor applications requiring good weathering, abrasion resistance, and many colors. These coatings are also used extensively for marine applications. Special primers must be used with cellulosics in order to obtain good adhesion.

Epoxy coatings are used wherever a hard, electrical insulating coating is required over a wide temperature range. Depending on the particular epoxy selected, the coatings can be used up to 150°C (300°F), or higher, with dielectric strengths of up to 1200 volts per mil, in thicknesses over 10 mils. No primers are required and adhesion is excellent. Chemical resistance to most solvents and mild acids and bases is good. The major limitation is a tendency toward brittleness if the coating is over a few mils thick.

Fluidized-Bed Coating. Fluidized-bed coating consists of immersing a heated part into a bed of powder which is suspended or fluidized by means of an air stream sent through a porous plate on the bottom of the bed. The fluidized powder acts at though it were a liquid, distributing itself evenly over the part, penetrating into crevices, corners, etc. Since the powder acts like a thin liquid, or very low-density fluid, any shaped part can be easily immersed (see Figure 19-40).

When the powder particles contact the heated part they

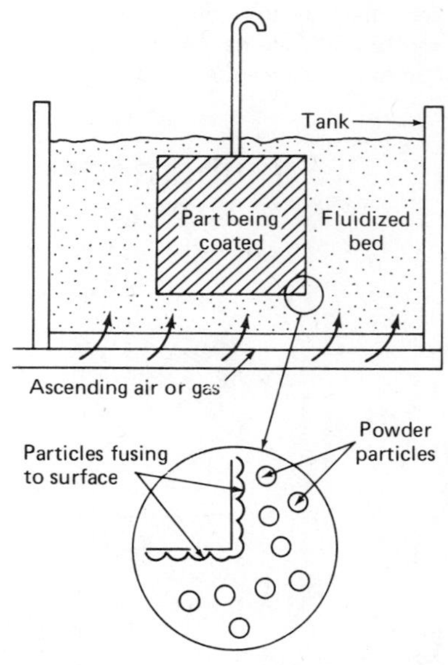

Fig. 19-40. Fluidized-bed coating and closeup of particles fusing on part surface. (Reprinted by permission from *Materials Engineering*, February 1960)

fuse to the surface to form a continuous film on the surface of the part. Depending on the powder, a post-heating treatment may or may not be required to complete the fusion of the coating. Usually a second heating is done to completely coalesce the coating and to produce a smoother surface on the coating.

The thickness of the coating depends on the initial temperature of the part, the length of time it is maintained in the powder bed, and, to a slight extent, on how much the part is moved in the bed. Usually the part is kept in a light to-and-fro motion in the bed which helps penetration of the powder into crevices and corners. With this process, coatings from 5 mils to 50 mils are commonly produced in one dip. Typical fluidized-bed-coated parts are shown in Figures 19-41, 19-42, 19-43, and 19-44.

Fluidized-bed coating is not confined to batch processing. The technique is well adapted for continuous coating of wire products or parts on a conveyor belt. Figure 19-45 illustrates a typical continuous process for wire, fencing, webs, etc. Figure 19-46 shows a similar continuous setup for treating individual parts, using a rising fluidized bed.

Table 19-17 summarizes the major types of plastics used in fluidized-bed coating, the preheating temperatures required, the fusion temperatures, and the characteristics of the resulting coatings.

Electrostatic Deposition. The most commonly used technique of powder spraying is electrostatic deposition. The powder is sprayed from a special gun which imparts a high-voltage electrostatic charge to the particles just as they leave the gun. The part to be coated is grounded, thus the potential difference is great enough so the particles are attracted to the workpiece with considerable force. This force is great

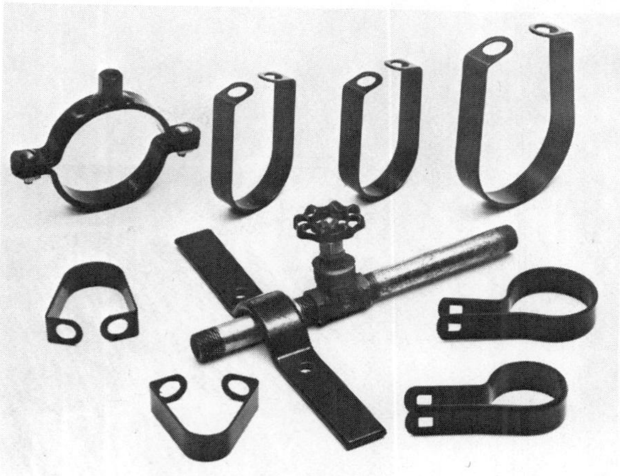

Fig. 19-41. Vinyl-coated metal hangars and clamps. (*Courtesy: Plastonics, Inc.*)

Fig. 19-42. Nylon-coated aluminum castings. (*Courtesy: Plastonics, Inc.*)

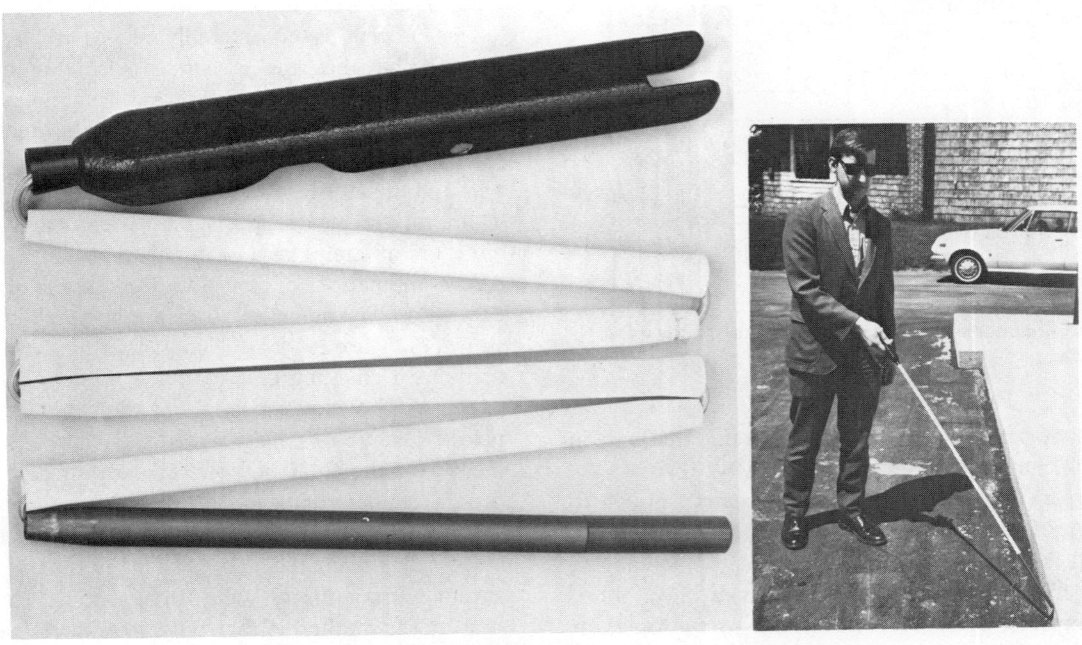

Fig. 19-43. Vinyl-coated metal can assembly. (*Courtesy: Plastonics, Inc.*)

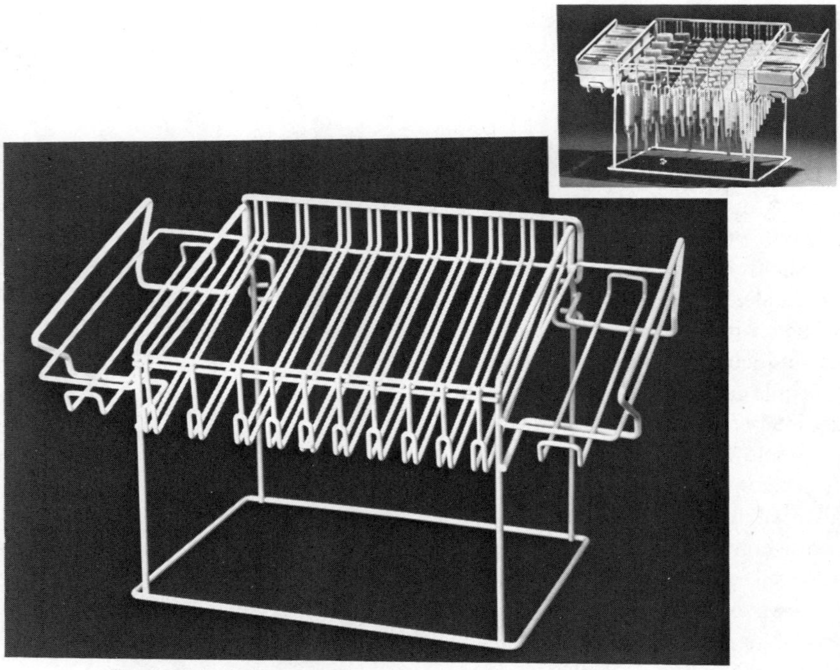

Fig. 19-44. Nylon-coated syringe rack. (*Courtesy: Plastonics, Inc.*)

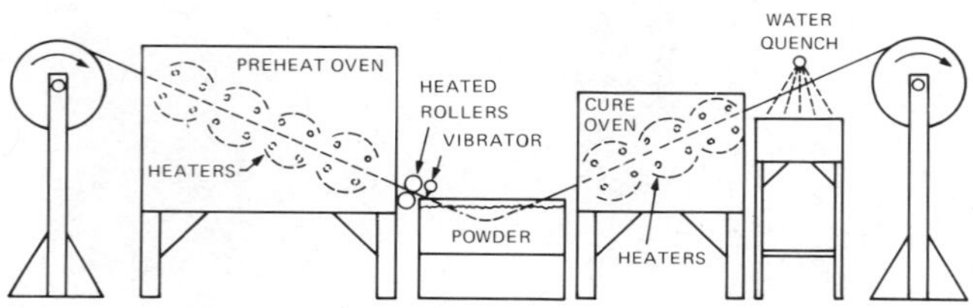

Fig. 19-45. Continuous process for fluidized-bed coating of wire, fencing, webs, etc.

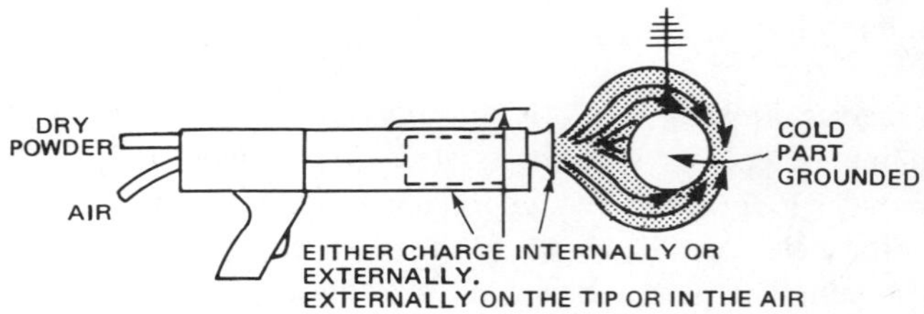

Fig. 19-47. Electrostatic spraying process.

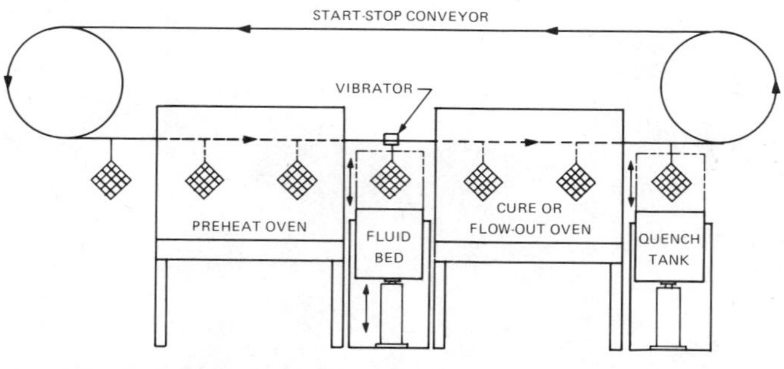

Fig. 19-46. Continuous setup for treating individual parts using a rising fluidized bed.

enough so that the powder particles surround the part in front, in back, and on all sides (see Figure 19-47).

After the part has been sprayed, it is heated enough to cause fusion of the coating (in the same manner as in fluidized-bed coating). The voltages used, while ranging from 50 to 150 kilovolts, involve currents of only 200 to 500 microamperes, so there is no hazard to the operator from the high voltages.

The process is considerably slower than fluidized-bed coating, but has the advantage that thin coatings from 1 to 3 mils, can be easily applied. The major limitation is the fact that the powder cannot be deposited reliably into small openings or tight angles. Also, the distance from the part and the air pressure must be controlled very accurately. Thus, very large, irregularly shaped parts may present difficulties in obtaining of uniform coatings.

A combination of the fluidized bed and electrostatic deposition is known as an electrostatic fluidized bed. As shown in Figure 19-48, this consists of a fluidized bed which contains a charging grid to put a potential difference between the powder particles and the part. The part is then placed just above the powder bed, instead of being immersed in the fluid powder. The advantage of this system over a regular fluidized bed or electrostatic deposition is that thinner coatings are possible, and small parts may be more quickly coated. The system is thus well suited for automated coating of many small parts. The main limitation of the method is that only small parts can be coated, since the attractive force is not great enough to cause the

Table 19-17. Summary of Plastics Suitable for Coatings to be Produced by the Fluidized-Bed Process.[a,*]

	Fluidizing Conditions				Fluidized-Bed Powder Coating Properties					
	Metal Preheat		Cure or Fusion		Maximum Operating					
Resin	Temp. °F.	Primer	Temp. °F.	Time Min.	Temp. °F.	Adhesion[b]	Weather Resistance	Chemical Resistance	Electrical Properties	Main Uses
Epoxy	250 to 450	No	250 to 450	5 to 60	200 to 400	Excellent	Good	Good	Excellent	Electrical Corrosion
Vinyl	450 to 550	Yes	400 to 600	1 to 3	225	Poor	Good	Fair[c]	Good	Decorative
Cellulose acetate butyrate	500 to 600	Yes	400 to 550	1 to 3	225	Poor	Good	Fair[c]	Excellent	Decorative
Nylon	650 to 800	Yes	650 to 700	1	300	Poor	Fair	Good	Good	Lubricity
Polyethylene	500 to 600	No	400 to 600	1 to 5	225	Fair	Good	Good	Excellent	Decorative Corrosion
Polypropylene	500 to 700	No	400 to 600	1 to 3	260	Poor	Good	Good	Excellent	Decorative Corrosion
Penton	500 to 650	Yes	450 to 600	1 to 10	350	Poor	Good	Excellent	Good	Corrosion
Teflon	800 to 1000	No	800 to 900	1 to 3	500	Poor	Good	Excellent	Excellent	Lubricity Corrosion

[a]Excluded are the ranges of major process conditions used for each and a comparison of the physical properties of the finished coatings.
[b]Without primer.
[c]No solvent resistance.
*Reprinted by permission from *Modern Plastics*, McGraw-Hill, Inc.

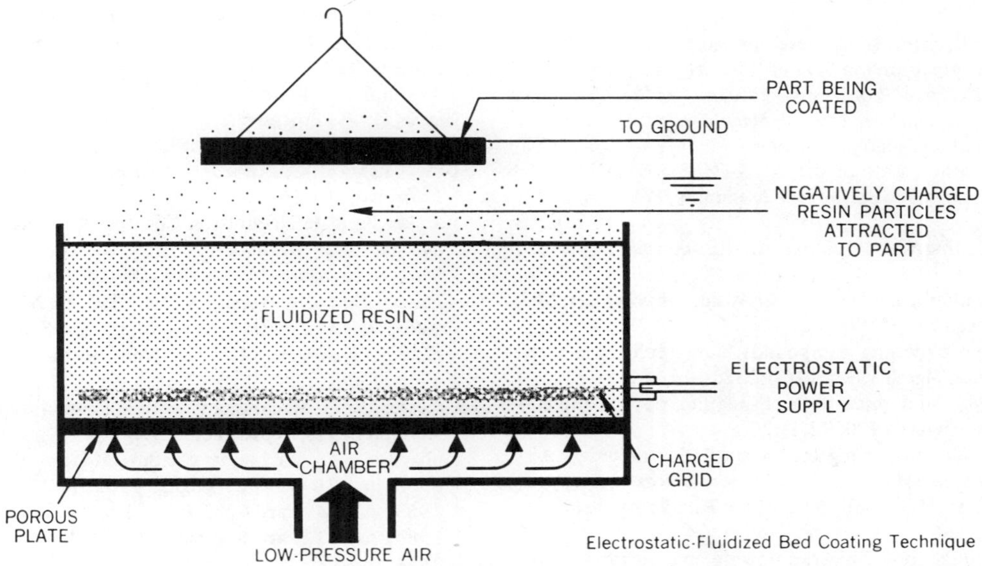

Fig. 19-48. Electrostatic fluidized bed.

powder particles to travel long distances without the propelling force of the air stream.

There are a number of other processes which have been or are being used to deposit powder coatings; however, at this time, none are used to the same extent as are fluidized beds or electrostatic spraying. Some of these techniques include flame spraying of powders using a special gun known as a Shori gun, spraying with a plasma gun, airless hot spray (AHS), and coating by electrophoretic deposition.

REFERENCES AND BIBLIOGRAPHY

Casting and Encapsulation

Abshier, C. S., Berry, J., and Maget, J. R. Toughening agents improve epoxy encapsulants. *Insulation/Circuits*, October 1977, **23**(10):27.

Antonen, R. C., Michael, K. W., and Engleman, J. H. New hybrid electronic molding compounds. *Journal of Electronic Materials*, 1977, **6**(1):49.

Antonen, R. C., and Meako, F. L. Why silicone encapsulants now? *Plastics Engineering*, March 1974, **30**(3):34.

Blaich, C. F. Jr. How to cast urethanes. *Plastics Technology*, November 1968, **14**(11):55.

Boot, R. J., Murphy, C. B., and Smith, O. L. The development of potting compounds for cryogenic applications. *Plastics Design & Processing*, October 1962, **2**(10):28.

Cope, D. E. Hydrophobic encapsulation: wetting agents facilitate filler wetout and dispersion. *Plastics Compounding*, March/April 1979, p. 64.

Delmonte, J. Urethane or epoxy? Which encapsulant to use. *Insulation/Circuits*, October 1977, **23**(10):37.

——. Epoxy encapsulation of electronic components. *SPE Journal*, November 1964, **20**(11):1223.

Du Bois, J. H. Permanent packaging by direct encapsulation. *Plastics World*, April 1962, **20**(4):34.

Feuchtbaum, R. B. Embedment selection. *Product Engineering*, January 3, 1966, **37**(1):44.

Firth, F. G. Potting and encapsulation. *Modern Plastics*, April 1956, **33**(8):125.

Harper, C. A. Molds and housings for producing embedded units. *SPE Journal*, February 1963, **19**(2):197.

——. Equipment for embedment processes. *Modern Plastics*, April 1961, **38**(8):105.

——. Urethanes, polysulfides, phenolics, polystyrenes, and polybutadienes for use as embedding materials. *Insulation*, April 1963, p. 23.

Hull, J. L. Encapsulation moves towards automation. *Plastics Engineering*, April 1978, **34**(4):23.

——. Tooling for encapsulation by transfer molding. *SPE Journal*, January 1966, **22**(1):25.

Insulation/Circuits. Techniques of encapsulation. May 1973, **19**(5):24.

Kessel, A. A., and Marcotte, G. M. Development of epoxy materials and techniques for precision molding of "floating" aerospace components. *Plastics Design & Processing*, February 1964, **4**(2):18.

C. A. Lawton Co. *Encapsulation Molding Machines.*

——. *Polyester Injection Molding Machines.*

Lee, S. M. Encapsulation, state of the art. *SAMPE Journal.* Part 1: November/December, 1978, p. 5. Part 2: January/February 1979, p. 4.

Liebowitz, L. A. Casting with liquid resins. *SPE Journal*, May 1968, **24**(5):59.

Loeb, W. E. Encapsulation by vacuum deposition of polymers. *SPE Journal*, September 1971, **27**(9):46.

Loeser, T. S. Epoxy troweling compounds—new application techniques. *SPE Journal*, March 1961, **17**(3):252.

Materials Engineering. Mini guide to encapsulating, potting and embedding materials. August 1970, **72**(2):20.

Lundberg, C. V. A guide to potting and encapsulation materials. *Materials in Design Engineering.* Part 1: Selection factors, seven types and their properties. May 1960, **51**(5):123. Part 2: Applications. April 1960, **51**(6):166.

McFaden, M. F. Utilization of plastics potting and casting in aerospace industries. *Insulation*, March 1961, p. 45.

Mermel, H. Plastics encapsulation: which way? *Plastics Technology*, June 1971, **17**(6):31.

Moylan, J. J., and Long, J. T. How to encapsulate with alkyds. *Modern Plastics*, March 1960, **37**(7):124.

Nelson, B. W., Morrisey, E. J., and Marcey, S. D. Casting to close tolerances with unfilled epoxy resins. *SPE Journal*, March 1961, **17**(3):257.

Pasadena Presses. *Transfer Molding and Encapsulation Presses.* Bulletin No. 1011.

Plastics Engineering. Precatalyzed LRM system promises lower cost encapsulation. July 1976, **32**(7):25.

Plastics World. Slush molding encapsulation. January 1962, **20**(1):16.

Rejda, S. New developments in epoxy resins, new lost-cost method for encapsulating motors. *Insulation*, March 1961, p. 33.

Rugg, G. B. Molding vs. casting for epoxy encapsulation. *Modern Plastics*, September 1961, **38**(1):109.

Salensky, G. Epoxy resin systems for electrical/electronic encapsulation. *Insulation/Circuits*, May 1972, **18**(5):19.

Sanders, M. A. Encapsulation through molding. *SPE Journal*, October 1961, **17**(10):1079.

Shaer, L. S. Processing plastics—spin casting thermoset parts. *Plastics Engineering*, February 1977, **33**(2):48.

——. Rubber mold spin-casting. *Machine Design*, January 20, 1977, **49**(2):113.

Stedfeld, R. A quick dip produces plastic or rubber parts. *Materials Engineering*, June 1979, **89**(6):30.

Sussman, V. Packaging electronic components. *Plastics World*, June 1961, **19**(6):28.

Tanner, W. C. Castable conductive plastics. *SPE Journal*, March 1959, **15**(3):126.

Uglione, H. L. Jr., and Bell, A. R. Jr. Epoxy molding powders: new method of encapsulation for welded modules. *Modern Plastics*, August 1964, **41**(8):133.

Zecher, R. F. Encapsulation of electrical components utilizing liquid resin molding techniques. *SPE Journal*, February 1969, **25**(2):40.

Zukor, L. J. Casting powdered resins. *Plastics Technology*, July 1961, **7**(7):47.

Machining and Finishing

Ashmore, J. Tumbling thermoplastics. *Modern Plastics*, March 1960, **18**(3).

Carlyon, G. "Machining Thermoplastics." In, *Modern Plastics Encyclopedia*, McGraw-Hill, Inc., New York, 1960.

Celanese Plastics Co. *Celcon Acetal Copolymer Fabrication and Machining.* Bulletin C3D, January 1967.

Dickie, J. F. Processor's guide to mechanical deflashing. *Plastics World*, August 1978, **36**(8):52.

Dow Chemical Co. *Styron—Machining and Mechanical Finishing.* Form No. 171-16D.

E.I. Du Pont de Nemours & Co., Inc. *A Guide to Cutting and Machining Kevlar Aramid.*

Elkins, S. High pressure water knife cuts tough composite materials. *Plastics Design & Processing*, June 1979, **19**(6):38.

Frados, J. (ed.) *Plastics Engineering Handbook of the Society of Plastics Engineers.* 4th Ed. Van Nostrand Reinhold Co., New York, 1976.

Graesser, E. G. Machining glass-base laminates. *Aircraft and Missiles Manufacturing*, September 1959.

Kobayashi, A. Machining plastics. *Modern Plastics.* Part 1: July 1963, p. 110. Part 2: August 1963, p. 109. Part 3: September 1963, p. 144. Part 4: October 1963, p. 229. Part 5: November 1963, p. 121. Part 6: December 1963, p. 126. Part 7: January 1964, p. 163.

Machining Data Handbook. 2nd Ed. Machinability Data Center, Metcut Research Associates Inc.

Mackey, B. A. Jr. How to drill precision holes in reinforced plastics in a hurry. *Plastics Engineering*, February 1980, **36**(2):22.

Mackrin, G. S. How to saw plastics. *Modern Plastics*, November 1959, **36**(11).

Oliver, L. R. Tough cutting materials handle FRP abrasion. *Plastics Design & Processing*, September 1978, **18**(9).

Plastics Design & Processing. Vapor polishing refinishes scratched telephone dials. August 1977, **17**(8):21.

——. Machining the engineering plastics. September 1968, **8**(9).

Plastics World. Molded plastics product finishing now. August 1964, **22**(8):16.

Reid, R. J. Jr. Mechanical finishing of plastics with coated abrasives. *Plastics World*, August 1960, **18**(8):24.

Silverman, M. N., and Richard, R. T. Modified phenylene oxide machines easily. *Modern Plastics*, May 1968, **45**(5):128.

Society of the Plastics Industry, Inc., Committee on Plastics Education. *Finishing and Decorating.* 1959.

Sprouse, W. A. Tumbling glamour frames. *Plastics World*, March 1960, **18**(3).

Summers, P. H. Fluidized bed–safe, efficient method for cleaning dies, molds and molding equipment. *Plastics Design & Processing*, December 1974, **14**(12).

Thomas, A. D. How to machine polycarbonate resins. *Modern Plastics*, December 1965, **42**(12):125.

Vasudeva, B. R. Here's how to eliminate fumes during fluidized bed cleaning. *Plastics Design & Processing*, February 1977, **17**(2):18.

Walker, V. Machining glass-reinforced plastics. *Plastics Technology*, June 1968, **14**(6):39.

Coloring and Colorants

Ahmed, M. *Coloring of Plastics, Theory and Practice.* Van Nostrand Reinhold Co., New York, 1979.

Benemelis, R. L. The pigmenting of plastics. *Plastics Technology*, August 1962, **8**(8):40.

Billmeyer, F. W. Jr. Color measurement in the computer age. *Plastics Engineering*, December 1978, **34**(12):35.

Bowman, J. E. Color matching by computer. *Plastics Compounding*, November/December 1978, p. 18.

Celanese Corporation of America. *Coloring Fortiflex A.* Product Data Bulletin P3C, September 15, 1958.

Daubs, D. Quality decoration of plastics. *Plastics World*, September 22, 1975, **33**(9):24.

Edwards, H. R. Variables influencing the performance of gel-coated composites. *Modern Plastics*, August 1979.

Elzy, F. T., and Rose, T. E. Additives for plastics–colorants. *Plastics Engineering*, July 1976, **32**(7):49.

Felsher, H. C., and Hanau, W. J. Exotic effects through the internal coloring of plastics. *Plastics Design & Processing*, October 1967, **7**(10):19.

Freemott, J. H. Take a look at liquid colorants. *Plastics Design & Processing*, March 1977, **17**(3):39.

Hackney, H. U. Reconsider liquid coloring for plastics. *Plastics Engineering*, March 1974, **30**(3):61.

Hawley, R. I., Cheves, J. H., and Hill, R. O. Jr. The art of coloring thermoplastics. *SPE Journal*, September 1965, **21**(9):1064.

Jarrett, W. In-house color program for profile extrusions. *Plastics Design & Processing*, August 1978, **18**(8):48.

Johnston, R. M., and Park, R. E. Coloring of unsaturated polyester resin laminates and gel coats. *SPE Journal*, November 1964, **20**(11):1211.

Kadykowski, R. Liquid colorants can keep you in the black. *Plastics World*, March 1977, **35**(3):52.

Keane, T. J. Small computer scales down the cost of color matching. *Plastics Engineering*, February 1976, **32**(2):42.

Manzione, J., and Danko, S. Color concentrates: avoiding hang-ups. *Plastics Technology*, July 1975, **21**(7):37.

Miller, M. C. Coloring systems for foamed plastics. *SPE Journal*, January 1966, **22**(1):44.

Modern Plastics. There are lots of color systems to choose from these days . . . . July 1976, **53**(7):34.

——. Coloring by direct feed with liquid color. August 1972, **49**(8):55.

——. How to choose the right colorant for the right resin. December 1962, **40**(4):90.

——. How to choose the correct colorant. April 1960, **37**(8):81.

Muller, C. P. Optimizing pigment blending. *Plastics Engineering*, June 1979, **35**(6):21.

Osmer, D. The opacity parameter in color matching. *Plastics Engineering*, September 1978, **34**(9):33.

Palmer, J. R. Avoiding pitfalls in the use of color concentrates. *Plastics Compounding*, January/February 1980, p. 35.

Plastics World. Does in-plant color blending pay? July 1972, **30**(7):44.

Reeve, T. B. Additives for plastics–colorants. *Plastics Engineering*, August 1977, **33**(8):31.

Richter, J. Color it azo. *Plastics Engineering*, April 1979, **35**(4):37.

Scheiner, L. L. Process profitably: do your own color blending. *Plastics Engineering*, July 1971, **17**(7):38.

Simpson, J. E. Know your colorant characteristics. *Modern Plastics*, March 1963, **40**(7):88.

——. Coloring thermoplastic resins. *Modern Plastics*, May 1963, **40**(9):107.

——. Coloring thermosetting resins. Modern Plastics, June 1963, **40**(10):89.

——. Color in plastics. *Materials in Design Engineering*, July 1963, 58:(1):61.

——. Choosing the right coloring method. *Modern Plastics*, November 1962, **40**(3):94.

——. A formula for specifying color concentrates. *Modern Plastics*, November 1962, **40**(3):142.

Snow, J. P. Jr., and Arnheim, W. M. Using liquid colors to best advantage. *Plastics Engineering*, June 1977, **33**(6):17.

Sorice, R. G. Colorants–the more they stay the same the more different they become. *Plastics Engineering*, August 1978, **34**(8):37.

Tomkins, D. M. Computer color matching makes sense even for small processors. *Plastics Engineering*, June 1979, **35**(6):29.

Weinrich, L. I., and Rotherham, H. G. Basic colorants for plastics. *Plastics Design & Processing*, September 1968, 8(9):12.

Wich, E. A. Choosing colorants for polystyrene. *Modern Plastics*, June 1961, **38**(10):86.

Zukor, L. J. Pigments, dyes and plastics. *Plastics Technology*, January 1962, 8(1):27.

Painting and Decorating

Adams, J. R. The techniques of spray painting plastics. *Plastics Design & Processing*, December 1967, 7(12):20.

Ayres, R. L., and Shofner, D. L. Designing with plastics–painted polyethylene parts. *Plastics Engineering*, September 1976, **32**(9):26.

Booth, G. L. Coating methods. *Modern Plastics.* Part 1: September 1958, **36**(1). Part 2: October 1958, **36**(2).

Bowes, H. D. Increase profits by paint-decorating plastic parts. *Plastics Engineering*, February 1978, **34**(2):31.

Bragole, R. A. "How to Paint Non-polar Plastics." Paper presentation, *Rubber Division Meeting, American Chemical Society*, Detroit, Michigan, May 1–4, 1973.

Brueggemann, W. H. Blistering of gel coat laminates. *Modern Plastics,* June 1959, **56**(6):80.

Celanese Plastics Co. *Decorating Celcon Acetal Copolymer.* Bulletin C3E, February 1966.

Davis, D. R. Coating with airless hot-spray. *Plastics Technology*, March 1962, **8**(3):45.

Davis, E. Insulative and conductive coatings. *SPE Journal*, May 1958, **14**(5):19.

Delia, M. A. Economical multi-color decorating. *Plastics World*, May 1967, **25**(5):32.

Dickinson, L. A. How to decorate plastics with spray and wipe techniques. *Plastics Design & Processing*, February 1978, **18**(2):46.

Dow Chemical Co. *Decorating Dow Plastics.* Form No. 171-401-75C-467, 1967.

Eastman Chemical Products, Inc. *Fabricating and Decorating Tech-*

niques for Tenite Polyterephthalate Plastic Formulations. Technical Report TR-38C, ECP 463, March 1974.

——. *Cementing and Decorating Articles Manufactured from Tenite Polyolefin Plastics.* Technical Report TR-21B, ECP 465, May 1968.

Felsher, H. C., and Hanau, W. J. Brilliance for plastics. *Modern Plastics*, September 1963, **40**(1):96.

——. Special decorative effects. *Modern Plastics*, December 1961, **39**(4):94.

Gorrell, J. H. New way to coat plastics and metals. *Plastics Technology*, October 1964, **10**(10).

Grant, R. R. Coating and laminating techniques for films and webs. *Plastics Machinery & Equipment*, April 1978, 8(4):16.

Hall, A. Painting plastics: the alternatives to solvents. *Plastics World*, January 20, 1975, **33**(1):46.

Hayden, T. E., and Scharnberg, J. Decorating of plastics. *SPE Journal*, June 1961, **17**(6):576.

Herman, R. M. Coating systems for molded polypropylene. *Modern Plastics*, May 1962, **39**(9):153.

La Bounty, T. J. Masks for decorating plastics. *Plastics Design & Processing.* Part 1: May 1962, **2**(5):22. Part 2: June 1962, **2**(6):29.

Lamp, R. W. Painting electrical conductivity onto structural foam plastic housings. *Plastics Design & Processing*, January 1976, **16**(1):17.

——. Guide to spraying plastics with electrically conductive paints. *Plastics Engineering*, June 1975, **31**(6):43.

Lau, G. Processors' guide to fluorocarbon coating. *Plastics Technology*, June 1969, **15**(6).

Lederman, B. E. Single paint system bonds to plastics by chemical grafting. *Plastics Design & Processing*, May 1975, **15**(5):20.

Materials Engineering. Take a look at polyurethane coatings to save energy, improve appearance. June 1978, **87**(6):39.

——. Coated plastic foams live a lot longer than uncoated. September 1973, 78(3):56.

——. Coatings and finishes for plastics. August 1971, **74**(2):42.

——. Polyurethane coatings: where, why and how to use them. February 1970, **71**(2).

Mock, J. A. Take a look at polyurethane coatings to save energy, improve performance. *Materials Engineering*, June 1978, **87**(6):39.

——. Jumping the heat hurdle with organic, inorganic coatings. *Materials Engineering*, May 1977, **85**(5):51.

——. Fluorocarbon finishes tackle tough severe environments. *Materials Engineering*, October 1971, **74**(4).

Modern Plastics. Continuous high-speed foiling of sheet opens new vistas in decorating. April 1978, **55**(4):56.

——. A fresh look at paints for plastics. March 1977, **54**(3):53.

——. New ways to decorate. March 1964, **41**(7):100.

——. Decorating molded products. April 1959, **36**(8):91.

Murray, B. G., and Andren, A. J. How to work with gel coats. *Modern Plastics*, June 1962, **39**(10):119.

Palkie, C. R. Organic and plastisol coatings. *Plastics World*, June 1965, **23**(6):50.

Plastics Machinery & Equipment. Coating and laminating techniques for films and webs. April 1978, 8(4).

Plastics Technology. Printing and decorating. November 1976, **22**(11).

Petrick, P. J. A guide to polyester gel coat application techniques. *Plastics Technology*, December 1968, **14**(12):44.

Plastics World. Molded plastics product finishing now. August 1964, **22**(8).

Product Engineering. Fire protection coatings work by forming foam-like chars. February 1976, **47**(2).

Skeels, A. N. Jr. Guide to plastic bottle decoration. *SPE Journal*, May 1971, **27**(5):30.

Smith, E. A. Printing and decorating polyolefin surfaces. *SPE Journal*, Part 1: February 1962, **18**(2):157. Part 2: March 1962, **18**(3):280.

Storms, C. D. Functional primers for plastics. *Plastics Design & Processing*, June 1977, **17**(6):57.

——. Solving plastic furniture finishing problems. *SPE Journal*, November 1970, **26**(11):43.

Tompkins, J. H., and Smith, K. R. Hi-fi textures for plastics. *Modern Plastics*, December 1961, **39**(4):109.

Thorson, M. Z. The art of plastics decoration. *SPE Journal*, December 1961, **17**(12):1301.

Trauernicht, J. O. Printing and decorating plastics: what's new, what's available. *Plastics Technology*, September 1969, **15**(9):41.

Tyler, B. J. Flame retardant coatings for plastics. *Materials in Design Engineering*, November 1960, **52**(6):16.

U.S. Industrial Chemicals Co. *Decorating Polyolefin Molded Items.* Bulletin PL 17-1264.

Surface Treatments

Ayres, R. L., and Shofner, D. L. Preparing polyolefin surfaces for inks and adhesives. *SPE Journal*, December 1972, **28**(12):51.

Bruner, W. M., and Baranano, C. M. How to make coatings stick to acetal resin. *Modern Plastics*, December 1961, **38**(4).

Dow Chemical Co. *Treating Polyethylene Surfaces.* Form No. 171-185B-68.

Doyle, D. J. Plastics and static buildup. *Plastics World*, October 1964, **22**(10):26.

E.I. Du Pont de Nemours & Co., Inc. *Delrin Acetal Resin Post Molding Operations: "Satinizing" Surface Treatment.* Bulletin No. A-22794-10M.

Eastman Chemical Products, Inc. *Electronic Treating of Polyethylene for Printing.* Technical Information Bulletin No. 1, July 1959.

Hall, A. Keeping static out of plastics processing. *Plastics World*, March 17, 1975, **33**(3):40.

Kent, G. Here's how to pretreat plastic parts for painting. *Plastics Technology*, June 1977, **23**(6):69.

Levitzky, J. J., Lindsey, F. J., and Kaghan, W. S. Corona discharge treatment of polypropylene films. *SPE Journal*, December 1964, **20**(12):1305.

Levy, W. W. Static electricity—cause, effect, and elimination. *SPE Journal*, October 1962, **18**(10):1288.

Modern Plastics. Electron radiation: fast new way to finish plastics. December 1969, **46**(12):84.

Perrone, L. Plasma: low-temperature surface treatment for plastics. *Plastics Machinery & Equipment*, September 1978, 8(9).

Rauhut, H. W. Pretreating polyethylene for optimum structural-adhesive joints. *Adhesives Age*, December 1969, **12**(12):28.

Sonkin, R. M. Corona-discharge treatment of polyolefin films. *Plastics Engineering*, February 1977, **33**(2):50.

Stiefling, H. Electronic surface treating of polyethylene film. *Plastics World*, November 1962, **20**(11):28.

Tremain, H. E. How to make plastics static-free. *Plastics Technology*, Part 1: March 1963, 9(3):38. Part 2: April 1963, 9(4):47.

Von der Heide, J. C., and Wilson, H. L. Guide to corona film treatment. *Modern Plastics*, May 1961, **38**(9):199.

Wechsberg, H. E., and Webber, J. B. Surface treatment of PE film by electrical discharge. *Modern Plastics*, June 1959, **36**(10):101.

Ziccarelli, J. J. Polyolefin surface treating methods. *Modern Plastics*, November 1962, **40**(3):126.

Hot Stamping

Alexander, M. J., and Dalgnault, A. P. Decorating plastics by hot stamping. *Plastics Design & Processing*, April 1967, 7(4):12.

Enlow, H. Hot stamping technology develops in-mold methods for wood decorating plastic parts. *Plastics Design & Processing*, October 1976, **16**(10):16.

Fuchs, G. For the right decorative touch try new hot stamping foils, machinery. *Plastics Design & Processing*, August 1978, **18**(8):53.

Gould, G. W. Decorating technology advances with hot stamping machinery and materials. *Plastics Design & Processing*, August 1979, **19**(8):16.

Heine, W. A. "Hot Stamping." In, *Modern Plastics Encyclopedia*, McGraw-Hill, Inc., 1973.

Hochner, W. L. In-mold decoration. *Plastics World*, February 1963, **21**(2):24.

Mainville, G. R. How to decorate plastics by hot die stamping. *Plastics Design & Processing*, November 1962, **2**(11):16.

Modern Plastics. Woodgraining brings new dimension to plastics design. November 1966, **44**(3):84.

Mysiewicz, T. G. News in hot-stamping systems. *Plastics Technology*, June 1977, **23**(6):61.

Plastics Design & Processing. Testing the permanance of hot stamping on plastics. June 1965, **15**(6):14.

Plastics World. Hot stamping–the look gets classier. October 1979, **37**(10):72.

——. Hot stamping–higher speeds, new finishes. February 1978, **36**(2):52.

——. In-mold decorating larger parts, complex shapes. February 1978, **36**(2):54.

——. If you can't color it, paint it or plate it–try hot foil stamping. June 1973, **31**(6):62.

——. Hot-stamping developments keep the lid on decorating costs. April 1972, **30**(4):54.

Richards, A. E. The variables of silicone rubber and how they affect hot stamping quality. *Plastics Design & Processing*, May 1977, **17**(5):39.

Shepherd, R. G. Jr., and Hirschey, M. E. Techniques for hot-stamping multicolor heat transfers. *Modern Plastics*, April 1965, **42**(8):132.

Metallization and Electroplating

Anis, N. The ABC's of electroplating ABS. *Plastics Engineering*, January 1977, **33**(1):14.

Bender, R. J. Future bright for vacuum metallization. *Plastics Design & Processing*, March 1962, **2**(3):14.

Clough, P. J., and Farrow, H. M. Thick vacuum metallized coatings are durable, corrosion resistant. *Materials in Design Engineering*, September 1960, **52**(3).

Courduvelis, C. J., and Wynschenk, J. There are new things happening in the field of decorative plastics plating. *Plastics Engineering*, January 1978, **34**(1):39.

Dunleavy, J. T. To give plastics low-cost mirror finishes–vacuum metallize. *Plastics Design & Processing*, May 1973, **13**(5):8.

Eastman Chemical Products, Inc. *High-Vacuum Metallizing of Tenite Butyrate.* Bulletin 54-1142, 1954.

Felscher, H. C., and Hanau, W. J. Brilliance for plastics through metallics. *Modern Plastics*, September 1966, **41**(1):96.

Goldie, W. Methods of preparing plastics for electroplating. *Plastics Design & Processing*, February 1968, **8**(2).

Himmler, D. W., and Alzapiedi, J. A. Preparing plastics for plating. *SPE Journal*, June 1973, **29**(6):57.

Kraus, J. J. Preplate systems. *SPE Journal*, April 1969, **25**(4):94.

Luch, D. Carbon filled polypropylene eliminates preplating in electroplating process. *Plastics Design & Processing*, September 1978, **18**(9):30.

Materials Engineering. Try metallizing to cut costs for bright, shiny plastic parts. March 1975, **81**(3):42.

——. How are plated plastics doing? November 1972, **76**(6):30.

Miller, B. How to design for electroplating–and make it stick! *Plastics World*, June 1974, **32**(6):44.

Mock, J. A. Take a look at vacuum metallizing for low cost, bright metallic finishes. *Materials Engineering*, April 1978, **87**(4).

——. Electroplated plastic parts: some new innovations. *Materials Engineering*, November 1978, **88**(6):55.

——. New materials, fast plating boost plated plastic parts. *Materials Engineering*, May 1967, **65**(5):88.

Modern Plastics. New metallizing process offers better adhesion and faster deposition. February 1980, **57**(2):22.

——. Sputter chroming is a new way to metallized flexible parts. February 1979, **56**(2):49.

——. It isn't so tough to plate SMC these days. August 1980, **57**(8):55.

——. Take another look at electroplating: it's a whole new decorating option. May 1978, **55**(5):45.

——. The new appeal of electroplating: costs are down, performance is up. December 1976, **53**(12).

——. Lower-cost electroplated PP. February 1971, **48**(2):46.

Mohler, J. B. Primer on electrodeposited coatings. *Materials Engineering*, January 1972, **75**(1):38.

Plastics Design & Processing. Portable metallizing system provides EMI shielding using thermal spraying technique. April 1978, **18**(4):49.

——. Focus on preplate system to reduce plating rejects. July 1977, **17**(7).

——. Tailor your part designs for quality–electroplated plastics. August 1977, **17**(8).

——. Electroplating-on-plastics: a plater's gamble on automation pays off. March 1966, **6**(3).

——. Flame-spray methods can be used to metal-coat plastics. October 1963, **3**(10):12.

——. Metallized plastics–new precision engineering materials. June 1963, **13**(6).

Plastics Engineering. Three ways to plate plastics. November 1973, **29**(11).

Plastics Technology. Why not ion plate plastics? December 1976, **22**(12):11.

——. Better plated-ABS parts. April 1964, **10**(4).

Plastics World. Sputtered chrome–new route to "plated" elastomers. September 1978, **36**(9):56.

——. Directly platable polypropylene. June 1978, **36**(6):66.

——. Chrome sputtering: high-volume technique for metallizing of plastics. September 1977, **35**(9):35.

——. Preplate process opens new doors for plastics in tough applications. September 1972, **30**(9):54.

——. Proper part design: key to profitable plating. February 1971, **29**(2):43.

Riley, M. W. Plating on plastics. *Plastics Technology*, September 1964, **10**(9):48.

Shaw, A. Designing plastic parts for metallizing. *Modern Plastics*, March 1961, **38**(7):114.

Sidders, J., and Storms, C. C. Resistance fired and sputtered chrome for vacuum metallizing exterior components. *Plastics Design & Processing*, November 1977, **17**(11).

Steurer, A. Tailor your part designs for quality electroplated plastics. *Plastics Design & Processing*, August 1977, **17**(8):43.

Storms, C. D. Three ways to plate plastics. *Plastics Engineering*, November 1973, **29**(11):35.

Weiss, A. Broader horizons for vacuum metallizing. *Plastics World*, July 1964, **22**(7).

Young, A. C. Why plated plastics work . . . or fail. *Plastics Design & Processing*, February 1972, **12**(2).

Powder Coating

Azzam, H. T. No-solvent, no-mess coating process. *SPE Journal*, March 1973, **29**(3):18.

Borsh, R. J. Fluidized bed coating with vinyl. *Modern Plastics*, July 1960, **37**(7):124.

Damm, E. P. Jr., and Faigenbaum, M. A. Coating polymers by electrophoretic deposition. *SPE Journal*, August 1971, **27**(8):58.

Davis, D. R. Coating with electrostatic dry-spray. *Plastics Technology*, June 1962, 8(6):37.

Denison, E. E. Application of coatings with buytrate powder. *SPE Journal*, June 1966, **22**(6):91.

Elbling, I. N., and Saunders, H. E. Powder coating–a live-wire alternative. *Plastics Engineering*, March 1976, **32**(3):39.

Forger, G. R. Powder coatings IV: a report. *Plastics Engineering*, November 1976, **32**(11):40.

Fossati, C. The fluidized bed process. *Plastics World*, June 1960, **18**(1):30.

Materials Engineering. Polyurethane coatings–where, why and how to use them. February 1970, **71**(2):47.

McSharry, J. J., Howell, S. G., and Memering, L. J. How to work with PE dispersions. *Modern Plastics*, April 1966, p. 147.

Mock, J. A. Powder coatings for more than a good appearance. *Materials Engineering*, January 1979, **89**(1):50.

——. Take a powder for clean air. *Materials Engineering*, May 1972, **75**(5):52.

——. Plastic powder coatings cut costs, up performance. *Materials Engineering*, March 1969, **69**(3):50.

Modern Plastics. In the new drift of powder coating, there's something for almost everyone. December 1975, **52**(12):34.

Mumma, R. H. *Coating with Powders*. Hercules, Inc.

Neumann, J. A., and Bockhoff, F. J. Fluidized plastic coatings for corrosion resistance. *Product Engineering*, January 1957, **28**(1): 140.

Pascoe, W. R. Fluidized bed: heavy coatings in one dip. *Materials in Design Engineering*, February 1960, **51**(2):91.

Pegg, F. E. Applying plastic coatings with the fluidized bed process. Part 1: Coating materials, September 1970, **10**(9):38. Part 2: Design and Processing, October 1970, **10**(10):22.

Pettigrew, C. K. Fluidized bed coating. *Modern Plastics.* Part 1: August 1966, **43**(12):111. Part 2: September 1966, **44**(1):150.

Plastics Engineering. Electrostatic powder coatings attract the electrical market. October 1973, **29**(10):60.

Product Engineering. Powder coatings offer low cost answer to product finishing problems. November 1977, **48**(11):21.

——. Cooler flame spraying widens choice of plastics. January 16, 1967, **38**(1):67.

Reif, R. B. A progress report on electrostatic coating processes. *Plastics Design & Processing*, May 1963, **3**(5):12.

Richart, D. S. A report on the fluidized bed coating system. Part 2: Plastics for coating and their selection. *Plastics Design & Processing*, July 1962, **2**(7):26.

Rothschild, M. V. Production setups for fluidized-bed coating. *Modern Plastics*, January 1962, **39**(1):113.

Savage, D. R. Look where dry powder finishing is heading. *Plastics Design & Processing*, August 1972, **12**(8):12.

Scharfenberger, J. A. New high-solids coating equipment offers energy/ecology advantages. *Modern Plastics*, February 1976, **53**(2):52.

Sharetts, R. R. A report on the fluidized bed coating system. Part 1: Present status, probable future directions. *Plastics Design & Processing*, June 1962, **2**(6):15.

Shaw, A. W., and Scharfenberger, J. Electrostatic spraying of thermosetting materials. *Modern Plastics*, May 1977, **54**(5):78.

SPE Journal. Plasma gun spray-coats PTFE commercially. October 1970, **26**(10):37.

20
Mold Fabrication and Design

The mold determines the shape of a part, or raw material, and in most cases, is also responsible for the final finish of the part. In almost every plastics process, the plastics material is at some stage softened or liquified so it will assume the form of the mold, after which it hardens by cooling and/or by completion of a chemical reaction. The ease or difficulty in assuming the shape of the mold will largely be determined by the design of the mold, and the final finish will be a reflection of the care which has gone into its fabrication. The speed with which parts are made and the percentage of waste generated are only two of the many variables in the process which depend on optimum mold design and fabrication.

The suggestions given in this chapter are, for the most part, simply general guidelines. Each type of mold and any individual mold, or die, must be analyzed as a separate case in order to assure an optimum design for the application. The references given at the end of the chapter, and the textbooks cited in particular, can be of great value in furnishing data for individual situations.

Molds used in plastics processing range from massive steel parts machined from large blocks of metal made to withstand tons of pressure, to thin sheet metal or plastic molds which have no pressure capabilities—and almost everything in between. Matching the diversity of materials for molds are the production methods used to fabricate the molds: machining, casting, hobbing, electroforming, metal spraying, plastics molding, etc. Thus the subject of mold fabrication techniques is also seen as a very large and complex one, which can be treated only fundamentally in this text.

Mold Materials

With the large variety of plastics processes, it is natural that there should also be a variety of molds and mold materials. Figure 20-1 lists the major molding processes and representative types of materials used to make the various molds. The materials shown in Figure 20-1 are those in most general use and are not necessarily the only materials which could be used for each application. For example, electroformed cavities are frequently used for injection molding of pen barrels and similar items. It should also be noted that Figure 20-1 is not intended as a listing of materials which might be used as substitutes for more costly materials. The actual cost of the material in a mold is usually minor—possibly from 5% to 10% of the total cost of the mold. The small savings in material costs which might be obtained by substituting a cheaper material must always be considered in light of the possible consequences if a mold broke due to fatigue or showed excessive wear in the midst of a long run.

Steel Molds

As shown in Figure 20-1, steel is used more than any other one material for mold fabrication. While a number of different types are shown, all the steel materials should have the following characteristics, emphasized to a greater or less degree, depending on the type:

1. Strength. This is one of the major requirements.
2. Toughness. Since molds can have fairly thin sections which are subject to sudden high stresses, due to injection and/or mold clamping, toughness is a must.
3. Wear resistance. This is particularly important in fabrication of compression molds. Frequently the resins used contain abrasive fillers and are subjected to high compressive loads while still in the powder state.
4. Corrosion resistance. This is of considerable importance when the resin being molded is a vinyl which, when degraded, can liberate corrosive hydrogen chloride gas. This gas can combine with moisture to produce hydrochloric acid.

There are, of course, a number of other characteristics which must be considered in the selection of a steel for any given application. Some of these are: hot hardness, machinability, annealed hardness, dimensional change on hardening, depth of hardening, and polishability.

The process used to make the steel can have an effect on the final finish. The cheapest steels are known as air-melted materials, and are apt to have pockets containing foreign materials, porosity, etc. A better grade of steel is vacuum-degassed steel in which the hydrogen, oxygen, and nitrogen are removed while the steel is molten, resulting in a material with virtually no porosity. Such a steel has considerably improved polishability. The highest grade of steel is known as vacuum-melted steel, or consumable electrode steel. In this process a steel billet is used as an electrode in a vacuum chamber. The vacuum-melted steels show the highest degree of polishability and are increasingly used in fabrication of molds for production of transparent parts such as automo-

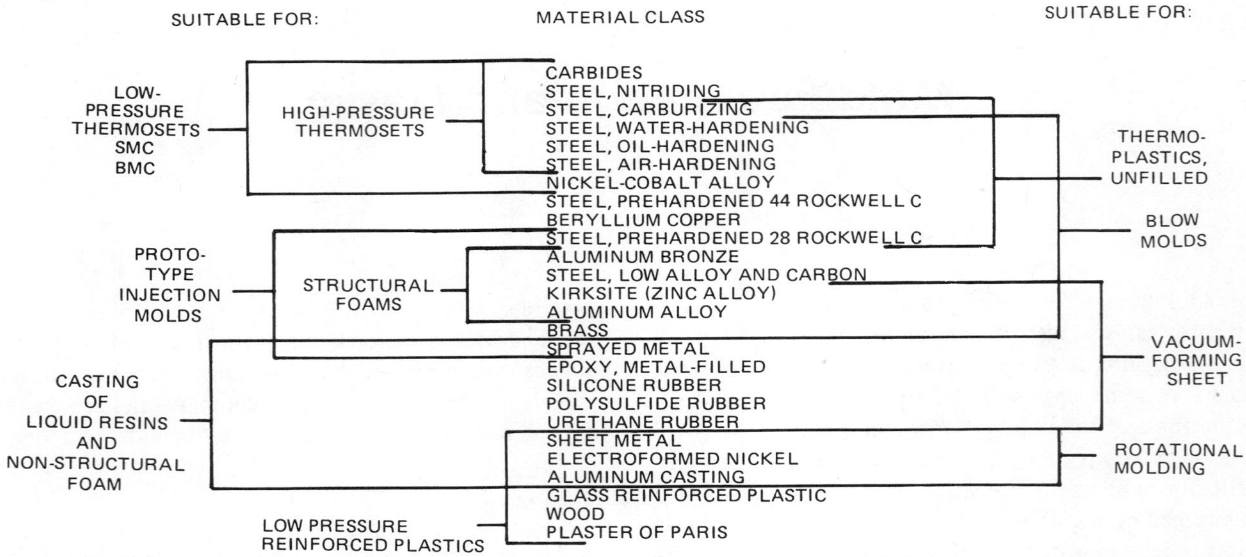

Fig. 20-1. Applications for mold-making materials. (Reprinted, with permission, from *Plastics Machinery & Equipment*, June 1978. Copyright, 1978, Industry Media, Inc.)

bile taillights which require the optimum in surface finish and polish on the mold.

Regardless of the process used to make the steel, there are four main steel types, roughly divided by the carbon content, heat treatment, and alloy content. The four types of steel used for mold making are:

1. Cold-hobbing or case-hardening types. These are soft materials used in making hobbed molds (explained below) which can be carburized to result in a hard surface.
2. Pre-heat-treated or prehardened steels. Cavities can be machined and no further heat treatment is required.
3. Deep hardening steels. Where the mold must withstand high compressive loads and must show minimum deformation on heat treatment, this type is recommended.
4. Special-purpose steels. Steels which must withstand corrosive conditions, excessive abrasion, etc., are in this category.

Table 20-1 lists thirteen of the most commonly used steels in the making of molds. There are a number of other steels, which for one reason or another may be preferred by a particular mold maker. However, by standardizing the most common types, the inventory can be kept down, and, possibly of more importance, the various heat treatment schedules can be minimized.

In Table 20-1 the carbon contents of the various steels are shown as being from 0.10% to as high as 1.0%. Since the carbon content of a non-alloyed steel to a large extent determines the final hardness of the steel, after an appropriate heat treatment, it is readily apparent that the P-1, P-5, and P-6 steels cannot be made very hard. They can, however, be surface hardened by a process known as carburizing. In this process the part is packed in carbon and is heated at a temperature and for a period of time which varies for each steel type. The result is a part with a much higher carbon content on the surface and for as much as 50 to 60 mils below the surface. Depending on the steel, the part is quenched in water or oil which then results in a hard, wear-resistant surface. A similar surface-hardening process, using nitrogen gas or ammonium salts in a molten bath instead of carbon, is called nitriding. The obvious advantage of easy machining in the soft state followed by the surface-hardening capability accounts for the popularity of these steels.

After the quenching operation, the part is in its maximum hardness state, whether surface hardened or deeply hardened, as might be with a high carbon-content steel. In almost all cases, the part is then reheated to a temperature ranging from 175°C to 680°C (350–1250°F) and held at this temperature for one hour per inch of thickness. This operation is known as tempering, and results in a hardness somewhat below the maximum, but with increased toughness. For optimum results, the process is repeated, in which case the mold is said to have been double tempered. The final hardness can then range from 30 to 65 Rockwell C, depending on the steel and the tempering treatment chosen.

The tempering process, while used primarily to obtain the final hardness desired, is also used as a means of relieving the stresses induced in the part by the quenching operation. By double tempering, maximum stress relief is obtained. Stresses are also set up in a block of steel by hard machining, grinding, etc. These stresses must also be relieved to prevent distortion or breakage when the mold is in use. The usual method is to rough machine to within $\frac{1}{8}$ to $\frac{1}{4}$ inch of the final size, and to stress relieve the part by heating it at 680–735°C (1250–1350°F) for 1 hour per inch of thickness in the largest dimension. Thus, an 8-inch-thick mold would be held for 8 hours at the stress-relief temperature, cooled

Table 20-1. Characteristics of Major Steel Types Used For Mold Making.

USE THESE AISI STEELS	FOR THESE REASONS / AND CONSIDER THESE PROPERTIES →	WEAR RESISTANCE	TOUGHNESS	HOT HARDNESS	CORROSION RESISTANCE	MACHINABILITY ANNEALED	AS PURCHASED HARDNESS, BRINELL	USUAL WORKING HARDNESS, ROCKWELL C	AGE OR QUENCHING MEDIUM	DEPTH OF HARDENING	DIMENSIONAL CHANGE ON HARDENING
FOR HOBBED AND CARBURIZED CAVITIES P-1 (.10C) [1]	EASIEST TO HOB AND MACHINE . . . DON'T USE FOR COMPRESSION MOLDS.	FAIR	EXCELLENT	POOR	VERY POOR	EXCELLENT	81-100	58-64	WATER QUENCH	SHALLOW	HIGH
P-5 (.10C)	HIGHER CORE STRENGTH AND WEAR RESISTANCE THAN P-1 . . . HOBBING NO PROBLEM.	GOOD	EXCELLENT	POOR	VERY POOR	EXCELLENT	103-123	56-60	OIL QUENCH	SHALLOW	MEDIUM
P-6 (.12C)	HIGHEST CORE STRENGTH. . . ESPECIALLY GOOD FOR HIGH OPERATIONAL PRESSURES.	GOOD	EXCELLENT	POOR	VERY POOR	GOOD	180-207	56-60	OIL QUENCH	MEDIUM	MEDIUM
MACHINED CAVITIES P-20 CARBURIZED (.35C)	SURFACE HARDNESS CAN BE AS HIGH AS ROCKWELL C62-65.	GOOD	GOOD	POOR	VERY POOR	EXCELLENT	150-180	56-60	OIL QUENCH	MEDIUM	MEDIUM
P-20 PREHARDENED	CAN BE HARDENED TO 300 BRINELL.	FAIR	GOOD	POOR	VERY POOR	GOOD	280-320	30	–	DEEP	–
P-21 PREHARDENED	CAN BE PREHARDENED TO 28 TO 30 RC. FOR HIGHER SURFACE HARDNESS NITRIDING IS RECOMMENDED.	FAIR	GOOD	GOOD	FAIR	GOOD	290-310	30	AIR AGE	DEEP	–
H-13 (.37C) [2]	EXCELLENT RESISTANCE TO TEMPERING . . . HIGH CORE STRENGTH . . . NO DISTORTION.	FAIR	GOOD	EXCEL.	POOR	FAIR	183-207	48-54	AIR QUENCH	VERY DEEP	LOW
420-STAINLESS (.27C) [2]	BEST POSSIBLE BALANCE OF POLISHABILITY, WEAR AND CORROSION RESISTANCE.	GOOD	GOOD	GOOD	EXCELLENT	FAIR	187-228	46-50	AIR OIL QUENCH [3]	MEDIUM DEEP	MEDIUM LOW
18% NICKEL MARAGING (.030)	PARTICULARLY SUITED FOR CLOSE TOLERANCES . . . GOOD MACHINABILITY, POLISHABILITY.	FAIR	GOOD	GOOD	FAIR	FAIR	283-297	50-54	AIR AGE	VERY DEEP	VERY LOW
FOR INSERTED CAVITIES A-2 (1.00C)	BEST FOR SMALL MOLDS . . . LONG WEARING . . . HARDENS TO ROCKWELL C-62.	GOOD	FAIR	GOOD	POOR	FAIR	202-217	59-61	AIR	DEEP	LOW
D-2	FOR SMALL MOLDS WHERE MAXIMUM RESISTANCE TO ABRASION IS REQUIRED. ALSO FOR MOLDS USED UP TO 750°F.	VERY GOOD	POOR	GOOD	POOR	POOR	217-255	58-60	AIR QUENCH	DEEP	VERY LOW
S-1	FOR SMALL AND MEDIUM SIZED MOLDS. TOUGHER THAN D2. CAN BE OIL HARDENED TO 53-56RC.	GOOD	VERY GOOD	GOOD	POOR	FAIR	199-225	53-56	OIL QUENCH	DEEP	MEDIUM
S-7	FOR MEDIUM AND LARGE MOLDS. TOUGH AND STABLE, HARDENS TO 53-56 RC.	GOOD	VERY GOOD	FAIR	FAIR	FAIR	190-225	53-56	AIR OIL QUENCH	MEDIUM OR LESS	MEDIUM LOW

1. (.10) – CARBON CONTENT 2. – WILL TAKE NITRIDING IF DESIRED 3. MUST BE OIL QUENCHED FOR SECTIONS OVER 4 IN. THICK.

slowly, and then finish machined. The final heat treatment would then be applied.

Another technique to minimize the possibility of distortion due to heat-treatment stresses is to use a prehardened steel such as the AISI* P-20. This steel can be machined and polished in the prehardened state (28 to 30 Rockwell C) and used without additional heat treatment. For complicated molds, a stress-relieving treatment, however, is usually recommended. If a harder surface is required for increased wear resistance then nitriding is usually used.

Two other terms used in heat treatment of steels are normalizing and annealing. Normalizing refers to a heat treatment which is given to the steel to relieve forging or machining stresses. It usually consists of simply heating the part to approximately 900°C (1650°F) (depending on the steel), followed by an air cool. Annealing is usually done when the steel is to be brought to its minimum hardness state in preparation for machining, or if maximum stress relief is desired. It also consists of heating to the recommended temperature, and then allowing the steel to cool slowly in the furnace to approximately 650°C (1200°F), after which the part may be air cooled.

Nonferrous Metal Molds

Several types of nonferrous metals are used for molds, principally where short runs are concerned, or where the pressures or wear conditions are not severe, or where the metal has some special characteristic such as high thermal conductivity, ease of casting, etc. The most commonly used nonferrous alloys are beryllium copper, aluminum, zinc, electroformed nickel, and low-melting types.

Beryllium Copper Alloys. Beryllium copper alloys with an approximate composition of 2.5% beryllium, 96.5% copper, and small amounts of cobalt and silica, have been used for many years as cast injection molds. Molds can be cast to give excellent surface reproduction, and can be hardened to as high as Rockwell C42, depending on the particular alloy and heat treatment. While these molds are expensive, their high thermal conductivity allows faster heat extraction during molding, thus resulting in a lower cycle time.

Aluminum. Aluminum alloy injection molds are mainly used for short-run production or for prototype fabrication. They are, however, used extensively for low-pressure applications such as rotational molding, blow molding, casting, or in structural and nonstructural foam processing. The use of aluminum in molds offers the very definite advantages of light weight, ease of machining, high thermal conductivity, and moderate costs. For molds to be fabricated by machining, the usual choices are the heat-treatable aircraft grades such as 6061 or 24S alloys. There are a number of casting alloys, the particular ones used are usually dependent on the type of mold to be made, the size, the application, etc.

Aluminum-bronze alloys have been used in molds, not for cavities or cores, because of softness, but as wear plates, bushings, plungers, etc., where steel on steel might present a galling problem. These alloys are available as plate or bar stock and as rods and tubes.

Zinc Alloys. Zinc alloys, the most common of which is Kirksite, are used extensively as castings, where they can equal aluminum in surface hardness and compressive strength. The use of a steel frame around the casting is recommended if the mold is used in any high-pressure applications.

Nickel. Nickel molds are usually made by an electroforming process, somewhat similar to electroplating. In general, electroformed nickel is used to produce molds such as pen barrel cavities and similar parts which might be very difficult to fabricate by normal machining. In some cases the cavity is built up to a thickness of approximately $\frac{1}{16}$ inch using a hard nickel deposit, and then the mold is further built up using a soft nickel, which is more rapidly deposited. The hard nickel is deposited at about 500 Brinell and the soft nickel at about 150 Brinell. In other cases the hard nickel is reinforced by a copper deposit, which also builds up at a more rapid rate than does the hard nickel. Nickel molds are also produced by a vapor deposition technique which is said to result in better tolerances, more uniformity of the shell thickness, higher strength, and better heat transfer. More details on the processes used to form these molds are included in the section on Mold Fabrication Techniques.

Low-Melting Alloys. Principally based on bismuth, low-melting alloys are used to produce molds for room-temperature-cure casting resins. After the resins are cured, the molds are either broken away or are melted off from the casting. Such materials are available under tradenames such as Cerrobend and Cerrotrue (Cerro Sales Corporation).

Nonmetallic Mold Materials

There are a number of nonmetallic materials which are also being used to produce molds. In general these materials are confined to usage where the pressures are not high such as casting, rotational moldings, nonstructural foam, thermoforming, etc. These nonmetallic mold materials are divided into two classes: elastomeric or flexible materials and hard, rigid materials.

The elastomeric materials include silicone room-temperature vulcanizables (RTVs) in many different hardnesses, and somewhat similar urethane, rubberlike materials, vinyl plastisols, gelatin, polysulfide, and various rubber latices. All of these materials share the common property of being castable or brushable around a master, and then hardening to form a mold. The main advantage in the use of the elas-

*AISI: American Institute of Steel and Iron.

tomeric materials is that parts can be made with undercuts and still be removable from the mold. The disadvantage is that, in general, only plastics castings can be made using such molds, since any pressure will cause deformation. An exception is an elastomeric male master used as a punch in a female cavity. If the mold is designed to almost fully confine the male, then appreciable pressures can be developed. A special characteristic of the silicone RTV mold is that usually mold release is not required even when casting epoxies or urethanes.

The rigid nonmetallics include such materials as plaster of paris, wood, wax, and cast plastics or glass-fiber-reinforced epoxy or polyesters. In all cases, except the wax, mold release is required on the mold. Cast epoxy reinforced with metal powder can be made strong enough to be used as an injection mold for small quantities of parts, if maximum pressures are not employed and temperatures are kept below 200–230°C (400–450°F). Backing the mold up with a metal frame is recommended.

Mold Surface Treatments

There are a number of surface treatments which can be applied to the molds to improve lubricity, wear qualities, corrosion resistance, etc. In general, these treatments are applicable only to metal molds; however, in some cases they can also be applied to the nonmetallic molds. Table 20-2 lists these treatments, the surfaces they are applied to, and the results which may be obtained.

Mold-Making Techniques

The particular method of making a mold depends to a great extent on the type of material of which the mold is made

Table 20-2. Mold Surface Treatments.*

Process	Material	Applied to	Purpose
Coating by impingement, molecularly bound	Tungsten disulfide	Any metal or hard nonmetallics	Reduce friction and metal-to-metal wear with dry film. Nonmigrating.
Coating by impingement, organically bound	Graphite	Any metal or hard nonmetallics	Reduce sticking of plastics to mold surface. Can migrate.
Electrolytic plating	Hard chrome	Steel, nickel and copper alloys	Protect polish, reduce wear and corrosion (except for chlorine or fluorine plastics).
	Gold	Nickel and brass	Corrosion only.
	Nickel	Steel and copper alloys	Resist corrosion except sulfur-bearing compounds, improves bond under chrome. Build up and repair worn or undersized molds.
Electroless plating	Nickel and nickel–silicon carbide	Steel, aluminum and beryllium copper	Protect nonmolding surfaces from rusting. Abrasion resistance.
	Phosphor nickel	Steel and copper	Resist wear and corrosion.
Blasting	Fine-mesh abrasives (liquid honing)	Any metal or hard nonmetallic	Remove scale after heat-treating. Provide a fine matte finish. Improves release of some thermoplastics.
	Sand	Any metal	Provide a coarse matte finish.
	Steel shot	Steel	Peen, to improve crack and fatigue resistance.
	Glass beads	Any metal	Peen, to remove fine scratches and tool marks.
Photo etch	Chemicals	Steel and aluminum	Decorate or texture mold surfaces.
Nitriding	Nitrogen gas or ammonia	Certain steel alloys	Improve corrosion resistance, reduces wear and galling. Alternative to chrome and nickel plating.
Liquid nitriding	Patented bath	All ferrous alloys	Improve lubricity and minimize galling.
Anodizing	Electrolytic oxidizing	Aluminum	Harden surface, improve wear and corrosion resistance.
Spraying and heat curing	Polytetrafluoroethylene	Aluminum, steel, nickel and copper alloys	Reduce sticking to mold surface.

*Reprinted, with permission, from *Plastics Machinery & Equipment*, June 1978. Copyright 1978, Industry Media, Inc.

and, to a lesser extent, on the application for which it is intended. Thus, if a mold is to be used for injection molding of a large quantity of parts it would probably be made of steel shaped by any one of a variety of methods. If only a few prototype parts are required, an epoxy mold would probably be chosen which could be made by a simple casting technique. It is extremely important that prior to designing or building a mold all the factors be considered; that is, pressures, temperatures, number of parts, finish required, production rate desired, and last, but not least, the anticipated cost of the mold and its amortization. Since the material cost is probably the least expensive item in the total mold cost, the best material should be chosen, consistent with a fabricating technique which will result in a satisfactory mold at the lowest cost.

Injection molds far outnumber all other types of molds. Thus, the following discussion is mainly related to the production of injection molds, unless otherwise noted. The majority of injection molds are made of steel, consisting of a cavity portion, termed the female half of the mold, and a punch or male half of the mold, both of which are responsible for the shape of the part. Associated with these two components are a number of other plates, pins, bushings, etc., all of which require very little in the way of special machining or shaping.

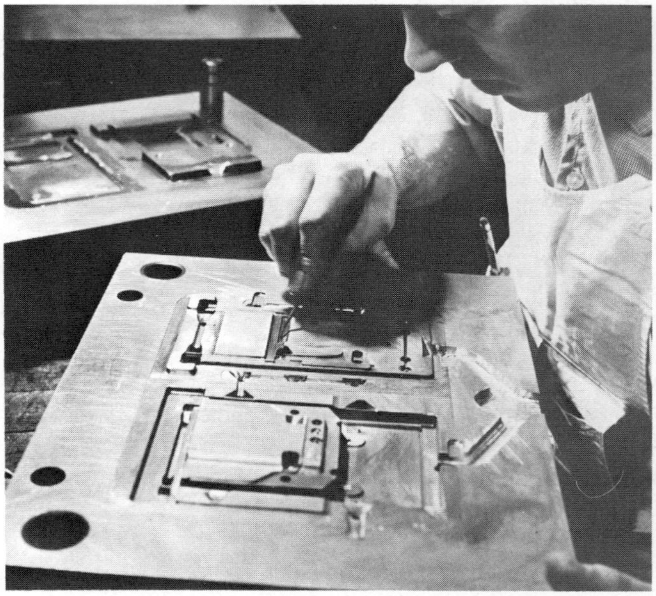

Fig. 20-2. Hand finishing of an injection mold.

Machining. Machining is metal removal by a cutting tool. The earliest technique used for forming a cavity, or the male portion of the mold, was by machining away the unwanted areas or sections of metal until the desired configuration was obtained. Some of the machines used to do this include drill presses for hole boring, end mills for cavity production, and lathes for production of circular male or female parts, for boring holes and cavities, and for grinding and polishing. In this respect it should be noted that the lathe, within its capabilities, is one of the most versatile of all machine tools, and also that lathe work is possibly the most inexpensive precision work available. While most lathes are manually operated, there are copy lathes which can very accurately machine out a metal piece by copying the contours of a pattern made of wood, plaster, or metal. Such equipment can be used to make multiple cavities in a block or make a number of identical male sections of a mold. Other tools used in the metal removal process include shapers and planers, which are used to square up a raw metal blank or remove a portion of a surface in an even plane. Grinders and polishers are used for final finishing, in many cases as a hand operation (see Figure 20-2).

Electric Discharge Machining (EDM). The production of small, intricately shaped cavities and cores is one of the major requirements of many injection molds. Simple, deep holes can be easily made by use of a drill press, an end mill, lathe, or a similar machine tool. However, if the cavity is an irregularly shaped hole or perhaps shaped like a star or a gear with sharp corners or points, it would require a great deal of time and effort to produce it by the conventional machining techniques described above.

In the late 1950s and early 1960s another method was developed for production of these "difficult-to-machine" cavities or core pieces. This method is called "electric discharge machining" or EDM. It is based on the erosive effect which a spark exhibits in jumping from one electrode to another. EDM takes advantage of the fact that erosion is much greater on the positive electrode than on the negative electrode by making the electrode negative and the workpiece positive. Figure 20-3 illustrates the minute cratering pro-

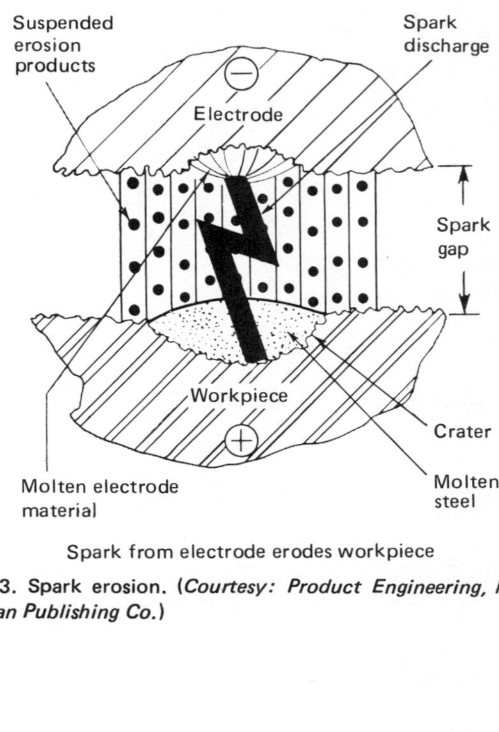

Fig. 20-3. Spark erosion. *(Courtesy: Product Engineering, Morgan Grampian Publishing Co.)*

duced by individual sparks and the large cumulative crater made by the high-frequency spark discharge. Also shown is the disparity in size of the positive and negative craters, which may vary as much as 8 to 1 in volume.

In the actual electric discharge machine, the electrode is made in the precise shape of the desired cavity, except slightly smaller on the sides to allow for the cratering effect. The electrode is advanced vertically into the workpiece to form the cavity. As it moves, the eroded metal particles are removed by a hydrocarbon flushing fluid, and since it is a special dielectric fluid, it regulates the sparking to cause even metal erosion. Cavity depth can be increased continually, with some corresponding electrode wear; however, the side erosion depth will reach a maximum and will not increase, since this is also the maximum length which the spark will travel in the dielectric fluid. This gap is called the spark overcut, and varies with the frequency and current being employed to form the cavity. By using a low frequency and a high current flow, the metal removal rate can be appreciably high; however, a rough surface results. A high-frequency spark and a low current will result in a smooth matte finish. Table 20-3 lists typical work removal rates and finishes obtained using various frequencies and current flows.

An electric discharge machine consists of the following units:

1. A power supply capable of delivering direct current at voltages up to 125 volts, at currents up to 400 amperes, and at variable frequencies from as low as 5 kilohertz to as high as 1000 kilohertz.
2. A means of holding the electrode and advancing it vertically under servo-control at a rate determined by the operator.
3. A container to hold the dielectric fluid and the workpiece. The workpiece holder must also allow for precise positioning in three planes.
4. A pump and filter mechanism to circulate the fluid and trap the eroded metal particles.

Figure 20-4 shows the schematic of such a machine and Figure 20-5 illustrates a typical machine.

EDM machines of the type pictured are capable of producing cavities (or cores using hollow electrodes) as large as the machine capacity or as small as 0.004 inch in diameter.

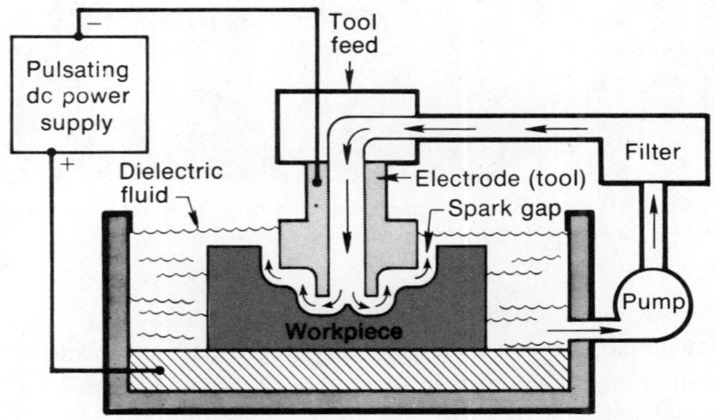

Fig. 20-4. Schematic of EDM (ram or plunger type). (*Courtesy: Machine Design*)

Table 20-3. Typical EDM Surface Textures with Copper Electrodes.*

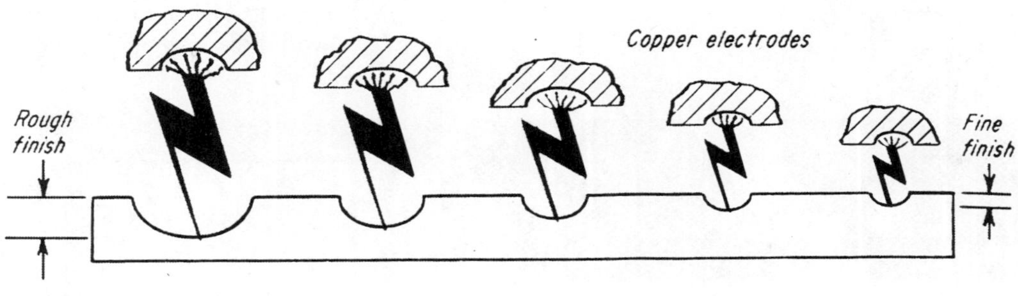

Power					
Frequency, kc	5	10	20	450	1000
Current, amp	1—20	5—17	4—12	3—9	0.5—3
Surface finish, microin.	200-500	150-200	100-150	50-75	25-10
Crater size, mils					
Workpiece, depth	1.9—4.0	1.5—1.9	1.0—1.5	0.5—0.6	0.1
width	5.3	2.4	2.0	0.65	0.15
Electrode, depth	1.2	0.57	0.45	0.18	0.03
width	1.3	0.58	0.46	0.2	0.04
Work removal rate, cu in./hr	1.7	0.5	0.25	0.1	0.005 to 0.0005

*Courtesy: *Product Engineering*, Grampian Publishing Co.

Fig. 20-5. Typical EDM machine showing operating chamber and controls. (*Courtesy: Charmilles Corporation of America*)

The cavities can be through or blind holes in relatively hard-to-machine shapes as shown in Figure 20-6a. Furthermore, the holes can be cut on surfaces at various angles, which are also normally difficult to machine with a drill bit or end mill (see Figure 20-6b and 20-6c).

A modification of the conventional ram- or plunger-type EDM is wire EDM; however, this method can only be used

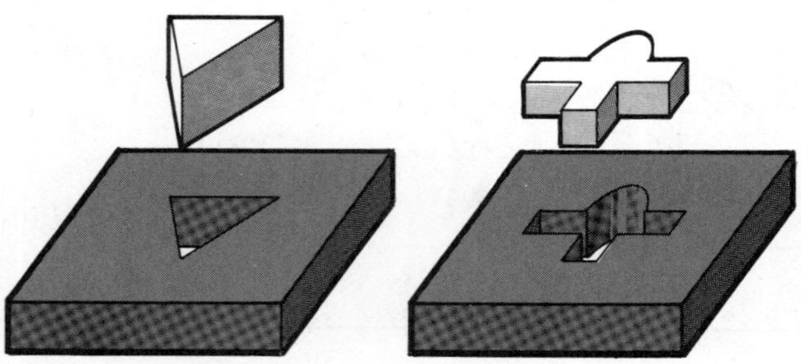

Fig. 20-6a. EDM hole cutting. (*Courtesy: Machine Design*)

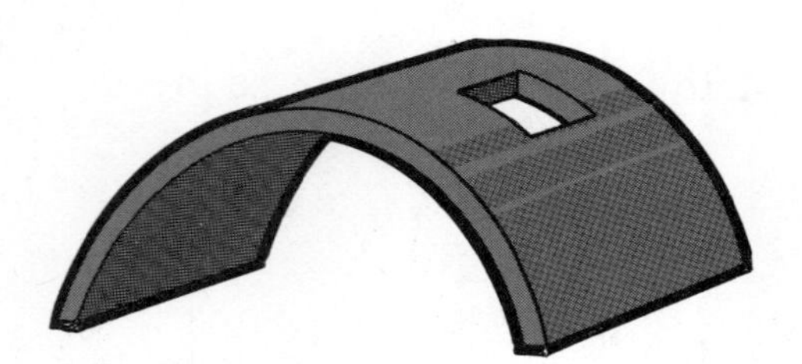

Fig. 20-6b. Delicate parts can be machined without damage or distortion since EDM does not involve any tool pressure. Also, hardened materials can be machined as easily as soft or annealed metals. (*Courtesy: Machine Design*)

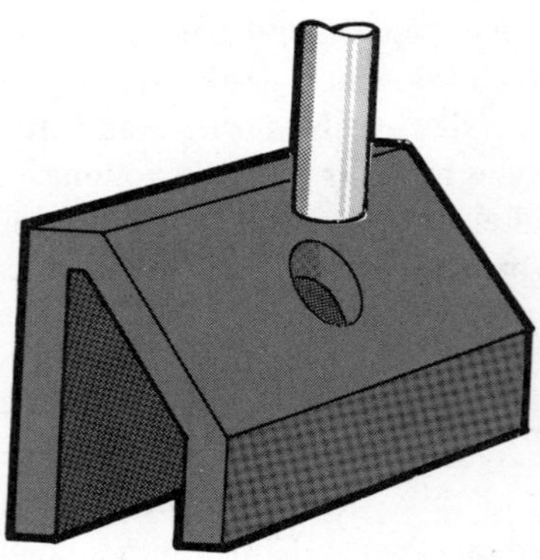

Fig. 20-6c. Holes can be cut on surfaces at various angles with the EDM process. (*Courtesy: Machine Design*)

for through holes. In this technique, a thin copper wire serves as the electrode. Accuracy is said to be within ±0.0002 inch. The finish on the cut is a matte texture in the range of 30 to 50 microinches. A schematic of this type of machine showing how the wire is advanced from reel to reel to take up the effects of erosion is illustrated in Figure 20-7.

Electrodes. Electrodes for the EDM are made from any good electrically conducting metal or graphite. The most commonly used materials are copper, copper-tungsten alloy, and graphite. The usual technique to produce the electrodes is by machining to shape. Copper-tungsten electrodes can also be produced by casting, using a proprietary process patented by the 3M Co. of St. Paul, Minnesota. Other techniques for production of graphite electrodes are described below. A silver-tungsten alloy is also used occasionally where its higher cost can be justified on the basis of its high workpiece erosion-to-wear ratio, as shown in Table 20-4.

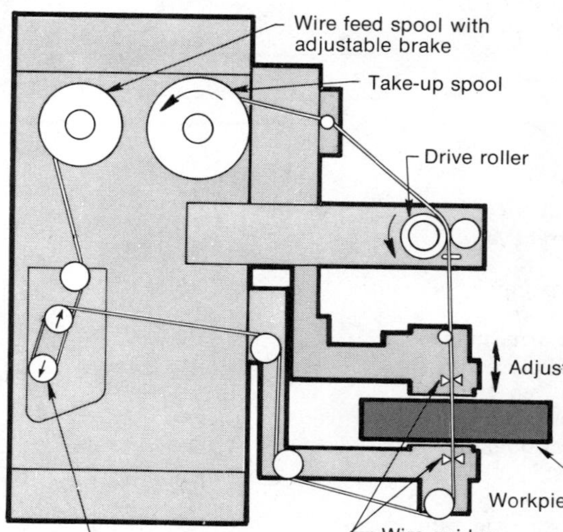

Fig. 20-7. Wire EDM. (*Courtesy: Machine Design*)

Table 20-4. Electrode Wear-to-Cut Ratios.*

Wear-to-cut ratios, (erosion volume, workpiece to electrode), for high-carbon, high-chrome steel workpiece and machinable electrodes:

Copper (99% conductivity)	4:1
Copper-tungsten	8:1
Graphite (commercial)	8:1
Silver-tungsten	15:1
Steel (finishing electrode only)	15:1

*Courtesy: *Product Engineering*, Grampian Publishing Co.

Conventional metal removal techniques, such as sawing, milling, lathe cutting, etc., are many times faster in terms of cubic inches of metal removed per minute than is normal EDM work. However, the convenience and accuracy of using EDM in many cases justify its use, particularly when a rough cut, a semi-finish cut, and a finish cut can all be made with one electrode. Such a setup is shown in Figure 20-8, which illustrates the use of a three-stage electrode, and the varying process conditions which are used with the electrode. For greatest saving in machine time and labor with such a setup, the cavity is initially rough cut, by conventional methods, to provide a flushing hole and to remove some of the metal. The three-stage electrode is then put through as shown in Figure 20-8.

In addition to the use of the metals described above as electrodes, steel is also used for special applications. One such use is when an irregular parting line between mold halves is required. Using conventional machining methods, the two halves of the mold would have to be made with extreme accuracy in order to avoid excessive flash. With EDM, one side of the mold is completely finished, as is the cavity in the second half of the mold. The parting line is then rough cut on the second half of the mold which is then installed as the workpiece in the EDM. The finished first half is then mounted to act as the electrode. The spark interaction between the two halves then results in a parting line that matches perfectly. Usually the time to form the parting line by this method is considerably less than if the same operation was done completely by machining. Typical electrodes used to produce various complicated cavities are shown in Figures 20-9 and 29-10.

Orbital-Action EDM. One of the problems in using the same electrode to produce a number of cavities is that the electrode wear may make it unsuitable for use after only a few operations. A technique known as orbital-action EDM offers a method of using the electrode regardless of the wear. The technique consists of moving either the workpiece or the electrode in a circular path which increases the extent of the cut by the amount of orbital motion (see Figure 20-11). Thus, as the electrode wears, the orbit is simply increased in size. Other advantages claimed for this

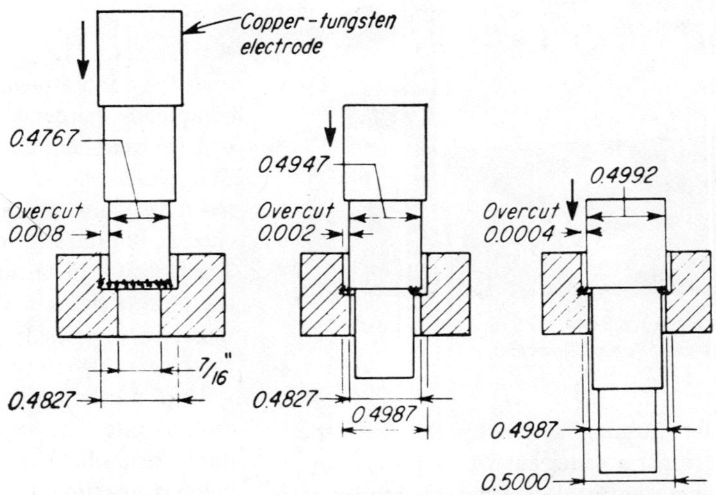

	Rough cut	Semi-finish	Finish cut
Current, amp	12	6	0.5
Frequency, kc	5	450	1000
Overcut, in.	0.008	0.002	0.0004
Finish, microin.	300	60	12 to 15
Tolerance, in.	0.001	0.0005	0.0002
Metal removed, cu in.	0.028 (88.5%)	0.003 (9.5%)	0.00065 (2%)
Erosion time, min.	3	11	58

Fig. 20-8. Use of a three-stage electrode to rough cut, semi-finish, and finish a cavity. (*Courtesy: Product Engineering, Morgan Grampian Publishing Co.*)

Fig. 20-9. Typical electrodes and dies made by ram and wire EDM. (*Courtesy: Charmilles Corporation of America*)

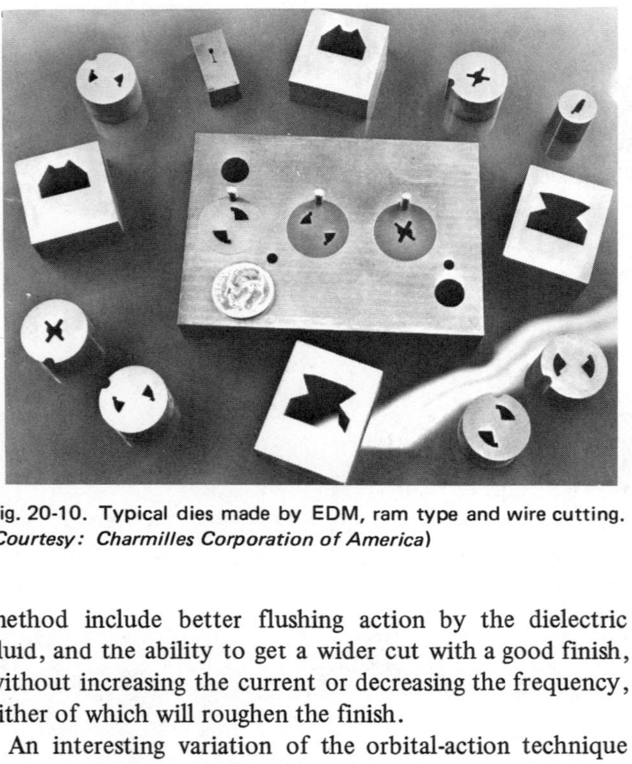

Fig. 20-10. Typical dies made by EDM, ram type and wire cutting. (*Courtesy: Charmilles Corporation of America*)

method include better flushing action by the dielectric fluid, and the ability to get a wider cut with a good finish, without increasing the current or decreasing the frequency, either of which will roughen the finish.

An interesting variation of the orbital-action technique is the use of the oscillatory action to produce graphite EDM electrodes. In this, the electrode is made by first making a model of the electrode in wood, plaster, or metal. The model is used to produce an abrading die, which is made of a silicon-carbide-filled epoxy casting. The abrading die is then forced into the graphite block, which is at the same time being moved with the rotary oscillatory motion. Close-tolerance electrodes are said to be made by this process without the use of machine tools. Figure 20-12 shows the sequence of operations and a machine used for the process. Many electrodes can be made using the same abrading die. The time is claimed to be considerably faster than electrode fabrication by conventional machining techniques.

Workpieces. The workpiece can be any conductive metal used for a mold or die. If the workpiece is steel, it can be completely hardened prior to EDM, since the condition of heat treatment has no effect on the spark erosion action. This would eliminate the side effects of hardening after machining—the possible distortion, cracking, surface heat checks, and the stresses induced by conventional machining. The fact that EDM can be applied to hardened molds makes the process particularly attractive for finished molds which must be reworked; for example, to open up a gate or machine a ridge in a mold to act as a strengthening rib.

While the hardness of a steel part has no effect on the erosion rate, the melting point of the material definitely does. Bismuth, which melts at 271°C (520°F) will form a crater four times wider and deeper than tungsten whose melting point is 3370°C (6098°F). The erosion rate is also directly dependent on the electrical conductivity characteristics of the workpiece material. Thus, aluminum can be machined by EDM at a more rapid rate than steel. To check out a new electrode then, it would be much cheaper, in terms of time and energy, to use a piece of scrap aluminum than to use steel.

Disadvantages. The major disadvantage of EDM is its relatively slow rate of metal removal as compared to most

Orbital-Action EDM is Different

In conventional EDM, the gap between the electrode (tool) and the work is created only by the overcut. Thus the electrode must be made to the exact size (minus two overcuts) of the required cavity. If a number of parts are required, electrode wear may make the electrode unsuitable (too small) after it has been used several times.

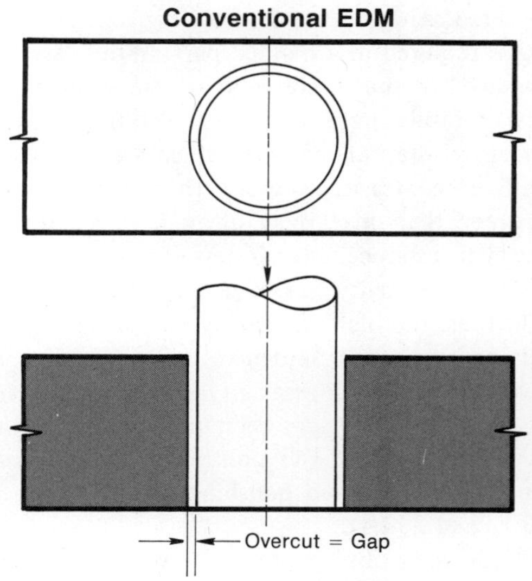

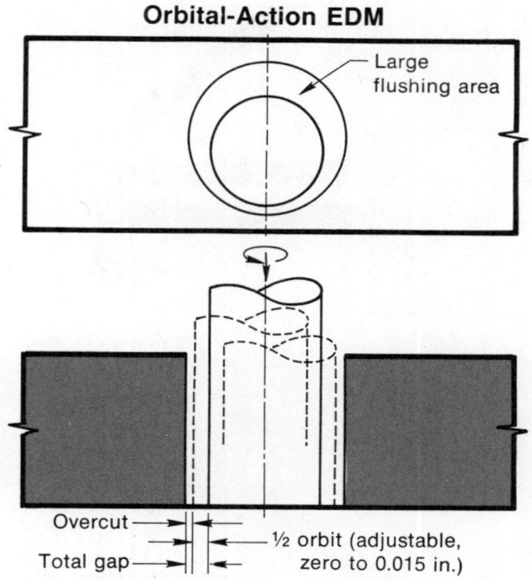

In an orbital-action system, the electrode is made smaller than the cavity to be machined. The difference is made up by the size of the orbit. And, as the electrode wears, the orbit can be increased to compensate.

The dielectric fluid, which serves as coolant and flushing medium as well as spark conductor, is much more effective in an orbit-action system than in a conventional EDM setup. Since the orbiting electrode is smaller than the cavity in the workpiece, the flushing area is much larger than in conventional systems.

Also, the continuous motion of the orbiting action actually pumps the fluid around the perimeter of the cut, just ahead of the spark, providing the agitation needed to carry away the debris. Flushing action is usually improved by increasing the size of the orbit and/or increasing orbital speed. At some combination of these variables, optimum flushing conditions are reached, however, and further increases in orbit size or speed have no effect.

Another advantage of orbital-action EDM becomes evident when the dielectric fluid is pumped down through the electrode, as shown in *Brief Review of Conventional EDM.* In such a configuration, flushing is restricted (in conventional EDM) by the non-eroded post left in the center, which reduces fluid flow.

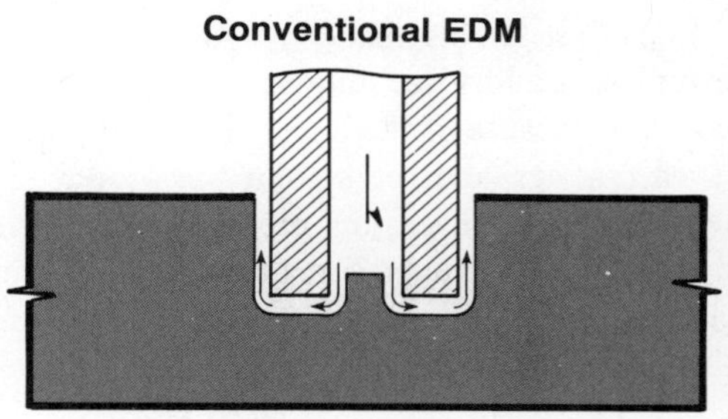

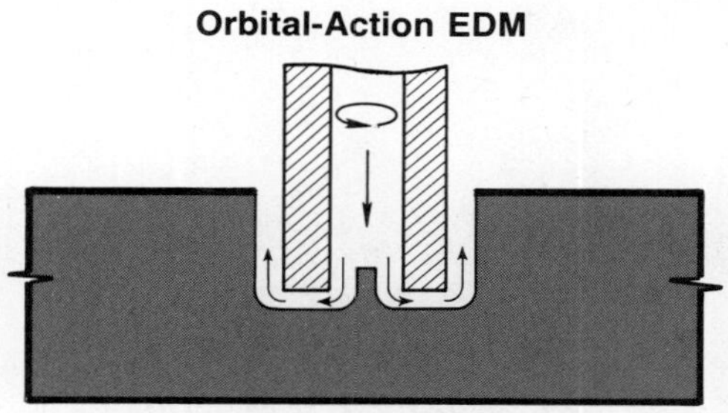

Orbital-action reduces the post diameter, increasing the fluid flow appreciably. In fact, if the fluid hole is small enough, the orbital motion can eliminate the post entirely. For example, machining 0.030 in. diameter holes, using a 0.030 in. orbit, would leave no posts.

Fig. 20-11. Conventional and orbital-action EDM. (*Courtesy: Machine Design*)

conventional methods of machining. The cost of equipment is considerably higher than other types of metal removal machines. Electrode wear can seriously limit production if many cavities are required, unless orbital-action is used. The cost of retrofitting ordinary EDM machines with orbital-action equipment can also be quite expensive. If maximum high currents are used to obtain high erosion rates, then localized heating can increase the workpiece surface hardness from 1 to 4 points (Rockwell C), with the possibility of surface cracking. Finishing at a lower current rate will usually eliminate both effects, however.

Despite the disadvantages mentioned, the many advantages of EDM, including its capacity to do jobs virtually impossible by other methods, have caused this method to be adopted by the great majority of mold makers and tooling shops.

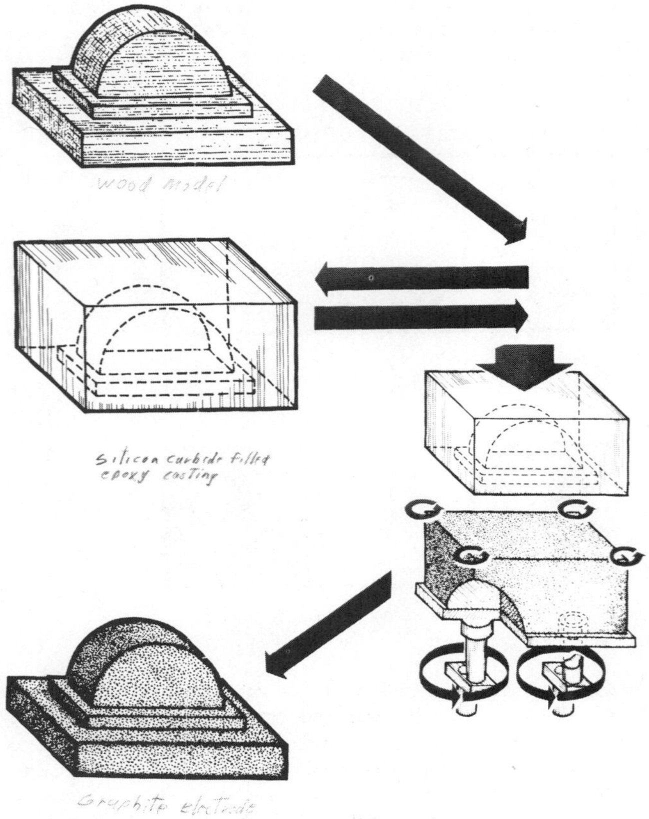

Fig. 20-12a. HAPCO process for production of graphite electrodes. (*Courtesy: Hausermann Abrading, Addison, Ill.*)

Fig. 20-12b. HAPCO machine for production of graphite electrodes. (*Courtesy: Hausermann Abrading, Addison, Ill.*)

Hobbing. Hobbing is a method of forming cavities by use of a hardened and polished steel master which is hydraulically forced into a soft steel blank. The process is particularly applicable when a number of duplicate cavities are required. Another use for hobbing is when depressed lettering is required in a molded part. In this case the lettering is engraved *in* the master hob which, when pressed into the cavity, results in raised letters. The final result is depressed letters in the molded part, which are then usually filled in with a contrasting color. This method is considerably cheaper than machining away most of the surface of a cavity to produce raised letters or figures.

After the cavity has been made, at pressures as high as 300 tons, the blank is usually cut to size to fit within a retaining ring, then hardened and polished. Case hardening gives a hard, wear-resistant surface and a strong, ductile core. If the master has been highly polished, the cavity will in most cases be well polished and require only a small amount of touch-up polishing, if any at all. (See Figures 20-13 and 20-14).

While the hobbing process is well suited for production of multiple cavities at the lowest costs, in many cases the cavities require several annealing operations, as well as the final heat treatment, which can add to the total cost. Just as important, the multiple heat treatments increase the possibility of warping and cracking, particularly in complicated molds having sections of widely varying thicknesses. Because of the specialized procedures, and the very high pressures required, most mold makers are not equipped to do hobbing and so the work is sent out to firms which specialize in this process (see Figure 20-15).

Casting. Cast molds made by gravity casting in sand by the typical foundry techniques have not proved to be very successful, because of the low density and porosity inherent in such castings, and the inability to produce fine surfaces. There are, however, casting processes which can be used to produce cast molds for use in a number of plastics processes, including injection molding, where high pressures must be resisted.

Shaw Process. One such casting method, a proprietary licensed process known as the Shaw process, utilizes a fired ceramic master against which any one of the common molten metals can be cast. The general procedure for preparing the ceramic master, in this and other proprietary techniques, is to prepare a pattern of wood, plaster, metal, or plastic, with the proper shrink allowance for the metal to be cast. A special, quick-setting rubber compound is then cast around the pattern. The rubber will very accurately reproduce even the grain structure, if the pattern is wood, which is frequently the case if furniture components are to be made in the mold.

After the rubber is stripped from the pattern, a ceramic mixture is poured into the rubber cavity and allowed to

Fig. 20-13. Hobbed molds, master hob, and plastics parts. (*Courtesy: L. W. Reinhold Plastics, Inc.*)

Fig. 20-14. Finely detailed and polished hobbed mold. (*Courtesy: L. W. Reinhold Plastics, Inc.*)

harden. The ceramic is then fired in an oven. The resulting master will show good surface reproduction with a small amount of porosity. This porosity will allow the molten metal, which is poured over the master, to degas, thus resulting in a more dense, sound casting with good surface reproduction.

Fig. 20-15. Special press used for hobbing. (*Courtesy: L. W. Reinhold Plastics, Inc.*)

This technique of preparing the ceramic master can be employed with any of the conventional casting metals; however, it is probably used more with beryllium copper than any other metal. The reasons for the use of this metal, as previously discussed, are the good surface reproduction in castings, the hardness which can be equal to steel (Rockwell C 45), and the high thermal conductivity which can appreciably decrease cycle times in injection molding.

In making the casting, a long, vertical riser, or sprue, is used to act as a pressure source, while the molten metal is being added and while it is cooling. This use of pressure helps produce a part which is uniform in density and porosity. The major disadvantage of the process is the fact that the master is rather fragile and only one casting can be made from each mold. The Shaw process, which was developed in England, is licensed to a number of fabricators in the United States. These specialists should be consulted on the applicability of the process for any given mold as compared to other methods of mold production.

Pressure Casting. Another method of preparing high-quality castings is the "pressure casting technique." A casting of beryllium copper, aluminum, or zinc alloy is cast around a polished steel master, which in this process is also called a hob. Immediately after pouring, the molten metal and the hob, enclosed in a yoke, are subjected to a hydraulic pressure of up to 100 tons to produce a dense, void-free casting (see Figure 20-16). This process is readily used for molds up to several hundred pounds in weight. Other precision castings are made by variations of this process, such as die casting and centrifugal casting, all of which are done by companies specializing in such processes. Small molds are also made by the "lost wax process," in which the master is a consumable material such as wax or polystyrene. This process has the advantage of being adaptable to very intricate masters, but has the disadvantage that a new master is required for each mold. The process has been used for many years in the production of fine jewelry and dental work.

Electroforming. As previously discussed, electroforming is a method of forming a mold by a deposition process. It is a considerably slower process for mold construction than casting or machining. However, it is one technique for fabricating certain cavities which would be extremely difficult or impossible to make by any other method. Examples would be a mold for injection molding of doll heads, or a pen barrel cavity with a decorative finish.

In electroforming, the master may be any material which can be made into a completely finished pattern. These could be metal masters or patterns of wood, wax, glass, plaster, or plastics parts which are to be duplicated. In the case of metals, electrical conductivity is no problem; nonconductive materials are sprayed with a silver nitrate solution, which is later reduced to pure silver, or are sprayed with a conductive paint. Masters that can be removed from the cavity are called permanent masters and can be used over and over to form a number of cavities. Masters containing undercuts which would prevent removal from the cavity are called expendable masters and are usually made

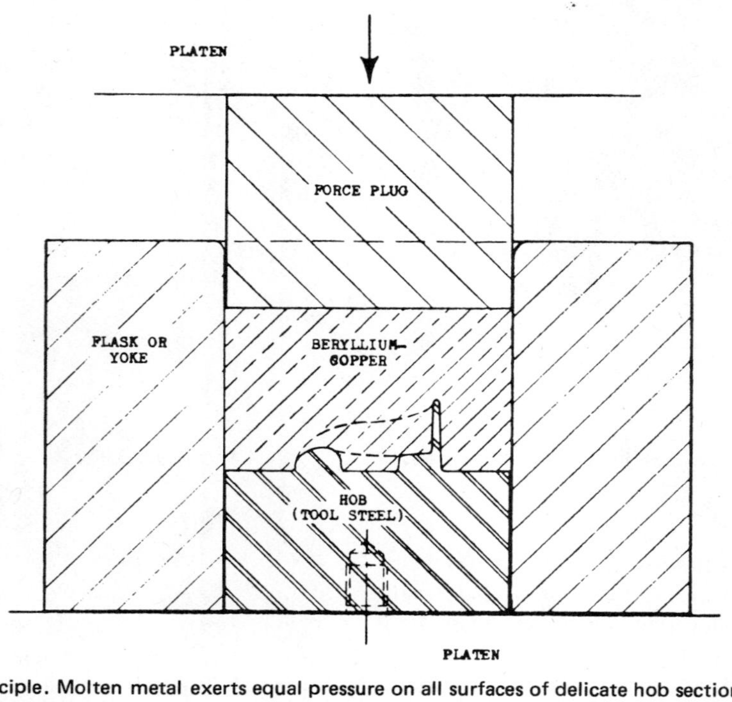

Fig. 20-16. Pressure casting principle. Molten metal exerts equal pressure on all surfaces of delicate hob section. (From *Plastics Mold Engineering Handbook* by DuBois and Pribble)

from materials such as low-melting lead, zinc and tin alloys, waxes, soluble or meltable plastics, soluble or breakable plasters, or dissolvable alloys such as aluminum and brass. A technique frequently used to prepare expendable masters is to make one permanent master, then cast an elastomeric cavity around it. The permanent master is then withdrawn and a number of expendable masters are then cast in the elastomer cavity, which is usually made using a silicone room-temperature-vulcanizing rubber (RTV). The use of the silicone RTV will give excellent reproduction of the original master, and if the expendable masters are cast in epoxy or wax, there is very little loss of detail.

The master is placed in the electroforming bath, nickel sulfamate, with certain proprietary additives. It is made the cathode in a circuit in which nickel bars are the anodes, as shown in Figure 20-17. Initial deposition is at a rate of 1 to 3 mils per hour, at current densities of 20 to 30 amps per square foot at a usual temperature of 35–50°C (100–125°F). If wax masters are used, the deposition may be done at room temperature. Deposition continues for 1 to 3 weeks or until a thickness of $\frac{1}{8}$ inch (or thicker if required) is obtained. The nickel that is deposited is a hard (Rockwell C 45 to 50), stress-free coating which has good elongation.

After deposition of a coating which is thick enough so the cavity can be handled without damage, the master is usually transferred to a second bath for deposition of a softer nickel or a copper backing so it will have sufficient strength and thickness for machining. Usually the cavity at this point is machined to fit tightly into a steel holder which can then be mounted into the mold base. Alternatively, cooling coils can be wound around the part at this time and then anchored firmly in place by the backing, after which it is machined on the exterior to fit into the mold base. Other backing materials which can be used include concrete and metal-filled epoxy; both of these materials will lower the cost of the molds. However, nonmetallic backed molds have considerably lower thermal conductivity which, in many cases, could result in increased cycle times. Thus, the nonmetallic backed molds would probably be suitable only for prototype work where the cycle times were inconsequential.

While it is a fact that a well-polished master of almost any masterial will result in a well-polished cavity without any additional polishing, this does not mean that any master can be satisfactorily used as a basis for a cavity. There are several design limitations which must be accommodated if an optimum part is to be secured.

1. Recesses in the master should be as shallow as possible, since the electroforming process has limited throwing power.
2. Exterior corners and angles should be rounded if possible to minimize burned spots and the development of "trees."
3. Undercuts should be avoided unless expendable masters are to be used.

Prior to the design of a master, it is advisable to consult with the electroformer to make sure the design being planned is practical in terms of the electroforming process and the cavity to be used. Figures 20-18, 20-19, 20-20, and 20-21 show some representative intricate injection molding cavities and molded parts.

Miscellaneous Metal Mold Fabrication Techniques. There are several other metal mold-making processes which are not used extensively, but for special applications have proved to be the best technique, the cheapest, or possibly a combination of both.

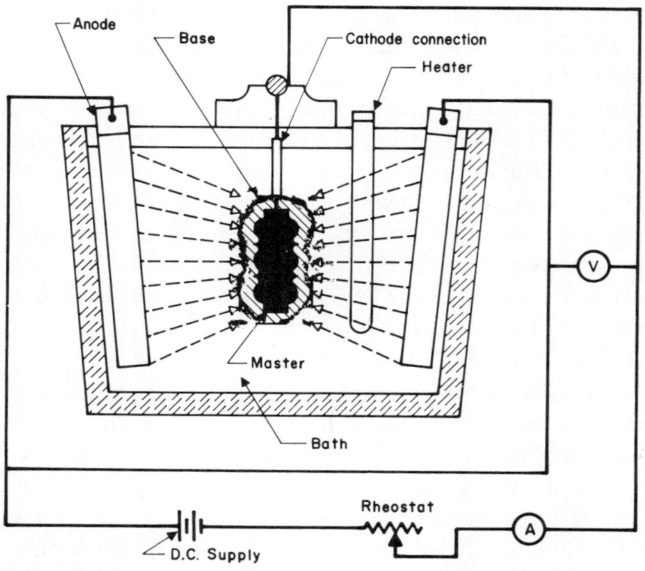

Fig. 20-17. Basic nickel electroforming process. (*Courtesy: Plastics Engineering*, February 1978)

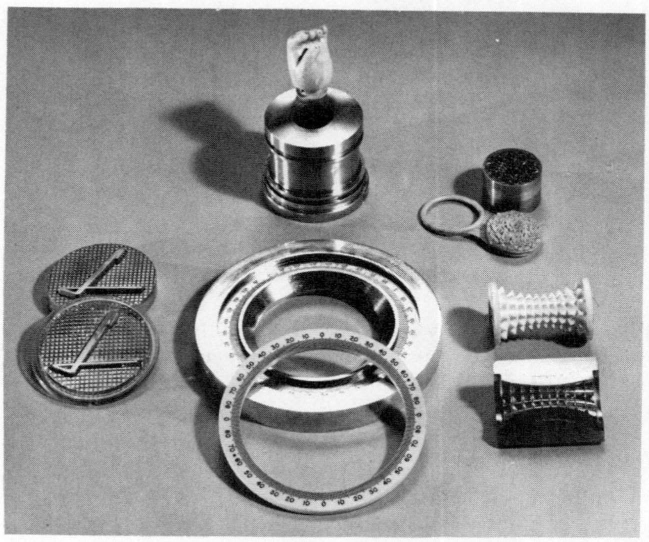

Fig. 20-18. Electroformed cavities and molded parts. (*Courtesy: Electromold Corporation*)

Fig. 20-19. Electroformed cavity and molded part. (*Courtesy: Electromold Corporation*)

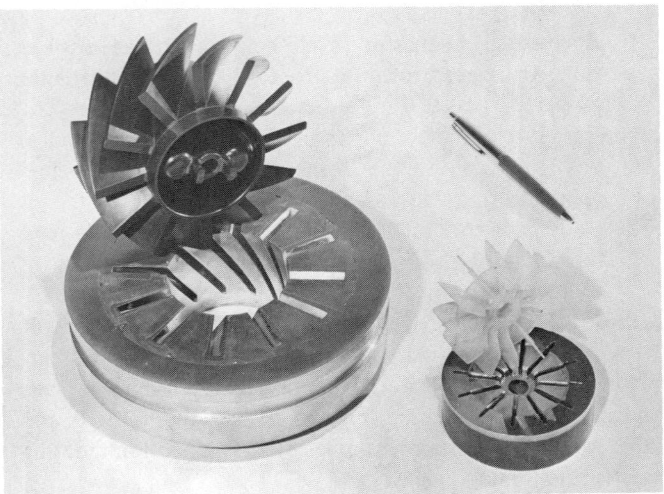

Fig. 20-20. Intricate parts molded in electroformed molds. (Reprinted, with permission, from *Plastics Machinery & Equipment*, June 1976. Copyright, 1976, Industry Media, Inc.)

Fig. 20-21. Multi-cavity electroformed mold. (Reprinted, with permission, from *Plastics Machinery & Equipment*, June 1976. Copyright, 1976, Industry Media, Inc.)

Nickel Vapor Deposition. A nickel shell can be built up on a master by deposition from a nickel carbonyl vapor. This process is consequently also called chemical vapor deposition (CVD), or gas plating. The process consists of heating a metal or ceramic master to 175°C (350°F) and then passing the nickel carbonyl gas over the part. The vapor is decomposed on the heated surface of the master and deposits nickel at a rate of approximately 0.010 inch per hour. The time for deposition of a thin shell mold is thus considerably faster than a similar mold made by electroforming. Other advantages claimed for this process are more even corner deposition and generally more uniform thickness deposition as compared to electroformed molds. The major disadvantages are the considerably higher prices for the molds and the necessity for masters which can withstand the 175°C (350°F) temperature without distortion.

Metal Spray. Metal spraying, also known as metallizing, has long been used as a method of building up worn shafts, abraded metal parts, etc. The same process using low-melting alloys is also being used to make molds using wood, plastics, and metal masters. In practice, the pattern, if wood, plastic, or plaster, is dried thoroughly and is then coated with a thin layer of a mold-release agent. The metal is then sprayed onto the master. Excellent surface reproduction is usually obtained. This process offers a quick, simple, inexpensive method of making molds which can be used for production of cast plastic parts, and other processes not involving high pressures. For protype injection molds, the use of a high-melting metal such as copper or aluminum is recommended.

Dipped Metal Molds. Dipped metal (slush cast) molds are made by simply dipping a polished metal master into a molten bath of a low-temperature melting eutectic alloy (see Figures 20-22, 20-23, 20-24, and 20-25). One such material, Cerrotru (Cerro Sales Corporation), with a melting point of 38°C (281°F), has the unique characteristic of expanding slightly as it solidifies, which greatly facilitates removal from the master. At the same time, the melting point of the alloy is high enough so that many plastics materials can undergo exothermic reactions or can be cured at elevated temperatures, at least for a preliminary cure, after which the casting is removed and the final cure is done in an oven. This technique is by far one of the fastest and cheapest methods of making a large quantity of molds for such applications as castings, encapsulation of electronic parts, prototype work, etc. Since the material can be used over and over, the cost of the alloy becomes negligible if a number of parts are made.

Nonmetallic Mold Fabrication. Nonmetallic molds are used in a number of plastics processes which include injection molding, thermoforming, low-pressure reinforced-plastics processing, foam processing, and casting. There are a number of reasons for the use of nonmetallic molds instead of metallic molds, which originally were the only types of molds employed.

Fig. 20-22. Use of a water-cooled aluminum master to form a number of nonpermanent casting molds. (*Courtesy: Cerro Metal Products, a division of The Marmon Group, Inc.*)

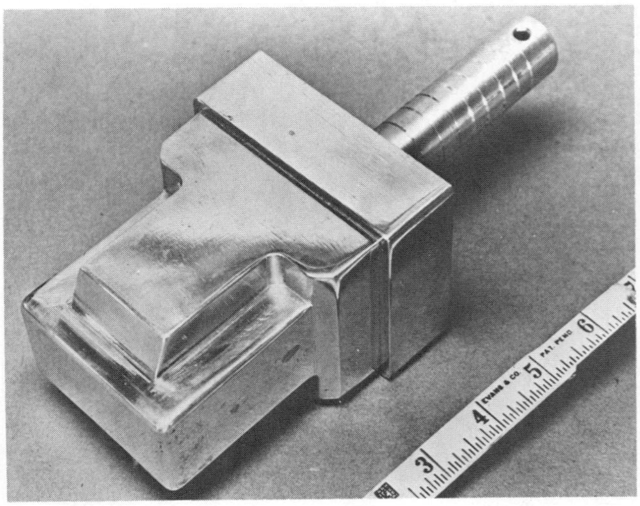

Fig. 20-23. Polished aluminum master. (*Courtesy: Cerro Metal Products, a division of The Marmon Group, Inc.*)

1. In most cases the cost is considerably less than a similar metal mold. This is particularly true if a small quantity of parts or simply prototypes are required.
2. Production time is usually much less than making a metal mold by machining, EDM, or electroforming.
3. The weight is usually much less than a metal mold.
4. In most cases repair work is very simple and can be done quickly.

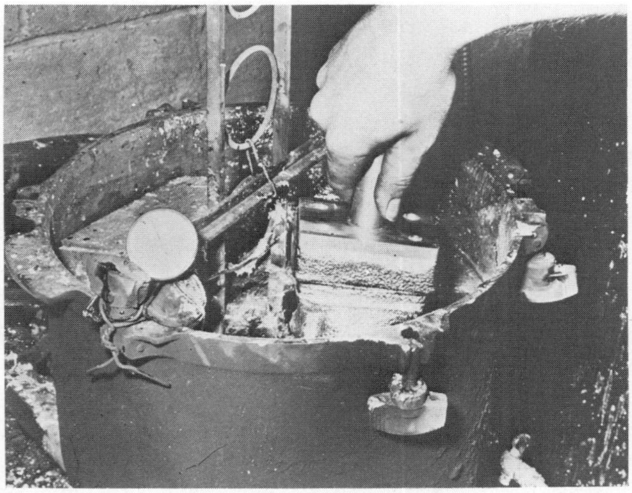

Fig. 20-24. Forming the mold by dipping the master into molten alloy. (*Courtesy: Cerro Metal Products, a division of The Marmon Group, Inc.*)

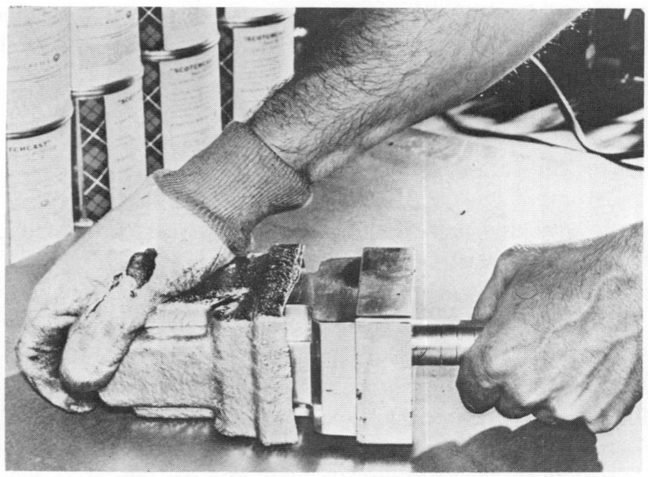

Fig. 20-25. Removing the dipped or slush-cast mold from the master. (*Courtesy: Cerro Metal Products, a division of The Marmon Group, Inc.*)

5. Nonmetallic molds are much more corrosion resistant than metal molds.
6. Room-temperature or low-temperature cures allow the use of wood, wax, or plastics patterns.
7. Transparent molds can be made, and for some applications such molds allow better control of the process.

There are also some definite shortcomings to the use of nonmetallic molds. These are mainly concerned with the lower temperature resistance of the nonmetallic materials, lower thermal conductivity, and the lower pressure resistance as compared to metal molds. However, with proper consideration of these characteristics, these materials are being used very satisfactorily for many applications. Nonmetallic mold materials and molds include rigid and heat-resistant, elastomeric, and miscellaneous types described in the next sections.

Rigid Molds. The rigid nonmetallic materials used to fabricate molds include phenolic casting resins, polyesters, and epoxy resins, all of which are used as casting materials and as laminates. Because of superior mechanical properties and low shrinkage, epoxies are by far the most used of the resins. In the form of castings, the epoxies are sometimes used in the unfilled state, but more often are filled with metal particles, glass, or possibly graphite. Aluminum- and iron-filled formulations are available from companies specializing in such materials, and these formulations are used where the maximum thermal conductivity is required; for example, in an injection mold, a blow mold, or a mold for expanded polystyrene foam production. The aluminum-filled mold will have the best thermal conductivity, but iron-filled formulations are used whenever highly abrasive materials are being molded. In either case, the mold is usually made by simply mixing the hardener with the filled resin and then casting over a parting agent-coated pattern, which is held in some type of container. After the epoxy has cured thoroughly, either at room temperature or at an elevated temperature, the pattern is removed and the surface of the casting is milled smooth. An alternative technique, where two mold halves are required, is to cast the epoxy over one half of the pattern, supported in plaster, then a second casting is made over the other half of the pattern. Numerous variations of this method are also used.

Epoxy and polyester resins are also employed to make fiberglass-reinforced molds, principally for applications in thermoforming and low-pressure reinforced plastics production. The general technique is to coat the pattern (wood, plaster, wax, plastics, etc.) with a gel coat of the resin. Several layers of impregnated fiberglass are then used to reinforce the mold, which is cured at room temperature or at a high temperature, as required.

Molds used for thermoforming or blow molding, frequently will incorporate cooling coils after the first one or two layers of fabric. The cooling coils are necessary since the low thermal conductivity of the mold will soon lead to heat buildup. The cooling coils, of course are held in place by additional layers of fiberglass. Vacuum holes for thermoforming can be made very simply in the mold by inserting greased piano wire rods into the mold, prior to cure, and then pulling them out after cure. Figure 20-26 illustrates a cast male mold with internal vacuum holes. In this case the pattern could well be a wooden male, suitably sealed. It could then be covered with plaster to form the female, into which the epoxy is later cast.

Composite molds can also be made using metal spray in conjunction with an epoxy-fiberglass mold. Such a mold might be made to incorporate the higher heat conductivity of the metal with the ease of construction of the reinforced epoxy. Figure 20-27 illustrates the concepts.

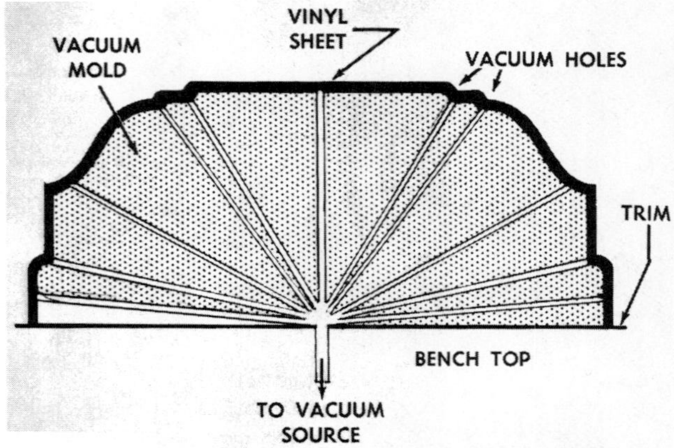

Fig. 20-26. Cross-sectional view of mold constructed to dissipate heat quickly, without incorporation of cooling coils. (*Courtesy: Plastics Design & Processing*)

Elastomeric (Flexible) Molds. Elastomeric molds provide intricately shaped cavities without the expense of complex, segmented molds. Such molds are used mainly in the production of cast parts for carved furniture sections, parts containing undercuts, or for production of small quantities of parts such as prototypes used to check the appearance or the functioning of a part later to be injection molded.

To make the mold, the liquid elastomer is simply cast over a pattern. The cured elastomer is then a flexible, elastic mold which, on removal of the pattern, can then be used as a cavity for production of parts using epoxy, polyester, polyurethane foam, plaster, or, with the right mold material, low-melting alloys (see Figure 20-28).

All of the elastomeric materials which are available as

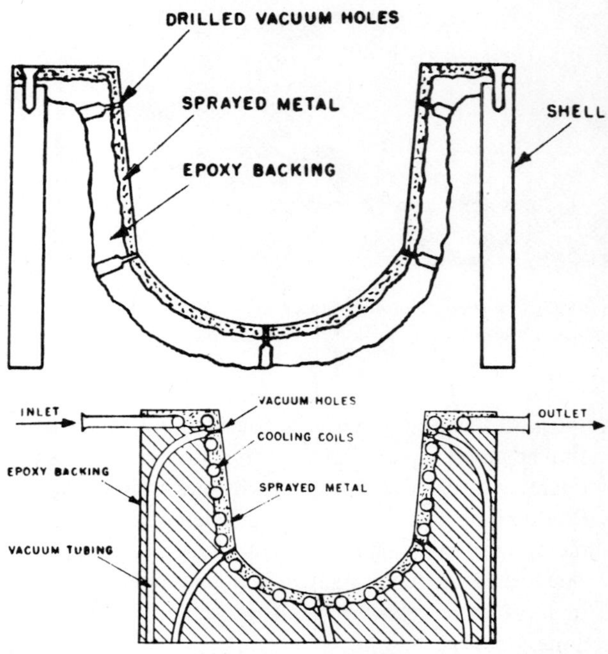

Fig. 20-27. Composite metal-plastics mold. (Reprinted by permission from *Plastics Technology*, September 1961)

Fig. 20-28. Silicone rubber mold used to make a "carved" polyurethane foam furniture part. (*Courtesy: Dow Corning Corporation*)

flowable liquids have been used to make such molds. The actual choice of the appropriate material to use is usually dictated by the mold or part to be cast, the economics of the process, and, occasionally, by the pattern material. Table 20-5 lists the major materials which are used and the advantages and disadvantages of each.

Table 20-5 indicates that the silicone RTV mold materials, while not the least expensive, are probably the best all-around materials. They are considered the best for most applications since they give excellent reproduction of any surface—wood grains, fabric weave patterns, phonograph record grooves, etc. With a room-temperature cure, with no exotherm, there is no heat to distort a wax or other temperature-sensitive pattern or mandrel. After the RTV is thoroughly cured, the silicone molds can be used with heat-curing epoxies or polyesters or low-melting alloys.

The major restriction to the use of the silicone RTV materials is that they should *not* be cured in the presence of any sulfur-bearing materials. Thus, patterns of some rubber compounds might be unusable, as would some modeling clay compounds and some plastics materials. If in doubt, the molder should try a small amount in contact with the pattern material before making an entire cast.

The silicone mold-making RTVs are two-part materials, as are most of the other mold-making elastomeric materials. Usually the resin and catalyst are colored differently, so that incomplete mixing can be easily detected by streaks in the mix. When casting the RTV, the pattern (wood, metal, plaster, etc.) and the container are usually coated with a mold release to insure easy parting of the silicone from the container and the pattern. When casting a resin into a silicone mold, however, mold release is usually *not* required, even with good adhesive materials such as epoxies. When making a two-piece silicone mold, where the RTV is cast against an already cured RTV section, it is *necessary* that mold release be used, since the uncured silicone will adhere very well to the cured silicone section. A satisfactory release agent may be made by simply dissolving approximately 5% petroleum jelly in 95% methylene chloride or about the same amount of paraffin wax in xylene.

There are a number of silicone RTV mold-making materials ranging from low-viscosity liquids, which can be easily poured, to high-viscosity materials which must be troweled

Table 20-5. Elastomeric Molding Materials.

Material	Form	Cure	Advantages	Disadvantages	Relative Cost
PVC melt or plastisol	One part	Melt or fusion @ approx. 175°C (350°F)	Withstands all casting materials. No release usually required.	Poor flexibility and elongation. Must be heated. Limited pattern materials.	Inexpensive
Rubber latex	One part	Warm, flowing air	Good flexibility and tear strength. Fair reproduction of details.	Very laborious process. Approx. 3–5 mils per coat. Very high shrinkage.	Inexpensive
Polysulfide	Two part	Room temperature	Good detail reproduction. Good release. Good tear strength.	Cannot tolerate exotherms over approx. 65°C (150°F).	Moderate
Polyurethane	Two part	Room temperature	Tough, good abrasion resistance. Can withstand exotherms to 120°C (250°F).	Must be protected from moisture in air or pattern. Short pot life.	Moderate
Silicone RTV	Two part	Room temperature	Withstands high temperatures. Release agents usually not required. Compatible with all casting materials. Excellent reproduction. Very low shrinkage.	Cure inhibition by sulfur-bearing compounds. High cost.	Expensive

Table 20-6. Typical Properties of Silicone Mold Making Elastomers.

Before Catalyzing	Silastic[1] A RTV Rubber	Silastic B RTV Rubber	Silastic C RTV Rubber	Silastic D RTV Rubber	Silastic E RTV Rubber	Silastic G RTV Rubber	Silastic J RTV Rubber
Viscosity, poises[2]	140	300	8,000	300	1,200	1,400	800
Color	White	White	Red	Red	White	White	Beige
Specific gravity	1.14	1.38	1.47	1.50	1.12	1.1	1.29
Non-volatile content, percent	98	98	98	98	98	97	–
Flow (sag or slump) MIL-S-7502	Pourable	Pourable	Nil	Pourable	Pourable	Pourable	Pourable
Working time, min or hours	–	–	10 min to 8 hr[2]	10 min to 4 hr[2]	2 hr	2 hr	$2\frac{1}{2}$
After vulcanizing for 72 hrs at 77°F							
ASTM D 176, Durometer hardness, shore A scale	45	60	65	60	35	35	60
ASTM D 412, Tensile strength, psi	400	600	650	600	700	850	750
ASTM D 412, Elongation, percent	180	120	180	100	400	350	175
ASTM D 621, Tear strength, die B, ppi	15	30	40	30	90	110	70
Temperature stability, upper limit, °F	500	600	600	600	500	500	500
Shrink, linear percent after 24 hrs at 77°F	0.3	0.2	0.3	0.2	Nil	0.2	Nil
After 7 days at 77°F	0.6	0.4	0.6	0.4	0.1	0.6	–
Thermal conductivity cal per [(cm) (degrees C) (sec)]	0.52×10^{-3}	7.5×10^{-4}	7.5×10^{-4}	7.5×10^{-4}	–	–	–
Vol. coefficient of thermal expansion per degree C (0-100)	7.50×10^{-4}	3.25×10^{-4}	3.25×10^{-4}	3.25×10^{-4}	–	–	–

Brookfield Viscometer, No. 5 Spindle, 5RPM.
[1] Silastic is a registered Trademark at the Dow Corning Corporation.
[2] Depending on catalyst used.

Properties of Typical Urethane Mold Making Materials.*

Before Catalyzing	8059 A & B	X-87144 A/B & B-40	87645 A/B & B-40	X-87647
Viscosity, poises*	2,600	3,500	7,000	16,000
Color	?	Clear, black & amber	Clear, amber & black	Amber, clear & black
Specific gravity	1.2	1.1	1.1	–
Non-volatile content, percent	–	–	–	–
Flow (sag or slump) MIL-S-7502	–	–	–	–
After vulcanizing for 72 hrs at 77°F				
ASTM D 176, Durometer hardness, shore A scale	46	41**	77	68D(61A @ 200°F)
ASTM D 412, Tensile strength, psi	600	530	5,290	6,000
ASTM D 412, Elongation, percent	300	330	490	360
ASTM D 624, Tear strength, die B, ppi	?	70	490	970
Temperature stability, upper limit, °F	?	200°F	200°F	?
Shrink, linear percent after 24 hrs at 77°F	–	–	–	–
After 7 days at 77°F	–	–	–	–
Thermal conductivity cal per [(cm) (degrees C) (sec)]	–	–	–	–
Vol. coefficient of thermal expansion per degree C(0–100)	–	–	–	–
Pot, life, min	40–66	50	25	10 min
Time to develop full properties at R.T. days	2–3	6–7	3	–
Time to develop full properties at 250°F, hr.	24	–	–	–

*Courtesy of Furane Plastics Co.
**Values after accelerated heat cure.

onto the pattern surface. The choice of the specific material to be used is usually dependent on the type of pattern. A pattern with intricate slots and crevices would require a material with a low viscosity. A large, deep pattern with vertical sides might require a high-viscosity RTV, but if the pattern is difficult to remove, it might be necessary to use a material with high tear strength and good elongation. Table 20-6 lists the properties of some of the RTV mold-

Fig. 20-29. Uralane 5715 figure cast into a mold of Uralane 8059 demonstrates versatility of these urethane products. (*Courtesy: Furane Plastics, Inc.*)

making materials available from one manufacturer. Also listed in Table 20-6 are properties of some urethane compounds used to make molds. Figure 20-29 shows a urethane part cast in a urethane mold. While the urethane compounds are less costly than the silicones and can be made stronger and more tear resistant, they are also very moisture sensitive, have shorter working times, and may be considerably harder. Prior to deciding on any one material to make an elastomeric mold then, it is well to consult with one or more of the material manufacturers for advice and instruction. Figure 20-30 illustrates some of the steps in making a silicone mold, but the method is applicable in general to any of the other two-part elastomers.

While most silicone and other elastomeric molds are used

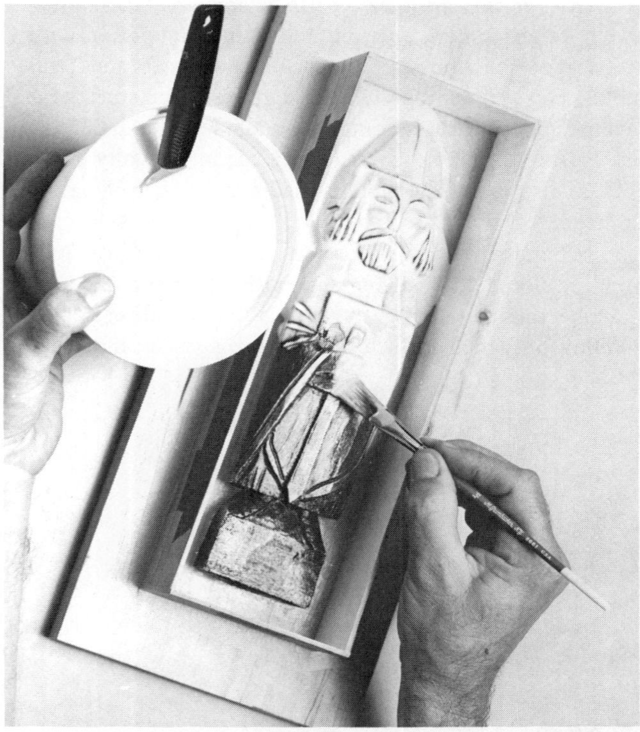

Fig. 20-30b. Coating master with thin brush coat of catalyzed silicone RTV. Brushing is used to minimize bubble formation which would occur if RTV were cast in place.

Fig. 20-30a. Applying mold release to wooden master.

Fig. 20-30c. Filling cavity with catalyzed RTV.

as cavities in preparing cast parts, there is one technique in which an elastomeric section is used in conjunction with a metal mold section to form a matched male and female mold. This rubber-metal mold can result in considerable savings in costs over a comparable all-metal mold. The rubber half of the mold is made of a casting, usually silicone RTV. This type of the mold is used mainly in forming resin-impregnated fiberglass fabric, where only one side of the part is required to have a smooth, polished surface. As the mold is closed, the rubber portion is deformed and forces the fiberglass lay-up tightly against the metal section. Depending on which surface of the part is to be smooth, the rubber can be either a male or female casting. Figures 20-31 and 20-32 show how the mold is made and used.

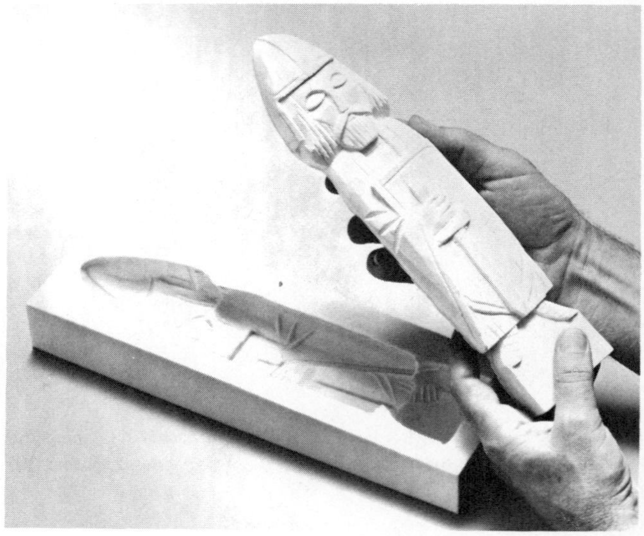

Fig. 20-30f. Cast plaster replica of original wood master. (Figs. 20-30a through 20-30f are *Courtesy: Dow Corning Corporation*)

Fig. 20-30d. Removal of master from cast mold.

Miscellaneous Nonmetallic Mold Materials. Depending on the particular material to be molded, or the pattern to be made, a great variety of nonmetallic materials have been used to make either patterns or molds or both. One of the materials used very extensively for pattern making is gypsum cement, or plaster of paris. Special formulations of this type of material under the trade names of Hydrocal and Ultracal (United States Gypsum Co.) are stronger and have higher heat resistance than ordinary plaster products, thus making them more useful for tooling purposes.

Tooling for hollow parts such as fiberglass ducting presents a special problem, since in many cases the duct is curved or in several sections so the mandrel cannot be easily removed.

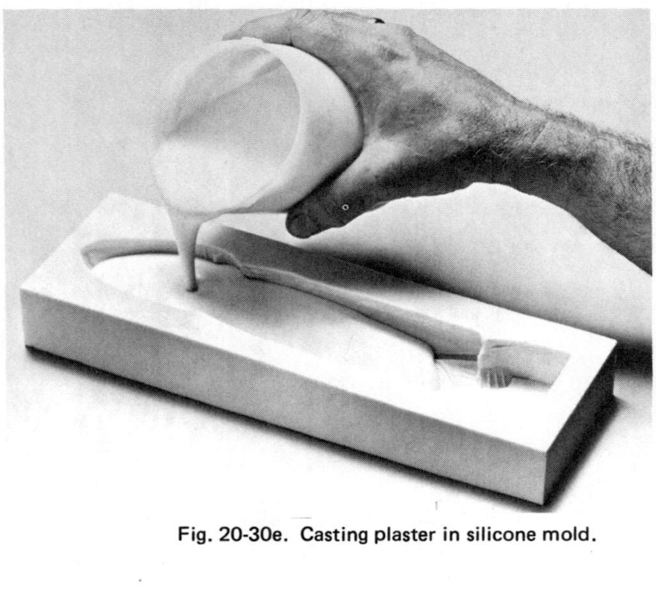

Fig. 20-30e. Casting plaster in silicone mold.

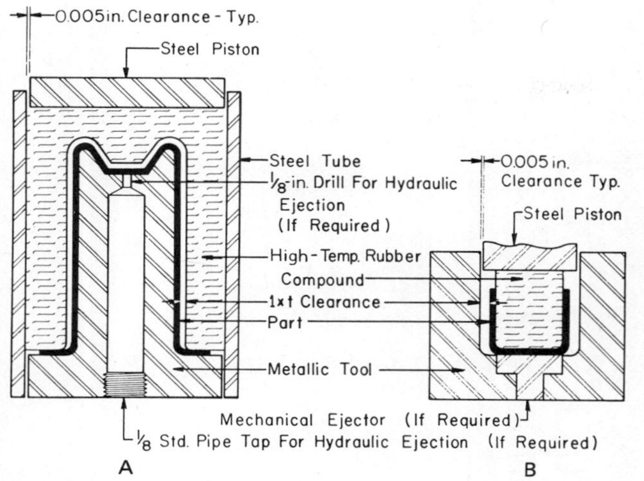

Fig. 20-31. Rubber-metal mold and fiberglass lay-up prior to pressurization. (A) Rubber used as deforming female cavity. (B) Rubber used as deforming male mold. (Thompson, 1957. Reprinted by permission from *Modern Plastics*, McGraw-Hill, Inc.)

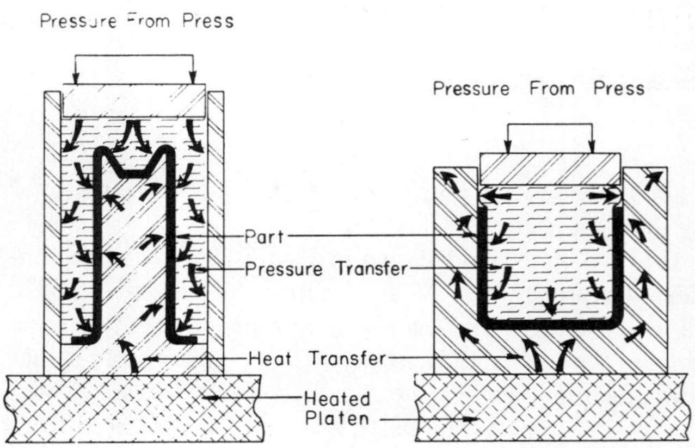

Fig. 20-32. Pressurization by deformation of rubber section of rubber-metal mold. (Thompson, 1957. Reprinted by permission from *Modern Plastics*, McGraw-Hill, Inc.)

In such a case a fragile type of plaster might be used. Such materials are called "breakaway" plasters since they can be broken loose by tapping the part with a rubber mallet and shaking the pieces out. In some cases even the breakaway plasters cannot be easily removed. In such cases, "washout" materials are used. These are plaster-type materials which can be removed by soaking in hot water. The use of these types of materials are largely confined to polyester and epoxy-fiberglass fabrication.

Another material used for molds is polytetrafluoroethylene (Teflon TFE) which is usually machined into the desired shape, PTFE is chosen for its lack of adhesion to any casting resin and its high-temperature resistance. When high-temperature resistance is not required, wax has been employed, since it too will release from almost anything. Wax also has the desirable characteristic of being meltable so it can be easily cast to shape over a pattern, or removed by melting.

Polysulfides and urethanes, previously mentioned as elastomeric mold materials, can also be obtained as semi-rigid materials, or can be filled with fillers which will both harden the polymers and at the same time reduce the weight and cost of the mold. Fillers such as vermiculite, phenolic microballoons, and glass beads are used, along with sand, wood flour, and similar materials.

Mold Design

Injection Molds. The number of molds used in injection processing far exceeds the number of molds used in any other plastics processing technique, and the design of these molds is, on the whole, considerably more complex than molds used for any other process. Thus, it is not surprising that the major portion of literature on mold design and fabrication is devoted to injection molds. At the same time, much of the design factors relating to injection molds can be used in design of transfer molds for thermosets, RIM molds, and molds used for casting and encapsulation.

The injection mold, like any other mold, is expected to form the plastics material into the desired shape in a minimum amount of time. To do this for a thermoplastic material means that the hot, liquid resin must be introduced into the mold at a high temperature and pressure, be uniformly distributed throughout the mold cavity, then be cooled rapidly enough so the mold may be opened and the part can be ejected without any significant distortion. The same function is required of a mold used for thermosetting resins, except that the resin is heated to a cured state in the mold before ejection. In order to accomplish all these operations in the minimum time, it is necessary that the various mold components be specifically designed for the particular part, the resin, and the molding press in which the mold will be mounted. Figure 20-33 illustrates a two-plate mold, the simplest type of mold in which all the required components are present. While there are obviously more than two plates in the mold, its name derives from the fact that there is only one parting line where the mold is opened, so it could be made with only two plates.

The sequence of operations in the two-plate mold is as follows:

1. Liquid material is introduced into the mold through the sprue bushing, forming the sprue.
2. From the sprue additional material is introduced to form the runner(s).
3. Small openings from the runners to the cavity, called gates, are filled next.
4. Last, material from the gates enters the cavities, filling them to form the part.
5. After a short period of time, the material changes from a liquid to a solid, due to the cooling action of

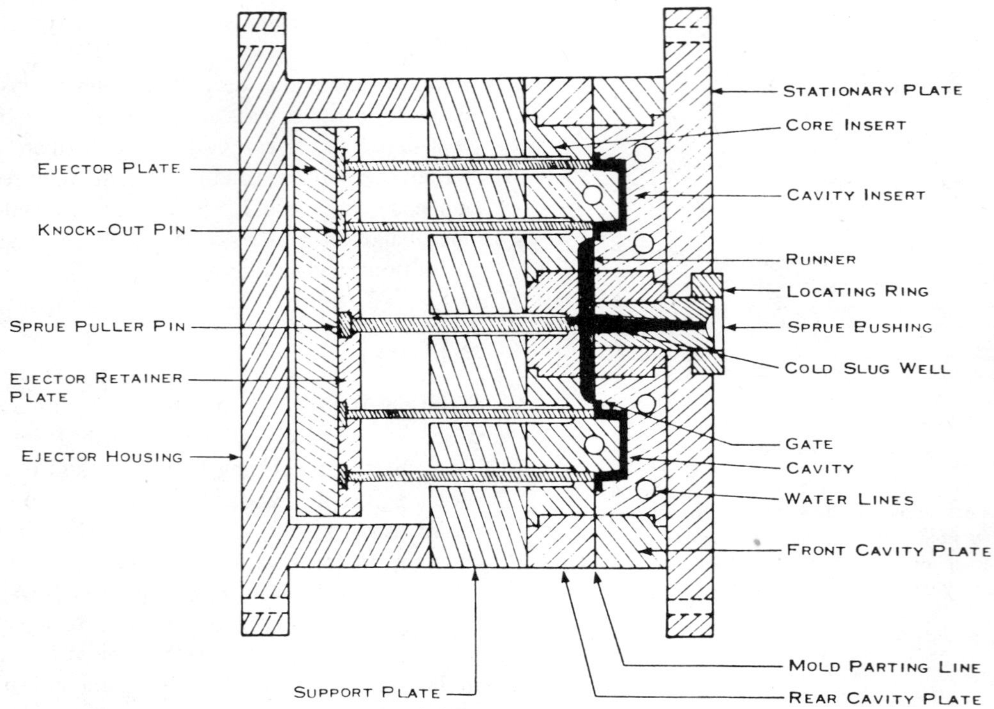

Fig. 20-33. Two-plate injection mold. (*Courtesy: Gulf Oil Chemicals Co.*)

the water lines. (If the material is a thermoset, hardening takes place by heating.)

6. When the part has hardened sufficiently, the mold is opened at the parting line and the ejector plate forces the knockout pins against the part(s), causing them to be ejected along with the runner(s) and sprue.
7. The mold is then closed and the entire cycle is repeated.
8. The runner system and the sprue are removed from the part, which after trimming and inspection, is ready for shipment. The runners and sprues, in the case of thermoplastics, are repellitized in a scrap grinder and then, mixed with virgin material, are replaced in the injection press hopper for another cycle.

In designing a mold such as the one in Figure 20-33, it is obvious that the sprue and runner system should be no larger than that required to feed the cavities in the shortest period of time. The gates should also be of minimum size in order to leave minimum markoff on the part at the point where the gate is cut away. The location, form, and size of the gate are important variables the designer must consider. The size and location of the water lines in the mold (which may be used for both heating and cooling) are important, since they must maintain a constant mold temperature, while at the same not unduly weaken the mold, which may have to withstand pressures up to 20,000 psi or greater. So the ejection system will work properly, the designer has to insure that the design is one in which the part will remain on the moving portion of the mold, later to be removed by the ejector pins.

The two-plate mold shown in Figure 20-33 may be considered representative of all injection (or transfer) molds; all other designs are merely improvements on this basic design. The main disadvantage of the two-plate mold is that the runner and sprue systems are produced and are ejected at the same time as the parts, and must be separated from them, usually in a manual operation (see Figure 10-33b). The runners and sprues then must be reground and then recycled in the press, again an extra operation requiring time, labor and energy.

The first major improvement in the two-plate mold was a design in which the runner and sprue systems are automatically removed from the parts as the mold is opened Such a mold, known as a three-plate mold, is shown in Figure 20-34. This mold has two parting lines, so that the runner and sprue systems are produced in one section of the mold, the runner plate, and the parts are produced in another, separate section, the cavity plate. On opening the mold, ejection takes place as two operations, illustrated in Figure 20-35 (see also Figure 10-34b). This type of mold is well suited for automation, as the press can be equipped with conveyor belts so the parts drop on one belt and the runners and sprues on another, which then drops them into a scrap grinder. While the three-plate mold has the definite advantage of producing automatically degated parts, it still has the disadvantage of requiring the reworking of the sprues and runners.

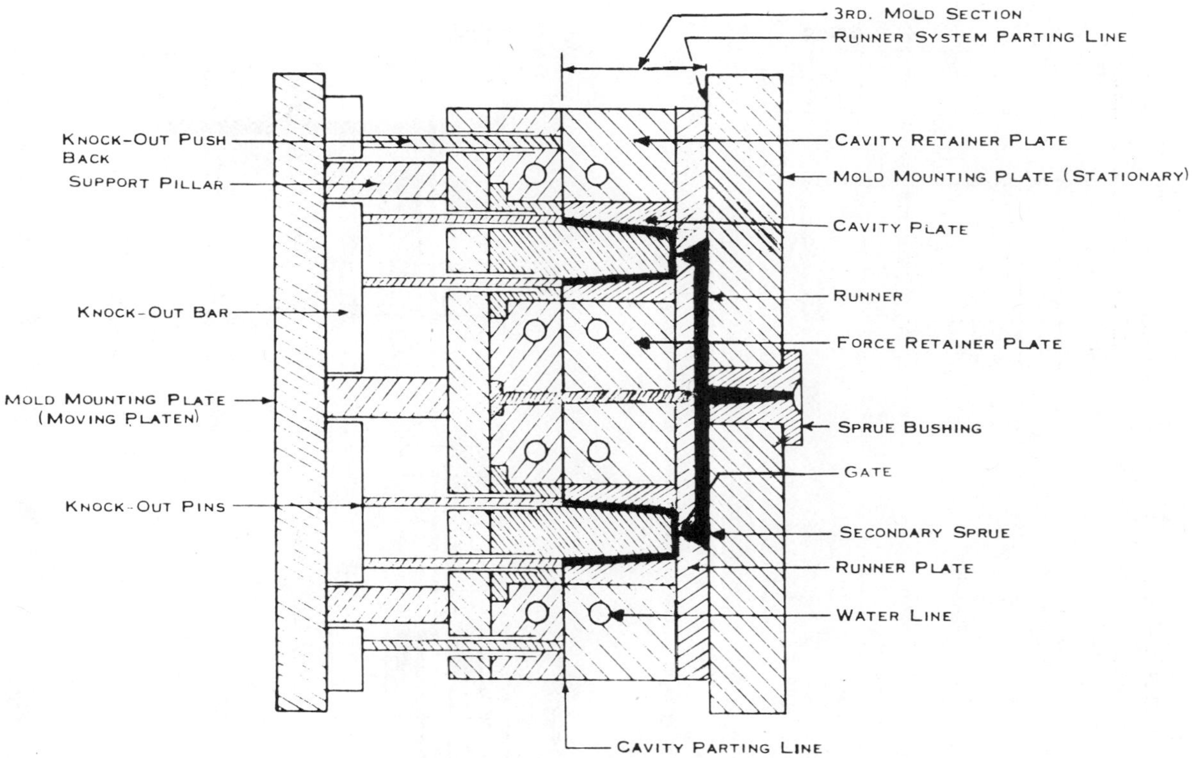

Fig. 20-34. Three-plate injection mold. (*Courtesy: Gulf Oil Chemicals Co.*)

The next major improvement in injection mold design was the elimination of the runner system entirely in the so-called runnerless molds, also described in Chapter 10. These molds, of which there are a number of variations, in all cases also include elimination of the sprue system. The basic idea behind all the runnerless mold designs is that, while sprue and runner channels are included in the mold, the sprue and runners are kept molten and so are never formed as items which are ejected. The resin is maintained in a fluid state, by various techniques which include several types of supplementary heaters in the runner plate, so it acts as a hot manifold. Special heat probes at the gate section and the use of the heat insulating properties of the resin itself help to maintain the inner core of the runner in a fluid state. The outer portion is allowed to harden and then acts as the insulation. The runner in this "insulated runner" mold is usually made considerably larger in diameter than the usual runner (approximately 1 inch; $\frac{1}{4}$ inch is a usual size).

The major disadvantage of the insulated runner technique is that in the event of any shutdown, the entire runner will "freeze" and so will have to be taken out. The advantage is that no extra power is required, and the mold, if heated probes are not used at the gate, is simpler than other runnerless molds. Figure 20-36 shows a section of a typical insulated runner mold, with the solidified resin shell encircling the molten portion. Also shown is the "emergency" parting line in the mold in case of freeze-up. The air insulation around the inner sprue or nozzle leading to the part cavity is the most critical section of the mold from the standpoint of temperature control. If the temperature at the gate becomes too high, then the gate drools and strings; if it is allowed to become too low, the gate will freeze and interfere with the next shot.

Figure 20-37 illustrates the three major types of runnerless molds. The first type was developed in the late 1950s and used cartridge heaters in the runner plate. It was called a hot runner mold or a heated or hot manifold mold. Insulated runner molds were developed shortly afterward. Internally heated runners were developed in the 1970s as a means of saving energy, since they use the insulating properties of the resin as a means of minimizing heat loss to the runner plate (see also Figure 10-56).

The major problem with all the runnerless systems is that of controlled flow at the gate. As with the insulated runner mold, the gate must be kept under a very tight temperature control if the system is to operate satisfactorily. The solution to the problem was the introduction of electrically heated probes or bushings at the end of the runner, as shown at the right side of Figures 20-37b and 20-37c. With these electrically heated and electronically controlled probes or bushings, it is now possible to fabricate molds which satisfactorily produce parts completely free of runners and/or sprues at high production rates on automatic equipment. The control system for each probe, however, is no longer a simple thermocouple and a relay. Figure

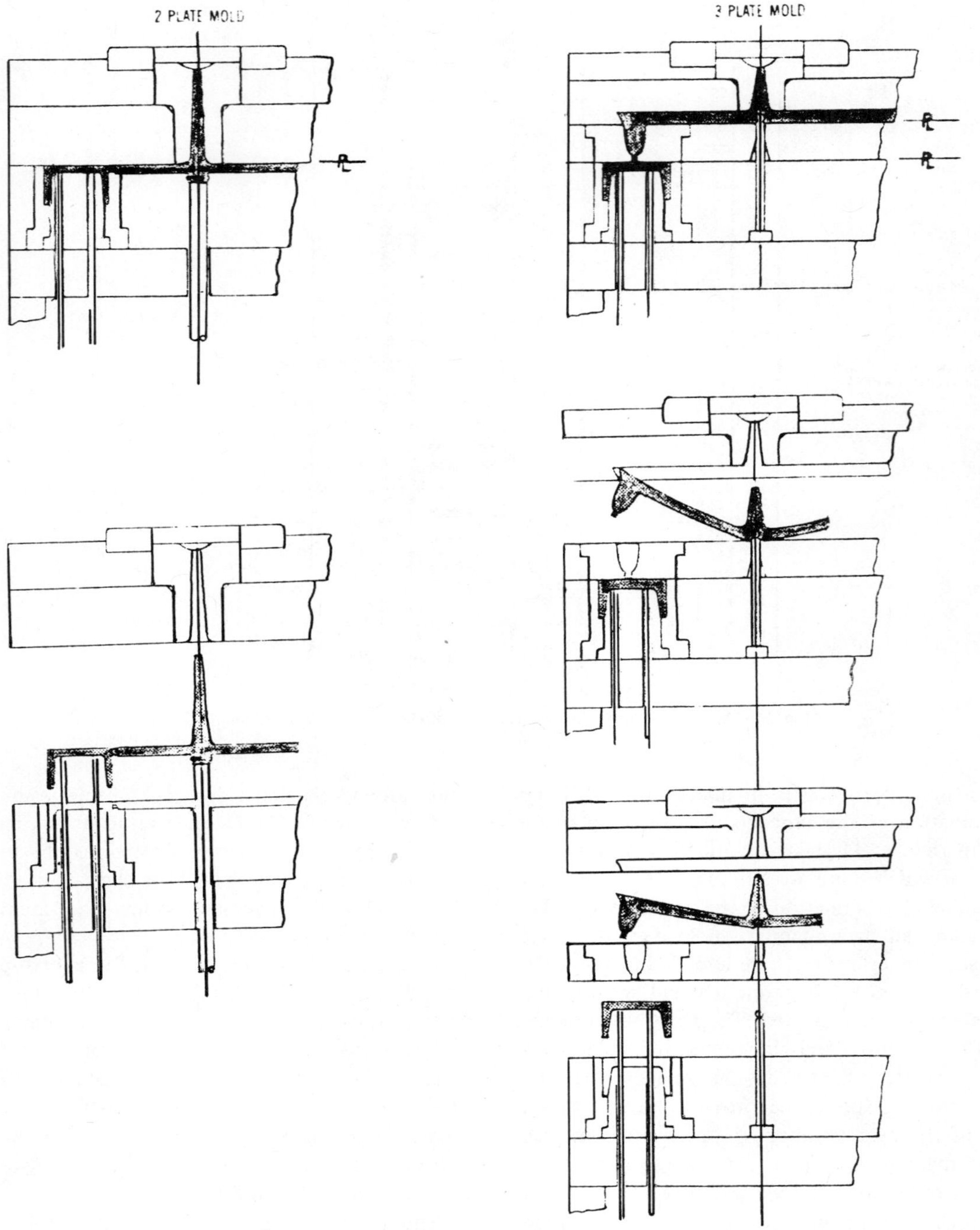

Fig. 20-35. Ejection action: two-plate versus three-plate molds. (*Courtesy: E.I. Du Pont de Nemours & Co., Inc.*)

20-38 illustrates the controls for a dual-heat "spear" probe (see also Chapter 10 for information on the "spear" and other types of heated probes).

Just as mold bases, cavity blocks, ejection pins, etc., are available on an "off-the-shelf" basis for two- and three-plate molds, the runnerless mold components are almost as readily available. The several different "hot tip" bushings or heated probes are readily obtained in different lengths, diameters, and current ratings. Heated manifolds, while necessarily custom made, are available from the sprue bushing or probe manufacturers.

Figures 20-39 and 20-40 show several types of custom-made hot manifolds, equipped with hot tip bushings, and either cartridge or tubular heaters. In the design of the manifold-probe assembly, there are a number of factors which must be considered in order to obtain optimum operation. Since each manifold is different, not all the factors pertaining to every manifold can be discussed; however,

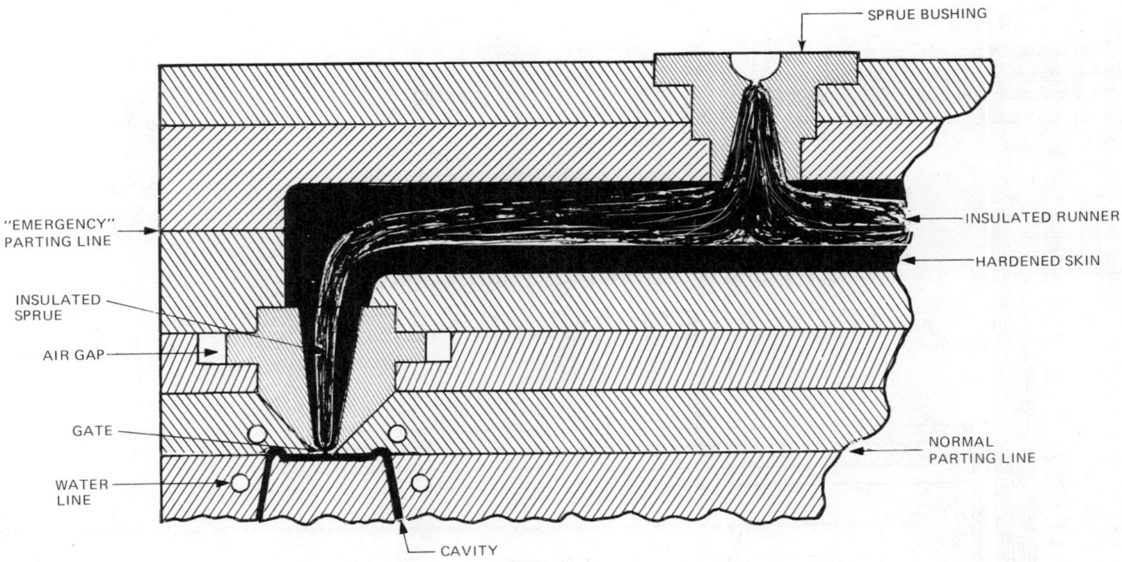

Fig. 20-36. Insulated runner system.

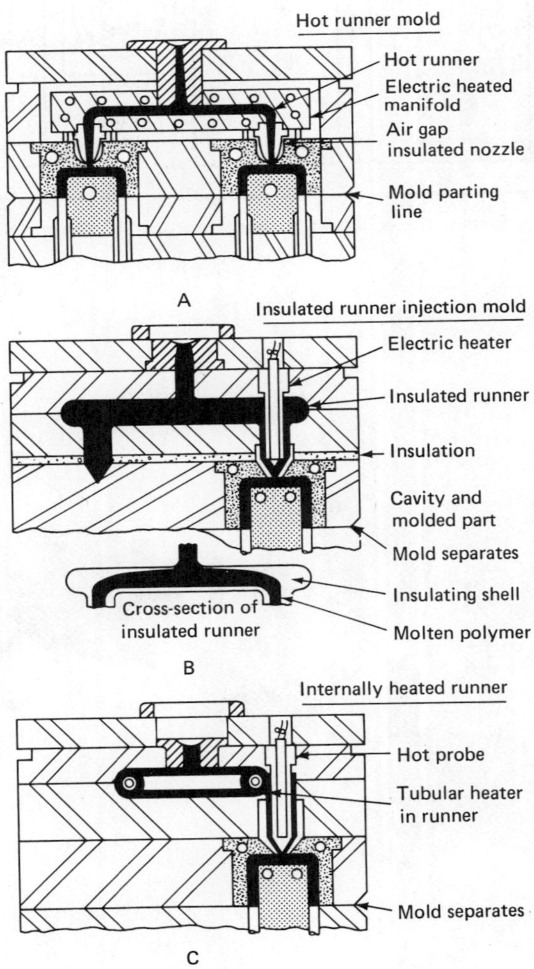

Fig. 20-37. Runnerless mold types. (From *Plastics Product Design* by Beck)

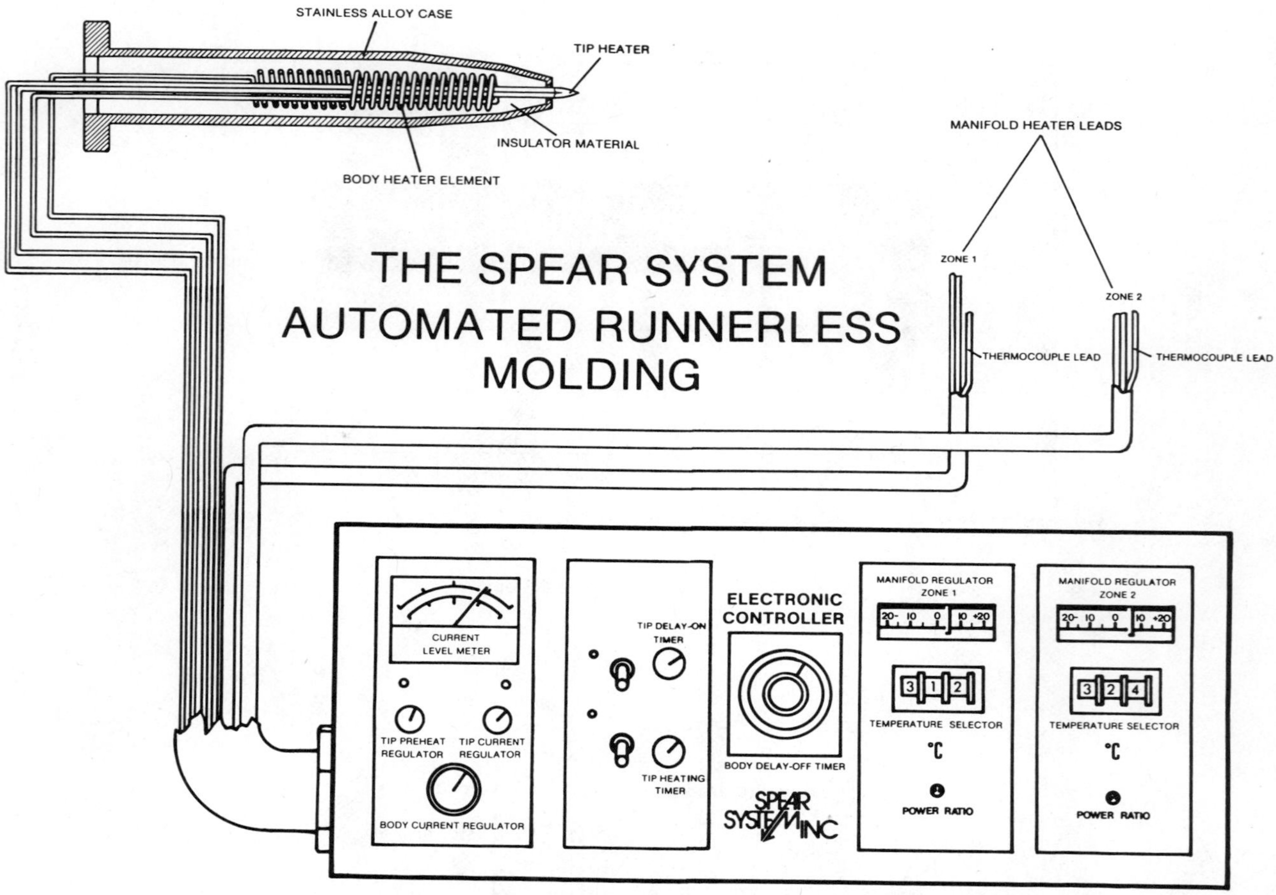

Fig. 20-38. Control system for dual-heated "spear" probe. (*Courtesy: Spear System, Inc.*)

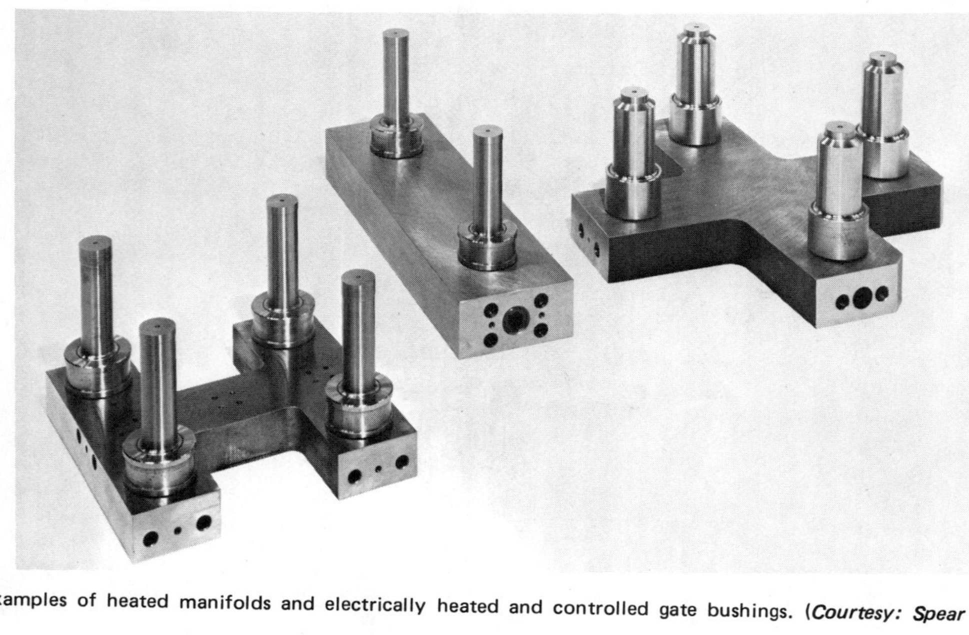

Fig. 20-39. Examples of heated manifolds and electrically heated and controlled gate bushings. (*Courtesy: Spear System, Inc.*)

Fig. 20-40. Heated manifold and electrically heated and controlled gate bushings.

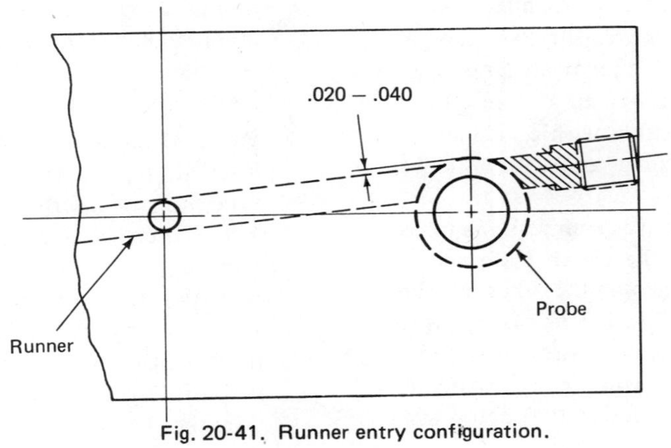

Fig. 20-41. Runner entry configuration.

the major design factors common to all manifolds are as follows:*

1. Material flow. The runner network should be designed for even flow and balance where possible. If balanced runners are not possible then the runner diameter may be varied to help balance the flow. Final balancing may be done by varying the operating temperatures of the heated probes.
2. Runner entry into the probe chamber. The runner should enter the probe flow chamber in an offset manner, (see Figure 20-41). This configuration will allow the resin to flow around the probe body in a swirling action, reducing the chance of dead spots in the flow. The design also prevents the possibility of injection pressure bending or deflecting the probe.
3. Runner size. The size of runners should be matched to the cavity size. Thus, different-sized runners should be used in a family mold, where the cavities might differ widely. This will minimize the time the resin resides in the manifold. In general, the manifold runner size is greater for general-purpose materials and smaller for the engineering plastics.
4. Heater capacity and arrangement. It is very important the the manifold be heated as evenly as possible to eliminate potential hot spots, especially near the runners. Care should be taken so that if cartridge heaters are used, they are not placed very close to a runner or cross a runner. The electrical current requirements can be calculated as follows:

Step 1. 108 × manifold weight (lbs.) × manifold differential operating temp. °F, (manifold temp.–room temp.).

Step 2. 3412 (BTU/KW Hrs.) × 0.5 (hrs to operating temp.) × manifold efficiency (from 0.3 to 0.35 depending on efficiency) 0.3 = Poor efficiency. 0.35 = High efficiency. (See below for efficiency calculations.)

Step 3. Divide the product of step 1 by the product of step 2. This will provide the total number of watts required to properly heat the manifold.

*The following material is largely based on *A Guide to Spear System Mold Design*. 2nd Ed. Spear System, Inc., Chatsworth, California.

Step 4. Divide the result of step 3 by the number of heater elements required to evenly heat the entire manifold. Always use heating elements equal to or greater than the number obtained in step 4.

Weight Calculation:

Volume (W × L × H) × Density (0.284 for mild steel)

Example: 3.45 W × 6.10 L × 2.56 H × 0.284 = 15.301 lbs. Manifold (M.F.) weight

$$\frac{108 \times 15.3 \text{ lbs (M.F. Wt.)} \times 378°\text{F (M.F. Temp.–Room Temp.)} = 624607.2}{3412 \text{ (BTU/KW Hrs.)} \times 0.5 \text{ (Heating Time)} \times 0.32 \text{ (Efficiency)} = 545.92} = 1144.1368$$

$$\frac{1144 \text{ Total watts required}}{8 \text{ Heating elements}} = 143 \text{ Watts per Heater}$$

5. Manifold efficiency. The efficiency of the manifold is determined by estimating the amount of heat loss to surrounding mold components. A manifold with a relatively *low surface contact area* through riser pads and runner bushings and with a relatively *large air gap* on all sides of the manifold will be a more efficient manifold than the same manifold with a high number of riser pads and runner bushings and with a small air gap.

   An average manifold will have an air gap of approximately 0.800 inch between the clamp plate and the top of the manifold. The air gap on the other five sides will be around 0.400 inch. The efficiency factor, if the manifold has normal placement of riser pads and an average number of probes, will be approximately 0.325. Poorer efficiency will lower the factor while better efficiency will raise it.
6. Thermocouple positioning. Thermocouple positioning is extremely important if uniformity of manifold temperature is to be obtained. Positions to be avoided are where heat transfer is high (such as near a riser pad), or remote from the heaters, or on the outer surface of the manifold.

   The appropriate position in manifolds of symmetric configuration is to put the tip of the thermocouple into the manifold approximately 0.200 to 0.400 inch from the heater wall at a mid-point from each side of the manifold.
7. Manifold spacer blocks. Steel blocks are used on all four sides of the hot manifold to connect the back plate to the fixed platen or clamping plate, as shown in Figures 20-42 and 20-43. These blocks should be heavy enough to hold both plates in position with no bowing or warpage under clamp pressure. The manifold is placed within the perimeter of the spacer blocks and between the plates (using the risers) so that an insulating air gap is formed on all six sides of the manifold. The thermocouples and manifold heater retainers are mounted through the spacer blocks (see Figure 20-20-43). Wires or plugs for the probes are also mounted through the spacer blocks are required.
8. Manifold riser pads. Pads are used as upper and lower supports for the manifold. Normally these pads are approximately 0.800 inch in diameter and height, and are made of mild steel or reinforced asbestos. Since these pads allow direct heat loss from the manifold to the plate surfaces, their area should be kept to a minimum. Upper riser pads should have a clearance of about 0.004 inch to allow for thermal expansion of the manifold as it is heated to operating temperature (see Figure 20-44).
9. Probe installation. An example of one method of probe installation is shown in Figure 20-42, where the "spear" type probe is inserted into a hole in the manifold, and a steel riser pad and a stainless steel O ring are used to prevent leakage when the manifold is heated to the operating temperature. A slot in the riser pad is used to allow exit of the heater wires, and the riser pad itself is made with a clearance of approximately 0.002 inch to allow to thermal expansion of the manifold.

   In the design of the manifold-probe assembly, it is extremely important that allowance be made for the linear expansion of the manifold when it is brought to the operating temperature. This expansion must be compensated for in order to insure proper alignment of the probe tip with the gate centerline. It is of equal importance that the probe tip not interfere with the side wall of the gate when the mold is at room temperature (see Figure 20-45).

Expansion calculations are usually made from a fixed point such as the sprue bushing centerline. The distance from the sprue bushing centerline to a gate centerline at room temperature is known as the gate pitch. Similarly, the distance to the probe tip centerline is called the probe pitch or the manifold hole pitch (see Figure 20-46). *At room temperature the probe pitch must always be less than the gate pitch.* The calculation for the room-temperature location of the probe (probe pitch), in relation to the gate pitch is as follows:

Assume the following conditions:

Gate pitch = 3.200 inch

Manifold operating temperature = 448°F

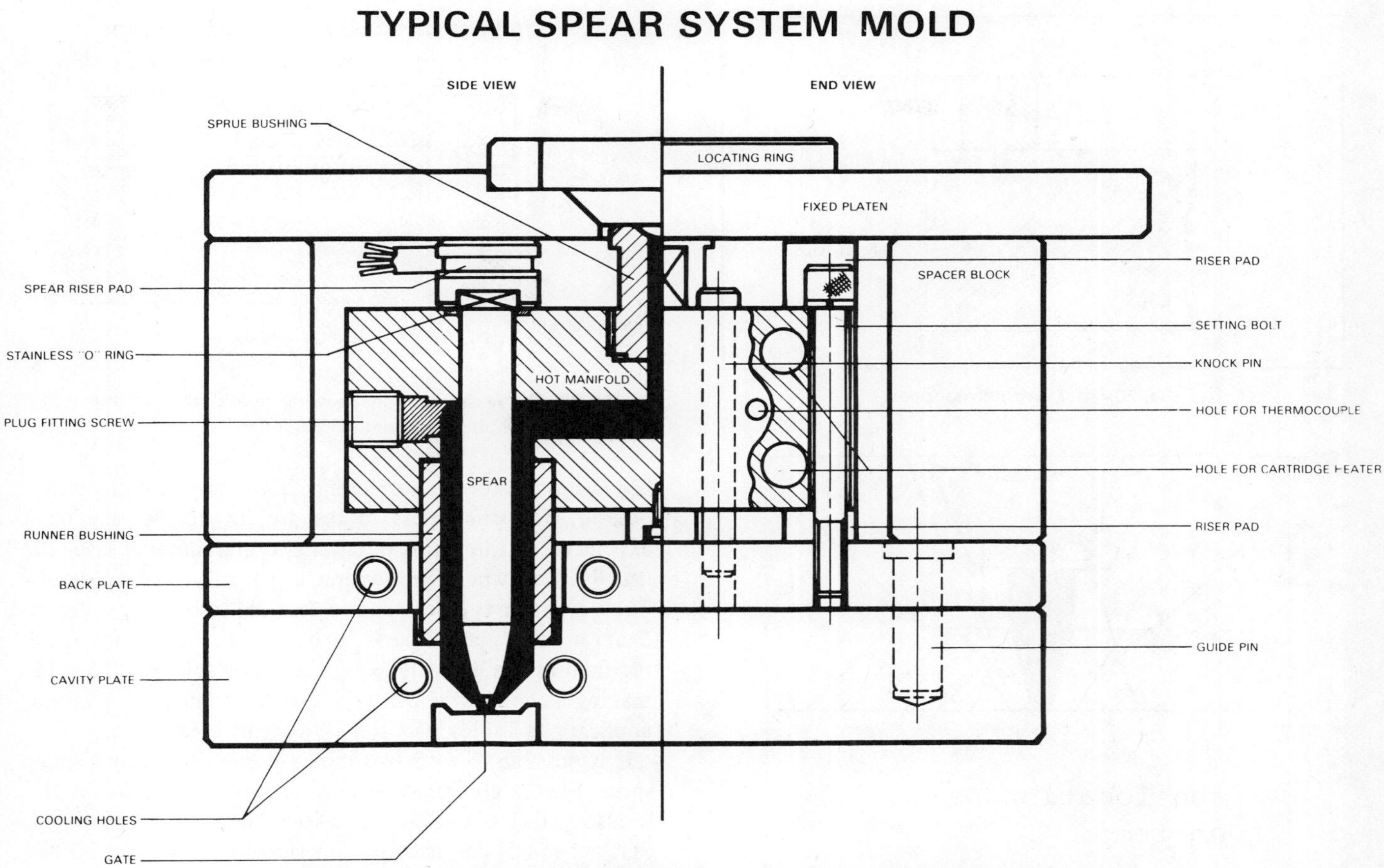

Fig. 20-42. Typical runnerless mold assembly using a hot manifold and hot tip probes.

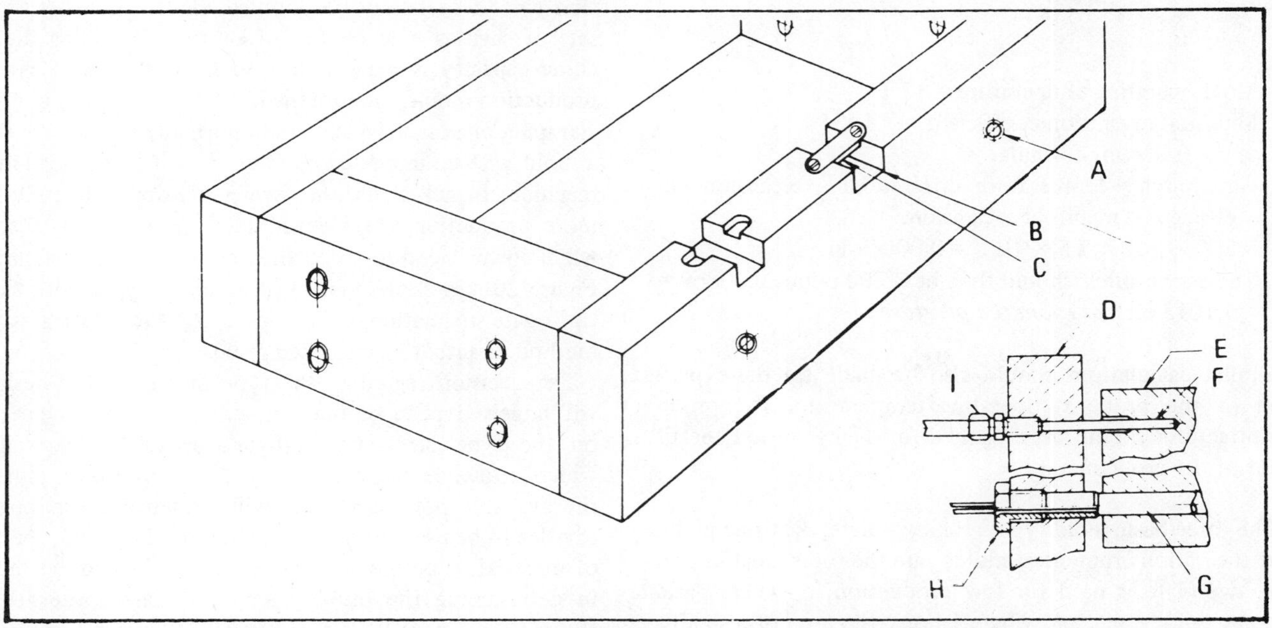

Ⓐ Hole for thermocouple
Ⓑ Channel for Spear lead wire
Ⓒ Fixing strap for Spear lead wire
Ⓓ Spacer block
Ⓔ Manifold
Ⓕ Thermocouple
Ⓖ Cartridge heater
Ⓗ Cartridge heater spacer
Ⓘ Thermocouple holder

Fig. 20-43. Manifold spacer blocks.

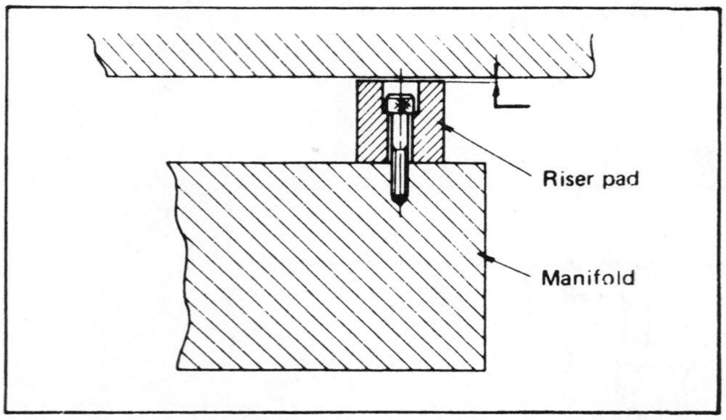

Fig. 20-44. Manifold riser pads.

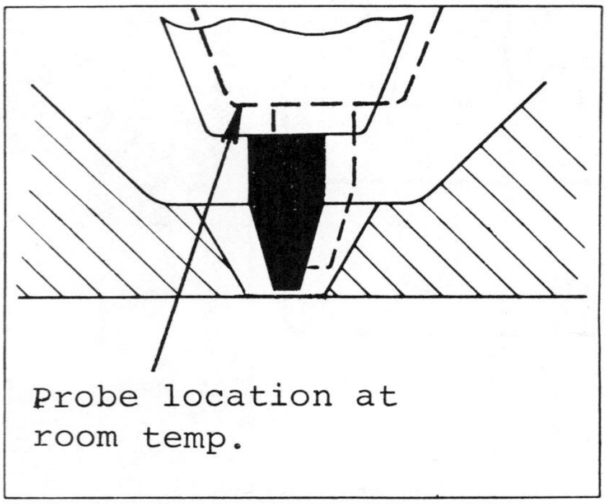

Fig. 20-45. Probe movement caused by thermal expansion of the manifold.

Mold operating temperature = 86°F
Manifold expansion coefficient = $7.5 \times 10^{-6}/°F$*
Use the following formula:
Gate pitch × temperature differential × expansion coefficient = manifold expansion.
$3.200 \times 362 \times 7.5 \times 10^{-6} = 0.00869$ in.
The probe pitch should then be 3.200 minus 0.00869 = 3.1013 inch *at room temperature.*

Similar calculations should also be made for the expansion of the probe at operating temperature. The probe manufacturer should be consulted for the expansion coefficient of the probe metal.

The three basic mold types—the two-plate, the three-plate, and the various runnerless molds—are the tools most in use. The two-plate is used for low production, or where parts require thick gates, and where labor costs are not a significant item. The three-plate mold is used where production is high and where the expense of a runnerless mold is not desired. The runnerless molds are, of course, the most expensive, but in terms of labor cost and material saved are usually used where production can justify the mold cost. The particular type of runnerless mold to be used is dependent on the type of parts to be made, the surface finish required at the gate, the equipment available, etc. To aid in making a choice of which type should be chosen for various applications Tables 20-7 and 20-8 are included.

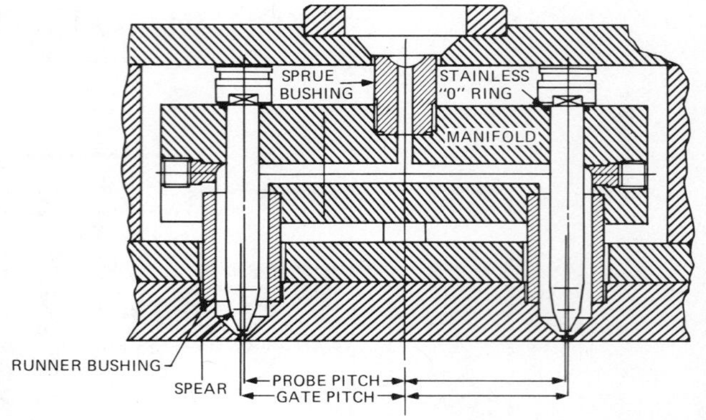

Fig. 20-46. Probe installation showing relationship of the probe pitch to the gate pitch, at room temperature.

It is possible to feed two molds at one time from a single sprue. Such a combination mold is termed a stacked mold; it also called a two-tier, two-level, or a sandwich mold. Stacked molds are usually runnerless molds. Figure 20-47a illustrates a stacked mold, in this case with insulated runners. (Also see Figure 10-59.) In this type of mold, production can theoretically be doubled, since it consists of two sets of cavities, one on top of another. Providing the machine capacity is high enough to fill both sets of cavities, production could be doubled, without requiring higher clamping pressures, which would normally be the case when a mold with an increased number of cavities is placed in the machine. In actual use, however, only approximately 80% more production is achieved, since the mold has a somewhat slower ejection speed than does a conventional mold. Figure 20-47b shows a stacked mold equipped with "spear tip" gate regulation, and Figure 20-47c illustrates one method of actuating a stacked mold.

The decision regarding the type of mold to be designed will largely depend on the number of parts to be made, and on the type, size, and configuration of the part—all of which influence the complexity of the mold. In addition, the available press capacities will influence the number of cavities to be used in the mold. On the other hand, the type of material to be molded will be one of the decisive factors in determining the primary and secondary sprue designs, the runner system layout, the gating, the vent system, and the method to use in cooling and regulating the mold temperature. All of these factors must be carefully considered in order to obtain the optimum design, regardless of the mold type. To aid in making the decisions detailed descrip-

*The coefficient of expansion (7.5×10^{-6}) applies to 4130 steel only. For other steels the coefficient may be different.

Table 20-7. Advantages of Various Runnerless-Molding Systems.*

Types of Runners	Advantages	Disadvantages
Insulated-runner, no heat	Simplest in design Fast color changes No controls Runs heat-sensitive materials	Impractical on small parts on long cycles Freezes up on delay Requires runner removal to start Leaves largest gate vestige
Insulated runner, heated probes	Fewer controls than full hot runner Smaller gate vestige than insulated runner Most efficient for frequency color changes	Same as above, except will tolerate somewhat longer delays and has less gate vestige
Hot manifold, center gate	Fewer freeze-up problems Easier start-up than insulated runner	More controls Longer warm-up time Longer time to change colors Tends to undermine mold-support similar to ejector box
Hot-manifold, hot-edge gate	Good for parts not suited to center gate Only one control for small molds Least vestige of any system	Overheat on delay Uneven cooling around cavities Most delicate at gate
Internally heated hot runner	Starts up most quickly and at normal operating settings No overheating or freeze-up on delay Less energy required than hot manifold Most efficient type for high production of one color	Most prone to blocked gates Requires more controls than other types Takes longest time for color changes

*Reprinted, with permission, from *Plastics Machinery & Equipment,* October 1978. Copyright 1978, Industry Media, Inc.

Table 20-8. Comparison of Various Runnerless-Molding Systems.*

Criteria	Insulated-Runner	Insulated-Runner, Hot Tip	Hot-Manifold	Hot-Edge Gate	Internally Heated Hot Runner	Internally Heated Hot Runner and Flow Controlled Gate
Gate vestige	D	B	B	A	B	B
Frequency of gate freeze-off	C	B	A	B	A	A
Ease of freeing frozen gate	C	B	A	A	A	A
Ease of freeing blocked gate	A	A	C	C	C	A
Can shut off one or more cavities	X	C	A	X	A	A
Start-up time	C	C	B	B	A	A
Waste of plastics materials	C	B	B	A	A	A
Simplicity	A	B	C	B	C	D
Power consumption	A(=0)	B	C	B	B	C
Number of controls	A(=0)	C	C	B	C	D
Usable with closed-loop control	X	C	B	A	C	B
Run small, light-weight parts on slow cycle	X	C	A	B	A	A
Run large, heavy parts on long cycle	X	C	A	B	A	A
Overall efficiency*	D	C	B	A	A	A

A = best B = good C = fair D = poor X = impractical

*Depends on application, but assumes high-volume parts.
*Reprinted, with permission, from *Plastics Machinery & Equipment,* October 1978. Copyright 1978, Industry Media, Inc.

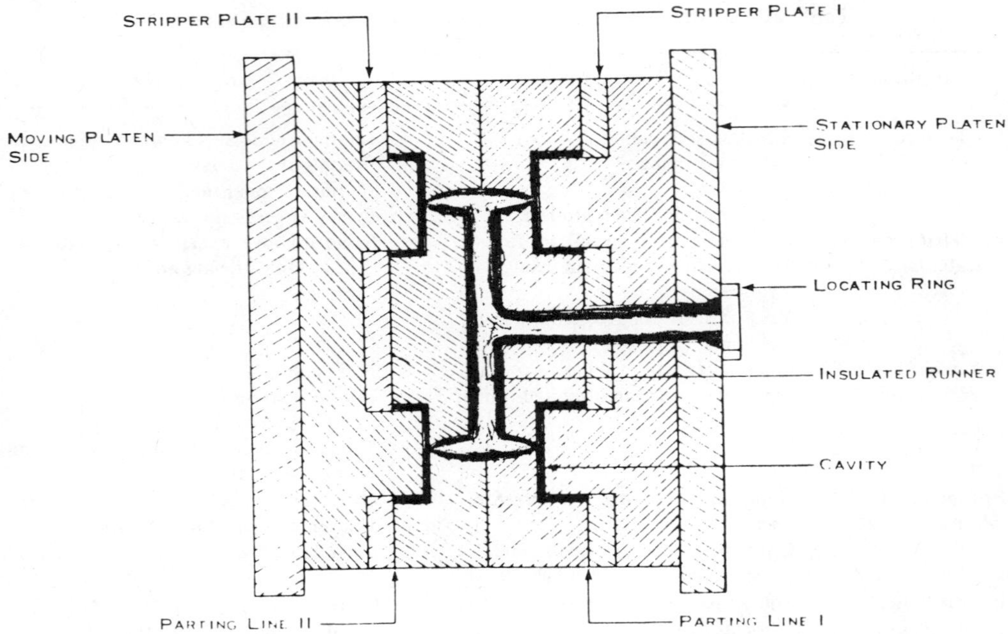

Fig. 20-47a. Stacked mold with insulated runners. (*Courtesy: Gulf Oil Chemicals Co.*)

Fig. 20-47b. Stacked mold equipped with "spear tip" gate regulation.

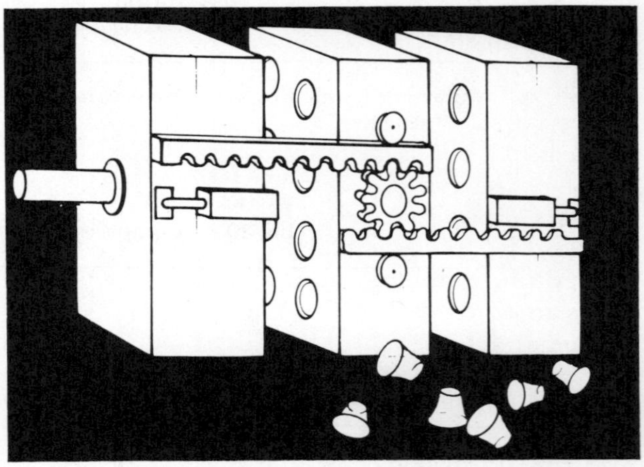

Fig. 20-47c. One method of actuation of a stacked mold.

tions of individual mold components—sprues, gates, runners, vents, etc.—are provided in the next sections. In general, the information given is applicable to all type of molds, cases where it is not will be readily apparent.

Sprues. The sprue bushing is that component of the mold which conveys the molten plastic from the nozzle of the press to the runner system. The design of the sprue, its shape, and length must be such as to result in minimum restriction to the flow of resin. The taper shown in Figure 20-48 is 2° to 6° so that the sprue can be easily pulled away from the nozzle* and bushing when the mold is opened. To insure that the sprue will be pulled away, the end of the

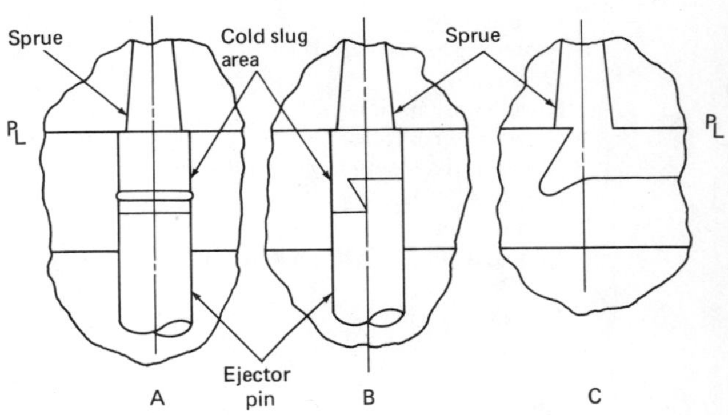

Fig. 20-48. Sprue pullers and anchoring techniques.

*See Chapter 10 for a detailed description of nozzles.

sprue is locked either into the movable cavity plate or onto the end of the ejector pin as shown in **A** and **B** of Figure 20-48. If no pin is used, the sprue can be locked into the plate as shown in **C**. Another use for the sprue ending is as a repository for the cold slug—the small amount of hardened material which may have formed at the end of the nozzle between shots. Depositing it here minimizes the possibility of the cold slug interfering with a runner flow or being incorporated into the cavity as a partially fused portion of a part.

In some cases the sprue can lead directly to the part, if a large gate vestige can be tolerated and a cold slug would be inconsequential (see Figure 20-49). In other cases, a secondary sprue can lead through a pinpoint gate directly into a cavity (see Figures 20-36 and 20-50).

Runners. Runners are the channels by which the melt is conveyed to the cavities through very short openings known as gates. It is very important that the runners be as short as possible to minimize pressure drops. At the same time, if at all possible, the runners should be of equal lengths to each cavity so the pressure drops will be the same for each cavity. Such a condition (equal length runners and the same type of cavity at the end of each runner) is known as a balanced runner system. A balanced runner system will minimize the possibility of "short shots" (partially filled cavities) where the pressure is low, and/or "flashes" where the pressure is so high as to cause the mold to open slightly and produce flash at certain cavities. Figure 20-51 illustrates balanced runner systems of several configurations. Also shown is an example of an unbalanced system, where the runners to some cavities are distinctly longer than others.

There are times when runners of different sizes, or diameters, are used in the same mold, to even out the flow in a mold containing cavities of widely differing sizes. In designing such a mold, the runners should initially all be machined to the same diameter. This will save machining costs, unless the cavities are so disproportionate as to conclusively indicate otherwise. Past experience must also be taken into account.

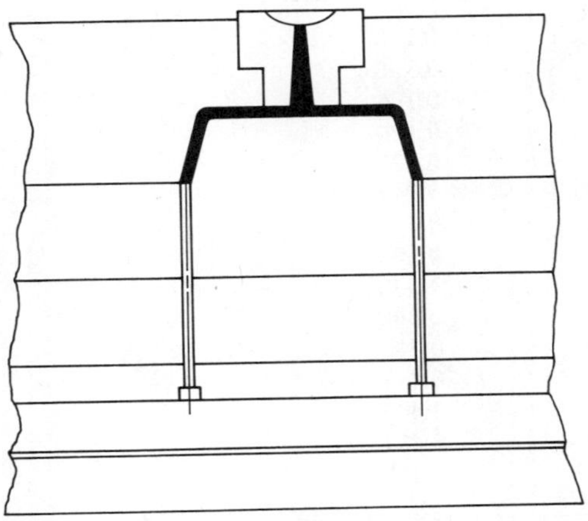

Fig. 20-49. Sprue gate.

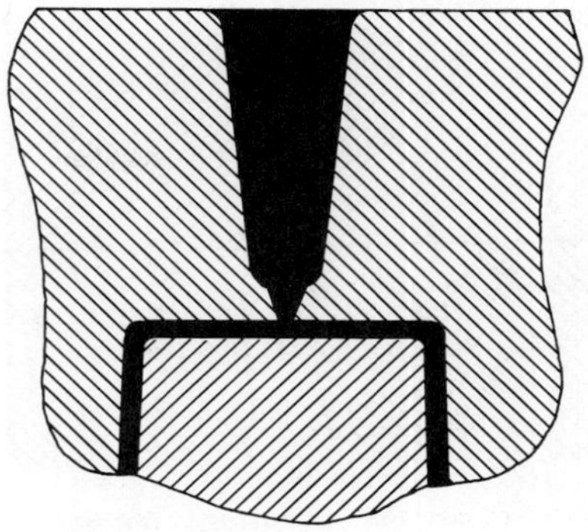

Fig. 20-50. Sprue pinpoint gate.

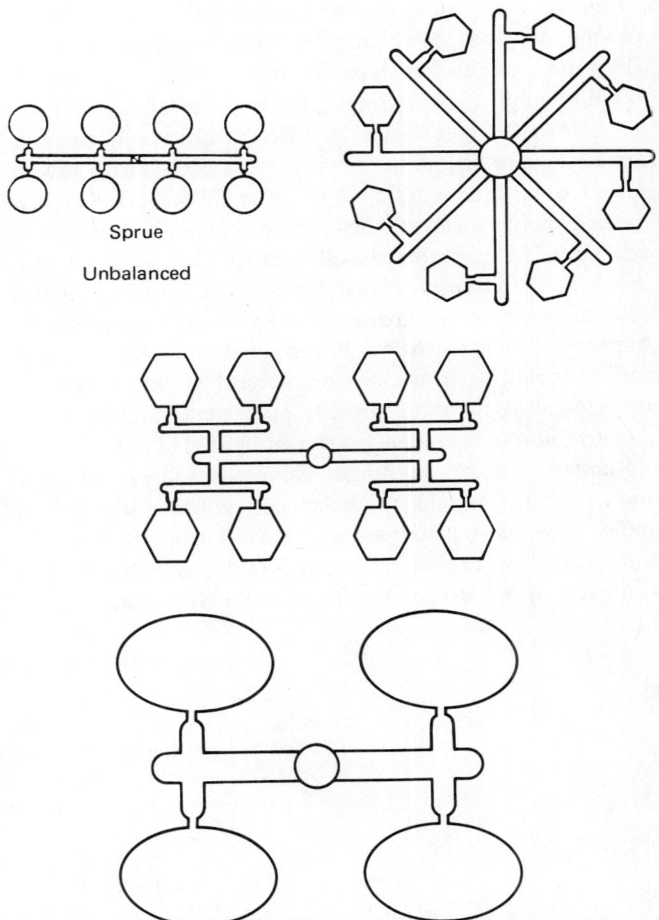

Fig. 20-51. Balanced versus unbalanced runners. (*Courtesy: Abtec Chemical Co.*)

One type of unbalanced mold is known as a "family" mold. In this mold configuration there are usually several cavities of differing sizes for a number of parts which must be assembled together, or which must be the same color. Runners of different sizes and/or diameters may also be used here to balance flow. Initial trials should be made with runners of the same diameter, and if poor performance results then the responsible runners can be enlarged.

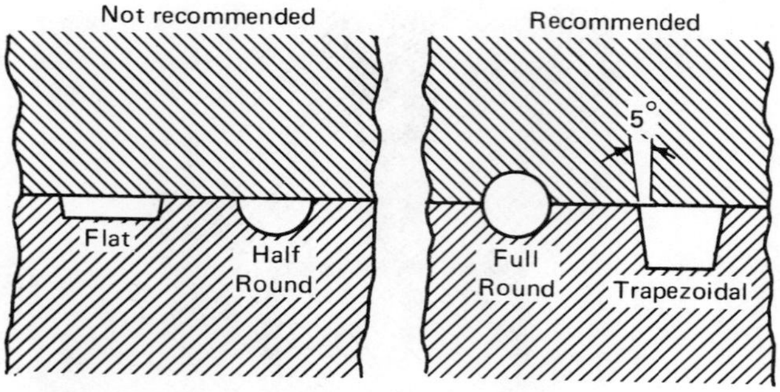

Fig. 20-52. Runner designs. (*Courtesy: Monsanto Plastics and Resins Co., St. Louis, Mo.*)

Runners should not only be of minimum length but should also be of a shape which offers least resistance to flow. There are basically four types of runners which can be used: (1) full round, (2) half round, (3) trapezoidal, and (4) rectangular or flat (see Figure 20-52). The full-round runner is the best since it offers the highest capacity with lowest surface area, thus resulting in the least chilling effect on the moving melt. However, this type of runner has the disadvantage of being extremely difficult to machine, since both halves of the mold must match perfectly, thus making this type of runner very expensive. The best compromise is the trapezoid which, as shown, is machined into only one mold surface, resulting in a considerable cost saving. Depths of trapezoidal runners to provide equivalent cross-sectional area to full-round runners are shown in Table 20-9.

Runner channels in runnerless molds follow the same rules regarding lengths and balanced layouts as are used for molds in which the runners are actually made. However, in most cases the runners are considerably larger than in an ordinary three-plate or two-plate mold. Insulated runners in particular can be very large, up to 1 inch in diameter. Polyethylene and other materials with low thermal conductivity and high specific heats are well suited for such insulated runners—provided fast, uninterrupted cycles are maintained.

Gates. Gates are the small openings between the runners or sprues and the mold cavities. The size, location and type of gate are all important factors in determining the success or failure of a mold, or, at the least, the attainment of the optimum cycle time. There are a number of different types of gate designs which are largely dependent on the type of mold, the size and type of part, and the type of material. Regardless of the various gate designs, the gates may be broadly classified as either round or rectangular.

The round gates are circular in cross-sectional area and may or may not be tapered. Rectangular cross-section gates are preferred because the width and thickness of the gate are independent variables. Thickness largely controls the freezing time of the gate (except in runnerless molds equipped with flow control gates), and also determines the time available for filling and packing the cavity. The width of the gate mainly controls the amount of flow for any given time and thickness. Thus, by simply increasing the gate width, the flow can be increased without affecting the freezing time. A second classification of gates is "impinging" and "nonimpinging." Impinging types are preferred, as explained below. Round gates are usually impinging, while rectangular gates may be of either type.

While there are a great many gate designs possible, they all share certain common design factors which should always be considered before a final decision is made. These factors, which apply to conventional molds and runnerless molds, are:

Table 20-9. Trapezoidal Runner Dimensions.*

Diameter of Round Runner	Decimal Equivalent	Cross-Section Area, Sq. Ins.	Depth of Equivalent Trapezoidal Runner
1/8	.125	0.0123	0.110
5/32	.1563	0.0192	0.138
3/16	.1875	0.0276	0.166
7/32	.2188	0.0376	0.194
1/4	.250	0.0491	0.221
9/32	.2813	0.0622	0.249
5/16	.3125	0.0767	0.277
11/32	.3438	0.0928	0.305
3/8	.375	0.1104	0.332
13/32	.4063	0.1296	0.360
7/16	.4375	0.1503	0.388
15/32	.4688	0.1725	0.415
1/2	.500	0.1964	0.443
17/32	.5313	0.2219	0.471
9/16	.5625	0.2484	0.498

*Courtesy: Monsanto Plastics and Resins Co.

1. Because of the surface imperfection left by the gate, it should be located in a surface that will not be visible, if possible.
2. The gate should be located as close to the center of the part as possible, to minimize the flow distance. If a central location cannot be used because of design or appearance requirements, or because of warpage due to unequal flow lengths, then multiple gating should be used, or one large gate (see Figure 20-53).
3. Gate into the thickest section of the part, if possible, since this section freezes last and therefore is most apt to show sink marks. Continuous pressure from the gate will minimize the markoff.
4. Parts using two gates, with *long* flows between the gates, will show weld marks, which are also a line of weakness. Three or four gates should be considered. Multiple gating, besides giving stronger weld lines and lowered stresses, can also allow faster cycles because the gates can be thinner.
5. If possible, flow from the gate should impinge on a wall or core pin to minimize "jetting." These are surface imperfections arising from a small amount of resin emerging from the gate as a fine jet, which freezes in the air as a curled mass. It is later pushed to a surface by the succeeding hot melt to result in a surface imperfection and a point of high stress and possible failure.

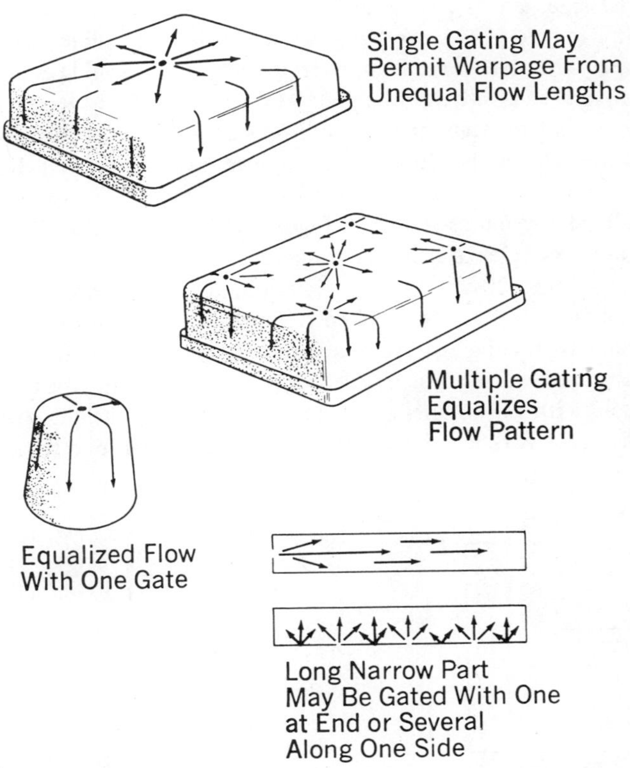

Fig. 20-53. Gating for equalized flow patterns. (*Courtesy: Celanese Plastics Co.*)

6. While a small gate is desirable to obtain rapid freezing, the gate size should also be as large as feasible to minimize the pressure drop across the gate.
7. The gate should be thin enough to cause fast freezing and prevent reverse flow and drool.
8. The gate "land" (the length of metal between the runner and the cavity) should be as short as possible, consistent with the clamping pressures and the injection pressures. Common lengths are 0.020–0.030 inch. Less than 0.020 inch may result in metal failure at the gate due to runner pressure. More than 0.030 inch may cause excessive pressure drops, necessitating higher injection pressures, higher melt temperatures, or both, with possible longer cycle times, material degradation, etc.

The principal types of gating systems used for various part configurations and their size calculations are shown in Figures 20-54 thorugh 20-58. It should be noted that the suggested gate dimensions are for acetal resin. However, the dimensions shown should be satisfactory for initial trials with other resins, bearing in mind that a smaller gate is much easier to enlarge, and cheaper than trying to make a large gate smaller. Inasmuch as acetal is considered one of the fairly high-flow resins, the gate dimensions shown are all on the small (conservative) side for most resins. Probably only nylon would be molded with smaller gates.

Round gates, also called center gating or direct gating (except the tunnel type), are used mainly for large molded parts or for single-cavity molds (see Figure 20-54). Center gating has the advantage of eliminating weld marks and of allowing for fast, uniform fill. The point gates, also called pinpoint or restricted gates, are usually not too satisfactory when used with highly viscous resins or with heat-sensitive polymers. It should be noted that for pinpoint or tunnel

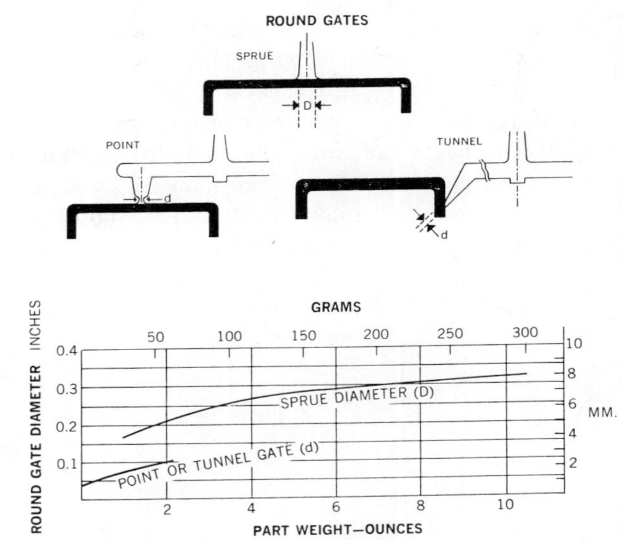

Fig. 20-54. Round gates. (*Courtesy: E.I. Du Pont de Nemours & Co., Inc.*)

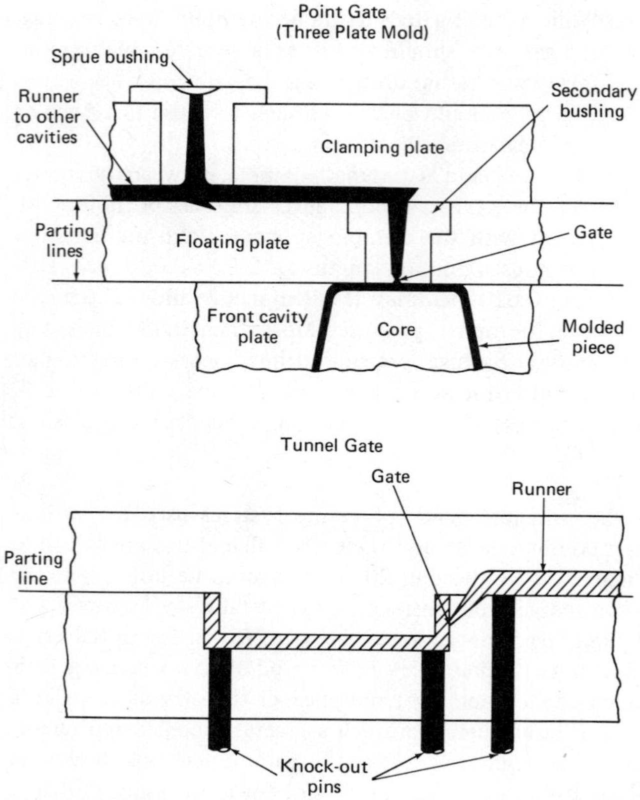

Fig. 20-55. Point and tunnel gates. (*Courtesy: E.I. Du Pont de Nemours & Co., Inc.*)

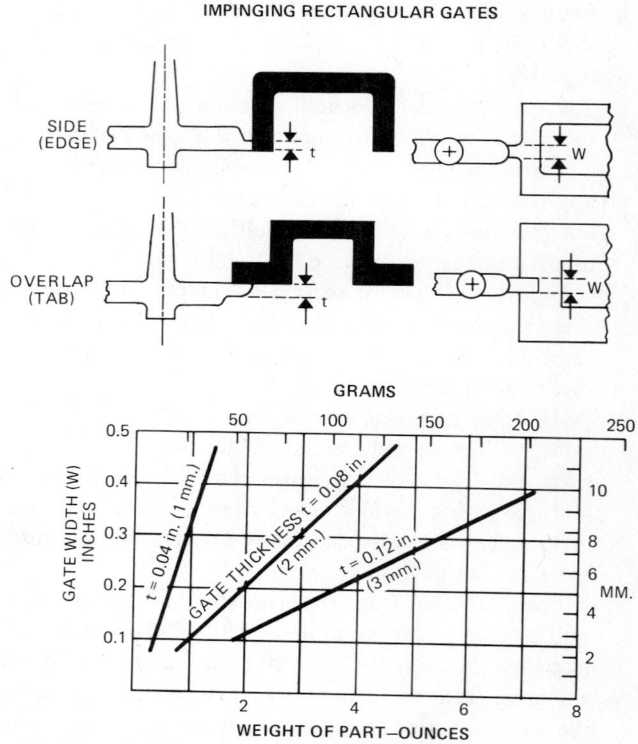

Fig. 20-56. Impinging rectangular gates. (*Courtesy: E.I. Du Pont de Nemours & Co., Inc.*)

gates, the gate diameter should not exceed the wall thickness. The tunnel gate, also called a submarine gate, is used when automatic degating is desired. In this gate, the melt is introduced into the cavity at a point below the parting line of the cavity. During the ejection cycle the gate is sheared off. Figure 20-55 details the use of a pinpoint gate and the tunnel gate.

Rectangular gates may be either impinging or nonimpinging types, with the former being preferred. Regardless of the type, however, the gate thickness is usually based on the part thickness. Thus, a gate for a high-flow resin might be started at a thickness of 50% of the average part thickness, with somewhat greater thicknesses for more highly viscous resins. If the initial trials show too rapid freeze-up or sink marks, then the thickness of the gate should be increased.

Figure 20-56 shows two types of directly impinging gates. The side gate is also known as an edge gate. A somewhat similar gate is the overlap gate, which is also called a tab gate. This gate, as shown, overlaps the cavity with a small tab in such a way that the melt jet must impinge on a cavity wall. Under certain design conditions this tab-runner combination cannot be directed to impinge on a cavity wall. In this case, a pinpoint gate is used to direct the jet against the side of the tab as shown in Figure 20-57, which also illustrates examples of balanced runner molds. It should be noted that Figure 20-56 allows calculation of the gate width assuming a given gate thickness. For the same gate width, using an impinging type tab, the gate thickness would in most cases have to be a little larger. Fortunately, trying the mold with the thinner gate would present very little in the way of rework problems if the thinner gate was found to be inadequate.

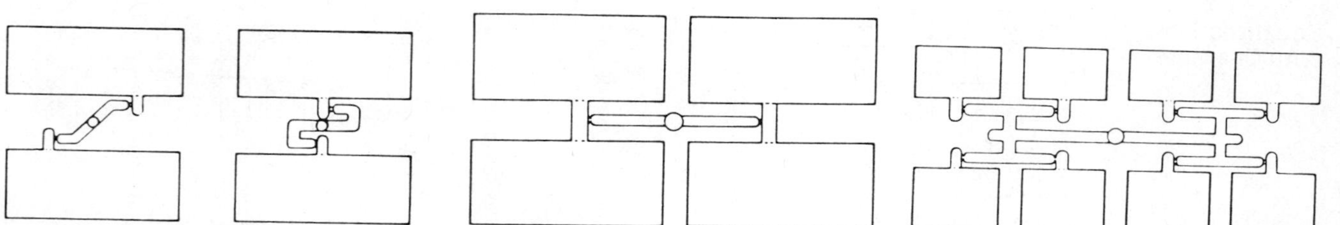

Two-Cavity Mold for Tab-Gated Parts   Four-Cavity Mold with Bridge Tab Gates   Eight-Cavity Tab-Gated Mold

Fig. 20-57. Nonimpinging rectangular gates.

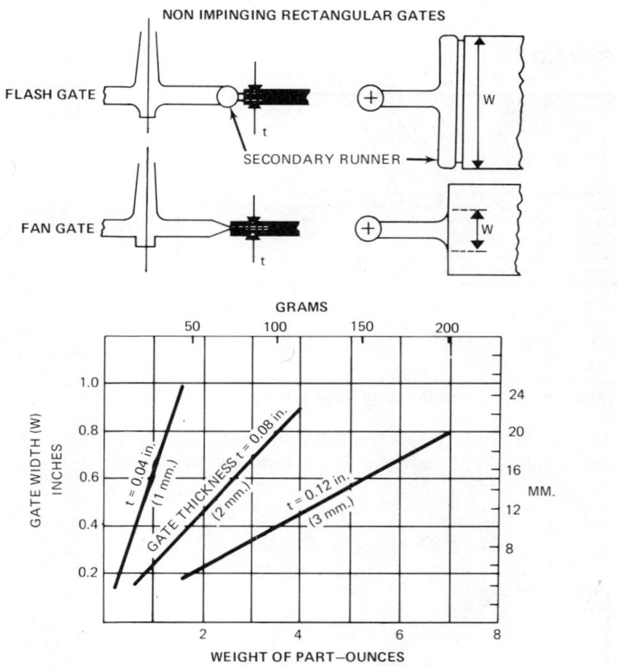

Fig. 20-58. Tab impinging gates. (*Courtesy: Monsanto Plastics and Resins Co., St. Louis, Mo.*)

Flash and fan gates are special nonimpinging gates always used for edge-gating applications (see Figure 20-58). These gates differ from the previously described edge gates in that they are used mainly for large-area parts having thin, flat sections, such as dials, clock faces, etc. The usual thickness of the gate is approximately one half of the part thickness. It should be noted that the flash gate has a secondary runner parallel to the cavity wall. This extra runner enables the flash gate to fill considerably faster than would a comparable fan gate. The disadvantage of the flash gate is the large added runner which must be reprocessed, unless runnerless molding is used. If the runnerless molding is used, the mold becomes considerably more complicated. In some cases, the best approach would be a compromise between a fan gate and a flash gate. Figure 20-58 illustrates the two concepts and, as shown, allows for width calculations of the fan gate for varying gate thicknesses from 0.04 inch to 0.12 inch.

Cylindrical and circular parts have special types of gates for their particular configuration. One gate design, the ring gate, is an adaptation of the flash gate used for flat moldings. In this gating system, shown in Figure 20-59, a secondary runner is employed in a manner similar to that used in the flash gate. This system is particularly useful in that it provides uniform flow around a tubular part and helps prevent core shift, due to the high injection pressure, which could result in uneven wall thickness.

The second gate design peculiar to cylindrical parts is the diaphragm or disc gate, also illustrated in Figure 20-59. This system is used because of its uniformity of molding and absence of weld lines. For large parts, over approximately 4 inches in diameter, the disc gate becomes expensive in material. One system used for large parts is a series of tunnel or submarine gates feeding studs placed around the part shown in Figure 20-60.

Valve gating. Valve gating to control the flow of polymer at the gate can be accomplished by several different techniques. In general, all the methods use a heated probe at the gate which either automatically or electronically shuts the flow. Figure 20-61 shows a mechanically actuated valve which, as the pressurized melt is forced around the probe, retracts allowing resin to enter the cavity until injection is complete. When the pressure diminishes, due to the screw retraction, the spring in the probe forces the valve to seal the gate. An automatically trimmed gate is the result. Another gate valve is the "spear" system (described in Chapter 10) in which the probe, using electronic controls, alternately melts and freezes the gate as required by the cycle. Both techniques, and other similar ones, are particularly suitable for use in runnerless molding, and should be more widely adopted in the future.

Vents. Venting of the mold is extremely important for successful parts production. Compression of the entrapped air by the rapidly entering resin can cause discoloration or charring of the resin front, particularly at the last corner filled. The vent should be located, if possible, directly opposite a gate, or at the point of a corner, etc. (see Figure 20-62). Typical vents on one mold half are 0.25–0.50 inch wide by 0.002–0.005 inch deep.

In some cases, the vent need be no more than a scratch made on the parting-line surface of the mold using a diamond-pointed tool. Other techniques used for venting include very shallow, flat areas machined on the surface of several ejection pins, or, for extreme cases, the use of vacuum

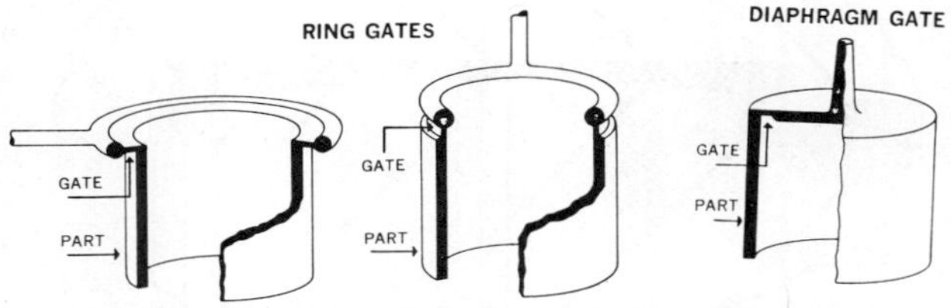

Fig. 20-59. Ring gates and diaphragm gate. (*Courtesy: E.I. Du Pont de Nemours & Co., Inc.*)

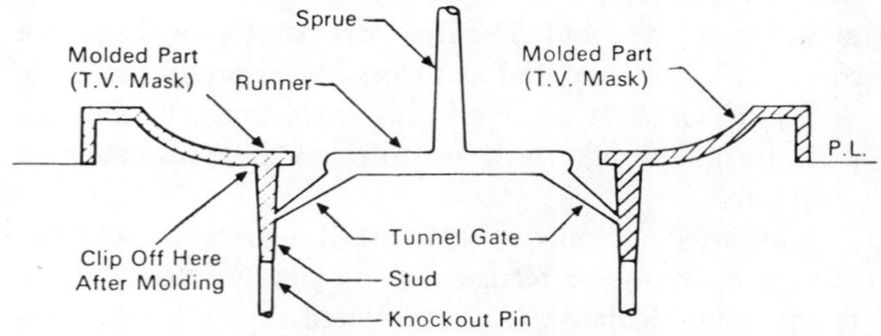

Fig. 20-60. Tunnel gate into stud. (*Courtesy: E.I. Du Pont de Nemours & Co., Inc.*)

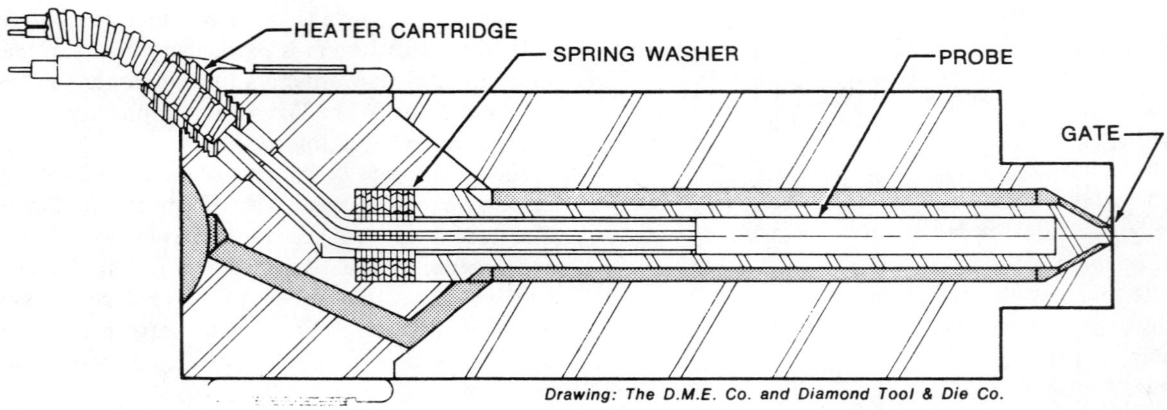

Fig. 20-61. Typical mechanical valve-gating system. Plastic is injected into the nozzle, forcing the probe rearward. The gate opens, and the cavity fills. (Reprinted, with permission, from *Plastics Machinery & Equipment*, May 1978. Copyright, 1978, Industry Media, Inc.)

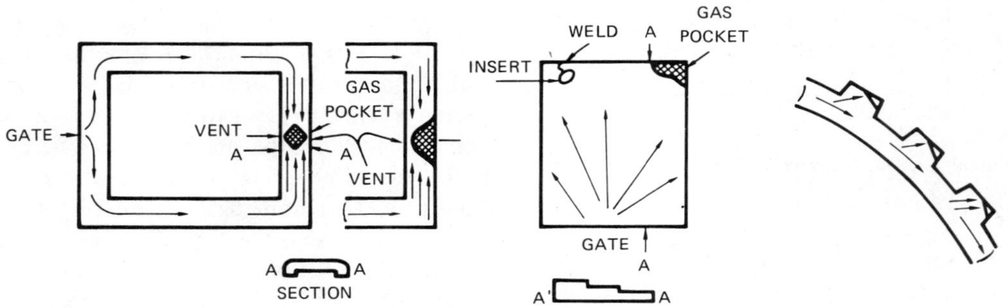

Fig. 20-62. Typical vent locations in a mold cavity.

to exhaust the air just prior to injection. If vent holes or channels are made in the mold they should not be over a few thousandths of an inch in diameter, or deep, in order to prevent the resin from entering and clogging the vent (see Figure 20-63).

Insufficient venting can be noticed by indications other than charring or black spots. Voids or apparent sink marks, due to air bubbles, may also be found. A simple test to determine if additional venting is required is to squirt a small amount of a hydrocarbon, such as kerosene, into the mold. During injection the air-kerosene mixture may ignite (the "diesel effect"), leaving a small deposit of soot. If no soot is seen, then venting may be presumed to be adequate.

Cooling. Cooling or, more properly, temperature regulation, is required so that uniformly rigidized parts can be removed from the mold. The first parts shot may freeze in a short period of time, but as successive injections are made the mold will heat up, increasing cycle times, or allowing parts to be ejected which are soft and show distortion or warpage.

Molds are commonly cooled with flowing water, directly from the supply, or refrigerated or sent through a mold-temperature regulator. If a refrigerated supply is used, it is important that the mold *not* be cooled below room temperature, since this could cause atmospheric moisture to con-

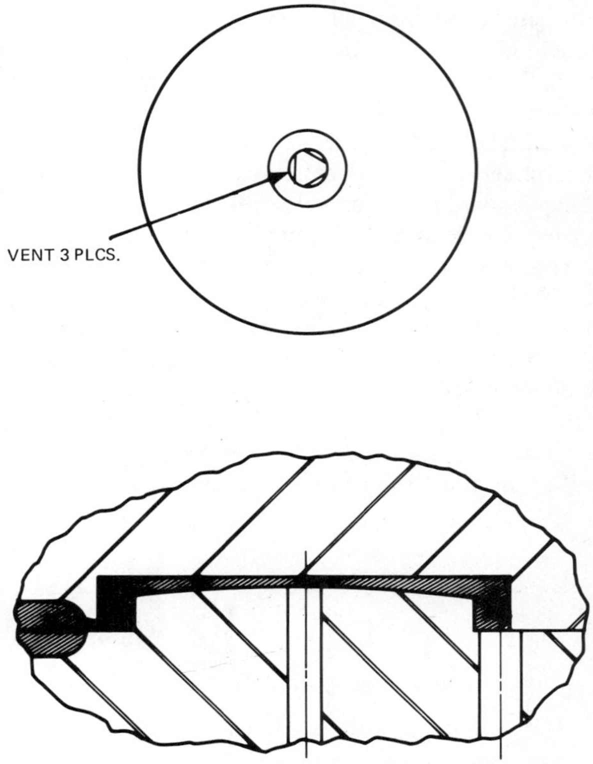

Fig. 20-63. Ejection pin venting.

dense on the mold. The use of a mold at the lowest temperature obtainable is not always the best policy, since, although this will result in the shortest cycle time, it will not always result in the highest gloss on the surface. Many molds, not connected to a temperature regulator, have the two halves of the mold connected in series. The cold water enters the part of the mold which could make the poorest surface; then, through a flexible hose, the warmed water enters the other half of the mold, which is thus kept at a higher temperature. The second half of the mold, for most resins such as polyethylene or styrene, will then produce a surface with considerably higher gloss. The actual mold temperatures range from approximately 35°C (100°F) to as high as 120°C (250°F), depending on the resin, the mold, the cycle, etc. The important factor is that the mold be designed with sufficient cooling capacity so that once a temperature or temperatures have been determined to be satisfactory, there will be very little variation during continuous operation.

The design of the cooling system will, of course, vary depending on the cavity shape and depth. In general, multicavity molds should have parallel coolant feed to the cavities, rather than series feed, for the reasons outlined above. Actual cooling systems include water channeling and a bubbler system (see Figure 20-64), flat plate spirals (see Figure 20-65), and special ring coolers used for one small location (see Figure 20-66). When using water channeling, it is important that the channels be located in relation to the cavity, not in random locations, or simply where it is convenient to drill the channels. Random locations can lead to hot and cold regions which can result in nonuniform cooling with consequent warpage, distortion, locked-in thermal stresses, and/or sink marks. Areas around gates, in particular, should be adequately cooled to assure minimum freezing time. In borderline cases of inadequate cooling, sometimes increasing the coolant velocity, or simply increasing the coolant line diameters (in or outside the mold), may result in sufficient cooling.

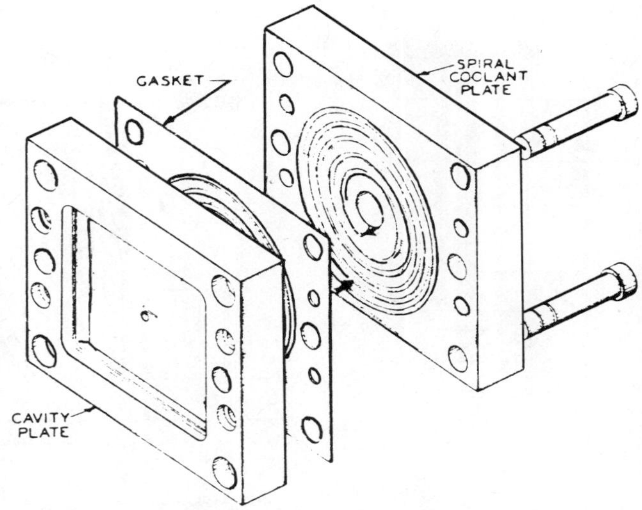

Fig. 20-65. Cavity plate with spiral-type cooling channel. Cold water enters near gate where melt is hottest, warms as it reaches cavity boundaries where melt has already somewhat cooled. This helps to reduce the difference in piece temperature from one part of the cavity to another. (*Courtesy: Phillips Chemical Co., a division of Phillips Petroleum Co., Bartlesville, Okla.*)

Ejection. Ejection of the molded part is done after the part has cooled sufficiently so it can be pushed off the core or out of the cavity. The ejector mechanism, of whatever type, is almost always actuated by the movement of the

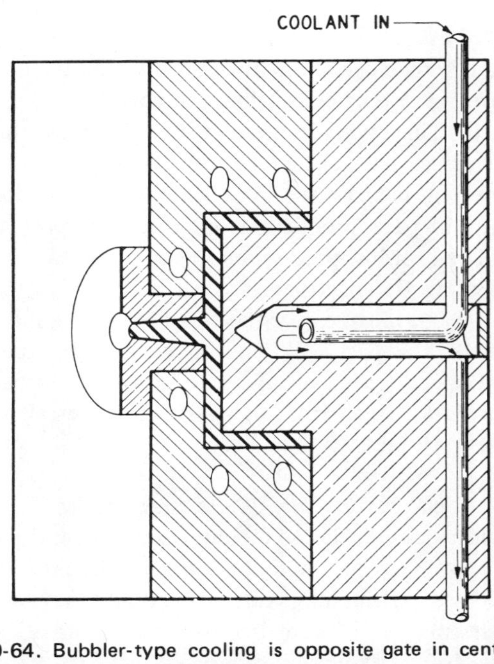

Fig. 20-64. Bubbler-type cooling is opposite gate in center-gated cavity. (*Courtesy: Phillips Chemical Co., a division of Phillips Petroleum Co., Bartlesville, Okla.*)

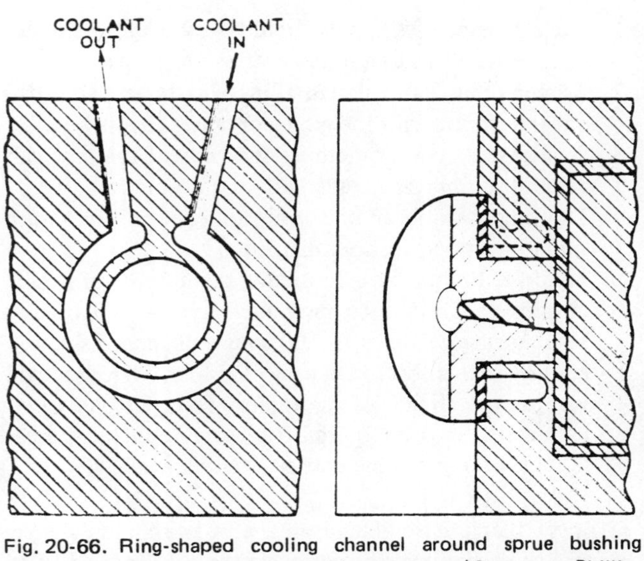

Fig. 20-66. Ring-shaped cooling channel around sprue bushing helps prevent overheating of metal in gate area. (*Courtesy: Phillips Chemical Co., a division of Phillips Petroleum Co., Bartlesville, Okla.*)

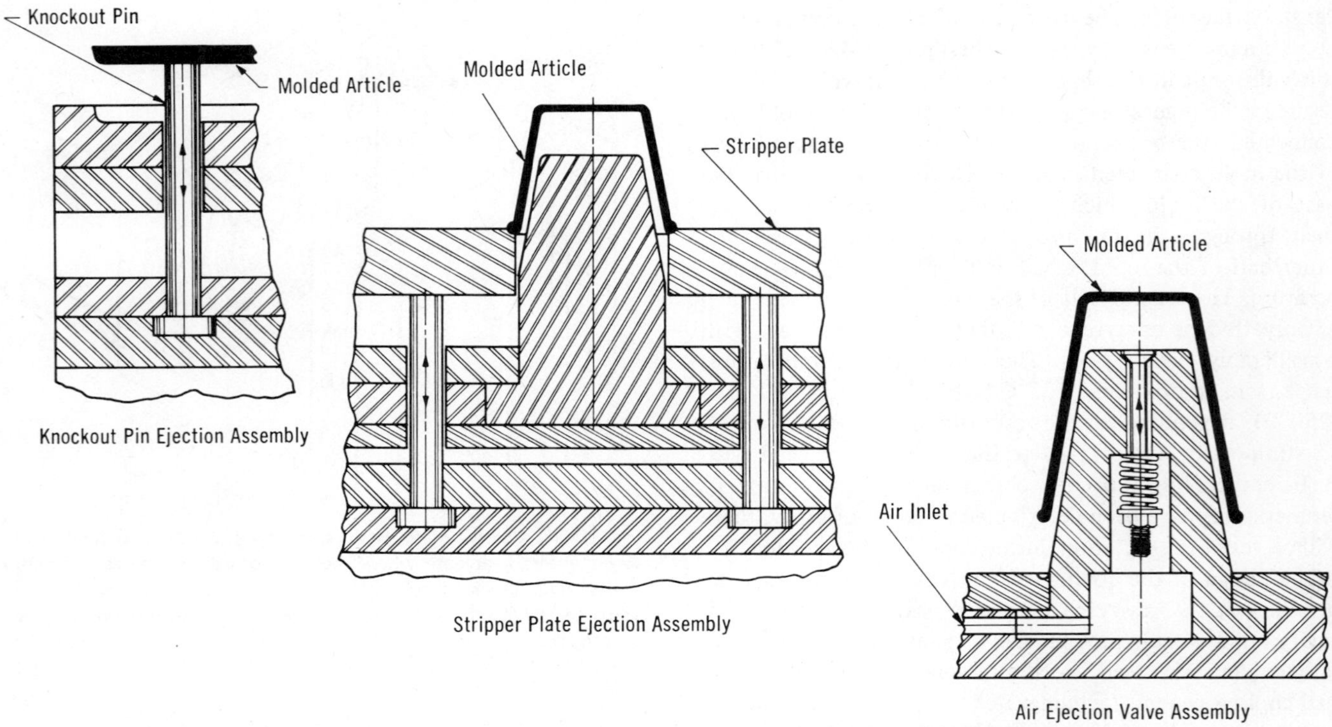

Fig. 20-67. Three methods of part ejection. (*Courtesy: Eastman Chemical Products, Inc.*)

mold during opening. There are three main methods of removing parts from cores or cavities: removal by pins or sleeves, by use of a stripper plate, and by means of compressed air. Frequently two or three methods will be used in combination (see Figure 20-67).

The use of knockout pins is the most common technique. The pins contact the part, usually at an edge, and push it off. In designing pin ejection, the following should be considered:

1. The diameter of the pins should be as large as the design allows, so as to minimize stress on the part.
2. As many pins as the design allows should be used, without interfering with the gating or water channeling.
3. Pins leave marks. Sometimes the ends of the pins can be used to impress a trademark or a decoration.
4. The pins must bear uniformly on the part to assure smooth ejection without distortion.
5. A rule of thumb for determining the number of knockoutpins required is to provide 1 square inch of pin-bearing surface for each 100 square inches of sidewall contact area (note: *not* projected area). Pin area will vary as the square of the radius. Thus, 82 pins of $\frac{1}{8}$-inch-diameter are required for 1 square inch, whereas 20 pins of $\frac{1}{4}$-inch-diameter give the same area.

Stripper plates have the advantage of bearing on the entire surface of the part, or parts, and so can be used to remove the part with less stresses developing, than with pin removal, and with negligible markoff. However, stripper plates have the disadvantage of being necessarily loose around cores, or cavities, and so the possibility exists that plastic will be forced between the stripper plate and the core plate. This material can interfere with proper closing of the mold if it is not removed.

Compressed air can be used to remove parts from shallow cavities by using a special air-ejection valve assembly. Air blown under the edge of a part can be used by the operator to remove parts, thus eliminating the need for any ejection mechanism. This saving in mold cost, which might be considerable, must be weighed against the fact that such a procedure will increase the cycle time, and may affect the uniformity of parts production. It should, however, always be considered for low-production runs.

In designing the ejection system it is important that the ejection pins be located near the core or male section of the mold, since the part will tend to be on the core after the mold opens, due to shrinkage. Conversely, the part will generally be loosened from the walls of the female cavity, again due to shrinkage.

Side-Pulls and Unscrewing Techniques. Holes in a part which run perpendicular to the direction of opening always present a problem to the mold designer. Should the holes be drilled as a postmold operation, or should they be put in during the molding? There is obviously no direct answer, since the drilling expense per part will have to be equated against the cost of installing side-pulls. One simple method of actuating side-pulls is the use of a canted pin cam which is fixed to one half of the mold and slides in an opening in the other half. As it moves, with the movable portion of the mold, it also moves a sliding plate in a direction perpendicular to the mold-opening axis. The sliding plate can then be attached to one or more pins, which extend into the mold when it is closed (see Figure 20-68).

Other methods of actuating side-pulls include hydraulic and compressed air cylinders, and electric solenoids and motors. The cam-actuated side-pulls, the cylinders, and solenoids, using rack and pinion mechanisms, are also used for automatic unscrewing. Electric motors, of course, can be well adapted for such tasks.

Cavities. A discussion of cavities is to some extent irrelevant in an abbreviated description of mold design, such as this, since all cavities are necessarily different; however,

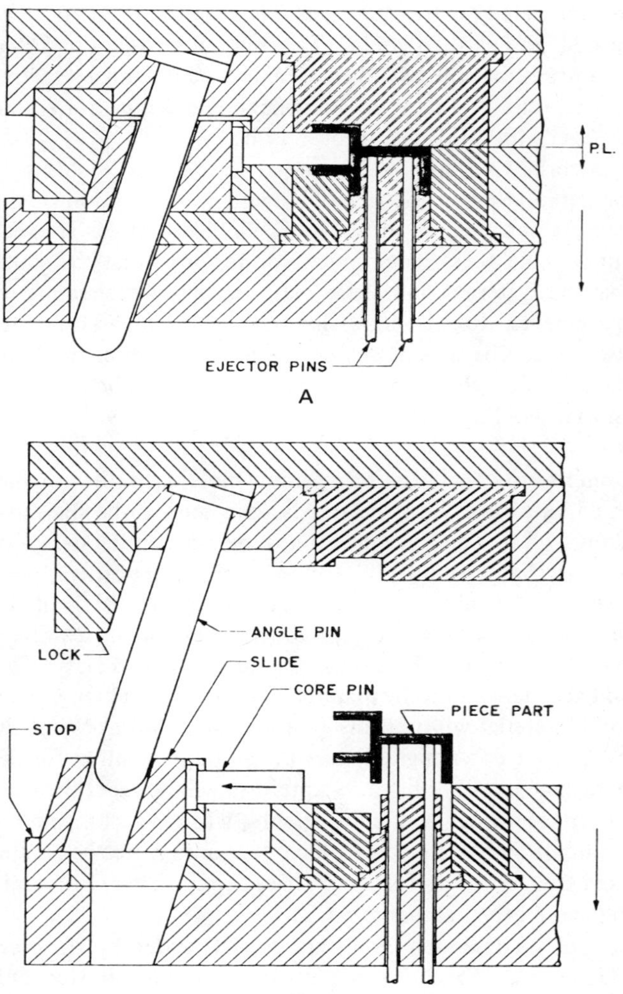

Fig. 20-68. (A) Mechanical cam in molding position. (B) Cam in ejection position. Note withdrawal of insert from part. (*Injection Molding* by Irvin Rubin. Copyright, 1972, John Wiley & Sons, Inc. Reprinted by permission of John Wiley & Sons, Inc.)

there are a few factors that are more or less common to all cavity designs.

1. In dimensioning cavities, shrink allowance must always be included. In almost every case, this is an increment which must be added to the final part dimension. The allowance will depend on the type of material, the molding conditions, and the thickness of the part. Shrink allowance information is always available from the material supplier.
2. Always assume that the part will tend to stick to the core section of the mold on opening and design the ejection system with this in mind.
3. Use the largest dimensional tolerances commensurate with proper part functioning, since this will produce the cheapest molds and the least rejects for improper dimensions.
4. Do not overspecify finish requirements for nonvisible sections of the part. Elimination of honing and polishing can save costs in certain molds. Consideration must be given to difficulties in part removal. If parts have too rough a mold surface, they may be difficult to eject. A light sand blast or hone can sometimes make polyethylene EVA, or nylon surfaces easier to eject than a highly polished surface. Runners though, should always be given a high polish to minimize flow resistance.
5. Draft allowance should always be as large as feasible for the part design, since this will aid in part removal. Depending on the resin and the mold surface, the draft may range from a minimum of $\frac{1}{2}°$ to as high as 5°. The more flexible resins will normally require the higher draft allowances.
6. Vents must always be provided—designed in—not as afterthoughts. Normally the vent should be at a location furthest from the gate, or at a point which might be expected to trap air.
7. Whenever possible, standard mold bases and cavity inserts should be used, since in most cases these will prove cheaper than "starting from scratch."

Structural Foam and RIM Molds. Molds for structural foam and RIM molding are similar to conventional injection molds, previously discussed, except that usually the parts are much larger—in most cases, they are quoted in pounds instead of in ounces. Another difference is that there are two main processes in structural foam molding—low-pressure molding and high-pressure molding. In both types, the pressures are substantially lower then in conventional injection molding. Consequently the molds may be made of cast aluminum, beryllium copper, Kirksite, steel, or epoxy and wood, in that order of acceptability. The epoxy and wooden molds obviously are for short production runs, since they cannot be temperature regulated with much success. The high-pressure molding process is further distinguished by the fact that the molds are of expanding types, which makes them somewhat more complicated than normal injection molds.

Structural foam molds and RIM molds are similar to conventional injection molds in that both have sprues, runners, gates, ejection pins, and cooling channels. Because of the size differences, the sprues, runners, and gates are much larger, but the same principles of short and balanced runners and gates are adhered to. Vents are extremely important in foam and RIM molding to get rid of the large amounts of air, but also to allow the generated gases to escape.

Because of the size of the parts and the necessity to inject the resin in the mold as rapidly as possible, the molds for structural foam may have as many as 40 or more nozzles. Other molds may use extra-large nozzles to accomplish the same result. In the injection of resin in both structural foaming and the RIM process, it is usually found best to introduce the material at the lowest point in the mold, thus assuring the escape of air at some higher point. In many cases the mold is made so it can be tilted or oriented to bring the nozzle to the lowest position on the mold. Flash gating is also used very frequently to assure fast filling.

Cooling, and in some cases heating, of the molds must be done on a larger scale than with conventional molds. In order to have more time for cooling, and still not slow production too much, in many structural molds which are not too large, a number of molds can be assembled on a rotary table. For the same reason, many molds are made as light as possible, even though aluminum is used, so that cooling can be accomplished as rapidly as possible.

Swirl marks present one problem in low-pressure structural molding which has not been entirely solved for all types of resins or parts. Some of the techniques to minimize swirl marks are slow injection (some say fast), heating of the mold to some experimentally derived temperature, and then cooling. Low-mass molds are used here to decrease the entire heating-cooling cycle. Overflow vents, an increase in the blowing agent content, and the use of solvents are also said to assist.

Transfer Molds. Transfer molds, used only for thermosets, are somewhat similar to injection molds, and so have many of the problems and solutions found in injection molding. Transfer molding was actually developed before injection molding, but was eclipsed by the much larger usage of thermoplastics.

One of the major differences between transfer molding and injection molding, is the necessity of precise temperature control in the mold; in transfer molds, the resin must be heated enough to liquify but not enough to cause setting up in the transfer chamber. This temperature and time relationship must usually be experimentally determined for each resin and charge weight. Another source of heat for the resin is the frictional heat developed at the gate, which must also be taken into account so premature curing will not take place. The frictional heat at the gate can also be used as part of the cure cycle, if the molding is done in a repetitive manner using automatic equipment, such as in a thermoset injection molding press. In general, transfer mold runners and gates are somewhat larger than in injection molds, due to the heavier filler loading of most thermosetting resins.

A special decision must be made in the design of a transfer mold as to whether it should be a "pot" mold or a plunger mold (see Chapter 16). The type of press available will, of course, influence the decision since any press can be used with a "pot" mold, but the press must have an auxiliary plunger to be used with a plunger mold. Another decision is whether to design the mold with replaceable inserts at the runners and gates. These are sometimes necessary if the resin is one with an abrasive filler which could cause rapid wear on runners and gates. Replaceable inserts are frequently made of carbides for greatest resistance to wear.

An obvious difference in transfer molding of thermosets, as contrasted to thermoplastics, is that the mold must be heated, rather than cooled. For minimum cycle times it is important that the heating system be tied in with a sensitive temperature regulator so that maximum temperatures can be used without danger of overcuring.

Injection molding of thermosets is done in molds very similar to thermoplastic molds, with the exception of the heating and cooling differences. While the molds are very similar, the press requirements do differ in that the screw may be different and there may be special preheating requirements; for example, lower barrel temperatures are used, etc. The advantages of thermoset injection molding include the elimination of preforms or preheaters and shorter molding cycles.

Compression Molds. Compression molding is the original molding process and the molds used tend to be the most simple of all. The first decision to be made in the design of a compression mold is the type—flash, semi-positive, positive, etc. Usually this decision is made on the basis of the number of parts required and the quality desired (see Chapter 16). The next decision is the depth of the cavity. This will also depend on the material to be used. If a high-impact loose material will possibly be used, then the cavity should be made deep enough to take the amount required for the part. On the other hand, if a preform will be used, then the cavity can be reduced considerably, with an accompanying saving in machining costs which may or may not be greater than the cost of preforming. These decisions will ultimately depend on the number of parts.

Small compression molds are usually heated by the press platens with possibly some auxiliary electrical heaters embedded in the mold. On the other hand, large compression molds, used for SMC, BMC, or fiberglass structures, are commonly made with self-contained heaters in the form of coils which can be used for steam or hot oil circulation.

Cooling provisions for such molds are not normally provided. The small molds are usually of steel. The large molds may be made of aluminum, or even epoxy-fiberglass or concrete backing up a sprayed or electroformed metal shell. The important consideration is that the mold be able to resist whatever the molding pressure is during the process. Normally this pressure is considerably higher when the mold is used for phenolics, melamines, and ureas than when it is used for epoxies or polyesters. Thus, the lightweight molds are mainly for use with the latter resins. Regardless of the resin used, the cavity should always be dimensioned with the recommended shrink allowance and draft angles.

Blow Molds. Blow molding consists of simply expanding a hot thermoplastic against a mold which then cools it to shape. The major consideration is have a mold with good heat conduction, built-in cooling coils, and freedom from wear at the neck and pinch-off points.

Blow molds are made usually of aluminum, either cast or machined, depending on the finish and quality required. Other materials include beryllium copper, where the highest-quality parts are required, or Kirksite which may not produce such high surface or dimensional quality, but is considerably cheaper. Cooling coils are either drilled in the mold as water channels, or, somewhat better, are installed as separate coils at the time of mold casting. The separate coils are considered better since they can be designed to result in maximum cooling in certain areas and less cooling in other areas. This differential cooling is used to obtain uniformity in the temperature of thick sections such as a neck of a bottle and a thin section such as the side. Another technique to attain some differential cooling is to always have the water inlet at the thickest mold section, so the coldest water will arrive there first.

Molds to be used for long production runs should be fitted with wear-resistant inserts at points such as the neck and the pinch-off area. These inserts can be of heat-treated steel, press fitted or bolted into the appropriate area. Vents should be provided, usually along the parting line, which should be in an inconspicuous area if possible. Since there will be a tendency to shrink away from the mold, the parting line can even be located in an area of slight undercut. To assure easy part removal, the finish should be a slight sandblast or vapor hone, which will also aid in venting.

Thermoforming Molds. Depending on the size of the part, thermoforming molds may be of cast aluminum, Kirksite, or epoxy-fiberglass. If the latter is used, the part is laid up on a form, first using a gel coat to assure a smooth finish. The lay-up is usually vacuum bagged for curing. Vacuum holes can either be drilled in after cure or, using greased piano wire, can be made at the time of cure, if a vacuum bag is not used. In many cases the mold consists of a thin shell of epoxy-fiber-glass fabric backed up by an epoxy filled with aluminum diamonds, or a lightweight filler such as vermiculite. Paper-base honeycomb is also used to obtain rigidity with minimum weight.

Cooling coils of copper tubing are usually embedded next to the shell, either with the backing, or as a separate operation prior to installing the backing. To prevent drilling into the coil, a low-voltage lamp and battery can be connected to the coil assembly and the drill body. In use, the operator cautiously drills into the shell and, if the lamp glows, another spot is picked to drill (see Figure 20-69).

Reinforced Plastics and Composites Molds. By reinforced plastics and composites, one usually means parts made with low pressures, from zero pressure to approximately 100 psi. For such molds, virtually anything can, and has been, used: plaster, plastics, concrete, sheet metal, wood, etc. In general, the mold need only withstand the slight forming pressure and the temperature used to cure the resin. The major factors are a dry material and a good surface. This means thorough drying of plaster and sealing of wood. The surface in most cases is simply sanded, sealed if necessary, then coated with mold release.

Cooling coils are not usually used with these kinds of molds, although coils can be embedded against the inside of a mold to act as heating coils. Parts to be cured at temperatures above 120°C (250°F) are usually made on metal molds.

In most cases, two-piece molds are not required for these processes since a vacuum bag, or an autoclave bag, is all that is required for one surface. When two finished surfaces are required, a conventional two-piece mold is used. The two-piece mold is also used if the part is required to be molded at pressures somewhat above that obtainable from a vacuum bag or an autoclave. A simple method of making such a "medium-pressure" mold is to use a cast or machined fe-

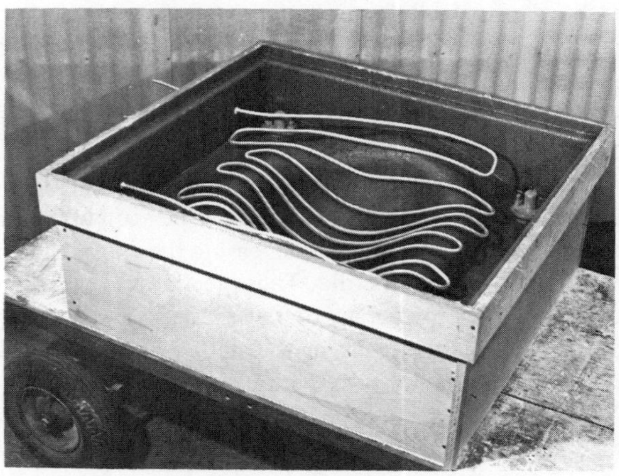

Fig. 20-69. Cooling-heating coil installation on an epoxy-fiberglass mold. Such a mold might be used for SMC molding, thermoforming, or as a casting form, etc. (*Courtesy: Owens-Corning Fiberglas Corporation*)

male and an elastomer male. By proper design, the male can be forced into the lay-up in the female half and, if deformed slightly, can exert pressure all over the part.

Extrusion Dies. Extrusions are produced in special types of "molds" called dies. The dies do the same job as molds; that is, they are responsible for the shape of the part. There is, however, one distinct difference between the two types of tools. In a mold, the part comes out in the exact shape and size of the mold (except for shrinkage and inadvertent distortion). In a die, the extrudate may be considerably different from the die shape. For example, a 4-inch diameter extruded polyethylene tube may be finally made into a 3-foot-diameter film. Another example is a die used to extrude a square bar which is made of four intersecting curves. For this reason, design of extrusion dies is as much an art as a science, and the prior experience of the designer in making similar parts is extremely important in assuring success in making a new die.

Film, sheeting, tubes, and pipes are all made in dies extruding material with a uniform cross-section. In designing these dies the important factors are the minimization of turbulence inside the die and the elimination of any points where stagnation of the resin might occur. If the die is appreciably larger than the nozzle of the extruder, then it is advisable to equip the die with auxiliary heaters which can be controlled separately from the barrel heater controls. Thus, variations in heat at the mouth of the die can be smoothed out to preserve the uniformity of the extrudate.

Profile extrusions are the main types which are not uniform in cross-section. These include extrusions such as table edges, decorative profiles, extruded shapes which are cut up to make small individual parts, etc. For such dies the same precautions concerning turbulence and avoiding stagnation usually hold, except where they might be deliberately put in to obtain some special effect. Profile dies, depending on the shape, may call for even more of the designer's past experience and skill than in designing uniform cross-section dies.

In many profile dies there are thick and thin sections. Flow in the nonuniform sections will also be nonuniform, in that the thick walls will tend to extrude faster than the thin sections. Equalization of flow can be accomplished by increasing the land length, or decreasing the gap slightly just before the mouth of the die. These techniques will help to prevent curling or warping of the extrusion as it emerges. Individual heater controls in the die can also be used to control flow in nonuniform sections.

Whether a uniform or nonuniform shape is being extruded, thickness can be increased by simply slowing the take-off equipment or by increasing the extruder RPM, assuming it is not working at its maximum. Conversely, the extrudate can be reduced in thickness by speeding up the take-off or reducing the extruder RPM.

REFERENCES AND BIBLIOGRAPHY

Mold Design

Bainbridge, R. W. Thermoset mold-gate design. *Durez Molder*, Issue 2, 1970, p. 4.

Baldwin, N. C. How to avoid warpage problems in injection molded parts. Part I: Part and mold design. *Plastics Design & Processing*, January 1974, **14**(1):25.

Ballentine, C. E. The importance of good fundamental mold design. *SPE Journal*, June 1965, **21**(6):241.

Bassett, H. D., Fazzari, A. M., and Staub, R. B. Tips on tooling for polysulfone resin. *Modern Plastics*, October 1965, **41**(2):117.

Beck, R. D. *Plastic Product Design.* Van Nostrand Reinhold Co., New York, 1970.

Benjamin, W. P. *Plastics Tooling.* McGraw-Hill Book Co., New York, 1972.

Bodner, S. S. Good design input = good production output. *Plastics Machinery & Equipment*, July 1973, **3**(7):18.

Borro, E. F., and Bainbridge, R. W. Mold design for transfer molding; *Plastics Technology*, August 1965, **11**(8):50.

Carley, J. F. Rheology and die design. *SPE Journal*, September 1963, **19**(9):977.

Dagniaux, S. D. J. Tailoring molds to the high speed production of thin-walled PP containers. *Plastics Engineering*, April 1979, **35**(4):25.

Davis, R. L., and Ashley, H. W. Factors influencing the design of molded plastics parts: Parts I–VIII. *Plastics World*, May 1962–December 1962: **20**(5):44, **20**(6):36, **20**(7):28, **20**(8):58, **20**(9):48, **20**(10):38, **20**(11):32, **20**(12):40.

Driscoll, J. R. Gating the mold right. *Plastics Machinery & Equipment*, May 1978, **8**(5):20.

Dow Corning Corporation. *How to Make a Multi-Piece Mold with Silastic RTV Silicone Rubber.* Dow Corning Bulletin 61-265a, March 1976.

DuBois, J. H. Sprue bushings and hot runner systems for thermoplastics. *Plastics Machinery & Equipment*, October 1978, **8**(10):43.

DuBois, J. H., and Pribble, W. I. *Plastics Mold Engineering Handbook.* 3rd Ed. Van Nostrand Reinhold Co., New York, 1978.

Engman, J. C. Gates for injection molding. *Plastics Technology*, October 1962, **8**(10):23.

Farris, R. N., and Meeks, P. J. Importance of mold rigidity in injection molding. *Plastics Technology*, May 1957, **3**(5).

Frizelle, W. G. What you need to know about mold, runner and gate design. *Plastics Technology*, June 1972, **18**(6):42.

Frados, J. (ed.) *Plastics Engineering Handbook of the Society of Plastics Engineers.* 4th Ed. Van Nostrand Reinhold Co., New York, 1976.

General Electric Co. *Noryl Thermoplastic Resins, Design.* Bulletin CDX-83.

Greninger, E. G. Designing and building compression and transfer molds. *Plastics World*, January 1965, **23**(1):30.

Hadden, C. W., and Talbot, H. M. Designing molds for polycarbonate. *Plastics Technology*, January 1975, **21**(1):38.

Kaplan, T., and Gordon, I. B. Designing molds for blow molding. *Plastics Technology*, February 1962, **8**(2):42.

Kelch, A., and Vickerson, H. Simple device gives controlled gating and improves quality of injected molded parts. *Plastics Design & Processing*, February 1962, **5**(2):316.

Levy, S. Maximizing flow through profile-extrusion dies. *Plastics Machinery & Equipment*, April 1978, **8**(4):24.

——. Designing plastics parts to withstand static loads. *Plastics World*, May 1964, **22**(5):18.

Martin, L. D. Injection molded plastics gears, Part I. Design considerations. *Plastics Design & Processing*, August 1968, **8**(8).

May, E. M. Mold design for high production. *Western Plastics*, Technical Report No. 74, July 1961.

Messenger, D. L. Design concepts for phenolics. *Plastics World*, February 1965, **23**(2):44.

Messenger, D. L. and Papoojian, R. S. Designing inlet channels of thermoset injection molds. *Plastics Engineering*, September 1974, **30**(9):38.

Meyer, W. H. Mold and part design for polypropylene. *Plastics Technology*, December 1962, **8**(12):36.

Misitano, G. Guidelines for effective design of RIM molds. *Plastics Engineering*, February 1979, **35**(2):27.

Naitove, M. H. Injection molding BMC: basics of mold design. *Plastics Technology*, October 1976, **22**(10):59.

Nuchtrob, W. C. Design and construction of tooling for sheet-milling compound. *Plastics Machinery & Equipment*, August 1978, **8**(8):34.

Olson, R. Structural foam tooling. *Plastics Technology*, June 1976, **22**(6).

Paulson, D. C. Care and design of injection molds. *Plastics Technology*, February 1969, **15**(2):53.

Perras, H. A. A molder's guide to mold design. *Plastics Technology*, January 1968, **14**(1):43.

Pickering, C. E. Molds or blow molding. *SPE Journal*, October 1959, **15**(10):891.

Plastics Design & Processing. Injection mold oversights reduce efficiency, up construction costs. March 1976, **16**(3):15.

Plastics Design Forum. How to pick the right toolmaker. March/April 1978, p. 71.

Plastics Machinery & Equipment. Cooling and heating systems for injection molds. October 1977, **7**(10):12.

——. Designing injection molds for thermosets, June 1973, **3**(6):18.

——. Mold design for injection molding with polypropylene. December 1973, **13**(12):18.

Prasad, A. Injection mold coolant lone design can determine mold cycling efficiency. *Plastics Design & Processing*, August 1976, **16**(8):9.

——, Molds that cool it. *Plastics Engineering*, June 1974, **30**(6):36.

Riley, M. W. *Plastics Tooling.* Reinhold Publishing Corporation, New York, 1955.

Ross, M. I. Mold design for automatic runnerless molding, *SPE Journal*, June 1965, **21**(6):559.

Sanders, M. A. Is this tolerance necessary? *SPE Journal*, January 1961, **17**(1):51.

Schaer, L. S. Rubber mold spin-casting for low cost prototype or production parts. *Machine Design*, January 20, 1977, **49**(2):113.

Schoeller, R. E. Blow mold designs change to meet new demands. *Plastics Technology*, May 1971, **17**(5):48.

Schroeder, K. Designing for automation in thermoset molds. *Plastics Machinery & Equipment*, January 1978, **8**(1):12.

Sisco, P. E. Design and construction of dies for blown and cast film, *Plastics Machinery & Equipment*, March 1978, **8**(3):37.

Thayer, G. Hot-tip gates save $. *Plastics Technology*, January 1976, **22**(1).

——. Mold design problems, *SPE Journal*, October 1957, **14**(10).

Tinkham, S. E. The design of transfer molds. *Plastics Machinery & Equipment*, July 1978, **8**(7):36.

——. New ideas in molds for thermosets. *Modern Plastics*, March 1967, **44**(7):147.

Tusch, R. L. Blow molding die design. *SPE Journal*, July 1963, **19**(7):632.

Vaill, E. W. Essentials of mold design in production planning. *SPE Journal*, March 1965, **21**(3):274.

Vansickle, R. W. Mold design technology for polyethylene. *Plastics Technology*, January 1958, **5**(1):35.

Wenninger, B. P. Tooling for thermoset molding today. *Plastics World*, August 1966, **24**(8):56.

Young, W. Treat molds right and cut downtime. *Plastics Design & Processing*, May 1972, **12**(5).

——. Hints for better mold performance. *SPE Journal*, December 1965, **21**(12):1362.

Mold Fabrication

Austin, M. How to use short-run tooling. *SPE Journal*, March 1971, **27**(3):29.

Bainbridge, W. W. Selective plate your molds to reduce corrosion. *Plastics Technology*, February 1967, **13**(2):51.

Barriner, F. Cut costs with disposable mandrels. *Plastics Technology*, October 1961, **7**(10):40.

Bender, L. Tooling with hydrophobic epoxies. *Plastics World*, June 1962, **20**(6):14.

Benjamin, W. P. *Plastics Tooling.* McGraw-Hill Book Co., New York, 1972.

Biery, J. Molds and plugs for production thermoforming. *Plastics Technology*, September 1965, **11**(9):41.

Birch, K. Spark erosion machining of mold cavities. *SPE Journal*, December 1960, **16**(12):1311.

Bogart, F. L. Plastics tooling for plastics production. *Plastics World*, June 1960, **18**(6):18.

Bonales, A. Electrical discharge machining. *The Western Electric Engineer*, January 1965, **IX**(1).

Bonales, A. The spark that carves. *Production Engineering*, September 27, 1965, **36**(9):53.

Bostwick, R., and Griff, A. L. Porous metal thermoforming molds. *Modern Plastics*, May 1962, **39**(9):139.

Brown, A. B. How to save on mold costs for prototype precision parts. *Plastics World*, November 18, 1974, **32**(11):38.

Cantillon, W. H. Berylium-copper molds for blow molding. *SPE Journal*, February 1966, **22**(2):75.

Carley, J. F. Rheology and die design. *SPE Journal*, September 1962, **18**(9).

Cerro Metal Products. *Mold Making with Cerro Alloys.* Bulletin HQ-763-1 G-2.

Cerro Sales Corporation. *How to Slush Cast Molds for Epoxy Encapsulation.*

Chastain, C. E. The perils of prototypes for injection molded plastic parts. *Machine Design*, November 15, 1973, **45**(21):137.

Cocci, A. J. Materials for moldmaking and their special treatment. *SPE Journal*, May 1965, **21**(5):462.

Colt Industries, Crucible Specialty Metals Division. *Crucible Data Sheet*, Issue #10, Bulletin DSS 131-5m-171.

Csaszar, E. J. Building injection molds by machining. *Plastics Design & Processing*, August 1975, **15**(8):10.

Danien, W. R. Choosing the right RIM molding material. *Plastics Design Forum*, September/October 1978, p. 51.

Delmonte, J. Porous, high temperature plastics tools for vacuum forming. *Plastics Design & Processing*, November 1965, **5**(11):12.

——. What cast urethanes can do. *Modern Plastics*, December 1965, **42**(4).

Delong, R. D. Design and construction of injection-blow molds, *Plastics Machinery & Equipment*, February 1978, **8**(2):24.

Dow Corning Corporation. *Silastic RTV*, Form #08-010, 1962.

Dreger, D. R. EDM goes into orbit. *Machine Design*, April 21, 1977, **49**(8):218.

E. I. Du Pont de Nemours & Co., Inc. *Molding Manual for Delrin Acetal Resins*, Bulletin A-34149, February 1964.

Dubois, J. H., and Pribble, W. I. *Plastics Mold Engineering Handbook*, 3rd Ed. Van Nostrand Reinhold Co., New York, 1978.

Dutt, D. E., and Mather, H. S. Low cost molds for short runs. *Modern Plastics*, November 1962, **40**(3):138.

Eastman Chemical Products, Inc. *Injection Molding Tenite Polypropylene and Tenite Polyallomer*, Bulletin TR-8A ECP 463, June 1970.

Electro-Coatings, Inc. *The Basics of Industrial Hard Chrome Plating.* 1975.

Falkin, S. How to choose molds for thermoforming. *Modern Plastics*, March 1967, **44**(7):138.

Fee, J. D. Guide to mold patterns and model materials. *SPE Journal*, March 1971, **27**(3).

Freeman Chemical Corporation. *Mold Making Guide.* Form 611, January 16, 1970.

Garner, P. J. New mold making technique. *SPE Journal*, May 1971, **27**(5):18.

General American Research Division. *Kanigen Electroless Nickel Alloy Coating*, March 1972.

General Electric Co. *Liquid Silicone Rubber for Flexible Molding Requirements.* Technical Data Book S-34 Silicone Products Dept., Waterford, N.Y.

——. *Ultra-High Strength RTV Silicone for Mold Making.* Technical Data Book S-29C, Silicone Products Dept. Waterford, N.Y.

Geronemo, V. F. Making a polyester mold. *Reinforced plastics*, May/June 1966, p. 24.

Grafton, P. Design and construction of compression molds. *Plastics Machinery & Equipment*, November 1978, 8(11).

Gregory, R. B. Advances in extrusion die design. *Modern Plastics*, March 1967, **44**(7):132.

Gulf Oil Co. *A Short Course in Polyethylene Molding.* Bulletin SP 12101 CD, April 1969.

Hall, F. E. Plastic parts in 24 hours. *Product Engineering*, February 14, 1966, **37**(2):87.

Hall, C. V. Surface quality; key to RP contact molds, *Plastics Technology*, July 1965, **11**(7).

Hankins, N. K., and Prideaux, D. G. Wash-away mandrels for filament winding. *Modern Plastics*, November 1963, **40**(3):115.

Harper, C. A. Molds for encapsulation. *Modern Plastics*, August 1961, **38**(12):103.

Harper, Jack R. Thin-wall silicone rubber molds now offer low tooling costs and quantity production rates. *Plastics Design & Processing*, October 1977, **17**(10):55.

Hartung, R. E. Molds for blow molding. *Modern Plastics*, March 1967, **44**(7):157.

Hauck, J. E. Fast method for making matched-die molds. *Modern Plastics*, March 1971, **48**(3).

Horvaka, J. P/M technology moves into mold construction. *Plastics Design & Processing*, March 1974, **14**(3):16.

Hubbell, F. H. How to protect molds against abrasion and corrosion. *Plastics Engineering*, December 1976, **32**(12):30.

Husky Injection Molding Systems, Ltd. Automation and short runs. Newsletter, March 1974.

Jaques, F., and Schmidt, J. Breakthrough in moldmaking—electroerosion. *Modern Plastics*, December 1960, **38**(4):109.

Jameson, E. C. Another look at electrical discharge machining. *Modern Plastics*, May 1965, **42**(9):158.

Junker, W. E. How to make low cost molds of paraffin, *Materials in Design Engineering*, March 1960, **37**(7):148.

Kadar, I. Guide to prototype molding. *Plastics Design & Processing.* Part I: Building the mold, August 1978, **18**(8):31. Part II: Material selection, casting techniques, September, **18**(9):37.

Kasmala, J., and Tierney, J. How to make molds for plastic furniture. *SPE Journal*, November 1970, **26**(11):39.

Kenny, M. E. Reinforced plastic molds for large blow-molding structures, *SPE Journal*, December 1965, **21**(12):1384.

Kish Industries, Inc. *Plastic Die Fabrication Procedure.*

Kroha, C. E. Selective plating: a simple way to replate molds on the press. *Plastics Engineering*, June 1977, **33**(16):24.

Lawry, W. A. Basic considerations in building epoxy molds for vacuum forming operations. *Plastics Design & Processing*, September 1961, 7(9):24.

Lombardo, M. D. A new concept in thermally controlled plastic molds. *Plastics Design & Processing*, May 1967, 7(5):29.

Levy, S. Maximizing flow through profile-extrusion dies. *Plastics Machinery & Equipment*, April 1978, 8(4):24.

——. Diamond wire saw cuts precision profile dies. *Plastics Machinery & Equipment*, August 1977, 7(8):22.

Lundy, D. W. Are your mold making methods up to today's tolerance requirements? *Plastics Machinery & Equipment*, November 1977, 7(11):25.

Martino, R. R. Mold construction for automatic molding of thermosetting materials. *SPE Journal*, February 1965, **21**(2):165.

Mast, H. Berylium copper molds: when, where and why to use them. *Plastics Technology*, October 1969. **15**(10):52.

McArthur, J. M. Steel for plastics mold cavities and cores. *SPE Journal*, June 1966, **22**(6):65.

Metco, Inc. *Wear Resistant Coatings Applied by Metco Flame Spraying.* Bulletin 207A, 1972.

——. *How to Cut Machine Downtime, Reduce Maintenance Costs & Increase Part Service Life with Metco Flame Spraying.* Bulletin 51K, 1966.

Metrocavich, S. Hobbing for raised and undercut sections. *SPE Journal*, November 1964, **20**(11):1191.

Miller, B. S. 97% yield in molding thermosets? Cold-manifold molds can do it. *Plastics World*, June 17, 1974, **32**(6):52.

Misch, D. Epoxies and elastomers frequently selected for making polyurethane foam molds. *Plastics Technology*, May 1971, **17**(5):50.

Misitano, G. Cooling most important for structural foam tooling. *Plastics Technology*, May 1971, **17**(5):43.

Mobay Chemical Co. *Injection Molding Merlon Polycarbonate, Thermoplastics Processing*, 83-M7.

Modern Plastics. 60 . . . 100 . . . 300 lb. injection molded parts? Easy. June 1970, **47**(6):70.

Monsanto Co. *Injection Molding of Lustrex and Lustran Styrene Polymers.* Product Information Bulletin No. 1104B.

Morse, G. T. Flexible urethane molds for the furniture industry. *SPE Journal*, June 1969, **25**(6):53.

Nachtrob, W. C. Design and construction of tooling for sheet-molding compound. *Plastics Machinery & Equipment*, August 1978, 8(8):34.

Newman, S., Pink, F., and Pesko, D. Processing plastics—make a production part from a test mold. *Plastics Engineering*, January 1977, **33**(1):22.

Noren, D. W. A new cure for mold-cooling headaches. *Plastics Technology*, April 1978, **24**(4):122.

O'Connor, R. T. Making low-cost molds with an epoxy-aluminum formulation. *Plastics World*, September 1960, **18**(9):26.

Owens-Corning Fiberglas Corporation. *Spray Metal Prototype Tooling.* Publication #5-EL-8365-A, September 1978.

Pawlak, J. J. Electroforming: an easy way to low-cost tooling. *Plastics Technology*, June 1968, **14**(6):45.

Peters, D. L., Morgan, B. P., and Cooper, W. E. Why it pays to build low-cost prototype blow molds. *Plastics Technology*, September 1968, **14**(9):34.

Peterson, P. S. Fast method for making matched die molds. *Modern Plastics*, March 1971, **48**(3):72.

Phillips Chemical Co. *Improving Surface Finish by Mold Venting.* Bulletin CWO 6016 26C, Vol. 3, No. 2, March/April 1962.

Plastics Design & Processing. Cut lead times on mold prototypes with spray metal tooling system. March 1978, **15**(3):31.

——. Low cost EDM electrodes produce intricate molds. June 1978, **36**(6):36.

——. New molding technique permits cost and time saving in production of EDM electrodes. October 1977, **17**(10):44.

——. Silicone carbide particles/nickel alloy coating protects molds against abrasive filled resin systems. August 1976, **16**(8).

——. Fundamentals of mold design–key elements in tooling for FRP molding. August 1976, **16**(8).

——. Extending mold life plus improving mold release. October 1971, **11**(10:27.

——. Heat curing thermoforming molds without an oven. May 1970, **10**(5):25.

——. Electroformed molds for plastics processing. July 1970, **10**(7):24.

——. Epoxy molds–easy way to get injection molded prototypes. July 1965, **5**(7):12.

Plastics Design Forum. Choosing the right RIM tooling material. September/October, 1978.

Plastics Engineering. Molds that cool it. June 1974, **30**(6):36.

Plastics Technical Evaluation Center. A review of plastics for tooling. Picatinny Arsenal, Dover, N.J., March, 1964.

Plastics Technology. Make molds with built-in release surface. May 1978, **24**(5):11.

——. Electroforming + RTV silicone rubber = new process for plastics tooling. August 1964, **10**(8):16.

——. Matched epoxy cast dies. November 1964, **10**(11):15.

——. Silicone rubber molds save time, cut costs. May 1962, **8**(5):23.

Plastics World. Mold making for short run parts. January 1979, **37**(1):56.

——. Produce molds fast, at low cost and with accuracy. August 1978, **36**(8):36.

——. Making molds by gas plating–old approach nets new cost benefits. November 1978, **28**(11):70.

——. How to buy an injection mold . . . that works. September 1972, **30**(9).

——. Instant molds. December 1964, **22**(12):34.

——. Polysulfide molds. August 1960. **18**(8):40.

Product Engineering. Precision metal, make EDM electrodes faster and for less. December 1978.

——. Molds for plastics and glass are easy to make with spray process. October 1973, **44**(10):14.

——. Epoxies hold the key to making low cost prototype parts. February 2, 1970, **41**(2):56.

——. Giant molds for plastics made from electroformed composite. May 25, 1970, **41**(5):82.

Reinforced Plastics. Mold Making for Large Pieces, March-April, 1963.

Ren Plastics, Inc. *The Techniques of Using Epoxy Plastic Tooling Materials.* Lansing, Michigan, 1964.

Rezolin, Inc. *Manual of Plastic Tooling Materials*, February 1963.

Robson, A. A. Modern electroformed nickel molds: fabrication techniques and application, *Plastics Design & Processing*, February 1979, **19**(2).

Rubber Age. What's new in vaporforming processes? November 1970, p. 72.

Rubino, A. G. Casting molds. *SPE Journal*, September 1962, **18**(9):1159.

Savla, M. What does non-ferrous tooling cost? *SPE Journal*, March 1971, **27**(3):24.

Schaeffer, and Diez, D. E. Prototype and mold fabrication. *Plastics Technology*, September 1961, **7**(9):23.

Schettig, J. R. How to select mold steels. *Plastics Technology*, November 1964, **10**(11):36.

Scheiner, L. L. Small refinements bring high quality injection tooling. *Plastics Technology*, May 1971, **17**(5):40.

Schick, H. P. How to build molds for thermoforming. *Plastics Technology*, September 1965, **11**(9):56.

Schoeller, R. Building injection molds by casting. *Plastics Design & Processing*, August 1975, **15**(8):13.

Sisco, P. E. Jr. Design and construction of dies for blown and cast film. *Plastics Machinery & Equipment*, March 1978, **8**(3):37.

Smith, H. A. Here's a guide to making flexible molds for foam furniture. *Plastics Technology*, September 1968, **14**(9):38.

——. A guide to making flexible molds with RTV silicone rubber. *Plastics Design & Processing*, August 1962, **2**(8):14.

Smith, R. W. Sophistication comes to thermoform tooling. *Plastics Technology*, May 1971, **17**(5):52.

Spier, I. M. Copper: the forgotten tool for mold construction. *Plastics Machinery & Equipment*, September 1977, **7**(9):16.

——. Ejection systems for injection molds. *Plastics Machinery & Equipment*, December 1977, **7**(12):13.

——. Salvaging old molds with improved cooling. *Plastics Machinery & Equipment*, July 1976, **6**(7):16.

Stokes, W. J. B. Design, construction, and use of electroformed molds. *Plastics Engineering*, February 1978, **34**(2):25.

——. Repairing molds by electroforming. *Modern Plastics*, December 1958, **35**(4):121.

Sullivan, J. W. Engineer's report: tool steels for plastics mold dies. *Plastics World*, May 1965, **23**(5):50.

Summers, P. H. Fluidized bed–safe, efficient method for cleaning dies, molds and molding equipment. *Plastics Design & Processing*, December 1974, **14**(12):20.

Thompson, R. Gate and vent design for structural foam molds. *Modern Plastics*, July 1957, **34**(11):115.

Throne, J. L. Gate and vent design for structural foam molds. *Plastics Design & Processing*, June 1977, **17**(6).

Tinkham, S. E. Streamlining the moldmaking process. *Plastics Machinery & Equipment*, October 1976, **6**(10):12.

United States Gypsum Co. *Tooling Techniques with USG Industrial Plasters and Gypsum Cements.* Bulletin TAC-164, July 1975.

——. *Tooling with Gypsum Plasters for Reinforced Plastics*, IGL Bulletin-312.

U.S. Industrial Co. *Polyolefin Injection Molding . . . . An Operating Manual.* Bulletin P2-1277.

Western Plastics. Aluminum molds key to injection molding large parts in short runs. November 1964.

Wright, D. H. Templectroforming of mold cavities, *SPE Journal*, October 1968, **24**(10):113.

Young, W. Getting the best performance from your molds: Part III, Durez Division, Hooker Chemicals & Plastics Corporation.

Young, W. Treat molds right and cut downtime. *Plastics Design & Processing*, May 1972, **12**(5):16.

21
Part Design

PRELIMINARY DESIGN CONSIDERATIONS

Before the design of any part is begun, the usual procedure is to first determine the function of the part and the environment in which it is to operate. This environmental survey should include the thermal conditions–high and low temperatures, and the mechanical stresses on the part–tensile load, impact, compression, etc. Depending on the anticipated use, these may be the major environmental factors, or the only factors which need be considered. In other types of parts, the major factors may be electrical stresses–high voltages or frequencies, dielectric losses, etc. Some parts may combine thermal stresses with chemical stresses–strong acids, solvents, oils, high humidity, etc. Operational time must also be considered.

Having prepared the list of environmental factors, the next step is usually to prepare a list of the requirements of the part from the standpoint of user acceptance. Such a list might include items such as color, lack of color, transparency, specific gravity (must it float or should it have a heavy filler), and cost. Included in this list should be any hard-to-define items such as the feel of the part, the sound it must make if struck, etc. This part of the list of requirements may indeed be the most difficult to determine; however, by proper consideration, the designer may avoid the pitfalls of having an unacceptable part after the material decision has been made, or worse after the first part has been made.

After all the requirements have been determined, the next step is to decide on the material. In selecting the optimum material, the most important factor should be used as the first screening guide; for example, must the material be transparent? In this case, since only transparent materials need be considered, the list of candidate materials is considerably shortened. The final choice is usually a material which will give the optimum performance at the lowest cost; in practice, this usually turns out to be a compromise.

With the selection of the material, the next choice the designer must make is the selection of the fabrication process. In many cases, the choice is dictated by the number of parts or the material chosen. For example, machining and casting are obvious choices when twelve prototypes are required, and injection molding if one million parts are required. In other cases, the fabrication process is dictated by the type of part or, occasionally, by the equipment available.

Regardless of the type of part or the overall design, there are certain fundamental design rules which are applicable to almost all types of parts–plastics or not. These include avoidance of stress concentrators, proper use of ribs and fillets, hole spacing, etc. The following discussion, while mainly dealing with injection-molded parts, has relevance to parts made by all processes and, to some extent, to parts made of all materials.*

Wall Thickness

The first rule of good design is to use a uniform wall thickness whenever possible. A thick section next to thin sections will cool last and shrink more away from the mold, resulting in a "sink mark." The differential cooling will also result in internal stresses, warpage, or distortion, and may even cause cracking. A corollary to this rule is to always use the minimum wall thickness which will result in satisfactory parts, since this will save material and molding time and reduce the part cost. Figure 21-1 shows the possible effects of unequal wall thickness, and designs to eliminate these effects. Note that sharp corners are avoided.

If a housing, shroud, or similar part is being designed, it may be thermoformed starting with a sheet of uniform thickness. Similarly, parts made using impregnated fabrics are automatically made with uniform wall thickness.

In the event that nonuniform wall thicknesses cannot be avoided, then the different wall sizes should be gradually blended, as shown in Figure 21-2. In any event, the wall thicknesses should not vary by more than a ratio of 3 : 1, if at all possible.

In addition to uniform thickness, it is desirable to design parts with wall sections that are well within normal molding capability. Table 21-1 gives suggested wall thicknesses for common thermoplastic injection molding materials. Wall thickness recommendations for the newer, heat-resistant materials similar to polycarbonate and polysulfone can be greater than the conventional materials.

Wall thicknesses for thermoset materials depend on the type of resin and the type of molding operation; for example, because epoxies, alkyds, and polyester molding powders and compounds are very fluid when heated, very thin walls (0.010 to 0.015 inch) are feasible. Phenolics have been molded up to 3 inches and more in thickness. Similar thicknesses in thermoplastics are possible, but impractical because

*The discussion and illustrations are largely based on *Plastic Product Design*, by R. D. Beck, Van Nostrand Reinhold Co., 1970. This work should be consulted for a fuller treatment of the topics discussed.

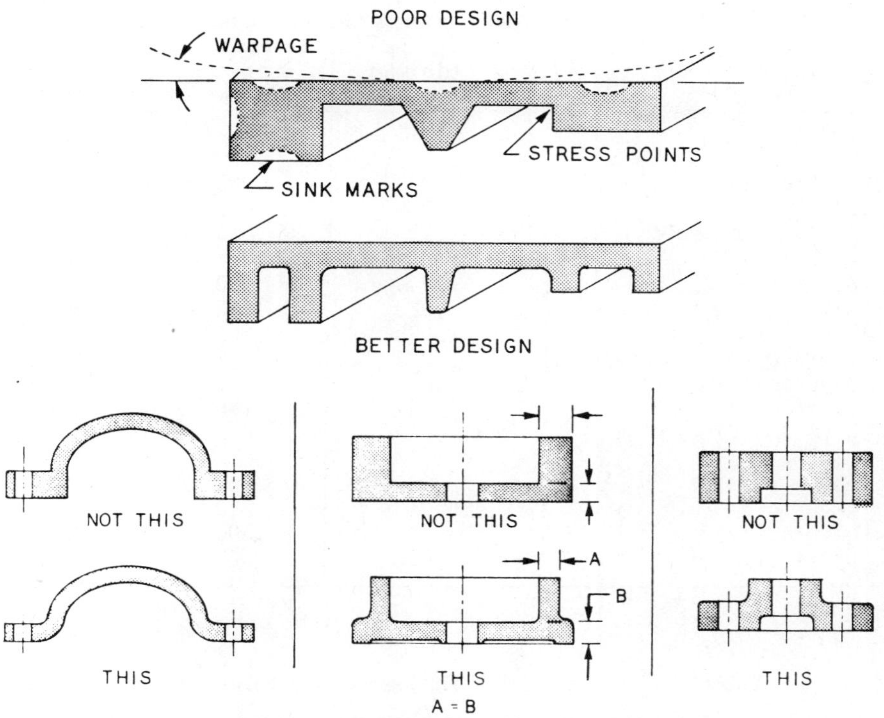

Fig. 21-1. Uniform wall thickness designs. (From *Plastic Product Design* by Beck)

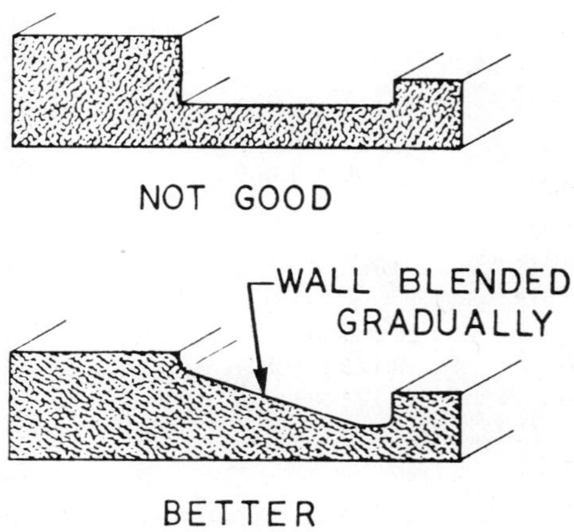

Fig. 21-2. Thick-thin wall blending. (From *Plastic Product Design* by Beck)

of the excessive cooling times required. Table 21-2 shows suggested wall thicknesses for common thermosetting materials.

Wall thicknesses for epoxy or polyester fabric-reinforced, low-pressure molded parts may range from a low of approximately 0.005 inch to a high of 1 inch, providing the processing excludes air bubbles, and the part is thoroughly cured.

Reinforcing Ribs and Bosses

There are two ways to increase the stiffness of a part. The wall thickness can be increased throughout, which results in an increase in the cross-sectional moment of inertia. However, a rectangular cross-section is relatively inefficient in bending. The second method of increasing stiffness is to provide reinforcing ribs or gussets. The "T" section which results when gussets are added provides stiffness equivalent to a rectangular section, with less material and therefore less weight. Ribs increase stiffness locally, as required, without adding superfluous material and weight.

Ribs can be used efficiently around attaching holes and bosses to transfer loads from the bolts to the primary structure. Ribs can also stiffen unsupported flat plates to inhibit buckling or increase the natural frequency of vibration. They can be used effectively around openings to increase radial stiffness and maintain concentricity, and to aid in filling bosses, as illustrated in Figure 21-3.

It is important to design ribs properly to minimize the possibility of warping and sink marks after cooling or curing. The suggestions given for wall thickness are applicable to the design of reinforcing ribs. Gradual thickness transitions should be provided with generous radii. In general, the thickness of a rib should be less than the thickness of the wall to which it is attached. When structural requirements demand rib thicknesses greater than the thickness of the adjacent wall, it is better to add additional ribs and maintain a maximum thickness limitation. Alternatively, the rib can be "split" as shown in Figure 21-4. When the part requires more than one rib, the spacing between ribs should not be less than two times the thickness of the wall to which the ribs are attached. It is more efficient (in terms of stiffness) to increase the height of a rib than to increase

Table 21-1. Suggested Wall Thickness for Thermoplastic Parts.*

Thermoplastic Materials	Minimum Inches	Average Inches	Maximum Inches
Acetal	.015	.062	.125
ABS	.030	.090	.125
Acrylic	.025	.093	.250
Cellulosics	.025	.075	.187
FEP fluoroplastic	.010	.035	.500
Nylon	.015	.062	.125
Polycarbonate	.040	.093	.375
Polyethylene (L.D.)	.020	.062	.250
Polyethylene (H.D.)	.035	.062	.250
Ethylene vinyl acetate	.020	.062	.125
Polypropylene	.025	.080	.300
Polysulfone	.040	.100	.375
Noryl (modified PPO)	.030	.080	.375
Polystyrene	.030	.062	.250
SAN	.030	.062	.250
PVC--rigid	.040	.093	.375
Polyurethane	.025	.500	1.50
Surlyn (ionomer)	.025	.062	.75

*Beck, 1970.

Table 21-2. Suggested Wall Thicknesses for Thermoset Parts.

Thermosetting	Minimum for Any Article	For Small Articles	Average for Most Articles	Large to Maximum Articles
Phenolics				
general-purpose and flock-filled	0.050	0.062	0.125	0.187 to 1.000
fabric-filled	0.062	0.125	0.187	0.187 to 0.375
mineral-filled	0.125	0.125	0.187	0.200 to 1.000
Alkyd				
glass-filled	0.040	0.093	0.125	0.187 to 0.500
mineral-filled	0.040	0.125	0.187	0.187 to 0.375
Ureas and Melamines				
cellulose-filled	0.035	0.062	0.100	0.125 to 0.187
fabric-filled	0.050	0.125	0.125	0.125 to 0.187
mineral-filled	0.040	0.093	0.187	0.187 to 0.375

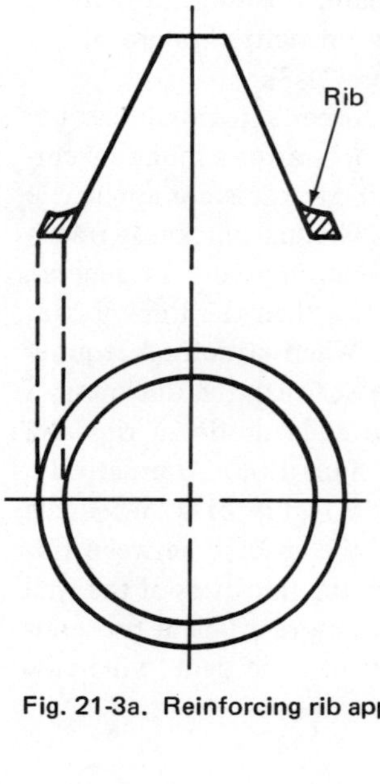

Fig. 21-3a. Reinforcing rib application.

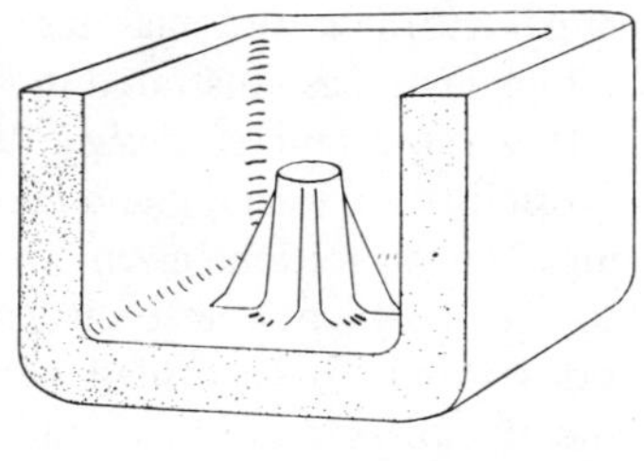

Fig. 21-3b. Ribs placed on sides of bosses aid in filling them.

its thickness, since stiffness is related to the cube of the height and is proportional to the thickness. However, the height should not be increased to such a degree that the rib is unstable and will buckle under a load.

Rib Design for Thermoplastic and Thermosetting Materials. The design of ribs using thermoplastic materials differs somewhat from thermoset rib design. Figure 21-5 illustrates the rib design and proportions used in most thermoplastic materials. The recommended proportions for thermoset ribs are shown in Figure 21-6.

Bosses

Bosses are protruding pads that are commonly used as mounting surfaces for the molded part. A boss is used to distribute bearing loads caused by attachment bolt preloads, and to transfer bolt loads to the main structure. Excessively

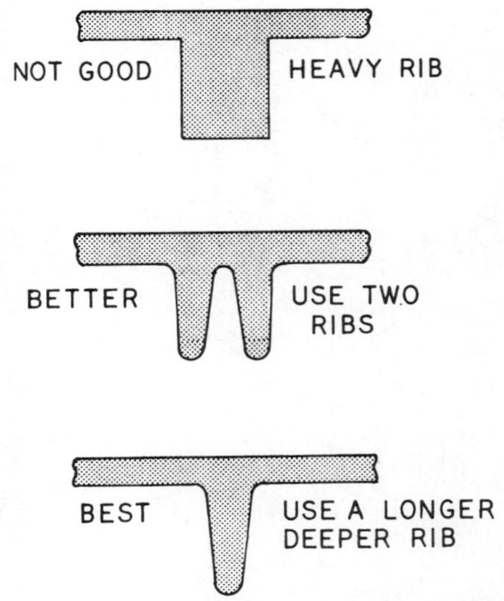

Fig. 21-4. General rib designs. (From *Plastic Product Design* by Beck)

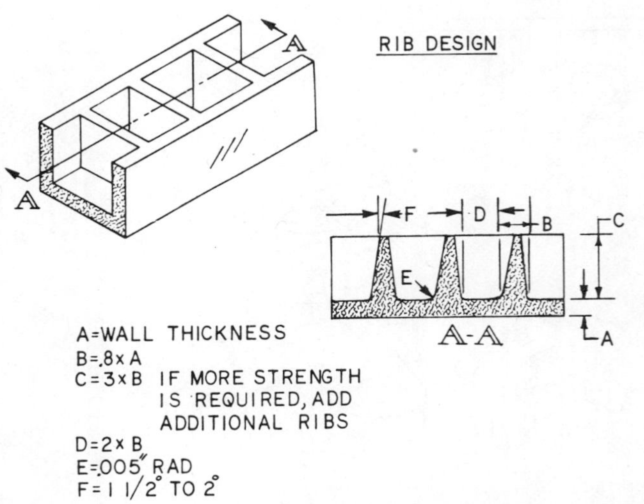

Fig. 21-5. Thermoplastic rib design. (*Courtesy: The Dow Chemical Co.*, reprinted from *Plastic Product Design* by Beck)

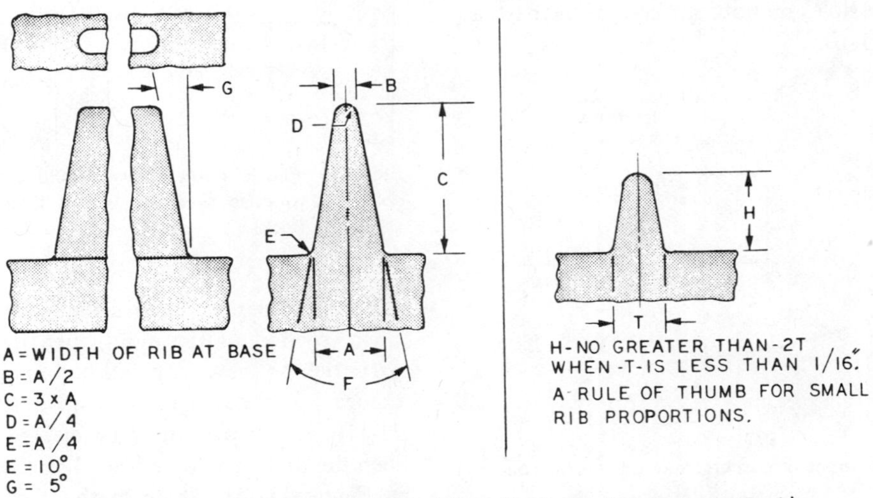

Fig. 21-6. Rib design for thermosets. (From *Plastic Product Design* by Beck)

thick bosses should be avoided. When the attachment loads are sufficiently high, ribs can be used to connect the bosses to the adjacent wall. Bosses must be large enough in diameter to carry the required load without failure. Figure 21-7 illustrates the proper design of bosses. In general, the height should not be more than twice the diameter.

Radii and Fillets

Radii and fillets promote the distribution of internal loads and stresses and help relieve stress concentrations associated with changes in wall thickness and changes in direction of internal loads. Radii and fillets also promote the even flow of material during molding, so that lower pressures can be used, or the cycle time may be lessened.

Stresses tend to concentrate at the junction of two sections such as corners or ribs, as shown in Figure 21-8. The curve in Figure 21-8 also illustrates the effectiveness of radii in reducing stress concentrations, and indicates the proper radius to be used for a given wall thickness. Stress concentrations are also present at external corners. When the external corner is sharp, good design dictates that the section at the corner should be increased by about 30%. Rounding the corner with an external radius tends to maintain a uniform thickness and therefore relieves a stress concentration (see Figure 21-9).

When stress concentrations are minimized, the strength of the part is increased. Furthermore, cracks are less likely to occur when generous radii are used at the junction of the two walls. Whenever possible, a minimum radius of $\frac{1}{64}$ inch (0.015 inch) should be provided.

Taper or Draft

Taper or draft is almost always required in order to assure easy removal of the part from the mold. The taper or draft is necessary because the molded material will tend to shrink onto the male section of the mold. However, sticking can

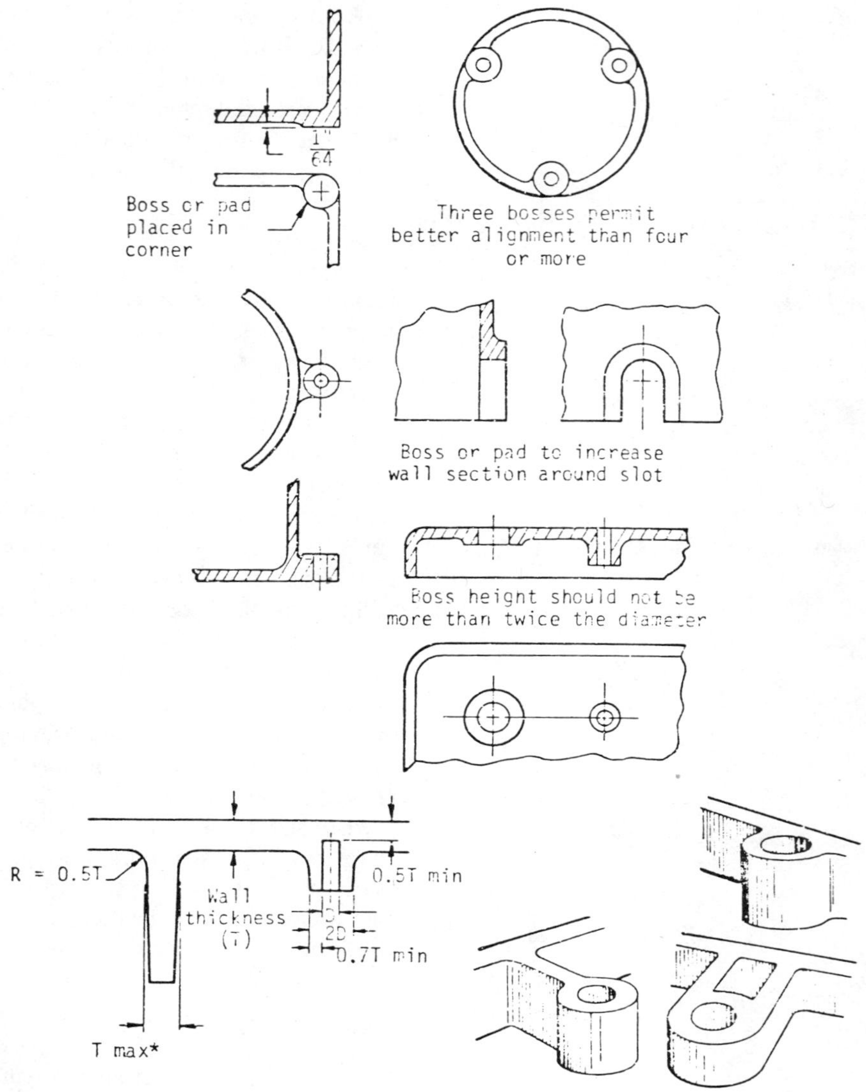

Fig. 21-7. Boss design.

take place on both male (plunger) or female (cavity) sections of a mold. Recommended drafts are a minimum of $\frac{1}{2}°$ to as high as 3° (if possible) on both surfaces of the part, as illustrated in Figure 21-10.

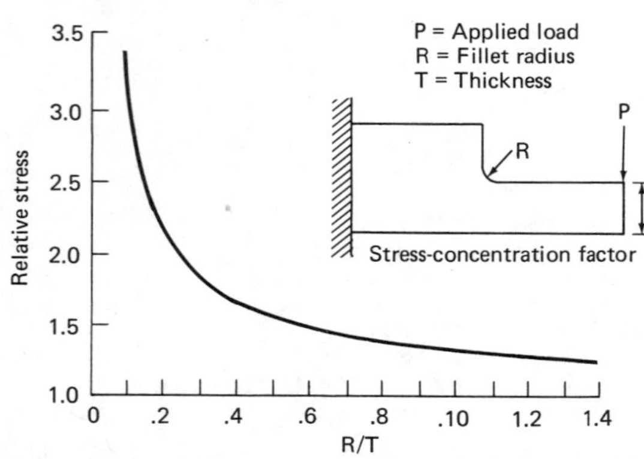

Fig. 21-8. Stress concentrations at corners, and radii versus concentration factor.

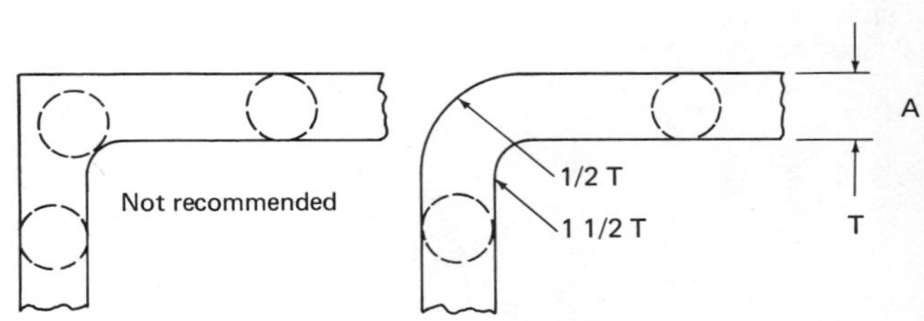

Fig. 21-9. Plastic corner flow should present no problem if the corner is rounded as shown in this figure. (From *Plastic Product Design* by Beck)

If a mold design requires the knockout pins to be in the cavity section of the mold, then it is imperative that the part stick to the cavity and not the plunger. In such a case the cavity is designed with no draft, as shown in Figure 21-11a. If the part must be ejected when on the plunger, then the draft would be placed on the cavity walls, as shown in Figure 21-11b. Other methods of assuring that the part

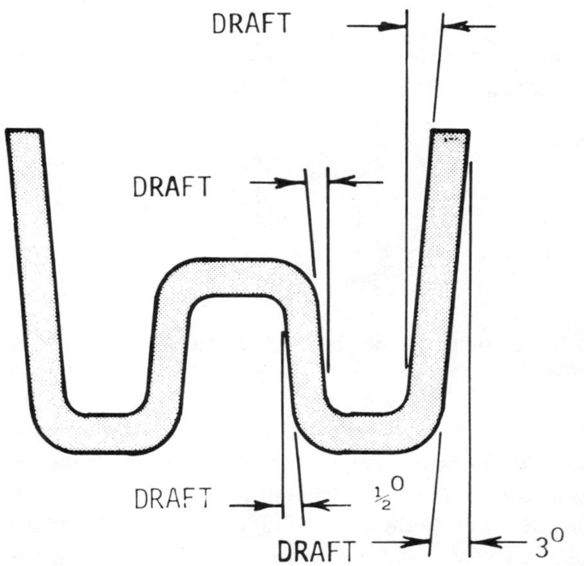

Fig. 21-10. Recommended draft angles. (From *Plastic Product Design* by Beck)

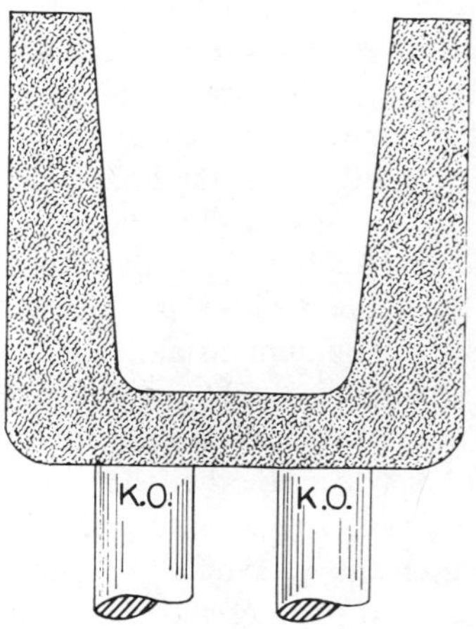

Fig. 21-11a. When draft is placed in the plunger only, the part will stay in the cavity. Knockout pins will then be required in the cavity. (From *Plastic Product Design* by Beck)

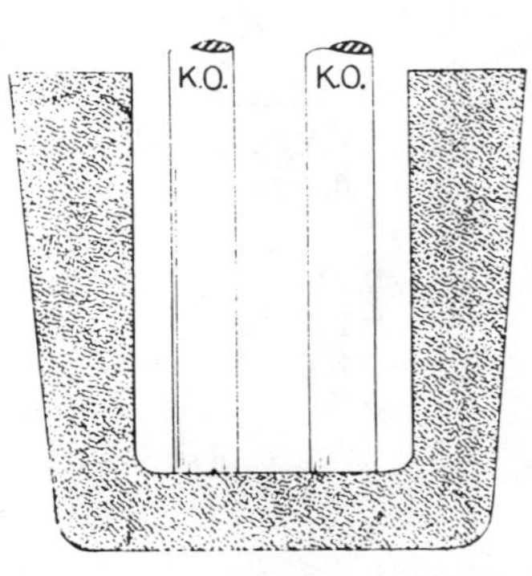

Fig. 21-11b. If draft is placed on the outside of the part only, the part will stay in the plunger. Knockout pins will then be required in the plunger. (From *Plastic Product Design* by Beck)

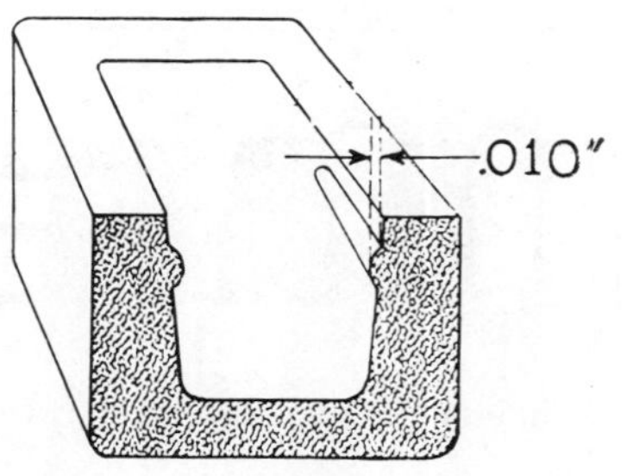

Fig. 21-11c. Pick-up ribs are frequently used to make the molded part remain in the cavity or on the punch. (From *Plastic Product Design* by Beck)

will stay in the cavity or on the plunger are to mold pick-up ribs on the part (see Figure 21-11c), or the desired surface can be liquid honed to increase sticking.

Molded Holes

Molded holes usually cost less than machined holes if large quantities of parts are to be made, provided the holes have optimum geometry and size. In general, molded holes in the direction of the mold closure (perpendicular to parting line) are less expensive than a hole at right angles or at an angle to mold closure. (The latter two holes require special retracting mechanisms for automatic molding. In these cases post-mold drilling should be considered.) Figure 21-12 illustrates the different hole orientations.

Molded holes may be through holes or blind holes, or if two or more mandrels are used to form the hole, they may be stepped, recessed, or intersecting. The various types of holes, illustrated in Figure 21-13 can be round, square, elliptical, etc. Round holes are always the cheapest since simple dowel pins can be used as mandrels in the mold to form the hole.

The pins which form the holes are sometimes subject to severe bending forces, and the pins may be distorted or broken off if the length-to-diameter ratio is excessive. The ratio of the mean diameter of a pin to its length is called the "slenderness ratio." Recommended slenderness ratios for pins supported by one hole and two holes are shown in Figures 21-14a and 21-14b.

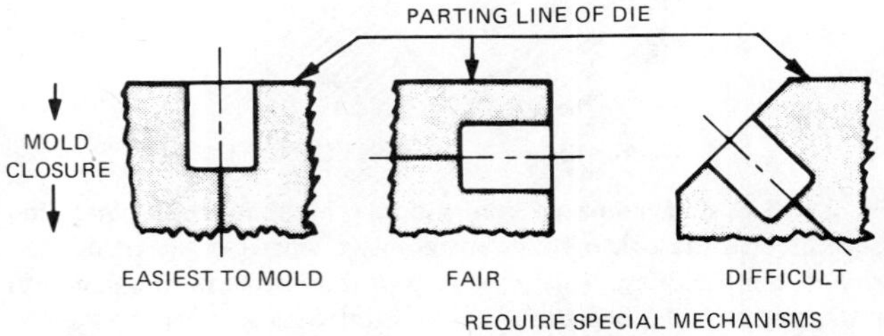

Fig. 21-12. Hole orientations. (From *Plastic Product Design* by Beck)

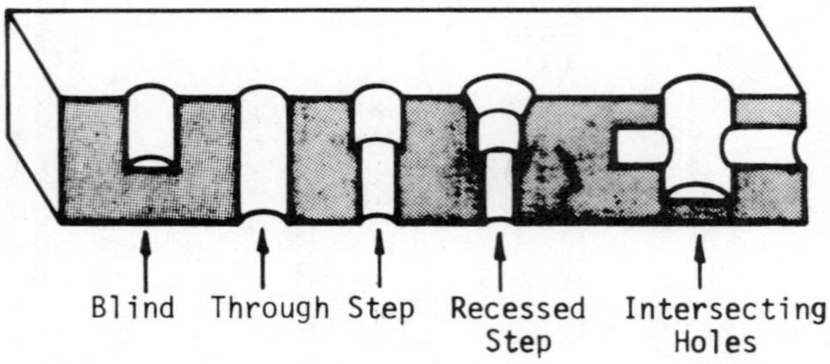

Fig. 21-13. Hole types. (From *Plastic Product Design* by Beck)

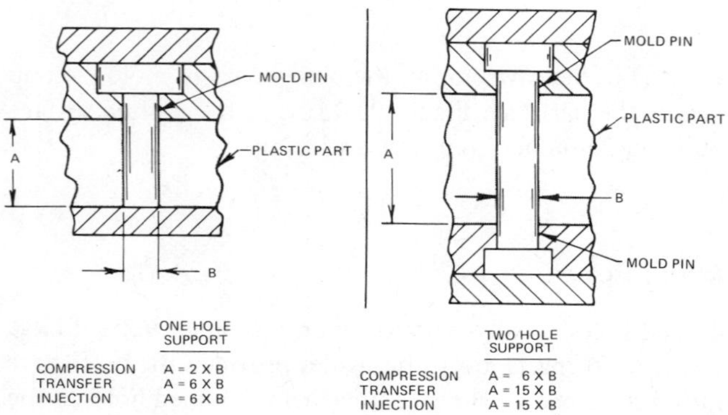

Fig. 21-14. Slenderness ratios for various pin types. (From *Plastic Product Design* by Beck)

In many cases, particularly for compression molding, a small amount of flash will be entrapped in the bottom pin support hole when a two-hole pin support is used (see Figure 21-15). On the next shot this can cause the pin entering the hole to bend or break. Better design is to use two pins butted together in the center, as shown in Figure 21-16. In this design, as the mold opens, any flash on either pin can be seen and easily removed. The two pins are shown as having different diameters, since this will minimize the

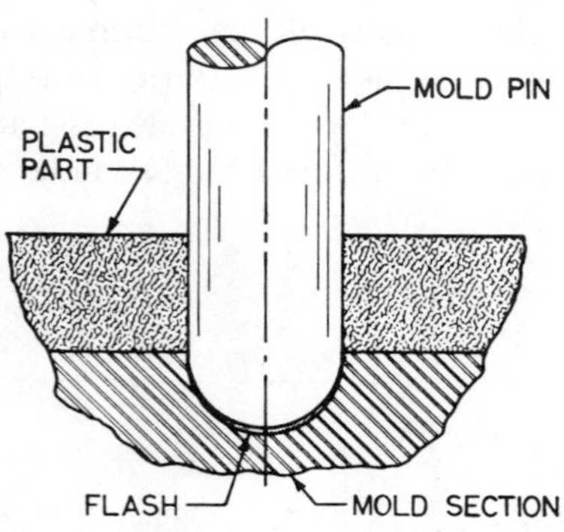

Fig. 21-15. Using two-hole pin support in compression molding could involve the risk of flash entrapment, since resin is introduced prior to mold closing. This possibility is minimized during injection or transfer molding because the resin is introduced after the mold is closed. (From *Plastic Product Design* by Beck)

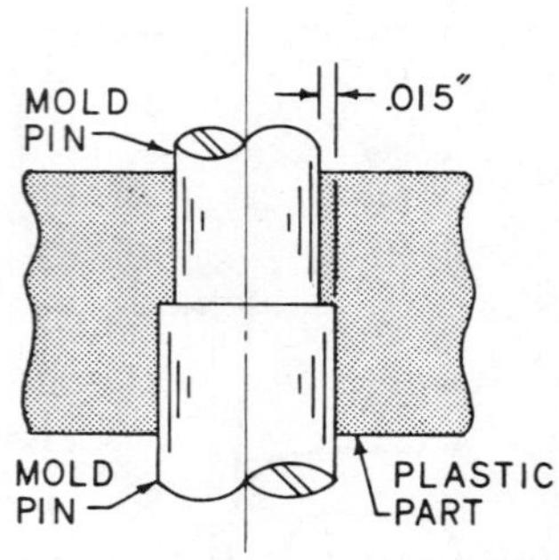

Fig. 21-16. Hole molding using stepped pins concept. (From *Plastic Product Design* by Beck)

possibility of pin misalignment interfering with later assembly with a shaft or a through bolt.

In the usual method of molding holes, the material is injected into the mold and flows around the pin which forms the hole. Thus, the material joins itself somewhere behind the pin to form a weld line, an area of potential weakness. If this proves detrimental to the operation of the part, the part designer and the mold designer should jointly decide how and where the molten polymer is to be injected in order to minimize possible weak areas. These areas of weakness might be in a side wall, the edge, or the area between holes.

One method of eliminating the major portion of the weld line weakness and still mark the hole location, is to use a short pin to mold a very shallow hole. This hole is then used as a drill marker hole for a later drilling operation. Whether the holes are molded in or are drilled in, they should be placed a minimum distance apart or from an edge or a side wall, to minimize weld line weaknesses. Table 21-3 lists recommended distances and Figure 21-17 shows the relationship of holes to each other and to the sidewall.

Holes to be tapped or to be used with self-tapping screws should be countersunk as shown in Figure 21-18. Feather edges should be avoided since they can lead to cracking at the edges of the part or mold.

Table 21-3. Hole Spacing.

Recommended Distances between Holes and Distances of Hole from Side Wall

Hole Diameter Inch	Minimum Distance From Edge, Inch	Minimum Distance From Each Other, Inch
0.062	0.093	0.140
0.093	0.109	0.187
0.125	0.156	0.250
0.187	0.218	0.312
0.250	0.250	0.437
0.312	0.312	0.562
0.375	0.343	0.875
0.500	0.437	0.875

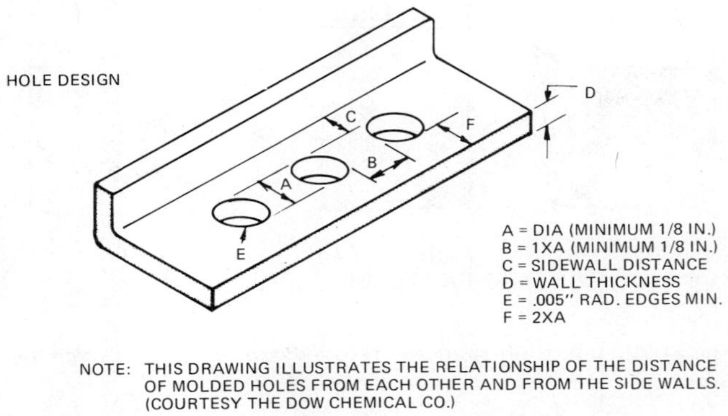

Fig. 21-17. Edge distance for molded holes. (*Courtesy: The Dow Chemical Co.*, reprinted from *Plastic Product Design* by Beck)

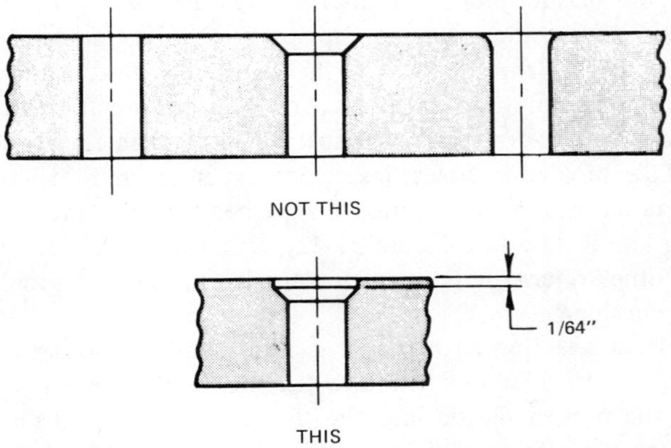

Fig. 21-18. Hole countersinking. (From *Plastic Product Design* by Beck)

Undercuts

Undercuts are indentations or projections on the walls of a plastics part which interfere with simple ejection from a two-part mold. *Undercuts should be avoided whenever possible.* They can unduly complicate the mold design and considerably increase the cost, since special split molds are required, retracting mechanisms may be needed, etc.

Designs with internal undercuts require more expensive tooling than parts with external undercuts. In some cases, the undercut may be postmold machined at less cost than molding it in. This is particularly true of external undercuts. Figure 21-19 shows various types of undercuts. The circular undercut is the type which lends itself well to a postmold machining operation.

Threads

Threads in molded plastics parts are commonly made in the American Standard Class I or II design, either by molding in or by tapping. The American Standard type is not a fine-pitched thread (maximum 32 threads/inch) and should be specified for all general-purpose threads. Molded threads can

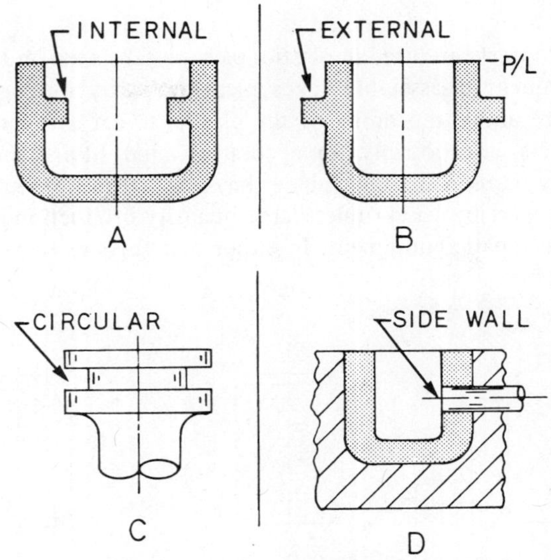

Fig. 21-19. Internal and external undercuts. (From *Plastic Product Design* by Beck)

be held to somewhat better tolerances than tapped threads of the same size, since tapped threads tend to have minute chips and cracks on the surface. When showing internal or external threads in a design, the threads should be shown and dimensioned as illustrated in Figure 21-20. Molded threads, in all cases, should start and stop abruptly. Feather edges should never be specified, since they are areas of weakness in parts or molds.

External threads can be molded using a split mold. However, the possibility exists that excessive flash may be present which will interfere with subsequent assembly. This can be remedied by a chase operation, although this is a costly extra step.

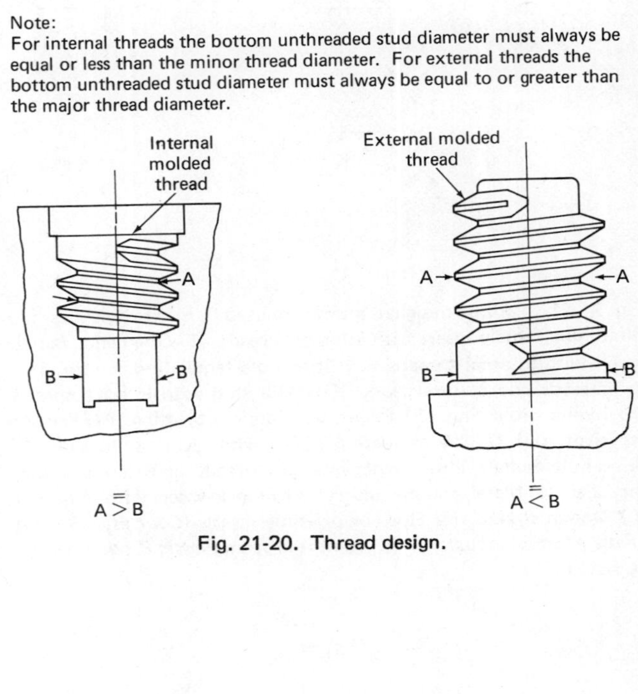

Fig. 21-20. Thread design.

Inserts

Metal inserts are used in plastics parts to take wear and tear if frequent disassembly takes place, to carry mechanical stresses above the limits of the plastic, to act as electrical contacts, and possibly for decorative value. Inserts should always be used sparingly since they are placed in the mold or the part by hand, unless large-quantity production justifies automatic equipment. In either case their use becomes expensive.

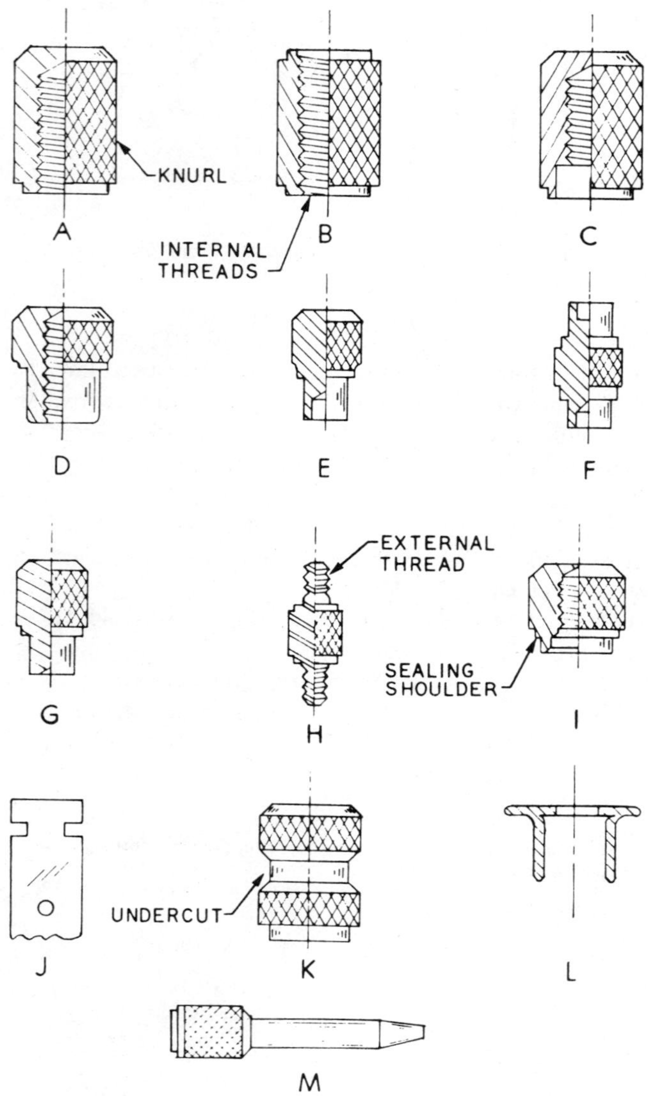

Fig. 21-21. Various designed inserts are used in molded plastics. (A) Blind-hole female insert with internal threads. (B) Open ends female insert with internal threads. (C) Blind-hole female insert with internal threads and counter bore. (D) Male stud with internal thread. (E) Eyelet projecting. (F) Eyelet both ends projecting. (G) Projecting rivet. (H) Double-projecting insert with external threads. (I) Blind-hole female insert with internal threads and double-sealing shoulder. (J) Metal stamped insert. (K) Female insert with undercut. (L) Drawn eyelet. (M) Rod- or pin-type insert. (*Courtesy: Society of the Plastics Industry, Inc.*, reprinted from *Plastic Product Design* by Beck)

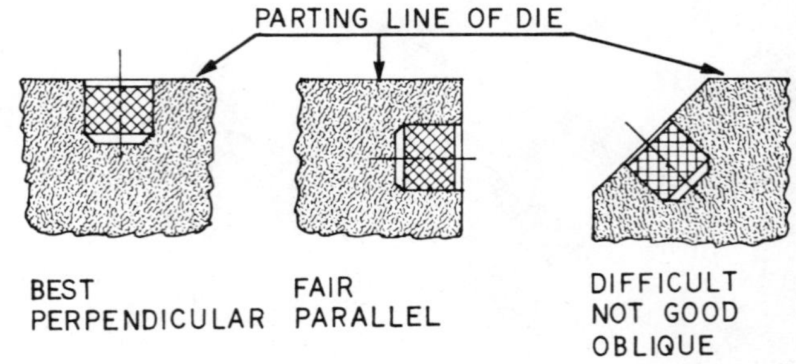

Fig. 21-22. Insert orientations. (From *Plastic Product Design* by Beck)

There are basically two types of inserts: molded-in inserts and postmolded inserts. In both types the insert is locked in by the plastic. Molded-in inserts have a medium or coarse knurl to provide adequate gripping power. Postmolded inserts may be threaded or may be installed by heat or ultrasonically, and may also be coarse knurled. Figure 21-21 shows some of the molded-in insert types available.

Like molded-in holes, inserts are expensive and difficult to mold in, if an orientation not perpendicular to the parting line is used (see Figure 21-22). If an insert is required at other orientations, then postmolded installation should be considered.

When selecting an insert, it is important to choose one with a shoulder that fits into the mold to prevent entry of the molten plastic into the threads. Figure 21-23 illustrates the effects of the insert shoulders to prevent thread fouling.

Like hole placements, the location of an insert in relation to the edge of a part or an adjacent wall is extremely important. In addition to the mechanical stress, which the insert is usually designed to take, a thermal stress is imposed on the plastic because of its higher coefficient of thermal expansion. Thus, when the molded part is cooled, the plastic tends to shrink more than the metal, causing stresses in the plastic which can, at some later date, result in cracks around the insert. The remedy, of course, is to use a sufficient amount of plastic around the insert, either in the form of a boss, or an increased distance from walls, edges, or other inserts. Table 21-4 shows the minimum thickness of plastic which should be molded around an insert to avoid the possibility of cracking. Inserts with sharp corners should also be avoided to minimize cracking (see Figure 21-24).

Tolerances

The tolerances specified for any molded or machined plastics part should be as loose as possible. Extremely tight tolerances call for more expensive molds, special controls during the molding operation, and possibly the use of shrink fixtures during cooling–all of which result in significantly more expensive parts. It is also important to note that dimensional changes in many plastics, due to temperature

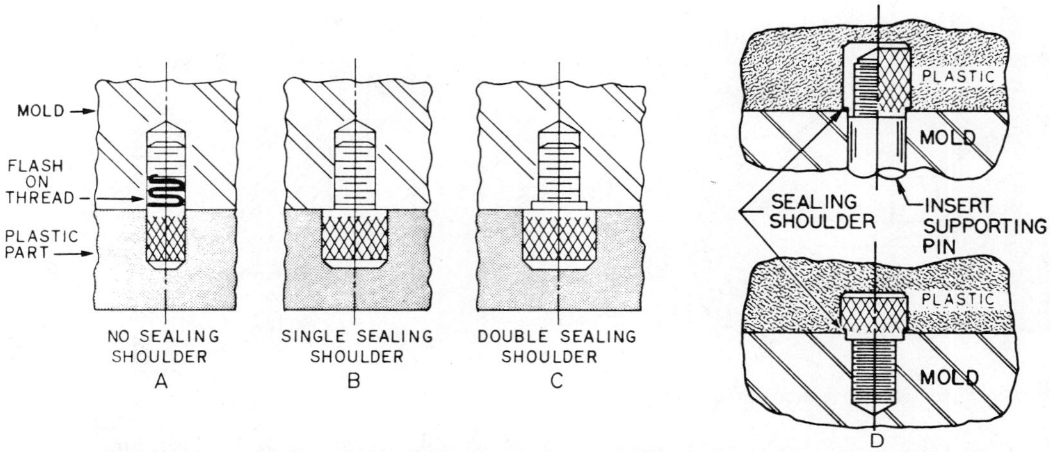

Fig. 21-23. Insert shoulder design. Avoid using an insert that has not been provided with a shoulder as shown in (A). A single-sealing shoulder (B) is better. A double-sealing shoulder is the best (C) but is more expensive. D is a good design.

variations from -55°C to +120°C (-65-+250°F) may be considerably more than the allowable tolerances.

For thermoplastic materials, injection molding generally results in parts with the best dimensional tolerances. (This also applies to compression molding of thermoplastics; however, the method is seldom used except for very limited production.) The dimensional tolerances for thermoplastic parts made by any method, however, will vary from part to part and from material to material.

The tolerances for compression-molded thermoset parts will depend on the type of resin, the filler, the molding conditions, and the type of mold used. These same factors (except perhaps the mold type) will also hold for transfer and/or injection molding. As a rule, parts made of thermosets (e.g., phenolics, epoxies, polyesters, alkyds, etc.) can be designed with tighter tolerances than corresponding thermoplastic parts. Furthermore, thermosets are more dimensionally stable under thermal conditions than are the same parts made of unfilled thermoplastics. Filled thermoplastics, depending on the amount and type of filler, can approach or be equal to the dimensional stability of thermosets.

Normally, the tolerance stated by the resin manufacturer

Table 21-4. Suggested Minimum Plastics Wall Thickness for Inserts of Various Diameters.

DIAMETER OF INSERTS INCHES	.125	.250	.375	.500	.750	1.00
PLASTIC MATERIALS						
ABS	.125	.250	.375	.500	.750	1.00
Acetal	.062	.125	.187	.250	.375	.500
Acrylics	.093	.125	.187	.250	.375	.500
Cellulosics	.125	.250	.375	.500	.750	1.00
Ethylene vinyl acetate	.040	.085	N.R.	N.R.	N.R.	N.R.
F.E.P. (fluorocarbon)	.025	.060	N.R.	N.R.	N.R.	N.R.
Nylon	.125	.250	.375	.500	.750	1.00
Noryl (modified PPO)	.062	.125	.187	.250	.375	.500
Polyallomers	.125	.250	.375	.500	.750	1.00
Polycarbonate	.062	.125	.187	.250	.375	.500
Polyethylene (H. D.)	.125	.250	.375	.500	.750	1.00
Polypropylene	.125	.250	.375	.500	.750	1.00
Polystyrene	Not Recommended					
Polysulfone	Not Recommended					
Surlyn (ionomer)	.062	.093	.125	.187	.250	.312
Phenolic G. P.	.093	.156	.187	.218	.312	.343
Phenolic (medium impact)	.078	.140	.156	.203	.281	.312
Phenolic (high impact)	.062	.125	.140	.187	.250	.281
Urea	.093	.156	.187	.218	.312	.343
Melamine	.125	.187	.218	.312	.343	.375
Epoxy	.020	.030	.040	.050	.060	.070
Alkyd	.125	.187	.187	.312	.343	.375
Diallyl phthalate	.125	.187	.250	.312	.343	.375
Polyester (premix)	.093	.125	.140	.187	.250	.281

*Beck, 1970.

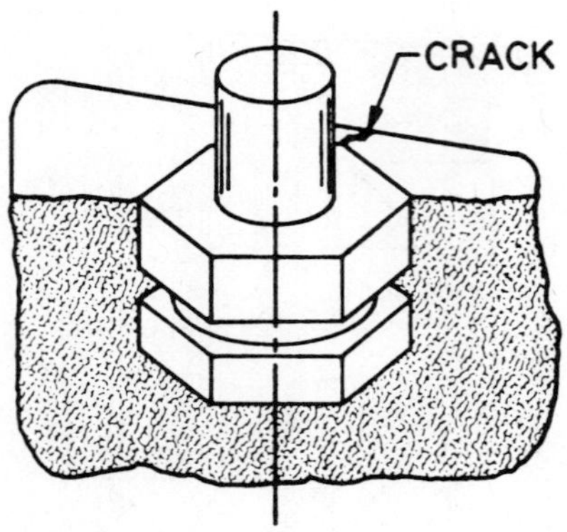

Fig. 21-24. Cracks caused by sharp-cornered inserts.

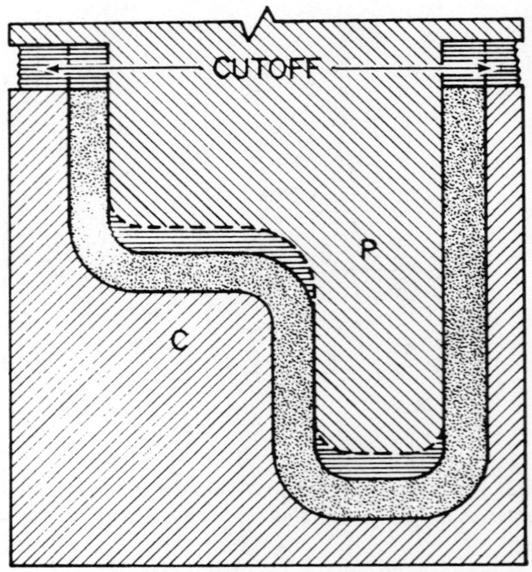

Fig. 21-26. Dimensions affected by flash. (From *Plastic Product Design* by Beck)

for a material is given in inches/inch and applies to measurements made at ambient temperature and 50% relative humidity. If the dimensions are extremely critical, the storage time at a particular temperature and relative humidity, before measuring, should be stated on the drawing. This information is mainly applicable to 6/6 nylon, used for precision gears or similar parts, which can change as much as 8% in volume going from "bone dry" to a 70-80% relative humidity.

Tolerances can vary depending on the location of the parting line and the depth of the flash. When designing a mold, the molder will, as a rule lay it out to give the best overall tolerances. Therefore, *the parting line location should not be shown on the part drawing*. This will give the molder freedom of choice for best producibility. If one dimension is extremely critical, however, this should be noted on the drawing as a guide to the mold designer. Alternatively, if a part is designed with a clearly defined parting plane, then dimensions shall be shown that will not be affected by the flash, which may be appreciable, depending on the mold type, amount of wear, etc. Figure 21-25 shows two methods of dimensioning, one of which is affected by the flash thickness in four dimensions, and the other in only two.

It is also important to note that flash can affect dimensions in one plane and not in another, as shown in Figure 21-26. This figure also illustrates how dimensions controlled by the cavity can be unaffected by thick flash, whereas dimensions controlled by the plunger can be seriously affected.

Typical molding tolerances for many molding compounds, both thermoplastics and thermosets, have been published by the Society of the Plastics Industry. (Table 21-5 shows examples of such data sheets.) In addition, all materials manufacturers publish data on the shrinkage of their polymers. The data published by the Society of the Plastics Industry and most commercial sources relate to injection-molded parts, unless otherwise stated. The most notable exceptions are the thermosets, for which data may be given for compression, transfer, and injection molding.

Shrinkage of parts made by thermoforming, blow molding, rotational molding, etc., (except compression and transfer molding) will show considerably higher tolerance variances than injection molding. These variations will have

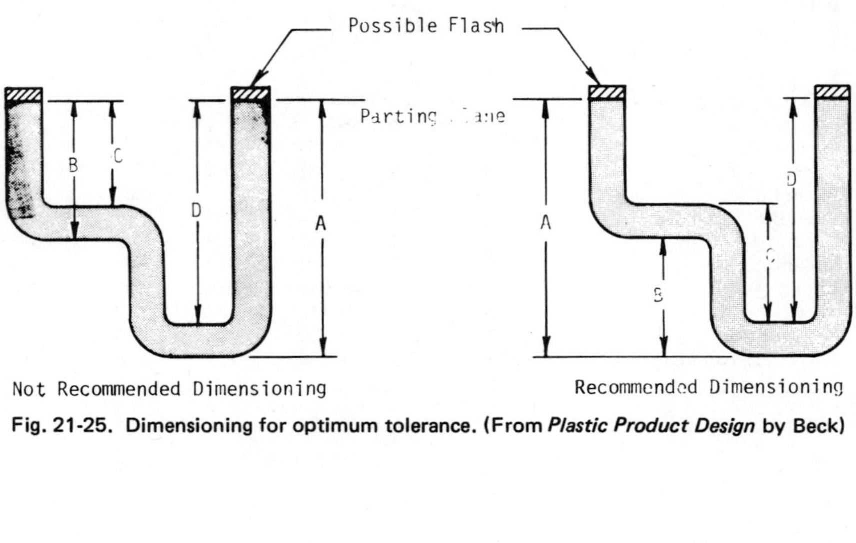

Fig. 21-25. Dimensioning for optimum tolerance. (From *Plastic Product Design* by Beck)

Table 21-5. Typical Tolerance Standards for Various Materials.*

STANDARDS AND PRACTICES OF PLASTICS CUSTOM MOLDERS — Engineering and Technical Standards

ABS

NOTE: The Commercial values shown below represent common production tolerances at the most economical level. The Fine values represent closer tolerances that can be held but at a greater cost.

Drawing Code	Dimensions (Inches)	Plus or Minus in Thousands of an Inch
A = Diameter (see Note #1)	0.000, 0.500, 1.000, 2.000	
B = Depth (see Note #3)	3.000, 4.000	
C = Height (see Note #3)	5.000, 6.000	

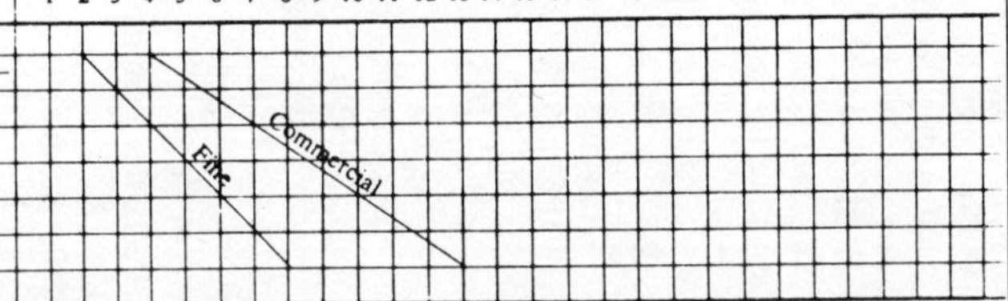

Drawing Code	Dimensions (Inches)	Comm. ±	Fine ±
	6.000 to 12.000 for each additional inch add (inches)	.003	.002
D=Bottom Wall (see Note #3)		.004	.002
E = Side Wall (see Note #4)		.003	.002
F = Hole Size Diameter (see Note #1)	0.000 to 0.125	.002	.001
	0.125 to 0.250	.002	.001
	0.250 to 0.500	.003	.002
	0.500 & Over	.004	.002
G = Hole Size Depth (see Note#5)	0.000 to 0.250	.003	.002
	0.250 to 0.500	.004	.002
	0.500 to 1.000	.005	.003
Draft Allowance per side (see Note #5)		2°	1°
Flatness (see Note #4)	0.000 to 3.000	.015	.010
	3.000 to 6.000	.030	.020
Thread Size (class)	Internal	1	2
	External	1	2
Concentricity (see Note #4)	(T.I.R.)	.009	.005
Fillets, Ribs, Corners (see Note #6)		.025	.015
Surface Finish	(see Note #7)		
Color Stability	(see Note #7)		

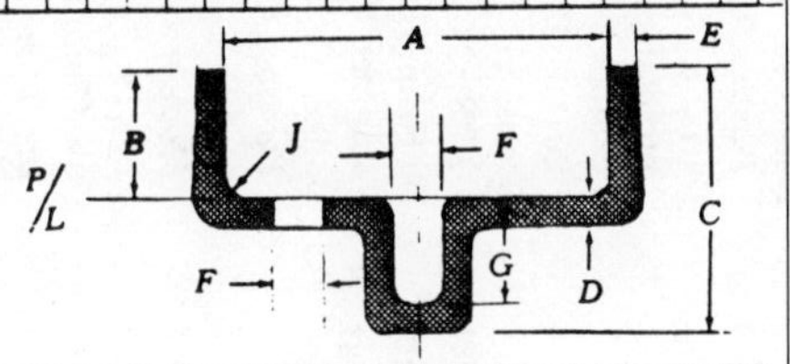

REFERENCE NOTES

1 – These tolerances do not include allowance for aging characteristics of material.
2 – Tolerances based on ⅛″ wall section.
3 – Parting line must be taken into consideration.
4 – Part design should maintain a wall thickness as nearly constant as possible. Complete uniformity in this dimension is impossible to achieve.
5 – Care must be taken that the ratio of the depth of a cored hole to its diameter does not reach a point that will result in excessive pin damage.
6 – These values should be increased whenever compatible with desired design and good molding technique.
7 – Customer-Molder understanding necessary prior to tooling.

STANDARDS AND PRACTICES OF PLASTICS CUSTOM MOLDERS — Engineering and Technical Standards

HIGH DENSITY POLYETHYLENE

NOTE: The Commercial values shown below represent common production tolerances at the most economical level. The Fine values represent closer tolerances that can be held but at a greater cost.

Drawing Code	Dimensions (Inches)	Plus or Minus in Thousands of an Inch
A = Diameter (see Note #1)	0.000, 0.500, 1.000, 2.000	
B = Depth (see Note #3)	3.000, 4.000	
C = Height (see Note #3)	5.000, 6.000	

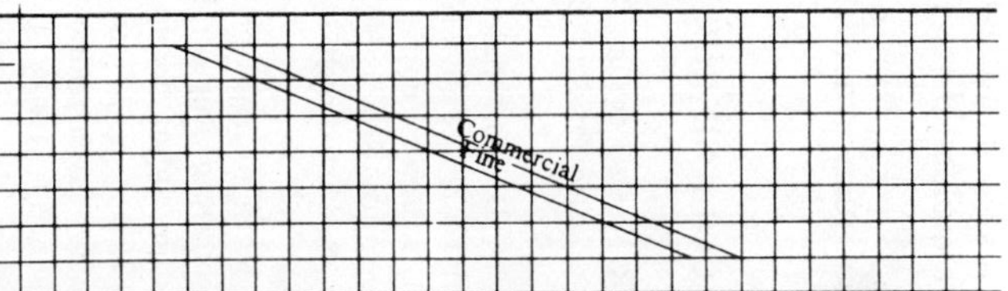

Drawing Code	Dimensions (Inches)	Comm. ±	Fine ±
	6.000 to 12.000 for each additional inch add (inches)	.006	.003
D=Bottom Wall (see Note #3)		.006	.004
E = Side Wall (see Note #4)		.006	.004
F = Hole Size Diameter (see Note #1)	0.000 to 0.125	.003	.002
	0.125 to 0.250	.005	.003
	0.250 to 0.500	.006	.004
	0.500 & Over	.008	.005
G = Hole Size Depth (see Note#5)	0.000 to 0.250	.005	.003
	0.250 to 0.500	.007	.004
	0.500 to 1.000	.009	.006
Draft Allowance per side (see Note #5)		2°	¾°
Flatness (see Note #4)	0.000 to 3.000	.023	.015
	3.000 to 6.000	.037	.022
Thread Size (class)	Internal	1	2
	External	1	2
Concentricity (see Note #4)	(T.I.R.)	.027	.010
Fillets, Ribs, Corners (see Note #6)		.025	.010
Surface Finish	(see Note #7)		
Color Stability	(see Note #7)		

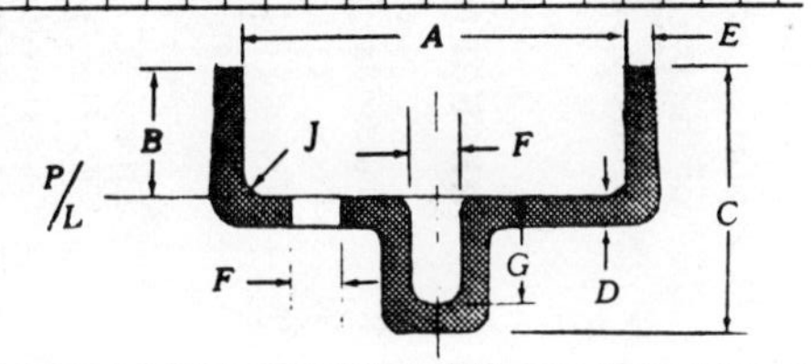

REFERENCE NOTES

1 – These tolerances do not include allowance for aging characteristics of material.
2 – Tolerances based on ⅛″ wall section.
3 – Parting line must be taken into consideration.
4 – Part design should maintain a wall thickness as nearly constant as possible. Complete uniformity in this dimension is impossible to achieve.
5 – Care must be taken that the ratio of the depth of a cored hole to its diameter does not reach a point that will result in excessive pin damage.
6 – These values should be increased whenever compatible with desired design and good molding technique.
7 – Customer-Molder understanding necessary prior to tooling.

*Courtesy: Society of the Plastics Industry, Inc.

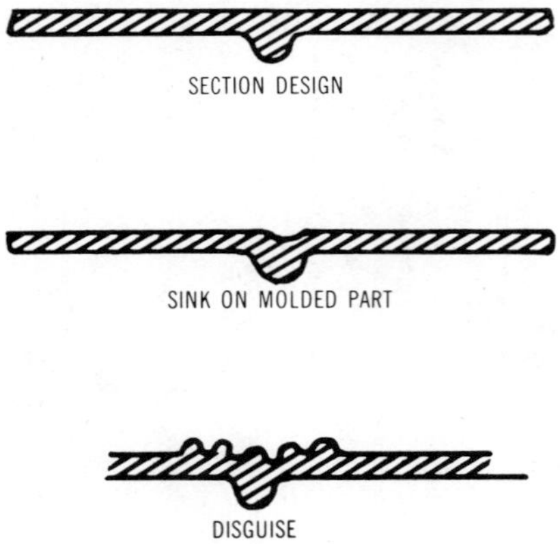

Fig. 21-27. Sink mark disguise. (*Courtesy: E.I. Du Pont de Nemours & Co., Inc.*)

to be determined empirically. In general, the tolerances for glass-reinforced thermoplastic parts can be considerably tighter than for parts made with unreinforced materials. Slight changes in the finished dimensions of a molded part can sometimes be made by the molder, without the cost and trouble of remachining, re-heat treating, and possibly replating the mold. The increase or decrease in the dimensions, or prevention of warpage, can, to some extent, be made by the molder. For example, an increase in molding pressure or a decrease in temperature can sometimes decrease shrinkage. The opposite changes can increase shrinkage. Warpage can be prevented by use of "shrink fixtures." These process changes can sometimes solve the problems caused by small out-of-tolerance dimensions.

Decorating

A large portion of molded plastic parts are sold as consumer items. Appearance, in many cases, may be the decisive factor in the marketing of the part. Thus, any designs that improve part appearance and do not affect the part functioning should be considered.

An example of the use of decorative design to disguise a defect is the incorporation of a corrugated section to include a sink mark, as shown in Figure 21-27.

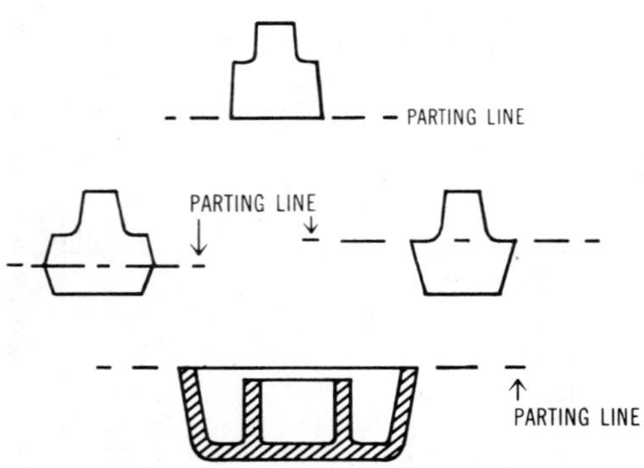

Fig. 21-28. Parting line locations. (*Courtesy: E.I. Du Pont de Nemours & Co., Inc.*)

Another idea is locating a parting line in an inconspicuous portion of a molded part. This location at the top or bottom of the part will also usually prove to be the cheapest method of mold construction as well (see Figure 21-28).

REFERENCES AND BIBLIOGRAPHY

Bainbridge, R. W. Part design. *Durez Molder*, Issue 3, 1973, p. 2.

Beck, R. D. *Plastic Product Design*. Van Nostrand Reinhold Co., New York, 1970.

Benkelman, W. D. To mold or to machine: a new look. *Modern Plastics*, September 1966, **43**(1).

Brown, W. E. Structural design criterial for plastic materials. *Western Plastics*, Technical Report No. 93, February 1963, p. 79.

Buszkiewicz, B. Part design by the numbers. *Plastics Engineering*, June 1979, **35**(6):37.

Carlyon, G. When to make plastic parts by machining. *Materials in Design Engineering*, August 1965, **62**(2):104.

——. Plastics in sliding bearings. *Research/Development Magazine*, February 1962.

——. To mold or to machine, that is the question. *Modern Plastics*, October 1960, **37**(2):90.

Celanese Plastics Co. *Fortiflex Polyethylene Injection Molding*. Bulletin P3A, NB7-22-071050, June 1971.

——. *Engineering Plastics in Design*. Bulletin G1B, 351-107100.

Clary, J. F. Design engineers: the next voice you hear may be the FCC. *Plastics World*, March 22, 1976, **34**(3):30.

Crate, J. F. "Structural Design with Plastics." Paper presentation, *Design News Seminar*, Los Angeles, June 16, 1965.

——. *Spring Rate Chart for Cantilever Springs of Delrin Acetal Resin.* E.I. Du Pont de Nemours & Co., Inc., Form A-59958.

Darvas, R. How to design foam core sandwich laminates. *Modern Plastics*, June 1964, **41**(10):125.

Davis, R. L., and Ashley, H. W. Factors influencing the design of molded plastics parts: Parts I–VIII. *Plastics World*, May 1962–December 1962: **20**(5):44, **20**(6):36, **20**(7):28, **20**(8), 58, **20**(9):48, **20**(10):38, **20**(11):32, **20**(12):40.

Deans, H. Designing for ultrasonics. *Plastics World*, August 1965, **23**(8):24.

De Haan, H. P. Designing the part. *Plaskon Engineering Plastic News*, summer 1974.

Dodge, G. P. Designing for use of post-molding inserts. *Modern Plastics*, April 1962, **39**(8):131.

Donaldson, A. L., and Velleu, R. B. Directionally reinforced plastics. *Modern Plastics*. Part I: October 1957, **34**(2):133. Part II: November 1957, **34**(3):143.

Dow Chemical Co. *Designing for Static and Rapid Mechanical Loading*. Form No. 153-132-63, 1963.

Dow, R. L. Tips for engineering phenolic parts. *Plastics Engineering*, February 1977, **33**(2):45.

Dreger, D. R. Plastics for load bearing parts. *Machine Design*, May 10, 1979, **51**(9):60.

DuBois, J. H. Plastics materials selection. *Plastics World*, October 1965, **23**(10):52.

——. Selecting materials to suit design. *Plastics World*, July 1962, **20**(7):26.

E.I. Du Pont de Nemours & Co., Inc. *Delrin Design Handbook*. Bulletin A-54208, April 1967.

——. *A Guide Book on Designing with Du Pont Engineering Plastics*. Bulletin A-36912, June 1964.

——. *Application Design*. Bulletin A-23013, December 1961.

——. *Teflon Fluorocarbon Resins*. Bulletin A-40844.

——. *Cold Heading with Du Pont Thermoplastics*. Form A-53178.

——. *Zytel Nylon Resins*. Bulletin A-55164.

——. *Stripping of Molded Undercuts*. Form A-68120.

Erley, W. I. Avoiding pitfalls in material selection. *Plastics Engineering*, April 1977, **33**(4):40.

Fenelon, P. J., and Fucella, D. C. Combining reinforcements for more balanced cost/performance. *Plastics Design Forum*, May/June 1979, p. 92.

Filbert, W. C. Sr. Some notes on sinks and warpage in molded parts—what they are, how to eliminate or minimize them. *Plastics Design & Processing*, March 1964, **4**(3):13.

Gabriel, L. What designers should know about mold finishes. *Plastics Design & Processing*, April 1965, **5**(4):10.

General Electric Co. *Noryl Thermoplastics Resins Design*. Bulletin CDX-83.

B.F. Goodrich Chemical Co. *Designing With Elastomers*. Bulletin E-3, 7303-SPM.

Hall, R. C. Pull strength of inserts ultrasonically installed. *SAMPE Journal*, September/October 1976, p. 10.

Hanna, R. D. The molded-in hinge in polypropylene components. *Plastics World*, May 1961, **19**(5):14.

Harley, R. E. Designing reinforced plastics for automotive applications. *Plastics World*, February 1965, **23**(2):22.

Hauser, C. M., Hauser, R. L., and Norling, J. F. Designing strength and stiffness into GRP panels. *Plastics Engineering*, November 1978, **34**(11):37.

Hazenfeld, R. C. Key design parameters for pultruded products. *Plastics Design Forum*, May/June 1977, p. 88.

Henry, D. W. Checklist for designing molded plastics parts. *Machine Design*, August 10, 1978, **50**(15):114.

Johnston, J. E. Designing for compression and transfer moldings. *Materials in Design Engineering*, November 1957, **47**(6):126.

Kent, J. R. Designing for injection molded plastics. *Materials in Design Engineering*, December 1957, **47**(7):106.

Kolb, D. S. Designing plastic parts for ultrasonic assembly. *Machine Design*, March 16, 1967, **39**(5).

Leunig, C. V. Mechanical design with polycarbonate resin. *SPE Journal*, December 1962, **18**(12).

Levy, S. Engineer's report: designing optics in plastics. *Plastics World*, December 1965, **23**(12):44.

——. Designing plastics parts to withstand static loads. *Plastics World.* Part 1: May 1964, **22**(5):18. Part 2: June 1964, **22**(6):32.

——. Designing plastics to withstand varying loads. *Plastics World*, August 1964, **22**(8):32.

——. Designing plastics parts for electrical and electronic uses. *Plastics World*, October 1964, **22**(10):40.

——. Designing for environmental resistance. *Plastics World*, May 1962, **20**(5):22.

Levy, S., and DuBois, J. H. *Plastic Product Design Engineering Handbook*. Van Nostrand Reinhold Co., New York, 1977.

Lund, H. J., and Zimmerman, R. L. ABS plastics. Part 1: Designing with ABS. *Plastics Design & Processing*, April 1968, **8**(4):12.

Martin, L. D. Injection molded plastic gears. *Plastics Design & Processing*. Part 1: Design considerations, August 1968, **8**(8):38. Part 2: **8**(9):33.

Materials Engineering. Pick the right thermoplastic for low temperature. May 1968, **67**(5):52.

McDermott, J. J. Seven ways to improve plastic moldings. *Materials in Design Engineering*, June 1965, **61**(6):100.

McGrew, F. C. Designing polymers for use as plastics. *Modern Plastics*, November 1957, **34**(3):55.

Mengason, J. *New Design Versatility with Integral Hinges*. E.I. Du Pont de Nemours & Co., Inc. Bulletin Form A-57987.

Messenger, D. L. Practical guide to engineering parts in phenolics. *Plastics Design & Processing*, June 1971, **11**(6):14.

——. Design concepts for phenolics. *Plastics World*, February 1965, **23**(2):44.

Miller, R. L. *Pull-Out Force and Strip Torque for Self-Tapping Screws*. E.I. Du Pont de Nemours & Co., Inc. Bulletin Form A-61369.

Misitano, G. Guidelines for effective design of RIM Molds. *Plastics Engineering*, February 1979, **35**(2).

Miska, K. H. *Honeycomb Structures Maintain Strength While Cutting Weight*. *Materials Engineering*, November 1973, **78**(6):36.

Modern Plastics. To mold or machine: a new look. September 1966, **44**(1):118.

Nagy, A. F., and Platt, C. A. Molded-in or post-molding inserts? *Plastics World*, January 1963, **21**(1):16.

Owens-Corning Fiberglas Corporation. Fiberglas/Plastics Design Guide, Publication #5-AU5164-B, 1972.

——. *Designing Matched Die Molded Fiberglas Reinforced Plastic Parts*. Publication #1-PL-2383, 1963.

Persak, K. J. Designing acetal parts. *Plastics Design & Processing*, June 1969, **9**(6):14.

Plastics Design Forum. Designing ribbed parts. March/April 1980, p. 59.

——. Shrinkage: the hidden troublemaker. March/April 1979, p. 61.

——. Shielding against electromagnetic interference. March/April 1979, p. 16.

——. Designing your way out of high temperature problems. November/December 1977, p. 18.

——. Gate marks—and what you can do about them. November/December 1977, p. 37.

——. The limitations of injection molding on product design. January/February 1977, p. 20.

——. What to watch out for in load bearing applications. July/August 1977, p. 75.

——. For large, hollow, stress-free parts, design for rotomolding. May/June 1977, p. 66.

Plastics Design & Processing. Designing with CAB sheet. February 1978, **18**(2):17.

——. Practical guidelines to designing parts in reinforced thermosets. January 1971, **11**(1):16.

Plastics World. SMC properties design study offers computer design information. December 1978, **36**(12):22.

——. Design tips on preventing warpage in molded parts. May 19, 1975, **33**(5):15.

PPG Fiberglass. *Trends in Designing with Reinforced Plastics*. Form F-305R2, October 1964.

Prout, E. C. Designing with the toughest of all plastics—UHMWPE. *Plastics Design Forum*, May/June 1978, p. 26.

Reinforced Plastics. RP design hints. November/December 1964.

Russell, G. P. Watch for pitfalls in designing for cost/performance. *Plastics World*, October 1978, **36**(10):47.

Scott, J. N., Smith, J. V., and Alexander, D. L. Design tips for polyolefin parts. *Modern Plastics*, June 1970, **47**(6):130.

Sheffler, F. W. Estimating for reinforced plastics. *Modern Plastics*, May 1957, **34**(9):135.

Theberge, J. E., Arkles, B., and Carreno, C. A. "Designing with Fiber Reinforced and Internally Lubricated Thermoplastic Elastomers." Paper presentation, *31st Annual Technical Conference*, Reinforced Plastics/Composites Institute, 1976.

Thomas, A. D., and Leunig, C. V. Mechanical design with polycarbonate resin. *SPE Journal*, December 1962, **18**(12):1464.

Thorkildsen, R. L., and Schmitz, J. V. Designing with plastics. *Plastics Technology*, December 1961, **7**(12):26.

Wegman, R. F., Bodner, M. J., and Taglier, K. J. How to design for adhesive bonding. *Materials in Design Engineering*, August 1961, **54**(2):120.

Wheeler, M. E. What to use: glass reinforced thermoplastics or thermosets. *Plastics Engineering*, November 1977, **33**(11):41.

Yates, D. N. Designing with plastics—the need is for lighter vehicles by 1985. *Plastics Engineering*, July 1977, **33**(7):35.

Zecher, R. F. Designing electrical components for transfer molding. *Plastics Design & Processing*, December 1968, **8**(12):22.

22
Plastics Identification

It is not the purpose of this chapter to provide a comprehensive methodology for the identification of every plastics material. As any analytical chemist will aver, proper identification is based on sample purity. The more extraneous ingredients there are in a sample, the more complex and difficult is the analysis. Throughout this book we have stressed the adaptability of base polymers to modification, alteration, and formulation. To say a material is a vinyl is to do no more than broadly categorize the material.

There is an adage in the plastics industry which affirms that given enough time, enough money and enough people, there is no plastics composition that would not survive disclosure. Modern, sophisticated techniques such as atomic absorption, nuclear magnetic resonance, spectral and thermal analysis, X-ray diffraction, electronic microscopy, and many others are all brought to bear in the identification of adhesives, coatings, molded parts, composite materials, etc.

For most purposes, however, identification of a type or class is all that is necessary. It may only be important to know whether a material is thermoplastic or thermosetting, or whether it is sharply or broadly melting. By using a combination of common sense, some knowledge of the plastics end use, and some relatively simple organoleptic techniques, it is possible to sufficiently "narrow the field" to achieve a reasonable identification. Organoleptic analysis emphasizes sensory clues as a method of identification. The sound emitted when struck, part flexibility or rigidity, flame characteristics (odor, color, solvent generation, etc.) when burned (if flammable), solubility, visual appearance, etc., are all mechanisms utilized in the organoleptic approach.

It is important to stress that these methods are qualitative and are meant only to reduce the list of possibilities. The more tests run, the higher the chances are that an accurate identification can be made. Misleading interpretations of the tests are all too common. For example, ABS will give the same characteristic response to burning as styrene, a flammable epoxy may have a flame retardant incorporated into the formula and thus not burn, included fillers may so distort a system's specific gravity that density measurements are useless, and flammable plasticizers may be added to PVC at such a level that the nonflammable nature of the base polymer will not be apparent.

On the other hand, the tests to be described below, in conjunction with the analyst's visual acuity, may be all that are necessary to extract the needed information. For example, if the sample is a thin film, all non-film formers are automatically excluded; if it is transparent, the list is rapidly narrowed to those plastics that are transparent; if the sample floats on water and is not a foam, it is probably a polyolefin, and thus almost 90% of the other plastics groups are eliminated.

With these precautions in mind, the following groups of simple test procedures are presented to provide the reader with a starting point for identification of plastics materials. We will examine flame tests, solution tests, density measurements, and some tests specifically for plastics films.

Flame Tests

One of the best simple identification tests requires burning the unknown material, thereby forcing it to reveal more identifying clues. Holding a flame under the edge of the unknown material will make possible the following observations:

1. Does the material burn?
2. If it burns, what color is the flame?
3. Does the material smoke?
4. If it smokes, what color is the smoke?
5. Is the smoke sooty (unsaturation) or clear?
6. Does the material drip?
7. Does it continue to burn after removal of flame?
8. What odor is given off?

The ease of igniting the material will depend upon the flame source. A gas (Bunsen) burner is ideal; but a cigarette lighter or even a match may be used if nothing better is available (with consideration given for the lower temperatures from the latter sources). Characteristic odors will be given off regardless of the flame used. Table 22-1 is useful for classifying plastics materials by their burning and odor characteristics. Many odors, while distinguishable, are difficult to describe. The accuracy of this test is greatly improved if known materials are available for direct comparison with the unknown material. Fillers, plasticizers, and other additives may present odors themselves and make identification more difficult. Therefore, the results of this test should not be considered conclusive.

Some caution is required both in operation and interpretation of this test: Flames should be extinguished before smelling the material. The odors should be sniffed cautiously. Large quantities of fumes may prove unpleasant

Table 22-1. Plastics Subjected to Flame.

Won't Burn

RETAINS ITS SHAPE
- Formaldehyde odor
  - Urinous odor–urea formaldehyde
  - Fishlike odor–melamine formaldehyde
  - Phenolic odor–phenol formaldehyde
- Surface char forms
  - Mild sulfurous odor, little smoke–polyphenylene sulfide
  - No smoke, glows–thermoplastic polyimide

DEFORMS
- Thin sections curl; chars slightly; weak burned hair odor–polytetrafluoroethylene
- Melts, swells; spurts and crackles; has strong burned hair odor–nylon
- Slight melting; weak acetic odor–polychlorotrifluoroethylene
- Cracks badly; phenolic odor–cast phenolic

Extinguishes Itself

ORANGE FLAME
- Black smoke; chars; drips; stress cracks in CCl_4; phenolic odor–polycarbonate

YELLOW FLAME
- Green mantle; spurts green and yellow; softens; white smoke; acrid odor–polyvinyl chloride
- Smoky; softens; chlorine odor–poly(vinyl chloride co acetate)
- Green mantle; spurts green; heavy black ash; sweet odor–polyvinylidene chloride
- Blue mantle; melts; spurts and crackles; burned hair odor–nylon
- No odor; dense white smoke; residual white ash; even if filled with red iron oxide–silicones
- No smoke; no drip; SO_2 odor (pungent, hot cinders); does not soften in CCl_4; does not soften in trifluoroethane (TFE)–polyether sulfone
- No smoke; no drip; pungent SO_2 odor; stress cracks in CCl_4; softens in trifluoroethane (TFE)–polysulfone
- Peculiar odor–poly(phenylene oxide) types

Burns after Flame Removal

ORANGE-YELLOW FLAME*
- Very much soot; sweet odor; metallic ring when struck–polystyrene
- Some soot; mild to heavy phenolic odor–epoxy
- Black smoke; drips; no char; N_xO odor (4th of July "snakes")–polyurethane
- Soot; sweet odor; no melt–styrene-based unsaturated polyester
- Soot; drips; strong characteristic odor–polybutylene terephthalate
- Soot; mixed floral/sweet odor–ABS

DARK YELLOW FLAME, BLACK SMOKE
- Some soot; spurts, acetic odor–polyvinyl acetate
- Drips; drippings burn; acetic odor–cellulose acetate
- Slight blue mantle; melts; drips; drippings burn; rancid butter odor–cellulose acetate butyrate
- Melts; drips; drippings burn; propionic (foul) odor–cellulose propionate

YELLOW FLAME*
- Blue mantle
  - Melts; drips; rancid butter odor–polyvinyl butyral
  - Melts; drips; burning area swells and is transparent; candle wax odor–polyethylene
  - Spurts; some black smoke; stress cracks in ketones; sweet odor–acrylic
  - Melts; drips; drippings burn; sweet odor–ethyl cellulose
  - Black smoke; melts; breaks up; mild phenolic odor–allyls
  - Burning area is transparent; swells; drips; heavy sweet odor–polypropylene
  - Weak sweet odor–polyvinyl formal
  - Melts; drips; strong formaldehyde odor–acetal
  - Sooty smoke; distinctive pungent odor; somewhat like PE and PP; no neck down when flexed; constant draw when stretched–polybutylene
- No mantle
  - Stiff char; does not melt; low smoke; almond oil odor–nitrile copolymers
  - Sooty smoke; floral odor; no melt–acrylic-based unsaturated polyester
  - Sooty smoke; drips; strong characteristic odor–polyethylene terephthalate
  - Sooty smoke; strong characteristic odor; strings out–thermoplastic polyester copolymer

INTENSE WHITE FLAME
- Camphor odor–plasticized cellulose nitrate
- No camphor odor–cellulose nitrate

*New thermoplastic elastomers and ethylenic copolymers will behave in a flame much like their base polymers: for example, olefinics: see polyethylene, polypropylene, etc. SBRs: see polystyrenes, ABS, etc. olefin/styrenes: see PE, PP, PS, ABS, etc.

and temporarily reduce the sense of smell. Repeated inhalation of small amounts of fumes possibly may be injurious.

The Beilstein test for the presence of chlorine may also be helpful. The tip of a copper wire should be heated to a red-hot temperature in a flame. The hot wire tip should then be drawn across the surface of the unknown material in order to pick up a small quantity of the plastic. The tip of the wire is returned to the flame. A bright green-colored flame indicates the presence of chlorine in the plastic. The copper wire can also be held just above the burning material. If it is heated red hot (by the burning plastic, or by the plastic and the ignition source), the flame will also show a greenish coloration due to formation of copper chloride.

Additional identification tests can be made if some laboratory equipment is available. A test requiring the use of only an oven or electrical hot plate capable of 260°C (500°F) temperatures is useful for classifying the material as thermosetting or thermoplastic. The unknown material should be placed on an unheated hot plate which subsequently is turned on. A thermoplastic material will soften and then melt as its temperature rises; a thermoset material normally will not soften, although a few of the newer thermosetting materials will soften under some conditions.

Solution Tests

The solubility of plastics can provide further clues for identifying an unknown. Ideally, each plastic would be soluble in only one solvent which would be unique to that plastic. Unfortunately, this is not true. Thermosetting plastics are virtually insoluble. And soluble plastics rarely have only one solvent. The solvents listed in Table 22-2 will, in most cases, affect only the plastics listed. Most plastics are attacked slowly by solvents. While the rate of solvent attack can be hastened by the use of small, very thin pieces of plastic, immersion of the material in the solvent for 12 hours or more is often required. Solvents can be dangerous. Some are flammable; contact with the skin or eyes can be injurious. Some solvent fumes can be disagreeable, harmful, or deadly in small amounts. It is recommended that all solvents be handled with care.

Density Measurements

Specific gravity measurements can provide help in identification of unknown, unfilled materials. These measurements may be made easily using the following equipment: an analytical balance of the overhead beam type, a beaker or wide-mouth jar, about 12 inches of fine wire, and methanol (methyl alcohol) or a solution of distilled water and detergent. (One part detergent to 100 parts water is sufficient to reduce the collection of air bubbles on the surface of the plastic material when immersed.) The distilled water is used if the unknown's specific gravity is greater than 1.0; methanol (with 0.7917 specific gravity at 23.0°C, 73.4°F) should be used if the unknown has a specific gravity below 1.0. The test liquids and the unknown sample should be at normal room temperatures while the test is conducted. The closer to 23.0°C (73.4°F), the more accurate the results obtained.

Table 22-2. Solvents for Plastics.

Only These Plastics Are:		In These:
Soluble	Partially Soluble	Solvents
Polyethylene Cellulose nitrate	Cellulose acetate	Dipropylene glycol
Vinyl butyral	–	Sec-undecyl alcohol
Vinyl chloride acetate	Vinyl chloride	Diisobutyl ketone
Vinyl acetate	–	Sec-hexyl alcohol
Vinyl acetate	Cellulose acetate	Cyclohexanol
Polymethyl methacrylate	–	Methyl formate Ethyl formate
Polystyrene	Vinyl acetate Vinyl butyral	Diethyl benzene
Ethylcellulose	–	Sec-amyl alcohol
Cellulose nitrate	–	Diethylene glycol
Cellulose acetate	Cellulose nitrate Cellulose acetate butyrate Ethylcellulose Polystyrene Polymethyl methacrylate	Furfuryl alcohol
Cellulose acetate Cellulose acetate butyrate Cellulose nitrate Vinyl acetate	–	Acetonitrile Acetone MEK

*Johnson and Sauber, 1966.
Note: This chart is useful for differentiation only between the plastics listed.

To determine the material's specific gravity, the following procedure can be used:

1. Cut a piece of the unknown material about $\frac{1}{2}'' \times \frac{1}{2}'' \times \frac{1}{2}''$. The sides must be smooth; any roughness might cause air bubbles to cling to the material.
2. Attach one end of the wire to the hook on one of the scale's pan supports. The other end of the wire must hang freely, yet must be long enough to support the unknown material while immersed in the liquid-filled beaker. Cut the wire to length as required.
3. Weigh the wire; note this weight as A.
4. Tie the plastic to the free end of the wire and weigh; this weight is noted as B.
5. $B - A = C$, the weight of plastic in air.
6. Fill the beaker (or jar) with either water or methanol. (Water is the odds-on choice; most plastics have a specific gravity greater than 1.0.)
7. Position the beaker under the suspended plastic. Raise the breaker so that the plastic is completely submerged in the liquid. The beaker should be supported without touching any moving part of the scale. Neither the plastic nor the supporting wire must touch any part of the beaker. Mark the point at which the wire enters the liquid.
8. Weigh the plastic piece to the nearest 0.001 gram while it is submerged and note this weight as D.
9. Remove the plastic from the liquid and untie it from the wire.
10. Replace the wire in the liquid to the same depth it was previously. Again, the wire must not touch the beaker's sides or bottom.
11. Weigh the wire alone in the liquid. Note this weight as E. $A - E = F$, the loss of weight of the wire in liquid.

The material's specific gravity is computed from one of the following formulas, depending on the liquid used:

$$\text{For water: Specific gravity} = \frac{(1.0)C}{C - D - F}$$

$$\text{For methanol: Specific gravity} = \frac{(0.7917)C}{C - D - F}$$

Table 22-3. Specific Gravity of Thermoplastics.

Specific Gravity Range	Materials in This Range	Material's Specific Gravity
0.80–0.89	Poly (methyl pentene)	0.83–0.84
	Polypropylene	0.85–0.90
	Thermoplastic rubber (olefinic types)	0.88–0.90
0.90–0.99	Thermoplastic rubber (olefinic-styrenic types)	0.90–1.20
	Polybutylene	0.908–0.917
	Low-density polyethylene	0.91–0.93
	High-density polyethylene	0.93–0.96
	Ionomer	0.93–0.96
	Thermoplastic rubber (styrene-butadiene types)	0.93–1.10
1.00–1.09	Chlorinated polyethylene	1.00–2.40
	Nylon 11 and 12	1.01–1.05
	ABS	1.01–1.21
	Thermoplastic polyurethane	1.05–1.25
	Polystyrene	1.05–1.08
	Nylon 6, 6/6, 6/10	1.06–1.17
	Poly (phenylene oxides)	1.06–1.10
	Poly (vinyl butyral)	1.07–1.20
1.10–1.19	Ethyl cellulose	1.10–1.16
	Aromatic nylon	1.12
	Poly (aryl ether)	1.14
	Poly (acrylonitrile)	1.15
	Poly (vinyl chloride)	1.15–1.65
	Cellulose acetate butyrate	1.17–1.24
	Acrylic	1.18–1.19
	Poly (vinyl acetate)	1.19
1.20–1.29	Poly (vinyl formal)	1.20
	Polycarbonate	1.20
	Cellulose acetate propionate	1.20–1.24
	Poly (phenylene oxides)–filled	1.21–1.36
	Nylon 11 and 12–glass filled	1.23–1.26
	ABS–glass filled	1.23–1.36
	Polysulfone	1.24
	Cellulose acetate	1.27–1.34
1.30–1.39	Allyl diglycol carbonate	1.30–1.40
	Aromatic nylon–filled	1.30–1.37
	Thermoplastic polyester	1.31–1.39
	Cellulose nitrate	1.35–1.40
	Nylon 6, 6/6, 6/10–filled	1.35–1.50
	Polyethersulfone	1.37
1.40–1.49	Acetal	1.41–1.425
	Poly (amide-imide)	1.41–1.45
	Thermoplastic polyester	1.41–1.73
	Thermoplastic polyimide	1.43
	Polysulfone–filled	1.46
1.50–1.59	Saran	1.58–1.75
1.60–1.69	Poly (ether sulfone)–filled	1.68
	PECTFE	1.68–1.69
1.70–1.79	PETFE	1.7–1.8
	Poly (vinylidene fluoride)	1.75–1.78
1.80+	Poly (phenylene sulfide)	1.8
	PTFE	2.10–2.30
	CTFE	2.10–2.30
	PFA	2.12–2.17
	FEP	2.12–2.17

Table 22-4. Specific Gravity of Thermosets.

Specific Gravity Range	Materials in This Range	Material's Specific Gravity
0.90–0.99	Silicone	0.99–1.50
1.00–1.29	Epoxy	1.00–2.40
	Polyurethane	1.05–1.50
	Polyester	1.15–1.46
1.30–1.39	Allylics	1.34–1.85
1.40–1.49	Polyimides	1.42–1.55
	Polyimides–filled	1.42–1.90
	Ureas	1.47–1.52
	Melamines	1.47–1.78
1.50–1.59	Polyester–filled	1.50–2.20
	Phenolics–filled	1.55–1.90
1.60+	Furan	1.75
	Silicone–filled	1.80–1.90
	Alkyd–filled	1.80–2.24

Tables 22-3 and 22-4 show the specific gravity of most plastics. It can be seen that the gravities of many of the materials overlap, and some have broad specific gravity ranges. Thus, identification of an unknown plastics material by means of a specific gravity test alone is difficult. However, once the specific gravity of an unknown plastic is determined, the chart is useful for reducing the list of possibilities.

Identifying Plastics Films

Plastics films may be divided into natural and synthetic materials. Cellophane and cellulosic films constitute the types of naturally derived films used mainly by the packaging industry. Synthetic films constitute a wide range of polymeric-based materials.

Cellophane will burn the same as paper. It will continue to burn after it is withdrawn from a flame. It is also insoluble in the commonly used organic solvents. Acetone will dissolve a nitrocellulose or saran coating; however, the base film is insoluble in acetone. The infrared spectrum of cellophane contains strong absorption bands at 2.9 and between 9-10.

Cellulose acetate is a strong and clear film which is easy to tear. It does not become opaque upon flexing and burns with an acetic acid odor. A commonly used spot test consists of adding 1 ml water and 4 ml anthrone solution in sulfuric acid to the film. A green to blue color indicates the presence of cellulose acetate. Absorption peaks are located at 5.6 and 8.0 on the infrared spectrum.

Cellulose propionate is essentially similar to its acetate and butyrate relatives. The one important physical test is infrared spectroscopy. The propionate group has its absorption peak at 8.0.

Polyethylene has a waxy feel, stretches easily, and floats in water. The odor, when burning, is paraffinic. Polyethylene is readily soluble in hot toluene and benzene. Characteristic absorption occurs at 3.4, 6.8, 7.4, and 14.

Polypropylene film is tougher and has better heat resistance than polyethylene and floats in water. It melts and drips while burning. A distinct increase in clarity and rigidity characterizes polypropylene as compared to polyethylene. Infrared examination shows absorption bands at 3.4, 7, 8, and 12. Flexing a thick film specimen will show a white opaque neck down that demonstrates the molecular orientation not present in other olefinic homopolymers.

Polyvinyl alcohol film is a translucent material which burns with a sweet nonacrid odor. It forms a gel with cold water, dissolves in hot water fairly readily, and is insoluble in chloroform and ethanol. Infrared spectra reveals the clear presence of hydroxyl groups at 3.0 and at 9.1. An additional peak occurs at 6.1.

Polyvinyl chloride may be used in either the rigid or flexible state. Rigid PVC is a hard material which becomes white at the bend upon flexing. It is soluble in dioxane, tetrahydrofuran, and cyclohexanone. A broad band exists at 14.5 on the infrared spectrum.

Polyvinylidene chloride films (saran) are clear and transparent. The material shrinks violently upon heating and is not soluble in dioxane. Absorption occurs at 9.3 and 9.6.

Polystyrene is a transparent sparkling brittle film which burns with a yellow sooty flame. A simple spot test for polystyrene consists of the following: Mix a few drops of the sample in a test tube with a few milliliters of carbon tetrachloride. A yellow precipitate will form that will dissolve by adding more of a standard bromine solution. After cooling in ice, an additional yellow precipitate indicates the presence of styrene. Polystyrene absorbs at 7.0 and 9.5 in the infrared range.

Polyester films are very strong and transparent. They burn slowly with a mildly sweet odor and melt into pearls. They are insoluble in all organic solvents. Infrared studies reveal absorption at 8.0 and 9.0.

Polyamide (nylon) films may be found in several different varieties. Nylon 6 may be identified by a color test employing *o*-nitrobenzaldehyde. The film is heated in a test tube. The vapors that arise contact a strip of filter paper freshly moistened with *o*-nitrobenzaldehyde in 2N sodium hydroxide. A deep mauve-black color identifies nylon 6. Nylon 11 film dissolves in concentrated formic acid and melts into pearls upon heating. Absorption occurs at 3.0 and 6.0.

Polycarbonate film is a smooth, sparkling nonstretchable material. It burns, moderately, producing a phenolic odor. The film is soluble in acetone and insoluble in heptane. Bands occur at 5.6 and 8.0 on the infrared spectrum. Table 22-5 summarizes the methods of analysis for various flexible films.

Spectroscopic Analysis

Spectroscopic analysis is required for precise qualitative and quantitative identification of the unknown's contents.

Table 22-5. Guide to Identification of Flexible Plastics Films.

Film	Burning	Odor	Density	Solubility	Chemical Test	Absorption Bands, Infrared Method	Comments
Cellophane Uncoated	Like paper	Like paper	1.45	Insolvent in organic solvents	Molisch test	2.9μ; 10.μ	Chemical test is probably more valuable than infrared techniques
Saran-coated	Like paper	Like paper	1.44	Coating soluble in acetone	Pyridine	2.9μ; 10μ	
Nitrocellulose-coated	Like paper	Like paper	1.48	Coating soluble in acetone	Diphenylamine	2.9μ; 10μ	
Cellulose acetate	Slow burning, beads at edge of burnt area	Acetic odor	1.25–1.35	Soluble in acetone	Molisch test	8.1μ; 5.78μ–5.83μ	Only soluble in acetone as an organic solvent
Cellulose acetate butyrate	Slow burning, beads at edge of burnt area	Rancid butter	1.25–1.35	Soluble in acetone	Molisch test	8.6μ; 5.8μ; 8.1μ	Best method to identify is by infrared

Cellulose propionate	Slow burning, beads at edge of burnt area	Rancid butter	1.25–1.35	Soluble in acetone	Molisch test	8.5μ; 12.4μ	Infrared will identify the three different cellulose esters fairly well
PVDC	Green edge to film, moderate rate	Rubbery	1.4	Soluble in hot toluene	Copper wire–green flame	3.4μ; 6.9μ; 7.3μ; 8.3μ	Classical copper wire flame test will yield green color due to presence of chlorine
Polyethylene							
Low density	Rapid, drips	Candle wax	0.914–0.921	Soluble in hot toluene	Inert	3.4μ; 6.8μ; 7.4μ	Easy to identify by feel and density, and cloudy appearance
Medium density	Like wax	Candle wax	0.921–0.935	Soluble in hot toluene	Inert	14.0μ; 6.8μ; 7.4μ	
High density	Like wax	Candle wax	0.935–0.960	Soluble in hot toluene	Inert	14.0μ; 6.8μ; 7.4μ	
Polypropylene							
Cast	Rapid, waxy	Wax	0.80–0.91	Soluble in hot toluene	Inert	3.4μ; 7.0μ; 8.0μ; 12.0μ	Clear and more rigid than polyethylene
Oriented	Rapid, waxy	Wax	0.80–0.91	Soluble in hot toluene	Inert	3.4μ; 7.0μ; 8.0μ; 12.0μ	Tensile strength of value in determining orientation of film
Polyvinyl chloride	Slowly, self-extinguishing	Pungent, irritating	1.2–1.4	Soluble in acetone	Morpholine	14.5μ	Must be cautious in detecting PVC as opposed to PVDC
Polyvinylidene choride	Green flame, slow	Pungent, resin like	1.68	Soluble in hot toluene and ether	Pyridine test	9.35μ; 9.55μ	Film soft to feel
Polystyrene	Moderately, melts into clear liquid	Flowers	1.1	Insoluble in most organic solvents	Bromine test	3.5μ; 6.9μ; 13.5μ	Very crisp and metallic-like sound
Polyester	Slowly upon removal from flame	Sweet	1.3	Insoluble in organic solvents	*o*-nitrobenzaldehyde	8.0μ; 9.0μ	Very clear and strong film in all gauges
Polyamide	Slow burning	Burning hair	1.09–1.4	Soluble in phenol	*o*-nitrobenzaldehyde	3.05μ; 3.45μ; 6.1μ; & 6.85μ	Various types may be detected by paper chromatography
Polycarbonate	Generally fast	–	–	Insoluble in most solvents	Iodophenol, ferric chloride for phenol content	Phenolic content to be determined	New film and in development
Polytetrafluorocarbon	Inert	–	2.2–2.3	Sensitive to certain organic solvents	Fluorine detection	8.0μ–11.0μ	Difficult to detect qualitatively

Several spectroscopic methods currently are used for analysis of materials, each having certain advantages.

Infrared spectroscopy is generally the surest and most rapid method of identification of organic compounds in the solid or liquid state. Emission spectroscopy rapidly determines the individual elements comprising organic and inorganic materials in a liquid or solid. X-ray diffraction is useful for identification of solid, crystalline compounds, both organic and inorganic; while mass spectroscopy provides analysis of all low-boiling materials—usually in gaseous or liquid forms.

The reader who wishes a more detailed study of analytic techniques in the plastics industry is urged to refer to the references as well as standard texts and reference books on analytical methods.

REFERENCES AND BIBLIOGRAPHY

Cain, D. S., and Stimler, S. S., "Infrared Spectra of Plastics and Resins, Part 3—Related Polymeric Materials (Elastomers)." Naval Research Laboratory Report 6503, February 1967.

E.I. Du Pont de Nemours & Co., Inc. *A Guide to the Identification of Thermoplastic Materials*. Bulletin A-22845, December 1961.

Feigl, F., and Anger, V. Spot tests for plastics. *Modern Plastics*, May 1960, **37**(5):151, 191–196.

Hummel, D. O., and Scholl, F. *Infrared Analysis of Polymers, Resins, and Additives, An Atlas,* Wiley-Interscience, New York, 1969.

Johnson, R. C., and Sauber, W. F. How to identify plastics. *Western Plastics*, March 1966.

Nyquist, R. A. *Infrared Spectra of Polymers and Resins*. 2nd Ed. Dow Chemical Co., Midland, Michigan, May 1961.

Sacharow, S. Analyzing packaging materials. *Adhesives Age*. Part I: January 1970, **13**(1):8. Part II: February 1970, **13**(2):8.

23
Materials and Processes Selection

This chapter is intended to be a summation of all the information given in the book. It is for the use of the designer, engineer, cost estimator, student, or anyone else who is responsible for the selection of the optimum material, process, and/or tooling for any given part. It is *not* intended that the information given in this chapter will be used as a substitute for the more detailed descriptions given elsewhere, but rather that the concise listings will act as a comparative guide to help in the decision making.

The selection of the optimum material for any given part should be based on an analysis of the requirements for the part. This analysis should, at the minimum, include the following:

1. The physical properties required: high or low tensile, impact, or compressive strength, etc.
2. The thermal requirements: strength at an elevated temperature, flexibility at a low temperature, etc.
3. The chemical and/or environmental requirements.
4. Electrical requirements: low loss materals for high-frequency applications, high dielectric strength, arc resistance, etc.
5. Miscellaneous properties: transparency, colors, flexibility, low costs, low density, etc.

The Materials Properties Charts included in this chapter are given whenever possible in the form of bar graphs, since this form of presentation shows at a glance what the maximum and minimum values for any given set of data are. Thus, perusal of many columns of data is eliminated. For those data not readily shown by bar graphs (e.g., lists of transparencies), the materials are shown in an order of apparent superiorities—*to the writer.* Others may have constructed the order of ranking in another way, because of a different emphasis on a certain characteristic. Such non-quantitative lists should then be examined very critically from the user's point of view.

The Processes Selection Guides cannot be presented in the form of bar graphs, since the process used to make any part is dependent on the type of part, the material used, and the quantity required. Thus, the same part may be fabricated by two different processes, depending on the quantity required. Another criterion for process selection is, in

Table 23-1. Material Selection Guide.

MECHANICAL PROPERTIES

- **HIGH TENSILE STRENGTH $>5\times10^3$ PSI**
  - **GLASS-REINFORCED $>15\times10^3$ PSI**
    THERMOSET POLYESTER
    ALKYD
    VINYL ESTER
    NYLON (INCLUDING AROMATICS)
    THERMOPLASTIC POLYESTERS (INCLUDING AROMATICS)
    POLYCARBONATES / ALLOYS
    POLYSULFONES
    POLYSTYRENE / COPOLYMERS (EXCEPT ABS)
    EPOXY
    POLYPHENYLENE SULFIDE
    ABS
    POLYACETAL
    PHENOLIC
    PPO BASED
    POLYIMIDES
    MELAMINE
  - **STANDARD AND NON-GLASS FILLED $>10\times10^3$ PSI**
    THERMOSET POLYESTER
    ALKYD
    VINYL ESTER
    NYLON (INCLUDING AROMATICS)
    POLYIMIDES
    POLYAMIDE–IMIDES
    THERMOPLASTIC POLYESTERS (INCLUDING AROMATICS)
    POLYSULFONES
    POLYSTYRENE / COPOLYMERS (EXCEPT ABS)
    EPOXY
    POLYPHENYLENE SULFIDE
    PHENOLIC
    PPO BASED
    UREAS
    PVC & COPOLYMERS
- **LOW TENSILE STRENGTH $<5\times10^3$ PSI**
  THERMOPLASTIC ELASTOMERS
  POLYMETHYLPENTENE
  POLYBUTYLENE
  FURAN
  SILICONE
  POLYETHYLENE & COPOLYMERS
  CELLULOSICS
  POLYURETHANE
  PVC & COPOLYMERS
  FLUOROPLASTIC / COPOLYMERS
  POLYPROPYLENE
  PHENOLIC
  ABS
  EPOXY
  POLYSTYRENE
  THERMOPLASTIC POLYESTERS
  POLYAMIDE-IMIDES
  POLYIMIDES
  ALKYD
  VINYL ESTER
- **IMPACT STRENGTH <8FT-LBS/IN. OF NOTCH**
  POLYBUTYLENES
  THERMOPLASTIC ELASTOMERS
  FLUOROPLASTICS
  NYLONS
  CELLULOSICS
  POLYTHYLENE & COPOLYMERS
  POLYURETHANE
  PVC & COPOLYMERS
  EPOXY
  THERMOSETTING POLYESTERS
  ALKYDS
  VINYL ESTERS
  POLYPROPYLENE
  POLYCARBONATE
  MELAMINES
  PHENOLIC
  POLYIMIDE
  ALLYLS
  ACRYLIC
  ABS
  PPO BASED
  POLYARYL ETHOR
  GLASS-REINFORCED SILICONE
  POLYSTYRENE COPOLYMERS

Table 23-2. Tensile Strengths of Some Plastics Materials.

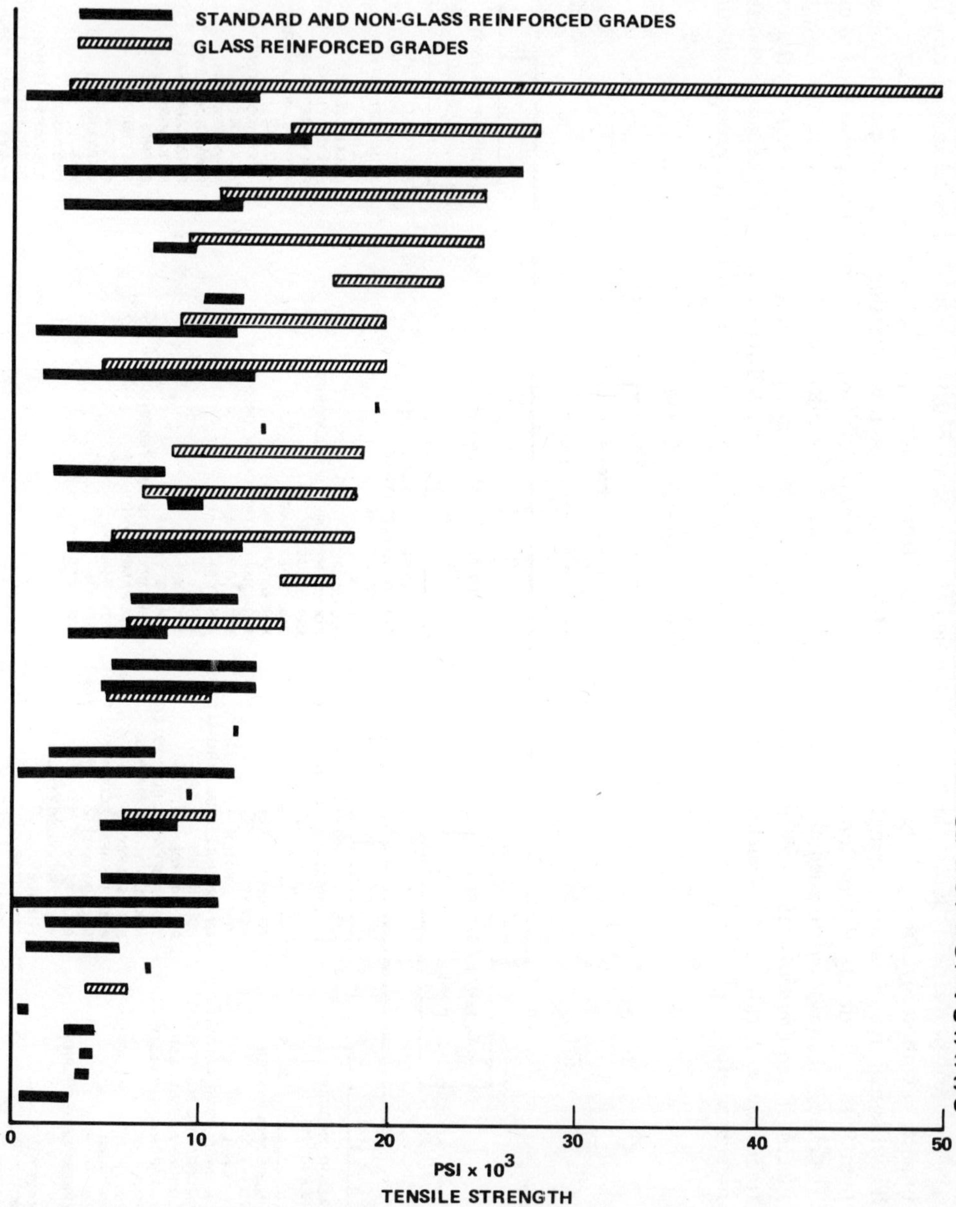

3–50 0.5–13	THERMOSET POLYESTER, ALKYD, VINYL ESTER
1.3–28 7.4–14	NYLONS INCLUDING AROMATICS
2.6–26	POLYIMIDES AND POLY (AM IDE-IMIDES)
10.8–25 2.5–12	THERMO PLASTIC POLYESTERS INCLUDING AROMATICS
9.6–25 7.5–9.5	POLYCARBONATE AND ALLOYS
17–23	GLASS REINFORCED POLYSUFONES
10.2–12.2	POLYSULFONES
9–20 1.5–12.0	POLYSTYRENE AND COPOLYMERS (EXCEPT ABS)
5–20 2–13	EPOXY
19.5	GLASS REINFORCED POLYPHENYLENE SULFIDE
13.4	POLYPHENYLENE SULFIDE
8.5–19	GLASS REINFORCED ABS
2.5–8	ABS
6.9–18.5 8.5–10	POLYACETAL
5.0–18 3–12	PHENOLIC
14.5–17	GLASS REINFORCED PPO BASED
6.5–12	PPO BASED
6.0–14.57 2.8–8.25	POLYPROPYLENE
5.5–13.6	UREA–FORMALDEHYDE
5.0–10.5 5.0–13.0	MELAMINE–FORMALDEHYDE
12	GLASS REINFORCED FLUOROPLASTICS
2.0 - 7.4	FLUOROPLASTICS AND COPOLYMERS
0.5–12	PVC AND COPOLYMERS
9.5	GLASS REINFORCED PVC
6.0–11 5.0–8.7	ALLYLS
5.0–11	ACRYLIC
0.2–11	POLYURETHANE
1.9–9	CELLULOSICS
9.0	GLASS REINFORCED POLYETHYLENE
0.6–5.5	POLYETHYLENE AND COPOLYMERS
7.5	POLYARYL ETHER
4–6.5	GLASS REINFORCED SILICONE
0.35–1.0	SILICONE
3.0–4.5	FURAN
3.8–4.4	POLYBUTYLENE
3.5–4.0	POLYMETHYLPENTENE
0.65–3.0	THERMOPLASTIC RUBBER

Table 23-3. Compressive Strengths of Some Plastics Materials.

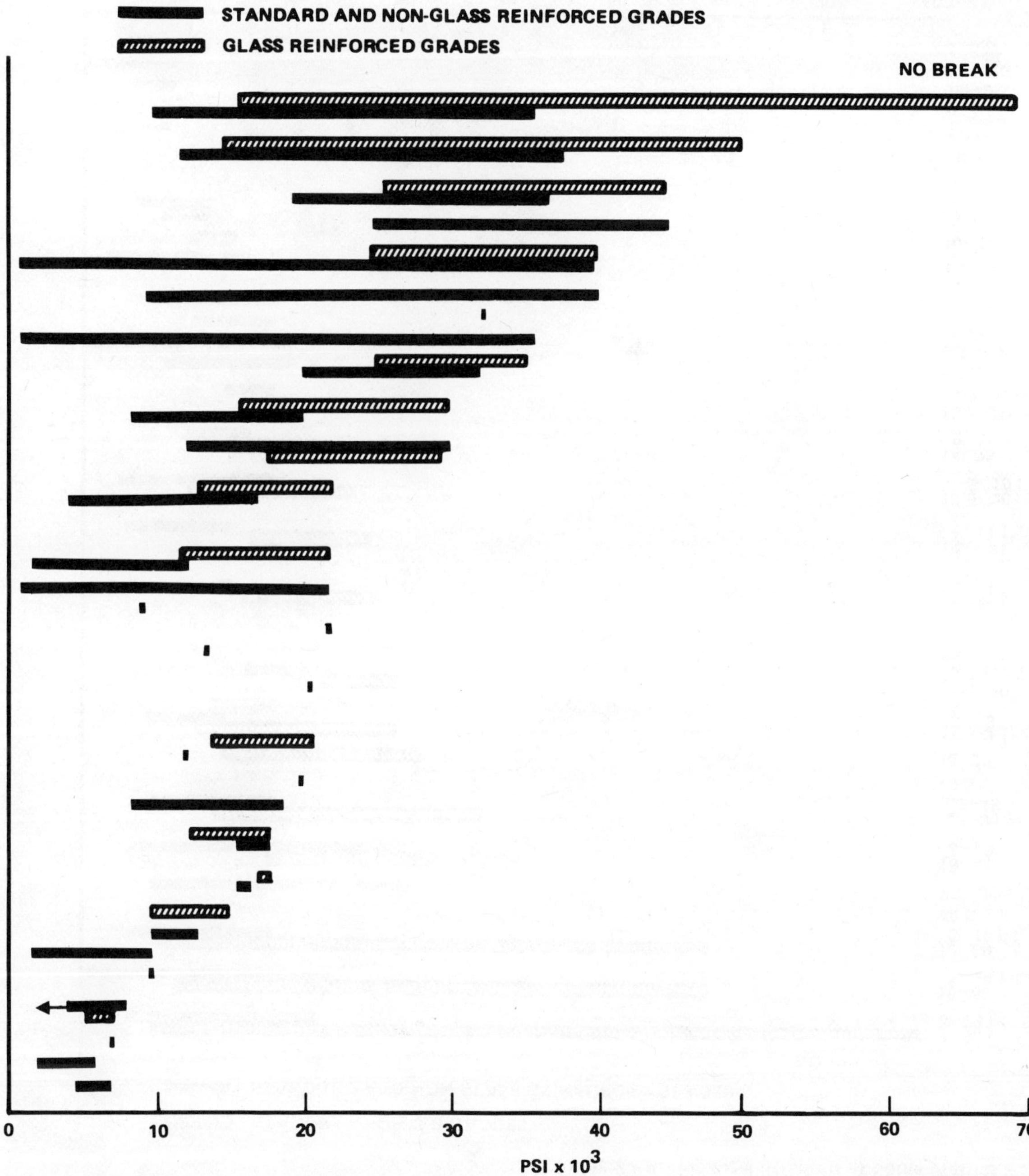

	THERMOPLASTIC RUBBER
16–70 10–36	PHENOLIC
15–50 12–38	THERMOSET POLYESTER ALKYD, VINYL ESTER
20–35 26–45	MELAMINE-FORMALDEHYDE
25–45	UREA-FORMALDEHYDE
25–40 1–40	EPOXY
9.7–40	POLYIMIDE & POLYAMIDE-IMIDE
32.5	GLASS RFD POLYIMIDE
1.1–36	CELLULOSICS
25–35 20–32	ALLYLS
16.2–30 8.6–20	THERMOPLASTIC POLYESTER
12.5–30 18–29.4	NYLONS INCLUDING AROMATICS
13–22 4–17	POLYSTYRENE & COPOLYMERS (EXCEPT ABS)
12–22 1.8–12.5	ABS
0.9–22	PVC & COPOLYMERS
9	GLASS RFD PVC
22.0	SULFONES (YIELD)
13.9	GLASS RFD SULFONES (YIELD)
21.0	GLASS RFD POLY-PHENYLENE SULFIDE
14–21	GLASS RFD POLYCARBONATE
12.5	POLYCARBONATE
20.0	POLYURETHANE
8.4–19	ACRYLIC
13–18 16–18	POLYACETAL (10% DEF)
17.6–17.9 16–16.4	PPO BASED
10–15	GLASS RFD SILICONE
10–13	FURAN
1.7–10	FLUOROPLASTICS & COPOLYMERS
10	GLASS RFD FLUOROPLASTICS
<3.7–8 5.5–7	POLYPROPYLENE
7	GLASS RFD POLYETHYLENE & COPOLYMERS
2.0–5.5	POLYETHYLENE & COPOLYMERS
5–6.6	POLYMETHYLPENTENE

Table 23-4. Flexural Strength of Some Plastics Materials.

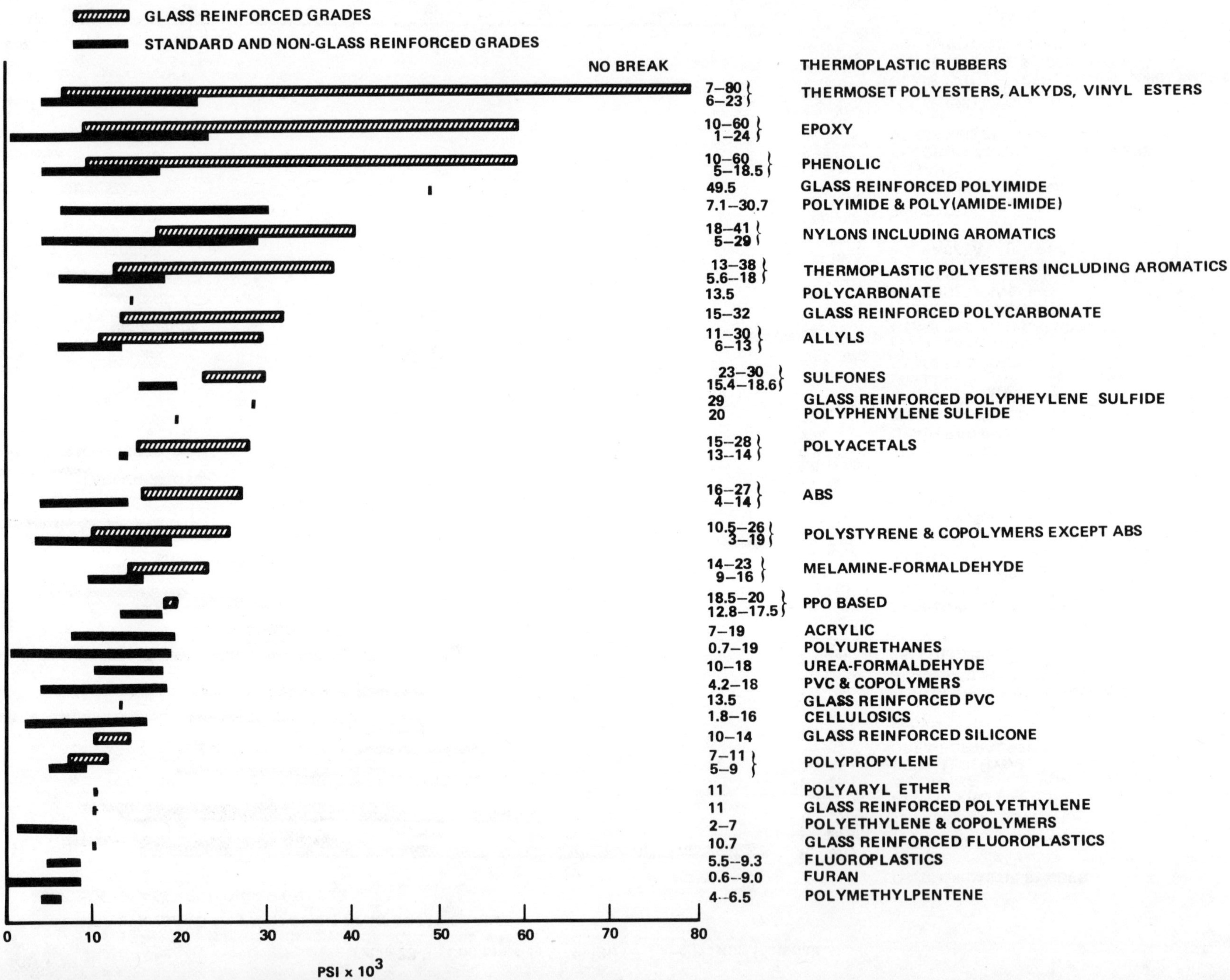

Table 23-5. Impact Strength of Plastics Materials.

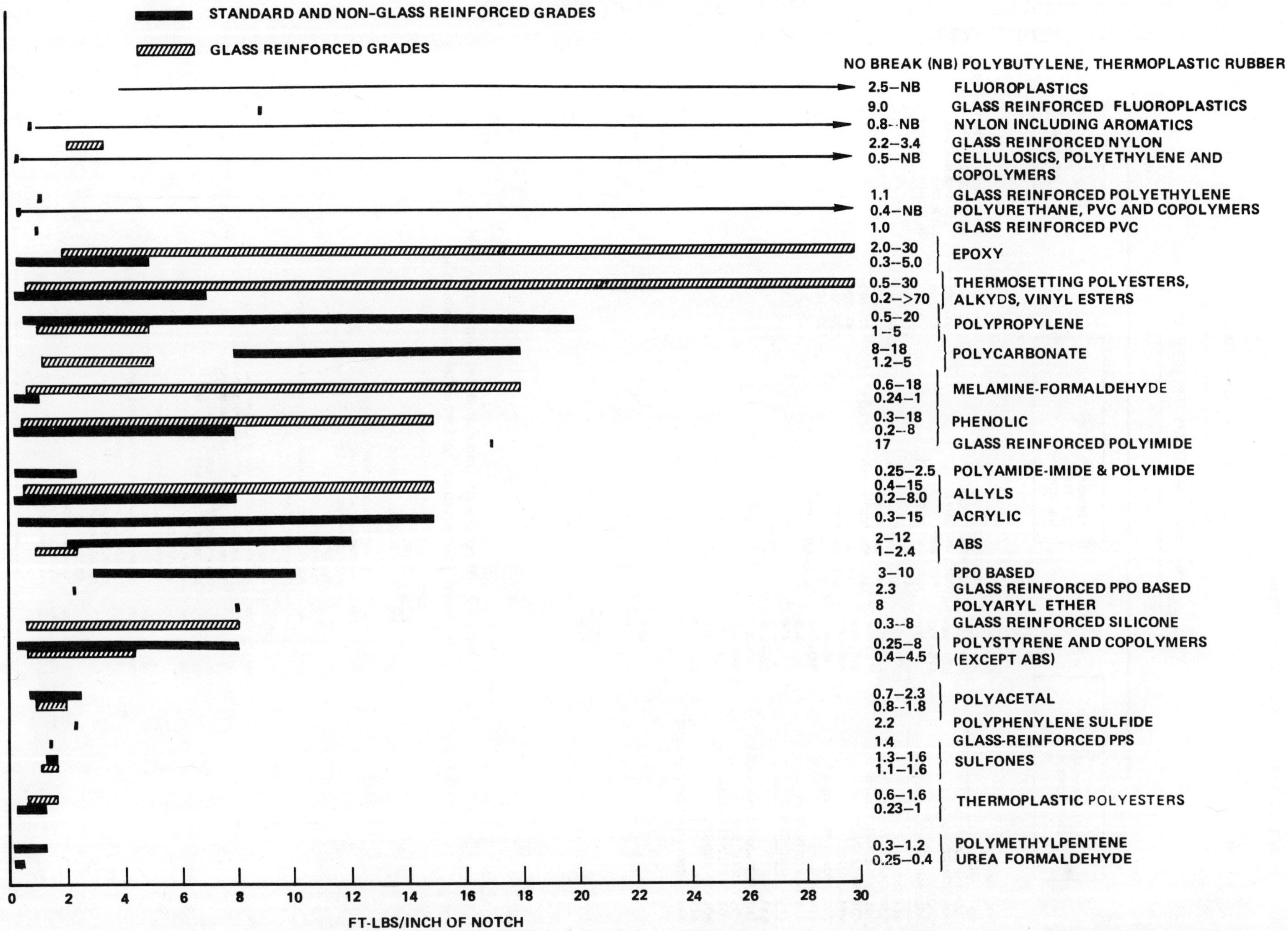

Table 23-6. Material Selection Guide.

CONTINUOUS OPERATING TEMPERATURE

HIGH TEMPERATURE >300°F (149°C)

	°F
POLYIMIDES	500–800
SILICONES	400–600
FLUOROPLASTICS	300–550
POLYAMIDE-IMIDE	520–545
EPOXY	175–500
POLYPHENYLENE SULFIDE	500
ALLYLS	300–450
PHENOLICS	212–450
POLYETHERSULFONES	400
POLYSULFONES	340–400
MELAMINES	300–400
THERMOSET POLYESTER	150–400
UREAS	200–350
ACRYLATE RUBBER	400
FLUOROSILICONES	400
NYLONS	225–350
THERMOPLASTIC POLYESTER	310–320
POLYARYLETHER	250–320
POLYMETHYLPENTENE	250–320
POLYCARBONATE	180–310
ALKYDS	250–300
CHLORINATED POLYETHYLENE	300

STANDARD TEMPERATURE 75 – 300°F (25 – 149°F)

	°F
PPO BASED	170–260
POLYPROPYLENE	205–250
POLYURETHANE	190–250
VINYLS	150–230
POLYBUTYLENE	225
ACETALS	195–220
ABS & SAN	160–220
POLYSTYRENE	125–205
ABS/POLYCARBONATE ALLOY	190–200
ACRYLICS	140–200
CELLULOSICS	120–200
POLYETHYLENE & COPOLYMER	120–180

LOW TEMPERATURE < 75°F (< 25°C)

	°F
CHLORINATED POLYETHYLENE	−60
POLYURETHANE	−65
FLUOROSILICONE	−90
SILICONE	−180
FLUOROPLASTICS	−300

Table 23-7. Approximate Recommended Continuous Heat Resistant Temperatures for Some Plastics Materials.

C°	F°	
316	600	MOST PLASTICS BURN OR CHAR
		POLYIMIDES
		POLYAMIDE-IMIDES, EPOXIES,
260	500	POLYPHENYLENE SULFIDES
		ALLYLS, PHENOLICS
		SILICONES
		FLUOROPLASTICS
204	400	POLYSULFONES, MELAMINES, POLYESTER (TS), ACRYLATE RUBBER
		FLUOROSILICONES
		NYLONS,
		POLYESTER (TP)
149	300	ALKYDS, CHLORINATED POLYETHYLENE
		PPO BASED, POLYPROPYLENE, POLYURETHANE
		POLYMETHYL PENTENE, POLYCARBONATE
		VINYLS, POLYBUTYLENE, ACETALS
100	212	BOILING POINT OF WATER, ABS, SAN
93	200	ACRYLICS
		CELLULOSICS
		POLYTHYLENE AND COPOLYMERS
38	100	
25	77	STANDARD AMBIENT ROOM TEMPERATURE
−0	32	FREEZING POINT OF WATER
		ETHYLENE-ACRYLATE RUBBER
		CHLORINATED POLYETHYLENE
		POLYURETHANE
−73	−100	FLUOROSILICONE
−129	−200	SILICONE
−185	−300	FLUOROPLASTICS

many cases, the size of the part; that is, some parts, because of their large or small size, are impossible to make by one process or another. The Process Selection Guide in Table 23-14, therefore, breaks down the processes into large and small parts, and then further breaks down the processes for thermosets and thermoplastics. The Guide in Table 23-15 uses a breakdown based on volume of parts to be fabricated, etc.

For any given part, the most important properties should be determined first, and then the bar charts or the other materials guides should be consulted to find the most likely materials or material. Then, an appropriate process should be chosen, again using the factors outlined above. Thus, by a process of elimination, the optimum material and process can be selected. Last, the tooling should be chosen, based mainly on the number of parts to be made. Prior to making a final decision, therefore, it is recommended that the detailed descriptions of the materials and processes be once again perused in order to assure that the best have been chosen.

Table 23-8. Material Selection Guide.

MISCELLANEOUS PROPERTIES

- **ELECTRICAL PROPERTIES**
  - **DC, LOW FREQUENCY LOW VOLTAGE**
    - MOST PLASTIC MATERIALS
  - **HIGH FREQUENCY HIGH VOLTAGE**
    - POLYETHYLENE AND COPOLYMERS
    - CHLORINATED PVC (HIGH VOLTAGE)
    - POLYSTYRENE AND COPOLYMERS
    - POLYPROPYLENE
    - OLEFINIC THERMOPLASTIC RUBBERS
    - AROMATIC POLYESTERS
    - FLUOROPLASTICS
  - **ARC RESISTANCE >200 SECS**

ACRYLICS	NO TRACK
ALKYDS	75–420
FLUOROPLASTICS	70-360
CELLULOSICS	50-310
EPOXY	45-300
ALLYLS	115-250
SILICONE	60-250
ACETAL	158-240
POLYESTER(TS)	100-240
POLYETHYLENE & COPOLYMERS	135-235
POLYIMIDE	230
MELAMINES	95–200
POLYPHENLENE SULFIDE	200

- **CHEMICAL RESISTANCE**
  - **EXCELLENT**
    - FLUOROPLASTICS
    - POLYIMIDES
    - POLYOLEFINS
    - ACETALS
    - POLYPHENYLENE SULFIDE
    - ALLYLS
    - EPOXIES
    - IONOMERS
    - POLYAMIDE-IMIDE
  - **GOOD**
    - POLYESTERS (TP & TS)
    - SILICONES
    - PHENOLICS
    - POLYSULFONES
    - NYLONS
    - VINYLS
    - POLYURETHANES
    - ACRYLICS
    - ALKYDS
    - STYRENE-ACRYLONITILE
    - AMINOPLASTICS
    - POLYARYL ETHER
    - POLYARYL SULFONE
  - **POOR**
    - STYRENICS
    - CELLULOSICS
    - POLYCARBONATE

Table 23-9. Guide to the Effect of Chemical Environments on Plastics.*

PLASTIC MATERIAL	Aromatic solvents 77	Aromatic solvents 200	Aliphatic solvents 77	Aliphatic solvents 200	Chlorinated solvents 77	Chlorinated solvents 200	Weak bases and salts 77	Weak bases and salts 200	Strong bases 77	Strong bases 200	Strong acids 77	Strong acids 200	Strong oxidants 77	Strong oxidants 200	Esters and ketones 77	Esters and ketones 200	24-hr. Water absorption % change by weight
	temperature F																
Acetals*	1-4	2-4	1	2	1-2	4	1-3	2-5	1-5	2-5	5	5	5	5	1	2-3	0.22 - 0.25
Acrylics	5	5	2	3	5	5	1	3	2	5	4	4-5	5	5	5	5	0.2 - 0.4
Acrylonitrile-Butadiene-Styrenes (ABS)	4	5	2	3-5	3-5	5	1	2-4	1	2-4	1-4	5	1-5	5	3-5	5	0.1 - 0.4
Aramids (aromatic polyamide)	1	1	1	1	1	1	2	3	4	5	3	4	2	5	1	2	0.6
Cellulose Acetates (CA)	2	3	2	3	3	4	2	3	3	5	3	5	3	5	5	5	2 - 7
Cellulose Acetate Butyrates (CAB)	4	5	1	3	3	4	2	4	3	5	3	5	3	5	5	5	0.9 - 2.0
Cellulose Acetate Propionates (CAP)	4	5	1	3	3	4	1	2	3	5	3	5	3	5	5	5	1.3 - 2.8
Diallyl Phthalates (DAP, filled)	1-2	2-4	2	3	2	4	2	3	2	4	1-2	2-3	2	4	3-4	4-5	0.2 - 0.7
Epoxies	1	2	1	2	1-2	3-4	1	1-2	1	2	2-3	3-4	4	4-5	2	3-4	0.01 - 0.10
Ethylene Copolymers (EVA) (Ethylene-Vinyl Acetates)	5	5	5	5	5	5	1	2	1	5	1	5	1	5	2	5	0.05 - 0.13
Fluorocarbons: Ethylene/Tetrafluoro-ethylene Copolymers (ETFE)	1	1	1	1	1	1	1	1	1	1	1	1	1	1	1	1	<0.03
Fluorocarbons: Fluorinated Ethylene Propylenes (FEP)	1	1	1	1	1	1	1	1	1	1	1	1	1	1	1	1	<0.01
Fluorocarbons: Perfluoroalkoxies (PFA)	1	1	1	1	1	1	1	1	1	1	1	1	1	1	1	1	<0.03
Fluorocarbons: Polychlorotrifluoro-ethylenes (CTFE)	1	1	1	1	3	4	1	1	1	1	1	1	1	1	1	1	0.01 - 0.10
Fluorocarbons: Polytetrafluoroethylenes (TFE)	1	1	1	1	1	1	1	1	1	1	1	1	1	1	1	1	0
Furans	1	1	1	1	1	1	2	2	2	2	1	1	5	5	1	1	0.01 - 0.20
Ionomers	2	4	1	4	4	4	1	4	1	4	2	4	1	5	1	4	0.1 - 1.4
Melamines (filled)	1	1	1	1	1	1	2	3	2	3	2	3	2	3	1	2	0.01 - 1.30
Nitriles (high barrier alloys of ABS or SAN)	1	4	1	2-4	1-4	2-5	1	2-4	1	2-4	2-5	5	3-5	5	1-5	5	0.2 - 0.5
Nylons	1	1	1	1	1	2	1	2	2	3	5	5	5	5	1	1	0.2 - 1.9
Phenolics (filled)	1	1	1	1	1	1	2	3	3	5	1	1	4	5	2	2	0.1 - 2.0
Polyallomers	2	4	2	4	4	5	1	1	1	1	1	3	1	4	1	3	<0.01

RATING CODE: 1. no effect or inert, **2.** slight effect, **3.** mild effect, **4.** softening or swelling, **5.** severe degradation.

MATERIALS

Some plastics incorporate a broad family of materials with widely differing properties in these environments. Special grades are also available with properties outside the ranges listed.

ENVIRONMENT

Chemical effects in plastics are defined by conditions, which include stress cracking, etching or chemical attack (ASTM D-543), staining (ASTM D-2299) and swelling or solvating (ASTM D-543). These conditions are influenced by the time of exposure, temperature and concentrations or combination of these.

*Courtesy: *Plastics World* and Shell Chemical Co.

Table 23-9. *(Continued)*

PLASTIC MATERIAL	Aromatic solvents		Aliphatic solvents		Chlorinated solvents		Weak bases and salts		Strong bases		Strong acids		Strong oxidants		Esters and ketones		24-hr. Water absorption
	temperature F																
	77	200	77	200	77	200	77	200	77	200	77	200	77	200	77	200	% change by weight
Polyamide-imides	1	1	1	1	2	3	1	1	3	4	2	3	2	3	1	1	0.22 - 0.28
Polyarylsulfones (PAS)	4	5	2	3	4	5	1	2	2	2	1	1	2	4	3	4	1.2 - 1.8
Polybutylenes (PB)	3	5	1	5	4	5	1	2	1	3	1	3	1	4	1	3	<0.01 - 0.3
Polycarbonates (PC)	5	5	1	1	5	5	1	5	5	5	1	1	1	1	5	5	0.15 - 0.35
Polyesters (thermoplastic)	2	5	1	3-5	3	5	1	3-4	2	5	3	4-5	2	3-5	2	3-4	0.06 - 0.09
Polyesters (thermoset-glass fiber filled)	1-3	3-5	2	3	2	4	2	3	3	5	2	3	2	4	3-4	4-5	0.01 - 2.50
Polyethylenes (LDPE-HDPE — low-density to high-density)	4	5	4	5	4	5	1	1	1	1	1-2	1-2	1-3	3-5	2	3	0.00 - 0.01
Polyethylenes (UHMWPE — ultra high molecular weight)	3	4	3	4	3	4	1	1	1	1	1	1	1	1	3	4	<0.01
Polyimides	1	1	1	1	1	1	2	3	4	5	3	4	2	5	1	1	0.3 - 0.4
Polyphenylene Oxides (PPO) (modified)	4	5	2	3	4	5	1	1	1	1	1	2	1	2	2	3	0.06 - 0.07
Polyphenylene Sulfides (PPS)	1	2	1	1	1	2	1	1	1	1	1	1	1	2	1	1	<0.05
Polyphenylsulfones	4	4	1	1	5	5	1	1	1	1	1	1	1	1	3	4	0.5
Polypropylenes (PP)	2	4	2	4	2-3	4-5	1	1	1	1	1	2-3	2-3	4-5	2	4	0.01 - 0.03
Polystyrenes (PS)	4	5	4	5	5	5	1	5	1	5	4	5	4	5	4	5	0.03 - 0.60
Polysulfones	4	4	1	1	5	5	1	1	1	1	1	1	1	1	3	4	0.2 - 0.3
Polyurethanes (PUR)	3	4	2	3	4	5	2-3	3-4	2-3	3-4	2-3	3-4	4	4	4	5	0.02 - 1.50
Polyvinyl Chlorides (PVC)	4	5	1	5	5	5	1	5	1	5	1	5	2	5	4	5	0.04 - 1.00
Polyvinyl Chlorides-Chlorinated (CPVC)	4	4	1	2	5	5	1	2	1	2	1	2	2	3	4	5	0.04 - 0.45
Polyvinylidene Fluorides (PVDF)	1	1	1	1	1	1	1	1	1	2	1	2	1	2	3	5	0.04
Silicones	4	4	2	3	4	5	1	2	4	5	3	4	4	5	2	4	0.1 - 0.2
Styrene Acrylonitriles (SAN)	4	5	3	4	3	5	1	3	1	3	1	3	3	4	4	5	0.20 - 0.35
Ureas (filled)	1	3	1	3	1	3	2	3	2	3	4	5	2	3	1	2	0.4 - 0.8
Vinyl Esters (glass fiber filled)	1	3	1-2	2-4	1-2	4	1	3	1	3	1	2	2	3	3-4	4-5	0.01 - 2.50

TEMPERATURE

In many cases the difference between the 77F and 200F ratings is simply a heat distortion effect with no degradation.

WATER ABSORPTION

In ordinary tap and salt water there is almost no chemical attack, however, ion-free water (such as distilled or deionized water) can show varying degrees of attack. Water absorption values have been included to quantify the most common effect of water.

*Widespread ratings exist since acetal copolymers are more inert than homopolymers in many "corrosive" environments.

Table 23-10. Comparison Table of Unmodified Thermoset Resin Systems.

PROPERTY	POLYESTERS	EPOXIES	PHENOLICS	MELAMINES	SILICONES	POLYIMIDES
SPECIFIC GRAVITY	1.10-1.46	1.10-1.4	1.30-1.86	1.40-1.48	1.30-1.34	1.42-1.90
COLOR POSSIBILITIES	VERY GOOD	GOOD	LIMITED	VERY GOOD	GOOD	–
BY-PRODUCTS OF CURE	NONE	NONE	H_2O	H_2O	H_2O, RCOOH	H_2O
MOLDING PRESSURES	0-HIGH	0-HIGH	LOW-HIGH	MEDIUM-HIGH	LOW-HIGH	LOW-HIGH
MOLDING TEMP., °F	<70-320	<70-330	270-320	270-320	290-350	330-660
SHRINKAGE, %	2-8	0.1-0.4	1-1.2	1-1.5	1-1.5	0.1-0.3
TENSILE STRENGTH, 10^3 PSI	6-10	4-13	6-9	7-9	4-5	5-27
ELONGATION, %	5	2.6	1.5-2.0	–	–	< 1 – 10
MODULUS OF ELASTICITY TENSION, 10^3 PSI	3-6.5	4-6	4-5	7.5-10	–	1.9-28.5
COMPRESSIVE STRENGTH, 10^3 PSI	13-36.5	15-21.5	12-15	40-45	9-15	4.7-40.0
FLEXURAL STRENGTH 10^3 PSI	8.5-18.5	13-21	11-17	11-14	9-14	7.1-49.5
IMPACT STRENGTH (IZOD)	0.2-4	0.2-1.0	0.25-0.40	0.24-0.40	0.25-6.0	0.25-0.8
HEAT RESISTANCE °F (CONTINUOUS)	250-500	250-500	160.250	210	>600	>600
HEAT DISTORTION, °F	140-400	115-550	165-260	298	>900	270-680
WATER ABSORPTION, %, (24 HRS.)	0.15-0.60	0.08-0.15	0.03-0.04	0.3-0.5	0.1-0.5	0.11-0.60
DIELECTRIC STRENGTH V PER MIL	200-420	400-500	350-400	300-400	200-500	250-725
DIELECTRIC CONSTANT ($60\text{-}10^6$ CPS)	2.8-5.2	3.3-5.0	4.0-7.5	4.3-7.6	3.2-5.2	3.2-131.5
DISSIPATION (POWER FACTOR $60\text{-}10^6$ CPS)	0.003-.028	0.002-.050	0.01-.15	0.015-0.080	0.0008-0.01	0.0018-0.013
ARC RESISTANCE, SEC.	125	45-120	TRACKS	100-145	250-350	230
BURNING RATE	SLOW TO SELF EXTINGUISHING	SLOW TO SELF EXTINGUISHING	VERY LOW	SELF-EXTINGUISHING	NONE TO SLOW	–
PRICE RANGE, $/LB	0.30-.50	0.65-1.30	0.12-0.40	0.40-0.60	5.00-7.50	–
EFFECT OF SUNLIGHT	YELLOWS	NONE	DARKENS	SLIGHT COLOR CHANGE	NONE TO SLIGHT	–
EFFECT OF WEAK ACIDS	NONE	NONE	NONE	NONE	NONE TO SLIGHT	RESISTANT
EFFECT OF STRONG ACIDS	NONE TO CONSIDERABLE	ATTACKED BY SOME	DECOMPOSED BY OXID ACIDS. NONE TO SLIGHT WITH REGULAR ORG. ACIDS	DECOMPOSES	SLIGHT	SLOWLY ATTACKED – RESISTANT
EFFECT OF WEAK ALKALIES	NONE TO SLIGHT	NONE	SLIGHT TO MARKED FUNCTION OF pH	NONE	NONE TO SLIGHT	SLOWLY ATTACKED
EFFECT OF STRONG ALKALIES	ATTACKED	SLIGHT	DECOMPOSES	ATTACKED	SLIGHT	ATTACKED
EFFECT OF ORGANIC SOLVENTS	ATTACKED (KETONES & CHLORINATED SOLVENTS)	GENERALLY RESISTANT	ATTACKED BY SOME	NONE	ATTACKED BY SOME	VERY RESISTANT
MACHINING QUALITIES	GOOD	GOOD	EXCELLENT	FAIR	FAIR TO GOOD	–
MAJOR LIMITATIONS	CURE SHRINKAGE	DERMATITIS, DIFFICULT TO MOLD RELEASE	COLORS LIMITED	HIGH COST	VERY HIGH COST	HIGH COST
MAJOR ADVANTAGES	CASE OF FABRICATION, CLARITY WITH FLAME RETARDANCY, MODERATE DISSIPATION (POWER) FACTOR	LOW SHRINKAGE, EXCELLENT ADHESION	GOOD GENERAL PROPERTIES, LOW COST	ARC RESISTANCE, SCRATCH RESISTANCE, COLORABILITY	HEAT RESISTANCE, LOW DISSIPATION (POWER) FACTOR	HEAT RESISTANCE

Table 23-11. Material Selection Guide.

MISCELLANEOUS PROPERTIES

LOW FRICTION < 0.8 ON STEEL
- ABS
- POLYACETALS
- SOME NYLONS
- POLYSULFONES
- THERMOPLASTIC POLYESTERS
- FLUOROPLASTICS
- GRAPHITE, MOLYBDENUM DISULFIDE AND FLUOROPLASTIC FILLED COMPOUNDS

COLORABILITY

UNLIMITED
- ABS
- ACETAL
- ACRYLIC
- ALKYD
- ALLYL
- CELLULOSIC
- FLUOROPLASTICS
- MELAMINES
- NYLONS
- PPO BASED
- POLYARYLETHER
- POLYCARBONATES
- THERMOPLASTIC & THERMOSETTING POLYESTER
- POLOLEFINS
- POLYSTYRENES
- POLYURETHANES
- THERMOPLASTIC ELASTOMERS
- UREAS
- VINYLS
- VINYL ESTERS

LIMITED
- EPOXY
- FURAN
- PHENOLIC
- POLYAMIDE-IMIDE
- POLYIMIDE
- POLYPHENYLENE SULFIDE
- SILICONES

ABRASION RESISTANCE < 10,000 MG/KC
- NYLONS
- POLYURETHANES
- POLYACETALS
- ACRYLIC/PVC ALLOY
- ABS/POLYURETHANE ALLOY
- POLYSULFONES
- PPO BASED
- THERMOPLASTIC POLYESTER
- POLYIMIDES
- SOME POLYSTYRENES
- FLUOROPLASTIC COPOLYMERS
- SOME FLUOROPLASTICS
- UHMWPE
- GRAPHITE FILLED COMPOUNDS

THERMOPLASTIC ELASTOMERS
- STYRENE BUTADENE
- POLYESTER
- OLEFINIC
- URETHANE
- VINYL COPOLYMERS

NON-FLAMMABILITY

INHERENTLY-V-O
- FLUOROPLASTICS
- IONOMER
- MELAMINES
- POLYPHENYLENE SULFIDES
- POLYSULFONES
- VINYLS
- POLYIMIDES
- POLYAMIDE-IMIDES

WITH MODIFICATION OR COMPOUNDING
- ABS
- ACRYLIC
- ALKYD
- CELLULOSE ACETATE
- ALLYLS
- EPOXY
- NYLONS
- PHENOLICS
- PPO BASED
- POLYCARBONATE
- POLYESTER (TP & TS)
- POLYOLEFINS
- STYRENICS
- POLYURETHANES
- SILICONES
- UREAS

LOWEST DENSITIES < 1.0 UNFILLED

POLYMETHYLPENTENE	0.83–0.84
POLYOLEFIN TPR	0.88–0.90
POLYPROPYLENE	0.89–0.91
POLY (ETHYLENE CO PROPYLENE)	0.896–0.899
POLYBUTYLENE	0.908–0.917
POLYETHYLENE (INCLUDING UHMWPE)	0.910–0.940
POLY (EHTYLENE CO VINYL ACETATE)	0.92–0.95
POLY (ETHYENE CO ETHYL ACRYLATE)	0.93
IONOMER	0.93–0.96

TRANSPARENCY (COLORLESS)

THICK SECTION > 0.100 IN.
- ACRYLIC
- POLYCARBONATE
- POLYSTYRENE
- CELLULOSSICS
- SAN
- POLYESTERS (TP)
- NYLON
- POLYSULFONE & POLYETHER SULFONE

THIN SECTION (0.001 TO 0.050 IN.)
- POLYETHYLENE
- POLYPROPYLENE
- POLYVINYLS
- POLYBUTYLENES
- IONOMER
- POLYMETHYLPENTENE
- FLUROPLASTICS

Table 23-12. Process Selection Guide.

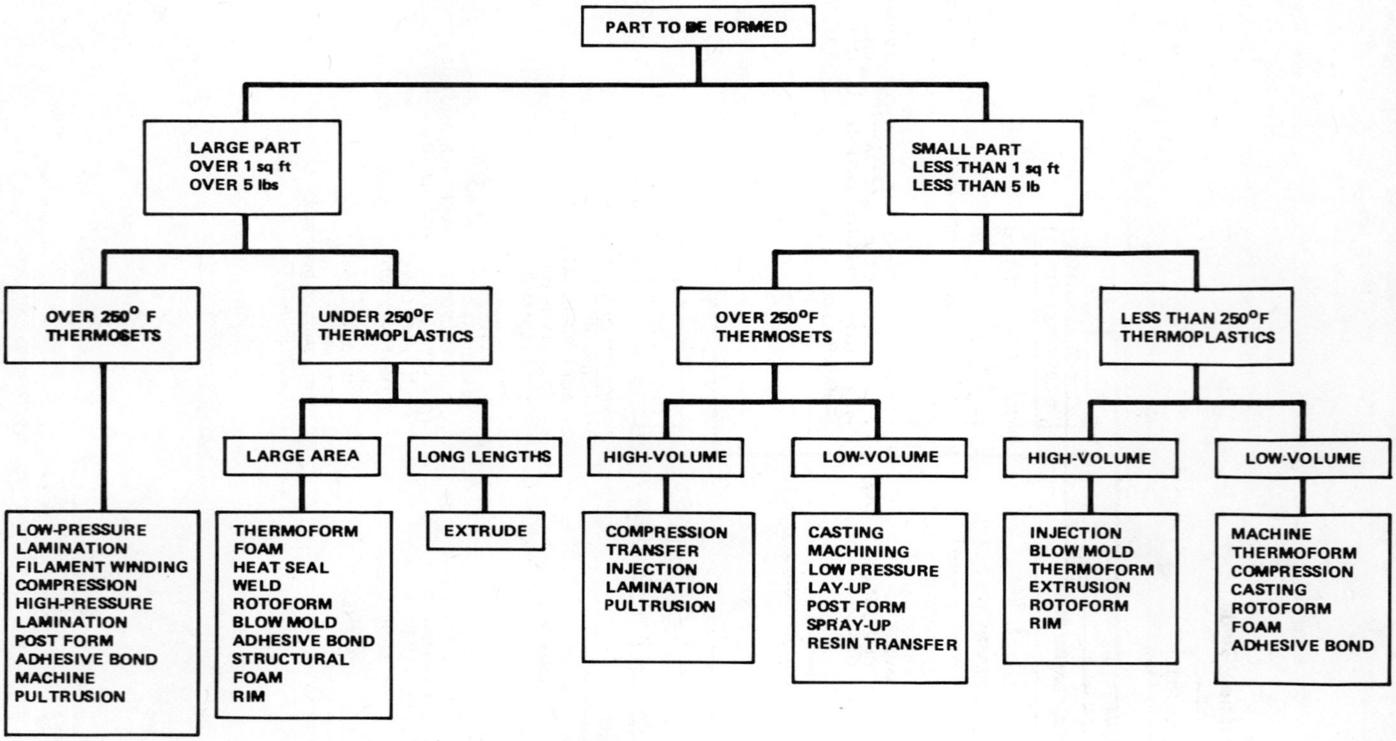

Table 23-13. Tooling Selection Guide.

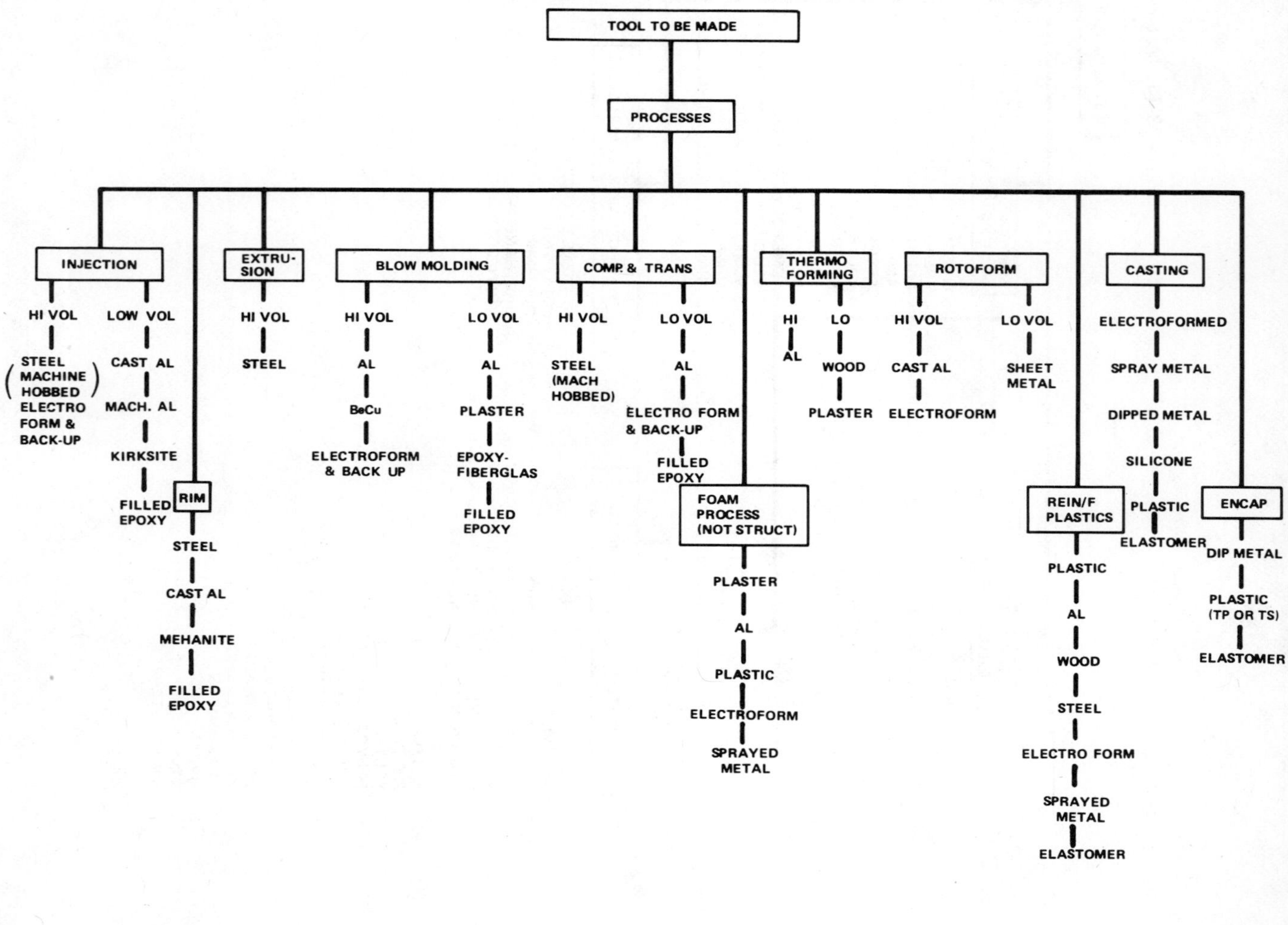

Table 23-14. Process Selection Guide.

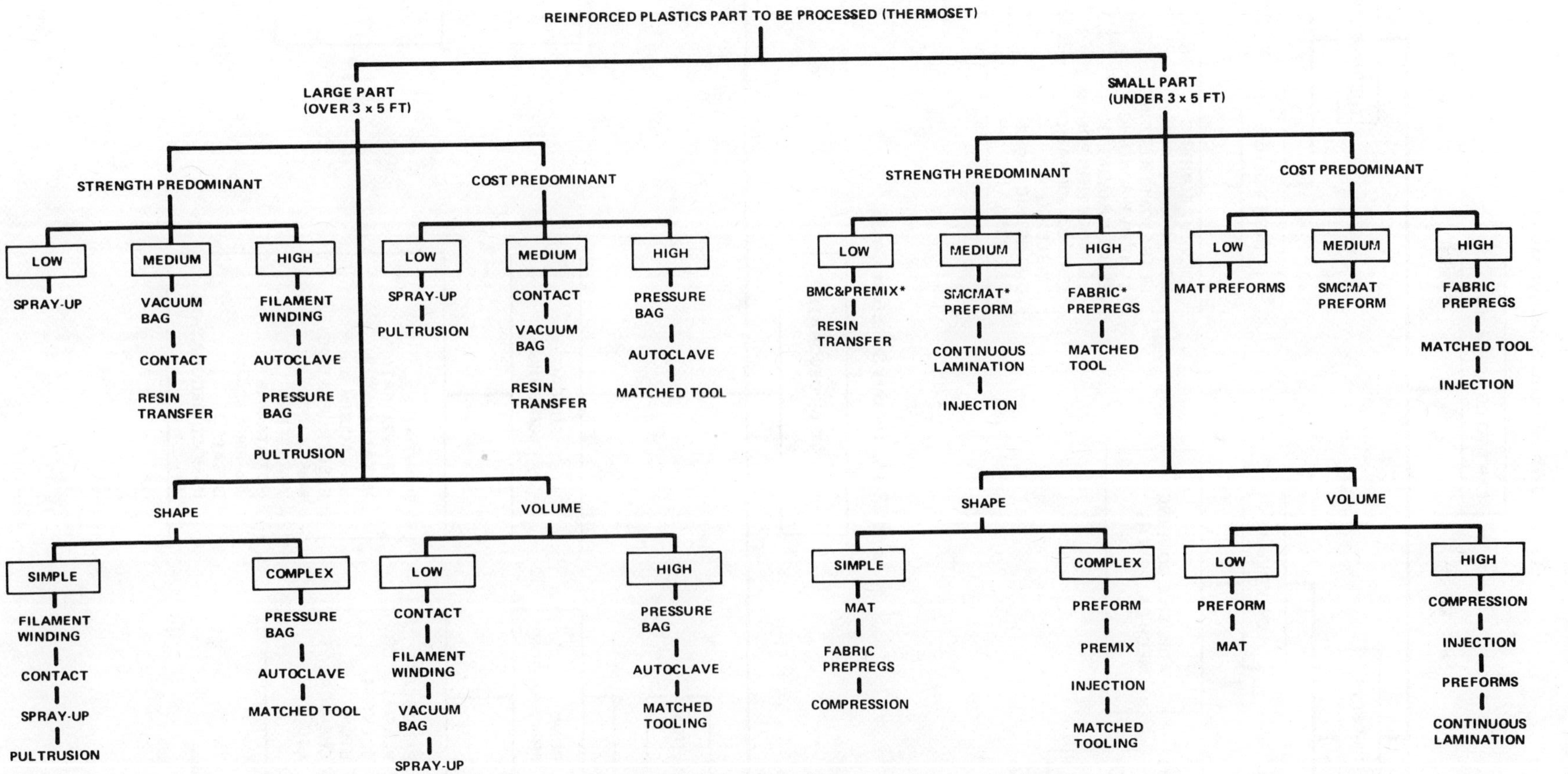

*BMC, PREMIX PREFORMS AND PREPREGS - ALL MATCHED TOOL COMPRESSION MOLDING

Table 23-15. Process Selection Guide.

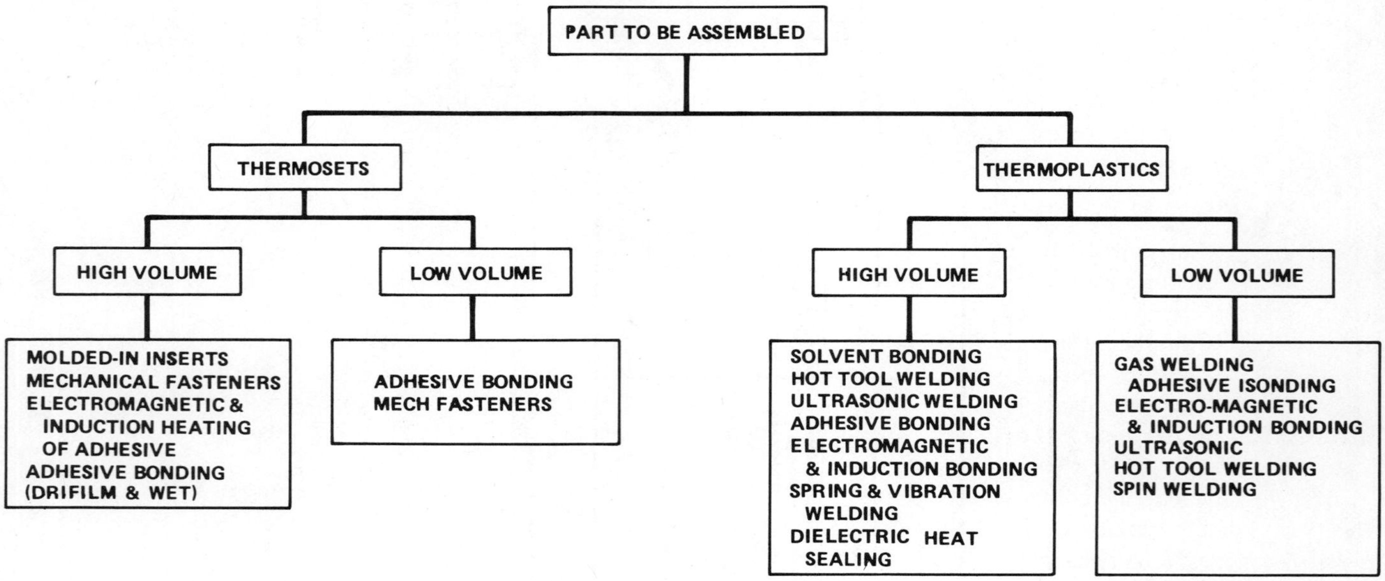

Table 23-16. Process Selection Guide.

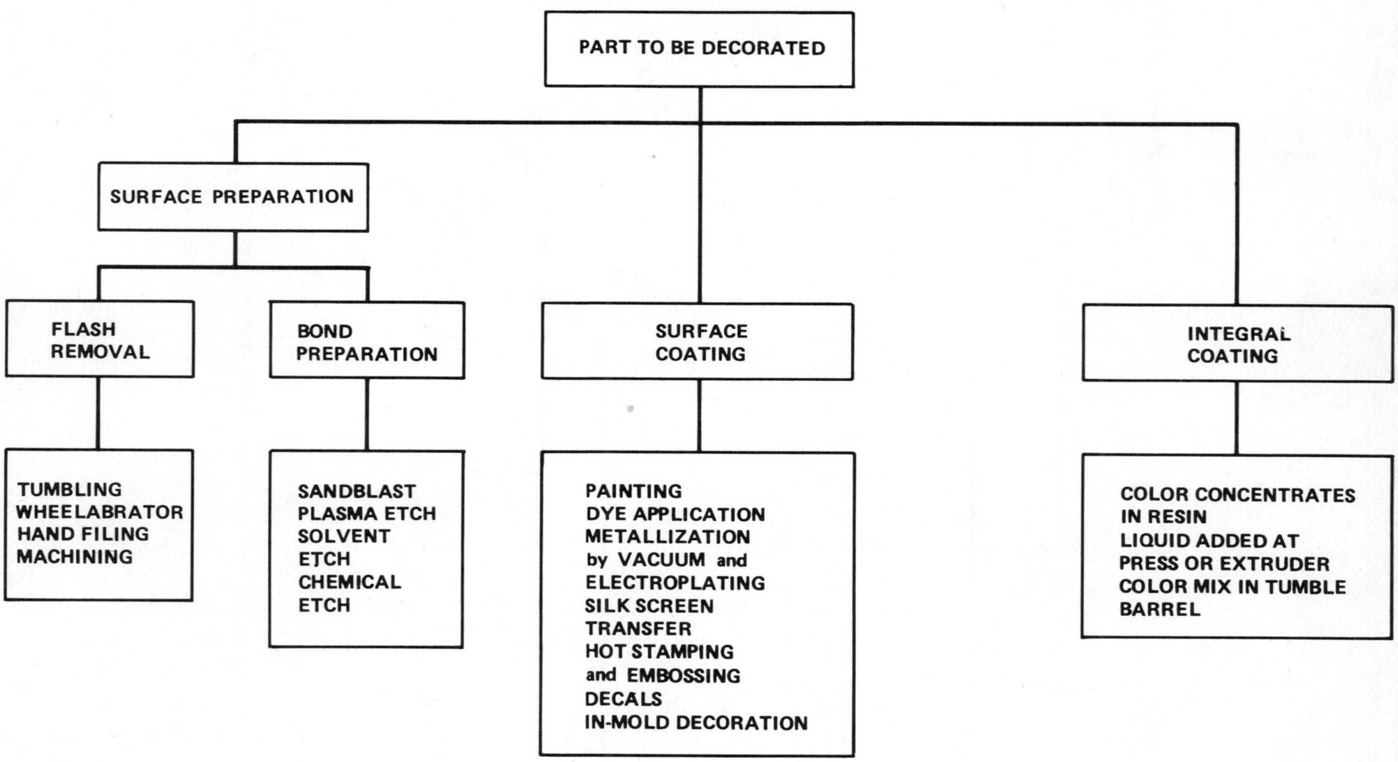

24
Safety in the Plastics Industry

The plastics industry, whether a plant is a producer of plastics raw materials or a processor of plastics into consumer and/or industrial items, is a hybrid industry. On one hand, the industry is deeply involved in chemical processing or the handling of various chemicals; on the other hand, it is also an industry employing many massive machines, with all the attendent dangers of moving parts, flying chips, hot metal, etc.

Even with the extremely diverse nature of the plastics industry, the safety hazards can be grouped essentially into three basic types: (1) fire hazards arising from the use of inflammable solvents, resins, dusts, and oils; (2) chemical hazards from the various chemical substances processed, in many cases at high temperature, including toxic or irritant fumes, lachrymators, dermatitis producers, etc.; and (3) mechanical or equipment hazards, all the "normal" hazards of working around moving machinery, thermal hazards from hot molds, hot oils, steam leaks, etc. Included in the equipment hazards are the problems of occasionally having to lift heavy weights; hazards of electrical shock—since virtually all the equipment is electrified; the possibility of flying particles, dust, etc.

The mere fact that hazards in the plastics industry appear to encompass those of several industries should not in itself lead to the conclusion that the industry is particularly hazardous. On the contrary, the knowledge and experience gained in the chemical industries and in industries using many types of electrified machinery can be, and is, used to good effect in exposing the potential safety hazards and, more important, in minimizing or eliminating these dangers. Information of specific hazards in the plastics plant and methods of reducing these hazards are outlined in the following sections. For detailed information the reader is advised to consult the *Plastics Industry Safety Handbook*, published by the Society of the Plastics Industry in cooperation with the National Safety Council. In addition, the regulations of the Occupational Safety and Health Act (OSHA) and the Toxic Substances Control Act (TSCA) must also be considered whenever possibly dangerous practices or toxic substances may be encountered in the plastics plant.

Fire Hazards

The fire hazards associated with the plastics industry are a consequence of the fact that many of the basic materials used in processing are flammable or explosive. Solvents and thinners such as acetone and styrene burn rapidly under most conditions, and their vapors are both explosive and toxic to personnel. Catalysts, especially of the peroxide type, are highly unstable; heat or shock may result in violent explosion. Mixtures or combinations of chemicals such as accelerators and catalysts generate exothermic reactions so rapidly they can ignite or—when confined, as in bottles—explode.

We have seen in the early chapters of this book that resins and polymers themselves may be flammable. Once combustion begins, the by-products of combustion may create dense, acrid smoke which makes combating the fire difficult. The by-products may also be highly toxic or corrosive.

Fillers, particularly fine metal powders, may also be combustible, more so if they are processed in equipment or environments where static discharge may occur.

Processing oils, extenders, and modifiers must be treated as potential fire hazards. In most cases, the greater hazard occurs, as with resins, after combustion has begun.

Chemical Hazards

Many of the basic raw materials used in the plastics industry fall into the category of allergens. Responses to the allergenic properties of chemicals used with plastics include lachrymal and dermal reactions, respiratory complications, corrosion, and carcinogenic effects. During processing some polymers reach temperatures that cause toxic fumes to be released.

Because of recent legislation, identification of potential hazards of plastics raw materials as well as the plastics themselves is becoming more commonplace. Although debate continues as to the specific, quantitatively acceptable exposure levels for toxic materials, the qualitative effects are becoming well known. The biggest problem is the fact that such a wide diversity of materials are used or handled in the plastics industry, it is almost impossible to completely categorize in detail all the available data. Information can be obtained from individual suppliers, the National Institutes of Health, Department of Labor, and many library references.

Flammability. The flammability of various plastics has been addressed in Chapter 3 and Chapter 22, as well as in the individual discussions of each plastic. Tables of fire hazard properties of other chemicals used in the plastics industry can be obtained from the National Board of Fire Underwriters, Boston, Massachusetts.

Fire prevention measures must begin in the original design of the plant. Relief valves and explosion vents should be supplied wherever required. Dangerous fumes or gases should always be vented so that the building and immediate outside environment will not be contaminated. Piping, valves, etc., must be leakproof and properly assembled.

Storage of flammables should adhere to local code requirements. Fire-resistant construction, such as fire doors and walls, wired glass, etc., should be compatible with the flammable nature of materials to be handled. Approved storage containers should be provided, with bulk storage kept away from normal working areas. Procedures for handling spills should be standardized. Particular attention must be paid to contamination of sewers and drains.

Obviously, fire fighting equipment such as sprinkler systems, fire extinguishers, etc., must be strategically located and properly maintained. Attention to fire prevention is part of an overall good housekeeping policy in any facility.

Potential Hazards

A complete listing of all the potential chemical hazards encountered in the plastics industry is too extensive to be attempted here. However, a brief summary of some of the more common problems can be presented by considering each of the most widely used plastics individually.

General Purpose Thermoplastics

Cellulosics. No particular health problem has been associated with cellulosics relative to ingestion, skin contact, or dust inhalation.

Polyolefins. Polyolefins are considered to be physiologically inert. Sometimes additives are used, such as antioxidants, which might be extractable from the polymer by body fluids or ambient liquids (e.g., in food or cosmetic containers). At sufficiently high temperatures, polyolefins will pyrolyze, yielding materials which could irritate the eyes and respiratory tract.

Fluorocarbons. Experience has demonstrated that no inhalation hazard exists when fluorocarbons are handled below 150-200°C (302-392°F). Above this temperature, pyrolysis can generate fumes which are believed to cause an influenza-like response.

Vinyls. There are two major problems associated with vinyls. The first is that vinyl chloride monomer has been shown to be carcinogenic. PVC is made today with unmeasureable residues of vinyl chloride in the polymer. Consequently the hazards reside only with the handling of the monomer.

The second problem area lies within the formulating ingredients added to vinyl polymers and copolymers. Stabilizers of lead, cadmium, and barium compounds and coloring pigments of cadmium, barium, and selenium compounds can induce toxic responses. Dust and vapor inhalation and skin contact must be prevented. The vapors of some commercial plasticizers may be irritants. Solvents used for organosols or solution systems must be treated appropriately.

Acrylics. The main health hazard from acrylic resins appears to be dermatitis, although pyrolysis can generate toxic fumes. Acrylonitrile is known to be toxic, but once polymerized it is relatively innocuous. Recent use of acrylonitrile in barrier resins in soft drink bottles was prohibited by the Food and Drug Administration, even though the amount of residual acrylonitrile monomer was not measureable.

Styrenics. Although the styrene odor is ubiquitous, the toxicity of the monomer is not great. Once polymerized the styrenics are innocuous. Copolymers with acrylonitrile should be treated similarly to polyacrylonitriles.

Engineering Thermoplastics

As a rule, most engineering thermoplastics are innocuous. Some, like nylon 6, may have trace residual monomer present which could produce toxic or allergenic responses. Fuming from pyrolysis is the greatest hazard of these materials.

Thermosetting Plastics

Phenolics and Amino Resins. The health hazards of the amino resins are basically caused by the formaldehyde fraction. This material is a potent primary irritant and may also cause dermatitis by sensitization. The phenol used in phenolics is caustic and causes burns. Also, the hexamethylene tetramine used as a curing agent is dermatitic.

The elevated temperatures and pressures used to process the B-staged, as-supplied resins can cause liberation of some free formaldehyde and, in the case of phenolics, some furfural. Machining and cutting operations produce resin dust, which may also cause a dermatitis problem. In the fully cured condition, these resins are innocuous.

Polyesters and Alkyds. If curing is properly carried out, the resulting resin will be biologically inert, because there will be no unreacted monomer content. In sufficient concentration, styrene vapor is highly irritating to the eyes and respiratory passages. However, the characteristic odor of styrene gives good warning for appropriate preventive measures to be taken.

The acids and alcohols used in these resins exhibit a low order of toxicity. Most of the amines that are used as promotors are strongly alkaline and may produce injury to skin and mucous membranes upon contact.

Since many alkyds are supplied as solutions, the main health hazards are those associated with the solvents rather than the resins themselves.

Epoxies. The main health problem in the use of epoxy resins is caused by the dermatitic potential of the amine, and some select anhydride hardeners. Recently, carcinogenicity has been found in a few of the resins and in occasional reactive diluents. These products have been removed from the marketplace or are supplied with precautionary labeling advising of the danger.

Polyurethanes. The monomeric isocyanates from which the resins are prepared are rather strong irritants to the skin and respiratory tract. They may cause allergic skin sensitization as well as an asthmatic reaction, which may or may not have an allergic basis.

The cured polymer will pyrolyze at sufficiently high temperatures and toxic by-products can be generated which should not be inhaled.

Silicones. Cured silicone systems are considered nontoxic. They cannot cause silicosis since the silica is chemically bound in the polymer and is not in its normal powdery, crystalline form.

The basis of prevention and protection against the chemical hazards in the plastics industry is the same as in any other industry—*good housekeeping.* Adequate ventilation, protective outerwear, eye protection, clean and orderly work areas, precautionary signs, labeling appropriate containers and storage facilities, adequate lighting, etc—all act in consort to provide the optimum in safe working conditions. Employee education about potential hazards and continuing supervision of work habits minimize the danger level in every case.

Machinery and Equipment Hazards

Regardless of the particular type of plastics processing done, whether in a chemical plant producing plastics raw stock, or in a processing plant making plastics articles, there are a number of hazards and potentially dangerous pieces of equipment which are common to all types of processors.

Moving Parts. All moving parts of equipment that are exposed and are large enough to cause injury should be considered hazardous. These would include all types of gears, belts, chains, connecting rods, etc. The possibilities include being struck by the part as it moves, being caught in or between parts, and being struck by a broken part or by a foreign object which may have fallen into the moving parts.

Nonmoving Parts. There are a number of nonmoving areas of machinery which also present hazards to the operator. These include any part which has a sharp edge, or point, or which may become hot during operation. Loss of lubricating oil or hydraulic fluid can create a dangerous or slippery surface.

Electrical hazards include high voltages which may be present in any part of the machine as part of the normal wiring. The insulation may be broken down because of abrasion, vibration, and/or heat. High voltage from static charges may build up around any type of rotating equipment that is not properly grounded. Inadequate lighting and insufficient workspace can cause a dangerous situation to arise simply by forcing the operator too close to the machine to be safe.

A source of hazard to the operator, not always taken into account, is noise. This is particularly insidious since it is a cumulative danger, and one that tends to be ignored "once the operator gets used to it."

Preventive Measures. Recognition of the various hazards is half the battle. The other half is the implementation of measures to eliminate or minimize these hazards. In almost all cases this can be accomplished by simple safety procedures, by strict adherence to local, state and Federal ordinances, by application of common sense and avoidance of what are usually considered to be unsafe practices.

Moving Parts. All moving parts that could possibly catch, strike, pinch, or squeeze the operator should be covered with some type of wire mesh or transparent plastic or safety glass guard. The operator working around such machinery should not wear loose-fitting clothing, a tie, or rings, watches, and jewelery. Gloves should *not* be worn around moving machinery unless the operator is specially instructed to do so.

The machine itself should be painted in a light color with the hazardous areas clearly visible or painted in a contrasting color. The start switches should be recessed at or below the control panel level. The start switches should also be fitted with springs that require firm pressure to actuate them. Stop switches, on the other hand, should be above the control panel and built in such a way that only a slight pressure will actuate them. If the machine is equipped with movable guards, such as on injection machines, these machines should be equipped with interlock switches such that when the guard is not in its normal position, the switch will be thrown, and the machine will not operate.

Nonmoving parts. Nonmoving parts with sharp edges or corners should be equipped with guards or painted so as to be conspicuous. All electrical wiring should be in conduit, except where attached to a moving part. The machines should always be adequately grounded. High-voltage circuits leading to switches or other controls that might be contacted by the operator should be actuated by low-voltage relay circuits, if at all possible.

General Considerations. There should be an organized safety program in every plant. Such a program should include regular inspections, a safety team with machine operators and not just management as members, and posters, if possible, to remind workers of the need for safety on the job. Good housekeeping should be stressed at all times by having adequate storage for tools and loose equipment, and containers for scraps, chips, and waste particles. A safety monitor should be appointed for each area to oversee all potentially dangerous operations and to see that all routine operations are performed in a safe manner. The safety monitor should be responsible for checking on unsafe machine operations, housekeeping, and such items as excessive noise, use of safety glasses, and gloves, where these are permitted.

Safety in Specific Processes

While the foregoing discussion applies to all types of plastics processes in general, there are certain hazards related to some processes much more than to others. The following descriptions, while labeled for specific processes, should be considered applicable in some degree to all.

Injection Molding. Possibly the greatest number of accidents in injection molding are caused by the moving parts of the press. For this reason, a movable safety guard with an interlock switch must be installed in the area of the mold. Fixed guards should be in place over moving rams, toggles, etc. In addition to the interlock switch, a mechanical stop should be provided to prevent mold closure if the safety door is not fully closed. All guards and safety switches should be inspected at regular intervals.

The machine barrel should be covered so the operator cannot come in contact with any hot sections. All electrical connections and leads should be covered so the operator cannot at any time come in contact with "live" wiring. A lockout procedure should be established so that all machine controls are locked in the "off" position when the machine is being worked on. Only the operator and the maintenance foreman should have the key to unlock the machine.

Extrusion. The same precautions given for injection molding regarding heat and electrical wiring should also be observed for the extrusion process. In addition, it should be noted that pressures at the die can be very high, so a pressure gauge should always be consulted when starting up, or when a new material is being run. A sheared bolt on a die caused by excessive pressure can act like a lethal weapon. Hands should *never* be placed into the throat of a machine to clear it or to remove a foreign object unless the machine is *stopped.* Saws or flying shears used to cut moving extrudate should always be provided with guards to reduce hazards of flying chips, etc.

Purging of the machine, when the hot melt is not put through a puller mechanism, should only be done when the operator is wearing adequate protective clothing and a face mask, since hot extrudate, particularly purging materials, can exit the machine at high velocity and force. A bucket of water should be kept at the machine to be used for immersing a hand or glove that might have come in contact with hot plastic. It should be noted that polyethylene, a common purging material, has a very high specific heat; this means that a small amount of hot polyethylene can inflict a severe burn.

When starting the extruder, it is important that the die end of the machine be heated first. Then the rest of the machine can be heated up to the feed end of the barrel in stages, to insure that the material at the exit end is thoroughly fluid before the rest of the material in the barrel heats up, or perhaps overheats and cannot be removed. The temperature and pressure at the die should be monitored when starting to make sure it conforms to previous runs, or is not too high if a new material is being run.

Compression and Transfer Molding. The principal hazards in compression and transfer molding are similar to those in injection molding, in that the operator is exposed to hot dies and closing molds. In addition, however, there may be irritating fumes from some molding compounds and the possibility of flash being ejected from a mold as it closes. Until the action of any given mold is thoroughly checked out, the operator should wear gloves and a face shield.

Compressed air, which is used generally to clean molds, must be treated cautiously since it can cause flash to fly in all directions. Serious accidents can be caused by its indiscriminate use. Oil leaks from accumulators can be a hazard around molding presses if good housekeeping is not maintained. Dielectric preheaters, if used, should never have the interlock switches disconnected so the unit can be operated with the door open. Such a practice is dangerous to the operator, and also can cause interference to television and radios in the vicinity, This can be the cause of a fine by the Federal Communications Commission.

Miscellaneous Processes. Processes such as thermoforming, rotoforming, blow molding, and reinforced plastics processing have no particular hazards other than the usual ones of fire hazards due to inflammable solvents, or chemical problems due to possibly toxic substances. The usual precautions of good housekeeping and the avoidance of loose clothing, jewelry, etc., should, of course, always be exercised.

REFERENCES AND BIBLIOGRAPHY

Delmonte, J. Safety considerations in handling encapsulants. *Insulation*, July 1960, p. 41.

Furane Plastics, Inc. *Safety Precautions to be Observed in Handling Epoxy Resins*. Technical Bulletin EP-8 (mod. 1), February 15, 1956.

Hallas, R. S. Safety and efficiency for plastics machinery. *Plastics Engineering*, June 1975, **31**(6):25.

Harper, C. A. Caution: handle with care. *SPE Journal*, March 1961, **17**(3):225.

Hauck, J. E. Flame retardant plastics. *Materials in Design Engineering*, November 1964, **60**(6):83.

Hersch, P. Handling epoxies safely. *Plastics Technology*, August 1968, **14**(8):55.

Houghton Laboratories Inc. *Hazards in Handling Epoxide Compounds*. Technical Data Bulletin, September 1956.

Lillie, R. W. Fire hazards and plastics. *Western Plastics*, January 1963, p. 9.

Manufacturing Chemists Association, Inc. *Tolylene Diisocyanate*. Chemical Safety Data Sheet SD-73, adopted 1959.

Materials in Design Engineering. How dangerous are plastics in fires? March 1961, **53**(3):180.

Mohr, J. G., and Oleesky, S. *SPI Handbook of Technology and Engineering of Reinforced Plastics/Composites*. 2nd Ed. Van Nostrand Reinhold Co., New York, 1973.

Naturman, L. I. Fires in plastics fabricating plants. *SPE Journal*, Spetember 1962, **18**(9):1133.

Shell Chemical Corporation. *Recommendations for Handling Epon Resins and Auxiliary Chemicals in Manufacturing Operations*. Technical Bulletin SC:57-15.

Society of the Plastics Industry, Inc. *Plastics Industry Safety Handbook*. Cahners Books, Boston, Mass., 1973.

Stoltz, G. H., and Gleed, J. S. "Safety first" should be the credo in your molding shop. *Plastics Technology*, April 1970, **16**(4):64.

Woolrich, P. F. Precautions in the use of urethane materials. *Plastics Compounding*, May/June 1979, **2**(3).

Appendix
Ancillary Materials and Equipment

Webster defines "ancillary" as subordinate or auxiliary. The materials and equipment described below are subordinate or auxiliary in the sense that any process could be done without them, using only the major pieces of equipment listed for the process. For example, injection molding can be done using only a press and a mold. However, using various types of ancillary equipment or ancillary materials, the parts can be made cheaper, quicker, or with better quality—or with all three improvements in the process.

The listings of the various materials and equipment will, in most cases, be very familiar to those persons who are engaged in a particular process. However, they are listed in some detail for the benefit of students, those who might want to venture into a new field, and for the "professionals" who might not be aware of all the "ancillaries" available from their own or other plastics fields.

The first list of ancillaries are those which might be used in most plastics processes. This is then followed by a listing of ancillaries for each process by type; for example, injection, extrusion, blow molding, etc. In each case, the major pieces of equipment will not be listed or described since this will have been done previously in the applicable chapters.

Equipment in Common Use

Process Control Equipment
- Temperature measuring equipment—surface pyrometers, (IR) optical pyrometers, temperature recorders—mechanical and temperature-recording paints and crayons.
- Humidity measuring equipment—wet and dry bulb indicators, dial indicators, and color change devices.
- Storage tanks and silos.
- Conveyors—belts, vacuum, auger feed, vibratory, etc.
- Decorating means—hot stamping, foil marking, branding, etc.
- Testing equipment—balances, tensile testing machine, viscosity tester, ovens, impact testers, thermal analyzers for various properties.

Elastomer Processing
- Durometers—Shore A and D. Scott tester or other machine capable of 20 in/min movement. Compression set test fixture, ovens, impact tester capable of operation at sub-zero temperatures.

Film Processing
- Film thickness measuring equipment—Beta ray or other type capable of measurements during operation. Film impact tester, rotatable blow molds, special film scrap compactors, shredders and granulators, pinhole detectors. Special testing equipment for measuring blocking, slip, machinability, stiffness, curl, running quality, haze, gloss, and see-through properties.

Injection Molding
- Storage silos and containers connected to various types of conveying systems—vacuum, air, auger, vibratory.
- Metering and blending equipment for colorants.
- Vacuum dryer-hoppers, drying ovens, desiccant regeneration systems for dehumidifiers.
- Automatic temperature, pressure, and cycle timers, with cathode ray tube display and/or recording equipment for noting resin used, conditions, number of parts made, etc.
- Mold temperature regulators. Rooftop water chillers.
- Mold-release agents—internal and external.
- Hopper magnets and metal detectors.
- Nozzle-attached melt homogenizers.
- Weigh feed equipment with displays and/or recorders.
- Gate cutters (many types).
- Anti-scale water treatments.
- Automatic part removers (mold sweeps or robots).
- Automatic conveying, stacking, and counting equipment.
- Automatic trimming and/or marking equipment.
- Annealing ovens or oil-filled tanks (for nylons and polystyrene).
- Fluidized-bed mold cleaners.
- Balances for checking part weight.
- Scrap granulators and grinders.
- Nozzle and sprue bushing refacing tools.
- Nozzle orifice gauges (wear checking).
- Spiral molds for resin flow tests.

Extrusion
- Storage bins and silos connected to conveyor equipment.
- Automatic hopper loaders.
- Hopper drier systems and desiccant regeneraters.
- Hopper magnets and metal detectors.
- Hopper color concentrate mixers, powder and/or liquid. Crammer feeders.
- Automatic barrel temperature and pressure controllers and recorders. Optical pyrometers.
- Screen changer equipment—slide or roll type.
- Die and screen cleaning equipment—fluidized bed or equivalent.
- Stainless steel diaphragm pressure gauges.
- Take-off equipment. Coilers, winders for sheet and film.
- Slitting equipment. Edge guides and edge trimmers for sheet and film.
- Thickness gauges—Beta ray, air, etc.
- Strand pelletizers and scrap grinders.
- Surface treaters—flame and corona discharge.
- Automatic cut-off and stackers for sheets—shears and saws.
- Vacuum sizing tanks—differential pressure apparatus.
- Radiation equipment for crosslinking film.
- Orientation (centering) equipment for film.
- Rotation or oscillating dies for film production.
- Static elimination equipment (film and sheet).

Blow Molding
- Automatic parison control.
- Air chiller for part cooling.
- Automatic trimming equipment.
- Surface treatment equipment for decorating or labeling.
- Automatic decorating or labeling equipment.
- Automatic conveyors and stacking equipment.

Thermoforming
- Automatic trimming equipment.
- Static elimination equipment.
- Automatic conveying equipment.
- Surface treating equipment—flame or corona.
- Decorating equipment—hot stamping, silk screen, foil, etc.
- Sealing equipment—ultrasonic, dielectric, or heat.
- Automatic stacking and counting equipment.

- Steel rule dies and "clicker" presses.
- Resin granulators.

Rotoforming
- Automatic loading and unloading equipment.
- Precise charge weighing equipment.
- Precise temperature and cycle controls.
- Resin drying equipment.
- Resin pulverizers (35 mesh).

Compression and Transfer Molding
- Automatic charge weighing equipment.
- Resin preformers and preheaters (ovens, IR heaters, or dielectric heaters).
- Automatic charge loading and parts removal (robots).
- Deflashing equipment–tumbling barrels, grit blasts, etc.
- Warm or cold runner molds.
- Shuttle molds for fast transfer.
- Spiral or cup molds for resin flowability tests.

Reinforced Plastics
- Humidity indicators.
- Bung hole mixers for catalysts, colorants, thinners, etc.
- Metering, mixing, and dispensing machines.
- Various types of ribbon or Z blade batch mixers.
- Spot heating apparatus–hand held heaters, heating tapes, IR heaters, ovens.
- Chopper guns, with resin transfer molding attachments, color blocks for rapid color changes in gun.
- Ultrasonic flaw detectors.
- Roving guides for spray chopper guns–prevent tangling.
- Various types of release agents.
- Thickness gauges–one side operation.
- Resin viscosity test apparatus–Brookfield viscometer, Zahn cup, etc.
- Hardness testers–Barcol, Shore D.
- Balances for hand mixing.
- Resin spreaders, rollers, tongue depressors, rags, etc.
- Solvent containers, flip-top solvent tank (safety types only).

Joining and Fastening
- Adhesive applicators.
- Mixing, metering, and dispensing equipment.
- Hand-held heat sealers.
- Viscosity testers–Zahn cup, Brookfield viscometer.
- Varying types of hot gas weld tips.
- Varying horn designs for ultrasonic sealers.

Miscellaneous Processes
- Vacuum deaerator for encapsulation.
- Rapid mixers for foam preparation.
- Foam cutters–band saws, hot wire, or fabric cutting machines.

Index